HL

PEARSON BACCALAUREATE

HIGHER LEVEL

Mathematics

2012 edition

DEVELOPED SPECIFICALLY FOR THE

IB DIPLOMA

IBRAHIM WAZIR • TIM GARRY

PETER ASHBOURNE • PAUL BARCLAY • PETER FLYNN • KEVIN FREDERICK • MIKE WAKEFORD

ALWAYS LEARNING

PEARSON

Published by Pearson Education Limited, Edinburgh Gate, Harlow, Essex, CM20 2JE.

www.pearsonglobalschools.com

Edited by Mary Nathan and Maggie Rumble
Designed by Tony Richardson
Typeset by TechType

First published 2009

This edition published 2012

20 19 18
IMP 10 9 8

British Library Cataloguing in Publication Data
A catalogue record for this book is available from the British Library

ISBN 978 0 435 07496 8

The authors and publisher would like to thank the following individuals and organisations for permission to reproduce photographs:
(Key: b-bottom; c-centre; l-left; r-right; t-top)
Alamy Images: 413t, 518br; **Art Directors and TRIP Photo Library:** 26tl; **Corbis:** 220t, 399tr; **Fotolia.com:** 190c, 279br, 700br, 707br, 745br; **Glow Images:** 48tl; **Pearson Education Ltd:** 1t, 429b; **Science Photo Library Ltd:** 225tr, 247br, 288t, 428bc, 956br, 962t; **Shutterstock.com:** 398b, 516cl, 605t, 762br, 832tl, 854bl, 965cr, 968-969bc, inside front cover: Dmitry Lobanov

The publisher would like to thank the International Baccalaureate Organization for permission to reproduce its intellectual property. This material has been developed independently by the publisher and the content is in no way connected with nor endorsed by the International Baccalaureate (IB). International Baccalaureate® is a registered trademark of the International Baccalaureate Organization.

Printed in Slovakia by Neografia

Websites
There are links to relevant websites in this book. In order to ensure that the links are up to date and that the links work we have made the links available on our website at www.pearsonhotlinks.co.uk. Search for this title or ISBN 9780435074968.

Contents

All accessed through the online e-book (see page ix)

Acknowledgements

We wish to again extend our sincere and heartfelt thanks to Jane Mann for her dedication and encouragement through all the hard work of the 1st and 2nd editions. We also wish to thank Maggie Rumble, Gwen Burns (1st edition), and Mary Nathan (2nd edition) for their highly skilled and attentive work as editors.

The authors and publisher would like to thank Ric Sims for writing the TOK chapter. The authors would also like to thank Douglas Butler, Simon Woodhead, Mark Hatsell and all at Autograph for superb dynamic mathematics software – and for making it possible for the authors to utilise Autograph's interactive and visual features in the e-book.

Autograph

The publishers would also like to thank David Harris for his professional guidance, Nicholas Georgiou for checking the answers, Texas Instruments for providing the TI-Smart View program.

Dedications

I dedicate this work to the memory of my parents.

My special thanks go to my wife Lody for standing beside me throughout writing this book. She has been my inspiration and motivation for continuing to improve my knowledge and move my career forward. She is my rock, and I dedicate this book to her.

My appreciation and thanks also go to my friend and teacher Ram Mohapatra for his help with the Options section and to Peter Ashbourne for his help with the complex numbers chapter.

My thanks go to all the students and teachers who used the 1st edition and sent us their comments and corrections.

Ibrahim Wazir

My gratitude and deepest love go to my wonderful family – Val, Bethany, Neil and Rhona – for your support, patience and good humour. Some of the considerable time and energy that went into writing and revising two textbooks was borrowed from precious family time. Please forgive me for that. It is time with you, my family, which I most cherish in life.

I also wish to thank my good friend Marty Kehoe for his help and friendly advice; and to all the students that have passed through my classrooms since 1983 – especially students in the past four years who have provided constructive feedback on the first edition.

Tim Garry

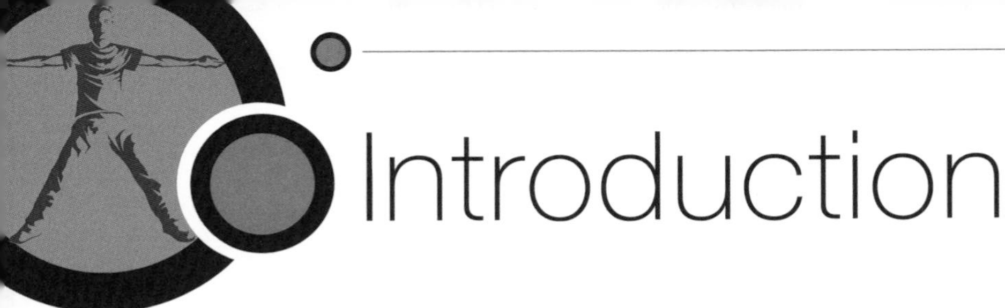

Introduction

This textbook comprehensively covers all of the material in the **core syllabus** for the two-year **Mathematics Higher Level** course in the **International Baccalaureate (IB) Diploma Programme**. A new syllabus for each of the IB mathematics courses was issued in early 2012 for which students will first take exams in May 2014. This second edition is specifically designed for the 2014 Higher Level syllabus. Students will first be taught the course with this syllabus in the autumn of 2012. All of the material for the **option syllabus** is contained on a Pearson website and is password protected (see below for more information). Your teacher will specify which one of the four option topics you will study.

Content

As you will see when you look at the table of contents, the six **core syllabus topics** (see margin) are fully covered, though some are split over different chapters in order to group the information as logically as possible. The textbook has been designed so that the chapters proceed in a manner that supports effective learning of the necessary concepts and skills. Thus – although not absolutely necessary – it is recommended that you read and study the chapters in numerical order. It is particularly important that all of the content in the first chapter, ***Fundamentals***, is thoroughly reviewed and understood before studying any of the other chapters. It covers most of the *presumed knowledge* for the course, including essential terminology, notation and techniques that are essential for successful completion of the Mathematics HL course.

IB Mathematics Higher Level Core syllabus topics

1 Algebra
2 Functions and equations
3 Circular functions and trigonometry
4 Vectors
5 Statistics and probability
6 Calculus

The previous syllabus for Mathematics HL contained a topic on matrices in the core syllabus. This topic is not in the 2014 syllabus, resulting in most of the content on matrices being removed. Matrices is an interesting and practical area of mathematics – so we decided to keep the chapter Matrix Algebra (Chapter 6) from the 1st edition. However, you could skip Chapter 6 and still cover the entire core syllabus.

Other than the final three chapters, each chapter has a set of **exercises** at the end of each section. Also, at the end of each of these chapters (except for Chapter 1) there is a set of **practice questions,** which are designed to give students practice with exam-like questions. Many of the end-of-chapter practice questions are taken from past IB exam papers. Near the end of the book, just before the index, you will find answers to all of the exercises and practice questions that appear in this textbook.

There are numerous **worked examples** throughout the textbook, showing you how to apply the concepts and skills you are studying.

This example appears in Section 5 of Chapter 3 Algebraic Functions, Equations and Inequalities.

```
68+48√2
                 135.882251
68-48√2
               0.1177490061
MAT
```

Example 31 – Another equation in quadratic form

Find all solutions, expressed exactly, to the equation $w^{\frac{1}{2}} = 4w^{\frac{1}{4}} - 2$.

Solution

$w^{\frac{1}{2}} - 4w^{\frac{1}{4}} + 2 = 0$ Set the equation to zero.

$\left(w^{\frac{1}{4}}\right)^2 - 4\left(w^{\frac{1}{4}}\right) + 2 = 0$ Attempt to write in quadratic form: $at^2 + bt + c = 0$

$t^2 - 4t + 2 = 0$ Make appropriate substitution; in this case, let $w^{\frac{1}{4}} = t$.

$t = \dfrac{-(-4) \pm \sqrt{(-4)^2 - 4(1)(2)}}{2}$ Trinomial does not factorize; apply quadratic formula.

$t = \dfrac{4 \pm \sqrt{8}}{2} = \dfrac{4 \pm 2\sqrt{2}}{2}$

$t = 2 \pm \sqrt{2}$

$w^{\frac{1}{4}} = 2 \pm \sqrt{2}$ Substituting $w^{\frac{1}{4}}$ back in for t; raise both sides to 4th power.

$w = (2 + \sqrt{2})^4$ or $w = (2 - \sqrt{2})^4$

$w = \left((2 + \sqrt{2})^2\right)^2$ or $w = \left((2 - \sqrt{2})^2\right)^2$

$w = (6 + 4\sqrt{2})^2$ or $w = (6 - 4\sqrt{2})^2$

$w = 68 + 48\sqrt{2} \approx 135.882$ or $w = 68 - 48\sqrt{2} \approx 0.117749$ (approx. values found with GDC)

Chapter 19 contains three full-length Paper 1 and Paper 2 **sample exams**. Solution keys for these exams are available from the authors' website.

Finally, you will find a **Theory of Knowledge** chapter, which should stimulate you to think more deeply and critically about the nature of knowledge in mathematics and the relationship between mathematics and other subject areas.

Website support

At www.pearsonbacconline.com you will find a selection of free online learning resources supporting the material in this book. More comprehensive support for teachers who adopt the textbook will be available at the authors' website: www.wazir-garry-math.org, which will be regularly updated. You will be required to register before gaining access to materials on the authors' website.

The following will be available from the authors' website:

1 Further practice/mock exams and mark schemes
2 Additional exercises with solutions
3 Internal Assessment ('Mathematical Exploration') notes and guidance
4 Graphing calculators and other technology
5 Instructional activities for students
6 Chapter tests and quizzes.

Worked solutions

Worked solutions for all exercises and practice questions can be accessed from the online e-book for this textbook (more on the e-book below).

HL Option topics

Over 600 textbook pages of material covering the four Higher Level Options can be accessed via the online e-book. All four Options will be presented as e-books when you log in to your e-book account and you can use whichever of them you need. The Options are covered comprehensively, with thorough explanations, worked examples, exercises and practice exam questions. Please note that unauthorized circulation of this material is not permitted.

Online e-book

Included with this textbook is an e-book that contains a digital copy of the textbook. To access this e-book, please follow the instructions on the inside front cover of this book. The textbook on the e-book offers far more than just another copy of the textbook. There are many interactive features on the e-book, which can be accessed by clicking on active links embedded in the pages of the digital version of the textbook. These features include:

1 Additional explanations and examples

2 Practice quizzes for each chapter

3 Dynamic demonstrations of key concepts

4 Audio-video graphing calculator support with activities and tips

5 Worked Solutions for all exercises and practice questions

6 All four Options chapters

7 Software illustrations and simulations.

These interactive resources are designed to support and enhance students' understanding of essential concepts and skills throughout the course. We are profoundly indebted to Peter Ashbourne, Paul Barclay, Peter Flynn, Kevin Frederick and Mike Wakeford – the team of highly experienced and gifted mathematics teachers who created these supplementary student resources on the e-book.

Overview of syllabus changes

As a result of the IB's cyclical curriculum review process, the IB Mathematics HL core syllabus for first exams in May 2014 differs from the previous syllabus in some ways. The following is an overview of the most important changes.

Topic 1 ***Algebra*** remains Topic 1 and has the following addition: *solution of systems of linear equations (maximum of three equations in three unknowns).*

Topic 2 ***Functions and equations*** remains Topic 2 and has the following addition: *sum and product of the roots of polynomial equations.*

Topic 3 ***Circular functions and trigonometry*** remains Topic 3 and is unchanged.

Topic 4 ***Matrices*** has been removed. *Solution of systems of linear equations* is in the *Algebra* topic and *row reduction* for finding the intersection of three planes is still in the *Vectors* topic.

Topic 5 ***Vectors*** is now Topic 4 and the *determinant representation of the vector product* has been removed.

Topic 6 ***Statistics and probability*** is now Topic 5 and *estimation of mean and variance of a population from a sample* has been removed.

Topic 7 ***Calculus*** is now Topic 6 and has the following additions: *informal idea of continuity* and *total distance travelled equals* $\int_a^b |v(t)|\, dt$. Also the *solution of first order differential equations by separation of variables* has been removed from this topic and is now in the **Calculus** Option topic.

Certainly, there is a great deal of useful mathematics that cannot 'fit' into the syllabus. We have decided to include a few non-syllabus items in the textbook and have clearly identified any such items as optional.

Internal assessment

This textbook, the online e-book, and the two supporting websites (from Pearson and the authors) provide comprehensive support for the new Internal Assessment component (*Mathematical Exploration*). There is a brief chapter near the end of the textbook on *Mathematical Exploration* in the context of the IA programme for Mathematics HL. Further in-depth information and guidance for teachers adopting the textbook will be provided on the authors' website. We will be updating teacher support and advice for Internal Assessment on our website regularly to address the latest developments, so teachers are encouraged to check from time to time for updates.

Information boxes

As you read this textbook, you will encounter numerous boxes of different colours containing a wide range of helpful information.

Assessment statements

3.6 Solution of triangles.

The cosine rule: $c^2 = a^2 + b^2 - 2ab\cos C$.

The sine rule: $= \frac{a}{\sin A} = \frac{b}{\sin B} = \frac{c}{\sin C}$.

Area of a triangle as $\frac{1}{2}ab\sin C$.

You will find a box like the one at the bottom of page x at the start of each chapter. They outline the components of the HL core syllabus (indicating syllabus section and sub-section numbers) that will be covered in that chapter.

The green box at right is an example (from Chapter 9) of a 'key' fact drawn out of the main text and highlighted. This makes them useful for quick reference and they also enable you to identify the core learning points within a section.

The process of 'breaking-up' the vector into its components, as we did in the example, is called **resolving** the vector into its components. Notice that the process of resolving a vector is not unique. That is, you can resolve a vector into several pairs of directions.

Beige boxes, like the one below (from Chapter 5), contain interesting information which will add to your wider knowledge but which does not fit within the main body of the text.

Radioactive carbon (carbon-14 or C-14), produced when nitrogen-14 is bombarded by cosmic rays in the atmosphere, drifts down to Earth and is absorbed from the air by plants. Animals eat the plants and take C-14 into their bodies. Humans in turn take C-14 into their bodies by eating both plants and animals. When a living organism dies, it stops absorbing C-14, and the C-14 that is already in the object begins to decay at a slow but steady rate, reverting to nitrogen-14. The half-life of C-14 is 5730 years. Half of the original amount of C-14 in the organic matter will have disintegrated after 5730 years; half of the remaining C-14 will have been lost after another 5730 years, and so forth. By measuring the ratio of C-14 to N-14, archaeologists are able to date organic materials. However, after about 50 000 years, the amount of C-14 remaining will be so small that the organic material cannot be dated reliably.

Margin hints (like the one at right) can be found alongside questions, exercises and worked examples, providing insight into how best to analyze and/or answer a question. They also identify common errors and pitfalls, and suggest approaches that IB examiners like to see.

• **Hint:** Notice here that P(B or C) is *not* the sum of P(B) and P(C) because B and C are not disjoint.

Blue boxes (like the one below) in the main body of the text have important facts, definitions, rules and theorems.

Inequality properties

For three real numbers a, b and c:

1 If $a > b$, and $b > c$, then $a > c$.
2 If $a > b$, and $c > 0$, then $ac > bc$.
3 If $a > b$, and $c < 0$, then $ac < bc$.
4 If $a > b$, then $a + c > b + c$.

Approach

This textbook is designed to be read by you – the student. It is important that you read this textbook *carefully*. Developing your ability to read and understand mathematical explanations will prove to be valuable in your long-term intellectual development, while also helping you to understand the mathematics necessary to be successful in your Mathematics

Higher Level course. You should always read a section thoroughly *before* attempting any of the exercises at the end of the section. In preparing this textbook, we have endeavoured to write clear and thorough explanations supported by suitable worked examples. Our primary goal was to present sound mathematics with sufficient rigour and detail at a level appropriate for a student of Higher Level Mathematics.

The positive feedback and constructive comments on the 1st edition, which we received from numerous teachers and students, was very much appreciated. Your comments assisted us greatly in being able to make many improvements and corrections in this 2nd edition. Thank you. We welcome your feedback with regard to any aspects of the textbook and the online e-book. We encourage teachers who adopt the textbook to register at our authors' website and make use of the materials available on it.

Email: info@wazir-garry-math.org
Website: www.wazir-garry-math.org

Ibrahim Wazir and *Tim Garry*

1 Fundamentals

Introduction

This first chapter reviews some of the *presumed knowledge* for the course – that is, mathematical knowledge that you *must* be familiar with before delving fully into the Mathematics Higher Level course (Chapter 2 and beyond). It is not necessary to work through each section in detail; however, it is very important that you read the entire chapter carefully in order to find out what is in it, and to become familiar with terminology, notations, and algebraic techniques used regularly in the course.

1.1 Sets, inequalities, absolute value and properties of real numbers

The language and notation of sets is often convenient for expressing results to a variety of problems in mathematics. We will review basic concepts, some important sets and useful notation. Some set concepts and notation will be applied again to probability problems in Chapter 12.

Sets of numbers and set notation

A set is a collection of objects or **elements**. Typically in mathematics and in this course the elements of a set will be numbers that can be defined by a list or a mathematical rule. Sets are usually denoted by capital letters. The elements, or members, of a set are listed between braces { }. For example, if the set A consists of the numbers 4, 5 and 6, we write $A = \{4, 5, 6\}$ where 4, 5 and 6 are the elements of set A. Symbolically, we write $4 \in A$, $5 \in A$ and $6 \in A$; read as '4 is an element of set A', or '4 is a member of set A' etc. To express that the number 3 is not an element of set A, we write $3 \notin A$.

The three dots seen in the set {1, 2, 3, ...} are an **ellipsis** and can have two different interpretations when used as a mathematical notation. When used in set notation, or raised up to show a repeated operation (e.g. $2 + 4 + 6 + \cdots + 48 + 50$), an ellipsis indicates that the numbers continue indefinitely in the same pattern. It should only be used in this way if the pattern is clear. Alternatively, an ellipsis can also be used to indicate that the decimal representation of an irrational number continues indefinitely and does *not* have a repeating pattern. For example, $\pi = 3.141\,592\,65\ldots$.

Sets whose number of elements can be counted are **finite**. If the number of elements in a set cannot be given a specific number then it is **infinite**. When we count objects, we start with the number 1, then 2, 3, etc; that is, the set {1, 2, 3, ...}. This is the set of positive integers (also known as the set of counting numbers) which is given the special symbol $\mathbb{Z}^+$. The number of elements in the set $A = \{4, 5, 6\}$ is three so it is a **finite set**. Even though we can define the set of positive integers in the form of a list, $\mathbb{Z}^+ = \{1, 2, 3, \ldots\}$, it is an **infinite set** because it is not possible to specify how many members are in the set.

Rather than defining a finite set by listing all the elements, we can specify the elements using a rule. For example, the set $B = \{x \mid 4 \leqslant x \leqslant 10, x \in \mathbb{Z}^+\}$ is read as 'B is the set of all x-values such that x is a positive integer between 4 and 10, inclusive'. This is an alternative way of writing $B = \{4, 5, 6, 7, 8, 9, 10\}$. Set notation using a mathematical rule is particularly useful when defining an infinite set, for which it is not possible to list all the elements, or a finite set with a large number of elements with a continuing pattern.

Example 1 – Defining sets

Using set notation and an appropriate mathematical rule, define each of following sets. Also indicate whether the set is finite or infinite.

a) The set of all integers between −8 and 6, not including −8 and 6 (i.e. exclusive).

b) The set of all integer multiples of $\frac{\pi}{4}$ greater than zero and less than or equal to 2π.

c) The set of positive odd integers.

Solution

a) $\{x \mid -8 < x < 6, x \in \mathbb{Z}\}$ finite set

b) $\left\{0 < n \cdot \frac{\pi}{4} \leqslant 2\pi, n \in \mathbb{Z}\right\}$ or $\left\{x \mid x = n \cdot \frac{\pi}{4}, 0 < n \leqslant 8, n \in \mathbb{Z}\right\}$ finite set

c) $\{2k - 1, k \in \mathbb{Z}^+\}$ or $\{2k + 1, k = 0, 1, 2, \ldots\}$ infinite set

Symbol	Set name	Set notation
$\mathbb{C}$	set of complex numbers	$\{a + bi \mid a, b \in \mathbb{R}\}$ where $i^2 = -1$
$\mathbb{R}$	set of real numbers	$\{x \in \mathbb{R}\}$
$\mathbb{R}^+$	set of positive real numbers	$\{x \mid x > 0, x \in \mathbb{R}\}$
$\mathbb{Q}$	set of rational numbers	$\left\{\frac{p}{q} \mid p, q \in \mathbb{Z}, q \neq 0\right\}$
$\mathbb{Q}^+$	set of positive rational numbers	$\{x \mid x > 0, x \in \mathbb{Q}\}$
$\mathbb{Z}$	set of integers	$\{\ldots, -3, -2, -1, 0, 1, 2, 3, \ldots\}$
$\mathbb{N}$	set of natural numbers (or whole numbers)	$\{0, 1, 2, 3, \ldots\}$
$\mathbb{Z}^+$	set of positive integers (or counting numbers)	$\{1, 2, 3, \ldots\}$

Table 1.1 Some important infinite sets are listed here, indicating their special symbols and how to express them with set notation, if possible.

There is some disagreement in the mathematics community about whether the number zero should be included in the natural numbers. So do not be confused if you see other textbooks indicate that the set of natural numbers, $\mathbb{N}$, does *not* include zero – and is defined as $\mathbb{N} = \{1, 2, 3, \ldots\}$. In IB mathematics the set $\mathbb{N}$ is defined to be the set of positive integers *and* zero, $\mathbb{N} = \{0, 1, 2, 3, \ldots\}$.

Now that we have the symbol $\mathbb{N}$ for the set of natural numbers $\{0, 1, 2, 3, \ldots\}$, we can also write the answer to Example 1, part c), the set of positive odd integers, as $\{2k + 1, k \in \mathbb{N}\}$.

A **real number** is any number that can be represented by a point on the real number line (Figure 1.1). Each point on the real number line corresponds to one unique real number, and conversely each real number corresponds to one unique point on the real number line. This kind of relationship is called a **one-to-one correspondence**. The number associated with a point on the real number line is called the **coordinate** of the point.

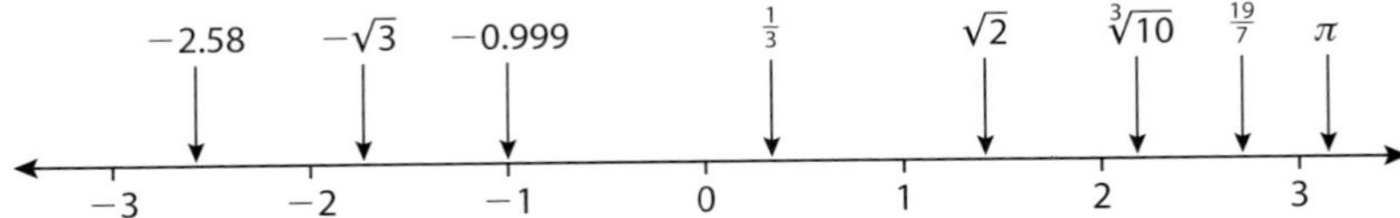

Figure 1.1 The real number line.

The real numbers are a subset of the **complex numbers.** It is likely that you will have limited or no experience with complex numbers or **imaginary numbers.** We will encounter complex and imaginary numbers in Chapter 3 and study them thoroughly in Chapter 10. However, it is worth saying a few introductory words about them at this point. The complex numbers, $\mathbb{C}$, involve a combination of real and imaginary numbers. Any complex number can be written in the form $a + bi$ where a and b are real numbers and $\boldsymbol{i}$ is the imaginary number defined such that $i^2 = -1$. For a complex number $a + bi$, if $b = 0$ then the complex number is a real number (e.g. $5 = 5 + 0i$, and $\sqrt{2} = \sqrt{2} + 0i$), and if $b \neq 0$ then the complex number is an imaginary number (e.g. $5 - 3i$, and $0 + 2i = 2i$). Hence, any complex number is either a real number or an imaginary number (see Figure 1.2).

We will see in Chapter 3 that some polynomial equations will have solutions that are imaginary numbers. For the quadratic equation $x^2 + 1 = 0$, we must find x such that $x^2 = -1$. A value for x will not be a real number. The symbol i was invented such that $i^2 = -1$. Hence, $x^2 + 1 = 0$ has two imaginary solutions, $x = i$ and $x = -i$. We define the imaginary number i as $i^2 = -1$ but we are allowed to write $i = \sqrt{-1}$. We will study complex numbers in greater depth in Chapter 10.

• **Hint:** The answer for Example 1, part a), was the set $\{x \mid -8 < x < 6, x \in \mathbb{Z}\}$. The definition for the elements of the set specified that all elements must be integers. If the definition of a set does not specify to which set the elements belong, it is assumed to be the set of real numbers $\mathbb{R}$. For example, the set $\{x \mid -8 < x < 6\}$ would contain all of the real numbers between -8 and 6 exclusive. This is an infinite set, whereas the set $\{x \mid -8 < x < 6, x \in \mathbb{Z}\}$ is finite.

Similarly, any real number is either rational or irrational, with the rational numbers and irrational numbers being subsets of the real numbers (Figure 1.2). We construct the **rational numbers** $\mathbb{Q}$ by taking ratios of integers. Thus, a real number is rational if it can be written as the ratio $\frac{p}{q}$ of any two integers, where $q \neq 0$. The decimal representation of a rational number either repeats or terminates. For example, $\frac{5}{7} = 0.714\,285\,714\,285\ldots = 0.\overline{714\,285}$ (the block of six digits repeats) or $\frac{3}{8} = 0.375$ (the decimal 'terminates' at 5, or alternatively has a repeating zero after the 5).

A real number that cannot be written as the ratio of two integers, such as π and $\sqrt{2}$, is called **irrational**. Irrational numbers have infinite non-repeating decimal representations. For example, $\sqrt{2} \approx 1.414\,213\,5623\ldots$ and $\pi \approx 3.141\,592\,653\,59\ldots$. There is no special symbol for the set of irrational numbers.

Figure 1.2 The diagram depicts the relationships between the different subsets of the complex numbers. The real numbers combined with the imaginary numbers make up the entire set of complex numbers. The rational numbers combined with the irrational numbers make up the entire set of real numbers.

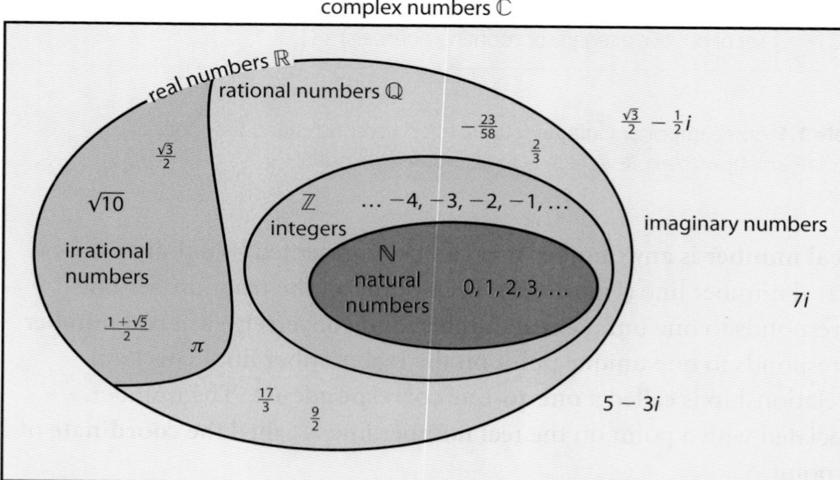

The earliest known use of irrational numbers was in India between 800–500 BCE. The first mathematical proof that a number could not be expressed as the ratio of two integers (i.e. irrational) is usually attributed to the Pythagoreans. The revelation that not all numbers were rational was a great shock to Pythagoras and his followers, given that their mathematics and theories about the physical world were based completely on positive integers and their ratios. Euclid (ca. 325–265 BCE) wrote a proof of the irrationality of $\sqrt{2}$ in his *Elements*, one of the most famous books in mathematics. Euclid's proof is considered to be an elegant proof because it is both simple and powerful. Euclid used a method called **proof by contradiction**, or in Latin, *reductio ad absurdum*. Here is a condensed version of his proof that $\sqrt{2}$ cannot be written as the ratio of two integers. This is equivalent to saying that there is no rational number $\frac{p}{q}$ whose square is 2 where $p, q \in \mathbb{Z}$. The proof begins by assuming that the statement to be proved is false – that is, we assume that there ***is*** a rational number $\frac{p}{q}$ completely simplified (i.e. p and q have no common factor) whose square is 2. Then $\left(\frac{p}{q}\right)^2 = 2$, and it follows that $p^2 = 2q^2$. Hence, p^2 has a factor of 2 which means that p must be an even number. Since p is even, then let's replace p with $2k$, where k is an integer, giving $4k^2 = 2q^2$ leading to $2k^2 = q^2$. Therefore, q^2 has a factor of 2 and so q is also even. This means that p and q both have a factor of 2. But this contradicts the assumption that p and q have no common factors. Therefore, the initial assumption that there is a rational number $\frac{p}{q}$ whose square is 2 leads to a contradiction. It logically follows then that this assumption must be false, i.e. there is no rational number whose square is 2.

Example 2 – Expressing a repeating decimal as a rational number

Express each as a rational number completely simplified.

a) $1.416\,6666\ldots = 1.41\overline{6}$

b) $38.245\,3453\ldots = 38.2\overline{453}$

Solution

a) Let $N = 1.416\,6666\ldots$

Then $1000N = 1416.666\,66\ldots$ and $100N = 141.666\,66\ldots$

Now subtract $100N$ from $1000N$:

$$\begin{aligned} 1000N &= 1416.666\,66\ldots \\ -100N &= -141.666\,66\ldots \\ \hline 900N &= 1275 \end{aligned}$$

This gives $N = \frac{1275}{900} = \frac{25 \times 51}{25 \times 36} = \frac{51}{36} = \frac{3 \times 17}{3 \times 12} = \frac{17}{12}$

Therefore, $1.41\overline{6} = \frac{17}{12}$.

b) Let $N = 38.245\,3453\ldots$

Then $10\,000N = 382\,453.453\,453\ldots$ and $10N = 382.453\,453\ldots$

Now subtract $10N$ from $10\,000N$:

$$\begin{aligned} 10\,000N &= 382\,453.453\,453\ldots \\ -10N &= -382.453\,453\ldots \\ \hline 9990N &= 382\,071 \end{aligned}$$

This gives $N = \frac{382\,071}{9990} = \frac{3 \times 127\,357}{3 \times 3330} = \frac{127\,357}{3330}$

Therefore, $38.2\overline{453} = \frac{127\,357}{3300}$.

Note: 382 071 is divisible by 3 because the sum of its digits (21) is divisible by 3. The fraction $\frac{127\,357}{3330}$ cannot be simplified because 127 357 and 3330 share no common factors; $3330 = 2 \times 3 \times 3 \times 5 \times 37$ (prime factorization) and 2, 3, 5 and 37 are not factors of 127 357

Another approach to expressing a repeating decimal as a rational number appears in Chapter 4.

Set relations, operations and diagrams

If every element of a set C is also an element of a set D, then C is a **subset** of set D, and is written symbolically as $C \subseteq D$. If two sets are equal (i.e. they have identical elements), they satisfy the definition of a subset and each would be a subset of the other. For example, if $C = \{2, 4, 6\}$ and $D = \{2, 4, 6\}$, then $C = D$, $C \subseteq D$ and $D \subseteq C$. What is more common is that a subset is a set that is contained in a larger set and does not contain at least one element of the larger set. Such a subset is called a **proper subset** and is denoted with the symbol $\subset$. For example, if $D = \{2, 4, 6\}$ and $E = \{2, 4\}$, then E is a proper subset of D and is written $E \subset D$, but $C \not\subset D$. Other than the set of complex numbers itself, all of the sets listed in Table 1.1 are proper subsets of the complex numbers.

The set of all elements under consideration for a particular situation or problem is called the **universal set**, usually denoted by the symbol U. The

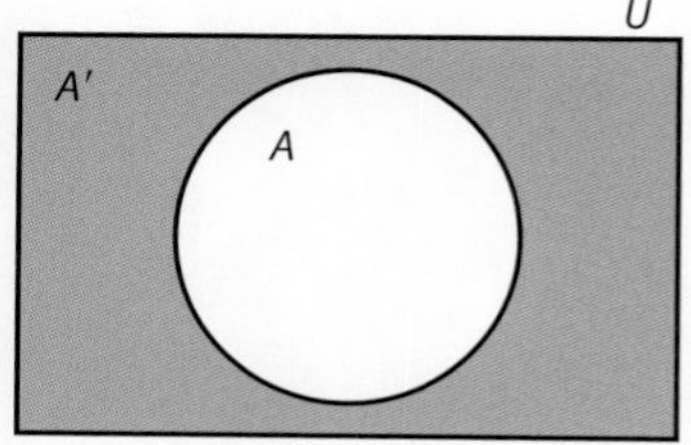

Figure 1.3 Venn diagram for the universal set U, set A, and the complement of A, A' (shaded region).

complement of a given set A is the set of all elements in the universal set that are not elements of set A, and is denoted by the symbol A'. **Venn diagrams** are used to pictorially represent the relationship of sets within a universal set. The universal set, U, is represented by a rectangle and any subset of U is represented by the interior of a circle within the rectangle (see Figure 1.3).

If for a certain problem the universal set is the complex numbers $\mathbb{C}$, then the complement of the real numbers is the imaginary numbers. For problems in secondary school mathematics, and in this course, the universal set will often be a subset of the complex numbers – commonly the real numbers $\mathbb{R}$. If the universal set is the real numbers, then the set of irrational numbers is the complement of the rational numbers $\mathbb{Q}$. See Figure 1.2.

The **intersection** of sets A and B, denoted by $A \cap B$ and read 'A intersection B', is the set of all elements that are in both set A *and* set B. The **union** of two sets A and B, denoted by $A \cup B$ and read 'A union B', is the set of all elements that are in set A *or* in set B (or in both). The set that contains no elements is called the **empty set** (or null set) and is denoted by $\varnothing$. Sets whose intersection is the empty set, i.e. they have no elements in common, are **disjoint sets.**

Although the set $\{2, 3\}$ is equal to the set $\{3, 2\}$, the ordered pairs $(2, 3)$ and $(3, 2)$ are not the same. Hence, for the Cartesian product of two sets A and B, in general, $A \times B \neq B \times A$.

The **Cartesian product** of two sets A and B is the set of all **ordered pairs** $\{(a, b)\}$, where $a \in A$ and $b \in B$. It is written as $A \times B = \{(a, b) \mid a \in A, b \in B\}$. For example, if $X = \{1, 2\}$ and $Y = \{3, 4, 5\}$,

then $X \times Y = \{(1,3), (1,4), (1,5), (2,3), (2,4), (2,5)\}$

and $Y \times X = \{(3,1), (3,2), (4,1), (4,2), (5,1), (5,2)\}$

• **Hint:** The symbol for the union of two sets, ∪, can be remembered by connecting it with the first letter in the word 'union'.

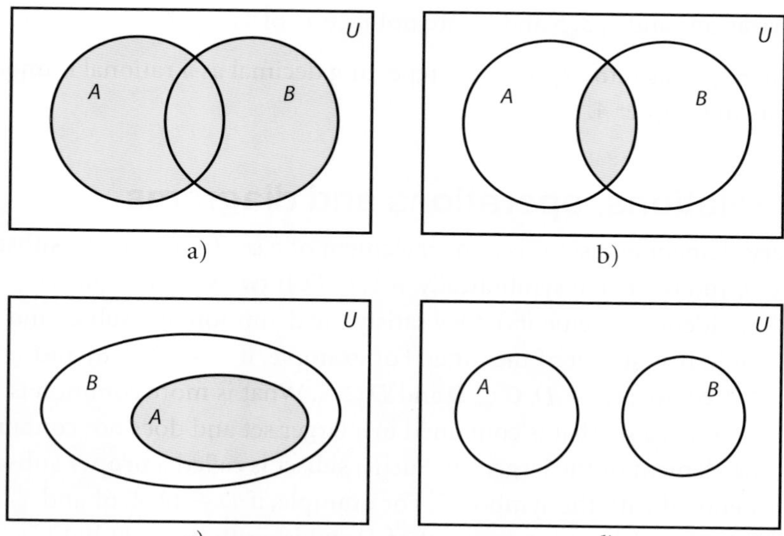

Figure 1.4 a) Union of sets A and B, $A \cup B$
b) Intersection of sets A and B, $A \cap B$
c) Proper subset, $A \subset B$ d) Two disjoint sets, A and B

Venn diagrams are named after the British mathematician, philosopher and writer John Venn (1834–1923). Although he was not the first to use diagrams as an aid to problems in set theory and logic, he was the first to formalize their usage and popularized them in his writings such as in his first book *Symbolic Logic* published in 1881.

Set relations and operations

Subset: $A \subseteq B$ means that A is a subset of B

Proper subset: $A \subset B$ means that $A \subseteq B$ but $A \neq B$

Intersection: $A \cap B = \{x \mid x \in A \text{ and } x \in B\}$

Union: $A \cup B = \{x \mid x \in A \text{ or } x \in B \text{ or both}\}$

Complement: $A' = \{x \mid x \notin A\}$

Empty set: $\varnothing$, the set with no elements

Cartesian product: $A \times B = \{(a, b) \mid a \in A, b \in B\}$

Example 3 – Set operations

Consider that the universal set U is defined to be $U = \{1, 2, 3, 4, 5, 6, 7, 8, 9, 10, 11, 12, 13\}$, and $A = \{2, 5, 8, 11\}$, $B = \{2, 4, 6, 8, 10, 12\}$, $C = \{2, 3, 5, 7, 11, 13\}$.

a) Find the following:

(i) $A \cap B$ (ii) $A \cup B$ (iii) A'

(iv) $A \cap C$ (v) $A \cap B \cap C$ (vi) $(B \cup C)'$

(vii) $A \cap (B \cup C)'$ (viii) $A \cup B \cup C$

b) Draw a Venn diagram to illustrate the relationship between the sets A, B and C.

Solution

a) (i) $A \cap B = \{2, 8\}$ (ii) $A \cup B = \{2, 4, 5, 6, 8, 10, 11, 12\}$

(iii) $A' = \{1, 3, 4, 6, 7, 9, 10, 12, 13\}$ (iv) $A \cap C = \{2, 5, 11\}$

(v) $A \cap B \cap C = \{2\}$ (vi) $(B \cup C)' = \{1, 9\}$

(vii) $A \cap (B \cup C)' = \varnothing$

(viii) $A \cup B \cup C = \{2, 3, 4, 5, 6, 7, 8, 10, 11, 12, 13\}$

b)

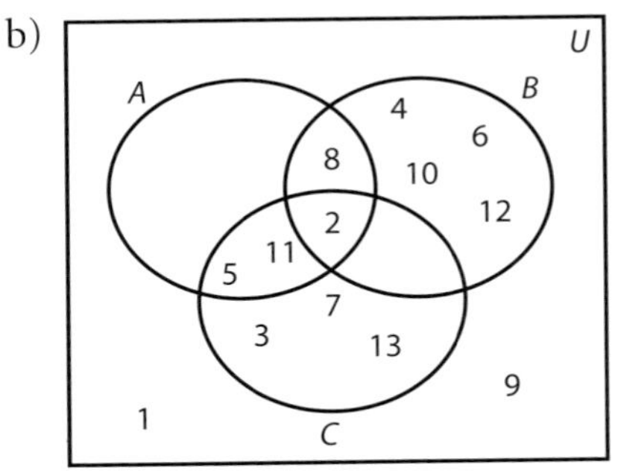

• **Hint:** When we list the elements of a set we never repeat an element. For example, for $A \cup B$ in Example 3 the numbers 2 and 8 are in both A and B but they are each written once when listing the elements in $A \cup B$.

In Example 3, instead of defining sets U, A, B and C using lists, we could have defined each of the sets using a rule. For example,
$U = \{x \mid 1 \leqslant x \leqslant 13, x \in \mathbb{Z}\}$,
$A = \{x \mid x = 3n - 1, n = 1, 2, 3, 4\}$,
$B = \{x \mid x = 2n, 1 \leqslant n \leqslant 6 \text{ and } n \in \mathbb{Z}\}$,
and
$C = \{x \mid x \leqslant 13, x \text{ is a prime number}\}$.

Inequalities (order relations)

An inequality is a statement involving one of four symbols that indicates an **order relation** between two numbers or algebraic expressions on either side of the symbol. The symbols are

$<$ (less than)

$>$ (greater than)

$\leqslant$ (less than or equal to)

$\geqslant$ (greater than or equal to).

The relation $a > b$ is read 'a is greater than b' and in the geometric representation of numbers on the real number line it means that a lies to the right of b. Since -2 lies to the right of -3 on the number line then $-2 > -3$. The inequality $a > b$ is equivalent to the inequality $b < a$ (b lies to the left of a on the number line), and similarly $-2 > -3$ is equivalent to $-3 < -2$.

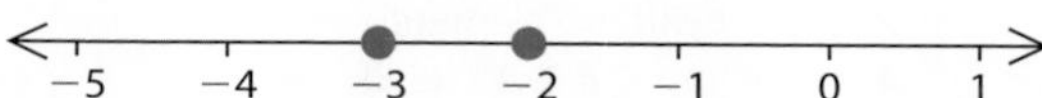

Working with inequalities is very important for many of the topics in this course. There are four basic properties for inequalities.

Inequality properties
For three real numbers a, b and c:
1 If $a > b$, and $b > c$, then $a > c$.
2 If $a > b$, and $c > 0$, then $ac > bc$.
3 If $a > b$, and $c < 0$, then $ac < bc$.
4 If $a > b$, then $a + c > b + c$.

The first property is sometimes referred to as the transitive property. The second property for inequalities expresses the fact that an inequality that is multiplied on both sides by a positive number does not change the inequality symbol. For example, given that $x > 6$ then multiplying both sides by $\frac{1}{2}$ gives $\frac{x}{2} > 3$.
The third property tells us that if we multiply both sides of an inequality by a negative number then the inequality symbol is reversed. For example, if $-3x \leqslant 12$ then multiplying both sides by $-\frac{1}{3}$ gives $x \geqslant -4$. The fourth property means that the same quantity being added to both sides will produce an equivalent inequality.

When you solve an inequality the result will be a range of possible values of the variable. The inequalities in the next example are solved by applying the properties for inequalities (stated above) and basic rules for solving linear equations with which you are familiar.

Example 4 – Solving inequalities

Solve each inequality.

a) $6x + 1 > x - 5$ b) $9 - 4x \leqslant 2x - 3$

c) $3(1 - 2x) < 15$ d) $-3 \leqslant 2x - 1 < 9$ e) $-2 \leqslant 4 - 3x < 13$

Solution

a) $6x + 1 > x - 5 \Rightarrow 6x > x - 6 \Rightarrow 5x > -6 \Rightarrow x > -\frac{6}{5}$

b) $9 - 4x \leqslant 2x - 3 \Rightarrow 12 - 4x \leqslant 2x \Rightarrow 12 \leqslant 6x \Rightarrow 2 \leqslant x$ or $x \geqslant 2$
Alternatively, $9 - 4x \leqslant 2x - 3 \Rightarrow -4x \leqslant 2x - 12 \Rightarrow -6x \leqslant -12 \Rightarrow x \geqslant 2$

c) $3(1 - 2x) < 15 \Rightarrow 1 - 2x < 5 \Rightarrow -2x < 4 \Rightarrow x > -2$

d) The inequality $-3 \leqslant 2x - 1 < 9$ is a 'double inequality' containing two separate inequalities $-3 \leqslant 2x - 1$ and $2x - 1 < 9$; we can solve each separately or simultaneously as shown here.
$-3 \leqslant 2x - 1 < 9 \Rightarrow -2 \leqslant 2x < 10 \Rightarrow -1 \leqslant x < 5$

This solution set is read 'x is any real number that is greater than or equal to -1 **and** less than 5'.

e) $-2 < 4 - 3x < 13 \Rightarrow -6 < -3x < 9 \Rightarrow 2 > x > -3 \Rightarrow -3 < x < 2$

• **Hint:** It is improper to write the solution to Example 4 e) as $2 > x > -3$. A double inequality should be written with the lesser quantity on the left and greater on the right, i.e. $-3 < x < 2$ for Example 4 e). A double inequality is the intersection of two sets. For example, the expression $-3 < x < 2$ represents the **intersection** of $x > -3$ **and** $x < 2$; i.e. the numbers greater than -3 **and** less than 2. The **union** of two sets **cannot** be written as a double inequality. Using inequalities to represent the numbers less than 4 **or** greater than 7 must be written as two separate inequalities, $x < 4$ **or** $x > 7$.

In Chapter 3, we will be solving further inequalities involving linear, quadratic and rational (fractional) expressions.

Intervals on the real number line

Except when studying complex numbers in Chapter 10 and solving certain polynomial equations in Chapter 3, problems that we encounter in this course will be in the context of the real numbers. For example, the solution set for the inequality in Example 4 c) is the set of all real numbers greater than negative three. Such a set can be represented geometrically by a part, or an **interval**, of the real number line and corresponds to a line segment or a ray. It can be written symbolically by an **inequality** or by **interval notation**. For example, the set of all real numbers x between 2 and 5 inclusive, can be expressed by the inequality $2 \leqslant x \leqslant 5$ or by the interval notation $x \in [2, 5]$. This is an example of a **closed interval** (i.e. both endpoints are included in the set) and corresponds to the line segment with endpoints of $x = 2$ and $x = 5$.

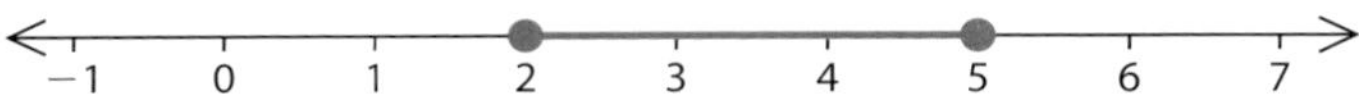

• **Hint:** Unless indicated otherwise, if interval notation is used, we assume that it indicates an infinite set containing any real number within the indicated range. For example, the expression $x \in [-4, 2]$ is read 'x is any real number between -4 and 2 inclusive.'

An example of an **open interval** is $-3 < x < 1$, also written as $x \in]-3, 1[$, where both endpoints are *not* included in the set. This set corresponds to a line segment with 'open dots' on the endpoints indicating they are excluded.

If an interval, such as $-4 \leqslant x < 2$, also written as $x \in [-4, 2[$, includes one endpoint but not the other, it is referred to as a **half-open** interval.

• **Hint:** The symbols ∞ (positive infinity) and $-\infty$ (negative infinity) do not represent real numbers. They are simply symbols used to indicate that an interval extends indefinitely in the positive or negative direction.

The three examples of intervals on the real number line given above are all considered **bounded** intervals in that they are line segments with two endpoints (regardless whether included or excluded). The set of all real numbers greater than 2 is an open interval because the one endpoint is excluded and can be expressed by the inequality $x > 2$, also written as $x \in [2, \infty[$. This is also an example of an **unbounded** interval and corresponds to a part of the real number line that is a ray.

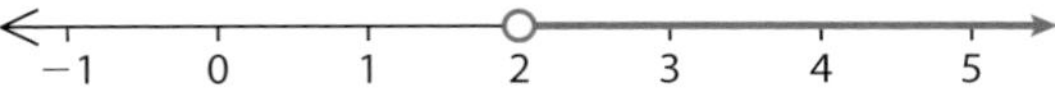

Table 1.2 The nine possible types of intervals – both bounded and unbounded. For all of the examples given, we assume that $a < b$.

Interval notation	Inequality	Interval type	Graph
$x \in [a, b]$	$a \leqslant x \leqslant b$	closed bounded	a b
$x \in]a, b[$	$a < x < b$	open bounded	a b
$x \in [a, b[$	$a \leqslant x < b$	half-open bounded	a b
$x \in]a, b]$	$a < x \leqslant b$	half-open bounded	a b
$x \in [a, \infty[$	$x \geqslant a$	half-open unbounded	a
$x \in]a, \infty[$	$x > a$	open unbounded	a
$x \in]-\infty, b]$	$x \leqslant b$	half-open unbounded	b
$x \in]-\infty, b[$	$x < b$	open unbounded	b
$x \in]-\infty, \infty[$	real number line		

Absolute value (or modulus)

The **absolute value** of a number a, denoted by $|a|$, is the distance from a to 0 on the real number line. Since a distance must be positive or zero, then the absolute value of a number is never negative. Note that if a is a negative number then $-a$ will be positive.

Definition of absolute value

If a is a real number, the **absolute value** of a is

$$|a| = \begin{cases} a & \text{if } a \geqslant 0 \\ -a & \text{if } a < 0 \end{cases}$$

Here are four useful properties of absolute value.

Given that a and b are real numbers, then:

1. $|a| \geqslant 0$ **2.** $|-a| = |a|$ **3.** $|ab| = |a||b|$ **4.** $\left|\frac{a}{b}\right| = \frac{|a|}{|b|}, b \neq 0$

Absolute value is used to define the distance between two numbers on the real number line. The distance between -4 and 0 on the number line is clearly seen to be 4 units in the figure below.

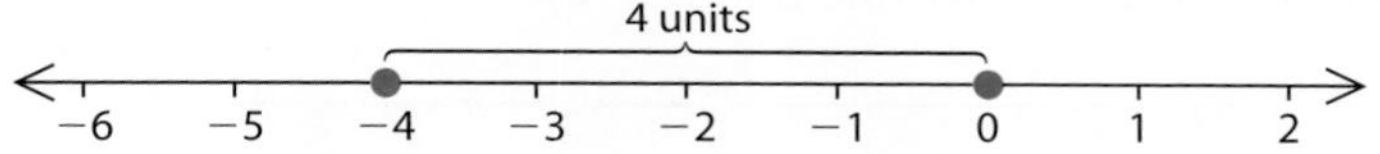

Also the distance between 4 and 0 is 4 units. Note that $|-4| = 4$ gives the distance between -4 and 0, and $|-4| = 4$ gives the distance between 4 and 0. These observations lead to the geometrical interpretation of absolute value as a distance: $|a|$ is the distance between a and 0 on the real number line.

Now consider the distance between two non-zero numbers on the number line. The distance between -5 and 2 is 7 units as shown below.

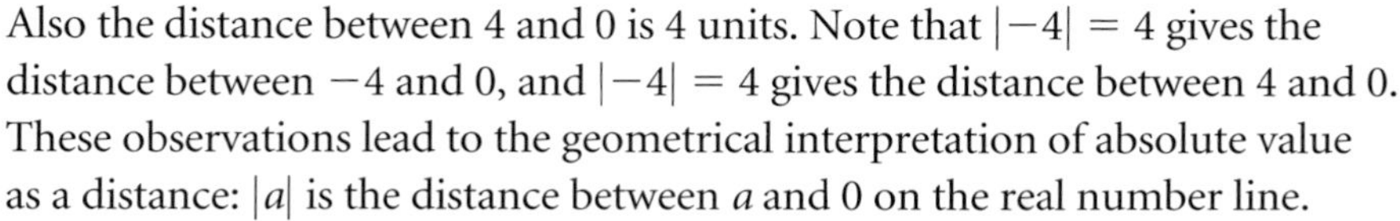

Note that $|-5 - 2| = 7$ gives the distance between -5 and 2. It is also true that $|2 - (-5)| = 7$ is the distance between -5 and 2. The examples suggest that the distance between two numbers on the number line is always given by the absolute value of their difference.

Distance between two points on the real number line

Given that a and b are real numbers, the distance between the points with coordinates a and b on the real number line is $|b-a|$, which is equivalent to $|a-b|$.

The geometric statement that 'the distance between k and 8 is $\frac{5}{2}$' can be expressed algebraically as $|k - 8| = \frac{5}{2}$, or as $|8 - k| = \frac{5}{2}$. The geometric interpretation of absolute value as a distance leads to a helpful method for solving equations involving absolute value expressions.

Example 5 – Absolute value as a distance

How many real numbers have an absolute value equal to 3? Since absolute value can be interpreted as the distance from a number on the real number line to 0 (the origin), then this question is equivalent to asking which real numbers are a distance of 3 units from 0 on the number line? See the figure below.

There are just two numbers whose distance to 0 is 3 units, namely, 3 and -3. In other words the equation $|x| = 3$ has two solutions: $x = 3$ or $x = -3$. This reasoning leads to a general method for solving a linear equation containing the absolute value of a variable expression.

If $|x| = c$, where x is an unknown and c is a positive real number, then $x = c$ or $x = -c$.

Example 6 – Solving absolute value linear equations

Solve for x in each equation.

a) $|x - 5| = 8$ b) $4|x + \frac{3}{2}| = 9$ c) $\left|\frac{3x - 4}{2}\right| - 7 = 9$

Solution

a) $|x - 5| = 8$

$x - 5 = 8$ or $x - 5 = -8$ Applying the property that if $|x| = c$, then $x = c$ or $x = -c$.

$x = 13$ or $x = -3$

b) $4|x + \frac{3}{2}| = 9 \Rightarrow |x + \frac{3}{2}| = \frac{9}{4}$

$x + \frac{3}{2} = \frac{9}{4}$ or $x + \frac{3}{2} = -\frac{9}{4}$

$x = \frac{3}{4}$ or $x = -\frac{15}{4}$

c) $\left|\frac{3x - 4}{2}\right| - 7 = 9 \Rightarrow \left|\frac{3x - 4}{2}\right| = 16$

$\frac{3x - 4}{2} = 16$ or $\frac{3x - 4}{2} = -16$

$3x - 4 = 32$ or $3x - 4 = -32$

$x = 12$ or $x = -\frac{28}{3}$

We will encounter more sophisticated equations involving absolute value in Chapter 3.

Properties of real numbers

There are four arithmetic operations with real numbers: addition, multiplication, subtraction and division. Since subtraction can be written as addition ($a - b = a + (-b)$), and division can be written as multiplication $\left(\frac{a}{b} = a\left(\frac{1}{b}\right), b \neq 0\right)$, then the properties of the real numbers are defined in terms of addition and multiplication only. In these definitions, $-a$ is the **additive inverse** (or opposite) of a, and $\frac{1}{a}$ is the **multiplicative inverse** (or reciprocal) of a.

Table 1.3 Properties of real numbers.

Property	Rule	Example
commutative property of addition:	$a + b = b + a$	$2x^3 + y = y + 2x^3$
commutative property of multiplication:	$ab = ba$	$(x - 2)3x^2 = 3x^2(x - 2)$
associative property of addition:	$(a + b) + c = a + (b + c)$	$(1 + x) - 5x = 1 + (x - 5x)$
associative property of multiplication:	$(ab)c = a(bc)$	$(3x \cdot 5y)\left(\frac{1}{y}\right) = (3x)\left(5y \cdot \frac{1}{y}\right)$
distributive property:	$a(b + c) = ab + ac$	$x^2(x - 2) = x^2 \cdot x + x^2(-2)$
additive identity property:	$a + 0 = a$	$4y + 0 = 4y$
multiplicative identity property:	$1 \cdot a = a$	$\frac{2}{3} = 1 \cdot \frac{2}{3} = \frac{4}{4} \cdot \frac{2}{3} = \frac{8}{12}$
additive inverse property:	$a + (-a) = 0$	$6y^2 + (-6y^2) = 0$
multiplicative inverse property:	$a \cdot \frac{1}{a} = 1, a \neq 0$	$(y - 3)\left(\frac{1}{y - 3}\right) = 1$

Note: These properties can be applied in either direction as shown in the 'rules' above.

Exercise 1.1

In questions 1–6, use the symbol $\subset$ (proper subset) to write a correct statement involving the two sets.

1 $\mathbb{Z}$ and $\mathbb{Q}$ **2** $\mathbb{N}$ and $\mathbb{Q}$ **3** $\mathbb{C}$ and $\mathbb{R}$

4 $\mathbb{Z}$ and $\mathbb{N}$ **5** $\mathbb{Z}$ and $\mathbb{Z}^+$ **6** $\mathbb{N}$ and $\mathbb{R}$

In questions 7–9, express each repeating decimal as a completely simplified fraction.

7 2.151515... **8** 11.913333... **9** $8.\overline{714285}$

In questions 10–15, state the indicated set given that $A = \{1, 2, 3, 4, 5, 6, 7, 8\}$, $B = \{1, 3, 5, 7, 9\}$ and $C = \{2, 4, 6\}$.

10 $A \cap B$ **11** $A \cup B$ **12** $B \cap C$

13 $A \cup C$ **14** $A \cap C$ **15** $A \cup B \cup C$

16 Consider that the universal set U is defined to be $U = \{1, 2, 3, 4, 5, 6, 7, 8, 9, 10\}$, and $A = \{2n, n \in \mathbb{Z}^+\}$, $B = \{2n - 1, n \in \mathbb{Z}^+\}$, $C = \{3n, n \in \mathbb{Z}^+\}$, $D = \{1, 6, 7, 9\}$.

a) Describe in words the elements of sets A, B and C.

b) Find the following:
(i) $A \cap B$ (ii) $A \cup B$ (iii) A' (iv) B' (v) $A \cap D$
(vi) $B \cap C$ (vii) $B \cap C \cap D$ (viii) $(C \cup D)'$ (ix) $A \cap (C \cap D)'$

c) Draw a Venn diagram to illustrate the relationship between the sets A, C and D.

17 P and Q are subsets of the universal set U with the three sets defined as follows.
$U = \{x \mid -2 \leqslant x \leqslant 4\}$, $P = \{x \mid -2 \leqslant x \leqslant 1\}$, $Q = \{x \mid 0 < x < 3\}$
(Remember: If not specified, the elements in an interval will be an infinite set of real numbers.)
Being careful with the inclusion or exclusion of endpoints, list the following sets.
a) $P \cap Q$ b) $P \cup Q$ c) P' d) Q' e) $(P \cup Q)'$ (f) $(P \cap Q)'$

In questions 18–23, solve the inequality.

18 $\frac{x}{5} > -2$ **19** $3 + 4x \leqslant -9$ **20** $7 - 3x < -3$

21 $6(2 - x) < 2x + 15$ **22** $9 \leqslant 8x - 3 < 11$ **23** $-4 \leqslant 1 - 5x \leqslant 16$

In questions 24–31, determine whether each statement is true for all real numbers x. If the statement is false, then indicate one counterexample, i.e. a value of x for which the statement is false.

24 $2x \geqslant x$ **25** $x^3 + 1 > x^3$ **26** $x^3 + x > x^3$

27 $x^2 \geqslant x$ **28** $x^2 \geqslant 0$ **29** $\sqrt{x} \geqslant 0$

30 $-x \leqslant 0$ **31** $\frac{1}{x} \leqslant x$

In questions 32–37, plot the two real numbers on the real number line, and then find the exact distance between their coordinates.

32 -7 and $\frac{15}{2}$ **33** -2 and -11 **34** 27.4 and 19.2

35 π and 3 **36** -3π and $\frac{2\pi}{3}$ **37** $\frac{61}{7}$ and $-\frac{23}{11}$

In questions 38–43, write an inequality to represent the given interval and state whether the interval is closed, open or half-open. Also state whether the interval is bounded or unbounded.

38 $[-5, 3]$ **39** $]-10, -2]$ **40** $[1, \infty[$

41 $]-\infty, 4]$ **42** $[0, 2\pi[$ **43** $[a, b]$

In questions 44–49, use interval notation to represent the subset of real numbers that is indicated by the inequality.

44 $x > -3$

45 $-4 < x < 6$

46 $x \leqslant 10$

47 $0 \leqslant x < 12$

48 $x < \pi$

49 $-3 \leqslant x \leqslant 3$

In questions 50–53, use both inequality and interval notation to represent the given subset of real numbers.

50 x is at least 6.

51 x is greater than or equal to 4 and less than 10.

52 x is negative.

53 x is any positive number less than 25.

In questions 54–57, express the inequality, or inequalities, using absolute value.

54 $-6 < x < 6$

55 $x \leqslant -4$ or $x \geqslant 4$

56 $-\pi \leqslant x \leqslant \pi$

57 $x < -1$ or $x > 1$

In questions 58–63, evaluate each absolute value expression.

58 $|-13|$

59 $|7 - 11|$

60 $-5|-5|$

61 $|-3|-|-8|$

62 $|\sqrt{3} - 3|$

63 $\dfrac{-1}{|-1|}$

In questions 64–71, find all values of x that make the equation true.

64 $|x| = 5$

65 $|x - 3| = 4$

66 $|6 - x| = 10$

67 $|x + 5| = -2$

68 $|3x + 5| = 1$

69 $\frac{1}{2}|x - \frac{2}{3}| = 5$

70 $\left|\dfrac{6 - 2x}{3}\right| + \frac{2}{5} = 8$

71 $2\left|\dfrac{x + 2}{2}\right| = 2$

72 For each of the following statements, find at least one counterexample that confirms the statement is false.

a) $|x + y| = |x| + |y|$

b) $|x - y| = |x| - |y|$

73 Using properties of inequalities, prove each of the statements.

a) If $x < y$ and $x > 0$, then $\frac{1}{y} < \frac{1}{x}$.

b) If $x < 0 < y$, then $\frac{1}{y} > \frac{1}{x}$.

1.2 Roots and radicals (surds)

Roots

If a number can be expressed as the product of two equal factors, then that factor is called the **square root** of the number. For example, 7 is the square root of 49 because $7 \times 7 = 49$. Now 49 is also equal to -7×-7, so -7 is also a square root of 49. Every positive real number will have two real number square roots, one positive and one negative. However, there are many instances where we only want the positive square root. The symbol $\sqrt{\ }$ (called the **radical sign**) indicates only the positive square root, referred to as the **principal square root**. Because $4^2 = 16$ and $(-4)^2 = 16$ the square roots of 16 are 4 and -4; but the principal square root of 16 is only positive four, that is $\sqrt{16} = 4$. The negative square root of 16 is written as $-\sqrt{16} = -4$, and when both square roots are wanted we write $\pm\sqrt{16}$. In the real numbers, every positive number has two square roots (one positive and the other negative) but only one principal square root (positive) denoted with the radical sign.

When a number can be expressed as the product of three equal factors, then that factor is called the **cube root** of the number. For example, -4 is the cube root of -64 because $(-4)^3 = -64$. With the radical sign this is written as $\sqrt[3]{-64} = -4$. In the real numbers, every number (positive or negative) has just one cube root. In the notation $\sqrt[n]{a}$, a is called the **radicand** and n is a positive integer called the **index**. The index indicates which root (square root or cube root or 4th root, etc.) is to be extracted. If no index is written it is assumed to be a 2, thereby indicating a square root.

In general, if a real number a can be expressed as the factor b multiplied n times, i.e. $b^n = a$, then that factor b is called the ***n*th root** of a. In the set of real numbers, if n is an even number (e.g. square root, 4th root, 6th root, etc.) then a has two nth roots (positive and negative) with the positive root being the **principal *n*th root**. Because $2^4 = 16$ and $(-2)^4 = 16$, then both 2 and -2 are 4th roots of 16. However, the principal 4th root of 16 is 2, written $\sqrt[4]{16} = 2$. If the index n is an odd number (e.g. cube root, 5th root, etc.) then the sign ($+$ or $-$) of the nth root of a will be the same as the sign of a. For example, the 5th root of 32 is 2, and the 5th root of -32 is -2. With the radical sign these results are written as $\sqrt[5]{32} = 2$ and $\sqrt[5]{-32} = -2$.

Our discussion here on roots and radicals is limited to the real numbers. We will learn in Chapter 10 that if we broaden our consideration to the complex numbers, then any number will have exactly n different nth roots. For example, the number 16 has four 4th roots: 2, -2, $2i$ and $-2i$. Your GDC may have the imaginary number i. Try taking the 4th power of $2i$ and $-2i$ (could also be entered as $2\sqrt{-1}$ and $-2\sqrt{-1}$) on your GDC (see calculator screen images below). You may need to change the mode of your calculator from real to complex.

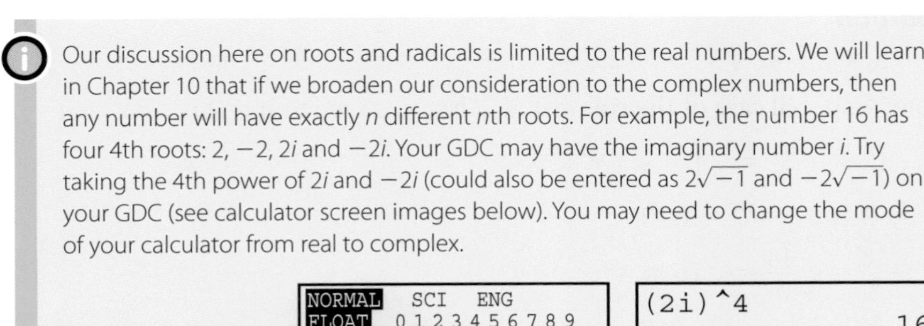

• **Hint:** There are many words that have more than one meaning in mathematics. The correct interpretation of a word will depend on the situation (context) in which it is being applied. The word *root* is not only used for square root, cube root, nth root, etc. but can also mean the solution of an equation. For example, $x = 3$ and $x = -1$ are roots of the equation $x^2 - 2x - 3 = 0$ (see Section 3.5).

3, x, A, 4 — y, 4, B, 8

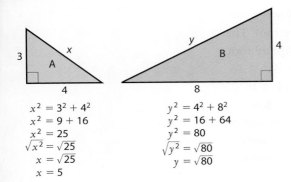

$x^2 = 3^2 + 4^2$
$x^2 = 9 + 16$
$x^2 = 25$
$\sqrt{x^2} = \sqrt{25}$
$x = \sqrt{25}$
$x = 5$

$y^2 = 4^2 + 8^2$
$y^2 = 16 + 64$
$y^2 = 80$
$\sqrt{y^2} = \sqrt{80}$
$y = \sqrt{80}$

• **Hint:** The solution for the hypotenuse of triangle A involves the equation $x^2 = 25$. Because x represents a length that must be positive, we want only the positive square root when taking the square root of both sides of the equation – i.e. $\sqrt{25}$. However, if there were no constraints on the value of x, we must remember that a positive number will have two square roots and we would write $\sqrt{x^2} = |x| = 5 \Rightarrow x = \pm 5$.

Radicals (surds)

Some roots are rational and some are irrational. Consider the two right triangles on the left. By applying Pythagoras' theorem, we find the length of the hypotenuse for triangle A to be exactly 5 (an integer and rational number) and the hypotenuse for triangle B to be exactly $\sqrt{80}$ (an irrational number). An irrational root – e.g. $\sqrt{80}, \sqrt{3}, \sqrt{10}, \sqrt[3]{4}$ – is called a **radical** or **surd**. The only way to express irrational roots exactly is in radical, or surd, form.

It is not immediately obvious that the following expressions are all equivalent.

$$\sqrt{80}, 2\sqrt{20}, \frac{16\sqrt{5}}{\sqrt{16}}, 2\sqrt{2}\sqrt{10}, \frac{10\sqrt{8}}{\sqrt{10}}, 4\sqrt{5}, 5\sqrt{\frac{16}{5}}$$

Square roots occur frequently in several of the topics in this course, so it will be useful for us to be able to simplify radicals and recognise equivalent radicals. Two useful rules for manipulating expressions with radicals are given below.

Simplifying radicals

For $a \geq 0, b \geq 0$ and $n \in \mathbb{Z}^+$, the following rules can be applied:

1 $\sqrt[n]{a} \times \sqrt[n]{b} = \sqrt[n]{ab}$ **2** $\frac{\sqrt[n]{a}}{\sqrt[n]{b}} = \sqrt[n]{\frac{a}{b}}$

Note: Each rule can be applied in either direction.

Example 7 – Simplifying radicals I

Simplify completely:

a) $\sqrt{5} \times \sqrt{5}$ b) $\sqrt{12} \times \sqrt{21}$ c) $\frac{\sqrt{48}}{\sqrt{3}}$

d) $\sqrt[3]{12} \times \sqrt[3]{18}$ e) $7\sqrt{2} - 3\sqrt{2}$ f) $\sqrt{5} + 2\sqrt{25} - 3\sqrt{5}$

g) $\sqrt{3}(2 - 2\sqrt{3})$ h) $(1 + \sqrt{2})(1 - \sqrt{2})$

Solution

a) $\sqrt{5} \times \sqrt{5} = \sqrt{5 \cdot 5} = \sqrt{25} = 5$

Note: A special case of the rule $\sqrt[n]{a} \times \sqrt[n]{b} = \sqrt[n]{ab}$ when $n = 2$ is $\sqrt{a} \times \sqrt{a} = a$.

b) $\sqrt{12} \times \sqrt{21} = \sqrt{4} \times \sqrt{3} \times \sqrt{3} \times \sqrt{7} = \sqrt{4} \times (\sqrt{3} \times \sqrt{3}) \times \sqrt{7}$

$= 2 \times 3 \times \sqrt{7} = 6\sqrt{7}$

c) $\frac{\sqrt{48}}{\sqrt{3}} = \sqrt{\frac{48}{3}} = \sqrt{16} = 4$

d) $\sqrt[3]{12} \times \sqrt[3]{18} = \sqrt[3]{12 \cdot 18} = \sqrt[3]{216} = 6$

e) $7\sqrt{2} - 3\sqrt{2} = 4\sqrt{2}$

f) $\sqrt{5} + 2\sqrt{25} - 3\sqrt{5} = 10 - 2\sqrt{5}$

g) $\sqrt{3}(2 - 2\sqrt{3}) = 2\sqrt{3} - 2\sqrt{3}\sqrt{3} = 2\sqrt{3} - 2 \cdot 3 = 2\sqrt{3} - 6$ or $-6 + 2\sqrt{3}$

h) $(1 + \sqrt{2})(1 - \sqrt{2}) = 1 - \sqrt{2} + \sqrt{2} - \sqrt{2}\sqrt{2} = 1 - 2 = -1$

The radical $\sqrt{24}$ can be simplified because one of the factors of 24 is 4, and the square root of 4 is rational (i.e. 4 is a perfect square).

$$\sqrt{24} = \sqrt{4 \cdot 6} = \sqrt{4}\sqrt{6} = 2\sqrt{6}$$

Rewriting 24 as the product of 3 and 8 (rather than 4 and 6) would not help simplify $\sqrt{24}$ because neither 3 nor 8 are perfect squares, i.e. there is no integer whose square is 3 or 8.

For any real number a, it would first appear that the rule $\sqrt{a^2} = a$ would be correct, but it is not. What if $a = -3$? Then $\sqrt{(-3)^2} = \sqrt{9} = 3$, not -3. The correct rule that is true for any real number a is $\sqrt{a^2} = |a|$. Generalizing for any index where n is a positive integer, we need to consider whether n is even or odd. If n is even, then $\sqrt[n]{a^n} = |a|$; and if n is odd, then $\sqrt[n]{a^n} = a$.
For example,
$\sqrt[6]{(-3)^6} = \sqrt[6]{729} = \sqrt[6]{3^6} = 3$;
and
$\sqrt[3]{(-5)^3} = \sqrt[3]{-125} = -5$.

Example 8 – Simplifying radicals II

Express each in terms of the simplest possible radical.

a) $\sqrt{80}$ b) $\sqrt{\frac{14}{81}}$ c) $\sqrt[3]{24}$ d) $5\sqrt{128}$

e) $\sqrt{x^2}$ f) $\sqrt{20a^4b^2}$ g) $\sqrt[3]{81}$ h) $\sqrt{4+9}$

Solution

a) $\sqrt{80} = \sqrt{16 \cdot 5} = \sqrt{16}\sqrt{5} = 4\sqrt{5}$

Note: 4 is a factor of 80 and is a perfect square, but 16 is the *largest* factor that is a perfect square

b) $\sqrt{\frac{14}{81}} = \frac{\sqrt{14}}{\sqrt{81}} = \frac{\sqrt{14}}{9}$

c) $\sqrt[3]{24} = \sqrt[3]{8} \times \sqrt[3]{3} = 2\sqrt[3]{3}$

d) $5\sqrt{128} = 5\sqrt{64}\sqrt{2} = 5 \cdot 8\sqrt{2} = 40\sqrt{2}$

e) $\sqrt{x^2} = |x|$

f) $\sqrt{20a^4b^2} = \sqrt{4}\sqrt{5}\sqrt{a^4}\sqrt{b^2} = 2a^2|b|\sqrt{5}$

g) $\sqrt[3]{81} = \sqrt[3]{27}\sqrt[3]{3} = 3\sqrt[3]{3}$

h) $\sqrt{4+9} = \sqrt{13}$

● **Hint:** Note that in Example 8 h) the square root of a sum is *not* equal to the sum of the square roots. That is, avoid the error $\sqrt{a+b} \neq \sqrt{a} + \sqrt{b}$.

In many cases we prefer not to have radicals in the denominator of a fraction. Recall from Example 7, part a), the special case of the rule $\sqrt[n]{a} \times \sqrt[n]{b} = \sqrt[n]{ab}$ when $n = 2$ is $\sqrt{a} \times \sqrt{a} = a$, assuming $a > 0$. The process of eliminating irrational numbers from the denominator is called **rationalizing the denominator.**

Example 9 – Rationalizing the denominator I

Rationalize the denominator of each expression.

a) $\frac{2}{\sqrt{3}}$ b) $\frac{\sqrt{7}}{4\sqrt{10}}$

Solution

a) $\frac{2}{\sqrt{3}} = \frac{2}{\sqrt{3}} \cdot \frac{\sqrt{3}}{\sqrt{3}} = \frac{2\sqrt{3}}{3}$

b) $\frac{\sqrt{7}}{4\sqrt{10}} = \frac{\sqrt{7}}{4\sqrt{10}} \cdot \frac{\sqrt{10}}{\sqrt{10}} = \frac{\sqrt{70}}{4 \cdot 10} = \frac{\sqrt{70}}{40}$

Changing a fraction from having a denominator that is irrational to an equivalent fraction where the denominator is rational (rationalizing the denominator) is not always a necessity. For example, expressing the cosine ratio of 45° as $\frac{1}{\sqrt{2}}$ rather than the equivalent value of $\frac{\sqrt{2}}{2}$ is mathematically correct. However, there will be instances where a fraction with a rational denominator will be preferred. It is a useful skill for simplifying some more complex fractions and for recognizing that two expressions are equivalent. For example, $\frac{1}{\sqrt{3}} = \frac{\sqrt{3}}{3}$, or a little less obvious, $\frac{3}{2+\sqrt{5}} = -6 + 3\sqrt{5}$. There are even situations where it might be useful to rationalize the numerator (see Example 11 below).

Recall the algebraic rule $(a+b)(a-b) = a^2 - b^2$. Any pair of expressions fitting the form of $a + b$ and $a - b$ are called a pair of **conjugates**. The result of multiplying a pair of conjugates is always a **difference of two squares**, $a^2 - b^2$, and this can be helpful in some algebraic manipulations – as we will see in the next example.

Example 10 – Rationalizing the denominator II

Express the quotient $\frac{2}{4-\sqrt{3}}$ so that the denominator is a rational number.

Solution

Multiply numerator and denominator by the conjugate of the denominator, $4+\sqrt{3}$, and simplify:

$$\frac{2}{4-\sqrt{3}} \cdot \frac{4+\sqrt{3}}{4+\sqrt{3}} = \frac{8+2\sqrt{3}}{4^2-(\sqrt{3})^2} = \frac{8+2\sqrt{3}}{16-3} = \frac{8+2\sqrt{3}}{13} \text{ or } \frac{8}{13} + \frac{2\sqrt{3}}{13}$$

Example 11 – Rationalizing the numerator

We will encounter the following situation in our study of calculus.

We are interested to analyze the behaviour of the quotient $\frac{\sqrt{x+h}-\sqrt{x}}{h}$ as the value of h approaches zero. It is not possible to directly substitute zero in for h in the present form of the quotient because that will give an undefined result of $\frac{0}{0}$. Perhaps we can perform the substitution if we rationalize the numerator. We will assume that x and $x + h$ are positive.

Solution

Multiplying numerator and denominator by the conjugate of the numerator and simplifying:

$$\frac{(\sqrt{x+h}-\sqrt{x})}{h} \cdot \frac{(\sqrt{x+h}+\sqrt{x})}{(\sqrt{x+h}+\sqrt{x})} = \frac{(\sqrt{x+h})^2-(\sqrt{x})^2}{h(\sqrt{x+h}+\sqrt{x})}$$

$$= \frac{x+h-x}{h(\sqrt{x+h}+\sqrt{x})}$$

$$= \frac{h}{h(\sqrt{x+h} + \sqrt{x})}$$

$$= \frac{1}{\sqrt{x+h} + \sqrt{x}}$$

Substituting zero for h into this expression causes no problems. Therefore, as h approaches zero, the expression $\frac{\sqrt{x+h} - \sqrt{x}}{h}$ would appear to approach the expression $\frac{1}{\sqrt{x+0} + \sqrt{x}} = \frac{1}{2\sqrt{x}}$.

Exercise 1.2

In questions 1–15, express each in terms of the simplest possible radical.

1 $\sqrt{h^2} \times \sqrt{h^2}$ **2** $\frac{\sqrt{45}}{\sqrt{5}}$ **3** $\sqrt{18} \times \sqrt{10}$

4 $\sqrt{\frac{28}{49}}$ **5** $\sqrt[3]{4} \times \sqrt[3]{16}$ **6** $\sqrt{\frac{15}{20}}$

7 $\sqrt{5}(3 + 4\sqrt{5})$ **8** $(2 + \sqrt{6})(2 - \sqrt{6})$ **9** $\sqrt{98}$

10 $4\sqrt{1000}$ **11** $\sqrt[3]{48}$ **12** $\sqrt{12x^3y^3}$

13 $\sqrt[5]{m^5}$ **14** $\sqrt{\frac{27}{6}}$ **15** $\sqrt{x^{16}(1+x)^2}$

In questions 16–18, completely simplify the expression.

16 $13\sqrt{7} - 10\sqrt{7}$ **17** $\sqrt{72} - 8\sqrt{3} + 3\sqrt{48}$ **18** $\sqrt{500} + 5\sqrt{20} - \sqrt{45}$

In questions 19–30, rationalize the denominator, simplifying if possible.

19 $\frac{1}{\sqrt{5}}$ **20** $\frac{2}{5\sqrt{2}}$ **21** $\frac{6\sqrt{7}}{\sqrt{3}}$

22 $\frac{4}{\sqrt{32}}$ **23** $\frac{2}{1+\sqrt{5}}$ **24** $\frac{1}{3+2\sqrt{5}}$

25 $\frac{\sqrt{3}}{2-\sqrt{3}}$ **26** $\frac{4}{\sqrt{2}+\sqrt{5}}$ **27** $\frac{x-y}{\sqrt{x}+\sqrt{y}}$

28 $\frac{1+\sqrt{3}}{2+\sqrt{3}}$ **29** $\sqrt{\frac{1}{x^2} - 1}$ **30** $\frac{h}{\sqrt{x+h} - \sqrt{x}}$

In questions 31–33, rationalize the numerator, simplifying if possible.

31 $\frac{\sqrt{a}-3}{a-9}$ **32** $\frac{\sqrt{x}-\sqrt{y}}{x-y}$ **33** $\frac{\sqrt{m}-\sqrt{7}}{7-x}$

1.3 Exponents (indices)

As we've already seen with roots in the previous section, repeated multiplication of identical numbers can be written more efficiently by using exponential notation.

Exponential notation

If b is any real number ($b \in \mathbb{R}$) and n is a positive integer ($n \in \mathbb{Z}^+$), then

$$b^n = \underbrace{b \cdot b \cdot b \cdot \ldots \cdot b}_{n \text{ factors}}$$

where n is the **exponent**, b is the **base** and b^n is called the ***n*th power** of b.
Note: n is also called the **power** or **index** (plural: indices).

Integer exponents

We now state seven laws of integer exponents (or indices) that you will have learned in a previous mathematics course. Familiarity with these rules is essential for work throughout this course.

Let a and b be real numbers ($a, b \in \mathbb{R}$) and let m and n be integers ($m, n \in \mathbb{Z}$). Assume that all denominators and bases are not equal to zero. All of the laws can be applied in either direction.

Table 1.4 Laws of exponents (indices) for integer exponents.

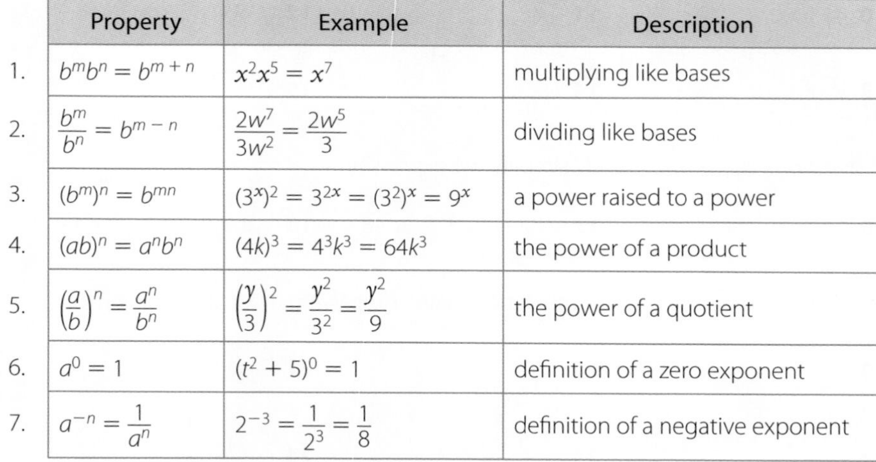

	Property	Example	Description
1.	$b^m b^n = b^{m+n}$	$x^2x^5 = x^7$	multiplying like bases
2.	$\frac{b^m}{b^n} = b^{m-n}$	$\frac{2w^7}{3w^2} = \frac{2w^5}{3}$	dividing like bases
3.	$(b^m)^n = b^{mn}$	$(3^x)^2 = 3^{2x} = (3^2)^x = 9^x$	a power raised to a power
4.	$(ab)^n = a^n b^n$	$(4k)^3 = 4^3k^3 = 64k^3$	the power of a product
5.	$\left(\frac{a}{b}\right)^n = \frac{a^n}{b^n}$	$\left(\frac{y}{3}\right)^2 = \frac{y^2}{3^2} = \frac{y^2}{9}$	the power of a quotient
6.	$a^0 = 1$	$(t^2 + 5)^0 = 1$	definition of a zero exponent
7.	$a^{-n} = \frac{1}{a^n}$	$2^{-3} = \frac{1}{2^3} = \frac{1}{8}$	definition of a negative exponent

• **Hint:** If the base of an exponential expression is negative, then it is necessary to write it in brackets. The expression such as -3^2 is equivalent to $-(3)^2$. Hence, $(-3)^2 = 9$ but $-3^2 = -9$.

Negative integers and fractions were first used as exponents in the modern conventional notation (as raised numbers, such as 5^{-2}, $x^{\frac{3}{2}}$) by Isaac Newton in a letter in 1676 to a fellow scientist in which he described his derivation of the binomial theorem (Chapter 4 in this book).

The last two laws of exponents listed above – the definition of a zero exponent and the definition of a negative exponent – are often assumed without proper explanation. The definition of a^n as repeated multiplication, i.e. n factors of a, is easily understood when n is a positive integer. So how do we formulate appropriate definitions for a^n when n is negative or zero? These definitions will have to be compatible with the laws for positive integer exponents. If the law stating $b^m b^n = b^{m+n}$ is to hold for a zero exponent, then $b^n b^0 = b^{n+0} = b^n$. Since the number 1 is the identity element for multiplication (multiplicative identity property) then $b^n \cdot 1 = b^n$. Therefore, we must define b^0 as the number 1. If the law $b^m b^n = b^{m+n}$ is to also hold for negative integer exponents, then

$b^n b^{-n} = b^{n-n} = b^0 = 1$. Since the product of b^n and b^{-n} is 1, then they must be reciprocals (multiplicative inverse property). Therefore, we must define b^{-n} as $\frac{1}{b^n}$.

Rational exponents (fractional exponents)

We know that $4^3 = 4 \times 4 \times 4$ and $4^0 = 1$ and $4^{-2} = \frac{1}{4^2} = \frac{1}{4 \times 4}$, but what meaning are we to give to $4^{\frac{1}{2}}$? In order to carry out algebraic operations with expressions having exponents that are rational numbers, it will be very helpful if they follow the laws established for integer exponents. From the law $b^m b^n = b^{m+n}$, it must follow that $4^{\frac{1}{2}} \times 4^{\frac{1}{2}} = 4^{\frac{1}{2}+\frac{1}{2}} = 4^1$. Likewise, from the law $(b^m)^n = b^{mn}$, it follows that $(4^{\frac{1}{2}})^2 = 4^{\frac{1}{2}\cdot 2} = 4^1$. Therefore, we need to define $4^{\frac{1}{2}}$ as the square root of 4 or, more precisely, as the principal (positive) square root of 4, that is, $\sqrt{4}$. We are now ready to use radicals to define a rational exponent of the form $\frac{1}{n}$, where n is a positive integer. If the rule $(b^m)^n = b^{mn}$ is to apply when $m = \frac{1}{n}$, it must follow that $(b^{\frac{1}{n}})^n = b^{\frac{n}{n}} = b^1$. This means that the nth power of $b^{\frac{1}{n}}$ is b and, from the discussion of nth roots in Section 1.2, we define $b^{\frac{1}{n}}$ as the principal nth root of b.

Definition of $b^{\frac{1}{n}}$

If $n \in \mathbb{Z}^+$, then $b^{\frac{1}{n}}$ is the principal nth root of b. Using a radical, this means

$$b^{\frac{1}{n}} = \sqrt[n]{b}$$

This definition allows us to evaluate exponential expressions such as the following:

$$36^{\frac{1}{2}} = \sqrt{36} = 6;\ (-27)^{\frac{1}{3}} = \sqrt[3]{-27} = -3;\ \left(\frac{1}{81}\right)^{\frac{1}{4}} = \sqrt[4]{\frac{1}{81}} = \frac{1}{3}$$

Now we can apply the definition of $b^{\frac{1}{n}}$ and the rule $(b^m)^n = b^{mn}$ to develop a rule for expressions with exponents of the form not just $\frac{1}{n}$ but of the more general form $\frac{m}{n}$.

$$b^{\frac{m}{n}} = b^{m\frac{1}{n}} = (b^m)^{\frac{1}{n}} = \sqrt[n]{b^m};\text{ or, equivalently, } b^{\frac{m}{n}} = b^{\frac{1}{n}\cdot m} = (b^{\frac{1}{n}})^m = (\sqrt[n]{b})^m$$

This will allow us to evaluate exponential expressions such as $9^{\frac{3}{2}}$, $(-8)^{\frac{5}{3}}$ and $64^{\frac{5}{6}}$.

Definition of a rational exponent

If m and n are positive integers with no common factors, then

$$b^{\frac{m}{n}} = \sqrt[n]{b^m} \text{ or } (\sqrt[n]{b})^m$$

If n is an even number, then we must have $b \geqslant 0$.

The numerator of a rational exponent indicates the power to which the base of the exponential expression is raised, and the denominator indicates the root to be taken. With this definition for rational exponents, we can conclude that all of the laws of exponents stated for integer exponents in Table 1.4 also hold true for rational exponents.

Example 12 – Applying laws of exponents

Evaluate and/or simplify each of the following expressions. Leave only positive exponents.

a) $(3a^2b)^3$ b) $3(a^2b)^3$ c) $(-2)^{-3}$

d) $(x+y)^0$ e) $(3^3)^{\frac{1}{2}} \cdot 9^{\frac{3}{4}}$ f) $\frac{m^2n^{-3}}{m^{-5}n^3}$

g) $(-27)^{-\frac{2}{3}}$ h) $8^{\frac{2}{3}}$ i) $(2^x)(2^{3-x})$

j) $(0.04)^{-2}$ k) $\frac{\sqrt{a}\sqrt{a^3}}{a^3}$ $(a>0)$ l) $\frac{x^{-2}y^3z^{-4}}{(2x^2)^3} \times \frac{8}{y^{-2}z^4}$

- **Hint for (o):** apply $b^mb^n = b^{m+n}$ in other direction.

m) $\sqrt[4]{81a^8b^{12}}$ n) $\frac{x^{\frac{3}{2}} + x^{\frac{1}{2}}}{x^{\frac{1}{2}}}$ $(x>0)$ o) $2^{n+3} - 2^{n+1}$

p) $\frac{\sqrt{a+b}}{a+b}$ q) $\frac{(x+y)^2}{(x+y)^{-2}}$ r) $\frac{x^2 + 2^{\frac{3}{2}} - 2(x^2+2)^{\frac{1}{2}}}{x^2}$

Solution

a) $(3a^2b)^3 = 3^3(a^2)^3b^3 = 27a^6b^3$

b) $3(a^2b)^3 = 3(a^2)^3b^3 = 3a^6b^3$

c) $(-2)^{-3} = \frac{1}{(-2)^3} = -\frac{1}{8}$

d) $(x+y)^0 = 1$

e) $(3^3)^{\frac{1}{2}} \cdot 9^{\frac{3}{4}} = 3^{\frac{3}{2}}(3^2)^{\frac{3}{4}} = 3^{\frac{3}{2}} \cdot 3^{\frac{3}{2}} = 3^{\frac{6}{2}} = 3^3 = 27$

f) $\frac{m^2n^{-3}}{m^{-5}n^3} = \frac{m^2}{m^{-5}} \cdot \frac{n^{-3}}{n^3} = \frac{m^{2-(-5)}}{1} \cdot \frac{1}{n^{3-(-3)}} = \frac{m^7}{n^6}$

g) $(-27)^{-\frac{2}{3}} = [(-3)^3]^{-\frac{2}{3}} = (-3)^{3(-\frac{2}{3})} = (-3)^{-2} = \frac{1}{(-3)^2} = \frac{1}{9}$

h) $8^{\frac{2}{3}} = \sqrt[3]{8^2} = \sqrt[3]{64} = 4$ or $8^{\frac{2}{3}} = (\sqrt[3]{8})^2 = (2)^2 = 4$ or $8^{\frac{2}{3}} = (2^3)^{\frac{2}{3}} = 2^2 = 4$

i) $(2^x)(2^{3-x}) = 2^{x+3-x} = 2^3 = 8$

j) $(0.04)^{-2} = \left(\frac{4}{100}\right)^{-2} = \left(\frac{1}{25}\right)^{-2} = \left(\frac{25}{1}\right)^2 = 625$

k) $\frac{\sqrt{a}\sqrt{a^3}}{a^3} = \frac{a^{\frac{1}{2}} \cdot a^{\frac{3}{2}}}{a^3} = \frac{a^{\frac{1}{2}+\frac{3}{2}}}{a^3} = \frac{a^2}{a^3} = \frac{1}{a}$

l) $\frac{x^{-2}y^3z^{-4}}{(2x^2)^3} \times \frac{8}{y^{-2}z^4} = \frac{x^{-2}y^3z^{-4}}{8x^6} \times \frac{8}{y^{-2}z^4} = \frac{y^3}{x^2x^6z^4} \times \frac{y^2}{z^4} = \frac{y^5}{x^8z^8}$

m) $\sqrt[4]{81a^8b^{12}} = \sqrt[4]{81} \cdot \sqrt[4]{a^8} \cdot \sqrt[4]{b^{12}} = 3a^{\frac{8}{4}}b^{\frac{12}{4}} = 3a^2b^3$

n) $\frac{x^{\frac{3}{2}} + x^{\frac{1}{2}}}{x^{\frac{1}{2}}} = \frac{x^{\frac{3}{2}}}{x^{\frac{1}{2}}} + \frac{x^{\frac{1}{2}}}{x^{\frac{1}{2}}} = \frac{x^{\frac{3}{2}-\frac{1}{2}}}{1} + 1 = x + 1$

o) $2^{n+3} - 2^{n+1} = (2^n)(2^3) - (2^n)(2^1) = 8(2^n) - 2(2^n) = 6(2^n)$

p) $\dfrac{\sqrt{a+b}}{a+b} = \dfrac{(a+b)^{\frac{1}{2}}}{(a+b)^1} = \dfrac{1}{(a+b)^{1-\frac{1}{2}}} = \dfrac{1}{(a+b)^{\frac{1}{2}}} = \dfrac{1}{\sqrt{a+b}}$

q) $\dfrac{(x+y)^2}{(x+y)^{-2}} = (x+y)^{2-(-2)} = (x+y)^4$

Although $(x+y)^4 = x^4 + 4x^3y + 6x^2y^2 + 4xy^3 + y^4$, merely expanding is not 'simplifying'.

r) $\dfrac{(x^2+2)^{\frac{3}{2}} - 2(x^2+2)^{\frac{1}{2}}}{x^2} = \dfrac{(x^2+2)^{\frac{1}{2}}[(x^2+2)^1 - 2]}{x^2} = \dfrac{(x^2+2)^{\frac{1}{2}}[x^2]}{x^2}$

$= (x^2+1)^{\frac{1}{2}}$ or $\sqrt{x^2+1}$

• **Hint:** Note that in Example 12 q) that the square of a sum is **not** equal to the sum of the squares. That is, avoid the error $(x+y)^2 \neq x^2 + y^2$, and in general $(x+y)^n \neq x^n + y^n$.

Exercise 1.3

In questions 1–6, simplify (without your GDC) each expression to a single integer.

1 $16^{\frac{1}{4}}$ **2** $9^{\frac{3}{2}}$ **3** $64^{\frac{2}{3}}$

4 $8^{\frac{4}{3}}$ **5** $32^{\frac{3}{5}}$ **6** $(\sqrt{2})^6$

In questions 7–9, simplify each expression (without your GDC) to a quotient of two integers.

7 $\left(\frac{8}{27}\right)^{\frac{2}{3}}$ **8** $\left(\frac{9}{16}\right)^{\frac{1}{2}}$ **9** $\left(\frac{25}{4}\right)^{\frac{3}{2}}$

In questions 10–13, evaluate (without your GDC) each expression.

10 $(-3)^{-2}$ **11** $(13)^0$ **12** $\dfrac{4 \cdot 3^{-2}}{2^{-2} \cdot 3^{-1}}$ **13** $\left(-\frac{3}{4}\right)^{-3}$

In questions 14–34, simplify each exponential expression (leave only positive exponents).

14 $(-xy^3)^2$ **15** $-(xy^3)^2$ **16** $(-2xy^3)^3$

17 $(2x^3y^{-5})(2x^{-1}y^3)^4$ **18** $(4m^2)^{-3}$ **19** $\dfrac{3k^3p^4}{(3k^3)^2p^2}$

20 $(-32)^{\frac{3}{5}}$ **21** $(125)^{\frac{2}{3}}$ **22** $\dfrac{x\sqrt{x}}{\sqrt[3]{x}}$

23 $\dfrac{4a^3b^5}{(2a^2b)^4} \cdot \dfrac{b^{-1}}{a^{-3}}$ **24** $\dfrac{(\sqrt[3]{x})(\sqrt[3]{x^4})}{\sqrt[3]{x^2}}$ **25** $\dfrac{6(a-b)^2}{3a-3b}$

26 $\dfrac{(x+4y)^{\frac{1}{2}}}{2(x+4y)^{-1}}$ **27** $\dfrac{p^2+q^2}{\sqrt{p^2+q^2}}$ **28** $\dfrac{5^{3x+1}}{25}$

29 $\dfrac{x^{\frac{1}{3}} + x^{\frac{1}{4}}}{x^{\frac{1}{2}}}$ **30** $3^{n+1} - 3^{n-2}$ **31** $\dfrac{8^{k+2}}{2^{3k+2}}$

32 $\sqrt[3]{24x^6y^{12}}$ **33** $\frac{1}{n}\sqrt{n^2+n^4}$ **34** $\dfrac{x+\sqrt{x}}{1+\sqrt{x}}$

• **Hint:** In question 34 it is incorrect to 'cancel' the term of $\sqrt{x}$ from the numerator and denominator. That is, remember $\dfrac{a+b}{c+b} \neq \dfrac{a}{c}$.

1.4 Scientific notation (standard form)

Exponents provide an efficient way of writing and calculating with very large or very small numbers. The need for this is especially great in science. For example, a light year (the distance that light travels in one year) is 9 460 730 472 581 kilometres and the mass of a single water molecule is 0.000 000 000 000 000 000 000 0056 grams. It is far more convenient and useful to write such numbers in **scientific notation** (also called **standard form**).

> **Scientific notation**
> A positive number N is written in scientific notation if it is expressed in the form:
> $$N = a \times 10^k, \text{ where } 1 \leqslant a < 10 \text{ and } k \text{ is an integer.}$$

In scientific notation, a light year is about 9.46×10^{12} kilometres. This expression is determined by observing that when a number is multiplied by 10^k and k is **positive**, the decimal point will move k places to the **right**. Therefore, $9.46 \times 10^{12} = 9\underbrace{460\,000\,000\,000}_{12 \text{ decimal places}}$. Knowing that when a number is multiplied by 10^k and k is **negative** the decimal point will move k places to the **left** helps us to express the mass of a water molecule as 5.6×10^{-24} grams. This expression is equivalent to $0.\underbrace{000\,000\,000\,000\,000\,000\,000\,0056}_{24 \text{ decimal places}}$.

Scientific notation is also a very convenient way of indicating the number of **significant figures** (digits) to which a number has been approximated. A light year expressed to an accuracy of 13 significant figures is 9 460 730 472 581 kilometres. However, many calculations will not require such a high degree of accuracy. For a certain calculation it may be more appropriate to have a light year approximated to 4 significant figures, which could be written as 9 461 000 000 000 kilometres, or more efficiently and clearly in scientific notation as 9.461×10^{12} kilometres.

Not only is scientific notation conveniently compact, it also allows a quick comparison of the magnitude of two numbers without the need to count zeros. Moreover, it enables us to use the laws of exponents to perform otherwise unwieldy calculations.

Example 13 – Scientific notation

Use scientific notation to calculate each of the following.

a) $64\,000 \times 2\,500\,000\,000$

b) $\dfrac{0.000\,000\,78}{0.000\,000\,0012}$

c) $\sqrt[3]{27\,000\,000\,000}$

Solution

a) $64\,000 \times 2\,500\,000\,000 = (6.4 \times 10^4)(2.5 \times 10^9)$
$= 6.4 \times 2.5 \times 10^4 \times 10^9$
$= 16 \times 10^{4+9}$
$= 1.6 \times 10^1 \times 10^{13} = 1.6 \times 10^{14}$

b) $\dfrac{0.000\,000\,78}{0.000\,000\,0012} = \dfrac{7.8 \times 10^{-7}}{1.2 \times 10^{-9}} = \dfrac{7.8}{1.2} \times \dfrac{10^{-7}}{10^{-9}} = 6.5 \times 10^{-7-(-9)}$
$= 6.5 \times 10^2$ or 650

c) $\sqrt[3]{27\,000\,000\,000} = (2.7 \times 10^{10})^{\frac{1}{3}} = (27 \times 10^9)^{\frac{1}{3}} = (27)^{\frac{1}{3}}(10^9)^{\frac{1}{3}}$
$= 3 \times 10^3$ or 3000

Your GDC will automatically express numbers in scientific notation when a large or small number exceeds its display range. For example, if you use your GDC to compute 2 raised to the 64th power, the display (depending on the GDC model) will show the approximation

`1.844674407E19` or `1.844674407 19`

The final digits indicate the power of 10, and we interpret the result as $1.844\,674\,407 \times 10^{19}$. ($2^{64}$ is exactly 18 446 744 073 709 551 616.)

The wheat and chessboard problem is a mathematical question that is posed as part of a story that has been told in many variations over the centuries. In any version of the story, the question is: If one grain of wheat is placed on the first square of an 8 by 8 chessboard, then two grains of wheat on the second square, four grains on the third square, and so on – each time doubling the grains of rice – then exactly how many grains of wheat in total are on the board after grains are placed on the last square?

Exercise 1.4

In questions 1–10, write each number in scientific notation, rounding to three significant figures.

1 253.8 **2** 0.007 81 **3** 7 405 239

4 0.000 001 0448 **5** 4.9812 **6** 0.001 991

7 Land area of Earth: 148 940 000 square kilometres

8 Relative density of hydrogen: 0.000 0899 grams per cm^3

9 Mean distance from the Earth to the Sun (a unit of length referred to as the Astronomical Unit, *AU*): 149 597 870.691 kilometres

10 Mass of an electron 0.000 000 000 000 000 000 000 000 000 000 910 938 15 kg

In questions 11–14, write each number in ordinary decimal notation.

11 2.7×10^{-3} **12** 5×10^7

13 9.035×10^{-8} **14** 4.18×10^{12}

In questions 15–22, use scientific notation and the laws of exponents to perform the indicated operations. Give the result in scientific notation rounded to two significant figures.

15 $(2.5 \times 10^{-3})(10 \times 10^5)$ **16** $\dfrac{3.2 \times 10^6}{1.6 \times 10^2}$

17 $\dfrac{(1 \times 10^{-3})(3.28 \times 10^6)}{4 \times 10^7}$ **18** $(2 \times 10^3)^4(3.5 \times 10^5)$

19 $(0.000\,000\,03)\,(6\,000\,000\,000\,000)$ **20** $\dfrac{(1\,000\,000)^2\sqrt{0.000\,000\,04}}{(8\,000\,000\,000)^{\frac{2}{3}}}$

21 $\dfrac{4 \times 10^4}{(6.4 \times 10^2)\,(2.5 \times 10^{-5})}$ **22** $(5.4 \times 10^2)^5\,(-1.1 \times 10^{-6})^2$

1.5 Algebraic expressions

Examples of algebraic expressions are

$$5a^3b^2 \qquad 2x^2 + 7x - 8 \qquad \frac{Gm_1m_2}{r^2} \qquad \frac{t}{\sqrt{1 - \left(\frac{v}{c}\right)^2}}$$

Algebraic expressions are formed by combining variables and constants using addition, subtraction, multiplication, division, exponents and radicals.

The word *algebra* comes from the 9th-century Arabic book *Hisâb al-Jabr w'al-Muqabala*, written by the Islamic mathematician and astronomer Abu Ja'far Muhammad ibn Musa al-Khwarizmi (c. 778–850). The book title refers to transposing and combining terms, two processes used in solving equations. In Latin translations, the title was shortened to *Aljabr*, from which we get the word *algebra*. Al-Khwarizmi worked as a scholar in Baghdad studying and writing about mathematics and science. Some of his works were later translated into Latin, thus helping to establish Hindu-Arabic numerals and algebra concepts into Europe. The word *algorithm* comes from a Latinized version of his name.

Polynomials

An algebraic expression that has only non-negative powers of one or more variables and contains no variable in a denominator is called a **polynomial**.

Definition of a polynomial in the variable x

Given $a_0, a_1, a_2, \ldots, a_n \in \mathbb{R}$, $a_n \neq 0$ and $n \geqslant 0$, $n \in \mathbb{Z}^+$, then a **polynomial in x** is a sum of distinct **terms** in the form

$$a_nx^n + a_{n-1}x^{n-1} + \ldots + a_1x + a_0$$

where $a_1, a_2, \ldots, a_n$ are the **coefficients**, a_0 is the **constant term** and n (the greatest exponent) is the **degree** of the polynomial.

• **Hint:** Polynomials with one, two and three terms are called **monomials**, **binomials** and **trinomials**, respectively. A polynomial of: degree 1 is '**linear**'; degree 2 is '**quadratic**'; degree 3 is '**cubic**'; degree 4 is '**quartic**' and degree 5 is '**quintic**'. Beyond degree 5 there are no generally accepted names for polynomials. Quadratic polynomials are studied in depth in Chapter 3.

Polynomials are added or subtracted using the properties of real numbers that were discussed in Section 1.1 of this chapter. We do this by combining **like terms** – terms containing the same variable(s) raised to the same power(s) – and applying the distributive property.

For example,

$2x^2y + 6x^2 - 7x^2y = 2x^2y - 7x^2y + 6x^2$ Rearrange terms so the like terms are together.

$= (2 - 7)x^2y + 6x^2$ Apply distributive property: $ab + ac = (b + c)a$.

$= -5x^2y + 6x^2$ No like terms remain, so polynomial is simplified.

Expanding and factorizing polynomials

We apply the distributive property in the other direction, i.e. $a(b + c) = ab + ac$, in order to multiply polynomials. For example,

$$\begin{aligned}(2x - 3)(x + 5) &= 2x(x + 5) - 3(x + 5) \\ &= 2x^2 + 10x - 3x - 15 && \text{Combining like terms } 10x \text{ and } -3x. \\ &= 2x^2 + 7x - 15 && \text{Terms written in descending order of the exponents.}\end{aligned}$$

The process of multiplying polynomials is often referred to as **expanding**. Especially in the case of a polynomial being raised to a power, the number of terms in the resulting polynomial, after applying the distributive property and combining like terms, has increased (expanded) compared to the original number of terms. For example,

$$\begin{aligned}(x + 3)^2 &= (x + 3)(x + 3) && \text{Squaring a 1st degree (linear) binomial.} \\ &= x(x + 3) + 3(x + 3) \\ &= x^2 + 3x + 3x + 9 \\ &= x^2 + 6x + 9 && \text{The result is a 2nd degree (quadratic) trinomial.}\end{aligned}$$

and

$$\begin{aligned}(x + 1)^3 &= (x + 1)(x + 1)(x + 1) && \text{Cubing a 1st degree binomial.} \\ &= (x + 1)(x^2 + x + x + 1) \\ &= x(x^2 + 2x + 1) + 1(x^2 + 2x + 1) && \text{Distributive property.} \\ &= x^3 + 2x^2 + x + x^2 + 2x + 1 \\ &= x^3 + 3x^2 + 3x + 1 && \text{Result is a 3rd degree (cubic) polynomial with four terms.}\end{aligned}$$

As stated in Section 1.2, pairs of binomials of the form $a + b$ and $a - b$ are called **conjugates**. In most instances, the product of two binomials produces a trinomial. However, the product of a pair of conjugates produces a binomial such that both terms are squares and the second term is negative – referred to as a **difference of two squares**. For example,

$$\begin{aligned}(x + 5)(x - 5) &= x(x - 5) + 5(x - 5) && \text{Multiplying two conjugates;} \\ &= x^2 - 5x + 5x - 25 && \text{distributive property.} \\ &= x^2 - 25 && x^2 - 25 \text{ is a difference of two squares.}\end{aligned}$$

The inverse (or 'undoing') of multiplication (expansion) is **factorization.** If it is helpful for us to rewrite a polynomial as a product, then we need to factorize it – i.e. apply the distributive property in the *reverse* direction $(ab + ac = (b + c)a)$. The previous four examples can be used to illustrate equivalent pairs of factorized and expanded polynomials.

Factorized		**Expanded**
$(2x - 3)(x + 5)$	=	$2x^2 + 7x - 15$
$(x + 3)^2$	=	$x^2 + 6x + 9$
$(x + 1)^3$	=	$x^3 + 3x^2 + 3x + 1$
$(x + 5)(x - 5)$	=	$x^2 - 25$

Certain polynomial expansions (products) and factorizations occur so frequently you should be able to quickly recognize and apply them. Here is a list of some of the more common ones. You can verify these identities by performing the multiplication (expanding).

Common polynomial expansion and factorization patterns

Expanding →

Product of two binomials	$(x+a)(x+b)$	$=x^2+(a+b)x+ab$	Factorizing a trinomial
Product of two binomials	$(ax+b)(cx+d)$	$=acx^2+(ad+bc)x+bd$	Factorizing a trinomial
Product of two conjugates	$(a+b)(a-b)$	$=a^2-b^2$	Difference of two squares
Square of sum of 2 terms	$(a+b)^2$	$=a^2+2ab+b^2$	Trinomial perfect square
Square of difference of 2 terms	$(a-b)^2$	$=a^2-2ab+b^2$	Trinomial perfect square
Cube of a sum of 2 terms	$(a+b)^3$	$=a^3+3a^2b+3ab^2+b^3$	Perfect cube
Cube of difference of 2 terms	$(a-b)^3$	$=a^3-3a^2b+3ab^2-b^3$	Perfect cube
	$(a+b)(a^2-ab+b^2)$	$=a^3+b^3$	Sum of two cubes
	$(a-b)(a^2+ab+b^2)$	$=a^3-b^3$	Difference of two cubes

← Factorizing

These identities are useful patterns into which we can substitute any number or algebraic expression for *a*, *b* or *x*. This allows us to efficiently find products and powers of polynomials and also to factorize many polynomials.

Example 14 – Multiplying polynomials

Find each product.

a) $(x+2)(x-7)$
b) $(3x-4)(4x+1)$
c) $(6x+y)(6x-y)$
d) $(4h-5)^2$
e) $(a+2)^3$
f) $(3x+2\sqrt{5})(3x-2\sqrt{5})$
g) $(x^2-y)^3$
h) $(1+3m)^2$
i) $(x+2i)(x-2i)$
j) $(x+y+4)(x+y-4)$
k) $(-6-15w)(w+2)$
l) $(a-b+c)^2$

Solution

a) This product fits the pattern $(x+a)(x+b)=x^2+(a+b)x+ab$.

$$(x+2)(x-7)=x^2+(2-7)x+(2)(-7)=x^2-5x-14$$

You should be able to perform the middle step 'mentally' without writing it.

b) This product fits the pattern $(ax+b)(cx+d)=acx^2+(ad+bc)x+bd$.

$$(3x-4)(4x+1)=12x^2+(3-16)x-4=12x^2-13x-4$$

c) This fits the pattern $(a+b)(a-b)=a^2-b^2$ where the result is a difference of two squares.

$$(5x^3+3y)(5x^3-3y)=(5x^3)^2-(3y)^2=25x^6-9y^2$$

d) This fits the pattern $(a - b)^2 = a^2 - 2ab + b^2$.

$$(4h - 5)^2 = (4h)^2 - 2(4h)(5) + (5)^2 = 16h^2 - 40h + 25$$

e) This fits the pattern $(a + b)^3 = a^3 + 3a^2b + 3ab^2 + b^3$.

$$(a + 2)^3 = (a)^3 + 3(a)^2(2) + 3(a)(2)^2 + (2)^3 = a^3 + 6a^2 + 12a + 8$$

f) This is a pair of conjugates, so they fit the pattern $(a + b)(a - b) = a^2 - b^2$.

$$(3x + 2\sqrt{5})(3x - 2\sqrt{5}) = (3x)^2 - (2\sqrt{5})^2 = 9x^2 - (4 \cdot 5) = 9x - 20$$

Note: As we have observed earlier, the product of two **irrational** conjugates is a single **rational** number. We used this result to simplify fractions with irrational denominators in Section 1.2.

g) This fits the pattern $(a - b)^3 = a^3 - 3a^2b + 3ab^2 - b^3$.

$$(x^2 - 4y)^3 = (x^2)^3 - 3(x^2)^2(4y) + 3(x^2)(4y)^2 - (4y)^3$$
$$= x^6 - 12x^4y + 48x^2y^2 - 64y^3$$

h) This fits the pattern $(a + b)^2 = a^2 + 2ab + b^2$.

$$(1 + 3m^2)^2 = (1)^2 + 2(1)(3m^2) + (3m^2)^2$$
$$= 1 + 6m^2 + 9m^4 \text{ or } 9m^4 + 6m^2 + 1$$

i) This fits the pattern $(a + b)(a - b) = a^2 - b^2$.

$$(x + 2i)(x - 2i) = x^2 - (2i)^2 = x^2 - 4i^2 = x^2 - 4(-1) = x^2 + 4$$

Remember from Section 1.1, that the imaginary number i is defined such that $i^2 = -1$.

j) Initially the product does not seem to fit a pattern and we can find the product simply by applying the distributive property.

$$(x + y + 4)(x + y - 4) = x^2 + xy - 4x + xy + y^2 - 4y + 4x + 4y - 16$$
$$= x^2 + 2xy + y^2 - 16$$

However, upon closer inspection we see that this is a product of two conjugates. This can be made clear with the insertion of brackets.

$$[(x + y) + 4][(x + y) - 4] = (x + y)^2 - 4^2 = x^2 + 2xy + y^2 - 16$$

k) Factor out GCF of -3 from the first factor, and then multiply.

$$(-6 - 9w)(3w + 2) = -3(2 + 3w)(3w + 2) = -3(3w + 2)^2$$
$$= -3(9w^2 + 12w + 4) = -27w^2 - 36w - 12$$

l) By inserting a pair of brackets, this product can be considered as the square of a binomial.

$$(a - b + c)^2 = [(a - b) + c]^2 = (a - b)^2 + 2(a - b)c + c^2$$
$$= a^2 - 2ab + b^2 + 2ac - 2bc + c^2$$

or $a^2 + b^2 + c^2 - 2ab + 2ac - 2bc$

Note: It would be incorrect to insert brackets to write $(a - b + c)^2 = [a - (b + c)]^2$ for (l). Why?

• **Hint:** The result in Example 14 i), $(x + 2i)(x - 2i) = x^2 + 4$, shows that imaginary numbers could be used to factorize certain polynomials. However, when we factorize a polynomial in this course we will only look for factors that contain coefficients and/or constants that are rational numbers. For example, we consider both of the polynomials $x^2 - 5$ and $x^2 + 9$ *not* to be factorable, even though $x^2 - 5 = (x + \sqrt{5})(x - \sqrt{5})$, and $x^2 + 9 = (x + 3i)(x - 3i)$.

Example 15 – Factorizing polynomials

Completely factorize the following expressions.

a) $2x^2 - 14x + 24$ b) $2x^2 + x - 15$ c) $8x^7 - 18x$
d) $3y^3 + 24y^2 + 48y$ e) $(x + 3)^2 - y^2$ f) $5x^3y + 20xy^3$
g) $c^3 + 27$ h) $1 - 8h^6$ i) $a^4 - \frac{1}{16}$
j) $15 - x^2 - 2x$ k) $3x^2 + 20x - 7$ l) $y^2 + 5y + \frac{25}{4}$

Solution

a) $2x^2 - 14x + 24 = 2(x^2 - 7x + 12)$ Factor out the greatest common factor (GCF).
$= 2[x^2 + (-3 - 4)x + (-3)(-4)]$ Fits the pattern $(x + a)(x + b) = x^2 + (a + b)x + ab$.
$= 2(x - 3)(x - 4)$ 'Trial and error' to find $-3 - 4 = -7$ and $(-3)(-4) = 12$.

b) The terms have no common factor and the leading coefficient is not equal to one. This factorization requires a logical 'trial and error' approach. There are eight possible factorizations.

$(2x - 1)(x + 15)$ $(2x - 3)(x + 5)$ $(2x - 5)(x + 3)$ $(2x - 15)(x + 1)$
$(2x + 1)(x - 15)$ $(2x + 3)(x - 5)$ $(2x + 5)(x - 3)$ $(2x + 15)(x - 1)$

Testing the middle term in each, you find that the correct factorization is $2x^2 + x - 15 = (2x - 5)(x + 3)$.

c) Factor out GCF then write as difference of two squares in the form $a^2 - b^2 = (a + b)(a - b)$.

$8x^7 - 18x = 2x(4x^6 - 9) = 2x[(2x^3)^2 - 3^2] = 2x(2x^3 + 3)(2x^3 - 3)$

d) $3y^3 + 24y^2 + 48y = 3y(y^2 + 8y + 16)$ Factor out the greatest common factor.
$= 3y(y^2 + 2 \cdot 4y + 4^2)$ Fits the pattern $a^2 + 2ab + b^2 = (a + b)^2$.
$= 3y(y + 4)^2$

e) Fits the difference of two squares pattern: $a^2 - b^2 = (a + b)(a - b)$ with $a = x + 3$ and $b = y$.
Therefore, $(x + 3)^2 - y^2 = [(x + 3) + y][(x + 3) - y]$
$= (x + y + 3)(x - y + 3)$

f) $5x^3y + 20xy^3 = 5xy(x^2 + 4y^2)$ We can only factor out the greatest common factor of $5xy$. Although both of the terms x^2 and $4y^2$ are perfect squares, the expression $x^2 + 4y^2$ is not a *difference* of squares – and, hence, it cannot be factorized. The sum of two squares, $a^2 + b^2$, cannot be factorized.

g) This binomial is the sum of two cubes, fitting the pattern $a^3 + b^3 = (a + b)(a^2 - ab + b^2)$.

$c^3 + 27 = c^3 + 3^3 = (c + 3)(c^2 - 3c + 9)$

h) This binomial is the difference of two cubes, fitting the pattern $a^3 - b^3 = (a - b)(a^2 + ab + b^2)$.

$1 - 8h^6 = 1^3 - (2h^2)^3 = (1 - 2h)(1 + 2h^2 + 4h^4)$

i) This binomial is the difference of two squares – but be sure to factorize completely.

$a^4 - \frac{1}{16} = (a^2 - \frac{1}{4})(a^2 + \frac{1}{4}) = (a + \frac{1}{2})(a - \frac{1}{2})(a^2 + \frac{1}{4})$

j) Write the terms in order of descending exponents and then factor out the -1 so that the leading coefficient is positive.

$15 - x^2 - 2x = -x^2 - 2x + 15 = -(x^2 + 2x - 15) = -(x + 5)(x - 3)$

k) When searching for factors of a quadratic like $3x^2 + 20x - 7$ we restrict our search to factors with coefficients and constants that are integers. Since 3 is a prime number, then we can start the factorizing by writing $3x^2 + 20x - 7 = (3x + ?)(x + ?)$. We know the two missing numbers have a product of -7, and since 7 is a prime number then the two missing numbers are either -7 and 1, or -1 and 7. With trial and error, it can be determined that $3x^2 + 20x - 7 = (3x - 1)(x + 7)$.

l) This fits the factoring pattern of $a^2 + 2ab + b^2 = (a + b)^2$ (trinomial perfect square). Consider the pattern written as $a^2 + (2b)a + b^2$ and substitute y for a, then $y^2 + (2b)y + b^2$. The last term, $\frac{25}{4}(b^2)$, is the square of $\frac{5}{2}$ which is one-half of 5, the coefficient of the middle term $(2b)$. Thus, $y^2 + 5y + \frac{25}{4} = (y + \frac{5}{2})^2$.

Guidelines for factoring polynomials

1. Factor out the greatest common factor (GCF), if one exists.
2. Determine if the polynomial, or any factors, fit any of the special polynomial patterns – and factor accordingly.
3. Any quadratic trinomial of the form $ax^2 + bx + c$ will require a logical trial and error approach, if it factorizes.

Most polynomials cannot be factored into a product of polynomials with integer or rational coefficients. In fact, factorizing is often difficult even when possible for polynomials with degree 3 or higher. Nevertheless, factorizing is a powerful algebraic technique that can be applied in many situations.

Algebraic fractions

An **algebraic fraction** (or rational expression) is a quotient of two algebraic expressions or two polynomials. Given a certain algebraic fraction, we must assume that the variable can only have values so that the denominator is not zero. For example, for the algebraic fraction $\frac{x + 3}{x^2 - 4}$, x cannot be 2 or -2. Most of the algebraic fractions that we will encounter will have numerators and denominators that are polynomials.

• **Hint:** Only common **factors** can be cancelled between the numerator and denominator of a fraction. For example, $\frac{5\times 3}{3}=\frac{5}{1}\times\frac{3}{3}=5\times 1=5$ where the common factors of 3 cancel; that is, $\frac{5\times \cancel{3}}{\cancel{3}}=5$. However, a common error is cancelling common terms that are *not* factors. For example, avoid the following common error: $\frac{5+3}{3}=\frac{5+\cancel{3}}{\cancel{3}}=5$. This is clearly incorrect, because $\frac{5+3}{3}=\frac{8}{3}\neq 5$.

Simplifying algebraic fractions

When trying to simplify algebraic fractions we need to completely factor the numerator and denominator and cancel any common factors.

Example 16 – Cancelling common factors in fractions

Simplify:

a) $\frac{2a^2-2ab}{6ab-6b^2}$ b) $\frac{1-x^2}{x^2+x-2}$ c) $\frac{(x+h)^2-x^2}{h}$

Solution

a) $\frac{2a^2-2ab}{6ab-6b^2}=\frac{2a(\cancel{a-b})}{6b(\cancel{a-b})}=\frac{{}^1\cancel{2}a}{{}_3\cancel{6}b}=\frac{a}{3b}$

b) $\frac{1-x^2}{x^2+x-2}=\frac{(1-x)(1+x)}{(x-1)(x+2)}=\frac{-(-1+x)(1+x)}{(x-1)(x+2)}=\frac{-\cancel{(x-1)}(x+1)}{\cancel{(x-1)}(x+2)}$

$=-\frac{x+1}{x+2}$ or $-\frac{x-1}{x+2}$

c) $\frac{(x+h)^2-x^2}{h}=\frac{x^2+2hx+h^2-x^2}{h}=\frac{2hx+h^2}{h}=\frac{\cancel{h}(2x+h)}{\cancel{h}}=2x+h$

Adding and subtracting algebraic fractions

Before any fractions – numerical or algebraic – can be added or subtracted they must be expressed with the same denominator, preferably the least common denominator. Then the numerators can be added or subtracted according to the rule: $\frac{a}{b}+\frac{c}{d}=\frac{ad}{bd}+\frac{bc}{bd}=\frac{ad+bc}{bd}$.

Example 17 – Working with algebraic fractions

Perform the indicated operation and simplify.

a) $x-\frac{1}{x}$ b) $\frac{2}{a+b}+\frac{3}{a-b}$ c) $\frac{2}{x+2}-\frac{x-4}{2x^2+x-6}$

Solution

a) $x-\frac{1}{x}=\frac{x}{1}-\frac{1}{x}=\frac{x^2}{x}-\frac{1}{x}=\frac{x^2-1}{x}$ or $\frac{(x+1)(x-1)}{x}$

b) $\frac{2}{a+b}+\frac{3}{a-b}=\frac{2}{a+b}\cdot\frac{a-b}{a-b}+\frac{3}{a-b}\cdot\frac{a+b}{a+b}=\frac{2(a-b)+3(a+b)}{(a+b)(a-b)}$

$=\frac{2a-2b+3a+3b}{a^2-b^2}=\frac{5a+b}{a^2-b^2}$

c) $\frac{2}{x+2}-\frac{x-4}{2x^2+x-6}=\frac{2}{x+2}-\frac{x-4}{(2x-3)(x+2)}$

$=\frac{2}{x+2}\cdot\frac{2x-3}{2x-3}-\frac{x-4}{(2x-3)(x+2)}$

$=\frac{2(2x-3)-(x-4)}{(2x-3)(x+2)}$

$=\frac{4x-6-x+4}{(2x-3)(x+2)}$

$=\frac{3x-2}{(2x-3)(x+2)}$ or $\frac{3x-2}{2x^2+x-6}$

• **Hint:** Although it is true that $\frac{a+b}{c}=\frac{a}{c}+\frac{b}{c}$, be careful to avoid an error here: $\frac{a}{b+c}\not\equiv\frac{a}{b}+\frac{a}{c}$. Also, be sure to only cancel common *factors* between numerator and denominator. It is true that $\frac{ac}{bc}=\frac{a}{b}$ (with the common factor of c cancelling) because $\frac{ac}{bc}=\frac{a}{b}\cdot\frac{c}{c}=\frac{a}{b}\cdot 1=\frac{a}{b}$; but, in general, it is *not* true that $\frac{a+c}{b+c}=\frac{a}{b}$. The term c is not a common factor of the numerator and denominator.

Simplifying a compound fraction

Fractional expressions with fractions in the numerator or denominator, or both, are usually referred to as compound fractions. A compound fraction is best simplified by first simplifying both its numerator and denominator into single fractions, and then multiplying numerator and denominator by the reciprocal of the denominator, i.e. $\dfrac{\frac{a}{b}}{\frac{c}{d}} = \dfrac{\frac{a}{b}\cdot\frac{d}{c}}{\frac{c}{d}\cdot\frac{d}{c}} = \dfrac{\frac{ad}{bc}}{1} = \dfrac{ad}{bc}$; thereby expressing the compound fraction as a single fraction.

Example 18 – Simplifying compound fractions

Simplify:

a) $\dfrac{\frac{1}{x+h} - \frac{1}{x}}{h}$ b) $\dfrac{\frac{a}{b}+1}{1-\frac{a}{b}}$ c) $\dfrac{x(1-2x)^{-\frac{3}{2}} + (1-2x)^{-\frac{1}{2}}}{1-x}$

Solution

a) $\dfrac{\frac{1}{x+h} - \frac{1}{x}}{h} = \dfrac{\frac{x}{x(x+h)} - \frac{x+h}{x(x+h)}}{\frac{h}{1}} = \dfrac{\frac{x-(x+h)}{x(x+h)}}{\frac{h}{1}} = \dfrac{x-x-h}{x(x+h)}\cdot\dfrac{1}{h}$

$= \dfrac{-\cancel{h}}{x(x+h)}\cdot\dfrac{1}{\cancel{h}} = -\dfrac{1}{x(x+h)}$

b) $\dfrac{\frac{a}{b}+1}{1-\frac{a}{b}} = \dfrac{\frac{a}{b}+\frac{b}{b}}{\frac{b}{b}-\frac{a}{b}} = \dfrac{\frac{a+b}{b}}{\frac{b-a}{b}} = \dfrac{a+b}{\cancel{b}}\cdot\dfrac{\cancel{b}}{b-a} = \dfrac{a+b}{b-a}$

c) $\dfrac{x(1-2x)^{-\frac{3}{2}} + (1-2x)^{-\frac{1}{2}}}{1-x} = \dfrac{(1-2x)^{-\frac{3}{2}}[\,x + (1-2x)^1]}{1-x}$ Factor out the power of $1-2x$ with the *smallest* exponent.

$= \dfrac{(1-2x)^{-\frac{3}{2}}[\,x+1-2x]}{1-x}$

$= \dfrac{(1-2x)^{-\frac{3}{2}}\cancel{(1-x)}}{\cancel{1-x}}$

$= \dfrac{1}{(1-2x)^{\frac{3}{2}}}$

With rules for rational exponents and radicals we can rewrite the result from c) above, but it's not any *simpler*…

$$\frac{1}{(1-2x)^{\frac{3}{2}}} = \frac{1}{\sqrt{(3x-2)^3}} = \frac{1}{\sqrt{(3x-2)^2}\sqrt{3x-2}} = \frac{1}{|3x-2|\sqrt{3x-2}}$$

Rationalizing the denominator

Recall Example 9 from Section 1.2, where we rationalized the denominator of the numerical fractions $\dfrac{2}{\sqrt{3}}$ and $\dfrac{\sqrt{7}}{4\sqrt{10}}$. Also recall that expressions of the form $a+b$ and $a-b$ are called conjugates and their product is a^2-b^2 (difference of two squares). If a fraction has an irrational denominator of the form $a+b\sqrt{c}$, we can change it to a rational expression by multiplying numerator and denominator by its conjugate $a-b\sqrt{c}$, given that $(a+b\sqrt{c})(a-b\sqrt{c}) = a^2-(b\sqrt{c})^2 = a^2-b^2c$.

Example 19 – Rationalizing the denominator

Rationalize the denominator of each fractional expression.

a) $\dfrac{1}{1-\sqrt{x}}\ x \geqslant 0, x \neq 1$ b) $\dfrac{x-2}{x+3\sqrt{2}}$

Solution

a) $\dfrac{1}{1-\sqrt{x}} = \dfrac{1}{1-\sqrt{x}} \cdot \dfrac{1+\sqrt{x}}{1+\sqrt{x}} = \dfrac{1+\sqrt{x}}{1-(\sqrt{x})^2} = \dfrac{1+\sqrt{x}}{1-x}$

b) $\dfrac{x-\sqrt{2}}{x+3\sqrt{2}} = \dfrac{x-\sqrt{2}}{x+3\sqrt{2}} \cdot \dfrac{x-3\sqrt{2}}{x-3\sqrt{2}}$

$$= \frac{x^2-(3\sqrt{2})x-(\sqrt{2})x+3\cdot 2}{x^2-(3\sqrt{2})^2} = \frac{x^2-(4\sqrt{2})x+6}{x-18}$$

Exercise 1.5

In questions 1–16, expand and simplify.

1 $(x-4)(x+5)$
2 $(3h-1)(2h-3)$
3 $(y+9)(y-9)$
4 $(4x+2)^2$
5 $(2n-5)^2$
6 $(2y-5)^3$
7 $(6a-7b)(6a+7b)$
8 $(2x+3+y)(2x+3-y)$
9 $(ax+b)^3$
10 $(ax+b)^4$
11 $(2+x\sqrt{5})(2-x\sqrt{5})$
12 $(2x-1)(4x^2+2x+1)$
13 $(x+y-z)^2$
14 $(x+yi)(x-yi)$
15 $(m+3)(3-m)$
16 $(1-\sqrt{x^2+1})^2$

In questions 17–36, completely factorize the expression.

17 $12x^2-48$
18 x^3-6x^2
19 x^2+x-12
20 $7-6m-m^2$
21 $x^2-10x+16$
22 y^2+7y+6
23 $3n^2-21n+30$
24 $2x^3+20x^2+18x$
25 a^2-16
26 $3y^2-14y-5$
27 $25n^4-4$
28 $ax^2+6ax+9a$
29 $2n(m+1)^2-(m+1)^2$
30 x^4-1
31 $9-(y-3)^2$
32 $4y^4-10y^3-96y^2$
33 $4x^2-20x+25$
34 $(2x+3)^{-2}+2x(2x+3)^{-3}$
35 $(n-2)^4-(n-2)^3(2n-3)$
36 $m^3-\frac{4}{3}m^2+\frac{4}{9}m$

In questions 37–46, simplify the algebraic fraction.

37 $\dfrac{x+4}{x^2+5x+4}$
38 $\dfrac{3n-3}{6n^2-6n}$
39 $\dfrac{a^2-b^2}{5a-5b}$
40 $\dfrac{x^2+4x+4}{x+2}$

41 $\dfrac{2a-5}{5-2a}$

42 $\dfrac{(2x+h)^2-4x^2}{h}$

43 $\dfrac{(x+1)^3(3x-5)-(x+1)^2(8x+3)}{(x-4)(x+1)^3}$

44 $\dfrac{3y(y+3)-2(2y+1)}{(y+2)^2}$

45 $\dfrac{a-\dfrac{a^2}{b}}{\dfrac{a^2}{b}-a}$

46 $\dfrac{1+\dfrac{1}{1+\dfrac{1}{x-1}}}{1-\dfrac{1}{x-1}}$

In questions 47–60, perform the indicated operation and simplify.

47 $\frac{1}{n}-1$

48 $\dfrac{2}{2x-1}-4$

49 $\frac{x}{5}-\frac{x-1}{3}$

50 $\frac{1}{a}-\frac{1}{b}$

51 $\dfrac{1}{(x-3)^2}-\dfrac{3}{x-3}$

52 $\dfrac{x}{x+3}+\dfrac{1}{x}$

53 $\dfrac{1}{x+y}+\dfrac{1}{x-y}$

54 $\dfrac{3}{x-2}+\dfrac{5}{2-x}$

55 $\dfrac{2x-6}{x}\cdot\dfrac{3x}{x-3}$

56 $\dfrac{2x+6}{7}\times\dfrac{1}{x^2-9}$

57 $\dfrac{a+b}{b}\cdot\dfrac{1}{a^2-b^2}$

58 $\dfrac{3x^2-3}{6x}\times\dfrac{5x^2}{1-x}$

59 $\dfrac{3}{y+2}+\dfrac{5}{y^2-3y-10}$

60 $\dfrac{8}{9-x^2}\div\dfrac{2x}{x^3-x^2-6x}$

In questions 61–64, rationalize the denominator of each fractional expression.

61 $\dfrac{1}{x-\sqrt{2}}$

62 $\dfrac{5}{2+x\sqrt{3}}$

63 $\dfrac{\sqrt{x}+\sqrt{y}}{\sqrt{x}-\sqrt{y}}$

64 $\dfrac{1}{\sqrt{x+h}+\sqrt{x}}$

1.6 Equations and formulae

Equations, identities and formulae

We will encounter a wide variety of equations in this course. Essentially an equation is a statement equating two algebraic expressions that may be true or false depending upon what value(s) are substituted for the variable(s). The value(s) of the variable(s) that make the equation true are called the **solutions** or **roots** of the equation. All of the solutions to an equation comprise the **solution set** of the equation. An equation that is true for all possible values of the variable is called an **identity**. All of the common polynomial expansion and factorization patterns shown in Section 1.5 are identities. For example, $(a+b)^2 = a^2 + 2ab + b^2$ is true for all values of a and b. The following are also examples of identities.

$$3(x-5) = 2(x+3) + x - 21 \qquad (x+y)^2 - 2xy = x^2 + y^2$$

One of the most famous equations in the history of mathematics, $x^n + y^n = z^n$, is associated with Pierre Fermat (1601–1665), a French lawyer and amateur mathematician. Writing in the margin of a French translation of *Arithmetica*, considered to be the first book of algebra, written by the 3rd-century BC Greek mathematician Diophantus, Fermat conjectured that the equation $x^n + y^n = z^n$ $(x, y, z, n \in \mathbb{Z})$ has no non-zero solutions for the variables x, y and z when the parameter n is greater than two. When $n = 2$ the equation is equivalent to Pythagoras' theorem for which there are an infinite number of integer solutions – Pythagorean triples, such as $3^2 + 4^2 = 5^2$ and $5^2 + 12^2 = 13^2$, and their multiples. Fermat claimed to have a proof for his conjecture but that he could not fit it in the margin. All the other margin conjectures in Fermat's copy of *Arithmetica* were proven by the start of the 19th century but this one remained unproven for over 350 years, until the English mathematician Andrew Wiles proved it in 1994.

An equation may be referred to as a **formula** (plural: formulae). These typically contain more than one variable and, often, other symbols that represent specific constants or **parameters** (constants that may change in value but do not alter the properties of the expression). Formulae with which you may be familiar include:

$A = \pi r^2$, $d = rt$, $d = \sqrt{(x_1 - x_2)^2 + (y_1 - y_2)^2}$ and $V = \frac{4}{3}\pi r^3$

Whereas most equations that we will encounter have numerical solutions, we can solve a formula for a certain variable in terms of other variables – sometimes referred to as changing the subject of a formula.

Example 20 – Changing the subject of a formula

Solve for the indicated variable in each formula.

a) $a^2 + b^2 = c^2$ Solve for *b*.

b) $T = 2\pi\sqrt{\frac{l}{g}}$ Solve for *l*.

c) $I = \frac{nR}{R + r}$ Solve for *R*.

Solution

a) $a^2 + b^2 = c^2 \Rightarrow b^2 = c^2 - a^2 \Rightarrow b = \pm\sqrt{c^2 - a^2}$
If b is a length then $b = \sqrt{c^2 - a^2}$.

b) $T = 2\pi\sqrt{\frac{l}{g}} \Rightarrow \sqrt{\frac{l}{g}} = \frac{T}{2\pi} \Rightarrow \frac{l}{g} = \frac{T^2}{4\pi^2} \Rightarrow l = \frac{T^2 g}{4\pi^2}$

c) $I = \frac{nR}{R + r} \Rightarrow I(R + r) = nR \Rightarrow IR + Ir = nR \Rightarrow IR - nR = -Ir$

$\Rightarrow R(I - n) = -Ir \Rightarrow R = \frac{-Ir}{I - n}$

Note that factorization was required in solving for R in Example 20 c).

Equations and graphs

Two important characteristics of any equation are the number of variables (unknowns) and the type of algebraic expressions it contains (e.g. polynomials, rational expressions, trigonometric, exponential, etc.). Nearly all of the equations in this course will have either one or two variables, and in this introductory chapter we will only discuss equations with algebraic expressions that are polynomials. Solutions for equations with a single variable will consist of individual numbers that can be *graphed* as points on a number line. The **graph** of an equation is a visual representation of the equation's solution set. For example, the solution set of the one-variable equation containing quadratic and linear polynomials $x^2 = 2x + 8$ is $x \in \{-2, 4\}$. The graph of this one-variable equation (Figure 1.5) is depicted below on a one-dimensional coordinate system, i.e. the real number line.

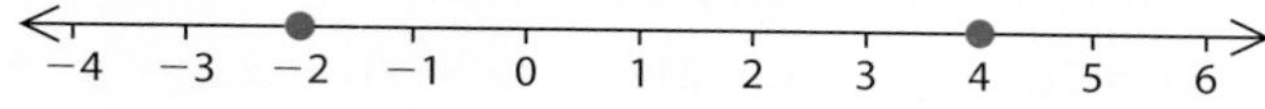

Figure 1.5 Graph of the solution set for the equation $x^2 = 2x + 8$.

The solution set of a two-variable equation will be an **ordered pair** of numbers. An ordered pair corresponds to a location indicated by a point on a two-dimensional coordinate system, i.e. a **coordinate plane**. For example, the solution set of the two-variable **quadratic equation** $y = x^2$ will be an infinite set of ordered pairs (x, y) that satisfy the equation. Four ordered pairs in the solution set are graphed in Figure 1.6 in red. The graph of all the ordered pairs in the solution set form a curve as shown in blue. (Quadratic equations will be covered in detail in Chapter 3.)

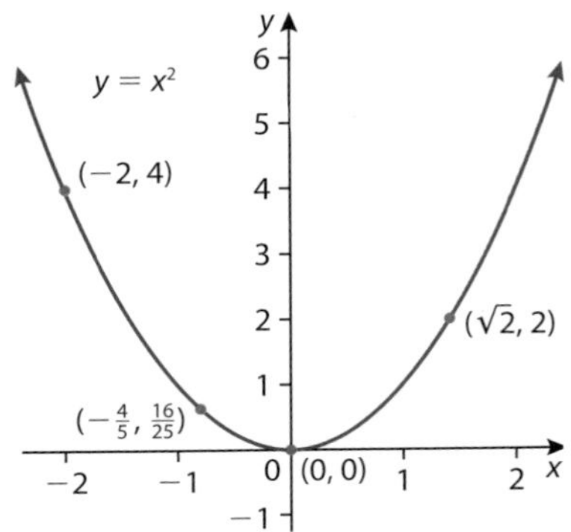

Figure 1.6 Four ordered pairs in the solution set of $y = x^2$ are graphed in red. The graph of all the ordered pairs in the solution set form a curve, as shown in blue.

Equations of lines

A one-variable **linear equation** in x can always be written in the form $ax + b = 0$, $a \neq 0$ and it will have exactly one solution, $x = -\frac{b}{a}$.

An example of a two-variable **linear equation in x and y** is $x - 2y = 2$. The graph of this equation's solution set (an infinite set of ordered pairs) is a **line** (Figure 1.7).

The **slope** m, or **gradient**, of a non-vertical line is defined by the formula $m = \frac{y_2 - y_1}{x_2 - x_1} = \frac{\text{vertical change}}{\text{horizontal change}}$. Because division by zero is undefined, the slope of a vertical line is undefined. Using the two points $(1, -\frac{1}{2})$ and $(4, 1)$, we compute the slope of the line with equation $x - 2y = 2$ to be

$$m = \frac{1 - \left(-\frac{1}{2}\right)}{4 - 1} = \frac{\frac{3}{2}}{\frac{3}{1}} = \frac{1}{2}.$$

If we solve for y, we can rewrite the equation in the form $y = \frac{1}{2}x - 1$. Note that the coefficient of x is the slope of the line and the constant term is the y-coordinate of the point at which the line intersects the y-axis, i.e. the y-intercept. There are several forms in which to write linear equations whose graphs are lines.

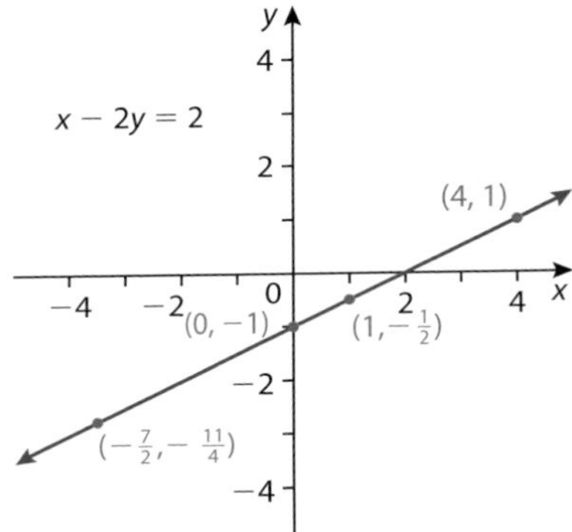

Figure 1.7 The graph of $x - 2y = 2$, ordered pairs shown in red.

Form	Equation	Characteristics
general form	$ax + by + c = 0$	every line has an equation in this form if both a and $b \neq 0$
slope-intercept form	$y = mx + c$	m is the slope; $(0, c)$ is the y-intercept
point-slope form	$y - y_1 = m(x - x_1)$	m is the slope; (x_1, y_1) is a known point on the line
horizontal line	$y = c$	slope is zero; $(0, c)$ is the y-intercept
vertical line	$x = c$	slope is undefined; unless line is y-axis, no y-intercept

Table 1.5 Forms of equations of lines.

Most problems involving equations and graphs fall into two categories: (1) given an equation, determine its graph; and (2) given a graph, or some information about it, find its equation. For lines, the first type of problem is often best solved by using the slope-intercept form, whereas for the second type of problem the point-slope form is usually most useful.

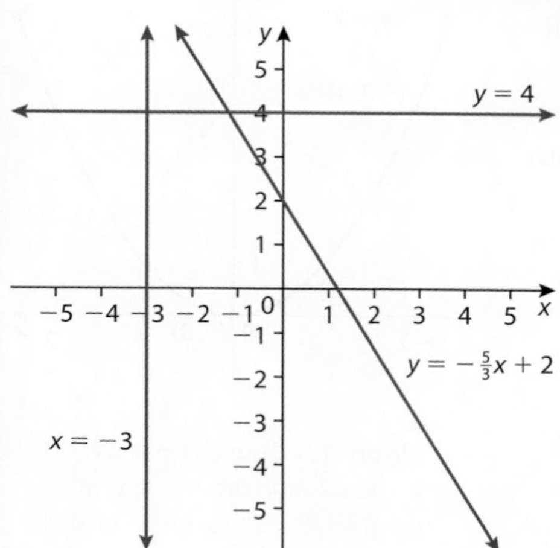

Example 21 – Sketching the graphs of linear equations

Without using a GDC, sketch the line that is the graph of each of the following linear equations written here in general form.

a) $5x + 3y - 6 = 0$

b) $y - 4 = 0$

c) $x + 3 = 0$

Solution

a) Solve for y to write the equation in slope-intercept form. $5x + 3y - 6 = 0 \Rightarrow 3y = -5x + 6 \Rightarrow y = -\frac{5}{3}x + 2$. The line has a y-intercept of $(0, 2)$ and a slope of $-\frac{5}{3}$.

b) The equation $y - 4 = 0$ is equivalent to $y = 4$, whose graph is a horizontal line with a y-intercept of $(0, 4)$.

c) The equation $x + 3 = 0$ is equivalent to $x = -3$, whose graph is a vertical line with no y-intercept; but, it has an x-intercept of $(-3, 0)$.

Example 22 – Finding the equation of a line

a) Find the equation of the line that passes through the point $(3, 31)$ and has a slope of 12. Write the equation in slope-intercept form.

b) Find the linear equation in C and F knowing that when $C = 10$ then $F = 50$, and when $C = 100$ then $F = 212$. Solve for F in terms of C.

Solution

a) Substitute into the point-slope form $y - y_1 = m(x - x_1)$; $x_1 = 3$, $y_1 = 31$ and $m = 12$.

$$y - y_1 = m(x - x_1) \Rightarrow y - 31 = 12(x - 3) \Rightarrow y = 12x - 36 + 31 \Rightarrow y = 12x - 5$$

b) The two points, ordered pairs (C, F), that are known to be on the line are $(10, 50)$ and $(100, 212)$. The variable C corresponds to the variable x and F corresponds to y in the definitions and forms stated above. The slope of the line is $m = \frac{F_2 - F_1}{C_2 - C_1} = \frac{212 - 50}{100 - 10} = \frac{162}{90} = \frac{9}{5}$. Choose one of the points on the line, say $(10, 50)$, and substitute it and the slope into the point-slope form.

$$F - F_1 = m(C - C_1) \Rightarrow F - 50 = \frac{9}{5}(C - 10) \Rightarrow F = \frac{9}{5}C - 18 + 50 \Rightarrow F = \frac{9}{5}C + 32$$

The slope of a line is a convenient tool for determining whether two lines are parallel or perpendicular. The two lines graphed in Figure 1.8 suggests the following property: Two distinct non-vertical lines are **parallel** if and only if their slopes are equal, $m_1 = m_2$.

The two lines graphed in Figure 1.9 suggests another property: Two non-vertical lines are perpendicular if and only if their slopes are negative reciprocals – that is, $m_1 = -\frac{1}{m_2}$, which is equivalent to $m_1 \cdot m_2 = -1$.

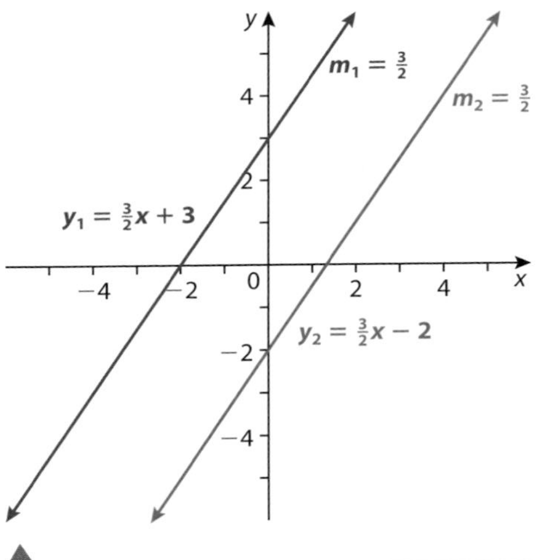

Figure 1.8 Parallel lines.

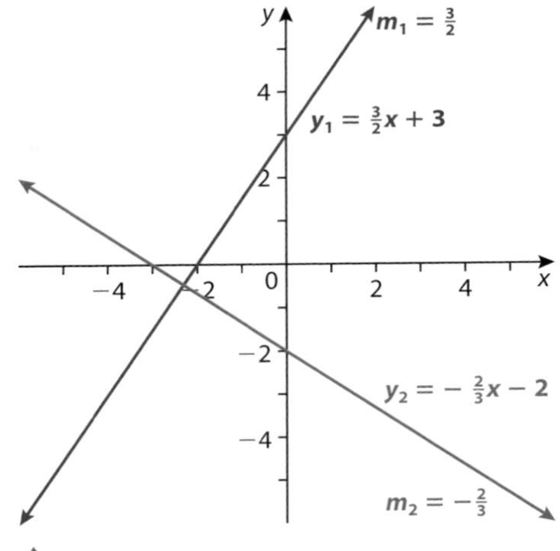

Figure 1.9 Perpendicular lines.

Distances and midpoints

Recall from Section 1.1 that absolute value (modulus) is used to define the **distance** (always positive) between two points on the real number line.

The distance between the points A and B on the real number line is $|B - A|$, which is equivalent to $|A - B|$.

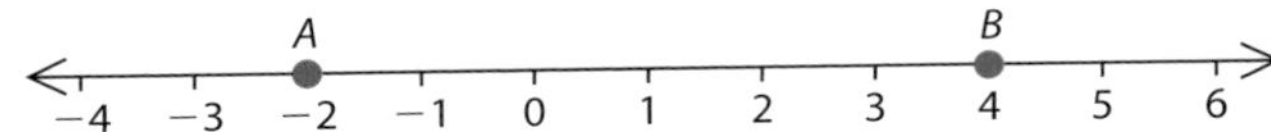

Figure 1.10 The length of the line segment $[AB]$ is AB.

The points A and B are the endpoints of a line segment that is denoted with the notation $[AB]$ and the length of the line segment is denoted AB. In Figure 1.10, the distance between A and B is $AB = |4 - (-2)| = |-2 - 4| = 6$.

The distance between two general points (x_1, y_1) and (x_2, y_2) on a coordinate plane can be found using the definition for distance on a number line and Pythagoras' theorem. For the points (x_1, y_1) and (x_2, y_2), the horizontal distance between them is $|x_1 - x_2|$ and the vertical distance is $|y_1 - y_2|$. As illustrated in Figure 1.11, these distances are the lengths of two legs of a right-angled triangle whose hypotenuse is the distance between the points. If d represents the distance between (x_1, y_1) and (x_2, y_2), then by Pythagoras' theorem $d^2 = |x_1 - x_2|^2 + |y_1 - y_2|^2$. Because the square of any number is positive, the absolute value is not necessary, giving us the **distance formula** for two-dimensional coordinates.

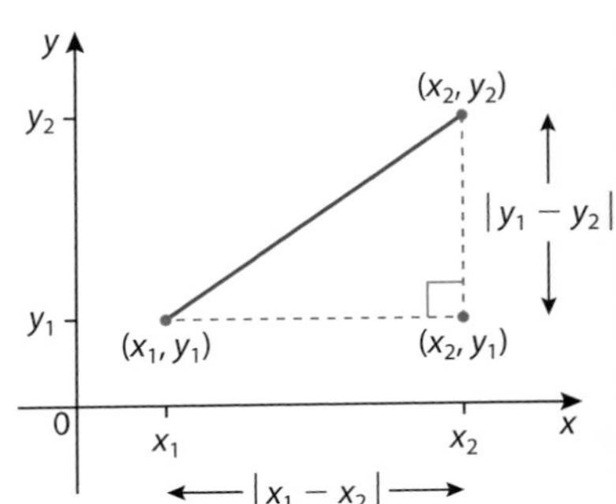

Figure 1.11 Distance between two points on a coordinate plane.

The distance formula

The distance d between the two points (x_1, y_1) and (x_2, y_2) in the coordinate plane is

$$d = \sqrt{(x_1 - x_2)^2 + (y_1 - y_2)^2}$$

The coordinates of the **midpoint** of a line segment are the average values of the corresponding coordinates of the two endpoints.

The midpoint formula

The midpoint of the line segment joining the points (x_1, y_1) and (x_2, y_2) in the coordinate plane is

$$\left(\frac{x_1 + x_2}{2}, \frac{y_1 + y_2}{2}\right)$$

Example 23 – Using the distance and midpoint formulae

a) Show that the points $P(1, 2)$, $Q(3, 1)$ and $R(4, 8)$ are the vertices of a right triangle.

b) Find the midpoint of the hypotenuse.

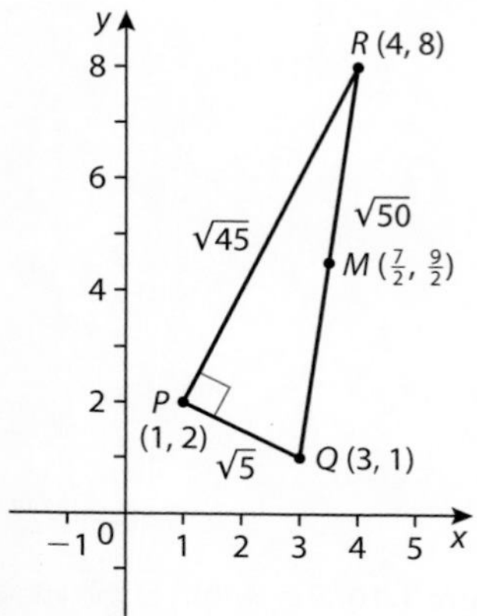

Figure 1.12 Diagram for Example 23.

Solution

a) The three points are plotted and the line segments joining them are drawn in Figure 1.12. Applying the distance formula, we can find the exact lengths of the three sides of the triangle.

$PQ = \sqrt{(1-3)^2 + (2-1)^2} = \sqrt{4+1} = \sqrt{5}$

$QR = \sqrt{(3-4)^2 + (1-8)^2} = \sqrt{1+49} = \sqrt{50}$

$PR = \sqrt{(1-4)^2 + (2-8)^2} = \sqrt{9+36} = \sqrt{45}$

$PQ^2 + PR^2 = QR^2$ because $(\sqrt{5})^2 + (\sqrt{45})^2 = 5 + 45 = 50 = (\sqrt{50})^2$. The lengths of the three sides of the triangle satisfy Pythagoras' theorem, confirming that the triangle is a right-angled triangle.

b) QR is the hypotenuse. Let the midpoint of QR be point M. Using the midpoint formula, $M = \left(\frac{3+4}{2}, \frac{1+8}{2}\right) = \left(\frac{7}{2}, \frac{9}{2}\right)$. This point is plotted in Figure 1.12.

Example 24 – Using the distance formula

Find x so that the distance between the points $(1, 2)$ and $(x, -10)$ is 13.

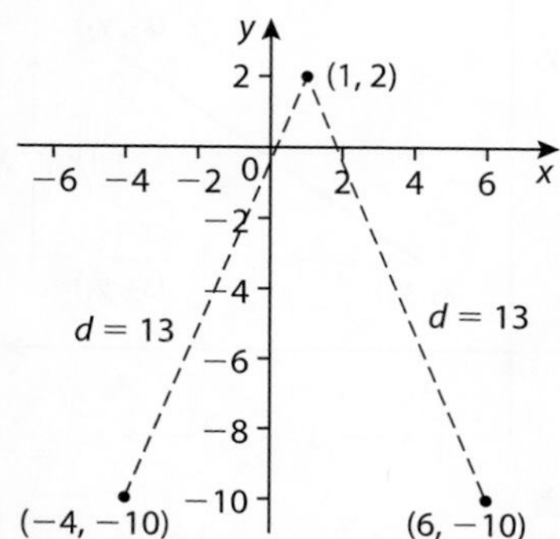

Figure 1.13 The graph shows the two points that are both a distance of 13 from (1, 2).

Solution

$d = 13 = \sqrt{(x-1)^2 + (-10-2)^2} \Rightarrow 13^2 = (x-1)^2 + (-12)^2$

$\Rightarrow 169 = x^2 - 2x + 1 + 144 \Rightarrow x^2 - 2x - 24 = 0$

$\Rightarrow (x-6)(x+4) = 0 \Rightarrow x - 6 = 0 \text{ or } x + 4 = 0$

$\Rightarrow x = 6 \text{ or } x = -4$

Simultaneous equations

Many problems that we solve with algebraic techniques involve sets of equations with several variables, rather than just a single equation with one or two variables. Such a set of equations is called a set of **simultaneous**

equations because we find the values for the variables that solve all of the equations simultaneously. In this section, we consider only the simplest set of simultaneous equations – a pair of linear equations in two variables. We will take a brief look at three methods for solving simultaneous linear equations. They are:

1. Graphical method
2. Elimination method
3. Substitution method

Although we will only look at pairs of linear equations in this section, it is worthwhile mentioning that the graphical and substitution methods are effective for solving sets of equations where not all of the equations are linear, e.g. one linear and one quadratic equation.

Graphical method

The graph of each equation in a system of two linear equations in two unknowns is a line. The graphical interpretation of the solution of a pair of simultaneous linear equations corresponds to determining what point, or points, lies on both lines. Two lines in a coordinate plane can only relate to one another in one of three ways: (1) intersect at exactly one point, (2) intersect at all points on each line (i.e. the lines are identical), or (3) the two lines do not intersect (i.e. the lines are parallel). These three possibilities are illustrated in Figure 1.14.

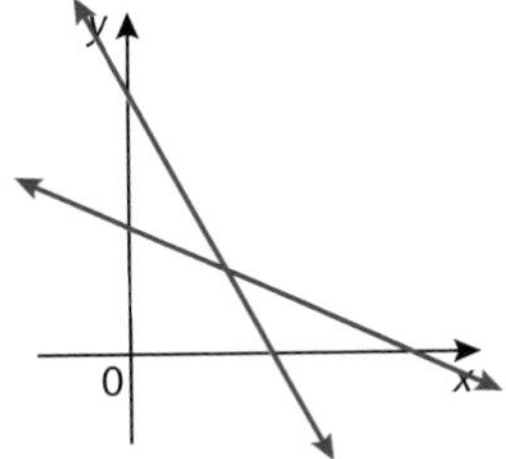

Intersect at exactly one point; exactly one solution

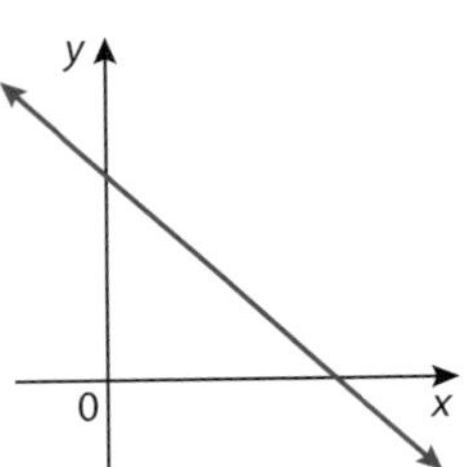

Identical–coincident lines; infinite solutions

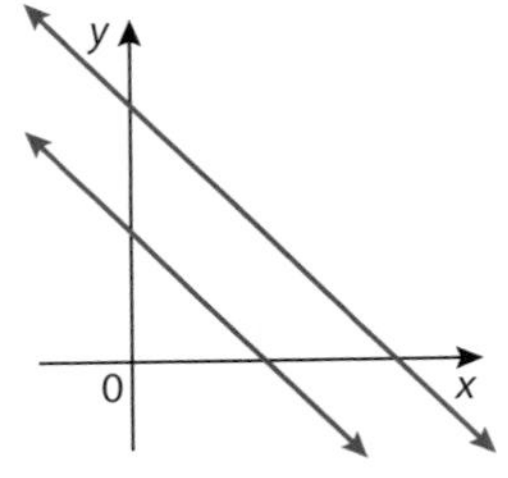

Never intersect–parallel lines; no solution

Figure 1.14 Possible relationship between two lines in a coordinate plane.

Although a graphical approach to solving simultaneous linear equations provides a helpful visual picture of the number and location of solutions, it can be tedious and inaccurate if done by hand. The graphical method is far more efficient and accurate when performed on a graphical display calculator (GDC).

Example 25 – Solving simultaneous equations with a GDC

Use the graphical features of a GDC to solve each pair of simultaneous equations.

a) $2x + 3y = 6$
$2x - y = -10$

b) $7x - 5y = 20$
$3x + y = 2$

Solution

a) First, we will rewrite each equation in slope-intercept form, i.e. $y = mx + c$. This is a necessity if we use our GDC, and is also very useful for graphing by hand (manual).

$2x + 3y = 6 \Rightarrow 3y = -2x + 6 \Rightarrow y = -\frac{2}{3}x + 2$ and $2x - y = -10 \Rightarrow y = 2x + 10$

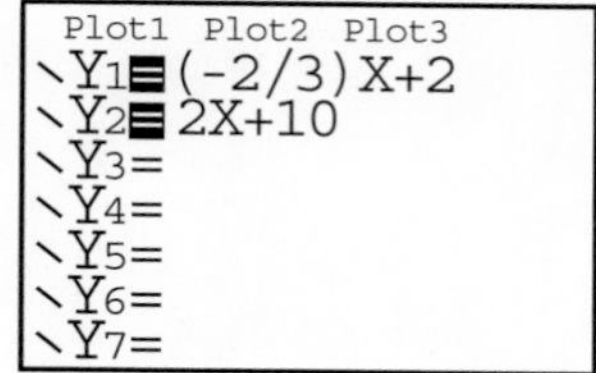

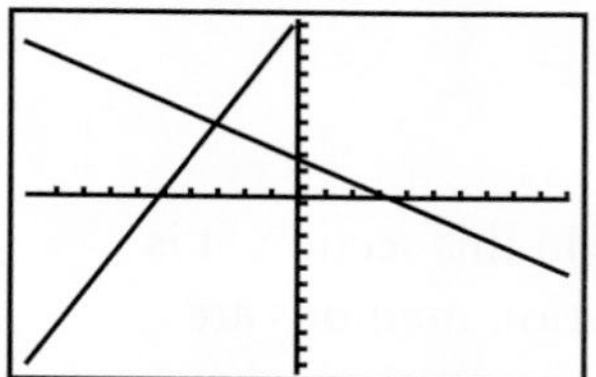

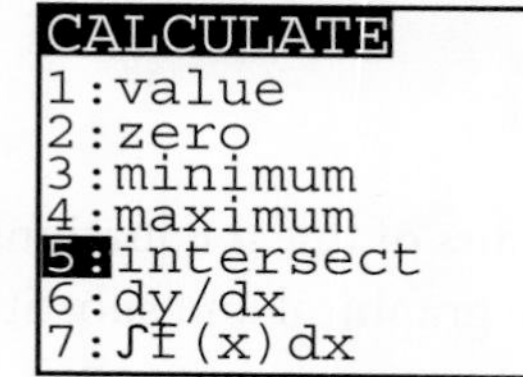

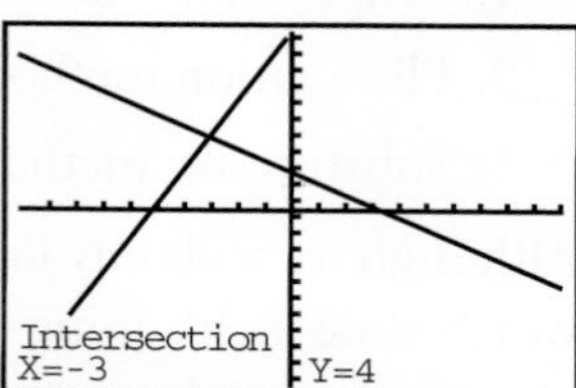

The intersection point and solution to the simultaneous equations is $x = -3$ and $y = 4$, or $(-3, 4)$. If we manually graphed the two linear equations in a) very carefully using graph paper, we may have been able to determine the exact coordinates of the intersection point. However, using a graphical method without a GDC to solve the simultaneous equations in b) would only allow us to crudely approximate the solution.

b) $7x - 5y = 20 \Rightarrow 5y = 7x - 20 \Rightarrow y = \frac{7}{5}x - 4$ and
$3x + y = 2 \Rightarrow y = -3x + 2$

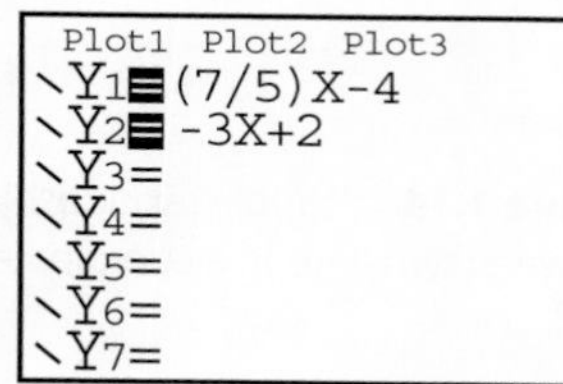

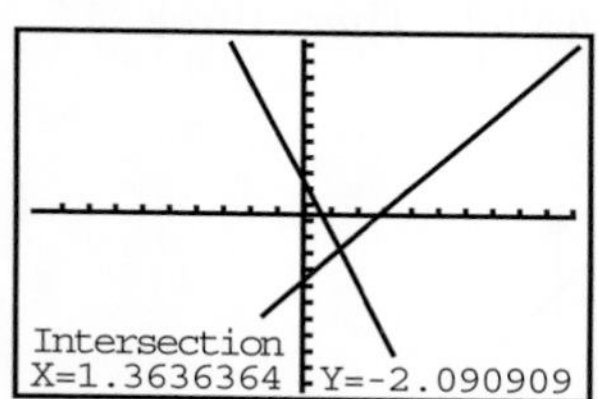

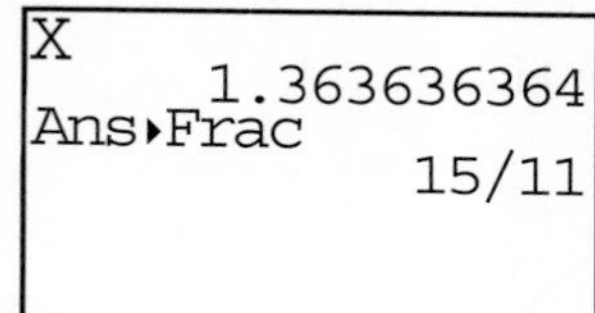

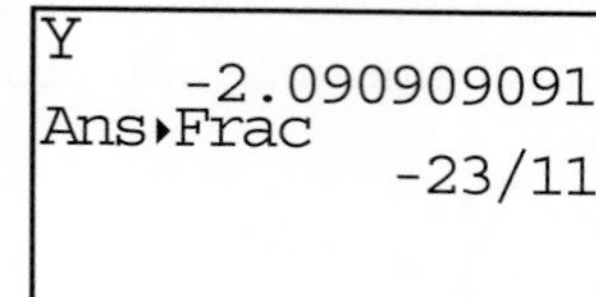

The solution to the simultaneous equations is $x = \frac{15}{11}$ and $y = -\frac{23}{11}$, or $\left(\frac{15}{11}, -\frac{23}{11}\right)$.

The full power and efficiency of the GDC is used in this example to find the exact solution.

Elimination method

To solve a system using the **elimination method**, we try to combine the two linear equations using sums or differences in order to eliminate one of the variables. Before combining the equations, we need to multiply one or both of the equations by a suitable constant to produce coefficients for one of the variables that are equal (then subtract the equations), or that differ only in sign (then add the equations).

Example 26 – Elimination method

Use the elimination method to solve each pair of simultaneous equations.

a) $5x + 3y = 9$
$2x - 4y = 14$

b) $x - 2y = 7$
$2x - 4y = 5$

Solution

a) We can obtain coefficients for y that differ only in sign by multiplying the first equation by 4 and the second equation by 3. Then we add the equations to eliminate the variable y.

$$\begin{array}{rcl} 5x + 3y = 9 & \rightarrow & 20x + 12y = 36 \\ 2x - 4y = 14 & \rightarrow & 6x - 12y = 42 \\ \hline & & 26x = 78 \\ & & x = \frac{78}{26} \\ & & x = 3 \end{array}$$

By substituting the value of 3 for x in either of the original equations we can solve for y.

$5x + 3y = 9 \Rightarrow 5(3) + 3y = 9 \Rightarrow 3y = -6 \Rightarrow y = -2$

The solution is $(3, -2)$.

b) To obtain coefficients for x that are equal, we multiply the first equation by 2 and then subtract the equations to eliminate the variable x.

$$\begin{array}{rcl} x - 2y = 7 & \rightarrow & 2x - 4y = 14 \\ 2x - 4y = 5 & \rightarrow & 2x - 4y = 5 \\ \hline & & 0 = 9 \end{array}$$

Because it is not possible for 0 to equal 9, there is no solution. The lines that are the graphs of the two equations are parallel. To confirm this we can rewrite each of the equations in the form $y = mx + c$.

$$x - 2y = 7 \Rightarrow 2y = x - 7 \Rightarrow y = \tfrac{1}{2}x - \tfrac{7}{2} \text{ and}$$
$$2x - 4y = 5 \Rightarrow 4y = 2x - 5 \Rightarrow y = \tfrac{1}{2}x - \tfrac{5}{2}$$

Both equations have a slope of $\frac{1}{2}$, but different y-intercepts. Therefore, the lines are parallel. This confirms that this pair of simultaneous equations has no solution.

Substitution method

The algebraic method that can be applied effectively to the widest variety of simultaneous equations, including non-linear equations, is the **substitution method**. Using this method, we choose one of the equations and solve for one of the variables in terms of the other variable. We then substitute this expression into the other equation to produce an equation with only one variable, which we can solve directly.

Example 27 – Substitution method

Use the substitution method to solve each pair of simultaneous equations.

a) $3x - y = -9$
$6x + 2y = 2$

b) $-2x + 6y = 4$
$3x - 9y = -6$

Solution

a) Solve for y in the top equation, $3x - y = -9 \Rightarrow y = 3x + 9$, and substitute $3x + 9$ in for y in the bottom equation:
$6x + 2(3x + 9) = 2 \Rightarrow 6x + 6x + 18 = 2 \Rightarrow 12x = -16 \Rightarrow x = -\frac{16}{12} = -\frac{4}{3}$.
Now substitute $-\frac{4}{3}$ for x in either equation to solve for y.
$3\left(-\frac{4}{3}\right) - y = -9 \Rightarrow y = -4 + 9 \Rightarrow y = 5$.
The solution is $x = -\frac{4}{3}, y = 5$, or $\left(-\frac{4}{3}, 5\right)$.

b) Solve for x in the top equation,
$-2x + 6y = 4 \Rightarrow 2x = 6y - 4 \Rightarrow x = 3y - 2$, and substitute $3y - 2$ in for x in the bottom equation:
$3(3y - 2) - 9y = -6 \Rightarrow 9y - 6 - 9y = -6 \Rightarrow 0 = 0$.
The resulting equation $0 = 0$ is true for any values of x and y. The two equations are equivalent, and their graphs will produce identical lines – i.e. coincident lines. Therefore, the solution set consists of all points (x, y) lying on the line $-2x + 6y = 4$ $\left(\text{or } y = \frac{1}{3}x + \frac{2}{3}\right)$.

Exercise 1.6

In questions 1–8, solve for the indicated variable in each formula.

1 $m(h - x) = n$ solve for x

2 $v = \sqrt{ab - t}$ solve for a

3 $A = \frac{h}{2}(b_1 + b_2)$ solve for b_1

4 $A = \frac{1}{2}r^2\theta$ solve for r

5 $\frac{f}{g} = \frac{h}{k}$ solve for k

6 $at = x - bt$ solve for t

7 $V = \frac{1}{3}\pi r^3 h$ solve for r

8 $F = \dfrac{g}{m_1 k + m_2 k}$ solve for k

In questions 9–12, find the equation of the line that passes through the two given points. Write the line in slope-intercept form ($y = mx + c$), if possible.

9 $(-9, 1)$ and $(3, -7)$

10 $(3, -4)$ and $(10, -4)$

11 $(-12, -9)$ and $(4, 11)$

12 $\left(\frac{7}{3}, -\frac{1}{2}\right)$ and $\left(\frac{7}{3}, \frac{5}{2}\right)$

13 Find the equation of the line that passes through the point $(7, -17)$ and is parallel to the line with equation $4x + y - 3 = 0$. Write the line in slope-intercept form ($y = mx + c$).

14 Find the equation of the line that passes through the point $\left(-5, \frac{11}{2}\right)$ and is perpendicular to the line with equation $2x - 5y - 35 = 0$. Write the line in slope-intercept form ($y = mx + c$).

In questions 15–18, a) find the exact distance between the points, and b) find the midpoint of the line segment joining the two points.

15 $(-4, 10)$ and $(4, -5)$

16 $(-1, 2)$ and $(5, 4)$

17 $\left(\frac{1}{2}, 1\right)$ and $\left(-\frac{5}{2}, \frac{4}{3}\right)$

18 $(12, 2)$ and $(-10, 9)$

In questions 19 and 20, find the value(s) of k so that the distance between the points is 5.

19 (5, −1) and (k, 2)

20 (−2, −7) and (1, k)

In questions 21–23, show that the given points form the vertices of the indicated polygon.

21 Right-angled triangle: (4, 0), (2, 1) and (−1, −5)

22 Isosceles triangle: (1, −3), (3, 2) and (−2, 4)

23 Parallelogram: (0, 1), (3, 7), (4, 4) and (1, −2)

In questions 24–29, use the elimination method to solve each pair of simultaneous equations.

24 $x + 3y = 8$
$x - 2y = 3$

25 $x - 6y = 1$
$3x + 2y = 13$

26 $6x + 3y = 6$
$5x + 4y = -1$

27 $x + 3y = -1$
$x - 2y = 7$

28 $8x - 12y = 4$
$-2x + 3y = 2$

29 $5x + 7y = 9$
$-11x - 5y = 1$

In questions 30–35, use the substitution method to solve each pair of simultaneous equations.

30 $2x + y = 1$
$3x + 2y = 3$

31 $3x - 2y = 7$
$5x - y = -7$

32 $2x + 8y = -6$
$-5x - 20y = 15$

33 $\frac{x}{5} + \frac{y}{2} = 8$
$x + y = 20$

34 $2x - y = -2$
$4x + y = 5$

35 $0.4x + 0.3y = 1$
$0.25x + 0.1y = -0.25$

In questions 36–38, solve the pair of simultaneous equations using any method – elimination, substitution or the graphical features of your GDC.

36 $3x + 2y = 9$
$7x + 11y = 2$

37 $3.62x - 5.88y = -10.11$
$0.08x - 0.02y = 0.92$

38 $2x - 3y = 4$
$5x + 2y = 1$

2 Functions

Assessment statements

2.1 Concept of a function $f: x \to f(x)$; domain, range, image (value).
One-to-one and many-to-one functions.
Composite functions $f \circ g$; identity function. Inverse function f^{-1} including domain restriction.

2.2 The graph of a function; its equation $y = f(x)$.
Investigation of key features of graphs such as intercepts, horizontal and vertical asymptotes, symmetry and consideration of domain and range.
The graphs of the absolute value functions, $y = |f(x)|$ and $y = f(|x|)$.
The graph of $y = \frac{1}{f(x)}$ from $y = f(x)$.

2.3 Transformations of graphs: translations, stretches, reflections in the axes.
The graph of $y = f^{-1}(x)$ as the reflection in the line $y = x$ of the graph $y = f(x)$.

2.4 The reciprocal function $x \to \frac{1}{x}$, $x \neq 0$: its graph; its self-inverse nature.

Introduction

The relationship between two quantities – how the value of one quantity depends on the value of another quantity – is the key behind the concept of a function. Functions and how we use them are at the very foundation of many topics in mathematics, and are essential to our understanding of much of what will be covered later in this book. This chapter will look at some general characteristics and properties of functions. We will consider composite and inverse functions, and investigate how the graphs of functions can be transformed by means of translations, stretches and reflections.

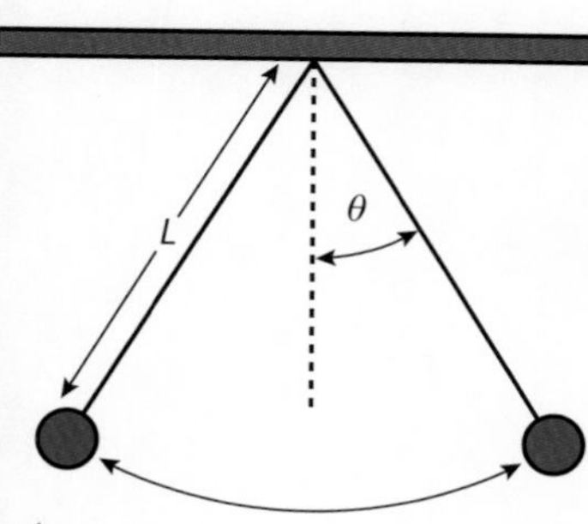

Figure 2.1 A simple pendulum.

2.1 Definition of a function

A simple pendulum consists of a heavy object hanging from a string of length L (in metres) and fixed at a pivot point (Figure 2.1). If you displace the suspended object to one side by a certain angle θ from the vertical and release it, the object will swing back and forth under the force of gravity. The period T (in seconds) of the pendulum is the time for the object to return to the point of release and, for a small angle θ, the two variables T and L are related by the formula $T = 2\pi\sqrt{\frac{L}{g}}$ where g is the gravitational field strength (acceleration due to gravity). Therefore, assuming that the force of gravity is constant at a given elevation ($g \approx 9.81\ \mathrm{m\,s^{-2}}$ at sea level), the formula can be used to calculate the value of T for any value of L.

As with the period T and the length L for a pendulum, many mathematical relationships concern how the value of one variable determines the value of a second variable. Other examples include:

Area of a circle determined by its radius:
$A = \pi r^2$ (π is a constant)

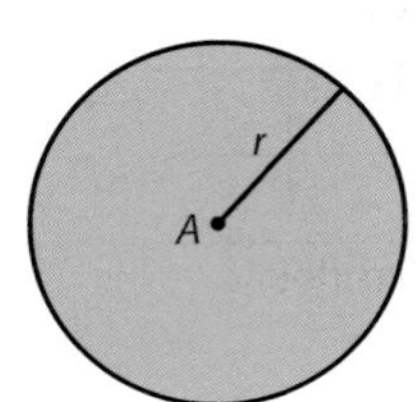

Converting degrees Celsius to degrees Fahrenheit:
$F = \frac{9}{5}C + 32$

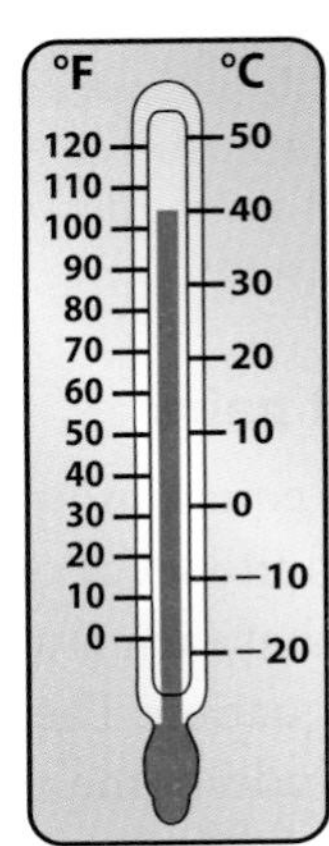

Distance that a number is from the origin determined by its absolute value:

In general, suppose that the values of a particular **independent variable**, for example x, determine the values of a **dependent variable** y in such a way that for a specific value of x, a single value of y is determined. Then we say that y is a **function** of x and we write $y = f(x)$ (read 'y equals f of x'), or $y = g(x)$ etc., where the letters f and g represent the name of the function. For the four mathematical relationships that were described above, we have:

Period T is a function of length L: $T = 2\pi\sqrt{\frac{L}{g}}$, or $f(L) = 2\pi\sqrt{\frac{L}{g}}$ where $T = f(L)$.

Area A is a function of radius r: $A = \pi r^2$, or $g(r) = \pi r^2$ where $A = g(r)$.

°F (degrees Fahrenheit) is a function of °C: $F = \frac{9}{5}C + 32$, or $t(C) = \frac{9}{5}C + 32$ where $F = t(C)$.

Distance y from origin is a function of x: $y = |x|$, or $f(x) = |x|$ where $y = f(x)$.

Along with equations, other useful ways of representing a function include a graph of the equation on a **Cartesian coordinate system** (also called

René Descartes

The Cartesian coordinate system is named in honour of the French mathematician and philosopher René Descartes (1596–1650). Descartes stimulated a revolution in the study of mathematics by merging its two major fields – algebra and geometry. With his coordinate system utilizing ordered pairs (*Cartesian coordinates*) of real numbers, geometric concepts could be formulated analytically and algebraic concepts (e.g. relationships between two variables) could be viewed graphically. Descartes initiated something that is very helpful to all students of mathematics – that is, considering mathematical concepts from multiple perspectives: graphical (visual) and analytical (algebraic).

a **rectangular coordinate system**), a **table**, a **set of ordered pairs**, or a **mapping**. These are illustrated below for the absolute value function $y = |x|$.

Graph

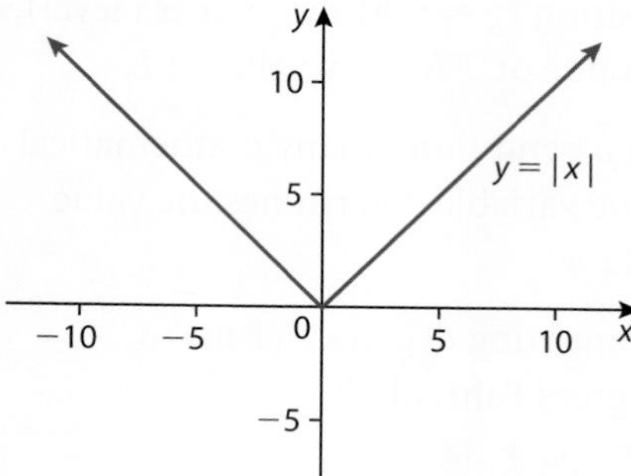

• **Hint:** The coordinate system for the graph of an equation has the independent variable on the horizontal axis and the dependent variable on the vertical axis.

Table $y = |x|$

x	y
-10	10
$-\frac{15}{2}$	$\frac{15}{2}$
-5	5
-3.6	3.6
0	0
$\sqrt{2}$	$\sqrt{2}$
5	5
8.3	8.3
10	10

Set of ordered pairs

The graph of the equation $y = |x|$ consists of an infinite set of ordered pairs (x, y) such that each is a solution of the equation. The following set includes some of the ordered pairs on the line: $\{(-23, 23), (-10, 10), (-\sqrt{7}, \sqrt{7}), (0, 0), (5, 5)\}$.

Mapping

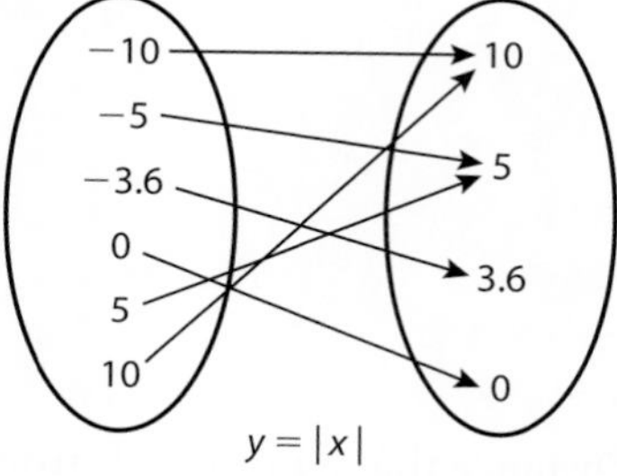

The largest possible set of values for the independent variable (the **input** set) is called the **domain** – and the set of resulting values for the dependent variable (the **output** set) is called the **range**. In the context of a mapping, each value in the domain is mapped to its **image** in the range.

All of the various ways of representing a mathematical function illustrate that its defining characteristic is that it is a rule by which each number in the domain determines a unique number in the range.

Definition of a function

A **function** is a correspondence (**mapping**) between two sets *X* and *Y* in which each element of set *X* corresponds to (maps to) exactly one element of set *Y*. The **domain** is set *X* (**independent variable**) and the **range** is set *Y* (**dependent variable**).

Not all equations represent a function. The solution set for the equation $x^2 + y^2 = 1$ is the set of ordered pairs (x, y) on the circle of radius equal to 1 and centre at the origin (see Figure 2.2). If we solve the equation for y, we get $y = \pm\sqrt{1 - x^2}$. It is clear that any value of x between -1 and 1 will produce two different values of y (opposites). Since at least one value in the domain (x) determines more than one value in the range (y), then the equation does not represent a function. A correspondence between two sets that does not satisfy the definition of a function is called a **relation**.

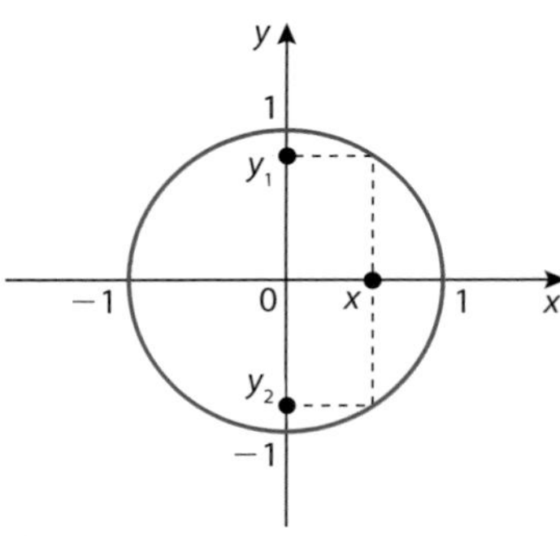

Figure 2.2 Graph of $x^2 + y^2 = 1$.

Alternative definition of a function

A **function** is a **relation** in which no two different ordered pairs have the same first coordinate.

A vertical line intersects the graph of a function at no more than one point (vertical line test).

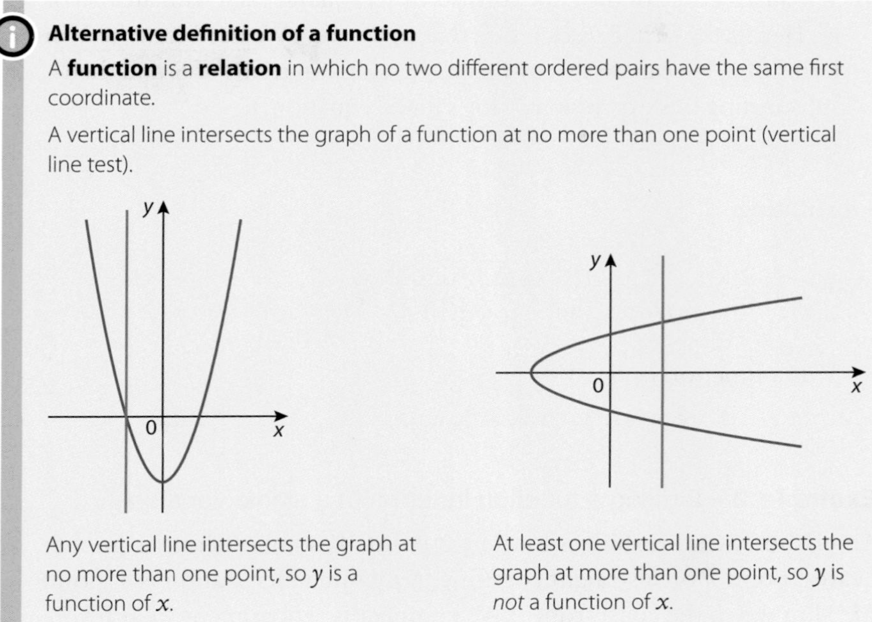

Any vertical line intersects the graph at no more than one point, so y is a function of x.

At least one vertical line intersects the graph at more than one point, so y is *not* a function of x.

Not only are functions important in the study of mathematics and science, we encounter and use them routinely – often in the form of tables. Examples include height and weight charts, income tax tables, loan payment schedules, and time and temperature charts. The importance of functions in mathematics is evident from the many functions that are installed on your GDC.

For example, the keys labelled SIN, x^{-1}, LN, $\sqrt{\ }$ each represent a function, because for each input (entry) there is only one output (answer). The calculator screen image shows that for the function $y = \ln x$, the input of $x = 10$ has only one output of $y \approx 2.302\,585\,093$.

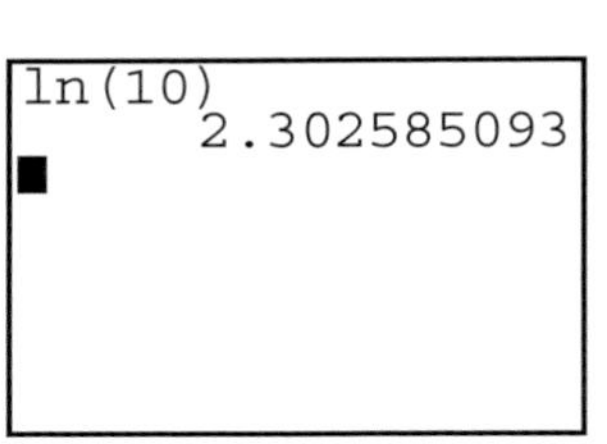

For many physical phenomena, we observe that one quantity depends on another. The word **function** is used to describe this dependence of one quantity on another – i.e. how the value of an independent variable determines the value of a dependent variable. A common mathematical task is to find how to express one variable as a function of another variable.

Example 1

a) Express the volume V of a cube as a function of the length e of each edge.

b) Express the volume V of a cube as a function of its surface area S.

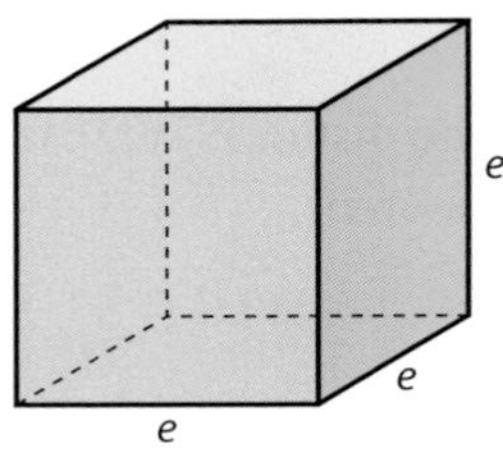

Solution

a) V as a function of e is $V = e^3$.

b) The surface area of the cube consists of six squares each with an area of e^2. Hence, the surface area is $6e^2$; that is, $S = 6e^2$. We need to write V in terms of S. We can do this by first expressing e in terms of S, and then substituting this expression in for e in the equation $V = e^3$.

$S = 6e^2 \Rightarrow e^2 = \frac{S}{6} \Rightarrow e = \sqrt{\frac{S}{6}}$.

Substituting,

$$V = \left(\sqrt{\frac{S}{6}}\right)^3 = \frac{(S^{\frac{1}{2}})^3}{(6^{\frac{1}{2}})^3} = \frac{S^{\frac{3}{2}}}{6^{\frac{3}{2}}} = \frac{S^1 \cdot S^{\frac{1}{2}}}{6^1 \cdot 6^{\frac{1}{2}}} = \frac{S}{6}\sqrt{\frac{S}{6}}$$

V as a function of S is $V = \frac{S}{6}\sqrt{\frac{S}{6}}$.

Example 2 – Finding a function in terms of a single variable

An offshore wind turbine is located at point W, 4 km offshore from the nearest point P on a straight coastline. A maintenance station is at point M, 3 km down the coast from P. An engineer is returning by boat from the wind turbine. He decides to row to a dock at point D that is located between P and M at an unknown distance x km from point P. The engineer can row 3 km/hr and walk 6 km/hr. Express the total time T (hours) for the trip from the wind turbine to the maintenance station as a function of x (km).

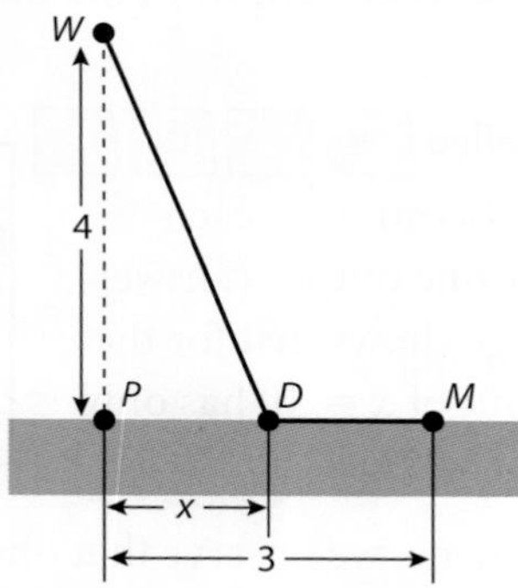

Solution

To get an equation for T in terms of x, we use the fact that time $= \frac{\text{distance}}{\text{rate}}$. We then have

$$T = \frac{\text{distance } WD}{3} + \frac{\text{distance } DM}{6}$$

The distance WD can be expressed in terms of x by using Pythagoras' theorem.

$$WD^2 = x^2 + 4^2 \Rightarrow WD = \sqrt{x^2 + 16}$$

To express T in terms of only the single variable x, we note that $DM = 3 - x$.

Then the total time T can be written in terms of x by the equation:

$$T = \frac{\sqrt{x^2 + 16}}{3} + \frac{3 - x}{6} \text{ or } T = \frac{1}{3}\sqrt{x^2 + 16} + \frac{1}{2} - \frac{x}{6}$$

Using our graphic display calculator (GDC) to graph the equation gives a helpful picture showing how T changes when x changes. In function graphing mode on a GDC, the independent variable is always x and the dependent variable is always y.

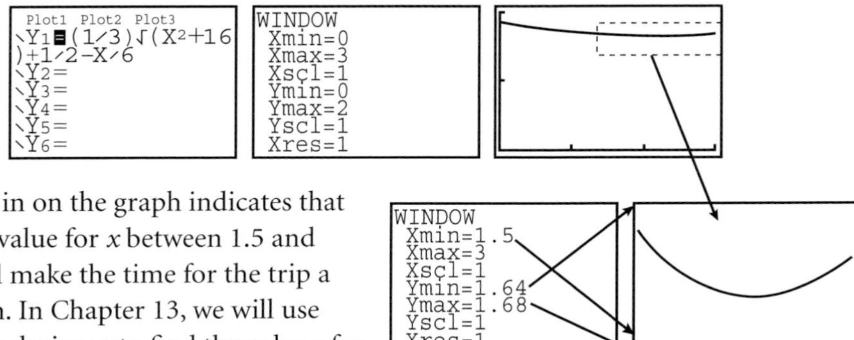

Zooming in on the graph indicates that there is a value for x between 1.5 and 3 that will make the time for the trip a minimum. In Chapter 13, we will use calculus techniques to find the value of x that gives a minimum time for the trip.

Domain and range of a function

The domain of a function may be stated explicitly, or it may be implied by the expression that defines the function. Except in Chapter 10, where we will encounter functions for which the variables can have values that are imaginary numbers, we can assume that any functions that we will work with are **real-valued functions of a real variable.** That is, the domain and range will only contain real numbers or some subset of the real numbers. Therefore, if not explicitly stated otherwise, the domain of a function is the set of all real numbers for which the expression is defined as a real number. For example, if a certain value of x is substituted into the algebraic expression defining a function and it causes division by zero or the square root of a negative number (both undefined in the real numbers) to occur, that value of x cannot be in the domain. The domain of a function may also be implied by the physical context or limitations that exist in a problem. For example, for both functions derived in Example 1 $\left(V = \frac{S}{6}\sqrt{\frac{S}{6}} \text{ and } V = e^3\right)$ the domain is the set of positive real numbers (symbolized by $\mathbb{R}^+$) because neither a length (edge of a cube) nor a surface area (face of a cube) can have a value that is negative or zero. In Example 2 the domain for the function is $0 < x < 3$ because of the constraints given in the problem. Usually the range of a function is not given explicitly and is determined by analyzing the output of the function for all values of the input (domain). The range of a function is often more difficult to find than the domain, and analyzing the graph of a function is very helpful in determining it. A combination of algebraic and graphical analysis is very useful in determining the domain and range of a function.

Example 3 – Domain of a function

Find the domain of each of the following functions.

a) $\{(-6,-3),(-1,0),(2,3),(3,0),(5,4)\}$

b) Volume of a sphere: $V=\frac{4}{3}\pi r^3$

c) $y=\dfrac{5}{2x-6}$

d) $y=\sqrt{3-x}$

Solution

a) The function consists of a set of ordered pairs. The domain of the function consists of all first coordinates of the ordered pairs. Therefore, the domain is the set $x \in \{-6,-1,2,3,5\}$.

b) The physical context tells you that a sphere cannot have a radius that is negative or zero. Therefore, the domain is the set of all real numbers r such that $r > 0$.

c) Since division by zero is not defined for real numbers then $2x-6 \neq 0$. Therefore, the domain is the set of all real numbers x such that $x \in \mathbb{R}, x \neq 3$.

d) Since the square root of a negative number is not real, then $3-x \geqslant 0$. Therefore, the domain is all real numbers x such that $x \leqslant 3$.

Example 4 – Domain and range of a function I

What is the domain and range for the function $y=x^2$?

Solution

- *Algebraic analysis*: Squaring any real number produces another real number. Therefore, the domain of $y=x^2$ is the set of all real numbers ($\mathbb{R}$). What about the range? Since the square of any positive or negative number will be positive and the square of zero is zero, the range is the set of all real numbers greater than or equal to zero.
- *Graphical analysis*: For the domain, focus on the x-axis and *horizontally* scan the graph from $-\infty$ to $+\infty$. There are no 'gaps' or blank regions in the graph and the parabola will continue to get 'wider' as x goes to either $-\infty$ or $+\infty$. Therefore, the domain is all real numbers. For the range, focus on the y-axis and *vertically* scan from $-\infty$ or $+\infty$. The parabola will continue 'higher' as y goes to $+\infty$, but the graph does not go below the x-axis. The parabola has no points with negative y-coordinates. Therefore, the range is the set of real numbers greater than or equal to zero. See Figure 2.3.

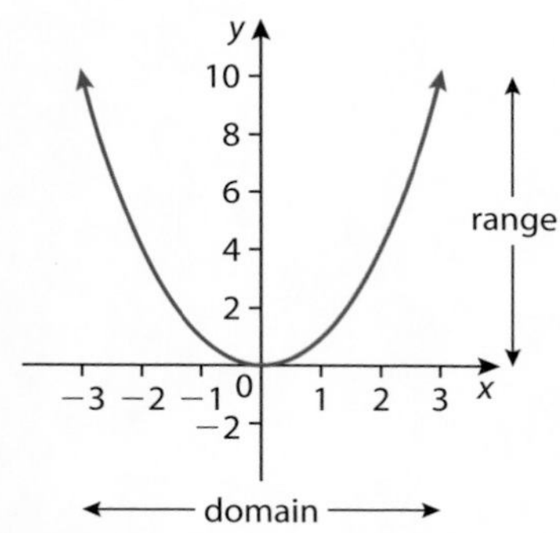

Figure 2.3 The graph of $y=x^2$.

Table 2.1 Different ways of expressing the domain and range of $y=x^2$.

Description in words	Interval notation (both formats)
domain is any real number	domain is $\{x : x \in \mathbb{R}\}$, or domain is $x \in]-\infty, \infty[$
range is any real number greater than or equal to zero	range is $\{y : y \geqslant 0\}$, or range is $y \in [0, \infty[$

Function notation

It is common practice to name a function using a single letter, with f, g and h being the most common. Given that the domain variable is x and the range variable is y, the symbol $f(x)$ denotes the unique value of y that is generated by the value of x. Another notation – sometimes referred to as **mapping notation** – is based on the idea that the function f is the rule that maps x to $f(x)$ and is written $f: x \mapsto f(x)$. For each value of x in the domain, the corresponding unique value of y in the range is called the **function value** at x, or the **image** of x under f. The image of x may be written as $f(x)$ or as y. For example, for the function $f(x) = x^2$: '$f(3) = 9$'; or 'if $x = 3$ then $y = 9$'.

• **Hint:** When asked to determine the domain and range of a function, it is wise for you to conduct both algebraic and graphical analysis – and not rely too much on either approach. For graphical analysis of a function, producing a *comprehensive graph* on your GDC is essential, i.e. a graph that shows all important features of the graph.

Notation	Description in words
$f(x) = x^2$	'the function f, in terms of x, is x^2'; or, simply, 'f of x equals x^2'
$f: x \mapsto x^2$	'the function f maps x to x^2'
$f(3) = 9$	'the value of the function f when $x = 3$ is 9'; or, simply, 'f of 3 equals 9'
$f: 3 \mapsto 9$	'the image of 3 under the function f is 9'

Table 2.2 Function notation.

• **Hint:** It is common to write $y = f(x)$ and call it a function but this can be considered a misuse of the notation. If we were to be very precise, we would call f the function and $f(x)$ the value of the function at x. But this is often overlooked and we accept writing expressions such as $y = x^2$ or $y = \sin x$ and calling them functions.

Example 5 – Domain and range of a function II

Find the domain and range of the function $h: x \mapsto \frac{1}{x-2}$.

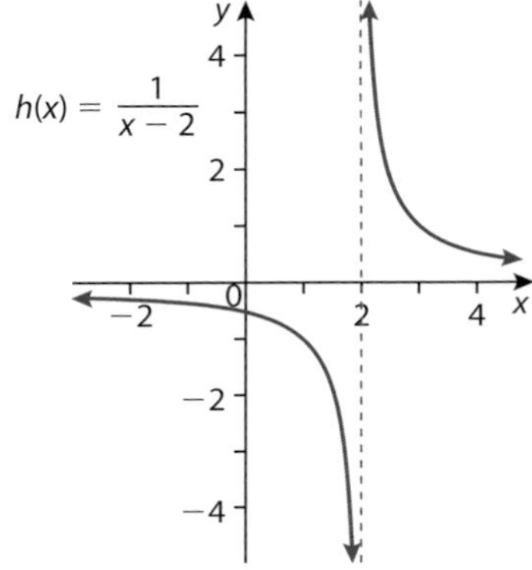

Solution

- *Algebraic analysis*: The function produces a real number for all x, except for $x = 2$ when division by zero occurs. Hence, $x = 2$ is the only real number not in the domain. Since the numerator of $\frac{1}{x-2}$ can never be zero, the value of y cannot be zero. Hence, $y = 0$ is the only real number not in the range.
- *Graphical analysis*: A horizontal scan shows a 'gap' at $x = 2$ dividing the graph of the equation into two branches that both continue indefinitely, with no other 'gaps' as $x \to \pm\infty$. Both branches are **asymptotic** (approach but do not intersect) to the vertical line $x = 2$. This line is a **vertical asymptote** and is drawn as a dashed line (it is *not* part of the graph of the equation). A vertical scan reveals a 'gap' at $y = 0$ (x-axis) with both branches of the graph continuing indefinitely, with no other 'gaps' as $y \to \pm\infty$. Both branches are also asymptotic to the x-axis. The x-axis is a **horizontal asymptote**.

  Both approaches confirm the following for $h: x \mapsto \frac{1}{x-2}$:

  The domain is $\{x: x \in \mathbb{R}, x \neq 2\}$ or $x \in]-\infty, 2[\cup]2, \infty[$

  The range is $\{y: y \in \mathbb{R}, y \neq 0\}$ or $y \in]-\infty, 0[\cup]0, \infty[$

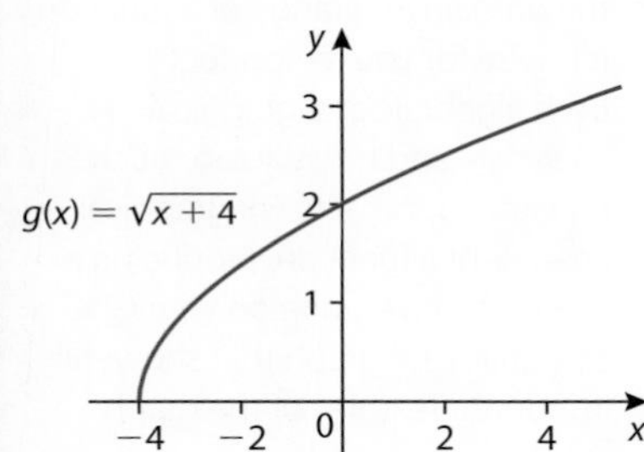

Example 6 – Domain and range of function II

Consider the function $g(x) = \sqrt{x+4}$.

a) Find: (i) $g(7)$ (ii) $g(32)$ (iii) $g(-4)$

b) Find the values of x for which g is undefined.

c) State the domain and range of g.

Solution

a) (i) $g(7) = \sqrt{7+4} = \sqrt{11} \approx 3.32$ (3 significant figures)
(ii) $g(32) = \sqrt{32+4} = \sqrt{36} = 6$
(iii) $g(-4) = \sqrt{-4+4} = \sqrt{0} = 0$

b) $g(x)$ will be undefined (square root of a negative) when $x + 4 < 0$.
$x + 4 < 0 \Rightarrow x < -4$. Therefore, $g(x)$ is undefined when $x < -4$.

c) It follows from the result in b) that the domain of g is $\{x : x \geqslant -4\}$. The symbol $\sqrt{\ }$ stands for the **principal square root** that, by definition, can only give a result that is positive or zero. Therefore, the range of g is $\{y : y \geqslant 0\}$. The domain and range are confirmed by analyzing the graph of the function.

Example 7 – Domain and range of a function III

Find the domain and range of the function
$f(x) = \dfrac{1}{\sqrt{9 - x^2}}$.

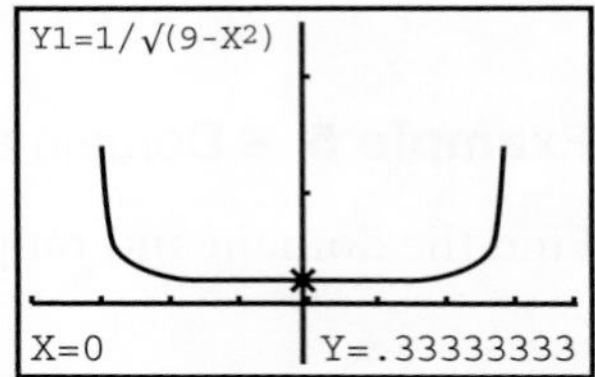

Solution

The graph of $y = \dfrac{1}{\sqrt{9 - x^2}}$ on a GDC, shown above, agrees with algebraic analysis indicating that the expression $\dfrac{1}{\sqrt{9 - x^2}}$ will be positive for all x, and is defined only for $-3 < x < 3$.

• **Hint:** As Example 7 illustrates, it is dangerous to completely trust graphs produced on a GDC without also doing some algebraic thinking. It is important to mentally check that the graph shown is comprehensive (shows all important features of the graph), and that the graph agrees with algebraic analysis of the function – e.g. where should the function be zero, positive, negative, undefined, increasing/decreasing without bound, etc.

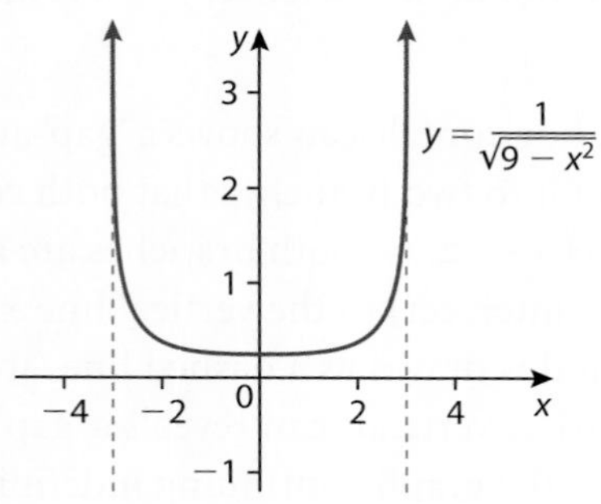

Further analysis and tracing the graph reveals that $f(x)$ has a minimum at $\left(0, \frac{1}{3}\right)$. The graph on the GDC (next page) is misleading in that it appears to show that the function has a maximum value (y) of approximately 2.803 7849. Can this be correct? A lack of algebraic thinking and over-reliance on your GDC could easily lead to a mistake. The graph abruptly stops its curve upwards because of low screen resolution.

Function values should get quite large for values of x a little less than 3, because the value of $\sqrt{9-x^2}$ will be small, making the fraction $\dfrac{1}{\sqrt{9-x^2}}$ large. Using your GDC to make a table for $f(x)$, or evaluating the function for values of x very close to -3 or 3, confirms that as x approaches -3 or 3, y increases without bound, i.e. y goes to $+\infty$. Hence, $f(x)$ has vertical asymptotes of $x = -3$ and $x = 3$. This combination of graphical and algebraic analysis leads to the conclusion that the domain of $f(x)$ is $\{x: -3 < x < 3\}$, and the range of $f(x)$ is $\{y: y \geqslant \frac{1}{3}\}$.

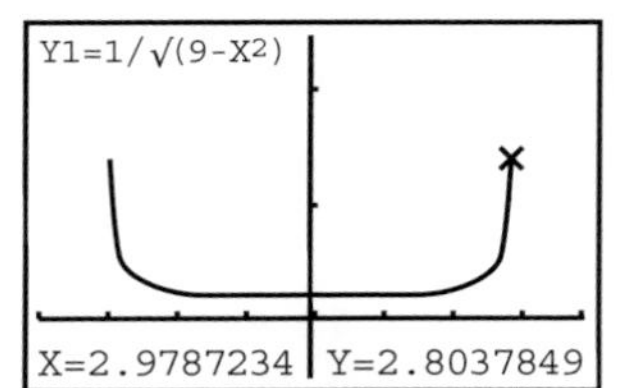

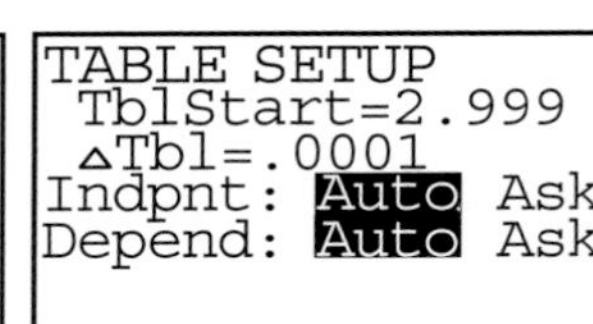

X	Y1
2.9994	16.668
2.9995	18.258
2.9996	20.413
2.9997	23.571
2.9998	28.868
2.9999	40.825
3	ERROR

X=2.9994

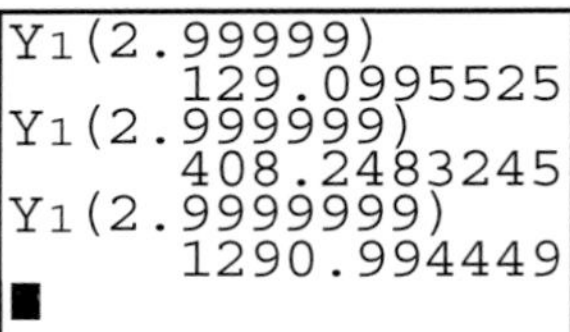

Exercise 2.1

For each equation 1–9, a) match it with its graph (choices are labelled A to L), and b) state whether or not the equation represents a function – with a justification. Assume that x is the independent variable and y is the dependent variable.

1 $y = 2x$ **2** $y = -3$ **3** $x - y = 2$

4 $x^2 + y^2 = 4$ **5** $y = 2 - x$ **6** $y = x^2 + 2$

7 $y^3 = x$ **8** $y = \frac{2}{x}$ **9** $x^2 + y = 2$

A
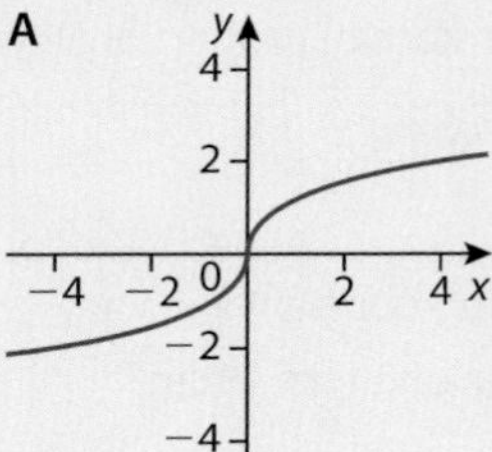

B
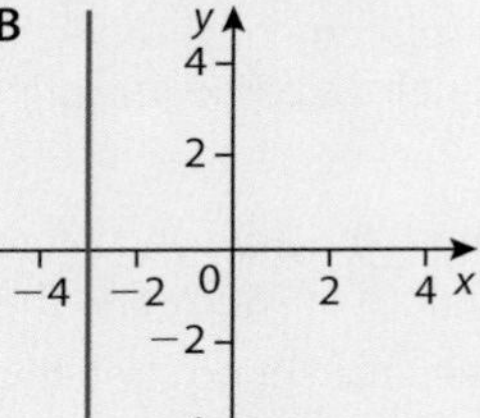

C
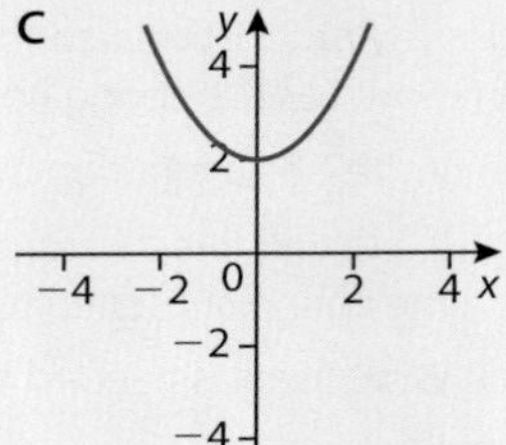

D
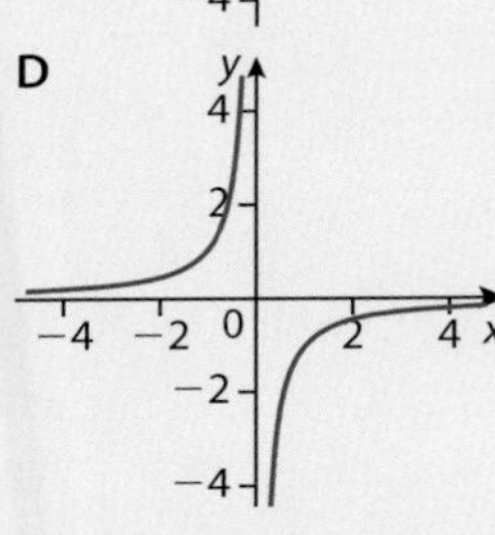

E
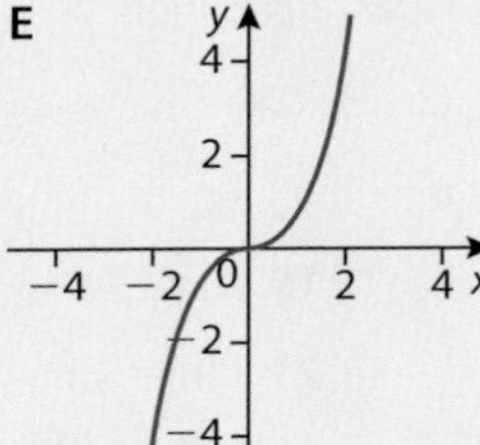

F
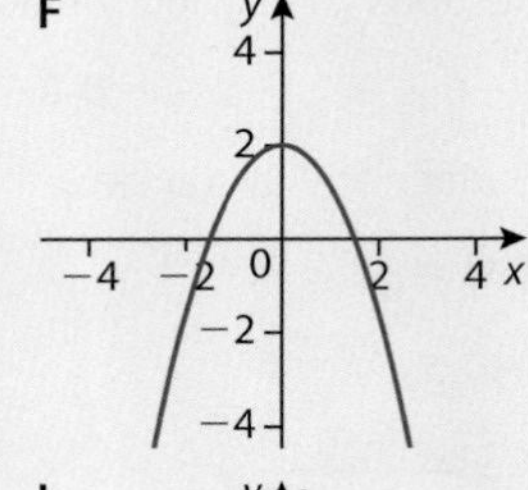

G
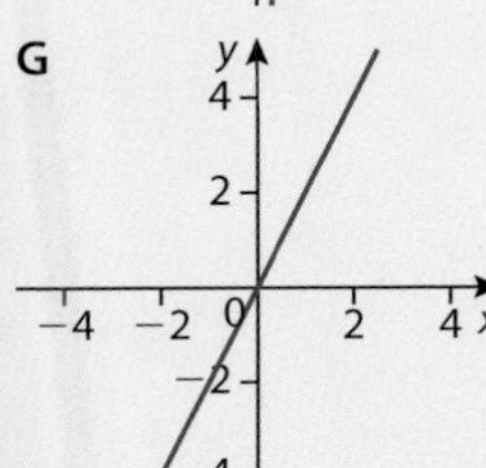

H
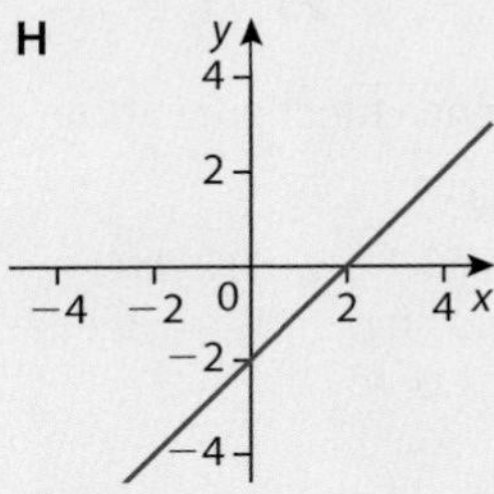

I
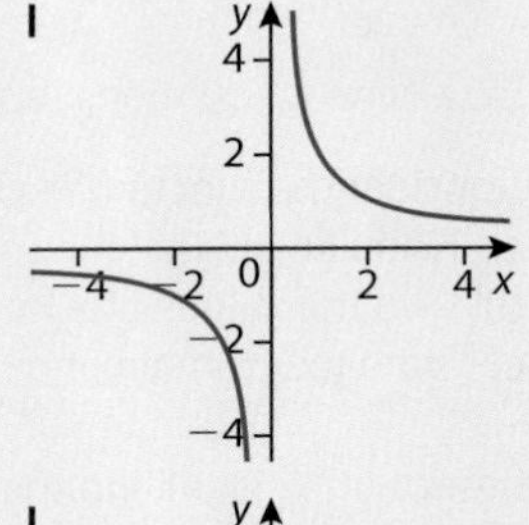

J
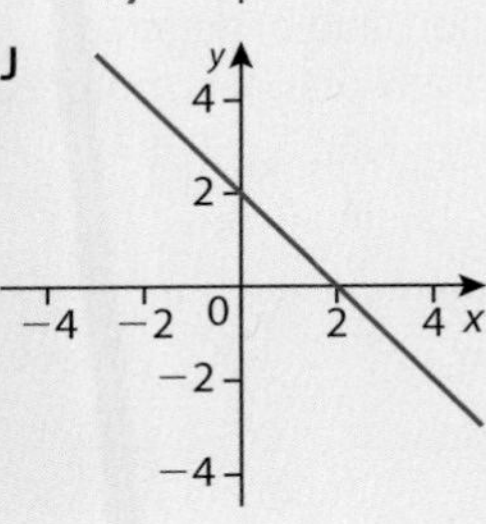

K
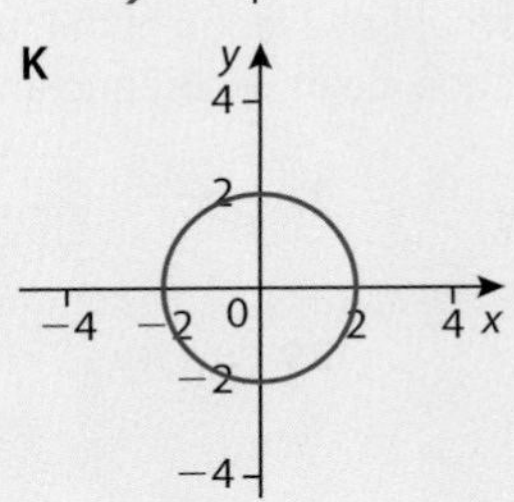

L
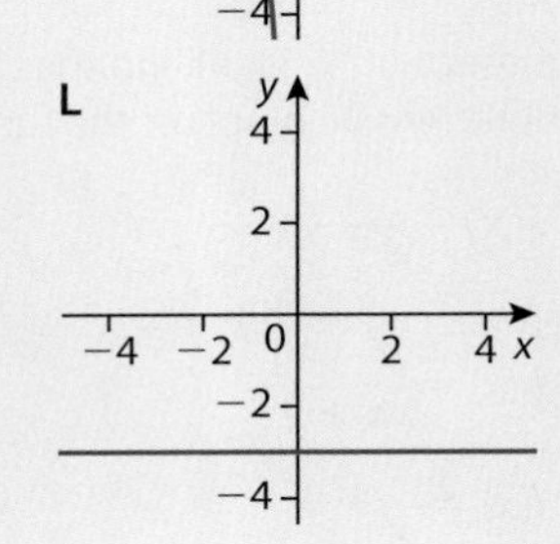

10 Express the area, A, of a circle as a function of its circumference, C.

11 Express the area, A, of an equilateral triangle as a function of the length, ℓ, of each of its sides.

12 A rectangular swimming pool with dimensions 12 metres by 18 metres is surrounded by a pavement of uniform width x metres. Find the area of the pavement, A, as a function of x.

13 In a right isosceles triangle, the two equal sides have length x units and the hypotenuse has length h units. Write h as a function of x.

14 The pressure P (measured in kilopascals, kPa) for a particular sample of gas is directly proportional to the temperature T (measured in kelvin, K) and inversely proportional to the volume V (measured in litres, ℓ). With k representing the constant of proportionality, this relationship can be written in the form of the equation $P = k\frac{T}{V}$.

a) Find the constant of proportionality, k, if 150 ℓ of gas exerts a pressure of 23.5 kPa at a temperature of 375 K.

b) Using the value of k from part a) and assuming that the temperature is held constant at 375 K, write the volume V as a function of pressure P for this sample of gas.

15 In physics, Hooke's law states that the force F (measured in newtons, N) needed to keep a spring stretched a displacement of x units beyond its natural length is directly proportional to the displacement x. Label the constant of proportionality k (known as the spring constant for a particular spring).

a) Write F as a function of x.

b) If a spring has a natural length of 12 cm and a force of 25 N is needed to keep the spring stretched to a length of 16 cm, find the spring constant k.

c) What force is needed to keep the spring stretched to a length of 18 cm?

In questions 16–23, find the domain of the function.

16 $\{(-6.2, -7), (-1.5, -2), (0.7, 0), (3.2, 3), (3.8, 3)\}$

17 Surface area of a sphere: $S = 4\pi r^2$

18 $f(x) = \frac{2}{5}x - 7$

19 $h: x \mapsto x^2 - 4$

20 $g(t) = \sqrt{3 - t}$

21 $h(t) = \sqrt[3]{t}$

22 $f: x \mapsto \dfrac{6}{x^2 - 9}$

23 $f(x) = \sqrt{\dfrac{1}{x^2} - 1}$

24 Do all linear equations represent a function? Explain.

25 Consider the function $h(x) = \sqrt{x - 4}$.

a) Find: (i) $h(21)$ (ii) $h(53)$ (iii) $h(4)$

b) Find the values of x for which h is undefined.

c) State the domain and range of h.

In questions 26–30, a) find the domain and range of the function, and b) sketch a comprehensive graph of the function clearly indicating any intercepts or asymptotes.

26 $f: x \mapsto \dfrac{1}{x - 5}$

27 $g(x) = \dfrac{1}{\sqrt{x^2 - 9}}$

28 $h(x) = \dfrac{2x - 1}{x + 2}$

29 $p: x \mapsto \sqrt{5 - 2x^2}$

30 $f(x) = \dfrac{1}{x} - 4$

2.2 Composite functions

Composition of functions

Consider the function in Example 6 in the previous section, $f(x) = \sqrt{x + 4}$. When you evaluate $f(x)$ for a certain value of x in the domain (for example, $x = 5$) it is necessary for you to perform computations in two separate steps in a certain order.

$f(5) = \sqrt{5 + 4} \Rightarrow f(5) = \sqrt{9}$ Step 1: compute the sum of 5 + 4.

$\Rightarrow f(5) = 3$ Step 2: compute the principal square root of 9.

Given that the function has two separate evaluation 'steps', $f(x)$ can be seen as a combination of two 'simpler' functions that are performed in a specified order. According to how $f(x)$ is evaluated (as shown above), the simpler function to be performed first is the rule of 'adding 4' and the second is the rule of 'taking the square root'. If $h(x) = x + 4$ and $g(x) = \sqrt{x}$, we can create (compose) the function $f(x)$ from a combination of $h(x)$ and $g(x)$ as follows:

$f(x) = g(h(x))$

$= g(x + 4)$ Step 1: substitute $x + 4$ for $h(x)$, making $x + 4$ the argument of $g(x)$.

$= \sqrt{x + 4}$ Step 2: apply the function $g(x)$ on the argument $x + 4$.

We obtain the rule $\sqrt{x + 4}$ by first applying the rule $x + 4$ and then applying the rule $\sqrt{x}$. A function that is obtained from 'simpler' functions by applying one after another in this way is called a **composite function**. In the example above, $f(x) = \sqrt{x + 4}$ is the **composition** of $h(x) = x + 4$ followed by $g(x) = \sqrt{x}$. In other words, f is obtained by substituting h into g, and can be denoted in function notation by $g(h(x))$ – read 'g of h of x'.

From the explanation on how f is the composition (or composite) of g and h, you can see why a composite function is sometimes referred to as a 'function of a function'. Also, note that in the notation $g(h(x))$ the function h that is applied first is written 'inside', and the function g that is applied second is written 'outside'.

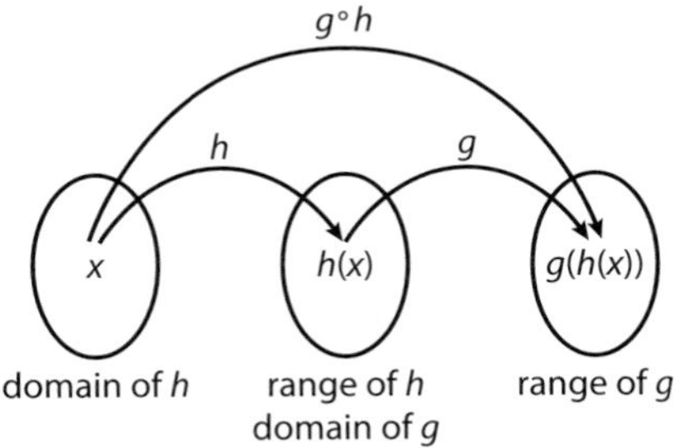

Figure 2.4 Mapping for composite function $g(h(x))$.

We start with a number x in the domain of h and find its image $h(x)$. If this number $h(x)$ is in the domain of g, we then compute the value of $g(h(x))$. The resulting composite function is denoted as $(g \circ h)(x)$. See mapping illustration in Figure 2.4.

Definition of the composition of two functions

The composition of two functions, g and h, such that h is applied first and g second is given by

$$(g \circ h)(x) = g(h(x))$$

The domain of the composite function $g \circ h$ is the set of all x in the domain of h such that $h(x)$ is in the domain of g.

• **Hint:** The notations $(g \circ h)(x)$ and $g(h(x))$ are both commonly used to denote a composite function where h is applied first and then followed by applying g. Since we are reading this from left to right, it is easy to apply the functions in the incorrect order. It may be helpful to read $g \circ h$ as 'g following h', or as 'g composed with h' to emphasize the order in which the functions are applied. Also, in either notation, $(g \circ h)(x)$ or $g(h(x))$, the function applied first is closest to the variable x.

Example 8 – Forming a composition of two functions I

If $f(x) = 3x$ and $g(x) = 2x - 6$, find:

a) $(f \circ g)(5)$ b) Express $(f \circ g)(x)$ as a single function rule (expression).

c) $(g \circ f)(5)$ d) Express $(g \circ f)(x)$ as a single function rule (expression).

e) $(g \circ g)(5)$ f) Express $(g \circ g)(x)$ as a single function rule (expression).

Solution

a) $(f \circ g)(5) = f(g(5)) = f(2 \cdot 5 - 6) = f(4) = 3 \cdot 4 = 12$

b) $(f \circ g)(x) = f(g(x)) = f(2x - 6) = 3(2x - 6) = 6x - 18$
Therefore, $(f \circ g)(x) = 6x - 18$.
Check with result from a): $(f \circ g)(5) = 6 \cdot 5 - 18 = 30 - 18 = 12$

c) $(g \circ f)(5) = g(f(5)) = g(3 \cdot 5) = g(15) = 2 \cdot 15 - 6 = 24$

d) $(g \circ f)(x) = g(f(x)) = g(3x) = 2(3x) - 6 = 6x - 6$
Therefore, $(g \circ f)(x) = 6x - 6$.
Check with result from c): $(g \circ f)(5) = 6 \cdot 5 - 6 = 30 - 6 = 24$

e) $(g \circ g)(5) = g(g(5)) = g(2 \cdot 5 - 6) = g(4) = 2 \cdot 4 - 6 = 2$

f) $(g \circ g)(x) = g(g(x)) = g(2x - 6) = 2(2x - 6) - 6 = 4x - 18$
Therefore, $(g \circ g)(x) = 4x - 18$.
Check with result from e): $(g \circ g)(5) = 4 \cdot 5 - 18 = 20 - 18 = 2$

It is important to notice that in parts b) and d) in Example 8, $f \circ g$ is *not* equal to $g \circ f$. At the start of this section, it was shown how the two functions $h(x) = x + 4$ and $g(x) = \sqrt{x}$ could be combined into the composite function $(g \circ h)(x)$ to create the single function $f(x) = \sqrt{x + 4}$. However, the composite function $(h \circ g)(x)$ – the functions applied in reverse order – creates a different function: $(h \circ g)(x) = h(g(x)) = h(\sqrt{x}) = \sqrt{x} + 4$. Since $\sqrt{x} + 4 \neq \sqrt{x + 4}$, then again $f \circ g$ is *not* equal to $g \circ f$. Is it always true that $f \circ g \neq g \circ f$? The next example will answer that question.

Example 9 – Forming a composition of two functions II

Given $f: x \mapsto 3x - 6$ and $g: x \mapsto \frac{1}{3}x + 2$, find the following:

a) $(f \circ g)(x)$ b) $(g \circ f)(x)$

Solution

a) $(f \circ g)(x) = f(g(x)) = f\left(\frac{1}{3}x + 2\right) = 3\left(\frac{1}{3}x + 2\right) - 6 = x + 6 - 6 = x$

b) $(g \circ f)(x) = g(f(x)) = g(3x - 6) = \frac{1}{3}(3x - 6) + 2 = x - 2 + 2 = x$

Example 9 shows that it is possible for $f \circ g$ to be equal to $g \circ f$. We will learn in the next section that this occurs in some cases where there is a 'special' relationship between the pair of functions. However, in general, $f \circ g \neq g \circ f$.

Decomposing a composite function

In Examples 8 and 9, we created a single function by forming the composition of two functions. As we did with the function $f(x) = \sqrt{x + 4}$ at the start of this section, it is also important for you to be able to identify two functions that *make up* a composite function, in other words, for you to *decompose* a function into two simpler functions. When you are doing this it is very useful to think of the function which is applied first as the 'inside' function, and the function that is applied second as the 'outside' function. In the function $f(x) = \sqrt{x + 4}$, the 'inside' function is $h(x) = x + 4$ and the 'outside' function is $g(x) = \sqrt{x}$.

• **Hint:** Decomposing composite functions – identifying the component functions that form a composite function – is an important skill when working with certain functions in the topic of calculus. For the composite function $f(x) = (g \circ h)(x)$, g and h are the component functions.

Example 10 – Decomposing a composite function

Each of the following functions is a composite function of the form $(f \circ g)(x)$. For each, find the two component functions f and g.

a) $h: x \mapsto \dfrac{1}{x + 3}$ b) $k: x \mapsto 2^{4x+1}$ c) $p(x) = \sqrt[3]{x^2 - 4}$

Solution

a) If you were to evaluate the function $h(x)$ for a certain x in the domain, you would first evaluate the expression $x + 3$, and then evaluate the expression $\frac{1}{x}$. Hence, the 'inside' function (applied first) is $y = x + 3$, and the 'outside' function (applied second) is $y = \frac{1}{x}$. Then, with $g(x) = x + 3$ and $f(x) = \frac{1}{x}$, it follows that $h: x \mapsto (f \circ g)(x)$.

b) Evaluating $k(x)$ requires you to first evaluate the expression $4x + 1$, and then evaluate the expression 2^x. Hence, the 'inside' function is $y = 4x + 1$, and the 'outside' function is $y = 2^x$. Then, with $g(x) = 4x + 1$ and $f(x) = 2^x$, it follows that $k: x \mapsto (f \circ g)(x)$.

c) Evaluating $p(x)$ requires you to perform three separate evaluation 'steps': (1) squaring a number, (2) subtracting four, and then (3) taking the cube root. Hence, it is possible to decompose $p(x)$ into three component functions: if $h(x) = x^2$, $g(x) = x - 4$ and $f(x) = \sqrt[3]{x}$, then $p(x) = (f \circ g \circ h)(x) = f(g(h(x)))$. However, for our purposes it is best to decompose the composite function into only two component functions: if $g(x) = x^2 - 4$ and $f(x) = \sqrt[3]{x}$, then $p(x) = (f \circ g)(x)$.

Finding the domain of a composite function

Referring back to Figure 2.4 (shown again here as Figure 2.5), it is important to note that in order for a value of x to be in the domain of the composite function $g \circ h$, two conditions must be met:

(1) x must be in the domain of h, and (2) $h(x)$ must be in the domain of g.

Likewise, it is also worth noting that $g(h(x))$ is in the range of $g \circ h$ only if x is in the domain of $g \circ h$. The next example illustrates these points – and also that, in general, the domains of $g \circ h$ and $h \circ g$ are not necessarily the same.

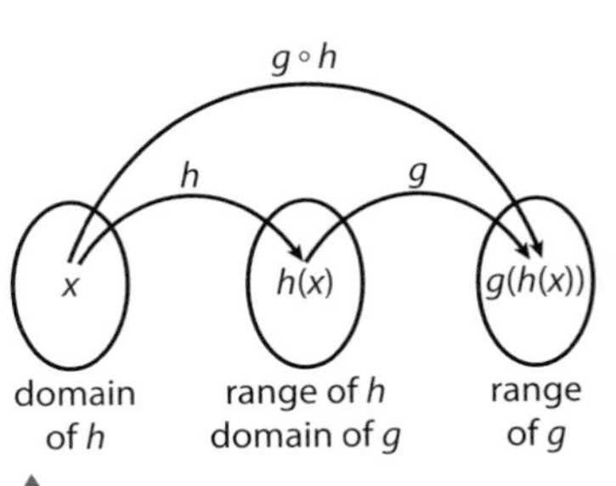

Figure 2.5 Mapping for composite function $g(h(x))$.

Example 11 – Domain and range of a composite function

Let $g(x) = x^2 - 4$ and $h(x) = \sqrt{x}$. Find:

a) $(g \circ h)(x)$ and its domain and range

b) $(h \circ g)(x)$ and its domain and range.

Solution

Firstly, establish the domain and range for both g and h. For $g(x) = x^2 - 4$, the domain is $x \in \mathbb{R}$ and the range is $y \geqslant -4$. For $h(x) = \sqrt{x}$, the domain is $x \geqslant 0$ and the range is $y \geqslant 0$.

a) $(g \circ h)(x) = g(h(x))$

$= g(\sqrt{x})$ — To be in the domain of $g \circ h$, $\sqrt{x}$ must be defined for $x \Rightarrow x \geqslant 0$.

$= (\sqrt{x})^2 - 4$ — Therefore, the domain of $g \circ h$ is $x \geqslant 0$.

$= x - 4$ — Since $x \geqslant 0$, the range for $y = x - 4$ is $y \geqslant -4$.

Therefore, $(g \circ h)(x) = x - 4$, and its domain is $x \geqslant 0$, and its range is $y \geqslant -4$.

b) $(h \circ g)(x) = h(g(x))$ — $g(x) = x^2 - 4$ must be in the domain of h $x^2 - 4 \geqslant 0 \Rightarrow x^2 \geqslant 4$

$= h(x^2 - 4)$ — Therefore, the domain of $h \circ g$ is $x \leqslant -2$ or $x \geqslant 2$

$= \sqrt{x^2 - 4}$ — and, with $x \leqslant -2$ or $x \geqslant 2$, the range for $y = \sqrt{x^2 - 4}$ is $y \geqslant 0$.

Therefore, $(h \circ g)(x) = \sqrt{x^2 - 4}$, and its domain is $x \leqslant -2$ or $x \geqslant 2$, and its range is $y \geqslant 0$.

Exercise 2.2

1 Let $f(x) = 2x$ and $g(x) = \frac{1}{x - 3}$, $x \neq 0$.

a) Find the value of (i) $(f \circ g)(5)$ and (ii) $(g \circ f)(5)$.

b) Find the function rule (expression) for (i) $(f \circ g)(x)$ and (ii) $(g \circ f)(x)$.

2 Let $f: x \mapsto 2x - 3$ and $g: x \mapsto 2 - x^2$.

In a)-f), evaluate:

a) $(f \circ g)(0)$ b) $(g \circ f)(0)$ c) $(f \circ f)(4)$

d) $(g \circ g)(-3)$ e) $(f \circ g)(-1)$ f) $(g \circ f)(-3)$

In g)-j), find the expression:

g) $(f \circ g)(x)$ h) $(g \circ f)(x)$

i) $(f \circ f)(x)$ j) $(g \circ g)(x)$

For each pair of functions in questions 3–12, find $(f \circ g)(x)$ and $(g \circ f)(x)$ and state the domain for each.

3 $f(x) = 4x - 1$, $g(x) = 2 + 3x$

4 $f(x) = x^2 + 1$, $g(x) = -2x$

5 $f(x) = \sqrt{x + 1}$, $g(x) = 1 + x^2$

6 $f(x) = \frac{2}{x + 4}$, $g(x) = x - 1$

7 $f(x) = 3x + 5, g(x) = \frac{x-5}{3}$

8 $f(x) = x^2 - 2x, g(x) = -x^2 - 2x$

9 $f(x) = \frac{2x}{4-x}, g(x) = \frac{1}{x^2}$

10 $f(x) = 2 - x^3, g(x) = \sqrt[3]{1-x^2}$

11 $f(x) = \frac{2}{x+3} - 3, g(x) = \frac{2}{x+3} - 3$ $[f = g]$

12 $f(x) = \frac{x}{x-1}, g(x) = x^2 - 1$

13 Let $g(x) = \sqrt{x-1}$ and $h(x) = 10 - x^2$. Find:

a) $(g \circ h)(x)$ and its domain and range, and

b) $(h \circ g)(x)$ and its domain and range.

14 Let $f(x) = \frac{1}{x}$ and $g(x) = 10 - x^2$. Find:

a) $(f \circ g)(x)$ and its domain and range, and

b) $(g \circ f)(x)$ and its domain and range.

In questions 15–22, determine functions g and h so that $f(x) = g(h(x))$.

15 $f(x) = (x+3)^2$

16 $f(x) = \sqrt{x-5}$

17 $f(x) = 7 - \sqrt{x}$

18 $f(x) = \frac{1}{x+3}$

19 $f(x) = 10^{x+1}$

20 $f(x) = \sqrt[3]{x-9}$

21 $f(x) = |x^2 - 9|$

22 $f(x) = \frac{1}{\sqrt{x-5}}$

In questions 23–26, find the domain for a) the function f, b) the function g, and c) the composite function $f \circ g$.

23 $f(x) = \sqrt{x}, g(x) = x^2 + 1$

24 $f(x) = \frac{1}{x}, g(x) = x + 3$

25 $f(x) = \frac{3}{x^2 - 1}, g(x) = x + 1$

26 $f(x) = 2x + 3, g(x) = \frac{x}{2}$

2.3 Inverse functions

Pairs of inverse functions

If we choose a number and cube it (raise it to the power of 3), and then take the cube root of the result, the answer is the original number. The same result would occur if we applied the two rules in the reverse order. That is, first take the cube root of a number and then cube the result – and again the answer is the original number. Let's write each of these rules as a function with function notation. Write the cubing function as $f(x) = x^3$, and the cube root function as $g(x) = \sqrt[3]{x}$. Now using what we know about composite functions and operations with radicals and exponents, we can write what was described above in symbolic form.

1. Cube a number and then take the cube root of the result:

$g(f(x)) = \sqrt[3]{x^3} = (x^3)^{\frac{1}{3}} = x^1 = x$

For example, $g(f(-2)) = \sqrt[3]{(-2)^3} = \sqrt[3]{-8} = -2$

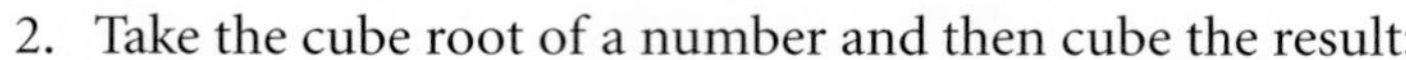

2. Take the cube root of a number and then cube the result:

$f(g(x)) = (\sqrt[3]{x})^3 = (x^{\frac{1}{3}})^3 = x^1 = x$

For example, $f(g(27)) = (\sqrt[3]{27})^3 = (3)^3 = 27$

Because function g has this reverse (inverse) effect on function f, we call function g the **inverse** of function f. Function f has the same inverse effect on function g [$g(27) = 3$ and then $f(3) = 27$], making f the inverse function of g. The functions f and g are inverses of each other. The cubing and cube root functions are an example of a pair of **inverse functions.** The mapping diagram for functions f and g in Figure 2.6 illustrates the relationship for a pair of inverse functions where the domain of one is the range for the other.

You are already familiar with pairs of **inverse operations**. Addition and subtraction are inverse operations. For example, the rule of 'adding six' ($x + 6$), and the rule of 'subtracting six' ($x - 6$), *undo* each other. Accordingly, the functions $f(x) = x + 6$ and $g(x) = x - 6$ are a pair of inverse functions. Multiplication and division are also inverse operations.

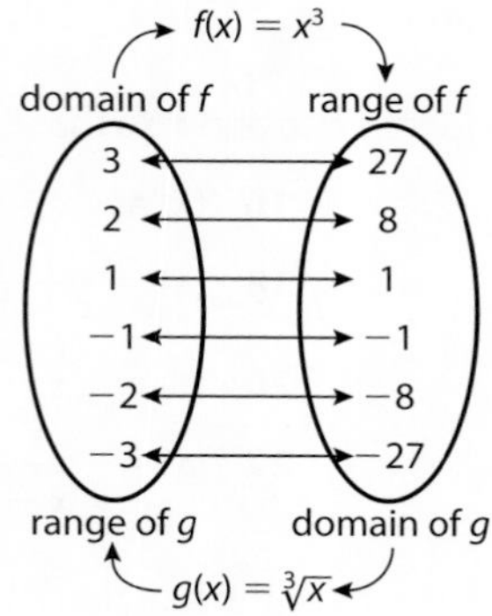

Figure 2.6 A mapping diagram for the cubing and cube root functions.

The composite of two inverse functions is the function that always produces the same number that was first substituted into the function. This function is called the **identity function** because it assigns each number in its domain to itself, and is denoted by $I(x) = x$.

Definition of the inverse of a function

If f and g are two functions such that $(f \circ g)(x) = x$ for every x in the domain of g and $(g \circ f)(x) = x$ for every x in the domain of f, the function g is the *inverse* of the function f. The notation to indicate the function that is the 'inverse of function f' is f^{-1}. Therefore,

$$(f \circ f^{-1})(x) = x \text{ and } (f^{-1} \circ f)(x) = x$$

The domain of f must be equal to the range of f^{-1}, and the range of f must be equal to the domain of f^{-1}.

Figure 2.7 shows a mapping diagram for a pair of inverse functions.

It follows from the definition that if g is the inverse of f, it must also be true that f is the inverse of g.

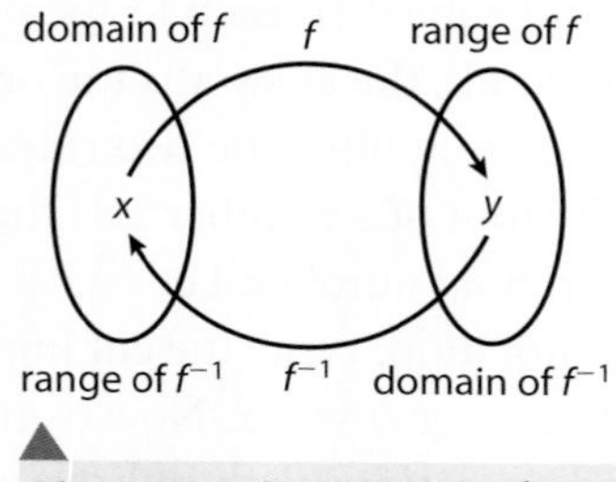

Figure 2.7 $f(x) = y$ and $f^{-1}(y) = x$.

For a pair of inverse functions, f and g, the composite functions $f(g(x))$ and $g(f(x))$ are equal, a 'special' relationship that we learned last section is not generally true for an arbitrary pair of functions.

Note: Remember that the notation $(f \circ g)(x)$ is equivalent to $f(g(x))$.

• **Hint:** Do not mistake the -1 in the notation f^{-1} for an exponent. It is *not* an exponent. If a superscript of -1 is applied to the name of a function, as in f^{-1} or $\sin^{-1}$, then it denotes the function that is the inverse of the named function (e.g. f or sin). If a superscript of -1 is applied to an expression, as in 7^{-1} or $(2x + 5)^{-1}$, then it is an exponent and denotes the reciprocal of the expression.

In general, the functions $f(x)$ and $g(x)$ are a pair of inverse functions if the following two statements are true:

1. $g(f(x)) = x$ for all x in the domain of f.
2. $f(g(x)) = x$ for all x in the domain of g.

Example 12 – Verifying a pair of functions are inverses

Given $h(x) = \frac{x-3}{2}$ and $p(x) = 2x + 3$, show that h and p are a pair of inverse functions.

Solution

Since the domain and range of both $h(x)$ and $p(x)$ is the set of all real numbers, then:

1. For any real number x, $p(h(x)) = p\left(\frac{x-3}{2}\right) = 2\left(\frac{x-3}{2}\right) + 3 = x - 3 + 3 = x$
2. For any real number x, $h(p(x)) = h(2x+3) = \frac{(2x+3)-3}{2} = \frac{2x}{2} = x$

Since $p(h(x)) = h(p(x)) = x$ then h and p are a pair of inverse functions.

Returning to our initial example, it is clear that both $f(x) = x^3$ and $g(x) = \sqrt[3]{x}$ satisfy the definition of a function because for both f and g every number in its domain determines exactly one number in its range. Since they are a pair of inverse functions then the 'reverse' is also true for both – that is, every number in its range is determined by exactly one number in its range. Such a function is called a **one-to-one function**. The phrase 'one-to-one' is appropriate because each value in the domain corresponds to exactly **one** value in the range, and each value in the range corresponds to exactly **one** value in the domain.

• **Hint:** The mapping diagram for f and g in Figure 2.6 nicely illustrates this 'one-to-one correspondence' between the domain and range for each function.

A one-to-one function
A function is **one-to-one** if each element y in the range is the image of exactly one element x in the domain.

The existence of an inverse function

Determining whether a function is one-to-one is very useful because the inverse of a one-to-one function will also be a function. Analyzing the graph of a function is the most effective way to determine whether a function is one-to-one. Let's look at the graph of the one-to-one function $f(x) = x^3$ shown in Figure 2.8. It is clear that as the values of x increase over the domain (i.e. from $-\infty$ to ∞) that the function values are always increasing. A function that is always increasing, or always decreasing, throughout its domain is one-to-one and has an inverse function.

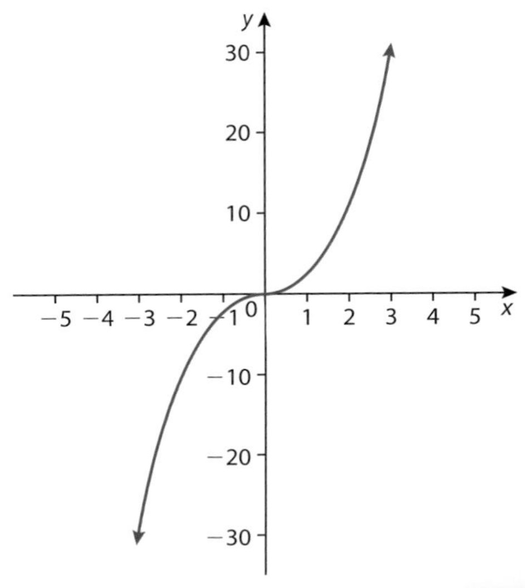

Figure 2.8 Graph of $f(x) = x^3$ which is increasing as x goes from $-\infty$ to ∞.

A function f is an **increasing function** if $x_1 < x_2$ implies $f(x_1) < f(x_2)$, and it is a **decreasing function** if $x_1 < x_2$ implies $f(x_1) > f(x_2)$. If a function is either increasing or decreasing, it is said to be **monotonic**.

Example 13 shows that a function that is not one-to-one (always increasing or always decreasing) can be made so by restricting its domain.

Example 13 – Restricting the domain so that a function is one-to-one

The function $f(x) = x^2$ (Figure 2.9) is not one-to-one for all real numbers. However, the function $g(x) = x^2$ with domain $x \geqslant 0$ (Figure 2.10) is always increasing (one-to-one), and the function $h(x) = x^2$ with domain $x \leqslant 0$ (Figure 2.11) is always decreasing (one-to-one).

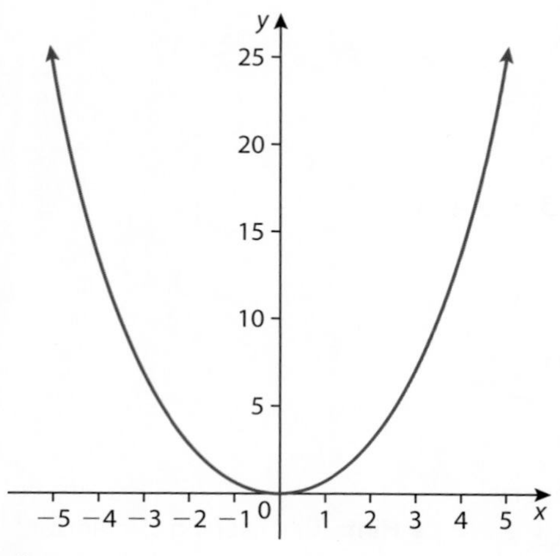

Figure 2.9 $f(x) = x^2$

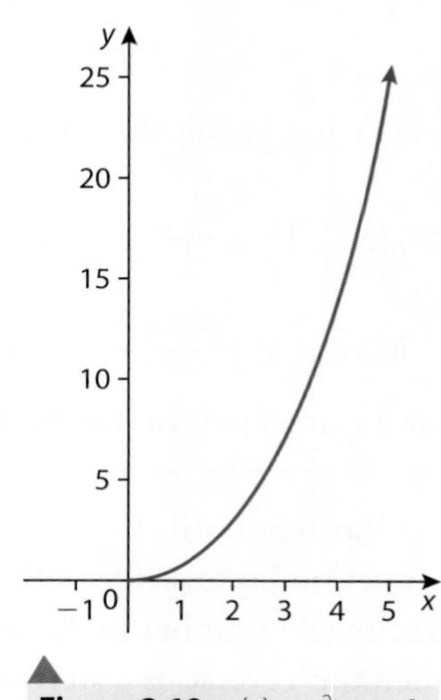

Figure 2.10 $g(x) = x^2, x \geqslant 0$

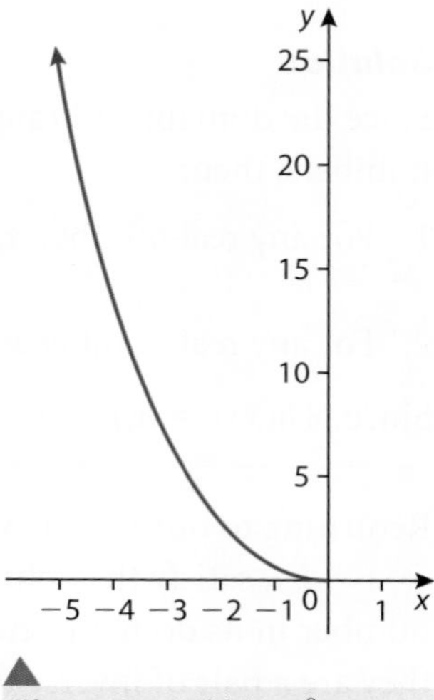

Figure 2.11 $h(x) = x^2, x \leqslant 0$

If a function f is always increasing or always decreasing in its domain (i.e. it is monotonic), then f has an inverse f^{-1}.

No horizontal line can pass through the graph of a one-to-one function at more than one point.

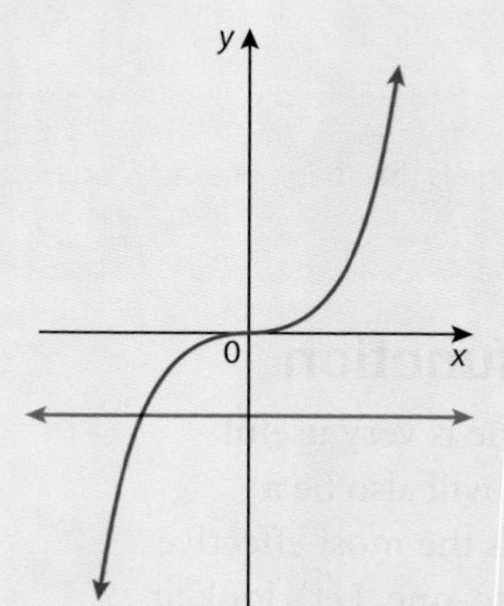

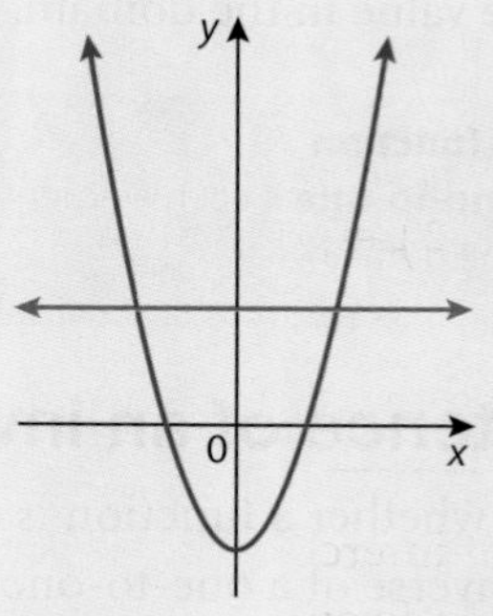

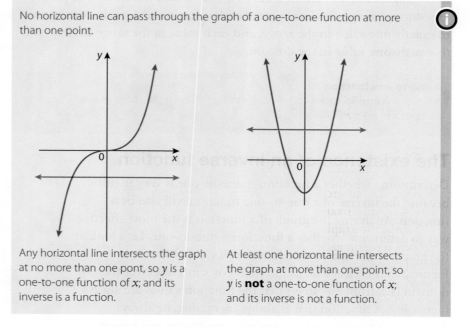

Any horizontal line intersects the graph at no more than one pont, so y is a one-to-one function of x; and its inverse is a function.

At least one horizontal line intersects the graph at more than one point, so y is **not** a one-to-one function of x; and its inverse is not a function.

A function for which at least one element y in the range is the image of more than one element x in the domain is called a **many-to-one function.** Examples of many-to-one functions that we have already encountered are $y = x^2, x \in \mathbb{R}$ and $y = |x|, x \in \mathbb{R}$. As Figure 2.12 illustrates for $y = |x|$, a horizontal line exists that intersects a many-to-one function at more than one point. Thus, the inverse of a many-to-one function will not be a function.

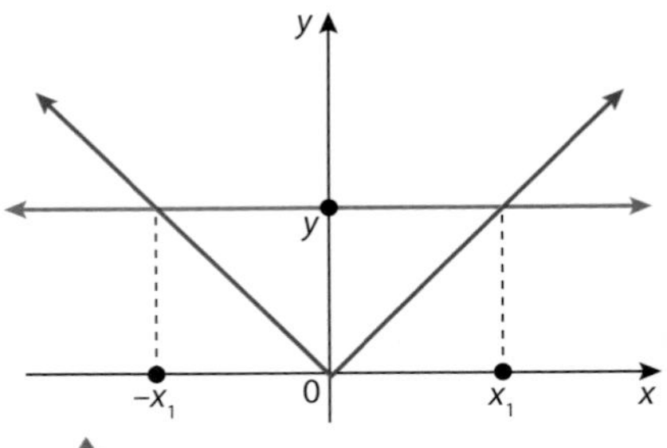

Figure 2.12 Graph of $y = |x|$; an example of a many-to-one function.

Finding the inverse of a function

Example 14 – Finding an inverse function I

The function f is defined for $x \in \mathbb{R}$ by $f(x) = 4x - 8$. Determine if f has an inverse f^{-1}. If not, restrict the domain of f in order to find an inverse function f^{-1}. Verify the result by showing that $(f \circ f^{-1})(x) = x$ and $(f^{-1} \circ f)(x) = x$. Graph f and its inverse function f^{-1} on the same set of axes.

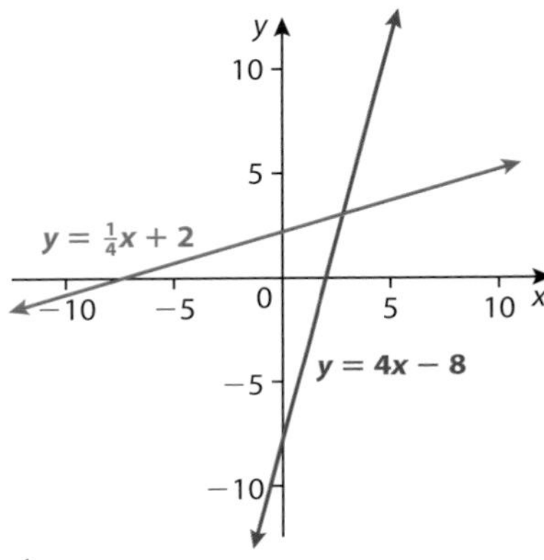

Figure 2.13 Graph of pair of inverse functions for Example 14.

Solution

Firstly, we recognize that f is an increasing function for $(-\infty, \infty)$ because the graph of $f(x) = 4x - 8$ is a straight line with a constant slope of 4. Therefore, f is a one-to-one function and it has an inverse f^{-1}. To find the equation for f^{-1}, we start by switching the domain (x) and range (y) since the domain of f becomes the range of f^{-1} and the range of f becomes the domain of f^{-1}, as stated in the definition and depicted in Figure 2.7. Also, recall that $y = f(x)$.

$f(x) = 4x - 8$

$y = 4x - 8$ Write $y = f(x)$.

$x = 4y - 8$ Interchange x and y (i.e. switch the domain and range).

$4y = x + 8$ Solve for y (dependent variable) in terms of x (independent variable).

$y = \frac{1}{4}x + 2$

$f^{-1}(x) = \frac{1}{4}x + 2$ Resulting equation is $y = f^{-1}(x)$.

Verify that f and f^{-1} are inverses by showing that $f(f^{-1}(x)) = x$ and $f^{-1}(f(x)) = x$.

$$f\left(\tfrac{1}{4}x + 2\right) = 4\left(\tfrac{1}{4}x + 2\right) - 8 = x + 8 - 8 = x$$

$$f^{-1}(4x - 8) = \tfrac{1}{4}(4x - 8) + 2 = x - 2 + 2 = x$$

This confirms that $y = 4x - 8$ and $y = \frac{1}{4}x + 2$ are inverses of each other.

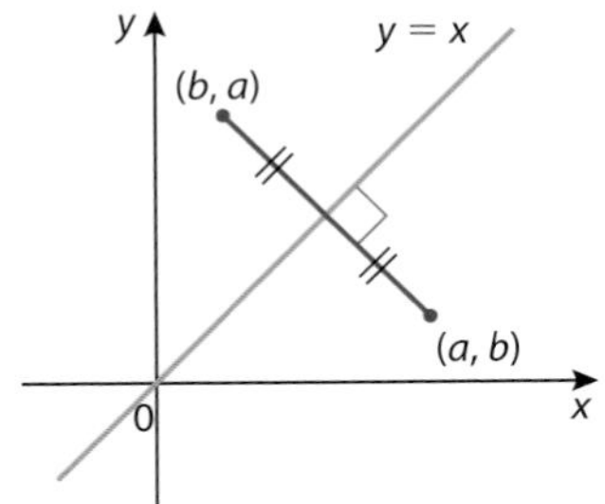

Figure 2.14 The point (b, a) is a reflection over the line $y = x$ of the point (a, b).

The method of interchanging domain (x) and range (y) to find the inverse function used in Example 14 also gives us a way for obtaining the graph of f^{-1} from the graph of f. Given the reversing effect that a pair of inverse functions have on each other, if $f(a) = b$ then $f^{-1}(b) = a$. Hence, if the ordered pair (a, b) is a point on the graph of $y = f(x)$, then the 'reversed' ordered pair (b, a) must be on the graph of $y = f^{-1}(x)$. Figure 2.14 shows that the point (b, a) can be found by reflecting the point (a, b) about the line $y = x$. Therefore, as Figure 2.15 illustrates, the following statement can be made about the graphs of a pair of inverse functions.

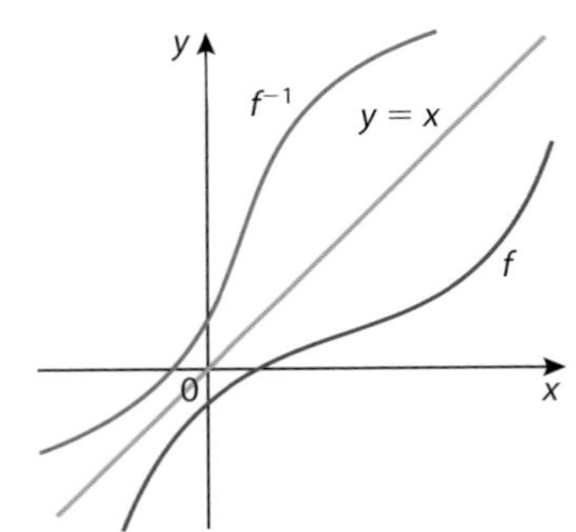

Figure 2.15 Graphs of f and f^{-1} are symmetrical about the line $y = x$.

Graphical symmetry of inverse functions

The graph of f^{-1} is a reflection of the graph of f about the line $y = x$.

Example 15 – Finding an inverse function II

The function f is defined for $x \in \mathbb{R}$ by $f: x \mapsto \dfrac{x^2+3}{x^2+1}$. Determine if f has an inverse f^{-1}. If not, restrict the domain of f in order to find an inverse function f^{-1}. Graph f and its inverse f^{-1} on the same set of axes.

Solution

A graph of f produced on a GDC reveals that it is not monotonic over its domain $(-\infty, \infty)$. It is increasing for $(-\infty, 0]$, and decreasing for $[0, \infty)$. Therefore, f does not have an inverse f^{-1} for $x \in \mathbb{R}$. It is customary to restrict the domain to the 'largest' set possible. Hence, we can choose to restrict the domain to either $x \in (-\infty, 0]$ (making f an increasing function), or $x \in [0, \infty)$ (making f a decreasing function). Let's change the domain from $x \in \mathbb{R}$ to $x \in [0, \infty)$.

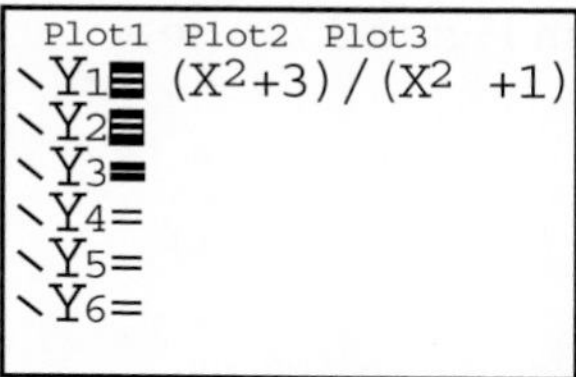

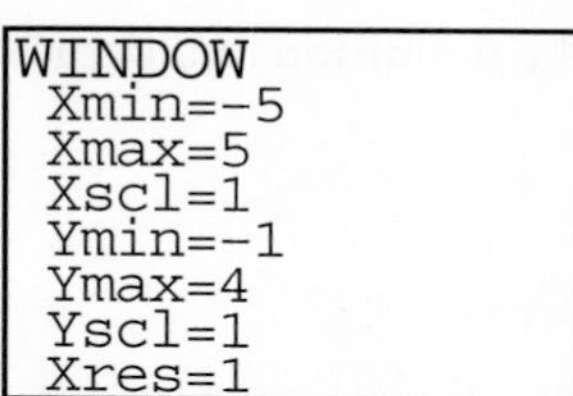

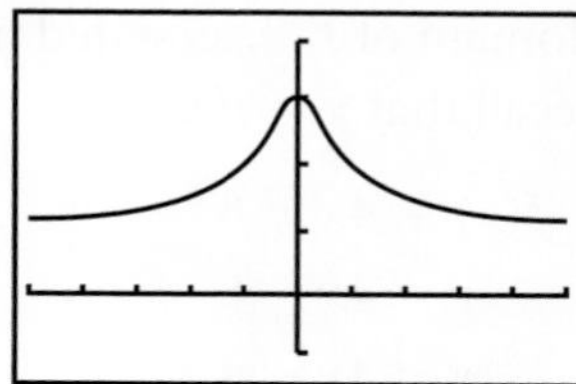

We use a method similar to that in Example 14 to find the equation for f^{-1}. First solve for x in terms of y and then interchange the domain (x) and range (y).

$$f: x \mapsto \frac{x^2+3}{x^2+1} \Rightarrow y = \frac{x^2+3}{x^2+1} \Rightarrow x^2y + y = x^2 + 3 \Rightarrow x^2y - x^2 = 3 - y$$

$$\Rightarrow x^2(y-1) = 3 - y \Rightarrow x^2 = \frac{3-y}{y-1} \Rightarrow x = \pm\sqrt{\frac{3-y}{y-1}} \Rightarrow y = \pm\sqrt{\frac{3-x}{x-1}}$$

Since we chose to restrict the domain of f to $x \in [0, \infty)$, then the range of f^{-1} will be $y \in [0, \infty)$. Therefore, from the working above, the resulting inverse function is $f^{-1}(x) = \sqrt{\dfrac{3-x}{x-1}}$.

Figure 2.16 Graphs of f and f^{-1} for Example 15 show symmetry about the line $y = x$.

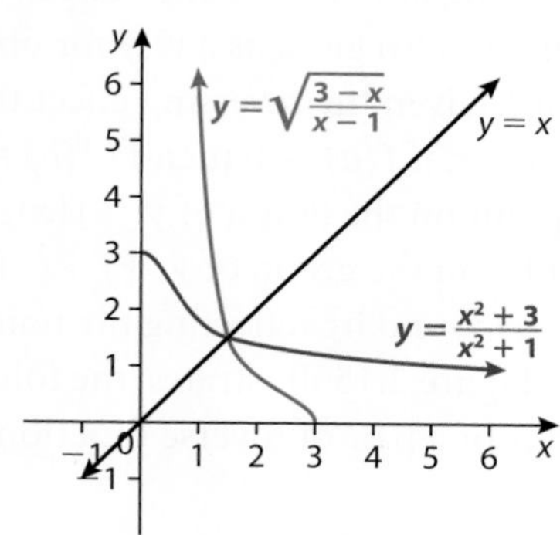

Finding the inverse of a function

To find the inverse of a function f, use the following steps:

1 Confirm that f is one-to-one (although, for this course, you can assume this).

2 Replace $f(x)$ with y.

3 Interchange x and y.

4 Solve for y.

5 Replace y with $f^{-1}(x)$.

6 The domain of f^{-1} is equal to the range of f, and the range of f^{-1} is equal to the domain of f.

Example 16

Consider the function $f: x \mapsto \sqrt{x+3}, x \geqslant -3$.

a) Determine the inverse function f^{-1}. b) What is the domain of f^{-1}?

Solution

a) Following the steps for finding the inverse of a function gives:

$y = \sqrt{x+3}$	Replace $f(x)$ with y.
$x = \sqrt{y+3}$	Interchange x and y.
$x^2 = y + 3$	Solve for y (squaring both sides here).
$y = x^2 - 3$	Solved for y.
$f^{-1}: x \mapsto x^2 - 3$	Replace y with $f^{-1}(x)$.

b) The domain explicitly defined for f is $x \geqslant -3$ and since the $\sqrt{\ }$ symbol stands for the principal square root (positive), then the range of f is all positive real numbers, i.e. $y \geqslant 0$. The domain of f^{-1} is equal to the range of f; therefore, the domain of f^{-1} is $x \geqslant 0$.

Graphing $y = \sqrt{x+3}$ and $y = x^2 - 3$ from Example 16 on your GDC visually confirms these results. Note that since the calculator would have automatically assumed that the domain is $x \in \mathbb{R}$, the domain for the equation $y = x^2 - 3$ has been changed to $x \geqslant 0$. In order to show that f and f^{-1} are reflections about the line $y = x$, the line $y = x$ has been graphed and a viewing window has been selected to ensure that the scales are equal on each axis. Using the trace feature of your GDC, you can explore a characteristic of inverse functions – that is, if some point (a, b) is on the graph of f, the point (b, a) must be on the graph of f^{-1}.

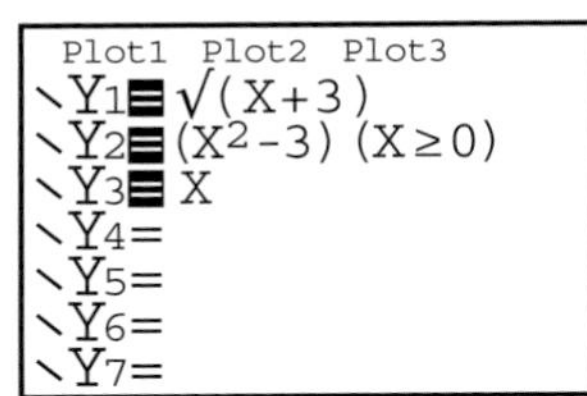

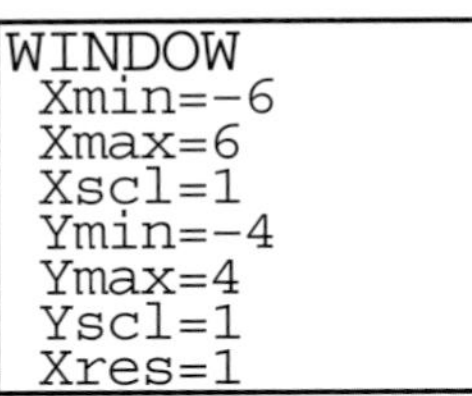

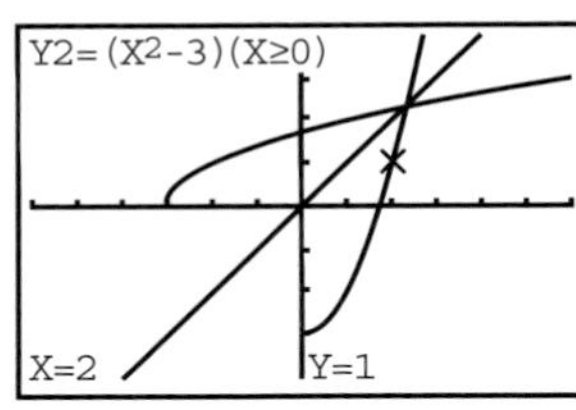

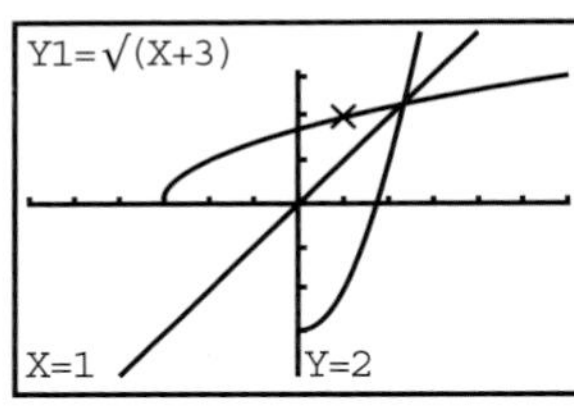

Example 17

Consider the function $f(x) = 2(x+4)$ and $g(x) = \frac{1-x}{3}$.

a) Find g^{-1} and state its domain and range.

b) Solve the equation $(f \circ g^{-1})(x) = 2$.

Solution

a)
$$y = \frac{1-x}{3}$$ Replace $f(x)$ with y.

$$x = \frac{1-y}{3}$$ Interchange x and y.

$$3x = 1 - y$$ Solve for y.

$$y = -3x + 1$$ Solved for y.

$$g^{-1}(x) = -3x + 1$$ Replace y with $g^{-1}(x)$.

g is a linear function and its domain is $x \in \mathbb{R}$ and its range is $y \in \mathbb{R}$; therefore, for g^{-1} the domain is $x \in \mathbb{R}$ and range is $y \in \mathbb{R}$.

b)
$$(f \circ g^{-1})(x) = f(g^{-1}(x)) = f(-3x+1) = 2$$
$$2[(-3x+1)+4] = 2$$
$$-6x + 2 + 8 = 2$$
$$-6x = -8$$
$$x = \tfrac{4}{3}$$

Exercise 2.3

In questions 1–4, assume that f is a one-to-one function.

1 a) If $f(2) = -5$, what is $f^{-1}(-5)$?

b) If $f^{-1}(6) = 10$, what is $f(10)$?

2 a) If $f(-1) = 13$, what is $f^{-1}(13)$?

b) If $f^{-1}(b) = a$, what is $f(a)$?

3 If $g(x) = 3x - 7$, what is $g^{-1}(5)$?

4 If $h(x) = x^2 - 8x$, with $x \geqslant 4$, what is $h^{-1}(-12)$?

In questions 5–14, show a) algebraically and b) graphically that f and g are inverse functions by verifying that $(f \circ g)(x) = x$ and $(g \circ f)(x) = x$, and by sketching the graphs of f and g on the same set of axes, with equal scales on the x- and y-axes. Use your GDC to assist in making your sketches on paper.

5 $f: x \mapsto x + 6;\ g: x \mapsto x - 6$

6 $f: x \mapsto 4x;\ g: x \mapsto \frac{x}{4}$

7 $f: x \mapsto 3x + 9;\ g: x \mapsto \frac{1}{3}x - 3$

8 $f: x \mapsto \frac{1}{x};\ g: x \mapsto \frac{1}{x}$

9 $f: x \mapsto x^2 - 2, x \geqslant 0;\ g: x \mapsto \sqrt{x+2}, x \geqslant -2$

10 $f: x \mapsto 5 - 7x;\ g: x \mapsto \frac{5-x}{7}$

11 $f: x \mapsto \frac{1}{1+x};\ g: x \mapsto \frac{1-x}{x}$

12 $f: x \mapsto (6 - x)^{\frac{1}{2}};\ g: x \mapsto 6 - x^2, x \geqslant 0$

13 $f: x \mapsto x^2 - 2x + 3, x \geqslant 1;\ g: x \mapsto 1 + \sqrt{x-2}, x \geqslant 2$

14 $f: x \mapsto \sqrt[3]{\frac{x+6}{2}};\ g: x \mapsto 2x^3 - 6$

In questions 15–24, find the inverse function f^{-1} and state its domain.

15 $f(x) = 2x - 3$

16 $f(x) = \dfrac{x + 7}{4}$

17 $f(x) = \sqrt{x}$

18 $f(x) = \dfrac{1}{x + 2}$

19 $f(x) = 4 - x^2, x \geqslant 0$

20 $f(x) = \sqrt{x - 5}$

21 $f(x) = ax + b, a \neq 0$

22 $f(x) = x^2 + 2x, x \geqslant -1$

23 $f(x) = \dfrac{x^2 - 1}{x^2 + 1}, x \leqslant 0$

24 $f(x) = x^3 + 1$

In questions 25–28, determine if f has an inverse f^{-1}. If not, restrict the domain of f in order to find an inverse function. Graph f and its inverse f^{-1} on the same set of axes.

25 $f(x) = \dfrac{2x + 3}{x - 1}$

26 $f(x) = (x - 2)^2$

27 $f(x) = \dfrac{1}{x^2}$

28 $f(x) = 2 - x^4$

29 Use your GDC to graph the function $f(x) = \dfrac{2x}{1 + x^2}, x \in \mathbb{R}$. Find three intervals for which f is a one-to-one function (monotonic) and hence will have an inverse f^{-1} on the interval. The union of all three intervals is all real numbers.

In questions 30–37, use the functions $g(x) = x + 3$ and $h(x) = 2x - 4$ to find the indicated value or the indicated function.

30 $(g^{-1} \circ h^{-1})(5)$

31 $(h^{-1} \circ g^{-1})(9)$

32 $(g^{-1} \circ g^{-1})(2)$

33 $(h^{-1} \circ h^{-1})(2)$

34 $g^{-1} \circ h^{-1}$

35 $h^{-1} \circ g^{-1}$

36 $(g \circ h)^{-1}$

37 $(h \circ g)^{-1}$

38 The reciprocal function in question 8, $f(x) = \dfrac{1}{x}$, is its own inverse (self-inverse). Show that any function in the form $f(x) = \dfrac{a}{x + b} - b, a \neq 0$ is its own inverse.

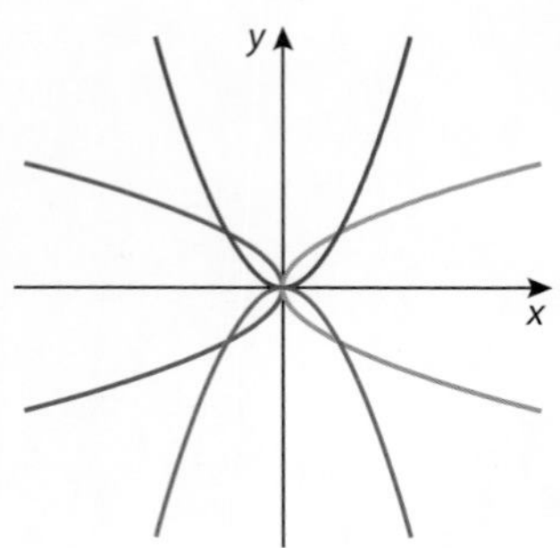

2.4 Transformations of functions

Even when you use your GDC to sketch the graph of a function, it is helpful to know what to expect in terms of the location and shape of the graph – and even more so if you're not allowed to use your GDC for a particular question. In this section, we look at how certain changes to the equation of a function can affect, or **transform**, the location and shape of its graph. We will investigate three different types of **transformations** of functions that include how the graph of a function can be **translated**, **reflected** and **stretched** (or shrunk). Studying graphical transformations gives us a better understanding of how to efficiently sketch and visualize many different functions. We will also take a closer look at two specific functions: the absolute value function, $y = |x|$, and the reciprocal function, $y = \frac{1}{x}$.

• **Hint:** When analyzing the graph of a function, it is often convenient to express a function in the form $y = f(x)$. As we have done throughout this chapter, we often refer to a function such as $f(x) = x^2$ by the equation $y = x^2$.

Graphs of common functions

It is important for you to be familiar with the location and shape of a certain set of common functions. For example, from your previous knowledge about linear equations, you can determine the location of the linear function $f(x) = ax + b$. You know that the graph of this function is a line whose slope is a and whose y-intercept is $(0, b)$.

The eight graphs in Figure 2.17 represent some of the most commonly used functions in algebra. You should be familiar with the characteristics of the graphs of these common functions. This will help you predict and analyze the graphs of more complicated functions that are derived from applying one or more transformations to these simple functions. There are other important basic functions with which you should be familiar – for

Figure 2.17 Graphs of common functions.

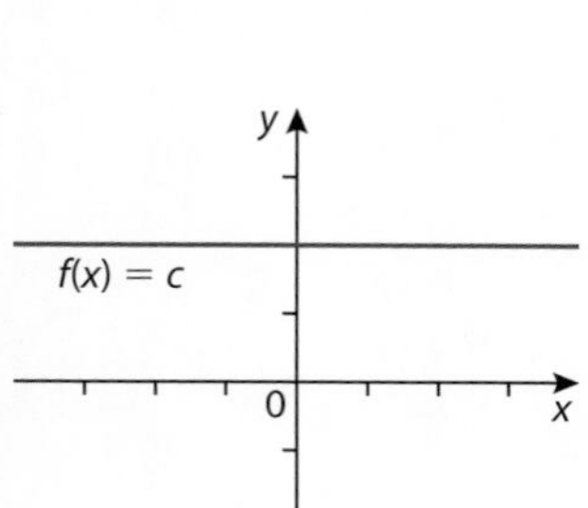

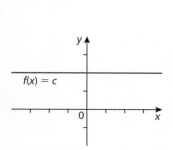

a) Constant function

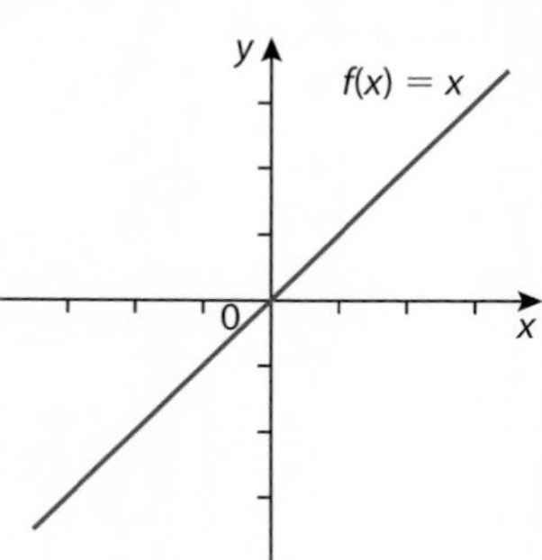

b) Identity function

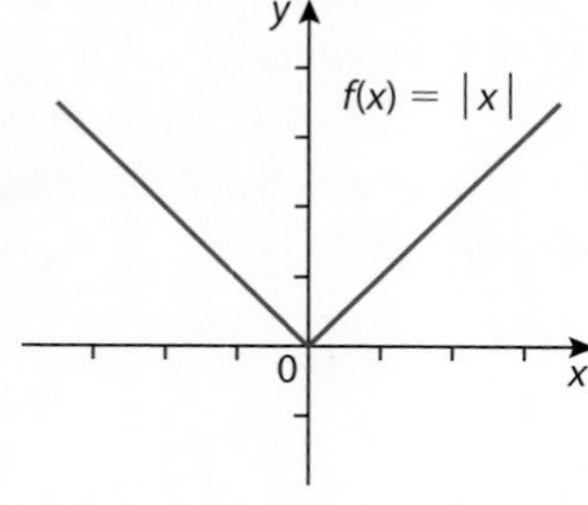

c) Absolute value function

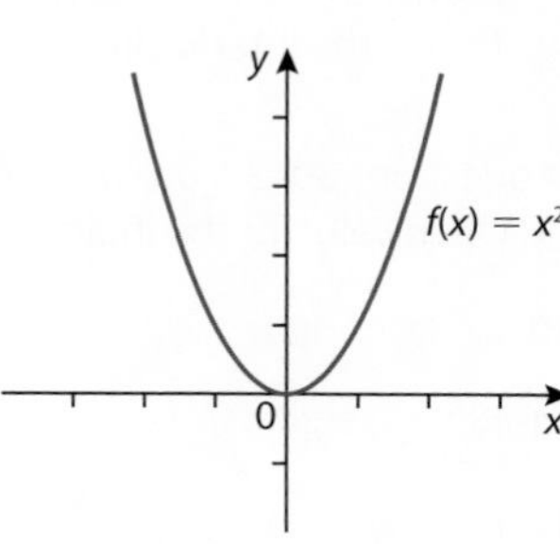

d) Squaring function

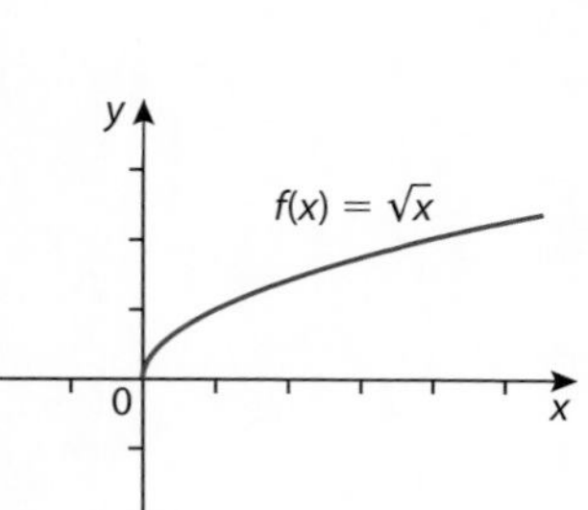

e) Square root function

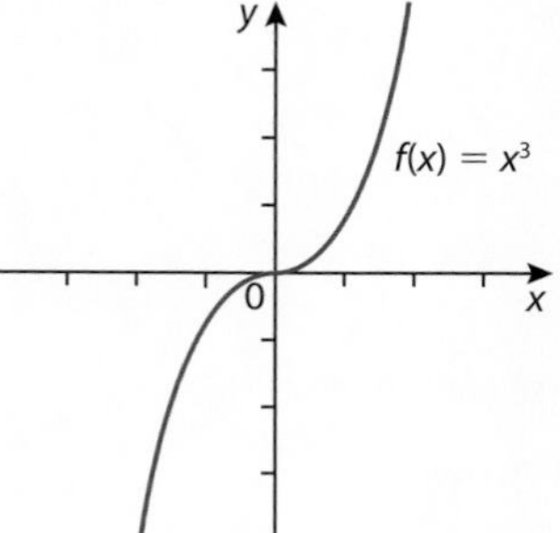

f) Cubing function

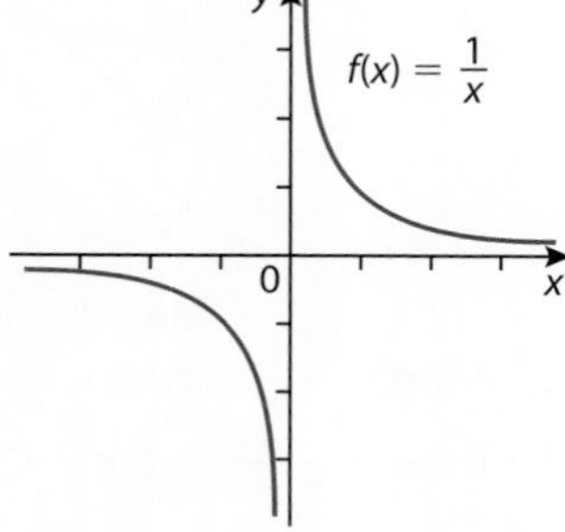

g) Reciprocal function

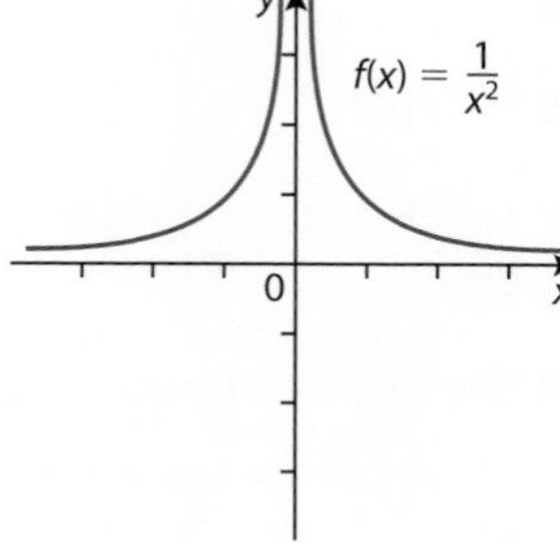

h) Inverse square function

example, exponential, logarithmic and trigonometric functions – but we will encounter these in later chapters.

We will see that many functions have graphs that are a transformation (translation, reflection or stretch), or a combination of transformations, of one of these common functions.

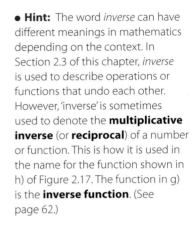

• **Hint:** The word *inverse* can have different meanings in mathematics depending on the context. In Section 2.3 of this chapter, *inverse* is used to describe operations or functions that undo each other. However, 'inverse' is sometimes used to denote the **multiplicative inverse** (or **reciprocal**) of a number or function. This is how it is used in the name for the function shown in h) of Figure 2.17. The function in g) is the **inverse function**. (See page 62.)

Vertical and horizontal translations

Use your GDC to graph each of the following three functions: $f(x) = x^2$, $g(x) = x^2 + 3$ and $h(x) = x^2 - 2$. How do the graphs of g and h compare with the graph of f that is one of the common functions displayed in Figure 2.17? The graphs of g and h both appear to have the same shape – it's only the location, or position, that has changed compared to f. Although the curves (parabolas) appear to be getting closer together, their vertical separation at every value of x is constant.

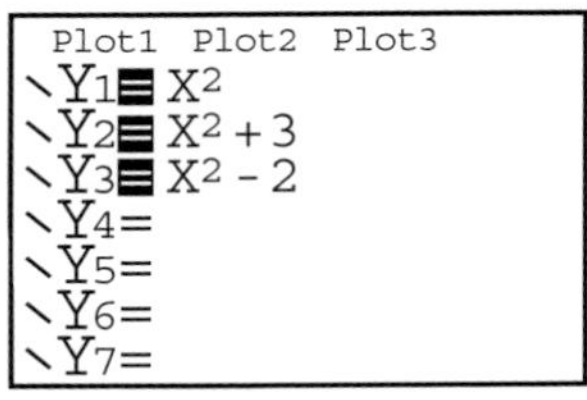

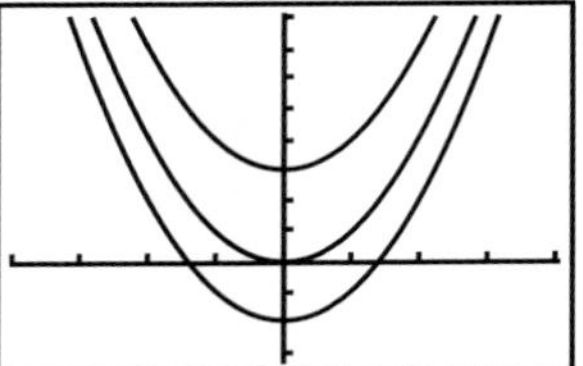

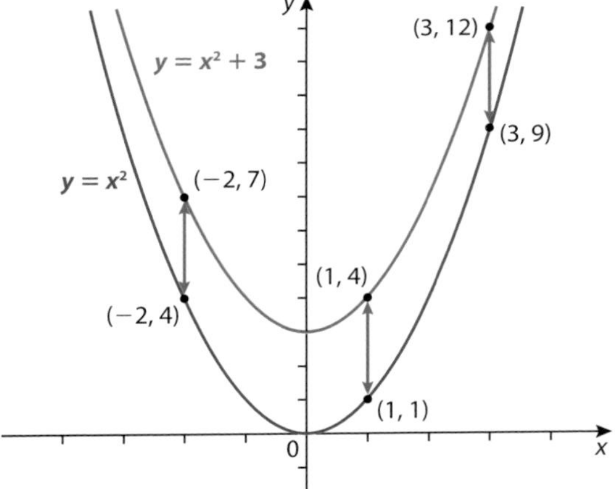

Figure 2.18 Translating $f(x) = x^2$ up.

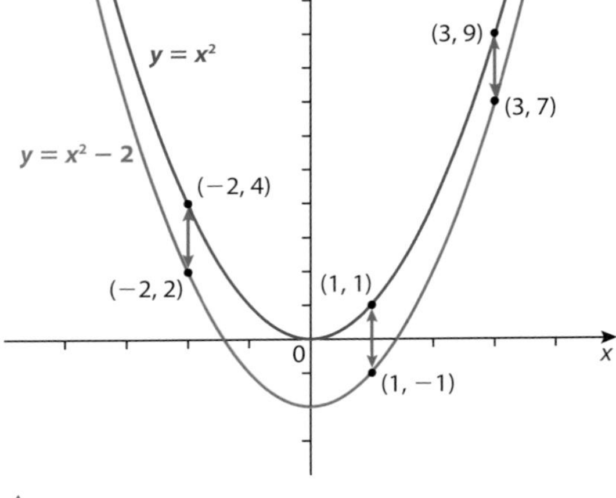

Figure 2.19 Translating $f(x) = x^2$ down.

As Figures 2.18 and 2.19 clearly show, you can obtain the graph of $g(x) = x^2 + 3$ by translating (shifting) the graph of $f(x) = x^2$ *up* three units, and you can obtain the graph of $h(x) = x^2 - 2$ by translating the graph of $f(x) = x^2$ *down* two units.

Vertical translations of a function

Given $k > 0$, then:

I. The graph of $y = f(x) + k$ is obtained by translating *up* k units the graph of $y = f(x)$.

II. The graph of $y = f(x) - k$ is obtained by translating *down* k units the graph of $y = f(x)$.

Change function g to $g(x) = (x + 3)^2$ and change function h to $h(x) = (x - 2)^2$. Graph these two functions along with the 'parent' function

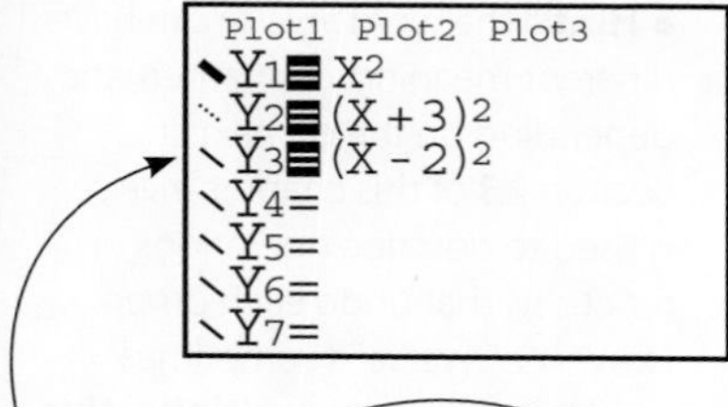

Note that a different graphing style is assigned to each equation on the GDC.

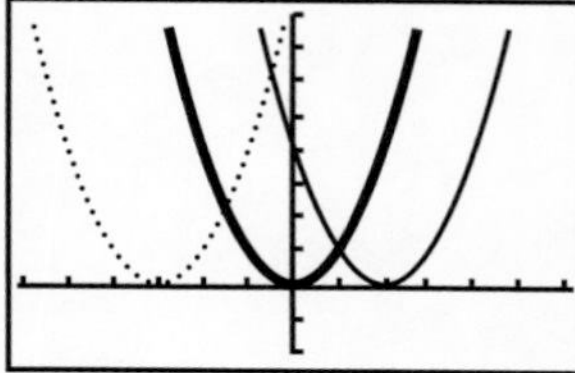

$f(x) = x^2$ on your GDC. This time we observe that functions g and h can be obtained by a horizontal translation of f.

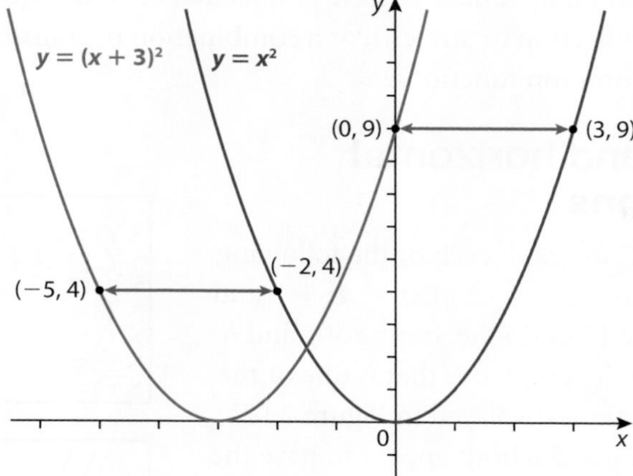

Figure 2.20 Translate $y = x^2$ left 3 units to produce graph of $y = (x + 3)^2$.

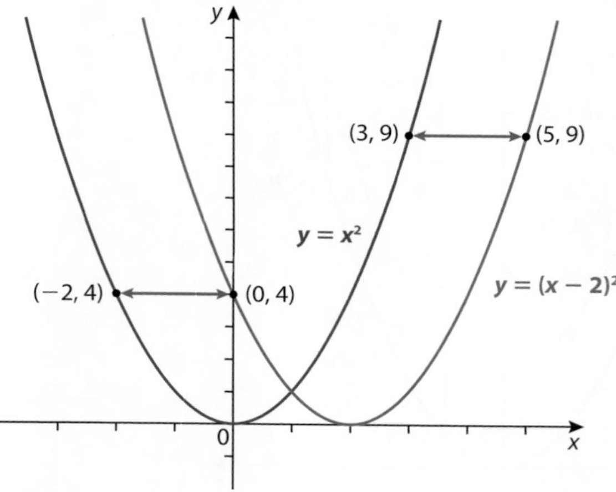

Figure 2.21 Translate $y = x^2$ right 2 units to produce graph of $y = (x - 2)^2$.

As Figures 2.20 and 2.21 clearly show, you can obtain the graph of $g(x) = (x + 3)^2$ by translating the graph of $f(x) = x^2$ three units to the *left*, and you can obtain the graph of $h(x) = (x - 2)^2$ by translating the graph of $f(x) = x^2$ two units to the *right*.

Horizontal translations of a function

Given $h > 0$, then:

I. The graph of $y = f(x - h)$ is obtained by translating the graph of $y = f(x)$ h units to the *right*.

II. The graph of $y = f(x + h)$ is obtained by translating the graph of $y = f(x)$ h units to the *left*.

• **Hint:** An alternative (and more consistent) approach to vertical and horizontal translations is to think of what number is being added directly to the x- or y-coordinate. For example, the equation for the graph obtained by translating the graph of $y = x^2$ three units up is $y = x^2 + 3$, which can also be written as $y - 3 = x^2$. In this form, negative three is added to the y-coordinate (vertical coordinate), which causes a vertical translation in the *upward* (or positive) direction. Likewise, the equation for the graph obtained by translating the graph of $y = x^2$ two units to the right is $y = (x - 2)^2$. Negative two is added to the x-coordinate (horizontal coordinate), which causes a horizontal translation to the right (or positive direction). There is consistency between vertical and horizontal translations. Assuming that movement up or to the right is considered positive, and that movement down or to the left is negative, then the direction for either type of translation is opposite to the sign ($\pm$) of the number being added to the vertical (y) or horizontal (x) coordinate. In fact, what is actually being translated is the y-axis or the x-axis. For example, the graph of $y - 3 = x^2$ can also be obtained by not changing the graph of $y = x^2$ but instead translating the y-axis three units down – which creates exactly the same effect as translating the graph of $y = x^2$ three units up.

Example 18 – Translations of a graph

The diagrams show how the graph of $y = \sqrt{x}$ is transformed to the graph of $y = f(x)$ in three steps. For each diagram, a) and b), give the equation of the curve.

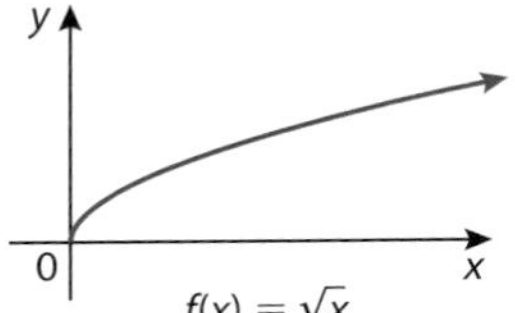

a) 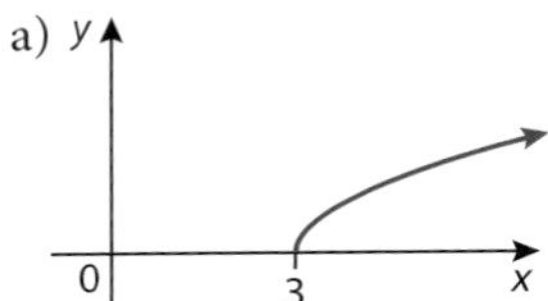

b) 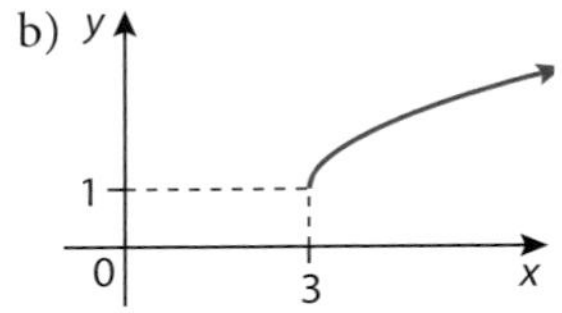

Solution

To obtain graph a), the graph of $y = \sqrt{x}$ is translated three units to the right. To produce the equation of the translated graph, -3 is added *inside* the argument of the function $y = \sqrt{x}$. Therefore, the equation of the curve graphed in a) is $y = \sqrt{x - 3}$.

To obtain graph b), the graph of $y = \sqrt{x - 3}$ is translated up one unit. To produce the equation of the translated graph, $+1$ is added *outside* the function. Therefore, the equation of the curve graphed in b) is $y = \sqrt{x - 3} + 1$ (or $y = 1 + \sqrt{x - 3}$).

Note that in Example 18, if the transformations had been performed in reverse order – that is, the vertical translation followed by the horizontal translation – it would produce the same final graph (in part b)) with the same equation. In other words, when applying both a vertical and horizontal translation on a function it does not make any difference which order they are applied (i.e. they are commutative). However, as we will see further on in the chapter, it *can* make a difference to how other sequences of transformations are applied. In general, transformations are *not* commutative.

Example 19

Write the equation of the absolute value function whose graph is shown on the right.

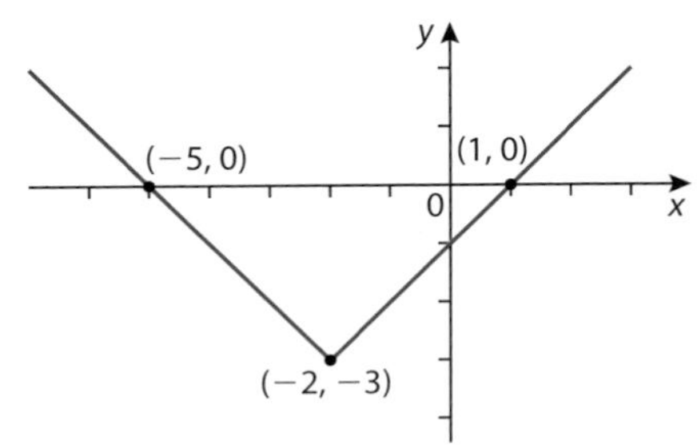

Solution

The graph shown is exactly the same shape as the graph of the equation $y = |x|$ but in a different position. Given that the vertex is $(-2, -3)$, it is clear that this graph can be obtained by translating $y = |x|$ two units left and then three units down. When we move $y = |x|$ two units left we get the graph of $y = |x + 2|$. Moving the graph of $y = |x + 2|$ three units down gives us the graph of $y = |x + 2| - 3$. Therefore, the equation of the graph shown is $y = |x + 2| - 3$. (Note: The two translations applied in reverse order produce the same result.)

Reflections

Use your GDC to graph the two functions $f(x) = x^2$ and $g(x) = -x^2$. The graph of $g(x) = -x^2$ is a reflection in the x-axis of $f(x) = x^2$. This certainly makes sense because g is formed by multiplying f by -1, causing the y-coordinate of each point on the graph of $y = -x^2$ to be the negative of the y-coordinate of the point on the graph of $y = x^2$ with the same x-coordinate.

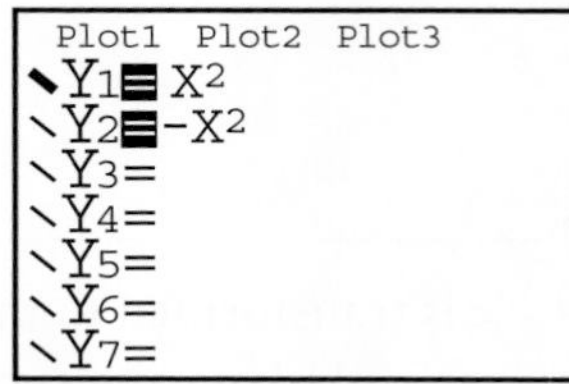

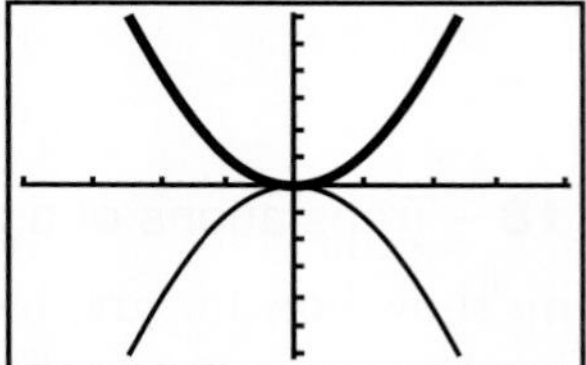

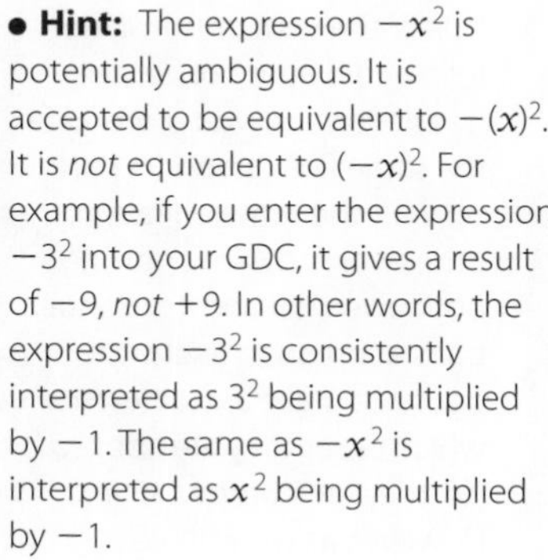

• **Hint:** The expression $-x^2$ is potentially ambiguous. It is accepted to be equivalent to $-(x)^2$. It is *not* equivalent to $(-x)^2$. For example, if you enter the expression -3^2 into your GDC, it gives a result of -9, *not* $+9$. In other words, the expression -3^2 is consistently interpreted as 3^2 being multiplied by -1. The same as $-x^2$ is interpreted as x^2 being multiplied by -1.

Figures 2.22 and 2.23 illustrate that the graph of $y = -f(x)$ is obtained by reflecting the graph of $y = f(x)$ in the x-axis.

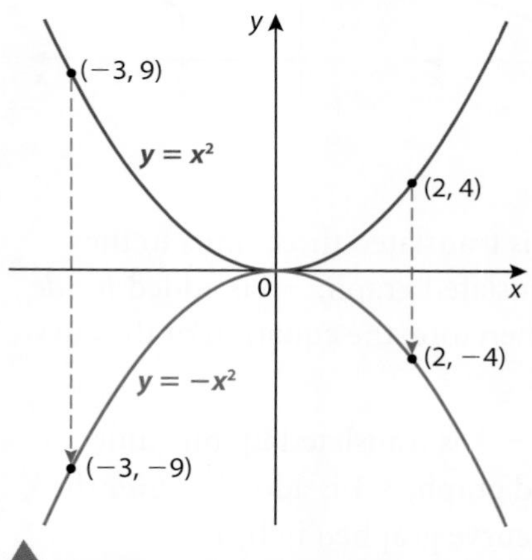

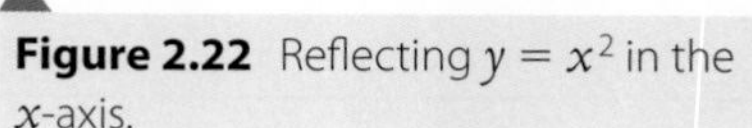

Figure 2.22 Reflecting $y = x^2$ in the x-axis.

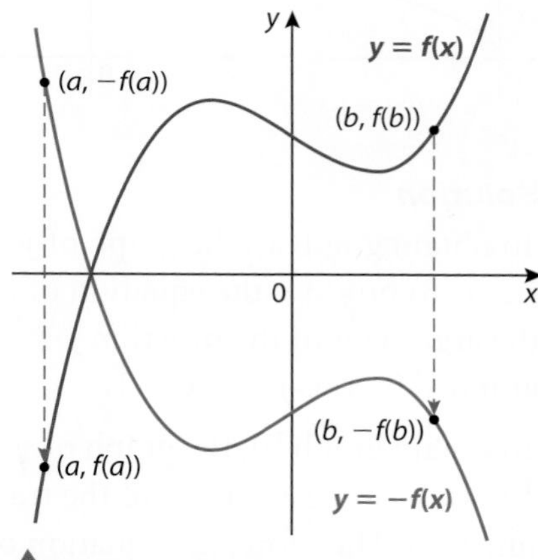

Figure 2.23 Reflecting $f(x)$ in the x-axis.

Graph the functions $f(x) = \sqrt{x - 2}$ and $g(x) = \sqrt{-x - 2}$. Previously, with $f(x) = x^2$ and $g(x) = -x^2$, g was formed by multiplying the entire function f by -1. However, for $f(x) = \sqrt{x - 2}$ and $g(x) = \sqrt{-x - 2}$, g is formed by multiplying the variable x by -1. In this case, the graph of $g(x) = \sqrt{-x - 2}$ is a reflection in the y-axis of $f(x) = \sqrt{x - 2}$. This makes sense if you recognize that the y-coordinate on the graph of $y = \sqrt{-x}$ will be the same as the y-coordinate on the graph of $y = \sqrt{x}$, if the value substituted for x in $y = \sqrt{-x}$ is the opposite of the value of x in $y = \sqrt{x}$. For example, if $x = 9$ then $y = \sqrt{9} = 3$; and, if $x = -9$ then $y = \sqrt{-(-9)} = \sqrt{9} = 3$. Opposite values of x in the two functions produce the same y-coordinate for each.

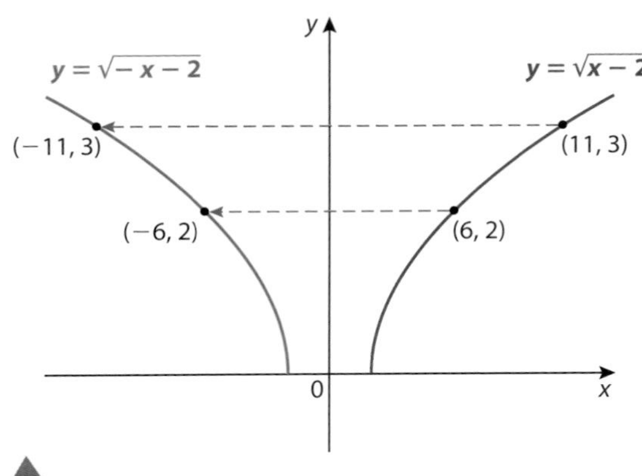

Figure 2.24 Reflecting $y = \sqrt{x-2}$ in the y-axis.

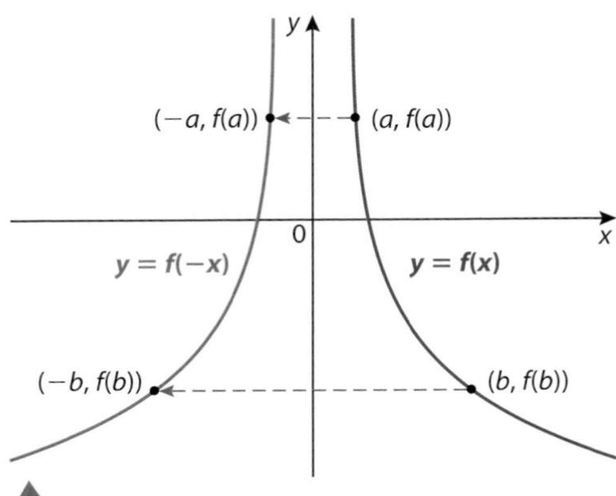

Figure 2.25 Reflecting $f(x)$ in the y-axis.

Figures 2.24 and 2.25 illustrate that the graph of $y = f(-x)$ is obtained by reflecting the graph of $y = f(x)$ in the y-axis.

Reflections of a function in the coordinate axes

I. The graph of $y = -f(x)$ is obtained by reflecting the graph of $y = f(x)$ in the x-axis.
II. The graph of $y = f(-x)$ is obtained by reflecting the graph of $y = f(x)$ in the y-axis.

Example 20 – Reflections in the coordinate axes

For $g(x) = 2x^3 - 6x^2 + 3$, find:

a) the function $h(x)$ that is the reflection of $g(x)$ in the x-axis

b) the function $p(x)$ that is the reflection of $g(x)$ in the y-axis.

Solution

a) Knowing that $y = -f(x)$ is the reflection of $y = f(x)$ in the x-axis, then $h(x) = -g(x) = -(2x^3 - 6x^2 + 3) \Rightarrow h(x) = -2x^3 + 6x^2 - 3$ will be the reflection of $g(x)$ in the x-axis. We can verify the result on the GDC – graphing the original equation $y = 2x^3 - 6x^2 + 3$ in bold style.

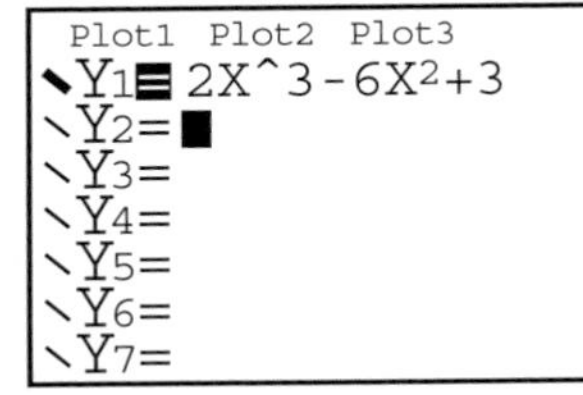

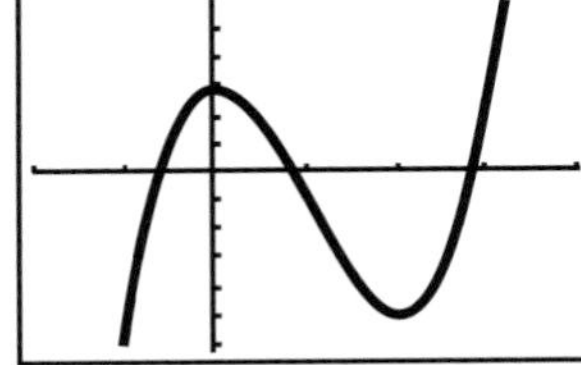

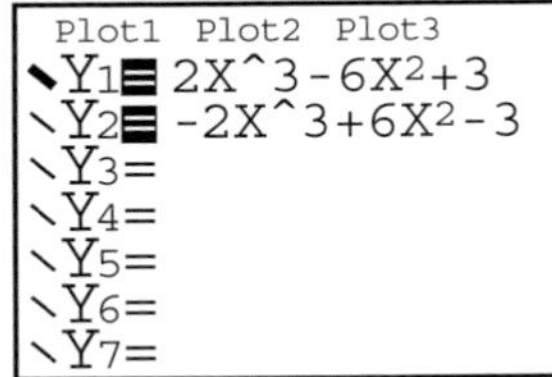

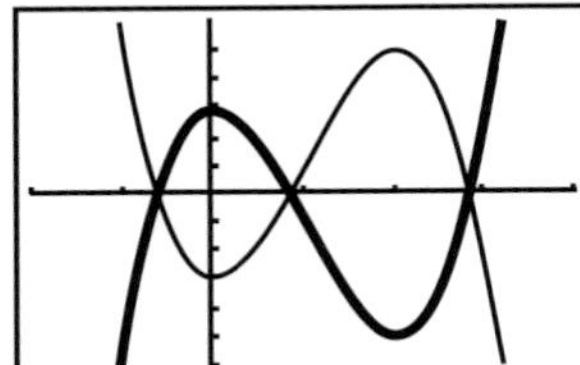

b) Knowing that $y = f(-x)$ is the reflection of $y = f(x)$ in the y-axis, we need to substitute $-x$ for x in $y = g(x)$. Thus, $p(x) = g(-x) = 2(-x)^3 - 6(-x)^2 + 3 \Rightarrow p(x) = -2x^3 - 6x^2 + 3$ will be the reflection of $g(x)$ in the y-axis. Again, we can verify the result on the GDC – graphing the original equation $y = 2x^3 - 6x^2 + 3$ in bold style.

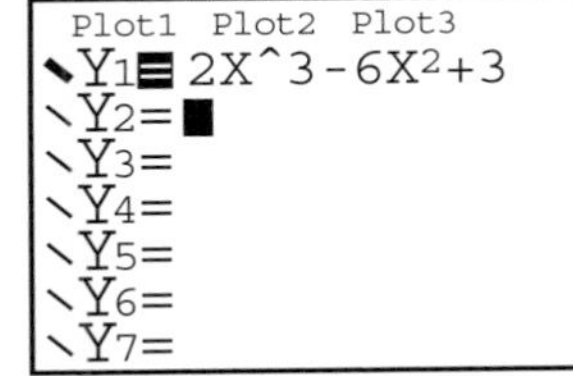

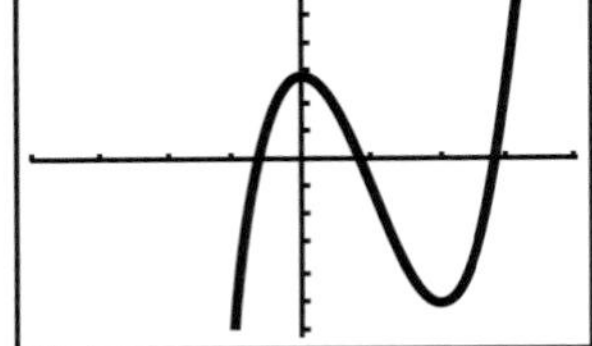

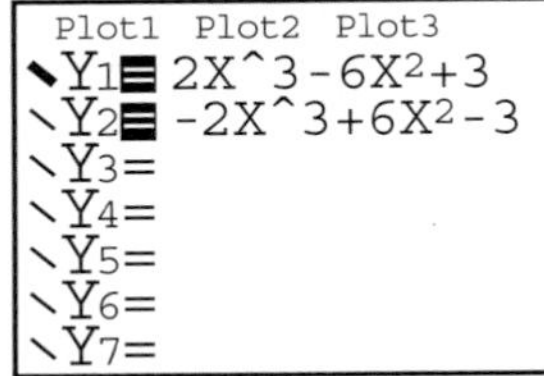

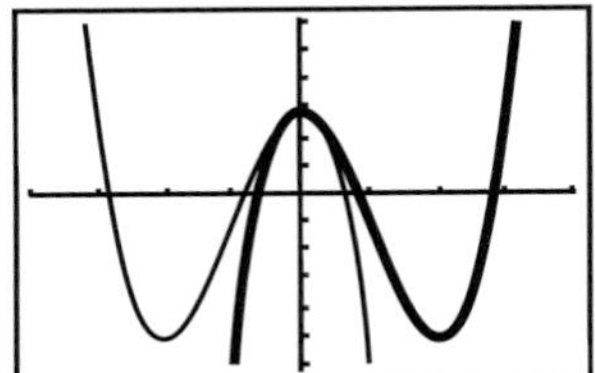

Non-rigid transformations: stretching and shrinking

Horizontal and vertical translations, and reflections in the x- and y-axes are called **rigid transformations** because the shape of the graph does not change – only its position is changed. **Non-rigid transformations** cause the shape of the original graph to change. The non-rigid transformations that we will study cause the shape of a graph to *stretch* or *shrink* in either the vertical or horizontal direction.

Vertical stretch or shrink

Graph the following three functions: $f(x) = x^2$, $g(x) = 3x^2$ and $h(x) = \frac{1}{3}x^2$. How do the graphs of g and h compare to the graph of f? Clearly, the shape of the graphs of g and h is not the same as the graph of f. Multiplying the function f by a positive number greater than one, or less than one, has distorted the shape of the graph. For a certain value of x, the y-coordinate of $y = 3x^2$ is three times the y-coordinate of $y = x^2$. Therefore, the graph of $y = 3x^2$ can be obtained by *vertically stretching* the graph of $y = x^2$ by a factor of 3 (**scale factor 3**). Likewise, the graph of $y = \frac{1}{3}x^2$ can be obtained by *vertically shrinking* the graph of $y = x^2$ by **scale factor $\frac{1}{3}$.**

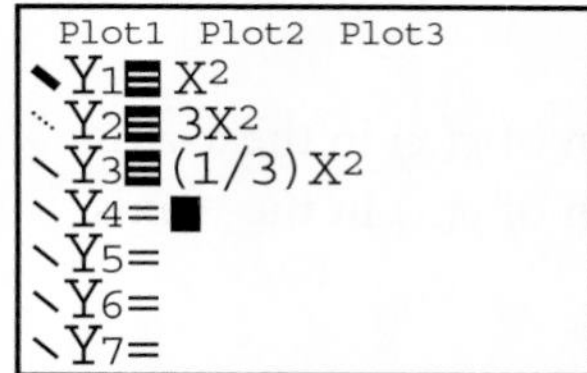

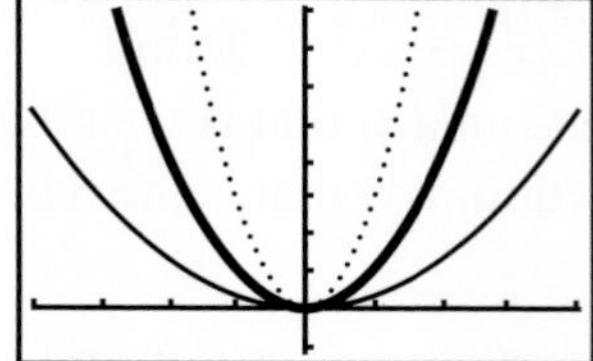

Figures 2.26 and 2.27 illustrate how multiplying a function by a positive number, a, *greater than one* causes a transformation by which the function *stretches* vertically by scale factor a. A point (x, y) on the graph of $y = f(x)$ is transformed to the point (x, ay) on the graph of $y = af(x)$.

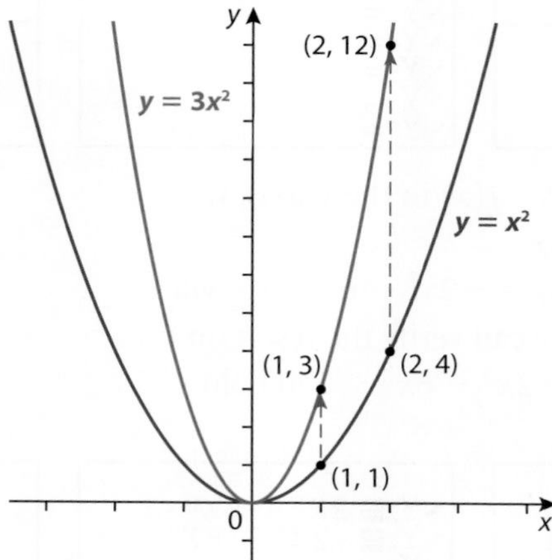

Figure 2.26 Vertical stretch of $y = x^2$ by scale factor 3.

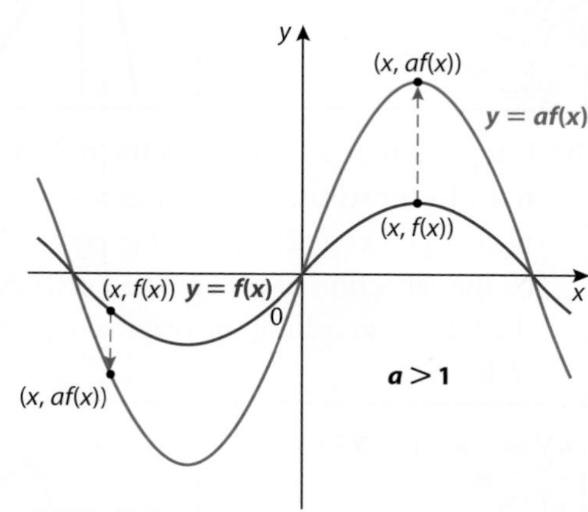

Figure 2.27 Vertical stretch of $f(x)$ by scale factor a.

Figures 2.28 and 2.29 illustrate how multiplying a function by a positive number, a, *greater than zero and less than one* causes the function to *shrink* vertically by scale factor a. A point (x, y) on the graph of $y = f(x)$ is transformed to the point (x, ay) on the graph of $y = af(x)$.

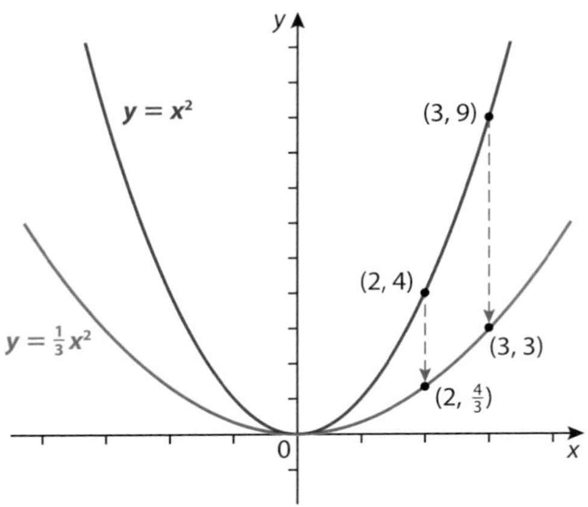

Figure 2.28 Vertical shrink of $y = x^2$ by scale factor $\frac{1}{3}$.

Figure 2.29 Vertical shrink of $f(x)$ by scale factor a.

Vertical stretching and shrinking of functions

I. If $a > 1$, the graph of $y = af(x)$ is obtained by *vertically stretching* the graph of $y = f(x)$.

II. If $0 < a < 1$, the graph of $y = af(x)$ is obtained by *vertically shrinking* the graph of $y = f(x)$.

Horizontal stretch or shrink

Let's investigate how the graph of $y = f(ax)$ is obtained from the graph of $y = f(x)$. Given $f(x) = x^2 - 4x$, find another function, $g(x)$, such that $g(x) = f(2x)$. We substitute $2x$ for x in the function f, giving $g(x) = (2x)^2 - 4(2x)$. For the purposes of our investigation, let's leave $g(x)$ in this form. On your GDC, graph these two functions, $f(x) = x^2 - 4x$ and $g(x) = (2x)^2 - 4(2x)$, using the indicated viewing window and graphing f in bold style.

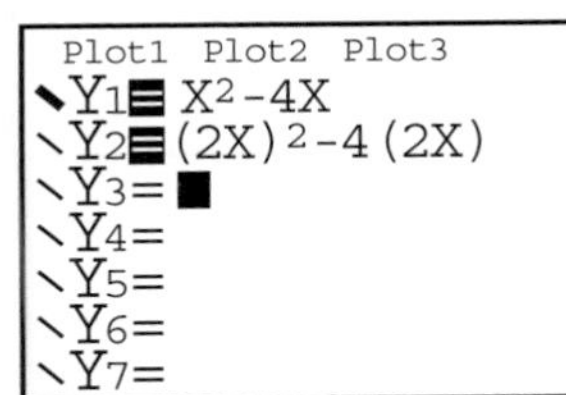

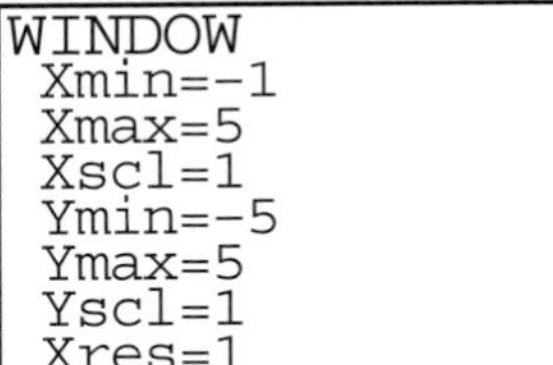

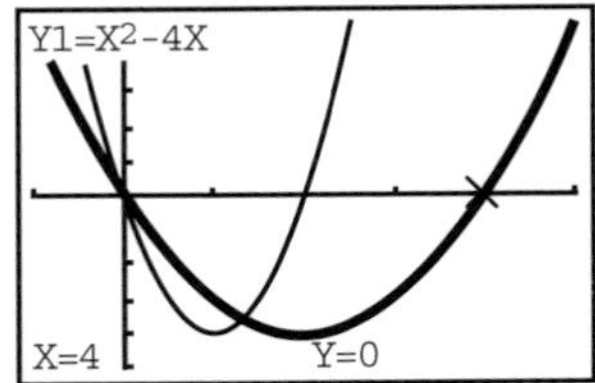

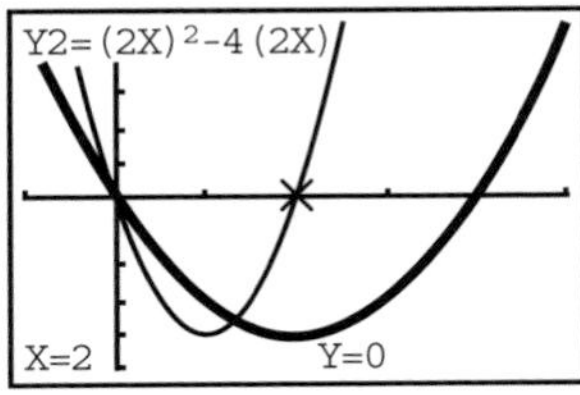

Comparing the graphs of the two equations, we see that $y = g(x)$ is *not* a translation or a reflection of $y = f(x)$. It is similar to the *shrinking* effect that occurs for $y = af(x)$ when $0 < a < 1$, except, instead of a vertical shrinking, the graph of $y = g(x) = f(2x)$ is obtained by *horizontally* shrinking the graph of $y = f(x)$. Given that it is a shrinking – rather than a stretching – the scale factor must be less than one. Consider the point $(4, 0)$ on the graph of $y = f(x)$. The point on the graph of $y = g(x) = f(2x)$ with the same y-coordinate and on the same side of the parabola is $(2, 0)$. The x-coordinate of the point on $y = f(2x)$ is the x-coordinate of the point on $y = f(x)$ multiplied by $\frac{1}{2}$. Use your GDC to confirm this for other pairs of corresponding points on $y = x^2 - 4x$ and $y = (2x)^2 - 4(2x)$ that have the same y-coordinate. The graph of $y = f(2x)$ can be obtained by *horizontally shrinking* the graph of $y = f(x)$ by scale factor $\frac{1}{2}$. This makes sense because if $f(2x_2) = (2x_2)^2 - 4(2x_2)$ and $f(x_1) = x_1^2 - 4x_1$ are to produce the same y-value then $2x_2 = x_1$; and, thus, $x_2 = \frac{1}{2}x_1$. Figures 2.30 and 2.31 illustrate how multiplying the x-variable of a function by a positive number, a, *greater than one* causes the function to *shrink* horizontally by scale factor $\frac{1}{a}$. A point (x, y) on the graph of $y = f(x)$ is transformed to the point $\left(\frac{1}{a}x, y\right)$ on the graph of $y = f(ax)$.

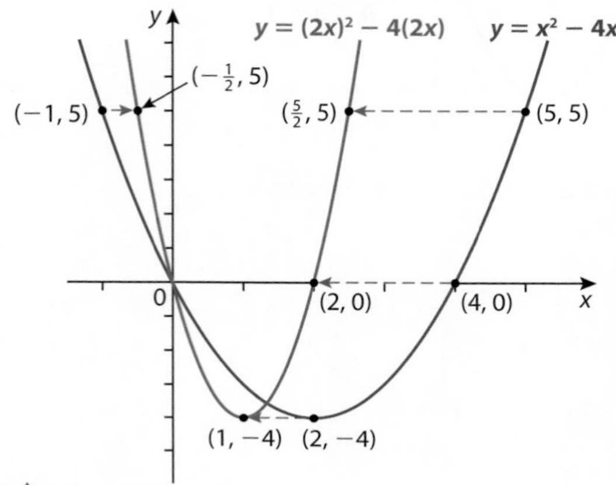

Figure 2.30 Horizontal shrink of $y = x^2 - 4x$ by scale factor $\frac{1}{2}$.

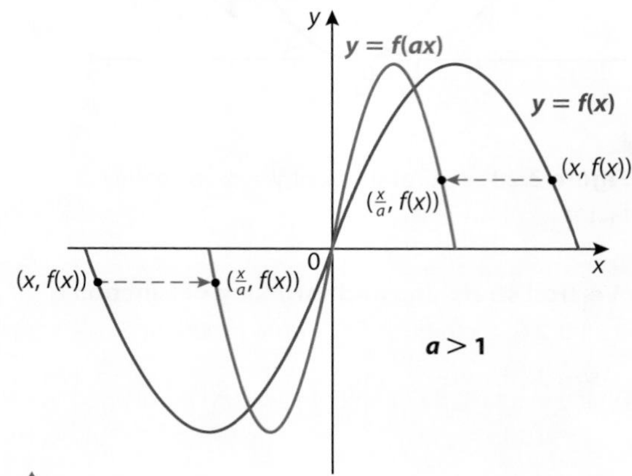

Figure 2.31 Horizontal shrink of $f(x)$ by scale factor $\frac{1}{a}$, $a > 1$.

If $0 < a < 1$, the graph of the function $y = f(ax)$ is obtained by a *horizontal stretching* of the graph of $y = f(x)$ – rather than a shrinking – because the scale factor $\frac{1}{a}$ will be a value greater than 1 if $0 < a < 1$. Now, letting $a = \frac{1}{2}$ and, again using the function $f(x) = x^2 - 4x$, find $g(x)$, such that $g(x) = f\left(\frac{1}{2}x\right)$. We substitute $\frac{x}{2}$ for x in f, giving $g(x) = \left(\frac{x}{2}\right)^2 - 4\left(\frac{x}{2}\right)$. On your GDC, graph the functions f and g using the indicated viewing window with f in bold.

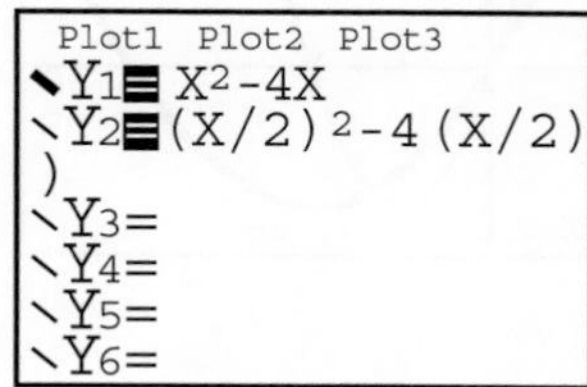

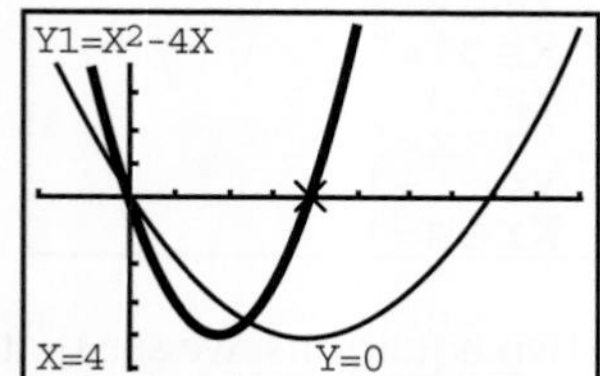

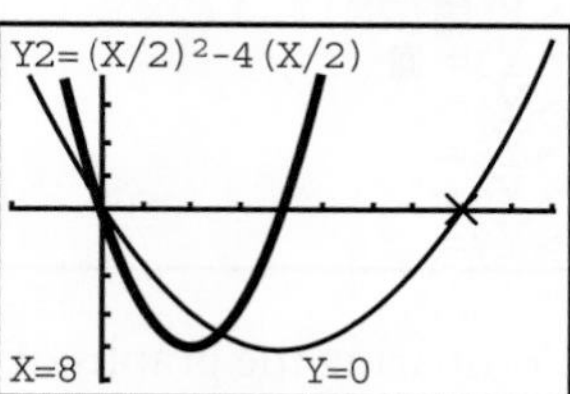

WINDOW
Xmin=⁻2
Xmax=10
Xscl=1
Ymin=⁻5
Ymax=5
Yscl=1
Xres=1

The graph of $y = \left(\frac{x}{2}\right)^2 - 4\left(\frac{x}{2}\right)$ is a horizontal stretching of the graph of $y = x^2 - 4x$ by scale factor $\frac{1}{a} = \frac{1}{\frac{1}{2}} = 2$. For example, the point $(4, 0)$ on $y = f(x)$ has been moved horizontally to the point $(8, 0)$ on $y = g(x) = f\left(\frac{x}{2}\right)$.

Figures 2.32 and 2.33 illustrate how multiplying the x-variable of a function by a positive number, a, *greater than zero and less than one* causes the function to *stretch* horizontally by scale factor $\frac{1}{a}$. A point (x, y) on the graph of $y = f(x)$ is transformed to the point $\left(\frac{1}{a}x, y\right)$ on the graph of $y = f(ax)$.

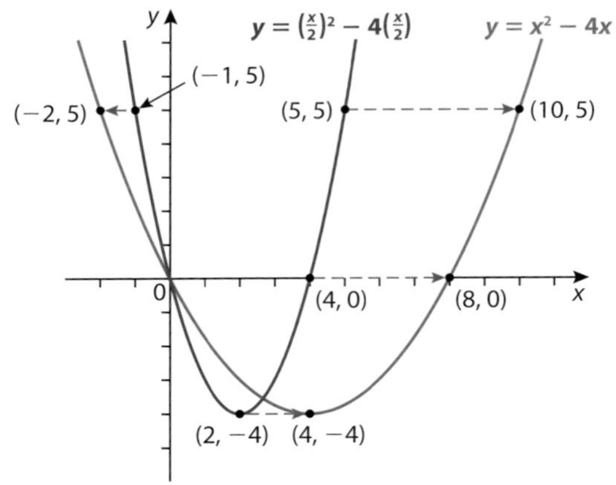

Figure 2.32 Horizontal stretch of $y = x^2 - 4x$ by scale factor 2.

Figure 2.33 Horizontal stretch of $f(x)$ by scale factor $\frac{1}{a}$, $0 < a < 1$.

Horizontal stretching and shrinking of functions

I. If $a > 1$, the graph of $y = f(ax)$ is obtained by *horizontally shrinking* the graph of $y = f(x)$.

II. If $0 < a < 1$, the graph of $y = f(ax)$ is obtained by *horizontally stretching* the graph of $y = f(x)$.

Example 21

The graph of $y = f(x)$ is shown. Sketch the graph of each of the following two functions.

a) $y = 3f(x)$

b) $y = \frac{1}{3}f(x)$

c) $y = f(3x)$

d) $y = f\left(\frac{1}{3}x\right)$

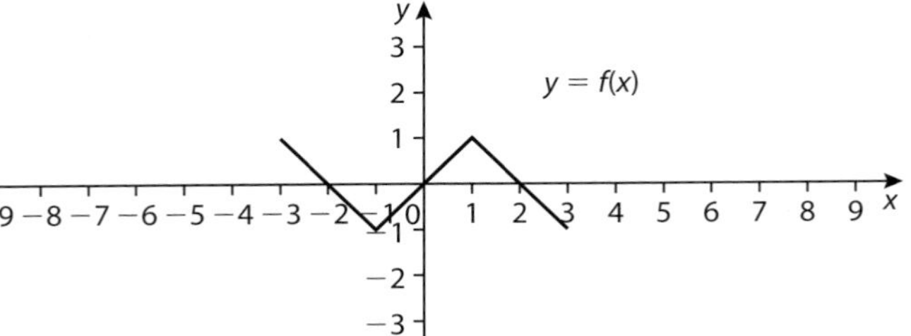

Solution

a) The graph of $y = 3f(x)$ is obtained by vertically stretching the graph of $y = f(x)$ by scale factor 3.

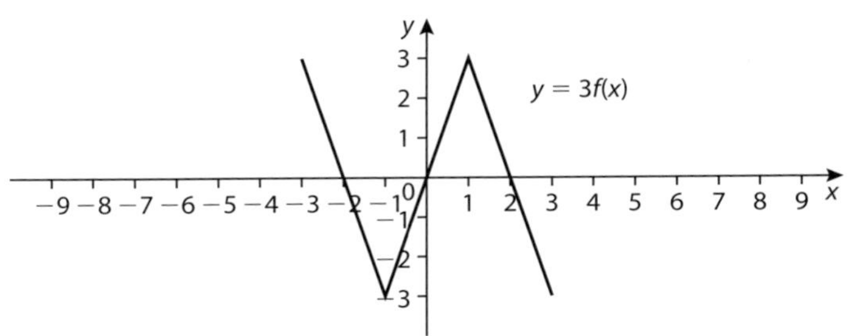

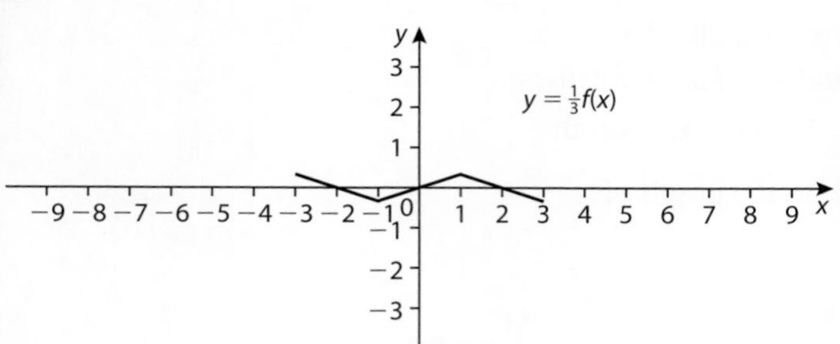

b) The graph of $y = \frac{1}{3}f(x)$ is obtained by vertically shrinking the graph of $y = f(x)$ by scale factor $\frac{1}{3}$.

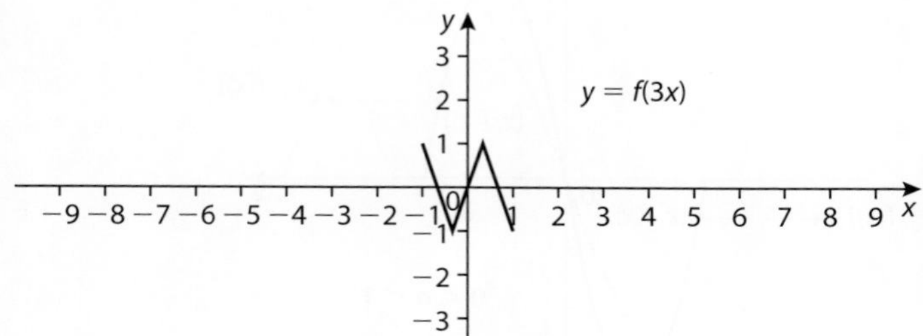

c) The graph of $y = f(3x)$ is obtained by horizontally shrinking the graph of $y = f(x)$ by scale factor $\frac{1}{3}$.

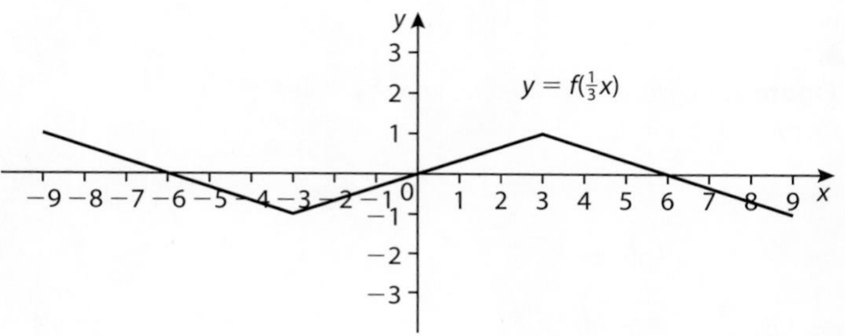

d) The graph of $y = f\left(\frac{1}{3}x\right)$ is obtained by horizontally stretching the graph of $y = f(x)$ by scale factor 3.

Example 22

Describe the sequence of transformations performed on the graph of $y = x^2$ to obtain the graph of $y = 4x^2 - 3$.

Solution

Step 1: Start with the graph of $y = x^2$.

Step 2: Vertically stretch $y = x^2$ by scale factor 4.

Step 3: Vertically translate $y = 4x^2$ three units down.

Step 1:

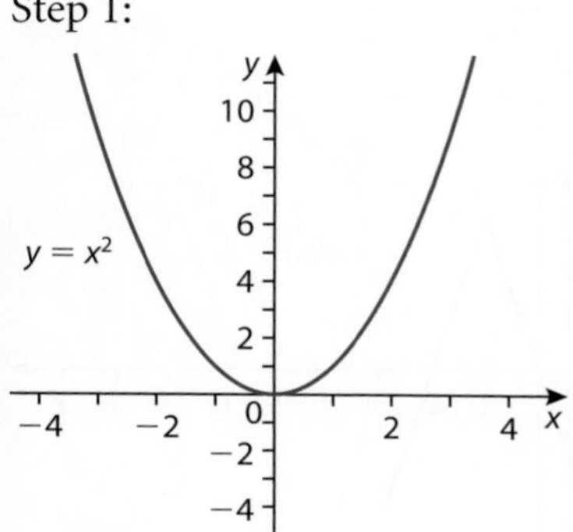

Step 2:

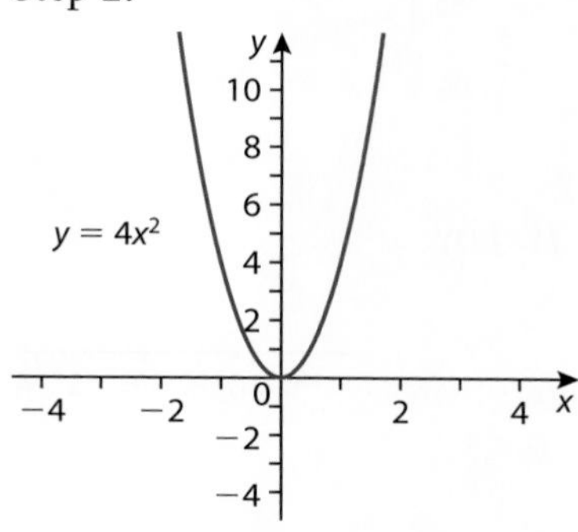

Step 3:

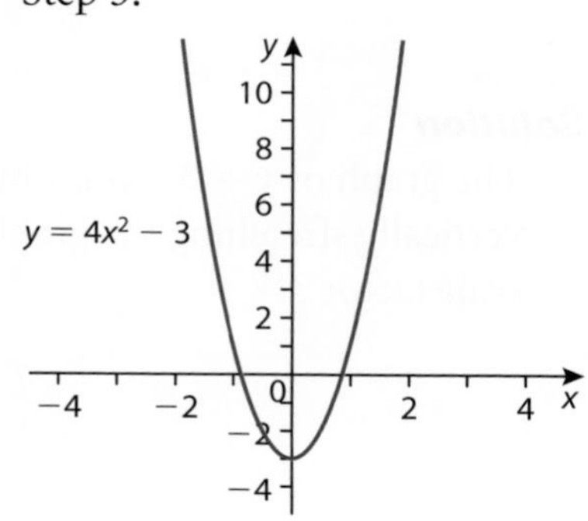

Note that in Example 22, a vertical stretch followed by a vertical translation does not produce the same graph if the two transformations are performed in reverse order. A vertical translation followed by a vertical stretch would generate the following sequence of equations:

Step1: $y = x^2$ Step 2: $y = x^2 - 3$ Step 3: $y = 4(x^2 - 3) = 4x^2 - 12$

This final equation is not the same as $y = 4x^2 - 3$.

When combining two or more transformations, the order in which they are performed can make a difference. In general, when a sequence of transformations includes a vertical/horizontal stretch or shrink, or a reflection through the x-axis, the order may make a difference.

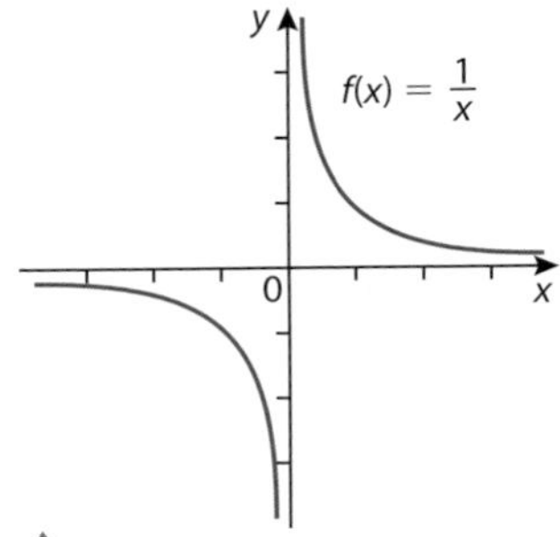

Figure 2.34 The reciprocal function $y = \frac{1}{x}$.

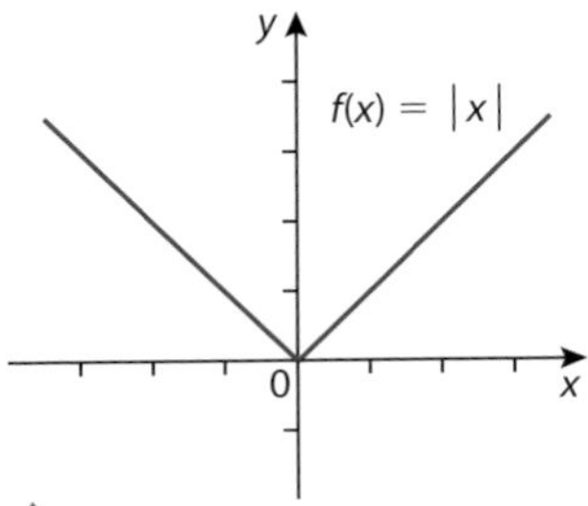

Figure 2.35 The absolute value function $y = |x|$.

Reciprocal and absolute value graphs

Two of the functions that appeared in the set of common functions in Figure 2.17 at the start of this section were the reciprocal function, $f(x) = \frac{1}{x}$, and the absolute value function (Figures 2.34 and 2.35).

Lets investigate how the graph of a given function, say $g(x)$, compares to that of a composite function $f(g(x))$, where the function f is either the reciprocal function or the absolute value function.

Example 23 – Graph of the reciprocal of a function

Given $f(x) = \frac{1}{x}$, $g(x) = -2x + 4$ and $h(x) = x^2 + 2x - 3$, sketch the graphs of the composite functions $f(g(x))$ and $f(h(x))$. Discuss the characteristics of each graph.

Solution

$$f(g(x)) = \frac{1}{g(x)} \Rightarrow y = \frac{1}{-2x + 4}$$

Clearly the reciprocal of g will be undefined wherever $g(x) = 0$ making the domain of $\frac{1}{g(x)}$ to be $\{x: x \in \mathbb{R}, x \neq 2\}$. Consequently the graph of $\frac{1}{g(x)}$ will have a **vertical asymptote** with equation $x = 2$. The graph of g illustrates that as x approaches the value of 2 ($x \to 2$) from the left side, the value of $g(x)$ is always positive but is converging to zero. Therefore, as $x \to 2$ from the left (or, $x \to 2^-$), the values of $\frac{1}{g(x)}$ become increasingly large in the positive direction. We can express this behaviour symbolically by writing, 'as $x \to 2^-$, $\frac{1}{g(x)} \to +\infty$'. Similarly, as $x \to 2^+$, $\frac{1}{g(x)} \to -\infty$.

Also, the x-axis ($y = 0$) is a **horizontal asymptote** for the

Figure 2.36 Graph of $g(x)$ and its reciprocal.

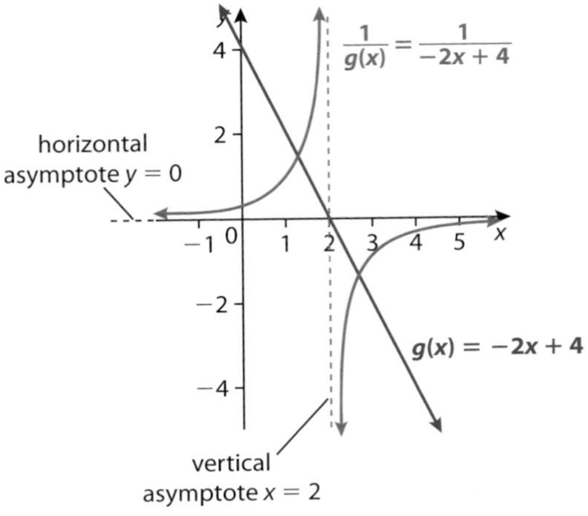

graph of $\frac{1}{g(x)}$ because as the value of $g(x)$ becomes very large (either positive or negative), the value of $\frac{1}{g(x)}$ converges to zero; or, symbolically, as $x \to \pm\infty$, $\frac{1}{g(x)} \to 0$.

$$f(h(x)) = \frac{1}{h(x)} = \frac{1}{x^2 + 2x - 3} = \frac{1}{(x+3)(x-1)}$$

Domain for $\frac{1}{h(x)}$ is $\{x : x \in \mathbb{R}, x \neq -3, x \neq 1\}$.

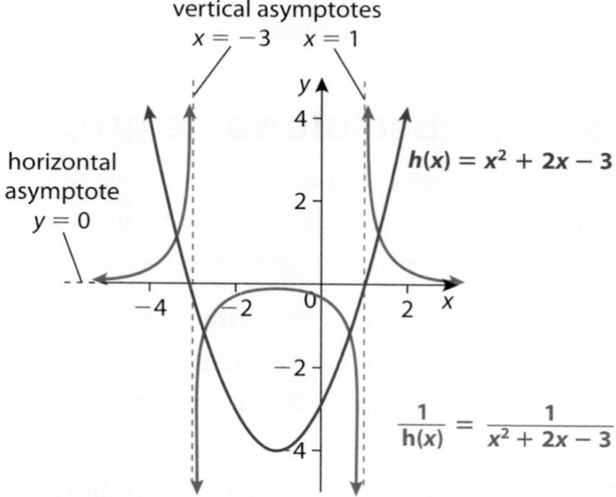

Figure 2.37 Graph of $h(x)$ and its reciprocal.

Since $h(x) = 0$ for $x = -3$ and $x = 1$ we anticipate that the graph of its reciprocal, $\frac{1}{h(x)}$, will have vertical asymptotes of $x = -3$ and $x = 1$. This is confirmed by the fact that as $x \to -3^-$, $\frac{1}{h(x)} \to +\infty$; as $x \to -3^+$, $\frac{1}{h(x)} \to -\infty$; and as $x \to 1^-$, $\frac{1}{h(x)} \to -\infty$; as $x \to 1^+$, $\frac{1}{h(x)} \to +\infty$.
The graph of $\frac{1}{h(x)}$ will also have a horizontal asymptote of $y = 0$ (x-axis) because as $x \to \pm\infty$, $\frac{1}{h(x)} \to 0$.

Vertical and horizontal asymptotes
In general, the line $x = c$ is a vertical asymptote of the graph of f if $f(x) \to \infty$ or $f(x) \to -\infty$ as x approaches c from either the left or the right. The line $y = c$ is a horizontal asymptote of the graph of f if $f(x)$ approaches c as $x \to \infty$ or $x \to -\infty$.

Example 24 – Graphs of composites with absolute value function

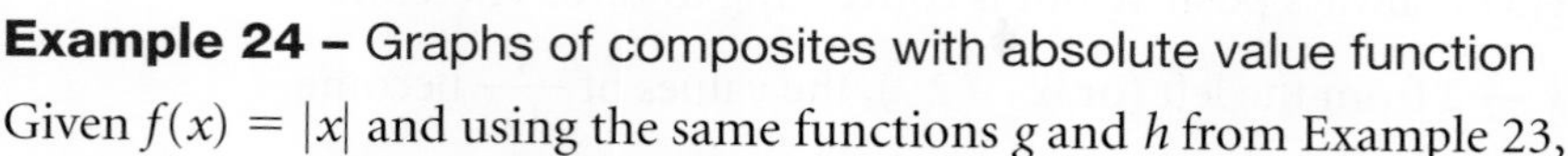

Given $f(x) = |x|$ and using the same functions g and h from Example 23,

a) graph the composite functions $f \circ g$ and $f \circ h$; and

b) graph the composite functions $g \circ f$ and $h \circ f$.

Solution

a) $(f \circ g)(x) = f(-2x + 4) = |-2x + 4|$

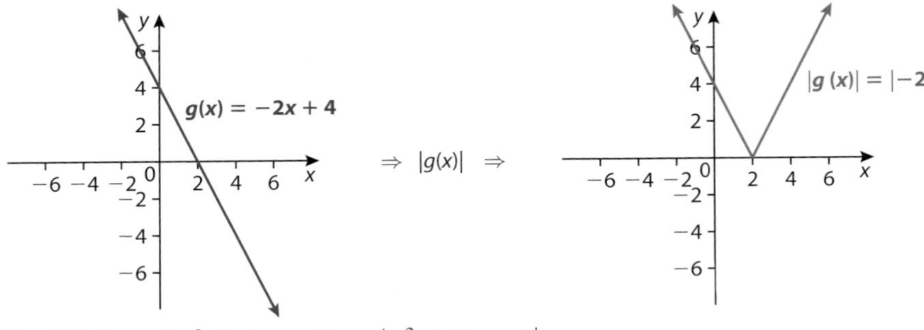

$(f \circ h)(x) = f(x^2 + 2x - 3) = |x^2 + 2x - 3|$

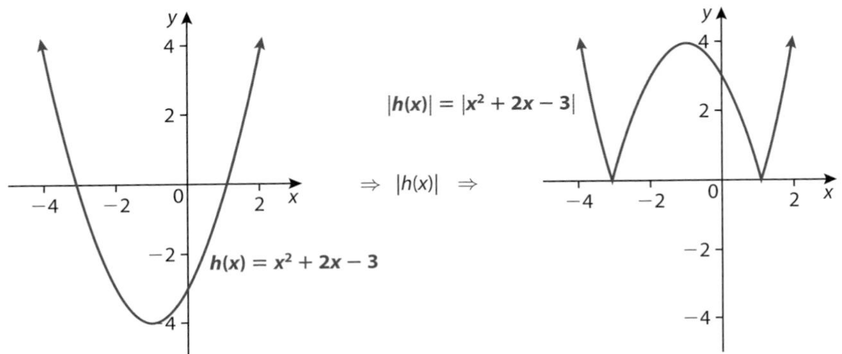

From these two examples with functions $g(x)$ and $h(x)$, we see the change that occurs from the graph of a function to the graph of the **absolute value of the function**. Any portion of the graph of $g(x)$ or $h(x)$ that was below the x-axis gets reflected above the x-axis.

b) $(g \circ f)(x) = g(|x|) = -2|x| + 4$

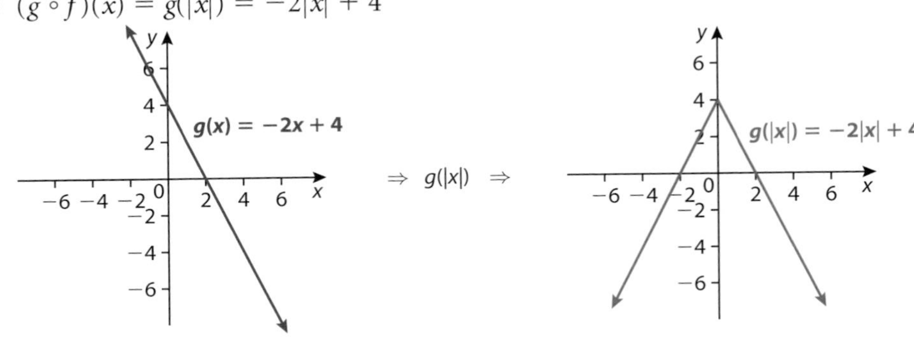

$(h \circ f)(x) = h(|x|) = |x|^2 + 2|x| - 3$

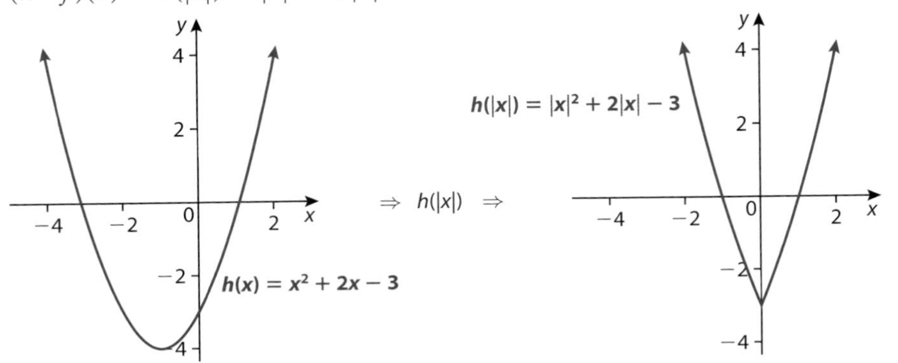

Similarly to part a) we can see a change from the graph of a function to the graph of the **function of the absolute value.** Any portion of the graph of $g(x)$ or $h(x)$ that was left of the y-axis is eliminated, and any portion that was to the right of the y-axis is reflected to the left of the y-axis. Since the portion that was right of the y-axis remains, the resulting graph is always symmetric about the y-axis.

Summary of transformations on the graphs of functions

Assume that a, h and k are positive real numbers.

Transformed function	Transformation performed on $y = f(x)$
$y = f(x) + k$	vertical translation k units up
$y = f(x) - k$	vertical translation k units down
$y = f(x - h)$	horizontal translation h units right
$y = f(x + h)$	horizontal translation h units left
$y = -f(x)$	reflection in the x-axis
$y = f(-x)$	reflection in the y-axis
$y = af(x)$	vertical stretch ($a > 1$) or shrink ($0 < a < 1$)
$y = f(ax)$	horizontal stretch ($0 < a < 1$) or shrink ($a > 1$)
$y = \lvert f(x) \rvert$	portion of graph of $y = f(x)$ below x-axis is reflected above x-axis
$y = f(\lvert x \rvert)$	symmetric about y-axis; portion right of y-axis is reflected over y-axis

Exercise 2.4

In questions 1–14, sketch the graph of f, without a GDC or by plotting points, by using your knowledge of some of the basic functions shown in Figure 2.17.

1 $f: x \mapsto x^2 - 6$

2 $f: x \mapsto (x - 6)^2$

3 $f: x \mapsto |x| + 4$

4 $f: x \mapsto |x + 4|$

5 $f: x \mapsto 5 + \sqrt{x - 2}$

6 $f: x \mapsto \dfrac{1}{x - 3}$

7 $f: x \mapsto \dfrac{1}{(x + 5)^2} + 2$

8 $f: x \mapsto -x^3 - 4$

9 $f: x \mapsto -|x - 1| + 6$

10 $f: x \mapsto \sqrt{-x + 3}$

11 $f: x \mapsto 3\sqrt{x}$

12 $f: x \mapsto \frac{1}{2}x^2$

13 $f: x \mapsto \left(\frac{1}{2}x\right)^2$

14 $f: x \mapsto (-x)^3$

In questions 15–18, write the equation for the graph that is shown.

15

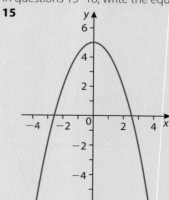

16

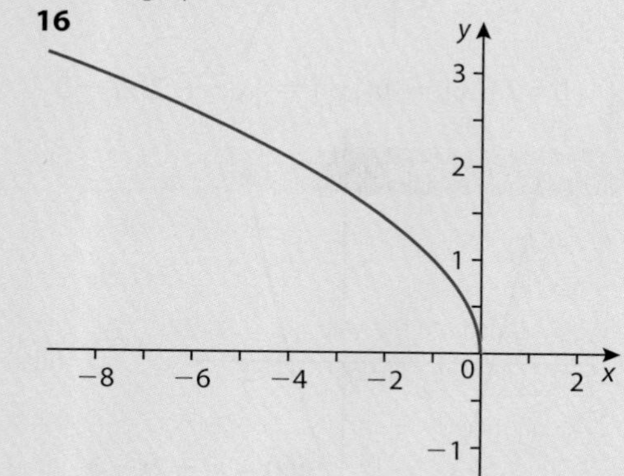

17

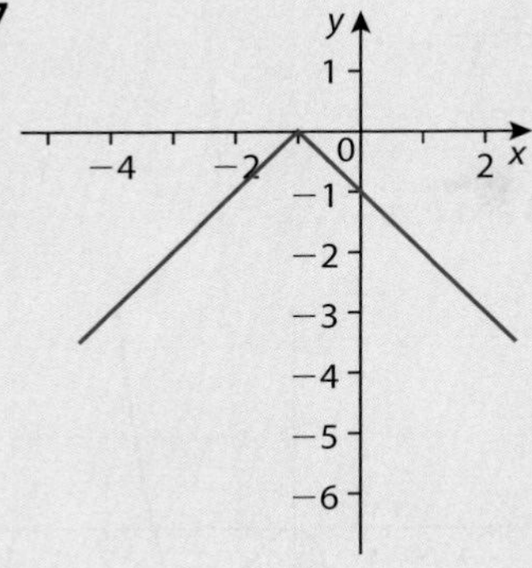

18 Vertical and horizontal asymptotes shown:

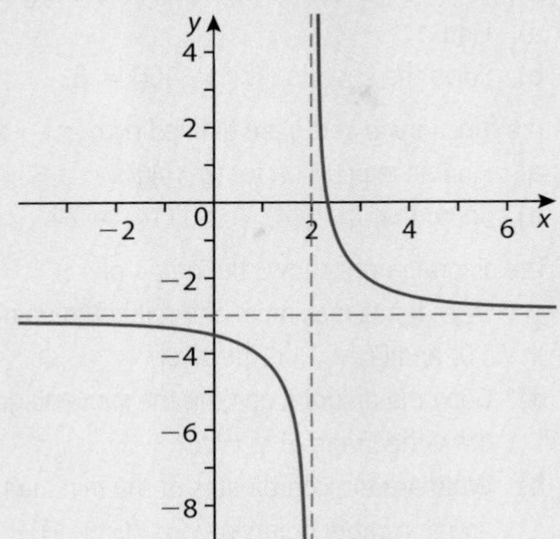

19 The graph of f is given. Sketch the graphs of the following functions.

a) $y = f(x) - 3$
b) $y = f(x - 3)$
c) $y = 2f(x)$
d) $y = f(2x)$
e) $y = -f(x)$
f) $y = f(-x)$
g) $y = 2f(x) + 4$

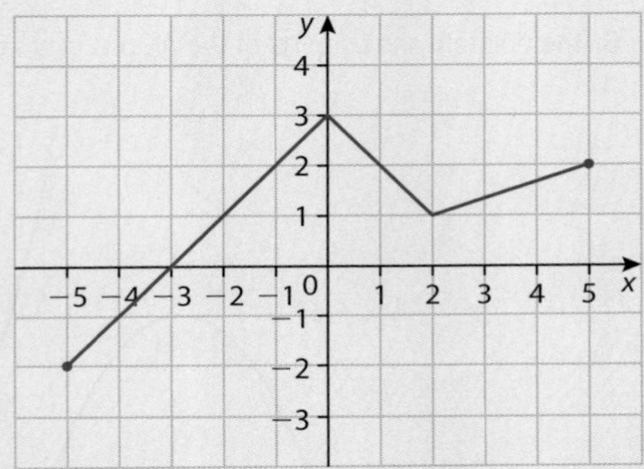

In questions 20–23, specify a sequence of transformations to perform on the graph of $y = x^2$ to obtain the graph of the given function.

20 $g: x \mapsto (x - 3)^2 + 5$

21 $h: x \mapsto -x^2 + 2$

22 $p: x \mapsto \frac{1}{2}(x + 4)^2$

23 $f: x \mapsto [3(x - 1)]^2 - 6$

Without using your GDC, for each function $f(x)$ in questions 24–26 sketch the graph of a) $\frac{1}{f(x)}$, b) $|f(x)|$ and c) $f(|x|)$. Clearly label any intercepts or asymptotes.

24 $f(x) = \frac{1}{2}x - 4$

25 $f(x) = (x - 4)(x + 2)$

26 $f(x) = x^3$

Practice questions

1 Let $f: x \mapsto \sqrt{x - 3}$ and $g: x \mapsto x^2 + 2x$. The function $(f \circ g)(x)$ is defined for all $x \in \mathbb{R}$ **except** for the interval $]a, b[$.

a) Calculate the values of a and b.

b) Find the range of $f \circ g$.

2 Two functions g and h are defined as $g(x) = 2x - 7$ and $h(x) = 3(2 - x)$.

Find: **a)** $g^{-1}(3)$

b) $(h \circ g)(6)$

3 Consider the functions $f(x) = 5x - 2$ and $g(x) = \dfrac{4 - x}{3}$.

a) Find g^{-1}.

b) Solve the equation $(f \circ g^{-1})(x) = 8$.

4 The functions g and h are defined by $g: x \mapsto x - 3$ and $h: x \mapsto 2x$.

a) Find an expression for $(g \circ h)(x)$.

b) Show that $g^{-1}(14) + h^{-1}(14) = 24$.

5 The diagram right shows the graph of $y = f(x)$. It has maximum and minimum points at (0, 0) and (1, −1), respectively.

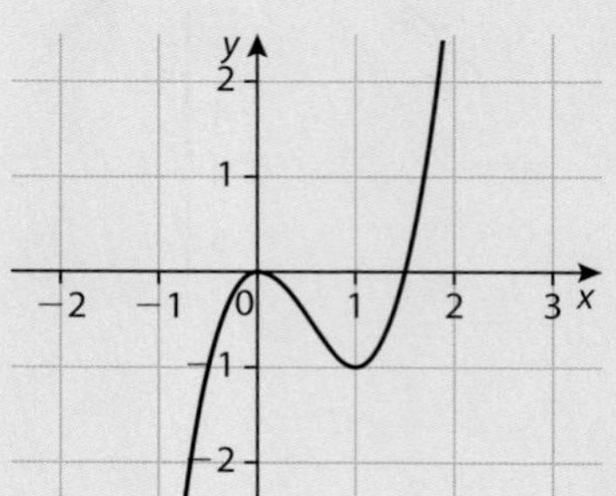

a) Copy the diagram and, on the same diagram, draw the graph of $y = f(x + 1) - \frac{1}{2}$.

b) What are the coordinates of the minimum and maximum points of $y = f(x + 1) - \frac{1}{2}$?

6 The diagram shows parts of the graphs of $y = x^2$ and $y = -\frac{1}{2}(x + 5)^2 + 3$.

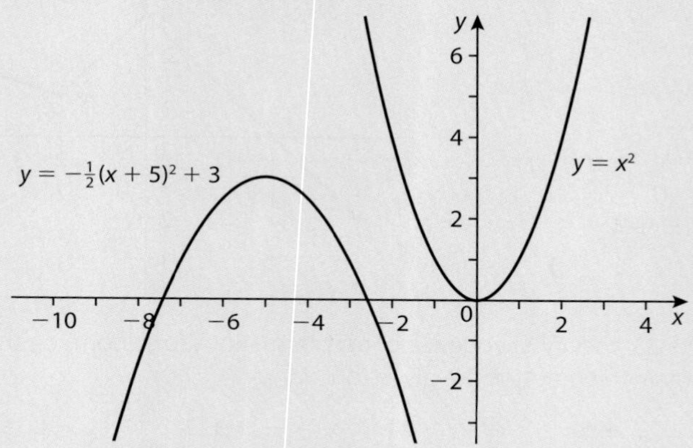

The graph of $y = x^2$ may be transformed into the graph of $y = -\frac{1}{2}(x + 5)^2 + 3$ by these transformations.

A reflection in the line $y = 0$, followed by
a vertical stretch by scale factor k, followed by
a horizontal translation of p units, followed by
a vertical translation of q units.

Write down the value of

a) k **b)** p **c)** q.

7 The function f is defined by $f(x) = \dfrac{4}{\sqrt{16 - x^2}}$, for $-4 < x < 4$.

a) Without using a GDC, sketch the graph of f.

b) Write down the equation of each vertical asymptote.

c) Write down the range of the function f.

8 Let $g: x \mapsto \dfrac{1}{x}, x \neq 0$.

a) Without using a GDC, sketch the graph of g.

The graph of g is transformed to the graph of h by a translation of 4 units to the left and 2 units down.

b) Find an expression for the function h.

c) **(i)** Find the x- and y-intercepts of h.
(ii) Write down the equations of the asymptotes of h.
(iii) Sketch the graph of h.

9 Consider $f(x) = \sqrt{x + 3}$.

a) Find:
(i) $f(8)$ **(ii)** $f(46)$ **(iii)** $f(-3)$

b) Find the values of x for which f is undefined.

c) Let $g: x \mapsto x^2 - 5$. Find $(g \circ f)(x)$.

10 Let $g(x) = \frac{x - 8}{2}$ and $h(x) = x^2 - 1$.

a) Find $g^{-1}(-2)$.

b) Find an expression for $(g^{-1} \circ h)(x)$.

c) Solve $(g^{-1} \circ h)(x) = 22$.

11 Given the functions $f: x \mapsto 3x - 1$ and $g: x \mapsto \frac{4}{x}$, find the following:

a) f^{-1} **b)** $f \circ g$ **c)** $(f \circ g)^{-1}$ **d)** $g \circ g$

12 **a)** The diagram shows part of the graph of the function $h(x) = \frac{a}{x - b}$. The curve passes through the point $A\,(-4, -8)$. The vertical line (MN) is an asymptote. Find the value of: **(i)** a **(ii)** b.

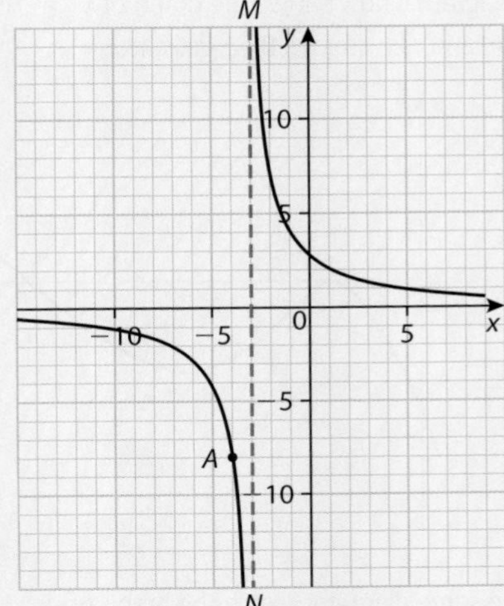

b) The graph of $h(x)$ is transformed as shown in the diagram right. The point A is transformed to $A'(-4, 8)$. Give a full geometric description of the transformation.

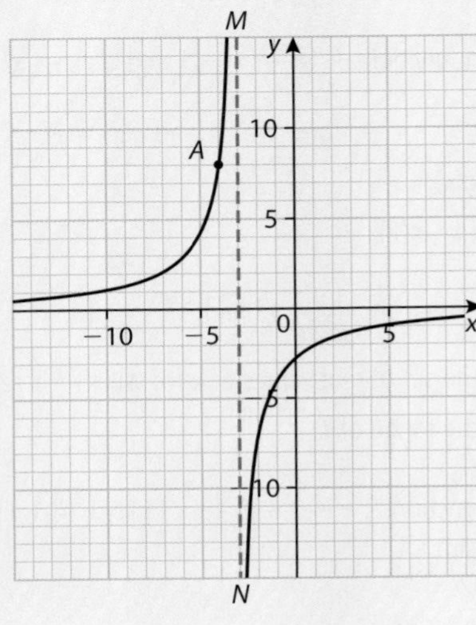

13 The graph of $y = f(x)$ is shown in the diagram.

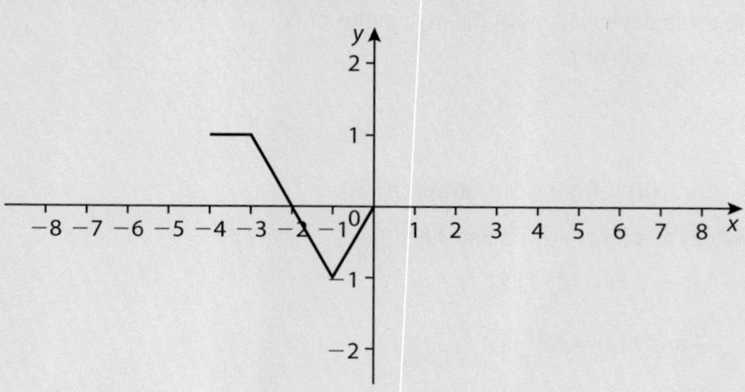

a) Make two copies of the coordinate system as shown in the diagram but without the graph of $y = f(x)$. On the first diagram sketch a graph of $y = 2f(x)$, and on the second diagram sketch a graph of $y = f(x - 4)$.

b) The point $A(-3, 1)$ is on the graph of $y = f(x)$. The point A' is the corresponding point on the graph of $y = -f(x) - 1$. Find the coordinates of A'.

14 The diagram below shows the graph of $y_1 = f(x)$. The x-axis is a tangent to $f(x)$ at $x = m$ and $f(x)$ crosses the x-axis at $x = n$.

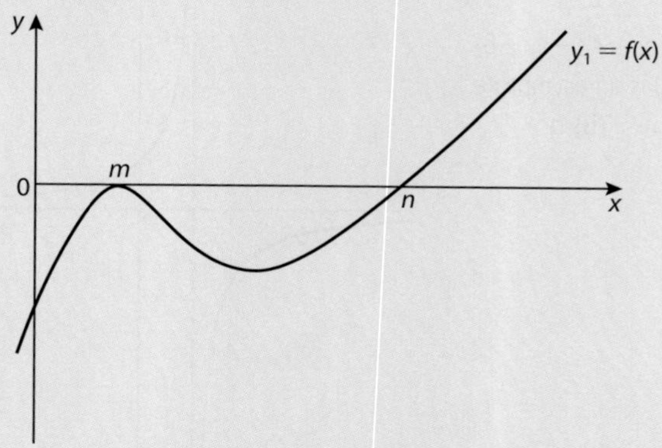

On the same diagram, sketch the graph of $y_2 = f(x - k)$, where $0 < k < n - m$ and indicate the coordinates of the points of intersection of y_2 with the x-axis.

15 Given functions $f : x \mapsto x + 1$ and $g : x \mapsto x^3$, find the function $(f \circ g)^{-1}$.

16 If $f(x) = \dfrac{x}{x + 1}$ for $x \neq -1$ and $g(x) = (f \circ f)(x)$, find

a) $g(x)$

b) $(g \circ g)(2)$.

17 Let $f : x \mapsto \sqrt{\dfrac{1}{x^2} - 2}$. Find

a) the set of real values of x for which f is real and finite;

b) the range of f.

18 The function $f : x \mapsto \dfrac{2x + 1}{x - 1}$, $x \in \mathbb{R}, x \neq 1$. Find the inverse function, f^{-1}, clearly stating its domain.

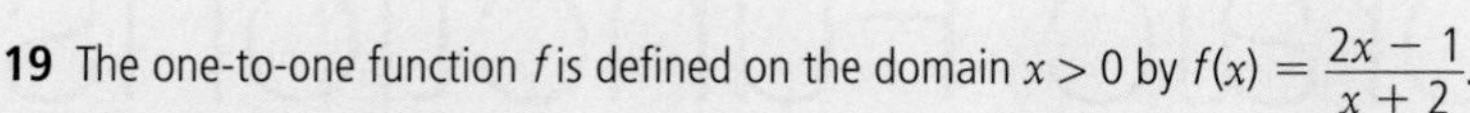

19 The one-to-one function f is defined on the domain $x > 0$ by $f(x) = \dfrac{2x - 1}{x + 2}$.

a) State the range, A, of f.

b) Obtain an expression for $f^{-1}(x)$, for $x \in A$.

20 The function f is defined by $f: x \mapsto x^3$.

Find an expression for $g(x)$ in terms of x in each of the following cases

a) $(f \circ g)(x) = x + 1$;

b) $(g \circ f)(x) = x + 1$.

21 **a)** Find the largest set S of values of x such that the function $f(x) = \dfrac{1}{\sqrt{3 - x^2}}$ takes real values.

b) Find the range of the function f defined on the domain S.

22 Let f and g be two functions. Given that $(f \circ g)(x) = \dfrac{x + 1}{2}$ and $g(x) = 2x - 1$, find $f(x - 3)$.

23 The diagram below shows the graph of $y = f(x)$ which passes through the points A, B, C and D.

Sketch, indicating clearly the images of A, B, C and D, the graphs of

a) $y = f(x - 4)$;

b) $y = f(-3x)$.

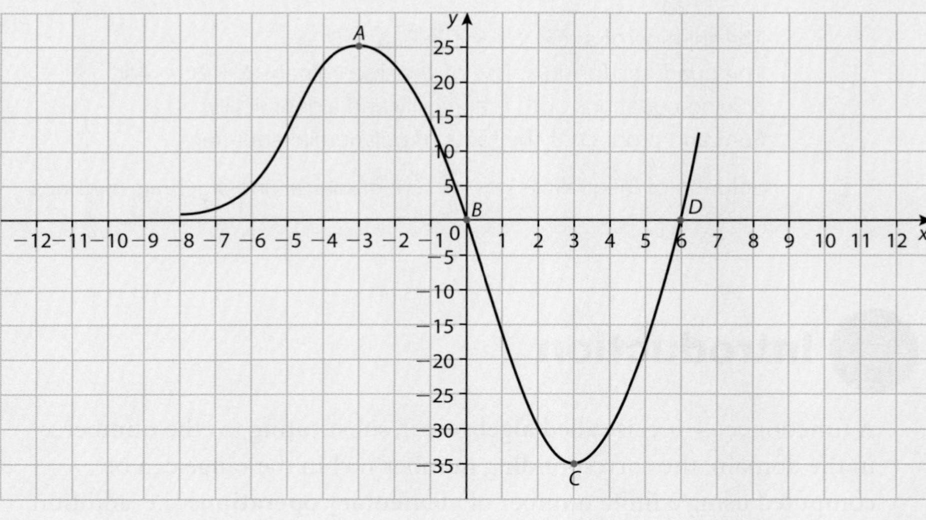

3 Algebraic Functions, Equations and Inequalities

Assessment statements

2.1 Odd and even functions (also see Chapter 7).

2.4 The rational function $x \mapsto \frac{ax + b}{cx + d}$ and its graph.

2.5 Polynomial functions.
The factor and remainder theorems.
The fundamental theorem of algebra.

2.6 The quadratic function $x \mapsto ax^2 + bx + c$: its graph, axis of symmetry $x = -\frac{b}{2a}$.
The solution of $ax^2 + bx + c = 0$, $a \neq 0$.
The quadratic formula. Use of the discriminant $\Delta = b^2 - 4ac$.
Solving equations both graphically and algebraically.
Sum and product of the roots of polynomial equations.

2.7 Solution of inequalities $g(x) \geqslant f(x)$; graphical and algebraic methods.

Introduction

A function $x \mapsto f(x)$ is called **algebraic** if, substituting for the number x in the domain, the corresponding number $f(x)$ in the range can be computed using a finite number of **elementary operations** (i.e. addition, subtraction, multiplication, division, and extracting a root). For example, $f(x) = \frac{x^2 + \sqrt{9 - x}}{2x - 6}$ is algebraic. For our purposes in this course, functions can be organized into three categories:

1. Algebraic functions
2. Exponential and logarithmic functions (Chapter 5)
3. Trigonometric and inverse trigonometric functions (Chapter 7)

The focus of this chapter is algebraic functions of a single variable which – given the definition above – are functions that contain polynomials, radicals (surds), rational expressions (quotients), or a combination of these. The chapter will begin by looking at polynomial functions in general and then moves onto a closer look at 2nd degree polynomial functions (quadratic functions). Solving equations containing polynomial functions is an important skill that will be covered. We will also study rational functions, which are quotients of polynomial functions and the associated topic of partial fractions (optional). The chapter will close with methods of solving inequalities and absolute value functions, and strategies for solving various equations.

The concept of a function is a fairly recent development in the history of mathematics. Its meaning started to gain some clarity about the time of René Descartes (1596–1650) when he defined a function to be any positive integral power of x (i.e. x^2, x^3, x^4, etc.). Leibniz (1646–1716) and Johann Bernoulli (1667–1748) developed the concept further. It was Euler (1707–1783) who introduced the now standard function notation $y = f(x)$.

3.1 Polynomial functions

The most common type of algebraic function is a polynomial function where, not surprisingly, the function's rule is given by a polynomial. For example,

$$f(x) = x^3, \quad h(t) = -2t^2 + 16t - 24, \quad g(y) = y^5 + y^4 - 11y^3 + 7y^2 + 10y - 8$$

Recalling the definition of a polynomial, we define a polynomial function.

Definition of a polynomial function in the variable x

A **polynomial function** P is a function that can be expressed as

$$P(x) = a_n x^n + a_{n-1}x^{n-1} + \ldots + a_1 x + a_0, \quad a_n \neq 0$$

where the non-negative integer n is the **degree** of the polynomial function. The numbers $a_0, a_1, a_2, \ldots, a_n$, are real numbers and are the **coefficients** of the polynomial. a_n is the **leading coefficient**, $a_n x^n$ is the **leading term** and a_0 is the **constant term**.

It is common practice to use subscript notation for coefficients of general polynomial functions, but for polynomial functions of low degree, the following simpler forms are often used.

Degree	Function form	Function name	Graph
Zero	$P(x) = a$	Constant function	Horizontal line
First	$P(x) = ax + b$	Linear function	Line with slope a
Second	$P(x) = ax^2 + bx + c$	Quadratic function	Parabola (U-shape, 1 turn)
Third	$P(x) = ax^3 + bx^2 + cx + d$	Cubic function	N-shape (2 or no turns)

Table 3.1 Features of polynomial functions of low degree.

To identify an individual term in a polynomial function, we use the function name correlated with the power of x contained in the term. For example, the polynomial function $f(x) = x^3 - 9x + 4$ has a *cubic* term of x^3, no *quadratic* term, a *linear* term of $-9x$, and a *constant* term of 4.

• **Hint:** When working with a polynomial function, such as $f(x) = x^3 - 9x + 4$, it is common to refer to it in a couple of different ways – either as 'the polynomial $f(x)$', or as 'the function $x^3 - 9x + 4$.'

For each polynomial function $P(x)$ there is a corresponding **polynomial equation** $P(x) = 0$. When we solve polynomial equations, we often refer to solutions as **roots**.

• **Hint:** The use of the word '**root**' here to denote the solution of a polynomial equation should not be confused with the use of the word in the context of square root, cube root, fifth root, etc.

Zeros and roots
If P is a function and c is a number such that $P(c) = 0$, then c is a **zero** of the function P (or of the polynomial P) and $x = c$ is a **root** of the equation $P(x) = 0$.

Approaches to finding zeros of various polynomial functions will be considered in the first three sections of this chapter.

Graphs of polynomial functions

As we reviewed in Section 1.6, the graph of a first-degree polynomial function (linear function), such as $P(x) = 2x - 5$, is a line (Figure 3.1a). The graph of every second-degree polynomial function (quadratic function) is a parabola (Figure 3.1b). A thorough review and discussion of quadratic functions and their graphs is in the next section.

The simplest type of polynomial function is one whose rule is given by a power of x. In Figure 3.1, the graphs of $P(x) = x^n$ for $n = 1, 2, 3, 4, 5$ and 6 are shown. As the figure suggests, the graph of $P(x) = x^n$ has the same general ∪-shape as $y = x^2$ when n is even, and the same general ∿ shape as $y = x^3$ when n is odd. However, as the degree n increases, the graphs of polynomial functions become flatter near the origin and steeper away from the origin.

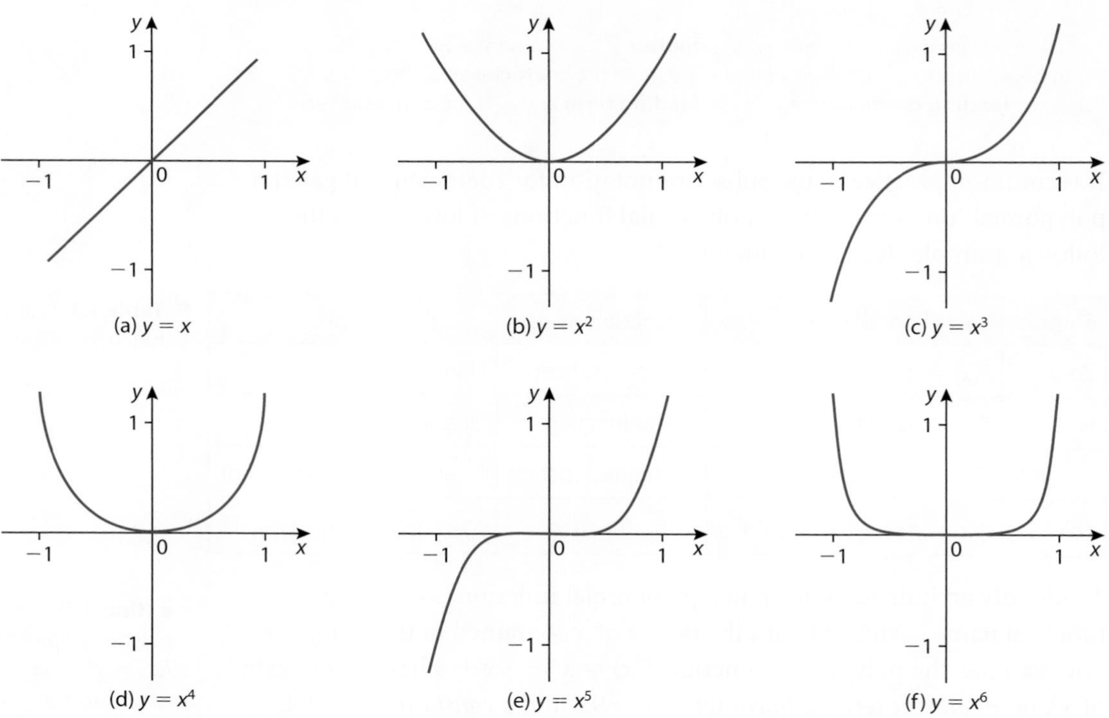

Figure 3.1 Graphs of $P(x) = x^n$ for increasing n.

Another interesting observation is that, depending on the degree of the polynomial function, its graph displays a certain type of symmetry. The graph of $P(x) = x^n$ is symmetric with respect to the origin when n is odd. Such a function is aptly called an **odd function**. The graph of $P(x) = x^n$ is symmetric with respect to the y-axis when n is even. Accordingly any such function is called an **even function**. Formal definitions for odd and even functions will be presented in Chapter 7 when we investigate the graphs of the sine and cosine functions.

Not all polynomial functions are even or odd – that is, not all polynomial functions display rotation symmetry about the origin or reflection symmetry about the y-axis. For example, the graph of the polynomial function $y = x^2 + x + 1$ is neither even nor odd. It has line symmetry, but the line of symmetry is not the y-axis.

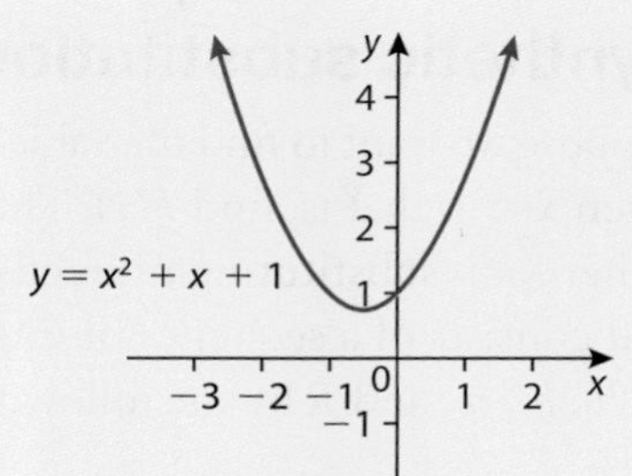

Note that the graph of an **even function** may or may not intersect the x-axis (x-intercept). As we will see, where and how often the graph of a function intersects the x-axis is helpful information when trying to determine the value and nature of the roots of a polynomial equation $P(x) = 0$.

The graphs of polynomial functions that are not in the form $P(x) = x^n$ are more difficult to sketch. However, the graphs of all polynomial functions share these properties:

1. It is a smooth curve (i.e. it has no sharp, pointed turns – only smooth, rounded turns).
2. It is continuous (i.e. it has no breaks, gaps or holes).
3. It rises ($P(x) \to \infty$) or falls ($P(x) \to -\infty$) without bound as $x \to +\infty$ or $x \to -\infty$.
4. It extends on forever both to the left ($-\infty$) and to the right ($+\infty$); domain is $\mathbb{R}$.
5. The graph of a polynomial function of degree n has at most $n - 1$ turning points.

The property that is listed third of the five properties of the graphs of polynomial functions is referred to as the **end behaviour** of the function because it describes how the curve *behaves* at the left and right *ends* (i.e. as $x \to +\infty$ and as $x \to -\infty$). The end behaviour of a polynomial function is determined by its degree and by the sign of its leading coefficient. See Exercise 3.1, Q11.

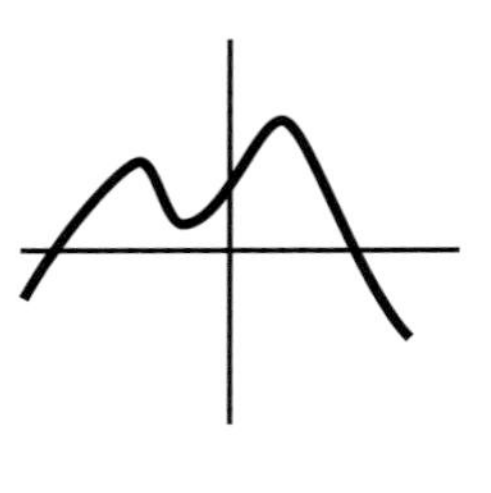

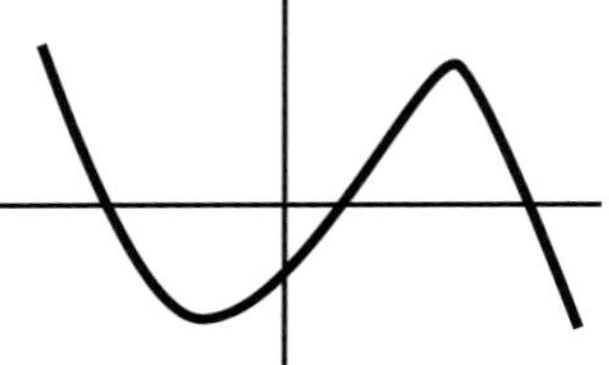

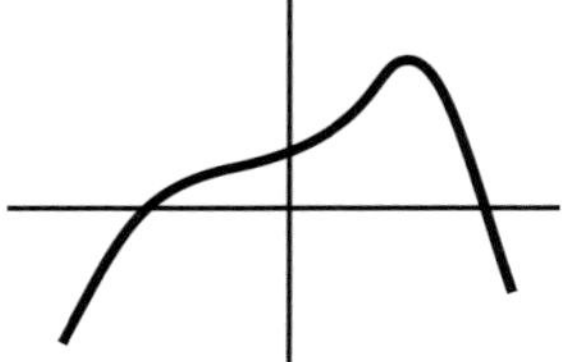

Figure 3.2 The graph of a polynomial function is a smooth, unbroken, continuous curve, such as the ones shown here.

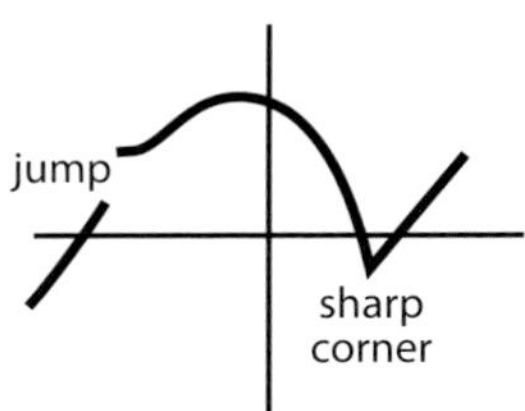

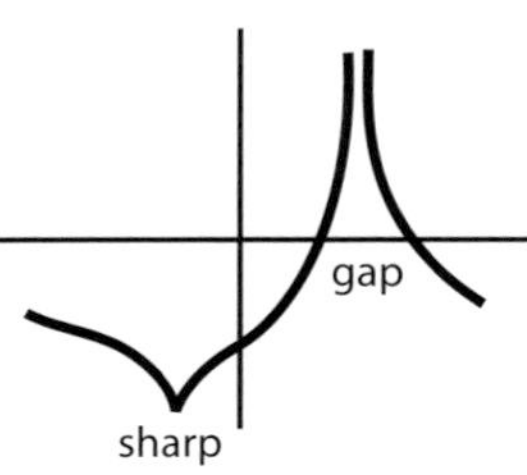

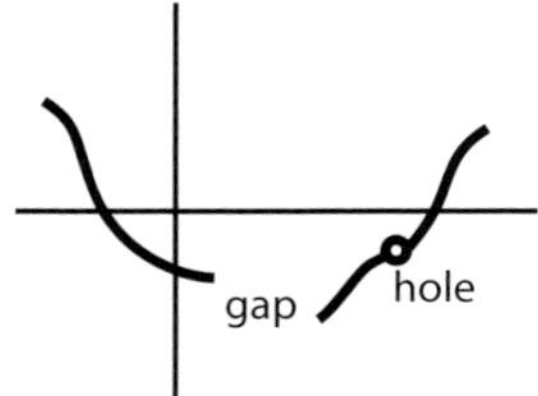

Figure 3.3 There can be no jumps, gaps, holes or sharp corners on the graph of a polynomial function. Thus none of the functions whose graphs are shown here are polynomial functions.

If we wish to sketch the graph of a polynomial function without a GDC, we need to compute some function values in order to locate a few points on the graph. This could prove to be quite tedious if the polynomial function has a high degree. We will now develop a method that provides an efficient procedure for evaluating polynomial functions. It will also be useful in the third section of this chapter for some situations when we divide polynomials. For simplicity, we give the method for a fourth-degree polynomial, but it is applicable to any nth degree polynomial.

Synthetic substitution (Optional)

Suppose we want to find the value of $P(x) = a_4x^4 + a_3x^3 + a_2x^2 + a_1x + a_0$ when $x = c$, that is, find $P(c)$. The computation of c^4 may be tricky, so rather than substituting c directly into $P(x)$ we will take a gradual approach that consists of a sequence of multiplications and additions. We define b_4, b_3, b_2, b_1, and R by the following equations.

$$b_4 = a_4 \quad (1)$$

$$b_3 = b_4c + a_3 \quad (2)$$

$$b_2 = b_3c + a_2 \quad (3)$$

$$b_1 = b_2c + a_1 \quad (4)$$

$$R = b_1c + a_0 \quad (5)$$

Our goal is to show that the value of $P(c)$ is equivalent to the value of R. Firstly, we substitute the expression for b_3 given by equation (2) into equation (3), and also use equation (1) to replace b_4 with a_4, to produce

$$\begin{aligned} b_2 &= (a_4c + a_3)c + a_2 \\ &= a_4c^2 + a_3c + a_2 \quad (6) \end{aligned}$$

We now substitute this expression for b_2 in (6) into (4) to give

$$\begin{aligned} b_1 &= (a_4c^2 + a_3c + a_2)c + a_1 \\ &= a_4c^3 + a_3c^2 + a_2c + a_1 \quad (7) \end{aligned}$$

To complete our goal we substitute this expression for b_1 in (7) into (5) to give

$$\begin{aligned} R &= (a_4c^3 + a_3c^2 + a_2c + a_1)c + a_0 \\ &= a_4c^4 + a_3c^3 + a_2c^2 + a_1c + a_0 \quad (8) \end{aligned}$$

This is the value of $P(x)$ when $x = c$. If we condense (6), (7) and (8) into one expression, we obtain

$$\begin{aligned} R &= \{[(a_4c + a_3)c + a_2]c + a_1\}c + a_0 \\ &= a_4c^4 + a_3c^3 + a_2c^2 + a_1c + a_0 = P(c) \quad (9) \end{aligned}$$

Carrying out the computations for equation (9) can be challenging. However, a nice pattern can be found if we closely inspect the expression $\{[(a_4c + a_3)c + a_2]c + a_1\}c + a_0$. Each nested computation involves finding the product of c and one of the coefficients, a_n, (starting with the leading coefficient) and then adding the next coefficient – and repeating this process until the constant term is used. Hence, the actual computation of R is quite straightforward if we arrange the nested computations required for (9) in the following systematic manner.

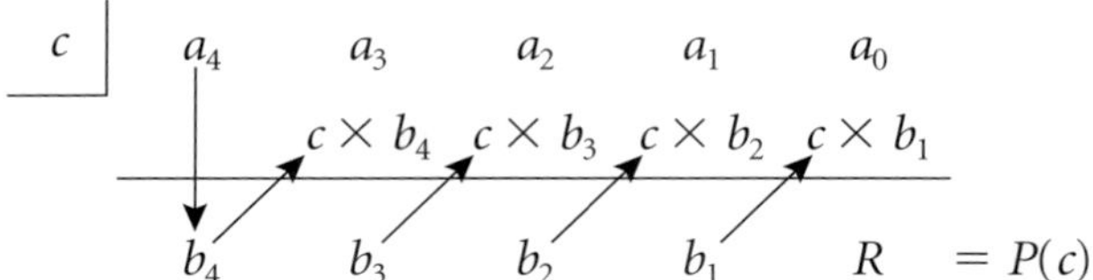

In this procedure we place c in a small box to the upper left. The coefficients of the polynomial function $P(x)$ are placed in the first line. We start by simply rewriting the leading coefficient below the horizontal line (remember $b_4 = a_4$). The diagonal arrows indicate that we multiply the number in the row below the line by c to obtain the next number in the second row above the line. Each b_n after the leading coefficient is obtained by adding the two numbers in the first and second rows directly above b_n. At the end of the procedure, the last such sum is $R = P(c)$. This method of computing the value of $P(x)$ when $x = c$ is called **synthetic substitution**.

Example 1 – Using synthetic substitution to find function values

Given $P(x) = 2x^4 + 6x^3 - 5x^2 + 7x - 12$, find the value of $P(x)$ when $x = -4, -1$ and 2.

Solution

We use the procedure for synthetic substitution just described.

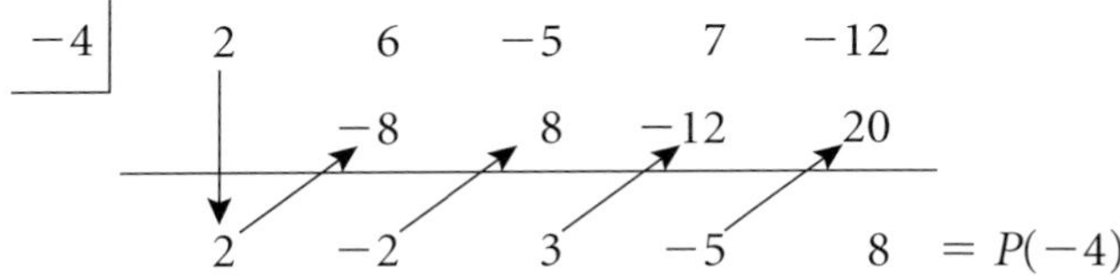

Therefore, $P(-4) = 8$.

Note: Contrast using synthetic substitution to evaluate $P(-4)$ with using direct substitution.

$$\begin{aligned} P(-4) &= 2(-4)^4 + 6(-4)^3 - 5(-4)^2 + 7(-4) - 12 \\ &= 2(256) + 6(-64) - 5(16) - 28 - 12 \\ &= 512 - 384 - 80 - 28 - 12 \\ &= 128 - 108 - 12 \\ &= 8 \end{aligned}$$

-1	2	6	-5	7	-12	
		-2	-4	9	-16	
	2	4	-9	16	-28	$= P(-1)$

Therefore, $P(-1) = -28$.

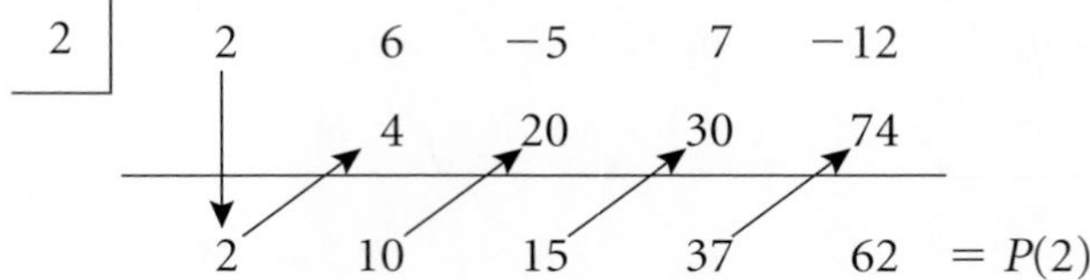

Therefore, $P(2) = 62$.

Since the graphs of all polynomial functions are continuous (no gaps or holes), then the function values we computed for the quartic polynomial function in Example 1 can give us information about the location of its zeros (i.e. x-intercepts of the graph). Since $P(-4) = 8$ and $P(-1) = -28$, then the graph of $P(x)$ must cross the x-axis ($P(x) = 0$) at least once between $x = -4$ and $x = -1$. Also, with $P(-1) = -28$ and $P(2) = 62$ there must be at least one x-intercept between $x = -1$ and $x = 2$. Hence, the polynomial equation $P(x) = 2x^4 + 6x^3 - 5x^2 + 7x - 12 = 0$ has at least one real root between -4 and -1, and at least one real root between -1 and 2. In Section 3.3 we will investigate real zeros of polynomial functions and then we will extend the investigation to include imaginary zeros, thereby extending the universal set for solving polynomial equations from the real numbers to complex numbers.

Graphing $P(x) = 2x^4 + 6x^3 - 5x^2 + 7x - 12$ on our GDC, we observe that the graph of $P(x)$ does indeed intersect the x-axis between -4 and -1 (just slightly greater than $x = -4$), and again between -1 and 2 (near $x = 1$).

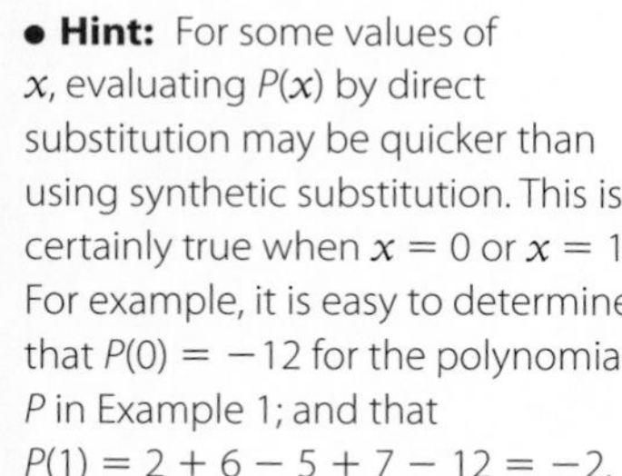
• **Hint:** For some values of x, evaluating $P(x)$ by direct substitution may be quicker than using synthetic substitution. This is certainly true when $x = 0$ or $x = 1$. For example, it is easy to determine that $P(0) = -12$ for the polynomial P in Example 1; and that $P(1) = 2 + 6 - 5 + 7 - 12 = -2$.

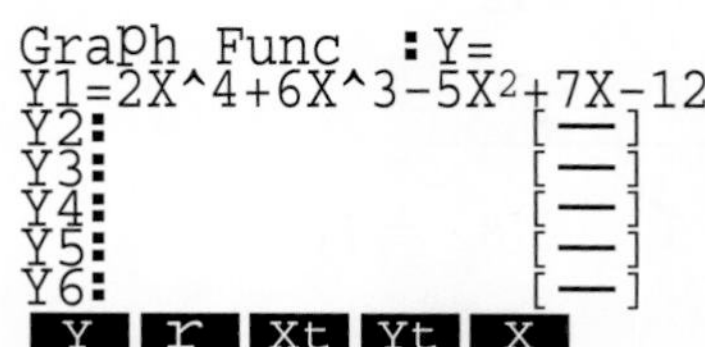

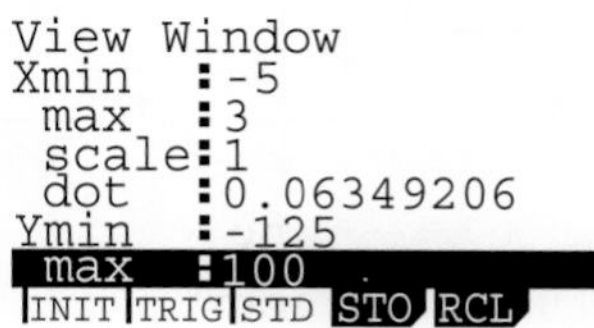

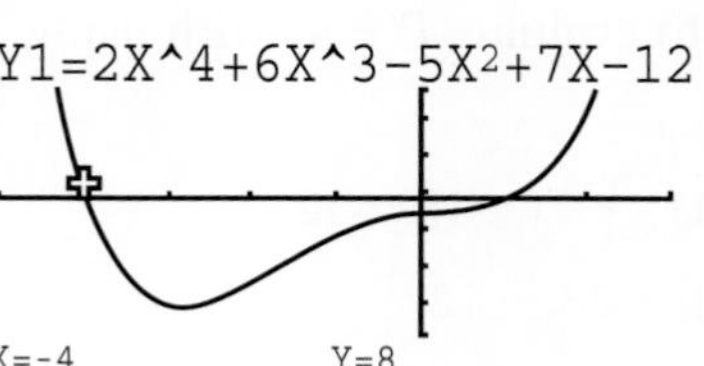

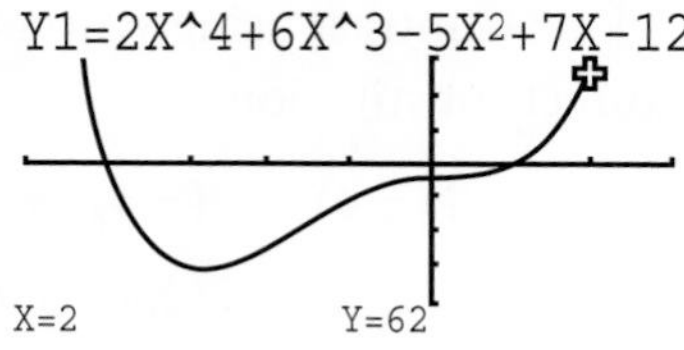

Example 2

Use synthetic substitution to find the y-coordinates of the points on the graph of $f(x) = x^3 - 4x^2 + 24$ for $x = -3, -1, 1, 3$ and 5. Sketch the graph of f for $-4 \leqslant x \leqslant 6$.

Solution

Important: In order for the method of synthetic substitution to work properly it is necessary to insert 0 for any 'missing' terms in the polynomial. The polynomial $x^3 - 4x^2 + 24$ has no linear term so the top row in the set-up for synthetic substitution must be 1 −4 0 24.

$$\begin{array}{r|rrrr} -3 & 1 & -4 & 0 & 24 \\ & & -3 & 21 & -63 \\ \hline & 1 & -7 & 21 & -39 \end{array} \qquad \begin{array}{r|rrrr} -1 & 1 & -4 & 0 & 24 \\ & & -1 & 5 & -5 \\ \hline & 1 & -5 & 5 & 19 \end{array} \qquad \begin{array}{r|rrrr} 1 & 1 & -4 & 0 & 24 \\ & & 1 & -3 & -3 \\ \hline & 1 & -3 & -3 & 21 \end{array}$$

$$\begin{array}{r|rrrr} 3 & 1 & -4 & 0 & 24 \\ & & 3 & -3 & -9 \\ \hline & 1 & -1 & -3 & 15 \end{array} \qquad \begin{array}{r|rrrr} 5 & 1 & -4 & 0 & 24 \\ & & 5 & 5 & 25 \\ \hline & 1 & 1 & 5 & 49 \end{array}$$

Therefore, the points $(-3, -39)$, $(-1, 19)$, $(1, 21)$, $(3, 15)$ and $(5, 49)$ are on the graph of f and have been plotted in the coordinate plane below.

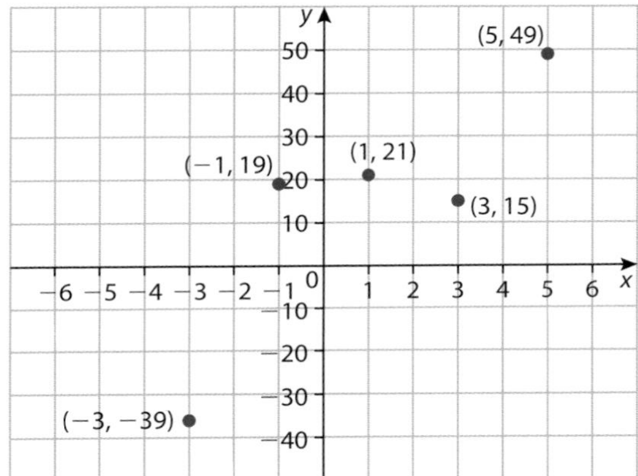

Recall that the end behaviour of a polynomial function is determined by its degree and by the sign of its leading coefficient. Since the leading term of f is x^3 then its graph will fall ($y \to -\infty$) as $x \to -\infty$ and will rise ($y \to \infty$) as $x \to +\infty$. Also a polynomial function of degree n has at most $n - 1$ turning points; therefore, the graph of f has at most two turning points. Given the coordinates of the five points found with the aid of synthetic substitution, there will clearly be exactly two turning points. The graph of f can now be accurately sketched.

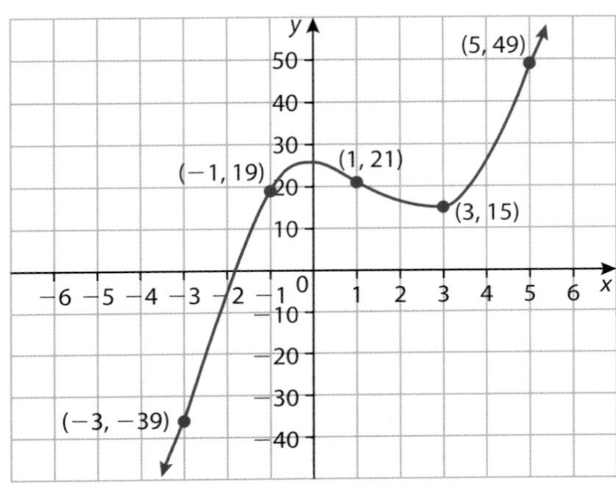

Exercise 3.1

In questions 1–4, use synthetic substitution to evaluate $P(x)$ for the given values of x.

1 $P(x) = x^4 + 2x^3 - 3x^2 - 4x - 20, \quad x = 2,\ x = -3$

2 $P(x) = 2x^5 - x^4 + 3x^3 - 15x - 9, \quad x = -1,\ x = 2$

3 $P(x) = x^5 + 5x^4 + 3x^3 - 6x^2 - 9x + 11, \quad x = -2,\ x = 4$

4 $P(x) = x^3 - (c + 3)x^2 + (3c + 5)x - 5c, \quad x = c,\ x = 2$

5 Given $P(x) = kx^3 + 2x^2 - 10x + 3$, for what value of k is $P(-2) = 15$?

6 Given $P(x) = 3x^4 - 2x^3 - 10x^2 + 3kx + 3$, for what value of k is $x = -\frac{1}{3}$ a zero of $P(x)$?

For questions 7 and 8, do not use your GDC.

7 a) Given $y = 2x^3 + 3x^2 - 5x - 4$, determine the y-value for each value of x such that $x \in \{-3, -2, -1, 0, 1, 2, 3\}$.

b) How many times must the graph of $y = 2x^3 + 3x^2 - 5x - 4$ cross the x-axis?

c) Sketch the graph of $y = 2x^3 + 3x^2 - 5x - 4$.

8 a) Given $y = x^4 - 4x^2 - 2x + 1$, determine the y-value for each value of x such that $x \in \{-3, -2, -1, 0, 1, 2, 3\}$.

b) How many times must the graph of $y = x^4 - 4x^2 - 2x + 1$ cross the x-axis?

c) Sketch the graph of $y = x^4 - 4x^2 - 2x + 1$.

9 Given $f(x) = x^3 + ax^2 - 5x + 7a$, find a so that $f(2) = 10$.

10 Given $f(x) = bx^3 - 5x^2 + 2bx + 10$, find b so that $f(\sqrt{3}) = -20$.

11 There are four possible end behaviours for a polynomial function $P(x)$. These are:

as $x \to \infty$, $P(x) \to \infty$ and as $x \to -\infty$, $P(x) \to \infty$ or symbolically (↖, ↗)

as $x \to \infty$, $P(x) \to -\infty$ and as $x \to -\infty$, $P(x) \to \infty$ or symbolically (↖, ↘)

as $x \to \infty$, $P(x) \to -\infty$ and as $x \to -\infty$, $P(x) \to -\infty$ or symbolically (↙, ↘)

as $x \to \infty$, $P(x) \to \infty$ and as $x \to -\infty$, $P(x) \to -\infty$ or symbolically (↙, ↗)

a) By sketching a graph on your GDC, state the type of end behaviour for each of the polynomial functions below.

(i) $P(x) = 2x^4 - 6x^3 + x^2 + 4x - 1$

(ii) $P(x) = -2x^4 - 6x^3 + x^2 + 4x - 1$

(iii) $P(x) = -6x^3 + x^2 + 4x - 1$

(iv) $P(x) = 6x^3 + x^2 - 4x - 1$

(v) $P(x) = x^2 - 4x - 1$

(vi) $P(x) = -2x^6 + x^5 + 2x^4 - 3x^3 + 4x^2 - x + 1$

(vii) $P(x) = x^5 + 2x^4 - x^3 + x^2 - x + 1$

(viii) $P(x) = -x^5 + 2x^4 - x^3 + x^2 - x + 1$

b) Use your results from a) to write a general statement about how the leading term of a polynomial function, $a_n x^n$, determines what type of end behaviour the graph of the function will display. Be specific about how the characteristics of the coefficient, a_n, and the power, n, of the leading term affect the function's end behaviour.

3.2 Quadratic functions

A **linear function** is a polynomial function of degree one that can be written in the general form $f(x) = ax + b$ where $a \neq 0$. Linear equations were briefly reviewed in Section 1.6. It is clear that any linear function will have a single solution (root) of $x = -\frac{b}{a}$. In essence, this is a formula that gives the zero of any linear polynomial.

In this section, we will focus on **quadratic functions** – functions consisting of a second-degree polynomial that can be written in the form $f(x) = ax^2 + bx + c$ such that $a \neq 0$. You are probably familiar with the quadratic formula that gives the zeros of any quadratic polynomial. We will also investigate other methods of finding zeros of quadratics and consider important characteristics of the graphs of quadratic functions.

The word *quadratic* comes from the Latin word *quadratus* that means four-sided, to make square, or simply a square. *Numerus quadratus* means a square number. Before modern algebraic notation was developed in the 17th and 18th centuries, the geometric figure of a square was used to indicate a number multiplying itself. Hence, raising a number to the power of two (in modern notation) is commonly referred to as the operation of squaring. *Quadratic* then came to be associated with a polynomial of degree two rather than being associated with the number four, as the prefix quad often indicates (e.g. quadruple).

Definition of a quadratic function
If a, b and c are real numbers, and $a \neq 0$, the function $f(x) = ax^2 + bx + c$ is a **quadratic function**. The graph of f is the graph of the equation $y = ax^2 + bx + c$ and is called a **parabola**.

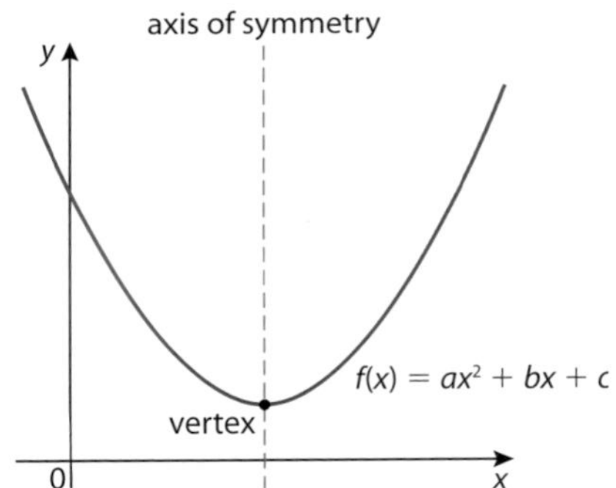

If $a > 0$ then the parabola opens upward.

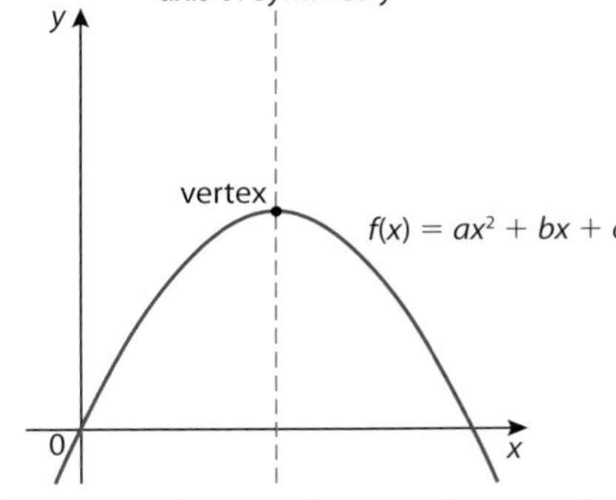

If $a < 0$ then the parabola opens downward.

Figure 3.4 'Concave up' and 'concave down' parabolas.

Each parabola is symmetric about a vertical line called its **axis of symmetry.** The axis of symmetry passes through a point on the parabola called the **vertex** of the parabola, as shown in Figure 3.4. If the leading coefficient, a, of the quadratic function $f(x) = ax^2 + bx + c$ is positive, the parabola opens upward (concave up) – and the y-coordinate of the vertex will be a **minimum value** for the function. If the leading coefficient, a, of $f(x) = ax^2 + bx + c$ is negative, the parabola opens downward (concave down) – and the y-coordinate of the vertex will be a **maximum value** for the function.

The graph of $f(x) = a(x - h)^2 + k$

From the previous chapter, we know that the graph of the equation $y = (x + 3)^2 + 2$ can be obtained by translating $y = x^2$ three units to the left and two units up. Being familiar with the shape and position of the graph of $y = x^2$, and knowing the two translations that transform $y = x^2$ to

$y = (x + 3)^2 + 2$, we can easily visualize and/or sketch the graph of $y = (x + 3)^2 + 2$ (see Figure 3.5). We can also determine the axis of symmetry and the vertex of the graph. Figure 3.6 shows that the graph of $y = (x + 3)^2 + 2$ has an axis of symmetry of $x = -3$ and a vertex at $(-3, 2)$. The equation $y = (x + 3)^2 + 2$ can also be written as $y = x^2 + 6x + 11$. Because we can easily identify the vertex of the parabola when the equation is written as $y = (x + 3)^2 + 2$, we often refer to this as the **vertex form** of the quadratic equation, and $y = x^2 + 6x + 11$ as the **general form**.

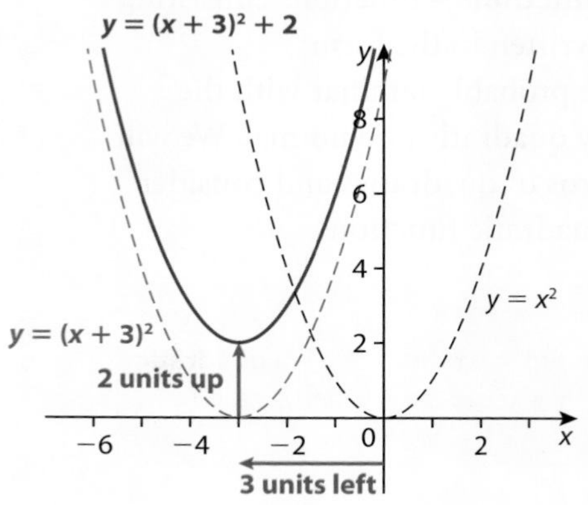

Figure 3.5 Translating $y = x^2$ to give $y = (x + 3)^2 + 2$.

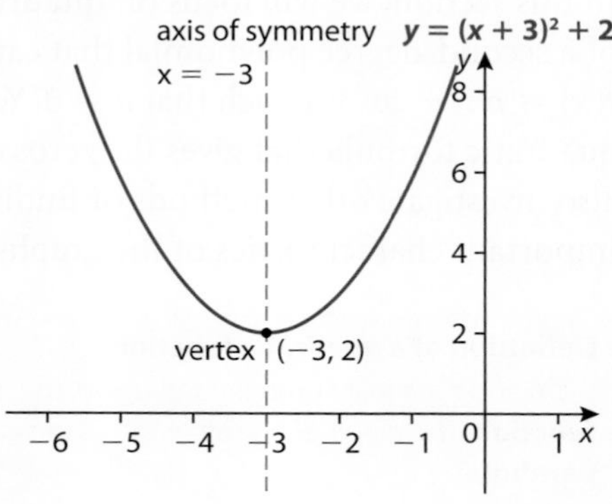

Figure 3.6 The axis of symmetry and the vertex.

• **Hint:** $f(x) = a(x - h)^2 + k$ is sometimes referred to as the **standard form** of a quadratic function.

Vertex form of a quadratic function

If a quadratic function is written in the form $f(x) = a(x - h)^2 + k$, with $a \neq 0$, the graph of f has an axis of symmetry of $x = h$ and a vertex at (h, k).

Completing the square

For visualizing and sketching purposes, it is helpful to have a quadratic function written in vertex form. How do we rewrite a quadratic function written in the form $f(x) = ax^2 + bx + c$ (general form) into the form $f(x) = a(x - h)^2 + k$ (vertex form)? We use the technique of **completing the square.**

For any real number p, the quadratic expression $x^2 + px + \left(\frac{p}{2}\right)^2$ is the square of $\left(x + \frac{p}{2}\right)$. Convince yourself of this by expanding $\left(x + \frac{p}{2}\right)^2$. The technique of *completing the square* is essentially the process of adding a constant to a quadratic expression to make it the square of a binomial. If the coefficient of the quadratic term (x^2) is positive one, the coefficient of the linear term is p, and the constant term is $\left(\frac{p}{2}\right)^2$, then $x^2 + px + \left(\frac{p}{2}\right)^2 = \left(x + \frac{p}{2}\right)^2$ and the square is completed.

Remember that the coefficient of the quadratic term (leading coefficient) must be equal to positive one before completing the square.

Example 3

Find the equation of the axis of symmetry and the coordinates of the vertex of the graph of $f(x) = x^2 - 8x + 18$ by rewriting the function in the form $x^2 + px + \left(\frac{p}{2}\right)^2$.

Solution

To complete the square and get the quadratic expression $x^2 - 8x + 18$ in the form $x^2 + px + \left(\frac{p}{2}\right)^2$, the constant term needs to be $\left(\frac{-8}{2}\right)^2 = 16$.

We need to add 16, but also subtract 16, so that we are adding zero overall and, hence, not changing the original expression.

$f(x) = x^2 - 8x + 16 - 16 + 18$	Actually adding zero $(-16 + 16)$ to the right side.
$f(x) = x^2 - 8x + 16 + 2$	$x^2 - 8x + 16$ fits the pattern $x^2 + px + \left(\frac{p}{2}\right)^2$ with $p = -8$.
$f(x) = (x - 4)^2 + 2$	$x^2 - 8x + 16 = (x - 4)^2$

The axis of symmetry of the graph of f is the vertical line $x = 4$ and the vertex is at $(4, 2)$. See Figure 3.7.

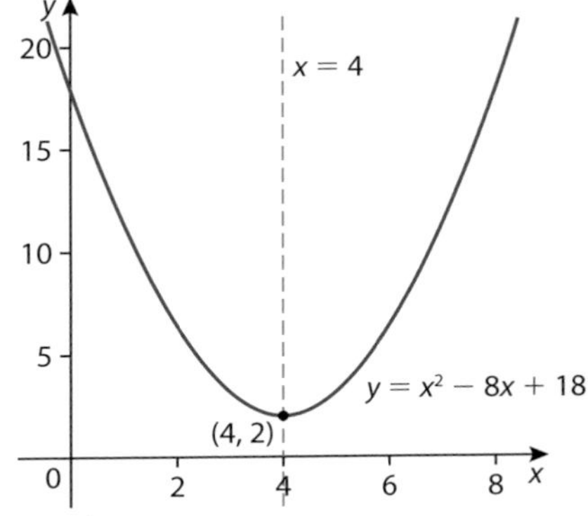

Figure 3.7

Example 4 – Properties of a parabola

For the function $g: x \mapsto -2x^2 - 12x + 7$,

a) find the axis of symmetry and the vertex of the graph
b) indicate the transformations that can be applied to $y = x^2$ to obtain the graph
c) find the minimum or maximum value.

Solution

a) $g: x \mapsto -2\left(x^2 + 6x - \frac{7}{2}\right)$	Factorize so that the coefficient of the quadratic term is $+1$.
$g: x \mapsto -2\left(x^2 + 6x + 9 - 9 - \frac{7}{2}\right)$	$p = 6 \Rightarrow \left(\frac{p}{2}\right)^2 = 9$; hence, add $+9 - 9$ (zero)
$g: x \mapsto -2\left[(x + 3)^2 - \frac{18}{2} - \frac{7}{2}\right]$	$x^2 + 6x + 9 = (x + 3)^2$
$g: x \mapsto -2\left[(x + 3)^2 - \frac{25}{2}\right]$	
$g: x \mapsto -2(x + 3)^2 + 25$	Multiply through by -2 to remove outer brackets.
$g: x \mapsto -2(x - (-3))^2 + 25$	Express in vertex form: $g: x \mapsto a(x - h)^2 + k$

The axis of symmetry of the graph of g is the vertical line $x = -3$ and the vertex is at $(-3, 25)$. See Figure 3.8.

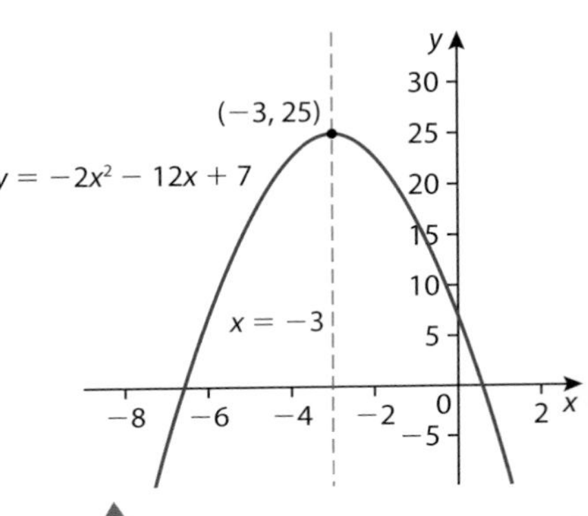

Figure 3.8

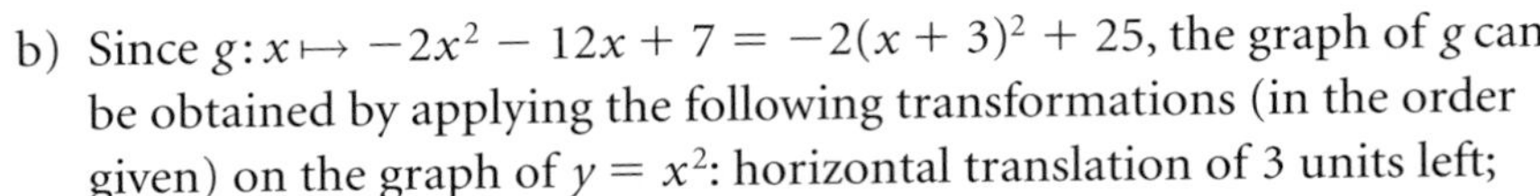

b) Since $g: x \mapsto -2x^2 - 12x + 7 = -2(x + 3)^2 + 25$, the graph of g can be obtained by applying the following transformations (in the order given) on the graph of $y = x^2$: horizontal translation of 3 units left;

reflection in the x-axis (parabola opening down); vertical stretch of factor 2; and a vertical translation of 25 units up.

c) The parabola opens down because the leading coefficient is negative. Therefore, g has a maximum and no minimum value. The maximum value is 25 (y-coordinate of vertex) at $x = -3$.

The technique of completing the square can be used to derive the quadratic formula. The following example derives a general expression for the axis of symmetry and vertex of a quadratic function in the general form $f(x) = ax^2 + bx + c$ by completing the square.

Example 5 – Graphical properties of general quadratic functions

Find the axis of symmetry and the vertex for the general quadratic function $f(x) = ax^2 + bx + c$.

Solution

$f(x) = a\left(x^2 + \frac{b}{a}x + \frac{c}{a}\right)$ Factorize so that the coefficient of the x^2 term is $+1$.

$f(x) = a\left[x^2 + \frac{b}{a}x + \left(\frac{b}{2a}\right)^2 - \left(\frac{b}{2a}\right)^2 + \frac{c}{a}\right]$ $p = \frac{b}{a} \Rightarrow \left(\frac{p}{2}\right)^2 = \left(\frac{b}{2a}\right)^2$

$f(x) = a\left[\left(x + \frac{b}{2a}\right)^2 - \frac{b^2}{4a^2} + \frac{c}{a}\right]$ $x^2 + \frac{b}{a}x + \left(\frac{b}{2a}\right)^2 = x + \left(\frac{b}{2a}\right)^2$

$f(x) = a\left(x + \frac{b}{2a}\right)^2 - \frac{b^2}{4a} + c$ Multiply through by a.

$f(x) = a\left(x - \left(-\frac{b}{2a}\right)\right)^2 + c - \frac{b^2}{4a}$ Express in vertex form: $f(x) = a(x - h)^2 + k$

This result leads to the following generalization.

Symmetry and vertex of $f(x) = ax^2 + bx + c$

For the graph of the quadratic function $f(x) = ax^2 + bx + c$, the axis of symmetry is the vertical line with the equation $x = -\frac{b}{2a}$ and the vertex has coordinates $\left(-\frac{b}{2a}, c - \frac{b^2}{4a}\right)$.

Check the results for Example 4 using the formulae for the axis of symmetry and vertex. For the function $g: x \mapsto -2x^2 - 12x + 7$:

$$x = -\frac{b}{2a} = -\frac{-12}{2(-2)} = -3 \Rightarrow \text{ axis of symmetry is the vertical line } x = -3$$

$$c - \frac{b^2}{4a} = 7 - \frac{(-12)^2}{4(-2)} = \frac{56}{8} + \frac{144}{8} = 25 \Rightarrow \text{ vertex has coordinates } (-3, 25)$$

These results agree with the results from Example 4.

Zeros of a quadratic function

A specific value for x is a **zero** of a quadratic function $f(x) = ax^2 + bx + c$ if it is a solution (or **root**) to the equation $ax^2 + bx + c = 0$.

As we will observe, every quadratic function will have two zeros although it is possible for the same zero to occur twice (double zero, or double root). The x-coordinate of any point(s) where f crosses the x-axis (y-coordinate is zero) is a **real zero** of the function. A quadratic function can have one, two or no real zeros as Figure 3.9 illustrates. To find non-real zeros we need to extend our search to the set of complex numbers and we will see that a quadratic function with no real zeros will have two distinct **imaginary zeros.** Finding all zeros of a quadratic function requires you to solve quadratic equations of the form $ax^2 + bx + c = 0$. Although $a \neq 0$, it is possible for b or c to be equal to zero. There are five general methods for solving quadratic equations as outlined in Table 3.2 below.

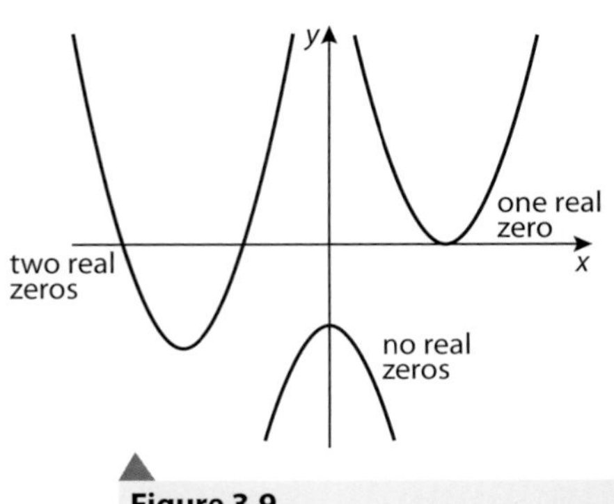

Figure 3.9

Table 3.2 Methods for solving quadratic equations.

Square root Examples	If $a^2 = c$ and $c > 0$, then $a = \pm\sqrt{c}$. $x^2 - 25 = 0$ $(x + 2)^2 = 15$ $x^2 = 25$ $x + 2 = \pm\sqrt{15}$ $x = \pm 5$ $x = -2 \pm\sqrt{15}$
Factorizing Examples	If $ab = 0$, then $a = 0$ or $b = 0$. $x^2 + 3x - 10 = 0$ $x^2 - 7x = 0$ $(x + 5)(x - 2) = 0$ $x(x - 7) = 0$ $x = -5$ or $x = 2$ $x = 0$ or $x = 7$
Completing the square Example	If $x^2 + px + q = 0$, then $x^2 + px + \left(\frac{p}{2}\right)^2 = -q + \left(\frac{p}{2}\right)^2$ which leads to $\left(x + \frac{p}{2}\right)^2 = -q + \frac{p^2}{4}$ and then the square root of both sides (as above). $x^2 - 8x + 5 = 0$ $x^2 - 8x + 16 = -5 + 16$ $(x - 4)^2 = 11$ $x - 4 = \pm\sqrt{11}$ $x = 4 \pm \sqrt{11}$
Quadratic formula Example	If $ax^2 + bx + c = 0$, then $x = \frac{-b \pm \sqrt{b^2 - 4ac}}{2a}$. $2x^2 - 3x - 4 = 0$ $x = \frac{-(-3) \pm \sqrt{(-3)^2 - 4(2)(-4)}}{2(2)}$ $x = \frac{3 \pm \sqrt{41}}{4}$
Graphing Example	Graph the equation $y = ax^2 + bx + c$ on your GDC. Use the calculating features of your GDC to determine the x-coordinates of the point(s) where the parabola intersects the x-axis. *Note*: This method works for finding real solutions, but **not** imaginary solutions. $2x^2 - 5x - 7 = 0$ GDC calculations reveal that the zeros are at $x = \frac{7}{2}$ and $x = -1$

Plot1 Plot2 Plot3
\Y1=2X²-5X-7
\Y2=
\Y3=
\Y4=
\Y5=
\Y6=
\Y7=

CALCULATE
1:value
2:zero
3:minimum
4:maximum
5:intersect
6:dy/dx
7:∫f(x)dx

Y1=2X²-5X-7 Left Bound? X=2.787234 Y=-5.398823

Y1=2X²-5X-7 Right Bound? X=3.8085106 Y=2.9669535

Y1=2X²-5X-7 Guess? X=3.6382979 Y=1.2829335

Zero X=3.5 Y=0

Y1=2X²-5X-7 Left Bound? X=-1.297872 Y=2.8583069

Y1=2X²-5X-7 Right Bound? X=-.6170213 Y=-3.153463

Y1=2X²-5X-7 Guess? X=-.8723404 Y=-1.116342

Zero X=-1 Y=0

Sum and product of the roots of a quadratic equation

In the next section, the Factor Theorem formally states the relationship between linear factors of the form $x - \alpha$ and the zeros for *any* polynomial.

Consider the quadratic equation $x^2 + 5x - 24 = 0$. This equation can be solved using factorization as follows.

$$x^2 + 5x - 24 = (x + 8)(x - 3) = 0 \Rightarrow x = -8 \text{ or } x = 3$$

Clearly, if $x - \alpha$ is a factor of the quadratic polynomial $ax^2 + bx + c$, then $x = \alpha$ is a root (solution) of the quadratic equation $ax^2 + bx + c = 0$.

Now let us consider the general quadratic equation $ax^2 + bx + c = 0$, whose roots are $x = \alpha$ and $x = \beta$. Given our observation from the previous paragraph, we can write the quadratic equation with roots α and β as:

$$\begin{aligned} ax^2 + bx + c = (x - \alpha)(x - \beta) &= 0 \\ x^2 - \alpha x - \beta x + \alpha\beta &= 0 \\ x^2 - (\alpha + \beta)x + \alpha\beta &= 0 \end{aligned}$$

Since the equation $ax^2 + bx + c = 0$ can also be written as $x^2 + \frac{b}{a}x + \frac{c}{a} = 0$, then:

$$x^2 - (\alpha + \beta)x + \alpha\beta = x^2 + \frac{b}{a}x + \frac{c}{a}$$

Equating coefficients of both sides, gives the following results.

$$\alpha + \beta = -\frac{b}{a} \text{ and } \alpha\beta = \frac{c}{a}$$

Sum and product of the roots of a quadratic equation
For any quadratic equation in the form $ax^2 + bx + c = 0$, the **sum of the roots** of the equation is $-\frac{b}{a}$ and the **product of the roots** is $\frac{c}{a}$. (In the next section, this result is extended to polynomial equations of any degree.)

Example 6

If α and β are the roots of each equation, find the sum, $\alpha + \beta$, and product, $\alpha\beta$, of the roots.

a) $x^2 - 5x + 3 = 0$ b) $3x^2 + 4x - 7 = 0$

Solution

a) For the equation $x^2 - 5x + 3 = 0$, $a = 1$, $b = -5$ and $c = 3$.
Therefore, $\alpha + \beta = -\frac{b}{a} = -\frac{-5}{1} = 5$ and $\alpha\beta = \frac{c}{a} = \frac{3}{1} = 3$.

b) For the equation $3x^2 + 4x - 7 = 0$, $a = 3$, $b = 4$ and $c = -7$.
Therefore, $\alpha + \beta = -\frac{b}{a} = -\frac{4}{3}$ and $\alpha\beta = \frac{c}{a} = \frac{-7}{3}$.

If the sum and product of the roots of a quadratic equation are known, then the equation can be written in the following form: $x^2 -$ (sum of roots)$x +$ (product of roots) $= 0$

Example 7

If α and β are the roots of the equation $2x^2 + 6x - 5 = 0$, find a quadratic equation whose roots are:

a) $2\alpha, 2\beta$ b) $\frac{1}{\alpha + 1}, \frac{1}{\beta + 1}$

Solution

For the equation $2x^2 + 6x - 5 = 0$, $a = 2$, $b = 6$ and $c = -5$.

Thus, $\alpha + \beta = -\frac{b}{a} = -\frac{6}{2} = -3$ and $\alpha\beta = \frac{c}{a} = \frac{-5}{2}$.

a) Sum of the new roots $= 2\alpha + 2\beta = 2(\alpha + \beta) = 2(-3) = -6$.

Thus for the new equation, $-\frac{b}{a} = -6$.

Product of the new roots $= 2\alpha \cdot 2\beta = 4\alpha\beta = 4\left(-\frac{5}{2}\right) = -10$.

Thus for the new equation, $\frac{c}{a} = -10$.

The new equation we are looking for can be written as $ax^2 + bx + c = 0$ or $x^2 + \frac{b}{a}x + \frac{c}{a} = 0$.

Therefore, the quadratic equation with roots $2\alpha, 2\beta$ is $x^2 - (-6)x - 10 = 0$
$\Rightarrow x^2 + 6x - 10 = 0$

b) Sum of the new roots $\frac{1}{\alpha + 1} + \frac{1}{\beta + 1} = \frac{\beta + 1 + \alpha + 1}{(\alpha + 1)(\beta + 1)}$

$$= \frac{\alpha + \beta + 2}{\alpha\beta + \alpha + \beta + 1} = \frac{-3 + 2}{-\frac{5}{2} - 3 + 1} = \frac{-1}{-\frac{9}{2}} = \frac{2}{9}.$$

Thus for the new equation, $-\frac{b}{a} = \frac{2}{9}$.

Product of the new roots $\left(\frac{1}{\alpha + 1}\right)\left(\frac{1}{\beta + 1}\right) = \frac{1}{\alpha\beta + \alpha + \beta + 1}$

$$= \frac{1}{-\frac{5}{2} - 3 + 1} = \frac{1}{-\frac{9}{2}} = -\frac{2}{9}.$$

Thus for the new equation, $\frac{c}{a} = -\frac{2}{9}$.

The new equation we are looking for can be written as $x^2 + \frac{b}{a}x + \frac{c}{a} = 0$.

Therefore, the quadratic equation with roots $\frac{1}{\alpha + 1}, \frac{1}{\beta + 1}$ is $x^2 - \frac{2}{9}x - \frac{2}{9} = 0$ or $9x^2 - 2x - 2 = 0$.

Example 8

Given that the roots of the equation $x^2 - 4x + 2 = 0$ are α and β, find the values of the following expressions.

a) $\alpha^2 + \beta^2$ b) $\frac{1}{\alpha^2} + \frac{1}{\beta^2}$

Solution

With $x^2 - 4x + 2 = 0$, $\alpha + \beta = -\frac{b}{a} = -\frac{-4}{1} = 4$ and $\alpha\beta = \frac{c}{a} = \frac{2}{1} = 2$.

Both of the expressions $\alpha^2 + \beta^2$ and $\frac{1}{\alpha^2} + \frac{1}{\beta^2}$ need to be expressed in terms of $\alpha + \beta$ and $\alpha\beta$.

a) $\alpha^2 + \beta^2 = \alpha^2 + 2\alpha\beta + \beta^2 - 2\alpha\beta = (\alpha + \beta)^2 - 2\alpha\beta$

Substituting the values for $\alpha + \beta$ and $\alpha\beta$ from above, gives
$\alpha^2 + \beta^2 = 4^2 - 2 \cdot 2 = 16 - 4 = 12$.

b) $\frac{1}{\alpha^2} + \frac{1}{\beta^2} = \frac{\beta^2}{\alpha^2\beta^2} + \frac{\alpha^2}{\alpha^2\beta^2} = \frac{\alpha^2 + \beta^2}{(\alpha\beta)^2}$

From part a) we know that $\alpha^2 + \beta^2 = (\alpha + \beta)^2 - 2\alpha\beta$. Substituting this into the numerator gives:

$\frac{1}{\alpha^2} + \frac{1}{\beta^2} = \frac{(\alpha + \beta)^2 - 2\alpha\beta}{(\alpha\beta)^2}$ Then substituting the values for $\alpha + \beta$ and $\alpha\beta$ from above, gives:

$= \frac{4^2 - 2 \cdot 2}{2^2} = \frac{12}{4} = 3$

Therefore, $\frac{1}{\alpha^2} + \frac{1}{\beta^2} = 3$.

The quadratic formula and the discriminant

The expression that is beneath the radical sign in the quadratic formula, $b^2 - 4ac$, determines whether the zeros of a quadratic function are real or imaginary. Because it acts to 'discriminate' between the types of zeros, $b^2 - 4ac$ is called the **discriminant**. It is often labelled with the Greek letter Δ (delta). The value of the discrimant can also indicate if the zeros are equal and if they are rational.

The discriminant and the nature of the zeros of a quadratic function

For the quadratic function $f(x) = ax^2 + bx + c$, $(a \neq 0)$ where a, b and c are real numbers:

If $\Delta = b^2 - 4ac > 0$, then f has two distinct real zeros, and the graph of f intersects the x-axis twice.

If $\Delta = b^2 - 4ac = 0$, then f has one real zero (double root), and the graph of f intersects the x-axis once (i.e. it is tangent to the x-axis).

If $\Delta = b^2 - 4ac < 0$, then f has two conjugate imaginary zeros, and the graph of f does not intersect the x-axis.

In the special case when a, b and c are integers and the discriminant is the square of an integer (a *perfect square*), the polynomial $ax^2 + bx + c$ has two distinct **rational zeros**.

When the discriminant is zero then the solution of a quadratic function is $x = \frac{-b \pm \sqrt{b^2 - 4ac}}{2a} = \frac{-b \pm \sqrt{0}}{2a} = -\frac{b}{2a}$. As mentioned, this solution of $-\frac{b}{2a}$ is called a double zero (or root) which can also be described as a **zero of multiplicity of 2**. If a and b are integers then the zero $-\frac{b}{2a}$ will be rational.

When we solve polynomial functions of higher degree later this chapter, we will encounter zeros of higher multiplicity.

• **Hint:** Remember that the **roots** of a polynomial equation are those values of x for which $P(x) = 0$. These values of x are called the **zeros** of the polynomial P.

Factorable quadratics

If the zeros of a quadratic polynomial are rational – either two distinct zeros or two equal zeros (double zero/root) – then the polynomial is factorable. That is, if $ax^2 + bx + c$ has rational zeros then $ax^2 + bx + c = (mx + n)(px + q)$ where m, n, p and q are rational numbers.

Example 9 – Using discriminant to determine the nature of the roots of a quadratic equation

Use the discriminant to determine how many real roots each equation has. Visually confirm the result by graphing the corresponding quadratic function for each equation on your GDC.

a) $2x^2 + 5x - 3 = 0$ b) $4x^2 - 12x + 9 = 0$ c) $2x^2 - 5x + 6 = 0$

Solution

a) The discriminant is $\Delta = 5^2 - 4(2)(-3) = 49 > 0$. Therefore, the equation has two distinct real roots. This result is confirmed by the graph of the quadratic function $y = 2x^2 + 5x - 3$ that clearly shows it intersecting the x-axis twice. Also since $\Delta = 49$ is a perfect square then the two roots are also rational and the quadratic polynomial $2x^2 + 5x - 3 = 0$ is factorable: $2x^2 + 5x - 3 = (2x - 1)(x + 3) = 0$. Thus, the two rational roots are $x = \frac{1}{2}$ and $x = -3$.

$y = 2x^2 + 5x - 3$

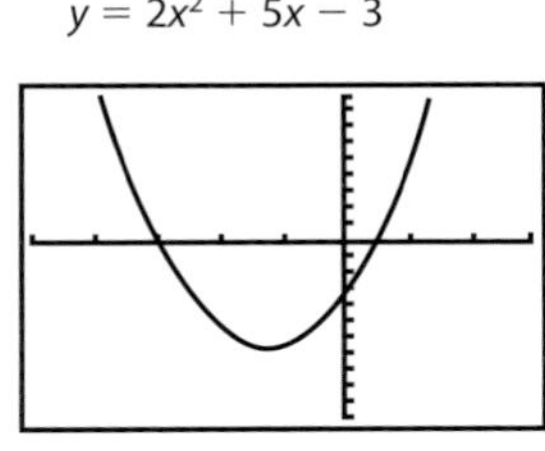

b) The discriminant is $\Delta = (-12)^2 - 4(4)(9) = 0$. Therefore, the equation has one rational root (a double root). The graph on the GDC of $y = 4x^2 - 12x + 9$ appears to intersect the x-axis at only one point. We can be more confident with this conclusion by investigating further – for example, tracing or looking at a table of values on the GDC.

$y = 4x^2 - 12x + 9$

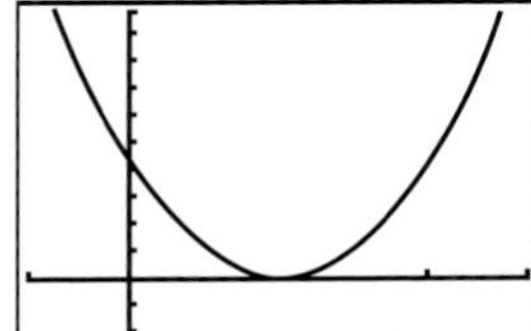

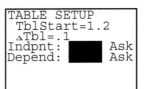

```
TABLE SETUP
 TblStart=1.2
 ΔTbl=.1
Indpnt: Auto Ask
Depend: Auto Ask
```

X	Y1
1.2	.36
1.3	.16
1.4	.04
1.5	0
1.6	.04
1.7	.16
1.8	.36

Y1=0

Also, since the root is rational ($\Delta = 0$), the polynomial $4x^2 - 12x + 9$ must be factorable.

$4x^2 - 12x + 9 = (2x - 3)(2x - 3) = \left[2\left(x - \frac{3}{2}\right)2\left(x - \frac{3}{2}\right)\right] = 4\left(x - \frac{3}{2}\right)^2 = 0$

There are two equal linear factors which means there are two equal rational zeros – both equal to $\frac{3}{2}$ in this case.

c) The discriminant is $\Delta = (-5)^2 - 4(2)(6) = -23 < 0$. Therefore, the equation has no real roots. This result is confirmed by the graph of the quadratic function $y = 2x^2 - 5x + 6$ that clearly shows that the graph does not intersect the x-axis. The equation will have two imaginary roots.

$y = 2x^2 - 5x + 6$

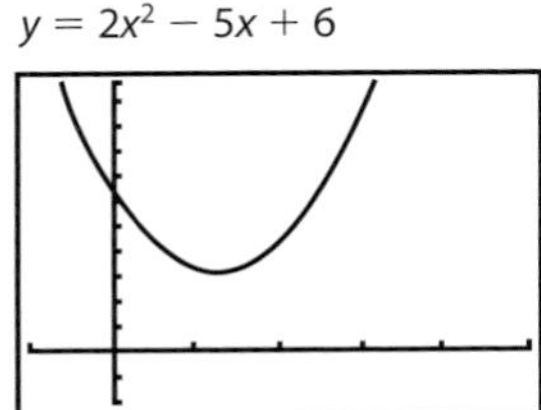

• **Hint:** If a quadratic polynomial has a zero of multiplicity 2($\Delta = 0$), as in Example 6 b), then not only is the polynomial factorable but its factorization will contain two equal linear factors. In such a case then $ax^2 + bx + c = a(x - p)^2$ where $x - p$ is the linear factor and $x = p$ is the rational zero.

Example 10 – The discriminant and number of real zeros

For $4x^2 + 4kx + 9 = 0$, determine the value(s) of k so that the equation has: a) one real zero, b) two distinct real zeros, and c) no real zeros.

Solution

a) For one real zero $\Delta = (4k)^2 - 4(4)(9) = 0 \Rightarrow 16k^2 - 144 = 0$

$\Rightarrow 16k^2 = 144 \Rightarrow k^2 = 9 \Rightarrow k = \pm 3$

b) For two distinct real zeros $\Delta = (4k)^2 - 4(4)(9) > 0 \Rightarrow 16k^2 > 144$
$\Rightarrow k^2 > 9 \Rightarrow k < -3$ or $k > 3$

c) For no real zeros $\Delta = (4k)^2 - 4(4)(9) < 0 \Rightarrow 16k^2 < 144 \Rightarrow k^2 < 9$
$\Rightarrow k > -3$ and $k < 3 \Rightarrow -3 < k < 3$

Example 11 – Conjugate imaginary solutions

Find the zeros of the function $g: x \rightarrow 2x^2 - 4x + 7$.

Solution

Solve the equation $2x^2 - 4x + 7 = 0$ using the quadratic formula with $a = 2$, $b = -4$, $c = 7$.

$$x = \frac{-(-4) \pm \sqrt{(-4)^2 - 4(2)(7)}}{2(2)} = \frac{4 \pm \sqrt{-40}}{4} = \frac{4 \pm \sqrt{4}\sqrt{-1}\sqrt{10}}{4}$$

$$= \frac{4 \pm 2i\sqrt{10}}{4} = 1 \pm \frac{i\sqrt{10}}{2}$$

The two zeros of g are $1 + \frac{\sqrt{10}}{2}i$ and $1 - \frac{\sqrt{10}}{2}i$.

Note that the imaginary zeros are written in the form $a + bi$ (introduced in Section 1.1) and that they clearly are a pair of conjugates, i.e. fitting the pattern $a + bi$ and $a - bi$.

Number of complex zeros of a quadratic polynomial
Every quadratic polynomial has exactly two complex zeros, provided that a zero of multiplicity 2 (two equal zeros) is counted as two zeros.

• **Hint:** Recall from Section 1.1 that the real numbers and the imaginary numbers are distinct subsets of the complex numbers. A complex number can be either real $\left(\text{e.g. } -7, \frac{\pi}{2}, 3 - \sqrt{2}\right)$ or imaginary (e.g. $4i$, $2 + i\sqrt{5}$).

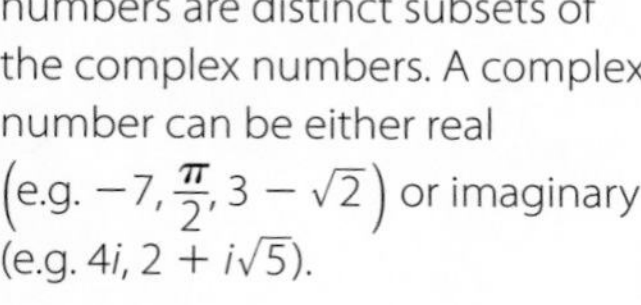

The graph of $f(x) = a(x - p)(x - q)$

If a quadratic function is written in the form $f(x) = a(x - p)(x - q)$ then we can easily identify the x-intercepts of the graph of f. Consider that $f(p) = a(p - p)(p - q) = a(0)(p - q) = 0$ and that $f(q) = a(q - p)(q - q) = a(q - p)(0) = 0$. Therefore, the quadratic function $f(x) = a(x - p)(x - q)$ will intersect the x-axis at the points $(p, 0)$ and $(q, 0)$. We need to factorize in order to rewrite a quadratic function in the form $f(x) = ax^2 + bx + c$ to the form $f(x) = a(x - p)(x - q)$. Hence, $f(x) = a(x - p)(x - q)$ can be referred to as the **factorized** form of a quadratic function. Recalling the symmetric nature of a parabola, it is clear that the x-intercepts $(p, 0)$ and $(q, 0)$ will be equidistant from the axis of symmetry (see Figure 3.10). As a result, the equation of the axis of symmetry and the x-coordinate of the vertex of the parabola can be found from finding the average of p and q.

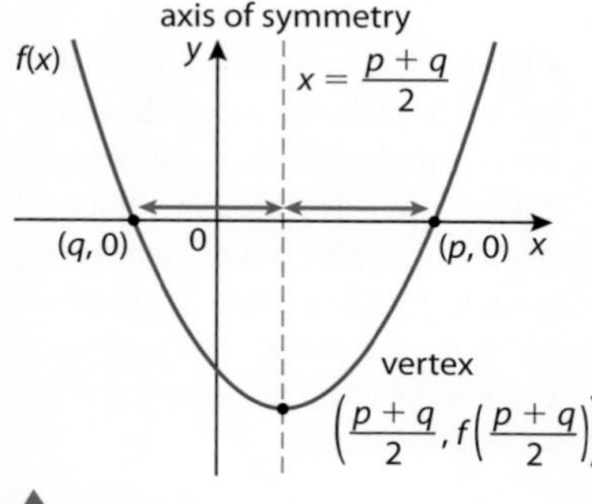

Figure 3.10

Factorized form of a quadratic function
If a quadratic function is written in the form $f(x) = a(x - p)(x - q)$, with $a \neq 0$, the graph of f has x-intercepts at $(p, 0)$ and $(q, 0)$, an axis of symmetry with equation $x = \frac{p + q}{2}$, and a vertex at $\left(\frac{p + q}{2}, f\left(\frac{p + q}{2}\right)\right)$.

Example 12

Find the equation of each quadratic function from the graph in the form $f(x) = a(x - p)(x - q)$ and also in the form $f(x) = ax^2 + bx + c$.

a) 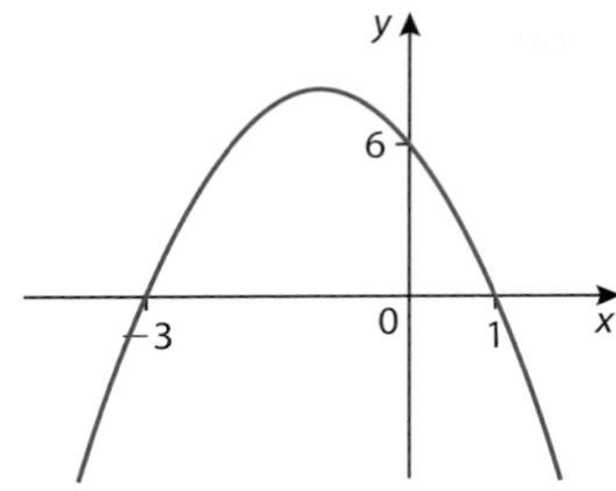

b) 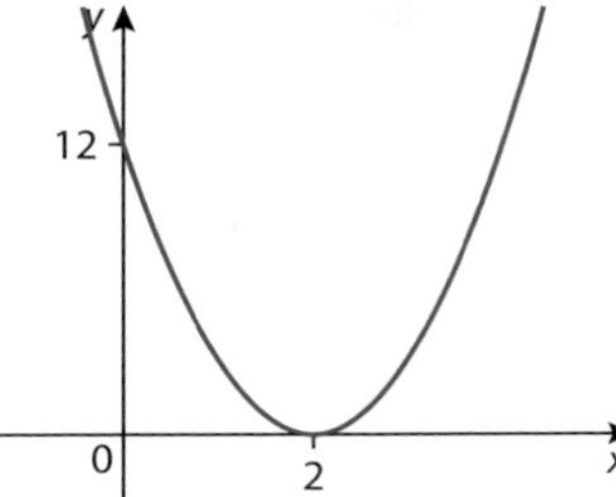

Solution

a) Since the x-intercepts are -3 and 1 then $y = a(x + 3)(x - 1)$. The y-intercept is 6, so when $x = 0$, $y = 6$. Hence, $6 = a(0 + 3)(0 - 1) = -3a \Rightarrow a = -2$ ($a < 0$ agrees with the fact that the parabola is opening down). The function is $f(x) = -2(x + 3)(x - 1)$, and expanding to remove brackets reveals that the function can also be written as $f(x) = -2x^2 - 4x + 6$.

b) The function has one x-intercept at 2 (double root), so $p = q = 2$ and $y = a(x - 2)(x - 2) = a(x - 2)^2$. The y-intercept is 12, so when $x = 0$, $y = 12$. Hence, $12 = a(0 - 2)^2 = 4a \Rightarrow a = 3$ ($a > 0$ agrees with the parabola opening up). The function is $f(x) = 3(x - 2)^2$. Expanding reveals that the function can also be written as $f(x) = 3x^2 - 12x + 12$.

Example 13

The graph of a quadratic function intersects the x-axis at the points $(-6, 0)$ and $(-2, 0)$ and also passes through the point $(2, 16)$. a) Write the function in the form $f(x) = a(x - p)(x - q)$. b) Find the vertex of the parabola. c) Write the function in the form $f(x) = a(x - h)^2 + k$.

Solution

a) The x-intercepts of -6 and -2 gives $f(x) = a(x + 6)(x + 2)$. Since f passes through $(2, 16)$, then $f(2) = 16 \Rightarrow f(2) = a(2 + 6)(2 + 2) = 16 \Rightarrow 32a = 16 \Rightarrow a = \frac{1}{2}$. Therefore, $f(x) = \frac{1}{2}(x + 6)(x + 2)$.

b) The x-coordinate of the vertex is the average of the x-intercepts. $x = \dfrac{-6 - 2}{2} = -4$, so the y-coordinate of the vertex is $y = f(-4) = \frac{1}{2}(-4 + 6)(-4 + 2) = -2$. Hence, the vertex is $(-4, -2)$.

c) In vertex form, the quadratic function is $f(x) = \frac{1}{2}(x + 4)^2 - 2$.

Table 3.3 Review of properties of quadratics.

Quadratic function, $a \neq 0$	Graph of function	Results
General form $f(x) = ax^2 + bx + c$ $\Delta = b^2 - 4ac$ (discriminant)	Parabola opens up if $a > 0$ Parabola opens down if $a < 0$ $x = -\frac{b}{2a}$; $\frac{-b-\sqrt{\Delta}}{2a}$; $\frac{-b+\sqrt{\Delta}}{2a}$; $\left(-\frac{b}{2a}, c - \frac{b^2}{4a}\right)$ If $\Delta \geqslant 0$, f has x-intercept(s) If $\Delta < 0$, f has no x-intercept(s)	Axis of symmetry is $x = -\dfrac{b}{2a}$ If $\Delta \geqslant 0$, f has x-intercept(s): $\left(\dfrac{-b \pm \sqrt{\Delta}}{2a}, 0\right)$ Vertex is: $\left(-\dfrac{b}{2a}, c - \dfrac{b^2}{4a}\right)$
Vertex form $f(x) = a(x - h)^2 + k$	$x = h$; (h, k)	Axis of symmetry is $x = h$ Vertex is (h, k)
Factorized form (two distinct rational zeros) $f(x) = a(x - p)(x - q)$	$x = \frac{p+q}{2}$; $(q, 0)$; $(p, 0)$; $\left(\frac{p+q}{2}, f\left(\frac{p+q}{2}\right)\right)$	Axis of symmetry is $x = \dfrac{p+q}{2}$ x-intercepts are: $(p, 0)$ and $(q, 0)$
Factorized form (one rational zero) $f(x) = a(x - p)^2$	$x = p$; $(p, 0)$	Axis of symmetry is $x = p$ Vertex and x-intercept is $(p, 0)$

Exercise 3.2

For each of the quadratic functions f in questions 1–5, find the following:

a) the axis of symmetry and the vertex, by algebraic methods

b) the transformation(s) that can be applied to $y = x^2$ to obtain the graph of $y = f(x)$

c) the minimum or maximum value of f.

Check your results using your GDC.

1 $f: x \mapsto x^2 - 10x + 32$

2 $f: x \mapsto x^2 + 6x + 8$

3 $f: x \mapsto -2x^2 - 4x + 10$

4 $f: x \mapsto 4x^2 - 4x + 9$

5 $f: x \mapsto \frac{1}{2}x^2 + 7x + 26$

In questions 6–13, solve the quadratic equation using factorization.

6 $x^2 + 2x - 8 = 0$

7 $x^2 = 3x + 10$

8 $6x^2 - 9x = 0$

9 $6 + 5x = x^2$

10 $x^2 + 9 = 6x$

11 $3x^2 + 11x - 4 = 0$

12 $3x^2 + 18 = 15x$

13 $9x - 2 = 4x^2$

In questions 14–19, use the method of completing the square to solve the quadratic equation.

14 $x^2 + 4x - 3 = 0$

15 $x^2 - 4x - 5 = 0$

16 $x^2 - 2x + 3 = 0$

17 $2x^2 + 16x + 6 = 0$

18 $x^2 + 2x - 8 = 0$

19 $-2x^2 + 4x + 9 = 0$

20 Let $f(x) = x^2 - 4x - 1$. a) Use the quadratic formula to find the zeros of the function. b) Use the zeros to find the equation for the axis of symmetry of the parabola. c) Find the minimum or maximum value of f.

In questions 21–24, determine the number of real solutions to each equation.

21 $x^2 + 3x + 2 = 0$

22 $2x^2 - 3x + 2 = 0$

23 $x^2 - 1 = 0$

24 $2x^2 - \frac{9}{4}x + 1 = 0$

25 Find the value(s) of p for which the equation $2x^2 + px + 1 = 0$ has one real solution.

26 Find the value(s) of k for which the equation $x^2 + 4x + k = 0$ has two distinct real solutions.

27 The equation $x^2 - 4kx + 4 = 0$ has two distinct real solutions. Find the set of all possible values of k.

28 Find all possible values of m so that the graph of the function $g: x \mapsto mx^2 + 6x + m$ does not touch the x-axis.

29 Find the range of values of k such that $3x^2 - 12x + k > 0$ for all real values of x. (Hint: Consider what must be true about the zeros of the quadratic equation $y = 3x^2 - 12x + k$.)

30 Prove that the expression $x - 2 - x^2$ is negative for all real values of x.

In questions 31 and 32, find a quadratic function in the form $y = ax^2 + bx + c$ that satisfies the given conditions.

31 The function has zeros of $x = -1$ and $x = 4$ and its graph intersects the y-axis at $(0, 8)$.

32 The function has zeros of $x = \frac{1}{2}$ and $x = 3$ and its graph passes through the point $(-1, 4)$.

33 Find the range of values for k in order for the equation $2x^2 + (3 - k)x + k + 3 = 0$ to have two imaginary solutions.

34 For what values of m does the function $f(x) = 5x^2 - mx + 2$ have two distinct real zeros?

35 The graph of a quadratic function passes through the points $(-3, 10)$, $(\frac{1}{4}, -\frac{9}{16})$ and $(1, 6)$. Express the function in the form $f(x) = ax^2 + bx + c$, where $a, b, c \in \mathbb{R}$.

36 The maximum value of the function $f(x) = ax^2 + bx + c$ is 10.
Given that $f(3) = f(-1) = 2$, find $f(2)$.

37 Find the values of x for which $4x + 1 < x^2 + 4$.

38 Show that there is no real value t for which the equation $2x^2 + (2 - t)x + t^2 + 3 = 0$ has real roots.

39 Show that the two roots of $ax^2 - a^2x - x + a = 0$ are reciprocals of each other.

40 Find the sum and product of the roots for each of the following quadratic equations.

a) $2x^2 + 6x - 5 = 0$ b) $x^2 = 1 - 3x$ c) $4x^2 - 6 = 0$

d) $x^2 + ax - 2a = 0$ e) $m(m - 2) = 4(m + 1)$ f) $3x - \frac{2}{x} = 1$

41 The roots of the equation $2x^2 - 3x + 6 = 0$ are α and β. Find a quadratic equation with integral coefficients whose roots are $\frac{\alpha}{\beta}$ and $\frac{\beta}{\alpha}$.

42 If α and β are the roots of the equation $3x^2 + 5x + 4 = 0$, find the values of the following expressions.

a) $\alpha^2 + \beta^2$ b) $\frac{\alpha}{\beta} + \frac{\beta}{\alpha}$

c) $\alpha^3 + \beta^3$ [Hint: factorise $\alpha^3 + \beta^3$ into a product of a binomial and a trinomial.]

43 Consider the quadratic equation $x^2 + 8x + k = 0$ where k is a constant.

a) Find both roots of the equation given that one root of the equation is three times the other.

b) Find the value of k.

44 The roots of the equation $x^2 + x + 4 = 0$ are α and β.

a) Without solving the equation, find the value of the expression $\frac{1}{\alpha} + \frac{1}{\beta}$.

b) Find a quadratic equation whose roots are $\frac{1}{\alpha}$ and $\frac{1}{\beta}$.

45 If α and β are roots of the quadratic equation $5x^2 - 3x - 1 = 0$, find a quadratic equation with integral coefficients which have the roots:

a) $\frac{1}{\alpha^2}$ and $\frac{1}{\beta^2}$ b) $\frac{\alpha^2}{\beta}$ and $\frac{\beta^2}{\alpha}$

3.3 Zeros, factors and remainders

Finding the zeros of polynomial functions is a feature of many problems in algebra, calculus and other areas of mathematics. In our analysis of quadratic functions in the previous section, we saw the connection between the graphical and algebraic approaches to finding zeros. Information obtained from the graph of a function can be used to help find its zeros and, conversely, information about the zeros of a polynomial function can be used to help sketch its graph. Results and observations from the last section lead us to make some statements about real zeros of all polynomial functions. Later in this section we will extend our consideration to imaginary zeros. The following box summarizes what we have observed thus far about the zeros of polynomial functions.

Real zeros of polynomial functions

If P is a polynomial function and c is a real number, then the following statements are equivalent.

- $x = c$ is a zero of the function P.
- $x = c$ is a solution (or root) of the polynomial equation $P(x) = 0$.
- $x - c$ is a linear factor of the polynomial P.
- $(c, 0)$ is an x-intercept of the graph of the function P.

Polynomial division

As with integers, finding the factors of polynomials is closely related to dividing polynomials. An integer n is **divisible** by another integer m if m is a factor of n. If n is not divisible by m we can use the process of **long division** to find the quotient of the numbers and the remainder. For example, let's use long division to divide 485 by 34.

$$\begin{array}{r} 14 \\ 34\overline{)485} \\ \underline{34} \\ 145 \\ \underline{136} \\ 9 \end{array} \qquad \text{check:} \quad \begin{array}{rl} 14 & \text{quotient} \\ \underline{\times\ 34} & \text{divisor} \\ 56 & \\ \underline{420} & \\ 476 & \\ \underline{+\ \ 9} & \text{remainder} \\ 485 & \text{dividend} \end{array}$$

The number 485 is the **dividend**, 34 is the **divisor**, 14 is the **quotient** and 9 is the **remainder**. The long division process (or algorithm) stops when a remainder is less than the divisor. The procedure shown above for checking the division result may be expressed as

$$485 = 34 \times 14 + 9$$

or in words as

$$\text{dividend} = \text{divisor} \times \text{quotient} + \text{remainder}$$

The process of division for polynomials is similar to that for integers. If a polynomial $D(x)$ is a factor of polynomial $P(x)$, then $P(x)$ is divisible by $D(x)$. However, if $D(x)$ is not a factor of $P(x)$ then we can use a **long division algorithm for polynomials** to find a quotient polynomial $Q(x)$ and a remainder polynomial $R(x)$ such that $P(x) = D(x) \cdot Q(x) + R(x)$. In the same way that the remainder must be less than the divisor when dividing integers, the remainder must be a polynomial of a lower degree than the divisor when dividing polynomials. Consequently, when the divisor is a linear polynomial (degree of 1) the remainder must be of degree 0, i.e. a constant.

• **Hint:** A common error when performing long division with polynomials is to add rather than subtract during each cycle of the process.

Example 14

Find the quotient $Q(x)$ and remainder $R(x)$ when $P(x) = 2x^3 - 5x^2 + 6x - 3$ is divided by $D(x) = x - 2$.

Solution

$$\begin{array}{r|l} & 2x^2 - x + 4 \\ x-2 & 2x^3 - 5x^2 + 6x - 3 \\ & 2x^3 - 4x^2 \quad \leftarrow 2x^2(x-2) \\ & -x^2 + 6x \quad \leftarrow \text{Subtract} \\ & -x^2 + 2x \quad \leftarrow -x(x-2) \\ & 4x - 3 \quad \leftarrow \text{Subtract} \\ & 4x - 8 \quad \leftarrow 4(x-2) \\ & 5 \quad \leftarrow \text{Subtract} \end{array}$$

Thus, the quotient $Q(x)$ is $2x^2 - x + 4$ and the remainder is 5. Therefore, we can write

$$2x^3 - 5x^2 + 6x - 3 = (x - 2)(2x^2 - x + 4) + 5$$

This equation provides a means to check the result by expanding and simplifying the right side and verifying it is equal to the left side.

$$\begin{aligned} 2x^3 - 5x^2 + 6x - 3 &= (x - 2)(2x^2 - x + 4) + 5 \\ &= (2x^3 - x^2 + 4x - 4x^2 + 2x - 8) + 5 \\ &= 2x^3 - 5x^2 + 6x - 3 \end{aligned}$$

Taking the identity $P(x) = D(x) \cdot Q(x) + R(x)$ and dividing both sides by $D(x)$ produces the equivalent identity $\frac{P(x)}{D(x)} = Q(x) + \frac{R(x)}{D(x)}$.

Hence, the result for Example 14 could also be written as

$\frac{2x^3 - 5x^2 + 6x - 3}{x - 2} = 2x^2 - x + 4 + \frac{5}{x - 2}$.

Note that writing the result in this manner is the same as rewriting $17 = 5 \times 3 + 2$ as $\frac{17}{5} = 3 + \frac{2}{5}$, which we commonly write as the 'mixed number' $3\frac{2}{5}$.

• **Hint:** When performing long division with polynomials it is necessary to write all polynomials so that the powers (exponents) of the terms are in descending order. Example 12 illustrates that if there are any 'missing' terms then they have a coefficient of zero and a zero must be included in the appropriate location in the division scheme.

Example 15

Divide $f(x) = 4x^3 - 31x - 15$ by $2x + 5$, and use the result to factor $f(x)$ completely.

Solution

$$\begin{array}{r|l} & 2x^2 - 5x - 3 \\ 2x+5 & 4x^3 + 0x^2 - 31x - 15 \\ & 4x^3 + 10x^2 \\ & -10x^2 - 31x \\ & -10x^2 - 25x \\ & -6x - 15 \\ & -6x - 15 \\ & 0 \end{array}$$

Thus $f(x) = 4x^3 - 31x - 15 = (2x + 5)(2x^2 - 5x - 3)$

... and factorizing the quadratic quotient (also a factor of $f(x)$), gives

$$f(x) = 4x^3 - 31x - 15 = (2x + 5)(2x^2 - 5x - 3)$$
$$= (2x + 5)(2x + 1)(x - 3)$$

This factorization would lead us to believe that the three zeros of $f(x)$ are $x = -\frac{5}{2}$, $x = -\frac{1}{2}$ and $x = 3$. Graphing $f(x)$ on our GDC and using the 'trace' feature confirms that all three values are zeros of the cubic polynomial.

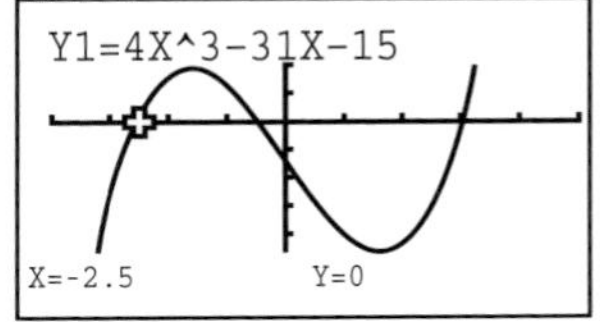

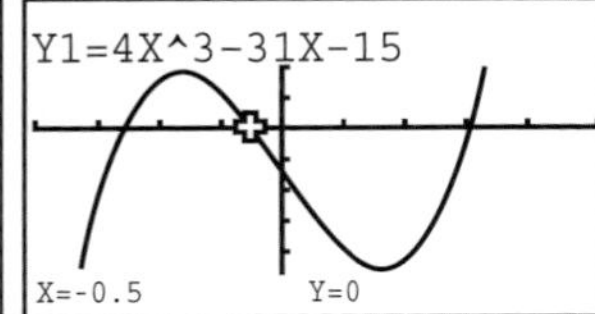

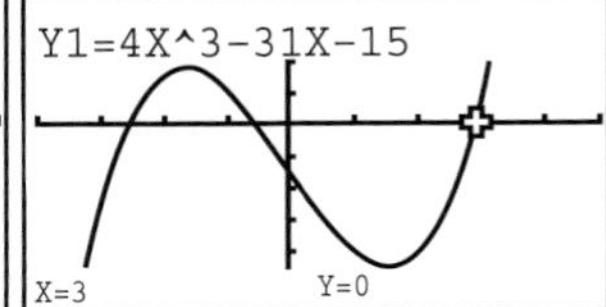

Division algorithm for polynomials

If $P(x)$ and $D(x)$ are polynomials such that $D(x) \neq 0$, and the degree of $D(x)$ is less than or equal to the degree of $P(x)$, then there exist unique polynomials $Q(x)$ and $R(x)$ such that

$$P(x) = D(x) \cdot Q(x) + R(x)$$

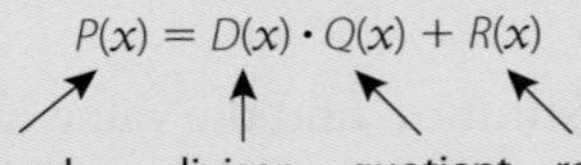

dividend divisor quotient remainder

and where $R(x)$ is either zero or of degree less than the degree of $D(x)$.

Remainder and factor theorems

As illustrated by Examples 14 and 15, we commonly divide polynomials of higher degree by linear polynomials. By doing so we can often uncover zeros of polynomials as occurred in Example 15. Let's look at what happens to the division algorithm when the divisor $D(x)$ is a linear polynomial of the form $x - c$. Since the degree of the remainder $R(x)$ must be less than the degree of the divisor (degree of one in this case) then the remainder will be a constant, simply written as R. Then the division algorithm for a linear divisor is the identity:

$$P(x) = (x - c) \cdot Q(x) + R$$

If we evaluate the polynomial function P at the number $x = c$, we obtain

$$P(c) = (c - c) \cdot Q(c) + R = 0 \cdot Q(c) + R = R$$

Thus the remainder R is equal to $P(c)$, the value of the polynomial P at $x = c$. Because this is true for any polynomial P and any linear divisor $x - c$, we have the following theorem.

The remainder theorem

If a polynomial function $P(x)$ is divided by $x - c$, then the remainder is the value $P(c)$.

Example 16

What is the remainder when $g(x) = 2x^3 + 5x^2 - 8x + 3$ is divided by $x + 4$?

Solution

The linear polynomial $x + 4$ is equivalent to $x - (-4)$. Applying the remainder theorem, the required remainder is equal to the value of $g(-4)$.

$$g(-4) = 2(-4)^3 + 5(-4)^2 - 8(-4) + 3 = 2(-64) + 5(16) + 32 + 3$$
$$= -128 + 80 + 35 = -13$$

Therefore, when the polynomial function $g(x)$ is divided by $x + 4$ the remainder is -13.

Figure 3.11 Connection between synthetic substitution and long division.

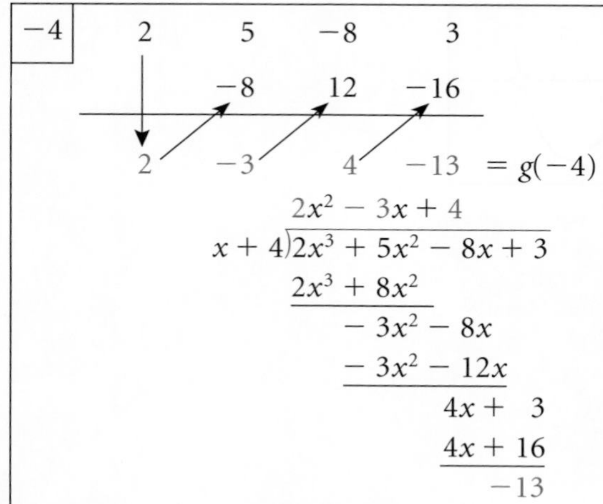
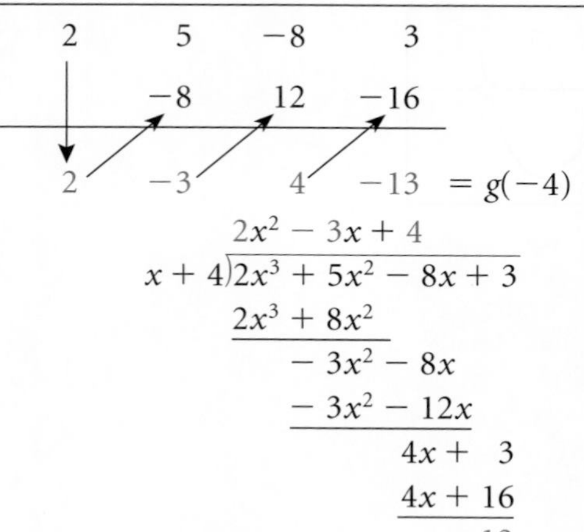

We found the value of $g(-4)$ in Example 16 by directly substituting -4 into $g(x)$. Alternatively, we could have used the efficient method of synthetic substitution that we developed in Section 3.1 to evaluate $g(-4)$.

We could also have found the remainder by performing long division, which is certainly the least efficient method. However, there is a very interesting and helpful connection between the process of long division with a linear divisor and synthetic substitution.

Not only does synthetic substitution find the value of the remainder, but the numbers in the bottom row preceding the remainder (shown in red in Figure 3.11) are the same as the coefficients of the quotient (also in red) found from the long division process. Clearly, synthetic substitution is the most efficient method for finding the remainder *and* quotient when dividing a polynomial by a linear polynomial in the form $x - c$. When this method is used to find a quotient and remainder we refer to it as **synthetic division.**

The numbers in the last row of the synthetic substitution process give both the remainder and the coefficients of the quotient when a polynomial is divided by a linear polynomial in the form $x - c$.

A consequence of the remainder theorem is the factor theorem, which also follows intuitively from our discussion in the previous section about the zeros and factors of quadratic functions. It formalizes the relationship between zeros and linear factors of all polynomial functions with real coefficients.

It is important to understand that the factor theorem is a **biconditional** statement of the form 'A if and only if B'. Such a statement is true in either 'direction'; that is, 'If A then B', and also 'If B then A' – usually abbreviated A → B and B → A, respectively.

The factor theorem

A polynomial function $P(x)$ has a factor $x - c$ if and only if $P(c) = 0$.

To illustrate the efficiency of synthetic division, let's answer the same problem posed in Example 14 (solution reproduced in Figure 3.12) in Example 17.

Example 17

Find the quotient $Q(x)$ and remainder $R(x)$ when $P(x) = 2x^3 - 5x^2 + 6x - 3$ is divided by $D(x) = x - 2$.

Solution

Using synthetic division

$$\begin{array}{r|rrrr} 2 & 2 & -5 & 6 & -3 \\ & & 4 & -2 & 8 \\ \hline & 2 & -1 & 4 & 5 \end{array} \leftarrow \text{remainder}$$

coefficients of the quotient ($2, -1, 4$)

$$\begin{array}{rll} & 2x^2 - x + 4 & \\ x-2 \overline{)} & 2x^3 - 5x^2 + 6x - 3 & \\ & 2x^3 - 4x^2 & \leftarrow 2x^2(x-2) \\ & -x^2 + 6x & \leftarrow \text{Subtract} \\ & -x^2 + 2x & \leftarrow -x(x-2) \\ & 4x - 3 & \leftarrow \text{Subtract} \\ & 4x - 8 & \leftarrow 4(x-2) \\ & 5 & \leftarrow \text{Subtract} \end{array}$$

The quotient $Q(x)$ is $2x^2 - x + 4$ and the remainder is 5.

Figure 3.12 Solution for Example 14.

Since a divisor of degree 1 is dividing a polynomial of degree 3 then the quotient must be of degree 2 and, with all polynomials written so that their terms are descending in powers (exponents), we know that the numbers in the bottom row of the synthetic division scheme are the coefficients of a quadratic polynomial. Hence, the quotient is $2x^2 - x + 4$ and the remainder is 5.

When one or more zeros of a given polynomial are known, applying the factor theorem and synthetic division is a very effective strategy to aid in finding factors and zeros of the polynomial.

Example 18

Given that $x = -\frac{1}{2}$ and $x = 8$ are zeros of the polynomial function $h(x) = x^4 - \frac{15}{2}x^3 - 30x - 16$, find the other two zeros of $h(x)$.

Solution

From the factor theorem, it follows that $x + \frac{1}{2}$ and $x - 8$ are factors of $h(x)$. Dividing the 4th degree polynomial by the two linear factors in succession will yield a quadratic factor. We can find the zeros of this quadratic factor by using known factorizing techniques or by applying the quadratic formula.

$$\begin{array}{r|rrrrr} -\frac{1}{2} & 1 & -\frac{15}{2} & 0 & -30 & -16 \\ & & -\frac{1}{2} & 4 & -2 & 16 \\ \hline 8 & 1 & -8 & 4 & -32 & 0 \\ & & 8 & 0 & 32 & \\ \hline & 1 & 0 & 4 & 0 & \end{array}$$

This row shows that $x^4 - \frac{15}{2}x^3 - 30x - 16 = (x + \frac{1}{2})(x^3 - 8x^2 + 4x - 32)$.

This row shows that $x^3 - 8x^2 + 4x - 32 = (x - 8)(x^2 + 4)$.

● Hint: Example 18 indicates that if we divide the quartic polynomial $x^4 - \frac{15}{2}x^3 - 30x - 16$ by $x^2 + 4$ the remainder will be zero, since $x^2 + 4$ is a factor. Synthetic division *only* works for linear divisors of the form $x - c$ so this division could only be done by using the long division process.

Hence, $x^4 - \frac{15}{2}x^3 - 30x - 16 = (x + \frac{1}{2})(x - 8)(x^2 + 4)$.

The zeros of the quadratic factor $x^2 + 4$ must also be zeros of $h(x)$.

$$x^2 + 4 = 0 \Rightarrow x^2 = -4 \Rightarrow x = \pm\sqrt{-4} \Rightarrow x = \pm\sqrt{4}\,\sqrt{-1} \Rightarrow x = \pm 2i$$

Therefore, the other two remaining zeros of $h(x)$ are $x = 2i$ and $x = -2i$.

Note that the two imaginary zeros, $x = 2i$ and $x = -2i$, of the polynomial in Example 18 are a pair of conjugates. In the previous section we asserted that imaginary zeros of a quadratic polynomial always come in conjugate pairs. Although it is beyond the scope of this book to prove it, we will accept that this is true for imaginary zeros of any polynomial.

Conjugate zeros

If a polynomial P has real coefficients, and if the complex number $z = a + bi$ is a zero of P, then its conjugate $z^* = a - bi$ is also a zero of P.

Example 19

Given that $2 - 3i$ is a zero of the polynomial $5x^3 - 19x^2 + 61x + 13$, find all remaining zeros of the polynomial.

Solution

Firstly, we need to consider what is the maximum number of zeros that the cubic polynomial can have. In the previous section we stated that every quadratic polynomial has exactly two complex zeros. It is reasonable to conjecture that a cubic will have three complex zeros. Since $2 - 3i$ is a zero, then $2 + 3i$ must also be a zero; and the third zero must be a real number. Although not explicitly stated in the remainder and factor theorems, both theorems are true for linear polynomials $x - c$ where the number c is real *or* imaginary, i.e. it can be any complex number. Therefore, the cubic polynomial has factors $x - (2 - 3i)$ and $x - (2 + 3i)$. Rather than attempting to divide the cubic polynomial by one of these factors, let's find the product of these factors and use it as a divisor.

$$\begin{aligned}
[x - (2 - 3i)][x - (2 + 3i)] &= [x - 2 + 3i][x - 2 - 3i] \\
&= [(x - 2) + 3i][(x - 2) - 3i] \\
&= (x - 2)^2 - (3i)^2 \\
&= x^2 - 4x + 4 - 9i^2 \\
&= x^2 - 4x + 4 + 9 \\
&= x^2 - 4x + 13
\end{aligned}$$

We can only use synthetic division with linear divisors, so we will need to divide $5x^3 - 19x^2 + 61x + 13$ by $x^2 - 4x + 13$ using long division.

$$\begin{array}{r}
5x + 1 \phantom{{}+ 61x + 13} \\
x^2 - 4x + 13 \overline{)\,5x^3 - 19x^2 + 61x + 13} \\
\underline{5x^3 - 20x^2 + 65x} \phantom{{}+ 13} \\
x^2 - 4x + 13 \\
\underline{x^2 - 4x + 13} \\
0
\end{array}$$

Thus, $5x^3 - 19x^2 + 61x + 13$ also has a linear factor of $5x + 1$ and therefore has a zero of $x = -\frac{1}{5}$.

The zeros of the cubic polynomial are:
$x = 2 - 3i$, $x = 2 + 3i$ and $x = -\frac{1}{5}$.

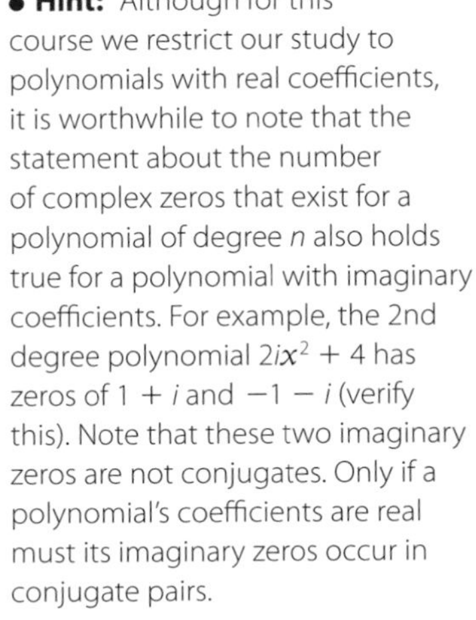

• **Hint:** Although for this course we restrict our study to polynomials with real coefficients, it is worthwhile to note that the statement about the number of complex zeros that exist for a polynomial of degree n also holds true for a polynomial with imaginary coefficients. For example, the 2nd degree polynomial $2ix^2 + 4$ has zeros of $1 + i$ and $-1 - i$ (verify this). Note that these two imaginary zeros are not conjugates. Only if a polynomial's coefficients are real must its imaginary zeros occur in conjugate pairs.

The cubic polynomial in Example 19 had three complex zeros – one real and two imaginary. The quartic polynomial in Example 18 had four complex zeros – two real and two imaginary. In Example 15, we factored a cubic polynomial into a product of three linear polynomials, so the factor theorem says it will have three real zeros. And in the previous section we concluded that, provided we take into account the multiplicity of a zero (e.g. double root), all quadratic polynomials have two complex zeros – either two real zeros or two imaginary zeros. These examples are illustrations of the following useful fact.

Zeros of polynomials of degree *n*
A polynomial of degree $n > 0$ with complex coefficients has exactly n complex zeros, provided that each zero is counted as many times as its multiplicity.

Since imaginary zeros always exist in conjugate pairs then if a polynomial with real coefficients has any imaginary zeros there can only be an even number of them. It logically follows then that a polynomial with an odd degree has at least one real zero. One consequence of this fact is that the graph of an odd-degree polynomial function must intersect the x-axis at least once. This agrees with our claim in Section 3.1 that the end behaviour of a polynomial function is influenced by its degree. Odd-degree polynomial functions will rise as $x \to \infty$ and fall as $x \to -\infty$ (or the other way around if the leading coefficient is negative) producing the same general ∿ shape as $y = x^3$, and hence will cross the x-axis at least once.

Example 20

Given that $2x + 1$ is a factor of the cubic function $f(x) = 2x^3 - 15x^2 + 24x + 16$

a) completely factorize the polynomial

b) find all of the zeros and their multiplicities

c) sketch its graph for the interval $-1 \leq x \leq 6$, given that the graph of the function has a turning point at $x = 1$

Solution

a) Remember that synthetic division can only be used for linear divisors of the form $x - c$. Because $2x + 1 = 2\left(x + \frac{1}{2}\right)$, then if $2x + 1$ is a factor $x + \frac{1}{2}$ is also a factor. So we can set up synthetic division with a divisor of $x + \frac{1}{2}$, but we must take the following into account.

$$\begin{aligned} 2x^3 - 15x^2 + 24x + 16 &= (2x + 1)\cdot Q(x) \\ &= 2(x + \tfrac{1}{2})\cdot Q(x) \\ &= \left(x + \tfrac{1}{2}\right)\cdot 2Q(x) \end{aligned}$$

$$\frac{2x^3 - 15x^2 + 24x + 16}{x + \frac{1}{2}} = 2Q(x)$$

When the polynomial is divided by $x + \frac{1}{2}$, the quotient will be two times the quotient from dividing by $2x + 1$. Dividing by two will give us the quotient that we want.

$$\begin{array}{r|rrrr} -\frac{1}{2} & 2 & -15 & 24 & 16 \\ & & -1 & 8 & -16 \\ \hline & 2 & -16 & 32 & 0 \end{array}$$

Hence, $2x^3 - 15x^2 + 24x + 16 = \left(x + \frac{1}{2}\right)(2x^2 - 16x + 32)$

and $2x^3 - 15x^2 + 24x + 16 = 2\left(x + \frac{1}{2}\right)\frac{1}{2}(2x^2 - 16x + 32)$

$= (2x + 1)(x^2 - 8x + 16)$ Factorize the quadratic factor.

$= (2x + 1)(x - 4)(x - 4)$ $x^2 - 8x + 16$ fits the pattern $x^2 + 2ax + a^2 = (x + a)^2$

$= (2x + 1)(x - 4)^2$

b) The zeros of $2x^3 - 15x^2 + 24x + 16$ are $x - \frac{1}{2}$ and $x = 4$ (multiplicity of two).

c) Because the polynomial is of degree 3 and its leading coefficient is positive, the end behaviour of the graph will be such that the graph rises as $x \to \infty$ and falls as $x \to -\infty$. That means the general shape of the graph will be a ∿ shape with one maximum and one minimum as shown right.

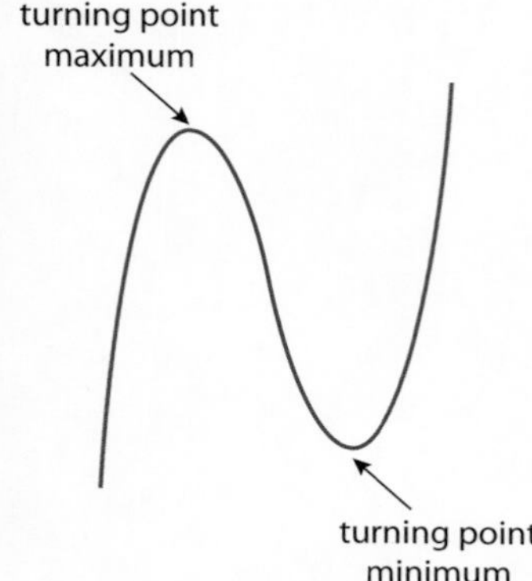

Find the coordinates of the given turning point by evaluating $f(1)$ using synthetic substitution.

$$\begin{array}{r|rrrr} 1 & 2 & -15 & 24 & 16 \\ & & 2 & -13 & 11 \\ \hline & 2 & -13 & 11 & 27 \end{array}$$

$\Rightarrow f(1) = 27$. Hence, the point $(1, 27)$ is on the graph.

Since $f(0) = 16$ then the y-intercept is $(0, 16)$, which means that $(1, 27)$ is a maximum point. Because the zero $x = 4$ has a multiplicity of two, then we know from the previous chapter on quadratic functions that the graph will be tangent to the x-axis at the point $(4, 0)$. The other x-intercept is $\left(-\frac{1}{2}, 0\right)$. We can now make a very accurate sketch of the function.

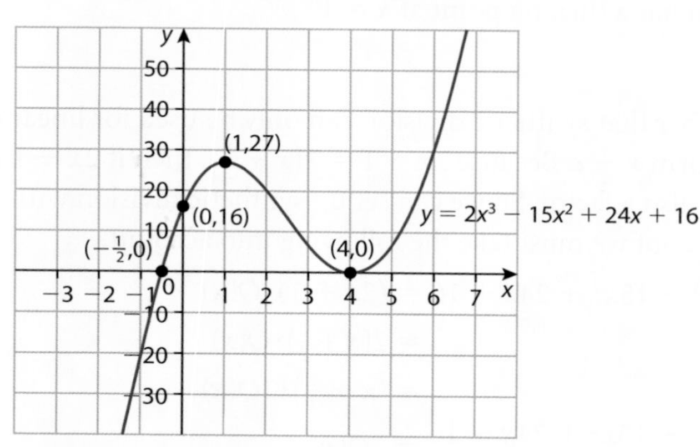

We know how to find the exact zeros of linear and quadratic functions. The quadratic formula is a general rule that gives the *exact* values of *all* complex zeros of *any* quadratic polynomial using radicals and the coefficients of the polynomial. We also know how to use our GDC to approximate real zeros. In this chapter, we have gained techniques to search for, or verify, the zeros of polynomial functions of degree 3 or higher. This leads us to an important question: Can we find exact values of all complex zeros of any polynomial function of 3rd degree and higher? This question was answered for cubic and quartic polynomials in the 16th century when the Italian mathematician Girolamo Cardano (1501–1576) presented a 'cubic formula' and a 'quartic formula'. These formulae were methods for finding all complex zeros of 3rd degree and 4th degree polynomials using only radicals and coefficients. Cardano's presentation of the formulae depended heavily on the work of other Italian mathematicians. Scipione del Ferro (1465–1526) is given credit as the first to find a general algebraic solution to cubic equations. Cardano's method of solving any cubic was obtained from Niccolo Fontana (1500–1557) known as 'Tartaglia'. Similarly, Cardano solved quartic equations using a method that he learned from his own student Lodovico Ferrari (1522–1565). The methods for solving cubic and quartic equations are quite complicated and are not part of this course. The question of finding formulae for exact zeros of polynomials of degree 5 (quintic) and higher was not resolved until the early 19th century. In 1824, a young Norwegian mathematician, Niels Henrik Abel (1802–1829), proved that it was impossible to find an algebraic formula for a general quintic equation. An even more remarkable discovery was made by the French mathematician Evariste Galois (1811–1832) who died in a pistol duel before turning 21. Galois proved that for any polynomial of degree 5 or greater, it is not possible, except in special cases, to find the exact zeros by using only radicals and the polynomial's coefficients. Mathematicians have developed sophisticated methods of approximating the zeros of polynomial equations of high degree and other types of equations for which there are no algebraic solution methods. These are studied in a branch of advanced mathematics called **numerical analysis**.

There is a theorem called the **fundamental theorem of algebra** that guarantees that *every* polynomial function of non-zero degree with complex coefficients has at least one complex zero. The theorem was first proved by the famous German mathematician Carl Friedrich Gauss (1777–1855). Many of the results in this section on the zeros of polynomials are directly connected with this important theorem.

Example 21

Find a polynomial P with integer coefficients of least degree having zeros of $x = 2$, $x = -\frac{1}{3}$ and $x = 1 - i$.

Solution

Given that $1 - i$ is a zero then its conjugate $1 + i$ must also be a zero. Thus, the required polynomial has four complex zeros, and four corresponding factors. The four factors are:

$x - 2,\ x + \frac{1}{3},\ x - (1 - i)$ and $x - (1 + i)$

$$\begin{aligned} P(x) &= (x-2)\left(x+\tfrac{1}{3}\right)[x-(1-i)][x-(1+i)] \\ &= \left(x^2-\tfrac{5}{3}x-\tfrac{2}{3}\right)[(x-1)+i][(x-1)-i] \\ &= (3x^2-5x-2)[(x-1)^2-i^2] \\ &= (3x^2-5x-2)(x^2-2x+1+1) \\ &= (3x^2-5x-2)(x^2-2x+2) \\ &= 3x^4-6x^3+6x^2-5x^3+10x^2-10x-2x^2+4x-4 \\ P(x) &= 3x^4-11x^3+14x^2-6x-4 \end{aligned}$$

Multiplying by 3 does not change the zeros ...

... but does guarantee integer coefficients.

Sum and product of the roots of any polynomial equation

In the previous section, we found a way to express the sum and product of the roots of a quadratic equation, $ax^2 + bx + c = 0$, in terms of a, b and c. It is natural to wonder whether a similar method could be found for polynomial equations of degree greater than two.

Using the same approach as in the previous section for quadratic equations, let's consider the general cubic equation $ax^3 + bx^2 + cx + d = 0$ whose roots are $x = \alpha$, $x = \beta$ and $x = \gamma$. It follows that this general cubic equation can be written in the form $x^3 + \frac{b}{a}x^2 + \frac{c}{a}x + \frac{d}{a} = 0$. Applying the Factor Theorem, it can also be written in the form $(x - \alpha)(x - \beta)(x - \gamma) = 0$. Expanding the brackets gives:

$$(x - \alpha)(x - \beta)(x - \gamma) = x^3 - \alpha x^2 - \beta x^2 - \gamma x^2 + \alpha\beta x + \beta\gamma x + \alpha\gamma x - \alpha\beta\gamma = 0$$

$$x^3 - (\alpha + \beta + \gamma)\,x^2 + (\alpha\beta + \beta\gamma + \alpha\gamma)x - \alpha\beta\gamma = 0$$

Equating coefficients for $x^3 + \frac{b}{a}x^2 + \frac{c}{a}x + \frac{d}{a} = 0$ and $x^3 - (\alpha + \beta + \gamma)x^2 + (\alpha\beta + \beta\gamma + \alpha\gamma)x - \alpha\beta\gamma = 0$ gives us the following results for the sum and product of the roots for any cubic equation.

$$\alpha + \beta + \gamma = -\frac{b}{a} \text{ and } \alpha\beta\gamma = -\frac{d}{a}$$

This result for the sum and product of the roots of any cubic equation looks very similar to that for any quadratic equation. The only difference is that the product of the roots, $\alpha\beta\gamma$, is the opposite of the quotient $\frac{\text{constant term}}{\text{leading coefficient}}$.

For the general quartic equation $ax^4 + bx^3 + cx^2 + dx + e = 0$ with roots α, β, γ and δ, the factored form of the equation expands as follows:

$$(x - \alpha)(x - \beta)(x - \gamma)(x - \delta) =$$
$$x^4 - (\alpha + \beta + \gamma + \delta)x^2 + (\alpha\beta + \alpha\gamma + \alpha\delta + \beta\gamma + \beta\delta + \gamma\delta)x -$$
$$(\alpha\beta\gamma + \alpha\beta\delta + \alpha\gamma\delta + \beta\gamma\delta) + \alpha\beta\gamma\delta = 0$$

Since this is equivalent to $x^4 + \frac{b}{a}x^3 + \frac{c}{a}x^2 + \frac{d}{a}x + \frac{e}{a} = 0$, then the sum and product of the roots for any quartic equation are:

$$\alpha + \beta + \gamma + \delta = -\frac{b}{a} \text{ and } \alpha\beta\gamma\delta = \frac{e}{a}.$$

These results for the sum and product of roots for polynomial equations of degree 2 (quadratic), degree 3 (cubic) and degree 4 (quartic) lead to the following result for any polynomial function of degree n that we state without a formal proof.

Sum and product of the roots (zeros) of any polynomial equation

For the **polynomial equation of degree *n*** given by $P(x) = a_n x^n + a_{n-1}x^{n-1} + \ldots + a_1 x + a_0 = 0$, $a_n \neq 0$ the **sum of the roots** is $-\frac{a_{n-1}}{a_n}$ and the **product of the roots** is $\frac{(-1)^n a_0}{a_n}$.

Example 22

Two of the roots of the equation $x^3 - 3x^2 + kx + 75 = 0$ are opposites. Find the values of all the roots and the constant k.

Solution

Let the three unknown roots be represented by α, $-\alpha$ and β.

Then $\alpha - \alpha + \beta = 3 \Rightarrow \beta = 3$ and $\alpha(-\alpha)\beta = -75 \Rightarrow \alpha(-\alpha)(3) = -75 \Rightarrow -3\alpha^2 = -75 \Rightarrow \alpha^2 = 25 \Rightarrow \alpha = \pm 5$

Therefore, the three roots are 5, -5 and 3.

To find the value of k, write the cubic in factored form and expand.

$$(x-3)(x+5)(x-5) = 0 \Rightarrow (x-3)(x^2-25) = 0$$
$$\Rightarrow x^3 - 3x^2 - 25x + 75 = 0$$

Therefore, $k = -25$.

Example 23

Consider the equation $2x^4 - x^3 - 4x^2 + 10x - 4 = 0$. Given that one of the zeros of the equation is $r_1 = 1 + i$, find the other three zeros r_2, r_3 and r_4.

Solution

There are other strategies (e.g. using factors and polynomial division) but it is more efficient to apply what we know about the sum and product of the roots (zeros) of a polynomial equation.

Firstly, since $r_1 = 1 + i$ is a zero, then its conjugate must also be a zero; hence $r_2 = 1 - i$.

From the fact that the sum of the roots is $-\frac{a_{n-1}}{a_n}$, then $r_1 + r_2 + r_3 + r_4 = -\frac{a_3}{a_4}$.

Substituting in known values gives $1 + i + 1 - i + r_3 + r_4 = -\frac{-1}{2}$
$\Rightarrow 2 + r_3 + r_4 = \frac{1}{2} \Rightarrow r_3 + r_4 = -\frac{3}{2}$

Also, since the product of the roots is $\frac{(-1)^n a_0}{a_n}$, then $r_1r_2r_3r_4 = \frac{(-1)^n a_0}{a_n}$.

Substituting gives:

$$(1+i)(1-i)r_3r_4 = \frac{(-1)^4(-4)}{2} \Rightarrow (1 - i^2)r_3r_4 = -2$$
$$\Rightarrow 2r_3r_4 = -2$$
$$\Rightarrow r_3r_4 = -1$$

To find r_3 and r_3, we need to use the pair of equations $\begin{cases} r_3 + r_4 = -\frac{3}{2} \\ r_3r_4 = -1 \end{cases}$

Solving for r_3 in the first equation gives $r_3 = -r_4 - \frac{3}{2}$.

Substituting into the other equation gives: $\left(-r_4 - \frac{3}{2}\right) r_4 = -1$

$$\Rightarrow r_4^2 + \frac{3}{2}r_4 - 1 = 0$$
$$\Rightarrow 2r_4^2 + 3r_4 - 2 = 0$$
$$\Rightarrow (2r_4 - 1)(r_4 + 2) = 0$$
$$\Rightarrow r_4 = \frac{1}{2} \text{ or } r_4 = -2$$

If $r_4 = \frac{1}{2}$, then $r_3 = -\frac{1}{2} - \frac{3}{2} = -2$. $\left[\text{And if } r_4 = -2, \text{ then } r_3 = \frac{1}{2}\right]$

Therefore the other three zeros are $1 - i, \frac{1}{2}$ and -2.

Exercise 3.3

In questions 1–5, two polynomials P and D are given. Use either synthetic division or long division to divide $P(x)$ by $D(x)$, and express $P(x)$ in the form $P(x) = D(x) \cdot Q(x) + R(x)$.

1 $P(x) = 3x^2 + 5x - 5,\ D(x) = x + 3$

2 $P(x) = 3x^4 - 8x^3 + 9x + 5,\ D(x) = x - 2$

3 $P(x) = x^3 - 5x^2 + 3x - 7,\ D(x) = x - 4$

4 $P(x) = 9x^3 + 12x^2 - 5x + 1,\ D(x) = 3x - 1$

5 $P(x) = x^5 + x^4 - 8x^3 + x + 2,\ D(x) = x^2 + x - 7$

6 Given that $x - 1$ is a factor of the function $f(x) = 2x^3 - 17x^2 + 22x - 7$ factorize f completely.

7 Given that $2x + 1$ is a factor of the function $f(x) = 6x^3 - 5x^2 - 12x - 4$ factorize f completely.

8 Given that $x + \frac{2}{3}$ is a factor of the function $f(x) = 3x^4 + 2x^3 - 36x^2 + 24x + 32$ factorize f completely.

In questions 9–12, find the quotient and the remainder.

9 $\dfrac{x^2 - 5x + 4}{x - 3}$

10 $\dfrac{x^3 + 2x^2 + 2x + 1}{x + 2}$

11 $\dfrac{9x^2 - x + 5}{3x^2 - 7x}$

12 $\dfrac{x^5 + 3x^3 - 6}{x - 1}$

In questions 13–16, use synthetic division and the remainder theorem to evaluate $P(c)$.

13 $P(x) = 2x^3 - 3x^2 + 4x - 7, c = 2$

14 $P(x) = x^5 - 2x^4 + 3x^2 + 20x + 3, c = -1$

15 $P(x) = 5x^4 + 30x^3 - 40x^2 + 36x + 14, c = -7$

16 $P(x) = x^3 - x + 1, c = \frac{1}{4}$

17 Given that $x = -6$ is a zero of the polynomial $x^3 + 2x^2 - 19x + 30$ find all remaining zeros of the polynomial.

18 Given that $x = 2$ is a double root of the polynomial $x^4 - 5x^3 + 7x^2 - 4$ find all remaining zeros of the polynomial.

19 Find the values of k such that -3 is a zero of $f(x) = x^3 - x^2 - k^2x$.

20 Find the values of a and b such that 1 and 4 are zeros of $f(x) = 2x^4 - 5x^3 - 14x^2 + ax + b$.

In questions 21–23, find a polynomial with real coefficients satisfying the given conditions.

21 Degree of 3; and zeros of -2, 1 and 4

22 Degree of 4; and zeros of -1, 3 (multiplicity of 2) and -2

23 Degree of 3; and 2 is the only zero (multiplicity of 3)

In questions 24–26, find a polynomial of lowest degree with real coefficients and the given zeros.

24 $x = -1$ and $x = 1 - i$

25 $x = 2, x = -4$ and $x = -3i$

26 $x = 3 + i$ and $x = 1 - 2i$

27 Given that $x = 2 - 3i$ is a zero of $f(x) = x^3 - 7x^2 + 25x - 39$ find the other remaining zeros.

28 The polynomial $6x^3 + 7x^2 + ax + b$ has a remainder of 72 when divided by $x - 2$ and is exactly divisible (i.e. remainder is zero) by $x + 1$.

a) Calculate a and b.

b) Show that $2x - 1$ is also a factor of the polynomial and, hence, find the third factor.

29 The polynomial $p(x) = (ax + b)^3$ leaves a remainder of -1 when divided by $x + 1$, and a remainder of 27 when divided by $x - 2$. Find the values of the real numbers a and b.

30 The quadratic polynomial $x^2 - 2x - 3$ is a factor of the quartic polynomial function $f(x) = 4x^4 - 6x^3 - 15x^2 - 8x - 3$. Find all of the zeros of the function f. Express the zeros exactly and completely simplified.

31 $x - 2$ and $x + 2$ are factors of $x^3 + ax^2 + bx + c$, and it leaves a remainder of 10 when divided by $x - 3$. Find the values of a, b and c.

32 Let $P(x) = x^3 + px^2 + qx + r$. Two of the zeros of $P(x) = 0$ are 3 and $1 + 4i$. Find the value of p, q and r.

33 When divided by $(x + 2)$ the expression $5x^3 - 3x^2 + ax + 7$ leaves a remainder of R. When the expression $4x^3 + ax^2 + 7x - 4$ is divided by $(x + 2)$ there is a remainder of $2R$. Find the value of the constant a.

34 The polynomial $x^3 + mx^2 + nx - 8$ is divisible by $(x + 1 + i)$. Find the value of m and n.

35 Given that the roots of the equation $x^3 - 9x^2 + bx - 216 = 0$ are consecutive terms in a geometric sequence, find the value of b and solve the equation.

36 a) Prove that when a polynomial $P(x)$ is divided by $ax - b$ the remainder is $P\left(\frac{b}{a}\right)$.

b) Hence, find the remainder when $9x^3 - x + 5$ is divided by $3x + 2$.

37 Find the sum and product of the roots of the following equations.

a) $x^4 - \frac{2}{3}x^3 + 3x^2 - 2x + 5 = 0$

b) $(x - 2)^3 = x^4 - 1$

c) $\dfrac{3}{x^2 + 2} = \dfrac{2x^2 - x}{2x^5 + 1}$

38 If α, β and γ are the three roots of the cubic equation $ax^3 + bx^2 + cx + d = 0$, show that $\alpha\beta + \alpha\gamma + \beta\gamma = \frac{c}{a}$.

39 One of the zeros of the equation $x^3 - 63x + 162 = 0$ is double another zero. Find all three zeros.

40 Find the three zeros of the equation $x^3 - 6x^2 - 24x + 64 = 0$ given that they are consecutive terms in a geometric sequence. [Hint: let the zeros be represented by $\frac{\alpha}{r}$, α, αr where r is the common ratio.]

41 Consider the equation $x^5 - 12x^4 + 62x^3 - 166x^2 + 229x - 130 = 0$. Given that two of the zeros of the equation are $x = 3 - 2i$ and $x = 2$, find the remaining three zeros.

42 Find the value of k such that the zeros of the equation $x^3 - 6x^2 + kx + 10 = 0$ are in arithmetic progression, that is, they can be represented by α, $\alpha + d$ and $\alpha + 2d$ for some constant d. [Hint: use the result from question 38.]

43 Find the value of k if the roots of the equation $x^3 + 3x^2 - 6x + k = 0$ are in geometric progression.

3.4 Rational functions

Another important category of algebraic functions is rational functions, which are functions in the form $R(x) = \frac{f(x)}{g(x)}$ where f and g are polynomials and the domain of the function R is the set of all real numbers except the real zeros of polynomial g in the denominator. Some examples of rational functions are

$$p(x) = \frac{1}{x-5}, \quad q(x) = \frac{x+2}{(x+3)(x-1)}, \quad \text{and} \quad r(x) = \frac{x}{x^2+1}$$

The domain of p excludes $x = 5$, and the domain of q excludes $x = -3$ and $x = 1$. The domain of r is all real numbers because the polynomial $x^2 + 1$ has no real zeros.

Example 24

Find the domain and range of $h(x) = \frac{1}{x-2}$. Sketch the graph of h.

Solution

Because the denominator is zero when $x = 2$, the domain of h is all real numbers except $x = 2$, i.e. $x \in \mathbb{R}, x \neq 2$. Determining the range of the function is a little less straightforward. It is clear that the function could never take on a value of zero because that will only occur if the numerator is zero. And since the denominator can have any value except zero it seems that the function values of h could be any real number except zero. To confirm this and to determine the behaviour of the function (and shape of the graph), some values of the domain and range (pairs of coordinates) are displayed in the tables below.

• **Hint:** A fraction is only zero if its numerator is zero.

x approaches 2 from the left

x	$h(x)$
−98	−0.01
−8	−0.1
0	−0.5
1	−1
1.5	−2
1.9	−10
1.99	−100
1.999	−1000

x approaches 2 from the right

x	$h(x)$
102	0.01
12	0.1
4	0.5
3	1
2.5	2
2.1	10
2.01	100
2.001	1000

The values in the tables provide clear evidence that the range of h is all real numbers except zero, i.e. $h(x) \in \mathbb{R}, h(x) \neq 0$. The values in the tables also show that as $x \to -\infty$, $h(x) \to 0$ from below (sometimes written $h(x) \to 0^-$) and as $x \to +\infty$, $h(x) \to 0$ from above ($h(x) \to 0^+$). It follows that the line with equation $y = 0$ (the x-axis) is a horizontal asymptote for the graph of h. As $x \rightarrow 2$ from the left (sometimes written $x \rightarrow 2^-$), $h(x)$ appears to decrease without bound, whereas as $x \rightarrow 2$ from the right ($x \rightarrow 2^+$), $h(x)$ appears to increase without bound. This indicates that the graph of h will have a vertical asymptote at $x = 2$. This behaviour is confirmed by the graph at left.

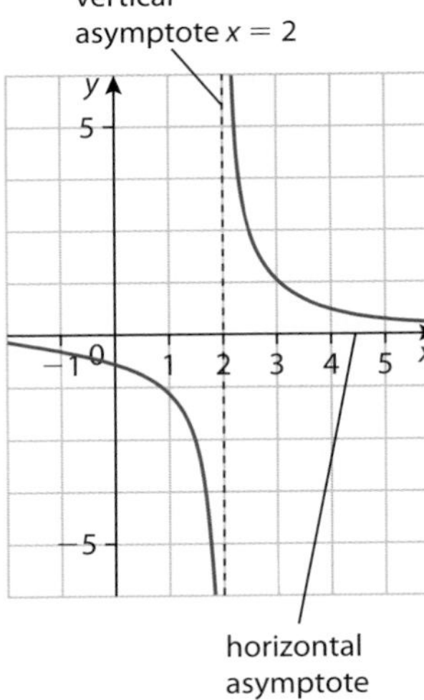

Horizontal and vertical asymptotes

The line $y = c$ is a **horizontal asymptote** of the graph of the function f if at least one of the following statements is true:

- as $x \rightarrow +\infty$, then $f(x) \rightarrow c^+$
- as $x \rightarrow -\infty$, then $f(x) \rightarrow c^+$
- as $x \rightarrow +\infty$, then $f(x) \rightarrow c^-$
- as $x \rightarrow -\infty$, then $f(x) \rightarrow c^-$

The line $x = d$ is a **vertical asymptote** of the graph of the function f if at least one of the following statements is true:

- as $x \rightarrow d^+$, then $f(x) \rightarrow +\infty$
- as $x \rightarrow d^+$, then $f(x) \rightarrow -\infty$
- as $x \rightarrow d^-$, then $f(x) \rightarrow +\infty$
- as $x \rightarrow d^-$, then $f(x) \rightarrow -\infty$

Example 25

Consider the function $f(x) = \dfrac{3x^2 - 12}{x^2 + 3x - 4}$. Sketch the graph of f and identify any asymptotes and any x- or y-intercepts. Use the sketch to confirm the domain and range of the function.

Solution

Firstly, let's completely factorize both the numerator and denominator.

$$f(x) = \frac{3x^2 - 12}{x^2 + 3x - 4} = \frac{3(x + 2)(x - 2)}{(x - 1)(x + 4)}$$

Axis intercepts:

The x-intercepts will occur where the numerator is zero. Hence, the x-intercepts are $(-2, 0)$ and $(2, 0)$. A y-intercept will occur when $x = 0$.
$f(0) = \dfrac{3(2)(-2)}{(-1)(4)} = 3$, so the y-intercept is $(0, 3)$.

Vertical asymptote(s):

Any vertical asymptote will occur where the denominator is zero, that is, where the function is undefined. From the factored form of f we see that the vertical asymptotes are $x = 1$ and $x = -4$. We need to determine if the graph of f falls ($f(x) \rightarrow -\infty$) or rises ($f(x) \rightarrow \infty$) on either side of each vertical asymptote. It's easiest to do this by simply analyzing what the sign of h will be as x approaches 1 and -4 from both the left and right. For example, as $x \rightarrow 1^-$ we can use a test value close to and to the left of 1 (e.g. $x = 0.9$) to check whether $f(x)$ is positive or negative to the left of 1.

$$f(x) = \frac{3(0.9 + 2)(0.9 - 2)}{(0.9 - 1)(0.9 + 4)} \Rightarrow \frac{(+)(-)}{(-)(+)} \Rightarrow f(x) > 0 \Rightarrow \text{as } x \rightarrow 1^-,$$

then $f(x) \rightarrow +\infty$ (rises)

As $x \rightarrow 1^+$ we use a test value close to and to the right of 1 (e.g. $x = 1.1$) to check whether $f(x)$ is positive or negative to the right of 1.

• **Hint:** The farther the number n is from 0, the closer the number $\frac{1}{n}$ is to 0. Conversely, the closer the number n is to 0, the farther the number $\frac{1}{n}$ is from 0. These facts can be expressed simply as:

$$\frac{1}{\text{BIG}} = \text{little and } \frac{1}{\text{little}} = \text{BIG}$$

They can also be expressed more mathematically using the concept of a limit expressed in limit notation as: $\lim_{n\to\infty}\frac{1}{n} = 0$ and $\lim_{n\to 0}\frac{1}{n} = \infty$.
Note: Infinity is not a number, so $\lim_{n\to 0}\frac{1}{n}$ actually does not exist, but writing $\lim_{n\to 0}\frac{1}{n} = \infty$ expresses the idea that $\frac{1}{n}$ increases without bound as n approaches 0.

$$f(x) = \frac{3(1.1+2)(1.1-2)}{(1.1-1)(1.1+4)} \Rightarrow \frac{(+)(-)}{(+)(+)} \Rightarrow f(x) < 0 \Rightarrow \text{as } x \to 1^+,$$
then $f(x) \to -\infty$ (falls)

Conducting similar analysis for the vertical asymptote of $x = -4$, produces:

$$f(x) = \frac{3(-4.1+2)(-4.1-2)}{(-4.1-1)(-4.1+4)} \Rightarrow \frac{(-)(-)}{(-)(-)} \Rightarrow f(x) > 0 \Rightarrow \text{as } x \to 4^-,$$
then $f(x) \to +\infty$ (rises)

$$f(x) = \frac{3(-3.9+2)(-3.9-2)}{(-3.9-1)(-3.9+4)} \Rightarrow \frac{(-)(-)}{(-)(+)} \Rightarrow f(x) < 0 \Rightarrow \text{as } x \to 4^+,$$
then $f(x) \to -\infty$ (falls)

Horizontal asymptote(s):

A horizontal asymptote (if it exists) is the value that $f(x)$ approaches as $x \to \pm\infty$. To find this value, we divide both the numerator and denominator by the highest power of x that appears in the denominator (x^2 for function f).

$$f(x) = \frac{\frac{3x^2}{x^2} - \frac{12}{x^2}}{\frac{x^2}{x^2} + \frac{3x}{x^2} - \frac{4}{x^2}} \text{ then, as } x \to \pm\infty, f(x) = \frac{3-0}{1+0-0} = 3$$

Hence, the horizontal asymptote is $y = 3$.

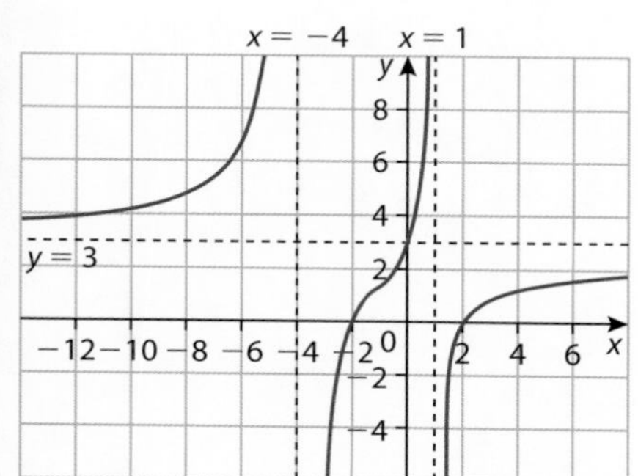

Sketch of graph:

Now we know the behaviour (rising or falling) of the function on either side of each vertical asymptote and that the graph will approach the horizontal asymptote as $x \to \pm\infty$, an accurate sketch of the graph can be made as shown right.

Domain and range:

Because the zeros of the polynomial in the denominator are $x = 1$ and $x = -4$, the domain of f is all real numbers except 1 and -4. From our analysis and from the sketch of the graph, it is clear that between $x = -4$ and $x = 1$ the function takes on all values from $-\infty$ to $+\infty$, therefore the range of f is all real numbers.

We are in the habit of cancelling factors in algebraic expressions (Section 1.5), such as

$$\frac{x^2-1}{x-1} = \frac{(x+1)\cancel{(x-1)}}{\cancel{x-1}} = x+1$$

However, the function $f(x) = \frac{x^2-1}{x-1}$ and the function $g(x) = x + 1$ are **not** the same function. The difference occurs when $x = 1$.
$f(1) = \frac{1^2-1}{1-1} = \frac{0}{0}$, which is undefined, and $g(1) = 1 + 1 = 2$. So, 1 is not in the domain of f but it is in the domain of g. As we might expect the

graphs of the two functions appear identical, but upon closer inspection it is clear that there is a 'hole' in the graph of f at the point $(1, 2)$. Thus, f is a *discontinuous* function but the polynomial function g is continuous. f and g are different functions.

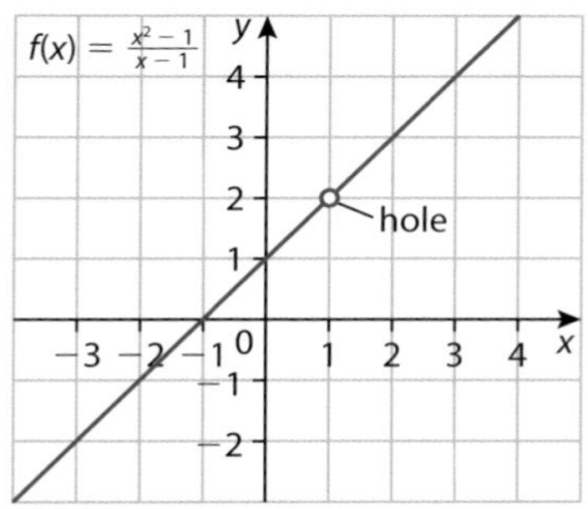

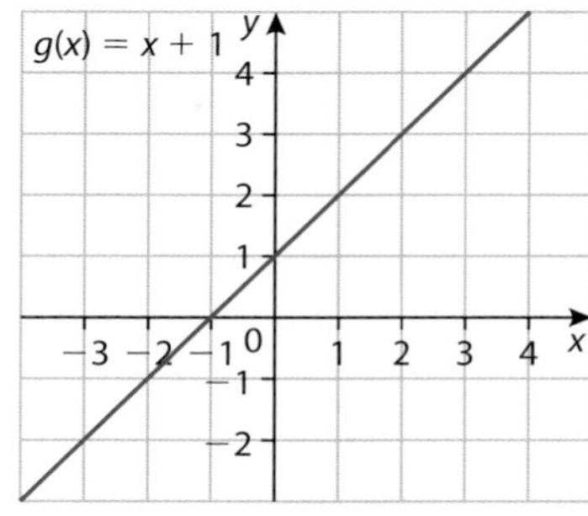

• **Hint:** Try graphing $\frac{x^2 - 1}{x - 1}$ on your GDC and zooming in closely to the region around the point $(1, 2)$. Can you see the 'hole'?

In working with rational functions, we often assume that every linear factor that appears in both the numerator and in the denominator has been cancelled. Therefore, for a rational function in the form $\frac{f(x)}{g(x)}$, we can usually assume that the polynomial functions f and g have no common factors.

Example 26

Find any asymptotes for the function $p(x) = \frac{x^2 - 9}{x - 4}$.

Solution

The denominator is zero when $x = 4$, thus the line with equation $x = 4$ is a vertical asymptote. Although the numerator $x^2 - 9$ is not divisible by $x - 4$, it does have a larger degree. Some insight into the behaviour of function p may be gained by dividing $x - 4$ into $x^2 - 9$. Since the degree of the numerator is one greater than the degree of the denominator, the quotient will be a linear polynomial. Recalling from the previous section that $\frac{P(x)}{D(x)} = Q(x) + \frac{R(x)}{D(x)}$, where Q and R are the quotient and remainder, we can rewrite $p(x)$ as a linear polynomial plus a fraction.

Since the denominator is in the form $x - c$ we can carry out the division efficiently by means of synthetic division.

$$\begin{array}{r|rrr} 4 & 1 & 0 & -9 \\ & & 4 & 16 \\ \hline & 1 & 4 & 7 \end{array}$$

Hence, $p(x) = \frac{x^2 - 9}{x - 4} = x + 4 + \frac{7}{x - 4}$.

As $x \to \pm\infty$, the fraction $\frac{7}{x - 4} \to 0$. This tells us about the end behaviour of function p, namely that the graph of p will get closer and closer to the line $y = x + 4$ as the values of x get further away from the origin. Symbolically, this can be expressed as follows: as $x \to \pm\infty$, $p(x) \to x + 4$.

We can graph both the rational function $p(x)$ and the line $y = x + 4$ on our GDC to visually confirm our analysis.

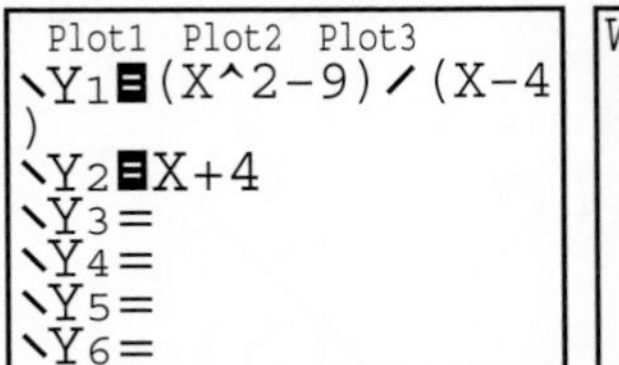

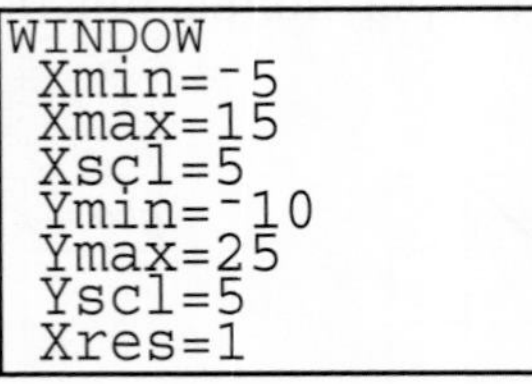

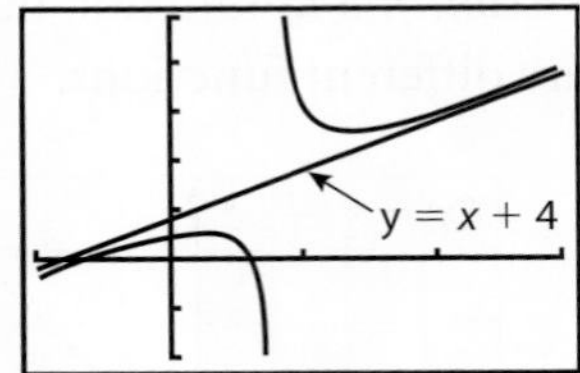

If a line is an asymptote of a graph but it is neither horizontal nor vertical, it is called an **oblique asymptote** (sometimes called a slant asymptote). The graph of any rational function of the form $\frac{f(x)}{g(x)}$, where the degree of function f is one more than the degree of function g will have an oblique asymptote.

Using Example 25 as a model, we can set out a general procedure for analyzing a rational function leading to a sketch of its graph and determining its domain and range.

Analyzing a rational function $R(x) = \frac{f(x)}{g(x)}$ given functions f and g have no common factors

1. **Factorize**: Completely factorize both the numerator and denominator.
2. **Intercepts**: A zero of f will be a zero of R and hence an x-intercept of the graph of R. The y-intercept is found by evaluating $R(0)$.
3. **Vertical asymptotes**: A zero of g will give the location of a vertical asymptote (if any). Then perform a sign analysis to see if $R(x) \to +\infty$ or $R(x) \to -\infty$ on either side of each vertical asymptote.
4. **Horizontal asymptote**: Find the horizontal asymptote (if any) by dividing both f and g by the highest power of x that appears in g, and then letting $x \to \pm\infty$.
5. **Oblique asymptotes**: If the degree of f is one more than the degree of g, then the graph of R will have an oblique asymptote. Divide g into f to find the quotient $Q(x)$ and remainder. The oblique asymptote will be the line with equation $y = Q(x)$.
6. **Sketch of graph**: Start by drawing dashed lines where the asymptotes are located. Use the information about the intercepts, whether $Q(x)$ falls or rises on either side of a vertical asymptote, and additional points as needed to make an accurate sketch.
7. **Domain and range**: The domain of R will be all real numbers except the zeros of g. You need to study the graph carefully in order to determine the range. Often, but not always (as in Example 25), the value of the function at the horizontal asymptote will not be included in the range.

End behaviour of a rational function

Let R be the rational function given by

$$R(x) = \frac{f(x)}{g(x)} = \frac{a_nx^n + a_{n-1}x^{n-1} + \ldots + a_1x + a_0}{b_mx^m + b_{m-1}x^{m-1} + \ldots + b_1x + b_0}$$

where functions f and g have no common factors. Then the following holds true:

1. If $n < m$, then the x-axis (line $y = 0$) is a horizontal asymptote for the graph of R.
2. If $n = m$, then the line $y = \frac{a_n}{b_m}$ is a horizontal asymptote for the graph of R.
3. If $n > m$, then the graph of R has no horizontal asymptote. However, if the degree of f is one more than the degree of g, then the graph of R will have an oblique asymptote.

Exercise 3.4

In questions 1–10, sketch the graph of the rational function without the aid of your GDC. On your sketch clearly indicate any x- or y-intercepts and any asymptotes (vertical, horizontal or oblique). Use your GDC to verify your sketch.

1 $f(x) = \dfrac{1}{x+2}$

2 $g(x) = \dfrac{3}{x-2}$

3 $h(x) = \dfrac{1-4x}{1-x}$

4 $R(x) = \dfrac{x}{x^2-9}$

5 $p(x) = \dfrac{2}{x^2+2x-3}$

6 $M(x) = \dfrac{x^2+1}{x}$

7 $f(x) = \dfrac{x}{x^2+4x+4}$

8 $h(x) = \dfrac{x^2+2x}{x-1}$

9 $g(x) = \dfrac{2x+8}{x^2-x-12}$

10 $C(x) = \dfrac{x-2}{x^2-4x}$

In questions 11–14, use your GDC to sketch a graph of the function, and state the domain and range of the function.

11 $f(x) = \dfrac{2x^2+5}{x^2-4}$

12 $g(x) = \dfrac{x+4}{x^2+3x-4}$

13 $h(x) = \dfrac{6}{x^2+6}$

14 $r(x) = \dfrac{x^2-2x+1}{x-1}$

In questions 15–18, use your GDC to sketch a graph of the function. Clearly label any x- or y-intercepts and any asymptotes.

15 $f(x) = \dfrac{2x-5}{2x^2+9x-18}$

16 $g(x) = \dfrac{x^2+x+1}{x-1}$

17 $h(x) = \dfrac{3x^2}{x^2+x+2}$

18 $g(x) = \dfrac{1}{x^3-x^2-4x+4}$

19 If a, b and c are all positive, sketch the curve $y = \dfrac{x-a}{(x-b)(x-c)}$ for each of the following conditions:

a) $a < b < c$ b) $b < a < c$ c) $b < c < a$

20 A drug is given to a patient and the concentration of the drug in the bloodstream is carefully monitored. At time $t \geqslant 0$ (in minutes after patient receiving the drug), the concentration, in milligrams per litre (mg/l) is given by the following function.

$$C(t) = \frac{25t}{t^2+4}$$

a) Sketch a graph of the drug concentration (mg/l) versus time (min).

b) When does the highest concentration of the drug occur, and what is it?

c) What eventually happens to the concentration of the drug in the bloodstream?

d) How long does it take for the concentration to drop below 0.5 mg/l?

3.5 Other equations and inequalities

We have studied some approaches to analyzing and solving polynomial equations in this chapter. Some problems lead to equations with expressions that are not polynomials, for example, expressions with radicals, fractions, or absolute value. Problems in mathematics often do not involve equations but inequalities. We need to be familiar with effective methods for solving inequalities involving polynomials – and again, radicals, fractions, or absolute value.

Equations involving a radical

Example 27 – Solving an equation with a single radical expression

Solve for x: $\sqrt{3x+6} = 2x + 1$

Solution

Squaring both sides gives $3x + 6 = (2x+1)^2$

$$3x + 6 = 4x^2 + 4x + 1$$

$$4x^2 + x - 5 = 0$$

Factorizing: $(4x+5)(x-1) = 0$

$$x = -\tfrac{5}{4} \text{ or } x = 1$$

Check both solutions in the original equation:

When $x = -\frac{5}{4}$, $\sqrt{3\left(-\frac{5}{4}\right) + 6} = 2\left(-\frac{5}{4}\right) + 1 \Rightarrow \sqrt{\frac{9}{4}} = -\frac{3}{2} \Rightarrow \frac{3}{2} \neq -\frac{3}{2}$

Therefore, $x = -\frac{5}{4}$ is *not* a solution.

When $x = 1$, $\sqrt{3(1) + 6} = 2(1) + 1 \Rightarrow \sqrt{9} = 3 \Rightarrow 3 = 3$

Therefore, $x = 1$ is the only solution.

If two quantities are equal, for example $a = b$, then it is certainly true that $a^2 = b^2$, and $a^3 = b^3$, etc. However, the converse is not necessarily true. A simple example can illustrate this.

Consider the trivial equation $x = 3$. There is only one value of x that makes the equation true – and that is 3. Now if we take this original equation and square both sides we transform it to the equation $x^2 = 9$. This transformed equation has two solutions, 3 and -3, so it is not equivalent to the original equation. By squaring both sides we gained an extra solution, often called an **extraneous solution**, that satisfies the transformed equation but not the original equation as occurred in Example 27. Whenever you raise both sides of an equation by a power it is imperative that you check all solutions in the original equation.

Every solution of the equation $a = b$ is also a solution of the equation $a^n = b^n$, but it is not necessarily true that every solution of $a^n = b^n$ is a solution of $a = b$.

Example 28 – Solving an equation with two radical expressions

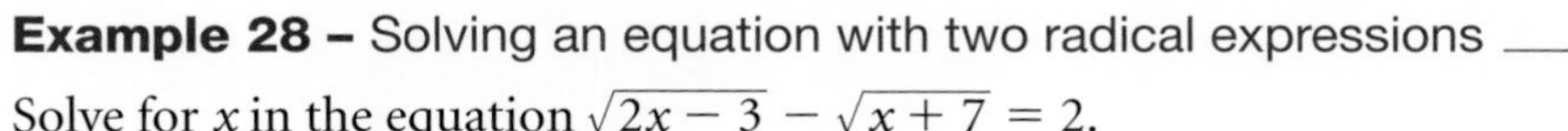

Solve for x in the equation $\sqrt{2x-3} - \sqrt{x+7} = 2$.

Solution

Squaring both sides of the original equation will produce a messy expression on the left side, so it is better to rearrange the terms so that one side of the equation contains only a single radical term.

$$\sqrt{2x - 3} = 2 + \sqrt{x + 7}$$

$$(\sqrt{2x - 3})^2 = (2 + \sqrt{x + 7})^2$$

$$2x - 3 = 4 + 4\sqrt{x + 7} + x + 7$$

$$x - 14 = 4\sqrt{x + 7}$$

$$(x - 14)^2 = (4\sqrt{x + 7})^2$$

Squaring both sides again to eliminate the radical.

$$x^2 - 28x + 196 = 16(x + 7)$$

$$x^2 - 44x + 84 = 0$$

$$(x - 2)(x - 42) = 0$$

$$x = 2 \text{ or } x = 42$$

Check both solutions in the original equation:

When $x = 2$, $\sqrt{2(2) - 3} \stackrel{?}{=} 2 + \sqrt{2 + 7} \Rightarrow \sqrt{1} \stackrel{?}{=} 2 + \sqrt{9} \Rightarrow 1 \neq 5$

Thus, $x = 2$ is *not* a solution.

When $x = 42$, $\sqrt{2(42) - 3} \stackrel{?}{=} 2 + \sqrt{42 + 7} \Rightarrow \sqrt{81} \stackrel{?}{=} 2 + \sqrt{49} \Rightarrow 9 = 2 + 7$

Thus, $x = 42$ is a solution.

We can verify the single solution of $x = 42$ using our GDC by graphing the equation $y = \sqrt{2x - 3} - \sqrt{x + 7} - 2$ and looking for x-intercepts (zeros). Since we are restricted to real number solutions then the smallest possible value for x that can be substituted into the equation is $\frac{3}{2}$. This helps determine a suitable viewing window for the graph on our GDC.

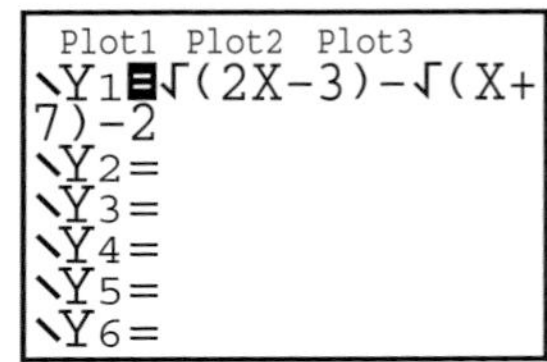


```
WINDOW
 Xmin=-5
 Xmax=60
 Xscl=5
 Ymin=-5
 Ymax=2
 Yscl=1
 Xres=1
```

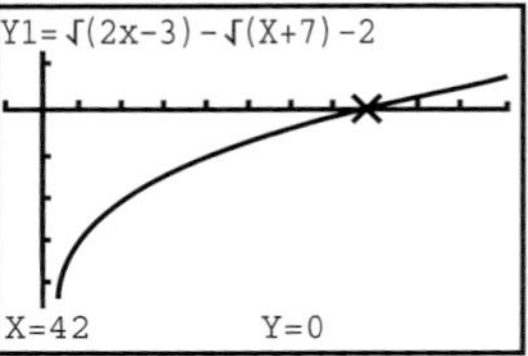

This verifies that $x = 42$ is the only solution to the equivalent equation $\sqrt{2x - 3} = 2 + \sqrt{x + 7}$.

Equations involving fractions

It is also possible for extraneous solutions to appear when solving equations with fractions.

Example 29 – An extraneous root in an equation with fractions

Find all real solutions of the equation $\frac{2x}{4 - x^2} + \frac{1}{x + 2} = 3$ and verify solution(s) with a GDC.

Solution

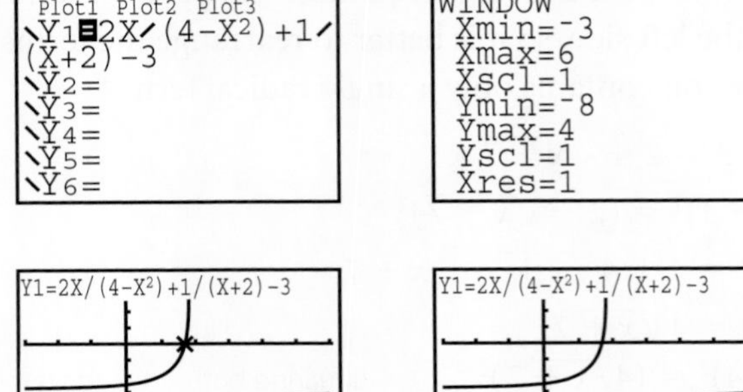

Multiply both sides of the equation by the least common denominator of the fractions, $4 - x^2$.

$$\frac{4 - x^2}{1} \cdot \frac{2x}{4 - x^2} + \frac{(2 - x)(2 + x)}{1} \cdot \frac{1}{x + 2} = 3(4 - x^2)$$

Factorizing $4 - x^2$ gives $(2 - x)(2 + x)$.

$$2x + 2 - x = 12 - 3x^2$$

$$3x^2 + x - 10 = 0$$

$$(3x - 5)(x + 2) = 0$$

$$x = \tfrac{5}{3} \text{ or } x = -2$$

Clearly $x = -2$ cannot be a solution because that would cause division by zero in the original equation.

The GDC images show that the equation $y = \dfrac{2x}{4 - x^2} + \dfrac{1}{x + 2} - 3$ has an x-intercept at $\left(\frac{5}{3}, 0\right)$, confirming the solution $x = \frac{5}{3}$.

• **Hint:** Not only is it possible to *gain* an extraneous solution when solving certain equations, it is also possible to *lose* a correct solution by incorrectly dividing both sides of an equation by a common factor. For example, solve for x in the equaton $4(x + 2)^2 = 3x(x + 2)$. Dividing both sides by $(x + 2)$, gives $4(x + 2) = 3x \Rightarrow 4x + 8 = 3x \Rightarrow x = -8$. However, there are two solutions, $x = -8$ *and* $x = -2$. The solution of $x = -2$ was lost because a factor of $x + 2$ was eliminated from both sides of the original equation. This is a common error to be avoided.

Equations in quadratic form

In Section 3.2 we covered methods of solving quadratic equations. As the three previous examples illustrate, quadratic equations commonly appear in a range of mathematical problems. The methods of solving quadratics can sometimes be applied to other equations. An equation in the form $at^2 + bt + c = 0$, where t is an algebraic expression, is an equation in **quadratic form**. We can solve such equations by substituting for the algebraic expression and then apply an appropriate method for solving a quadratic equation.

Example 30 – A 4th degree polynomial equation in quadratic form

Find all real solutions of the equation $2m^4 - 5m^2 + 2 = 0$.

Solution

The equation can be written as $2(m^2)^2 - 5(m^2) + 2 = 0$ showing it is quadratic in terms of m^2. Let $t = m^2$, and substituting gives $2t^2 - 5t + 2 = 0$. Solve for t, substitute m^2 back in for t, and then solve for m.

$2m^4 - 5m^2 + 2 = 0$

Substitute t for m^2 $\quad 2t^2 - 5t + 2 = 0$

$(2t - 1)(t - 2) = 0$

$t = \frac{1}{2}$ or $t = 2$

Substituting m^2 for t $\quad m^2 = \frac{1}{2}$ or $m^2 = 2$

$m = \pm\sqrt{\frac{1}{2}} = \pm\frac{\sqrt{2}}{2}$ or $m = \pm\sqrt{2}$

These four solutions – which are two pairs of opposites – can be checked by substituting them directly into the original equation. A value for m will be raised to the 4th and 2nd powers, thus we only need to check one value from each pair of opposites.

When $m = \frac{\sqrt{2}}{2}$, $2\left(\frac{\sqrt{2}}{2}\right)^4 - 5\left(\frac{\sqrt{2}}{2}\right)^2 + 2 = 0 \Rightarrow 2\left(\frac{1}{4}\right) - 5\left(\frac{1}{2}\right) + 2 = 0$

$\Rightarrow \frac{1}{2} - \frac{5}{2} + 2 = 0 \Rightarrow 0 = 0$

When $m = \sqrt{2}$, $2(\sqrt{2})^4 - 5(\sqrt{2})^2 + 2 = 0 \Rightarrow 2(4) - 5(2) + 2 = 0$

$\Rightarrow 8 - 10 + 2 = 0 \Rightarrow 0 = 0$

Therefore, the solutions to the equation are $m = \frac{\sqrt{2}}{2}, -\frac{\sqrt{2}}{2}, \sqrt{2}$ and $-\sqrt{2}$.

Example 31 – Another equation in quadratic form

Find all solutions, expressed exactly, to the equation $w^{\frac{1}{2}} = 4w^{\frac{1}{4}} - 2$.

Solution

$w^{\frac{1}{2}} - 4w^{\frac{1}{4}} + 2 = 0$ — Set the equation to zero.

$\left(w^{\frac{1}{4}}\right)^2 - 4\left(w^{\frac{1}{4}}\right) + 2 = 0$ — Attempt to write in quadratic form: $at^2 + bt + c = 0$

$t^2 - 4t + 2 = 0$ — Make appropriate substitution; in this case, let $w^{\frac{1}{4}} = t$.

$t = \frac{-(-4) \pm \sqrt{(-4)^2 - 4(1)(2)}}{2}$ — Trinomial does not factorize; apply quadratic formula.

$t = \frac{4 \pm \sqrt{8}}{2} = \frac{4 \pm 2\sqrt{2}}{2}$

$t = 2 \pm \sqrt{2}$

$w^{\frac{1}{4}} = 2 \pm \sqrt{2}$ Substituting $w^{\frac{1}{4}}$ back in for t; raise both sides to 4th power.

```
68+48√2
           135.882251
68-48√2
         0.1177490061
▸MAT
```

$w = (2 + \sqrt{2})^4$ or $w = (2 - \sqrt{2})^4$

$w = ((2 + \sqrt{2})^2)^2$ or $w = ((2 - \sqrt{2})^2)^2$

$w = (6 + 4\sqrt{2})^2$ or $w = (6 - 4\sqrt{2})^2$

$w = 68 + 48\sqrt{2} \approx 135.882$ or $w = 68 - 48\sqrt{2} \approx 0.117749$ (approx. values found with GDC)

It will be difficult to check these two solutions by substituting them directly into the original equation as we did in the previous example. It will be more efficient to use our GDC.

• **Hint:** We will encounter equations in later chapters – for example, equations with logarithms and trigonometric functions – that will be in quadratic form.

Most GDC models have an equation 'solver'. The main limitation of this GDC feature is that it will usually return only approximate solutions. However, even if exact solutions are required, approximate solutions from a GDC are still very helpful as a check of the exact solutions obtained algebraically.

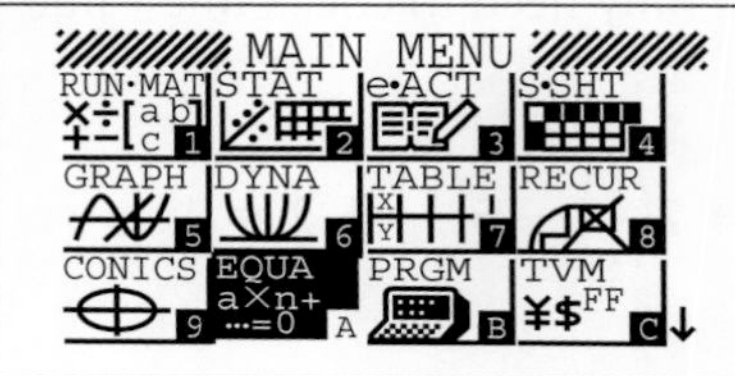

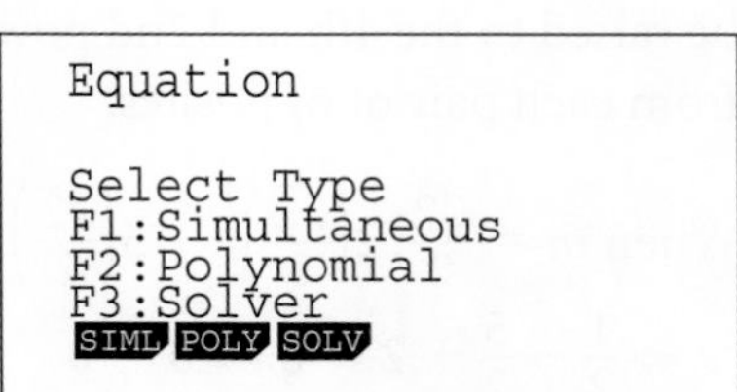


```
Eq:X^(1÷2)-4X^(1÷4)+2
  X=0.1177490061
Lft=0
Rgt=0

REPT
```

```
Eq:X^(1÷2)-4X^(1÷4)+2
  X=135.882251
Lft=0
Rgt=0

REPT
```

Equations involving absolute value

Equations involving absolute value occur in a range of different topics in mathematics. To solve an equation containing one or more absolute value expressions, we apply the definition from Section 1.1, which states that the absolute value of a real number a, denoted by $|a|$, is given by

$$|a| = \begin{cases} a & \text{if } a \geqslant 0 \\ -a & \text{if } a < 0 \end{cases}$$

Also recall that in Section 1.1 we stated that $|a|$ is the distance between the coordinate a and the origin on the real number line.

Example 32 – Equation with an absolute value expression

Use an algebraic approach to solve the equation $|2x + 7| = 13$. Check any solution(s) on a GDC.

Solution

The expression inside the absolute value symbols must be either 13 or -13, so $2x + 7$ equals 13 or -13. Hence, the given equation is satisfied if either

$$2x + 7 = 13 \quad \text{or} \quad 2x + 7 = -13$$
$$2x = 6 \quad\quad 2x = -20$$
$$x = 3 \quad\quad x = -10$$

The solutions are $x = 3$ and $x = -10$.

To check the solutions on a GDC, graph the equation $y = |2x + 7| - 13$ and confirm that $x = 3$ and $x = -10$ are the x-intercepts of the graph.

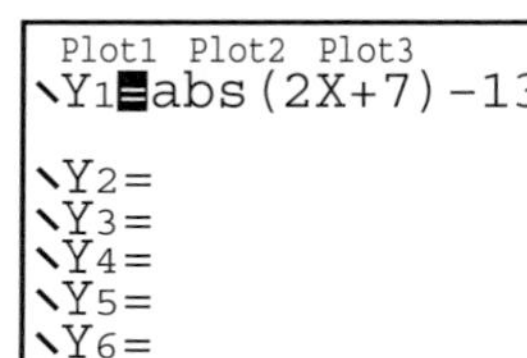

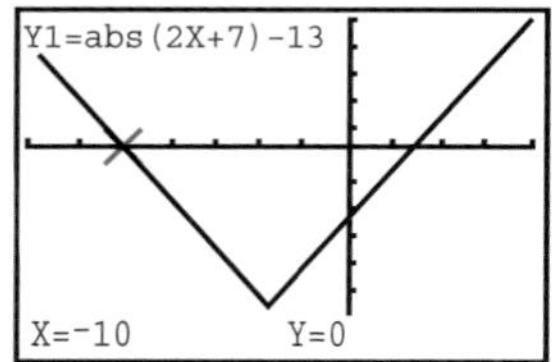

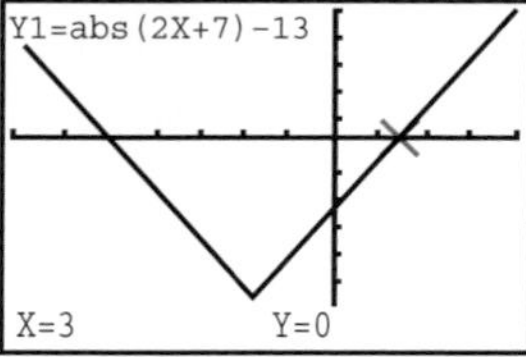

The x-intercepts of the graph of $y = |2x + 7| - 13$ agree with the solutions to the equation.

Example 33 – Equation with two absolute value expressions

Find algebraically the solution(s) to the equation $|2x - 1| = |7 - 3x|$. Check the solution(s) graphically.

Solution

There are four possibilities:

$$2x - 1 = 7 - 3x \text{ or } 2x - 1 = -(7 - 3x) \text{ or } -(2x - 1) = 7 - 3x$$
$$\text{or } -(2x - 1) = -(7 - 3x)$$

The first and last equations are equivalent, and the second and third equations are also equivalent. Thus, it is only necessary to solve the first two equations.

$$2x - 1 = 7 - 3x \quad \text{or} \quad 2x - 1 = -(7 - 3x)$$
$$5x = 8 \quad\quad 2x - 1 = -7 + 3x$$
$$x = \tfrac{8}{5} \quad\quad 6 = x \Rightarrow x = 6$$

To check, we can graph the equations $y_1 = |2x - 1|$ and $y_2 = |7 - 3x|$, and confirm that the x-coordinates of their points of intersection agree with the solutions to the given equation.

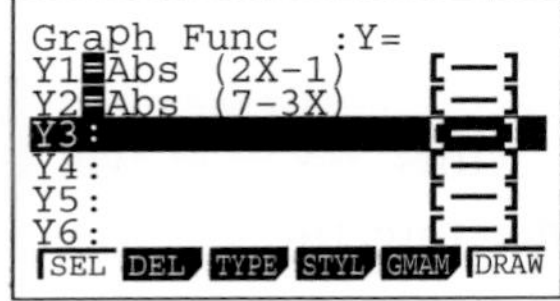

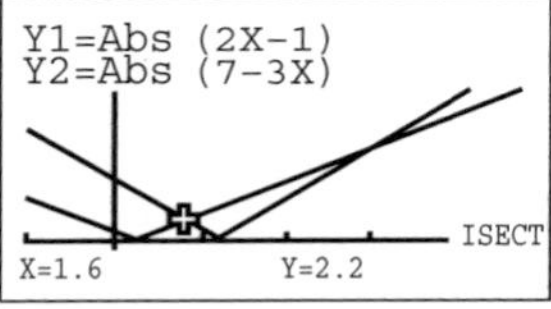

Solving inequalities

Working with inequalities is very important for many of the topics in this course. Inequalities were covered in Section 1.1 in the context of order on the real number line. Recall the four important properties for inequalities.

Properties of inequalities

For three real numbers *a*, *b* and *c*:

1. If $a > b$ and $b > c$, then $a > c$.
2. If $a > b$ and $c > 0$, then $ac > bc$.
3. If $a > b$ and $c < 0$, then $ac < bc$.
4. If $a > b$, then $a + c > b + c$.

Quadratic inequalities

In the topics covered in this course, you will need to be as proficient with solving inequalities as with solving equations. We solved some simple linear inequalities in Section 1.1. Here we will consider strategies for other inequalities – particularly involving quadratic and absolute value expressions.

Example 34 – A quadratic inequality

Find the values of x that solve the inequality $x^2 > x$.

Solution

It is possible to determine the solution set to this inequality by a method of trial and error, or simply using a mental process. That may be successful but generally speaking it is a good idea to attempt to find the solution set by some algebraic method and then check, usually by means of a GDC. For this example, it is tempting to consider dividing both sides by x, but that cannot be done because it is not known whether x is positive or negative. Recall that when multiplying or dividing both sides of an inequality by a negative number it is necessary to reverse the inequality sign (3rd property of inequalities listed above). Instead a better approach is to place all terms on one side of the inequality (with zero on the other side) and then try to factorize.

$$x^2 > x$$
$$x^2 - x > 0$$
$$x(x - 1) > 0$$

Now analyze the signs of the two different factors in a 'sign chart'.

sign chart

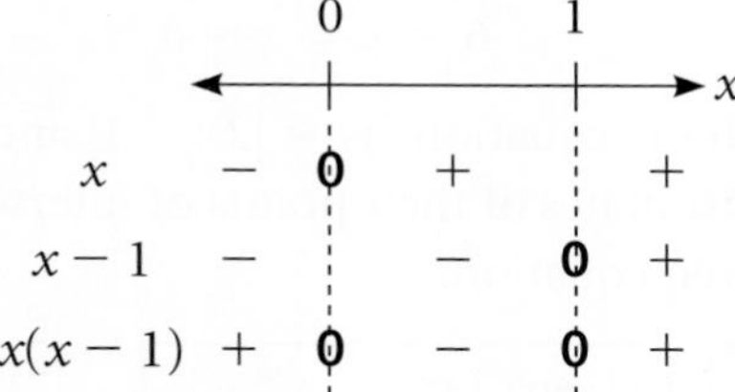

• **Hint:** The solution set, $x < 0$ or $x > 1$, for Example 34 comprises two intervals that do not intersect (disjoint). It is incorrect to write the solution as $0 > x > 1$, or as $1 < x < 0$. Both of these formats imply that the solution set consists of the values of x *between* 0 and 1, but that is not the case. Only write the 'combined' inequality $a < x < b$ if $x > a$ *and* $x < b$ where the two intervals are intersecting *between a* and *b*.

The sign chart indicates that the product of the two factors, $x(x - 1)$, will be positive when x is less than 0 or greater than 1. Therefore, the solution set is $x < 0$ or $x > 1$.

Inequalities with quadratic polynomials arise in many different contexts. Problems in which we need to analyze the value of the discriminant of a quadratic equation will usually require us to solve a quadratic inequality, as the next example illustrates.

Example 35 – A quadratic from evaluating a discriminant

Given $f(x) = 3kx^2 - (k + 3)x + k - 2$, find the range of values of k for which f has no real zeros.

Solution

The quadratic function f will have no real zeros when its discriminant is negative. Since f is written in the form $ax^2 + bx + c = 0$ then, in terms of the parameter k, $a = 3k$, $b = -(k + 3)$ and $c = k - 2$. Substituting these values into the discriminant, we have the inequality

$(-(k + 3))^2 - 4(3k)(k - 2) < 0$

$k^2 + 6k + 9 - 12k^2 + 24k < 0$

$-11k^2 + 30k + 9 < 0$ Easier to factorize if leading coefficient is positive.

$11k^2 - 30k - 9 > 0$ Multiply both sides by -1; reverse inequality sign.

$$k = \frac{-(-30) \pm \sqrt{(-30)^2 - 4(11)(-9)}}{2(11)} = \frac{30 \pm \sqrt{1296}}{22} = \frac{30 \pm 36}{22}$$

$$k = \frac{30 + 36}{22} = \frac{66}{22} = 3 \text{ or } k = \frac{30 - 36}{22} = -\frac{6}{22} = -\frac{3}{11}$$

The two rational zeros indicate $11k^2 - 30k - 9$ could have been factorized into $(11k + 3)(k - 3)$:

$$(11k + 3)(k - 3) > 0$$

The results of the sign chart indicate that the solution set to the inequality is $k < -\frac{3}{11}$ or $k > 3$. Therefore, any value of k such that $k < -\frac{3}{11}$ or $k > 3$ will cause the function f to have no real zeros.

sign chart

	$k < -\frac{3}{11}$	$-\frac{3}{11}$		3	$k > 3$
$11k + 3$	$-$	0	$+$		$+$
$k - 3$	$-$		$-$	0	$+$
$(11k + 3)(k - 3)$	$+$	0	$-$	0	$+$

Absolute value inequalities

In Section 1.1 we described how absolute value is used to indicate distance on the number line. For example, the equation $|x| = 3$ means that some number x is a distance of 3 units from the origin. The two solutions to this equation are $x = 3$ and $x = -3$. Consequently, the inequality $|x| < 3$ means that x lies *at most* 3 units from the origin, as shown in Figure 3.13.

Figure 3.13

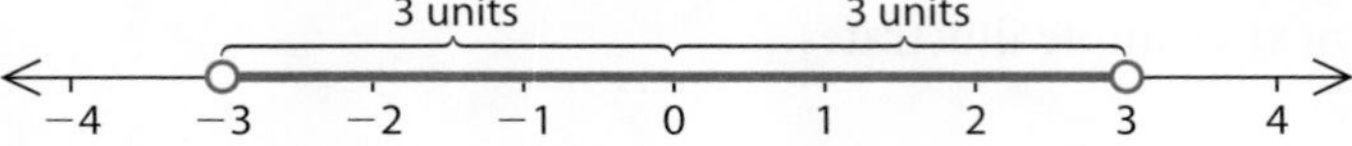

This means that x lies *between* -3 and 3, that is, $-3 < x < 3$. Similarly, the inequality $|x| > 3$ means that x lies 3 *or more* units from the origin. This occurs if x is to the left of -3 (that is, $x < -3$) or if x lies to the right of 3 (that is, $x > 3$).

Properties of absolute value inequalities

For any real numbers x and c such that $c > 0$:

1. $|x| < c$ if and only if $-c < x < c$.
2. $|x| > c$ if and only if $x < -c$ or $x > c$.

Example 36 – Absolute value inequality I

Solve for x: $|3x - 7| \geqslant 8$

Solution

Applying the second property for absolute value inequalities, we have

$$3x - 7 \leqslant -8 \text{ or } 3x - 7 \geqslant 8$$
$$3x \leqslant -1 \text{ or } 3x \geqslant 15$$
$$x \leqslant -\tfrac{1}{3} \text{ or } x \geqslant 5$$

Therefore, the solution set is the union of two half-open intervals $x \leqslant -\frac{1}{3}$ or $x \geqslant 5$, which can also be written in interval notation as $\left]-\infty, -\frac{1}{3}\right] \cup [5, \infty[$.

Example 37 – Absolute value inequality II

Find the values of x which satisfy the inequality $\left|\dfrac{x}{x + 4}\right| < 2$.

Solution

Applying the first property for absolute value inequalities gives

$$-2 < \frac{x}{x + 4} < 2$$

We cannot multiply both sides by $x + 4$ unless we take into account the two different cases: (1) when $x + 4$ is positive (inequality is *not* reversed), and (2) when $x + 4$ is negative (inequality sign *is* reversed). Instead, let's solve the two inequalities in the 'combined' inequality separately by rearranging so that zero is on one side and then analyze where the expression on the other side is zero, positive and negative. This is similar to the approach used in Example 34.

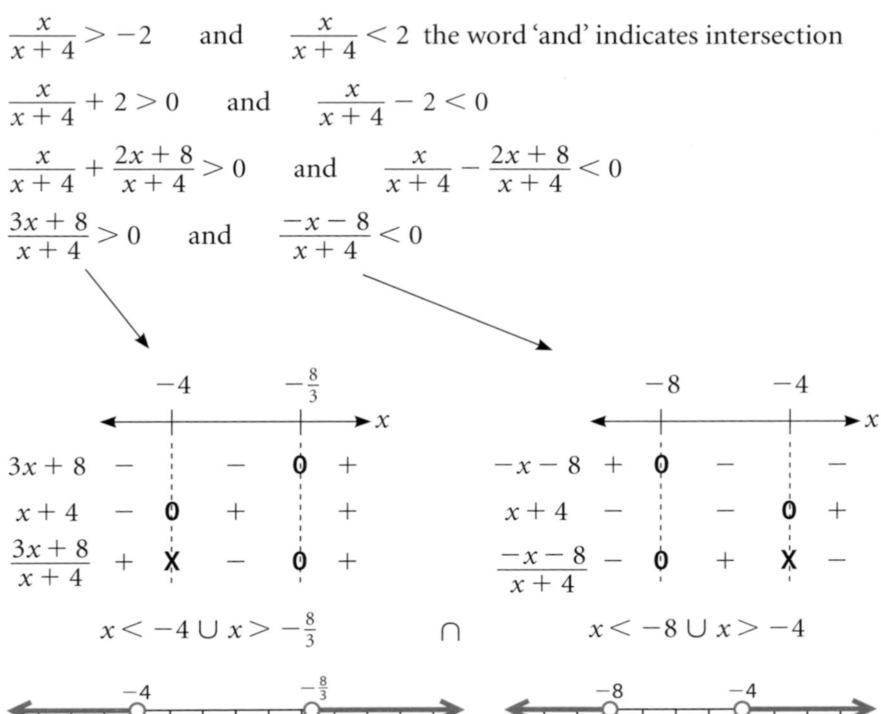

The solution set for the original 'combined' inequality, $-2 < \frac{x}{x+4} < 2$, will be the intersection of the solution sets of the two separate inequalities graphed above on the number line. Thus, the solution set is $x < -8$ or $x > -\frac{8}{3}$.

A graphical check using a GDC can be effectively performed by graphing the equation $y = \left|\frac{x}{x+4}\right| - 2$ and observing where the graph is below the x-axis. The values of x for which this is true will correspond to the solution set for the inequality $\left|\frac{x}{x+4}\right| < 2$.

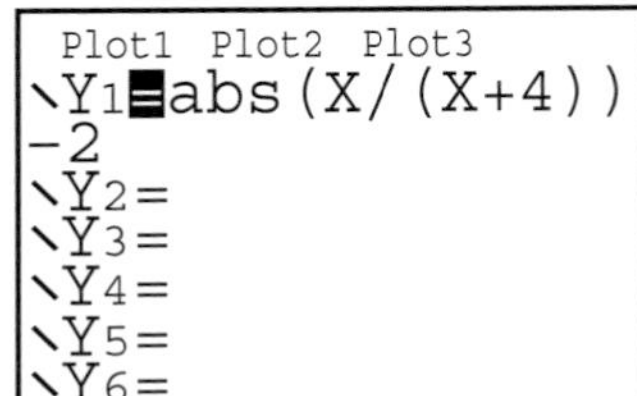


```
WINDOW
 Xmin=-12
 Xmax=2
 Xscl=1
 Ymin=-3
 Ymax=3
 Yscl=1
 Xres=1
```

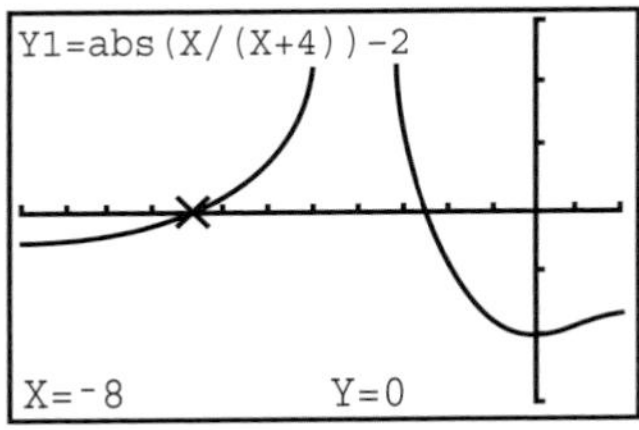

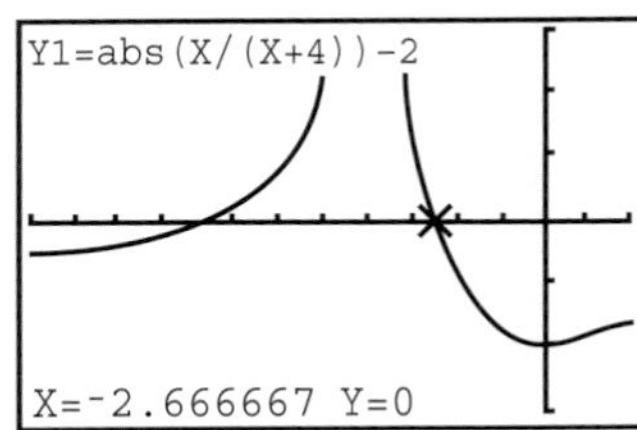

Example 38 – Algebraic and graphical methods

Solve the inequality $|x - 4| > 2|x - 7|$.

Solution

Method 1 – Algebraic

If $a > 0$, $b > 0$ and $a = b$, then $a^2 = b^2$. Since the expressions on both sides must be positive then we can square both sides and remove the absolute value signs.

$$(x - 4)^2 > (2(x - 7))^2$$
$$x^2 - 8x + 16 > 4(x^2 - 14x + 49)$$
$$x^2 - 8x + 16 > 4x^2 - 56x + 196$$
$$0 > 3x^2 - 48x + 180$$
$$0 > x^2 - 16x + 60$$
$$(x - 10)(x - 6) < 0$$

	$x < 6$	6	$6 < x < 10$	10	$x > 10$
$x - 10$	−		−	0	+
$x - 6$	−	0	+		+
$(x - 10)(x - 6)$	+	0	−	0	+

Therefore, the solution set is the open interval $6 < x < 10$.

Method 2 - Graphical

We can graph the two equations $y_1 = |x - 4|$ and $y_2 = 2|x - 7|$ and use our GDC to determine for what values of x the graph of y_1 is above the graph of y_2.

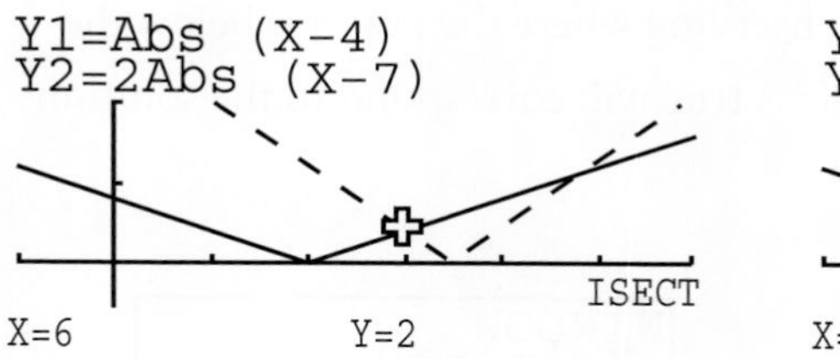

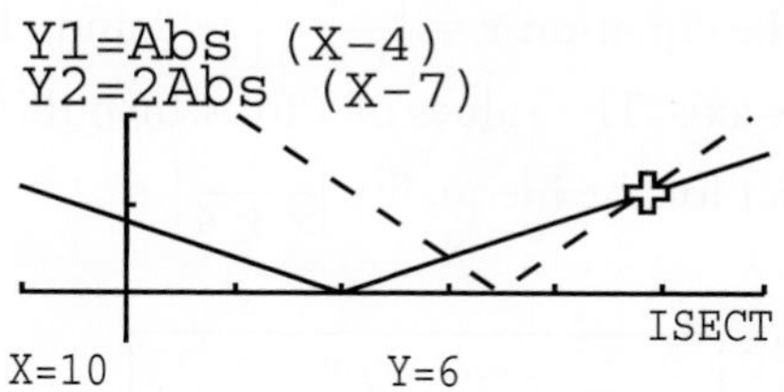

The equation $y_2 = 2|x - 7|$ has been graphed in a dashed style. By using the 'intersect' command on the GDC we find that the graph of y_1 is above the graph of y_2 for $6 < x < 10$. Therefore, the solution set is the open interval $6 < x < 10$.

Example 39 – Inequality involving rational expressions

For what values of x is $\frac{x}{x + 8} \leqslant \frac{1}{x - 1}$? Solve algebraically.

Solution

As applied in previous examples, an effective algebraic approach is to rearrange the inequality so that both fractions are on the same side with zero on the other side. Then combine the two fractions into one fraction and analyze where the fraction is zero, positive and negative.

		-8		-2		1		4		x
$x+2$	$-$		$-$	0	$+$		$+$		$+$	
$x-4$	$-$		$-$		$-$		$-$	0	$+$	
$x+8$	$-$	0	$+$		$+$		$+$		$+$	
$x-1$	$-$		$-$		$-$	0	$+$		$+$	
$\frac{(x+2)(x-4)}{(x+8)(x-1)}$	$+$	X	$-$		$+$	X	$-$		$+$	

$$\frac{x}{x+8} - \frac{1}{x-1} \leqslant 0$$

$$\frac{x(x-1)-(x+8)}{(x+8)(x-1)} \leqslant 0$$

$$\frac{x^2-2x-8}{(x+8)(x-1)} \leqslant 0$$

$$\frac{(x+2)(x-4)}{(x+8)(x-1)} \leqslant 0$$

Therefore, $\frac{x}{x+8} \leqslant \frac{1}{x-1}$ when $-8 < x \leqslant -2$ or $1 < x \leqslant 4$, which can also be expressed in interval notation as $]-8,-2] \cup]1,4]$.

Exercise 3.5

In questions 1–22, solve for x in the equation. If possible, find all real solutions and express them exactly. If this is not possible, then solve using your GDC and approximate any solutions to three significant figures. Be sure to check answers and to recognize any extraneous solutions.

1 $\sqrt{x+6} + 2x = 9$

2 $\sqrt{x+7} + 5 = x$

3 $\sqrt{7x+14} - 2 = x$

4 $\sqrt{2x+3} - \sqrt{x-2} = 2$

5 $\frac{5}{x+4} - \frac{4}{x} = \frac{21}{5x+20}$

6 $\frac{x+1}{2x+3} = \frac{5x-1}{7x+3}$

7 $\frac{1}{x} - \frac{1}{x+1} = \frac{1}{x+4}$

8 $\frac{2x}{1-x^2} + \frac{1}{x+1} = 2$

9 $x^4 - 2x^2 - 15 = 0$

10 $2x^{\frac{2}{3}} - x^{\frac{1}{3}} - 15 = 0$

11 $x^6 - 35x^3 + 216 = 0$

12 $5x^{-2} - x^{-1} - 2 = 0$

13 $|3x+4| = 8$

14 $|x+6| = |3x-24|$

15 $|5x+1| = 2x$

16 $|x-1| + |x| = 3$

17 $\left|\frac{x+1}{x-1}\right| = 3$

18 $\sqrt{x} - \frac{6}{\sqrt{x}} = 1$

19 $\sqrt{4-x} - \sqrt{6+x} = \sqrt{14+2x}$

20 $\frac{6}{x^2+1} = \frac{1}{x^2} + \frac{10}{x^2+4}$

21 $x - \sqrt{x+10} = 0$

22 $6x - 37\sqrt{x} + 56 = 0$

In questions 23–30, find the values of x that solve the inequality.

23 $3x^2 - 4 < 4x$

24 $\frac{2x-1}{x+2} \geqslant 1$

25 $2x^2 + 8x \leqslant 120$

26 $|1-4x| > 7$

27 $|x-3| > |x-14|$

28 $\left|\frac{x^2-4}{x}\right| \leqslant 3$

29 $\frac{x}{x-2} > \frac{1}{x+1}$

30 $\frac{4x-1}{x^2-2x-3} < 3$

31 Find the values of p for which the equation $px^2 - 3x + 1 = 0$ has a) one real solution, b) two real solutions, and c) no real solutions.

32 Given $f(x) = x^2 + x(k - 1) + k^2$, find the range of values of k so that $f(x) > 0$ for all real values of x.

33 Show that both of the following inequalities are true for all real numbers m and n such that $m > n > 0$.

a) $m + \frac{1}{n} > 2$ b) $(m + n)\left(\frac{1}{m} + \frac{1}{n}\right) > 4$

34 Find all of the exact solutions to the equation $(x^2 + x)^2 = 5x^2 + 5x - 6$.

35 If a, b and c are positive and unequal, show that $(a + b + c)^2 < 3(a^2 + b^2 + c^2)$.

36 Find the values of x that solve each inequality.

a) $\left|\frac{2x - 3}{x}\right| < 1$ b) $\frac{3}{x - 1} - \frac{2}{x + 1} < 1$

37 Provide a geometric or algebraic argument to show that $|a + b| \leq |a| + |b|$ for all $a, b \in \mathbb{R}$.

3.6 Partial fractions (Optional)

In arithmetic, when we add fractions we find the least common denominator. Then we multiply both the numerator and denominator of each term by what is needed to complete the common denominator. For example:

$$\frac{2}{3} + \frac{5}{7} = \frac{2}{3} \cdot \frac{7}{7} + \frac{5}{7} \cdot \frac{3}{3} = \frac{14 + 15}{21} = \frac{29}{21}$$

$$\frac{2}{3} + \frac{5}{9} + \frac{1}{27} = \frac{2}{3} \cdot \frac{9}{9} + \frac{5}{9} \cdot \frac{3}{3} + \frac{1}{27} = \frac{18 + 15 + 1}{27} = \frac{34}{27}$$

Reversing the process is called expressing each compound fraction as *partial fractions*. That is, given for example the fraction $\frac{29}{21} = \frac{29}{3 \times 7}$, we express it as a sum of two fractions. One fraction has denominator 3 and the other has denominator 7. Hence, we have the name *partial fractions*.

The process of finding the *partial fractions* is a straightforward process. We write:

$\frac{29}{3 \times 7} = \frac{a}{3} + \frac{b}{7}$ and then we solve for two integers a and b.

$$\frac{29}{3 \times 7} = \frac{a}{3} + \frac{b}{7} = \frac{7a + 3b}{21} \Rightarrow 7a + 3b = 29$$

Now by trial and error we can find that $a = 2$ and $b = 5$. Other answers are also possible $(-1, 12)$, $(8, -9)$ …

Notice the situation in the second example. The L.C.M. contains different powers of the same number. Consequently, when finding the partial fractions decomposition you need to consider that all powers less than or equal to the highest one may be present. That is, when we set up the process of decomposing $\frac{24}{27}$ we set it up in the following manner:

$$\frac{24}{27} = \frac{a}{27} + \frac{b}{9} + \frac{c}{3}$$

Then we attempt to find the values of a, b, and c.

In algebra, we carry out that process on the addition of rational expressions. Once again we multiply the numerator and denominator of each term by what was missing from the denominator of that term.

Partial fractions decomposition (PFD)

With partial fractions decomposition, we are going to reverse the process and decompose a rational expression into two or more simpler proper rational expressions. This is a very useful skill in which a single fraction with a factorable denominator is split into the sum of two or more fractions (partial fractions) whose denominators are the factors of the original denominator.

The method of partial fractions decomposition is extremely helpful in evaluating certain integrals as you will see in Section 16.5 (optional).

For example: $\frac{12x - 1}{2x^2 - 5x - 3} = \frac{2}{2x + 1} + \frac{5}{x - 3}$

Example 40

Find the partial fraction decomposition of $\frac{x + 1}{x^2 + 5x + 6}$.

Solution

$\frac{x + 1}{x^2 + 5x + 6} \equiv \frac{x + 1}{(x + 2)(x + 3)}$, and hence we will attempt to find two numbers a and b such that:

$$\frac{x + 1}{x^2 + 5x + 6} \equiv \frac{a}{x + 2} + \frac{b}{x + 3}$$

(Notice that we wrote this as an identity rather than equality because it has to be true for all values of x and not only for a few.)

$$\frac{x + 1}{x^2 + 5x + 6} \equiv \frac{a}{x + 2} + \frac{b}{x + 3} \equiv \frac{a(x + 3) + b(x + 2)}{(x + 2)(x + 3)}$$

Since the denominators of these identical fractions are the same, their numerators must also be the same. That is

$$x + 1 \equiv a(x + 3) + b(x + 2).$$

We have two methods of solution here.

First method

$$x + 1 \equiv a(x + 3) + b(x + 2) \Leftrightarrow x + 1 \equiv (a + b)x + (3a + 2b)$$

For two polynomials to be identical, the coefficients of the same powers must be the same, that is, the coefficient of x on the left must be the same as the coefficient of x on the right and similarly the constant terms. Hence:

$$1 = a + b \text{ and } 1 = 3a + 2b$$

$1 = a + b$ and $1 = 3a + 2b$

Now, solving the system with two equations will yield:

$a = -1$ and $b = 2$

Hence, $\frac{x+1}{x^2+5x+6} \equiv \frac{-1}{x+2} + \frac{2}{x+3}$.

This is also called the 'cover-up' method. This method allows the choice of numbers that are not initially in the domain of the original rational expression.

Second method

$x + 1 \equiv a(x+3) + b(x+2)$

Again, since this is an identity, the two sides must be the same for any choice of x. Hence, we can substitute any two numbers for x to get the value of each of a and b, specifically replacing x with -3 yields:

$x + 1 \equiv a(x+3) + b(x+2) \Rightarrow -2 = -b \Rightarrow b = 2.$

Notice how the choice of -3 eliminated the term with a and allowed us to find b directly. Replacing x with -2 yields:

$x + 1 \equiv a(x+3) + b(x+2) \Rightarrow -1 = a.$

This is of course the same result as above. Also notice here how the choice of -2 eliminated the term with b and allowed us to find a directly.

Note: This method is helpful in cases where there are no repeated factors.

The second method is faster whenever applicable. (We will discuss this in more detail later.)

Example 41

Find the PFD for $\frac{5x^2+16x+17}{2x^3+9x^2+7x-6}$.

Solution

$$\frac{5x^2+16x+17}{2x^3+9x^2+7x-6} \equiv \frac{5x^2+16x+17}{(2x-1)(x+2)(x+3)} \equiv \frac{a}{2x-1} + \frac{b}{x+2} + \frac{c}{x+3}$$

First method

$$5x^2+16x+17 \equiv a(x+2)(x+3) + b(2x-1)(x+3) + c(2x-1)(x+2)$$
$$\equiv (a+2b+2c)x^2 + (5a+5b+3c)x + 6a - 3b - 2c$$

This leads to this system: $\begin{cases} a + 2b + 2c = 5 \\ 5a + 5b + 3c = 16 \\ 6a - 3b - 2c = 17 \end{cases}$

Using any method of your choice for solving systems of equations, you should have:

$a = 3, b = -1, c = 2$ and hence:

$$\frac{5x^2+16x+17}{2x^3+9x^2+7x-6} \equiv \frac{3}{2x-1} - \frac{1}{x+2} + \frac{2}{x+3}$$

Second method

$5x^2 + 16x + 17 \equiv a(x+2)(x+3) + b(2x-1)(x+3) + c(2x-1)(x+2)$

$x = -2 \Rightarrow 5 = -5b \Rightarrow b = -1$

$x = -3 \Rightarrow 14 = 7c \Rightarrow c = 2$

$x = \frac{1}{2} \Rightarrow \frac{105}{4} = \frac{35}{4}a \Rightarrow a = 3$

Properties

1 Partial fractions decomposition only works for proper rational expressions, that is, the degree of the numerator must be less than the degree of the denominator. If it is not, then you must perform long division first, and then perform the partial fractions decomposition on the rational part (the remainder over the divisor). After you've done the partial fraction decomposition, just add back in the quotient part from the long division.

2 **Linear factors**: We can only decompose the partial fractions into proper rational expressions. Hence, in each partial fraction, when the denominator is linear, only a constant can be in the numerator. So, for every linear factor in the denominator, you will need a constant in the numerator. See Examples 40 and 41 above.

3 **Repeated linear factors**: If the denominator of the rational expression contains repeated linear factors, then following our discussion in the introduction, the process is as follows.

We need to include a factor in the expansion for each power possible. For example, if we have $(x-1)^3$, we will need to include $(x-1)$, an $(x-1)^2$, and $(x-1)^3$. Each of those $(x-1)$ factors would have a constant term in the numerator because $x-1$ is linear, no matter what power it is raised to.

For example: $\frac{13x^3 - 62x^2 + 101x - 58}{(x-1)^3(2x-5)} \equiv \frac{a}{(x-1)^3} + \frac{b}{(x-1)^2} + \frac{c}{x-1} + \frac{d}{2x-5}$

4 **Irreducible quadratic factors:** If the rational expression we are decomposing contains irreducible quadratic factors in the denominator, then the numerator could have a linear term and/or a constant term. So, for every irreducible quadratic factor in the denominator, you will need a linear term and a constant term in the numerator.

For example: $\frac{-8x^3 + 15x^2 - 26x + 33}{(x-1)^2(2x^2+5)} \equiv \frac{a}{(x-1)^2} + \frac{b}{x-1} + \frac{cx+d}{2x^2+5}$

Note: It may turn out that any of the numbers a, b, c, or d is zero.

Example 42

Write $\frac{3x-1}{x^2+4x+4}$ as the sum of partial fractions.

Solution

The first step is to factorise the denominator.

$x^2 + 4x + 4 = (x+2)^2$

Here the denominator has a repeated linear factor: $\frac{3x-1}{x^2+4x+4} = \frac{3x-1}{(x+2)^2}$

Because there are two (i.e. repeated) linear factors of $x + 2$ in the denominator of the original rational expression then it *must* have a partial fraction with a denominator of $(x + 2)^2$, and it *may* also have a partial fraction with a denominator of $x + 2$.

Thus, we are looking for constants A and B such that:

$$\frac{3x-1}{(x+2)^2} \equiv \frac{A}{x+2} + \frac{B}{(x+2)^2}$$

Multiplying both sides of the equation by $(x + 2)^2$ gives:

$$3x - 1 \equiv A(x + 2) + B$$

Essentially, the task is to find the unique values of A and B such that this equation is an identity, i.e. it is true for all values of x for which the original fraction is defined (in this case $x \neq -2$). However, as you recall, the 'cover-up' method allows us to choose 'helpful' values of x including such numbers. For example, in this case, if $x = -2$ then A is eliminated and the value of B can be found directly.

Let $x = -2$: $3x - 1 \equiv A(x + 2) + B \Rightarrow 3(-2) - 1 = A \cdot 0 + B$
$\Rightarrow B = -7$

Let $x = 0$: $3x - 1 \equiv A(x + 2) + B \Rightarrow 3 \cdot 0 - 1 = 2A - 7$
$\Rightarrow 2A = 6$
$\Rightarrow A = 3$

Therefore, $\dfrac{3x-1}{x^2+4x+4} = \dfrac{3}{x+2} - \dfrac{7}{(x+2)^2}$

Example 43

Write $\dfrac{2}{x^3+3x^2+2x}$ as the sum of partial fractions.

Solution

We first factorize the denominator and discover that one of the factors is an irreducible quadratic factor:

$$\frac{2}{x^3+3x^2+2x} = \frac{2}{x(x^2+2x+2)} \equiv \frac{a}{x} + \frac{bx+c}{x^2+2x+2}$$

Simplifying the expression gives:

$$2 \equiv a(x^2 + 2x + 2) + x(bx + c) \Rightarrow 2 \equiv (a + b)x^2 + (2a + c)x + 2a \Rightarrow$$

$$\begin{cases} a+b=0 \\ 2a+c=0 \\ 2a=2 \end{cases} \Rightarrow \begin{cases} a=1 \\ b=-1 \\ c=-2 \end{cases}$$

Therefore $\dfrac{2}{x^3+2x^2+2x} = \dfrac{1}{x} - \dfrac{x+2}{x^2+2x+2}$.

Exercise 3.6

Decompose each of the following rational expressions into partial fractions.

1 $\dfrac{5x+1}{x^2+x-2}$

2 $\dfrac{x+4}{x^2-2x}$

3 $\dfrac{x+2}{x^2+4x+3}$

4 $\dfrac{5x^2+20x+6}{x^3+2x^2+x}$

5 $\dfrac{2x^2+x-12}{x^3+5x^2+6x}$

6 $\dfrac{4x^2+2x-1}{x^3+x^2}$

7 $\dfrac{3}{x^2+x-2}$

8 $\dfrac{5-x}{2x^2+x-1}$

9 $\dfrac{3x+4}{(x+2)^2}$

10 $\dfrac{12}{x^4-x^3-2x^2}$

11 $\dfrac{2}{x^3+x}$

12 $\dfrac{x+2}{x^3+3x}$

13 $\dfrac{3x+2}{x^3+6x}$

14 $\dfrac{2x+3}{x^3+8x}$

15 $\dfrac{x+5}{x^3-4x^2-5x}$

Practice questions

1 Solve for x in the equation $x^2-(a+3b)x+3ab=0$.

2 Find the values of x that solve the following inequality.

$$\frac{3x-2}{5}+3 \geqslant \frac{4x-1}{3}$$

3 For what value of c is the vertex of the parabola $y=3x^2-8x+c$ at $\left(\frac{4}{3}, -\frac{1}{3}\right)$?

4 The quadratic function $f(x)=ax^2+bx+c$ has the following characteristics:
(i) passes through the point (2, 4); **(ii)** has a maximum value of 6 when $x=4$; and **(iii)** has a zero of $x=4+2\sqrt{3}$

Find the values of a, b and c.

5 If the roots of the equation $x^3+5x^2+px+q=0$ are ω, 2ω and $\omega+3$, find the values of ω, p and q.

6 Find all values of m such that the equation $mx^2-2(m+2)x+m+2=0$ has **a)** two real roots; **b)** two real roots (one positive and one negative).

7 $x-1$ and $x+1$ are factors of the polynomial x^3+ax^2+bx+c, and the polynomial has a remainder of 12 when divided by $x-2$. Find the values of a, b and c.

8 Solve the inequality $|x|<5|x-6|$.

9 Find the range of values for k in order for the equation $2x^2+(3-k)x+k+3=0$ to have two imaginary solutions.

10 Consider the rational function $f(x)=\dfrac{2x^2+8x+7}{x^2+4x+5}$. Do not use your GDC for this question.

a) Write $f(x)$ in the form $a-\dfrac{b}{(x+c)^2+d}$.

b) State the values of **(i)** $\lim_{x\to+\infty} f(x)$, and **(ii)** $\lim_{x\to-\infty} f(x)$.

c) State the coordinates of the minimum point on the graph of $f(x)$.

11 Find the values of k so that the equation $(k-2)x^2+4x-2k+1=0$ has two distinct real roots.

12 When the function $f(x)=6x^4+11x^3-22x^2+ax+6$ is divided by $(x+1)$ the remainder is -20. Find the value of a.

13 The polynomial $p(x)=(ax+b)^3$ leaves a remainder of -1 when divided by $(x+1)$, and a remainder of 27 when divided by $(x-2)$. Find the values of the real numbers a and b.

14 The polynomial $f(x)=x^3+3x^2+ax+b$ leaves the same remainder when divided by $(x-2)$ as when divided by $(x+1)$. Find the value of a.

15 When the polynomial x^4+ax+3 is divided by $(x-1)$, the remainder is 8. Find the value of a.

16 The polynomial x^3+ax^2-3x+b is divisible by $(x-2)$ and has a remainder 6 when divided by $(x+1)$. Find the value of a and of b.

17 The polynomial x^2-4x+3 is a factor of $x^3+(a-4)x^2+(3-4a)x+3$. Calculate the value of the constant a.

18 Consider $f(x)=x^3-2x^2-5x+k$. Find the value of k if $(x+2)$ is a factor of $f(x)$.

19 Find the real number k for which $1+ki\,(i=\sqrt{-1})$ is a zero of the polynomial z^2+kz+5.

20 The equation $kx^2-3x+(k+2)=0$ has two distinct real roots. Find the set of possible values of k.

21 Consider the equation $(1+2k)x^2-10x+k-2=0$, $k\in\mathbb{R}$. Find the set of values of k for which the equation has real roots.

22 Find the range of values of m such that for all x

$$m(x+1)\leqslant x^2.$$

23 Find the values of x for which $|5-3x|\leqslant|x+1|$.

24 Solve the inequality $x^2-4+\dfrac{3}{x}<0$.

25 Solve the inequality $|x-2|\geqslant|2x+1|$.

26 Let $f(x)=\dfrac{x+4}{x+1}$, $x\neq-1$ and $g(x)=\dfrac{x-2}{x-4}$, $x\neq4$.

Find the set of values of x such that $f(x)\leqslant g(x)$.

27 Solve the inequality $\left|\dfrac{x+9}{x-9}\right|\leqslant2$.

28 Given that $2+i$ is a root of the equation $x^3-6x^2+13x-10=0$ find the other two roots.

29 Find all values of x that satisify the inequality $\dfrac{2x}{|x-1|}<1$.

4 Sequences and Series

Assessment statements

1.1 Arithmetic sequences and series; sum of finite arithmetic sequences; geometric sequences and series; sum of finite and infinite geometric series.
Sigma notation.

1.3 Counting principles, including permutations and combinations. The binomial theorem: expansion of $(a + b)^n$, $n \in \mathbb{N}$.

1.4 Proof by mathematical induction.

Introduction

The heights of consecutive bounds of a ball, compound interest, and Fibonacci numbers are only a few of the applications of sequences and series that you have seen in previous courses. In this chapter you will review these concepts, consolidate your understanding and take them one step further.

4.1 Sequences

Take the following pattern as an example:

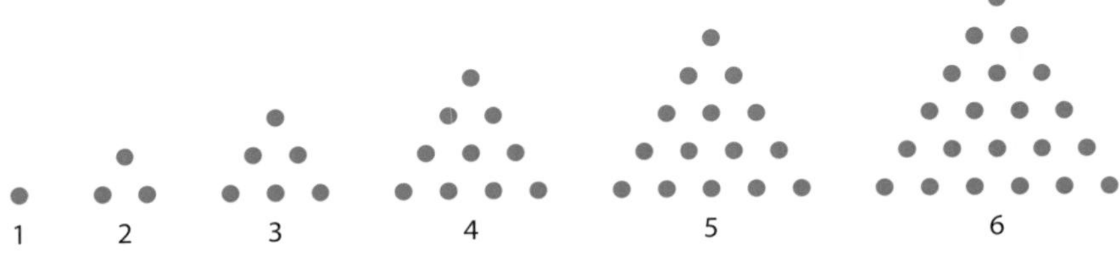

The first figure represents 1 dot, the second represents 3 dots, etc. This pattern can also be described differently. For example, in function notation:

$$f(1) = 1, f(2) = 3, f(3) = 6, \text{etc., where the domain is } \mathbb{Z}^+$$

Here are some more examples of sequences:

1 6, 12, 18, 24, 30

2 $3, 9, 27, \ldots, 3^k, \ldots$

3 $\left\{\frac{1}{i^2}; i = 1, 2, 3, \ldots, 10\right\}$

4 $\{b_1, b_2, \ldots, b_n, \ldots\}$, sometimes used with an abbreviation $\{b_n\}$

The first and third sequences are **finite** and the second and fourth are **infinite**. Notice that, in the second and third sequences, we were able to define a rule that yields the nth number in the sequence (called the nth term) as a function of n, the term's number. In this sense, a sequence is a **function** that assigns a **unique** number (a_n) to each positive integer n.

Example 1

Find the first five terms and the 50th term of the sequence $\{b_n\}$ such that $b_n = 2 - \frac{1}{n^2}$.

Solution

Since we know an *explicit* expression for the nth term as a *function* of its number n, we only need to find the value of that function for the required terms:

$$b_1 = 2 - \frac{1}{1^2} = 1;\ b_2 = 2 - \frac{1}{2^2} = 1\frac{3}{4};\ b_3 = 2 - \frac{1}{3^2} = 1\frac{8}{9};\ b_4 = 2 - \frac{1}{4^2} = 1\frac{15}{16};$$

$$b_5 = 2 - \frac{1}{5^2} = 1\frac{24}{25};\ \text{and}\ b_{50} = 2 - \frac{1}{50^2} = 1\frac{2499}{2500}.$$

So, informally, **a sequence is an ordered set of real numbers**. That is, there is a first number, a second, and so forth. The notation used for such sets is shown above. The way we defined the function in Example 1 is called the **explicit** definition of a sequence. There are other ways to define sequences, one of which is the **recursive** definition. The following example will show you how this is used.

Example 2

Find the first five terms and the 20th term of the sequence $\{b_n\}$ such that $b_1 = 5$ and $b_n = 2(b_{n-1} + 3)$.

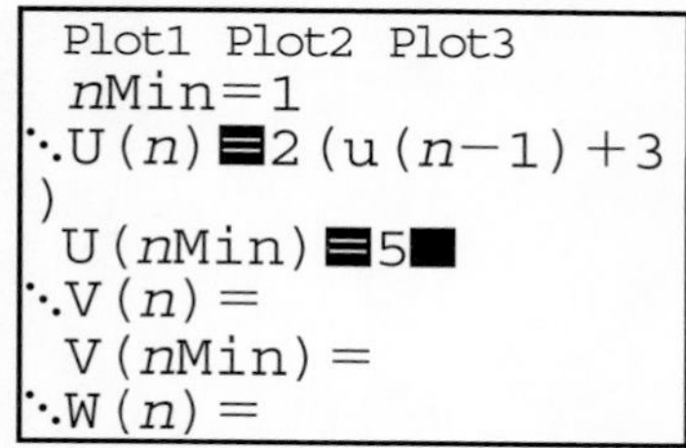


```
U(5)
                170
U(20)
            5767162
```

Solution

The defining formula for this sequence is recursive. It allows us to find the nth term b_n if we know the preceding term b_{n-1}. Thus, we can find the second term from the first, the third from the second, and so on. Since we know the first term, $b_1 = 5$, we can calculate the rest:

$$b_2 = 2(b_1 + 3) = 2(5 + 3) = 16$$
$$b_3 = 2(b_2 + 3) = 2(16 + 3) = 38$$
$$b_4 = 2(b_3 + 3) = 2(38 + 3) = 82$$
$$b_5 = 2(b_4 + 3) = 2(82 + 3) = 170$$

Thus, the first five terms of this sequence are 5, 16, 38, 82, 170. However, to find the 20th term, we must first find all 19 preceding terms. This is one of the drawbacks of the recursive definition, unless we can change the definition into explicit form. This can easily be done using a GDC.

Example 3

A Fibonacci sequence is defined recursively as

$$F_n = \begin{cases} 1 & n = 1 \\ 1 & n = 2 \\ F_{n-1} + F_{n-2} & n > 2 \end{cases}$$

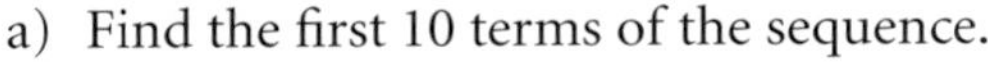

a) Find the first 10 terms of the sequence.

b) Evaluate $S_n = \sum_{i=1}^{n} F_i$ for $n = 1, 2, 3, \ldots, 10$.

c) By observing that $F_1 = F_3 - F_2$, $F_2 = F_4 - F_3$, and so on, derive a formula for the sum of the first n Fibonacci numbers.

Fibonacci numbers are a sequence of numbers named after Leonardo of Pisa, known as Fibonacci (a short form of filius Bonaccio, 'son of Bonaccio').

Solution

a) 1, 1, 2, 3, 5, 8, 13, 21, 34, 55

b) $S_1 = 1, S_2 = 2, S_3 = 4, S_4 = 7, S_5 = 12, S_6 = 20, S_7 = 33, S_8 = 54,$
$S_9 = 88, S_{10} = 143$

c) Since $F_3 = F_2 + F_1$, then

$$\begin{aligned} F_1 &= F_3 - F_2 \\ F_2 &= F_4 - F_3 \\ F_3 &= F_5 - F_4 \\ F_4 &= F_6 - F_5 \\ &\vdots \\ F_n &= F_{n+2} - F_{n+1} \\ \hline S_n &= F_{n+2} - F_2 \end{aligned}$$

Notice that $S_5 = 12 = F_7 - F_2 = 13 - 1$ and $S_8 = 54 = F_{10} - F_2 = 55 - 1$.

Note: parts a) and b) can be made easy by using a spreadsheet. Here is an example:

	A	B	C	D
1	F(n)	S(n)		
2	1	1		
3	1	2		
4	2	4		
5	3	7		
6	5	12		
7	8	20		
8	13	33		
9	21	54		
10	34	88		
11	55	143		Let this cell be A2 + A3 Then copy it down
12	89	232		
13	144	376		
14	233	609		
15	377	986		Let this cell be B10 + A11 Then copy it down
16	610	1596		
17	987	2583		

Notice that not all sequences have formulae, either recursive or explicit. Some sequences are given only by listing their terms. Among the many kinds of sequences that there are, two types are of interest to us: arithmetic and geometric sequences, which we will discuss in the next two sections.

Exercise 4.1

Find the first five terms of each infinite sequence defined in questions 1–6.

1 $s(n) = 2n - 3$

2 $g(k) = 2^k - 3$

3 $f(n) = 3 \times 2^{-n}$

4 $\begin{cases} a_1 = 5 \\ a_n = a_{n-1} + 3; \text{ for } n > 1 \end{cases}$

5 $a_n = (-1)^n(2^n) + 3$

6 $\begin{cases} b_1 = 3 \\ b_n = b_{n-1} + 2n; \text{ for } n \geqslant 2 \end{cases}$

Find the first five terms and the 50th term of each infinite sequence defined in questions 7–14.

7 $a_n = 2n - 3$

8 $b_n = 2 \times 3^{n-1}$

9 $u_n = (-1)^{n-1}\dfrac{2n}{n^2 + 2}$

10 $a_n = n^{n-1}$

11 $a_n = 2a_{n-1} + 5$ and $a_1 = 3$

12 $u_{n+1} = \dfrac{3}{2u_n + 1}$ and $u_1 = 0$

13 $b_n = 3 \cdot b_{n-1}$ and $b_1 = 2$

14 $a_n = a_{n-1} + 2$ and $a_1 = -1$

Suggest a recursive definition for each sequence in questions 15–17.

15 $\frac{1}{3}, \frac{1}{12}, \frac{1}{48}, \frac{1}{192}, \ldots$

16 $\frac{1}{2}a, \frac{2}{3}a^3, \frac{8}{9}a^5, \frac{32}{27}a^7, \ldots$

17 $a - 5k, 2a - 4k, 3a - 3k, 4a - 2k, 5a - k, \ldots$

In questions 18–21, write down a possible formula that gives the nth term of each sequence.

18 $4, 7, 12, 19, \ldots$

19 $2, 5, 8, 11, \ldots$

20 $1, \frac{3}{4}, \frac{5}{9}, \frac{7}{16}, \frac{9}{25}, \ldots$

21 $\frac{1}{4}, \frac{3}{5}, \frac{5}{6}, 1, \frac{9}{8}, \ldots$

22 Define $a_n = \dfrac{F_{n+1}}{F_n}, n > 1$, where F_n is a member of a Fibonacci sequence.

a) Write the first 10 terms of a_n.

b) Show that $a_n = 1 + \dfrac{1}{a_{n-1}}$

23 Define the sequence

$$F_n = \frac{1}{\sqrt{5}}\left(\frac{(1 + \sqrt{5})^n - (1 - \sqrt{5})^n}{2^n}\right)$$

a) Find the first 10 terms of this sequence and compare them to Fibonacci numbers.

b) Show that $3 \pm \sqrt{5} = \dfrac{(1 \pm \sqrt{5})^2}{2}$.

c) Use the result in b) to verify that F_n satisfies the recursive definition of Fibonacci sequences.

4.2 Arithmetic sequences

Examine the following sequences and the most likely recursive formula for each of them.

7, 14, 21, 28, 35, 42, ... $\quad a_1 = 7$ and $a_n = a_{n-1} + 7$, for $n > 1$

2, 11, 20, 29, 38, 47, ... $\quad a_1 = 2$ and $a_n = a_{n-1} + 9$, for $n > 1$

48, 39, 30, 21, 12, 3, −6, ... $\quad a_1 = 48$ and $a_n = a_{n-1} - 9$, for $n > 1$

Note that in each case above, every term is formed by adding a constant number to the preceding term. Sequences formed in this manner are called **arithmetic sequences**.

Definition of an arithmetic sequence

A sequence $a_1, a_2, a_3, \ldots$ is an **arithmetic sequence** if there is a constant d for which

$$a_n = a_{n-1} + d$$

for all integers $n > 1$. d is called the **common difference** of the sequence, and $d = a_n - a_{n-1}$ for all integers $n > 1$.

So, for the sequences above, 7 is the common difference for the first, 9 is the common difference for the second and −9 is the common difference for the third.

This description gives us the recursive definition of the arithmetic sequence. It is possible, however, to find the explicit definition of the sequence.

Applying the recursive definition repeatedly will enable you to see the expression we are seeking:

$$a_2 = a_1 + d;\ a_3 = a_2 + d = a_1 + d + d = a_1 + 2d;$$
$$a_4 = a_3 + d = a_1 + 2d + d = a_1 + 3d;\ \ldots$$

So, as you see, you can get to the nth term by adding d to a_1, $(n - 1)$ times, and therefore:

***n*th term of an arithmetic sequence**

The general (nth) term of an arithmetic sequence, a_n, with first term a_1 and common difference d, may be expressed explicitly as

$$a_n = a_1 + (n - 1)d$$

This result is useful in finding any term of the sequence without knowing all the previous terms.

Note: The arithmetic sequence can be looked at as a linear function as explained in the introduction to this chapter, i.e. for every increase of one unit in n, the value of the term will increase by d units. As the first term is a_1, the point $(1, a_1)$ belongs to this function. The constant increase d can be considered to be the gradient (slope) of this linear model; hence, the nth term, the dependent variable in this case, can be found by using the *point-slope* form of the equation of a line:

$$y - y_1 = m(x - x_1)$$
$$a_n - a_1 = d(n - 1) \Leftrightarrow a_n = a_1 + (n - 1)d$$

This agrees with our definition of an arithmetic sequence.

Example 4

Find the nth and the 50th terms of the sequence 2, 11, 20, 29, 38, 47, …

Solution

This is an arithmetic sequence whose first term is 2 and common difference is 9. Therefore,

$$a_n = a_1 + (n - 1)d = 2 + (n - 1) \times 9 = 9n - 7$$

$$\Rightarrow a_{50} = 9 \times 50 - 7 = 443$$

Example 5

Find the recursive and the explicit forms of the definition of the following sequence, then calculate the value of the 25th term.

13, 8, 3, −2, …

Solution

This is clearly an arithmetic sequence, since we observe that −5 is the common difference.

Recursive definition: $a_1 = 13$

$$a_n = a_{n-1} - 5$$

Explicit definition: $a_n = 13 - 5(n - 1) = 18 - 5n$, and

$$a_{25} = 18 - 5 \times 25 = -107$$

Example 6

Find a definition for the arithmetic sequence whose first term is 5 and fifth term is 11.

Solution

Since the fifth term is given, using the explicit form, we have

$$a_5 = a_1 + (5 - 1)d \Rightarrow 11 = 5 + 4d \Rightarrow d = \frac{3}{2}$$

This leads to the general term,

$a_n = 5 + \frac{3}{2}(n - 1)$, or, equivalently, the recursive form

$$\begin{cases} a_1 = 5 \\ a_n = a_{n-1} + \frac{3}{2}, n > 1 \end{cases}$$

• **Hint:** Definition: In a finite arithmetic sequence $a_1, a_2, a_3, \ldots, a_k$, the terms $a_2, a_3 \ldots, a_{k-1}$ are called **arithmetic means** between a_1 and a_k.

Example 7

Insert four arithmetic means between 3 and 7.

Solution

Since there are four means between 3 and 7, the problem can be reduced to a situation similar to Example 6 by considering the first term to be 3 and the sixth term to be 7. The rest is left as an exercise for you!

Exercise 4.2

1 Insert four arithmetic means between 3 and 7.

2 Say whether each given sequence is an arithmetic sequence. If yes, find the common difference and the 50th term; if not, say why not.

a) $a_n = 2n - 3$

b) $b_n = n + 2$

c) $c_n = c_{n-1} + 2$, and $c_1 = -1$

d) $u_n = 3u_{n-1} + 2$

e) 2, 5, 7, 12, 19, ...

f) 2, −5, −12, −19, ...

For each arithmetic sequence in questions 3–8, find:

a) the 8th term

b) an explicit formula for the nth term

c) a recursive formula for the nth term.

3 −2, 2, 6, 10, ...

4 29, 25, 21, 17, ...

5 −6, 3, 12, 21, ...

6 10.07, 9.95, 9.83, 9.71, ...

7 100, 97, 94, 91, ...

8 $2, \frac{3}{4}, -\frac{1}{2}, -\frac{7}{4}, \ldots$

9 Find five arithmetic means between 13 and −23.

10 Find three arithmetic means between 299 and 300.

11 In an arithmetic sequence, $a_5 = 6$ and $a_{14} = 42$. Find an explicit formula for the nth term of this sequence.

12 In an arithmetic sequence, $a_3 = -40$ and $a_9 = -18$. Find an explicit formula for the nth term of this sequence.

In each of questions 13–17, the first 3 terms and the last term of an arithmetic sequence are given. Find the number of terms.

13 3, 9, 15, ..., 525

14 9, 3, −3, ..., −201

15 $3\frac{1}{8}, 4\frac{1}{4}, 5\frac{3}{8}, \ldots, 14\frac{3}{8}$

16 $\frac{1}{3}, \frac{1}{2}, \frac{2}{3}, \ldots, 2\frac{5}{6}$

17 $1 - k, 1 + k, 1 + 3k, \ldots, 1 + 19k$

18 Find five arithmetic means between 15 and −21.

19 Find three arithmetic means between 99 and 100.

20 In an arithmetic sequence, $a_3 = 11$ and $a_{12} = 47$. Find an explicit formula for the nth term of this sequence.

21 In an arithmetic sequence, $a_7 = -48$ and $a_{13} = -10$. Find an explicit formula for the nth term of this sequence.

22 The 30th term of an arithmetic sequence is 147 and the common difference is 4. Find a formula for the nth term.

23 The first term of an arithmetic sequence is −7 and the common difference is 3. Is 9803 a term of this sequence? If so, which one?

24 The first term of an arithmetic sequence is 9689 and the 100th term is 8996. Show that the 110th term is 8926. Is 1 a term of this sequence? If so, which one?

25 The first term of an arithmetic sequence is 2 and the 30th term is 147. Is 995 a term of this sequence? If so, which one?

4.3 Geometric sequences

Examine the following sequences and the most likely recursive formula for each of them.

7, 14, 28, 56, 112, 224, … $a_1 = 7$ and $a_n = a_{n-1} \times 2$, for $n > 1$

2, 18, 162, 1458, 13 122, … $a_1 = 2$ and $a_n = a_{n-1} \times 9$, for $n > 1$

48, −24, 12, −6, 3, −1.5, … $a_1 = 48$ and $a_n = a_{n-1} \times -0.5$, for $n > 1$

Note that in each case above, every term is formed by multiplying a constant number with the preceding term. Sequences formed in this manner are called **geometric sequences.**

Definition of a geometric sequence

A sequence $a_1, a_2, a_3, \ldots$ is a **geometric sequence** if there is a constant r for which

$$a_n = a_{n-1} \times r$$

for all integers $n > 1$. r is called the **common ratio** of the sequence, and $r = a_n \div a_{n-1}$ for all integers $n > 1$.

So, for the sequences above, 2 is the common ratio for the first, 9 is the common ratio for the second and −0.5 is the common ratio for the third.

This description gives us the recursive definition of the geometric sequence. It is possible, however, to find the explicit definition of the sequence.

Applying the recursive definition repeatedly will enable you to see the expression we are seeking:

$$a_2 = a_1 \times r;\ a_3 = a_2 \times r = a_1 \times r \times r = a_1 \times r^2;$$
$$a_4 = a_3 \times r = a_1 \times r^2 \times r = a_1 \times r^3; \ldots$$

So, as you see, you can get to the nth term by multiplying a_1 with r, $(n - 1)$ times, and therefore:

***n*th term of geometric sequence**

The general (nth) term of a geometric sequence, a_n, with common ratio r and first term a_1, may be expressed explicitly as

$$a_n = a_1 \times r^{(n-1)}$$

This result is useful in finding any term of the sequence without knowing all the previous terms.

Example 8

a) Find the geometric sequence with $a_1 = 2$ and $r = 3$.

b) Describe the sequence 3, −12, 48, −192, 768, …

c) Describe the sequence $1, \frac{1}{2}, \frac{1}{4}, \frac{1}{8}, \ldots$

d) Graph the sequence $a_n = \frac{1}{4} \cdot 3^{n-1}$

Solution

a) The geometric sequence is 2, 6, 18, 54, …, $2 \times 3^{n-1}$. Notice that the ratio of a term to the preceding term is 3.

b) This is a geometric sequence with $a_1 = 3$ and $r = -4$. The nth term is $a_n = 3 \times (-4)^{n-1}$. Notice that, when the common ratio is negative, the terms of the sequence alternate in sign.

c) The nth term of this sequence is $a_n = 1 \cdot \left(\frac{1}{2}\right)^{n-1}$. Notice that the ratio of any two consecutive terms is $\frac{1}{2}$. Also, notice that the terms decrease in value.

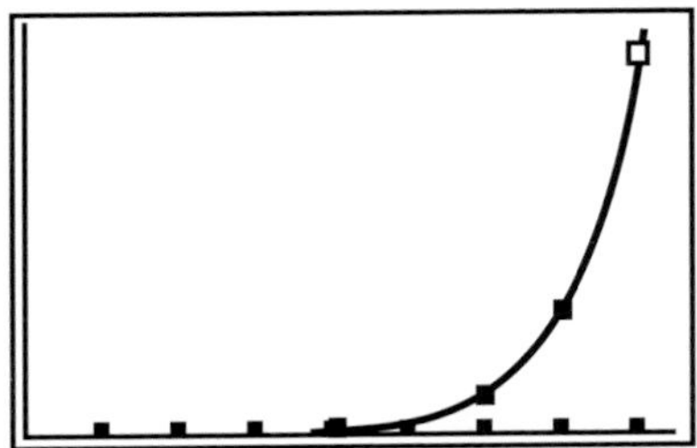

d) The graph of the geometric sequence is shown on the left. Notice that the points lie on the graph of the function $y = \frac{1}{4} \cdot 3^{x-1}$.

Example 9

At 8:00 a.m., 1000 mg of medicine is administered to a patient. At the end of each hour, the concentration of medicine is 60% of the amount present at the beginning of the hour.

a) What portion of the medicine remains in the patient's body at noon if no additional medication has been given?

b) If a second dosage of 1000 mg is administered at 10:00 a.m., what is the total concentration of the medication in the patient's body at noon?

Solution

a) We use the geometric model, as there is a constant multiple by the end of each hour. Hence, the concentration at the end of any hour after administering the medicine is given by:

$a_n = a_1 \times r^{(n-1)}$, where n is the number of hours

Thus, at noon $n = 5$, and $a_5 = 1000 \times 0.6^{(5-1)} = 129.6$.

b) For the second dosage, the amount of medicine at noon corresponds to $n = 3$, and $a_3 = 1000 \times 0.6^{(3-1)} = 360$.

So, the concentration of medicine is $129.6 + 360 = 489.6$ mg.

Compound interest

Interest compounded annually

When we borrow money we pay interest, and when we invest money we receive interest. Suppose an amount of €1000 is put into a savings account that bears an annual interest of 6%. How much money will we have in the bank at the end of four years?

It is important to note that the 6% interest is given annually and is added to the savings account, so that in the following year it will also earn interest, and so on.

Time in years	Amount in the account
0	1000
1	$1000 + 1000 \times 0.06 = 1000(1 + 0.06)$
2	$1000(1 + 0.06) + (1000(1 + 0.06)) \times 0.06 = 1000(1 + 0.06)(1 + 0.06) = 1000(1 + 0.06)^2$
3	$1000(1 + 0.06)^2 + (1000(1 + 0.06)^2) \times 0.06 = 1000(1 + 0.06)^2(1 + 0.06) = 1000(1 + 0.06)^3$
4	$1000(1 + 0.06)^3 + (1000(1 + 0.06)^3) \times 0.06 = 1000(1 + 0.06)^3(1 + 0.06) = \mathbf{1000(1 + 0.06)^4}$

Table 4.1 Compound interest.

This appears to be a geometric sequence with five terms. You will notice that the number of terms is five, as both the beginning and the end of the first year are counted. (Initial value, when time = 0, is the first term.)

In general, if a **principal** of P euros is invested in an account that yields an interest rate r (expressed as a decimal) annually, and this interest is added at the end of the year, every year, to the principal, then we can use the geometric sequence formula to calculate the **future value** A, which is accumulated after t years.

If we repeat the steps above, with

$$A_0 = P = \text{initial amount}$$
$$r = \text{annual interest rate}$$
$$t = \text{number of years}$$

it becomes easier to develop the formula:

Table 4.2 Compound interest formula.

Time in years	Amount in the account
0	$A_0 = P$
1	$A_1 = P + Pr = P(1 + r)$
2	$A_2 = A_1(1 + r) = P(1 + r)^2$
⋮	
t	$A_t = P(1 + r)^t$

Notice that since we are counting from 0 to t, we have $t + 1$ terms, and hence using the geometric sequence formula,

$$a_n = a_1 \times r^{(n-1)} \Rightarrow A_t = A_0 \times (1 + r)^t$$

Interest compounded *n* times per year

Suppose that the principal P is invested as before but the interest is paid n times per year. Then $\frac{r}{n}$ is the interest paid every compounding period. Since every year we have n periods, for t years, we have nt periods. The amount A in the account after t years is

$$A = P\left(1 + \frac{r}{n}\right)^{nt}$$

Example 10

€1000 is invested in an account paying compound interest at a rate of 6%. Calculate the amount of money in the account after 10 years if

a) the compounding is annual

b) the compounding is quarterly

c) the compounding is monthly.

Solution

a) The amount after 10 years is

$A = 1000(1 + 0.06)^{10} = €1790.85.$

b) The amount after 10 years quarterly compounding is

$A = 1000\left(1 + \frac{0.06}{4}\right)^{40} = €1814.02.$

c) The amount after 10 years monthly compounding is

$A = 1000\left(1 + \frac{0.06}{12}\right)^{120} = €1819.40.$

Example 11

You invested €1000 at 6% compounded quarterly. How long will it take this investment to increase to €2000?

Solution

Let $P = 1000$, $r = 0.06$, $n = 4$ and $A = 2000$ in the compound interest formula:

$$A = P\left(1 + \frac{r}{n}\right)^{nt}$$

Then solve for t:

$$2000 = 1000\left(1 + \frac{0.06}{4}\right)^{4t} \Rightarrow 2 = 1.015^{4t}$$

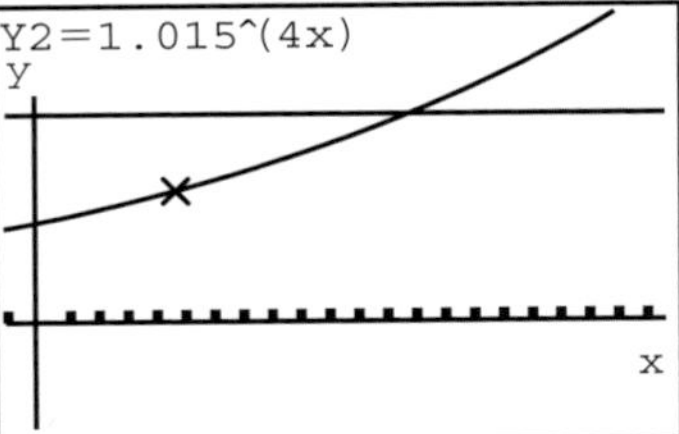

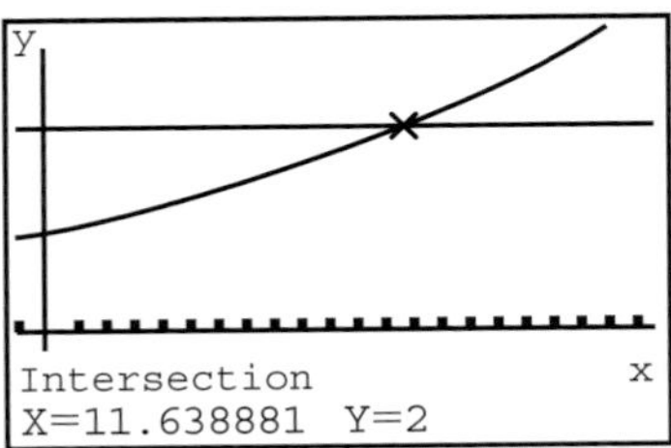

Using a GDC, we can graph the functions $y = 2$ and $y = 1.015^{4t}$ and then find the intersection between their graphs.

As you can see, it will take the €1000 investment 11.64 years to double to €2000. This translates into approximately 47 quarters.

You can check your work to see that this is accurate by using the compound interest formula:

$$A = 1000\left(1 + \frac{0.06}{4}\right)^{47} = €2013.28$$

Later in the book, you will learn how to solve the problem algebraically.

Example 12

You want to invest €1000. What interest rate is required to make this investment grow to €2000 in 10 years if interest is compounded quarterly?

Solution

Let $P = 1000$, $n = 4$, $t = 10$ and $A = 2000$ in the compound interest formula:

$$A = P\left(1 + \frac{r}{n}\right)^{nt}$$

Now solve for r:

$$2000 = 1000\left(1 + \frac{r}{4}\right)^{40} \Rightarrow 2 = \left(1 + \frac{r}{4}\right)^{40} \Rightarrow 1 + \frac{r}{4} = \sqrt[40]{2} \Rightarrow r = 4(\sqrt[40]{2} - 1)$$
$$= 0.0699$$

So, at a rate of 7% compounded quarterly, the €1000 investment will grow to at least €2000 in 10 years.

You can check to see whether your work is accurate by using the compound interest formula:

$$A = 1000\left(1 + \frac{0.07}{4}\right)^{40} = €2001.60$$

Population growth

The same formulae can be applied when dealing with population growth.

Example 13

The city of Baden in Lower Austria grows at an annual rate of 0.35%. The population of Baden in 1981 was 23 140. What is the estimate of the population of this city for 2013?

Solution

This situation can be modelled by a geometric sequence whose first term is 23 140 and whose common ratio is 1.0035. Since we count the population of 1981 among the terms, the number of terms is 33.

2013 is equivalent to the 33rd term in this sequence. The estimated population for Baden is, therefore,

$$\text{Population (2013)} = a_{31} = 23\,140(1.0035)^{32} = 25\,877$$

Note: Later in the book, more realistic population growth models will be explored and more efficient methods will be developed, as well as the ability to calculate interest that is continuously compounded.

Exercise 4.3

In each of questions 1–15 determine whether the sequence in each question is arithmetic, geometric, or neither. Find the common difference for the arithmetic ones and the common ratio for the geometric ones. Find the common difference or ratio and the 10th term for each arithmetic or geometric one as appropriate.

1 $3, 3^{a+1}, 3^{2a+1}, 3^{3a+1}, \ldots$

2 $a_n = 3n - 3$

3 $b_n = 2^{n+2}$

4 $c_n = 2c_{n-1} - 2$, and $c_1 = -1$

5 $u_n = 3u_{n-1}, u_1 = 4$

6 2, 5, 12.5, 31.25, 78.125, …

7 2, −5, 12.5, −31.25, 78.125, …

8 2, 2.75, 3.5, 4.25, 5, …

9 $18, -12, 8, -\frac{16}{3}, \frac{32}{9}, \ldots$

10 52, 55, 58, 61, …

11 −1, 3, −9, 27, −81, …

12 0.1, 0.2, 0.4, 0.8, 1.6, 3.2, …

13 3, 6, 12, 18, 21, 27, …

14 6, 14, 20, 28, 34, …

15 2.4, 3.7, 5, 6.3, 7.6, …

For each arithmetic or geometric sequence in questions 16–32 find

a) the 8th term
b) an explicit formula for the nth term
c) a recursive formula for the nth term.

16 $-3, 2, 7, 12, \ldots$

17 $19, 15, 11, 7, \ldots$

18 $-8, 3, 14, 25, \ldots$

19 $10.05, 9.95, 9.85, 9.75, \ldots$

20 $100, 99, 98, 97, \ldots$

21 $2, \frac{1}{2}, -1, -\frac{5}{2}, \ldots$

22 $3, 6, 12, 24, \ldots$

23 $4, 12, 36, 108, \ldots$

24 $5, -5, 5, -5, \ldots$

25 $3, -6, 12, -24, \ldots$

26 $972, -324, 108, -36, \ldots$

27 $-2, 3, -\frac{9}{2}, \frac{27}{4} \ldots$

28 $35, 25, \frac{125}{7}, \frac{625}{49}, \ldots$

29 $-6, -3, -\frac{3}{2}, -\frac{3}{4}, \ldots$

30 $9.5, 19, 38, 76, \ldots$

31 $100, 95, 90.25, \ldots$

32 $2, \frac{3}{4}, \frac{9}{32}, \frac{27}{256}, \ldots$

33 Insert 4 geometric means between 3 and 96.

- **Hint:** Definition: In a finite geometric sequence $a_1, a_2, a_3, \ldots, a_k$, the terms $a_2, a_3, \ldots, a_{k-1}$ are called *geometric means* between a_1 and a_k.

34 Find 3 geometric means between 7 and 4375.

35 Find a geometric mean between 16 and 81.

- **Hint:** This is also called the *mean proportional*.

36 Find 4 geometric means between 7 and 1701.

37 Find a geometric mean between 9 and 64.

38 The first term of a geometric sequence is 24 and the fourth term is 3, find the fifth term and an expression for the nth term.

39 The first term of a geometric sequence is 24 and the third term is 6, find the fourth term and an expression for the nth term.

40 The common ratio in a geometric sequence is $\frac{2}{7}$ and the fourth term is $\frac{14}{3}$. Find the third term.

41 Which term of the geometric sequence 6, 18, 54, … is 118 098?

42 The 4th term and the 7th term of a geometric sequence are 18 and $\frac{729}{8}$. Is $\frac{59049}{128}$ a term of this sequence? If so, which term is it?

43 The 3rd term and the 6th term of a geometric sequence are 18 and $\frac{243}{4}$. Is $\frac{19683}{64}$ a term of this sequence? If so, which term is it?

44 Jim put €1500 into a savings account that pays 4% interest compounded semiannually. How much will his account hold 10 years later if he does not make any additional investments in this account?

45 At her daughter Jane's birth, Charlotte set aside £500 into a savings account. The interest she earned was 4% compounded quarterly. How much money will Jane have on her 16th birthday?

46 How much money should you invest now if you wish to have an amount of €4000 in your account after 6 years if interest is compounded quarterly at an annual rate of 5%?

47 In 2007, the population of Switzerland was estimated to be 7554 (in thousands). How large would the Swiss population be in 2012 if it grows at a rate of 0.5% annually?

48 The common ratio in a geometric sequence is $\frac{3}{7}$ and the fourth term is $\frac{14}{3}$. Find the third term.

49 Which term of the geometric sequence 7, 21, 63, … is 137 781?

50 Tim put €2500 into a savings account that pays 4% interest compounded semiannually. How much will his account hold 10 years later if he does not make any additional investments in this account?

51 At her son William's birth, Jane set aside £1000 into a savings account. The interest she earned was 6% compounded quarterly. How much money will William have on his 18th birthday?

4.4 Series

The word 'series' in common language implies much the same thing as 'sequence'. But in mathematics when we talk of a series, we are referring in particular to sums of terms in a sequence, e.g. for a sequence of values a_n, the corresponding series is the sequence of S_n with

$$S_n = a_1 + a_2 + \ldots + a_{n-1} + a_n$$

If the terms are in an arithmetic sequence, we call the sum an **arithmetic series.**

Sigma notation

Most of the series we consider in mathematics are **infinite** series. This name is used to emphasize the fact that the series contain infinitely many terms. Any sum in the series S_k will be called a partial sum and is given by

$$S_k = a_1 + a_2 + \ldots + a_{k-1} + a_k$$

For convenience, this partial sum is written using the sigma notation:

$$S_k = \sum_{i=1}^{i=k} a_i = a_1 + a_2 + \ldots + a_{k-1} + a_k$$

Sigma notation is a concise and convenient way to represent long sums. Here, the symbol Σ is the Greek capital letter *sigma* that refers to the initial letter of the word 'sum'. So, the expression $\sum_{i=1}^{i=k} a_i$ means the sum of all the terms a_i, where i takes the values from 1 to k. We can also write $\sum_{i=m}^{n} a_i$ to mean the sum of the terms a_i, where i takes the values from m to n. In such a sum, m is called the lower limit and n the upper limit.

Example 14

Write out what is meant by:

a) $\sum_{i=1}^{5} i^4$ b) $\sum_{r=3}^{7} 3^r$ c) $\sum_{j=1}^{n} x_j p(x_j)$

Solution

a) $\sum_{i=1}^{5} i^4 = 1^4 + 2^4 + 3^4 + 4^4 + 5^4$

b) $\sum_{r=3}^{7} 3^r = 3^3 + 3^4 + 3^5 + 3^6 + 3^7$

c) $\sum_{j=1}^{n} x_j p(x_j) = x_1 p(x_1) + x_2 p(x_2) + \ldots + x_n p(x_n)$

Example 15

Evaluate $\sum_{n=0}^{5} 2^n$

Solution

$$\sum_{n=0}^{5} 2^n = 2^0 + 2^1 + 2^2 + 2^3 + 2^4 + 2^5 = 63$$

Example 16

Write the sum $\frac{1}{2} - \frac{2}{3} + \frac{3}{4} - \frac{4}{5} + \ldots + \frac{99}{100}$ in sigma notation.

Solution

We notice that each term's numerator and denominator are consecutive integers, so they take on the absolute value of $\frac{k}{k+1}$ or any equivalent form. We also notice that the signs of the terms alternate and that we have 99 terms. To take care of the sign, we use some power of (-1) that will start with a positive value. If we use $(-1)^k$, the first term will be negative, so we can use $(-1)^{k+1}$ instead. We can, therefore, write the sum as

$$(-1)^{1+1}\frac{1}{2} + (-1)^{2+1}\frac{2}{3} + (-1)^{3+1}\frac{3}{4} + \ldots + (-1)^{99+1}\frac{99}{100} = \sum_{k=1}^{99} (-1)^{k+1}\frac{k}{k+1}$$

Properties of the sigma notation

There are a number of useful results that we can obtain when we use sigma notation.

1 For example, suppose we had a sum of constant terms

$$\sum_{i=1}^{5} 2$$

What does this mean? If we write this out in full, we get

$$\sum_{i=1}^{5} 2 = 2 + 2 + 2 + 2 + 2 = 5 \times 2 = 10.$$

In general, if we sum a constant n times then we can write

$$\sum_{i=1}^{n} k = k + k + \ldots + k = n \times k = nk.$$

2 Suppose we have the sum of a constant times i. What does this give us? For example,

$$\sum_{i=1}^{5} 5i = 5 \times 1 + 5 \times 2 + 5 \times 3 + 5 \times 4 + 5 \times 5 = 5 \times (1 + 2 + 3 + 4 + 5) = 75.$$

However, this can also be interpreted as follows

$$\sum_{i=1}^{5} 5i = 5 \times 1 + 5 \times 2 + 5 \times 3 + 5 \times 4 + 5 \times 5 = 5 \times (1 + 2 + 3 + 4 + 5) = 5\sum_{i=1}^{5} i$$

which implies that

$$\sum_{i=1}^{5} 5i = 5\sum_{i=1}^{5} i$$

In general, we can say

$$\begin{aligned}\sum_{i=1}^{n} ki &= k \times 1 + k \times 2 + \ldots + k \times n \\ &= k \times (1 + 2 + \ldots + n) \\ &= k\sum_{i=1}^{n} i\end{aligned}$$

3 Suppose that we need to consider the summation of two different functions, such as

$$\begin{aligned}\sum_{k=1}^{n} (k^2 + k^3) &= (1^2 + 1^3) + (2^2 + 2^3) + \ldots + n^2 + n^3 \\ &= (1^2 + 2^2 + \ldots + n^2) + (1^3 + 2^3 + \ldots + n^3) \\ &= \sum_{k=1}^{n} (k^2) + \sum_{k=1}^{n} (k^3)\end{aligned}$$

In general,

$$\sum_{k=1}^{n} (f(k) + g(k)) = \sum_{k=1}^{n} f(k) + \sum_{k=1}^{n} g(k)$$

Arithmetic series

In arithmetic series, we are concerned with adding the terms of arithmetic sequences. It is very helpful to be able to find an easy expression for the partial sums of this series.

Let us start with an example:

Find the partial sum for the first 50 terms of the series

$$3 + 8 + 13 + 18 + \ldots$$

We express S_{50} in two different ways:

$$\begin{aligned}S_{50} &= 3 + 8 + 13 + \ldots + 248, \text{ and} \\ S_{50} &= 248 + 243 + 238 + \ldots + 3 \\ \hline 2S_{50} &= 251 + 251 + 251 + \ldots + 251\end{aligned}$$

There are 50 terms in this sum, and hence

$$2S_{50} = 50 \times 251 \Rightarrow S_{50} = \frac{50}{2}(251) = 6275.$$

This reasoning can be extended to any arithmetic series in order to develop a formula for the nth partial sum S_n.

Let $\{a_n\}$ be an arithmetic sequence with first term a_1 and a common difference d. We can construct the series in two ways: Forward, by adding d to a_1 repeatedly, and backwards by subtracting d from a_n repeatedly. We get the following two expressions for the sum:

$$S_n = a_1 + a_2 + a_3 + \ldots + a_n = a_1 + (a_1 + d) + (a_1 + 2d) + \ldots + (a_1 + (n-1)d)$$

and

$$S_n = a_n + a_{n-1} + a_{n-2} + \ldots + a_1 = a_n + (a_n - d) + (a_n - 2d) + \ldots + (a_n - (n-1)d)$$

By adding, term by term vertically, we get

$$\begin{array}{rcl} S_n &=& a_1 + (a_1 + d) + (a_1 + 2d) + \ldots + (a_1 + (n-1)d) \\ S_n &=& a_n + (a_n - d) + (a_n - 2d) + \ldots + (a_n - (n-1)d) \\ \hline 2S_n &=& (a_1 + a_n) + (a_1 + a_n) + (a_1 + a_n) + \ldots + (a_1 + a_n) \end{array}$$

Since we have n terms, we can reduce the expression above to

$2S_n = n(a_1 + a_n)$, which can be reduced to

$S_n = \frac{n}{2}(a_1 + a_n)$, which in turn can be changed to give an interesting perspective of the sum,

i.e. $S_n = n\left(\frac{a_1 + a_n}{2}\right)$ is n times the average of the first and last terms!

If we substitute $a_1 + (n-1)d$ for a_n then we arrive at an alternative formula for the sum:

$$S_n = \frac{n}{2}(a_1 + a_1 + (n-1)d) = \frac{n}{2}(2a_1 + (n-1)d)$$

Sum of an arithmetic series

The sum, S_n, of n terms of an arithmetic series with common difference d, first term a_1, and nth term a_n is:

$$S_n = \frac{n}{2}(a_1 + a_n) \quad \text{or} \quad S_n = \frac{n}{2}(2a_1 + (n-1)d)$$

Example 17

Find the partial sum for the first 50 terms of the series

$$3 + 8 + 13 + 18 + \ldots$$

Solution

Using the second formula for the sum, we get

$$S_{50} = \frac{50}{2}(2 \times 3 + (50 - 1)5) = 25 \times 251 = 6275.$$

Using the first formula requires that we know the nth term. So, $a_{50} = 3 + 49 \times 5 = 248$, which now can be used:

$$S_{50} = 25(3 + 248) = 6275.$$

Geometric series

As is the case with arithmetic series, it is often desirable to find a general expression for the nth partial sum of a geometric series.

Let us start with an example:

Find the partial sum for the first 20 terms of the series

$$3 + 6 + 12 + 24 + \dots$$

We express S_{20} in two different ways and subtract them:

$$\begin{aligned} S_{20} &= 3 + 6 + 12 + \dots + 1\,572\,864 \\ 2S_{20} &= \quad\; 6 + 12 + \dots + 1\,572\,864 + 3\,145\,728 \\ \hline -S_{20} &= 3 \qquad\qquad\qquad\qquad\qquad - 3\,145\,728 \\ \Rightarrow S_{20} &= 3\,145\,725 \end{aligned}$$

This reasoning can be extended to any geometric series in order to develop a formula for the nth partial sum S_n.

Let $\{a_n\}$ be a geometric sequence with first term a_1 and a common ratio $r \neq 1$. We can construct the series in two ways as before and using the definition of the geometric sequence, i.e. $a_n = a_{n-1} \times r$, then

$$\begin{aligned} S_n &= a_1 + a_2 + a_3 + \dots + a_{n-1} + a_n, \text{ and} \\ rS_n &= ra_1 + ra_2 + ra_3 + \dots + ra_{n-1} + ra_n \\ &= a_2 + a_3 + \dots + a_{n-1} + a_n + ra_n \end{aligned}$$

Now, we subtract the first and last expressions to get

$$S_n - rS_n = a_1 - ra_n \Rightarrow S_n(1-r) = a_1 - ra_n \Rightarrow S_n = \frac{a_1 - ra_n}{1-r}; r \neq 1.$$

This expression, however, requires that r, a_1, as well as a_n be known in order to find the sum. However, using the nth term expression developed earlier, we can simplify this sum formula to

$$S_n = \frac{a_1 - ra_n}{1-r} = \frac{a_1 - ra_1r^{n-1}}{1-r} = \frac{a_1(1-r^n)}{1-r}; r \neq 1.$$

Sum of a geometric series

The sum, S_n, of n terms of a geometric series with common ratio r ($r \neq 1$) and first term a_1, is:

$$S_n = \frac{a_1(1-r^n)}{1-r} \quad \left[\text{equivalent to } S_n = \frac{a_1(r^n-1)}{r-1}\right]$$

Example 18

Find the partial sum for the first 20 terms of the series $3 + 6 + 12 + 24 + \dots$ in the opening example for this section.

Solution

$$S_{20} = \frac{3(1-2^{20})}{1-2} = \frac{3(1-1\,048\,576)}{-1} = 3\,145\,725$$

Infinite geometric series

Consider the series

$$\sum_{k=1}^{n} 2\left(\tfrac{1}{2}\right)^{k-1} = 2 + 1 + \tfrac{1}{2} + \tfrac{1}{4} + \tfrac{1}{8} + \dots$$

Consider also finding the partial sums for 10, 20 and 100 terms. The sums we are looking for are the partial sums of a geometric series. So,

$$\sum_{k=1}^{10} 2\left(\tfrac{1}{2}\right)^{k-1} = 2 \times \frac{1 - \left(\tfrac{1}{2}\right)^{10}}{1 - \tfrac{1}{2}} \approx 3.996$$

$$\sum_{k=1}^{20} 2\left(\tfrac{1}{2}\right)^{k-1} = 2 \times \frac{1 - \left(\tfrac{1}{2}\right)^{20}}{1 - \tfrac{1}{2}} \approx 3.999\,996$$

$$\sum_{k=1}^{100} 2\left(\tfrac{1}{2}\right)^{k-1} = 2 \times \frac{1 - \left(\tfrac{1}{2}\right)^{100}}{1 - \tfrac{1}{2}} \approx 4$$

As the number of terms increases, the partial sum appears to be approaching the number 4. This is no coincidence. In the language of limits,

$$\lim_{n\to\infty} \sum_{k=1}^{n} 2\left(\tfrac{1}{2}\right)^{k-1} = \lim_{n\to\infty} 2 \times \frac{1 - \left(\tfrac{1}{2}\right)^{k}}{1 - \tfrac{1}{2}} = 2 \times \frac{1 - 0}{\tfrac{1}{2}} = 4, \text{since } \lim_{n\to\infty}\left(\tfrac{1}{2}\right)^{n} = 0.$$

This type of problem allows us to extend the usual concept of a 'sum' of a **finite** number of terms to make sense of sums in which an **infinite** number of terms is involved. Such series are called **infinite series**.

One thing to be made clear about infinite series is that they are not true sums! The associative property of addition of real numbers allows us to extend the definition of the sum of two numbers, such as $a + b$, to three or four or n numbers, but not to an infinite number of numbers. For example, you can add any specific number of 5s together and get a real number, but if you add an *infinite* number of 5s together, you cannot get a real number! The remarkable thing about infinite series is that, in some cases, such as the example above, the sequence of partial sums (which are true sums) approach a finite limit L. The limit in our example is 4. This we write as

$$\lim_{n\to\infty} \sum_{k=1}^{n} a_k = \lim_{n\to\infty} (a_1 + a_2 + \ldots + a_n) = L.$$

We say that the series **converges** to L, and it is convenient to define L as the **sum of the infinite series**. We use the notation

$$\sum_{k=1}^{\infty} a_k = \lim_{n\to\infty} \sum_{k=1}^{n} a_k = L.$$

We can, therefore, write the limit above as

$$\sum_{k=1}^{\infty} 2\left(\tfrac{1}{2}\right)^{k-1} = \lim_{n\to\infty} \sum_{k=1}^{n} 2\left(\tfrac{1}{2}\right)^{k-1} = 4.$$

If the series does not have a limit, it **diverges** and does not have a sum.

We are now ready to develop a general rule for **infinite geometric series**. As you know, the sum of the geometric series is given by

$$S_n = \frac{a_1 - ra_n}{1 - r} = \frac{a_1 - ra_1r^{n-1}}{1 - r} = \frac{a_1(1 - r^n)}{1 - r}; r \neq 1.$$

If $|r| < 1$, then $\lim_{n\to\infty} r^n = 0$ and

$$S_n = S = \lim_{n\to\infty} \frac{a_1(1 - r^n)}{1 - r} = \frac{a_1}{1 - r}.$$

We will call this **the sum of the infinite geometric series.** In all other cases the series diverges. The proof is left as an exercise.

$$\sum_{k=1}^{\infty} 2\left(\frac{1}{2}\right)^{k-1} = \frac{2}{1-\frac{1}{2}} = 4, \text{ as already shown.}$$

Sum of an infinite geometric series

The sum, S_∞, of an infinite geometric series with first term a_1, such that the common ratio r satisfies the condition $|r| < 1$ is given by:

$$S_\infty = \frac{a_1}{1-r}$$

Example 19

A rational number is a number that can be expressed as a quotient of two integers. Show that $0.\overline{6} = 0.666\ldots$ is a rational number.

Solution

$$0.\overline{6} = 0.666\ldots = 0.6 + 0.06 + 0.006 + 0.0006 + \ldots$$
$$= \frac{6}{10} + \frac{6}{10}\cdot\frac{1}{10} + \frac{6}{10}\cdot\left(\frac{1}{10}\right)^2 + \frac{6}{10}\cdot\left(\frac{1}{10}\right)^3 + \ldots$$

This is an infinite geometric series with $a_1 = \frac{6}{10}$ and $r = \frac{1}{10}$; therefore,

$$0.\overline{6} = \frac{\frac{6}{10}}{1-\frac{1}{10}} = \frac{6}{10}\cdot\frac{10}{9} = \frac{2}{3}$$

Example 20

If a ball has elasticity such that it bounces up 80% of its previous height, find the total vertical distances travelled down and up by this ball when it is dropped from an altitude of 3 metres. Ignore friction and air resistance.

Solution

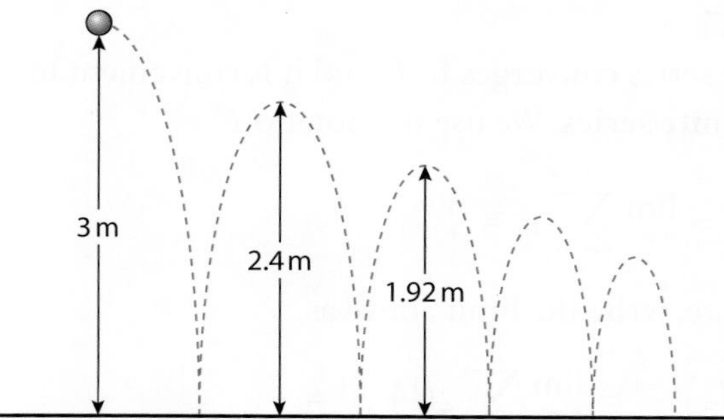

After the ball is dropped the initial 3 m, it bounces up and down a distance of 2.4 m. Each bounce after the first bounce, the ball travels 0.8 times the previous height twice – once upwards and once downwards. So, the total vertical distance is given by

$$h = 3 + 2(2.4 + (2.4 \times 0.8) + (2.4 \times 0.8^2) + \ldots) = 3 + 2 \times l$$

The amount in parenthesis is an infinite geometric series with $a_1 = 2.4$ and $r = 0.8$. The value of that quantity is

$$l = \frac{2.4}{1-0.8} = 12.$$

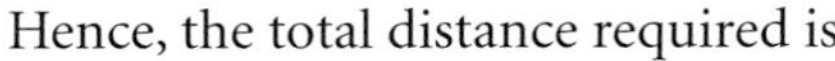

Hence, the total distance required is

$$h = 3 + 2(12) = 27 \text{ m}.$$

Applications of series to compound interest calculations

Annuities

An **annuity** is a sequence of equal periodic payments. If you are saving money by depositing the same amount at the end of each compounding period, the annuity is called **ordinary annuity**. Using geometric series you can calculate the **future value (FV)** of this annuity, which is the amount of money you have after making the last payment.

You invest €1000 at the end of each year for 10 years at a fixed annual interest rate of 6%. See table below.

Year	Amount invested	Future value
10	1000	1000
9	1000	$1000(1 + 0.06)$
8	1000	$1000(1 + 0.06)^2$
⋮		
1	1000	$1000(1 + 0.06)^9$

Table 4.3 Calculating the future value.

The future value of this investment is the sum of all the entries in the last column, so it is

$$FV = 1000 + 1000(1 + 0.06) + 1000(1 + 0.06)^2 + \ldots + 1000(1 + 0.06)^9$$

This sum is a partial sum of a geometric series with $n = 10$ and $r = 1 + 0.06$. Hence,

$$FV = \frac{1000(1 - (1 + 0.06)^{10})}{1 - (1 + 0.06)} = \frac{1000(1-(1 + 0.06)^{10})}{-0.06} = 13\,180.79.$$

```
Plot1 Plot2 Plot3
nMin=1
\u(n)■u(n−1)*(1+
0.06)
 u(nMin)■1000
\v(n)=
 v(nMin)=
\w(n)=
```

```
sum(seq(u(n),n,1,
10)
            13180.79494
```

This result can also be produced with a GDC, as shown.

We can generalize the previous formula in the same manner. Let the periodic payment be R and the periodic interest rate be i, i.e. $i = \frac{r}{n}$. Let the number of periodic payments be m.

Period	Amount invested	Future value
m	R	R
$m - 1$	R	$R(1 + i)$
$m - 2$	R	$R(1 + i)^2$
⋮		
1	R	$R(1 + i)^{m-1}$

Table 4.4 Calculating the future value – formula.

The future value of this investment is the sum of all the entries in the last column, so it is

$$FV = R + R(1 + i) + R(1 + i)^2 + \ldots + R(1 + i)^{m-1}$$

This sum is a partial sum of a geometric series with m terms and $r = 1 + i$. Hence,

$$FV = \frac{R(1 - (1 + i)^m)}{1 - (1 + i)} = \frac{R(1 - (1 + i)^m)}{-i} = R\left(\frac{(1 + i)^m - 1}{i}\right)$$

Note: If the payment is made at the beginning of the period rather than at the end, the annuity is called **annuity due** and the future value after m periods will be slightly different. The table for this situation is given below.

Table 4.5 Calculating the future value (annuity due).

Period	Amount invested	Future value
m	R	$R(1 + i)$
$m - 1$	R	$R(1 + i)^2$
$m - 2$	R	$R(1 + i)^3$
$\vdots$		
1	R	$R(1 + i)^m$

The future value of this investment is the sum of all the entries in the last column, so it is

$$FV = R(1 + i) + R(1 + i)^2 + \ldots + R(1 + i)^{m-1} + R(1 + i)^m$$

This sum is a partial sum of a geometric series with m terms and $r = 1 + i$. Hence,

$$FV = \frac{R(1 + i(1 - (1 + i)^m))}{1 - (1 + i)} = \frac{R(1 + i - (1 + i)^{m+1})}{-i} = R\left(\frac{(1 + i)^{m+1} - 1}{i} - 1\right)$$

If the previous investment is made at the beginning of the year rather than at the end, then in 10 years we have

$$FV = R\left(\frac{(1 + i)^{m+1} - 1}{i} - 1\right) = 1000\left(\frac{(1 + 0.06)^{10+1} - 1}{0.006} - 1\right) = 13\,971.64.$$

Exercise 4.4

1 Find the sum of the arithmetic series $11 + 17 + \ldots + 365$.

2 Find the sum:
$2 - 3 + \frac{9}{2} - \frac{27}{4} + \ldots - \frac{177\,147}{1024}$

3 Evaluate $\sum_{k=0}^{13} (2 - 0.3k)$.

4 Evaluate $2 - \frac{4}{5} + \frac{8}{25} - \frac{16}{125} + \ldots$

5 Evaluate $\frac{1}{3} + \frac{\sqrt{3}}{12} + \frac{1}{16} + \frac{\sqrt{3}}{64} + \frac{3}{256} + \ldots$

6 Express each repeating decimal as a fraction:
a) $0.\overline{52}$ b) $0.4\overline{53}$ c) $3.01\overline{37}$

7 At the beginning of every month, Maggie invests £150 in an account that pays 6% annual rate. How much money will there be in the account after six years?

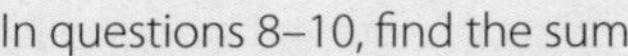

In questions 8–10, find the sum.

8 $9 + 13 + 17 + \ldots + 85$

9 $8 + 14 + 20 + \ldots + 278$

10 $155 + 158 + 161 + \ldots + 527$

11 The kth term of an arithmetic sequence is $2 + 3k$. Find, in terms of n, the sum of the first n terms of this sequence.

12 How many terms should we add to exceed 678 when we add $17 + 20 + 23 \ldots$?

13 How many terms should we add to exceed 2335 when we add $-18 - 11 - 4 \ldots$?

14 An arithmetic sequence has a as first term and $2d$ as common difference, i.e., a, $a + 2d, a + 4d, \ldots$. The sum of the first 50 terms is T. Another sequence, with first term $a + d$, and common difference $2d$, is combined with the first one to produce a new arithmetic sequence. Let the sum of the first 100 terms of the new combined sequence be S. If $2T + 200 = S$, find d.

15 Consider the arithmetic sequence $3, 7, 11, \ldots, 999$.

a) Find the number of terms and the sum of this sequence.

b) Create a new sequence by removing every third term, i.e., $11, 23, \ldots$. Find the sum of the terms of the remaining sequence.

16 The sum of the first 10 terms of an arithmetic sequence is 235 and the sum of the second 10 terms is 735. Find the first term and the common difference.

In questions 17–19, use your GDC or a spreadsheet to evaluate each sum.

17 $\displaystyle\sum_{k=1}^{20}(k^2 + 1)$

18 $\displaystyle\sum_{i=3}^{17}\frac{1}{i^2 + 3}$

19 $\displaystyle\sum_{n=1}^{100}(-1)^n\frac{3}{n}$

20 Find the sum of the arithmetic series

$13 + 19 + \ldots + 367$

21 Find the sum

$2 - \frac{4}{3} + \frac{8}{9} - \frac{16}{27} + \ldots - \frac{4096}{177\,147}$

22 Evaluate $\displaystyle\sum_{k=0}^{11}(3 + 0.2k)$.

23 Evaluate $2 - \frac{4}{3} + \frac{8}{9} - \frac{16}{27} + \ldots$

24 Evaluate $\frac{1}{2} + \frac{\sqrt{2}}{2\sqrt{3}} + \frac{1}{3} + \frac{\sqrt{2}}{3\sqrt{3}} + \frac{2}{9} + \ldots$

In questions 25–27, find the first four partial sums and then the nth partial sum of each sequence.

25 $u_n = \frac{3}{5^n}$

26 $v_n = \frac{1}{n^2 + 3n + 2}$ Hint: Show that $v_n = \frac{1}{n+1} - \frac{1}{n+2}$

27 $u_n = \sqrt{n+1} - \sqrt{n}$

28 A ball is dropped from a height of 16 m. Every time it hits the ground it bounces 81% of its previous height.

a) Find the maximum height it reaches *after* the 10th bounce.

b) Find the total distance travelled by the ball till it rests. (Assume no friction and no loss of elasticity).

29

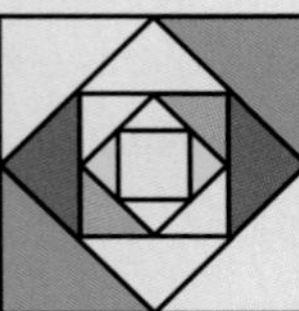

The sides of a square are 16 cm in length. A new square is formed by joining the midpoints of the adjacent sides and two of the resulting triangles are coloured as shown.

a) If the process is repeated 6 more times, determine the total area of the shaded region.

b) If the process is repeated indefinitely, find the total area of the shaded region.

30

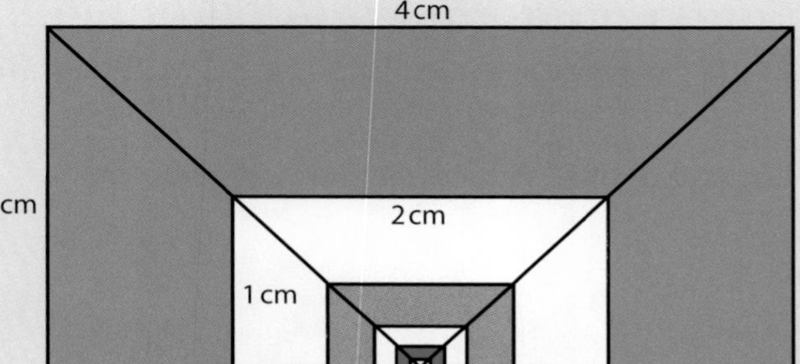

The largest rectangle has dimensions 4 by 2, as shown; another rectangle is constructed inside it with dimensions 2 by 1. The process is repeated. The region surrounding every other inner rectangle is shaded, as shown.

a) Find the total area for the three regions shaded already.

b) If the process is repeated indefinitely, find the total area of the shaded regions.

In questions 31–34, find each sum.

31 $7 + 12 + 17 + 22 + \dots + 337 + 342$

32 $9486 + 9479 + 9472 + 7465 + \dots + 8919 + 8912$

33 $2 + 6 + 18 + 54 + \dots + 3\,188\,646 + 9\,565\,938$

34 $120 + 24 + \frac{24}{5} + \frac{24}{25} + \dots + \frac{24}{78\,125}$

4.5 Counting principles

Simple counting problems

This section will introduce you to some of the basic principles of counting. In Section 4.6 you will apply some of this in justifying the binomial theorem and in Chapter 12 you will use these principles to tackle many probability problems. We will start with two examples.

Example 21

Nine paper chips each carrying the numerals 1–9 are placed in a box. Two chips are chosen such that the first chip is chosen, the number is recorded and the chip is put back in the box, then the second chip is drawn. The numbers on the chips are added. In how many ways can you get a sum of 8?

Solution

To solve this problem, count the different number of ways that a total of 8 can be obtained:

1st chip	1	2	3	4	5	6	7
2nd chip	7	6	5	4	3	2	1

From this list, it is clear that you can have 7 different ways of receiving a sum of 8.

Example 22

Suppose now that the first chip is chosen, the number is recorded and the chip is *not* put back in the box, then the second chip is drawn. In how many ways can you get a sum of 8?

Solution

To solve this problem too, count the different number of ways that a total of 8 can be obtained:

1st chip	1	2	3	5	6	7
2nd chip	7	6	5	3	2	1

From this list, it is clear that you can have 6 different ways of receiving a sum of 8.

The difference between the two situations is described by saying that the first random selection is done **with replacement**, while the second is **without replacement**, which ruled out the use of two 4s.

Fundamental principle of counting

The above examples show you simple counting principles in which you can list each possible way that an event can happen. In many other cases, listing the ways an event can happen may not be feasible. In such cases we need to rely on counting principles. The most important of which is the **fundamental principle of counting**, also known as the multiplication principle. Consider the following situations:

Example 23

You can make a sandwich from one of three types of bread and one of four kinds of cheese, with or without pickles. How many different kinds of sandwiches can be made?

Solution

With each type of bread you can have 4 sandwiches. There are 12 possible sandwiches altogether. These are without pickles; if you want sandwiches with pickles, then you have 24 possible ones. That is, there are $3 \times 4 \times 2 = 24$ possible sandwiches.

Example 24

How many 3-digit even numbers are there?

Solution

The first digit cannot be zero, since the number has to be a 3-digit number, so there are 9 ways the hundred's digit can be. There is no condition on what the ten's digit should be, so we have 10 possibilities, and to be even, the number must end with 0, 2, 4, 6, or 8. Therefore, we have $9 \times 10 \times 5 = 450$ 3-digit even numbers.

Examples 23 and 24 are examples of the following principle:

Fundamental principle of counting

If there are m ways an event can occur followed by n ways a second event can occur, then there are a total of $(m)(n)$ ways that the two can occur.

This principle can be extended to more than two events or processes:

If there are k events than can happen in $n_1, n_2, \ldots, n_k$ ways, then the whole sequence can happen in

$$n_1 \times n_2 \times \ldots \times n_k \text{ ways.}$$

Example 25

A large school issues special coded identification cards that consist of two letters of the alphabet followed by three numerals. For example, AB 737 is such a code. How many different ID cards can be issued if the letters or numbers can be used more than once?

Solution

As the letters can be used more than once, then each letter position can be filled in 26 different ways, i.e. the letters can be filled in $26 \times 26 = 676$ ways. Each number position can be filled in 10 different ways; hence, the numerals can be filled in $10 \times 10 \times 10 = 1000$ different ways. So, the code can be formed in $676 \times 1000 = 676\,000$ different ways.

Permutations

One major application of the fundamental principle is in determining the number of ways the n objects can be arranged. Consider the following situation for example. You have 5 books you want to put on a shelf: maths (M), physics (P), English (E), biology (B), and history (H). In how many ways can you do this?

To find this out, number the positions you want to place the books in as shown

If we decide to put the maths book in position 1, then there are four different ways of putting a book in position 2.

M	P				M	E				M	B				M	H			
1	2	3	4	5	1	2	3	4	5	1	2	3	4	5	1	2	3	4	5

Since we can put any of the 5 books in the first position, then there will be $5 \times 4 = 20$ ways of shelving the first two books. Once you place the books in positions 1 and 2, the third book can be any one of three books left.

M	P	E			M	P	B			M	P	H		
1	2	3	4	5	1	2	3	4	5	1	2	3	4	5

Once you use three books, there are two books for the fourth position and only one way of placing the fifth book. So, the number of ways of arranging all 5 books is

$$5 \times 4 \times 3 \times 2 \times 1 = 120 = 5!$$

Factorial notation

The product of the first n positive integers is denoted by $n!$ and is called ***n* factorial:**

$$n! = 1 \times 2 \times 3 \times 4 \ldots (n-2) \times (n-1) \times n$$

We also define $0! = 1$.

Permutations

An arrangement is called a **permutation**. It is the reorganization of objects or symbols into distinguishable sequences. When we place things in order, we say we have made an arrangement. When we change the order, we say we have changed the arrangement. So each of the arrangements that can be made by taking *some* or *all* of a number of things is known as a **permutation**.

• **Hint:** A permutation of n different objects can be understood as an ordering (arrangement) of the objects such that one object is first, one is second, one is third, and so on.

Number of permutations of *n* objects

The previous set up can be applied to n objects rather than only 5. The number of ways of filling in the first position can be done in n ways.

n	$n-1$	$n-2$	$n-3$	…	1
1	2	3	4		n

Once the first position is filled, the second position can be filled by any of the $n-1$ objects left, and hence using the fundamental principle there will be $n \cdot (n-1)$ different ways for filling the first two positions. Repeating the same procedure till the nth position is filled is therefore

$$n \cdot (n-1) \cdot (n-2) \ldots 2 \cdot 1 = n!$$

Frequently, we are engaged in arranging a **subset** of the whole collection rather than the entire collection. For example, suppose we want to shelve 3 of the books rather than all 5 of them. The discussion will be analogous to the previous situation. However, we have to limit our search to the first three positions only, i.e. the number of ways we can shelve three out of the 5 books is

$$5 \times 4 \times 3 = 60$$

To change this product into factorial notation, we do the following:

$$5 \times 4 \times 3 = 5 \times 4 \times 3 \times \frac{2!}{2!} = \frac{5 \times 4 \times 3 \times 2 \times 1}{2!} = \frac{5!}{2!} = \frac{5!}{(5-3)!}$$

This leads us to the following general result.

Number of permutations of *n* objects taken *r* at a time

The number of permutations of n objects taken r at a time is

$$^nP_r = {}_nP_r = P_r^n = P(n, r) = \frac{n!}{(n-r)!};\ n \geq r$$

To verify the formula above, you can proceed in the same manner as with the permutation of n objects.

n	$n-1$	$n-2$	$n-3$		$n-(r-1)$
↓	↓	↓	↓	…	↓
1	2	3	4		r

When you arrive to the rth position, you would have used $r - 1$ objects already, and hence you are left with $n - (r - 1) = n - r + 1$ objects to fill this position. So, the number of ways of arranging n objects taken r at a time is

$$^nP_r = n \cdot (n-1) \cdot (n-2) \ldots (n-r+1)$$

Here again, to make the expression more manageable, we can write it in factorial notation:

$$\begin{aligned} ^nP_r &= n \cdot (n-1) \cdot (n-2) \ldots (n-r+1) \\ &= n \cdot (n-1) \cdot (n-2) \ldots (n-r+1)\frac{(n-r)!}{(n-r)!} \\ &= \frac{n \cdot (n-1) \cdot (n-2) \ldots (n-r+1) \cdot (n-r)!}{(n-r)!} = \frac{n!}{(n-r)!} \end{aligned}$$

Example 26

15 drivers are taking part in a Formula 1 car race. In how many different ways can the top 6 positions be filled?

Solution

Since the drivers are all different, this is a permutation of 15 'objects' taken 6 at a time.

$$^{15}P_6 = \frac{15!}{(15-6)!} = 3\,603\,600$$

This can also be easily calculated using a GDC.

```
15 nPr 6
              3603600
15!/9!
              3603600
```

Combinations

A **combination** is a selection of some or all of a number of different objects. It is an unordered collection of unique sizes. In a permutation, the order of occurrence of the objects or the arrangement is important, but in combination the order of occurrence of the objects is not important. In that sense, a combination of r objects out of n objects is a subset of the set of n objects.

For example, there are 24 permutations of three letters out of ABCD, while there are only 4 combinations! Here is why:

ABC	ABD	ACD	BCD
ACB	ADB	ADC	BDC
BAC	BAD	CAD	CBD
BCA	BDA	CDA	CDB
CAB	DAB	DAC	DBC
CBA	DBA	DCA	DCB

For one combination, ABC for example, there are $3! = 6$ permutations. This is true for all combinations. So, the number of permutations is 6 times the number of combinations, i.e.

$$^4P_3 = 3!\,^4C_3$$

where $^4C_3$ is the number of combinations of the 4 letters taken 3 at a time.

According to the previous result, we can write

$$^4C_3 = \frac{^4P_3}{3!} = \frac{\frac{4!}{(4-3)!}}{3!} = \frac{4!}{3!(4-3)!}$$

The last result can also be generalized to n elements combined r at a time. (The ISO notation for this quantity, which is also used by the IB is $\binom{n}{r}$. In this book, we will follow the ISO notation.)

Every subset of r objects (combination), gives rise to $r!$ permutations. So, if you have $\binom{n}{r}$ combinations, these will result in $r!\binom{n}{r}$ permutations. Therefore,

$$^nP_r = r!\binom{n}{r} \Leftrightarrow \binom{n}{r} = \frac{^nP_r}{r!} = \frac{\frac{n!}{(n-r)!}}{r!} = \frac{n!}{(n-r)!r!}$$

$\binom{n}{r} = \frac{n!}{r!(n-r)!} = \binom{n}{n-r}$. This symmetry is obvious as when we pick r objects, we leave $n-r$ objects behind, and hence the number of ways of choosing r objects is the same as the number of ways of $n-r$ objects not chosen.

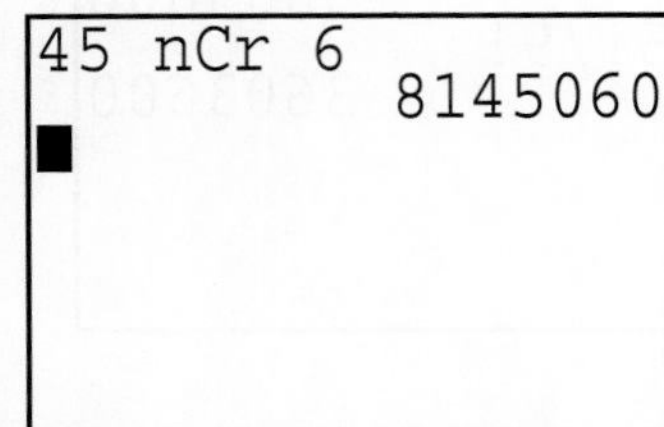

Example 27

A lottery has 45 numbers. If you buy a ticket, then you choose 6 of these numbers. How many different choices does this lottery have?

Solution

Since 6 numbers will have to be chosen and order is not an issue here, this is a combination case. The number of possible choices is

$$\binom{45}{6} = 8\ 145\ 060.$$

This can also be calculated using a GDC.

Example 28

In poker, a deck of 52 cards is used, and a 'hand' is made up of 5 cards.

a) How many hands are there?

b) How many hands are there with 3 diamonds and 2 hearts?

Solution

a) Since the order is not important, as a player can reorder the cards after receiving them, this is a combination of 52 cards taken 5 at a time: $\binom{52}{5} = 2\ 598\ 960$.

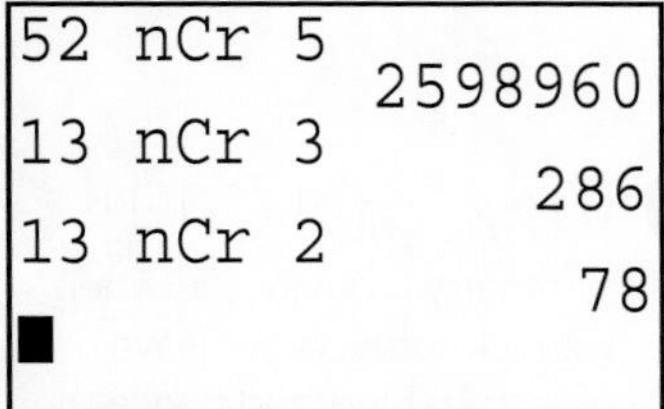

b) Since there are 13 diamonds and we want 3 of them, there are $\binom{13}{3} = 286$ ways to get the 3 diamonds. Since there are 13 hearts and we want 2 of them, there are $\binom{13}{2} = 78$ ways to get the 2 hearts. Since we want them both to occur at the same time, we use the fundamental counting principle and multiply 286 and 78 together to get 22 308 possible hands.

Example 29

A code is made up of 6 different digits. How many possible codes are there?

Solution

Since there are 10 digits and we are choosing 6 of them, and since the order we use these digits makes a difference in the code, then this is a permutation case. The number of possible codes is

$$^{10}P_6 = 151\ 200.$$

Exercise 4.5

1 Evaluate each of the following expressions.

a) $^5P_5$ b) $5!$ c) $^{20}P_1$ d) $^8P_3$

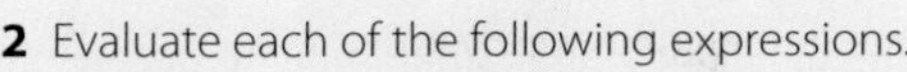

2 Evaluate each of the following expressions.

a) $\binom{5}{5}$ b) $\binom{5}{0}$ c) $\binom{10}{3}$ d) $\binom{10}{7}$

3 Evaluate each of the following expressions.

a) $\binom{7}{3} + \binom{7}{4}$ b) $\binom{8}{4}$ c) $\binom{10}{6} + \binom{10}{7}$ d) $\binom{11}{7}$

4 Evaluate each of the following expressions.

a) $\binom{8}{5} - \binom{8}{3}$ b) $11 \cdot 10!$ c) $\binom{10}{3} - \binom{10}{7}$ d) $\binom{10}{1}$

5 Tell whether each of the following expressions is true.

a) $\frac{10!}{5!} = 2!$ b) $(5!)^2 = 25!$ c) $\binom{101}{8} = \binom{101}{93}$

6 You are buying a computer and have the following choices: three types of HD, two types of DVD players, four types of graphic cards. How many different systems can you choose from?

7 You are going to a restaurant with a set menu. They have three starters, four main meals, two drinks, and three deserts. How many different choices are available for you to choose your meal from?

8 A school is in need of three teachers: PE, maths, and English. They have 8 applicants for the PE position, 3 applicants for the maths position and 13 applicants for English. How many different combinations of choices do they have?

9 You are given a multiple choice test where each question has four possible answers. The test is made up of 12 questions and you are guessing at random. In how many ways can you answer all the questions on the test?

10 The test in question 9 is divided into two parts, the first six are true/false questions and the last six are multiple choice as described. In how many different ways can you answer all questions on that test?

11 Passwords on a network are made up of two parts. One part consists of three letters of the alphabet, not necessarily different, and five digits, also not necessarily different. How many passwords are possible on this network?

12 How many 5-digit numbers can be made if the units digit cannot be 0?

13 Four couples are to be seated in a theatre row. In how many different ways can they be seated if

a) no restrictions are made

b) every two members of each couple like to sit together?

14 Five girls and three boys should go through a doorway in single file. In how many orders can they do that if

a) there are no constraints

b) the girls must go first?

15 Write all the permutations of the letters in JANE.

16 Write all the permutations of the letters in MAGIC taken three at a time.

17 A computer code is made up of three letters followed by four digits.

a) In how many ways is the code possible?

b) If 97 of the three-letter combinations cannot be used because they are offensive, how many codes are still possible?

18 A local bridge club has 17 members, 10 females and 7 males. They have to elect three officers: president, deputy, and treasurer. In how many ways is this possible if

a) there are no restrictions

b) the president is a male

c) the deputy must be a male, the president can be any gender, but the treasurer must be a female

d) the president and deputy are of the same gender

e) all three officers are not the same gender.

19 The research and development department for a computer manufacturer has 26 employees: 8 mathematicians, 12 computer scientists, and 6 electrical engineers. They need to select three employees to be leaders of the group. In how many ways can they do this if

a) the three officers are of the same specialization

b) at least one of them must be an engineer

c) two of them must be mathematicians?

20 A 'combination' lock has three numbers, each in the range 1 to 50.

a) How many different combinations are possible?

b) How many combinations do not have duplicates?

c) How many have the first and second numbers matching?

d) How many have exactly two of the numbers matching?

21 In how many ways can five married couples be seated around a circle so that spouses sit together?

22 a) How many subsets of $\{1, 2, 3, \ldots, 9\}$ have two elements?

b) How many subsets of $\{1, 2, 3, \ldots, 9\}$ have an odd number of elements?

23 Nine seniors and 12 juniors make up the maths club at a school. They need four members for an upcoming competition.

a) How many 4-member teams can they form?

b) How many of these 4-member teams have the same number of juniors and seniors?

c) How many of these 4-member teams have more juniors than seniors?

24 This problem uses the same data as question 23 above. Tim, a junior, is the strongest 'mathlete' among his group while senior Gwen is the strongest among her group. Either Tim or Gwen must be on the team, but they cannot both be on the team. Answer the same questions as above.

25 A shipment of 100 hard disks contains 4 defective disks. We choose a sample of 6 disks for inspection.

a) How many different possible samples are there?

b) How many samples could contain all 4 defective disks? What percentage of the total is that?

c) How many samples could contain at least 1 defective disk? What percentage of the total is that?

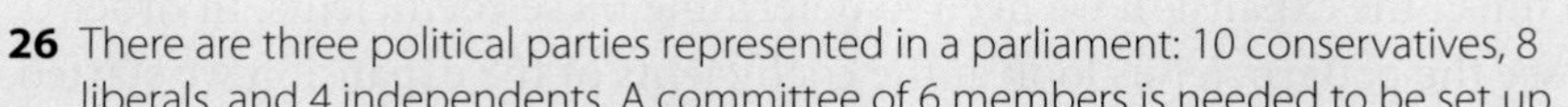

26 There are three political parties represented in a parliament: 10 conservatives, 8 liberals, and 4 independents. A committee of 6 members is needed to be set up.

a) How many different committees are possible?

b) How many committees with equal representation are possible?

27 How many ways are there for 9 boys and 6 girls to stand in a line so that no two girls stand next to each other?

4.6 The binomial theorem

A binomial is a polynomial with two terms. For example, $x + y$ is a binomial. In principle, it is easy to raise $x + y$ to any power; but raising it to high powers would be tedious. We will find a formula that gives the expansion of $(x + y)^n$ for any positive integer n. The proof of the binomial theorem is given in Section 4.7.

Let us look at some special cases of the expansion of $(x + y)^n$:

$(x + y)^0 = 1$

$(x + y)^1 = x + y$

$(x + y)^2 = x^2 + 2xy + y^2$

$(x + y)^3 = x^3 + 3x^2y + 3xy^2 + y^3$

$(x + y)^4 = x^4 + 4x^3y + 6x^2y^2 + 4xy^3 + y^4$

$(x + y)^5 = x^5 + 5x^4y + 10x^3y^2 + 10x^2y^3 + 5xy^4 + y^5$

$(x + y)^6 = x^6 + 6x^5y + 15x^4y^2 + 20x^3y^3 + 15x^2y^4 + 6xy^5 + y^6$

There are several things that you will have noticed after looking at the expansion:

- There are $n + 1$ terms in the expansion of $(x + y)^n$.
- The degree of each term is n.
- The powers on x begin with n and decrease to 0.
- The powers on y begin with 0 and increase to n.
- The coefficients are symmetric.

For instance, notice how the exponents of x and y behave in the expansion of $(x + y)^5$.

The exponents of x decrease:

$(x + y)^5 = x^{\boxed{5}} + 5x^{\boxed{4}}y + 10x^{\boxed{3}}y^2 + 10x^{\boxed{2}}y^3 + 5x^{\boxed{1}}y^4 + x^{\boxed{0}}y^5$

The exponents of y increase:

$(x + y)^5 = x^5y^{\boxed{0}} + 5x^4y^{\boxed{1}} + 10x^3y^{\boxed{2}} + 10x^2y^{\boxed{3}} + 5xy^{\boxed{4}} + y^{\boxed{5}}$

Using this pattern, we can now proceed to expand any binomial raised to power n: $(x + y)^n$. For example, leaving a blank for the missing coefficients, the expansion for $(x + y)^7$ can be written as

$(x + y)^7$

$= \square x^7 + \square x^6y + \square x^5y^2 + \square x^4y^3 + \square x^3y^4 + \square x^2y^5 + \square xy^6 + \square y^7$

To finish the expansion we need to determine these coefficients. In order to see the pattern, let us look at the coefficients of the expansion we started the section with.

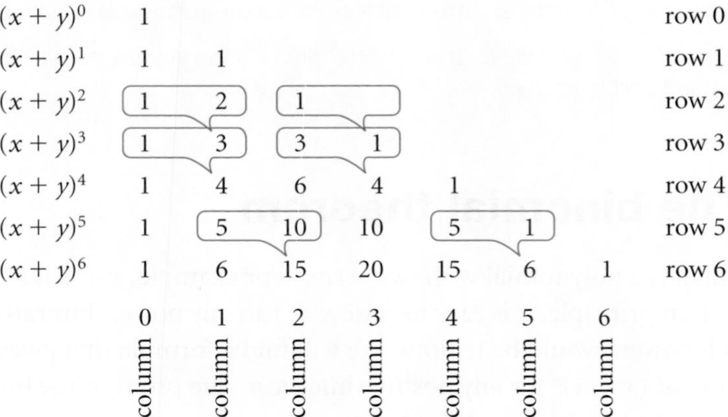

$(x + y)^0$ 1 row 0
$(x + y)^1$ 1 1 row 1
$(x + y)^2$ 1 2 1 row 2
$(x + y)^3$ 1 3 3 1 row 3
$(x + y)^4$ 1 4 6 4 1 row 4
$(x + y)^5$ 1 5 10 10 5 1 row 5
$(x + y)^6$ 1 6 15 20 15 6 1 row 6
column 0
column 1
column 2
column 3
column 4
column 5
column 6

A triangle like the one above is known as Pascal's triangle. Notice how the first and **second** terms in row **3** give you the **second** term in row **4**; the third and **fourth** terms in row **3** give you the **fourth** term of row **4**; the second and **third** terms in row **5** give you the **third** term in row **6**; and the fifth and **sixth** terms in row **5** give you the **sixth** term in row **6**, and so on. So now we can state the key property of Pascal's triangle.

Pascal's triangle was known to Persian and Chinese mathematicans in the 13th century.

Pascal's triangle

Every entry in a row is the sum of the term directly above it and the entry diagonally above and to the left of it. When there is no entry, the value is considered zero.

Take the last entry in row 5, for example; there is no entry directly above it, so its value is $0 + 1 = 1$.

From this property it is easy to find all the terms in any row of Pascal's triangle from the row above it. So, for the expansion of $(x + y)^7$, the terms are found from row 6 as follows:

$$\begin{array}{ccccccccc} 0 \rightarrow & 1 \rightarrow & 6 \rightarrow & 15 \rightarrow & 20 \rightarrow & 15 \rightarrow & 6 \rightarrow & 1 \rightarrow & 0 \\ & \downarrow & \downarrow & \downarrow & \downarrow & \downarrow & \downarrow & \downarrow & \downarrow \\ & 1 & 7 & 21 & 35 & 35 & 21 & 7 & 1 \end{array}$$

So, $(x + y)^7 = x^7 + \boxed{7}x^6y + \boxed{21}x^5y^2 + \boxed{35}x^4y^3 + \boxed{35}x^3y^4 + \boxed{21}x^2y^5 + \boxed{7}xy^6 + y^7$.

Note: Several sources use a slightly different arrangement for Pascal's triangle. The common usage considers the triangle as isosceles and uses the principle that every two entries add up to give the entry diagonally below them, as shown in the following diagram.

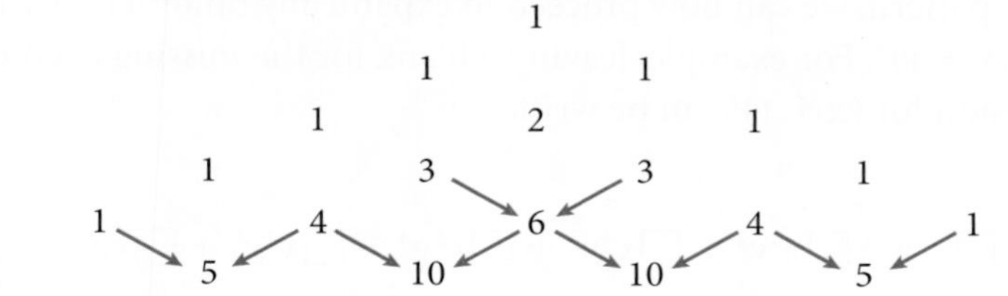

Example 30

Use Pascal's triangle to expand $(2k - 3)^5$.

Solution

We can find the expansion above by replacing x by $2k$ and y by -3 in the binomial expansion of $(x + y)^5$.

Using the fifth row of Pascal's triangle for the coefficients will give us the following:

$$\mathbf{1}(2k)^5 + \mathbf{5}(2k)^4(-3) + \mathbf{10}(2k)^3(-3)^2 + \mathbf{10}(2k)^2(-3)^3 + \mathbf{5}(2k)(-3)^4 + \mathbf{1}(-3)^5 = 32k^5 - 240k^4 + 720k^3 - 1080k^2 + 810k - 243.$$

Pascal's triangle is an easy and useful tool in finding the coefficients of the binomial expansion for relatively small values of n. It is not very efficient doing that for large values of n. Imagine you want to evaluate $(x + y)^{20}$. Using Pascal's triangle, you will need the terms in the 19th row and the 18th row and so on. This makes the process tedious and not practical.

Luckily, we have a formula that can find the coefficients of any Pascal's triangle row. This formula is the binomial formula, whose proof is beyond the scope of this book. Every entry in Pascal's triangle is denoted by $\binom{n}{r}$, which is also known as the binomial coefficient.

In $\binom{n}{r}$, n is the row number and r is the column number.

The factorial notation makes many formulae involving the multiplication of consecutive positive integers shorter and easier to write. That includes the binomial coefficient.

The binomial coefficient

With n and r as non-negative integers such that $n \geqslant r$, the binomial coefficient $\binom{n}{r}$ is defined by

$$\binom{n}{r} = \frac{n!}{r!(n-r)!}$$

Example 31

Find the value of a) $\binom{7}{3}$ b) $\binom{7}{4}$ c) $\binom{7}{0}$ d) $\binom{7}{7}$

Solution

a) $\binom{7}{3} = \frac{7!}{3!(7-3)!} = \frac{7!}{3!4!} = \frac{\cancel{1 \cdot 2 \cdot 3 \cdot 4} \cdot 5 \cdot 6 \cdot 7}{(1 \cdot 2 \cdot 3)(\cancel{1 \cdot 2 \cdot 3 \cdot 4})} = \frac{5 \cdot 6 \cdot 7}{1 \cdot 2 \cdot 3} = 35$

b) $\binom{7}{4} = \frac{7!}{3!(7-4)!} = \frac{7!}{4!3!} = \frac{\cancel{1 \cdot 2 \cdot 3 \cdot 4} \cdot 5 \cdot 6 \cdot 7}{(\cancel{1 \cdot 2 \cdot 3 \cdot 4})(1 \cdot 2 \cdot 3)} = \frac{5 \cdot 6 \cdot 7}{1 \cdot 2 \cdot 3} = 35$

c) $\binom{7}{0} = \frac{7!}{0!(7-0)!} = \frac{\cancel{7!}}{0!\cancel{7!}} = \frac{1}{1} = 1$

d) $\binom{7}{7} = \frac{7!}{7!(7-7)!} = \frac{\cancel{7!}}{\cancel{7!}0!} = \frac{1}{1} = 1$

● **Hint:** Your calculator can do the tedious work of evaluating the binomial coefficient. If you have a TI, the binomial coefficient appears as ${}_nC_r$, which is another notation frequently used in mathematical literature.

```
7 nCr 3
                35
7 nCr 4
                35
7 nCr 0
                 1
■
```

Although the binomial coefficient $\binom{n}{r}$ appears as a fraction, all its results where n and r are non-negative integers are positive integers. Also, notice the **symmetry** of the coefficient in the previous examples. This is a property that you are asked to prove in the exercises:

$$\binom{n}{r} = \binom{n}{n-r}$$

Example 32

Calculate the following:

$\binom{6}{0}, \binom{6}{1}, \binom{6}{2}, \binom{6}{3}, \binom{6}{4}, \binom{6}{5}, \binom{6}{6}$

Solution

$\binom{6}{0} = 1, \binom{6}{1} = 6, \binom{6}{2} = 15, \binom{6}{3} = 20, \binom{6}{4} = 15, \binom{6}{5} = 6, \binom{6}{6} = 1$

The values we calculated above are precisely the entries in the sixth row of Pascal's triangle.

We can write Pascal's triangle in the following manner:

$$\begin{matrix} \binom{0}{0} \\ \binom{1}{0} & \binom{1}{1} \\ \binom{2}{0} & \binom{2}{1} & \binom{2}{2} \\ \binom{3}{0} & \binom{3}{1} & \binom{3}{2} & \binom{3}{3} \\ \cdots & \cdots & \cdots & \cdots \\ \binom{n}{0} & \binom{n}{1} & \cdots & \cdots & \cdots & \cdots & \binom{n}{n} \end{matrix}$$

Example 33

Calculate $\binom{n}{r-1} + \binom{n}{r}$.

• **Hint:** You will be able to provide reasons for the steps after you do the exercises!

This is called Pascal's rule.

Solution

$$\begin{aligned} \binom{n}{r-1} + \binom{n}{r} &= \frac{n!}{(r-1)!(n-r+1)!} + \frac{n!}{r!(n-r)!} \\ &= \frac{n! \cdot r}{r \cdot (r-1)!(n-r+1)!} + \frac{n! \cdot (n-r+1)}{r!(n-r)! \cdot (n-r+1)} \\ &= \frac{n! \cdot r}{r!(n-r+1)!} + \frac{n! \cdot (n-r+1)}{r!(n-r+1)!} \\ &= \frac{n! \cdot r + n! \cdot (n-r+1)}{r!(n-r+1)!} = \frac{n!(r+n-r+1)}{r!(n-r+1)!} \\ &= \frac{n!(n+1)}{r!(n-r+1)!} = \frac{(n+1)!}{r!(n+1-r)!} = \binom{n+1}{r} \end{aligned}$$

If we read the result above carefully, it says that the sum of the terms in the nth row $(r - 1)$th and rth columns is equal to the entry in the $(n + 1)$th row and rth column. That is, the two entries on the left are adjacent entries in the nth row of Pascal's triangle and the entry on the right is the entry in the $(n + 1)$th row directly below the rightmost entry. This is precisely the principle behind Pascal's triangle!

Using the binomial theorem

We are now prepared to state the binomial theorem. The proof of the theorem is optional and will require mathematical induction. We will develop the proof in Section 4.7.

$$(x+y)^n = \binom{n}{0} x^n + \binom{n}{1} x^{n-1}y + \binom{n}{2} x^{n-2}y^2 + \binom{n}{3} x^{n-3}y^3 + \ldots + \binom{n}{n-1} xy^{n-1} + \binom{n}{n} y^n$$

In a compact form, we can use sigma notation to express the theorem as follows:

$$(x+y)^n = \sum_{i=0}^{n} \binom{n}{i} x^{n-i} y^i$$

Example 34

Use the binomial theorem to expand $(x + y)^7$.

Solution

$$(x+y)^7 = \binom{7}{0}x^7 + \binom{7}{1}x^{7-1}y + \binom{7}{2}x^{7-2}y^2 + \binom{7}{3}x^{7-3}y^3 + \binom{7}{4}x^{7-4}y^4$$
$$+ \binom{7}{5}x^{7-5}y^5 + \binom{7}{6}xy^6 + \binom{7}{7}y^7$$
$$= x^7 + 7x^6y + 21x^5y^2 + 35x^4y^3 + 35x^3y^4 + 21x^2y^5 + 7xy^6 + y^7$$

Note: Why is the binomial theorem related to the number of combinations of n elements taken r at a time?

Consider evaluating $(x + y)^n$. In doing so, you have to multiply $(x + y)$ n times by itself. As you know, one term has to be x^n. How to get this term? x^n is the result of multiplying x in each of the n factors $(x + y)$ and that can only happen in one way. However, consider the term containing x^r. To have a power of r over the x, means that the x in each of r factors has to be multiplied, and the rest will be the $n - r$ y-terms. This can happen in $\binom{n}{r}$ ways. Hence, the coefficient of the term x^ry^{n-r} is $\binom{n}{r}$.

Example 35

Find the expansion for $(2k - 3)^5$.

Solution

$$(2k-3)^5 = \binom{5}{0}(2k)^5 + \binom{5}{1}(2k)^4(-3) + \binom{5}{2}(2k)^3(-3)^2 + \binom{5}{3}(2k)^2(-3)^3$$
$$+ \binom{5}{4}(2k)(-3)^4 + \binom{5}{5}(-3)^5$$
$$= 32k^5 - 240k^4 + 720k^3 - 1080k^2 + 810k - 243$$

Example 36

Find the term containing a^3 in the expansion $(2a - 3b)^9$.

Solution

To find the term, we do not need to expand the whole expression.

Since $(x + y)^n = \sum_{i=0}^{n} \binom{n}{i} x^{n-i} y^i$, the term containing a^3 is the term where $n - i = 3$, i.e. when $i = 6$. So, the required term is

$$\binom{9}{6}(2a)^{9-6}(-3b)^6 = 84 \cdot 8a^3 \cdot 729b^6 = 489\,888a^3b^6.$$

Example 37

Find the term independent of x in $\left(4x^3 - \frac{2}{x^2}\right)^5$.

Solution

The phrase 'independent of x' means the term with no x variable, i.e. the constant term. A constant is equivalent to the product of a number and x^0, since $x^0 = 1$. We are looking for the term in the expansion such that the resulting power is zero. In terms of i, each term in the expansion is given by

$$\binom{5}{i}(4x^3)^{5-i}(-2x^{-2})^i$$

Thus, for the constant term:

$$3(5 - i) - 2i = 0 \Rightarrow 15 - 5i = 0 \Rightarrow i = 3$$

Therefore, the term independent of x is:

$$\binom{5}{3}(4x^3)^2(-2x^{-2})^3 = 10 \cdot 16x^6(-8x^{-6}) = -1280$$

Example 38

Find the coefficient of b^6 in the expansion of $\left(2b^2 - \frac{1}{b}\right)^{12}$.

Solution

The general term is

$$\binom{12}{i}(2b^2)^{12-i}\left(-\frac{1}{b}\right)^i = \binom{12}{i}(2)^{12-i}(b^2)^{12-i}\left(-\frac{1}{b}\right)^i$$
$$= \binom{12}{i}(2)^{12-i}b^{24-2i}b^{-i}(-1)^i = \binom{12}{i}(2)^{12-i}b^{24-3i}(-1)^i$$

$24 - 3i = 6 \Rightarrow i = 6$. So, the coefficient in question is $\binom{12}{6}(2)^6(-1)^6 = 59\,136$.

Exercise 4.6

1 Use Pascal's triangle to expand each binomial.

a) $(x + 2y)^5$ b) $(a - b)^4$ c) $(x - 3)^6$

d) $(2 - x^3)^4$ e) $(x - 3b)^7$ f) $\left(2n + \frac{1}{n^2}\right)^6$

g) $\left(\frac{3}{x} - 2\sqrt{x}\right)^4$

2 Evaluate each expression.

a) $\binom{8}{3}$ b) $\binom{18}{5} - \binom{18}{13}$ c) $\binom{7}{4}\binom{7}{3}$

d) $\binom{5}{0} + \binom{5}{1} + \binom{5}{2} + \binom{5}{3} + \binom{5}{4} + \binom{5}{5}$

e) $\binom{6}{0} - \binom{6}{1} + \binom{6}{2} - \binom{6}{3} + \binom{6}{4} - \binom{6}{5} + \binom{6}{6}$

3 Use the binomial theorem to expand each of the following.

a) $(x + 2y)^7$ b) $(a - b)^6$ c) $(x - 3)^5$

d) $(2 - x^3)^6$ e) $(x - 3b)^7$ f) $\left(2n + \frac{1}{n^2}\right)^6$

g) $\left(\frac{3}{x} - 2\sqrt{x}\right)^4$ h) $(1 + \sqrt{5})^4 + (1 - \sqrt{5})^4$

i) $(\sqrt{3} + 1)^8 - (\sqrt{3} - 1)^8$ j) $(1 + i)^8$, where $i^2 = -1$

k) $(\sqrt{2} - i)^6$, where $i^2 = -1$

4 Consider the expression $\left(x - \frac{2}{x}\right)^{45}$.

a) Find the first three terms of this expansion.

b) Find the constant term if it exists or justify why it does not exist.

c) Find the last three terms of the expansion.

d) Find the term containing x^3 if it exists or justify why it does not exist.

5 Prove that $\binom{n}{k} = \binom{n}{n-k}$ for all $n, k \in \mathbb{N}$ and $n \geqslant k$.

6 Prove that for any positive integer n,

$\binom{n}{1} + \binom{n}{2} + \ldots + \binom{n}{n-1} + \binom{n}{n} = 2^n - 1$ • **Hint:** $2^n = (1 + 1)^n$

7 Consider all $n, k \in \mathbb{N}$ and $n \geqslant k$.

a) Verify that $k! = k(k - 1)!$

b) Verify that $(n - k + 1)! = (n - k + 1)(n - k)!$

c) Justify the steps given in the proof of $\binom{n}{r-1} + \binom{n}{r} = \binom{n+1}{r}$ in the examples.

8 Find the value of the expression:

$\binom{6}{0}\left(\frac{1}{3}\right)^6 + \binom{6}{1}\left(\frac{1}{3}\right)^5\left(\frac{2}{3}\right) + \binom{6}{2}\left(\frac{1}{3}\right)^4\left(\frac{2}{3}\right)^2 + \ldots + \binom{6}{6}\left(\frac{2}{3}\right)^6$

9 Find the value of the expression:

$\binom{8}{0}\left(\frac{2}{5}\right)^8 + \binom{8}{1}\left(\frac{2}{5}\right)^7\left(\frac{3}{5}\right) + \binom{8}{2}\left(\frac{2}{5}\right)^6\left(\frac{3}{5}\right)^2 + \ldots + \binom{8}{8}\left(\frac{3}{5}\right)^8$

10 Find the value of the expression:

$\binom{n}{0}\left(\frac{1}{7}\right)^n + \binom{n}{1}\left(\frac{1}{7}\right)^{n-1}\left(\frac{6}{7}\right) + \binom{n}{2}\left(\frac{1}{7}\right)^{n-2}\left(\frac{6}{7}\right)^2 + \ldots + \binom{n}{n}\left(\frac{6}{7}\right)^n$

11 Find the term independent of x in the expansion of $\left(x^2 - \frac{1}{x}\right)^6$.

12 Find the term independent of x in the expansion of $\left(3x - \frac{2}{x}\right)^8$.

13 Find the term independent of x in the expansion of $\left(2x - \frac{3}{x^3}\right)^8$.

14 Find the first three terms of the expansion of $(1 + x)^{10}$ and use them to find an approximation to

a) 1.01^{10} b) 0.99^{10}

15 Show that $\binom{n}{r-1} + 2\binom{n}{r} + \binom{n}{r+1} = \binom{n+2}{r+1}$ and interpret your result on the entries in Pascal's triangle.

16 Express each repeating decimal as a fraction:

a) $0.\overline{7}$ b) $0.3\overline{45}$ c) $3.21\overline{29}$

17 Find the coefficient of x^6 in the expansion of $(2x - 3)^9$.

18 Find the coefficient of x^3b^4 in $(ax + b)^7$.

19 Find the constant term of $\left(\frac{2}{z^2} - z\right)^{15}$.

20 Expand $(3n - 2m)^5$.

21 Find the coefficient of r^{10} in $(4 + 3r^2)^9$.

4.7 Mathematical induction

Domino effect

In addition to playing games of strategy, another familiar activity using dominoes is to place them on edge in lines, then topple the first tile, which falls on and topples the second, which topples the third, etc., resulting in all of the tiles falling. Arrangements of millions of tiles have been made that have taken many minutes to fall.

The Netherlands has hosted an annual domino toppling competition called *Domino Day* since 1986. The record, achieved in 2006, is 4 079 381 dominoes.

Similar phenomena of chains of small events each causing similar events leading to an eventual grand result, by analogy, are called *domino effects*. The phenomenon also has some theoretical bearing to familiar applications like the amplifier, digital signals, or information processing.

Induction

In mathematics, we have a parallel in **mathematical induction**, which is a method for proving a statement that is maintained about every natural number. For example,

$$1 + 2 + 3 + \ldots + n = \frac{n(n + 1)}{2}$$

This claims that the sum of consecutive numbers from 1 to n is half the product of the last term, n, and the integer after it.

We want to prove that this will be true for $n = 1$, $n = 2$, $n = 3$, and so on. Now we can test the formula for any given number, say $n = 3$:

$1 + 2 + 3 = \frac{1}{2} \cdot 3 \cdot 4 = 6$, which is true.

It is also true for $n = 4$:

$$1 + 2 + 3 + 4 = \frac{1}{2} \cdot 4 \cdot 5 = 10$$

But how are we to prove this rule for every value of n?

The method of proof is shown to the right. It is called the principle of mathematical induction.

Note: The order of the steps varies from one source to the other. We present you with both arrangements.

When the statement is true for $n = 1$, then according to 1), it will also be true for $n = 2$. But that implies it will be true for $n = 3$; which implies it will be true for $n = 4$. And so on. It will be true for every natural number.

To prove a statement by induction, then, we must prove parts 1) and 2) above.

The hypothesis of Step 1) – 'The statement is true for $n = k$' – is called the **induction assumption**, or the **induction hypothesis**. It is what we assume when we prove a theorem by induction.

Mathematical induction

1) When a statement is true for the natural number $n = k$, then it is also true for its successor, $n = k + 1$; and
2) the statement is true for $n = 1$; then the statement is true for every natural number n.

Example 39

Prove that the sum of the first n natural numbers is given by this formula:

$$1 + 2 + 3 + \ldots + n = \frac{n(n + 1)}{2}$$

We will call this statement $S(n)$, because it depends on n.

Proof

We will do Steps 1) and 2) above. First, we will assume that the statement is true for $n = k$; that is, we will assume that $S(k)$ is true:

$$S(k): 1 + 2 + 3 + \ldots + k = \frac{k(k + 1)}{2} \qquad (1)$$

This is the induction assumption. Assuming this, we must prove that $S(k+1)$ is also true. That is, we must show:

$$S(k+1): 1 + 2 + 3 + \ldots + (k+1) = \frac{(k+1)((k+1)+1)}{2} \qquad (2)$$

To do that, we will simply add the next term $(k+1)$ to both sides of the induction assumption, equation (1), and then simplify:

$$\begin{aligned} S(k+1): 1 + 2 + 3 + \ldots + k + (k+1) &= \frac{k(k+1)}{2} + (k+1) \\ &= \frac{k(k+1) + 2(k+1)}{2} \\ &= \frac{(k+1)(k+2)}{2} \\ &= \frac{(k+1)((k+1)+1)}{2} \end{aligned}$$

This is equation (2), which is the first thing we wanted to show.

It is extremely important to note that mathematical induction can be used to prove results obtained in some other way. It is *not* a tool for discovering formulae or theorems.

Next, we must show that the statement is true for $n = 1$. We have

$$S(1): 1 = \frac{1(1+1)}{2}$$

The formula therefore is true for $n = 1$. We have now fulfilled both conditions of the principle of mathematical induction. $S(n)$ is therefore true for every natural number.

Example 40

In an investigation to find the sum of the first n positive *odd* integers, we can do the following: Investigate the sums of the first few odd integers and then try to come up with a conjecture. Then mathematical induction will provide us with a tool to prove the conjecture.

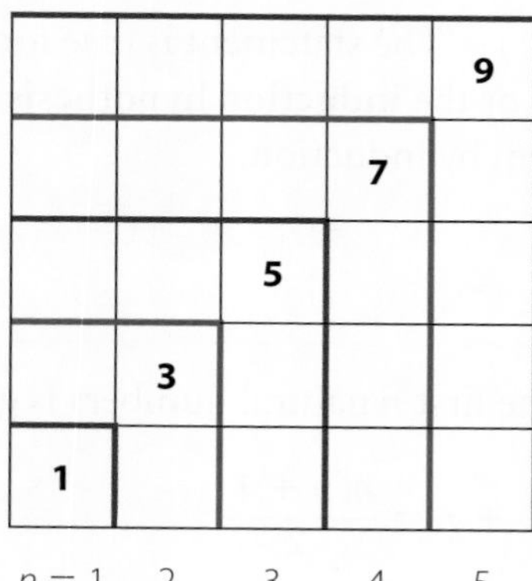

For $n = 1$, the sum is $1 = 1$.

For $n = 2$, the sum is $1 + 3 = 4$.

For $n = 3$, the sum is $1 + 3 + 5 = 9$.

For $n = 4$, the sum is $1 + 3 + 5 + 7 = 16$.

For $n = 5$, the sum is $1 + 3 + 5 + 7 + 9 = 25$.

It is clear that the number of integers you add, and the sum, are related, i.e. the sum of n such integers is n^2.

n	1	2	3	4	5	6	...	n
SUM	1	4	9	16	25	36	...	n^2

Solution

Let $S(n)$ denote the statement that the sum of the first n odd positive integers is n^2.

First, we must complete the basis step, i.e. we must show that $S(1)$ is true. Then we must carry out the inductive step, i.e. we have to *show* that $S(k+1)$ is true whenever $S(k)$ is assumed true.

Basis step: $S(1)$, which means that the sum of the first odd integer is 1^2. This is obvious as the sum of 1 is 1!

Inductive step: We must show that the implication $S(k) \Rightarrow S(k+1)$ is true, regardless of the choice of k. To that end, we start with an assumption that $S(k)$ is true for any choice of k; i.e.

$$1 + 3 + 5 + \ldots + (2k-1) = k^2.$$

Now, we must show that $S(k+1)$ is true.

$$S(k+1)\text{: } 1 + 3 + 5 + \ldots + (2k+1) = (k+1)^2,$$

(the $(k+1)$th odd integer is $2(k+1) - 1 = 2k + 1$)

The left-hand side can be written as

$$1 + 3 + 5 + \ldots + (2k-1) + (2k+1) = k^2 + (2k+1) = k^2 + 2k + 1 = (k+1)^2,$$

Therefore, $1 + 3 + 5 + \ldots + (2k+1) = (k+1)^2$, which is nothing but $S(k+1)$.

This shows that $S(k+1)$ follows from $S(k)$. Since $S(1)$ is true, and the implication $S(k) \Rightarrow S(k+1)$ is true for all positive integers k, the mathematical induction principle shows that $S(n)$ is true for all positive integers n.

Example 41

Prove that $3^n < n!$ for all integers $n > 6$.

Solution

Let $S(n)$ be the statement that $3^n < n!$

Basis step: To prove this inequality the basis step must be $S(7)$. Note that $S(6)$: $3^6 = 729 < 6! = 720$ is not true! $S(7)$: $3^7 = 2187 < 7! = 5040$ is true.

In general, a proof by mathematical induction that a statement $S(n)$ is true for every positive integer $n \geqslant 1$ consists of two steps:

BASIS STEP: The statement $S(1)$ is *shown* to be true.

INDUCTIVE STEP: The implication $S(k) \Rightarrow S(k+1)$ is *shown* to be true for any positive integer k.

Not all statements are true for all positive integers $n \geqslant 1$. In such cases, a variation of the mathematical induction principle is used: A statement $S(n)$ is true for every positive integer $n \geqslant n_0$ consists of two steps:

BASIS STEP: The statement $S(n_0)$ is shown to be true.

INDUCTIVE STEP: The implication $S(k) \Rightarrow (k+1)$ is shown to be true for any positive integer $k \geqslant n_0$. For example, $2^n < n!$ can only be true for $n \geqslant 4$.

Note: The nth odd positive integer is $2n - 1$. This is so because we are adding '2' a total of $n - 1$ times to 1; i.e. $1 + 2(n-1) = 2n - 1$.

Note: In a proof by mathematical induction, we do not assume that $S(k)$ is true for all positive integers! We only show that if it is assumed that $S(k)$ is true, then $S(k+1)$ is also true.

Inductive step: Assume $S(k)$ is true, i.e. assume that $3^k < k!$ is true. We must show that $S(k+1)$ is also true, i.e. we must show that $3^{k+1} < (k+1)!$

On the assumption that $3^k < k!$, multiply both sides of this inequality by 3.

$3 \cdot 3^k < 3 \cdot k!$, and since $k > 6$, then $3 < k + 1$; hence,

$$\begin{aligned} 3 \cdot 3^k &< 3 \cdot k! \\ &< (k+1) \cdot k! \\ &= (k+1)! \\ \Rightarrow 3^{k+1} &< (k+1)! \end{aligned}$$

This shows that $S(k+1)$ is true whenever $S(k)$ is true. This completes the inductive step of the proof.

Therefore, $3^n < n!$ for all integers $n > 6$.

Note: When we use mathematical induction to prove a statement $S(n)$, we show that $S(1)$ is true. Then we know that $S(2)$ is true, since $S(1) \Rightarrow S(1 + 1)$.

Further, we know that $S(3)$ is true, since $S(2) \Rightarrow S(2 + 1)$. Continuing along these lines, we see that $S(n)$ is true for every positive integer n.

Example 42

Show that in an arithmetic sequence where $a_n = a_{n-1} + d$, the nth term can be given by the formula

$$a_n = a_1 + (n-1)d.$$

Solution

Let $S(n)$ be the statement that $a_n = a_1 + (n-1)d$.

Basis step: To prove this formula the basis step must be $S(1)$.
$S(1)$: $a_1 = a_1 + (1-1)d = a_1$ is true.

Inductive step: Assume $S(k)$ is true, i.e. assume that $a_k = a_1 + (k-1)d$ is true. We must show that $S(k+1)$ is also true, i.e. we must show that $a_{k+1} = a_1 + (k+1-1)d = a_1 + kd$.

On the assumption that $a_k = a_1 + (k-1)d$:

$a_{k+1} = a_k + d$ by definition of an arithmetic sequence; hence,

$$a_{k+1} = \underbrace{a_k} + d = \underbrace{a_1 + (k-1)d} + d = a_1 + kd$$

This shows that $S(k+1)$ is true whenever $S(k)$ is true. This completes the inductive step of the proof.

Therefore, $a_k = a_1 + (k-1)d$ for all integers n.

Example 43

Show that in an arithmetic series: $S_n = \frac{n}{2}(2a_1 + (n-1)d)$.

Solution

Let $P(n)$ be the statement that $S_n = \frac{n}{2}(2a_1 + (n-1)d)$.

Basis step: To prove this formula the basis step must be $P(1)$.
$P(1)$: $S_1 = \frac{1}{2}(2a_1 + (1-1)d) = a_1$ is true. $(S_1 = a_1)$

Notice here that we are using *P*(*n*) rather than *S*(*n*). The use of the name does not influence the method!

Inductive step: Assume $P(k)$ is true, i.e. assume that $S_k = \frac{k}{2}(2a_1 + (k-1)d)$ is true. We must show that $P(k+1)$ is also true, i.e. we must show that

$$S_{k+1} = \frac{k+1}{2}(2a_1 + (k+1-1)d) = \frac{k+1}{2}(2a_1 + kd).$$

On the assumption that $S_k = \frac{k}{2}(2a_1 + (k-1)d)$:

$S_{k+1} = S_k + a_{k+1}$ by definition of an arithmetic series; hence,

$$S_{k+1} = S_k + a_{k+1} = \frac{k}{2}(2a_1 + (k-1)d) + a_1 + kd$$

By combining like terms and simplifying, the expression (page 194) can be reduced to

$$S_{k+1} = \frac{k}{2}\cdot 2a_1 + \frac{k}{2}(k-1)d + a_1 + kd = (k+1)a_1 + \frac{k}{2}(k-1)d + kd$$

$$= \frac{(k+1)}{2}\cdot 2a_1 + \frac{k(k+1)}{2}d = \frac{k+1}{2}(2a_1 + kd)$$

This shows that $P(k+1)$ is true whenever $P(k)$ is true. This completes the inductive step of the proof.

Therefore, $S_n = \frac{n}{2}(2a_1 + (n-1)d)$ for all integers n.

Example 44

Show that 3 divides $n^3 + 2n$ for all non-negative integers n.

Solution

Let $P(n)$ be the statement that '3 divides $n^3 + 2n$'.

Basis step: To prove this formula the basis step must be $P(0)$.

$P(0)$: is true since $0^3 + 2(0) = 0$ is a multiple of 3. (If you are not convinced, you can try $P(1)$: $1^3 + 2(1) = 3$ is a multiple of 3.)

Inductive step: Assume $P(k)$ is true, i.e. assume that 3 divides $k^3 + 2k$. We must prove that

$P(k+1)$ is true, i.e. 3 divides $(k+1)^3 + 2(k+1)$.

Note that

$(k+1)^3 + 2(k+1)$

$= k^3 + 3k^2 + 3k + 1 + 2k + 2 = (k^3 + 2k) + 3k^2 + 3k + 1 + 2$

$= (k^3 + 2k) + 3(k^2 + k + 1)$

Since both terms in this sum are multiples of 3 – the first by the induction hypothesis and the second because it is 3 times an integer – it follows that the sum is a multiple of 3.

Hence, $(k+1)^3 + 2(k+1)$ is a multiple of 3.

This shows that $P(k+1)$ is true whenever $P(k)$ is true. This completes the inductive step of the proof.

Therefore, 3 divides $n^3 + 2n$ for all non-negative integers n.

Example 45

Show, using mathematical induction, that for all non-negative integers n

$$\binom{n}{0} + \binom{n}{1} + \binom{n}{2} + \dots + \binom{n}{n-1} + \binom{n}{n} = 2^n$$

Solution

Let $P(n)$ be the statement that $\binom{n}{0} + \binom{n}{1} + \binom{n}{2} + \dots + \binom{n}{n-1} + \binom{n}{n} = 2^n$.

Basis step: To prove this formula, the basis step must be $P(0)$.

$P(0)$: is true since $\binom{0}{0} = 2^0 = 1$ is true. Moreover, $P(1)$ is also true since $\binom{1}{0} + \binom{1}{1} = 2^1$ is true!

Inductive step: Assume $P(k)$ is true, i.e. assume that

$$\binom{k}{0} + \binom{k}{1} + \binom{k}{2} + \dots + \binom{k}{k-1} + \binom{k}{k} = 2^k.$$

Recall from Section 4.5 that $\binom{n}{r-1} + \binom{n}{r} = \binom{n+1}{r}$ which we claim to be the basis of Pascal's triangle. Using this fact, we can perform the following addition:

$$\begin{array}{rl}
& \binom{k}{0} + \binom{k}{1} + \binom{k}{2} + \dots + \binom{k}{k-2} + \binom{k}{k-1} + \binom{k}{k} = 2^k \\
\binom{k}{0} + & \binom{k}{1} + \binom{k}{2} + \dots + \binom{k}{k-1} + \binom{k}{k} = 2^k \\
\hline
\binom{k}{0} + & \binom{k+1}{1} + \binom{k+1}{2} + \dots + \binom{k+1}{k-1} + \binom{k+1}{k} + \binom{k}{k} = 2 \cdot 2^k
\end{array}$$

However, $\binom{k}{0} = \binom{k+1}{0} = \binom{k}{k} = \binom{k+1}{k+1} = 1$, so the last result can be written as

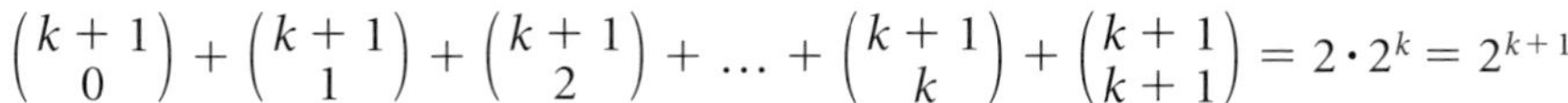

$$\binom{k+1}{0} + \binom{k+1}{1} + \binom{k+1}{2} + \ldots + \binom{k+1}{k} + \binom{k+1}{k+1} = 2 \cdot 2^k = 2^{k+1}$$

This shows that $P(k+1)$ is true whenever $P(k)$ is true. This completes the inductive step of the proof.

Therefore, $\binom{n}{0} + \binom{n}{1} + \binom{n}{2} + \ldots + \binom{n}{n-1} + \binom{n}{n} = 2^n$ for all non-negative integers n.

Proof of the binomial theorem (optional)

Before we get into the proof, we need to state a few properties of the summation notation.

1. *Change of limits* property: If $f(i)$ is an expression used in the summation process, then the following is true:

   $$\sum_{i=k}^{m} f(i) = \sum_{i=k+r}^{m+r} f(i-r)$$

   For example, suppose we need to find $10^2 + 11^2 + \ldots + 49^2$ using summation notation. We can either write it as

   $$\sum_{i=10}^{49} i^2 \text{ or } \sum_{i=1}^{40} (i+9)^2. \text{ Here } r = -9.$$

2. Another useful property is the following:

   $$\begin{aligned}\sum_{i=k}^{m} f(i) &= f(k) + f(k+1) + \ldots + f(m) \\ &= f(k) + \sum_{i=k+1}^{m} f(i)\end{aligned}$$

   Or

   $$\begin{aligned}\sum_{i=k}^{m} f(i) &= f(k) + f(k+1) + \ldots + f(m-1) + f(m) \\ &= \sum_{i=k}^{m-1} f(i) + f(m)\end{aligned}$$

The binomial theorem

Let $P(n)$ be the statement that $(a+b)^n = \sum_{i=0}^{n} \binom{n}{i} a^{n-i} b^i; \forall n \geqslant 0$.

Basis step: To prove this formula the basis step must be $P(0)$.

$P(0)$ is true since $(a+b)^0 = 1 = \sum_{i=0}^{0} \binom{0}{i} a^{n-i} b^i = \binom{0}{0} a^{0-0} b^0 = 1 \cdot 1 \cdot 1 = 1$.

Also, $P(1)$ is true since

$(a+b)^1 = a + b$

● **Hint:** The symbol ∀ stands for the universal quantifier: 'For all n'.

$$= \sum_{i=0}^{1} \binom{1}{i} a^{n-i} b^{i}$$

$$= \binom{1}{0} a^{1-0} b^{0} + \binom{1}{1} a^{1-1} b^{1}$$

$$= 1 \cdot a \cdot 1 + 1 \cdot 1 \cdot b$$

$$= a + b.$$

Inductive step: Assume $P(k)$ is true, i.e. assume that $(a+b)^k = \sum_{i=0}^{k} \binom{k}{i} a^{k-i} b^{i}$. We must prove that $P(k+1)$ is true, i.e. $(a+b)^{k+1} = \sum_{i=0}^{k+1} \binom{k+1}{i} a^{k+1-i} b^{i}$.

$$(a+b)^{k+1} = (a+b)(a+b)^k = (a+b) \sum_{i=0}^{k} \binom{k}{i} a^{k-i} b^{i}$$

and using the distributive property, we get

$$RHS = a \sum_{i=0}^{k} \binom{k}{i} a^{k-i} b^{i} + b \sum_{i=0}^{k} \binom{k}{i} a^{k-i} b^{i}$$

$$= \sum_{i=0}^{k} \binom{k}{i} a \cdot a^{k-i} b^{i} + \sum_{i=0}^{k} \binom{k}{i} b \cdot a^{k-i} b^{i}$$

$$= \sum_{i=0}^{k} \binom{k}{i} a^{k+1-i} b^{i} + \sum_{i=0}^{k} \binom{k}{i} a^{k-i} b^{i+1}$$

Now, using property 2 on page 197,

$$RHS = \binom{k}{0} a^{k+1} + \sum_{i=1}^{k} \binom{k}{i} a^{k+1-i} b^{i} + \sum_{i=0}^{k-1} \binom{k}{i} a^{k-i} b^{i+1} + \binom{k}{k} b^{k+1}$$

Moreover, using property 1, we have

$$RHS = \binom{k}{0} a^{k+1} + \sum_{i=1}^{k} \binom{k}{i} a^{k+1-i} b^{i} + \sum_{i=1}^{k} \binom{k}{i-1} a^{k-(i-1)} b^{(i-1)+1} + \binom{k}{k} b^{k+1}$$

$$= \binom{k}{0} a^{k+1} + \left\{ \sum_{i=1}^{k} \binom{k}{i} a^{k+1-i} b^{i} + \sum_{i=1}^{k} \binom{k}{i-1} a^{k+1-i} b^{i} \right\} + \binom{k}{k} b^{k+1}$$

Now, you observe that the terms inside the brackets have a common factor, so

$$RHS = \binom{k}{0} a^{k+1} + \sum_{i=1}^{k} \left\{ \binom{k}{i} + \binom{k}{i-1} \right\} a^{k+1-i} b^{i} + \binom{k}{k} b^{k+1}$$

Finally, using Pascal's property along with the fact that

$$\binom{k}{0} = \binom{k+1}{0} = \binom{k}{k} = \binom{k+1}{k+1} = 1, \text{ we have}$$

$$RHS = \binom{k+1}{0}a^{k+1} + \sum_{i=1}^{k}\binom{k+1}{i}a^{k+1-i}b^i + \binom{k+1}{k+1}b^{k+1}$$

$$= \sum_{i=0}^{k+1}\binom{k+1}{i}a^{k+1-i}b^i$$

This shows that $P(k + 1)$ is true whenever $P(k)$ is true. This completes the inductive step of the proof.

Therefore, $(a + b)^n = \sum_{i=0}^{n}\binom{n}{i}a^{n-i}b^i; \forall n \geq 0$.

Exercise 4.7

1 Find a formula for the sum of the first n even positive integers and prove it using mathematical induction.

2 Let $a_1, a_2, a_3, \ldots$ be a sequence defined by

$a_1 = 1, a_n = 3a_{n-1}; n \geq 1$

Show that $a_n = 3^{n-1}$ for all positive integers n.

3 Let $a_1, a_2, a_3, \ldots$ be a sequence defined by

$a_1 = 1, a_n = a_{n-1} + 4; n \geq 2$

Show that $a_n = 4n - 3$ for all positive integers $n > 1$.

4 Let $a_1, a_2, a_3, \ldots$ be a sequence defined by

$a_1 = 1, a_n = 2a_{n-1} + 1; n \geq 2$

Show that $a_n = 2^n - 1$ for all positive integers $n > 1$.

5 Let $a_1, a_2, a_3, \ldots$ be a sequence defined by

$a_1 = \frac{1}{2}, a_n = a_{n-1} + \frac{1}{n(n+1)}; n \geq 2$

Show that $a_n = \frac{n}{n+1}$ for all positive integers $n > 1$.

6 Find a formula for $\frac{1}{2} + \frac{1}{4} + \frac{1}{8} + \ldots + \frac{1}{2^n}$ and then use mathematical induction to prove your formula.

7 Show that $1 + 2 + 2^2 + \ldots + 2^n = 2^{n+1} - 1$ for all non-negative integers n.

8 Show, using mathematical induction, that in a geometric sequence $a_n = a_1r^{n-1}$.

9 Show, using mathematical induction, that in a geometric series $S_n = \frac{a - ar^n}{1 - r}$.

10 Prove that $2^n < n!$ for all positive integers larger than 3.

11 Prove that $2^n > n^2$ for all positive integers larger than 4.

12 Show that $1 \cdot 1! + 2 \cdot 2! + 3 \cdot 3! + \ldots n \cdot n! = (n + 1)! - 1$.

13 Show that $\frac{1}{1 \cdot 2} + \frac{1}{2 \cdot 3} + \frac{1}{3 \cdot 4} + \ldots + \frac{1}{n \cdot (n+1)} = \frac{n}{n+1}$ for all positive integers n.

14 Show that $n^3 - n$ is divisible by 3 for all positive integers n.

15 Show that $n^5 - n$ is divisible by 5 for all positive integers n.

16 Show that $n^3 - n$ is divisible by 6 for all positive integers n.

17 Show that $n^2 + n$ is an even number for all integers n.

18 Show that $5^n - 1$ is divisible by 4 for all integers n.

19 Show that $\begin{pmatrix} a & 0 \\ 0 & b \end{pmatrix}^n = \begin{pmatrix} a^n & 0 \\ 0 & b^n \end{pmatrix}$ for every positive integer n and where a and b are real numbers.

20 Prove each of the following statements.

a) $\sum_{i=1}^{n} (2i + 4) = n^2 + 5n$ for each positive integer n.

b) $\sum_{i=1}^{n} (2 \cdot 3^{i-1}) = 3^{n-1}$ for each positive integer n.

c) $\sum_{i=1}^{n} \frac{1}{(2i-1)(2i+1)} = \frac{n}{2n+1}$ for each positive integer n.

Practice questions

1 In an arithmetic sequence, the first term is 4, the 4th term is 19 and the nth term is 99. Find the common difference and the number of terms n.

2 How much money should you invest now if you wish to have an amount of €3000 in your account after 6 years if interest is compounded quarterly at an annual rate of 6%?

3 Two students, Nick and Charlotte, decide to start preparing for their IB exams 15 weeks ahead of the exams. Nick starts by studying for 12 hours in the first week and plans to increase the amount by 2 hours per week. Charlotte starts with 12 hours in the first week and decides to increase her time by 10% every week.

a) How many hours did each student study in week 5?

b) How many hours in total does each student study for the 15 weeks?

c) In which week will Charlotte exceed 40 hours per week?

d) In which week does Charlotte catch up with Nick in the number of hours spent on studying per week?

4 Two diet schemes are available for relatively overweight people to lose weight. Plan A promises the patient an initial weight loss of 1000 g the first month, with a steady loss of an additional 80 g every month after the first. So, the second month the patient will lose 1080 g and so on for a maximum duration of 12 months.

Plan B starts with a weight loss of 1000 g the first month and an increase in weight loss by 6% more every following month.

a) Write down the amount of grams lost under Plan B in the second and third months.

b) Find the weight lost in the 12th month for each plan.

c) Find the total weight loss during a 12-month period under

(i) Plan A **(ii)** Plan B.

5 Planning on buying your first car in 10 years, you start a savings plan where you invest €500 at the beginning of the year for 10 years. Your investment scheme offers a fixed rate of 6% per year compounded annually.

Calculate, giving your answers to the nearest euro (€),

(a) how much the first €500 is worth at the end of 10 years

(b) the total value your investment will give you at the end of the 10 years.

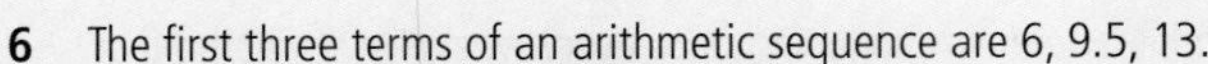

6 The first three terms of an arithmetic sequence are 6, 9.5, 13.

a) What is the 40th term of the sequence?

b) What is the sum of the first 103 terms of the sequence?

7 $\{a^n\}$ is defined as follows

$a_n = \sqrt[3]{(8 - a^3_{n-1})}$

a) Given that $a_1 = 1$, evaluate a_2, a_3, a_4. Describe $\{a_n\}$.

b) Given that $a_1 = 2$, evaluate a_2, a_3, a_4. Describe $\{a_n\}$.

8 A marathon runner plans her training programme for a 20 km race. On the first day she plans to run 2 km, and then she wants to increase her distance by 500 m on each subsequent training day.

a) On which day of her training does she first run a distance of 20 km?

b) By the time she manages to run the 20 km distance, what is the total distance she would have run for the whole training programme?

9 In the nation of Telefonica, cellular phones were first introduced in the year 2000. During the first year, the number of people who bought a cellular phone was 1600. In 2001, the number of new participants was 2400, and in 2002 the new participants numbered 3600.

a) You notice that the trend is a geometric sequence; find the common ratio.

Assuming that the trend continues,

b) how many participants will join in 2012?

c) in what year would the number of new participants first exceed 50 000?

Between 2000 and 2002, the total number of participants reaches 7600.

d) What is the total number of participants between 2000 and 2012?

During this period, the total adult population of Telefonica remains at approximately 800 000.

e) Use this information to suggest a reason why this trend in growth would not continue.

10 In an arithmetic sequence, the first term is 25, the fourth term is 13 and the nth term is $-11\,995$. Find the common difference d and the number of terms n.

11 The midpoints M, N, P, Q of the sides of a square of side 1 cm are joined to form a new square.

a) Show that the side of the second square $MNPQ$ is $\frac{\sqrt{2}}{2}$.

b) Find the area of square $MNPQ$.

A new third square $RSTU$ is constructed in the same manner.

c) **(i)** Find the area of the third square just constructed.

(ii) Show that the areas of the squares are in a geometric sequence and find its common ratio.

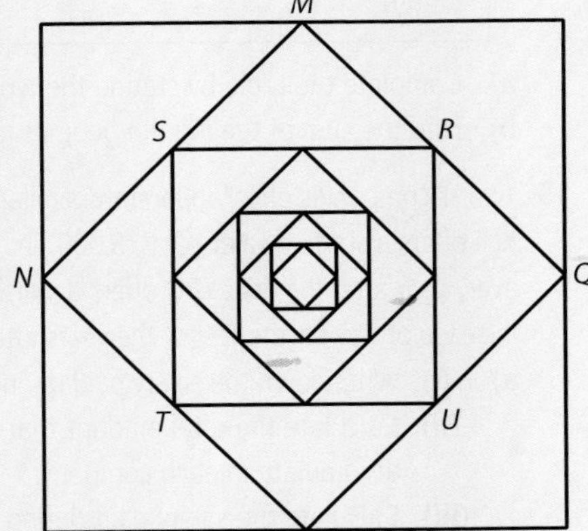

The procedure continues indefinitely.

d) **(i)** Find the area of the tenth square.

(ii) Find the sum of the areas of all the squares.

12 Tim is a dedicated swimmer. He goes swimming once every week. He starts the first week of the year by swimming 200 metres. Each week after that he swims 20 m more than the previous week. He does that all year long (52 weeks).

a) How far does he swim in the final week?

b) How far does he swim altogether?

13 The diagram below shows three iterations of constructing squares in the following manner: A square of side 3 units is given, then it is divided into nine smaller squares as shown and the middle square is shaded. Each of the unshaded squares is in turn divided into nine squares and the process is repeated. The area of the first shaded square is 1 unit.

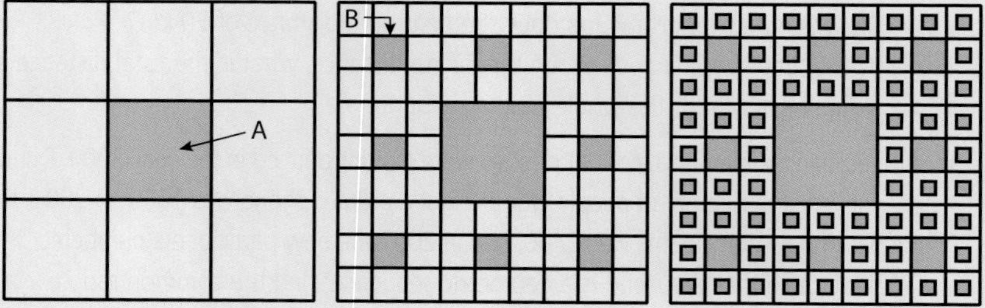

a) Find the area of each of the squares A and B.

b) Find the area of any small square in the third diagram.

c) Find the area of the shaded regions in the second and third iterations.

d) If the process is continued indefinitely, find the area left unshaded.

14 The table below shows four series of numbers. One series is an arithmetic one, one is a converging geometric series, one is a diverging geometric series and the fourth is neither geometric nor arithmetic.

Series		Type of series
(i)	$2 + 22 + 222 + 2222 + \dots$	
(ii)	$2 + \frac{4}{3} + \frac{8}{9} + \frac{16}{27} + \dots$	
(iii)	$0.8 + 0.78 + 0.76 + 0.74 + \dots$	
(iv)	$2 + \frac{8}{3} + \frac{32}{9} + \frac{128}{27} + \dots$	

a) Complete the table by stating the type of each series.

b) Find the sum of the infinite geometric series above.

15 Two IT companies offer 'apparently' similar salary schemes for their new appointees. Kell offers a starting salary of €18 000 per year and then an annual increase of €400 every year after the first. YBO offers a starting salary of €17 000 per year and an annual increase of 7% for the rest of the years after the first.

a) **(i)** Write down the salary paid during the second and third years for each company.

(ii) Calculate the total amount that an employee working for 10 years will accumulate in each company.

(iii) Calculate the salary paid during the tenth year for each company.

b) Tim works at Kell and Merijayne works at YBO.

(i) When would Merijayne start earning more than Tim?

(ii) What is the minimum number of years that Merijayne requires so that her total earnings exceed Tim's total earnings?

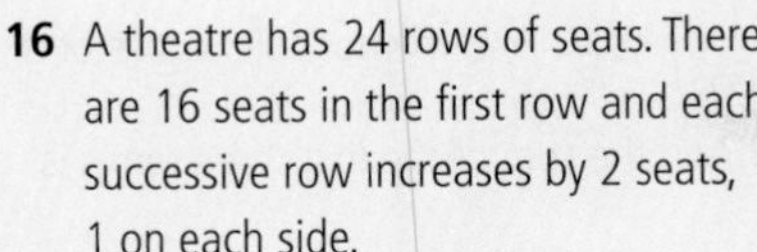

16 A theatre has 24 rows of seats. There are 16 seats in the first row and each successive row increases by 2 seats, 1 on each side.

a) Calculate the number of seats in the 24th row.

b) Calculate the number of seats in the whole theatre.

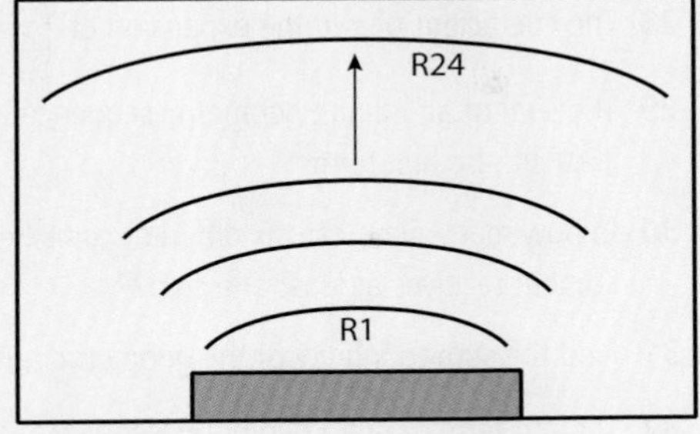

17 The amount of €7000 is invested at 5.25% annual compound interest.

a) Write down an expression for the value of this investment after t full years.

b) Calculate the minimum number of years required for this amount to become €10 000.

c) For the same number of years as in part b), would an investment of the same amount be better if it were at a 5% rate compounded quarterly?

18 With S_n denoting the sum of the first n terms of an arithmetic sequence, we are given that $S_1 = 9$ and $S_2 = 20$.

a) Find the second term.

b) Calculate the common difference of the sequence.

c) Find the fourth term.

19 The second term of an arithmetic sequence is 7. The sum of the first four terms of the arithmetic sequence is 12. Find the first term, a, and the common difference, d, of the sequence.

20 Given that

$(1 + x)^5 (1 + ax)^6 \equiv 1 + bx + 10x^2 + \dots\dots\dots + a^6 x^{11}$,

find the values of $a, b \in \mathbb{Z}$, where $a \neq 0$.

21 The ratio of the fifth term to the twelfth term of a sequence in an arithmetic progression is $\frac{6}{13}$. If each term of this sequence is positive, and the product of the first term and the third term is 32, find the sum of the first 100 terms of this sequence.

22 Using mathematical induction, prove that the number $2^{2n} - 3n - 1$ is divisible by 9, for $n = 1, 2, \dots$.

23 An arithmetic sequence has 5 and 13 as its first two terms respectively.

a) Write down, in terms of n, an expression for the nth term, an.

b) Find the number of terms of the sequence which are less than 400.

24 Find the coefficient of x^7 in the expansion of $(2 + 3x)^{10}$, giving your answer as a whole number.

25 The sum of the first n terms of an arithmetic sequence is $S_n = 3n^2 - 2n$. Find the nth term u_n.

26 Mr Blue, Mr Black, Mr Green, Mrs White, Mrs Yellow and Mrs Red sit around a circular table for a meeting. Mr Black and Mrs White must not sit together.

Calculate the number of different ways these six people can sit at the table without Mr Black and Mrs White sitting together.

27 Find the sum of the positive terms of the arithmetic sequence 85, 78, 71, ….

28 The coefficient of x in the expansion of $\left(x + \frac{1}{a(x)^2}\right)^7$ is $\frac{7}{3}$. Find the possible values of a.

29 The sum of an infinite geometric sequence is $\frac{27}{2}$, and the sum of the first three terms is 13. Find the first term.

30 In how many ways can six different coins be divided between two students so that each student receives at least one coin?

31 Find the sum to infinity of the geometric series $-12 + 8 - \frac{16}{3}$.

32 The nth term, u_n, of a geometric sequence is given by $u_n = 3(4)^{n+1}$, $n \in \mathbb{Z}^+$.

a) Find the common ratio r.

b) Hence, or otherwise, find S_n, the sum of the first n terms of this sequence.

33 Consider the infinite geometric series

$$1 + \left(\frac{2x}{3}\right) + \left(\frac{2x}{3}\right)^2 + \left(\frac{2x}{3}\right)^3 + \ldots$$

a) For what values of x does the series converge?

b) Find the sum of the series if $x = 1.2$.

34 How many four-digit numbers are there which contain at least one digit 3?

35 Consider the arithmetic series $2 + 5 + 8 + \ldots$.

a) Find an expression for S_n, the sum of the first n terms.

b) Find the value of n for which $S_n = 1365$.

36 Find the coefficient of x^3 in the binomial expansion of $\left(1 - \frac{1}{2}x\right)^8$.

37 Find $\sum_{r=1}^{50} \ln(2^r)$, giving the answer in the form $a \ln 2$, where $a \in \mathbb{Q}$.

38 A sequence $\{u_n\}$ is defined by $u_0 = 1$, $u_1 = 2$, $u_{n+1} = 3u_n - 2u_{n-1}$ where $n \in \mathbb{Z}^+$.

a) Find u_2, u_3, and u_4.

b) **(i)** Express u_n in terms of n.

(ii) Verify that your answer to part b)(i) satisfies the equation $u_{n+1} = 3u_n - 2u_{n-1}$.

39 A geometric sequence has all positive terms. The sum of the first two terms is 15 and the sum to infinity is 27. Find the value of

a) the common ratio;

b) the first term.

40 The first four terms of an arithmetic sequence are 2, $a - b$, $2a + b + 7$, and $a - 3b$, where a and b are constants. Find a and b.

41 A committee of four children is chosen from eight children. The two oldest children cannot both be chosen. Find the number of ways the committee may be chosen.

42 The three terms a, 1, b are in arithmetic progression. The three terms 1, a, b are in geometric progression. Find the value of a and of b given that $a \neq b$.

43 The diagram on the following page shows a sector AOB of a circle of radius 1 and centre O, where $\hat{AOB} = \theta$.

The lines (AB_1), (A_1B_2), (A_2B_3) are perpendicular to OB. A_1B_1, A_2B_2 are all arcs of circles with centre O.

Calculate the sum to infinity of the arc lengths

$AB + A_1B_1 + A_2B_2 + A_3B_3 + \ldots$

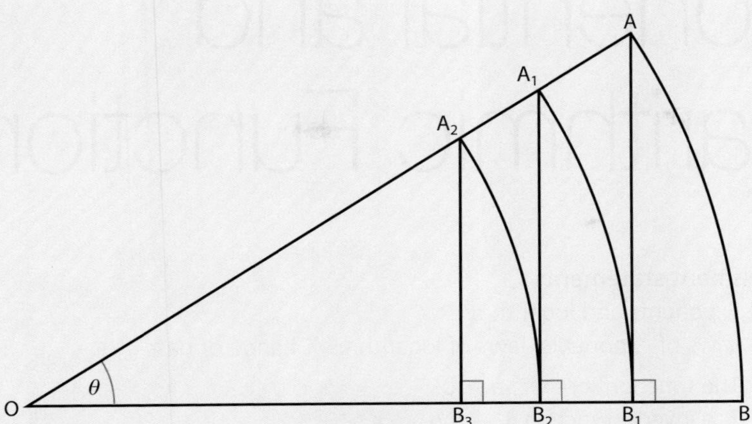

44 The sum of the first n terms of a series is given by

$S_n = 2n^2 - n$, where $n \in \mathbb{Z}^+$.

a) Find the first three terms of the series.

b) Find an expression for the nth term of the series, giving your answer in terms of n.

45 a) Find the expansion of $(2 + x)^5$, giving your answer in ascending powers of x.

b) By letting $x = 0.01$ or otherwise, find the exact value of 2.01^5.

46 A sum of \$5000 is invested at a compound interest rate of 6.3% per annum.

a) Write down an expression for the value of the investment after n full years.

b) What will be the value of the investment at the end of five years?

c) The value of the investment will exceed \$10 000 after n full years.

(i) Write an inequality to represent this information.

(ii) Calculate the minimum value of n.

47 Use mathematical induction to prove that $5^n + 9^n + 2$ is divisible by 4, for $n \in \mathbb{Z}^+$.

48 The sum of the first n terms of an arithmetic sequence $\{u_n\}$ is given by the formula $S_n = 4n^2 - 2n$. Three terms of this sequence, u_2, u_m and u_{32}, are consecutive terms in a geometric sequence. Find m.

5 Exponential and Logarithmic Functions

Assessment statements

1.2 Exponents and logarithms.
Laws of exponents; laws of logarithms. Change of base.

2.4 The function $x \mapsto a^x$, $a > 0$.
The inverse function $x \mapsto \log_a x$, $x > 0$.
Graphs of $y = a^x$ and $y = \log_a x$.
The exponential function $x \mapsto e^x$.
The logarithmic function $x \mapsto \ln x$, $x > 0$.

2.6 Solutions of $a^x = b$ using logarithms.

Introduction

A variety of functions have already been considered in this text (see Figure 2.17 in Section 2.4): polynomial functions (e.g. linear, quadratic and cubic functions), functions with radicals (e.g. square root function), rational functions (e.g. inverse and inverse square functions) and the absolute value function. This chapter examines exponential and logarithmic functions.

Exponential functions help us model a wide variety of physical phenomena. The natural exponential function (or simply, *the* exponential function), $f(x) = e^x$, is one of the most important functions in calculus. Exponential functions and their applications – especially to situations involving growth and decay – will be covered at length.

Logarithms, which were originally invented as a computational tool, lead to logarithmic functions. These functions are closely related to exponential functions and play an equally important part in calculus and a range of applications. We will learn that certain exponential and logarithmic functions are inverses of each other.

5.1 Exponential functions

Characteristics of exponential functions

• **Hint:** Another word for exponent is **index** (plural: **indices**).

We begin our study of exponential functions by comparing two algebraic expressions that represent two seemingly similar but very different functions. The two expressions $y = x^2$ and $y = 2^x$ are similar in that they both contain a **base** and an **exponent** (or power). In $y = x^2$, the base is the variable x and the exponent is the constant 2. In $y = 2^x$, the base is the constant 2 and the exponent is the variable x.

The quadratic function $y = x^2$ is in the form 'variable base$^{\text{constant power}}$', where the base is a variable and the exponent is an integer greater than or equal to zero (non-negative integer). Any function in this form is called a **power function**.

The function $y = 2^x$ is in the form 'constant base$^{\text{variable power}}$', where the base is a positive real number (not equal to one) and the exponent is a variable. Any function in this form is called an **exponential function**.

To illustrate a fundamental difference between exponential functions and power functions, consider the function values for $y = x^2$ and $y = 2^x$ when x is an integer from 0 to 10. Table 5.1 showing these results displays clearly how the values for the exponential function eventually increase at a significantly faster rate than the power function.

x	$y = x^2$	$y = 2^x$
0	0	1
1	1	2
2	4	4
3	9	8
4	16	16
5	25	32
6	36	64
7	49	128
8	64	256
9	81	512
10	100	1024

Table 5.1 Contrast between power function and exponential function.

Another important point to make is that power functions can easily be defined (and computed) for any real number. For any power function $y = x^n$, where n is any positive integer, y is found by simply taking x and repeatedly multiplying it n times. Hence, x can be any real number. For example, for the power function $y = x^3$, if $x = \pi$, then $y = \pi^3 \approx 31.006\,276\,68\ldots$. Since a power function like $y = x^3$ is defined for all real numbers, we can graph it as a continuous curve so that every real number is the x-coordinate of some point on the curve. What about the exponential function $y = 2^x$? Can we compute a value for y for any real number x? Before we try, let's first consider x being any rational number and recall the following laws of exponents (indices) that were covered in Section 1.3.

Laws of exponents

For $b > 0$ and $m, n \in \mathbb{Q}$ (rational numbers):

$b^m \cdot b^n = b^{m+n}$ $\quad \frac{b^m}{b^n} = b^{m-n}$ $\quad (b^m)^n = b^{mn}$ $\quad b^0 = 1$ $\quad b^{-m} = \frac{1}{b^m}$

Also, in Section 1.3, we covered the definition of a rational exponent.

Rational exponent

For $b > 0$ and $m, n \in \mathbb{Z}$ (integers):

$$b^{\frac{m}{n}} = \sqrt[n]{b^m} = (\sqrt[n]{b})^m$$

From these established facts, we are able to compute b^x ($b > 0$) when x is any rational number. For example, $b^{4.7} = b^{\frac{47}{10}}$ represents the 10th root of b raised to the 47th power, i.e. $\sqrt[10]{b^{47}}$. Now, we would like to define b^x when x is any real number such as π or $\sqrt{2}$. We know that π has a non-terminating, non-repeating decimal representation that begins $\pi = 3.141\,592\,653\,589\,793\ldots$. Consider the sequence of numbers

$$b^3, b^{3.1}, b^{3.14}, b^{3.141}, b^{3.1415}, b^{3.14159}, \ldots$$

To demonstrate just how quickly $y = 2^x$ increases, consider what would happen if you were able to repeatedly fold a piece of paper in half 50 times. A typical piece of paper is about five thousandths of a centimetre thick. Each time you fold the piece of paper the thickness of the paper doubles, so after 50 folds the thickness of the folded paper is the height of a stack of 2^{50} pieces of paper. The thickness of the paper after being folded 50 times would be $2^{50} \times 0.005$ cm – which is more than 56 million kilometres (nearly 35 million miles)! Compare that with the height of a stack of 50^2 pieces of paper that would be a meagre $12\frac{1}{2}$ cm – only 0.000 125 km.

Every term in this sequence is defined because each has a rational exponent. Although it is beyond the scope of this text, it can be proved that each number in the sequence gets closer and closer to a certain real number – defined as b^{π}. Similarly, we can define other irrational exponents in such a way that the laws of exponents hold for all real exponents. Table 5.2 shows a sequence of exponential expressions approaching the value of 2^{π}.

Table 5.2 Approaching the value of 2^{π}.

x	2^x (12 s.f.)
3	8.000 000 000 00
3.1	8.574 187 700 29
3.14	8.815 240 927 01
3.141	8.821 353 304 55
3.1415	8.824 411 082 48
3.141 59	8.824 961 595 06
3.141 592	8.824 973 829 06
3.141 5926	8.824 977 499 27
3.141 592 65	8.824 977 805 12

Your GDC will give an approximate value for 2^{π} to at least 10 significant figures, as shown below.

```
2^π
            8.824977827
```

Graphs of exponential functions

Using this definition of irrational powers, we can now construct a complete graph of any exponential function $f(x) = b^x$ such that b is a number greater than zero ($b \neq 1$) and x is any real number.

Example 1

Graph each exponential function by plotting points.

a) $f(x) = 3^x$ b) $g(x) = \left(\frac{1}{3}\right)^x$

Solution

We can easily compute values for each function for integral values of x from -3 to 3. Knowing that exponential functions are defined for all real numbers – not just integers – we can sketch a smooth curve in Figure 5.1, filling in between the ordered pairs shown in the table.

x	$f(x) = 3^x$	$g(x) = \left(\frac{1}{3}\right)^x$
-3	$\frac{1}{27}$	27
-2	$\frac{1}{9}$	9
-1	$\frac{1}{3}$	3
0	1	1
1	3	$\frac{1}{3}$
2	9	$\frac{1}{9}$
3	27	$\frac{1}{27}$

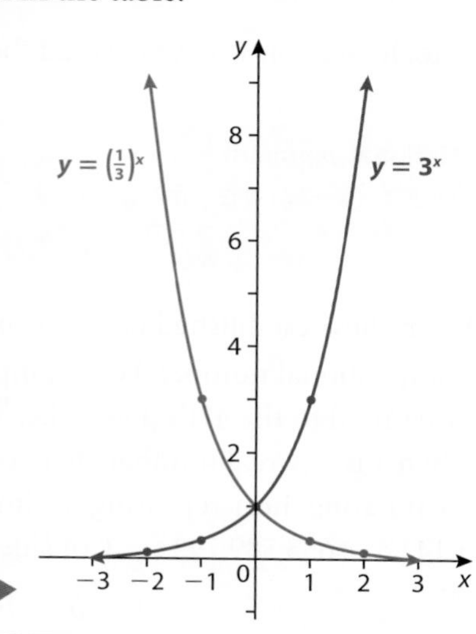

Figure 5.1

Remember that in Section 2.4 we established that the graph of $y = f(-x)$ is obtained by reflecting the graph of $y = f(x)$ in the y-axis. It is clear from the table and the graph in Figure 5.1 that the graph of function g is a reflection of function f about the y-axis. Let's use some laws of exponents to show that $g(x) = f(-x)$.

$$g(x) = \left(\frac{1}{3}\right)^x = \frac{1^x}{3^x} = \frac{1}{3^x} = 3^{-x} = f(-x)$$

It is useful to point out that both of the graphs, $y = 3^x$ and $y = \left(\frac{1}{3}\right)^x$, pass through the point $(0, 1)$ and have a horizontal asymptote of $y = 0$ (x-axis). The same is true for the graph of all exponential functions in the form $y = b^x$ given that $b \neq 1$. If $b = 1$, then $y = 1^x = 1$ and the graph is a horizontal line rather than a constantly increasing or decreasing curve.

Exponential functions

If $b > 0$ and $b \neq 1$, the **exponential function** with base b is the function defined by

$$f(x) = b^x$$

The **domain** of f is the set of real numbers ($x \in \mathbb{R}$) and the **range** of f is the set of positive real numbers ($y > 0$). The graph of f passes through $(0, 1)$, has the x-axis as a **horizontal asymptote**, and, depending on the value of the base of the exponential function b, will either be a continually increasing **exponential growth curve** or a continually decreasing **exponential decay curve**.

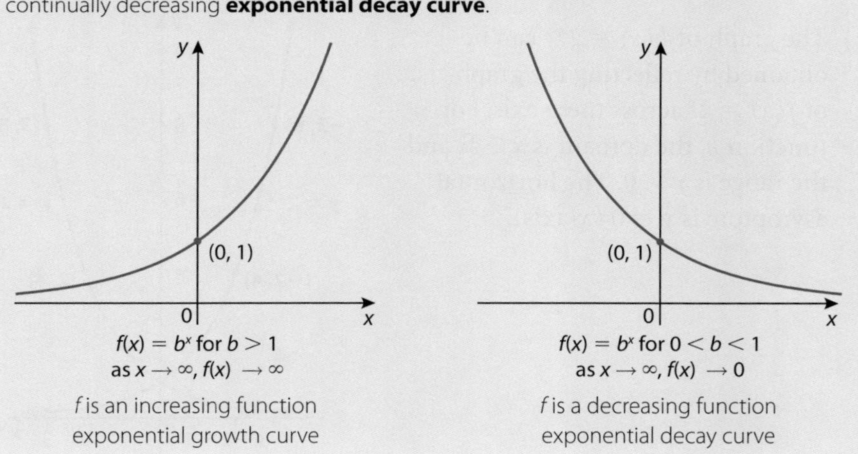

$f(x) = b^x$ for $b > 1$
as $x \to \infty$, $f(x) \to \infty$

f is an increasing function
exponential growth curve

$f(x) = b^x$ for $0 < b < 1$
as $x \to \infty$, $f(x) \to 0$

f is a decreasing function
exponential decay curve

The graphs of all exponential functions will display a characteristic growth or decay curve. As we shall see, many natural phenomena exhibit exponential growth or decay. Also, the graphs of exponential functions behave **asymptotically** for either very large positive values of x (decay curve) or very large negative values of x (growth curve). This means that there will exist a horizontal line that the graph will approach, but not intersect, as either $x \to \infty$ or as $x \to -\infty$.

Transformations of exponential functions

Recalling from Section 2.4 how the graphs of functions are translated and reflected, we can efficiently sketch the graph of many exponential functions.

Example 2

Using the graph of $f(x) = 2^x$, sketch the graph of each function. State the domain and range for each function and the equation of its horizontal asymptote.

a) $g(x) = 2^x + 3$ b) $h(x) = 2^{-x}$ c) $p(x) = -2^x$

d) $r(x) = 2^{x-4}$ e) $v(x) = 3(2^x)$

Solution

a) The graph of $g(x) = 2^x + 3$ can be obtained by translating the graph of $f(x) = 2^x$ vertically three units up. For function g, the domain is x is any real number ($x \in \mathbb{R}$) and the range is $y > 3$. The horizontal asymptote for g is $y = 3$.

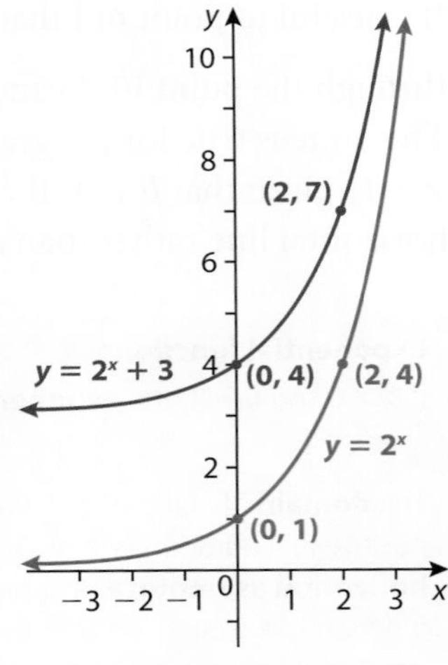

b) The graph of $h(x) = 2^{-x}$ can be obtained by reflecting the graph of $f(x) = 2^x$ across the y-axis. For function h, the domain is $x \in \mathbb{R}$ and the range is $y > 0$. The horizontal asymptote is $y = 0$ (x-axis).

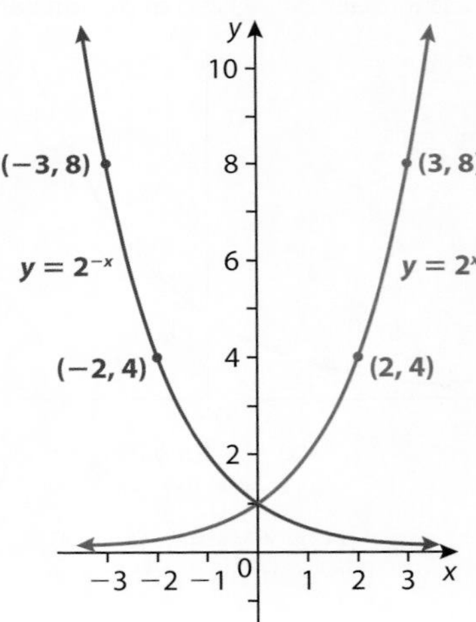

c) The graph of $p(x) = -2^x$ can be obtained by reflecting the graph of $f(x) = 2^x$ across the x-axis. For function p, the domain is $x \in \mathbb{R}$ and the range is $y < 0$. The horizontal asymptote is $y = 0$ (x-axis).

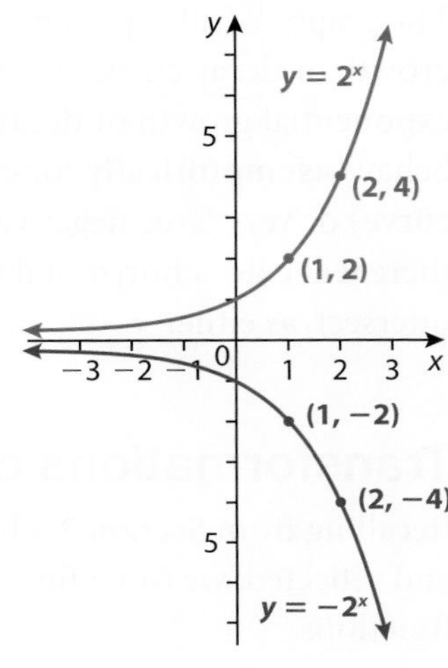

d) The graph of $r(x) = 2^{x-4}$ can be obtained by translating the graph of $f(x) = 2^x$ four units to the right. For function r, the domain is $x \in \mathbb{R}$ and the range is $y > 0$. The horizontal asymptote is $y = 0$ (x-axis).

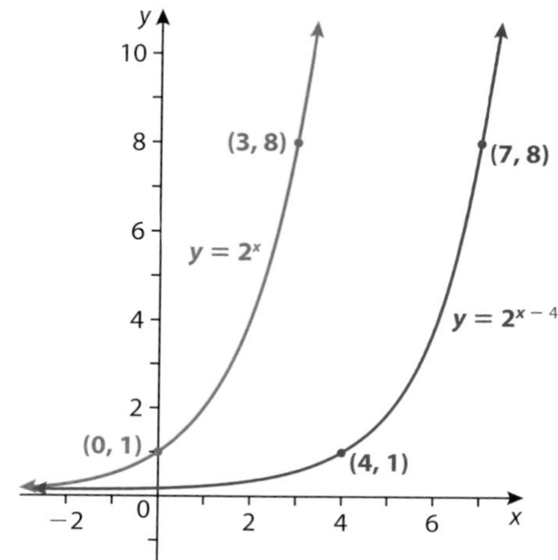

e) The graph of $v(x) = 3(2^x)$ can be obtained by a vertical stretch of the graph of $f(x) = 2^x$ by scale factor 3. For function v, the domain is $x \in \mathbb{R}$ and the range is $y > 0$. The horizontal asymptote is $y = 0$ (x-axis).

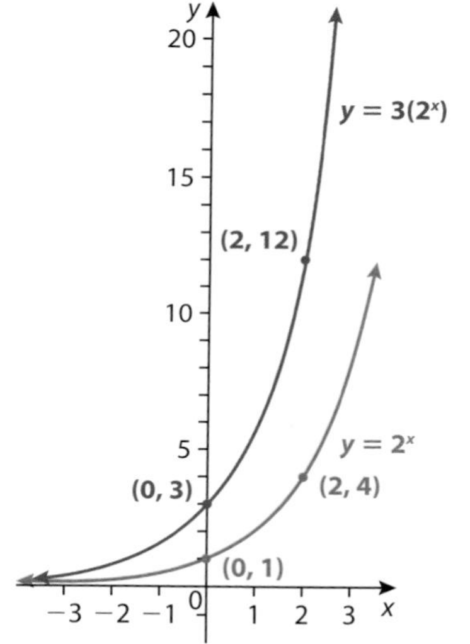

Note that for function p in part c) of Example 2 the horizontal asymptote is an **upper bound** (i.e. no function value is equal to or greater than $y = 0$). Whereas, in parts a), b), d) and e) the horizontal asymptote for each function is a **lower bound** (i.e. no function value is equal to or less than the y-value of the asymptote).

5.2 Exponential growth and decay

Mathematical models of growth and decay

Exponential functions are well suited as a mathematical model for a wide variety of steadily increasing or decreasing phenomena of many kinds, including population growth (or decline), investment of money with compound interest and radioactive decay. Recall from the previous chapter that the formula for finding terms in a geometric sequence (repeated multiplication by common ratio r) is an exponential function. Many instances of growth or decay occur geometrically (repeated multiplication by a growth or decay factor).

Exponential models

Exponential models are equations of the form $A(t) = A_0b^t$, where $A_0 \neq 0, b > 0$ and $b \neq 1$. $A(t)$ is the **amount after time *t***. $A(0) = A_0b^0 = A_0(1) = A_0$, so A_0 is called the **initial amount** or value (often the value at time $(t) = 0$). If $b > 1$, then A(t) is an **exponential growth model**. If $0 < b < 1$, then $A(t)$ is an **exponential decay model**. The value of b, the base of the exponential function, is often called the **growth or decay factor**.

Example 3

A sample count of bacteria in a culture indicates that the number of bacteria is doubling every hour. Given that the estimated count at 15:00 was 12 000 bacteria, find the estimated count three hours earlier at 12:00 and write an exponential growth function for the number of bacteria at any hour t.

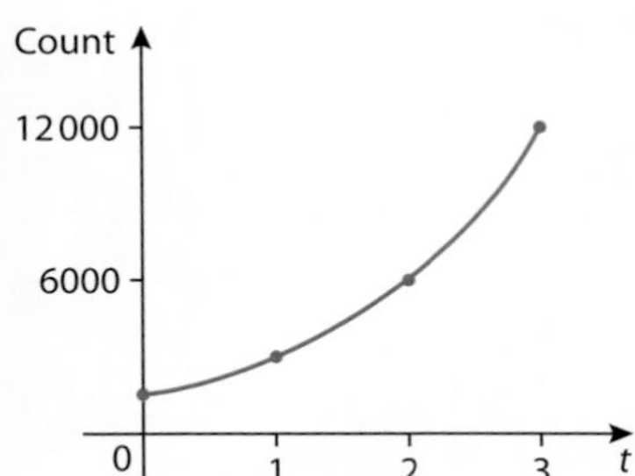

Solution

Consider the time at 12:00 to be the starting, or initial, time and label it $t = 0$ hours. Then the time at 15:00 is $t = 3$. The amount at any time t (in hours) will double after an hour so the growth factor, b, is 2. Therefore, $A(t) = A_0(2)^t$. Knowing that $A(3) = 12\,000$, compute A_0: $12\,000 = A_0(2)^3 \Rightarrow 12\,000 = 8A_0 \Rightarrow A_0 = 1500$. Therefore, the estimated count at 12:00 was 1500, and the growth function for number of bacteria at time t is $A(t) = 1500(2)^t$.

Radioactive carbon (carbon-14 or C-14), produced when nitrogen-14 is bombarded by cosmic rays in the atmosphere, drifts down to Earth and is absorbed from the air by plants. Animals eat the plants and take C-14 into their bodies. Humans in turn take C-14 into their bodies by eating both plants and animals. When a living organism dies, it stops absorbing C-14, and the C-14 that is already in the object begins to decay at a slow but steady rate, reverting to nitrogen-14. The half-life of C-14 is 5730 years. Half of the original amount of C-14 in the organic matter will have disintegrated after 5730 years; half of the remaining C-14 will have been lost after another 5730 years, and so forth. By measuring the ratio of C-14 to N-14, archaeologists are able to date organic materials. However, after about 50 000 years, the amount of C-14 remaining will be so small that the organic material cannot be dated reliably.

Radioactive material decays at exponential rates. The **half-life** is the amount of time it takes for a given amount of material to decay to half of its original amount. An exponential function that models decay with a known value for the half-life, h, will be of the form $A(t) = A_0\left(\frac{1}{2}\right)^h$, where the decay factor is $\frac{1}{2}$ and h represents the number of half-lives that have occurred (i.e. the number of times that A_0 is multiplied by $\frac{1}{2}$). If t represents the amount of time, the number of half-lives will be $\frac{t}{h}$. For example, if the half-life of a certain material is 25 days and the amount of time that has passed since measuring the amount A_0 is 75 days, then the number of half-lives is

$k = \frac{t}{h} = \frac{75}{25} = 3$, and the amount of material remaining is equal to

$A_0\left(\frac{1}{2}\right)^3 = \frac{A_0}{8}$.

Half-life formula

If a certain initial amount, A_0, of material decays with a half-life of h, the amount of material that remains at time t is given by the exponential decay model $A(t) = A_0\left(\frac{1}{2}\right)^{\frac{t}{h}}$. The time units (e.g. seconds, hours, years) for h and t must be the same.

Example 4

The half-life of radioactive carbon-14 is approximately 5730 years. How much of a 10 g sample of carbon-14 remains after 15 000 years?

Solution

The exponential decay model for the carbon-14 is $A(t) = A_0\left(\frac{1}{2}\right)^{\frac{t}{5730}}$.

What remains of 10 g after 15 000 years is given by

$A(15\,000) = 10\left(\frac{1}{2}\right)^{\frac{15\,000}{5730}} \approx 1.63$ g.

Compound interest

Recall from Chapter 4 that exponential functions occur in calculating compound interest. If an initial amount of money P, called the **principal**, is invested at an interest rate r per time period, then after one time period the amount of interest is $P \times r$ and the total amount of money is $A = P + Pr = P(1 + r)$. If the interest is added to the principal, the new principal is $P(1 + r)$, and the total amount after another time period is $A = P(1 + r)(1 + r) = P(1 + r)^2$. In the same way, after a third time period the amount is $A = P(1 + r)^3$. In general, after k periods the total amount is $A = P(1 + r)^k$, an exponential function with growth factor $1 + r$. For example, if the amount of money in a bank account is earning interest at a rate of 6.5% per time period, the growth factor is $1 + 0.065 = 1.065$. Is it possible for r to be negative? Yes, if an amount (not just money) is decreasing. For example, if the population of a town is decreasing by 12% per time period, the decay factor is $1 - 0.12 = 0.88$.

For compound interest, if the annual interest rate is r and interest is compounded (number of times added in) n times per year, then each time period the interest rate is $\frac{r}{n}$, and there are $n \times t$ time periods in t years.

Compound interest formula

The exponential function for calculating the amount of money after t years, $A(t)$, where P is the initial amount or principal, the annual interest rate is r and the number of times interest is compounded per year is n, is given by

$$A(t) = P\left(1 + \frac{r}{n}\right)^{nt}$$

Example 5

An initial amount of 1000 euros is deposited into an account earning $5\frac{1}{4}\%$ interest per year. Find the amounts in the account after eight years if interest is compounded annually, semi-annually, quarterly, monthly and daily.

Solution

We use the exponential function associated with compound interest with values of $P = 1000$, $r = 0.0525$ and $t = 8$ to complete the results in Table 5.3.

Compounding	n	Amount after 8 years
Annual	1	$1000\left(1 + \frac{0.0525}{1}\right)^{8} = 1505.83$
Semi-annual	2	$1000\left(1 + \frac{0.0525}{2}\right)^{2(8)} = 1513.74$
Quarterly	4	$1000\left(1 + \frac{0.0525}{4}\right)^{4(8)} = 1517.81$
Monthly	12	$1000\left(1 + \frac{0.0525}{12}\right)^{12(8)} = 1520.57$
Daily	365	$1000\left(1 + \frac{0.0525}{365}\right)^{365(8)} = 1521.92$

Table 5.3 Compound interest calculations.

Example 6

A new car is purchased for \$22 000. If the value of the car decreases (depreciates) at a rate of approximately 15% per year, what will be the approximate value of the car to the nearest whole dollar in $4\frac{1}{2}$ years?

Solution

The decay factor for the exponential function is $1 - r = 1 - 0.15 = 0.85$. In other words, after each year the car's value is 85% of what it was one year before. We use the exponential decay model $A(t) = A_0 b^t$ with values $A_0 = 22\,000$, $b = 0.85$ and $t = 4.5$.

$$A(4.5) = 22\,000(0.85)^{4.5} \approx 10\,588$$

The value of the car will be approximately \$10 588.

Exercise 5.1 and 5.2

1 a) Write the equation for an exponential equation with base $b > 0$.

b) Given $b \neq 1$, state the domain and range of this function.

c) Sketch the general shape of the graph of this exponential function for each of two cases:

(i) $b > 1$ (ii) $0 < b < 1$

For questions 2–7, sketch a graph of the function and state its domain, range, y-intercept and the equation of its horizontal asymptote.

2 $f(x) = 3^{x+4}$

3 $g(x) = -2^x + 8$

4 $h(x) = 4^{-x} - 1$

5 $p(x) = \dfrac{1}{2^x - 1}$

6 $q(x) = 3(3^{-x}) - 3$

7 $k(x) = 2^{-|x-2|} + 1$

8 If a general exponential function is written in the form $f(x) = a(b)^{x-c} + d$, state the domain, range, y-intercept and the equation of the horizontal asymptote in terms of the parameters a, b, c and d.

9 Using your GDC and a graph-viewing window with Xmin $= -2$, Xmax $= 2$, Ymin $= 0$ and Ymax $= 4$, sketch a graph for each exponential equation on the same set of axes.

a) $y = 2^x$ b) $y = 4^x$ c) $y = 8^x$

d) $y = 2^{-x}$ e) $y = 4^{-x}$ f) $y = 8^{-x}$

10 Write equations that are equivalent to the equations in 9 d), e) and f) but have an exponent of positive x rather than negative x.

11 If $1 < a < b$, which is steeper: the graph of $y = a^x$ or $y = b^x$?

12 The population of a city triples every 25 years. At time $t = 0$, the population is 100 000. Write a function for the population $P(t)$ as a function of t. What is the population after:

a) 50 years b) 70 years c) 100 years?

13 An experiment involves a colony of bacteria in a solution. It is determined that the number of bacteria doubles approximately every 3 minutes and the initial number of bacteria at the start of the experiment is 10^4. Write a function for the number of bacteria $N(t)$ as a function of t (in minutes). Approximately how many bacteria are there after:

a) 3 minutes b) 9 minutes c) 27 minutes d) one hour?

14 A bank offers an investment account that will double your money in 10 years.

a) Express $A(t)$, the amount of money in the account after t years, in the form $A(t) = A_0(r)^t$.

b) If interest was added into the account just once at the end of each year (simple interest), then find the annual interest rate for the account (to 3 significant figures).

15 If \$10 000 is invested at an annual interest rate of 11%, compounded quarterly, find the value of the investment after the given number of years.

a) 5 years b) 10 years c) 15 years

16 A sum of \$5000 is deposited into an investment account that earns interest at a rate of 9% per year compounded monthly.

a) Write the function $A(t)$ that computes the value of the investment after t years.

b) Use your GDC to sketch a graph of $A(t)$ with values of t on the horizontal axis ranging from $t = 0$ years to $t = 25$ years.

c) Use the graph on your GDC to determine the minimum number of years (to the nearest whole year) for this investment to have a value greater than \$20 000.

17 If \$10 000 is invested at an annual interest rate of 11% for a period of five years, find the value of the investment for the following compounding periods.

a) annually b) monthly c) daily d) hourly

18 Imagine a bank account that has the fantastic annual interest rate of 100%. If you deposit \$1 into this account, how much will be in the account exactly one year later, for the following compounding periods?

a) annually b) monthly c) daily d) hourly e) every minute

19 Each year for the past eight years, the population of deer in a national park increases at a steady rate of 3.2% per year. The present population is approximately 248 000.

a) What was the approximate number of deer one year ago?

b) What was the approximate number of deer eight years ago?

20 Radioactive carbon-14 has a half-life of 5730 years. The remains of an animal are found 20 000 years after it died. About what percentage (to 3 significant figures) of the original amount of carbon-14 (when the animal was alive) would you expect to find?

21 Once a certain drug enters the bloodstream of a human patient, it has a half-life of 36 hours. An amount of the drug, A_0, is injected in the bloodstream at 12:00 on Monday. How much of the drug will be in the bloodstream of the patient five days later at 12:00 on Friday?

22 An open can is filled with 1000 ml of fluid that evaporates at a rate of 30% per week.

a) Write a function, $A(w)$, that gives the amount of fluid after w weeks.

b) Use your GDC to find how many weeks (whole number) it will take for the volume of fluid to be less than 1 ml.

23 Why are exponential functions of the form $f(x) = b^x$ defined so that $b > 0$?

24 You are offered a highly paid job that lasts for just one month – exactly 30 days. Which of the following payment plans, I or II, would give you the largest salary? How much would you get paid?

I One dollar on the first day of the month, two dollars on the second day, three dollars on the third day, and so on (getting paid one dollar more each day) until the end of the 30 days. (You would have a total of \$55 after 10 days.)

II One cent (\$0.01) on the first day of the month, two cents (\$0.02) on the second day, four cents on the third day, eight cents on the fourth day, and so on (each day getting paid double from the previous day) until the end of the 30 days. (You would have a total of \$10.23 after 10 days.)

25 Each exponential function graphed below can be written in the form $f(x) = k(a)^x$. Find the value of a and k for each.

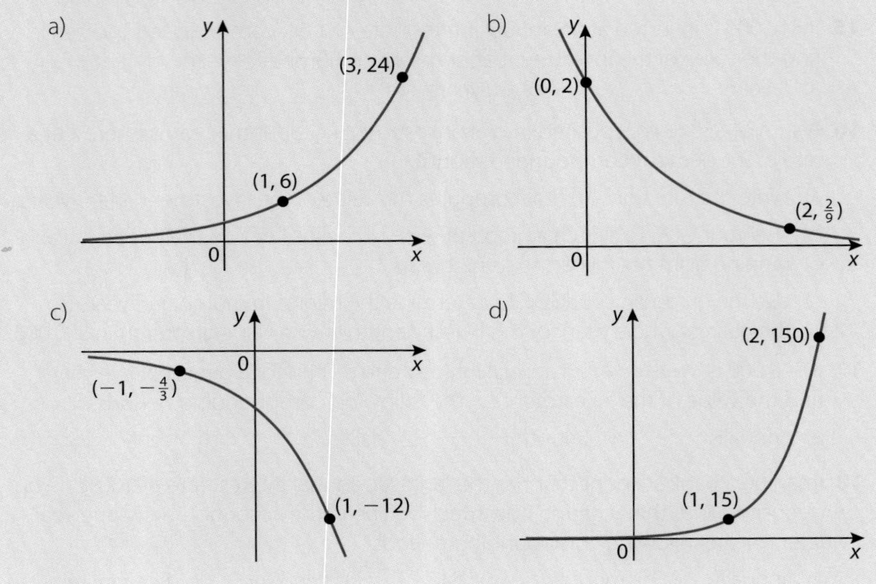

5.3 The number e

Recalling the definition of an exponential function, $f(x) = b^x$, we recognize that b is any positive constant and x is any real number. Graphs of $y = b^x$ for a few values where $b \geqslant 1$ are shown in Figure 5.2. As noted in the first section of this chapter, all the graphs pass through the point (0, 1).

Figure 5.2 Graphs of $y = b^x$ for some values when $b \geqslant 1$.

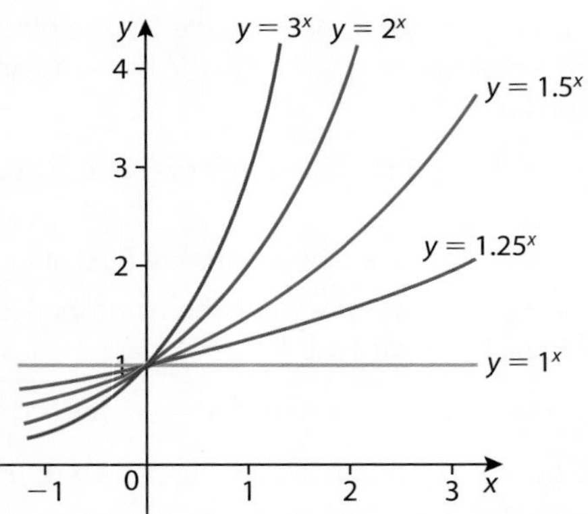

The question arises: what is the *best* number to choose for the base b? There is a good argument for $b = 10$ since we most commonly use a base 10 number system. Your GDC will have the expression 10^x as a built-in

command. The base $b = 2$ is also plausible because a binary number system (base 2) is used in many processes, especially in computer systems. However, the most important base is an irrational number that is denoted with the letter e. As we will see, the value of e approximated to six significant figures is 2.718 28. The importance of e will be clearer when we get to calculus topics. The number π – another very useful irrational number – has a natural geometric significance as the ratio of circumference to diameter for any circle. The number e also occurs in a 'natural' manner. We will illustrate this two different ways: first, by considering the **rate of change** of an exponential function, and secondly, by revisiting compound interest and considering **continuous change** rather than **incremental change**.

The 'discovery' of the constant e is attributed to Jakob Bernoulli (1654–1705). He was a member of the famous Bernoulli family of distinguished mathematicians, scientists and philosophers. This included his brother Johann (1667–1748), who made important developments in calculus, and his nephew Daniel (1700–1782), who is most well known for Bernoulli's principle in physics. The constant e is of enormous mathematical significance – and it appears 'naturally' in many mathematical processes. Jakob Bernoulli first observed e when studying sequences of numbers in connection to compound interest problems.

Rate of change (slope) of an exponential function

Since exponential functions (and associated logarithmic functions) are very important in calculus, the criteria we will use to determine the best value for b will be based on considering the slope of the curve $y = b^x$. In calculus we are interested in the rate of change (i.e. slope of the graph) of functions. Our goal to is to find a value for b such that the slope of the graph of $y = b^x$ at any value of x is equal to the function value y. We could investigate this by trial and error – and with a GDC this might prove fruitful – but it would not guarantee us an exact value and it could prove inefficient. Let's narrow our investigation to studying the slope of the curves at the point (0, 1) which is convenient because it is shared by all the curves.

To obtain a good estimate for the value of e we will use the diagram in Figure 5.3 where the scale on the x- and y-axes are equal and $P(0, 1)$ is the y-intercept of the graph of $y = e^x$. Q is a point on $y = e^x$ close to point P with coordinates (h, e^h). PR and RQ are parallel to the x- and y-axes, respectively, and they intersect at point $R(h, 1)$. The slope of the curve is always changing. It is not constant as with a straight line. As we will justify more thoroughly in our study of differential calculus in Chapter 13, the slope of a curve at a point will be equal to the slope of the line tangent to the curve at that point. PS is the tangent line to the curve at P, intersecting RQ at S. Thus, we are looking for the value e such that the slope of the tangent line PS is equal to 1. It follows that $\frac{RS}{PR} = 1$ and because $PR = h$ then $RS = h$. Since we have set Q close to P then we can assume that h is very small. Therefore, $RS \approx RQ$ and $\frac{RQ}{RS} \approx 1$. The value of $\frac{RQ}{RS}$ will get closer and closer to the value of 1 as h gets smaller (i.e. as Q gets closer to P). Since the y-coordinate of R is 1, then $RQ = e^h - 1$. Substituting h for RS and $e^h - 1$ for RQ into $\frac{RQ}{RS} \approx 1$, gives $\frac{e^h - 1}{h} \approx 1$. We wish to obtain an estimate for e so we multiply through by h to get $e^h - 1 \approx h$ leading to $e^h \approx h + 1$. To isolate e we raise both sides to the $\frac{1}{h}$ power, finally producing, $e \approx (1 + h)^{\frac{1}{h}}$.

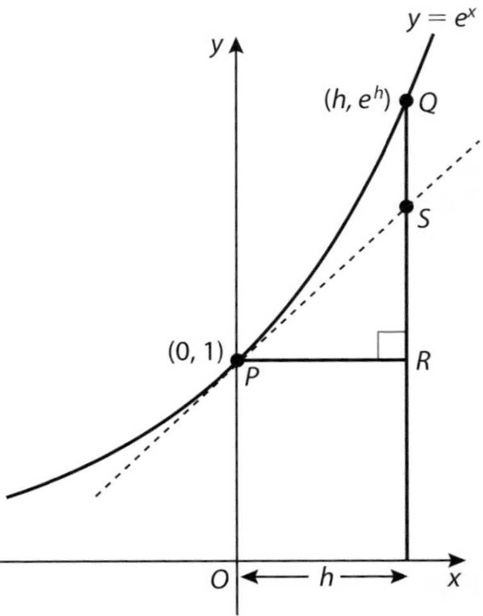

Figure 5.3 Graph of $y = e^x$; slope of the tangent line PS is equal to 1.

h	$e \approx (1+h)^{\frac{1}{h}}$
0.1	2.593742...
0.01	2.704814...
0.001	2.716924...
0.0001	2.718146...
0.00001	2.718268...
0.000001	2.718280...
0.0000001	2.718282...

Table 5.4 Values for $e \approx (1+h)^{\frac{1}{h}}$ as h approaches zero (accuracy to 7 significant figures).

Given that h is made small enough, the expression above should give a good estimation of the value of e. Using the approximation $e \approx (1+h)^{\frac{1}{h}}$, Table 5.4 shows values for e as h approaches zero.

To an accuracy of six significant figures, it appears that the value of e is approximately 2.718 28.

Definition of e (I) $e = \lim_{h \to 0}(1+h)^{\frac{1}{h}}$

The definition is read 'e equals the limit of $(1+h)^{\frac{1}{h}}$ as n goes to zero.'

Geometrically speaking, as point Q gets closer to point P ($h \to 0$), and also closer to point S, we wanted the slope of the tangent line at $(0, 1)$, $\frac{RS}{PR}$, to be equal to 1. This is the same as saying that we wanted $\frac{e^h - 1}{h} \to 1$ as $h \to 0$ (see coloured triangle in Figure 5.4). The value of e approximated to increasing accuracy in Table 5.4 is the number that makes this happen. A non-geometrical way of describing this feature of the graph is to say that the **rate of change** (slope) of the function $y = e^x$ at $x = 0$ is equal to 1.

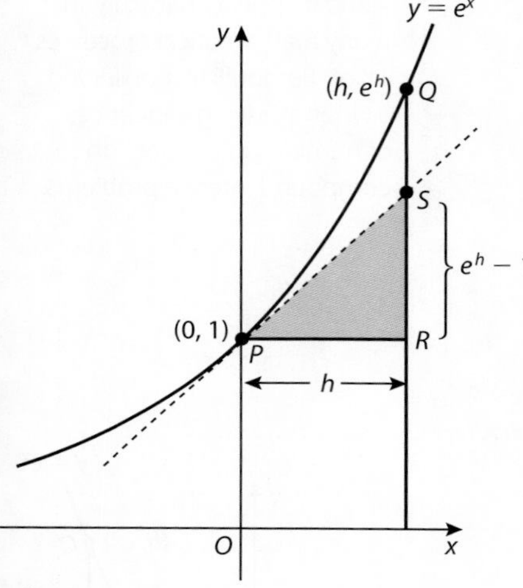

Figure 5.4 At $x = 0$, the rate of change of $y = e^x$ is equal to 1.

The rate of change of $y = e^x$ at a *general* value of x can be similarly obtained by fixing point P on the curve with coordinates (x, e^x) and a nearby point Q with coordinates $(x + h, e^{x+h})$. See Figure 5.5 below. Then the rate of change of the function at point P is $\frac{e^{x+h} - e^x}{h}$ as $h \to 0$. We cannot evaluate the limit of $\frac{e^{x+h} - e^x}{h}$ as $h \to 0$ directly by substituting 0 for h. By applying some algebra and knowing that $\frac{e^h - 1}{h} \to 1$ as $h \to 0$, we can evaluate the required limit.

As $h \to 0$, $\frac{e^{x+h} - e^x}{h} = \frac{e^x e^h - e^x}{h} = \frac{e^x(e^h - 1)}{h} = e^x\left[\frac{e^h - 1}{h}\right] \to e^x \cdot 1 = e^x$

Therefore, for any value of x, the rate of change of the function $y = e^x$ is e^x. In other words, the rate of change of the function at any value in the domain (x) is equal to the corresponding value of the range (y). This is the amazing feature of $y = e^x$ that makes e the most useful and 'natural' choice for the base of an exponential function, and the irrational number $e \approx 2.718\,28\ldots$ is the only base for which this is true.

Figure 5.5 The rate of change of $y = e^x$ at a general value of x.

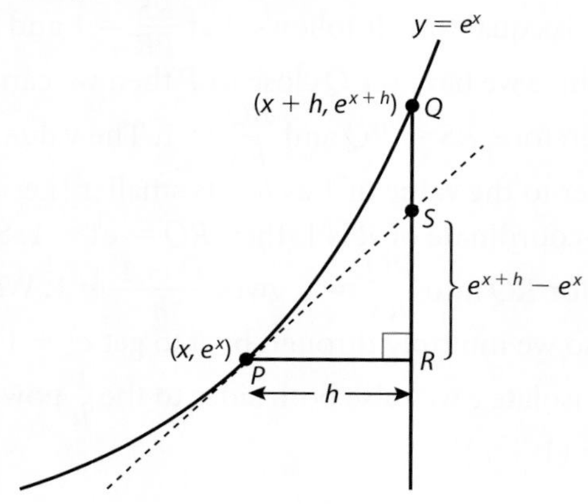

Continuously compounded interest

In the previous section and in Chapter 4, we computed amounts of money resulting from an initial amount (principal) with interest being compounded (added in) at discrete intervals (e.g. yearly, monthly, daily). In the formula that we used, $A(t) = P\left(1 + \frac{r}{n}\right)^{nt}$, n is the number of times that interest is compounded per year. Instead of adding interest only at discrete intervals, let's investigate what happens if we try to add interest continuously – that is, let the value of n increase without bound ($n \to \infty$).

Consider investing just \$1 at a very generous annual interest rate of 100%. How much will be in the account at the end of just one year? It depends on how often the interest is compounded. If it is only added at the end of the year ($n = 1$), the account will have \$2 at the end of the year. Is it possible to compound the interest more often to get a one-year balance of \$2.50 or of \$3.00? We use the compound interest formula with $P = \$1$, $r = 1.00$ (100%) and $t = 1$, and compute the amounts for increasing values of n. $A(1) = 1\left(1 + \frac{1}{n}\right)^{n \cdot 1} = \left(1 + \frac{1}{n}\right)^n$. This can be done very efficiently on your GDC by entering the equation $y = \left(1 + \frac{1}{x}\right)^x$ to display a table showing function values of increasing values of x.

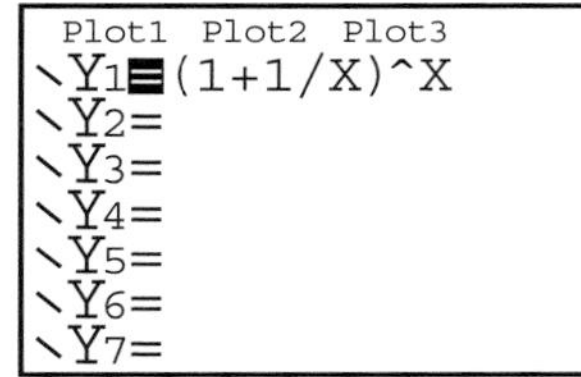


```
TABLE SETUP
 TblStart=1
 ΔTbl=1
Indpnt: Auto Ask
Depend: Auto Ask
```

```
X     | Y1
1     | 2
2     | 2.25
4     | 2.4414
Y1=2.44140625
```

```
X     | Y1
1     | 2
2     | 2.25
4     | 2.4414
12    | 2.613
Y1=2.61303529022
```

```
X     | Y1
1     | 2
2     | 2.25
4     | 2.4414
12    | 2.613
365   | 2.7146
Y1=2.71456748202
```

```
X     | Y1
1     | 2
2     | 2.25
4     | 2.4414
12    | 2.613
365   | 2.7146
8760  | 2.7181
Y1=2.71812669063
```

```
X      | Y1
1      | 2
2      | 2.25
4      | 2.4414
12     | 2.613
365    | 2.7146
8760   | 2.7181
525600 | 2.7183
Y1=2.7182792154
```

```
X      | Y1
2      | 2.25
4      | 2.4414
12     | 2.613
365    | 2.7146
8760   | 2.7181
525600 | 2.7183
3.15E7 | 2.7183
Y1=2.71828247254
```

As the number of compounding periods during the year increases, the amount at the end of the year appears to approach a limiting value.

As $n \to \infty$, the quantity of $\left(1 + \frac{1}{n}\right)^n$ approaches the number e. To 13 decimal places, e is approximately 2.718 281 828 4590.

Table 5.5

Compounding	n	$A(1) = \left(1 + \frac{1}{n}\right)^n$
Annual	1	2
Semi-annual	2	2.25
Quarterly	4	2.441 406 25...
Monthly	12	2.613 035 290 22...
Daily	365	2.714 567 482 02...
Hourly	8 760	2.718 126 690 63...
Every minute	525 600	2.718 279 2154...
Every second	31 536 000	2.718 282 472 54...

Leonhard Euler (1701–1783) was the dominant mathematical figure of the 18th century and is one of the most influential and prolific mathematicians of all time. Euler's collected works fill over 70 large volumes. Nearly every branch of mathematics has significant theorems that are attributed to Euler.

Euler proved mathematically that the limit of $\left(1+\frac{1}{n}\right)^n$ as n goes to infinity is precisely equal to an irrational constant which he labelled e. His mathematical writings were influential not just because of the content and quantity but also because of Euler's insistence on clarity and efficient mathematical notation. Euler introduced many of the common algebraic notations that we use today. Along with the symbol e for the base of natural logarithms (1727), Euler introduced $f(x)$ for a function (1734), i for the square root of negative one (1777), π for pi, Σ for summation (1755), and many others. His introductory algebra text, written originally in German (Euler was Swiss), is still available in English translation. Euler spent most of his working life in Russia and Germany. Switzerland honoured Euler by placing his image on the 10 Swiss franc banknote.

Definition of *e* (II)

$$e = \lim_{n\to\infty}\left(1+\frac{1}{n}\right)^n$$

The definition is read as 'e equals the limit of $\left(1+\frac{1}{n}\right)^n$ as n goes to infinity'.

Note that the two definitions that we have provided for the number e are equivalent. Take our first limit definition for e: $e = \lim_{h\to 0}(1+h)^{\frac{1}{h}}$.
Let $\frac{1}{h} = n$, it follows that $h = \frac{1}{n}$ and as $h \to 0$ then $n \to \infty$. Substituting $\frac{1}{n}$ for h, n for $\frac{1}{h}$, and evaluating the limit as $n \to \infty$ transforms $\lim_{h\to 0}(1+h)^{\frac{1}{h}}$ to $\lim_{n\to\infty}\left(1+\frac{1}{n}\right)^n$, which is our second limit definition for e.

As the number of compoundings, n, increase without bound, we approach continuous compounding – where interest is being added continuously. In the formula for calculating amounts resulting from compound interest, letting $m = \frac{n}{r}$ produces

$$A(t) = P\left(1+\frac{r}{n}\right)^{nt} = P\left(1+\frac{1}{m}\right)^{mrt} = P\left[\left(1+\frac{1}{m}\right)^m\right]^{rt}$$

Now if $n \to \infty$ and the interest rate r is constant, then $\frac{n}{r} = m \to \infty$. From the limit definition of e, we know that if $m \to \infty$, then $\left(1+\frac{1}{m}\right)^m \to e$.
Therefore, for continuous compounding, it follows that
$A(t) = P\left[\left(1+\frac{1}{m}\right)^n\right]^{rt} = P[e]^{rt}$.

This result is part of the reason that e is the best choice for the base of an exponential function modelling change that occurs continuously (e.g. radioactive decay) rather than in discrete intervals.

Continuous compound interest formula

An exponential function for calculating the amount of money after t years, $A(t)$, for interest compounded continuously, where P is the initial amount or principal and r is the annual interest rate, is given by $A(t) = Pe^{rt}$.

Example 7

An initial investment of 1000 euros earns interest at an annual rate of $5\frac{1}{4}\%$. Find the total amount after 10 years if the interest is compounded:
a) annually (simple interest), b) quarterly, and c) continuously.

Solution

a) $A(t) = P(1 + r)^t = 1000(1 + 0.0525)^{10} = 1669.10$ euros

b) $A(t) = P\left(1 + \frac{r}{n}\right)^{nt} = 1000\left(1 + \frac{0.0525}{4}\right)^{4(10)} = 1684.70$ euros

c) $A(t) = Pe^{rt} = 1000e^{0.0525(10)} = 1690.46$ euros

The natural exponential function and continuous change

For many applications involving continuous change, the most suitable choice for a mathematical model is an exponential function with a base having the value of e.

The natural exponential function

The natural exponential function is the function defined as

$$f(x) = e^x$$

As with other exponential functions, the domain of the natural exponential function is the set of all real numbers ($x \in \mathbb{R}$), and its range is the set of positive real numbers ($y > 0$). The natural exponential function is often referred to as <u>the</u> exponential function.

The formula developed for continuously compounded interest does not apply only to applications involving adding interest to financial accounts. It can be used to model growth or decay of a quantity that is changing *geometrically* (i.e. repeated multiplication by a constant ratio, or growth/decay factor) and the change is continuous, or approaching continuous. Another version of a formula for continuous change, which we will learn more about in calculus, is stated below:

Continuous exponential growth/decay

If an initial quantity C (when $t = 0$) increases or decreases continuously at a rate r over a certain time period, the amount $A(t)$ after t time periods is given by the function $A(t) = Ce^{rt}$. If $r > 0$, the quantity is increasing (growing). If $r < 0$, the quantity is decreasing (decaying).

Example 8

The cost of the new Boeing 787 Dreamliner airplane will be 150 million US dollars when purchased new. The airplane will lose value at a continuous rate. This is modeled by the continuous decay function $C(t) = 150e^{-0.053t}$ where $A(t)$ is the value of the airplane (in millions) after t years.

a) How much (to the nearest million dollars) would a Dreamliner jet be worth precisely five years after being purchased?

b) If a Dreamliner jet is purchased in 2010, what would be the first year that the jet is worth less than half of its original cost?

c) Find the value of b (to 4 s.f.) for a discrete decay model, $D(t) = 150b^t$, so that $D(t)$ is a suitable model to describe the same decay as $C(t)$.

Solution

a) $C(5) = 150e^{-0.053(5)} \approx 115$. The value is approximately \$115 million after five years.

b) Using a GDC, we graph the decay equation $y = 150e^{-0.053x}$ and the horizontal line $y = 75$ and determine the intersection point.

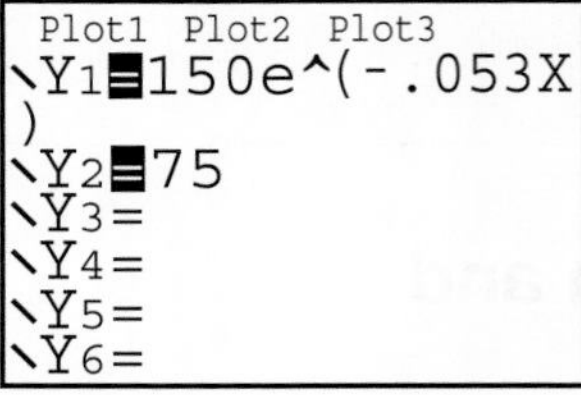

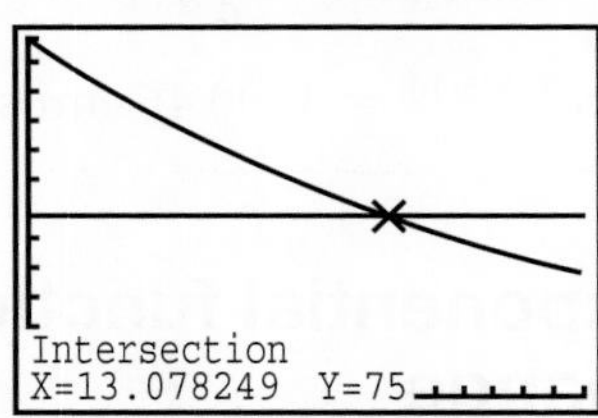

The x-coordinate of the intersection point is approximately 13.08. At the start of 2013, the jet's value is not yet half of its original value. Therefore, the first year that the jet is worth less than half of its original cost is 2014.

c) One way to find the value of b so that $D(t) = 150b^t$ serves as a reasonable substitute for $C(t) = 150e^{-0.053t}$ is to compute some function values for $C(t)$ and use them to compute the relative change from one year to the next.

$$C(1) = 150e^{-0.053(1)} \approx 142.2570$$
$$C(2) = 150e^{-0.053(2)} \approx 134.9137$$
$$C(3) = 150e^{-0.053(3)} \approx 127.9495$$

Relative change from year 1 to year 2: $\dfrac{134.9137 - 142.2570}{142.2570} \approx -0.051\,62$

Compute relative change from year 2 to year 3 to make sure it agrees with result above.

Relative change from year 2 to year 3: $\dfrac{127.9495 - 134.9137}{134.9137} \approx -0.051\,62$

The annual rate of decay, b, is the fraction of what remains after each year. Thus, $b = 1 - 0.051\,62 = 0.948\,38$; and to 4 s.f. the annual rate of decay is $b \approx 0.9484$. Therefore, the discrete decay model is $D(t) = 150(0.9484)^t$.

```
Plot1 Plot2 Plot3
\Y1=150e^(-.053X)
\Y2=150(.9484)^X
\Y3=
\Y4=
\Y5=
```

X	Y1	Y2
0	150	150
1	142.26	142.26
2	134.91	134.92
3	127.95	127.96
4	121.34	121.34
5	115.08	115.09
6	109.14	109.15

X=0

To check that the two decay models give similar results for each year, we can use a GDC to display a table of values for both models side by side for easy comparison.

Exercise 5.3

For questions 1–6, sketch a graph of the function and state its: a) domain and range; b) coordinates of any x-intercept(s) and y-intercept; c) and the equation of any asymptote(s).

1 $f(x) = e^{2x-1}$ **2** $g(x) = e^{-x} + 1$

3 $h(x) = -2e^x$

4 $p(x) = e^{x^2} - e$

5 $h(x) = \frac{1}{1 - e^x}$

6 $h(x) = e^{|x+2|} - 1$

7 a) State a definition of the number *e* as a limit.

b) Evaluate $\left(1 - \frac{1}{n}\right)^n$ for $n = 100$, $n = 10\,000$ and $n = 1\,000\,000$.

c) To 5 significant figures, what appears to be the value of $\lim_{n\to\infty}\left(1 - \frac{1}{n}\right)^n$? How does this number relate to the number *e*?

8 Use your GDC to graph the curve $y = \left(1 + \frac{1}{x}\right)^x$ and the horizontal line $y = 2.72$. Use a graph window so that x ranges from 0 to 20 and y ranges from 0 to 3. Describe the behaviour of the graph of $y = \left(1 + \frac{1}{x}\right)^x$. Will it ever intersect the graph of $y = 2.72$? Explain.

9 Two different banks, Bank A and Bank B, offer accounts with exactly the same annual interest rate of 6.85%. However, the account from Bank A has the interest compounded monthly whereas the account from Bank B compounds the interest continuously. To decide which bank to open an account with, you calculate the **amount of interest** you would earn after three years from an initial deposit of 500 euros in each bank's account. It is assumed that you make no further deposits and no withdrawals during the three years. How much interest would you earn from each of the accounts? Which bank's account earns more – and how much more?

10 Dina wishes to deposit \$1000 into an investment account and then withdraw the total in the account in five years. She has the choice of two different accounts. *Blue Star account*: interest is earned at an annual interest rate of 6.13% compounded weekly (52 weeks in a year). *Red Star account*: interest is earned at an annual interest rate of 5.95% compounded continuously. Which investment account – *Blue Star* **or** *Red Star* – will result in the greatest total at the end of five years? What is the total after five years for this account? How much more is it than the total for the other account?

11 Strontium-90 is a radioactive isotope of strontium. Strontium-90 decays according to the function $A(t) = Ce^{-0.0239t}$, where t is time in years and C is the initial amount of strontium-90 when $t = 0$. If you have 1 kilogram of strontium-90 to start with, how much (approximated to 3 significant figures) will you have after:

a) 1 year?

b) 10 years?

c) 100 years?

d) 250 years?

12 A radioactive substance decays in such a way that the mass (in kilograms) remaining after t days is given by the function $A(t) = 5e^{-0.0347t}$.

a) Find the mass (i.e. initial mass) at time $t = 0$.

b) What **percentage** of the initial mass remains after 10 days?

c) On your GDC and then on paper, draw a graph of the function $A(t)$ for $0 \leqslant t \leqslant 50$.

d) Use one of your graphs to approximate, to the nearest whole day, the half-life of the radioactive substance.

13 Which of the given interest rates and compounding periods would provide the better investment?

a) $8\frac{1}{2}\%$ per year, compounded semi-annually

b) $8\frac{1}{4}\%$ per year, compounded quarterly

c) 8% per year, compounded continuously

14 In certain conditions the bacterium that causes cholera, *Vibrios cholerae*, can grow rapidly in number. In a laboratory experiment a culture of *Vibrios cholerae* is started with 20 bacterium. The bacterium's growth is modeled with the following continuous growth model $A(t) = 20e^{0.068t}$ where $A(t)$ is the number of bacteria after t minutes.

a) Determine the value of r for the discrete growth model $B(t) = 20(r)^t$, so that $B(t)$ is equivalent to $A(t)$.

b) For both of these models, by what percentage does the number of bacteria grow each minute?

15 By comparing the graph of each of the following equations to the graph of $y = e^x$, determine if the slope of the tangent line at the point (0, 1) for the graph of each equation is less than or greater than 1.

a) $y = 2^x$

b) $y = \left(\frac{5}{2}\right)^x$

c) $y = \left(\frac{11}{4}\right)^x$

d) $y = 3^x$

16 Consider that £1000 is invested at 4.5% interest compounded continuously.

a) How much money is in the account after 10 years? After 20 years?

b) Use your GDC to determine how many years (to nearest tenth of a year) it takes for the initial investment to double to £2000.

c) If £5000 is invested at the same rate of interest also compounded continuously, how many years (to nearest tenth) would it take to double?

d) Are the answers to b) and c) the same or different? Why?

5.4 Logarithmic functions

In Example 7 of the previous section, we used the equation $A(t) = 1000e^{0.0525t}$ to compute the amount of money in an account after t years. Now suppose we wish to determine how much time, t, it takes for the initial investment of 1000 euros to double. To find this we need to solve the following equation for t: $2000 = 1000e^{0.0525t} \Rightarrow 2 = e^{0.0525t}$. The unknown t is in the exponent. At this point in the book, we do not have an algebraic method to solve such an equation, but developing the concept of a **logarithm** will provide us with the means to do so.

John Napier (1550–1617) was a Scottish landowner, scholar and mathematician who 'invented' logarithms – a word he coined which derives from two Greek words: *logos* – meaning ratio, and *arithmos* – meaning number. Logarithms made numerical calculations much easier in areas such as astronomy, navigation, engineering and warfare. English mathematician Henry Briggs (1561–1630) came to Scotland to work with Napier and together they perfected logarithms, which included the idea of using the base ten. After Napier died in 1617, Briggs took over the work on logarithms and published a book of tables in 1624. By the second half of the 17th century, the use of logarithms had spread around the world. They became as popular as electronic calculators in our time. The great French mathematician Pierre-Simon Laplace (1749–1827) even suggested that the logarithms of Napier and Briggs doubled the life of astronomers, because it so greatly reduced the labours of calculation. In fact, without the invention of logarithms it is difficult to imagine how Kepler and Newton could have made their great scientific advances. In 1621, an English mathematician and clergyman, William Oughtred (1574–1660) used logarithms as the basis for the invention of the slide rule. The slide rule was a very effective calculation tool that remained in common use for over three hundred years.

The inverse of an exponential function

For $b > 1$, an exponential function with base b is increasing for all x, and for $0 < b < 1$ an exponential function is decreasing for all x. It follows from this that all exponential functions must be one-to-one. Recall from Section 2.3 that a one-to-one function passes both a vertical line test and a horizontal line test. We demonstrated that an inverse function would exist for any one-to-one function. Therefore, an exponential function with base b such that $b > 0$ and $b \neq 1$ will have an inverse function, which is given in the following definition. Also recall from Section 2.3 that the domain of a function f is the range of its inverse function f^{-1}, and, similarly, the range of f is the domain of f^{-1}. The domain and range are switched around for a function and its inverse.

Definition of a logarithmic function

For $b > 0$ and $b \neq 1$, the **logarithmic function** $y = \log_b x$ (read as 'logarithm with base b of x') is the inverse of the exponential function with base b.

$$y = \log_b x \text{ if and only if } x = b^y$$

The domain of the logarithmic function $y = \log_b x$ is the set of positive real numbers ($x > 0$) and its range is all real numbers ($y \in \mathbb{R}$).

Logarithmic expressions and equations

When evaluating logarithms, note that *a logarithm is an exponent.* This means that the value of $\log_b x$ is the exponent to which b must be raised to obtain x. For example, $\log_2 8 = 3$ because 2 must be raised to the power of 3 to obtain 8 – that is, $\log_2 8 = 3$ if and only if $2^3 = 8$.

We can use the definition of a logarithmic function to translate a logarithmic equation into an exponential equation and vice versa. When doing this, it is helpful to remember, as the definition stated, that in either form – logarithmic or exponential – the base is the same.

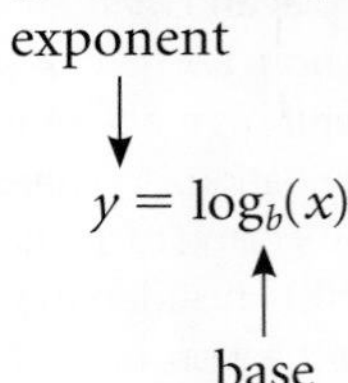

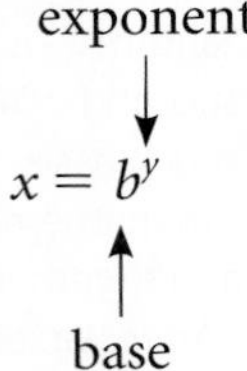

Example 9

Find the value of each of the following logarithms.

a) $\log_7 49$ b) $\log_5(\frac{1}{5})$ c) $\log_6 \sqrt{6}$ d) $\log_4 64$ e) $\log_{10} 0.001$

Solution

For each logarithmic expression in a) to e), we set it equal to y and use the definition of a logarithmic function to obtain an equivalent equation in exponential form. We then solve for y by applying the logical fact that if $b > 0$, $b \neq 1$ and $b^y = b^k$ then $y = k$.

a) Let $y = \log_7 49$ which is equivalent to the exponential equation $7^y = 49$. Since $49 = 7^2$, then $7^y = 7^2$. Therefore, $y = 2 \Rightarrow \log_7 49 = 2$.

b) Let $y = \log_5(\frac{1}{5})$ which is equivalent to the exponential equation $5^y = \frac{1}{5}$. Since $\frac{1}{5} = 5^{-1}$, then $5^y = 5^{-1}$. Therefore, $y = -1 \Rightarrow \log_5(\frac{1}{5}) = -1$.

c) Let $y = \log_6 \sqrt{6}$ which is equivalent to the exponential equation $6^y = \sqrt{6}$. Since $\sqrt{6} = 6^{\frac{1}{2}}$, then $6^y = 6^{\frac{1}{2}}$. Therefore, $y = \frac{1}{2} \Rightarrow \log_6 \sqrt{6} = \frac{1}{2}$.

d) Let $y = \log_4 64$ which is equivalent to the exponential equation $4^y = 64$. Since $64 = 4^3$, then $4^y = 4^3$. Therefore, $y = 3 \Rightarrow \log_4 64 = 3$.

e) Let $y = \log_{10} 0.001$ which is equivalent to the exponential equation $10^y = 0.001$. Since $0.001 = \frac{1}{1000} = \frac{1}{10^3} = 10^{-3}$, then $10^y = 10^{-3}$. Therefore, $y = -3 \Rightarrow \log_{10} 0.001 = -3$.

Example 10

Find the domain of the function $f(x) = \log_2(4 - x^2)$.

Solution

From the definition of a logarithmic function the domain of $y = \log_b x$ is $x > 0$, thus for $f(x)$ it follows that $4 - x^2 > 0 \Rightarrow (2 + x)(2 - x) > 0 \Rightarrow -2 < x < 2$.

Hence, the domain is $-2 < x < 2$.

Properties of logarithms

As with all functions and their inverses, their graphs are reflections of each other over the line $y = x$. Figure 5.6 illustrates this relationship for exponential and logarithmic functions, and also confirms the domain and range for the logarithmic function stated in the previous definition.

Notice that the points (0, 1) and (1, 0) are mirror images of each other over the line $y = x$. This corresponds to the fact that since $b^0 = 1$ then $\log_b 1 = 0$. Another pair of mirror image points, $(1, b)$ and $(b, 1)$, highlight the fact that $\log_b b = 1$.

Notice also that since the x-axis is a horizontal asymptote of $y = b^x$, the y-axis is a vertical asymptote of $y = \log_b x$.

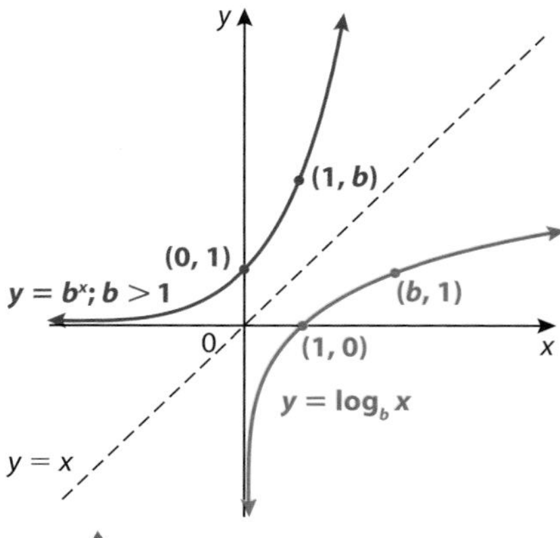

Figure 5.6 Reflection of $y = \log_b x$ over the line $y = x$.

In Section 2.3, we established that a function f and its inverse function f^{-1} satisfy the equations

$f^{-1}(f(x)) = x$ for x in the domain of f

$f(f^{-1}(x)) = x$ for x in the domain of f^{-1}

When applied to $f(x) = b^x$ and $f^{-1}(x) = \log_b x$, these equations become

$\log_b(b^x) = x$ $x \in \mathbb{R}$

$b^{\log_b x} = x$ $x > 0$

Properties of logarithms I

For $b > 0$ and $b \neq 1$, the following statements are true:

1. $\log_b 1 = 0$ (*because* $b^0 = 1$)
2. $\log_b b = 1$ (*because* $b^1 = b$)
3. $\log_b(b^x) = x$ (*because* $b^x = b^x$)
4. $b^{\log_b x} = x$ (*because* $\log_b x$ *is the power to which* b *must be raised to get* x)

The logarithmic function with base 10 is called the **common logarithmic function**. On calculators and on your GDC, this function is denoted by **log**. The value of the expression $\log_{10} 1000$ is 3 because 10^3 is 1000. Generally, for common logarithms (i.e. base 10) we omit writing the base of 10. Hence, if **log** is written with no base indicated, it is assumed to have a base of 10. For example, $\log 0.01 = -2$.

Common logarithm: $\log_{10} x = \log x$

As with exponential functions, the most widely used logarithmic function – and the other logarithmic function supplied on all calculators – is the logarithmic function with the base of e. This function is known as the **natural logarithmic function** and it is the inverse of the natural exponential function $y = e^x$. The natural logarithmic function is denoted by the symbol **ln**, and the expression $\ln x$ is read as 'the natural logarithm of x'.

Natural logarithm: $\log_e x = \ln x$

Example 11

Evaluate the following expressions:

a) $\log\left(\frac{1}{10}\right)$ b) $\log(\sqrt{10})$ c) $\log 1$ d) $10^{\log 47}$ e) $\log 50$

f) $\ln e$ g) $\ln\left(\frac{1}{e^3}\right)$ h) $\ln 1$ i) $e^{\ln 5}$ j) $\ln 50$

Solution

a) $\log\left(\frac{1}{10}\right) = \log(10^{-1}) = -1$

b) $\log(\sqrt{10}) = \log(10^{\frac{1}{2}}) = \frac{1}{2}$

c) $\log 1 = \log(10^0) = 0$

d) $10^{\log 47} = 47$

e) $\log 50 \approx 1.699$ (using GDC)

f) $\ln e = 1$

g) $\ln\left(\frac{1}{e^3}\right) = \ln(e^{-3}) = -3$

h) $\ln 1 = \ln(e^0) = 0$

i) $e^{\ln 5} = 5$

j) $\ln 50 \approx 3.912$ (using GDC)

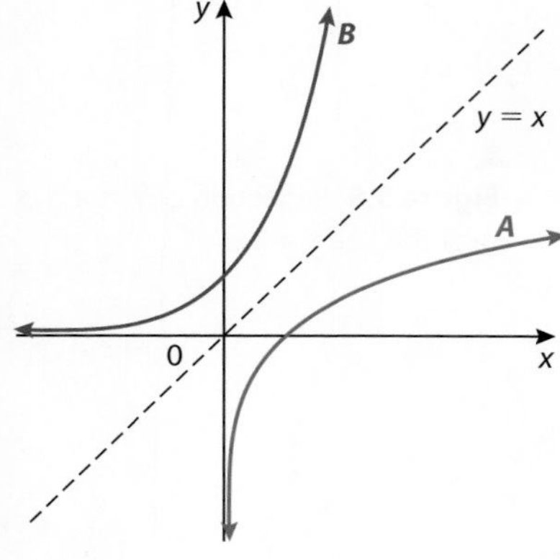

Example 12

The diagram shows the graph of the line $y = x$ and two curves. Curve A is the graph of the equation $y = \log x$. Curve B is the reflection of curve A in the line $y = x$.

a) Write the equation for curve B.

b) Write the coordinates of the y-intercept of curve B.

Solution

a) Curve A is the graph of $y = \log x$, the common logarithm with base 10, which could also be written as $y = \log_{10} x$. Curve B is the inverse of $y = \log_{10} x$, since it is the reflection of it in the line $y = x$. Hence, the equation for curve B is the exponential equation $y = 10^x$.

b) The y-intercept occurs when $x = 0$. For curve B, $y = 10^0 = 1$. Therefore, the y-intercept for curve B is $(0, 1)$.

The logarithmic function with base b is the inverse of the exponential function with base b. Therefore, it makes sense that the laws of exponents (Section 1.3) should have corresponding properties involving logarithms. For example, the exponential property $b^0 = 1$ corresponds to the logarithmic property $\log_b 1 = 0$. We will state and prove three further important logarithmic properties that correspond to the following three exponential properties.

1. $b^m \cdot b^n = b^{m+n}$
2. $\frac{b^m}{b^n} = b^{m-n}$
3. $(b^m)^n = b^{mn}$

Properties of logarithms II

Given $M > 0$, $N > 0$ and k is any real number, the following properties are true for logarithms with $b > 0$ and $b \neq 1$.

Property		Description
1. $\log_b(MN)$	$= \log_b M + \log_b N$	the log of a product is the sum of the logs of its factors
2. $\log_b\left(\frac{M}{N}\right)$	$= \log_b M - \log_b N$	the log of a quotient is the log of the numerator minus the log of the denominator
3. $\log_b(M^k)$	$= k \log_b M$	the log of a number raised to an exponent is the exponent times the log of the number

Any of these properties can be applied in either direction.

Proofs

Property 1: Let $x = \log_b M$ and $y = \log_b N$.

The corresponding exponential forms of these two equations are

$$b^x = M \text{ and } b^y = N$$

Then, $\log_b(MN) = \log_b(b^x b^y) = \log_b(b^{x+y}) = x + y$.

It's given that $x = \log_b M$ and $y = \log_b N$; hence, $x + y = \log_b M + \log_b N$.

Therefore, $\log_b(MN) = \log_b M + \log_b N$.

Property 2: Again, let $x = \log_b M$ and $y = \log_b N \Rightarrow b^x = M$ and $b^y = N$.

Then, $\log_b\left(\frac{M}{N}\right) = \log_b\left(\frac{b^x}{b^y}\right) = \log_b(b^{x-y}) = x - y$.

With $x = \log_b M$ and $y = \log_b N$, then $x - y = \log_b M - \log_b N$.

Therefore, $\log_b\left(\frac{M}{N}\right) = \log_b M - \log_b N$.

Property 3: Let $x = \log_b M \Rightarrow b^x = M$.

Now, let's take the logarithm of M^k and substitute b^x for M:

$\log_b(M^k) = \log_b[(b^x)^k] = \log_b(b^{kx}) = kx$

It's given that $x = \log_b M$; hence, $kx = k\log_b M$.

Therefore, $\log_b(M^k) = k\log_b M$.

• **Hint:** The notation $f(x)$ uses brackets *not* to indicate multiplication but to indicate the argument of the function f. The symbol f is the name of a function, not a variable – it is not multiplying the variable x. Therefore, $f(x + y)$ is NOT equal to $f(x) + f(y)$. Likewise, the symbol **log** is also the name of a function. Therefore, $\log_b(x + y)$ is not equal to $\log_b(x) + \log_b(y)$. Other mistakes to avoid include incorrectly simplifying quotients or powers of logarithms. Specifically, $\frac{\log_b x}{\log_b y} \neq \log\left(\frac{x}{y}\right)$ and $(\log_b x)^k \neq k(\log_b x)$.

Example 13

Use the properties of logarithms to write each logarithmic expression as a sum, difference, and/or constant multiple of simple logarithms (i.e. logarithms without sums, products, quotients or exponents).

a) $\log_2(8x)$ b) $\ln\left(\frac{3}{y}\right)$ c) $\log(\sqrt{7})$

d) $\log_b\left(\frac{x^3}{y^2}\right)$ e) $\ln(5e^2)$ f) $\log\left(\frac{m+n}{n}\right)$

Solution

a) $\log_2(8x) = \log_2 8 + \log_2 x = 3 + \log_2 x$

b) $\ln\left(\frac{3}{y}\right) = \ln 3 - \ln y$

c) $\log(\sqrt{7}) = \log(7^{\frac{1}{2}}) = \frac{1}{2}\log 7$

d) $\log_b\left(\frac{x^3}{y^2}\right) = \log_b(x^3) - \log_b(y^2) = 3\log_b x - 2\log_b y$

e) $\ln(5e^2) = \ln 5 + \ln(e^2) = \ln 5 + 2\ln e = \ln 5 + 2(1) = 2 + \ln 5$
  ($2 + \ln 5 \approx 3.609$ using GDC)

f) $\log\left(\frac{m+n}{m}\right) = \log(m + n) - \log m$
(Remember: $\log(m + n) \neq \log m + \log n$)

Example 14

Write each expression as the logarithm of a single quantity.

a) $\log 6 + \log x$　　b) $\log_2 5 + 2\log_2 3$

c) $\ln y - \ln 4$　　d) $\log_b 12 - \frac{1}{2}\log_b 9$

e) $\log_3 M + \log_3 N - 2\log_3 P$　　f) $\log_2 80 - \log_2 5$

Solution

a) $\log 6 + \log x = \log(6x)$

b) $\log_2 5 + 2\log_2 3 = \log_2 5 + \log_2(3^2) = \log_2 5 + \log_2 9 = \log_2(5 \cdot 9)$
$= \log_2 45$

c) $\ln y - \ln 4 = \ln\left(\frac{y}{4}\right)$

d) $\log_b 12 - \frac{1}{2}\log_b 9 = \log_b 12 - \log_b(9^{\frac{1}{2}}) = \log_b 12 - \log_b(\sqrt{9})$
$= \log_b 12 - \log_b 3 = \log_b\left(\frac{12}{3}\right) = \log_b 4$

e) $\log_3 M + \log_3 N - 2\log_3 P = \log_3(MN) - \log_3(P^2) = \log_3\left(\frac{MN}{P^2}\right)$

f) $\log_2 80 - \log_2 5 = \log_2\left(\frac{80}{5}\right) = \log_2 16 = 4$　　(because $2^4 = 16$)

Change of base

The answer to part f) of Example 14 was $\log_2 16$ which we can compute to be exactly 4 because we know that $2^4 = 16$. The answer to part e) of Example 13 was $2 + \ln 5$ which we approximated to 3.609 using the natural logarithm function key (**ln**) on our GDC. But, what if we wanted to compute an approximate value for $\log_2 45$, the answer to part b) of Example 14? Our GDC can only evaluate common logarithms (base 10) and natural logarithms (base *e*). To evaluate logarithmic expressions and graph logarithmic functions to other bases we need to apply a **change of base formula**.

Change of base formula

Let a, b and x be positive real numbers such that $a \neq 1$ and $b \neq 1$. Then $\log_b x$ can be expressed in terms of logarithms to any other base a as follows:

$$\log_b x = \frac{\log_a x}{\log_a b}$$

Proof

$y = \log_b x \Rightarrow b^y = x$　　Convert from logarithmic form to exponential form.

$\log_a x = \log_a(b^y)$　　If $b^y = x$, then log of each with same bases must be equal.

$\log_a x = y\log_a b$　　Applying the property $\log_b(M^k) = k\log_b M$.

$y = \frac{\log_a x}{\log_a b}$　　Divide both sides by $\log_a b$.

$\log_b x = \frac{\log_a x}{\log_a b}$　　Substitute $\log_b x$ for y.

To apply the change of base formula, let $a = 10$ or $a = e$. Then the logarithm of any base b can be expressed in terms of either common logarithms or natural logarithms. For example:

$$\log_2 x = \frac{\log x}{\log 2} \quad \text{or} \quad \frac{\ln x}{\ln 2}$$

$$\log_5 x = \frac{\log x}{\log 5} \quad \text{or} \quad \frac{\ln x}{\ln 5}$$

$$\log_2 45 = \frac{\log 45}{\log 2} = \frac{\ln 45}{\ln 2} \approx 5.492 \quad \text{(using GDC)}$$

Example 15

Use the change of base formula and common or natural logarithms to evaluate each logarithmic expression. Start by making a rough mental estimate. Approximate your answer to 4 significant figures.

a) $\log_3 30$

b) $\log_9 6$

Solution

a) The value of $\log_3 30$ is the power to which 3 is raised to obtain 30. Because $3^3 = 27$ and $3^4 = 81$, the value of $\log_3 30$ is between 3 and 4, and will be much closer to 3 than 4 – perhaps around 3.1. Using the change of base formula and common logarithms, we obtain $\log_3 30 = \frac{\log 30}{\log 3} \approx 3.096$. This agrees well with the mental estimate.

After computing the answer on your GDC, use your GDC to also check it by raising 3 to the answer and confirming that it gives a result of 30.

```
log(30)/log(3)
          3.095903274
3^Ans
                   30
■
```

b) The value of $\log_9 6$ is the power to which 9 is raised to obtain 6. Because $9^{\frac{1}{2}} = \sqrt{9} = 3$ and $9^1 = 9$, the value of $\log_9 6$ is between $\frac{1}{2}$ and 1 – perhaps around 0.75. Using the change of base formula and natural logarithms, we obtain $\log_9 6 = \frac{\ln 6}{\ln 9} \approx 0.815$. This agrees well with the mental estimate.

```
ln(6)/ln(9)
          .8154648768
9^Ans
                    6
■
```

Exercise 5.4

In questions 1–9, express each logarithmic equation as an exponential equation.

1 $\log_2 16 = 4$ **2** $\ln 1 = 0$ **3** $\log 100 = 2$

4 $\log 0.01 = -2$ **5** $\log_7 343 = 3$ **6** $\ln\left(\frac{1}{e}\right) = -1$

7 $\log 50 = y$ **8** $\ln x = 12$ **9** $\ln(x + 2) = 3$

In questions10–18, express each exponential equation as a logarithmic equation.

10 $2^{10} = 1024$ **11** $10^{-4} = 0.0001$ **12** $4^{-\frac{1}{2}} = \frac{1}{2}$

13 $3^4 = 81$ **14** $10^0 = 1$ **15** $e^x = 5$

16 $2^{-3} = 0.125$ **17** $e^4 = y$ **18** $10^{x+1} = y$

In questions 19–38, find the exact value of the expression without using your GDC.

19 $\log_2 64$ **20** $\log_4 64$ **21** $\log_2\left(\frac{1}{8}\right)$ **22** $\log_3(3^5)$

23 $\log_{16} 8$ **24** $\log_{27} 3$ **25** $\log_{10} 0.001$ **26** $\ln e^{13}$

27 $\log_8 1$ **28** $10^{\log 6}$ **29** $\log_3\left(\frac{1}{27}\right)$ **30** $e^{\ln\sqrt{2}}$

31 $\log 1000$ **32** $\ln(\sqrt{e})$ **33** $\ln\left(\frac{1}{e^2}\right)$ **34** $\log_3(81^{22})$

35 $\log_4 2$ **36** $3^{\log_3 18}$ **37** $\log_5(\sqrt[3]{5})$ **38** $10^{\log \pi}$

In questions 39–46, use a GDC to evaluate the expression, correct to 4 significant figures.

39 $\log 50$ **40** $\log\sqrt{3}$ **41** $\ln 50$ **42** $\ln\sqrt{3}$

43 $\log 25$ **44** $\log\left(\frac{1+\sqrt{5}}{2}\right)$ **45** $\ln 100$ **46** $\ln(100^3)$

In questions 47–52, find the domain of each function.

47 $f(x) = \log(x - 2)$ **48** $g(x) = \ln(x^2)$ **49** $h(x) = \log(x) - 2$

50 $y = \log_7(8 - 5x)$ **51** $y = \sqrt{x + 2} - \log_3(9 - 3x)$

52 $y = \sqrt{\ln(1 - x)}$

In 53–55, find the domain *and* range of each function.

53 $y = \frac{1}{\ln x}$ **54** $y = |\ln(x - 1)|$ **55** $y = \frac{x}{\log x}$

For questions 56–59, find the equation of the function that is graphed in the form $f(x) = \log_b x$.

56

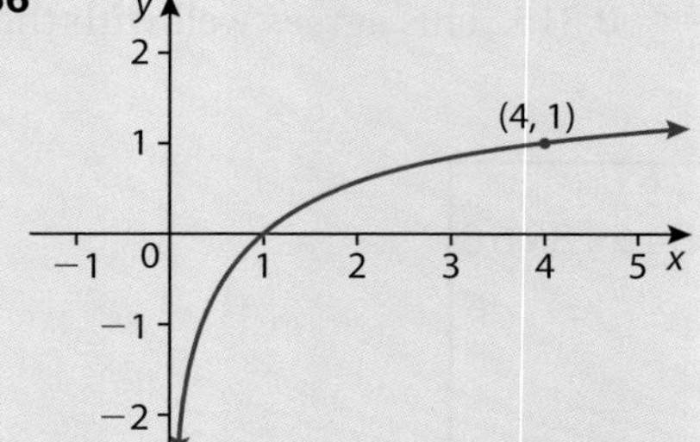

57

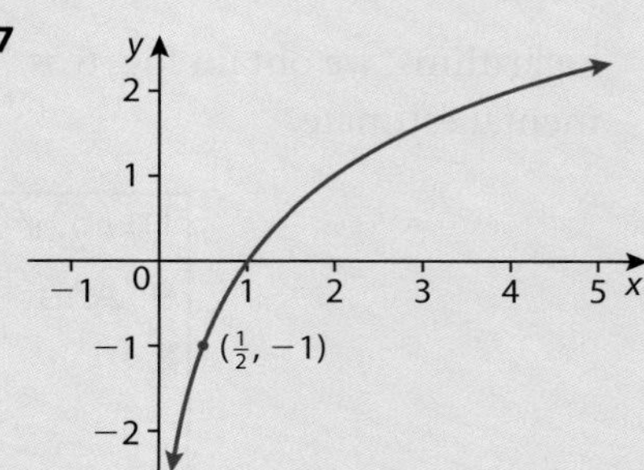

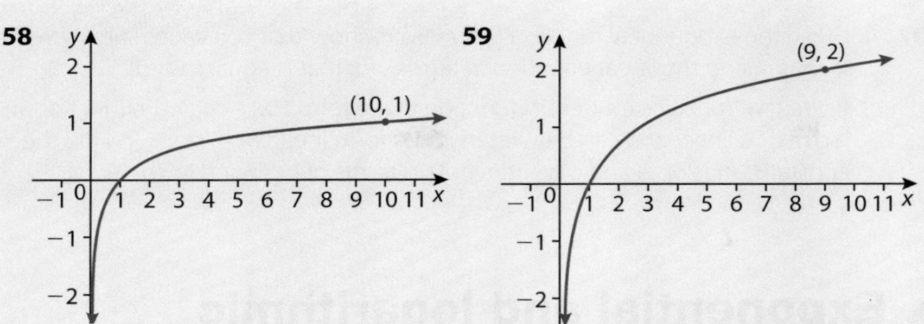

In questions 60–65, use properties of logarithms to write each logarithmic expression as a sum, difference and/or constant multiple of simple logarithms (i.e. logarithms without sums, products, quotients or exponents).

60 $\log_2(2m)$

61 $\log\left(\frac{9}{x}\right)$

62 $\ln\sqrt[5]{x}$

63 $\log_3(ab^3)$

64 $\log[10x(1+r)^t]$

65 $\ln\left(\frac{m^3}{n}\right)$

In 66–71, write each expression in terms of $\log_b p$, $\log_b q$ and $\log_b r$.

66 $\log_b pqr$

67 $\log_b\left(\frac{p^2q^3}{r}\right)$

68 $\log_b\sqrt[4]{pq}$

69 $\log_b\sqrt{\frac{qr}{p}}$

70 $\log_b\frac{p\sqrt{q}}{r}$

71 $\log_b\frac{(pq)^3}{\sqrt{r}}$

In 72–77, write each expression as the logarithm of a single quantity.

72 $\log(x^2)+\log\left(\frac{1}{x}\right)$

73 $\log_3 9 + 3\log_3 2$

74 $4\ln y - \ln 4$

75 $\log_b 12 - \frac{1}{2}\log_b 9$

76 $\log x - \log y - \log z$

77 $2\ln 6 - 1$ • **Hint:** $\ln(?) = 1$

In questions 78–81, use the change of base formula and common or natural logarithms to evaluate each logarithmic expression. Approximate your answer to 3 significant figures.

78 $\log_2 1000$ **79** $\log_{\frac{1}{2}} 40$ **80** $\log_6 40$ **81** $\log_5(0.75)$

In questions 82 and 83, use the change of base formula to evaluate $f(20)$.

82 $f(x) = \log_2 x$

83 $f(x) = \log_5 x$

84 Use the change of base formula to prove the following statement.

$$\log_b a = \frac{1}{\log_a b}$$

85 Show that $\log e = \frac{1}{\ln 10}$.

86 The relationship between the number of decibels dB (one variable) and the intensity I of a sound (in watts per square metre) is given by the formula $dB = 10\log\left(\frac{I}{10^{-16}}\right)$. Use properties of logarithms to write the formula in simpler form. Then find the number of decibels of a sound with an intensity of 10^{-4} watts per square metre.

87 a) Given the exponential function $f(x) = 5(2)^x$, show that $f(x)$ varies linearly with x; that is, find the linear equation in terms of x that is equal to $f(x)$.

b) Prove that for any exponential function in the form $f(x) = ab^x$, the function $\log(f(x))$ is linear and can be written in the form $\log(f(x)) = mx + c$. Find the constants m (slope) and c (y-intercept) in terms of $\log a$ and $\log b$.

5.5 Exponential and logarithmic equations

Solving exponential equations

At the start of the previous section, we wanted to find a way to determine how much time t (in years) it would take for an investment of 1000 euros to double, if the investment earns interest at an annual rate of $5\frac{1}{4}\%$. Since the interest is compounded continuously, we need to solve this equation: $2000 = 1000e^{0.0525t} \Rightarrow 2 = e^{0.0525t}$. The equation has the variable t in the exponent. With the properties of logarithms established in the previous section, we now have a way to algebraically solve such equations. Along with these properties, we need to apply the logic that if two expressions are equal then their logarithms must also be equal. That is, if $m = n$, then $\log_b m = \log_b n$.

Example 16

Solve the equation for the variable t. Give your answer accurate to 3 significant figures.

$$2 = e^{0.0525t}$$

Solution

$$2 = e^{0.0525t}$$

$$\ln 2 = \ln(e^{0.0525t})$$ Take natural logarithm of both sides.

$$\ln 2 = 0.0525t$$ Apply the property $\log_b(b^x) = x$ and $\ln e = 1$.

$$t = \frac{\ln 2}{0.0525} \approx 13.2$$

With interest compounding continuously at an annual interest rate of $5\frac{1}{4}\%$, it takes about 13.2 years for the investment to double.

This example serves to illustrate a general strategy for solving exponential equations. To solve an exponential equation, first isolate the exponential expression and take the logarithm of both sides. Then apply a property of logarithms so that the variable is no longer in the exponent and it can be isolated on one side of the equation. By taking the logarithm of both sides of an exponential equation, we are making use of the inverse relationship between exponential and logarithmic functions. Symbolically, this method can be represented as follows – solving for x:

(i) If $b = 10$ or e: $y = b^x \Rightarrow \log_b y = \log_b b^x \Rightarrow \log_b y = x$

(ii) If $b \neq 10$ or e:

$$y = b^x \Rightarrow \log_a y = \log_a b^x \Rightarrow \log_a y = x \log_a b \Rightarrow x = \frac{\log_a y}{\log_a b}$$

Example 17

Solve for x in the equation $3^{x-4} = 24$. Approximate the answer to 3 significant figures.

Solution

$3^{x-4} = 24$

$\log(3^{x-4}) = \log 24$ Take common logarithm of both sides.

$(x-4)\log 3 = \log 24$ Apply the property $\log_b(M^k) = k \log_b M$.

$x - 4 = \dfrac{\log 24}{\log 3}$ Divide both sides by log 3. $\left[\text{Note: } \dfrac{\log 24}{\log 3} \neq \log 8\right]$

$x = \dfrac{\log 24}{\log 3} + 4$

$x \approx 6.89$ Using GDC .

• **Hint:** We could have used natural logarithms instead of common logarithms to solve the equation in Example 17. Using the same method but with natural logarithms, we get

$$x = \frac{\ln 24}{\ln 3} + 4 \approx 6.89.$$

Recall Example 11 in Section 4.3 in which we solved an exponential equation graphically, because we did not yet have the tools to solve it algebraically. Let's solve it now using logarithms.

Example 18

You invested €1000 at 6% compounded quarterly. How long will it take this investment to increase to €2000?

Solution

Using the compound interest formula from Section 4.3, $A(t) = P\left(1 + \frac{r}{n}\right)^{nt}$, with $P = €1000$, $r = 0.06$ and $n = 4$, we need to solve for t when $A(t) = 2P$.

$2P = P\left(1 + \dfrac{0.06}{4}\right)^{4t}$ Substitute $2P$ for $A(t)$.

$2 = 1.015^{4t}$ Divide both sides by P.

$\ln 2 = \ln(1.015^{4t})$ Take natural logarithm of both sides.

$\ln 2 = 4t \ln 1.015$ Apply the property $\log_b(M^k) = k \log_b M$.

$t = \dfrac{\ln 2}{4 \ln 1.015}$

$t \approx 11.6389$ Evaluated on GDC.

The investment will double in 11.64 years – about 11 years and 8 months.

```
ln(2)/(4ln(1.015
))
        11.63888141
```

• **Hint:** Be sure to use brackets appropriately when entering the expression $\frac{\ln 2}{4 \ln 1.015}$ on your GDC. Following the rules for order of operations, your GDC will give an incorrect result if entered as shown here.

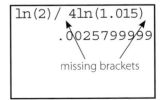

```
ln(2)/ 4ln(1.015)
          .0025799999
```

missing brackets

Example 19

The bacteria that cause 'strep throat' will grow in number at a rate of about 2.3% per minute. To the nearest whole minute, how long will it take for these bacteria to double in number?

Solution

Let t represent time in minutes and let A_0 represent the number of bacteria at $t = 0$.

Using the exponential growth model from Section 5.2, $A(t) = A_0b^t$, the growth factor, b, is $1 + 0.023 = 1.023$ giving $A(t) = A_0(1.023)^t$. The same equation would apply to money earning 2.3% annual interest with the money being added (compounded) once per year rather than once per minute. So, our mathematical model assumes that the number of bacteria increase incrementally, with the number increasing by 2.3% at the end of each minute. To find the doubling time, find the value of t so that $A(t) = 2A_0$.

$2A_0 = A_0(1.023)^t$ Substitute $2A_0$ for $A(t)$.

$2 = 1.023^t$ Divide both sides by A_0.

$\ln 2 = \ln(1.023^t)$ Take natural logarithm of both sides.

$\ln 2 = t \ln 1.023$ Apply the property $\log_b(M^k) = k \log_b M$.

$$t = \frac{\ln 2}{\ln 1.023} \approx 30.482$$

The number of bacteria will double in about 30 minutes.

Alternative solution

What if we assumed continuous growth instead of incremental growth? We apply the continuous exponential growth model from Section 5.3: $A(t) = Ce^{rt}$ with initial amount C and $r = 0.023$.

$2C = Ce^{0.023t}$ Substitute $2C$ for $A(t)$.

$2 = e^{0.023t}$ Divide both sides by C.

$\ln 2 = \ln(e^{0.023t})$ Take natural logarithm of both sides.

$\ln 2 = 0.023t$ Apply the property $\log_b(b^x) = x$.

$$t = \frac{\ln 2}{0.023} \approx 30.137$$

Continuous growth has a slightly shorter doubling time, but rounded to the nearest minute it also gives an answer of 30 minutes.

Example 20

\$1000 is invested in an investment account that earns interest at an annual rate of 10% compounded monthly. Calculate the minimum number of years needed for the amount in the account to exceed \$4000.

Solution

We use the exponential function associated with compound interest, $A(t) = P\left(1 + \frac{r}{n}\right)^{nt}$ with $P = 1000$, $r = 0.1$ and $n = 12$.

$$4000 = 1000\left(1 + \frac{0.1}{12}\right)^{12t} \Rightarrow 4 = (1.008\overline{3})^{12t} \Rightarrow \log 4 = \log[(1.008\overline{3})^{12t}] \Rightarrow$$

$$\log 4 = 12t\log(1.008\overline{3}) \Rightarrow t = \frac{\log 4}{12\log(1.008\overline{3})} \approx 13.92 \text{ years}$$

The minimum number of years needed for the account to exceed \$4000 is 14 years.

Example 21

A 20 g sample of radioactive iodine decays so that the mass remaining after t days is given by the equation $A(t) = 20e^{-0.087t}$, where $A(t)$ is measured in grams. After how many days (to the nearest whole day) is there only 5 g remaining?

Solution

$$5 = 20e^{-0.087t} \Rightarrow \frac{5}{20} = e^{-0.087t} \Rightarrow \ln 0.25 = \ln(e^{-0.087t}) \Rightarrow$$

$$\ln 0.25 = -0.087t \Rightarrow t = \frac{\ln 0.25}{-0.087} \approx 15.93$$

After about 16 days there is only 5 g remaining.

Example 22 – An equation in quadratic form

Solve for x in the equation $3^{2x} - 18 = 3^{x+1}$. Express any answers *exactly*.

Solution

The key to solving this equation is recognizing that it can be written in *quadratic form*. In Section 3.5, we solved equations of the form $at^2 + bt + c = 0$, where t is an algebraic expression. This is not immediately clear for this equation. We need to apply some laws of exponents to show that the equation is quadratic for the expression 3^x.

$3^{2x} - 18 = 3^{x+1}$

$(3^x)^2 - 3^1 \cdot 3^x - 18 = 0$ Applying rules $b^{mn} = (b^m)^n$ and $b^{m+n} = b^m b^n$.

Substituting a single variable, for example y, for the expression 3^x clearly makes the equation quadratic in terms of 3^x. We solve first for y and then solve for x after substituting 3^x back for y.

$$y^2 - 3y - 18 = 0$$
$$(y + 3)(y - 6) = 0$$
$$y = -3 \text{ or } y = 6$$
$$3^x = -3 \text{ or } 3^x = 6$$

$3^x = -3$ has no solution. Raising a positive number to a power cannot produce a negative number.

$$3^x = 6$$
$$\ln(3^x) = \ln 6 \quad \text{Take logarithm of both sides.}$$
$$x\ln 3 = \ln 6$$

Therefore, the one solution to the equation is exactly $x = \frac{\ln 6}{\ln 3}$.

- **Hint:** There are a couple of common algebra errors to avoid in the working for Example 22.
- If $3^x = -3$, then it does not follow that $x = -1$. An exponent of -1 indicates reciprocal.
- If $x = \frac{\ln 6}{\ln 3}$, it does ***not*** follow that $x = \ln 2$. The rule $\log m - \log n = \log\left(\frac{m}{n}\right)$ does not apply to the expression $\frac{\ln 6}{\ln 3}$.

Solving logarithmic equations

A logarithmic equation is an equation where the variable appears within the argument of a logarithm. For example, $\log x = \frac{1}{2}$ or $\ln x = 4$. We can solve both of these logarithmic equations directly by applying the definition of a logarithmic function (Section 5.4):

$$y = \log_b x \text{ if and only if } x = b^y$$

The logarithmic equation $\log x = \frac{1}{2}$ is equivalent to the exponential equation $x = 10^{\frac{1}{2}} = \sqrt{10}$, which leads directly to the solution. Likewise, the equation $\ln x = 4$ is equivalent to $x = e^4 \approx 54.598$. Both of these equations could have been solved by means of another method that makes use of the following two facts:

$$\text{(i) if } a = b \text{ then } n^a = n^b; \quad \text{and (ii) } b^{\log_b x} = x$$

To understand (ii) above, remember that a **logarithm is an exponent.** The value of $\log_b x$ is the exponent to which b is raised to give x. And b is being raised to this value; hence, the expression $b^{\log_b x}$ is equivalent to x. Therefore, another method for solving the logarithmic equation $\ln x = 4$ is to **exponentiate** both sides, i.e. use the expressions on either side of the equal sign as exponents for exponential expressions with equal bases. The base needs to be the base of the logarithm. For example,

$$\ln x = 4 \Rightarrow e^{\ln x} = e^4 \Rightarrow x = e^4$$

Example 23

Solve for x: $\log_3(2x - 5) = 2$

Solution

$$\log_3(2x-5) = 2 \Rightarrow 3^{\log_3(2x-5)} = 3^2 \quad \text{Exponentiate both side with base} = 3.$$
$$2x - 5 = 9 \quad \text{Applying property } b^{\log_b x} = x.$$
$$2x = 14$$
$$x = 7$$

Example 24

Solve for x in terms of k: $\log_2(5x) = 3 + k$

Solution

$\log_2(5x) = 3 + k \Rightarrow 2^{\log_2(5x)} = 2^{3+k}$ Exponentiate both sides with base = 2.

$5x = 2^3 \cdot 2^k$ Law of exponents $b^m \cdot b^n = b^{m+n}$ used 'in reverse'.

$x = \frac{8}{5}(2^k)$

For some logarithmic equations, it is necessary to first apply a property, or properties, of logarithms to simplify combinations of logarithmic expressions before solving.

Example 25

Solve for x: $\log_2 x + \log_2(10 - x) = 4$

Solution

$\log_2 x + \log_2(10 - x) = 4$

$\log_2[x(10 - x)] = 4$ Property of logarithms: $\log_b M + \log_b N = \log_b(MN)$.

$10x - x^2 = 2^4$ Changing from logarithmic form to exponential form.

$x^2 - 10x + 16 = 0$

$(x - 2)(x - 8) = 0$

$x = 2$ or $x = 8$

When solving logarithmic equations, you should be careful to always check if the *original* equation is a true statement when any solutions are substituted in for the variable. For Example 25, both of the solutions $x = 2$ and $x = 8$ produce true statements when substituted into the original equations. Sometimes 'extra' (extraneous) invalid solutions (met in Chapter 3) are produced, as illustrated in the next example.

Example 26

Solve for x: $\ln(x - 2) + \ln(2x - 3) = 2\ln x$

Solution

$\ln(x - 2) + \ln(2x - 3) = 2\ln x$

$\ln[(x - 2)(2x - 3)] = \ln x^2$ Properties of logarithms.

$\ln(2x^2 - 7x + 6) = \ln x^2$

$e^{\ln(2x^2 - 7x + 6)} = e^{\ln x^2}$ Exponentiate both sides.

$2x^2 - 7x + 6 = x^2$

$x^2 - 7x + 6 = 0$

$(x - 6)(x - 1) = 0$ Factorize.

$x = 6$ or $x = 1$

Substituting these two *possible* solutions indicates that $x = 1$ is not a valid solution. The reason is that if you try to substitute 1 for x into the original equation, we are not able to evaluate the expression $\ln(2x - 3)$ because we can only take the logarithm of a positive number. Therefore, $x = 6$ is the only solution. $x = 1$ is an extraneous solution that is not valid.

Solving, or checking the solutions to, a logarithmic equation on your GDC will help you avoid, or determine, extraneous solutions. To solve Example 26 on your GDC, a useful approach is to first set the equation equal to zero. Then graph the expression (after setting it equal to y) and observe where the graph intersects the x-axis (i.e. $y = 0$).

Graphical solution for Example 26:

$\ln(x - 2) + \ln(2x - 3) = 2\ln x \Rightarrow \ln(x - 2) + \ln(2x - 3) - 2\ln x = 0$

Graph the equation $y = \ln(x - 2) + \ln(2x - 3) - 2\ln x$ on your GDC and find x-intercepts.

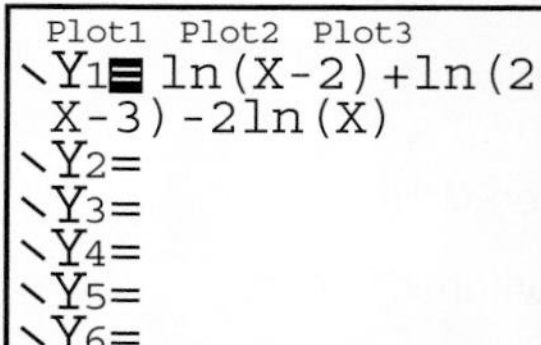

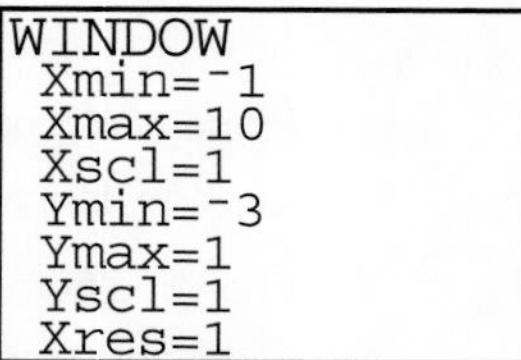

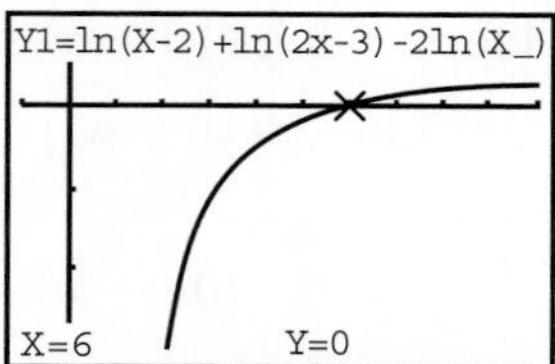

The graph only intersects the x-axis at $x = 6$ and not at $x = 1$. Hence, $x = 6$ is the only valid solution and $x = 1$ is an extraneous solution.

Exponential and logarithmic inequalities

In Section 3.5, we covered methods of solving a variety of inequalities. These methods can also be applied to solving inequalities involving exponential and logarithmic functions. It is important to consider the domain of any functions in the inequality, and to check any solutions in the original inequality in case any extraneous solutions occur.

Example 27

Find the solution set to the inequality: $2\log_3 x - 1 < 0$.

Solution

Due to the domain of the logarithmic function, all solutions must be positive.

Method 1 (algebraic solution)

Solve the equation $2\log_3 x - 1 = 0$ and find the exact solution.

$$2\log_3 x = 1 \Rightarrow \log_3 x = \tfrac{1}{2} \Rightarrow x = 3^{\frac{1}{2}} = \sqrt{3}$$

Substitute 'test' values, x_1 and x_2, into the original inequality such that $0 < x_1 < \sqrt{3}$ and $x_2 > \sqrt{3}$.

Let $x_1 = 1$: $\log_3 1 - 1 = 0 - 1 = -1 < 0$ (true)
Let $x_2 = 9$: $\log_3 9 - 1 = 2 - 1 = 1 \times 0$ (false)

Therefore, the solution set is $0 < x < \sqrt{3}$.

Method 2 (graphical solution)

Graph the equation $y = 2\log_3 x - 1$ on your GDC and use it to determine the portion of the graph that is less than zero (i.e. below the x-axis). But, how do we input the expression $\log_3 x$ on the GDC? We can use the change of base formula to write $\log_3 x = \frac{\log x}{\log 3}$.

```
Plot1 Plot2 Plot3
\Y1=2(log(X)/log
(3))-1
\Y2=
\Y3=
\Y4=
\Y5=
\Y6=
```

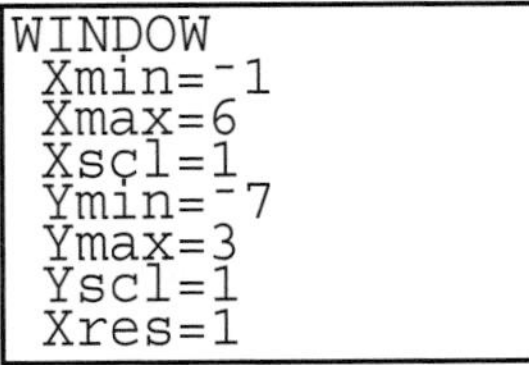

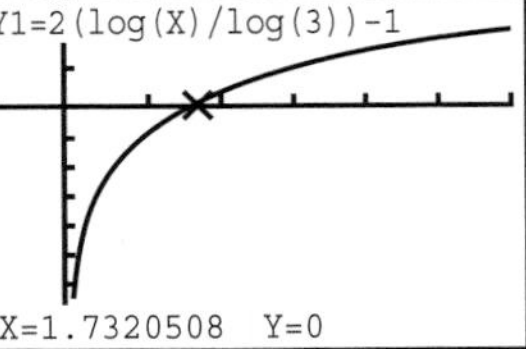

The y-axis is a vertical asymptote. The graph indicates that the solution set is $0 < x < 1.732\,0508$. Although the graphical method is efficient and effective it does not give an exact result.

Example 28

Solve: $(e^x - 2)(e^x + 6) \leqslant 3e^x$

Solution

The fact that the left side is factorized is not helpful because the other side of the inequality is not zero. So we need to expand the left side and rearrange terms to get zero on the right side.

$$(e^x - 2)(e^x + 6) \leqslant 3e^x$$
$$(e^x)^2 + 4e^x - 12 \leqslant 3e^x$$
$$e^{2x} + e^x - 12 \leqslant 0$$
Now factorize this expression.
$$(e^x - 3)(e^x + 4) \leqslant 0$$
Find where each factor is zero and construct a sign chart.

$e^x - 3 = 0 \Rightarrow e^x = 3 \Rightarrow x = \ln 3$
and $e^x + 4 = 0 \Rightarrow e^x = -4 \Rightarrow$ no solution

Since $x = \ln 3$ is the only zero of the expression $(e^x - 3)(e^x + 4)$ we only need to test x-values on either side of $x = \ln 3$. The factor $e^x + 4$ will always be positive.

x		$\ln 3$	
$e^x - 3$	$-$	0	$+$
$e^x + 4$	$+$		$+$
$(e^x - 3)(e^x + 4)$	$-$	0	$+$

Therefore, the solution set is $x \leqslant \ln 3$.

Exercise 5.5

In questions 1–12, solve for x. Give x accurate to 3 significant figures.

1 $10^x = 5$

2 $4^x = 32$

3 $8^{x-6} = 60$

4 $2^{x+3} = 100$

5 $\left(\frac{1}{5}\right)^x = 22$

6 $e^x = 15$

7 $10^x = e$

8 $3^{2x-1} = 35$

9 $2^{x+1} = 3^{x-1}$

10 $2e^{10x} = 19$

11 $6^{\frac{x}{2}} = 5^{1-x}$

12 $\left(1 + \frac{0.05}{12}\right)^{12x} = 3$

In questions 13–16, solve for x. Give answers **exactly**.

13 $4^x - 2^{x+1} = 48$ • **Hint:** write 4 as 2^2

14 $2^{2x+1} - 2^{x+1} + 1 = 2^x$

15 $6^{2x+1} - 17(6^x) + 12 = 0$

16 $3^{2x+1} + 3 = 10(3^x)$

17 \$5000 is invested in an account that pays 7.5% interest per year, compounded quarterly.

a) Find the amount in the account after three years.

b) How long will it take for the money in the account to double? Give the answer to the nearest quarter of a year.

18 How long will it take for an investment of €500 to triple in value if the interest is 8.5% per year, compounded continuously. Give the answer in number of years accurate to 3 significant figures.

19 A single bacterium begins a colony in a laboratory dish. If the colony doubles every hour, after how many hours does the colony first have more than one million bacteria?

20 Find the least number of years for an investment to double if interest is compounded annually with the following interest rates.

a) 3% b) 6% c) 9%

21 A new car purchased in 2005 decreases in value by 11% per year. When is the first year that the car is worth less than one-half of its original value?

22 Uranium-235 is a radioactive substance that has a half-life of 2.7×10^5 years.

a) Find the amount remaining from a 1 g sample after a thousand years.

b) How long will it take a 1 g sample to decompose until its mass is 700 milligrams (i.e. 0.7 g)? Give the answer in years accurate to 3 significant figures.

23 The stray dog population in a town is growing exponentially with about 18% more stray dogs each year. In 2008, there are 16 stray dogs.

a) Find the projected population of stray dogs after five years.

b) When is the first year that the number of stray dogs is greater than 70?

24 Initially a water tank contains one thousand litres of water. At the time $t = 0$ minutes, a tap is opened and water flows out of the tank. The volume, V litres, which remains in the tank after t minutes is given by the following exponential function: $V(t) = 1000(0.925)^t$.

a) Find the value of V after 10 minutes.

b) Find how long, to the nearest second, it takes for half of the initial amount of water to flow out of the tank.

c) The tank is considered 'empty' when only 5% of the water remains. From when the tap is first opened, how many whole minutes have passed before the tank can first be considered empty?

25 The mass m kilograms of a radioactive substance at time t days is given by $m = 5e^{-0.13t}$.

a) What is the initial mass?

b) How long does it take for the substance to decay to 0.5 kg? Give the answer in days accurate to 3 significant figures.

In questions 26–36, solve for x in the logarithmic equation. Give exact answers and be sure to check for extraneous solutions.

26 $\log_2(3x - 4) = 4$

27 $\log(x - 4) = 2$

28 $\ln x = -3$

29 $\log_{16} x = \frac{1}{2}$

30 $\log \sqrt{x + 2} = 1$

31 $\ln(x^2) = 16$

32 $\log_2(x^2 + 8) = \log_2 x + \log_2 6$

33 $\log_3(x - 8) + \log_3 x = 2$

34 $\log 7 - \log(4x + 5) + \log(2x - 3) = 0$

35 $\log_3 x + \log_3(x - 2) = 1$

36 $\log x^8 = (\log x)^4$

In questions 37–40, solve each inequality.

37 $5 \log x + 2 > 0$

38 $2\log x^2 - 3\log x < \log 8x - \log 4x$

39 $(e^x - 2)(e^x - 3) < 2e^x$

40 $3 + \ln x > e^x$

Practice questions

1 A portion of the graph $y = 2 - \log_3(x + 1)$ is shown. It intersects the x-axis at point P, the y-axis at point Q and the line $y = 3$ at point R. Find the following:

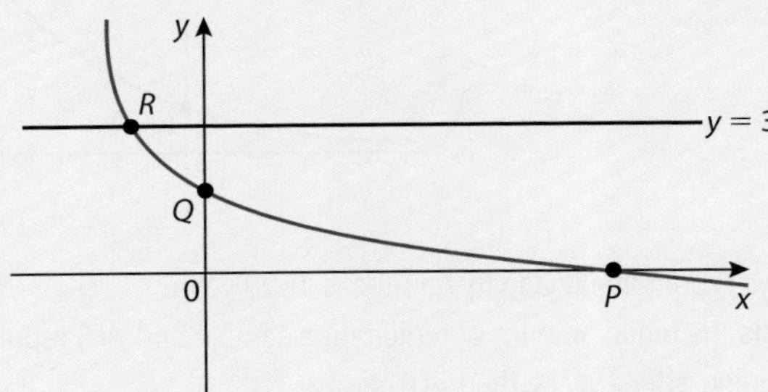

a) The x-coordinate of point P.

b) The y-coordinate of point Q.

c) The coordinates of point R.

2 The amount $A(t)$, in grams, of a certain radioactive substance remaining after t years decays by the formula $A(t) = A_0e^{-0.0045t}$, where A_0 is the initial amount.

a) If 5 grams are left after 800 years, how many grams were present initially?

b) What is the half-life of the substance?

3 a) Find expressions for the nth term and the sum to n terms of the following arithmetic series, $\ln y + \ln y^2 + \ln y^3 + \dots$ where $y > 0$.

b) Hence, find expressions for the nth term and the sum to n terms of the following arithmetic series, $\ln(xy) + \ln(xy^2) + \ln(xy^3) + \dots$ where $x > 0$ and $y > 0$.

4 Solve, for x, the equation $\log_2(5x^2 - x - 2) = 2 + 2\log_2 x$.

5 If $\log_2 4\sqrt{2} = x$, $\log_z y = 4$, and $y = 4x^2 - 2x - 6 + z$, find y.

6 Find the **exact** values of t for which $2e^{3t} - 7e^{2t} + 7e^t = 2$.

7 Find the **exact** solution(s) to the equation $8e^2 - 2e\ln x = (\ln x)^2$.

8 Find the exact value of x for each equation.

a) $\log_3 x - 4\log_x 3 + 3 = 0$

b) $\log_2(x - 5) + \log_2(x + 2) = 3$

9 Express each as a single logarithm.

a) $2\log a + 3\log b - \log c$

b) $3\ln x - \frac{1}{2}\ln y + 1$

10 A piece of wood is recovered from an ancient building during an archaeological excavation. The formula $A(t) = A_0e^{-0.000\,124t}$ is used to determine the age of the wood, where A_0 is the amount of carbon in any living tree, $A(t)$ is the amount of carbon in the wood being dated and t is the age of the wood in years. For the ancient piece of wood it is found that $A(t)$ is 79% of the amount of the carbon in a living tree. How old is the piece of wood, to the nearest 100 years?

11 The graph of the equation $y = \log_3(2x - 3) - 4$ intersects the x-axis at the point $(c, 0)$. Without using your GDC, find the exact value of c.

12 The graph of $y = b^x$, $b > 1$ is shown. On separate coordinate planes, sketch the graphs of

a) $y = b^{-x}$

b) $y = b^{1-x}$

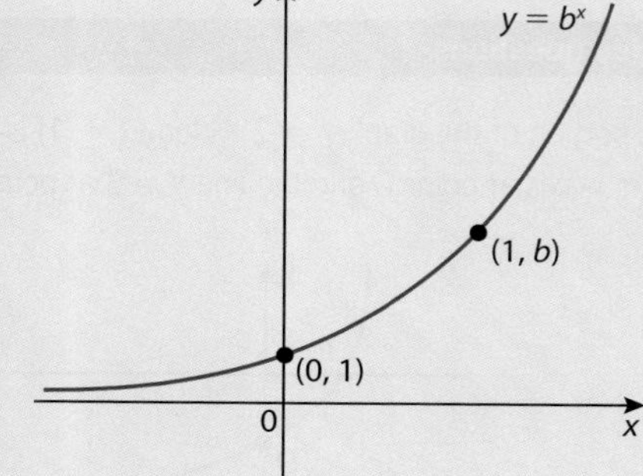

13 Radium decays exponentially and its half-life is 1600 years.
If A_0 represents the initial amount of radium in a sample and $A(t)$ represents the amount remaining after t years, then $A(t) = A_0e^{-kt}$.

a) Find the value of k approximated to four significant figures.

b) Find what percentage of the original amount of radium will be remaining after 4000 years.

14 Solve the equation $e^{-x} - x + 1 = 0$.

15 Find the set of values of x for which $|0.1x^2 - 2x + 3| < \log_{10} x$.

16 Determine the values of x that satisfy the inequality $\dfrac{xe^x}{x^2 - 1} \geqslant 1$.

17 a) Solve the equation $2(4^x) + 4^{-x} = 3$.

b) (i) Solve the equation $a^x = e^{2x+1}$ where $a > 0$, giving your answer for x in terms of a.

(ii) For what value of a does the equation have no solution?

18 The solution of $2^{2x+3} = 2^{x+1} + 3$ can be expressed in the form $a + \log_2 b$ where $a, b \in \mathbb{Z}$. Find the value of a and of b.

19 Solve $2(\ln x)^2 = 3\ln x - 1$ for x. Give your answers in **exact** form.

20 A sum of \$100 is invested.

a) If the interest is compounded annually at a rate of 5% per year, find the total value V of the investment after 20 years.

b) If the interest is compounded monthly at a rate of $\frac{5}{12}$% per month, find the minimum number of months for the value of the investment to exceed V.

21 Solve the equation $9\log_5 x = 25\log_x 5$ expressing your answer in the form $5^{\frac{p}{q}}$, where $p, q \in \mathbb{Z}$.

22 Solve $|\ln(x + 3)| = 1$. Give your answers in **exact** form.

23 Solve the equation $\left|e^{2x} - \frac{1}{x+2}\right| = 2$.

24 An experiment is carried out in which the number n of bacteria in a liquid, is given by the formula $n = 650e^{kt}$, where t is the time in minutes after the beginning of the experiment and k is a constant. The number of bacteria doubles every 20 minutes. Find the exact value of k.

25 The function f is defined for $x > 2$ by $f(x) = \ln x + \ln(x - 2) - \ln(x^2 - 4)$.

a) Express $f(x)$ in the form $\ln\left(\frac{x}{x+a}\right)$.

b) Find an expression for $f^{-1}(x)$.

26 **a)** The function f is defined by $f: x \mapsto e^x - 1 - x$.

(i) Use your GDC to find the minimum value of f.

(ii) Prove that $e^x \geqslant 1 + x$ for all real values of x.

b) Use mathematical induction to prove that

$$(1 + 1)\left(1 + \frac{1}{2}\right)\left(1 + \frac{1}{3}\right) \cdots \left(1 + \frac{1}{n}\right) = n + 1 \text{ for all integers } n \geqslant 1$$

c) Use the results of parts a) and b) to prove that

$$e^{\left(1 + \frac{1}{2} + \frac{1}{3} + \ldots + \frac{1}{n}\right)} > n$$

d) Find a value of n for which

$$1 + \frac{1}{2} + \frac{1}{3} + \cdots + \frac{1}{n} > 100$$

6 Matrix Algebra

Assessment statement

1.9 Solutions of systems of linear equations (a maximum of three equations in three unknowns), including cases when there is a unique solution, an infinity of solutions or no solution.

Note: Sections 6.1 to 6.3 are not required for examinations. However, it is highly recommended that you review these sections because of their important applications. Sections 6.1 and 6.2 can be omitted. Special attention must be paid to the determinant concept in Section 6.3 because it will be used later in the book.

In Section 6.4 the Gauss-Jordan elimination method is required in its 'raw' form, i.e. using equations. However, for reasons of efficiency, and if you were to use a GDC to solve a system of equations, the matrix form is more appropriate. Even though it is not required for examination purposes, in exams, any 'mathematically sound' method is accepted.

Introduction

Ever since their first emergence, matrices have been and remain significant mathematical tools. Uses of matrices span several areas from simply solving systems of simultaneous linear equations, to describing atomic structure, designing computer game graphics, analyzing relationships, coding, and operations research, to mention a few. If you have ever used a spreadsheet such as Excel or Lotus, or have ever created a table, then you have used a matrix. Matrices make the presentation of data understandable and help make calculations easy to perform. For example, your teacher's grade book may look something like this:

Student	Quiz 1	Quiz 2	Test 1	Test 2	Homework	Grade
Tim	70	80	86	82	95	A
Maher	89	56	80	60	55	C
...	...	...	...	...	...	...

If we want to know Tim's grade on Test 2, we simply follow along the row 'Tim' to the column 'Test 2' and find that he received a score of 82. Take a look at the matrix below about the sale of cameras in a store according to location and type.

	City	Donau	Neubau	Moedling
Nikon	153	98	74	56
Canon	211	120	57	29
Olympus	82	31	12	5
Other	308	242	183	107

If we want to know how many Canon cameras were sold in the Neubau shop, we follow along the row 'Canon' to the column 'Neubau' and find that 57 Canons were sold.

6.1 Basic definitions

What is a matrix?

A matrix is a rectangular array of elements. The elements can be symbolic expressions or numbers.

Matrix $[A]$ is denoted by

$$A = \begin{pmatrix} a_{11} & a_{12} & \cdots & a_{1n} \\ a_{21} & a_{22} & \cdots & a_{2n} \\ \vdots & \vdots & \vdots & \vdots \\ a_{m1} & a_{m2} & \cdots & a_{mn} \end{pmatrix} \begin{matrix} \leftarrow \\ \leftarrow \\ \vdots \\ \leftarrow \end{matrix} \Bigg\} \ m \text{ rows}$$

$$\underbrace{\uparrow \quad \uparrow \quad \cdots \quad \uparrow}_{n \text{ columns}}$$

Row i of A has n elements and is $(a_{i1} \quad a_{i2} \quad \ldots \quad a_{in})$.

Column j of A has m elements and is $\begin{pmatrix} a_{1j} \\ a_{2j} \\ \vdots \\ a_{mj} \end{pmatrix}$.

The number of rows and columns of the matrix define its size (order). So, a matrix that has m rows and n columns is said to have an $m \times n$ (m by n) size (order). A matrix A with $m \times n$ order (size) is sometimes denoted as $[A]_{m \times n}$ or $[A]_{mn}$ to show that A is a matrix with m rows and n columns. (Some authors use $[a_{ij}]$ to represent a matrix.) The sales matrix has a 4×4 order. When $m = n$, the matrix is said to be a **square matrix** with order n, so the sales matrix is a square matrix of order 4.

Every entry in the matrix is called an **entry** or **element** of the matrix, and is denoted by a_{ij}, where i is the row number and j is the column number of that element. The ordered pair (i, j) is also called the **address** of the element. So, in the grades matrix example, the entry (2, 4) is 60, the student Maher's grade on Test 2, while (2, 4) in the sales matrix example is 29, Canon's sales in the Moedling shop.

Arthur Cayley (1821–1895)

Arthur Cayley entered Trinity College, Cambridge in 1838. While still an undergraduate, he published three papers in the *Cambridge Mathematical Journal*. Cayley graduated as Senior Wrangler in 1842 and won the first Smith's prize. Winning a fellowship enabled him to teach for four years at Cambridge. He published 28 papers in the *Cambridge Mathematical Journal* during these years. Since a fellowship had limited tenure, Cayley needed to find a profession. He spent 14 years as a lawyer but, although very skilled in his legal specialty, he always considered it as a means to make money so that he could pursue mathematics. During these 14 years as a lawyer he published around 250 mathematical papers.

His published work comprises over 900 papers and notes covering several fields of modern mathematics. The most important aspect of his work was in developing the algebra of matrices.

Vectors

A vector is a matrix that has only one row or one column. There are two types of vectors – row vectors and column vectors.

Row vector

If a matrix has one row, it is called a row vector.

$B = (b_1\ b_2\ \dots\ b_m)$ is a row vector with dimension m.

$B = (1\ 2)$ could represent the position of a point in a plane and is an example of a row vector of dimension 2.

Column vector

If a matrix has one column, it is called a column vector.

$C = \begin{pmatrix} c_1 \\ c_2 \\ \vdots \\ c_n \end{pmatrix}$ is a column vector with dimension n.

$C = \begin{pmatrix} 1 \\ 2 \end{pmatrix}$ again could represent the position of a point in a plane and is an example of a column vector of dimension 2.

As you see, vectors can be represented by row or column matrices.

Submatrix

If some row(s) and/or column(s) of a matrix A are deleted, the remaining matrix is called a submatrix of A.

For example, if we are interested in the sales of the three main types of cameras in the central part of the city, we can represent them with the following *submatrix* of the original matrix:

$$\begin{pmatrix} 153 & 98 \\ 211 & 120 \\ 82 & 31 \end{pmatrix}$$

Zero matrix

A matrix for which all entries are equal to zero ($a_{ij} = 0$ for all i and j).

$$(0\quad 0), \begin{pmatrix} 0 & 0 \\ 0 & 0 \end{pmatrix}, \begin{pmatrix} 0 & 0 & 0 \\ 0 & 0 & 0 \end{pmatrix} \text{ are zero matrices.}$$

Diagonal

A square matrix where all entries except the diagonal entries are zero is called a **diagonal matrix.**

In a square matrix, the entries $a_{11}, a_{22}, \dots, a_{nn}$ are called the **diagonal elements** of the matrix. Sometimes the diagonal of the matrix is also called the **principal** or **main** of the matrix.

$$\begin{pmatrix} 153 & 0 & 0 & 0 \\ 0 & 120 & 0 & 0 \\ 0 & 0 & 12 & 0 \\ 0 & 0 & 0 & 107 \end{pmatrix}$$

What is the diagonal in our sales matrix? Here, $a_{11} = 153$, $a_{22} = 120$, $a_{33} = 12$ and $a_{44} = 107$.

Triangular matrix

You can use a matrix to present data showing distances between different cities.

	Graz	Salzburg	Innsbruck	Linz
Vienna	191	298	478	185
Graz		282	461	220
Salzburg			188	135
Innsbruck				320

◀ **Table 6.1**

The data in Table 6.1 can be represented by a triangular matrix (upper triangular in this case).

$$\begin{pmatrix} 191 & 298 & 478 & 185 \\ 0 & 282 & 461 & 220 \\ 0 & 0 & 188 & 135 \\ 0 & 0 & 0 & 320 \end{pmatrix}$$

In a triangular matrix, the entries on one side of its diagonal are all zero.

Definition of a triangular matrix

A triangular matrix is a square matrix with order n for which $a_{ij} = 0$ when $i > j$ (upper triangular) or, alternatively, when $i < j$ (lower triangular).

Another way of representing the distance data is given by the following matrix:

	Vienna	*Graz*	*Salzburg*	*Innsbruck*	*Linz*
Vienna	0	191	298	478	185
Graz	191	0	282	461	220
Salzburg	298	282	0	188	135
Innsbruck	478	461	188	0	320
Linz	185	220	1325	320	0

Again the data in the table can be represented by a matrix called a **symmetric** matrix. In such matrices, $a_{ij} = a_{ji}$ for all i and j. All symmetric matrices are square!

$$\begin{pmatrix} 0 & 191 & 298 & 478 & 185 \\ 191 & 0 & 282 & 461 & 220 \\ 298 & 282 & 0 & 188 & 135 \\ 478 & 461 & 188 & 0 & 320 \\ 185 & 220 & 135 & 320 & 0 \end{pmatrix}$$

6.2 Matrix operations

When are two matrices considered to be equal?

Two matrices A and B are equal if the size of A and B is the same (number of rows and columns are the same for A and B) and $a_{ij} = b_{ij}$ for all i and j.

For example, $\begin{pmatrix} 2 & 3 \\ 5 & 7 \end{pmatrix}$ and $\begin{pmatrix} 2 & x \\ x^2 - 4 & 7 \end{pmatrix}$ can only be equal if $x = 3$ and $x^2 - 4 = 5$, which can only be true if $x = 3$.

How do you add/subtract two matrices?

Two matrices A and B can be added only if they have the *same size.* If C is the sum of the two matrices, then we write

$$C = A + B$$

where $c_{ij} = a_{ij} + b_{ij}$, i.e. we add 'corresponding' terms, one by one.

For example,

$$\begin{pmatrix} 2 & 3 \\ 5 & 7 \end{pmatrix} + \begin{pmatrix} x & y \\ a & b \end{pmatrix} = \begin{pmatrix} 2+x & 3+y \\ 5+a & 7+b \end{pmatrix}$$

Subtraction is done similarly:

$$\begin{pmatrix} 2 & 3 & 1 \\ 5 & 7 & 0 \end{pmatrix} - \begin{pmatrix} x & y & 8 \\ a & b & 2 \end{pmatrix} = \begin{pmatrix} 2-x & 3-y & -7 \\ 5-a & 7-b & -2 \end{pmatrix}$$

The operations of addition and subtraction of matrices obey all rules of addition and subtraction of real numbers. That is,

$A + B = B + A$; $A + (B + C) = (A + B) + C$; $A - (B + C) = A - B - C$.

How do we multiply a scalar by a matrix?

A scalar is any object that is not a matrix. The multiplication by a scalar is straightforward. You multiply each term of the matrix by the scalar.

If A is an $m \times n$ matrix, and c is a scalar, the scalar product of c and A is another matrix $B = cA$ such that every entry b_{ij} of B is a multiple of its corresponding A entry, i.e. $b_{ij} = c \times a_{ij}$.

It is often convenient to rewrite the scalar multiple cA by factoring c out of every entry in the matrix. For instance, in the following example, the scalar $\frac{1}{2}$ has been factored out of the matrix.

$$\begin{pmatrix} \frac{1}{2} & -\frac{3}{2} \\ \frac{5}{2} & \frac{1}{2} \end{pmatrix} = \frac{1}{2}\begin{pmatrix} 1 & -3 \\ 5 & 1 \end{pmatrix}$$

Matrix multiplication

At first glance, the following definition may seem unusual. You will see later, however, that this definition of the product of two matrices has many practical applications.

Matrix multiplication

If $A = (a_{ij})$ is an $m \times n$ matrix and $B = (b_{ij})$ is an $n \times p$ matrix, the product AB is an $m \times p$ matrix, $AB = (c_{ij})$, where

$$c_{ij} = \sum_{k=1}^{n} a_{ik}b_{kj} = a_{i1}b_{1j} + a_{i2}b_{2j} + \ldots + a_{in}b_{nj}$$

for each $i = 1, 2, \ldots, m$ and $j = 1, 2, \ldots, n$.

This definition means that each entry with an address ij AB is obtained by multiplying the entries in the ith row of A by the *corresponding* entries in the jth column of B and then adding the results. The following shows the process in detail:

$$c_{ij} = \begin{pmatrix} a_{i1} & a_{i2} & \ldots & a_{in} \end{pmatrix} \begin{pmatrix} b_{1j} \\ b_{2j} \\ \vdots \\ b_{nj} \end{pmatrix} = a_{i1}b_{1j} + a_{i2}b_{2j} + \ldots + a_{in}b_{nj}$$

Example 1

Find $C = AB$ if $A = \begin{pmatrix} 3 & -5 & 2 \\ 2 & 1 & 7 \end{pmatrix}$ and $B = \begin{pmatrix} 3 & -2 & 1 & 5 \\ 5 & 8 & -4 & 0 \\ -9 & 10 & 5 & 3 \end{pmatrix}$.

Solution

A is a **2** × **3** matrix and B is a **3** × **4** matrix, so the product must be a **2** × **4** matrix. Every entry in the product is the result of multiplying the entries in the rows of A and columns of B. For example:

$$c_{12} = \sum_{k=1}^{3} a_{1k}b_{k2} = (a_{11} \quad a_{12} \quad a_{13})\begin{pmatrix} b_{12} \\ b_{22} \\ b_{32} \end{pmatrix} = (3 \quad -5 \quad 2)\begin{pmatrix} -2 \\ 8 \\ 10 \end{pmatrix}$$

$$= 3 \times (-2) - 5 \times 8 + 2 \times 10 = -26$$

or

$$c_{23} = \sum_{k=1}^{3} a_{2k}b_{k3} = (a_{21} \quad a_{22} \quad a_{23})\begin{pmatrix} b_{13} \\ b_{23} \\ b_{33} \end{pmatrix} = (2 \quad 1 \quad 7)\begin{pmatrix} 1 \\ -4 \\ 5 \end{pmatrix}$$

$$= 2 \times 1 + 1 \times (-4) + 7 \times 5 = 33$$

The operation is repeated eight times to get

$$C = AB = \begin{pmatrix} -34 & -26 & 33 & 21 \\ -52 & 74 & 33 & 31 \end{pmatrix}$$

This product can also be found using a GDC.

```
[A][B]
[[-34 -26 33 21...
 [-52 -74 33 31...
```

For the product of two matrices to be defined, the number of columns in the first matrix should be the same as the number of rows in the second matrix.

$$\underset{m \times n}{A} \quad \underset{n \times p}{B} = \underset{m \times p}{AB}$$

equal (the inner dimensions n and n)

order of AB (the outer dimensions m and p)

Examples – matrix multiplication

a) $\begin{pmatrix} 5 & 0 & 3 \\ -2 & 1 & 2 \end{pmatrix}\begin{pmatrix} -2 & 4 \\ 1 & -1 \\ 3 & -2 \end{pmatrix} = \begin{pmatrix} -1 & 14 \\ 11 & -13 \end{pmatrix}$

2 × 3   3 × 2   2 × 2

b) $\begin{pmatrix} 4 & -5 \\ 1 & 7 \end{pmatrix}\begin{pmatrix} 1 & 0 \\ 0 & 1 \end{pmatrix} = \begin{pmatrix} 4 & -5 \\ 1 & 7 \end{pmatrix}$

2 × 2   2 × 2   2 × 2

c) $\begin{pmatrix} 5 & 0 & 3 \\ -2 & 1 & 2 \\ 2 & 1 & 3 \end{pmatrix} \begin{pmatrix} -\frac{1}{7} & -\frac{3}{7} & \frac{3}{7} \\ -\frac{10}{7} & -\frac{9}{7} & \frac{16}{7} \\ \frac{4}{7} & \frac{5}{7} & -\frac{5}{7} \end{pmatrix} = \begin{pmatrix} 1 & 0 & 0 \\ 0 & 1 & 0 \\ 0 & 0 & 1 \end{pmatrix}$

$3 \times 3 \qquad 3 \times 3 \qquad 3 \times 3$

As you see from part b) above, the matrix $\begin{pmatrix} 1 & 0 \\ 0 & 1 \end{pmatrix}$ does not create a new value when it is multiplied by another matrix. This is why it is called the **identity** matrix of order 2.

The identity matrix

A $n \times n$ diagonal matrix where $a_{ij} = 1$ and $i = j$ is called the identity matrix of order n.

Examples – identity matrices

a) $\begin{pmatrix} a & b & c \\ d & e & f \\ g & h & i \end{pmatrix} \begin{pmatrix} 1 & 0 & 0 \\ 0 & 1 & 0 \\ 0 & 0 & 1 \end{pmatrix} = \begin{pmatrix} a & b & c \\ d & e & f \\ g & h & i \end{pmatrix}$

b) $\begin{pmatrix} 1 & 0 & 0 \\ 0 & 1 & 0 \\ 0 & 0 & 1 \end{pmatrix} \begin{pmatrix} a & b & c \\ d & e & f \\ g & h & i \end{pmatrix} = \begin{pmatrix} a & b & c \\ d & e & f \\ g & h & i \end{pmatrix}$

c) $\begin{pmatrix} a & b & c & m \\ d & e & f & n \\ g & h & i & p \\ j & k & l & q \end{pmatrix} \begin{pmatrix} 1 & 0 & 0 & 0 \\ 0 & 1 & 0 & 0 \\ 0 & 0 & 1 & 0 \\ 0 & 0 & 0 & 1 \end{pmatrix} = \begin{pmatrix} a & b & c & m \\ d & e & f & n \\ g & h & i & p \\ j & k & l & q \end{pmatrix}$

Sometimes, the identity matrix is denoted by I_n, where n is the order. So, in parts a) and b) above, the identity is I_3, and in c) it is I_4.

Examples – comparing AB with BA

a) $\begin{pmatrix} 2 & -1 & 3 \end{pmatrix} \begin{pmatrix} 2 \\ 5 \\ 4 \end{pmatrix} = (11)$

$1 \times 3 \qquad 3 \times 1 \quad 1 \times 1$

b) $\begin{pmatrix} 2 \\ 5 \\ 4 \end{pmatrix} \begin{pmatrix} 2 & -1 & 3 \end{pmatrix} = \begin{pmatrix} 4 & -2 & 6 \\ 10 & -5 & 15 \\ 8 & -4 & 12 \end{pmatrix}$

$3 \times 1 \qquad 1 \times 3 \qquad 3 \times 3$

Notice the difference between the products in parts a) and b). Matrix multiplication, in general, is **not commutative**. It is usually not true that $AB = BA$.

Let $A = \begin{pmatrix} 3 & 6 \\ 5 & 2 \end{pmatrix}$ and $B = \begin{pmatrix} -2 & 3 \\ 1 & 5 \end{pmatrix}$, then $AB = \begin{pmatrix} 3 & 6 \\ 5 & 2 \end{pmatrix} \begin{pmatrix} -2 & 3 \\ 1 & 5 \end{pmatrix} = \begin{pmatrix} 0 & 39 \\ -8 & 25 \end{pmatrix}$

but

$$BA = \begin{pmatrix} -2 & 3 \\ 1 & 5 \end{pmatrix} \begin{pmatrix} 3 & 6 \\ 5 & 2 \end{pmatrix} = \begin{pmatrix} 9 & -6 \\ 28 & 16 \end{pmatrix} \Rightarrow AB \neq BA$$

However, if we let

$$A = \begin{pmatrix} 3 & 6 \\ 5 & 2 \end{pmatrix} \text{ and } B = \begin{pmatrix} 2 & 6 \\ 5 & 1 \end{pmatrix}, \text{ then } AB = \begin{pmatrix} 3 & 6 \\ 5 & 2 \end{pmatrix}\begin{pmatrix} 2 & 6 \\ 5 & 1 \end{pmatrix} = \begin{pmatrix} 36 & 24 \\ 20 & 32 \end{pmatrix} \text{ and}$$

$$BA = \begin{pmatrix} 2 & 6 \\ 5 & 1 \end{pmatrix}\begin{pmatrix} 3 & 6 \\ 5 & 2 \end{pmatrix} = \begin{pmatrix} 36 & 24 \\ 20 & 32 \end{pmatrix} \Rightarrow AB = BA$$

Thus, in general, $AB \neq BA$. However, for some matrices A and B, it may happen that $AB = BA$.

Example 2

Find the average sales in each of the regions (City, Donau, Neubau and Moedling), given the following information.

	City	Donau	Neubau	Moedling
Nikon	153	98	74	56
Canon	211	120	57	29
Olympus	82	31	12	5
Other	308	242	183	107

The average selling price for each make of camera is as follows:
Nikon €1200, Canon €1100, Olympus €900, Other €600

Solution

We set up a matrix multiplication in which the individual camera sales are multiplied by the corresponding price. Since the rows represent the sales of the different makes of camera, create a row matrix of the different prices and perform the multiplication.

$$\begin{pmatrix} 1200 & 1100 & 900 & 600 \end{pmatrix} \begin{pmatrix} 153 & 98 & 74 & 56 \\ 211 & 120 & 57 & 29 \\ 82 & 31 & 12 & 5 \\ 308 & 242 & 183 & 107 \end{pmatrix} = \begin{pmatrix} 674\,300 & 422\,700 & 272\,100 & 167\,800 \end{pmatrix}$$

So, the regions' sales are:

	City	Donau	Neubau	Moedling
Sales	674 300	422 700	272 100	167 800

Remember that we are multiplying a 1×4 matrix with a 4×4 matrix and hence we get a 1×4 matrix.

Exercise 6.1 and 6.2

1 Consider the following matrices

$$A = \begin{pmatrix} -2 & x \\ y-1 & 3 \end{pmatrix}, B = \begin{pmatrix} x+1 & -3 \\ 4 & y-2 \end{pmatrix}$$

a) Evaluate each of the following
(i) $A + B$ (ii) $3A - B$.

b) Find x and y such that $A = B$.

c) Find x and y such that $A + B$ is a diagonal matrix.

d) Find AB and BA.

2 Solve for the variables.

a) $\begin{pmatrix} 3 & 0 \\ 4 & 2 \end{pmatrix}\begin{pmatrix} x \\ y \end{pmatrix} = \begin{pmatrix} 6 \\ -12 \end{pmatrix}$ b) $\begin{pmatrix} 2 & p \\ 3 & q \end{pmatrix}\begin{pmatrix} 4 \\ 5 \end{pmatrix} = \begin{pmatrix} 18 \\ -8 \end{pmatrix}$

3 The diagram below shows the major highways connecting some European cities: Vienna (V), Munich (M), Frankfurt (F), Stuttgart (S), Zurich (Z), Milano (L) and Paris (P).

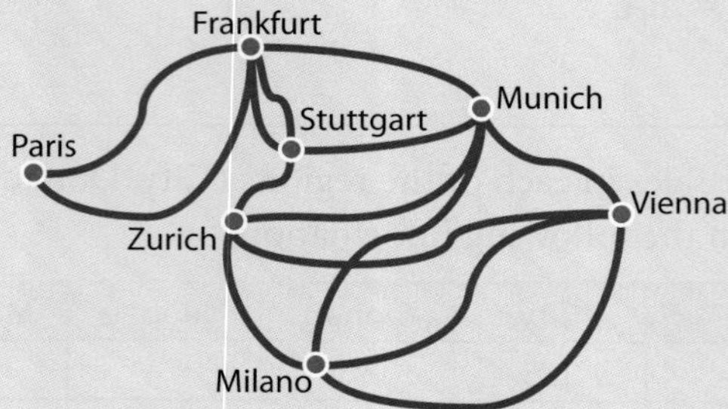

a) Write the number of *direct* routes between each pair of cities into a matrix as started below:

$$\begin{array}{c} \\ V \\ M \\ F \\ S \\ Z \\ L \\ P \end{array}\begin{array}{c} \begin{array}{ccccccc} V & M & F & S & Z & L & P \end{array} \\ \begin{bmatrix} 0 & 1 & 0 & 0 & 1 & 2 & 0 \\ & & & & & & \\ & & & & & & \\ & & & & & & \\ & & & & & & \\ & & & & & & \\ & & & & & & \end{bmatrix} \end{array}$$

b) Multiply the matrix from part a) by itself and interpret what it signifies.

4 Consider the following matrices:

$A = \begin{pmatrix} 2 & 5 & 1 \\ 0 & -3 & 2 \\ 7 & 0 & -1 \end{pmatrix}, B = \begin{pmatrix} m & -2 \\ 3m & -1 \\ 2 & 3 \end{pmatrix}, C = \begin{pmatrix} x-1 & 5 & y \\ 0 & -x & y+1 \\ 2x+y & x-3y & 2y-x \end{pmatrix}$

a) Find $A + C$. b) Find AB. c) Find BA.

d) Solve for x and y if $A = C$. e) Find $B + C$.

f) Solve for m if $3B + 2\begin{pmatrix} -1 & m^2 \\ -5 & 2 \\ 1 & -1 \end{pmatrix} = \begin{pmatrix} 7 & 12 \\ 17 & 1 \\ 2m+2 & 7 \end{pmatrix}$.

5 Find a, b and c so that the following equation is true:

$2 \cdot \begin{pmatrix} a-1 & b \\ c+2 & 3 \end{pmatrix} + \begin{pmatrix} 3 & -1 \\ 0 & 5 \end{pmatrix} = \begin{pmatrix} -5 & 5 \\ 8 & c+9 \end{pmatrix}$

6 Find x and y such that:

$\begin{pmatrix} 2 & -3 \\ -5 & 7 \end{pmatrix}\begin{pmatrix} x-11 & 1-x \\ -5 & x+2y \end{pmatrix} = \begin{pmatrix} 1 & 0 \\ 0 & 1 \end{pmatrix}$

7 Find m and n if

$\begin{pmatrix} m^2-1 & m+2 \\ 5 & -2 \end{pmatrix} = \begin{pmatrix} 3 & n+1 \\ 5 & n-5 \end{pmatrix}$.

8 There are two supermarkets in your area. Your shopping list consists of 2 kg of tomatoes, 500 g of meat and 3 litres of milk. Prices differ between the different shops, and it is difficult to switch between stores to make certain you are paying the least amount of money. A better strategy is to check and see where you pay less on *average*! The prices of the different items are given below. Which shop should you go to?

Product	Price in shop A	Price in shop B
Tomato	€1.66/kg	€1.58/kg
Meat	€2.55/100 g	€2.6/100 g
Milk	€0.90/litre	€0.95/litre

9 Consider the matrices

$$A = \begin{pmatrix} 2 & 0 \\ -5 & 1 \end{pmatrix},\ B = \begin{pmatrix} 3 & -1 \\ 1 & 4 \end{pmatrix} \text{ and } C = \begin{pmatrix} -3 & 5 \\ 2 & 7 \end{pmatrix}.$$

a) Find $A + (B + C)$ and $(A + B) + C$.

b) Make a conjecture about the addition of 2×2 matrices observed in a) above and prove it.

c) Find $A(BC)$ and $(AB)C$.

d) Make a conjecture about the multiplication of 2×2 matrices observed in c) above and prove it.

10 A company stores and sells air conditioning units, electric heaters and humidifiers. Row matrix A represents the number of each unit sold last year, and matrix B represents the profit margin for each unit. Find AB and describe what the product represents.

$$A = (235 \quad 562 \quad 117),\ B = \begin{pmatrix} €120 \\ €95 \\ €56 \end{pmatrix}$$

11 Find r and s such that the following equation is true: $rA + B = A$, where

$$A = \begin{pmatrix} 2 & 3 \\ 5 & 7 \end{pmatrix} \text{ and } B = \begin{pmatrix} -4 & -6 \\ s-8 & -14 \end{pmatrix}.$$

12 Let $A = \begin{pmatrix} 1 & 1 \\ 0 & 1 \end{pmatrix}$.

a) Find:
(i) A^2 (ii) A^3 (iii) A^4 (iv) A^n

Let $B = \begin{pmatrix} 3 & 3 \\ 0 & 3 \end{pmatrix}$.

b) Find:
(i) B^2 (ii) B^3 (iii) B^4 (iv) B^n

13 Solve for x and y such that $\boldsymbol{AB} = \boldsymbol{BA}$ if $A = \begin{pmatrix} 2 & 3 \\ 4 & 1 \end{pmatrix}$ and $B = \begin{pmatrix} x & 2 \\ y & 3 \end{pmatrix}$.

14 Solve for x and y such that $\boldsymbol{AB} = \boldsymbol{BA}$ if $A = \begin{pmatrix} 3 & x \\ -2 & 1 \end{pmatrix}$ and $B = \begin{pmatrix} 5 & 2 \\ y & 1 \end{pmatrix}$.

15 Solve for x such that $\boldsymbol{AB} = \boldsymbol{BA}$ if $A = \begin{pmatrix} 1 & 2 & 3 \\ x & 2 & -3 \\ 1 & 0 & 4 \end{pmatrix}$ and $B = \begin{pmatrix} -8 & x+3 & 12 \\ 23 & x-6 & -18 \\ 2 & -2 & 8 \end{pmatrix}$.

16 Solve for x and y such that $\boldsymbol{AB} = \boldsymbol{BA}$ if $A = \begin{pmatrix} y & 2 & y+2 \\ x & 2 & -3 \\ 1 & y-1 & 4 \end{pmatrix}$ and $B = \begin{pmatrix} -8 & x+3 & 12 \\ 23 & x-6 & -18 \\ 2 & -2 & 8 \end{pmatrix}$.

6.3 Applications to systems

There is a wide range of applications of matrices in solving systems of equations. Recall from your algebra that the equation of a straight line can take the form

$$ax + by = c$$

where a, b and c are constants and x and y are variables. We call this equation a **linear equation in two variables.** Similarly, the equation of a plane in three-dimensional space has the form

$$ax + by + cz = d$$

where a, b, c and d are constants. We call this equation a **linear equation in three variables.**

A **solution** of a linear equation in n variables (in this case two or three) is an ordered set of real numbers (x_0, y_0, z_0) so that the equation in question is satisfied when these values are substituted for the corresponding variables. For example, the equation

$$x + 2y = 4$$

is satisfied when $x = 2$ and $y = 1$. Some other solutions are $x = -4$ and $y = 4$, $x = 0$ and $y = 2$, and $x = -2$ and $y = 3$.

The set of all solutions of a linear equation is its **solution set**, and when this set is found, the equation is said to have been **solved.** To describe the entire solution set we often use a **parametric representation** as illustrated in the following examples.

Example 3

Solve the linear equation $x + 2y = 4$.

Solution

To find the solution set of an equation in two variables, we solve for one variable in terms of the other. For instance, if we solve for x, we obtain

$$x = 4 - 2y.$$

In this form, y is **free**, in the sense that it can take on any real value, while x is not free, since its value depends on that of y. To represent this solution set in general terms, we introduce a third variable, for example, t, called a **parameter**, and by letting $y = t$ we represent the solution set as

$$x = 4 - 2t,\ y = t,\ t \text{ is any real number.}$$

Particular solutions can then be obtained by assigning values to the parameter t. For instance, $t = 1$ yields the solution $x = 2$ and $y = 1$, and $t = 3$ yields the solution $x = -2$ and $y = 3$.

Note that the solution set of a linear equation can be represented parametrically in several ways. For instance, in this example, if we solve for y in terms of x, the parametric representation would take the following form:

$$x = m,\ y = 2 - \tfrac{1}{2}m,\ m \text{ is a real number.}$$

Also, by choosing $m = 2$, one particular solution would be $(x, y) = (2, 1)$, and by choosing $m = -2$, another particular solution would be $(-2, 3)$.

Example 4

Solve the linear equation $3x + 2y - z = 3$.

Solution

Choosing x and y as the *free* variables, we solve for z.

$$z = 3x + 2y - 3$$

Letting $x = p$ and $y = q$, we obtain the parametric representation:

$$x = p,\ y = q,\ z = 3x + 2y - 3,\ p \text{ and } q \text{ any real numbers.}$$

A particular solution $(x, y, z) = (1, 1, 2)$.

Parametric representation is very important when we study vectors and lines later on in the book.

Systems of linear equations – refresher

A system of k equations in n variables is a set of k linear equations in the same n variables. For example,

$$\begin{aligned} 2x + 3y &= 3 \\ x - y &= 4 \end{aligned}$$

is a system of two linear equations in two variables, while

$$\begin{aligned} x - 2y + 3z &= 9 \\ x - 3y &= 4 \end{aligned}$$

is a system with two equations and three variables, and

$$\begin{aligned} x - 2y + 3z &= 9 \\ x - 3y &= 4 \\ 2x - 5y + 5z &= 17 \end{aligned}$$

is a system with three equations and three variables.

A **solution** of a system of equations is an ordered set of numbers x_0, y_0, ... which satisfy every equation in the system. For example, $(3, -1)$ is a solution of

$$\begin{aligned} 2x + 3y &= 3 \\ x - y &= 4 \end{aligned}$$

Both equations in the system are satisfied when $x = 3$ and $y = -1$ are substituted into the equations. On the contrary, $(0, 1)$ is not a solution of the system, even though it satisfies the first equation, as it does not satisfy the second.

As you already know, there are several ways of finding solutions to systems. In this chapter, we will consider using matrix methods to solve systems of equations.

Taking our example above, notice how we can write the system of equations in matrix form:

$$\begin{cases} 2x + 3y = 3 \\ x - y = 4 \end{cases} \Rightarrow \begin{pmatrix} 2 & 3 \\ 1 & -1 \end{pmatrix}\begin{pmatrix} x \\ y \end{pmatrix} = \begin{pmatrix} 3 \\ 4 \end{pmatrix}$$

The representation of the system of equations in this way enables us to use matrix operations in solving systems. This matrix equation can be written as

$$\begin{pmatrix} 2 & 3 \\ 1 & -1 \end{pmatrix}\begin{pmatrix} x \\ y \end{pmatrix} = \begin{pmatrix} 3 \\ 4 \end{pmatrix} \Rightarrow AX = C$$

where A is the coefficient matrix, X is the variables matrix and C is the constants matrix. However, to solve this equation, the inverse of a matrix has to be defined as the solution of the system in the form

$$X = A^{-1}C$$

where A^{-1} is the inverse of the matrix A.

Matrix inverse (Optional)

To solve the equation $2x = 6$ for x, we need to multiply both sides of the equation by $\frac{1}{2}$:

$\frac{1}{2} \times 2x = \frac{1}{2} \times 6 \Rightarrow x = 3$. This is so, because $\frac{1}{2} \times 2 = 2 \times \frac{1}{2} = 1$.

$\frac{1}{2}$ is called the multiplicative inverse of 2. The inverse of a matrix is defined in a similar manner and plays a similar role in solving a matrix equation, such as $AX = C$.

Inverse of a matrix

A square matrix B is the inverse of a square matrix A if $AB = BA = I$, where I is the identity matrix.

The notation A^{-1} is used to denote the inverse of a matrix A. Thus, $B = A^{-1}$. Note that only square matrices can have multiplicative inverses.

Example – matrix inverse

$A = \begin{pmatrix} 7 & 5 \\ 4 & 3 \end{pmatrix}$ and $B = \begin{pmatrix} 3 & -5 \\ -4 & 7 \end{pmatrix}$ are multiplicative inverses since

$$AB = \begin{pmatrix} 7 & 5 \\ 4 & 3 \end{pmatrix}\begin{pmatrix} 3 & -5 \\ -4 & 7 \end{pmatrix} = \begin{pmatrix} 21 - 20 & -35 + 35 \\ 12 - 12 & -20 + 21 \end{pmatrix} = \begin{pmatrix} 1 & 0 \\ 0 & 1 \end{pmatrix}$$

$$BA = \begin{pmatrix} 3 & -5 \\ -4 & 7 \end{pmatrix}\begin{pmatrix} 7 & 5 \\ 4 & 3 \end{pmatrix} = \begin{pmatrix} 21 - 20 & 15 - 15 \\ -28 + 28 & -20 + 21 \end{pmatrix} = \begin{pmatrix} 1 & 0 \\ 0 & 1 \end{pmatrix}$$

Finding the inverse can also be achieved using a GDC.

```
[A]-1
          [[3   -5]
           [-4  7 ]]
[A]-1[A]
            [[1 0]
             [0 1]]
```

There are a few methods available for finding the inverse of a 2 × 2 matrix. We will be using the following method only, since the other methods are beyond the scope of this textbook.

Let $A = \begin{pmatrix} a & b \\ c & d \end{pmatrix}$ and assume $A^{-1} = \begin{pmatrix} e & f \\ g & h \end{pmatrix}$ and then solve the following matrix equation for e, f, g and h in terms of a, b, c and d.

$$\begin{pmatrix} a & b \\ c & d \end{pmatrix}\begin{pmatrix} e & f \\ g & h \end{pmatrix} = \begin{pmatrix} 1 & 0 \\ 0 & 1 \end{pmatrix} \Rightarrow \begin{pmatrix} ae + bg & af + bh \\ ce + dg & cf + dh \end{pmatrix} = \begin{pmatrix} 1 & 0 \\ 0 & 1 \end{pmatrix}$$

Now we can set up two systems to solve for the required variables, i.e.:

$$\begin{pmatrix} ae + bg & af + bh \\ ce + dg & cf + dh \end{pmatrix} = \begin{pmatrix} 1 & 0 \\ 0 & 1 \end{pmatrix}$$

$$\left.\begin{matrix} ae + bg = 1 \\ ce + dg = 0 \end{matrix}\right\} \Rightarrow \left.\begin{matrix} dae + \mathbf{dbg} = d \\ bce + \mathbf{bdg} = 0 \end{matrix}\right\} \Rightarrow e = \frac{d}{ad - bc},\ g = \frac{-c}{ad - bc}$$

$$\left.\begin{matrix} af + bh = 0 \\ cf + dh = 1 \end{matrix}\right\} \Rightarrow \left.\begin{matrix} daf + \mathbf{dbh} = 0 \\ bcf + \mathbf{bdh} = b \end{matrix}\right\} \Rightarrow f = \frac{-b}{ad - bc},\ h = \frac{a}{ad - bc}$$

Therefore, $A^{-1} = \begin{pmatrix} \frac{d}{ad - bc} & \frac{-b}{ad - bc} \\ \frac{-c}{ad - bc} & \frac{a}{ad - bc} \end{pmatrix}$ or $A^{-1} = \frac{1}{ad - bc}\begin{pmatrix} d & -b \\ -c & a \end{pmatrix}$.

Example 5

Find the inverse of $\begin{pmatrix} 4 & 7 \\ 3 & 5 \end{pmatrix}$.

Solution

Here $a = 4$, $b = 7$, $c = 3$ and $d = 5$, so $ad - bc = -1$. Thus,

$$A^{-1} = \frac{1}{ad - bc}\begin{pmatrix} d & -b \\ -c & a \end{pmatrix} = \frac{1}{-1}\begin{pmatrix} 5 & -7 \\ -3 & 4 \end{pmatrix} = \begin{pmatrix} -5 & 7 \\ 3 & -4 \end{pmatrix}.$$

```
[A]
                [[4 7]
                 [3 5]]
[A]-1
               [[-5 7]
                [3 -4]]
```

The determinant

The number $ad - bc$ is called the **determinant** of the 2×2 matrix $A = \begin{pmatrix} a & b \\ c & d \end{pmatrix}$.

The notation we will use for this number is **det *A***, so $\det A = ad - bc$.

The determinant plays an important role in determining whether a matrix has an inverse or not.

If the determinant is zero, i.e. $ad - bc = 0$, the matrix does not have an inverse. If a matrix has no inverse, it is called a **singular matrix**; if it is invertible, it is called **non-singular**.

Example 6

Solve the system of equations.

$$\begin{aligned} 2x + 3y &= 3 \\ x - y &= 4 \end{aligned}$$

Solution

In matrix form, the system can be written as

$$\begin{pmatrix} 2 & 3 \\ 1 & -1 \end{pmatrix}\begin{pmatrix} x \\ y \end{pmatrix} = \begin{pmatrix} 3 \\ 4 \end{pmatrix} \Rightarrow \begin{pmatrix} x \\ y \end{pmatrix} = \begin{pmatrix} 2 & 3 \\ 1 & -1 \end{pmatrix}^{-1}\begin{pmatrix} 3 \\ 4 \end{pmatrix}$$

$$\Rightarrow \begin{pmatrix} x \\ y \end{pmatrix} = \frac{1}{-5}\begin{pmatrix} -1 & -3 \\ -1 & 2 \end{pmatrix}\begin{pmatrix} 3 \\ 4 \end{pmatrix}$$

$$\Rightarrow \begin{pmatrix} x \\ y \end{pmatrix} = \frac{1}{-5}\begin{pmatrix} -15 \\ 5 \end{pmatrix} = \begin{pmatrix} 3 \\ -1 \end{pmatrix}$$

```
[A]-1 [C]
                 [[3 ]
                  [-1]]
```

Solving systems of equations in three variables follows similar procedures. However, finding the inverse of a 3 × 3 matrix will be delegated to the GDC at this level. As in the case of a 2 × 2 matrix, the existence of an inverse for a 3 × 3 matrix depends on the value of its determinant.

The determinant of a 3 × 3 matrix A can be achieved in one of two ways:

1. $A = \begin{pmatrix} a & b & c \\ d & e & f \\ g & h & i \end{pmatrix} \Rightarrow \det A = a(ei - fh) - b(di - fg) + c(dh - eg)$

   For example, if

   $$A = \begin{pmatrix} 5 & 1 & -4 \\ 2 & -3 & -5 \\ 7 & 2 & -6 \end{pmatrix} \Rightarrow \det A = 5(18 + 10) - 1(-12 + 35) - 4(4 + 21) = 17$$

```
[A]
        [[5 1  -4]
         [2 -3 -5]
         [7 2  -6]]
det([A])
                17
```

2. A practical method is to use a 'special' set up as follows:

   $$\det A = \begin{vmatrix} a & b & c & a & b \\ d & e & f & d & e \\ g & h & i & g & h \end{vmatrix} = aei + bfg + cdh - gec - hfa - idb$$

   This is done by 'copying' the first two columns and adding them to the end of the matrix, multiplying down the main diagonals and adding the products, and then multiplying up the second diagonals and subtracting them from the previous product, as shown. In the example above:

   $$\begin{vmatrix} 5 & 1 & -4 & 5 & 1 \\ 2 & -3 & -5 & 2 & -3 \\ 7 & 2 & -6 & 7 & 2 \end{vmatrix}$$

$= 5(-3)(-6) + 1(-5)(7) + (-4) \cdot 2 \cdot 2 - 7(-3)(-4) - 2(-5) \cdot 5 - (-6) \cdot 2 \cdot 1$

$= 90 - 35 - 16 - 84 + 50 + 12$

$= 152 - 135$

$= 17$

   In fact, this arrangement is simply a reordering of the calculations involved in the previous method.

Example 7

Solve the system of equations.

$$5x + y - 4z = 5$$
$$2x - 3y - 5z = 2$$
$$7x + 2y - 6z = 5$$

Solution

We write this system in matrix form:

$$\begin{pmatrix} 5 & 1 & -4 \\ 2 & -3 & -5 \\ 7 & 2 & -6 \end{pmatrix}\begin{pmatrix} x \\ y \\ z \end{pmatrix} = \begin{pmatrix} 5 \\ 2 \\ 5 \end{pmatrix}$$

Since det $A \neq 0$, we can find the solution in the same way we did for the 2×2 matrix, i.e.

$$\begin{pmatrix} 5 & 1 & -4 \\ 2 & -3 & -5 \\ 7 & 2 & -6 \end{pmatrix}\begin{pmatrix} x \\ y \\ z \end{pmatrix} = \begin{pmatrix} 5 \\ 2 \\ 5 \end{pmatrix} \Rightarrow \begin{pmatrix} x \\ y \\ z \end{pmatrix} = \begin{pmatrix} 5 & 1 & -4 \\ 2 & -3 & -5 \\ 7 & 2 & -6 \end{pmatrix}^{-1}\begin{pmatrix} 5 \\ 2 \\ 5 \end{pmatrix}$$

Using a GDC:

```
[A]-1 [C]
              [[3 ]
               [-2]
               [2 ]]
```

To check your work, you can store the answer matrix as D and then substitute the values into the system:

$$\begin{pmatrix} 5 & 1 & -4 \\ 2 & -3 & -5 \\ 7 & 2 & -6 \end{pmatrix}\begin{pmatrix} 3 \\ -2 \\ 2 \end{pmatrix} = \begin{pmatrix} 15 - 2 - 8 \\ 6 + 6 - 10 \\ 21 - 4 - 12 \end{pmatrix} = \begin{pmatrix} 5 \\ 2 \\ 5 \end{pmatrix}, \text{ or}$$

```
[A] [D]
              [[5]
               [2]
               [5]]
```

Area of a triangle

An interesting application of determinants that you may find helpful is finding the area of a triangle whose vertices are given as points in a coordinate plane. The following result will become obvious as you study Chapter 14.

Area of a triangle

The area of a triangle with vertices (x_1, y_1), (x_2, y_2), and (x_3, y_3) is equal to $\left|\frac{1}{2}|A|\right|$ where $A = \begin{pmatrix} x_1 & y_1 & 1 \\ x_2 & y_2 & 1 \\ x_3 & y_3 & 1 \end{pmatrix}$.

Example 8

Find the area of triangle ABC whose vertices are $A(1, 3)$, $B(5, -1)$ and $C(-2, 5)$.

Solution

We let $(x_1, y_1) = (1, 3)$, $(x_2, y_2) = (5, -1)$, and $(x_3, y_3) = (-2, 5)$. To find the area, we evaluate the determinant:

$$\begin{vmatrix} x_1 & y_1 & 1 \\ x_2 & y_2 & 1 \\ x_3 & y_3 & 1 \end{vmatrix} = \begin{vmatrix} 1 & 3 & 1 \\ 5 & -1 & 1 \\ -2 & 5 & 1 \end{vmatrix} = -4.$$

Using this value, we can conclude that the area of the triangle is given by:

$$\text{Area} = \left| \frac{1}{2} \begin{vmatrix} 1 & 3 & 1 \\ 5 & -1 & 1 \\ -2 & 5 & 1 \end{vmatrix} \right| = \left| \frac{1}{2} \cdot -4 \right| = 2$$

• **Hint:** Try using determinants to find the area of triangle *ABC* with *A*(2, 3), *B*(12, 3), and *C*(12, 9). Confirm your answer by using the usual area formula of a triangle, $\frac{1}{2}$(*base* × *height*).

Lines in planes

In our previous discussion, what if the three points are collinear? The answer is very simple. The triangle would collapse into a line segment and the area becomes zero. This fact helps us develop two techniques that are very helpful in dealing with questions of collinearity and equations of lines.

For example, take the points $A(-2, -3)$, $B(1, 3)$ and $C(3, 7)$. Find the area of 'triangle' *ABC*.

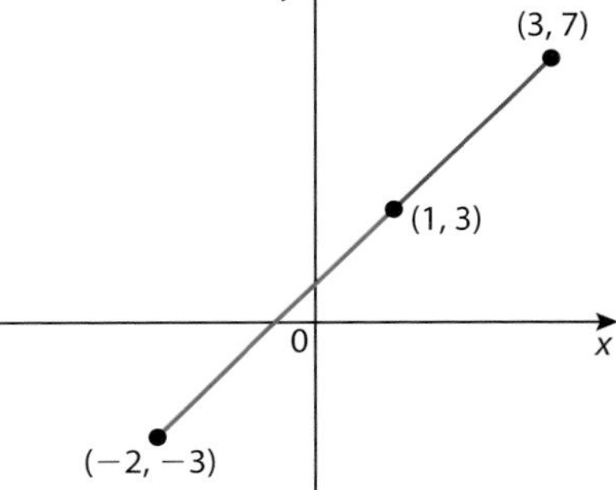

$$\text{Area} = \left| \frac{1}{2} \begin{vmatrix} -2 & -3 & 1 \\ 1 & 3 & 1 \\ 3 & 7 & 1 \end{vmatrix} \right| = \left| \frac{1}{2} \cdot -0 \right| = 0$$

This result can be stated in general as given below:

Test for collinearity

The three points (x_1, y_1), (x_2, y_2), and (x_3, y_3) are collinear if and only if

$$\begin{vmatrix} x_1 & y_1 & 1 \\ x_2 & y_2 & 1 \\ x_3 & y_3 & 1 \end{vmatrix} = 0.$$

Example 9

Determine whether the points $(-2, 3)$, $(2, 5)$ and $(5, 7)$ lie on the same line.

Solution

By setting up the matrix as suggested by the rule above, we have

$$\begin{vmatrix} -2 & 3 & 1 \\ 2 & 5 & 1 \\ 5 & 7 & 1 \end{vmatrix} = 2 \neq 0.$$

Because the value of the determinant is not equal to zero, the points cannot lie on a line.

Two-point equation of a line

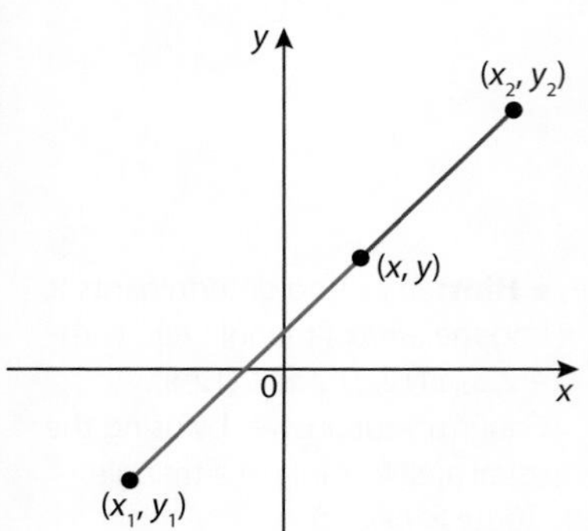

The test for collinearity leads us to the following result, which enables us to find the equation of a line containing two points. Consider two points (x_1, y_1), (x_2, y_2) which lie on a given line. To find the equation of the line through these two points, we introduce a general point (x, y) on the line. These three points (x_1, y_1), (x_2, y_2) and (x, y) are collinear, and hence they satisfy the determinant equation

$$\begin{vmatrix} x & y & 1 \\ x_1 & y_1 & 1 \\ x_2 & y_2 & 1 \end{vmatrix} = 0$$

which gives us the equation of the line in the form:

$$(y_1 - y_2)x + (x_2 - x_1)y + (x_1y_2 - y_1x_2) = 0$$

which in turn is of the form: $Ax + By + C = 0$.

Example 10

Find the equation of the line through $(-2, 3)$ and $(3, 7)$.

Solution

Applying the determinant formula for the equation of a line produces

$$\begin{vmatrix} x & y & 1 \\ -2 & 3 & 1 \\ 3 & 7 & 1 \end{vmatrix} = (3 - 7)x + (3 + 2)y + (-14 - 9) = 0$$

$$-4x + 5y - 23 = 0$$

Exercise 6.3

1 Consider the matrix M which satisfies the matrix equation

$$\begin{pmatrix} 3 & 7 \\ -4 & -9 \end{pmatrix} M = \begin{pmatrix} 2 & 1 \\ 3 & 5 \end{pmatrix}.$$

a) Write out the inverse of matrix $\begin{pmatrix} 3 & 7 \\ -4 & -9 \end{pmatrix}$.

b) Hence, write M as a product of two matrices.

c) Evaluate M.

d) Now consider the equation containing the matrix N:

$$N\begin{pmatrix} 3 & 7 \\ -4 & -9 \end{pmatrix} = \begin{pmatrix} 2 & 1 \\ 3 & 5 \end{pmatrix}$$

(i) Write N as a product of two matrices.

(ii) Evaluate N.

e) Write a short paragraph describing your work on this problem.

2 Find the matrix E in the following equation:

$$\begin{pmatrix} 1 & 3 \\ 3 & 4 \end{pmatrix} = \begin{pmatrix} 1 & 0 \\ 3 & 1 \end{pmatrix} E \begin{pmatrix} 1 & 0 \\ 0 & -5 \end{pmatrix}$$

3 a) Prove that the matrix $A = \begin{pmatrix} 2 & -3 & 1 \\ 1 & 1 & -3 \\ 3 & -2 & -3 \end{pmatrix}$ should have an inverse.

b) Write out A^{-1}.

c) Hence, solve the system of equations:

$$\begin{cases} 2x - 3y + z = 4.2 \\ x + y - 3z = -1.1 \\ 3x - 2y - 3z = 2.9 \end{cases}$$

4 Find the inverse for each matrix.

a) $A = \begin{pmatrix} \frac{\sqrt{3}}{2} & -\frac{1}{2} \\ \frac{1}{2} & \frac{\sqrt{3}}{2} \end{pmatrix}$ b) $B = \begin{pmatrix} a & 1 \\ a+2 & \frac{3}{a}+1 \end{pmatrix}$

5 For what values of x is the following matrix singular?

$A = \begin{pmatrix} x+1 & 3 \\ 3x-1 & x+3 \end{pmatrix}$

6 Find n such that $\begin{pmatrix} 2 & -1 & 4 \\ 2n & 2 & 0 \\ 2 & 1 & 4n \end{pmatrix}$ is the inverse of $\begin{pmatrix} -2 & -3 & 4 \\ 1 & 2 & -2 \\ 3n & 2 & -5n \end{pmatrix}$.

7 Consider the two matrices $A = \begin{pmatrix} 4 & 2 \\ 0 & -3 \end{pmatrix}$ and $B = \begin{pmatrix} 2 & 1 \\ 3 & 5 \end{pmatrix}$.

a) Find X such that $XA = B$.

b) Find Y such that $AY = B$.

c) Is $X = Y$? Explain.

8 Consider the two matrices

$P = \begin{pmatrix} 2 & 0 & -1 \\ 3 & 5 & 4 \\ 1 & 0 & -1 \end{pmatrix}$ and $Q = \begin{pmatrix} 3 & -1 & 1 \\ 4 & 0 & 0 \\ 1 & 2 & -1 \end{pmatrix}$.

a) Find PQ and QP.

b) Find P^{-1}, Q^{-1}, $P^{-1}Q^{-1}$, $Q^{-1}P^{-1}$ $(PQ)^{-1}$, and $(QP)^{-1}$.

c) Write a few sentences about your observations in parts a) and b).

9 Consider the matrices A and B.

$A = \begin{pmatrix} 3 & -2 & 1 \\ -4 & 1 & -3 \\ 1 & -5 & 1 \end{pmatrix}$; $B = \begin{pmatrix} -29 \\ 37 \\ -24 \end{pmatrix}$

a) Find the matrix C if $AC = B$.

b) Solve the system of equations:

$$\begin{cases} 3x - 2y + z = -29 \\ 4x - y + 3z = -37 \\ -x + 5y - z = 24 \end{cases}$$

10 Solve the matrix equation

$$\begin{pmatrix} 2 & 2+x \\ 5 & 4+x \end{pmatrix}\begin{pmatrix} 3 & x \\ x-4 & 2 \end{pmatrix} = \begin{pmatrix} 3 & x \\ x-4 & 2 \end{pmatrix}\begin{pmatrix} 2 & 2+x \\ 5 & 4+x \end{pmatrix}$$

11 Consider the matrices A and B below. Find x and y such that $AB = BA$.

$$A = \begin{pmatrix} 2 & 1 \\ 5 & 3 \end{pmatrix};\ B = \begin{pmatrix} 2 - x & 1 \\ 5x & y \end{pmatrix}$$

12 Consider the matrices A and B below. Find x and y such that $AB = BA$.

$$A = \begin{pmatrix} 3 & 1 \\ -5 & 2 \end{pmatrix};\ B = \begin{pmatrix} 1 - x & x \\ 5x & y \end{pmatrix}$$

13 Consider the matrices A and B below. Find x and y such that $AB = BA$.

$$A = \begin{pmatrix} 3 + x & 1 \\ -5 & 2 \end{pmatrix};\ B = \begin{pmatrix} y - x & x \\ 5x - y + 1 & y + x \end{pmatrix}$$

14 In each case, you are given two points in the plane. Use matrix methods to find an equation of a line that contains the given points.

a) $A(-5, -6)$, $B(3, 11)$

b) $A(5, -2)$, $B(3, -2)$

c) $A(-5, 3)$, $B(-5, 8)$

15 Find the area of the parallelogram with the given points as three of its vertices:

a) $A(-5, -6)$, $B(3, 11)$, $C(8, 1)$

b) $A(3, -5)$, $B(3, 11)$, $C(8, 11)$

c) $A(4, -6)$, $B(-3, 9)$, $C(7, 7)$

16 Find x such that the area of triangle ABC is 10 square units.

a) $A(x, -6)$, $B(3, 11)$, $C(8, 3)$

b) $A(-5, x)$, $B(3, x + 2)$, $C(x^2 + 2x - 3, 1)$

17 Find the value of k such that the points P, Q, and R are collinear.

a) $P(2, -5)$, $Q(4, k)$, $R(5, -2)$

b) $P(-6, 2)$, $Q(-5, k)$, $R(-3, 5)$

18 Exploration:

Consider the matrix $A = \begin{pmatrix} 2 & 7 \\ 5 & 5 \end{pmatrix}$. Define $f(x) = \det(xI - A)$ where x is any real number and I is the identity matrix.

a) Find $\det(A)$.

b) Expand $f(x)$ and compare the constant term to your answer in a).

c) How is the coefficient of x in the expansion of $f(x)$ related to A?

d) Find $f(A)$ and simplify it.

e) Now repeat parts a)–d) with matrix $B = \begin{pmatrix} a & b \\ c & d \end{pmatrix}$.

• **Hint:** $f(x)$ is called the characteristic polynomial of A.

19 Exploration:

Consider the matrix $A = \begin{pmatrix} 2 & 7 & 1 \\ -1 & 3 & 2 \\ 5 & 5 & -4 \end{pmatrix}$. Define $f(x) = \det(xI - A)$ where x is any real number and I is the identity matrix.

a) Find $\det(A)$.

b) Expand $f(x)$ and compare the constant term to your answer in a).

c) How is the coefficient of x^2 in the expansion of $f(x)$ related to A?

d) Find $f(A)$ and simplify it.

e) Now repeat parts a)–d) with matrix $B = \begin{pmatrix} a & b & c \\ d & e & f \\ g & h & i \end{pmatrix}$.

6.4 Further properties and applications

Pages 267–269 are optional material. You can choose not to work on them. However, starting with Gauss-Jordan elimination (on page 269) the material is required in examinations.

In question 8 of Exercise 6.3, you were asked to make some observations concerning the answers to parts a) and b). The purpose is for you to discover some properties of inverse matrices.

Let us take the following matrices, for example:

Consider the two matrices A and B, where $A = \begin{pmatrix} -1 & 1 & 2 \\ 3 & 2 & 1 \\ 1 & -2 & -1 \end{pmatrix}$,

$B = \begin{pmatrix} 1 & 2 & 3 \\ 1 & 3 & 3 \\ 2 & 4 & 3 \end{pmatrix}$.

Find A^{-1}, B^{-1}, AB, BA, $(AB)^{-1}$, $A^{-1}B^{-1}$, $B^{-1}A^{-1}$, and $(BA)^{-1}$.

As shown below,

$$A^{-1} = \begin{pmatrix} 0 & \frac{1}{4} & \frac{1}{4} \\ -\frac{1}{3} & \frac{1}{12} & -\frac{7}{12} \\ \frac{2}{3} & \frac{1}{12} & \frac{5}{12} \end{pmatrix}, B^{-1} = \begin{pmatrix} 1 & -2 & 1 \\ -1 & 1 & 0 \\ \frac{2}{3} & 0 & -\frac{1}{3} \end{pmatrix}$$

```
[[0.0  .3  .3 ]
 [-.3  .1  -.6]
 [.7   .1  .4 ]]
Ans▸Frac
[[0.0   1/4  1/4...
 [-1/3  1/12 -7/...
 [2/3   1/12 5/1...
```

```
[[1.0  -2.0 1.0...
 [-1.0 1.0  0.0...
 [.7   0.0  -.3...
Ans▸Frac
[[1.0  -2.0 1.0...
 [-1.0 1.0  0.0...
 [2/3  0.0  -1/...
```

Also,

$$AB = \begin{pmatrix} 4 & 9 & 6 \\ 7 & 16 & 18 \\ -3 & -8 & -6 \end{pmatrix}, BA = \begin{pmatrix} 8 & -1 & 1 \\ 11 & 1 & 2 \\ 13 & 4 & 5 \end{pmatrix}$$

```
[[4.0   9.0  6.0...
 [7.0   16.0 18....
 [-3.0  -8.0 -6....
[B][A]
[[8.0   -1.0 1.0...
 [11.0  1.0  2.0...
 [13.0  4.0  5.0...
```

```
([A][B])-1
[[1.3  .2   1.8]
 [-.3  -.2  -.8]
 [-.2  .1   .0 ]]
```

```
Ans▸Frac
[[4/3   1/6   11/...
 [-1/3  -1/6  -5/...
 [-2/9  5/36  1/3...
```

$$(AB)^{-1} = \begin{pmatrix} \frac{4}{3} & \frac{1}{6} & \frac{11}{6} \\ -\frac{1}{3} & -\frac{1}{6} & -\frac{5}{6} \\ -\frac{2}{9} & \frac{5}{36} & \frac{1}{36} \end{pmatrix}, \text{ also}$$

```
[A]-1[B]-1
[[-.1  .3    -.1]
 [-.8  .7    -.1]
 [.9   -1.3  .5 ]]
```

```
[[-.1  .3    -.1]
 [-.8  .7    -.1]
 [.9   -1.3  .5 ]]
Ans▸Frac
[[-1/12   1/4   -...
 [-29/36  3/4   -...
 [31/36   -5/4  1...
```

$$A^{-1}B^{-1} = \begin{pmatrix} -\frac{1}{12} & \frac{1}{4} & -\frac{1}{12} \\ -\frac{29}{36} & \frac{3}{4} & -\frac{5}{36} \\ \frac{31}{36} & -\frac{5}{4} & \frac{19}{36} \end{pmatrix}.$$

This last result shows that $(AB)^{-1} \neq A^{-1}B^{-1}$. However, as you notice below $(AB)^{-1} = B^{-1}A^{-1}$:

$$B^{-1}A^{-1} = \begin{pmatrix} \frac{4}{3} & \frac{1}{6} & \frac{11}{6} \\ -\frac{1}{3} & -\frac{1}{6} & -\frac{5}{6} \\ -\frac{2}{9} & \frac{5}{36} & \frac{1}{36} \end{pmatrix}.$$

```
[B]-1[A]-1
 [[1.3  .2   1.8]
   [-.3 -.2 -.8]
   [-.2 .1   .0 ]]
```

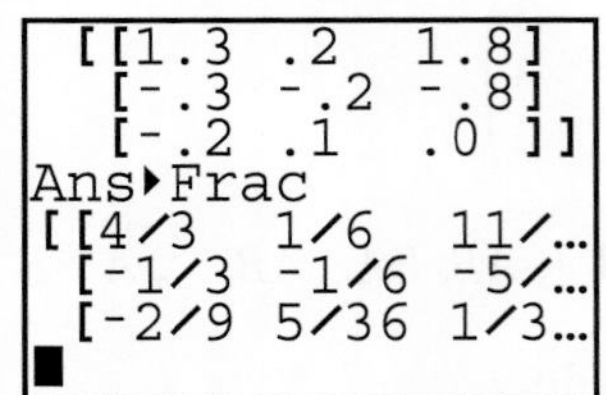

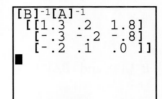

Finally, we also have

$$(BA)^{-1} = \begin{pmatrix} -\frac{1}{12} & \frac{1}{4} & -\frac{1}{12} \\ -\frac{29}{36} & \frac{3}{4} & -\frac{5}{36} \\ \frac{31}{36} & -\frac{5}{4} & \frac{19}{36} \end{pmatrix}.$$

This in turn is nothing but $A^{-1}B^{-1}$.

So, in general we have the following result:

If A and B are non-singular matrices of order n, then AB is also non-singular and $(AB)^{-1} = B^{-1}A^{-1}$.

The proof of this theorem is straightforward:

To show that $B^{-1}A^{-1}$ is the inverse of AB, we need only show that it conforms to the definition of an inverse matrix. That is,

$$(AB)(B^{-1}A^{-1}) = (B^{-1}A^{-1})\,(AB) = I.$$

Now, $(AB)(B^{-1}A^{-1}) = A(B\,B^{-1})A^{-1} = A(I)A^{-1} = AA^{-1} = I$.

Similarly, $(B^{-1}A^{-1})\,(AB) = B^{-1}(A^{-1}A)B = B^{-1}(I)B = B^{-1}B = I$.

Hence, AB is non-singular (invertible) and its inverse is $B^{-1}A^{-1}$.

The following properties will be listed without proof:

$(A^{-1})^{-1} = A$

$(cA)^{-1} = \frac{1}{c}A^{-1}; c \neq 0$

$\det(AB) = \det A \cdot \det B$

This last result is helpful in proving the following property.

If A is non-singular, then $\det A^{-1} = \frac{1}{\det A}$.

Proof: Since $AA^{-1} = I$, then

$$\det(AA^{-1}) = \det I \Rightarrow \det A \cdot \det A^{-1} = 1 \Rightarrow \det A^{-1} = \frac{1}{\det A}.$$

In the previous section, we solved a system of equations using inverse matrices. However, that method works as long as the system is consistent with a unique solution. In many cases, the solution either has an infinite number of solutions or is inconsistent. There is another method of solution which we want to introduce you to.

Some terminology

As we have seen before, it is usual to represent a system of equations using matrix notation. In the previous section you learned how to solve a system of equations by writing the system in matrix form. For example, to solve the system

$$\begin{cases} 2x + 3y - 4z = 8 \\ \quad\quad 2y + 4z = -3 \\ x \quad\quad - 2z = 4 \end{cases}$$

we wrote

$$\begin{pmatrix} 2 & 3 & -4 \\ 0 & 2 & 4 \\ 1 & 0 & -2 \end{pmatrix}\begin{pmatrix} x \\ y \\ z \end{pmatrix} = \begin{pmatrix} 8 \\ -3 \\ 4 \end{pmatrix}$$

The first matrix is called the **coefficient matrix** (or **matrix of coefficients**) and the matrix on the right is called the **constants matrix** or the **answers matrix**. If the system has a unique solution then it can be solved. As you see, the method is limited and it has a strict constraint. Thanks to a slightly different arrangement, we can use matrices to arrive at our solution regardless of whether it is unique, has an infinite number of solutions, or simply no solution. To that end we need to write the system as follows:

$$\left(\begin{array}{ccc|c} 2 & 3 & -4 & 8 \\ 0 & 2 & 4 & -3 \\ 1 & 0 & -2 & 4 \end{array}\right)$$

This is called the **augmented** matrix of the system. It is customary to put a bar between the coefficients and the answers. However, this bar is not necessary and we will not be using it in this book. Just remember that the last column is the answers column!

Gauss-Jordan elimination

The idea behind this method is very simple. We successively apply certain simple operations to the system of equations reducing them into a special form that is easy to solve. The operations are called **elementary row**

operations and they can be applied to the system without changing the solution to the system. That is, the solution to the reduced system (**reduced row echelon form**) is the same as that for the original system. We can apply the operations either to the system itself or to its augmented matrix. Since the latter is easier to work with, we recommend that you first write the augmented matrix, reduce it, and then write the equivalent system to read the solution from.

There are three types of **elementary row operations.**

1. **Multiply any row by non-zero real number.**
2. **Interchange any two rows.**
3. **Add a multiple of one row to another row.**

Note: The order with which we apply the operations is not unique!

We will demonstrate the method with an example.

Consider the following system and its associated matrix:

$$\begin{cases} 2x + y - z = 2 \\ x + 3y + 2z = 1 \\ 2x + 4y + 6z = 6 \end{cases} \Leftrightarrow \left(\begin{array}{ccc|c} 2 & 1 & -1 & 2 \\ 1 & 3 & 2 & 1 \\ 2 & 4 & 6 & 6 \end{array}\right)$$

Switch row 1 and row 2 – type 2 operation:

$$\begin{cases} x + 3y + 2z = 1 \\ 2x + y - z = 2 \\ 2x + 4y + 6z = 6 \end{cases} \Leftrightarrow \left(\begin{array}{ccc|c} 1 & 3 & 2 & 1 \\ 2 & 1 & -1 & 2 \\ 2 & 4 & 6 & 6 \end{array}\right)$$

Multiply row 3 by $\frac{1}{2}$ – type 1 operation:

$$\begin{cases} x + 3y + 2z = 1 \\ 2x + y - z = 2 \\ x + 2y + 3z = 3 \end{cases} \Leftrightarrow \left(\begin{array}{ccc|c} 1 & 3 & 2 & 1 \\ 2 & 1 & -1 & 2 \\ 1 & 2 & 3 & 3 \end{array}\right)$$

Multiply row 1 by -2 and add it to row 2, and multiply row 1 by -1 and add it to row 3 – type 3 operations:

$$\begin{cases} x + 3y + 2z = 1 \\ -5y - 5z = 0 \\ -y + z = 2 \end{cases} \Leftrightarrow \left(\begin{array}{ccc|c} 1 & 3 & 2 & 1 \\ 0 & -5 & -5 & 0 \\ 0 & -1 & 1 & 2 \end{array}\right)$$

Notice here that row 1 did not change and rows 2 and three were replaced with the result of the elementary operation.

Multiply row 2 by $-\frac{1}{5}$:

$$\begin{cases} x + 3y + 2z = 1 \\ y + z = 0 \\ -y + z = 2 \end{cases} \Leftrightarrow \left(\begin{array}{ccc|c} 1 & 3 & 2 & 1 \\ 0 & 1 & 1 & 0 \\ 0 & -1 & 1 & 2 \end{array}\right)$$

Now, add row 2 to row 3, and multiply row 2 by -3 and add it to row 1:

$$\begin{cases} x - z = 1 \\ y + z = 0 \\ 2z = 2 \end{cases} \Leftrightarrow \left(\begin{array}{ccc|c} 1 & 0 & -1 & 1 \\ 0 & 1 & 1 & 0 \\ 0 & 0 & 2 & 2 \end{array}\right)$$

Now multiply row 3 by $\frac{1}{2}$:

$$\begin{cases} x \quad\quad - z = 1 \\ \quad\quad y + z = 0 \\ \quad\quad\quad\quad z = 1 \end{cases} \Leftrightarrow \left(\begin{array}{ccc|c} 1 & 0 & -1 & 1 \\ 0 & 1 & 1 & 0 \\ 0 & 0 & 1 & 1 \end{array}\right)$$

Lastly, add row 3 to row 1, and multiply row 3 by -1 and add it to row 2:

$$\begin{cases} x \quad\quad\quad = 2 \\ \quad\quad y \quad = -1 \\ \quad\quad\quad z = 1 \end{cases} \Leftrightarrow \left(\begin{array}{ccc|c} 1 & 0 & 0 & 2 \\ 0 & 1 & 0 & -1 \\ 0 & 0 & 1 & 1 \end{array}\right)$$

As you notice, from this last system it is easy to read the solution of $(2, -1, 1)$. You can verify that this solution is also the solution to the original system.

The simplified matrix is in its reduced row echelon form (to be defined later).

Of course, when we do the work, we do not have to show the processes in parallel. We just perform the operation on the matrix and then translate it into the equation form.

Note: This whole operation can easily be performed using a GDC.

```
[A]
     [[2 1 -1 2]
      [1 3 2  1]
      [2 4 6  6]]
```

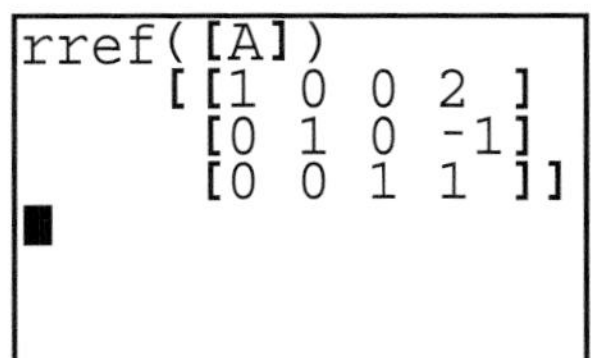

Example 11

Solve the following system:

$$\begin{cases} x + y + 2z = 1 \\ x + \quad\quad z = 2 \\ \quad y + \quad z = 0 \end{cases}$$

Solution

The augmented matrix is:

$$\begin{cases} x + y + 2z = 1 \\ x + y \quad\quad z = 2 \\ \quad y + z = 0 \end{cases} \Leftrightarrow \begin{pmatrix} 1 & 1 & 2 & 1 \\ 1 & 0 & 1 & 2 \\ 0 & 1 & 1 & 0 \end{pmatrix}$$

Multiply row 1 with -1 and add to row 2:

$$\begin{cases} x + y + 2z = 1 \\ \quad -y - z = 1 \\ \quad\quad y + z = 0 \end{cases} \Leftrightarrow \begin{pmatrix} 1 & 1 & 2 & 1 \\ 0 & -1 & -1 & 1 \\ 0 & 1 & 1 & 0 \end{pmatrix}$$

Add row 2 to row 1 and row 2 to row 3:

$$\begin{cases} x \quad\quad + z = 2 \\ \quad -y - z = 1 \\ \quad\quad\quad 0 = 1 \end{cases} \Leftrightarrow \begin{pmatrix} 1 & 0 & 1 & 2 \\ 0 & -1 & -1 & 1 \\ 0 & 0 & 0 & 1 \end{pmatrix}$$

At this stage, work can stop because if you write the last row as an equation, it reads

$$0x + 0y + 0z = 1.$$

This statement cannot be true for any value, and hence the system is inconsistent.

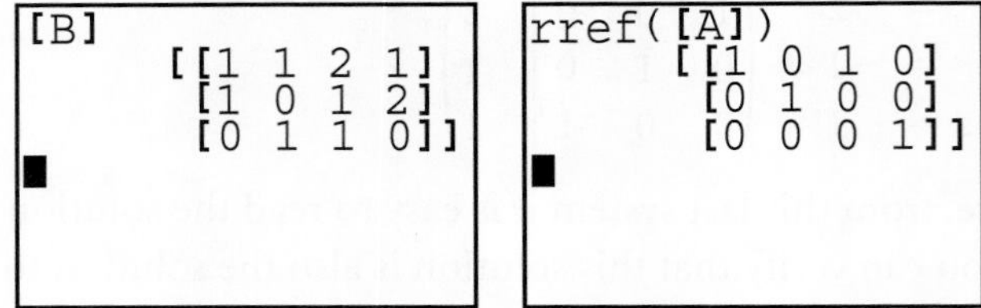

Example 12

Solve the following system:

$$\begin{cases} 2x + y - z = 4 \\ x + 3y + 7z = 7 \\ 2x + 4y + 8z = 10 \end{cases}$$

Solution

The augmented matrix is:

$$\begin{cases} 2x + y - z = 4 \\ x + 3y + 7z = 7 \\ 2x + 4y + 8z = 10 \end{cases} \Leftrightarrow \begin{pmatrix} 2 & 1 & -1 & 4 \\ 1 & 3 & 7 & 7 \\ 2 & 4 & 8 & 10 \end{pmatrix}$$

$$\begin{cases} x + 3y + 7z = 7 \\ 2x + y - z = 4 \\ 2x + 4y + 8z = 10 \end{cases} \Leftrightarrow \begin{pmatrix} 1 & 3 & 7 & 7 \\ 2 & 1 & -1 & 4 \\ 2 & 4 & 8 & 10 \end{pmatrix} \quad R_1 \Leftrightarrow R_2$$

$$\begin{cases} x + 3y + 7z = 7 \\ -5y - 15z = -10 \\ 3y + 9z = 6 \end{cases} \Leftrightarrow \begin{pmatrix} 1 & 3 & 7 & 7 \\ 0 & -5 & -15 & -10 \\ 0 & 3 & 9 & 6 \end{pmatrix} \quad \begin{cases} -R_2 + R_3 \\ -2R_1 + R_2 \end{cases}$$

$$\begin{cases} x + 3y + 7z = 7 \\ y + 3z = 2 \\ y + 3z = 2 \end{cases} \Leftrightarrow \begin{pmatrix} 1 & 3 & 7 & 7 \\ 0 & 1 & 3 & 2 \\ 0 & 1 & 3 & 2 \end{pmatrix} \quad \begin{cases} -\frac{1}{5}R_2 \\ \frac{1}{3}R_3 \end{cases}$$

$$\begin{cases} x \qquad - 2z = 1 \\ y + 3z = 2 \\ 0 = 0 \end{cases} \Leftrightarrow \begin{pmatrix} 1 & 0 & -2 & 1 \\ 0 & 1 & 3 & 2 \\ 0 & 0 & 0 & 0 \end{pmatrix} \quad \begin{cases} -R_2 + R_3 \\ -3R_2 + R_1 \end{cases}$$

Since the last row is all zeros, there is not much that we can do. The conclusion is that this last row is true for any choice of values for the variables. Now we are left with a system of two equations and three variables.

$$\begin{cases} x - 2z = 4 \\ y + 3z = 2 \end{cases}$$

We need to solve for two of the variables in terms of the third. A wise choice here would be to solve for x and y in terms of z. That is,

$$x = 1 + 2z, y = 2 - 3z.$$

This means that for every choice of a value for z, we have a corresponding solution for the system. For example, if $z = 0$, then the solution would be $(1, 2, 0)$, for $z = 2$, the solution is $(5, -4, 2)$, and so on. This means that we have an infinite number of solutions. It is customary to present the solution in terms of a parameter, t for example. We let $z = t$, and our general solution would then be

$$(1 + 2t, 2 - 3t, t).$$

So, what is a **reduced row echelon form** (rref)?

We are confident that by now, you have a feel for what it is:

A matrix is in rref if it satisfies the following properties:

1. If there are any rows consisting entirely of zeros, they appear at the bottom of the matrix.
2. In any non-zero row, the first non-zero entry is 1. This entry is called the **pivot** of the row.
3. For any consecutive rows, the pivot of the lower row must be to the right of the pivot of the preceding row.
4. Any column that contains a pivot, has zeros everywhere else.

See the demonstration below; A is in rref while B is not.

$$A = \begin{pmatrix} \boxed{1} & 0 & 3 & 0 & 5 & 8 \\ 0 \rightarrow & \boxed{1} & 4 & 0 & 4 & 2 \\ 0 & 0 \rightarrow & 0 \rightarrow & \boxed{1} & 5 & 2 \\ 0 & 0 & 0 & 0 & 0 & 0 \end{pmatrix} \quad B = \begin{pmatrix} 1 & 0 & 0 & 2 & 3 & 4 & 5 \\ 0 & 0 & 0 & 1 & 3 & 6 & 7 \\ 0 & 0 & 1 \leftarrow & 0 & 2 & 2 & 1 \\ 0 & 0 & 0 & 0 & 0 & 0 & 0 \\ 0 & 0 & 0 & 0 & 0 & 0 & 0 \end{pmatrix}$$

Curve fitting

Another application of matrices (systems) is to help fit specific models to sets of points.

Example 13

Fit a quadratic model to pass through the points $(-1, 10)$, $(2, 4)$, and $(3, 14)$.

Solution

The problem is to find parameters a, b, and c that will force the curve representing the function $f(x) = ax^2 + bx + c$ to contain the given points. This means

$$f(-1) = 10, f(2) = 4, \text{ and } f(3) = 14.$$

Since we need to find the three unknown parameters, we need three equations which are offered by the conditions above:

$$f(x) = ax^2 + bx + c$$
$$f(-1) = a - b + c = 10$$
$$f(2) = 4a + 2b + c = 4$$
$$f(3) = 9a + 3b + c = 14$$

This is clearly a system of three equations which can be solved using matrix methods, among other methods of course.

Using *rref*, we get the following result:

$$\left(\begin{array}{ccc|c} 1 & -1 & 1 & 10 \\ 4 & 2 & 1 & 4 \\ 9 & 3 & 1 & 14 \end{array}\right) \Leftrightarrow \left(\begin{array}{ccc|c} 1 & 0 & 0 & 3 \\ 0 & 1 & 0 & -5 \\ 0 & 0 & 1 & 2 \end{array}\right)$$

Which means that $a = 3$, $b = -5$, and $c = 2$; so the function is $f(x) = 3x^2 - 5x + 2$.

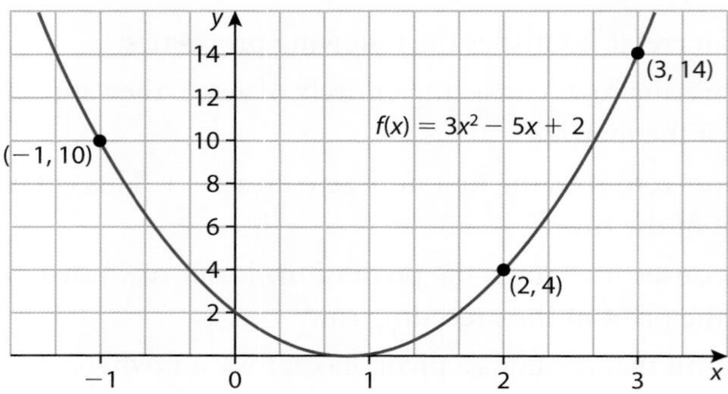

Equivalently, we can use the inverse matrix directly:

$$\begin{pmatrix} 1 & -1 & 1 \\ 4 & 2 & 1 \\ 9 & 3 & 1 \end{pmatrix}\begin{pmatrix} a \\ b \\ c \end{pmatrix} = \begin{pmatrix} 10 \\ 4 \\ 14 \end{pmatrix} \Leftrightarrow \begin{pmatrix} a \\ b \\ c \end{pmatrix} = \begin{pmatrix} 1 & -1 & 1 \\ 4 & 2 & 1 \\ 9 & 3 & 1 \end{pmatrix}^{-1}\begin{pmatrix} 10 \\ 4 \\ 14 \end{pmatrix} = \begin{pmatrix} 3 \\ -5 \\ 2 \end{pmatrix}$$

```
    [[1 -1 1 10]
     [4 2  1 4 ]
     [9 3  1 14]]
rref([A]
    [[1 0 0 3 ]
     [0 1 0 -5]
     [0 0 1 2 ]]
```

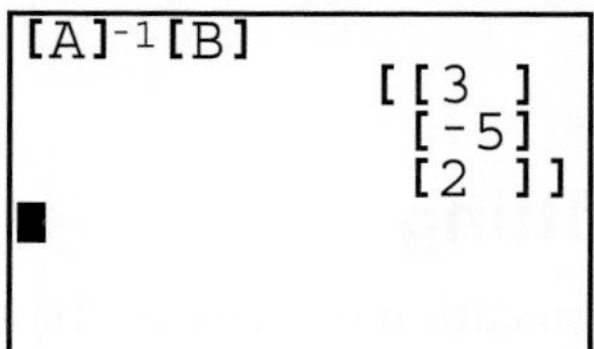

Exercise 6.4

1 Given the matrix $A = \begin{pmatrix} 5 & 6 \\ -1 & 0 \end{pmatrix}$ find the value of the real number m such that $\det(A - mI) = 0$, where I is the 2×2 multiplication identity matrix.

2 a) Find the values of a and b, given that the matrix $A = \begin{pmatrix} a & -4 & -6 \\ -8 & 5 & 7 \\ -5 & 3 & 4 \end{pmatrix}$ is the inverse of the matrix $B = \begin{pmatrix} 1 & 2 & -2 \\ 3 & b & 1 \\ -1 & 1 & -3 \end{pmatrix}$.

b) For the values of a and b found in part a), solve the system of linear equations:

$x + 2y - 2z = 5$

$3x + by + z = 0$

$-x + y - 3z = a - 1$

3 Find the value(s) of m so that the matrix $\begin{pmatrix} 1 & m & 1 \\ 3 & 1-m & 2 \\ m & -3 & m-1 \end{pmatrix}$ is singular.

4 Solve each system of equations. If a solution does not exist, justify why not.

a) $\begin{cases} 4x - y + z = -5 \\ 2x + 2y + 3z = 10 \\ 5x - 2y + 6z = 1 \end{cases}$

b) $\begin{cases} 4x - 2y + 3z = -2 \\ 2x + 2y + 5z = 16 \\ 8x - 5y - 2z = 4 \end{cases}$

c) $\begin{cases} 5x - 3y + 2z = 2 \\ 2x + 2y - 3z = 3 \\ x - 7y + 8z = -4 \end{cases}$

d) $\begin{cases} 3x - 2y + z = -29 \\ -4x + y - 3z = 37 \\ x - 5y + z = -24 \end{cases}$

e) $\begin{cases} 2x + 3y + 5z = 4 \\ 3x + 5y + 9z = 7 \\ 5x + 9y + 17z = 13 \end{cases}$

f) $\begin{cases} 2x + 3y + 5z = 4 \\ 3x + 5y + 9z = 7 \\ 5x + 9y + 17z = 1 \end{cases}$

g) $\begin{cases} -x + 4y - 2z = 12 \\ 2x - 9y + 5z = -25 \\ -x + 5y - 4z = 10 \end{cases}$

h) $\begin{cases} x - 3y - 2z = 8 \\ -2x + 7y + 3z = -19 \\ x - y - 3z = 3 \end{cases}$

5 a) Find the values of k such that the following matrix is not singular

$$A = \begin{pmatrix} 1 & 1 & k-1 \\ k & 0 & -1 \\ 6 & 2 & -3 \end{pmatrix}.$$

b) Find the value(s) of k such that A is the inverse of B, where

$$B = \begin{pmatrix} k-3 & -3 & k \\ 3 & k+2 & -1 \\ -2 & -4 & 1 \end{pmatrix}.$$

c) For the value of k found in b), apply elementary row operations to reduce the matrix $\begin{pmatrix} 1 & 1 & k-1 & 1 & 0 & 0 \\ k & 0 & -1 & 0 & 1 & 0 \\ 6 & 2 & -3 & 0 & 0 & 1 \end{pmatrix}$ into $\begin{pmatrix} 1 & 0 & 0 & a & b & c \\ 0 & 1 & 0 & d & e & f \\ 0 & 0 & 1 & g & h & i \end{pmatrix}$ where $a, \ldots, i$ are to be determined.

6 a) Find the values of k such that the following matrix is not singular.

$$A = \begin{pmatrix} \frac{2}{5} & -\frac{17}{5} & \frac{k+9}{5} \\ -\frac{1}{5} & \frac{21}{5} & -\frac{13}{5} \\ k-2 & 3 & -2 \end{pmatrix}$$

b) Find the value(s) of k such that A is the inverse of B, where

$$B = \begin{pmatrix} k+1 & 1 & k \\ 2 & k+2 & -3 \\ 3 & 6 & -5 \end{pmatrix}.$$

c) For the value of k found in b), apply elementary row operations to reduce the matrix $\begin{pmatrix} 2 & -17 & k+9 & 1 & 0 & 0 \\ -1 & 21 & -13 & 0 & 1 & 0 \\ 5(k-2) & 15 & -10 & 0 & 0 & 1 \end{pmatrix}$ into $\begin{pmatrix} 1 & 0 & 0 & a & b & c \\ 0 & 1 & 0 & d & e & f \\ 0 & 0 & 1 & g & h & i \end{pmatrix}$ where $a, \ldots, i$ are to be determined.

7 Use elementary row operations to transform the matrix $[A \vdots I]$ to a matrix in the form $[I \vdots B]$. Comment on the relationship between A and B and support your conclusion.

a) $\left(\begin{array}{ccc|ccc} 2 & 0 & 3 & 1 & 0 & 0 \\ -1 & 1 & 1 & 0 & 1 & 0 \\ 2 & -2 & 1 & 0 & 0 & 1 \end{array}\right)$ b) $\left(\begin{array}{ccc|ccc} 1 & 4 & 5 & 1 & 0 & 0 \\ 2 & -3 & 1 & 0 & 1 & 0 \\ -1 & 8 & 6 & 0 & 0 & 1 \end{array}\right)$

8 Determine the function f so that the curve representing it contains the indicated points.

a) $f(x) = ax^2 + bx + c$ to contain $(-1, 5)$, $(2, -1)$, and $(4, 35)$.

b) $f(x) = ax^2 + bx + c$ to contain $(-1, 12)$ and $(2, -3)$.

• **Hint:** there is more than one curve!

c) $f(x) = ax^3 + bx^2 + cx + d$ to contain the points $(-1, 5)$, $(1, -3)$, $(2, 5)$, and $(3, 45)$. [optional material]

d) $f(x) = ax^3 + bx^2 + cx + d$ to contain the points $(-3, 4)$, $(-1, 4)$, and $(2, 4)$.

9 Consider the following system of equations:

$$\begin{cases} 2x + y + 3z = -5 \\ 3x - y + 4z = 2 \\ 5x + 7z = m - 5 \end{cases}$$

Find the value(s) of m for which this system is consistent. For the value of m found, find the most general solution of the system.

10 Consider the following system of equations:

$$\begin{cases} -3x + 2y + 3z = 1 \\ 4x - y - 5z = -5 \\ x + y - 2z = m - 3 \end{cases}$$

Find the value(s) of m for which this system is consistent. For the value of m found, find the most general solution of the system.

11 Consider the matrix $A = \begin{pmatrix} 3 & -4 & -6 \\ -8 & 5 & 7 \\ -5 & 3 & 4 \end{pmatrix}$.

a) Find $\det(A)$.

b) Use the third elementary row operation to transform the matrix A into matrix B in triangular form (i.e. **add a multiple of one row to another row**).

c) Find $\det(B)$.

d) Use a GDC to find $\det(C)$ for $C = \begin{pmatrix} 2 & 1 & -3 & 5 \\ 4 & 3 & -4 & -6 \\ 6 & -8 & 5 & 7 \\ -6 & -5 & 3 & 4 \end{pmatrix}$.

e) Repeat b) and c) for C.

Practice questions

1 If $\begin{pmatrix} 2x & 3 \\ -4x & x \end{pmatrix}$ and det $A = 14$, find x.

2 Let $M = \begin{pmatrix} a & 2 \\ 2 & -1 \end{pmatrix}$, where $a \in \mathbb{Z}$.

a) Find M^2 in terms of a.

b) If M^2 is equal to $\begin{pmatrix} 5 & -4 \\ -4 & 5 \end{pmatrix}$, find the value of a.

Using this value of a, find M^{-1} and hence solve the system of equations:

$$-x + 2y = -3$$
$$2x - y = 3$$

3 Two matrices are given, where $A = \begin{pmatrix} 5 & 2 \\ 2 & 0 \end{pmatrix}$ and $BA = \begin{pmatrix} 11 & 2 \\ 44 & 8 \end{pmatrix}$. Find B.

4 The matrices A, B, and X are given, where

$A = \begin{pmatrix} 3 & 1 \\ -5 & 6 \end{pmatrix}$, $B = \begin{pmatrix} 4 & 8 \\ 0 & -3 \end{pmatrix}$, $X = \begin{pmatrix} a & b \\ c & d \end{pmatrix}$ with $a, b, c, d \in \mathbb{R}$.

Find the values of a, b, c and d such that $AX + X = B$.

5 $A = \begin{pmatrix} 5 & -2 \\ 7 & 1 \end{pmatrix}$ is a 2×2 matrix.

a) Write out A^{-1}.

b) **(i)** If $XA + B = C$, where B, C, and X are 2×2 matrices, express X in terms of A^{-1}, B, and C.

(ii) Find X if $B = \begin{pmatrix} 6 & 7 \\ 5 & -2 \end{pmatrix}$ and $C = \begin{pmatrix} -5 & 0 \\ -8 & 7 \end{pmatrix}$.

6 Given $A = \begin{pmatrix} a & b \\ c & 1 \end{pmatrix}$ and $B = \begin{pmatrix} 1 & 2 \\ d & c \end{pmatrix}$,

a) write out $A + B$;

b) find AB.

7 **a)** Write out the inverse of the matrix $\begin{pmatrix} 1 & -3 & 1 \\ 2 & 2 & -1 \\ 1 & -5 & 3 \end{pmatrix}$.

b) Hence, solve the system of simultaneous equations:

$$x - 3y + z = 1$$
$$2x + 2y - z = 2$$
$$x - 5y + 3z = 3$$

8 Given the two matrices C and D, where

$C = \begin{pmatrix} -2 & 4 \\ 1 & 7 \end{pmatrix}$ and $D = \begin{pmatrix} 5 & 2 \\ -1 & a \end{pmatrix}$,

the matrix Q is given such that $3Q = 2C - D$.

b) Find Q.

b) Find CD.

c) Find D^{-1}.

9 a) Find the values of a and b given that the matrix $A = \begin{pmatrix} a & -4 & -6 \\ -8 & 5 & 7 \\ -5 & 3 & 4 \end{pmatrix}$ is the inverse of the matrix $B = \begin{pmatrix} 1 & 2 & -2 \\ 3 & b & 1 \\ -1 & 1 & -3 \end{pmatrix}$.

b) For the values of a and b found in part a), solve the system of linear equations:

$$x + 2y - 2z = 5$$
$$3x + by + z = 0$$
$$-x + y - 3z = a - 1$$

10 a) Given matrices A, B, C for which $AB = C$ and $\det A \neq 0$, express B in terms of A and C.

b) Let $A = \begin{pmatrix} 1 & 2 & 3 \\ 2 & -1 & 2 \\ 3 & -3 & 2 \end{pmatrix}$, $D = \begin{pmatrix} -4 & 13 & -7 \\ -2 & 7 & -4 \\ 3 & -9 & 5 \end{pmatrix}$ and $C = \begin{pmatrix} 5 \\ 7 \\ 10 \end{pmatrix}$.

(i) Find the matrix DA.

(ii) Find B if $AB = C$.

c) Find the coordinates of the point of intersection of the planes $x + 2y + 3z = 5$, $2x - y + 2z = 7$ and $3x - 3y + 2z = 10$. (This can be answered after Chapter 14.)

11 a) Find the determinant of the matrix $\begin{pmatrix} 1 & 1 & 2 \\ 1 & 2 & 1 \\ 2 & 1 & 5 \end{pmatrix}$.

b) Find the value of λ for which the following system of equations can be solved.

$$\begin{pmatrix} 1 & 1 & 2 \\ 1 & 2 & 1 \\ 2 & 1 & 5 \end{pmatrix}\begin{pmatrix} x \\ y \\ z \end{pmatrix} = \begin{pmatrix} 3 \\ 4 \\ \lambda \end{pmatrix}$$

c) For this value of λ, find the general solution to the system of equations.

12 The square matrix X is such that $X^3 = 0$. Show that the inverse of the matrix $(I - X)$ is $I + X + X^2$.

7 Trigonometric Functions and Equations

Assessment statements

2.1 Odd and even functions (also see Chapter 3).

3.1 The circle: radian measure of angles; length of an arc; area of a sector.

3.2 The circular functions $\sin x$, $\cos x$ and $\tan x$: their domains and ranges; their periodic nature; and their graphs.
Definition of $\cos\theta$ and $\sin\theta$ in terms of the unit circle.
Definition of $\tan\theta$ as $\frac{\sin\theta}{\cos\theta}$.
Exact values of sin, cos and tan of $0, \frac{\pi}{6}, \frac{\pi}{4}, \frac{\pi}{3}, \frac{\pi}{2}$ and their multiples.
Definition of the reciprocal trigonometric ratios $\sec\theta$, $\csc\theta$ and $\cot\theta$.
Pythagorean identities: $\cos^2\theta + \sin^2\theta = 1$; $1 + \tan^2\theta = \sec^2\theta$; $1 + \cot^2\theta = \csc^2\theta$.

3.3 Compound angle identities.
Double angle identities.

3.4 Composite functions of the form $f(x) = a\sin(b(x + c)) + d$.

3.5 The inverse functions $x \mapsto \arcsin x$, $x \mapsto \arccos x$, $x \mapsto \arctan x$; their domains and ranges; their graphs.

3.6 Algebraic and graphical methods of solving trigonometric equations in a finite interval including the use of trigonometric identities and factorization.

Introduction

The word *trigonometry* comes from two Greek words, *trigonon* and *metron*, meaning 'triangle measurement'. Trigonometry developed out of the use and study of triangles, in surveying, navigation, architecture and astronomy, to find relationships between lengths of sides of triangles and measurement of angles. As a result, trigonometric functions were initially defined as functions of angles – that is, functions with angle measurements as their domains. With the development of calculus in the seventeenth century and the growth of knowledge in the sciences, the application of trigonometric functions grew to include a wide variety of periodic (repetitive) phenomena such as wave motion, vibrating strings, oscillating pendulums, alternating electrical current and biological cycles. These applications of trigonometric functions require their domains to be sets of real numbers without reference to angles or triangles. Hence, trigonometry can be approached from two different perspectives – **functions**

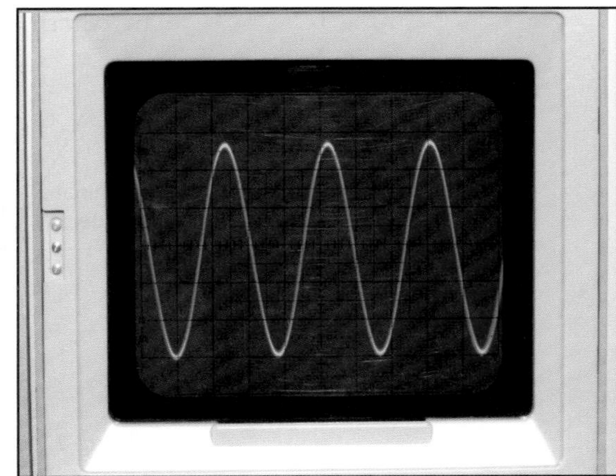

The oscilloscope shows the graph of pressure of sound wave versus time for a high-pitched sound. The graph is a repetitive pattern that can be expressed as the sum of different 'sine' waves. A sine wave is any transformation of the graph of the trigonometric function $y = \sin x$ and takes the form $y = a\sin[b(x + c)] + d$.

of angles, or **functions of real numbers.** The first perspective is the focus of the next chapter where trigonometric functions will be defined in terms of the **ratios of sides of a right triangle.** The second perspective is the focus of this chapter, where trigonometric functions will be defined in terms of a real number that is the **length of an arc along the unit circle.** While it is possible to define trigonometric functions in these two different ways, they assign the same value (interpreted as an angle, an arc length, or simply a real number) to a particular real number. Although this chapter will not refer much to triangles, it seems fitting to begin by looking at angles and arc lengths – geometric objects indispensable to the two different ways of viewing trigonometry.

7.1 Angles, circles, arcs and sectors

Angles

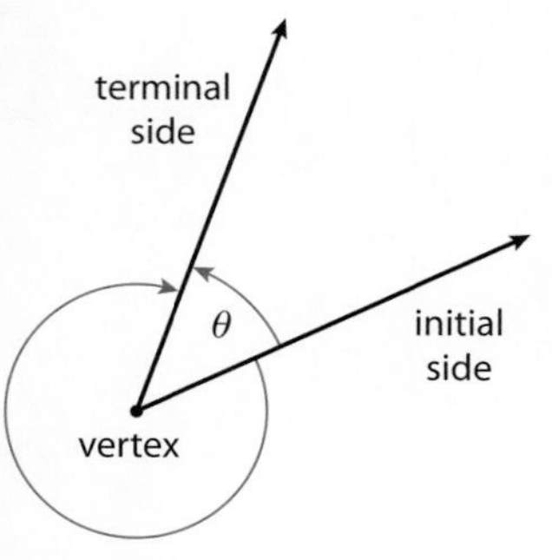

Figure 7.1 Components of an angle.

An **angle** in a plane is made by rotating a ray about its endpoint, called the **vertex** of the angle. The starting position of the ray is called the **initial side** and the position of the ray after rotation is called the **terminal side** of the angle (Figure 7.1). An angle having its vertex at the origin and its initial side lying on the positive x-axis is said to be in **standard position** (Figure 7.2a). A **positive angle** is produced when a ray is rotated in an anticlockwise direction, and a **negative angle** when a ray is rotated in a clockwise direction. Two angles in standard position whose terminal sides are in the same location – regardless of the direction or number of rotations – are called **coterminal angles.** Greek letters are often used to represent angles, and the direction of rotation is indicated by an arc with an arrow at its endpoint. The x- and y-axes divide the coordinate plane into four quadrants (numbered with Roman numerals). Figure 7.2b shows a positive angle α (alpha) and a negative angle β (beta) that are coterminal in quadrant III.

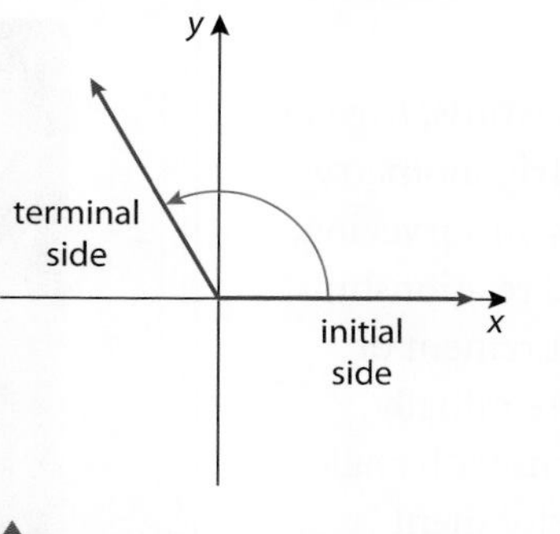

Figure 7.2a Standard position of an angle.

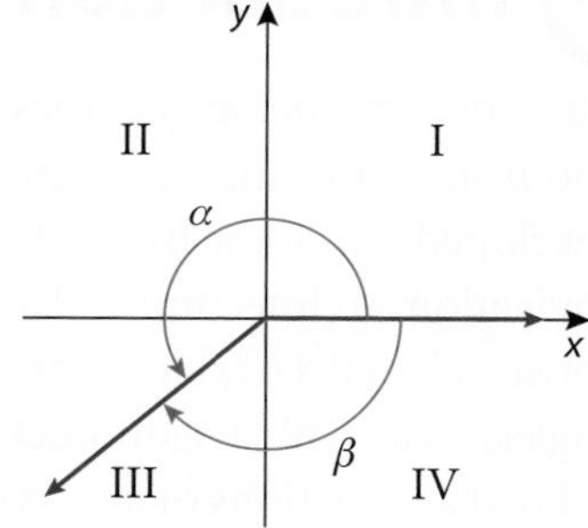

Figure 7.2b Coterminal angles.

Measuring angles: degree measure and radian measure

Perhaps the most natural unit for measuring large angles is the **revolution.** For example, most cars have an instrument (a tachometer) that indicates the number of revolutions per minute (rpm) at which the engine is operating. However, to measure smaller angles, we need a smaller unit. A common unit for measuring angles is the **degree**, of which there are 360 in one revolution. Hence, the unit of one degree (1°) is defined to be 1/360 of one anticlockwise revolution about the vertex.

The convention of having 360 degrees in one revolution can be traced back around 4000 years to ancient Babylonian civilizations. The number system most widely used today is a base 10, or **decimal**, system. Babylonian mathematics used a base 60, or **sexagesimal**, number system. Although 60 may seem to be an awkward number to have as a base, it does have certain advantages. It is the smallest number that has 2, 3, 4, 5 and 6 as factors – and it also has factors of 10, 12, 15, 20 and 30. But why 360 degrees? We're not certain but it may have to do with the Babylonians assigning 60 divisions to each angle in an equilateral triangle and exactly six equilateral triangles can be arranged around a single point. That makes $6 \times 60 = 360$ equal divisions in one full revolution. There are few numbers as small as 360 that have so many different factors. This makes the degree a useful unit for dividing one revolution into an equal number of parts. 120 degrees is $\frac{1}{3}$ of a revolution, 90 degrees is $\frac{1}{4}$ of a revolution, 60 degrees is $\frac{1}{6}$, 45 degrees is $\frac{1}{8}$, and so on.

There is another method of measuring angles that is more natural. Instead of dividing a full revolution into an arbitrary number of equal divisions (e.g. 360), consider an angle that has its vertex at the centre of a circle (a **central angle**) and subtends (or intercepts) a part of the circle, called an **arc of the circle**. Figure 7.3 shows three circles with radii of different lengths ($r_1 < r_2 < r_3$) and the same central angle θ (theta) subtending (intercepting) the arc lengths s_1, s_2 and s_3. Regardless of the size of the circle (i.e. length of the radius), the ratio of arc length (s) to radius (r) for a given circle will be constant. For the angle θ in Figure 7.3, $\frac{s_1}{r_1} = \frac{s_2}{r_2} = \frac{s_3}{r_3}$. Because this ratio is an arc length divided by another length (radius), it is just an ordinary real number and has no units.

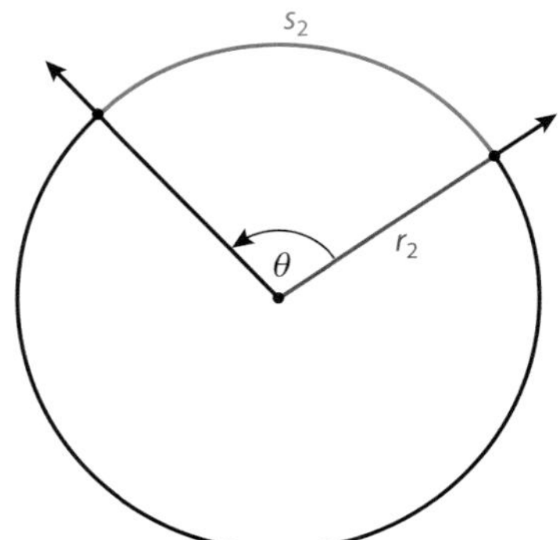

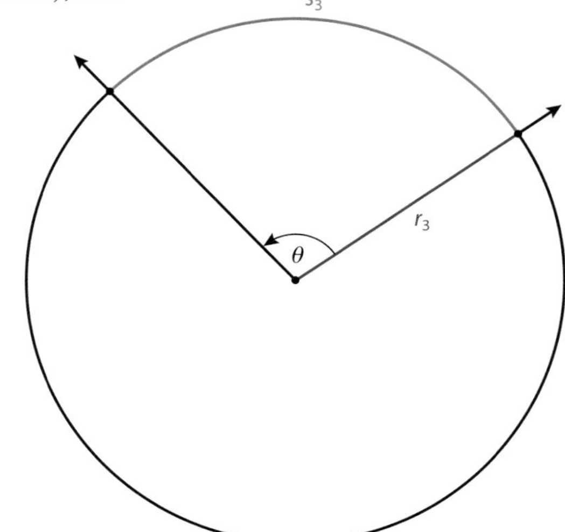

Figure 7.3 Different circles with the same central angle θ subtending different arcs, but the ratio of arc length to radius remains constant.

Minor and major arcs

If a central angle is **less** than 180°, the subtended arc is referred to as a **minor arc**. If a central angle is **greater** than 180°, the subtended arc is referred to as a **major arc**.

The ratio $\frac{s}{r}$ indicates how many radius lengths, r, fit into the length of the arc s. For example, if $\frac{s}{r} = 2$, the length of s is equal to two radius lengths. This accounts for the name **radian** and leads to the following definition.

When the measure of an angle is, for example, 5 radians, the word 'radians' does not indicate units (as when writing centimetres, seconds or degrees) but indicates the *method* of angle measurement. If the measure of an angle is in units of degrees, we must indicate this by word or symbol. For example, $\theta = 5$ degrees or $\theta = 5°$. However, when radian measure is used it is customary to write no units or symbol. For example, a central angle θ that subtends an arc equal to five radius lengths (radians) is simply given as $\theta = 5$.

Radian measure

One **radian** is the measure of a central angle θ of a circle that subtends an arc s of the circle that is exactly the same length as the radius r of the circle. That is, when $\theta = 1$ radian, arc length = radius.

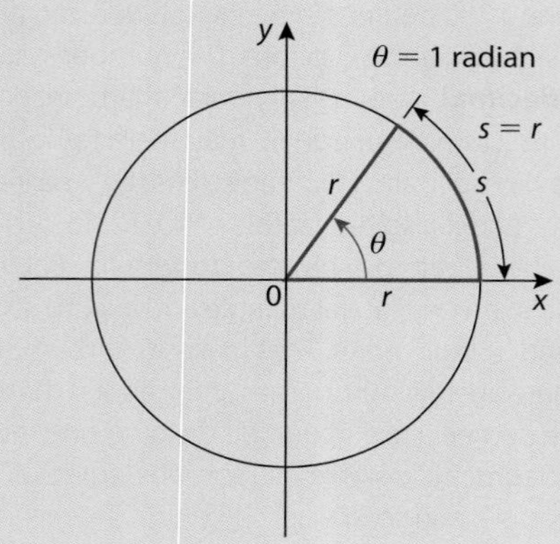

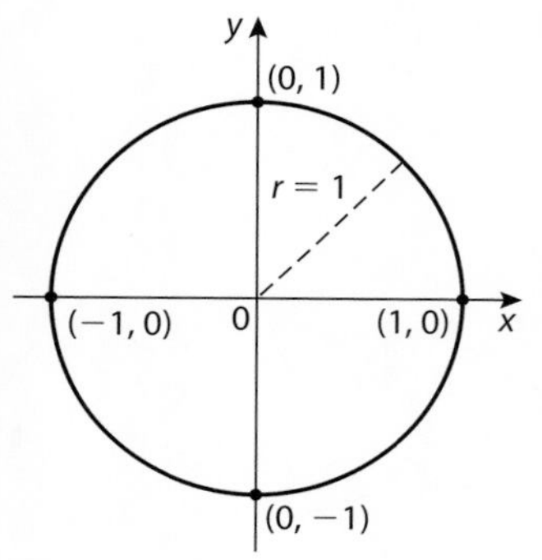

Figure 7.4 The unit circle.

The unit circle

When an angle is measured in radians it makes sense to draw it, or visualize it, so that it is in standard position. It follows that the angle will be a central angle of a circle whose centre is at the origin, as shown above. As Figure 7.3 illustrated, it makes no difference what size circle is used. The most practical circle to use is the circle with a radius of one unit so the radian measure of an angle will simply be equal to the length of the subtended arc.

Radian measure: $\theta = \frac{s}{r}$ If $r = 1$, then $\theta = \frac{s}{1} = s$.

The circle with a radius of one unit and centre at the origin $(0, 0)$ is called the **unit circle** (Figure 7.4). The equation for the unit circle is $x^2 + y^2 = 1$. Because the circumference of a circle with radius r is $2\pi r$, a central angle of one full anticlockwise revolution (360°) subtends an arc on the unit circle equal to 2π units. Hence, if an angle has a degree measure of 360°, its radian measure is exactly 2π. It follows that an angle of 180° has a radian measure of exactly π. This fact can be used to convert between degree measure and radian measure, and vice versa.

Conversion between degrees and radians

Because $180° = \pi$ radians, $1° = \frac{\pi}{180}$ radians, and 1 radian $= \frac{180°}{\pi}$. An angle with a radian measure of 1 has a degree measure of approximately 57.3° (to 3 significant figures).

Example 1

The angles of 30° and 45°, and their multiples, are often encountered in trigonometry. Convert 30° and 45° to radian measure and sketch the corresponding arc on the unit circle. Use these results to convert 60° and 90° to radian measure.

Solution

(Note that the 'degree' units cancel.)

$$30° = 30°\left(\frac{\pi}{180°}\right) = \frac{30°}{180°}\pi = \frac{\pi}{6}$$

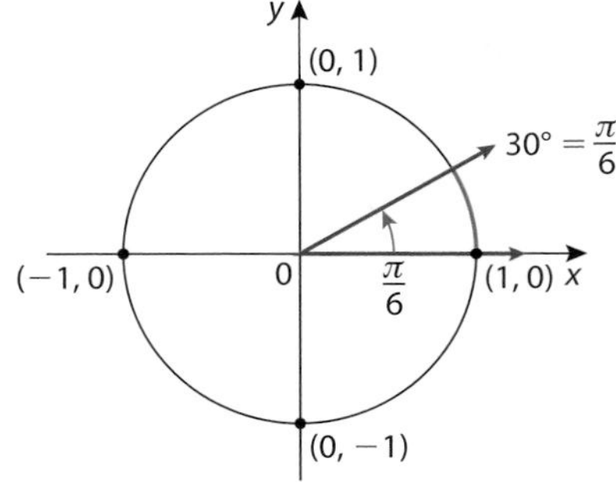

$$45° = 45°\left(\frac{\pi}{180°}\right) = \frac{45°}{180°}\pi = \frac{\pi}{4}$$

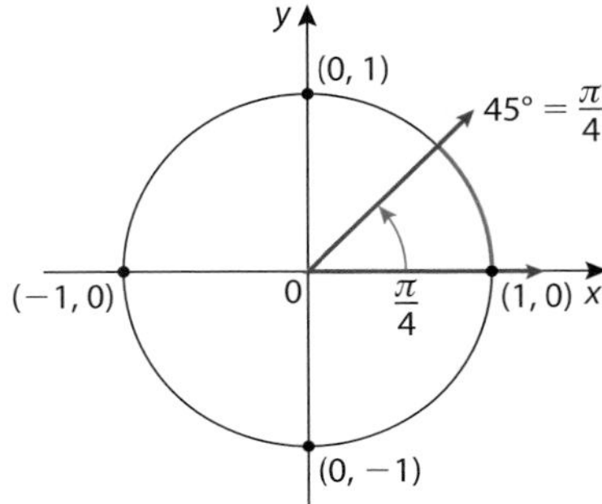

Since $60° = 2(30°)$ and $30° = \frac{\pi}{6}$, then $60° = 2\left(\frac{\pi}{6}\right) = \frac{\pi}{3}$. Similarly, $90° = 2(45°)$ and $45° = \frac{\pi}{4}$, so $90° = 2\left(\frac{\pi}{4}\right) = \frac{\pi}{2}$.

● **Hint:** It is very helpful to be able to quickly recall the results from Example 1: $30° = \frac{\pi}{6}$, $45° = \frac{\pi}{4}$, $60° = \frac{\pi}{3}$ and $90° = \frac{\pi}{2}$. Of course, not all angles are multiples of 30° or 45° when expressed in degrees, and not all angles are multiples of $\frac{\pi}{6}$ or $\frac{\pi}{4}$ when expressed in radians. However, these 'special' angles often appear in problems and applications. Knowing these four facts can help you to quickly convert mentally between degrees and radians for many common angles. For example, to convert 225° to radians, apply the fact that $225° = 5(45°)$. Since $45° = \frac{\pi}{4}$, then $225° = 5(45°) = 5\left(\frac{\pi}{4}\right) = \frac{5\pi}{4}$.

As another example, convert $\frac{11\pi}{6}$ to degrees: $\frac{11\pi}{6} = 11\left(\frac{\pi}{6}\right) = 11(30°) = 330°$.

Example 2

a) Convert the following radian measures to degrees. Express exactly, if possible. Otherwise, express accurate to 3 significant figures.

(i) $\frac{4\pi}{3}$ (ii) $-\frac{3\pi}{2}$ (iii) 5 (iv) 1.38

b) Convert the following degree measures to radians. Express exactly, if possible. Otherwise, express accurate to 3 significant figures.

(i) 135° (ii) −150° (iii) 175° (iv) 10°

Solution

a) (i) $\frac{4\pi}{3} = 4\left(\frac{\pi}{3}\right) = 4(60°) = 240°$

(ii) $-\frac{3\pi}{2} = -\frac{3}{2}(\pi) = -\frac{3}{2}(180°) = -270°$

(iii) $5\left(\frac{180°}{\pi}\right) \approx 286.479° \approx 286°$

(iv) $1.38\left(\frac{180°}{\pi}\right) \approx 79.068° \approx 79.1°$

● **Hint:** All GDCs will have a degree mode and a radian mode. Before doing any calculations with angles on your GDC, be certain that the mode setting for angle measurement is set correctly. Although you may be more familiar with degree measure, as you progress further in mathematics – and especially in calculus – radian measure is far more useful.

b) (i) $135° = 3(45°) = 3\left(\frac{\pi}{4}\right) = \frac{3\pi}{4}$

(ii) $-150° = -5(30°) = -5\left(\frac{\pi}{6}\right) = -\frac{5\pi}{6}$

(iii) $175°\left(\frac{\pi}{180°}\right) \approx 3.0543 \approx 3.05$

(iv) $10°\left(\frac{\pi}{180°}\right) \approx 0.174\,53 \approx 0.175$

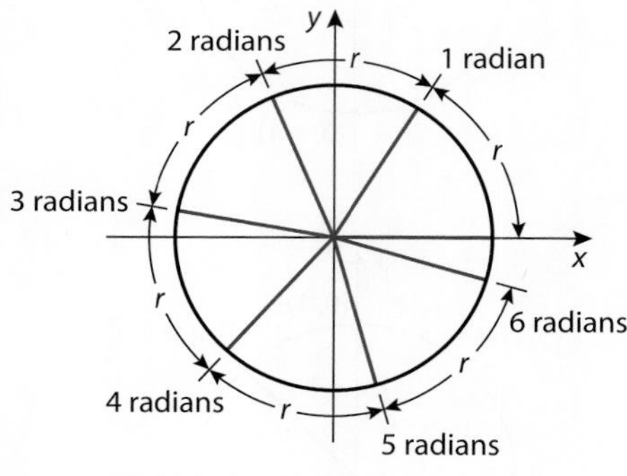

Figure 7.5 Arcs with lengths equal to the radius placed along circumference of a circle.

Because 2π is approximately 6.28 (3 significant figures), there are a little more than six radius lengths in one revolution, as shown in Figure 7.5.

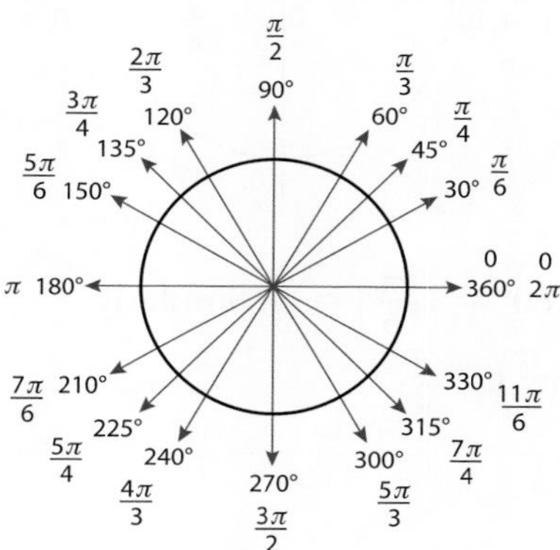

Figure 7.6 Degree measure and radian measure for common angles.

Figure 7.6 shows all of the angles between 0° and 360° inclusive, that are multiples of 30° or 45°, and their equivalent radian measure. You will benefit by being able to convert quickly between degree measure and radian measure for these common angles.

Arc length

For any angle θ, its radian measure is given by $\theta = \frac{s}{r}$. Simple rearrangement of this formula leads to another formula for computing arc length.

Arc length

For a circle of radius r, a central angle θ subtends an arc of the circle of length s given by

$$s = r\theta$$

where θ is in radian measure.

Example 3

A circle has a radius of 10 cm. Find the length of the arc of the circle subtended by a central angle of 150°.

Solution

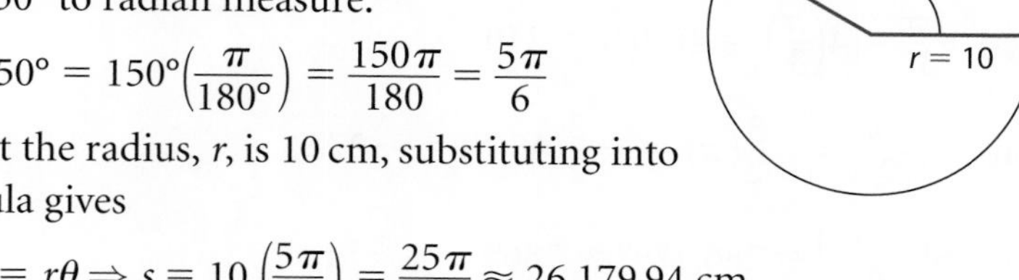

To use the formula $s = r\theta$, we must first convert 150° to radian measure.

$$150° = 150°\left(\frac{\pi}{180°}\right) = \frac{150\pi}{180} = \frac{5\pi}{6}$$

Given that the radius, r, is 10 cm, substituting into the formula gives

$$s = r\theta \Rightarrow s = 10\left(\frac{5\pi}{6}\right) = \frac{25\pi}{3} \approx 26.179\,94 \text{ cm}$$

The length of the arc is approximately 26.18 cm (4 significant figures).

Note that the units of the product $r\theta$ are the same as the units of r because in radian measure θ has no units.

Example 4

The diagram shows a circle of centre O with radius $r = 6$ cm. Angle AOB subtends the minor arc AB such that the length of the arc is 10 cm. Find the measure of angle AOB in degrees to 3 significant figures.

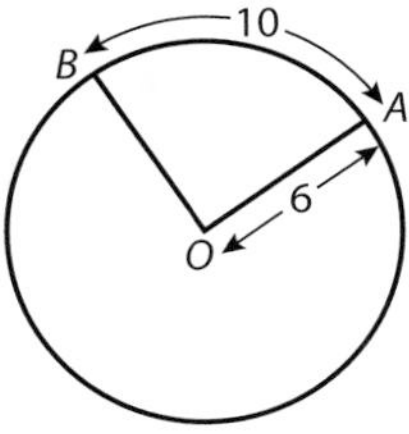

Solution

From the arc length formula, $s = r\theta$, we can state that $\theta = \frac{s}{r}$. Remember that the result for θ will be in radian measure. Therefore, angle $AOB = \frac{10}{6} = \frac{5}{3}$ or $1.\overline{6}$ radians. Now, we convert to degrees: $\frac{5}{3}\left(\frac{180°}{\pi}\right) \approx 95.49297°$. The degree measure of angle AOB is approximately 95.5°.

Geometry of a circle

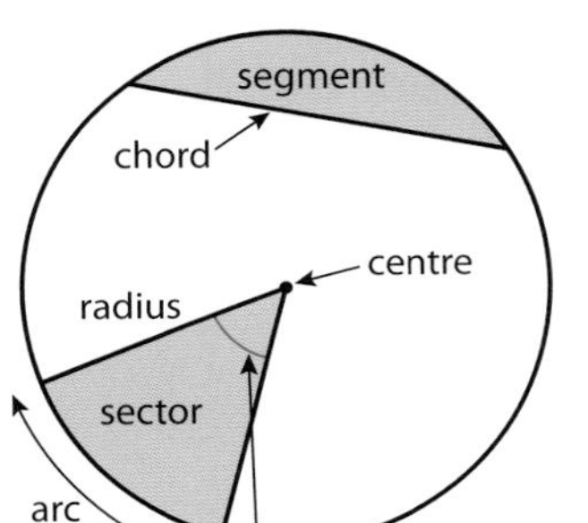

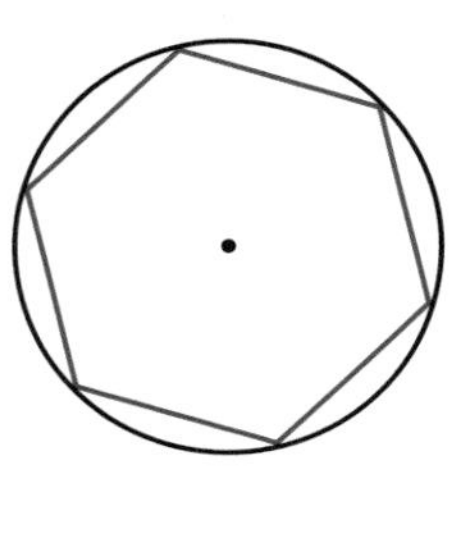

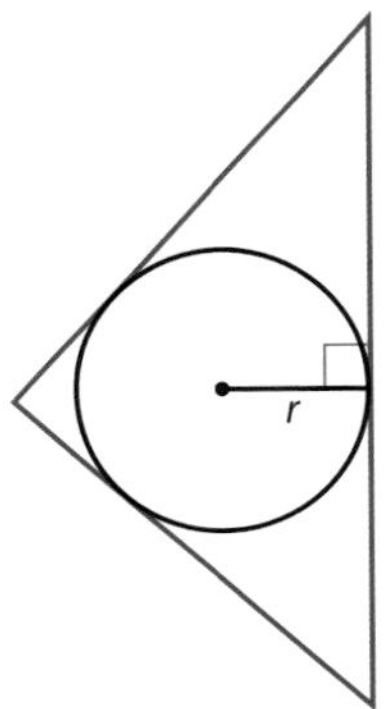

Figure 7.7 Circle terminology.

Sector of a circle

A **sector of a circle** is the region bounded by an arc of the circle and the two sides of a central angle (Figure 7.8). The ratio of the area of a sector to the area of the circle (πr^2) is equal to the ratio of the length of the subtended arc to the circumference of the circle ($2\pi r$). If s is the arc length and A is the area of the sector, we can write the following proportion: $\frac{A}{\pi r^2} = \frac{s}{2\pi r}$. Solving for A gives $A = \frac{\pi r^2 s}{2\pi r} = \frac{1}{2}rs$. From the formula for arc length we have $s = r\theta$, with θ the radian measure of the central angle. Substituting $r\theta$ for s gives the area of a sector to be $A = \frac{1}{2}rs = \frac{1}{2}r(r\theta) = \frac{1}{2}r^2\theta$. This result makes sense because, if the sector is the entire circle, $\theta = 2\pi$ and area $A = \frac{1}{2}r^2\theta = \frac{1}{2}r^2(2\pi) = \pi r^2$, which is the formula for the area of a circle.

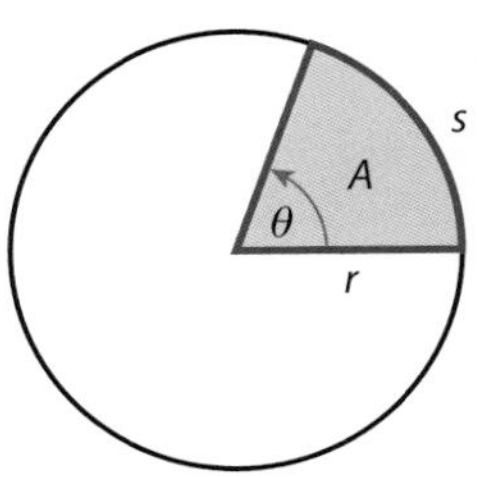

Figure 7.8 Sector of a circle.

Area of a sector

In a circle of radius r, the area of a sector with a central angle θ measured in radians is

$$A = \frac{1}{2}r^2\theta$$

Example 5

A circle of radius 9 cm has a sector whose central angle has radian measure $\frac{2\pi}{3}$. Find the exact values of the following: a) the length of the arc subtended by the central angle, and b) the area of the sector.

Solution

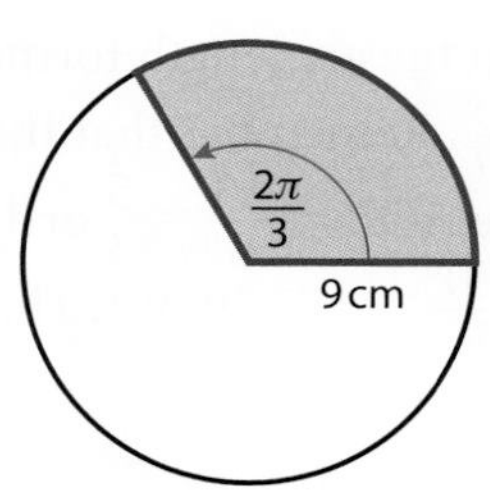

a) $s = r\theta \Rightarrow s = 9\left(\frac{2\pi}{3}\right) = 6\pi$

The length of the arc is exactly 6π cm.

b) $A = \frac{1}{2}r^2\theta \Rightarrow A = \frac{1}{2}(9)^2\left(\frac{2\pi}{3}\right) = 27\pi$

The area of the sector is exactly $27\pi\,\text{cm}^2$.

• **Hint:** The formula for arc length, $s = r\theta$, and the formula for area of a sector, $A = \frac{1}{2}r^2\theta$, are true only when θ is in radians.

Exercise 7.1

In questions 1–9, find the exact radian measure of the angle given in degree measure.

1 60° **2** 150° **3** −270°

4 36° **5** 135° **6** 50°

7 −45° **8** 400° **9** −480°

In questions 10–18, find the degree measure of the angle given in radian measure. If possible, express exactly. Otherwise, express accurate to 3 significant figures.

10 $\frac{3\pi}{4}$ **11** $-\frac{7\pi}{2}$ **12** 2

13 $\frac{7\pi}{6}$ **14** −2.5 **15** $\frac{5\pi}{3}$

16 $\frac{\pi}{12}$ **17** 1.57 **18** $\frac{8\pi}{3}$

In questions 19–24, the measure of an angle in standard position is given. Find two angles – one positive and one negative – that are coterminal with the given angle. If no units are given, assume the angle is in radian measure.

19 30° **20** $\frac{3\pi}{2}$ **21** 175°

22 $-\frac{\pi}{6}$ **23** $\frac{5\pi}{3}$ **24** 3.25

In questions 25 and 26, find the length of the arc s in the figure.

25

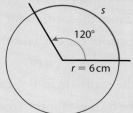

s

120°

$r = 6$ cm

26

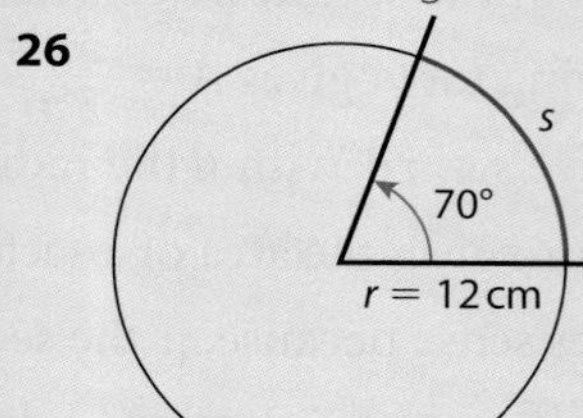

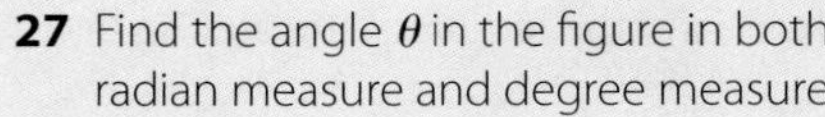

27 Find the angle θ in the figure in both radian measure and degree measure.

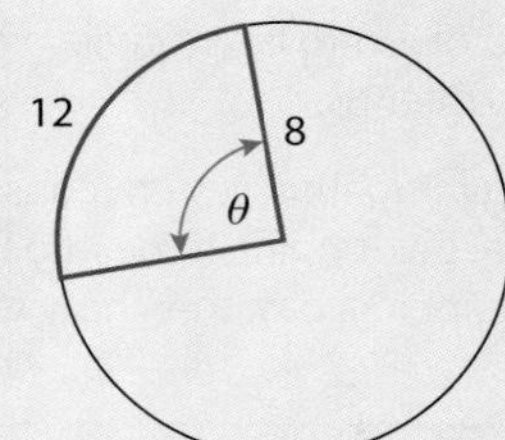

28 Find the radius r of the circle in the figure.

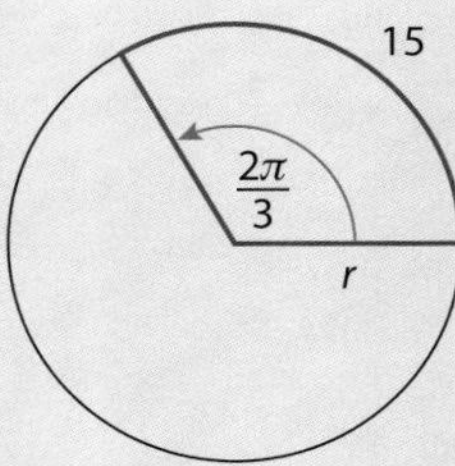

In questions 29 and 30, find the area of the sector in each figure.

29

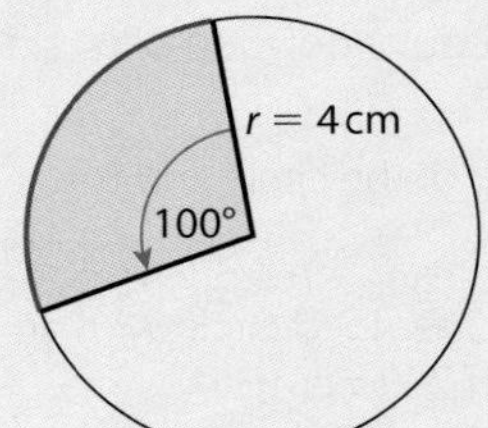

30

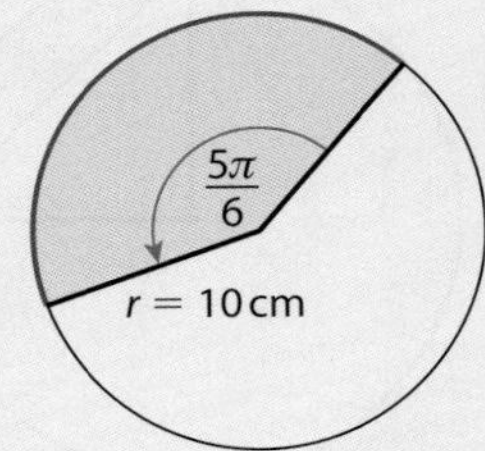

31 An arc of length 60 cm subtends a central angle α in a circle of radius 20 cm. Find the measure of α in both degrees and radians, approximate to 3 significant figures.

32 Find the length of an arc that subtends a central angle with radian measure of 2 in a circle of radius 16 cm.

33 The area of a sector of a circle with a central angle of 60° is 24 cm^2. Find the radius of the circle.

34 A bicycle with tyres 70 cm in diameter is travelling such that its tyres complete one and a half revolutions every second. That is, the **angular velocity** of a wheel is 1.5 revolutions per second.

a) What is the angular velocity of a wheel in radians per second?

b) At what speed (in km/hr) is the bicycle travelling along the ground? (This is the **linear velocity** of the bicycle.)

35 A bicycle with tyres 70 cm in diameter is travelling along a road at 25 km/hr. What is the angular velocity of a wheel of the bicycle in radians per second?

36 Given that ω is the angular velocity in radians/second of a point on a circle with radius r cm, express the linear velocity, v, in cm/second, of the point as a function in terms of ω and r.

37 A chord of 26 cm is in a circle of radius 20 cm. Find the length of the arc the chord subtends.

38 A circular irrigation system consists of a 400 metre pipe that is rotated around a central pivot point. If the irrigation pipe makes one full revolution around the pivot point in a day, then how much area, in square metres, does it irrigate each hour?

39 a) Find the radius of a circle circumscribed about a regular polygon of 64 sides if one side is 3 cm.

b) What is the difference between the circumference of the circle and the perimeter of the polygon?

40 What is the area of an equilateral triangle that has an inscribed circle with an area of $50\pi\,\text{cm}^2$, and a circumscribed circle with an area of $200\pi\,\text{cm}^2$?

41 In the diagram, the sector of a circle is subtended by two perpendicular radii. If the area of the sector is **A** square units, then find an expression for the area of the circle in terms of **A**.

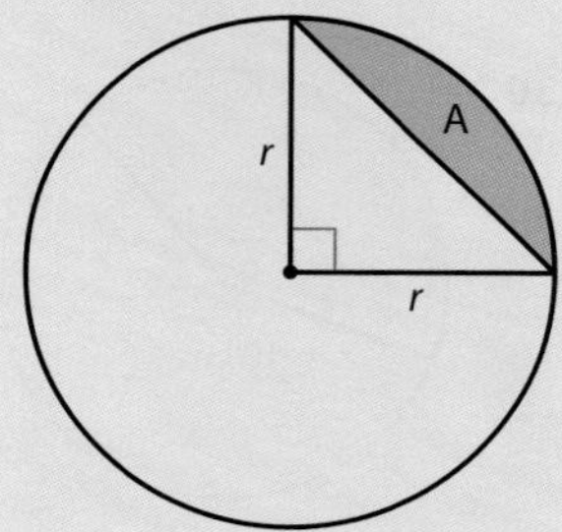

7.2 The unit circle and trigonometric functions

Several important functions can be described by mapping the coordinates of points on the real number line onto the points of the unit circle. Recall from the previous section that the unit circle has its centre at $(0, 0)$, it has a radius of one unit and its equation is $x^2 + y^2 = 1$.

A wrapping function: the real number line and the unit circle

Suppose that the real number line is tangent to the unit circle at the point $(1, 0)$ – and that zero on the number line matches with $(1, 0)$ on the circle, as shown in Figure 7.9. Because of the properties of circles, the real number line in this position will be perpendicular to the x-axis. The scales on the number line and the x- and y-axes need to be the same. Imagine that the real number line is flexible like a string and can wrap around the circle, with zero on the number line remaining fixed to the point $(1, 0)$ on the unit circle. When the top portion of the string moves along the circle, the wrapping is anticlockwise ($t > 0$), and when the bottom portion of the string moves along the circle, the wrapping is clockwise ($t < 0$). As the string wraps around the unit circle, each real number t on the string is mapped onto a point (x, y) on the circle. Hence, the real number line from 0 to t makes an arc of length t starting on the circle at $(1, 0)$ and ending at the point (x, y) on the circle. For example, since the circumference of the unit circle is 2π, the number $t = 2\pi$ will be wrapped anticlockwise around the circle to the point $(1, 0)$. Similarly, the number $t = \pi$ will be wrapped anticlockwise halfway around the circle to the point $(-1, 0)$ on the circle. And the number $t = -\frac{\pi}{2}$ will be wrapped clockwise one-quarter of the way around the circle to the point $(0, -1)$ on the circle. Note that each number t on the real number line is mapped (corresponds) to *exactly one* point on the unit circle, thereby satisfying the definition of a function (Section 2.1) – consequently this mapping is called a **wrapping function**.

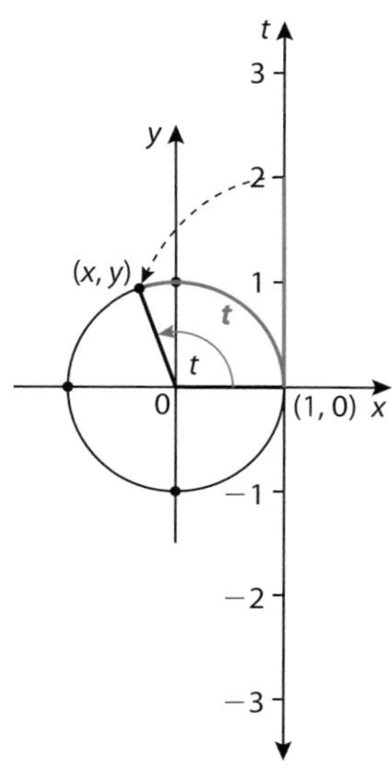

Figure 7.9 The wrapping function.

Before we leave our mental picture of the string (representing the real number line) wrapping around the unit circle, consider any pair of points on the string that are exactly 2π units from each other. Let these two points represent the real numbers t_1 and $t_1 + 2\pi$. Because the circumference of the unit circle is 2π, these two numbers will be mapped to the same point on the unit circle. Furthermore, consider the infinite number of points whose distance from t_1 is any integer multiple of 2π, i.e. $t_1 + k \cdot 2\pi$, $k \in \mathbb{Z}$, and again all of these numbers will be mapped to the same point on the unit circle. Consequently, the wrapping function is not a one-to-one function as defined in Section 2.3. Output for the function (points on the unit circle) are unchanged by the addition of any integer multiple of 2π to any input value (a real number). Functions that behave in such a repetitive (or cyclic) manner are called **periodic**.

We are surrounded by periodic functions. A few examples include: the average daily temperature variation during the year; sunrise and the day of the year; animal populations over many years; the height of tides and the position of the Moon; and an electrocardiogram, which is a graphic tracing of the heart's electrical activity.

Definition of a periodic function

A function f such that $f(x) = f(x + p)$ is a **periodic function**. If p is the least positive constant for which $f(x) = f(x + p)$ is true, p is called the **period** of the function.

Trigonometric functions

From our discussions about functions in Chapter 2, any function will have a domain (input) and range (output) that are sets having individual numbers as elements. We use the individual coordinates x and y of the points on the unit circle to define six **trigonometric functions**: the **sine**, **cosine**, **tangent**, **cosecant**, **secant** and **cotangent** functions. The names of these functions are often abbreviated in writing (but not speaking) as **sin**, **cos**, **tan**, **csc**, **sec**, **cot**, respectively.

When the real number t is wrapped to a point (x, y) on the unit circle, the value of the y-coordinate is assigned to the sine function; the x-coordinate is assigned to the cosine function; and the ratio of the two coordinates $\frac{y}{x}$ is assigned to the tangent function. Sine, cosine and tangent are often referred to as the **basic trigonometric functions.** The other three, cosecant, secant and cotangent, are each a reciprocal of one of the basic trigonometric functions and thus, are often referred to as the **reciprocal trigonometric functions.** All six are defined by means of the length of an arc on the unit circle as follows.

• **Hint:** To help you remember these definitions, note that the functions in the bottom row are the reciprocals of the function directly above in the top row.

Definition of the trigonometric functions

Let t be any real number and (x, y) a point on the unit circle to which t is mapped. Then the function definitions are:

$\sin t = y$	$\cos t = x$	$\tan t = \frac{y}{x}, x \neq 0$
$\csc t = \frac{1}{y}, y \neq 0$	$\sec t = \frac{1}{x}, x \neq 0$	$\cot t = \frac{x}{y}, y \neq 0$

• **Hint:** Most calculators do not have keys for cosecant, secant and cotangent. You have to use the sine, cosine or tangent keys and the appropriate quotient. Because cosecant is the reciprocal of sine, to evaluate $\csc\frac{\pi}{3}$, for example, you need to evaluate $\frac{1}{\sin\frac{\pi}{3}}$. There is a key on your GDC labelled $\sin^{-1}$. It is **not** the reciprocal of sine but represents the inverse of the sine function, also denoted as the arcsine function (abbreviated arcsin). This is the same for $\cos^{-1}$ and $\tan^{-1}$. We will learn about these three inverse trigonometric functions in the last section of this chapter.

Figure 7.10 Signs of the trigonometric functions depend on the quadrant where the arc t terminates.

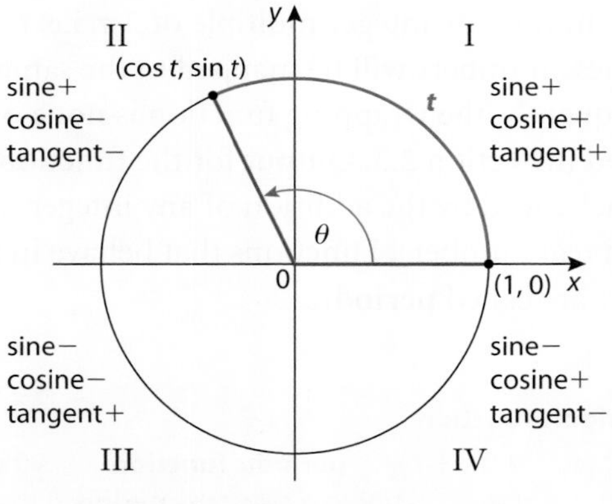

On the unit circle: $x = \cos t, y = \sin t$.

• **Hint:** When sine, cosine and tangent are defined as circular functions based on the unit circle, radian measure is used. The values for the domain of the sine and cosine functions are real numbers that are arc lengths on the unit circle. As we know from the previous section, the arc length on the unit circle subtends an angle in standard position, whose radian measure is equivalent to the arc length (see Figure 7.10).

Because the definitions for the sine, cosine and tangent functions given here do not refer to triangles or angles, but rather to a real number representing an arc length on the unit circle, the name **circular functions** is also given to them. In fact, from this chapter's perspective that these functions are *functions of real numbers* rather than *functions of angles*, 'circular' is a more appropriate adjective than 'trigonometric'. Nevertheless, trigonometric is the more common label and will be used throughout the book.

Let's use the definitions for these three trigonometric, or circular, functions to evaluate them for some 'easy' values of t.

Example 6

Evaluate the sine, cosine and tangent functions for the following values of t.

a) $t = 0$ b) $t = \frac{\pi}{2}$ c) $t = \pi$

d) $t = \frac{3\pi}{2}$ e) $t = 2\pi$

Solution

Evaluating the sin, cos and tan functions for any value of t involves finding the coordinates of the point on the unit circle where the arc of length t will 'wrap to' (or terminate), starting at the point $(1, 0)$. It is useful to remember that an arc of length π is equal to one-half of the circumference of the unit circle. All of the values for t in this example are positive, so the arc length will wrap along the unit circle in an anticlockwise direction.

a) An arc of length $t = 0$ has no length so it 'terminates' at the point $(1, 0)$. By definition:

$\sin 0 = y = 0$ $\cos 0 = x = 1$

$\tan 0 = \frac{y}{x} = \frac{0}{1} = 0$ $\csc 0 = \frac{1}{y} = \frac{1}{0}$ is undefined

$\sec 0 = \frac{1}{x} = \frac{1}{1} = 1$ $\cot 0 = \frac{x}{y} = \frac{1}{0}$ is undefined

b) An arc of length $t = \frac{\pi}{2}$ is equivalent to one-quarter of the circumference of the unit circle (Figure 7.11) so it terminates at the point $(0, 1)$.
By definition:

$\sin \frac{\pi}{2} = y = 1$ $\cos \frac{\pi}{2} = x = 0$

$\tan \frac{\pi}{2} = \frac{y}{x} = \frac{1}{0}$ is undefined $\csc \frac{\pi}{2} = \frac{1}{y} = 1$

$\sec \frac{\pi}{2} = \frac{1}{x}$ is undefined $\cot \frac{\pi}{2} = \frac{x}{y} = 0$

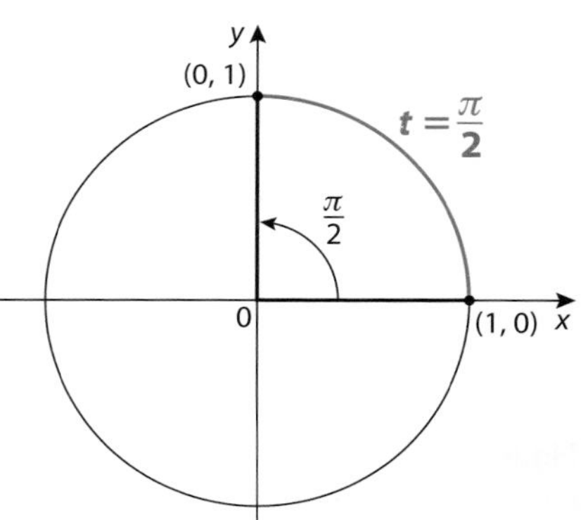

Figure 7.11 Arc length of $\frac{\pi}{2}$ or one-quarter of an anticlockwise revolution.

c) An arc of length $t = \pi$ is equivalent to one-half of the circumference of the unit circle (Figure 7.12) so it terminates at the point $(-1, 0)$. By definition:

$\sin \pi = y = 0$ $\cos \pi = x = -1$

$\tan \pi = \frac{y}{x} = \frac{0}{-1} = 0$ $\csc \pi = \frac{1}{y}$ is undefined

$\sec \pi = \frac{1}{-1} = -1$ $\cot \pi = \frac{x}{y}$ is undefined

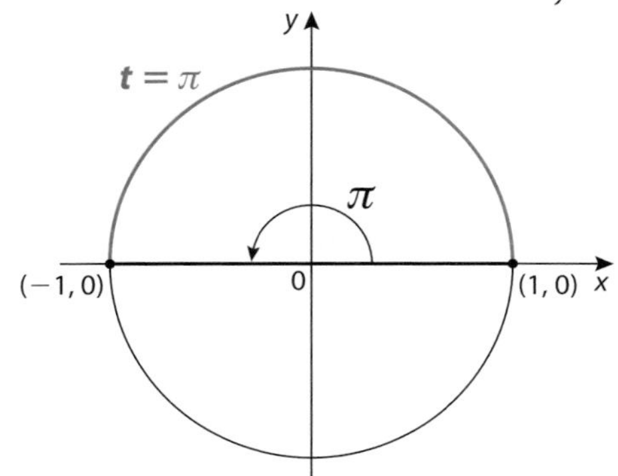

Figure 7.12 Arc length of π, one-half of an anticlockwise revolution.

d) An arc of length $t = \frac{3\pi}{2}$ is equivalent to three-quarters of the circumference of the unit circle (Figure 7.13), so it terminates at the point $(0, -1)$. By definition:

$\sin\frac{3\pi}{2} = y = -1$ $\cos\frac{3\pi}{2} = x = 0$

$\tan\frac{3\pi}{2} = \frac{y}{x} = \frac{-1}{0}$ is undefined $\csc\frac{3\pi}{2} = \frac{1}{y} = -1$

$\sec\frac{3\pi}{2} = \frac{1}{x}$ is undefined $\cot\frac{3\pi}{2} = \frac{x}{y} = 0$

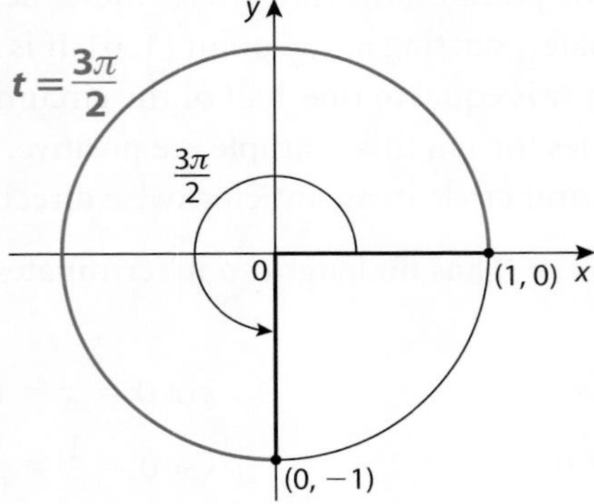

Figure 7.13 Arc length of $\frac{3\pi}{2}$, three-quarters of an anticlockwise revolution.

e) An arc of length $t = 2\pi$ terminates at the same point as arc of length $t = 0$ (Figure 7.14), so the values of the trigonometric functions are the same as found in part a):

$\sin 0 = y = 0$ $\cos 0 = x = 1$

$\tan 0 = \frac{y}{x} = \frac{0}{1} = 0$ $\csc 0 = \frac{1}{y}$ is undefined

$\sec 0 = \frac{1}{x} = 1$ $\cot 0 = \frac{x}{y}$ is undefined

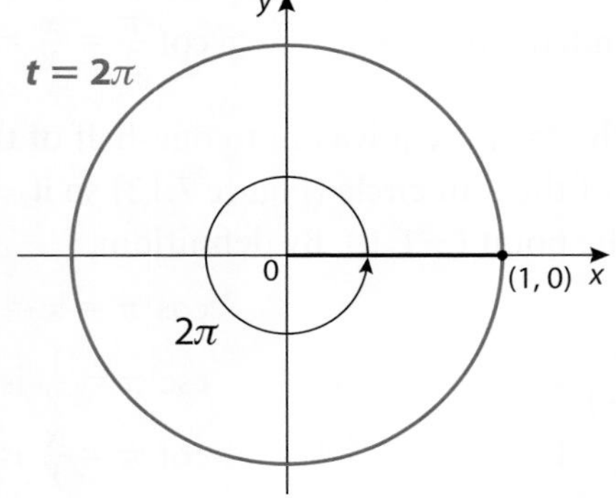

Figure 7.14 Arc length of 2π, one full anticlockwise revolution.

Domain and range of trigonometric functions

If s and t are coterminal arcs (i.e. terminate at the same point), then the trigonometric functions of s are equal to those of t. That is, $\sin s = \sin t$, $\cos s = \cos t$, etc.

Because every real number t corresponds to exactly one point on the unit circle, the domain for both the sine function and the cosine function is the set of all real numbers. In Example 6, the tangent function and the three reciprocal trigonometric functions were sometimes undefined. Hence, the domain for these functions cannot be all real numbers. From the definitions of the functions, it is clear that the tangent and secant functions will be undefined when the x-coordinate of the arc's terminal point is zero. Therefore, the domain of the tangent and secant functions is all real numbers but **not** including the infinite set of numbers generated by adding any integer multiple of π to $\frac{\pi}{2}$. For example, $\frac{\pi}{2} + \pi = \frac{3\pi}{2}$ and $\frac{\pi}{2} - \pi = -\frac{\pi}{2}$ (see Figure 7.15), thus the tangent and secant of $\frac{3\pi}{2}$ and $-\frac{\pi}{2}$ are undefined. Similarly, the cotangent and cosecant functions will be undefined when the y-coordinate of the arc's terminal point is zero. Therefore, the domain of the cotangent and cosecant functions is all real numbers but **not** including all of the integer multiples of π.

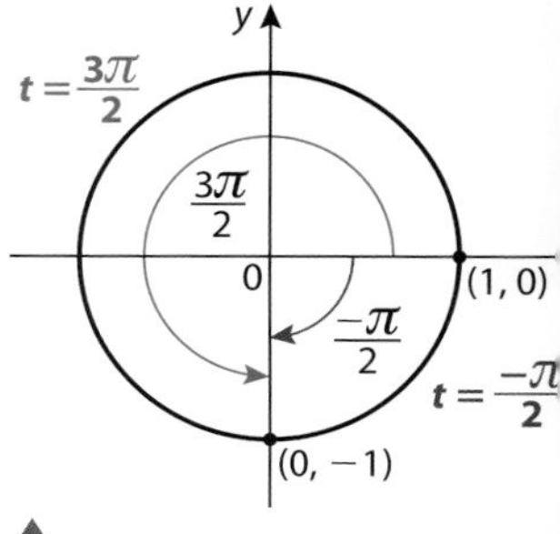

Figure 7.15

Domains of the six trigonometric functions

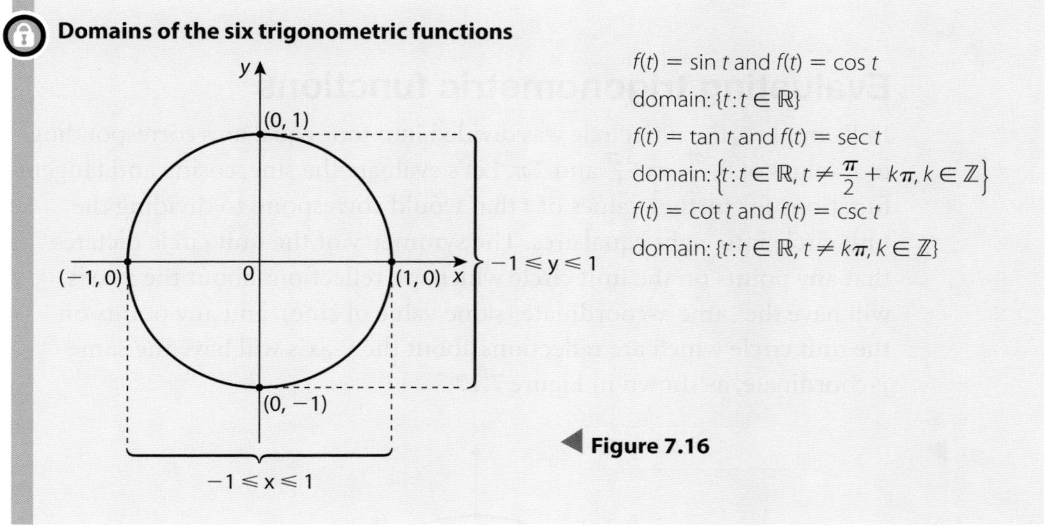

$f(t) = \sin t$ and $f(t) = \cos t$

domain: $\{t : t \in \mathbb{R}\}$

$f(t) = \tan t$ and $f(t) = \sec t$

domain: $\left\{t : t \in \mathbb{R}, t \neq \frac{\pi}{2} + k\pi, k \in \mathbb{Z}\right\}$

$f(t) = \cot t$ and $f(t) = \csc t$

domain: $\{t : t \in \mathbb{R}, t \neq k\pi, k \in \mathbb{Z}\}$

Figure 7.16

To determine the range of the sine and cosine functions, consider the unit circle shown in Figure 7.16. Because $\sin t = y$ and $\cos t = x$ and (x, y) is on the unit circle, we can see that $-1 \leq y \leq 1$ and $-1 \leq x \leq 1$. Therefore, $-1 \leq \sin t \leq 1$ and $-1 \leq \cos t \leq 1$. The range for the tangent function will not be bounded as for sine and cosine. As t approaches values where $x = \cos t = 0$, the value of $\frac{y}{x} = \tan t$ will become very large – either negative or positive, depending on which quadrant t is in. Therefore, $-\infty < \tan t < \infty$; or, in other words, $\tan t$ can be any real number.

Domain and range of sine, cosine and tangent functions

$f(t) = \sin t$	domain: $\{t : t \in \mathbb{R}\}$	range: $-1 \leq f(t) \leq 1$
$f(t) = \cos t$	domain: $\{t : t \in \mathbb{R}\}$	range: $-1 \leq f(t) \leq 1$
$f(t) = \tan t$	domain: $\left\{t : t \in \mathbb{R}, t \neq \frac{\pi}{2} + k\pi, k \in \mathbb{Z}\right\}$	range: $f(t) \in \mathbb{R}$

From our previous discussion of periodic functions, we can conclude that all three of these trigonometric functions are periodic. Given that the sine and cosine functions are generated directly from the wrapping function, the period of each of these functions is 2π. That is,

$$\sin t = \sin(t + k \cdot 2\pi),\ k \in \mathbb{Z} \text{ and } \cos t = \cos(t + k \cdot 2\pi),\ k \in \mathbb{Z}$$

Since the cosecant and secant functions are reciprocals, respectively, of sine and cosine, the period of cosecant and secant will also be 2π.

Initial evidence from Example 6 indicates that the period of the tangent function is π. That is,

$$\tan t = \tan(t + k \cdot \pi), k \in \mathbb{Z}$$

We will establish these results graphically in the next section. Also note that since these functions are periodic then they are not one-to-one functions. This is an important fact with regard to establishing inverse trigonometric functions (Section 7.6).

Evaluating trigonometric functions

In Example 6, the unit circle was divided into four equal arcs corresponding to t values of $0, \frac{\pi}{2}, \pi, \frac{3\pi}{2}$ and 2π. Let's evaluate the sine, cosine and tangent functions for further values of t that would correspond to dividing the unit circle into eight equal arcs. The symmetry of the unit circle dictates that any points on the unit circle which are reflections about the x-axis will have the same x-coordinate (same value of sine), and any points on the unit circle which are reflections about the y-axis will have the same y-coordinate, as shown in Figure 7.17.

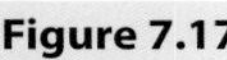
Figure 7.17

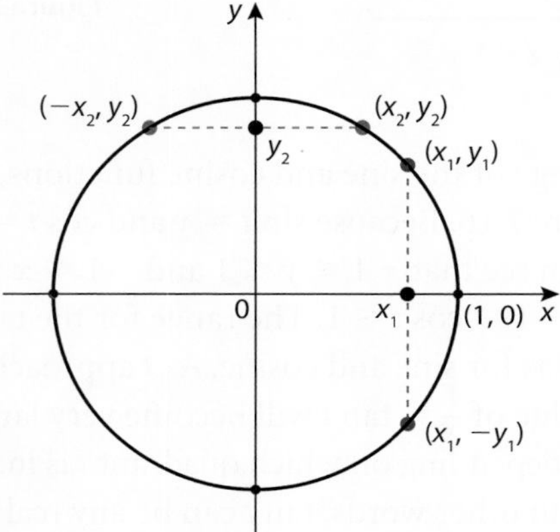

Example 7

Evaluate the sine, cosine and tangent functions for $t = \frac{\pi}{4}$, and then use that result to evaluate the same functions for $t = \frac{3\pi}{4}$, $t = \frac{5\pi}{4}$ and $t = \frac{7\pi}{4}$.

Solution

When an arc of length $t = \frac{\pi}{4}$ is wrapped along the unit circle starting at $(1, 0)$, it will terminate at a point (x_1, y_1) in quadrant I that is equidistant from $(1, 0)$ and $(0, 1)$. Since the line $y = x$ is a line of symmetry for the unit circle, (x_1, y_1) is on this line. Hence, the point (x_1, y_1) is the point of intersection of the unit circle $x^2 + y^2 = 1$ with the line $y = x$. Let's find the coordinates of the intersection point by solving this pair of simultaneous equations by substituting x for y into the equation $x^2 + y^2 = 1$.

$$x^2 + y^2 = 1 \Rightarrow x^2 + x^2 = 1 \Rightarrow 2x^2 = 1 \Rightarrow x^2 = \tfrac{1}{2} \Rightarrow x = \pm\sqrt{\frac{1}{2}} = \pm\frac{1}{\sqrt{2}}$$

Rationalizing the denominator gives $x = \pm\frac{\sqrt{2}}{2}$ and, since the point is in the first quadrant, $x = \frac{\sqrt{2}}{2}$. Given that the point is on the line $y = x$ then $y = \frac{\sqrt{2}}{2}$. Therefore, the arc of length $t = \frac{\pi}{4}$ will terminate at the point $\left(\frac{\sqrt{2}}{2}, \frac{\sqrt{2}}{2}\right)$ on the unit circle. Using the symmetry of the unit circle, we can also determine the points on the unit circle where arcs of length $t = \frac{3\pi}{4}$, $t = \frac{5\pi}{4}$ and $t = \frac{7\pi}{4}$ terminate. These arcs and the coordinates of their terminal points are given in Figure 7.18.

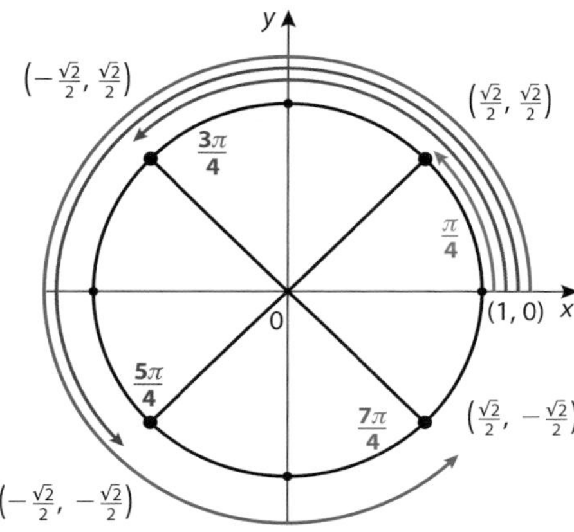

Figure 7.18

Using the coordinates of these points, we can now evaluate the trigonometric functions for $t = \frac{\pi}{4}, \frac{3\pi}{4}, \frac{5\pi}{4}$ and $\frac{7\pi}{4}$. By definition:

$$t = \frac{\pi}{4}: \quad \sin\frac{\pi}{4} = y = \frac{\sqrt{2}}{2} \qquad \cos\frac{\pi}{4} = x = \frac{\sqrt{2}}{2} \qquad \tan\frac{\pi}{4} = \frac{y}{x} = \frac{\frac{\sqrt{2}}{2}}{\frac{\sqrt{2}}{2}} = 1$$

$$t = \frac{3\pi}{4}: \sin\frac{3\pi}{4} = y = \frac{\sqrt{2}}{2} \qquad \cos\frac{3\pi}{4} = x = -\frac{\sqrt{2}}{2} \qquad \tan\frac{3\pi}{4} = \frac{y}{x} = \frac{\frac{\sqrt{2}}{2}}{-\frac{\sqrt{2}}{2}} = -1$$

$$t = \frac{5\pi}{4}: \sin\frac{5\pi}{4} = y = -\frac{\sqrt{2}}{2} \qquad \cos\frac{5\pi}{4} = x = -\frac{\sqrt{2}}{2} \qquad \tan\frac{5\pi}{4} = \frac{y}{x} = \frac{-\frac{\sqrt{2}}{2}}{-\frac{\sqrt{2}}{2}} = 1$$

$$t = \frac{7\pi}{4}: \sin\frac{7\pi}{4} = y = -\frac{\sqrt{2}}{2} \qquad \cos\frac{7\pi}{4} = x = \frac{\sqrt{2}}{2} \qquad \tan\frac{7\pi}{4} = \frac{y}{x} = \frac{-\frac{\sqrt{2}}{2}}{\frac{\sqrt{2}}{2}} = -1$$

We can use a method similar to that of Example 7 to find the point on the unit circle where an arc of length $t = \frac{\pi}{6}$ terminates in the first quadrant. Then we can again apply symmetry about the line $y = x$ and the y- and x-axes to find points on the circle corresponding to arcs whose lengths are multiples of $\frac{\pi}{6}$, e.g. $\frac{2\pi}{6} = \frac{\pi}{3}, \frac{4\pi}{6} = \frac{2\pi}{3}$, etc. Arcs whose lengths are multiples of $\frac{\pi}{4}$ and $\frac{\pi}{6}$ correspond to eight equally spaced points and twelve equally spaced points, respectively, around the unit circle, as shown in Figures 7.19 and 7.20. The coordinates of these points give us the sine, cosine and tangent values for common values of *t*.

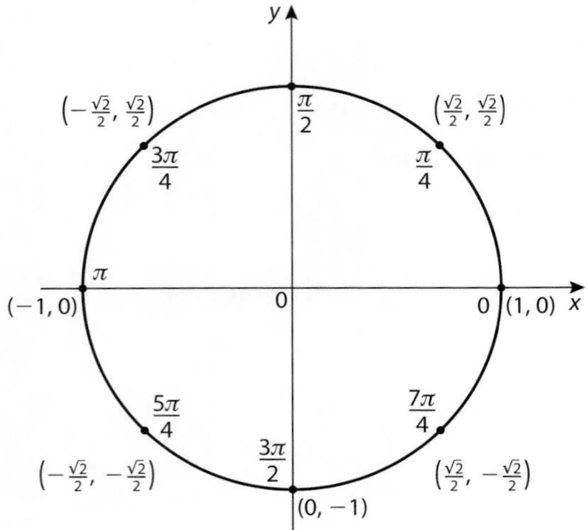

Figure 7.19 Arc lengths that are multiples of $\frac{\pi}{4}$ divide the unit circle into eight equally spaced points.

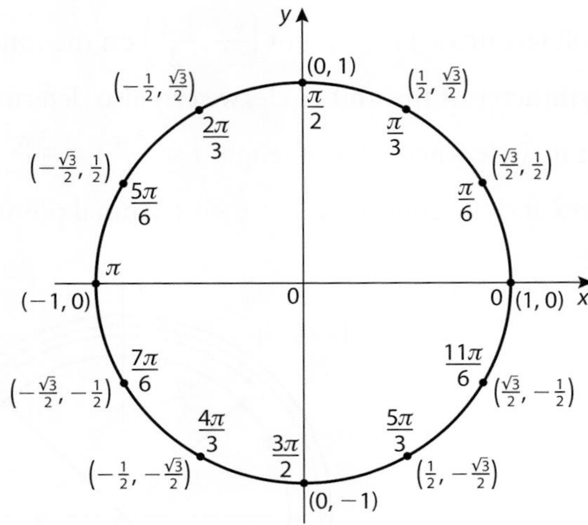

Figure 7.20 Arc lengths that are multiples of $\frac{\pi}{6}$ divide the unit circle into twelve equally spaced points.

The tangent, cosecant, secant and cotangent functions can all be expressed in terms of the sine and/or cosine functions. The following four identities follow directly from the definitions for the trigonometric functions.

$\tan t = \frac{\sin t}{\cos t}$ $\quad \csc t = \frac{1}{\sin t}$

$\sec t = \frac{1}{\cos t}$ $\quad \cot t = \frac{\cos t}{\sin t}$

You will find it very helpful to know from memory the exact values of sine and cosine for numbers that are multiples of $\frac{\pi}{6}$ and $\frac{\pi}{4}$. Use the unit circle diagrams shown in Figures 7.19 and 7.20 as a guide to help you do this and to visualize the location of the terminal points of different arc lengths. With the symmetry of the unit circle and a point's location in the coordinate plane telling us the sign of *x* and *y* (see Figure 7.10), we only need to remember the sine and cosine of common values of *t* in the first quadrant and on the positive *x*- and *y*-axes. These are organized in Table 7.1.

Table 7.1 The trigonometric functions evaluated for special values of *t*.

t	$\sin t$	$\cos t$	$\tan t$	$\csc t$	$\sec t$	$\cot t$
0	0	1	0	undefined	1	undefined
$\frac{\pi}{6}$	$\frac{1}{2}$	$\frac{\sqrt{3}}{2}$	$\frac{\sqrt{3}}{3}$	2	$\frac{2\sqrt{3}}{3}$	$\sqrt{3}$
$\frac{\pi}{4}$	$\frac{\sqrt{2}}{2}$	$\frac{\sqrt{2}}{2}$	1	$\sqrt{2}$	$\sqrt{2}$	1
$\frac{\pi}{3}$	$\frac{\sqrt{3}}{2}$	$\frac{1}{2}$	$\sqrt{3}$	$\frac{2\sqrt{3}}{3}$	2	$\frac{\sqrt{3}}{3}$
$\frac{\pi}{2}$	1	0	undefined	1	undefined	0

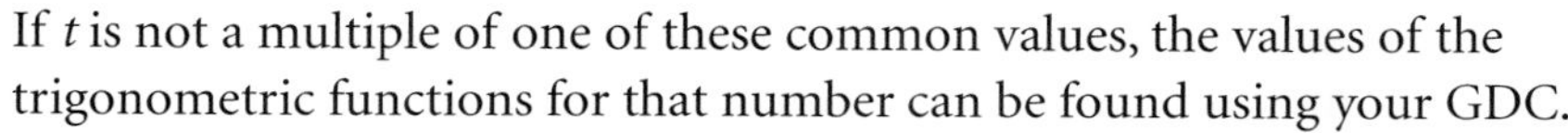

If t is not a multiple of one of these common values, the values of the trigonometric functions for that number can be found using your GDC.

For any arc s on the unit circle ($r = 1$) the arc length formula from the previous section, $s = r\theta$, shows us that each real number t not only measures an arc along the unit circle but also measures a central angle in radians. That is, $t = r\theta = 1 \cdot \theta = \theta$ in radian measure. Therefore, when you are evaluating a trigonometric function it does not make a difference whether the argument of the function is considered to be a real number (i.e. length of an arc) or an angle in radians.

• **Hint:** Memorize the values of sin t and cos t for the values of t that are highlighted in the red box in Table 7.1. These values can be used to derive the values of all six trigonometric functions for any multiple of $\frac{\pi}{6}, \frac{\pi}{4}, \frac{\pi}{3}$ or $\frac{\pi}{2}$.

Example 8

Find the following function values. Find the exact value, if possible. Otherwise, find the approximate value accurate to 3 significant figures.

a) $\sin\frac{2\pi}{3}$ b) $\cos\frac{5\pi}{4}$ c) $\tan\frac{11\pi}{6}$

d) $\csc\frac{13\pi}{6}$ e) $\sec 3.75$

Solution

a) The terminal point for $\frac{2\pi}{3}$ is in the second quadrant and is the reflection in the y-axis of the terminal point for $\frac{\pi}{3}$, whose y-coordinate is $\frac{\sqrt{3}}{2}$. Therefore, $\sin\frac{2\pi}{3} = \frac{\sqrt{3}}{2}$.

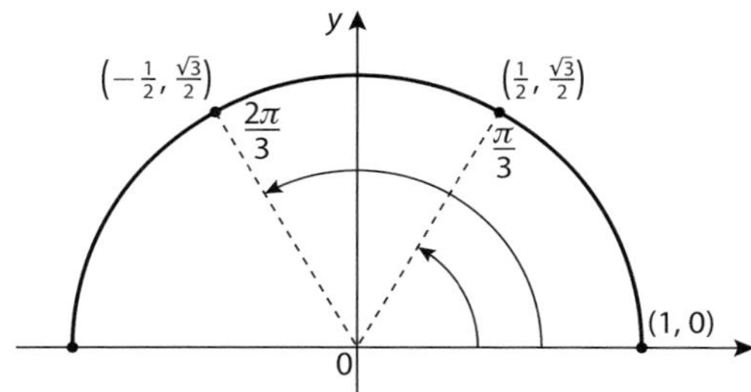

b) $\frac{5\pi}{4}$ is in the third quadrant. Hence, its x-coordinate and cosine must be negative. All of the odd multiples of $\frac{\pi}{4}$ have terminal points with x- and y-coordinates of $\pm\frac{\sqrt{2}}{2}$. Therefore, $\cos\frac{5\pi}{4} = -\frac{\sqrt{2}}{2}$.

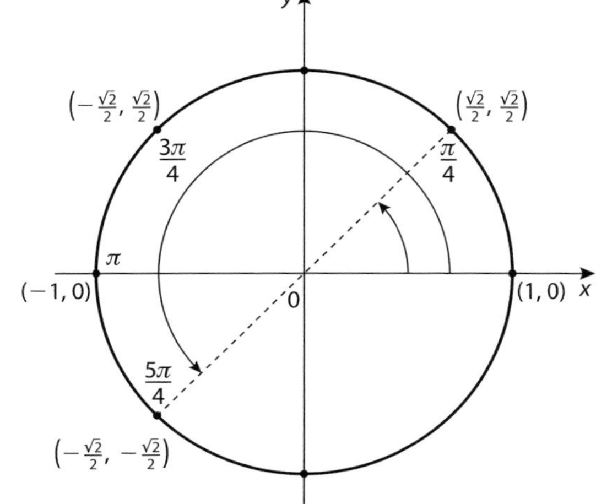

c) $\frac{11\pi}{6}$ is in the fourth quadrant, so its tangent will be negative. Its terminal point is the reflection in the x-axis of the terminal point for $\frac{\pi}{6}$, whose coordinates are $\left(\frac{\sqrt{3}}{2}, \frac{1}{2}\right)$. Therefore, $\tan\frac{11\pi}{6} = \frac{y}{x} = \frac{-\frac{1}{2}}{\frac{\sqrt{3}}{2}} = -\frac{1}{\sqrt{3}} = -\frac{\sqrt{3}}{3}$.

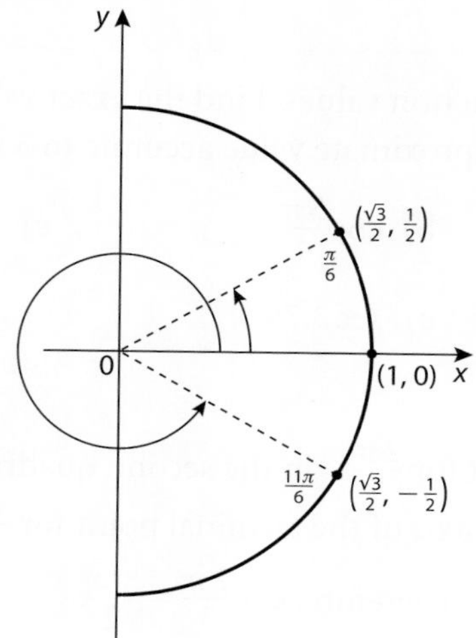

d) $\frac{13\pi}{6}$ is more than one revolution. Because $\frac{13\pi}{6} = \frac{\pi}{6} + 2\pi$ and the period of the cosecant function is 2π [i.e. $\csc t = \csc(t + k \cdot 2\pi), k \in \mathbb{Z}$], then $\csc\frac{13\pi}{6} = \csc\frac{\pi}{6} = \frac{1}{\sin\frac{\pi}{6}} = \frac{1}{\frac{1}{2}} = 2$.

e) To evaluate sec 3.75 you must use your GDC. An arc of length 3.75 will have its terminal point in the third quadrant since $\pi \approx 3.14$ and $\frac{3\pi}{2} \approx 4.71$, meaning $\pi < 3.75 < \frac{3\pi}{2}$. Hence, cos 3.75 must be negative, and because the secant function is the reciprocal of cosine, then sec 3.75 is also negative. This fact indicates that the result in the second GDC image below must be incorrect with the GDC wrongly set to 'degree' mode. Changing to 'radian' mode allows for the correct result to be computed. To an accuracy of three significant figures, $\sec 3.75 \approx -1.22$.

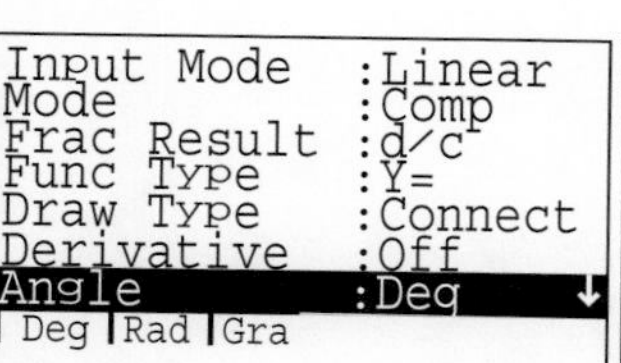

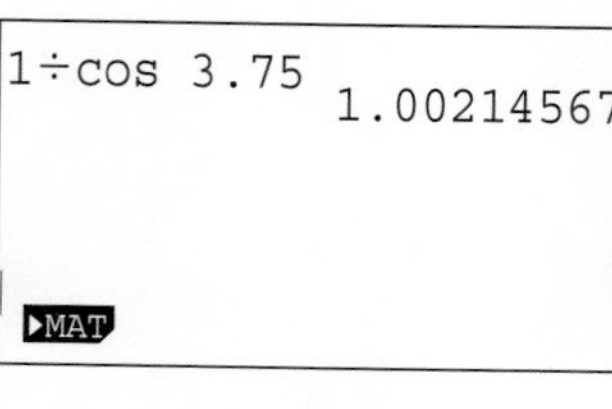

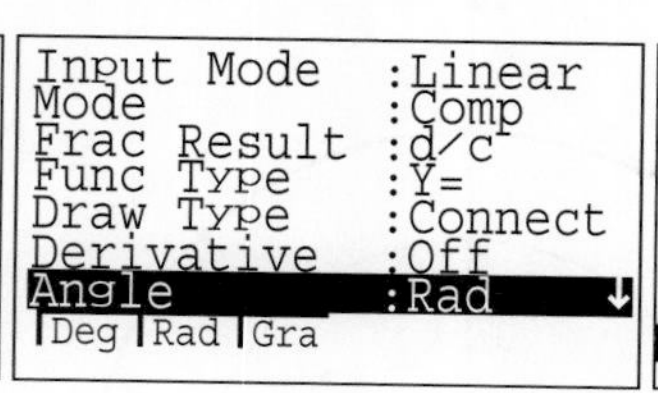

1÷cos 3.75 1.002145671
1÷cos 3.75 -1.21868088
MAT

Have you ever wondered how your calculator computes a value for a trigonometric function – such as cos 0.75? Evaluating an algebraic function (Chapter 3) is relatively straightforward because, by definition, it consists of a finite number of elementary operations (i.e. addition, subtraction, multiplication, division, and extracting a root). It is not so straightforward to evaluate non-algebraic functions like exponential, logarithmic and trigonometric functions and efforts by mathematicians to do so have led to some sophisticated approximation techniques using **power series** that

are studied in further calculus. A power series is an infinite series that can be thought of as a polynomial with an infinite number of terms. You will learn about the theory and application of power series if your Mathematics HL class covers the *Option: Infinite series and differential equations*. If you look in the Mathematics HL Information (Formulae) Booklet in the Topic 10 section (for series and differential equations) you will see the power series (infinite polynomial) approximation for some functions including the cosine function.

$\cos x = 1 - \frac{x^2}{2!} + \frac{x^4}{4!} - \ldots$ where $n! = 1 \cdot 2 \cdot 3 \ldots n$ [$n!$ is read '*n* factorial']

Exploiting the fact that polynomial functions are easy to evaluate, we can easily program a calculator to compute enough terms of the power series to obtain a result to the required accuracy. For example, if we use the first three terms of the power series for cosine to find cos 0.75, we get

$\cos 0.75 = 1 - \frac{0.75^2}{2!} + \frac{0.75^4}{4!} = 0.731\,933\,593\,75$. Compare this to the value obtained using your GDC.

Several important mathematicians in the 17th and 18th centuries, including Isaac Newton, James Gregory, Gottfried Leibniz, Leonhard Euler and Joseph Fourier, contributed to the development of using power series to represent non-algebraic functions. However, the two names most commonly associated with power series are the English mathematician Brook Taylor (1685–1731) and the Scottish mathematician Colin Maclaurin (1698–1746).

Exercise 7.2

1 a) By knowing the ratios of sides in any triangle with angles measuring 30°, 60° and 90° (see figure), find the coordinates of the points on the unit circle where an arc of length $t = \frac{\pi}{6}$ and $t = \frac{\pi}{3}$ terminate in the first quadrant.

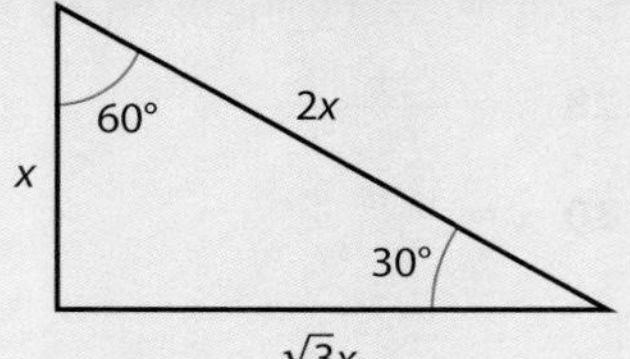

b) Using the result from a) and applying symmetry about the unit circle, find the coordinates of the points on the unit circle corresponding to arcs whose lengths are $\frac{2\pi}{3}, \frac{5\pi}{6}, \frac{7\pi}{6}, \frac{4\pi}{3}, \frac{5\pi}{3}, \frac{11\pi}{6}$.

Draw a large unit circle and label all of these points with their coordinates and the measure of the arc that terminates at each point.

Questions 2–9

The figure of quadrant I of the unit circle shown right indicates angles in intervals of 10 degrees and also indicates angles in radian measure of 0.5, 1 and 1.5. Use the figure and the definitions of the sine and cosine functions to approximate the function values to one decimal place in questions 2–9. Check your answers with your GDC (be sure to be in the correct angle measure mode).

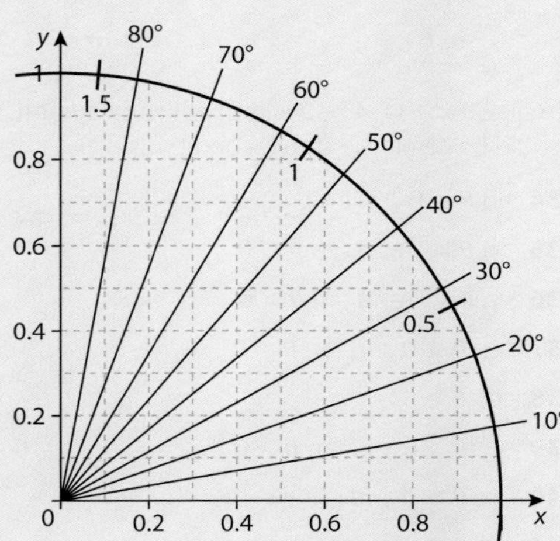

2 $\cos 50°$ **3** $\sin 80°$ **4** $\cos 1$

5 $\sin 0.5$ **6** $\tan 70°$ **7** $\cos 1.5$

8 $\sin 20°$ **9** $\tan 1$

In questions 10–18, t is the length of an arc on the unit circle starting from $(1, 0)$. a) State the quadrant in which the terminal point of the arc lies. b) Find the coordinates of the terminal point (x, y) on the unit circle. Give exact values for x and y, if possible. Otherwise, approximate values to 3 significant figures.

10 $t = \frac{\pi}{6}$ **11** $t = \frac{5\pi}{3}$ **12** $t = \frac{7\pi}{4}$

13 $t = \frac{3\pi}{2}$ **14** $t = 2$ **15** $t = -\frac{\pi}{4}$

16 $t = -1$ **17** $t = -\frac{5\pi}{4}$ **18** $t = 3.52$

In questions 19–27, state the exact value of the sine, cosine and tangent of the given real number.

19 $\frac{\pi}{3}$ **20** $\frac{5\pi}{6}$ **21** $-\frac{3\pi}{4}$

22 $\frac{\pi}{2}$ **23** $-\frac{4\pi}{3}$ **24** 3π

25 $\frac{3\pi}{2}$ **26** $-\frac{7\pi}{6}$ **27** $t = 1.25\pi$

In questions 28–31, use the periodic properties of the sine and cosine functions to find the exact value of $\sin x$ and $\cos x$.

28 $x = \frac{13\pi}{6}$ **29** $x = \frac{10\pi}{3}$

30 $x = \frac{15\pi}{4}$ **31** $x = \frac{17\pi}{6}$

32 Find the exact function values, if possible. Do not use your GDC.

a) $\cos \frac{5\pi}{6}$ b) $\sin 315°$ c) $\tan \frac{3\pi}{2}$

d) $\sec \frac{5\pi}{3}$ e) $\csc 240°$

33 Find the exact function values, if possible. Otherwise, use your GDC to find the approximate value accurate to three significant figures.

a) $\sin 2.5$ b) $\cot 120°$ c) $\cos \frac{5\pi}{4}$

d) $\sec 6$ e) $\tan \pi$

In questions 34–41, specify in which quadrant(s) an angle θ in standard position could be given the stated conditions.

34 $\sin \theta > 0$

35 $\sin \theta > 0$ and $\cos \theta < 0$

36 $\sin \theta < 0$ and $\tan \theta > 0$

37 $\cos \theta < 0$ and $\tan \theta < 0$

38 $\cos \theta > 0$

39 $\sec \theta > 0$ and $\tan \theta > 0$

40 $\cos \theta > 0$ and $\csc \theta < 0$

41 $\cot \theta < 0$

7.3 Graphs of trigonometric functions

The graph of a function provides a useful visual image of its behaviour. For example, from the previous section we know that trigonometric functions are periodic, i.e. their values repeat in a regular manner. The graphs of the trigonometric functions should provide a picture of this periodic behaviour. In this section, we will graph the sine, cosine and tangent functions and transformations of the sine and cosine functions.

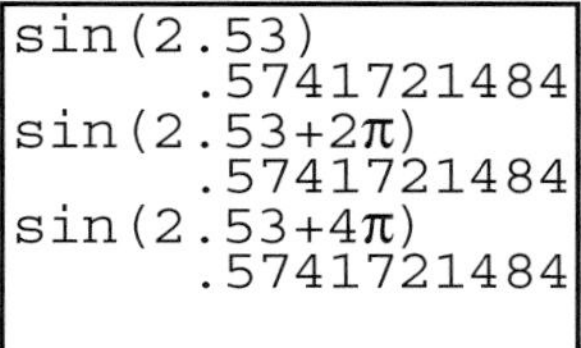

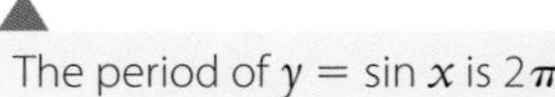

The period of $y = \sin x$ is 2π.

Graphs of the sine and cosine functions

Since the period of the sine function is 2π, we know that two values of t (domain) that differ by 2π (e.g. $\frac{\pi}{6}$ and $\frac{13\pi}{6}$ in Example 8d) will produce the same value for y (range). This means that any portion of the graph of $y = \sin t$ with a t-interval of length 2π (called one **period** or **cycle** of the graph) will repeat. Remember that the domain of the sine function is all real numbers, so one period of the graph of $y = \sin t$ will repeat indefinitely in the positive and negative direction. Therefore, in order to construct a complete graph of $y = \sin t$, we need to graph just one period of the function, that is, from $t = 0$ to $t = 2\pi$, and then repeat the pattern in both directions.

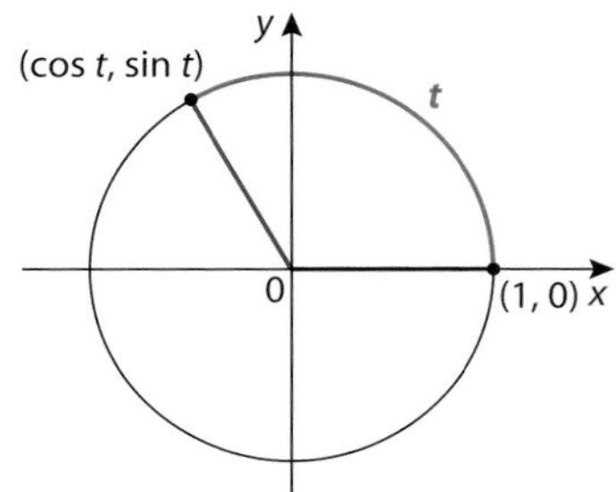

Figure 7.21 Coordinates of terminal point of arc t gives the values of $\cos t$ and $\sin t$.

We know from the previous section that $\sin t$ is the y-coordinate of the terminal point on the unit circle corresponding to the real number t (Figure 7.21). In order to generate one period of the graph of $y = \sin t$, we need to record the y-coordinates of a point on the unit circle and the corresponding value of t as the point travels anticlockwise one revolution, starting from the point $(1, 0)$. These values are then plotted on a graph with t on the horizontal axis and y (i.e. $\sin t$) on the vertical axis. Figure 7.22 illustrates this process in a sequence of diagrams.

Figure 7.22 Graph of the sine function for $0 \leqslant t \leqslant 2\pi$ generated from a point travelling along the unit circle.

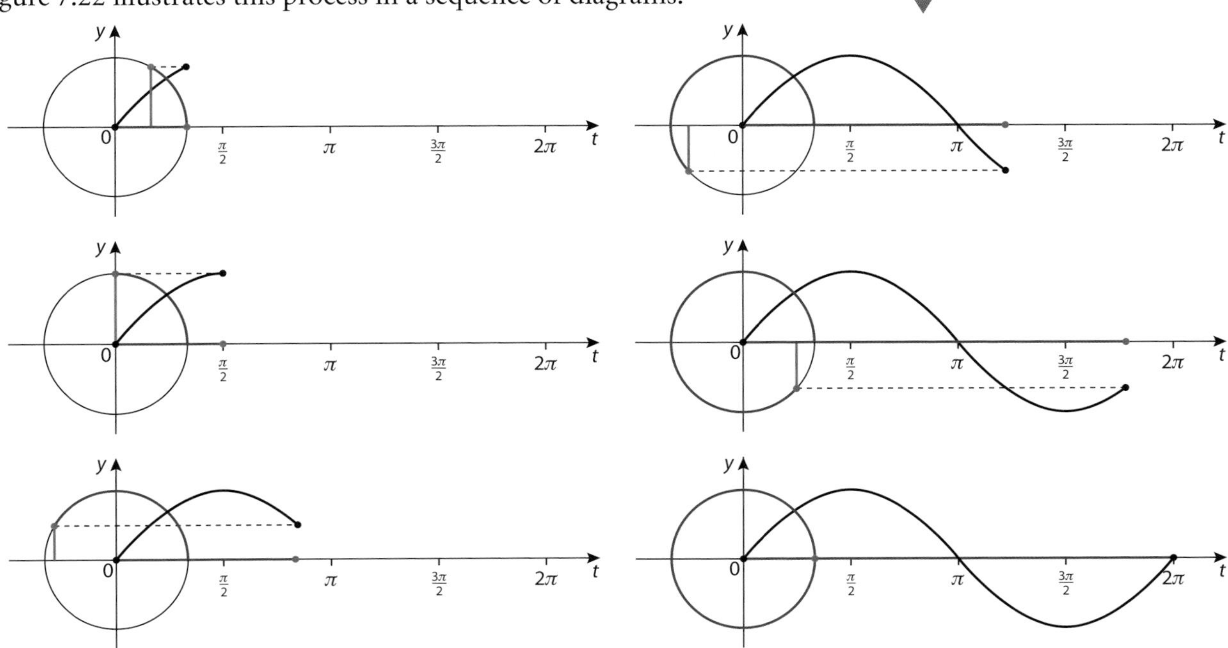

As the point $(\cos t, \sin t)$ travels along the unit circle, the x-coordinate (i.e. $\cos t$) goes through the same cycle of values as the y-coordinate ($\sin t$). The only difference is that the x-coordinate begins at a different value in the cycle – when $t = 0$, $y = 0$, but $x = 1$. The result is that the graph of $y = \cos t$ is the exact same shape as $y = \sin t$ but it has been shifted to the left $\frac{\pi}{2}$ units. The graph of $y = \cos t$ for $0 \leq t \leq 2\pi$ is shown in Figure 7.23.

Figure 7.23 Graph of $y = \cos t$ for $0 \leq t \leq 2\pi$.

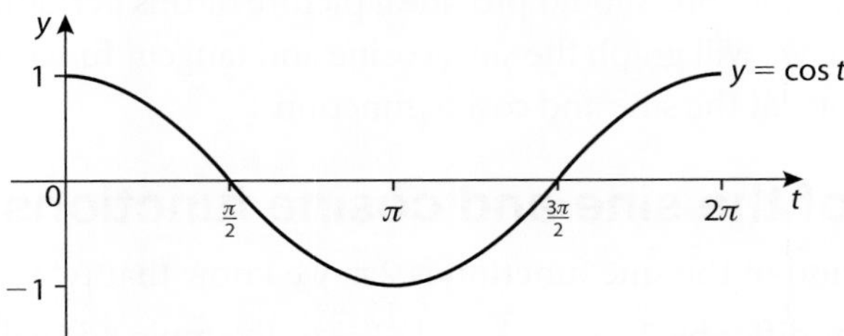

The convention is to use the letter x to denote the variable in the domain of the function. Hence, we will use the letter x rather than t and write the trigonometric functions as $y = \sin x$, $y = \cos x$ and $y = \tan x$.

Because the period for both the sine function and cosine function is 2π, to graph $y = \sin x$ and $y = \cos x$ for wider intervals of x we simply need to repeat the shape of the graph that we generated from the unit circle for $0 \leq x \leq 2\pi$ (Figures 7.22 and 7.23). Figure 7.24 shows the graphs of $y = \sin x$ and $y = \cos x$ for $-4\pi \leq x \leq 4\pi$.

Figure 7.24 $y = \sin x$ and $y = \cos x$, $0 \leq x \leq 4\pi$.

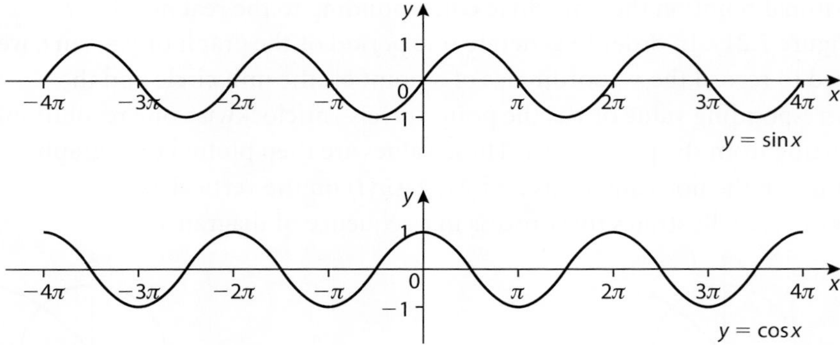

Aside from their periodic behaviour, these graphs reveal further properties of the graphs of $y = \sin x$ and $y = \cos x$. Note that the sine function has a maximum value of $y = 1$ for all $x = \frac{\pi}{2} + k \cdot 2\pi$, $k \in \mathbb{Z}$, and has a minimum value of $y = -1$ for all $x = -\frac{\pi}{2} + k \cdot 2\pi$, $k \in \mathbb{Z}$. The cosine function has a maximum value of $y = 1$ for all $x = k \cdot 2\pi$, $k \in \mathbb{Z}$, and has a minimum value of $y = -1$ for all $x = \pi + k \cdot 2\pi$, $k \in \mathbb{Z}$. This also confirms – as established in the previous section – that both functions have a domain of all real numbers and a range of $-1 \leq y \leq 1$.

Closer inspection of the graphs, in Figure 7.24, shows that the graph of $y = \sin x$ has rotational symmetry about the origin – that is, it can be rotated one-half of a revolution about $(0, 0)$ and it remains the same. This graph symmetry can be expressed with the identity: $\sin(-x) = -\sin x$. For example, $\sin\left(-\frac{\pi}{6}\right) = -\frac{1}{2}$ and $-\left[\sin\left(\frac{\pi}{6}\right)\right] = -\left[\frac{1}{2}\right] = -\frac{1}{2}$. A function that is symmetric about the origin is called an **odd function**. The graph of $y = \cos x$ has line symmetry in the y-axis – that is, it can be reflected in the line $x = 0$ and it remains the same. This graph symmetry can be expressed with the identity: $\cos(-x) = \cos x$. For example, $\cos\left(-\frac{\pi}{6}\right) = \frac{\sqrt{3}}{2}$ and $\cos\frac{\pi}{6} = \frac{\sqrt{3}}{2}$. A function that is symmetric about the y-axis is called an **even function**.

Recall that odd and even functions were first discussed in Section 3.1.

Odd and even functions

A function is **odd** if, for each x in the domain of f, $f(-x) = -f(x)$.

The graph of an odd function is symmetric with respect to the origin (rotational symmetry).

A function is **even** if, for each x in the domain of f, $f(-x) = f(x)$.

The graph of an even function is symmetric with respect to the y-axis (line symmetry).

Graphs of transformations of the sine and cosine functions

In Section 2.4, we learned how to transform the graph of a function by horizontal and vertical translations, by reflections in the coordinate axes, and by stretching and shrinking – both horizontal and vertical. The following is a review of these transformations.

Review of transformations of graphs of functions

Assume that a, b, c and d are real numbers.

To obtain the graph of:	From the graph of $y = f(x)$:
$y = f(x) + d$	Translate d units up for $d > 0$, d units down for $d < 0$.
$y = f(x + c)$	Translate c units left for $c > 0$, c units right for $c < 0$.
$y = -f(x)$	Reflect in the x-axis.
$y = af(x)$	Vertical stretch ($a > 1$) or shrink ($0 < a < 1$) of factor a.
$y = f(-x)$	Reflect in the y-axis.
$y = f(bx)$	Horizontal stretch ($0 < b < 1$) or shrink ($b > 1$) of factor $\frac{1}{b}$.

In this section, we will look at the composition of sine and cosine functions of the form

$$f(x) = a\sin[b(x + c)] + d \quad \text{and} \quad f(x) = a\cos[b(x + c)] + d$$

Example 9

Sketch the graph of each function on the interval $-\pi \leqslant x \leqslant 3\pi$.

a) $f(x) = 2\cos x$

b) $g(x) = \cos x + 3$

c) $h(x) = 2\cos x + 3$

d) $p(x) = \frac{1}{2}\sin x - 2$

Solution

a) Since $a = 2$, the graph of $y = 2\cos x$ is obtained by vertically stretching the graph of $y = \cos x$ by a factor of 2.

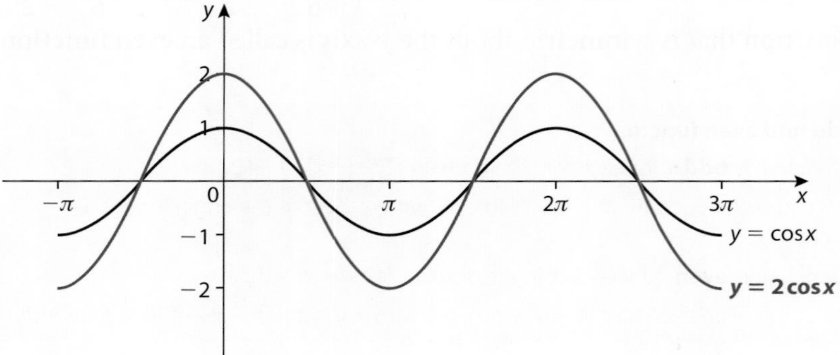

b) Since $d = 3$, the graph of $y = \cos x + 3$ is obtained by translating the graph of $y = \cos x$ three units up.

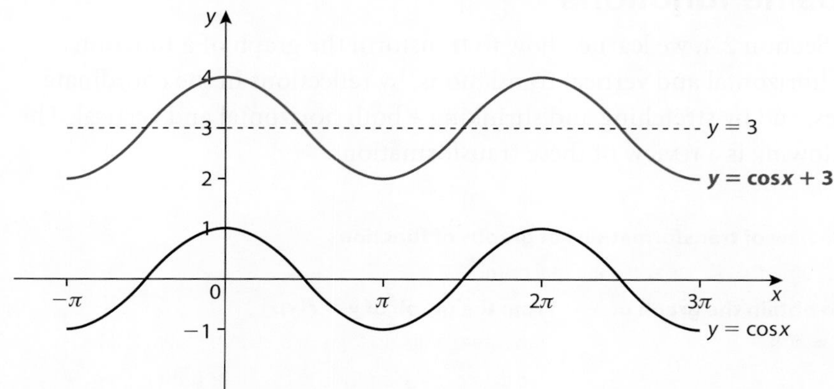

c) We can obtain the graph of $y = 2\cos x + 3$ by combining both of the transformations to the graph of $y = \cos x$ performed in parts a) and b) – namely, a vertical stretch of factor 2 and a translation three units up.

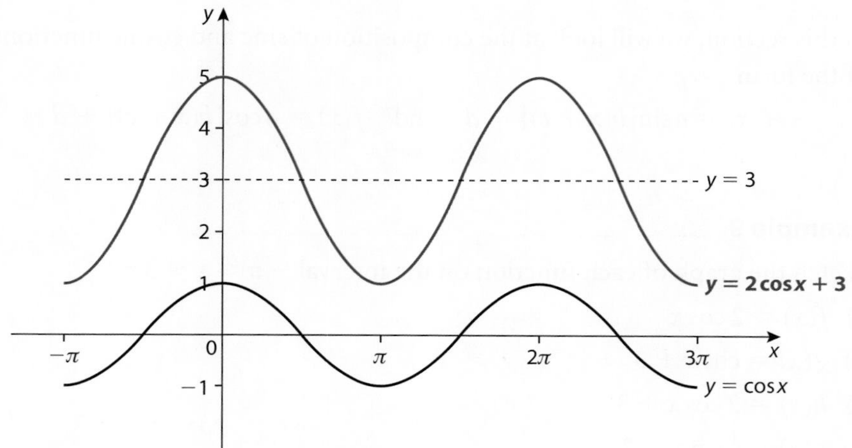

d) The graph of $y = \frac{1}{2}\sin x - 2$ can be obtained by vertically shrinking the graph of $y = \sin x$ by a factor of $\frac{1}{2}$ and then translating it down two units.

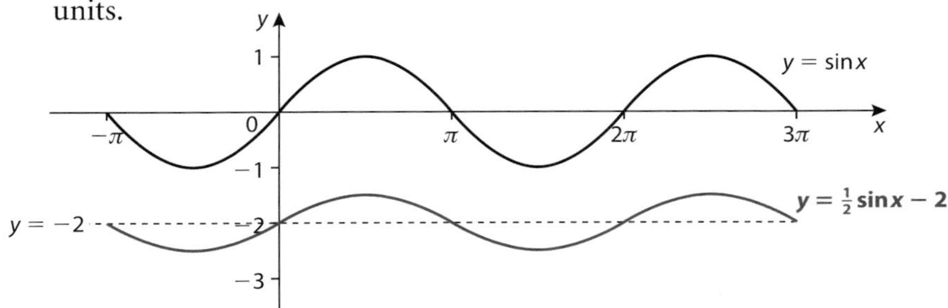

In part a), the graph of $y = 2\cos x$ has many of the same properties as the graph of $y = \cos x$: same period, and the maximum and minimum values occur at the same values of x. However, the graph ranges between -2 and 2 instead of -1 and 1. This difference is best described by referring to the **amplitude** of each graph. The amplitude of $y = \cos x$ is 1 and the amplitude of $y = 2\cos x$ is 2. The amplitude of a sine or cosine graph is not always equal to its maximum value. In part b), the amplitude of $y = \cos x + 3$ is 1; in part c), the amplitude of $y = 2\cos x + 3$ is 2; and the amplitude of $y = \frac{1}{2}\sin x - 2$ is $\frac{1}{2}$. For all three of these, the graphs oscillate about the horizontal line $y = d$. How *high* and *low* the graph oscillates with respect to the mid-line, $y = d$, is the graph's amplitude. With respect to the general form $y = af(x)$, changing the amplitude is equivalent to a vertical stretching or shrinking. Thus, we can give a more precise definition of amplitude in terms of the parameter a.

Amplitude of the graph of sine and cosine functions

The graphs of $f(x) = a\sin[b(x + c)] + d$ and $f(x) = a\cos[b(x + c)] + d$ have an **amplitude** equal to $|a|$.

Example 10

Waves are produced in a long tank of water. The depth of the water, d metres, at t seconds, at a fixed location in the tank, is modelled by the function $d(t) = M\cos\left(\frac{\pi}{2}t\right) + K$, where M and K are positive constants. On the right is the graph of $d(t)$ for $0 \leqslant t \leqslant 12$ indicating that the point (2, 5.1) is a minimum and the point (8, 9.7) is a maximum.

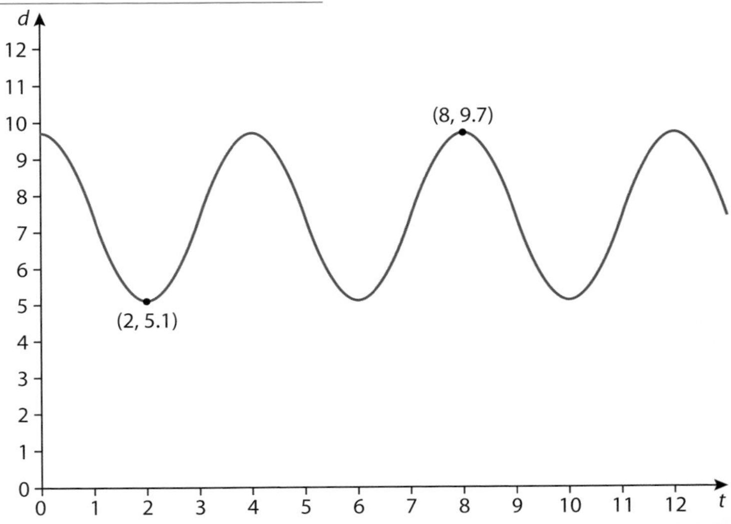

a) Find the value of K and the value of M.

b) After $t = 0$, find the first time when the depth of the water is 9.7 metres.

Solution

a) The constant K is equivalent to the constant d in the general form of a cosine function: $f(x) = a\cos[b(x + c)] + d$. To find the value of K and the equation of the horizontal mid-line, $y = K$, find the average of the function's maximum and minimum value: $K = \frac{9.7 + 5.1}{2} = 7.4$.

The constant M is equivalent to the constant a whose absolute value is the amplitude. The amplitude is the difference between the function's maximum value and the mid-line: $|M| = 9.7 - 7.4 = 2.3$. Thus, $M = 2.3$ or $M = -2.3$. Try $M = 2.3$ by evaluating the function at one of the known values:

$d(2) = 2.3\cos\left(\frac{\pi}{2}(2)\right) + 7.4 = 2.3\cos\pi + 7.4 = 2.3(-1) + 7.4 = 5.1.$

This agrees with the point (2, 5.1) on the graph. Therefore, $M = 2.3$.

b) Maximum values of the function ($d(8) = 9.7$) occur at values of t that differ by a value equal to the period. From the graph, we can see that the difference in t-values from the minimum (2, 5.1) to the maximum (8, 9.7) is equivalent to one-and-a-half periods. Therefore, the period is 4 and the first time after $t = 0$ at which $d = 9.7$ is $t = 4$.

All four of the functions in Example 9 had the same period of 2π, but the function in Example 10 had a period of 4. Because $y = \sin x$ completes one period from $x = 0$ to $x = 2\pi$, it follows that $y = \sin bx$ completes one period from $bx = 0$ to $bx = 2\pi$. This implies that $y = \sin bx$ completes one period from $x = 0$ to $x = \frac{2\pi}{b}$. This agrees with the period for the function $d(t) = 2.3\cos\left(\frac{\pi}{2}t\right) + 7.4$ in Example 10: period $= \frac{2\pi}{b} = \frac{2\pi}{\frac{\pi}{2}} = \frac{2\pi}{1} \cdot \frac{2}{\pi} = 4$.

Note that the change in amplitude and vertical translation had no effect on the period. We should also expect that a horizontal translation of a sine or cosine curve should not affect the period. The next example looks at a function that is horizontally translated (shifted) and has a period different from 2π.

Example 11

Sketch the function $f(x) = \sin\left(2x + \frac{2\pi}{3}\right)$.

Solution

To determine how to transform the graph of $y = \sin x$ to obtain the graph of $y = \sin\left(2x + \frac{2\pi}{3}\right)$, we need to make sure the function is written in the form $f(x) = a\sin[b(x + c)] + d$. Clearly, $a = 1$ and $d = 0$, but we will need to factorize a 2 from the expression $2x + \frac{2\pi}{3}$ to get $f(x) = \sin\left[2\left(x + \frac{\pi}{3}\right)\right]$. According to our general transformations from Chapter 2, we expect that the graph of f is obtained by first performing a horizontal shrinking of factor $\frac{1}{2}$ to the graph of $y = \sin x$ and then a translation to the left $\frac{\pi}{3}$ units (see Section 2.4).

Transformations of the graphs of trigonometric functions follow the same rules as for other functions. The rules were established in Section 2.4 and summarized on page 84.

The graphs on the next page illustrate the two-stage sequence of transforming $y = \sin x$ to $y = \sin\left[2\left(x + \frac{\pi}{3}\right)\right]$.

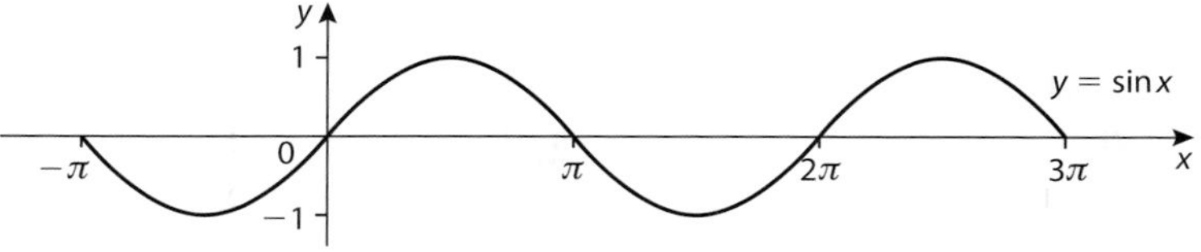

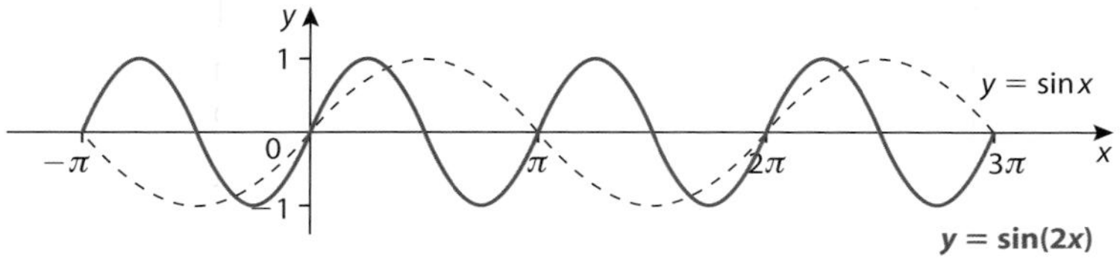

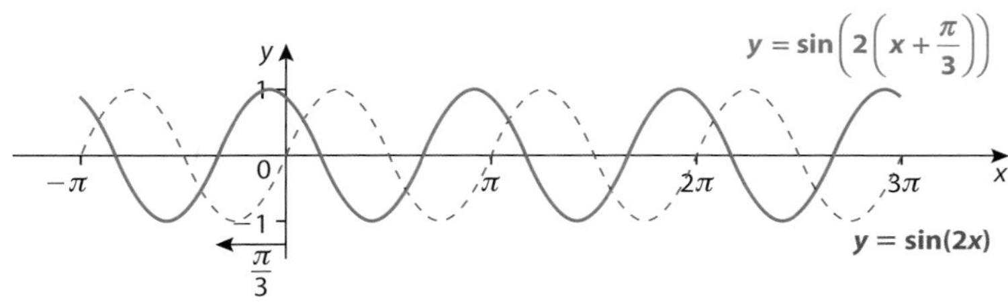

Note: A horizontal translation of a sine or cosine curve is often referred to as a **phase shift**. The equations $y = \sin\left(x + \frac{\pi}{3}\right)$ and $y = \sin\left[2\left(x + \frac{\pi}{3}\right)\right]$ both underwent a phase shift of $-\frac{\pi}{3}$.

Period and horizontal translation (phase shift) of sine and cosine functions

Given that b is a positive real number, $y = a\sin[b(x + c)] + d$ and $y = a\cos[b(x + c)] + d$ have a **period** of $\frac{2\pi}{b}$ and a horizontal translation (**phase shift**) of $-c$.

Example 12

The graph of a function in the form $y = a\cos bx$ is given in the diagram right.

a) Write down the value of a.

b) Calculate the value of b.

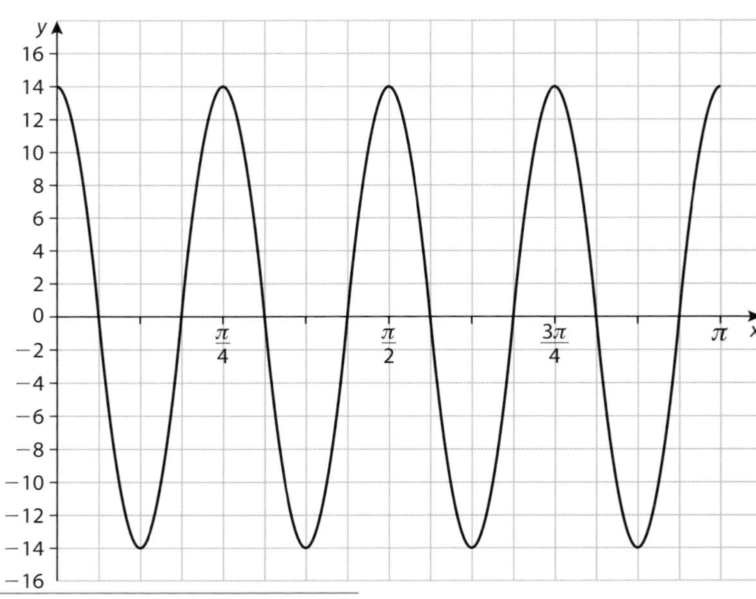

Solution

a) The amplitude of the graph is 14. Therefore, $a = 14$.

b) From inspecting the graph we can see that the period is $\frac{\pi}{4}$.

$$\text{Period} = \frac{2\pi}{b} = \frac{\pi}{4}$$
$$b\pi = 8\pi \Rightarrow b = 8.$$

Example 13

For the function $f(x) = 2\cos\left(\frac{x}{2}\right) - \frac{3}{2}$:

a) Sketch the function for the interval $-\pi \leqslant x \leqslant 5\pi$. Write down its amplitude and period.

b) Determine the domain and range for $f(x)$.

c) Write $f(x)$ as a trigonometric function in terms of sine rather than cosine.

Solution

a) $a = 2 \Rightarrow$ amplitude $= 2$; $b = \frac{1}{2} \Rightarrow$ period $= \frac{2\pi}{\frac{1}{2}} = 4\pi$. To obtain the graph of $y = 2\cos\left(\frac{x}{2}\right) - \frac{3}{2}$, we perform the following transformations on $y = \cos x$: (i) a horizontal stretch by factor $\frac{1}{\frac{1}{2}} = 2$, (ii) a vertical stretch by factor 2, and (iii) a vertical translation down $\frac{3}{2}$ units.

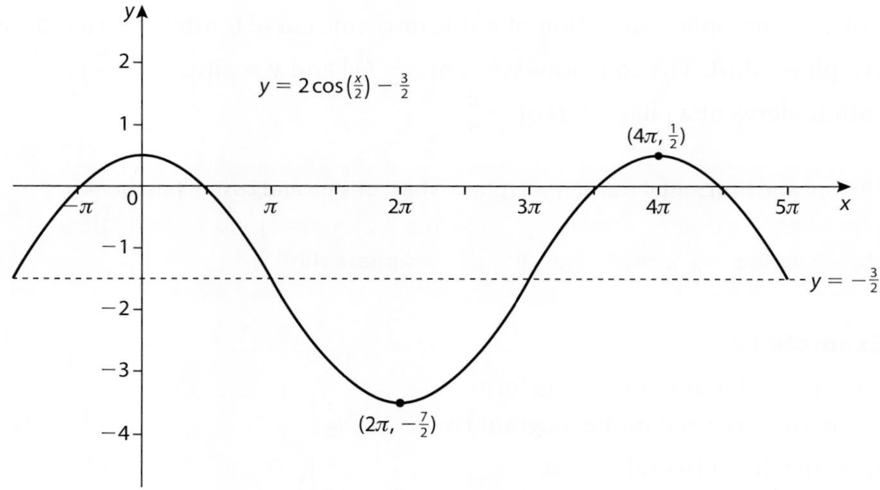

b) The domain is all real numbers. The function will reach a maximum value of $d + a = -\frac{3}{2} + 2 = \frac{1}{2}$, and a minimum value of $d - a = -\frac{3}{2} - 2 = -\frac{7}{2}$.

Hence, the range is $-\frac{7}{2} \leqslant y \leqslant \frac{1}{2}$.

c) The graph of $y = \cos x$ can be obtained by translating the graph of $y = \sin x$ to the left $\frac{\pi}{2}$ units. Thus, $\cos x = \sin\left(x + \frac{\pi}{2}\right)$, or, in other words, any cosine function can be written as a sine function with a phase shift $= -\frac{\pi}{2}$. Therefore, $f(x) = 2\cos\left(\frac{x}{2}\right) - \frac{3}{2} = 2\sin\left(\frac{x}{2} + \frac{\pi}{2}\right) - \frac{3}{2}$.

Horizontal translation (phase shift) identities

The following are true for all values of x:

$$\cos x = \sin\left(x + \frac{\pi}{2}\right) \qquad \sin x = \cos\left(x - \frac{\pi}{2}\right)$$

$$\cos x = \sin\left(\frac{\pi}{2} - x\right) \qquad \sin x = \cos\left(\frac{\pi}{2} - x\right)$$

The identity $\cos x = \sin\left(x + \frac{\pi}{2}\right)$ is equivalent to the identity $\cos x = \sin\left(\frac{\pi}{2} - x\right)$ because $\sin\left(\frac{\pi}{2} - x\right) = \sin\left[-\left(x - \frac{\pi}{2}\right)\right]$ and the graph of $y = \sin\left[-\left(x - \frac{\pi}{2}\right)\right]$ can be obtained by first translating $y = \sin x$ to the right $\frac{\pi}{2}$ units, and then reflecting the graph in the y-axis. This produces the same graph as $y = \cos x$. This can be confirmed nicely on your GDC as shown.
Therefore, $\cos x = \sin\left(\frac{\pi}{2} - x\right)$. In fact, it is also true that $\sin x = \cos\left(\frac{\pi}{2} - x\right)$. Clearly, $x + \left(\frac{\pi}{2} - x\right) = \frac{\pi}{2}$. If the domain ($x$) values were being treated as angles, then x and $\frac{\pi}{2} - x$ would be complementary angles.
This is why cosine is considered the co-function of sine.
Two trigonometric functions f and g are co-functions if the following are true for all x: $f(x) = g\left(\frac{\pi}{2} - x\right)$ and $f\left(\frac{\pi}{2} - x\right) = g(x)$.

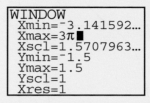

WINDOW
Xmin=⁻3.141592…
Xmax=3π
Xscl=1.5707963…
Ymin=⁻1.5
Ymax=1.5
Yscl=1
Xres=1

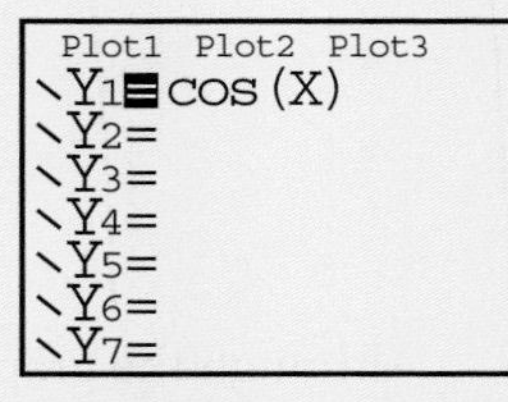

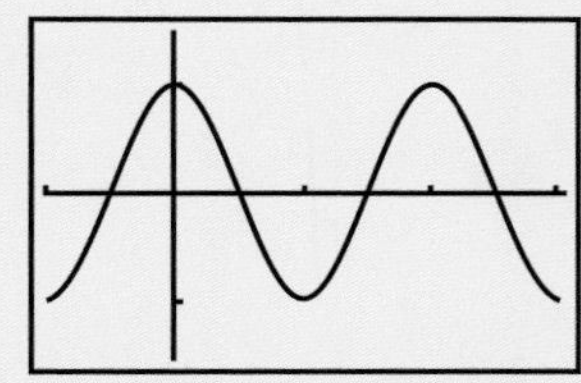

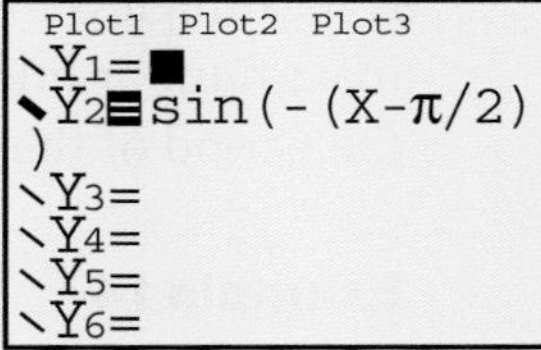

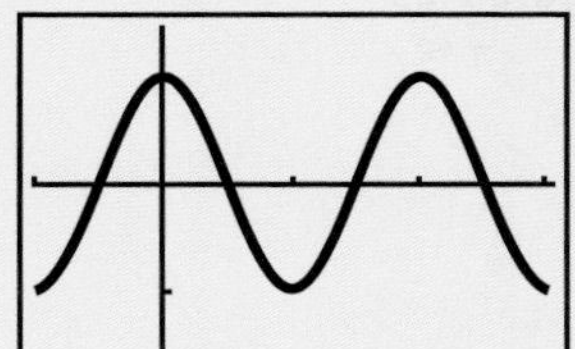

Graph of the tangent function

From work done earlier in this chapter, we expect that the behaviour of the tangent function will be significantly different from that of the sine and cosine functions. In Section 7.2, we concluded that the function $f(x) = \tan x$ has a domain of all real numbers such that $x \neq \frac{\pi}{2} + k\pi$, $k \in \mathbb{Z}$, and that its range is all real numbers. Also, the results for Example 6 in Section 7.2 led us to speculate that the period of the tangent function is π. This makes sense since the identity $\tan x = \frac{\sin x}{\cos x}$ informs us that $\tan x$ will be zero whenever $\sin x = 0$, which occurs at values of x that differ by π (visualize arcs on the unit circle whose terminal points are either $(1, 0)$ or $(-1, 0)$). The values of x for which $\cos x = 0$ cause $\tan x$ to be undefined ('gaps' in the domain) also differ by π (the points $(0, 1)$ or $(0, -1)$ on the unit circle). As x approaches these values where $\cos x = 0$, the value of $\tan x$ will become very large – either very large negative or very large positive.
Thus, the graph of $y = \tan x$ has vertical asymptotes at $x = \frac{\pi}{2} + k\pi$, $k \in \mathbb{Z}$.
Consequently, the graphical behaviour of the tangent function will not be a wave pattern such as that produced by the sine and cosine functions, but rather a series of separate curves that repeat every π units. Figure 7.25 shows the graph of $y = \tan x$ for $-2\pi \leqslant x \leqslant 2\pi$.

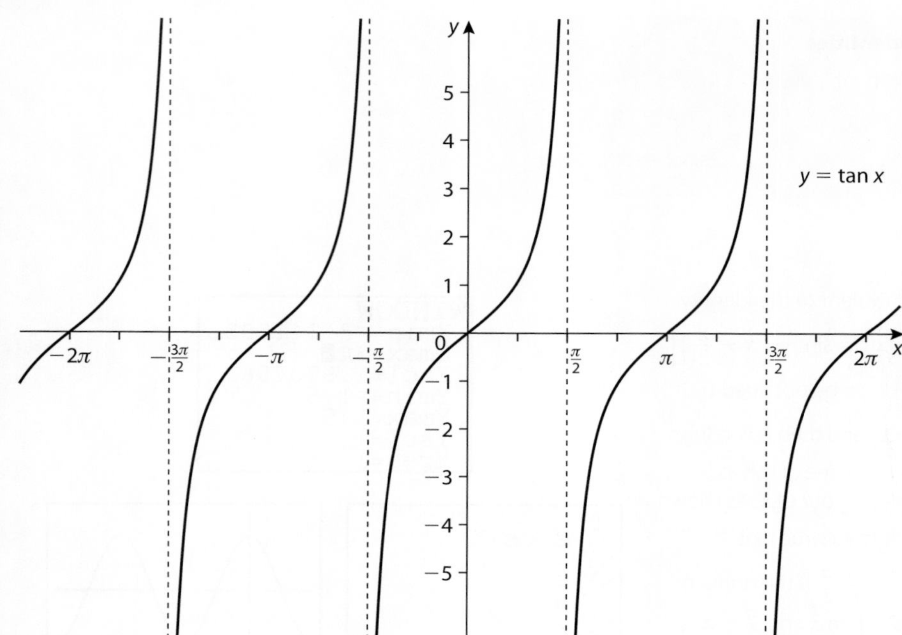

▲ **Figure 7.25** $y = \tan x$ for $-2\pi \leq x \leq 2\pi$.

The graph gives clear confirmation that the period of the tangent function is π, that is, $\tan x = \tan(x + k \cdot \pi)$, $k \in \mathbb{Z}$.

The graph of $y = \tan x$ has rotational symmetry about the origin – that is, it can be rotated one-half of a revolution about (0, 0) and it remains the same. Hence, like the sine function, tangent is an odd function and $\tan(-x) = -\tan x$.

Although the graph of $y = \tan x$ can undergo a vertical stretch or shrink, it is meaningless to consider its amplitude since the tangent function has no maximum or minimum value. However, other transformations can affect the period of the tangent function.

Example 14

Sketch each function.

a) $f(x) = \tan 2x$ b) $g(x) = \tan\left[2\left(x - \frac{\pi}{4}\right)\right]$

Solution

a) An equation in the form $y = f(bx)$ indicates a horizontal shrinking of $f(x)$ by a factor of $\frac{1}{b}$. Hence, the period of $y = \tan 2x$ is $\frac{1}{2} \cdot \pi = \frac{\pi}{2}$.

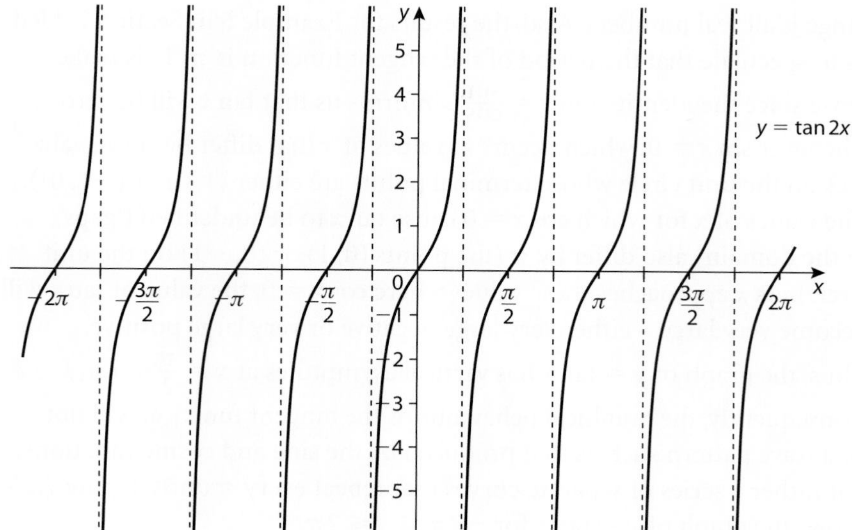

b) The graph of $y = \tan\left[2\left(x - \frac{\pi}{4}\right)\right]$ is obtained by first performing a horizontal shrinking of the graph of $y = \tan x$ by a factor of $\frac{1}{2}$ and then translating the graph to the right $\frac{\pi}{4}$ units. As for $f(x) = \tan 2x$ in part a), the period of $g(x) = \tan\left[2\left(x - \frac{\pi}{4}\right)\right]$ is $\frac{\pi}{2}$.

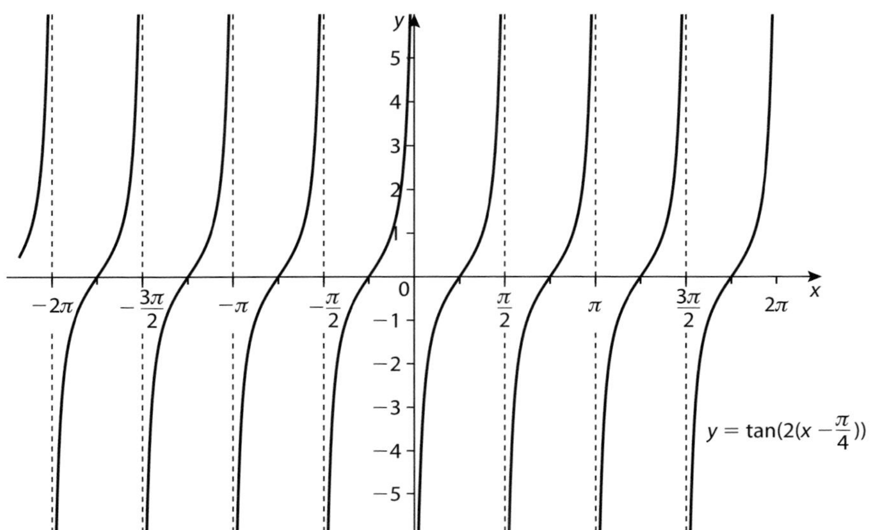

Exercise 7.3

In questions 1–9, without using your GDC, sketch a graph of each equation on the interval $-\pi \leqslant x \leqslant 3\pi$.

1 $y = 2\sin x$

2 $y = \cos x - 2$

3 $y = \frac{1}{2}\cos x$

4 $y = \sin\left(x - \frac{\pi}{2}\right)$

5 $y = \cos(2x)$

6 $y = 1 + \tan x$

7 $y = \sin\left(\frac{x}{2}\right)$

8 $y = \tan\left(x + \frac{\pi}{2}\right)$

9 $y = \cos\left(2x - \frac{\pi}{4}\right)$

For each function in questions 10–12:

a) Sketch the function for the interval $-\pi \leqslant x \leqslant 5\pi$. Write down its amplitude and period.

b) Determine the domain and range for $f(x)$.

10 $f(x) = \frac{1}{2}\cos x - 3$

11 $g(x) = 3\sin(3x) - \frac{1}{2}$

12 $g(x) = 1.2\sin\left(\frac{x}{2}\right) + 4.3$

In questions 13 and 14, a graph of a trigonometric equation is shown, on the interval $0 \leqslant x \leqslant 12$, that can be written in the form $y = A\sin\left(\frac{\pi}{4}x\right) + B$. Two points – one a minimum and the other a maximum – are indicated on the graph. Find the value of A and B for each.

13

14

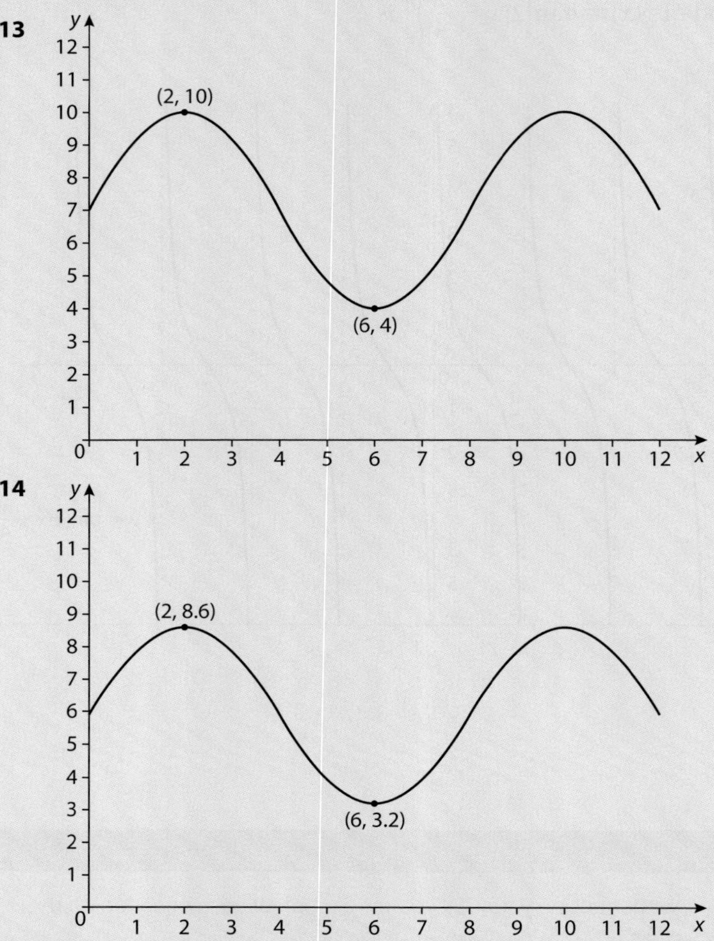

15 A graph of a trigonometric equation is shown below, on the interval $0 \leqslant x \leqslant 12$, that can be written in the form $y = A\cos\left(\frac{\pi}{4}x\right) + B$. Two points – one a minimum and the other a maximum – are indicated on the graph. Find the value of A and of B for each.

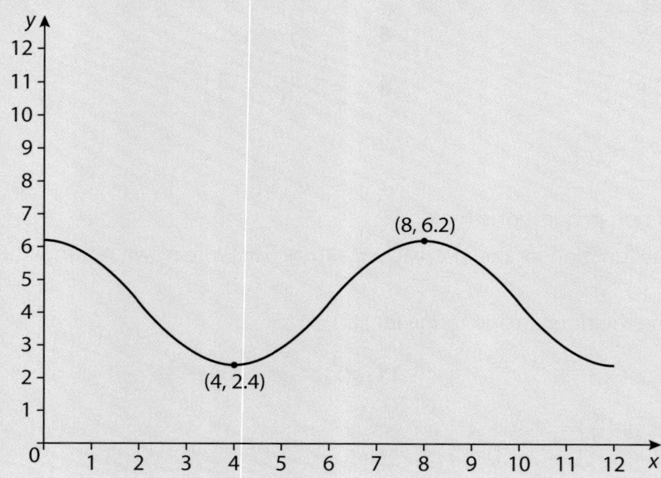

16 The graph of a function in the form $y = p\cos qx$ is given in the diagram below.

a) Write down the value of p.

b) Calculate the value of q.

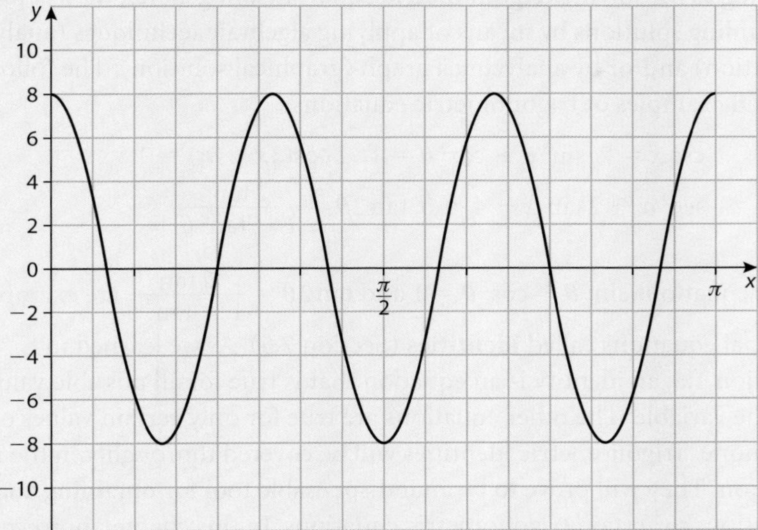

17 a) With help from your GDC, sketch the graphs of the three reciprocal trigonometric functions $y = \csc x$, $y = \sec x$ and $y = \cot x$ for the interval $0 \leqslant x \leqslant 2\pi$. Include any vertical asymptotes as dashed lines.

b) The domain of all of the trigonometric functions is stated in Section 7.2. State the range for each of the three reciprocal trigonometric functions.

18 The diagram shows part of the graph of a function whose equation is in the form $y = a\sin(bx) + c$.

a) Write down the values of a, b and c.

b) Find the exact value of the x-coordinate of the point P, the point where the graph crosses the x-axis as shown in the diagram.

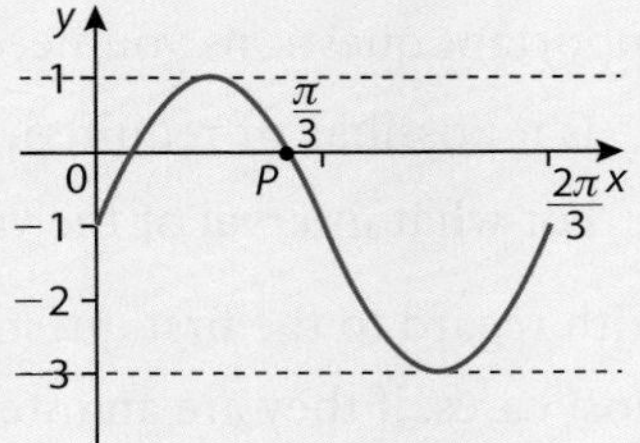

19 The graph below represents $y = a\sin(x + b) + c$, where a, b, and c are constants. Find values for a, b, and c.

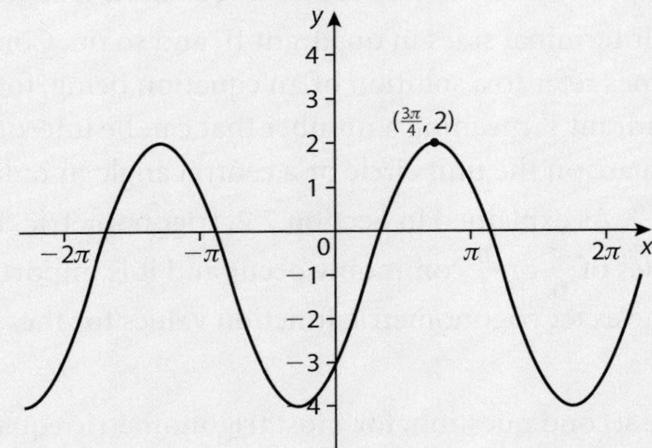

7.4 Trigonometric equations

The primary focus of this section is to give an overview of concepts and strategies for solving **trigonometric equations**. In general, we will look at finding solutions by means of applying algebraic techniques (analytic solution) and/or by analyzing a graph (graphical solution). The following are all examples of trigonometric equations:

$$\csc x = 2,\ \sin^2\theta + \cos^2\theta = 1,\ 2\cos(3x - \pi) = 1,$$

$$\sec^2\alpha - 2\tan\alpha - 4 = 0,\ \tan 2\theta = \frac{2\tan}{1 - \tan^2\theta}$$

The mathematical symbol ≡ is used to indicate that an equation has the special property of being an **identity**. It is not consistently used. You will notice that it is not used in the identities listed in the IB Information (Formulae) Booklet for Mathematics HL. The trigonometric identities required for this course are covered in the next section of this chapter.

The equations $\sin^2\theta + \cos^2\theta = 1$ and $\tan 2\theta = \frac{2\tan}{1 - \tan^2\theta}$ are examples of special equations called **identities** (Section 7.5). As we learned in Section 1.6, an identity is an equation that is true for all possible values of the variable. The other equations are true for only certain values or for none. Trigonometric identities will be covered thoroughly in the next section. They will prove to be an indispensable tool for obtaining analytic solutions to certain trigonometric equations. In this chapter, however, we will be applying methods similar to that used to solve equations encountered earlier in this book

The unit circle and exact solutions to trigonometric equations

When you are asked to solve a trigonometric equation, there are two important questions you need to consider:

1. Is it possible, or required, to express any solution(s) exactly?
2. For what interval of the variable are all solutions to be found?

With regard to the first question, exact solutions are only attainable, in most cases, if they are an integer multiple of $\frac{\pi}{6}$ or $\frac{\pi}{4}$. Although we are primarily interested in finding numerical solutions (rather than angles in degrees), the language of angles is convenient. Recall from the first section of this chapter that if angles are given using radian measure, then angles between 0 and $\frac{\pi}{2}$ have their terminal sides in quadrant I, angles between $\frac{\pi}{2}$ and π have their terminal sides in quadrant II, and so on. Consequently, we will sometimes refer to a solution of an equation being, for example, a 'number in quadrant I', meaning a number that can be interpreted as either the length of an arc on the unit circle or a central angle in radian measure between 0 and $\frac{\pi}{2}$. As explained in Section 7.2, trigonometric domain values that are multiples of $\frac{\pi}{6}$ or $\frac{\pi}{4}$ commonly occur and it is important to be familiar with the exact trigonometric function values for these numbers (Table 7.1).

Concerning the second question, for most trigonometric equations there are infinitely many solutions. For example, the solutions to the equation

$\sin x = \frac{1}{2}$ are any number (arc or central angle) in quadrants I or II positioned so that the terminal point on the unit circle has a y-coordinate of $\frac{1}{2}$ (Figure 7.26). There are an infinite set of numbers that do this, being $\frac{\pi}{6}$ plus any multiple of 2π (quadrant I) or $\frac{5\pi}{6}$ plus any multiple of 2π (quadrant II). This infinite set is concisely written as $x = \frac{\pi}{6} + k \cdot 2\pi$ or $x = \frac{5\pi}{6} + k \cdot 2\pi$, $k \in \mathbb{Z}$. However, for this course the number of solutions to any trigonometric equation will be limited to a finite set by the fact that the solution set will always be restricted to a specified interval. For the equation $\sin x = \frac{1}{2}$, if the solution set is restricted to the interval $0 \leqslant x < 2\pi$, then the solutions are $\frac{\pi}{6}$ and $\frac{5\pi}{6}$. If the solution set is restricted to the interval $-2\pi < x < 2\pi$, then the solutions are $-\frac{11\pi}{6}$, $-\frac{7\pi}{6}$, $\frac{\pi}{6}$ and $\frac{5\pi}{6}$. If the solution set is restricted to the interval $0 \leqslant x < 4\pi$, then the solutions are $\frac{\pi}{6}$, $\frac{5\pi}{6}$, $\frac{13\pi}{6}$ and $\frac{17\pi}{6}$. Figure 7.27 illustrates how the graph of $y = \sin x$ can be used to locate the solutions for the equation $\sin x = \frac{1}{2}$ for different intervals of x. When asked to solve a trigonometric equation, a solution interval will always be given, as in the example below.

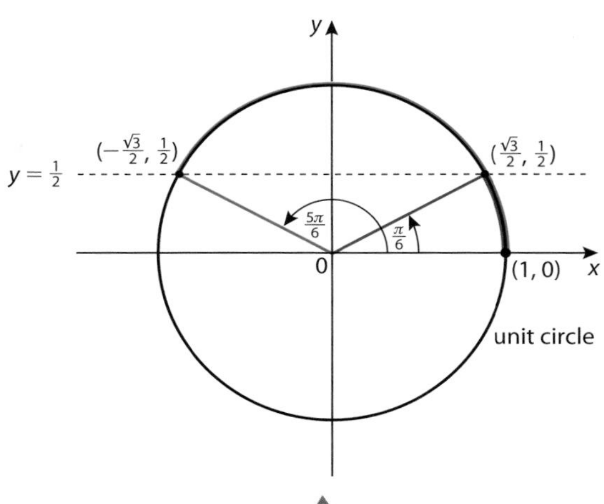

Figure 7.26 Solution to $\sin x = \frac{1}{2}$, $0 \leqslant x < 2\pi$.

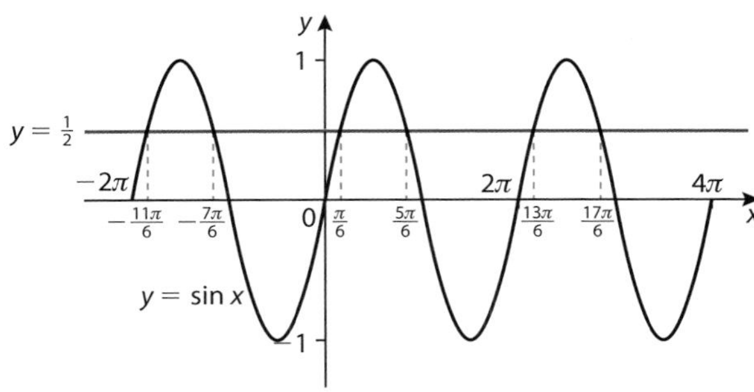

Figure 7.27 Points of intersection between $y = \sin x$ and $y = \frac{1}{2}$.

• **Hint:** As explained here, if the solution set for the equation $\sin x = \frac{1}{2}$ is not restricted, then the **general solution** is $x = \frac{\pi}{6} + k \cdot 2\pi$ or $x = \frac{5\pi}{6} + k \cdot 2\pi$, $k \in \mathbb{Z}$. This infinite solution corresponds to all of the points of intersection between the graphs of $y = \sin x$ and $y = \frac{1}{2}$ as they will repeatedly intersect as the graphs extend indefinitely in both directions (Figure 7.27). It is recommended that you are familiar with how to use a parameter (k in this case) to write the general solution for an equation with an infinite solution set, though it is not required for this course.

Example 15

Find the exact solution(s) to the equation $\sin x \cos x = 2 \cos x$ for $-\pi < x < \pi$.

Solution

There is a temptation to divide both sides by $\cos x$, but as pointed out in Section 3.5, this can result in losing a solution to the equation. In fact, for this equation, both solutions would be lost. Instead, set the equation equal to zero and factorize out the common factor of $\cos x$.

$$\sin x \cos x - 2 \cos x = 0$$
$$\cos x(\sin x - 2) = 0$$
$$\cos x = 0 \quad \text{or} \quad \sin x = 2$$

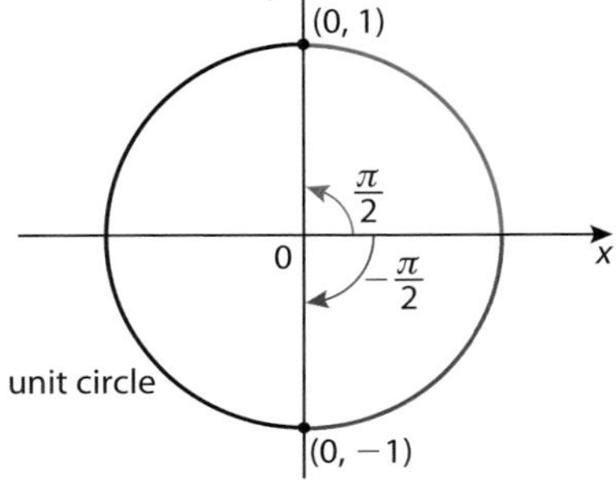

2 is outside the range of the sine function so there is no solution to $\sin x = 2$. Solutions to $\cos x = 0$ occur for arcs (angles) that terminate where the x-coordinate is 0. For the solution interval $-\pi < x < \pi$, this

occurs where the unit circle intersects the y-axis as shown in the diagram. Therefore this analytic solution gives the exact solutions of $x = \frac{\pi}{2}$ and $x = -\frac{\pi}{2}$.

Your GDC can be a very effective tool for searching for solutions graphically. However, it can be limited when exact solutions are requested. The sequence of GDC images below show a graphical solution for the equation in Example 15.

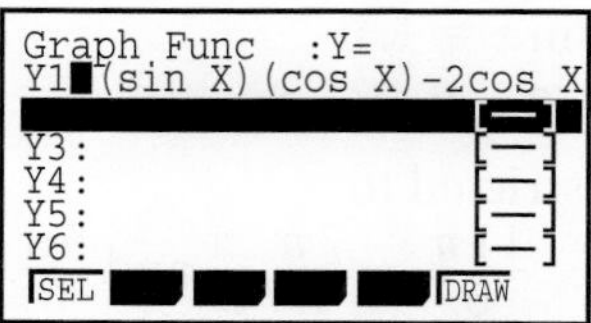

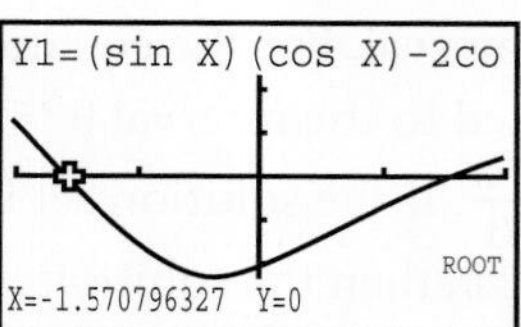

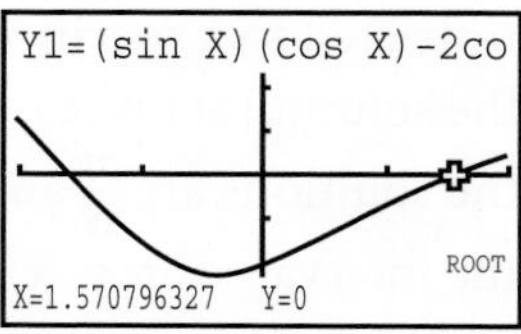

The GDC gives the two solutions in the interval $-\pi < x < \pi$ as $x = -1.570\,796\,327$ and $x = 1.570\,796\,327$. These values are approximations (to 10 significant figures) of the irrational numbers, $x = -\frac{\pi}{2}$ and $x = \frac{\pi}{2}$, and confirms that they are the correct solutions. If exact solutions are required then you need to first attempt an analytic solution, and then a graphical confirmation can be performed.

Example 16

Find the exact solution(s) to the equation $\tan(\theta) + 1 = 0$ for $0 \leqslant x < 360°$.

• **Hint:** The expression $\tan x + 1$ is not equivalent to $\tan(x + 1)$. In the first expression, x alone is the argument of the function, and in the second expression, $x + 1$ is the argument of the function. It is a good habit to use brackets to make it absolutely clear what is, or is not, the argument of a function. For example, there is no ambiguity if $\tan x + 1$ is written as $\tan(x) + 1$, or as $1 + \tan x$.

Solution

Since the solution interval is expressed in terms of degrees, it is necessary to give any solution as an angle in degree measure. Solutions to this equation are values of θ such that $\tan\theta = -1$. Applying the identity $\tan\theta = \frac{\sin\theta}{\cos\theta}$, we have $\frac{\sin\theta}{\cos\theta} = -1$. We need to find any angles θ such that $\sin\theta$ and $\cos\theta$ have opposite signs. This occurs in quadrant II at $\theta = 135°$ and in quadrant IV at $\theta = 315°$ as shown in the diagram.

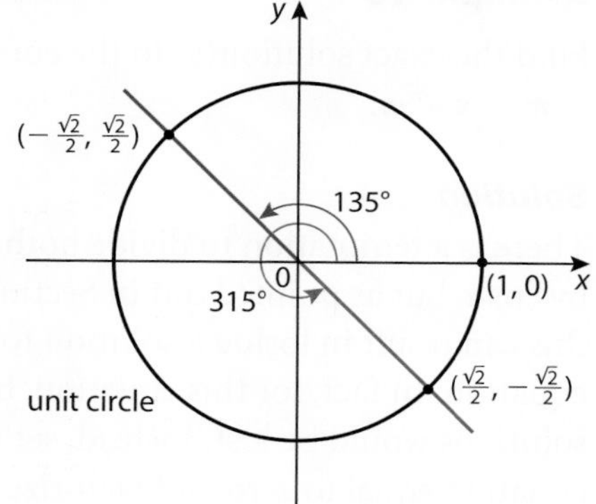

It is possible to arrive at exact answers that are not multiples of $\frac{\pi}{6}$ or $\frac{\pi}{4}$, as the next example illustrates.

Example 17

Find the exact solution(s) to the equation $\cos^2\left(x - \frac{\pi}{3}\right) = \frac{1}{2}$ for $0 \leqslant x < 2\pi$.

Solution

The expression $\cos^2\left(x - \frac{\pi}{3}\right)$ can also be written as $\left[\cos\left(x - \frac{\pi}{3}\right)\right]^2$. The first step is to take the square root of both sides – remembering that every positive number has two square roots – which gives $\cos\left(x - \frac{\pi}{3}\right) = \pm\sqrt{\frac{1}{2}} = \pm\frac{1}{\sqrt{2}} = \pm\frac{\sqrt{2}}{2}$. All of the odd integer multiples of $\frac{\pi}{4}\left(\ldots -\frac{3\pi}{4}, -\frac{\pi}{4}, 0, \frac{\pi}{4}, \frac{3\pi}{4}, \ldots\right)$ have a cosine equal to either $\frac{\sqrt{2}}{2}$ or $-\frac{\sqrt{2}}{2}$.

That is, $x - \frac{\pi}{3} = \frac{\pi}{4} + k\cdot\frac{\pi}{2}$. Now, solve for x.

$x = \frac{\pi}{3} + \frac{\pi}{4} + k\cdot\frac{\pi}{2} = \frac{7\pi}{12} + k\cdot\frac{6\pi}{12}$. The last step is to substitute in different integer values for k to generate all the possible values for x so that $0 \leqslant x < 2\pi$.

When $k = 0$: $x = \frac{7\pi}{12}$; when $k = 1$: $x = \frac{7\pi}{12} + \frac{6\pi}{12} = \frac{13\pi}{12}$;

when $k = 2$: $x = \frac{7\pi}{12} + \frac{12\pi}{12} = \frac{19\pi}{12}$;

when $k = 3$: $x = \frac{7\pi}{12} + \frac{18\pi}{12} = \frac{25\pi}{12}$; however, $\frac{25\pi}{12} > 2\pi$ … but,

when $k = -1$: $x = \frac{7\pi}{12} - \frac{6\pi}{12} = \frac{\pi}{12}$.

Therefore, there are four exact solutions in the interval $0 \leqslant x < 2\pi$, and they are: $x = \frac{\pi}{12}, \frac{7\pi}{12}, \frac{13\pi}{12}$ or $\frac{19\pi}{12}$.

• **Hint:** As we did at the end of Example 15, check the solutions to trigonometric equations with your GDC. The sequence of GDC images here verifies that $x = \frac{\pi}{12}$ is the first solution to the equation in Example 17.

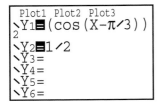

```
Plot1 Plot2 Plot3
\Y1■(cos(X-π/3))²
\Y2■1/2
\Y3=
\Y4=
\Y5=
\Y6=
```

```
WINDOW
 Xmin=0
 Xmax=6.2831853...
 Xscl=1.5707963...
 Ymin=-1.5
 Ymax=1.5
 Yscl=1
 Xres=1
```

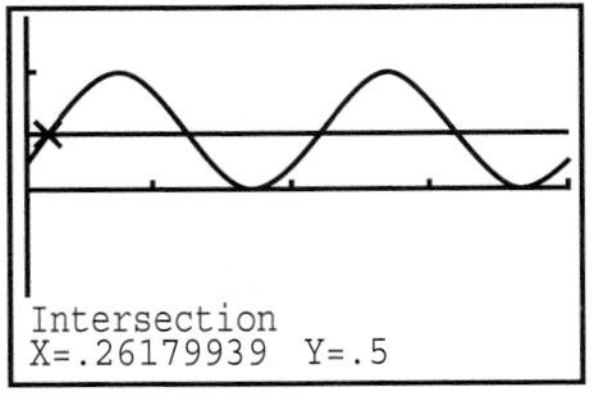


```
π/12
               .2617993878
```

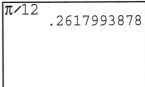

When entering the equation $x = \cos^2\left(x - \frac{\pi}{3}\right)$ into your GDC (as shown in the first GDC image), you will have to enter it in the form $y = \left[\cos\left(x - \frac{\pi}{3}\right)\right]^2$. Be aware that $\cos^2\left(x - \frac{\pi}{3}\right)$ is not equivalent to $\cos\left(x - \frac{\pi}{3}\right)^2$. The expression $\cos\left(x - \frac{\pi}{3}\right)^2$ indicates that the quantity $x - \frac{\pi}{3}$ is squared first and then the cosine of the resulting value is found. However, the expression $y = \cos\left(x - \frac{\pi}{3}\right)$. indicates that the cosine of $x - \frac{\pi}{3}$ is found first and then that value is squared.

Graphical solutions to trigonometric equations

If exact solutions are not required then a graphical solution using your GDC is a very effective way to find approximate solutions to trigonometric equations. Unless instructed to do otherwise, you should give approximate solutions to an accuracy of three significant figures.

Example 18

Find all solutions to the equation $3\tan x = 2\cos x$ in the interval $0 \leqslant x < 2\pi$.

Solution

Graph the equation $y = 3\tan x - 2\cos x$ and find all of its zeros (x-intercepts) in the interval $0 \leqslant x < 2\pi$. Because the domain of the tangent function is $\left\{x : x \in \mathbb{R}, x \neq \frac{\pi}{2} + k\pi, k \in \mathbb{Z}\right\}$, then we expect there to be 'gaps' (and vertical asymptotes) in the graph at $x = \frac{\pi}{2}$ and at $x = \frac{3\pi}{2}$.

It is possible to solve the equation in Example 18 analytically. See Exercise 7.4, question 30.
The exact solutions are $x = \frac{\pi}{6}$ and $x = \frac{5\pi}{6}$. The GDC image shows their approximate values agree with the solutions found in the example.

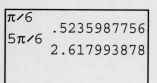

π/6
.5235987756
5π/6
2.617993878

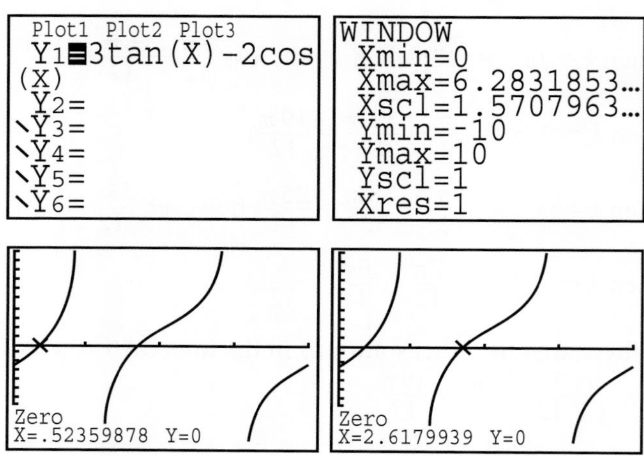

This sequence of GDC images indicates approximate solutions of $x \approx 0.524$ and $x \approx 2.62$ to an accuracy of three significant figures.

A graphical approach is effective and appropriate when it is very difficult, or not possible, to find exact solutions.

Example 19

The peak height, h metres, of ocean waves during a storm is given by the equation $h = 9 + 4\sin\left(\frac{t}{2}\right)$, where t is the number of hours after midnight.
A tsunami alarm is triggered when the peak height goes above 12.5 metres. Find the value of t when the alarm first sounds.

Solution

Graph the equations $y = 9 + 4\sin\left(\frac{x}{2}\right)$ and $y = 12.5$ and find the first point of intersection for $x > 0$.

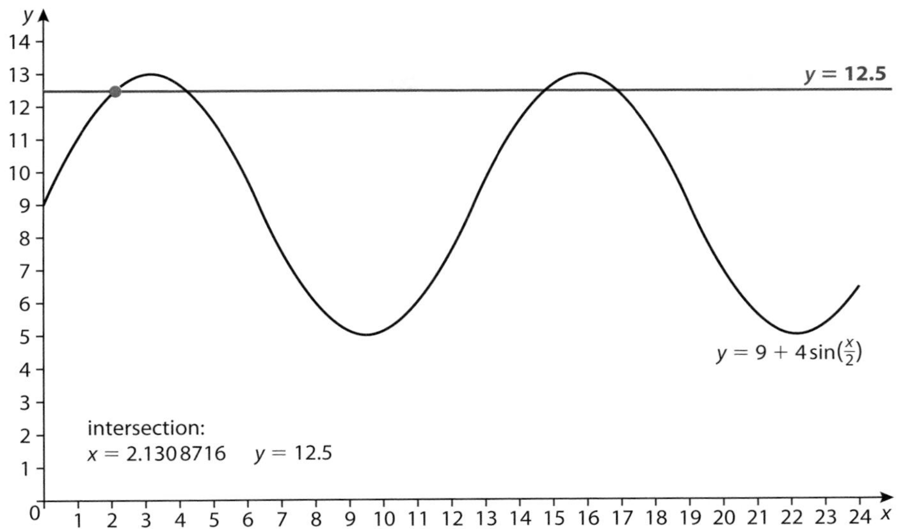

Using the Intersect command on the GDC indicates that the first point of intersection has an x-coordinate of approximately 2.13. Therefore, the alarm will first sound when $t \approx 2.13$ hours.

Analytic solutions to trigonometric equations

An analytical approach requires you to devise a solution strategy utilizing algebraic methods that you have applied to other types of equations – such as quadratic equations. Trigonometric equations that demand an analytic approach will often, but not always, result in exact solutions. Although our approach for equations in this section focuses on algebraic techniques, it is important to use graphical methods to support or confirm our analytical solutions.

Example 20

Solve $2\sin^2 x + \sin x = 0$ for $0 \leqslant x < 2\pi$.

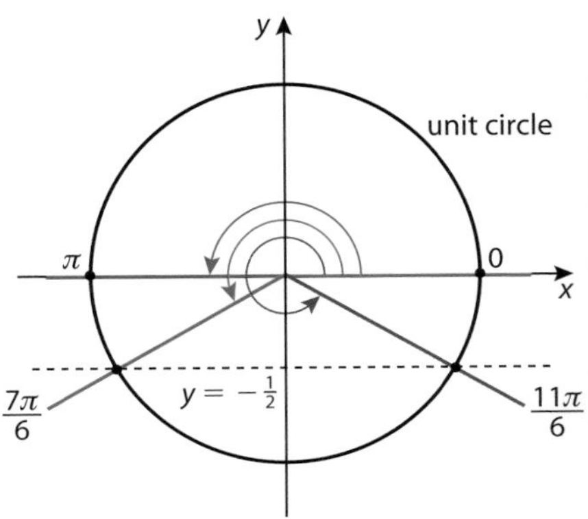

Solution

Factorizing gives $\sin x(2\sin x + 1) = 0$

$\sin x = 0$ or $\sin x = -\frac{1}{2}$

Solutions to $\sin x = 0$ are where the angle is on the x-axis; and solutions to $\sin x = -\frac{1}{2}$ are angles in quadrant III and IV such that their intersection point with the unit circle has y-coordinate of $-\frac{1}{2}$.

for $\sin x = 0$: $x = 0, \pi$ for $\sin x = -\frac{1}{2}$: $x = \frac{7\pi}{6}, \frac{11\pi}{6}$

Therefore, the solutions are $x = 0, \pi, \frac{7\pi}{6}, \frac{11\pi}{6}$.

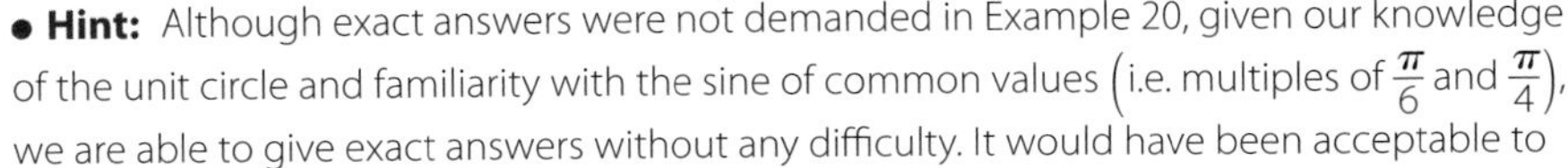

• **Hint:** Although exact answers were not demanded in Example 20, given our knowledge of the unit circle and familiarity with the sine of common values $\left(\text{i.e. multiples of } \frac{\pi}{6} \text{ and } \frac{\pi}{4}\right)$, we are able to give exact answers without any difficulty. It would have been acceptable to give approximate solutions using your GDC, but it is worth recognizing that this would have required considerable more effort than providing exact solutions. Entering and graphing the equation $y = 2\sin^2 x + \sin x$ on your GDC (see GDC images) would not be the most efficient or appropriate solution method, but if sufficient time is available it is an effective way to confirm your exact solutions. [Note that $\sin^2 x$ must be entered in a GDC as $(\sin x)^2$.]

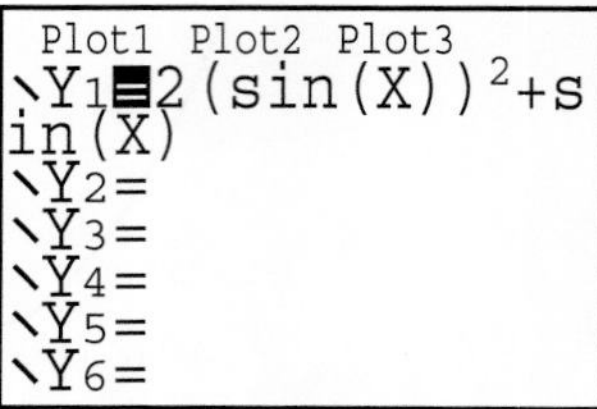

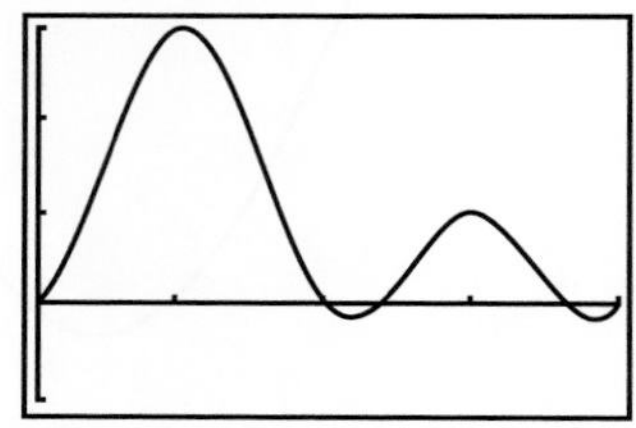

• **Hint:** As we will see in the next section, it is often the case that an analytic solution is not possible unless a substitution is made using a suitable trigonometric identity.

The next example illustrates how the application of a trigonometric identity can be helpful to rewrite the equation in a way that allows us to solve it algebraically. The next section will introduce many further trigonometric identities and examples of using them to assist in solving trigonometric equations.

Example 21

Solve $3\cos x + \cot x = 0$ for $0 \leqslant x \leqslant 2\pi$.

Solution

Since the structure of this equation is such that an expression is set equal to zero, it would be nice to be able to use the same algebraic technique as the previous example – that is, factorize and solve for when each factor is zero. However, it is not possible to factorize the expression $3\cos x + \cot x$, and rewriting the equation as $3\cos x = -\cot x$ does not help. Are there any expressions in the equation for which we can substitute an equivalent expression that will make the equation accessible to an algebraic solution? We do not have any equivalent expressions for $\cos x$, but we do have an identity for $\cot x$. Since $\cot x$ is the reciprocal of $\tan x$ we know that $\cot x = \frac{\cos x}{\sin x}$. Let's see what happens when we substitute $\frac{\cos x}{\sin x}$ for $\cot x$.

$3\cos x + \frac{\cos x}{\sin x} = 0$ Now, get a common denominator.

$\frac{3\sin x\cos x}{\sin x} + \frac{\cos x}{\sin x} = 0$

$\frac{3\sin x\cos x + \cos x}{\sin x} = 0$ Noting that $\sin x \neq 0$, multiply both sides by $\sin x$. A fraction equals zero when the denominator equals zero.

$3\sin x\cos x + \cos x = 0$

$\cos x(3\sin x + 1) = 0$ Factorize.

$\cos x = 0$ or $\sin x = -\frac{1}{3}$

For $\cos x = 0$: $x = \frac{\pi}{2}, \frac{3\pi}{2}$

We know that solutions to $\cos x = 0$ are angles on the y-axis giving the two exact solutions of $\frac{\pi}{2}$ and $\frac{3\pi}{2}$. Although we know solutions to $\sin x = -\frac{1}{3}$ are angles in quadrants III and IV, we do not know their exact values. So, we will need to use our GDC to find approximate solutions to $\sin x = -\frac{1}{3}$ for $0 \leqslant x \leqslant 2\pi$.

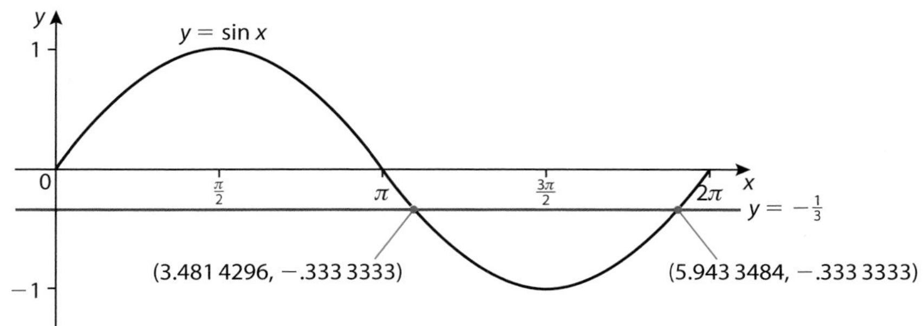

Thus, for $\sin x = -\frac{1}{3}$: $x \approx 3.48$ or $x \approx 5.94$ (3 significant figures)

Therefore, the full solution set for the equation is $x = \frac{\pi}{2}, \frac{3\pi}{2}$; $x \approx 3.48, 5.94$.

• **Hint:** A strategy that often proves fruitful is to try and rewrite a trigonometric equation in terms of just one trigonometric function. If that is not possible, then try and rewrite it in terms of only the sine and cosine functions. This strategy was used in Example 21.

Exercise 7.4

In questions 1–12, find the exact solution(s) for $0 \leqslant x < 2\pi$. Verify your solution(s) with your GDC.

1 $\cos x = \frac{1}{2}$

2 $2\sin x + 1 = 0$

3 $1 - \tan x = 0$

4 $\sqrt{3} = 2\sin x$

5 $2\sin^2 x = 1$

6 $4\cos^2 x = 3$

7 $\tan^2 x - 1 = 0$

8 $4\cos^2 x = 1$

9 $\tan x(\tan x + 1) = 0$

10 $\sin x \cos x = 0$

11 $5 - \sec x = 3$

12 $\csc^2 x = 2$

In questions 13–20, use your GDC to find approximate solution(s) for $0 \leqslant x < 2\pi$. Express solutions accurate to 3 significant figures.

13 $\sin x = 0.4$

14 $3\cos x + 1 = 0$

15 $\tan x = 2$

16 $\sec 2x = 3.46$

17 $\cos(x - 1) = -0.38$

18 $3\tan^2 x = 1$

19 $\csc(2x - 3) = \frac{3}{2}$

20 $3\cot x = 10$

In questions 21–24, given that k is any integer, list all of the possible values for x that are in the specified interval.

21 $\frac{\pi}{2} + k \cdot \pi, -3\pi \leqslant x \leqslant 3\pi$

22 $\frac{\pi}{6} + k \cdot 2\pi, -2\pi \leqslant x \leqslant 2\pi$

23 $\frac{7\pi}{12} + k \cdot \pi, 0 \leqslant x < 2\pi$

24 $\frac{\pi}{4} + k \cdot \frac{\pi}{4}, 0 \leqslant x < 4\pi$

In questions 25–32, find the **exact** solutions for the indicated interval. The interval will also indicate whether the solutions are given in degree or radian measure. Write a complete analytic solution.

25 $\cos\left(x - \frac{\pi}{6}\right) = -\frac{1}{2}, 0 \leqslant x < 2\pi$

26 $\tan(\theta + \pi) = 1, -\pi \leqslant \theta \leqslant \pi$

27 $\sin 2x = \frac{\sqrt{3}}{2}, 0 \leqslant x < 360°$

28 $\sin^2\left(\alpha + \frac{\pi}{2}\right) = \frac{3}{4}, -\frac{\pi}{2} \leqslant \alpha \leqslant \frac{\pi}{2}$

29 $2\cos^2\theta - 5\cos\theta - 3 = 0, 0 \leqslant \theta < 2\pi$

30 $3\tan x = 2\cos x, 0 \leqslant x < 2\pi$

31 $2\cos(x + 90°) = \sqrt{2}, 0 \leqslant x < 360°$

32 $9\sec^2\theta = 12, 0 \leqslant \theta < \pi$

33 The number, N, of empty birds' nests in a park is approximated by the function $N = 74 + 42\sin\left(\frac{\pi}{12}t\right)$, where t is the number of hours after midnight. Find the value of t when the number of empty nests first equals 90. Approximate the answer to 1 decimal place.

34 In Edinburgh, the number of hours of daylight on day D is modelled by the function $H = 12 + 7.26\sin\left[\frac{2\pi}{365}(D - 80)\right]$, where D is the number of days after December 31 (e.g. January 1 is $D = 1$, January 2 is $D = 2$, and so on). Do not use your GDC on part a).

a) Which days of the year have 12 hours of daylight?

b) Which days of the year have about 15 hours of daylight?

c) How many days of the year have more than 17 hours of daylight?

In questions 35–42, solve the equation for the stated solution interval. Find exact solutions when possible, otherwise give solutions to three significant figures. Verify solutions with your GDC.

35 $2\cos^2 x + \cos x = 0, 0 \leqslant x < 2\pi$

36 $2\sin^2\theta - \sin\theta - 1 = 0, 0 \leqslant \theta < 2\pi$

37 $\tan^2 x - \tan x = 2, -90° \leqslant x \leqslant 90°$

38 $3\cos^2 x - 6\cos x = 2, -\pi < x \leqslant \pi$

39 $2\sin\beta = 3\cos\beta, 0 \leqslant \beta \leqslant 180°$

40 $\sin^2 x = \cos^2 x, 0 \leqslant x \leqslant \pi$

41 $\sec^2 x + 2\sec x + 4 = 0, 0 \leqslant x < 2\pi$

42 $\sin x \tan x = 3\sin x, 0 \leqslant x < 360°$

7.5 Trigonometric identities

You will recall that an identity is an equation that is true for all values of the variable for which the expressions in the equation are defined. Several trigonometric identities have been introduced earlier in this chapter. They are reviewed here (Table 7.2) and a number of important new identities are presented and proved in this section.

Trigonometric identities are used in a variety of ways. For example, one of the reciprocal identities is applied whenever the cosecant, secant or cotangent function is evaluated on a calculator. The following uses of trigonometric identities will be illustrated in this section.

1. Evaluate trigonometric functions.

The **co-function identities** for sine and cosine were established in Section 7.3 by means of investigating horizontal shifts of graphs of the sine and cosine functions. Similarly we can prove co-function identities for secant and cosecant, and for tangent and cotangent. These appear in Table 7.2 on the next page.

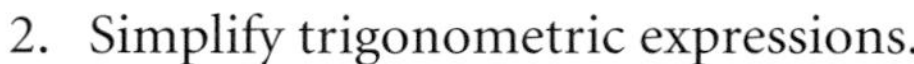

2. Simplify trigonometric expressions.
3. Prove other trigonometric identities.
4. Solve trigonometric equations.

The first portion of this section is devoted to developing some further trigonometric identities that are organized into three groups: Pythagorean identities, compound angle identities, and double angle identities.

Table 7.2 Summary of fundamental trigonometric identities.

Reciprocal identities:		
$\csc x = \frac{1}{\sin x}$	$\sec x = \frac{1}{\cos x}$	$\cot x = \frac{1}{\tan x}$
Tangent and cotangent identities:		
$\tan x = \frac{\sin x}{\cos x}$	$\cot x = \frac{\cos x}{\sin x}$	
Odd/even function identities:		
$\sin(-x) = -\sin x$	$\cos(-x) = \cos x$	$\tan(-x) = -\tan x$
$\csc(-x) = -\csc x$	$\sec(-x) = \cos x$	$\cot(-x) = -\tan x$
Co-function identities:		
$\sin\left(\frac{\pi}{2} - x\right) = \cos x$	$\sec\left(\frac{\pi}{2} - x\right) = \csc x$	$\tan\left(\frac{\pi}{2} - x\right) = \cot x$
$\cos\left(\frac{\pi}{2} - x\right) = \sin x$	$\csc\left(\frac{\pi}{2} - x\right) = \sec x$	$\cot\left(\frac{\pi}{2} - x\right) = \tan x$

It was confirmed in Section 7.3 that sine and tangent are **odd functions** and that cosine is an **even function**. We will accept without proof that if a function is odd, then its reciprocal is also odd; and the same is true for even functions. Therefore, cosecant and cotangent are odd functions, and secant is an even function.

Pythagorean identities

At the start of the previous section, it was stated that the equation $\sin^2\theta + \cos^2\theta = 1$ is an identity; that is, it's true for all possible values of θ. Let's prove that this is the case.

Recall from Section 7.1 that the equation for the unit circle is $x^2 + y^2 = 1$. That is, the coordinates (x, y) of any point on the circle satisfy the equation $x^2 + y^2 = 1$. As we learned in Section 7.2, if θ is any real number that represents a central angle (in radian measure) of the unit circle that terminates at (x, y), then $x = \cos\theta$ and $y = \sin\theta$. Substituting directly into the equation for the circle gives $\sin^2\theta + \cos^2\theta = 1$. Therefore, the equation $\sin^2\theta + \cos^2\theta = 1$ is true for any real number x.

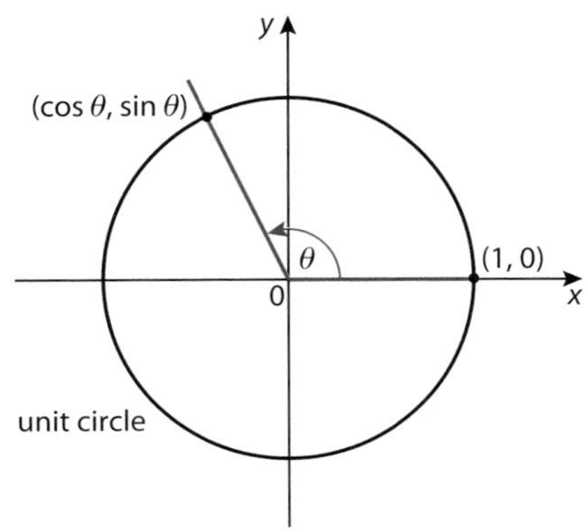

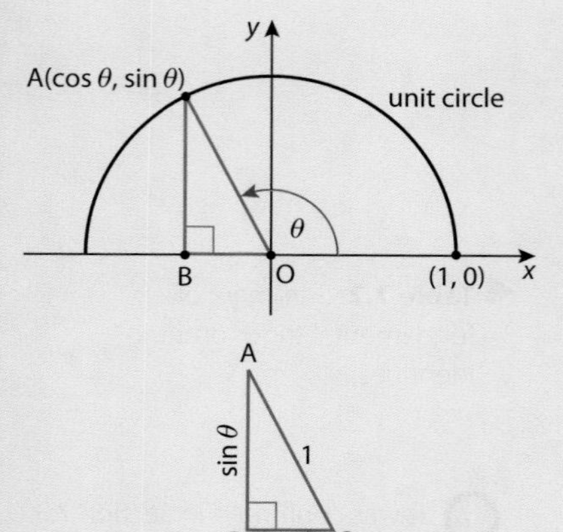

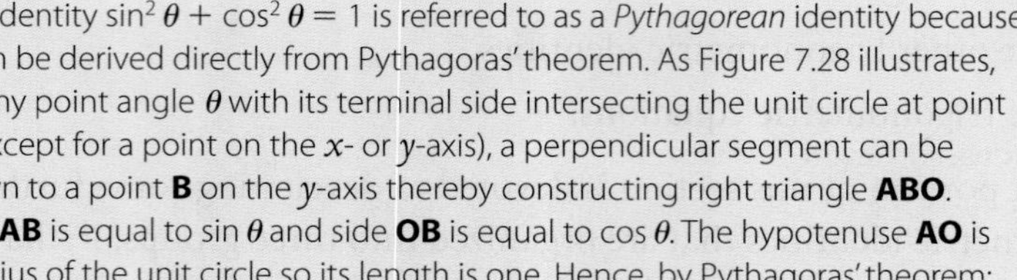
The identity $\sin^2\theta + \cos^2\theta = 1$ is referred to as a *Pythagorean* identity because it can be derived directly from Pythagoras' theorem. As Figure 7.28 illustrates, for any point angle θ with its terminal side intersecting the unit circle at point **A** (except for a point on the x- or y-axis), a perpendicular segment can be drawn to a point **B** on the y-axis thereby constructing right triangle **ABO**. Side **AB** is equal to $\sin\theta$ and side **OB** is equal to $\cos\theta$. The hypotenuse **AO** is a radius of the unit circle so its length is one. Hence, by Pythagoras' theorem: $\sin^2\theta + \cos^2\theta = 1$.

Figure 7.28

• **Hint:** Graph the equation $y = \sin^2 x + \cos^2 x$ on your GDC with the y-axis ranging from -2 to 2 and the x-axis ranging from -2π to 2π (radian mode) or $-360°$ to $360°$ (degree mode). What do you observe?

Phrases such as 'prove the identity' and 'verify the identity' are often used. Both mean, 'prove that the given equation is an identity'. We do this by performing a series of algebraic manipulations to show that the expression on one side of the equation can be transformed into the expression on the other side, or that both expressions can be transformed into some third expression. When verifying that an equation is an identity, you should not perform an operation to both sides of the equation; for example, multiplying both sides of the equation by a quantity. This can only be done if it is known that the two sides of the equation are equal, but this is exactly what we are trying to verify in the process of 'proving an identity.'

Example 22

Prove that $1 + \tan^2\theta = \sec^2\theta$ is an identity.

Solution

There is more of an opportunity to perform algebraic manipulations on the left side than the right side. Thus, our task is to transform the expression $1 + \tan^2\theta$ into the expression $\sec^2\theta$.

$1 + \tan^2\theta = \sec^2\theta$ Using the identity $\tan\theta = \frac{\sin\theta}{\cos\theta}$ substitute $\frac{\sin^2\theta}{\cos^2\theta}$ for $\tan^2\theta$.

$1 + \frac{\sin^2\theta}{\cos^2\theta} =$ Find a common denominator.

$\frac{\cos^2\theta}{\cos^2\theta} + \frac{\sin^2\theta}{\cos^2\theta} =$

$\frac{\cos^2\theta + \sin^2\theta}{\cos^2\theta} =$ Apply the Pythagorean identity $\sin^2\theta + \cos^2\theta = 1$.

$\frac{1}{\cos^2\theta} =$ Because $\frac{1}{\cos\theta} = \sec\theta$, then $\frac{1}{\cos^2\theta} = \sec^2\theta$.

$\sec^2\theta = \sec^2\theta$ Q.E.D.

Q.E.D. is an abbreviation for the Latin phrase '*quod erat demonstrandum*' which means 'that which was to be proved (or demonstrated)'. It is often written at the end of a proof to indicate that its conclusion has been reached.

Another identity than can be proved in a manner similar to the identity in Example 22 is $1 + \cot^2\theta = \csc^2\theta$.

Pythagorean identities

$\sin^2\theta + \cos^2\theta = 1$ $\qquad$ $1 + \tan^2\theta = \sec^2\theta$ $\qquad$ $1 + \cot^2\theta = \csc^2\theta$

The Pythagorean identities are sometimes used in radical forms such as $\sin\theta = \pm\sqrt{1 - \cos^2\theta}$ or $\tan\theta = \pm\sqrt{\sec^2\theta - 1}$ where the sign (+ or −) depends on θ (which quadrant it is in).

Example 23

a) Express $2\cos^2 x + \sin x$ in terms of $\sin x$ only.

b) Solve the equation $2\cos^2 x + \sin x = -1$ for x in the interval $0 \leqslant x \leqslant 2\pi$, expressing your answer(s) exactly.

Solution

a) $2\cos^2 x + \sin x = 2(1 - \sin^2 x) + \sin x$ $\quad$ Using Pythagorean identity:

$= 2 - 2\sin^2 x + \sin x$ $\quad$ $\cos^2 x = 1 - \sin^2 x$.

b) $2\cos^2 x + \sin x = -1$

$2 - 2\sin^2 x + \sin x = -1$ $\quad$ Substitute result from a).

$2\sin^2 x - \sin x - 3 = 0$ $\quad$ (Alternatively: let $\sin x = y$, then $2y^2 - y - 3 = 0$)

$(2\sin x - 3)(\sin x + 1) = 0$ $\quad$ Factorize. (alt: $(2y - 3)(y + 1) = 0$)

$\sin x = \frac{3}{2}$ or $\sin x = -1$ $\quad$ (Alt: $y = \frac{3}{2}$ or $y = -1 \Rightarrow \sin x = \frac{3}{2}$ or $\sin x = -1$)

For $x = \frac{3}{2}$: no solution because $\frac{3}{2}$ is not in the range of the sine function.

For $\sin x = -1$: $x = \frac{3\pi}{2}$.

Therefore, there is only one solution in $0 \leqslant x \leqslant 2\pi : x = \frac{3\pi}{2}$.

Use your GDC to check this result by rewriting $2\cos^2 x + \sin x = -1$ as $2\cos^2 x + \sin x + 1 = 0$ and then graph $y = 2\cos^2 x + \sin x + 1$; confirming a single zero at $x = \frac{3\pi}{2}$ in the interval $x \in [0, 2\pi]$.

```
Plot1 Plot2 Plot3
\Y1■2(cos(X))²+s
in(X)+1
\Y2=
\Y3=
\Y4=
\Y5=
\Y6=
```

```
WINDOW
 Xmin=0
 Xmax=6.2831853...
 Xscl=π/2■
 Ymin=-1
 Ymax=4
 Yscl=1
 Xres=1
```

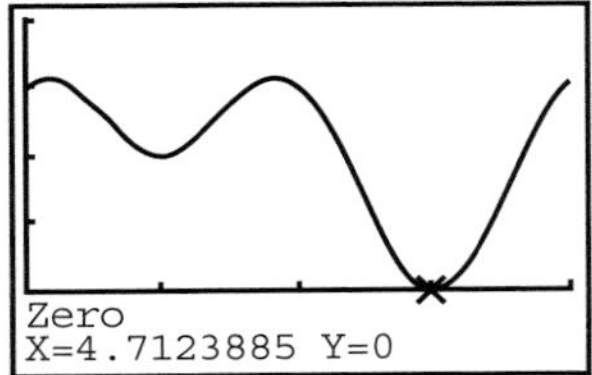


```
X
        4.712388457
3π/2
         4.71238898
```

Compound angle identities (sum and difference identities)

In this section we develop trigonometric identities known as the compound angle identities for sine, cosine and tangent. These contain the expressions $\sin(\alpha + \beta)$, $\sin(\alpha - \beta)$, $\cos(\alpha + \beta)$, $\cos(\alpha - \beta)$, $\tan(\alpha + \beta)$ and $\tan(\alpha - \beta)$. We first find a formula for $\cos(\alpha + \beta)$.

On first reaction you might wonder whether $\cos(\alpha + \beta) = \cos\alpha + \cos\beta$. Often it is easier to prove a mathematical statement false than to prove it true. One counter-example is sufficient to prove a statement false. Let $\alpha = \frac{\pi}{3}$ and $\beta = \frac{\pi}{6}$. Does $\cos\left(\frac{\pi}{3} + \frac{\pi}{6}\right) = \cos\frac{\pi}{3} + \cos\frac{\pi}{6}$?

• **Hint:** As will occur in Chapter 8, Greek letters such as α (alpha), β (beta), or θ (theta) are frequently used to name angles. In the development of the formula for $\cos(\alpha + \beta)$, α and β are arcs along the unit circle, but they could just as well be representing the central angle (in radian measure) that cuts off (subtends) the arc.

$$\cos\left(\frac{\pi}{3} + \frac{\pi}{6}\right) = \cos\left(\frac{2\pi}{6} + \frac{\pi}{6}\right) = \left(\frac{3\pi}{6}\right) = \cos\left(\frac{\pi}{2}\right) = 0$$

and $\cos\frac{\pi}{3} + \cos\frac{\pi}{6} = \frac{1}{2} + \frac{\sqrt{3}}{2} = \frac{1 + \sqrt{3}}{2}$.

Thus, the answer is 'no'; $\cos\left(\frac{\pi}{3} + \frac{\pi}{6}\right) \neq \cos\frac{\pi}{3} + \cos\frac{\pi}{6}$.

Although $\cos(\alpha + \beta) = \cos\alpha + \cos\beta$ may be true for some values (e.g. it's true for $\alpha = \frac{\pi}{2}$ and $\beta = \frac{3\pi}{4}$), it's not true for **all** possible values of α and β, and therefore, it is **not** an identity.

Derivation of identity for the cosine of the sum of two numbers

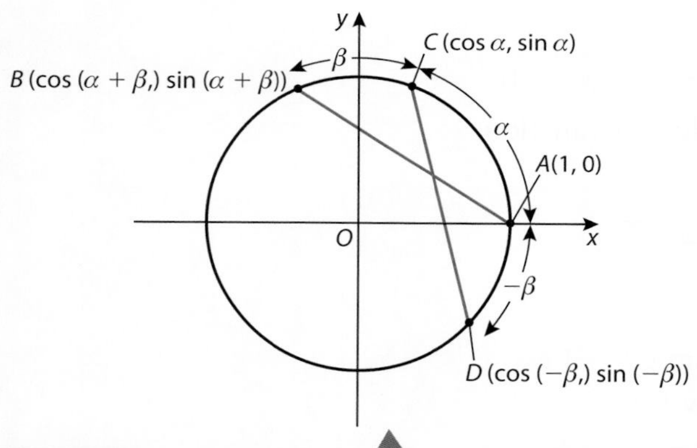

Figure 7.29

To find a formula for $\cos(\alpha + \beta)$, we use Figure 7.29 showing the four points A, B, C and D on the unit circle and the two chords AB and CD. The arc lengths α, β and $-\beta$ have been marked. The coordinates of A, B, C and D in terms of sines and cosines of the arcs are also indicated. The coordinates of point D are $(\cos(-\beta), \sin(-\beta))$, but we can apply the odd/even identities to write the coordinates of D more simply as $(\cos\beta, -\sin\beta)$. Observe that the arc length from A to B is equal to the arc length from D to C because they both have a length equal to $\alpha + \beta$. Since equal arcs on a circle determine equal chords, it must follow that $AB = CD$. Using the respective coordinates for A, B, C and D, we can express $AB = CD$ using the distance formula as

$$\sqrt{(\cos(\alpha + \beta) - 1)^2 + \sin^2(\alpha + \beta)} = \sqrt{(\cos\alpha - \cos\beta)^2 + (\sin\alpha + \sin\beta)^2}$$

Squaring both sides and expanding, gives

$$\cos^2(\alpha + \beta) - 2\cos(\alpha + \beta) + 1 + \sin^2(\alpha + \beta)$$
$$= \cos^2\alpha - 2\cos\alpha\cos\beta + \cos^2\beta + \sin^2\alpha + 2\sin\alpha\sin\beta + \sin^2\beta$$
$$[\cos^2(\alpha + \beta) + \sin^2(\alpha + \beta)] - 2\cos(\alpha + \beta) + 1$$
$$= (\cos^2\alpha + \sin^2\alpha) + (\sin^2\beta + \cos^2\beta) - 2\cos\alpha\cos\beta + 2\sin\alpha\sin\beta$$

Applying the Pythagorean identity $\sin^2\theta + \cos^2\theta = 1$, we can replace three expressions with 1:

$$1 - 2\cos(\alpha + \beta) + 1 = 1 + 1 - 2\cos\alpha\cos\beta + 2\sin\alpha\sin\beta$$

Subtracting 2 from each side and dividing both sides by -2, gives

$$\cos(\alpha + \beta) = \cos\alpha\cos\beta - \sin\alpha\sin\beta$$

This is the **identity for the cosine of the sum of two numbers.**

Previously we were only able to find exact values of a trigonometric function for certain 'special' numbers, i.e. multiples of $\frac{\pi}{6}$ or $\frac{\pi}{4}$.

Example 24 – Using the sum identity for cosine

Find the exact values for a) $\cos\frac{5\pi}{12}$, and b) $\cos 75°$.

Solution

a) $\frac{5\pi}{12} = \frac{\pi}{4} + \frac{\pi}{6}$

Applying the identity $\cos(\alpha + \beta) = \cos\alpha\cos\beta - \sin\alpha\sin\beta$ with

$\alpha = \frac{\pi}{4}$ and $\beta = \frac{\pi}{6}$, gives $\cos\left(\frac{\pi}{4} + \frac{\pi}{6}\right) = \cos\frac{\pi}{4}\cos\frac{\pi}{6} - \sin\frac{\pi}{4}\sin\frac{\pi}{6}$

$$= \left(\frac{\sqrt{2}}{2}\right)\left(\frac{\sqrt{3}}{2}\right) - \left(\frac{\sqrt{2}}{2}\right)\left(\frac{1}{2}\right)$$

$$= \frac{\sqrt{6}}{4} - \frac{\sqrt{2}}{4} = \frac{\sqrt{6} - \sqrt{2}}{4}.$$

Therefore, $\cos\frac{5\pi}{12} = \frac{\sqrt{6} - \sqrt{2}}{4}$.

Derivation of identity for the cosine of the difference of two numbers

We can use the identity for the cosine of the sum of two numbers and the fact that cosine is an even function and sine is an odd function to derive the formula for $\cos(\alpha + \beta)$.

Let's replace β with $-\beta$ in $\cos(\alpha + \beta) = \cos\alpha\cos\beta - \sin\alpha\sin\beta$.

$$\cos[\alpha + (-\beta)] = \cos\alpha\cos(-\beta) - \sin\alpha\sin(-\beta)$$

Substituting $-\sin\beta$ for $\sin(-\beta)$, and $\cos\beta$ for $\cos(-\beta)$, gives

$$\cos(\alpha - \beta) = \cos\alpha\cos\beta + \sin\alpha\sin\beta$$

This is the **identity for the cosine of the difference of two numbers.**

Example 25 – Using the sum and difference identities for cosine

Given that A and B are numbers representing arcs or angles that are in the first quadrant, and $\sin A = \frac{4}{5}$ and $\cos B = \frac{12}{13}$, find the exact values of a) $\cos(A + B)$ and b) $\cos(A - B)$.

Solution

We are given the exact values for $\sin A$ and $\cos B$, but we also need exact values for $\sin B$ and $\cos A$ in order to use the sum and difference identities for cosine.

Since B is in the first quadrant then $B > 0$ and re-arranging one of the Pythagorean identities, we have

$$\sin B = \sqrt{1 - \cos^2 B} = \sqrt{1 - \left(\frac{12}{13}\right)^2} = \sqrt{\frac{25}{169}} = \frac{5}{13}.$$

Similarly, $\cos A = \sqrt{1 - \sin^2 A} = \sqrt{1 - \left(\frac{4}{5}\right)^2} = \sqrt{\frac{9}{25}} = \frac{3}{5}.$

a) Substituting into the identity for the cosine of the sum of two numbers, gives

$$\cos(A + B) = \cos A \cos B - \sin A \sin B = \left(\frac{3}{5}\right)\left(\frac{12}{13}\right) - \left(\frac{4}{5}\right)\left(\frac{5}{13}\right) = \frac{16}{65}.$$

Therefore, $\cos(A + B) = \frac{16}{65}.$

b) Substituting into the identity for the cosine of the difference of two numbers, gives

$$\cos(A - B) = \cos A \cos B + \sin A \sin B = \left(\frac{3}{5}\right)\left(\frac{12}{13}\right) + \left(\frac{4}{5}\right)\left(\frac{5}{13}\right) = \frac{56}{65}.$$

Therefore, $\cos(A - B) = \frac{56}{65}.$

• **Hint:** Notice that in Example 25, we obtained $\cos(A + B)$ and $\cos(A - B)$ without finding the actual values of A and B.

Derivation of identities for the sine of the sum/difference of two numbers

The identity $\cos(\alpha - \beta) = \cos\alpha\cos\beta + \sin\alpha\sin\beta$ can be used to derive an identity for $\sin(\alpha + \beta)$. Substituting $\frac{\pi}{2}$ for α and $(\alpha + \beta)$ for β, gives

$$\cos\left[\frac{\pi}{2} - (\alpha + \beta)\right] = \cos\left[\left(\frac{\pi}{2} - \alpha\right) - \beta\right]$$
$$= \cos\left(\frac{\pi}{2} - \alpha\right)\cos\beta + \sin\left(\frac{\pi}{2} - \alpha\right)\sin\beta$$

Now using the co-function identities $\cos\left(\frac{\pi}{2} - x\right) = \sin x$ and $\sin\left(\frac{\pi}{2} - x\right) = \cos x$, we have,

$$\sin(\alpha + \beta) = \sin\alpha\cos\beta + \cos\alpha\sin\beta$$

This is the **identity for the sine of the sum of two numbers.**

By replacing β with $-\beta$, in the identity $\sin(\alpha + \beta) = \sin\alpha\cos\beta + \cos\alpha\sin\beta$, we get

$$\sin(\alpha - \beta) = \sin\alpha\cos(-\beta) + \cos\alpha\sin(-\beta)$$

Applying the odd/even identities for $\cos(-\beta)$ and $\sin(-\beta)$, produces

$$\sin(\alpha - \beta) = \sin\alpha\cos\beta - \cos\alpha\sin\beta$$

This is the **identity for the sine of the difference of two numbers.**

Derivation of identities for the tangent of the sum/difference of two numbers

To produce an identity for $\sin(\alpha + \beta)$ in terms of $\tan\alpha$ and $\tan\beta$, we start with the fundamental identity that the tangent is the quotient of sine and cosine. We have

$$\tan(\alpha + \beta) = \frac{\sin(\alpha + \beta)}{\cos(\alpha + \beta)} \quad \text{given } \cos(\alpha + \beta) \neq 0$$
$$= \frac{\sin\alpha\cos\beta + \cos\alpha\sin\beta}{\cos\alpha\cos\beta - \sin\alpha\sin\beta}$$

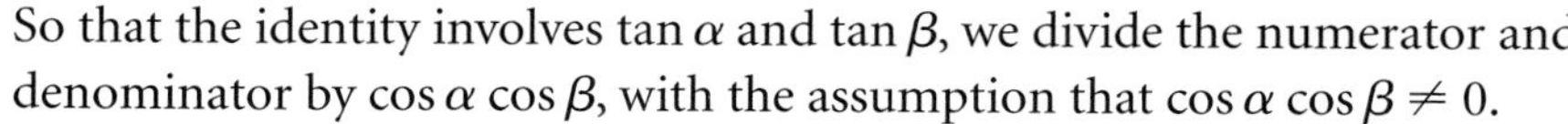

So that the identity involves $\tan\alpha$ and $\tan\beta$, we divide the numerator and denominator by $\cos\alpha\cos\beta$, with the assumption that $\cos\alpha\cos\beta \neq 0$.

$$= \frac{\dfrac{\sin\alpha\cos\beta}{\cos\alpha\cos\beta} + \dfrac{\cos\alpha\sin\beta}{\cos\alpha\cos\beta}}{\dfrac{\cos\alpha\cos\beta}{\cos\alpha\cos\beta} - \dfrac{\sin\alpha\sin\beta}{\cos\alpha\cos\beta}}$$

$$\tan(\alpha + \beta) = \frac{\tan\alpha + \tan\beta}{1 - \tan\alpha\tan\beta}$$

This is the **identity for the tangent of the sum of two numbers.**

If in this identity β is replaced with $-\beta$, we get

$$\tan[\alpha + (-\beta)] = \frac{\tan\alpha + \tan(-\beta)}{1 - \tan\alpha\tan(-\beta)}$$

Tangent is an odd function, so $\tan(-\beta) = -\tan\beta$. Making this substitution, gives

$$\tan(\alpha - \beta) = \frac{\tan\alpha - \tan\beta}{1 + \tan\alpha\tan\beta}$$

This is the **identity for the tangent of the difference of two numbers.**

Compound angle identities

$\cos(\alpha + \beta) = \cos\alpha\cos\beta - \sin\alpha\sin\beta$ $\quad$ $\cos(\alpha - \beta) = \cos\alpha\cos\beta + \sin\alpha\sin\beta$

$\sin(\alpha + \beta) = \sin\alpha\cos\beta + \cos\alpha\sin\beta$ $\quad$ $\sin(\alpha - \beta) = \sin\alpha\cos\beta - \cos\alpha\sin\beta$

$\tan(\alpha + \beta) = \dfrac{\tan\alpha + \tan\beta}{1 - \tan\alpha\tan\beta}$ $\quad$ $\tan(\alpha - \beta) = \dfrac{\tan\alpha - \tan\beta}{1 + \tan\alpha\tan\beta}$

• **Hint:** The compound angle identities are also referred to as the 'sum and difference identities', or the 'addition and subtraction identities'.

Example 26 – Using the sum identity for tangent

If $\tan(A + B) = \frac{1}{7}$ and $\tan A = 3$, find the value of $\tan B$.

Solution

Using the identity for the tangent of the sum of two numbers, we write

$$\tan(A + B) = \frac{\tan A + \tan B}{1 - \tan A\tan B}$$

Substituting $\frac{1}{7}$ for $\tan(A + B)$, and 3 for $\tan A$.

$$\frac{1}{7} = \frac{3 + \tan B}{1 - 3\tan B}$$

Cross-multiply and solve for $\tan B$.

$$21 + 7\tan B = 1 - 3\tan B$$

$$10\tan B = -20$$

$$\tan B = -2$$

Note that, similar to Example 25, we found the exact value of $\tan B$ without finding the actual value of B. In fact, we're not even certain which quadrant B is in, only that it must be in either quadrant II or IV since $\tan B < 0$.

Double angle identities

Is $\sin 2\theta = 2\sin\theta$ an identity? Clearly, it is not – as the counter-example $\theta = \frac{\pi}{6}$ shows.

$$\sin\left(2\cdot\frac{\pi}{6}\right) = \sin\left(\frac{\pi}{3}\right) = \frac{\sqrt{3}}{2}, \text{ and } 2\sin\left(\frac{\pi}{6}\right) = 2\left(\frac{1}{2}\right) = 1$$

A direct consequence of the compound angle identities developed in the past few pages are formulas for $\sin 2\theta$, $\cos 2\theta$ and $\tan 2\theta$, that is, **double angle identities**. For example, the formula for $\sin 2\theta$ can be derived by taking the identity for the sine of two numbers and by letting $\alpha = \beta = \theta$.

$$\sin 2\theta = \sin(\theta + \theta) = \sin\theta\cos\theta + \cos\theta\sin\theta = 2\sin\theta\cos\theta$$

Similarly, for $\cos 2\theta$ we have,

$$\cos 2\theta = \cos(\theta + \theta) = \cos\theta\cos\theta - \sin\theta\sin\theta = \cos^2\theta - \sin^2\theta$$

By applying the Pythagorean identity $\sin^2\theta + \cos^2\theta = 1$, we can write the double angle identity for $\cos 2\theta$ in two other useful ways.

$$\cos 2\theta = \cos^2\theta - \sin^2\theta = \cos^2\theta - (1 - \cos^2\theta) = 2\cos^2\theta - 1$$

$$\cos 2\theta = \cos^2\theta - \sin^2\theta = (1 - \sin^2\theta) - \sin^2\theta = 1 - 2\sin^2\theta$$

• **Hint:** The double angle identity for the tangent function does not hold if $\theta = \frac{\pi}{4} + k\cdot\frac{\pi}{2}$, where k is any integer, because for these values of θ the denominator is zero. The identity also does not hold if $\theta = \frac{\pi}{2} + k\cdot\pi$, where k is any integer, because for these values $\tan\theta$ does not exist. Nevertheless, the equation is still an identity because it is true for all values of θ for which both sides are defined.

To derive the formula for expressing $\tan 2\theta$ in terms of $\tan\theta$, we take the same approach and start with the identity for the tangent of the sum of two numbers and let $\alpha = \beta = \theta$.

$$\tan(\theta + \theta) = \frac{\tan\theta + \tan\theta}{1 - \tan\theta\tan\theta} = \frac{2\tan\theta}{1 - \tan^2\theta}$$

We now have a useful set of identities for the sine, cosine and tangent of twice an angle (or number).

Double angle identities

$\sin 2\theta = 2\sin\theta\cos\theta$

$$\cos 2\theta = \begin{cases} \cos^2\theta - \sin^2\theta \\ 2\cos^2\theta - 1 \\ 1 - 2\sin^2\theta \end{cases}$$

$$\tan 2\theta = \frac{2\tan\theta}{1 - \tan^2\theta}$$

Now let's look at some further applications of the trigonometric identities we have established, especially for solving more sophisticated equations.

Example 27

Solve the equation $\cos 2x + \cos x = 0$ for $0 \leqslant x \leqslant 2\pi$.

Solution

Taking an initial look at the graph of $y = \cos 2x + \cos x$ suggests that there are possibly three solutions in the interval $x \in [0, 2\pi]$. Although the expression $\cos 2x + \cos x$ contains terms with only the cosine function, it is not possible to perform any algebraic operations on them because they have different arguments. In order to solve algebraically, we need both cosine functions to have arguments of x (rather than $2x$). There are three different double angle identities for $\cos 2x$. It is best to have the equation in terms of one trigonometric function, so we choose to substitute $2\cos^2 x - 1$ for $\cos 2x$.

$$\cos 2x + \cos x = 0 \Rightarrow 2\cos^2 x - 1 + \cos x = 0 \Rightarrow 2\cos^2 x + \cos x - 1 = 0$$

$$(2\cos x - 1)(\cos x + 1) = 0 \Rightarrow \cos x = \tfrac{1}{2} \text{ or } \cos x = -1$$

For $\cos x = \frac{1}{2}$: $x = \frac{\pi}{3}, \frac{5\pi}{3}$; for $\cos x = -1$: $x = \pi$.

Therefore, all of the solutions in the interval $0 \leqslant x \leqslant 2\pi$ are: $x = \frac{\pi}{3}, \pi, \frac{5\pi}{3}$.

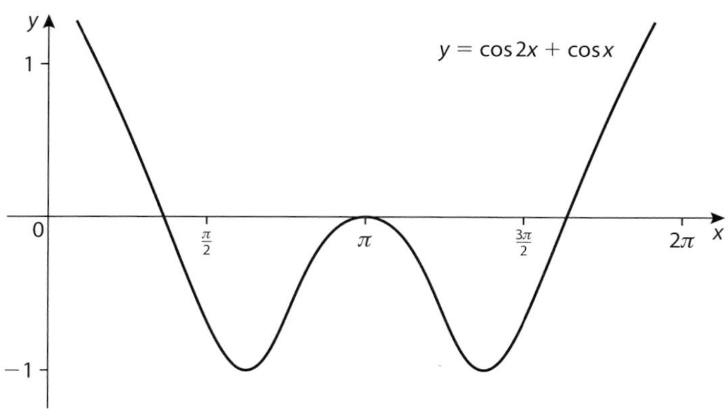

Example 28

Solve the equation $2 \sin 2x = 3 \cos x$ for $0 \leqslant x \leqslant \pi$.

Solution

$2 \sin 2x = 3 \cos x$

$2(2 \sin x \cos x) = 3 \cos x$ — Using double angle identity for sine.

$4 \sin x \cos x = 3 \cos x$ — Do not divide by $\cos x$; solution(s) may be eliminated.

$4 \sin x \cos x - 3 \cos x = 0$ — Set equal to zero to prepare for solving by factorization.

$\cos x(4 \sin x - 3) = 0$ — Factorize.

$\cos x = 0$ or $\sin x = \frac{3}{4}$

For $\cos x = 0$: $x = \frac{\pi}{2}$.

For $\sin x = \frac{3}{4}$: $x \approx 0.848$ or 2.29.

Approximate solutions are found using the Intersect command on the GDC.
All solutions in interval $0 \leqslant x \leqslant \pi$ are: $x = \frac{\pi}{2}$; $x \approx 0.848, 2.29$.

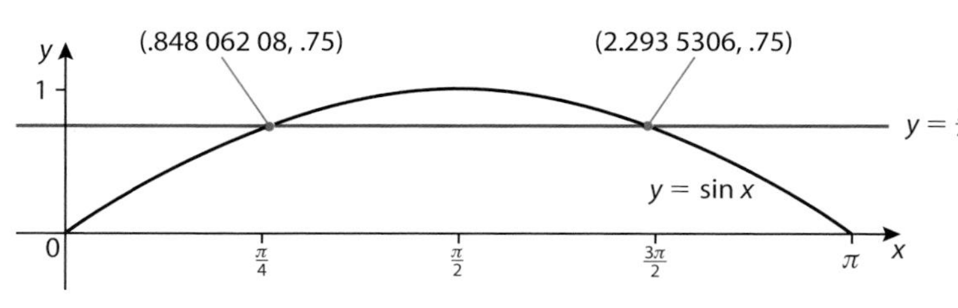

The next example illustrates how trigonometric identities can be applied to find exact values to trigonometric expressions.

Example 29

Given that $\cos x = \frac{1}{4}$ and that $0 < x < \frac{\pi}{2}$, find the *exact* values of

a) $\sin x$ b) $\sin 2x$

Solution

a) Given $0 < x < \frac{\pi}{2}$ it follows that $\sin x > 0$, because the arc with length x will terminate in the first quadrant. The Pythagorean identity is useful when relating $\sin x$ and $\cos x$.

$$\sin^2 x = 1 - \cos^2 x \Rightarrow \sin x = \sqrt{1 - \cos^2 x}$$

$$\Rightarrow \sin x = \sqrt{1 - \left(\frac{1}{4}\right)^2} = \sqrt{\frac{15}{16}} = \frac{\sqrt{15}}{4}$$

b) $\sin 2x = 2 \sin x \cos x = 2\left(\frac{\sqrt{15}}{4}\right)\left(\frac{1}{4}\right) = \frac{\sqrt{15}}{8}$

Example 30

Prove the following identity.

$$\frac{\cos A}{\cos A - \sin A} + \frac{\sin A}{\cos A + \sin A} = 1 + \tan 2A$$

• **Hint:** An effective approach to proving identities is to try and work exclusively on one side of the equation. Choosing the side that has an expression that is more 'complicated' is often an efficient path to transform the expression to the one on the other side by means of algebraic manipulations and substitutions. If you do choose to simplify both sides, be careful to work on each side independent of the other. In other words, as mentioned previously, do not perform an operation to both sides (e.g. multiplying both sides by the same quantity). This is only valid if it is known that both sides are equal but this is precisely what you are trying to prove.

Solution

Although we could apply a double angle identity to $\tan 2A$ on the right side it would not help to simplify the expression. The left side appears riper for simplification given that the common denominator of the two fractions is $\cos^2 A - \sin^2 A$ which is equivalent to $\cos 2A$.

$$\frac{\cos A}{\cos A - \sin A} \cdot \frac{\cos A + \sin A}{\cos A + \sin A} + \frac{\sin A}{\cos A + \sin A} \cdot \frac{\cos A - \sin A}{\cos A - \sin A} = \text{RHS}$$

Find a common denominator.

$$\frac{\cos^2 A + \sin A \cos A}{\cos^2 A - \sin^2 A} + \frac{\sin A \cos A - \sin^2 A}{\cos^2 A - \sin^2 A} = \text{RHS}$$

Multiply conjugates $(a + b)(a - b) = a^2 - b^2$.

$$\frac{\cos^2 A - \sin^2 A + 2 \sin A \cos A}{\cos^2 A - \sin^2 A} = \text{RHS}$$

$$\frac{\cos 2A + 2 \sin A \cos A}{\cos 2A} = \text{RHS}$$ Substitute $\cos 2A$ for $\cos^2 A - \sin^2 A$.

Observing that the right-hand side (RHS) has a term equal to 1 directs us to split the left side into two fractions since one of the terms in the numerator is equal to the denominator.

$$\frac{\cos 2A}{\cos 2A} + \frac{2 \sin A \cos A}{\cos 2A} = \text{RHS}$$

$$1 + \frac{\sin 2A}{\cos 2A} = \text{RHS}$$ Substitute $\sin 2A$ for $2 \sin A \cos A$.

$1 + \tan 2A = 1 + \tan 2A$ Q.E.D. Apply tangent identity $\tan x = \frac{\sin x}{\cos x}$.

Table 7.3 Summary of trigonometric identities.

Reciprocal identities		
$\csc\theta = \dfrac{1}{\sin\theta}$	$\sec\theta = \dfrac{1}{\cos\theta}$	$\cot\theta = \dfrac{1}{\tan\theta}$
Tangent and cotangent identities		
$\tan\theta = \dfrac{\sin\theta}{\cos\theta}$	$\cot\theta = \dfrac{\cos\theta}{\sin\theta}$	
Odd/even function identities		
$\sin(-\theta) = -\sin\theta$	$\cos(-\theta) = \cos\theta$	$\tan(-\theta) = -\tan\theta$
$\csc(-\theta) = -\csc\theta$	$\sec(-\theta) = \cos\theta$	$\cot(-\theta) = -\tan\theta$
Co-function identities		
$\sin\left(\frac{\pi}{2} - \theta\right) = \cos\theta$	$\sec\left(\frac{\pi}{2} - \theta\right) = \csc\theta$	$\tan\left(\frac{\pi}{2} - \theta\right) = \cot\theta$
$\cos\left(\frac{\pi}{2} - \theta\right) = \sin\theta$	$\csc\left(\frac{\pi}{2} - \theta\right) = \sec\theta$	$\cot\left(\frac{\pi}{2} - \theta\right) = \tan\theta$
Pythagorean identities		
$\sin^2\theta + \cos^2\theta = 1$	$1 + \tan^2\theta = \sec^2\theta$	$1 + \cot^2\theta = \csc^2\theta$
Compound angle identities		
$\sin(\alpha \pm \beta) = \sin\alpha\cos\beta \pm \cos\alpha\sin\beta$ $\cos(\alpha \pm \beta) = \cos\alpha\cos\beta \mp \sin\alpha\sin\beta$ $\tan(\alpha \pm \beta) = \dfrac{\tan\alpha \pm \tan\beta}{1 \mp \tan\alpha\tan\beta}$		
Double angle identities		
$\sin 2\theta = 2\sin\theta\cos\theta$ $\cos 2\theta = \begin{cases}\cos^2\theta - \sin^2\theta \\ 2\cos^2\theta - 1 \\ 1 - 2\sin^2\theta\end{cases}$ $\tan 2\theta = \dfrac{2\tan\theta}{1 - \tan^2\theta}$		

Exercise 7.5

In questions 1–6, use a compound angle identity to find the **exact** value of the expression.

1 $\cos\frac{7\pi}{12}$ **2** $\sin 165°$

3 $\tan\frac{\pi}{12}$ **4** $\sin\left(-\frac{5\pi}{12}\right)$

5 $\cos 255°$ **6** $\cot 75°$

7 a) Find the **exact** value of $\cos\frac{\pi}{12}$.

b) By writing $\cos\frac{\pi}{12}$ as $\cos\left(2\cdot\frac{\pi}{24}\right)$ and using a double angle identity for cosine, find the **exact** value of $\cos\frac{\pi}{24}$.

In questions 8–10, prove the co-function identity using the compound angle identities.

8 $\tan\left(\frac{\pi}{2} - \theta\right) = \cot\theta$ **9** $\sin\left(\frac{\pi}{2} - \theta\right) = \cos\theta$ **10** $\csc\left(\frac{\pi}{2} - \theta\right) = \sec\theta$

11 Given that $\sin x = \frac{3}{5}$ and that $0 < x < \frac{\pi}{2}$, find the exact values of

a) $\cos x$ b) $\cos 2x$ c) $\sin 2x$

12 Given that $\cos x = -\frac{2}{3}$ and that $\frac{\pi}{2} < x < \pi$, find the exact values of

a) $\sin x$ b) $\sin 2x$ c) $\cos 2x$

In questions 13–16, find the exact values of $\sin 2\theta$, $\cos 2\theta$ and $\tan 2\theta$ subject to the given conditions.

13 $\sin\theta = \frac{2}{3}, \frac{\pi}{2} < \theta < \pi$

14 $\cos\theta = -\frac{4}{5}, \pi < \theta < \frac{3\pi}{2}$

15 $\tan\theta = 2, 0 < \theta < \frac{\pi}{2}$

16 $\sec\theta = -4, \csc\theta > 0$

In questions 17–20, use a compound angle identity to write the given expression as a function of x alone.

17 $\cos(x - \pi)$

18 $\sin\left(x - \frac{\pi}{2}\right)$

19 $\tan(x + \pi)$

20 $\cos\left(x + \frac{\pi}{2}\right)$

In questions 21–24, use identities to find an equivalent expression involving only sines and cosines, and then simplify it.

21 $\sec\theta + \sin\theta$

22 $\dfrac{\sec\theta\csc\theta}{\tan\theta\sin\theta}$

23 $\dfrac{\sec\theta + \csc\theta}{2}$

24 $\dfrac{1}{\cos^2\theta} + \dfrac{1}{\cot^2\theta}$

In questions 25–32, simplify each expression.

25 $\cos\theta - \cos\theta\sin^2\theta$

26 $\dfrac{1 - \cos^2\theta}{\sin^2\theta}$

27 $\cos 2\theta + \sin^2\theta$

28 $\dfrac{\sin^2\theta}{\cos^2\theta} + \dfrac{1}{\cot^2\theta}$

29 $\sin(\alpha + \beta) + \sin(\alpha - \beta)$

30 $\dfrac{1 + \cos 2A}{2}$

31 $\cos(\alpha + \beta) + \cos(\alpha - \beta)$

32 $2\cos^2\theta - \cos 2\theta$

In questions 33–46, prove each identity.

33 $\dfrac{\cos 2\theta}{\cos\theta + \sin\theta} = \cos\theta - \sin\theta$

34 $(1 - \cos\alpha)(1 + \sec\alpha) = \sin\alpha\tan\alpha$

35 $\dfrac{1 - \tan^2 x}{1 + \tan^2 x} = \cos 2x$

36 $\cos^4\theta - \sin^4\theta = \cos 2\theta$

37 $\cot\theta - \tan\theta = 2\cot 2\theta$

38 $\dfrac{\cos\beta - \sin\beta}{\cos\beta + \sin\beta} = \dfrac{\cos 2\beta}{1 + \sin 2\beta}$

39 $\dfrac{1}{\sec\theta(1-\sin\theta)} = \sec\theta + \tan\theta$

40 $(\tan A - \sec A)^2 = \dfrac{1-\sin A}{1+\sin A}$

41 $\dfrac{\tan 2x \tan x}{\tan 2x - \tan x} = \sin 2x$

42 $\dfrac{\sin 2\theta - \cos 2\theta + 1}{\sin 2\theta + \cos 2\theta + 1} = \tan\theta$

43 $\dfrac{1+\cos\alpha}{\sin\alpha} = 2\csc\alpha - \dfrac{\sin\alpha}{1+\cos\alpha}$

44 $\dfrac{1+\cos\beta}{\sin\beta} + \dfrac{\sin\beta}{1+\cos\beta} = 2\csc\beta$

45 $\dfrac{\cot x - 1}{1-\tan x} = \dfrac{\csc x}{\sec x}$

46 $\sin\left(\dfrac{\theta}{2}\right) = \pm\sqrt{\dfrac{1-\cos\theta}{2}}$

• **Hint:** For question 46, first prove that $\sin^2 x = 1 - \dfrac{\cos 2x}{2}$, then make a suitable substitution for x. This identity is called the **half-angle identity** for sine. Can you find the corresponding half-angle identity for cosine?

47 Given the figure shown right, find an expression in terms of x for the value of $\tan\theta$.

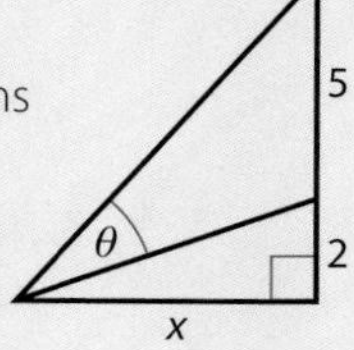

In questions 48–57, solve each equation for x in the given interval. Give answers exactly, if possible. Otherwise, give answers accurate to three significant figures.

48 $2\sin^2 x - \cos x = 1, 0 \leqslant x < 2\pi$

49 $\sec^2 x = 8\cos x, -\pi < x \leqslant \pi$

50 $2\cos x + \sin 2x = 0, -180° < x \leqslant 180°$

51 $2\sin x = \cos 2x, 0 \leqslant x < 2\pi$

52 $\cos 2x = \sin^2 x, 0 \leqslant x < 2\pi$

53 $2\sin x\cos x + 1 = 0, 0 \leqslant x < 2\pi$

54 $\cos^2 x - \sin^2 x = -\frac{1}{2}, 0 \leqslant x \leqslant \pi$

55 $\sec^2 x - \tan x - 1 = 0, 0 \leqslant x < 2\pi$

56 $\tan 2x + \tan x = 0, 0 \leqslant x < 2\pi$

57 $2\sin 2x\cos 3x + \cos 3x = 0, 0 \leqslant x \leqslant 180°$

58 Find an identity for $\sin 3x$ in terms of $\sin x$.

59 a) By squaring $\sin^2 x + \cos^2 x$, prove that $\sin^4 x + \cos^4 x = \frac{1}{4}(\cos 4x + 3)$.

b) Hence, or otherwise, solve the equation $\sin^4 x + \cos^4 x = \frac{1}{2}$ for $0 \leqslant x < 2\pi$.

 7.6 Inverse trigonometric functions

In Section 2.3, we learned that if a function f is one-to-one then f has an inverse f^{-1}. A defining characteristic of a one-to-one function is that it is always increasing or always decreasing in its domain. Also, recall that no horizontal line can pass through the graph of a one-to-one function at more than one point. It is evident that none of the trigonometric functions are one-to-one functions given their periodic nature. Therefore, the inverse of any of the trigonometric functions over their domain is not a function.

Defining the inverse sine function

Recall that the domain of $y = \sin x$ is all real numbers ($\mathbb{R}$) and its range is the set of all real numbers in the closed interval $-1 \leqslant y \leqslant 1$. The sine function is not one-to-one and hence its inverse is not a function, since more than one value of x corresponds to the same value of y. For example, $\sin\frac{\pi}{6} = \sin\frac{5\pi}{6} = \sin\frac{13\pi}{6} = \frac{1}{2}$. That is, for $y = \sin x$ there are an infinite number of ordered pairs with a y-coordinate of $\frac{1}{2}$ (see Figure 7.30).

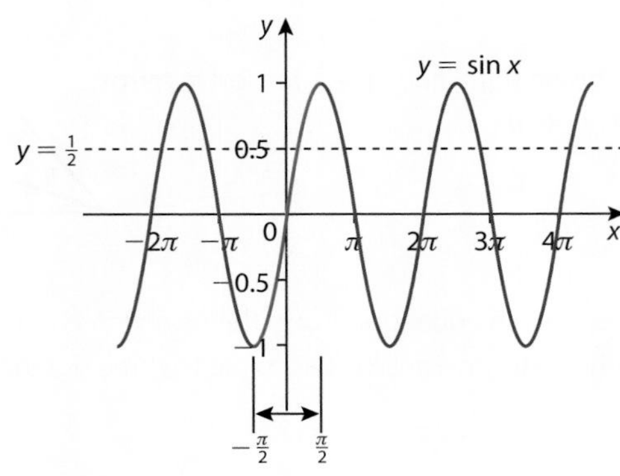

Figure 7.30 A horizontal line, $y = \frac{1}{2}$ shown here, can intersect the graph of $y = \sin x$ more than once, thus indicating that the inverse of $y = \sin x$ is not a function. The portion of the graph (in red) from $-\frac{\pi}{2}$ to $-\frac{\pi}{2}$ is used to define the inverse and only intersects a horizontal line once.

Examples 13 and 15 in Section 2.3, showed us that a function that is not one-to-one can often be made so by restricting its domain. Consequently, even though there is no inverse function for the sine function for all $\mathbb{R}$, we can define the inverse sine function if we restrict its domain so that it is one-to-one (and passes the horizontal line test). We have an unlimited number of ways of restricting the domain but it seems sensible to select an interval of x including zero, and it's standard to restrict the domain to the 'largest' set possible. Consider restricting the domain of $y = \sin x$ to the interval $-\frac{\pi}{2} \leqslant x \leqslant \frac{\pi}{2}$. In this interval, $y = \sin x$ is always increasing and takes on every value from -1 to 1 exactly once. Thus, the function $y = \sin x$ with domain $-\frac{\pi}{2} \leqslant x \leqslant \frac{\pi}{2}$ is one-to-one and its inverse is a function. We have the following definition:

The equation $y = \arcsin x$ is interpreted, 'y is the arc whose sine is x', or 'y is the angle whose sine is x,' or 'y is the real number whose sine is x.' Any GDC labels the inverse sine function as $\sin^{-1} x$. The symbols $y = \arcsin x$ and $y = \sin^{-1} x$ are both commonly used to indicate the inverse sine function, but a disadvantage of writing $y = \sin^{-1} x$ is that it can be confused with $y = (\sin x)^{-1} = \frac{1}{\sin x} = \csc x$.

Inverse sine function

The inverse sine function, denoted by $x = \arcsin x$ or $y = \sin^{-1} x$, is the function with a domain of $-1 \leqslant x \leqslant 1$ and a range of $-\frac{\pi}{2} \leqslant y \leqslant \frac{\pi}{2}$ defined by

$$y = \arcsin x \quad \text{if and only if} \quad x = \sin y$$

Thus, $\arcsin x$ (or $\sin^{-1} x$) is the number in the closed interval $\left[-\frac{\pi}{2}, \frac{\pi}{2}\right]$ whose sine is x. For example, $\arcsin\frac{1}{2} = \frac{\pi}{6}$ because the one number in the interval $\left[-\frac{\pi}{2}, \frac{\pi}{2}\right]$ whose sine is $\frac{1}{2}$ is $\frac{\pi}{6}$. Your GDC is programmed such that it will give the same result. If your GDC is in radian mode it will give the approximate value of $\frac{\pi}{6}$ to several significant figures, and if it is in degree mode, it will give the exact result of 30°. See the GDC images on the next page.

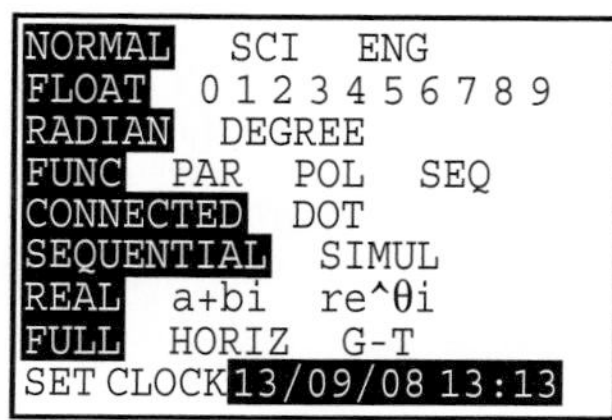

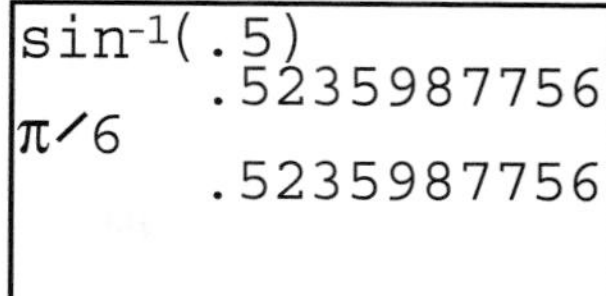

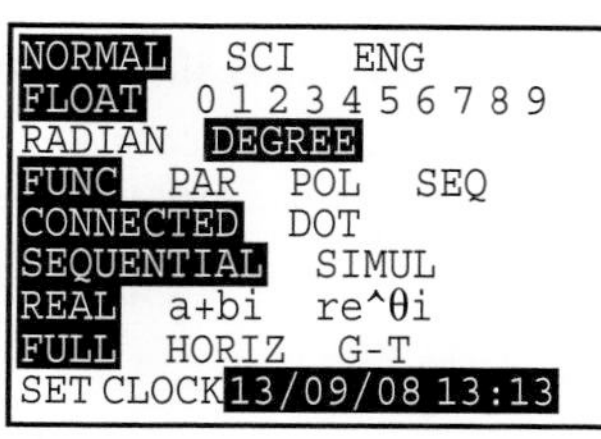

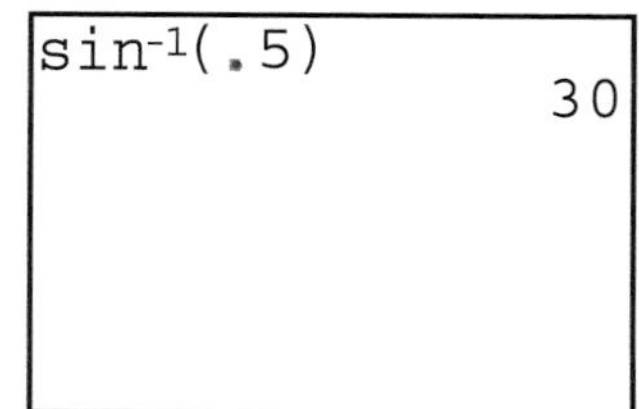

From the graphical symmetry of inverse functions, the graph of $y = \arcsin x$ is a reflection of $y = \sin x$ about the line $y = x$, as shown in Figures 7.31 and 7.32.

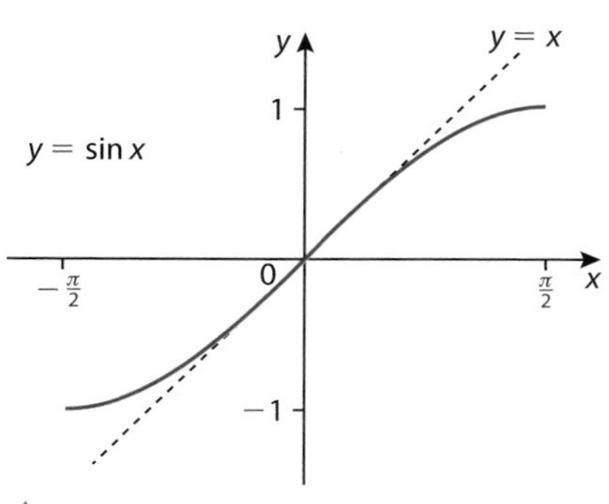

Figure 7.31 The graph of $y = \sin x$ with domain restricted to $-\frac{\pi}{2} \leq x \leq \frac{\pi}{2}$.

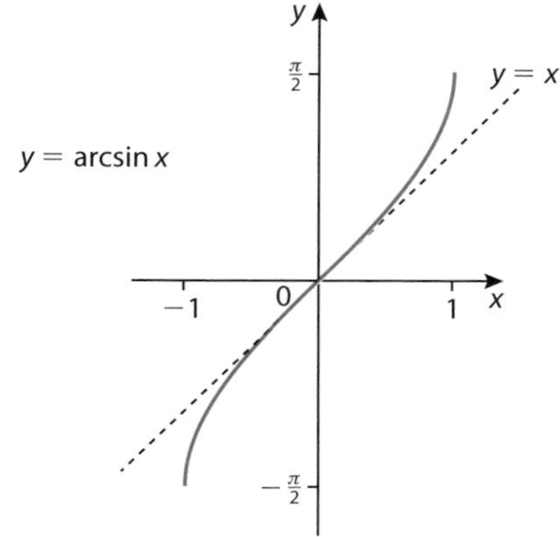

Figure 7.32 The graph of $y = \arcsin x$.

Defining the inverse cosine and inverse tangent functions

The inverse cosine function and inverse tangent function can be defined by following a parallel procedure to that used for defining the inverse sine function. The graphs of $y = \cos x$ and $y = \tan x$ (Figures 7.33 and 7.34) clearly show that neither function is one-to-one and consequently their inverses are not functions. Consider restricting the domain of the cosine function to the closed interval $0 \leq x \leq \pi$ (Figure 7.33) and restricting the domain of the tangent function to the open interval $-\frac{\pi}{2} < x < \frac{\pi}{2}$ (Figure 7.34). The interval for tangent cannot include the endpoints, $-\frac{\pi}{2}$ and $\frac{\pi}{2}$, because tangent is undefined for these values. For these domain restrictions cosine and tangent will attain each of its function values exactly once. Hence, with these restrictions, both cosine and tangent will be one-to-one and their inverses will be functions.

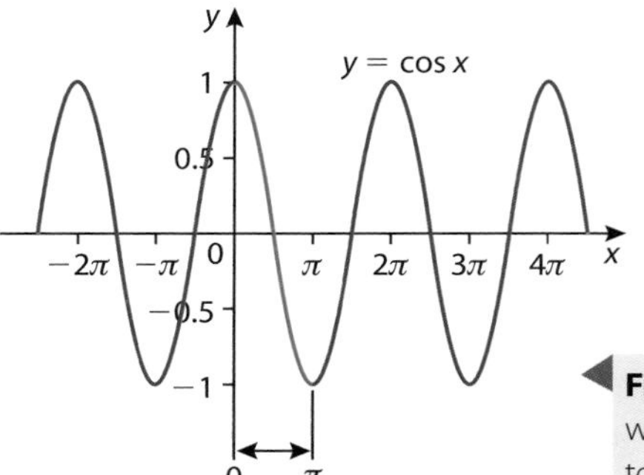

Figure 7.33 The graph of $y = \cos x$ with portion of the graph (in red) from 0 to π (inclusive) used to define its inverse.

Figure 7.34 The graph of $y = \tan x$ with the portion of the graph (in red) from $-\frac{\pi}{2}$ to $\frac{\pi}{2}$ (exclusive) used to define its inverse.

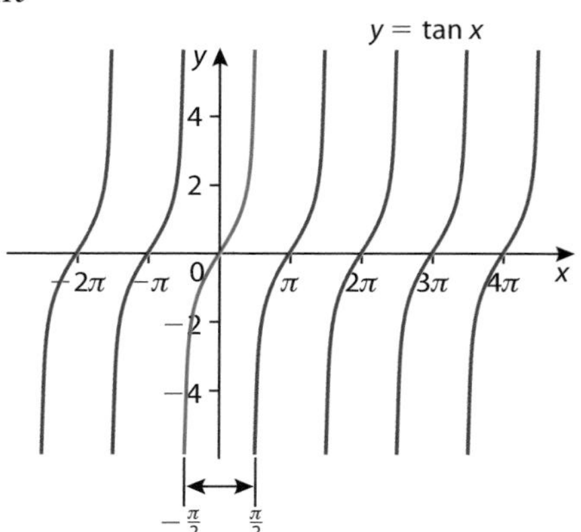

Inverse cosine function

The inverse cosine function, denoted by $y = \arccos x$, or $y = \cos^{-1} x$, is the function with a domain of $-1 \leqslant x \leqslant 1$ and a range of $0 \leqslant y \leqslant \pi$ defined by

$$y = \arccos x \quad \text{if and only if} \quad x = \cos y$$

Inverse tangent function

The inverse tangent function, denoted by $y = \arctan x$, or $y = \tan^{-1} x$, is the function with a domain of $\mathbb{R}$ and a range of $-\frac{\pi}{2} < y < \frac{\pi}{2}$ defined by

$$y = \arctan x \quad \text{if and only if} \quad x = \tan y$$

The graphs of $y = \cos x$ (for the appropriate interval) and $y = \arccos x$ are shown in Figures 7.35 and 7.36.

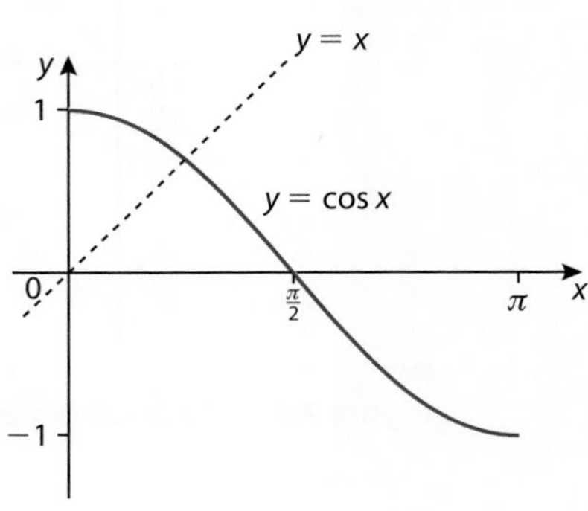

Figure 7.35 The graph of $y = \cos x$ with domain restricted to $0 \leqslant x \leqslant \pi$.

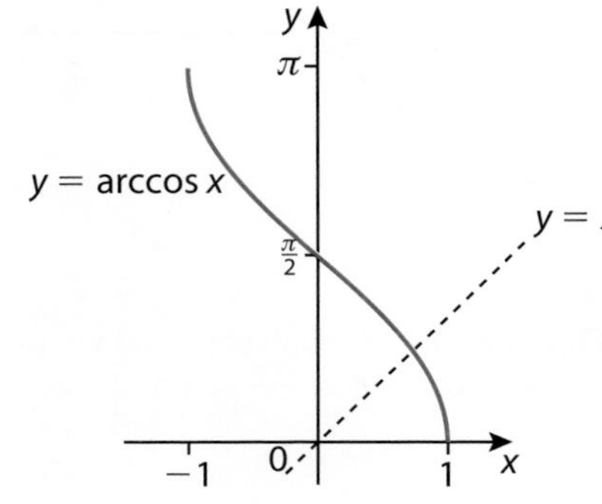

Figure 7.36 The graph of $y = \arccos x$.

The inverse cotangent, secant and cosecant functions are rarely used (and are not in the Maths Higher Level syllabus) so definitions will not be given for them here.

The graphs of $y = \tan x$ (for the appropriate interval) and $y = \arctan x$ are shown in Figures 7.37 and 7.38.

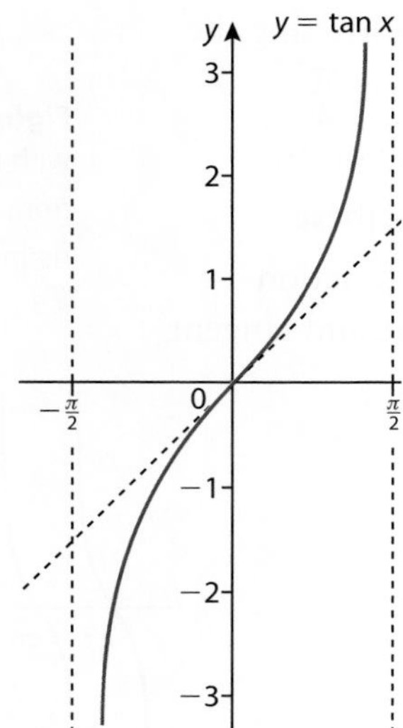

Figure 7.37 The graph of $y = \tan x$ with domain restricted to $-\frac{\pi}{2} < x < \frac{\pi}{2}$.

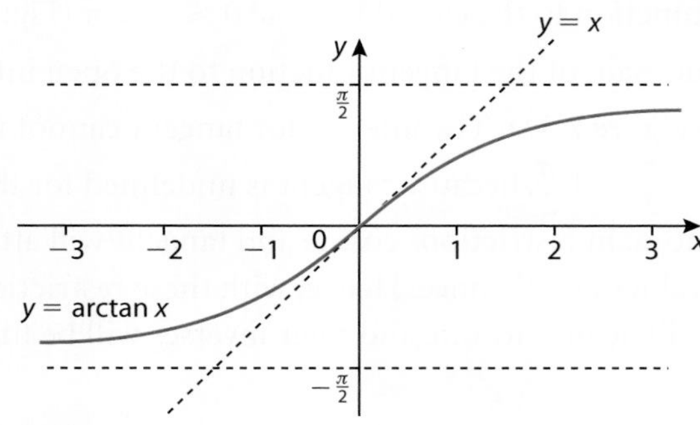

Figure 7.38 The graph of $y = \arctan x$.

• **Hint:** Unless specifically instructed otherwise, we will assume that the result of evaluating an inverse trigonometric function will be a real number that can be interpreted as either an arc length on the unit circle or an angle in radian measure. If the result is to be an angle in degree measure then the instructions will explicitly request this.

Example 30

Without using your GDC, find the exact value of each expression.

a) $\arcsin\left(-\frac{\sqrt{3}}{2}\right)$ b) $\arccos 1$ c) $\arctan\sqrt{3}$ d) $\arcsin\frac{3}{2}$

Solution

a) The expression $\arcsin\left(-\frac{\sqrt{3}}{2}\right)$ can be interpreted as 'the number y such that $-\frac{\pi}{2} \leqslant y \leqslant \frac{\pi}{2}$ whose sine is $-\frac{\sqrt{3}}{2}$' or 'the number in quadrant I or IV whose sine is $-\frac{\sqrt{3}}{2}$.' We know sine function values are negative in quadrants III and IV, so the number we are looking for is in quadrant IV. The diagram shows that the required number is $-\frac{\pi}{3}$. An angle of $-\frac{\pi}{3}$ in standard position will intersect the unit circle at a point whose y-coordinate is $-\frac{\sqrt{3}}{2}$.

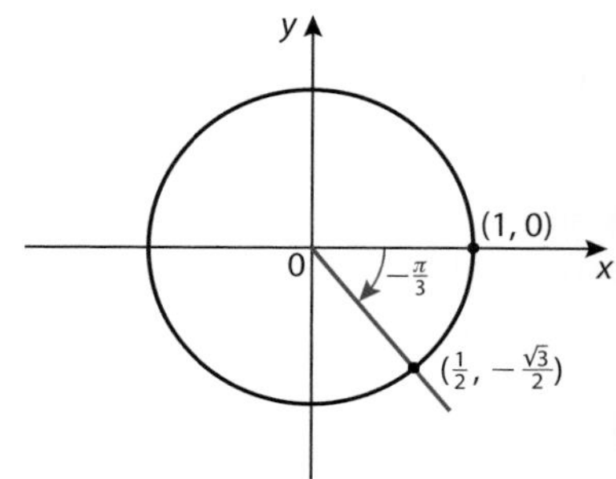

Therefore, $\arcsin\left(-\frac{\sqrt{3}}{2}\right) = -\frac{\pi}{3}$.

b) The range of the function $y = \arccos x$ is $0 \leqslant y \leqslant \pi$. Thus we are looking for a number in quadrant I or II whose cosine is 1. The number we are looking for is 0, because an angle of measure 0 in standard position will intersect the unit circle at a point whose x-coordinate is 1. Therefore, $\arccos 1 = 0$.

c) The range of the function $y = \arctan x$ is $-\frac{\pi}{2} < y < \frac{\pi}{2}$. Thus we are looking for a number in quadrant I or IV for which the ratio $\frac{\text{sine}}{\text{cosine}}$ is equal to $\sqrt{3}$. It must be in quadrant I because in quadrant IV tangent values are negative. Familiarity with the sine and cosine values for common angles covered earlier in this chapter helps us to recognize that the required ratio will be $\frac{\frac{\sqrt{3}}{2}}{\frac{1}{2}}$. The required number is $\frac{\pi}{3}$ because it is in the first quadrant with $\sin\frac{\pi}{3} = \frac{\sqrt{3}}{2}$ and $\cos\frac{\pi}{3} = \frac{1}{2}$.

Therefore, $\arctan\sqrt{3} = \frac{\pi}{3}$.

d) The domain of the function $y = \arccos x$ is $-1 \leqslant x \leqslant 1$, but $\frac{3}{2}$ is not in this interval. There is no number whose sine is $\frac{3}{2}$. Therefore, $\arcsin\frac{3}{2}$ is not defined.

Compositions of trigonometric and inverse trigonometric functions

Recall from Chapter 2 that for a pair of inverse functions the following two properties hold true.

$f(f^{-1}(x)) = x$ for all x in the domain of f^{-1}; and $f^{-1}(f(x)) = x$ for all x in the domain of f.

It follows that the following properties hold true for the inverse sine, cosine and tangent functions.

• **Hint:** Note that the inverse property $\arcsin(\sin\beta) = \beta$ does **not** hold true when $\beta = \frac{3\pi}{4}$.

$\arcsin\left(\sin\frac{3\pi}{4}\right) = \arcsin\left(\frac{\sqrt{2}}{2}\right) = \frac{\pi}{4}$

and

$\arcsin\left(\sin\frac{5\pi}{4}\right) = \arcsin\left(-\frac{\sqrt{2}}{2}\right) = -\frac{\pi}{4}$.

The property $\arcsin(\sin\beta) = \beta$ is not valid for values of β outside the interval $-\frac{\pi}{2} \leqslant \beta \leqslant \frac{\pi}{2}$. Similarly, the property $\arccos(\cos\beta) = \beta$ is not valid for values of β outside the interval $0 \leqslant \beta \leqslant \pi$; and $\arctan(\tan\beta) = \beta$ is not valid for values of β outside the interval $-\frac{\pi}{2} < \beta < \frac{\pi}{2}$.

Inverse properties

If $-1 \leqslant \alpha \leqslant 1$, then $\sin(\arcsin\alpha) = \alpha$; and if $-\frac{\pi}{2} \leqslant \beta \leqslant \frac{\pi}{2}$, then $\arcsin(\sin\beta) = \beta$.

If $-1 \leqslant \alpha \leqslant 1$, then $\cos(\arccos\alpha) = \alpha$; and if $0 \leqslant \beta \leqslant \pi$ then $\arccos(\cos\beta) = \beta$.

If $\alpha \in \mathbb{R}$, then $\tan(\arctan\alpha) = \alpha$; and if $-\frac{\pi}{2} < \beta < \frac{\pi}{2}$, then $\arctan(\tan\beta) = \beta$.

Example 31

Find the exact values, if possible, for the following expressions.

a) $\cos^{-1}\left(\cos\frac{4\pi}{3}\right)$ b) $\tan(\arctan(-7))$ c) $\sin(\arcsin\sqrt{3})$

Solution

a) $\frac{4\pi}{3}$ is not in the range of the $\cos^{-1}$, or arccos, function $0 \leqslant \beta \leqslant \pi$. However, using the symmetry of the unit circle we know that $\frac{4\pi}{3}$ has the same cosine as $\frac{2\pi}{3}$ (see figure) which is in the interval $0 \leqslant \beta \leqslant \pi$. Thus, $\cos^{-1}\left(\cos\frac{4\pi}{3}\right) = \cos^{-1}\left(\cos\frac{2\pi}{3}\right) = \frac{2\pi}{3}$.

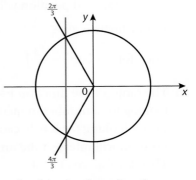

b) -7 is in the range of the tangent function (and in the domain of the arctangent function), so the inverse property applies. Therefore, $\tan(\arctan(-7)) = -7$.

c) $\sqrt{3}$ is not in the range of the sine function $-1 \leqslant \alpha \leqslant 1$, so $\arcsin\sqrt{3}$ is not defined. It follows that $\sin(\arcsin\sqrt{3})$ is not defined.

All of the results in Example 31 can be quickly verified on your GDC as shown below. Be sure to be in radian mode.

```
cos⁻¹(cos(4π/3))
        2.094395102
2π/3
        2.094395102
```

```
tan(tan⁻¹(-7))
                 -7
```

```
sin(sin⁻¹(√(3))
```

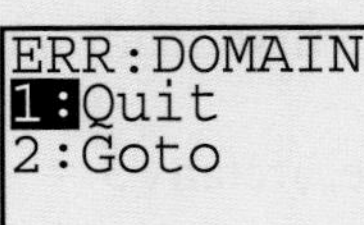

Example 32

Without using your GDC, find the exact value of each expression.

a) $\cos\left[\sin^{-1}\left(-\frac{8}{17}\right)\right]$

b) $\arcsin\left(\tan\frac{3\pi}{4}\right)$

c) $\sec\left[\arctan\left(\frac{3}{5}\right)\right]$

Solution

a) If we let $\theta = \sin^{-1}\left(-\frac{8}{17}\right)$, then $\sin\theta = -\frac{8}{17}$. Because $\sin\theta$ is negative, then θ must be an angle (arc) in quadrant IV. From a simple sketch of an appropriately labeled triangle in quadrant IV, we can determine $\cos\theta = \cos\left(\sin^{-1}\left(-\frac{8}{17}\right)\right)$.

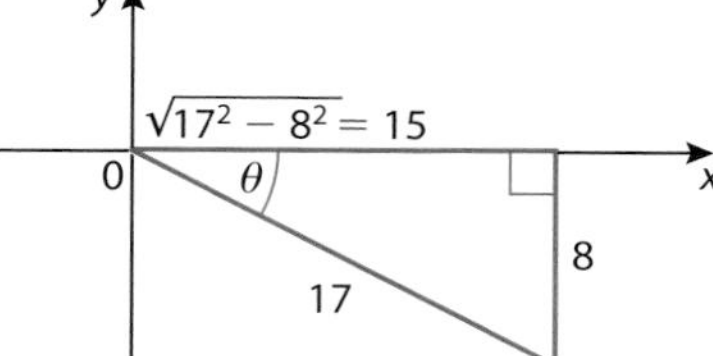

Therefore, $\cos\left(\sin^{-1}\left(-\frac{8}{17}\right)\right) = \frac{15}{17}$.

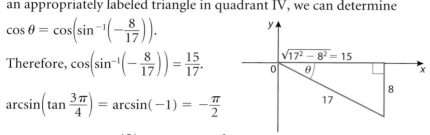

b) $\arcsin\left(\tan\frac{3\pi}{4}\right) = \arcsin(-1) = -\frac{\pi}{2}$

c) If we let $\theta = \arctan\left(\frac{3}{5}\right)$ then $\tan\theta = \frac{3}{5}$. Because $\tan\theta > 0$ then θ must be in quadrant I. Consequently, we can construct a right triangle containing θ in quadrant I by drawing a line from the origin to the point (5, 3), as shown in the diagram. The hypotenuse is $\sqrt{25 + 9} = \sqrt{34}$.

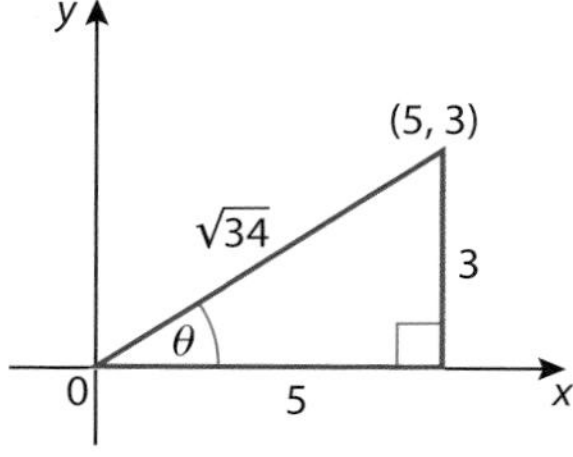

Therefore, $\sec\left[\arctan\left(\frac{3}{5}\right)\right] = \sec\theta = \frac{1}{\cos\theta} = \frac{1}{\frac{5}{\sqrt{34}}} = \frac{\sqrt{34}}{5}$.

Example 33

If $C = \arctan 3 + \arcsin\left(\frac{5}{13}\right)$, find the exact value of $\cos C$.

Solution

Let $A = \arctan 3$ and $B = \arcsin\left(\frac{5}{13}\right)$. Thus, $C = A + B$ and a strategy for finding $\cos C$ is to use the following compound angle identity: $\cos C = \cos(A + B) = \cos A \cos B - \sin A \sin B$. We know that $\sin B = \frac{5}{13}$. We need to find exact values for $\cos A$, $\cos B$ and $\sin A$. The range for $\arctan x$ is $-\frac{\pi}{2} < x < \frac{\pi}{2}$ and the range for $\arcsin x$ is $-\frac{\pi}{2} \leqslant x \leqslant \frac{\pi}{2}$, and since $\tan A = 3 > 0$ and $\sin B = \frac{5}{13} > 0$, both A and B are in quadrant I.

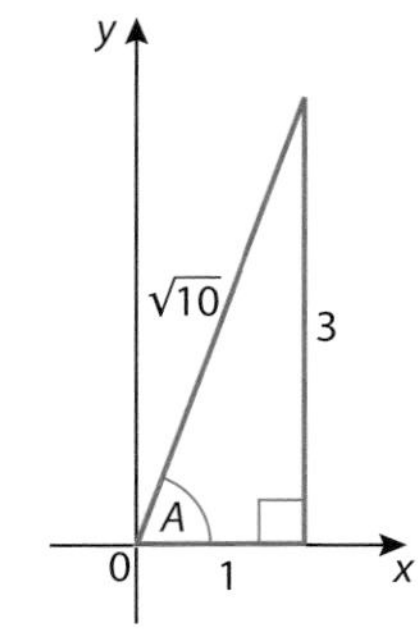

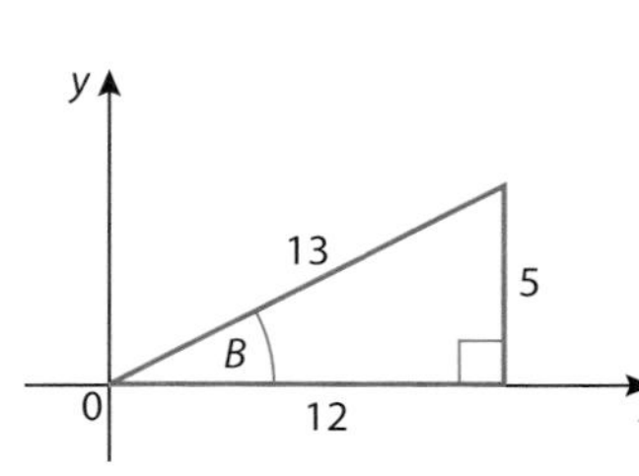

$\sin A = \frac{3}{\sqrt{10}} = \frac{3\sqrt{10}}{10}$ $\qquad$ $\sin B = \frac{5}{13}$

$\cos A = \frac{1}{\sqrt{10}} = \frac{\sqrt{10}}{10}$ $\qquad$ $\cos B = \frac{12}{13}$

Hence, $\cos C = \cos(A + B) = \cos A \cos B - \sin A \sin B$

$$= \left(\frac{\sqrt{10}}{10}\right)\left(\frac{12}{13}\right) - \left(\frac{3\sqrt{10}}{10}\right)\left(\frac{5}{13}\right)$$

$$= \frac{(12 - 15)\sqrt{10}}{130}$$

$$= \frac{-3\sqrt{10}}{130}$$

Therefore, $\cos C = \dfrac{-3\sqrt{10}}{130}$.

Example 34

Find all solutions, accurate to three significant figures, to the equation $3 \sin 2\theta = 1$ in the interval $0 \leqslant \theta < 2\pi$.

Solution

A reasonable idea is to apply a double angle identity and substitute $2 \sin \theta \cos \theta$ for $\sin 2\theta$. Although a substitution like this proved to be an effective technique in the previous section, it is not always the best strategy. In this case, the transformed equation becomes $6 \sin \theta \cos \theta = 1$ which would prove difficult to solve. A better approach is

$$3 \sin 2\theta = 1$$

$$\sin 2\theta = \tfrac{1}{3}$$

$$2\theta = \arcsin(\tfrac{1}{3})$$

$$\theta = \frac{1}{2}\arcsin(\tfrac{1}{3})$$

There is one angle in quadrant I with a sine equal to $\frac{1}{3}$ and one angle in quadrant II with a sine equal to $\frac{1}{3}$ (see figure). None of the common angles has a sine equal to $\frac{1}{3}$, so we will need to use the inverse sine ($\sin^{-1}$) on our GDC to obtain an approximate answer. Since the range of the inverse sine function, $\sin^{-1}$, is $-\frac{\pi}{2} \leqslant y \leqslant \frac{\pi}{2}$ your GDC's computation of $\sin^{-1}(\frac{1}{3})$ will only give the angle (arc) in quadrant I. From the symmetry of the unit circle, we can obtain the angle in quadrant II by subtracting the angle in quadrant I from π. The GDC images below show the computation to find both answers – and a check of the two answers.

```
sin⁻¹(1/3)
        .3398369095
.5*Ans
        .1699184547
Ans→A
        .1699184547
```

```
sin⁻¹(1/3)
        .3398369095
.5(π-Ans)
        1.400877872
Ans→B
        1.400877872
```

```
3sin(2A)
                  1
3sin(2B)
                  1
```

Therefore, $\theta \approx 0.170$ or $\theta \approx 1.40$ accurate to 3 significant figures.

To an observer, the apparent size of an object depends on the distance from the observer to the object. The farther an object is from an observer, the smaller its apparent size. For example, although the Sun's diameter is 400 times wider than our Moon's diameter, the two objects appear to have the same diameter as viewed from the Earth (see Figure 7.39). Thus, during a total solar eclipse, the Moon blocks out the Sun. Also, if an object is sufficiently above or below the horizontal position of the observer, the apparent size of the object will also decrease if you move close to the object. Thus for this situation, there will be a distance for which the angle subtended at the eye of the observer is a maximum (Example 35).

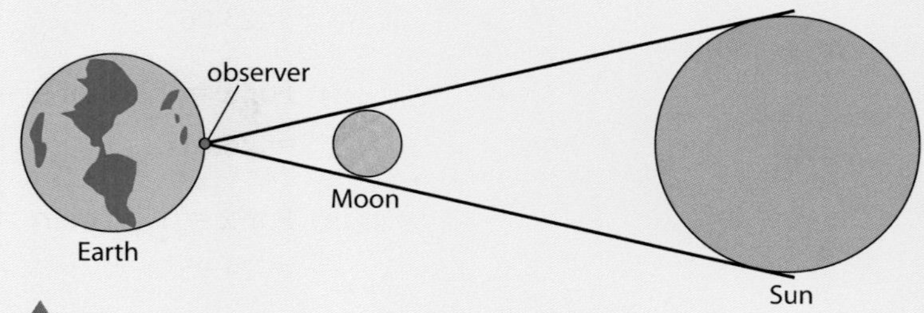

Figure 7.39

On the surface of the Earth the angle subtended by the moon and the Sun is nearly the same. It is approximately 0.54 degrees for the Moon and 0.52 degrees for the Sun. The Sun is 400 times wider than the Moon and coincidentally 400 times further from the Earth than the Moon.

Example 35

A painting that is 125 cm from top to bottom is hanging on the wall of a gallery such that it's base is 250 cm from the floor. Pablo is standing x cm from the wall from which the painting is hung. Pablo's eyes are 170 cm from the floor and from where he stands the painting subtends an angle α degrees. a) Write a function for α in terms of x. b) Find α, accurate to four significant figures, for the following values of x: (i) $x = 75$ cm; (ii) $x = 125$ cm; and (iii) $x = 175$ cm. c) Using a GDC, approximate to the nearest cm, how far Pablo should stand from the wall so that the subtended angle α is a maximum.

Solution

a) The figure shows α, the angle subtended by the painting, and β, the angle subtended by the part of the wall above eye level and below the painting. Let θ be the sum of these two angles. Hence, $\theta = \alpha + \beta$ and $\alpha = \theta - \beta$. From the compound angle identity for tangent, we have

$$\tan\alpha = \frac{\tan\theta - \tan\beta}{1 + \tan\theta\tan\beta}$$

From the right triangles in the figure, we can determine that

$$\tan\beta = \frac{80}{x} \quad \text{and} \quad \tan\theta = \frac{205}{x}$$

Substituting these into the expression for $\tan\alpha$, gives

$$\tan\alpha = \frac{\frac{205}{x} - \frac{80}{x}}{1 + \left(\frac{205}{x}\right)\left(\frac{80}{x}\right)}$$

$$\tan\alpha = \frac{\frac{125}{x}}{1 + \left(\frac{205}{x}\right)\left(\frac{80}{x}\right)} \cdot \frac{x^2}{x^2}$$

$$\tan\alpha = \frac{125x}{x^2 + 16\,400}$$

Therefore, $\alpha = \tan^{-1}\left(\frac{125x}{x^2 + 16\,400}\right)$.

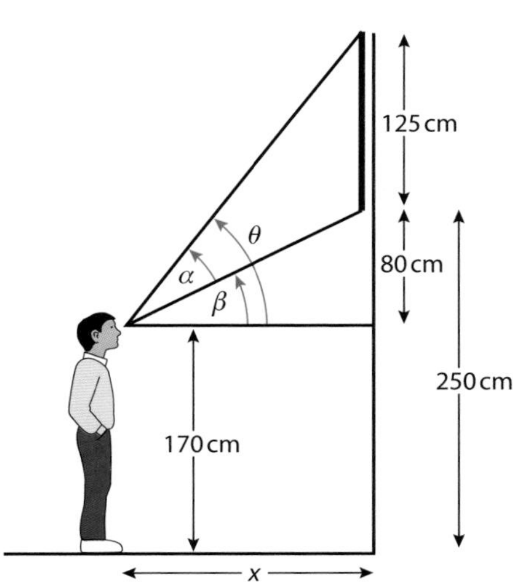

b) (i) For $x = 75$ cm: $\alpha = \tan^{-1}\left(\dfrac{125 \cdot 75}{75^2 + 16\,400}\right) \approx \tan^{-1}(0.425\,6527)$ $\approx 23.06°$.

(ii) For $x = 125$ cm: $\alpha = \tan^{-1}\left(\dfrac{125 \cdot 125}{125^2 + 16\,400}\right) \approx \tan^{-1}(0.487\,9001)$ $\approx 26.01°$.

(iii) For $x = 175$ cm: $\alpha = \tan^{-1}\left(\dfrac{125 \cdot 175}{175^2 + 16\,400}\right) \approx \tan^{-1}(0.465\,1781)$ $\approx 24.95°$.

c) Graph the function found in a). On the GDC, it will be entered as $y = \tan^{-1}\left(\dfrac{125x}{x^2 + 16\,400}\right)$. Find the value of x that gives the maximum value for y (subtended angle α) by either tracing or using a 'maximum' command on the calculator. See the GDC images below.

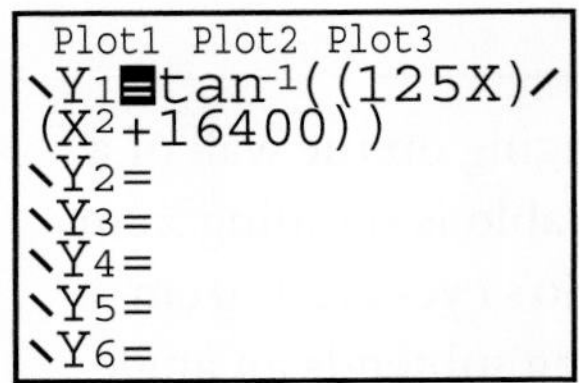

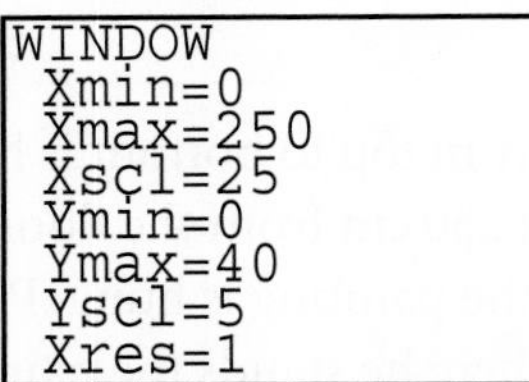

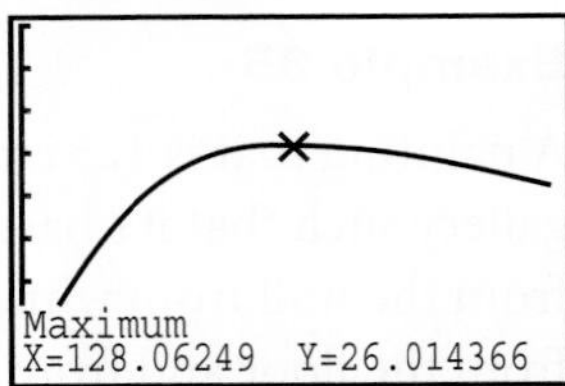

Therefore, if Pablo stands 128 cm away from the wall the painting will subtend the widest possible angle at his eye – or, in other words, give him the 'best' view of the painting.

Exercise 7.6

In questions 1–6, find the exact value (in radian measure) of each expression without using your GDC.

1 $\arcsin 1$

2 $\arccos\left(\dfrac{1}{\sqrt{2}}\right)$

3 $\arctan(-\sqrt{3})$

4 $\arccos\left(-\dfrac{1}{2}\right)$

5 $\arctan 0$

6 $\arcsin\left(\dfrac{-\sqrt{3}}{2}\right)$

In questions 7–20, without using your GDC, find the exact value, if possible, for each expression. Verify your result with your GDC.

7 $\sin^{-1}\left(\sin\dfrac{2\pi}{3}\right)$

8 $\cos^{-1}\left(\cos\dfrac{3}{2}\right)$

9 $\tan(\arctan 12)$

10 $\cos\left(\arccos\dfrac{2\pi}{3}\right)$

11 $\arctan\left(\tan\left(-\dfrac{3\pi}{4}\right)\right)$

12 $\sin(\arcsin \pi)$

13 $\sin\left(\arctan\dfrac{3}{4}\right)$

14 $\cos\left(\arcsin\left(\dfrac{7}{25}\right)\right)$

15 $\arcsin\left(\tan\dfrac{\pi}{3}\right)$

16 $\tan^{-1}\left(2\sin\dfrac{\pi}{3}\right)$

17 $\cos\left(\arctan\left(\dfrac{1}{2}\right)\right)$

18 $\cos(\sin^{-1}(0.6))$

19 $\sin\left(\arccos\left(\dfrac{3}{5}\right) + \arctan\left(\dfrac{5}{12}\right)\right)$

20 $\cos\left(\tan^{-1}3 + \sin^{-1}\left(\dfrac{1}{3}\right)\right)$

In questions 21–26, rewrite the expression as an algebraic expression in terms of x.

21 $\cos(\arcsin x)$

22 $\tan(\arccos x)$

23 $\cos(\tan^{-1} x)$

24 $\sin(2\cos^{-1} x)$

25 $\tan\left(\frac{1}{2}\arccos x\right)$

26 $\sin(\arcsin x + 2\arctan x)$

27 Show that $\arcsin\frac{4}{5} + \arcsin\frac{5}{13} = \arccos\frac{16}{65}$.

28 Show that $\arctan\frac{1}{2} + \arctan\frac{1}{3} = \frac{\pi}{4}$.

29 Find x if $\tan^{-1} x + \tan^{-1}(1 - x) = \tan^{-1}\frac{4}{3}$.

In questions 30–37, solve for x in the indicated interval.

30 $5\cos(2x) = 2, 0 \leq x \leq \pi$

31 $\tan\left(\frac{x}{2}\right) = 2, 0 < x \leq 2\pi$

32 $2\cos x - \sin x = 0, 0 < x \leq 2\pi$

33 $3\sec^2 x = 2\tan x + 4, 0 < x \leq 2\pi$

34 $2\tan^2 x - 3\tan x + 1 = 0, 0 \leq x \leq \pi$

35 $\tan x \csc x = 5, 0 < x \leq 2\pi$

36 $\tan 2x + 3\tan x = 0, 0 < x \leq 2\pi$

37 $2\cos^2 x - 3\sin 2x = 2, 0 \leq x \leq \pi$

38 An offshore lighthouse is located 2 km from a straight coastline. The lighthouse has a revolving light. Let θ be the angle that the beam of light from the lighthouse makes with the coastline; and P is the point on the coast the shortest distance from the lighthouse (see figure). If d is the distance in km from P to the point B where the beam of light is hitting the coast, express θ as a function of d. Sketch a complete graph of this function and indicate the portion of the graph that sufficiently represents the given situation.

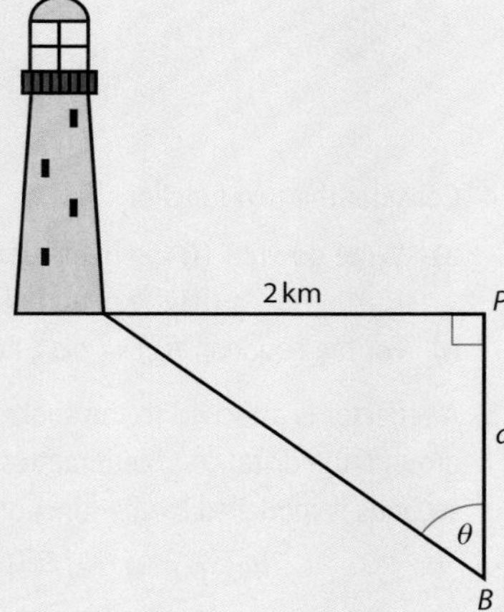

39 The screen in a movie cinema is 7 metres from top to bottom and is positioned 3 metres above the horizontal floor of the cinema. The first row of seats is 2.5 metres from the wall that the screen is on and the rows are each 1 metre apart. You decide to sit in the row where you get the 'best' view, that is, where the angle subtended at your eyes by the screen is a maximum. When you are sitting in one of the cinema's seats your eyes are 1.2 metres above the horizontal floor.

a) Let x be the distance that you are from the wall that the screen is on, and θ is the angle subtended at your eyes by the screen.
   (i) Draw a clear diagram to represent all the information given.
   (ii) Find a function for θ in terms of x.
   (iii) Sketch a graph of the function.
   (iv) Use your GDC to find the value of x that gives a maximum for θ. In which row should you sit?

b) Suppose that, starting with the first row of seats, the floor of the cinema is sloping upwards at an angle of 20° above the horizontal. Again, the first row of seats is 2.5 metres from the wall that the screen is on and the rows are each 1 metre apart measured along the sloping floor. Let x be the distance from where the first row starts and your seat in the cinema.
   (i) Draw a clear diagram to represent all the information given.
   (ii) Find a function for θ in terms of x.
   (iii) Sketch a graph of the function.
   (iv) Use your GDC to find the value of x that gives a maximum for θ. In which row should you sit?

Practice questions

1 A toy on an elastic string is attached to the top of a doorway. It is pulled down and released, allowing it to bounce up and down. The length of the elastic string, L centimetres, is modelled by the function $L = 110 + 25\cos(2\pi t)$, where t is time in seconds after release.

a) Find the length of the elastic string after 2 seconds.

b) Find the minimum length of the string.

c) Find the first time after release that the string is 85 cm.

d) What is the period of the motion?

2 Find the exact solution(s) to the equation $2\sin^2 x - \cos x + 1 = 0$ for $0 \leqslant x \leqslant 2\pi$.

3 The diagram shows a circle of radius 6 cm.
The perimeter of the shaded sector is 25 cm.
Find the radian measure of the angle θ.

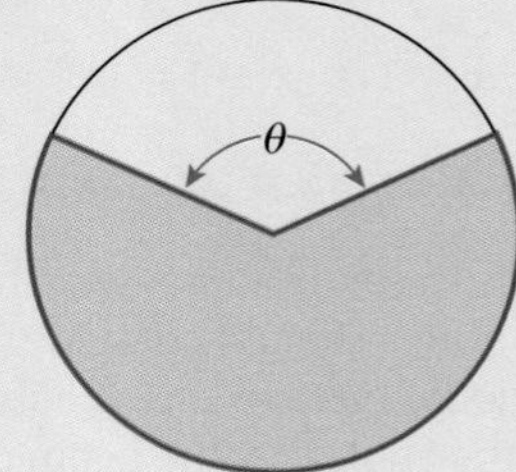

4 Consider the two functions $f(x) = \cos 4x$ and $g(x) = \cos\left(\frac{x}{2}\right)$.

a) Write down: **(i)** the minimum value of the function f

(ii) the period of g.

b) For the equation $f(x) = g(x)$, find the number of solutions in the interval $0 \leqslant x \leqslant \pi$.

5 A reflector is attached to the spoke of a bicycle wheel. As the wheel rolls along the ground, the distance, d centimetres, that the reflector is above the ground after t seconds is modelled by the function

$$d = p + q\cos\left(\frac{2\pi}{m}t\right), \text{ where } p, q \text{ and } m \text{ are constants.}$$

The distance d is at a maximum of 64 cm at $t = 0$ seconds and at $t = 0.5$ seconds, and is at a minimum of 6 cm at $t = 0.25$ seconds and at $t = 0.75$ seconds. Write down the value of:

a) p **b)** q **c)** m.

6 Find all solutions to $1 + \sin 3x = \cos(0.25x)$ such that $x \in [0, \pi]$.

7 Find all solutions to both trigonometric equations in the interval $x \in [0, 2\pi]$. Express the solutions exactly.

a) $2\cos^2 x + 5\cos x + 2 = 0$ **b)** $\sin 2x - \cos x = 0$

8 The value of x is in the interval $\frac{\pi}{2} < x < \pi$ and $\cos^2 x = \frac{8}{9}$. Without using your GDC, find the exact values for the following:

a) $\sin x$ **b)** $\cos 2x$ **c)** $\sin 2x$

9 The depth, d metres, of water in a harbour varies with the tides during each day. The first high (maximum) tide after midnight occurs at 5:00 a.m. with a depth of 5.8 m. The first low (minimum) tide occurs at 10:30 a.m. with a depth of 2.6 m.

a) Find a trigonometric function that models the depth, d, of the water t hours after midnight.

b) Find the depth of the water at 12 noon.

c) A large boat needs at least 3.5 m of water to dock in the harbour. During what time interval after 12 noon can the boat dock safely?

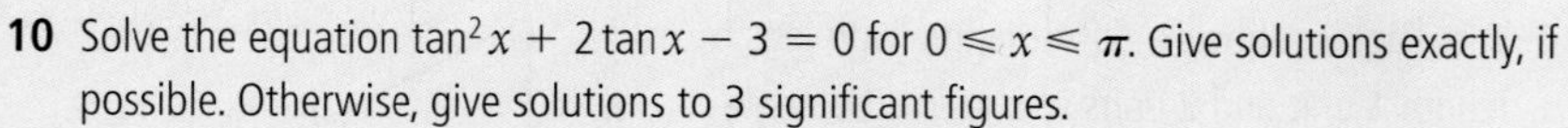

10 Solve the equation $\tan^2 x + 2\tan x - 3 = 0$ for $0 \leqslant x \leqslant \pi$. Give solutions exactly, if possible. Otherwise, give solutions to 3 significant figures.

11 The following diagram shows a circle of centre O and radius 10 cm. The arc ABC subtends an angle of $\frac{3}{2}$ radians at the centre O.

a) Find the length of the arc ACB.

b) Find the area of the shaded region.

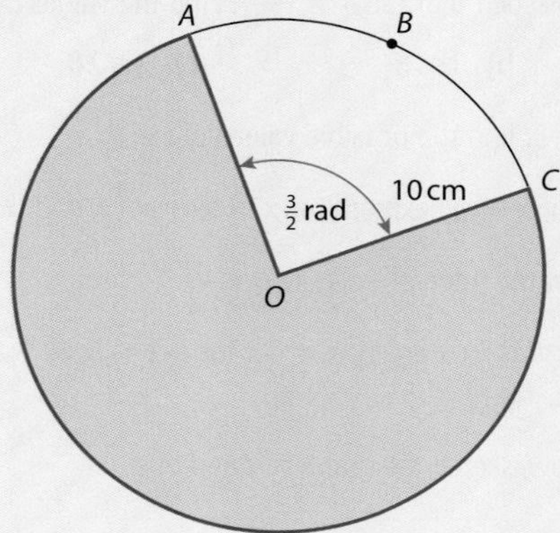

12 Consider the function $f(x) = \frac{5}{2}\cos\left(2x - \frac{\pi}{2}\right)$. For what values of k will the equation $f(x) = k$ have no solutions?

13 A portion of the graph of $y = k + a\sin x$ is shown below. The graph passes through the points (0, 1) and $\left(\frac{3\pi}{2}, 3\right)$. Find the value of k and a.

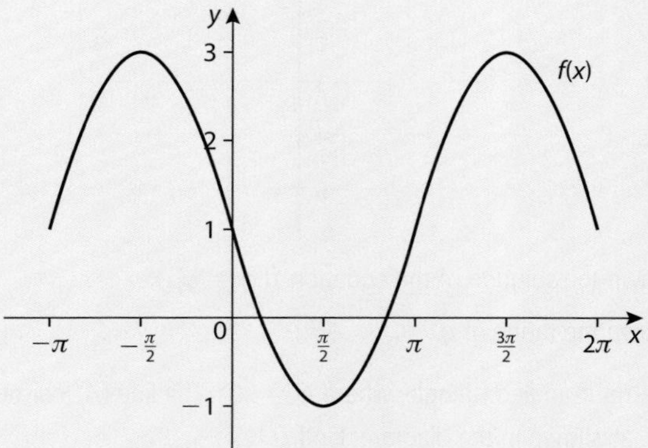

14 The angle α satisfies the equation $2\tan^2\alpha - 5\sec\alpha - 10 = 0$ where α is in the second quadrant. Find the **exact** value of $\sec\alpha$.

15 Triangles PTS and RTS are right-angled at T with angles α and β as shown in the diagram. Find the exact values of the following:

a) $\sin(\alpha + \beta)$

b) $\cos(\alpha + \beta)$

c) $\tan(\alpha + \beta)$

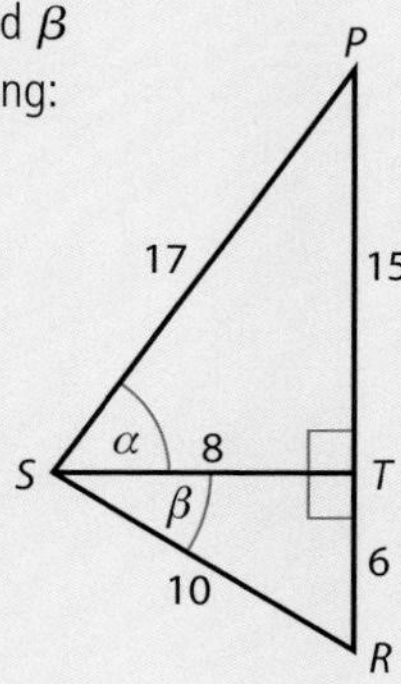

16 The diagram shows a right triangle with legs of length 1 unit and 2 units as shown. The angle at vertex P has a degree measure of $p°$. Find the exact values of $\sin 2p°$ and $\sin 3p°$.

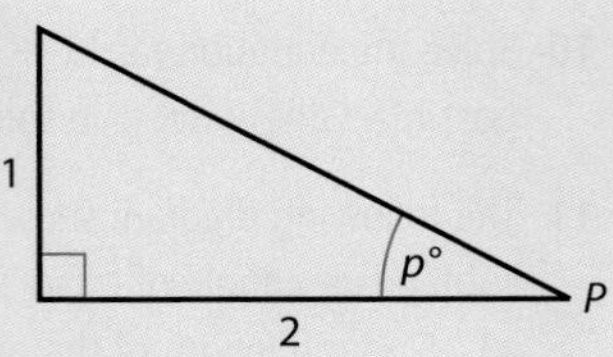

17 The obtuse angle B is such that $\tan B = -\frac{5}{12}$. Find the values of

a) $\sin B$ **b)** $\cos B$ **c)** $\sin 2B$ **d)** $\cos 2B$

18 Given that $\tan 2\theta = \frac{3}{4}$, find the possible values of $\tan\theta$.

19 If $\sin(x - \alpha) = k\sin(x + \alpha)$ express $\tan x$ in terms of k and α.

20 Solve $\tan^2 2\theta = 1$, in the interval $-\frac{\pi}{2} \leq \theta \leq \frac{\pi}{2}$.

21 Let f be the function $f(x) = x\arccos x + \frac{1}{2}x$ for $-1 \leq x \leq 1$ and g the function $g(x) = \cos 2x$ for $-1 \leq x \leq 1$.

a) On the grid below, sketch the graph of f and of g.

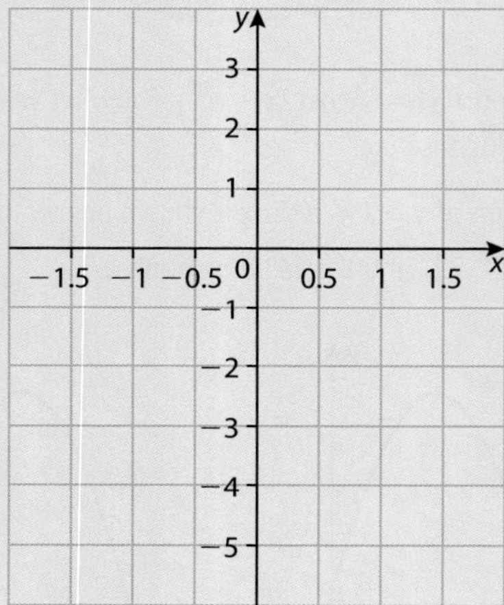

b) Write down the solution of the equation $f(x) = g(x)$.

c) Write down the range of g.

22 Let ABC be a right-angled triangle, where $\hat{C} = 90°$. The line (AD) bisects $B\hat{A}C$, $BD = 3$, and $DC = 2$, as shown in the diagram. Find $D\hat{A}C$.

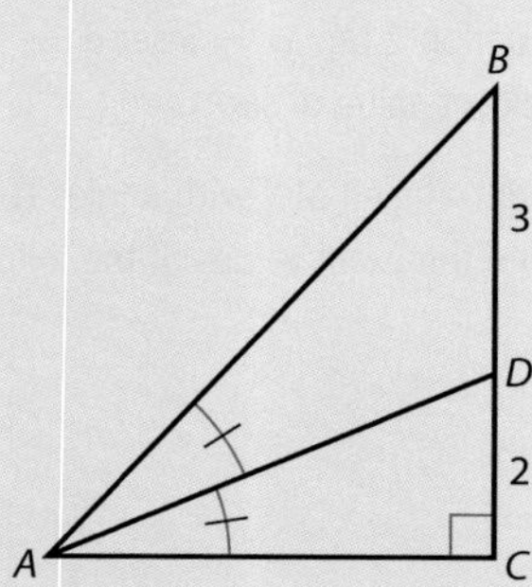

23 The diagram below shows the boundary of the cross section of a water channel.

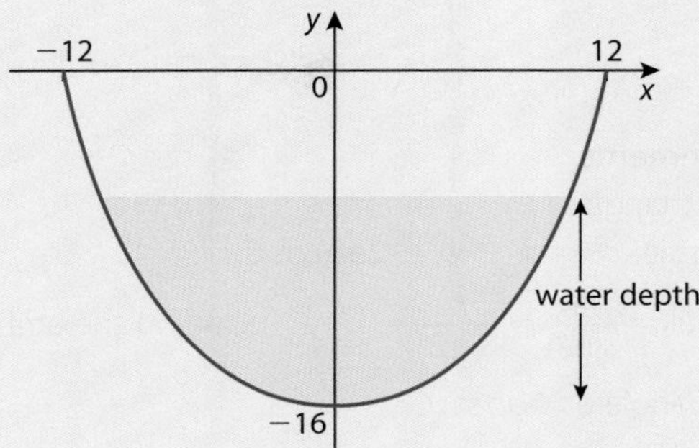

The equation that represents this boundary is $y = 16\sec\left(\frac{\pi x}{36}\right) - 32$ where x and y are both measured in cm. The top of the channel is level with the ground and has a width of 24 cm. The maximum depth of the channel is 16 cm. Find the width of the water surface in the channel when the water depth is 10 cm. Give your answer in the form $a \arccos b$, where $a, b \in \mathbb{R}$.

8 Triangle Trigonometry

Assessment statements

3.6 Solution of triangles.

The cosine rule: $c^2 = a^2 + b^2 - 2ab\cos C$.

The sine rule: $= \frac{a}{\sin A} = \frac{b}{\sin B} = \frac{c}{\sin C}$, including the ambiguous case.

Area of a triangle as $\frac{1}{2}ab\sin C$.

Applications in two and three dimensions.

Introduction

In this chapter, we approach trigonometry from a **right triangle** perspective where trigonometric functions will be defined in terms of the **ratios of sides of a right triangle**. Over two thousand years ago, the Greeks developed trigonometry to make helpful calculations for surveying, navigating, building and other practical pursuits. Their calculations were based on the angles and lengths of sides of a right triangle. The modern development of trigonometry, based on the length of an arc on the unit circle, was covered in the previous chapter. We begin a more classical approach by introducing some terminology regarding right triangles.

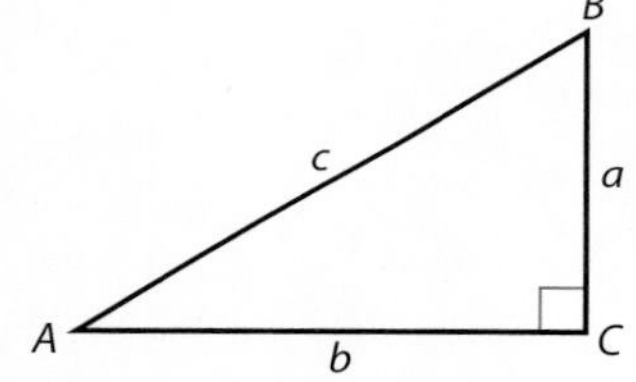

Figure 8.1 Conventional triangle notation.

• **Hint:** In IB notation, $[AC]$ denotes the line segment connecting points A and C. The notation AC represents the *length* of this line segment. Also, the notation $A\hat{B}C$ denotes the angle with its vertex at point B, with one side of the angle containing the point A and the other side containing point C.

8.1 Right triangles and trigonometric functions of acute angles

Right triangles

The conventional notation for triangles is to label the three vertices with capital letters, for example A, B and C. The same capital letters can be used to represent the measure of the angles at these vertices. However, we will often use a Greek letter, such as α (alpha), β (beta) or θ (theta) to do so. The corresponding lower-case letters, a, b and c, represent the lengths of the sides opposite the vertices. For example, b represents the length of the side opposite angle B, that is, the line segment AC, or $[AC]$ (Figure 8.1).

In a right triangle, the longest side is opposite the right angle (i.e. measure of 90°) and is called the **hypotenuse**, and the two shorter sides adjacent to the right angle are often called the **legs** (Figure 8.2). Because the sum of the three angles in any triangle in plane geometry is 180°, then the two non-right angles are both **acute angles** (i.e. measure between 0 and 90 degrees). It also follows that the two acute angles in a right triangle are a pair of **complementary angles** (i.e. have a sum of 90°).

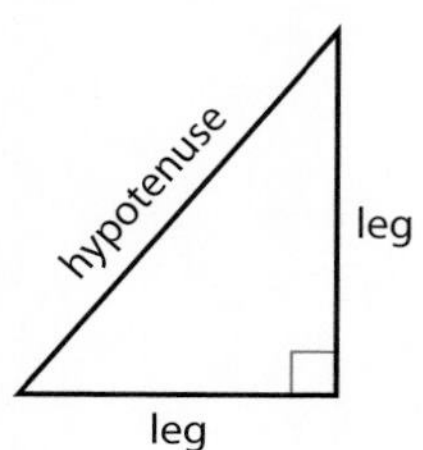

Figure 8.2 Right triangle terminology.

Trigonometric functions of an acute angle

We can use properties of similar triangles and the definitions of the sine, cosine and tangent functions from Chapter 7 to define these functions in terms of the sides of a right triangle.

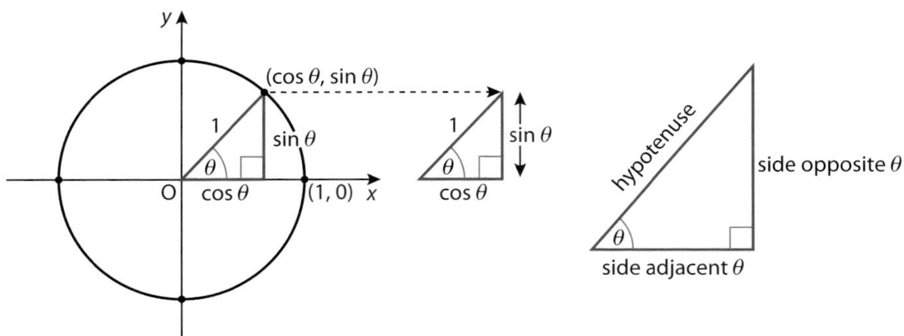

Figure 8.3 Trigonometric functions defined in terms of sides of similar triangles.

The right triangles shown in Figure 8.3 are **similar triangles** because corresponding angles have equal measure – each has a right angle and an acute angle of measure θ. It follows that the ratios of corresponding sides are equal, allowing us to write the following three proportions involving the sine, cosine and tangent of the acute angle θ.

$$\frac{\sin\theta}{1} = \frac{\text{opposite}}{\text{hypotenuse}} \qquad \frac{\cos\theta}{1} = \frac{\text{adjacent}}{\text{hypotenuse}} \qquad \frac{\tan\theta}{1} = \frac{\sin\theta}{\cos\theta} = \frac{\text{opposite}}{\text{adjacent}}$$

The definitions of the trigonometric functions in terms of the sides of a right triangle follow directly from these three equations.

Right triangle definition of the trigonometric functions

Let θ be an **acute angle** of a right triangle, then the sine, cosine and tangent functions of the angle θ are defined as the following ratios in the right triangle:

$$\sin\theta = \frac{\text{side opposite angle }\theta}{\text{hypotenuse}}$$

$$\cos\theta = \frac{\text{side adjacent angle }\theta}{\text{hypotenuse}}$$

$$\tan\theta = \frac{\text{side opposite angle }\theta}{\text{side adjacent angle }\theta}$$

It follows that the sine, cosine and tangent of an acute angle are positive.

It is important to understand that properties of similar triangles are the foundation of right triangle trigonometry. Regardless of the size (i.e. lengths of sides) of a right triangle, so long as the angles do not change, the ratio of any two sides in the right triangle will remain *constant*. All the right triangles in Figure 8.4 have an acute angle with a measure of 30° (thus, the other acute angle is 60°). For each triangle, the ratio of the side opposite the 30° angle to the hypotenuse is exactly $\frac{1}{2}$. In other words, the sine of 30° is always $\frac{1}{2}$. This agrees with results from the previous chapter, knowing that an angle of 30° is equivalent to $\frac{\pi}{6}$ in radian measure.

Thales of Miletus (circa 624–547) was the first of the Seven Sages, or wise men of ancient Greece, and is considered by many to be the first Greek scientist, mathematician and philosopher. Thales visited Egypt and brought back knowledge of astronomy and geometry. According to several accounts, Thales, with no special instruments, determined the height of Egyptian pyramids. He applied formal geometric reasoning. Diogenes Laertius, a 3rd-century biographer of ancient Greek philosophers, wrote: 'Hieronymus says that [Thales] even succeeded in measuring the pyramids by observation of the length of their shadow at the moment when our shadows are equal to our own height.' Thales used the geometric principle that the ratios of corresponding sides of similar triangles are equal.

Figure 8.4 Corresponding ratios of a pair of sides for similar triangles are equal.

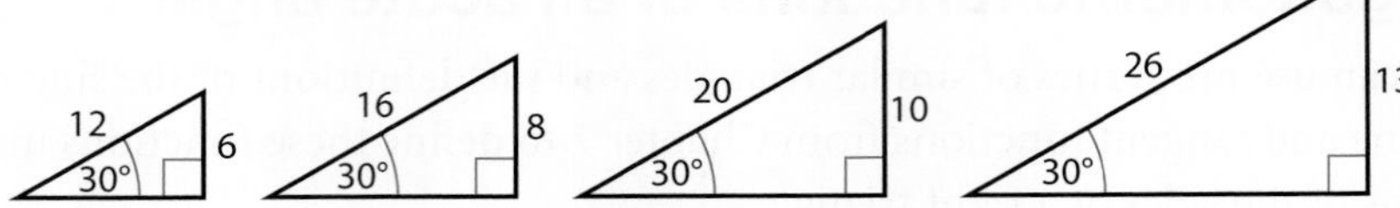

For any right triangle, the sine ratio for 30° is always $\frac{1}{2}$: $\sin 30° = \frac{1}{2}$. The trigonometric functions of acute angles are not always rational numbers such as $\frac{1}{2}$. We will see in upcoming examples that the sine of 60° is exactly $\frac{\sqrt{3}}{2}$.

Geometric derivation of trigonometric functions for 30°, 45° and 60°

We can use Pythagoras' theorem and properties of triangles to find the exact values for the most common acute angles: 30°, 45° and 60°.

Sine, cosine and tangent values for 45°

Derivation

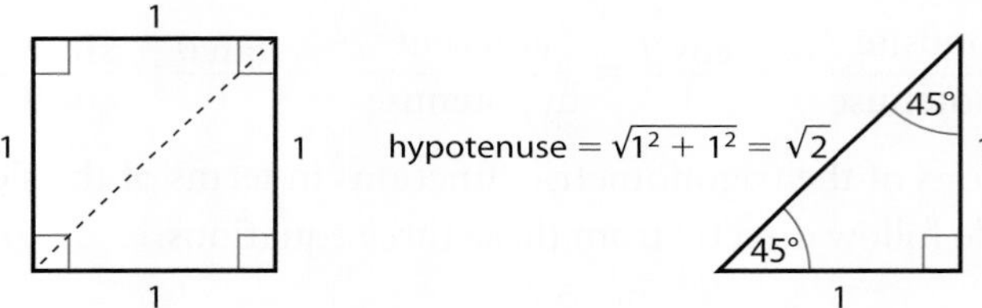

Consider a square with each side equal to one unit. Draw a diagonal of the square, forming two isosceles right triangles. From geometry, we know that the diagonal will bisect each of the two right angles forming two isosceles right triangles, each with two acute angles of 45°. The isosceles right triangles have legs of length one unit and, from Pythagoras' theorem, a hypotenuse of exactly $\sqrt{2}$ units. The trigonometric functions are then calculated as follows:

$$\sin 45° = \frac{\text{opposite}}{\text{hypotenuse}} = \frac{1}{\sqrt{2}} = \frac{\sqrt{2}}{2}$$

(Multiplying by $\frac{\sqrt{2}}{\sqrt{2}}$ to rationalize the denominator.)

$$\cos 45° = \frac{\text{adjacent}}{\text{hypotenuse}} = \frac{1}{\sqrt{2}} = \frac{\sqrt{2}}{2}$$

$$\tan 45° = \frac{\text{opposite}}{\text{adjacent}} = \frac{1}{1} = 1$$

Sine, cosine and tangent values for 30° and 60°

Derivation

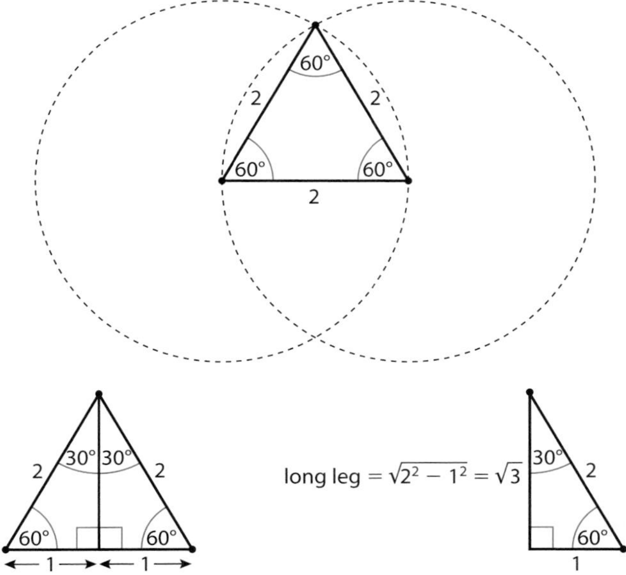

Start with a line segment of length two units. Using each endpoint as a centre and the segment as a radius, construct two circles. The endpoints of the original line segment and the point of intersection of the two circles are the vertices of an equilateral triangle. Each side has a length of two units and the measure of each angle is 60°. From geometry, the altitude drawn from one of the vertices bisects the angle at that vertex and also bisects the opposite side to which it is perpendicular. Two right triangles are formed that have acute angles of 30° and 60°, a hypotenuse of two units, and a short leg of one unit. Using Pythagoras' theorem, the long leg is $\sqrt{3}$ units. The trigonometric functions of 30° and 60° are then calculated as follows:

$$\sin 60° = \frac{\text{opposite}}{\text{hypotenuse}} = \frac{\sqrt{3}}{2} \qquad \sin 30° = \frac{\text{opposite}}{\text{hypotenuse}} = \frac{1}{2}$$

$$\cos 60° = \frac{\text{adjacent}}{\text{hypotenuse}} = \frac{1}{2} \qquad \cos 30° = \frac{\text{adjacent}}{\text{hypotenuse}} = \frac{\sqrt{3}}{2}$$

$$\tan 60° = \frac{\text{opposite}}{\text{adjacent}} = \frac{\sqrt{3}}{1} = \sqrt{3} \qquad \tan 30° = \frac{\text{opposite}}{\text{adjacent}} = \frac{1}{\sqrt{3}} = \frac{\sqrt{3}}{3}$$ (Rationalizing the denominator.)

The geometric derivation of the values of the sine, cosine and tangent functions for the 'special' acute angles 30°, 45° and 60° agree with the results from the previous chapter. The results for these angles – in both degree and radian measure – are summarised in the box below.

Values of sine, cosine and tangent for common acute angles

$\sin 30° = \sin\frac{\pi}{6} = \frac{1}{2}$ $\quad \cos 30° = \cos\frac{\pi}{6} = \frac{\sqrt{3}}{2}$ $\quad \tan 30° = \tan\frac{\pi}{6} = \frac{\sqrt{3}}{3}$

$\sin 45° = \sin\frac{\pi}{4} = \frac{\sqrt{2}}{2}$ $\quad \cos 45° = \cos\frac{\pi}{4} = \frac{\sqrt{2}}{2}$ $\quad \tan 45° = \tan\frac{\pi}{4} = 1$

$\sin 60° = \sin\frac{\pi}{3} = \frac{\sqrt{3}}{2}$ $\quad \cos 60° = \cos\frac{\pi}{3} = \frac{1}{2}$ $\quad \tan 60° = \tan\frac{\pi}{3} = \sqrt{3}$

• Hint: It is important that you are able to recall – without a calculator – the exact trigonometric values for these common angles.

Observe that $\sin 30° = \cos 60° = \frac{1}{2}$, $\sin 60° = \cos 30° = \frac{\sqrt{3}}{2}$ and $\sin 45° = \cos 45° = \frac{\sqrt{2}}{2}$. Complementary angles (sum of 90°) have equal function values for sine and cosine. That is, for all angles x measured in degrees, $\sin x = \cos(90° - x)$ or $\sin(90° - x) = \cos x$. As noted in Chapter 7, it is for this reason that sine and cosine are called co-functions.

Solution of right triangles

Every triangle has three sides and three angles – six different parts. The ancient Greeks knew how to solve for all of the unknown angles and sides in a right triangle given that either the length of two sides, or the length of one side and the measure of one angle, were known. To **solve a right triangle** means to find the measure of any unknown sides or angles. We can accomplish this by applying Pythagoras' theorem and trigonometric functions. We will utilize trigonometric functions in two different ways when solving for missing parts in right triangles – to find the length of a side, and to find the measure of an angle. Solving right triangles using the sine, cosine and tangent functions is essential to finding solutions to problems in fields such as astronomy, navigation, engineering and architecture. In Sections 8.3 and 8.4, we will see how trigonometry can also be used to solve for missing parts in triangles that are not right triangles.

Angles of depression and elevation

An imaginary line segment from an observation point O to a point P (representing the location of an object) is called the **line of sight** of P. If P is above O, the acute angle between the line of sight of P and a horizontal line passing through O is called the **angle of elevation** of P. If P is below O, the angle between the line of sight and the horizontal is called the **angle of depression** of P. This is illustrated in Figure 8.5.

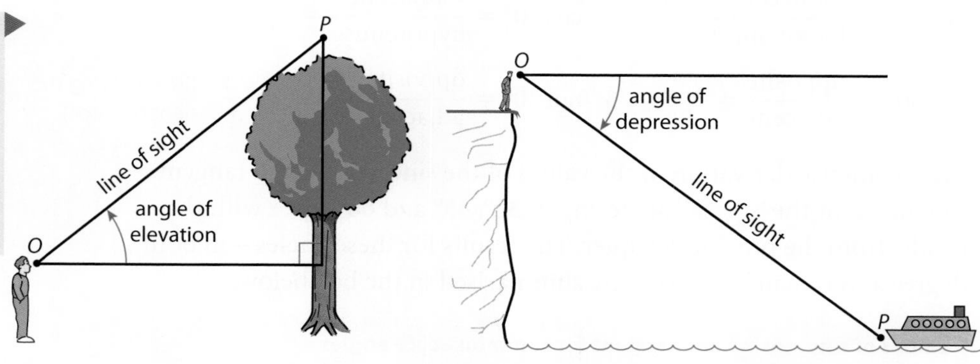

Figure 8.5 An angle of elevation or depression is always measured from the horizontal. Also, note that for each diagram, the angle of elevation from O to P is equal to the angle of depression from P to O.

Example 1

Solve triangle ABC given $c = 8.76$ cm and angle $A = 30°$, where the right angle is at C. Give exact answers when possible, otherwise give to an accuracy of 3 significant figures.

Solution

Knowing that the conventional notation is to use a lower-case letter to represent the length of a side opposite the vertex denoted with the corresponding upper-case letter, we sketch triangle ABC indicating the known measurements.

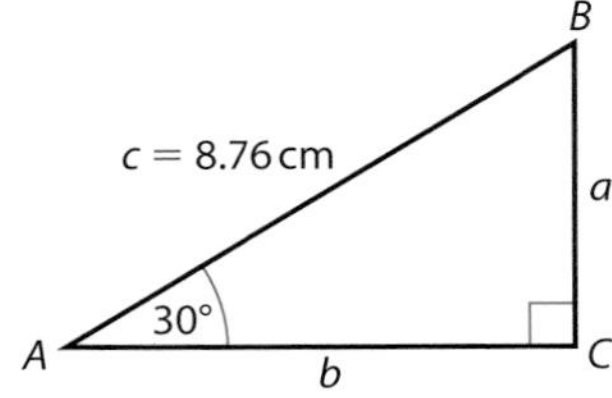

From the definition of sine and cosine functions, we have

$$\sin 30° = \frac{\text{opposite}}{\text{hypotenuse}} = \frac{a}{8.76} \qquad \cos 30° = \frac{\text{adjacent}}{\text{hypotenuse}} = \frac{b}{8.76}$$

$$a = 8.76 \sin 30° \qquad b = 8.76 \cos 30°$$

$$a = 8.76\left(\frac{1}{2}\right) = 4.38 \qquad b = 8.76\left(\frac{\sqrt{3}}{2}\right) \approx 7.586\,382\,537 \approx 7.59$$

Therefore, $a = 4.38$ cm, $b \approx 7.59$ cm, and it's clear that angle $B = 60°$.

We can use Pythagoras' theorem to check our results for a and b.

$$a^2 + b^2 = c^2 \Rightarrow \sqrt{a^2 + b^2} = 8.76$$

Be aware that the result for a is exactly 4.38 cm (assuming measurements given for angle A and side c are exact), but the result for b can only be approximated. To reduce error when performing the check, we should use the most accurate value (i.e. most significant figures) possible for b. The most effective way to do this on our GDC is to use results that are stored to several significant figures, as shown in the GDC screen image.

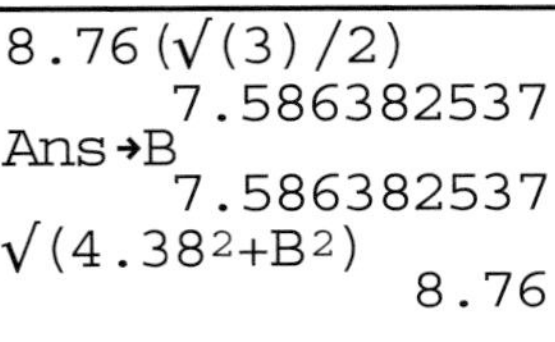

Example 2

A man who is 183 cm tall casts a 72 cm long shadow on the horizontal ground. What is the angle of elevation of the sun to the nearest tenth of a degree?

Solution

In the diagram, the angle of elevation of the sun is labelled θ.

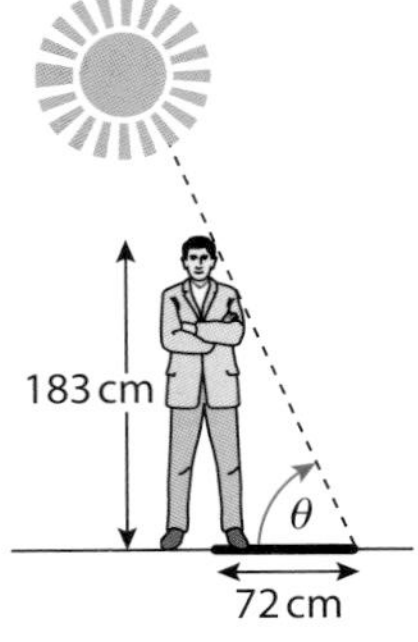

$$\tan\theta = \frac{183}{72}$$

$$\theta = \tan^{-1}\left(\frac{183}{72}\right)$$

$$\theta \approx 68.5°$$

```
tan-1(183/72)
        68.52320902
```

GDC computation in degree mode

The angle of elevation of the sun is approximately 68.5°.

• **Hint:** As noted earlier, the notation for indicating the inverse of a function is a superscript of negative one. For example, the inverse of the cosine function is written as $\cos^{-1}$. The negative one is *not* an exponent, so it does not denote reciprocal. Do not make this error: $\cos^{-1} x \neq \frac{1}{\cos x}$.

Example 3

During a training exercise, an air force pilot is flying his jet at a constant altitude of 1200 metres. His task is to fire a missile at a target. At the moment he fires his missile he is able to see the target at an angle of depression of 18.5°. Assuming the missile travels in a straight line, what distance will the missile cover (to the nearest metre) from the jet to the target?

Solution

Draw a diagram to represent the information and let x be the distance that the missile travels from the plane to the target. A right triangle can be 'extracted' from the diagram with one leg 1200 metres, the angle opposite that leg is 18.5°, and the hypotenuse is x. Applying the sine ratio, we can write the equation $\sin 18.5° = \frac{1200}{x}$.

Then $x = \frac{1200}{\sin 18.5°} \approx 3781.85$. Hence, the missile travels approximately 3782 metres.

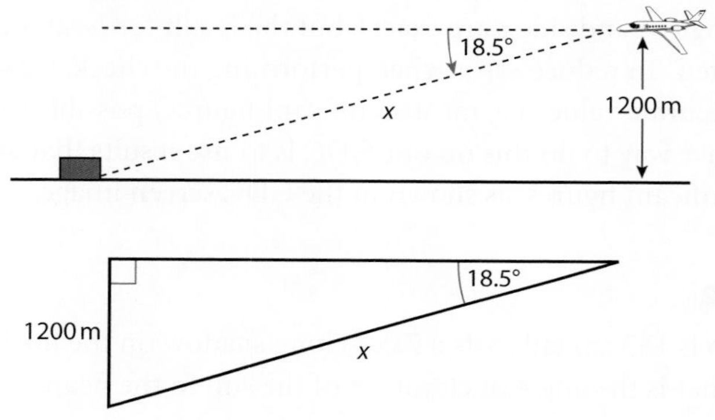

Example 4

A boat is sailing directly towards a cliff. The angle of elevation of a point on the top of the cliff and straight ahead of the boat increases from 10° to 15° as the ship sails a distance of 50 metres. Find the height of the cliff.

Solution

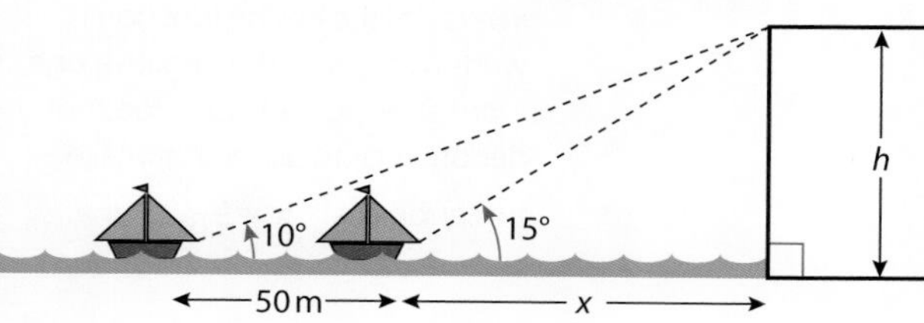

Draw a diagram that accurately represents the information with the height of the cliff labelled h metres and the distance from the base of the cliff to the later position of the boat labelled x metres. There are two right triangles that can be 'extracted' from the diagram. From the smaller right triangle, we have

$$\tan 15° = \frac{h}{x} \Rightarrow h = x \tan 15°$$

From the larger right triangle, we have

$$\tan 10° = \frac{h}{x + 50} \Rightarrow h = (x + 50)\tan 10°$$

We can solve for x by setting the two expressions for h equal to each other.

Then we can solve for h by substitution.

$$x\tan 15° = (x + 50)\tan 10°$$
$$x\tan 15° = x\tan 10° + 50\tan 10°$$
$$x(\tan 15° - \tan 10°) = 50\tan 10°$$
$$x = \frac{50\tan 10°}{\tan 15° - \tan 10°} \approx 96.225$$

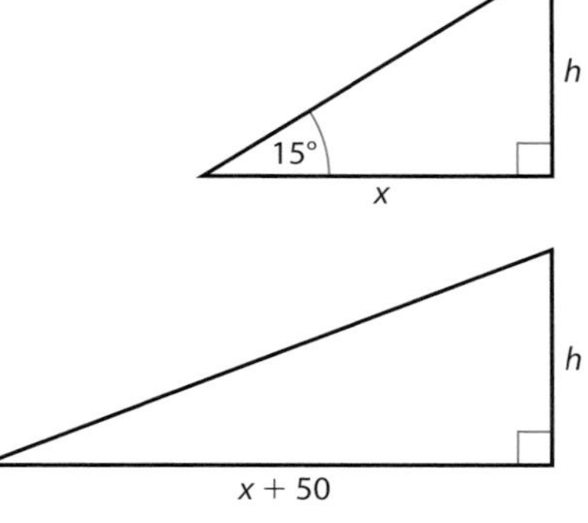

Substituting this value for x into $h = x\tan 15°$, gives

$$h \approx 96.225\tan 15° \approx 25.783$$

Therefore, the height of the cliff is approximately 25.8 metres.

Example 5

Using a suitable right triangle, find the exact minimum distance from the point (8, 3) to the line with the equation $2x - y + 2 = 0$.

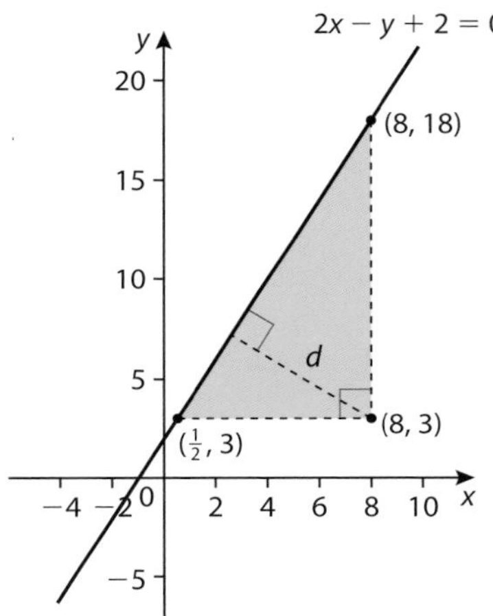

Solution

Graph the line with equation $2x - y + 2 = 0$. The minimum distance from the point (8, 3) to the line is the length of the line segment drawn from the point *perpendicular* to the line. This minimum distance is labelled d in the diagram. d is also the height of the large yellow triangle formed by drawing vertical and horizontal line segments from (8, 3) to the line.

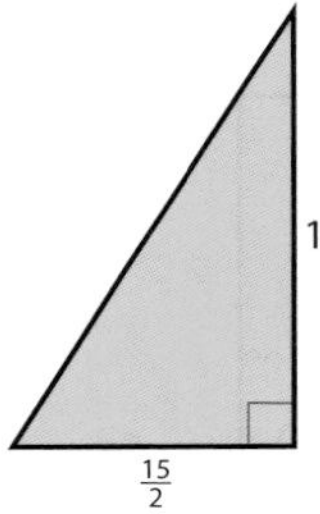

The area of the right triangle is
$A = \frac{1}{2}\left(\frac{15}{2}\right)(15) = \frac{225}{4}$.

The area of the triangle can also be found by using the hypotenuse as the base and the distance d as the height. By Pythagoras' theorem, we have

$$\text{hypotenuse} = \sqrt{\left(\frac{15}{2}\right)^2 + 15^2} = \sqrt{\frac{1125}{4}} = \frac{\sqrt{225}\sqrt{5}}{\sqrt{4}} = \frac{15\sqrt{5}}{2}$$

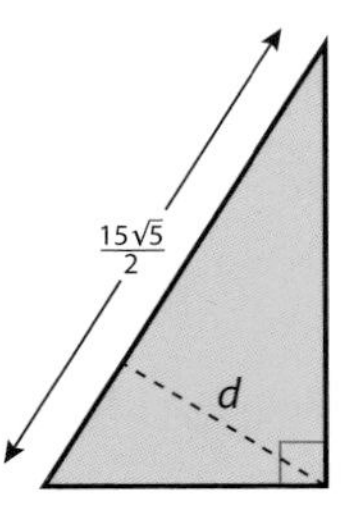

Thus the area can also be expressed as $A = \frac{1}{2}\left(\frac{15\sqrt{5}}{2}\right)d$. We can solve for d by equating the two results for the area of the triangle.

$$\frac{1}{2}\left(\frac{15\sqrt{5}}{2}\right)d = \frac{225}{4}$$
$$\frac{15\sqrt{5}}{4}d = \frac{225}{4}$$
$$d = \frac{225}{4} \cdot \frac{4}{15\sqrt{5}}$$
$$d = \frac{15}{\sqrt{5}} = \frac{15}{\sqrt{5}} \cdot \frac{\sqrt{5}}{\sqrt{5}} = \frac{15\sqrt{5}}{5} = 3\sqrt{5}$$

Therefore, the minimum distance from the point (8, 3) to the line with equation $2x - y + 2 = 0$ is $3\sqrt{5}$ units.

Exercise 8.1

For each question 1–9, a) sketch a right triangle corresponding to the given trigonometric function of the acute angle θ, b) find the exact value of the other five trigonometric functions, and c) use your GDC to find the degree measure of θ and the other acute angle (approximate to 3 significant figures).

1 $\sin\theta = \frac{3}{5}$ **2** $\cos\theta = \frac{5}{8}$ **3** $\tan\theta = 2$

4 $\cos\theta = \frac{7}{10}$ **5** $\cot\theta = \frac{1}{3}$ **6** $\sin\theta = \frac{\sqrt{7}}{4}$

7 $\sec\theta = \frac{11}{\sqrt{61}}$ **8** $\tan\theta = \frac{9}{10}$ **9** $\csc\theta = \frac{4\sqrt{65}}{65}$

In questions 10–15, find the exact value of θ in degree measure ($0 < \theta < 90°$) and in radian measure $\left(0 < \theta < \frac{\pi}{2}\right)$ without using your GDC.

10 $\cos\theta = \frac{1}{2}$ **11** $\sin\theta = \frac{\sqrt{2}}{2}$ **12** $\tan\theta = \sqrt{3}$

13 $\csc\theta = \frac{2\sqrt{3}}{3}$ **14** $\cot\theta = 1$ **15** $\cos\theta = \frac{\sqrt{3}}{2}$

In questions 16–21, solve for x and y. Give your answer exact or to 3 s.f.

16

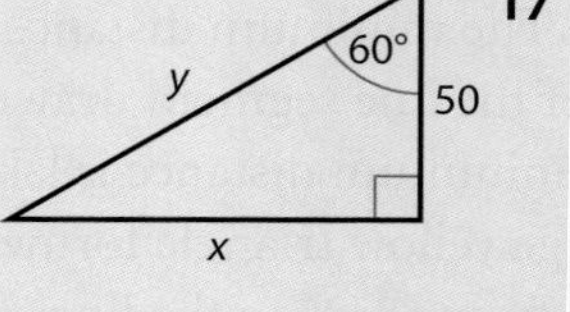

17

15
y
55°
x

18

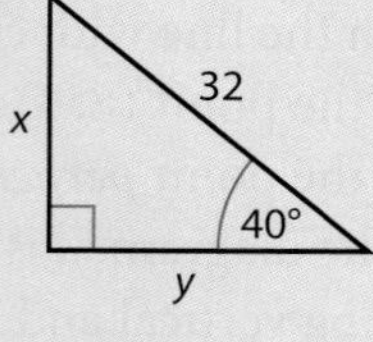

19

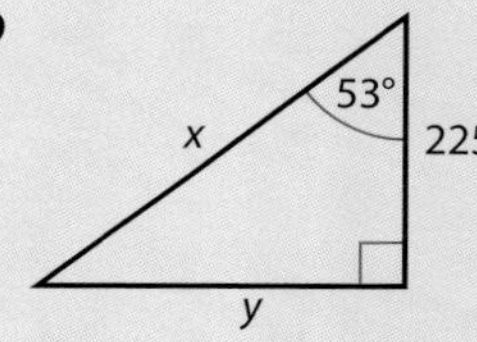

20

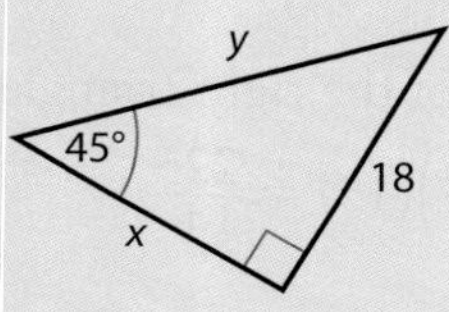

21

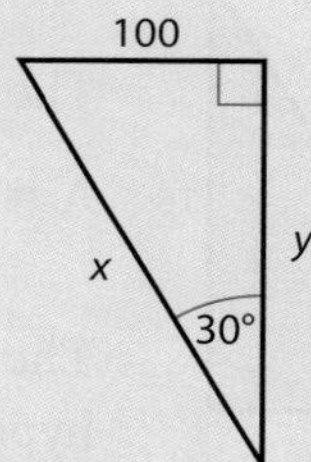

In questions 22–25, find the degree measure of the angles α and β. If possible, give an exact answer – otherwise, approximate to three significant figures.

22

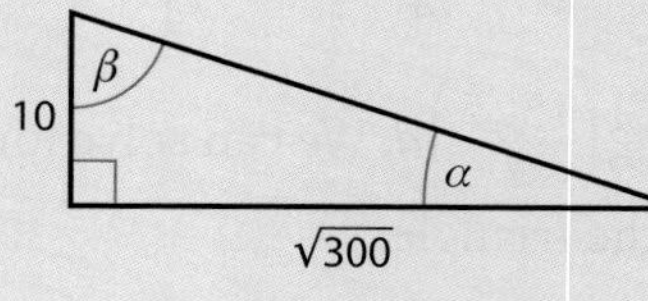

23

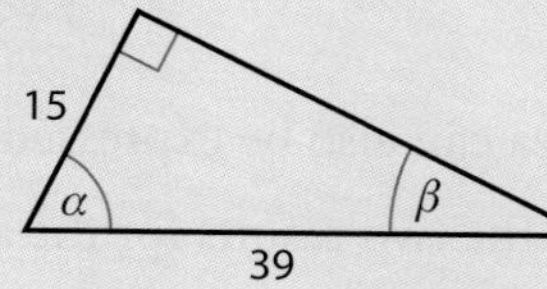

24

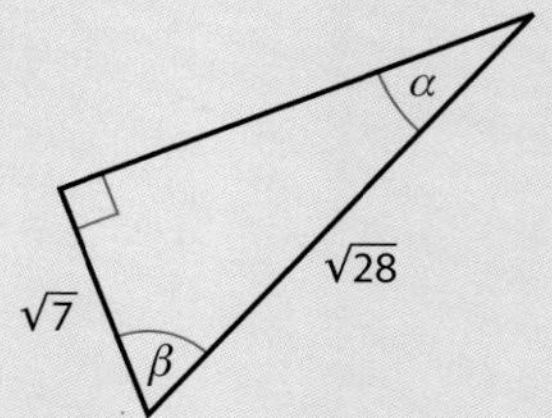

25

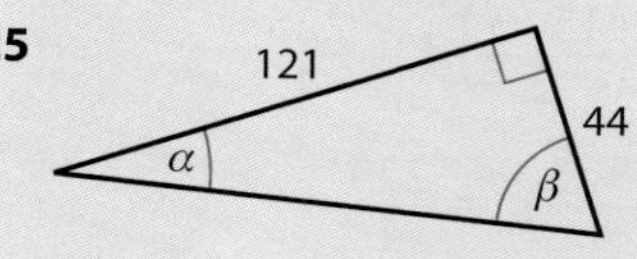

26 The tallest tree in the world is reputed to be a giant redwood named *Hyperion* located in Redwood National Park in California, USA. At a point 41.5 metres from the centre of its base and on the same elevation, the angle of elevation of the top of the tree is 70°. How tall is the tree? Give your answer to three significant figures.

27 The Eiffel Tower in Paris is 300 metres high (not including the antenna on top). What will be the angle of elevation of the top of the tower from a point on the ground (assumed level) that is 125 metres from the centre of the tower's base?

28 A 1.62-metre tall woman standing 3 metres from a streetlight casts a 2-metre long shadow. What is the height of the streetlight?

29 A pilot measures the angles of depression to two ships to be 40° and 52° (see the figure). If the pilot is flying at an elevation of 10 000 metres, find the distance between the two ships.

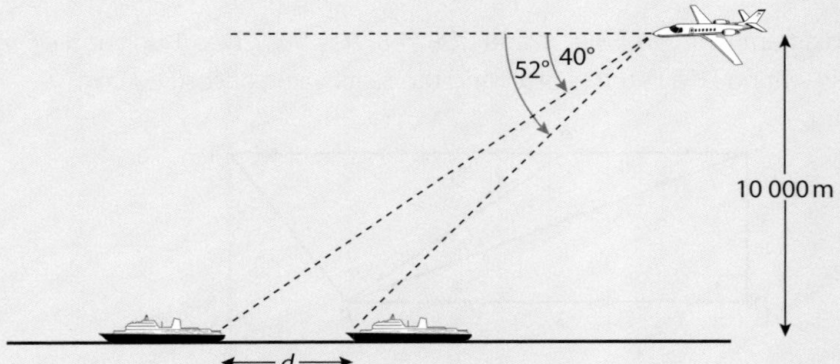

30 Find the measure of all the angles in a triangle with sides of length 8 cm, 8 cm and 6 cm.

31 From a 50-metre observation tower on the shoreline, a boat is sighted at an angle of depression of 4° moving directly toward the shore at a constant speed. Five minutes later the angle of depression of the boat is 12°. What is the speed of the boat in kilometres per hour?

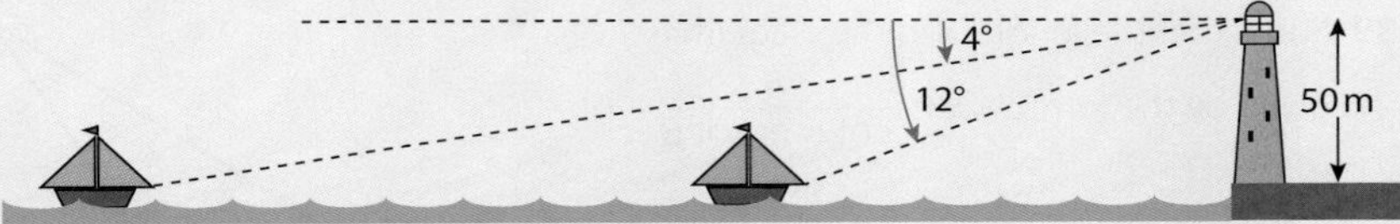

32 Find the length of x indicated in the diagram. Approximate your answer to 3 significant figures.

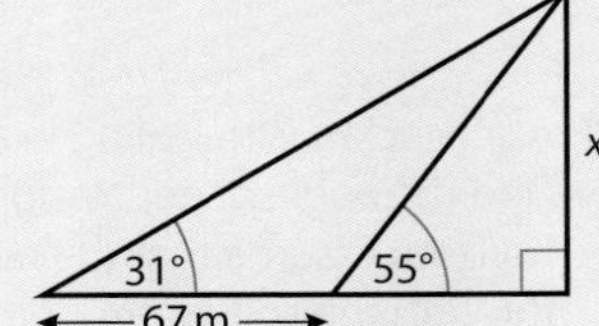

33 A support wire for a tower is connected from an anchor point on level ground to the top of the tower. The straight wire makes a 65° angle with the ground at the anchor point. At a point 25 metres farther from the tower than the wire's anchor point and on the same side of the tower, the angle of elevation to the top of the tower is 35°. Find the wire length to the nearest tenth of a metre.

34 A 30-metre high building sits on top of a hill. The angles of elevation of the top and bottom of the building from the same spot at the base of the hill are measured to be 55° and 50° respectively. Relative to its base, how high is the hill to the nearest metre?

35 The angle of elevation of the top of a vertical pole as seen from a point 10 metres away from the pole is double its angle of elevation as seen from a point 70 metres from the pole. Find the height (to the nearest tenth of a metre) of the pole above the level of the observer's eyes.

36 Angle ABC of a right triangle is bisected by segment BD. The lengths of sides AB and BC are given in the diagram. Find the exact length of BD, expressing your answer in simplest form.

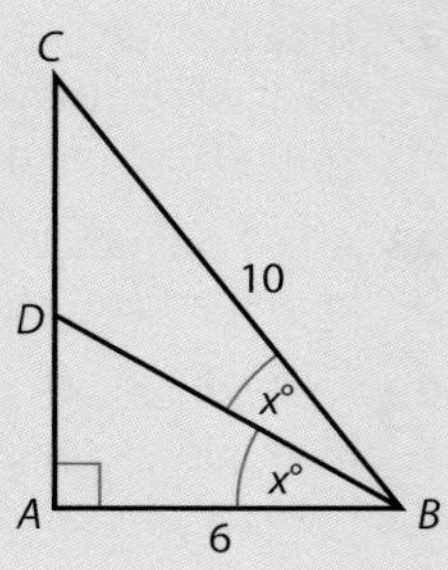

37 In the diagram, $D\hat{E}C = C\hat{E}B = x°$ and $C\hat{D}E = B\hat{E}A = 90°$, $CD = 1$ unit and $DE = 3$ units. By writing $D\hat{E}A$ in terms of $x°$, find the exact value of $\cos(D\hat{E}A)$.

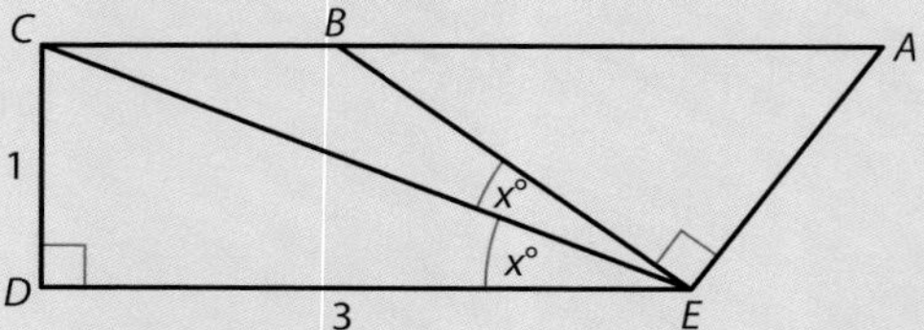

38 For any point with coordinates (p, q) and any line with equation $ax + by + c = 0$, find a formula in terms of a, b, c, p and q that gives the minimum (perpendicular) distance, d, from the point to the line.

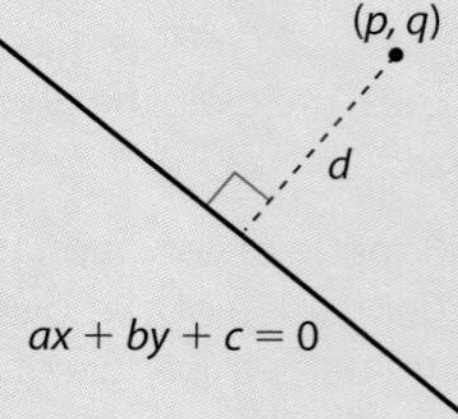

39 Show that the length x in the diagram is given by the formula $x = \dfrac{d}{\cot \alpha - \cot \beta}$.

• **Hint:** First try expressing the formula using the tangent ratio.

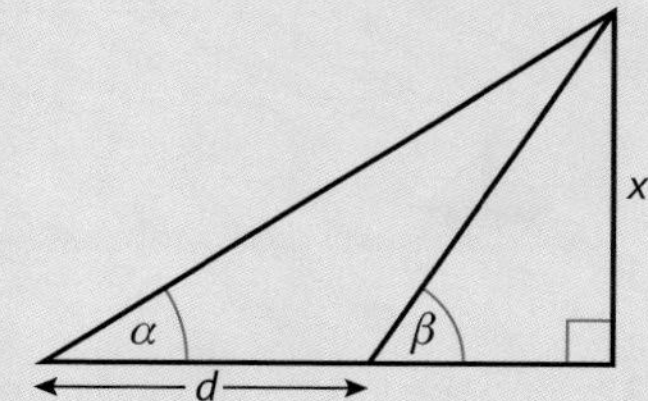

40 A spacecraft is travelling in a circular orbit 200 km above the surface of the Earth. Find the angle of depression (to the nearest degree) from the spacecraft to the horizon. Assume that the radius of the Earth is 6400 km. The 'horizontal' line through the spacecraft from which the angle of depression is measured will be parallel to a line tangent to the surface of the Earth directly below the spacecraft.

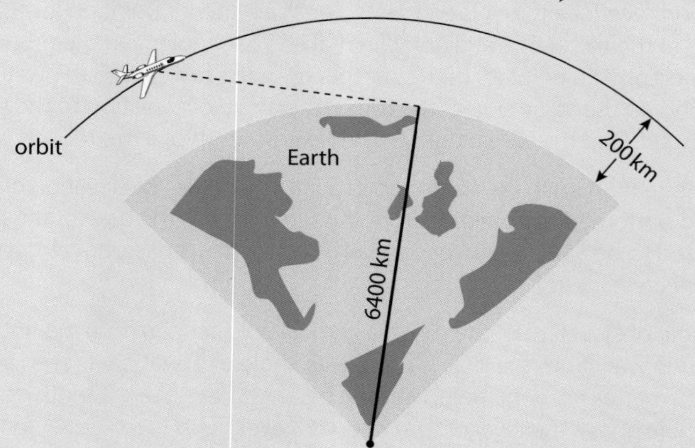

8.2 Trigonometric functions of any angle

In this section, we will extend the trigonometric ratios to all angles allowing us to solve problems involving any size angle.

Defining trigonometric functions for any angle in standard position

Consider the point $P(x, y)$ on the terminal side of an angle θ in standard position (Figure 8.6) such that r is the distance from the origin O to P. If θ is an acute angle then we can construct a right triangle POQ (Figure 8.7) by dropping a perpendicular from P to a point Q on the x-axis, and it follows that:

$$\sin\theta = \frac{y}{r} \qquad \cos\theta = \frac{x}{r} \qquad \tan\theta = \frac{y}{x}\ (x \neq 0)$$

$$\csc\theta = \frac{r}{y}\ (y \neq 0) \qquad \sec\theta = \frac{r}{x}\ (x \neq 0) \qquad \cot\theta = \frac{x}{y}\ (y \neq 0)$$

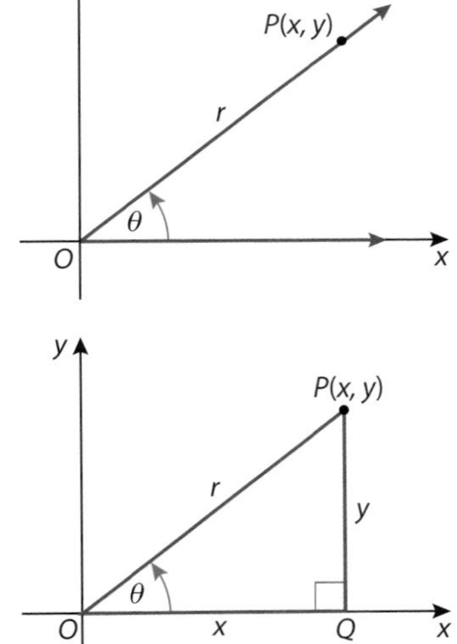

Figure 8.6

Figure 8.7

Extending this to angles other than acute angles allows us to define the trigonometric functions for any angle – positive or negative. It is important to note that the values of the trigonometric ratios do not depend on the choice of the point $P(x, y)$. If $P'(x', y')$ is any other point on the terminal side of angle θ, as in Figure 8.8, then triangles POQ and $P'OQ'$ are similar and the trigonometric ratios for corresponding angles are equal.

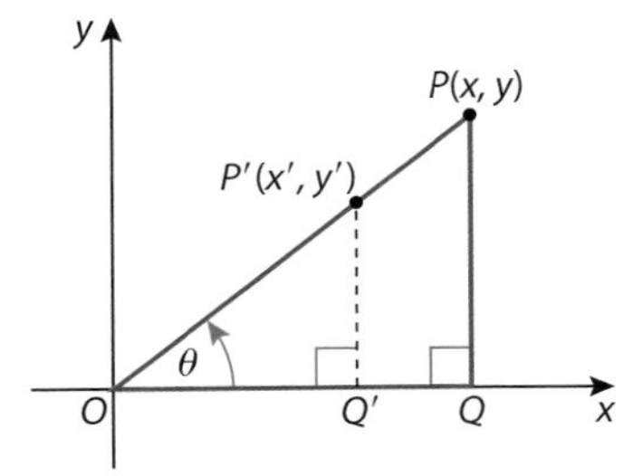

Figure 8.8

Definition of trigonometric functions

Let θ be any angle (in degree or radian measure) in standard position, with (x, y) any point on the terminal side of θ, and $r = \sqrt{x^2 + y^2}$, the distance from the origin to the point (x, y), as shown below.

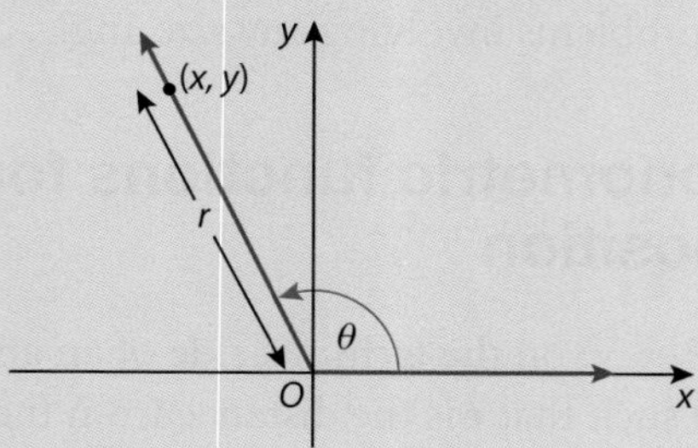

Then the trigonometric functions are defined as follows:

$$\sin\theta = \frac{y}{r} \qquad \cos\theta = \frac{x}{r} \qquad \tan\theta = \frac{y}{x}\ (x \neq 0)$$

$$\csc\theta = \frac{r}{y}\ (y \neq 0) \qquad \sec\theta = \frac{r}{x}\ (x \neq 0) \qquad \cot\theta = \frac{x}{y}\ (y \neq 0)$$

Example 6

Find the sine, cosine and tangent of an angle α that contains the point $(-3, 4)$ on its terminal side when in standard position.

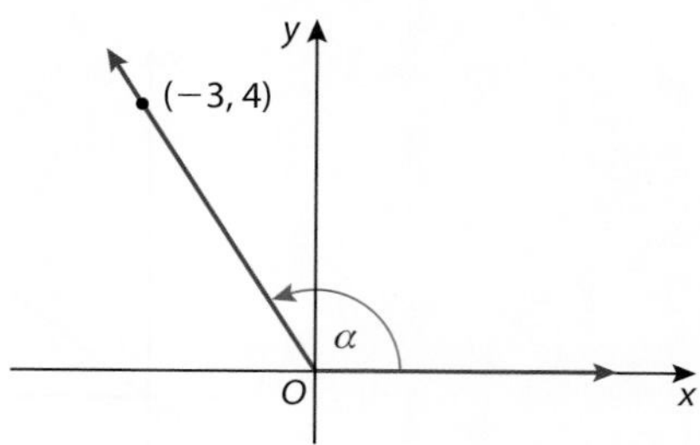

Solution

$$r = \sqrt{x^2 + y^2} = \sqrt{(-3)^2 + 4^2} = \sqrt{25} = 5$$

Then, $\sin\alpha = \frac{y}{r} = \frac{4}{5}$

$$\cos\alpha = \frac{x}{r} = \frac{-3}{5} = -\frac{3}{5}$$

$$\tan\alpha = \frac{y}{x} = \frac{4}{-3} = -\frac{4}{3}$$

Note that for the angle α in Example 6, we can form a right triangle by constructing a line segment from the point $(-3, 4)$ perpendicular to the x-axis, as shown in Figure 8.9. Clearly, $\theta = 180° - \alpha$. Furthermore, the values of the sine, cosine and tangent of the angle θ are the same as that for the angle α, except that the *sign* may be different.

Figure 8.9

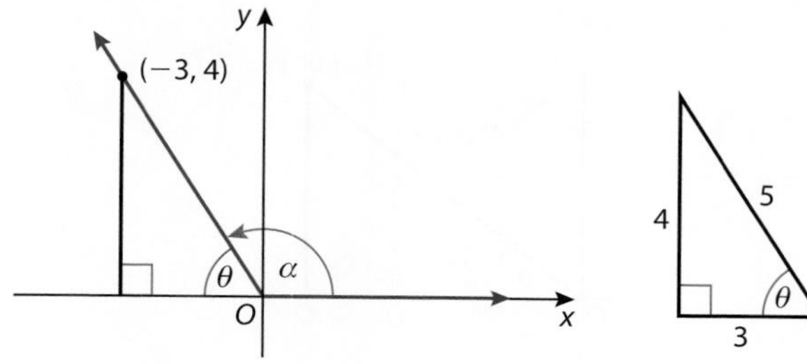

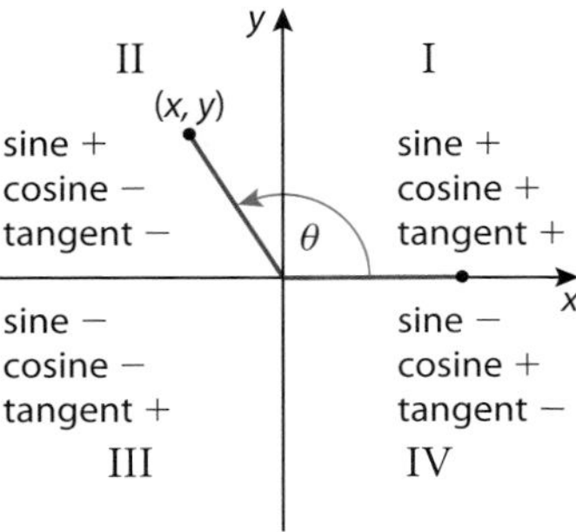

Figure 8.10 Sign of trigonometric function values depends on the quadrant in which the terminal side of the angle lies.

Whether the trigonometric functions are defined in terms of the length of an arc or in terms of an angle, the signs of trigonometric function values are determined by the quadrant in which the arc or angle lies, when in standard position (Figure 8.10).

Example 7

Find the sine, cosine and tangent of the obtuse angle that measures 150°.

Solution

The terminal side of the angle forms a 30° angle with the x-axis. The sine values for 150° and 30° will be exactly the same, and the cosine and tangent values will be the same but of opposite sign. We know that

$\sin 30° = \frac{1}{2}$, $\cos 30° = \frac{\sqrt{3}}{2}$ and $\tan 30° = \frac{\sqrt{3}}{3}$.

Therefore, $\sin 150° = \frac{1}{2}$, $\cos 150° = -\frac{\sqrt{3}}{2}$ and $\tan 150° = -\frac{\sqrt{3}}{3}$.

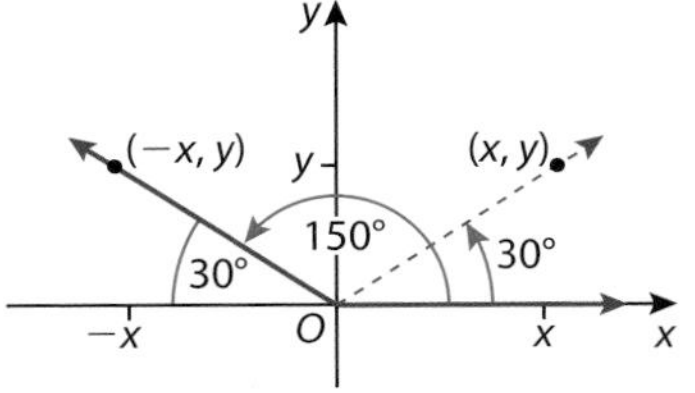

Example 7 illustrates three trigonometric identities for angles whose sum is 180° (i.e. a pair of supplementary angles). The following are true for any acute angle θ:

$\sin(180° - \theta) = \sin \theta$

$\cos(180° - \theta) = -\cos \theta$

$\tan(180° - \theta) = -\tan \theta$

$\csc(180° - \theta) = \csc \theta$

$\sec(180° - \theta) = -\sec \theta$

$\cot(180° - \theta) = -\cot \theta$

Example 8

Given that $\sin \theta = \frac{5}{13}$ and $90° < \theta < 180°$, find the exact values of $\cos \theta$ and $\tan \theta$.

Solution

θ is an angle in the second quadrant. It follows from the definition $\sin \theta = \frac{y}{r}$ that with θ in standard position there must be a point on the terminal side of the angle that is 13 units from the origin (i.e. $r = 13$) and which has a y-coordinate of 5, as shown in the diagram.

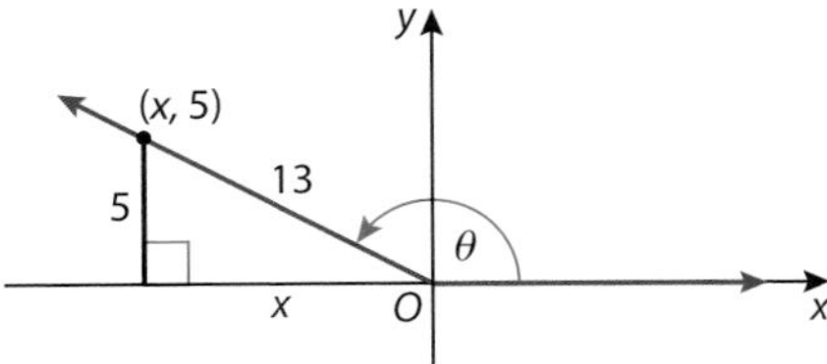

Using Pythagoras' theorem, $|x| = \sqrt{13^2 - 5^2} = \sqrt{144} = 12$. Because θ is in the second quadrant, the x-coordinate of the point must be negative, thus $x = -12$.

Therefore, $\cos \theta = \frac{-12}{13} = -\frac{12}{13}$, and $\tan \theta = \frac{5}{-12} = -\frac{5}{12}$.

Example 9

a) Find the acute angle with the same sine ratio as (i) 135°, and (ii) 117°.
b) Find the acute angle with the same cosine ratio as (i) 300°, and (ii) 342°.

Solution

a) (i) Angles in the first and second quadrants have the same sine ratio. Hence, the identity $\sin(180° - \theta) = \sin\theta$. Since $180° - 135° = 45°$, then $\sin 135° = \sin 45°$.

(ii) Since $180° - 117° = 63°$, then $\sin 117° = \sin 63°$.

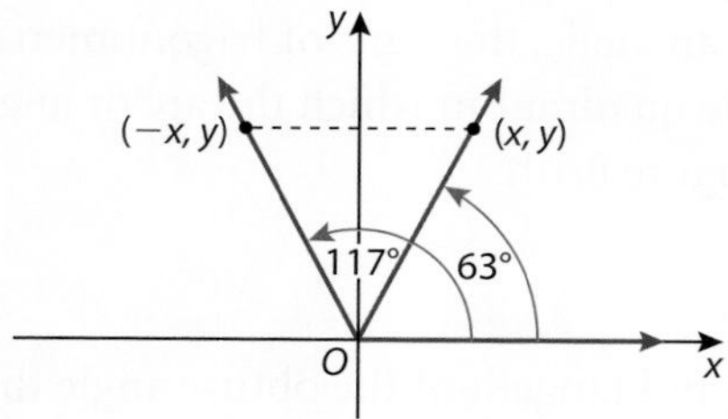

b) (i) Angles in the first and fourth quadrants have the same cosine ratio. Hence, the identity $\cos(360° - \theta) = \cos\theta$. Since $360° - 300° = 60°$, then $\cos 300° = \cos 60°$.

(ii) Since $360° - 342° = 18°$, then $\cos 342° = \cos 18°$.

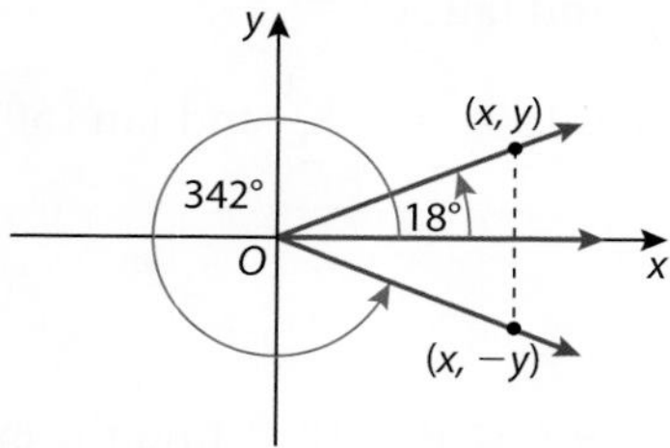

Areas of triangles

You are familiar with the standard formula for the area of a triangle, area $= \frac{1}{2} \times$ base $\times$ height (or area $= \frac{1}{2}bh$), where the base, b, is a side of the triangle and the height, h, (or altitude) is a line segment perpendicular to the base (or the line containing it) and drawn to the vertex opposite to the base, as shown in Figure 8.11.

Figure 8.11

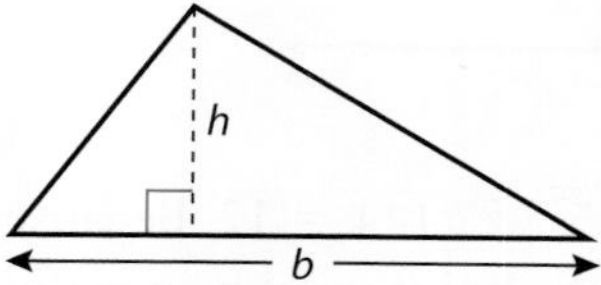

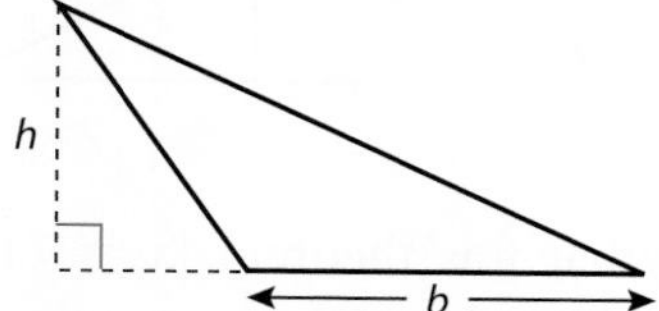

If the lengths of two sides of a triangle and the measure of the angle between these sides (often called the included angle) are known, then the triangle is unique and has a fixed area. Hence, we should be able to calculate the area from just these measurements, i.e. from knowing two sides and the included angle. This calculation is quite straightforward if the triangle is a right triangle (Figure 8.12) and we know the lengths of the two legs on either side of the right angle.

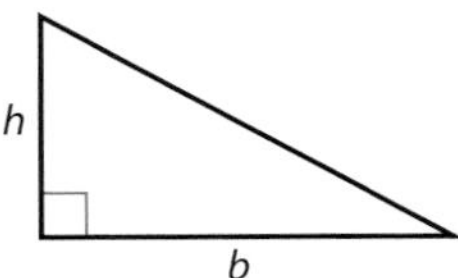

Figure 8.12 A right triangle.

Let's develop a general area formula that will apply to any triangle – right, acute or obtuse. For triangle ABC shown in Figure 8.13, suppose we know the lengths of the two sides a and b and the included angle C. If the length of the height from B is h, the area of the triangle is $\frac{1}{2}bh$. From right triangle trigonometry, we know that $\sin C = \frac{h}{a}$, or $h = a\sin C$. Substituting $a\sin C$ for h, area $= \frac{1}{2}bh = \frac{1}{2}b(a\sin C) = \frac{1}{2}ab\sin C$.

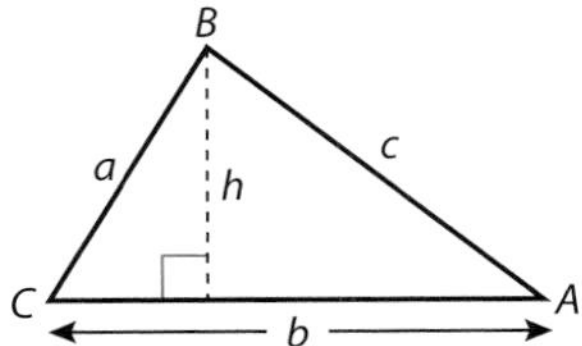

Figure 8.13 An acute triangle.

If the angle C is obtuse, then from Figure 8.14 we see that $\sin(180° - C) = \frac{h}{a}$. So, the height is $h = a\sin(180° - C)$. However, $\sin(180° - C) = \sin C$. Thus, $h = a\sin C$ and, again, area $= \frac{1}{2}ab\sin C$.

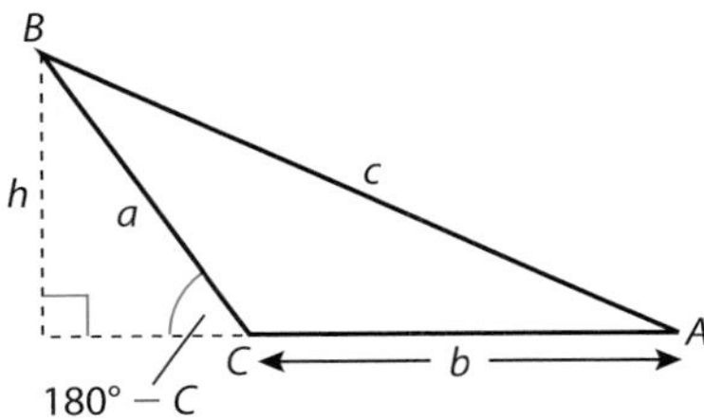

Figure 8.14 An obtuse triangle.

Area of a triangle

For a triangle with sides of lengths a and b and included angle C,

$$\text{Area of } \triangle = \frac{1}{2}ab\sin C$$

• **Hint:** Note that the procedure for finding the area of a triangle from a pair of sides and the included angle can be performed three different ways. For any triangle labelled in the manner of the triangles in Figures 8.13 and 8.14, its area is expressed by any of the following three expressions.

$$\begin{aligned}\text{Area of } \triangle &= \tfrac{1}{2}ab\sin C\\ &= \tfrac{1}{2}ac\sin B\\ &= \tfrac{1}{2}bc\sin A\end{aligned}$$

These three equivalent expressions will prove to be helpful for developing an important formula for solving non-right triangles in the next section.

Example 10

The circle shown has a radius of 1 cm and the central angle θ subtends an arc of length of $\frac{2\pi}{3}$ cm. Find the area of the shaded region.

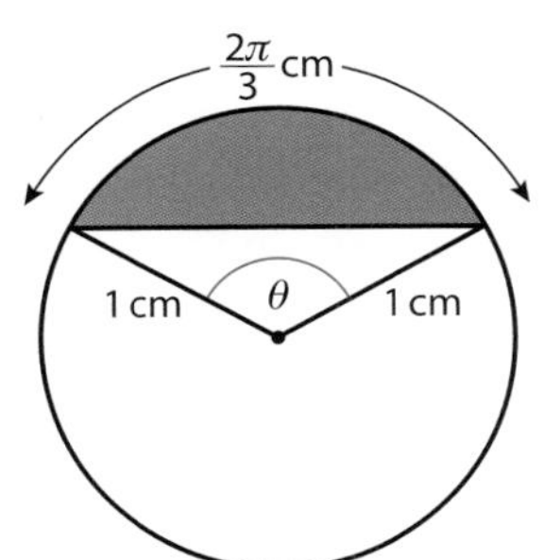

The region bounded by an arc of a circle and the chord connecting the endpoints of the arc is called a **segment** of the circle (see figure for Example 10).

Solution

The formula for the area of a sector is $A = \frac{1}{2}r^2\theta$ (Section 7.1), where θ is the central angle in radian measure. Since the radius of the circle is one, the length of the arc subtended by θ is the same as the radian measure of θ.

Thus, area of sector $= \frac{1}{2}(1)^2\left(\frac{2\pi}{3}\right) = \frac{\pi}{3}$ cm^2.

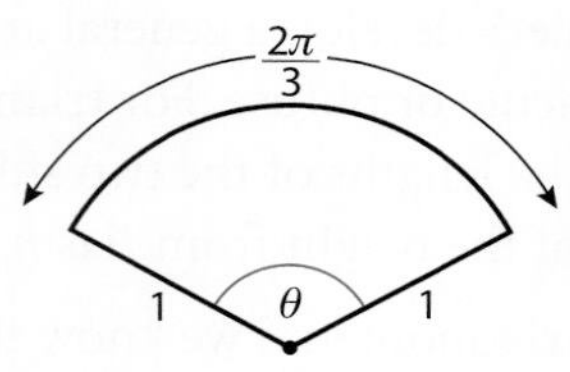

The area of the triangle formed by the two radii and the chord is equal to

$\frac{1}{2}(1)(1)\sin\left(\frac{2\pi}{3}\right) = \frac{1}{2}\left(\frac{\sqrt{3}}{2}\right) = \frac{\sqrt{3}}{4}$ cm^2.

$\left[\sin\frac{2\pi}{3} = \sin\left(\pi - \frac{2\pi}{3}\right) = \sin\frac{\pi}{3} = \frac{\sqrt{3}}{2}\right]$

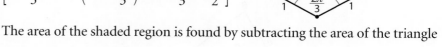

The area of the shaded region is found by subtracting the area of the triangle from the area of the sector.

Area $= \frac{\pi}{3} - \frac{\sqrt{3}}{4}$ or $\frac{4\pi - 3\sqrt{3}}{12}$ or approximately 0.614 cm^2 (3 s.f.).

Example 11

Show that it is possible to construct two different triangles with an area of 35 cm^2 that have sides measuring 8 cm and 13 cm. For each triangle, find the measure of the (included) angle between the sides of 8 cm and 13 cm to the nearest tenth of a degree.

Solution

We can visualize the two different triangles with equal areas – one with an acute included angle (α) and the other with an obtuse included angle (β).

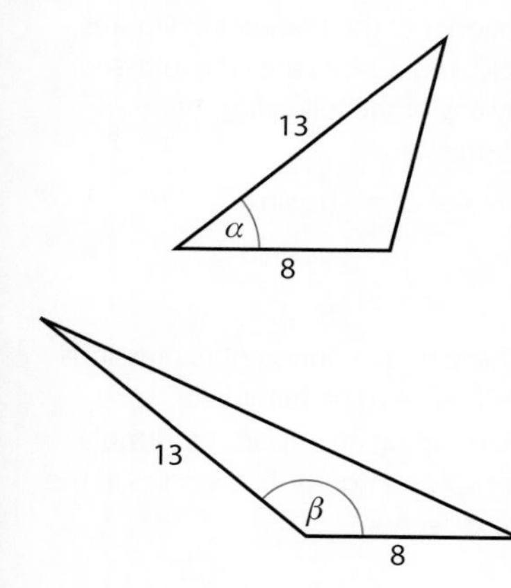

$$\text{Area} = \tfrac{1}{2}(\text{side})(\text{side})(\text{sine of included angle}) = 35 \text{ cm}^2$$

$$= \tfrac{1}{2}(8)(13)(\sin\alpha) = 35$$

$$52\sin\alpha = 35$$

$$\sin\alpha = \frac{35}{52}$$

$$\alpha = \sin^{-1}\left(\frac{35}{52}\right)$$ Recall that the GDC will only give the acute angle with sine ratio of $\frac{35}{52}$.

$$\alpha \approx 42.3°$$ Round to the nearest tenth.

Knowing that $\sin(180° - \alpha) = \sin\alpha$, the obtuse angle β is equal to $180° - 42.3° = 137.7°$.

Check this answer by computing on your GDC:

$\frac{1}{2}(8)(13)(\sin 137.7°) \approx 34.997 \approx 35$ cm^2.

Therefore, there are two different triangles with sides 8 cm and 13 cm and area of 35 cm^2 – one with an included angle of 42.3° and the other with an included angle of 137.7°.

Exercise 8.2

In questions 1–4, find the exact value of the sine, cosine and tangent functions of the angle θ.

1

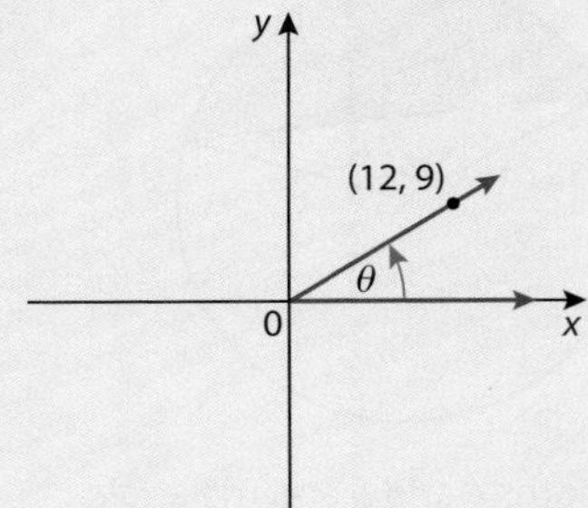

2

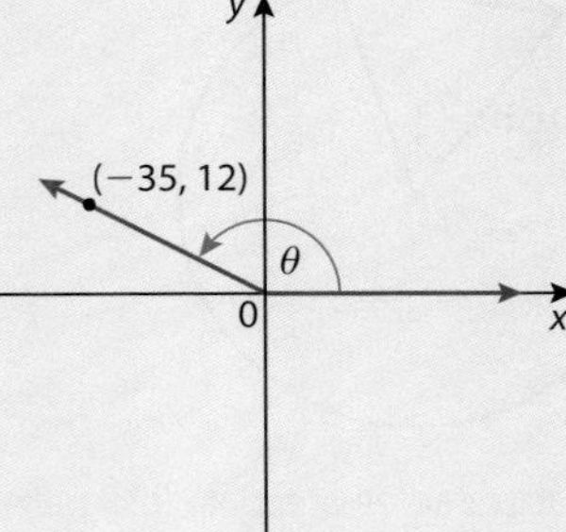

3

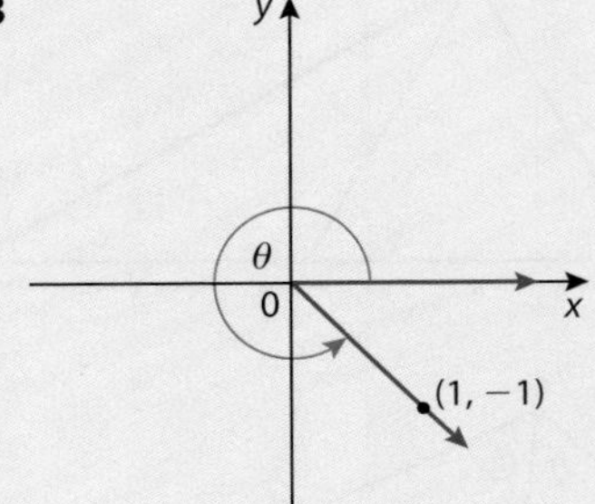

4

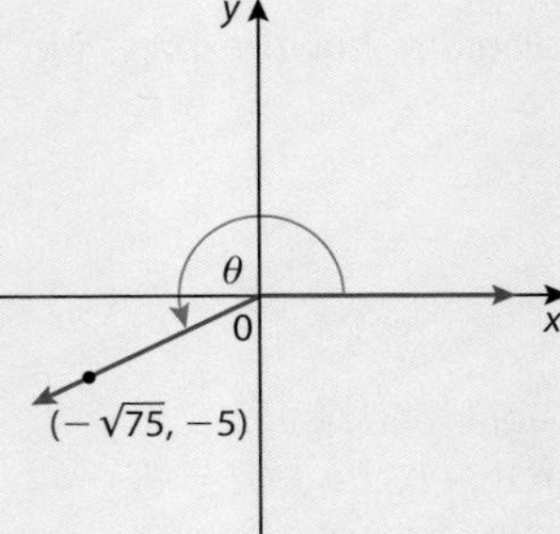

5 Without using your GDC, determine the exact values of all six trigonometric functions for the following angles.

a) $120°$ b) $135°$ c) $330°$ d) $270°$ e) $240°$
f) $\frac{5\pi}{4}$ g) $-\frac{\pi}{6}$ h) $\frac{7\pi}{6}$ i) $-60°$ j) $-\frac{3\pi}{2}$
k) $\frac{5\pi}{3}$ l) $-210°$ m) $-\frac{\pi}{4}$ n) π o) 4.25π

6 Given that $\cos\theta = \frac{8}{17}$ and $0° < \theta < 90°$, find the exact values of the other five trigonometric functions.

7 Given that $\tan\theta = -\frac{6}{5}$ and $\sin\theta < 0$, find the exact values of $\sin\theta$ and $\cos\theta$.

8 Given that $\sin\theta = 0$ and $\cos\theta < 0$, find the exact values of the other five trigonometric functions.

9 If $\sec\theta = 2$ and $\frac{3\pi}{2} < \theta < 2\pi$, find the exact values of the other five trigonometric functions.

10 a) Find the acute angle with the same sine ratio as (i) 150°, and (ii) 95°.
b) Find the acute angle with the same cosine ratio as (i) 315°, and (ii) 353°.
c) Find the acute angle with the same tangent ratio as (i) 240°, and (ii) 200°.

11 Find the area of each triangle. Express the area exactly, or, if not possible, express it accurate to 3 s.f.

a)

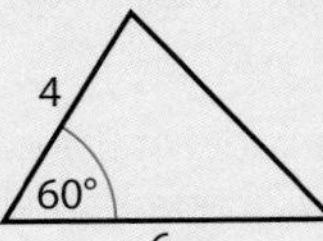

b)

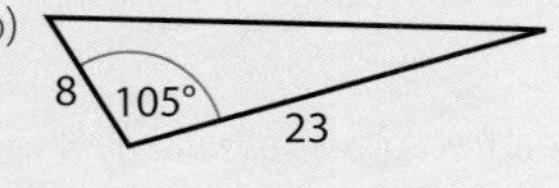

c)

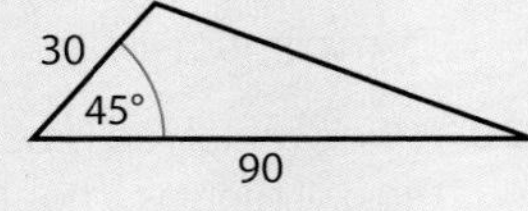

12 Triangle ABC has an area of 43 cm^2. The length of side AB is 12 cm and the length of side AC is 15 cm. Find the degree measure of angle A.

13 A chord AB subtends an angle of 120° at O, the centre of a circle with radius 15 cm. Find the area of a) the sector AOB, and b) the triangle AOB.

14 Find the area of the shaded region (called a *segment*) in each circle.

a)

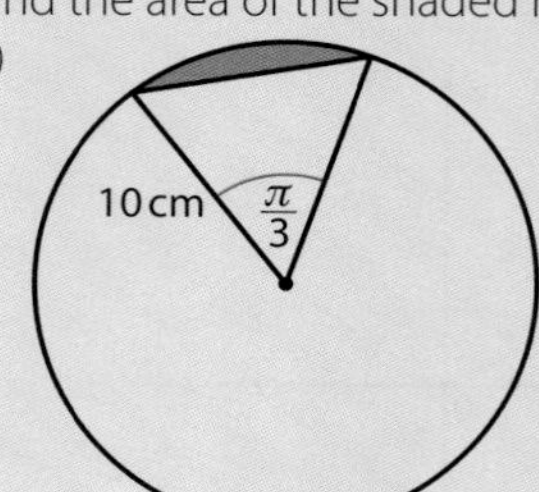

b)

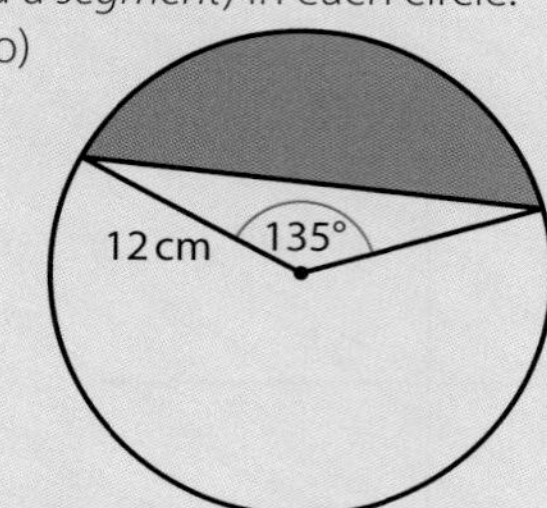

15 Two adjacent sides of a parallelogram have lengths a and b and the angle between these two sides is θ. Express the area of the parallelogram in terms of a, b and θ.

16 For the diagram shown, express y in terms of x.

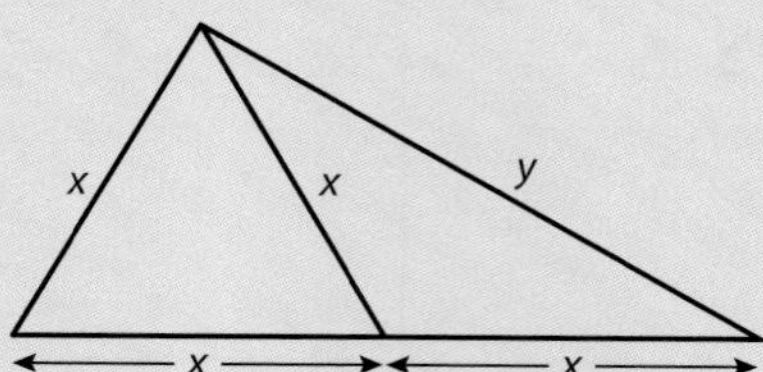

17 In the diagram, GJ bisects $F\hat{G}H$ such that $F\hat{G}J = H\hat{G}J = \theta$. Express x in terms of h, f and $\cos\theta$.

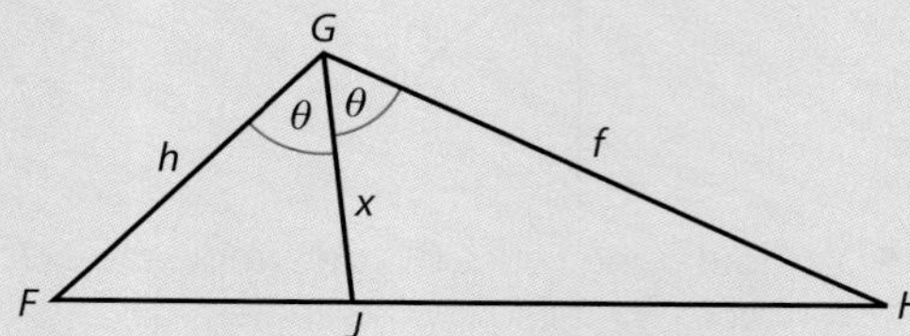

18 If s is the length of each side of a regular polygon with n sides and r is the radius of the circumscribed circle, show that $s = 2r\sin\left(\frac{180°}{n}\right)$. (Note: A *regular* polygon has all sides equal.)

The figure shows a regular pentagon ($n = 5$) with each side of length s circumscribed by a circle with radius r.

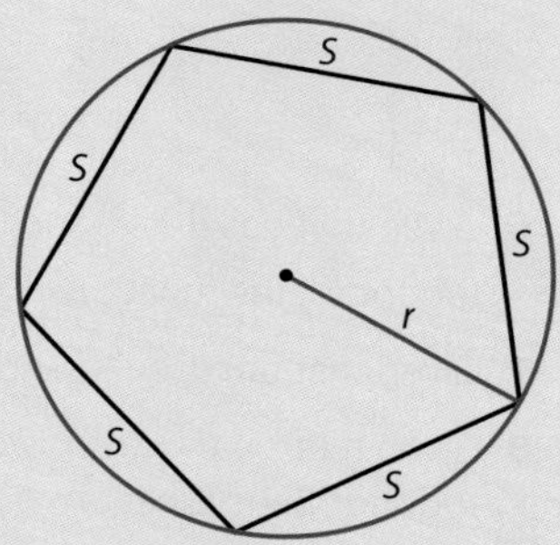

19 Suppose a triangle has two sides of lengths 6 cm and 8 cm and an included angle x.

a) Express the area of the triangle as a function of x.

b) State the domain and range of the function and sketch its graph for a suitable interval of x.

c) Find the exact coordinates of the maximum point of the function. What type of triangle corresponds to this maximum? Explain why this triangle gives a maximum area.

20 A long metal rod is being carried down a hallway 3 metres wide. At the end of the hall there is a right-angled turn into a narrower hallway 2 metres wide. The angle that the rod makes with the outer wall is θ (see figure on the next page).

a) Show that the length, L, of the rod is given by the function $L(\theta) = 3\csc\theta + 2\sec\theta$.

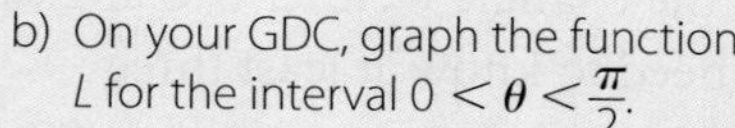

b) On your GDC, graph the function L for the interval $0 < \theta < \frac{\pi}{2}$.

c) Using the built-in features of your GDC, find the minimum value of the function L. Explain why this is the length of the longest rod that can be carried around the corner.

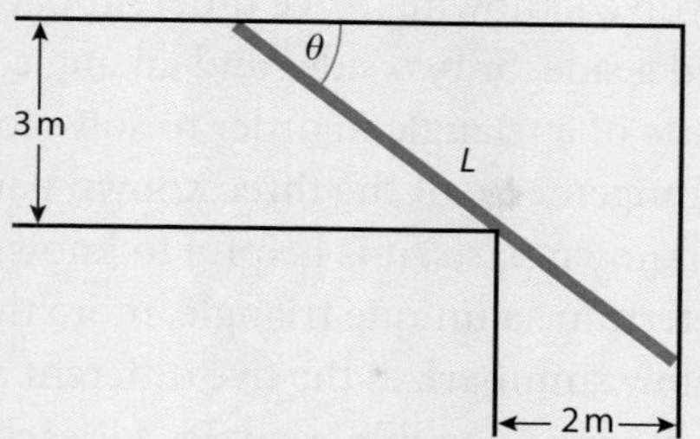

21 As viewed from the surface of the Earth (A), the angle subtended by the full Moon ($D\hat{A}E$) is 0.5182°. Given that the distance from the Earth's surface to the Moon's surface (AB) is approximately 383 500 kilometres, find the radius, r, of the Moon to three significant figures.

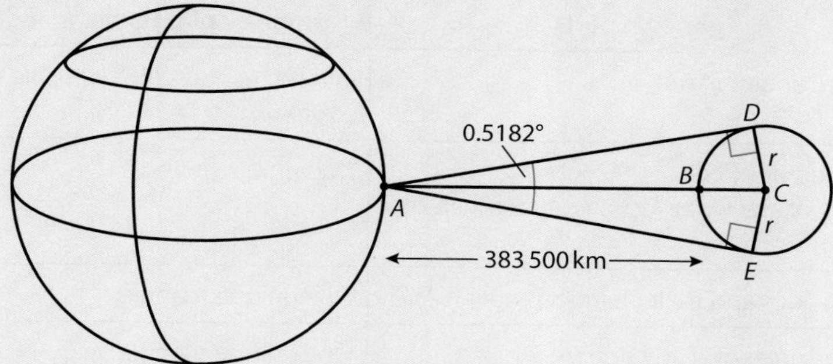

22 a) Given that $\sin\theta = x$, find $\sec\theta$ in terms of x.

b) Given that $\tan\beta = y$, find $\sin\beta$ in terms of y.

23 The figure shows the unit circle with angle θ in standard position. Segment BC is tangent to the circle at P and $B\hat{O}C$ is a right angle. Each of the six trigonometric functions of θ is equal to the length of a line segment in the figure. For example, we know from the previous section (and previous chapter) that $\sin\theta = AP$. For each of the five other trigonometric functions, find a line segment in the figure whose length equals the function value of θ.

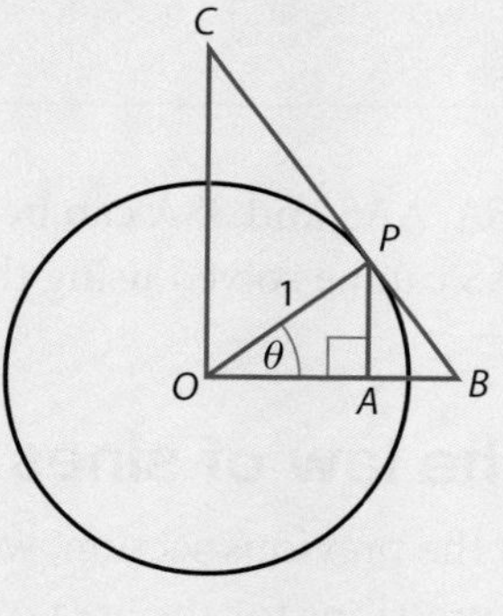

The law of sines

In Section 8.1 we used techniques from right triangle trigonometry to solve right triangles when an acute angle and one side are known, or when two sides are known. In this section and the next, we will study methods for finding unknown lengths and angles in triangles that are not right triangles. These general methods are effective for solving problems involving any kind of triangle – right, acute or obtuse.

Possible triangles constructed from three given parts

As mentioned in the previous paragraph, we've solved right triangles by either knowing an acute angle and one side, or knowing two sides. Since the triangles also have a right angle, each of those two cases actually involved knowing three different parts of the triangle – either two angles and a side, or two sides and an angle. We need to know at least three parts of a triangle in order to solve for other unknown parts. Different arrangements of the three known parts can be given. Before solving for unknown parts, it is helpful to know whether the three known parts determine a unique triangle, more than one triangle, or none. The table below summarizes the five different arrangements of three parts and the number of possible triangles for each. You are encouraged to confirm these results on your own with manual or computer generated sketches.

Possible triangles formed with three known parts

Known parts	Number of possible triangles
Three angles (AAA)	Infinite triangles (not possible to solve)
Three sides (SSS) (sum of any two must be greater than the third)	One unique triangle
Two sides and their included angle (SAS)	One unique triangle
Two angles and any side (ASA or AAS)	One unique triangle
Two sides and a non-included angle (SSA)	No triangle, one triangle or two triangles

ASA, AAS and SSA can be solved using the **law of sines**, whereas SSS and SAS can be solved using the **law of cosines** (next section).

The law of sines (or sine rule)

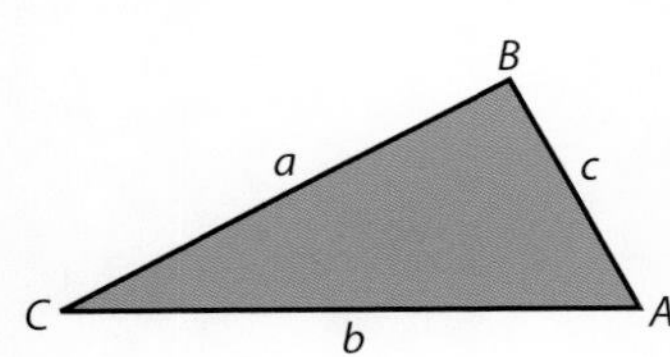

In the previous section, we showed that we can write three equivalent expressions for the area of any triangle for which we know two sides and the included angle.

$$\text{Area of } \triangle = \tfrac{1}{2}ab\sin C = \tfrac{1}{2}ac\sin B = \tfrac{1}{2}bc\sin A$$

If each of these expressions is divided by $\frac{1}{2}abc$,

$$\frac{\frac{1}{2}ab\sin C}{\frac{1}{2}abc} = \frac{\frac{1}{2}ac\sin B}{\frac{1}{2}abc} = \frac{\frac{1}{2}bc\sin A}{\frac{1}{2}abc}$$

we obtain three equivalent ratios – each containing the sine of an angle divided by the length of the side opposite the angle.

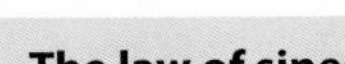

The law of sines

If A, B and C are the angle measures of any triangle and a, b and c are, respectively, the lengths of the sides opposite these angles, then

$$\frac{\sin A}{a} = \frac{\sin B}{b} = \frac{\sin C}{c}$$

Alternatively, the law of sines can also be written as $\frac{a}{\sin A} = \frac{b}{\sin B} = \frac{c}{\sin C}$.

Solving triangles given two angles and any side (ASA or AAS)

If we know two angles and any side of a triangle, we can use the law of sines to find any of the other angles or sides of the triangle.

Example 12

Find all of the unknown angles and sides of triangle DEF shown in the diagram. Approximate all measurements to 1 decimal place.

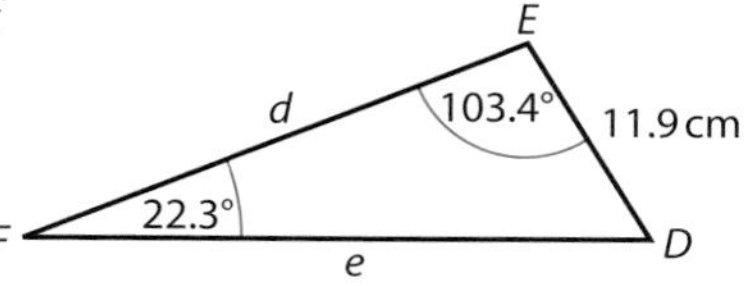

Solution

The third angle of the triangle is

$$D = 180° - E - F = 180° - 103.4° - 22.3° = 54.3°.$$

Using the law of sines, we can write the following proportion to solve for the length e:

$$\frac{\sin 22.3°}{11.9} = \frac{\sin 103.4°}{e}$$

$$e = \frac{11.9 \sin 103.4°}{\sin 22.3°} \approx 30.507 \text{ cm}$$

We can write another proportion from the law of sines to solve for d:

$$\frac{\sin 22.3°}{11.9} = \frac{\sin 54.3°}{d}$$

$$d = \frac{11.9 \sin 54.3°}{\sin 22.3°} \approx 25.467 \text{ cm}$$

Therefore, the other parts of the triangle are $D = 54.3°$, $e \approx 30.5$ cm and $d \approx 25.5$ cm.

• **Hint:** When using your GDC to find angles and lengths with the law of sines (or the law of cosines), remember to store intermediate answers on the GDC for greater accuracy. By not rounding until the final answer, you reduce the amount of round-off error.

Example 13

A tree on a sloping hill casts a shadow 45 m along the side of the hill. The gradient of the hill is $\frac{1}{5}$ (or 20%) and the angle of elevation of the sun is 35°. How tall is the tree to the nearest tenth of a metre?

Solution

α is the angle that the hill makes with the horizontal. Its measure can be found by computing the inverse tangent of $\frac{1}{5}$.

$$\alpha = \tan^{-1}\left(\frac{1}{5}\right) \approx 11.3099°$$

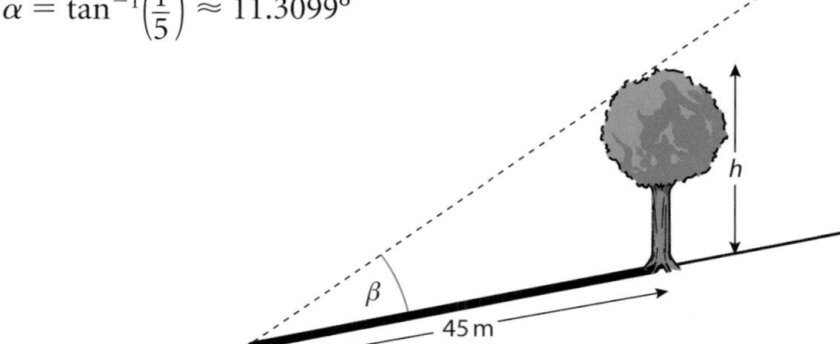

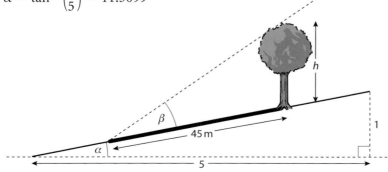

The height of the tree is labelled h. The angle of elevation of the sun is the angle between the sun's rays and the horizontal. In the diagram, this angle of elevation is the sum of α and β. Thus, $\beta \approx 35° - 11.3099° \approx 23.6901°$. For the larger right triangle with $\alpha + \beta = 35°$ as one of its acute angles, the other acute angle – and the angle in the obtuse triangle opposite the side of 45 m – must be 55°. Now we can apply the law of sines for the obtuse triangle to solve for h.

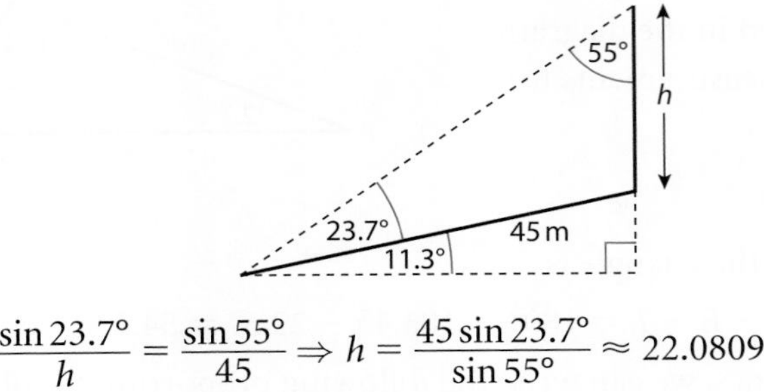

$$\frac{\sin 23.7°}{h} = \frac{\sin 55°}{45} \Rightarrow h = \frac{45 \sin 23.7°}{\sin 55°} \approx 22.0809$$

Therefore, the tree is approximately 22.1 m tall.

Two sides and a non-included angle (SSA) – the ambiguous case

The arrangement where we are given the lengths of two sides of a triangle and the measure of an angle not between those two sides can produce three different results: no triangle, one unique triangle or two different triangles. Let's explore these possibilities with the following example.

Example 14

Find all of the unknown angles and sides of triangle ABC where $a = 35$ cm, $b = 50$ cm and $A = 30°$. Approximate all measurements to 1 decimal place.

Solution

Figure 8.15 shows the three parts we have from which to try and construct a triangle.

Figure 8.15

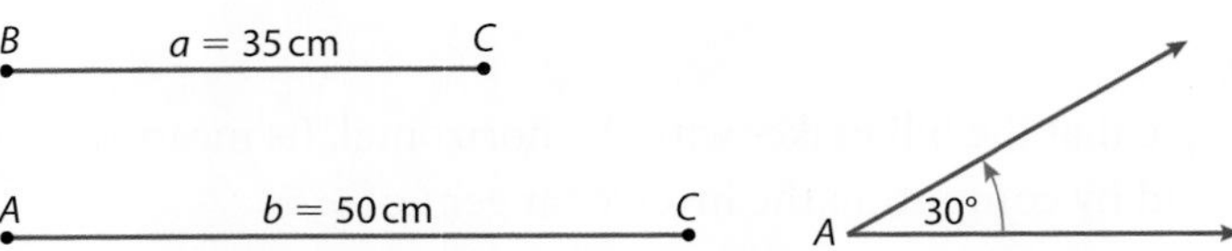

We attempt to construct the triangle, as shown in Figure 8.16. We first draw angle A with its initial side (or base line of the triangle) extended. We then measure off the known side $b = AC = 50$. To construct side a (opposite angle A), we take point C as the centre and with radius $a = 35$ we draw an arc of a circle. The points on this arc are all possible positions for vertex B – one of the endpoints of side a, or BC. Point B must be on the base line, so B can be located at any point of intersection of the circular arc and the base line. In this instance, with these particular measurements for the two sides and non-included angle, there are two points of intersection, which we label B_1 and B_2.

Figure 8.16

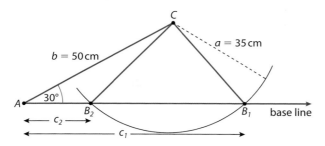

Therefore, we can construct two different triangles, triangle AB_1C (Figure 8.17) and triangle AB_2C (Figure 8.18). The angle B_1 will be acute and angle B_2 will be obtuse. To complete the solution of this problem, we need to solve each of these triangles.

- Solve triangle AB_1C:

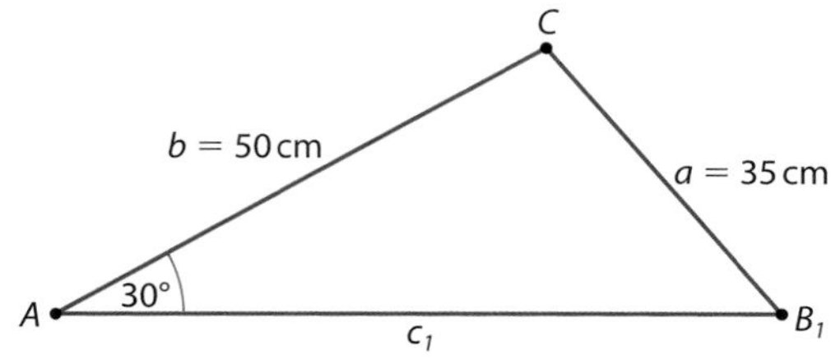

Figure 8.17

We can solve for acute angle B_1 using the law of sines:

$$\frac{\sin 30°}{35} = \frac{\sin B_1}{50}$$

$$\sin B_1 = \frac{50 \sin 30°}{35} = \frac{50(0.5)}{35}$$

$$B_1 = \sin^{-1}\left(\frac{5}{7}\right) \approx 45.5847°$$

Then, $C \approx 180° - 30° - 45.5847° \approx 104.4153°$.

With another application of the law of sines, we can solve for side c_1:

$$\frac{\sin 30°}{35} = \frac{\sin 104.4153°}{c_1}$$

$$c_1 = \frac{35 \sin 104.4153°}{\sin 30°} \approx \frac{35(0.96852)}{0.5} \approx 67.7964 \text{ cm}$$

Therefore, for triangle AB_1C, $B_1 \approx 45.6°$, $C \approx 104.4°$ and $c_1 \approx 67.8$ cm.

- Solve triangle AB_2C:

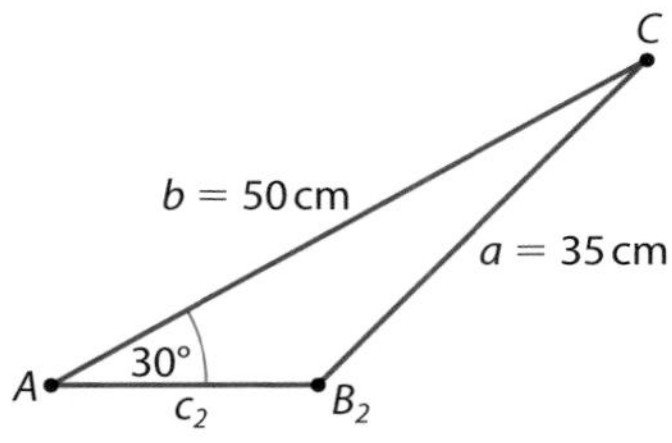

Figure 8.18

Solving for obtuse angle B_2, using the law of sines, gives the same result as above, except we know that $90° < B_2 < 180°$.

We also know that $\sin(180° - \theta) = \sin\theta$.

Thus, $B_2 = 180° - B_1 \approx 180° - 45.5847° \approx 134.4153°$.

Then, $C \approx 180° - 30° - 134.4153° \approx 15.5847°$.

With another application of the law of sines, we can solve for side c_2:

$$\frac{\sin 30°}{35} = \frac{\sin 15.5847°}{c_2}$$

$$c_2 \approx \frac{35 \sin 15.5847°}{\sin 30°} \approx \frac{35(0.26866)}{0.5} \approx 18.8062 \text{ cm}$$

Therefore, for triangle AB_2C, $B_2 \approx 134.4°$, $C \approx 15.6°$ and $c_2 \approx 18.8$ cm.

Now that we have solved this specific example, let's take a more general look and examine all the possible conditions and outcomes for the SSA arrangement. In general, we are given the lengths of two sides – call them a and b – and a non-included angle – for example, angle A that is opposite side a. From these measurements, we can determine the number of different triangles. Figure 8.19 shows the four different possibilities (or cases) when angle A is acute. The number of triangles depends on the length of side a.

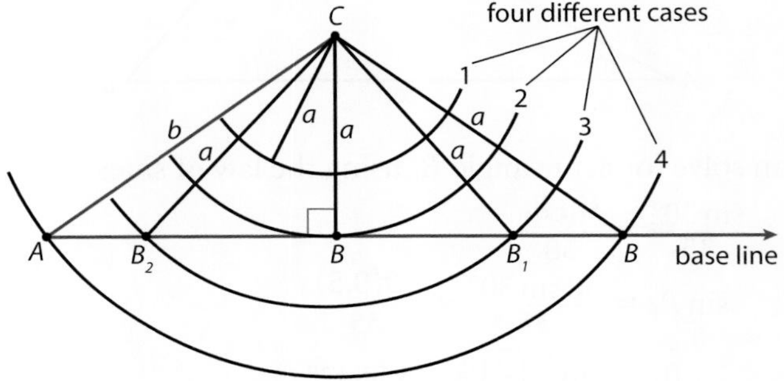

Figure 8.19 Four distinct cases for SSA when angle A is acute.

In case 2, side a is perpendicular to the base line resulting in a single right triangle, shown in Figure 8.20. In this case, clearly $\sin A = \frac{a}{b}$ and $a = b\sin A$. In case 1, the length of a is shorter than it is in case 2, i.e $b\sin A$. In case 3, which occurred in Example 14, the length of a is longer than $b\sin A$, but less than b. And, in case 4, the length of a is greater than b. These results are summarized in the table below. Because the number of triangles may be none, one or two, depending on the length of a (the side opposite the given angle), the SSA arrangement is called the ambiguous case.

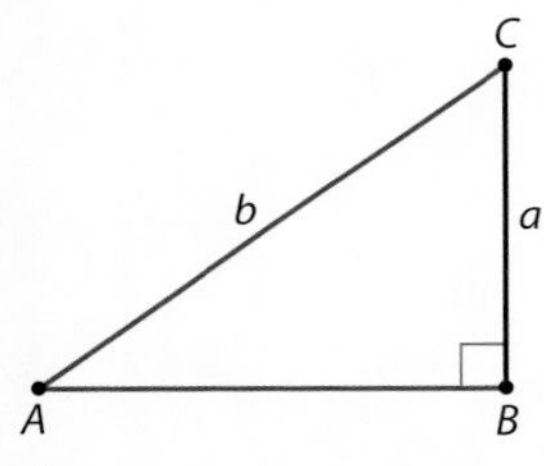

Figure 8.20 Case 2 for SSA: $a = b \sin A$, one right angle.

The ambiguous case (SSA)

Given the lengths of sides a and b and the fact that the non-included angle A is acute, the following four cases and resulting triangles can occur.

Length of a	Number of triangles	Case in Figure 8.19
$a < b\sin A$	No triangle	1
$a = b\sin A$	One right triangle	2
$b\sin A < a < b$	Two triangles	3
$a \geqslant b$	One triangle	4

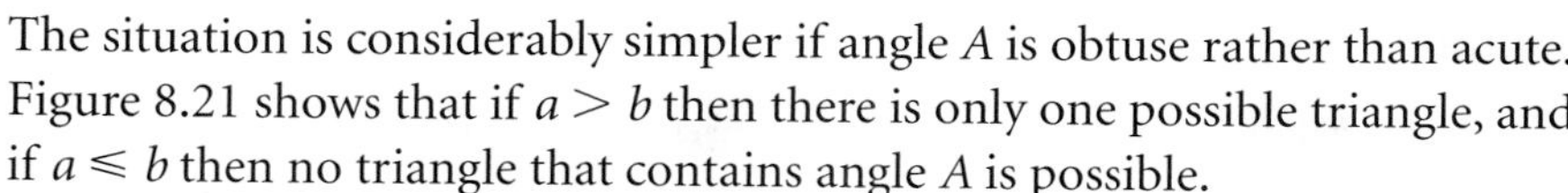

The situation is considerably simpler if angle A is obtuse rather than acute. Figure 8.21 shows that if $a > b$ then there is only one possible triangle, and if $a \leqslant b$ then no triangle that contains angle A is possible.

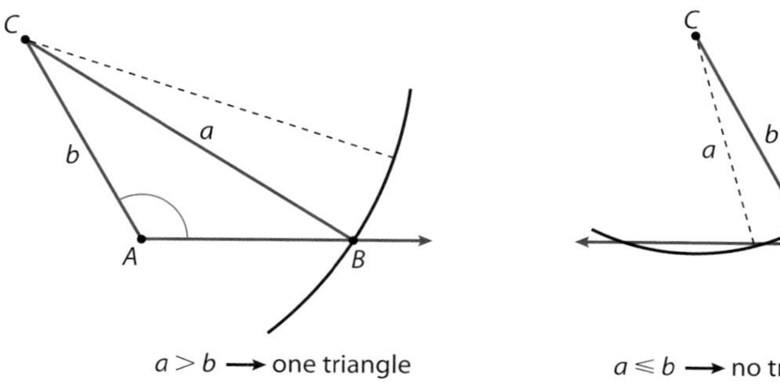

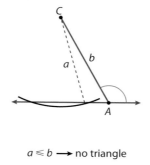

Figure 8.21 Angle A is obtuse.

Example 15

For triangle ABC, if side $b = 50$ cm and angle $A = 30°$, find the values for the length of side a that will produce: a) no triangle, b) one triangle, c) two triangles. This is the same SSA information given in Example 14 with the exception that side a is not fixed at 35 cm, but is allowed to vary.

Solution

Because this is a SSA arrangement and given A is an acute angle, then the number of different triangles that can be constructed is dependent on the length of a. First calculate the value of $b \sin A$:

$$b \sin A = 50 \sin 30° = 50(0.5) = 25 \text{ cm}$$

Thus, if a is exactly 25 cm then triangle ABC is a right triangle, as shown in the figure.

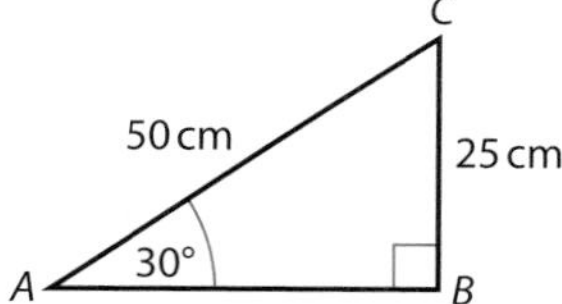

a) If $a < 25$ cm, there is no triangle.
b) If $a = 25$ cm, or $a \geqslant 50$ cm, there is one unique triangle.
c) If 25 cm $< a <$ 50 cm, there are two different possible triangles.

• **Hint:** It is important to be familiar with the notation for line segments and angles commonly used in IB exam questions. For example, the line segment labelled b in the diagram (below) is denoted as $[AC]$ in IB notation. Angle A, the angle between $[BA]$ and $[AC]$, is denoted as $B\hat{A}C$. Also, the line containing points A and B is denoted as (AB).

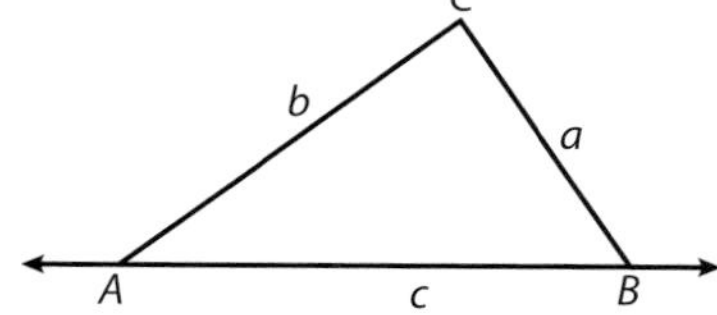

Example 16

The diagrams below show two different triangles both satisfying the conditions: $HK = 18$ cm, $JK = 15$ cm, $J\hat{H}K = 53°$.

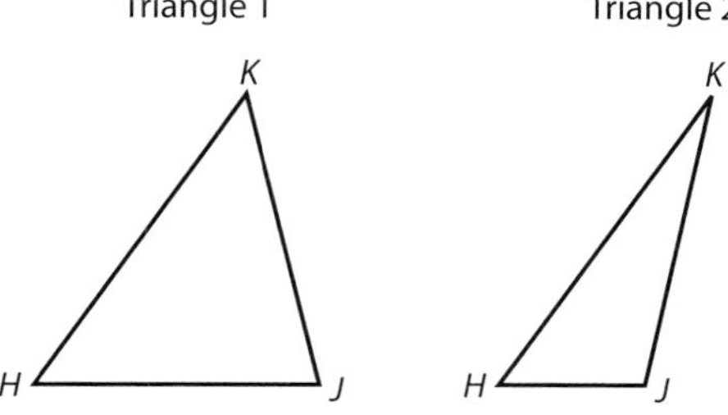

a) Calculate the size of $H\hat{J}K$ in Triangle 2.
b) Calculate the area of Triangle 1.

Solution

a) From the law of sines, $\frac{\sin(H\hat{J}K)}{18} = \frac{\sin 53°}{15} \Rightarrow \sin(H\hat{J}K) = \frac{18 \sin 53°}{15}$

$\approx 0.958\,36 \Rightarrow \sin^{-1}(0.958\,36) \approx 73.408°$

However, $H\hat{J}K > 90° \Rightarrow H\hat{J}K \approx 180° - 73.408° \approx 106.592°$.

Therefore, in Triangle 2 $H\hat{J}K \approx 107°$ (3 s.f.).

b) In Triangle 1, $H\hat{J}K < 90° \Rightarrow H\hat{J}K \approx 73.408°$

$\Rightarrow H\hat{K}J \approx 180° - (73.408° + 53°) \approx 53.592°$

Area $= \frac{1}{2}(18)(15)\sin(53.592°) \approx 108.649\text{ cm}^2$.

Therefore, the area of Triangle 1 is approximately 109 cm^2 (3 s.f.).

8.4 The law of cosines

Two cases remain in our list of different ways to arrange three known parts of a triangle. If three sides of a triangle are known (SSS arrangement), or two sides of a triangle and the angle between them are known (SAS arrangement), then a unique triangle is determined. However, in both of these cases, the law of sines cannot solve the triangle.

Figure 8.22

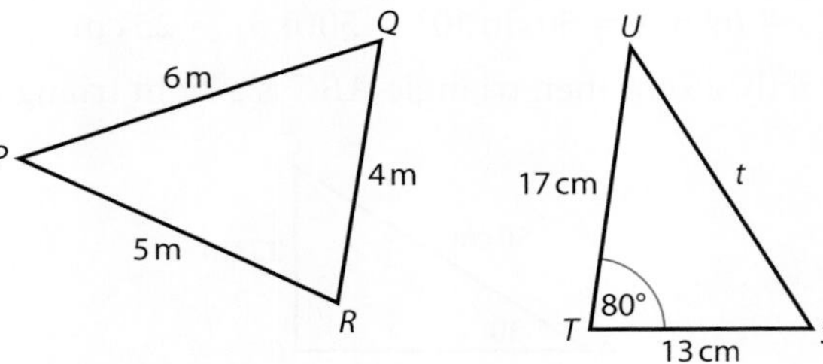

For example, it is not possible to set up an equation using the law of sines to solve triangle *PQR* or triangle *STU* in Figure 8.22.

- Trying to solve $\triangle PQR$: $\frac{\sin P}{4} = \frac{\sin R}{6} \Rightarrow$ two unknowns; cannot solve for angle *P* or angle *R*.
- Trying to solve $\triangle STU$: $\frac{\sin 80°}{t} = \frac{\sin U}{13} \Rightarrow$ two unknowns; cannot solve for angle *U* or side *t*.

The law of cosines (or cosine rule)

We will need the **law of cosines** to solve triangles with these kinds of arrangements of sides and angles. To derive this law, we need to place a general triangle *ABC* in the coordinate plane so that one of the vertices is at the origin and one of the sides is on the positive *x*-axis. Figure 8.23 shows both an acute triangle *ABC* and an obtuse triangle *ABC*. In either case, the coordinates of vertex *C* are $x = b\cos C$ and $y = b\sin C$. Because *c* is the distance from *A* to *B*, then we can use the distance formula to write

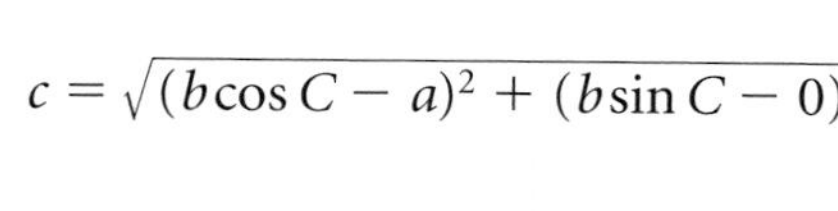

$$c = \sqrt{(b\cos C - a)^2 + (b\sin C - 0)^2}$$ Distance between $(b\cos C, b\sin C)$ and $(a, 0)$.

$$c^2 = (b\cos C - a)^2 + (b\sin C - 0)^2$$ Squaring both sides.

$$c^2 = b^2\cos^2 C - 2ab\cos C + a^2 + b^2\sin^2 C$$ Expand.

$$c^2 = b^2(\cos^2 C + \sin^2 C) - 2ab\cos C + a^2$$ Factor out b^2 from two terms.

$$c^2 = b^2 - 2ab\cos C + a^2$$ Apply trigonometric identity $\cos^2\theta + \sin^2\theta = 1$.

$$c^2 = a^2 + b^2 - 2ab\cos C$$ Rearrange terms.

This equation gives one form of the law of cosines. Two other forms are obtained in a similar manner by having either vertex A or vertex B, rather than C, located at the origin.

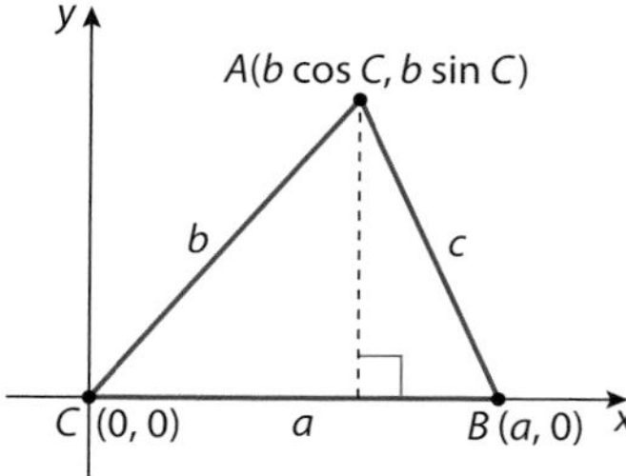

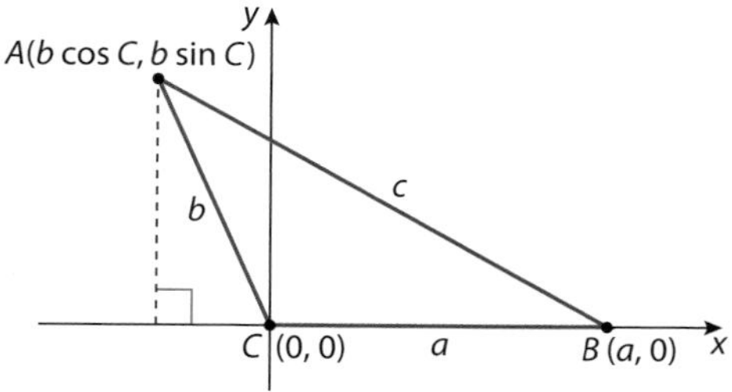

Figure 8.23 Deriving the cosine rule.

The law of cosines

In any triangle ABC with corresponding sides a, b and c:

$$c^2 = a^2 + b^2 - 2ab\cos C$$
$$b^2 = a^2 + c^2 - 2ac\cos B$$
$$a^2 = b^2 + c^2 - 2bc\cos A$$

It is helpful to understand the underlying pattern of the law of cosines when applying it to solve for parts of triangles. The pattern relies on choosing one particular angle of the triangle and then identifying the two sides that are adjacent to the angle and the one side that is opposite to it. The law of cosines can be used to solve for the chosen angle or the side opposite the chosen angle.

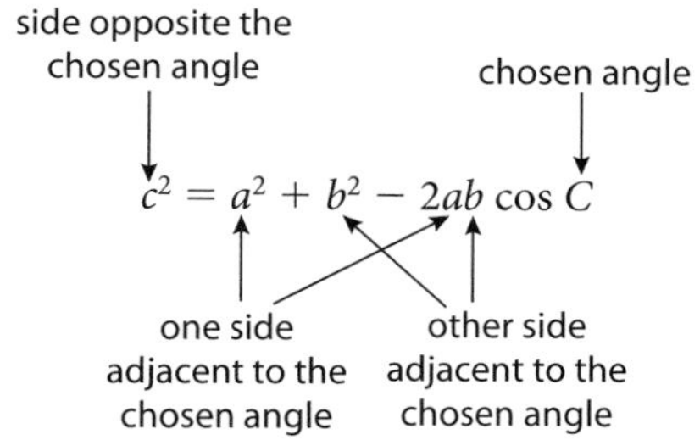

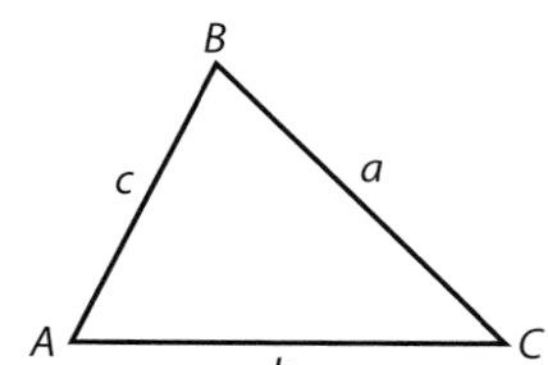

Solving triangles given two sides and the included angle (SAS)

If we know two sides and the included angle, we can use the law of cosines to solve for the side opposite the given angle. Then it is best to solve for one of the two remaining angles using the law of sines.

Example 17

Find all of the unknown angles and sides of triangle *STU*, one of the triangles shown earlier in Figure 8.22. Approximate all measurements to 1 decimal place.

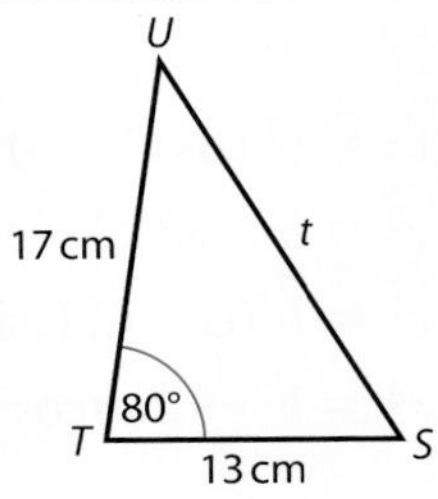

Solution

We first solve for side *t*, opposite the known angle $S\hat{T}U$, using the law of cosines:

$$t^2 = 13^2 + 17^2 - 2(13)(17)\cos 80°$$

$$t = \sqrt{13^2 + 17^2 - 2(13)(17)\cos 80°}$$

$$t \approx 19.5256$$

Now use the law of sines to solve for one of the other angles, say $S\hat{T}U$:

$$\frac{\sin T\hat{S}U}{17} = \frac{\sin 80°}{19.5256}$$

$$\sin T\hat{S}U = \frac{17\sin 80°}{19.5256}$$

$$T\hat{S}U = \sin^{-1}\left(\frac{17\sin 80°}{19.5256}\right)$$

$$T\hat{S}U \approx 59.0288°$$

Then, $S\hat{U}T \approx 180° - (80° + 59.0288°) \approx 40.9712°$.

Therefore, the other parts of the triangle are $t \approx 19.5$ cm, $T\hat{S}U \approx 59.0°$ and $S\hat{U}T \approx 41.0°$.

You may have noticed that the formula for the law of cosines looks similar to the formula for Pythagoras' theorem. In fact, Pythagoras' theorem can be considered a special case of the law of cosines. When the chosen angle in the law of cosines is 90°, and since $\cos 90° = 0$, the law of cosines becomes Pythagoras' theorem.

If angle $C = 90°$, then
$c^2 = a^2 + b^2 - 2ab\cos C$
$\Rightarrow c^2 = a^2 + b^2 - 2ab\cos 90°$
$\Rightarrow c^2 = a^2 + b^2 - 2ab(0)$
$\Rightarrow c^2 = a^2 + b^2$ or $a^2 + b^2 = c^2$

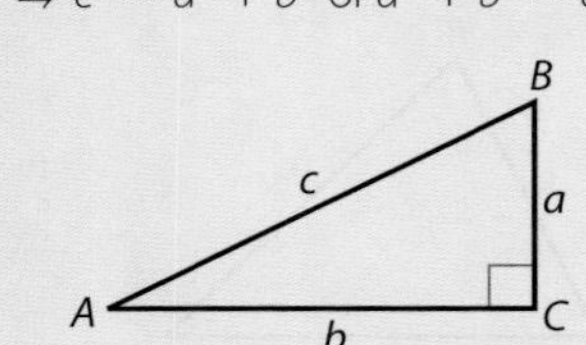

• **Hint:** As previously mentioned, remember to store intermediate answers on the GDC for greater accuracy. By not rounding until the final answer, you reduce the amount of round-off error. The GDC screen images below show the calculations in the solution for Example 17 above.

```
√(13²+17-2(13)(
17)cos(80))
        19.52556031
Ans→T
        19.52556031
```

```
Ans→T
        19.52556031
sin⁻¹(17sin(80)/T
)
        59.02884098
Ans→S
        59.02884098
```

```
sin⁻¹(17sin(80)/T
)
        59.02884098
Ans→S
        59.02884098
180-(80+S)
        40.97115902
```

Example 18

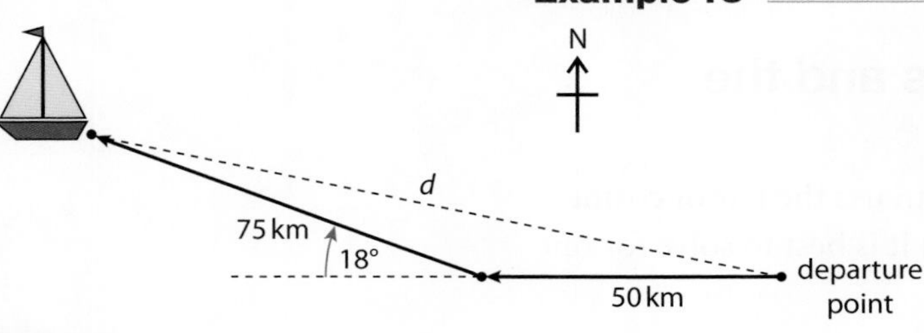

A ship travels 50 km due west, then changes its course 18° northward, as shown in the diagram. After travelling 75 km in that direction, how far is the ship from its point of departure? Give your answer to the nearest tenth of a kilometre.

Solution

Let d be the distance from the departure point to the position of the ship. A large obtuse triangle is formed by the three distances of 50 km, 75 km and d km. The angle opposite side d is $180° - 18° = 162°$. Using the law of cosines, we can write the following equation to solve for d:

$$d^2 = 50^2 + 75^2 - 2(50)(75)\cos 162°$$

$$d = \sqrt{50^2 + 75^2 - 2(50)(75)\cos 162°} \approx 123.523$$

Therefore, the ship is approximately 123.5 km from its departure point.

Solving triangles given three sides (SSS)

Given three line segments such that the sum of the lengths of any two is greater than the length of the third, then they will form a unique triangle. Therefore, if we know three sides of a triangle we can solve for the three angle measures. To use the law of cosines to solve for an unknown angle, it is best to first rearrange the formula so that the chosen angle is the subject of the formula.

Solve for angle C in:

$$c^2 = a^2 + b^2 - 2ab\cos C \Rightarrow 2ab\cos C = a^2 + b^2 - c^2 \Rightarrow \cos C = \frac{a^2 + b^2 - c^2}{2ab}$$

Then, $C = \cos^{-1}\left(\frac{a^2 + b^2 - c^2}{2ab}\right)$.

Example 19

Find all of the unknown angles of triangle PQR, the second triangle shown earlier in Figure 8.22. Approximate all measurements to 1 decimal place.

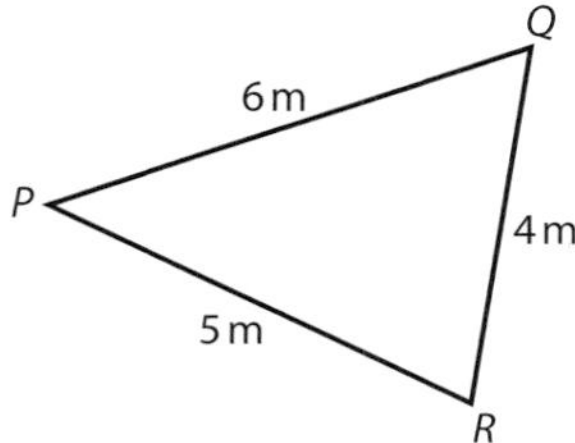

Solution

Note that the smallest angle will be opposite the shortest side. Let's first solve for the smallest angle – thus, writing the law of cosines with chosen angle P:

$$P = \cos^{-1}\left(\frac{5^2 + 6^2 - 4^2}{2(5)(6)}\right) \approx 41.4096°$$

Now that we know the measure of angle P, we have two sides and a non-included angle (SSA), and the law of sines can be used to find the other non-included angle. Consider the sides $QR = 4$, $RP = 5$ and the angle $P \approx 41.4096°$. Substituting into the law of sines, we can solve for angle Q that is opposite RP.

$$\frac{\sin Q}{5} = \frac{\sin 41.4096°}{4}$$

$$\sin Q = \frac{5 \sin 41.4096°}{4}$$

$$Q = \sin^{-1}\left(\frac{5 \sin 41.4096°}{4}\right) \approx 55.7711°$$

Then, $R \approx 180° - (41.4096° + 55.7711°) \approx 82.8192°$.

Therefore, the three angles of triangle PQR are $P \approx 41.4°$, $Q \approx 55.8°$ and $R \approx 82.8°$.

Example 20

A ladder that is 8 m long is leaning against a non-vertical wall that slopes away from the ladder. The foot of the ladder is 3.5 m from the base of the wall, and the distance from the top of the ladder down the wall to the ground is 5.75 m. To the nearest tenth of a degree, what is the acute angle at which the wall is inclined to the horizontal?

Solution

Let's start by drawing a diagram that accurately represents the given information. θ marks the acute angle of inclination of the wall. Its supplement is $F\hat{B}T$. From the law of cosines:

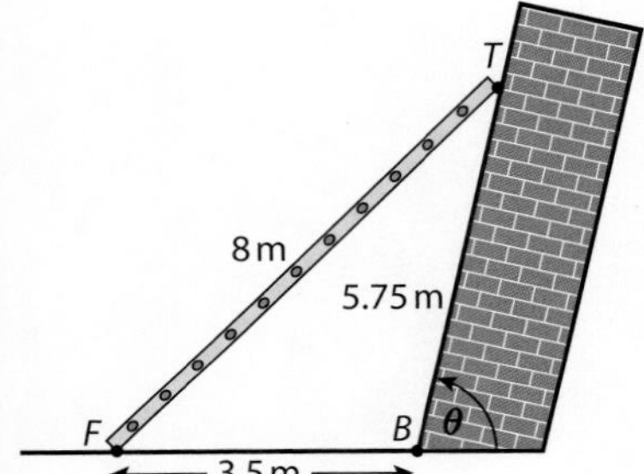

$$\cos F\hat{B}T = \frac{3.5^2 + 5.75^2 - 8^2}{2(3.5)(5.75)}$$

$$F\hat{B}T = \cos^{-1}\left(\frac{3.5^2 + 5.75^2 - 8^2}{2(3.5)(5.75)}\right) \approx 117.664°$$

$$\theta \approx 180° - 117.664° \approx 62.336°$$

Therefore, the angle of inclination of the wall is approximately 62.3°.

Exercise 8.3 and 8.4

In questions1–6, state the number of distinct triangles (none, one, two or infinite) that can be constructed with the given measurements. If the answer is one or two triangles, provide a sketch of each triangle.

1 $A\hat{C}B = 30°$, $A\hat{B}C = 50°$ and $B\hat{A}C = 100°$
2 $A\hat{C}B = 30°$, $AC = 12$ cm and $BC = 17$ cm
3 $A\hat{C}B = 30°$, $AB = 7$ cm and $AC = 14$ cm
4 $A\hat{C}B = 47°$, $BC = 20$ cm and $A\hat{B}C = 55°$
5 $B\hat{A}C = 25°$, $AB = 12$ cm and $BC = 7$ cm
6 $AB = 23$ cm, $AC = 19$ cm and $BC = 11$ cm

In questions 7–15, solve the triangle. In other words, find the measurements of all unknown sides and angles. If two triangles are possible, solve for both.

7 $B\hat{A}C = 37°$, $A\hat{B}C = 28°$ and $AC = 14$
8 $A\hat{B}C = 68°$, $A\hat{C}B = 47°$ and $AC = 23$
9 $B\hat{A}C = 18°$, $A\hat{C}B = 51°$ and $AC = 4.7$
10 $A\hat{C}B = 112°$, $A\hat{B}C = 25°$ and $BC = 240$
11 $BC = 68$, $A\hat{C}B = 71°$ and $AC = 59$
12 $BC = 16$, $AC = 14$ and $AB = 12$
13 $BC = 42$, $AC = 37$ and $AB = 26$
14 $BC = 34$, $A\hat{B}C = 43°$ and $AC = 28$
15 $AC = 0.55$, $B\hat{A}C = 62°$ and $BC = 0.51$

16 Find the lengths of the diagonals of a parallelogram whose sides measure 14 cm and 18 cm and which has one angle of 37°.

17 Find the measures of the angles of an isosceles triangle whose sides are 10 cm, 8 cm and 8 cm.

18 Given that for triangle DEF, $E\hat{D}F = 43°$, $DF = 24$ and $FE = 18$, find the two possible measures of $D\hat{F}E$.

19 A tractor drove from a point A directly north for 500 m, and then drove north-east (i.e. bearing of 45°) for 300 m, stopping at point B. What is the distance between points A and B?

20 Find the measure of the smallest angle in the triangle shown.

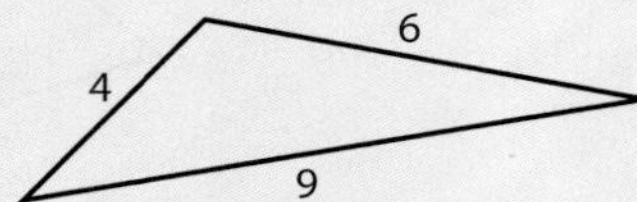

21 Find the area of triangle PQR.

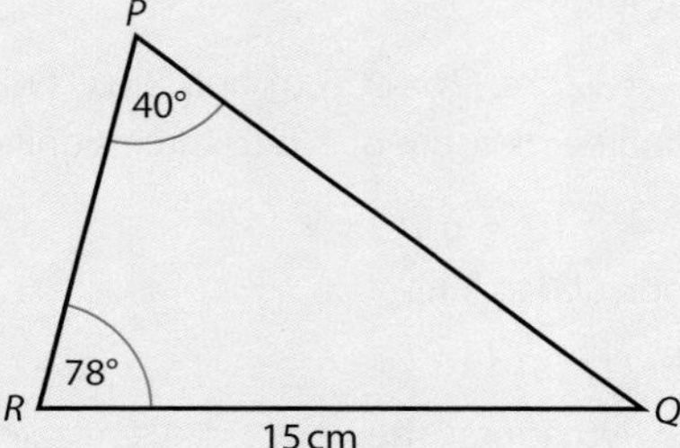

In questions 22 and 23, find a value for the length of BC so that the number of possible triangles is: a) one, b) two and c) none.

22 $B\hat{A}C = 36°$, $AB = 5$

23 $B\hat{A}C = 60°$, $AB = 10$

24 A 50 m vertical pole is to be erected on the side of a sloping hill that makes a 8° angle with the horizontal (see diagram). Find the length of each of the two supporting wires (x and y) that will be anchored 35 m uphill and downhill from the base of the pole.

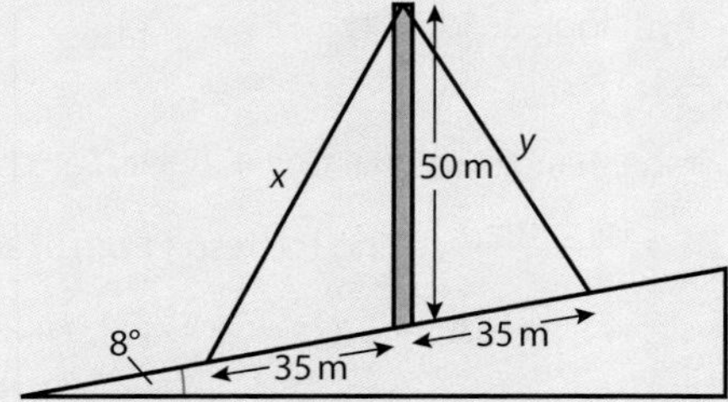

25 The lengths of the sides of a triangle ABC are $x - 2$, x and $x + 2$. The largest angle is 120°.

a) Find the value of x.

b) Show that the area of the triangle is $\frac{15\sqrt{3}}{4}$.

c) Find $\sin A + \sin B + \sin C$ giving your answer in the form $\frac{p\sqrt{q}}{r}$ where $p, q, r \in \mathbb{R}$.

26 Find the area of a triangle that has sides of lengths 6, 7 and 8 cm.

27 Let a, b and c be the sides of a triangle where c is the longest side.

a) If $c^2 > a^2 + b^2$, then what is true about triangle ABC?

b) If $c^2 < a^2 + b^2$, then what is true about triangle ABC?

c) Use the cosine rule to prove each of your conclusions for a) and b).

28 Consider triangle DEF with $E\hat{D}F = 43.6°$, $DE = 19.3$ and $EF = 15.1$. Find DF.

29 In the diagram, $WX = x$ cm, $XY = 3x$ cm, $YZ = 20$ cm, $\sin\theta = \frac{4}{5}$ and $W\hat{X}Y = 120°$.

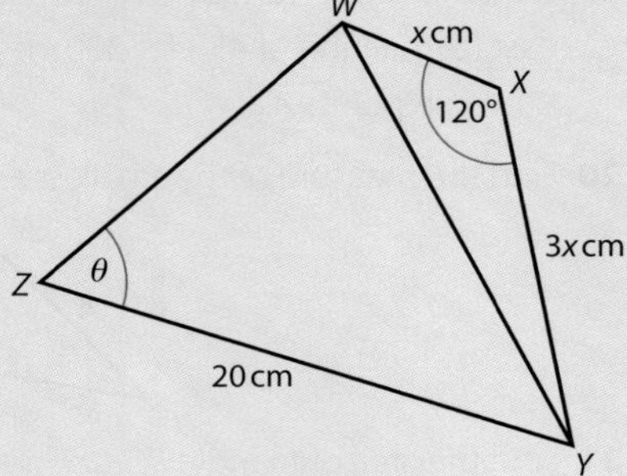

a) If the area of triangle WZY is 112 cm^2, find the length of $[WZ]$.

b) Given that θ is an acute angle, state the value of $\cos\theta$ and hence find the length of $[WY]$.

c) Find the **exact** value of x.

d) Find the degree measure of $X\hat{Y}Z$ to three significant figures.

30 In triangle FGH, $FG = 12$ cm, $FH = 15$ cm, and $\angle G$ is twice the size of $\angle H$. Find the approximate degree measure of $\angle H$ to three significant figures.

31 In triangle PQR, $QR = p$, $PR = q$, $PQ = r$ and $[QS]$ is perpendicular to $[PR]$.

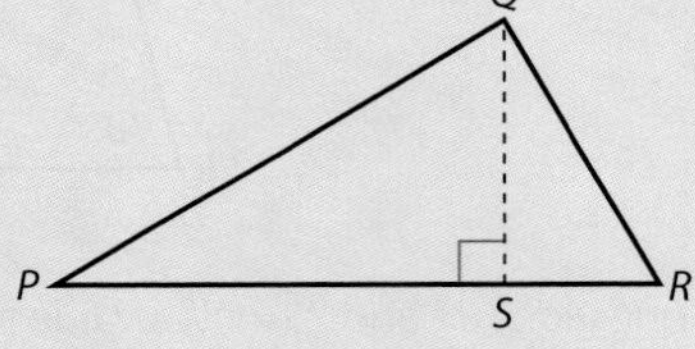

a) Show that $RS = q - r\cos P$.

b) Hence, by using Pythagoras' theorem in the triangle QRS, prove the cosine rule for the triangle PQR.

c) If $P\hat{Q}R = 60°$, use the cosine rule to show that $p = \frac{1}{2}\left(r \pm \sqrt{4q^2 - 3r^2}\right)$.

32 For triangle ABC we can express its area, **A**, as $\mathbf{A} = \frac{1}{2}ab\sin C$. The cosine rule can be used to write the expression $c^2 = a^2 + b^2 - 2ab\cos C$.

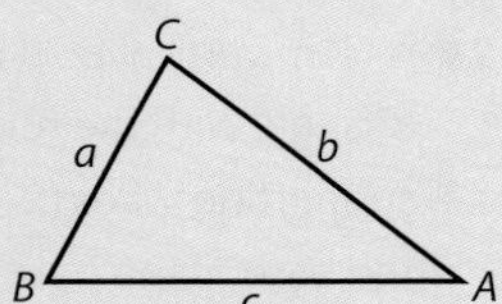

a) Using these two expressions show that $16\mathbf{A}^2 = 4a^2b^2 - (a^2 + b^2 - c^2)$.

• **Hint:** use the Pythagorean identity $\sin^2 C + \cos^2 C = 1$.

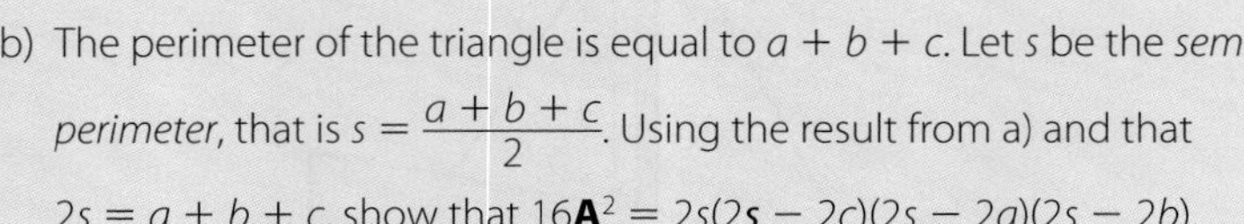

b) The perimeter of the triangle is equal to $a + b + c$. Let s be the *semi-perimeter*, that is $s = \dfrac{a + b + c}{2}$. Using the result from a) and that $2s = a + b + c$, show that $16\mathbf{A}^2 = 2s(2s - 2c)(2s - 2a)(2s - 2b)$.

c) Finally, show that the result in b) gives $\mathbf{A} = \sqrt{s(s-a)(s-b)(s-c)}$. This notable result expresses the area of a triangle in terms of *only* the length of its three sides. Although quite possibly known before his time, the formula is attributed to the ancient Greek mathematician and engineer, Heron of Alexandria (ca. 10–70 AD) and is thus called **Heron's formula**. The first written reference to the formula is Heron's proof of it in his book *Metrica*, written in approximately 60 AD.

8.5 Applications

There are some additional applications of triangle trigonometry – both right triangles and non-right triangles – that we should take some time to examine.

Equations of lines and angles between two lines

Recall from Section 1.6, the slope m, or gradient, of a non-vertical line is defined as $m = \frac{y_2 - y_1}{x_2 - x_1} = \frac{\text{vertical change}}{\text{horizontal change}}$.

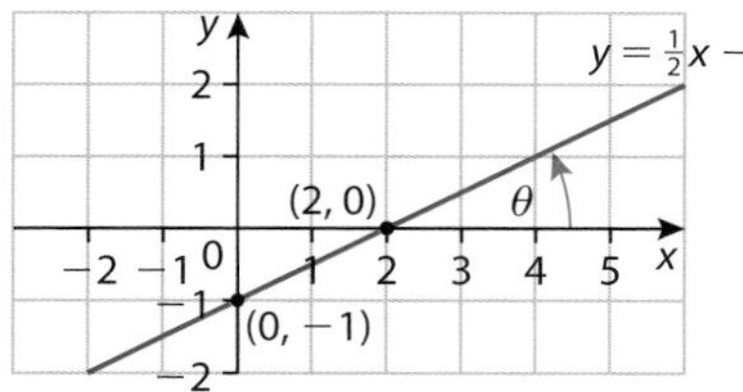

Figure 8.24

The equation of the line shown in Figure 8.24 has a slope $m = \frac{1}{2}$ and a y-intercept of $(0, -1)$. So, the equation of the line is $y = \frac{1}{2}x - 1$. We can find the measure of the acute angle θ between the line and the x-axis by using the tangent function (Figure 8.25).

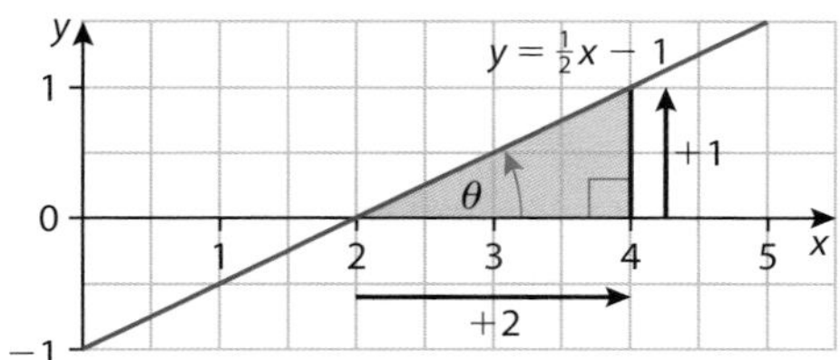

Figure 8.25

$\theta = \tan^{-1}(m) = \tan^{-1}\left(\frac{1}{2}\right) \approx 26.6°$.

Clearly, the slope, m, of this line is equal to $\tan\theta$. If we know the angle between the line and the x-axis, and the y-intercept $(0, c)$, we can write the equation of the line in slope-intercept form ($y = mx + c$) as $y = (\tan\theta)x + c$.

Before we can generalize for any non-horizontal line, let's look at a line with a negative slope.

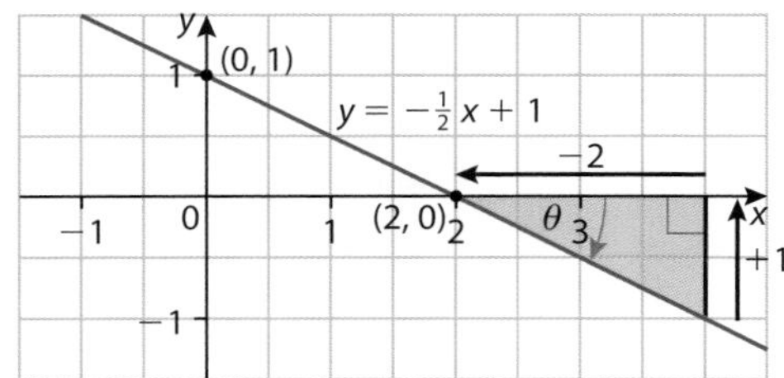

Figure 8.26

The slope of the line is $-\frac{1}{2}$. In order for $\tan\theta$ to be equal to the slope of the line, the angle θ must be the angle that the line makes with the x-axis in the positive direction, as shown in Figure 8.26. In this example, $\theta = \tan^{-1}(m) = \tan^{-1}\left(-\frac{1}{2}\right) \approx -26.6°$.

Remember, an angle with a negative measure indicates a clockwise rotation from the initial side to the terminal side of the angle.

Equations of lines intersecting the *x*-axis

If a line has a y-intercept of $(0, c)$ and makes an angle of θ with the positive direction of the x-axis, such that $-90° < \theta < 90°$, then the slope (gradient) of the line is $m = \tan\theta$ and the equation of the line is $y = (\tan\theta)x + c$. Note: The angle this line makes with any horizontal line will be θ.

Let's use triangle trigonometry to find the angle between any two intersecting lines – not just for a line intersecting the x-axis. Realize that any pair of intersecting lines that are not perpendicular will have both an acute angle and an obtuse angle between them. When asked for an angle between two lines, the convention is to give the acute angle.

Example 21

Find the acute angle between the lines $y = 3x$ and $y = -x$.

Solution

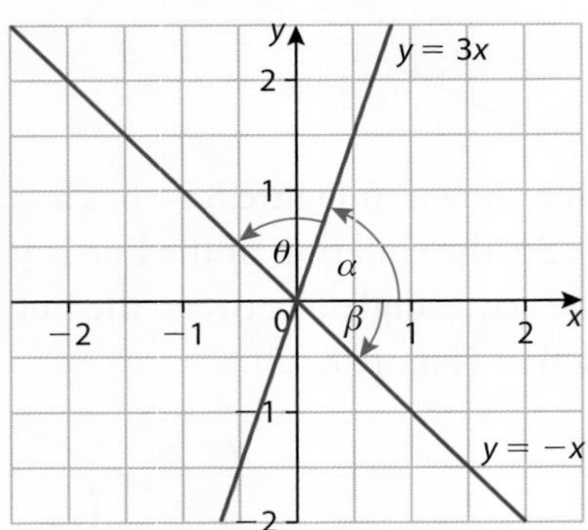

The angle between the line $y = 3x$ and the positive x-axis is α, and the angle between the line $y = -x$ and the positive x-axis is β.

$$\alpha = \tan^{-1}(3) \approx 71.565°$$
$$\beta = \tan^{-1}(-1) = -45°$$

The obtuse angle between the two lines is
$\alpha - \beta \approx 71.565° - (-45°) \approx 116.565°$.

Therefore, the acute angle θ between the two lines is
$\theta = 180° - 116.565° \approx 63.4°$.

Example 22

Find the acute angle between the lines $y = 5x - 2$ and $y = \frac{1}{3}x - 1$.

Solution

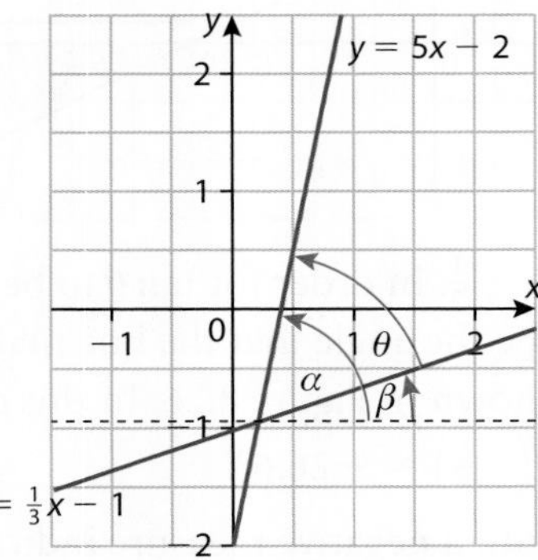

A horizontal line is drawn through the point of intersection.

The angle between $y = 5x - 2$ and this horizontal line is α, and the angle between $y = \frac{1}{3}x - 1$ and this horizontal line is β.

$$\alpha = \tan^{-1}(5) \approx 78.690° \quad \text{and} \quad \beta = \tan^{-1}\left(\tfrac{1}{3}\right) = 18.435°$$

The acute angle θ between the two lines is
$\theta = \alpha - \beta \approx 78.690° - 18.435° \approx 60.3°$.

We can generalize the procedure for finding the angle between two lines as follows.

Angle between two lines

Given two non-vertical lines with equations of $y_1 = m_1x + c_1$ and $y_2 = m_2x + c_2$, the angle between the two lines is $|\tan^{-1}(m_1) - \tan^{-1}(m_2)|$. Note: This angle may be acute or obtuse.

Example 23

a) Find the exact equation of line L_1 that passes through the origin and makes an angle of $-60°$ with the positive direction of the x-axis (or $120°$).

b) The equation of line L_2 is $7x + y + 1 = 0$. Find the acute angle between the lines L_1 and L_2.

Solution

a) The equation of the line is given by $y = (\tan\theta)x$

$$\Rightarrow y = [\tan(-60°)]x = \left[\frac{\sin(-60°)}{\cos(-60°)}\right]x = \left[\frac{-\frac{\sqrt{3}}{2}}{\frac{1}{2}}\right]x = (-\sqrt{3})x$$

Therefore, the equation of L_1 is $y = -\sqrt{3}x$ or $y = -x\sqrt{3}$.

Note: $\tan(-60°) = \tan 120° = (-\sqrt{3})$.

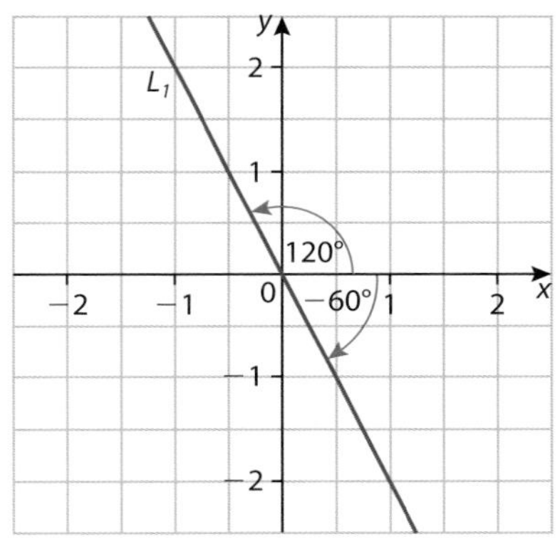

b) L_2: $7x + y + 1 = 0 \Rightarrow y = -7x - 1$

θ is the acute angle between the lines L_1 and L_2.

$\theta = |\tan^{-1}(m_1) - \tan^{-1}(m_2)| = |\tan^{-1}(-\sqrt{3}) - \tan^{-1}(-7)|$

$\Rightarrow \theta \approx |-60° - (-81.870°)| \approx |-21.87°|$

Therefore, the acute angle between the lines is approximately $21.9°$ (3 s.f.).

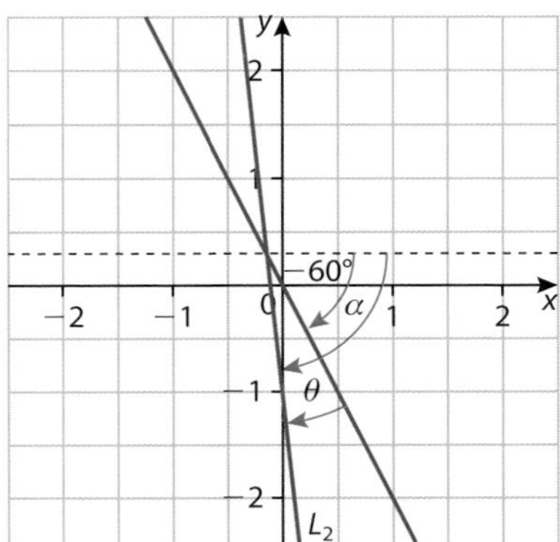

Further applications involving the solution of triangles

Many problems that involve distances and angles are represented by diagrams with multiple triangles – right and otherwise. These diagrams can be confusing and difficult to interpret correctly. In these situations, it is important to carry out a careful analysis of the given information and diagram – this will usually lead to drawing additional diagrams. Often we can extract a triangle, or triangles, for which we have enough information to allow us to solve the triangle(s).

Example 24

Two boats, J and K, are 500 m apart. A lighthouse is on top of a 470 m cliff. The base, B, of the cliff is in line horizontally with $[JK]$. From the top, T, of the lighthouse, the angles of depression of J and K are, respectively, 25° and 40°. Find, correct to the nearest metre, the height, h, of the lighthouse from its base on the clifftop ground to the top T.

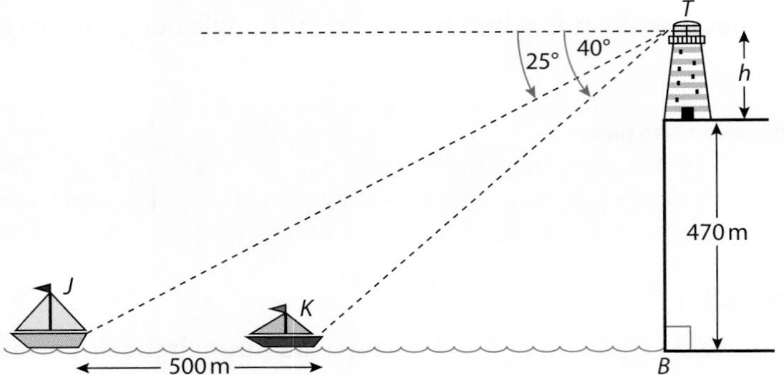

Solution

First, extract obtuse triangle JKT and apply the law of sines to solve for the side KT, which is also the hypotenuse of the right triangle KBT.

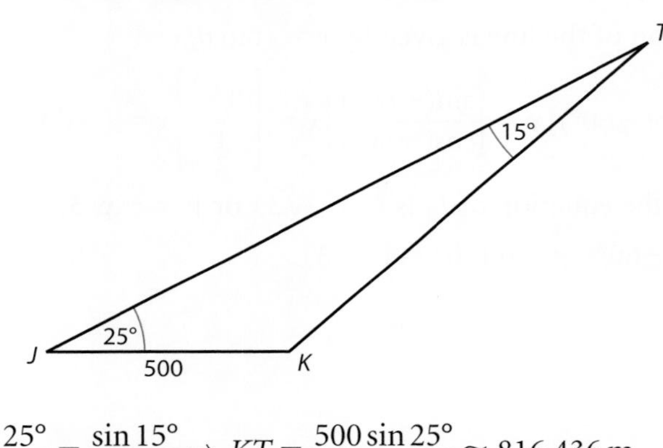

$$\frac{\sin 25°}{KT} = \frac{\sin 15°}{500} \Rightarrow KT = \frac{500 \sin 25°}{\sin 15°} \approx 816.436 \text{ m}$$

We can now use the right triangle KBT to find the side BT – which is equal to the height of the cliff plus the height of the lighthouse.

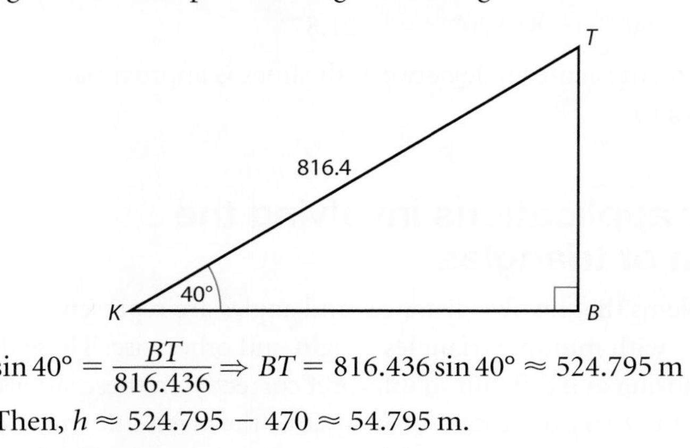

$$\sin 40° = \frac{BT}{816.436} \Rightarrow BT = 816.436 \sin 40° \approx 524.795 \text{ m}$$

Then, $h \approx 524.795 - 470 \approx 54.795$ m.

Therefore, the height of the lighthouse is 54.8 m.

Example 25

The diagram shows a point P that is 10 km due south of a point D. A straight road PQ is such that the (compass) bearing of Q from P is 045°. A and B are two points on this road which are both 8 km from D. Find the bearing of B from D, approximated to 3 s.f.

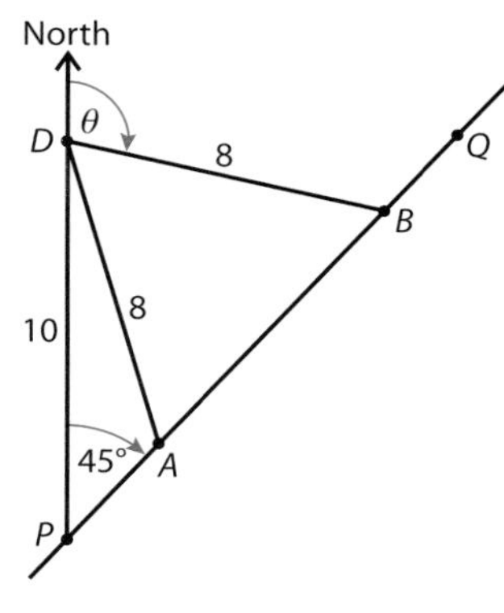

Solution

The angle θ in the diagram is the bearing of B from D. A strategy that will lead to finding θ is:

(1) Extract triangle PDB and use the law of sines to solve for $D\hat{B}P$.

(2) Triangle ADB is isosceles (two sides equal), so $D\hat{A}B = D\hat{B}P$; and since the sum of angles in triangle ADB is 180°, we can solve for $A\hat{D}B$.

(3) We can solve for $D\hat{A}P$ because it is supplementary to $D\hat{A}B$, and then we can find the third angle in triangle APD.

(4) Since $\theta + A\hat{D}B + A\hat{D}P = 180°$, we can solve for θ.

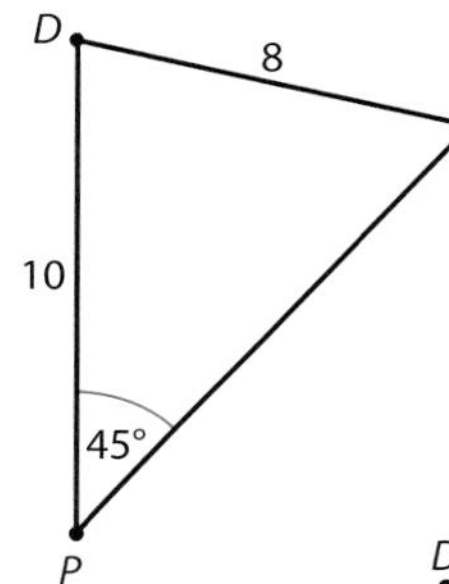

$$\frac{\sin D\hat{B}P}{10} = \frac{\sin 45°}{8}$$

$$\sin D\hat{B}P = \frac{10 \sin 45°}{8}$$

$$D\hat{B}P = \sin^{-1}\left(\frac{10 \sin 45°}{8}\right) \approx 62.11°$$

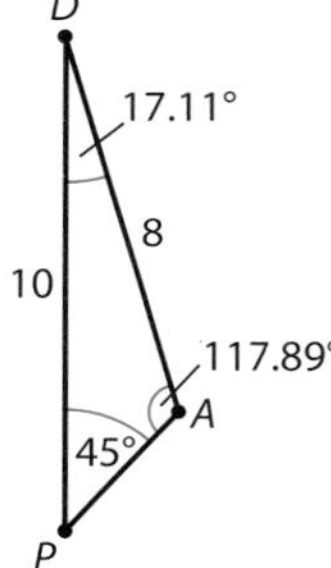

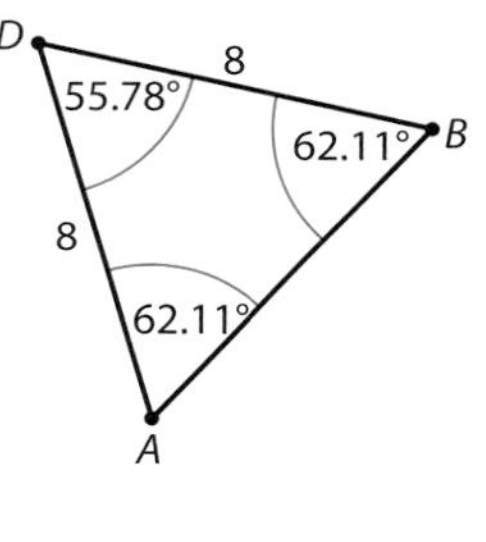

$$D\hat{A}B = D\hat{B}P \approx 62.11°$$

$$A\hat{D}B \approx 180° - 2(62.11°) \approx 55.78°$$

$$P\hat{A}D \approx 180° - 62.11° \approx 117.89°$$

$$A\hat{D}P \approx 180° - (45° + 117.89°) \approx 17.11°$$

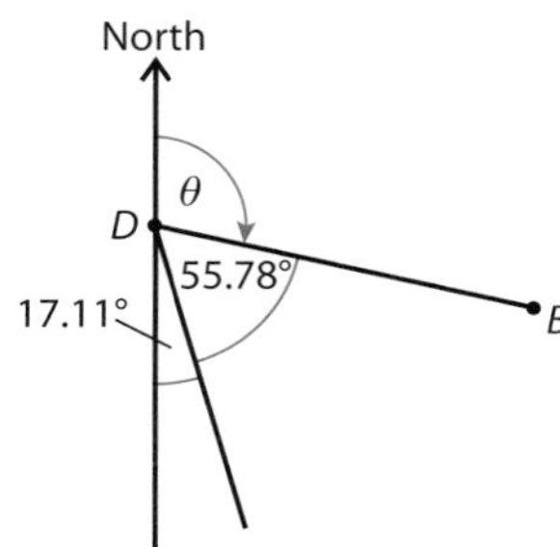

Compass bearings are measured **clockwise** from north.

$$\theta \approx 180° - (17.11° + 55.78°) \approx 107.11°$$

Therefore, the bearing of B from D is approximately 107° to an accuracy of 3 s.f.

Three-dimensional trigonometry problems

Of course, not all applications of triangle trigonometry are restricted to just two dimensions. In many problems, it is necessary to calculate lengths and angles in three-dimensional structures. As in the preceding section, it is very important to carefully analyze the three-dimensional diagram and to extract any relevant triangles in order to solve for the necessary angle or length.

Example 26

The diagram shows a vertical pole GH that is supported by two wires fixed to the horizontal ground at C and D. The following measurements are indicated in the diagram: $CD = 50$ m, $G\hat{D}H = 32°$, $H\hat{D}C = 26°$ and $H\hat{C}D = 80°$.

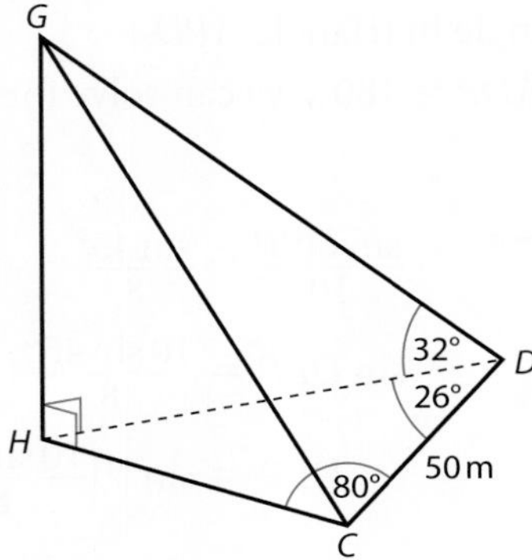

Find a) the distance between H and D, and b) the height of the pole GH.

Solution

a) In triangle HDC: $D\hat{H}C = 180° - (80° + 26°) = 74°$.

Now apply the law of sines:

$$\frac{\sin 80°}{HD} = \frac{\sin 74°}{50} \Rightarrow HD = \frac{50 \sin 80°}{\sin 74°} \approx 51.225\text{ m}$$

Therefore, the distance from H to D is 51.2 m accurate to 3 s.f.

b) Using the right triangle GHD:

$$\tan 32° = \frac{GH}{51.225} \Rightarrow GH = 51.225 \tan 32° \approx 32.009\text{ m}$$

Therefore, the height of the pole is 32.0 m accurate to 3 s.f.

Example 27

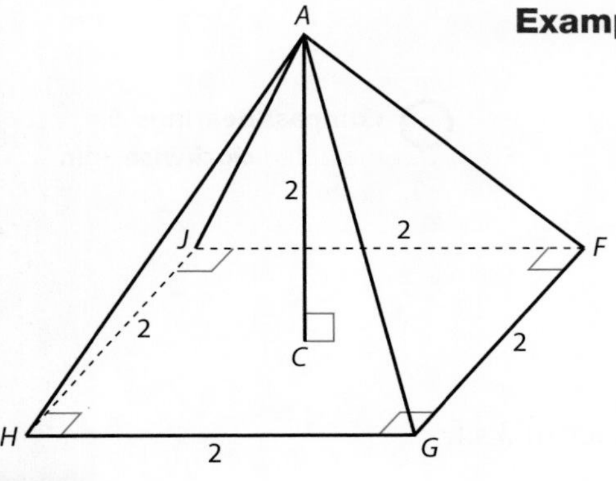

The figure shown is a pyramid with a square base. It is a *right* pyramid, so the line segment (i.e. the height) drawn from the top vertex A perpendicular to the base will intersect the square base at its centre C. If each side of the square base has a length of 2 cm and the height of the pyramid is also 2 cm, find:

a) the measure of $A\hat{G}F$

b) the total surface area of the pyramid.

Solution

a) Label the midpoint of $[GF]$ as point M and draw two line segments, $[CM]$ and $[AM]$. Since C is the centre of the square base then $CM = 1$ cm. Extract right triangle ACM and use Pythagoras' theorem to find the length of $[AM]$.

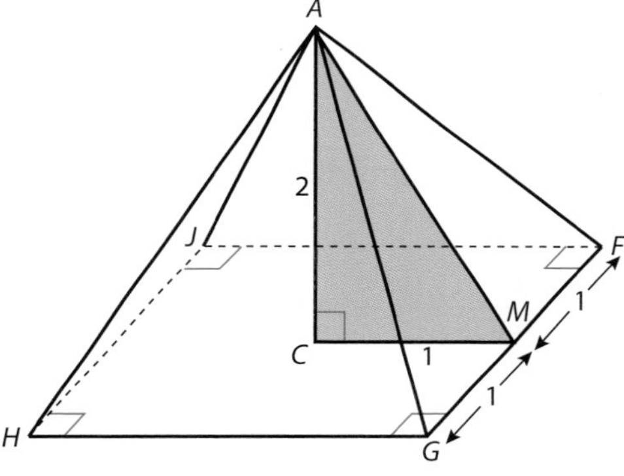

$AM = \sqrt{1^2 + 2^2} = \sqrt{5}$ $\quad$ $[AM]$ is perpendicular to $[GF]$.

Extract right triangle AMG and use the tangent ratio to find $A\hat{G}M$ (same as $A\hat{G}F$):

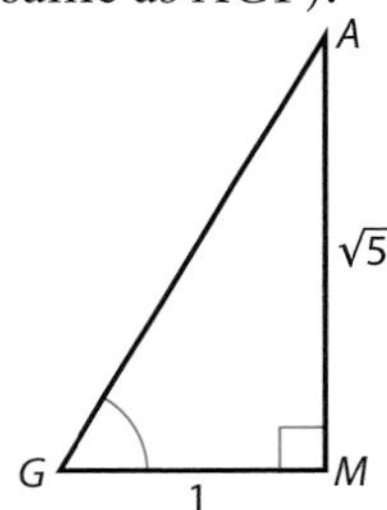

$$\tan(A\hat{G}M) = \frac{\sqrt{5}}{1}$$

$$A\hat{G}M = \tan^{-1}(\sqrt{5}) \approx 65.905°$$

Therefore, $A\hat{G}M = A\hat{G}F \approx 65.9°$.

b) The total surface area comprises the square base plus four identical lateral faces that are all isosceles triangles. Triangle AGM is one-half the area of one of these triangular faces.

$$\text{Area of triangle } AGM = \frac{1}{2}(1)(\sqrt{5}) = \frac{\sqrt{5}}{2}$$

$$\Rightarrow \text{area of triangle } AGF = 2\left(\frac{\sqrt{5}}{2}\right) = \sqrt{5}$$

$$\text{Surface area} = \text{area of square base} + \text{area of four lateral faces}$$
$$= 2^2 + 4\sqrt{5} = 4 + 4\sqrt{5} \approx 12.94 \text{ cm}^2$$

Example 28

For the rectangular box shown, find a) the measure of $A\hat{B}C$, and b) the area of triangle ABC.

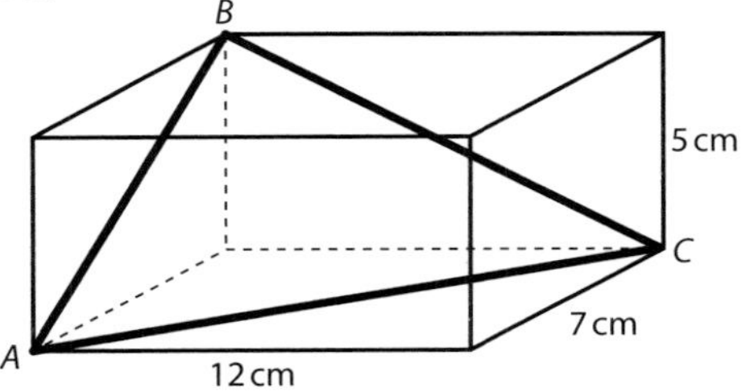

Solution

a) Each of the three sides of triangle ABC is the hypotenuse of a right triangle. Using Pythagoras' theorem:

$$AC = \sqrt{7^2 + 12^2} = \sqrt{49 + 144} = \sqrt{193} = 13.892$$

$$AB = \sqrt{5^2 + 7^2} = \sqrt{25 + 49} = \sqrt{74} \approx 8.602$$

$$BC = \sqrt{5^2 + 12^2} = \sqrt{25 + 144} = \sqrt{169} = 13$$

Apply the law of cosines to find $A\hat{B}C$, using exact lengths of the sides of the triangle.

$$\cos A\hat{B}C = \frac{(\sqrt{74})^2 + 13^2 - (\sqrt{193})^2}{2(\sqrt{74})(13)} \Rightarrow A\hat{B}C = \cos^{-1}\left[\frac{74 + 169 - 193}{2(\sqrt{74})(13)}\right] \approx 77.082°$$

Therefore, the measure of $A\hat{B}C$ is approximately 77.1° to 3 s.f.

b) Area of triangle $= \frac{1}{2}(AB)(BC)\sin A\hat{B}C = \frac{1}{2}(\sqrt{74})(13)\sin(77.082°)$
$\approx 54.499\,96\ \text{cm}^2$

Therefore, the area of triangle ABC is approximately 54.5 cm^2.

Exercise 8.5

In questions 1–4, determine:

a) the slope (gradient) of the line (approximate to 3 s.f. if not exact)

b) the equation of the line.

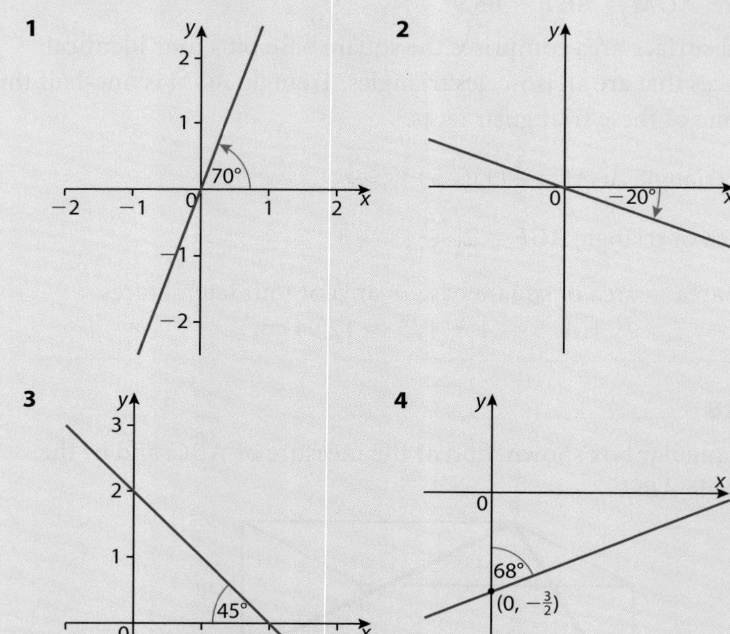

In questions 5–7, find the acute angle that the line through the given pair of points makes with the x-axis.

5 $(1, 4)$ and $(-1, 2)$

6 $(-3, 1)$ and $(6, -5)$

7 $\left(2, \frac{1}{2}\right)$ and $(-4, -10)$

In questions 8 and 9, find the acute angle between the two given lines.

8 $y = -2x$ and $y = x$

9 $y = -3x + 5$ and $y = 2x$

10 a) Find the exact equation of line L_1 that passes through the origin and makes an angle of 30° with the positive direction of the x-axis.

b) The equation of line L_2 is $x + 2y = 6$. Find the acute angle between L_1 and L_2.

11 Calculate AB given $CD = 30$ cm, and the angle measures given in the diagram.

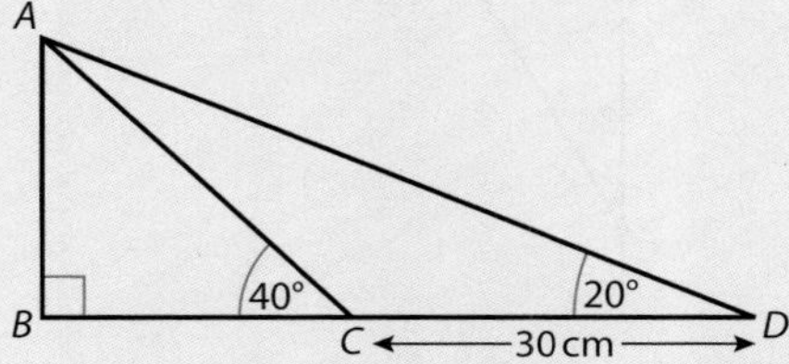

12 The circle with centre O and radius of 8 cm has two chords PR and RS, such that $PR = 5$ cm and $RS = 10$ cm. Find each of the angles $P\hat{R}O$ and $S\hat{R}O$, and then calculate the area of the triangle PRS.

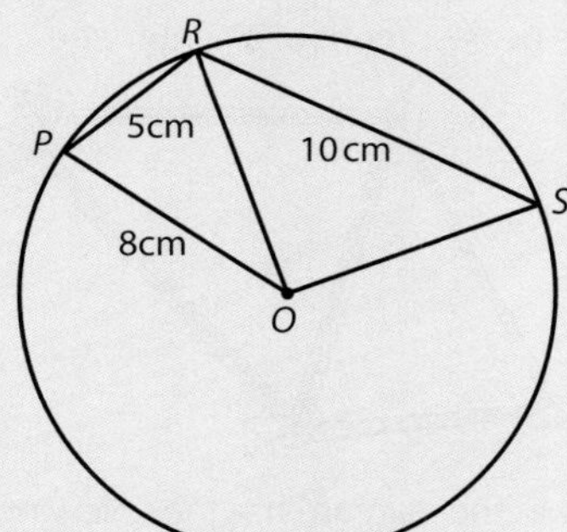

13 A forester was conducting a survey of a tropical jungle that was mostly inaccessible on foot. The points F and G indicate the location of two rare trees. To find the distance between points F and G, a line AB of length 250 m is measured out so that F and G are on opposite sides of AB. The angles between the line segment AB and the line of sight from each endpoint of AB to each tree are measured, and are shown in the diagram. Calculate the distance between F and G.

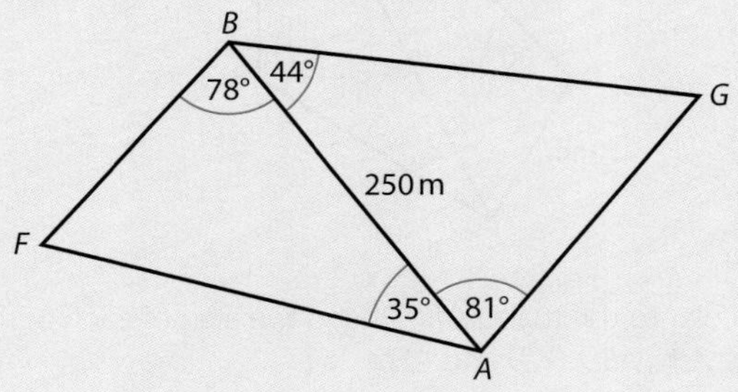

14 Calculate the distance between the tips of the hands of a large clock on a building at 10 o'clock if the minute hand is 3 m long and the hour hand is 2.25 m long.

15 An airplane takes off from point *A*. It flies 850 km on a bearing of 030°. It then changes direction to a bearing of 065° and flies a further 500 km and lands at point *B*.

a) What is the straight line distance from *A to B*?

b) What is the bearing from *A* to *B*?

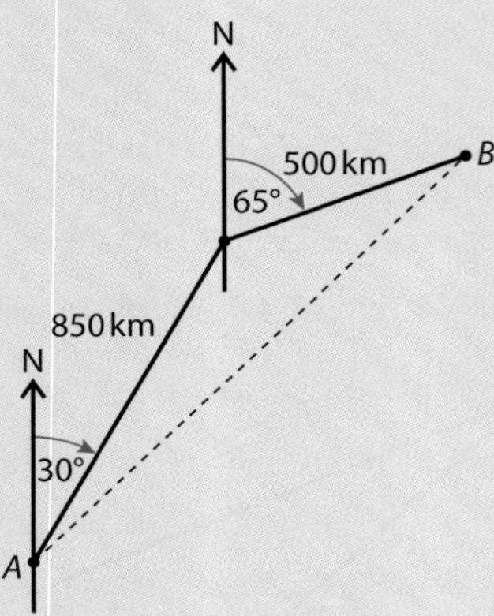

16 The traditional bicycle frame consists of tubes connected together in the shape of a triangle and a quadrilateral (four-sided polygon). In the diagram, *AB*, *BC*, *CD* and *AD* represent the four tubes of the quadrilateral section of the frame. A frame maker has prepared three tubes such that $AD = 53$ cm, $AB = 55$ cm and $BC = 11$ cm. If $D\hat{A}B = 76°$ and $A\hat{B}C = 97°$, what must be the length of tube *CD*? Give your answer to the nearest tenth of a centimetre.

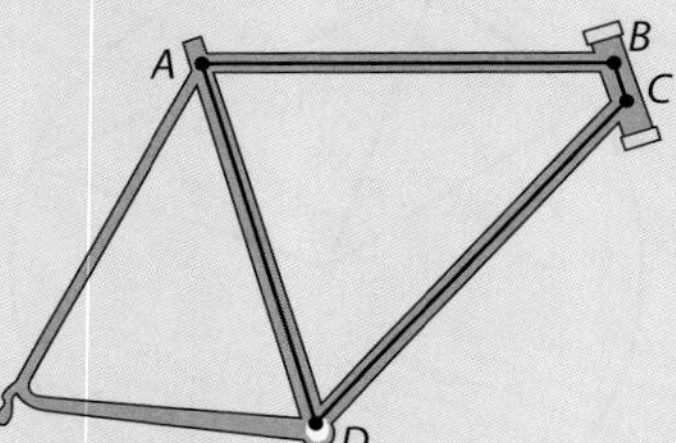

17 The tetrahedron shown in the diagram has the following measurements.

$AB = 12$ cm, $DC = 10$ cm, $A\hat{C}B = 45°$ and $A\hat{D}B = 60°$

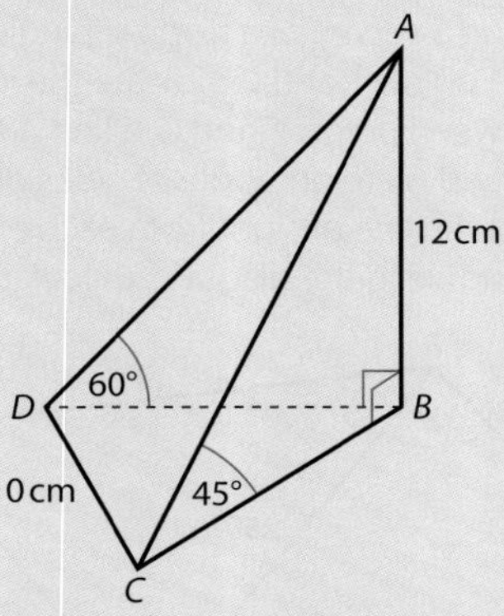

AB is perpendicular to the triangle *BCD*. Find the area of each of the four triangular faces: *ABC*, *ABD*, *BCD* and *ACD*.

18 Find the measure of angle *DEF* in the rectangular box.

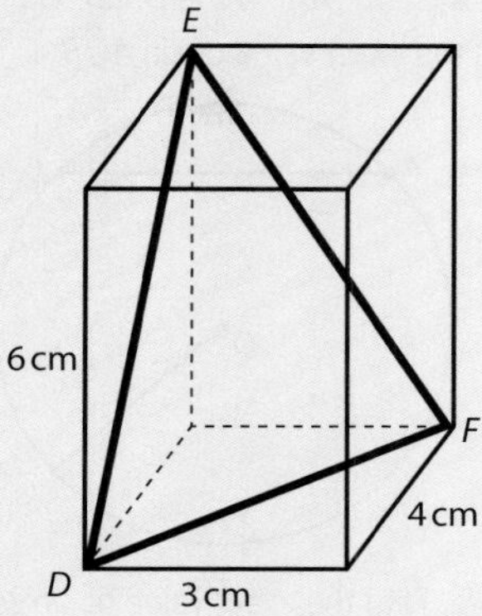

19 At a point *A*, due south of a building, the angle of elevation from the ground to the top of a building is 58°. At a point *B* (on level ground with *A*), 80 m due west of *A*, the angle of elevation to the top of the building is 27°. Find the height of the building.

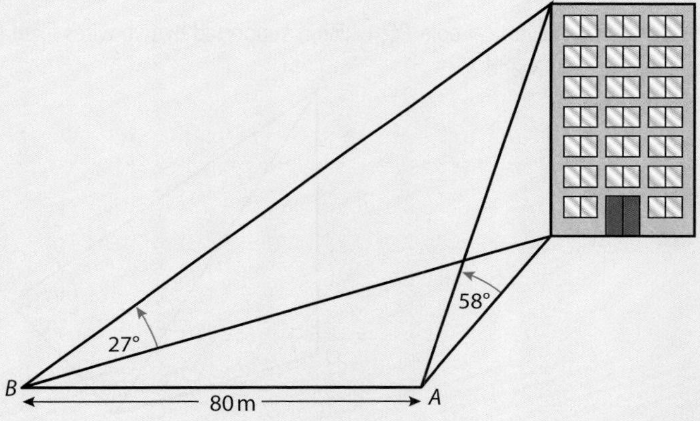

20 A right pyramid has a square base with sides of length 8 cm. The height of the pyramid is 10 cm. Calculate the angle between two adjacent lateral faces. In other words, find the dihedral angle between two planes each containing one of two adjacent lateral faces. There are four lateral faces that are isosceles triangles and one square base. Two adjacent lateral faces are shaded in the diagram.

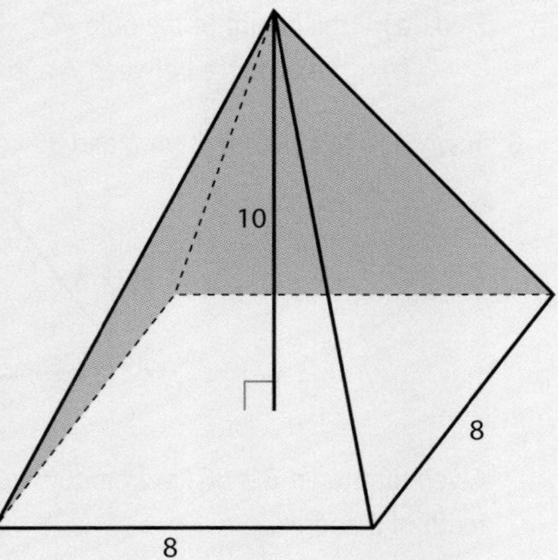

• **Hint:** *AB* lies in the plane P_1 and *AC* lies in a second plane P_2 (see Figure 8.27). If *AB* and *AC* are both perpendicular to the line of intersection of the planes, then $B\hat{A}C$ is the angle between the planes. This angle is often called the dihedral angle of the planes.

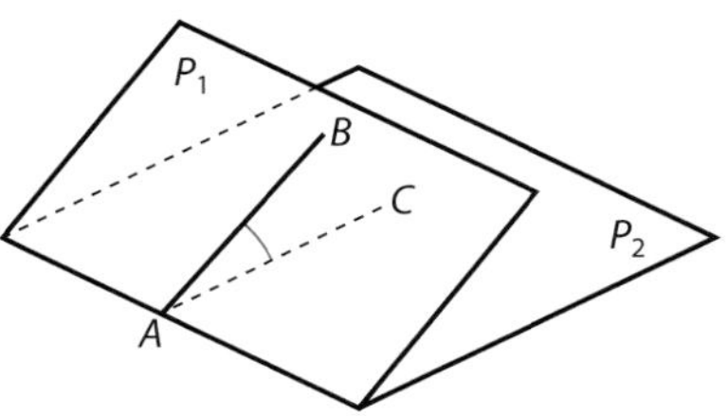

Figure 8.27 Dihedral angle *BAC* of planes P_1 and P_2.

Practice questions

1 The shortest distance from a chord [AB] to the centre O of a circle is 3 units. The radius of the circle is 5 units. Find the exact value of $\sin A\hat{O}B$.

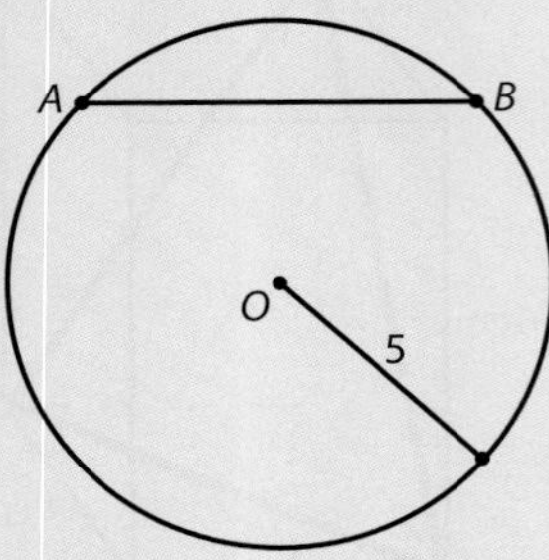

2 In a right triangle, $\tan\theta = \frac{3}{7}$. Find the exact value of $\sin 2\theta$ and $\cos 2\theta$.

3 A triangle has sides of length 4, 5 and 7 units. Find, to the nearest tenth of a degree, the size of the largest angle.

4 If A is an obtuse angle in a triangle and $\sin A = \frac{5}{13}$, calculate the exact value of $\sin 2A$.

5 The diagram shows a vertical pole PQ, which is supported by two wires fixed to the horizontal ground at A and B.

$BQ = 40$ m

$P\hat{B}Q = 36°$

$B\hat{A}Q = 70°$

$A\hat{B}Q = 30°$

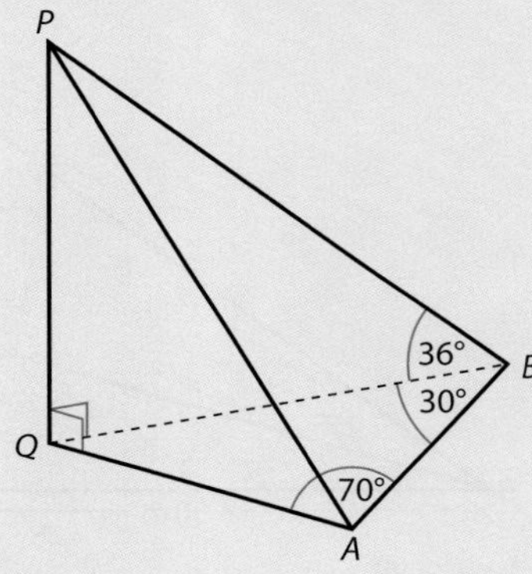

Find: **a)** the height of the pole PQ

b) the distance between A and B.

6 Town A is 48 km from town B and 32 km from town C, as shown in the diagram.

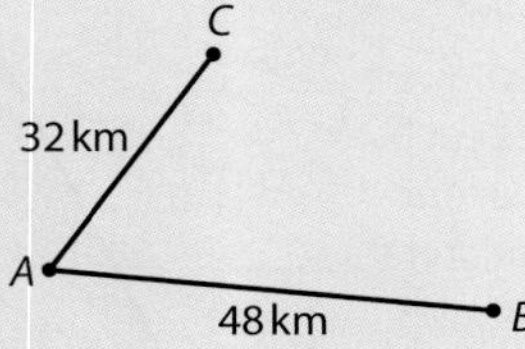

Given that town B is 56 km from town C, find the size of the angle $C\hat{A}B$ to the nearest tenth of a degree.

7 The following diagram shows a triangle with sides 5 cm, 7 cm and 8 cm.

Find: **a)** the size of the smallest angle, in degrees

b) the area of the triangle.

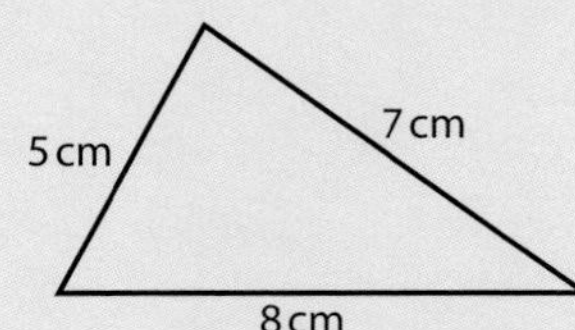

8 The diagrams below show two different triangles, both satisfying the conditions: $AB = 20$ cm, $AC = 17$ cm, $A\hat{B}C = 50°$.

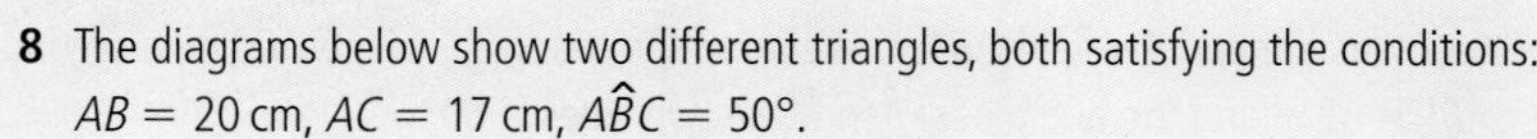

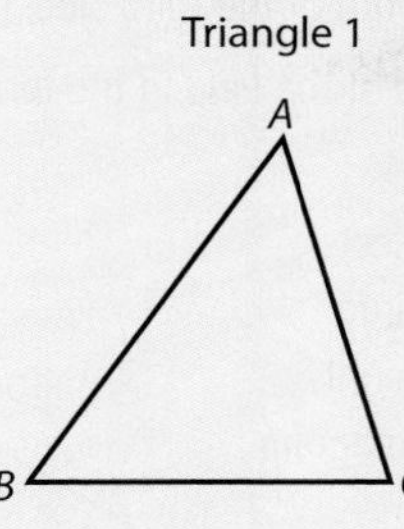

a) Calculate the size of $A\hat{C}B$ in Triangle 2.

b) Calculate the area of Triangle 1.

9 Two boats *A* and *B* start moving from the same point *P*. Boat *A* moves in a straight line at 20 km/h and boat *B* moves in a straight line at 32 km/h. The angle between their paths is 70°. Find the distance between the two boats after 2.5 hours.

10 In triangle *JKL*, $JL = 25$, $KL = 38$ and $K\hat{J}L = 51°$, as shown in the diagram.

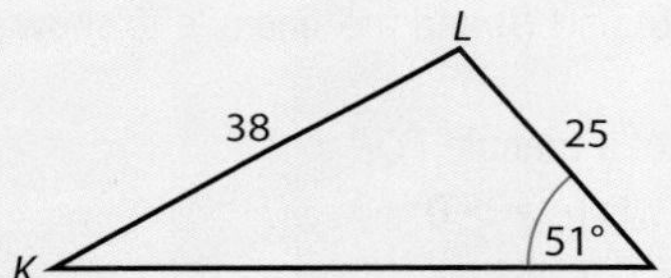

Find $J\hat{K}L$, giving your answer correct to the nearest degree.

11 The following diagram shows a triangle *ABC*, where $BC = 5$ cm, $A\hat{B}C = 60°$ and $A\hat{C}B = 40°$.

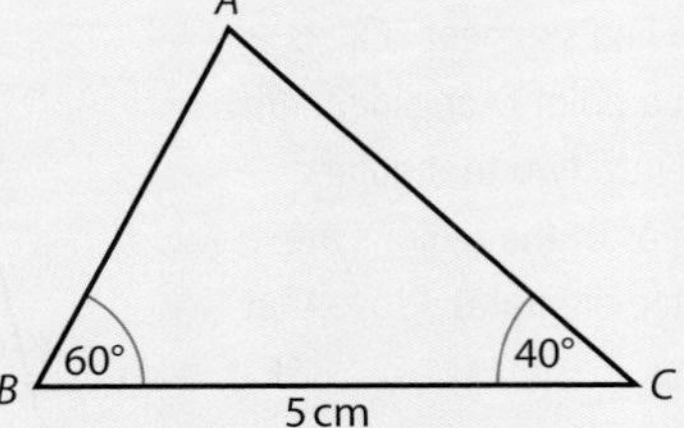

a) Calculate *AB*. **b)** Find the area of the triangle.

12 Find the measure of the acute angle between a pair of diagonals of a cube.

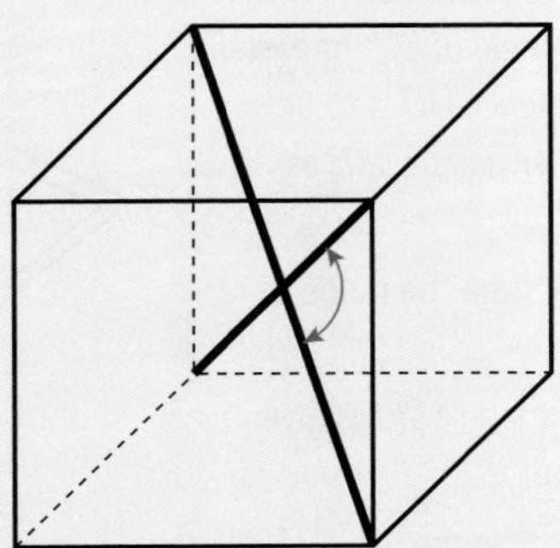

13 A farmer owns a triangular field *ABC*. One side of the triangle, [*AC*], is 104 m, a second side, [*AB*], is 65 m and the angle between these two sides is 60°.

a) Use the cosine rule to calculate the length of the third side, [*BC*], of the field.

b) Given that $\sin 60° = \frac{\sqrt{3}}{2}$, find the area of the field in the form $p\sqrt{3}$, where *p* is an integer.

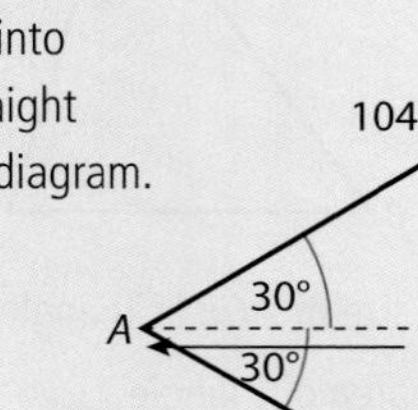

Let *D* be a point on [*BC*] such that [*AD*] bisects the 60° angle. The farmer divides the field into two parts, A_1 and A_2, by constructing a straight fence [*AD*] of length *x* m, as shown in the diagram.

c) **(i)** Show that the area of A_1 is given by $\frac{65x}{4}$.

(ii) Find a similar expression for the area of A_2.

(iii) Hence, find the value of *x* in the form $q\sqrt{3}$, where *q* is an integer.

d) **(i)** Explain why $\sin A\hat{D}C = \sin A\hat{D}B$.

(ii) Use the result of part **(i)** and the sine rule to show that $\frac{BD}{DC} = \frac{5}{8}$.

14 The lengths of the sides of a triangle *PQR* are $x - 2$, x and $x + a$ where $a > 0$.

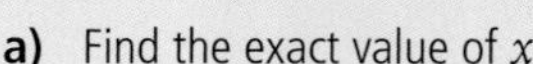

Angle *P* is 30° and angle *Q* is 45°, as shown in the diagram.

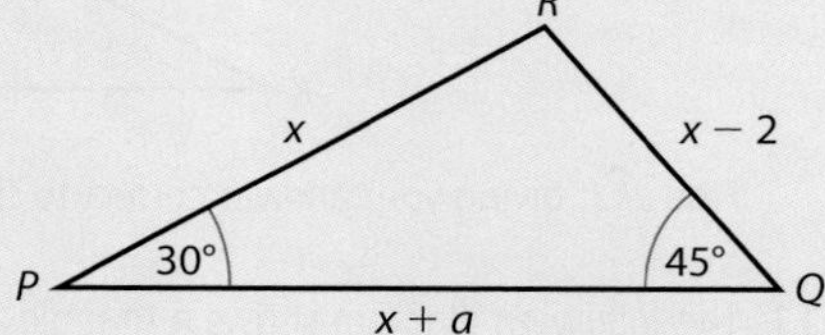

a) Find the exact value of *x*.

b) Find the exact area of triangle *PQR*.

15 Given a triangle *ABC*, a line segment [*CD*] is drawn from vertex *C* to a point D on side [*AB*]. Triangle *ABC* is divided into two triangular regions by [*CD*]. The areas of the regions are denoted as T_1 and T_2 (see diagram). Prove that for any triangle *ABC* the ratio of the areas $\frac{T_1}{T_2}$ is equal to the ratio of the lengths $\frac{BD}{AD}$.

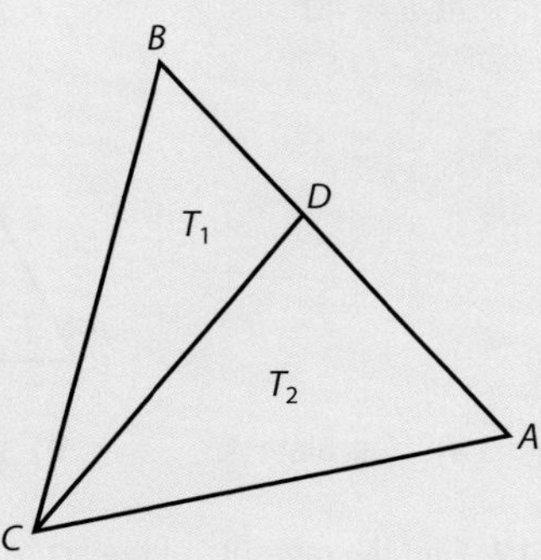

16 One corner, *K*, of a field consists of two stone walls, [*KJ*] and [*KL*], at an angle of 60° to each other. A 30-metre wooden fence [*JL*] is to be built to create a triangular enclosure *JKL*, as shown in the diagram.

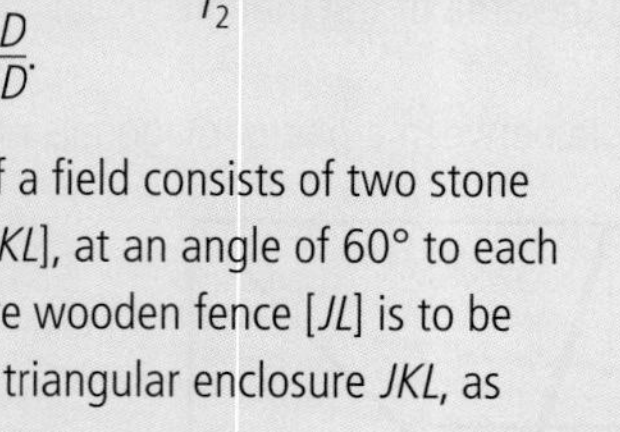

a) If $K\hat{J}L$ is denoted by θ, state the range of possible values for θ.

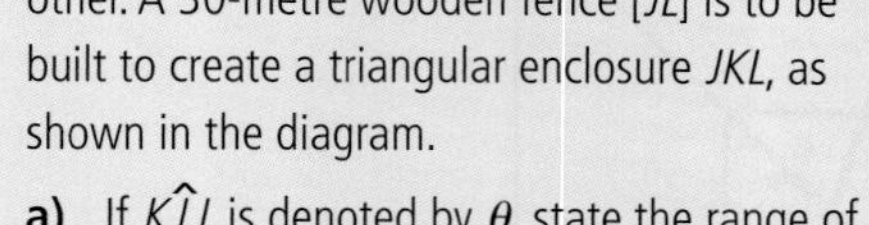

b) Show that the area of triangle *JKL* is given by $300\sqrt{3}\sin\theta\sin(\theta + 60°)$.

c) Use your GDC to determine the value of θ that gives the maximum area for the enclosure.

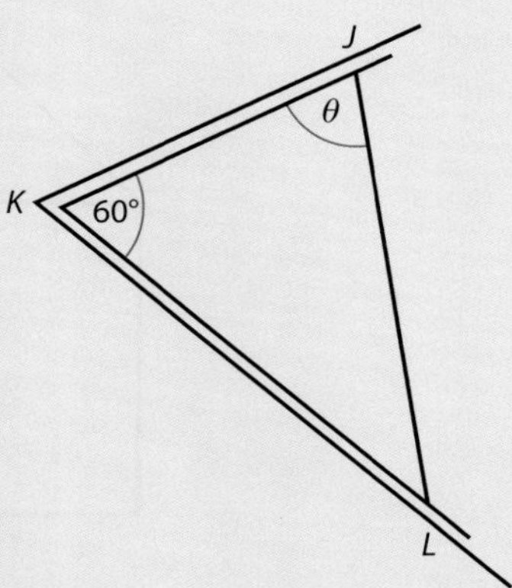

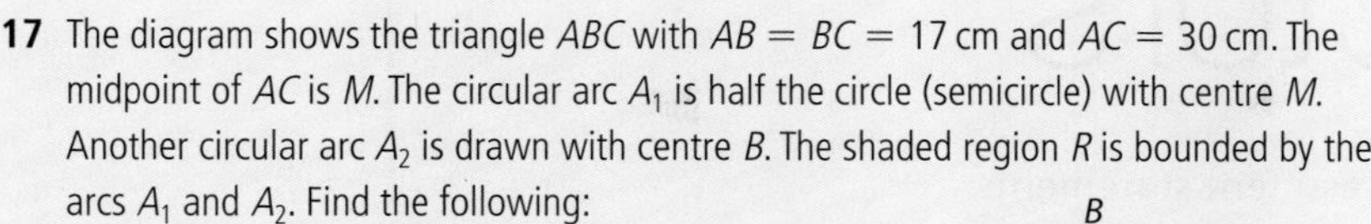

17 The diagram shows the triangle ABC with $AB = BC = 17$ cm and $AC = 30$ cm. The midpoint of AC is M. The circular arc A_1 is half the circle (semicircle) with centre M. Another circular arc A_2 is drawn with centre B. The shaded region R is bounded by the arcs A_1 and A_2. Find the following:

a) the area of triangle ABC

b) the measure of $A\hat{B}C$ in radians

c) the area of the shaded region R.

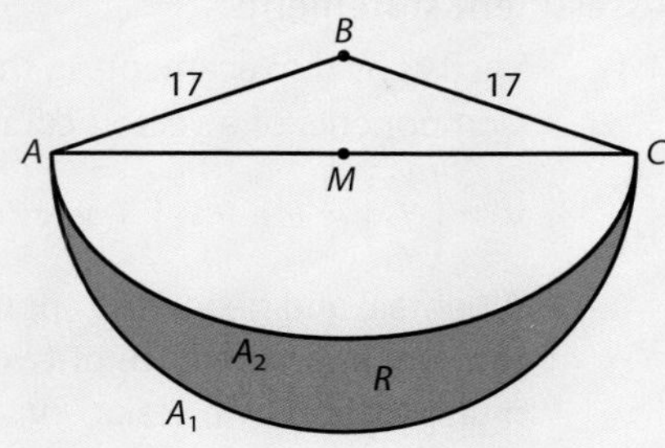

18 a) In the diagram, radii drawn to endpoints of a chord of the unit circle determine a central angle α. Show that the length of the chord is equal to $L = \sqrt{2 - 2\cos\alpha}$.

b) By using the substitution $\theta = \frac{\alpha}{2}$ in the double angle formula $\cos 2\theta = 1 - 2\sin^2\theta$, derive a formula for $\sin\frac{\alpha}{2}$, that is a half-angle formula for the sine function.

c) Use the result in **a)** and your result in **b)** to show that the length of the chord is equal to $L = 2\sin\left(\frac{\alpha}{2}\right)$.

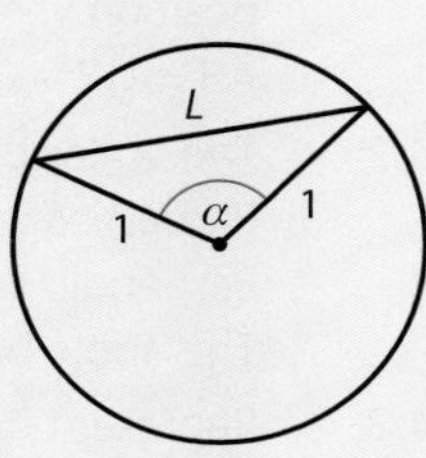

19 In triangle ABC, $A\hat{B}C = 2\theta$ and $B\hat{A}C = \theta$. Determine an expression for $\cos\theta$ in terms of a and b.

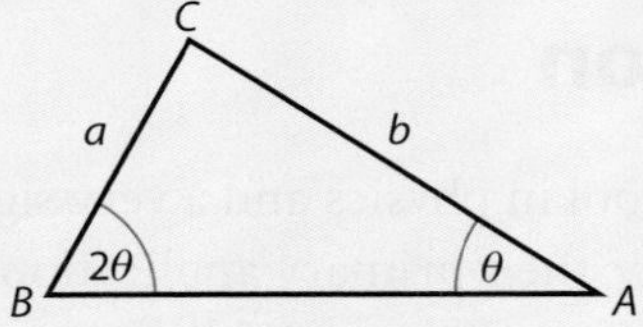

9 Vectors

Assessment statements

4.1 Vectors as displacements in the plane.
Components of a vector; column representation.

$$\mathbf{v} = \begin{pmatrix} v_1 \\ v_2 \\ v_3 \end{pmatrix} = v_1\mathbf{i} + v_2\mathbf{j} + v_3\mathbf{k}$$

Algebraic and geometric approaches to the following topics:
the sum and difference of two vectors; the zero vector; the vector $-\mathbf{v}$;
multiplication by a scalar, $k\mathbf{v}$;
magnitude of a vector, $|\mathbf{v}|$;
unit vectors; base vectors, $\mathbf{i}$, $\mathbf{j}$ and $\mathbf{k}$;
position vectors $\overrightarrow{OA} = \mathbf{a}$;
$\overrightarrow{AB} = \overrightarrow{OB} - \overrightarrow{OA} = \mathbf{b} - \mathbf{a}$.

4.2 The scalar product of two vectors.
Properties of the scalar product.
Perpendicular vectors; parallel vectors.
The angle between two vectors.

4.3 Representation of a line as $\mathbf{r} = \mathbf{a} + t\mathbf{b}$.
The angle between two lines.
(See also Chapter 14.)

Introduction

Vectors are an essential tool in physics and a very significant part of mathematics. Historically, their primary application was to represent forces, and the operation called '**vector addition**' corresponds to the combining of various forces. Many other applications in physics and other fields have been found since. In this chapter, we will discuss what vectors are and how to add, subtract and multiply them by scalars; we will also examine why vectors are useful in everyday life and how they are used in real-life applications. Then we will discuss scalar products.

Control panel of a passenger jet cockpit.

9.1 Vectors as displacements in the plane

We can represent physical quantities like temperature, distance, area, speed, density, pressure and volume by a single number indicating magnitude or size. These are called **scalar quantities.** Other physical quantities possess the properties of magnitude and direction. We define the force needed to pull a truck up a 10° slope by its **magnitude** and **direction**. Force, displacement, velocity, acceleration, lift, drag, thrust and weight are quantities that cannot be described by a single number. These are called **vector quantities**. Distance and displacement, for example, have distinctly different meanings; so do speed and velocity. Speed is a scalar quantity that refers to 'how fast an object is moving'.

Velocity is a vector quantity that refers to 'the rate at which an object *changes its position*'. When evaluating the velocity of an object, we must keep track of direction. It would not be enough to say that an object has a velocity of 55 km/h; we must include direction information in order to fully describe the velocity of the object. For instance, you must describe the object's velocity as being 55 km/h east. This is one of the essential differences between speed and velocity. Speed is a **scalar** quantity and does not keep track of direction; velocity is a **vector** quantity and is direction-conscious.

Thus, an aeroplane moving westward with a speed of 600 km/h has a velocity of 600 km/h west. Note that speed has no direction (it is scalar) and velocity, at any instant, is simply the speed with a direction.

We represent vector quantities with **directed line segments** (Figure 9.1). The directed line segment $\overrightarrow{AB}$ has **initial point** A and **terminal point** B. We use the notation $\overrightarrow{AB}$ to indicate that the line segment represents a vector quantity. We use $|\overrightarrow{AB}|$ to represent the **magnitude** of the directed line segment. The terms **size, length** or **norm** are also used. The direction of $\overrightarrow{AB}$ is from A to B. $\overrightarrow{BA}$ has the same length but the opposite direction to $\overrightarrow{AB}$ and hence cannot be equal to it.

Two directed line segments that have the same magnitude and direction are equivalent. For example, the directed line segments in Figure 9.2 are all equivalent.

We call the set of all directed line segments equivalent to a given directed line segment $\overrightarrow{AB}$ a **vector v**, and write $\mathbf{v} = \overrightarrow{AB}$. We denote vectors by lower-case, boldface letters such as **a**, **u**, and **v**.

We say that two vectors **a** and **b** are equal if their corresponding directed line segments are equivalent.

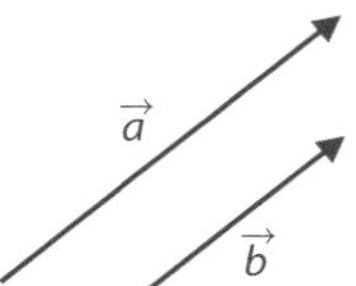

Vectors $\vec{a}$ and $\vec{b}$ have the same direction but different magnitudes $\Rightarrow \vec{a} \neq \vec{b}$.

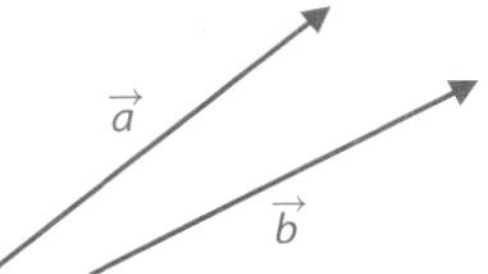

Vectors $\vec{a}$ and $\vec{b}$ have equal magnitudes but different directions $\Rightarrow \vec{a} \neq \vec{b}$.

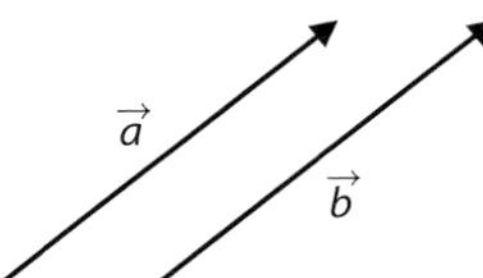

Vectors $\vec{a}$ and $\vec{b}$ have equal magnitudes and the same direction $\Rightarrow \vec{a} = \vec{b}$.

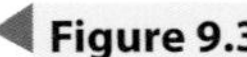

Figure 9.3

The notion of vector, as presented here, is due to the mathematician-physicist J. Williard Gibbs (1839–1903) of Yale University. His book *Vector Analysis* (1881) made these ideas accessible to a wide audience.

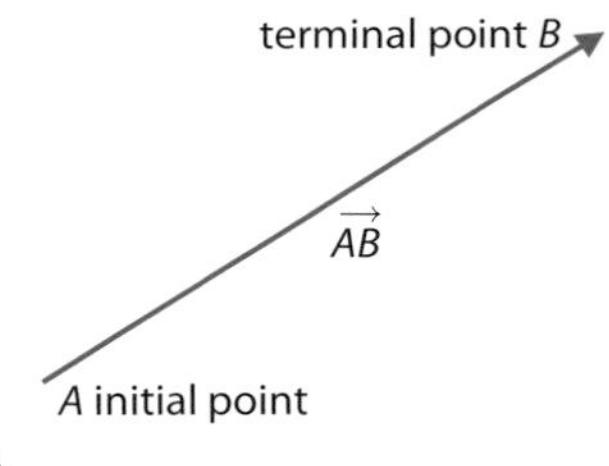

Figure 9.1

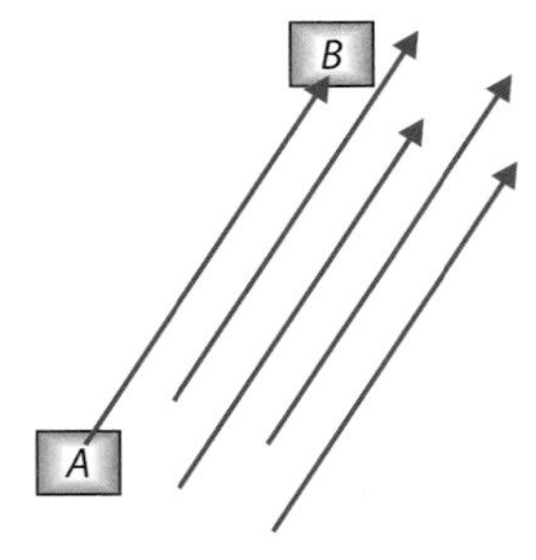

Figure 9.2

• **Hint:** Note: When we handwrite vectors, we cannot use boldface, so the convention is to use the arrow notation.

Definition 1: Two vectors **u** and **v** are equal if they have the same magnitude and the same direction.

Definition 2: The negative of a vector **u**, denoted by $-\mathbf{u}$, is a vector with the same magnitude but opposite direction.

Example 1

Marco walked around the park as shown in the diagram. What is Marco's displacement at the end of his walk?

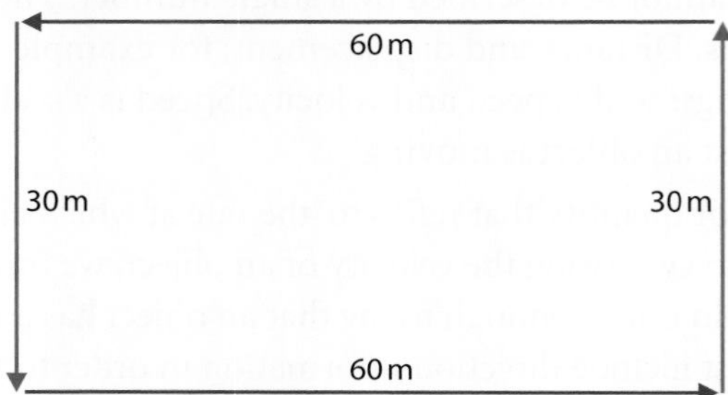

Solution

Even though he walked a total distance of 180 m, his displacement is zero since he returned to his original position. So, his displacement is **0**.

This is a displacement and hence direction is also important, not only magnitude. The 30 m south 'cancelled' the 30 m north, and the 60 m east is cancelled by the 60 m west.

Vectors can also be looked at as displacement/translation in the plane. Take, for example, the directed segments PQ and RS as representing the vectors **u** and **v**, respectively. The points $P(0, 0)$, $Q(2, 5)$, $R(3, 1)$ and $S(5, 6)$ are shown in Figure 9.4.

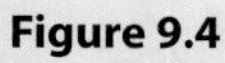
Figure 9.4

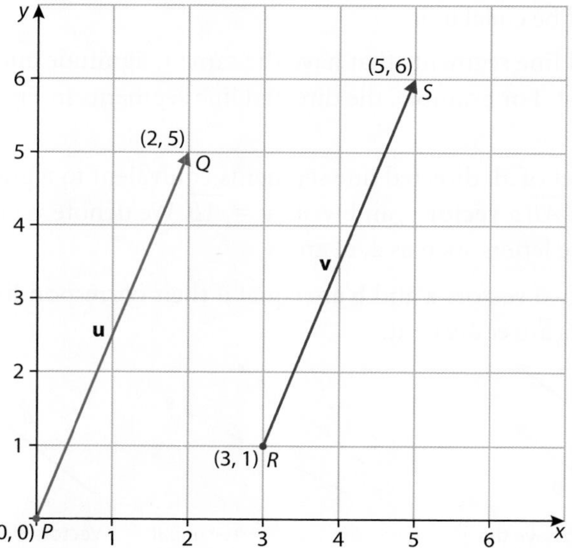

We can prove that these two vectors are equal.

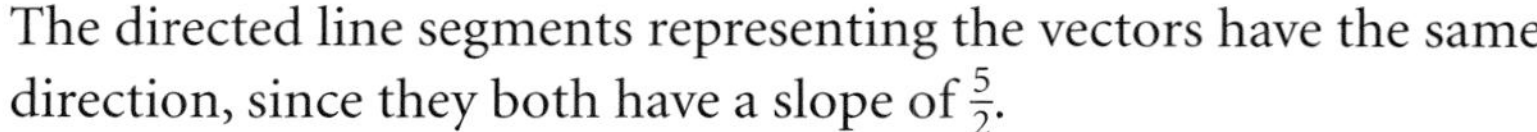

The directed line segments representing the vectors have the same direction, since they both have a slope of $\frac{5}{2}$.

They also have the same magnitude, as:

$$|\overrightarrow{PQ}| = \sqrt{5^2 + 2^2} = \sqrt{29} \text{ and}$$
$$|\overrightarrow{RS}| = \sqrt{(5-3)^2 + (6-1)^2} = \sqrt{29}$$

Component form

The directed line segment with the origin as its initial point is the most convenient way of representing a vector. This representation of the vector is said to be in **standard position**. In Figure 9.4, **u** is in standard position. A vector in standard position can be uniquely represented by the coordinates of its terminal point (u_1, u_2). This is called the **component form of a vector u**, written as $\mathbf{u} = (u_1, u_2)$.

The coordinates u_1 and u_2 are the **components** of the vector **u**. In Figure 9.4, the components of the vector **u** are 2 and 5.

If the initial and terminal points of the vector are the same, the vector is a **zero vector** and is denoted by $\mathbf{0} = (0, 0)$.

If **u** is a vector in the plane with initial point (0, 0) and terminal point (u_1, u_2), the **component form** of **u** is $\mathbf{u} = (u_1, u_2)$.
Note: The component form is also written as $\begin{pmatrix} u_1 \\ u_2 \end{pmatrix}$.

So, a vector in the plane is also an ordered pair (u_1, u_2) of real numbers. The numbers u_1 and u_2 are the components of **u**. The vector $\mathbf{u} = (u_1, u_2)$ is also called the **position vector** of the point (u_1, u_2).

If the vector **u** is not in standard position and is represented by a directed segment AB, then it can be written in its component form, observing the following fact:

$\mathbf{u} = (u_1, u_2) = (x_2 - x_1, y_2 - y_1)$, where $A(x_1, y_1)$ and $B(x_2, y_2)$ (Figure 9.5).

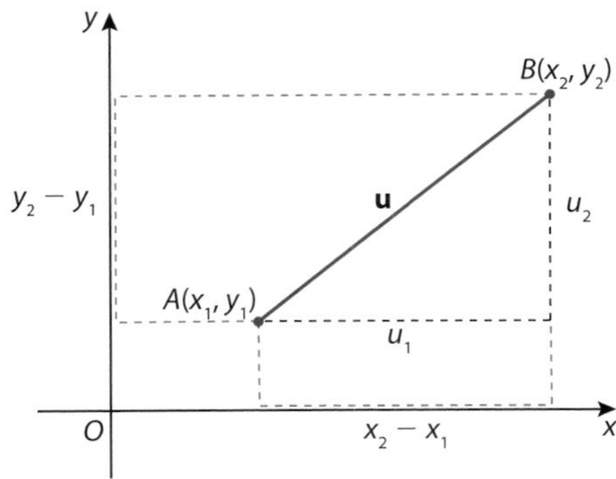

Figure 9.5

The length of vector **u** can be given using Pythagoras' theorem and/or the distance formula:

$$|\mathbf{u}| = \sqrt{u_1^2 + u_2^2} = \sqrt{(x_2 - x_1)^2 + (y_2 - y_1)^2}$$

Example 2

a) Find the components and the length of the vector between the points $P(-2, 3)$ and $Q(4, 7)$.

b) $\overrightarrow{RS}$ is another representation of the vector **u** where $R(7, -3)$. Find the coordinates of S.

Solution

a) $\overrightarrow{PQ} = (4 - (-2), 7 - 3) = (6, 4)$

$|\overrightarrow{PQ}| = \sqrt{36 + 16} = \sqrt{52} = 2\sqrt{13}$

b) Let S have coordinates (x, y). Therefore,

$\overrightarrow{RS} = (x - 7, y + 3)$.

But,

$\overrightarrow{RS} = \overrightarrow{PQ} \Rightarrow x - 7 = 6$ and $y + 3 = 4 \Rightarrow x = 13, y = 1$.

So, S has coordinates $(13, 1)$.

Example 3

The directed segment from $(-1, 2)$ to $(3, 5)$ represents a vector **v**. Find the length of vector **v**, draw the vector in standard position and find the opposite of the vector in component form.

Solution

The length of vector **v** can be found using the distance formula:

$$|\mathbf{v}| = \sqrt{(3 + 1)^2 + (5 - 2)^2} = 5$$

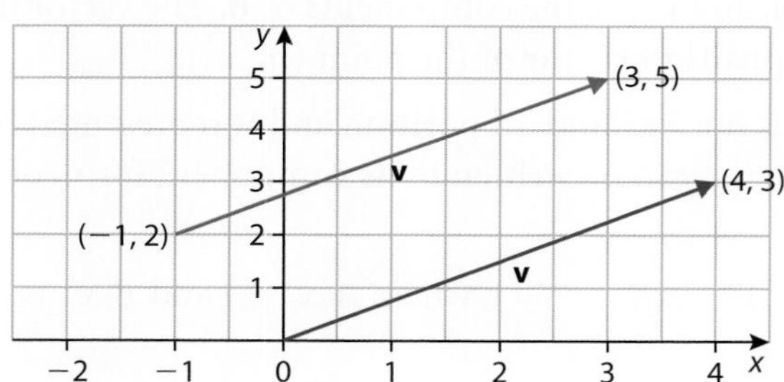

The opposite of this vector can be represented by $-\mathbf{v} = (-4, -3)$.

9.2 Vector operations

Two of the most basic and important operations are scalar multiplication and vector addition.

Scalar multiplication

In working with vectors, numbers are considered scalars. In this discussion, scalars will be limited to real numbers only. Geometrically, the product of a vector **u** and a scalar k, $\mathbf{v} = k\mathbf{u}$, is a vector that is $|k|$ times as long as **u**. If

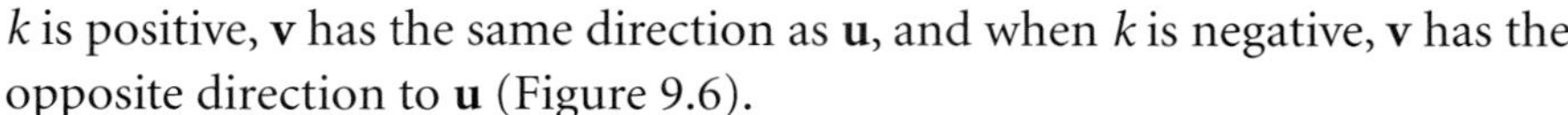

k is positive, **v** has the same direction as **u**, and when k is negative, **v** has the opposite direction to **u** (Figure 9.6).

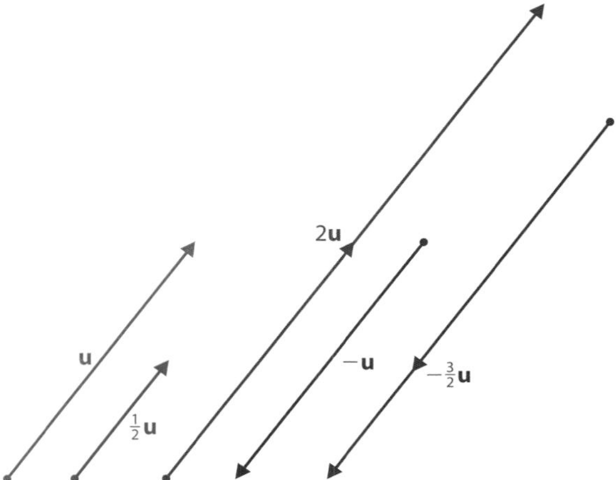

Figure 9.6

Consequence: It becomes clear from this discussion that for two vectors to be parallel, it is necessary and sufficient that one of them is a scalar multiple of the other. That is, if **v** and **u** are parallel, then $\mathbf{v} = k\mathbf{u}$; and vice versa, if $\mathbf{v} = k\mathbf{u}$, then **v** and **u** are parallel**.**

In terms of their components, the operation of scalar multiplication is straightforward.

If $\mathbf{u} = (u_1, u_2)$ then $\mathbf{v} = k\mathbf{u} = k(u_1, u_2) = (ku_1, ku_2)$.

Example 4

Find the magnitude of each vector.

a) $\mathbf{u} = (3, -4)$ b) $\mathbf{v} = (6, -8)$ c) $\mathbf{w} = (7, 0)$ d) $\mathbf{z} = \left(\frac{1}{2}, \frac{\sqrt{3}}{2}\right)$

Solution

a) $|\mathbf{u}| = \sqrt{3^2 + 4^2} = 5$

b) $|\mathbf{v}| = \sqrt{6^2 + (-8)^2} = 10$ Notice that $\mathbf{v} = 2\mathbf{u}$ and so $|\mathbf{v}| = 2|\mathbf{u}|$.

c) $|\mathbf{w}| = \sqrt{7^2 + 0^2} = 7$

d) $|\mathbf{z}| = \sqrt{\left(\frac{1}{2}\right)^2 + \left(\frac{\sqrt{3}}{2}\right)^2} = 1$ This is also called a unit vector as you will see later.

Vector addition

There are two equivalent ways of looking at the addition of vectors geometrically. One is the triangular method and the other is the parallelogram method.

Let **u** and **v** denote two vectors. Draw the vectors such that the terminal point of **u** and initial point of **v** coincide. The vector joining the initial point of **u** to the terminal point of **v** is the sum (resultant) of vectors **u** and **v** and is denoted by $\mathbf{u} + \mathbf{v}$ (Figure 9.7).

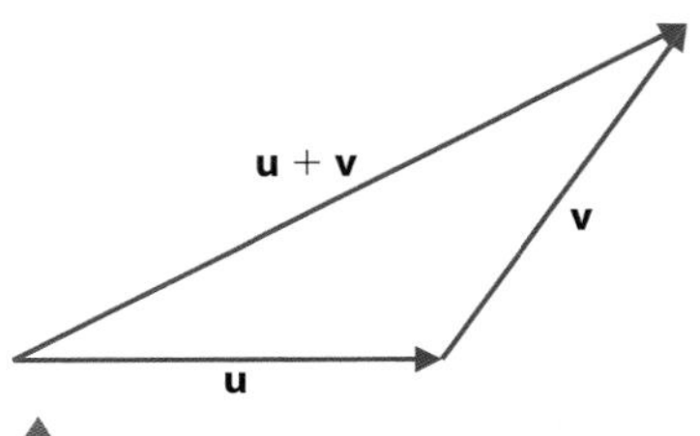

Figure 9.7

Another equivalent way of looking at the sum also gives us the grounds to say that vector addition is commutative.

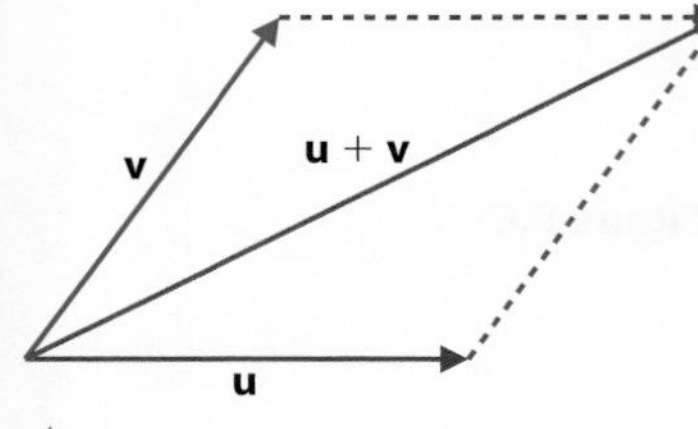

Figure 9.8

Let **u** and **v** denote two vectors. Draw the vectors such that the initial point of **u** and initial point of **v** coincide. The vector joining the common initial point of **u** and **v** to the opposite corner of the parallelogram, formed by the vectors as its adjacent sides, is the sum (resultant) of vectors **u** and **v** and is denoted by $\mathbf{u} + \mathbf{v}$ (Figure 9.8).

The difference of two vectors is an extremely important rule that will be used later in the chapter.

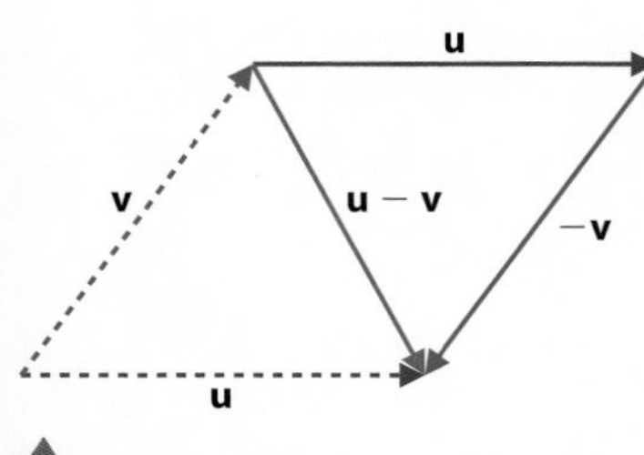

Figure 9.9

As Figure 9.9 shows, it is an extension of the addition rule. An easy way of looking at it is through a combination of the parallelogram rule and the triangle rule. We draw the vectors **u** and **v** in the usual way, then we draw $-\mathbf{v}$ starting at the terminal point of **u** and we add $\mathbf{u} + (-\mathbf{v})$ to get the difference $\mathbf{u} - \mathbf{v}$. As it turns out, the difference of the two vectors **u** and **v** is the diagonal of the parallelogram with its initial point the terminal of **v** and its terminal point the terminal point of **u**.

Example 5

Consider the vectors $\mathbf{u} = (2, -3)$ and $\mathbf{w} = (1, 3)$.

a) Write down the components of $\mathbf{v} = 2\mathbf{u}$.

b) Find $|\mathbf{u}|$ and $|\mathbf{v}|$ and compare them.

c) Draw the vectors $\mathbf{u}, \mathbf{v}, \mathbf{w}, 2\mathbf{w}, \mathbf{u} + \mathbf{w}, \mathbf{v} + 2\mathbf{w}, \mathbf{u} - \mathbf{w}, \mathbf{v} - 2\mathbf{w}$.

d) Comment on the results of c) above.

Solution

a) $\mathbf{v} = 2(2, -3) = (4, -6)$

b) $|\mathbf{u}| = \sqrt{4 + 9} = \sqrt{13}$, $|\mathbf{v}| = \sqrt{16 + 36} = \sqrt{52} = 2\sqrt{13}$. Clearly, $|\mathbf{v}| = 2|\mathbf{u}|$.

c)

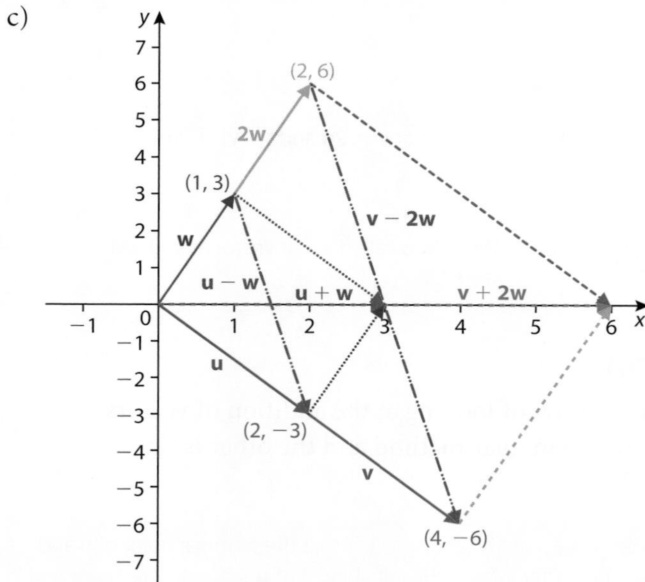

d) We observe that $\mathbf{u} + \mathbf{w} = (3, 0)$ which turns out to be $(1 + 2, 3 - 3)$, the sum of the corresponding components. We observe the same for $\mathbf{v} + 2\mathbf{w} = (6, 0)$, which in turn is $(2 + 4, 6 - 6)$.

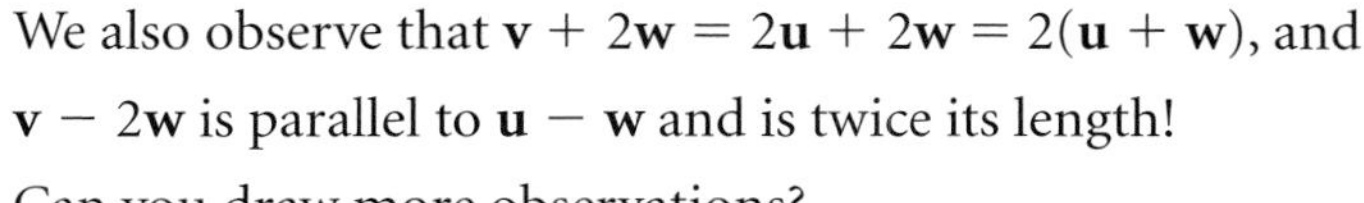

We also observe that $\mathbf{v} + 2\mathbf{w} = 2\mathbf{u} + 2\mathbf{w} = 2(\mathbf{u} + \mathbf{w})$, and
$\mathbf{v} - 2\mathbf{w}$ is parallel to $\mathbf{u} - \mathbf{w}$ and is twice its length!
Can you draw more observations?

Example 6

ABCD is a quadrilateral with vertices that have position vectors **a**, **b**, **c**, and **d** respectively. *P*, *Q*, *R*, and *S* are the midpoints of the sides.

a) Express each of the following in terms of **a**, **b**, **c**, and **d**:
$\overrightarrow{AB}$, $\overrightarrow{CD}$, $\overrightarrow{AP}$, and $\overrightarrow{OP}$

b) Prove that *PQRS* is a parallelogram using vector methods.

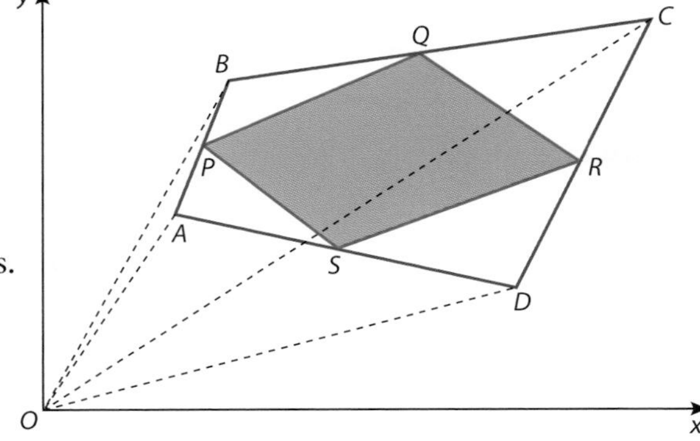

Solution

a) $\overrightarrow{AB} = \overrightarrow{OB} - \overrightarrow{OA} = \mathbf{b} - \mathbf{a}$

$\overrightarrow{CD} = \overrightarrow{OD} - \overrightarrow{OC} = \mathbf{d} - \mathbf{c}$

$\overrightarrow{AP} = \frac{1}{2}\overrightarrow{AB} = \frac{1}{2}(\mathbf{b} - \mathbf{a})$

$\overrightarrow{OP} = \overrightarrow{OA} + \overrightarrow{AP} = \mathbf{a} + \frac{1}{2}(\mathbf{b} - \mathbf{a}) = \frac{1}{2}(\mathbf{b} + \mathbf{a})$

b) One way of proving *PQRS* is a parallelogram is to show a pair of opposite sides parallel and congruent.

You can show that $\overrightarrow{OQ} = \frac{1}{2}(\mathbf{b} + \mathbf{c})$, $\overrightarrow{OR} = \frac{1}{2}(\mathbf{d} + \mathbf{c})$, and $\overrightarrow{OS} = \frac{1}{2}(\mathbf{d} + \mathbf{a})$ as we did for $\overrightarrow{OP}$.

Now, $\overrightarrow{PQ} = \overrightarrow{OQ} - \overrightarrow{OP} = \frac{1}{2}(\mathbf{b} + \mathbf{c}) - \frac{1}{2}(\mathbf{b} + \mathbf{a}) = \frac{1}{2}(\mathbf{c} - \mathbf{a})$, and
$\overrightarrow{SR} = \overrightarrow{OR} - \overrightarrow{OS} = \frac{1}{2}(\mathbf{d} + \mathbf{c}) - \frac{1}{2}(\mathbf{d} + \mathbf{a}) = \frac{1}{2}(\mathbf{c} - \mathbf{a})$.

Therefore, $\overrightarrow{PQ} = \overrightarrow{SR}$, and since they are opposite sides of the quadrilateral, so it is a parallelogram.

Base vectors in the coordinate plane

As you have seen before, vectors can also be represented in a coordinate system using their component form. This is a very useful tool that helps make many applications of vectors simple and easy. At the heart of the component approach to vectors we find the 'base' vectors **i** and **j**.

i is a vector of magnitude 1 with the direction of the positive *x*-axis and **j** is a vector of magnitude 1 with the direction of the positive *y*-axis. These vectors and any vector that has a magnitude of 1 are called **unit vectors.** Since vectors of same direction and length are equal, each vector **i** and **j** may be drawn at any point in the plane, but it is usually more convenient to draw them at the origin, as shown in Figure 9.10.

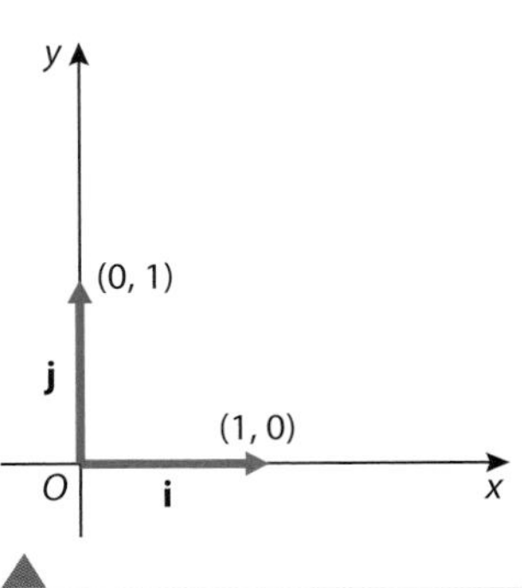

Figure 9.10

Now, the vector $k\mathbf{i}$ has magnitude k and is parallel to the vector **i**. Similarly, the vector $m\mathbf{j}$ has magnitude m and is parallel to **j**.

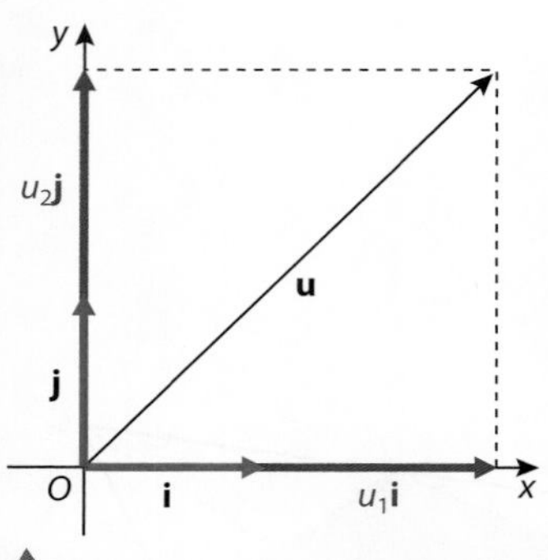

Figure 9.11

Consider the vector $\mathbf{u} = (u_1, u_2)$. This vector, in standard position, has an ***x*-component** u_1 and ***y*-component** u_2 (Figure 9.11).

Since the vector **u** is the diagonal of the parallelogram with adjacent sides $u_1\mathbf{i}$ and $u_2\mathbf{j}$, then it is the sum of the two vectors, i.e. $\mathbf{u} = u_1\mathbf{i} + u_2\mathbf{j}$. It is customary to say that $u_1\mathbf{i}$ is the **horizontal component** and $u_2\mathbf{j}$ is the **vertical component** of **u**.

The previous discussion shows that it is always possible to express any vector in the plane as a linear combination of the unit vectors **i** and **j**.

This form of representation of vectors opens the door to a rich world of vector applications.

Vector addition and subtraction in component form

If vector **u** has components (u_1, u_2), then its component form is: $\mathbf{u} = u_1\mathbf{i} + u_2\mathbf{j}$

Consider the two vectors $\mathbf{u} = u_1\mathbf{i} + u_2\mathbf{j}$ and $\mathbf{v} = v_1\mathbf{i} + v_2\mathbf{j}$.

(i) Vector sum $\mathbf{u} + \mathbf{v}$

$$\mathbf{u} + \mathbf{v} = (u_1\mathbf{i} + u_2\mathbf{j}) + (v_1\mathbf{i} + v_2\mathbf{j}) = (u_1\mathbf{i} + v_1\mathbf{i}) + (u_2\mathbf{j} + v_2\mathbf{j})$$
$$= (u_1 + v_1)\mathbf{i} + (u_2 + v_2)\mathbf{j}$$

For example, to add the two vectors $\mathbf{u} = 2\mathbf{i} + 4\mathbf{j}$ and $\mathbf{v} = 5\mathbf{i} - 3\mathbf{j}$, it is enough to add the corresponding components:

$$\mathbf{u} + \mathbf{v} = (2 + 5)\mathbf{i} + (4 - 3)\mathbf{j} = 7\mathbf{i} + \mathbf{j}$$

(ii) Vector difference $\mathbf{u} - \mathbf{v}$

$$\mathbf{u} - \mathbf{v} = (u_1\mathbf{i} + u_2\mathbf{j}) - (v_1\mathbf{i} + v_2\mathbf{j}) = (u_1\mathbf{i} - v_1\mathbf{i}) + (u_2\mathbf{j} - v_2\mathbf{j})$$
$$= (u_1 - v_1)\mathbf{i} + (u_2 - v_2)\mathbf{j}$$

For example, to subtract the two vectors $\mathbf{u} = 2\mathbf{i} + 4\mathbf{j}$ and $\mathbf{v} = 5\mathbf{i} - 3\mathbf{j}$, it is enough to subtract the corresponding components:

$$\mathbf{u} - \mathbf{v} = (2 - 5)\mathbf{i} + (4 + 3)\mathbf{j} = -3\mathbf{i} + 7\mathbf{j}$$

This interpretation of the difference gives us another way of finding the components of any vector in the plane, even if it is not in standard position (Figure 9.12).

Figure 9.12

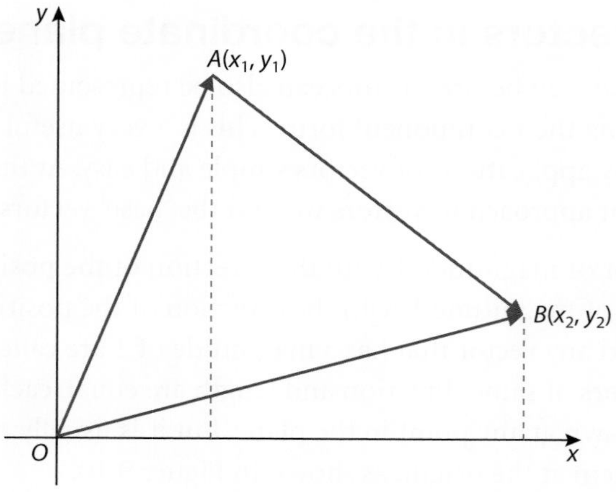

Consider the vector $\overrightarrow{AB}$ where the position vectors of its endpoints are given by the vectors $\overrightarrow{OA} = x_1\mathbf{i} + y_1\mathbf{j}$ and $\overrightarrow{OB} = x_2\mathbf{i} + y_2\mathbf{j}$.

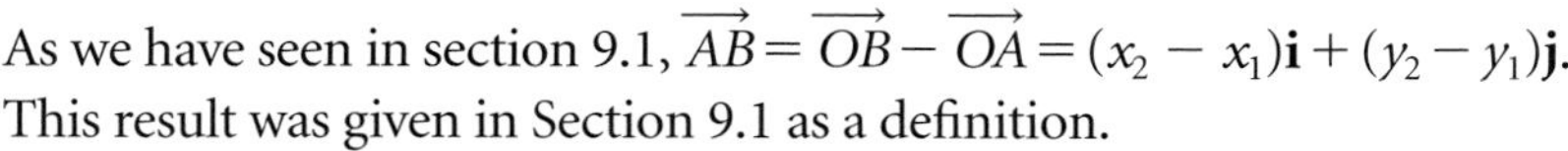

As we have seen in section 9.1, $\overrightarrow{AB} = \overrightarrow{OB} - \overrightarrow{OA} = (x_2 - x_1)\mathbf{i} + (y_2 - y_1)\mathbf{j}$.
This result was given in Section 9.1 as a definition.

- Many of the laws of ordinary algebra are also valid for vector algebra. These laws are:
  - Commutative law for addition: $\mathbf{a} + \mathbf{b} = \mathbf{b} + \mathbf{a}$
  - Associative law for addition: $(\mathbf{a} + \mathbf{b}) + \mathbf{c} = \mathbf{a} + (\mathbf{b} + \mathbf{c})$

    The verification of the associative law is shown in Figure 9.13.

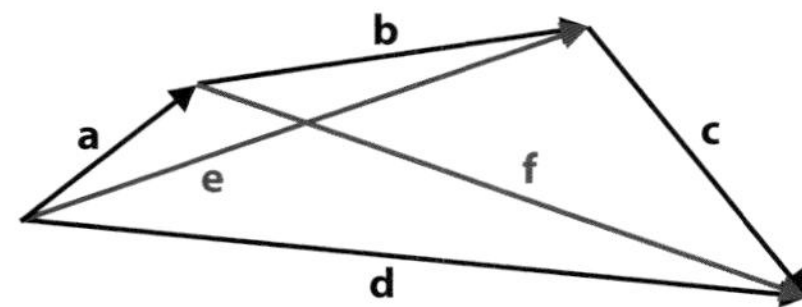

Figure 9.13

    If we add **a** and **b** we get a vector **e**. And similarly, if **b** is added to **c**, we get **f**.

    Now $\mathbf{d} = \mathbf{e} + \mathbf{c} = \mathbf{a} + \mathbf{f}$. Replacing **e** with $(\mathbf{a} + \mathbf{b})$ and **f** with $(\mathbf{b} + \mathbf{c})$, we get $(\mathbf{a} + \mathbf{b}) + \mathbf{c} = \mathbf{a} + (\mathbf{b} + \mathbf{c})$ and we see that the law is verified.
  - Commutative law for multiplication: $m\mathbf{a} = \mathbf{a}m$
  - Distributive law (1): $(m + n)\mathbf{a} = m\mathbf{a} + n\mathbf{a}$, where m and n are two different scalars.
  - Distributive law (2): $m(\mathbf{a} + \mathbf{b}) = m\mathbf{a} + m\mathbf{b}$

These laws allow the manipulation of vector quantities in much the same way as ordinary algebraic equations.

Two vectors **u** and **v** are parallel iff $\mathbf{v} = k\mathbf{u}$. This also means that in component form:

$$\frac{v_1}{u_1} = \frac{v_2}{u_2} = k$$

Exercise 9.1 and 9.2

1 Consider the vectors **u** and **v** given.
Sketch each indicated vector.

a) 2**u**
b) −**v**
c) **u** + **v**
d) **2u** − **v**
e) **v** − 2**u**

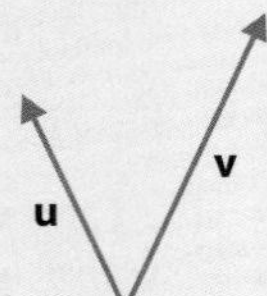

For questions 2–5, consider the points A and B given and answer the following questions:

a) Find $|\overrightarrow{AB}|$.
b) Find the components of the vector $\mathbf{u} = \overrightarrow{AB}$ and sketch it in standard position.
c) Write the vector $\mathbf{v} = \frac{1}{|\overrightarrow{AB}|} \cdot \mathbf{u}$ in component form.
d) Find $|\mathbf{v}|$.
e) Sketch the vector **v** and compare it to **u**.

2 $A(3, 4)$ and $B(7, -1)$

3 $A(-2, 3)$ and $B(5, 1)$

4 $A(3, 5)$ and $B(0, 5)$

5 $A(2, -4)$ and $B(2, 1)$

6 Consider the vector shown.

a) Write down the component representation of the vector.
b) Find the length of the vector.
c) Sketch the vector in standard position.
d) Find a vector equal to this one with initial point $(-1, 1)$.

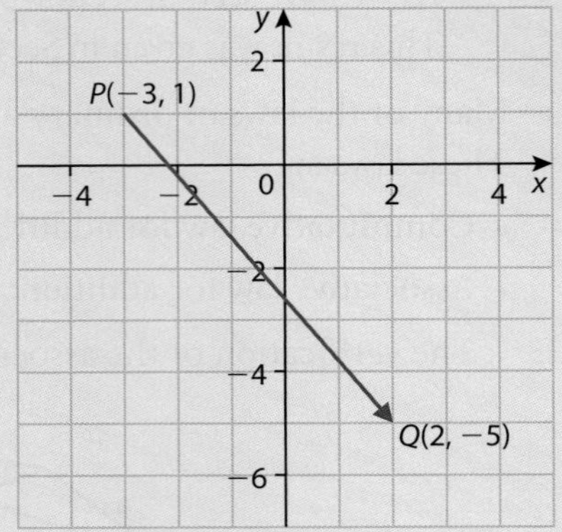

For questions 7–9, the initial point P and terminal point Q are given. Answer the same questions as in question 6.

7 $P(3, 2), Q(7, 8)$

8 $P(2, 2), Q(7, 7)$

9 $P(-6, -8), Q(-2, -2)$

10 Which of the vectors **a**, **b**, or **c** in the figure shown right is equivalent to **u** − **v**? Which is equivalent to **v** + **u**?

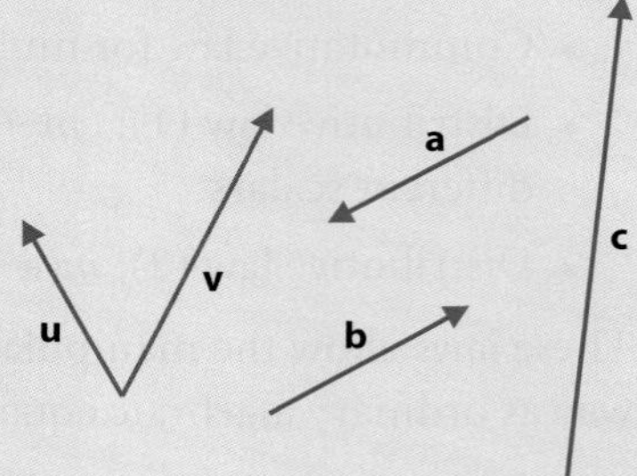

11 Find the terminal point of $\mathbf{v} = 3\mathbf{i} - 2\mathbf{j}$ if the initial point is $(-2, 1)$.

12 Find the initial point of $\mathbf{v} = (-3, 1)$ if the terminal point is $(5, 0)$.

13 Find the terminal point of $\mathbf{v} = (6, 7)$ if the initial point is $(-2, 1)$.

14 Find the initial point of $\mathbf{v} = 2\mathbf{i} + 7\mathbf{j}$ if the terminal point is $(-3, 2)$.

15 Consider the vectors $\mathbf{u} = 3\mathbf{i} - \mathbf{j}$ and $\mathbf{v} = -\mathbf{i} + 3\mathbf{j}$.
a) Find $\mathbf{u} + \mathbf{v}$, $\mathbf{u} - \mathbf{v}$, $2\mathbf{u} + 3\mathbf{v}$ and $2\mathbf{u} - 3\mathbf{v}$.
b) Find $|\mathbf{u} + \mathbf{v}|$, $|\mathbf{u} - \mathbf{v}|$, $|\mathbf{u}| + |\mathbf{v}|$ and $|\mathbf{u}| - |\mathbf{v}|$.
c) Find $|2\mathbf{u} + 3\mathbf{v}|$, $|2\mathbf{u} - 3\mathbf{v}|$, $2|\mathbf{u}| + 3|\mathbf{v}|$ and $2|\mathbf{u}| - 3|\mathbf{v}|$.

16 Let $\mathbf{u} = (1, 5)$ and $\mathbf{v} = (3, -4)$. Find the vector $\mathbf{x}$ such that $2\mathbf{u} - 3\mathbf{x} + \mathbf{v} = 5\mathbf{x} - 2\mathbf{v}$.

17 Find **u** and **v** if $\mathbf{u} - 2\mathbf{v} = 2\mathbf{i} - 3\mathbf{j}$ and $\mathbf{u} + 3\mathbf{v} = \mathbf{i} + \mathbf{j}$.

18 Find the lengths of the diagonals of the parallelogram whose sides are the vectors $2\mathbf{i} - 3\mathbf{j}$ and $\mathbf{i} + \mathbf{j}$.

19 Vectors **u** and **v** form two sides of parallelogram $PQRS$, as shown. Express each of the following vectors in terms of **u** and **v**.
a) $\overrightarrow{PR}$
b) $\overrightarrow{PM}$, where M is the midpoint of $[RS]$
c) $\overrightarrow{QS}$
d) $\overrightarrow{QN}$

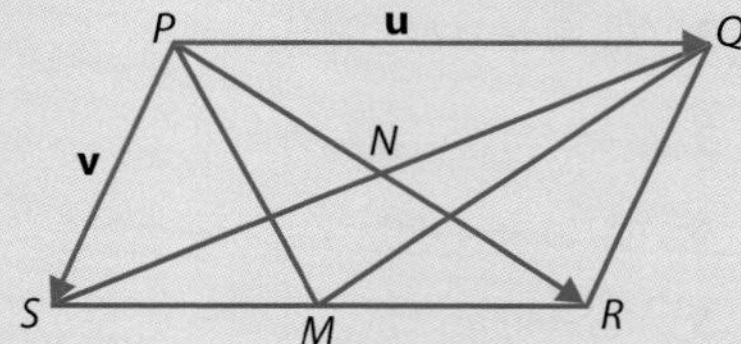

20 Find (x, y) so that the diagram at the right is a parallelogram.

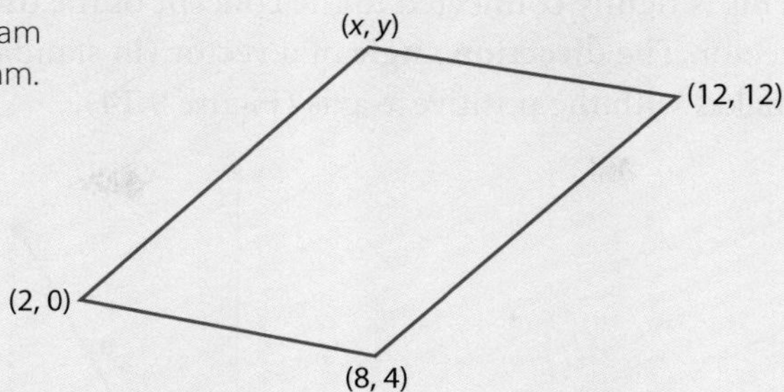

21 Find x and y in the parallelogram shown right.

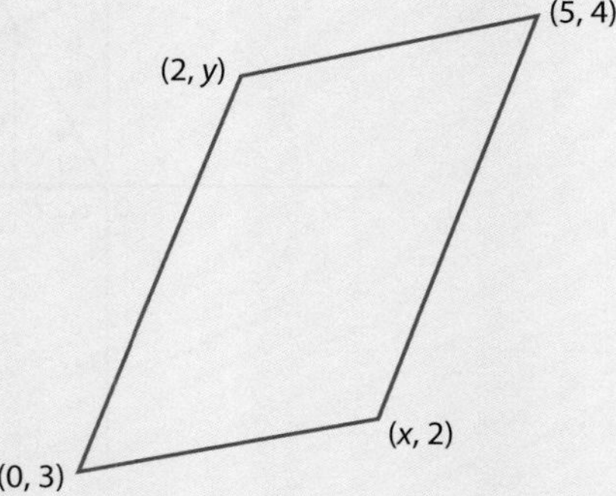

22 Find the scalars r and s such that

$\begin{pmatrix} 8 \\ 46 \end{pmatrix} = r\begin{pmatrix} 1 \\ 9 \end{pmatrix} + s\begin{pmatrix} 1 \\ -4 \end{pmatrix}$.

Note: $\begin{pmatrix} 8 \\ 46 \end{pmatrix}$ is said to be written as a linear combination of $\begin{pmatrix} 1 \\ 9 \end{pmatrix}$ and $\begin{pmatrix} 1 \\ -4 \end{pmatrix}$.

23 Write (4, 7) as a linear combination of (2, 3) and (2, 1).

24 Write (5, −5) as a linear combination of (1, −1) and (−1, 1).

25 Write (−11, 0) as a linear combination of (2, 5) and (3, 2).

26 Let $\mathbf{u} = \mathbf{i} + \mathbf{j}$ and $\mathbf{v} = -\mathbf{i} + \mathbf{j}$. Show that, if $\mathbf{w}$ is any vector in the plane, then it can be written as a linear combination of $\mathbf{u}$ and $\mathbf{v}$. (You can generalize the result to any two non-zero, non-parallel vectors $\mathbf{u}$ and $\mathbf{v}$.)

9.3 Unit vectors and direction angles

Consider the vector $\mathbf{u} = 3\mathbf{i} + 4\mathbf{j}$. To find the magnitude of this vector, $|\mathbf{u}|$, we use the distance formula:

$$|\mathbf{u}| = \sqrt{3^2 + 4^2} = 5$$

If we divide the vector $\mathbf{u}$ by $|\mathbf{u}| = 5$, i.e. we multiply the vector $\mathbf{u}$ by the reciprocal of its magnitude, we get another vector that is parallel to $\mathbf{u}$, since they are scalar multiples of each other. The new vector is

$$\frac{\mathbf{u}}{5} = \frac{3}{5}\mathbf{i} + \frac{4}{5}\mathbf{j}$$

This vector is a unit vector in the same direction as $\mathbf{u}$, because

$$\left|\frac{\mathbf{u}}{5}\right| = \sqrt{\left(\frac{3}{5}\right)^2 + \left(\frac{4}{5}\right)^2} = 1$$

Therefore, to find a unit vector in the same direction as a given vector, we divide that vector by its own magnitude.

This is tightly connected to the concept of the **direction angle** of a given vector. The **direction angle** of a vector (in standard position) is the angle it makes with the positive x-axis (Figure 9.14).

Figure 9.14

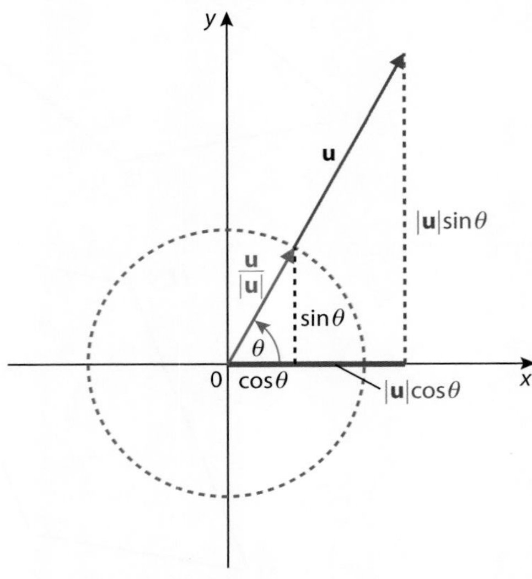

To find a unit vector parallel to a vector **u**, we simply find the vector $\frac{\mathbf{u}}{|\mathbf{u}|}$:

$$\frac{\mathbf{u}}{|\mathbf{u}|} = \frac{\mathbf{u}}{\sqrt{u_1^2 + u_2^2}} = \left(\frac{u_1}{\sqrt{u_2^1 + u_2^2}}, \frac{u_2}{\sqrt{u_2^1 + u_2^2}}\right)$$

So, the vector **u** can be expressed in terms of the unit vector parallel to it in the following manner:

$\mathbf{u} = u_1\mathbf{i} + u_2\mathbf{j} = (|\mathbf{u}|\cos\theta)\mathbf{i} + (|\mathbf{u}|\sin\theta)\mathbf{j} = |\mathbf{u}|(\cos\theta\mathbf{i} + \sin\theta\mathbf{j})$ where $u_1 = |\mathbf{u}|\cos\theta$ and $u_2 = |\mathbf{u}|\sin\theta$. This fact implies two important tools that help us:

1. find the direction of a given vector
2. find vectors of any magnitude parallel to a given vector.

Applications of unit vectors and direction angles

Given a vector $\mathbf{u} = u_1\mathbf{i} + u_2\mathbf{j}$, find the direction angle of this vector and another vector, whose magnitude is m, that is parallel to the vector **u.**

1. To help determine the direction angle, we observe the following:

   $u_1 = |\mathbf{u}|\cos\theta$ and $u_2 = |\mathbf{u}|\sin\theta$

   This implies that $\frac{u_2}{u_1} = \frac{|\mathbf{u}|\sin\theta}{|\mathbf{u}|\cos\theta} = \tan\theta.$

   So, $\tan^{-1}\theta$ is the reference angle for the direction angle in question. To know what the direction angle is, it is best to look at the numbers u_1 and u_2 in order to determine which quadrant the vector is in. The following example (Example 6) will clarify this point.

2. To find a vector of magnitude m parallel to **u**, we must first find the unit vector in the direction of **u** and then we multiply it by the scalar m.

   The unit vector in the direction of **u** is $\frac{\mathbf{u}}{|\mathbf{u}|} = \frac{1}{|\mathbf{u}|}(u_1\mathbf{i} + u_2\mathbf{j})$, and the vector of magnitude m in this direction will be

   $$m\frac{\mathbf{u}}{|\mathbf{u}|} = \frac{m}{\sqrt{u_1^2 + u_2^2}}(u_1\mathbf{i} + u_2\mathbf{j}).$$

Example 7

Find the direction angle (to the nearest degree) of each vector, and find a vector of magnitude 7 that is parallel to each.

a) $\mathbf{u} = 2\mathbf{i} + 2\mathbf{j}$

b) $\mathbf{v} = -3\mathbf{i} + 3\mathbf{j}$

c) $\mathbf{w} = 3\mathbf{i} - 4\mathbf{j}$

Solution

a) The direction angle for **u** is θ, as shown in Figure 9.15.

$\tan\theta = \frac{2}{2} = 1 \Rightarrow \theta = 45°$

A vector of magnitude 7 that is parallel to **u** is

$$7\frac{\mathbf{u}}{|\mathbf{u}|} = \frac{7}{\sqrt{2^2 + 2^2}}(2\mathbf{i} + 2\mathbf{j}) = \frac{7}{2\sqrt{2}}(2\mathbf{i} + 2\mathbf{j}) = \frac{7}{\sqrt{2}}(\mathbf{i} + \mathbf{j}).$$

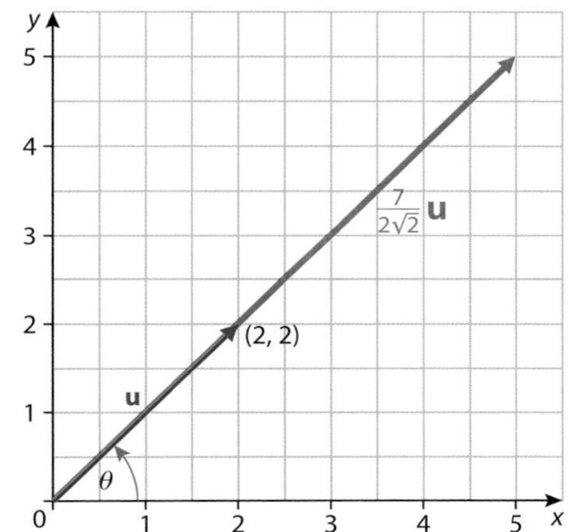

Figure 9.15

b) The direction angle for **v** is $180° - \theta$, as shown in Figure 9.16.

$\tan\theta = \frac{-3}{3} = -1 \Rightarrow \theta = 180° - 45° = 135°$

A vector of magnitude 7 that is parallel to **v** is

$$7\frac{\mathbf{v}}{|\mathbf{v}|} = \frac{7}{\sqrt{3^2 + 3^2}}(-3\mathbf{i} + 3\mathbf{j}) = \frac{7}{3\sqrt{2}}(-3\mathbf{i} + 3\mathbf{j}) = \frac{7}{\sqrt{2}}(-\mathbf{i} + \mathbf{j}).$$

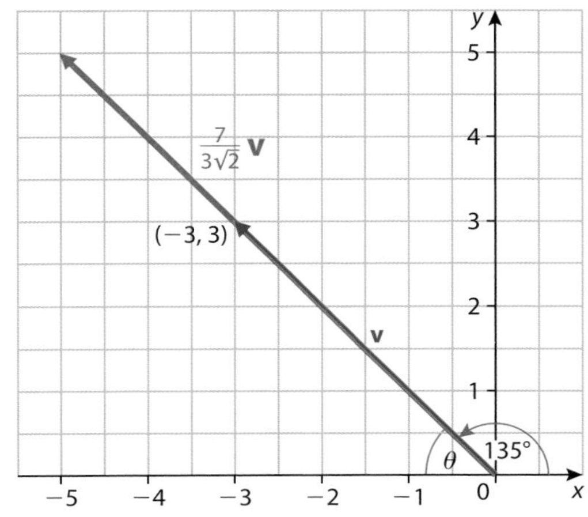

Figure 9.16

c) The direction angle for **w** is θ, as shown in Figure 9.17.

$\tan\theta = \frac{-4}{3} \Rightarrow \theta \approx -53°$

A vector of magnitude 7 that is parallel to **w** is

$7\frac{\mathbf{u}}{|\mathbf{u}|} = \frac{7}{\sqrt{3^2 + (-4)^2}}(3\mathbf{i} - 4\mathbf{j}) = \frac{7}{5}(3\mathbf{i} - 4\mathbf{j}).$

Figure 9.17

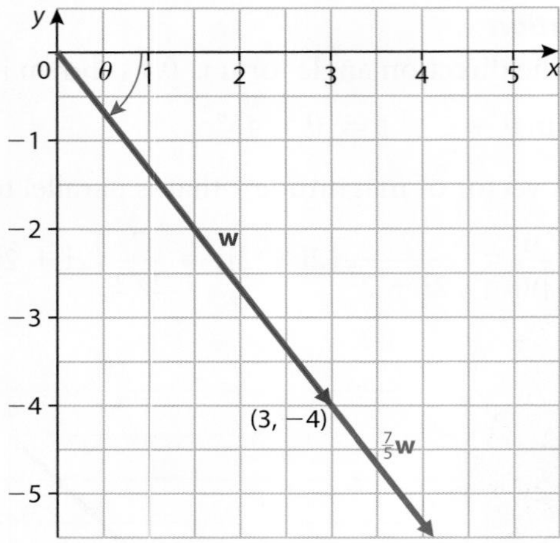

Using vectors to model force, displacement and velocity

The force on an object can be represented by a vector. We can think of the force as a push or pull on an object such as a person pulling a box along a plane or the weight of a truck which is a downward pull of the Earth's gravity on the truck. If several forces act on an object, the **resultant** force experienced by the object is the vector sum of the forces.

Force

Example 8

What force is required to pull a boat of 800 N up a ramp inclined at 15° from the horizontal? Friction is ignored in this case.

Solution

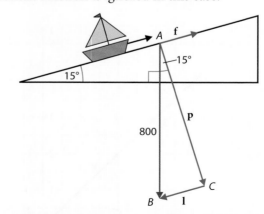

The process of 'breaking-up' the vector into its components, as we did in the example, is called **resolving** the vector into its components. Notice that the process of resolving a vector is not unique. That is, you can resolve a vector into several pairs of directions.

The situation can be shown on a diagram. The weight is represented by the vector $\overrightarrow{AB}$. The weight of the boat has two components – one perpendicular to the ramp, which is the force responsible for keeping the boat on the ramp and preventing it from tumbling down (**p**). The other force is parallel to the ramp, and is the force responsible for pulling the boat down the ramp (**l**). Therefore, the force we need, **f**, must counter **l**.

In triangle ABC:

$$\sin \angle A = |\mathbf{l}|/800 \Rightarrow |\mathbf{l}| = 800 \sin \angle A = 800 \sin 15° = 207.06.$$

We need an upward force of 207.06 N along the ramp to move the boat.

Example 9

In many countries, it is a requirement that disabled people have access to all places without needing the help of others. Consider an office building whose entrance is 40 cm above ground level. Assuming, on average, that the weight of a person including the equipment used is 1200 N, answer the following questions:

a) At what angle should the ramp designed for disabled persons be set if, on average, the force that a person can apply using their hands is 300 N?

b) How long should the ramp be?

Solution

a)

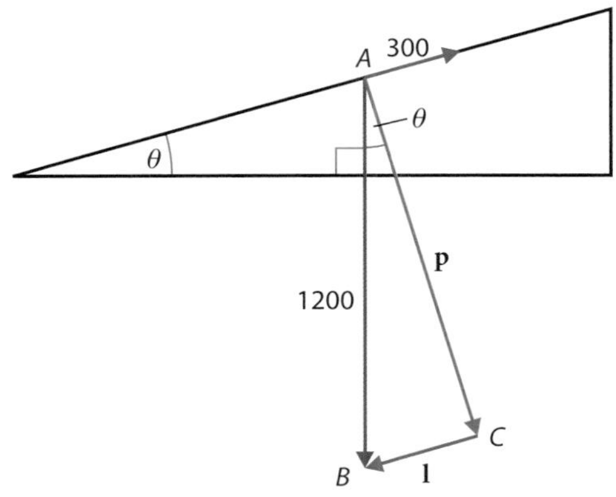

As the diagram above shows, $|\mathbf{l}| = 300$, and

$$\sin \angle A = \frac{|\mathbf{l}|}{1200} = \frac{300}{1200} \Rightarrow \angle A = \sin^{-1} 0.25 \approx 14.47°.$$

b) The length d of the ramp can be found using right triangle trigonometry:

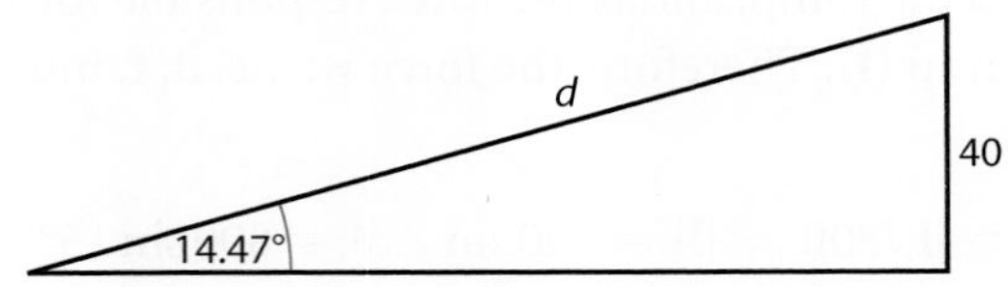

$$\sin 14.47 = \frac{40}{d} \Rightarrow d = \frac{40}{\sin 14.47} \approx \frac{40}{0.25} = 160 \text{ cm}$$

Resultant force

Two forces $\mathbf{F}_1$ with magnitude 20 N and $\mathbf{F}_2$ with magnitude 40 N are acting on an object at equilibrium as shown in the diagram. Find the force $\mathbf{F}$ required to keep the object at equilibrium.

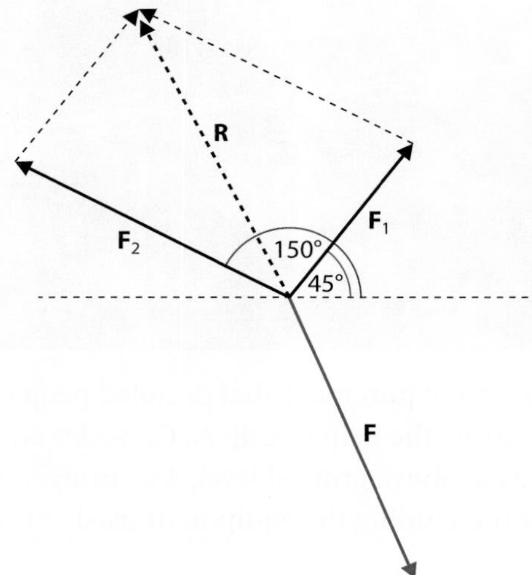

We will write the vectors for $\mathbf{F}_1$ and $\mathbf{F}_2$ in component form:

$$\mathbf{F}_1 = (20\cos 45°)\mathbf{i} + (20\sin 45°)\mathbf{j} = 10\sqrt{2}\mathbf{i} + 10\sqrt{2}\mathbf{j}$$

$$\mathbf{F}_1 = (40\cos 150°)\mathbf{i} + (40\sin 150°)\mathbf{j} = -20\sqrt{3}\mathbf{i} + 20\mathbf{j}$$

Now, the resultant force $\mathbf{R}$ is

$$\mathbf{R} = (10\sqrt{2}\mathbf{i} + 10\sqrt{2}\mathbf{j}) + (-20\sqrt{3}\mathbf{i} + 20\mathbf{j})$$

$$= (10\sqrt{2} - 20\sqrt{3})\mathbf{i} + (10\sqrt{2} + 20)\mathbf{j}$$

Finally, the force $\mathbf{F}$ required to keep the object at equilibrium is

$$\mathbf{F} = -\mathbf{R} = (-10\sqrt{2} + 20\sqrt{3})\mathbf{i} - (10\sqrt{2} + 20)\mathbf{j}$$

Vectors can be used to help tackle displacement situations. For example, an object at a position defined by the position vector (**a**, **b**) and a velocity vector (**c**, **d**) has a position vector (**a**, **b**) + t(**c**, **d**) after time t.

Displacement and velocity

Note: In navigation, the convention is that the **course** or **bearing** of a moving object is the angle that its direction makes with the north direction measured clockwise. So, for example, a ship going east has a bearing of 90°.

The velocity of an object can be represented by a vector whose direction is the direction of motion and whose magnitude is the speed of the object.

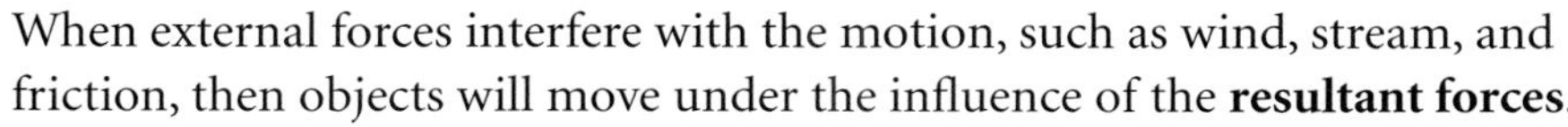

When external forces interfere with the motion, such as wind, stream, and friction, then objects will move under the influence of the **resultant forces**.

Example 10

An aeroplane heads in a northerly direction with a speed of 450 km/h. The wind is blowing in the direction of N 60° E with a speed of 60 km/h.

a) Write down the component forms of the plane's air velocity and the wind velocity.

b) Find the true velocity of the plane.

c) Find the true speed and direction of the plane.

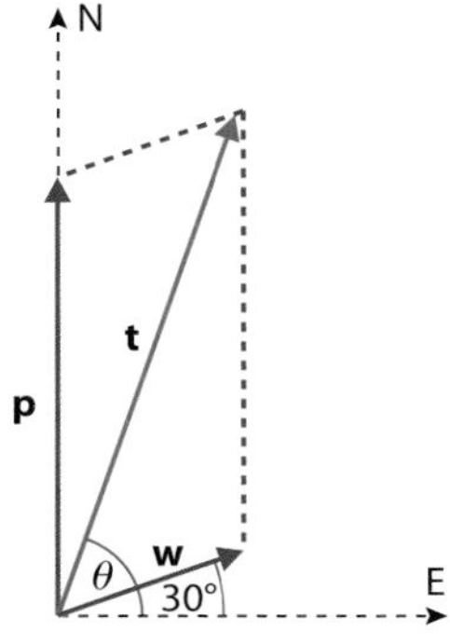

Solution

Let **p** be the vector for the plane's air velocity, **w** the wind's velocity, and **t** the true velocity.

a) $\mathbf{p} = 0\mathbf{i} + 450\mathbf{j}$

$$\mathbf{w} = (60\cos 30°)\mathbf{i} + (60\sin 30°)\mathbf{j} = 30\sqrt{3}\mathbf{i} + 30\mathbf{j}$$

b) The true velocity of the plane is the resultant of the two forces above, therefore

$$\mathbf{t} = \mathbf{p} + \mathbf{w} = (0\mathbf{i} + 450\mathbf{j}) + (30\sqrt{3}\mathbf{i} + 30\mathbf{j}) = 30\sqrt{3}\mathbf{i} + 480\mathbf{j}.$$

c) The true speed is given by the magnitude of **t**,

$$|\mathbf{t}| = \sqrt{(30\sqrt{3})^2 + 480^2} \approx 482.8 \text{ km/h}.$$

The direction is determined by the angle θ that the true velocity makes with the horizontal. From our discussion earlier, this can be found by using the property that $\tan\theta = \dfrac{480}{30\sqrt{3}} \approx 9.24$, and so $\theta \approx 83.8°$. So, we can now give the true direction of the plane as N 6.2° E.

Example 11

The position vector of a ship (MB) from its starting position at a port RJ is given by $\begin{pmatrix} x \\ y \end{pmatrix} = \begin{pmatrix} 5 \\ 20 \end{pmatrix} + t\begin{pmatrix} 12 \\ 16 \end{pmatrix}$. Distances are in kilometres and speeds are in km/h. t is time after 00 hour.

a) Find the position of the MB after 2 hours.

b) What is the speed of the MB?

c) Another ship (LW) is at sea in a location $\begin{pmatrix} 41 \\ 68 \end{pmatrix}$ relative to the same port. LW has stopped for some reason. Show that if LW does not start to move, the two ships will collide. Find the time of the potential collision.

d) To avoid collision, LW is ordered to leave its position and start moving at a velocity of $\begin{pmatrix} 15 \\ -36 \end{pmatrix}$ one hour after MB started. Find the position vector of LW.

e) How far apart are the two ships after two hours since the start of MB?

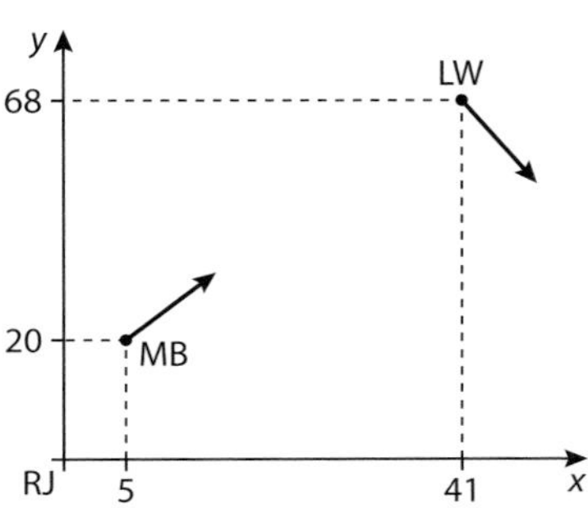

Solution

a) MB is at a position with vector $\begin{pmatrix} x \\ y \end{pmatrix} = \begin{pmatrix} 5 \\ 20 \end{pmatrix} + 2\begin{pmatrix} 12 \\ 16 \end{pmatrix} = \begin{pmatrix} 29 \\ 52 \end{pmatrix}$.

b) Since the velocity of the ship is $\begin{pmatrix} 12 \\ 16 \end{pmatrix}$, the speed is $\left|\begin{pmatrix} 12 \\ 16 \end{pmatrix}\right| = \sqrt{12^2 + 16^2}$ $= 20\,\text{km/h}$.

c) The collision can happen if the position vectors of the two ships are equal: $\begin{pmatrix} 5 \\ 20 \end{pmatrix} + t\begin{pmatrix} 12 \\ 16 \end{pmatrix} = \begin{pmatrix} 41 \\ 68 \end{pmatrix} \Rightarrow 5 + 12t = 41$ and $20 + 16t = 68 \Rightarrow 12t = 36$ and $16t = 48 \Rightarrow t = 3$. After 3 hours, at 03:00, a collision could happen.

d) Since LW started one hour later, its position vector is $\begin{pmatrix} x \\ y \end{pmatrix} = \begin{pmatrix} 41 \\ 68 \end{pmatrix} + (t-1)\begin{pmatrix} 15 \\ -36 \end{pmatrix}, t \geq 1$.

e) MB is at $\begin{pmatrix} 29 \\ 52 \end{pmatrix}$ and LW is at $\begin{pmatrix} 41 \\ 68 \end{pmatrix} + (2-1)\begin{pmatrix} 15 \\ -36 \end{pmatrix} = \begin{pmatrix} 56 \\ 32 \end{pmatrix}$. The distance between them is $\sqrt{(56-29)^2 + (32-52)^2} = \sqrt{1129} = 33.6\,\text{km}$.

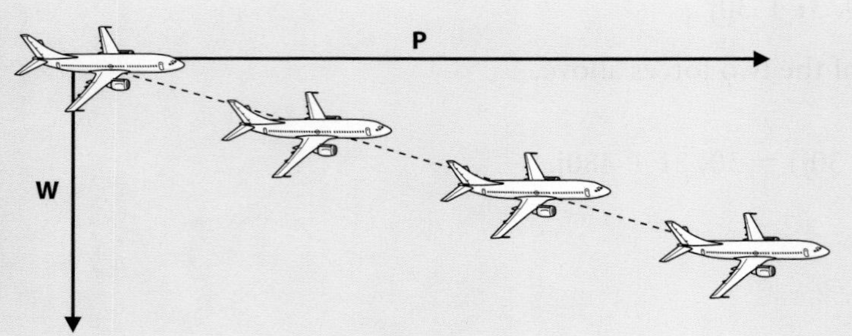

When the wind is strong and is acting in a direction different from that of the airplane and if you watch the plane from the ground you will notice that the 'nose' of the plane is in a direction (air velocity) different from the motion of the plane's 'true' velocity.

Exercise 9.3

1 Find the direction angle for each vector.

a) $\mathbf{u} = (2, 0)$

b) $\mathbf{v} = (0, 3)$

c) $\mathbf{w} = (-3, 0)$

d) $\mathbf{u} + \mathbf{v}$

e) $\mathbf{v} + \mathbf{w}$

2 Find the magnitude and direction angle for each vector.

a) $\mathbf{u} = (3, 2)$ b) $\mathbf{v} = (-3, -2)$ c) $2\mathbf{u}$

d) $3\mathbf{v}$ e) $2\mathbf{u} + 3\mathbf{v}$ f) $2\mathbf{u} - 3\mathbf{v}$

3 Find the magnitude and direction angle for each vector.

a) $\mathbf{u} = (-4, 7)$ b) $\mathbf{v} = (2, 5)$ c) $3\mathbf{u}$

d) $-2\mathbf{v}$ e) $3\mathbf{u} + 2\mathbf{v}$ f) $\mathbf{u} - \mathbf{v}$

4 Write each of the following vectors in component form. θ is the angle that the vector makes with the positive horizontal axis.

a) $|\mathbf{u}| = 310, \theta = 62°$ b) $|\mathbf{u}| = 43.2, \theta = 19.6°$

c) $|\mathbf{u}| = 12, \theta = 135°$ d) $|\mathbf{u}| = 240, \theta = 300°$

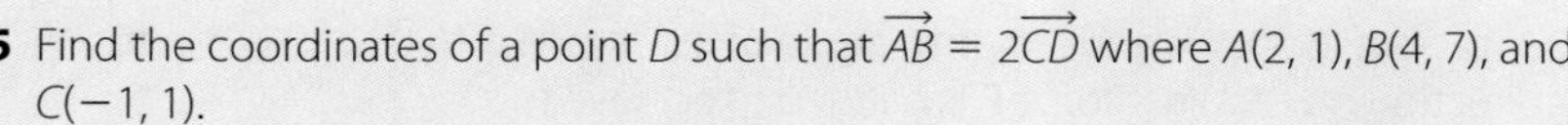

5 Find the coordinates of a point D such that $\overrightarrow{AB} = 2\overrightarrow{CD}$ where $A(2, 1)$, $B(4, 7)$, and $C(-1, 1)$.

6 Find the unit vector in the same direction as **u** in each of the following cases.

a) $\mathbf{u} = (3, 4)$

b) $\mathbf{u} = 2\mathbf{i} - 5\mathbf{j}$

7 Find a unit vector in the plane making an angle θ with the positive x-axis where

a) $\theta = 150°$

b) $\theta = 315°$

8 Find a vector of magnitude 7 that is parallel to $\mathbf{u} = 3\mathbf{i} - 4\mathbf{j}$.

9 Find a vector of magnitude 3 that is parallel to $\mathbf{u} = 2\mathbf{i} + 3\mathbf{j}$.

10 Find a vector of magnitude 7 that is perpendicular to $\mathbf{u} = 3\mathbf{i} - 4\mathbf{j}$.

11 Find a vector of magnitude 3 that is perpendicular to $\mathbf{u} = 2\mathbf{i} + 3\mathbf{j}$.

12 A plane is flying on a bearing of 170° at a speed of 840 km/h. The wind is blowing in the direction N 120° E with a strength of 60 km/h.

a) Find the vector components of the plane's still-air velocity and the wind's velocity.

b) Determine the true velocity (ground) of the plane in component form.

c) Write down the true speed and direction of the plane.

Note: In navigation, the convention is that the **course** or **bearing** of a moving object is the angle that its direction makes with the north direction measured clockwise. So, for example, a ship going east has a bearing of 090°.

13 A plane is flying on a compass heading of 340° at 520 km/h. The wind is blowing with the bearing 320° at 64 km/h.

a) Find the component form of the velocities of the plane and the wind.

b) Find the actual ground speed and direction of the plane.

14

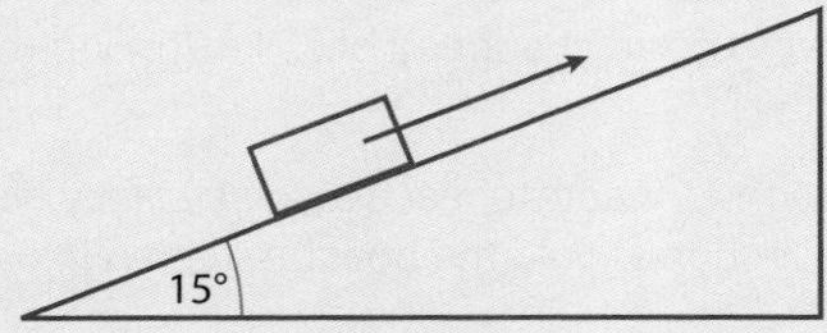

A box is being pulled up a 15° inclined plane. The force needed is 25 N. Find the horizontal and vertical components of the force vector and interpret each of them.

15 A motor boat with the power to steer across a river at 30 km/h is moving such that the bow is pointed in a northerly direction. The stream is moving eastward at 6 km/h. The river is 1 km wide. Where on the opposite side will the boat meet the land?

16 A force of 2500 N is applied at an angle of 38° to pull a 10 000 N ship in the direction given. What force **F** is needed to achieve this?

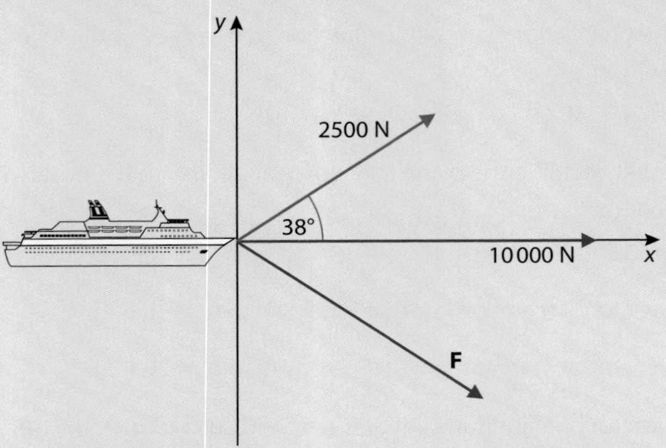

17 A boat is observed to have a bearing of 072°. The speed of the boat relative to still water is 40 km/h. Water is flowing directly south. The boat appears to be heading directly east.

a) Express the velocity of the boat with respect to the water in component form.

b) Find the speed of the water stream and the true speed of the boat.

18 A 50 N weight is suspended by two strings as shown. Find the tensions **T** and **S** in the strings.

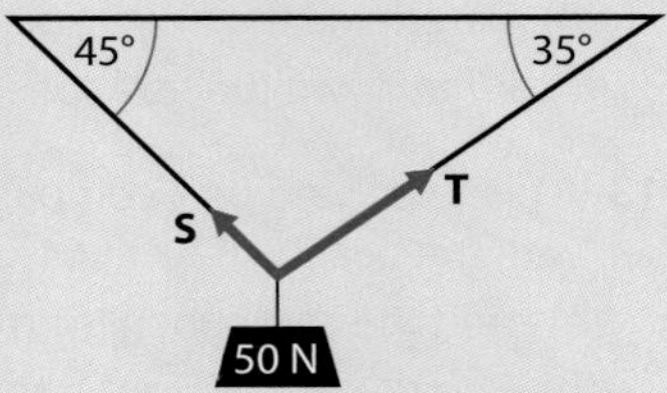

19 A runner runs in a westerly direction on the deck of a cruise ship at 8 km/h. The cruise ship is moving north at a speed of 35 km/h. Find the velocity of the runner relative to the water.

20 The boat in question 15 wants to reach a point exactly north of the starting point. In which direction should the boat be steered in order to achieve this objective?

21 Forces **F** = (−10, 3), **G** = (−4, 1) and **H** = (4, −10) act on a point **P**. Find the additional force required to keep the system in equilibrium.

22 A wind is blowing due west at 60 km/h. A small plane with air speed of 300 km/h is trying to maintain a course due north. In what direction should the pilot steer the plane to keep the targeted course? How fast is the plane moving?

23 The points $P(2, 2)$, $Q(10, 2)$ and $R(12, 6)$ are three vertices of a parallelogram. Find the fourth vertex S if

a) P and R are vertices of the same diagonal

b) P and R are vertices of a common side.

24 Show, using vector operations, that the diagonals of a parallelogram intersect each other.

25 Show, using vector operations, that the line segment joining the midpoints of two sides of a triangle is parallel to the third side and has half its length.

26 Prove that the midpoints of the sides of any quadrilateral are the vertices of a parallelogram.

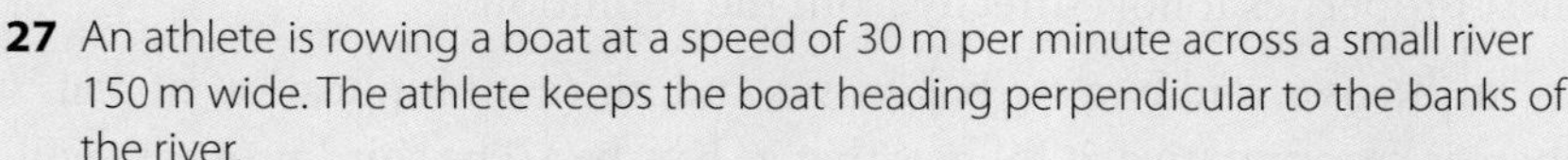

27 An athlete is rowing a boat at a speed of 30 m per minute across a small river 150 m wide. The athlete keeps the boat heading perpendicular to the banks of the river.

a) How far down the river does the boat reach the opposite side if the river is flowing at a rate of 10 m/minute?

b) How long does the trip last?

c) At what angle must the athlete steer the boat in order to reach a point directly opposite the starting point on the other side of the river? How long does the trip take?

28 A jet heads in the direction N 30° E at a speed of 400 km/h. The jet experiences a 20 km/h crosswind flowing due east. Find

a) the true velocity **p** of the jet,

b) the true speed and direction of the jet.

29 A box is carried by two strings F and G as shown right. The string F makes an angle of 45° with the horizontal while G makes an angle of 30°. The forces in F and G have a magnitude of 200 N each. The weight of the box is 300 N. What is the magnitude of the resultant force on the box and in which direction does it move?

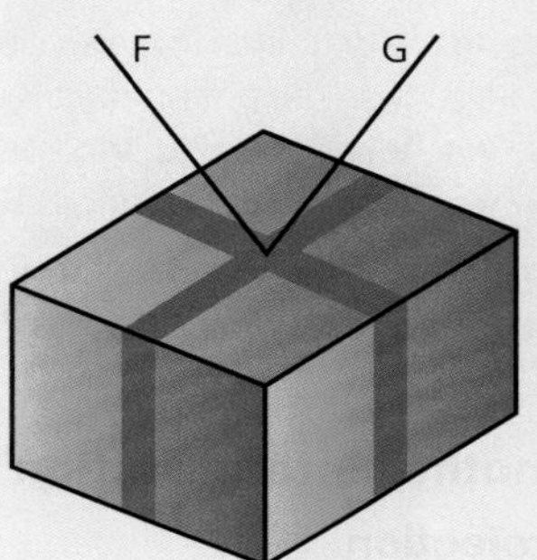

9.4 Scalar product of two vectors

The multiplication of two vectors is not uniquely defined: in other words, it is unclear whether the product will be a vector or not. For this reason there are two types of vector multiplication:

The **scalar** or **dot product** of two vectors, which results in a scalar; and the **vector** or **cross product** of two vectors, which results in a vector.

In this chapter, we shall discuss only the scalar or dot product. We will discuss the vector product in Chapter 14.

The **scalar product of two vectors**, **a** and **b** denoted by $\mathbf{a} \cdot \mathbf{b}$, is defined as the product of the magnitudes of the vectors times the cosine of the angle between them:
$\mathbf{a} \cdot \mathbf{b} = |\mathbf{a}||\mathbf{b}|\cos\theta$

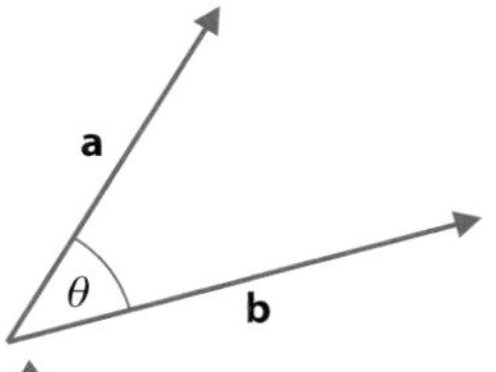

Figure 9.18

This is illustrated in Figure 9.18.

Note that the result of a dot product is a scalar, not a vector. The rules for scalar products are given in the following list:

$$\mathbf{a} \cdot \mathbf{b} = \mathbf{b} \cdot \mathbf{a}$$
$$0 \cdot \mathbf{a} = \mathbf{a} \cdot 0 = 0$$
$$\mathbf{a} \cdot (\mathbf{b} + \mathbf{c}) = \mathbf{a} \cdot \mathbf{b} + \mathbf{a} \cdot \mathbf{c}$$
$$\mathbf{a} \cdot \mathbf{a} = |\mathbf{a}|^2$$
$$k(\mathbf{a} \cdot \mathbf{b}) = k\mathbf{a} \cdot \mathbf{b} = \mathbf{a} \cdot k\mathbf{b}, \text{ with } k \text{ any scalar.}$$

The first properties follow directly from the definition:

$\mathbf{a}\cdot\mathbf{b} = |\mathbf{a}|\,|\mathbf{b}|\cos\theta$, and $\mathbf{b}\cdot\mathbf{a} = |\mathbf{b}|\,|\mathbf{a}|\cos\theta$, and, since multiplication of real numbers is commutative, it follows that $\mathbf{a}\cdot\mathbf{b} = \mathbf{b}\cdot\mathbf{a}$ The third property will be proved later in this section. Proofs of the rest of the properties are left as exercises.

Using the definition, it is immediately clear that for two non-zero vectors **u** and **v**, if **u** and **v** are perpendicular, the dot product is zero. This is so, because
$\mathbf{u}\cdot\mathbf{v} = |\mathbf{u}||\mathbf{v}|\cos\theta = |\mathbf{u}||\mathbf{v}|\cos 90° = |\mathbf{u}||\mathbf{v}| \times 0 = 0.$

The converse is also true: if $\mathbf{u}\cdot\mathbf{v} = 0$, the vectors are perpendicular,
$\mathbf{u}\cdot\mathbf{v} = 0 \Rightarrow |\mathbf{u}||\mathbf{v}|\cos\theta = 0 \Rightarrow \cos\theta = 0 \Rightarrow \theta = 90°.$

Using the definition, it is also clear that for two non-zero vectors **u** and **v**, if **u** and **v** are parallel then the dot product is equal to $\pm\,|\mathbf{u}||\mathbf{v}|$. This is so, because
$\mathbf{u}\cdot\mathbf{v} = |\mathbf{u}||\mathbf{v}|\cos\theta = |\mathbf{u}||\mathbf{v}|\cos 0° = |\mathbf{u}||\mathbf{v}| \times 1 = |\mathbf{u}||\mathbf{v}|$, or
$\mathbf{u}\cdot\mathbf{v} = |\mathbf{u}||\mathbf{v}|\cos\theta = |\mathbf{u}||\mathbf{v}|\cos 180° = |\mathbf{u}||\mathbf{v}| \times (-1) = -|\mathbf{u}||\mathbf{v}|.$

The converse is also true: if $\mathbf{u}\cdot\mathbf{v} = \pm\,|\mathbf{u}||\mathbf{v}|$, the vectors are parallel, since
$\mathbf{u}\cdot\mathbf{v} = |\mathbf{u}||\mathbf{v}|\cos\theta \Rightarrow |\mathbf{u}||\mathbf{v}|\cos\theta = \pm\,|\mathbf{u}||\mathbf{v}| \Rightarrow \cos\theta = \pm 1 \Rightarrow \theta = 0°$ or $\theta = 180°.$

Another interpretation of the dot product

Projection

(This subsection is optional – it is beyond the scope of the IB syllabus, but very helpful in clarifying the concept of dot products.)

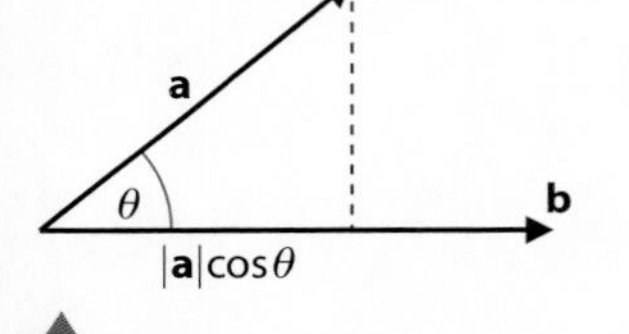

Figure 9.19

The quantity $|\mathbf{a}|\cos\theta$ is called the projection of the vector **a** on vector **b** (Figure 9.19). So, the dot product $\mathbf{b}\cdot\mathbf{a} = |\mathbf{b}||\mathbf{a}|\cos\theta = |\mathbf{b}|(|\mathbf{a}|\cos\theta)$
$= |\mathbf{b}| \times$ (the projection of **a** on **b**).

This fact is used in proving the third property on the list on page 419.

If we let B and C stand for the projections of **b** and **c** on **a**, we have
$\mathbf{a}(\mathbf{b} + \mathbf{c}) = |\mathbf{a}|(B + C) = |\mathbf{a}|B + |\mathbf{a}|C = \mathbf{a}\cdot\mathbf{b} + \mathbf{a}\cdot\mathbf{c}.$
This is called the **distributive property** of scalar products over vector addition. See Figure 9.20.

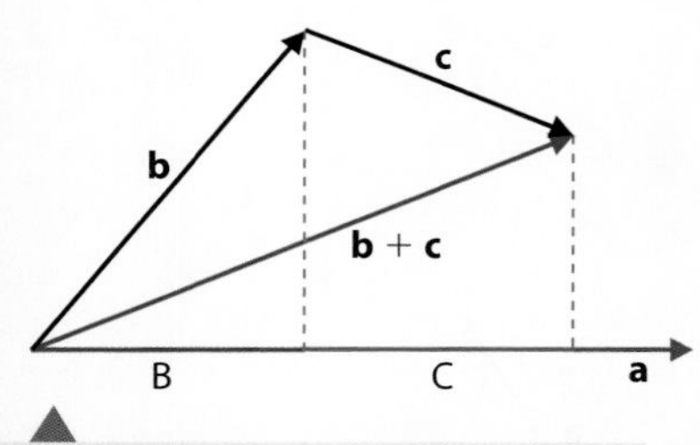

Figure 9.20

With this result, we can develop another definition for the dot product that is more useful in the calculation of this product.

Theorem

If vectors are expressed in component form, $\mathbf{u} = u_1\mathbf{i} + u_2\mathbf{j}$ and $\mathbf{v} = v_1\mathbf{i} + v_2\mathbf{j}$, then $\mathbf{u}\cdot\mathbf{v} = (u_1\mathbf{i} + u_2\mathbf{j})\cdot(v_1\mathbf{i} + v_2\mathbf{j}) = u_1v_1 + u_2v_2.$

Proof

$\mathbf{u}\cdot\mathbf{v} = (u_1\mathbf{i} + u_2\mathbf{j})\cdot(v_1\mathbf{i} + v_2\mathbf{j}) = u_1v_1\mathbf{i}^2 + u_1v_2\mathbf{ij} + u_2v_1\mathbf{ji} + u_2v_2\mathbf{j}^2$
However, $\mathbf{i}^2 = \mathbf{j}^2 = 1$ and $\mathbf{ij} = \mathbf{ji} = 0$. (Proof is left as an exercise for you.)

Therefore, $\mathbf{u}\cdot\mathbf{v} = (u_1\mathbf{i} + u_2\mathbf{j})\cdot(v_1\mathbf{i} + v_2\mathbf{j}) = u_1v_1 + u_2v_2.$

For example, to find the scalar product of the two vectors $\mathbf{u} = 2\mathbf{i} + 4\mathbf{j}$ and $\mathbf{v} = 5\mathbf{i} - 3\mathbf{j}$, it is enough to add the products' corresponding components:

$$\mathbf{u}\cdot\mathbf{v} = 2 \times 5 + 4 \times (-3) = -2$$

If we start the definition of the scalar product as $\mathbf{u} \cdot \mathbf{v} = u_1v_1 + u_2v_2$, we can deduce the other definition.

Start with the law of cosines which you learned in Chapter 8. Consider the diagram opposite and apply the law to finding *BC* in triangle *ABC*.

$$|\mathbf{u} - \mathbf{v}|^2 = |\mathbf{u}|^2 + |\mathbf{v}|^2 - 2|\mathbf{u}|\,|\mathbf{v}|\cos\theta$$

Using the fact that $\mathbf{u} \cdot \mathbf{u} = u_1u_1 + u_2u_2 = \mathbf{u}^2$,

$$\begin{aligned}|\mathbf{u} - \mathbf{v}|^2 &= (\mathbf{u} - \mathbf{v})^2 = (\mathbf{u} - \mathbf{v}) \cdot (\mathbf{u} - \mathbf{v})\\ &= u^2 - \mathbf{u} \cdot \mathbf{v} - \mathbf{v} \cdot \mathbf{u} + v^2 = u^2 - \mathbf{u} \cdot \mathbf{v} - \mathbf{u} \cdot \mathbf{v} + v^2\\ &= |\mathbf{u}|^2 - 2(\mathbf{u} \cdot \mathbf{v}) + |\mathbf{v}|^2\end{aligned}$$

Now, comparing the two results

$$|\mathbf{u} - \mathbf{v}|^2 = |\mathbf{u}|^2 - 2(\mathbf{u} \cdot \mathbf{v}) + |\mathbf{v}|^2 = |\mathbf{u}|^2 + |\mathbf{v}|^2 - 2|\mathbf{u}|\,|\mathbf{v}|\cos\theta$$

$$\Rightarrow -2(\mathbf{u} \cdot \mathbf{v}) = -2|\mathbf{u}|\,|\mathbf{v}|\cos\theta \Rightarrow \mathbf{u} \cdot \mathbf{v} = |\mathbf{u}|\,|\mathbf{v}|\cos\theta$$

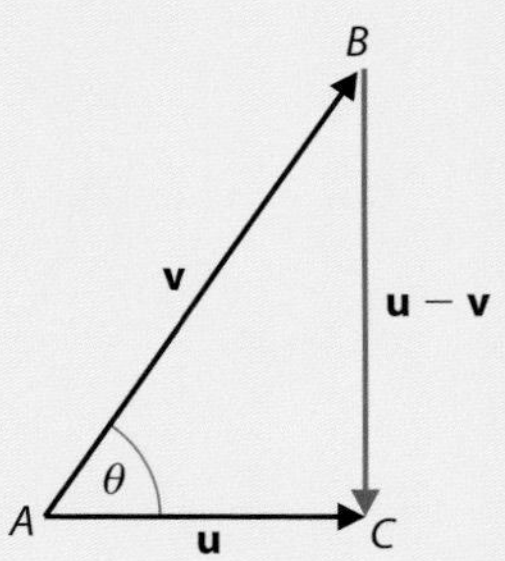

Example 12

Find the dot product of $\mathbf{u} = 2\mathbf{i} - 3\mathbf{j}$ and $\mathbf{v} = 3\mathbf{i} + 2\mathbf{j}$.

Solution

$$\mathbf{u} \cdot \mathbf{v} = 2 \times 3 - 3 \times 2 = 0$$

What does this tell us about the two vectors?

The angle between two vectors

The basic definition of the scalar product offers us a method for finding the angle between two vectors.

Since $\mathbf{u} \cdot \mathbf{v} = |\mathbf{u}||\mathbf{v}|\cos\theta$, then $\cos\theta = \dfrac{\mathbf{u} \cdot \mathbf{v}}{|\mathbf{u}||\mathbf{v}|}$.

Note: When the vectors **u** and **v** are given in component form, then the angle cosine can be directly calculated with

$$\cos\theta = \frac{\mathbf{u} \cdot \mathbf{v}}{|\mathbf{u}||\mathbf{v}|} = \frac{u_1v_1 + u_2v_2}{\sqrt{u_1^2 + u_2^2}\sqrt{v_1^2 + v_2^2}}$$

Example 13

Find the angle between the following two vectors:

$$\mathbf{v} = -3\mathbf{i} + 3\mathbf{j} \text{ and } \mathbf{w} = 2\mathbf{i} - 4\mathbf{j}$$

Solution

$$\cos\theta = \frac{\mathbf{v} \cdot \mathbf{w}}{|\mathbf{v}||\mathbf{w}|} = \frac{-3 \times 2 + 3 \times -4}{\sqrt{(-3)^2 + 3^2} \times \sqrt{2^2 + 4^2}} = \frac{-18}{\sqrt{18}\sqrt{20}} \Rightarrow \theta = 161.57°$$

Example 14

Consider the segment $[AB]$ with $A(-2, -3)$ and $B(3, 1)$. Use dot products to find the equation of the circle whose diameter is AB.

Solution

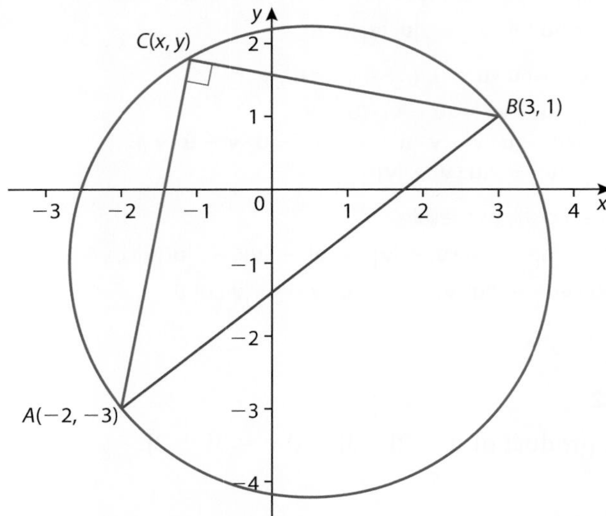

Consider any point $C(x, y)$ on the graph. Find the vectors $\overrightarrow{AC}$ and $\overrightarrow{BC}$. For the point C to be on the circle, the angle at C must be a right angle. Hence, the vectors $\overrightarrow{AC}$ and $\overrightarrow{BC}$ are perpendicular.

For perpendicular vectors, the dot product must be zero.

$$\overrightarrow{AC} = (x + 2, y + 3), \overrightarrow{BC} = (x - 3, y - 1)$$

$$\overrightarrow{AC} \cdot \overrightarrow{BC} = 0 \Rightarrow (x + 2)(x - 3) + (y + 3)(y - 1) = 0$$

$$\Rightarrow x^2 - x + y^2 + 2y = 9$$

Example 15

Show that the vector $\mathbf{n} = a\mathbf{i} + b\mathbf{j}$ is orthogonal (perpendicular) to the line l with equation $ax + by + c = 0$.

Solution

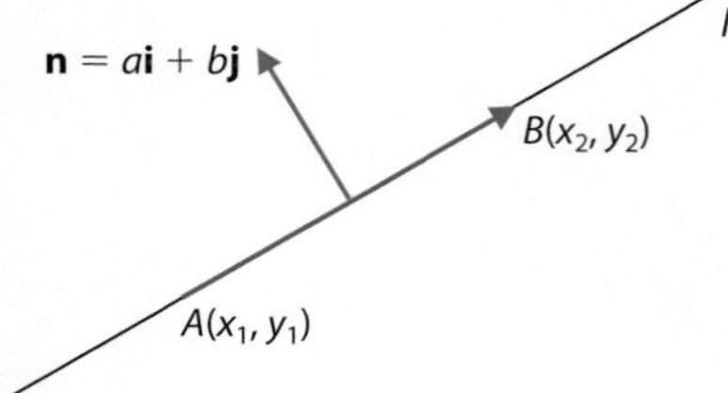

Consider two points A and B on the line with the coordinates as shown.

$\overrightarrow{AB} = (x_2 - x_1, y_2 - y_1)$ and

$\mathbf{n} \cdot \overrightarrow{AB} = (a, b) \cdot (x_2 - x_1, y_2 - y_1) = (ax_2 + by_2) - (ax_1 + by_1)$, but A and B are on the line, so

$ax_2 + by_2 = -c$ and $ax_1 + by_1 = -c \Rightarrow \mathbf{n} \cdot \overrightarrow{AB} = -c + c = 0.$

Example 16

Find the distance from the point $P(x_0, y_0)$ to the line l with equation $ax + by + c = 0$.

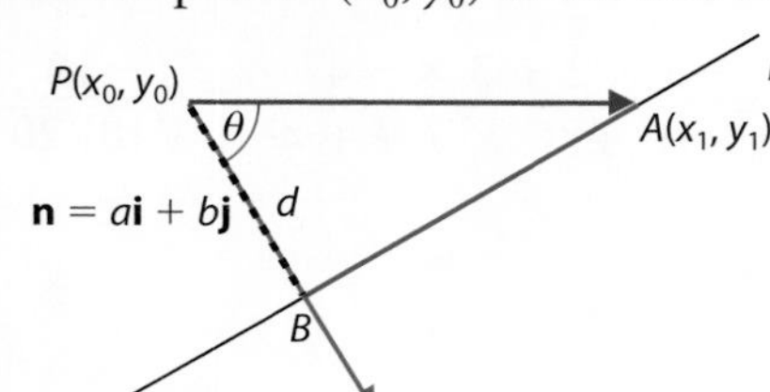

Solution

The required distance, d, can be found using triangle PAB.

$$d = \left| |\overrightarrow{PA}| \cos\theta \right| = \left| |\overrightarrow{PA}| \frac{\overrightarrow{PA} \cdot \mathbf{n}}{|\overrightarrow{PA}||\mathbf{n}|} \right| = \left| \frac{\overrightarrow{PA} \cdot \mathbf{n}}{|\mathbf{n}|} \right|, \left(\frac{\overrightarrow{PA} \cdot \mathbf{n}}{|\mathbf{n}|} \text{ is called the component of } \overrightarrow{PA} \text{ along } \mathbf{n}. \right)$$

Now,

$$\overrightarrow{PA} = (x_1 - x_0, y_1 - y_0) \Rightarrow \overrightarrow{PA} \cdot \mathbf{n} = a(x_1 - x_0) + b(y_1 - y_0)$$

$$\Rightarrow \overrightarrow{PA} \cdot \mathbf{n} = ax_1 + by_1 - ax_0 - by_0 = -c - ax_0 - by_0$$

Therefore, $d = \left| \frac{\overrightarrow{PA} \cdot \mathbf{n}}{|\mathbf{n}|} \right| = \left| \frac{-c - ax_0 - by_0}{\sqrt{a^2 + b^2}} \right| = \frac{|ax_0 + by_0 + c|}{\sqrt{a^2 + b^2}}$.

So, for example, the distance from $A(2, -3)$ to the line with equation $5x + 3y = 2$ is

$$d = \frac{|5(2) + 3(-3) - 2|}{\sqrt{5^2 + 3^2}} = \frac{1}{\sqrt{34}} = \frac{\sqrt{34}}{34}.$$

Example 17

The instrument panel in a plane indicates that its airspeed (the speed of the plane relative to the surrounding air) is 200 km/h and that its compass heading (the direction in which the plane's nose is pointing) is N 45° E. There is a steady wind blowing from the west at 50 km/h. Because of the wind, the plane's *true* velocity is different from the panel reading. Find the true velocity of the plane. Also, find its true speed and direction.

Solution

A diagram can help clarify the situation.

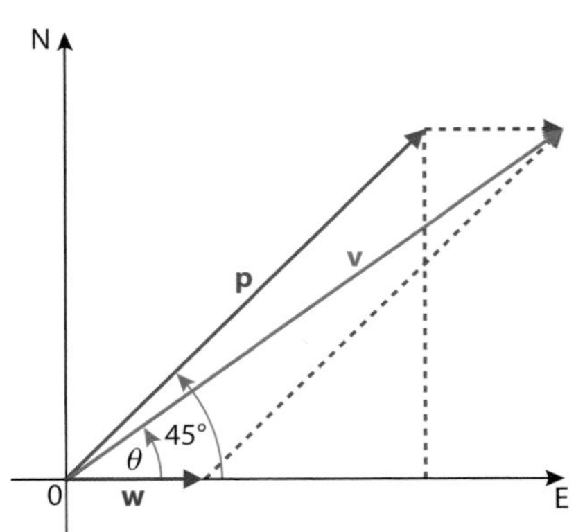

The plane velocity $\mathbf{p}$ can be expressed in its component form:

$$x = |\mathbf{p}|\cos 45° = 200\cos 45° = 100\sqrt{2},$$

$$y = |\mathbf{p}|\sin 45° = 200\sin 45° = 100\sqrt{2},$$

so $\mathbf{p}$ can be written as $\mathbf{p} = (100\sqrt{2}, 100\sqrt{2})$.

The wind velocity $\mathbf{w}$ can also be expressed in component form:

$$\mathbf{w} = (50, 0)$$

So, the true velocity, $\mathbf{v} = (100\sqrt{2} + 50, 100\sqrt{2})$.

To find the true speed, we find the magnitude of the resultant found above:

$$|\mathbf{v}| = \sqrt{(100\sqrt{2} + 50)^2 + (100\sqrt{2})^2} \approx 238 \text{ km/h}$$

To find the true direction, we find θ and calculate the *heading* of the plane:

$$\tan\theta = \frac{100\sqrt{2}}{100\sqrt{2} + 50} \approx 0.739 \Rightarrow \theta \approx 36.5°,$$

so the true direction is N 53.5° E.

Exercise 9.4

1 Find (i) $\mathbf{u} \cdot \mathbf{v}$ and (ii) the angle between $\mathbf{u}$ and $\mathbf{v}$ to the nearest degree.

a) $\mathbf{u} = \mathbf{i} + \sqrt{3}\,\mathbf{j}$, $\mathbf{v} = \sqrt{3}\,\mathbf{i} - \mathbf{j}$

b) $\mathbf{u} = (2, 5)$, $\mathbf{v} = (4, 1)$

c) $\mathbf{u} = 2\mathbf{i} - 3\mathbf{j}$, $\mathbf{v} = 4\mathbf{i} - \mathbf{j}$

d) $\mathbf{u} = 2\mathbf{j}$, $\mathbf{v} = -\mathbf{i} + \sqrt{3}\,\mathbf{j}$

e) $\mathbf{u} = (-3, 0)$, $\mathbf{v} = (0, 7)$

f) $\mathbf{u} = (3, 0)$, $\mathbf{v} = (\sqrt{3}, 1)$

g) $\mathbf{u} = -6\mathbf{j}$, $\mathbf{v} = -2\mathbf{i} + 2\sqrt{3}\mathbf{j}$

h) $\mathbf{u} = 2\mathbf{i} + 2\mathbf{j}$, $\mathbf{v} = -4\mathbf{i} - 4\mathbf{j}$

2 Using the vectors $\mathbf{u} = 3\mathbf{i} - 2\mathbf{j}$, $\mathbf{v} = \mathbf{i} + 3\mathbf{j}$ and $\mathbf{w} = 4\mathbf{i} + 5\mathbf{j}$, find each of the indicated results.

a) $\mathbf{u} \cdot (\mathbf{v} + \mathbf{w})$

b) $\mathbf{u} \cdot \mathbf{v} + \mathbf{u} \cdot \mathbf{w}$

c) $\mathbf{u}(\mathbf{v} \cdot \mathbf{w})$

d) $(\mathbf{u} \cdot \mathbf{v})\mathbf{w}$

e) $(\mathbf{u} \cdot \mathbf{v})(\mathbf{u} \cdot \mathbf{w})$

f) $(\mathbf{u} + \mathbf{v}) \cdot (\mathbf{u} - \mathbf{v})$

g) Looking at a)–d) write one paragraph to summarize what you learned!

3 Determine whether $\mathbf{u}$ is orthogonal, parallel or neither to $\mathbf{v}$:

$$\mathbf{u} = \begin{pmatrix} -\frac{1}{2} \\ 2 \end{pmatrix}, \mathbf{v} = \begin{pmatrix} -2 \\ \frac{1}{2} \end{pmatrix}$$

$$\mathbf{u} = \begin{pmatrix} 8 \\ 4 \end{pmatrix}, \mathbf{v} = \begin{pmatrix} 6 \\ -12 \end{pmatrix}$$

$$\mathbf{u} = \begin{pmatrix} 2\sqrt{3} \\ 2 \end{pmatrix}, \mathbf{v} = \begin{pmatrix} 1 \\ -\sqrt{3} \end{pmatrix}$$

• **Hint:** The work done by any force is defined as the product of the force multiplied by the distance it moves a certain object. In other words, it is the product of the force multiplied by the displacement of the object. As such, work is the dot product between the force and displacement $\mathbf{W} = \mathbf{F} \cdot \mathbf{D}$.

4 Find the work done by the force $\mathbf{F}$ in moving an object between points M and N.

a) $\mathbf{F} = 400\mathbf{i} - 50\mathbf{j}$, $M(2, 3)$, $N(12, 43)$

b) $\mathbf{F} = 30\mathbf{i} + 150\mathbf{j}$, $M(0, 30)$, $N(15, 70)$

c) $\mathbf{F} = \begin{pmatrix} 5 \\ 25 \end{pmatrix}$, $M(0, 0)$, $N(1, 6)$

5 Find the interior angles of the triangle ABC.

a) $A(1, 2)$, $B(3, 4)$, $C(2, 5)$

b) $A(3, 4)$, $B(-1, -7)$, $C(-8, -2)$

c) $A(3, -5)$, $B(1, -9)$, $C(-7, -9)$

6 Find a vector perpendicular to $\mathbf{u}$ in each case below. (Answers are not unique!)

a) $\mathbf{u} = (3, 5)$

b) $\mathbf{u} = \frac{1}{2}\mathbf{i} - \frac{3}{4}\mathbf{j}$

7 Use the dot product to find the equation of a circle whose diameter is [AB].

a) $A(1, 2)$, $B(3, 4)$

b) $A(3, 4)$, $B(-1, -7)$

8 Decide whether the triangle ABC is right-angled using vector algebra: $A(1, -3)$, $B(2, 0)$, $C(6, -2)$

9 Find t such that $\mathbf{a} = t\mathbf{i} - 3\mathbf{j}$ is perpendicular to $\mathbf{b} = 5\mathbf{i} + 7\mathbf{j}$.

10 For what value(s) of b are the vectors $(-6, b)$ and (b, b^2) perpendicular?

11 Find a unit vector that makes an angle of $60°$ with $\mathbf{u} = (3, 4)$.

12 Find t such that $\mathbf{a} = t\mathbf{i} - \mathbf{j}$ and $\mathbf{b} = \mathbf{i} + \mathbf{j}$ make an angle of $\frac{3}{4}\pi$ radians.

13 Use the dot product to prove that the diagonals of a rhombus are perpendicular to each other.

14 Find the component of **u** along **v** if

a) $\mathbf{u} = (0, 7), \mathbf{v} = (6, 8)$

b) $\mathbf{u} = \begin{pmatrix} -\frac{1}{2} \\ 2 \end{pmatrix}, \mathbf{v} = \begin{pmatrix} -2 \\ \frac{1}{2} \end{pmatrix}$

15 A young man pulls a sled horizontally by exerting a force of 16 N on the rope that is tied to its front end. The rope makes an angle of 45° with the horizontal. Find the work done in pulling the sled 55 m.

16 Find the distance from the point P to the line l in each case:

a) $P(0, 0), l: 3x - 4y + 5 = 0$

b) $P(2, 2), l: 3x - 2y = 2$

c) $P(1, 5), l: 5x - 3y = 11$

17 Given three points in the plane P, Q, and R such that $\overrightarrow{OP} \perp \overrightarrow{QR}$ and $\overrightarrow{OQ} \perp \overrightarrow{PR}$, use scalar product to show that $\overrightarrow{OR} \perp \overrightarrow{PQ}$.

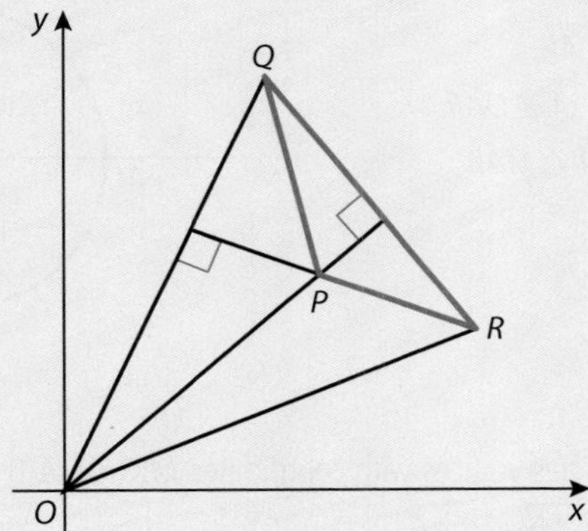

18 Two vectors $\begin{pmatrix} 3 \\ 4 \end{pmatrix}$ and $\begin{pmatrix} x \\ 1 \end{pmatrix}$ have an angle of 30° between them. Find the possible values of x.

19 A weight of 1000 N is supported by two forces $\mathbf{a} = (-200, 400)$ and $\mathbf{b} = (200, 600)$. The weight is in equilibrium. Find the angles α, β, and θ.

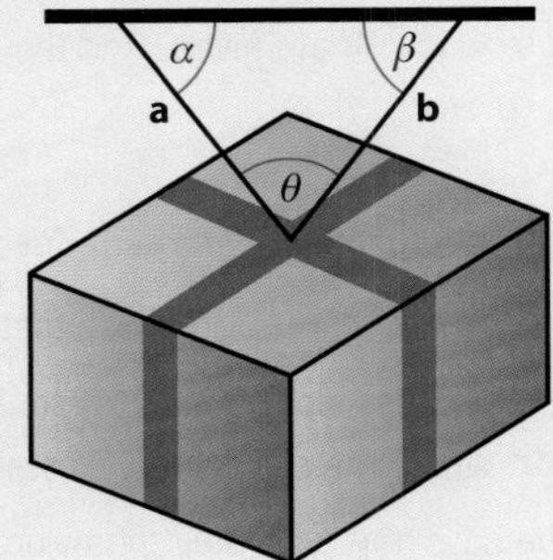

20 Show that the vector $|\mathbf{a}|\mathbf{b} + |\mathbf{b}|\mathbf{a}$ bisects the angle between the two vectors **a** and **b**.

Practice questions

1 $ABCD$ is a rectangle with M the midpoint of $[AB]$. $\mathbf{u}$ and $\mathbf{v}$ represent the vectors joining M to D and C respectively. Express each of the following vectors in terms of $\mathbf{u}$ and $\mathbf{v}$.

a) $\overrightarrow{DC}$

b) $\overrightarrow{AM}$

c) $\overrightarrow{BC}$

d) $\overrightarrow{AC}$

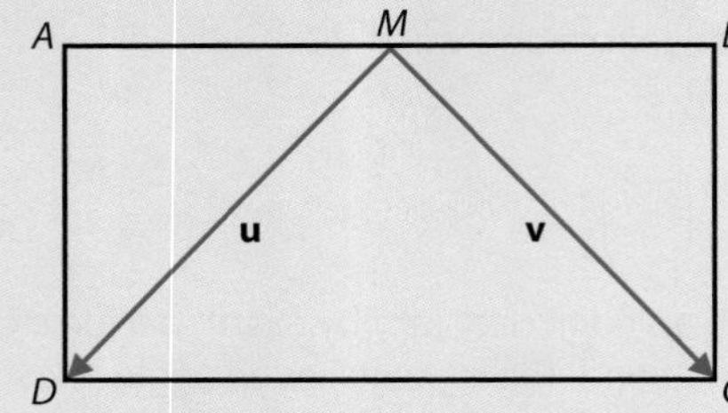

2 Consider the vectors $\mathbf{u} = \mathbf{i} - 2\mathbf{j}$ and $\mathbf{v} = 4\mathbf{i} + 3\mathbf{j}$.

a) Find the component form of the vector $\mathbf{w} = 2\mathbf{u} + \mathbf{v}$.

b) Find the vector $\mathbf{z}$ which has a magnitude of 6 units and same direction as $\mathbf{w}$.

3 M and A are the ends of the diameter of a circle with centre at the origin. The radius of the circle is 15 cm and $\overrightarrow{OR} = \begin{pmatrix} 10 \\ 5\sqrt{5} \end{pmatrix}$.

a) Verify that R lies on the circle.

b) Find the vector $\overrightarrow{AR}$.

c) Find the cosine of $\angle OAR$.

d) Find the area of $\triangle MAR$.

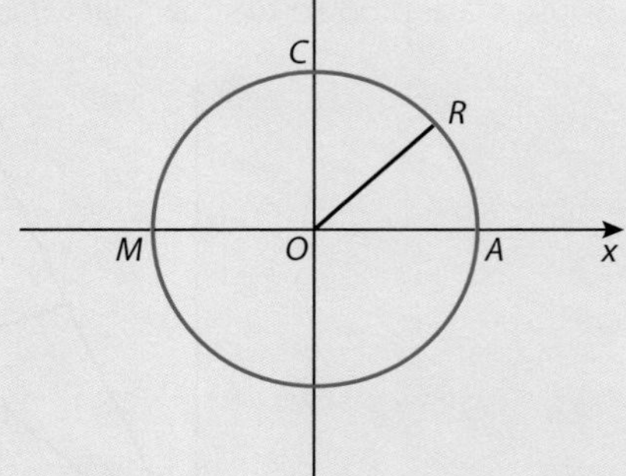

4 Quadrilateral $MARC$ has vertices with coordinates $M(0, 0)$, $A(6, 2)$, $R(11, 4)$ and $C(3, 8)$.

a) Find the vectors $\overrightarrow{MR}$ and $\overrightarrow{AC}$.

b) Find the angle between the diagonals of quadrilateral $MARC$.

c) Let the vector $\mathbf{u}$ be the vector joining the midpoints of $[MA]$ and $[AR]$, and $\mathbf{v}$ be the vector joining the midpoints of $[RC]$ and $[CM]$. Compare $\mathbf{u}$ and $\mathbf{v}$ to $\overrightarrow{MR}$, and hence show that the quadrilateral connecting the midpoints of the sides of $MARC$ form a parallelogram.

5 Vectors $\mathbf{u} = 5\mathbf{i} + 3\mathbf{j}$ and $\mathbf{v} = \mathbf{i} - 4\mathbf{j}$ are given. Find the scalars m and n such that $m(\mathbf{u} + \mathbf{v}) - 5\mathbf{i} + 7\mathbf{j} = n(\mathbf{u} - \mathbf{v})$.

6 Vector $\begin{pmatrix} 1 \\ 0 \end{pmatrix}$ represents a displacement in the eastern direction while vector $\begin{pmatrix} 0 \\ 1 \end{pmatrix}$ represents a displacement north. Distances are in kilometres.

Two crews of workers are laying gas pipes in a north-south direction across the North Sea. Consider the base port where the crews leave to start work as the origin (0, 0).

At 07:00 the crews left the base port with their motor boats to two different locations. The crew called 'Marco' travel at a velocity of $\begin{pmatrix} 9 \\ 12 \end{pmatrix}$ and the crew called 'Tony' travel at a velocity of $\begin{pmatrix} 18 \\ -8 \end{pmatrix}$. Speeds are in km/h.

a) Find the speed of each boat.

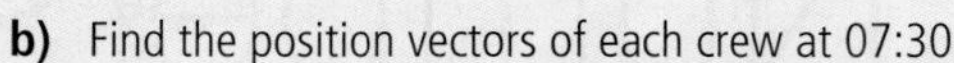

b) Find the position vectors of each crew at 07:30.

c) Hence, or otherwise, find the distance between the vehicles at 07:30.

d) At 07:30 'Tony' stops and the crew begins laying pipes towards the north. 'Marco' continues travelling in the same direction at the same speed until it is exactly north of 'Tony'. At this point, 'Marco' stops and the crew then begins laying pipes towards the south. At what time does 'Marco' start work?

e) Each crew lays an average of 400 m of pipe in an hour. If they work non-stop until their lunch break at 12:30, what is the distance between them at this time?

f) How long would 'Marco' take to return to base port from its lunchtime position, assuming it travelled in a straight line and with the same average speed as on the morning journey? (Give your answer to the nearest minute.)

7 Triangle *TRI* is defined as follows:

$\overrightarrow{OT} = \begin{pmatrix} 3 \\ -1 \end{pmatrix}$, $\overrightarrow{TR} = \begin{pmatrix} 5 \\ 6 \end{pmatrix}$, $\overrightarrow{TR} \cdot \overrightarrow{IR} = 0$, and $\overrightarrow{TI} = k\mathbf{j}$ where k is a scalar and $\mathbf{j}$ is the unit vector in the y-direction.

a) Draw an accurate diagram of $\triangle TRI$.

b) Write the vector $\overrightarrow{IR}$.

8 Vector $\begin{pmatrix} 1 \\ 0 \end{pmatrix}$ represents a displacement in the eastern direction while vector $\begin{pmatrix} 0 \\ 1 \end{pmatrix}$ represents a displacement north. Distances are in kilometres.

The position vector of a plane for AUA airlines from its starting position in Vienna is given by $\begin{pmatrix} x \\ y \end{pmatrix} = \begin{pmatrix} 25 \\ 40 \end{pmatrix} + t\begin{pmatrix} 360 \\ 480 \end{pmatrix}$. Speeds are in km/h and t is time after 00 hour.

a) Find the position of the AUA plane after 2 hours.

b) What is the speed of the plane?

c) A plane for LH airline started at the same time from a location $\begin{pmatrix} -155 \\ 1300 \end{pmatrix}$ relative to Vienna and moving with a velocity vector $\begin{pmatrix} 480 \\ -360 \end{pmatrix}$, flying at the same height as the AUA plane. Show that if the LH plane does not change route, the two planes will collide. Find the time of the potential collision.

d) To avoid collision, the LH plane is ordered to leave its position and start moving at a velocity of $\begin{pmatrix} 450 \\ -390 \end{pmatrix}$ one hour after it started. Find the position vector of the LH plane at that time.

e) How far apart are the two planes after two hours?

9 For what value(s) of n are the vectors $\begin{pmatrix} 3n \\ 2n+3 \end{pmatrix}$ and $\begin{pmatrix} 2n-1 \\ 4-2n \end{pmatrix}$ perpendicular. Otherwise, show that it is not possible.

10 Let α be the angle between the vectors **a** and **b**, where

$$\mathbf{a} = (\cos\theta)\mathbf{i} + (\sin\theta)\mathbf{j},\ \mathbf{b} = (\sin\theta)\mathbf{i} + (\cos\theta)\mathbf{j} \text{ and } 0 < \theta < \frac{\pi}{4}.$$

Express α in terms of θ.

11 Given two non-zero vectors **a** and **b** such that $|\mathbf{a} + \mathbf{b}| = |\mathbf{a} - \mathbf{b}|$, find the value of $\mathbf{a} \cdot \mathbf{b}$.

10 Complex Numbers

Assessment statements

1.5 Complex numbers: the number $i = \sqrt{-1}$; the term's real part, imaginary part, conjugate, modulus and argument.
Cartesian form $z = a + ib$.
Sums, products and quotients of complex numbers.

1.6 Modulus–argument (polar) form $z = r(\cos\theta + i\sin\theta) = r\text{cis}(\theta) = re^{i\theta}$.
The complex plane.

1.7 De Moivre's theorem.
Powers and roots of a complex number.

1.8 Conjugate roots of polynomial equations with real coefficients.

Introduction

You have already met complex numbers in Chapters 1 and 3. This chapter will broaden your understanding to include trigonometric representation of complex numbers and some applications.

Fractals can be generated using complex numbers.

Solving a linear equation of the form

$$ax + b = 0, \text{ with } a \neq 0$$

is a straightforward procedure if we are using the set of real numbers. The situation, as you already know, is different with quadratic equations. For example, as you have seen in Chapter 3, solving the quadratic equation

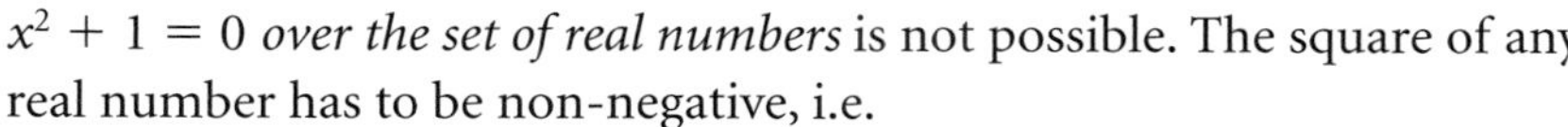

$x^2 + 1 = 0$ *over the set of real numbers* is not possible. The square of any real number has to be non-negative, i.e.

$(x^2 \geqslant 0 \Leftrightarrow x^2 + 1 \geqslant 1) \Rightarrow x^2 + 1 > 0$ for any choice of a real number x.

This means that $x^2 + 1 = 0$ is impossible for every real number x. This forces us to introduce a new set where such a solution is possible.

Numbers such as $\sqrt{-1}$ are not intuitive and many mathematicians in the past resisted their introduction, so they are called **imaginary numbers**.

The situation with finding a solution to $x^2 + 1 = 0$ is analogous to the following scenario: For a child in the first or second grade, a question such as $5 + ? = 9$ is manageable. However, a question such as $5 + ? = 2$ is impossible because the student's knowledge is *restricted* to the set of positive integers.

However, at a later stage when the same student is faced with the same question, he/she can solve it because their scope has been *extended* to include negative numbers too.

Also, at early stages an equation such as

$$x^2 = 5$$

cannot be solved till the student's knowledge of sets is extended to include irrational numbers where he/she can recognize numbers such as $x = \pm\sqrt{5}$.

The situation is much the same for $x^2 + 1 = 0$. We *extend* our number system to include numbers such as $\sqrt{-1}$; i.e. a number whose square is -1.

Thanks to Euler's (1707–1783) seminal work on imaginary numbers, they now feature prominently in the number system. Euler skilfully employed them to obtain many interesting results. Later, Gauss (1777–1855) represented them as points in the plane and renamed them as **complex numbers**, using them to obtain various significant results in number theory.

10.1 Complex numbers, sums, products and quotients

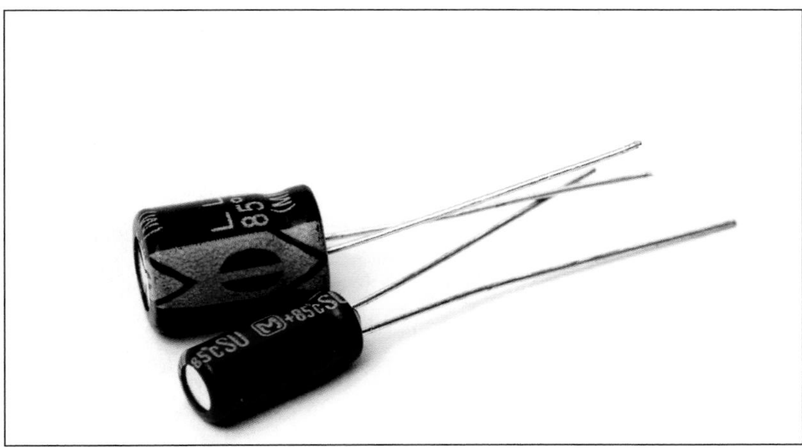

Electronic components like capacitors are used in AC circuits. Their effects are represented using complex numbers.

As you have seen in the introduction, the development of complex numbers had its origin in the search for methods of solving polynomial equations. The quadratic formula

$$x = \frac{-b}{2a} \pm \frac{\sqrt{b^2 - 4ac}}{2a}$$

had been used earlier than the 16th century to solve quadratic equations – in more primitive notations, of course. However, mathematicians stopped short of using it for cases where $b^2 - 4ac$ was negative. The use of the formula in cases where $b^2 - 4ac$ is negative depends on two principles (in addition to the other principles inherent in the set of real numbers, such as associativity and commutativity of multiplication).

1. $\sqrt{-1} \cdot \sqrt{-1} = -1$
2. $\sqrt{-k} = \sqrt{k} \cdot \sqrt{-1}$ **for any real number** $k > 0$

Example 1

Multiply $\sqrt{-36} \cdot \sqrt{-49}$.

Solution

First we simplify each square root using rule 2.

$$\sqrt{-36} = \sqrt{36} \cdot \sqrt{-1} = 6 \cdot \sqrt{-1}$$
$$\sqrt{-49} = \sqrt{49} \cdot \sqrt{-1} = 7 \cdot \sqrt{-1}$$

And hence using rule 1 with the other obvious rules:

$$\sqrt{-36} \cdot \sqrt{-49} = 6 \cdot \sqrt{-1} \cdot 7 \cdot \sqrt{-1} = 42 \cdot \sqrt{-1} \cdot \sqrt{-1} = -42$$

To deal with the quadratic formula expressions that consist of combinations of real numbers and square roots of negative numbers, we can apply the rules of binomials to numbers of the form

$$a + b\sqrt{-1}$$

where a and b are real numbers. For example, to add $5 + 7\sqrt{-1}$ to $2 - 3\sqrt{-1}$ we combine 'like' terms as we do in polynomials:

$$\begin{aligned}(5 + 7\sqrt{-1}) + (2 - 3\sqrt{-1}) &= 5 + 2 + 7\sqrt{-1} - 3\sqrt{-1} \\ &= (5 + 2) + (7 - 3)\sqrt{-1} = 7 + 4\sqrt{-1}\end{aligned}$$

Similarly, to multiply these numbers we use the binomial multiplication procedures:

$$\begin{aligned}(5 + 7\sqrt{-1}) \cdot (2 - 3\sqrt{-1}) &= 5 \cdot 2 + (7\sqrt{-1}) \cdot (-3\sqrt{-1}) + 5 \cdot (-3\sqrt{-1}) \\ &\quad + (7\sqrt{-1}) \cdot 2 \\ &= 10 - 21 \cdot (\sqrt{-1})^2 - 15 \cdot \sqrt{-1} + 14 \cdot \sqrt{-1} \\ &= 10 - 21 \cdot (-1) + (-15 + 14)\sqrt{-1} \\ &= 31 - \sqrt{-1}\end{aligned}$$

Euler introduced the symbol i for $\sqrt{-1}$.

A **pure imaginary number** is a number of the form ki, where k is a real number and i, the **imaginary unit**, is defined by $i^2 = -1$.

Note: In some cases, especially in engineering sciences, the number i is sometimes denoted as j.

Note: With this definition of i, a few interesting results are immediately apparent. For example,

$i^3 = i^2 \cdot i = -1 \cdot i = -i$, and

$i^4 = i^2 \cdot i^2 = (-1) \cdot (-1) = 1$, and so

$i^5 = i^4 \cdot i = 1 \cdot i = i$, and also

$i^6 = i^4 \cdot i^2 = i^2 = -1$; $i^7 = -i$, and finally $i^8 = 1$.

This leads you to be able to evaluate any positive integer power of i using the following property:

$$i^{4n+k} = i^k,\ k = 0, 1, 2, 3.$$

So, for example $i^{2122} = i^{2120+2} = i^2 = -1$.

Example 2

Simplify

a) $\sqrt{-36} + \sqrt{-49}$ b) $\sqrt{-36} \cdot \sqrt{-49}$

Solution

a) $\sqrt{-36} + \sqrt{-49} = \sqrt{36}\sqrt{-1} + \sqrt{49}\sqrt{-1}$
$= 6i + 7i = 13i$

b) $\sqrt{-36} \cdot \sqrt{-49} = 6i \cdot 7i = 42i^2$
$= 42(-1) = -42$

We do not define $i = \sqrt{-1}$ for a reason. It is the convention in mathematics that when we write $\sqrt{9}$ then we mean the non-negative square root of 9, namely 3. We do not mean -3! i does not belong to this category since we cannot say that i is the positive square root of -1, i.e. $i > 0$. If we do, then $-1 = i \cdot i > 0$, which is false, and if we say $i < 0$, then $-i > 0$, and $-1 = -i \cdot -i > 0$, which is also false. Actually $-i$ is also a square root of -1 because $-i \cdot -i = i^2 = -1$.

With this in mind, we can use a 'convention' which calls i the **principal** square root of -1 and write $i = \sqrt{-1}$.

Gauss introduced the idea of complex numbers by giving them the following definition.

A **complex number** is a number that can be written in the form $a + bi$ where a and b are real numbers and $i^2 = -1$. a is called the **real part** of the number and b is the **imaginary part**.

Notation

It is customary to denote complex numbers with the variable z.

$z = 5 + 7i$ is the complex number with real part 5 and imaginary part 7 and $z = 2 - 3i$ has 2 as real part and -3 as imaginary.

It is usual to write **Re**(z) for the real part of z and **Im**(z) for the imaginary part. So, Re$(2 + 3i) = 2$ and Im$(2 + 3i) = 3$.

Note that both the real and imaginary parts are real numbers!

A GDC can be set up to do basic complex number operations. For example, if you have a TI-84 Plus, the set up is as follows.

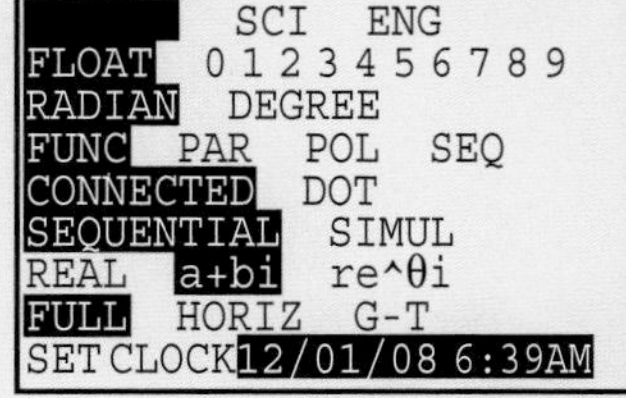

Algebraic structure of complex numbers

Gauss' definition of the complex numbers triggers the following understanding of the set of complex numbers as an extension to our number sets in algebra.

The set of *complex numbers* $\mathbb{C}$ is the set of ordered pairs of real numbers $\mathbb{C} = \{z = (x, y): x, y \in \mathbb{R}\}$, with the following additional structure:

Equality

Two complex numbers $z_1 = (x_1, y_1)$ and $z_2 = (x_2, y_2)$ are equal if their corresponding components are equal: $(x_1, y_1) = (x_2, y_2)$ if $x_1 = x_2$ and $y_1 = y_2$. That is, *two complex numbers are equal if and only if their real parts are equal and their imaginary parts are equal.*

This is equivalent to saying: $a + bi = c + di \Leftrightarrow a = c$ and $b = d$.

For example, if $2 - (y - 2)i = x + 3 + 5i$, then x must be -1 and y must be -3. **Explain why.**

An interesting application of the way equality works is in finding the square roots of complex numbers without a need for the trigonometric forms developed later in the chapter.

Find the square root(s) of $z = 5 + 12i$. Let the square root of z be $x + yi$, then $(x + yi)^2 = 5 + 12i \Rightarrow x^2 - y^2 + 2xyi = 5 + 12i \Rightarrow x^2 - y^2 = 5$ and $2xy = 12 \Rightarrow xy = 6 \Rightarrow y = \frac{6}{x}$, and when we substitute this value in $x^2 - y^2 = 5$, we have $x^2 - \left(\frac{6}{x}\right)^2 = 5$. This simplifies to $x^4 - 5x^2 - 36 = 0$ which yields $x^2 = -4$ or $x^2 = 9, \Rightarrow x = \pm 3$. This leads to $x = \pm 2i$, that is, the two square roots of $5 + 12i$ are $3 + 2i$ or $-3 - 2i$.

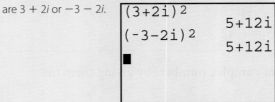

```
(3+2i)²
                5+12i
(-3-2i)²
                5+12i
```

Addition and subtraction for complex numbers are defined as follows:

Addition

$(x_1, y_1) + (x_2, y_2) = (x_1 + x_2, y_2 + y_2)$

This is equivalent to saying: $(a + bi) + (c + di) = (a + c) + (b + d)i$.

Multiplication

$(x_1, y_1)(x_2, y_2) = (x_1x_2 - y_1y_2, x_1y_2 + x_2y_1)$

This is equivalent to using the binomial multiplication on $(a + bi)(c + di)$:

$$(a + bi) \cdot (c + di) = ac + bdi^2 + adi + bci = ac - bd + (ad + bc)i$$

Addition and multiplication of complex numbers inherit most of the properties of addition and multiplication of real numbers:

$z + w = w + z$ and $zw = wz$ (Commutativity)

$z + (u + v) = (z + u) + v$ and $z(uv) = (zu)v$ (Associativity)

$z(u + v) = zu + zv$ (Distributive property)

A number of complex numbers take up unique positions. For example, the number $(0, 0)$ has the properties of 0:

$$(x, y) + (0, 0) = (x, y) \text{ and } (x, y)(0, 0) = (0, 0).$$

It is therefore normal to identify it with 0. The symbol is exactly the same symbol used to identify the 'real' 0. So, the real and complex zeros are the same number.

Another complex number of significance is $(1, 0)$. This number plays an important role in multiplication that stems from the following property:

$$(x, y)\,(1, 0) = (x \cdot 1 - y \cdot 0, x \cdot 0 + y \cdot 1) = (x, y)$$

For complex numbers, $(1, 0)$ behaves like the identity for multiplication for real numbers. Again, it is normal to write $(1, 0) = 1$.

The third number of significance is $(0, 1)$. It has the notable characteristic of having a negative square, i.e.

$$(0, 1)(0, 1) = (0 \cdot 0 - 1 \cdot 1, 0 \cdot 1 + 1 \cdot 0) = (-1, 0)$$

Using the definition above, $(0, 1) = 0 + 1i = i$. So, the last result should be no surprise to us since we know that

$$i \cdot i = -1 = (-1, 0).$$

Since (x, y) represents the complex number $x + yi$, then every real number x can be written as $x + 0i = (x, 0)$. The set of real numbers is therefore a subset of the set of complex numbers. They are the complex numbers whose imaginary part is 0. Similarly, pure imaginary numbers are of the form $0 + yi = (0, y)$. They are the complex numbers whose real part is 0.

Notation

So far, we have learned how to represent a complex number in two forms:

(x, y) and $x + yi$.

Now, from the properties above

$$(x, y) = (x, 0) + (0, y) = (x, 0) + (y, 0)(0, 1)$$

(Check the truth of this equation.)

This last equation justifies why we can write $(x, y) = x + yi$.

Example 3

Simplify each expression.

a) $(4 - 5i) + (7 + 8i)$

b) $(4 - 5i) - (7 + 8i)$

c) $(4 - 5i)(7 + 8i)$

Solution

a) $(4 - 5i) + (7 + 8i) = (4 + 7) + (-5 + 8)i = 11 + 3i$

b) $(4 - 5i) - (7 + 8i) = (4 - 7) + (-5 - 8)i = -3 - 13i$

c) $(4 - 5i)(7 + 8i) = (4 \cdot 7 - (-5) \cdot 8) + (4 \cdot 8 + (-5) \cdot 7)i = 68 - 3i$

```
(4-5i)/(8i)
          -.625-.5i
Ans▸Frac
          -5/8-1/2i
(4-5i)*(7+8i)
              68-3i
```

Division

Multiplication can be used to perform division of complex numbers.

The **division** of two complex numbers, $\frac{a + bi}{c + di}$, involves finding a complex number $(x + yi)$ satisfying $\frac{a + bi}{c + di} = x + yi$; hence, it is sufficient to find the unknowns x and y.

Example 4

Find the quotient $\frac{2+3i}{1+2i}$.

Solution

Let $\frac{2+3i}{1+2i} = x + iy$. Hence, using multiplication and the equality of complex numbers,

$2 + 3i = (1 + 2i)(x + iy) \Leftrightarrow 2 + 3i = x - 2y + i(2x + y)$

$$\Leftrightarrow \begin{cases} 2 = x - 2y \\ 3 = 2x + y \end{cases} \Rightarrow x = \frac{8}{5}, y = \frac{1}{5}$$

Thus, $\frac{2+3i}{1+2i} = \frac{8}{5} - \frac{1}{5}i$.

```
(2+3i)/(1+2i)
        1.6-.2i
Ans▸Frac
       8/5-1/5i
■
```

Now, in general, $\frac{a+bi}{c+di} = x + yi \Leftrightarrow a + bi = (x + yi)(c + di)$.

With the multiplication as described above:

$$a + bi = (cx - dy) + (dx + cy)i$$

Again by applying the equality of complex numbers property above we get a system of two equations that can be solved.

$$\begin{cases} cx - dy = a \\ dx + cy = b \end{cases} \Rightarrow x = \frac{ac + bd}{c^2 + d^2}; y = \frac{bc - ad}{c^2 + d^2}$$

The denominator $c^2 + d^2$ resulted from multiplying $c + di$ by $c - di$, which is its conjugate.

Conjugate

With every complex number $(a + bi)$ we associate another complex number $(a - bi)$ which is called its conjugate. The conjugate of number z is most often denoted with a bar over it, sometimes with an asterisk to the right of it, occasionally with an apostrophe and even less often with the plain symbol Conj as in

$$\bar{z} = z^* = z' = \text{Conj}(z).$$

In this book, we will use z^* for the conjugate.

> Although the conjugate notation z^* will be used in the book, in your own work you can use any notation you feel comfortable with. You just need to understand that the IB questions use this one.

The importance of the conjugate stems from the following property

$$(a + bi)(a - bi) = a^2 - b^2i^2 = a^2 + b^2$$

which is a non-negative real number. So the product of a complex number and its conjugate is always a real number.

Example 5

Find the conjugate of z and verify the property mentioned above.

a) $z = 2 + 3i$

b) $z = 5i$

c) $z = 11$

Solution

a) $z^* = 2 - 3i$, and $(2 + 3i)(2 - 3i) = 4 - 9i^2 = 4 + 9 = 13$.

b) $z^* = -5i$, and $(5i)(-5i) = -5i^2 = (-5)(-1) = 5$.

c) $z^* = 11$, and $11 \cdot 11 = 121$.

So, the method used in dividing two complex numbers can be achieved by multiplying the quotient by a fraction whose numerator and denominator are the conjugate $c - di$.

$$\frac{a + bi}{c + di} = \frac{a + bi}{c + di} \cdot \frac{c - di}{c - di} = \frac{(a + bi)(c - di)}{c^2 + d^2} = \frac{ac + bd}{c^2 + d^2} + \frac{bc - ad}{c^2 + d^2}i$$

Example 6

Find each quotient and write your answer in standard form.

a) $\dfrac{4 - 5i}{7 + 8i}$

b) $\dfrac{4 - 5i}{8i}$

c) $\dfrac{4 - 5i}{7}$

Solution

a) $\dfrac{4 - 5i}{7 + 8i} = \dfrac{4 - 5i}{7 + 8i} \cdot \dfrac{7 - 8i}{7 - 8i} = \dfrac{28 - 40 + (-32 - 35)i}{49 + 64} = -\dfrac{12}{113} - \dfrac{67}{113}i$

b) $\dfrac{4 - 5i}{8i} = \dfrac{4 - 5i}{8i} \cdot \dfrac{-8i}{-8i} = \dfrac{-32i - 40}{64} = -\dfrac{5}{8} - \dfrac{1}{2}i$

c) $\dfrac{4 - 5i}{7} = \dfrac{4}{7} - \dfrac{5}{7}i$

```
(4-5i)/(7+8i)
-.1061946903-.5...
Ans▶Frac
 -12/113-67/113i
```

```
(4-5i)/(8i)
       -.625-.5i
Ans▶Frac
       -5/8-1/2i
```

Example 7

Solve the system of equations and express your answer in Cartesian form.

$$(1 + i)z_1 - iz_2 = -3$$
$$2z_1 + (1 - i)z_2 = 3 - 3i$$

Solution

Multiply the first equation by 2, and the second equation by $(1 + i)$.

$$2(1 + i)z_1 - 2iz_2 = -6 \qquad \textbf{(1)}$$

$$2(1 + i)z_1 + (1 + i)(1 - i)z_2 = (1 + i)(3 - 3i)$$

$$2(1 + i)z_1 + 2z_2 = 6 \qquad \textbf{(2)}$$

By subtracting **(2)** from **(1)**, we get

$$(-2 - 2i)z_2 = -12$$

And hence

$$z_2 = \frac{-12}{-2 - 2i} = 3 - 3i$$

$$z_1 = \frac{-3 + i(3 - 3i)}{1 + i} = \frac{3}{2} + \frac{3}{2i}$$

Properties of conjugates

Here is a theorem that lists some of the important properties of conjugates. In the next section, we will add a few more to the list.

Theorem

Let z, z_1 and z_2 be complex numbers, then

(1) $(z^*)^* = z$

(2) $z^* = z$ if and only if z is real.

(3) $(z_1 + z_2)^* = z_1^* + z_2^*$ The conjugate of the sum is the sum of conjugates.

(4) $(-z)^* = -z^*$

(5) $(z_1 \cdot z_2)^* = z_1^* \cdot z_2^*$ The conjugate of the product is the product of conjugates.

(6) $(z^{-1})^* = (z^*)^{-1}$, if $z \neq 0$.

The product can be extended to powers of complex numbers, i.e. $(z^2)^* = (z \cdot z)^* = z^* \cdot z^* = (z^*)^2$. This result can be generalized for any non-negative integer power n, i.e. $(z^n)^* = (z^*)^n$ and can be proved by mathematical induction.

The basis case, when $n = 0$, is obviously true: $(z^0)^* = 1 = (z^*)^0$.

Now assume $(z^k)^* = (z^*)^k$. $(z^{k+1})^* = (z^k z)^* = (z^k)^* z^* = (z^*)^k z^*$ (using the product rule).

Therefore, $(z^{k+1})^* = (z^*)^k z^* = (z^*)^{k+1}$.

So, since if the statement is true for $n = k$, it is also true for $n = k + 1$, then by the principle of mathematical induction it is true for all $n \geqslant 0$.

Proof

(1) and (2) are obvious. For (1), $((a + bi)^*)^* = (a - bi)^* = a + bi$, and for (2), $a - bi = a + bi \Rightarrow 2bi = 0 \Rightarrow b = 0$.

(3) is proved by straightforward calculation:

Let $z_1 = x_1 + iy_1$ and $z_2 = x_2 + iy_2$, then

$$\begin{aligned}(z_1 + z_2)^* &= ((x_1 + iy_1) + (x_2 + iy_2))^* = ((x_1 + x_2) + i(y_1 + y_2))^* \\ &= (x_1 + x_2) - i(y_1 + y_2) = (x_1 - iy_1) + (x_2 - iy_2) = z_1^* + z_2^*.\end{aligned}$$

(4) can now be proved using the above results:

$$(z + (-z))^* = 0^* = 0$$

but, $(z + (-z))^* = 0^* = z^* + (-z)^*$,

so $z^* + (-z)^* = 0$, and $(-z)^* = -z^*$.

Also (5) is proved by straightforward calculation:

$$\begin{aligned}(z_1 \cdot z_2)^* &= ((x_1 + iy_1) \cdot (x_2 + iy_2))^* = ((x_1x_2 - y_1y_2) + i(y_1x_2 + x_1y_2))^* \\ &= (x_1x_2 - y_1y_2) - i(y_1x_2 + x_1y_2) \\ &= (x_1 - iy_1) \cdot (x_2 - iy_2) = z_1^* \cdot z_2^*\end{aligned}$$

And finally, (6):

$(z(z^{-1}))^* = 1^* = 1$

but, $(z(z^{-1}))^* = z^*(z^{-1})^*$, so $z^*(z^{-1})^* = 1$,

and $(z^{-1})^* = \frac{1}{z^*} = (z^*)^{-1}$.

Conjugate zeros of polynomials

In Chapter 3, you used the following result without proof.

If c is a root of a polynomial equation with real coefficients, then c is also a root.*

Theorem: If c is a root of a polynomial equation with real coefficients, then c^* is also a root of the equation.

We give the proof for $n = 3$, but the method is general.

$$P(x) = ax^3 + bx^2 + dx + e$$

Since c is a root of $P(x) = 0$, we have

$ac^3 + bc^2 + dc + e = 0$

$\Rightarrow (ac^3 + bc^2 + dc + e)^* = 0$ — Since $0^* = 0$.

$\Rightarrow (ac^3)^* + (bc^2)^* + (dc)^* + e^* = 0$ — Sum of conjugates theorem.

$\Rightarrow a(c^*)^3 + b(c^*)^2 + d(c^*) + e = 0$ — Result of product conjugate.

$\Rightarrow (c^*)$ is a root of $P(x) = 0$.

Example 8

$1 + 2i$ is a zero of the polynomial $P(x) = x^3 - 5x^2 + 11x - 15$. Find all other zeros.

Solution

Since the polynomial has real coefficients, then $1 - 2i$ is also a zero. Hence, using the factor theorem, $P(x) = (x - (1 + 2i))(x - (1 - 2i))(x - c)$, where c is a real number to be found.

Now, $P(x) = (x^2 - 2x + 5)(x - c)$. c can either be found by division or by factoring by trial and error. In either case, $c = 3$.

Example 9[1]

$1 + 2i$ is a zero of the polynomial
$P(x) = x^3 + (i - 2)x^2 + (2i + 5)x + 8 + i$.
Find all other zeros.

[1] Not included in present IB syllabus.

Solution

Since the polynomial does not have real coefficients, then $1 - 2i$ is not necessarily also a zero. To find the other zeros, we can perform synthetic substitution

$$\begin{array}{r|cccc} 1+2i & 1 & i-2 & 2i+5 & 8+i \\ & & 1+2i & -7+i & -8-i \\ \hline & 1 & -1+3i & -2+3i & 0 \end{array}$$

This shows that $P(x) = (x - 1 - 2i)(x^2 + (-1 + 3i)x - 2 + 3i)$. The second factor can be factored into $(x + 1)(x - 2 + 3i)$ giving us the other two zeros as -1 and $2 - 3i$.

Note: $x^2 + (-1 + 3i)x - 2 + 3i = 0$ can be solved using the quadratic formula.

$$x = \frac{-b \pm \sqrt{b^2 - 4ac}}{2a} = \frac{1 - 3i \pm \sqrt{(-1 + 3i)^2 - 4(-2 + 3i)}}{2}$$

$$= \frac{1 - 3i \pm \sqrt{-8 - 6i + 8 - 12i}}{2} = \frac{1 - 3i \pm \sqrt{-18i}}{2}$$

To find $\sqrt{-18i}$ we let $(a + bi)^2 = -18i \Rightarrow a^2 - b^2 + 2abi = -18i$, then equating the real parts and imaginary parts to each other: $a^2 - b^2 = 0$ and $2ab = -18$ will yield $\sqrt{-18i} = \pm 3 \mp 3i$, and hence

$$x = \frac{1 - 3i \pm \sqrt{-18i}}{2} = \frac{1 - 3i \pm (\pm 3 \mp 3i)}{2}$$

which will yield $x = -1$ or $x = 2 - 3i$.

Exercise 10.1

Express each of the following numbers in the form $a + bi$.

1 $5 + \sqrt{-4}$ **2** $7 - \sqrt{-7}$ **3** -6

4 $-\sqrt{49}$ **5** $\sqrt{-81}$ **6** $-\sqrt{\frac{-25}{16}}$

Perform the following operations and express your answer in the form $a + bi$.

7 $(-3 + 4i) + (2 - 5i)$ **8** $(-3 + 4i) - (2 - 5i)$

9 $(-3 + 4i)(2 - 5i)$ **10** $3i - (2 - 4i)$

11 $(2 - 7i)(3 + 4i)$ **12** $(1 + i)(2 - 3i)$

13 $\frac{3 + 2i}{2 + 5i}$ **14** $\frac{2 - i}{3 + 2i}$

15 $\left(\frac{2}{3} - \frac{1}{2}i\right) + \left(\frac{1}{3} + \frac{1}{2}i\right)$ **16** $\left(\frac{2}{3} - \frac{1}{2}i\right)\left(\frac{2}{3} + \frac{1}{2}i\right)$

17 $\left(\frac{2}{3} - \frac{1}{2}i\right) \div \left(\frac{1}{3} + \frac{1}{2}i\right)$ **18** $(2 + i)(3 - 2i)$

19 $\frac{1}{i}(3 - 7i)$ **20** $(2 + 5i) - (-2 - 5i)$

21 $\frac{13}{5 - 12i}$ **22** $\frac{12i}{3 + 4i}$

23 $3i\left(3 - \frac{2}{3}i\right)$

24 $(3 + 5i)(6 - 10i)$

25 $\dfrac{39 - 52i}{24 + 10i}$

26 $(7 - 4i)^{-1}$

27 $(5 - 12i)^{-1}$

28 $\dfrac{3}{3 - 4i} + \dfrac{2}{6 + 8i}$

29 $\dfrac{(7 + 8i)(2 - 5i)}{5 - 12i}$

30 $\dfrac{5 - \sqrt{-144}}{3 + \sqrt{-16}}$

31 Let $z = a + bi$. Find a and b if $(2 + 3i)z = 7 + i$.

32 $(2 + yi)(x + i) = 1 + 3i$, where x and y are real numbers. Solve for x and y.

33 a) Evaluate $(1 + i\sqrt{3})^3$.

b) Prove that $(1 + i\sqrt{3})^{6n} = 8^{2n}$, *where* $n \in Z^+$.

c) Hence, find $(1 + i\sqrt{3})^{48}$.

34 a) Evaluate $(-\sqrt{2} + i\sqrt{2})^2$.

b) Prove that $(-\sqrt{2} + i\sqrt{2})^{4k} = (-16)^k$, *where* $k \in Z^+$.

c) Hence, find $(-\sqrt{2} + i\sqrt{2})^{46}$.

35 If z is a complex number such that $|z + 4i| = 2|z + i|$, find the value of $|z|$.
($|z| = \sqrt{x^2 + y^2}$ where $z = x + iy$.)

36 Find the complex number z and write it in the form $a + bi$ if $z = 3 + \dfrac{2i}{2 - i\sqrt{2}}$.

37 Find the values of the two real numbers x and y such that $(x + iy)(4 - 7i) = 3 + 2i$.

38 Find the complex number z and write it in the form $a + bi$ if $i(z + 1) = 3z - 2$.

39 Find the complex number z and write it in the form $a + bi$ if $\dfrac{2 - i}{1 + 2i}\sqrt{z} = 2 - 3i$.

40 Find the values of the two real numbers x and y such that $(x + iy)^2 = 3 - 4i$.

41 a) Find the values of the two real numbers x and y such that $(x + iy)^2 = -8 + 6i$.

b) Hence, solve the following equation

$$z^2 + (1 - i)z + 2 - 2i = 0.$$

42 If $z \in \mathbb{C}$, find all solutions to the equation $z^3 - 27i = 0$.

43 Given that $z = \frac{1}{2} + 2i$ is a zero of the polynomial $f(x) = 4x^3 - 16x^2 + 29x - 51$, find the other zeros.

44 Find a polynomial function with integer coefficients and lowest possible degree that has $\frac{1}{2}$, -1 and $3 + i\sqrt{2}$ as zeros.

45 Find a polynomial function with integer coefficients and lowest possible degree that has -2, -2 and $1 + i\sqrt{3}$ as zeros.

46 Given that $z = 5 + 2i$ is a zero of the polynomial $f(x) = x^3 - 7x^2 - x + 87$, find the other zeros.

47 Given that $z = 1 - i\sqrt{3}$ is a zero of the polynomial $f(x) = 3x^3 - 4x^2 + 8x + 8$, find the other zeros.

48 Let $z \in \mathbb{C}$. If $\dfrac{z}{z^*} = a + bi$, show that $|a + bi| = 1$.

49 Given that $z = (k + i)^4$ where k is a real number, find all values of k such that

a) z is a real number

b) z is purely imaginary.

50 Solve the system of equations.

$iz_1 + 2z_2 = 3 - i$
$2z_1 + (2 + i)z_2 = 7 + 2i$

51 Solve the system of equations.

$iz_1 - (1 + i)z_2 = 3$
$(2 + i)z_1 + iz_2 = 4$

10.2 The complex plane

Our definition of complex numbers as ordered pairs of real numbers enables us to look at them from a different perspective. Every ordered pair (x, y) determines a unique complex number $x + yi$, and vice versa. This correspondence is embodied in the geometric representation of complex numbers. Looking at complex numbers as points in the plane equipped with additional structure changes the plane into what we call **complex plane**, or **Gauss plane**, or **Argand plane** (**diagram**). The complex plane has two axes, the horizontal axis is called the **real axis**, and the vertical axis is the **imaginary axis**. Every complex number $z = x + yi$ is represented by a point (x, y) in the plane. The real part is measured along the real axis and the imaginary part along the imaginary axis.

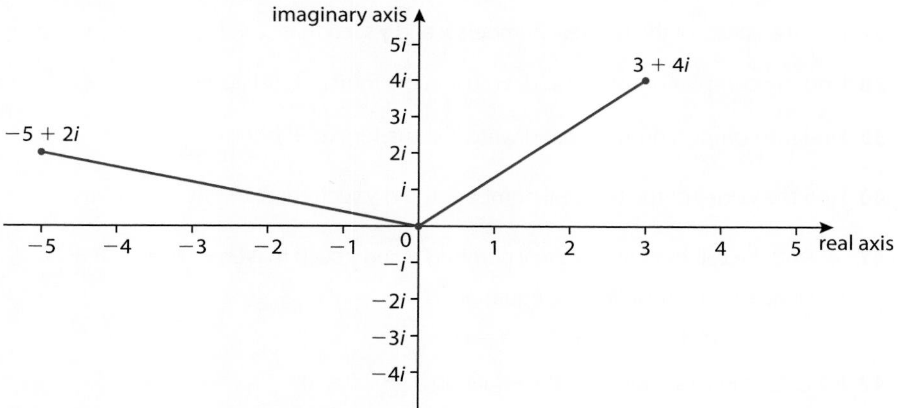

The diagram above illustrates how the two complex numbers $3 + 4i$ and $-5 + 2i$ are plotted in the complex plane.

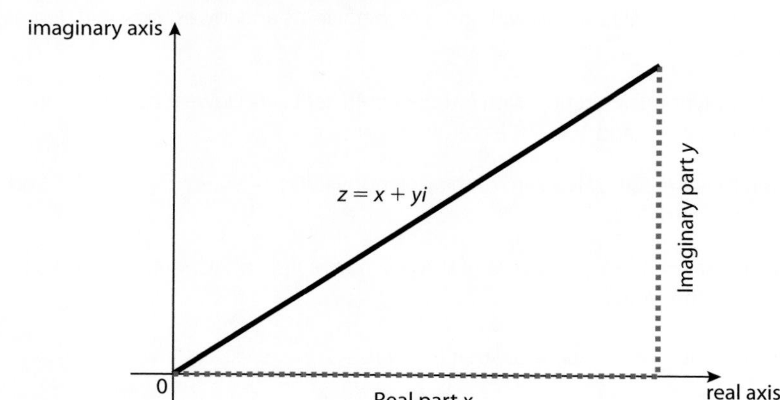

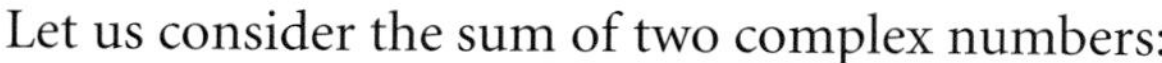

Let us consider the sum of two complex numbers:

$$z_1 = x_1 + y_1 i, \text{ and } z_2 = x_2 + y_2 i$$

As we have defined addition before:

$$z_1 + z_2 = (x_1 + x_2) + (y_1 + y_2)i$$

This suggests that we consider complex numbers as vectors; i.e. we regard the complex number $z = x + iy$ as a vector in standard form whose terminal point is the complex number (x, y).

Since we are representing the complex numbers by vectors, this results in some analogies between the two sets. So, adding two complex numbers or subtracting them, or multiplying by a scalar, are similar in both sets.

Example 10

Consider the complex numbers $z_1 = 3 + 4i$ and $z_2 = -5 + 2i$.
Find $z_1 + z_2$ and $z_1 - z_2$.

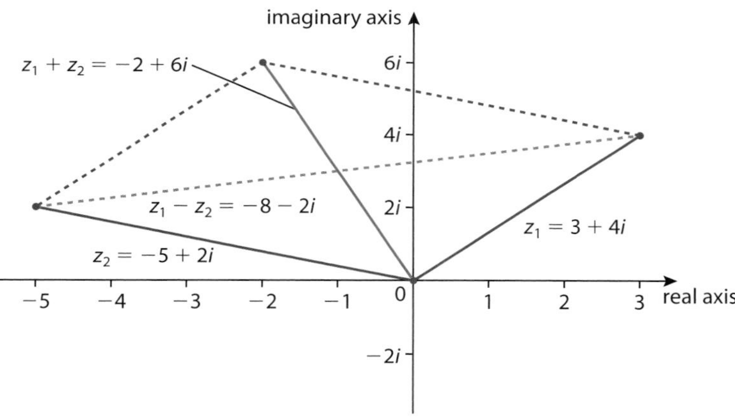

Note here that the vector representing the sum, $-2 + 6i$, is the diagonal of the parallelogram with sides representing $3 + 4i$ and $-5 + 2i$, while the vector representing the difference is the second diagonal of the parallelogram.

The length, norm, of a vector also has a parallel in complex numbers. You recall that for a vector $\mathbf{v} = (x, y)$ the length of the vector is

$$|\mathbf{v}| = \sqrt{x^2 + y^2}.$$

For complex numbers, the **modulus** or **absolute value** (or magnitude) of the complex number $z = x + yi$ is

$$|z| = \sqrt{x^2 + y^2}.$$

It follows immediately that since

$$z^* = x - yi \Rightarrow |z^*| = \sqrt{x^2 + (-y)^2} = \sqrt{x^2 + y^2}, \text{ then}$$

$$|z^*| = |z|.$$

Also of interest is the following result.

$z \cdot z^* = (x + iy)(x - iy) = x^2 + y^2$,
$|z|^2 = x^2 + y^2$, and $|z^*|^2 = x^2 + y^2$
$\Rightarrow z \cdot z^* = |z|^2 = |z^*|^2$

For example:

$(3 + 4i)(3 - 4i) = 9 + 16 = 25 = (\sqrt{3^2 + 4^2})^2$

Example 11

Calculate the moduli of the following complex numbers

a) $z_1 = 5 - 6i$ b) $z_2 = 12 + 5i$

Solution

a) $|z_1| = |5 - 6i| = \sqrt{5^2 + 6^2} = \sqrt{61}$

b) $|z_2| = |12 + 5i| = \sqrt{12^2 + 5^2} = \sqrt{169} = 13$

Example 12

Graph each set of complex numbers.

a) $A = \{z \mid |z| = 3\}$ b) $B = \{z \mid |z| \leqslant 3\}$

Solution

a) A is the set of complex numbers whose distance from the origin is 3 units. So, the set is a circle with radius 3 and centre (0, 0) as shown.

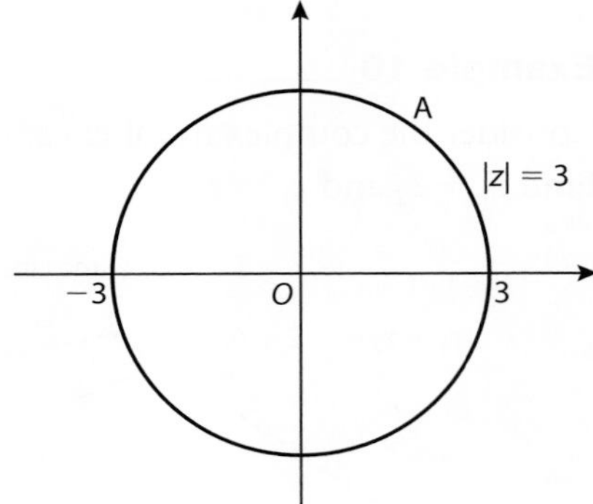

b) B is the set of complex numbers whose distance from the origin is less than or equal to 3. So, the set is a disk of radius 3 and centre at the origin.

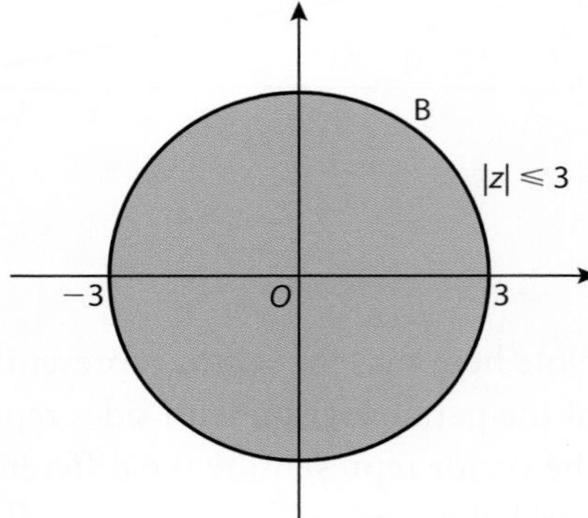

Another important property is the following result:

$$|z_1z_2| = |z_1||z_2|$$

Proof:

$$|z_1z_2| = |(x_1x_2 - y_1y_2) + (x_1y_2 + x_2y_1)i| = \sqrt{(x_1x_2 - y_1y_2)^2 + (x_1y_2 + x_2y_1)^2}$$
$$= \sqrt{(x_1x_2)^2 - 2x_1x_2y_1y_2 + (y_1y_2)^2 + (x_1y_2)^2 + 2x_1y_2x_2y_1 + (x_2y_1)^2}$$
$$= \sqrt{(x_1x_2)^2 + (y_1y_2)^2 + (x_1y_2)^2 + (x_2y_1)^2}$$

But,

$$|z_1||z_2| = \sqrt{x_1^2 + y_1^2} \cdot \sqrt{x_2^2 + y_2^2} = \sqrt{(x_1^2 + y_1^2)(x_2^2 + y_2^2)}$$
$$= \sqrt{(x_1x_2)^2 + (y_1y_2)^2 + (x_1y_2)^2 + (x_2y_1)^2}$$

And so the result follows.

Example 13

Evaluate $|(3 + 4i)(5 + 12i)|$.

Solution

$|(3 + 4i)(5 + 12i)| = |3 + 4i|\,|5 + 12i| = \sqrt{9 + 16}\,\sqrt{25 + 144} = 5 \times 13 = 65,$

or $|(3 + 4i)(5 + 12i)| = |-33 + 56i| = \sqrt{(-33)^2 + 56^2} = \sqrt{4255} = 65$

Trigonometric/polar form of a complex number

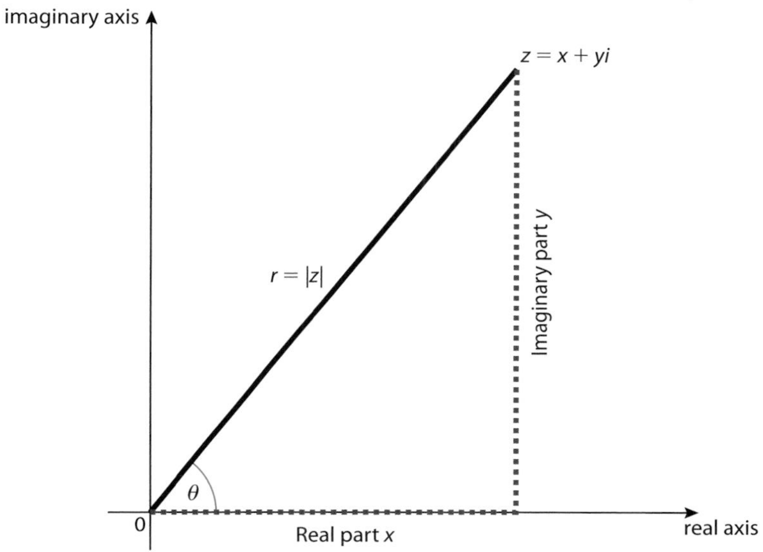

We know by now that every complex number $z = x + yi$ can be considered as an ordered pair (x, y). Hence, using our knowledge of vectors, we can introduce a new form for representing complex numbers – the trigonometric form (also known as polar form).

The trigonometric form uses the modulus of the complex number as its distance from the origin, $r \geqslant 0$, and θ the angle the 'vector' makes with the real axis.

Clearly $x = r\cos\theta$ and $y = r\sin\theta$; $r = \sqrt{x^2 + y^2}$; and $\tan\theta = \frac{y}{x}$.

Therefore, $z = x + yi = r\cos\theta + (r\sin\theta)i = r(\cos\theta + i\sin\theta)$.

The angle θ is called the **argument** of the complex number, $\arg(z)$.

$\text{Arg}(z)$ is not unique. However, all values differ by a multiple of 2π.

Note:

The trigonometric form is called 'modulus-argument' by the IB. Please keep that in mind. Also this trigonometric form is abbreviated, for ease of writing, as follows:

$z = x + yi = r(\cos\theta + i\sin\theta) = r\,\text{cis}\,\theta$. (cis θ stands for $\cos\theta + i\sin\theta$.)

Example 14

Write the following numbers in trigonometric form.

a) $z = 1 + i$ b) $z = \sqrt{3} - i$

c) $z = -5i$ d) $z = 17$

Solution

a) $r = \sqrt{1^2 + 1^2} = \sqrt{2}$; $\tan\theta = \frac{1}{1} = 1$.
Hence, by observing the real and imaginary parts being positive, we can conclude that the argument must be $\theta = \frac{\pi}{4}$.

$\therefore z = \sqrt{2}\left(\cos\frac{\pi}{4} + i\sin\frac{\pi}{4}\right) = \sqrt{2}\operatorname{cis}\frac{\pi}{4}$

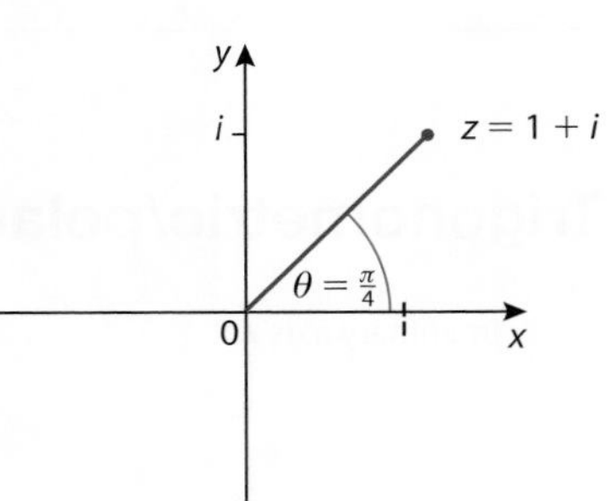

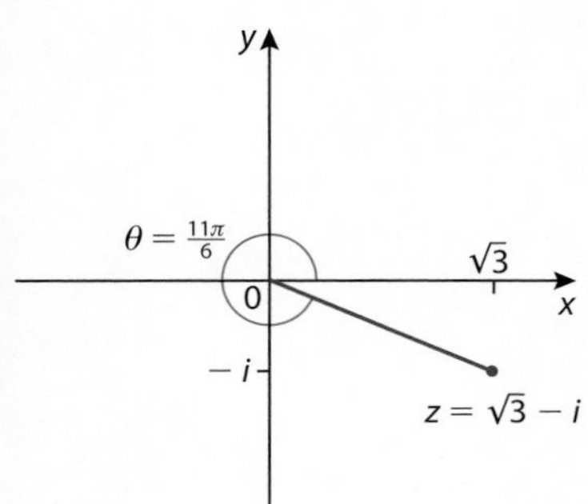

b) $r = \sqrt{(\sqrt{3})^2 + (-1)^2} = \sqrt{4} = 2$; $\tan\theta = \frac{-1}{\sqrt{3}}$. The real part is positive, the imaginary part is negative, and the point is therefore in the fourth quadrant, so $\theta = \frac{11\pi}{6}$.

$\therefore z = 2\left(\cos\frac{11\pi}{6} + i\sin\frac{11\pi}{6}\right) = 2\operatorname{cis}\frac{11\pi}{8}$

We can also use $\theta = -\frac{\pi}{6}$.

c) $r = 5$ and $\theta = \frac{3\pi}{2}$ since it is on the negative side of the imaginary axis.

$\therefore z = 5\left(\cos\frac{3\pi}{2} + i\sin\frac{3\pi}{2}\right)$

We can also use $\theta = -\frac{\pi}{2}$.

d) $r = 17$ and $\theta = 0$

$\therefore z = 17\,(\cos 0 + i\sin 0)$

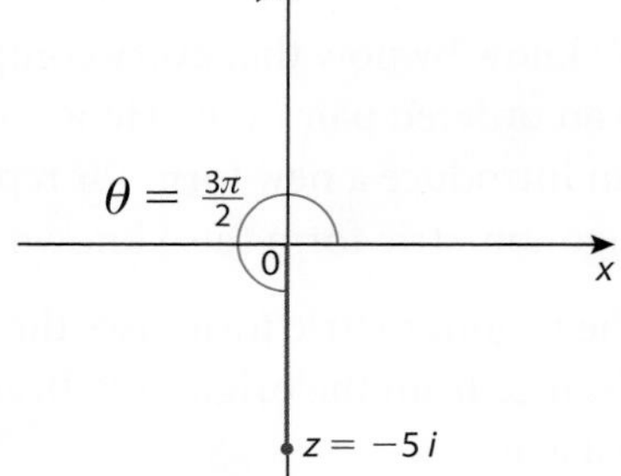

Example 15

Convert each complex number into its rectangular form.

a) $z = 3\cos 150° + 3i\sin 150°$ b) $z = 12\operatorname{cis}\frac{4\pi}{3}$

c) $z = 6(\cos 50° + i\sin 50°)$ d) $z = 15\left(\cos\frac{\pi}{2} + i\sin\frac{\pi}{2}\right)$

Solution

a) $z = 3\left(\frac{-\sqrt{3}}{2}\right) + 3i\left(\frac{1}{2}\right) = \frac{-3\sqrt{3}}{2} + \frac{3i}{2}$

b) $z = 12\cos\frac{4\pi}{3} + 12i\sin\frac{4\pi}{3} = 12 \cdot \frac{-1}{2} + 12i \cdot -\frac{\sqrt{3}}{2} = -6 - \frac{6i}{\sqrt{3}}$

c) $z = 6\cos 50° + 6i\sin 50° = 6 \cdot 0.643 + 6i \cdot 0.766 = 3.857 + 4.596i$

d) $z = 15(0 + i) = 15i$

Multiplication

The trigonometric form of the complex number offers a very interesting and efficient method for multiplying complex numbers.

Let

$$z_1 = r_1(\cos\theta_1 + i\sin\theta_1) \text{ and } z_2 = r_2(\cos\theta_2 + i\sin\theta_2)$$

be two complex numbers written in trigonometric form. Then

$$z_1z_2 = (r_1(\cos\theta_1 + i\sin\theta_1))(r_2(\cos\theta_2 + i\sin\theta_2)$$
$$= r_1r_2[(\cos\theta_1\cos\theta_2 - \sin\theta_1\sin\theta_2) + i(\sin\theta_1\cos\theta_2 + \sin\theta_2\cos\theta_1)].$$

Now, using the addition formulae for sine and cosine, we have

$$\boxed{z_1z_2 = r_1r_2[(\cos(\theta_1 + \theta_2)) + i(\sin(\theta_1 + \theta_2))]}$$

This formula says: *To multiply two complex numbers written in trigonometric form, we multiply the moduli and add the arguments.*

The analogy between complex numbers and vectors stops at multiplication. As you recall, multiplication of vectors is not 'well defined' in the sense that there are two products – the scalar product which is a scalar, not a vector, and the vector product (discussed later) which is a vector but is not in the plane! Complex number products are complex numbers!

Example 16

Let $z_1 = 2 + 2i\sqrt{3}$ and $z_2 = -1 - i\sqrt{3}$.

a) Evaluate z_1z_2 by using their standard forms (rectangular or Cartesian).

b) Evaluate z_1z_2 by using their trigonometric forms and verify that the two results are the same.

Solution

a) $z_1z_2 = (2 + 2i\sqrt{3})(-1 - i\sqrt{3}) = (-2 + 6) + (-2\sqrt{3} - 2\sqrt{3})i = 4 - 4i\sqrt{3}$

b) Converting both to trigonometric form, we get

$z_1 = 4\,\text{cis}\frac{\pi}{3}$ and $z_2 = 2\,\text{cis}\frac{4\pi}{3}$, then

$$z_1z_2 = 4 \cdot 2\left(\text{cis}\left(\frac{\pi}{3} + \frac{4\pi}{3}\right)\right) = 8\,\text{cis}\left(\frac{5\pi}{3}\right) = 8\left(\cos\frac{5\pi}{3} + i\sin\frac{5\pi}{3}\right)$$
$$= 8\left(\frac{1}{2} + i\left(\frac{-\sqrt{3}}{2}\right)\right) = 4 - 4i\sqrt{3}.$$

Note: You may observe here that multiplying z_1 by z_2 resulted in a new number whose magnitude is twice that of z_1 and is rotated by an angle of $\frac{4\pi}{3}$. Alternatively, you can see it as multiplying z_2 by z_1 which results in a complex number whose magnitude is 4 times that of z_2 and is rotated by an angle of $\frac{\pi}{3}$.

Example 17

Let $z_1 = -2 + 2i$ and $z_2 = 3\sqrt{3} - 3i$.

Convert to trigonometric form and multiply.

Solution

$$z_1 = 2\sqrt{2}\ \text{cis}\frac{3\pi}{4} \text{ and } z_2 = 6\ \text{cis}\frac{11\pi}{6}, \text{ then}$$

$$z_1z_2 = 12\sqrt{2}\left(\text{cis}\left(\frac{3\pi}{4} + \frac{11\pi}{6}\right)\right) = 12\sqrt{2}\ \text{cis}\left(\frac{31\pi}{12}\right) = 12\sqrt{2}\ \text{cis}\left(\frac{7\pi}{12} + 2\pi\right)$$

$$= 12\sqrt{2}\ \text{cis}\left(\frac{7\pi}{12}\right) = 12\sqrt{2}\left(\cos\frac{7\pi}{12} + i\sin\frac{7\pi}{12}\right)$$

Note: You can simplify this answer further to get an exact rectangular form.

$$z_1z_2 = 12\sqrt{2}\left(\cos\frac{7\pi}{12} + i\sin\frac{7\pi}{12}\right) = 12\sqrt{2}\left(\cos\left(\frac{3\pi + 4\pi}{12}\right) + i\sin\frac{3\pi + 4\pi}{12}\right)$$

$$= 12\sqrt{2}\left(\cos\left(\frac{\pi}{4} + \frac{\pi}{3}\right) + i\sin\left(\frac{\pi}{4} + \frac{\pi}{3}\right)\right)$$

$$= 12\sqrt{2}\left(\left(\frac{\sqrt{2}}{2}\cdot\frac{1}{2} - \frac{\sqrt{2}}{2}\cdot\frac{\sqrt{3}}{2}\right) + i\left(\frac{\sqrt{2}}{2}\cdot\frac{1}{2} + \frac{\sqrt{2}}{2}\cdot\frac{\sqrt{3}}{2}\right)\right)$$

$$= 12\sqrt{2}\left(\frac{\sqrt{2} - \sqrt{6}}{4} + i\frac{\sqrt{2} + \sqrt{6}}{4}\right) = (6 - 6\sqrt{3}) + i(6 + 6\sqrt{3})$$

Note: By comparing the Cartesian form of the product to the polar form, i.e. $12\sqrt{2}\left(\cos\frac{7\pi}{12} + i\sin\frac{7\pi}{12}\right)$ and $12\sqrt{2}\left(\frac{\sqrt{2} - \sqrt{6}}{4} + i\frac{\sqrt{2} - \sqrt{6}}{4}\right)$, we can conclude that $\cos\frac{7\pi}{12} = \frac{\sqrt{2} - \sqrt{6}}{4}$ and $\sin\frac{7\pi}{12} = \frac{\sqrt{2} - \sqrt{6}}{4}$.

This observation gives us a way of using complex number multiplication in order to find exact values of some trigonometric functions.

You may have noticed that the conjugate of a complex number $z = r(\cos\theta + i\sin\theta)$ is $z^* = r(\cos\theta - i\sin\theta) = r(\cos(-\theta) + i\sin(-\theta))$.

Also, $z \cdot z^* = r(\cos\theta + i\sin\theta)\cdot r(\cos\theta - i\sin\theta)$
$= r^2(\cos^2\theta + \sin^2\theta)$
$= r^2$.

Graphically, a complex number and its conjugate are reflections of each other in the real axis. See the figure opposite.

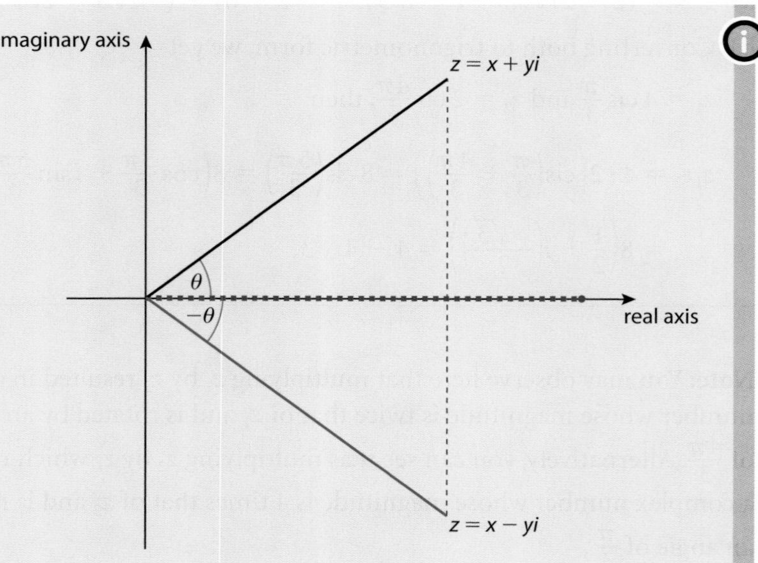

Division of complex numbers

A similar approach gives us the rules for division of complex numbers.

Let

$z_1 = r_1(\cos\theta_1 + i\sin\theta_1)$ and $z_2 = r_2(\cos\theta_2 + i\sin\theta_2)$

be two complex numbers written in trigonometric form. Then

$$\frac{z_1}{z_2} = \frac{r_1(\cos\theta_1 + i\sin\theta_1)}{r_2(\cos\theta_2 + i\sin\theta_2)} \cdot \frac{\cos\theta_2 - i\sin\theta_2}{\cos\theta_2 - i\sin\theta_2}$$

$$= \frac{r_1}{r_2}\left(\frac{(\cos\theta_1\cos\theta_2 + \sin\theta_1\sin\theta_2) + i(\sin\theta_1\cos\theta_2 - \sin\theta_2\cos\theta_1)}{\cos^2\theta_2 + \sin^2\theta_2)}\right)$$

$$= \frac{r_1}{r_2}\left(\frac{(\cos\theta_1\cos\theta_2 + \sin\theta_1\sin\theta_2) + i(\sin\theta_1\cos\theta_2 - \sin\theta_2\cos\theta_1)}{1}\right).$$

Now, using the subtraction formulas for sine and cosine, we have

$$\boxed{\frac{z_1}{z_2} = \frac{r_1}{r_2}[(\cos(\theta_1 - \theta_2)) + i(\sin(\theta_1 - \theta_2))]}$$

This formula says: *To divide two complex numbers written in trigonometric form, we divide the moduli and subtract the arguments.*

In particular, if we take $z_1 = 1$ and $z_2 = z$ (i.e. $\theta_1 = 0$ and $\theta_2 = \theta$), we will have the following result.

If $z = r(\cos\theta + i\sin\theta)$ then $\boxed{\frac{1}{z} = \frac{1}{r}(\cos(-\theta) + i\sin(-\theta)) = \frac{1}{r}(\cos(\theta) - i\sin(\theta))}$

Example 18

Let $z_1 = 1 + i$ and $z_2 = \sqrt{3} - i$.

a) Convert into trigonometric form.

b) Evaluate $\frac{1}{z_2}$.

c) Evaluate $\frac{z_1}{z_2}$.

d) Use the results above to find the exact values of $\sin\frac{5\pi}{12}$ and $\cos\frac{5\pi}{12}$.

Solution

a) $z_1 + \sqrt{2}\operatorname{cis}\frac{\pi}{4}$; $z_2 = 2\operatorname{cis}\frac{11\pi}{6} = 2\operatorname{cis}\frac{-\pi}{6}$

b) $\frac{1}{z_2} = \frac{1}{2}\operatorname{cis}\left(-\frac{-\pi}{6}\right) = \frac{1}{2}\operatorname{cis}\frac{\pi}{6}$

c) $\frac{z_1}{z_2}$ can be found by either multiplying z_1 by $\frac{1}{z_2}$, or by using division as shown above.

$$\frac{z_1}{z_2} = z_1 \cdot \frac{1}{z_2} = \left(\sqrt{2}\operatorname{cis}\frac{\pi}{4}\right)\cdot\left(\frac{1}{2}\operatorname{cis}\frac{\pi}{6}\right) = \frac{\sqrt{2}}{2}\operatorname{cis}\left(\frac{\pi}{4} + \frac{\pi}{6}\right) = \frac{\sqrt{2}}{2}\operatorname{cis}\left(\frac{5\pi}{12}\right), \text{ or}$$

$$\frac{z_1}{z_2} = \frac{\sqrt{2}\operatorname{cis}\frac{\pi}{4}}{2\operatorname{cis}\frac{-\pi}{6}} = \frac{\sqrt{2}}{2}\operatorname{cis}\left(\frac{\pi}{4} - \frac{-\pi}{6}\right) = \frac{\sqrt{2}}{2}\operatorname{cis}\left(\frac{5\pi}{12}\right)$$

d) $\frac{z_1}{z_2} = \frac{1+i}{\sqrt{3}-i} \cdot \frac{\sqrt{3}+i}{\sqrt{3}+i} = \frac{\sqrt{3}-1+(\sqrt{3}+1)i}{4}$

Comparing this to part c).

$\frac{\sqrt{3}-1}{4} = \frac{\sqrt{2}}{2}\cos\frac{5\pi}{12} \Rightarrow \cos\frac{5\pi}{12} = \frac{\sqrt{3}-1}{4} \cdot \frac{2}{\sqrt{2}} = \frac{\sqrt{6}-\sqrt{2}}{4}$.

Also, $\frac{\sqrt{3}+1}{4} = \frac{\sqrt{2}}{2}\sin\frac{5\pi}{12} \Rightarrow \sin\frac{5\pi}{12} = \frac{\sqrt{3}+1}{4} \cdot \frac{2}{\sqrt{2}} = \frac{\sqrt{6}+\sqrt{2}}{4}$.

Exercise 10.2

In questions 1–14, write the complex number in polar form with argument θ, such that $0 \leq \theta < 2\pi$.

1 $2 + 2i$

2 $\sqrt{3} + i$

3 $2 - 2i$

4 $\sqrt{6} - i\sqrt{2}$

5 $2 - 2i\sqrt{3}$

6 $-3 + 3i$

7 $4i$

8 $-3\sqrt{3} - 3i$

9 $i + 1$

10 -15

11 $(4 + 3i)^{-1}$

12 $i(3 + 3i)$

13 π

14 ei

In questions 15–24, find z_1z_2 and $\frac{z_1}{z_2}$.

15 $z_1 = \cos\frac{\pi}{2} + i\sin\frac{\pi}{2}, z_2 = \cos\frac{\pi}{3} + i\sin\frac{\pi}{3}$

16 $z_1 = \cos\frac{5\pi}{6} + i\sin\frac{5\pi}{6}, z_2 = \cos\frac{7\pi}{6} + i\sin\frac{7\pi}{6}$

17 $z_1 = \cos\frac{\pi}{6} + i\sin\frac{\pi}{6}, z_2 = \cos\frac{2\pi}{3} + i\sin\frac{2\pi}{3}$

18 $z_1 = \cos\frac{13\pi}{12} + i\sin\frac{13\pi}{12}, z_2 = \cos\frac{5\pi}{12} + i\sin\frac{5\pi}{12}$

19 $z_1 = 3\left(\cos\frac{3\pi}{4} + i\sin\frac{3\pi}{4}\right), z_2 = \frac{2}{3}\left(\cos\frac{4\pi}{3} + i\sin\frac{4\pi}{3}\right)$

20 $z_1 = 3\sqrt{2}\left(\cos\frac{5\pi}{4} + i\sin\frac{5\pi}{4}\right), z_2 = 2\left(\cos\frac{5\pi}{3} + i\sin\frac{5\pi}{3}\right)$

21 $z_1 = \cos 135° + i\sin 135°, z_2 = \cos 90° + i\sin 90°$

22 $z_1 = 3(\cos 120° + i\sin 120°), z_2 = 2(\cos 240° + i\sin 240°)$

23 $z_1 = \frac{5}{8}(\cos 225° + i\sin 225°), z_2 = \frac{\sqrt{3}}{2}(\cos 330° + i\sin 330°)$

24 $z_1 = 3\sqrt{2}(\cos 315° + i\sin 315°), z_2 = 2(\cos 300° + i\sin 300°)$

In questions 25–30, write z_1 and z_2 in polar form, and then find the reciprocals $\frac{1}{z_1}, \frac{1}{z_2}$, the product z_1z_2, and the quotient $\frac{z_1}{z_2}$ $(-\pi < \theta < \pi)$.

25 $z_1 = \sqrt{3} + i$ and $z_2 = 2 - 2i\sqrt{3}$

26 $z_1 = \sqrt{6} + i\sqrt{2}$ and $z_2 = 2\sqrt{3} - 6i$

27 $z_1 = 4\sqrt{3} + 4i$ and $z_2 = -3 - 3i$

28 $z_1 = i\sqrt{3}$ and $z_2 = -\sqrt{2} - i\sqrt{6}$

29 $z_1 = \sqrt{5} + i\sqrt{5}$ and $z_2 = 2i\sqrt{2}$

30 $z_1 = 1 + i\sqrt{3}$ and $z_2 = 2\sqrt{3}$

31 Consider the complex number z where $|z - i| = |z + 2i|$.

a) Show that $\text{Im}(z) = -\frac{1}{2}$.

b) Let z_1 and z_2 be the two possible values of z, such that $|z| = 1$.

(i) Sketch a diagram to show the points which represent z_1 and z_2 in the complex plane.

(ii) Find $\arg(z_1)$ and $\arg(z_2)$.

32 Use the Argand diagram to show that $|z_1 + z_2| \leqslant |z_1| + |z_2|$.

33 If $z = \sqrt{3}\left(\cos\frac{2\pi}{3} + i\sin 2\frac{\pi}{3}\right)$, express each of the following complex numbers in Cartesian form.

a) $\dfrac{3}{\sqrt{3} + z}$ b) $\dfrac{2z}{3 + z^2}$ c) $\dfrac{3 - z^2}{3 + z^2}$

34 Find the modulus and argument (amplitude) of each of the complex numbers

$$z_1 = 2\sqrt{3} - 2i,\ z_2 = 2 + 2i \text{ and } z_3 = (2\sqrt{3} - 2i)(2 + 2i).$$

35 If the numbers in question 34 represent the vertices of a triangle in the Argand diagram, find the area of that triangle.

36 Identify, in the complex plane, the set of points that correspond to the following equations.

a) $|z| = 3$

b) $z^* = -z$

c) $z + z^* = 8$

d) $|z - 3| = 2$

e) $|z - 1| + |z - 3| = 2$

37 Identify, in the complex plane, the set of points that correspond to the following inequations.

a) $|z| \leqslant 3$

b) $|z - 3i| \geqslant 2$

10.3 Powers and roots of complex numbers

The formula established for the product of two complex numbers can be applied to derive a special formula for the nth power of a complex number.

Let $z = r(\cos\theta + i\sin\theta)$, now

$$\begin{aligned} z^2 &= (r(\cos\theta + i\sin\theta))(r(\cos\theta + i\sin\theta)) \\ &= r^2((\cos\theta\cos\theta - \sin\theta\sin\theta) + i(\sin\theta\cos\theta + \cos\theta\sin\theta)) \\ &= r^2((\cos^2\theta - \sin^2\theta) + i(2\sin\theta\cos\theta)) = r^2(\cos 2\theta + i\sin 2\theta). \end{aligned}$$

Similarly,

$$\begin{aligned} z^3 &= z \cdot z^2 = (r(\cos\theta + i\sin\theta))(r^2(\cos 2\theta + i\sin 2\theta)) \\ &= r^3(\cos(\theta + 2\theta) + i\sin(\theta + 2\theta)) = r^3(\cos 3\theta + i\sin 3\theta). \end{aligned}$$

In general, we obtain the following theorem, named after the French mathematician A. De Moivre (1667–1754).

Note: As a matter of fact, de Moivre stated 'his' formula only implicitly. Its standard form is due to Euler and was generalized by him to any real n.

De Moivre's theorem

If $z = r(\cos\theta + i\sin\theta)$ and n is a positive integer, then

$$z^n = (r(\cos\theta + i\sin\theta))^n = r^n(\cos n\theta + i\sin n\theta).$$

The theorem: To find the nth power of any complex number written in trigonometric form, we take the nth power of the modulus and multiply the argument with n.

Proof

The proof of this theorem follows as an application of mathematical induction.

Let $P(n)$ be the statement $z^n = r^n(\cos n\theta + i\sin n\theta)$.

Basis step:

To prove this formula the basis step must be $P(1)$.

$P(1)$: is true since

$z^1 = r^1(\cos\theta + i\sin\theta)$, which is given!

[If you are not convinced, you can try

$P(2)$: $z^2 = r^2(\cos 2\theta + i\sin 2\theta)$, which we showed above.]

Inductive step:

Assume that $P(k)$ is true, i.e.

$z^k = r^k(\cos k\theta + i\sin k\theta)$. We need to show that $P(k+1)$ is also true.

So we have to show that $z^{k+1} = r^{k+1}(\cos(k+1)\theta + i\sin(k+1)\theta)$.

Now,

$$\begin{aligned}
z^{k+1} &= z^k \cdot z = (\cos k\theta + i\sin k\theta)(r(\cos\theta + i\sin\theta)) \text{ by assumption}\\
&= r^k r[(\cos k\theta\cos\theta - \sin k\theta\sin\theta) + i(\sin k\theta\cos\theta + \cos k\theta\sin\theta)]\\
&= r^{k+1}[\cos(k\theta + \theta) + i\sin(k\theta + \theta)] \text{ by addition formulae for sine and cosine}\\
&= r^{k+1}(\cos(k+1)\theta + i\sin(k+1)\theta)
\end{aligned}$$

Therefore, by the principle of mathematical induction, since the theorem is true for $n = 1$, and whenever it is true for $n = k$, it was proved true for $n = k + 1$, then the theorem is true for positive integers n.

Note: In fact the theorem is valid for all real numbers n. However, the proof is beyond the scope of this course and this book and therefore we will consider the theorem true for all real numbers without proof at the moment.

Example 19

Find $(1 + i)^6$.

Solution

We convert the number into polar form first.

$(1 + i) = \sqrt{2}\left(\cos\frac{\pi}{4} + i\sin\frac{\pi}{4}\right)$

Now we can apply De Moivre's theorem.

$$(1 + i)^6 = \left[\sqrt{2}\left(\cos\frac{\pi}{4} + i\sin\frac{\pi}{4}\right)\right]^6 = (\sqrt{2})^6\left(\cos\left(6\cdot\frac{\pi}{4}\right) + i\sin\left(6\cdot\frac{\pi}{4}\right)\right)$$
$$= 8\left(\cos\frac{3\pi}{2} + i\sin\frac{3\pi}{2}\right) = 8(-i) = -8i$$

Imagine you wanted to use the binomial theorem to evaluate the power.

$$(1 + i)^6 = 1 + 6i + 15i^2 + 20i^3 + 15i^4 + 6i^5 + i^6$$
$$= 1 + 6i - 15 - 20i + 15 + 6i - 1 = 8i$$

When the powers get larger, we are sure you will appreciate De Moivre!

Applications of De Moivre's theorem

Several applications of this theorem prove very helpful in dealing with trigonometric identities and expressions.

For example, when $n = -1$, the theorem gives the following result.

$z^{-1} = r^{-1}(\cos(-\theta) + i\sin(-\theta)) = \frac{1}{r}(\cos\theta - i\sin\theta)$

Also,

$z^{-n} = (z^{-1})^n = (r^{-1}(\cos(-\theta) + i\sin(-\theta)))^n = r^{-n}(\cos(-n\theta) + i\sin(-n\theta))$.

If we take the case when $r = 1$, then

$z^n = \cos n\theta + i\sin n\theta$ and $z^{-n} = \cos(-n\theta) + i\sin(-n\theta) = \cos n\theta - i\sin n\theta$

$\Rightarrow z^n + z^{-n} = 2\cos n\theta$ and $z^n - z^{-n} = 2i\sin n\theta$.

These relationships are quite helpful in allowing us to write powers of $\cos\theta$ and $\sin\theta$ in terms of cosines and sines of multiples of θ.

Example 20

Find $\cos^3\theta$ in terms of first powers of the cosine function.

Solution

Starting with

$$\left(z + \frac{1}{z}\right)^3 = (2\cos\theta)^3$$

and expanding the left-hand side, we get

$$z^3 + 3z + \frac{3}{z} + \frac{1}{z^3} = 8\cos^3\theta \Rightarrow z^3 + \frac{1}{z^3} + 3\left(z + \frac{1}{z}\right) = 8\cos^3\theta$$
$$\Updownarrow \qquad \Updownarrow$$
$$\Rightarrow 2\cos 3\theta + 3(2\cos\theta) = 8\cos^3\theta$$
$$\Rightarrow \cos^3\theta = \frac{1}{8}(2\cos 3\theta + 3(2\cos\theta))$$
$$= \frac{1}{4}(\cos 3\theta + 3\cos\theta)$$

Example 21

Simplify the following expression:

$$\frac{(\cos 6\theta + i\sin 6\theta)(\cos 3\theta + i\sin 3\theta)}{\cos 4\theta + i\sin 4\theta}$$

Solution

$$\frac{(\cos 6\theta + i\sin 6\theta)(\cos 3\theta + i\sin 3\theta)}{\cos 4\theta + i\sin 4\theta}$$
$$= \frac{(\cos\theta + i\sin\theta)^6(\cos\theta + i\sin\theta)^3}{(\cos\theta + i\sin\theta)^4}$$

Using the laws of exponents, we have

$$\frac{(\cos\theta + i\sin\theta)^6(\cos\theta + i\sin\theta)^3}{(\cos\theta + i\sin\theta)^4} = (\cos\theta + i\sin\theta)^5$$
$$= \cos 5\theta + i\sin 5\theta.$$

*n*th roots of a complex number

De Moivre's theorem is an essential tool for finding nth roots of complex numbers.

An nth root of a given number z is a number w that satisfies the following relation

$$w^n = z.$$

For example, $w = 1 + i$ is a 6th root of $z = -8i$ because, as you have seen above,

$$(1 + i)^6 = -8i, \text{ or}$$

$w = -\sqrt{3} + i$ is a 10th root of $512 + 512i\sqrt{3}$.

This is also because $w^{10} = (-\sqrt{3} + i)^{10} = 512 + 512i\sqrt{3}$.

How to find the *n*th roots:

To find them, we apply the definition of an nth root as mentioned above.

Let $w = s(\cos\alpha + i\sin\alpha)$ be an nth root of $z = r(\cos\theta + i\sin\theta)$. This means that $w^n = z$, i.e.

$$(s(\cos\alpha + i\sin\alpha))^n = r(\cos\theta + i\sin\theta) \Rightarrow$$
$$s^n(\cos n\alpha + i\sin n\alpha) = r(\cos\theta + i\sin\theta)$$

However, two complex numbers are equal if their moduli are equal, that is,

$$s^n = r \Leftrightarrow s = \sqrt[n]{r} = r^{\frac{1}{n}}.$$

Also,

$$\cos n\alpha = \cos\theta \text{ and } \sin n\alpha = \sin\theta.$$

From your trigonometry chapters, you recall that both sine and cosine functions are periodic of period 2π each; hence,

$$\begin{cases} \cos n\alpha = \cos\theta \\ \sin n\alpha = \sin\theta \end{cases} \Rightarrow n\alpha = \theta + 2k\pi, k = 0, 1, 2, \ldots$$

This leads to

$$\alpha = \frac{\theta + 2k\pi}{n} = \frac{\theta}{n} + \frac{2k\pi}{n};\ k = 0, 1, 2, 3, ..., n - 1.$$

Notice that we stop the values of k at $n - 1$. This is so because for values larger than or equal to n, principal arguments for these roots will be identical to those for $k = 0$ till $n - 1$.

nth roots of a complex number

Let $z = r(\cos\theta + i\sin\theta)$ and let n be a positive integer, then z has n distinct nth roots

$$z_k = \sqrt[n]{r}\left(\cos\left(\frac{\theta}{n} + \frac{2k\pi}{n}\right) + i\sin\left(\frac{\theta}{n} + \frac{2k\pi}{n}\right)\right)$$

where $k = 1, 2, 3, \ldots, n - 1$.

Note: Each of the n nth roots of z has the same modulus $\sqrt[n]{r} = r^{\frac{1}{n}}$. Thus all these roots lie on a circle in the complex plane whose radius is $\sqrt[n]{r} = r^{\frac{1}{n}}$. Also, since the arguments of consecutive roots differ by $\frac{2\pi}{n}$, then the roots are also equally spaced on this circle.

Example 22

Find the cube roots of $z = -8 + 8i$.

Solution

$r = 8\sqrt{2}$ and $\theta = \frac{3\pi}{4}$, so the roots are

$$w = \sqrt[n]{r}\left(\cos\left(\frac{\theta}{n} + \frac{2k\pi}{n}\right) + i\sin\left(\frac{\theta}{n} + \frac{2k\pi}{n}\right)\right) = \sqrt[3]{(8\sqrt{2})}\left(\cos\left(\frac{\frac{3\pi}{4}}{3} + \frac{2k\pi}{3}\right) + i\sin\left(\frac{\frac{3\pi}{4}}{3} + \frac{2k\pi}{3}\right)\right)$$

$$= 2(\sqrt[6]{2})\left(\cos\left(\frac{\pi}{4} + \frac{2k\pi}{3}\right) + i\sin\left(\frac{\pi}{4} + \frac{2k\pi}{3}\right)\right);\ k = 0, 1, 2$$

$$w_1 = 2(\sqrt[6]{2})\left(\cos\left(\frac{\pi}{4}\right) + i\sin\left(\frac{\pi}{4}\right)\right)$$

$$w_2 = 2(\sqrt[6]{2})\left(\cos\left(\frac{\pi}{4} + \frac{2\pi}{3}\right) + i\sin\left(\frac{\pi}{4} + \frac{2\pi}{3}\right)\right) = 2\sqrt[6]{2}\left(\cos\left(\frac{11\pi}{12}\right) + i\sin\left(\frac{11\pi}{12}\right)\right)$$

$$w_3 = 2(\sqrt[6]{2})\left(\cos\left(\frac{\pi}{4} + \frac{4\pi}{3}\right) + i\sin\left(\frac{\pi}{4} + \frac{4\pi}{3}\right)\right) = 2\sqrt[6]{2}\left(\cos\left(\frac{19\pi}{12}\right) + i\sin\left(\frac{19\pi}{12}\right)\right)$$

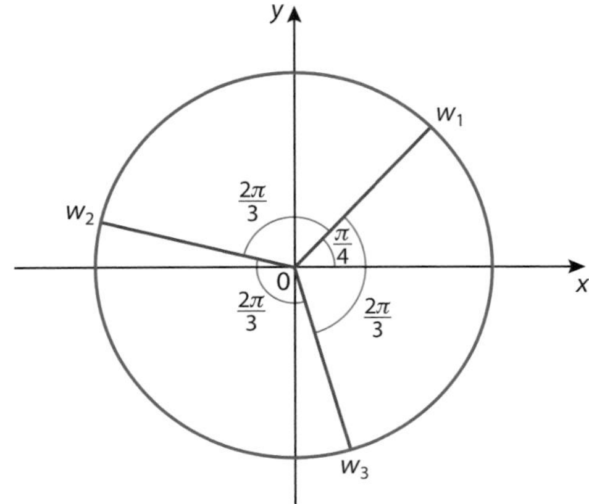

Notice how the arguments are distributed equally around a circle with radius $2(\sqrt[6]{2})$. The difference between any two arguments is $\frac{2\pi}{3}$.

Notice that if you try to go beyond $k = 2$, then you get back to w_1.

$$w_4 = 2\sqrt[6]{2}\left(\cos\left(\frac{\pi}{4} + \frac{6\pi}{3}\right) + i\sin\left(\frac{\pi}{4} + \frac{6\pi}{3}\right)\right) = 2\sqrt[6]{2}\left(\cos\left(\frac{\pi}{4} + 2\pi\right) + i\sin\left(\frac{\pi}{4} + 2\pi\right)\right)$$

$$= 2\sqrt[6]{2}\left(\cos\left(\frac{\pi}{4}\right) + i\sin\left(\frac{\pi}{4}\right)\right) = w_1$$

Also, if you raise any of the roots to the third power, you will eventually get z; for example,

$$(w_2)^3 = \left[2\sqrt[6]{2}\left(\cos\left(\frac{11\pi}{12}\right) + i\sin\left(\frac{11\pi}{12}\right)\right)\right]^3 = 8\sqrt{2}\left(\cos\left(\frac{33\pi}{12}\right) + i\sin\left(\frac{33\pi}{12}\right)\right)$$

$$= 8\sqrt{2}\left(\cos\left(\frac{11\pi}{4}\right) + i\sin\left(\frac{11\pi}{4}\right)\right) = 8\sqrt{2}\left(\cos\left(\frac{3\pi}{4}\right) + i\sin\left(\frac{3\pi}{4}\right)\right) = z$$

Example 23

Find the six sixth roots of $z = -64$ and graph these roots in the complex plane.

Solution

Here $r = 64$ and $\theta = \pi$. So the roots are

$$w = s\left(\cos\left(\frac{\theta}{n} + \frac{2k\pi}{n}\right) + i\sin\left(\frac{\theta}{n} + \frac{2k\pi}{n}\right)\right)$$

$$= \sqrt[6]{64}\left(\cos\left(\frac{\pi}{6} + \frac{2k\pi}{6}\right) + i\sin\left(\frac{\pi}{6} + \frac{2k\pi}{6}\right)\right)$$

$$= 2\left(\cos\left(\frac{\pi}{6} + \frac{k\pi}{3}\right) + i\sin\left(\frac{\pi}{6} + \frac{k\pi}{3}\right)\right);\ k = 0, 1, 2, 3, 4, 5$$

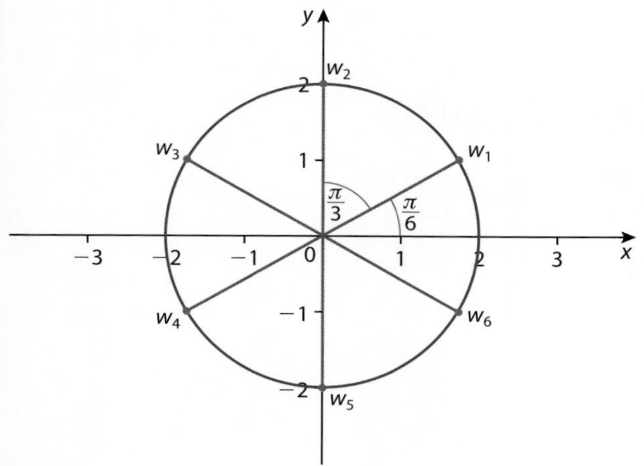

$$w_1 = 2\left(\cos\left(\frac{\pi}{6}\right) + i\sin\left(\frac{\pi}{6}\right)\right)$$

$$w_2 = 2\left(\cos\left(\frac{\pi}{6} + \frac{\pi}{3}\right) + i\sin\left(\frac{\pi}{6} + \frac{\pi}{3}\right)\right) = 2\left(\cos\left(\frac{\pi}{2}\right) + i\sin\left(\frac{\pi}{2}\right)\right)$$

$$w_3 = 2\left(\cos\left(\frac{\pi}{6} + \frac{2\pi}{3}\right) + i\sin\left(\frac{\pi}{6} + \frac{2\pi}{3}\right)\right) = 2\left(\cos\left(\frac{5\pi}{6}\right) + i\sin\left(\frac{5\pi}{6}\right)\right)$$

$$w_4 = 2\left(\cos\left(\frac{\pi}{6} + \frac{3\pi}{3}\right) + i\sin\left(\frac{\pi}{6} + \frac{3\pi}{3}\right)\right) = 2\left(\cos\left(\frac{7\pi}{6}\right) + i\sin\left(\frac{7\pi}{6}\right)\right)$$

$$w_5 = 2\left(\cos\left(\frac{\pi}{6} + \frac{4\pi}{3}\right) + i\sin\left(\frac{\pi}{6} + \frac{4\pi}{3}\right)\right) = 2\left(\cos\left(\frac{3\pi}{2}\right) + i\sin\left(\frac{3\pi}{2}\right)\right)$$

$$w_6 = 2\left(\cos\left(\frac{\pi}{6} + \frac{5\pi}{3}\right) + i\sin\left(\frac{\pi}{6} + \frac{5\pi}{3}\right)\right) = 2\left(\cos\left(\frac{11\pi}{6}\right) + i\sin\left(\frac{11\pi}{6}\right)\right)$$

*n*th roots of unity

The rules we established can be applied to finding the *n*th roots of 1 (unity). Since 1 is a real number, then in polar/trigonometric form it has a modulus of 1 and an argument of 0. We can write it as

$$1 = 1(\cos 0 + i\sin 0).$$

Now applying the rules above, 1 has *n* distinct *n*th roots given by

$$\begin{aligned} z_k &= \sqrt[n]{r}\left(\cos\left(\frac{\theta}{n} + \frac{2k\pi}{n}\right) + i\sin\left(\frac{\theta}{n} + \frac{2k\pi}{n}\right)\right) \\ &= \sqrt[n]{1}\left(\cos\left(\frac{0}{n} + \frac{2k\pi}{n}\right) + i\sin\left(\frac{0}{n} + \frac{2k\pi}{n}\right)\right) \\ &= \cos\left(\frac{2k\pi}{n}\right) + i\sin\left(\frac{2k\pi}{n}\right);\ k = 0, 1, 2, \ldots, n-1 \end{aligned}$$

Or in degrees,

$$z_k = \cos\left(\frac{360k}{n}\right) + i\sin\left(\frac{360k}{n}\right);\ k = 0, 1, 2, \ldots, n-1$$

Example 24

Find

a) the square roots of unity

b) the cube roots of unity.

Solution

a) Here $k = 2$, and therefore the two roots are

$$z_k = \cos\left(\frac{360k}{2}\right) + i\sin\left(\frac{360k}{2}\right);\ k = 0, 1$$

$$z_0 = \cos\left(\frac{0}{2}\right) + i\sin\left(\frac{0}{2}\right) = 1$$

$$z_1 = \cos\left(\frac{360}{2}\right) + i\sin\left(\frac{360}{2}\right) = \cos 180 + i\sin 180 = -1$$

b) Here $k = 3$, and the three roots are

$$z_k = \cos\left(\frac{2k\pi}{3}\right) + i\sin\left(\frac{2k\pi}{3}\right);\ k = 0, 1, 2, 3$$

$$z_0 = \cos\left(\frac{0}{3}\right) + i\sin\left(\frac{0}{3}\right) = 1$$

$$z_1 = \cos\left(\frac{2\pi}{3}\right) + i\sin\left(\frac{2\pi}{3}\right) = -\frac{1}{2} + i\frac{\sqrt{3}}{2}$$

$$z_2 = \cos\left(\frac{4\pi}{3}\right) + i\sin\left(\frac{4\pi}{3}\right) = -\frac{1}{2} - i\frac{\sqrt{3}}{2}$$

Euler's formula

The material in this part depends on work that you will do in the Analysis option. Otherwise, you will have to accept the result without proof.

In the options section on infinite series, we have the following results.

Taylor's (Maclaurin's) series expansion for $\sin x$, $\cos x$ and e^x are

$$\sin x = x - \frac{x^3}{3!} + \frac{x^5}{5!} - \frac{x^7}{7!} + \ldots = \sum_0^{\infty}(-1)^n \frac{x^{2n+1}}{(2n+1)!}$$

$$\cos x = 1 - \frac{x^2}{2!} + \frac{x^4}{4!} - \frac{x^6}{6!} + \ldots = \sum_0^{\infty}(-1)^n \frac{x^{2n}}{(2n)!}$$

$$e^x = 1 + x + \frac{x^2}{2!} + \frac{x^3}{3!} + \frac{x^4}{4!} + \ldots = \sum_0^{\infty} \frac{x^n}{n!}$$

Now if you add

$$\sin x + \cos x = 1 + x - \frac{x^2}{2!} - \frac{x^3}{3!} + \frac{x^4}{4!} + \frac{x^5}{5!} - \frac{x^6}{6!} - \frac{x^7}{7!} + \ldots$$

and compare the result to e^x expansion, we notice a stark similarity in the terms, except for the 'discrepancy' in the signs! The signs in the sum alternate in a way where pairs of terms alternate! This property is typical of powers of i.

Look at $i, i^2, i^3, i^4, i^5, i^6, i^7, i^8, \ldots = i, \boxed{-1, -i}, \boxed{1, i}, \boxed{-1, -i}, 1, \ldots$

This suggests expanding e^{ix}

$$e^{ix} = 1 + ix + \frac{i^2x^2}{2!} + \frac{i^3x^3}{3!} + \frac{i^4x^4}{4!} + \frac{i^5x^5}{5!} + \frac{i^6x^6}{6!} + \ldots$$

$$= 1 + \frac{i^2x^2}{2!} + \frac{i^4x^4}{4}! + \frac{i^6x^6}{6!} + ix + \frac{i^3x^3}{3!} + \frac{i^5x^5}{5!} + \ldots$$

$$= 1 - \frac{x^2}{2!} + \frac{x^4}{4!} - \frac{x^6}{6!} + \ldots + i\left(x + \frac{i^2x^3}{3!} + \frac{i^4x^5}{5!} + \ldots\right)$$

$$= 1 - \frac{x^2}{2!} + \frac{x^4}{4!} - \frac{x^6}{6!} + \ldots + i\left(x - \frac{x^3}{3!} + \frac{x^5}{5!} + \ldots\right)$$

$$= \cos x + i \sin x$$

Since, for any complex number

$z = x + iy = r(\cos\theta + i\sin\theta)$ and since $e^{i\theta} = \cos\theta + i\sin\theta$, then
$z = r(\cos\theta + i\sin\theta) = re^{i\theta}$.

This is known as Euler's formula.

Example 25

Evaluate each of the following

a) $e^{i\pi}$ b) $e^{i\frac{\pi}{2}}$

Solution

a) $e^{i\pi} = \cos\pi + i\sin\pi = -1$

b) $e^{i\frac{\pi}{2}} = \cos\frac{\pi}{2} + i\sin\frac{\pi}{2} = i$

Example 26

Use Euler's formula to prove DeMoivre's theorem.

Solution

$$(r(\cos\theta + i\sin\theta))^n = (re^{i\theta})^n = r^n e^{in\theta}$$
$$= r^n(\cos n\theta + i\sin n\theta)$$

Example 27

Find the real and imaginary parts of the complex numbers:

a) $z = 3e^{i\frac{\pi}{6}}$ b) $z = 7e^{2i}$

Solution

a) Since $|z| = 3$ and $\arg(z) = \frac{\pi}{6}$, $\text{Re}(z) = 3\cos\frac{\pi}{6} = \frac{3\sqrt{3}}{2}$ and $\text{Im}(z) = 3\sin\frac{\pi}{6} = \frac{3}{2}$.

b) Since $|z| = 7$ and $\arg(z) = 2$, $\text{Re}(z) = 7\cos 2$ and $\text{Im}(z) = 7\sin 2$.

Example 28

Express $z = 5 + 5i$ in exponential form.

Solution

$|z| = 5\sqrt{2}$ and $\tan\theta = \frac{5}{5} = 1 \Rightarrow \theta = \frac{\pi}{4}$, therefore $z = 5\sqrt{2}\,e^{i\frac{\pi}{4}}$.

Example 29

Evaluate $(5 + 5i)^6$ and express your answer in rectangular form.

Solution

Let $z = 5 + 5i$. From the example above, $z = 5\sqrt{2}\,e^{i\frac{\pi}{4}}$; hence,

$$z^6 = \left(5\sqrt{2}\,e^{i\frac{\pi}{4}}\right)^6 = (5\sqrt{2})^6 e^{i\frac{\pi}{4}\times 6} = 125\,000\,e^{i\frac{3\pi}{2}} = -125\,000i.$$

Alternatively,

$$(5 + 5i)^6 = \left(5\sqrt{2}\left(\cos\frac{\pi}{4} + i\sin\frac{\pi}{4}\right)\right)^6 = (5\sqrt{2})^6\left(\cos\frac{6\pi}{4} + i\sin\frac{6\pi}{4}\right) = -125\,000i.$$

Example 30

Simplify the following expression:

$$\frac{(\cos 6\theta + i\sin 6\theta)(\cos 3\theta + i\sin 3\theta)}{\cos 4\theta + i\sin 4\theta}$$

Solution

$$\frac{(\cos 6\theta + i\sin 6\theta)(\cos 3\theta + i\sin 3\theta)}{\cos 4\theta + i\sin 4\theta} = \frac{e^{6i\theta}\cdot e^{3i\theta}}{e^{4i\theta}} = e^{5i\theta} = \cos 5\theta + i\sin 5\theta$$

Example 31

Use Euler's formula to find the cube roots of i.

Solution

$i = e^{i\left(\frac{\pi}{2} + 2k\pi\right)} \Rightarrow i^{\frac{1}{3}} = \left(e^{i\left(\frac{\pi}{2} + 2k\pi\right)}\right)^{\frac{1}{3}} = e^{i\left(\frac{\pi}{6} + \frac{2k\pi}{3}\right)}$; $k = 0, 1, 2$

Therefore,

$$z_0 = e^{i\left(\frac{\pi}{6}\right)} = \cos\frac{\pi}{6} + i\sin\frac{\pi}{6} = \frac{\sqrt{3}}{2} + \frac{i}{2}$$

$$z_1 = e^{i\left(\frac{\pi}{6} + \frac{2\pi}{3}\right)} = e^{i\left(\frac{5\pi}{6}\right)} = \cos\frac{5\pi}{6} + i\sin\frac{5\pi}{6} = -\frac{\sqrt{3}}{2} + \frac{i}{2}$$

$$z_2 = e^{i\left(\frac{\pi}{6} + \frac{4\pi}{3}\right)} = e^{i\left(\frac{3\pi}{2}\right)} = \cos\frac{3\pi}{2} + i\sin\frac{3\pi}{2} = -i$$

As you notice here, Euler's formula provides us with a very powerful tool to perform otherwise extremely laborious calculations.

Exercise 10.3

In questions 1–6, write the complex number in Cartesian form.

1 $z = 4e^{-i\frac{2\pi}{3}}$

2 $z = 3e^{2\pi i}$

3 $z = 3e^{0.5\pi i}$

4 $z = 4\,\text{cis}\left(\frac{7\pi}{12}\right)$ (exact value)

5 $z = 13e^{\frac{\pi i}{3}}$

6 $z = 3e^{1+\frac{\pi}{3}i}$

In questions 7–16, write each complex number in exponential form.

7 $2 + 2i$

8 $\sqrt{3} + i$

9 $\sqrt{6} - i\sqrt{2}$

10 $2 - 2i\sqrt{3}$

11 $-3 + 3i$

12 $4i$

13 $-3\sqrt{3} - 3i$

14 $i(3 + 3i)$

15 π

16 ei

In questions 17–25, find each complex number. Express in exact rectangular form when possible.

17 $(1 + i)^{10}$

18 $(\sqrt{3} - i)^6$

19 $(3 + 3i\sqrt{3})^9$

20 $(2 - 2i)^{12}$

21 $(\sqrt{3} - i\sqrt{3})^8$

22 $(-3 + 3i)^7$

23 $(\sqrt{3} - i\sqrt{3})^{-8}$

24 $(-3\sqrt{3} - 3i)^{-7}$

25 $2(\sqrt{3} + i)^7$

In questions 26–30, find each root and graph them in the complex plane.

26 The square roots of $4 + 4i\sqrt{3}$.

27 The cube roots of $4 + 4i\sqrt{3}$.

28 The fourth roots of -1.

29 The sixth roots of i.

30 The fifth roots of $-9 - 9i\sqrt{2}$.

In questions 31–36, solve each equation.

31 $z^5 - 32 = 0$

32 $z^8 + i = 0$

33 $z^3 + 4\sqrt{3} - 4i = 0$

34 $z^4 - 16 = 0$

35 $z^5 + 128 = 128i$

36 $z^6 - 64i = 0$

In questions 37–40, use De Moivre's theorem to simplify each of the following expressions.

37 $(\cos(9\beta) + i\sin(9\beta))(\cos(5\beta) - i\sin(5\beta))$

38 $\dfrac{(\cos(6\beta) + i\sin(6\beta))(\cos(4\beta) + i\sin(4\beta))}{(\cos(3\beta)) + i\sin(3\beta))}$

39 $(\cos(9\beta) + i\sin(9\beta))^{\frac{1}{3}}$

40 $\sqrt[n]{(\cos(2n\beta) + i\sin(2n\beta))}$

41 Use $e^{i\theta}$ to prove that $\cos(\alpha + \beta) = \cos\alpha\cos\beta - \sin\alpha\sin\beta$.

42 Use De Moivre's theorem to show that $\cos 4\alpha = 8\cos^4\alpha - 8\cos^2\alpha + 1$.

43 Use De Moivre's theorem to show that $\cos 5\alpha = 16\cos^5\alpha - 20\cos^3\alpha + 5\cos\alpha$.

44 Use De Moivre's theorem to show that $\cos^4\alpha = \frac{1}{8}(\cos 4\alpha + 4\cos 2\alpha + 3)$.

45 Let $z = \cos 2\alpha + i\sin 2\alpha$.

a) Show that $z + \frac{1}{z} = 2\cos 2\alpha$ and that $2i\sin 2\alpha = z - \frac{1}{z}$.

b) Find an expression for $\cos 2n\alpha$ and $\sin 2n\alpha$ in terms of z.

46 Let the cubic roots of 1 be 1, ω and ω^2. Simplify $(1 + 3\omega)(1 + 3\omega^2)$.

47 a) Show that the fourth roots of unity can be written as 1, β, β^2, and β^3.

b) Simplify $(1 + \beta)(1 + \beta^2 + \beta^3)$.

c) Show that $\beta + \beta^2 + \beta^3 = -1$.

48 a) Show that the fifth roots of unity can be written as 1, α, α^2, α^3 and α^4.

b) Simplify $(1 + \alpha)(1 + \alpha^4)$.

c) Show that $1 + \alpha + \alpha^2 + \alpha^3 + \alpha^4 = 0$.

49 Show that $(1 + i\sqrt{3})^n + (1 - i\sqrt{3})^n$ is real and find its value for $n = 18$.

50 Given that $z = (2a + 3i)^3$, and $a \in \mathbb{R}^+$, find the values of a such that $\arg z = 135°$.

Practice questions

1 Let $z = x + yi$. Find the values of x and y if $(1 - i)z = 1 - 3i$.

2 Let x and y be real numbers, and ω be one of the complex solutions of the equation $z^3 = 1$. Evaluate:

a) $1 + \omega + \omega^2$

b) $(\omega x + \omega^2 y)(\omega y + \omega^2 x)$

3 a) Evaluate $(1 + i)^2$ where $i = \sqrt{-1}$.

b) Prove, by mathematical induction, that $(1 + i)^{4n} = (-4)^n$, where $n \in \mathbb{N}^+$.

c) Hence or otherwise, find $(1 + i)^{32}$.

4 Let $z_1 = \dfrac{\sqrt{6} - i\sqrt{2}}{2}$ and $z_2 = 1 - i$.

a) Write z_1 and z_2 in the form $r(\cos\theta + i\sin\theta)$, where $r > 0$ and $-\frac{\pi}{2} \leq \theta \leq \frac{\pi}{2}$.

b) Show that $\frac{z_1}{z_2} = \cos\frac{\pi}{12} + i\sin\frac{\pi}{12}$.

c) Find the value of $\frac{z_1}{z_2}$ in the form $a + bi$, where a and b are to be determined exactly in radical (surd) form. Hence or otherwise, find the exact values of $\cos\frac{\pi}{12}$ and $\sin\frac{\pi}{12}$.

5 Let $z_1 = a\left(\cos\frac{\pi}{4} + i\sin\frac{\pi}{4}\right)$ and $z_2 = b\left(\cos\frac{\pi}{3} + i\sin\frac{\pi}{3}\right)$.

Express $\left(\frac{z_1}{z_2}\right)^3$ in the form $z = x + yi$.

6 If z is a complex number and $|z + 16| = 4|z + 1|$, find the value of $|z|$.

7 Find the values of a and b, where a and b are real, given that $(a + bi)(2 - i) = 5 - i$.

8 Given that $z = (b + i)^2$, where b is real and positive, find the exact value of b when $\arg z = 60°$.

9 The complex number z satisfies $i(z + 2) = 1 - 2z$, where $i = \sqrt{-1}$. Write z in the form $z = a + bi$, where a and b are real numbers.

10 a) Express $z^5 - 1$ as a product of two factors, one of which is linear.

b) Find the zeros of $z^5 - 1$, giving your answers in the form

$$r(\cos\theta + i\sin\theta) \text{ where } r > 0 \text{ and } -\pi < \theta \leq \pi.$$

c) Express $z^4 + z^3 + z^2 + z + 1$ as a product of two real quadratic factors.

11 a) Express the complex number $8i$ in polar form.

b) The cube root of $8i$ which lies in the first quadrant is denoted by z. Express z

(i) in polar form

(ii) in Cartesian form.

12 Consider the complex number $z = \dfrac{\left(\cos\frac{\pi}{4} - i\sin\frac{\pi}{4}\right)^2\left(\cos\frac{\pi}{3} + i\sin\frac{\pi}{3}\right)^3}{\left(\cos\frac{\pi}{24} - i\sin\frac{\pi}{24}\right)^4}$.

a) (i) Find the modulus of z.

(ii) Find the argument of z, giving your answer in radians.

b) Using De Moivre's theorem, show that z is a cube root of one, i.e. $z = \sqrt[3]{1}$.

c) Simplify $(1 + 2z)(2 + z^2)$, expressing your answer in the form $a + bi$, where a and b are exact real numbers.

13 The complex number z satisfies the equation $\sqrt{z} = \frac{2}{1 - i} + 1 - 4i$.

Express z in the form $x + iy$ where $x, y \in \mathbb{Z}$.

14 a) Prove, using mathematical induction, that for a positive integer n,

$$(\cos\theta + i\sin\theta)^n = \cos n\theta + i\sin n\theta \text{ where } i^2 = -1.$$

b) The complex number z is defined by $z = \cos\theta + i\sin\theta$.

(i) Show that $\frac{1}{z} = \cos(-\theta) + i\sin(-\theta)$.

(ii) Deduce that $z^n + z^{-n} = 2\cos n\theta$.

c) (i) Find the binomial expansion of $(z + z^{-1})^5$.

(ii) Hence, show that $\cos^5\theta = \frac{1}{16}(a\cos 5\theta + b\cos 3\theta + c\cos\theta)$, where a, b and c are positive integers to be found.

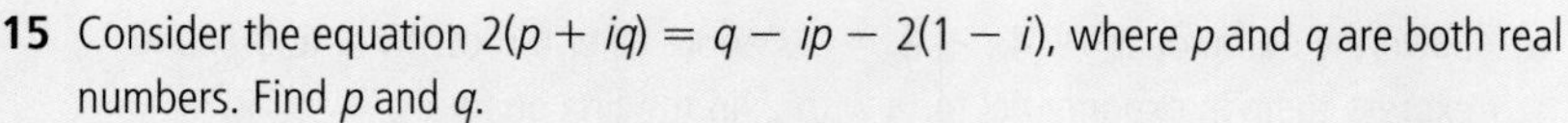

15 Consider the equation $2(p + iq) = q - ip - 2(1 - i)$, where p and q are both real numbers. Find p and q.

16 Consider $z^5 - 32 = 0$.

(i) Show that $z_1 = 2\left(\cos\frac{2\pi}{5} + i\sin\frac{2\pi}{5}\right)$ is one of the complex roots of this equation.

(ii) Find z_1^2, z_1^3, z_1^4 and z_1^5 giving your answer in the modulus argument form.

(iii) Plot the points that represent z_1, z_1^2, z_1^3, z_1^4 and z_1^5 in the complex plane.

(iv) The point $z^n{}_1$ is mapped to $z^{n+1}{}_1$ by a composition of two linear transformations, where $n = 1, 2, 3, 4$. Give a full geometric description of the two transformations.

17 A complex number z is such that $|z| = |z - 3i|$.

a) Show that the imaginary part of z is $\frac{3}{2}$.

b) Let z_1 and z_2 be the two possible values of z, such that $|z| = 3$.

(i) Sketch a diagram to show the points which represent z_1 and z_2 in the complex plane, where z_1 is in the first quadrant.

(ii) Show that $\arg(z_1) = \frac{\pi}{6}$.

(iii) Find $\arg(z_2)$.

c) Given that $\arg\left(\frac{z_1^k z_2}{2i}\right) = \pi$, find a value for k.

18 Given that $(a + i)(2 - bi) = 7 - i$, find the value of a and of b, where $a, b \in \mathbb{Z}$.

19 Consider the complex number $z = \cos\theta + i\sin\theta$.

a) Using De Moivre's theorem show that

$$z^n + \frac{1}{z^n} = 2\cos n\theta.$$

b) By expanding $\left(z + \frac{1}{z}\right)^4$ show that

$$\cos^4\theta = \tfrac{1}{8}(\cos 4\theta + 4\cos 2\theta + 3).$$

20 Consider the complex geometric series $e^{i\theta} + \frac{1}{2}e^{2i\theta} + \frac{1}{4}e^{3i\theta} + \ldots$

a) Find an expression for z, the common ratio of this series.

b) Show that $|z| < 1$.

c) Write down an expression for the sum to infinity of this series.

d) **(i)** Express your answer to part **c)** in terms of $\sin\theta$ and $\cos\theta$.

(ii) Hence, show that

$$\cos\theta + \tfrac{1}{2}\cos 2\theta + \tfrac{1}{4}\cos 3\theta + \ldots = \frac{4\cos\theta - 2}{5 - 4\cos\theta}.$$

21 Let $P(z) = z^3 + az^2 + bz + c$, where a, b and $c \in \mathbb{R}$. Two of the roots of $P(z) = 0$ are -2 and $(-3 + 2i)$. Find the value of a, of b and of c.

22 Given that $|z| = 2\sqrt{5}$, find the complex number z that satisfies the equation

$$\frac{25}{z} - \frac{15}{z^*} = 1 - 8i.$$

23 Solve the simultaneous system of equations giving your answers in $x + iy$ form:

$$iz_1 + 2z_2 = 3$$

$$z_1 + (1 - i)z_2 = 4$$

24 a) Solve the equation $x^2 - 4x + 8 = 0$. Denote its two roots by z_1 and z_2 and express them in exponential form with z_1 in the first quadrant.

b) Find the value of $\frac{z_1^4}{z_2^2}$ and write it in the form $x + yi$.

c) Show that $z_1^4 = z_2^4$.

d) Find the value of $\frac{z_1}{z_2} + \frac{z_2}{z_1}$.

e) For what values of n is z_1^n real?

25 a) Show that $z = \cos\frac{2\pi}{7} + i\sin\frac{2\pi}{7}$ is a root of the equation $x^7 - 1 = 0$.

b) Show that $z^7 - 1 = (z - 1)(z^6 + z^5 + z^4 + z^3 + z^2 + z + 1)$ and deduce that $z^6 + z^5 + z^4 + z^3 + z^2 + z + 1 = 0$.

c) Show that $\cos\frac{2\pi}{7} + \cos\frac{4\pi}{7} + \cos\frac{6\pi}{7} = -\frac{1}{2}$.

11 Statistics

Assessment statements

5.1 Concepts of population, sample, random sample and frequency distribution of discrete and continuous data.
Grouped data: use of mid-interval values, interval width, upper and lower interval boundaries.
Mean, variance, standard deviation.

Introduction

You will almost inevitably encounter statistics in one form or another on a daily basis. Here is an example:

The World Health Organization (WHO) collects and reports data pertaining to worldwide population health on all 192 UN member countries. Among the indicators reported is the **health-adjusted life expectancy** (HALE), which is based on life expectancy at birth, but includes an adjustment for time spent in poor health. It is most easily understood as the equivalent number of years in full health that a newborn can expect to live, based on current rates of ill-health and mortality. According to WHO rankings, lost years due to disability are substantially higher in poorer countries. Several factors contribute to this trend including injury, blindness, paralysis, and the debilitating effects of tropical disease.

More information on HALE can be found by visiting www.pearsonhotlinks.com, enter the ISBN or title of this book and select weblink 1.

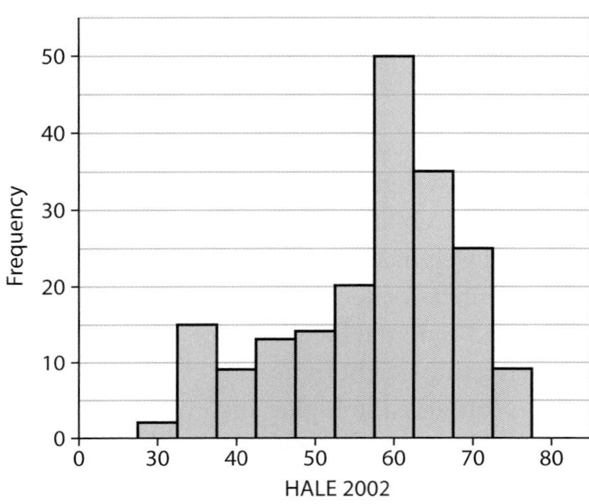

Of the 192 countries ranked by WHO, Japan has the highest life expectancy (75 years) and the lowest ranking country is Sierra Leone (29 years).

Reports similar to this one are commonplace in publications of several organizations, newspapers and magazines, and on the internet.

Questions that come to mind as we read such a report include: How did the researchers collect the data? How can we be sure that these results are reliable? What conclusions should be drawn from this report? The increased frequency with which statistical techniques are used in all fields, from business to agriculture to social and natural sciences, leads to the need for statistical literacy – familiarity with the goals and methods of these techniques – to be a part of any well-rounded educational programme.

Since statistical methods for summary and analysis provide us with powerful tools for making sense out of the data we collect, in this chapter we will first start by introducing two basic components of most statistical problems – population and sample – and then delve into the methods of presenting and making sense of data.

In the language of statistics, one of the most basic concepts is sampling. In most statistical problems, we draw a specified number of measurements or data – a sample – from a much larger body of measurements, called the population. On the basis of our observation of the data in the well-chosen sample, we try to describe or predict the behaviour of the population.

A **population** is any entire collection of people, animals, plants or things from which we may collect data. It is the entire group we are interested in, which we wish to describe or draw conclusions about. In order to make any generalizations about a population, a **sample**, that is meant to be representative of the population, is often studied. For each population there are many possible samples.

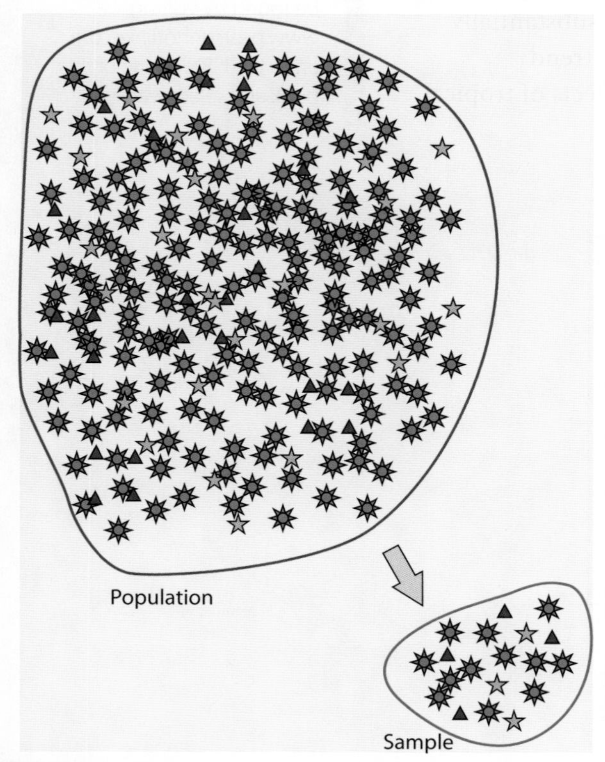

For example, a report on the effect the economic status (ES) has on healthy children's postures stated that:

'…ES, independent of overt malnutrition, affects height, weight, … with some gender differences in healthy children. Influence of income on height and weight show sexual dimorphism, a slight but significant effect is observed only in boys. MPH (mid-parental height) is the most prominent variable effecting height in healthy children. Higher height … observed in higher income groups suggest that secular trend in growth still exists, at least in boys, in a country of favorable economic development.'

Source: *European Journal of Clinical Nutrition* (2007) **61**, 752–758

The population is the 3-tuple measurement (economic status, height, weight) of all children of age 3–18 in Turkey. The sample is the set of measurements of the 428 boys and 386 girls that took part in the study. Notice that the population and sample are the measurements and not the people! The boys and girls are 'experimental units' or subjects in this study.

In this chapter we will present some basic techniques in **descriptive statistics** – the branch of statistics concerned with describing sets of measurements, both samples and populations.

11.1 Graphical tools

Once you have collected a set of measurements, how can you display this set in a clear, understandable and readable form? First, you must be able to define what is meant by measurement or 'data' and to categorize the types of data you are likely to encounter. We begin by introducing some definitions of the new terms in the statistical language that you need to know.

A **variable** is a characteristic that changes or varies over time and/or for different objects under consideration.

For example, if you are measuring the height of adults in a certain area, the height is a variable that changes with time for an individual and from person to person. When a variable is actually measured, a set of measurements or **data** will result. So, if you gather the heights of the students at your school, the set of measurements you get is a **data set**.

As the process of data collection begins, it becomes clear that often the number of data collected is so large that it is difficult for the statistician to see the findings of the data. The statistician's objective is to summarize succinctly, bringing out the important characteristics of the numbers and values in such a way that a clear and accurate picture emerges.

There are several ways of summarizing and describing data. Among them are tables and graphs and numerical measures.

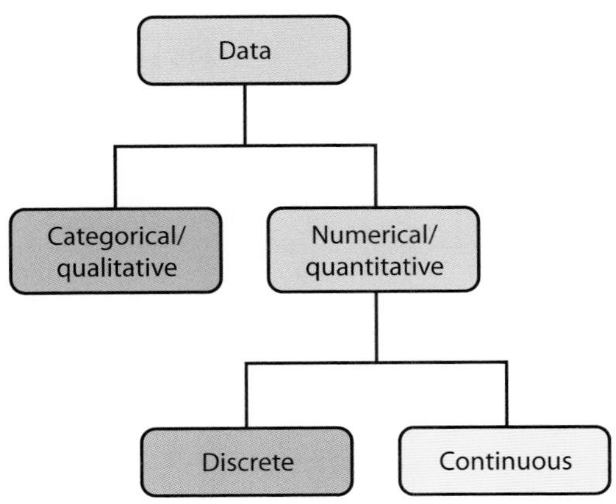

Frequently we use pie charts as a way of summarizing a set of categorical data or displaying the different values of a given variable (e.g., percentage distribution). This type of chart is a circle divided into a series of segments. Each segment represents a particular category. The area of each segment is the same proportion of a circle as the category is of the total data set.

Pie charts usually show the component parts of a whole. Often you will see a segment of the drawing separated from the rest of the pie in order to emphasize an important piece of information.

For example, in a large school, there are 230 students in the Maths Studies class, 180 students in the Standard Level maths class and 90 students in the HL mathematics class.

The pie chart for this data is given below.

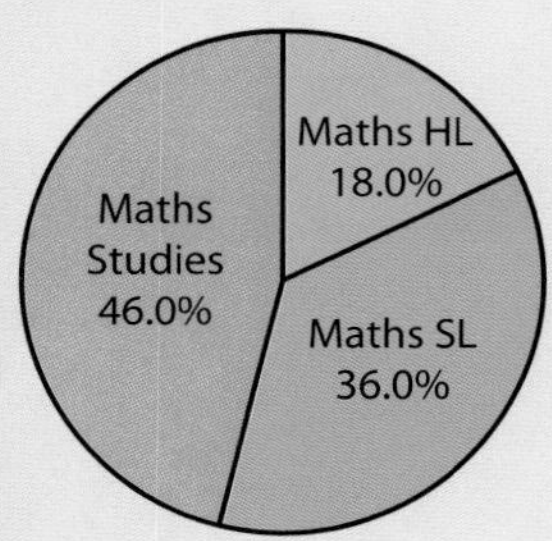

Classification of variables

Numerical or categorical

When classifying data, there are two major classifications: numerical or categorical data.

NUMERICAL (QUANTITATIVE) DATA – Quantitative variables measure a numerical quantity or amount on each experimental unit. Quantitative data yields a numerical response.

Examples: Yearly income of company presidents, the heights of students at school, the length of time it takes students to finish their lunch at school, and the total score you receive on exams, are all numerical.

Moreover, there are two types of numerical data:

DISCRETE – responses which arise from counting.

Example: Number of courses students take in a day.

CONTINUOUS – responses which arise from measuring.

Example: Time it takes a student to travel from home to school.

CATEGORICAL (QUALITATIVE) DATA – Qualitative variables measure a quality or characteristic of the experimental unit. Categorical data yields a qualitative response, i.e. data is kind or type rather than quantity.

Examples: Categorizing students into first year IB or second year IB; into Maths Studies SL, Maths SL, Further Maths SL, or Maths HL; or political affiliation, will result in qualitative variables and data.

Bar graphs are one of the many techniques also used to present data in a visual form so that the reader may readily recognize patterns or trends. A bar graph may be either horizontal or vertical. The important point to note about bar graphs is their bar length or height – the greater their length or height, the greater their value.

Bar graphs usually present categorical and numeric variables grouped in class intervals. They consist of an axis and a series of labelled horizontal or vertical bars. The bars depict frequencies of different values of a variable or simply the different values themselves. The student data in the previous box can be represented by a bar graph as shown below.

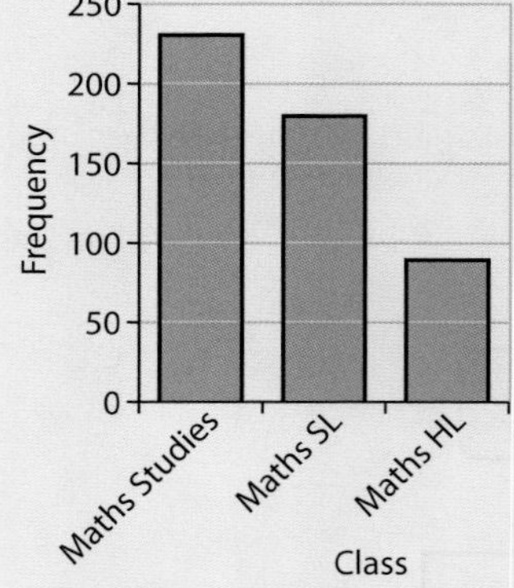

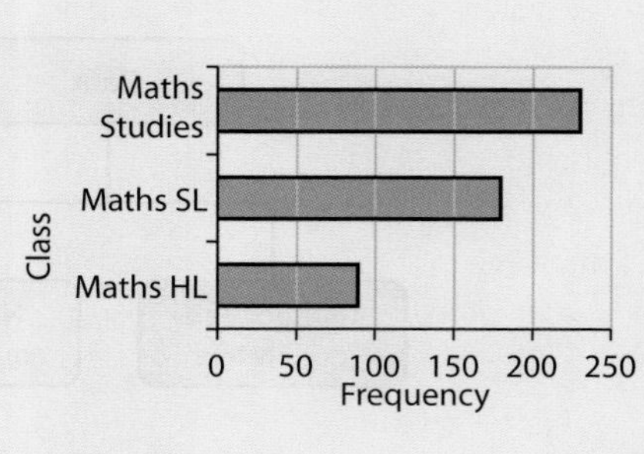

Notice here that the parts do not need to show the component parts of a whole. The key is to show their relative heights.

When data is first collected, there are some simple ways of beginning to organize the data. These include an ordered array and the stem-and-leaf display – not required.

- Data in raw form (as collected):
  24, 26, 24, 21, 27, 27, 30, 41, 32, 38
- Data in ordered array from smallest to largest (an ordered array is an arrangement of data in either ascending or descending order):
  21, 24, 24, 26, 27, 27, 30, 32, 38, 41

Suppose a consumer organization was interested in studying weekly food and living expenses of college students. A survey of 80 students yielded the following expenses to the nearest euro:

38	50	55	60	46	51	58	64	50	49	48	65	58	61	65	53
39	51	56	61	48	53	59	65	54	54	54	59	65	66	47	49
40	51	56	62	47	55	60	63	60	59	59	50	46	45	54	47
41	52	57	64	50	53	58	67	67	66	65	58	54	52	55	52
44	52	57	64	51	55	61	68	67	54	55	48	57	57	66	66

Table 11.1

The first step in the analysis is a summary of the data, which should show the following information:

- What values of the variable have been measured?
- How often has each value occurred?

A stem-and-leaf plot, or stem plot, is a technique used to classify and organize data as they are collected.

225	250	213	216	183
211	200	246	243	231
209	209	225	200	217
224	230	237	185	235
258	225	232	216	227
216	256	226	271	217
196	243	232	230	246
228	200	216	219	
200	224	209	191	

A stem-and-leaf plot looks something like a bar graph. Each number in the data is broken down into a stem and a leaf, thus the name. Here is a set of data representing the lives of 43 light bulbs of a certain type.

The stem of the number, in this case, consists of the multiples of 10. For example, 183, 18 is the stem, and 3 is the leaf. The leaf of the number will always be a single digit.

The stem-and-leaf plot shows how the data are spread–that is, highest number, lowest number, most common number and outliers and it preserves the individual values.

Once you have decided that a stem-and-leaf plot is the best way to show your data, draw it as follows:

On the left-hand side, write down the thousands, hundreds or tens (all digits except the last one). These will be your stems.

Draw a line to the right of these stems.

On the other side of the line, write down the ones (the last digit of a number). These will be your leaves.

Stem-and-leaf display	
18	3 5
19	1 6
20	0 0 0 0 9 9 9
21	1 3 6 6 6 6 7 7 9
22	4 4 5 5 5 6 7 8
23	0 0 1 2 2 5 7
24	3 3 6 6
25	0 6 8
26	
27	1

For example, if the observed value is 25, then the stem is 2 and the leaf is the 5. If the observed value is 369, then the stem is 36 and the leaf is 9. Where observations are accurate to one or more decimal places, such as 23.7, the stem is 23 and the leaf is 7. If the range of values is too great, the number 23.7 can be rounded up to 24 to limit the number of stems.

Such summaries can be done in many ways. The most useful are the frequency distribution and the histogram. There are other methods of presenting data, some of which we will discuss later. The rest are not within the scope of this book.

Frequency distribution (table)

A **frequency distribution** is a table used to organize data. The left column (called classes or groups) includes numerical intervals on a variable being studied. The right column is a list of the frequencies, or number of observations, for each class. Intervals normally are of equal size, must cover the range of the sample observations, and are non-overlapping (Table 11.2).

There are some general rules for preparing frequency distributions that make it easier to summarize data and to communicate results.

Construction of a frequency distribution (table)

If classes are described with discrete limits such as '30–34', '35–39', '40–44'..., then the boundaries are midway between the neighbouring class limits / end points. That is, the classes above will be considered as '29.5, but less than 34.5', '34.5, but less than 39.5', '39.5, but less than 44.5' etc. Here, the boundaries are 29.5, 34.5, 39.5, 44.5. Each class width is 5. See Example 3.

Rule 1: Intervals (classes) must be inclusive and non-overlapping; each observation must belong to one and only one class interval. Consider a frequency distribution for the living expenses of the 80 college students. If the frequency distribution contains the intervals '35–40' and '40–45', to which of these two classes would a person spending €40 belong?

The boundaries, or endpoints, of each class must be clearly defined. For our example, appropriate intervals would be '35 but less than 40' and '40 but less than 45'.

Rule 2: Determine k, the number of classes. Practice and experience are the best guidelines for deciding on the number of classes. In general, the number of classes could be between 5 and 10. But this is not an absolute rule. Practitioners use their judgement in these issues. If the number of classes is too few, some characteristics of the distribution will be hidden, and if too many, some characteristics will be lost with the detail.

In some cases, we do not necessarily create intervals with the same width. Look at the end of this section for an example.

Rule 3: Intervals should be the same width, w. The width is determined by the following:

$$\text{interval width} = \frac{\text{largest number} - \text{smallest number}}{\text{number of intervals}}$$

Both the number of intervals and the interval width should be rounded upward, possibly to the next largest integer. The above formula can be used when there are no natural ways of grouping the data. If this formula is used, the interval width is generally rounded to a convenient whole number to provide for easy interpretation.

In the example of the weekly living expenses of students, a reasonable grouping with nice round numbers was that of '35 but less than 40' and '40 but less than 45', etc.

Living expenses (€)	Number of students	Percentage of students
35 but < 40	2	2.50
40 but < 45	3	3.75
45 but < 50	11	13.75
50 but < 55	21	26.25
55 but < 60	19	23.75
60 but < 65	11	13.75
65 but < 70	13	16.25
Total	**80**	**100.00**

Table 11.2 Frequency and percentage frequency distributions of the weekly expenses of 80 students.

Grouping the data in a table like this one enables us to see some of its characteristics. For example, we can observe that there are few students who spend as little as 35 to 45 euros, while the majority of the students spend more than €45. Grouping the data will also cause some loss of detail, as we do not see from the table what the real values in each class are.

In the table above, the impression we get is that the class midpoint, also known as the mid-interval value, will represent the data in that interval. For example, 37.5 will represent the data in the first class, while 62.5 will represent the data in the 60 to 65 class. 35 and 40 are known as the **interval boundaries**.

Graphically, we have a tool that helps visualize the distribution. This tool is the **histogram**.

Histogram

A histogram is a graph that consists of vertical bars constructed on a horizontal line that is marked off with intervals for the variable being displayed. The intervals correspond to those in a frequency distribution table. The height of each bar is proportional to the number of observations in that interval. The number of observations can also be displayed above the bars.

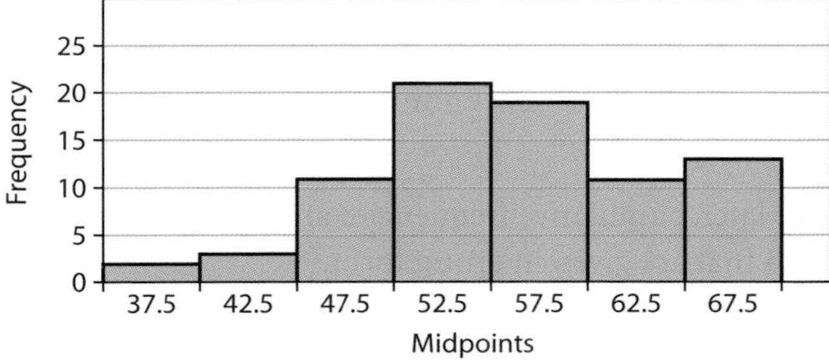

By looking at the histogram, it becomes visually clear that our observations above are true. From the histogram we can also see that the distribution is not symmetric.

To get a histogram on your GDC:

- Enter your data into a list
- Go to **StatPlot** and change it as shown below
- Graph

L1	L2	L3 1
38	-------	-------
39		
40		
41		
44		
45		
48		

L1(1)=38

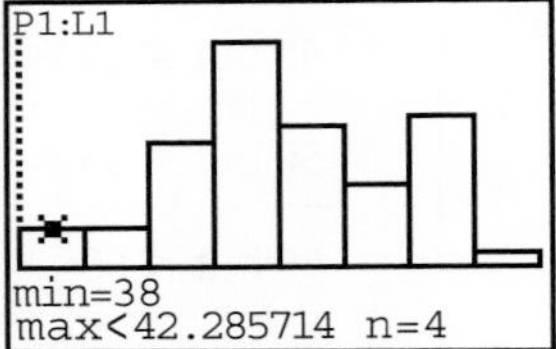

Cumulative and relative cumulative frequency distributions

A **cumulative frequency distribution** contains the total number of observations whose values are less than the upper limit for each interval. It is constructed by adding the frequencies of all frequency distribution intervals up to and including the present interval. A **relative cumulative frequency distribution** converts all cumulative frequencies to cumulative percentages.

In our example above, the following is a cumulative distribution and a relative (percentage) cumulative distribution.

Table 11.3 Cumulative frequency and cumulative relative frequency distributions of the weekly expenses of 80 students.

Living expenses (€)	Number of students	Cumulative number of students	Percentage of students	Cumulative percentage of students
35 but < 40	2	2	2.50	2.50
40 but < 45	3	5	3.75	6.25
45 but < 50	11	16	13.75	20.00
50 but < 55	21	37	26.25	46.25
55 but < 60	19	56	23.75	70.00
60 but < 65	11	67	13.75	83.75
65 but < 70	13	80	16.25	100.00
	80		**100.00**	

Notice how every cumulative frequency is added to the frequency in the next interval to give you the next cumulative frequency. The same is true for the relative frequencies.

As we will see later, cumulative frequencies and their graphs help in analyzing data that are given in group form.

Cumulative line graph/cumulative frequency graph

Sometimes called an **ogive**, this is a line that connects points that are the cumulative percentage of observations below the upper limit of each class in a cumulative frequency distribution.

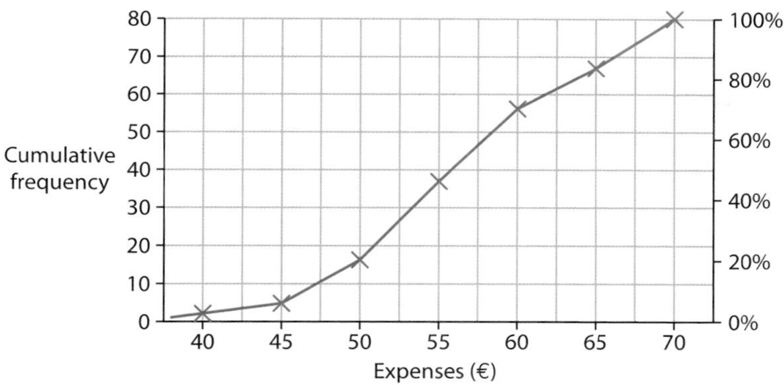

Notice how the height of each line at the upper boundary represents the cumulative frequency for that interval. For example, at 50 the height is 16 and at 60 it is 56.

Example 1

Here is the WHO data in raw form.

29	36	40	44	48	52	54	56	59	60	61	61	62	63	64	66	68	71	72	73	63	64	66	68
31	36	41	44	49	52	54	57	59	60	61	62	62	64	64	66	68	71	72	75	63	64	66	68
33	36	41	44	49	52	55	57	59	60	61	62	62	64	65	66	69	71	72	35	38	43	47	71
34	37	41	45	49	53	55	58	59	60	61	62	63	64	65	66	69	71	73	36	40	44	48	71
34	37	42	45	50	53	55	58	59	60	61	62	63	64	65	67	70	71	73	50	54	56	59	72
35	37	42	45	50	53	55	58	59	60	61	62	63	64	65	67	70	71	73	51	54	56	59	72
35	37	43	46	50	54	55	58	59	60	61	62	63	64	65	67	70	71	73	60	60	61	62	73
35	38	43	46	50	54	55	58	59	60	61	62	63	64	65	67	70	72	73	60	61	61	62	73

Prepare a frequency table starting with 25 and with a class interval of 5. Then draw a histogram of the data and a cumulative frequency graph.

Solution

We first sort the data and then make sure we count every number in one class only.

Life expectancy (years)[1]	Number of countries	Life expectancy (years)	Number of countries
25–30	1	55–60	26
30–35	4	60–65	54
35–40	14	65–70	22
40–45	14	70–75	27
45–50	11	75–80	1
50–55	18		

[1] 25–30 contains all observations larger than or equal to 25 but less than 30.

The histogram created by Excel is shown on the next page. Since we have classes of equal width, the height and the area give the same impression about the frequency of the class interval. For example, the class of 60–65 contains almost twice as much as the class of 55–60, and the height of the histogram is also twice as high. So is the area. Similarly, the height of the 65–70 class is double that of the 45–50 class.

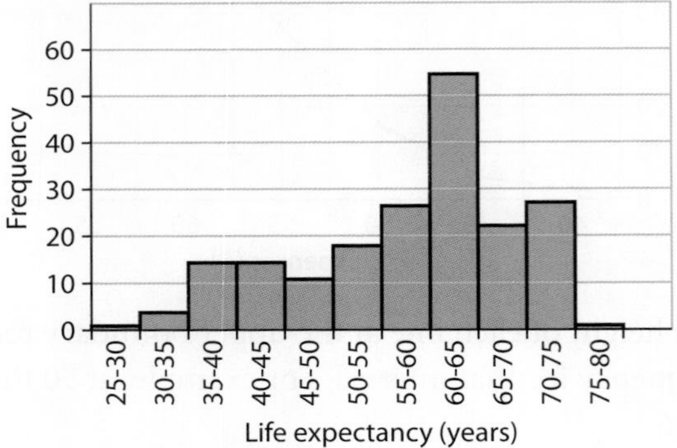

Life expectancy (years)	Number of countries	Cumulative number of countries	Life expectancy (years)	Number of countries	Cumulative number of countries
25–30	1	1	55–60	26	88
30–35	4	5	60–65	54	142
35–40	14	19	65–70	22	164
40–45	14	33	70–75	27	191
45–50	11	44	75–80	1	192
50–55	18	62			

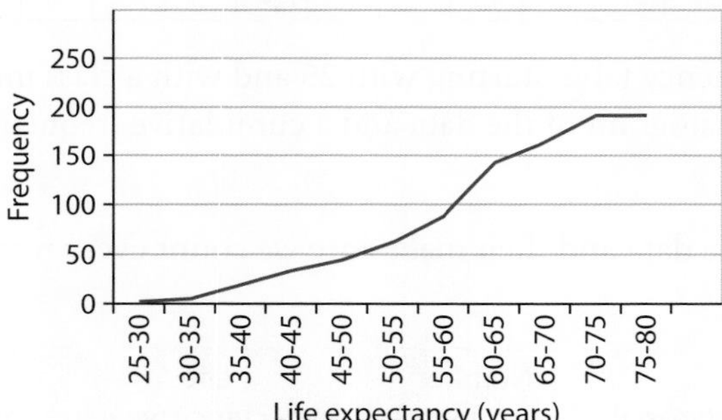

Histograms when class widths are unequal

In some cases, the class widths are not equal. The basic idea behind the histogram is that the area of each 'bar' reflects the frequency of the class. Hence, using the frequency along the vertical axis is a practical thing. However, when the classes have different widths, this practice will be misleading. An alternative for the usual representation is to use the 'frequency density'. The idea behind it is simple: the area of each bar must represent the frequency of the class. So, the height of each bar is measured by its density, which is the frequency of the class per unit of the class size.

This can be found by taking the height of each bar as:

$$\textit{Class height} = \textit{frequency density} = \frac{\textit{frequency}}{\textit{class width}}.$$

This means that the area of each bar $= \textit{width} \times \textit{height}$

$$= \cancel{\textit{width}} \times \frac{\textit{frequency}}{\cancel{\textit{width}}}$$
$$= \textit{frequency}$$

Note: The modal class in a grouped frequency distribution is the class with the largest frequency density.

Example 2

The following table gives the weights (in Newtons) of young children visiting a pediatrician's practice in a certain week.

Weight	5–10	10–12	12–14	14–16	16–20	20–30
Frequency	10	18	24	30	22	16

To draw a meaningful histogram, we find the frequency density for each class.

Weight	5–10	10–12	12–14	14–16	16–20	20–30
Class width	5	2	2	2	4	10
Frequency	10	18	24	30	22	16
Frequency density	2	9	12	15	5.5	1.6

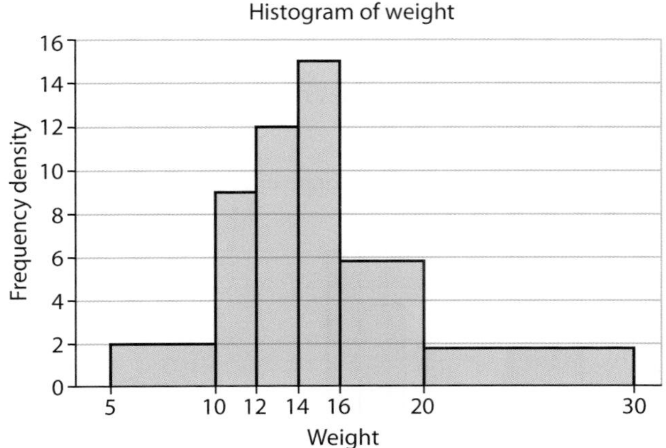

The modal class here is the class 14–16 as it has the largest frequency density of 15.

Look at the histogram below. Notice that if we were to draw the histogram using the frequency itself, the histogram would have given us the wrong representation of the relative size of the classes 5–10, 16–20, and 20–30.

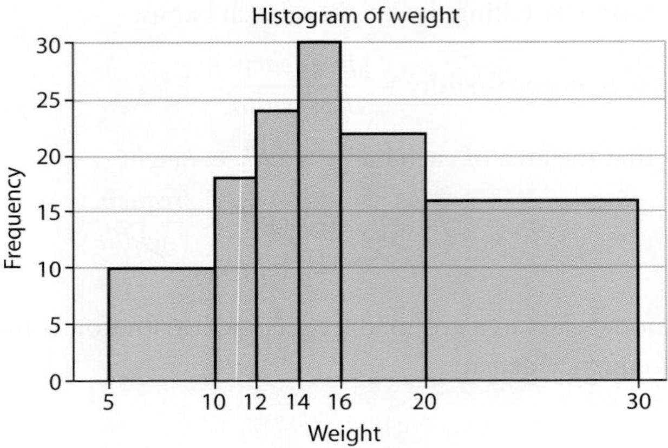

Example 3

The ages (to the nearest year) of visitors to the Prater (Amusement park) in Vienna on a Sunday in July are given in the table below.

Age	6–10	11–15	16–18	19–20	21–25	26–30	31–40
Frequency	120	265	390	320	240	100	45

Draw a histogram of the data in the table.

We first represent classes by their boundaries and change the frequencies into densities.

Age	5.5–10.5	10.5–15.5	15.5–18.5	18.5–20.5	20.5–25.5	25.5–30.5	30.5–40.5
Class width	5	5	3	2	5	5	10
Frequency	120	265	390	320	240	100	45
Density	24	53	130	160	48	20	4.5

Here is the histogram. The modal class here is the class 18.5–20.5 with a frequency density of 160.

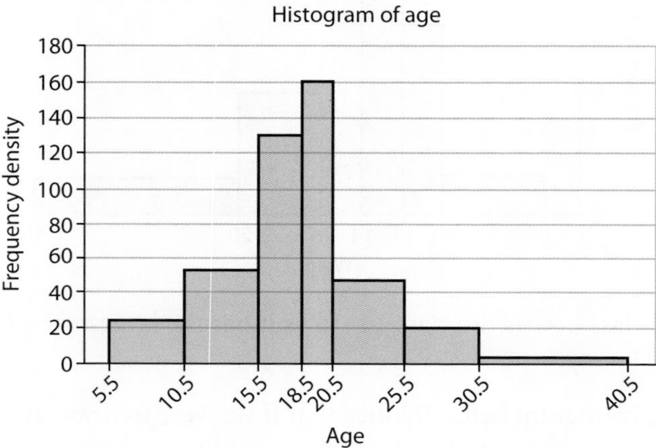

Note: The frequency of each class is given by:

$$width \times height = width \times density = \cancel{width} \times \frac{frequency}{\cancel{width}} = frequency$$

Example 4

The histogram below represents the heights of students (in centimetres) in a high school. Write out the frequency table for this distribution.

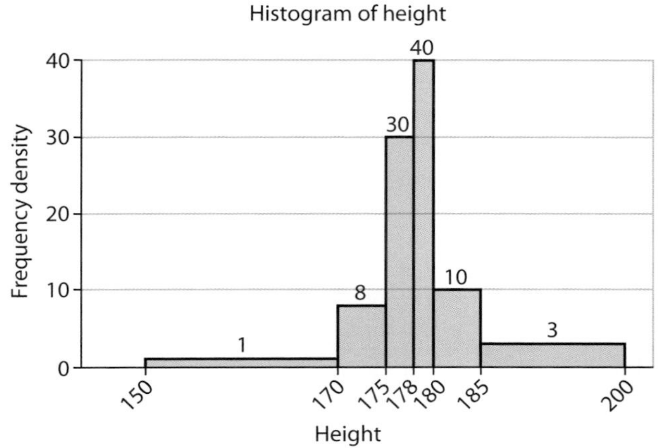

Remember that the frequency is equal to the product of the class width and the frequency density.

Hence, for the 150–170 class the frequency is $20 \times 1 = 20$, and for the 170–175 class the frequency is $5 \times 8 = 40$. Therefore, the frequency distribution is given by:

Height	150–170	170–175	175–178	178–180	180–185	185–200
Frequency	20	40	90	80	50	45

Exercise 11.1

1 Identify the experimental units, sensible population and sample on which each of the following variables is measured. Then indicate whether the variable is quantitative or qualitative.
a) Gender of a student
b) Number of errors on a final exam for 10th-grade students
c) Height of a newborn child
d) Eye colour for children aged less than 14
e) Amount of time it takes to travel to work
f) Rating of a country's leader: excellent, good, fair, poor
g) Country of origin of students at international schools

2 State what you expect the shapes of the distributions of the following variables to be: uniform, unimodal, bimodal, symmetric, etc. Explain why.
a) Number of goals shot by football players during last season.
b) Weights of newborn babies in a major hospital during the course of 10 years.
c) Number of countries visited by a student at an international school.
d) Number of emails received by a high school student at your school per week.

3 Identify each variable as quantitative or qualitative:
a) Amount of time to finish your extended essay.
b) Number of students in each section of IB Maths HL.

c) Rating of your textbook as excellent, good, satisfactory, terrible.
d) Country of origin of each student on Maths HL courses.

4 Identify each variable as discrete or continuous:
a) Population of each country represented by HL students in your session of the exam.
b) Weight of IB Maths HL exams printed every May since 1976.
c) Time it takes to mark an exam paper by an examiner.
d) Number of customers served at a bank counter.
e) Time it takes to finish a transaction at a bank counter.
f) Amount of sugar used in preparing your favourite cake.

5 Grade point averages (GPA) in several colleges are on a scale of 0–4. Here are the GPAs of 45 students at a certain college.

1.8	1.9	1.9	2.0	2.1	2.1	2.1	2.2	2.2	2.3	2.3	2.4	2.4	2.4	2.5
2.5	2.5	2.5	2.5	2.5	2.6	2.6	2.6	2.6	2.6	2.7	2.7	2.7	2.7	2.7
2.8	2.8	2.8	2.9	2.9	2.9	3.0	3.0	3.0	3.1	3.1	3.1	3.2	3.2	3.4

Prepare a frequency histogram, a relative frequency histogram and a cumulative frequency graph. Describe the data in two to three sentences.

6 The following are the grades of an IB course with 40 students (two sections) on a 100-point test. Use the graphical methods you have learned so far to describe the grades.

61	62	93	94	91	92	86	87	55	56
63	64	86	87	82	83	76	77	57	58
94	95	89	90	67	68	62	63	72	73
87	88	68	69	65	66	75	76	84	85

7 The length of time (months) between repeated speeding violations of 50 young drivers are given in the table below:

2.1	1.3	9.9	0.3	32.3	8.3	2.7	0.2	4.4	7.4
9	18	1.6	2.4	3.9	2.4	6.6	1	2	14.1
14.7	5.8	8.2	8.2	7.4	1.4	16.7	24	9.6	8.7
19.2	26.7	1.2	18	3.3	11.4	4.3	3.5	6.9	1.6
4.1	0.4	13.5	5.6	6.1	23.1	0.2	12.6	18.4	3.7

a) Construct a histogram for the data.
b) Would you describe the shape as symmetric?
c) The law in this country requires that the driving licence be taken away if the driver repeats the violation within a period of 10 months. Use a cumulative frequency graph to estimate the fraction of drivers who may lose their licence.

8 To decide on the number of counters needed to be open during rush hours in a supermarket, the management collected data from 60 customers for the time they spent waiting to be served. The times, in minutes, are given in the following table.

3.6	0.7	5.2	0.6	1.3	0.3	1.8	2.2	1.1	0.4
1	1.2	0.7	1.3	0.7	1.6	2.5	0.3	1.7	0.8
0.3	1.2	0.2	0.9	1.9	1.2	0.8	2.1	2.3	1.1
0.8	1.7	1.8	0.4	0.6	0.2	0.9	1.8	2.8	1.8
0.4	0.5	1.1	1.1	0.8	4.5	1.6	0.5	1.3	1.9
0.6	0.6	3.1	3.1	1.1	1.1	1.1	1.4	1	1.4

a) Construct a relative frequency histogram for the times.
b) Construct a cumulative frequency graph and estimate the number of customers who have to wait 2 minutes or more.

9 The histogram below shows the number of days spent by heart patients in Austrian hospitals in the 2003–2005 period.

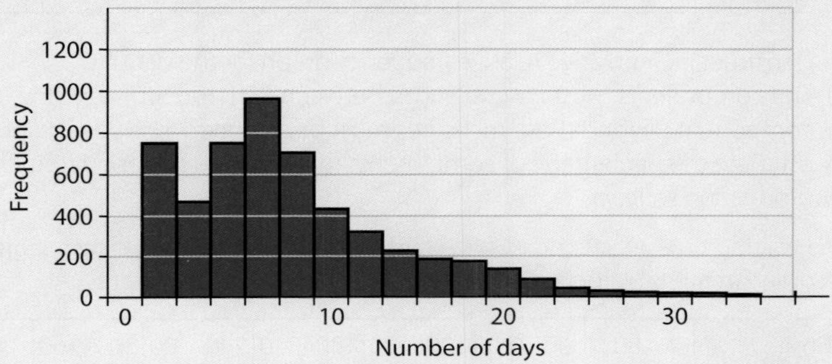

a) Describe the data in a few sentences.
b) Draw a cumulative frequency graph for the data.
c) What percentage of the patients stayed less than 6 days?

10 One of the authors exercises on almost a daily basis. He records the length of time he exercises on most of the days. Here is what he recorded for 2006.

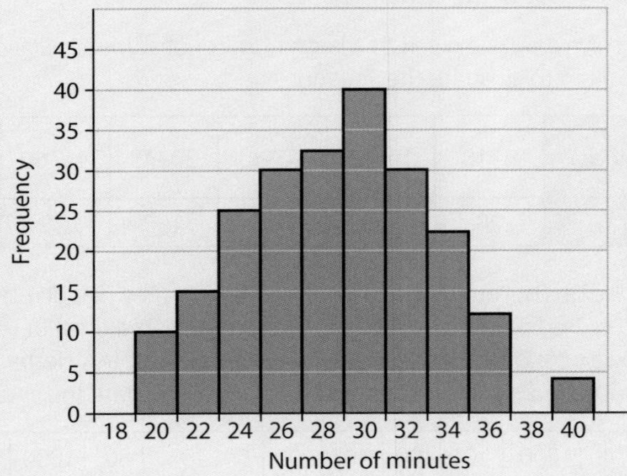

a) What is the longest time he has spent doing his exercises?
b) What percentage of the time did he exercise more than 30 minutes?
c) Draw a cumulative frequency graph for his exercise time.

11 Radar devices are installed at several locations on a main highway. Speeds, in km/h, of 400 cars travelling on that highway are measured and summarized in the following table.

Speed	60–75	75–90	90–105	105–120	120–135	Over 135
Frequency	20	70	110	150	40	10

a) Construct a frequency table for the data.
b) Draw a histogram to illustrate the data.
c) Draw a cumulative frequency graph for the data.
d) The speed limit in this country is 130 km/h. Use your graph in c) to estimate the percentage of the drivers driving faster than this limit.

12 Electronic components used in the production of CD players are manufactured in a factory and their measures must be very accurate. Here are the measures of a sample of 400 such components.

Length (mm)	Less than 5.00	5.00–5.05	5.05–5.10	5.10–5.15	5.15–5.20	More than 5.20
Frequency	16	100	123	104	48	9

a) Construct a cumulative relative frequency graph for the data.
b) The components must have a length between 5.01 and 5.18, and any component beyond these measures must be scrapped. Use your graph to estimate the percentage of components that must be scrapped from this production facility.

13 The waiting time, in seconds, of 300 customers at a supermarket cash register are recorded in the table below.

Time	<60	60–120	120–180	180–240	240–300	300–360	>360
Frequency	12	15	42	105	66	45	15

a) Draw a histogram of the data.
b) Construct a cumulative frequency graph of the data.
c) Use the cumulative frequency graph to estimate the waiting time that is exceeded by 25% of the customers.

14 The time to solve a puzzle given to a large number of students is given below. Draw a histogram to illustrate the situation.

Time (seconds)	5–10	10–20	20–30	30–45	45–60	More than 60
Frequency	20	120	70	150	20	0

15 Post offices weigh the letters customers send before they decide on the amount of postage required. The table below lists the masses (in grams) of letters processed by a post office in a large city on a certain day. (Any letter heavier than 2000 g is considered a parcel.) Draw a histogram to illustrate the situation.

Mass	1–200	201–400	401–600	601–800	801–1000	1001–2000
Frequency	3220	450	130	96	54	40

16 In a study to determine the relative frequency of delays at a major airport, the following histogram has been produced. Develop the frequency distribution of the study. Flights more than 2 hours late were considered as atypical.

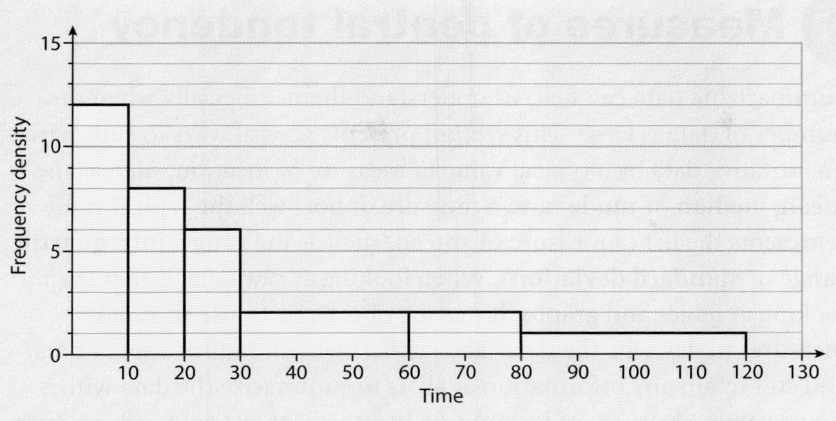

17 Copy and complete the frequency distribution for the data represented by the frequency polygon below.

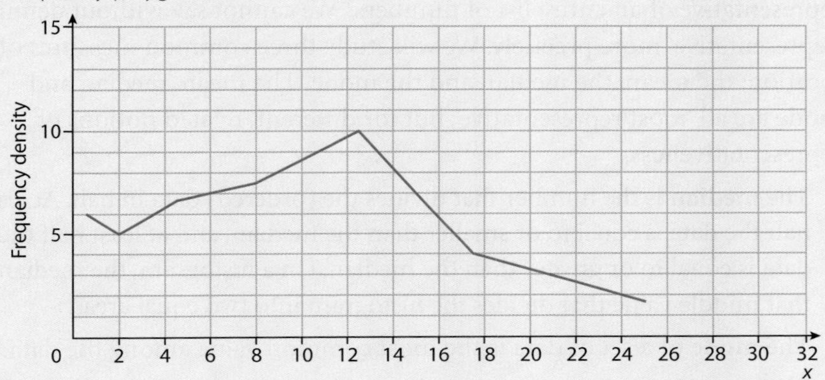

x	$0 \leqslant x < 1$	$1 \leqslant x < 3$	$3 \leqslant x < 6$	$6 \leqslant x < 10$	$10 \leqslant x < 15$	$15 \leqslant x < 20$	$20 \leqslant x < 30$
Frequency	6						

18 Write out the frequency distribution for the data represented by the frequency polygon below. The lowest boundary is 0.

19 Find the value of m, n, p, and q in the frequency density calculation table below.

x	20–24	25–34	35–42	43–49	50–60
Frequency	m	n	20	21	p
Frequency density	2	4.5	q	3	3

11.2 Measures of central tendency

Summarizing data can help us understand them, especially when the number of data is large. This section presents several ways to summarize quantitative data by a typical value (a measure of location, such as the **mean**, **median** or **mode**) and a measure of how well the typical value represents the list (a measure of spread, such as the **range**, **interquartile range** or **standard deviation**). When looking at raw data, rather than looking at tables and graphs, it may be of interest to use summary measures to describe the data. The farthest we can reduce a set of data, and still retain any information at all, is to summarize the data with a single value. Measures of location do just that – they try to capture with a single number what is typical of the data. What single number is most representative of an entire list of numbers? We cannot say without defining 'representative' more precisely. We will study three common measures of location: the mean, the median and the mode. The mean, median and mode are all 'most representative', but for different, related notions of representativeness.

- The **median** is the number that divides the (ordered) data in half. At least half the data is equal to or smaller than the median, and at least half the data is equal to or greater than the median. (In a histogram, the median is that middle value that divides the histogram into two equal areas.)
- The **mode** of a set of data is the most common value among the data.
- The **mean** (more precisely, the arithmetic mean) is commonly called the average. It is the sum of the data, divided by the number of data:

$$\text{mean} = \frac{\text{sum of data}}{\text{number of data}} = \frac{\text{total}}{\text{number of data}}$$

When these measures are computed for a population, they are called **parameters**. When they are computed for a sample, they are called **statistics**.

Statistic and parameter

A statistic is a descriptive measure computed from a sample of data. A parameter is a descriptive measure computed from an entire population of data.

Measures of central tendency provide information about a 'typical' observation in the data, or locate the data set.

The mean and the median

The most common measure of central tendency is the arithmetic mean, usually referred to simply as the 'mean' or the 'average.'

Example 5

The following are the five closing prices of the NASDAQ Index for the first business week in November 2007. This is a sample of size $n = 5$ for the closing prices from the entire 2007 population: 2794.83, 2810.38, 2795.18, 2825.18, 2748.76.

What is the average closing price?

Solution

$$\text{Average} = \frac{2794.83 + 2810.38 + 2795.18 + 2825.18 + 2748.76}{5} = 2794.87.$$

This is called the sample mean. A second measure of central tendency is the median, which is the value in the middle position when the measurements are ordered from smallest to largest. The median of this data can only be calculated if we first sort them in ascending order:

2748.76   2794.83   **2795.18**   2810.38   2825.18

The **arithmetic mean** or **average** of a set of n measurements (data set) is equal to the sum of the measurements divided by n.

Notation

The sample mean: $\bar{x} = \frac{\sum_{i=1}^{n} x_i}{n} = \frac{x_1 + x_2 + x_3 + \ldots + x_n}{n}$, where n is the sample size.

This is a **statistic**.

The population mean: $\mu = \frac{\sum_{i=1}^{N} x_i}{N} = \frac{x_1 + x_2 + x_3 + \ldots + x_N}{N}$, where N is the population size. This is a **parameter**.

It is important to observe that you normally do not know the mean of the population μ and that you usually estimate it with the sample mean $\bar{x}$.

The **median** of a set of n measurements is the value of x that falls in the middle position when the data is sorted in ascending order.

In the previous example, we calculated the sample median by finding the third measurement to be in the middle position. However, in a different situation, where the number of measurements is even, the process is slightly different.

Let us assume that you took six tests last term and your marks were, in ascending order, 52, 63, 74, 78, 80, 89.

52   63   [74   78]   80   89

There are two 'middle' observations here. To find the median, choose a value halfway between the two middle observations:

$$m = \frac{74 + 78}{2} = 76$$

Note: The position of the median can be given by $\frac{n+1}{2}$. If this number ends with a decimal, you need to average the adjacent values.

In the NASDAQ Index case, we have five observations, the position of the median is then at $\frac{5+1}{2} = 3$, which we found. In the grades example, the position of the median score is at $\frac{6+1}{2} = 3.5$, and hence we average the numbers at positions **three** and **four**.

Although both the mean and median are good measures for the centre of a distribution, the median is less sensitive to extreme values or **outliers**. For example, the value 52 in the previous example is lower than all your test scores and is the only failing score you have. The median, 76, is not affected by this outlier even if it were much lower than 52. Assume, for example, that your lowest score is 12 rather than 52. The median calculation

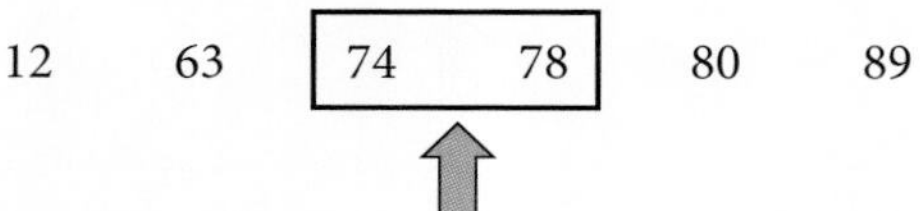

still gives the same median of 76. If we were to calculate the mean of the original set, we would get

$$\bar{x} = \frac{\sum x}{6} = \frac{436}{6} = 72.\bar{6}.$$

While the new mean, with 12 as the lowest score, is

$$\bar{x} = \frac{\sum x}{6} = \frac{396}{6} = 66.$$

Clearly, the low outlier 'pulled' the mean towards it while leaving the median untouched. However, because the mean depends on every observation and uses all the information in the data, it is generally, wherever possible, the preferred measure of central tendency.

A third way to locate the centre of a distribution is to look for the value of x that occurs with the highest frequency. This measure of the centre is called the **mode**.

Example 6

Here is a table listing the frequency distribution of 25 families in Lower Austria that were polled in a marketing survey to list the number of litres of milk consumed during a particular week.

Number of litres	Frequency	Relative frequency
0	2	0.08
1	5	0.20
2	9	0.36
3	5	0.20
4	3	0.12
5	1	0.04

Find the frequency histogram.

Solution

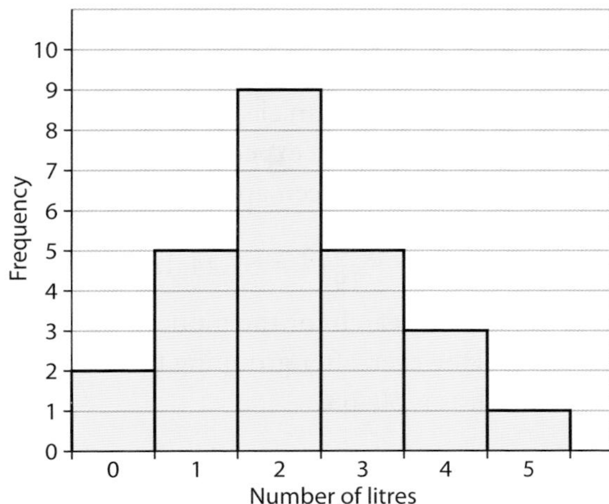

The histogram (Example 3) shows a relatively symmetric shape with a modal class at $x = 2$. Apparently, the mean and median are not far from each other. The median is the 13th observation, which is 2, and the mean is calculated to be 2.2.

For lists, the **mode** is the most common (frequent) value. A list can have more than one mode. For histograms, the mode is a relative maximum.

Shape of the distribution

An examination of the shape of a distribution will illustrate how the distribution is centred around the mean. Distributions are either symmetric or they are not symmetric, in which case the shape of the distribution is described as asymmetric or skewed.

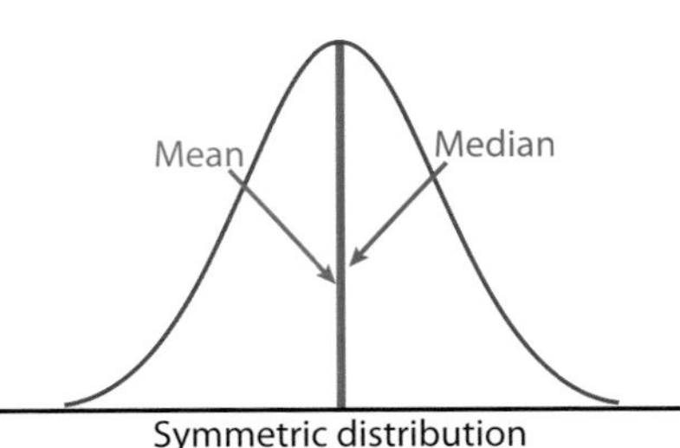

Symmetric distribution

Symmetry

The shape of a distribution is said to be **symmetric** if the observations are balanced, or evenly distributed, about the mean. In a symmetric distribution, *the mean and the median are equal.*

Skewness

A distribution is **skewed** if the observations are not symmetrically distributed above and below the mean.

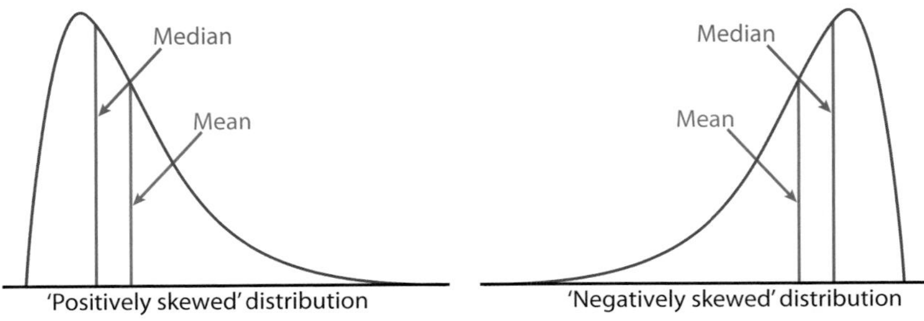

'Positively skewed' distribution 'Negatively skewed' distribution

A **positively skewed** (or skewed to the right) distribution has a tail that extends to the right in the direction of positive values. A **negatively skewed** (or skewed to the left) distribution has a tail that extends to the left in the direction of negative values.

Looking back at the WHO data, we can clearly see that the data is skewed to the left. Few countries have low life expectancies. The bulk of the countries have life expectancies between approximately 50 and 65 years.

The average HALE is $\mu = \frac{\sum x}{n} = \frac{11028}{192} = 57.44$. Looking at the raw data, it does not appear sensible to search for the mode, as there are very few of them (61, 59, 60 or 62). However, after grouping the data into classes, we can see that the modal class is 60–65.

As there are 192 observations, which means that the median is at $\frac{n+1}{2} = \frac{192+1}{2} = 96.5$, we take the average of the 96th and 97th observations, which are Palau and Moldova with 60 each. So, the median is 60!

Knowing the median, we could say that a typical life expectancy is 60 years. How much does this really tell us? How well does this median describe the real situation? After all, not all countries have the same 60 years HALE. Whenever we find the centre of data, the next step is always to ask how well it actually summarizes the data.

When we describe a distribution numerically, we always report a measure of its **spread** along with its centre.

Exercise 11.2

1 You are given eight measurements: 5, 4, 7, 8, 6, 6, 5, 7.

a) Find $\bar{x}$. b) Find the median.

c) Based on the previous results, is the data symmetric or skewed? Explain and support your conclusion with an appropriate graph.

2 You are given ten measurements: 5, 7, 8, 6, 12, 7, 8, 11, 4, 10.

a) Find $\bar{x}$. b) Find the median. c) Find the mode.

3 The following table gives the number of DVD players owned by a sample of 50 typical families in a large city in Germany.

Number of DVD players	0	1	2	3
Number of households	12	24	8	6

Find the average and the median number of DVD players. Which measure is more appropriate here? Explain.

4 Ten of the Fortune 500 large businesses that lost money in 2006 are listed below:

Company	Loss ($ million)	Company	Loss ($ million)
Vodafone	39 093	General Motors	10 567
Kodak	1362	Japan Airlines	417
UAL	21 167	Japan Post	3
Mitsubishi Motors	814	AMR	861
Visteon	270	Karstadt Quelle	393

Calculate the mean and median of the losses. Which measure is more appropriate in this case? Explain.

5 Even on a crucial examination, students tend to lose focus while writing their tests. In a psychology experiment, 20 students were given a 10-minute quiz and the number of seconds they spent 'on task' were recorded. Here are the results:

350	380	500	460	480	400	370	380	450	530
520	460	390	360	410	470	470	490	390	340

Find the mean and median of the time spent on task. If you were writing a report to describe these times, which measure of central tendency would you use and why?

6 At 5:30 p.m. during the holiday season, a toy shop counted the number of items sold and the revenue collected for that day. The result was $n = 90$ toys with a total revenue of $\sum x = €4460$.

a) Find the average amount spent on each toy that day.

Shortly before the shop closed at 6 p.m., two new purchases of €74 and €60 were made.

b) Calculate the new mean of the sales per toy that day.

7 A farmer has 144 bags of new potatoes weighing 2.15 kg each. He also has 56 bags of potatoes from last year with an average weight of 1.80 kg. Find the mean weight of a bag of potatoes available from this farmer.

8 The following are the grades earned by 25 students on a 50-mark test in statistics.
26, 27, 36, 38, 23, 26, 20, 35, 19, 24, 25, 27, 34, 27, 26, 42, 46, 18, 22, 23, 24, 42, 46, 33, 40.

a) Calculate the mean of the grades.

b) Draw a stem plot of the grades. Use the plot to estimate where the median is.

c) Draw a histogram of the grades.

d) Develop a cumulative frequency graph of the grades. Use your graph to estimate the median.

9 The following are data concerning the injuries in road accidents in the UK classified by severity.

Year	Fatal	Serious	Slight
1970	758	7860	13 515
1975	699	6912	13 041
1980	644	7218	13 926
1985	550	6507	13 587
1990	491	5237	14 443
1995	361	4071	12 102
2000	297	3007	11 825
2005	264	2250	10 922

a) Draw bar graphs for the total number of injuries and describe any patterns you observe.

b) Draw pie charts for the different types of injuries for the years 1970, 1990 and 2005.

10 The following data report the car driver casualties in Durham county in the UK in 2006.

a) Draw a histogram of the data.

b) Estimate the mean of the data.

c) Develop a cumulative frequency graph and use it to estimate the median of the data.

Age	Number
15–19	103
20–24	125
25–29	103
30–34	80
35–39	88
40–44	96
45–49	78
50–54	60
55–59	45
60–64	33
65–69	17
70–74	13
75–79	26

11 Use the data in question 9 of Exercise 11.1 to estimate the median and the mean of the number of days in hospital by heart patients.

12 Use the data in question 10 of Exercise 11.1 to estimate the median and the mean of the exercise time of the author for 2006.

13 Use the data in question 11 of Exercise 11.1 to estimate the median and the mean speed of cars on the highway.

14 Use the data in question 12 of Exercise 11.1 to estimate the median and the mean length of components at this facility.

15 Use the data in question 13 of Exercise 11.1 to estimate the median and the mean of the waiting time for customers at this supermarket.

16 a) Given that $\sum_{i=1}^{40} x_i = 1664$, find $\overline{x}$.

b) Given that $\sum_{i=1}^{20} (x_i - 20) = 1664$, find $\overline{x}$.

17 For a large class of 60 students, 12 points are added to each grade to boost the student's score on a relatively difficult test.

a) Knowing that $\sum(x + 12) = 4404$, find the mean score (without the 12 points) of this group of 60 students.

b) Another section of the class has 40 students and their average score is 67.4. Find the average of the whole group of 100 students.

11.3 Measures of variability

Measures of location summarize what is typical of elements of a list, but not every element is typical. Are all the elements close to each other? Are most of the elements close to each other? What is the biggest difference between elements? On average, how far are the elements from each other? The answer lies in the measures of spread or variability.

It is possible that two data sets have the same mean, but the individual observations in one set could vary more from the mean than do the observations in the second set. It takes more than the mean alone to describe data. Measures of variability (also called measures of dispersion or spread), which include the range, the variance, the standard deviation, interquartile range and the coefficient of variation, will help to summarize the data.

Range

The range in a set of data is the difference between the largest and smallest observations.

Consider the expense data given at the beginning of this chapter. Also consider the same data when the largest value of 68 is replaced by 120. What is the range for these two sets of data?

	Expense data	Expense data with outlier
Minimum	38	38
Maximum	68	120
Range	30	82

Table 11.4

Notice that the range is a *single number*, not an interval of values as you might think from its use in common speech. The maximum of the HALE data is 79 and the minimum is 29, so the range is 50.

Range doesn't take into account how the data is distributed and is, of course, affected by extreme values (outliers) as you see above.

Variance and standard deviation

Note: For an SL treatment of this topic, see our SL book. In this chapter, we will gear the discussion to HL notation.

The most comprehensive measures of dispersion are those in terms of the average deviation from some location parameter.

Variance

The sample variance, s^2, is the sum of the squared differences between each observation and the sample mean divided by the sample size minus 1.

$$s^2 = \frac{\sum_{i=1}^{n}(x_i - \overline{x})^2}{n-1}$$

• **Hint:** Discussing the reason we define the sample variance in this manner is beyond the scope of this book. The use of $n - 1$ in the denominator has to deal with the use of the sample variance as an estimate of the population variance. Such an estimate has to be unbiased, and this sample variance is the most unbiased estimate of the population variance. However, the IB syllabus uses a different definition of this variance.

The IB variance is listed as s_n^2 and is evaluated as follows:

$$s_n^2 = \frac{\sum_{i=1}^{n}(x_i - \overline{x})^2}{n}$$

s^2 is called the *unbiased estimate* of the population variance σ^2 and is denoted as s_{n-1}^2. However, it is not required for the current IB syllabus.

It is obvious that $s_{n-1}^2 = \frac{\sum_{i=1}^{n}(x_i - \overline{x})^2}{n-1} = \frac{n}{n-1} \cdot \frac{\sum_{i=1}^{n}(x_i - \overline{x})^2}{n} = \frac{n}{n-1} \cdot$

Or $s_n^2 = \frac{n-1}{n} \cdot s_{n-1}^2$ in case you want to use s_x^2 from your GDC.

With your calculators you should also be careful as the listed s_x in TI and Casio GDCs corresponds to s_{n-1}^2. So, when you use your GDC, make sure you use what is called σ_x.

The population variance, σ^2, is the sum of the squared differences between each observation and the population mean divided by the population size, N.

$$\sigma^2 = \frac{\sum_{i=1}^{n}(x_i - \mu)^2}{N}$$

The variance is a measure of the variation about the mean squared. In order to bring the measure down to the data measurements, the square root is taken and the measure looked at is the standard deviation.

The standard deviation measures the **standard amount of deviation** or **spread** around the mean.

Standard deviation

The sample standard deviation, s_n, is the (positive) square root of the variance, and is defined as:

$$s_n = \sqrt{s_n^2} = \sqrt{\frac{\sum_{i=1}^{n}(x_i - \overline{x})^2}{n}} \quad \text{(IB)}$$

$$s_{n-1} = \sqrt{s_{n-1}^2} = \sqrt{\frac{\sum_{i=1}^{n}(x_i - \overline{x})^2}{n-1}} = \sqrt{\frac{n}{n-1}} \cdot s_n$$

$$\text{or, } s_n = \sqrt{\frac{n-1}{n}} \cdot s_{n-1}$$

When does $s = 0$? Answer: When all the data take on the same value and there is no variability about the mean.

When is s large? Answer: When there is a large amount of variability about the mean.

The population standard deviation is:

$$\sigma = \sqrt{\sigma^2} = \sqrt{\frac{\sum_{i=1}^{n}(x_i - \mu)^2}{N}}$$

These are measures of variation about the mean.

Consider the following example:

In business, investors invest their money in stocks whose prices fluctuate with market conditions. Stocks are considered risky if they have high fluctuations. Here are the closing prices of two stocks traded on Vienna's stock market for the first seven business days in September 2007:

Stock A	Stock B
4	1
4.25	3
5	2.5
4.75	5
5.75	7
5.25	6.5
6	10
$\overline{x}_A = 5$ Median (A) = 5	$\overline{x}_B = 5$ Median (B) = 5

Even though the two stocks have similar central values, they do behave differently. It is obvious that stock B is more variable and it becomes more obvious when we calculate the standard deviations.

We will calculate the standard deviation manually in this example to demonstrate the process. You do not have to do this manually all the time!

$$s_A^2 = \frac{\sum_{i=1}^{7}(x_i - 5)^2}{7} = \frac{(4-5)^2 + (4.25-5)^2 + (5-5)^2 + (4.75-5)^2 + (5.75-5)^2 + (5.25-5)^2 + (6-5)^2}{7} = 0.464$$

$$s_B^2 = \frac{\sum_{i=1}^{7}(x_i - 5)^2}{7} = \frac{(1-5)^2 + (3-5)^2 + (2.5-5)^2 + (5-5)^2 + (7-5)^2 + (6.5-5)^2 + (10-5)^2}{7} = 8.21$$

This means that the standard deviations are $s_A = 0.681$ and $s_B = 2.866$. Stock B is 4.2 times as variable as stock A.

Note: When computing s_n^2 or s_{n-1}^2 manually, you may find the following shortcut of some use:

$$s_n^2 = \frac{\sum_{i=1}^{n}(x_i - \overline{x})^2}{n} = \frac{\sum_{i=1}^{n}(x_i^2 - 2x_i\overline{x} + \overline{x}^2)}{n} = \frac{\sum_{i=1}^{n} x_i^2 - 2\sum_{i=1}^{n} x_i\overline{x} + \sum_{i=1}^{n}\overline{x}^2}{n}$$

$$= \frac{\sum_{i=1}^{n} x_i^2}{n} - \frac{2\overline{x}\sum_{i=1}^{n} x_i}{n} + \frac{\sum_{i=1}^{n}\overline{x}^2}{n} = \frac{\sum_{i=1}^{n} x_i^2}{n} - 2\overline{x}\sum_{i=1}^{n}\frac{x_i}{n} + \frac{n\overline{x}^2}{n} = \frac{\sum_{i=1}^{n} x_i^2}{n} - \overline{x}^2$$

$$s_{n-1}^2 = \left(\frac{\sum_{i=1}^{n} x_i^2}{n} - \overline{x}^2\right) \cdot \frac{n}{n-1} = \frac{\sum_{i=1}^{n} x_i^2 - n\overline{x}^2}{n-1}$$

However, remember that once you have a good understanding of the standard deviation, you will rely on a GDC or software to do most of the calculation for you.

Here is how you can use your TI GDC:

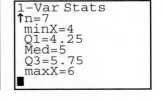

```
EDIT CALC TESTS
1:Edit...
2:SortA(
3:SortD(
4:ClrList
5:SetUpEditor
```

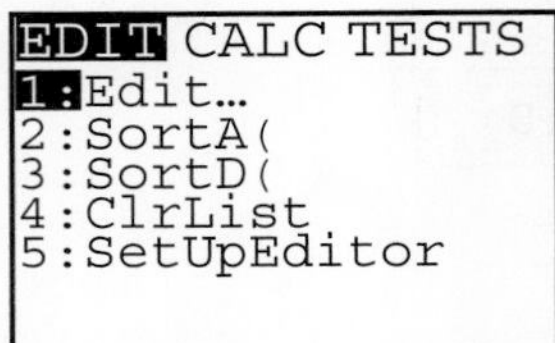

```
1-Var Stats L1
```

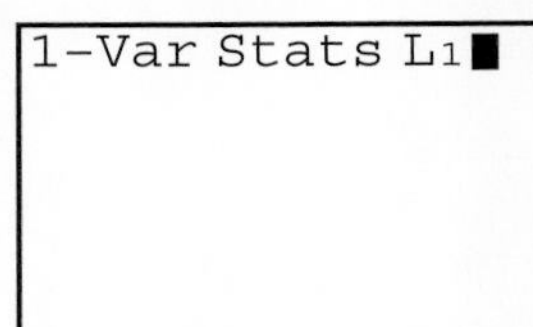


```
1-Var Stats
↑n=7
 minX=4
 Q1=4.25
 Med=5
 Q3=5.75
 maxX=6
```

The s_x used by your GDC gives $s_{n-1} = \sqrt{\dfrac{\sum_{i=1}^{n}(x_i - \overline{x})^2}{n-1}}$ and $\sigma = \sqrt{\dfrac{\sum_{i=1}^{n}(x_i - \overline{x})^2}{n}}$ which is the s_n used in IB exams.

The screenshots also show you that the GDC gives you $\sum x^2$, which can be used if you want to find the variance by hand.

$$s_n^2 = \frac{\sum_{i=1}^{n} x_i^2}{n} - \overline{x}^2 = \frac{178.25}{7} - 5^2 = 0.464 \Rightarrow s_n = 0.681$$

$$s_{n-1} = \sqrt{\frac{7}{6}} \times 0.681 = 0.736, \text{ or}$$

$$s_{n-1}^2 = \frac{\sum_{i=1}^{n} x_i^2 - n\overline{x}^2}{n-1} = \frac{178.25 - 7.(5)^2}{6} = \frac{3.25}{6} = 0.542$$

$$\Rightarrow s_{n-1} = 0.736$$

The interquartile range and measures of non-central tendency

To understand another measure of spread known as the **interquartile range,** it is first necessary to define percentiles and quartiles.

Percentiles and quartiles

Data must first be in ascending order.

Percentiles separate large ordered data sets into hundredths. The *p*th percentile is a number such that *p* per cent of the observations are at or below that number.

Quartiles are descriptive measures that separate large ordered data sets into four quarters.

To score in the 90th percentile indicates 90% of the test scores were less than or equal to your score. An excellent performance! You scored in the upper 10% of all persons taking the test.

- **First quartile, Q_1**
  The first quartile, Q_1, is another name for the 25th percentile. The first quartile divides the ordered data such that 25% of the observations are at or below this value. Q_1 is located in the $0.25(n + 1)$st position when the data is in ascending order. That is,

$$Q_1 = \frac{n+1}{4} \text{ ordered observation}$$

- **Third quartile, Q_3**
  The third quartile, Q_3, is another name for the 75th percentile. The third quartile divides the ordered data such that 75% of the observations are at or below this value. Q_3 is located in the $0.75(n + 1)$st position when the data is in ascending order. That is,

$$Q_3 = \frac{3(n+1)}{4} \text{ ordered observation}$$

- **The median**
  The median is the 50th percentile, or the second quartile, Q_2.

A measure which helps to measure variability and is not affected by extreme values is the interquartile range. It avoids the problem of extreme values by just looking at the range of the middle 50% of the data.

Hint: The first quartile is also called the lower quartile. The third quartile is also called the upper quartile.

A practical method to calculate the quartiles is to split the data into two halves at the median. (When n is odd, include the median in both halves!) The first quartile is the median of the first half and the third quartile is the median of the second half. For example, with the stocks data, {4, 4.25, 4.75, 5, 5.25, 5.75, 6}, $n = 7$, the median is the fourth observation, 5. The first quartile is then the median of {4, 4.25, 4.75, 5}, which is 4.5, and the third quartile is the median of {5, 5.25, 5.75, 6}, which is 5.5.

Interquartile range

The interquartile range (IQR) measures the spread in the middle 50% of the data; it is the difference between the observations at the 25th and the 75th percentiles:

$$\textbf{IQR} = \mathbf{Q_3 - Q_1}$$

If we consider the student expense data in Table 11.1 (on page 467) and once again look at that same data with the outlier 120 replacing the largest value 68, we have the following results:

	Expense data	Expense data with outlier
Minimum	38	38
Q_1	50	50
Median	55	55
Q_3	61	61
Maximum	68	120
Range	30	82
IQR	11	11

Range doesn't take into account how the data is distributed and is, of course, affected by extreme values. We clearly saw that in Table 11.4 (page 487). However, the IQR evidently does not have that problem.

Five-number summary

Five-number summary refers to the five descriptive measures: minimum, first quartile, median, third quartile, maximum.
Clearly, $X_{minimum} < Q_1 < \text{Median} < Q_3 < X_{maximum}$.

Box-and-whisker plot

Whenever we have a five-number summary, we can put the information together in one graphical display called a **box plot**, also known as a **box-and-whisker** plot. In the student expenditure data, the IQR is €11. This is evident in the box plot below, where the IQR is the difference between 50 and 61.

Let us make a box plot with the student expense data.

- Draw an axis spanning the range of the data. Mark the numbers corresponding to the median, minimum, maximum, and the lower and upper quartiles.
- Draw a rectangle with lower end at Q1 and upper end at Q3, as shown below.
- To help us consider outliers, mark the points corresponding to lower and upper fences. Mark them with a dotted line since they are not part of the box. The fences are constructed at the following positions:
  - Lower fence: $Q_1 - 1.5 \times \text{IQR}$ (Here it is $50 - 1.5(11) = 33.5$.)
  - Upper fence: $Q_3 + 1.5 \times \text{IQR}$ (Here it is $61 + 1.5(11) = 77.5$.)

  Any point beyond the lower or upper fence is considered an **outlier**.
- Mark any outlier with an asterisk (*) on the graph (shown below).
- Extend horizontal lines called 'whiskers' from the ends of the box to the smallest and largest observations that are not outliers. In the first case these are 38 and 68, while in the second they are 38 and 67.

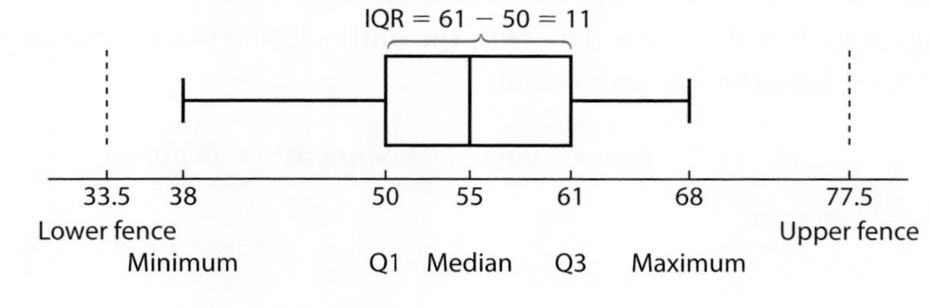

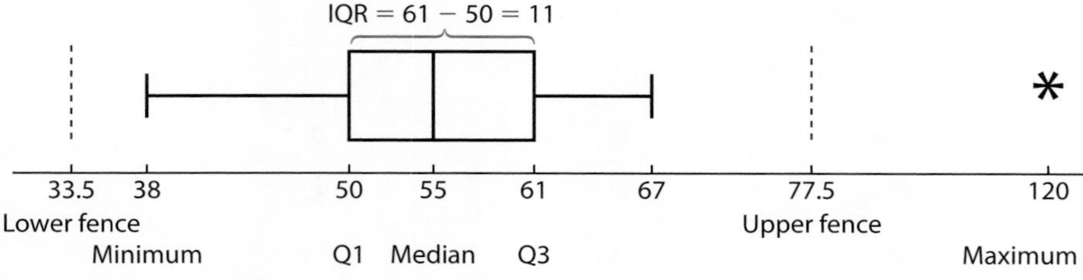

An outlier is an *unusual* observation. It lies at an abnormal distance from the rest of the data. There is no unique way of describing what an outlier is. A common practice is to consider any observation that is further than 1.5 IQR from the first or third quartile as an outlier. Outliers are important in statistical analysis. Outliers may contain important information not shared with the rest of the data. Statisticians look very carefully at outliers because of their influence on the shapes of distributions and their effect on the values of the other statistics, such as the mean and standard deviation.

Here is a box plot of the data done by a software package.

Box plot of student expense data

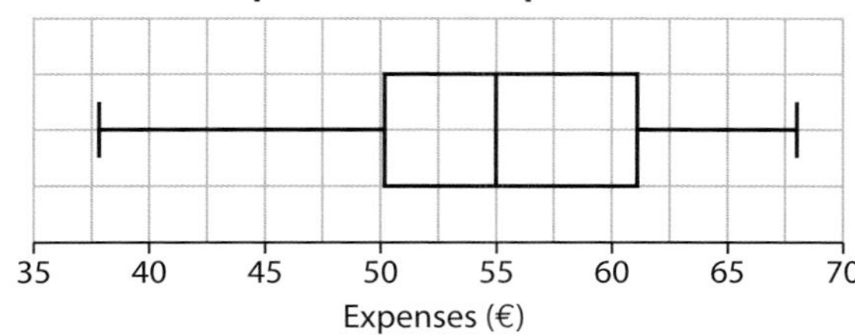

As you can see, the box contains the middle 50% of the data. The width of the box is nothing but the IQR! Now we know that the middle 50% of the students' expenditure is €11. This seems, at times, as a reasonable summary of the spread of the distribution, as you can see in the histogram below.

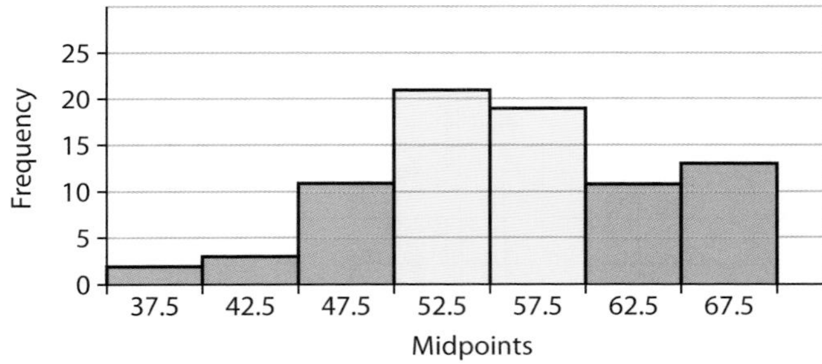

If you locate the IQR on the histogram, you can also get another visual indication of the spread of the data.

How to use your GDC for histograms and box plots:

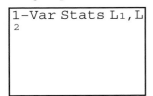

```
L1    L2    L3    2
38    1     --------
39    1
40    1
41    1
44    1
45    1
46    2
L2(1)=1
```

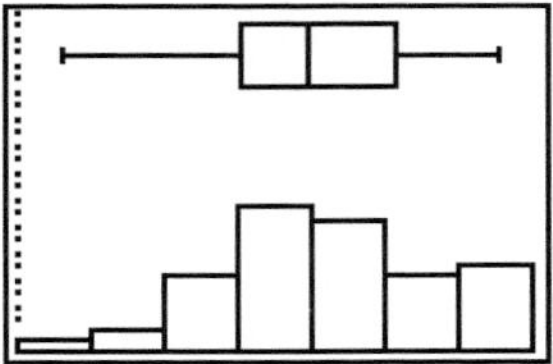

For grouped data:

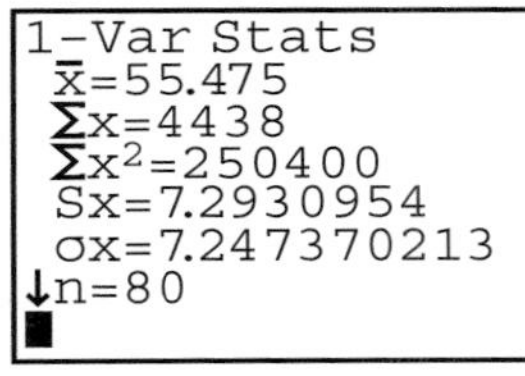


```
1-Var Stats
↑n=80
 minX=38
 Q1=50.5
 Med=55
 Q3=61
 maxX=68
```

An ogive can also be produced:

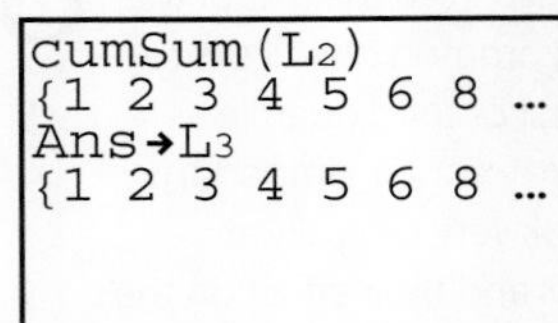

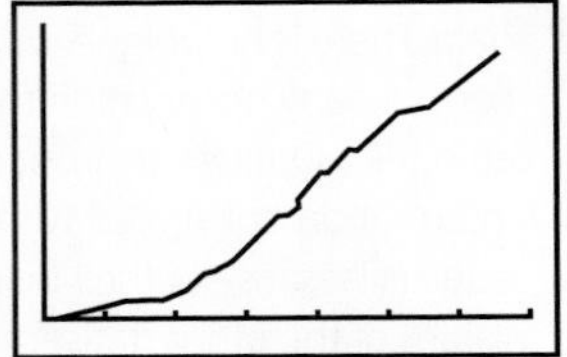

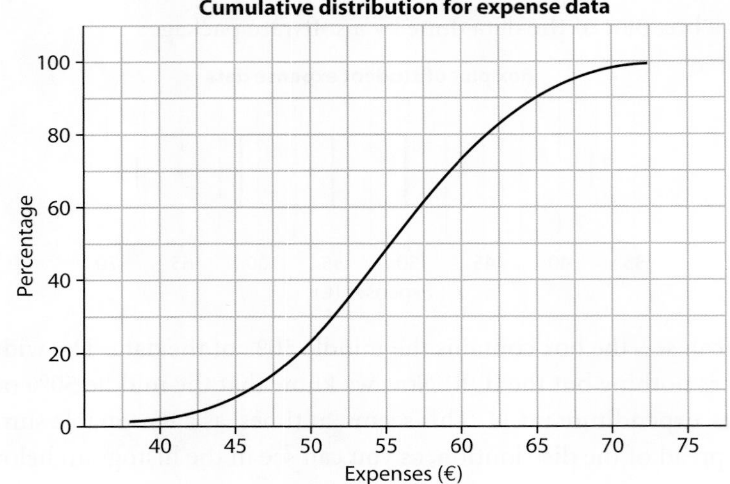

This is a realistic ogive.

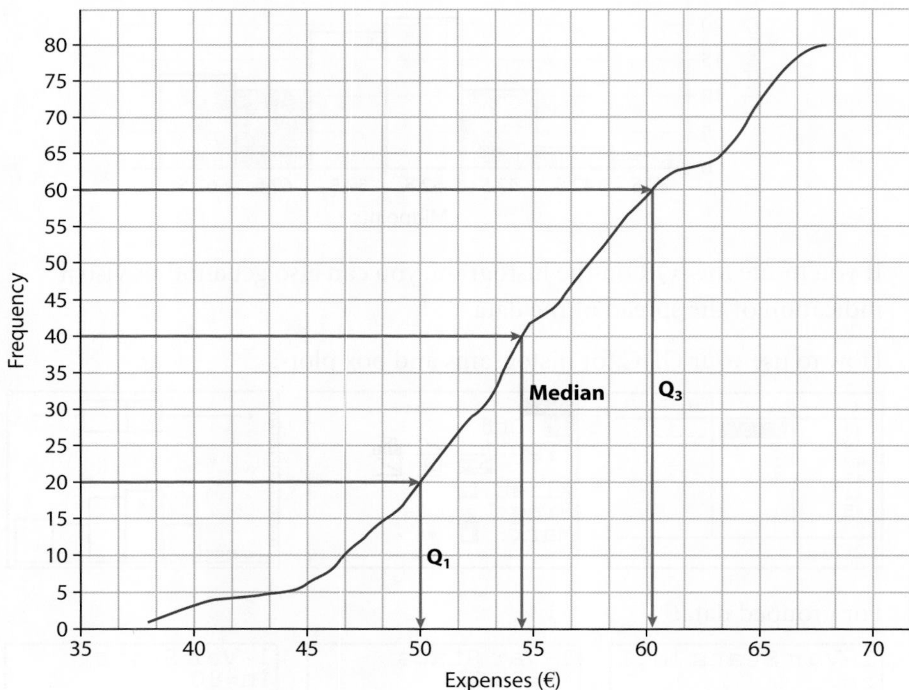

Notice how we locate the first quartile. Since there are 80 observations, the first quartile is approximately at the $\frac{n+1}{4} = \frac{81}{4} \approx 20$th position, which appears to be around 50.

The median is at the $\frac{n+1}{2} = \frac{81}{2} \approx$ 40th–41st position, i.e. approximately at 55.

Similarly, the third quartile is at $\frac{3(n+1)}{4} = \frac{243}{4} \approx$ 61st, which happens here at approximately 61!

The calculation of the mean and variance for grouped data is essentially the same as for raw data. The difference lies in the use of frequencies to save typing (writing) all numbers. Here is a comparison:

Statistic	Raw data	Grouped data	Grouped data with intervals
$\overline{x}$	$\overline{x} = \frac{\sum_{all\ x} x}{n}$	$\overline{x} = \frac{\sum_{all\ x} x_i \cdot f(x_i)}{n} = \frac{\sum_{all\ x} x_i \cdot f(x_i)}{\sum f(x_i)}$	$\overline{x} = \frac{\sum_{all\ x} m_i \cdot f(m_i)}{n} = \frac{\sum_{all\ x} m_i \cdot f(m_i)}{\sum f(m_i)}$
s_n^2	$s_n^2 = \frac{\sum_{all\ x} (x_i - \overline{x})^2}{n}$	$s_n^2 = \frac{\sum_{all\ x} (x_i - \overline{x})^2 \cdot f(x_i)}{n} = \frac{\sum_{all\ x} (x_i - \overline{x})^2 \cdot f(x_i)}{\sum f(x_i)}$	$s_n^2 = \frac{\sum_{all\ x} (m_i - x)^2 \cdot f(m_i)}{n} = \frac{\sum_{all\ x} (m_i - \overline{x})^2 \cdot f(m_i)}{\sum f(m_i)}$

where

x_i = data point

$f(x_i)$ = frequency of x_i

m_i = interval midpoint (mid mark or mid value)

$f(m_i)$ = frequency of interval i

$\sum f(x_i), \sum f(m_i)$ = total number of data points

For the grouped data reproduced here, this is how we estimate the mean and variance:

Living expenses	Midpoint m	Number of students $f(m)$	$m_i \times f(m_i)$	$(m_i - \overline{x})^2$	$(m_i - \overline{x})^2 \times f(m_i)$
35 but <40	37.5	2	75	344.5	688.9
40 but <45	42.5	3	127.5	183.9	551.6
45 but <50	47.5	11	522.5	73.3	806.0
50 but <55	52.5	21	1102.5	12.7	266.1
55 but <60	57.5	19	1092.5	2.1	39.4
60 but <65	62.5	11	687.5	41.5	456.2
65 but <70	67.5	13	877.5	130.9	1701.4
Totals		$\sum f(m_i) = 80$	$\sum_{all\ x} m_i \cdot f(m_i) = 4485$	$\sum_{all\ x} (m_i - \overline{x})^2 \cdot f(m_i) = 4509.6$	
		Mean	$\frac{4485}{80} = 56.06$	**Variance**	$\frac{4509.6}{80} = 56.37$
				Standard deviation	7.51

The numbers here are estimates of the mean and the variance and eventually the standard deviation. As you will notice, they are not equal to the values we calculated earlier, but are close. The reason for this is that, with grouping, we lost the detail in each interval. For example, the interval between 45 and 50 is represented by the midpoint 47.5. In essence, we are assuming that every number in the interval is equal to 47.5.

Example 7

Speed limits in some European cities are set to 50 km/h. Drivers in various cities react to such limits differently. The results of the survey to compare drivers' behaviour in Brussels, Vienna and Stockholm are given in the accompanying table. Use box plots to compare the results.

Vienna	62	60	59	50	61	63	53	46	58	49	51	37	47	51	63	52	44	50	45	44
Brussels	64	61	63	57	49	49	46	58	45	60	51	36	65	45	47	46				
Stockholm	43	44	34	35	31	34	29	33	36	38	45	47	29	48	51	49	48			

Solution

Parallel box plots may be an appropriate tool to enable a comparison between the three data sets.

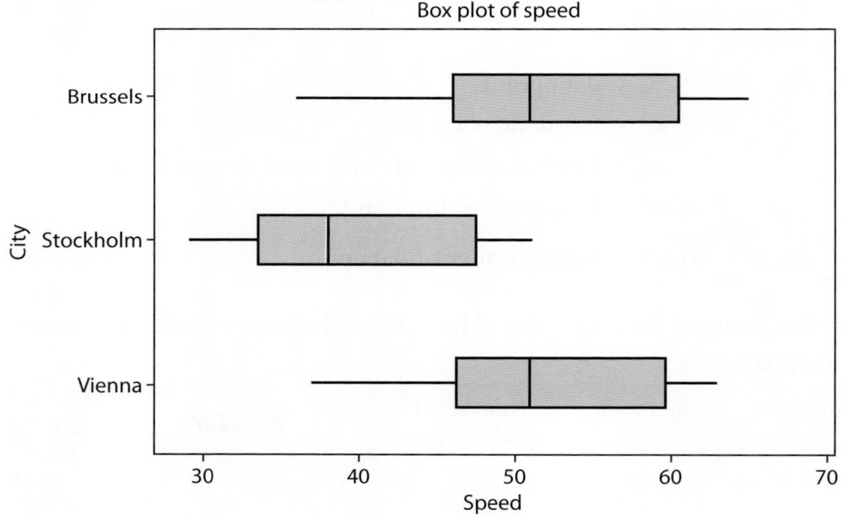

It appears that, on average, drivers in Brussels and Vienna tend to be on the 'speedy' side. The median in both cities is higher than 50, which means than more than 50% of the drivers in the two cities tend not to respect the speed limit. The variation in both cities is comparable with Brussels having a slightly wider range than Vienna. Almost all drivers in Stockholm appear to adhere to the 50 km/h limit. The median is around 40 km/h and the third quartile about 47, which means that more than 75% of the drivers in this city drive at a speed less than the 50 km/h limit.

Shape, centre and spread

Statistics is about variation, so spread is an important fundamental concept. Measures of spread help us to precisely analyze what we do not know! If the values we are looking at are scattered very far from the centre, the IQR and the standard deviation will be large. If these are large, our central values will not represent the data well. That is why we always report spread with any central value.

A practical way of seeing the significance of the standard deviation can be demonstrated with the following (optional) observations:

Empirical rule:

If the data is close to being symmetrical, as in the figure left, the following is true:

- The interval $\mu \pm \sigma$ contains approximately 68% of the measurements.
- The interval $\mu \pm 2\sigma$ contains approximately 95% of the measurements.
- The interval $\mu \pm 3\sigma$ contains approximately 99.7% of the measurements.

The empirical rule usually indicates if an observation is very far from the expected or not. Take the following example:

Symmetric distribution

I have recorded my car's fuel efficiency over the last 98 times that I have filled the tank with gasoline. Opposite is the data expressing how many kilometeres per litre the car travelled:

km/litre	Frequency	km/litre	Frequency
6.0	1	10.0	14
7.0	1	10.5	7
7.5	4	11.0	9
8.0	8	11.5	5
8.5	14	12.0	1
9.0	21	12.5	2
9.5	11		

The summary measures are:

Mean	9.454
σ	1.223
Median	9.25
Q_1	8.5
Q_3	10.125
IQR	1.625

The histogram shows that the distribution is almost symmetric. The possible outlier has little effect on the mean and standard deviation. That is why the mean and median are almost the same. Looking at the box plot, you can see that there is one outlier.

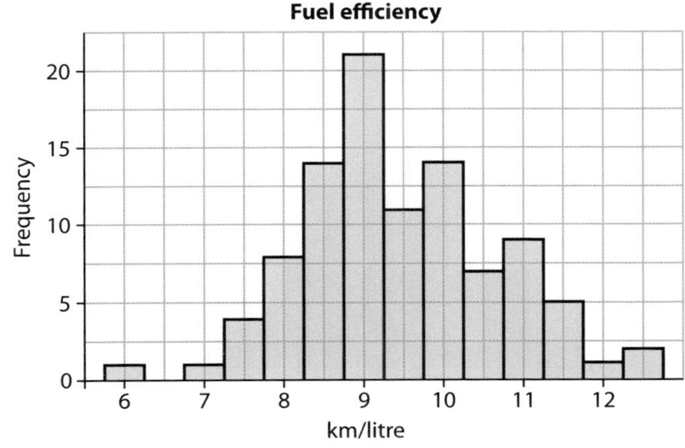

Chebyshev's rule:
This rule is similar to the empirical rule in the sense that it tries to interpret the standard deviation. Contrary to the empirical rule, it applies to any data set, regardless of the shape of the distribution.

Rule: For any number $k > 1$, at least $\left(1 - \frac{1}{k^2}\right)$ of the measurements will fall within k standard deviations of the mean. [i.e. within the interval $(\bar{x} - k\sigma, \bar{x} + k\sigma)$.]

a) No useful information is provided on the fraction of measurements that fall within 1 standard deviation from the mean – $(\bar{x} - \sigma, \bar{x} + \sigma)$.
b) At least $\frac{8}{9}$ will fall within 3 standard deviations from the mean – $(\bar{x} - 3\sigma, \bar{x} + 3\sigma)$.
c) At least $\frac{3}{4}$ will fall within 2 standard deviations from the mean – $(\bar{x} - 2\sigma, \bar{x} + 2\sigma)$.

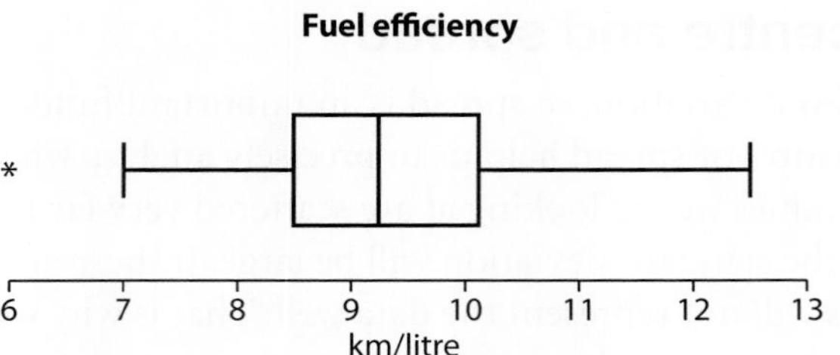

The confirmation is below:

$9.25 - 1.5 \times 1.625 = 6.8$, which is why 6 is considered as an outlier.

$10.125 + 1.5 \times 1.625 = 12.6$, and hence no outliers on this side.

If we use the empirical rule, we can expect about 99.7% of the data to lie within three standard deviations of the mean, i.e. $9.454 - 3 \times 1.223 = 5.8$ and $9.454 + 3 \times 1.223 = 13.1$. In fact, you see all the data is within the specified interval, including the potential outlier!

Question: What should you be able to tell about a quantitative variable?

Answer: Report the shape of its distribution, and include a centre and a spread.

Question: Which central measure and which measure of spread?

Answer: The rules are:

- If the shape is skewed, report the median and IQR. You *may* want to include the mean and standard deviation, but you should point out that the mean and median differ as this difference is a sign that the data is skewed. A histogram can help.
- If the shape is symmetrical, report the mean and standard deviation. You may report the median and IQR as well.
- If there are clear outliers, report the data with and without the outliers. The differences may be revealing.

Example 8

The records of a large high school show the heights of their students for the year 2006.

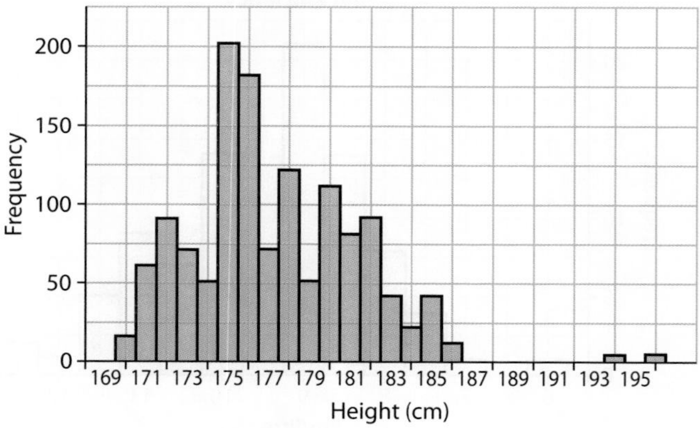

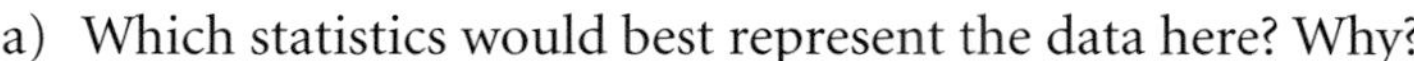

a) Which statistics would best represent the data here? Why?

b) Calculate the mean and standard deviation.

c) Develop a cumulative frequency graph of the data.

d) Use your result of c) above to estimate the median, Q_1, Q_3 and IQR.

e) Are there any outliers in the data? Why?

f) Write a few sentences describing the distribution.

Solution

a) The data appears to have outliers and is slightly skewed to the right. The most appropriate measure is the median, since the mean is influenced by the extreme values.

b) To calculate the mean and standard deviation, we will set up a table that will facilitate the calculation.

Height (cm) x_i	Number of students $f(x)$	$x_i \times f(x_i)$	$(x_i - \overline{x})^2$	$(x_i - \overline{x})^2 \times f(x_i)$
170	15	2550	51.84	777.6
171	60	10 260	38.44	2306.4
172	90	15 480	27.04	2433.6
173	70	12 110	17.64	1234.8
174	50	8700	10.24	512
175	200	35 000	4.84	968
176	180	31 680	1.44	259.2
177	70	12 390	0.04	2.8
178	120	21 360	0.64	76.8
179	50	8950	3.24	162
180	110	19 800	7.84	862.4
181	80	14 480	14.44	1155.2
182	90	16 380	23.04	2073.6
183	40	7320	33.64	1345.6
184	20	3680	46.24	924.8
185	40	7400	60.84	2433.6
186	10	1860	77.44	774.4
194	2	388	282.24	564.5
196	3	588	353.44	1060.3
Totals	$\sum f(x_i) = 1300$	$\sum_{all\ x} x_i \cdot f(x_i) = 230\,376$	$\sum_{all\ x} (x_i - \overline{x})^2 \cdot f(x_i) = 19\,927.6$	
	Mean	$\frac{230\,376}{1300} = 177.2$	**Variance**	$\frac{19\,927.4}{1300} = 15.33$
			Standard deviation	3.92

Note: Using the alternative formula for the variance will also give the same result. (Due to rounding, answers will differ slightly.)

$$s_n^2 = \frac{\sum_{i=1}^{n} x_i^2 \times f(x_i)}{n} - \overline{x}^2 = \frac{40\,845\,390}{1300} - 177.2123^2 = 15.3315 \Rightarrow$$

$$s_n = \sqrt{15.3315} = 3.92$$

c) To develop the cumulative frequency graph, we first need to develop the cumulative frequency table. This is done by accumulating the frequencies as shown below.

x	$f(x)$	Cum $f(x)$	x	$f(x)$	Cum $f(x)$
170	15	15	184	20	1245
171	60	75	185	40	1285
172	90	165	186	10	1295
173	70	235	187	0	1295
174	50	285	188	0	1295
175	200	485	189	0	1295
176	180	665	190	0	1295
177	70	735	191	0	1295
178	120	855	192	0	1295
179	50	905	193	0	1295
180	110	1015	194	2	1297
181	80	1095	195	0	1297
182	90	1185	196	3	1300
183	40	1225			

The cumulative frequency table is constructed such that the cumulative frequency corresponding to any measurement is the number of observations that are less than or equal to its value. So, for example, the cumulative frequency corresponding to a height of 174 cm is 285, which consists of the 50 observations with height 174 cm and the 235 observations for heights less than 174 cm.

The cumulative frequency graph plots the observations on the horizontal axis against their cumulative frequencies on the vertical axis, as shown below.

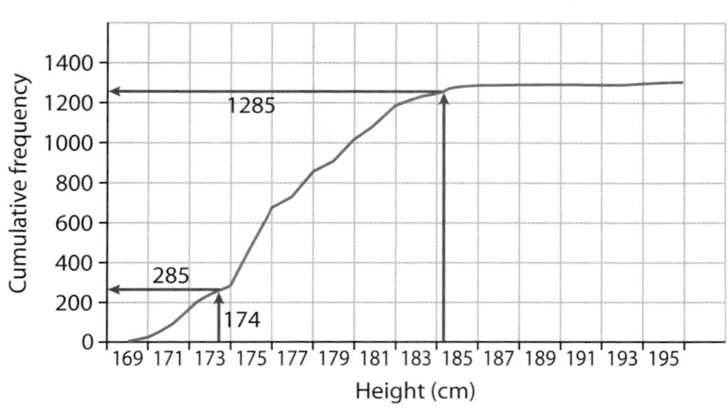

d) The median is the observation between $\frac{1300}{2} = 650$th and 651st observations, since the number is even. From the cumulative table, we can see that the median is in the 176 interval. So the median is 176.

Q_1 is at $\frac{1301}{4} \approx 325$th observation. From the table, as 174 has a cumulative frequency of 285, and 175 has a cumulative frequency of 485, then Q1 has to be **175**.

Also, Q_3 is at $\frac{3 \times 1301}{4} \approx 976$th observation. So, similarly, it is **180**.

IQR $= 180 - 175 =$ **5**.

e) To check for outliers, we can calculate the lengths of the whiskers.

Lower fence: $175 - 1.5 \times 5 = 167.5$, which is lower than the minimum value, so there are no outliers on the left.

Upper fence: $180 + 1.5 \times 5 = 187.5$. So we have five outliers, two at 194 cm and three at 196 cm.

f) The distribution appears to be bimodal with two modes at 175 and 176. It is slightly skewed to the right with a few extreme values at 194 and 196. This is further confirmed by the fact that the mean of 177.2 is higher than the median of 176.

Note: Here are the calculations using a GDC:

L1	L2	L3 3
170	15	
171	60	
172	90	
173	70	
174	50	
175	200	
176	180	

L3(1) =

1-Var Stats
$\bar{x}$=177.2123077
Σx=230376
Σx²=40845390
Sx=3.916704232
σx=3.915197517
↓n=1300

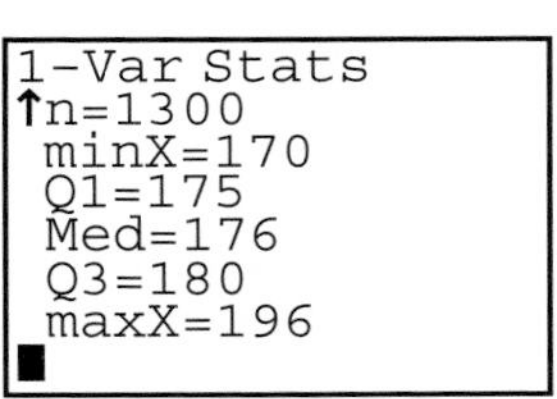

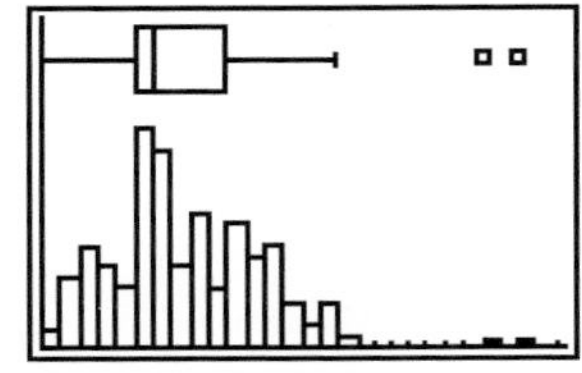

Exercise 11.3

1 The pulse rates of 15 patients chosen at random from visitors of a local clinic are given below:

72, 80, 67, 68, 80, 68, 80, 56, 76, 68, 71, 76, 60, 79, 71

a) Calculate the mean and standard deviation of the pulse rate of the patients at the clinic.

b) Draw a box plot of the data and indicate the values of the different parts of the box.

c) Check if there are any outliers.

2 The number of passengers on 50 flights from Washington to London on a commercial airline were:

165	173	158	171	177	156	178	210	160	164
141	127	119	146	147	155	187	162	185	125
163	179	187	174	166	174	139	138	153	142
153	163	185	149	154	154	180	117	168	182
130	182	209	126	159	150	143	198	189	218

a) Calculate the mean and standard deviation of the number of passengers on this airline between the two cities.

b) Set up a stem plot for the data and use it to find the median of the number of passengers.

c) Develop a cumulative frequency graph. Estimate the median, and first and third quartiles. Draw a box plot.

d) Find the IQR and use it to check whether there are any outliers.

e) Use the empirical rule to check for outliers.

3 At a school, 100 students took a 'mock' IB exam using paper 3. The paper was marked out of 60 marks. Here are the results

Marks	0–9	10–19	20–29	30–39	40–49	50–60
No. of students	5	9	16	24	27	19

a) Draw a cumulative frequency curve.

b) Estimate the median and quartiles.

4 130 first-year IB students were given a placement test to decide whether they go for SL or HL. The times for these students to finish the test are given in the table below:

Time	30–40	40–50	50–60	60–70	70–80	80–90	90–100	100–110	110–120
No. of students	8	12	24	29	19	16	12	8	2

a) Develop a cumulative frequency curve.

b) Estimate the median and the IQR.

c) 20 students did not manage to finish the test after 120 minutes and had to hand it in uncompleted. Estimate the median finishing time for all 130 students.

5 The mean score of 26 students on a 40-point paper is 22. The mean for another group of 84 other students is 32. Find the mean of the combined group of 110 students.

6 The scores on a 100-mark test of a sample of 80 students in a large school are given below.

Score	59–63	63–67	67–71	71–75	75–79	79–83	83–87
No. of students	6	10	18	24	10	8	4

a) Find the mean and standard deviation of the scores of all students.

b) A bonus of 13 points is to be added to these scores. What is the new value of the mean and standard deviation?

7 *Cats* is a famous musical. In a large theatre in Vienna (1744 capacity), during a period of 10 years, it played 1000 performances. The manager of the group kept a record of the empty seats on the days it played. Here is the table.

Number of empty seats	1–10	11–20	21–30	31–40	41–50	51–60	61–70	71–80	81–90	91–100
Days	15	50	100	170	260	220	90	45	30	20

a) Copy and complete the following cumulative frequency table for the above information.

Number of empty seats	$x \leq 10$	$x \leq 20$	$x \leq 30$	$x \leq 40$	$x \leq 50$	$x \leq 60$	$x \leq 70$	$x \leq 80$	$x \leq 90$	$x \leq 100$
Days	15		165			815				1000

b) Draw a cumulative frequency graph of this distribution. Use 1 unit on the vertical axis to represent the number of 100 days and 1 unit on the horizontal axis to represent every 10 seats.

c) Use the graph from b) to answer the following questions:

(i) Find an estimate of the median number of empty seats.

(ii) Find an estimate for the first quartile, third quartile and the IQR.

(iii) The days the number of empty seats was less than 35 seats were considered bumper days (lots of profit). How many days were considered bumper days?

(iv) The highest 15% of the days with empty seats were categorized as loss days. What is the number of empty seats above which a day is claimed as a loss?

8 Aptitude tests sometimes use jigsaw puzzles to test the ability of new applicants to perform precision assembly work in electronic instruments. One such company that produces the computerized parts of video and CD players gave the following results:

Time to finish the puzzle (nearest second)	Number of employees
30–35	16
35–40	24
40–45	22
45–50	26
50–55	38
55–60	36
60–65	32
65–70	18

a) Draw a histogram of the data.

b) Draw a cumulative frequency curve and estimate the median and IQR.

c) Calculate the estimates of the mean and standard deviation of all such participants.

9 The heights of football players at a given school are given in the table below:

Height	Frequency	Height	Frequency	Height	Frequency	Height	Frequency	Height	Frequency	Height	Frequency
152	2	160	7	168	18	175	5	183	9	191	4
155	6	163	5	170	7	178	11	185	4	193	1
157	9	165	20	173	12	180	8	188	2		

a) Find the five-number summary for this data.

b) Display the data with a box plot and a histogram.

c) Find the mean and standard deviation of the data.

d) Describe the data with a few sentences.

e) Draw a cumulative frequency graph and estimate the height of the player that is in the 90th percentile.

f) 10 players' data were missing when we collected the data. The average height of the 10 players is 182. Find the average height of all the players, including the last 10.

10 Consider 10 data measures.

a) If the mean of the first 9 measures is 12 and the 10th measure is 12, what is the mean of the 10 measures?

b) If the mean of the first 9 measures is 11, and the 10th measure is 21, what is the mean of the 10 measures?

c) If the mean of the first 9 measures is 11, and the mean of the 10 measures is 21, what is the value of the 10th measure?

11 Suppose that the mean of a set of 10 data points is 30.

a) It is discovered that a data point having a value of 25 was incorrectly entered as 15. What should be the revised value of the mean?

b) Suppose an additional point of value 32 was added. Will this increase or decrease the value of the mean?

12 Half the values of a sample are equal to 20, one-sixth are equal to 40, and one-third are equal to 60. What is the sample mean?

13 The seven numbers 7, 10, 12, 17, 21, x and y have a mean $\mu = 12$ and a variance $\sigma^2 = \frac{172}{7}$. Find x and y given that $x < y$.

14 A sample of 25 observations was taken out of a large population of measurements. If it is given that $\sum_{i=1}^{25} x_i = 278$ and $\sum_{i=1}^{n} x_i^2 = 3682$, estimate the mean and the variance of the population of measurements.

15 Use the data in question 9 of Exercise 11.1 to estimate the IQR and the standard deviation of the number of days in hospital by heart patients.

16 Use the data in question 10 of Exercise 11.1 to estimate the IQR and the standard deviation of the exercise time of the author for 2006.

17 Use the data in question 11 of Exercise 11.1 to estimate the IQR and the standard deviation of the speed of cars on the highway.

18 Use the data in question 12 of Exercise 11.1 to estimate the IQR and the standard deviation of the length of components at this facility.

19 Use the data in question 13 of Exercise 11.1 to estimate the IQR and the standard deviation of the waiting time for customers at this supermarket.

Practice questions

1 Given that μ is the mean of a data set $y_1, y_2, \ldots, y_{30}$, and you know that

$$\sum_{i=1}^{30} y_i = 360 \text{ and } \sum_{i=1}^{30} (y_i - \mu)^2 = 925, \text{ find}$$

a) the value of μ

b) the standard deviation of the set.

2 Laura made a survey of some students at school asking them about the time it takes each of them to come to school every morning. She scribbled the numbers on a piece of paper and, unfortunately, could not read the number of students who spend 40 minutes on their trip to school. The average number of minutes she had originally found was 34 minutes. Find out how many students spend 40 minutes on their trip.

Time in minutes	10	20	30	40	50
Number of students with this time	1	2	5	**?**	3

3 The following table gives 50 measurements of the time it took a certain reaction to be done in a laboratory experiment.

3.1	5.1	4.9	1.8	2.8	5.6	3.6	2.2	2.5	3.4
4.5	2.5	3.5	3.6	3.7	5.1	4.1	4.8	4.9	1.6
2.9	3.6	2.1	6.1	3.5	4.7	4	3.9	3.7	3.9
2.7	4.3	4	5.7	4.4	3.7	3.7	4.6	4.2	4
3.8	5.6	6.2	4.9	2.5	4.2	2.9	3.1	2.8	3.9

a) Construct a frequency table and histogram starting at 1.6 and with interval length of 0.5.

b) What fraction of the measurements is less than 5.1?

c) Estimate, from your histogram, the median of this data set.

d) Estimate the mean and standard deviation using your frequency table.

e) Construct a cumulative frequency graph.

f) From your cumulative frequency graph, estimate each of the five numbers in the five-number summary.

4 In large cities around the world, governments offer parking facilities for public use. The histogram below gives a picture of the number of parking sites available with the capacity of each, in a number of cities chosen at random.

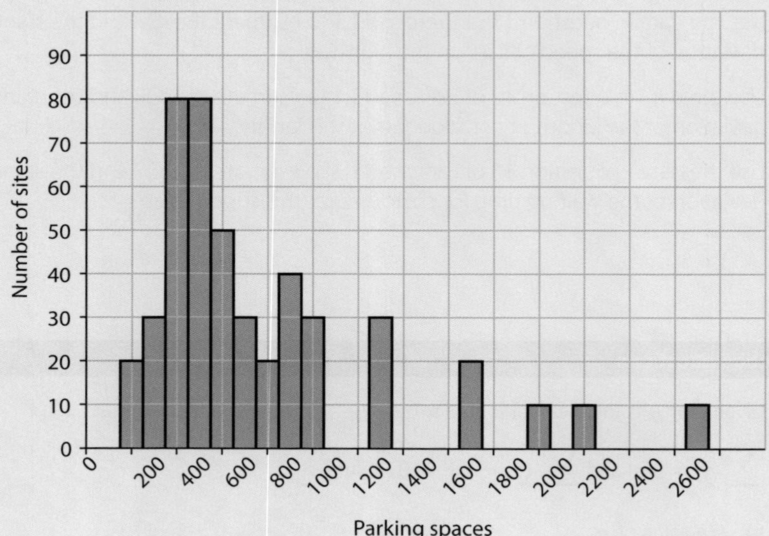

a) Which statistics would best represent the data here? Why?

b) Calculate the mean and standard deviation.

c) Develop a cumulative frequency graph of the data.

d) Use your result from **c)** above to estimate the median, Q_1, Q_3 and IQR.

e) Are there any outliers in the data? Why?

f) Write a few sentences describing the distribution.

5 The box plots display the case prices (in €) of red wines produced in France, Italy and Spain.

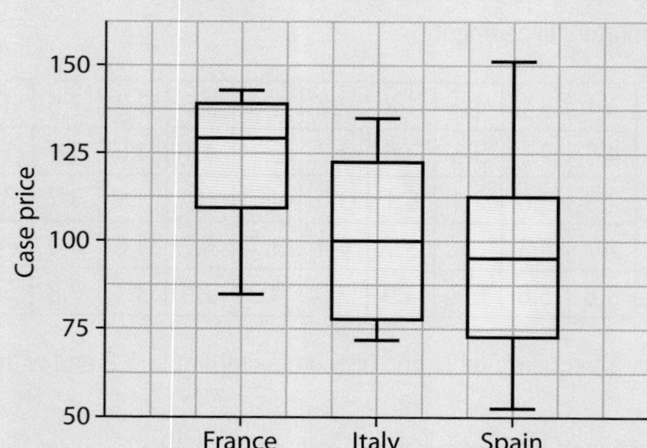

a) Which country appears to produce the most expensive red wine? The cheapest?

b) In which country are the red wines generally more expensive?

c) Write a few sentences comparing the pricing of red wines in the three countries.

6

112.72	53.55	54.12	54.33	58.79	59.26	60.39	62.45	52.22	52.52	52.58	52.85
54.06	51.34	51.93	52.09	52.14	52.24	52.24	52.53	53.5	51.82	51.93	52
52.78	52.82	50.28	50.49	51.28	51.28	51.52	51.62	52.4	52.43	49.83	50.46
50.95	51.07	51.11	49.45	49.45	49.73	49.76	49.93	50.19	50.32	50.63	48.64
49.79	50.19	50.62	50.96	49.09	49.16	49.29	49.74	49.74	49.75	49.84	49.76
52.9	52.91	53.4	52.18	52.57	52.72	50.56	50.87	50.9	49.32	49.7	

The table shows the record for the times (seconds) of the 71 male swimmers in the 100 m swim on the first day during the Summer Olympics 2000 in Sydney.

a) Calculate the mean time and the standard deviation.

b) Calculate the median and IQR.

c) Explain the differences between these two sets of measures.

7 In a survey of universities in major cities in the world, the percentage of first-year students who graduate on time (some require 4 years and some 5 years) was reported. The summary statistics are given below.

Number of universities surveyed	120	Mean percentage	69
Median percentage	70	Standard deviation	9.8
Minimum	42	Maximum	86
Range	44	Q_1	60.25
Q_3	75.75		

a) Is this distribution symmetric? Explain.

b) Check for outliers.

c) Create a box plot of the data.

d) Describe the data in a short paragraph.

8 The International Heart Association studies, among other factors, the influence of cholesterol level (in mg/dl) on the conditions of heart patients. In a study of 2000 subjects, the following cumulative relative frequency graph was recorded.

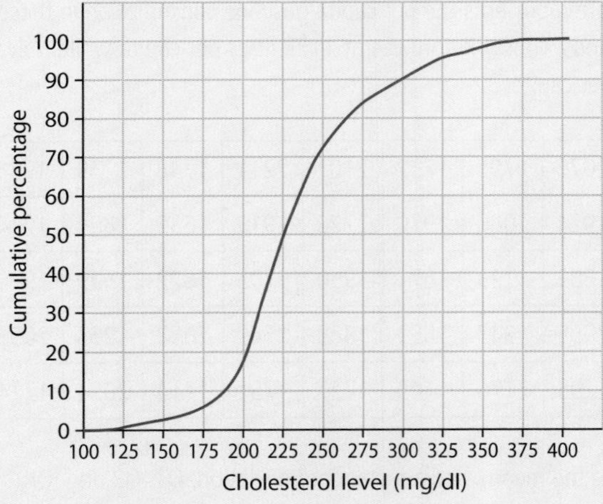

a) Estimate the median cholesterol level of heart patients in the study.

b) Estimate the first and third quartiles, and the 90th and 10th percentiles.

c) Estimate the IQR. Also estimate the number of patients in the middle 50% of this distribution.

d) Create a box plot of the data.

e) Give a short description of the distribution.

9 Many of the streets in Vienna, Austria have a speed limit of 30 km/h. On one Sunday evening the police registered the speed of cars passing an important intersection, in order to give speeding tickets when drivers exceeded the limit. Here is a random sample of 100 cars recorded that evening.

26	46	39	41	44	37	38	35	34	31
27	47	39	41	44	37	38	35	34	32
27	47	39	41	44	37	38	35	34	32
27	48	39	41	44	27	38	35	34	32
29	48	40	41	45	37	38	36	34	33
30	48	40	41	45	37	38	36	35	33
30	48	40	42	45	38	39	36	35	33
30	49	40	42	46	38	39	36	35	33
30	50	41	42	46	38	39	36	35	33
31	54	41	43	46	38	39	36	35	33

a) Prepare a frequency table for the data.

b) Draw a histogram of the data and describe the shape.

c) Calculate, showing all work, the mean and standard deviation of the data.

d) Prepare a cumulative frequency table of the data.

e) Find the median, Q_1, Q_3 and IQR.

f) Are there any outliers in the data? Explain using an appropriate diagram.

10 The following is the data collected from 50 industrial countries chosen at random in 2001. The data represents the per capita gasoline consumption in these countries. The Netherlands' consumption was at 1123 litres per capita while Italy stood at 2220 litres per capita.

2062	2076	1795	1732	2101	2211	1748	1239	1936	1658
1639	1924	2086	1970	2220	1919	1632	1894	1934	1903
1714	1689	1123	1671	1950	1705	1822	1539	1976	1999
2017	2055	1943	1553	1888	1749	2053	1963	2053	2117
1600	1795	2176	1445	1727	1751	1714	2024	1714	2133

a) Calculate the mean, median, standard deviation, Q1, Q3 and IQR.

b) Are there any outliers?

c) Draw a box plot.

d) What consumption levels are within 1 standard deviation from the mean?

e) Germany, with a consumption level of 2758 litres per capita, was not included in the sample. What effect on the different statistics calculated would adding Germany have? Do not recalculate the statistics.

11 90 students on a statistics course were given an experiment where each reported, to the nearest minute, the time, x, it took them to commute to school on a specific day. The teachers then reported back that the total travelling time for the course participants was $\sum x = 4460$ minutes.

a) Find the mean number of minutes the students spent travelling to school that day.

Four students who were absent when the data was first collected reported that they spent 35, 39, 28 and 32 minutes, respectively.

b) Calculate the new mean including these four students.

12 Two thousand students at a large university take the final statistics examination, which is marked on a 100-scale, and the distribution of marks received is given in the table below.

Marks	1–10	11–20	21–30	31–40	41–50	51–60	61–70	71–80	81–90	91–100
Number of candidates	30	100	200	340	520	440	180	90	60	40

a) Complete the table below so that it represents the cumulative frequency for each interval.

Marks	$\leqslant 10$	$\leqslant 20$	$\leqslant 30$	$\leqslant 40$	$\leqslant 50$	$\leqslant 60$	$\leqslant 70$	$\leqslant 80$	$\leqslant 90$	$\leqslant 100$
Number of candidates	30	130				1630				

b) Draw a cumulative frequency graph of the distribution, using a scale of 1 cm for 100 students on the vertical axis and 1 cm for 10 marks on the horizontal axis.

c) Use your graph from **b)** to answer parts **(i)–(iii)** below.

(i) Find an estimate for the median score.

(ii) Candidates who scored less than 35 were required to retake the examination. How many candidates had to retake the exam?

(iii) The highest-scoring 15% of candidates were awarded a distinction. Find the mark above for which a distinction was awarded.

13 At a conference of 100 mathematicians there are 72 men and 28 women. The men have a mean height of 1.79 m and the women have a mean height of 1.62 m. Find the mean height of the 100 mathematicians.

14 The mean of the population $x_1, x_2, \ldots, x_{25}$ is m. Given that $\sum_{i=1}^{25} x_i = 300$ and $\sum_{i=1}^{25} (x_i - m)^2 = 625$, find

a) the value of m

b) the standard deviation of the population.

15 The table shows the scores of competitors in a competition.

Score	10	20	30	40	50
Number of competitors with this score	1	2	5	k	3

The mean score is 34. Find the value of k.

16 A survey is carried out to find the waiting times for 100 customers at a supermarket.

a) Calculate an estimate for the mean of the waiting times, by using an appropriate approximation to represent each interval.

b) Construct a cumulative frequency table for this data.

c) Use the cumulative frequency table to draw, on graph paper, a cumulative frequency graph, using a scale of 1 cm per 20 seconds waiting time for the horizontal axis and 1 cm per 10 customers for the vertical axis.

d) Use the cumulative frequency graph to find estimates for the median and the lower and upper quartiles.

Waiting time (seconds)	Number of customers
0–30	5
30–60	15
60–90	33
90–120	21
120–150	11
150–180	7
180–210	5
210–240	3

17 The following diagram represents the lengths, in cm, of 80 plants grown in a laboratory.

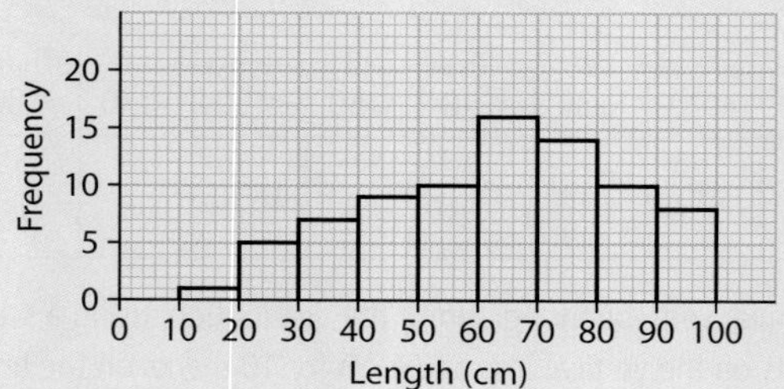

a) How many plants have lengths in cm between

(i) 50 and 60?

(ii) 70 and 90?

b) Calculate estimates for the mean and the standard deviation of the lengths of the plants.

c) Explain what feature of the diagram suggests that the median is different from the mean.

d) The following is an extract from the cumulative frequency table.

Length in cm less than	Cumulative frequency
.	.
50	22
60	32
70	48
80	62
.	.

Use the information in the table to estimate the median. Give your answer to 2 significant figures.

18 The table below represents the weights, W, in grams, of 80 packets of roasted peanuts.

Weight (W)	$80 < W \leqslant 85$	$85 < W \leqslant 90$	$90 < W \leqslant 95$	$95 < W \leqslant 100$	$100 < W \leqslant 105$	$105 < W \leqslant 110$	$110 < W \leqslant 115$
Number of packets	5	10	15	26	13	7	4

a) Use the midpoint of each interval to find an estimate for the standard deviation of the weights.

b) Copy and complete the following cumulative frequency table for the above data.

Weight (W)	$W \leqslant 85$	$W \leqslant 90$	$W \leqslant 95$	$W \leqslant 100$	$W \leqslant 105$	$W \leqslant 110$	$W \leqslant 115$
Number of packets	5	15					80

c) A cumulative frequency graph of the distribution is shown below, with a scale of 2 cm for 10 packets on the vertical axis and 2 cm for 5 grams on the horizontal axis.

Use the graph to estimate

(i) the median

(ii) the upper quartile (that is, the third quartile).

Give your answers to the nearest gram.

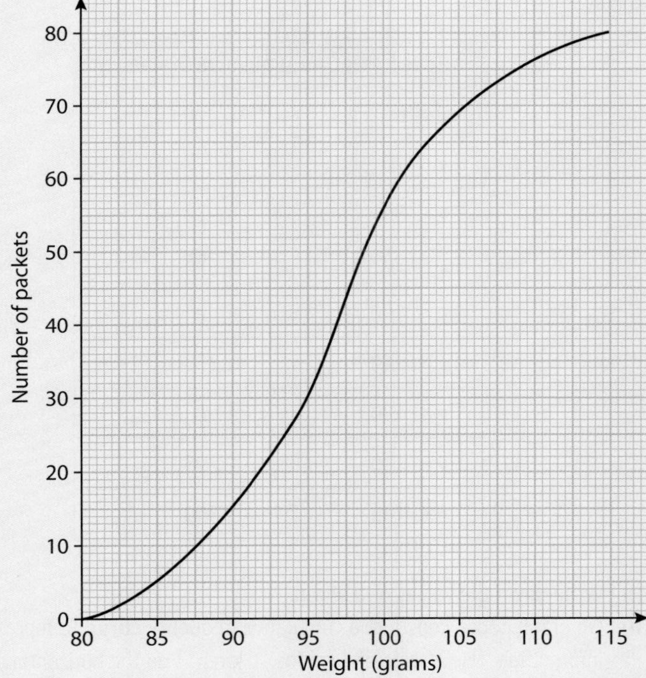

d) Let $W_1, W_2, \ldots, W_{80}$ be the individual weights of the packets, and let $\overline{W}$ be their mean. What is the value of the sum

$$(W_1 - \overline{W}) + (W_2 - \overline{W}) + (W_3 - \overline{W}) + \ldots + (W_{79} - \overline{W}) + (W_{80} - \overline{W})?$$

e) One of the 80 packets is selected at random. Given that its weight satisfies $85 < W \leqslant 110$, find the probability that its weight is greater than 100 grams.

19 The speeds, in km h^{-1}, of cars passing a point on a highway are recorded in the following table.

Speed v	Number of cars
$v \leq 60$	0
$60 < v \leq 70$	7
$70 < v \leq 80$	25
$80 < v \leq 90$	63
$90 < v \leq 100$	70
$100 < v \leq 110$	71
$110 < v \leq 120$	39
$120 < v \leq 130$	20
$130 < v \leq 140$	5
$v > 140$	0

a) Calculate an estimate of the mean speed of the cars.

b) The following table gives some of the cumulative frequencies for the information above.

Speed v	Cumulative frequency
$v \leq 60$	0
$v \leq 70$	7
$v \leq 80$	32
$v \leq 90$	95
$v \leq 100$	a
$v \leq 110$	236
$v \leq 120$	b
$v \leq 130$	295
$v \leq 140$	300

(i) Write down the values of a and b.

(ii) On graph paper, construct a cumulative frequency **curve** to represent this information. Use a scale of 1 cm for 10 km h^{-1} on the horizontal axis and a scale of 1 cm for 20 cars on the vertical axis.

c) Use your graph to determine

(i) the percentage of cars travelling at a speed in excess of 105 km h^{-1}

(ii) the speed which is exceeded by 15% of the cars.

20 A taxi company has 200 taxi cabs. The cumulative frequency curve below shows the fares in dollars ($) taken by the cabs on a particular morning.

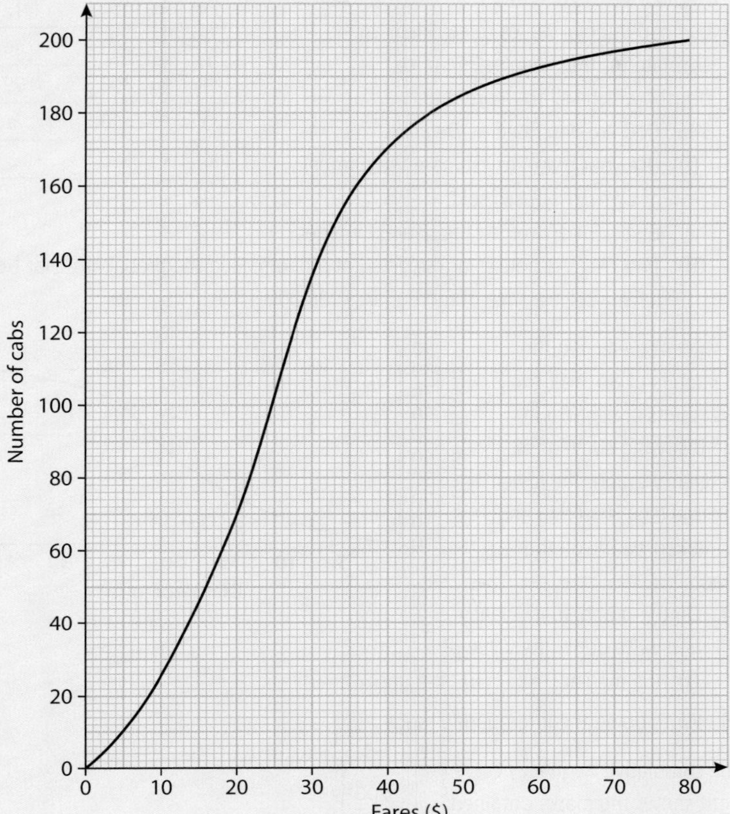

a) Use the curve to estimate

(i) the median fare

(ii) the number of cabs in which the fare taken is $35 or less.

The company charges 55 cents per kilometre for distance travelled. There are no other charges. Use the curve to answer the following.

b) On that morning, 40% of the cabs travel less than *a* km. Find the value of *a*.

c) What percentage of the cabs travel more than 90 km on that morning?

21 Three positive integers *a*, *b* and *c*, where $a < b < c$, are such that their median is 11, their mean is 9 and their range is 10. Find the value of *a*.

22 In a suburb of a large city, 100 houses were sold in a three-month period. The following **cumulative frequency table** shows the distribution of selling prices (in thousands of dollars).

Selling price *P* ($ thousand)	$P \leqslant 100$	$P \leqslant 200$	$P \leqslant 300$	$P \leqslant 400$	$P \leqslant 500$
Total number of houses	12	58	87	94	100

a) Represent this information on a cumulative frequency **curve**, using a scale of 1 cm to represent $50 000 on the horizontal axis and 1 cm to represent 5 houses on the vertical axis.

b) Use your curve to find the interquartile range.

The price information is represented in the following frequency distribution.

c) Find the values of a and b.

d) Use mid-interval values to calculate an estimate for the mean selling price.

e) Houses which sell for more than \$350 000 are described as *De Luxe*.

(i) Use your graph to estimate the number of *De Luxe* houses sold. Give your answer to the nearest integer.

(ii) Two *De Luxe* houses are selected at random. Find the probability that **both** have a selling price of more than \$400 000.

Selling price P (\$ thousand)	Total number of houses
$0 < P \leq 100$	12
$100 < P \leq 200$	46
$200 < P \leq 300$	29
$300 < P \leq 400$	a
$400 < P \leq 500$	b

23 A student measured the diameters of 80 snail shells. His results are shown in the following cumulative frequency graph. The lower quartile (LQ) is 14 mm and is marked clearly on the graph.

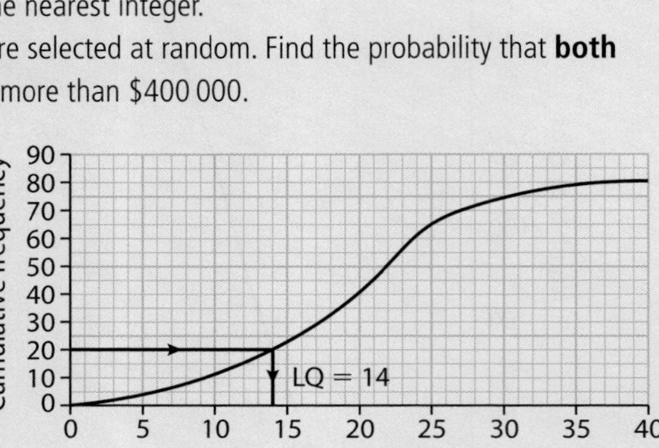

a) On the graph, mark clearly and write down the value of

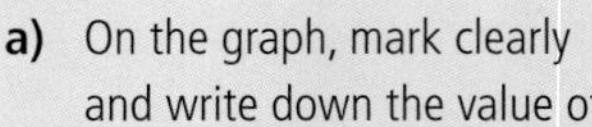

(i) the median **(ii)** the upper quartile.

b) Write down the interquartile range.

24 The cumulative frequency curve right shows the marks obtained in an examination by a group of 200 students.

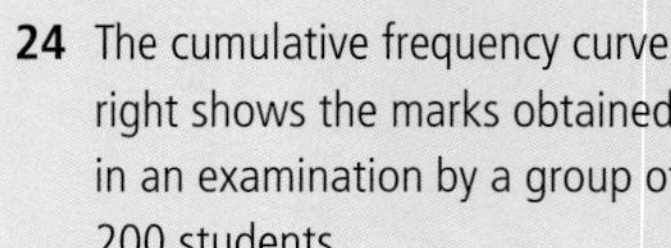

a) Use the cumulative frequency curve to complete the frequency table on the next page.

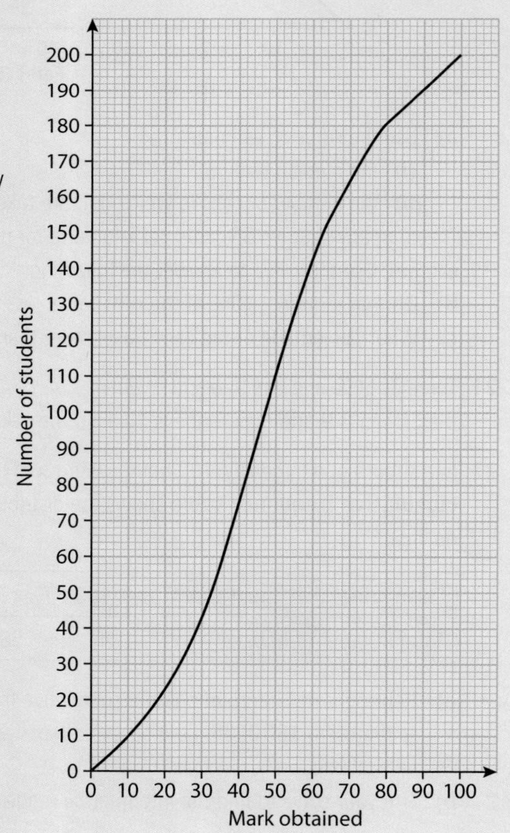

Mark (x)	$0 \leqslant x < 20$	$20 \leqslant x < 40$	$40 \leqslant x < 60$	$60 \leqslant x < 80$	$80 \leqslant x < 100$
Number of students	22				20

b) Forty per cent of the students fail. Find the pass mark.

25 The cumulative frequency curve right shows the heights (in centimetres) of 120 basketball players.
Use the curve to estimate

a) the median height

b) the interquartile range.

26 Let a, b, c and d be integers such that $a < b$, $b < c$ and $c = d$.
The mode of these four numbers is 11.
The range of these four numbers is 8.
The mean of these four numbers is 8.
Calculate the value of each of the integers a, b, c and d.

27 A test, to be marked out of 100, is completed by 800 students. The cumulative frequency graph for the marks is given below.

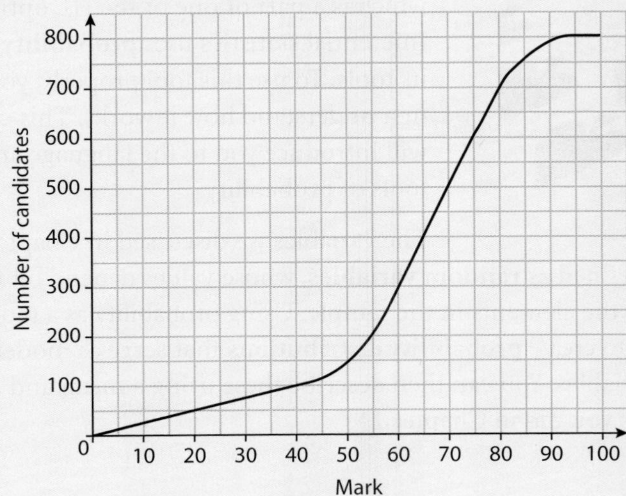

a) Write down the number of students who scored 40 marks or less on the test.

b) The middle 50% of test results lie between marks a and b, where $a < b$. Find a and b.

28 x and y are integers with $x < y$. The set of numbers $\{x, y, 10, 12, 16, 16, 18, 18\}$ have a mean of 13 and a variance σ^2 of 21. Find x and y.

12 Probability

Assessment statements

5.2 Concepts of trial, outcome, equally likely outcomes, sample space (U) and event.
The probability of an event A as $P(A) = n(A)/n(U)$.
The complementary events as A and A' (not A);
$P(A) + P(A') = 1$.
Use of Venn diagrams, tree diagrams and tables of outcomes to solve problems.

5.3 Combined events, the formula: $P(A \cup B) = P(A) + P(B) - P(A \cap B)$.
$P(A \cap B) = 0$ for mutually exclusive events.

5.4 Conditional probability; the definition: $P(A|B) = P(A \cap B)/P(B)$.
Independent events; the definition: $P(A|B) = P(A) = P(A|B')$.
Use of Bayes' theorem for a maximum of three events.

Introduction

Now that you have learned to describe a data set in Chapter 11, how can you use sample data to draw conclusions about the populations from which you drew your samples?

The techniques we use in drawing conclusions are part of what we call **inferential statistics**, which is a part of one of the HL options. Inferential statistics uses **probability** as one of its tools. To use this tool properly, you must first understand how it works. This chapter will introduce you to the language and basic tools of probability.

The variables we discussed in Chapter 11 can now be redefined as **random variables**, whose values depend on the chance selection of the elements in the sample. Using probability as a tool, you will be able to create **probability distributions** that serve as models for random variables. You can then describe these using a mean and a standard deviation as you did in Chapter 11.

12.1 Randomness

Probability is the study of randomness and uncertainty.

The reasoning in statistics rests on asking, 'How often would this method give a correct answer if I used it very many times?' When we produce data by random sampling or by experiments, the laws of probability enable us to answer the question, 'What would happen if we did this many times?'

What does 'random' mean? In ordinary speech, we use 'random' to denote things that are unpredictable. Events that are **random** are not perfectly predictable, but *they have long-term regularities* that we can describe and quantify using probability. In contrast, **haphazard** events *do not necessarily have long-term regularities.* Take, for example, the tossing of an unbiased coin and observing the number of heads that appear. This is random behaviour.

Please distinguish between random and haphazard (chaos). At first glance they might seem to be the same because neither of their outcomes can be anticipated with certainty.

When you throw the coin, there are only two outcomes, heads or tails. Figure 12.1 shows the results of the first 50 tosses of an experiment that tossed the coin 5000 times. Two sets of trials are shown. The red graph shows the result of the first trial: the first toss was a head followed by a tail, making the proportion of heads to be 0.5. The third toss was also a tail, so the proportion of heads is 0.33, then 0.25. On the other hand, the other set of trials, shown in green, starts with a series of tails, then a head, which raises the proportion to 0.2, etc.

The proportion of heads is quite variable at first. However, in the long run, and as the number of tosses increases, the proportion of heads stabilizes around 0.5. We say that 0.5 is the **probability** of a head.

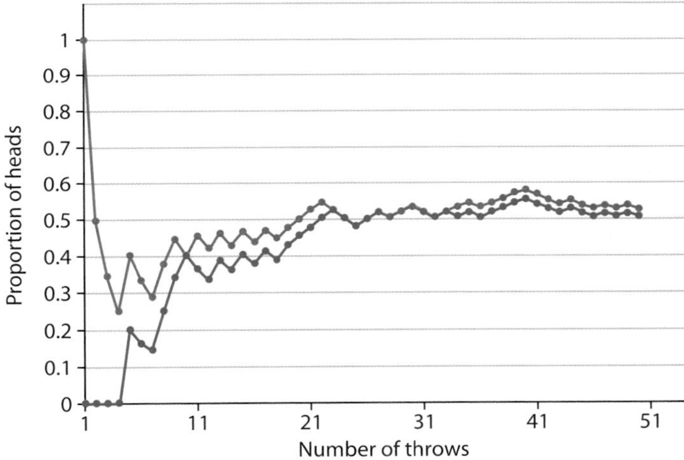

Figure 12.1

It is important that you know that the proportion of heads in a small number of tosses can be far from the probability. Probability describes only what happens in the long run. How a fair coin lands when it is tossed is an example of a random event. One cannot predict perfectly whether the coin will land heads or tails. However, in repeated tosses, the fraction of times the coin lands heads will tend to settle down to a limit of 50%. The outcome of an individual toss is not perfectly predictable, but the long-term average behaviour is predictable. Thus, it is reasonable to consider the outcome of tossing a fair coin to be random.

Imagine the following scenario:

I drive every day to school. Shortly before school, there is a traffic light. It appears that it is always red when I get there. I collected data over the course of one year (180 school days) and considered the green light to be a 'success'. Here is a partial table of the collected data.

Day	1	2	3	4	5	6	7	...
Light	red	green	red	green	red	red	red	...
Percentage green	0	50	33.3	50	40	33.3	28.6	...

The first day it was red, so the proportion of success is 0% (0 out of 1); the second day it was green, so the frequency is now 50% (1 out of 2); the third day it was red again, so 33.3% (1 out of 3), and so on. As we collect more data, the new measurement becomes a smaller and smaller fraction of the accumulated frequency, so, in the long run, the graph settles to the real chance of finding it green, which in this case is about 30%. The graph is shown below.

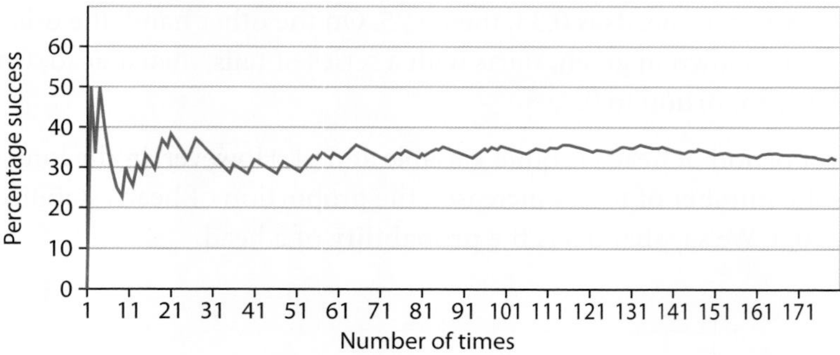

Actually, if you run a simulation for a longer period, you can see that it really stabilizes around 30%. See graph below.

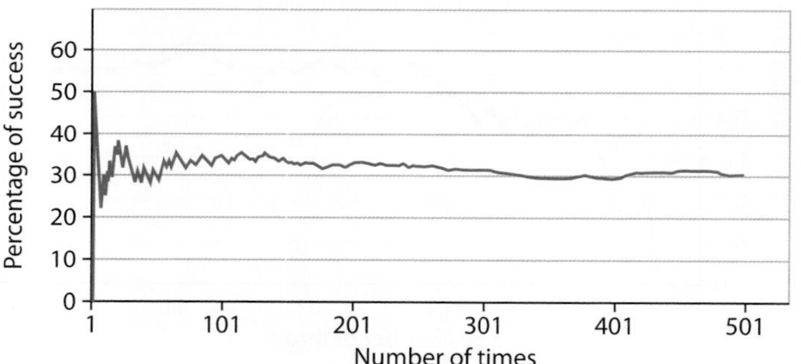

You have to observe here that the randomness in the experiment is not in the traffic light itself, as it is controlled by a timer. In fact, if the system works well, it may turn green at the same time every day. The randomness of the event is the time I arrive at the traffic light.

The French Count Buffon (1707–1788) tossed a coin 4040 times and received 2048 heads, i.e. a proportion of 50.69%. Also, the English statistician Karl Pearson (1857–1936) tossed a coin 24 000 times and received 12 012 heads, a 50.05% proportion for heads.

Count Buffon ▶

If we ask for the probability of finding the traffic light green in the above example, our answer will be about 30%. We base our answer on knowing that, in the long run, the fraction of time that the traffic light was green is 30%. We could also say that the **long-run relative frequency** of the green light settles down to about 30%.

12.2 Basic definitions

Data is obtained by observing either uncontrolled events in nature or controlled situations in a laboratory. We use the term **experiment** to describe either method of data collection.

An **experiment** is the process by which an observation (or measurement) is obtained. A **random** (chance) **experiment** is an experiment where there is uncertainty concerning which of two or more possible outcomes will result.

Tossing a coin, rolling a die and observing the number on the top surface, counting cars at a traffic light when it turns green, measuring daily rainfall in a certain area, etc. are a few experiments in this sense of the word.

A description of a random phenomenon in the language of mathematics is called a **probability model**. For example, when we toss a coin, we cannot know the outcome in advance. What *do* we know? We are willing to say that the outcome will be either heads or tails. Because the coin appears to be balanced, we believe that each of these outcomes has probability 0.50. This description of coin tossing has two parts:

- A list of possible outcomes.
- A probability for each outcome.

This two-part description is the starting point for a probability model. We will begin by describing the outcomes of a random phenomenon and learn how to assign probabilities to the outcomes by using one of the definitions of probability.

The notation for sample space could also be **U** (IB notation) or any other letter.

The **sample space S** of a random experiment (or phenomenon) is the set of all possible outcomes.

For example, for one toss of a coin, the sample space is

$$S = \{\text{heads, tails}\}, \text{ or simply } \{h, t\}$$

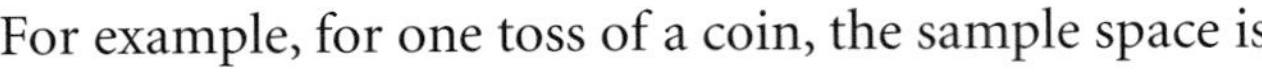

Example 1

Toss a coin twice (or two coins once) and record the results. What is the sample space?

Solution

$$S = \{hh, ht, th, tt\}$$

Example 2

Toss a coin twice (or two coins once) and count the number of heads showing. What is the sample space?

Solution

$$S = \{0, 1, 2\}$$

A **simple event** is the outcome we observe in a single repetition (trial) of the experiment.

For example, an experiment is throwing a die and observing the number that appears on the top face. The simple events in this experiment are $\{1\}$, $\{2\}$, $\{3\}$, $\{4\}$, $\{5\}$ and $\{6\}$. Of course, the set of all these simple events is the sample space of the experiment.

Set theory provides a foundation for all of mathematics. The language of probability is much the same as the language of set theory. Logical statements can be interpreted as statements about sets. This will enable us later to introduce a method of understanding how to set up probability problems that we need to tackle.

We are now ready to define an **event**. There are several ways of looking at it, which in essence are all the same.

An **event** is an **outcome** or a **set of outcomes** of a random experiment.

With this understanding, we can also look at the event as a subset of the sample space or as a collection of simple events.

Example 3

When rolling a standard six-sided die, what are the sets of event A 'observe an odd number', and event B 'observe a number less than 5'.

Solution

Event A is the set $\{1, 3, 5\}$. Event B is the set $\{1, 2, 3, 4\}$.

Sometimes it helps to visualize an experiment using some tools of set theory. Basically, there are several similarities between the ideas of set theory and probability, and it is very helpful when we see the connection. A simple but powerful diagram is the **Venn diagram**. The diagram shows the outcomes of the die rolling experiment.

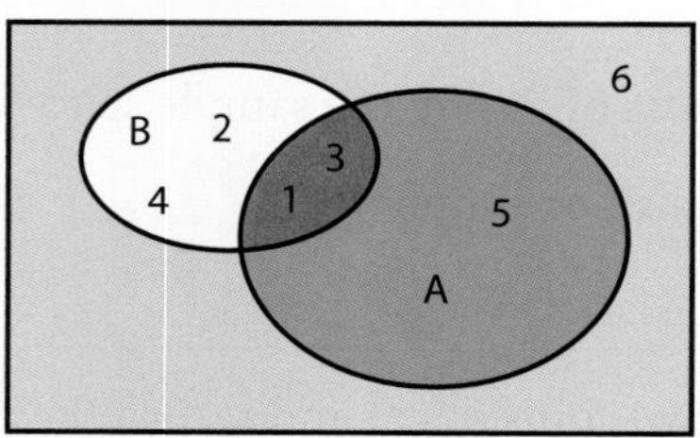

In general, in this book, we will use a rectangle to represent the sample space and closed curves to represent events, as shown in Example 3.

To understand the definitions more clearly, let's look at the following additional example.

Example 4

Suppose we choose one card at random from a deck of 52 playing cards, what is the sample space S?

Solution

S = {A♣, 2♣, … K♣, A♦, 2♦, … K♦, A♥, 2♥, … K♥, A♠, 2♠, … K♠}

Some events of interest:

K = event of king = {K♣, K♦, K♥, K♠}

H = event of heart = {A♥, 2♥, … K♥}

J = event of jack or better

= {J♣, J♦, J♥, J♠, Q♣, Q♦, Q♥, Q♠, K♣, K♦, K♥, K♠, A♣, A♦, A♥, A♠}

Q = event of queen = {Q♣, Q♦, Q♥, Q♠}

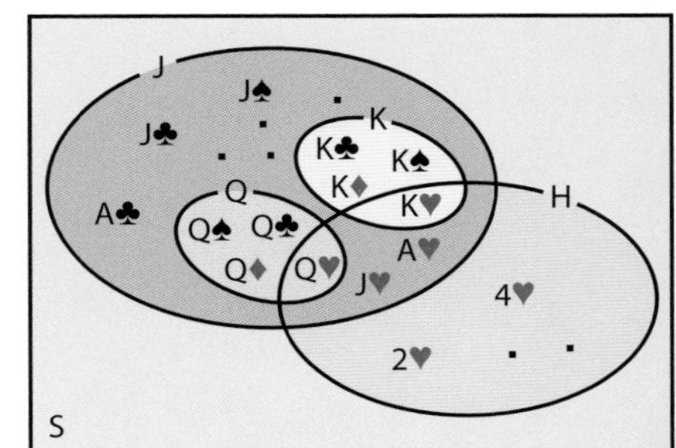

Some useful set theory results

Set operations have a number of properties, which are basic consequences of the definitions.

Some examples are:

$A \cup B = B \cup A$
$(A')' = A$
$A \cap S = A$
$A \cup S = S$
$A \cap A' = \varnothing$
$A \cup A' = S$

S is the sample space and $\varnothing$ is the empty set.

Two mainly valuable properties are known as De Morgan's laws, which state that:

$(A \cup B)' = A' \cap B'$

$(A \cap B)' = A' \cup B'$

And finally $A \cap (B \cup C) = (A \cap B) \cup (A \cap C)$

$A \cup (B \cap C) = (A \cup B) \cap (A \cup C)$

Example 5

Toss a coin three times and record the results. Show the event 'observing two heads' as a Venn diagram.

Solution

The sample space is made up of 8 possible outcomes such as hhh, hht, tht, etc.

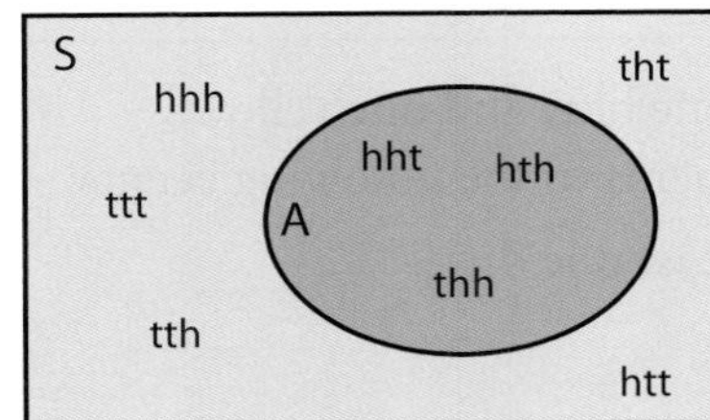

Observing exactly two heads is an event with three elements: {hht, hth, thh}.

Tree diagrams, tables and grids

In an experiment to check the blood types of patients, the experiment can be thought of as a two-stage experiment: first we identify the type of the blood and then we classify the Rh factor + or −.

The simple events in this experiment can be counted using another tool, the **tree diagram**, which is extremely powerful and helpful in solving probability problems.

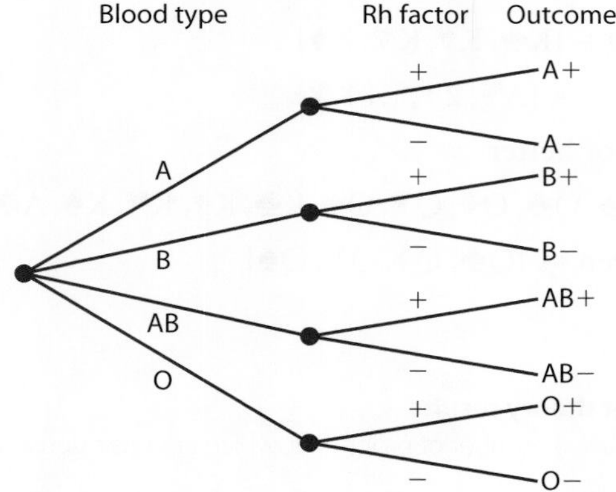

Our sample space in this experiment is the set {A+, A−, B+, B−, AB+, AB−, O+, O−} as we can read from the last column.

This data can also be arranged in a **probability table**:

	Blood type			
Rh factor	A	B	AB	O
Positive	A+	B+	AB+	O+
Negative	A−	B−	AB−	O−

Or using a 2-dimensional grid as shown right:

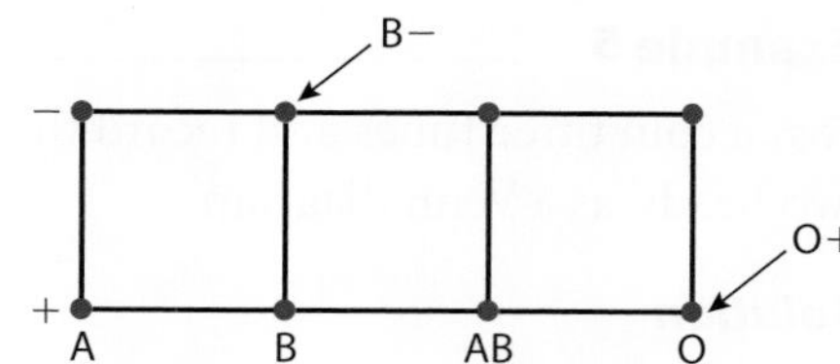

Example 6

Two tetrahedral dice, one blue and one yellow, are rolled. List the elements of the following events:

T = {3 appears on at least one die}

B = {the blue die is a 3}

S = {sum of the dice is a six}

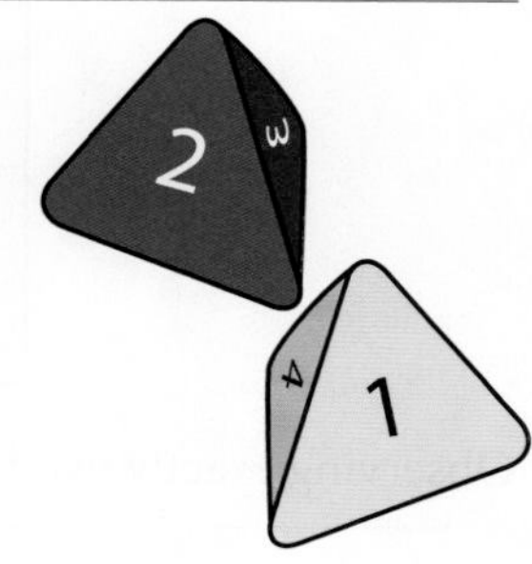

Solution

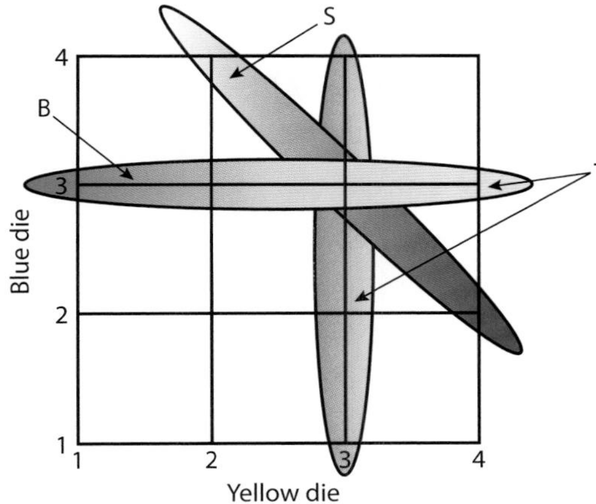

$T = \{(1, 3), (2, 3), (3, 3), (4, 3), (3, 4), (3, 2), (3, 1)\}$
$B = \{(1, 3), (2, 3), (3, 3), (4, 3)\}$
$S = \{(2, 4), (3, 3), (4, 2)\}$

Exercise 12.1 and 12.2

1 In a large school, a student is selected at random. Give a reasonable sample space for answers to each of the following questions:

a) Are you left-handed or right-handed?

b) What is your height in centimetres?

c) How many minutes did you study last night?

2 We throw a coin and a standard six-sided die and we record the number and the face that appear in that order. For example, (5, h) represents a 5 on the die and a head on the coin. Find the sample space.

3 We draw cards from a deck of 52 playing cards.

a) List the sample space if we draw one card at a time.

b) List the sample space if we draw two cards at a time.

c) How many outcomes do you have in each of the experiments above?

4 Tim carried out an experiment where he tossed 20 coins together and observed the number of heads showing. He repeated this experiment 10 times and got the following results:

11, 9, 10, 8, 13, 9, 6, 7, 10, 11

a) Use Tim's data to get the probability of obtaining a head.

b) He tossed the 20 coins for the 11th time. How many heads should he expect to get?

c) He tossed the coins 1000 times. How many heads should he expect to see?

5 In the game 'Dungeons and Dragons', a four-sided die with sides marked with 1, 2, 3 and 4 spots is used. The intelligence of the player is determined by rolling the die twice and adding 1 to the sum of the spots.

a) What is the sample space for rolling the die twice? (Record the spots on the 1st and 2nd throws.)

b) What is the sample space for the intelligence of the player?

6 A box contains three balls, blue, green and yellow. You run an experiment where you draw a ball, look at its colour and then replace it and draw a second ball.

a) What is the sample space of this experiment?

b) What is the event of drawing yellow first?

c) What is the event of drawing the same colour twice?

7 Repeat the same exercise as in question 6 above, without replacing the first ball.

8 Nick flips a coin three times and each time he notes whether it is heads or tails.

a) What is the sample space of this experiment?

b) What is the event that heads occur more often than tails?

9 Franz lives in Vienna. He and his family decided that their next vacation will be to either Italy or Hungary. If they go to Italy, they can fly, drive or take the train. If they go to Hungary, they will drive or take a boat. Letting the outcome of the experiment be the location of their vacation and their mode of travel, list all the points in the sample space. Also list the sample space of the event 'fly to destination.'

10 A hospital codes patients according to whether they have health insurance or no insurance, and according to their condition. The condition of the patient is rated as good (g), fair (f), serious (s), or critical (c). The clerk at the front desk marks 0, for non-insured patients, and 1 for insured, and uses one of the letters for the condition. So, (1, c) means an insured patient with critical condition.

a) List the sample space of this experiment.

b) What is the event 'not insured, in serious or critical condition'?

c) What is the event 'patient in good or fair condition'?

d) What is the event 'patient has insurance'?

11 A social study investigates people for different characteristics. One part of the study classifies people according to gender (G_1 = female, G_2 = male), drinking habits (K_1 = abstain, K_2 = drinks occasionally, K_3 = drinks frequently), and marital status (M_1 = married, M_2 = single, M_3 = divorced, M_4 = widowed).

a) List the elements of an appropriate sample space for observing a person in this study.

b) Define the following events:
A = the person is a male, B = the person drinks, and C = the person is single
List the elements of each A, B and C.

c) Interpret the following events in the context of this situation:

$$A \cup B; A \cap C; C'; A \cap B \cap C; A' \cap B.$$

12 Cars leaving the highway can take a right turn (R), left turn (L), or go straight (S). You are collecting data on traffic patterns at this intersection and you group your observations by taking four cars at a time every 5 minutes.

a) List a few outcomes in your sample space U. How many are there?

b) List the outcomes in the event that all cars go in the same direction.

c) List the outcomes that only two cars turn right.

d) List the outcomes that only two cars go in the same direction.

13 You are collecting data on traffic at an intersection for cars leaving a highway. Your task is to collect information about the size of the vehicle: truck (*T*), bus (*B*), car (*C*). You also have to record whether the driver has the safety belt on (*SY*) or no safety belt (*SN*), as well as whether the headlights are on (*O*) or off (*F*).

a) List the outcomes of your sample space, U.

b) List the outcomes of the event *SY* that the driver has the safety belt on.

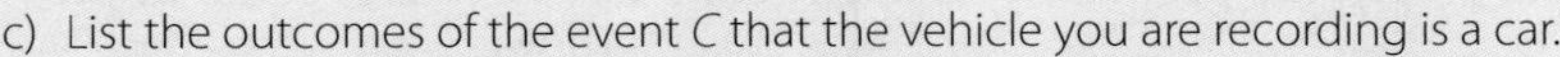

c) List the outcomes of the event C that the vehicle you are recording is a car.
d) List the outcomes of the event in $C \cap SY$, C', and $C \cup SY$.

14 Many electric systems use a built in 'back-up' system so that the equipment using the system will work even if some parts fail. Such a system is given in the diagram below.

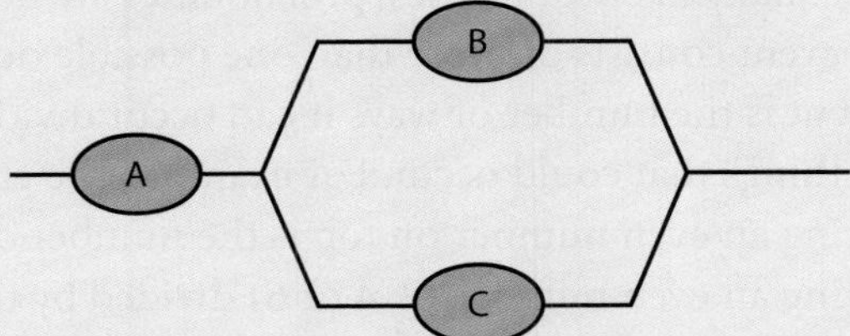

Two parts of this system are installed 'in parallel', so that the system will work if at least one of them works. If we code a working system by 1 and a failing system by 0, then one of the outcomes would be (1, 0, 1), which means parts A and C work while B failed.

a) List the outcomes of your sample space, U.
b) List the outcomes of the event X that exactly 2 of the parts work.
c) List the outcomes of the event Y that at least 2 of the parts work?
d) List the outcomes of the event Z that the system functions.
e) List the outcomes of the events: Z', $X \cup Z$, $X \cap Z$, $Y \cup Z$, and $Y \cap Z$.

15 Your school library has 5 copies of George Polya's *How To Solve It* book. Copies 1 and 2 are first-edition, and copies 3, 4 and 5 are second edition. You are searching for a first-edition book, and you will stop when you find a copy. For example, if you find copy 2 immediately, then the outcome is 2. Outcome 542 represents the outcome that a first edition was found on the third attempt.

a) List the outcomes of your sample space, U.
b) List the outcomes of the event A that two books must be searched.
c) List the outcomes of the event B that at least two books must be searched.
d) List the outcomes of the event C that copy 1 is found.

12.3 Probability assignments

There are a few theories of probability that assign meaning to statements like 'the probability that A occurs is $p\%$'. In this book, we will primarily examine only the **relative frequency theory**. In essence, we will follow the idea that probability is 'the long-run proportion of repetitions on which an event occurs'. This allows us to 'merge' two concepts into one.

- Equally likely outcomes
  In the theory of equally likely outcomes, probability has to do with symmetries and the indistinguishability of outcomes. If a given experiment or trial has n possible outcomes among which there is no preference, they are equally likely. The probability of each outcome is then $\frac{100\%}{n}$ or $\frac{1}{n}$. For example, if a coin is balanced well, there is no reason for it to land heads in preference to tails when it is tossed, so,

accordingly, the probability that the coin lands heads is equal to the probability that it lands tails, and both are $\frac{100\%}{2} = 50\%$. Similarly, if a die is fair, the chance that when it is rolled it lands with the side with 1 on top is the same as the chance that it shows 2, 3, 4, 5 or 6: $\frac{100\%}{6}$ or $\frac{1}{6}$.
In the theory of equally likely outcomes, probabilities are between 0% and 100%. If an event consists of more than one possible outcome, the chance of the event is the number of ways it can occur divided by the total number of things that could occur. For example, the chance that a die lands showing an even number on top is the number of ways it could land showing an even number (2, 4 or 6) divided by the total number of things that could occur (6, namely showing 1, 2, 3, 4, 5 or 6).

- Frequency theory
  In the frequency theory, probability is the limit of the relative frequency with which an event occurs in repeated trials. Relative frequencies are always between 0% and 100%. According to the frequency theory of probability, 'the probability that A occurs is $p\%$' means that if you repeat the experiment over and over again, independently and under essentially identical conditions, the percentage of the time that A occurs will converge to p. For example, to say that the chance a coin lands heads is 50% means that if you toss the coin over and over again, independently, the ratio of the number of times the coin lands heads to the total number of tosses approaches a limiting value of 50%, as the number of tosses grows. Because the ratio of heads to tosses is always between 0% and 100%, when the probability exists it must be between 0% and 100%.

In all theories, probability is on a scale of 0% to 100%. 'Probability' and 'chance' are synonymous.

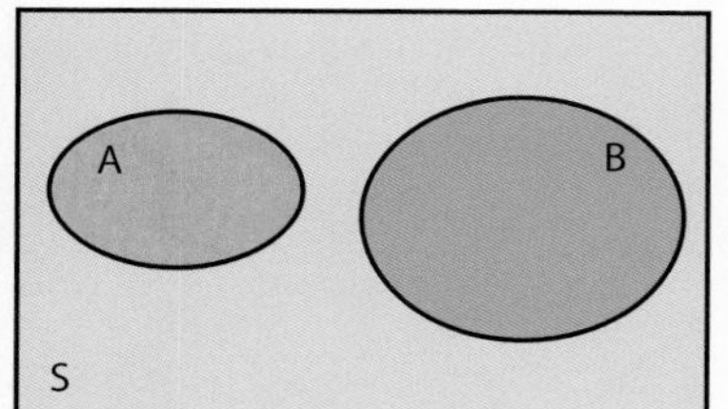

Using Venn diagrams and the 'equally likely' concept, we can say that the probability of any event is the number of elements in an event A divided by the total number of elements in the sample space **S**. This is equivalent to saying: $P(A) = \frac{n(A)}{n(S)}$, where $n(A)$ represents the number of outcomes in A and $n(S)$ represents the total number of outcomes. So, in Example 5, the probability of observing exactly two heads is: $P(2 \text{ heads}) = \frac{3}{8}$.

Probability rules

Regardless of which theory we subscribe to, the probability rules apply.

Rule 1

Any probability is a number between 0 and 1, i.e. the probability $P(A)$ of any event A satisfies $0 \leqslant P(A) \leqslant 1$. If the probability of any event is 0, the event *never* occurs. Likewise, if the probability is 1, it *always* occurs. In rolling a standard die, it is impossible to get the number 9, so $P(9) = 0$. Also, the probability of observing any integer between 1 and 6, inclusive, is 1.

No matter how little a chance you think an event has, there is no such thing as negative probability.

Rule 2

All possible outcomes together must have a probability of 1, i.e. the probability of the sample space **S** is 1: $P(S) = 1$. Informally, this is sometimes called the 'something has to happen rule'.

No matter how large a chance you think an event has, there is no such thing as a probability larger than 1!

Rule 3

If two events have no outcomes in common, the probability that one or the other occurs is the sum of their individual probabilities. Two events that have no outcomes in common, and hence can never occur together, are called **disjoint** events or **mutually exclusive** events.

$$P(A \text{ or } B) = P(A) + P(B)$$

This is the **addition rule for mutually exclusive events.**

For example, in tossing three coins, the events of getting exactly two heads or exactly two tails are disjoint, and hence the probability of getting exactly two heads or two tails is $\frac{3}{8} + \frac{3}{8} = \frac{6}{8} = \frac{3}{4}$.

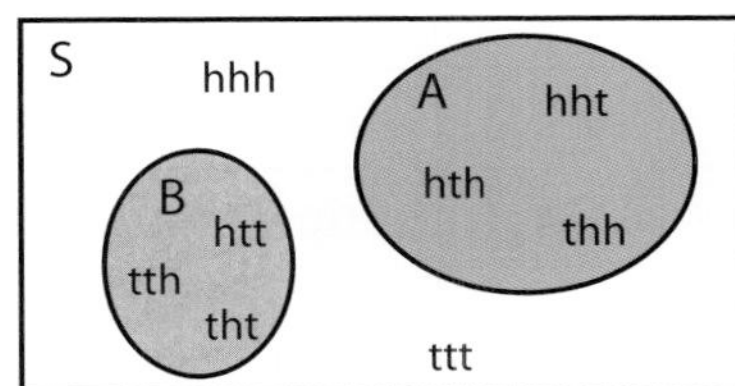

Additionally, we can always add the probabilities of **outcomes** because they are always disjoint. A trial cannot come out in two different ways at the same time. This will give you a way to check whether the probabilities you assigned are *legitimate.*

Rule 4

Suppose that the probability that you receive a 7 on your IB exam is 0.2, then the probability of *not* receiving a 7 on the exam is 0.8. The event that contains the outcomes **not in *A*** is called the **complement** of *A*, and is denoted by ***A*′**.

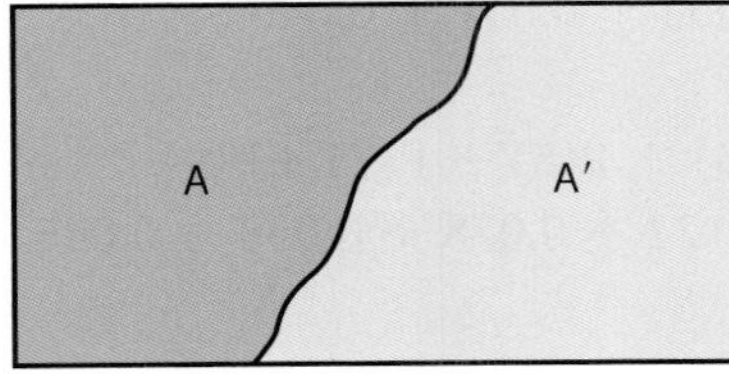

$P(A') = 1 - P(A)$, or $P(A) = 1 - P(A')$.

You have to be careful with these rules. By the 'something has to happen' rule, the total of the probabilities of all possible outcomes **must be 1**. This is so because they are disjoint, and their sum covers all the elements of the sample space. Suppose someone reports the following probabilities for students in your high school (4 years). If the probability that a grade 1, 2, 3 or 4 student is chosen at random from the high school is 0.24, 0.24, 0.25 and 0.19 respectively, with no other possibilities, you should know immediately that there is something wrong. These probabilities add up to 0.92. Similarly, if someone claims that these probabilities are 0.24, 0.28, 0.25, 0.26 respectively, there is also something wrong. These probabilities add up to 1.03, which is more than 1.

Example 7

Data for traffic violations was collected in a certain country and a summary is given below:

Age group	18–20 years	21–29 years	30–39 years	Over 40 years
Probability	0.06	0.47	0.29	0.18

What is the probability that the offender is a) in the youngest age group, b) between 21 and 40, and c) younger than 40?

Solution

Each probability is between 0 and 1, and the probabilities add up to 1. Therefore, this is a legitimate assignment of probabilities.

a) The probability that the offender is in the youngest group is 6%.

b) The probability that the driver is in the group 21 to 39 years is $0.47 + 0.29 = 0.76$.

c) The probability that a driver is younger than 40 years is $1 - 0.18 = 0.82$.

Example 8

It is a striking fact that when people create codes for their cellphones, the first digits follow distributions very similar to the following one:

First digit	0	1	2	3	4	5	6	7	8	9
Probability	0.009	0.300	0.174	0.122	0.096	0.078	0.067	0.058	0.051	0.045

a) Find the probabilities of the following three events:

$A = \{\text{first digit is } 1\}$

$B = \{\text{first digit is more than } 5\}$

$C = \{\text{first digit is an odd number}\}$

b) Find the probability that the first digit is (i)1 or greater than 5, (ii) not 1, and (iii) an odd number or a number larger than 5.

Solution

a) From the table:

$\mathrm{P}(A) = 0.300$

$$\begin{aligned}\mathrm{P}(B) &= \mathrm{P}(6) + \mathrm{P}(7) + \mathrm{P}(8) + \mathrm{P}(9) \\ &= 0.067 + 0.058 + 0.051 + 0.045 \\ &= 0.221\end{aligned}$$

$$\begin{aligned}\mathrm{P}(C) &= \mathrm{P}(1) + \mathrm{P}(3) + \mathrm{P}(5) + \mathrm{P}(7) + \mathrm{P}(9) \\ &= 0.300 + 0.122 + 0.078 + 0.058 + 0.045 \\ &= 0.603\end{aligned}$$

b) (i) Since A and B are mutually exclusive, by the addition rule, the probability that the first digit is 1 or greater than 5 is

$$\mathrm{P}(A \text{ or } B) = 0.300 + 0.\,221 = 0.521.$$

(ii) Using the complement rule, the probability that the first digit is not 1 is

$$\mathrm{P}(A') = 1 - \mathrm{P}(A) = 1 - 0.300 = 0.700.$$

(iii) The probability that the first digit is an odd number or a number larger than 5:

$$\begin{aligned}\mathrm{P}(B \text{ or } C) &= \mathrm{P}(1) + \mathrm{P}(3) + \mathrm{P}(5) + \mathrm{P}(6) + \mathrm{P}(7) + \mathrm{P}(8) + \mathrm{P}(9) \\ &= 0.300 + 0.122 + 0.078 + 0.067 + 0.058 + 0.051 \\ &\quad + 0.045 \\ &= 0.721\end{aligned}$$

• **Hint:** Notice here that P(B or C) is *not* the sum of P(B) and P(C) because B and C are not disjoint.

Equally likely outcomes

In some cases we are able to assume that individual outcomes are equally likely because of some balance in the experiment. Tossing a balanced coin renders heads or tails equally likely, with each having a probability of 50%, and rolling a standard balanced die gives the numbers from 1 to 6 as equally likely, with each having a probability of $\frac{1}{6}$.

Suppose in Example 8 we consider all the digits to be equally likely to happen, then our table would be:

First digit	0	1	2	3	4	5	6	7	8	9
Probability	0.1	0.1	0.1	0.1	0.1	0.1	0.1	0.1	0.1	0.1

$\mathrm{P}(A) = 0.1$

$\mathrm{P}(B) = \mathrm{P}(6) + \mathrm{P}(7) + \mathrm{P}(8) + \mathrm{P}(9) = 4 \times 0.1 = 0.4$

$\mathrm{P}(C) = \mathrm{P}(1) + \mathrm{P}(3) + \mathrm{P}(5) + \mathrm{P}(7) + \mathrm{P}(9) = 5 \times 0.1 = 0.5$

Also, by the complement rule, the probability that the first digit is not 1 is

$$\mathrm{P}(A') = 1 - \mathrm{P}(A) = 1 - 0.1 = 0.9.$$

2-dimensional grids are also very helpful tools that are used to visualize 2-stage or sequential probability models. For example, consider rolling a normal unbiased cubical die twice. Here are some events and how to use the grid in calculating their probabilities:

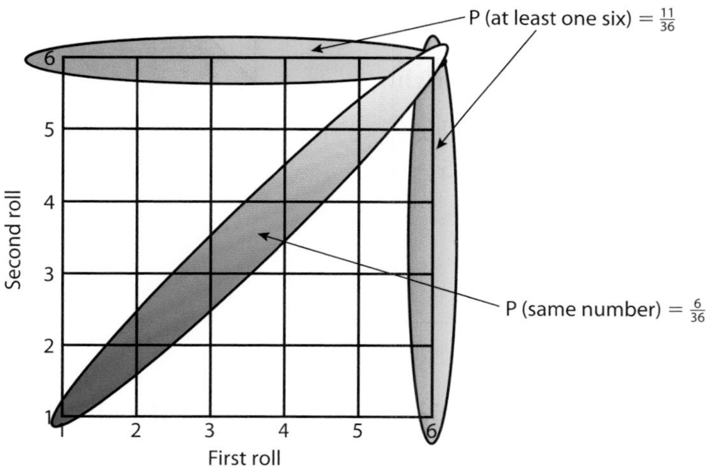

If we are interested in the probability that at least one roll shows a 6, we count the points on the column corresponding to 6 on the first roll and the points on the row corresponding to 6 on the second roll observing naturally that the point in the corner should not be counted twice.

If we are interested in the number showing on both rolls to be the same, then we count the points on the diagonal as shown.

Finally, if we are interested in the probability that the first roll shows a number larger than the second roll, then we pick the points below the diagonal.

Hence, P(first number > second number) $= \frac{15}{36}$.

Geometric probability

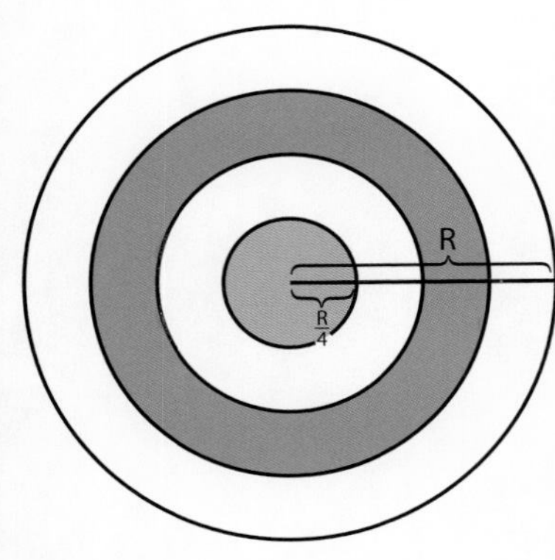

Some cases give rise to interpreting events as areas in the plane. Take for example shooting at a circular target at random. What is the probability of hitting the central part?

The probability of hitting the central part is given by

$$P = \frac{\pi\left(\frac{R}{4}\right)^2}{\pi R^2} = \frac{1}{16}.$$

Example 9

Lydia and Rania agreed to meet at the 'museum quarter' between 12:00 and 13:00. The first person to arrive will wait 15 minutes. If the second person does not show up, the first person will leave and they meet afterwards. Assuming that their arrivals are at random, what is the probability that they meet?

Solution

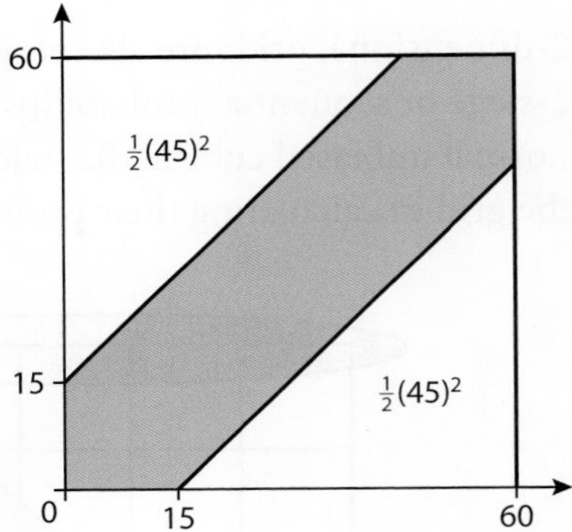

If Lydia arrives x minutes after 12:00 and Rania arrives y minutes after 12:00, then the condition for them to meet is $|x - y| \leqslant 15$, and $x \leqslant 60$, $y \leqslant 60$.

Geometrically, the outcomes of their 'encounter' region is given in the shaded region in the diagram right.

The area for each triangle is $\frac{1}{2}bh = \frac{1}{2}(45)^2$, so, the shaded area is $60^2 - 45^2$.

The probability they meet is therefore $\frac{60^2 - 45^2}{60^2} = \frac{7}{16}$.

Probability calculation for equally likely outcomes using counting principles

In an experiment where all outcomes are equally likely, the theoretical probability of an event A is given by

$$P(A) = \frac{n(A)}{n(S)}$$

where $n(A)$ is the number of outcomes that make up the event A, and $n(S)$ is the total number of outcomes in the sample space.

The new ideas we want to discuss here involve the calculation of $n(A)$ and $n(S)$. Such calculations will involve what you learned in Chapter 4 about counting principles.

Example 10

In a group of 18 students, eight are females. What is the probability of choosing five students

a) with all girls?

b) with three girls and two boys?

c) with at least one boy?

Solution

The total number of outcomes is the number of ways we can choose 5 out of the 18 students. So

$$n(S) = \binom{18}{5} = 8586.$$

a) This event will require that we pick our group from among the 8 girls. So,

$$n(A) = \binom{8}{5} = 56 \Rightarrow \mathrm{P}(A) = \frac{56}{8568} = 0.0065.$$

b) This event will require that we pick three out of the 8 girls, and at the same time, we pick 2 out of the 10 boys. So, using the multiplication principle,

$$n(B) = \binom{8}{3} \cdot \binom{10}{2} = 56.45 = 2520 \Rightarrow \mathrm{P}(B) = \frac{2520}{8568} = 0.294.$$

Note: Did you observe that $\binom{8}{5} = \binom{8}{3}$? Why?

c) This event can be approached in two ways:

- To have at least 1 boy means that we can have 1, 2, 3, 4 or 5 boys. These are mutually exclusive, so the probability in question is the sum

$$\mathrm{P}(C) = \frac{\binom{10}{1}\binom{8}{4} + \binom{10}{2}\binom{8}{3} + \binom{10}{3}\binom{8}{2} + \binom{10}{4}\binom{8}{1} + \binom{10}{5}\binom{8}{0}}{\binom{18}{5}} = \frac{8512}{8568} = 0.9935$$

- To recognize that at least 1 boy is the complement of no boys at all, i.e. 0 boys or all 5 girls.

$$\mathrm{P}(C) = 1 - \mathrm{P}(A) = 1 - 0.0065 = 0.9935.$$

Example 11

A deck of playing cards has 52 cards. In a game, the player is given five cards. Find the probability of the player having

a) three cards of one denomination and two cards of another (three 7s and two Js for example).

This game can be played at two stages, First, the player is given five cards, and then he/she can decide to exchange some of the cards. (The cards exchanged are discarded and not returned to the deck!)

A player was given the following hand: Q♠, Q♦, Q♥, 4♣, 9♠. She decided to change the last two cards. Find the probability of the player having

b) three cards of one denomination and two cards of another

c) four queens.

Solution

a) The sample space consists of all possible 5-card hands that can be given out:

$$n(S) = \binom{52}{5} = 2\,598\,960$$

Call the event of interest A.

As there are 13 denominations in the deck of cards then there are 13 choices for the first required denomination. Once a denomination is chosen, say 9,
then there are $\binom{4}{3}$ ways of choosing 3 cards out of the four. Using the multiplication rule, there are $13 \cdot \binom{4}{3}$ ways of choosing 3 cards of the first denomination. We are now left with 12 possible denominations for the second one, each can give us $\binom{4}{2}$ ways of getting two of the cards, and hence using the multiplication rule, there are $12 \cdot \binom{4}{2}$ ways of choosing the cards for the second denomination. Again using the multiplication rule we will have $\left[13 \cdot \binom{4}{3}\right]\left[12 \cdot \binom{4}{2}\right] = 3744$ ways of choosing the first and second denominations.

The requested probability is then

$$P(A) = \frac{3744}{2\,598\,960} \approx 0.001\,44.$$

b) Since we have 3 queens, then we need only look for 2 cards of a different denomination. Now, there are only 47 cards left in the deck because we had 5 already. So the sample space has $n(S) = \binom{47}{2} = 1081$ ways of getting the rest of the 5 cards. The other cards could be two 4's, two 9's or two of the rest of the 10 denominations.

We have $\binom{3}{2} = 3$ ways of getting two 4's since 4♣ is already discarded. We also have $\binom{3}{2} = 3$ ways to get two 9's. Or, for each of the other 10 denominations (no Q, no 4 and no 9), we have $\binom{4}{2} = 6$ different ways of getting two of them, i.e. we have $10 \cdot \binom{4}{2} = 60$ different ways of getting two cards of the same denomination other than Q, 4 or 9.

So, the total number of ways of getting two cards of the same denomination is $3 + 3 + 60 = 66$ ways.

So, the required probability is $P(A) = \dfrac{n(A)}{n(S)} = \dfrac{66}{1081} \approx 0.0611.$

c) To have 4 Q's we only have to look for one, and there is only one way of getting the missing Q♣. That leaves us with one card to be chosen from the 46 cards left. 46 ways!

Therefore, $P(A) = \frac{46}{1081} \approx 0.0426.$

Exercise 12.3

1 In a simple experiment, chips with integers 1–20 inclusive were placed in a box and one chip was picked at random.

a) What is the probability that the number drawn is a multiple of 3?

b) What is the probability that the number drawn is not a multiple of 4?

2 The probability an event *A* happens is 0.37.

a) What is the probability that it does not happen?

b) What is the probability that it may or may not happen?

3 You are playing with an ordinary deck of 52 cards by drawing cards at random and looking at them.

a) Find the probability that the card you draw is

(i) the ace of hearts

(ii) the ace of hearts or any spade

(iii) an ace or any heart

(iv) not a face card.

b) Now you draw the ten of diamonds, put it on the table and draw a second card. What is the probability that the second card is

(i) the ace of hearts?

(ii) not a face card?

c) Now you draw the ten of diamonds, return it to the deck and draw a second card. What is the probability that the second card is

(i) the ace of hearts?

(ii) not a face card?

4 On Monday morning, my class wanted to know how many hours students spent studying on Sunday night. They stopped schoolmates at random as they arrived and asked each, 'How many hours did you study last night?' Here are the answers of the sample they chose on Monday, 14 January, 2008.

Number of hours	0	1	2	3	4	5
Number of students	4	12	8	3	2	1

a) Find the probability that a student spent less than three hours studying Sunday night.

b) Find the probability that a student studied for two or three hours.

c) Find the probability that a student studied less than six hours.

5 We throw a coin and a standard six-sided die and we record the number and the face that appear. Find

a) the probability of having a number larger than 3

b) the probability that we receive a head and a 6.

6 A die is constructed in a way that a 1 has the chance to occur twice as often as any other number.

a) Find the probability that a 5 appears.

b) Find the probability an odd number will occur.

7 You are given two fair dice to roll in an experiment.

a) Your first task is to report the numbers you observe.
   (i) What is the sample space of your experiment?
   (ii) What is the probability that the two numbers are the same?
   (iii) What is the probability that the two numbers differ by 2?
   (iv) What is the probability that the two numbers are not the same?

b) In a second stage, your task is to report the sum of the numbers that appear.
   (i) What is the probability that the sum is 1?
   (ii) What is the probability that the sum is 9?
   (iii) What is the probability that the sum is 8?
   (iv) What is the probability that the sum is 13?

8 The blood types of people can be one of four types: O, A, B or AB. The distribution of people with these types differs from one group of people to another. Here are the distributions of blood types for randomly chosen people in the US, China and Russia.

Country \ Blood type	O	A	B	AB
US	0.43	0.41	0.12	?
China	0.36	0.27	0.26	0.11
Russia	0.39	0.34	?	0.09

a) What is the probability of type AB in the US?

b) Dirk lives in the US and has type B blood. What is the probability that a randomly chosen US citizen can donate blood to Dirk? (Type B can only receive from O and B.)

c) What is the probability of randomly choosing an American and a Chinese (independently) with type O blood?

d) What is the probability of randomly choosing an American, a Chinese and a Russian (independently) with type O blood?

e) What is the probability of randomly choosing an American, a Chinese and a Russian (independently) with the same blood type?

9 In each of the following situations, state whether or not the given assignment of probabilities to individual outcomes is legitimate. Give reasons for your answer.

a) A die is loaded such that the probability of each face is according to the following assignment (x is the number of spots on the upper face and $P(x)$ is its probability.)

x	1	2	3	4	5	6
$P(x)$	0	$\frac{1}{6}$	$\frac{1}{3}$	$\frac{1}{3}$	$\frac{1}{6}$	0

b) A student at your school categorized in terms of gender and whether they are diploma candidates or not.
P(female, diploma candidate) = 0.57, P(female, not a diploma candidate) = 0.23,
P(male, diploma candidate) = 0.43, P(male, not a diploma candidate) = 0.18.

c) Draw a card from a deck of 52 cards (x is the suit of the card and $P(x)$ is its probability).

x	Hearts	Spades	Diamonds	Clubs
$P(x)$	$\frac{12}{52}$	$\frac{15}{52}$	$\frac{12}{52}$	$\frac{13}{52}$

10 In Switzerland, there are three 'official' mother tongues, German, French and Italian. You choose a Swiss at random and ask, 'What is your mother tongue?' Here is the distribution of responses:

Language	German	French	Italian	Other
Probability	0.58	0.24	0.12	?

a) What is the probability that a Swiss person's mother tongue is not one of the official ones?

b) What is the probability that a Swiss person's mother tongue is not German?

c) What is the probability that you choose two Swiss independent of each other and they both have German mother tongue?

d) What is the probability that you choose two Swiss independent of each other and they both have the same mother tongue?

11 The majority of email messages are now 'spam.' Choose a spam email message at random. Here is the distribution of topics:

Topic	Adult	Financial	Health	Leisure	Products	Scams
Probability	0.165	0.142	0.075	0.081	0.209	0.145

a) What is the probability of choosing a spam message that does not concern these topics?

Parents are usually concerned with spam messages with 'adult' content and scams.

b) What is the probability that a randomly chosen spam email falls into one of the other categories?

12 Consider n to be a positive integer. Let $f(x) = \dfrac{\binom{n}{x+1}}{\binom{n}{x}}$, where x is also a positive integer. Determine the values of x (in terms of n) for which $f(x) < 1$.

13 Determine $\boldsymbol{n}$ in each of the following cases:

a) $\binom{n}{2} = 190$ b) $\binom{n}{4} = \binom{n}{8}$

14 An experiment involves rolling a pair of dice, 1 white and 1 red, and recording the numbers that come up. Find the probability

a) that the sum is greater than 8

b) that a number greater than 4 appears on the white die

c) that at most a total of 5 appears.

15 Three books are picked from a shelf containing 5 novels, 3 science books and a thesaurus. What is the probability that

a) the thesaurus is selected?

b) two novels and a science book are selected?

16 Five cards are chosen at random from a deck of 52 cards. Find the probability that the set contains

a) 3 kings

b) 4 hearts and 1 diamond.

17 A class consists of 10 girls and 12 boys. A team of 6 members is to be chosen at random. What is the probability that the team contains

a) one boy?

b) more boys than girls?

18 A committee of six people is to be chosen from a group of 15 people that contains two married couples.

a) What is the probability that the committee will include both married couples?

b) What is the probability that the committee will include the three youngest members in the group?

19 The computer department at your school has received a shipment of 25 printers, of which 10 are colour laser printers and the rest are black-and-white laser models. Six printers are selected at random to be checked for defects. What is the probability that

a) exactly 3 of them are colour lasers?

b) at least 3 are colour lasers?

20 The city of Graz just bought 30 buses to put in service in their public transport network. Shortly after the first week in service, 10 buses developed cracks on their instruments panel.

a) How many ways are there to select a sample of 6 buses for a thorough inspection?

b) What is the probability that half of the chosen buses have cracks?

c) What is the probability that at least half of the chosen buses have cracks?

d) What is the probability that at most half of the chosen buses have cracks?

21 A small factory has a 24-hour production facility. They employ 30 workers on the day shift (08:00–16:00), 22 workers on the evening shift (16:00–24:00) and 15 workers on the morning shift (00:00–08:00). A quality control consultant is to select 9 workers for in-depth interviews.

a) What is the probability that all 9 come from the day shift?

b) What is the probability that all 9 come from the same shift?

c) What is the probability that at least two of the shifts are represented?

d) What is the probability that at least one of the shifts is unrepresented?

22 a) A box contains 8 chips numbered 1 to 8. Two are chosen at random and their numbers are added together. What is the probability that their sum is 7?

b) A box contains 20 chips numbered 1 to 20. Two are chosen at random. What is the probability that the numbers on the two chips differ by 3?

c) A box contains 20 chips numbered 1 to 20. Two are chosen at random. What is the probability that the numbers on the two chips differ by more than 3?

23 Tim and Val want to meet for dinner at their favourite restaurant. They reserved a table for between 20:00 and 22:00. The first person to arrive will order a salad and spend 30 minutes before ordering the main meal. If the second person does not arrive within the 30 minutes, the first person will pay the bill and leave. What is the probability that they manage to eat dinner together at the restaurant?

24 Bus 48A and Tram 49 serve different routes in the city. They share one stop next to my house. They stop at this station every 20 minutes. Every stay is 3 minutes long. Assuming their arrivals at the hour are random, find the probability that both are at the stop at 12:00 on Monday.

25 During a dinner party, Magda plans on opening six bottles of wine. Her supply includes 8 French, 10 Australian and 12 Italian wines. She sends her sister Mara to choose the bottles. Mara has no knowledge of the wine types and picks the bottles at random.

a) What is the probability that two of each type get selected?

b) What is the probability that all served bottles are of the same type?

c) What is the probability of serving only Italian and French wines?

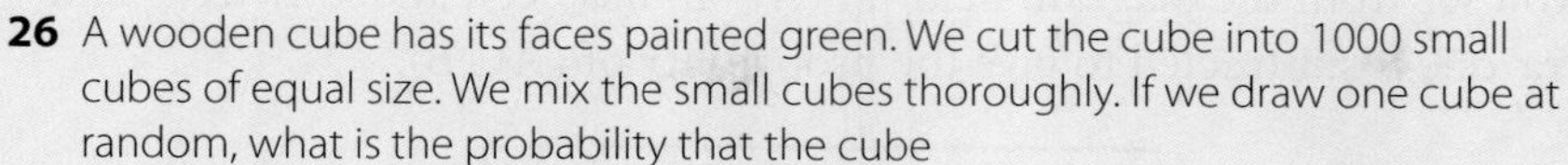

26 A wooden cube has its faces painted green. We cut the cube into 1000 small cubes of equal size. We mix the small cubes thoroughly. If we draw one cube at random, what is the probability that the cube

a) has two faces coloured green?

b) has three coloured faces?

c) does not have a coloured face at all?

12.4 Operations with events

In Example 8, we talked about the following events:

$B = \{\text{first digit is more than } 5\}$
$C = \{\text{first digit is an odd number}\}$

We also claimed that these two events are not disjoint. This brings us to another concept for looking at combined events.

The **intersection** of two events *B* and *C*, denoted by the symbol $B \cap C$ or simply *BC*, is the event containing all outcomes common to *B* and *C*.

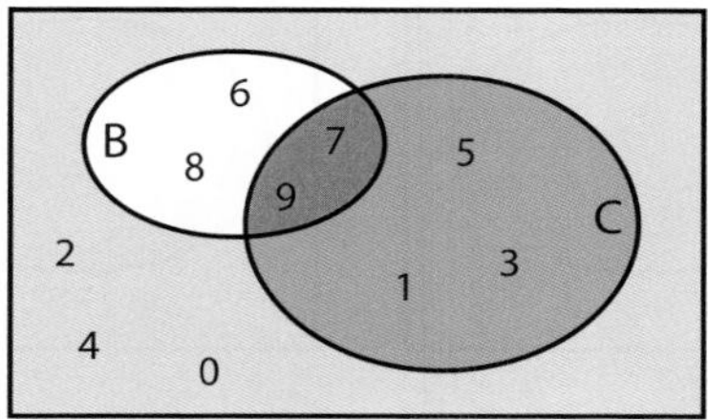

Here $B \cap C = \{7, 9\}$ because these outcomes are in both *B* and *C*. Since the intersection has outcomes common to the two events *B* and *C*, they are not mutually exclusive.

The probability of $B \cap C$ is $0.058 + 0.045 = 0.103$. Recall from Example 8 that we said that the probability of *B* or *C* is not simply the sum of the probabilities. That brings us to the next concept. How can we find the probability of *B* or *C* when they are not mutually exclusive? To answer this question, we need to define another operation.

The **union** of two events *B* and *C*, denoted by the symbol $B \cup C$, is the event containing all the outcomes that belong to *B* or to *C* or to both.

Here $B \cup C = \{1, 3, 5, 6, 7, 8, 9\}$. In calculating the probability of $B \cup C$, we observe that the outcomes 7 and 9 are counted twice. To remedy the situation, if we decide to add the probabilities of *B* and *C*, we subtract one of the incidents of double counting. So, $P(B \cup C) = 0.221 + 0.603 - 0.103 = 0.721$, which is the result we received with direct calculation. In general, we can state the following probability rule:

Rule 5

For any two events *A* and *B*, $P(A \cup B) = P(A) + P(B) - P(A \cap B)$.

As you see from the diagram below, $P(A \cap B)$ has been added twice, so the 'extra' one is subtracted to give the probability of $(A \cup B)$.

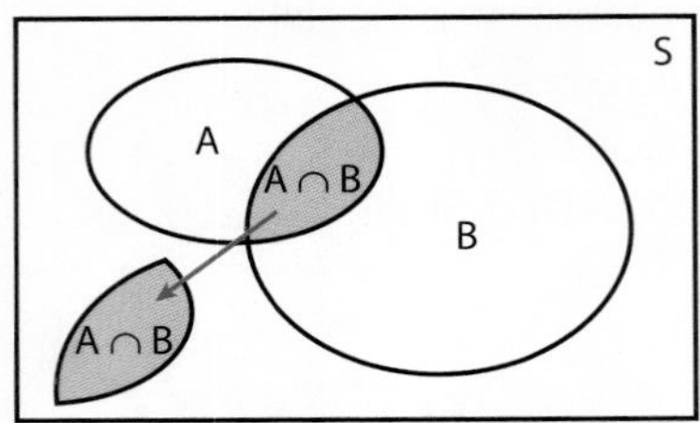

This general probability addition rule applies to the case of mutually exclusive events too. Consider any two events A and B. The probability of A or B is given by

$$\begin{aligned} P(A \cup B) &= P(A) + P(B) - P(A \cap B) \\ &= P(A) + P(B), \text{ since } P(A \cap B) = 0. \end{aligned}$$

Some useful results

1 $P(A \cap B) = P(B \cap A)$

2 $P(A) = P(A \cap B) + P(A \cap B')$

3 $P(B) = P(A \cap B) + P(A' \cap B)$

4 $P(A' \cap B') = 1 - P(A \cup B)$

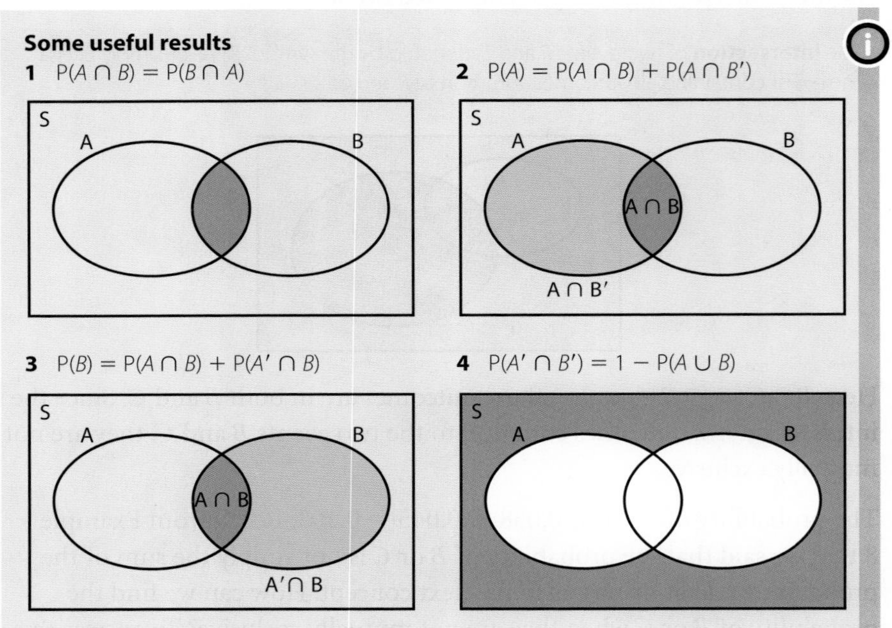

Rule 6

The simple multiplication rule.

Consider the following situation: In a large school, 55% of the students are male. It is also known that the percentage of smokers among males and females in this school is the same, 22%. What is the probability of selecting a student at random from this population and the student is a male smoker?

Applying common sense only, we can think of the problem in the following manner. Since the proportion of smokers is the same in both groups, smoking and gender are independent of each other in the sense that knowing that the student is a male does not influence the probability that he smokes!

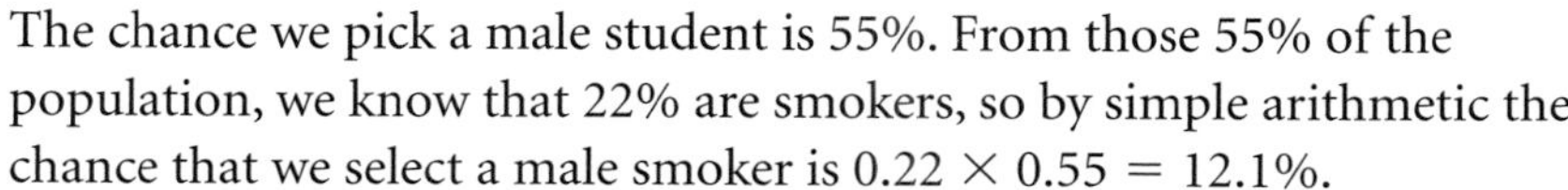

The chance we pick a male student is 55%. From those 55% of the population, we know that 22% are smokers, so by simple arithmetic the chance that we select a male smoker is $0.22 \times 0.55 = 12.1\%$.

This is an example of the multiplication rule for independent events.

Two events A and B are **independent** if knowing that one of them occurs does not change the probability that the other occurs.

The **multiplication rule for independent events:** If two events A and B are independent, then $P(A \cap B) = P(A) \times P(B)$.

Example 12

Reconsider the situation with the traffic light at the beginning of this chapter. The probability that I find the light green is 30%. What is the probability that I find it green on two consecutive days?

Solution

We will assume that my arrival and finding the light green is a random event, and that if it turns green on one day it does not influence how it turns the next day. In that case our calculation is very simple:

$$\begin{aligned} \text{P(green the first and second day)} &= \text{P(green first day)} \times \text{P(green second day)} \\ &= 0.30 \times 0.30 = 0.09. \end{aligned}$$

Do not confuse independent with disjoint. 'Disjoint' means that if one of the events occurs then the other does not occur; while 'independent' means that knowing one of the events occurs does not influence the probability of whether the other occurs or not!

This rule can also be extended to more than two independent events. For example, on the assumption of independence, what is the chance that I find the light green five days of the week?

$$\text{P(green on five days)} = 0.3 \times 0.3 \times 0.3 \times 0.3 \times 0.3 = 0.002\,43$$

Example 13

Computers bought from a well-known producer require repairs quite frequently. It is estimated that 17% of computers bought from the company require one repair job during the first month of purchase, 7% will need repairs twice during the first month, and 4% require three or more repairs.

a) What is the probability that a computer chosen at random from this producer will need
   - (i) no repairs?
   - (ii) no more than one repair?
   - (iii) some repair?

b) If you buy two such computers, what is the probability that
   - (i) neither will require repair?
   - (ii) both will need repair?

Solution

a) Since all of the events listed are disjoint, the addition rule can be used.

(i) P(no repairs) = 1 − P(some repairs) = 1 − (0.17 + 0.07 + 0.04)
= 1 − (0.28) = 0.72

(ii) P(no more than one repair) = P(no repairs or one repair)
= 0.72 + 0.17 = 0.89

(iii) P(some repairs) = P(one or two or three or more repairs)
= 0.17 + 0.07 + 0.04 = 0.28

b) Since repairs on the two computers are independent from one another, the multiplication rule can be used. Use the probabilities of events from part a) in the calculations.

(i) P(neither will need repair) = (0.72)(0.72) = 0.5184

(ii) P(both will need repair) = (0.28)(0.28) = 0.0784

Conditional probability

In probability, conditioning means incorporating new restrictions on the outcome of an experiment: updating probabilities to take into account new information. This section describes conditioning, and how conditional probability can be used to solve complicated problems. Let us start with an example.

Example 14

A public health department wanted to study the smoking behaviour of high school students. They interviewed 768 students from grades 10–12 and asked them about their smoking habits. They categorized the students into three categories: smokers (more than 1 pack of 20 cigarettes per week), occasional smokers (less than 1 pack per week), and non-smokers. The results are summarized below:

	Smoker	Occasional	Non-smoker	Total
Male	127	73	214	414
Female	99	66	189	354
Total	226	139	403	768

If we select a student at random from this study, what is the probability that we select a) a girl, b) a male smoker, and c) a non-smoker?

Solution

a) $\text{P(female)} = \frac{354}{768} = 0.461$

So, 46.1% of our sample are females.

b) Since we have 127 boys categorized as smokers, the chance of a male smoker will be

$\text{P(male smoker)} = \frac{127}{768} = 0.165.$

c) $P(\text{non-smoker}) = \frac{403}{768} = 0.525$

In the above example, what if we know that the selected student is a girl? Does that influence the probability that the selected student is a non-smoker? Yes, it does!

Knowing that the selected student is a female changes our choices. The 'revised' sample space is not made up of all students anymore. It is only the female students. The chance of finding a non-smoker among the females is $\frac{189}{354} = 0.534$, i.e. 53.4% of the females are non-smokers as compared to the 52.5% of non-smokers in the whole population.

This probability is called a conditional probability, and we write this as $P(\text{non-smoker} \mid \text{female}) = \frac{189}{354}$.

We read this as, '*Probability of selecting a non-smoker* **given that** *we have selected a female*'.

The conditional probability of A given B, $P(A \mid B)$, is the probability of the event A, updated on the basis of the knowledge that the event B occurred. Suppose that A is an event with probability $P(A) = p \neq 0$, and that $A \cap B = \emptyset$ (A and B are disjoint). Then if we learn that B occurred we know A did not occur, so we should revise the probability of A to be zero, $P(A \mid B) = 0$ (the conditional probability of A given B is zero).

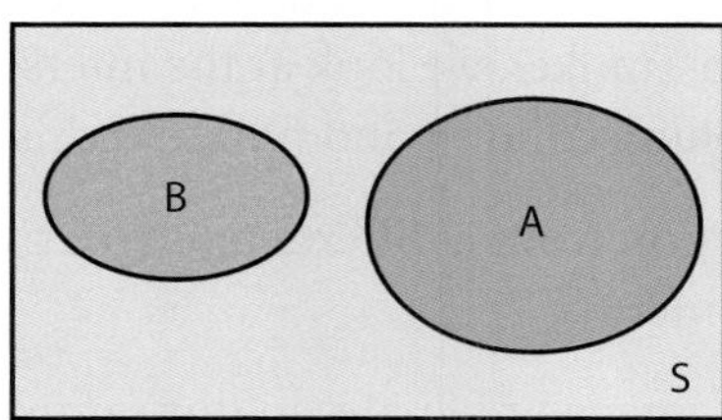

On the other hand, suppose that $A \cap B = B$ (B is a subset of A, so B implies A). Then if we learn that B occurred we know A must have occurred as well, so we should revise the probability of A to be 100%, $P(A \mid B) = 1$(the conditional probability of A given B is 100%).

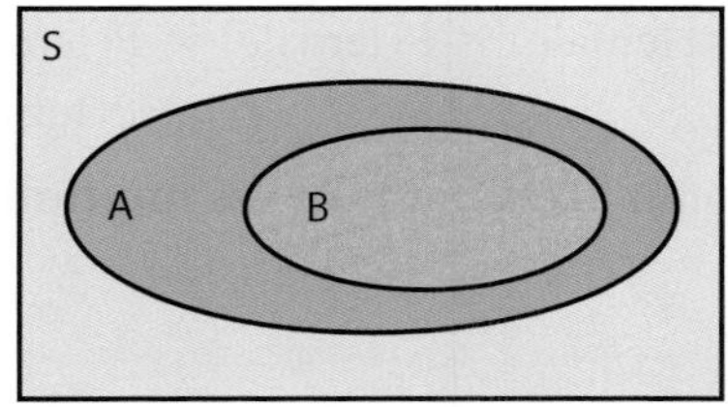

Remember that the probability we assign to an event can change if we know that some other event has occurred. This idea is the key to understanding conditional probability.

Imagine the following scenario:
You are playing cards and your opponent is about to give you a card. What is the probability that the card you receive is a queen?

As you know, there are 52 cards in the deck, 4 of these cards are queens. So, assuming that the deck was thoroughly shuffled, the probability of receiving a queen is

$$P(\text{queen}) = \frac{4}{52} = \frac{1}{13}.$$

This calculation assumes that you know nothing about any cards already dealt from the deck.

Suppose now that you are looking at the five cards you have in your hand, and one of them is a queen. You know nothing about the other 47 cards except that exactly three queens are among them. The probability of being given a queen as the next card, given what you know, is

$$P(\text{queen} \mid 1 \text{ queen in hand}) = \frac{3}{47} \neq \frac{1}{13}.$$

So, knowing that there is one queen among your five cards changes the probability of the next card being a queen.

Consider Example 14 again. We want to express the table frequencies as relative frequencies or probabilities. Our table will look like this:

	Smoker	Occasional	Non-smoker
Male	0.165	0.095	0.279
Female	0.129	0.086	**0.246**

To find the probability of selecting a student at random and finding that student is a female non-smoker, we look at the intersection of the female row with the non-smoking column and find that this probability is 0.246.

Looking at this calculation from a different perspective, we can think about it in the following manner:

We know that the percentage of females in our sample is 46.1, and among those females, in Example 14, we found that 53.4% of those are non-smokers. So, the percentage of female non-smokers in the population is the 53.4% of those 46.1% females, i.e. $0.534 \times 0.461 = 0.246$.

In terms of events, this can be read as:

$$\begin{aligned} P(\text{non-smoker} \mid \text{female}) \times P(\text{female}) &= P(\text{female and non-smoker}) \\ &= P(\text{female} \cap \text{non-smoker}). \end{aligned}$$

The previous discussion is an example of the **multiplication rule** of any two events A and B.

Multiplication rule

Given any events A and B, the probability that both events happen is given by

$$P(A \cap B) = P(A|B) \times P(B).$$

Example 15

In a psychology lab, researchers are studying the colour preferences of young children. Six green toys and four red toys (identical apart from colour) are placed in a container. The child is asked to select two toys at random. What is the probability that the child chooses two red toys?

Solution

To solve this problem, we will use a tree diagram.

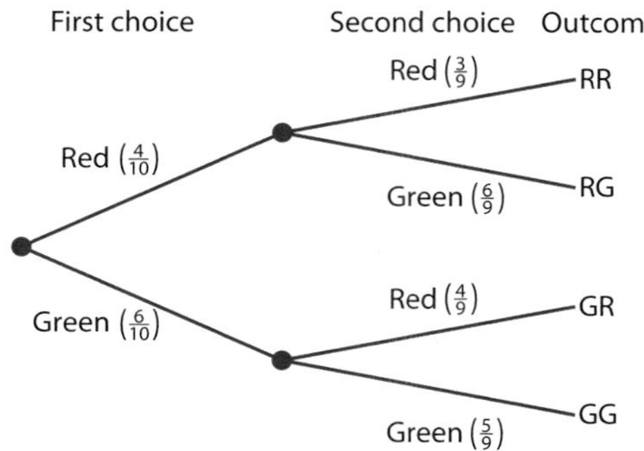

As you notice, every entry on the 'branches' has a conditional probability. So, Red on the second choice is actually either Red|Red or Red|Green. We are interested in RR, so the probability is

$$P(RR) = P(R) \times P(R|R) = \frac{4}{10} \times \frac{3}{9} = 13.3\%.$$

If $P(A \cap B) = P(A|B) \times P(B)$, as discussed above, and if $P(B) \neq 0$, we can rearrange the multiplication rule to produce a definition of the conditional probability $P(A|B)$ in terms of the 'unconditional' probabilities $P(A \cap B)$ and $P(B)$.

> When $P(B) \neq 0$, the **conditional probability** of A given B is $P(A|B) = \dfrac{P(A \cap B)}{P(B)}$.

Why does this formula make sense?

First of all, note that it does agree with the intuitive answers we found above. If $A \cap B = \emptyset$, $P(A \cap B) = 0$, so $P(A|B) = 0/P(B) = 0$;

and if $A \cap B = B$, $P(A|B) = P(B)/P(B) = 100\%$.

Now, if we learn that B occurred, we can restrict attention to just those outcomes that are in B, and disregard the rest of S, so we have a new sample space that is just B (see diagram below).

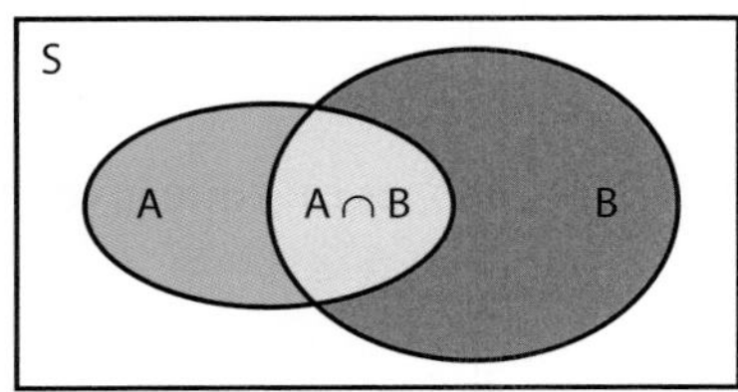

For A to have occurred in addition to B, requires that $A \cap B$ occurred, so the conditional probability of A given B is $P(A \cap B)/P(B)$, just as we defined it above.

Example 16

In an experiment to study the phenomenon of colour blindness, researchers collected information concerning 1000 people in a small town and categorized them according to colour blindness and gender. Here is a summary of the findings:

	Male	Female	Total
Colour-blind	40	2	42
Not colour-blind	470	488	958
Total	510	490	1000

What is the probability that a person is colour-blind given that the person is a woman?

Solution

To answer this question, we notice that we do not have to search the whole population for this event. We limit our search to the women. We have 490 women. As we only need to consider women, then when we search for colour blindness, we only look for the women who are colour-blind, i.e. the intersection. Here we only have two women. Therefore, the chance we get a colour-blind person given the person is a woman is

$$\mathrm{P}(C|W) = \frac{\mathrm{P}(C \cap W)}{\mathrm{P}(W)} = \frac{n(C \cap W)}{n(W)} = 0.004,$$ where C is for colour-blind and W for woman.

Notice here that we used the frequency rather than the probability. However, these are equivalent since dividing by $n(S)$ will transform the frequency into a probability.

$$\frac{n(C \cap W)}{n(W)} = \frac{\frac{n(C \cap W)}{n(S)}}{\frac{n(W)}{n(S)}} = \frac{\mathrm{P}(C \cap W)}{\mathrm{P}(W)} = \mathrm{P}(C|W)$$

Example 17

AUA, a national airline, is known for its punctuality. The probability that a regularly scheduled flight departs on time is $\mathrm{P}(D) = 0.83$, the probability that it arrives on time is $\mathrm{P}(A) = 0.92$, and the probability that it arrives and departs on time, $\mathrm{P}(A \cap D) = 0.78$. Find the probability that a flight

a) arrives on time given that it departed on time

b) departs on time given that it arrived on time.

Solution

a) The probability that a flight arrives on time given that it departed on time is

$$\mathrm{P}(A|D) = \frac{\mathrm{P}(A \cap D)}{\mathrm{P}(D)} = \frac{0.78}{0.83} = 0.94.$$

b) The probability that a flight departs on time given that it arrived on time

$$\mathrm{P}(D|A) = \frac{\mathrm{P}(D \cap A)}{\mathrm{P}(A)} = \frac{0.78}{0.92} = 0.85.$$

Independence

Two events are **independent** if learning that one occurred does not affect the chance that the other occurred. That is, if $P(A|B) = P(A)$, and vice versa.

This means that if we apply our definition to the general multiplication rule, then

$$P(A \cap B) = P(A|B) \times P(B) = P(A) \times P(B)$$

which is the multiplication rule for independent events we studied earlier.

These results give us some helpful tools in checking the independence of events.

> Two events are **independent** if and only if either $P(A \cap B) = P(A) \times P(B)$, or $P(A|B) = P(A)$. Otherwise, the events are **dependent**.

Example 18

Take another look at the AUA situation in Example 17. Are the events of arriving on time (A) and departing on time (D) independent?

Solution

We can answer this question in two different ways:

a) $P(A) = 0.92$ and we found that $P(A|D) = 0.94$. Since the two values are not the same, we can say that the two events are not independent.

b) Alternately, $P(A \cap D) = 0.78$ and
$P(A) \times P(D) = 0.92 \times 0.83 = 0.76 \neq P(A \cap D)$.

Example 19

In many countries, the police stop drivers on suspicion of drunk driving. The stopped drivers are given a breath test, a blood test or both. In a country where this problem is vigorously dealt with, the police records show the following:

81% of the drivers stopped are given a breath test, 40% a blood test, and 25% both tests.

a) What is the probability that a suspected driver is given
   (i) a test?
   (ii) exactly one test?
   (iii) no test?

b) Are giving the two tests independent?

Solution

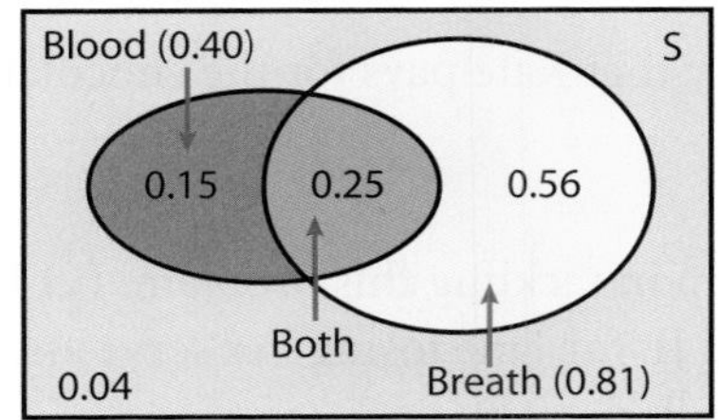

A Venn diagram can help explain the solution.

a) (i) The probability that a driver receives a test means that he/she receives either a blood test, a breath test or both tests. The probability as such can be calculated directly from the diagram, or by applying the addition rule. The diagram shows that if 81% receive the breath test and 25% are also given the blood test, then 56% do not receive a blood test. Similarly, 15% of the blood test receivers do not get a breath test. So, the probability of receiving a test is $0.56 + 0.25 + 0.15 = 0.96$.

Also, if we apply the addition rule,

$$\begin{aligned}\text{P(breath or blood)} &= \text{P(breath)} + \text{P(blood)} - \text{P(both)}\\ &= 0.81 + 0.40 - 0.25 = 0.96.\end{aligned}$$

(ii) To receive exactly one test is to receive a blood test or a breath test, but not both! So, from the Venn diagram it is clear that this probability is $0.15 + 0.56 = 0.71$. To approach it differently, since we know that the union of the two events still contains the intersection, we can subtract the probability of the intersection from that of the union: $0.96 - 0.25 = 0.71$.

(iii) To receive no test is equivalent to the complement of the union of the events. Hence, $\text{P(no test)} = 1 - \text{P(1 test)} = 1 - 0.96 = 0.04$.

b) To check for independence, we can use any of the two methods we tried before. Since all the necessary probabilities are given, we can use the product rule. If they were independent, then

$\text{P(both tests)} = \text{P(breath)} \times \text{P(blood)} = 0.81 \times 0.40 = 0.324$, but $\text{P(both tests)} = 0.25$. Therefore, the events of receiving a breath test and a blood test are not independent.

Example 20

Jane and Kate frequently play tennis with each other. When Jane serves first, she wins 60% of the time, and the same pattern occurs with Kate. They alternate the serve of course. They usually play for a prize, which is a chocolate bar. The first one who loses on her serve will have to buy the chocolate. Jane serves first.

a) Find the probability that Jane pays on her second serve.

b) Find the probability that Jane eventually pays for the chocolate.

c) Find the probability that Kate pays for the chocolate.

Solution

A tree diagram can help in tackling this problem. Let JW stand for Jane winning her serve, and JL for Jane losing her serve and hence paying. KW and KL are defined similarly.

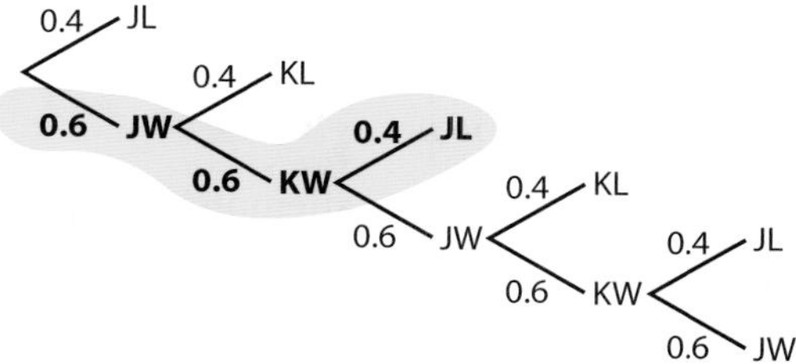

a) For Jane to pay on her second serve, she should win her first serve, Kate must also win her first serve, and then Jane loses her second serve. See diagram above. The probability this happens is

$$P(JW) \cdot P(KW) \cdot P(JL) = 0.6 \times 0.6 \times 0.4 = 0.4 \cdot (0.6)^2 = 0.144.$$

b) For Jane to pay, she needs to be the first one to lose on her serve. This means she loses on the first serve or the second or the third, and so on. So, the probability that she pays is

$$\begin{aligned} P(\text{J pays}) &= P(JL) + P(JW) \cdot P(KW) \cdot P(JL) \\ &\quad + P(JW) \cdot P(KW) \cdot P(JW) \cdot P(KW) \cdot P(JL) \\ &= 0.4 + 0.4 \cdot (0.6)^2 + 0.4 \cdot (0.6)^4 + \ldots \end{aligned}$$

This appears to be the sum of an infinite geometric series with $(0.6)^2$ as a common ratio; hence,

$$P(\text{J pays}) = 0.4 + 0.4 \cdot (0.6)^2 + 0.4 \cdot (0.6)^4 + \ldots = \frac{0.4}{1 - (0.6)^2} = 0.625.$$

c) Obviously,

$$P(\text{K pays}) = 1 - P(\text{J pays}) = 1 - 0.625 = 0.375.$$

This also gives us the opportunity to look at it differently. For Kate to pay, she needs to lose on her first serve, i.e. 0.6×0.4, or on her second, third, etc.

$$\begin{aligned} P(\text{K pays}) &= 0.6 \times 0.4 + 0.6 \times (0.6)^2 \times 0.4 + 0.6 \times (0.6)^4 \times 0.4 + \ldots \\ &= \frac{0.6 \times 0.4}{1 - (0.6)^2} = 0.375 \end{aligned}$$

Example 21

A target for a dart game is shown here. The radius of the board is 40 cm and it is divided into three regions as shown. You score 2 points if you hit the centre, 1 point for the middle region and 0 points for the outer region.

a) What is the probability of scoring a 1 in one attempt?

b) What is the probability of scoring a 2 in one attempt?

c) How many attempts are necessary so that the probability of scoring at least one 2 is at least 50%?

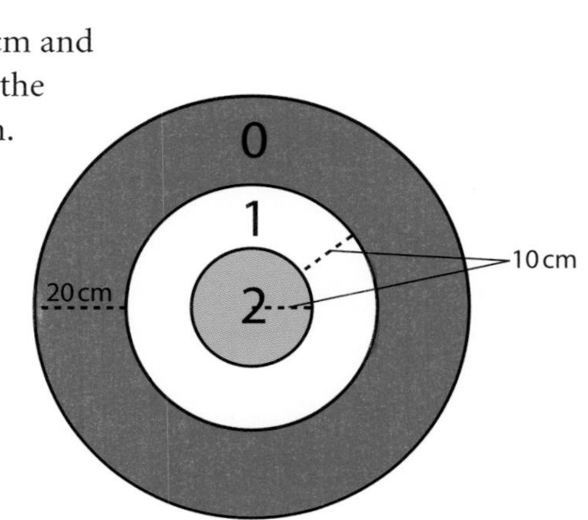

Solution

a) $P(1) = \dfrac{\pi(20^2 - 10^2)}{\pi(40^2)} = \dfrac{3}{16}$

b) $P(2) = \dfrac{\pi(10^2)}{\pi(40^2)} = \dfrac{1}{16}$

c) Let the number of attempts be n.

P(at least one 2) $= 1 - $ P(no 2 in n attempts) $= 1 - \left(\frac{15}{16}\right)^n$

For this probability to be at least 50%, then

$$1 - \left(\frac{15}{16}\right)^n \geqslant 0.5 \Leftrightarrow \left(\frac{15}{16}\right)^n \leqslant 0.5$$

$$\Rightarrow n \geqslant \ln\left(\frac{15}{16}\right) \leqslant \ln(0.5)$$

$$\Rightarrow n \geqslant \frac{\ln(0.5)}{\ln\left(\frac{15}{16}\right)} \quad \left\{\text{since } \ln\left(\frac{15}{16}\right) < 0\right\}$$

$$\Rightarrow n \leqslant 10.74$$

So, 11 attempts are required.

Exercise 12.4

1 Events A and B are given such that $P(A) = \frac{3}{4}$, $P(A \cup B) = \frac{4}{5}$ and $P(A \cap B) = \frac{3}{10}$. Find $P(B)$.

2 Events A and B are given such that $P(A) = \frac{7}{10}$, $P(A \cup B) = \frac{9}{10}$ and $P(A \cap B) = \frac{3}{10}$. Find

a) $P(B)$ b) $P(B' \cap A)$ c) $P(B \cap A')$

d) $P(B' \cap A')$ e) $P(B|A')$

3 If events A and B are given such that $P(A) = \frac{1}{3}$, $P(A \cup B) = \frac{4}{9}$ and $P(B) = \frac{2}{9}$, show that A and B are neither independent nor mutually exclusive.

4 Events A and B are given such that $P(A) = \frac{3}{7}$ and $P(A \cap B) = \frac{3}{10}$. If A and B are independent, find $P(A \cup B)$.

5 Driving tests in a certain city are not easy to pass the first time you take them. After going through training, the percentage of new drivers passing the test the first time is 60%. If a driver fails the first test, there is a chance of passing it on a second try, two weeks later. 75% of the second-chance drivers pass the test. Otherwise, the driver has to retrain and take the test after 6 months. Find the probability that a randomly chosen new driver will pass the test without having to wait 6 months.

6 People with O-negative blood type are universal donors, i.e. they can donate blood to individuals with any blood type. Only 8% of people have O-negative.

a) One person randomly appears to give blood. What is the probability that he/she does not have O-negative?

b) Two people appear independently to give blood. What is the probability that
  (i) both have O-negative?
  (ii) at least one of them has O-negative?
  (iii) only one of them has O-negative?

c) Eight people appear randomly to give blood. What is the probability that at least one of them has O-negative?

7 PIN numbers for cellular phones usually consist of four digits that are not necessarily different.

a) How many possible PINs are there?

b) You don't want to consider the pins that start with 0. What is the probability that a PIN chosen at random does not start with a zero?

c) What is the probability that a PIN contains at least one zero?

d) Given a PIN with at least one zero, what is the probability that it starts with a zero?

8 An urn contains six red balls and two blue ones. We make two draws and each time we put the ball back after marking its colour.

a) What is the probability that at least one of the balls is red?

b) Given that at least one is red, what is the probability that the second one is red?

c) Given that at least one is red, what is the probability that the second one is blue?

9 Two dice are rolled and the numbers on the top face are observed.

a) List the elements of the sample space.

b) Let x represent the sum of the numbers observed. Copy and complete the following table.

x	2	3	4	5	6	7	8	9	10	11	12
P(x)		$\frac{1}{18}$									

c) What is the probability that at least one die shows a 6?

d) What is the probability that the sum is at most 10?

e) What is the probability that a die shows 4 or the sum is 10?

f) Given that the sum is 10, what is the probability that one of the dice is a 4?

10 A large school has the following numbers categorized by class and gender:

Grade / Gender	Grade 9	Grade 10	Grade 11	Grade 12	Total
Male	180	170	230	220	800
Female	200	130	190	180	700

a) What is the probability that a student chosen at random will be a female?

b) What is the probability that a student chosen at random is a male grade 12 student?

c) What is the probability that a female student chosen at random is a grade 12 student?

d) What is the probability that a student chosen at random is a grade 12 or female student?

e) What is the probability that a grade 12 student chosen at random is a male?

f) Are gender and grade independent of each other? Explain.

11 Some young people do not like to wear glasses. A survey considered a large number of teenage students as to whether they needed glasses to correct their vision and whether they used the glasses when they needed to. Here are the results.

		Used glasses when needed	
		Yes	No
Need glasses for correct vision	Yes	0.41	0.15
	No	0.04	0.40

a) Find the probability that a randomly chosen young person from this group
   (i) is judged to need glasses
   (ii) needs to use glasses but does not use them.

b) From those who are judged to need glasses, what is the probability that he/she does not use them?

c) Are the events of using and needing glasses independent?

12 Fill in the missing entries in the following table.

P(A)	P(B)	Conditions for events A and B	P($A \cap B$)	P($A \cup B$)	P($A \mid B$)
0.3	0.4	Mutually exclusive			
0.3	0.4	Independent			
0.1	0.5			0.6	
0.2	0.5		0.1		

13 In a large graduating class, there are 100 students taking the IB examination. 40 students are doing Maths/SL, 30 students are doing Physics/SL and 12 are doing both.

a) A student is chosen at random. Find the probability that this student is doing Physics/SL given that he/she is doing Maths/SL.

b) Are doing Physics/SL and Maths/SL independent?

14 A market chain in Germany accepts only Mastercard and Visa. It estimates that 21% of its customers use Mastercard, 57% use Visa and 13% use both cards.

a) What is the probability that a customer will have an acceptable credit card?

b) What proportion of their customers has neither card?

c) What proportion of their customers has exactly one acceptable card?

15 132 of 300 patients at a hospital are signed up for a special exercise program which consists of a swimming class and an aerobics class. Each of these 132 patients takes at least one of the two classes. There are 78 patients in the swimming class and 84 in the aerobics class. Find the probability that a randomly chosen patient at this hospital is

a) not in the exercise program

b) enrolled in both classes.

16 An ordinary unbiased 6-sided die is rolled three times. Find the probability of rolling

a) three twos

b) at least one two

c) exactly one two.

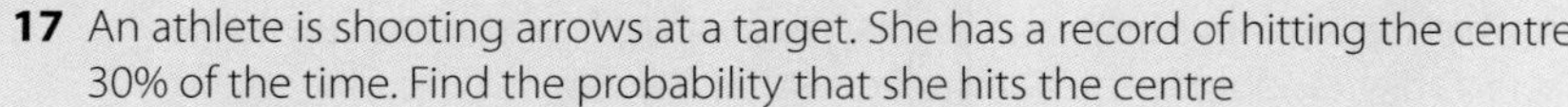

17 An athlete is shooting arrows at a target. She has a record of hitting the centre 30% of the time. Find the probability that she hits the centre

a) with her second shot

b) exactly once with her first three shots

c) at least once with her first three shots.

18 Two unbiased dodecahedral (12 faces) dice are thrown. The scores are the numbers on the top side.

Find the probability that

a) at least one twelve shows

b) a sum of 12 shows on both dice

c) a total score of at least 20 shows on both dice

d) given that a twelve shows on one die, a total score of at least 20 is achieved.

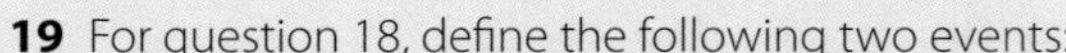

19 For question 18, define the following two events:

A = {at least one of the numbers is a 10}

B = {the sum of the numbers is at most 15}

Describe the following events, list their elements and find their probabilities:

a) $A \cap B$ b) $A \cup B$ c) $(A \cap B)'$

d) $(A \cup B)'$ e) $A' \cup B'$ f) $A' \cap B'$

g) $(A \cap B') \cup (A' \cap B)$

20 Consider any events A, B, and C. Prove each of the following:

a) $P(A \cap B) \geqslant P(A) + P(B) - 1$

b) $P(A \cup B \cup C) = P(A) + P(B) + P(C) - P(A \cap B) - P(A \cap C) - P(B \cap C) + P(A \cap B \cap C)$

21 Three fair 6-sided dice are rolled.

a) Find the probability that triples are rolled.

b) Given that the roll is a sum of 8 or less, find the probability that triples are rolled.

c) Find the probability that at least one six appears.

d) Given that the dice all have different numbers, find the probability that at least one six appears.

22 You are given four coins: one has two heads, one has two tails, and the other two are normal. You choose a coin at random and toss it. The result is tails. What is the probability that the opposite face is heads?

23 George and Kassanthra play a game in which they roll two unbiased cubical dice. The first one who rolls a sum of 6 wins. Kassanthra rolls the dice first.

a) What is the probability that Kassanthra wins on her second roll?

b) What is the probability that George wins on his second roll?

c) What is the probability that Kassanthra wins?

24 A small repair shop for washing machines has the following demand for their services:

On 10% of the days they have no requests, they have one request on 30% of the days, and two requests 50% of the time.

a) On Monday, what is the chance of more than two requests?

b) What is the chance of no requests for a whole (5 day) week?

25 A small class has five boys and six girls. A group of four students are to be selected at random to interview a new school director.

a) Find the probability that the group contains at least one boy.

b) Find the probability that the majority of the group is girls.

c) Given that the group contains at least one boy, what is the chance that the boys are in the majority?

26 A construction company is bidding on three projects: B_1, B_2 and B_3. From previous experience they have the following probabilities of winning the bids: $P(B_1) = 0.22$, $P(B_2) = 0.25$ and $P(B_3) = 0.28$. Winning the bids is not independent one from another. The joint probabilities are given below.

	B_1	B_2	B_3
B_1		0.11	0.05
B_2	0.11		0.07
B_3	0.05	0.07	

Also, $P(B_1 \cap B_2 \cap B_3) = 0.01$. Find the following probabilities:

a) $P(B_1 \cup B_2)$

b) $P(B'_1 \cap B'_2)$

c) $P(B'_1 \cap B'_2) \cup B_3$

d) $P(B'_1 \cap B'_2 \cap B_3)$

e) $P(B_2 \cap B_3|B_1)$

f) $P(B_2 \cup B_3|B_1)$

27 Circuit boards used in electronic equipment go through more than one layer of inspection. The process of finding faults in the solder joints on these boards is highly subjective and prone to disagreements among inspectors. In a batch of 20 000 joints, Nick found 1448 faulty joints while David found 1502 faulty ones. All in all, among both inspectors, 2390 joints were judged to be faulty. Find the probability that a randomly chosen joint is

a) judged to be faulty by neither of the two inspectors

b) judged to be defective by David but not Nick.

12.5 Bayes' theorem

Lie detectors (polygraphs), drug and alcohol tests, and disease screening tests are among the many applications where the results are frequently cautiously scrutinized. Tests with high precision rates are open to error. Bayes' theorem helps us understand and analyze the results of such tests.

To understand the theorem, we start this section with an example.

Suppose I have 9 indistinguishable boxes with blue and red balls as shown in the diagram (next page). They are of three types: Type A contain 2 balls each – one red and one blue, B contain 3 balls each – one blue and two red, and C contain 4 balls each – one blue and three red. I mix up the boxes and choose one at random, and then I pick a ball from that box, also at random. I show you the ball. What is the probability that you can guess the box type it was drawn from if the ball is red?

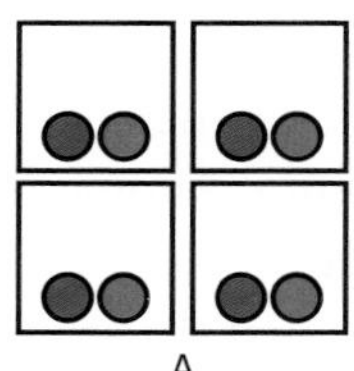

A

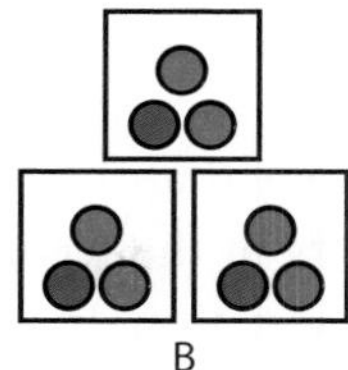

B

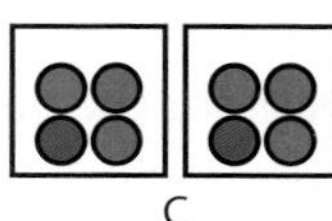

C

Letting R represent red ball and A, B or C representing box types, the question is to find

$$P(A|R), P(B|R), \text{ or } P(C|R).$$

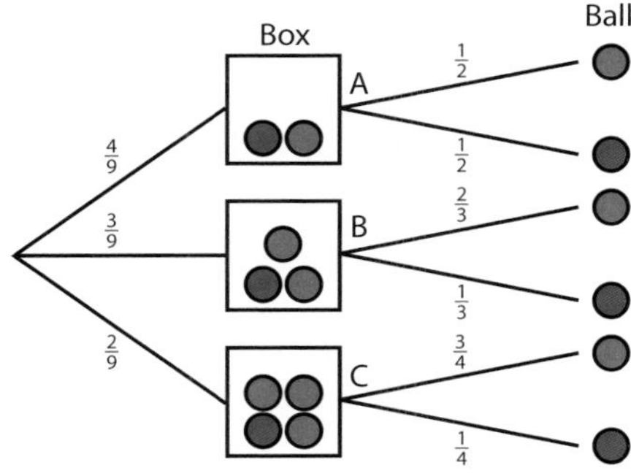

Recalling information from conditional probability, your task is to calculate

$$P(A|R) = \frac{P(A \cap R)}{P(R)}.$$

Since you already know $P(A)$ and $P(R|A)$, then

$$P(A \cap R) = P(R|A) \cdot P(A) = \tfrac{4}{9} \cdot \tfrac{1}{2} = \tfrac{2}{9}.$$

However, to find $P(R)$, we need to see what constitutes the event R.

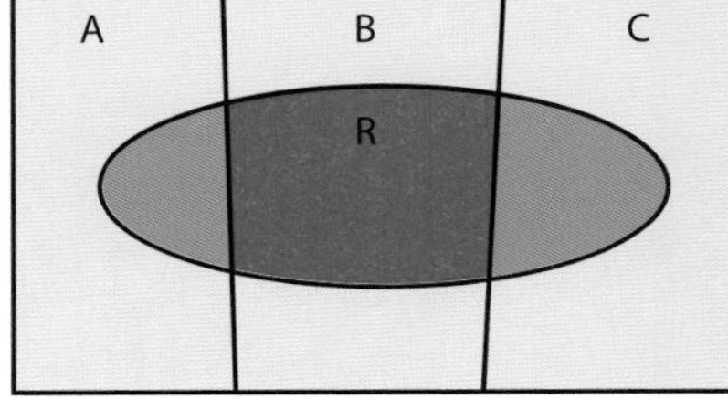

As you see in the diagram right, A, B and C are mutually exclusive, and hence

$$R = (R \cap A) \cup (R \cap B) \cup (R \cap C) \Rightarrow$$
$$P(R) = P(R \cap A) + P(R \cap B) + P(R \cap C), \text{ and so}$$
$$\begin{aligned} P(R) &= P(R \cap A) + P(R \cap B) + P(R \cap C) \\ &= P(R|A) \cdot P(A) + P(R|B) \cdot P(B) + P(R|C) \cdot P(C) \\ &= \tfrac{1}{2} \cdot \tfrac{4}{9} + \tfrac{2}{3} \cdot \tfrac{3}{9} + \tfrac{3}{4} \cdot \tfrac{2}{9} = \tfrac{11}{18} \end{aligned}$$

And now

$$P(A|R) = \frac{P(A \cap R)}{P(R)} = \frac{\frac{2}{9}}{\frac{11}{18}} = \tfrac{4}{11}, \text{ and similarly}$$

$$P(B|R) = \frac{P(B \cap R)}{P(R)} = \frac{\frac{2}{3} \cdot \frac{3}{9}}{\frac{11}{18}} = \tfrac{4}{11}, \text{ and finally}$$

$$P(C|R) = \frac{P(C \cap R)}{P(R)} = \frac{\frac{3}{4} \cdot \frac{2}{9}}{\frac{11}{18}} = \tfrac{3}{11}.$$

So, A or B could more likely be the source of a red ball than C.

Conditional probability typically deals with the probability of an event when you have information about something that happened earlier. In conditional probability, we assume that the first event is known (selecting a box) and ask for the probability of the second event (colour of ball). Bayes' rule deals with a reverse situation. It assumes that the second event is known (red ball) and asks for the probability of the first event.

Let us look at another example to see how we apply Bayes' theorem. We will then make a formal statement of the rule.

Example 22

60% of the students at a university are male and 40% are female. Records show that 30% of the males have IB diplomas while 75% of the females have IB diplomas. A student is selected and found to have a diploma. What is the probability that the student is a female?

Solution

Let the event female be called F, event male M and event IB diploma I. The question asks for $P(F|I)$.

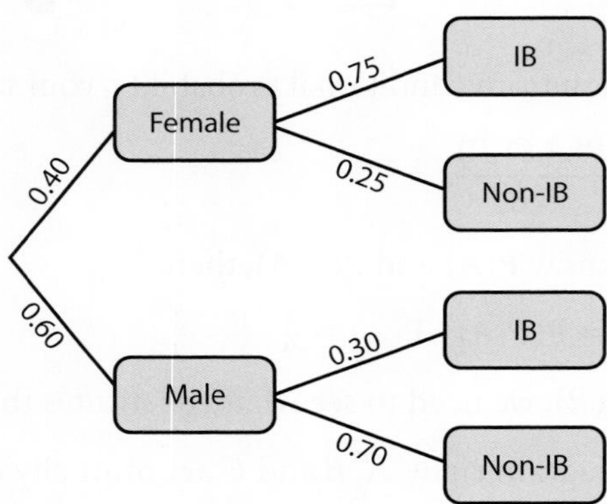

$$P(F|I) = \frac{P(F \cap I)}{P(I)} = \frac{P(F) \cdot P(I|F)}{P(I)}$$

Since $I = (F \cap I) \cup (M \cap I)$, then

$$P(I) = 0.40 \times 0.75 + 0.60 \times 0.30 = 0.48, \text{ and so}$$

$$P(F|I) = \frac{P(F \cap I)}{P(I)} = \frac{0.40 \times 0.75}{0.48} = 0.625.$$

Note: Another interpretation of this number is the proportion of females among the IB holders.

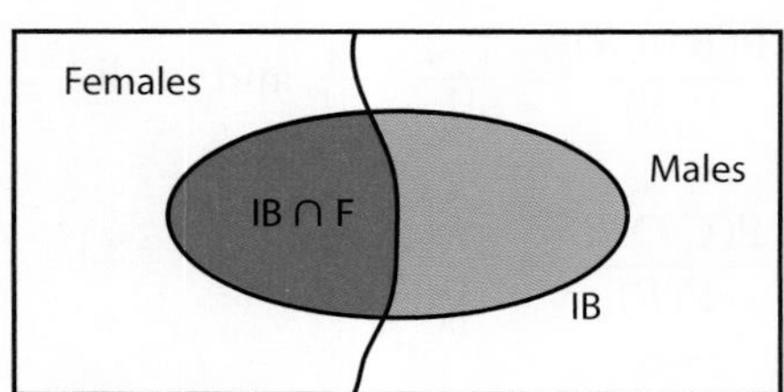

Bayes' theorem – simple case

Let S be a sample space with E_1 and E_2 mutually exclusive events that *partition* this sample space. Let F be a non-empty event in this sample space. Then

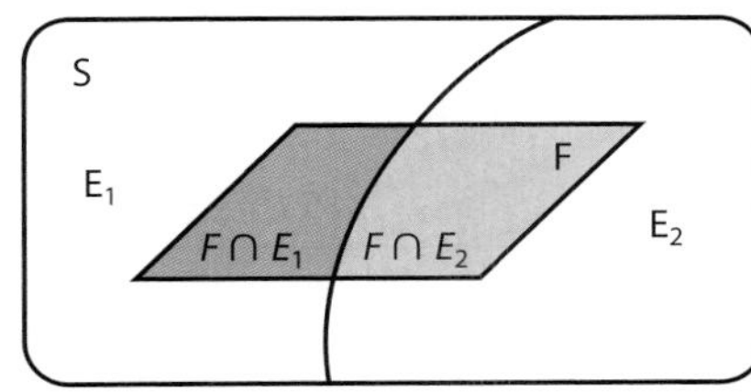

$$P(E_1|F) = \frac{P(E_1 \cap F)}{P(F)} = \frac{P(E_1 \cap F)}{P(E_1 \cap F) + P(E_2 \cap F)}$$

$$= \frac{P(E_1) \cdot P(F|E_1)}{P(E_1) \cdot P(F|E_1) + P(E_2) \cdot P(F|E_2)}$$

Note:
The probability of an event in terms of its mutually exclusive and exhaustive subsets is usually called the **total probability**:
$P(F) = P(E_1) \cdot P(F|E_1) + P(E_2) \cdot P(F|E_2)$

Sometimes, this rule is stated differently:

$$P(E|F) = \frac{P(E \cap F)}{P(F)} = \frac{P(E \cap F)}{P(E \cap F) + P(E' \cap F)}$$

$$= \frac{P(E_1) \cdot P(F|E)}{P(E) \cdot P(F|E) + P(E') \cdot P(F|E')}$$

where E' is the complement of E.

Example 23

Lie detectors (polygraphs) are not considered reliable in court. They are, however, administered to employees in sensitive positions. One such test gives a positive reading (person is lying) when a person is lying 88% of the time and a negative reading (person telling the truth) when a person is telling the truth 86% of the time. In a security-related question, the vast majority of subjects have no reason to lie so that 99% of the subjects will tell the truth. An employee produces a positive response on the polygraph. What is the probability that this employee is actually telling the truth?

Solution

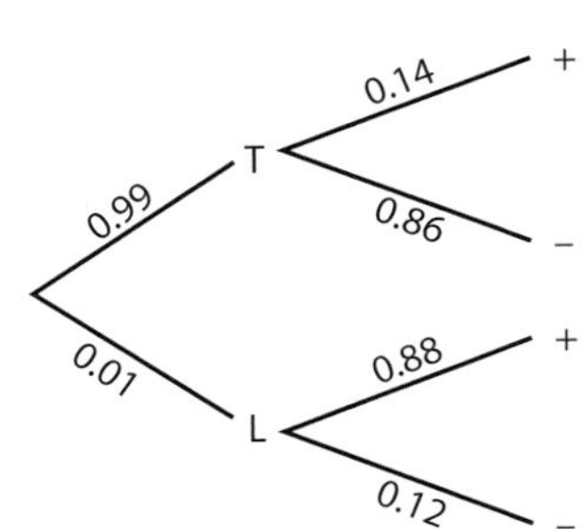

Let $+$ denote the event of the test being positive (subject is lying), and $-$ denote the event of the test being negative (subject telling the truth). Let L represent the person is lying, and T the person is telling the truth. Hence, we have

$$P(+|L) = 0.88, P(-|L) = 0.12, P(-|T) = 0.86, P(+|T) = 0.14$$

$$P(+) = P(+ \cap L) + P(+ \cap T) = P(+|L) \cdot P(L) + P(+|T) \cdot P(T)$$

$$= 0.88 \times 0.01 + 0.14 \times 0.99$$

And finally,

$$P(T|+) = \frac{P(+|T) \cdot P(T)}{P(+|L) \cdot P(L) + P(+|T) \cdot P(T)} = \frac{0.14 \times 0.99}{0.88 \times 0.01 + 0.14 \times 0.99} = 0.94.$$

Thus, in screening this population of mostly innocent people, 94% of the positive polygraph readings will be misleading!

Bayes' theorem – general rule

Let S be a sample space with $E_1, E_2, \ldots, E_n$ mutually exclusive events that *partition* this sample space. Let F be a non-empty event in this sample space. Then

$$\mathrm{P}(E_i|F) = \frac{\mathrm{P}(E_i \cap F)}{\mathrm{P}(F)} = \frac{\mathrm{P}(E_i \cap F)}{\mathrm{P}(E_1 \cap F) + \mathrm{P}(E_2 \cap F) + \ldots + \mathrm{P}(E_n \cap F)}$$

$$= \frac{\mathrm{P}(E_i) \cdot \mathrm{P}(F|E_i)}{\mathrm{P}(E_1) \cdot \mathrm{P}(F|E_1) + \mathrm{P}(E_2) \cdot \mathrm{P}(F|E_2) + \ldots + \mathrm{P}(E_n) \cdot \mathrm{P}(F|E_n)}$$

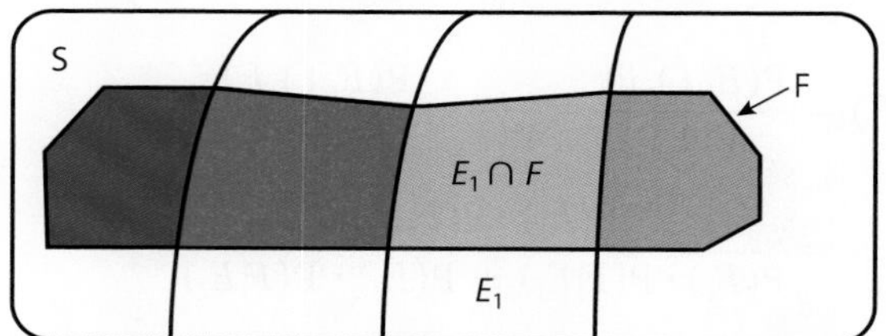

Example 24

A paper factory produces high-quality paper using two machines. Like any process, the final product is checked for quality. 98% of the time machine A produces paper that conforms to the accepted norms. 97.5% of machine B's paper conforms to norms. Machine A produces 60% of the paper in this company. You pick a paper at random and it does not conform to the norms, what is the chance it was produced by A?

Solution

Let the non-conforming paper be called N. Since the paper is produced by these two machines, then the non-conforming paper has to be from the output of one of these machines, i.e.

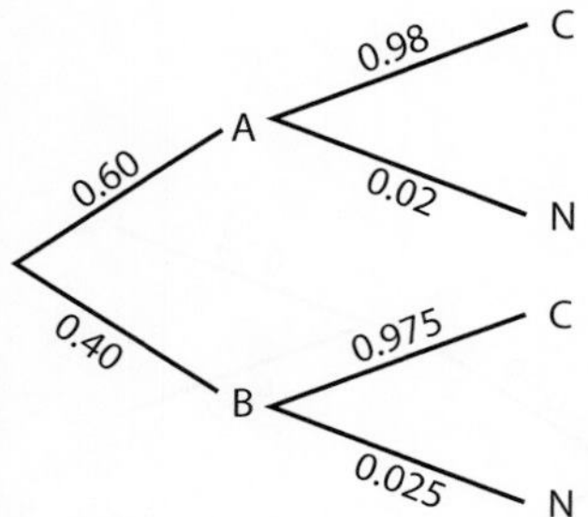

$$N = (N \cap A) \cup (N \cap B),$$

$$\mathrm{P}(N) = \mathrm{P}(N \cap A) + \mathrm{P}(N \cap B) = \mathrm{P}(N|A) \cdot \mathrm{P}(A) + \mathrm{P}(N|B) \cdot \mathrm{P}(B)$$

$$= 0.02 \times 0.60 + 0.025 \times 0.40 = 0.022$$

So, using Bayes' theorem,

$$\mathrm{P}(A|N) = \frac{\mathrm{P}(A \cap N)}{\mathrm{P}(N)} = \frac{\mathrm{P}(N|A) \cdot \mathrm{P}(A)}{\mathrm{P}(N)} = \frac{0.02 \times 0.60}{0.022} = 54.5\%.$$

Example 25

In many countries the law requires that a driver's licence is withdrawn if he/she is found to have more that 0.05% blood alcohol concentration (BAC). Suppose that the police use a test that will correctly identify a 'drunk' driver (testing positive) 99% of the time, and will correctly identify a sober driver (testing negative) 99% of the time. Let's assume that 0.5% of the drivers in your city drive under the influence of alcohol. We want to know the probability that given a positive test a driver is actually 'drunk'.

Solution

Let D be the event of being drunk and N indicate being not drunk. Let $+$ be the event of a positive test. We need to know $P(D|+)$.

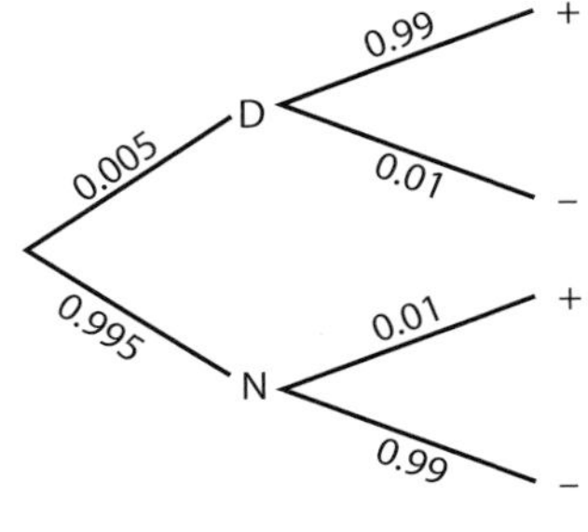

The question is to find the proportion of the drunk drivers who test positive out of all of those who test positive.

$P(D)$, or the probability that the driver is drunk, regardless of any other information, is 0.005, since 0.5% of the drivers drink and drive.

$P(N)$, or the probability that the driver is sober, is $1 - P(D)$, or 0.995.

$P(+|D)$, or the probability that the test is positive given that the driver is drunk, is 0.99, since the test is 99% accurate.

$P(+|N)$, or the probability that the test is positive, given that the driver is not drunk, is 0.01, since the test will produce a *false positive* for 1% of sober drivers.

$P(+)$, or the probability of a positive test event, regardless of other information, is 0.0149 or 1.49%, which is found by adding the probability that a *true positive* result will appear ($99\% \times 0.5\% = 0.495\%$) plus the probability that a *false positive* will appear ($1\% \times 99.5\% = 0.995\%$).

Given this information, we can compute the $P(D|+)$ of a driver who tested positive actually being a drunk driver:

$$P(D|+) = \frac{P(D \cap +)}{P(+)} = \frac{P(+|D)P(D)}{P(+|D)P(D) + P(+|N)\,P(N)}$$

$$= \frac{0.99 \times 0.005}{0.99 \times 0.005 + 0.01 \times 0.995} = 0.3322$$

Even with the high precision of the test, the probability that a driver who tested positive in fact did have a high BAC is only about 33%, so it is essentially more likely that the driver is not a drunk driver.

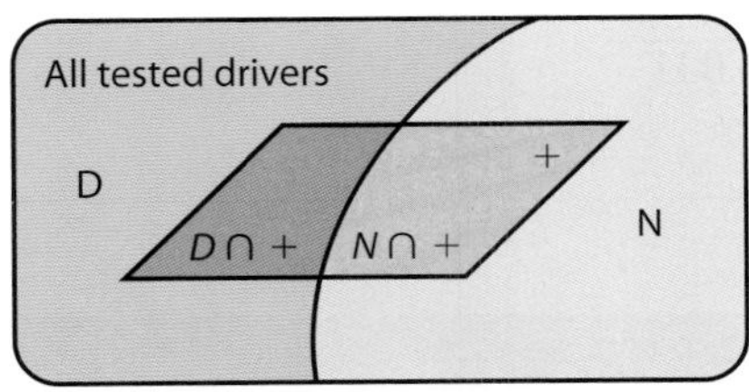

Example 26

A computer manufacturer receives hard disks from three different suppliers. Marco supplies 40% of the disks, Berto supplies 25%, while Lukas supplies the rest. Disks from Marco have a defective rate of 4%, those from Berto 3%, while Lukas' have a 5% rate.

a) A disk is checked at random. What is the probability that it is defective?

b) A disk is checked and found defective. What is the probability that it was supplied by Lukas?

Solution

Let M represent Marco, B represent Berto and L represent Lukas. Also let D represent defective items and G the non-defective ones. A tree diagram may be helpful here.

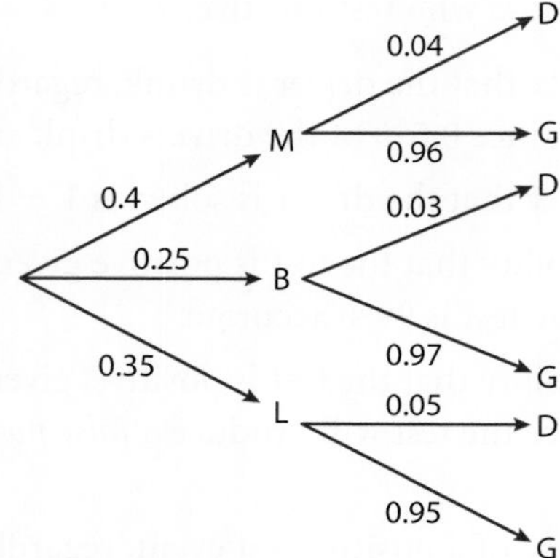

a) A defective disk could be supplied by any of the three suppliers, hence:

$$\begin{aligned} P(D) &= P(M \cap D) + P(B \cap D) + P(L \cap D) \\ &= P(M) \cdot P(D|M) + P(B) \cdot P(D|B) + P(L) \cdot P(D|L) \\ &= 0.4 \times 0.04 + 0.25 \times 0.03 + 0.35 \times 0.05 \\ &= 0.041 \end{aligned}$$

b) This is a Bayes theorem application. We are reversing the order of events:

$$\begin{aligned} P(L|D) &= \frac{P(L \cap D)}{P(D)} \\ &= \frac{P(L) \cdot P(D|L)}{P(M) \cdot P(D|M) + P(B) \cdot P(D|B) + P(L) \cdot P(D|L)} \\ &= \frac{0.35 \times 0.05}{0.041} \\ &= 0.427 \end{aligned}$$

Exercise 12.5

1 In a sample space S, we have the following events and some associated probabilities:

$$P(E) = \tfrac{2}{3},\ P(A|E) = \tfrac{3}{50} \text{ and } P(A|E') = \tfrac{1}{25}$$

a) Represent the information using a tree diagram.

b) Find $P(A)$.

c) Find $P(E|A)$.

2 The rate of prostate cancer among men in 2002 was approximately 26 cases per 100 000 people. Diagnosing this type of cancer saves the lives of about 70% of those treated. At a hospital, the probability of diagnosing a person with this cancer correctly is 78% and the probability of diagnosing a person without this cancer as having the disease is 6%.

a) What is the probability of diagnosing a person as having this cancer?
b) What percentage of those diagnosed with this cancer actually have it?

3 Police in a small town plan to enforce speed limits by installing 'radar traps' at the two main town entrances: east end and west end. Traffic statistics show that 40% of the cars entering the town use the east entrance and the rest use the west one. The east entrance traps are operated 40% of the time and the west entrance are operated 60% of the time. Assuming that the proportion of 'speeders' is the same at both entrances, what is the probability that:
a) a speeding driver is spotted passing one of the traps?
b) a speeding driver who got spotted has actually come from the west end?

4 Coloured balls are placed in three boxes as follows:

	Box		
	1	2	3
Green	4	8	6
Red	6	2	8
Blue	10	6	6

A box is selected at random from which a ball is randomly drawn.
a) What is the probability that the ball is green?
b) Given that the ball is green, what is the probability that it was drawn from box 2?

5 Two coins are in your pocket. When tossed, one of the coins is biased with 0.6 probability of landing heads, while the other coin is unbiased. You select one of the coins at random and toss it.
a) What is the probability it lands heads?
b) Given that it lands on tails, what is the probability that it was the unbiased coin?

6 When answering a question on a multiple choice test, a student is given 5 choices, one of which is correct. The test is so designed that the choices are very close and the probability of getting the correct answer, when you know the material, is 0.6. In a class where 70% of the students are well prepared, a randomly chosen student answers the question correctly. What is the probability that the student really knew the material?

7 Nigel is a student at Wigley College and lives in the dorms. To avoid coming late to his morning classes he usually sets his alarm clock. 85% of the time he manages to remember and set his alarm. When the alarm goes off he manages to go to his morning classes 90% of the time. If the alarm is not set, he still manages to get up and go to class on 60% of the days.
a) What percentage of the days does he manage to get to his morning classes?
b) He made it to class one day. What is the chance that he did that without having set the alarm?

8 Marco plays tennis. In this game, a player has two 'serves'. If the first serve is successful, the game continues. If the first serve is not successful, the player is given another chance. If the second serve fails, then the player loses the point.

Marco is successful with his first serve 60% of the time and 95% successful with his second serve. When his first serve is successful he goes on to win the point 75% of the time, and when it takes him two serves, he wins the point 50% of the time.

a) What is the probability that Marco wins the point?

b) If Marco wins a point, what is the probability that he succeeded with the first serve?

9 In Vienna, conventional wisdom has it that in February days are snowy or fine. 80% of the time a fine day follows a fine day. 40% of the time a snowy day is followed by a fine day. The forecast for the first of February to be a fine day is 0.75.

a) Find the probability that 2nd February is fine.

b) Given that 2nd February turns out to be snowy, what is the probability that the 1st of February was a fine day?

10 It is known that 33% of people over the age of 50 around the world have some kind of arthritis. A test has been developed to detect arthritis in individuals. This test was given to a large group of individuals with confirmed cases and a positive test result was achieved in 87% of the cases. That same test gave a positive test to 4% of individuals that do not have arthritis.

If this test is given to an individual at random and it tests positive, what is the probability that the individual has this disease?

11 (Relatively challenging!) A high school has a large graduating class. The table below shows how the students are categorized according to their college plans and gender.

	Local university	University abroad
Male	51%	4%
Female	16%	29%

5% of these students are in the IB mathematics/HL class. 72% of the HL class will attend local university and the rest are going abroad.

a) What percentage of the graduating class are mathematics/HL students planning on studying locally?

b) What percentage of the non-mathematics/HL students are going to study at a local university?

c) Among the students studying locally, what percentage are mathematics/HL students?

d) Among the students studying abroad, what percentage are mathematics/HL students?

12 When Olympic athletes are tested for illegal drug use (doping), the results of a single test are used to ban the athlete from competition. In an experiment on 1000 athletes, 100 were using the testosterone drug. During the medical examination, the available test would positively identify 50% of the users. It would also falsely identify 9% of the non-users as users.

If an athlete tests positive, what is the probability that he/she is really doping?

13 An engineering company employs three architects that are responsible for cost estimates of new projects. Antonio makes 30% of the estimates, Richard makes 20% and Marco 50%. Like all estimates, there are usually some errors in these estimates. The record of percentages of 'serious' errors that cost the company thousands of euros shows Antonio at 3%, Richard at 2%, and Marco at 1%. Which of the three engineers is probably responsible for most of the serious errors?

14 At a small airport, if an aircraft is present at 10 km distance from the runway, radar detects it and generates an alarm signal 99% of the time. If an aircraft is not present, the radar generates a (false) alarm, with probability 0.10. We assume that an aircraft is present with probability 0.05.

a) What is the probability that the radar gives an alarm signal?

b) Given that there is no alarm signal, what is the probability that an aircraft is there?

15 You enter a chess tournament where your probability of winning a game is 0.3 against half the players (novices), 0.4 against a quarter of the players (experienced) and 0.5 against the remaining quarter of the players (masters). You play a game against a randomly chosen opponent.

a) What is the probability of winning?

b) Given that you won, what is the probability that the game was against a master?

16 Driving tests in a certain city are relatively easy to pass the first time you take them. After going through training, the percentage of new drivers passing the test the first time is 80%. If a driver fails the first test, there is chance of passing it on a second harder test, two weeks later. 50% of the second-chance drivers pass the test. If the second test is unsuccessful, a third attempt, a week later, is given and 30% of the participants pass it. Otherwise, the driver has to retrain and take the test after 1 year.

a) Find the probability that a randomly chosen new driver will pass the test without having to wait one year.

b) Find the probability that a randomly chosen new driver that passed the test did so on the second attempt.

17 A school has 250 employees categorized by task and gender in the following table.

	Teaching	Administrative	Support
Male	84	14	52
Female	56	26	18

An employee is randomly selected. Let A be the event that he/she is an administrative staff member, T teaching staff, S support, M male, and F female.

a) Write down the following probabilities: $P(F)$, $P(F \cap T)$, $P(F \cup A')$, $P(F'|A)$.

b) Which events are independent of F, which are mutually exclusive to F. Justify your choices.

c) (i) Given that 90% of teachers, as well as 80% of the administrative staff and 30% of the support staff, own cars, find the probability that a staff member chosen at random owns a car.

(ii) Knowing that the randomly chosen staff member owns a car, find the probability that he/she is a teacher.

18 Car insurance companies categorize drivers as high risk, medium risk and low risk. (For your information only: Teens and seniors are considered high risk!)

20% of the drivers insured by 'First Insurance' are high risk, and 50% are medium risk driver. The company's actuaries estimate the chance that each class of driver will have at least one accident in the coming 12 months as follows: High risk 6%, medium 3% and low 1%.

a) Find the probability that a randomly chosen driver is a high-risk driver that will have an accident in a 12-month period.

b) Find the probability that a randomly chosen driver insured by this company will have an accident in the next 12 months.

c) A customer has a claim for an accident. What is the probability that he/she is a high-risk driver?

Practice questions

1 Two independent events A and B are given such that $P(A) = k$, $P(B) = k + 0.3$ and $P(A \cap B) = 0.18$

a) Find k.
b) Find $P(A \cup B)$.
c) Find $P(A' | B')$.

2 Many airport authorities test prospective employees for drug use, with the intent of improving efficiency and reducing accidents. This procedure has plenty of opponents who claim that it creates difficulties for some classes of people and that it prevents others from getting these jobs even if they were not drug users. The claim depends on the fact that these tests are not 100% accurate. To test this claim, let us assume that a test is 98% accurate in the sense that it identifies a person as a user or non-user 98% of the time. Each job applicant takes this test twice. The tests are done at separate times and are designed to be independent of each other. What is the probability that

a) a non-user fails both tests?
b) a drug user is detected (i.e. he/she fails at least one test)?
c) a drug user passes both tests?

3 Communications satellites are difficult to repair when something goes wrong. One satellite works on solar energy and has two systems that provide electricity: the main system with a probability of failure of 0.002, and a back-up system that works independently of the main one. It has a failure rate of 0.01. What is the probability that the systems do not fail at the same time?

4 In a group of 200 students taking the IB examination, 120 take Spanish, 60 take French and 10 take both.

a) If a student is selected at random, what is the probability that he/she
- **(i)** takes either French or Spanish?
- **(ii)** takes either French or Spanish but not both?
- **(iii)** does not take any French or Spanish?

b) Given that a student takes the Spanish exam, what is the chance that he/she takes French?

5 In a factory producing disk drives for computers, there are three machines that work independently to produce one of the components. In any production process, machines are not 100% fault free. The production after one 'run' from these machines is listed below.

	Defective	Non-defective
Machine I	6	120
Machine II	4	80
Machine III	10	150

a) A component is chosen at random from the produced lot. Find the probability that the chosen component is
- **(i)** from machine I
- **(ii)** a defective component from machine II
- **(iii)** non-defective or from machine I
- **(iv)** from machine I given that it is defective.

b) Is the quality of the component dependent on the machine used?

6 At a school, the students are organizing a lottery to raise money for the needy in their community. The lottery tickets they have consist of small coloured envelopes inside which there is a small note. The note says: 'You won a prize!' or 'Sorry, try another ticket.' The envelopes have several colours. They have 70 red envelopes that contain two prizes, and the rest (130 tickets) contain four other prizes.

a) You want to help this class and you buy a ticket hoping that it does not have a prize. Additionally, you don't like the red colour. You pick your ticket at random by closing your eyes. What is the probability that your wish comes true?

b) You are surprised – you picked a red envelope. What is the probability that you did not win a prize?

7 You are given two events A and B with the following conditions:

$$P(A \mid B) = 0.30, \; P(B \mid A) = 0.60, \; P(A \cap B) = 0.18$$

a) Find $P(B)$.

b) Are A and B independent? Why?

c) Find $P(B \cap A')$.

8 In several ski resorts in Austria and Switzerland, the local sports authorities use high school students as 'ski instructors' to help deal with the surge in demand during vacations. However, to become an instructor, you have to pass a test and be a senior at your school. Here are the results of a survey of 120 students in a Swiss school who are training to become instructors. In this group, there are 70 boys and 50 girls. 74 students took the test, 32 boys and 16 girls passed the test, and the rest, including 12 girls, failed the test. 10 of the students, including 6 girls, were too young to take the ski test.

a) Copy and complete the table.

	Boys	Girls
Passed the ski test	32	16
Failed the ski test		12
Training, but did not take the test yet		
Too young to take the test		

b) Find the probability that

(i) a student chosen at random has taken the test

(ii) a girl chosen at random has taken the test

(iii) a randomly chosen boy and randomly chosen girl have both passed the ski test.

9 Two events A and B are such that $P(A) = \frac{9}{16}$, $P(B) = \frac{3}{8}$, and $P(A \mid B) = \frac{1}{4}$. Find the probability that

a) both events will happen

b) only one of the events will happen

c) neither event will happen.

10 Martina plays tennis. When she serves, she has a 60% chance of succeeding with her first serve and continuing the game. She has a 95% chance on the second serve. Of course if both serves are not successful, she loses the point.

a) Find the probability that she misses both serves.

If Martina succeeds with the first serve, her chances of gaining the point against Steffy is

75%. If she is only successful with the second serve, her chances against Steffy for that point go down to 50%.

b) Find the probability that Martina wins a point against Steffy.

11 For the events A and B, $P(A) = 0.6$, $P(B) = 0.8$ and $P(A \cup B) = 1$.
Find

a) $P(A \cap B)$

b) $P(A' \cup B')$.

12 In a survey, 100 students were asked, 'Do you prefer to watch television or play sport?' Of the 46 boys in the survey, 33 said they would choose sport, while 29 girls made this choice.

	Boys	Girls	Total
Television			
Sport	33	29	
Total	46		100

By completing this table or otherwise, find the probability that

a) a student selected at random prefers to watch television

b) a student prefers to watch television given that the student is a boy.

13 Two ordinary, six-sided dice are rolled and the total score is noted.

a) Complete the tree diagram by entering probabilities and listing outcomes.

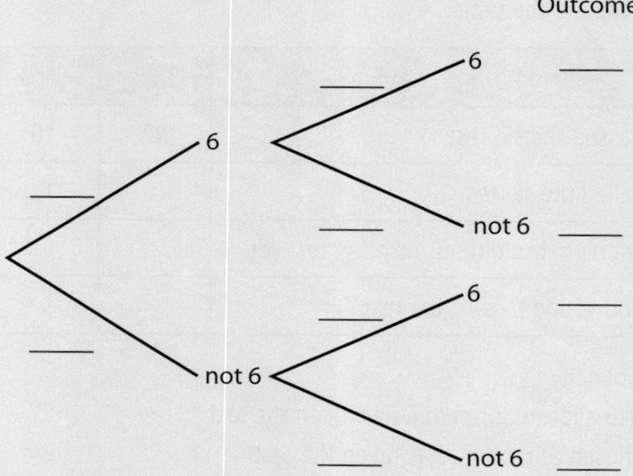

b) Find the probability of getting one or more sixes.

14 The Venn diagram right shows a sample space U and events A and B.

$n(U) = 36$, $n(A) = 11$, $n(B) = 6$ and $n(A \cup B)' = 21$.

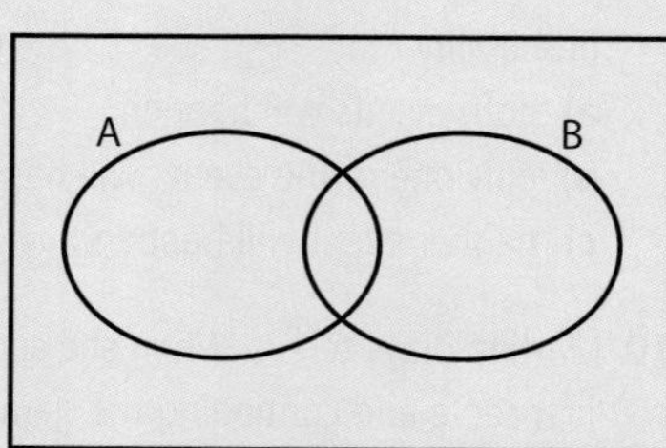

a) On the diagram, shade the region $(A \cup B)'$.

b) Find

(i) $n(A \cap B)$

(ii) $P(A \cap B)$.

c) Explain why events A and B are not mutually exclusive.

15 In a survey of 200 people, 90 of whom were female, it was found that 60 people were unemployed, including 20 males.

a) Using this information, complete the table below.

	Males	Females	Totals
Unemployed			
Employed			
Totals			200

b) If a person is selected at random from this group of 200, find the probability that this person is

(i) an unemployed female

(ii) a male given that the person is employed.

16 A bag contains 10 red balls, 10 green balls and 6 white balls. Two balls are drawn at random from the bag without replacement. What is the probability that they are of different colours?

17 The Venn diagram right shows the universal set U and the sets A and B.

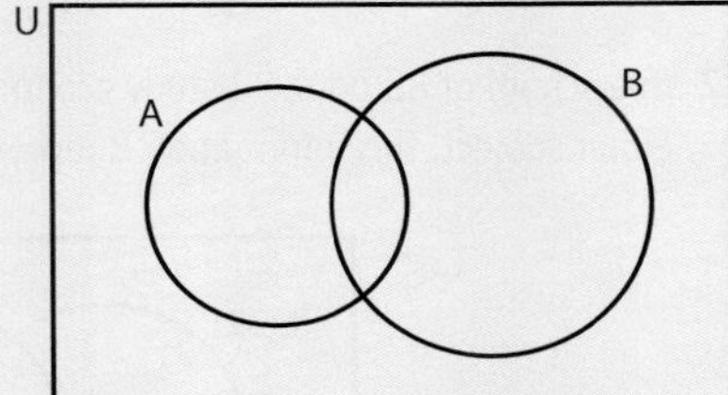

a) Shade the area in the diagram which represents the set $B \cap A'$.

$n(U) = 100,\ n(A) = 30,\ n(B) = 50,\ n(A \cup B) = 65.$

b) Find $n(B \cap A')$.

c) An element is selected at random from U. What is the probability that this element is in $B \cap A'$?

18 The events B and C are dependent, where C is the event 'a student takes chemistry', and B is the event 'a student takes biology'. It is known that

$P(C) = 0.4,\ P(B|C) = 0.6,\ P(B|C') = 0.5.$

a) Complete the following tree diagram.

b) Calculate the probability that a student takes biology.

c) Given that a student takes biology, what is the probability that the student takes chemistry?

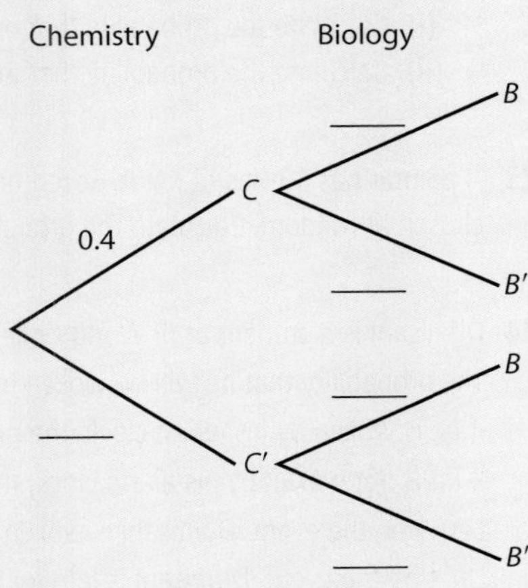

19 Two fair dice are thrown and the number showing on each is noted. The sum of these two numbers is S. Find the probability that

a) S is less than 8

b) at least one die shows a 3

c) at least one die shows a 3 given that S is less than 8.

20 For events A and B, the probabilities are $P(A) = \frac{3}{11}$ and $P(B) = \frac{4}{11}$.
Calculate the value of $P(A \cap B)$ if

a) $P(A \cup B) = \frac{6}{11}$

b) events A and B are independent.

21 Consider events A and B such that $P(A) \neq 0$, $P(A) \neq 1$, $P(B) \neq 0$ and $P(B) \neq 1$.
In each of the situations **a)**, **b)**, **c)** below, state whether A and B are mutually exclusive (M), independent (I), or neither (N).

a) $P(A|B) = P(A)$

b) $P(A \cap B) = 0$

c) $P(A \cap B) = P(A)$

22 In a school of 88 boys, 32 study economics (E), 28 study history (H) and 39 do not study either subject. This information is represented in the following Venn diagram.

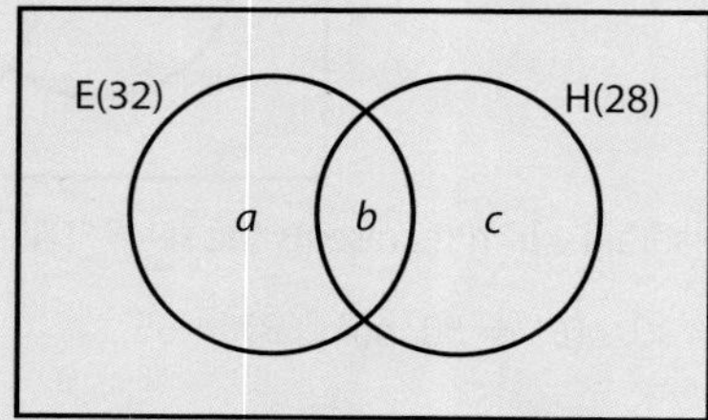

a) Calculate the values a, b, c.

b) A student is selected at random.

(i) Calculate the probability that he studies both economics and history.

(ii) Given that he studies economics, calculate the probability that he does not study history.

c) A group of three students is selected at random from the school.

(i) Calculate the probability that none of these students studies economics.

(ii) Calculate the probability that at least one of these students studies economics.

23 A painter has 12 tins of paint. Seven tins are red and five tins are yellow. Two tins are chosen at random. Calculate the probability that both tins are the same colour.

24 Dumisani is a student at IB World College.
The probability that he will be woken by his alarm clock is $\frac{7}{8}$.
If he is woken by his alarm clock, the probability he will be late for school is $\frac{1}{4}$.
If he is not woken by his alarm clock, the probability he will be late for school is $\frac{3}{5}$.
Let W be the event 'Dumisani is woken by his alarm clock'.
Let L be the event 'Dumisani is late for school'.

a) Copy and complete the tree diagram below.

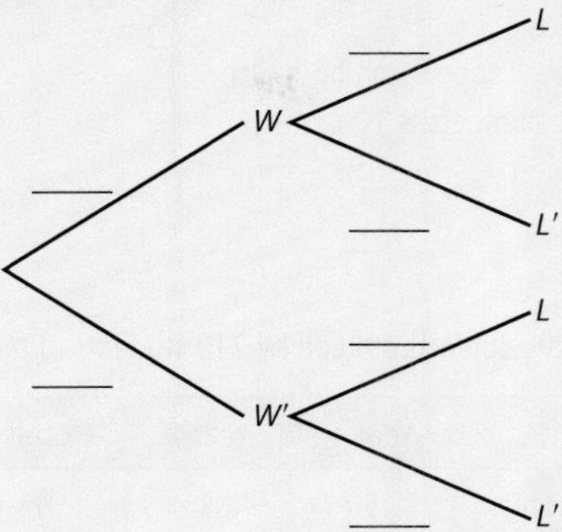

b) Calculate the probability that Dumisani will be late for school.

c) Given that Dumisani is late for school, what is the probability that he was woken by his alarm clock?

25 The diagram shows a circle divided into three sectors A, B and C. The angles at the centre of the circle are 90°, 120° and 150°. Sectors A and B are shaded as shown.

The arrow is spun. It cannot land on the lines between the sectors. Let A, B, C and S be the events defined by

A : Arrow lands in sector A
B : Arrow lands in sector B
C : Arrow lands in sector C
S : Arrow lands in a shaded region.

Find

a) $P(B)$ **b)** $P(S)$ **c)** $P(A|S)$.

26 A packet of seeds contains 40% red seeds and 60% yellow seeds. The probability that a red seed grows is 0.9, and that a yellow seed grows is 0.8. A seed is chosen at random from the packet.

a) Complete the probability tree diagram below.

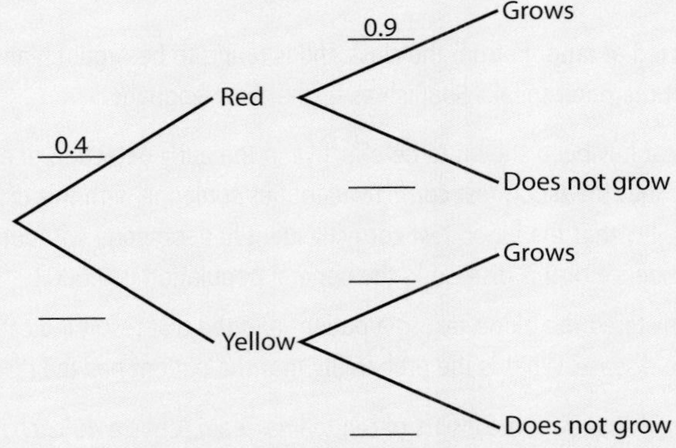

b) **(i)** Calculate the probability that the chosen seed is red and grows.
(ii) Calculate the probability that the chosen seed grows.
(iii) Given that the seed grows, calculate the probability that it is red.

27 Two unbiased six-sided dice are rolled, a red one and a black one. Let E and F be the events

E: the same number appears on both dice
F: the sum of the numbers is 10.

Find **a)** $\text{P}(E)$
b) $\text{P}(F)$
c) $\text{P}(E \cup F)$.

28 The table below shows the subjects studied by 210 students at a college.

	Year 1	Year 2	Totals
History	50	35	85
Science	15	30	45
Art	45	35	80
Totals	110	100	210

a) A student from the college is selected at random.

Let A be the event the student studies art.
Let B be the event the student is in year 2.

(i) Find $\text{P}(A)$.
(ii) Find the probability that the student is a year 2 art student.
(iii) Are the events A and B independent? Justify your answer.

b) Given that a history student is selected at random, calculate the probability that the student is in year 1.

c) Two students are selected at random from the college. Calculate the probability that one student is in year 1 and the other in year 2.

29 A bag contains 2 red balls, 3 blue balls and 4 green balls. A ball is chosen at random from the bag and is not replaced. A second ball is chosen. Find the probability of choosing one green ball and one blue ball in any order.

30 In a bilingual school there is a class of 21 pupils. In this class, 15 of the pupils speak Spanish as their first language and 12 of these 15 pupils are Argentinian. The other 6 pupils in the class speak English as their first language and 3 of these 6 pupils are Argentinian.

A pupil is selected at random from the class and is found to be Argentinian. Find the probability that the pupil speaks Spanish as his/her first language.

31 A new blood test has been shown to be effective in the early detection of a disease. The probability that the blood test correctly identifies someone with this disease is 0.99, and the probability that the blood test correctly identifies someone without that disease is 0.95. The incidence of this disease in the general population is 0.0001.

A doctor administered the blood test to a patient and the test result indicated that this patient had the disease. What is the probability that the patient has the disease?

32 The local Football Association consists of ten teams. Team A has a 40% chance of winning any game against a higher-ranked team, and a 75% chance of winning any game against a lower-ranked team. If A is currently in fourth position, find the probability that A wins its next game.

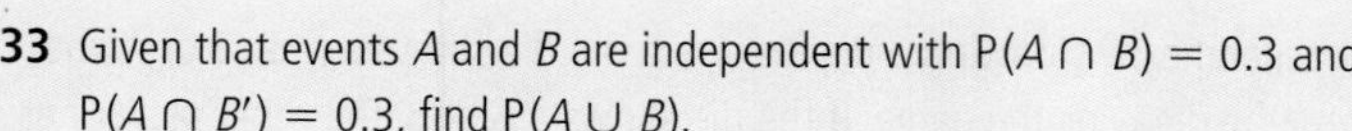

33 Given that events A and B are independent with $P(A \cap B) = 0.3$ and $P(A \cap B') = 0.3$, find $P(A \cup B)$.

34 A girl walks to school every day. If it is not raining, the probability that she is late is $\frac{1}{5}$. If it is raining, the probability that she is late is $\frac{2}{3}$. The probability that it rains on a particular day is $\frac{1}{4}$.

On one particular day the girl is late. Find the probability that it was raining on that day.

35 Given that $P(x) = \frac{2}{3}$, $P(y|x) = \frac{2}{5}$ and $P(y|x') = \frac{1}{4}$, find

a) $P(y')$ **b)** $P(x' \cup y')$.

36 The probability that a man leaves his umbrella in any shop he visits is $\frac{1}{3}$. After visiting two shops in succession, he finds he has left his umbrella in one of them. What is the probability that he left his umbrella in the second shop?

37 Two women, Ann and Bridget, play a game in which they take it in turns to throw an unbiased six-sided die. The first woman to throw a '6' wins the game. Ann is the first to throw.

a) Find the probability that

(i) Bridget wins on her first throw

(ii) Ann wins on her second throw

(iii) Ann wins on her nth throw.

b) Let p be the probability that Ann wins the game. Show that $p = \frac{1}{6} + \frac{25}{36}p$.

c) Find the probability that Bridget wins the game.

d) Suppose that the game is played six times. Find the probability that Ann wins more games than Bridget.

38 A box contains 22 red apples and 3 green apples. Three apples are selected at random, one after the other, without replacement.

a) The first two apples are green. What is the probability that the third apple is red?

b) What is the probability that exactly two of the three apples are red?

39 The probability that it rains during a summer's day in a certain town is 0.2. In this town, the probability that the daily maximum temperature exceeds 25 °C is 0.3 when it rains and 0.6 when it does not rain. Given that the maximum daily temperature exceeded 25 °C on a particular summer's day, find the probability that it rained on that day.

40 An integer is chosen at random from the first one thousand positive integers. Find the probability that the integer chosen is

a) a multiple of 4

b) a multiple of **both** 4 and 6.

41 The independent events A and B are such that $P(A) = 0.4$ and $P(A \cup B) = 0.88$. Find

a) $P(B)$

b) the probability that either A occurs or B occurs but **not both**.

42 Robert travels to work by train every weekday from Monday to Friday. The probability that he catches the 08:00 train on Monday is 0.66. The probability that he catches the 08:00 train on any other weekday is 0.75. A weekday is chosen at random.

a) Find the probability that he catches the train on that day.

b) Given that he catches the 08:00 train on that day, find the probability that the chosen day is Monday.

43 Jack and Jill play a game, by throwing a die in turn. If the die shows a 1, 2, 3 or 4, the player who threw the die wins the game. If the die shows a 5 or 6, the other player has the next throw. Jack plays first and the game continues until there is a winner.

a) Write down the probability that Jack wins on his first throw.

b) Calculate the probability that Jill wins on her first throw.

c) Calculate the probability that Jack wins the game.

44 Bag A contains 2 red and 3 green balls.

a) Two balls are chosen at random from the bag without replacement. Find the probability that 2 red balls are chosen.

Bag B contains 4 red and n green balls.

b) Two balls are chosen without replacement from this bag. If the probability that two red balls are chosen is $\frac{2}{15}$, show that $n = 6$.

A standard die with six faces is rolled. If a 1 or 6 is obtained, two balls are chosen from bag A, otherwise two balls are chosen from bag B.

c) Calculate the probability that two red balls are chosen.

d) Given that two red balls are chosen, find the probability that a 1 or a 6 was obtained on the die.

45 Given that $(A \cup B)' = \varnothing$, $\mathrm{P}(A'|B) = \frac{1}{3}$ and $\mathrm{P}(A) = \frac{6}{7}$, find $\mathrm{P}(B)$.

13 Differential Calculus I: Fundamentals

Assessment statements

6.1 Informal ideas of limit, continuity and convergence.

Definition of derivative from first principles: $f'(x) = \lim_{h \to 0} \frac{f(x+h) - f(x)}{h}$.

Derivative interpreted as gradient function and as rate of change.
Find equations of tangents and normals.
Identifying increasing and decreasing functions.
The second derivative.

6.2 Derivative of x^n.

6.3 Local maximum and minimum points.
Points of inflexion with zero and non-zero gradients.
Graphical behaviour of functions including the relationship between the graphs of f, f' and f''.

6.6 Kinematic problems involving displacement, s, velocity, v, and acceleration, a. (See also Chapter 15.)

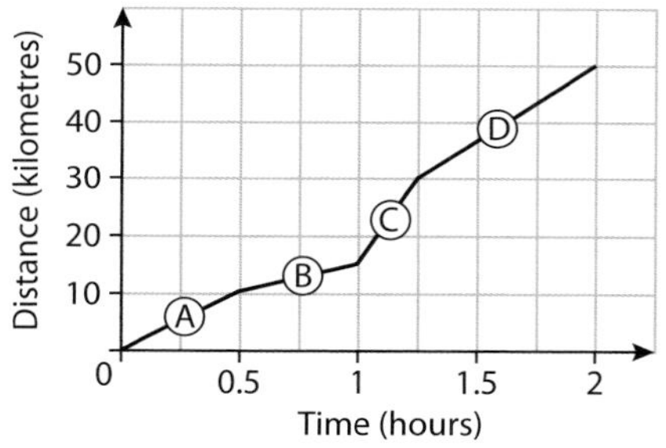

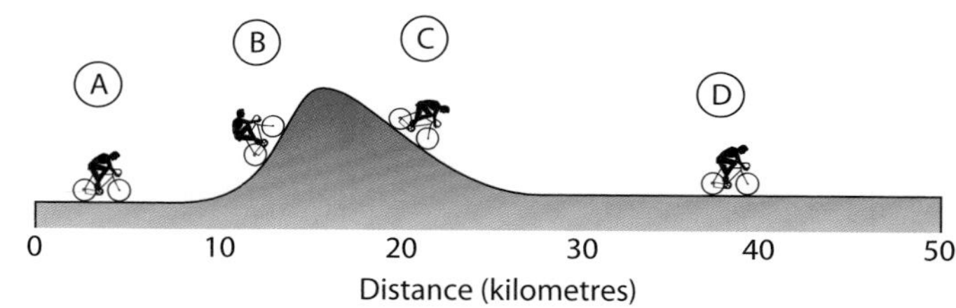

A bicycle ride over a hill: The graph above left shows distance (km) versus time (hrs) for a 50-kilometre bicycle ride that included going up and then down a steep hill. There are four time intervals labelled A, B, C and D. During which interval is the bicyclist's speed the least? the greatest? During which two intervals is the bicyclist's speed about the same? How does the shape of the distance-time graph give information about the speed of the bicyclist during a certain interval? or at a particular moment (instant) during the ride?

Introduction

Calculus is the branch of mathematics that was developed to analyze and model change – such as velocity and acceleration. We can also apply it to study change in the context of slope, area, volume and a wide range of other real-life phenomena. Although mathematical techniques that you have studied previously deal with many of these concepts, the ability to model change was restricted. For example, consider the curve in Figure 13.1. This shows the motion of an object by indicating the distance (y metres) travelled after a certain amount of time (t seconds). Pre-calculus mathematics will only allow us to compute the **average velocity** between two different times (Figure 13.2). With calculus – specifically, techniques of differential calculus – we will be able to find the velocity of the object at a particular instant, known as its **instantaneous velocity** (Figure 13.3). The starting point for our study of calculus is the idea of a limit.

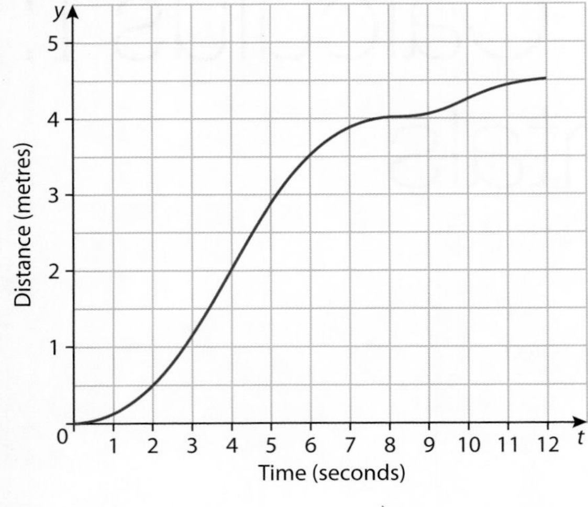

Figure 13.1 Distance–time graph for an object's motion.

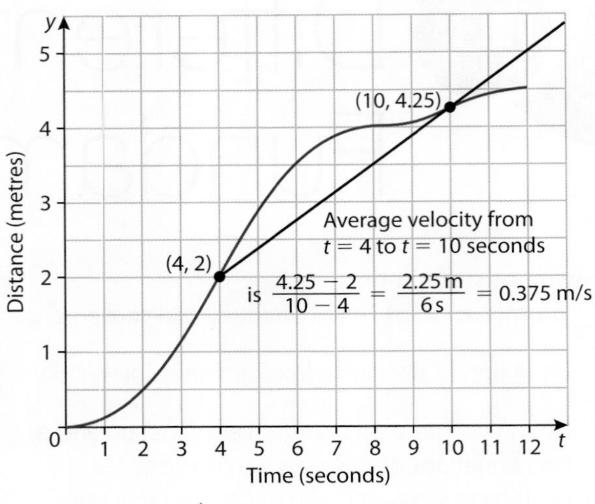

Figure 13.2 Computing average velocity from a distance–time graph.

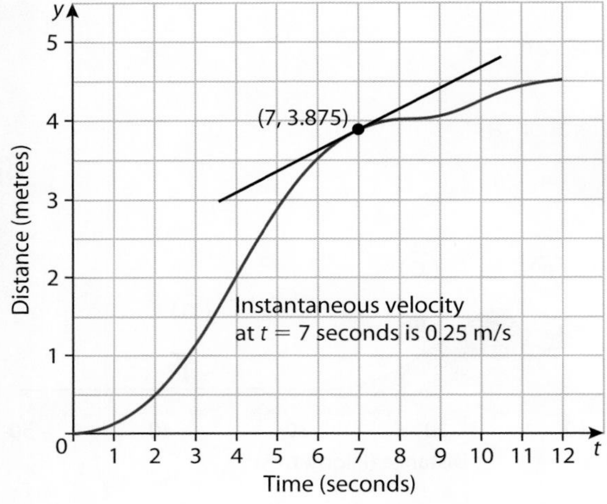

Figure 13.3 Instantaneous velocity from a distance–time graph.

13.1 Limits of functions

A **limit** is one of the ideas that distinguish calculus from algebra, geometry and trigonometry. The notion of a limit is a fundamental concept of calculus. Limits are not new to us. We often use the idea of a 'limit' in many non-mathematical situations. Mathematically speaking, we have encountered limits on at least two occasions previously in this book – finding the sum of an infinite geometric series (Section 4.4) and computing the irrational number e (Section 5.3).

Recall from Section 4.4 that we established that if the sequence of partial sums for an infinite series **converges** to a finite number L we say that the infinite series has a 'sum' of L. Further on in that section, we used limits to algebraically confirm that the infinite series $2 + 1 + \frac{1}{2} + \frac{1}{4} + \frac{1}{8} + \ldots$ has a sum of 4. As part of the algebra for this, we reasoned that as the value of n increases in the positive direction without bound (i.e. $n \to +\infty$) the expression $\left(\frac{1}{2}\right)^n$ converges to zero – in other words, the **limit** of $\left(\frac{1}{2}\right)^n$ as n goes to positive infinity is zero. We express this result more efficiently using limit notation, as we did in Chapter 4, by writing $\lim_{n \to \infty}\left(\frac{1}{2}\right)^n = 0$.

It is beyond the requirements of this course to establish a precise formal definition of a limit, but a closer look at justifying this limit and a couple of others can lead us to a useful informal definition.

Example 1

Evaluate $\lim_{n \to \infty} \left(\frac{1}{2}\right)^n$ by using your GDC to analyze the behaviour of the function $f(x) = \left(\frac{1}{2}\right)^x$ for large positive values of x.

Solution

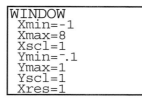

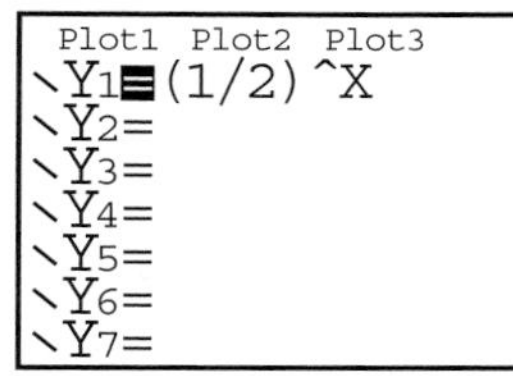

WINDOW
Xmin=-1
Xmax=8
Xscl=1
Ymin=-.1
Ymax=1
Yscl=1
Xres=1

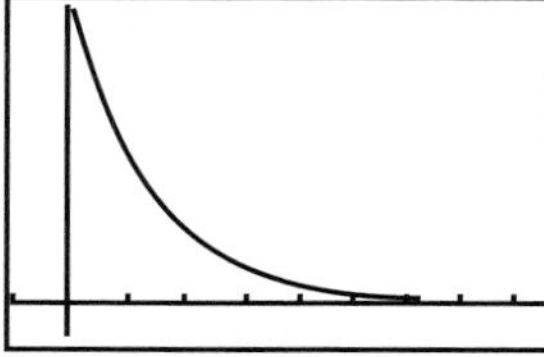

TABLE SETUP
TblStart=0
ΔTbl=1
Indpnt: Auto Ask
Depend: Auto Ask

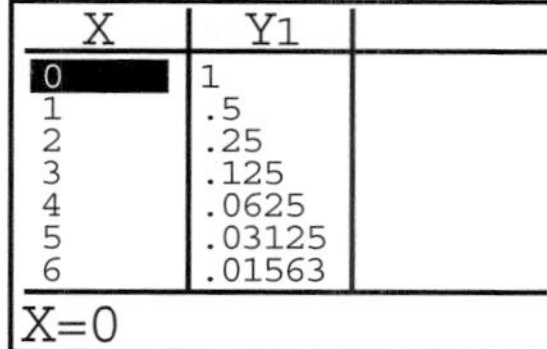

X	Y1
0	1
1	.5
2	.25
3	.125
4	.0625
5	.03125
6	.01563

X=0

X	Y1
7	.00781
8	.00391
9	.00195
10	9.8E-4
11	4.9E-4
12	2.4E-4
13	1.2E-4

Y1=1.22070313E-4

The GDC screen images show the graph and table of values for $y = \left(\frac{1}{2}\right)^x$. Clearly, the larger the value of x, the closer that y gets to zero. Although there is no value of x that will produce a value of y equal to zero, we can get as close to zero as we wish. For example, if we wish to produce a value of y within 0.001 of zero, then we could choose $x = 10$ and $y = \left(\frac{1}{2}\right)^{10} = \frac{1}{1024}$ $\approx 0.000\,976\,56$; and if we want a result within 0.000 0001 of zero, then we could choose $x = 24$ and $y = \left(\frac{1}{2}\right)^{24} = \frac{1}{16\,777\,216} \approx 0.000\,000\,059\,605$; and so on. Therefore, we can conclude that $\lim_{n \to \infty} \left(\frac{1}{2}\right)^n = 0$.

In calculus we are interested in limits of functions of real numbers. Although many of the limits of functions that we will encounter can only be approached and not actually reached (as in Example 1), this is not always the case. For example, if asked to evaluate the limit of the function $f(x) = \frac{x}{2} - 1$ as x approaches 6, we simply need to evaluate the function for $x = 6$. Since $f(6) = 2$, then $\lim_{x \to 6} \left(\frac{x}{2} - 1\right) = 2$.

However, it is more common that we are unable to evaluate the limit of $f(x)$ as x approaches some number c because $f(c)$ does not exist.

The line $y = c$ is a **horizontal asymptote** of the graph of a function $y = f(x)$ if either $\lim_{x \to \infty} f(x) = c$ or $\lim_{x \to -\infty} f(x) = c$. For example, the line $y = 0$ (x-axis) is a horizontal asymptote of the graph of $y = \left(\frac{1}{2}\right)^x$ because $\lim_{x \to \infty} \left(\frac{1}{2}\right)^x = 0$.

Example 2

Find the following two limits using your GDC to analyze the behaviour of the relevant function.

a) $\lim_{x \to 0} \frac{\sin x}{x}$ b) $\lim_{x \to 0} \frac{\cos x - 1}{x}$

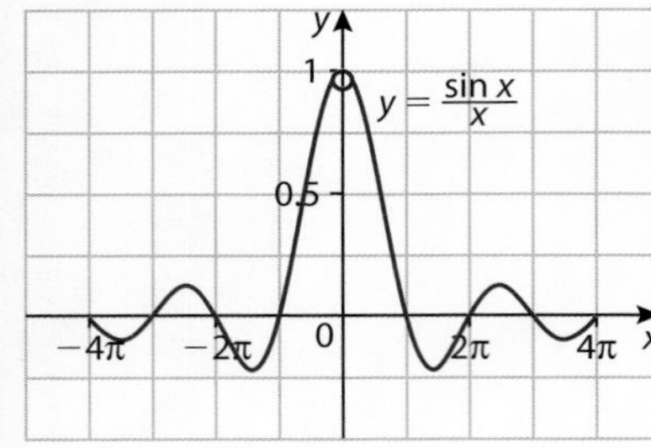


```
Plot1 Plot2 Plot3
\Y1■ sin(X)/X
\Y2=
\Y3=
\Y4=
\Y5=
\Y6=
\Y7=
```

```
Y1(-.05)
         .9995833854
Y1(.05)
         .9995833854
1-Ans
     4.166145864E-4
```

```
Y1(-.002)
         .9999993333
Y1(.002)
         .9999993333
1-Ans
       6.66666655E-7
```

```
Plot1 Plot2 Plot3
\Y1■ (cos(X)-1)/X
\Y2=
\Y3=
\Y4=
\Y5=
\Y6=
```

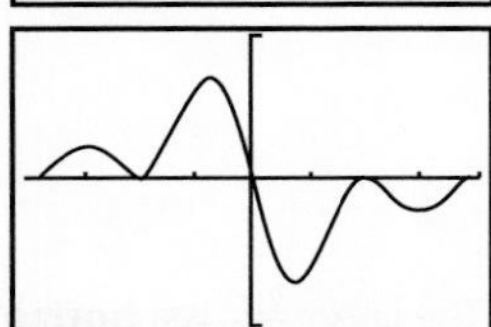

X	Y1
-.03	.015
-.02	.01
-.01	.005
0	ERROR
.01	-.005
.02	-.01
.03	-.015

X=-.03

Solution

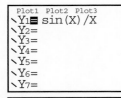

a) We are not able to evaluate this limit by direct substitution because when $x = 0$, $\frac{\sin x}{x} = \frac{0}{0}$ and is therefore undefined. Let's use our GDC again to analyze the behaviour of the function $f(x) = \frac{\sin x}{x}$ as x approaches zero from the right side and the left side.

Although there is no point on the graph of $y = \frac{\sin x}{x}$ corresponding to $x = 0$, it is clear from the graph that as x approaches zero (from either direction) the value of $\frac{\sin x}{x}$ converges to one. We can get the value of $\frac{\sin x}{x}$ arbitrarily close to 1 depending on our choice of x.

If we want $\frac{\sin x}{x}$ to be within 0.001 of 1, we choose $x = 0.05$ giving $\frac{\sin 0.05}{0.05} \approx 0.999\,583$ and $1 - 0.999\,583 = 0.000\,417 < 0.001$; and if we want $\frac{\sin x}{x}$ to be within 0.000 001 of 1, then we choose $x = 0.002$ giving $\frac{\sin 0.02}{0.02} \approx 0.999\,999\,3333$ and $1 - 0.999\,999\,3333$ $= 0.000\,000\,6667 < 0.000\,001$; and so on.

Therefore, $\lim_{x \to 0} \frac{\sin x}{x} = 1$.

b) As with $y = \frac{\sin x}{x}$, substituting $x = 0$ into the function $y = \frac{\cos x - 1}{x}$ produces the meaningless fraction $\frac{0}{0}$. The graph of $y = \frac{\cos x - 1}{x}$ (GDC images right) reveals that the function values approach 0 as x goes to 0. A table produced on a GDC also shows that the function values approach zero from both directions.

Therefore, $\lim_{x \to 0} \frac{\cos x - 1}{x} = 0$.

These two limits are confirmed analytically in the next section.

The analysis and result for Example 2 illustrates why it is preferred (and often necessary) that in calculus the argument of a trigonometric function be in radian measure rather than degrees. The limit of $\frac{\sin x}{x}$ as x goes to ∞ is not equal to 1 if x is in degrees. With your GDC in radian mode, duplicate the graph of $y = \frac{\sin x}{x}$ shown in the solution for Example 2. Now change the window dimensions on your GDC to Xmin = −720 and Xmax = 720 (equivalent to -4π and 4π) and graph $y = \frac{\sin x}{x}$ in degree mode. Explain why no graph appears.

It is interesting to note that if you ask your GDC to evaluate the function $y = \frac{\sin x}{x}$ for a sufficiently small value of x it will give a result of exactly 1. The function is undefined for $x = 0$ and can never give a result of exactly 1, so obviously the calculator is making an error. Due to memory restrictions the calculator has rounded off the result to 1. The GDC image below shows that for $x = 0.000\,01$ the result has been rounded to 1 when the actual value is $0.999\,999\,999\,98\overline{3}$ (digit 3 repeating). Even the Google calculator (see image below) gives an incorrect result.

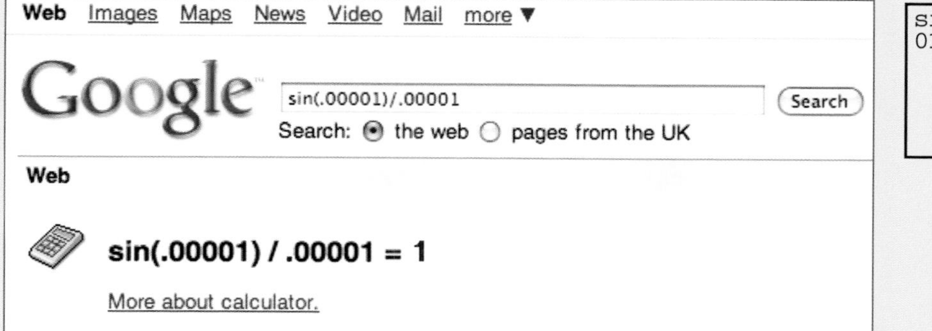


```
sin(.00001)/.000
01
               1
```

Functions do not necessarily converge to a finite value at every point – it's possible for a limit not to exist.

Example 3

Find $\lim\limits_{x\to 0}\frac{1}{x^2}$, if it exists.

Solution

As x approaches zero, the value of $\frac{1}{x^2}$ becomes increasingly large in the positive direction. The graph of the function (left) seems to indicate that we can make the values of $y = \frac{1}{x^2}$ arbitrarily large by choosing x close enough to zero. Therefore, the values of $y = \frac{1}{x^2}$ do not approach a finite number, so $\lim\limits_{x\to 0}\frac{1}{x^2}$ does not exist.

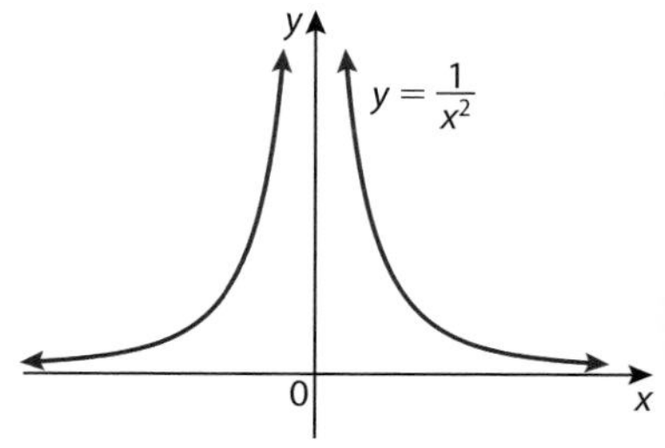

Although we can describe the behaviour of the function $y = \frac{1}{x^2}$ by writing $\lim\limits_{x\to 0}\frac{1}{x^2} = \infty$, this does not mean that we consider ∞ to represent a number – it does not. This notation is simply a convenient way to indicate in what manner the limit does not exist.

The line $x = c$ is a **vertical asymptote** of the graph of a function $y = f(x)$ if either $\lim\limits_{x\to c} f(x) = \infty$ or $\lim\limits_{x\to c} f(x) = -\infty$.
For example, the line $x = 0$ (y-axis) is a vertical asymptote of the graph of $y = \frac{1}{x^2}$ because $\lim\limits_{x\to 0}\frac{1}{x^2} = \infty$.

Limit of a function

If $f(x)$ becomes arbitrarily close to a unique finite number L as x approaches c from either side, then the **limit** of $f(x)$ as x approaches c is L. The notation for indicating this is $\lim\limits_{x\to c} f(x) = L$.
When a function $f(x)$ becomes *arbitrarily close* to a finite number L, we say that $f(x)$ **converges** to L.

Often when trying to determine the limit of a quotient by direct substitution, we may get a meaningless fraction such as $\frac{0}{0}$ or $\frac{\infty}{\infty}$. Such an expression is called an **indeterminate form** because we cannot use it to determine the desired limit. When confronted with an indeterminate form we need to perform some algebraic manipulation to the quotient to get it into a form so that the limit can be evaluated by direct substitution and/or applying known limits.

For our purposes in this course, it is also important to be able to apply some basic algebraic manipulation in order to evaluate the limits of some functions algebraically, rather than by conjecturing from a graph or table.

The following five properties of limits are also useful.

Properties of limits

Let a and b be real numbers, and let f and g be functions with the following limits.

$\lim_{x\to a} f(x) = L$ and $\lim_{x\to a} g(x) = K$

1 Constant: $\lim_{x\to a} b = b$

2 Scalar multiple: $\lim_{x\to a} [b \cdot f(x)] = b \cdot L$

3 Sum or difference: $\lim_{x\to a} [f(x) \pm g(x)] = L \pm K$

4 Product: $\lim_{x\to a} [f(x) \cdot g(x)] = L \cdot K$

5 Quotient: $\lim_{x\to a} \left[\frac{f(x)}{g(x)}\right] = \frac{L}{K} \quad K \neq 0$

Example 4

Evaluate each limit algebraically.

a) $\lim_{x\to\infty} \frac{5x-3}{x}$

b) $\lim_{p\to 0} (3x^2 - 4px + p^2)$

c) $\lim_{h\to 0} \frac{[(x+h)^2 - 6] - (x^2 - 6)}{h}$

d) $\lim_{x\to\infty} \frac{3x^2 + 5x - 1}{2x^2 + 1}$

Solution

a) $$\lim_{x\to\infty} \frac{5x-3}{x} = \lim_{x\to\infty}\left(\frac{5x}{x} - \frac{3}{x}\right)$$ Split the fraction into two terms and ...

$$= \lim_{x\to\infty} 5 - \lim_{x\to\infty} \frac{3}{x}$$ ... evaluate the limit of each term separately.

$$= 5 - 0 = 5$$ Therefore, $\lim_{x\to\infty} \frac{5x-3}{x} = 5$.

b) $$\lim_{p\to 0} (3x^2 - 4px + p^2) = \lim_{p\to 0} 3x^2 - \lim_{p\to 0} 4px + \lim_{p\to 0} p^2$$ Evaluate the limit of each term separately.

$$= 3x^2 - 0 + 0 = 3x^2$$ Therefore, $\lim_{p\to 0} (3x^2 - 4px + p^2) = 3x^2$.

c) $$\lim_{h\to 0} \frac{[(x+h)^2 - 6] - (x^2 - 6)}{h} = \lim_{h\to 0} \frac{x^2 + 2xh + h^2 - 6 - x^2 + 6}{h}$$

$$= \lim_{h\to 0} \frac{2xh + h^2}{h}$$

$$= \lim_{h\to 0} \frac{h(2x + h)}{h}$$

$$= \lim_{h\to 0} 2x + \lim_{h\to 0} h$$

$$= 2x + 0 = 2x$$

Therefore, $\lim_{h\to 0} \frac{[(x+h)^2 - 6] - (x^2 - 6)}{h} = 2x.$

d) $\lim_{x \to \infty} \frac{3x^2 + 5x - 1}{2x^2 + 1}$

Dividing numerator and denominator by largest power of x, i.e. x^2.

$= \lim_{x \to \infty} \frac{\frac{3x^2}{x^2} + \frac{5x}{x^2} - \frac{1}{x^2}}{\frac{2x^2}{x^2} + \frac{1}{x^2}}$

$= \lim_{x \to \infty} \frac{3 + \frac{5}{x} - \frac{1}{x^2}}{2 + \frac{1}{x^2}}$

$= \frac{\lim_{x \to \infty} 3 + \lim_{x \to \infty} \frac{5}{x} - \lim_{x \to \infty} \frac{1}{x^2}}{\lim_{x \to \infty} 2 + \lim_{x \to \infty} \frac{1}{x^2}}$

Applying $\lim_{x \to a} \left[\frac{f(x)}{g(x)}\right] = \frac{L}{K}$ and $\lim_{x \to a} [f(x) \pm g(x)] = L \pm K$.

$= \frac{3 + 0 - 0}{2 + 0}$

Therefore, $\lim_{x \to \infty} \frac{3x^2 + 5x - 1}{2x^2 + 1} = \frac{3}{2}$.

The limits in parts b) and c) of Example 4 show that in some cases the limit of a function is itself a function.

Connect the limit in Example 4 part d) with the end behaviour of rational functions covered in Section 3.4. Since $\lim_{x \to \infty} \frac{3x^2 + 5x - 1}{2x^2 + 1} = \frac{3}{2}$, the rational function $y = \frac{3x^2 + 5x - 1}{2x^2 + 1}$ will have a horizontal asymptote of $y = \frac{3}{2}$. In other words, as $x \to \infty$ or $x \to -\infty$ the function will approach the value of $y = \frac{3}{2}$ as illustrated in the graph shown.

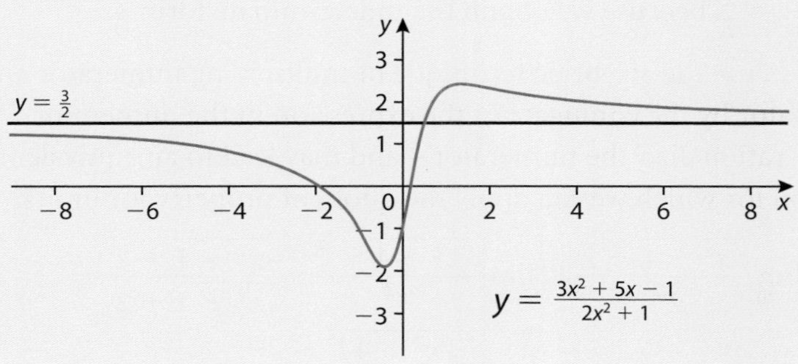

In this section we evaluated limits by guessing and checking with the help of our GDC. This process led us to conclude that $\lim_{x \to 0} \frac{\sin x}{x} = 1$ and $\lim_{x \to 0} \cos \frac{x - 1}{x} = 0$.

It was reasonable to take this approach since it is not possible to perform algebraic manipulations on these expressions as we did with the expressions in Example 4. However, if possible we should always try to use analytic methods to evaluate a limit as illustrated in the next example.

Example 5

a) Estimate the value of $\lim_{x \to 0} \frac{\sqrt{x^2 + 4} - 2}{x^2}$ by evaluating the function $f(x) = \frac{\sqrt{x^2 + 4} - 2}{x^2}$ for $x = \pm 0.5, \pm 0.01, \pm 0.0001, \pm 0.000005, \pm 0.000001$.

b) Using algebra and properties of limits, evaluate $\lim_{x\to 0}\frac{\sqrt{x^2+4}-2}{x^2}$.

c) Comment on the two results.

Solution

a) A GDC that displays results to an accuracy of ten significant figures gives the following results.

x	$f(x)=\frac{\sqrt{x^2+4}-2}{x^2}$
± 0.5	0.246 211 2512
± 0.01	0.249 998 438
± 0.0001	0.25
$\pm 0.000\,005$	0.248
$\pm 0.000\,003$	0.244 444 4444
$\pm 0.000\,001$	0

The GDC results in the table seem unusual. Initially as x approaches zero from either direction the function values appear to be approaching $\frac{1}{4}$, but then as the function is evaluated for values even closer to zero, the function values continue to decrease to zero.

Is $\lim_{x\to 0}\frac{\sqrt{x^2+4}-2}{x^2}$ equal to $\frac{1}{4}$ or 0?

If we trust our GDC, we may be tempted to conclude that $\lim_{x\to 0}\frac{\sqrt{x^2+4}-2}{x^2}=0$.

b) We cannot immediately apply the limit property for quotients, $\lim_{x\to a}\left[\frac{f(x)}{g(x)}\right]=\frac{L}{K}$ because we obtain the indeterminate form $\frac{0}{0}$.

We need to use the algebraic technique of multiplying numerator and denominator by the conjugate of the expression in the numerator. This will 'rationalize' the numerator – and may lead to an equivalent expression for which we can apply the quotient property for limits.

$$\begin{aligned}\lim_{x\to 0}\frac{\sqrt{x^2+4}-2}{x^2} &= \lim_{x\to 0}\frac{\sqrt{x^2+4}-2}{x^2}\cdot\frac{\sqrt{x^2+4}+2}{\sqrt{x^2+4}+2}\\ &= \lim_{x\to 0}\frac{\left(\sqrt{x^2+4}\right)^2-2^2}{x^2\left(\sqrt{x^2+4}+2\right)}\\ &= \lim_{x\to 0}\frac{x^2+4-4}{x^2\left(\sqrt{x^2+4}+2\right)}\\ &= \lim_{x\to 0}\frac{\cancel{x^2}}{\cancel{x^2}\left(\sqrt{x^2+4}+2\right)}\\ &= \lim_{x\to 0}\frac{1}{\sqrt{x^2+4}+2}\\ &= \frac{\lim_{x\to 0}1}{\lim_{x\to 0}\sqrt{x^2+4}+2}=\frac{1}{\sqrt{4}+2}=\frac{1}{4}\end{aligned}$$

Therefore, $\lim_{x\to 0}\frac{\sqrt{x^2+4}-2}{x^2}=\frac{1}{4}$.

c) Because of memory limitations a GDC will sometimes give a false value. Because $\sqrt{x^2 + 4}$ is very close to 2 when x is very small, a GDC will eventually consider $\sqrt{x^2 + 4}$ to be equal to 2.000 000 00 … (to as many digits as the GDC is capable of computing) when x is sufficiently small. Your GDC is a very powerful tool, but like any tool it does have its limitations.

Exercise 13.1

In questions 1–4, evaluate each limit algebraically and then confirm your result by means of a table or graph on your GDC.

1 $\lim_{n\to\infty} \frac{1 + 4n}{n}$ **2** $\lim_{h\to 0} (3x^2 + 2hx + h^2)$

3 $\lim_{d\to 0} \frac{(x + d)^2 - x^2}{d}$ **4** $\lim_{x\to 3} \frac{x^2 - 9}{x - 3}$

In questions 5–7, investigate the limit of the expression (if it exists) as $x \to \infty$ by evaluating the expression for the following values of x: 10, 50, 100, 1000, 10 000 and 1 000 000. Hence, make a conjecture for the value of each limit.

5 $\lim_{x\to\infty} \frac{3x + 2}{x^2 - 3}$ **6** $\lim_{x\to\infty} \frac{5x - 6}{2x + 5}$ **7** $\lim_{x\to\infty} \frac{3x^2 + 2}{x - 3}$

In questions 8–13, find the limit, if it exists.

8 $\lim_{x\to 4} \frac{x - 4}{x^2 - 16}$ **9** $\lim_{x\to 1} \frac{x^2 + x - 2}{x^2 - 1}$

10 $\lim_{x\to 0} \frac{\sqrt{2 + x} - \sqrt{2}}{x}$ • **Hint:** multiply numerator and denominator by conjugate of numerator

11 $\lim_{x\to\infty} \frac{x^3 - 1}{4x^3 - 3x + 1}$

12 $\lim_{x\to 0} \frac{\tan x}{x}$ • **Hint:** rewrite $\tan x$ as $\frac{\sin x}{\cos x}$ and apply property $\lim_{x\to 0} [f(x) \cdot g(x)] = L \cdot K$

13 $\lim_{\theta\to 0} \frac{\sin 3\theta}{\theta}$ • **Hint:** rewrite $\frac{\sin 3\theta}{\theta}$ as $\left(3\frac{\sin 3\theta}{3\theta}\right)$ and apply $\lim_{x\to 0} \frac{\sin x}{x} = 1$

14 Use the graphing or table capabilities of your GDC to investigate the values of the expression $\left(1 + \frac{1}{c}\right)^c$ as c increases without bound (i.e. $c \to \infty$). Explain the significance of the result.

15 If it is known that the line $y = 3$ is a horizontal asymptote for the function $f(x)$, state the value of each of the following two limits: $\lim_{x\to\infty} f(x)$ and $\lim_{x\to -\infty} f(x)$.

16 If it is known that the line $x = a$ is a vertical asymptote for the function $g(x)$ and $g(x) > 0$, what conclusion can be made about $\lim_{x\to a} g(x)$?

17 State the equations of all horizontal and vertical asymptotes for the following functions. Confirm using your GDC.

a) $f(x) = \frac{3x - 1}{1 + x}$ b) $g(x) = \frac{1}{(x - 2)^2}$ c) $g(x) = \frac{1}{x - a} + b$

d) $R(x) = \frac{2x^2 - 3}{x^2 - 9}$ e) $d(x) = \frac{5 - 3x}{x^2 - 5x}$ f) $p(x) = \frac{x^2 - 4}{x - 4}$

For questions 18 and 19, a) use your GDC to estimate the limit, and b) use analytic methods to evaluate the limit.

18 $\lim_{x \to 2} \frac{\sqrt{x^2+5}-3}{x^2-2x}$ **19** $\lim_{x \to +\infty} \frac{4x-1}{\sqrt{x^2+2}}$

20 Show that $\lim_{h \to 0} \frac{\sqrt{x+h}-\sqrt{x}}{h} = \frac{1}{2\sqrt{x}}$.

21 Show that $\lim_{h \to 0} \frac{\frac{1}{x+h}-\frac{1}{x}}{h} = -\frac{1}{x^2}$.

13.2 The derivative of a function: definition and basic rules

Tangent lines and the slope (gradient) of a curve

In Section 1.6, we reviewed linear equations in two variables. And, later in Section 2.1, we established that any non-vertical line represents a function for which we typically assign the variables x and y for values in the domain and range of the function, respectively. Any linear function can be written in the form $y = mx + c$. This is the slope-intercept form for a linear equation, where m is the slope (or gradient) of the graph and c is the y-coordinate of the point at which the graph intersects the y-axis (i.e. the y-intercept). The value of the slope m, defined as $m = \frac{y_2 - y_1}{x_2 - x_1} = \frac{\text{vertical change}}{\text{horizontal change}}$, will be the same for any pair of points, (x_1, y_1) and (x_2, y_2), on the line. An essential characteristic of the graph of a linear function is that it has a constant slope. This is not true for the graphs of non-linear functions.

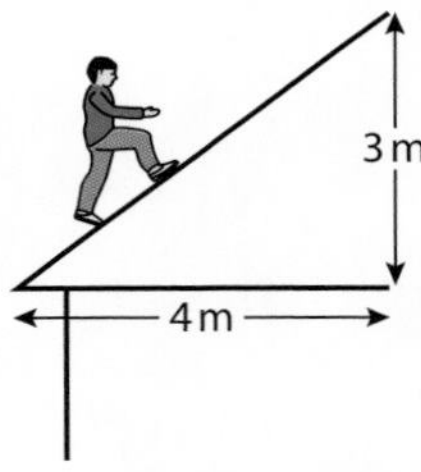

Figure 13.4 Slope of a straight line.

Consider a person walking up the side of a pitched roof as shown in Figure 13.4. At *any* point along the line segment PQ the person is experiencing a slope of $\frac{3}{4}$. Now consider someone walking up the curve shown in Figure 13.5, which passes through the three points A, B and C. As the person walks along the curve from A to C, he/she will experience a steadily increasing slope. The slope is continually changing from one point to the next along the curve. Therefore, it is incorrect to say that a non-linear function, whose graph is a curve, has ***a*** slope – it has *infinitely many* slopes. We need a means to determine the slope of a non-linear function *at a specific point* on its graph.

Figure 13.5 Slope of a curve.

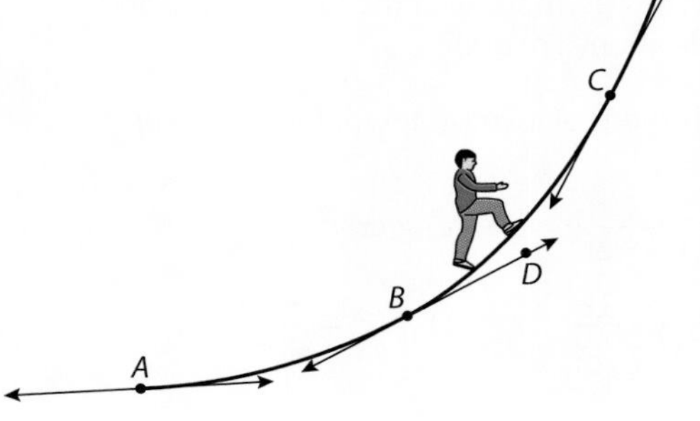

Imagine if the slope of the curve in Figure 13.5 stopped increasing (remained constant) after point B. From that point on, a person walking up the curve would move along a line with a slope equal to the slope of the curve at point B. This line – containing point D in the diagram – only 'touches'

the curve once at B. Line (BD) is **tangent** to the curve at point B. Therefore, finding the slope of the line that is tangent to a curve at a certain point will give us the slope of the curve at that point.

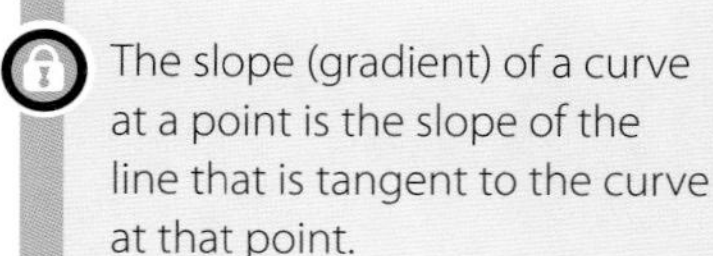
The slope (gradient) of a curve at a point is the slope of the line that is tangent to the curve at that point.

Finding the slope of a curve at a point – or better – finding a rule (function) that gives us the slope at any point on the curve is very useful information in many applications. The slope of a line, or of a curve at a point, is a measure of how fast variable y is changing as variable x changes. **The slope represents the rate of change of y with respect to x.** To find the slope of a tangent line, we first need to clarify what it means to say that a line is tangent to a curve at a point. Then we can establish a method to find the tangent line at a point.

• **Hint:** The word 'curve' can often mean the same as 'function', even if the function is linear.

The three graphs in Figure 13.6 show different configurations of tangent lines. A tangent line may cross or intersect the graph at one or more points.

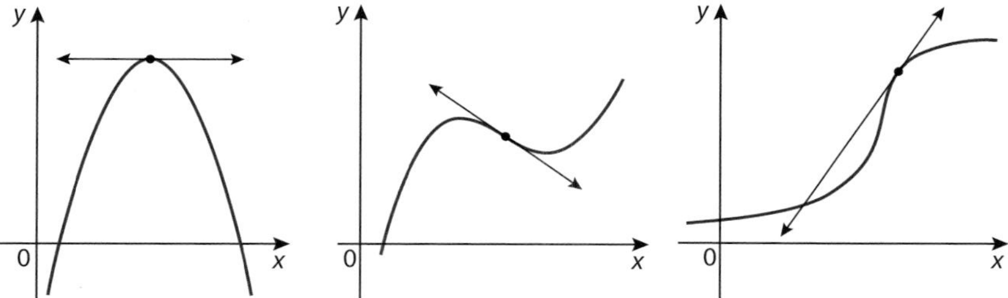

Figure 13.6 Different configurations of lines tangent to a curve.

For many functions, the graph has a tangent at *every* point. Informally, a function is said to be *smooth* if it has this property. Any linear function is certainly smooth, since the tangent at each point coincides with the original graph. However, some graphs are not smooth at every point. Consider the point $(0, 0)$ on the graph of the function $y = |x|$ (Figure 13.7). Zooming in on $(0, 0)$ will always produce a V-shape rather than smoothing out to appear more and more linear. Therefore, there is no tangent to the graph at this point.

Figure 13.7 $y = |x|$

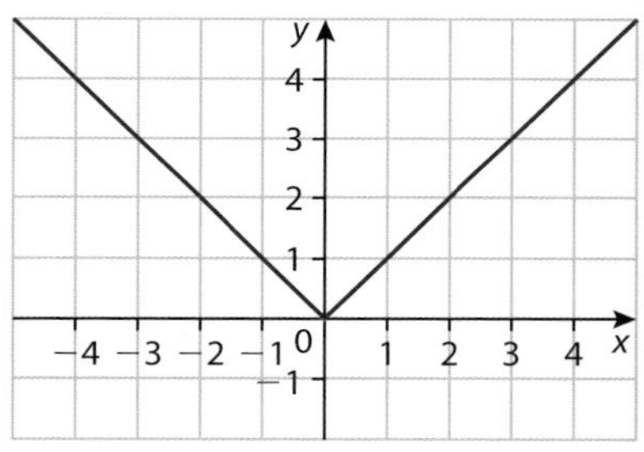

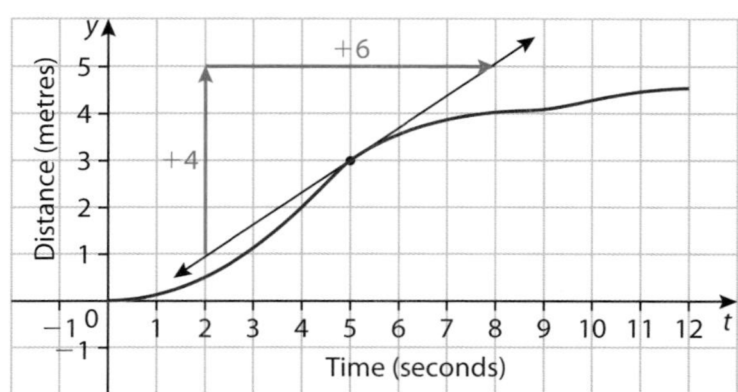

Figure 13.8 Estimating the slope of a tangent line.

One way to find the tangent line of a graph at a particular point is to make a visual estimate. Figure 13.8 reproduces the time-distance graph for an object's motion from the previous section (Figure 13.1). The slope at any point (t, y) on the curve will give us the rate of change of the distance y with respect to time t, in other words the object's **instantaneous velocity** at time t. In the figure, an estimate of the line tangent to the curve at $(5, 3)$ has been drawn. Reading from the graph, the slope appears to be $\frac{4}{6} = \frac{2}{3}$. Or, in other words, the object has a velocity of approximately 0.667 m/s at the instant when $t = 5$ seconds.

A more precise method of finding tangent lines makes use of a secant line and a limit process. Suppose that f is any smooth function, so the tangent to its graph exists at all points. A **secant line** (or chord) is drawn through the point for which we are trying to find a tangent to f and a second point on the graph of f, as shown in Figure 13.9a. If P is the point of tangency with coordinates $(x, f(x))$, choose a point Q to be horizontally some h units away. Hence, the coordinates of point Q are $(x + h, f(x + h))$. Then the slope of the secant line (PQ) is $m_{\text{sec}} = \dfrac{f(x+h) - f(x)}{(x+h) - x} = \dfrac{f(x+h) - f(x)}{h}$.
The right side of this equation is often referred to as a **difference quotient.** The numerator is the change in y, and the denominator h is the change in x. The limit process of achieving better and better approximations for the slope of the tangent at P consists of finding the slope of the secant (PQ) as Q moves ever closer to P, as shown in the graphs in Figure 13.9b and Figure 13.9c. In doing so, the value of h will approach zero.

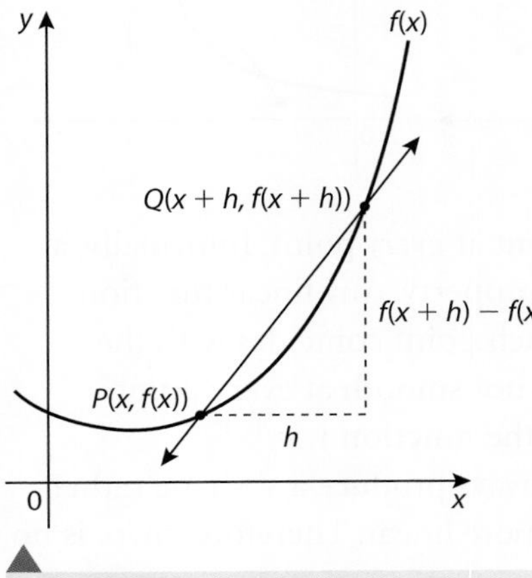

Figure 13.9a

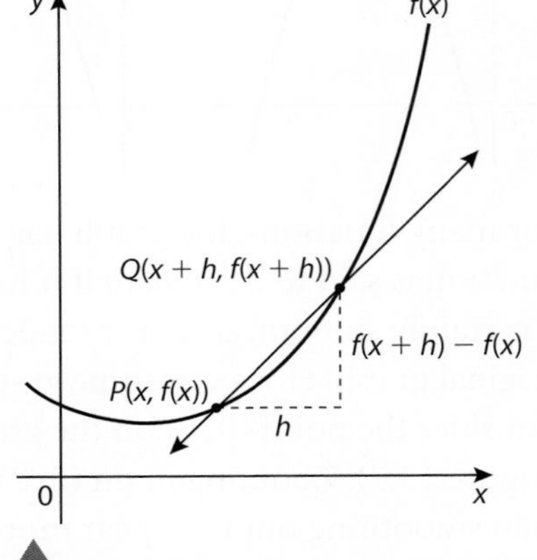

Figure 13.9b

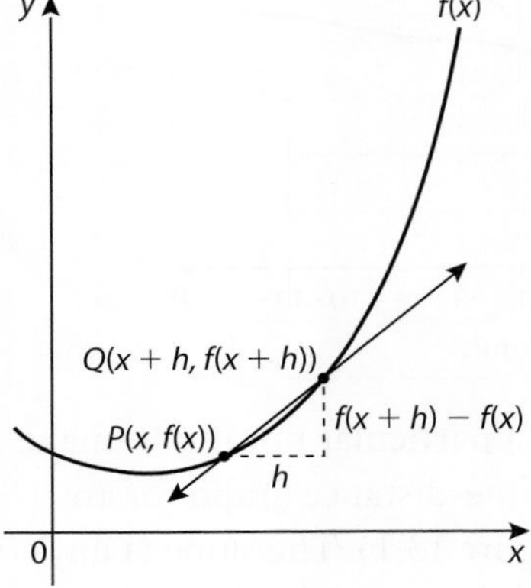

Figure 13.9c As h tends to zero, the secant line becomes a better approximation of the tangent line.

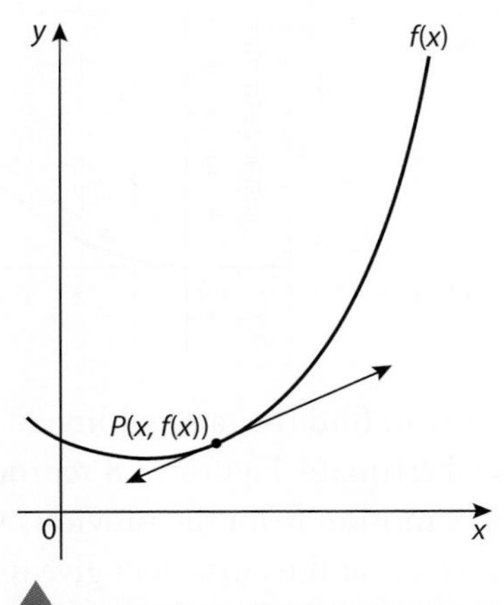

Figure 13.9d Tangent to f at point P.

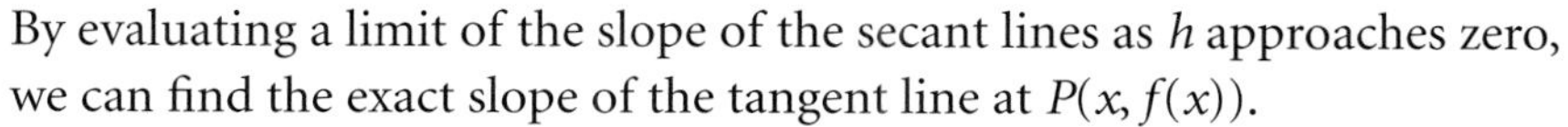

By evaluating a limit of the slope of the secant lines as h approaches zero, we can find the exact slope of the tangent line at $P(x, f(x))$.

> The word 'secant', as applied to a line, comes from the Latin word *secare*, meaning to cut. The word 'tangent' comes from the Latin verb *tangere*, meaning to touch.

The slope (gradient) of a curve at a point

The slope of the curve $y = f(x)$ at the point $(x, f(x))$ is equal to the slope of its tangent line at $(x, f(x))$, and is given by

$$m_{\tan} = \lim_{h \to 0} m_{\sec} = \lim_{h \to 0} \frac{f(x + h) - f(x)}{h}$$

provided that this limit exists.

Let's apply the definition of the slope of a curve at a point to find a rule, or function, for the slope of all of the tangent lines to a curve.

Example 6

Find a rule for the slopes of the tangent lines to the graph of $f(x) = x^2 + 1$. Use this rule to find the exact slope of the curve at the point where $x = 0$ and at the point where $x = 1$.

Solution

Let $(x, f(x))$ represent any point on the graph of f. By definition, the slope of the tangent line at $(x, f(x))$ is:

$$m = \lim_{h \to 0} \frac{f(x + h) - f(x)}{h} = \lim_{h \to 0} \frac{[(x + h)^2 + 1] - [x^2 + 1]}{h}$$

$$= \lim_{h \to 0} \frac{[x^2 + 2xh + h^2 + 1] - [x^2 + 1]}{h}$$

$$= \lim_{h \to 0} \frac{x^2 - x^2 + 2xh + h^2 + 1 - 1}{h}$$

$$= \lim_{h \to 0} \frac{h(2x + h)}{h}$$

$$= \lim_{h \to 0} (2x + h)$$

$$= 2x$$

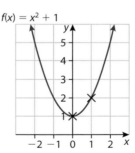

Therefore, the slope at any point $(x, f(x))$ on the graph of f is $2x$.

At the point where $x = 0$, the slope is $2(0) = 0$. This makes visual sense because the point $(0, 1)$ is the vertex of the parabola $y = x^2 + 1$, and we expect that the tangent at this point is a horizontal line with a slope of zero. At the point where $x = 1$, the slope is $2(1) = 2$. This also makes visual sense because moving along the curve from $(0, 1)$ to $(1, 2)$ the slope is steadily increasing.

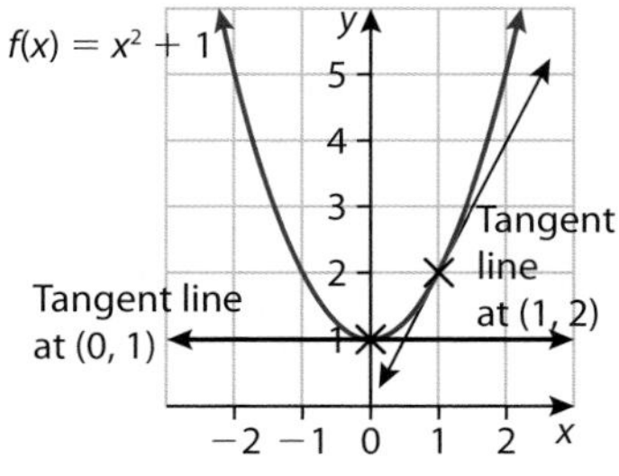

In Example 6, from the function $f(x) = x^2 + 1$ we used the limit process to derive another function with the rule $2x$. With this derived function we can compute the slope (gradient) of the graph of $f(x)$ at a point from simply inputting the x-coordinate of the point. This *derived* function is called the **derivative** of f at x. It is given the notation $f'(x)$, which is commonly read as 'f prime of x', or simply, 'the derivative of f of x.'

The derivative and differentiation

- The **derivative**, $f'(x)$, at a point x in the domain of f is the slope (gradient) of the graph of f at $(x, f(x))$, and is given by

$$f'(x) = \lim_{h \to 0} \frac{f(x+h) - f(x)}{h}$$

  provided that this limit exists.
- If the derivative exists at each point of the domain of f, we say that f is **smooth**.
- The process of finding the derivative, $f'(x)$, is called **differentiation**.
- If $y = f(x)$, then $f'(x)$ is a formula for the instantaneous **rate of change** of y with respect to x.

If finding the derivative of a function indicated with the function notation $f(x)$, then – as shown already – the derivative is usually denoted as $f'(x)$. However, there are two other notations with which you should be familiar. Commonly, if a function is given as y in terms of x, then the derivative is denoted as y', read as 'y prime.' The notation $\frac{dy}{dx}$ is also often used to indicate a derivative, and is read as 'the derivative of y with respect to x.' Note: $\frac{dy}{dx}$ is not a fraction. If, for example, $y = x^2 + 1$, the derivative can be denoted by writing $\frac{d}{dx}(x^2 + 1) = 2x$. This is read as 'the derivative of $x^2 + 1$ with respect to x is $2x$.'

Differentiating from first principles

Depending on the particular purpose that you have in differentiating a function, you can consider the derivative as giving the slope of the graph of the function *or* the rate of change of the dependent variable (commonly y) with respect to the independent variable (commonly x). Both interpretations are useful and widely applied.

Using the limit definition directly to find the derivative of a function (as we did in Example 6) is often called 'differentiating from first principles'.

Example 7

Differentiating from first principles, find the derivative of $f(x) = x^3$.

Solution

$$\begin{aligned} f'(x) = \lim_{h \to 0} \frac{f(x+h) - f(x)}{h} &= \lim_{h \to 0} \frac{(x+h)^3 - x^3}{h} \\ &= \lim_{h \to 0} \frac{(x+h)(x+h)^2 - x^3}{h} \\ &= \lim_{h \to 0} \frac{(x+h)(x^2 + 2hx + h^2) - x^3}{h} \\ &= \lim_{h \to 0} \frac{x^3 + 3hx^2 + 3h^2x + h^3 - x^3}{h} \\ &= \lim_{h \to 0} \frac{h(3x^2 + 3hx + h^2)}{h} \\ &= \lim_{h \to 0} (3x^2 + 3hx + h^2) \\ &= 3x^2 \end{aligned}$$

Therefore, the derivative of $f(x) = x^3$ is $f'(x) = 3x^2$.

As in Example 6, the result for Example 7 is a function that gives us the slope at any point on the graph of $y = x^3$. For example, the points $(1, 1)$ and $(-1, -1)$ both lie on $y = x^3$, and the slopes at these points are respectively $f'(1) = 3(1)^2 = 3$ and $f'(-1) = 3(-1)^2 = 3$. Hence, the tangents at these points will be parallel, as shown in Figure 13.10.

slope = 3
y
x
$y = x^3$
(1, 1)
(−1, −1)

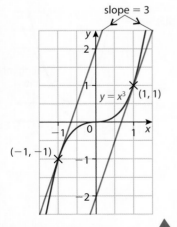

Figure 13.10 Two tangents to $y = x^3$ that are parallel.

Let's examine the relationship between the slopes of tangents to the curve $f(x) = x^2 + 1$ (Example 6) and slopes of tangents to $g(x) = x^2$. Recall that we found the derivative of $f(x)$ to be $f'(x) = 2x$. It appears from the graphs of f and g, in Figure 13.11, that the slopes of tangents at points with the same x-coordinate are equal. For example, the tangent to g at the point $(1, 2)$ looks parallel to the tangent to f at $(1, 1)$, as shown in Figure 13.11. This implies that the derivatives of the two functions are equal. Rather than confirming this conjecture by finding the derivative of $g(x) = x^2$ by first principles (i.e. using the limit definition), let's use the graphical and computing power of our GDC. Any GDC model is capable of computing the slope of a curve at a point – either on the GDC's 'home' screen, or its graphing screen. The screen images below show computing derivative values for $y = x^2$ on the 'home' screen.

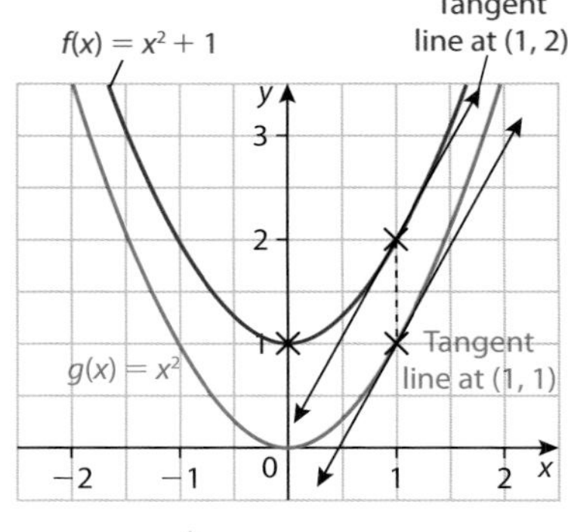

Figure 13.11

The exact command name and syntax for computing the value of a derivative at a point may vary from one GDC model to another.

This command finds the value of the derivative of $y = x^2$ in **terms of** x, at the point $x = 1$.

```
MATH NUM CPX PRB
1:▸Frac
2:▸Dec
3:³
4:³√(
5:ˣ√
6:fMin(
7↓fMax(
```

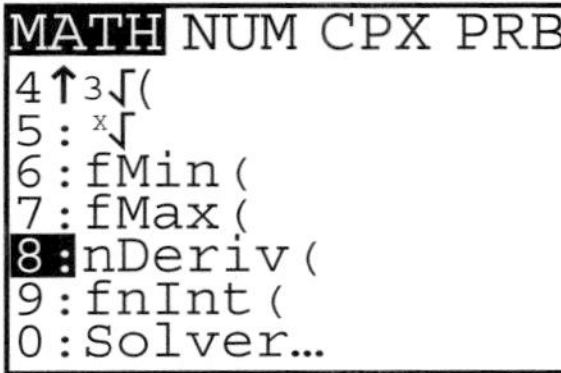

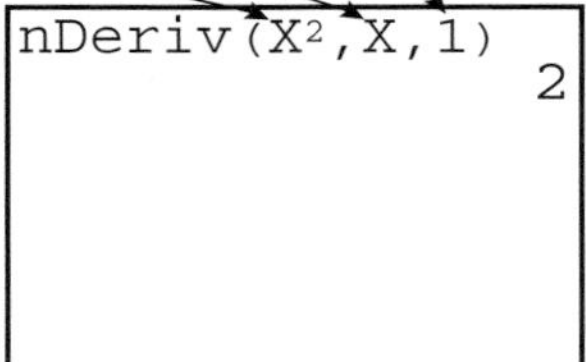


```
nDeriv(X²,X,1)
                2
nDeriv(X²,X,2)
                4
nDeriv(X²,X,3)
                6
```

```
                6
nDeriv(X²,X,-1)
               -2
nDeriv(X²,X,17)
               34
nDeriv(X²,X,-9)
              -18
```

Our GDC results confirm our conjecture that the derivative of $g(x) = x^2$ is $g'(x) = 2x$.

Example 8

From first principles, find:

a) y' given $y = 3x^2 + 2x$ b) $\frac{dy}{dx}$ given $y = \frac{1}{x}$

Solution

We will apply the definition of the derivative, $f'(x) = \lim_{h \to 0} \frac{f(x+h) - f(x)}{h}$, in both a) and b).

$$\text{a) } y' = \lim_{h \to 0} \frac{[3(x+h)^2 + 2(x+h)] - (3x^2 + 2x)}{h}$$

$$= \lim_{h \to 0} \frac{(3x^2 + 6hx + 3h^2 + 2x + 2h) - (3x^2 + 2x)}{h}$$

$$= \lim_{h \to 0} \frac{6hx + 3h^2 + 2h}{h}$$

$$= \lim_{h \to 0} (6x + 3h + 2) \Rightarrow y' = 6x + 2$$

b) $\dfrac{dy}{dx} = \dfrac{d}{dx}\left(\dfrac{1}{x}\right) = \lim\limits_{h\to 0} \dfrac{\dfrac{1}{x+h} - \dfrac{1}{x}}{h}$

$$= \lim_{h\to 0} \frac{\dfrac{x}{x(x+h)} - \dfrac{x+h}{x(x+h)}}{h}$$

$$= \lim_{h\to 0} \left(\frac{\dfrac{-h}{x(x+h)}}{\dfrac{h}{1}} \right)$$

$$= \lim_{h\to 0} \left(\frac{-h}{x(x+h)} \cdot \frac{1}{h} \right)$$

$$= \lim_{h\to 0} \left(\frac{-1}{x^2 + hx} \right) \Rightarrow \frac{d}{dx}\left(\frac{1}{x}\right) = -\frac{1}{x^2} \text{ or } \frac{d}{dx}(x^{-1}) = -x^{-2}$$

Basic differentiation rules

We have now established the following results:

- If $f(x) = x^2$, then $f'(x) = 2x$.
- If $f(x) = x^2 + 1$, then $f'(x) = 2x$.
- If $f(x) = 3x^2 + 2x$, then $f'(x) = 6x + 2$.
- If $f(x) = x^3$, then $f'(x) = 3x^2$.
- If $f(x) = x^{-1}$, then $f'(x) = -x^{-2}$.

In addition, we know that if $f(x) = x$, then $f'(x) = 1$, since the line $y = x$ has a constant slope equal to 1; and that if $f(x) = 1$, then $f'(x) = 0$ because the line $y = 1$ is horizontal and thus has a constant slope equal to 0. Furthermore, the graph of any function $f(x) = c$, where c is a constant, is a horizontal line, confirming that if $f(x) = c$, $c \in \mathbb{R}$, then $f'(x) = 0$. In other words, the derivative of a constant is zero. This leads to our first basic rule of differentiation.

The constant rule

The derivative of a constant function is zero. That is, given c is a real number, and if $f(x) = c$, then $f'(x) = 0$.

These following results:

$$\begin{aligned} f(x) &= x^{-1} &&\Rightarrow & f'(x) &= -x^{-2} \\ f(x) &= x^0 = 1 &&\Rightarrow & f'(x) &= 0 \\ f(x) &= x^1 = x &&\Rightarrow & f'(x) &= 1 \\ f(x) &= x^2 &&\Rightarrow & f'(x) &= 2x \\ f(x) &= x^3 &&\Rightarrow & f'(x) &= 3x^2 \end{aligned}$$

can be summarized in the single statement:

$$\text{if } f(x) = x^n \text{ then } f'(x) = nx^{n-1} \text{ for } n = -1, 0, 1, 2, 3$$

In fact, this statement is true not just for these values but for *any* value of n that is a rational number ($n \in \mathbb{Q}$). This leads to our second basic rule of differentiation.

Functions of the form $f(x) = x^n$ are called **power functions**, so the differentiation rule $\frac{d}{dx}(x^n) = nx^{n-1}$ gives the rule for differentiating power functions – and is often referred to as the **power rule**.

The derivative of x^n

Given n is a rational number, and if $f(x) = x^n$, then $f'(x) = nx^{n-1}$.

Recall from Chapter 4 the binomial theorem for positive integers

$$(a+b)^n = \sum_{r=0}^{n} \binom{n}{r} a^{n-r} b^r.$$

Applying this to the limit definition of the derivative gives,

$$\frac{d}{dx}(x^n) = \lim_{h\to 0}\frac{(x+h)^n - x^n}{h}$$

$$= \lim_{h\to 0}\frac{\left(\binom{n}{0}x^n + \binom{n}{1}x^{n-1}h + \binom{n}{2}x^{n-2}h^2 + \ldots + \binom{n}{n-1}xh^{n-1} + \binom{n}{n}h^n\right) - x^n}{h}$$

$$= \lim_{h\to 0}\frac{\left(x^n + nx^{n-1}h + \frac{1}{2}n(n-1)x^{n-2}h^2 + \ldots + nxh^{n-1} + h^n\right) - x^n}{h}$$

$$= \lim_{h\to 0} nx^{n-1} + \lim_{h\to 0}\frac{1}{2}n(n-1)x^{n-2}h + \ldots + \lim_{h\to 0} nxh^{n-2} + \lim_{h\to 0} h^{n-1}$$

$$= nx^{n-1} + 0 + \ldots + 0 + 0$$

$$= nx^{n-1}$$

Therefore, $\frac{d}{dx}(x^n) = nx^{n-1}$.

Another basic rule of differentiation is suggested by our result that the derivative of $f(x) = x^2 + 1$ is $f'(x) = 2x$. The derivative of a sum of a number of terms is obtained by differentiating each term separately – i.e. differentiating 'term-by-term'. That is,

$$\frac{d}{dx}(x^2+1) = \frac{d}{dx}(x^2) + \frac{d}{dx}(1) = 2x + 0 = 2x.$$

The sum and difference rule

If $f(x) = g(x) \pm h(x)$ then $f'(x) = g'(x) \pm h'(x)$.

The sum rule for derivatives can help us give a very convincing justification of our first differentiation rule: the constant rule. The fact that the derivative of a constant must be zero can be verified by considering the transformation of the graph of a function (Section 2.4). The graph of the function $f(x) + c$, where $c \in \mathbb{R}$, is a vertical translation by c units of the graph of $f(x)$. As Figure 13.12 illustrates, when the graph of a function is translated vertically its shape is preserved. Hence, the slope of the tangent line to the graph of $f(x) + c$ will be the same as that for $f(x)$ at a particular value of x. This means that the derivatives for the two functions must be equal. That is,

$$\frac{d}{dx}[f(x) + c] = \frac{d}{dx}[f(x)]$$

$$\frac{d}{dx}[f(x)] + \frac{d}{dx}(c) = \frac{d}{dx}[f(x)]$$

This is only true if $\frac{d}{dx}(c) = 0$.

Figure 13.12 Translating the graph of a function vertically does not alter the slope of the tangent line at a particular value of x. Hence the derivatives of the two functions are equal.

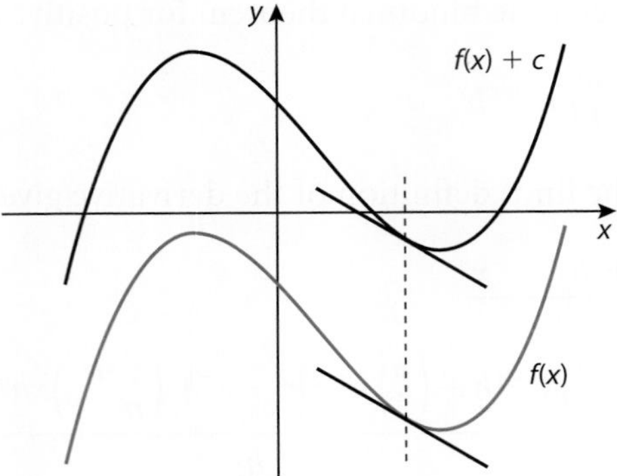

A fourth basic rule of differentiation is illustrated by our result that the derivative of $f(x) = 3x^2 + 2x$ is $f'(x) = 6x + 2$. Using the sum rule, $f'(x) = \frac{d}{dx}(3x^2 + 2x) = \frac{d}{dx}(3x^2) + \frac{d}{dx}(2x) = 6x + 2$. The fact that $\frac{d}{dx}(3x^2) = 6x$ suggests that $3 \cdot \frac{d}{dx}(x^2) = 3 \cdot 2x = 6x$. In other words, the derivative of a function being multiplied by a constant is equal to the constant multiplying the derivative of the function.

The constant multiple rule

If $f(x) = c \cdot g(x)$ then $f'(x) = c \cdot g'(x)$.

As mentioned before, and as you have seen, there are different notations used for indicating a derivative or differentiation. These can be traced back to the fact that calculus was first developed by Isaac Newton (1642–1727) and Gottfried Leibniz (1646–1716) independently of each other – and hence introduced different symbols for methods of calculus. The 'prime' notations y' and $f'(x)$ come from notations that Newton used for derivatives. The $\frac{dy}{dx}$ notation is similar to that used by Leibniz for indicating differentiation. Each has its advantages and disadvantages. For example, it is often easier to write our four basic rules of differentiation using Leibniz notation as shown below.

Constant rule: $\frac{d}{dx}(c) = 0, \ c \in \mathbb{R}$

Power rule: $\frac{d}{dx}(x^n) = nx^{n-1}, \ n \in \mathbb{Q}$

Sum and difference rule: $\frac{d}{dx}[g(x) + h(x)] = \frac{d}{dx}[g(x)] + \frac{d}{dx}[h(x)]$

Constant multiple rule: $\frac{d}{dx}[c \cdot f(x)] = c \cdot \frac{d}{dx}[f(x)], c \in \mathbb{R}$

Example 9

For each function: (i) find the derivative using the basic differentiation rules; (ii) find the slope of the graph of the function at the indicated points; and (iii) use your GDC to confirm your answer for (ii).

Function	Points
a) $f(x) = x^3 + 2x^2 - 15x - 13$	$(-3, 23), (3, -13)$
b) $f(x) = (2x - 7)^2$	$(2, 9), (\frac{7}{2}, 0)$
c) $f(x) = 3\sqrt{x} - 6$	$(4, 0), (9, 3)$
d) $f(x) = \frac{x^4}{4} - \frac{3x^3}{2} - 2x^2 + \frac{15x}{2} + \frac{3}{4}$	$(5, -43), (0, 0)$

Solution

a) (i) $\frac{d}{dx}(x^3 + 2x^2 - 15x - 13) = \frac{d}{dx}(x^3) + 2\cdot\frac{d}{dx}(x^2) - 15\cdot\frac{d}{dx}(x) - \frac{d}{dx}(13)$

$$= 3x^2 + 2(2x) - 15(1) - 0$$
$$= 3x^2 + 4x - 15$$

Therefore, the derivative of $f(x) = x^3 + 2x^2 - 15x - 13$ is $f'(x) = 3x^2 + 4x - 15$.

(ii) Slope of curve at $(-3, 23)$ is $f'(-3) = 3(-3)^2 + 4(-3) - 15 = 27 - 12 - 15 = 0$.

We should observe a horizontal tangent (slope $= 0$) to the curve at $(-3, 23)$.

Slope of curve at $(3, -13)$ is $f'(3) = 3(3)^2 + 4(3) - 15 = 27 + 12 - 15 = 24$.

We should observe a very steep tangent (slope $= 24$) to the curve at $(3, -13)$.

(iii) Not only can we use the GDC to compute the value of the derivative at a particular value of x on the 'home' screen, but we can also do it on the graph screen.

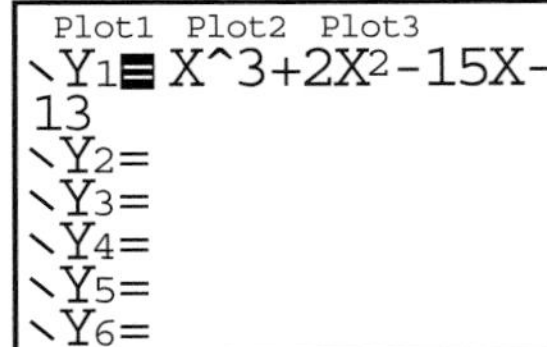

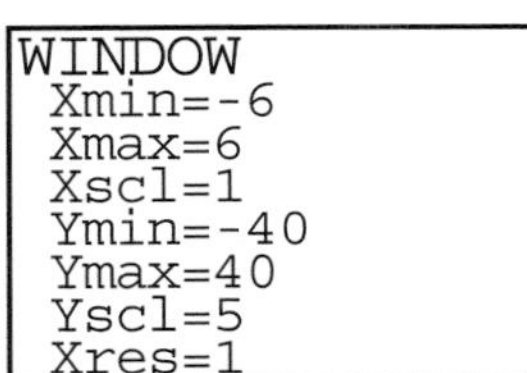

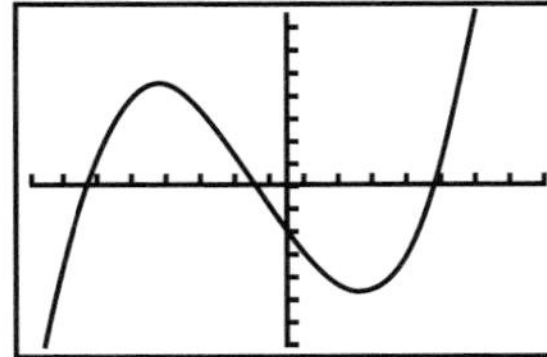

CALCULATE
1:value
2:zero
3:minimum
4:maximum
5:intersect
6:dy/dx
7:∫f(x)dx

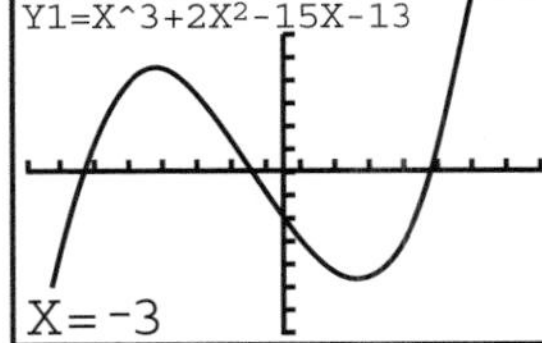

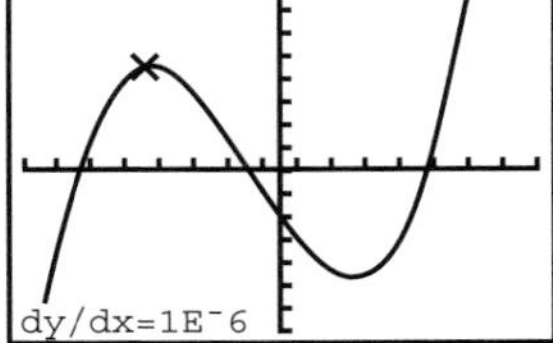

The GDC computes a slope of 1E⁻6 at the point $(-3, 23)$.
(1E⁻6 $= 1 \times 10^{-6} = 0.000\,001$)

Although the method the GDC uses is very accurate, sometimes there is a small amount of error in its calculation. This most commonly occurs when performing calculus computations (e.g. the value of the derivative at a point). 1E⁻6 $= 0.000\,001$ is very close to zero which is the exact value of the derivative. Observe that the graph of $y = x^3 + 2x^2 - 15x - 13$ appears to have a 'turning point' at $(-3, 23)$, confirming that a line tangent to the curve at that point would be horizontal.

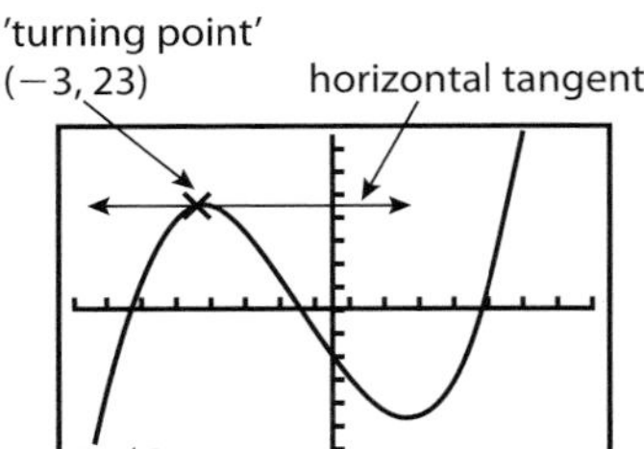

Let's check on our GDC that the slope of the curve is 24 at $(3, -13)$. Again, the GDC exhibits a small amount of error in its result.

Most GDCs are also capable of drawing a tangent at a point and displaying its equation as shown in the final screen image below.

```
CALCULATE
1:value
2:zero
3:minimum
4:maximum
5:intersect
6:dy/dx
7:∫f(x)dx
```

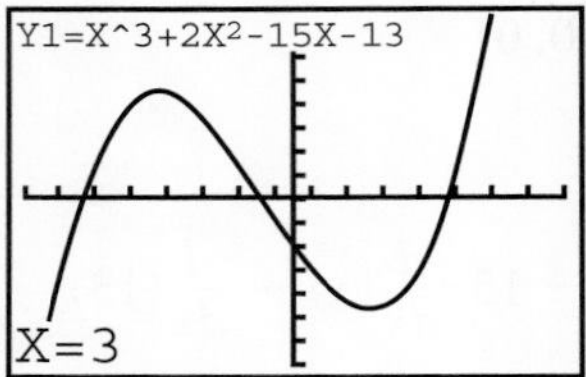

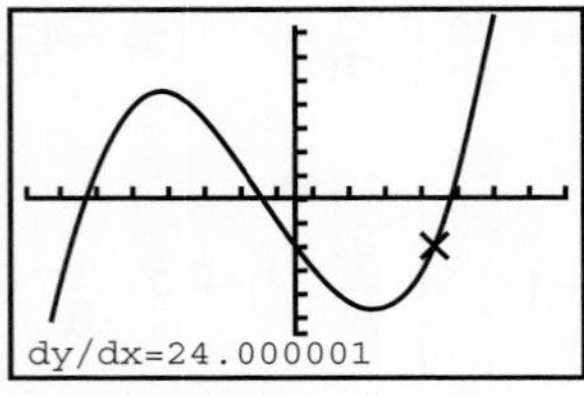

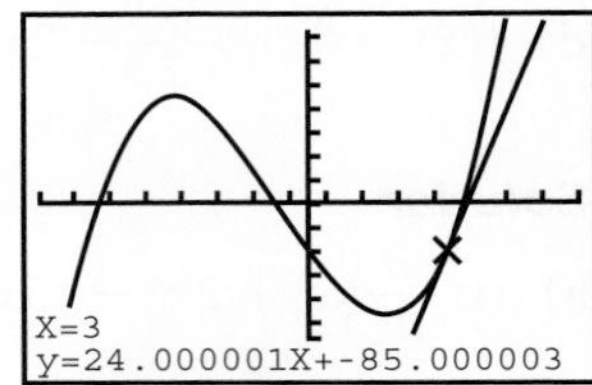

The equation of the tangent line at $(3, -13)$ is $y = 24x - 85$. We will look at finding the equations of tangent lines analytically in the last section of the chapter.

b) (i) $\frac{d}{dx}[(2x-7)^2] = \frac{d}{dx}[(2x-7)(2x-7)]$ Differentiate term-by-term after expanding.

$$= \frac{d}{dx}(4x^2 - 28x + 49)$$

$$= 4\frac{d}{dx}(x^2) - 28\frac{d}{dx}(x) + \frac{d}{dx}(49)$$

$$= 8x - 28 + 0$$

Therefore, the derivative of $f(x) = (2x-7)^2$ is $f'(x) = 8x - 28$.

(ii) Slope of curve at $(2, 9)$ is $f'(2) = 8(2) - 28 = -12$.

Slope of curve at $\left(\frac{7}{2}, 0\right)$ is $f'\left(\frac{7}{2}\right) = 8\left(\frac{7}{2}\right) - 28 = 0$.

Thus, we should observe a horizontal tangent to the curve at $\left(\frac{7}{2}, 0\right)$.

(iii)

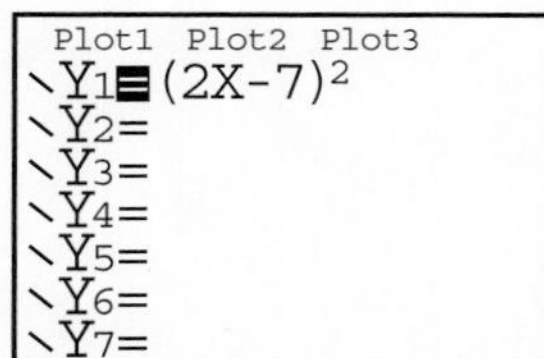

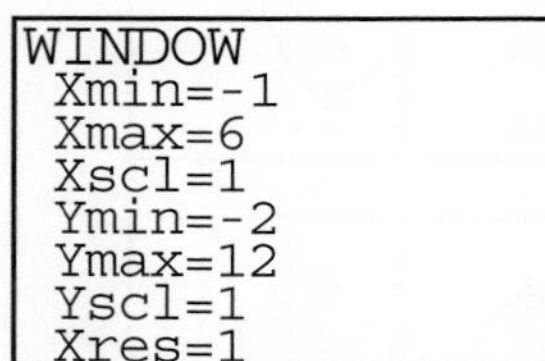

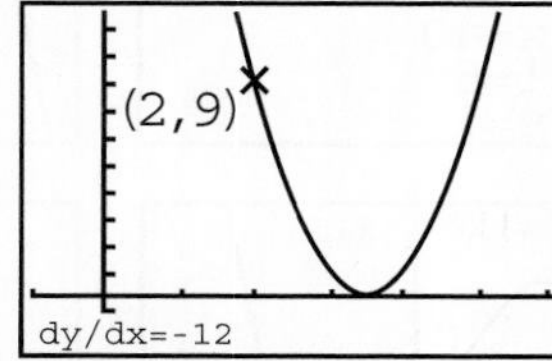

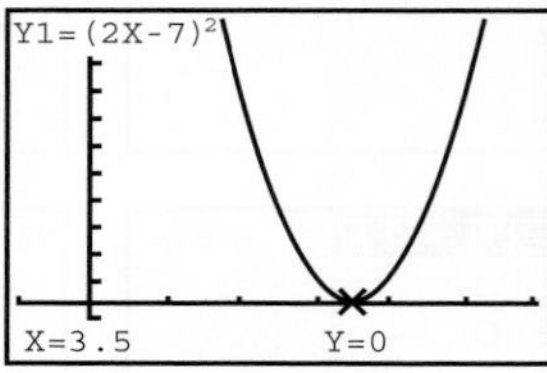

There's no error this time in the GDC's computation of the slope at $(2, 9)$. The vertex of the parabola is at $\left(\frac{7}{2}, 0\right)$, confirming that it has a horizontal tangent at that point.

c) (i) $\frac{d}{dx}(3\sqrt{x} - 6) = 3\frac{d}{dx}(x^{\frac{1}{2}}) - \frac{d}{dx}(6)$

$$= 3\left(\frac{1}{2}x^{-\frac{1}{2}}\right) - 0$$

$$= \frac{3}{2x^{\frac{1}{2}}}$$

Therefore, the derivative of $f(x) = 3\sqrt{x} - 6$ is $f'(x) = \frac{3}{2x^{\frac{1}{2}}}$ or $f'(x) = \frac{3}{2\sqrt{x}}$.

(ii) Slope of curve at (4, 0) is $f'(4) = \frac{3}{2\sqrt{4}} = \frac{3}{4}$.

Slope of curve at (9, 3) is $f'(9) = \frac{3}{2\sqrt{9}} = \frac{1}{2}$.

Thus, because the slope at $x = 9$ is less than that at $x = 4$, we should observe the graph of the equation becoming less steep as we move along the curve from $x = 4$ to $x = 9$.

(iii)

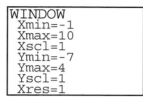

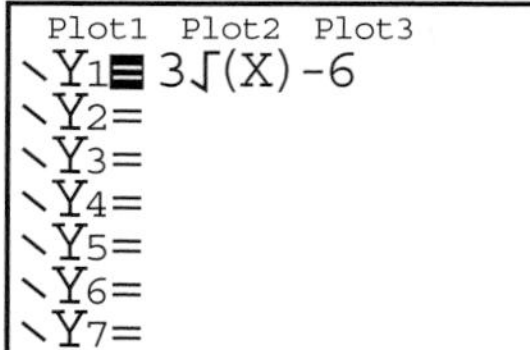


```
WINDOW
 Xmin=-1
 Xmax=10
 Xscl=1
 Ymin=-7
 Ymax=4
 Yscl=1
 Xres=1
```

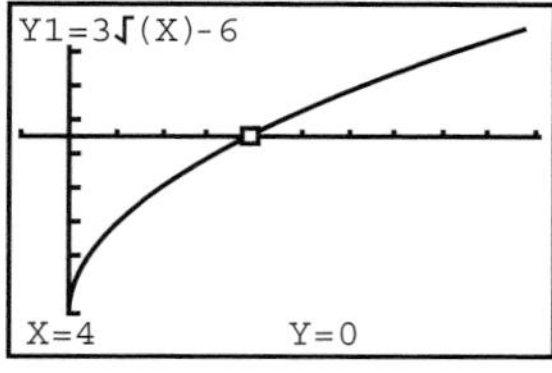


```
nDeriv(3√(X)-6,X
,4)
          .750000006
nDeriv(3√(X)-6,X
,9)
         .5000000009
```

The slope of the graph of $y = 3\sqrt{x} - 6$ appears to steadily decrease as x increases. Let's check the results for (ii) by evaluating the derivative at a point on the 'home' screen. The GDC confirms the slopes for the curve when $x = 4$ and $x = 9$, but again the GDC computations have incorporated a small amount of error.

d) (i) $\frac{d}{dx}\left(\frac{x^4}{4} - \frac{3x^3}{2} - 2x^2 + \frac{15x}{2} + \frac{3}{4}\right)$

$= \frac{1}{4}\frac{d}{dx}(x^4) - \frac{3}{2}\frac{d}{dx}(x^3) - 2\frac{d}{dx}(x^2) + \frac{15}{2}\frac{d}{dx}(x) + \frac{d}{dx}\left(\frac{3}{4}\right)$

$= \frac{1}{4}(4x^3) - \frac{3}{2}(3x^2) - 2\frac{d}{dx}(2x) + \frac{15}{2}(1) + 0$

$= x^3 - \frac{9x^2}{2} - 4x + \frac{15}{2}$

Therefore, the derivative of $f(x) = \frac{x^4}{4} - \frac{3x^3}{2} - 2x^2 + \frac{15x}{2} + \frac{3}{4}$ is $f'(x) = x^3 - \frac{9x^2}{2} - 4x + \frac{15}{2}$.

(ii) Slope of curve at (5, −43) is $f'(5) = 5^3 - \frac{9(5)^2}{2} - 4(5) + \frac{15}{2} = 0$.

Thus, there should be a horizontal tangent to the curve at (5, −43).

Slope of curve at (0, 0) is $f'(0) = \frac{15}{2}$.

(iii) Your GDC is not capable of computing the derivative function – only the specific value of the derivative for a given value of x. However, we can have the GDC graph the values of the derivative over a given *interval* of x. We can then graph the derivative function found from differentiation rules (result from (i)) and see if the two graphs match.

```
Plot1 Plot2 Plot3
\Y1■X^4/4-(3X^3)
/2-2X^2+15X/2+3/
4
\Y2=
\Y3=
\Y4=
\Y5=
```

```
WINDOW
 Xmin=-4
 Xmax=8
 Xscl=1
 Ymin=-50
 Ymax=50
 Yscl=10
 Xres=1
```

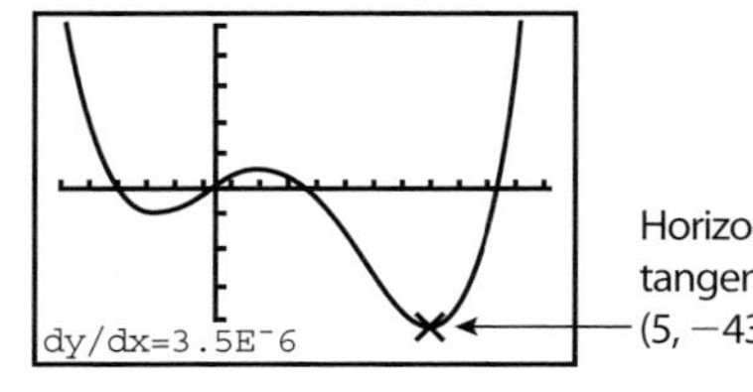

Horizontal tangent at (5, −43)

The command nDeriv(Y_1, X, X) computes the value of the derivative of function Y_1 in terms of x for all x.

```
Plot1 Plot2 Plot3
\Y1=X^4/4-(3X^3)/2-2X^2+15X/2+3/4
\Y2=nDeriv(Y1,X,X)
\Y3=X^3-(9X2)/2-4X+15/2
```

Values of the derivative of $f(x)$ will be graphed as Y_2, and the derivative function, $f'(x) = x^3 - \frac{9x^2}{2} - 4x + \frac{15}{2}$, determined by manual application of differentiation rules (part (i)), will be graphed as Y_3. Note that the graph of Y_3 will be in bold style to distinguish it from Y_2, and that the equation Y_1 has been turned 'off.'

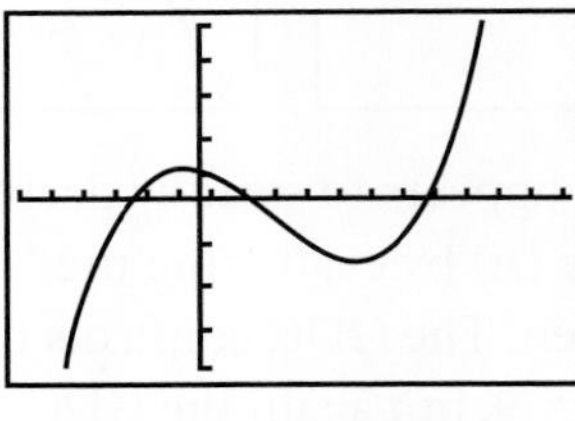

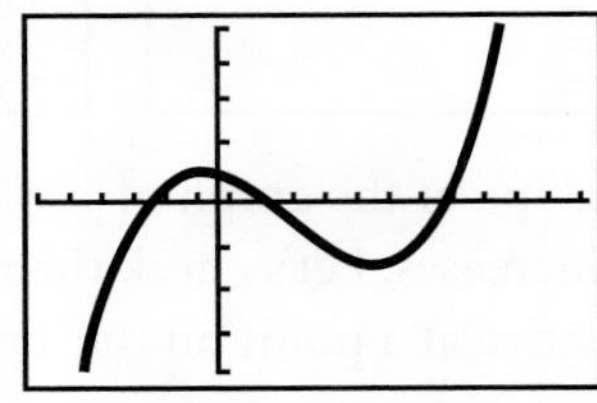

$Y_1 = \frac{x^4}{4} - \frac{3x^3}{2} - 2x^2 + \frac{15x}{2} + \frac{3}{4}$ $\quad$ $Y_2 = \text{nDeriv}(Y_1, X, X)$ $\quad$ $Y_3 = x^3 - \frac{9x^2}{2} - 4x + \frac{15}{2}$

Since the two graphs match, this confirms that the derivative found in part (i) using differentiation rules is correct.

Example 10

The curve $y = ax^3 + 7x^2 - 8x - 5$ has a turning point at the point where $x = -2$. Determine the value of a.

Solution

There must be a horizontal tangent, and a slope of zero, at the point where the graph has a turning point.

$$\frac{dy}{dx} = \frac{d}{dx}(ax^3 + 7x^2 - 8x - 5)$$

$$= a\frac{d}{dx}(x^3) + 7\frac{d}{dx}(x^2) - 8\frac{d}{dx}(x) + \frac{d}{dx}(-5) = 3ax^2 + 14x - 8$$

$$\frac{dy}{dx} = 0 \text{ when } x = -2\text{: } 3a(-2)^2 + 14(-2) - 8 = 0$$

$$\Rightarrow\ 12a - 28 - 8 = 0 \Rightarrow 12a = 36 \Rightarrow a = 3$$

Recall that the derivative of a function is a formula for the **instantaneous rate of change** of the dependent variable (commonly y) with respect to the dependent variable (x). In other words, as illustrated earlier in this section, the slope of the tangent at a point gives the slope, or rate of change, of the curve at that point. The slope of a **secant line** (that crosses the curve at two points) gives the **average rate of change** between the two points.

Example 11

Boiling water is poured into a cup. The temperature of the water in degrees Celsius, C, after t minutes is given by $C = 19 + \frac{182}{t^{\frac{3}{2}}}$, for times $t \geqslant 1$ minute.

a) Find the average rate of change of the temperature from $t = 2$ to $t = 6$.

b) Find the rate of change of the temperature at the instant that $t = 4$.

Solution

a)

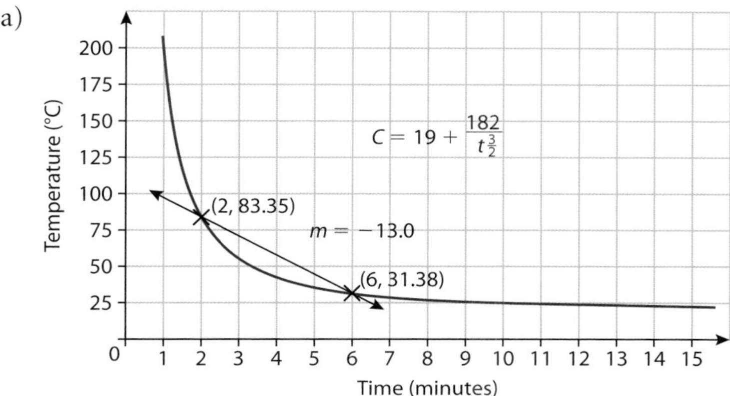

When $t = 2$, $C \approx 83.35°$ and when $t = 6$, $C \approx 31.38°$. The average rate of change from $t = 2$ to $t = 6$ is the slope of the line through the points (2, 83.35) and (6, 31.38).

Average rate of change $= \frac{83.35 - 31.38}{2 - 6} = \frac{51.97}{-4} = -12.9925.$

To an accuracy of 3 significant figures, the average rate of change from $t = 2$ to $t = 6$ is -13.0 °C per minute. During that period of time the water is, on average, becoming 13 degrees cooler every minute.

b) Let's compute the derivative $\frac{dC}{dt}$, i.e. the rate of change of degrees C with respect to time t, from which we can compute the rate the temperature is changing at the moment when $t = 4$.

$$\frac{dC}{dt} = \frac{d}{dt}\left(19 + \frac{182}{t^{\frac{3}{2}}}\right) = \frac{d}{dt}(19 + 182t^{-\frac{3}{2}}) = \frac{d}{dt}(19) + 182\frac{d}{dt}(t^{-\frac{3}{2}})$$

$$= 0 + 182\left(-\frac{3}{2}t^{-\frac{3}{2} - 1}\right) = -273t^{-\frac{5}{2}}$$

$$\frac{dC}{dt} = -\frac{273}{t^{\frac{5}{2}}} = -\frac{273}{\sqrt{t^5}}$$

At $t = 4$:

$$\frac{dC}{dt} = -\frac{273}{\sqrt{4^5}} = -\frac{273}{32} \approx -8.53$$

Therefore, the temperature's instantaneous rate of change at $t = 4$ minutes is -8.53 °C per minute.

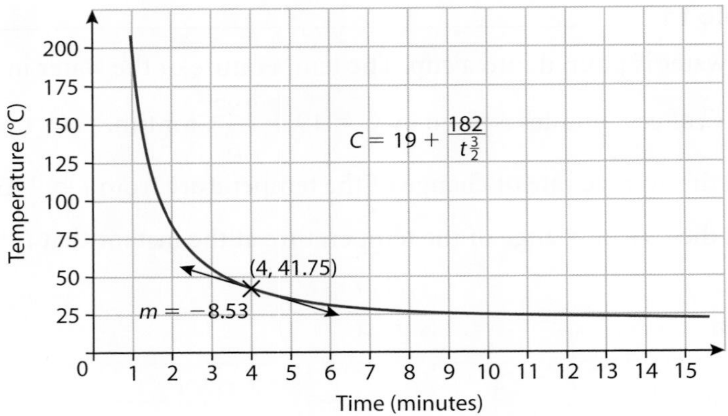

Differentiating sin *x* and cos *x* using limit definition for derivative

To add to our growing list of differentiation rules, we will now determine the derivatives for the sine and cosine functions. The results will help us determine the derivatives for the other trigonometric functions in Chapter 15.

The rigorous analytical method (applying limit definition of derivative) for finding these two derivatives requires two limit results that we found by decidedly non-rigorous methods in Example 2 in the previous section; namely that $\lim\limits_{x \to 0} \frac{\sin x}{x} = 1$ and $\lim\limits_{x \to 0} \frac{\cos x - 1}{x} = 0$. We conjectured the value of these limits after exploring the behaviour of the expressions on our GDC. Example 5 illustrated that estimating limits by such informal methods is not foolproof. Hence, we will now put these two limit results on firmer ground through a more rigorous approach.

We first state, without proof, an important theorem in mathematics.

The squeeze theorem
If $g(x) \leqslant f(x) \leqslant h(x)$ for all $x \neq c$ in some interval about c, and

$$\lim_{x \to c} g(x) = \lim_{x \to c} h(x) = L,$$

then

$$\lim_{x \to c} f(x) = L.$$

Figure 13.13 Squeezing *f* between *g* and *h* forces the limiting value of *f* to be between the limiting values of *g* and *h*.

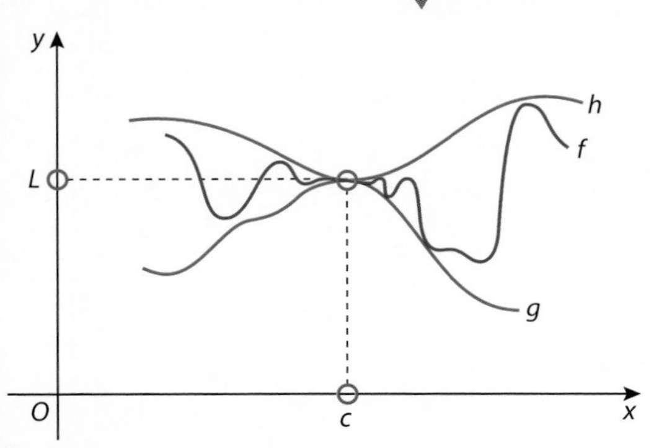

The theorem describes a function *f* whose values are 'squeezed' between the values of two other functions, *g* and *h*. If *g* and *h* have the same limit as $x \to c$, then *f* has the same limit, as suggested by Figure 13.13.

Consider a sector of a circle with centre O, central angle θ (in radian measure) and radius 1 (see Figure 13.14). Further consider right triangle AOC, sector AOB and triangle AOB. We know that point B has coordinates $(\cos\theta, \sin\theta)$ and point C has coordinates $(1, \tan\theta)$. From Section 7.1, we also know that the area of a sector with central angle θ is $\frac{1}{2}r^2\theta$. It is clear that the area of sector AOB must be between the area of $\triangle AOC$ and the area of $\triangle AOB$, that is, the sector is 'squeezed' between the two triangles (Figure 13.15).

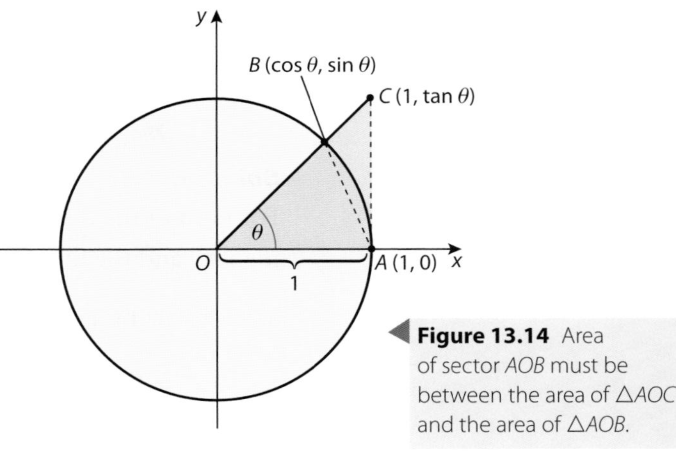

Figure 13.14 Area of sector AOB must be between the area of $\triangle AOC$ and the area of $\triangle AOB$.

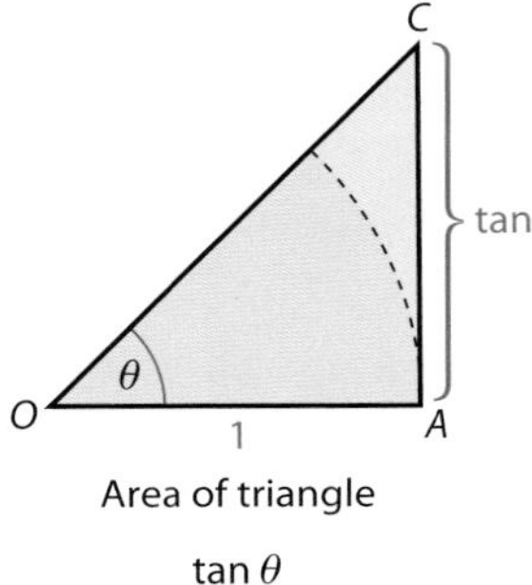

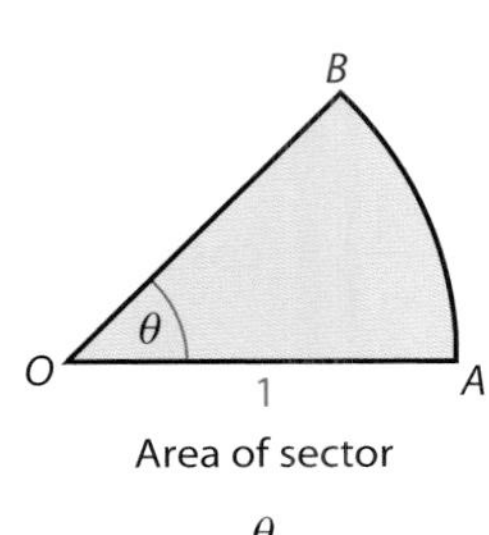

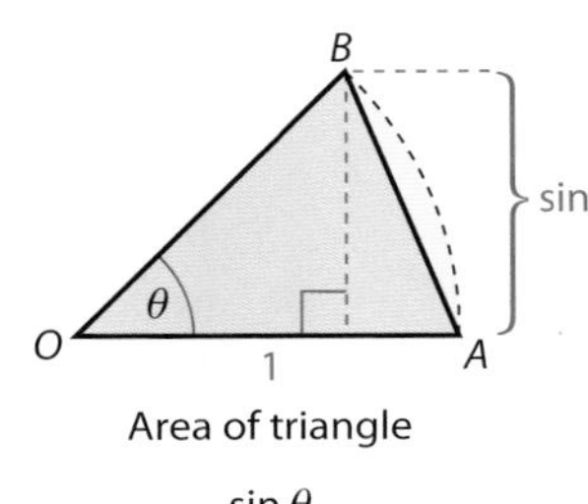

Area of triangle	$\geqslant$	Area of sector	$\geqslant$	Area of triangle
$\frac{\tan\theta}{2}$	$\geqslant$	$\frac{\theta}{2}$	$\geqslant$	$\frac{\sin\theta}{2}$

Figure 13.15 Area of sector AOB is squeezed between the two triangles.

Multiplying all the area expressions by $\frac{2}{\sin\theta}$ gives

$$\frac{1}{\cos\theta} \geqslant \frac{\theta}{\sin\theta} \geqslant 1.$$

Given the fact that if $\frac{a}{b} > \frac{c}{d}$, then $\frac{b}{a} < \frac{d}{c}$, we can write the reciprocals of the three expressions and reverse the inequality signs. This gives

$$\cos\theta \leqslant \frac{\sin\theta}{\theta} \leqslant 1.$$

It follows that

$$\lim_{\theta\to 0}\cos\theta \leqslant \lim_{\theta\to 0}\frac{\sin\theta}{\theta} \leqslant \lim_{\theta\to 0} 1.$$

From direct substitution, $\lim_{\theta\to 0}\cos\theta = 1$. Thus,

$$1 \leqslant \lim_{\theta\to 0}\frac{\sin\theta}{\theta} \leqslant 1.$$

We can now apply the squeeze theorem and conclude that $\lim_{\theta\to 0}\frac{\sin\theta}{\theta} = 1$.

Furthermore, because $\cos(-\theta) = \cos\theta$ and $\frac{\sin(-\theta)}{-\theta} = \frac{\sin(\theta)}{\theta}$, we can also conclude that this limit is true for all non-zero values of θ in the interval $-\frac{\pi}{2} < \theta < \frac{\pi}{2}$.

The above result, $\lim_{x\to 0}\frac{\sin x}{x} = 1$, can be used to algebraically deduce that $\lim_{x\to 0}\frac{\cos x - 1}{x} = 0$. This is saved for you to do in Exercise 13.2, question 26.

Example 12

Differentiate from first principles:

a) $f(x) = \sin x$ b) $f(x) = \cos x$

Solution

For both of the derivatives we will need to make use of a compound angle identity and the limit results $\lim_{x\to 0}\frac{\sin x}{x} = 1$ and $\lim_{x\to 0}\frac{\cos x - 1}{x} = 0$.

a) We start by substituting into the limit definition for the derivative.

$$f'(x) = \lim_{h\to 0}\frac{f(x+h) - f(x)}{h} = \lim_{h\to 0}\frac{\sin(x+h) - \sin x}{h}$$

Applying $\sin(A + B) = \sin A\cos B + \cos A\sin B$.

$$= \lim_{h\to 0}\frac{\sin x\cos h + \cos x\sin h - \sin x}{h}$$

Splitting argument into two fractions.

$$= \lim_{h\to 0}\left[\frac{\sin x\cos h - \sin x}{h} + \frac{\cos x\sin h}{h}\right]$$

Factorizing common factors in each fraction.

$$= \lim_{h\to 0}\left[\sin x\left(\frac{\cos h - 1}{h}\right) + \cos x\left(\frac{\sin h}{h}\right)\right]$$

Applying $\lim_{x\to a}[f(x)\cdot g(x)] = L\cdot K$.

$$= \lim_{h\to 0}\sin x\cdot\lim_{h\to 0}\left(\frac{\cos h - 1}{h}\right) + \lim_{h\to 0}\cos x\cdot\lim_{h\to 0}\left(\frac{\sin h}{h}\right)$$

Applying $\lim_{x\to 0}\frac{\sin x}{x} = 1$ and $\lim_{x\to 0}\frac{\cos x - 1}{x} = 0$.

$$= \sin x\cdot 0 + \cos x\cdot 1$$

$$= \cos x$$

Thus, if $f(x) = \sin x$ then $f'(x) = \cos x$, or using Leibniz notation $\frac{d}{dx}(\sin x) = \cos x$.

b) Again, we start by substituting into the limit definition for the derivative.

$$f'(x) = \lim_{h\to 0}\frac{f(x+h) - f(x)}{h} = \lim_{h\to 0}\frac{\cos(x+h) - \cos x}{h}$$

Applying $\cos(A + B) = \cos A\cos B - \sin A\sin B$.

$$= \lim_{h\to 0}\frac{\cos x\cos h - \sin x\sin h - \cos x}{h}$$

Splitting argument into two fractions.

$$= \lim_{h\to 0}\left[\frac{\cos x\cos h - \cos x}{h} - \frac{\sin x\sin h}{h}\right]$$

Factorizing common factors in each fraction.

$$= \lim_{h\to 0}\left[\cos x\left(\frac{\cos h - 1}{h}\right) - \sin x\left(\frac{\sin h}{h}\right)\right]$$

Applying $\lim_{x\to a}[f(x)\cdot g(x)] = L\cdot K$.

$$= \lim_{h\to 0}\sin x\cdot\lim_{h\to 0}\left(\frac{\cos h - 1}{h}\right) - \lim_{h\to 0}\sin x\cdot\lim_{h\to 0}\left(\frac{\sin h}{h}\right)$$

Applying $\lim_{x\to 0}\frac{\sin x}{x} = 1$ and $\lim_{x\to 0}\frac{\cos x - 1}{x} = 0$.

$$= \cos x\cdot 0 - \sin x\cdot 1$$

$$= -\sin x$$

Thus, if $f(x) = \cos x$ then $f'(x) = -\sin x$, or using Leibniz notation $\frac{d}{dx}(\cos x) = -\sin x$.

We will confirm these two results graphically at the start of Chapter 15.

Exercise 13.2

In questions 1–4, find the derivative of the function by applying the limit definition $f'(x) = \lim_{h \to 0} \frac{f(x+h) - f(x)}{h}$.

1 $f(x) = 1 - x^2$

2 $g(x) = x^3 + 2$

3 $h(x) = \sqrt{x}$

4 $r(x) = \frac{1}{x^2}$

5 Using your results from questions 1–4, find the slope of the graph of each function in 1–4 at the point where $x = 1$. Sketch each function and draw a line tangent to the graph at $x = 1$.

In questions 6–12, a) find the derivative of the function, and b) compute the slope of the graph of the function at the indicated point. Use a GDC to confirm your results.

6 $y = 3x^2 - 4x$ point (0, 0)

7 $y = 1 - 6x - x^2$ point (−3, 10)

8 $y = \frac{2}{x^3}$ point (−1, 2)

9 $y = x^5 - x^3 - x$ point (1, −1)

10 $y = (x + 2)(x - 6)$ point (2, −16)

11 $y = 2x + \frac{1}{x} - \frac{3}{x^3}$ point (1, 0)

12 $y = \frac{x^3 + 1}{x^2}$ point (−1, 0)

13 The slope of the curve $y = x^2 + ax + b$ at the point (2, −4) is −1. Find the value of a and the value of b.

In questions 14–17, find the coordinates of any points on the graph of the function where the slope is equal to the given value.

14 $y = x^2 + 3x$ slope = 3

15 $y = x^3$ slope = 12

16 $y = x^2 - 5x + 1$ slope = 0

17 $y = x^2 - 3x$ slope = −1

18 Use the graph of f to answer each of the following questions.

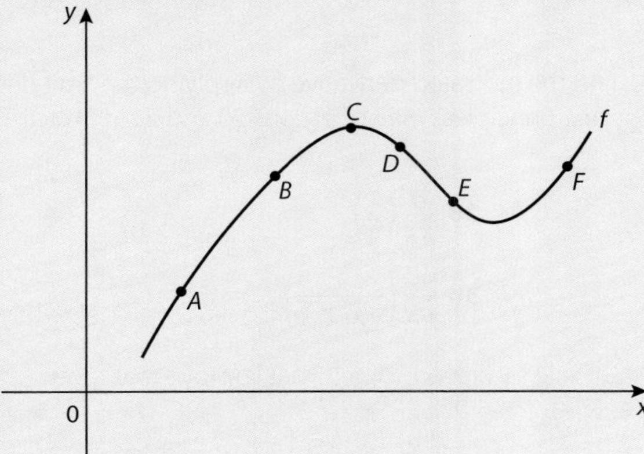

a) Between which two consecutive points is the average rate of change of the function greatest?

b) At what points is the instantaneous rate of change of f positive, negative and zero?

c) For which two pairs of points is the average rate of change approximately equal?

19 The slope of the curve $y = x^2 - 4x + 6$ at the point (3, 3) is equal to the slope of the curve $y = 8x - 3x^2$ at (a, b). Find the value of a and the value of b.

20 The graph of the equation $y = ax^3 - 2x^2 - x + 7$ has a slope of 3 at the point where $x = 2$. Find the value of a.

21 Find the coordinates of the point on the graph of $y = x^2 - x$ at which the tangent is parallel to the line $y = 5x$.

22 Let $f(x) = x^3 + 1$.

a) Evaluate $\frac{f(2 + h) - f(2)}{h}$ for $h = 0.1$.

b) What number does $\frac{f(2 + h) - f(2)}{h}$ approach as h approaches zero?

23 From first principles, find the derivative for the general quadratic function, $f(x) = ax^2 + bx + c$. Confirm your result by checking that it produces:
(i) the derivative of x^2 when $a = 1, b = 0, c = 0$
(ii) the derivative of $3x^2 - 4x + 2$ when $a = 3, b = -4, c = 2$.

24 A car is parked with the windows and doors closed for five hours. The temperature inside the car in degrees Celsius, C, is given by $C = 2\sqrt{t^3} + 17$ with t representing the number of hours since the car was first parked.

a) Find the average rate of change of the temperature from $t = 1$ to $t = 4$.

b) Find the function that gives the instantaneous rate of change of the temperature for any time t, $0 < t < 5$.

c) Find the time t at which the instantaneous rate of change of the temperature is equal to the average rate of change from $t = 1$ to $t = 4$.

25 A function f is even if $f(-x) = f(x)$ and a function g is odd if $g(-x) = -g(x)$.

a) If the function h is even, prove that the derivative of h is odd. In other words, if

$$h(-x) = h(x), \text{ then, } h'(-x) = -h'(x).$$

b) If the function p is odd, prove that the derivative of h is even. In other words, if

$$p(-x) = -p(x), \text{ then, } p'(-x) = p'(x).$$

26 Using algebraic manipulation and the proven result $\lim_{x \to 0} \frac{\sin x}{x} = 1$, prove that $\lim_{x \to 0} \frac{\cos x - 1}{x} = 0$.

In questions 27–30, find the indicated derivative by applying the limit definition of the derivative (i.e. by first principles). (See questions 20 and 21 in Exercise 13.1 for 27 and 28 below.)

27 $\frac{d}{dx}(\sqrt{x})$

28 $\frac{d}{dx}\left(\frac{1}{x}\right)$

29 $\frac{d}{dx}\left(\frac{2 + x}{3 - x}\right)$

30 $\frac{d}{dx}\left(\frac{1}{\sqrt{x + 2}}\right)$

31 Prove the constant rule by first principles. That is, prove that given a constant c, $c \in \mathbb{R}$, $\frac{d}{dx}(c) = 0$.

13.3 Maxima and minima – first and second derivatives

The relationship between a function and its derivative

The derivative, written in Newton notation as $f'(x)$ or in Leibniz notation as $\frac{dy}{dx}$, is a function derived from a function f that gives the slope of the graph of f at any x in the function's domain (given that the curve is 'smooth' at the value of x). The derivative is a slope, or rate of change, function. Knowing the slope of a function at different values in its domain tells us about properties of the function and the shape of its graph.

If the graph of a function is 'smooth' at a particular point, the function is considered to be *differentiable* at this point. In other words, a tangent line exists at this point. All functions that will be differentiated in this course will be differentiable at all values in the function's domain.

In the previous section, we observed that if a graph 'turns' at a particular point (for example, at the vertex of a parabola), then it has a horizontal tangent (slope = 0) at the point. Hence, the derivative will equal zero at a 'turning point'. In Section 3.2, we found the vertex of the graph of a quadratic function by using the technique of completing the square to write its equation in vertex form. We can also find the vertex by means of differentiation. As we look at the graph of a parabola moving from left to right (i.e. domain values increasing), it either turns from going down to going up (decreasing to increasing), or from going up to going down (increasing to decreasing) (Figure 13.16).

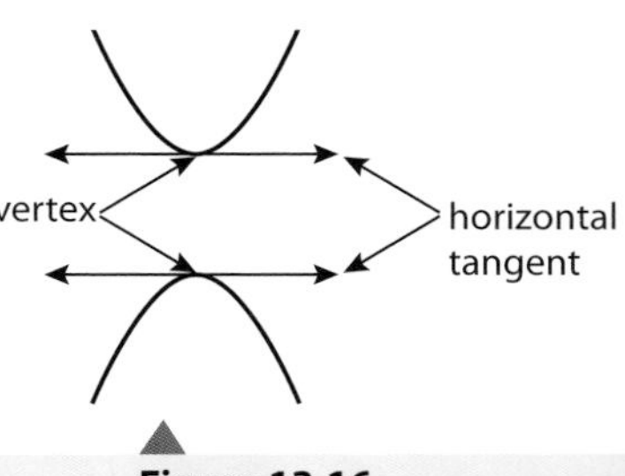

Figure 13.16

Example 13

Using differentiation, find the vertex of the parabola with the equation $y = x^2 - 8x + 14$.

Solution

Find the value of x for which the derivative, $\frac{dy}{dx}$, is zero.

$\frac{dy}{dx} = \frac{d}{dx}(x^2 - 8x + 14) = 2x - 8 = 0 \Rightarrow x = 4$

Thus, the x-coordinate of the vertex is 4.

To find the y-coordinate of the vertex, we substitute $x = 4$ into the equation, giving $y = 4^2 - 8(4) + 14 = -2$. Therefore, the vertex has coordinates $(4, -2)$.

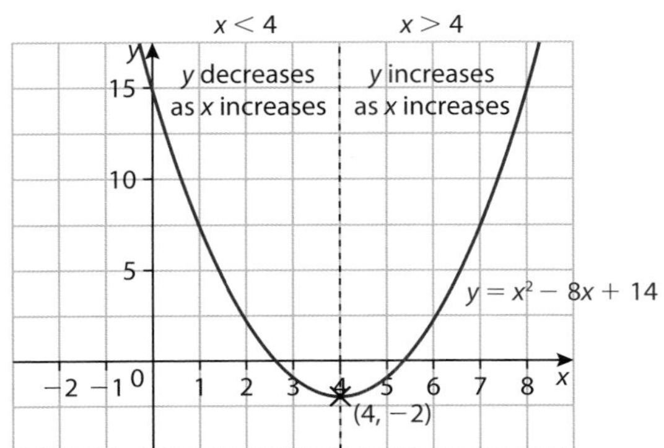

Figure 13.17 Slope changes from negative to positive as x increases.

We know that the parabola in Example 13 will 'open up' because the coefficient of the quadratic term, x^2, is positive. The parabola has a negative slope (decreasing) to the left of the vertex and a positive slope (increasing) to the right of the vertex (Figure 13.17). As the values of x increase, the derivative of $y = x^2 - 8x + 14$ will change from negative to zero to positive, accordingly.

$$\frac{dy}{dx} = 2x - 8 \Rightarrow \frac{dy}{dx} < 0 \text{ for } x < 4 \text{ and } \frac{dy}{dx} = 0 \text{ for } x = 4 \text{ and } \frac{dy}{dx} > 0 \text{ for } x > 4$$

In other words, the function $f(x) = x^2 - 8x + 14$ is decreasing for all $x < 4$; it is neither decreasing nor increasing at $x = 4$; and it is increasing for all $x > 4$. A point at which a function is neither increasing nor decreasing (i.e. there is a horizontal tangent) is called a **stationary point.** A convenient way to demonstrate where a function is increasing or decreasing and the location of any stationary points is with a **sign chart** for the function and its derivative, as shown in Figure 13.18 for $f(x) = x^2 - 8x + 14$. The derivative $f'(x) = 2x - 8$ is zero only at $x = 4$, thereby dividing the domain of f (i.e. $\mathbb{R}$) into two intervals: $x < 4$ and $x > 4$. $f'(x) = 2x - 8$ is a **continuous** function (i.e. no 'gaps' in the domain) so it is only necessary to test one point in each interval in order to determine the sign of all the values of the derivative in that interval. $f'(x)$ can only change sign at $x = 4$. For example, the fact that $f'(3) = 2(3) - 8 = -2 < 0$ means that $f'(x) < 0$ for all x when $x < 4$. Therefore, f is decreasing for all x in the open interval $(-\infty, 4)$.

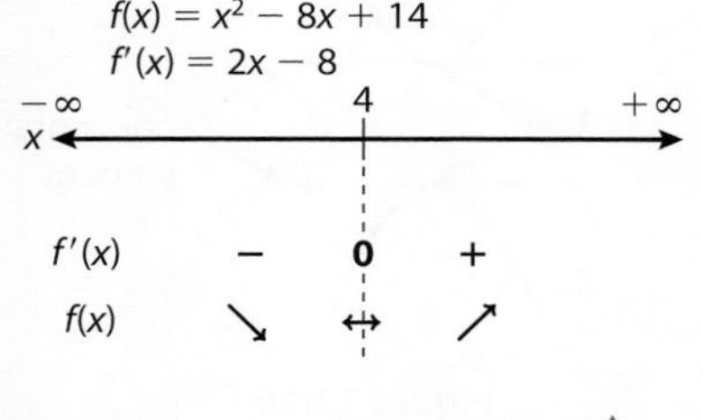

Figure 13.18 Sign chart for $f'(x)$ and $f(x)$.

Geometrically speaking, a function is **continuous** if there is no break in its graph; and a function is **differentiable** (i.e. a derivative exists) at any points where it is 'smooth'.

Increasing and decreasing functions and stationary points

If $f'(x) > 0$ for $a < x < b$, then $f(x)$ is **increasing** on the interval $a < x < b$.

If $f'(x) < 0$ for $a < x < b$, then $f(x)$ is **decreasing** on the interval $a < x < b$.

If $f'(x) = 0$ for $a < x < b$, then $f(x)$ is **constant** on the interval $a < x < b$.

If $f'(x) = 0$ for a single value $x = c$ on some interval $a < c < b$, then $f(x)$ has a **stationary point** at $x = c$. The corresponding point $(c, f(c))$ on the graph of f is called a stationary point.

It is at stationary points, or endpoints of the domain if the domain is not all real numbers, where a function may have a maximum or minimum value. These points at which extreme values of a function *may* occur are often referred to as **critical points.** Whether a function is increasing or decreasing on either side of a stationary point will indicate whether the stationary point is a maximum, minimum or neither.

Example 14

Consider the function $f(x) = 2x^3 + 3x^2 - 12x - 4,\ x \in \mathbb{R}$.

a) Find any stationary points of f.

b) Using the derivative of f, classify any stationary points as a maximum or minimum.

Solution

a) $f'(x) = 6x^2 + 6x - 12 = 0 \Rightarrow 6(x^2 + x - 2) = 0$
$\Rightarrow 6(x + 2)(x - 1) = 0 \Rightarrow x = -2$ or $x = 1$

With a domain of all real numbers there are no domain endpoints that may be an extreme value. Thus, f has two critical points: one at $x = -2$ and the other at $x = 1$.

When $x = -2$: $y = 2(-2)^3 + 3(-2)^2 - 12(-2) - 4 = 16 \Rightarrow f$ has a stationary point at $(-2, 16)$.

When $x = 1$: $y = 2(1)^3 + 3(1)^2 - 12(1) - 4 = -11 \Rightarrow f$ has a stationary point at $(1, -11)$.

b) Construct a sign chart for $f'(x)$ and $f(x)$ (left) to show where f is increasing or decreasing. The derivative $f'(x)$ has two zeros, at $x = -2$ and $x = 1$, thereby dividing the domain of f into three intervals that need to be tested. Since $f'(-3) = 6(-1)(-4) = 24 > 0$, then $f'(x) > 0$ for all $x < -2$. Likewise, since $f'(2) = 6(4)(1) = 24 > 0$, then $f'(x) > 0$ for all $x > 1$. Thus, f is increasing on the open intervals $(-\infty, -2)$ and $(1, \infty)$. Since $f'(0) = -12 < 0$, then $f'(x) < 0$ for all x such that $-2 < x < 1$. Thus, f is decreasing on the open interval $(-2, 1)$, i.e. $-2 < x < 1$. From this information, we can visualize for increasing values of x that the graph of f is going up for all $x < -2$, then turning down at $x = -2$, then going down for values of x from -2 to 1, then turning up at $x = 1$, and then going up for all $x > 1$. The basic shape of the graph of f will look something like the rough sketch shown left. Clearly, the stationary point $(-2, 16)$ is a maximum and the stationary point $(1, -11)$ is a minimum.

$f(x) = 2x^3 + 3x^2 - 12x - 4$

$f'(x) = 6(x + 2)(x - 1)$

		-2		1	
$f'(x)$	+	0	−	0	+
$f(x)$	↗	↔	↘	↔	↗
	increasing	stationary	decreasing	stationary	increasing

$x = -2$

f(x)

$x = 1$

The graph of $f(x) = 2x^3 + 3x^2 - 12x - 4$ from Example 14 (Figure 13.19) visually confirms the results acquired from analyzing the derivative of f.

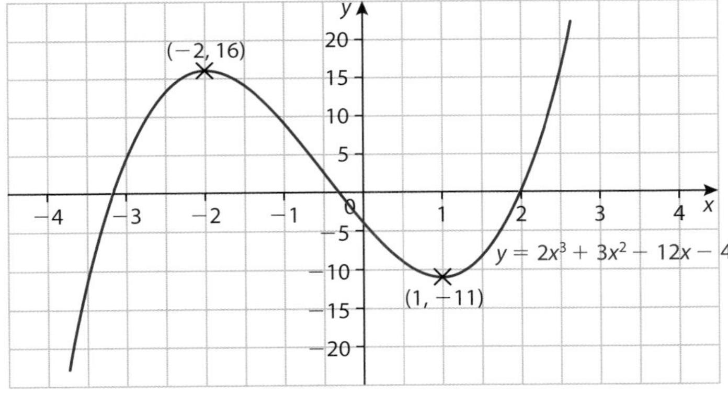

Figure 3.19

For Example 14, we can express the result for part b) most clearly by saying that $f(x)$ has a **relative maximum** value of 16 at $x = -2$, and $f(x)$ has a **relative minimum** value of -11 at $x = 1$. The reason that these *extreme* values are described as 'relative' (sometimes described as 'local') is because

The plural of 'maximum' is 'maxima', and the plural of 'minimum' is 'minima'. Maxima and minima are collectively referred to as 'extrema' – the plural of 'extremum' (extreme value). Extrema of a function that do not occur at domain endpoints will be 'turning points' of the graph of the function.

they are a maximum or minimum for the function in the immediate vicinity of the point, but not for the entire domain of the function. A point that is a maximum/minimum for the entire domain is called an **absolute**, or **global, maximum/minimum**.

The first derivative test

From Example 14, we can see that a function f has a maximum at some $x = c$ if $f'(c) = 0$ and f is *increasing* immediately to the left of $x = c$ and *decreasing* immediately to the right of $x = c$. Similarly, f has a minimum at some $x = c$ if $f'(c) = 0$ and f is *decreasing* immediately to the left of $x = c$ and *increasing* immediately to the right of $x = c$. It is important to understand, however, that not all stationary points are either a maximum or minimum.

Example 15

For the function $f(x) = x^4 - 2x^3$, find all stationary points and describe them completely.

Solution

$f'(x) = \dfrac{d}{dx}(x^4 - 2x^3) = 4x^3 - 6x^2 = 0 \Rightarrow 2x^2(2x - 3) = 0$

$\Rightarrow x = 0$ or $x = \dfrac{3}{2}$

The implied domain is all real numbers, so $x = 0$ and $x = \dfrac{3}{2}$ are the critical points of f.

When $x = 0$, $y = f(0) = 0$.

When $x = \dfrac{3}{2}$, $y = f\left(\dfrac{3}{2}\right) = \left(\dfrac{3}{2}\right)^4 - 2\left(\dfrac{3}{2}\right)^3 = \dfrac{81}{16} - \dfrac{54}{8} = -\dfrac{27}{16}$.

Therefore, f has stationary points at $(0, 0)$ and $\left(\dfrac{3}{2}, -\dfrac{27}{16}\right)$.

Because f has two stationary points, there are three intervals for which to test the sign of the derivative. We could use some form of a sign chart as shown previously, or we can use a more detailed table that summarizes the testing of the three intervals and the two critical points as shown below.

Interval/point	$x < 0$	$x = 0$	$0 < x < \frac{3}{2}$	$x = \frac{3}{2}$	$x > \frac{3}{2}$
Test value	$x = -1$		$x = 1$		$x = 2$
Sign of $f'(x)$	$f'(-1) = -10 < 0$	0	$f'(1) = -2 < 0$	0	$f'(2) = 8 > 0$
Conclusion	f decreasing ↘	none	f decreasing ↘	abs. min.	f increasing ↗

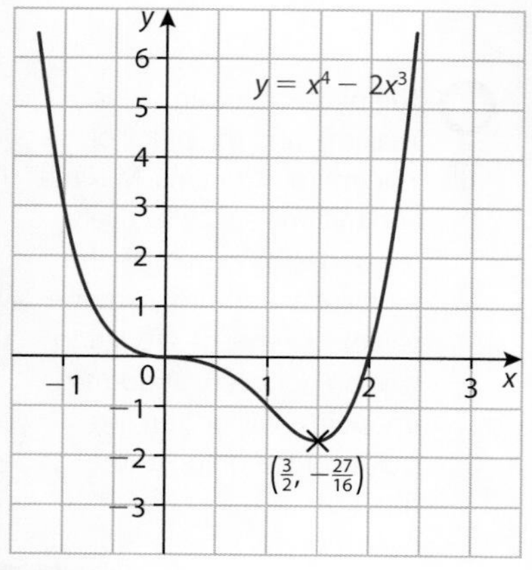

On either side of $x = 0$, f does not change from either decreasing to increasing or from increasing to decreasing. Although there is a horizontal tangent at $(0, 0)$, it is *not* an extreme value (turning point). The function steadily decreases as x approaches zero, then at $x = 0$ the function has a rate of change (slope) of zero for an instant and then continues on decreasing. As x approaches $\dfrac{3}{2}$, f is decreasing and then switches to increasing at $x = \dfrac{3}{2}$.

Therefore, the stationary point $(0, 0)$ is neither a maximum nor a minimum; and the stationary point $\left(\frac{3}{2}, -\frac{27}{16}\right)$ is an absolute minimum. Or, in other words, f has an absolute (global) minimum value of $-\frac{27}{16}$ at $x = \frac{3}{2}$.

The reason that an *absolute*, rather than a *relative*, minimum value occurs at $x = \frac{3}{2}$ is because for all $x < \frac{3}{2}$ the function f is either decreasing or constant (at $x = 0$) and for all $x < \frac{3}{2}$ f is increasing.

First derivative test for maxima and minima of a function

Suppose that $x = c$ is a critical point of a continuous and smooth function f. That is, $f(c) = 0$ and $x = c$ is a stationary point or $x = c$ is an endpoint of the domain.

I. At a stationary point $x = c$:

1. If $f'(x)$ changes sign from positive to negative as x increases through $x = c$, then f has a relative maximum at $x = c$.

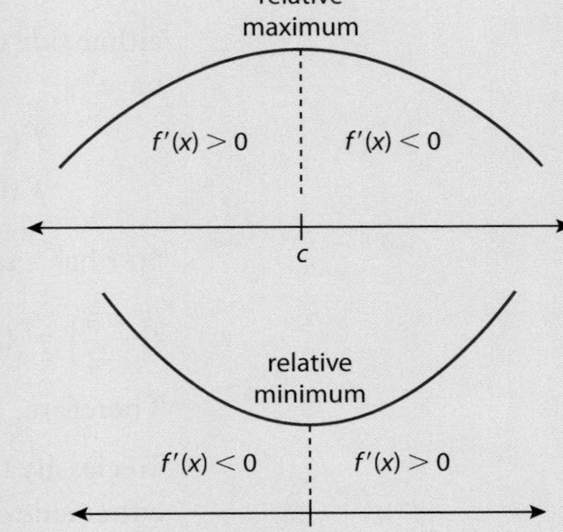

2. If $f'(x)$ changes sign from negative to positive as x increases through $x = c$, then f has a relative minimum at $x = c$.

3. If $f'(x)$ does not change sign as x increases through $x = c$, then f has neither a relative maximum nor a relative minimum at $x = c$.

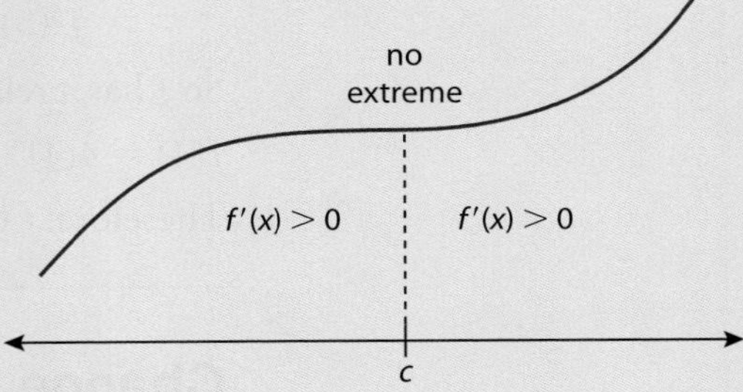

II. At a domain endpoint $x = c$:

If $x = c$ is an endpoint of the domain, then $x = c$ will be a relative maximum or minimum of f if the sign of $f'(x)$ is always positive or always negative for $x > c$ (at a left endpoint), or for $x < c$ (at a right endpoint), as illustrated below.

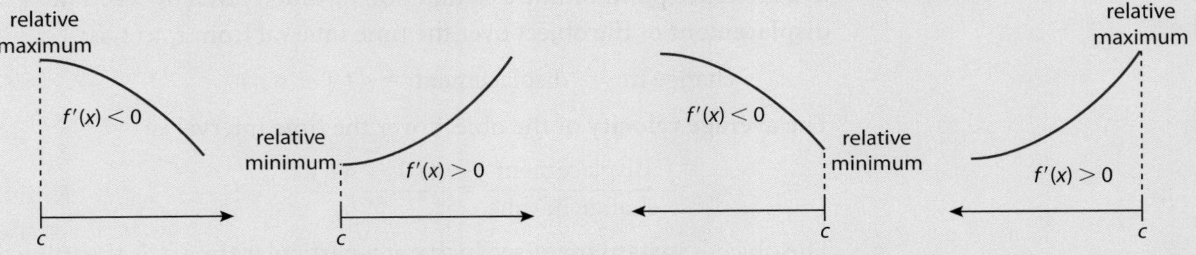

If it is possible to show that a relative maximum/minimum at $x = c$ is the greatest/least value for the entire domain of f, then it is classified as an absolute maximum/minimum.

Example 16

Apply the first derivative test to find any local extreme values for $f(x)$. Identify any absolute extrema.

$$f(x) = 4x^3 - 9x^2 - 120x + 25$$

Solution

$$f'(x) = \frac{d}{dx}(4x^3 - 9x^2 - 120x + 25) = 12x^2 - 18x - 120$$

$$f'(x) = 12x^2 - 18x - 120 = 0 \Rightarrow 6(2x^2 - 3x - 20) = 0$$
$$\Rightarrow 6(2x + 5)(x - 4) = 0$$

Thus, f has stationary points at $x = -\frac{5}{2}$ and $x = 4$.

To classify the stationary point at $x = -\frac{5}{2}$, we need to choose test points on either side of $-\frac{5}{2}$, for example, $x = -3$ (left) and $x = 0$ (right). Then we have

$$f'(-3) = 6(-1)(-7) = 42 > 0$$
$$f'(0) = 6(5)(-4) = -120 < 0$$

So f has a relative maximum at $x = -\frac{5}{2}$.

$$f\left(-\frac{5}{2}\right) = 4\left(-\frac{5}{2}\right)^3 - 9\left(-\frac{5}{2}\right)^2 - 120\left(-\frac{5}{2}\right) + 25 = 206.25$$

Therefore, f has a relative maximum value of 206.25 at $x = -\frac{5}{2}$.

To classify the stationary point at $x = 4$, we need to choose test points on either side of 4, for example, $x = 0$ (left) and $x = 5$ (right). Then we have

$$f'(0) = -120 < 0$$
$$f'(5) = 6(15)(1) = 90 > 0$$

So f has a relative minimum at $x = 4$.

$$f(4) = 4(4)^3 - 9(4)^2 - 120(4) + 25 = -343$$

Therefore, f has a relative minimum value of -343 at $x = 4$.

Change in displacement and velocity

Consider the motion of an object such that we know its position s relative to a reference point or line as a function of time t given by $s(t)$. The **displacement** of the object over the time interval from t_1 to t_2 is:

$$\text{change in } s = \text{displacement} = s(t_2) - s(t_1)$$

The **average velocity** of the object over the time interval is:

$$v_{\text{avg}} = \frac{\text{displacement}}{\text{change in time}} = \frac{s(t_2) - s(t_1)}{t_2 - t_1}$$

The object's **instantaneous velocity** at a particular time, t, is the value of the derivative of the position function, s, with respect to time at t.

$$\text{velocity} = \frac{ds}{dt} = s'(t)$$

Example 17

A rocket is launched upwards into the air. Its vertical position, s metres, above the ground at t seconds is given by
$s(t) = -5t^2 + 18t + 1$.

a) Find the average velocity over the time interval from $t = 1$ second to $t = 2$ seconds.

b) Find the instantaneous velocity at $t = 1$ second.

c) Find the maximum height reached by the rocket and the time at which this occurs.

Solution

a) $v_{avg} = \dfrac{s(2) - s(1)}{2 - 1} = \dfrac{[-5(2)^2 + 18(2) + 1] - [-5 + 18 + 1]}{1}$

$= 3$ metres per second (or m s^{-1})

b) $s'(t) = -10t + 18 \Rightarrow s'(1) = -10 + 18 = 8\ \text{m s}^{-1}$

c) $s'(t) = -10t + 18 = 0 \Rightarrow t = 1.8$

Thus, s has a stationary point at $t = 1.8$. t must be positive and ranges from time of launch ($t = 0$) to when the rocket hits the ground, i.e. $h = 0$.

$$s(t) = -5t^2 + 18t + 1 = 0 \Rightarrow t = \frac{-18 \pm \sqrt{18^2 - 4(-5)(1)}}{2(-5)}$$

$$\Rightarrow t \approx -0.5472 \text{ or } t \approx 3.655$$

So, the rocket hits the ground about 3.66 seconds after the time of launch. Hence, the domain for the position (s) and velocity (v) functions is $0 \leqslant t \leqslant 3.66$. Therefore, the function s has three critical points: $t = 0$, $t = 1.8$ and $t \approx 3.66$.

The maximum of the function, i.e. the maximum height, most likely occurs at the critical point $t = 1.8$. Let's confirm this.

Applying the first derivative test, we determine the sign of the derivative, $s'(t)$, for values on either side of $t = 1.8$, for example, $t = 0$ and $t = 2$. $s'(0) = 18 > 0$ and $s'(2) = -2 < 0$. Neither of the domain endpoints, $t = 0$ and $t \approx 3.66$, are at a maximum or minimum because the function is not constantly increasing or constantly decreasing before or after the endpoint. Since the function changes from increasing to decreasing at $t = 1.8$ and $s(1.8) = -5(1.8)^2 + 18(1.8) + 1 = 17.2$, then the rocket reaches a maximum height of 17.2 metres 1.8 seconds after it was launched.

The relationship between a function and its second derivative

You may have wondered why the strategy we are applying to locate and classify extrema for a function focuses on using the *first* derivative of the function. This implies that we are interested in using some other type of derivative, namely the *second* derivative. There is another useful test for the purpose of analyzing the stationary point of a function that makes use of the derivative of the derivative, i.e. the second derivative, of the function.

When we differentiate a function $y = f(x)$, we obtain the first derivative $f'(x)$ $\left(\text{also denoted as } \frac{dy}{dx}\right)$. Often this is a function that can also be differentiated. The result of doing so is the derivative of $f'(x)$, which is denoted in Newton notation as $f''(x)$ or in Leibniz notation as $\frac{d^2y}{dx^2}$ and called the second derivative of f with respect to x. For example, if $f(x) = x^3$, then $f'(x) = 3x^2$ and $f''(x) = 6x$.

Second derivatives, like first derivatives, occur often in methods of applying calculus. In Example 17, the function $s(t)$ gave the position, in metres above the ground, of a projectile (toy rocket) where t, in seconds, is the time since the projectile was launched. The function $s'(t)$, the first derivative of the position function, then gives the rate of change of the object's position, i.e. its velocity, in metres per second (m s^{-1}). Differentiation of this function gives the rate of change of the object's velocity, i.e. its *acceleration*, measured in metres per second per second (m s^{-2}).

The graphs of the position, velocity and acceleration functions for Example 17 aligned vertically (Figure 13.20) nicely illustrate the relationships between a function, its first derivative and its second derivative. The slope of the graph of $s(t)$ is initially a large positive value (graph is steep), but steadily decreases until it is zero (horizontal tangent) at $t = 1.8$ and then continues to decrease, becoming a large negative value (again, steep, but in the other direction). This corresponds to the real-life situation in which the rocket is launched with a high initial velocity ($v(0) = 18\ \text{m s}^{-1}$) and then its velocity decreases steadily due to gravity. The rocket's velocity is zero for just an instant when it reaches its maximum height at $t = 1.8$ and then its velocity becomes more and more negative because it has changed direction and is moving back (negative direction) to the ground. The rate of change of the velocity, $v'(t)$, is constant and it is negative because the velocity is decreasing from positive values to zero to negative values. This is clear from the fact that the graph of the velocity function, $v(t)$, is a straight line with a negative slope. It follows then that the acceleration function – the rate of change of velocity – is a negative constant, $a = -10$ in this case, and its graph is a horizontal line.

In Example 17, it is not possible to have a negative function value for $s(t)$ because the rocket's position is always above, or at, ground level. In many motion problems in calculus, we consider a simplified version by limiting

Position function:
$s(t) = -5t^2 + 18t + 1$

Velocity function:
$v(t) = s'(t) = -10t + 18$

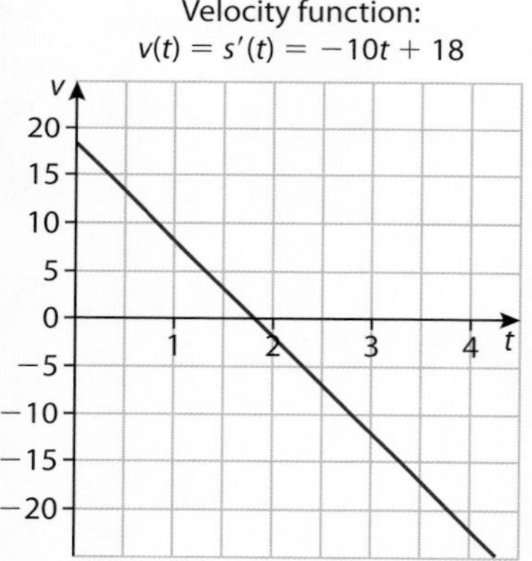

Acceleration function:
$a(t) = v'(t) = s''(t) = -10$

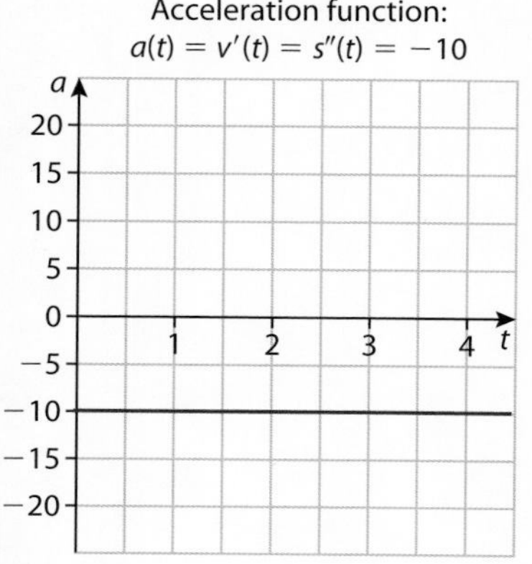

Figure 13.20 Position, velocity and acceleration functions for rocket.

an object's motion to a line with its position given as its **displacement** from a fixed point (usually the origin). At a position left of the fixed point, the object's displacement is negative, and at a position right of the fixed point, the displacement is positive. Velocity can also be positive or negative depending on the direction of travel (i.e. the sign of the rate of change of the object's displacement). Likewise, acceleration is positive if velocity is increasing (i.e. rate of change of velocity is positive) and negative if velocity is decreasing.

A common misconception is that acceleration is positive for motion in the positive direction (usually 'right' or 'up') and negative for motion in the negative direction (usually 'left' or 'down'). Acceleration indicates how velocity is changing. Even though an object may be moving in a positive direction (e.g. to the right) if it is slowing down, then its acceleration is acting in the opposite direction and would be negative. In Example 17, the rocket was always accelerating in the negative direction, -10 m s^{-2}, due to the force of gravity. Note: A more accurate value for the acceleration of a free-falling object due to gravity is -9.8 m s^{-2}.

It would be incorrect to graph a function and its first and/or second derivative on the same axes. For example, the position $s(t)$, velocity $v(t)$ and acceleration $a(t)$ functions graphed on separate axes in Figure 13.20 will have different units on each vertical axis: metres for $s(t)$, metres per second for $v(t)$ and metres per second per second for $a(t)$.

Motion along a line

If an object moves in a straight line such that at time t its displacement (position) from a fixed point is $s(t)$, then the first derivative $s'(t)$, also written as $\frac{ds}{dt}$, gives the velocity $v(t)$ at time t.

The second derivative $s''(t)$, also written as $\frac{d^2s}{dt^2}$, is the first derivative of $v(t)$. Hence, the second derivative of the displacement, or position, function is a measure of the rate at which the velocity is changing, i.e. it represents the acceleration of the object, which we express as

$$a(t) = v'(t) = s''(t) \quad \text{or} \quad a(t) = \frac{dv}{dt} = \frac{d^2s}{dt^2}.$$

Displacement can be negative, positive or zero. **Distance** is the absolute value of displacement. **Velocity** can be negative, positive or zero. **Speed** is the absolute value of velocity.

Example 18

An object moves along a straight line so that after t seconds its displacement from the origin is s metres. Given that $s(t) = -2t^3 + 6t^2$, answer the following:

a) Find expressions for the (i) velocity and (ii) acceleration at time t seconds.

b) Find the (i) initial velocity and (ii) initial acceleration of the object (i.e. at time when $t = 0$).

c) Find the (i) maximum displacement and (ii) maximum velocity for the interval $0 \leqslant t \leqslant 3$.

Solution

a) (i) $v(t) = \frac{ds}{dt} = \frac{d}{dt}(-2t^3 + 6t^2) = -6t^2 + 12t$

(ii) $a(t) = \frac{d^2s}{dt^2} = \frac{dv}{dt} = \frac{d}{dt}(-6t^2 + 12t) = -12t + 12$

b) (i) $v(0) = -6(0)^2 + 12(0) = 0 \Rightarrow$ The object's initial velocity is 0 m s^{-1}.

(ii) $a(0) = -12(0) + 12 = 12 \Rightarrow$ The object's initial acceleration is 12 m s^{-2}.

c) (i) To find the maximum displacement, we can apply the first derivative test to $s(t)$. Since the first derivative of displacement, $s(t)$, is velocity, $v(t)$, then the critical points of $s(t)$ are where the velocity is zero (stationary points) and domain endpoints.

$$s'(t) = v(t) = -6t^2 + 12t = 0 \Rightarrow 6t(-t+2) = 0$$
$$\Rightarrow v(t) = 0 \text{ when } t = 0 \text{ or } t = 2$$

For the interval $0 \leqslant t \leqslant 3$, the critical points to be tested for finding the maximum displacement are at $t = 0$, $t = 2$ and $t = 3$. Check whether the velocity is increasing or decreasing on either side of the stationary point at $t = 2$ by finding the sign of $v(t)$ for $t = 1$ and $t = 2.5$.

$v(1) = -6(1)^2 + 12(1) = 6$ and $v(2.5) = -6(2.5)^2 + 12(2.5) = -7.5$

Hence, the displacement s is increasing for $0 < t < 2$ and decreasing for $2 < t < 3$. This indicates that the stationary point at $t = 2$ must be an absolute maximum for s in the interval $0 \leqslant t \leqslant 3$.

$s(2) = -2(2)^3 + 6(2)^2 = 8$

Therefore, the object has a maximum displacement of 8 metres at $t = 2$ seconds.

(ii) To find the maximum velocity, we can apply the first derivative test to $v(t)$. The first derivative of $v(t)$ is acceleration $a(t)$, which is the *second* derivative of $s(t)$. Hence, where $s''(t) = 0$ (acceleration is zero) indicates critical points for $v(t)$, i.e. where velocity may change from increasing to decreasing, or vice versa.

$$s''(t) = a(t) = \frac{d}{dt}(-6t^2 + 12t) = -12t + 12$$
$$\Rightarrow 12(-t+1) = 0 \Rightarrow a(t) = 0 \text{ when } t = 1$$

For the interval $0 \leqslant t \leqslant 3$, the critical points to be tested for finding the maximum velocity are at $t = 0$, $t = 1$ and $t = 3$. Check whether the velocity is increasing or decreasing on either side of $t = 1$ by finding the sign of $a(t)$ for $t = 0.5$ and $t = 2$.

$a(0.5) = -12(0.5) + 12 = 6$ and $a(2) = -12(2) + 12 = -12$

Hence, the velocity v is increasing for $0 < t < 1$ and decreasing for $1 < t < 3$. This indicates that the point at $t = 1$ must be an absolute maximum for v in the interval $0 \leqslant t \leqslant 3$.

$v(1) = -6(1)^2 + 12(1) = 6$

Therefore, the object has a maximum velocity of 6 metres per second at $t = 1$ second.

The second derivative of a function tells us how the first derivative of the function changes. From this we can use the second derivative, as we did the first derivative, to reveal information about the shape of the graph of a function. Note in Example 18 that the object's velocity changed from increasing to decreasing when the object's acceleration was zero at $t = 1$.

Let's examine graphically the significance of the point where acceleration is zero (i.e. velocity changing from increasing to decreasing) in connection to the displacement graph for Example 18. In other words, what can the second derivative of a function tell us about the shape of the function's graph?

Figure 13.21 shows the graphs of the displacement, velocity and acceleration functions for the motion of the object in Example 18. A dashed vertical line highlights the nature of the three graphs where $t = 1$. At this point, velocity has a maximum value and acceleration is zero. It is also where velocity changes from increasing to decreasing, which has a corresponding effect on the shape of the displacement function $s(t)$.

At the point where $t = 1$, the graph of $s(t)$ changes from curving 'upwards' (*concave up*) to curving 'downwards' (*concave down*) because its slope (corresponding to velocity) changes from increasing to decreasing. This can only occur when velocity (first derivative) has a maximum and hence where acceleration (second derivative) is zero. We can see from this illustration that for a general function $f(x)$, finding intervals where the first derivative $f'(x)$ is increasing (positive acceleration) or decreasing (negative acceleration) can be used to determine where the graph of $f(x)$ is curving upward or curving downward. A point at which a function's curvature (concavity) changes – as at $t = 1$ for the graph of $s(t)$ left – is called a **point of inflexion**.

Figure 13.21

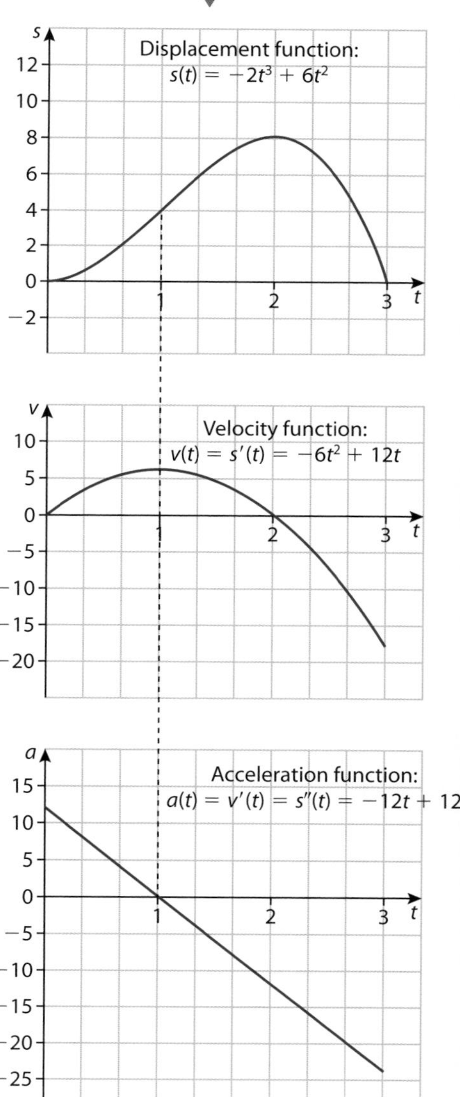

Concavity and the second derivative

The graph of $f(x)$ is **concave up** where $f'(x)$ is increasing and **concave down** where $f'(x)$ is decreasing. It follows that:

(i) if $f''(x) > 0$ for all x in some interval of the domain of f, the graph of f is concave up in the interval

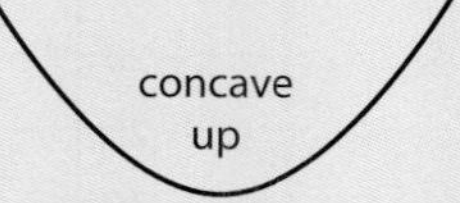

(ii) if $f''(x) < 0$ for all x in some interval of the domain of f, the graph of f is concave down in the interval.

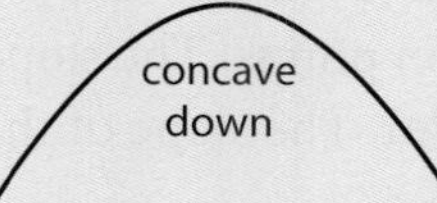

If $f(x)$ is a continuous function, its graph can only change concavity (up to down, or down to up) where $f''(x) = 0$. Hence, for a continuous function, an **inflexion point** may only occur where $f''(x) = 0$.

Note: Concavity is not defined for a line – it is neither concave up nor concave down.

Example 19

Determine the intervals on which the graph of $y = x^4 - 4x^3$ is concave up or concave down and identify any inflexion points.

Solution

We first note that the function is continuous for its domain of all real numbers. To locate points of inflexion, we then find for what value(s) the second derivative is zero.

$$\frac{dy}{dx} = \frac{d}{dx}(x^4 - 4x^3) = 4x^3 - 12x^2$$

$$\Rightarrow \frac{d^2y}{dx^2} = \frac{d}{dx}(4x^3 - 12x^2) = 12x^2 - 24x = 12x(x-2)$$

Setting $\frac{d^2y}{dx^2} = 0$, it follows that inflexion points may occur at $t = 0$ and $t = 2$. These two values divide the domain of the function into three intervals that we need to test. Let's choose $t = -1$, $t = 1$ and $t = 3$ as our test values. At $t = -1$, $\frac{d^2y}{dx^2} = 36 > 0$; at $t = 1$, $\frac{d^2y}{dx^2} = -12 < 0$; and at $t = 3$, $\frac{d^2y}{dx^2} = 36 > 0$. These results can be organized in a sign chart, illustrating that the graph of $y = x^4 - 4x^3$ is concave up for the open intervals $(-\infty, 0)$ and $(2, \infty)$, and concave down on the open interval $(0, 2)$.

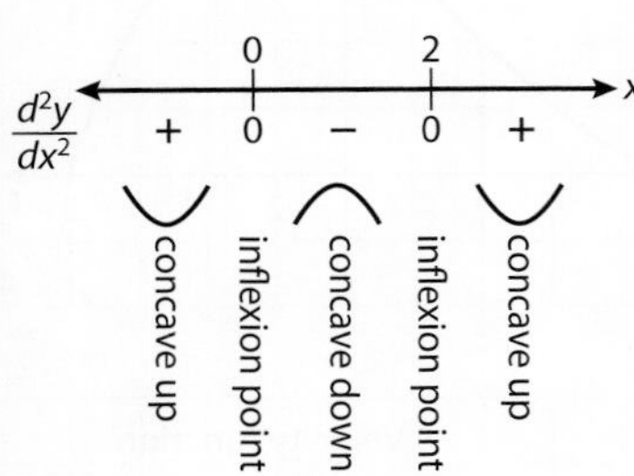

At $t = 0$, $y = 0$ and at $t = 2$, $y = 2^4 - 4(2)^3 = -16$. Therefore, $(0, 0)$ and $(2, -16)$ are inflexion points because it is at these points the concavity of the graph changes.

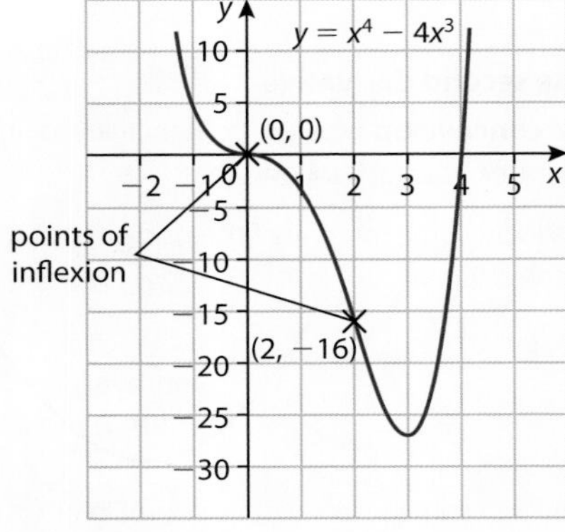

Figure 13.22 Inflexion points.

The graph of the function (Figure 13.22) from Example 19 reveals two different types of inflexion points. The slope of the curve at $(0, 0)$ is zero – i.e. it is a stationary point. The slope of the curve at the other inflexion point, $(2, -16)$, is negative.

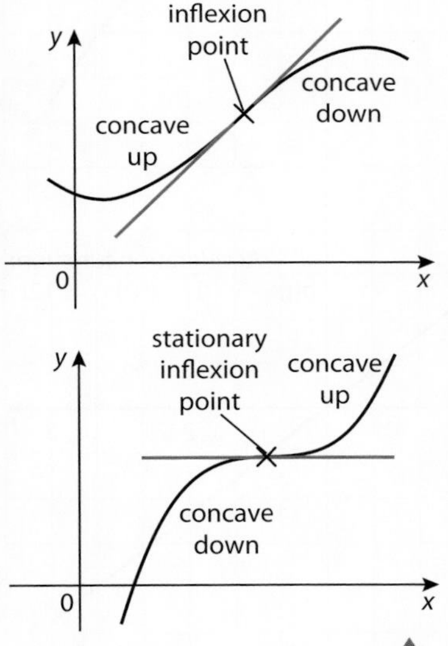

Figure 13.23 The concavity of a graph changes at a point of inflexion.

For either type of inflexion point, the graph crosses its tangent line at the point of inflexion, as shown in Figure 13.23.

The fact that the second derivative of a function is zero at a certain point does not guarantee that an inflexion point exists at the point.

The functions $y = x^3$ and $y = x^4$ will serve to illustrate that $\frac{d^2y}{dx^2} = 0$ is a necessary but not sufficient condition for the existence of an inflexion point.

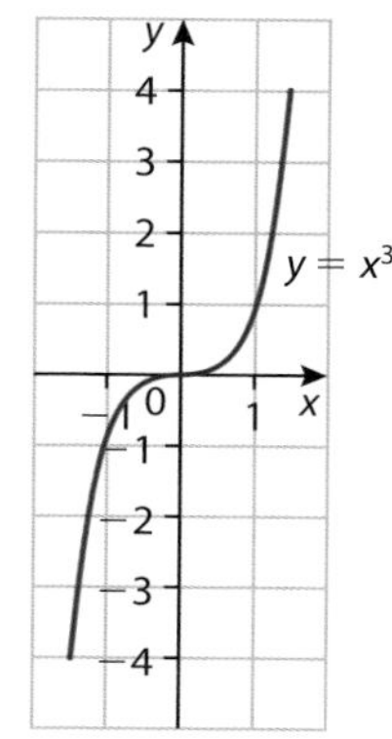

- For $y = x^3$: $\frac{dy}{dx} = \frac{d}{dx}(x^3) = 3x^2 \Rightarrow \frac{d^2y}{dx^2} = \frac{d}{dx}(3x^2) = 6x \Rightarrow \frac{d^2y}{dx^2} = 0$ at $x = 0$. We can conclude from this that there may be an inflexion point at $x = 0$. We need to investigate further by checking to see if $\frac{d^2y}{dx^2}$ changes sign at $x = 0$. At $x = -1$, $\frac{d^2y}{dx^2} = -6$ and at $x = 1$, $\frac{d^2y}{dx^2} = 6$.
  Thus, there is an inflexion point at $x = 0$ (confirmed by graph) because the second derivative changes sign at $x = 0$.

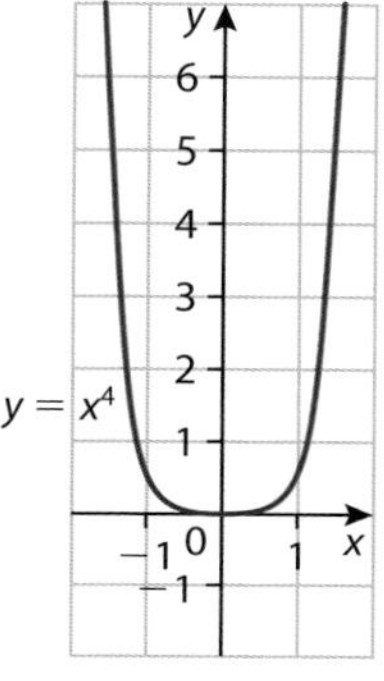

- For $y = x^4$: $\frac{dy}{dx} = \frac{d}{dx}(x^4) = 4x^3 \Rightarrow \frac{d^2y}{dx^2} = \frac{d}{dx}(4x^3) = 12x^2 \Rightarrow \frac{d^2y}{dx^2} = 0$ at $x = 0$. Again, we need to see if $\frac{d^2y}{dx^2}$ changes sign at $x = 0$.
  At $x = -1$, $\frac{d^2y}{dx^2} = 12$ and at $x = 1$, $\frac{d^2y}{dx^2} = 12$. Thus, there is *no* inflexion point at $x = 0$ (confirmed by graph) because the second derivative does *not* change sign at $x = 0$.

The second derivative test

Earlier in this section, we developed the first derivative test for locating maxima and minima of a function. Instead of using the first derivative to check whether a function changes from increasing to decreasing (maximum) or decreasing to increasing (minimum) at a stationary point, we can simply evaluate the second derivative at the stationary point. If the graph is concave up at the stationary point then it will be a minimum, and if it is concave down then it will be a maximum. If the second derivative is zero at a stationary point (as for $y = x^3$ and $y = x^4$), no conclusion can be made and we need to go back to the first derivative test. Using the second derivative in this way is a very efficient method for telling us whether a stationary point is a relative maximum or minimum.

The second derivative test

1. If $f'(c) = 0$ and $f''(c) < 0$, then f has a relative maximum at $x = c$.

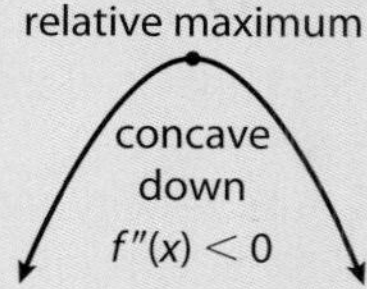

2. If $f'(c) = 0$ and $f''(c) > 0$, then f has a relative minimum at $x = c$.

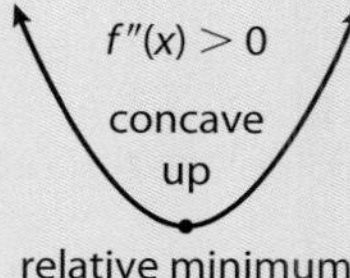

If $f''(c) = 0$, the test fails and the first derivative test should be applied.

Example 20

Find any relative extrema for $f(x) = 3x^5 - 25x^3 + 60x + 20$.

Solution

The implied domain of f is all real numbers. Solve $f'(x) = 0$ to obtain possible extrema.

$$\begin{aligned} f'(x) &= 15x^4 - 75x^2 + 60 = 0 \\ &15(x^4 - 5x^2 + 4) = 0 \\ &15(x^2 - 4)(x^2 - 1) = 0 \\ &15(x + 2)(x - 2)(x + 1)(x - 1) = 0 \end{aligned}$$

Therefore, f has four stationary points: $x = -2$, $x = -1$, $x = 1$ and $x = 2$.

Applying the second derivative test:

$$\begin{aligned} &f''(x) = 60x^3 - 150x = 30x(2x^2 - 5) \\ &f''(-2) = -180 < 0 \Rightarrow f \text{ has a relative maximum at } x = -2 \\ &f''(-1) = 90 > 0 \Rightarrow f \text{ has a relative minimum at } x = -1 \\ &f''(1) = -90 < 0 \Rightarrow f \text{ has a relative maximum at } x = 1 \\ &f''(2) = 180 > 0 \Rightarrow f \text{ has a relative minimum at } x = 2 \end{aligned}$$

Exercise 13.3

In questions 1–3, find the vertex of the parabola using differentiation.

1 $y = x^2 - 2x - 6$ **2** $y = 4x^2 + 12x + 17$ **3** $y = -x^2 + 6x - 7$

For questions 4–7, a) find the derivative, $f'(x)$, b) indicate the interval(s) for which $f(x)$ is increasing, and c) the interval(s) for which $f(x)$ is decreasing.

4 $y = x^2 - 5x + 6$ **5** $y = 7 - 4x - 3x^2$

6 $y = \frac{1}{3}x^3 - x$ **7** $y = x^4 - 4x^3$

For questions 8–13:

a) find the coordinates of any stationary points for the graph of the equation
b) state, with reasoning, whether each stationary point is a minimum, maximum or neither
c) sketch a graph of the equation and indicate the coordinates of each stationary point on the graph.

8 $y = 2x^3 + 3x^2 - 72x + 5$ **9** $y = \frac{1}{6}x^3 - 5$

10 $y = x(x - 3)^2$ **11** $y = x^4 - 2x^3 - 5x^2 + 6$

12 $y = x^3 - 2x^2 - 7x + 10$ **13** $y = x - \sqrt{x}$

14 An object moves along a line such that its displacement, s metres, from the origin O is given by $s(t) = t^3 - 4t^2 + t$.

a) Find expressions for the object's velocity and acceleration in terms of t.
b) For the interval $-1 \leqslant t \leqslant 3$, sketch the displacement-time, velocity-time, and acceleration-time graphs on separate sets of axes, vertically aligned as in Figure 13.21.
c) For the interval $-1 \leqslant t \leqslant 3$, find the time at which the displacement is a maximum and find its value.
d) For the interval $-1 \leqslant t \leqslant 3$, find the time at which the velocity is a minimum and find its value.
e) In words, accurately describe the motion of the object during the interval $-1 \leqslant t \leqslant 3$.

For each function $f(x)$ in questions 15–20, find any relative extrema and points of inflexion. State the coordinates of any such points. Use your GDC to assist you in sketching the function.

15 $f(x) = x^3 - 12x$

16 $f(x) = \frac{1}{4}x^4 - 2x^2$

17 $f(x) = x + \frac{4}{x}$

18 $y = x^2 - \frac{1}{x}$

19 $f(x) = -3x^5 + 5x^3$

20 $f(x) = 3x^4 - 4x^3 - 12x^2 + 5$

21 An object moves along a line such that its displacement, s metres, from a fixed point P is given by $s(t) = t(t - 3)(8t - 9)$.

a) Find the initial velocity and initial acceleration of the object.

b) Find the velocity and acceleration of the object at $t = 3$ seconds.

c) Find for what values of t the object changes direction. What significance do these times have in connection to the displacement of the object?

d) Find for what value of t the object's velocity is a minimum. What significance does this time have in connection to the acceleration of the object?

22 The delivery cost per tonne of bananas, D (in thousands of dollars), when x tonnes of bananas are shipped is given by $D = 3x + \frac{100}{x}$, $x > 0$. Find the value of x for which the delivery cost per tonne of bananas is a minimum, and find the value of the minimum delivery cost. Explain why this cost is a minimum rather than a maximum.

23 The curve $y = x^4 + ax^2 + bx + c$ passes through the point $(-1, -8)$ and at that point $\frac{d^2y}{dx^2} = \frac{dy}{dx} = 6$. Find the values of a, b and c and sketch the curve.

24 Find any maxima, minima or stationary points of inflexion of the function $f(x) = \frac{x^3 + 3x - 1}{x^2}$, stating, with explanation, the nature of each point.

Sketch the curve, indicating clearly what happens as $x \to \pm\infty$.

25 For each of the five functions graphed below sketch its derivative on a separate pair of axes. Do not use your GDC. It is helpful to use the result from question 25 in Exercise 13.2 – that the derivative of an even function is odd and the derivative of an odd function is even.

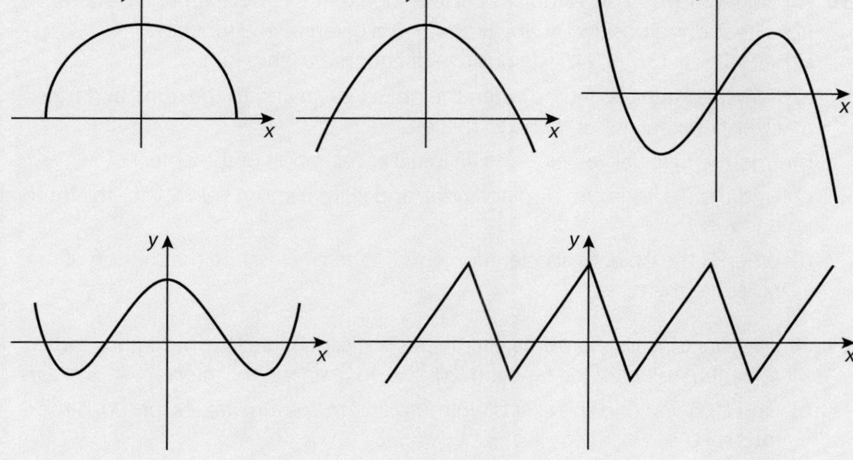

In questions 26 and 27, the graph of the **derivative** of a function f is shown.

a) On what intervals is f increasing or decreasing?

b) At what value(s) of x does f have a local maximum or minimum?

26

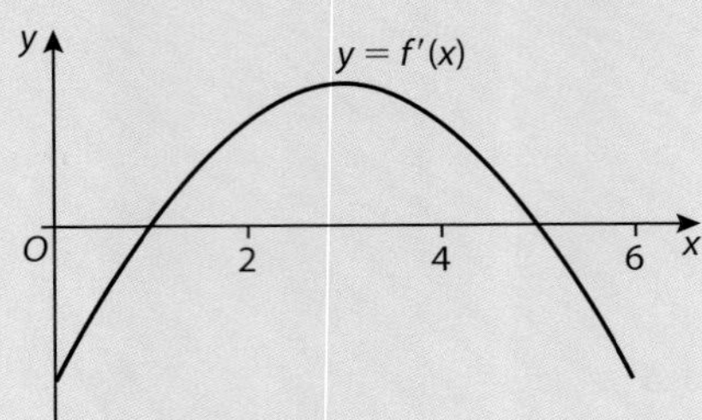

27

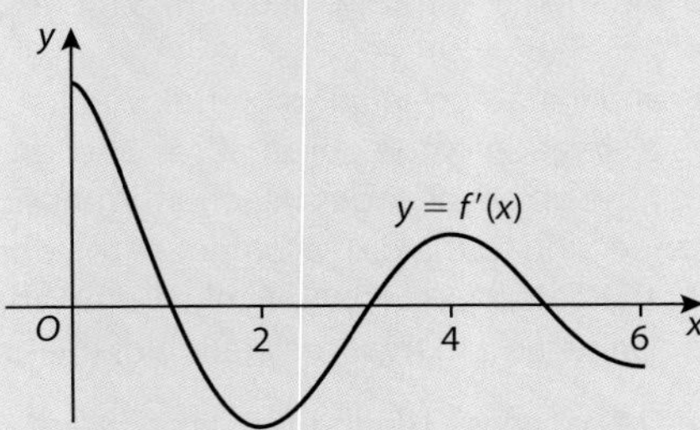

28 The graph of the **second derivative** f'' of a function f is shown. Approximate the x-coordinates of the inflexion points of f. Give reasons for your answers.

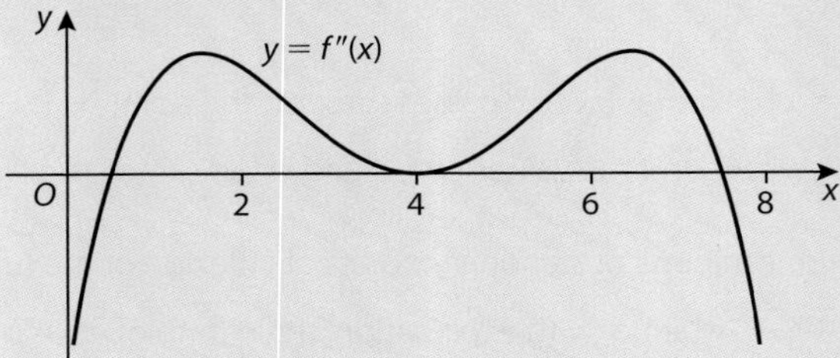

29 Sketch a continuous curve $y = f(x)$ with the following properties. Label coordinates where possible.

$f(-2) = 8$	$f(0) = 4$	$f(2) = 0$	$f'(2) = f'(-2) = 0$
$f'(x) > 0$ for $\|x\| > 2$	$f'(x) < 0$ for $\|x\| < 2$	$f''(x) < 0$ for $x < 0$	$f''(x) > 0$ for $x > 0$

30 An object moves along a horizontal line such that its displacement, s metres, from its starting position at any time $t \geqslant 0$ is given by the function $s(t) = -2t^3 + 15t^2 - 24t$. The positive direction is to the right.

a) Find the intervals of time when the object is moving to the right, and the intervals when it is moving to the left.

b) Find the (i) initial velocity, and (ii) initial acceleration of the object.

c) Find the (i) maximum displacement, and (ii) maximum velocity for the interval $0 \leqslant t \leqslant 5$.

d) When is the object's acceleration equal to zero? Describe the motion of the object at this time.

31 a) Use your GDC to approximate to three significant figures the maximum and minimum values of the function $f(x) = x - \sqrt{2}\sin x$ in the interval $0 \leqslant x \leqslant 2\pi$.

b) Find $f'(x)$ and find the exact minimum and maximum values for $f(x)$ in the interval $0 \leqslant x \leqslant 2\pi$.

13.4 Tangents and normals

In many areas of mathematics and physics, it is useful to have an accurate description of a line that is tangent or normal (perpendicular) to a curve. The most complete mathematical description we can obtain is to find the algebraic equation of such lines. In this chapter, much of our work has been in connection to the slopes of tangent lines, so this will be our starting point.

Finding equations of tangents

We now make use of the basic differentiation rules that we established earlier to determine the equation of lines that are tangent to a curve at a point. The first example shows how we can approximate the square root of a number quite accurately without a calculator by making use of a tangent line.

Example 21

a) Find the equation of the line tangent to $y = \sqrt{x}$ at $x = 9$.
b) Use this tangent line to approximate $\sqrt{10}$.

Solution

a) We can find the equation of any line if we know its slope and a point it passes through. Since $y = 3$ when $x = 9$, the point of tangency is $(9, 3)$. We differentiate to find the slope of the curve at $x = 9$, thus giving us the slope of the tangent line.

$$\frac{dy}{dx} = \frac{d}{dx}(\sqrt{x}) = \frac{d}{dx}(x^{\frac{1}{2}}) = \frac{1}{2}x^{-\frac{1}{2}} = \frac{1}{2\sqrt{x}}$$

At $x = 9$: $\frac{dy}{dx} = \frac{1}{2\sqrt{9}} = \frac{1}{6}$ $\Rightarrow$ The slope of the curve and tangent line at $x = 9$ is $\frac{1}{6}$.

Now that we have a point and a slope for the line we can substitute in the point-slope form for the equation of a line.

$$y - 3 = \tfrac{1}{6}(x - 9) \quad \Rightarrow \quad y = \tfrac{1}{6}x + \tfrac{3}{2}$$

The equation of the line tangent to $y = \sqrt{x}$ at $x = 9$ is $y = \frac{x}{6} + \frac{3}{2}$.

b) For values of x near 9, $y = \sqrt{x} \approx \frac{x}{6} + \frac{3}{2}$.

$$\sqrt{10} \approx \frac{10}{6} + \frac{3}{2} = \frac{19}{6} \qquad 6\overline{)19.00} = 3.1\overline{6}$$

The actual value of $\sqrt{10}$ to 4 significant figures is 3.162. Our approximation expressed to 3 significant figures is 3.167. The percentage error is less than 0.2%.

Figure 13.24

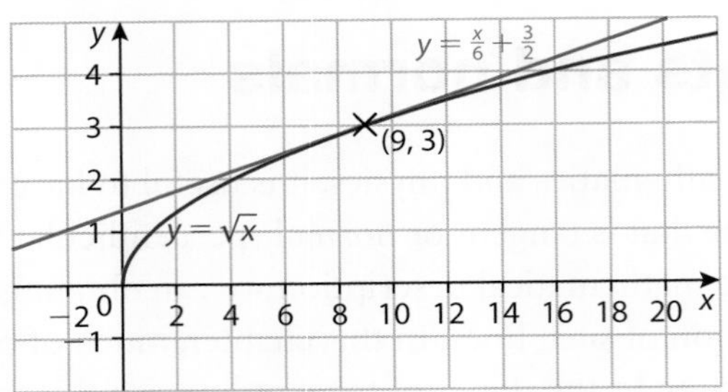

Finding the tangent to a curve was a challenge that motivated many of the initial developments of calculus in the 17th century. In one of his books on mathematics, Descartes wrote the following about the problem of how to find a tangent to a curve:

And I dare say that this is not only the most useful and most general problem in geometry that I know, but even that I have ever desired to know.

The graphs of $y = \sqrt{x}$ and its tangent at $x = 9$, $y = \frac{x}{6} + \frac{3}{2}$, in Figure 13.24 illustrate that the tangent is a very good approximation to the curve in the interval $5 < x < 13$ centred on the point of tangency $(9, 3)$.

Example 22

Find the equation of the tangent to $f(x) = x + \frac{1}{x}$ at the point $\left(\frac{1}{2}, \frac{5}{2}\right)$.

Solution

$$f(x) = x + \frac{1}{x} = x + x^{-1}$$
$$f'(x) = 1 - x^{-2} = 1 - \frac{1}{x^2}$$

When $x = \frac{1}{2}$, $f'\left(\frac{1}{2}\right) = 1 - \frac{1}{\left(\frac{1}{2}\right)^2} = -3.$ Hence, the slope of the tangent is -3.

$$y - \frac{5}{2} = -3\left(x - \frac{1}{2}\right) \Rightarrow y = -3x + \frac{3}{2} + \frac{5}{2} \Rightarrow y = -3x + 4$$

The equation of the line tangent to $f(x) = x + \frac{1}{x}$ at $x = \frac{1}{2}$ is $y = -3x + 4$.

Example 23

Consider the function $g(x) = x^2(x - 1)$.

a) Find the two points on the graph of g at which the slope of the curve is 8.

b) Find the equations of the tangents at both of these points.

Solution

a) In order to differentiate by applying the power rule term-by-term, we first need to write the equation for g in expanded form:
$g(x) = x^2(x - 1) = x^3 - x^2$

$$g'(x) = \frac{d}{dx}(x^3 - x^2) = 3x^2 - 2x$$
$$g'(x) = 3x^2 - 2x = 8 \Rightarrow 3x^2 - 2x - 8 = 0$$
$$(3x + 4)(x - 2) = 0 \Rightarrow x = -\frac{4}{3} \text{ or } x = 2$$
$$g\left(-\frac{4}{3}\right) = \left(-\frac{4}{3}\right)^3 - \left(-\frac{4}{3}\right)^2 = -\frac{112}{27} \text{ and } g(2) = 2^3 - 2^2 = 4$$

Thus, the slope of the curve is equal to 8 at the points $\left(-\frac{4}{3}, -\frac{112}{27}\right)$ and $(2, 4)$.

b) Tangent at $\left(-\frac{4}{3}, -\frac{112}{27}\right)$:

$$y - \left(-\frac{112}{27}\right) = 8\left[x - \left(-\frac{4}{3}\right)\right] \Rightarrow y = 8x + \frac{32}{3} - \frac{112}{27}$$

$$\Rightarrow y = 8x + \frac{176}{27}$$

Therefore, the equation of the tangent at $\left(-\frac{4}{3}, -\frac{112}{27}\right)$ is $y = 8x + \frac{176}{27}$.

Tangent at $(2, 4)$:

$$y - 4 = 8(x - 2) \Rightarrow y = 8x - 16 + 4 \Rightarrow y = 8x - 12$$

Therefore, the equation of the tangent at $(2, 4)$ is $y = 8x - 12$.

Figure 13.25 shows the results for Example 23 – the graph of the function g and the two tangent lines to the graph of the function that have a slope of 8. Note that the scales on the x- and y-axes are not equal which causes the slope of the tangent lines to appear less than 8 for this particular graph.

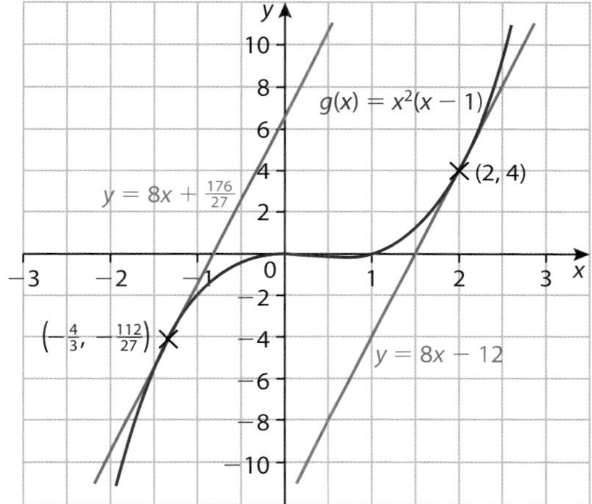

Figure 13.25

The normal to a curve at a point

Another line we often need to find is the line that is 'perpendicular' to a curve at a certain point, which we define to be the line that is perpendicular to the tangent at that point. In this particular context, we apply the adjective 'normal' rather than 'perpendicular' to denote that two lines are at right angles to one another.

A **normal** to a graph of a function at a point is the line through the point that is at a right angle to the tangent at the point. In other words, the tangent and normal to a curve at a certain point are perpendicular.

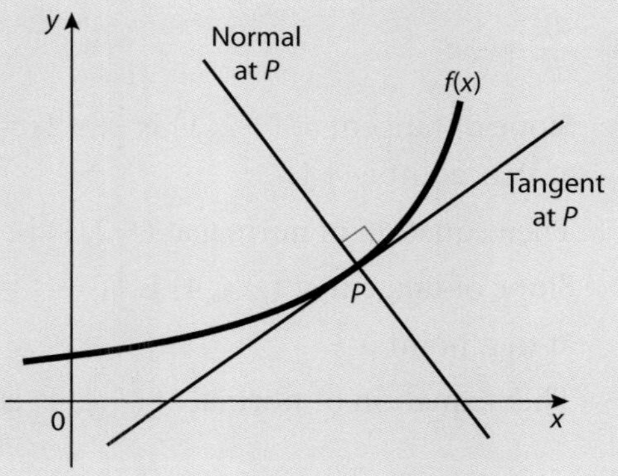

Recall that two perpendicular lines have slopes that are opposite reciprocals. If the slopes of two perpendicular lines are m_1 and m_2, then $m_1 = -\frac{1}{m_2}$ or $m_1m_2 = -1$. The exception is if one of the lines is horizontal (slope is zero) and the other is vertical (slope is undefined).

Example 24

Find the equation of the normal to the graph of $y = 2x^2 - 6x + 3$ at the point $(1, -1)$.

Solution

$$\frac{dy}{dx} = \frac{d}{dx}(2x^2 - 6x + 3) = 4x - 6$$

Slope of tangent at $(1, -1)$ is $4(1) - 6 = -2$. Hence, slope of normal is $+\frac{1}{2}$.
Equation of normal: $y - (-1) = \frac{1}{2}(x - 1) \Rightarrow y = \frac{1}{2}x - \frac{3}{2}$

Figure 13.26 shows the results for Example 24 with the curve at both its tangent and normal at the point $(1, -1)$. Please be aware that if you graph a function with its tangent and normal at a certain point, the normal will only appear perpendicular if the scales on both the x- and y-axes are equal. Regardless of whether the scales are equal or not, the tangent will always appear tangent to the curve.

Figure 13.26

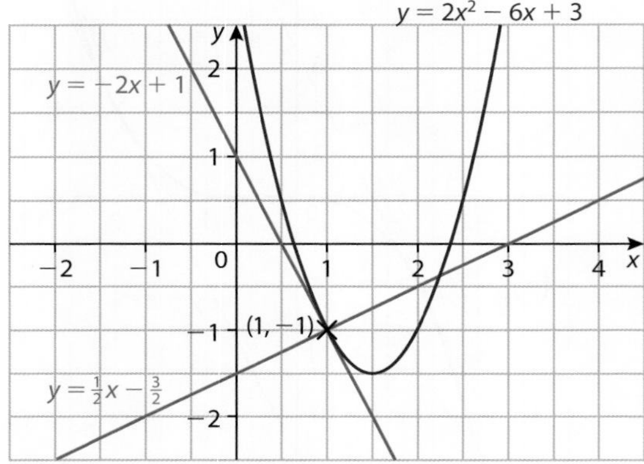

Example 25

Consider the parabola with equation $y = \frac{1}{4}x^2$.

a) Find the equation of the normals at the points $(-2, 1)$ and $(-4, 4)$.
b) Show that the point of intersection of these two normals lies on the parabola.

Solution

a) $\frac{dy}{dx} = \frac{1}{2}x$

Slope of tangent at $(-2, 1)$ is $\frac{1}{2}(-2) = -1$, so the slope of the normal at that point is $+1$.

Then equation of normal at $(-2, 1)$ is: $y - 1 = x - (-2) \Rightarrow y = x + 3$

Slope of tangent at $(-4, 4)$ is $\frac{1}{2}(-4) = -2$, so the slope of the normal at that point is $\frac{1}{2}$.

Then equation of normal at $(-4, 4)$ is: $y - 4 = \frac{1}{2}[x - (-4)]$

$$\Rightarrow y = \frac{1}{2}x + 6$$

b) Set the equations of the two normals equal to each other to find their intersection.

$$x + 3 = \tfrac{1}{2}x + 6 \quad \Rightarrow \quad \tfrac{1}{2}x = 3 \quad \Rightarrow \quad x = 6 \text{ then } y = 9$$
$$\Rightarrow \quad \text{intersection point is } (6, 9)$$

Substitute the coordinates of the points into the equation for the parabola.

$$y = \tfrac{1}{4}x^2 \quad \Rightarrow \quad 9 = \tfrac{1}{4}(6)^2 \quad \Rightarrow \quad 9 = \tfrac{1}{4}\cdot 36 \quad \Rightarrow \quad 9 = 9$$

This confirms that the intersection point, (6, 9), of the normals is also a point on the parabola.

Exercise 13.4

1 Find an equation of the tangent line to the graph of the equation at the indicated value of x.

a) $y = x^2 + 2x + 1$ $\quad x = -3$

b) $y = x^3 + x^2$ $\quad x = -\frac{2}{3}$

c) $y = 3x^2 - x + 1$ $\quad x = 0$

d) $y = 2x + \frac{1}{x}$ $\quad x = \frac{1}{2}$

2 Find the equations of the normal to the functions in question 1 at the indicated value of x.

3 Find the equations of the lines tangent to the curve $y = x^3 - 3x^2 + 2x$ at any point where the curve intersects the x-axis.

4 Find the equation of the tangent to the curve $y = x^2 - 2x$ that is perpendicular to the line $x - 2y = 1$.

5 Using your GDC for assistance, make accurate sketches of the curves $y = x^2 - 6x + 20$ and $y = x^3 - 3x^2 - x$ on the same set of axes. The two curves have the same slope at an integer value for x somewhere in the interval $0 \leqslant x \leqslant \frac{3}{2}$.

a) Find this value of x.

b) Find the equation for the line tangent to each curve at this value of x.

6 Find the equation of the normal to the curve $y = x^2 + 4x - 2$ at the point where $x = -3$. Find the coordinates of the other point where this normal intersects the curve again.

7 Consider the function $g(x) = \dfrac{1 - x^3}{x^4}$. Find the equation of both the tangent and the normal to the graph of g at the point (1, 0).

8 The normal to the curve $y = ax^{\frac{1}{2}} + bx$ at the point where $x = 1$ has a slope of 1 and intersects the y-axis at (0, −4). Find the value of a and the value of b.

9 a) Find the equation of the tangent to the function $f(x) = x^3 + \frac{1}{2}x^2 + 1$ at the point $\left(-1, \frac{1}{2}\right)$.

b) Find the coordinates of another point on the graph of f where the tangent is parallel to the tangent found in a).

10 Find the equation of both the tangent and the normal to the curve $y = \sqrt{x}(1 - \sqrt{x})$ at the point where $x = 4$.

11 Consider the function $f(x) = (1 + x)^2(5 - x)$.

a) Show that the line tangent to the graph of f where $x = 1$ does not intersect the graph of the function again.

b) Also show that the tangent line at (0, 5) intersects the graph of f at a turning point.

c) Sketch the graph of f and the two tangents from a) and b).

12 Find equations of both lines through the point (2, −3) that are tangent to the parabola $y = x^2 + x$.

13 Find all tangent lines through the origin to the graph of $y = 1 + (x - 1)^2$.

14 a) Find the equation of the tangent line to $y = \sqrt[3]{x}$ at $x = 8$.

b) Use the equation of this tangent line to approximate $\sqrt[3]{9}$ to three significant figures.

15 Find the equation of the tangent line for $f(x) = \frac{1}{\sqrt{x}}$ at $x = a$.

16 The tangent to the graph of $y = x^3$ at a point P intersects the curve again at another point Q.
Find the coordinates of Q in terms of the coordinates of P.

17 Two circles of radius r are tangent to each other. Two lines pass through the centre of one circle and are tangent to the other circle at points A and B as shown in the diagram. Find an expression for the distance between A and B.

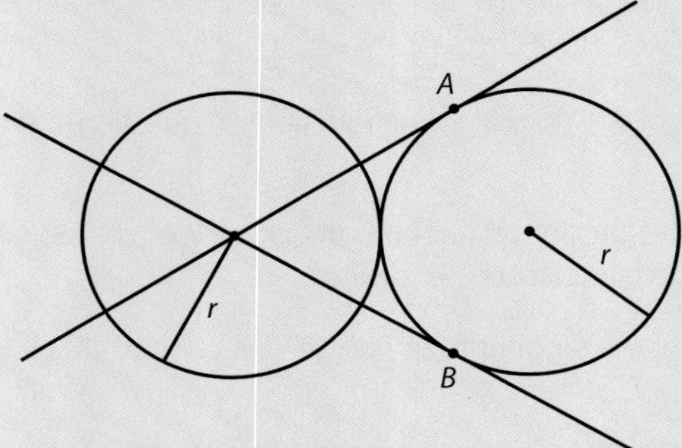

18 Prove that there is no line through the point (1, 2) that is tangent to the curve $y = 4 - x^2$.

Practice questions

1 The function f is defined as $f(x) = x^2$.

a) Find the gradient (slope) of f at the point P, where $x = 1.5$.

b) Find an equation for the tangent to f at the point P.

c) Draw a diagram to show clearly the graph of f and the tangent at P.

d) The tangent of part **b)** intersects the x-axis at the point Q and the y-axis at the point R. Find the coordinates of Q and R.

e) Verify that Q is the midpoint of $[PR]$.

f) Find an equation, in terms of a, for the tangent to f at the point $S(a, a^2)$, $a \neq 0$.

g) The tangent of part **f)** intersects the x-axis at the point T and the y-axis at the point U. Find the coordinates of T and U.

h) Prove that, whatever the value of a, T is the midpoint of SU.

2 The curve with equation $y = Ax + B + \frac{C}{x}$, $x \in \mathbb{R}, x \neq 0$, has a minimum at $P(1, 4)$ and a maximum at $Q(-1, 0)$. Find the value of each of the constants A, B and C.

3 Differentiate:

a) $x^2(2 - 3x^3)$

b) $\frac{1}{x}$

4 Consider the function $f(x) = \frac{8}{x} + 2x,\ x > 0$.

a) Solve the equation $f'(x) = 0$. Show that the graph of f has a turning point at $(2, 8)$.

b) Find the equations of the asymptotes to the graph of f, and hence sketch the graph.

5 Find the coordinates of the stationary point on the curve with equation $y = 4x^2 + \frac{1}{x}$.

6 The curve $y = ax^3 - 2x^2 - x + 7$ has a gradient (slope) of 3 at the point where $x = 2$. Determine the value of a.

7 If $f(2) = 3$ and $f'(2) = 5$, find an equation of **a)** the line tangent to the graph of f at $x = 2$, and **b)** the line normal to the graph of f at $x = 2$.

8 The function $g(x)$ is defined for $-3 \leqslant x \leqslant 3$. The behaviour of $g'(x)$ and $g''(x)$ is given in the tables below.

x	$-3 < x < -2$	-2	$-2 < x < 1$	1	$1 < x < 3$
$g'(x)$	negative	0	positive	0	negative

x	$-3 < x < -\frac{1}{2}$	$-\frac{1}{2}$	$-\frac{1}{2} < x < 3$
$g''(x)$	positive	0	negative

Use the information above to answer the following. In each case, justify your answer.

a) Write down the value of x for which g has a maximum.

b) On which intervals is the value of g decreasing?

c) Write down the value of x for which the graph of g has a point of inflexion.

d) Given that $g(-3) = 0$, sketch the graph of g. On the sketch, clearly indicate the position of the maximum point, the minimum point and the point of inflexion.

9 Given the function $f(x) = x^2 - 3bx + (c + 2)$, determine the values of b and c such that $f(1) = 0$ and $f'(3) = 0$.

10 **Figure 1** shows the graphs of the functions f_1, f_2, f_3, f_4. **Figure 2** includes the graphs of the derivatives of the functions shown in **Figure 1**.

Figure 1

f_1

f_2

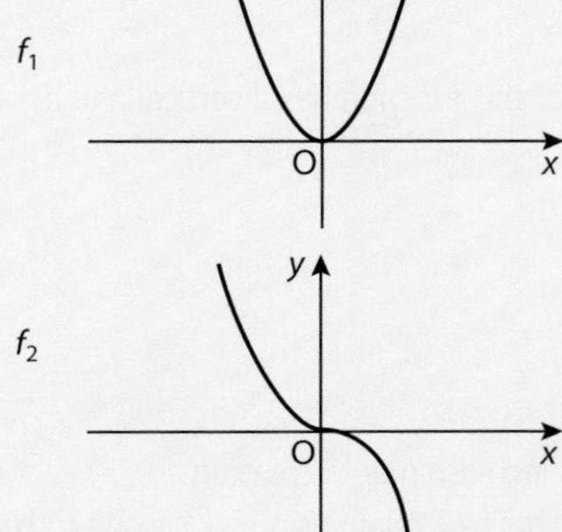

f_3

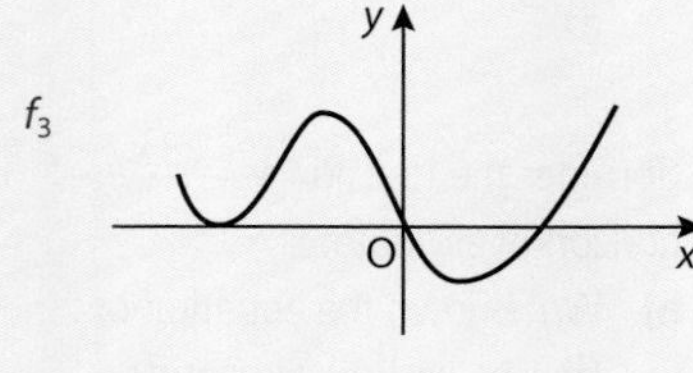

f_4

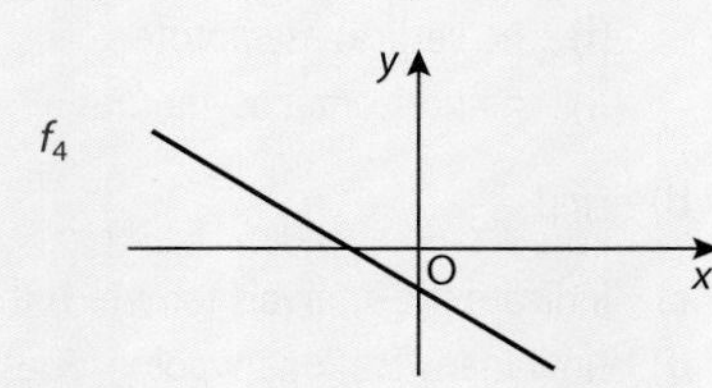

Figure 2

a)

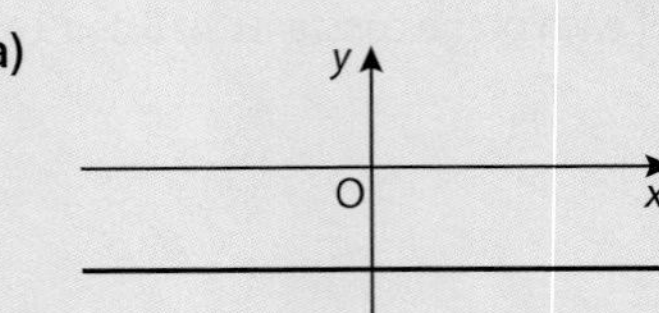

b)

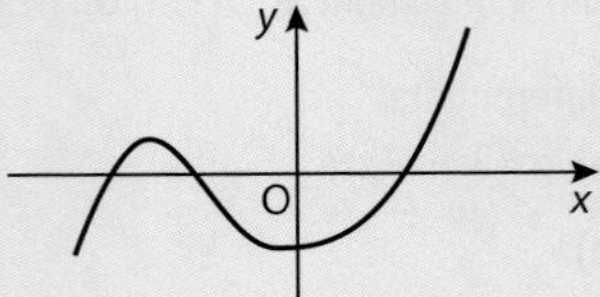

c)

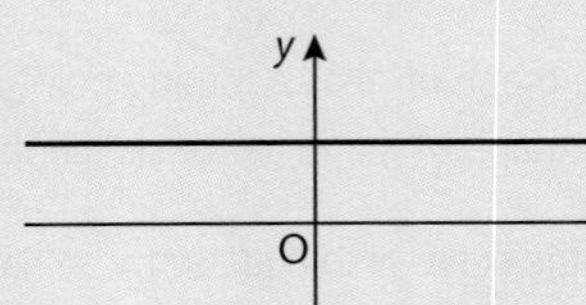

d)

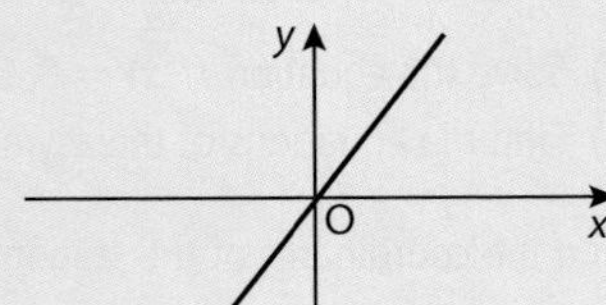

e)

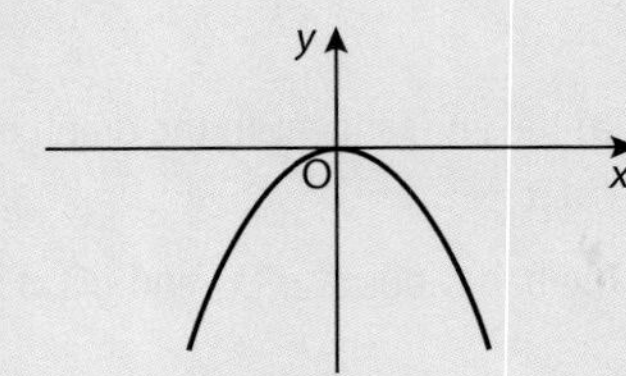

Complete the table below by matching each function with its derivative.

Function	Derivative diagram
f_1	
f_2	
f_3	
f_4	

11 Consider the function $f(x) = 1 + \sin x$.

a) Find the average rate of change of f from $x = 0$ to $x = \frac{\pi}{2}$.

b) Find the instantaneous rate of change of f at $x = \frac{\pi}{4}$.

c) At what value of x in the interval $0 < x < \frac{\pi}{2}$ is the instantaneous rate of change of f equal to the average rate of change of f from $x = 0$ to $x = \frac{\pi}{2}$ (answer to part **a)**)?

12 Consider the function $y = \frac{3x - 2}{x}$. The graph of this function has a vertical and a horizontal asymptote.

a) Write down the equation of

(i) the vertical asymptote

(ii) the horizontal asymptote.

b) Find $\frac{dy}{dx}$.

c) Indicate the intervals for which the curve is increasing or decreasing.

d) How many stationary points does the curve have? Explain using your result to **b)**.

13 Show that there are two points at which the function $h(x) = 2x^2 - x^4$ has a maximum value, and one point at which h has a minimum value. Find the coordinates of these three points, indicating whether it is a maximum or minimum.

14 The normal to the curve $y = x^{\frac{1}{2}} + x^{\frac{1}{3}}$ at the point (1, 2) meets the axes at $(a, 0)$ and $(0, b)$.
Find a and b.

15 The displacement, s metres, of a car, t seconds after leaving a fixed point A, is given by $s(t) = 10t - \frac{1}{2}t^2$.

a) Calculate the velocity when $t = 0$.
b) Calculate the value of t when the velocity is zero.
c) Calculate the displacement of the car from A when the velocity is zero.

16 A ball is thrown vertically upwards from ground level such that its height h metres at t seconds is given by $h = 14t - 4.9t^2$.

a) Write expressions for the ball's velocity and acceleration.
b) Find the maximum height the ball reaches and the time it takes to reach the maximum.
c) At the moment the ball reaches its maximum height, what is the ball's velocity and acceleration?

17 Find the exact coordinates of the inflexion point on the curve $y = x^3 + 12x^2 - x - 12$.

18 Consider the function $f(x) = 2\cos x - 3$. At the point on the curve where $x = \frac{\pi}{3}$, find:

a) the equation of the line tangent to f
b) the equation of the line normal to f.
Express both equations exactly.

19 A manufacturer produces closed cylindrical cans of radius r cm and height h cm. Each can has a total surface area of $54\pi\,\text{cm}^2$.

a) Solve for h in terms of r, and hence find an expression for the volume, $V\,\text{cm}^3$, of each can in terms of r.
b) Find the value of r for which the cans have their maximum possible volume.

20 The curve $y = ax^2 + bx + c$ has a maximum point at (2, 18) and passes through the point (0, 10). Find a, b and c.

21 For the function $f(x) = \frac{1}{2}x^2 - 5x + 3$, find:

a) the equation of the tangent line at $x = -2$
b) the equation of the normal line at $x = -2$.

22 Consider the function $f(x) = x^4 - x^3$.

a) Find the coordinates of any maximum or minimum points. Identify each as relative or absolute.
b) State the domain and range of f.
c) Find the coordinates of any inflexion point(s).
d) Sketch the function clearly indicating any maximum, minimum or inflexion points.

23 Evaluate each limit.

a) $\lim_{x\to\infty} \frac{2 - 3x + 5x^2}{8 - 3x^2}$

b) $\lim_{x\to 0} \frac{\sqrt{x + 4} - 2}{x}$

c) $\lim_{x\to 1} \frac{x^3 - 1}{x - 1}$

d) $\lim_{h\to 0} \frac{\sqrt{(x + h) + 2} - \sqrt{x + 2}}{h}$

24 Find the derivative $f'(x)$ for each function.

a) $f(x) = \dfrac{x^2 - 4x}{\sqrt{x}}$ **b)** $f(x) = x^3 - 3\sin x$

c) $f(x) = \dfrac{1}{x} + \dfrac{x}{2}$ **d)** $f(x) = \dfrac{7}{3x^{13}}$

25 A point (p, q) is on the graph of $y = x^3 + x^2 - 9x - 9$, and the line tangent to the graph at (p, q) passes through the point $(4, -1)$. Find p and q.

26 For what values of c, such that $c \geqslant 0$, is the line $y = -\frac{1}{12}x + c$ normal to the graph of $y = x^3 + \frac{1}{3}$?

27 Find the points on the curve $y = \frac{1}{3}x^3 - x$ where the tangent line is parallel to the line $y = 3x$.

28 At what point does the line that is normal to the graph of $y = x - x^2$ at the point $(1, 0)$ intersect the graph of the curve a second time?

29 If $f(x) = \sqrt{x + 2}$, find $f'(x)$ by first principles.

30 An object moves along a line according to the position function $s(t) = t^3 - 9t^2 + 24t$. Find the positions of the object when

a) its velocity is zero

b) its acceleration is zero.

31 A particle moves along a straight line in the time interval $0 \leqslant t \leqslant 2\pi$ such that its displacement from the origin O is s metres given by the function $s = t + \sin t$.

a) Find the value(s) of t in the interval $0 \leqslant t \leqslant 2\pi$ when the particle's direction changes.

b) Show that the particle always remains on the same side of the origin O.

c) Find the value(s) of t in the interval $0 \leqslant t \leqslant 2\pi$ when the particle's acceleration is zero.

d) Sketch a graph of the particle's displacement from O for $0 \leqslant t \leqslant 2\pi$, and state the maximum value of s in this interval.

32 The curve whose equation is $y = ax^3 + bx^2 + cx + d$ has a point of inflexion at $(-1, 4)$, a turning point when $x = 2$, and it passes through the point $(3, -7)$. Find the values of a, b, c and d, and the y-coordinate of the turning point.

33 Find the stationary values of the function $f(x) = 1 - \dfrac{9}{x^2} + \dfrac{18}{x^4}$ and determine their nature.

34 a) Find the equation of the tangent to the curve $y = \dfrac{1}{x}$ at the point $(1, 1)$.

b) Find the equation of the tangent to the curve $y = \cos x$ at the point $\left(\dfrac{\pi}{2}, 0\right)$.

c) Deduce that $\dfrac{1}{x} > \cos x$ for $0 \leqslant x \leqslant \dfrac{\pi}{2}$.

35 Show that there is just one tangent to the curve $y = x^3 - x + 2$ that passes through the origin.
Find its equation and the coordinates of the point of tangency.

36 The displacement, s metres, of a moving body B from a fixed point O, at time t seconds, is given by $s = 50t - 10t^2 + 1000$.

a) Find the velocity of B in m s^{-1}.

b) Find its maximum displacement from O.

37 The diagram shows a sketch of the graph of $y = f'(x)$ for $a \leqslant x \leqslant b$.

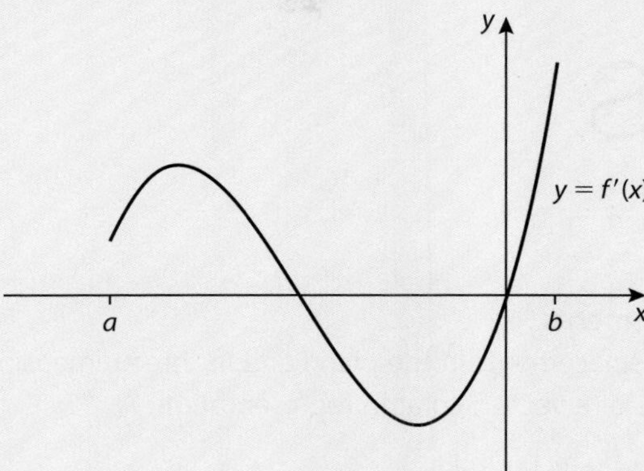

On the grid below, which has the same scale on the x-axis, draw a sketch of the graph of $y = f(x)$ for $a \leqslant x \leqslant b$, given that $f(0) = 0$ and $f(x) \geqslant 0$ for all x. On your graph you should clearly indicate any minimum or maximum points, or points of inflexion.

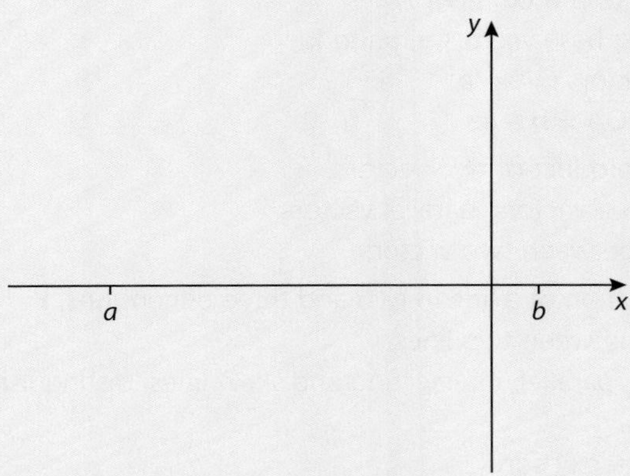

14 Vectors, Lines and Planes

Assessment statements

4.1 Vectors as displacements in the plane and in three dimensions.
Components of a vector; column representation.
$\mathbf{v} = \begin{pmatrix} v_1 \\ v_2 \\ v_3 \end{pmatrix} = v_1\mathbf{i} + v_2\mathbf{j} + v_3\mathbf{k}$
Algebraic and geometric approaches to the following topics:
the sum and difference of two vectors; the zero vector; the vector $-\mathbf{v}$;
multiplication by a scalar, $k\mathbf{v}$;
magnitude of a vector, $|\mathbf{v}|$;
unit vectors; base vectors, $\mathbf{i}$, $\mathbf{j}$ and $\mathbf{k}$;
position vectors $\overrightarrow{OA} = \mathbf{a}$;
$\overrightarrow{AB} = \overrightarrow{OB} - \overrightarrow{OA} = \mathbf{b} - \mathbf{a}$.

4.2 The scalar product of two vectors.
Perpendicular vectors; parallel vectors.
The angle between two vectors.

4.3 Vector equation of a line in two and three dimensions: $\mathbf{r} = \mathbf{a} + \lambda\mathbf{b}$.
The angle between two lines.

4.4 Coincident, parallel, intersecting and skew lines, distinguishing between these cases.
Points of intersection.

4.5 The vector product of two vectors, $\mathbf{v} \times \mathbf{w}$.
Properties of the vector product.
Geometric interpretation of $|\mathbf{v} \times \mathbf{w}|$.

4.6 Vector equation of a plane $\mathbf{r} = \mathbf{a} + \lambda\mathbf{b} + \mu\mathbf{c}$.
Use of normal vector to obtain the form $\mathbf{r} \cdot \mathbf{n} = \mathbf{a} \cdot \mathbf{n}$.
Cartesian equation of a plane $ax + by + cz = d$.

4.7 Intersections of: a line with a plane; two planes; three planes.
Angle between: a line and a plane; two planes.

Introduction

You have seen vectors in the plane in Chapter 9. We will limit our discussion to mainly three-dimensional space in this chapter. If you need to refresh your knowledge of the plane case, refer to Chapter 9.

Because we live in a three-dimensional world, it is essential that we study objects in three dimensions. To that end, we consider in this section a three-dimensional coordinate system in which points are determined by ordered triples. We construct the coordinate system in the following manner: Choose three mutually perpendicular axes, as shown in Figure 14.1, to serve as our reference. The orientation of the system is *right-handed* in the sense that if you hold your right hand so that the fingers curl from the positive x-axis towards the positive y-axis, your thumb points along the z-axis (see below). Looking at it in a different perspective, if you are looking straight at the system, the yz-plane is the plane facing you, and the xz-plane is perpendicular to it and extending out of the page towards you, and the xy-plane is the bottom part of that picture (Figure 14.2). The xy-, xz- and yz-planes are called the **coordinate planes**. Points in space are assigned coordinates in the same manner as in the plane. So, the point P (left) is assigned the ordered triple (x, y, z) to indicate that it is x, y and z units from the yz-, xz- and xy-planes.

Figure 14.1

In this chapter, we will extend our study of vectors to space. The good news is that many of the rules you know from the plane also apply to vectors in space. So, we will only have to introduce a few new concepts. Some of the material will either be a repeat of what you have learned for two-dimensional space or an extension.

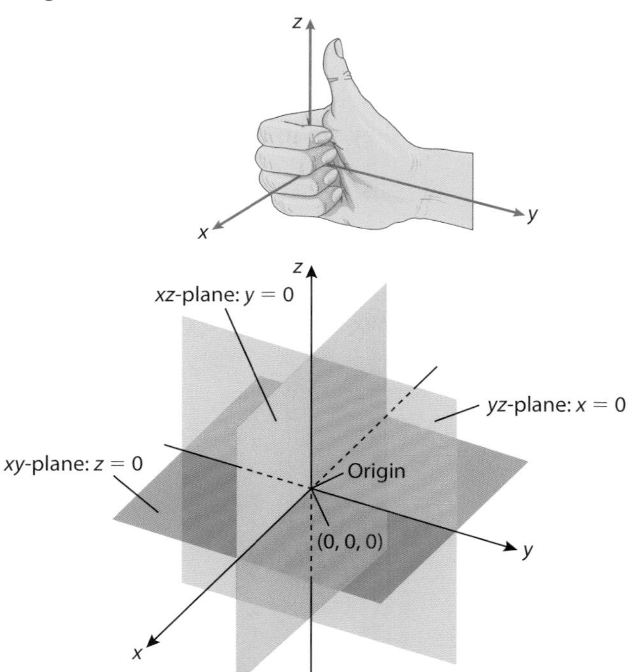

Figure 14.2 The coordinate planes divide space into 8 octants.

14.1 Vectors from a geometric viewpoint

Vectors can be represented geometrically by arrows in two- or three-dimensional space; the direction of the arrow specifies the direction of the vector, and the length of the arrow describes its magnitude. The first point on the arrow is called the **initial point** of the vector and the tip is called the **terminal point**. We shall denote vectors in lower-case boldface type, such as **v**, when using one letter to name the vector, and we will use $\overrightarrow{AB}$ to denote the vector from A to B. The handwritten notation will be the latter too.

If the initial point of a vector is at the origin, the vector is said to be in standard position. It is also called the **position vector** of point P. The terminal point will have coordinates of the form (x, y, z). We call these coordinates the **components** of **v** and we write $\mathbf{v} = (x, y, z)$ or $\mathbf{v} = \begin{pmatrix} x \\ y \\ z \end{pmatrix}$.

The length (magnitude) of a vector **v** is also known as its **modulus** or its **norm** and it is written as $|\mathbf{v}|$.

Look back at Figure 14.1. Using Pythagoras' theorem, we can show that the magnitude of a vector **v**, $|\mathbf{v}| = \sqrt{x^2 + y^2 + z^2}$.

Let $\overrightarrow{OP} = \mathbf{v}$, then

$|\mathbf{v}| = |\overrightarrow{OP}| = \sqrt{OB^2 + BP^2}$, since the triangle OBP is right-angled at B. Now, consider triangle OAB, which is right-angled at A:

$OB^2 = OA^2 + AB^2 = x^2 + y^2$, and, therefore,

$$|\mathbf{v}| = \sqrt{OB^2 + BP^2} = \sqrt{(x^2 + y^2) + z^2} = \sqrt{x^2 + y^2 + z^2}.$$

Two vectors like **v** and $\overrightarrow{AB}$ are equal (equivalent) if they have the same length (magnitude) and the same direction; we write $\mathbf{v} = \overrightarrow{AB}$. Geometrically, two vectors are equal if they are translations of one another as you see in Figures 14.3 and 14.4. Notice in Figure 14.4 that the four vectors are equal, even though they are in different positions.

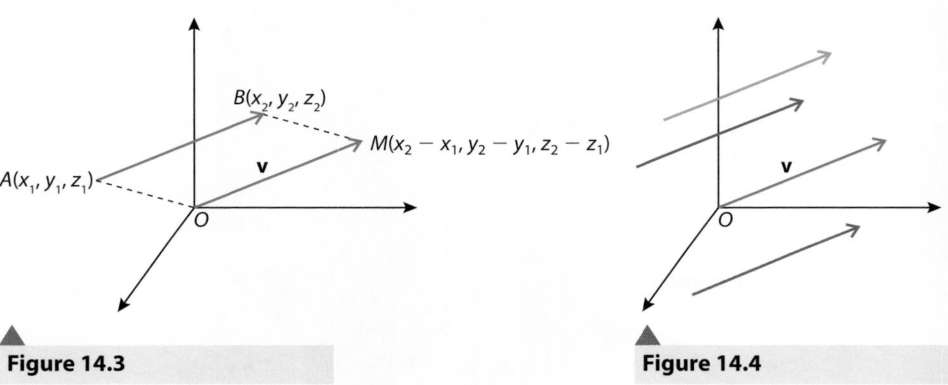

Figure 14.3

Figure 14.4

Because vectors are not affected by translation, the initial point of a vector **v** can be moved to any convenient position by making an appropriate translation.

Two vectors are said to be opposite if they have equal modulus but opposite direction (Figure 14.5).

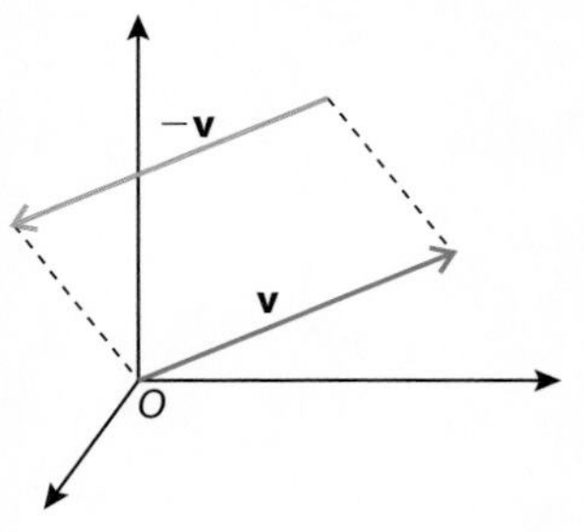

Figure 14.5

If the initial and terminal points of a vector coincide, the vector has length zero; we call this the **zero vector** and denote it by **0**.

The zero vector does not have a specific direction, so we will agree that it can be assigned any convenient direction in a specific problem.

Addition and subtraction of vectors

As you recall from Chapter 9, according to the **triangular rule**, if **u** and **v** are vectors, the sum **u** + **v** is the vector from the initial point of **u** to the terminal point of **v**, when the vectors are positioned so that the initial point of **v** is the terminal point of **u**, as shown in Figure 14.6.

Equivalently, **u** + **v** is also the diagonal of the parallelogram whose sides are **u** and **v**, as shown in Figure 14.7.

The difference of the two vectors **u** and **v** can be dealt with in the same manner. So, the vector **w** = **u** – **v** is a vector such that **u** = **v** + **w**.

In Figure 14.8, we can clearly see that the difference is along the diagonal joining the two terminal points of the vectors and in the direction from **v** to **u**.

If k is a real positive number, k**v** is a vector of magnitude $k|\mathbf{v}|$ and in the same direction as **v**. It follows that when k is negative, k**v** has magnitude $|k| \times |\mathbf{v}|$ and is in the opposite direction to **v** (Figure 14.9).

• **Hint:** When we discuss vectors, we will refer to real numbers as scalars.

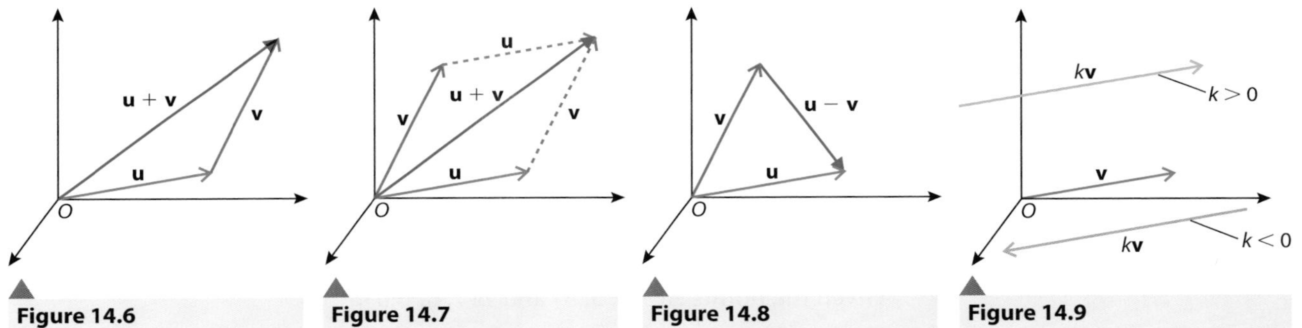

Figure 14.6

Figure 14.7

Figure 14.8

Figure 14.9

A result of the previous situation is the necessary and sufficient condition for two vectors to be parallel:

Two vectors are parallel if one of them is a scalar multiple of the other.
For example, the vector $(-3, 4, -2)$ is parallel to the vector $(4.5, -6, 3)$ since $(-3, 4, -2) = -\frac{2}{3}(4.5, -6, 3)$.

Components provide a simple way to algebraically perform several operations on vectors. First, by definition, we know that two vectors are equal if they have the same length and the same magnitude. So, if we choose to draw the two equal vectors $\mathbf{u} = (u_1, u_2, u_3)$ and $\mathbf{v} = (v_1, v_2, v_3)$ from the origin, their terminal points must coincide, and hence $u_1 = v_1$, $u_2 = v_2$ and $u_3 = v_3$. So, we showed that equal vectors have the same components. The converse is obviously true, i.e. if $u_1 = v_1$, $u_2 = v_2$ and $u_3 = v_3$, the two vectors are equal. The following results are also obvious from the simple geometry of similar figures:

If $\mathbf{u} = (u_1, u_2, u_3)$ and $\mathbf{v} = (v_1, v_2, v_3)$ and k is any real number, then

$$\mathbf{u} + \mathbf{v} = (u_1 + v_1, u_2 + v_2, u_3 + v_3) \text{ and } k\mathbf{u} = (ku_1, ku_2, ku_3).$$

If the initial point of the vector is not at the origin, the following theorem generalizes the previous notation to any position:

If $\overrightarrow{AB}$ is a vector with initial point $A(x_1, y_1, z_1)$ and terminal point $B(x_2, y_2, z_2)$, then $\overrightarrow{AB} = \overrightarrow{OB} - \overrightarrow{OA} = (x_2 - x_1, y_2 - y_1, z_2 - z_1)$, as you see in Figure 14.10.

Figure 14.10

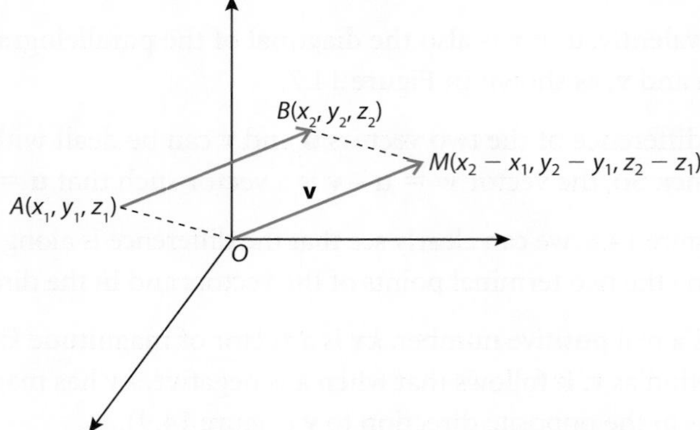

As illustrated in Figure 14.10, either by applying the distance formula or by using the equality of vectors **v** and $\overrightarrow{AB}$,

$$|\overrightarrow{AB}| = \sqrt{(x_2 - x_1)^2 + (y_2 - y_1)^2 + (z_2 - z_1)^2}$$

Additionally, the following results can follow easily from properties of real numbers: $\mathbf{u} + \mathbf{v} = \mathbf{v} + \mathbf{u}$; $(\mathbf{u} + \mathbf{v}) + \mathbf{w} = \mathbf{u} + (\mathbf{v} + \mathbf{w})$; $k(\mathbf{u} + \mathbf{v}) = k\mathbf{u} + k\mathbf{v}$; and the other obvious relationships.

Example 1

Given the points $A(-2, 3, 5)$ and $B(1, 0, -4)$,

a) find the components of vector $\overrightarrow{AB}$

b) find the components of vector $\overrightarrow{BA}$

c) find the components of vector $3\overrightarrow{AB}$

d) find the components of vector $\overrightarrow{OA} + \overrightarrow{OB}$

e) calculate $|\overrightarrow{AB}|$ and $|\overrightarrow{BA}|$

f) calculate $|3\overrightarrow{AB}|$ and $|\overrightarrow{OA} + \overrightarrow{OB}|$.

Solution

a) $\overrightarrow{AB} = \overrightarrow{OB} - \overrightarrow{OA} = (x_2 - x_1, y_2 - y_1, z_2 - z_1)$
$= (1 - (-2), 0 - 3, -4 - 5) = (3, -3, -9)$

b) Since $\overrightarrow{BA}$ is the opposite of $\overrightarrow{AB}$, then $\overrightarrow{BA} = (-3, 3, 9)$.

c) $3\overrightarrow{AB} = 3(3, -3, -9) = (9, -9, -27)$

d) $\overrightarrow{OA} + \overrightarrow{OB} = (-2 + 1, 3 + 0, 5 - 4) = (-1, 3, 1)$

e) $|\overrightarrow{AB}| = \sqrt{(x_2 - x_1)^2 + (y_2 - y_1)^2 + (z_2 - z_1)^2} = \sqrt{9 + 9 + 81} = 3\sqrt{11}$
$|\overrightarrow{BA}| = \sqrt{(x_2 - x_1)^2 + (y_2 - y_1)^2 + (z_2 - z_1)^2} = \sqrt{9 + 9 + 81} = 3\sqrt{11}$

f) $|3\overrightarrow{AB}| = \sqrt{(x_2 - x_1)^2 + (y_2 - y_1)^2 + (z_2 - z_1)^2} = \sqrt{81 + 81 + 729}$
$= \sqrt{891} = 9\sqrt{11}$

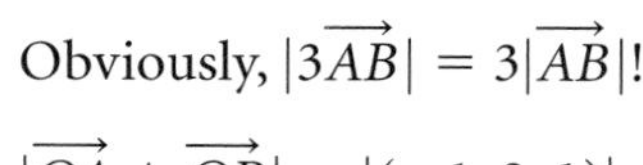

Obviously, $|3\overrightarrow{AB}| = 3|\overrightarrow{AB}|$!

$|\overrightarrow{OA} + \overrightarrow{OB}| = |(-1, 3, 1)| = \sqrt{1 + 9 + 1} = \sqrt{11}$

Notice that $|\overrightarrow{OA} + \overrightarrow{OB}| = \sqrt{11} \neq |\overrightarrow{OA}| + |\overrightarrow{OB}|$

$$= \sqrt{4 + 9 + 25} + \sqrt{1 + 0 + 16} = \sqrt{38} + \sqrt{17}.$$

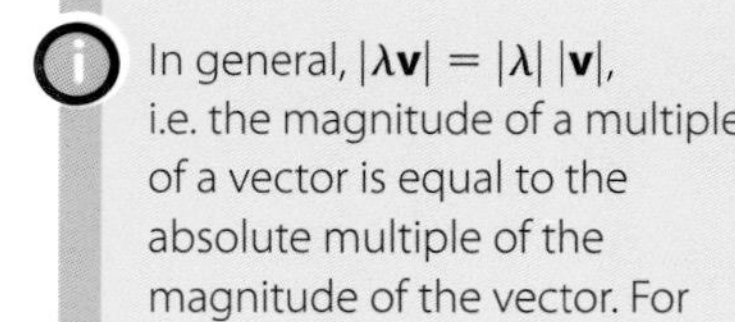

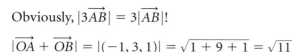

In general, $|\lambda\mathbf{v}| = |\lambda|\,|\mathbf{v}|$, i.e. the magnitude of a multiple of a vector is equal to the absolute multiple of the magnitude of the vector. For example, $|-3\mathbf{v}| = 3|\mathbf{v}|$.

Example 2

Determine the relationship between the coordinates of point $M(x, y, z)$ so that the points M, $A(0, -1, 5)$ and $B(1, 2, 3)$ are collinear.

Solution

For the points to be collinear, it is enough to make $\overrightarrow{AM}$ parallel to $\overrightarrow{AB}$.
If the two vectors are parallel, then one of them is a scalar multiple of the other. Say $\overrightarrow{AM} = t\overrightarrow{AB}$.

$$\overrightarrow{AM} = (x, y + 1, z - 5) = t(1, 3, -2) = (t, 3t, -2t)$$

So, $x = t$, $y + 1 = 3t$, and $z - 5 = -2t$.

Unit vectors

A vector of length 1 is called a **unit vector**. So, in two-dimensional space, the vectors $\mathbf{i} = (1, 0)$ and $\mathbf{j} = (0, 1)$ are unit vectors along the x- and y-axes, and in three-dimensional space, the unit vectors along the axes are $\mathbf{i} = (1, 0, 0)$, $\mathbf{j} = (0, 1, 0)$ and $\mathbf{k} = (0, 0, 1)$. The vectors **i**, **j** and **k** are called the **base vectors** of the 3-space.

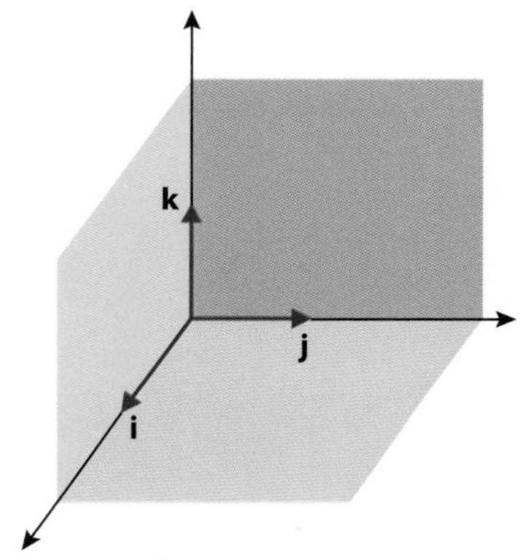

Figure 14.11

• **Hint:** The terms '2-space' and '3-space' are short forms for two-dimensional space and three-dimensional space respectively.

It follows immediately that each vector in 3-space can be expressed uniquely in terms of **i**, **j** and **k** as follows:

$$\begin{aligned}\mathbf{u} &= (x, y, z) = (x, 0, 0) + (0, y, 0) + (0, 0, z)\\ &= x(1, 0, 0) + y(0, 1, 0) + z(0, 0, 1) = x\mathbf{i} + y\mathbf{j} + z\mathbf{k}\end{aligned}$$

So, in Example 1, $\overrightarrow{AB} = (3, -3, -9) = 3\mathbf{i} - 3\mathbf{j} - 9\mathbf{k}$.

Unit vectors can be found in any direction, not only in the direction of the axes. For example, if we want to find the unit vector in the same direction as **u**, we need to find a vector parallel to **u**, which has a magnitude of 1. Since **u** has a magnitude of $|\mathbf{u}|$, it is enough to multiply this vector by $1/|\mathbf{u}|$ to 'normalize' it. So, the unit vector **v** in the same direction as **u** is

$\mathbf{v} = \dfrac{1}{|\mathbf{u}|}\mathbf{u} = \dfrac{\mathbf{u}}{|\mathbf{u}|}$. This is a unit vector since its length is 1. This is why:

Recall that $|\mathbf{u}|$ is a real number (scalar), and so is $1/|\mathbf{u}|$.

Let $1/|\mathbf{u}| = k \Rightarrow \mathbf{v} = \dfrac{1}{|\mathbf{u}|}\mathbf{u} = k\mathbf{u} \Rightarrow |\mathbf{v}| = |k\mathbf{u}| = k|\mathbf{u}| = \dfrac{1}{|\mathbf{u}|}\cdot|\mathbf{u}| = 1.$

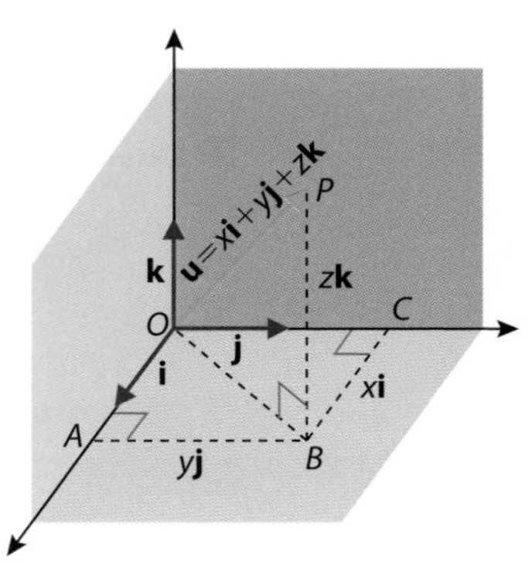

Figure 14.12

Example 3

Find a unit vector in the direction of $\mathbf{v} = \mathbf{i} - 2\mathbf{j} + 3\mathbf{k}$.

Solution

The length of the vector $\mathbf{v}$ is $\sqrt{1^2 + 2^2 + 3^2} = \sqrt{14}$, so the unit vector is

$$\frac{1}{\sqrt{14}}(\mathbf{i} - 2\mathbf{j} + 3\mathbf{k}) = \frac{\mathbf{i}}{\sqrt{14}} - \frac{2\mathbf{j}}{\sqrt{14}} + \frac{3\mathbf{k}}{\sqrt{14}}.$$

To verify that this is a unit vector, we find its length:

$$\sqrt{\left(\frac{1}{\sqrt{14}}\right)^2 + \left(\frac{2}{\sqrt{14}}\right)^2 + \left(\frac{3}{\sqrt{14}}\right)^2} = \sqrt{\frac{1}{14} + \frac{4}{14} + \frac{9}{14}} = 1$$

The unit vector plays another important role: it determines the direction of the given vector.

Recall from Chapter 9 that, in 2-space, we can write the vector in a form that gives us its direction (in terms of the angle it makes with the horizontal axis, called the direction angle) and its magnitude.

In the diagram below, θ is the angle with the horizontal axis.

The unit vector $\mathbf{v}$, in the same direction as $\mathbf{u}$, is:

$$\mathbf{v} = 1\cos\theta\,\mathbf{i} + 1\sin\theta\,\mathbf{j}$$

and from the results above,

$$\mathbf{v} = \frac{1}{|\mathbf{u}|}\mathbf{u} \Rightarrow$$

$$\mathbf{u} = |\mathbf{u}|\,(\mathbf{v})$$

$$= |\mathbf{u}|\cos\theta\,\mathbf{i} + |\mathbf{u}|\sin\theta\,\mathbf{j}$$

$$= |\mathbf{u}|(\cos\theta\,\mathbf{i} + \sin\theta\,\mathbf{j}).$$

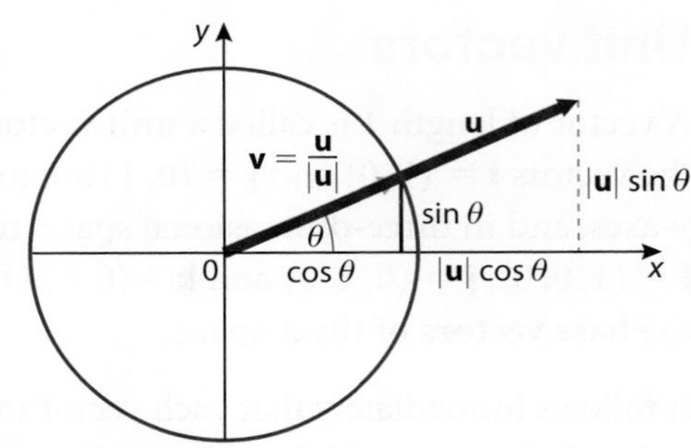

Example 4

Find the vector with magnitude 2 that makes an angle of 60° with the positive x-axis.

Solution

$$\mathbf{v} = |\mathbf{v}|\,(\cos 60°\,\mathbf{i} + \sin 60°\,\mathbf{j}) = 2\left(\frac{1}{2}\mathbf{i} + \frac{\sqrt{3}}{2}\mathbf{j}\right) = \mathbf{i} + \sqrt{3}\mathbf{j}$$

Example 5

Find the direction and magnitude of the vector $\mathbf{v} = 2\sqrt{3}\mathbf{i} - 2\mathbf{j}$.

Solution

$$|\mathbf{v}| = \sqrt{(2\sqrt{3})^2 + 4} = 4$$

$$\cos\theta = \frac{2\sqrt{3}}{4} = \frac{\sqrt{3}}{2},\ \sin\theta = \frac{-2}{4} = -\frac{1}{2} \Rightarrow \theta = -\frac{\pi}{6}$$

Example 6

a) Find the unit vector that has the same direction as $\mathbf{v} = \mathbf{i} + 2\mathbf{j} - 2\mathbf{k}$.
b) Find a vector of length 6 that is parallel to $\mathbf{v} = \mathbf{i} - 2\mathbf{j} + 3\mathbf{k}$.

Solution

a) The vector $\mathbf{v}$ has magnitude $|\mathbf{v}| = \sqrt{1 + 2^2 + (-2)^2} = 3$, so the unit vector $\boldsymbol{v}$ in the same direction as $\mathbf{v}$ is

$$\boldsymbol{v} = \frac{1}{3}\mathbf{v} = \frac{1}{3}\mathbf{i} + \frac{2}{3}\mathbf{j} - \frac{2}{3}\mathbf{k}.$$

b) Let $\mathbf{u}$ be the vector in question and $\boldsymbol{v}$ be the unit vector in the direction of $\mathbf{v}$.

$$\mathbf{u} = 6 \cdot \boldsymbol{v} = 6 \times \frac{1}{\sqrt{14}}(\mathbf{i} - 2\mathbf{j} + 3\mathbf{k}) = \frac{6\mathbf{i}}{\sqrt{14}} - \frac{12\mathbf{j}}{\sqrt{14}} + \frac{18\mathbf{k}}{\sqrt{14}}$$

Example 7

Note: This problem introduces you to the vector equation of a line, as we will see in Section 14.4.

If $\mathbf{r}_1$ and $\mathbf{r}_2$ are the position vectors of two points A and B in space, and λ is a real number, show that $\mathbf{r} = (1 - \lambda)\mathbf{r}_1 + \lambda\mathbf{r}_2$ is the position vector of a point C on the straight line joining A and B. Consider the cases where $\lambda = 0, 1, \frac{1}{2}, -1, 2$ and $\frac{2}{3}$.

Solution

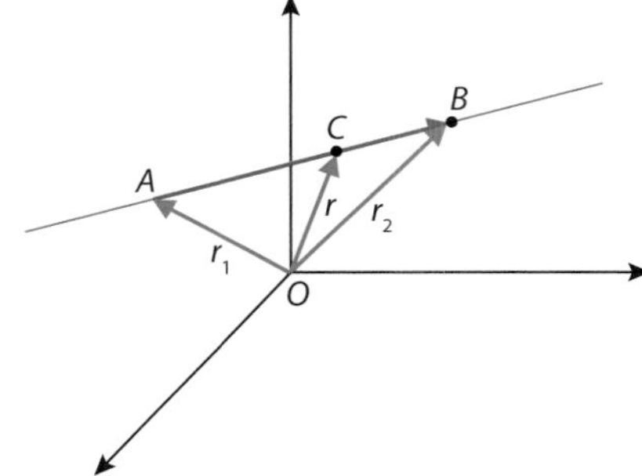

Rewrite the equation:

$$\mathbf{r} = (1 - \lambda)\mathbf{r}_1 + \lambda\mathbf{r}_2 = \mathbf{r}_1 + \lambda(\mathbf{r}_2 - \mathbf{r}_1)$$

Since $\mathbf{r}_2 - \mathbf{r}_1 = \overrightarrow{AB}$, then the position vector $\mathbf{r}$ of C, which is simply $\mathbf{r}_1 + \overrightarrow{AC}$ gives us $\mathbf{r} = \mathbf{r}_1 + \lambda\overrightarrow{AB} = \mathbf{r}_1 + \overrightarrow{AC}$, which in turn gives $\overrightarrow{AC} = \lambda\overrightarrow{AB}$. As you have seen before, this means that $\overrightarrow{AC}$ is parallel to $\overrightarrow{AB}$ and is a multiple of it.

If $\lambda = 0$, then $\mathbf{r} = \mathbf{r}_1$ and C is at A.

If $\lambda = 1$, then $\mathbf{r} = \mathbf{r}_1 + \overrightarrow{AB}$ and C is at B.

If $\lambda = \frac{1}{2}$, then $\mathbf{r} = \mathbf{r}_1 + \frac{1}{2}\overrightarrow{AB}$ and C is the midpoint of AB.

If $\lambda = -1$, then $\mathbf{r} = \mathbf{r}_1 + -\overrightarrow{AB}$ and A is the midpoint of CB.

If $\lambda = 2$, then $\mathbf{r} = \mathbf{r}_1 + 2\overrightarrow{AB}$ and B is the midpoint of AC.

If $\lambda = \frac{2}{3}$, then $\mathbf{r} = \mathbf{r}_1 + \frac{2}{3}\overrightarrow{AB}$ and C is $\frac{2}{3}$ the way between A and B.

Exercise 14.1

1 Write the vector $\overrightarrow{AB}$ in component form in each of the following cases.

a) $A\left(-\frac{3}{2}, -\frac{1}{2}, 1\right); B\left(1, -\frac{5}{2}, 1\right)$ b) $A\left(-2, -\sqrt{3}, -\frac{1}{2}\right); B\left(1, \sqrt{3}, -\frac{1}{2}\right)$

c) $A(2, -3, 5); B(1, -1, 3)$ d) $A(a, -a, 2a); B(-a, -2a, a)$

2 Given the coordinates of point P or Q and the components of $\overrightarrow{PQ}$, find the missing items.

a) $P\left(-\frac{3}{2}, -\frac{1}{2}, 1\right); \overrightarrow{PQ}\left(1, -\frac{5}{2}, 1\right)$ b) $\overrightarrow{PQ}\left(-\frac{3}{2}, -\frac{1}{2}, 1\right); Q\left(1, -\frac{5}{2}, 1\right)$

c) $P(a, -2a, 2a); \overrightarrow{PQ}(-a, -2a, a)$

3 Determine the relationship between the coordinates of point $M(x, y, z)$ so that the points M, A and B are collinear.

a) $A(0, 0, 5); B(1, 1, 0)$
b) $A(-1, 0, 1); B(3, 5, -2)$
c) $A(2, 3, 4); B(-2, -3, 5)$

4 Given the coordinates of the points A and B, find the symmetric image C of B with respect to A.

a) $A(3, -4, 0); B(-1, 0, 1)$
b) $A(-1, 3, 5); B\left(-1, \frac{1}{2}, \frac{1}{3}\right)$
c) $A(1, 2, -1); B(a, 2a, b)$

5 Given a triangle ABC and a point G such that $\overrightarrow{GA} + \overrightarrow{GB} + \overrightarrow{GC} = 0$, find the coordinates of G in each of the following cases.

a) $A(-1, -1, -1); B(-1, 2, -1); C(1, 2, 3)$
b) $A(2, -3, 1); B(1, -2, -5); C(0, 0, 1)$
c) $A(a, 2a, 3a); B(b, 2b, 3b); C(c, 2c, 3c)$

6 Determine the fourth vertex D of the parallelogram $ABCD$ having AB and BC as adjacent sides.

a) $A(\sqrt{3}, 2, -1); B(1, 3, 0); C(-\sqrt{3}, 2, -5)$
b) $A(\sqrt{2}, \sqrt{3}, \sqrt{5}); B(3\sqrt{2}, -\sqrt{3}, 5\sqrt{5}); C(-2\sqrt{2}, \sqrt{3}, -3\sqrt{5})$
c) $A\left(-\frac{1}{2}, \frac{1}{3}, 0\right); B\left(\frac{1}{2}, \frac{2}{3}, 5\right); C\left(\frac{7}{2}, -\frac{1}{3}, 1\right)$

7 Determine the values of m and n such that the vectors $\mathbf{v}(m - 2, m + n, -2m + n)$ and $\mathbf{w}(2, 4, -6)$ have the same direction.

8 Find a unit vector in the same direction as each vector.

a) $\mathbf{v} = 2\mathbf{i} + 2\mathbf{j} - \mathbf{k}$
b) $\mathbf{v} = 6\mathbf{i} - 4\mathbf{j} + 2\mathbf{k}$
c) $\mathbf{v} = 2\mathbf{i} - \mathbf{j} - 2\mathbf{k}$

9 Find a vector with the given magnitude and in the same direction as the given vector.

a) Magnitude 2, $\mathbf{v} = 2\mathbf{i} + 2\mathbf{j} - \mathbf{k}$
b) Magnitude 4, $\mathbf{v} = 6\mathbf{i} - 4\mathbf{j} + 2\mathbf{k}$
c) Magnitude 5, $\mathbf{v} = 2\mathbf{i} - \mathbf{j} - 2\mathbf{k}$

10 Let $\mathbf{u} = \mathbf{i} + 3\mathbf{j} - 2\mathbf{k}$ and $\mathbf{v} = 2\mathbf{i} + \mathbf{j}$. Find

a) $|\mathbf{u} + \mathbf{v}|$ b) $|\mathbf{u}| + |\mathbf{v}|$

c) $|-3\mathbf{u}| + |3\mathbf{v}|$ d) $\frac{1}{|\mathbf{u}|}\mathbf{u}$

e) $\left|\frac{1}{|\mathbf{u}|}\mathbf{u}\right|$

11 Find the terminal points for each vector.
a) $\mathbf{w} = 4\mathbf{i} + 2\mathbf{j} - 2\mathbf{k}$, given the initial point $(-1, 2, -3)$
b) $\mathbf{v} = 2\mathbf{i} - 3\mathbf{j} + \mathbf{k}$, given the initial point $(-2, 1, 4)$

12 Find vectors that satisfy the stated conditions:
a) opposite direction of $\mathbf{u} = (-3, 4)$ and third the magnitude of $\mathbf{u}$
b) length of 12 and same direction as $\mathbf{w} = 4\mathbf{i} + 2\mathbf{j} - 2\mathbf{k}$
c) of the form $x\mathbf{i} + y\mathbf{j} - 2\mathbf{k}$ and parallel to $\mathbf{w} = \mathbf{i} - 4\mathbf{j} + 3\mathbf{k}$

13 Let $\mathbf{u}$, $\mathbf{v}$ and $\mathbf{w}$ be the vectors from each vertex of a triangle to the midpoint of the opposite side. Find the value of $\mathbf{u} + \mathbf{v} + \mathbf{w}$.

14 Find the scalar t (or show that there is none) so that the vector $\mathbf{v} = t\mathbf{i} - 2t\mathbf{j} + 3t\mathbf{k}$ is a unit vector.

15 Find the scalar t (or show that there is none) so that the vector $\mathbf{v} = 2\mathbf{i} - 2t\mathbf{j} + 3t\mathbf{k}$ is a unit vector.

16 Find the scalar t (or show that there is none) so that the vector $\mathbf{v} = 0.5\mathbf{i} - t\mathbf{j} + 1.5t\mathbf{k}$ is a unit vector.

17 The diagram shows a cube of length 8 units.
a) Find the position vectors of all the vertices.
b) L, M and N are the midpoints of the respective edges. Find the position vectors of L, M and N.
c) Show that $\overrightarrow{LM} + \overrightarrow{MN} + \overrightarrow{NL} = \vec{0}$.

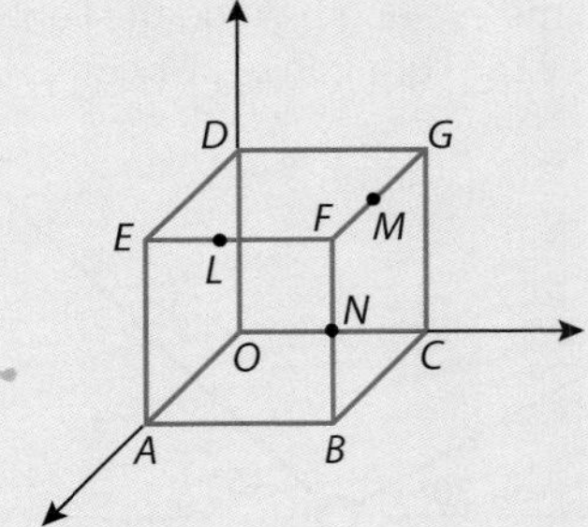

18 A triangular prism is given with the lengths of the sides $OA = 8$, $OB = 10$ and $OE = 12$.

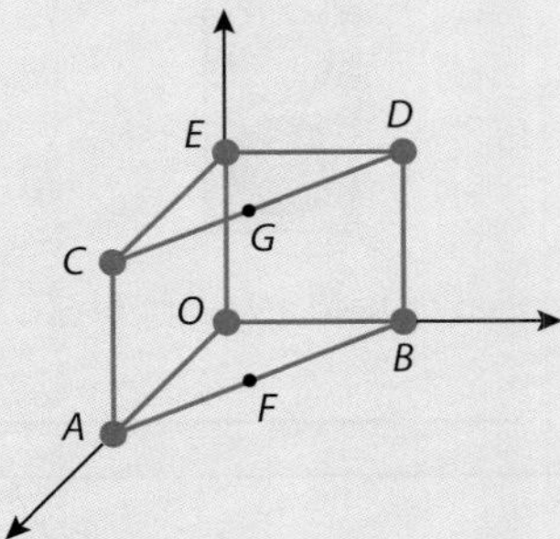

a) Find the position vectors of C and D.
b) F and G are the midpoints of the respective edges. Find their position vectors.
c) Find the vectors $\overrightarrow{AG}$ and $\overrightarrow{FD}$ and explain your results.

19 Find α such that $|\alpha\mathbf{i} + (\alpha - 1)\mathbf{j} + (\alpha + 1)\mathbf{k}| = 2$.

20 Let $\mathbf{a} = \begin{pmatrix} 4 \\ -2 \\ 1 \end{pmatrix}$, $\mathbf{b} = \begin{pmatrix} 1 \\ 1 \\ 1 \end{pmatrix}$, $\mathbf{c} = \begin{pmatrix} -1 \\ 3 \\ 2 \end{pmatrix}$, $\mathbf{d} = \begin{pmatrix} -3 \\ 0 \\ 1 \end{pmatrix}$.

Find the scalars α, β and μ (or show that they cannot exist) such that $\mathbf{a} = \alpha\mathbf{b} + \beta\mathbf{c} + \mu\mathbf{d}$.

21 Repeat question 20 for $\mathbf{a} = \begin{pmatrix} -1 \\ 1 \\ 5 \end{pmatrix}$, $\mathbf{b} = \begin{pmatrix} 1 \\ 0 \\ 1 \end{pmatrix}$, $\mathbf{c} = \begin{pmatrix} 3 \\ 2 \\ 0 \end{pmatrix}$, $\mathbf{d} = \begin{pmatrix} 0 \\ 1 \\ 1 \end{pmatrix}$.

22 Repeat question 20 for $\mathbf{a} = \begin{pmatrix} 2 \\ 1 \\ -1 \end{pmatrix}$, $\mathbf{b} = \begin{pmatrix} 1 \\ -1 \\ 0 \end{pmatrix}$, $\mathbf{c} = \begin{pmatrix} 3 \\ 0 \\ 1 \end{pmatrix}$, $\mathbf{d} = \begin{pmatrix} 4 \\ -1 \\ 1 \end{pmatrix}$.

23 Let $\mathbf{u}$ and $\mathbf{v}$ be non-zero vectors such that $|\mathbf{u} - \mathbf{v}| = |\mathbf{u} + \mathbf{v}|$.

a) What can you conclude about the parallelogram with $\mathbf{u}$ and $\mathbf{v}$ as adjacent sides?

b) Show that if

$$\mathbf{u} = \begin{pmatrix} u_1 \\ u_2 \\ u_3 \end{pmatrix} \text{ and } \mathbf{v} = \begin{pmatrix} v_1 \\ v_2 \\ v_3 \end{pmatrix},$$

then

$$u_1v_1 + u_2v_2 + u_3v_3 = 0.$$

24 A 125 N traffic light is hanging from two flexible cables. The magnitude of the force that each cable applies to the 'eye ring' holding the lights is called the cable tension. Find the cable tensions if the light is in equilibrium.

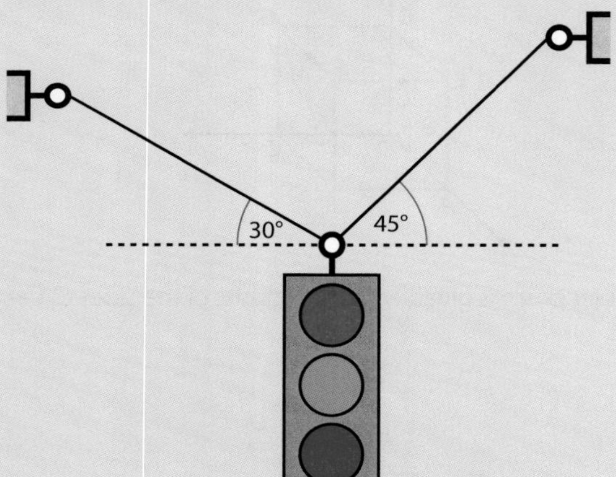

25 Find the tension in the cables used to hold a weight of 300 N as shown in the diagram.

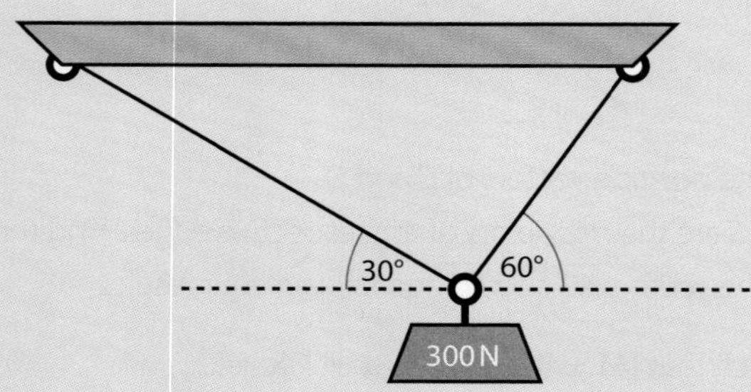

14.2 Scalar (dot) product

If $\mathbf{u} = (u_1, u_2, u_3)$ and $\mathbf{v} = (v_1, v_2, v_3)$ are two vectors, the dot product (scalar) is written as $\mathbf{u}\cdot\mathbf{v}$ and is defined as

$$\mathbf{u}\cdot\mathbf{v} = u_1v_1 + u_2v_2 + u_3v_3.$$

Result 1: $\mathbf{u}^2 = \mathbf{u}\cdot\mathbf{u} = u_1\cdot u_1 + u_2\cdot u_2 + u_3\cdot u_3 = u_1^2 + u_2^2 + u_3^2 = |\mathbf{u}|^2$

From this definition, we can deduce another geometric 'definition' of the dot product:

$\mathbf{u}\cdot\mathbf{v} = |\mathbf{u}||\mathbf{v}|\cos\theta$, where θ is the angle between the two vectors.

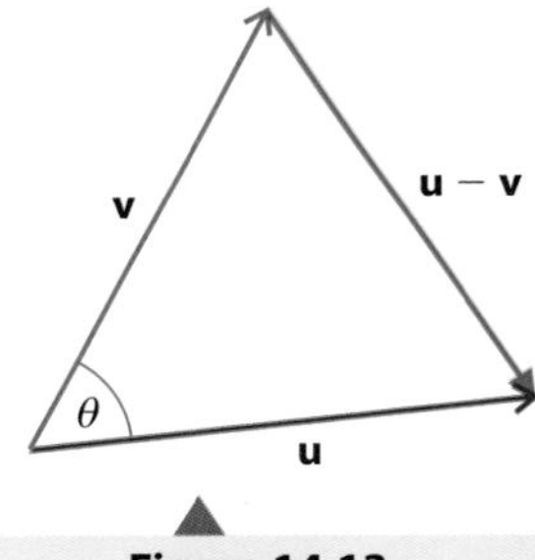

Figure 14.13

Proof:

Let **u** and **v** be drawn from the same point, as shown in Figure 14.13. Then

$$|\mathbf{u} - \mathbf{v}|^2 = (\mathbf{u} - \mathbf{v})\cdot(\mathbf{u} - \mathbf{v}) = \mathbf{u}^2 + \mathbf{v}^2 - 2\mathbf{u}\cdot\mathbf{v}$$
$$= |\mathbf{u}|^2 + |\mathbf{v}|^2 - 2\mathbf{u}\cdot\mathbf{v}.$$

Also, using the law of cosines,

$$|\mathbf{u} - \mathbf{v}|^2 = |\mathbf{u}|^2 + |\mathbf{v}|^2 - 2|\mathbf{u}|\cdot|\mathbf{v}|\cdot\cos\theta.$$

Conversely, using the law of cosines in the figure above gives

$$|\mathbf{u} - \mathbf{v}|^2 = |\mathbf{u}|^2 + |\mathbf{v}|^2 - 2|\mathbf{u}|\cdot|\mathbf{v}|\cdot\cos\theta,$$

which in turn will give

$$2|\mathbf{u}|\cdot|\mathbf{v}|\cdot\cos\theta = |\mathbf{u}|^2 + |\mathbf{v}|^2 - |\mathbf{u} - \mathbf{v}|^2$$
$$= (u_1^2 + u_2^2 + u_3^2) + (v_1^2 + v_2^2 + v_3^2) - [(u_1 - v_1)^2 + (u_2 - v_2)^2 + (u_3 - v_3)^2]$$
$$= 2(u_1v_1 + u_2v_2 + u_3v_3).$$

Thus, $|\mathbf{u}|\cdot|\mathbf{v}|\cdot\cos\theta = u_1v_1 + u_2v_2 + u_3v_3$ and $\mathbf{u}\cdot\mathbf{v} = |\mathbf{u}||\mathbf{v}|\cos\theta$.

By comparing the two results, we can conclude that $\mathbf{u}\cdot\mathbf{v} = |\mathbf{u}|\cdot|\mathbf{v}|\cos\theta$.

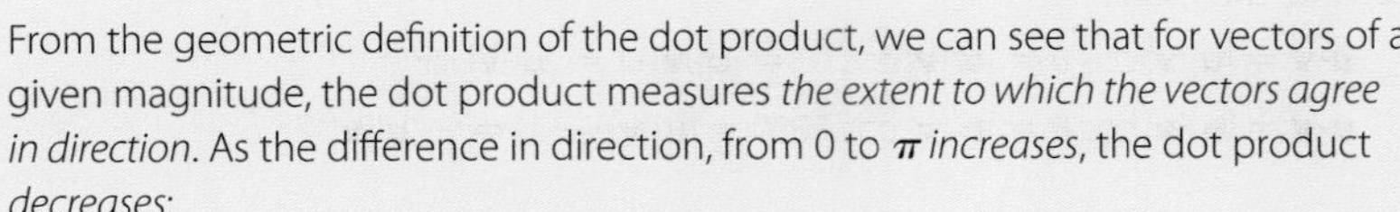

From the geometric definition of the dot product, we can see that for vectors of a given magnitude, the dot product measures *the extent to which the vectors agree in direction*. As the difference in direction, from 0 to π *increases*, the dot product *decreases*:

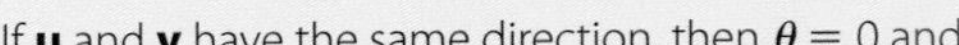

If **u** and **v** have the same direction, then $\theta = 0$ and

$$\mathbf{u}\cdot\mathbf{v} = |\mathbf{u}||\mathbf{v}|\cos\theta = |\mathbf{u}||\mathbf{v}|.$$

This is the largest possible value for $\mathbf{u}\cdot\mathbf{v}$.

If **u** and **v** are at right angles, then $\theta = \frac{\pi}{2}$ and

$$\mathbf{u}\cdot\mathbf{v} = 0.$$

If **u** and **v** have opposite directions, then $\theta = \pi$ and

$$\mathbf{u}\cdot\mathbf{v} = |\mathbf{u}||\mathbf{v}|\cos\pi = -|\mathbf{u}||\mathbf{v}|.$$

This is the least possible value for $\mathbf{u}\cdot\mathbf{v}$.

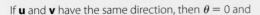

The scalar product can be used, among other things, to find angles between vectors:

$$\mathbf{u} \cdot \mathbf{v} = |\mathbf{u}||\mathbf{v}| \cos\theta \Leftrightarrow \cos\theta = \frac{\mathbf{u} \cdot \mathbf{v}}{|\mathbf{u}||\mathbf{v}|}.$$

$\cos\theta = \frac{\mathbf{u} \cdot \mathbf{v}}{|\mathbf{u}||\mathbf{v}|} = \frac{\mathbf{u}}{|\mathbf{u}|} \cdot \frac{\mathbf{v}}{|\mathbf{v}|} = u \cdot v$, where u and v are unit vectors in the direction of **u** and **v** respectively. That is, the cosine of the angle between two vectors is the dot product of the corresponding unit vectors.

Example 8

Find the angle between the vectors $\mathbf{u} = \mathbf{i} - 2\mathbf{j} + 2\mathbf{k}$ and $\mathbf{v} = -3\mathbf{i} + 6\mathbf{j} + 2\mathbf{k}$.

Solution

From the previous results, we have

$$\cos\theta = \frac{\mathbf{u} \cdot \mathbf{v}}{|\mathbf{u}| \cdot |\mathbf{v}|} = \frac{-3 - 12 + 4}{\sqrt{1 + 4 + 4}\sqrt{9 + 36 + 4}} = \frac{-11}{21}$$

$$\Rightarrow \theta = \cos^{-1}\left(\frac{-11}{21}\right) \approx 2.12 \text{ radians}$$

Result 2: A direct conclusion of the previous definitions is that if two vectors are *perpendicular*, the dot product is *zero*.

This is so because when the two vectors are perpendicular the angle between them is $\pm 90°$ and, therefore,

$$\mathbf{u} \cdot \mathbf{v} = |\mathbf{u}||\mathbf{v}| \cos\theta = |\mathbf{u}||\mathbf{v}| \cos 90° = |\mathbf{u}||\mathbf{v}| \cdot 0 = 0.$$

The base vectors of the coordinate system are obviously perpendicular: $\mathbf{i} \cdot \mathbf{j} = (1, 0, 0) \cdot (0, 1, 0) = 0$, and similarly, $\mathbf{i} \cdot \mathbf{k} = 0$ and $\mathbf{j} \cdot \mathbf{k} = 0$.

Result 3: If two vectors **u** and **v** are parallel, then $\mathbf{u} \cdot \mathbf{v} = \pm|\mathbf{u}||\mathbf{v}|$.

Again, this is so because when the vectors are parallel the angle between them is either 0° or 180° and, therefore,

$$\mathbf{u} \cdot \mathbf{v} = |\mathbf{u}||\mathbf{v}| \cos\theta = |\mathbf{u}||\mathbf{v}| \cos 0° = |\mathbf{u}||\mathbf{v}| \cdot 1 = |\mathbf{u}||\mathbf{v}|, \text{ or}$$

$$\mathbf{u} \cdot \mathbf{v} = |\mathbf{u}||\mathbf{v}| \cos\theta = |\mathbf{u}||\mathbf{v}| \cos 180° = |\mathbf{u}||\mathbf{v}| \cdot (-1) = -|\mathbf{u}||\mathbf{v}|.$$

Example 9

Determine which, if any, of the following vectors are orthogonal.

$$\mathbf{u} = 7\mathbf{i} + 3\mathbf{j} + 2\mathbf{k}, \mathbf{v} = -3\mathbf{i} + 5\mathbf{j} + 3\mathbf{k}, \mathbf{w} = \mathbf{i} + \mathbf{k}$$

Solution

$\mathbf{u} \cdot \mathbf{v} = 7(-3) + 3 \times 5 + 2 \times 3 = 0$; orthogonal vectors

$\mathbf{u} \cdot \mathbf{w} = 7 \times 1 + 3 \times 0 + 2 \times 1 = 9$; not orthogonal

$\mathbf{v} \cdot \mathbf{w} = -3 \times 1 + 5 \times 0 + 3 \times 1 = 0$; orthogonal vectors

Example 10

$A(1, 2, 3)$, $B(-3, 2, 4)$ and $C(1, -4, 3)$ are the vertices of a triangle. Show that the triangle is right-angled and find its area.

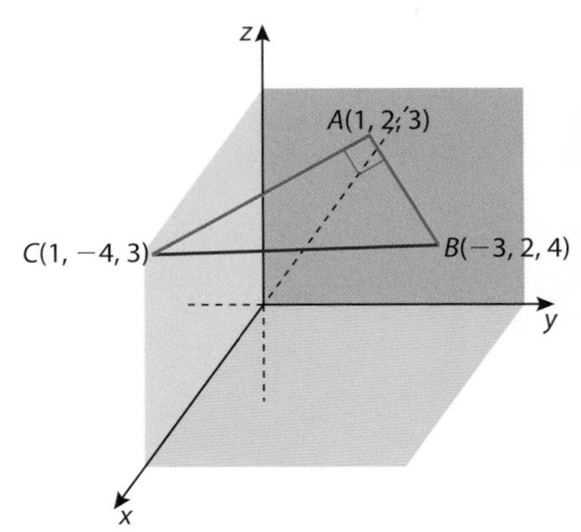

Solution

$$\overrightarrow{AB} = (-3 - 1)\mathbf{i} + (2 - 2)\mathbf{j} + (4 - 3)\mathbf{k} = -4\mathbf{i} + \mathbf{k}$$

$$\overrightarrow{AC} = (1 - 1)\mathbf{i} + (-4 - 2)\mathbf{j} + (3 - 3)\mathbf{k} = -6\mathbf{j}$$

$$\overrightarrow{BC} = (1 - (-3))\mathbf{i} + (-4 - 2)\mathbf{j} + (3 - 4)\mathbf{k} = 4\mathbf{i} - 6\mathbf{j} - \mathbf{k}$$

Since $\overrightarrow{AB} \cdot \overrightarrow{AC} = -4 \times 0 + 0 \times -6 + 1 \times 0 = 0$, the vectors are perpendicular. So the triangle is right-angled at A.

The area of this right triangle is half the product of the legs.

$$\text{Area} = \tfrac{1}{2}|\overrightarrow{AB}||\overrightarrow{AC}| = \tfrac{1}{2} \cdot \sqrt{(-4)^2 + 1} \cdot 6 = 3\sqrt{17}$$

Theorem

a) $|\mathbf{u} \cdot \mathbf{v}| \leqslant |\mathbf{u}|\,|\mathbf{v}|$

b) $|\mathbf{u} + \mathbf{v}| \leqslant |\mathbf{u}| + |\mathbf{v}|$

Proof

a) Since, $\mathbf{u} \cdot \mathbf{v} = |\mathbf{u}||\mathbf{v}|\cos\theta$, then $|\mathbf{u} \cdot \mathbf{v}| = ||\mathbf{u}|\,|\mathbf{v}|\cos\theta| = |\mathbf{u}|\,|\mathbf{v}||\cos\theta|$, and as $|\cos\theta| \leqslant 1$, then $|\mathbf{u} \cdot \mathbf{v}| \leqslant |\mathbf{u}||\mathbf{v}|$.

b) $$\left.\begin{aligned} |\mathbf{u} + \mathbf{v}|^2 &= (\mathbf{u} + \mathbf{v}) \cdot (\mathbf{u} + \mathbf{v}) = \mathbf{u} \cdot \mathbf{u} + \mathbf{u} \cdot \mathbf{v} + \mathbf{v} \cdot \mathbf{u} + \mathbf{v} \cdot \mathbf{v} \\ &= |\mathbf{u}|^2 + 2(\mathbf{u} \cdot \mathbf{v}) + |\mathbf{v}|^2 \end{aligned}\right\} \Leftarrow \mathbf{u} \cdot \mathbf{v} = \mathbf{v} \cdot \mathbf{u}$$

but, $\mathbf{u} \cdot \mathbf{v} \leqslant |\mathbf{u} \cdot \mathbf{v}|$, since $\mathbf{u} \cdot \mathbf{v}$ may also be negative while $|\mathbf{u} \cdot \mathbf{v}|$ is not.

Also, $|\mathbf{u} \cdot \mathbf{v}| \leqslant |\mathbf{u}|\,|\mathbf{v}|$, therefore

$$\begin{aligned} |\mathbf{u} + \mathbf{v}|^2 &= |\mathbf{u}|^2 + 2(\mathbf{u} \cdot \mathbf{v}) + |\mathbf{v}|^2 \leqslant |\mathbf{u}|^2 + 2|\mathbf{u} \cdot \mathbf{v}| + |\mathbf{v}|^2 \\ &\leqslant |\mathbf{u}|^2 + 2|\mathbf{u}||\mathbf{v}| + |\mathbf{v}|^2 = (|\mathbf{u}| + |\mathbf{v}|)^2. \end{aligned}$$

This then implies that $|\mathbf{u} + \mathbf{v}| \leqslant |\mathbf{u}| + |\mathbf{v}|$ (taking square roots).

Direction angles, direction cosines

Figure 14.14 shows a non-zero vector $\mathbf{v}$. The angles α, β and γ that the vector makes with the unit coordinate vectors are called the **direction angles** of $\mathbf{v}$, and $\cos\alpha$, $\cos\beta$ and $\cos\gamma$ are called the **direction cosines**.

Let $\mathbf{v} = x\mathbf{i} + y\mathbf{j} + z\mathbf{k}$. Considering the right triangles OAP, OCP and ODP, the hypotenuse in each of these triangles is OP, i.e. $|\mathbf{v}|$. From your trigonometry chapters, you know that the side adjacent to an angle θ in a right triangle is related to it by

$$\cos\theta = \frac{\text{adjacent}}{\text{hypotenuse}} \Leftrightarrow \text{adjacent} = \text{hypotenuse}\cdot\cos\theta, \text{ so in this case}$$

$x = |\mathbf{v}|\cos\alpha$, $y = |\mathbf{v}|\cos\beta$, $z = |\mathbf{v}|\cos\gamma$, and so

$$\mathbf{v} = (|\mathbf{v}|\cos\alpha)\,\mathbf{i} + (|\mathbf{v}|\cos\beta)\,\mathbf{j} + (|\mathbf{v}|\cos\gamma)\,\mathbf{k} = |\mathbf{v}|(\cos\alpha\,\mathbf{i} + \cos\beta\,\mathbf{j} + \cos\gamma\,\mathbf{k}).$$

Taking the magnitude of both sides,

$$|\mathbf{v}| = |\mathbf{v}|\sqrt{\cos^2\alpha + \cos^2\beta + \cos^2\gamma}.$$

Figure 14.14

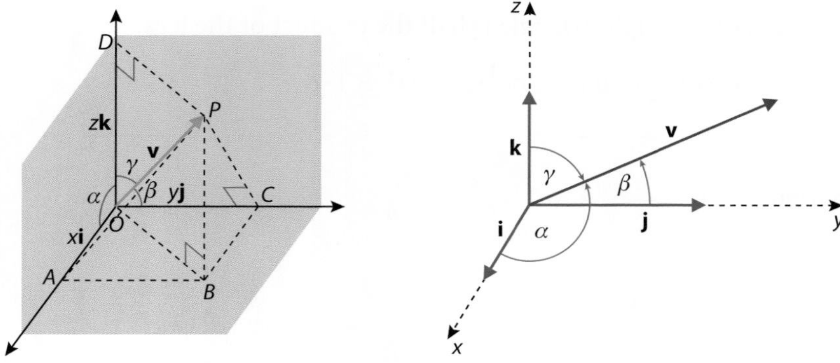

Therefore,
$\cos^2\alpha + \cos^2\beta + \cos^2\gamma = 1$, i.e. the sum of the squares of the direction cosines is always 1.
For a unit vector, the expression will be of the form

$$\mathbf{u} = |\mathbf{u}|(\cos\alpha\,\mathbf{i} + \cos\beta\,\mathbf{j} + \cos\gamma\,\mathbf{k}) = \cos\alpha\,\mathbf{i} + \cos\beta\,\mathbf{j} + \cos\gamma\,\mathbf{k}\ (|\mathbf{u}| = 1).$$

This means that for a unit vector its x-, y- and z-coordinates are its direction cosines.

It is also important that you remember that $\cos\alpha = \frac{x}{|\mathbf{v}|}$, $\cos\beta = \frac{y}{|\mathbf{v}|}$, $\cos\gamma = \frac{z}{|\mathbf{v}|}$.

Example 11

Find the direction cosines of the vector $\mathbf{v} = 4\mathbf{i} - 2\mathbf{j} + 4\mathbf{k}$, and then approximate the direction angles to the nearest degree.

Solution

$$|\mathbf{v}| = \sqrt{4^2 + (-2)^2 + 4^2} = 6 \Rightarrow \mathbf{v} = \frac{\mathbf{v}}{|\mathbf{v}|} = \frac{2}{3}\mathbf{i} - \frac{1}{3}\mathbf{j} + \frac{2}{3}\mathbf{k},$$

$$\text{thus } \cos\alpha = \frac{2}{3}, \cos\beta = -\frac{1}{3}, \cos\gamma = \frac{2}{3}$$

From your GDC you will obtain:

$$\alpha = \cos^{-1}\left(\frac{2}{3}\right) \approx 48°, \beta = \cos^{-1}\left(-\frac{1}{3}\right) \approx 109°, \gamma = \cos^{-1}\left(\frac{2}{3}\right) \approx 48°$$

Example 12

Find the angle that a main diagonal of a cube with side a makes with the adjacent edges.

Solution

We can place the cube in a coordinate system such that three of its adjacent edges lie on the coordinate axes as shown (right). The diagonal, represented by the vector **v** has a terminal point (a, a, a). Hence,

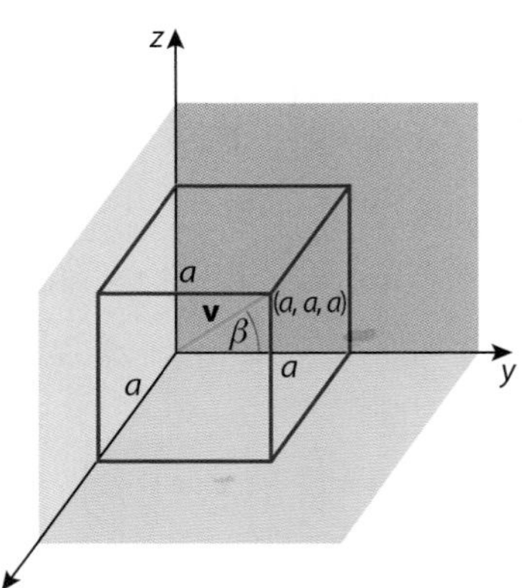

$$|\mathbf{v}| = \sqrt{a^2 + a^2 + a^2} = a\sqrt{3}.$$

Take angle β, for example:

$$\beta = \cos^{-1}\left(\frac{a}{a\sqrt{3}}\right) = \cos^{-1}\left(\frac{1}{\sqrt{3}}\right) \approx 54.7°$$

Example 13

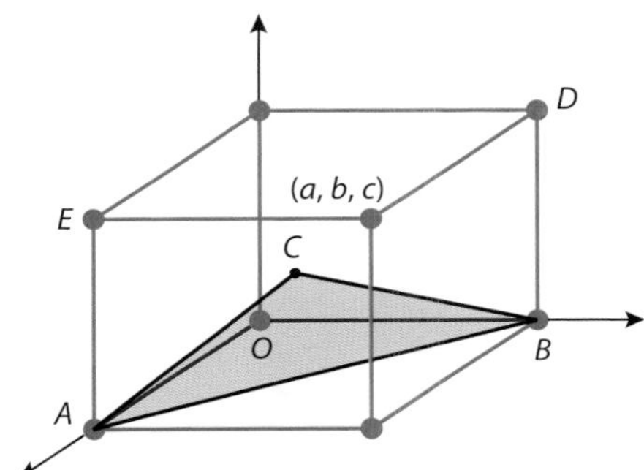

The point C is at the centre of the rectangular box whose edges have measures a, b and c. Find the measure of angle $A\hat{C}B$ in terms of a, b and c.

Solution

The point diagonally opposite to A is $D(0, b, c)$. So, $C\left(\frac{a}{2}, \frac{b}{2}, \frac{c}{2}\right)$. Consequently,

$$\overrightarrow{CA} = \left(a - \frac{a}{2}, 0 - \frac{b}{2}, 0 - \frac{c}{2}\right) = \left(\frac{a}{2}, -\frac{b}{2}, -\frac{c}{2}\right) \text{ and } \overrightarrow{CB}$$

$$= \left(0 - \frac{a}{2}, b - \frac{b}{2}, 0 - \frac{c}{2}\right) = \left(-\frac{a}{2}, \frac{b}{2}, -\frac{c}{2}\right)$$

$$\cos A\hat{C}B = \frac{\overrightarrow{CA} \cdot \overrightarrow{CB}}{|\overrightarrow{CA}||\overrightarrow{CB}|} = \frac{-\frac{a^2}{4} - \frac{b^2}{4} + \frac{c^2}{4}}{\sqrt{\frac{a^2}{4} + \frac{b^2}{4} + \frac{c^2}{4}}\sqrt{\frac{a^2}{4} + \frac{b^2}{4} + \frac{c^2}{4}}}$$

$$= \frac{c^2 - a^2 - b^2}{a^2 + b^2 + c^2}$$

Exercise 14.2

1 Find the dot product and the angle between the vectors.

a) $\mathbf{u} = (3, -2, 4)$, $\mathbf{v} = 2\mathbf{i} - \mathbf{j} - 6\mathbf{k}$

b) $\mathbf{u} = \begin{pmatrix} 2 \\ -6 \\ 0 \end{pmatrix}$, $\mathbf{v} = \begin{pmatrix} -1 \\ 3 \\ 5 \end{pmatrix}$

c) $\mathbf{u} = 3\mathbf{i} - \mathbf{j}$, $\mathbf{v} = 5\mathbf{i} + 2\mathbf{j}$

d) $\mathbf{u} = \mathbf{i} - 3\mathbf{j}$, $\mathbf{v} = 5\mathbf{j} + 2\mathbf{k}$

e) $|\mathbf{u}| = 3$, $|\mathbf{v}| = 4$, the angle between $\mathbf{u}$ and $\mathbf{v}$ is $\frac{\pi}{3}$

f) $|\mathbf{u}| = 3$, $|\mathbf{v}| = 4$, the angle between $\mathbf{u}$ and $\mathbf{v}$ is $\frac{2\pi}{3}$

• **Hint:** Orthogonal means 'at right angles to each other'.

2 State whether the following vectors are orthogonal. If not orthogonal, is the angle acute?

a) $\mathbf{u} = \begin{pmatrix} 2 \\ -6 \\ 4 \end{pmatrix}$, $\mathbf{v} = \begin{pmatrix} -1 \\ 3 \\ 5 \end{pmatrix}$

b) $\mathbf{u} = 3\mathbf{i} - 7\mathbf{j}$, $\mathbf{v} = 5\mathbf{i} + 2\mathbf{j}$

c) $\mathbf{u} = \mathbf{i} - 3\mathbf{j} + 6\mathbf{k}$, $\mathbf{v} = 6\mathbf{j} + 3\mathbf{k}$

3 a) Show that the vectors $\mathbf{v} = -y\mathbf{i} + x\mathbf{j}$ and $\mathbf{w} = y\mathbf{i} - x\mathbf{j}$ are both perpendicular to $\mathbf{u} = x\mathbf{i} + y\mathbf{j}$.

b) Find two unit vectors that are perpendicular to $\mathbf{u} = 2\mathbf{i} - 3\mathbf{j}$. Plot the three vectors in the same coordinate system.

4 (i) Find the direction cosines of $\mathbf{v}$.

(ii) Show that they satisfy $\cos^2\alpha + \cos^2\beta + \cos^2\gamma = 1$.

(iii) Approximate the direction angles to the nearest degree.

a) $\mathbf{v} = 2\mathbf{i} - 3\mathbf{j} + \mathbf{k}$

b) $\mathbf{v} = \mathbf{i} - 2\mathbf{j} + \mathbf{k}$

c) $\mathbf{v} = 3\mathbf{i} - 2\mathbf{j} + \mathbf{k}$

d) $\mathbf{v} = 3\mathbf{i} - 4\mathbf{k}$

5 Find a unit vector with direction angles $\frac{\pi}{3}, \frac{\pi}{4}, \frac{2\pi}{3}$.

6 Find a vector with magnitude 3 and direction angles $\frac{\pi}{4}, \frac{\pi}{4}, \frac{\pi}{2}$.

7 Determine m so that $\mathbf{u}$ and $\mathbf{v}$ are perpendicular.

a) $\mathbf{u} = (3, 5, 0)$; $\mathbf{v} = (m - 2, m + 3, 0)$

b) $\mathbf{u} = (2m, m - 1, m + 1)$; $\mathbf{v} = (m - 1, m, m - 1)$

8 Given the vectors $\mathbf{u} = (-3, 1, 2)$, $\mathbf{v} = (1, 2, 1)$, and $\mathbf{w} = \mathbf{u} + m\mathbf{v}$, determine the value of m so that the vectors $\mathbf{u}$ and $\mathbf{w}$ are orthogonal.

9 Given the vectors $\mathbf{u} = (-2, 5, 4)$ and $\mathbf{v} = (6, -3, 0)$, find, to the nearest degree, the measures of the angles between the following vectors.

a) $\mathbf{u}$ and $\mathbf{v}$

b) $\mathbf{u}$ and $\mathbf{u} + \mathbf{v}$

c) $\mathbf{v}$ and $\mathbf{u} + \mathbf{v}$

10 Consider the following three points: $A(1, 2, -3)$, $B(3, 5, -2)$ and $C(m, 1, -10m)$. Determine m so that

a) A, B and C are collinear

b) $\overrightarrow{AB}$ and $\overrightarrow{AC}$ are perpendicular.

11 Consider the triangle with vertices $A(4, -2, -1)$, $B(3, -5, -1)$ and $C(3, 1, 2)$. Find the vector equations of each of its medians and then find the coordinates of its centroid (i.e. where the medians meet).

12 Consider the tetrahedron $ABCD$ with vertices as shown in the diagram. Find, to the nearest degree, all the angles in the tetrahedron.

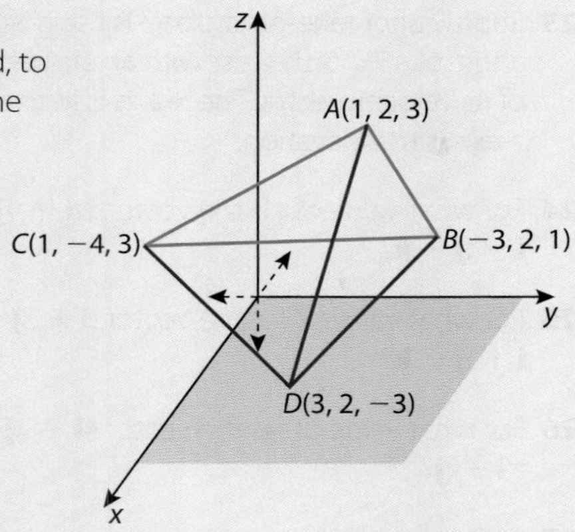

13 In question 12 above, use the angles you found to calculate the total surface area of the tetrahedron.

14 In question 12, what angles does $\overrightarrow{DC}$ make with each of the coordinate axes?

15 In question 12, find $(\overrightarrow{DA} - \overrightarrow{DB}) \cdot \overrightarrow{AC}$.

16 Find k such that the angle between the vectors $\begin{pmatrix} 3 \\ -k \\ -1 \end{pmatrix}$ and $\begin{pmatrix} 1 \\ -3 \\ k \end{pmatrix}$ is $\frac{\pi}{3}$.

17 Find k such that the angle between the vectors $\begin{pmatrix} k \\ 1 \\ 1 \end{pmatrix}$ and $\begin{pmatrix} 1 \\ k \\ 1 \end{pmatrix}$ is $\frac{\pi}{3}$.

18 Find x and y such that $\begin{pmatrix} 2 \\ x \\ y \end{pmatrix}$ is perpendicular to both $\begin{pmatrix} 3 \\ 1 \\ -1 \end{pmatrix}$ and $\begin{pmatrix} 4 \\ -1 \\ 2 \end{pmatrix}$.

19 Consider the vectors $\begin{pmatrix} 1 - x \\ 2x - 2 \\ 3 + x \end{pmatrix}$ and $\begin{pmatrix} 2 - x \\ 1 + x \\ 1 + x \end{pmatrix}$. Find the value(s) of x such that the two vectors are parallel.

20 In triangle ABC, $\overrightarrow{OA} = \begin{pmatrix} 2 \\ 3 \\ 1 \end{pmatrix}$, $\overrightarrow{OB} = \begin{pmatrix} 3 \\ 5 \\ 4 \end{pmatrix}$ and $\overrightarrow{BC} = \begin{pmatrix} -1 \\ 4 \\ 0 \end{pmatrix}$.

Find the measure of $A\hat{B}C$.
Find $\overrightarrow{AC}$ and use it to find the measure of $B\hat{A}C$.

21 Find the value(s) of b such that the vectors are orthogonal.

a) $(b, 3, 2)$ and $(1, b, 1)$

b) $(4, -2, 7)$ and $(b^2, b, 0)$

c) $\begin{pmatrix} b \\ 11 \\ -3 \end{pmatrix}$ and $\begin{pmatrix} 2b \\ -b \\ -5 \end{pmatrix}$

d) $\begin{pmatrix} 2 \\ 5 \\ 2b \end{pmatrix}$ and $\begin{pmatrix} 6 \\ 4 \\ -b \end{pmatrix}$

22 If two vectors $\mathbf{p}$ and $\mathbf{q}$ are such that $|\mathbf{p}| = |\mathbf{q}|$, show that $\mathbf{p} + \mathbf{q}$ and $\mathbf{p} - \mathbf{q}$ are perpendicular. (This proves that the diagonals of a rhombus are perpendicular to each other!)

23 Shortly after take-off, a plane is rising at a rate of 300 m/min. It is heading at an angle of 45° north-west with an airspeed of 200 km/h. Find the components of its velocity vector. The x-axis is in the east direction, the y-axis north and the z-axis is the elevation.

24 For what value of t is the vector $2t\mathbf{i} + 4\mathbf{j} - (10 + t)\mathbf{k}$ perpendicular to the vector $\mathbf{i} + t\mathbf{j} + \mathbf{k}$?

25 For what value of t is the vector $t\mathbf{i} + 3\mathbf{j} + 2\mathbf{k}$ perpendicular to the vector $\mathbf{i} + t\mathbf{j} + \mathbf{k}$?

26 For what value of t is the vector $4\mathbf{i} - 2\mathbf{j} + 7\mathbf{k}$ perpendicular to the vector $t^2\mathbf{i} + t\mathbf{j}$?

27 Find the angle between the diagonal of a cube and a diagonal of one of the faces. Consider all possible cases!

28 Show that the vector $|\mathbf{a}|\mathbf{b} + |\mathbf{b}|\mathbf{a}$ bisects the angle between the two vectors $\mathbf{a}$ and $\mathbf{b}$.

29 Let $\mathbf{u} = \mathbf{i} + m\mathbf{j} + \mathbf{k}$ and $\mathbf{v} = 2\mathbf{i} - \mathbf{j} + n\mathbf{k}$. Compute all values of m and n for which $\mathbf{u} \perp \mathbf{v}$ and $|\mathbf{u}| = |\mathbf{v}|$.

30 Show that $\frac{\pi}{4}, \frac{\pi}{6}, \frac{2\pi}{3}$ cannot be the direction angles of a vector.

31 If a vector has direction angles $\alpha = \frac{\pi}{3}$ and $\beta = \frac{\pi}{4}$, find the third direction angle γ.

32 If a vector has all its direction angles equal, what is the measure of each angle?

33 If the direction angles of a vector $\mathbf{u}$ are α, β and γ, then what are the direction angles of $-\mathbf{u}$?

34 Find all possible values of a unit vector $\mathbf{u}$ that will be perpendicular to both $\mathbf{i} + 2\mathbf{j} + \mathbf{k}$ and $3\mathbf{i} - 4\mathbf{j} + 2\mathbf{k}$.

14.3 Vector (cross) product

In several applications of vectors there is a need to find a vector that is orthogonal to two given vectors. In this section we will discuss a new type of vector multiplication that can be used for this purpose.

If $\mathbf{u} = (u_1, u_2, u_3)$ and $\mathbf{v} = (v_1, v_2, v_3)$ are two vectors, then the vector (cross) product is written as $\mathbf{u} \times \mathbf{v}$ and is defined as

$$\mathbf{u} \times \mathbf{v} = \begin{vmatrix} u_2 & u_3 \\ v_2 & v_3 \end{vmatrix} \mathbf{i} - \begin{vmatrix} u_1 & u_3 \\ v_1 & v_3 \end{vmatrix} \mathbf{j} + \begin{vmatrix} u_1 & u_2 \\ v_1 & v_2 \end{vmatrix} \mathbf{k},$$

or, using the properties of determinants, we can observe that this definition is equivalent to

$$\mathbf{u} \times \mathbf{v} = \begin{vmatrix} \mathbf{i} & \mathbf{j} & \mathbf{k} \\ u_1 & u_2 & u_3 \\ v_1 & v_2 & v_3 \end{vmatrix}.$$

Example 14

Given the vectors $\mathbf{u} = 2\mathbf{i} - 3\mathbf{j} + \mathbf{k}$ and $\mathbf{v} = \mathbf{i} + 3\mathbf{j} - 2\mathbf{k}$, find

a) $\mathbf{u} \times \mathbf{v}$ b) $\mathbf{v} \times \mathbf{u}$ c) $\mathbf{u} \times \mathbf{u}$

Solution

a) $$\mathbf{u} \times \mathbf{v} = \begin{vmatrix} \mathbf{i} & \mathbf{j} & \mathbf{k} \\ 2 & -3 & 1 \\ 1 & 3 & -2 \end{vmatrix} = \begin{vmatrix} -3 & 1 \\ 3 & -2 \end{vmatrix}\mathbf{i} - \begin{vmatrix} 2 & 1 \\ 1 & -2 \end{vmatrix}\mathbf{j} + \begin{vmatrix} 2 & -3 \\ 1 & 3 \end{vmatrix}\mathbf{k}$$

$$= 3\mathbf{i} + 5\mathbf{j} + 9\mathbf{k}.$$

You can also get the same result by simply evaluating the determinant using the short cut you learned in Chapter 6.

b) $$\mathbf{v} \times \mathbf{u} = \begin{vmatrix} \mathbf{i} & \mathbf{j} & \mathbf{k} \\ 1 & 3 & -2 \\ 2 & -3 & 1 \end{vmatrix} = \begin{vmatrix} 3 & -2 \\ -3 & 1 \end{vmatrix}\mathbf{i} - \begin{vmatrix} 1 & -2 \\ 2 & 1 \end{vmatrix}\mathbf{j} + \begin{vmatrix} 1 & 3 \\ 2 & -3 \end{vmatrix}\mathbf{k}$$

$$= -3\mathbf{i} - 5\mathbf{j} - 9\mathbf{k}.$$

Observe here that $\mathbf{u} \times \mathbf{v} = -(\mathbf{v} \times \mathbf{u})$!

c) $$\mathbf{u} \times \mathbf{u} = \begin{vmatrix} \mathbf{i} & \mathbf{j} & \mathbf{k} \\ 2 & -3 & 1 \\ 2 & -3 & 1 \end{vmatrix} = \begin{vmatrix} -3 & 1 \\ -3 & 1 \end{vmatrix}\mathbf{i} - \begin{vmatrix} 2 & 1 \\ 2 & 1 \end{vmatrix}\mathbf{j} + \begin{vmatrix} 2 & -3 \\ 2 & -3 \end{vmatrix}\mathbf{k} = 0.$$

Determinants have many useful applications when we are dealing with vector products. Here are some of the properties which we state without proof.

1 If two rows of a determinant are proportional, then the value of that determinant is zero.

So, for example, if $\mathbf{u} = a\mathbf{i} + b\mathbf{j} + c\mathbf{k}$ and $\mathbf{v} = ma\mathbf{i} + mb\mathbf{j} + mc\mathbf{k}$, then

$$\mathbf{u} \times \mathbf{v} = \begin{vmatrix} \mathbf{i} & \mathbf{j} & \mathbf{k} \\ a & b & c \\ ma & mb & mc \end{vmatrix} = 0.$$

This result leads to an important property of vector products:

Two non-zero vectors are parallel if their cross product is zero.

2 If two rows of a determinant are interchanged, then its value is multiplied by (-1).

So, for instance, if $\mathbf{u} = u_1\mathbf{i} + u_2\mathbf{j} + u_3\mathbf{k}$ and $\mathbf{v} = v_1\mathbf{i} + v_2\mathbf{j} + v_3\mathbf{k}$, then

$$\mathbf{u} \times \mathbf{v} = \begin{vmatrix} \mathbf{i} & \mathbf{j} & \mathbf{k} \\ u_1 & u_2 & u_3 \\ v_1 & v_2 & v_3 \end{vmatrix} = - \begin{vmatrix} \mathbf{i} & \mathbf{j} & \mathbf{k} \\ v_1 & v_2 & v_2 \\ u_1 & u_2 & u_3 \end{vmatrix} = -(\mathbf{v} \times \mathbf{u}).$$

Properties

The following results are important in future work and are straightforward to prove. Most of the proofs are left as exercises.

1 $\mathbf{u} \times (\mathbf{v} \pm \mathbf{w}) = (\mathbf{u} \times \mathbf{v}) \pm (\mathbf{u} \times \mathbf{w})$

2 $\mathbf{u} \times 0 = 0$

3 $\mathbf{u} \times \mathbf{u} = 0$

4 $\mathbf{i} \times \mathbf{j} = \mathbf{k}; \mathbf{j} \times \mathbf{k} = \mathbf{i}; \mathbf{k} \times \mathbf{i} = \mathbf{j}$

5 $\mathbf{u} \cdot (\mathbf{u} \times \mathbf{v}) = 0$ (i.e. $\mathbf{u} \times \mathbf{v}$ *is orthogonal to* $\mathbf{u}$.)

6 $\mathbf{v} \cdot (\mathbf{u} \times \mathbf{v}) = 0$ (i.e. $\mathbf{u} \times \mathbf{v}$ *is orthogonal to* $\mathbf{v}$.)

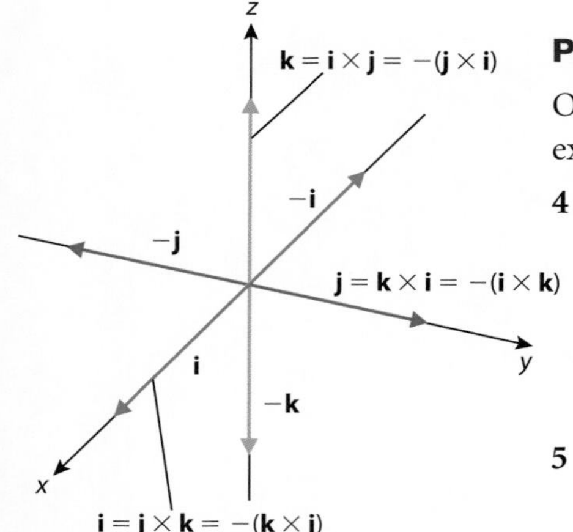

Proofs

Only properties 4 and 5 will be proved here, the rest will be left as an exercise.

4 To prove the first result, we simply apply the definition.

$$\mathbf{i} \times \mathbf{j} = \begin{vmatrix} \mathbf{i} & \mathbf{j} & \mathbf{k} \\ 1 & 0 & 0 \\ 0 & 1 & 0 \end{vmatrix} = \mathbf{k}, \text{ details are left as an exercise.}$$

5 $\mathbf{u} \cdot (\mathbf{u} \times \mathbf{v}) = (u_1, u_2, u_3) \cdot \left(\begin{vmatrix} u_2 & u_3 \\ v_2 & v_3 \end{vmatrix}, -\begin{vmatrix} u_1 & u_3 \\ v_1 & v_3 \end{vmatrix}, \begin{vmatrix} u_1 & u_2 \\ v_1 & v_2 \end{vmatrix} \right)$, so that

$$\begin{aligned} \mathbf{u} \cdot (\mathbf{u} \times \mathbf{v}) &= u_1 \begin{vmatrix} u_2 & u_3 \\ v_2 & v_3 \end{vmatrix} - u_2 \begin{vmatrix} u_1 & u_3 \\ v_1 & v_3 \end{vmatrix} + u_3 \begin{vmatrix} u_1 & u_2 \\ v_1 & v_2 \end{vmatrix} \\ &= u_1u_2v_3 - u_1u_3v_2 - u_2u_1v_3 + u_2u_3v_1 + u_3u_1v_2 - u_3u_2v_1 \\ &= 0 \end{aligned}$$

Properties 4 and 5 lead to an equivalent 'geometric' definition of the cross product:

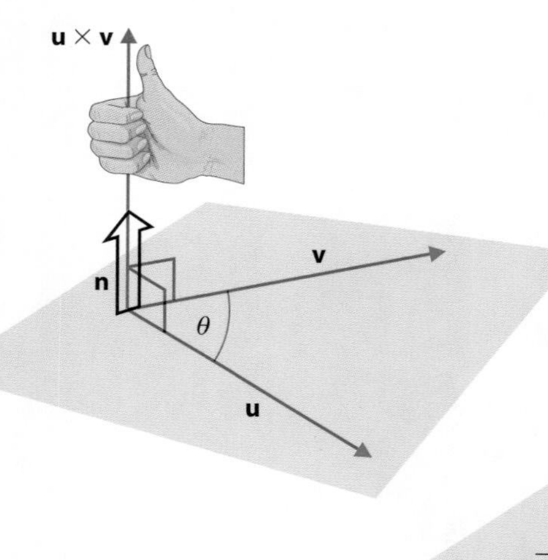

Definition of the cross product (Theorem):

($\mathbf{u} \times \mathbf{v}$) is a vector perpendicular to both $\mathbf{u}$ and $\mathbf{v}$, obeying the right-hand rule shown right, and has the magnitude:

$$|\mathbf{u} \times \mathbf{v}| = |\mathbf{u}||\mathbf{v}|\sin\theta.$$

Note: As we mentioned before, please do not forget that

$$\mathbf{u} \times \mathbf{v} = -(\mathbf{v} \times \mathbf{u}).$$

Proof

The algebraic manipulation required for the proof is tremendous, so we will keep the details away from this discussion:

$$|\mathbf{u} \times \mathbf{v}| = \sqrt{(u_2v_3 - u_3v_2)^2 + (u_1v_3 - u_3v_1)^2 + (u_1v_2 - u_2v_1)^2}$$

$$= \sqrt{(u_1^2 + u_2^2 + u_3^2)(v_1^2 + v_2^2 + v_3^2) - (u_1v_1 + u_2v_2 + u_3v_3)^2}$$

$$= \sqrt{|\mathbf{u}|^2|\mathbf{v}|^2 - (\mathbf{u} \cdot \mathbf{v})^2} = \sqrt{|\mathbf{u}|^2|\mathbf{v}|^2 - (|\mathbf{u}|\,|\mathbf{v}|\cos\theta)^2}$$

$$= |\mathbf{u}||\mathbf{v}|\sqrt{1 - \cos^2\theta} = |\mathbf{u}||\mathbf{v}|\sin\theta$$

• **Hint:** The vector product gives you another method for finding the angle between two vectors.
$|\mathbf{u} \times \mathbf{v}| = |\mathbf{u}||\mathbf{v}|\sin\theta$

$\Leftrightarrow \sin\theta = \dfrac{|\mathbf{u} \times \mathbf{v}|}{|\mathbf{u}||\mathbf{v}|}$

Example 15

Find a unit vector orthogonal to both vectors $\mathbf{u} = 2\mathbf{i} - 3\mathbf{j} + \mathbf{k}$ and $\mathbf{v} = \mathbf{i} + 3\mathbf{j} - 2\mathbf{k}$.

Solution

$\mathbf{u} \times \mathbf{v}$ is orthogonal to both vectors.

And $\mathbf{u} \times \mathbf{v} = 3\mathbf{i} + 5\mathbf{j} + 9\mathbf{k}$. A unit vector in the same direction as $\mathbf{u} \times \mathbf{v}$ will also be orthogonal to both vectors. Remembering that a unit vector is equal to the vector itself multiplied by the reciprocal of its magnitude, as we have seen in Chapter 9 and in Section 14.1, we find the magnitude of $\mathbf{u} \times \mathbf{v}$ first.

$$|\mathbf{u} \times \mathbf{v}| = \sqrt{9 + 25 + 81} = \sqrt{115},$$

and the required unit vector is therefore

$$\frac{\mathbf{u} \times \mathbf{v}}{|\mathbf{u} \times \mathbf{v}|} = \frac{3\mathbf{i} + 5\mathbf{j} + 9\mathbf{k}}{\sqrt{115}}.$$

Corollary: The last result leads to the conclusion that the magnitude of the cross product is the area of the parallelogram that has $\mathbf{u}$ and $\mathbf{v}$ as adjacent sides.

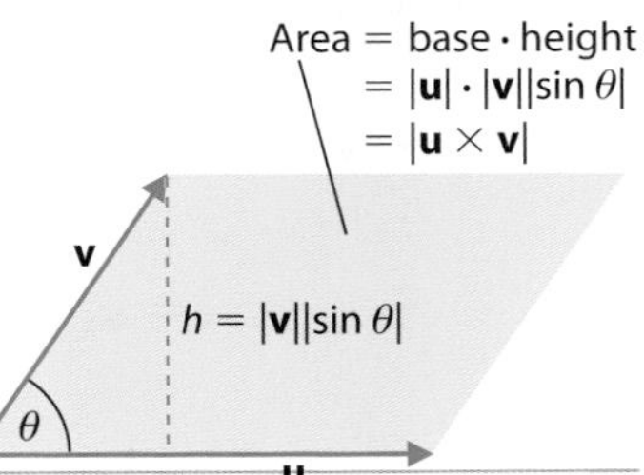

Example 16

Show that the quadrilateral $ABCD$ with its vertices at the following points is a parallelogram and find its area.
$A(3, 0, 2)$, $B(6, 2, 5)$, $C(1, 2, 2)$, $D(4, 4, 5)$

Solution

$$\overrightarrow{AB} = \begin{pmatrix} 3 \\ 2 \\ 3 \end{pmatrix}, \overrightarrow{BD} = \begin{pmatrix} -2 \\ 2 \\ 0 \end{pmatrix}, \overrightarrow{CD} = \begin{pmatrix} 3 \\ 2 \\ 3 \end{pmatrix}, \overrightarrow{AC} = \begin{pmatrix} -2 \\ 2 \\ 0 \end{pmatrix}$$

This implies that $\overrightarrow{AB} = \overrightarrow{CD}$ and $\overrightarrow{BD} = \overrightarrow{AC}$ which in turn means that the pairs of opposite sides of the quadrilateral are congruent and parallel. (You need only one pair.) Thus, *ABDC* is a parallelogram with *AB* and *BD* as adjacent sides.

Furthermore, since

$$\overrightarrow{AB} \times \overrightarrow{BD} = \begin{vmatrix} \mathbf{i} & \mathbf{j} & \mathbf{k} \\ 3 & 2 & 3 \\ -2 & 2 & 0 \end{vmatrix} = -6\mathbf{i} - 6\mathbf{j} + 10\mathbf{k}$$

the area of the parallelogram is

$$|\overrightarrow{AB} \times \overrightarrow{BD}| = \sqrt{36 + 36 + 100} = \sqrt{172} = 2\sqrt{43}.$$

Example 17

Find the area of the triangle determined by the points $A(2, 2, 0)$, $B(-1, 0, 2)$ and $C(0, 4, 3)$.

Solution

The area of the triangle *ABC* is half the area of the parallelogram formed with *AB* and *AC* as its adjacent sides.

But $\overrightarrow{AB} = (-3, -2, 2)$ and $\overrightarrow{AC} = (-2, 2, 3)$, so $\overrightarrow{AB} \times \overrightarrow{AC} = (-10, 5, -10)$, and hence area of triangle $ABC = \frac{1}{2}|\overrightarrow{AB} \times \overrightarrow{AC}| = \frac{15}{2}$.

The scalar triple product

(This product is very helpful in its geometric interpretation as it is of great help in finding the equation of a plane later in the chapter.)

If $\mathbf{u} = (u_1, u_2, u_3)$, $\mathbf{v} = (v_1, v_2, v_3)$ and $\mathbf{w} = (w_1, w_2, w_3)$ are three vectors, then the scalar triple product is $\mathbf{u} \cdot (\mathbf{v} \times \mathbf{w})$. The component expression of this product can be found by applying the above definition:

$$\mathbf{u} \cdot (\mathbf{v} \times \mathbf{w}) = \mathbf{u} \cdot \left(\begin{vmatrix} v_2 & v_3 \\ w_2 & w_3 \end{vmatrix} \mathbf{i} - \begin{vmatrix} v_1 & v_3 \\ w_1 & w_3 \end{vmatrix} \mathbf{j} + \begin{vmatrix} v_1 & v_2 \\ w_1 & w_2 \end{vmatrix} \mathbf{k} \right)$$

$$= u_1 \begin{vmatrix} v_2 & v_3 \\ w_2 & w_3 \end{vmatrix} - u_2 \begin{vmatrix} v_1 & v_3 \\ w_1 & w_3 \end{vmatrix} + u_3 \begin{vmatrix} v_1 & v_2 \\ w_1 & w_2 \end{vmatrix}$$

$$= \begin{vmatrix} u_1 & u_2 & u_3 \\ v_1 & v_2 & v_3 \\ w_1 & w_2 & w_3 \end{vmatrix}$$

Example 18

Calculate the scalar triple product of the vectors:

$$\mathbf{u} = 2\mathbf{i} - \mathbf{j} - 5\mathbf{k},\ \mathbf{v} = 2\mathbf{i} + 5\mathbf{j} - 5\mathbf{k},\ \mathbf{w} = \mathbf{i} + 4\mathbf{j} + 3\mathbf{k}$$

Solution

$$\mathbf{u} \cdot (\mathbf{v} \times \mathbf{w}) = \begin{vmatrix} 2 & -1 & -5 \\ 2 & 5 & -5 \\ 1 & 4 & 3 \end{vmatrix} = 66$$

Geometric interpretation

$|\mathbf{u} \cdot (\mathbf{v} \times \mathbf{w})|$ is the volume of the parallelepiped that has the three vectors as adjacent edges.

Proof

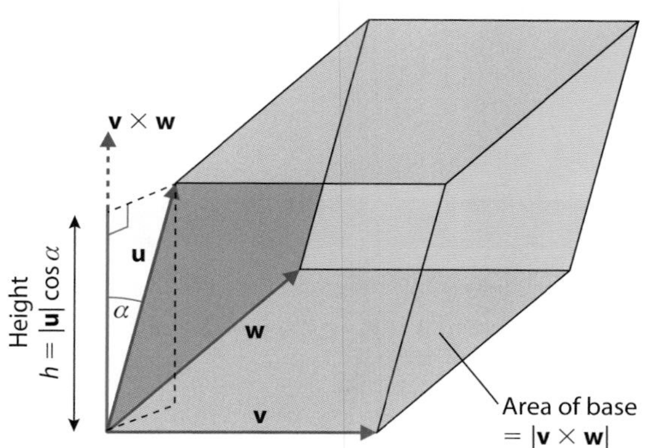

In the diagram above, $|\mathbf{v} \times \mathbf{w}|$ is the area of the parallelogram with sides $\mathbf{v}$ and $\mathbf{w}$, which is the base of the parallelepiped. Also,

$$|\mathbf{u} \cdot (\mathbf{v} \times \mathbf{w})| = |\mathbf{u}||\mathbf{v} \times \mathbf{w}|\cos\alpha$$
$$= |\mathbf{v} \times \mathbf{w}||\mathbf{u}|\cos\alpha.$$

But, $|\mathbf{u}|\cos\alpha = h$, the height of the parallelepiped, and $|\mathbf{v} \times \mathbf{w}|$ is the area of the base; therefore, the triple product's absolute value is the volume of the parallelepiped.

A direct consequence of this theorem is that the volume of the parallelepiped is 0 if and only if the three vectors are coplanar. That is:

If $\mathbf{u} = (u_1, u_2, u_3)$, $\mathbf{v} = (v_1, v_2, v_3)$ *and* $\mathbf{w} = (w_1, w_2, w_3)$ *are three vectors drawn from the same initial point, they lie in the same plane if:*

$$\mathbf{u} \cdot (\mathbf{v} \times \mathbf{w}) = \begin{vmatrix} u_1 & u_2 & u_3 \\ v_1 & v_2 & v_3 \\ w_1 & w_2 & w_3 \end{vmatrix} = 0.$$

The parentheses in the scalar triple product is unnecessary, i.e. $\mathbf{u} \cdot (\mathbf{v} \times \mathbf{w}) = \mathbf{u} \cdot \mathbf{v} \times \mathbf{w}$. Can you justify?

Example 19

Consider the three vectors

$$\mathbf{u} = 2\mathbf{i} + \mathbf{j} + m\mathbf{k},\ 3\mathbf{i} + 2\mathbf{j} + 3\mathbf{k} \text{ and } \mathbf{w} = m\mathbf{i} + 2\mathbf{j} + \mathbf{k}.$$

a) Find the volume of the parallelepiped that has these vectors as sides.

b) Show that these vectors can never be on the same plane.

Solution

a) The volume of the parallelepiped is given by the absolute value of their scalar triple product:

$$|\mathbf{u} \cdot (\mathbf{v} \times \mathbf{w})| = \left\| \begin{matrix} 2 & 1 & m \\ 3 & 2 & 3 \\ m & 2 & 1 \end{matrix} \right\| = |-2m^2 + 9m - 11|$$

b) For the vectors to be coplanar, their scalar triple product must be zero. That is, $-2m^2 + 9m - 11 = 0$.

However, since this is a quadratic equation, it can have real roots if $b^2 - 4ac \geqslant 0$, but $b^2 - 4ac = 81 - 88 = -7 < 0$, and thus the equation does not admit any real roots and the three vectors can therefore never be coplanar.

Exercise 14.3

1 a) Find the cross product using the definition: $\mathbf{i} \times (\mathbf{i} + \mathbf{j} + \mathbf{k})$.

b) Compare your answer to $(\mathbf{i} \times \mathbf{i}) + (\mathbf{i} \times \mathbf{j}) + (\mathbf{i} \times \mathbf{k})$.

2 Repeat question 1 for $\mathbf{j} \times (\mathbf{i} + \mathbf{j} + \mathbf{k})$.

3 Repeat question 1 for $\mathbf{k} \times (\mathbf{i} + \mathbf{j} + \mathbf{k})$.

4 Use the definition of vector products to verify $\mathbf{u} \times (\mathbf{v} \pm \mathbf{w}) = (\mathbf{u} \times \mathbf{v}) \pm (\mathbf{u} \times \mathbf{w})$. (This is the distributive property of vector product over addition and subtraction.)

In questions 5–8, find $\mathbf{u} \times \mathbf{v}$ and check that it is orthogonal to both $\mathbf{u}$ and $\mathbf{v}$.

5 $\mathbf{u} = (2, 3, -2)$, $\mathbf{v} = (-3, 2, 3)$

6 $\mathbf{u} = 4\mathbf{i} + 3\mathbf{j}$, $\mathbf{v} = -2\mathbf{j} + 2\mathbf{k}$

7 $\mathbf{u} = \begin{pmatrix} 1 \\ 2 \\ -1 \end{pmatrix}$, $\mathbf{v} = \begin{pmatrix} 4 \\ 1 \\ -3 \end{pmatrix}$

8 $\mathbf{u} = 5\mathbf{i} + \mathbf{j} + 2\mathbf{k}$, $\mathbf{v} = 3\mathbf{i} + \mathbf{k}$

9 Consider the following vectors:

$$\mathbf{u} = 2\mathbf{i} + \mathbf{j} + m\mathbf{k},\ \mathbf{v} = 3\mathbf{i} + 2\mathbf{j} + 3\mathbf{k},\ \mathbf{w} = m\mathbf{i} + 2\mathbf{j} + \mathbf{k}$$

Find

a) $\mathbf{u} \cdot (\mathbf{v} \times \mathbf{w})$

b) $\mathbf{w} \cdot (\mathbf{u} \times \mathbf{v})$

c) $\mathbf{v} \cdot (\mathbf{w} \times \mathbf{u})$

10 Consider the following vectors:

$$\mathbf{u} = (3, 0, 4),\ \mathbf{v} = (1, 2, 8),\ \mathbf{w} = (2, 5, 6)$$

Find

a) $\mathbf{u} \times (\mathbf{v} \times \mathbf{w})$

b) $(\mathbf{u} \times \mathbf{v}) \times \mathbf{w}$

c) $(\mathbf{u} \times \mathbf{v}) \times (\mathbf{v} \times \mathbf{w})$

d) $(\mathbf{v} \times \mathbf{w}) \times (\mathbf{u} \times \mathbf{v})$

e) $(\mathbf{u} \cdot \mathbf{w})\mathbf{v} - (\mathbf{u} \cdot \mathbf{v})\mathbf{w}$

f) $(\mathbf{w} \cdot \mathbf{u})\mathbf{v} - (\mathbf{w} \cdot \mathbf{v})\mathbf{u}$

11 Find a unit vector that is orthogonal to both

$$\mathbf{u} = -6\mathbf{i} + 4\mathbf{j} + \mathbf{k} \text{ and } \mathbf{v} = 3\mathbf{i} + \mathbf{j} + 5\mathbf{k}.$$

12 Find a unit vector that is normal (perpendicular) to the plane determined by the points $A(1, -1, 2)$, $B(2, 0, -1)$ and $C(0, 2, 1)$.

In questions 13–14, find the area of the parallelogram that has **u** and **v** as adjacent sides.

13 $\mathbf{u} = 2\mathbf{i} + 3\mathbf{k}, \mathbf{v} = \mathbf{i} + 4\mathbf{j} + 2\mathbf{k}$

14 $\mathbf{u} = 3\mathbf{i} + 4\mathbf{j} + \mathbf{k}, \mathbf{v} = 3\mathbf{j} - \mathbf{k}$

15 Verify that the points are the vertices of a parallelogram and find its area: $(2, -1, 1)$, $(5, 1, 4)$, $(0, 1, 1)$ and $(3, 3, 4)$.

16 Show that the points $P(1, -1, 2)$, $Q(2, 0, 1)$, $R(3, 2, 0)$ and $S(5, 4, -2)$ are coplanar.

17 For what value(s) of m are the following four points on the same plane? $A(m, 3, -2)$, $B(3, 4, m)$, $C(2, 0, -2)$ and $D(4, 8, 4)$.

In questions 18–19, find the area of the triangle with the given vertices.

18 $A(2, 6, -1)$, $B(1, 1, 1)$, $C(3, 5, 2)$

19 $A(3, 1, -2)$, $B(2, 5, 6)$, $C(6, 1, 8)$

In questions 20–22, find $\mathbf{u} \cdot (\mathbf{v} \times \mathbf{w})$.

20 $\mathbf{u} = 3\mathbf{i} - 2\mathbf{j} + 2\mathbf{k}, \mathbf{v} = 5\mathbf{i} + 2\mathbf{j} - 2\mathbf{k}, \mathbf{w} = \mathbf{i} + 2\mathbf{j} + 6\mathbf{k}$

21 $\mathbf{u} = (2, -1, 3), \mathbf{v} = (1, 4, 3), \mathbf{w} = (-3, 2, -2)$

22 $\mathbf{u} = \begin{pmatrix} 3 \\ 2 \\ 1 \end{pmatrix}, \mathbf{v} = \begin{pmatrix} 1 \\ -3 \\ 1 \end{pmatrix}, \mathbf{w} = \begin{pmatrix} 5 \\ 1 \\ 2 \end{pmatrix}$

In questions 23–24, find the volume of the parallelepiped with **u**, **v** and **w** as adjacent edges.

23 $\mathbf{u} = (3, -5, 3), \mathbf{v} = (1, 5, -1), \mathbf{w} = (3, 2, -3)$

24 $\mathbf{u} = 4\mathbf{i} + 2\mathbf{j} + 3\mathbf{k}, \mathbf{v} = 5\mathbf{i} + 6\mathbf{j} + 2\mathbf{k}, \mathbf{w} = 2\mathbf{i} + 3\mathbf{j} + 5\mathbf{k}$

In questions 25–26, determine whether the three given vectors are coplanar.

25 $\mathbf{u} = (2, -1, 2), \mathbf{v} = (4, 1, -1), \mathbf{w} = (6, -3, 1)$

26 $\mathbf{u} = (4, -2, -1), \mathbf{v} = (9, -6, -1), \mathbf{w} = (6, -6, 1)$

In questions 27–28, find m such that the following vectors are coplanar; otherwise, show that it is not possible.

27 $\mathbf{u} = (1, m, 1), \mathbf{v} = (3, 0, m), \mathbf{w} = (5, -4, 0)$

28 $\mathbf{u} = (2, -3, 2m), \mathbf{v} = (m, -3, 1), \mathbf{w} = (1, 3, -2)$

29 Consider the parallelepiped given in the diagram.

a) Find the volume.

b) Find the area of the face determined by **u** and **v**.

c) Find the height of the parallelepiped from vertex D to the base.

d) Find the angle that **w** makes with the plane determined by **u** and **v**.

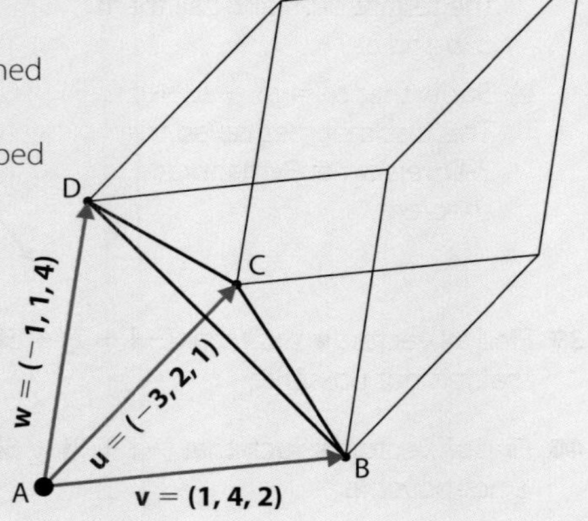

30 a) From geometry, you know that the volume of a tetrahedron is $\frac{1}{3}$(base)(height). Use the results from the previous problem to find the volume of tetrahedron *ABCD*. Compare this volume to the volume of the parallelepiped and make a general conjecture.

b) Use the results you have from part a) to find the volume of the tetrahedron whose vertices are

$$A(0, 3, 1), B(3, 2, -2), C(2, 1, 2) \text{ and } D(4, -1, 4).$$

31 What can you conclude about the angle between two non-zero vectors **u** and **v** if $\mathbf{u} \cdot \mathbf{v} = |\mathbf{u} \times \mathbf{v}|$?

32 Show that $|\mathbf{u} \times \mathbf{v}| = \sqrt{|\mathbf{u}|^2|\mathbf{v}|^2 - (\mathbf{u} \cdot \mathbf{v})^2}$.

• **Hint:** Use right triangle trigonometry to find *d* in terms of θ first.

33 Use the diagram on the right to show that the distance from a point *P* in space to a line *L* through two points *A* and *B* can be expressed as

$$d = \frac{|\overrightarrow{AP} \times \overrightarrow{AB}|}{|\overrightarrow{AB}|}.$$

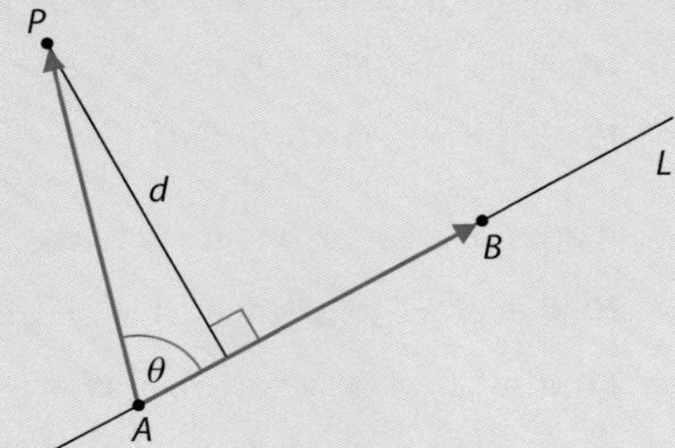

34 Use the result in the previous problem to find the distance from *A* to the line through the points *B* and *C*.

a) $A(-2, 2, 3), B(2, 2, 1), C(-1, 4, -3)$

b) $A(5, 4), B(3, 2), C(1, 3)$

c) $A(2, 0, 1), B(1, -2, 2), C(3, 0, 2)$

35 Express $(\mathbf{u} + \mathbf{v}) \times (\mathbf{v} - \mathbf{u})$ in terms of $(\mathbf{u} \times \mathbf{v})$.

36 Express $(2\mathbf{u} + 3\mathbf{v}) \times (4\mathbf{v} - 5\mathbf{u})$ in terms of $(\mathbf{u} \times \mathbf{v})$.

37 Express $(m\mathbf{u} + n\mathbf{v}) \times (p\mathbf{v} - q\mathbf{u})$ in terms of $(\mathbf{u} \times \mathbf{v})$, where *m*, *n*, *p* and *q* are scalars.

38 Refer to the diagram on the right. You are given a tetrahedron with vertex at the origin and base *ABC*.

a) Find the area of the base *ABC*. Call it *o*.

b) Find the area of each face of the tetrahedron and call them *a*, *b* and *c*.

c) Show that $o^2 = a^2 + b^2 + c^2$. This is sometimes called the 3-D version of Pythagoras' theorem.

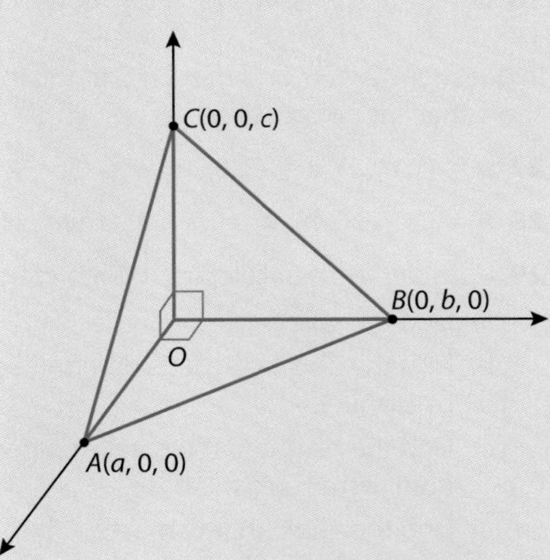

39 Find all vectors **v** such that $(-\mathbf{i} + 2\mathbf{j} + 3\mathbf{k}) \times \mathbf{v} = \mathbf{i} + 5\mathbf{j} - 3\mathbf{k}$; otherwise, show that it is not possible.

40 Find all vectors **v** such that $(-\mathbf{i} + 2\mathbf{j} + 3\mathbf{k}) \times \mathbf{v} = \mathbf{i} + 5\mathbf{j}$; otherwise, show that it is not possible.

14.4 Lines in space

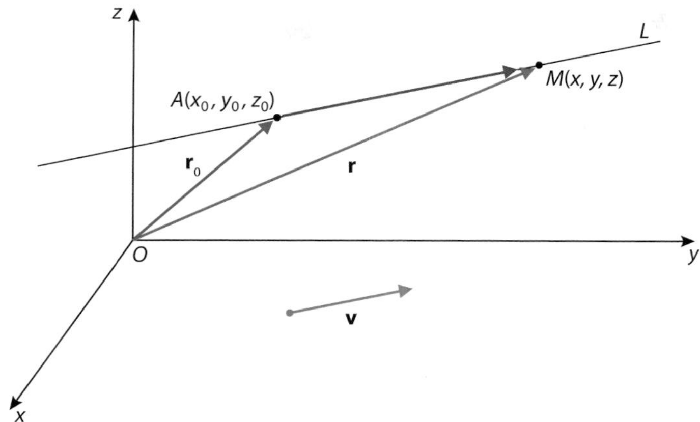

Similar to the plane, a straight line in space can be determined by any two points A and M that lie on it. Alternatively, the line can be determined by specifying a point on it and a direction given by a non-zero vector parallel to it. To investigate equations that describe lines in space, let us begin with a straight line L that passes through the point $A(x_0, y_0, z_0)$ and parallel to the vector $\mathbf{v} = a\mathbf{i} + b\mathbf{j} + c\mathbf{k}$ as shown in the diagram. Now, if L is the line that passes through A and is parallel to the non-zero vector $\mathbf{v}$, then L consists of all the points $M(x, y, z)$ for which the vector $\overrightarrow{AM}$ is parallel to $\mathbf{v}$.

In Section 14.1 we established that:

Two vectors are parallel if one of them is a scalar multiple of the other. That is, **v** is parallel to **u** if and only if $\mathbf{v} = t\mathbf{u}$ for some real number t.

This means that for the point M to be on L, $\overrightarrow{AM}$ must be a scalar multiple of $\mathbf{v}$, i.e. $\overrightarrow{AM} = t\mathbf{v}$, where t is a scalar.

This equation can be written in coordinate form as

$$(x - x_0, y - y_0, z - z_0) = t(a, b, c) = (ta, tb, tc).$$

For two vectors to be equal, their components must be the same, then

$$x - x_0 = ta, y - y_0 = tb, z - z_0 = tc.$$

This leads to the result:

$$x = x_0 + at, y = y_0 + bt, z = z_0 + ct.$$

The line that passes through the point $A(x_0, y_0, z_0)$ and parallel to the vector $\mathbf{v} = (a, b, c)$ has parametric equations:

$$x = x_0 + at, y = y_0 + bt, z = z_0 + ct$$

Example 20

a) Find parametric equations of the line through $A(1, -2, 3)$ and parallel to $\mathbf{v} = 5\mathbf{i} + 4\mathbf{j} - 6\mathbf{k}$.

b) Find parametric equations of the line through the points $A(1, -2, 3)$ and $B(2, 4, -2)$.

Solution

a) From the previous theorem, $x = 1 + 5t, y = -2 + 4t, z = 3 - 6t$.

b) We need to find a vector parallel to the given line. The vector $\overrightarrow{AB}$ provides a good choice: $\overrightarrow{AB} = (1, 6, -5)$. So the equations are

$$x = 1 + t, y = -2 + 6t, z = 3 - 5t.$$

Another set of equations could be

$$x = 2 + t, y = 4 + 6t, z = -2 - 5t.$$

Other sets are possible by considering any vector parallel to $\overrightarrow{AB}$.

Vector equation of a line

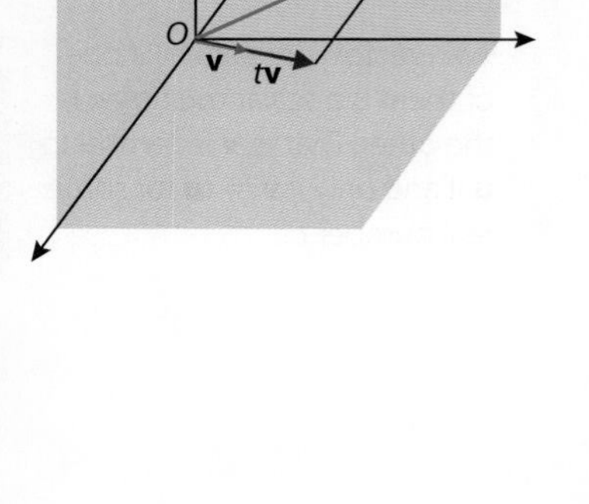

An alternative route to interpreting the equation

$$\overrightarrow{AM} = t\mathbf{v}$$

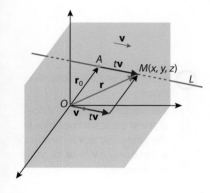

is to express it in terms of the position vectors $\mathbf{r}_0$ of the fixed point A, and $\mathbf{r}$, the position vector of M.

In Section 14.1, we discussed the difference of two vectors which can be of immediate use here.

$$\overrightarrow{AM} = \overrightarrow{OM} - \overrightarrow{OA}$$

$$\Rightarrow t\mathbf{v} = \mathbf{r} - \mathbf{r}_0$$

And hence we arrive at:

The **vector equation** of the line

$$\mathbf{r} = \mathbf{r}_0 + t\mathbf{v}$$

where $\mathbf{r}$ is the position vector of any point on the line, while $\mathbf{r}_0$ is the position vector of a fixed point (A in this case) on the line and $\mathbf{v}$ is the vector parallel to the given line.

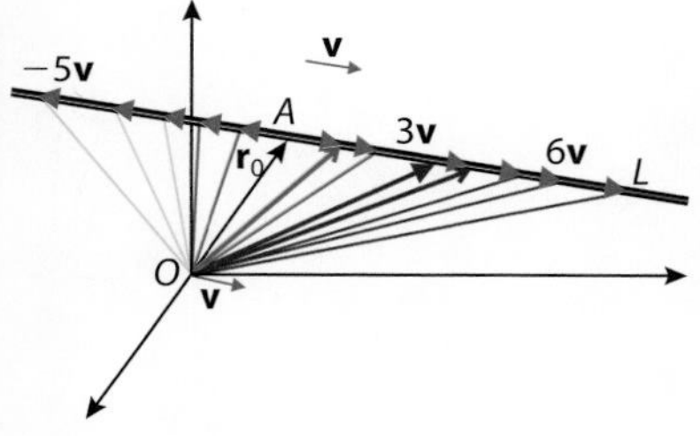

Figure 14.15 By observing Figure 14.15, you will notice, for example, that for each value of t you describe a point on the line. When $t > 0$, the points are in the same direction as $\mathbf{v}$. When $t < 0$, the points are in the opposite direction.

The two approaches are very closely related. We can even say that the parametric equations are a detailed form of the vector equation!

$$\mathbf{r} = \mathbf{r}_0 + t\mathbf{v}$$

$$\Leftrightarrow (x, y, z) = (x_0, y_0, z_0) + t(a, b, c)$$

$$\Leftrightarrow (x, y, z) = (x_0, y_0, z_0) + (ta, tb, tc) = (x_0 + ta, y_0 + tb, z_0 + tc)$$

$$\Leftrightarrow \begin{cases} x = x_0 + ta \\ y = y_0 + tb \\ z = z_0 + tc \end{cases}$$

You can interpret vector equations in several ways. One of these has to do with displacement. That is, to reach point M from point O, you first arrive at A, and then go towards M along the line a multiple of $\mathbf{v}$, $t\mathbf{v}$.

Example 21

Find a vector equation of the line that contains $(-1, 3, 0)$ and is parallel to $\mathbf{v} = 3\mathbf{i} - 2\mathbf{j} + \mathbf{k}$.

Solution

From the previous discussion,

$$\mathbf{r} = (-\mathbf{i} + 3\mathbf{j}) + t(3\mathbf{i} - 2\mathbf{j} + \mathbf{k}).$$

When $t = 0$, the equation gives the point $(-1, 3, 0)$. When $t = 1$, the equation yields

$\mathbf{r} = (-\mathbf{i} + 3\mathbf{j}) + (3\mathbf{i} - 2\mathbf{j} + \mathbf{k}) = 2\mathbf{i} + \mathbf{j} + \mathbf{k}$, a point shifted by $1\mathbf{v}$ down the line. Similarly, when $t = 3$,

$\mathbf{r} = (-\mathbf{i} + 3\mathbf{j}) + 3(3\mathbf{i} - 2\mathbf{j} + \mathbf{k}) = 8\mathbf{i} - 3\mathbf{j} + 3\mathbf{k}$, a point $3\mathbf{v}$ down the line, etc.

Alternatively, the equation can be written as

$$\mathbf{r} = (-1 + 3t)\mathbf{i} + (3 - 2t)\mathbf{j} + t\mathbf{k}.$$

This last form allows us to recognize the parametric equations of the line by simply reading the components of the vector on the right-hand side of the equation.

Example 22

Find a vector equation of the line passing through $A(2, 7)$ and $B(6, 2)$.

Solution

We let the vector $\overrightarrow{AB} = (6 - 2, 2 - 7) = (4, -5)$ be the vector giving the direction of the line, so

$$\mathbf{r} = (2, 7) + t(4, -5), \text{ or equivalently}$$

$$\mathbf{r} = 2\mathbf{i} + 7\mathbf{j} + t(4\mathbf{i} - 5\mathbf{j}).$$

Example 23

Find parametric equations for the line through $A(-1, 1, 3)$ and parallel to the vector $\mathbf{v} = 2\mathbf{i} + 3\mathbf{j} - \mathbf{k}$.

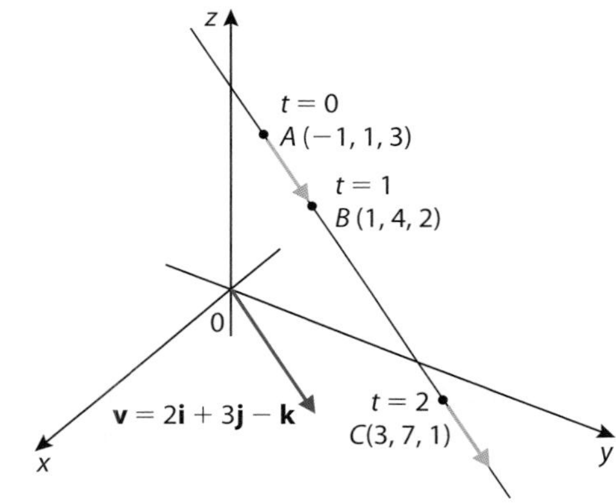

Solution

$$x = -1 + 2t, y = 1 + 3t, z = 3 - 1$$

If you select a few points with their parameter values you can see how the equation represents the line. For $t = 0$, as you expect, you are at point A; for $t = 1$, the point is B; and for $t = 2$, the point is C. The arrows show the direction of increasing values of t.

Line segments

Sometimes, we would like to 'parametrize' a line segment. That is, to write the equation so that it describes the points making up the segment. For example, to parametrize the line segment between $A(3, 7, 1)$ and $B(1, 4, 2)$, we first find the direction vector $AB = (-2, -3, 1)$, then we use point A as the fixed point on the line. Thus, the parametric equations are:

$$\begin{cases} x = 3 - 2t \\ y = 7 - 3t \\ z = 1 + t \end{cases}$$

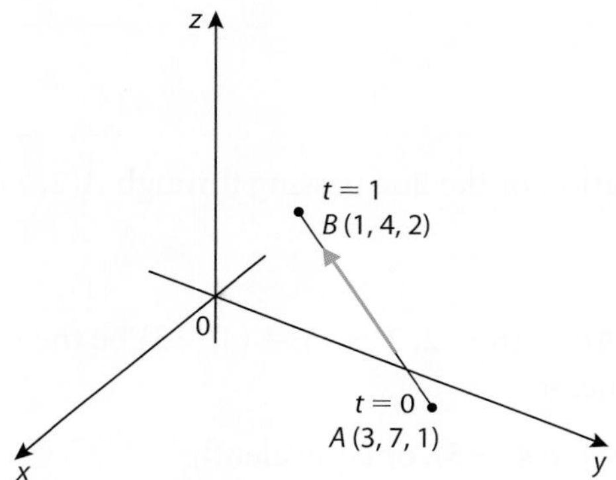

Notice that when $t = 0$, the line starts at the point $A(3, 7, 1)$; when $t = 1$, the line is at $B(1, 4, 2)$. Therefore, to parametrize this segment we restrict the values of t to $0 \leqslant t \leqslant 1$. The new equations are then

$$x = 3 - 2t, y = 7 - 3t, z = 1 + t, 0 \leqslant t \leqslant 1.$$

In general, to parametrize a line segment AB so that we represent the points included between the endpoints only, we can use the vector equation

$$\mathbf{r}(t) = (1 - t)\overrightarrow{OA} + t\overrightarrow{OB}, 0 \leqslant t \leqslant 1.$$

In this parametrization, when $t = 0, \mathbf{r} = \overrightarrow{OA}$, and when $t = 1, \mathbf{r} = \overrightarrow{OB}$. This way $\mathbf{r}$ traces the segment AB from A to B for $0 \leqslant t \leqslant 1$.

Note: The parametrization with the vector equation can be expressed differently if we want to use parametric equation.

If $A(x_1, y_1, z_1)$ and $B(x_2, y_2, z_2)$ are the endpoints of the segment, then

$$\begin{aligned}\mathbf{r}(t) &= (1-t)\overrightarrow{OA} + t\overrightarrow{OB} = \overrightarrow{OA} - t\overrightarrow{OA} + t\overrightarrow{OB}\\ &= \overrightarrow{OA} + t(\overrightarrow{OB} - \overrightarrow{OA})\\ &= (x_1, y_1, z_1) + t(x_2 - x_1, y_2 - y_1, z_2 - z_1)\\ &= (x_1 + t(x_2 - x_1), y_1 + t(y_2 - y_1), z_1 + t(z_2 - z_1)).\end{aligned}$$

So the parametric equations are

$$x = x_1 + t(x_2 - x_1), y = y_1 + t(y_2 - y_1), z = z_1 + t(z_2 - z_1),$$
$$0 \leqslant t \leqslant 1.$$

Example 24

Parametrize the segment through $A(2, -1, 5)$ and $B(4, 3, 2)$. Use the equation to find the midpoint of the segment.

Solution

$$\begin{aligned}\mathbf{r}(t) &= (1-t)\overrightarrow{OA} + t\overrightarrow{OB}, 0 \leqslant t \leqslant 1\\ &= (1-t)(2, -1, 5) + t(4, 3, 2)\\ &= (2 + 2t, -1 + 4t, 5 - 3t)\end{aligned}$$

For the midpoint, $t = \frac{1}{2}$, and hence its coordinates are

$$\mathbf{r}\left(\tfrac{1}{2}\right) = \left(2 + 2\left(\tfrac{1}{2}\right), -1 + 4\left(\tfrac{1}{2}\right), 5 - 3\left(\tfrac{1}{2}\right)\right) = \left(3, 1, \tfrac{7}{2}\right).$$

Note: This method can be used to find points that divide the segment in any ratio: $\frac{2}{3}$ the way from A to B, etc.

Equivalently, the parametric equations can be used.

$$x = 2 + t(4 - 2), y = -1 + t(3 + 1), z = 5 + t(2 - 5)$$
$$x = 2 + 2t, y = -1 + 4t, z = 5 - 3t, 0 \leqslant t \leqslant 1$$

Symmetric (Cartesian) equations of lines

Another set of equations for a line is obtained by eliminating the parameter from the parametric equation.

If $a \neq 0$, $b \neq 0$ and $c \neq 0$, then the set of parametric equations can be re-arranged to yield the set of Cartesian (symmetric) equations:

$$\left.\begin{aligned} x - x_0 = ta &\Leftrightarrow \frac{x - x_0}{a} = t\\ y - y_0 = tb &\Leftrightarrow \frac{y - y_0}{b} = t\\ z - z_0 = tc &\Leftrightarrow \frac{z - z_0}{c} = t \end{aligned}\right\} \Leftrightarrow \frac{x - x_0}{a} = \frac{y - y_0}{b} = \frac{z - z_0}{c}$$

Notice that the coordinates (x_0, y_0, z_0) of the fixed point A on L appear in the numerators of the fractions, and that the components a, b and c of a direction vector appear in the denominators of these fractions.

Example 25

Find the Cartesian equations of the line through $A(3, -7, 4)$ and $B(1, -4, -1)$.

Solution

In order to use the Cartesian equation, we find the vector **v** parallel to the line. Since A and B are two points that lie on the line, the vector $\overrightarrow{AB}$ will suffice. Thus, we let

$$\mathbf{v} = \overrightarrow{AB} = (1 - 3)\mathbf{i} + (-4 + 7)\mathbf{j} + (-1 - 4)\mathbf{k} = -2\mathbf{i} + 3\mathbf{j} - 5\mathbf{k}.$$

If we use A as the fixed point, then the Cartesian equations are

$$\frac{x - 3}{-2} = \frac{y + 7}{3} = \frac{z - 4}{-5}.$$

Similarly, if we use B as the fixed point, then

$$\frac{x - 1}{-2} = \frac{y + 4}{3} = \frac{z + 1}{-5}.$$

Example 26

Let L be the line with Cartesian equations

$$\frac{x - 2}{3} = \frac{y + 1}{-2} = z - 4.$$

Find a set of parametric equations for L.

Solution

Since the numbers in the denominators are the components of a vector parallel to L, then

$$\mathbf{v} = 3\mathbf{i} - 2\mathbf{j} + \mathbf{k}.$$

The point $(2, -1, 4)$ lies on L.

Thus, a set of parametric equations of L is

$$x = 2 + 3t, y = -1 - 2t, z = 4 + t.$$

A vector equation would be

$$\mathbf{r} = (2, -1, 4) + t(3, -2, 1).$$

Note: If any of the components a, b or c is zero, then the Cartesian equations are written in a mixed form. For example, if $c = 0$, then we write

$$\frac{x - x_0}{a} = \frac{y - y_0}{b}, z = z_0.$$

For example, the Cartesian set of equations for a line parallel to $2\mathbf{i} - 3\mathbf{j}$ through the point $(2, 1, -3)$ is

$$\frac{x - 2}{2} = \frac{y - 1}{-3}, z = -3.$$

Intersecting, parallel and skew straight lines

In the plane, lines can coincide, intersect or be parallel. This is not necessarily so in space. In addition to the three cases above, there is the case of skew straight lines. Although these lines are not parallel, they do not intersect either. They lie in different planes.

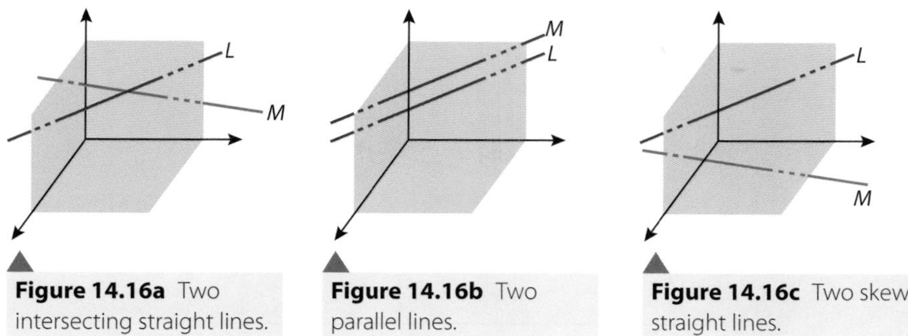

Figure 14.16a Two intersecting straight lines.

Figure 14.16b Two parallel lines.

Figure 14.16c Two skew straight lines.

How do we know whether two lines are parallel?

If the 'direction' vectors are parallel, then the lines are. Check to see if one of the vectors is a scalar multiple of the other. Alternatively, you can find the angle between them, and if it is 0° or 180°, the lines are either parallel or coincident. The case for coincidence is always there, and you need to check it by examining a point on one of the lines to see whether it is also on the other line.

Example 27

Show that the following two lines are parallel.

$$L_1: x = 2 - 3t, y = t, z = -1 + 2t$$

$$L_2: x = 1 + 6s, y = 2 - 2s, z = 2 - 4s$$

Solution

Let $\mathbf{l}_1$ be the vector parallel to L_1 and $\mathbf{l}_2$ be the vector parallel to L_2.

$$\mathbf{l}_1 = -3\mathbf{i} + \mathbf{j} + 2\mathbf{k} \text{ and } \mathbf{l}_2 = 6\mathbf{i} - 2\mathbf{j} - 4\mathbf{k}.$$

Now you can easily see that $\mathbf{l}_2 = -2\mathbf{l}_1$, and hence the vectors are parallel.

To check whether the lines coincide, we examine the point $(2, 0, -1)$, which is on the first line, and see whether it lies on the second line too.

If we choose $y = 0$, then $0 = 2 - 2s$, so $s = 1$; and when we substitute $s = 1$ into $x = 1 + 6s$ we find out that x must be 7 in order for the point $(2, 0, -1)$ to be on L_2. Therefore, the lines cannot intersect, and their 'direction' vectors are parallel, so they must be parallel.

Are the lines intersecting or skew?

If the direction vectors are not parallel, the lines either intersect or are skew. For the purposes of this course, the method starts by examining whether the lines intersect. If they do, we can find the coordinates of the point of intersection; if they do not intersect, we cannot find the coordinates of the point of intersection. Finding the coordinates of the point of intersection is a straightforward method that you already know: solving systems of equations. This can best be explained with an example.

The vector equation of a line has an interesting application in proving a result that you already know – the condition for two lines to be perpendicular.

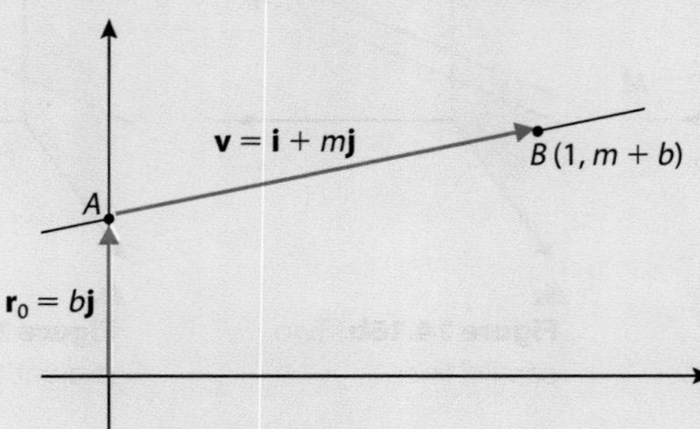

Consider the slope-intercept form of the equation of the line is $y = mx + b$. We can think of the fixed point on the line to be its y-intercept, i.e. $\mathbf{r}_0 = b\mathbf{j}$, and another point $(1, m + b)$ on the line $\Rightarrow \mathbf{r} = \mathbf{i} + (m + b)\mathbf{j}$. So, the direction vector of the line is $\mathbf{v} = \mathbf{r} - \mathbf{r}_0 = (\mathbf{i} + (m + b)\mathbf{j}) - b\mathbf{j} = \mathbf{i} + m\mathbf{j}$.

Now, if you have two lines with slopes m_1 and m_2, their direction vectors can be written as $\mathbf{v}_1 = \mathbf{i} + m_1\mathbf{j}$ and $\mathbf{v}_2 = \mathbf{i} + m_2\mathbf{j}$. For the two lines to be perpendicular, their direction vectors will also be perpendicular, and hence

$$\mathbf{v}_1 \cdot \mathbf{v}_2 = (\mathbf{i} + m_1\mathbf{j}) \cdot (\mathbf{i} + m_2\mathbf{j}) = 1 + m_1 m_2 = 0$$
$$\Rightarrow m_1 m_2 = -1.$$

Example 28

The lines L_1 and L_2 have the following equations:

$$L_1 : x = 1 + 4t, y = 5 - 4t, z = -1 + 5t$$
$$L_2 : x = 2 + 8s, y = 4 - 3s, z = 5 + s$$

Show that the lines are skew.

Solution

We first examine whether the lines are parallel. Since the vector parallel to L_1 is $\mathbf{l}_1 = (4, -4, 5)$ and the vector parallel to L_2 is $\mathbf{l}_2 = (8, -3, 1)$, they are not scalar multiples of each other and the vectors and consequently the lines are not parallel.

For the lines to intersect, there should be some point $M(x_0, y_0, z_0)$ which satisfies the equations of both lines for some values of t and s. That is,

$$x_0 = 1 + 4t = 2 + 8s; y_0 = 5 - 4t = 4 - 3s; z_0 = -1 + 5t = 5 + s.$$

This leads to a set of three simultaneous equations in two unknowns: s and t.

By solving the first two equations:

$$\left.\begin{aligned} 1 + 4t &= 2 + 8s \\ 5 - 4t &= 4 - 3s \end{aligned}\right\} \Rightarrow 6 = 6 + 5s \Rightarrow s = 0, t = \tfrac{1}{4}$$

For the system to be consistent, these values must satisfy the third equation, i.e. $-1 + \frac{5}{4} = 5 + 0$, which is false. Hence, the system is inconsistent and the lines are skew.

Example 29

The lines L_1 and L_2 have the following equations:

$$L_1 : x = 1 + 2t, y = 3 - 4t, z = -2 + 4t$$
$$L_2 : x = 4 + 3s, y = 4 + s, z = -4 - 2s$$

Show that the lines intersect.

Solution

We first examine whether the lines are parallel. Since the vector parallel to L_1 is $\mathbf{l}_1 = (2, -4, 4)$ and the vector parallel to L_2 is $\mathbf{l}_2 = (3, 1, -2)$, they are not scalar multiples of each other and the vectors and consequently the lines are not parallel.

For the lines to intersect, there should be some point $M(x_0, y_0, z_0)$ which satisfies the equations of both lines for some values of t and s. That is,

$$x_0 = 1 + 2t = 4 + 3s; y_0 = 3 - 4t = 4 + s; z_0 = -2 + 4t = -4 - 2s.$$

This leads to a set of three simultaneous equations in two unknowns: s and t.

By solving the first two equations:

$$\left.\begin{aligned} 1 + 2t &= 4 + 3s \\ 3 - 4t &= 4 + s \end{aligned}\right\} \Rightarrow 5 = 12 + 7s \Rightarrow s = -1, t = 0$$

For the system to be consistent, these values must satisfy the third equation, i.e. $-2 + 4(0) = -4 - 2(-1) \Rightarrow -2 = -2$, which is a correct statement. Hence, the two lines intersect.

The point of intersection can be found through substitution of the value of the parameter into the corresponding line equation:

$$L_1: (1, 3, -2) \text{ and } L_2: (4 - 3, 4 - 1, -4 - 2(-1)) = (1, 3, -2)$$

Note: In vector form, finding the point of intersection, if it exists, follows a similar approach. For example, the vector equations of lines L_1 and L_2 are

L_1: $\mathbf{r} = (1, 3, -2) + t(2, -4, 4)$

L_2: $\mathbf{r} = (4, 4, -4) + s(3, 1, -2)$

The condition of intersection is therefore

$(1, 3, -2) + t(2, -4, 4)$
$= (4, 4, -4) + s(3, 1, -2)$,

which leads to the same conclusion as in the case of parametric equations.

Application of lines to motion

The vector form of the equation of a line in space is more revealing when we think of the line as the path of an object, placed in an appropriate coordinate system and starting at position $A(x_0, y_0, z_0)$ and moving in the direction of $\mathbf{v}$.

Generally speaking, you find an object at an initial location A, represented by $\mathbf{r}_0$. The object moves on its path with a velocity vector $\mathbf{v} = (a, b, c)$. The object's position at any point in time after the start can then be described by $\mathbf{r} = \mathbf{r}_0 + t\mathbf{v}$.

Figure 14.17

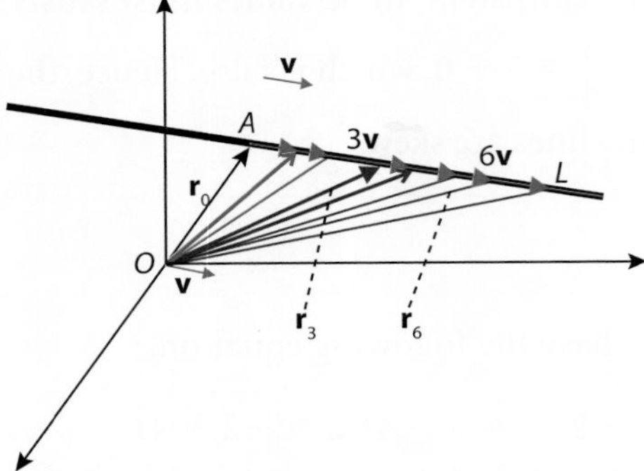

Assuming the unit of time is seconds, the equation tells us that for every second, the object moves a units in the x direction, b in the y direction and c in the z direction. So, for example, after 2 seconds you find the object at $\mathbf{r} = \mathbf{r}_0 + 2\mathbf{v}$.

The speed of the object is then $|\mathbf{v}|$ in the $\mathbf{v}$ direction.

In general, we can write the vector equation in a slightly modified form.

$$\mathbf{r}(t) = \mathbf{r}_0 + t\mathbf{v}$$

$$= \underbrace{\mathbf{r}_0}_{\text{initial position}} + \underbrace{t}_{\text{time}} \cdot \underbrace{|\mathbf{v}|}_{\text{speed}} \cdot \underbrace{\frac{\mathbf{v}}{|\mathbf{v}|}}_{\text{direction}}$$

In other words, the position of an object at time t is the *initial position* plus its *rate* × *time* (distance moved) in the *direction* $\mathbf{u} = \frac{\mathbf{v}}{|\mathbf{v}|}$ of its straight-line motion.

Example 30

A model plane is to fly directly from a platform at a reference point (2, 1, 1) toward a point (5, 5, 6) at a speed of 60 m/min. What is the position of the plane (to the nearest metre) after 10 minutes?

Solution

The unit vector in the direction of the flight is $\mathbf{u} = \frac{3}{5\sqrt{2}}\mathbf{i} + \frac{4}{5\sqrt{2}}\mathbf{j} + \frac{5}{5\sqrt{2}}\mathbf{k}$.

The position of the plane at any time t is

$$\mathbf{r}(t)_0 = \mathbf{r}_0 + t(\mathit{speed})(\mathbf{u})$$

$$= (2\mathbf{i} + \mathbf{j} + \mathbf{k}) + (10)(60)\left(\frac{3}{5\sqrt{2}}\mathbf{i} + \frac{4}{5\sqrt{2}}\mathbf{j} + \frac{5}{5\sqrt{2}}\mathbf{k}\right)$$

$$= (2\mathbf{i} + \mathbf{j} + \mathbf{k}) + \left(\frac{360}{\sqrt{2}}\mathbf{i} + \frac{480}{\sqrt{2}}\mathbf{j} + \frac{600}{\sqrt{2}}\mathbf{k}\right).$$

So, the plane is approximately at (257, 340, 425).

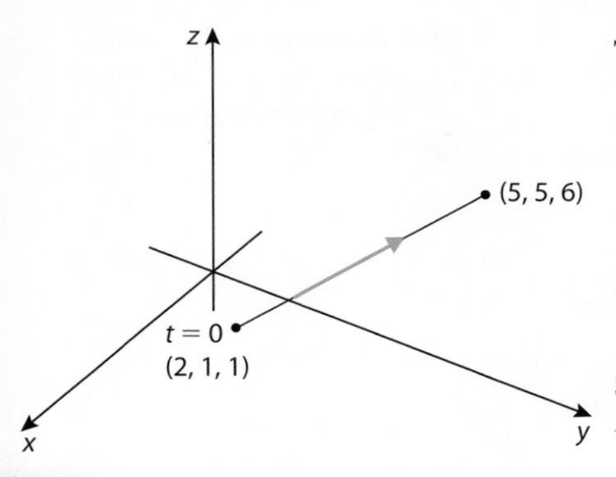

Example 31

An object is moving in the plane of an appropriately fitted coordinate system such that its position is given by

$$\mathbf{r} = (3, 1) + t(-2, 3),$$

where t stands for time in hours after start and distances are measured in km.

a) Find the initial position of the object.
b) Show the position of the object on a graph at start, 1 hour and 3 hours after start.
c) Find the velocity and speed of the object.

Solution

a) Initial position is when $t = 0$. This is the point (3, 1).
b) See graph.

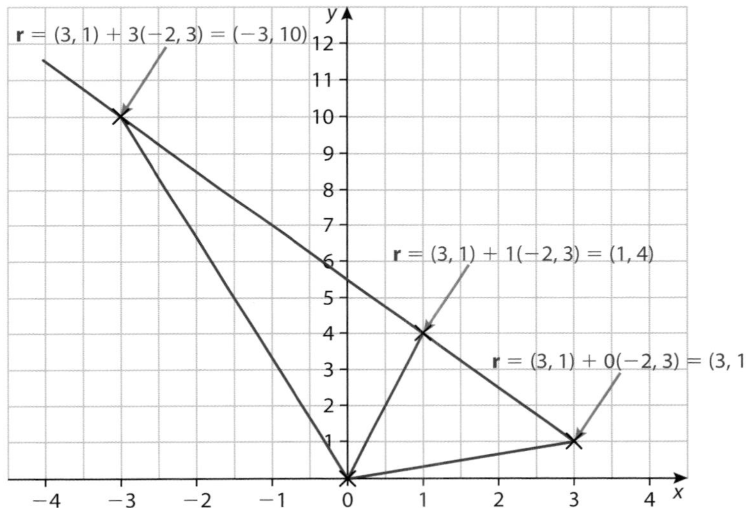

c) The velocity vector is $\mathbf{v} = (-2, 3)$, which means that every hour the object moves 2 units west and 3 units north.

   The speed is $|\mathbf{v}| = \sqrt{(-2)^2 + 3^2} = \sqrt{13}$ km/h.

   We can also express the velocity as $\sqrt{13}$ km/h in the direction of $(-2, 3)$.

Note: We can also express the direction in terms of the unit vector in the direction of $\mathbf{v}$ instead. That is, we can say that the speed is $\sqrt{13}$ km/h in the direction of $\left(\frac{-2}{\sqrt{13}}, \frac{3}{\sqrt{13}}\right)$, or, equivalently, at an angle of $\cos^{-1}\left(\frac{-2}{\sqrt{13}}\right) \approx 124°$ to the positive x-direction.

Example 32

At 12:00 midday a plane A is passing in the vicinity of an airport at a height of 12 km and a speed of 800 km/h. The direction of the plane is (4, 3, 0). [Consider that (1, 0, 0) is a displacement of 1 km due east, (0, 1, 0) due north, and (0, 0, 1) is an altitude of 1 km.]

a) Using the airport as the origin, find the position vector **r** of the plane t hours after midday.
b) Find the position of the plane 1 hour after midday.
c) Another plane B is heading towards the airport with velocity vector $(-300, -400, 0)$ from a location $(600, 480, 12)$. Is there a danger of collision?

Solution

a) The position vector at midday is $(0, 0, 12)$. The direction of the velocity vector is given by the unit vector $\frac{1}{5}(4, 3, 0)$. So, the velocity vector of this plane is $800 \cdot \frac{1}{5}(4, 3, 0) = (640, 480, 0)$.
The position vector of the plane is $\mathbf{r} = (0, 0, 12) + t(640, 480, 0)$.
b) $\mathbf{r} = (0, 0, 12) + (640, 480, 0) = (640, 480, 12)$
c) A collision can happen if the two planes pass the same point at the same time.
The position vector for the second plane is $\mathbf{r} = (600, 480, 12) + t(-300, -400, 0)$.
If the two paths intersect, they may intersect at instances corresponding to t_1 and t_2 and they should have the same position, i.e.

$$(0, 0, 12) + t_1(640, 480, 0) = (600, 480, 12) + t_2(-300, -400, 0).$$

This gives rise to a set of three equations in two variables:

$$\left.\begin{aligned} 640t_1 &= 600 - 300t_2 \\ 480t_1 &= 480 - 400t_2 \\ 12 &= 12 \end{aligned}\right\}$$

Solving the system of equations simultaneously will give $t_1 = \frac{6}{7}$ and $t_2 = \frac{6}{35}$.
This means that the planes' paths will cross at $(548.57, 411.43, 12)$. There is no collision though because plane A will pass that point at 12:51 while plane B will pass this point at 12.10!

Distance from a point to a line (optional)

2-space

Theorem: If the equation of a line l is written in the form $ax + by + c = 0$, then the distance from a point $P_0(x_0, y_0)$ to the line l is given by

$$d = \frac{|ax_0 + by_0 + c|}{\sqrt{a^2 + b^2}}.$$

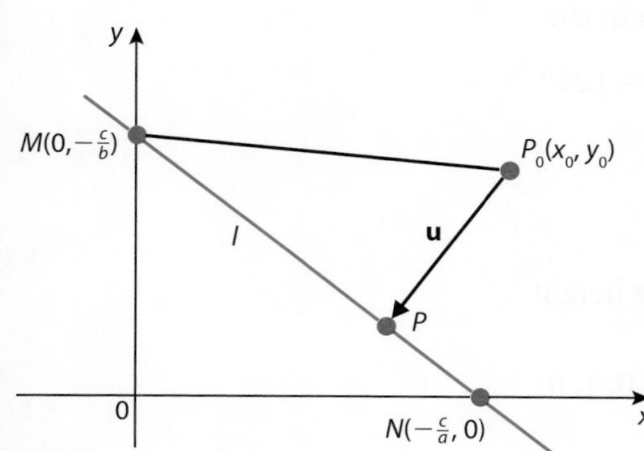

There are several methods of proving this theorem. We will follow a vector approach, leaving some other interesting methods for the website.

The x- and y-intercepts of the line l are

$$N\left(-\frac{c}{a}, 0\right) \text{ and } M\left(0, -\frac{c}{b}\right).$$

So, a vector parallel to l can be any vector in the direction of $\overrightarrow{NM} = \left(\frac{c}{a}, -\frac{c}{b}\right)$.

For convenience we will consider the vector $\mathbf{L} = \left(\frac{1}{a}, -\frac{1}{b}\right)$.

Consider a vector in the direction of $\overrightarrow{P_0P}$ perpendicular to l. Consider vector $\mathbf{u} = (a, b)$ which is perpendicular to l because $\mathbf{d} \cdot \mathbf{L} = 0$ and hence parallel to $\overrightarrow{P_0P}$, then in triangle MPP_0, the distance $|\overrightarrow{P_0P}|$ is

$$|\overrightarrow{MP_0}|\cos(MP_0P).$$

$$\overrightarrow{MP_0} = \left(x_0, y_0 + \frac{c}{b}\right)$$

$$|\overrightarrow{P_0P}| = \left||\overrightarrow{MP_0}| \cdot \cos M\hat{P_0}P\right| = \left|\frac{|\overrightarrow{MP_0}| \cdot \overrightarrow{MP_0} \cdot \overrightarrow{P_0P}}{|\overrightarrow{MP_0}| \cdot |\overrightarrow{P_0P}|}\right| = \left|\frac{\overrightarrow{MP_0} \cdot \mathbf{u}}{|\mathbf{u}|}\right|$$

$$= \left|\frac{\left(x_0, y_0 + \frac{c}{b}\right) \cdot (a, b)}{\sqrt{a^2 + b^2}}\right| = \frac{|ax_0 + by_0 + c|}{\sqrt{a^2 + b^2}}$$

3-space

$$d = |\overrightarrow{AP_0}| \cdot \sin(\theta) = |\overrightarrow{AP_0}| \cdot \frac{|\mathbf{L} \times \overrightarrow{AP_0}|}{|\mathbf{L}| \cdot |\overrightarrow{AP_0}|} = \frac{|\mathbf{L} \times \overrightarrow{AP_0}|}{|\mathbf{L}|}$$

where A is any point on line l and $\mathbf{L}$ is a vector parallel to l.

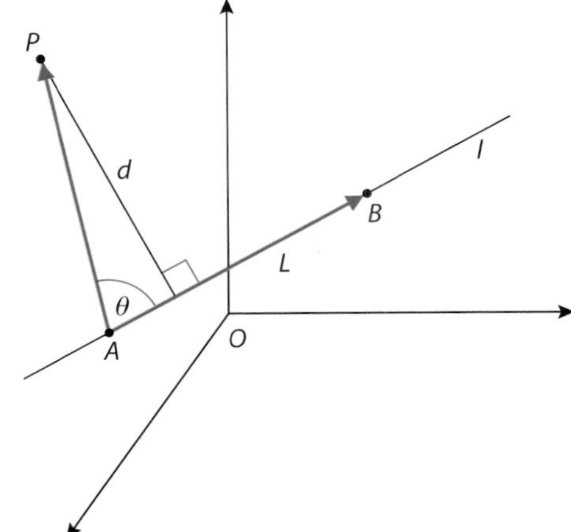

Example 33

Find the distance from the point $(1, 3)$ to the line with equation

$$2x - y = 7.$$

Solution

The equation can be written as $2x - y - 7 = 0$ and hence the distance is

$$d = \frac{|2(1) - 3 - 7|}{\sqrt{2^2 + 1}} = \frac{8}{\sqrt{5}}.$$

Example 34

Find the distance from the point $P(8, 1, -3)$ to the line containing $M(3, 0, 6)$ and $N(5, -2, 7)$.

Solution

We have:

$$\overrightarrow{MN} = (2, -2, 1) \text{ and } \overrightarrow{MP} = (5, 1, -9) \Rightarrow$$

$$d = \frac{|(2, -2, 1) \times (5, 1, -9)|}{|(2, -2, 1)|} = \frac{|(17, 23, 12)|}{\sqrt{2^2 + (-2)^2 + 1}}$$

$$= \frac{\sqrt{17^2 + 23^2 + 12^2}}{3} = \frac{\sqrt{962}}{3}$$

Distance between two skew straight lines

In the diagram below, two skew straight lines L_1 and L_2 are given with direction vectors $\mathbf{v}_1$ and $\mathbf{v}_2$ respectively. We need to find the distance d, defined as the length of the 'common perpendicular', between them.

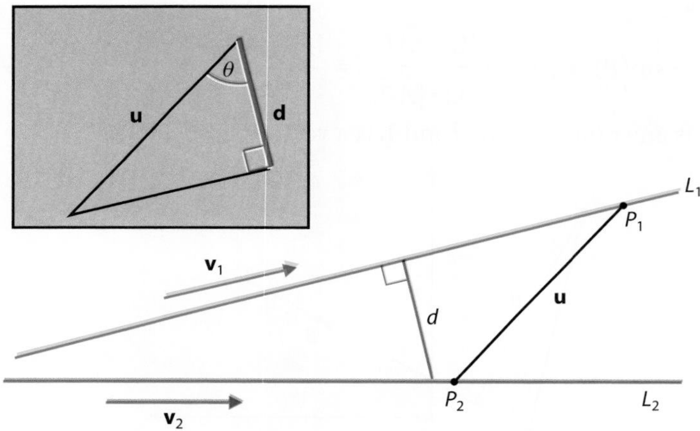

Consider any two fixed points P_1 on L_1 and P_2 on L_2. The distance d is the length of the orthogonal projection of vector $\mathbf{u} = \overrightarrow{P_1P_2}$ on a direction perpendicular to both $\mathbf{v}_1$ and $\mathbf{v}_2$. This direction can be determined by $\mathbf{v}_1 \times \mathbf{v}_2$. So,

$$d = \left| |\overrightarrow{P_1P_2}| \cos\theta \right| = \left| |\overrightarrow{P_1P_2}| \frac{\overrightarrow{P_1P_2} \cdot (\mathbf{v}_1 \times \mathbf{v}_2)}{|\overrightarrow{P_1P_2}||\mathbf{v}_1 \times \mathbf{v}_2|} \right| = \left| \frac{\overrightarrow{P_1P_2} \cdot (\mathbf{v}_1 \times \mathbf{v}_2)}{|\mathbf{v}_1 \times \mathbf{v}_2|} \right|.$$

Example 35

Find the distance between the following skew lines:

$$L_1: \mathbf{r} = (2, 3, 1) + t(1, 2, -3);\ L_2: \mathbf{r} = 4\mathbf{i} + 2\mathbf{j} + s(3\mathbf{i} - \mathbf{j} + \mathbf{k})$$

Solution

The two fixed points could be taken as $(2, 3, 1)$ and $(4, 2, 0)$ while the vectors are $\mathbf{v}_1 = (1, 2, -3)$ and $\mathbf{v}_2 = (3, -1, 1)$.

$$d = \left| \frac{(2, -1, -1)(-1, -10, -7)}{|(-1, -10, -7)|} \right|$$

$$= \left| \frac{-2 + 10 + 7}{\sqrt{1 + 100 + 49}} \right| = \frac{15}{\sqrt{150}} = \frac{\sqrt{6}}{2}$$

Note: The minimum distance could be found using other methods too. One of them would be to consider the line going from any point on L_1 to any point on L_2. This will give a parametric equation in s and t. Then considering that this line will be perpendicular to both L_1 and L_2, i.e. $\mathbf{u} \cdot \mathbf{v}_1 = 0, \mathbf{u} \cdot \mathbf{v}_2 = 0$, enables us to set up a system of two equations that could be solved for s and t. Lastly, we get the distance between the points corresponding to the specific values we just established.

Exercise 14.4

1 Find a vector equation, a set of parametric equations and a set of Cartesian equations of the line containing the point A and parallel to the vector $\mathbf{u}$.

a) $A(-1, 0, 2), \mathbf{u} = (1, 5, -4)$
b) $A(3, -1, 2), \mathbf{u} = (2, 5, -1)$
c) $A(1, -2, 6), \mathbf{u} = (3, 5, -11)$

2 Find all three forms of the equation of the line that passes through the points A and B.

a) $A(-1, 4, 2), B(7, 5, 0)$
b) $A(4, 2, -3), B(0, -2, 1)$
c) $A(1, 3, -3), B(5, 1, 2)$

3 a) Write the equation of the line through the points $(3, -2)$ and $(5, 1)$ in the form $\mathbf{r} = \mathbf{a} + t\mathbf{b}$.
b) Write the equation of the line through the points $(0, -2)$ and $(5, 0)$ in the form $\mathbf{r} = \mathbf{a} + t\mathbf{b}$.

4 The equation of a line in 2-space is given by $\mathbf{r} = (2, 1) + t(3, -2)$. Write the equation in the form $ax + by = c$.

5 Find the equation of a line through $(2, -3)$ that is parallel to the line with equation $\mathbf{r} = 3\mathbf{i} - 7\mathbf{j} + \lambda(4\mathbf{i} - 3\mathbf{j})$.

6 Find the equation of a line through $(-2, 1, 4)$ and parallel to the vector $3\mathbf{i} - 4\mathbf{j} + 7\mathbf{k}$.

7 In each of the following, find the point of intersection of the two given lines, and if they do not intersect, explain why.

a) $L_1: \mathbf{r} = (2, 2, 3) + t(1, 3, 1)$
$L_2: \mathbf{r} = (2, 3, 4) + t(1, 4, 2)$
b) $L_1: \mathbf{r} = (-1, 3, 1) + t(4, 1, 0)$
$L_2: \mathbf{r} = (-13, 1, 2) + t(12, 6, 3)$
c) $L_1: \mathbf{r} = (1, 3, 5) + t(7, 1, -3)$
$L_2: \mathbf{r} = (4, 6, 7) + t(-1, 0, 2)$

d) L_1: $\begin{pmatrix} x \\ y \\ z \end{pmatrix} = \begin{pmatrix} 3 \\ 4 \\ 6 \end{pmatrix} + t\begin{pmatrix} -2 \\ 1 \\ -1 \end{pmatrix}$

L_2: $\begin{pmatrix} x \\ y \\ z \end{pmatrix} = \begin{pmatrix} 5 \\ -2 \\ 7 \end{pmatrix} + s\begin{pmatrix} -4 \\ 2 \\ -2 \end{pmatrix}$

8 Find the vector and parametric equations of each line:

a) through the points $(2, -1)$ and $(3, 2)$

b) through the point $(2, -1)$ and parallel to the vector $\begin{pmatrix} -3 \\ 7 \end{pmatrix}$

c) through the point $(2, -1)$ and perpendicular to the vector $\begin{pmatrix} -3 \\ 7 \end{pmatrix}$

d) with y-intercept $(0, 2)$ and in the direction of $2\mathbf{i} - 4\mathbf{j}$

9 Consider the line with equation

$$\begin{pmatrix} x \\ y \\ z \end{pmatrix} = \begin{pmatrix} 3 \\ 4 \\ 6 \end{pmatrix} + t\begin{pmatrix} -2 \\ 1 \\ -1 \end{pmatrix}.$$

a) For what value of t does this line pass through the point $\left(0, \frac{11}{2}, \frac{9}{2}\right)$?

b) Does the point $(-1, 4, 6)$ lie on this line?

c) For what value of m does the point $\left(\frac{1-2m}{2}, 2m, 3\right)$ lie on the given line?

10 Consider the following equations representing the paths of cars after starting time $t \geqslant 0$, where distances are measured in km and time in hours. For each car, determine

(i) starting position

(ii) the velocity vector

(iii) the speed.

a) $\mathbf{r} = (3, -4) + t\begin{pmatrix} 7 \\ 24 \end{pmatrix}$

b) $\begin{pmatrix} x \\ y \end{pmatrix} = \begin{pmatrix} -3 \\ 1 \end{pmatrix} + t\begin{pmatrix} 5 \\ -12 \end{pmatrix}$

c) $(x, y) = (5, -2) + t(24, -7)$

11 Find the velocity vector of each of the following racing cars taking part in the Paris–Dakar rally:

a) direction $\begin{pmatrix} -3 \\ 4 \end{pmatrix}$ with a speed of 160 km/h

b) direction $\begin{pmatrix} 12 \\ -5 \end{pmatrix}$ with a speed of 170 km/h

12 After leaving an intersection of roads located at 3 km east and 2 km north of a city, a car is moving towards a traffic light 7 km east and 5 km north of the city at a speed of 30 km/h. (Consider the city as the origin for an appropriate coordinate system.)

a) What is the velocity vector of the car?

b) Write down the equation of the position of the car after t hours.

c) When will the car reach the traffic light?

13 Consider the vectors $\mathbf{u} = (1, a, b)$, $\mathbf{v} = \mathbf{i} - 3\mathbf{j} + 2\mathbf{k}$ and $\mathbf{w} = -2\mathbf{i} + \mathbf{j} - \mathbf{k}$.

a) Find a and b so that $\mathbf{u}$ is perpendicular to both $\mathbf{v}$ and $\mathbf{w}$.

b) If O is the origin, P a point whose position vector is $\mathbf{v}$ and Q is with position vector $\mathbf{w}$, find the cosine of the angle between $\mathbf{v}$ and $\mathbf{w}$.

c) Hence, find the sine of the angle and use it to find the area of the triangle OPQ.

14 The triangle ABC has vertices at the points $A(-1, 2, 3)$, $B(-1, 3, 5)$ and $C(0, -1, 1)$.

a) Find the size of the angle θ between the vectors $\overrightarrow{AB}$ and $\overrightarrow{AC}$.

b) Hence, or otherwise, find the area of triangle ABC.

Let L_1 be the line parallel to $\overrightarrow{AB}$ which passes through $D(2, -1, 0)$, and L_2 be the line parallel to $\overrightarrow{AC}$ which passes through $E(-1, 1, 1)$.

c) (i) Find the equations of the lines L_1 and L_2.

(ii) Hence, show that L_1 and L_2 do not intersect.

15 Consider the points $A(1, 3, -17)$ and $B(6, -7, 8)$ which lie on the line l.

a) Find an equation of line l, giving the answer in parametric form.

b) The point P is on l such that $\overrightarrow{OP}$ is perpendicular to l. Find the coordinates of P.

16 a) Starting with the equation of a line in the form $mx + ny = p$, find a vector equation of the line.

b) (i) Starting with a vector equation of a line where $\mathbf{r} = \mathbf{r}_0 + t\mathbf{v}$, with $\mathbf{r}_0 = \begin{pmatrix} x_0 \\ y_0 \end{pmatrix}$ and $\mathbf{v} = \begin{pmatrix} a \\ b \end{pmatrix}$, find an equation of the line in the form $mx + ny = p$.

(ii) What is the relationship between the components of the direction vector $\mathbf{v} = \begin{pmatrix} a \\ b \end{pmatrix}$ and the slope of the line?

17 Find a parametrization for the line segment between points A and B in each of the following questions.

(i) $A(0, 0, 0)$, $B(1, 1, 3)$

(ii) $A(-1, 0, 1)$, $B(1, 1, -2)$

(iii) $A(1, 0, -1)$, $B(0, 3, 0)$

18 Find a vector equation and a set of parametric equations of the line through the point $(0, 2, 3)$ and parallel to the line $\mathbf{r} = (\mathbf{i} - 2\mathbf{j}) + 2t\mathbf{k}$.

19 Find a vector equation and a set of parametric equations of the line through the point $(1, 2, -1)$ and parallel to the line $\mathbf{r} = t(2\mathbf{i} - 3\mathbf{j} + \mathbf{k})$.

20 Find a vector equation and a set of parametric equations of the line through the origin and the point $A(x_0, y_0, z_0)$.

21 Find a vector equation and a set of parametric equations of the line through $(3, 2, -3)$ and perpendicular to

a) the xz-plane

b) the yz-plane.

22 Write a set of symmetric equations for the line through the origin and the point $A(x_0, y_0, z_0)$, $x_0, y_0, z_0 \neq 0$.

In questions 23–29, determine whether the lines l_1 and l_2 are parallel, skew or intersecting. If they intersect, find the coordinates of the point of intersection.

23 l_1: $x - 3 = 1 - y = \dfrac{z-5}{2}$, l_2: $\mathbf{r} = \mathbf{i} + 4\mathbf{j} + 2\mathbf{k} + \lambda(\mathbf{j} + \mathbf{k})$

24 l_1: $\begin{cases} x = -1 + s \\ y = 2 - 3s \\ z = 1 + 2s \end{cases}$ l_2: $\begin{cases} x = 2 - 2m \\ y = -1 + 6m \\ z = -4m \end{cases}$

25 l_1: $\dfrac{x-3}{2} = \dfrac{1+y}{4} = 2 - z$, l_2: $\mathbf{r} = 3\mathbf{i} + 2\mathbf{j} - 2\mathbf{k} + \lambda(2\mathbf{i} + \mathbf{j} + 2\mathbf{k})$

26 l_1: $x - 1 = \dfrac{y-1}{3} = \dfrac{z+4}{2}$, l_2: $1 - x = -1 - y = \dfrac{z}{2}$

27 l_1: $\mathbf{r} = \mathbf{i} + 2\mathbf{j} + \lambda(-6\mathbf{i} + 9\mathbf{j} - 3\mathbf{k})$, l_2: $\mathbf{r} = 2\mathbf{i} + 3\mathbf{j} + m(2\mathbf{i} - 3\mathbf{j} + \mathbf{k})$

28 $\frac{x-2}{5} = y - 1 = \frac{z-2}{3}$ and $\frac{x+4}{3} = \frac{7-y}{3} = \frac{10-z}{4}$

29 $x = 1 + t, y = 2 - 2t, z = t + 5$ and $x = 2 + 2t, y = 5 - 9t, z = 2 + 6t$

30 Find the point on the line

$$\mathbf{r} = 2\mathbf{i} + 3\mathbf{j} + \mathbf{k} + t(-3\mathbf{i} + \mathbf{j} + \mathbf{k})$$

that is closest to the origin. (Hint: use the parametric form and the distance formula and minimize the distance using derivatives!)

31 Find the point on the line

$$\mathbf{r} = 4\mathbf{j} + 5\mathbf{k} + t(\mathbf{i} - 3\mathbf{j} + \mathbf{k})$$

that is closest to the origin.

32 Find the point on the line

$$\mathbf{r} = 5\mathbf{i} + 2\mathbf{j} + \mathbf{k} + t(\mathbf{i} - 3\mathbf{j} + \mathbf{k})$$

that is closest to the point $(-1, 4, 1)$.

14.5 Planes

To define/specify a plane is to identify it in a way that makes it unique. One way is to set up an equation in a frame that will identify every point that belongs to the plane. There are several ways of specifying a plane but we will only mention four of them here. The rest will be cases that we address in some problems later. For more helpful geometric concepts please refer to the book's website.

A plane can be defined

- by three non-collinear points
- by two intersecting straight lines
- to be perpendicular to a certain direction and at a specific distance from the origin (for example)
- by being drawn through a given point and perpendicular to a given direction.

A direction, for our purposes, can be defined by a vector. In the case of a plane, the vector determining the direction is perpendicular to the plane and is said to be **normal to the plane**.

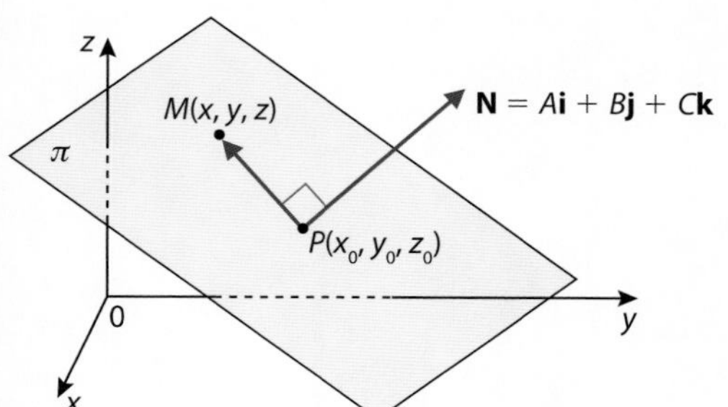

Equations of a plane

From the many ways of defining a plane above, the last two are mostly appropriate for deriving equations of a plane.

Cartesian (scalar) equation of a plane

Consider a plane π and a fixed point $P(x_0, y_0, z_0)$ on that plane. A vector $\mathbf{N} = A\mathbf{i} + B\mathbf{j} + C\mathbf{k}$, called the normal vector to the plane, is a vector perpendicular to the plane.

To find an equation for the plane, consider an arbitrary point $M(x, y, z)$ in space. Recalling that a line perpendicular to the plane is perpendicular to every line in the plane, we can conclude that for the point M to be on the plane, the vector **N** must be perpendicular to $\overrightarrow{PM}$.

Hence,

$$\mathbf{N} \cdot \overrightarrow{PM} = 0, \text{ but } \overrightarrow{PM} = (x - x_0)\mathbf{i} + (y - y_0)\mathbf{j} + (z - z_0)\mathbf{k}, \text{ and}$$

$$\mathbf{N} \cdot \overrightarrow{PM} = 0$$

$$\Leftrightarrow (A\mathbf{i} + B\mathbf{j} + C\mathbf{k}) \cdot ((x - x_0)\mathbf{i} + (y - y_0)\mathbf{j} + (z - z_0)\mathbf{k}) = 0$$

Using the scalar product definition this can be simplified to

$$A(x - x_0) + B(y - y_0) + C(z - z_0) = 0.$$

This is a Cartesian equation of a plane that passes through a point $P(x_0, y_0, z_0)$ and has a normal vector $\mathbf{N} = A\mathbf{i} + B\mathbf{j} + C\mathbf{k}$.

Note: If **N** is normal to a given plane, then any vector parallel to **N** will be normal to the plane. Suppose we have chosen 3**N** as our normal, then

$$3A(x - x_0) + 3B(y - y_0) + 3C(z - z_0) = 0$$

$$\Leftrightarrow A(x - x_0) + B(y - y_0) + C(z - z_0)$$

Specifically, the unit vector **n** in the same direction as **N** is of particular importance, as we will see soon.

Note: The above equation can be simplified further.
$A(x - x_0) + B(y - y_0) + C(z - z_0) = 0$
$\Leftrightarrow Ax + By + Cz = Ax_0 + By_0 + Cz_0$, and setting $Ax_0 + By_0 + Cz_0 = D$ will give us a more concise form of the equation,

$$Ax + By + Cz = D$$

which is similar to the equation of a line in the plane, i.e. $Ax + By = C$.

Note: In many sources, the equation of the plane is given in the form

$$Ax + By + Cz + D = 0.$$

This is the case when we set the quantity $Ax_0 + By_0 + Cz_0 = -D$. Each form has some advantage in using it. We will adhere to the previous form for reasons that will be clear in the following discussion.

Example 36

Write an equation for the plane that contains $(2, -3, 5)$ and has normal $\mathbf{N} = 2\mathbf{i} + \mathbf{j} - 3\mathbf{k}$.

Solution

A Cartesian equation for the plane is of the form:

$$A(x - x_0) + B(y - y_0) + C(z - z_0) = 0$$

$$\Rightarrow 2(x - 2) + (y + 3) - 3(z - 5) = 0$$

$$\Rightarrow 2x + y - 3z = -14$$

Alternatively, since $\mathbf{N} = 2\mathbf{i} + \mathbf{j} - 3\mathbf{k}$ is the normal to the plane, then $2x + y - 3z = D$, but the line contains the point $(2, -3, 5)$, and thus

$$2(2) - 3 - 3(5) = D \Rightarrow -14 = D.$$

And therefore

$$2x + y - 3z = -14 \text{ as before.}$$

Example 37

Show that every equation of the form $Ax + By + Cz = D$ with $A^2 + B^2 + C^2 \neq 0$ represents a plane in space.

Solution

The equation $Ax + By + Cz = D$ is a linear equation in 3 variables, x, y and z. This means that it has an infinite number of solutions, and hence we can be confident that there exist numbers x_0, y_0, z_0 such that $Ax_0 + By_0 + Cz_0 = D$.

Since the equation $Ax + By + Cz = D$ is also true, then

$$Ax + By + Cz = Ax_0 + By_0 + Cz_0 = D$$
$$\Leftrightarrow Ax + By + Cz - (Ax_0 + By_0 + Cz_0) = 0$$
$$\Leftrightarrow A(x - x_0) + B(y - y_0) + C(z - z_0) = 0$$

The last equation represents the equation of a plane through a fixed point $P(x_0, y_0, z_0)$ with a normal vector $\mathbf{N} = A\mathbf{i} + B\mathbf{j} + C\mathbf{k}$. The condition $A^2 + B^2 + C^2 \neq 0$ guarantees that $\mathbf{N} \neq \mathbf{0}$.

Vector equation of a plane

We can write the equation of the plane in vector notation. Using the same set up as before: the normal vector $\mathbf{N} = A\mathbf{i} + B\mathbf{j} + C\mathbf{k}$, a fixed point $P(x_0, y_0, z_0)$ with a position vector $\mathbf{r}_0 = x_0\mathbf{i} + y_0\mathbf{j} + z_0\mathbf{k}$, and an arbitrary point $M(x, y, z)$ with a position vector $\mathbf{r} = x\mathbf{i} + y\mathbf{i} + z\mathbf{k}$.

The equation $A(x - x_0) + B(y - y_0) + C(z - z_0) = 0$ can be interpreted as the scalar product $\mathbf{N} \cdot (\mathbf{r} - \mathbf{r}_0) = 0$ which you can also see in the diagram. The normal $\mathbf{N}$ is perpendicular to $PM = \mathbf{r} - \mathbf{r}_0$ and hence their dot product must be zero. Using the distributive property of the scalar product, we have

$$\mathbf{N} \cdot (\mathbf{r} - \mathbf{r}_0) = 0 \Leftrightarrow \mathbf{N} \cdot \mathbf{r} - \mathbf{N} \cdot r_0 = 0$$
$$\Leftrightarrow \mathbf{N} \cdot \mathbf{r} = \mathbf{N} \cdot \mathbf{r}_0$$

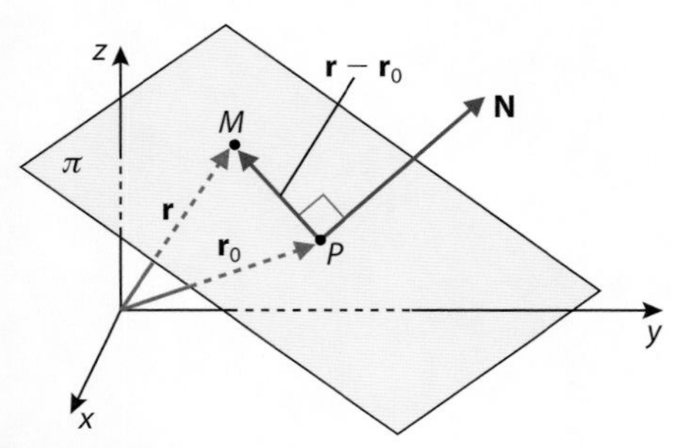

This is one form of the vector equation of a plane that passes through a point with position vector $\mathbf{r}_0$ and has a normal $\mathbf{N}$.

Note: Notice here that $\mathbf{N} \cdot \mathbf{r} = Ax + By + Cz$ and $\mathbf{N} \cdot \mathbf{r}_0 = Ax_0 + By_0 + Cz_0 = D$, which shows that $\mathbf{N} \cdot \mathbf{r} = \mathbf{N} \cdot \mathbf{r}_0$ is another way of stating $Ax + By + Cz = D$.

Unit vector equation of a plane

The diagram shows vector **N** as drawn from the origin O, along with the position vectors **r** and $\mathbf{r}_0$.

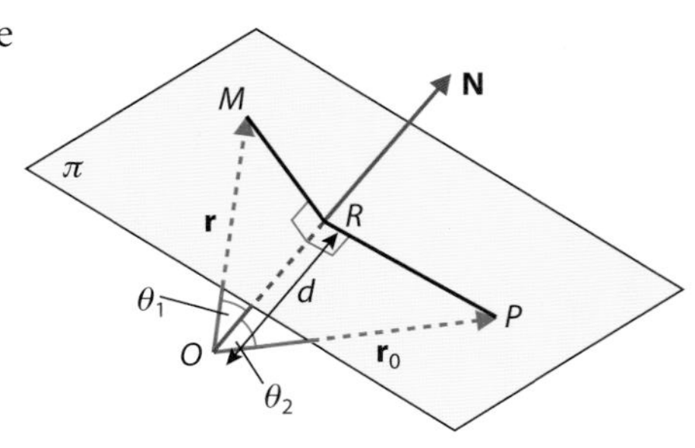

The vector equation $\mathbf{N} \cdot \mathbf{r} = \mathbf{N} \cdot \mathbf{r}_0$ can be investigated further.

$$\mathbf{N} \cdot \mathbf{r} = |\mathbf{N}||\mathbf{r}|\cos \theta_1 = |\mathbf{N}|OR = |\mathbf{N}|d$$

where OR $= d$ is the distance from the origin to the plane.

Also

$$\mathbf{N} \cdot \mathbf{r}_0 = |\mathbf{N}||\mathbf{r}_0|\cos \theta_2 = |\mathbf{N}|d.$$

In both cases, the result is of course the same. Both sides of the equation $\mathbf{N} \cdot \mathbf{r} = \mathbf{N} \cdot \mathbf{r}_0$ are equal to the same value: the magnitude of the normal multiplied by the distance from the origin.

Furthermore, if we divide each side by $|\mathbf{N}|$, we get the distance from the origin to the plane.

That is,

$$\mathbf{N} \cdot \mathbf{r} = |\mathbf{N}|d \Rightarrow d = \frac{\mathbf{N} \cdot \mathbf{r}}{|\mathbf{N}|}$$

as well as

$$\mathbf{N} \cdot \mathbf{r}_0 = |\mathbf{N}|d \Rightarrow d = \frac{\mathbf{N} \cdot \mathbf{r}_0}{|\mathbf{N}|}.$$

This last result gives us the basis for forming a new vector equation of the plane in terms of a unit vector perpendicular to it.

Let us call the unit vector normal to the plane **n**. So, using the results just established, we can write

$d = \dfrac{\mathbf{N} \cdot \mathbf{r}}{|\mathbf{N}|} = \dfrac{\mathbf{N} \cdot \mathbf{r}_0}{|\mathbf{N}|}$, which in turn can be simplified to

$\dfrac{\mathbf{N}}{|\mathbf{N}|} \cdot \mathbf{r} = \dfrac{\mathbf{N}}{|\mathbf{N}|} \cdot \mathbf{r}_0$, and since $\dfrac{\mathbf{N}}{|\mathbf{N}|} = \mathbf{n}$, then obviously

$$\mathbf{n} \cdot \mathbf{r} = \mathbf{n} \cdot \mathbf{r}_0 = d.$$

This equation is very practical when we need to find the distance from the origin to the plane. The distance from the origin to a plane is the scalar product between the unit normal and the position vector of any point on the plane.

This will be shown in the examples below.

Example 38

Write a vector equation for the plane that contains $(2, -3, 5)$ and has normal $\mathbf{N} = 2\mathbf{i} + \mathbf{j} - 3\mathbf{k}$.

Solution

We apply the results of the previous discussion:

$$\mathbf{N} \cdot \mathbf{r} = \mathbf{N} \cdot \mathbf{r}_0$$
$$\Rightarrow (2\mathbf{i} + \mathbf{j} - 3\mathbf{k}) \cdot \mathbf{r} = (2\mathbf{i} + \mathbf{j} - 3\mathbf{k}) \cdot (2\mathbf{i} - 3\mathbf{j} + 5\mathbf{k})$$
$$\Rightarrow (2\mathbf{i} + \mathbf{j} - 3\mathbf{k}) \cdot \mathbf{r} = -14$$

Notice that this result can easily transfer into Cartesian form by expanding the scalar product on the left.

$$(2\mathbf{i} + \mathbf{j} - 3\mathbf{k}) \cdot \mathbf{r} = -14$$
$$\Rightarrow (2\mathbf{i} + \mathbf{j} - 3\mathbf{k}) \cdot (x\mathbf{i} + y\mathbf{j} + z\mathbf{k}) = -14$$
$$\Rightarrow 2x + y - 3z = -14$$

Example 39

Show that the line l with equation $\mathbf{r} = 2\mathbf{i} - \mathbf{j} + 5\mathbf{k} + k(3\mathbf{i} + 2\mathbf{j} - 2\mathbf{k})$ is parallel to the plane P whose equation is $\mathbf{r} \cdot (2\mathbf{i} - 2\mathbf{j} + \mathbf{k}) = -3$ and find the distance between them.

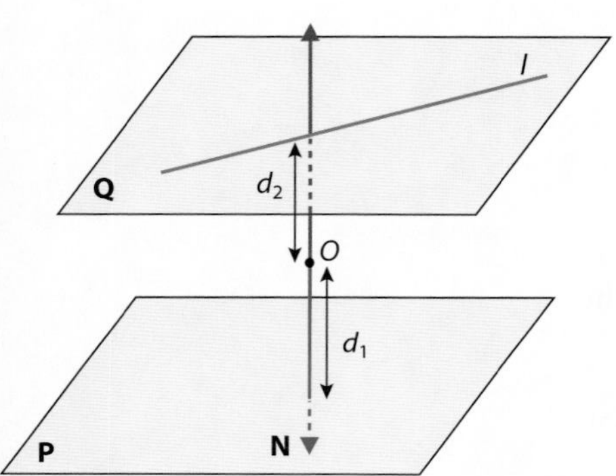

Solution

Since $\mathbf{v} = 3\mathbf{i} + 2\mathbf{j} - 2\mathbf{k}$ is the direction vector of the line l, and since this vector is perpendicular to $\mathbf{N} = 2\mathbf{i} - 2\mathbf{j} + \mathbf{k}$, the normal to P, as $(3\mathbf{i} + 2\mathbf{j} - 2\mathbf{k}) \cdot (2\mathbf{i} - 2\mathbf{j} + \mathbf{k}) = 6 - 4 - 2 = 0$, then the line l must be parallel to plane P.

To find the distance between the line and the plane P, we may find the distance between a point on l, $(2, -1, 5)$ for example, and plane P. One way would be to consider a plane Q containing the given point and parallel to P. The equation of plane Q can be found using the last result:

$$\mathbf{N} \cdot \mathbf{r} = \mathbf{N} \cdot \mathbf{r}_0$$
$$\Rightarrow (2\mathbf{i} - 2\mathbf{j} + \mathbf{k}) \cdot \mathbf{r} = (2\mathbf{i} - 2\mathbf{j} + \mathbf{k}) \cdot (2\mathbf{i} - \mathbf{j} + 5\mathbf{k})$$
$$\Rightarrow (2\mathbf{i} - 2\mathbf{j} + \mathbf{k}) \cdot \mathbf{r} = 11$$

The distance from the origin to P is given by

$$d = \left|\frac{\mathbf{N} \cdot \mathbf{r}_0}{|\mathbf{N}|}\right| = \frac{3}{\sqrt{4 + 4 + 1}} = 1$$

while the distance from the origin to Q is

$$d = \left|\frac{\mathbf{N} \cdot \mathbf{r}_0}{|\mathbf{N}|}\right| = \frac{11}{\sqrt{4 + 4 + 1}} = \frac{11}{3}.$$

The distance between the two planes is the sum of these two distances since the planes are on opposite sides of the origin (see note), and hence the required distance is $\frac{14}{3}$.

Note: The vector equation for P is $\mathbf{r} \cdot (2\mathbf{i} - 2\mathbf{j} + \mathbf{k}) = -3$, or $\mathbf{r} \cdot (-2\mathbf{i} + 2\mathbf{j} - \mathbf{k}) = 3$ and the equation for Q is $(2\mathbf{i} - 2\mathbf{j} + \mathbf{k}) \cdot \mathbf{r} = 11$. The normals to the two planes are opposite, and hence they are on opposite sides of the origin. If the two normals are in the same direction, then the distance between them will be the difference of the two distances from the origin.

Example 40

Show that the plane with vector equation $\mathbf{r} \cdot (2\mathbf{i} - 2\mathbf{j} + \mathbf{k}) = -3$ contains the line with equation $\mathbf{r} = \mathbf{i} + 3\mathbf{j} + \mathbf{k} + k(3\mathbf{i} + 2\mathbf{j} - 2\mathbf{k})$.

Solution

We have several methods available to us at this stage. One method is to check whether two points are common to the line and the plane. One point on the line is $(1, 3, 1)$. Since $(1, 3, 1) \cdot (2, -2, 1) = 2 - 6 + 1 = -3$, then the point is on the plane. Another point on the line can be found by choosing any value for k, say $k = 1$. Thus, another point has the position vector

$$\mathbf{r} = \mathbf{i} + 3\mathbf{j} + \mathbf{k} + 3\mathbf{i} + 2\mathbf{j} - 2\mathbf{k} = 4\mathbf{i} + 5\mathbf{j} - \mathbf{k}.$$

Since $(4\mathbf{i} + 5\mathbf{j} - \mathbf{k}) \cdot (2\mathbf{i} - 2\mathbf{j} + \mathbf{k}) = 8 - 10 - 1 = -3$, this point will also lie on the plane and therefore the plane will contain the whole line.

Another method would be to check only one point and prove that the line is parallel to the plane as in Example 39 above.

Example 41

Find the vector equation of the line through $(1, 2, 3)$ that is perpendicular to the plane with vector equation $\mathbf{r} \cdot (2\mathbf{i} - 2\mathbf{j} + \mathbf{k}) = -3$ and find their point of intersection.

Solution

A vector parallel to the required line must be parallel to the normal vector to the plane. Hence, a vector equation of the line is $\mathbf{r} = (1, 2, 3) + k(2, -2, 1)$.

To find the point of intersection, we consider any point on the line. Such a point would have the position vector $(1 + 2k, 2 - 2k, 3 + k)$. For this point to be on the plane, the following equation must be true:

$$(1 + 2k, 2 - 2k, 3 + k) \cdot (2, -2, 1) = -3;$$

i.e. $2 + 4k - 4 + 4k + 3 + k = -3$, so

$9k = -4$, and $k = -\frac{4}{9}$ giving the point of intersection of the line and the plane as

$$\left(1 + 2 \times -\tfrac{4}{9}, 2 - 2 \times -\tfrac{4}{9}, 3 - \tfrac{4}{9}\right) = \left(\tfrac{1}{9}, \tfrac{26}{9}, \tfrac{23}{9}\right).$$

Parametric form for the equation of a plane

We start this section with an example that demonstrates the following theorem:

Three coplanar vectors **u**, **v** *and* **w** *are given. If* **u** *and* **v** *are not parallel, then* **w** *can always be expressed as a linear combination of* **u** *and* **v**, *i.e. it is always possible to find two scalars s and t such that* $\mathbf{w} = s\mathbf{u} + t\mathbf{v}$. (The proof of the theorem is not included in this book.)

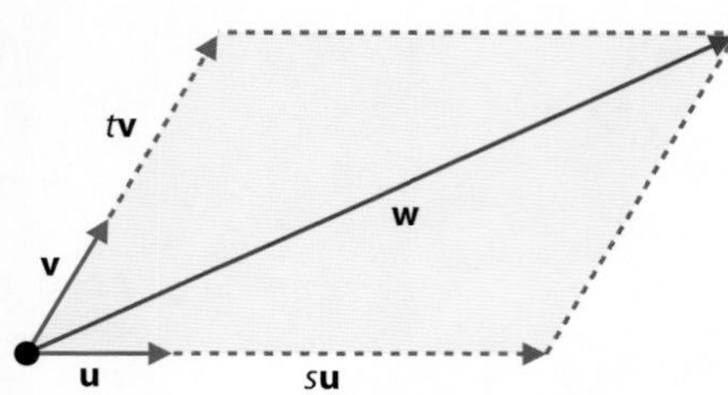

Using the diagram on the left, this means that it is always possible to construct a parallelogram whose diagonal is **w** and whose sides are the non-parallel vectors **u** and **v** or their multiples.

For example, given the two non-parallel vectors $\mathbf{u} = (1, 0, -2)$ and $\mathbf{v} = (3, 1, -9)$, then we can always find the scalars s and t so that vector $\mathbf{w} = (2, 1, -7)$ can be expressed as a linear combination of **u** and **v**. Thus,

$$(2, 1, -7) = s(1, 0, -2) + t(3, 1, -9).$$

To find s and t we solve the system of equations:

$$\begin{cases} s + 3t = 2 \\ t = 1 \\ -2s - 9t = -7 \end{cases}$$

This system is consistent and yields the solution $s = -1$ and $t = 1$. Thus,

$$\mathbf{w} = -\mathbf{u} + \mathbf{v}.$$

Now, consider the plane which is parallel to vectors **u** and **v** and which contains the point A whose position vector is $\mathbf{r}_0$. As the figure below shows, the two vectors determine the direction of the plane and A 'fixes' it in space. So, if M is any point in this plane then, according to the previous theorem, $\overrightarrow{AM} = s\mathbf{u} + t\mathbf{v}$ where s and t are two scalars.

If **r** is the position vector of M, then

$$\mathbf{r} = \mathbf{r}_0 + \overrightarrow{AM} = \mathbf{r}_0 + s\mathbf{u} + t\mathbf{v}.$$

Thus, any equation of the form $\mathbf{r} = \mathbf{r}_0 + s\mathbf{u} + t\mathbf{v}$, where s and t are independent scalars, represents the equation of a plane parallel to **u** and **v** and contains the point with position vector $\mathbf{r}_0$.

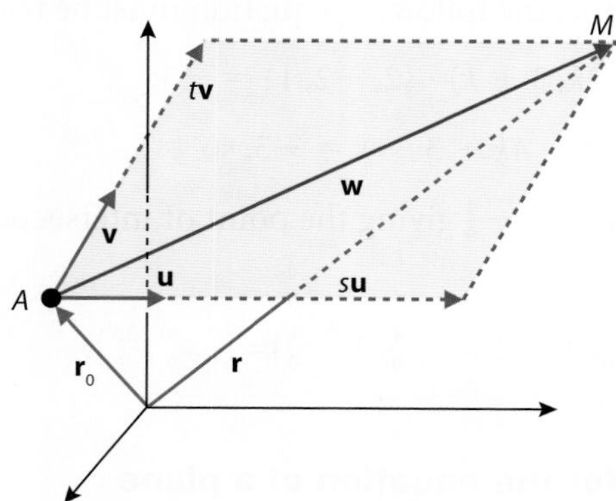

Note that this equation is not unique. This is because one can start at any other fixed point on the plane other than A and may choose any number of intersecting vectors in the plane other than **u** and **v**. The parametric form of the equation of the plane is seldom needed or used.

Example 42

Find an equation of a plane with normal $\mathbf{q} = 2\mathbf{i} - 3\mathbf{j} + \mathbf{k}$ and that contains the point $A(2, 1, 1)$. Use all forms you learned.

Solution

The Cartesian equation

Consider the equation:

$$A(x - x_0) + B(y - y_0) + C(z - z_0) = 0$$

$$\Rightarrow 2(x - 2) - 3(y - 1) + (z - 1) = 0$$

The equation would be: $2x - 3y + z - 2 = 0$.

Or, start the equation:

$$Ax + By + Cz = D$$

$$\Rightarrow 2x - 3y + z = D$$

Since the plane contains the point $(2, 1, 1)$, then

$2(2) - 3(1) + 1 = D$, and thus $D = 2$.

Vector equations

Finding the vector equation can also be achieved by applying:

$$\mathbf{N} \cdot \mathbf{r} = \mathbf{N} \cdot \mathbf{r}_0 \Rightarrow (2, -3, 1) \cdot (x, y, z) = (2, -3, 1) \cdot (2, 1, 1)$$

$$\Rightarrow (2, -3, 1) \cdot (x, y, z) = 2$$

Note: It is easy to transform the vector equation into Cartesian form by simply performing the dot product. The opposite is also true.

Parametric equation

A parametric equation of this plane is not as straightforward and may not be the most efficient way of doing this problem. However, for the sake of giving an example we present a way of doing it.

The parametric form requires that we have two vectors parallel to the plane. We may find the two vectors by considering that they have to be perpendicular to $(2, -3, 1)$. So, take a vector $(1, 1, z)$ and find z so that this vector is perpendicular to $(2, -3, 1)$:

$$\Rightarrow 2 - 3 + z = 0 \text{ and } z = 1$$

Do the same with $(1, 0, z)$, i.e. $2 + 0 + z = 0$ and $z = -2$. Therefore, two vectors that are perpendicular to $(2, -3, 1)$ are $(1, 1, 1)$ and $(1, 0, -2)$, and a parametric equation of the plane is

$$\mathbf{r} = \mathbf{r}_0 + s\mathbf{u} + t\mathbf{v} \Rightarrow \mathbf{r} = (2, 1, 1) + s(1, 1, 1) + t(1, 0, -2).$$

Observe that the choice of the vectors is arbitrary and hence the parametric form is not unique.

Example 43

Find the equation of the plane that contains the following three points:

$$A(1, 3, 0), B(-2, 1, 2) \text{ and } C(1, -2, -1).$$

Solution

Consider any point $M(x, y, z)$ on this plane. For this point to belong to the plane, the following vectors must be coplanar: $\overrightarrow{AM}$, $\overrightarrow{AB}$ and $\overrightarrow{AC}$.

This means that the parallelepiped with these vectors as edges is flat, i.e. with volume zero. Since we know that the volume of the parallelepiped is the absolute value of the scalar triple product, we equate that value to zero and get the equation. Here are the details:

$$\overrightarrow{AM} = (x - 1, y - 3, z), \overrightarrow{AB} = (-3, -2, 2), \overrightarrow{AC} = (0, -5, -1)$$

$$\Rightarrow \overrightarrow{AM} \cdot (\overrightarrow{AB} \times \overrightarrow{AC}) = \begin{vmatrix} x-1 & y-3 & z \\ -3 & -2 & 2 \\ 0 & -5 & -1 \end{vmatrix} = 0$$

$$\Rightarrow 3(4x - y + 5z - 1) = 0$$

So, the Cartesian equation of the plane is $4x - y + 5z - 1 = 0$.

We can also deduce the vector equation for this plane by writing it in scalar product form:

$$(4\mathbf{i} - \mathbf{j} + 5\mathbf{k}) \cdot (x\mathbf{i} + y\mathbf{j} + z\mathbf{k}) = 1.$$

The vector form could also be achieved if we think of the problem as a plane containing a fixed point and normal to a given vector.

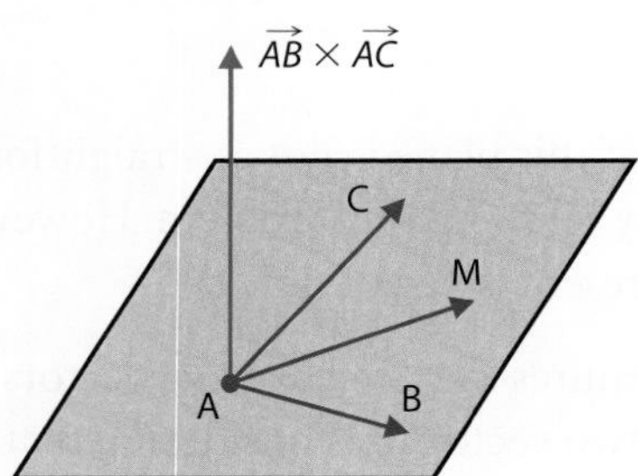

The normal can be found by computing the cross product of two vectors in the plane; in this case, we can take $\overrightarrow{AB}$ and $\overrightarrow{AC}$. So,

$$\overrightarrow{AB} \times \overrightarrow{AC} = \begin{vmatrix} \mathbf{i} & \mathbf{j} & \mathbf{k} \\ -3 & -2 & 2 \\ 0 & -5 & -1 \end{vmatrix} = 3(4\mathbf{i} - \mathbf{j} + 5\mathbf{k}).$$

The equation of the plane is then

$$(4\mathbf{i} - \mathbf{j} + 5\mathbf{k}) \cdot (x\mathbf{i} + y\mathbf{j} + z\mathbf{k}) = (4\mathbf{i} - \mathbf{j} + 5\mathbf{k}) \cdot (\mathbf{i} + 3\mathbf{j})$$

$$(4\mathbf{i} - \mathbf{j} + 5\mathbf{k}) \cdot (x\mathbf{i} + y\mathbf{j} + z\mathbf{k}) = 1.$$

This is the same as above.

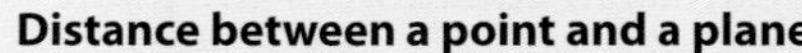

Distance between a point and a plane

The distance between a point $P(x_0, y_0, z_0)$ and a plane with equation $Ax + By + Cz = D$ is given by

$$d = \frac{|Ax_0 + By_0 + Cz_0 - D|}{\sqrt{A^2 + B^2 + C^2}}.$$

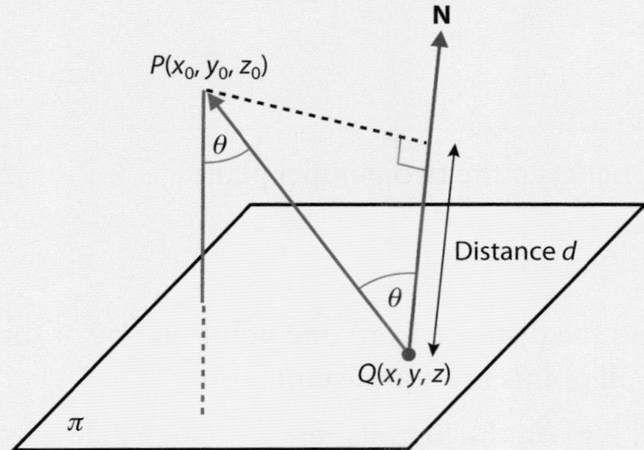

In the diagram above, let $Q(x, y, z)$ be any point on the plane and $\mathbf{N}(A, B, C)$ be a normal to the plane. The distance we are looking for is d. Then,

$$d = \left||\overrightarrow{QP}|\cos\theta\right| = \left||\overrightarrow{QP}|\frac{\overrightarrow{QP}\cdot\mathbf{N}}{|\overrightarrow{QP}||\mathbf{N}|}\right|$$

$$= \left|\frac{\overrightarrow{QP}\cdot\mathbf{N}}{|\mathbf{N}|}\right| = \left|\frac{(A, B, C)\cdot(x_0 - x, y_0 - y, z_0 - z)}{\sqrt{A^2 + B^2 + C^2}}\right|$$

$$= \frac{|A(x_0 - x) + B(y_0 - y) + C(z_0 - z)|}{\sqrt{A^2 + B^2 + C^2}}$$

$$= \frac{|Ax_0 + By_0 + Cz_0 - (Ax + Bx + Cz)|}{\sqrt{A^2 + B^2 + C^2}}$$

Since $Q(x, y, z)$ is on the plane, then $Ax + By + Cz = D$, so replacing this expression in the result above will yield

$$d = \frac{|Ax_0 + By_0 + Cz_0 - D|}{\sqrt{A^2 + B^2 + C^2}}.$$

This formula is similar to the distance between a point and a line in 2-space.

Example 44

Show that the line l with equation $\mathbf{r} = 2\mathbf{i} - \mathbf{j} + 5\mathbf{k} + k(3\mathbf{i} + 2\mathbf{j} - 2\mathbf{k})$ is parallel to the plane P whose equation is $\mathbf{r}\cdot(2\mathbf{i} - 2\mathbf{j} + \mathbf{k}) = -3$ and find the distance between them.

Solution

In a previous example we showed that the line is parallel to the plane because it is perpendicular to the normal of the plane. To find the distance, we used a relatively complex approach. At this moment we can utilize the distance formula just established to find the required distance.

A point on the line is $(2, -1, 5)$ and the Cartesian equation of the plane is simply

$$2x - 2y + z = -3.$$

Hence, the distance is

$$d = \frac{|2(2) - 2(-1) + 1(5) - (-3)|}{\sqrt{4 + 4 + 1}} = \frac{14}{3}.$$

Example 45

Find the distance between the two parallel planes: $x + 2y - 2z = 3$ and $2x + 4y - 4z = 7$.

Solution

It is enough to find the distance from one point on one of the planes to the other plane since all points are equidistant.

Take the point $(1, 1, z)$ on the first plane:

$$1 + 2 - 2z = 3, \text{ so } z = 0.$$

Thus, the point is $(1, 1, 0)$ and the distance between the planes is

$$d = \frac{|2(1) + 4(1) - 4(0) - 7|}{\sqrt{4 + 16 + 16}} = \frac{1}{6}.$$

Example 46

Find the distance between the two skew straight lines

$$L_1\colon \mathbf{r} = \begin{pmatrix} 1 \\ 5 \\ -1 \end{pmatrix} + \lambda \begin{pmatrix} 4 \\ -4 \\ 5 \end{pmatrix} \text{ and } L_2\colon \mathbf{r} = \begin{pmatrix} 2 \\ 4 \\ 5 \end{pmatrix} + \mu \begin{pmatrix} 8 \\ -3 \\ 1 \end{pmatrix}.$$

Solution

We can reduce this problem to the type in the previous example by creating two planes that contain the given lines and are parallel to each other.

For the two planes to be parallel, they must be perpendicular to the same vector. Hence, by finding the cross product of the direction vectors of the lines we would have found a vector perpendicular to both.

$$l_1 \times l_2 = \begin{vmatrix} \mathbf{i} & \mathbf{j} & \mathbf{k} \\ 4 & -4 & 5 \\ 8 & -3 & 1 \end{vmatrix} = 11\mathbf{i} + 36\mathbf{j} + 20\mathbf{k}$$

Considering the point $(2, 4, 5)$ on L_2, the plane containing this line will be

$$\begin{pmatrix} 11 \\ 36 \\ 20 \end{pmatrix} \cdot \begin{pmatrix} x \\ y \\ z \end{pmatrix} = \begin{pmatrix} 11 \\ 36 \\ 20 \end{pmatrix} \cdot \begin{pmatrix} 2 \\ 4 \\ 5 \end{pmatrix}$$

or $11x + 36y + 20z = 266$ and the distance between $(1, 5, -1)$ on L_1 to this plane will be

$$d = \frac{|11(1) + 36(5) + 20(-1) - 266|}{\sqrt{11^2 + 36^2 + 20^2}} = \frac{95}{\sqrt{1817}}.$$

The angle between two planes

The angle between two planes is defined to be the *acute* angle between them as you see in the figure below.

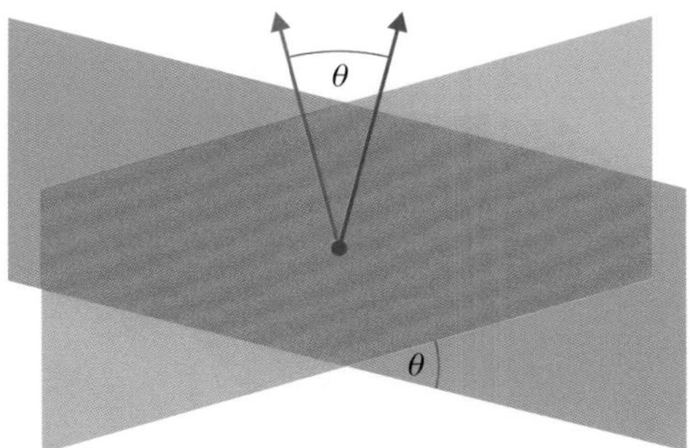

Consider two planes **P** and **Q** with unit normals $\mathbf{n}_1$ and $\mathbf{n}_2$ respectively. Their vector equations are of the form

$$\mathbf{r} \cdot \mathbf{n}_1 = d_1 \text{ and } \mathbf{r} \cdot \mathbf{n}_2 = d_2.$$

The angle between the planes is equal to the angle between the normal units $\mathbf{n}_1$ and $\mathbf{n}_2$ or the normals $\mathbf{N}_1$ and $\mathbf{N}_2$, and hence

$$\cos\theta = \mathbf{n}_1 \cdot \mathbf{n}_2 = \frac{\mathbf{N}_1 \cdot \mathbf{N}_2}{|\mathbf{N}_1|\,|\mathbf{N}_2|}.$$

For example, the angle between the planes with vector equations

$$\mathbf{r} \cdot (\mathbf{i} + \mathbf{j} - 2\mathbf{k}) = 3 \text{ and } \mathbf{r} \cdot (2\mathbf{i} - 2\mathbf{j} + \mathbf{k}) = 2$$

is given by

$$\cos\theta = \frac{(\mathbf{i} + \mathbf{j} - 2\mathbf{k}) \cdot (2\mathbf{i} - 2\mathbf{j} + \mathbf{k})}{\sqrt{1+1+4} \cdot \sqrt{4+4+1}} = -\frac{2}{3\sqrt{6}}.$$

This is the cosine of the obtuse angle between the two planes. The acute angle between them is $\cos^{-1}\left(\frac{2}{3\sqrt{6}}\right)$.

Example 47

Find the angle between the planes with equations

$$2x - 3y = 0 \text{ and } 3x + y - z = 4.$$

Solution

The two normals are $2\mathbf{i} - 3\mathbf{j}$ and $3\mathbf{i} + \mathbf{j} - \mathbf{k}$, and therefore the angle is given by

$$\cos\theta = \frac{(2\mathbf{i} - 3\mathbf{j}) \cdot (3\mathbf{i} + \mathbf{j} - \mathbf{k})}{\sqrt{13}\sqrt{11}} = \frac{3}{\sqrt{143}}.$$

So, the angle between the planes is $\cos^{-1}\left(\frac{3}{\sqrt{143}}\right)$.

The angle between a line and a plane

The angle between a line and a plane can be defined as the angle θ formed by the line and its projection on the plane, as shown in the figure below.

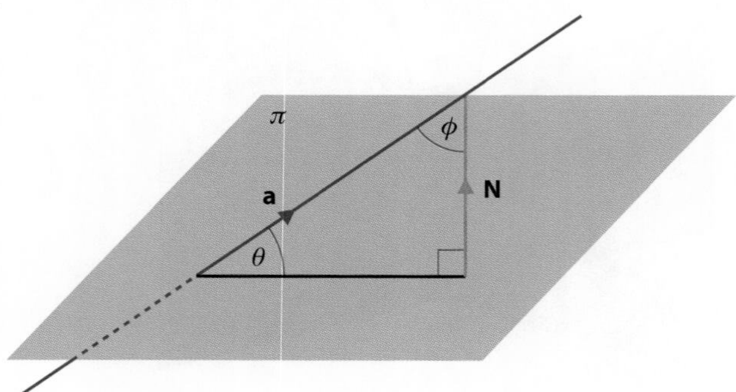

Consider the line l with equation $\mathbf{r} = \mathbf{r}_0 + \lambda\mathbf{a}$ and the plane with equation $\mathbf{r} \cdot \mathbf{N} = D$.

The acute angle ϕ between **N**, the normal to the plane, and the line l can be found by using the law of cosines:

$$\cos\phi = \left|\frac{\mathbf{a} \cdot \mathbf{N}}{|\mathbf{a}|\ |\mathbf{N}|}\right|$$

If θ is the acute angle between the line and the plane then

$$\theta = \frac{\pi}{2} - \phi.$$

Therefore, to find the angle between the line and the plane, we

- either find the angle ϕ first and then find its complement, or
- since ϕ and θ are complements, then $\sin\theta = \cos\phi = \left|\frac{\mathbf{a} \cdot \mathbf{N}}{|\mathbf{a}|\ |\mathbf{N}|}\right|$.

For example: to find the angle between the line with equation

$$\mathbf{r} = \mathbf{i} + 2\mathbf{j} - \mathbf{k} + \lambda(\mathbf{i} - \mathbf{j} + \mathbf{k})$$

and the plane with equation

$$\mathbf{r} \cdot (2\mathbf{i} - \mathbf{j} + \mathbf{k}) = 3$$

we find that

$$\sin\theta = \frac{(\mathbf{i} - \mathbf{j} + \mathbf{k}) \cdot (2\mathbf{i} - \mathbf{j} + \mathbf{k})}{\sqrt{3} \cdot \sqrt{4}} = \frac{4}{3\sqrt{2}}$$

$$\Rightarrow \theta = \sin^{-1}\frac{4}{3\sqrt{2}}.$$

Example 48

Find the angle between the line with equations

$\frac{x-1}{2} = \frac{y-2}{3} = \frac{z}{2}$, and the plane with equation

$2x - y - z = 7.$

Solution

The direction of the line is given by $\mathbf{a} = (2, 3, 2)$ and the normal to the plane by

$\mathbf{N} = (2, -1, -1)$, and the angle is given by

$$\sin\theta = \left|\frac{(2, 3, 2)\cdot(2, -1, -1)}{\sqrt{17}\cdot\sqrt{6}}\right| = \frac{1}{\sqrt{102}}$$

$$\Rightarrow \theta = \sin^{-1}\frac{1}{\sqrt{102}}.$$

Line of intersection of two planes

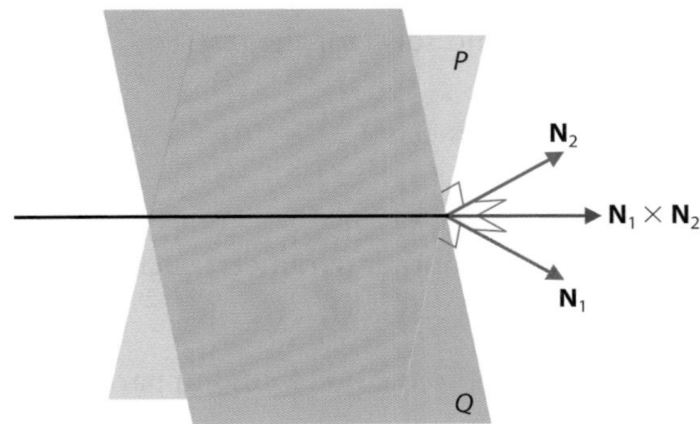

Unless two planes are parallel they will intersect along a straight line. Consider two planes P and Q that have l as their line of intersection. Also let the planes have the following vector equations:

$$\mathbf{r}\cdot\mathbf{N}_1 = D_1 \text{ and } \mathbf{r}\cdot\mathbf{N}_2 = D_2.$$

Since the line l lies in plane P then $\mathbf{N}_1$, the normal to this plane, must be perpendicular to it. This is also true for $\mathbf{N}_2$. Therefore, the direction of line l is perpendicular to both $\mathbf{N}_1$ and $\mathbf{N}_2$.

To find the line of intersection, we will demonstrate two methods:

1 Use the cross product of $\mathbf{N}_1$ and $\mathbf{N}_2$ as the direction of l and a specific point on the line.

2 Use the fact that all points on l must satisfy the equations of both planes; i.e. we solve a system of equations.

These methods are best demonstrated when we apply them to a particular situation.

Let the planes P and Q have the equations:

$$\mathbf{r}\cdot(\mathbf{i} + \mathbf{j} - 3\mathbf{k}) = 6 \text{ and } \mathbf{r}\cdot(2\mathbf{i} - \mathbf{j} + \mathbf{k}) = 4.$$

1 To find a vector equation of the line of intersection, we need first to find the cross product of the two normals and then find a point on the line l.

$$(\mathbf{i} + \mathbf{j} - 3\mathbf{k}) \times (2\mathbf{i} - \mathbf{j} + \mathbf{k}) = 2\mathrm{i} + 7\mathrm{j} + 3\mathbf{k}$$

To find a point on the line we use the fact that the points on that line must satisfy both equations. So, consider the points on both planes that have the x-coordinate zero; i.e.

$$(0, y, z) \cdot (\mathbf{i} + \mathbf{j} - 3\mathbf{k}) = 6 \text{ and } (0, y, z) \cdot (2\mathbf{i} - \mathbf{j} + \mathbf{k}) = 4$$

$$\Rightarrow \begin{cases} y - 3z = 6 \\ -y + z = 4 \end{cases} \Rightarrow z = -5 \text{ and } y = -9$$

So, the vector equation of the line is: $\mathbf{r} = (0, -9, -5) + t(2, 7, 3)$.

2 The second method uses a system of equations to find the equation. The equations of the planes in Cartesian form are:

$$x + y - 3z = 6 \text{ and } 2x - y + z = 4.$$

Since this system has to be solved simultaneously, and since there are two equations in three variables, we should consider one of these variables as a parameter and solve for the rest. So,

$$\begin{cases} x + y - 3z = 6 \\ 2x - y + z = 4 \end{cases} \Rightarrow 3x - 2z = 10 \Rightarrow z = -5 + \tfrac{3}{2}x;\ y = -9 + \tfrac{7}{2}x.$$

Therefore, we either consider x to be the parameter or, for convenience purposes, we replace it by another parameter such as the following:

$$x = 2\lambda,\ y = 7\lambda - 9,\ z = 3\lambda - 5$$

This equation is equivalent to the one found in part (1).

3 If the equations of the planes are in parametric form it may not be necessary to convert them into Cartesian form. However, from the example below, you may notice that it may be more straightforward to follow the Cartesian method.

Find the intersection between the two planes

$$P: \mathbf{r} = \mathrm{i} + \mathrm{j} + \lambda(2\mathbf{i} - \mathbf{k}) + \mu(\mathbf{i} - \mathbf{j} + \mathbf{k}), \text{ and}$$

$$Q: \mathbf{r} = 3\mathbf{i} - \mathbf{k} + s(\mathbf{i} - \mathbf{j} + 2\mathbf{k}) + t(\mathbf{i} + 2\mathbf{j} - \mathbf{k}).$$

A point on P will have the following coordinates: $(1 + 2\lambda + \mu, 1 - \mu, -\lambda + \mu)$, while a point on Q will have the coordinates $(3 + s + t, -s + 2t, -1 + 2s - t)$. For the collection of points on the intersection, we must have the coordinates satisfy both equations, and hence

$$1 + 2\lambda + \mu = 3 + s + t$$

$$1 - \mu = -s + 2t$$

$$-\lambda + \mu = -1 + 2s - t$$

This system of three equations in four unknowns must have an infinite number of solutions if the planes intersect – the set of points that belong to the line of intersection. We can solve this system best, after re-arranging terms, by Gaussian elimination:

$$\begin{pmatrix} 2 & 1 & -1 & -1 & 2 \\ 0 & -1 & 1 & -2 & -1 \\ -1 & 1 & -2 & 1 & -1 \end{pmatrix} \Rightarrow \begin{pmatrix} 1 & 0 & 0 & -\frac{3}{2} & \frac{1}{2} \\ 0 & 1 & 0 & \frac{9}{2} & \frac{5}{2} \\ 0 & 0 & 1 & \frac{5}{2} & \frac{3}{2} \end{pmatrix}$$

The last result expresses λ, μ and s in terms of t. To find the equation of the line we have to substitute these values in either of the two equations above. For example, it is easier to substitute for s in terms of t in the equation for Q. So, from the result above,

$$s = \tfrac{3}{2} - \tfrac{5}{2}t$$

and the line will have the vector equation

$$\begin{aligned}\mathbf{r} &= 3\mathbf{i} - \mathbf{k} + \left(\tfrac{3}{2} - \tfrac{5}{2}t\right)(\mathbf{i} - \mathbf{j} + 2\mathbf{k}) + t(\mathbf{i} + 2\mathbf{j} - \mathbf{k})\\ &= \left(\tfrac{9}{2}\mathbf{i} - \tfrac{3}{2}\mathbf{j} + 2\mathbf{k}\right) + t\left(-\tfrac{3}{2}\mathbf{i} + \tfrac{9}{2}\mathbf{j} - 6\mathbf{k}\right).\end{aligned}$$

If $\lambda = \frac{1}{2} + \frac{3}{2}t$ and $\mu = \frac{5}{2} - \frac{9}{2}t$ are substituted into the equation for P we will get the same result. (Try it!)

As we notice from the above discussion, the process is long and complex, even with many steps that are 'hidden' to save space. Alternatively, the Cartesian solution may be more efficient.

P: point $(1, 1, 0)$ is on the plane, and $(2, 0, -1) \times (1, -1, 1) = (-1, -3, -2)$ is perpendicular to the plane, so the Cartesian equation is

$$(x - 1) + 3(y - 1) + 2z = 0, \text{ or } x + 3y + 2z = 4.$$

Q: point $(3, 0, -1)$ is on the plane, and $(1, -1, 2) \times (1, 2, -1) = (-3, 3, 3)$ is perpendicular to the plane, so the Cartesian equation is

$$-3(x - 3) + 3y + 3(z + 1) = 0, \text{ or } x - y - z = 4.$$

The intersection between the planes is the result of solving the following system:

$$\begin{cases} x + 3y + 2z = 4 \\ x - y - z = 4 \end{cases} \Rightarrow x = 4 + m,\ y = -3m,\ z = 4m$$

This result compares to the previous one and appears to be more elegant!

Note:
Three planes can intersect in three lines as shown here.

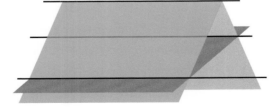

The three lines of intersection are parallel. Hence, the system of equations they represent is inconsistent.

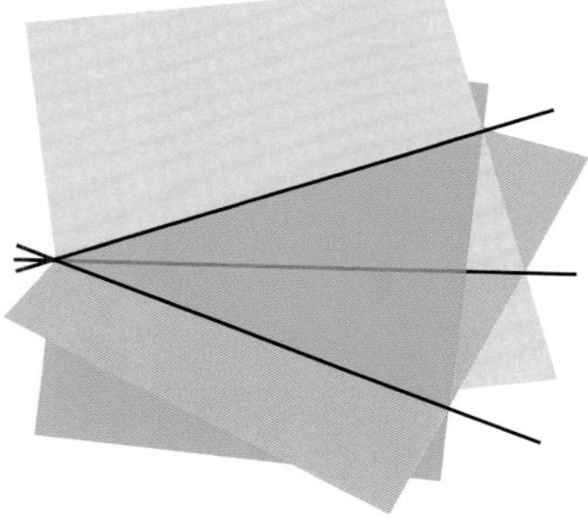

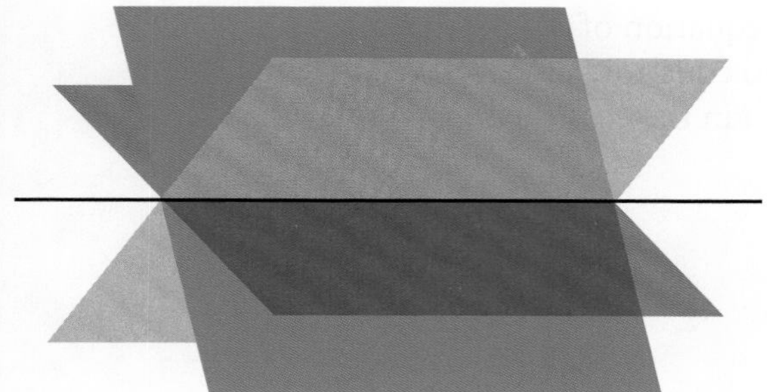

If the lines of intersection are not parallel, then the three planes meet at one point as shown. This system is consistent with a unique solution. The three planes can also all pass through one straight line. In that case, the system is consistent with an infinite number of solutions.

Exercise 14.5

1 Which of the points $A(3, -2, -1)$, $B(2, 1, -1)$, $C(1, 4, 0)$ lie in the plane
$3x + 2y - 3z = 11$?

2 Which of the points $A(3, 2, -3)$, $B(2, 1, -2)$, $C(1, 4, 0)$ lie in the plane
$(\mathbf{i} - 3\mathbf{j} + \mathbf{k}) \cdot (x\mathbf{i} + y\mathbf{j} + z\mathbf{k}) = -6$?

In questions 3–16, find an equation for the plane satisfying the given conditions. Give two forms for each equation out of the three forms: Cartesian, vector or parametric.

3 Contains the point $(3, -2, 4)$ and perpendicular to $2\mathbf{i} - 4\mathbf{j} + 3\mathbf{k}$

4 Contains the point $(-3, 2, 1)$ and perpendicular to $2\mathbf{i} + 3\mathbf{k}$

5 Contains the point $(0, 3, 1)$ and perpendicular to $3\mathbf{k}$

6 Contains the point $(3, -2, 4)$ and parallel to the plane $5x + y - 2z = 7$

7 Contains the point $(3, 0, 1)$ and parallel to the plane $y - 2z = 11$

8 Contains the point $(3, -2, 4)$ and the line $\mathbf{r} = \begin{pmatrix} 1 \\ -2 \\ 5 \end{pmatrix} + t\begin{pmatrix} 2 \\ 1 \\ 2 \end{pmatrix}$

9 Contains the lines $\mathbf{r} = \begin{pmatrix} 1 \\ -2 \\ 5 \end{pmatrix} + t\begin{pmatrix} 2 \\ 1 \\ 2 \end{pmatrix}$ and $\mathbf{r} = \begin{pmatrix} 4 \\ 0 \\ 7 \end{pmatrix} + t\begin{pmatrix} 3 \\ 2 \\ 2 \end{pmatrix}$

10 Contains the point $(1, -3, 2)$ and the line $x = 2t, y = 2 + t, z = -1 + 3t$

11 Contains the point $M(p, q, r)$ and perpendicular to the vector $\overrightarrow{OM}$

12 Contains the three points $(1, 2, 2)$, $(3, -1, 0)$ and $(7, 0, -2)$

13 Contains the three points $(2, -2, -2)$, $(3, -1, 3)$ and $(0, 1, 5)$

14 Contains the point $(1, -2, 3)$ and the line $x - 2 = y + 1 = \frac{z-5}{3}$

15 Contains the two parallel lines
$\mathbf{r} = (1, -1, 5) + t(3\mathbf{i} + 2\mathbf{j} + 4\mathbf{k})$ and $\mathbf{r} = (-3, 4, 0) + t(3\mathbf{i} + 2\mathbf{j} + 4\mathbf{k})$

16 Contains the point $(1, 1, 0)$ and parallel to the two lines
$x = 2 + t, y = -1, z = t$ and $x = s, y = 2 - s, z = -1 + s$

In questions 17–22, find the acute angle between the given lines or planes.

17 $3x + 4y - z = 1$ and $x - 2y = 3$

18 $4x - 7y + z = 3$ and $3x + 2y + 2z = 17$

19 $x = 4$ and $x + z = 4$

20 $x - 2y + 2z = 3$ and $x = 2 - 6t, y = 4 + 3t, z = 1 - 2t$

21 $(3\mathbf{i} - \mathbf{k}) \cdot (x, y, z) = 4$ and $\mathbf{r} = (2\mathbf{j} + 3\mathbf{k}) + \lambda(-\mathbf{i} + 2\mathbf{j} - \mathbf{k})$

22 $x + y + z = 7$ and $z = 0$

In questions 23–26 find the points of intersection of the given line and plane.

23 $\mathbf{r} = 5\mathbf{i} - 2\mathbf{k} + \lambda(\mathbf{i} - 3\mathbf{j} + 4\mathbf{k})$ and $(\mathbf{i} - 3\mathbf{j} + 2\mathbf{k}) \cdot (x\mathbf{i} + y\mathbf{j} + z\mathbf{k}) = -35$

24 $\mathbf{r} = \begin{pmatrix} 2 \\ 4 \\ 0 \end{pmatrix} + \mu \begin{pmatrix} 0 \\ -3 \\ 3 \end{pmatrix}$ and $4x - 2y + 3z - 30 = 0$

25 $x - 3 = \dfrac{y - 4}{5} = \dfrac{z - 6}{3}$ and $(2\mathbf{i} - 4\mathbf{j} + 6\mathbf{k}) \cdot (x\mathbf{i} + y\mathbf{j} + z\mathbf{k}) = 5$

26 $x = t, y = 4 - \frac{1}{3}t, z = 5 - \frac{5}{3}t$ and $3x - y + 2z = 6$

In questions 27–30, find the line of intersection between the given planes.

27 $x = 10$ and $x + y + z = 3$

28 $2x - y + z = 5$ and $x + y - z = 4$

29 $\begin{pmatrix} 1 \\ -1 \\ -2 \end{pmatrix} \cdot \begin{pmatrix} x \\ y \\ z \end{pmatrix} = 1$ and $x - y - 2z = 5$

30 $\mathbf{r} = \begin{pmatrix} 1 \\ 0 \\ 2 \end{pmatrix} + \lambda \begin{pmatrix} 1 \\ -2 \\ 0 \end{pmatrix} + \mu \begin{pmatrix} 3 \\ 2 \\ -8 \end{pmatrix}$ and $3x - y - z = 3$

31 Find a plane through $A(2, 1, -1)$ and perpendicular to the line of intersection of the planes $2x + y - z = 3$ and $x + 2y + z = 2$.

32 Find a plane through the points $A(1, 2, 3)$ and $B(3, 2, 1)$ and perpendicular to the plane $(4\mathbf{i} - \mathbf{j} + 2\mathbf{k}) \cdot (x\mathbf{i} + y\mathbf{j} + z\mathbf{k}) = 7$.

33 What point on the line through (1, 2, 5) and (3, 1, 1) is closest to the point $(2, -1, 5)$?

34 Find an equation of the plane that contains the line

$$\mathbf{r} = (-\mathbf{i} + 2\mathbf{j} + 3\mathbf{k}) + \lambda(\mathbf{i} - 2\mathbf{j} + \mathbf{k})$$

and is parallel to $\dfrac{x - 1}{3} = \dfrac{y + 2}{2} = \dfrac{z - 4}{4}$.

35 Find an equation of the plane that contains the line

$$x = 1 + 2t, y = 1 + 2t, z = 2 - t$$

and is parallel to $x - 1 = \dfrac{y - 2}{2} = z - 7$.

36 Show that the equation

$$\frac{x}{A} + \frac{y}{B} + \frac{z}{C} = 1$$

is the equation of a plane.

37 Find the equation of a plane that contains the point $(4, -3, -1)$ and is perpendicular to the planes $2x - 3y + 4z = 5$ and $4x - 3z = 5$.

38 Find the equation of a plane that contains the point (2, 3, 0) and is perpendicular to the plane $2x - 3y + 4z = 5$ and parallel to the line $\mathbf{r}(t) = (t - 3)\mathbf{i} + (4 - 2t)\mathbf{j} + (1 + t)\mathbf{k}$.

Review exercise

1 Briefly discuss how you test if two vectors are parallel or perpendicular. Use more than one approach.

2 Briefly discuss how you test if three vectors are coplanar.

3 Briefly discuss how you find the angle between two vectors.

4 Briefly discuss how you find the equation of a line.

5 Briefly discuss how you find the equation of a plane.

6 Briefly discuss how you find the angle between two planes.

7 Briefly discuss how you find the angle between a line and a plane.

Find vector, parametric and Cartesian equations for the lines in questions 8–15.

8 The line through the point $(4, -3, 0)$ parallel to the vector $\mathbf{i} + 2\mathbf{j} + \mathbf{k}$.

9 The line through $A(-1, 1, 4)$ and $B(4, 6, -1)$.

10 The line through $A(2, 3, 0)$ and $B(0, 1, 2)$.

11 The line through the origin parallel to the vector $\mathbf{j} + 2\mathbf{k}$.

12 The line through the point $(4, -1, 2)$ parallel to the line $x = 2 + 3t, y = 3 - t, z = 4t$.

13 The line through $(1, 2, 2)$ parallel to the y-axis.

14 The line through $(3, 5, 6)$ perpendicular to the plane $4x - 8y + 7z = 23$.

15 The line through $(3, 5, 6)$ perpendicular to the vectors $\mathbf{u} = 2\mathbf{i} + 3\mathbf{j} + 4\mathbf{k}$ and $\mathbf{v} = 4\mathbf{i} + 5\mathbf{j} + 6\mathbf{k}$.

Find equations for the planes in questions 16–20.

16 The plane through $A(1, 3, 0)$ normal to the vector $\mathbf{n} = 4\mathbf{i} - \mathbf{j} + \mathbf{k}$.

17 The plane through $P(2, 0, 5)$ parallel to the plane $2x + 3y - z = 11$.

18 The plane through $(2, 1, -1)$, $(3, -1, 0)$ and $(1, -1, 2)$.

19 The plane through $B(2, 5, 4)$ perpendicular to the line $x = 2 + 3t, y = 3 - t, z = 4t$.

20 The plane through $P(2, -1, 2)$ perpendicular to the vector from the origin to P.

21 Find the point of intersection of the lines $\mathbf{r} = \begin{pmatrix} 1 \\ 2 \\ 3 \end{pmatrix} + t\begin{pmatrix} 2 \\ 3 \\ 4 \end{pmatrix}$ and $x = 2 + s$, $y = 4 + 2s, z = -4s - 1$.

22 Find the equation of the plane determined by the straight lines in the previous question.

23 Find the point of intersection of the lines $x = 2 - y = z - 1$ and $\frac{x - 2}{2} = y - 3 = \frac{z - 6}{5}$.

24 Find the equation of the plane determined by the straight lines in the previous question.

25 Find the equation of the plane through $M(1, -2, 1)$ and perpendicular to the vector from the origin to M.

Practice questions

1 $ABCD$ is a rectangle and O is the midpoint of $[AB]$.
Express each of the following vectors in terms of $\overrightarrow{OC}$ and $\overrightarrow{OD}$

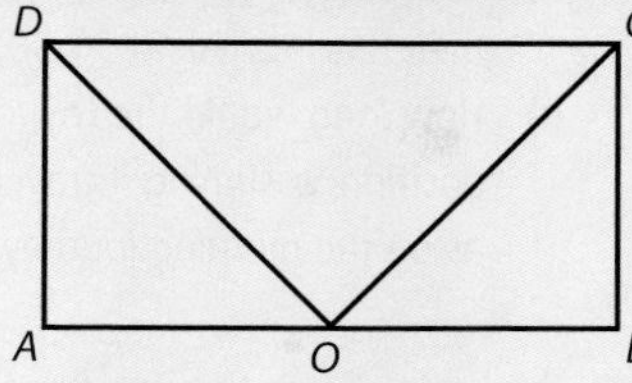

a) $\overrightarrow{CD}$ 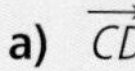 **b)** $\overrightarrow{OA}$ **c)** $\overrightarrow{AD}$

2 The vectors $\mathbf{i}$ and $\mathbf{j}$ are unit vectors along the x-axis and y-axis respectively.
The vectors $\mathbf{u} = -\mathbf{i} + \mathbf{j}$ and $\mathbf{v} = 3\mathbf{i} + 5\mathbf{j}$ are given.

a) Find $\mathbf{u} + 2\mathbf{v}$ in terms of $\mathbf{i}$ and $\mathbf{j}$.

A vector $\mathbf{w}$ has the same direction as $\mathbf{u} + 2\mathbf{v}$, and has a magnitude of 26.

b) Find $\mathbf{w}$ in terms of $\mathbf{i}$ and $\mathbf{j}$.

3 The circle shown has centre O and radius 6. $\overrightarrow{OA}$ is the vector $\begin{pmatrix} 6 \\ 0 \end{pmatrix}$, $\overrightarrow{OB}$ is the vector $\begin{pmatrix} -6 \\ 0 \end{pmatrix}$ and $\overrightarrow{OC}$ is the vector $\begin{pmatrix} 5 \\ \sqrt{11} \end{pmatrix}$.

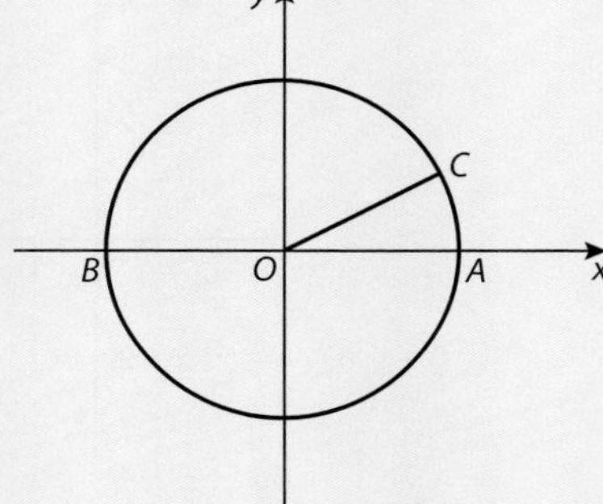

a) Verify that A, B and C lie on the circle.

b) Find the vector $\overrightarrow{AC}$.

c) Using an appropriate scalar product, or otherwise, find the cosine of angle OAC.

d) Find the area of triangle ABC, giving your answer in the form $a\sqrt{11}$, where $a \in \mathbb{N}$.

4 The quadrilateral $OABC$ has vertices with coordinates $O(0, 0)$, $A(5, 1)$, $B(10, 5)$ and $C(2, 7)$.

a) Find the vectors $\overrightarrow{OB}$ and $\overrightarrow{AC}$.

b) Find the angle between the diagonals of the quadrilateral $OABC$.

5 The vectors $\mathbf{u}$ and $\mathbf{v}$ are given by $\mathbf{u} = 3\mathbf{i} + 5\mathbf{j}$ and $\mathbf{v} = \mathbf{i} - 2\mathbf{j}$.
Find scalars a and b such that $a(\mathbf{u} + \mathbf{v}) = 8\mathbf{i} + (b - 2)\mathbf{j}$.

6 Find a vector equation of the line passing through $(-1, 4)$ and $(3, -1)$. Give your answer in the form $\mathbf{r} = \mathbf{p} + t\mathbf{d}$, where $t \in \mathbb{R}$.

7 In this question, the vector $\begin{pmatrix} 1 \\ 0 \end{pmatrix}$ represents a displacement due east and the vector $\begin{pmatrix} 0 \\ 1 \end{pmatrix}$ a displacement due north. Distances are in kilometres and time in hours.
Two crews of workers are laying an underground cable in a north-south direction across a desert. At 06:00 each crew sets out from their base camp, which is situated at the origin (0, 0). One crew is in a Toyundai vehicle and the other in a Chryssault vehicle.
The Toyundai has velocity vector $\begin{pmatrix} 18 \\ 24 \end{pmatrix}$ and the Chryssault has velocity vector $\begin{pmatrix} 36 \\ -16 \end{pmatrix}$.

a) Find the speed of each vehicle.

b) **(i)** Find the position vectors of each vehicle at 06:30.
(ii) Hence, or otherwise, find the distance between the vehicles at 06:30.

c) At this time (06:30) the Chryssault stops and its crew begin their day's work, laying cable in a northerly direction. The Toyundai continues travelling in the same direction, at the same speed, until it is exactly north of the Chryssault. The Toyundai crew then begin their day's work, laying cable in a southerly direction. At what time does the Toyundai crew begin laying cable?

d) Each crew lays an average of 800 m of cable in an hour. If they work non-stop until their lunch break at 11:30, what is the distance between them at this time?

e) How long would the Toyundai take to return to base camp from its lunchtime position, assuming it travelled in a straight line and with the same average speed as on the morning journey? (Give your answer to the nearest minute.)

8 The line L passes through the origin and is parallel to the vector $2\mathbf{i} + 3\mathbf{j}$.
Write down a vector equation for L.

9 The triangle ABC is defined by the following information:
$\overrightarrow{OA} = \begin{pmatrix} 2 \\ -3 \end{pmatrix}$, $\overrightarrow{AB} = \begin{pmatrix} 3 \\ 4 \end{pmatrix}$, $\overrightarrow{AB} \cdot \overrightarrow{BC} = 0$, $\overrightarrow{AC}$ is parallel to $\begin{pmatrix} 0 \\ 1 \end{pmatrix}$

a) On the grid below, draw an accurate diagram of triangle ABC.

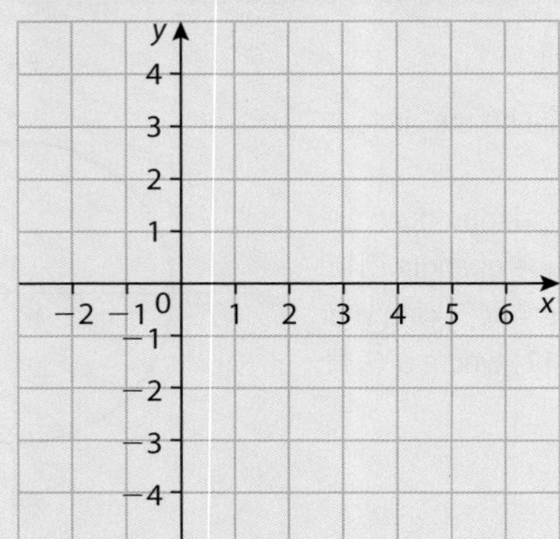

b) Write down the vector $\overrightarrow{OC}$.

10 In this question, the vector $\begin{pmatrix} 1 \\ 0 \end{pmatrix}$ represents a displacement due east and the vector $\begin{pmatrix} 0 \\ 1 \end{pmatrix}$ represents a displacement due north.
The point (0, 0) is the position of Shipple Airport. The position vector $\mathbf{r}_1$ of an aircraft, *Air One*, is given by

$$\mathbf{r}_1 = \begin{pmatrix} 16 \\ 12 \end{pmatrix} + t\begin{pmatrix} 12 \\ -5 \end{pmatrix},$$

where t is the time in minutes since 12:00.

a) Show that *Air One*

(i) is 20 km from Shipple Airport at 12:00

(ii) has a speed of 13 km/min.

b) Show that a Cartesian equation of the path of *Air One* is:

$$5x + 12y = 224.$$

The position vector $\mathbf{r}_2$ of an aircraft, *Air Two*, is given by

$$\mathbf{r}_2 = \begin{pmatrix} 23 \\ -5 \end{pmatrix} + t\begin{pmatrix} 2.5 \\ 6 \end{pmatrix},$$

where t is the time in minutes since 12:00.

c) Find the angle between the paths of the two aircraft.

d) **(i)** Find a Cartesian equation for the path of *Air Two*.

(ii) Hence, find the coordinates of the point where the two paths cross.

e) Given that the two aircraft are flying at the same height, show that they do not collide.

11 Find the size of the angle between the two vectors $\begin{pmatrix}1\\2\end{pmatrix}$ and $\begin{pmatrix}6\\-8\end{pmatrix}$. Give your answer to the nearest degree.

12 A line passes through the point $(4, -1)$ and its direction is perpendicular to the vector $\begin{pmatrix}2\\3\end{pmatrix}$. Find the equation of the line in the form $ax + by = p$, where a, b and p are integers to be determined.

13 In this question, the vector $\begin{pmatrix}1\\0\end{pmatrix}$ represents a displacement due east and the vector $\begin{pmatrix}0\\1\end{pmatrix}$ represents a displacement due north. Distances are in kilometres.
The diagram shows the path of the oil tanker *Aristides* relative to the port of Orto, which is situated at the point (0, 0).

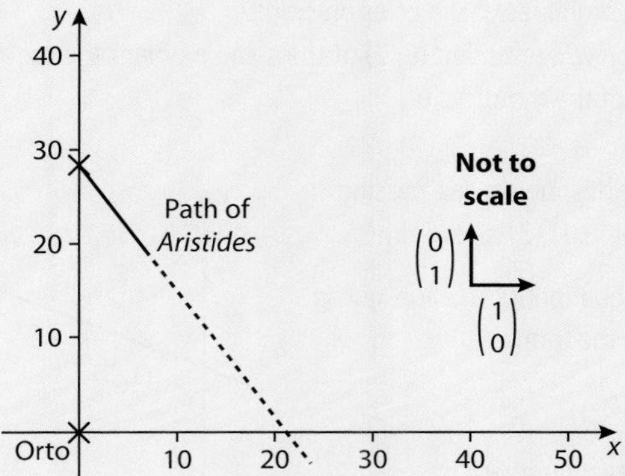

The position of the *Aristides* is given by the vector equation

$$\begin{pmatrix}x\\y\end{pmatrix} = \begin{pmatrix}0\\28\end{pmatrix} + t\begin{pmatrix}6\\-8\end{pmatrix}$$

at a time t hours after 12:00.

a) Find the position of the *Aristides* at 13:00.

b) Find

(i) the velocity vector

(ii) the speed of the *Aristides*.

c) Find a Cartesian equation for the path of the *Aristides* in the form $ax + by = g$.

Another ship, the cargo vessel *Boadicea*, is stationary, with position vector $\begin{pmatrix}18\\4\end{pmatrix}$.

d) Show that the two ships will collide, and find the time of collision.

To avoid collision, the *Boadicea* starts to move at 13:00 with velocity vector $\begin{pmatrix}5\\12\end{pmatrix}$.

e) Show that the position of the *Boadicea* for $t \geqslant 1$ is given by

$$\begin{pmatrix}x\\y\end{pmatrix} = \begin{pmatrix}13\\-8\end{pmatrix} + t\begin{pmatrix}5\\12\end{pmatrix}.$$

f) Find how far apart the two ships are at 15:00.

14 Find the angle between the following vectors **a** and **b**, giving your answer to the nearest degree.

$\mathbf{a} = -4\mathbf{i} - 2\mathbf{j}$
$\mathbf{b} = \mathbf{i} - 7\mathbf{j}$

15 In this question, a unit vector represents a displacement of 1 metre.
A miniature car moves in a straight line, starting at the point (2, 0). After t seconds, its position, (x, y), is given by the vector equation

$$\begin{pmatrix} x \\ y \end{pmatrix} = \begin{pmatrix} 2 \\ 0 \end{pmatrix} + t\begin{pmatrix} 0.7 \\ 1 \end{pmatrix}.$$

a) How far from the point (0, 0) is the car after 2 seconds?
b) Find the speed of the car.
c) Obtain the equation of the car's path in the form $ax + by = c$.

Another miniature vehicle, a motorcycle, starts at the point (0, 2) and travels in a straight line with constant speed. The equation of its path is

$$y = 0.6x + 2, \ x \geqslant 0.$$

Eventually, the two miniature vehicles collide.

d) Find the coordinates of the collision point.
e) If the motorcycle left point (0, 2) at the same moment the car left point (2 , 0), find the speed of the motorcycle.

16 The diagram right shows a line passing through the points (1, 3) and (6, 5).

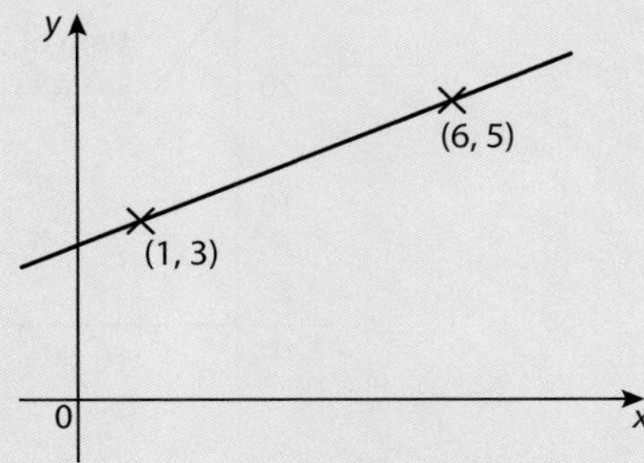

Find a vector equation for the line, giving your answer in the form

$$\begin{pmatrix} x \\ y \end{pmatrix} = \begin{pmatrix} a \\ b \end{pmatrix} + t\begin{pmatrix} c \\ d \end{pmatrix},$$

where t is any real number.

17 The vectors $\begin{pmatrix} 2x \\ x - 5 \end{pmatrix}$ and $\begin{pmatrix} x + 1 \\ 5 \end{pmatrix}$ are perpendicular for two values of x.

a) Write down the quadratic equation which the two values of x must satisfy.
b) Find the two values of x.

18 The diagram below shows the positions of towns O, A, B and X.

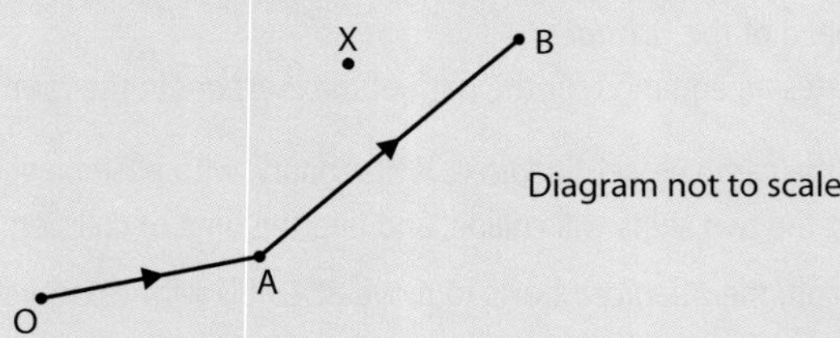

Town A is 240 km east and 70 km north of O.
Town B is 480 km east and 250 km north of O.
Town X is 339 km east and 238 km north of O.

A plane flies at a constant speed of 300 km h^{-1} from O towards A.

a) **(i)** Show that a unit vector in the direction of $\overrightarrow{OA}$ is $\begin{pmatrix} 0.96 \\ 0.28 \end{pmatrix}$.

(ii) Write down the velocity vector for the plane in the form $\begin{pmatrix} v_1 \\ v_2 \end{pmatrix}$.

(iii) How long does it take for the plane to reach A?

At A the plane changes direction so it now flies towards B. The angle between the original direction and the new direction is θ, as shown in the following diagram. This diagram also shows the point Y, between A and B, where the plane comes closest to X.

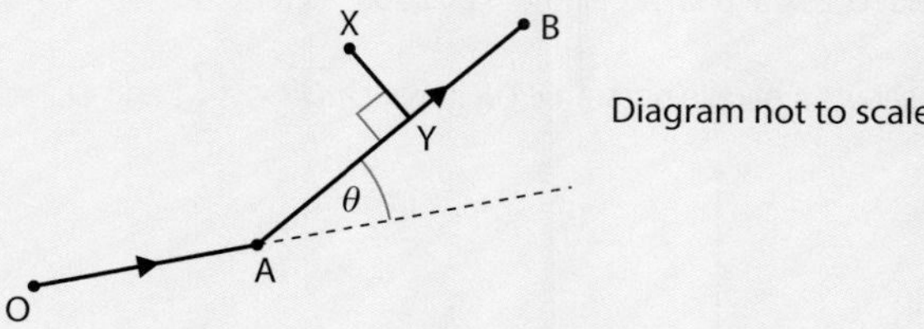

Diagram not to scale

b) Use the scalar product of two vectors to find the value of θ in degrees.

c) **(i)** Write down the vector $\overrightarrow{AX}$.

(ii) Show that the vector $\mathbf{n} = \begin{pmatrix} -3 \\ 4 \end{pmatrix}$ is perpendicular to $\overrightarrow{AB}$.

(iii) By finding the projection of $\overrightarrow{AX}$ in the direction of $\mathbf{n}$, calculate the distance XY.

d) How far is the plane from A when it reaches Y?

19 A vector equation of a line is $\begin{pmatrix} x \\ y \end{pmatrix} = \begin{pmatrix} 1 \\ 2 \end{pmatrix} + t\begin{pmatrix} -2 \\ 3 \end{pmatrix}$, $t \in \mathbb{R}$.
Find the equation of this line in the form $ax + by = c$, where a, b and $c \in \mathbb{Z}$.

20 Three of the coordinates of the parallelogram $STUV$ are $S(-2, -2)$, $T(7, 7)$ and $U(5, 15)$.

a) Find the vector $\overrightarrow{ST}$ and hence the coordinates of V.

b) Find a vector equation of the line (UV) in the form $\mathbf{r} = \mathbf{p} + \lambda\mathbf{d}$, where $\lambda \in \mathbb{R}$.

c) Show that the point E with position vector $\begin{pmatrix} 1 \\ 11 \end{pmatrix}$ is on the line (UV), and find the value of λ for this point.

The point W has position vector $\begin{pmatrix} a \\ 17 \end{pmatrix}$, $a \in \mathbb{R}$.

d) **(i)** If $\overrightarrow{EW} = 2\sqrt{13}$, show that one value of a is -3 and find the other possible value of a.

(ii) For $a = -3$, calculate the angle between $\overrightarrow{EW}$ and $\overrightarrow{ET}$.

21 Calculate the acute angle between the lines with equations
$\mathbf{r} = \begin{pmatrix} 4 \\ -1 \end{pmatrix} + s\begin{pmatrix} 4 \\ 3 \end{pmatrix}$ and $\mathbf{r} = \begin{pmatrix} 2 \\ 4 \end{pmatrix} + t\begin{pmatrix} 1 \\ -1 \end{pmatrix}$.

22 The diagram on the right shows the point O with coordinates (0, 0), the point A with position vector $\mathbf{a} = 12\mathbf{i} + 5\mathbf{j}$, and the point B with position vector $\mathbf{b} = 6\mathbf{i} + 8\mathbf{j}$. The angle between (OA) and (OB) is θ.

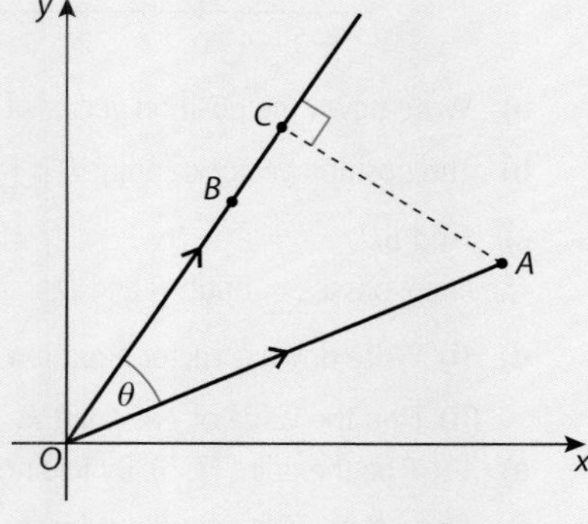

Find

a) $|\mathbf{a}|$

b) a unit vector in the direction of $\mathbf{b}$

c) the exact value of $\cos\theta$ in the form $\frac{p}{q}$, where $p, q \in \mathbb{Z}$.

23 The vector equations of two lines are given below.

$$\mathbf{r}_1 = \begin{pmatrix} 5 \\ 1 \end{pmatrix} + \lambda \begin{pmatrix} 3 \\ -2 \end{pmatrix}, \mathbf{r}_2 = \begin{pmatrix} -2 \\ 2 \end{pmatrix} + t\begin{pmatrix} 4 \\ 1 \end{pmatrix}$$

The lines intersect at the point P. Find the position vector of P.

24 The diagram shows a parallelogram $OPQR$ in which $\overrightarrow{OP} = \begin{pmatrix} 7 \\ 3 \end{pmatrix}$ and $\overrightarrow{OQ} = \begin{pmatrix} 10 \\ 1 \end{pmatrix}$.

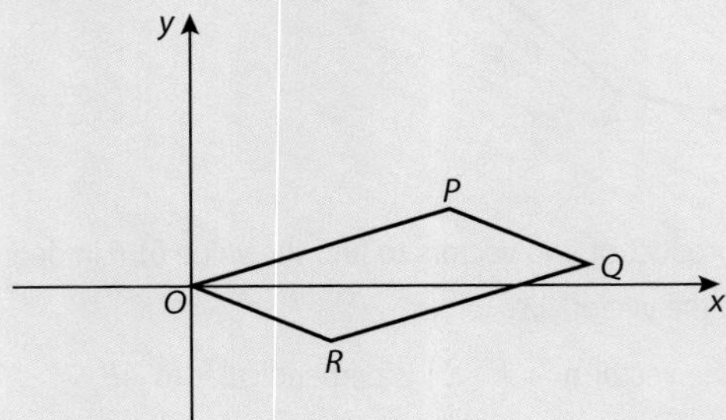

a) Find the vector $\overrightarrow{OR}$.

b) Use the scalar product of two vectors to show that $\cos O\hat{P}Q = -\frac{15}{\sqrt{754}}$.

c) **(i)** Explain why $\cos P\hat{Q}R = -\cos O\hat{P}Q$.

(ii) Hence, show that $\sin P\hat{Q}R = \frac{23}{\sqrt{754}}$.

(iii) Calculate the area of the parallelogram $OPQR$, giving your answer as an integer.

25 The diagram shows points A, B and C, which are three vertices of a parallelogram $ABCD$. The point A has position vector $\begin{pmatrix} 2 \\ 2 \end{pmatrix}$.

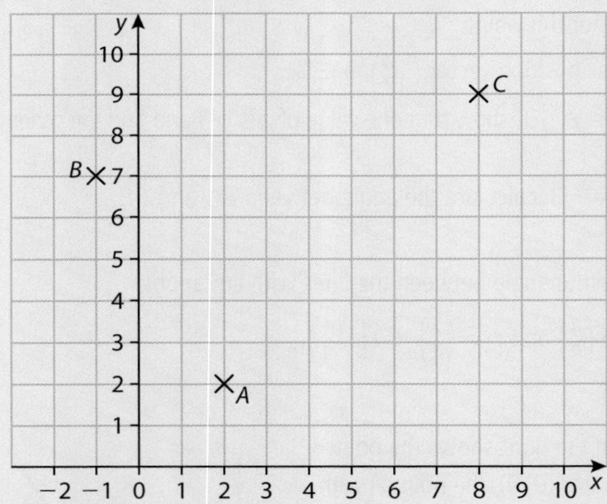

a) Write down the position vector of B and C.

b) The position vector of point D is $\begin{pmatrix} d \\ 4 \end{pmatrix}$. Find d.

c) Find $\overrightarrow{BD}$.

The line L passes through B and D.

d) **(i)** Write down a vector equation of L in the form $\begin{pmatrix} x \\ y \end{pmatrix} = \begin{pmatrix} -1 \\ 7 \end{pmatrix} + t\begin{pmatrix} m \\ n \end{pmatrix}$.

(ii) Find the value of t at point B.

e) Let P be the point (7, 5). By finding the value of t at P, show that P lies on the line L.

f) Show that $\overrightarrow{CP}$ is perpendicular to $\overrightarrow{BD}$.

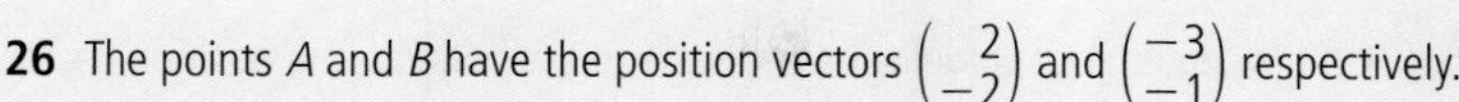

26 The points A and B have the position vectors $\begin{pmatrix} 2 \\ -2 \end{pmatrix}$ and $\begin{pmatrix} -3 \\ -1 \end{pmatrix}$ respectively.

a) **(i)** Find the vector $\overrightarrow{AB}$.

(ii) Find $|\overrightarrow{AB}|$.

The point D has position vector $\begin{pmatrix} d \\ 23 \end{pmatrix}$.

b) Find the vector $\overrightarrow{AD}$ in terms of d.

The angle $B\hat{A}D$ is 90°.

c) **(i)** Show that $d = 7$.

(ii) Write down the position vector of the point D.

The quadrilateral $ABCD$ is a rectangle.

d) Find the position vector of the point C.

e) Find the area of the rectangle $ABCD$.

27 Points A, B and C have position vectors $4\mathbf{i} + 2\mathbf{j}$, $\mathbf{i} - 3\mathbf{j}$ and $-5\mathbf{i} - 5\mathbf{j}$, respectively. Let D be a point on the x-axis such that $ABCD$ forms a parallelogram.

a) **(i)** Find $\overrightarrow{BC}$.

(ii) Find the position vector of D.

b) Find the angle between $\overrightarrow{BD}$ and $\overrightarrow{AC}$.

The line L_1 passes through A and is parallel to $\mathbf{i} + 4\mathbf{j}$. The line L_2 passes through B and is parallel to $2\mathbf{i} + 7\mathbf{j}$. A vector equation of L_1 is $\mathbf{r} = (4\mathbf{i} + 2\mathbf{j}) + s(\mathbf{i} + 4\mathbf{j})$.

c) Write down a vector equation of L_2 in the form $\mathbf{r} = \mathbf{b} + t\mathbf{q}$.

d) The lines L_1 and L_2 intersect at the point P. Find the position vector of P.

28 The diagram shows a cube, $OABCDEFG$, where the length of each edge is 5 cm. Express the following vectors in terms of $\mathbf{i}$, $\mathbf{j}$ and $\mathbf{k}$.

a) $\overrightarrow{OG}$

b) $\overrightarrow{BD}$

c) $\overrightarrow{EB}$

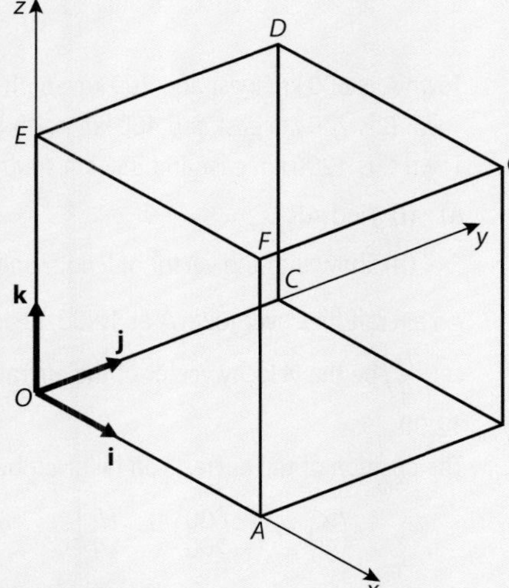

29 In this question, distance is in kilometres and time is in hours.

A balloon is moving at a constant height with a speed of 18 km h^{-1}, in the direction of the vector $\begin{pmatrix} 3 \\ 4 \\ 0 \end{pmatrix}$.

At time $t = 0$, the balloon is at point B with coordinates (0, 0, 5).

a) Show that the position vector $\mathbf{b}$ of the balloon at time t is given by

$$\mathbf{b} = \begin{pmatrix} x \\ y \\ z \end{pmatrix} = \begin{pmatrix} 0 \\ 0 \\ 5 \end{pmatrix} + \frac{18t}{5}\begin{pmatrix} 3 \\ 4 \\ 0 \end{pmatrix}.$$

At time $t = 0$, a helicopter goes to deliver a message to the balloon. The position vector $\mathbf{h}$ of the helicopter at time t is given by

$$\mathbf{h} = \begin{pmatrix} x \\ y \\ z \end{pmatrix} = \begin{pmatrix} 49 \\ 32 \\ 0 \end{pmatrix} + t\begin{pmatrix} -48 \\ -24 \\ 6 \end{pmatrix}.$$

b) (i) Write down the coordinates of the starting position of the helicopter.
(ii) Find the speed of the helicopter.

c) The helicopter reaches the balloon at point R.
(i) Find the time the helicopter takes to reach the balloon.
(ii) Find the coordinates of R.

30 In this question, the vector $\begin{pmatrix} 1 \\ 0 \end{pmatrix}$ represents a displacement due east and the vector $\begin{pmatrix} 0 \\ 1 \end{pmatrix}$ represents a displacement of 1 km north.

The diagram below shows the positions of towns A, B and C in relation to an airport O, which is at the point (0, 0). An aircraft flies over the three towns at a constant speed of 250 km h^{-1}.

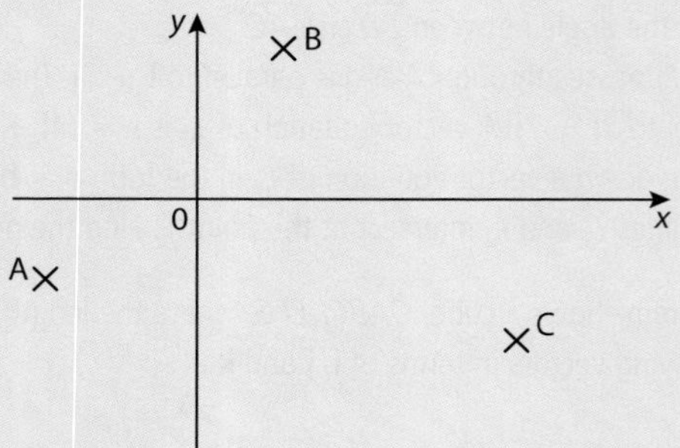

Town A is 600 km west and 200 km south of the airport.
Town B is 200 km east and 400 km north of the airport.
Town C is 1200 km east and 350 km south of the airport.

a) (i) Find $\overrightarrow{AB}$.
(ii) Show that the vector of length one unit in the direction of $\overrightarrow{AB}$ is $\begin{pmatrix} 0.8 \\ 0.6 \end{pmatrix}$.

An aircraft flies over town A at 12:00, heading towards town B at 250 km h^{-1}.

Let $\begin{pmatrix} p \\ q \end{pmatrix}$ be the velocity vector of the aircraft. Let t be the number of hours in flight after 12:00.

The position of the aircraft can be given by the vector equation

$$\begin{pmatrix} x \\ y \end{pmatrix} = \begin{pmatrix} -600 \\ -200 \end{pmatrix} + t\begin{pmatrix} p \\ q \end{pmatrix}.$$

b) (i) Show that the velocity vector is $\begin{pmatrix} 200 \\ 150 \end{pmatrix}$.
(ii) Find the position of the aircraft at 13:00.
(iii) At what time is the aircraft flying over town B?

Over town B the aircraft changes direction so it now flies towards town C. It takes five hours to travel the 1250 km between B and C. Over town A the pilot noted that she had 17 000 litres of fuel left. The aircraft uses 1800 litres of fuel per hour when travelling at 250 km h^{-1}. When the fuel gets below 1000 litres a warning light comes on.

c) How far from town C will the aircraft be when the warning light comes on?

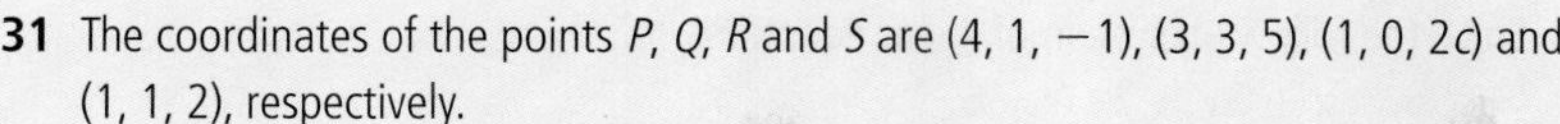

31 The coordinates of the points P, Q, R and S are $(4, 1, -1)$, $(3, 3, 5)$, $(1, 0, 2c)$ and $(1, 1, 2)$, respectively.

a) Find the value of c so that the vectors $\overrightarrow{OR}$ and $\overrightarrow{PR}$ are orthogonal.
For the remainder of the question, use the value of c found in part **a)** for the coordinate of the point R.

b) Evaluate $\overrightarrow{PS} \times \overrightarrow{PR}$.

c) Find an equation of the line l which passes through the point Q and is parallel to the vector PR.

d) Find an equation of the plane π which contains the line l and passes through the point S.

e) Find the shortest distance between the point P and the plane π.

32 Consider the points $A(1, 2, 1)$, $B(0, -1, 2)$, $C(1, 0, 2)$ and $D(2, -1, -6)$.

a) Find the vectors $\overrightarrow{AB}$ and $\overrightarrow{BC}$.

b) Calculate $\overrightarrow{AB} \times \overrightarrow{BC}$.

c) Hence, or otherwise, find the area of triangle ABC.

d) Find the equation of the plane P containing the points A, B and C.

e) Find a set of parametric equations for the line through the point D and perpendicular to the plane P.

f) Find the distance from the point D to the plane P.

g) Find a unit vector which is perpendicular to the plane P.

h) The point E is a reflection of D in the plane P. Find the coordinates of E.

33 a) If $\mathbf{u} = \mathbf{i} + 2\mathbf{j} + 3\mathbf{k}$ and $\mathbf{v} = 2\mathbf{i} - \mathbf{j} + 2\mathbf{k}$, show that $\mathbf{u} \times \mathbf{v} = 7\mathbf{i} + 4\mathbf{j} - 5\mathbf{k}$.

b) Let $\mathbf{w} = \lambda\mathbf{u} + \mu\mathbf{v}$ where λ and μ are scalars. Show that $\mathbf{w}$ is perpendicular to the line of intersection of the planes $x + 2y + 3z = 5$ and $2x - y + 2z = 7$ for all values of λ and μ.

34 Three points A, B and C have coordinates $(2, 1, -2)$, $(2, -1, -1)$ and $(1, 2, 2)$ respectively. The vectors $\overrightarrow{OA}$, $\overrightarrow{OB}$ and $\overrightarrow{OC}$, where O is the origin, form three concurrent edges of a parallelepiped $OAPBCQSR$ as shown in the following diagram.

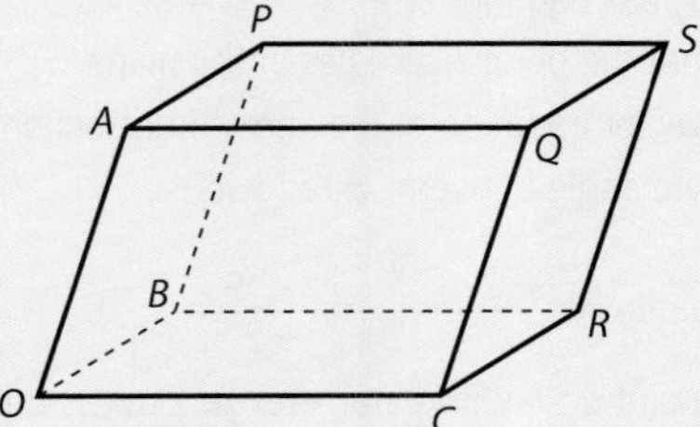

a) Find the coordinates of P, Q, R and S.

b) Find an equation for the plane $OAPB$.

c) Calculate the volume, V, of the parallelepiped given that $V = \overrightarrow{OA} \times \overrightarrow{OB} \cdot \overrightarrow{OC}$.

35 The triangle ABC has vertices at the points $A(-1, 2, 3)$, $B(-1, 3, 5)$ and $C(0, -1, 1)$.

a) Find the size of the angle θ between the vectors $\overrightarrow{AB}$ and $\overrightarrow{AC}$.

b) Hence, or otherwise, find the area of triangle ABC.

Let l_1 be the line parallel to $\overrightarrow{AB}$ which passes through $D(2, -1, 0)$ and l_2 be the line parallel to $\overrightarrow{AC}$ which passes through $E(-1, 1, 1)$.

c) **(i)** Find the equations of the lines l_1 and l_2.
(ii) Hence, show that l_1 and l_2 do not intersect.

d) Find the shortest distance between l_1 and l_2.

36 a) Solve the following system of linear equations:

$$x + 3y - 2z = -6$$
$$2x + y + 3z = 7$$
$$3x - y + z = 6$$

b) Find the vector $\mathbf{v} = (\mathbf{i} + 3\mathbf{j} - 2\mathbf{k}) \times (2\mathbf{i} + \mathbf{j} + 3\mathbf{k})$.

c) If $\mathbf{a} = \mathbf{i} + 3\mathbf{j} - 2\mathbf{k}$, $\mathbf{b} = 2\mathbf{i} + \mathbf{j} + 3\mathbf{k}$ and $\mathbf{u} = m\mathbf{a} + n\mathbf{b}$ where m, n are scalars, and $\mathbf{u} \neq 0$, show that $\mathbf{v}$ is perpendicular to $\mathbf{u}$ for all m and n.

d) The line l lies in the plane $3x - y + z = 6$, passes through the point $(1, -1, 2)$ and is perpendicular to $\mathbf{v}$. Find the equation of l.

37 The points A, B, C, D have the following coordinates: $A(1, 3, 1)$, $B(1, 2, 4)$, $C(2, 3, 6)$, $D(5, -2, 1)$.

a) (i) Evaluate the vector product $\overrightarrow{AB} \times \overrightarrow{AC}$, giving your answer in terms of the unit vectors $\mathbf{i}$, $\mathbf{j}$, $\mathbf{k}$.

(ii) Find the area of the triangle ABC.

The plane containing the points A, B, C is denoted by Π and the line passing through D perpendicular to Π is denoted by L. The point of intersection of L and Π is denoted by P.

b) (i) Find the Cartesian equation of Π.

(ii) Find the Cartesian equation of L.

c) Determine the coordinates of P.

d) Find the perpendicular distance of D from Π.

38 The point $A(2, 5, -1)$ is on the line L, which is perpendicular to the plane with equation $x + y + z - 1 = 0$.

a) Find the Cartesian equation of the line L.

b) Find the point of intersection of the line L and the plane.

c) The point A is reflected in the plane. Find the coordinates of the image of A.

d) Calculate the distance from the point $B(2, 0, 6)$ to the line L.

39 a) The point $P(1, 2, 11)$ lies in the plane π_1. The vector $3\mathbf{i} - 4\mathbf{j} + \mathbf{k}$ is perpendicular to π_1. Find the Cartesian equation of π_1.

b) The plane π_2 has equation $x + 3y - z = -4$.

(i) Show that the point P also lies in the plane π_2.

(ii) Find a vector equation of the line of intersection of π_1 and π_2.

c) Find the acute angle between π_1 and π_2.

40 A line l_1 has equation $\dfrac{x+2}{3} = \dfrac{y}{1} = \dfrac{z-9}{-2}$.

a) Let M be a point on l_1 with parameter μ. Express the coordinates of M in terms of μ.

b) The line l_2 is parallel to l_1 and passes through $P(4, 0, -3)$.

(i) Write down an equation for l_2.

(ii) Express $\overrightarrow{PM}$ in terms of μ.

c) The vector $\overrightarrow{PM}$ is perpendicular to l_1.

(i) Find the value of μ.

(ii) Find the distance between l_1 and l_2.

d) The plane π_1 contains l_1 and l_2. Find an equation for π_1, giving your answer in the form $Ax + By + Cz = D$.

e) The plane π_2 has equation $x - 5y - z = -11$. Verify that l_1 is the line of intersection of the planes π_1 and π_2.

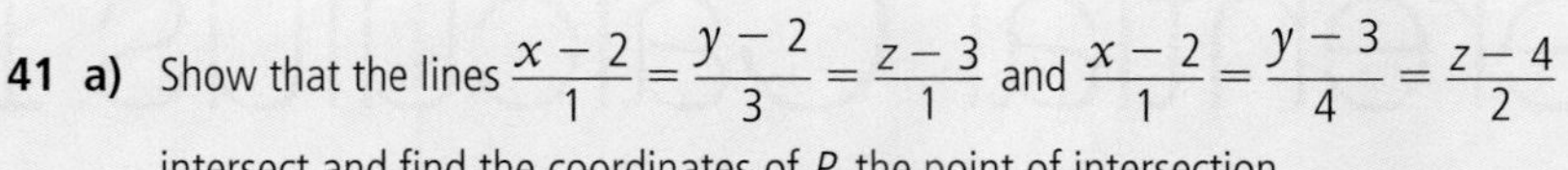

41 a) Show that the lines $\frac{x-2}{1} = \frac{y-2}{3} = \frac{z-3}{1}$ and $\frac{x-2}{1} = \frac{y-3}{4} = \frac{z-4}{2}$ intersect and find the coordinates of P, the point of intersection.

b) Find the Cartesian equation of the plane π that contains the two lines.

c) The point $Q(3, 4, 3)$ lies on π. The line L passes through the midpoint of $[PQ]$. Point S is on L such that $|\overrightarrow{PS}| = |\overrightarrow{QS}| = 3$, and the triangle PQS is normal to the plane π. Given that there are two possible positions for S, find their coordinates.

42 a) The plane π_1 has equation $\mathbf{r} = \begin{pmatrix} 2 \\ 1 \\ 1 \end{pmatrix} + \lambda\begin{pmatrix} -2 \\ 1 \\ 8 \end{pmatrix} + \mu\begin{pmatrix} 1 \\ -3 \\ -9 \end{pmatrix}$.

The plane π_2 has the equation $\mathbf{r} = \begin{pmatrix} 2 \\ 0 \\ 1 \end{pmatrix} + s\begin{pmatrix} 1 \\ 2 \\ 1 \end{pmatrix} + t\begin{pmatrix} 1 \\ 1 \\ 1 \end{pmatrix}$.

(i) For points which lie in π_1 and π_2, show that $\lambda = \mu$.

(ii) Hence, or otherwise, find a vector equation of the line of intersection of π_1 and π_2.

b) The plane π_3 contains the line $\frac{2-x}{3} = \frac{y}{-4} = z + 1$ and is perpendicular to $3\mathbf{i} - 2\mathbf{j} + \mathbf{k}$.

Find the Cartesian equation of π_3.

c) Find the intersection of π_1, π_2 and π_3.

15 Differential Calculus II: Further Techniques and Applications

Assessment statements

6.2 Derivative of x^n ($n \in \mathbb{Q}$), $\sin x$, $\cos x$, $\tan x$, e^x and $\ln x$.
Differentiation of a sum and a real multiple of a function.
The chain rule for composite functions.
Implicit differentiation.
Related rates of change.
The product and quotient rules.
Derivatives of $\sec x$, $\csc x$, $\cot x$, a^x, $\log_a x$, $\arcsin x$, $\arccos x$ and $\arctan x$.

6.3 Optimization problems.

Introduction

The primary purpose of the earlier chapter on calculus, Chapter 13, was to establish some fundamental concepts and techniques of differential calculus. Chapter 13 also introduced some applications involving the differentiation of functions: finding maxima and minima of a function; kinematic problems involving displacement, velocity and acceleration; and finding equations of tangents and normals. The focus of this chapter is to expand our set of differentiation rules and techniques and to deepen and extend the applications introduced in Chapter 13 – particularly using methods of finding extrema in the context of finding an 'optimum' solution to a problem and solving problems involving more than one rate of change. We start by investigating the derivatives of some important functions.

It is not an exaggeration to consider Isaac Newton (1642–1727) the most influential person in the development of modern science and mathematics. Newton was educated at Cambridge University and later was a professor of mathematics there. When Newton entered Cambridge in 1661, he did not know much mathematics but he learned quickly by reading works of Euclid and Descartes and attending lectures of Isaac Barrow, the first professor of mathematics at Cambridge. Cambridge was closed in 1665 and 1666 because of the Great Plague that swept through London and other parts of England. Studying and thinking on his own during these two years (and still not yet 25 years old), Newton discovered that white light can be decomposed into rays of different colours, how to represent functions using infinite series (including the binomial theorem), formulated the law of universal gravitation, and developed differential and integral calculus (several years before its independent discovery by Leibniz – see page 707). These great discoveries were all published much later because of Newton's fear of criticism and controversy. In 1687, Newton published his *Principia Mathematica*, one of the greatest scientific works ever written, in which he presented his version of calculus and applied it to investigate and explain a wide range of physical phenomena.

Newton's intellectual interests were not restricted to physics and mathematics. He left behind many papers dealing with theology and alchemy (attempting to change ordinary metals into gold). He was also a successful Warden of the Royal Mint (overseeing the production of official coins) and held political office, representing Cambridge University in Parliament several times.

15.1 Derivatives of composite functions, products and quotients

Derivatives of composite functions: the chain rule

We know how to differentiate functions such as $f(x) = x^3 + 2x - 3$ and $g(x) = \sqrt{x}$, but how do we differentiate the composite function $f(g(x)) = \sqrt{x^3 + 2x - 3}$? The rule for computing the derivative of the composite of two functions, i.e. the 'function of a function', is called the **chain rule**. Because most functions that we encounter in applications are composites of other functions, it can be argued that the chain rule is the most important, and most widely used, rule of differentiation.

Below are some examples of functions that we can differentiate with the rules that we have learned thus far in Chapter 13, and further examples of functions which are best differentiated with the chain rule.

Differentiate *without* the chain rule	Differentiate *with* the chain rule
$y = \cos x$	$y = \cos 2x$
$y = 3x^2 + 5x$	$x = \sqrt{3x^2 + 5x}$
$y = \sin x$	$y = \sin^2 x$
$y = \frac{1}{3x^2}$	$y = \frac{1}{3x^2 + x}$

The chain rule says, in a very basic sense, that given two functions, the derivative of their composite is the product of their derivatives – remembering that a derivative is a rate of change of one quantity (variable) with respect to another quantity (variable). For example, the function $y = 8x + 6 = 2(4x + 3)$ is the composite of the functions $y = 2u$ and $u = 4x + 3$. Note that the function y is in terms of u, and the function u is in terms of x. How are the derivatives of these three functions related?

Clearly, $\frac{dy}{dx} = 8$, $\frac{dy}{du} = 2$ and $\frac{du}{dx} = 4$. Since $8 = 2 \cdot 4$, the derivatives relate such that $\frac{dy}{dx} = \frac{dy}{du} \cdot \frac{du}{dx}$. In other words, rates of change multiply.

Again, if we think of derivatives as rates of change, the relationship $\frac{dy}{dx} = \frac{dy}{du} \cdot \frac{du}{dx}$ can be illustrated by a practical example. Consider the pair of levers in Figure 15.1 with lever endpoints U and U′ connected by a segment that can shrink and stretch but always remains horizontal. Hence, points U and U′ are always the same distance u from the ground.

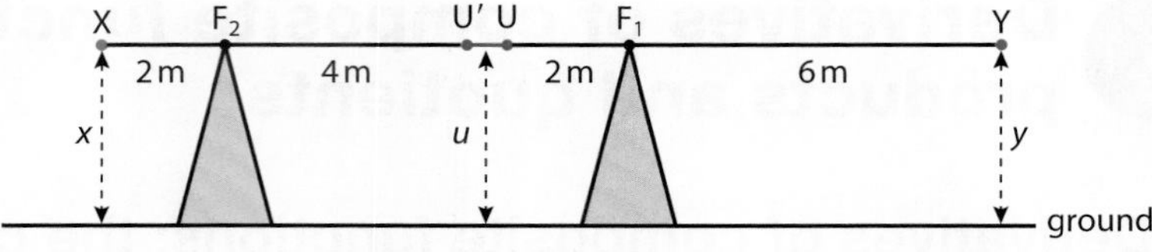

Figure 15.1 Two levers with horizontal connection between U′ and U.

As point Y moves down, points U and U′ move up, and point X moves down but at a rate different from that of Y. Let *dy*, *du* and *dx* represent the change in distance from the ground for the points Y, U and X, respectively. Because $YF_1 = 6$ and $UF_1 = 2$, if point Y moves such that $dy = 3$, then $du = 1$. Since $U'F_2 = 4$ and $XF_2 = 2$, if point U′ moves so that $du = 2$, then $dx = 1$.

Hence, $\frac{dy}{du} = 3$ and $\frac{du}{dx} = 2$.

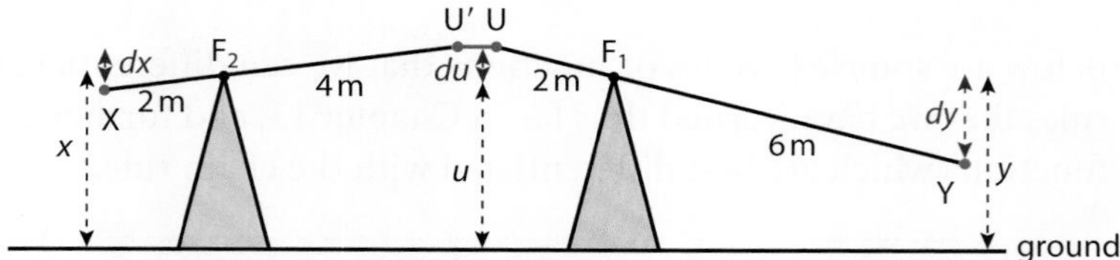

Figure 15.2 *dx*, *du* and *dy* represent the change in distance from the ground for X, U and Y.

Combining these two results, we can see that for every 6 units that Y's distance changes, X's distance will change 1 unit. That is, $\frac{dy}{dx} = 6$.

Therefore, we can write $\frac{dy}{dx} = \frac{dy}{du} \cdot \frac{du}{dx} = 3 \cdot 2 = 6$. In other words, the rate of change of *y* with respect to *x* is the product of the rate of change of *y* with respect to *u* and the rate of change of *u* with respect to *x*.

Example 1

The polynomial function $y = 16x^4 - 8x^2 + 1 = (4x^2 - 1)^2$ is the composite of $y = u^2$ and $u = 4x^2 - 1$. Use the chain rule to find $\frac{dy}{dx}$, the derivative of *y* with respect to *x*.

Solution

$$y = u^2 \Rightarrow \frac{dy}{du} = 2u$$

$$u = 4x^2 - 1 \Rightarrow \frac{du}{dx} = 8x$$

Applying the chain rule:
$$\frac{dy}{dx} = \frac{dy}{du} \cdot \frac{du}{dx} = 2u \cdot 8x$$
$$= 2(4x^2 - 1) \cdot 8x$$
$$= 64x^3 - 16x$$

In this particular case, we could have differentiated the function in expanded form by differentiating term-by-term rather than differentiating the factored form by the chain rule. $\frac{dy}{dx} = \frac{d}{dx}(16x^4 - 8x^2 + 1) = 64x^3 - 16x$;

confirming the result above. It is not always easier to differentiate powers of polynomials by expanding and then differentiating term-by-term.
For example, it is far better to find the derivative of $y = (3x + 5)^8$ by the chain rule.

In Section 2.2, we often wrote composite functions using nested function notation. For example, the notation $f(g(x))$ denotes a function composed of functions f and g such that g is the 'inside' function and f is the 'outside' function. For the composite function $y = (4x^2 - 1)^2$ in Example 1, the 'inside' function is $g(x) = 4x^2 - 1$ and the 'outside' function is $f(u) = u^2$. Looking again at the solution for Example 1, we see that we can choose to express and work out the chain rule in function notation rather than Leibniz notation.

For $y = f(g(x)) = (4x^2 - 1)^2$ and $y = f(u) = u^2$, $u = g(x) = 4x^2 - 1$,

Leibniz notation	**Function notation**
$\frac{dy}{dx} = \frac{dy}{du} \cdot \frac{du}{dx} = 2u \cdot 8x$	$\frac{d}{dx}[f(g(x))] = f'(u) \cdot g'(x) = 2u \cdot 8x$
$= 2(4x^2 - 1) \cdot 8x$	$= f'(g(x)) \cdot g'(x) = 2(4x^2 - 1) \cdot 8x$
$= 64x^3 - 16x$	$= 64x^3 - 16x$

This leads us to formally state the chain rule in two different notations.

The chain rule

If $y = f(u)$ is a function in terms of u and $u = g(x)$ is a function in terms of x, the function $y = f(g(x))$ is differentiated as follows:

$$\frac{dy}{dx} = \frac{dy}{du} \cdot \frac{du}{dx} \qquad \text{(Leibniz form)}$$

or, equivalently,

$$\frac{dy}{dx} = \frac{d}{dx}[f(g(x))] = f'(g(x)) \cdot g'(x) \qquad \text{(function notation form)}$$

Let Δu be the change in u corresponding to a change of Δx in x, that is, $\Delta u = g(x + \Delta x) - g(x)$. Then the corresponding change in y is $\Delta y = f(u + \Delta u) - f(u)$. It would be tempting to try to **prove the chain rule** by writing $\frac{\Delta y}{\Delta x} = \frac{\Delta y}{\Delta u} \cdot \frac{\Delta u}{\Delta x}$, which is a true statement if none of the denominators are zero. Recognizing that the definition of the derivative $f'(x) = \lim_{h \to 0} \frac{f(x + h) - f(x)}{h}$, is equivalent to $\frac{dy}{dx} = \lim_{\Delta x \to 0} \frac{\Delta y}{\Delta x}$, we could then proceed as follows:

$$\lim_{\Delta x \to 0} \frac{\Delta y}{\Delta x} = \lim_{\Delta x \to 0} \left(\frac{\Delta y}{\Delta u} \cdot \frac{\Delta u}{\Delta x} \right) = \lim_{\Delta x \to 0} \frac{\Delta y}{\Delta u} \cdot \lim_{\Delta x \to 0} \frac{\Delta u}{\Delta x}$$

$$= \lim_{\Delta u \to 0} \frac{\Delta y}{\Delta u} \cdot \lim_{\Delta x \to 0} \frac{\Delta u}{\Delta x} \text{ because if } \Delta x \to 0 \text{ then } \Delta u \to 0$$

$$= \frac{dy}{du} \cdot \frac{du}{dx}$$

This would work as a proof if we knew that Δu, the change in u, was non-zero – but we do not know this. It is possible that a small change in x could produce no change in u. Nonetheless, this reasoning does provide an intuitive justification relating the chain rule to the limit definition of the derivative. A properly rigorous proof can be constructed with a different approach, but we will not present it here.

The chain rule needs to be applied carefully. Consider the function notation form for the chain rule $\frac{d}{dx}[f(g(x))] = f'(g(x))\cdot g'(x)$. Although it is the product of two derivatives, it is important to point out that the first derivative involves the function f differentiated at $g(x)$ and the second is function g differentiated at x. The chain rule written in Leibniz form, $\frac{dy}{dx} = \frac{dy}{du}\cdot\frac{du}{dx}$, is easily remembered because it appears to be an obvious statement about fractions – but, they are *not* fractions. The expressions $\frac{dy}{dx}$, $\frac{dy}{du}$ and $\frac{du}{dx}$ are derivatives or, more precisely, limits and although du and dx essentially represent very small changes in the variables u and x, we cannot guarantee that they are non-zero.

The function notation form of the chain rule offers a very useful way of saying the rule 'in words', and, thus, a very useful structure for applying it.

f is 'outside' function   *g* is 'inside' function

$$\frac{dy}{dx} = \frac{d}{dx}[f(g(x))] = \underbrace{f'(g(x))}\cdot g'(x)$$

derivative of 'outside' function with 'inside' function unchanged × derivative of 'inside' function

The chain rule in words:

$$\begin{pmatrix}\text{derivative of}\\ \text{composite}\end{pmatrix} = \begin{pmatrix}\text{derivative of 'outside' function}\\ \text{with 'inside' function unchanged}\end{pmatrix} \times \begin{pmatrix}\text{derivative of}\\ \text{'inside' function}\end{pmatrix}$$

Although this is taking some liberties with mathematical language, the mathematical interpretation of the phrase "with 'inside' function unchanged" is that the derivative of the 'outside' function f is evaluated at $g(x)$, the 'inside' function.

• **Hint:** The chain rule is our most important rule of differentiation. It is an indispensable tool in differential calculus. Forgetting to apply the chain rule when it needs to be applied, or by applying it improperly, is a common source of errors in calculus computations. It is important to understand it, practise it and master it.

The chain rule acquired its name because we use it to take derivatives of composites of functions by 'chaining' together their derivatives. A function could be the composite of more than two functions. If a function were the composite of three functions, we would take the product of three derivatives 'chained' together. For example, if $y = f(u)$, $u = g(v)$ and $v = h(x)$, the derivative of the function $y = f(g(h(x)))$ is $\frac{dy}{dx} = \frac{dy}{du}\cdot\frac{du}{dv}\cdot\frac{dv}{dx}$.

Example 2

Differentiate each function by applying the chain rule. Start by 'decomposing' the composite function into the 'outside' function and the 'inside' function.

a) $y = \cos 3x$ b) $y = \sqrt{3x^2 + 5x}$

c) $y = \dfrac{1}{3x^2 + x}$ d) $y = \sin^2 x$

e) $y = \sin x^2$ f) $y = \sqrt[3]{(7 - 5x)^2}$

Solution

a) $y = f(g(x)) = \cos 3x \Rightarrow$ 'outside' function is $f(u) = \cos u$

$\Rightarrow$ 'inside' function is $g(x) = 3x$

In Leibniz form: $\dfrac{dy}{dx} = \dfrac{dy}{du} \cdot \dfrac{du}{dx} = (-\sin u) \cdot 3 = -3\sin(3x)$

Or, alternatively, in function notation form:

$$\frac{dy}{dx} = f'(g(x)) \cdot g'(x) = \underbrace{[-\sin(3x)]} \cdot 3 = -3\sin(3x)$$

derivative of 'outside' function with 'inside' function unchanged × derivative of 'inside' function

b) $y = f(g(x)) = \sqrt{3x^2 + 5x} \Rightarrow$

'outside' function is $f(u) = \sqrt{u} = u^{\frac{1}{2}}$

$f'(u) = \frac{1}{2}u^{-\frac{1}{2}} \Rightarrow$ 'inside' function is $g(x) = 3x^2 + 5x$

$$\frac{dy}{dx} = f'(g(x)) \cdot g'(x) = \tfrac{1}{2}(3x^2 + 5x)^{-\frac{1}{2}} \cdot (6x + 5)$$

$$\frac{dy}{dx} = \frac{6x + 5}{2(3x^2 + 5x)^{\frac{1}{2}}} \text{ or } \frac{6x + 5}{2\sqrt{3x^2 + 5x}}$$

c) $y = f(g(x)) = \dfrac{1}{3x^2 + x} \Rightarrow$

'outside' function is $f(u) = \dfrac{1}{u} = u^{-1}$

$f'(u) = -u^{-2} \Rightarrow$ 'inside' function is $g(x) = 3x^2 + x$

$$\frac{dy}{dx} = f'(g(x)) \cdot g'(x) = -(3x^2 + x)^{-2} \cdot (6x + 1)$$

$$\frac{dy}{dx} = -\frac{6x + 1}{(3x^2 + x)^2}$$

d) The expression $\sin^2 x$ is an abbreviated way of writing $(\sin x)^2$.

$y = f(g(x)) = \sin^2 x = (\sin x)^2 \Rightarrow$

'outside' function is $f(u) = u^2$

$f'(u) = 2u \Rightarrow$ 'inside' function is $g(x) = \sin x$

$$\frac{dy}{dx} = f'(g(x)) \cdot g'(x) = 2\sin x \cdot \cos x$$

$$\frac{dy}{dx} = 2\sin x \cos x$$

e) The expression $\sin x^2$ is equivalent to $\sin(x^2)$, and is **not** $(\sin x)^2$.

If $y = f(g(x)) = \sin(x^2)$, then the 'outside' function is $f(u) = \sin u$, and the 'inside' function is $g(x) = x^2$.

By the chain rule, $\frac{dy}{dx} = f'(g(x)) \cdot g'(x)$

$$= \cos(x^2) \cdot 2x$$

$$\frac{dy}{dx} = 2x\cos(x^2)$$

f) First change from radical (surd) form to rational exponent form.

$y = \sqrt[3]{(7-5x)^2} = (7-5x)^{\frac{2}{3}}$

$y = f(g(x)) = (7-5x)^{\frac{2}{3}} \Rightarrow$ 'outside' function $f(u) = u^{\frac{2}{3}}$

$\Rightarrow$ 'inside' function $g(x) = 7 - 5x$

By the chain rule, $\frac{dy}{dx} = f'(g(x)) \cdot g'(x)$

$$= \tfrac{2}{3}(7-5x)^{-\frac{1}{3}} \cdot (-5)$$

$$\frac{dy}{dx} = -\frac{10}{3(7-5x)^{\frac{1}{3}}} \text{ or } -\frac{10}{3(\sqrt[3]{7-5x})}$$

● **Hint:** Aim to write a function in a way that eliminates any confusion regarding the argument of the function. For example, write $\sin(x^2)$ rather than $\sin x^2$; $1 + \ln x$ rather than $\ln x + 1$; $5 + \sqrt{x}$ rather than $\sqrt{x} + 5$; $\ln(4 - x^2)$ rather than $\ln 4 - x^2$.

Example 3

Find the derivative of the function $y = (2x + 3)^3$ by:

a) expanding the binomial and differentiating term-by-term

b) the chain rule.

Solution

a) $y = (2x+3)^3 = (2x+3)(2x+3)^2$

$$= (2x+3)(4x^2 + 12x + 9)$$

$$= 8x^3 + 24x^2 + 18x + 12x^2 + 36x + 27$$

$$= 8x^3 + 36x^2 + 54x + 27$$

$$\frac{dy}{dx} = 24x^2 + 72x + 54$$

b) $y = f(g(x)) = (2x+3)^3 \Rightarrow y = f(u) = u^3;\ u = g(x) = 2x + 3$

$\Rightarrow f'(u) = 3u^2;\ g'(x) = 2$

$$\frac{dy}{dx} = \frac{dy}{du} \cdot \frac{du}{dx} = 3u^2 \cdot 2 = 6u^2$$

$$= 6(2x+3)^2$$

$$= 6(4x^2 + 12x + 9)$$

$$= 24x^2 + 72x + 54$$

The product rule

With the differentiation rules that we have learned thus far we can differentiate some functions that are products. For example, we can differentiate the function $f(x) = (x^2 + 3x)(2x - 1)$ by expanding and then differentiating the polynomial term-by-term. In doing so, we are applying the sum and difference, constant multiple and power rules from Section 13.2.

$$f(x) = (x^2 + 3x)(2x - 1) = 2x^3 + 5x^2 - 3x$$

$$f'(x) = 2\frac{d}{dx}(x^3) + 5\frac{d}{dx}(x^2) - 3\frac{d}{dx}(x)$$

$$f'(x) = 6x^2 + 10x - 3$$

The sum and difference rule states that the derivative of a sum/difference of two functions is the sum/difference of their derivatives. Perhaps the derivative of the product of two functions is the product of their derivatives. Let's try this with the above example.

$$f(x) = (x^2 + 3x)(2x - 1)$$

$$f'(x) = \frac{d}{dx}(x^2 + 3x) \cdot \frac{d}{dx}(2x - 1)?$$

$$f'(x) = (2x + 3) \cdot 2?$$

$$f'(x) = 4x + 6? \quad \text{However, } 4x + 6 \neq 6x^2 + 10x - 3.$$

Thus, one important fact we have learned from this example is that the derivative of a product of two functions is *not* the product of their derivatives. However, there are many products, such as $y = (4x - 3)^3(x - 1)^4$ and $f(x) = x^2 \sin x$, for which it is either difficult or impossible to write the function as a polynomial. In order to differentiate functions like this, we need a '**product**' **rule**.

Gottfried Wilhelm Leibniz (1646–1716)

Leibniz was a German philosopher, mathematician, scientist and professional diplomat – and, although self-taught in mathematics, was a major contributor to the development of mathematics in the 17th century. He developed the elementary concepts of calculus independent of, but slightly after, Newton. Nevertheless, the notation that Leibniz created for differential and integral calculus is still in use today. Leibniz' approach to the development of calculus was more purely mathematical, whereas Newton's was more directly connected to solving problems in physics. Leibniz created the idea of differentials (infinitely small differences in length), which he used to define the slope of a tangent, before the modern concept of limits was fully developed. Thus, Leibniz considered the derivative $\frac{dy}{dx}$ as the quotient of two differentials, dy and dx. Though it caused some confusion and consternation in his time (and to some extent still), Leibniz manipulated differentials algebraically to establish many of the important differentiation rules – including the product rule.

The product rule

If y is a function in terms of x that can be expressed as the product of two functions u and v that are also in terms of x, the product $y = uv$ can be differentiated as follows:

$$\frac{dy}{dx} = \frac{d}{dx}(uv) = u\frac{dv}{dx} + v\frac{du}{dx}$$

or, equivalently, if $y = f(x) \cdot g(x)$, then

$$\frac{dy}{dx} = \frac{d}{dx}[f(x) \cdot g(x)] = f(x) \cdot g'(x) + g(x) \cdot f'(x)$$

Proof of the product rule

Let $y = F(x) = f(x) \cdot g(x)$ where f and g are differentiable functions of x (i.e. derivative exists for all x) and their product is defined for all values of x in the domain.

We proceed by applying the limit definition of the derivative and properties of limits. Note that in the second line of the proof we have introduced the additional term, $f(x + h)g(x)$, and its opposite (thereby adding zero) in the numerator. The purpose of this is to allow us to analyze separately the changes in f and g as h goes to zero. Thus, in the fifth line we are eventually able to isolate limits that are the derivatives of f and g.

$$\begin{aligned}
F'(x) &= \lim_{h\to 0}\frac{f(x+h)g(x+h) - f(x)g(x)}{h} \\
&= \lim_{h\to 0}\frac{f(x+h)g(x+h) - f(x+h)g(x) + f(x+h)g(x) - f(x)g(x)}{h} \\
&= \lim_{h\to 0}\left[f(x+h)\frac{g(x+h) - g(x)}{h} + g(x)\frac{f(x+h) - f(x)}{h}\right] \\
&= \lim_{h\to 0}\left[f(x+h)\frac{g(x+h) - g(x)}{h}\right] + \lim_{h\to 0}\left[g(x)\frac{f(x+h) - f(x)}{h}\right] \\
&= \lim_{h\to 0}f(x+h) \cdot \lim_{h\to 0}\frac{g(x+h) - g(x)}{h} + \lim_{h\to 0}g(x) \cdot \lim_{h\to 0}\frac{f(x+h) - f(x)}{h} \\
&= f(x) \cdot g'(x) + g(x) \cdot f'(x)
\end{aligned}$$

A less formal but perhaps more intuitive justification can be provided by considering the product rule written in the form

$$\frac{dy}{dx} = \frac{d}{dx}(uv) = u\frac{dv}{dx} + v\frac{du}{dx}$$

and analyzing the relationship between the functions u, v and y when there is a small change in the variable x. Recall that the definition of the derivative (Section 13.2) is essentially the limit of $\frac{\text{change in } y}{\text{change in } x}$ as the 'change in x' goes to zero. Let δx (read 'delta x') and δy represent small changes in x and y, respectively. As $\delta x \to 0$, then $\frac{\delta y}{\delta x} \to \frac{dy}{dx}$, i.e. the derivative of y with respect to x.

Any small change in x, i.e. δx, will cause small changes, δu and δv, in the values of functions u and v respectively. Since $y = uv$, these changes will also cause a small change, δy, in the value of function y.

Now consider the rectangles in Figure 15.3. The area of the first smaller rectangle is $y = uv$. The values of u and v then increase by δu and δv respectively.

The area of the larger rectangle is $y + \delta y = uv + u\delta v + v\delta u + \delta u\delta v$.

The product uv changes by the amount $\delta y = u\delta v + v\delta u + \delta u\delta v$.

Dividing through by δx: $\frac{\delta y}{\delta x} = u\frac{\delta v}{\delta x} + v\frac{\delta u}{\delta x} + \delta u\frac{\delta v}{\delta x}$.

Let $\delta x \to 0$ and $\delta u \to 0$, then:

$$\frac{\delta y}{\delta x} = u\frac{\delta v}{\delta x} + v\frac{\delta u}{\delta x} + \delta u\frac{\delta v}{\delta x} \Rightarrow \frac{dy}{dx} = u\frac{dv}{dx} + v\frac{du}{dx} + 0 \cdot \frac{dv}{dx}$$

Giving $\frac{dy}{dx} = u\frac{dv}{dx} + v\frac{du}{dx}$, the product rule.

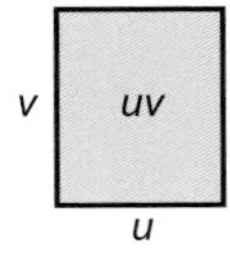

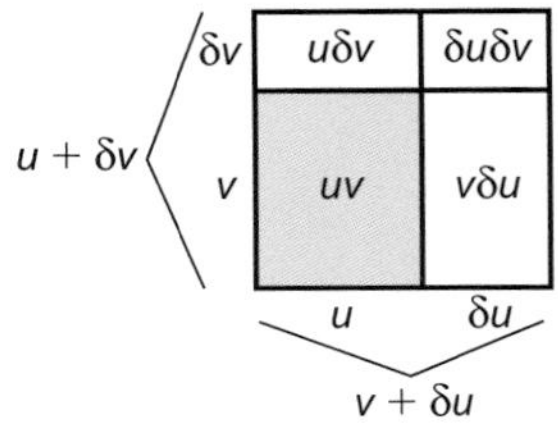

Figure 15.3

Example 4

Find the derivative of the function $y = (x^2 + 3x)(2x - 1)$ by:

a) expanding the binomial and differentiating term-by-term

b) the product rule.

Solution

a) Expanding gives $y = (x^2 + 3x)(2x - 1) = 2x^3 + 5x^2 - 3x$.

Therefore, $\frac{dy}{dx} = 6x^2 + 10x - 3$.

b) Let $u(x) = x^2 + 3x$ and $v(x) = 2x - 1$, then $y = u(x) \cdot v(x)$ or simply $y = uv$.

By the product rule (in Leibniz form),

$$\frac{dy}{dx} = \frac{d}{dx}(uv) = u\frac{dv}{dx} + v\frac{du}{dx} = (x^2 + 3x) \cdot 2 + (2x - 1) \cdot (2x + 3)$$
$$= (2x^2 + 6x) + (4x^2 + 4x - 3)$$
$$= 6x^2 + 10x - 3$$

This result agrees with the derivative we obtained earlier from differentiating the expanded polynomial.

Example 5

Given $y = x^2 \sin x$, find $\frac{dy}{dx}$.

Solution

Let $y = f(x) \cdot g(x) = x^2 \sin x \Rightarrow f(x) = x^2$ and $g(x) = \sin x$.

By the product rule (function notation form),

$$\frac{dy}{dx} = \frac{d}{dx}[f(x) \cdot g(x)] = f(x) \cdot g'(x) + g(x) \cdot f'(x)$$
$$= x^2 \cdot \cos x + (\sin x) \cdot 2x$$
$$\frac{dy}{dx} = x^2 \cos x + 2x \sin x$$

As with the chain rule, it is very helpful to remember the structure of the product rule in words.

$$\frac{dy}{dx} = \frac{d}{dx}\underbrace{[f(x) \cdot g(x)]}_{\text{product of two functions, i.e. factors}} = f(x) \cdot g'(x) + g(x) \cdot f'(x)$$

first factor, second factor (labels on $f(x)$ and $g(x)$)

product of two functions, i.e. factors = first factor × derivative of second factor + second factor × derivative of first factor

Example 6

Find an equation of the line tangent to the curve $y = \sin x \cos(2x)$ at the point where $x = \frac{\pi}{6}$.

Solution

To find the slope of the line tangent we need to find the derivative of $y = \sin x \cos(2x)$. To do this we will have to use more than one of the differentiation rules. Firstly, we need the product rule since the function consists of the two factors $\sin x$ and $\cos(2x)$. Secondly, the second factor is a composite of cosine and $2x$ so we need the chain rule. In essence the application of the chain rule will be 'nested' within the product rule.

$$\frac{dy}{dx} = \sin x\frac{d}{dx}(\cos(2x)) + \cos(2x)\frac{d}{dx}\sin x \quad \text{Product rule applied to entire function.}$$

$$\frac{dy}{dx} = \sin x(-2\sin(2x)) + \cos(2x)\cos x \quad \text{Chain rule for } \frac{d}{dx}(\cos(2x)).$$

$$\frac{dy}{dx} = -2\sin x \sin(2x) + \cos x \cos(2x)$$

At $x = \frac{\pi}{6}, \frac{dy}{dx} = -2\sin\left(\frac{\pi}{6}\right)\sin\left(2 \cdot \frac{\pi}{6}\right) + \cos\left(\frac{\pi}{6}\right)\cos\left(2 \cdot \frac{\pi}{6}\right)$

$$= -2\sin\left(\frac{\pi}{6}\right)\sin\left(\frac{\pi}{3}\right) + \cos\left(\frac{\pi}{6}\right)\cos\left(\frac{\pi}{3}\right) = -2\left(\frac{1}{2}\right)\left(\frac{\sqrt{3}}{2}\right) + \left(\frac{\sqrt{3}}{2}\right)\left(\frac{1}{2}\right) = -\frac{\sqrt{3}}{4}.$$

Hence, slope of the tangent line is $-\frac{\sqrt{3}}{4}$.

Find the y-coordinate of the tangent point:

At $x = \frac{\pi}{6}, y = \sin\left(\frac{\pi}{6}\right)\cos\left(2 \cdot \frac{\pi}{6}\right) = \sin\left(\frac{\pi}{6}\right)\cos\left(\frac{\pi}{3}\right) = \left(\frac{1}{2}\right)\left(\frac{1}{2}\right) = \frac{1}{4} \Rightarrow$ tangent point is $\left(\frac{\pi}{6}, \frac{1}{4}\right)$

Using point-slope form for a linear equation, gives

$$y - \frac{1}{4} = -\frac{\sqrt{3}}{4}\left(x - \frac{\pi}{6}\right) \Rightarrow y = -\frac{\sqrt{3}}{4}x + \frac{\pi\sqrt{3}}{24} + \frac{1}{4} \text{ or } y = -\frac{\sqrt{3}}{4}x + \frac{6 + \pi\sqrt{3}}{24}.$$

Therefore, an equation for the line tangent to $y = \sin x \cos(2x)$ at $x = \frac{\pi}{6}$ is

$$y = -\frac{\sqrt{3}}{4}x + \frac{6 + \pi\sqrt{3}}{24}.$$

Our GDC can give a quick visual check for this result. $\left[\frac{\pi}{6} \approx 0.523\,598\,78\right]$

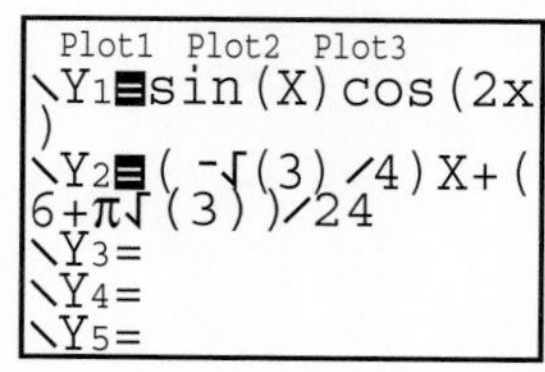

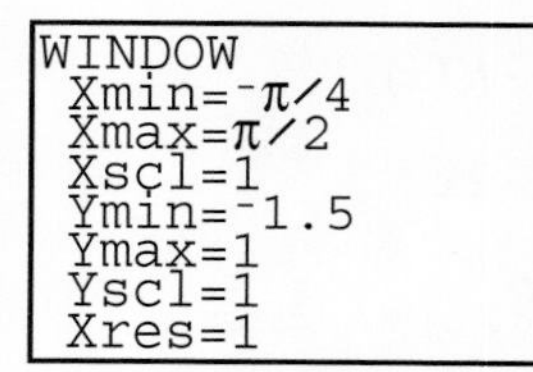

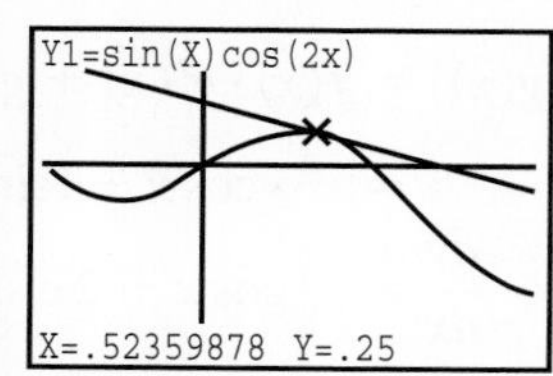

The quotient rule

Just as the derivative of the product of two functions is not the product of their derivatives, the derivative of a quotient of two functions is not the quotient of their derivatives. Let's derive a rule for the quotient of two functions by, once again, returning to the limit definition for the derivative. Let $y = F(x) = \dfrac{f(x)}{g(x)}$ where f and g are differentiable functions of x and their quotient is defined for all values of x in the domain.

As with the proof of the product rule we introduce a term, $f(g)g(x)$ in this case, and its opposite (thereby adding zero) in the numerator (in the 3rd line below). This allows us (in the 5th line) to isolate limits that are the derivatives of f and g.

$$F'(x) = \lim_{h \to 0} \frac{\dfrac{f(x+h)}{g(x+h)} - \dfrac{f(x)}{g(x)}}{h}$$

$$= \lim_{h \to 0} \frac{f(x+h)g(x) - f(x)g(x+h)}{h \cdot g(x)g(x+h)}$$

$$= \lim_{h \to 0} \frac{f(x+h)g(x) - f(x)g(x) + f(x)g(x) - f(x)g(x+h)}{h \cdot g(x)g(x+h)}$$

$$= \lim_{h \to 0} \frac{g(x)\dfrac{f(x+h) - f(x)}{h} - f(x)\dfrac{g(x+h) - g(x)}{h}}{g(x)g(x+h)}$$

$$= \frac{\lim\limits_{h \to 0} g(x) \cdot \lim\limits_{h \to 0} \dfrac{f(x+h) - f(x)}{h} - \lim\limits_{h \to 0} f(x) \cdot \lim\limits_{h \to 0} \dfrac{g(x+h) - g(x)}{h}}{\lim\limits_{h \to 0} g(x)g(x+h)}$$

$$= \frac{g(x) \cdot f'(x) - f(x) \cdot g'(x)}{g(x)g(x)}$$

$$= \frac{g(x) \cdot f'(x) - f(x) \cdot g'(x)}{[g(x)]^2}$$

The quotient rule

If y is a function in terms of x that can be expressed as the quotient of two functions u and v that are also in terms of x, the quotient $y = \frac{u}{v}$ can be differentiated as follows:

$$\frac{dy}{dx} = \frac{d}{dx}\left(\frac{u}{v}\right) = \frac{v\dfrac{du}{dx} - u\dfrac{dv}{dx}}{v^2}$$

or, equivalently, if $y = \dfrac{f(x)}{g(x)}$, then

$$\frac{dy}{dx} = \frac{d}{dx}\left[\frac{f(x)}{g(x)}\right] = \frac{g(x) \cdot f'(x) - f(x) \cdot g'(x)}{[g(x)]^2}$$

As with the chain rule and the product rule, it is helpful to recognize the structure of the quotient rule by remembering it in words:

$$\begin{pmatrix}\text{derivative}\\ \text{of quotient}\end{pmatrix} = \frac{(\text{denominator}) \times \begin{pmatrix}\text{derivative of}\\ \text{numerator}\end{pmatrix} - (\text{numerator})\begin{pmatrix}\text{derivative of}\\ \text{denominator}\end{pmatrix}}{(\text{denominator})^2}$$

• **Hint:** Since order is important in subtraction (subtraction is not commutative), be sure to set up the numerator of the quotient rule correctly.

• **Hint:** Note that we could have proved the quotient rule by writing the quotient $\frac{f(x)}{g(x)}$ as the product $f(x)\,[g(x)]^{-1}$ and apply the product rule and chain rule. As some of the examples here show, the derivative of a quotient can also be found by means of the product rule and/or the chain rule.

Example 7

For each function, find its derivative (i) by the quotient rule, and (ii) by another method.

a) $g(x) = \frac{5x-1}{3x^2}$ b) $h(x) = \frac{1}{2x-3}$ c) $f(x) = \frac{3x-2}{2x-5}$

Solution

a) (i) $g(x) = y = \frac{u}{v} = \frac{5x-1}{3x^2}$

$$g'(x) = \frac{dy}{dx} = \frac{v\frac{du}{dx} - u\frac{dv}{dx}}{v^2} = \frac{3x^2 \cdot 5 - (5x-1)\cdot 6x}{(3x^2)^2}$$

$$= \frac{15x^2 - 30x^2 + 6x}{9x^4}$$

$$= \frac{3x(-5x+2)}{9x^4}$$

$$g'(x) = \frac{-5x+2}{3x^3}$$

(ii) Using algebra, 'split' the numerator:

$$g(x) = \frac{5x-1}{3x^2} = \frac{5x}{3x^2} - \frac{1}{3x^2} = \frac{5}{3x} - \frac{1}{3x^2} = \frac{5}{3}x^{-1} - \frac{1}{3}x^{-2}$$

Now, differentiate term-by-term using the power rule.

$$g'(x) = \frac{5}{3}\frac{d}{dx}(-x^{-1}) - \frac{1}{3}\frac{d}{dx}(x^{-3})$$

$$= \frac{5}{3}(-x^{-2}) - \frac{1}{3}(-2x^{-3})$$

$$g'(x) = -\frac{5}{3x^2} + \frac{2}{3x^3}$$

[Results for (i) and (ii) are equivalent:

$$-\frac{5}{3x^2} + \frac{2}{3x^3} = -\frac{5}{3x^2}\cdot\frac{x}{x} + \frac{2}{3x^3} = -\frac{5x}{3x^3} + \frac{2}{3x^3} = \frac{-5x+2}{3x^3}$$]

b) (i) $y = \frac{f(x)}{g(x)} = \frac{1}{2x-3} \Rightarrow f(x) = 1$ and $g(x) = 2x - 3$

By the quotient rule (function notation form),

$$\frac{dy}{dx} = \frac{d}{dx}\left[\frac{f(x)}{g(x)}\right] = \frac{g(x)\cdot f'(x) - f(x)\cdot g'(x)}{[g(x)]^2}$$

$$= \frac{(2x-3)\cdot 0 - 1\cdot(2)}{(2x-3)^2}$$

$$\frac{dy}{dx} = -\frac{2}{(2x-3)^2}$$

(ii) $y = f(g(x)) = \frac{1}{2x-3} = (2x-3)^{-1} \Rightarrow$ 'outside' function is $f(u) = u^{-1}$

$\Rightarrow f'(u) = -u^{-2}$

$\Rightarrow$ 'inside' function is $g(x) = 2x - 3$

By the chain rule (function notation form),

$$\frac{dy}{dx} = f'(g(x))\cdot g'(x) = -(2x-3)^{-2}\cdot 2$$

$$\frac{dy}{dx} = -\frac{2}{(2x-3)^2}$$

c) (i) $f(x) = y = \frac{u}{v} = \frac{3x - 2}{2x - 5}$ $f'(x) = \frac{dy}{dx} = \frac{v\frac{du}{dx} - u\frac{dv}{dx}}{v^2}$

$$= \frac{(2x - 5) \cdot 3 - (3x - 2) \cdot 2}{(2x - 5)^2}$$

$$= \frac{6x - 15 - 6x + 4}{(2x - 5)^2}$$

$$f'(x) = \frac{-11}{(2x - 5)^2}$$

(ii) Rewrite $f(x)$ as a product and apply the product rule (with chain rule imbedded).

$f(x) = y = \frac{3x - 2}{2x - 5} = (3x - 2)(2x - 5)^{-1} \Rightarrow y = uv, u = 3x - 2$ and $v = (2x - 5)^{-1}$

Note: $v = (2x - 5)^{-1}$ is a composite function, so we'll need the chain rule to find $\frac{dv}{dx}$.

$$f'(x) = \frac{d}{dx}(uv) = u\frac{dv}{dx} + v\frac{du}{dx}$$

$$= (3x - 2) \cdot \frac{d}{dx}[(2x - 5)^{-1}] + (2x - 5)^{-1} \cdot 3$$

$$= (3x - 2)[-(2x - 5)^{-2} \cdot 2] + 3(2x - 5)^{-1}$$

Chain rule applied for $\frac{d}{dx}[(2x - 5)^{-1})]$.

$$= (-6x + 4)(2x - 5)^{-2} + 3(2x - 5)^{-1}$$

$$= (2x - 5)^{-2}[(-6x + 4) + 3(2x - 5)]$$

Factorizing out GCF of $(2x - 5)^2$.

$$= (2x - 5)^{-2}[-6x + 4 + 6x - 15]$$

$$f'(x) = \frac{-11}{(2x - 5)^2}$$

• **Hint:** The function $h(x) = \frac{3x^2}{5x - 1}$ initially looks similar to the function g in Example 7, part a) (they're reciprocals). However, it is *not* possible to 'split' the denominator and express as two fractions. Recognize that $\frac{3x^2}{5x - 1}$ is *not* equivalent to $\frac{3x^2}{5x} - \frac{3x^2}{1}$. Hence, in order to differentiate $h(x) = \frac{3x^2}{5x - 1}$ we would apply either the quotient rule, or the product rule with the function rewritten as $h(x) = 3x^2 (5x - 1)^{-1}$ and using the chain rule to differentiate the factor $(5x - 1)^{-1}$.

As Example 7 demonstrates, before differentiating a quotient it is worthwhile to consider if performing some algebra may allow other more efficient differentiation techniques to be used.

Higher derivatives

If $y = f(x)$ is a function of x then, in general, the derivative – expressed as either $\frac{dy}{dx}$ or $f'(x)$ – will be some other function of x. As we have learned the derivative indicates the rate of change of $f(x)$ with respect to x, as a function of x. In Section 13.3 we took the 'derivative of the derivative' of a function, that is, a function's second derivative, denoted by $\frac{d^2y}{dx^2}$ or $f''(x)$. The second derivative is an effective tool in verifying maximum, minimum and inflexion points on the graph of a function. In general, $\frac{d^2y}{dx^2}$ will also be a function of x and so may be differentiated to give the third derivative of y with respect to x, denoted by $\frac{d^3y}{dx^3}$. The nth derivative of y with respect to x is denoted by $\frac{d^ny}{dx^n}$. If the notation $f(x)$ is used, the first, second and third derivatives are written as $f'(x)$, $f''(x)$ and $f'''(x)$, respectively. The fourth derivative and higher is denoted using a superscript number rather than a 'prime' mark. For example, $f^{(4)}(x)$ represents the fourth derivative of the function f with respect to x.

The process of computing the nth derivative of a function can be very tedious and can only be achieved by computing the successive derivatives in turn. It is worthwhile to attempt to simplify the function $\frac{dy}{dx}$ before differentiating to find $\frac{d^2y}{dx^2}$, and in turn try to simplify this result before computing $\frac{d^3y}{dx^3}$, and so on.

Example 8

Given $y = \frac{1}{x}$, find a formula for the nth derivative $\frac{d^ny}{dx^n}$.

Solution

Let's take successive derivatives of the function until we can discern a pattern and then formulate a conjecture for the formula.

$$y = \frac{1}{x} = x^{-1}$$

$$\frac{dy}{dx} = -x^{-2} = \frac{-1}{x^2}$$

$$\frac{d^2y}{dx^2} = (-2)(-1)x^{-3} = \frac{2}{x^3}$$

$$\frac{d^3y}{dx^3} = (-3)(2)(1)x^{-4} = \frac{-6}{x^4}$$

$$\frac{d^4y}{dx^4} = (4)(3)(2)(1)x^{-5} = \frac{24}{x^5}$$

$$\frac{d^5y}{dx^5} = (-5)(4)(3)(2)(1)x^{-6} = \frac{-120}{x^6}$$

We observe that the sign of the result alternates: negative when n is odd, and positive when n is even. Thus, we need to incorporate the expression $(-1)^n$ into our formula since the successive values of $(-1)^n$ are $-1, 1, -1, 1, \ldots$. Another factor needs to be $n!$ (n factorial) because $n! = n(n-1)(n-2)\cdots 2 \cdot 1$. The last piece of the formula is that the power of x in the denominator is one more than the value of n.

Therefore, $\frac{d^{(n)}y}{dx^{(n)}} = \frac{(-1)^n n!}{x^{n+1}}$.

Exercise 15.1

1 Find the derivative of each function.

a) $y = (3x - 8)^4$ b) $y = \sqrt{1 - x}$ c) $y = \sin x \cos x$

d) $y = 2\sin\left(\frac{x}{2}\right)$ e) $y = (x^2 + 4)^{-2}$ f) $y = \frac{x + 1}{x - 1}$

g) $y = \frac{1}{\sqrt{x + 2}}$ h) $y = \cos^2 x$ i) $y = x\sqrt{1 - x}$

j) $y = \frac{1}{3x^2 - 5x + 7}$ k) $y = \sqrt[3]{2x + 5}$ l) $y = (2x - 1)^3(x^4 + 1)$

m) $y = \frac{\sin x}{x}$ n) $y = \frac{x^2}{x + 2}$ o) $y = \sqrt[3]{x^2}\cos x$

2 Find the equation of the line tangent to the given curve at the specified value of x. Express the equation exactly in the form $y = mx + c$.

a) $y = (2x^2 - 1)^3$ $x = -1$ b) $y = \sqrt{3x^2 - 2}$ $x = 3$

c) $y = \sin 2x$ $x = \pi$ d) $y = \frac{x^3 + 1}{2x}$ $x = 1$

3 An object moves along a line so that its position s relative to a starting point at any time $t \geqslant 0$ is given by $s(t) = \cos(t^2 - 1)$.

a) Find the velocity of the object as a function of t.

b) What is the object's velocity at $t = 0$?

c) In the interval $0 < t < 2.5$, find any times (values of t) for which the object is stationary.

d) Describe the object's motion during the interval $0 < t < 2.5$.

For questions 4–6, find the equation of a) the tangent, and b) the normal to the curve at the given point.

4 $y = \frac{2}{x^2 - 8}$ at (3, 2)

5 $y = \sqrt{1 + 4x}$ at (2, 3)

6 $y = \frac{x}{x + 1}$ at $(1, \frac{1}{2})$

7 Consider the trigonometric curve $y = \sin\left(2x - \frac{\pi}{2}\right)$.

a) Find $\frac{dy}{dx}$ and $\frac{d^2y}{dx^2}$.

b) Find the exact coordinates of any inflexion points for the curve in the interval $0 < x < \pi$.

8 A curve has equation $y = x(x - 4)^2$.

a) For this curve, find

(i) the x-intercepts

(ii) the coordinates of the maximum point

(iii) the coordinates of the point of inflexion.

b) Use your answers to part a) to sketch a graph of the curve for $0 \leqslant x \leqslant 4$, clearly indicating the features you have found in part a).

9 Consider the function $f(x) = \frac{x^2 - 3x + 4}{(x + 1)^2}$.

a) Show that $f'(x) = \frac{5x - 11}{(x + 1)^3}$.

b) Show that $f''(x) = \frac{-10x + 38}{(x + 1)^4}$.

c) Does the graph of f have an inflexion point at $x = 3.8$? Explain.

10 Find the first and second derivatives of the function $f(x) = \dfrac{x-a}{x+a}$.

11 Given $y = \dfrac{1}{1-x}$, find a formula for the nth derivative $\dfrac{d^ny}{dx^n}$.

12 The graph of the function $g(x) = \dfrac{8}{4+x^2}$ is called the *witch of Agnesi*.

a) Find the exact coordinates of any extreme values or inflexion points.

b) Determine all values of x for which (i) $g(x) < 0$, (ii) $g(x) = 0$, and (iii) $g(x) > 0$.

c) Find (i) $\lim\limits_{x \to -\infty} g(x)$, and (ii) $\lim\limits_{x \to +\infty} g(x)$.

d) Sketch the graph of g.

13 Use the product rule to prove the constant multiple rule for differentiation. That is, show that $\dfrac{d}{dx}(c \cdot f(x)) = c \cdot \dfrac{d}{dx}(f(x))$ for any constant c.

14 If $y = x^4 - 6x^2$, show that y, $\dfrac{dy}{dx}$ and $\dfrac{d^2y}{dx^2}$ are all negative on the interval $0 < x < 1$, but that $\dfrac{d^3y}{dx^3}$ is positive on the same interval.

15.2 Derivatives of trigonometric and exponential functions

Although it is important to provide formal justifications for any of our differentiation rules (as we did in the previous section), we should not forget that the derivative is a rule that gives us the slope of the line tangent to the graph of a function at a particular point. Thus, we can use a function's derivative to deduce the behaviour of its graph. Conversely, we can gain insight about the derivative of a function from the shape of its graph.

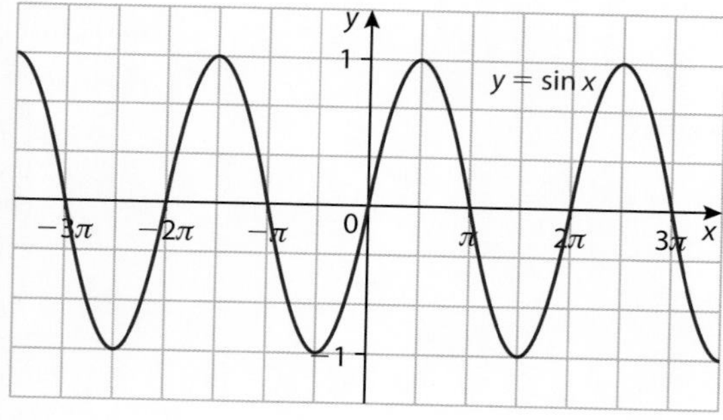

Figure 15.4

In Chapter 13, we formally determined that the derivative of $\sin x$ is $\cos x$ and that the derivative of $\cos x$ is $-\sin x$ by using the limit definition of the derivative. We could have made a very confident conjecture for the derivative of $\sin x$ by analyzing its graph as follows.

We start with the graph of $f(x) = \sin x$ (Figure 15.4). The graph of $y = \sin x$ is periodic, with period 2π, so the same will be true of its derivative that gives the slope at each point on the graph. Therefore, it's only necessary for us to consider the portion of the graph in the interval $0 \leqslant x \leqslant 2\pi$.

Figure 15.5 shows two pairs of axes having equal scales on the x- and y-axes and corresponding x-coordinates aligned vertically. On the top pair of axes $y = \sin x$ is graphed with tangent lines drawn at nine selected points. The points were chosen such that the slopes of the tangents at those points, in order, appear to be equal to $1, \frac{1}{2}, 0, -\frac{1}{2}, -1, -\frac{1}{2}, 0, \frac{1}{2}, 1$. The values of these slopes were then plotted in the bottom graph with the y-coordinate of each point indicating the slope of the curve for that particular x value. Hence, the points in the bottom pair of axes should be on the graph of the derivative of $y = \sin x$.

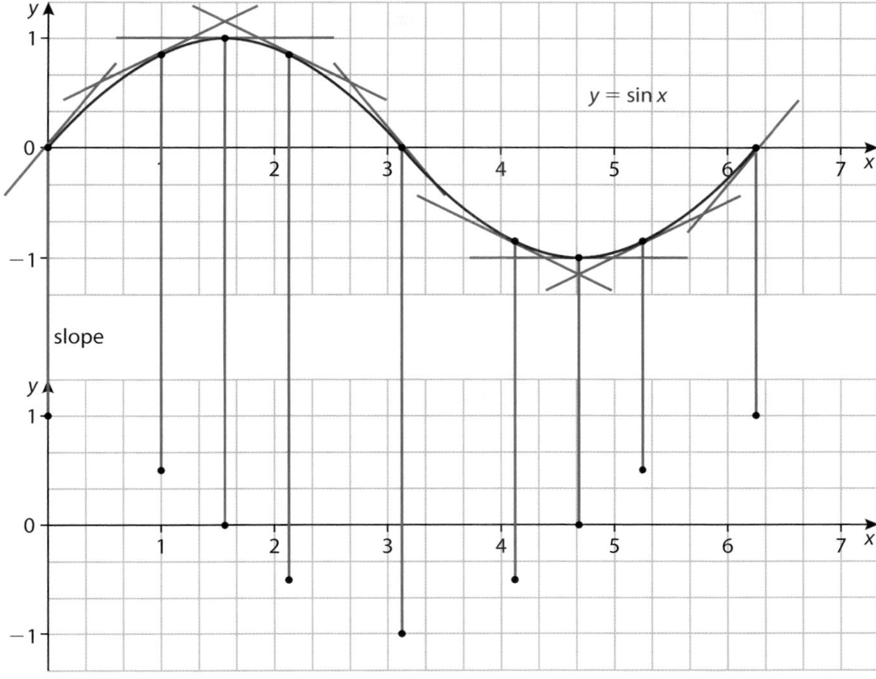

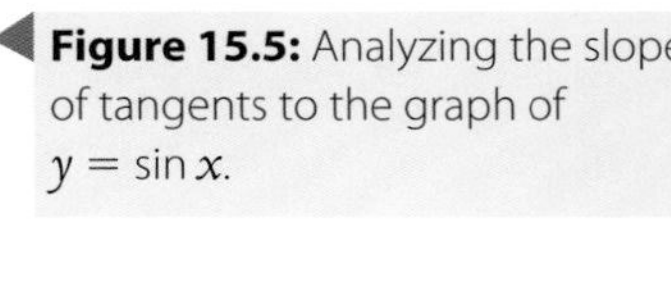
Figure 15.5: Analyzing the slope of tangents to the graph of $y = \sin x$.

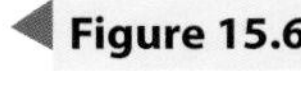
• **Hint:** Note that the graphs in Figures 15.4, 15.5, 15.6 and 15.7 have x in radians. As mentioned previously, we must only use radian measure when trigonometric functions are involved in calculus.

Figure 15.6 is the same as Figure 15.5 except with the graph of $y = \sin x$, the grid lines and the lines connecting points between the two graphs removed.

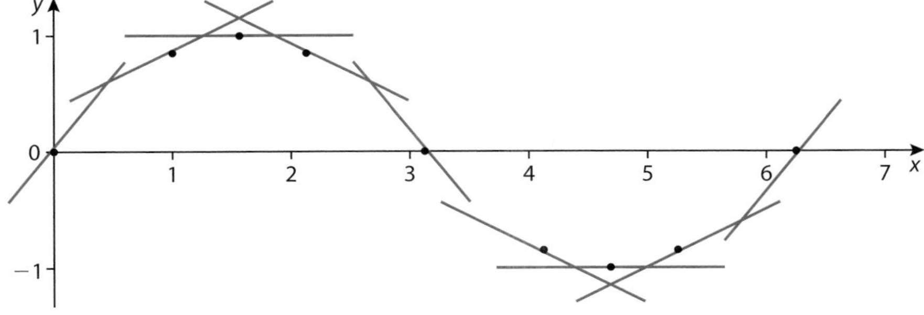

Figure 15.6

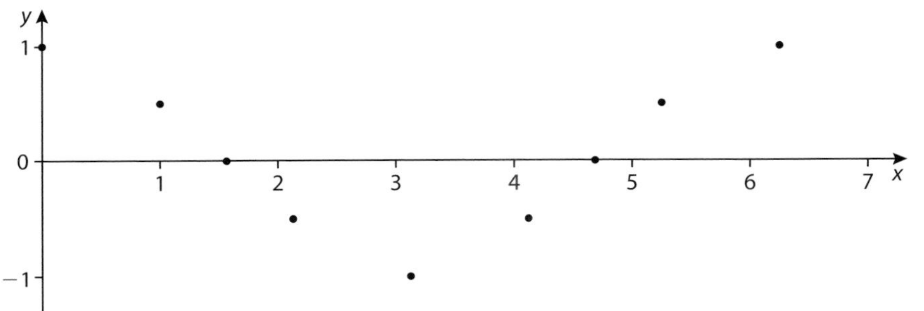

Figure 15.7

Clearly the points representing the slope of the tangents to $y = \sin x$ plotted in Figure 15.7 are tracing out the graph of $y = \cos x$.

Although we will use this informal approach to conjecture the derivatives for $y = e^x$ and $y = \ln x$, it does not always work so smoothly. For example, let's analyze the graph of $y = \tan x$ in an attempt to guess its derivative.

We can use our GDC command that evaluates the derivative of a function at a specified point to graph the value of the derivative at all points on a graph. We used this technique in Chapter 13 to confirm the result in Example 9 part d). The GDC screen images below show the graph of $y = \tan x$ and then the GDC graphing its derivative (in bold) on the same set of axes. Although, as pointed out in Section 13.3, in general it is incorrect to graph a function and its derivative on the same pair of axes (units on the vertical axis will not be the same), it is helpful in seeing the connection between the graph of a function and that of its derivative.

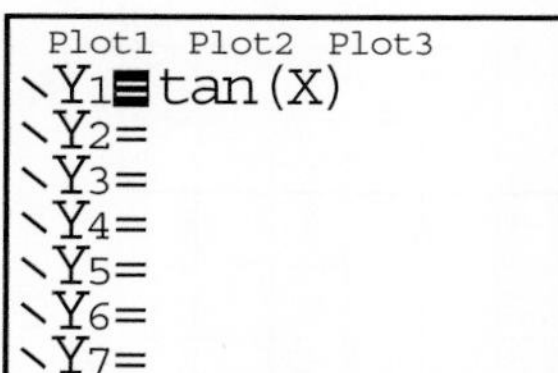

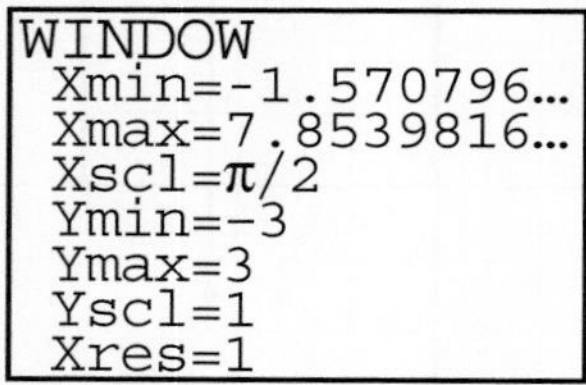

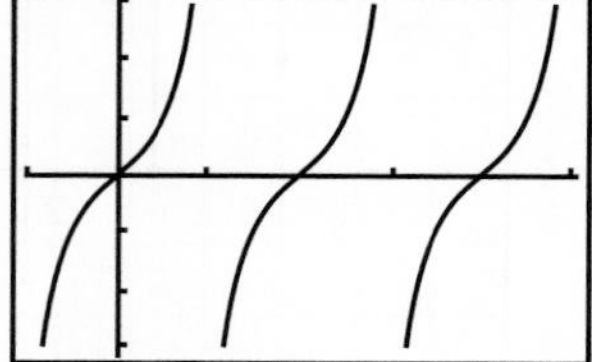

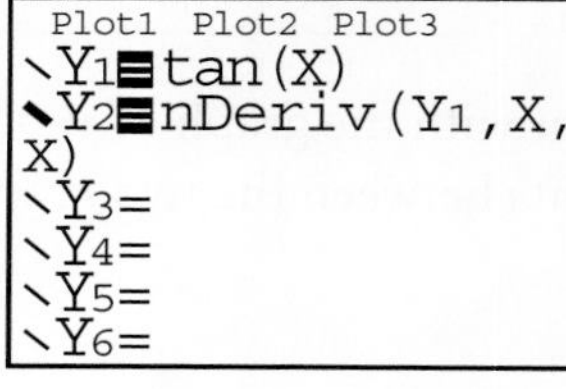

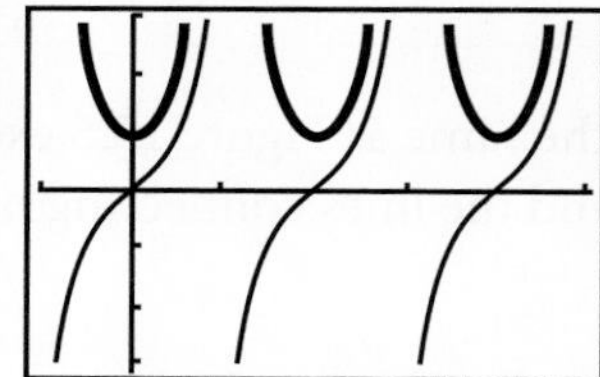

The graph of the derivative of $\tan x$ is always above the x-axis meaning that the derivative is always positive. This clearly agrees with the fact that the tangent function, except for where it is undefined, is always increasing (moving upwards) as the values of x increase. However, the shape of the graph does not bring to mind an easy conjecture for a rule for the derivative of $\tan x$.

Rather than use the limit definition for finding the derivative of $\tan x$ let's write $\tan x$ as $\frac{\sin x}{\cos x}$ and use the quotient rule.

$$\begin{aligned}
\frac{d}{dx}(\tan x) = \frac{d}{dx}\left(\frac{\sin x}{\cos x}\right) &= \frac{\cos x \frac{d}{dx}(\sin x) - \sin x \frac{d}{dx}(\cos x)}{\cos^2 x} \\
&= \frac{\cos x \cos x - \sin x(-\sin x)}{\cos^2 x} \\
&= \frac{\cos^2 x + \sin^2 x}{\cos^2 x} \\
&= \frac{1}{\cos^2 x} \\
&= \sec^2 x
\end{aligned}$$

Therefore, $\frac{d}{dx}(\tan x) = \sec^2 x$.

Similarly, it can be shown that $\frac{d}{dx}(\cot x) = -\csc^2 x$.

To find the derivative of sec x we can use the chain rule as follows.

$$\frac{d}{dx}(\sec x) = \frac{d}{dx}\left(\frac{1}{\cos x}\right) = \frac{d}{dx}[(\cos x)^{-1}]$$

$$= -(\cos x)^{-2}(-\sin x)$$ Applying chain rule.

$$= \frac{\sin x}{\cos^2 x}$$

$$= \frac{1}{\cos x} \cdot \frac{\sin x}{\cos x}$$

$$= \sec x \tan x$$

Therefore, $\frac{d}{dx}(\sec x) = \sec x \tan x$.

Similarly, it can be shown that $\frac{d}{dx}(\csc x) = -\csc x \cot x$.

The table below lists the derivatives of the six trigonometric functions.

$f(x)$	$f'(x)$
$\sin x$	$\cos x$
$\cos x$	$-\sin x$
$\tan x$	$\sec^2 x$
$\cot x$	$-\csc^2 x$
$\sec x$	$\sec x \tan x$
$\csc x$	$-\csc x \cot x$

Example 9

Find the derivative of each function.

a) $y = \cos(\sqrt{x})$

b) $y = \frac{x^3}{\sin x}$

c) $y = x^2 \tan(3x)$

d) $y = \sec^2(3x)$

Solution

a) $$\frac{dy}{dx} = \frac{d}{dx}[\cos(\sqrt{x})] = -\sin(\sqrt{x}) \cdot \frac{d}{dx}(\sqrt{x})$$ Applying chain rule.

$$= -\sin(\sqrt{x}) \cdot \frac{d}{dx}\left(x^{\frac{1}{2}}\right)$$

$$= -\sin(\sqrt{x}) \cdot \left(\frac{1}{2}x^{-\frac{1}{2}}\right)$$ Applying power rule.

Therefore, $\frac{dy}{dx} = -\frac{\sin(\sqrt{x})}{2\sqrt{x}}$.

b) Method 1 (quotient rule):

$$\frac{dy}{dx} = \frac{d}{dx}\left(\frac{x^3}{\sin x}\right) = \frac{\sin x \cdot \frac{d}{dx}(x^3) - x^3 \cdot \frac{d}{dx}(\sin x)}{\sin^2 x}$$ Applying quotient rule.

Therefore, $\frac{dy}{dx} = \frac{3x^2 \sin x - x^3 \cos x}{\sin^2 x}$.

Method 2 (product rule and chain rule):

$$\frac{dy}{dx} = \frac{d}{dx}\left(\frac{x^3}{\sin x}\right) = \frac{d}{dx}[x^3 \cdot (\sin x)^{-1}]$$ Rewriting as a product.

$$= x^3 \cdot \frac{d}{dx}[(\sin x)^{-1}] + (\sin x)^{-1} \cdot \frac{d}{dx}(x^3)$$ Applying product rule.

$$= x^3[-(\sin x)^{-2}\cos x] + (\sin x)^{-1}(3x^2)$$

$$= (\sin x)^{-2}[-x^3\cos x + 3x^2\sin x]$$ Factor out common factor of $(\sin x)^{-2}$.

Therefore, $\dfrac{dy}{dx} = \dfrac{3x^2\sin x - x^3\cos x}{\sin^2 x}$.

c) $\dfrac{dy}{dx} = \dfrac{d}{dx}[x^2\tan(3x)] = x^2 \cdot \dfrac{d}{dx}(\tan(3x)) + \tan(3x) \cdot \dfrac{d}{dx}(x^2)$

$$= x^2 \cdot \frac{d}{dx}(\tan(3x)) + \tan(3x) \cdot \frac{d}{dx}(x^2)$$ Applying product rule.

$$= x^2(3\sec^2(3x)) + (\tan(3x))(2x)$$ Applying chain rule for $\frac{d}{dx}(\tan(3x))$.

$$= 3x^2\sec^2(3x) + 2x\tan(3x)$$

d) $\dfrac{dy}{dx} = \dfrac{d}{dx}[\sec^2(3x)] = \dfrac{d}{dx}[(\sec(3x))^2]$

$$= 2\sec(3x) \cdot \frac{d}{dx}(\sec(3x))$$ Applying chain rule 1st time.

$$= 2\sec(3x) \cdot (\sec(3x)\tan(3x) \cdot \frac{d}{dx}(3x))$$ Applying chain rule 2nd time.

$$= 2\sec(3x) \cdot (\sec(3x)\tan(3x) \cdot 3)$$

$$= 6\sec^2(3x)\tan(3x)$$ Equivalent to $\frac{6\sin(3x)}{\cos^3(3x)}$.

Example 10

The motion of a particle moving along a straight line for the interval $0 < t < 12$ (t in seconds) is given by the function $s(t) = \sin\left(\frac{t}{2}\right) - \cos\left(\frac{t}{2}\right) + 1$, where s is the particle's displacement in centimetres from the origin O. The particle's displacement is negative when left of O and positive when right of O.

a) Find the exact time and displacement when the particle is (i) furthest to the right and (ii) furthest to the left during the interval $0 < t < 12$.

b) Find the particle's maximum speed to the right exactly and at what exact time it occurs.

Solution

For part a) displacement can only be a maximum or minimum when velocity is zero, i.e. $v(t) = 0$. Similarly for part b) velocity can only be a maximum or minimum when acceleration is zero, i.e. $a(t) = 0$. So we begin by finding the first and second derivatives of $s(t)$ giving us the velocity function, $v(t)$, and acceleration function, $a(t)$, respectively.

a) $v(t) = s'(t) = \frac{d}{dx}\left[\sin\left(\frac{t}{2}\right) - \cos\left(\frac{t}{2}\right) + 1\right] = \frac{1}{2}\cos\left(\frac{t}{2}\right) + \frac{1}{2}\sin\left(\frac{t}{2}\right)$

Solve $\frac{1}{2}\cos\left(\frac{t}{2}\right) + \frac{1}{2}\sin\left(\frac{t}{2}\right) = 0$:

$\sin\left(\frac{t}{2}\right) = -\cos\left(\frac{t}{2}\right)$

$\frac{\sin\left(\frac{t}{2}\right)}{\cos\left(\frac{t}{2}\right)} = \tan\left(\frac{t}{2}\right) = -1$ Given that $\cos\left(\frac{t}{2}\right) \neq 0$.

$\tan\left(\frac{t}{2}\right) = -1$ when $\frac{t}{2} = \frac{3\pi}{4} + k \cdot \pi, k \in \mathbb{Z}$

Thus, $t = \frac{3\pi}{2} + k \cdot 2\pi, k \in \mathbb{Z}$. For $0 < t < 12$, $t = \frac{3\pi}{2}$ or $t = \frac{7\pi}{2}$.

(i) Checking the sign (direction) of the particle's velocity just before and after these two times will show if they are maximum or minimum values. Test values are $t = \pi$ and 2π for $t = \frac{3\pi}{2}$.

$v(\pi) = \frac{1}{2}\cos\left(\frac{\pi}{2}\right) + \frac{1}{2}\sin\left(\frac{\pi}{2}\right) = 0 + \frac{1}{2} \cdot 1 = \frac{1}{2} > 0 \Rightarrow$ displacement increasing before $t = \frac{3\pi}{2}$

$v(2\pi) = \frac{1}{2}\cos\left(\frac{2\pi}{2}\right) + \frac{1}{2}\sin\left(\frac{2\pi}{2}\right) = \frac{1}{2}(-1) + 0 < 0 \Rightarrow$ displacement decreasing before $t = \frac{3\pi}{2}$

Hence, $s\left(\frac{3\pi}{2}\right) = \sin\left(\frac{3\pi}{4}\right) - \cos\left(\frac{3\pi}{4}\right) + 1 = \frac{\sqrt{2}}{2} - \left(-\frac{\sqrt{2}}{2}\right) + 1$
$= 1 + \sqrt{2}$ is a maximum.

Therefore, the particle is furthest to the right (maximum displacement) at $t = \frac{3\pi}{2}$ seconds when its displacement is $1 + \sqrt{2}$ cm.

(ii) Test values are $t = 3\pi$ and 4π for $t = \frac{7\pi}{2}$.

$v(3\pi) = \frac{1}{2}\cos\left(\frac{3\pi}{2}\right) + \frac{1}{2}\sin\left(\frac{3\pi}{2}\right) = 0 + \frac{1}{2}(-1) < 0 \Rightarrow$ displacement decreasing before $t = \frac{7\pi}{2}$

$v(4\pi) = \frac{1}{2}\cos(2\pi) + \frac{1}{2}\sin(2\pi) = \frac{1}{2}(1) + 0 > 0 \Rightarrow$ displacement increasing after $t = \frac{7\pi}{2}$

Hence, $s\left(\frac{7\pi}{2}\right) = \sin\left(\frac{7\pi}{4}\right) - \cos\left(\frac{7\pi}{4}\right) + 1 = -\frac{\sqrt{2}}{2} - \left(\frac{\sqrt{2}}{2}\right) +$
$1 = 1 - \sqrt{2}$ is a minimum.

Therefore, the particle is furthest to the left (minimum displacement) at $t = \frac{7\pi}{2}$ seconds when its displacement is $1 - \sqrt{2}$ cm.

b) $a(t) = v'(t) = \frac{d}{dx}\left[\frac{1}{2}\cos\left(\frac{t}{2}\right) + \frac{1}{2}\sin\left(\frac{t}{2}\right)\right] = -\frac{1}{4}\sin\left(\frac{t}{2}\right) + \frac{1}{4}\cos\left(\frac{t}{2}\right)$

Solve $\frac{1}{4}\cos\left(\frac{t}{2}\right) - \frac{1}{4}\sin\left(\frac{t}{2}\right) = 0$:

$\sin\left(\frac{t}{2}\right) = \cos\left(\frac{t}{2}\right)$

$\frac{\sin\left(\frac{t}{2}\right)}{\cos\left(\frac{t}{2}\right)} = \tan\left(\frac{t}{2}\right) = 1$ Given that $\cos\left(\frac{t}{2}\right) \neq 0$.

$\tan\left(\frac{t}{2}\right) = 1$ when $\frac{t}{2} = \frac{\pi}{4} + k \cdot \pi, k \in \mathbb{Z}$

Thus, $t = \frac{\pi}{2} + k \cdot 2\pi, k \in \mathbb{Z}$. For $0 < t < 12$, $t = \frac{\pi}{2}$ or $t = \frac{5\pi}{2}$.

To find maximum velocity (moving right, speed > 0), let's evaluate the velocity at all critical points, i.e. at endpoints for the time interval, $t = 0$ and $t = 12$, and where the acceleration is zero, $t = \frac{\pi}{2}$ and $t = \frac{5\pi}{2}$.

$v(0) = \frac{1}{2}\cos(0) + \frac{1}{2}\sin(0) = \frac{1}{2}$

$v\left(\frac{\pi}{2}\right) = \frac{1}{2}\cos\left(\frac{\pi}{4}\right) + \frac{1}{2}\sin\left(\frac{\pi}{4}\right) = \frac{\sqrt{2}}{4} + \frac{\sqrt{2}}{4} = \frac{\sqrt{2}}{2} \approx 0.707$

$v\left(\frac{5\pi}{2}\right) = \frac{1}{2}\cos\left(\frac{5\pi}{4}\right) + \frac{1}{2}\sin\left(\frac{5\pi}{4}\right) = -\frac{\sqrt{2}}{4} - \frac{\sqrt{2}}{4} = -\frac{\sqrt{2}}{2} \approx -0.707$

$v(12) = \frac{1}{2}\cos(6) + \frac{1}{2}\sin(6) \approx -0.424$

Therefore, the particle has a maximum velocity of $\frac{\sqrt{2}}{2}$ cm/sec when $t = \frac{\pi}{2}$ seconds.

A graph of the displacement function $s(t) = \sin\left(\frac{t}{2}\right) - \cos\left(\frac{t}{2}\right) + 1$ gives a good visual confirmation of our results.

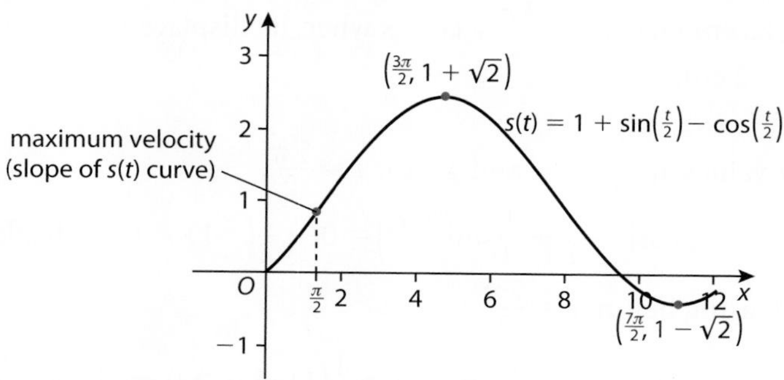

Derivatives of exponential functions

Let's review some important facts about exponential functions in general. An exponential function with base b is defined as $f(x) = b^x$, $b > 0$ and $b \neq 1$. The graph of f passes through (0, 1), has the x-axis as a horizontal asymptote, and, depending on the value of the base of the exponential function b, will either be a continually increasing exponential growth curve (Figure 15.8) or a continually decreasing exponential decay curve (Figure 15.9).

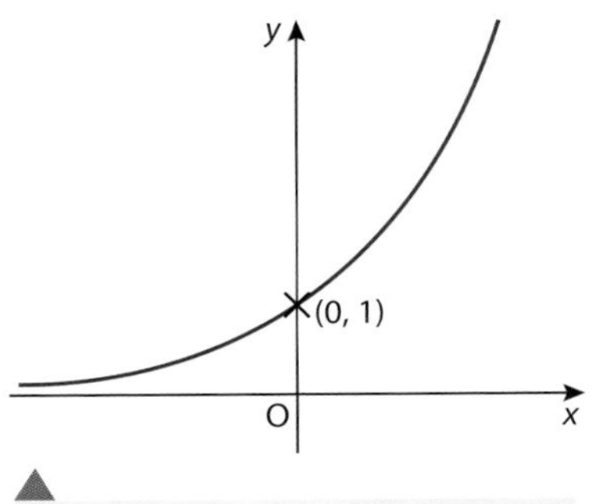

Figure 15.8

as $x \to \infty, f(x) \to \infty$

f is an increasing function

exponential growth curve

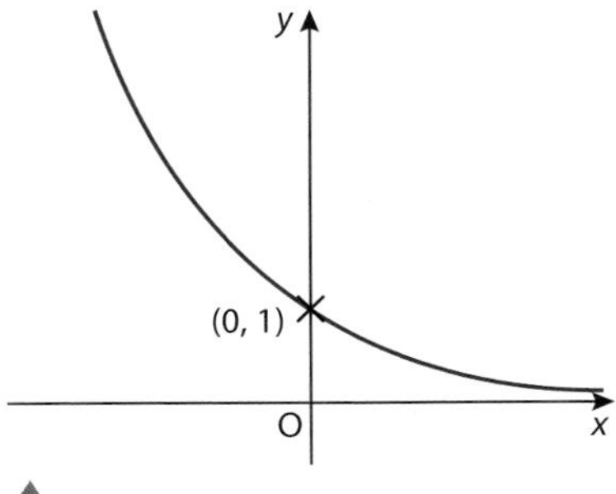

Figure 15.9

as $x \to \infty, f(x) \to 0$

f is a decreasing function

exponential decay curve

In Chapter 5 we learned that *the* exponential function e^x, sometimes written as 'exp x', is a particularly important function for modelling exponential growth and decay. The number e was defined in Section 5.3 as the limit of $\left(1 + \frac{1}{x}\right)^x$ as $x \to \infty$. Although the method was not successful in coming up with a conjecture for the derivative of the tangent function, let's try to guess the derivative of e^x by having our GDC graph its derivative.

- **Hint:** You may be tempted to find the derivative of e^x by applying the rule for differentiating powers $\frac{d}{dx}(x^n) = nx^{n-1}$ but this *only* applies if a variable is raised to a constant power. An exponential function, such as $y = e^x$, is a constant raised to a variable power, so the power rule does *not* apply.

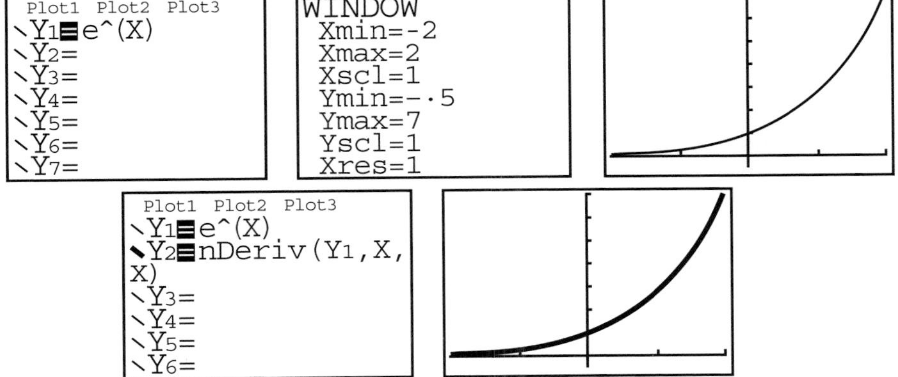

The graph of the derivative of e^x appears to be identical to e^x itself! This is a very interesting result, but one which we will see fits in exactly with the nature of exponential growth/decay.

Let's try to apply the limit definition of the derivative to provide a formal justification.

$$\frac{d}{dx}(e^x) = \lim_{h \to 0} \frac{e^{x+h} - e^x}{h}$$ Applying limit definition $f'(x) = \lim_{h \to 0} \frac{f(x+h) - f(x)}{h}$.

$$= \lim_{h \to 0} \frac{e^x \cdot e^h - e^x}{h}$$ Reverse of law of exponents: $a^m \cdot a^n = a^{m+n}$.

$$= \lim_{h \to 0} \frac{e^x(e^h - 1)}{h}$$ Factorizing.

$$= \lim_{h \to 0} e^x \cdot \lim_{h \to 0} \frac{(e^h - 1)}{h}$$ Applying properties of limits.

$$= e^x \cdot \lim_{h \to 0} \frac{e^h - 1}{h}$$ e^x is not affected by the value of h.

A closer look at the limit that is multiplying e^x reveals that it is equivalent to the slope of the graph of $y = e^x$ at $x = 0$: $\lim\limits_{h \to 0} \frac{e^{0+h} - e^0}{h} = \lim\limits_{h \to 0} \frac{e^h - 1}{h}$. To finish our differentiation of e^x by first principles, we need to evaluate this limit. It is beyond the scope of this course to give a formal algebraic proof for the limit. Nevertheless, we can provide a convincing informal justification by evaluating the expression $\frac{e^h - 1}{h}$ for values of h approaching zero, as shown in the table.

h	$\frac{e^h - 1}{h}$
0.1	1.051 709 181
0.01	1.005 016 708
0.0001	1.000 050 002
0.000 001	1.000 000 005

Thus, $\lim\limits_{h \to 0} \frac{e^h - 1}{h} = 1$ and we can complete our algebraic work for the derivative of e^x.

$$\frac{d}{dx}(e^x) = e^x \cdot \lim_{h \to 0} \frac{e^h - 1}{h} = e^x \cdot 1 = e^x$$

The derivative of the exponential function *is* the exponential function. More precisely, the slope of the graph of $f(x) = e^x$ at any point (x, e^x) is equal to the y-coordinate of the point.

The derivative of the exponential function

If $f(x) = e^x$, then $f'(x) = e^x$. Or, in Leibniz notation, $\frac{d}{dx}(e^x) = e^x$.

Example 11

Differentiate each of the following functions.

a) $y = e^{2x + \ln x}$ b) $y = \sqrt{x^2 + e^{4x}}$ c) $y = \frac{e^x - e^{-x}}{e^x + e^{-x}}$

Solution

a) Because $e^{2x + \ln x} = e^{2x}e^{\ln x}$ and $e^{\ln x} = x$, then $e^{2x + \ln x} = xe^{2x}$.

$$\frac{dy}{dx} = \frac{d}{dx}(e^{2x + \ln x}) = \frac{d}{dx}(xe^{2x})$$

$$= x \cdot \frac{d}{dx}(e^{2x}) + e^{2x} \cdot \frac{d}{dx}(x)$$ Applying the product rule.

Therefore, $\frac{dy}{dx} = 2xe^{2x} + e^{2x}$.

b) $$\frac{dy}{dx} = \frac{d}{dx}(\sqrt{x^2 + e^{4x}}) = \frac{d}{dx}\left[(x^2 + e^{4x})^{\frac{1}{2}}\right]$$

$$= \frac{1}{2}(x^2 + e^{4x})^{-\frac{1}{2}} \cdot \frac{d}{dx}(x^2 + e^{4x})$$ Applying power rule and chain rule.

$$= \frac{2x + 4e^{4x}}{2\sqrt{x^2 + e^{4x}}}$$

Therefore, $\frac{dy}{dx} = \frac{x + 2e^{4x}}{\sqrt{x^2 + e^{4x}}}$.

c) $\frac{dy}{dx} = \frac{d}{dx}\left(\frac{e^x - e^{-x}}{e^x + e^{-x}}\right)$

$$= \frac{(e^x + e^{-x}) \cdot \frac{d}{dx}(e^x - e^{-x}) - (e^x - e^{-x}) \cdot \frac{d}{dx}(e^x + e^{-x})}{(e^x + e^{-x})^2}$$ Quotient rule.

$$= \frac{(e^x + e^{-x})(e^x + e^{-x}) - (e^x - e^{-x})(e^x - e^{-x})}{(e^x + e^{-x})^2}$$

$$= \frac{(e^{2x} + 2e^xe^{-x} + e^{-2x}) - (e^{2x} - 2e^xe^{-x} + e^{-2x})}{(e^x + e^{-x})^2}$$

$$= \frac{4e^xe^{-x}}{(e^x + e^{-x})^2}$$

Therefore, $\frac{dy}{dx} = \frac{4}{(e^x + e^{-x})^2}$.

What about exponential functions with bases other than e? We now differentiate the general exponential function $f(x) = b^x$, $b > 1$, $b \neq 0$, repeating the same steps we did with $f(x) = e^x$.

$$\frac{d}{dx}(b^x) = \lim_{h \to 0} \frac{b^{x+h} - b^x}{h}$$ Definition of derivative.

$$= \lim_{h \to 0} \frac{b^x \cdot b^h - b^x}{h}$$ Reverse of $a^m \cdot a^n = a^{m+n}$.

$$= \lim_{h \to 0} \frac{b^x(b^h - 1)}{h}$$ Factorizing.

$$= b^x \cdot \lim_{h \to 0} \frac{b^h - 1}{h}$$ b^x is not affected by the value of h.

As with e^x, $\lim_{h \to 0} \frac{b^h - 1}{h}$ is equivalent to the slope of the graph of $f(x) = b^x$ at $x = 0$, i.e. $f'(0)$. Therefore, the derivative of the general exponential function $f(x) = b^x$ is $b^x \cdot f'(0)$. Although the value of $f'(0)$ will be a constant, it will depend on the value of the base b.

Application of the chain rule gives us the means to determine the value of $f'(0)$ in terms of b for the function $f(x) = b^x$. We can then state the rule for the derivative of the general exponential function $f(x) = b^x$.

We can use the laws of logarithms to write b^x in terms of e^x. Recall from Section 5.5 that $b^{\log_b x} = x$, and if $b = e$ then $e^{\ln x} = x$. Hence, $b^x = e^{x \ln b}$ because $e^{x \ln b} = e^{\ln(b^x)} = b^x$. We can now find the derivative of b^x by applying the chain rule to its equivalent expression $e^{x \ln b}$.

$y = f(g(x)) = e^{x \ln b}$ $\Rightarrow$ 'outside' function is $f(u) = e^u$

$f'(u) = e^u$ $\Rightarrow$ 'inside' function is $g(x) = x \ln b$

$g'(x) = \ln b$ [$\ln b$ is a constant]

$$\frac{dy}{dx} = f(g(x)) \cdot g'(x) = e^{x \ln b} \cdot \ln b$$

$$\frac{dy}{dx} = b^x \ln b$$

Therefore, $\frac{d}{dx}(b^x) = b^x \ln b$.

This result agrees with the fact that $\frac{d}{dx}(e^x) = e^x$. Using this 'new' general rule, $\frac{d}{dx}(b^x) = b^x \ln b$, then $\frac{d}{dx}(e^x) = e^x \ln e$. Since $\ln e = 1$ then $\frac{d}{dx}(e^x) = e^x$.

• **Hint:** Be careful to distinguish between the power rule, $\frac{d}{dx}(x^n) = nx^{n-1}$, where the base is a variable and the exponent is a constant, and the rule for differentiating exponential functions, $\frac{d}{dx}(b^x) = b^x \ln b$, where the base is a constant and the exponent is a variable.

The derivative of the general exponential function

For $b > 0$ and $b \neq 1$, if $f(x) = b^x$, then $f'(x) = b^x \ln b$. Or, in Leibniz notation, $\frac{d}{dx}(b^x) = b^x \ln b$.

Earlier we established that the derivative of the general exponential function $f(x) = b^x$ is $b^x \cdot f'(0)$, where $f'(0)$ is the slope of the graph at $x = 0$. From our result above, we can see that for a specific base b the slope of the curve $y = b^x$ when $x = 0$ is $\ln b$ because $b^0 \ln b = \ln b$. The first GDC screen image below shows the value of $f'(0)$ for $b = 2, 3$ and $\frac{1}{2}$. Evaluating $\ln 2$, $\ln 3$ and $\ln(\frac{1}{2})$ confirms that $f'(0)$ is equal to $\ln b$.

```
nDeriv(2^X,X,0)
          .6931472361
nDeriv(3^X,X,0)
           1.09861251
nDeriv((1/2)^X,X
,0)
         -.6931472361
```

```
ln(2)
          .6931471806
ln(3)
          1.098612289
ln(1/2)
         -.6931471806
```

Example 12

Find the equation of the line tangent to the curve $y = 2^x$ at the point where $x = 3$. Express the equation of the line exactly in the form $y = mx + c$.

Solution

We first find the derivative of $y = 2^x$ and then evaluate it at $x = 3$ to get the slope of the tangent.

$$y' = \frac{d}{dx}(2^x) = 2^x(\ln 2) \Rightarrow y'(3) = 2^3(\ln 2) = 8 \ln 2 = \ln 2^8$$
$$= \ln 256 \Rightarrow m = \ln 256$$

Finding the y-coordinate of the tangent point, $y(3) = 2^3 = 8 \Rightarrow$ point is $(3, 8)$

Substituting into the point-slope form for a linear equation, gives

$$y - y_1 = m(x - x_1) \Rightarrow y - 8 = \ln 256(x - 3)$$

Therefore, the equation of the tangent line is $y = (\ln 256)x + 8 - 3 \ln 256$.

The GDC images below nicely confirm the result.

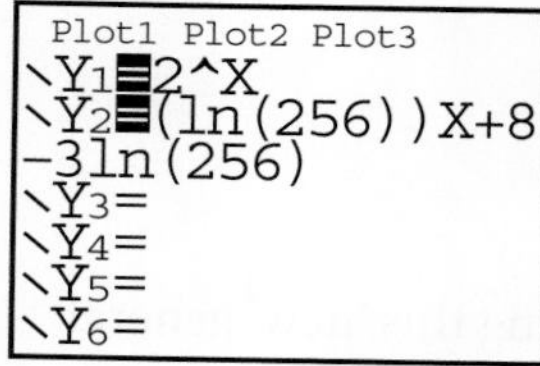


```
WINDOW
 Xmin=-1
 Xmax=5
 Xscl=1
 Ymin=-5
 Ymax=20
 Yscl=5
 Xres=1
```

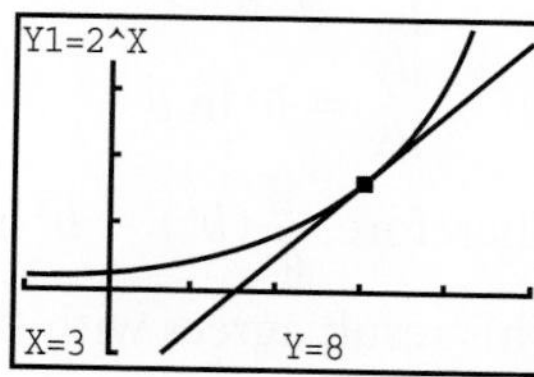

Example 13

Find the coordinates of the point P lying on the graph of $y = 5^x$ such that the line tangent to the curve at P passes through the origin.

Solution

Let $P = (x_0, y_0)$ be a point on the graph of $y = 5^x$. Since $\frac{dy}{dx} = 5^x(\ln 5)$ the slope of the tangent line to the curve at P is given by $\frac{dy}{dx} = 5^{x_0}(\ln 5)$. Substituting into the point-slope form for a linear equation gives,

$$y - y_0 = 5^{x_0}(\ln 5)(x - x_0)$$

If the line passes through the origin then $(0, 0)$ must satisfy the equation.

$$0 - y_0 = 5^{x_0}(\ln 5)(0 - x_0) \Rightarrow -y_0 = 5^{x_0}(\ln 5)(-x_0)$$

But $y_0 = 5^{x_0}$, so $-5^{x_0} = 5^{x_0}(\ln 5)(-x_0) \Rightarrow x_0 = \frac{5^{x_0}}{5^{x_0}\ln 5} = \frac{1}{\ln 5}$.

Then $y_0 = 5^{\frac{1}{\ln 5}} \Rightarrow (y_0)^{\ln 5} = \left(5^{\frac{1}{\ln 5}}\right)^{\ln 5} \Rightarrow (y_0)^{\ln 5} = 5 \Rightarrow y_0 = e$ because $e^{\ln x} = x$.

Therefore, the point P on the graph of $y = 5^x$ has coordinates $\left(\frac{1}{\ln 5}, e\right)$.

As a check let's find the equation of the tangent to $y = 5^x$ at this point.

Since $\frac{dy}{dx} = 5^{x_0}(\ln 5)$ the slope is $5^{\frac{1}{\ln 5}}(\ln 5)$, but we showed above that $5^{\frac{1}{\ln 5}} = e$. So the slope is equivalent to $e\ln 5$. Substituting in the point-slope form gives $y - e = e\ln 5\left(x - \frac{1}{\ln 5}\right) \Rightarrow y = e(\ln 5)x$. Clearly this line passes through $(0, 0)$.

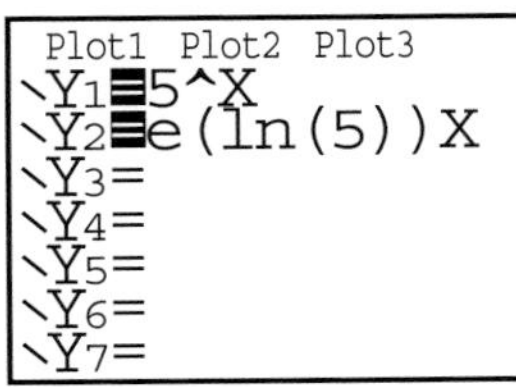

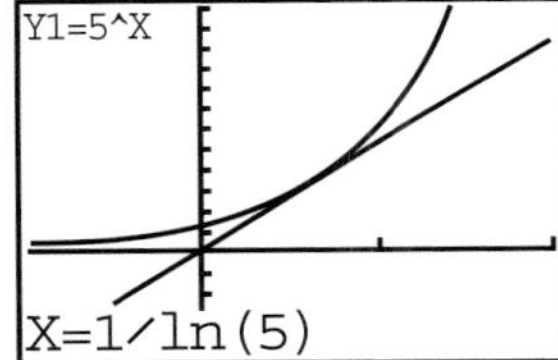

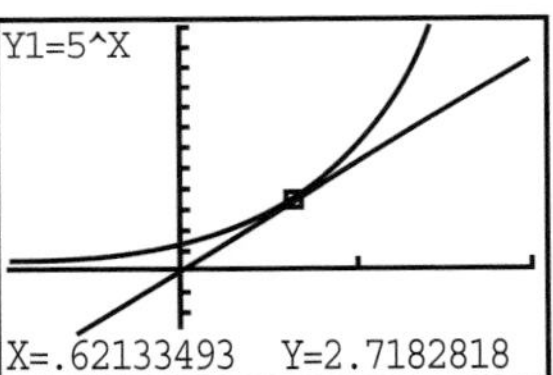

If $f(x) = b^x$, then $f'(x) = b^x \cdot f'(0)$. The value of $f'(0)$ is the slope of the graph of $f(x) = b^x$ at the point $(0, 1)$. Hence, this will be a particular constant for each value of b ($b > 1, b \neq 0$). Therefore, if $f(x) = b^x$, then $f'(x) = kb^x$ where k is a constant dependent on the value of b. If the amount of a quantity y at a time t is given by $y = b^t$ then $\frac{dy}{dt} = kb^t = ky$. In other words, the rate of change of the quantity y at time t is proportional to the amount of y at time t. This is the essential behaviour of exponential growth/decay. It is because of this property that exponential functions have so many applications to real-life phenomena. Here are some good examples:

1. The rate of population growth for many living organisms is proportional to the size of the population p: $\frac{dp}{dt} = kp$.
2. The rate at which a radioactive substance decays is proportional to the amount A of the substance present: $\frac{dA}{dt} = kA$.
3. Newton's law of cooling states that if a substance is placed in cooler surroundings then its temperature decreases at a rate proportional to the temperature difference T between the temperature of the substance and the temperature of its surroundings: $\frac{dT}{dt} = kT$.

Exercise 15.2

1 Find the derivative of each function.

a) $y = x^2e^x$ b) $y = 8^x$ c) $y = \tan e^x$

d) $y = \dfrac{x}{1+\cos x}$ e) $y = \dfrac{e^x}{x}$ f) $y = \frac{1}{3}\sec^3 2x - \sec 2x$

g) $y = 4^{-x}$ h) $y = \cos x \tan x$ i) $y = \dfrac{x}{e^x - 1}$

j) $y = 4\cos(\sin 3x)$ k) $y = 2^{x+1}$ l) $y = \dfrac{1}{\csc x - \sec x}$

2 Find the equation of the line tangent to the given curve at the specified value of x. Express the equation exactly in the form $y = mx + c$.

a) $y = \sin x$ $x = \frac{\pi}{3}$

b) $y = x + e^x$ $x = \frac{\pi}{3}$

c) $y = 4\tan 2x$ $x = \frac{\pi}{8}$

3 Consider the function $g(x) = x + 2\cos x$. For the interval $0 \leqslant x \leqslant 2\pi$.

a) find the exact x-coordinates of any stationary points

b) determine whether each stationary point is a maximum, minimum or neither and give a brief explanation.

4 Find the coordinates of any stationary points on the curve $y = x - e^x$. Classify any such points as a maximum, minimum or neither and explain.

5 Find the coordinates of any stationary points for each function on the interval $0 \leqslant x \leqslant 2\pi$. Indicate whether a stationary point is a maximum, minimum or neither.

a) $f(x) = 4\sin x - \cos 2x$ b) $g(x) = \tan x(\tan x + 2)$

6 Find the equation of the normal line to the curve $y = 3 + \sin x$ at the point where $x = \frac{\pi}{2}$.

7 Consider the function $f(x) = e^x - x^3$.

a) Find $f'(x)$ and $f''(x)$.

b) Find the x-coordinates (accurate to three significant figures) for any points where $f'(x) = 0$.

c) Indicate the intervals for which $f(x)$ is increasing, and indicate the intervals for which $f(x)$ is decreasing.

d) For the values of x found in part b), state whether that point on the graph of f is a maximum, minimum or neither.

e) Find the x-coordinate of any inflexion point(s) for the graph of f.

f) Indicate the intervals for which $f(x)$ is concave up, and indicate the intervals for which $f(x)$ is concave down.

8 Show that the curves $y = e^{-x}$ and $y = e^{-x}\cos x$ are tangent at each point common to both curves. Sketch the two curves over the interval $-\frac{\pi}{2} \leqslant x \leqslant \frac{3\pi}{2}$.

9 A particle moves in a straight line such that its displacement, s metres, is given by $s(t) = 4\cos t - \cos 2t$. If the particle comes to rest after T seconds, where $T > 0$, find:

a) the particle's acceleration at time T

b) the maximum speed of the particle for $0 < t < T$.

10 Find an equation for a line that is tangent to the graph of $y = e^x$ that passes through the origin.

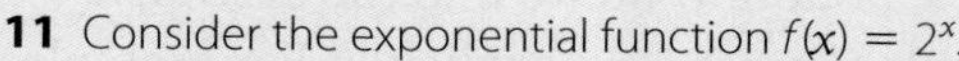

11 Consider the exponential function $f(x) = 2^x$.

a) Find $f'(x)$.

b) Find the equation of the tangent to the graph of f at the point (0, 1).

c) Explain why the graph of f has no stationary points.

12 Consider the function $h(x) = \dfrac{x^2 - 3}{e^x}$.

a) Find the exact coordinates of any stationary points.

b) Determine whether each stationary point is a maximum, minimum or neither.

c) What do the function values approach as (i) $x \to \infty$ and (ii) $x \to -\infty$.

d) Write down the equation of any asymptotes for the graph of $h(x)$.

e) Make an accurate sketch of the curve indicating any extrema and points where the graph intersects the x- and y-axis.

13 Given $y = \sin x$, and $\dfrac{dy}{dx} = \sin(x + a)$, $\dfrac{d^2y}{dx^2} = \sin(x + b)$ and $\dfrac{d^3y}{dx^3} = \sin(x + c)$, find:

a) the values of a, b and c

b) a formula for $\dfrac{d^{(n)}y}{dx^{(n)}}$.

14 a) Find the first three derivatives of $y = xe^x$.

b) Suggest a formula for $\dfrac{d^{(n)}}{dx^{(n)}}(xe^x)$ that is true for all positive integers n.

c) Prove that your formula is true by using mathematical induction.

15.3 Implicit differentiation, logarithmic functions and inverse trigonometric functions

Implicit differentiation

An equation such as $3x - 2y - 8 = 0$ is said to define y as a function of x because it satisfies the definition of a function in that each value of x (domain) determines (corresponds to) a unique value of y (range). We can manipulate the equation in order to solve for y in terms of x, giving $y = \frac{3}{2}x - 4$. In this form, in which y is alone on one side of the equation, the equation is said to define y **explicitly** as a function of x. In the original form of the equation, $x - 2y - 8 = 0$, the function is said to define y **implicitly** as a function of x. If we wish to find the derivative of y with respect to x, $\dfrac{dy}{dx}$, from an equation in which y is defined implicitly as a function of x we can often solve for y and then differentiate using one of the rules that we have established. For example, if we were asked to find $\dfrac{dy}{dx}$ for the equation $xy = 1$ we can write y explicitly as a function of x and then differentiate.

$$xy = 1 \Rightarrow y = \frac{1}{x} = x^{-1} \Rightarrow \frac{dy}{dx} = \frac{d}{dx}(x^{-1}) = -x^{-2} = -\frac{1}{x^2}$$

Most of the functions that we have encountered thus far can be described by expressing one variable explicitly in terms of another variable – for example, $y = \cos(2x)$ or $y = \sqrt{1 - x^2}$. But how do we find the derivative y for an equation where we are not able to solve for y explicitly? For example, if we have the equation

$$x^3 + y^3 - 9xy = 0 \text{ (Figure 15.10)}$$

we cannot solve for y in terms of x. However, there may exist one or more functions f such that if $y = f(x)$ then the equation

$$x^3 + [f(x)]^3 - 9x[f(x)] = 0$$

holds for all values of x in the domain of f. Hence, the function f is defined implicitly by the given equation.

With the assumption that the equation $x^3 + y^3 - 9xy = 0$ defines y as at least one differentiable function of x (see Figure 15.11), the derivative of y with respect to x, $\frac{dy}{dx}$, can be found by the technique of **implicit differentiation.**

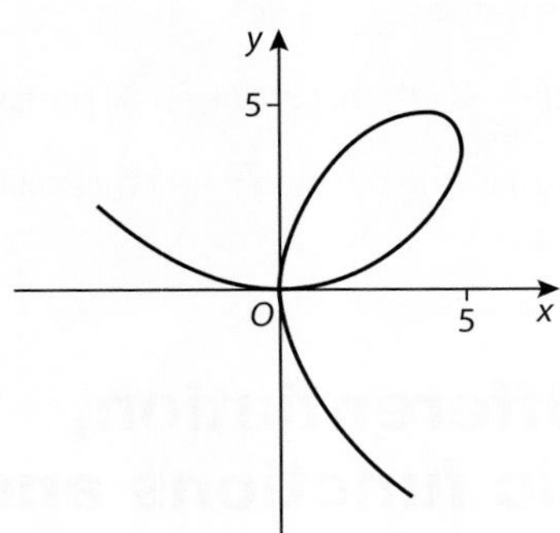

Figure 15.10 The graph of $x^3 + y^3 - 9xy = 0$ (called a *folium*, Latin for 'leaf'). This type of curve was first studied by Rene Descartes in 1638.

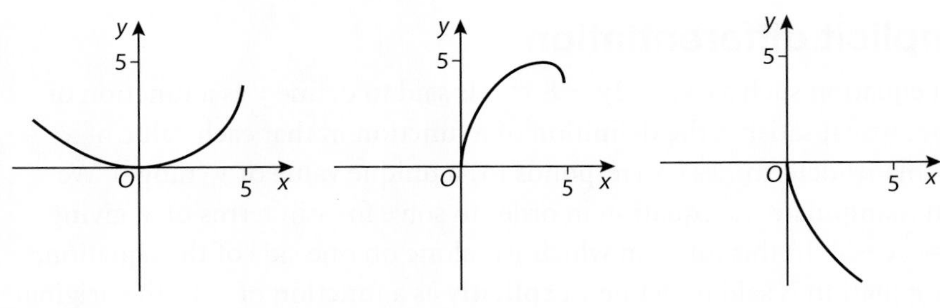

Figure 15.11 Although the equation $x^3 + y^3 - 9xy = 0$ is not a function, we can see that the graph of the equation can be separated into the graphs of three separate functions (they each pass the vertical line test for a function). This demonstrates that the equation implicitly defines y as three functions of x.

Initially we differentiate term-by-term, with respect to x, obtaining

$$\frac{d}{dx}(x^3) + \frac{d}{dx}(y^3) - \frac{d}{dx}(9xy) = \frac{d}{dx}(0).$$

The first and last terms are easily differentiated, and we can apply the constant rule to the third term, giving

$$3x^2 + \frac{d}{dx}(y^3) - 9\frac{d}{dx}(xy) = 0.$$

Differentiating the second and third terms is a little more complicated requiring the use of the chain rule (and also product rule for the third term). If y is defined implicitly as a function of x, then y^3 is also a (composite) function of x. Thus, applying the appropriate rules, we have

$$3x^2 + 3y^2 \cdot \frac{d}{dx}(y) - 9\left(x \cdot \frac{d}{dx}(y) + y \cdot \frac{d}{dx}(x)\right) = 0$$

$$3x^2 + 3y^2 \cdot \frac{dy}{dx} - 9\left(x \cdot \frac{dy}{dx} + y\right) = 0$$

$$3x^2 + 3y^2\frac{dy}{dx} - 9x\frac{dy}{dx} - 9y = 0$$

Now we solve the equation for $\frac{dy}{dx}$.

$$\frac{dy}{dx}(3y^2 - 9x) = -3x^2 + 9y \Rightarrow \frac{dy}{dx} = \frac{-3x^2 + 9y}{3y^2 - 9x}$$

Therefore, $\frac{dy}{dx} = \frac{-x^2 + 3y}{y^2 - 3x}$.

The process of implicit differentiation has given us a formula for $\frac{dy}{dx}$ that is the slope of the curve at any point (except where there is a vertical tangent and slope is undefined) and it is in terms of *both* x and y. This is not unexpected since we can see from the graph of the equation (Figure 15.10) that it is possible for two or three different points on the curve to have the same x-coordinate and the slope of the curve (given by $\frac{dy}{dx}$) will depend on the values of both x and y, and not only x as with functions where y is explicitly defined in terms of x.

In the examples and exercises of this section it is assumed that for any given equation y is implicitly defined as a differentiable function of x (or more than one differentiable function as in the above example) so that the technique of implicit differentiation can be applied.

Process of implicit differentiation

1. Differentiate, term-by-term, both sides of the equation **with respect to** x. The chain rule must be applied for any terms containing y.
2. Collect all terms containing $\frac{dy}{dx}$ on one side of the equation and all other terms on the other side.
3. Factor out $\frac{dy}{dx}$.
4. Solve for $\frac{dy}{dx}$ by dividing both sides by the factor multiplying $\frac{dy}{dx}$.
5. Simplify the result, if possible.

Example 14

Consider the equation for the unit circle $x^2 + y^2 = 1$ which is a relation (not a function).

a) Solve for y, and write all equations that express y as a function of x.
   Find $\frac{dy}{dx}$ for each of these functions.

b) Find $\frac{dy}{dx}$ by implicit differentiation.

c) Find the equation of the line tangent to the unit circle at the point $\left(-\frac{1}{2}, \frac{\sqrt{3}}{2}\right)$.

Solution

a) Solving for y produces two equations, each defining y as a function of x.

$$x^2 + y^2 = 1 \Rightarrow y^2 = 1 - x^2 \Rightarrow y = \sqrt{1 - x^2} \text{ and } y = -\sqrt{1 - x^2}$$

Differentiating each of these with respect to x gives,

$$\frac{dy}{dx} = \frac{d}{dx}(\sqrt{1 - x^2}) = \frac{d}{dx}\left[(1 - x^2)^{\frac{1}{2}}\right] = \frac{1}{2}(1 - x^2)^{-\frac{1}{2}}(-2x) \Rightarrow$$

$$\frac{dy}{dx} = \frac{-x}{\sqrt{1 - x^2}}$$

$$\frac{dy}{dx} = \frac{d}{dx}(-\sqrt{1 - x^2}) = \frac{d}{dx}\left[-(1 - x^2)^{\frac{1}{2}}\right]$$

$$= -\frac{1}{2}(1 - x^2)^{-\frac{1}{2}}(-2x) \Rightarrow \frac{dy}{dx} = \frac{x}{\sqrt{1 - x^2}}$$

For the function $y = \sqrt{1 - x^2}$ we have $\frac{dy}{dx} = \frac{-x}{\sqrt{1 - x^2}}$.

Since $y = \sqrt{1 - x^2}$, then $\frac{dy}{dx} = -\frac{x}{y}$.

For the function $y = -\sqrt{1 - x^2}$ we have $\frac{dy}{dx} = \frac{x}{\sqrt{1 - x^2}}$.

Since $y = -\sqrt{1 - x^2}, \Rightarrow -y = \sqrt{1 - x^2}$, then $\frac{dy}{dx} = -\frac{x}{y}$.

b) $\frac{d}{dx}(x^2) + \frac{d}{dx}(y^2) = \frac{d}{dx}(1)$ Differentiating both sides term-by-term.

$2x + 2y\frac{dy}{dx} = 0$ Chain rule applied to differentiate y^2.

$2y\frac{dy}{dx} = -2x$

$\frac{dy}{dx} = \frac{-2x}{2y}$ Solving for $\frac{dy}{dx}$.

Therefore, $\frac{dy}{dx} = -\frac{x}{y}$.

c) At the point $\left(-\frac{1}{2}, \frac{\sqrt{3}}{2}\right)$ the slope of the tangent line is $\frac{dy}{dx} = -\left(\frac{-\frac{1}{2}}{\frac{\sqrt{3}}{2}}\right)$

$= \frac{1}{\sqrt{3}} = \frac{\sqrt{3}}{3}$.

• **Hint:** Example 14 illustrates that even when it is possible to solve an equation explicitly for y in terms of x, it may be more efficient to find $\frac{dy}{dx}$ by implicit differentiation.

Substituting into the point-slope form gives,

$$y - \frac{\sqrt{3}}{2} = \frac{\sqrt{3}}{3}\left(x + \frac{1}{2}\right) \Rightarrow y = \frac{\sqrt{3}}{3}x + \frac{\sqrt{3}}{6} + \frac{\sqrt{3}}{2} \Rightarrow y = \frac{\sqrt{3}}{3}x + \frac{2\sqrt{3}}{3}$$

We can get a visual check by graphing the unit circle and the tangent line on our GDC. In order to graph the complete unit circle on our GDC we need to graph both functions found in part a).

```
Plot1 Plot2 Plot3
\Y1■√(1-X²)
\Y2■-√(1-X²)
\Y3■(√(3)/3)X+2√
(3)/3
\Y4=
\Y5=
\Y6=
```

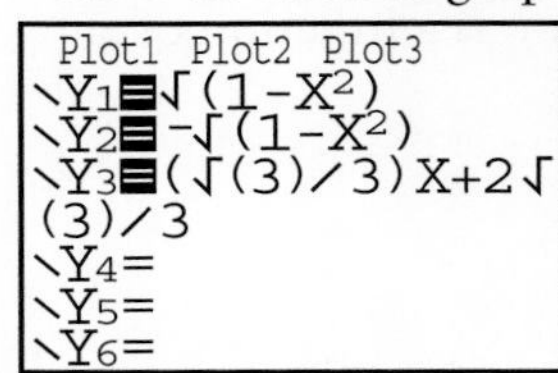

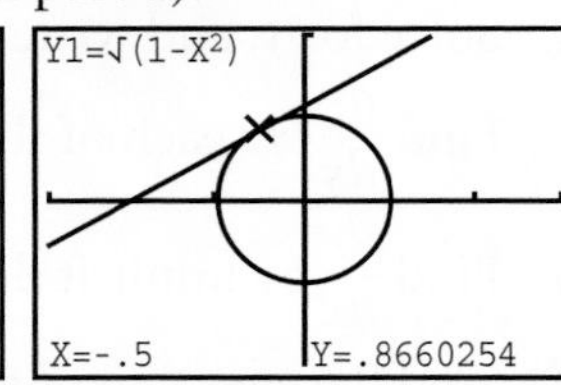

Example 15

a) Find the points on the graph of $x^2 + 4xy + 13y^2 = 9$ at which the tangent is horizontal.

b) Determine whether each point is a maximum, minimum or neither.

Solution

a) We need to find $\frac{dy}{dx}$ which we do by implicit differentiation.

$$\frac{d}{dx}(x^2) + 4\frac{d}{dx}(xy) + 13\frac{d}{dx}(y^2) = \frac{d}{dx}(9)$$ Differentiating both sides term-by-term.

$$2x + 4\left(x\frac{d}{dx}(y) + y\frac{d}{dx}(x)\right) + 13\left(2y\frac{d}{dx}(y)\right) = 0$$ Applying chain and product rules.

$$2x + 4x\frac{dy}{dx} + 4y + 26y\frac{dy}{dx} = 0$$ Collecting terms containing $\frac{dy}{dx}$ on one side.

$$\frac{dy}{dx}(4x + 26y) = -2x - 4y$$ Factor out $\frac{dy}{dx}$.

$$\frac{dy}{dx} = \frac{-2x - 4y}{4x + 26y} = \frac{-x - 2y}{2x + 13y}$$ Solving for $\frac{dy}{dx}$.

To find horizontal tangents, solve $\frac{dy}{dx} = 0$.

$$\frac{-x - 2y}{2x + 13y} = 0 \Rightarrow -x - 2y = 0 \Rightarrow y = -\frac{x}{2}$$

Of course, there are an infinite number of ordered pairs (x, y) that satisfy the equation $y = -\frac{x}{2}$. But the only ordered pairs that we want are ones that are on the curve $x^2 + 4xy + 13y^2 = 9$. So we substitute $-\frac{x}{2}$ for y and solve to find x-coordinates of points on the curve where $\frac{dy}{dx} = 0$.

$$x^2 + 4xy + 13y^2 = 9 \Rightarrow x^2 + 4x\left(-\frac{x}{2}\right) + 13\left(-\frac{x}{2}\right)^2 = 9$$

$$x^2 - 2x^2 + \frac{13}{4}x^2 = 9$$

$$4x^2 - 8x^2 + 13x^2 = 36$$ Multiplying both sides by 4.

$$9x^2 = 36$$

$$x^2 = 4 \Rightarrow x = 2 \text{ or } x = -2$$

y-coordinates: for $x = 2$, $y = -\frac{2}{2} = -1$; for $x = -2$, $y = -\left(\frac{-2}{2}\right) = 1$

Therefore, the tangents to the curve at $(2, -1)$ and $(-2, 1)$ are horizontal.

b) It is very difficult to determine the nature of the points $(2, -1)$ and $(-2, 1)$ by testing the sign of the derivative to either side of each point. Since $\frac{dy}{dx}$ is in terms of both x and y we need an explicit equation for y in terms of x to find the y-coordinate – but no explicit equation for y exists.

It is also impossible to graph the curve $x^2 + 4xy + 13y^2 = 9$ on our GDC to see its shape. Let's find the second derivative, $\frac{d^2y}{dx^2}$, and apply the second derivative test (Section 13.3).

$$\frac{d^2y}{dx^2} = \frac{d}{dx}\left(\frac{-x-2y}{2x+13y}\right)$$

$$= \frac{(2x+13y)\left[\frac{d}{dx}(-x-2y)\right] - (-x-2y)\left[\frac{d}{dx}(2x+13y)\right]}{(2x+13y)^2}$$

Applying quotient rule.

$$= \frac{(2x+13y)\left(-1-2\frac{dy}{dx}\right) + (x+2y)\left(2+13\frac{dy}{dx}\right)}{(2x+13y)^2}$$

$$= (2x+13y)\left(-1-2\left(\frac{-x-2y}{2x+13y}\right)\right) + (x+2y)\left(2+13\left(\frac{-x-2y}{2x+13y}\right)\right)$$

Substituting for $\frac{dy}{dx}$.

$$= \frac{2x+13y}{2x+13y} \cdot \frac{(2x+13y)\left(-1+\frac{2x+4y}{2x+13y}\right) + (x+2y)\left(2-\frac{13x+26y}{2x+13y}\right)}{(2x+13y)^2}$$

$$= \frac{(2x+13y)(-2x-13y+2x+4y) + (x+2y)(4x+26y-13x-26y)}{(2x+13y)^3}$$

$$= \frac{(2x+13y)(-9y) + (x+2y)(-9x)}{(2x+13y)^3}$$

$$\therefore \quad \frac{d^2y}{dx^2} = -\frac{9x^2+36xy+117y^2}{(2x+13y)^3} = \frac{-9(x^2+4xy+13y^2)}{(2x+13y)^3}$$

Now applying the second derivative test for both points where $\frac{dy}{dx} = 0$, we have

for $(2, -1)$, $\frac{d^2y}{dx^2} = \frac{-9(2^2+4(2)(-1)+13(-1)^2)}{(2(2)^2+13(-1))^3}$

$= \frac{81}{125} > 0 \Rightarrow (2, -1)$ is a minimum

for $(-2, 1)$, $\frac{d^2y}{dx^2} = \frac{-9(-2)^2+4(-2)(1)+13(1)^2)}{(2(-2)^2+13(1))^3}$

$= -\frac{3}{343} < 0 \Rightarrow (-2, 1)$ is a maximum

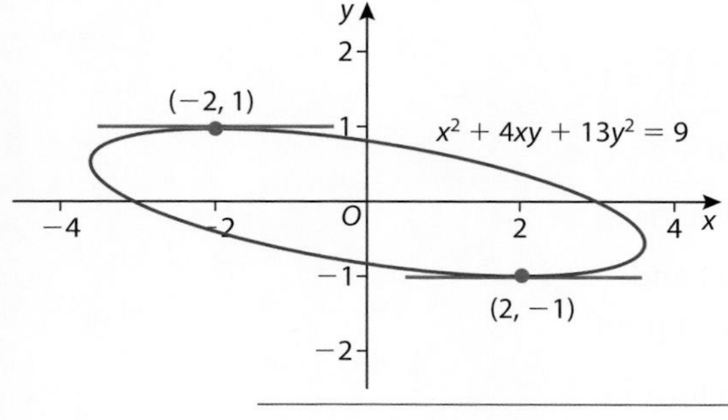

Even though it is not possible to graph the curve $x^2 + 4xy + 13y^2 = 9$ on our GDC, it is possible to find graphing software that can. The graph visually confirms our results for parts a) and b) of Example 15.

Previously we have established the rules for differentiating trigonometric functions and exponential functions. We still need to determine how to differentiate other important non-algebraic functions, namely logarithmic functions and inverse trigonometric functions.

Derivatives of logarithmic functions

At the start of the previous section we explored how we can often form a strong conjecture for the derivative of a function by analyzing the shape of the function's graph with the aid of some features of our GDC. Let's take this informal approach for finding the derivative for the natural logarithm function, $y = \ln x$, and then check our conjecture by deriving $\frac{d}{dx}(\ln x)$ by means of implicit differentiation.

The graph of $y = \ln x$ (Figure 15.12) is a particularly straightforward one. Its x-intercept is $(1, 0)$, and since its domain is all positive real numbers, it has no y-intercept. It is asymptotic to the y-axis, and the graph rises steadily, though less steeply as $x \to \infty$. There is neither an upper nor a lower bound, so its range is all real numbers.

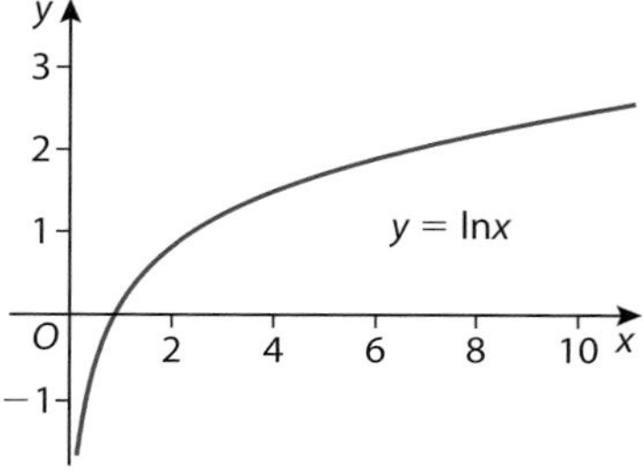

Figure 15.12

Let's cleverly use our GDC to view a graph of $y = \ln x$, a graph of its derivative, and to construct a table of ordered pairs with x and the value of the derivative at x (as computed by the GDC).

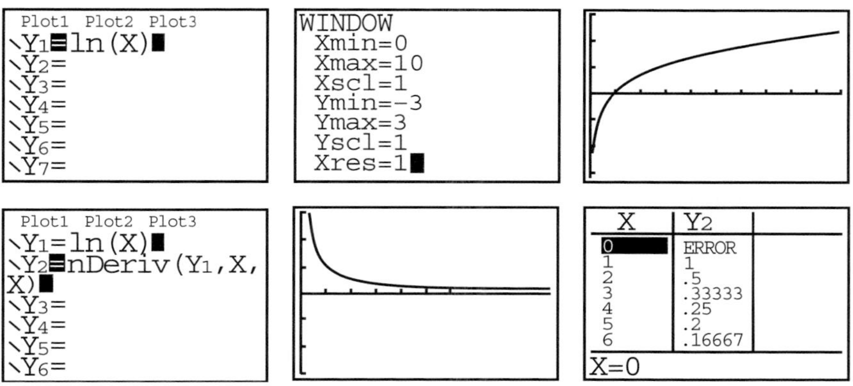

In the table, each value in the $\mathbf{Y}_2$ column is the slope of the curve (derivative) at the particular x value for $y = \ln x$. From the graph of the derivative and especially from the table, we conjecture that the derivative of $\ln x$ is $\frac{1}{x}$. This agrees with the fact that for $x > 0$, the slope of the graph of $y = \ln x$ is always positive and as x increases the slope decreases.

The inverse of $y = \ln x$ is $y = e^x$. Knowing this and that $\frac{d}{dx}(e^x) = e^x$, we can use implicit differentiation to confirm our conjecture.

$$y = \ln x$$

$$e^y = x \qquad \text{Inverse function relationship.}$$

$$\frac{d}{dx}(e^y) = \frac{d}{dx}(x) \qquad \text{Differentiate implicitly.}$$

$$e^y\frac{dy}{dx} = 1$$

$$\frac{dy}{dx} = \frac{1}{e^y}$$

$$\frac{dy}{dx} = \frac{1}{x} \qquad \text{Substituting } x \text{ for } e^y.$$

Therefore, $\frac{d}{dx}(\ln x) = \frac{1}{x}$.

The derivative of the natural logarithm function

If $f(x) = \ln x$, then $f'(x) = \frac{1}{x}$. Or, in Leibniz notation, $\frac{d}{dx}(\ln x) = \frac{1}{x}$.

It is interesting to note that that the derivative of the **non-algebraic** function $f(x) = \ln x$ is the **algebraic** function $f'(x) = \frac{1}{x}$. Non-algebraic functions, such as trigonometric, exponential and logarithmic functions are often referred to as 'transcendental' functions. A **transcendental function** is a function that is not algebraic – in other words, it cannot be composed of a finite number of the elementary operations of addition, subtraction, multiplication, division and extracting a root. A **transcendental number** is a real number that is not a root of any polynomial equation with rational coefficients. For example, π and e are transcendental numbers.

What about the derivative of a logarithmic function with a base, b, other than e; that is, logarithmic functions other than the natural logarithmic function?

To find the derivative of $\log_b x$ with any base $(b > 0, b \neq 1)$, we can use the change of base formula (Section 5.4) for logarithms to express $\log_b x$ in terms of the natural logarithm, $\ln x$, and then differentiate.

$$\log_b x = \frac{\ln x}{\ln b}$$ Applying change of base formula.

$$\frac{d}{dx}(\log_b x) = \frac{d}{dx}\left(\frac{\ln x}{\ln b}\right) = \frac{d}{dx}\left(\frac{1}{\ln b} \cdot \ln x\right)$$ Differentiating both sides.

$$= \frac{1}{\ln b} \cdot \frac{d}{dx}(\ln x)$$ $\frac{1}{\ln b}$ is a constant.

$$= \frac{1}{\ln b} \cdot \frac{1}{x}$$

Therefore, $\frac{d}{dx}(\log_b x) = \frac{1}{x \ln b}$.

The derivative of the general logarithm function

If $f(x) = \log_b x$ $(b > 0, b \neq 1)$, then $f'(x) = \frac{1}{x \ln b}$. Or, in Leibniz notation, $\frac{d}{dx}(\log_b x) = \frac{1}{x \ln b}$.

Example 16

a) Given $g(x) = \frac{1 + x}{1 - x}$, find $g'(x)$.

b) Hence, find $f'(x)$ for $f(x) = \ln\left(\frac{1 + x}{1 - x}\right)$.

c) (i) Show that $f(x)$ is an odd function.
(ii) Show that $f(x)$ has no stationary points.
(iii) Show that $f(x)$ has one point of inflexion, and give its coordinates.

Solution

a) $$g'(x) = \frac{(1-x)\frac{d}{dx}(1+x) - (1+x)\frac{d}{dx}(1-x)}{(1-x)^2}$$ Applying quotient rule.

$$= \frac{1-x+1+x}{(1-x)^2}$$

$$\therefore g'(x) = \frac{2}{(1-x)^2}$$

b) $$f'(x) = \frac{d}{dx}\left[\ln\left(\frac{1+x}{1-x}\right)\right] = \frac{1}{\frac{1+x}{1-x}} \cdot \frac{d}{dx}\left(\frac{1+x}{1-x}\right)$$ Applying $\frac{d}{dx}(\ln x) = \frac{1}{x}$ and chain rule.

$$= \left(\frac{1-x}{1+x}\right)\left(\frac{2}{(1-x)^2}\right)$$ Substituting result from part a).

$$= \frac{1}{1+x} \cdot \frac{2}{1-x}$$

$$\therefore f'(x) = \frac{2}{1-x^2}$$

c) (i) In Section 7.3 we stated that a function f is odd if, for each x in the domain of f, $f(-x) = -f(x)$ with its graph symmetric about the origin. This symmetry leads to the fact (see question 25 in Exercise 13.2) that the graph of the derivative of an odd function is symmetric about the y-axis, i.e. an even function. A function f is even if $f(-x) = f(x)$. Thus, it will suffice to show that $f'(x)$ is even in order to show that $f(x)$ is odd.

$$f'(-x) = \frac{2}{1-(-x)^2} = \frac{2}{1-x^2} = f(x)$$

Therefore, $f'(x)$ is even and it follows that $f(x)$ is odd.

(ii) A stationary point for a function can only occur where its derivative is zero.

Clearly, $f'(x) = \frac{2}{1-x^2} \neq 0$ because a rational expression can only equal zero when its numerator is zero. Therefore, $f(x)$ has no stationary points.

(iii) To find any inflexion points we start by finding where the second derivative is zero.

$$f''(x) = \frac{d}{dx}\left(\frac{2}{1-x^2}\right) = 2\frac{d}{dx}[(1-x^2)^{-1}]$$ Power and chain rules instead of quotient rule.

$$= 2[-(1-x^2)^{-2}(-2x)]$$

$$= f''(x) = \frac{4x}{(1-x^2)^2} = 0 \text{ when } x = 0$$

To confirm that an inflexion point does occur at $x = 0$ we need to show that the concavity of the graph of f changes at $x = 0$ ($f''(x)$ changes sign). Because $f(x)$ is defined only for $-1 < x < 1$, we choose $x = -\frac{1}{2}$ and $x = \frac{1}{2}$ as test points.

$$f''\left(-\frac{1}{2}\right) = \frac{4\left(-\frac{1}{2}\right)}{\left(1-\left(-\frac{1}{2}\right)^2\right)^2} = -\frac{32}{9} < 0 \text{ and}$$

$$f''\left(\frac{1}{2}\right) = \frac{4\left(\frac{1}{2}\right)}{\left(1-\left(\frac{1}{2}\right)^2\right)^2} = \frac{32}{9} > 0$$

Since $f''(x)$ changes sign (and $f(x)$ changes concavity) at $x = 0$, f has an inflexion point there. $f(0) = \ln\left(\frac{1+0}{1-0}\right) = \ln(1) = 0$. Therefore, the inflexion point is at $(0, 0)$. (See GDC images below).

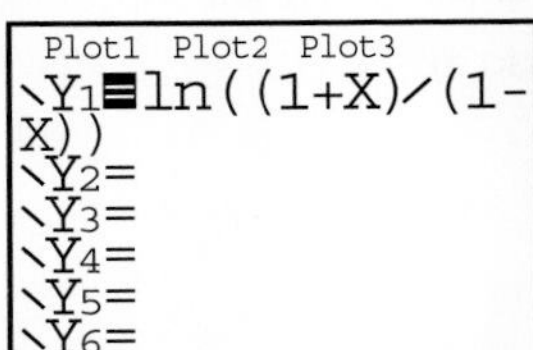

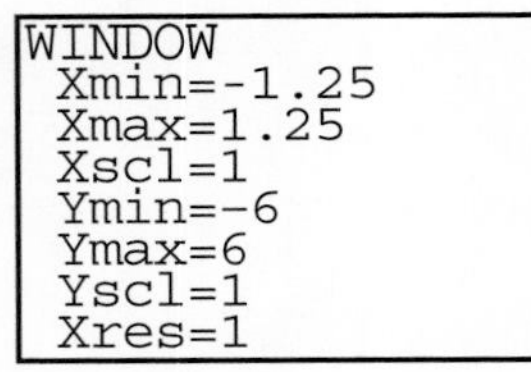

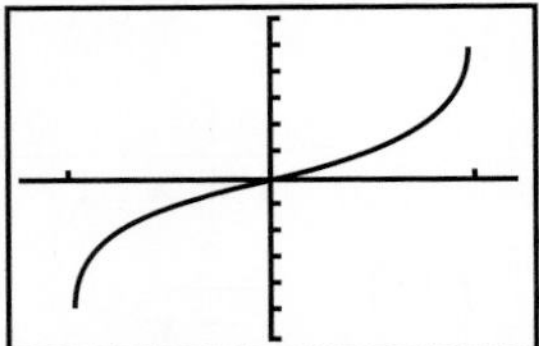

Example 17

Find the equation of the line tangent to the graph of $y = \log_{10}(x^3)$ at the point $x = 4$. Express the equation exactly with any logarithms being expressed as natural logarithms.

Solution

$$\frac{dy}{dx} = \frac{d}{dx}[\log_{10}(x^3)] = \frac{1}{x^3 \ln 10} \cdot \frac{d}{dx}(x^3) \quad \text{Applying } \frac{d}{dx}(\log_b x) = \frac{1}{x \ln b} \text{ and chain rule.}$$

$$= \frac{1}{x^3 \ln 10} \cdot 3x^2$$

$$\frac{dy}{dx} = \frac{3}{x \ln 10}$$

[Alternatively, we could have used laws of logarithms to write $y = \log_{10}(x^3) = 3\log_{10} x$ and then $\frac{dy}{dx} = 3\frac{d}{dx}(\log_{10} x) = \frac{3}{x \ln 10}$, avoiding use of the chain rule.]

When $x = 4$, $\frac{dy}{dx} = \frac{3}{4 \ln 10}$ and $y = \log_{10}(4^3) = \log_{10} 64 = \frac{\ln 64}{\ln 10}$ (using change of base formula). Thus, the tangent line intersects the curve at the point $\left(4, \frac{\ln 64}{\ln 10}\right)$ and has a slope of $\frac{3}{4 \ln 10}$. Substituting into the point-slope form for a linear equation gives:

$$y - \frac{\ln 64}{\ln 10} = \frac{3}{4 \ln 10}(x - 4) \quad \Rightarrow \quad y = \frac{3x}{4 \ln 10} - \frac{3}{\ln 10} + \frac{\ln 64}{\ln 10} \quad \Rightarrow$$

$$y = \frac{3x}{4 \ln 10} + \frac{-3 + \ln 64}{\ln 10}$$

Graphing the curve $y = \log_{10}(x^3)$ and the computed tangent line appears to give a good visual confirmation that the equation of the tangent line is correct.

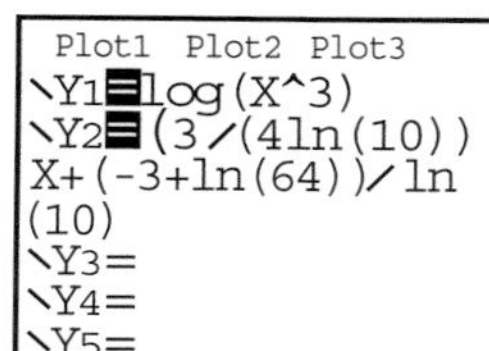

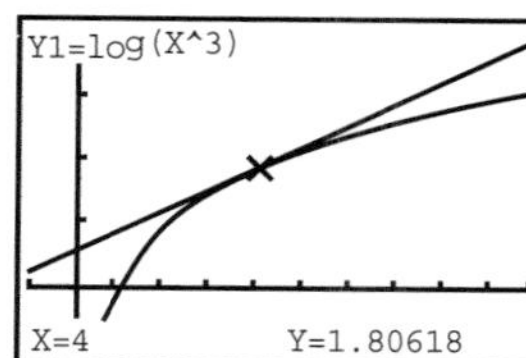

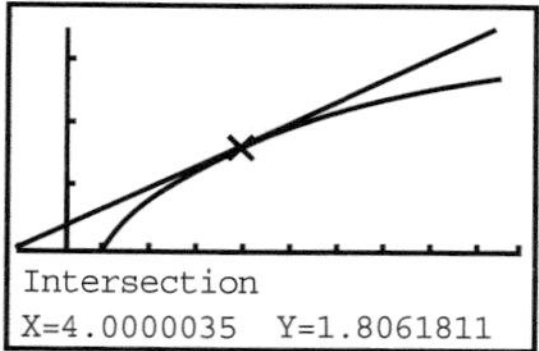

Derivatives of inverse trigonometric functions

In the preceding pages, we established that the derivative of the *non-algebraic* (transcendental) function $f(x) = \ln x$ is the *algebraic* function $f'(x) = \frac{1}{x}$. The same is true for the inverse trigonometric functions – they are transcendental but their derivatives are algebraic. The inverse trigonometric functions were discussed in Section 7.6. We will now use implicit differentiation to find the derivatives of the inverse functions for sine, cosine, and tangent functions – which are usually referred to as arcsin *x*, arccos *x* and arctan *x* respectively. Their graphs are shown again in Figure 15.13.

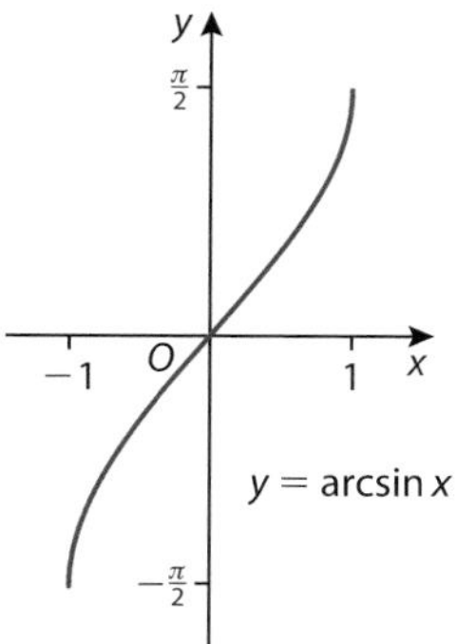

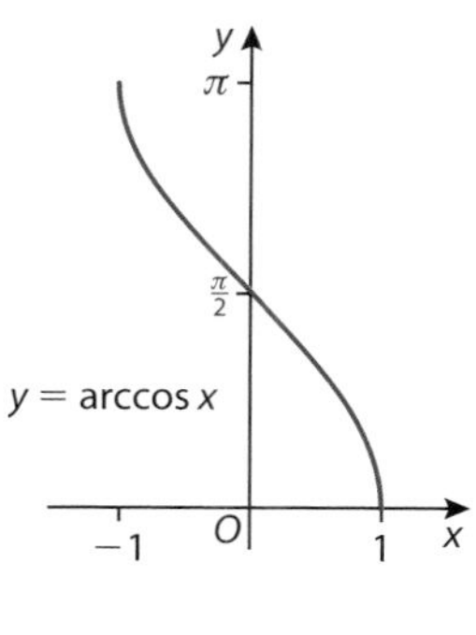

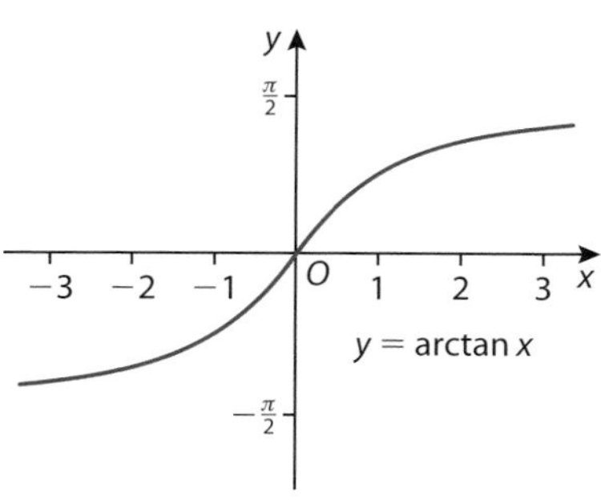

Figure 15.13

Given the smooth shape of their graphs we will assume that the functions $y = \arcsin x$, $y = \arccos x$ and $y = \arctan x$ are differentiable (i.e. the derivative exists) except where a vertical tangent exists. Since $y = \arcsin x$ and $y = \arccos x$ have vertical tangents at $x = -1$ and $x = 1$ they are differentiable throughout the interval $-1 < x < 1$. $y = \arctan x$ is differentiable for all real numbers.

• **Hint:** Recall from Chapter 7 that the notations $y = \arcsin x$ and $y = \sin^{-1} x$ are synonymous, but we will generally use $y = \arcsin x$.

Recall the definition of the arcsine function,

$$y = \arcsin x \Rightarrow \sin y = x \text{ for } -\frac{\pi}{2} \leqslant y \leqslant \frac{\pi}{2}.$$

Differentiating $\sin y = x$ implicitly with respect to *x* gives:

$$\frac{d}{dx}(\sin y) = \frac{d}{dx}(x) \quad \text{Differentiating both sides.}$$

$$(\cos y)\frac{dy}{dx} = 1 \quad \text{Implicit differentiation.}$$

$$\frac{dy}{dx} = \frac{1}{\cos y} \quad \text{Dividing by } \cos y.$$

That is, $\frac{d}{dx}(\arcsin x) = \frac{1}{\cos y}$.

Dividing by $\cos y$ in the last step is allowed because $\cos y \neq 0$ for the interval in which $y = \arcsin x$ is differentiable, i.e. $-\frac{\pi}{2} < y < \frac{\pi}{2}$ (quadrants I and IV). In fact, $\cos y > 0$ for $-\frac{\pi}{2} < y < \frac{\pi}{2}$. From the identity $\sin^2 x + \cos^2 x = 1$ we have $\cos x = \pm\sqrt{1 - \sin^2 x}$. Since $\cos y > 0$ we can replace $\cos y$ with $\sqrt{1 - \sin^2 y}$ and because $\sin y = x$ we get $\cos y = \sqrt{1 - x^2}$.
Therefore, $\frac{d}{dx}(\arcsin x) = \frac{1}{\sqrt{1 - x^2}}$.

We can apply a similar process to find the derivative of the arcos x function, obtaining the result

$$\frac{d}{dx}(\arccos x) = -\frac{1}{\sqrt{1 - x^2}}.$$

Although the domain for the inverse sine and inverse cosine functions is the fairly narrow closed interval $-1 \leqslant x \leqslant 1$ and they are differentiable on the open interval $-1 < x < 1$, the inverse tangent function is defined and differentiable for all real numbers. To find $\frac{d}{dx}(\arctan x)$, we follow a similar procedure to that for $\frac{d}{dx}(\arcsin x)$.

The definition of the inverse tangent (arctan) function is

$$y = \arctan x \Rightarrow \tan y = x \text{ for } -\frac{\pi}{2} \leqslant y \leqslant \frac{\pi}{2}.$$

Differentiating $\tan y = x$ implicitly with respect to x gives:

$$\frac{d}{dx}(\tan y) = \frac{d}{dx}(x)$$ Differentiating both sides.

$$(\sec^2 y)\frac{dy}{dx} = 1$$ Implicit differentiation.

$$\frac{dy}{dx} = \frac{1}{\sec^2 y}$$ Dividing by $\sec^2 y$.

$$\frac{dy}{dx} = \frac{1}{1 + \tan^2 y}$$ Applying identity $1 + \tan^2 y = \sec^2 y$.

Therefore, $\frac{d}{dx}(\arctan x) = \frac{1}{1 + x^2}$. $\tan y = x$.

The derivatives for the inverse secant, inverse cosecant and inverse cotangent functions can also be found by means of implicit differentiation. They are included in the list below but are not necessary for this course.

Derivatives of the inverse trigonometric functions

$\frac{d}{dx}(\arcsin x) = \frac{1}{\sqrt{1 - x^2}}$ $\frac{d}{dx}(\operatorname{arccsc} x) = -\frac{1}{x\sqrt{x^2 - 1}}$

$\frac{d}{dx}(\arccos x) = -\frac{1}{\sqrt{1 - x^2}}$ $\frac{d}{dx}(\operatorname{arcsec} x) = \frac{1}{x\sqrt{x^2 - 1}}$

$\frac{d}{dx}(\arctan x) = \frac{1}{1 + x^2}$ $\frac{d}{dx}(\operatorname{arccot} x) = -\frac{1}{1 + x^2}$

Example 18

Find the $\frac{dy}{dx}$ for each of the following.

a) $y = \cos^{-1}(e^{2x})$

b) $y = x\arcsin 2x + \frac{1}{2}\sqrt{1 - 4x^2}$

c) $\ln(x + y) = \arctan\left(\frac{x}{y}\right)$

Solution

a) $\dfrac{dy}{dx} = \dfrac{d}{dx}[\cos^{-1}(e^{2x})] = \dfrac{-1}{\sqrt{1-(e^{2x})^2}} \cdot \dfrac{d}{dx}(e^{2x})$ Chain rule and $\dfrac{d}{dx}(\arccos x) = \dfrac{-1}{\sqrt{1-x^2}}$.

$$= \frac{-1}{\sqrt{1-e^{4x}}} \cdot e^{2x} \cdot 2$$ Chain rule, again.

$$\frac{dy}{dx} = -\frac{2e^{2x}}{\sqrt{1-e^{4x}}}$$

b) $\dfrac{dy}{dx} = \dfrac{d}{dx}\left(x\arcsin 2x + \frac{1}{2}(1-4x^2)^{\frac{1}{2}}\right)$

$$= x\frac{d}{dx}(\arcsin 2x) + \arcsin 2x\frac{d}{dx}(x) + \frac{1}{2} \cdot \frac{1}{2}(1-4x^2)^{-\frac{1}{2}}\frac{d}{dx}(1-4x^2)$$

$$= x\left(\frac{1}{\sqrt{1-(2x)^2}}\frac{d}{dx}(2x)\right) + \arcsin 2x + \frac{-8x}{4\sqrt{1-4x^2}}$$

$$= \frac{2x}{\sqrt{1-4x^2}} + \arcsin 2x + \frac{-2x}{\sqrt{1-4x^2}}$$

$$\frac{dy}{dx} = \arcsin 2x$$

c) $\dfrac{d}{dx}[\ln(x+y)] = \dfrac{d}{dx}\left[\arctan\left(\dfrac{x}{y}\right)\right]$ Differentiating both sides implicitly.

$$\frac{1}{x+y}\left(1+\frac{dy}{dx}\right) = \frac{1}{1+\frac{x^2}{y^2}}\left(\frac{y - x\frac{dy}{dx}}{y^2}\right)$$ Chain rule, $\dfrac{d}{dx}(\arctan x) = \dfrac{1}{1+x^2}$, quotient rule.

$$\frac{1+\frac{dy}{dx}}{x+y} = \frac{y - x\frac{dy}{dx}}{x^2+y^2}$$

$$x^2 + y^2 + \frac{dy}{dx}x^2 + \frac{dy}{dx}y^2 = xy + y^2 - \frac{dy}{dx}x^2 - \frac{dy}{dx}xy$$

$$\frac{dy}{dx}(2x^2 + xy + y^2) = xy - x^2$$

$$\frac{dy}{dx} = \frac{xy - x^2}{2x^2 + xy + y^2}$$

Example 19

A painting that is 175 cm from top to bottom is hanging on the wall of a gallery such that it's base is 225 cm above the eye level of an observer. How far from the wall should the observer stand to get the best view of the painting, that is, so that the angle subtended at the observer's eye by the painting is a maximum? (This is similar to Example 34 in Section 7.6.)

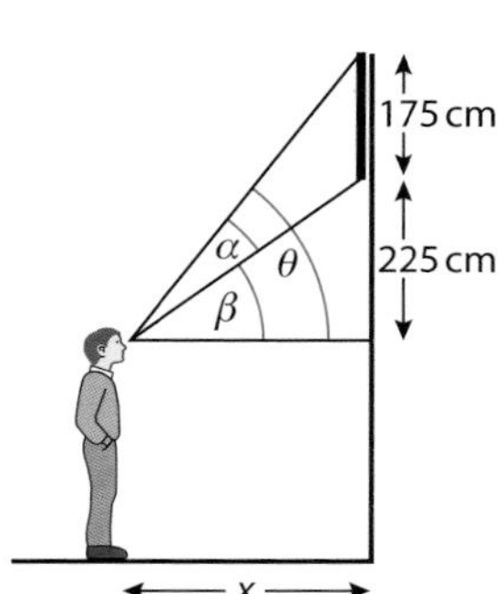

Solution

Change all lengths from centimetres to metres.

$\tan\theta = \dfrac{4}{x}$ and $\tan\beta = \dfrac{\frac{9}{4}}{x}$

Because $0 < \theta < \frac{\pi}{2}$ and $0 < \beta < \frac{\pi}{2}$, we have

$\theta = \arctan\frac{4}{x}$ and $\beta = \arctan\frac{\frac{9}{4}}{x}$.

Substituting these values of θ and β into the equation $\alpha = \theta - \beta$ gives

$\alpha = \arctan\frac{4}{x} - \arctan\frac{\frac{9}{4}}{x}$.

Differentiating with respect to x gives:

$$\frac{d\alpha}{dx} = \frac{d}{dx}\left[\arctan(4x^{-1}) - \arctan\left(\frac{9}{4}x^{-1}\right)\right]$$
$$= \frac{1}{1 + (4x^{-1})^2}(-4x^{-2}) - \frac{1}{1 + \left(\frac{9}{4}x^{-1}\right)^2}\left(-\frac{9}{4}x^{-2}\right)$$
$$= \frac{-4}{x^2 + 16} + \frac{\frac{9}{4}}{x^2 + \frac{81}{16}}$$
$$= \frac{-4}{x^2 + 16} + \frac{36}{16x^2 + 81}$$

Setting $\frac{d\alpha}{dx} = 0$, we get:

$$36(x^2 + 16) - 4(16x^2 + 81) = 0$$
$$-28x^2 + 252 = 0$$
$$x^2 = \tfrac{252}{28} = 9 \Rightarrow x = \pm 3, \text{ however } x \neq -3$$

We use the first derivative test to determine if the angle α is a maximum when $x = 3$, using test values of $x = 2$ and $x = 4$.

When $x = 2$, $\frac{d\alpha}{dx} = \frac{7}{145} > 0$ and when $x = 4$, $\frac{d\alpha}{dx} = -\frac{49}{2696} < 0$.

Hence, the angle α has an absolute maximum value at $x = 3$. Therefore, the observer should stand 3 metres away from the wall to get the 'best' view of the painting.

Summary of differentiation rules

Derivative of $f(x)$	$y = f(x)$	$\Rightarrow$	$f'(x) = \lim_{h \to 0} \frac{f(x+h) - f(x)}{h}$
Derivative of x^n	$f(x) = x^n$	$\Rightarrow$	$f'(x) = nx^{n-1}$
Derivative of $\sin x$	$f(x) = \sin x$	$\Rightarrow$	$f'(x) = \cos x$
Derivative of $\cos x$	$f(x) = \cos x$	$\Rightarrow$	$f'(x) = -\sin x$
Derivative of $\tan x$	$f(x) = \tan x$	$\Rightarrow$	$f'(x) = \sec^2 x$
Derivative of $\sec x$	$f(x) = \sec x$	$\Rightarrow$	$f'(x) = \sec x \tan x$
Derivative of $\csc x$	$f(x) = \csc x$	$\Rightarrow$	$f'(x) = -\csc x \cot x$
Derivative of $\cot x$	$f(x) = \cot x$	$\Rightarrow$	$f'(x) = -\csc^2 x$

Note: derivative rules for trigonometric functions only apply if x is in radian measure.

Derivative of e^x	$f(x) = e^x$	$\Rightarrow$	$f'(x) = e^x$
Derivative of b^x	$f(x) = b^x$	$\Rightarrow$	$f'(x) = b^x \ln b$
Derivative of $\ln x$	$f(x) = \ln x$	$\Rightarrow$	$f'(x) = \frac{1}{x}$
Derivative of $\log_b x$	$f(x) = \log_b x$	$\Rightarrow$	$f'(x) = \frac{1}{x \ln b}$

Derivative of arcsin x $\quad f(x) = \arcsin x \quad \Rightarrow \quad f'(x) = \frac{1}{\sqrt{1-x^2}}$

Derivative of arccos x $\quad f(x) = \arccos x \quad \Rightarrow \quad f'(x) = -\frac{1}{\sqrt{1-x^2}}$

Derivative of arctan x $\quad f(x) = \arctan x \quad \Rightarrow \quad f'(x) = \frac{1}{1+x^2}$

Derivative of arcsec x $\quad f(x) = \text{arcsec}\, x \quad \Rightarrow \quad f'(x) = \frac{1}{x\sqrt{x^2-1}}$

Derivative of arccsc x $\quad f(x) = \text{arccsc}\, x \quad \Rightarrow \quad f'(x) = -\frac{1}{x\sqrt{x^2-1}}$

Derivative of arccot x $\quad f(x) = \text{arccot}\, x \quad \Rightarrow \quad f'(x) = -\frac{1}{1+x^2}$

Chain rule for composite functions: $\frac{dy}{dx} = \frac{d}{dx}[f(g(x))] = f'(g(x)) \cdot g'(x)$

Product rule: $\frac{dy}{dx} = \frac{d}{dx}[f(x) \cdot g(x)] = f(x) \cdot g'(x) + g(x) \cdot f'(x)$

Quotient rule: $\frac{dy}{dx} = \frac{d}{dx}\left[\frac{f(x)}{g(x)}\right] = \frac{g(x) \cdot f'(x) - f(x) \cdot g'(x)}{[g(x)]^2}$

Exercise 15.3

In questions 1–12, find the derivative of y with respect to x, $\frac{dy}{dx}$, by implicit differentiation.

1 $x^2 + y^2 = 16$

2 $x^2y + xy^2 = 6$

3 $x = \tan y$

4 $x^2 - 3xy^2 + y^3x - y^2 = 2$

5 $\frac{x}{y} - \frac{y}{x} = 1$

6 $xy\sqrt{x+y} = 1$

7 $x + \sin y = xy$

8 $x^2y^3 = x^4 - y^4$

9 $xy + e^y = 0$

10 $(x+2)^2 + (y+3)^2 = 25$

11 $x = \tan y$

12 $y + \sqrt{xy} = 3x^3$

In questions 13–16, find the lines that are a) tangent and b) normal to the curve at the given point.

13 $x^3 - xy - 3y^2 = 0$, $(2, -2)$

14 $16x^4 + y^4 = 32$, $(1, 2)$

15 $2xy + \pi \sin y = 2\pi$, $\left(1, \frac{\pi}{2}\right)$

16 $\sqrt[3]{xy} = 14x + y$, $(2, -32)$

17 For the circle $x^2 + y^2 = r^2$ show that the tangent line at any point (x_1, y_1) on the circle is perpendicular to the line that passes through (x_1, y_1) and the centre of the circle.

18 Consider the equation $x^2 + xy + y^2 = 7$.

a) Find the two points where the curve intersects the x-axis. Show that the tangents to the curve at these two points are parallel.

b) Find any points where the tangent to the curve is parallel to the x-axis.

c) Find any points where the tangent to the curve is parallel to the y-axis.

19 The line that is normal to the curve $x^2 + 2xy - 3y^2 = 0$ at $(1, 1)$ intersects the curve at what other point?

In questions 20 and 21, find $\frac{dy}{dx}$ and $\frac{d^2y}{dx^2}$ for the given equation.

20 $4x^2 + 9y^2 = 36$

21 $xy = 2x - 3y$

22 Consider the equation $xy^3 = 1$. Find $\frac{dy}{dx}$ and $\frac{d^2y}{dx^2}$ by two different methods.

a) Solve for y in terms of x and differentiate explicitly.

b) Differentiate implicitly.

23 The graph (shown right) of the equation $x^2 + y^2 = 2x^2 + 2y^2 - x^2$ is a type of curve called a *cardioid*. A cardioid is a heart-shaped curve generated by a fixed point on a circle as it rolls around another circle having the same radius. Find the equation of the line tangent to this particular cardioid at the point $(0, \frac{1}{2})$.

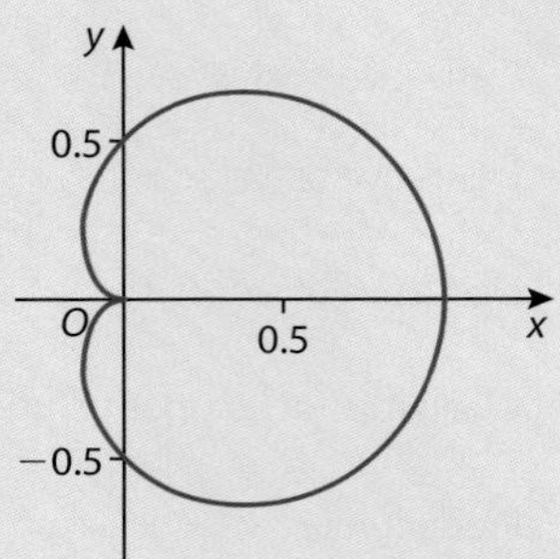

In questions 24–33, find the derivative of y with respect to x, $\frac{dy}{dx}$.

24 $y = \ln(x^3 + 1)$

25 $y = \ln(\sin x)$

26 $y = \log_5\sqrt{x^2 - 1}$

27 $y = \ln\sqrt{\frac{1+x}{1-x}}$

28 $y = \sqrt{\log_{10} x}$

29 $y = \ln\left(\frac{a-x}{a+x}\right)$

30 $y = \ln(e^{\cos x})$

31 $y = \frac{1}{\log_3 x}$

32 $y = x\ln(x) - x$

33 $y = \ln(ax) - (\ln b)\log_b x$

34 Find the equation of the line tangent to the graph of $y = \log_2 x$ at the point $x = 8$. Express the equation exactly. Can you find a way to graph $y = \log_2 x$ on your GDC in order to check your answer?

35 Given $y = \sqrt{\frac{x^2-1}{x^2+1}}$ we could find $\frac{dy}{dx}$ by applying the chain rule and the quotient rule. However, it is much easier to first take the natural logarithm of both sides, use the properties of logarithms to simplify as much as possible, and then differentiate implicitly to find $\frac{dy}{dx}$. This technique is called *logarithmic differentiation*. Use this technique to show that $\frac{dy}{dx} = \frac{2x}{(x^2-1)^{\frac{1}{2}}(x^2+1)^{\frac{3}{2}}}$.

36 Find the x-coordinate, between 0 and 1, of the point of inflexion on the graph of the function $f(x) = x^2\ln(x^2)$. Express your answer exactly.

37 a) Given $g(x) = \frac{\ln x}{x}$, find expressions for $g'(x)$ and $g''(x)$.

b) Show that g has an absolute maximum at $x = e$, and state the maximum value of g.

In questions 38–41, find the derivative of y with respect to x, $\frac{dy}{dx}$.

38 $y = \arctan(x + 1)$

39 $y = \sin^{-1}\left(\frac{x}{\sqrt{1+x^2}}\right)$

40 $y = \arccos\left(\frac{3}{x^2}\right)$

41 $\ln\sqrt{1+x^2} = x\tan^{-1}x$

42 Given that $f(x) = \arcsin x + \arccos x$, find $f'(x)$. What can you conclude about the function f?

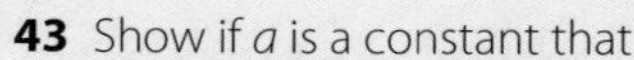

43 Show if a is a constant that

a) $\frac{d}{dx}\left[\arctan\left(\frac{x}{a}\right)\right] = \frac{a}{a^2 + x^2}$ b) $\frac{d}{dx}\left[\arcsin\left(\frac{x}{a}\right)\right] = \frac{1}{\sqrt{a^2 - x^2}}$

44 Find the equation of the line tangent to the curve $y = 4x \arctan 2x$ at the point on the curve where $x = \frac{1}{2}$. Express the equation exactly in the form $y = mx + c$, where m and c are constants.

45 Consider the function $f(x) = \arcsin(\cos x)$ with domain of $0 \leqslant x < \pi$.

a) Prove that f is a linear function.

b) Express the function exactly in the form $f(x) = ax + b$, where a and b are constants.

46 A 3-metre tall statue is on top of a column such that the bottom of the statue is 2 metres above the eye level of a person viewing the statue. How far from the base of the column should the person stand to get the best view of the statue, that is, so that the angle subtended at the observer's eye by the statue is a maximum?

47 A particle moves along the x-axis so that its displacement, s (in metres), from the origin at any time $t \geqslant 0$ (in seconds) is given by $s(t) = \arctan\sqrt{t}$.

a) Find the exact velocity of the particle at (i) $t = 1$ second, and at (ii) $t = 4$ seconds.

b) Find the exact acceleration of the particle at (i) $t = 1$ second, and at (ii) $t = 4$ seconds.

c) Describe the motion of the particle.

d) What is the limiting displacement of the particle as t approaches infinity?

15.4 Related rates

A claim was made in the first section of this chapter that 'the chain rule is the most important, and most widely used, rule of differentiation'. The chain rule has been repeatedly applied in all parts of this chapter thus far. Another important use of the chain rule is to find the rates of change of two or more variables that are changing with respect to time. Calculus provides us with the tools and techniques to solve problems where quantities (variables) are changing rather than static.

When a stone is thrown into a pond, a circular pattern of ripples is formed. In this situation we can observe an ever-widening circle moving across the water. As the circular ripple moves across the water, the radius r of the circle, its circumference C, and its area A all increase as a function of time t. Not only are these quantities (variables) functions of time, but their values at any particular time t are related to one another by familiar formulae such as $C = 2\pi r$ and $A = \pi r^2$. Thus their rates of change are also related to one another.

Example 20

A stone is thrown into a pond causing ripples in the form of concentric circles to move away from the point of impact at a rate of 20 cm per second. Find the following when a circular ripple has a radius of 50 cm and again when its radius is 100 cm.

a) the rate of change of the circle's circumference

b) the rate of change of the circle's area

Solution

In calculus, a derivative represents a rate of change of one variable with respect to another variable. If the circles are moving outward at a rate of 20 cm/sec, then the rate of change of the radius is 20 cm/sec, and in the notation of calculus we write

$$\frac{dr}{dt} = 20.$$

a) Knowing that the relationship between the radius, r, and the circumference, C, is $C = 2\pi r$, and that the rate of change of the radius with respect to time is $\frac{dr}{dt} = 20$, we can use the chain rule to find the rate of change of the circumference with respect to time, i.e. $\frac{dC}{dt}$.

$$\frac{dC}{dt} = \frac{dC}{dr} \cdot \frac{dr}{dt}$$

We need to find $\frac{dC}{dr}$, the rate of change (derivative) of the circumference with respect to the radius. This rate can be derived from the relationship between the variables.

$$C = 2\pi r$$

$$\frac{d}{dr}(C) = \frac{d}{dr}(2\pi r)$$ Differentiate both sides with respect to r.

$$\frac{dC}{dr} = 2\pi$$ Implicit differentiation on the left side.

Since the circumference C is a linear function of the radius r ($C = 2\pi r$), the derivative $\frac{dC}{dr}$ is a constant.

We now substitute in for $\frac{dC}{dr}$ and $\frac{dr}{dt}$ to find the rate of change of the circumference with respect to time, $\frac{dC}{dt}$.

$$\frac{dC}{dt} = \frac{dC}{dr} \cdot \frac{dr}{dt} \Rightarrow \frac{dC}{dt} = 2\pi \cdot 20 = 40\pi \text{ cm/sec}$$

The rate of change of a circular ripple's circumference is constant (40π). Therefore, the rate of change of the circumference is 40π cm/sec when the radius is 50 cm and also when its 100 cm.

• **Hint:** There is a slightly different method to determine $\frac{dC}{dt}$. We can find the rate by differentiating implicitly with respect to time, t, both sides of the equation, $C = 2\pi r$, that gives the relationship between the two changing quantities (variables).

$C = 2\pi r$

Differentiate both sides with respect to t:

$\frac{d}{dt}(C) = \frac{d}{dt}(2\pi r)$

Implicit differentiation:

$\frac{dC}{dt} = 2\pi\frac{dr}{dt}$

Substitute $\frac{dr}{dt} = 20$:

$\frac{dC}{dt} = 2\pi \cdot 20 = 40\pi$ cm/sec

b) Similarly, to find the rate of change of the area with respect to time, $\frac{dA}{dt}$, we can use the chain rule to write

$$\frac{dA}{dt} = \frac{dA}{dr} \cdot \frac{dr}{dt}.$$

Find $\frac{dA}{dr}$ from the formula, $A = \pi r^2$, that relates the variables A and r.

$$\frac{d}{dr}(A) = \frac{d}{dr}(\pi r^2)$$ Differentiate both sides with respect to r.

$$\frac{dA}{dr} = \pi(2r) = 2\pi r$$ Implicit differentiation on the left side.

Since the area A is a non-linear function of the radius r ($A = \pi r^2$), the derivative $\frac{dA}{dr}$ is not a constant but has different values depending on the value of r.

We substitute in for $\frac{dA}{dr}$ and $\frac{dr}{dt}$ to find the rate of change of the area with respect to time, $\frac{dA}{dt}$.

$$\frac{dA}{dt} = \frac{dA}{dr} \cdot \frac{dr}{dt} \Rightarrow \frac{dA}{dt} = 2\pi r \cdot 20 = 40\pi r$$

Thus, the rate of change of the circle's area with respect to time, $\frac{dA}{dt}$, is a linear function in terms of the radius r.

When the radius is 50 cm, $\frac{dA}{dt} = 40\pi \cdot 50 = 2000\pi\,\text{cm}^2/\text{sec}$
$\approx 6280\,\text{cm}^2/\text{sec}\ [\approx 0.628\,\text{m}^2/\text{sec}]$.

When the radius is 100 cm, $\frac{dA}{dt} = 40\pi \cdot 100 = 4000\pi\,\text{cm}^2/\text{sec}$
$\approx 12\,600\,\text{cm}^2/\text{sec}\ [\approx 1.26\,\text{m}^2/\text{sec}]$.

Note that when $r = 100$ cm the area is changing at twice the rate it was when $r = 50$ cm.

● **Hint:** It is important to include the appropriate units when giving a rate of change (derivative) answer. For example cm/sec, m²/hour, litres/sec, etc.

Example 21

A 4-metre ladder stands upright against a vertical wall. If the foot of the ladder is pulled away from the wall at a constant rate of 0.75 m/sec, how fast is the top of the ladder coming down the wall at the instant it is (i) 3 metres above the ground, and (ii) 1 metre above the ground. Give answers approximate to three significant figures.

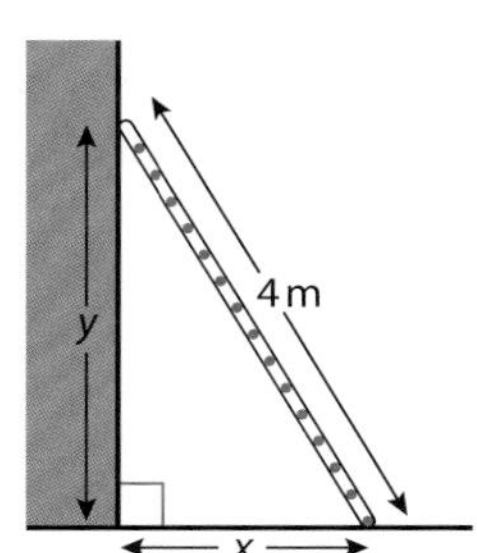

Solution

Let x and y represent the distances of the foot and top of the ladder, respectively, from the bottom of the wall. Then from Pythagoras' theorem, we have

$$x^2 + y^2 = 16.$$

Given that the ladder is being pulled away at a rate of 0.75 m/sec, then

$$\frac{dx}{dt} = 0.75 = \frac{3}{4}.$$

So we know the rate $\frac{dx}{dt}$, and we need to find $\frac{dy}{dt}$ when $y = 3$ and when $y = 1$.

Rather than starting with the chain rule and writing an equation relating the different rates, let's utilize the chain rule by differentiating implicitly with respect to time the equation relating the relevant variables x and y.

$$\frac{d}{dt}(x^2 + y^2) = \frac{d}{dt}(16)$$

$$2x\frac{dx}{dt} + 2y\frac{dy}{dt} = 0$$

$$\frac{dy}{dt} = -\frac{x}{y}\frac{dx}{dt}$$

(i) We know $\frac{dx}{dt} = \frac{3}{4}$, so to find $\frac{dy}{dt}$ when $y = 3$ m, we find the corresponding value for x.

$$x^2 + y^2 = 16 \Rightarrow x = \sqrt{16 - y^2}; \text{ for } y = 3: x = \sqrt{16 - 3^2} = \sqrt{7}$$

Hence, when $y = 3$: $\frac{dy}{dt} = -\frac{\sqrt{7}}{3} \cdot \frac{3}{4} = -\frac{\sqrt{7}}{4} \approx -0.661$ m/sec.

(ii) For $y = 1$: $x = \sqrt{16 - 1^2} = \sqrt{15}$

Hence, when $y = 1$: $\frac{dy}{dt} = -\frac{\sqrt{15}}{1} \cdot \frac{3}{4} = -\frac{3\sqrt{15}}{4} \approx -2.90$ m/sec.

It makes sense that $\frac{dy}{dt}$ is negative because the distance y decreases as the ladders slides down.

Example 22

In the preceding example, how fast is the angle between the ladder and the ground changing when $y = 2$ m?

Solution

We know $\frac{dx}{dt} = \frac{3}{4}$ and we seek to find $\frac{d\theta}{dt}$. We need a relationship, true at any instant, between the variables θ and x. Several trigonometric ratios could be used, but perhaps the most straightforward is

$$x = 4\cos\theta.$$

Now we differentiate implicitly with respect to t and solve for $\frac{d\theta}{dt}$.

$$\frac{d}{dt}(x) = \frac{d}{dt}(4\cos\theta)$$

$$\frac{dx}{dt} = -4\sin\theta\frac{d\theta}{dt}$$

$$\frac{d\theta}{dt} = -\frac{1}{4\sin\theta}\frac{dx}{dt}$$

When $y = 2$ we find that $\sin\theta = \frac{1}{2}$. Substituting appropriately for $\sin\theta$ and $\frac{d\theta}{dt}$, we have

$$\frac{d\theta}{dt} = -\frac{1}{4(\frac{1}{2})} \cdot \frac{3}{4} = -\frac{3}{8}.$$

Therefore, the angle is decreasing at a rate of $\frac{3}{8}$ radians/sec (or approximately 21.5°/sec).

The solution strategy used in the preceding two examples is summarized below.

Solving problems involving related rates

1. Identify any rate(s) of change you know and the rate of change to be found.
2. Draw a diagram with all of the important information clearly labelled.
3. Write an equation relating the variables whose rates of change are either known or are to be found.
4. Using the chain rule, differentiate the equation implicitly with respect to time. Solve for the rate to be found.
5. Substitute in all known values for any variables and any rates of change. Compute the required rate of change. Be sure to include appropriate units with the result.

Example 23

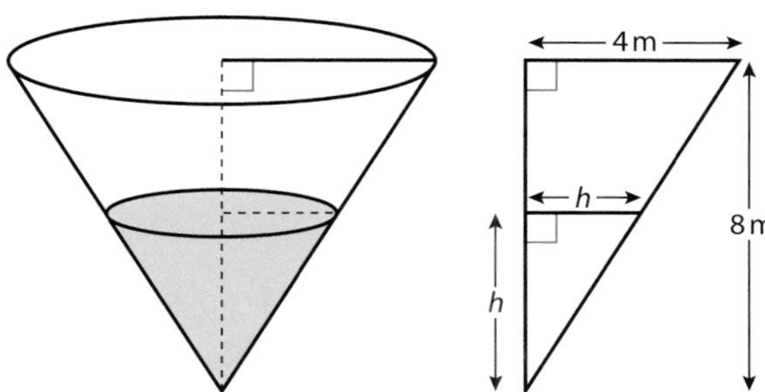

Consider a conical tank as shown in the diagram. Its radius at the top is 4 metres and its height is 8 metres. The tank is being filled with water at a rate of 2 m^3/min. How fast is the water level rising when it is 5 metres high?

Solution

We know the rate of change of the volume with respect to time, that is, $\frac{dV}{dt} = 2$ m^3/min and we seek to find the rate of change of the height of the water level with respect to time, call it $\frac{dh}{dt}$.

Not including t, there are three variables involved in this problem: V, r and h. The formula for the volume of a cone will give us an equation that relates all of these variables.

$$V = \tfrac{1}{3}\pi r^2 h$$

If we differentiate this equation now we will get the rate $\frac{dr}{dt}$ in our result. We need to either find $\frac{dr}{dt}$ (which is possible) or eliminate r from the equation by solving for it in terms of one of the other variables and substitute. By using similar triangles we can write a proportion involving r and h.

$$\frac{r}{h} = \frac{4}{8} \Rightarrow r = \frac{h}{2}$$

Hence, $V = \frac{1}{3}\pi\left(\frac{h}{2}\right)^2 h \Rightarrow V = \frac{\pi}{12}h^3$.

Differentiating implicitly with respect to t and solving for $\frac{dh}{dt}$:

$$\frac{dV}{dt} = \frac{\pi}{12} \cdot 3h^2\frac{dh}{dt} \Rightarrow \frac{dV}{dt} = \frac{\pi}{4}h^2\frac{dh}{dt} \Rightarrow \frac{dh}{dt} = \frac{4}{\pi h^2}\frac{dV}{dt}$$

Substituting $h = 5$ and $\frac{dV}{dt} = 2$ gives

$$\frac{dh}{dt} = \frac{4}{\pi(5)^2} \cdot 2 = \frac{8}{25\pi} \approx 0.102 \text{ m/min [or 10.2 cm/min]}.$$

Therefore, the water level is rising at a rate of 0.102 m/min when the water level is at 5 m.

• **Hint:** Be careful not to substitute in known quantities too early in the process of solving a related rates problem. Substitute the known values of any variables and any rates of change *after* differentiation. For example, in Example 23 h remained a variable (it is a quantity that is changing over time) until the last stage of the solution when we substituted $h = 5$. If we substituted earlier into $V = \frac{\pi}{12}h^3$, we would have obtained $\frac{dV}{dt} = 0$, which is obviously wrong.

The following example involves two rates of change.

Example 24

At 12 noon ship A is 65 km due north of a second ship, B. Ship A sails south at a rate of 14 km/hr, and ship B sails west at a rate of 16 km/hr.

a) How fast are the two ships approaching each other $1\frac{1}{2}$ hours later at 1:30?

b) At what time do the two ships stop approaching and begin moving away from each other?

Solution

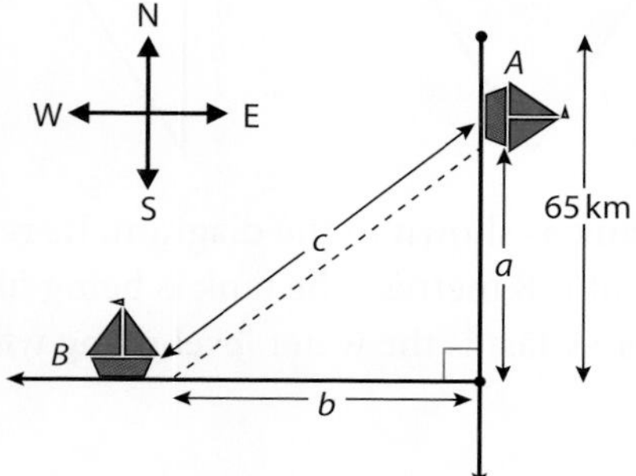

Let a and b be the distances that ships A and B, respectively, are from the intersection of the ships' paths (see diagram). Let c be the distance between the two ships. Since a is decreasing and b is increasing, we know that $\frac{da}{dt} = -14$ km/hr and $\frac{db}{dt} = 16$ km/hr.

a) The three variables are related by the equation

$$c^2 = a^2 + b^2.$$

Differentiating implicitly with respect to t gives

$$2c\frac{dc}{dt} = 2a\frac{da}{dt} + 2b\frac{db}{dt}.$$

The rate at which the ships are approaching is $\frac{dc}{dt}$. Solving for $\frac{dc}{dt}$:

$$\frac{dc}{dt} = \frac{a\frac{da}{dt} + b\frac{db}{dt}}{c}$$

Substituting $\frac{da}{dt} = -14$ and $\frac{db}{dt} = 16$:

$$\frac{dc}{dt} = \frac{-14a + 16b}{c}$$

The distances a and b are both functions of time; thus, they can be written in terms of t as

$$a = 65 - 14t \text{ and } b = 16t.$$

Evaluating these expressions when $t = 1\frac{1}{2}$, gives $a = 44$, $b = 24$ and $c = \sqrt{44^2 + 24^2} \approx 50.12$. Substituting these values into the expression for $\frac{dc}{dt}$ gives

$$\frac{dc}{dt} \approx \frac{-14(44) + 16(24)}{50.12} \approx -4.629.$$

Therefore, at 1:30 the distance between the two ships is decreasing at a rate of approximately -4.63 km/hr.

b) The time at which the two ships will stop approaching each other and begin to move away is when the value of $\frac{dc}{dt}$ changes from negative to positive. So we need to find when $\frac{dc}{dt} = 0$.

$$\frac{dc}{dt} = \frac{-14a + 16b}{c} = 0 \Rightarrow -14a + 16b = 0$$

Substituting in $a = 65 - 14t$ and $b = 16t$ gives:

$$-14(65 - 14t) + (16t) = 0 \Rightarrow 452t - 910 = 0 \Rightarrow t = \frac{910}{452} \approx 2.013$$

Therefore, just moments after 2:00 the two ships will stop approaching and start moving away from each other.

Exercise 15.4

1 A water tank is in the shape of an inverted cone. Water is being drained from the tank at a constant rate of 2 m³/min. (Since volume is decreasing, $\frac{dV}{dt}$ is negative.) The height of the tank is 8 m, and the diameter of the top of the tank is 6 m. When the height of the water is 5 m, find, in units of cm/min, the following:

a) the rate of change of the water level

b) the rate of change of the radius of the surface of the water.

2 A spherical balloon is being inflated at a constant rate of 240 cm³/sec. [$V = \frac{4}{3}\pi r^3$]

a) At what rate is the radius increasing when the radius is equal to 8 cm?

b) At what rate is the radius increasing 5 seconds after the start of inflation?

3 Oil is dripping from a car engine on to a garage floor, making a growing circular stain. The radius, r, of the stain is increasing at a constant rate of 1 cm/hr. When the radius is 4 cm, find:

a) the rate of change of the circumference of the stain

b) the rate of change of the area of the stain.

4 A hot air balloon is rising straight up from a level field at a constant rate of 50 m/min. An observer is standing 150 m from the point on the ground where the balloon was launched. Let θ be the angle between the ground and the observer's line of sight to the balloon from the point at which the observer is standing (angle of elevation of the balloon). What is the rate of change of θ (in radians/min) when the height of the balloon is 250 m?

5 Jenny is flying a kite at a constant height above level ground of 72 m. The wind carries the kite away horizontally at a rate of 6 m/sec. How fast must Jenny let out the string at the moment when the kite is 120 m away from her?

6 A 5-foot boy is walking toward a 20-foot lamp post at a constant rate of 6 ft/sec. The light from the lamp post causes the boy to cast a shadow. How fast is the tip of his shadow moving?

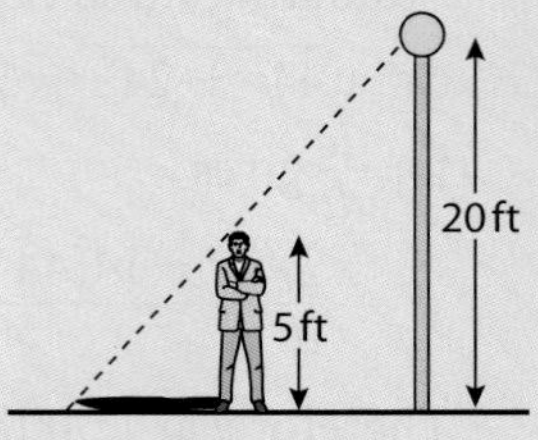

7 Two cars start from a point A at the same time. One travels west at 60 km/hr and the other travels north at 35 km/hr. How fast is the distance between them increasing 3 hours later?

8 A point moves along the curve $y = \sqrt{x^2 + 1}$ in such a way that $\frac{dx}{dt} = 4$. Find $\frac{dy}{dt}$ when $x = 3$.

9 A horizontal trough is 4 m long, 1.5 m wide and 1 m deep. Its cross-section is an isosceles triangle. Water is flowing into the trough at a constant rate of 0.03 m^3/sec. Find the rate at which the water level is rising 25 seconds after the water started flowing into the trough.

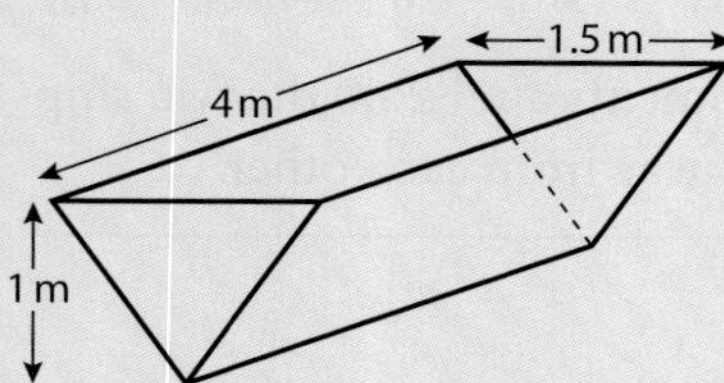

10 If the radius of a sphere is increasing at the constant rate of 3 mm/sec, how fast is the volume changing when the surface area is 10 mm^2? [Surface area $= 4\pi r^2$]

11 Two roads, A and B, intersect each other at an angle of 60°. Two cars, one on road A travelling at 40 km/hr and the other on road B travelling at 50 km/hr, are approaching the intersection. If, at a certain moment, the two cars are both 2 km from the intersection, how fast is the distance between them changing?

12 If the diagonal of a cube is increasing at a rate of 8 cm/sec, how fast is a side of the cube increasing?

13 A point P is moving along the circle with equation $x^2 + y^2 = 100$ at a constant rate of 3 units/sec. How fast is the projection of P on the x-axis moving when P is 5 units above the x-axis?

14 A jet is flying at a constant speed at an altitude of 10 000 m on a path that will take it directly over an observer on the ground. At a given instant the observer determines that the angle of elevation of the jet is $\frac{\pi}{3}$ radians and is increasing at a constant rate of $\frac{1}{60}$ radians/sec. Find the speed of the jet.

15 A television cameraman is filming an automobile race from a platform that is 40 metres from the racing track, following a car that is moving at 288 km/hr. How fast, in degrees per second, will the camera be turning when a) the car is directly in front of the camera and b) a half second later? Answer to the nearest whole degree.

16 A plane is flying due east at 640 km/hr and climbing vertically at a rate of 180 m/min. An airport tower is tracking it. Determine how fast the distance between the plane and the tower is changing when the plane is 5 km above the ground over a point exactly 6 km due west of the tower. Express the answer in km/hr.

15.5 Optimization

Many problems in science and mathematics involve finding the maximum or minimum value (**optimum** value) of a function over a specified or implied domain. The development of the calculus in the seventeenth century was motivated to a large extent by maxima and minima (**optimization**) problems. One such problem lead Pierre de Fermat (1601–1665) to develop his Principle of Least Time: a ray of light will follow the path that takes the least (or minimum) time. The solution to Fermat's principle lead to Snell's law, or law of refraction (see the investigation at the end of this section). The solution is found by applying techniques of differential calculus – which can also be used to solve other optimization problems involving ideas such as least cost, maximum profit, minimum surface area and greatest volume.

Previously, we learned the theory of how to use the derivative of a function to locate points where the function has a maximum or minimum (i.e. extreme) value. It is important to remember that if the derivative of a function is zero at a certain point it does not *necessarily* follow that the function has an extreme value (relative or absolute) at that point – it only ensures that the function has a horizontal tangent (stationary point) at that point. An extreme value *may* occur where the derivative is zero or at the endpoints of the function's domain.

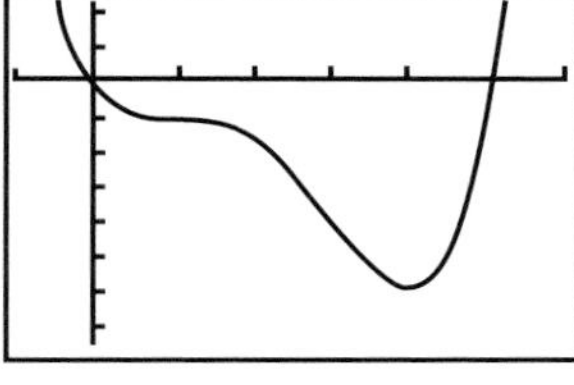

The graph of $f(x) = x^4 - 8x^3 + 18x^2 - 16x - 2$ is shown left. The derivative of $f(x)$ is $f'(x) = 4x^3 - 24x^2 + 36x - 16 = 4(x - 4)(x - 1)^2$. The function has horizontal tangents at both $x = 1$ and $x = 4$, since the derivative is zero at these points. However, an extreme value (absolute minimum) occurs only at $x = 4$. It is important to confirm – graphically (see GDC images) or algebraically – the precise nature of a point on a function where the derivative is zero. Some different algebraic methods for confirming that a value is a maximum or minimum will be illustrated in the examples that follow.

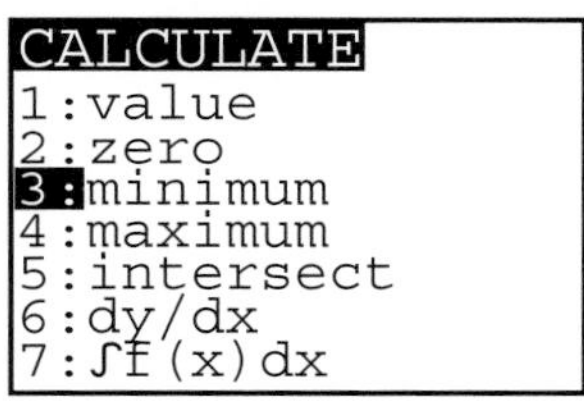

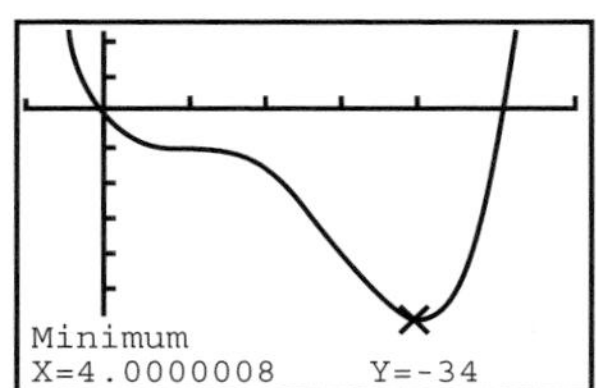

It is also useful to remember that one can often find extreme values (extrema) without calculus (e.g. using a 'minimum' command on a graphics calculator, as shown). Calculator or computer technology can be very helpful in modelling, solving or confirming solutions to optimization problems. However, it is important to learn how to apply algebraic methods of differentiation to optimization problems because it may be the only efficient way to obtain an accurate solution.

Let's start with a relatively straightforward example. We can use the steps in the solution to develop a general strategy that can be applied to more sophisticated problems.

Example 25 – Finding a maximum area (Developing a general strategy)

Find the maximum area of a rectangle inscribed in an isosceles right triangle whose hypotenuse is 20 cm long.

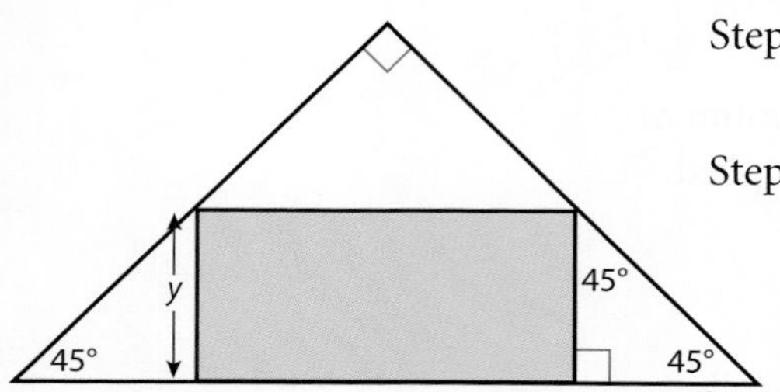

Solution

Step 1: Draw an accurate diagram. Let the base of the rectangle be x cm and the height y cm. Then the area of the rectangle is $A = xy\,\text{cm}^2$.

Step 2: Express area as a function in terms of only one variable.

It can be deduced from the diagram that $y = 10 - \frac{x}{2}$.

Therefore, $A(x) = x\left(10 - \frac{x}{2}\right) = 10x - \frac{x^2}{2}$.

x must be positive and from the diagram it is clear that x must be less than 20 (domain of A: $0 < x < 20$).

Step 3: Find the derivative of the area function and find for what value(s) of x it is zero.

$A'(x) = 10 - x \qquad A'(x) = 0$ when $x = 10$

Step 4: Analyze $A(x)$ at $x = 10$ and also at the endpoints of the domain, $x = 0$ and $x = 20$.

The second derivative test (Section 13.3) provides information about the concavity of a function. The second derivative is $A''(x) = -1$ and since $A''(x)$ is always negative then $A(x)$ is always concave down, indicating $A(x)$ has a maximum at $x = 10$.

$A(0) = 0$ and $A(20) = 0$, indicating $A(x)$ has an absolute maximum at $x = 10$.

Therefore, the rectangle has a maximum area equal to $A(10) = 10\left(10 - \frac{10}{2}\right) = 50\,\text{cm}^2$.

General strategy for solving optimization problems

Step 1: Draw a diagram that accurately illustrates the problem. Label all known parts of the diagram. Using variables, label the important unknown quantity (or quantities) (for example, x for base and y for height in Example 25).

Step 2: For the quantity that is to be optimized (area in Example 25), express this quantity as a function in terms of a single variable. From the diagram and/or information provided, determine the domain of this function.

Step 3: Find the derivative of the function from Step 2, and determine where the derivative is zero. This value (or values) of the derivative, along with any domain endpoints, are the **critical values** ($x = 0$, $x = 10$ and $x = 20$ in Example 25) to be tested.

Step 4: Using algebraic (e.g. second derivative test) or graphical (e.g. GDC) methods, analyze the nature (maximum, minimum, neither) of the points at the critical values for the optimized function. Be sure to answer the precise question that was asked in the problem.

Example 26 – Finding a minimum length – two posts problem

Two vertical posts, with heights of 7 m and 13 m, are secured by a rope going from the top of one post to a point on the ground between the posts and then to the top of the other post. The distance between the two posts is 25 m. Where should the point at which the rope touches the ground be located so that the least amount of rope is used?

Solution

Step 1: An accurate diagram is drawn. The posts are drawn as line segments PQ and TS and the point where the rope touches the ground is labelled R. The optimum location of point R can be given as a distance from the base of the shorter post, QR, or from the taller post, SR. It is decided to give the answer as the distance from the shorter post – and this is labelled x. There are two other important unknown quantities: the lengths of the two portions of the rope, PR and TR. These are labelled a and b, respectively.

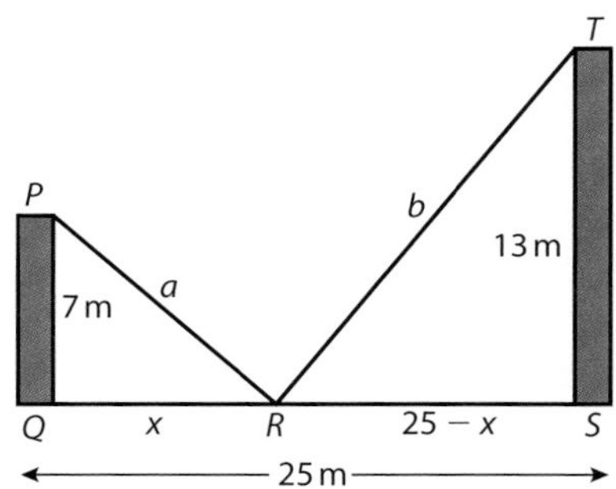

Step 2: The quantity to be minimized is the length L of the rope, which is the sum of a and b. From Pythagoras' theorem, $a = \sqrt{x^2 + 49}$ and $b = \sqrt{(25 - x)^2 + 169}$. Therefore, the function for length (L) can be expressed in terms of the single variable x as

$$L(x) = \sqrt{x^2 + 49} + \sqrt{(25 - x)^2 + 169}$$

$$= \sqrt{x^2 + 49} + \sqrt{x^2 - 50x + 625 + 169}$$

$$L(x) = \sqrt{x^2 + 49} + \sqrt{x^2 - 50x + 794}$$

From the given information and diagram, the domain of $L(x)$ is $0 \leqslant x \leqslant 25$.

Step 3: To facilitate differentiation, express $L(x)$ using fractional exponents:

$$L(x) = (x^2 + 49)^{\frac{1}{2}} + (x^2 - 50x + 794)^{\frac{1}{2}}$$

Then apply the chain rule for differentiation:

$$\frac{dL}{dx} = \tfrac{1}{2}(x^2 + 49)^{-\frac{1}{2}}(2x) + \tfrac{1}{2}(x^2 - 50x + 794)^{-\frac{1}{2}}(2x - 50) \Rightarrow$$

$$\frac{dL}{dx} = \frac{x}{\sqrt{x^2 + 49}} + \frac{x - 25}{\sqrt{x^2 - 50x + 794}}$$

By setting $\frac{dL}{dx} = 0$, we obtain

$$x\sqrt{x^2 - 50x + 794} = -(x - 25)\sqrt{x^2 + 49}$$

$$x^2(x^2 - 50x + 794) = (25 - x)^2(x^2 + 49)$$

$$x^4 - 50x^3 + 794x^2 = x^4 - 50x^3 + 674x^2 - 2450x + 30\,625$$

$$120x^2 + 2450x - 30\,625 = 0$$

$$5(4x - 35)(6x + 175) = 0$$

$$x = \frac{35}{4} \quad \text{or} \quad x = -\frac{175}{6}$$

Step 4: Since $x = -\frac{175}{6}$ is not in the domain for $L(x)$, then the critical values are $x = 0$, $x = \frac{35}{4}$ and $x = 25$. Simply evaluate $L(x)$ for these critical values.

$L(0) = 7 + \sqrt{794} \approx 35.18, \quad L(25) = \sqrt{674} + 13 \approx 38.96,$

$L\left(\frac{35}{4}\right) = 5\sqrt{41} \approx 32.02$

Therefore, the rope should touch the ground at a distance of $\frac{35}{4} = 8.75$ m from the base of the shorter post, to give a minimum rope length of approximately 32.02 m.

The minimum value could also be confirmed from the graph of $L(x)$, but it would be difficult to confirm using the second derivative test because of the algebra required. From this example, we can see that applied optimization problems can involve a high level of algebra. If you have access to suitable graphing technology, you could perform Steps 3 and 4 graphically rather than algebraically.

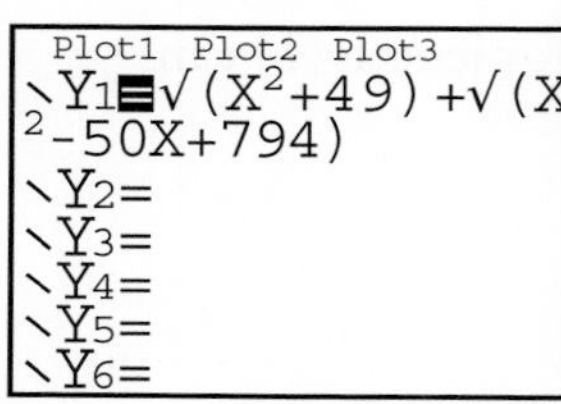

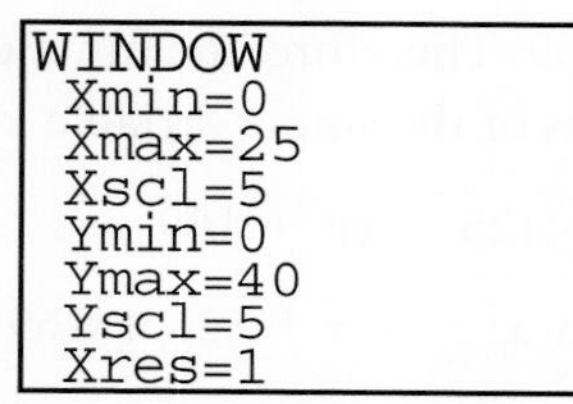

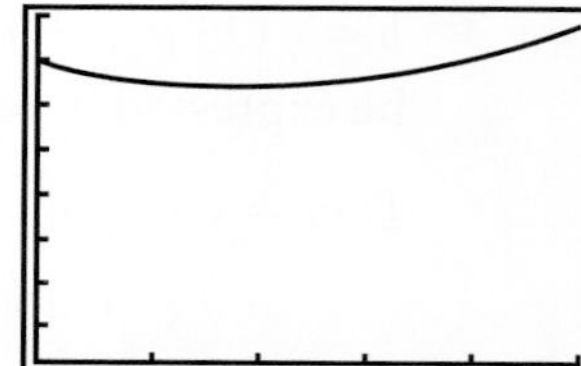

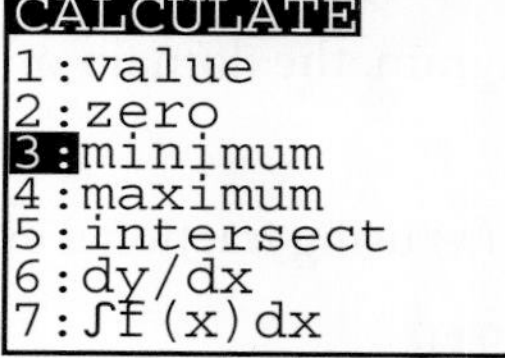

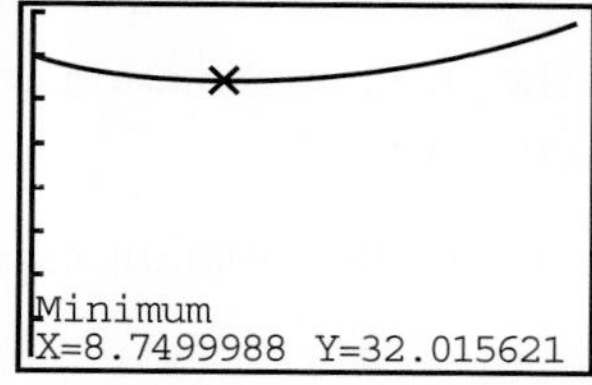

It is interesting to observe that the result for x produced by the calculator does not appear to be exact. Why is that? Algebraic techniques using differentiation give us the certainty of an exact solution while also allowing us to deal with the abstract nature of optimization problems involving parameters rather than fixed measurements (e.g. the heights of the posts).

In both Example 25 and 26, the extreme value occurred at a point where the derivative was zero. Although this often happens, an extreme value may occur at the endpoint of the domain.

Example 27 – An endpoint maximum

A supply of four metres of wire is to be used to form a square and a circle. How much of the wire should be used to make the square and how much should be used to make the circle in order to enclose the greatest amount of area? Guess the answer before looking at the following solution.

Solution

Step 1: Let x = length of each edge of the square and r = radius of the circle.

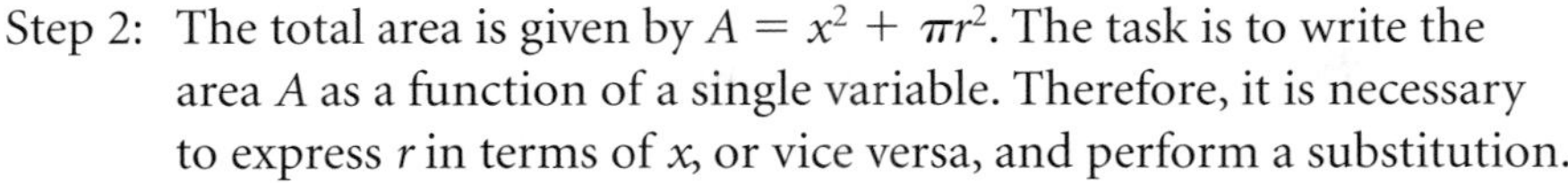

4 m

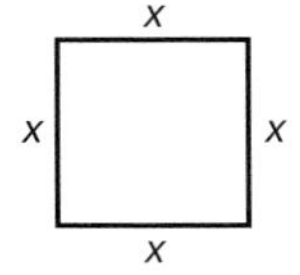

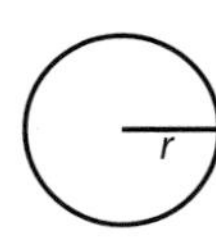

Step 2: The total area is given by $A = x^2 + \pi r^2$. The task is to write the area A as a function of a single variable. Therefore, it is necessary to express r in terms of x, or vice versa, and perform a substitution.

The perimeter of the square is $4x$ and the circumference of the circle is $2\pi r$. The total amount of wire is 4 m which gives

$$4 = 4x + 2\pi r \quad \Rightarrow \quad 2\pi r = 4 - 4x \quad \Rightarrow \quad r = \frac{2(1-x)}{\pi}$$

Substituting gives $A(x) = x^2 + \pi\left[\frac{2(1-x)}{\pi}\right]^2 = x^2 + \frac{4(1-x)^2}{\pi}$

$$= \frac{1}{\pi}[(\pi + 4)x^2 - 8x + 4]$$

Because the square's perimeter is $4x$, then the domain for $A(x)$ is $0 \leqslant x \leqslant 1$.

Step 3: Differentiate the function $A(x)$, set equal to zero, and solve.

$$\frac{d}{dx}\left(\frac{1}{\pi}[(\pi + 4)x^2 - 8x + 4]\right) = \frac{1}{\pi}[2(\pi + 4)x - 8] = 0$$

$$2(\pi + 4)x - 8 = 0 \Rightarrow (\pi + 4)x = 4 \Rightarrow x = \frac{4}{\pi + 4} \approx 0.5601$$

The critical values are $x = 0$, $x \approx 0.5601$ and $x = 1$.

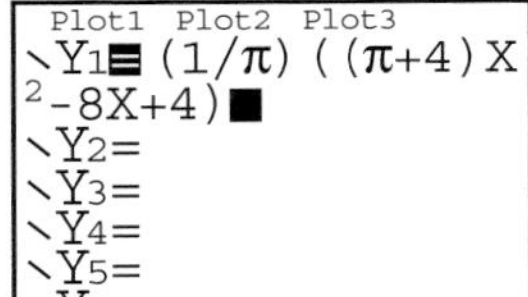

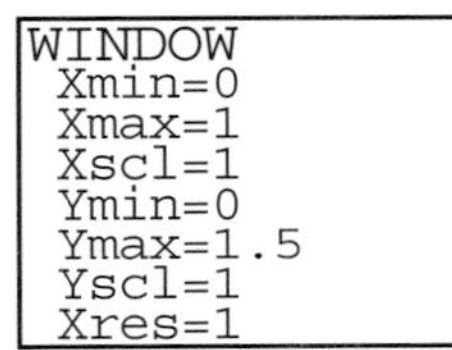

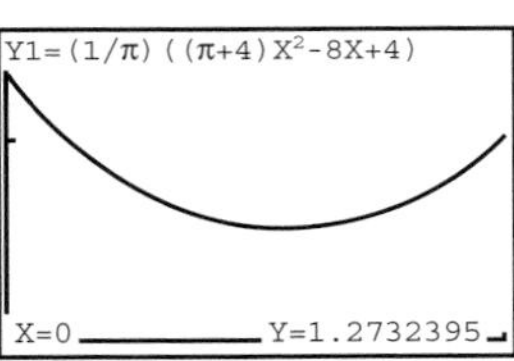

Step 4: Evaluating $A(x)$: $A(0) \approx 1.273$, $A(0.5601) \approx 0.5601$ and $A(1) = 1$. Therefore, the maximum area occurs when $x = 0$ which means <u>all</u> the wire is used for the circle.

What would the answer be if Example 27 asked for the dimensions of the square and circle to enclose the *least* total area?

Example 28 – Minimizing time

A pipeline needs to be constructed to link an offshore drilling rig to an onshore refinery depot. The oil rig is located at a distance (perpendicular to the coast) of 140 km from the coast. The depot is located inland at a distance (perpendicular) of 60 km from the coast. For modelling purposes, the coastline is assumed to follow a straight line. The point on the coastline nearest to the oil rig is 160 km from the point on the coastline nearest to the depot. The rate at which crude oil is pumped through the pipeline varies according to several variables, including pipe dimensions, materials, temperature, etc. On average, oil flows through the offshore section of the pipeline at a rate of 9 km per hour and 5 km per hour through the onshore section. Assume that both sections of pipeline can travel straight from one point to another. At what point should the pipeline intersect with the coastline in order for the oil to take a minimum amount of time to flow from the rig to the depot?

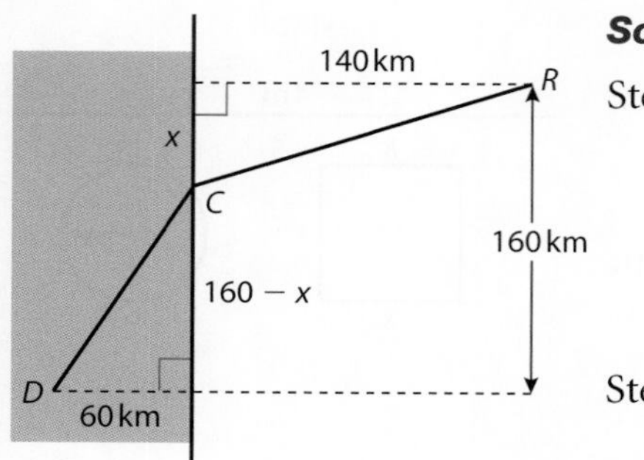

Solution

Step 1: The optimum location of the point, C, where the pipeline comes ashore will be designated by the distance, x, it is from the point on the coast that is a minimum distance (perpendicular) from the rig, R (140 km). The distance from R to C is $\sqrt{x^2 + 140^2}$ and the distance from D (depot) to C is $\sqrt{(160 - x)^2 + 60^2}$.

Step 2: The quantity to be minimized is time, so it is necessary to express the total time it takes the oil to flow from R to D in terms of a single variable.

$$\text{time} = \frac{\text{distance}}{\text{rate}} \Rightarrow \text{time (offshore)} = \frac{\sqrt{x^2 + 19\,600}\text{ km}}{9\text{ km/hr}};$$

$$\text{time (onshore)} = \frac{\sqrt{x^2 - 320x + 29\,200}\text{ km}}{5\text{ km/hr}}$$

The function for time T in terms of x is:

$$T(x) = \frac{\sqrt{x^2 + 19\,600}}{9} + \frac{\sqrt{x^2 - 320x + 29\,200}}{5}$$

and the domain for $T(x)$ is $0 \leqslant x \leqslant 160$.

Steps 3/4: The algebra for finding the derivative of $T(x)$ is similar to that of Step 3 in Example 26. Let's use graphing technology to find the value of x that produces a minimum for $T(x)$.

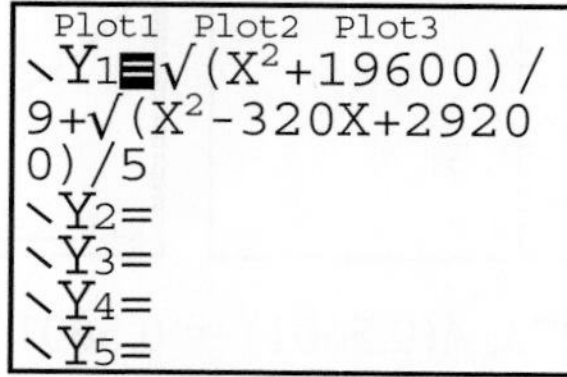

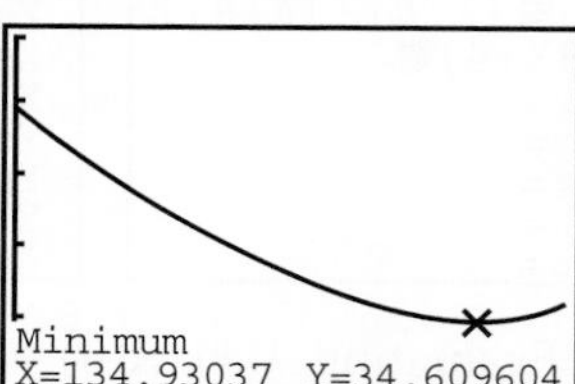

Therefore, the optimum point for the pipeline to intersect with the coast is approximately 134.9 km from the point on the coast nearest to the drilling rig.

The result could also be obtained by having a calculator or computer graph the derivative of $T(x)$ and compute any zeros for $T'(x)$ in the domain.

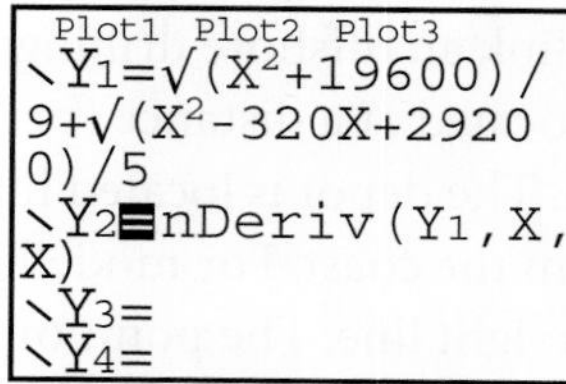

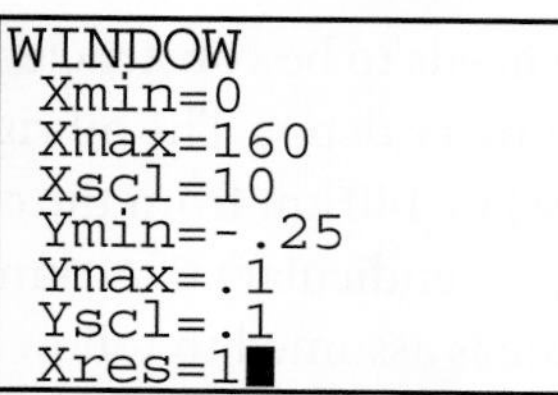


```
CALCULATE
1:value
2:zero
3:minimum
4:maximum
5:intersect
6:dy/dx
7:∫f(x)dx
```

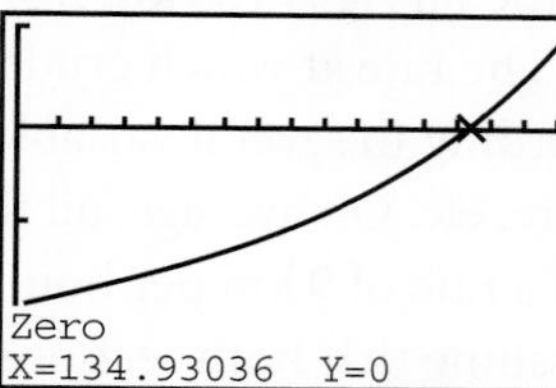

See the **Investigation** and how solving a problem similar to Example 28 derives Snell's law (or law of refraction).

Investigation – Snell's law

The speed of light depends on the medium through which light travels and is generally slower in denser media. The speed of light in a vacuum is an important physical constant and is exactly 299 792 458 m/s. A metre is defined to be the distance that light travels in a vacuum in $\frac{1}{299\,792\,458}$ of a second. Typically, the speed of light in a vacuum (denoted by the letter c) is given the approximate value of 3×10^8 m/s, but in the Earth's atmosphere light travels more slowly than that and even more slowly through glass and water.

Fermat's principle in optics states that light travels from one point to another along a path for which time is a minimum. Investigate the path that a ray of light will follow in going from a point A in a transparent medium, where the speed of light is c_1, to a point B in a different transparent medium, where its speed is c_2, as illustrated in the diagram left. Using algebra and differentiation, prove that for time to be a minimum the following relationship must hold: $\frac{\sin\theta_1}{c_1} = \frac{\sin\theta_2}{c_2}$. This equation is known as Snell's law or the law of refraction. Why is a graphics calculator not helpful?

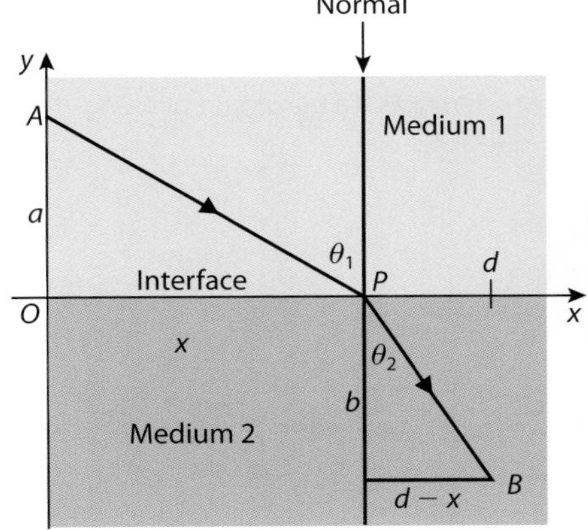

Assume that the two points, A and B, lie in the xy-plane and the x-axis (interface) separates the two media. A light ray is refracted (deflected) when it passes from one medium to another. θ_1 is the **angle of incidence** and θ_2 is the **angle of refraction** (both angles measured between ray and normal to the interface).

Exercise 15.5

1 Find the dimensions of the rectangle with maximum area that is inscribed in a semicircle with radius 1 cm. Two vertices of the rectangle are on the semicircle and the other two vertices are on the x-axis, as shown in the diagram.

2 A rectangular piece of aluminium is to be rolled to make a cylinder with open ends (a tube). Regardless of the dimensions of the rectangle, the perimeter of the rectangle must be 40 cm. Find the dimensions (length and width) of the rectangle that gives a maximum volume for the cylinder.

3 Find the minimum distance from the graph of the function $y = \sqrt{x}$ and the point $(\frac{3}{2}, 0)$.

4 A rectangular box has height h cm, width x cm and length $2x$ cm. It is designed to have a volume equal to 1 litre (1000 cm^3).

a) Show that $h = \frac{500}{x^2}$ cm.

b) Find an expression for the total surface area, S cm^2, of the box in terms of x.

c) Find the dimensions of the box that produces a minimum surface area.

5 The figure right consists of a rectangle *ABCD* and two semicircles on either end. The rectangle has an area of 100 cm². If x represents the length of the rectangle *AB*, find the value of x that makes the perimeter of the entire figure a minimum.

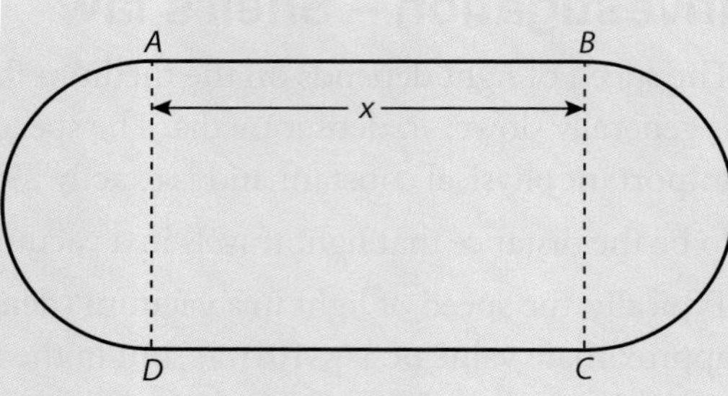

• **Hint:** Write an equation for θ in terms of x and find the value of x which makes θ a maximum by using your GDC.

6 Two vertical posts, with heights 12 metres and 8 metres, are 10 metres apart on horizontal ground. A rope that stretches is attached to the top of both posts and is stretched down so that it touches the ground at point *A* between the two posts. The distance from the base of the taller post to point *A* is represented by x and the angle between the two sections of rope is θ. What value of x makes θ a maximum?

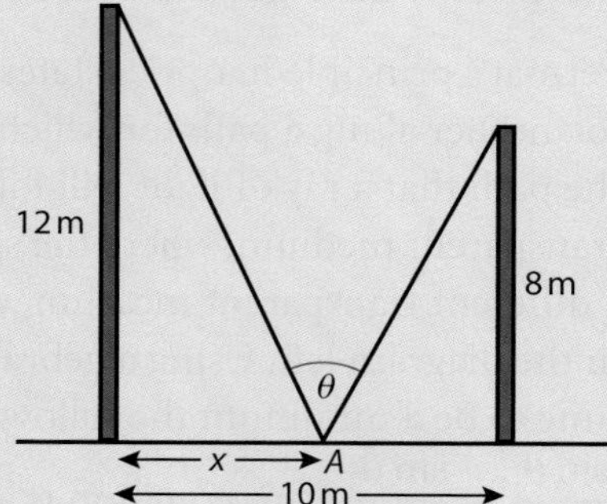

7 A ladder is to be carried horizontally down an L-shaped hallway. The first section of the hallway is 2 metres wide and then there is a right-angled turn into a 3-metre wide section. What is the longest ladder that can be carried around the corner?

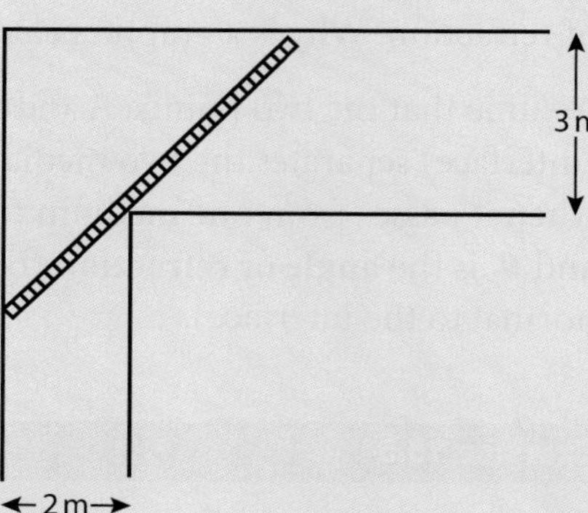

8 Charlie is walking from the wildlife observation tower (point *T*) to the Big Desert Park office (point *O*). The tower is 7 km due west and 10 km due south from the office. There is a road that goes to the office that Charlie can get to if she walks 10 km due north from the tower. Charlie can walk at a rate of 2 kilometres per hour (kph) through the sandy terrain of the park, but she can walk a faster rate of 5 kph on the road. To what point, *A*, on the road should Charlie walk to in order to take the least time to walk from the tower to the office? Find the value of d such that point *A* is d km from the office.

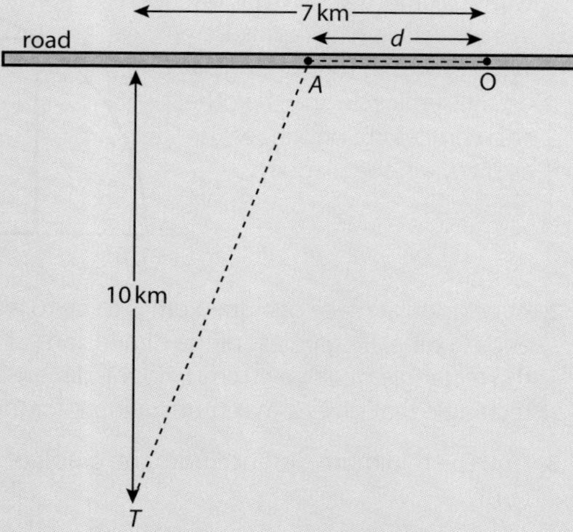

9 Two vertices of a rectangle are on the x-axis, and the other two vertices are on the curve $y = \frac{8}{x^2 + 4}$. (See Exercise 15.1, question 12.) Find the maximum area of the rectangle.

10 A ship sailing due south at 16 knots is 10 nautical miles north of a second ship going due west at 12 knots. Find the minimum distance between the two ships.

11 Find the height, h, and the base radius, r, of the largest right circular cylinder that can be made by cutting it away from a sphere with a radius of R.

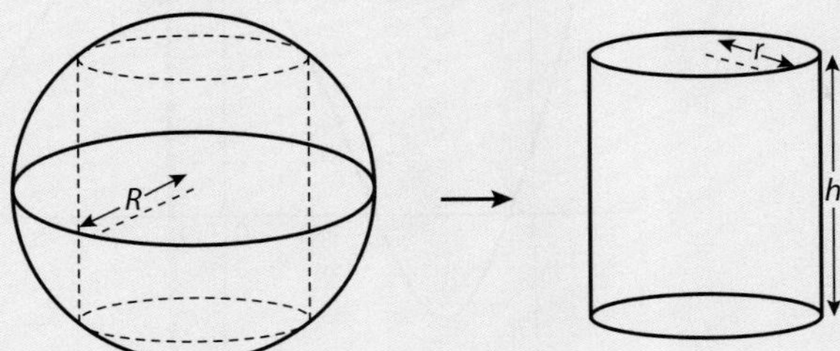

12 Nadia is standing at point A that is a km away in the countryside from a straight road XY (see diagram). She wishes to reach the point Y where the distance from X to Y is b km. Her speed on the road is r km/hr and her speed travelling across the countryside is c km/hr, such that $r > c$. If she wishes to reach Y as quickly as possible, find the position of point P where she joins the road.

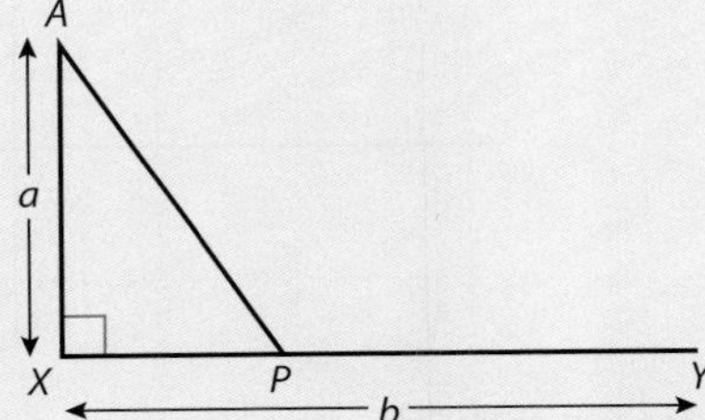

13 A cone of height h and radius r is constructed from a circle with radius 10 cm by removing a sector AOC of arc length x cm and then connecting the edges OA and OC. What arc length x will produce the cone of maximum volume, and what is the volume?

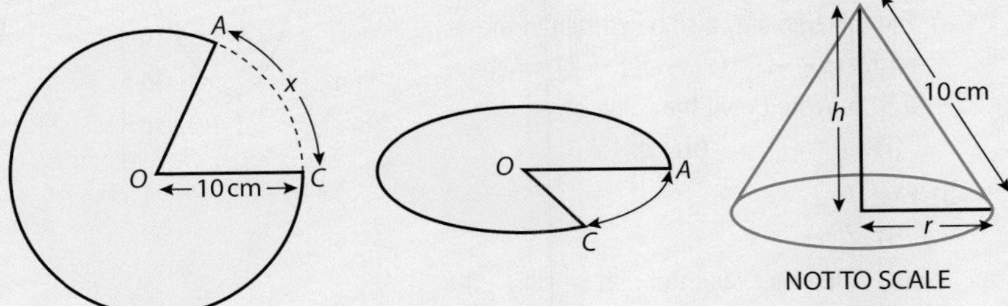

14 Point P is a units above the line AB, and point Q is b units below line AB (see diagram). The velocity of light is u units/second above AB and v units/second below AB, and $u > v$. The angles α and β are the angles that a ray of light makes with a perpendicular (normal) to line AB above and below AB, respectively. Show that the following relationship must hold true.

$$\frac{\sin \alpha}{\sin \beta} = \frac{u}{v}$$

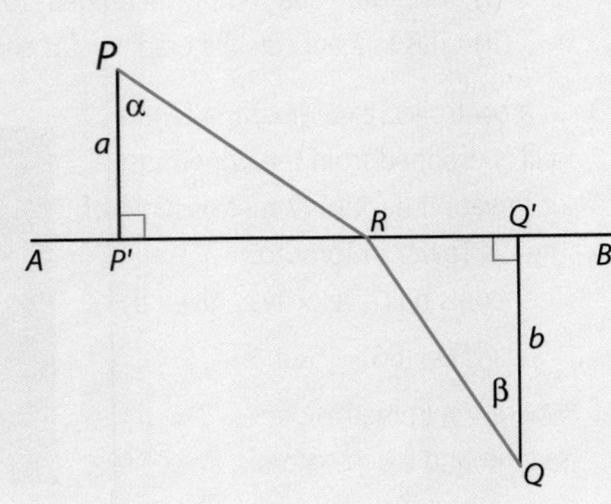

Practice questions

1 The diagram shows the graph of $y = f(x)$.

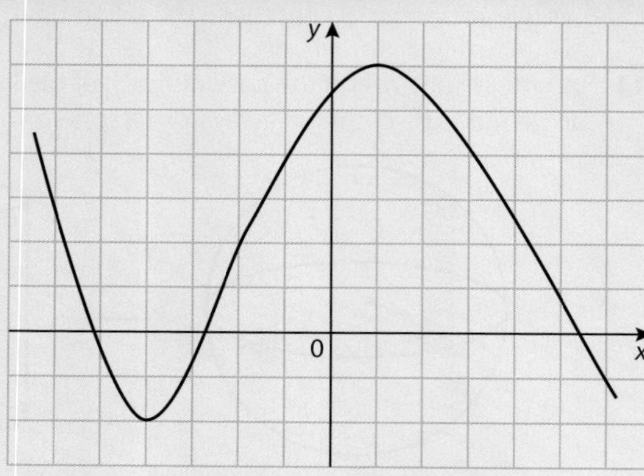

Copy the grid right and sketch the graph of $y = f'(x)$.

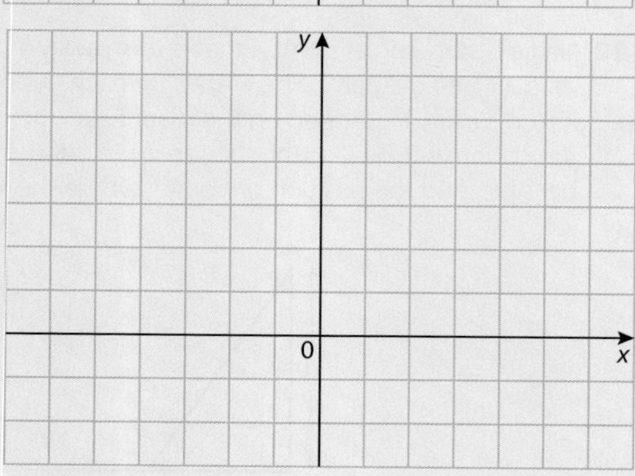

2 The diagram right shows part of the graph of the function $f: x \mapsto -x^3 - 2x^2 + 8x$.

The graph intersects the x-axis at $(-4, 0)$, $(0, 0)$ and $(2, 0)$. There is a minimum point at C and a maximum point at D.

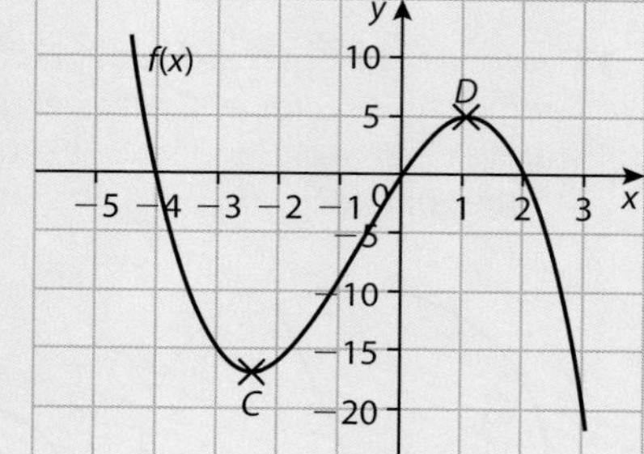

a) The function may also be written in the form $f: x \mapsto -x(x - a)(x - b)$, where $a < b$. Write down the value of

(i) a **(ii)** b.

b) Find

(i) $f'(x)$

(ii) the exact values of x at which $f'(x) = 0$

(iii) the value of the function at D.

c) **(i)** Find the equation of the tangent to the graph of f at $(0, 0)$.

(ii) This tangent cuts the graph of f at another point. Give the x-coordinate of this point.

3 In a controlled experiment, a tennis ball is dropped from the uppermost observation deck (447 metres high) of the CN Tower in Toronto.

The tennis ball's velocity is given by

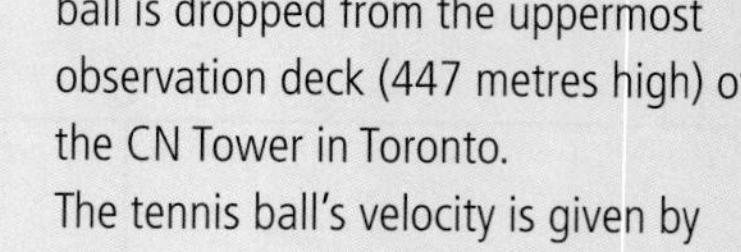

$$v(t) = 66 - 66e^{-0.15t}$$

where v is in metres per second and t is in seconds.

a) Find the value of v when
  (i) $t = 0$ **(ii)** $t = 10$.
b) **(i)** Find an expression for the acceleration, a, as a function of t.
  (ii) What is the value of a when $t = 0$?
c) **(i)** As t becomes large, what value does v approach?
  (ii) As t becomes large, what value does a approach?
  (iii) Explain the relationship between the answers to parts **c)(i)** and **(ii)**.

4 Given the function $f(x) = x^3 + 7x^2 + 8x - 3$,
a) identify any points as a relative maximum or minimum and find their exact coordinates
b) find the exact coordinates of any inflexion point(s).

5 Consider the function $g(x) = 2 + \frac{1}{e^{3x}}$.
a) **(i)** Find $g'(x)$.
  (ii) Explain briefly how this shows that $g(x)$ is a decreasing function for all values of x (i.e. that $g(x)$ always decreases in value as x increases).
Let P be the point on the graph of g where $x = -\frac{1}{3}$.
b) Find an expression in terms of e for
  (i) the y-coordinate of P
  (ii) the gradient of the tangent to the curve at P.
c) Find the equation of the tangent to the curve at P, giving your answer in the form $y = mx + c$.

6 Consider the function f given by $f(x) = \frac{2x^2 - 13x + 20}{(x - 1)^2}, x \neq 1$.
a) Show that $f'(x) = \frac{9x - 27}{(x - 1)^3}, x \neq 1$.

The second derivative is given by $f''(x) = \frac{72 - 18x}{(x - 1)^4}, x \neq 1$.

b) Using values of $f'(x)$ and $f''(x)$, explain why a minimum must occur at $x = 3$.
c) There is a point of inflexion on the graph of f. Write down the coordinates of this point.

7 Differentiate with respect to x:
a) $y = \frac{1}{(2x + 3)^2}$
b) $y = e^{\sin 5x}$
c) $y = \tan^2(x^2)$

8 The curve with equation $y = Ax + B + \frac{C}{x}, x \in \mathbb{R}, x \neq 0$, has a minimum at $P(1, 4)$ and a maximum at $Q(-1, 0)$. Find the value of each of the constants A, B and C.

9 Find $\frac{dy}{dx}$ and $\frac{d^2y}{dx^2}$ at the point (1, 1) on the curve $x^3 + y^3 = 2$.

10 Differentiate with respect to x:
a) $y = \frac{x}{e^x - 1}$ **b)** $y = e^x \sin 2x$ **c)** $y = (x^2 - 1)\ln(3x)$

11 The normal to the curve $y = x^2 - 4x$ at the point $(3, -3)$ intersects the x-axis at point P and the y-axis at point Q. Find the equation of the normal and the coordinates of P and Q.

12 Let $y = h(x)$ be a function of x for $0 \leqslant x \leqslant 6$. The graph of h has an inflexion point at P and a maximum point at M.
Partial sketches of the curves of $h'(x)$ and $h''(x)$ are shown below.

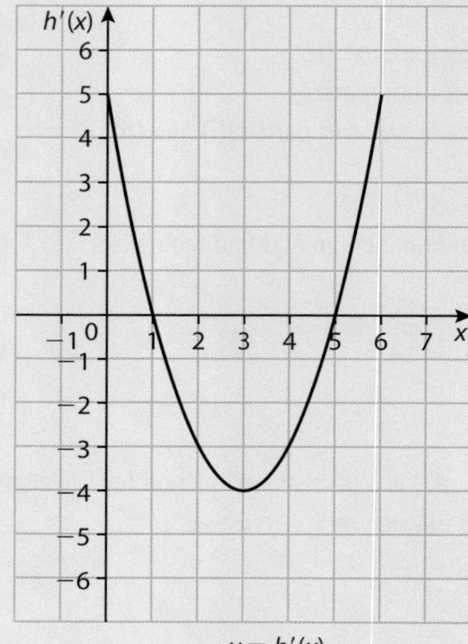

$y = h'(x)$

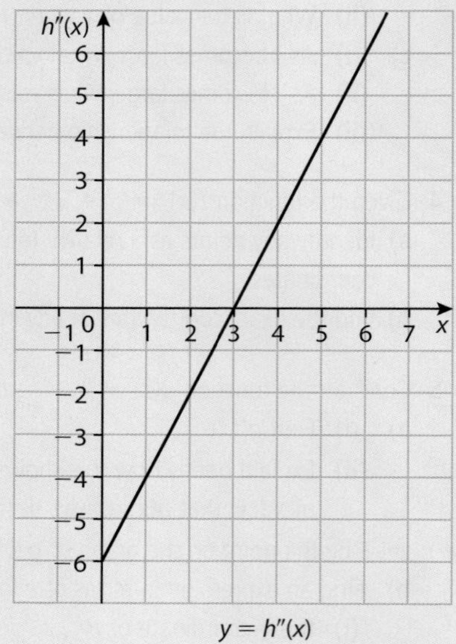

$y = h''(x)$

Use the above information to answer the following.

a) Write down the x-coordinate of P and justify your answer.

b) Write down the x-coordinate of M and justify your answer.

c) Given that $h(3) = 0$, sketch the graph of h. On the sketch, mark the points P and M.

13 Find the equation of the normal to the curve $x^2 + xy + y^2 - 3y = 10$ at the point (2, 3).

14 A cylinder is to be made with an exact volume of $128\pi \text{ cm}^3$. What should be the height h and the radius r of the cylinder's base so that the cylinder's surface area is a minimum?

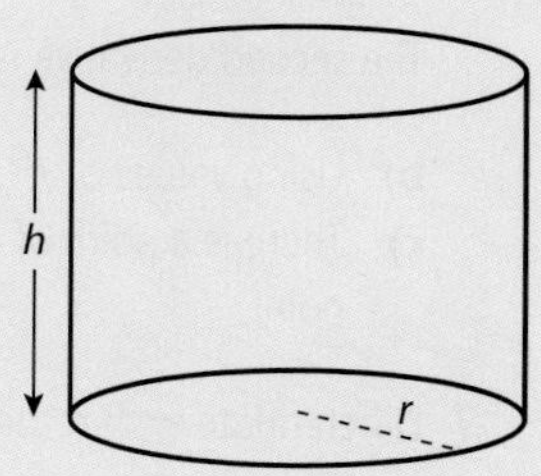

15 A rectangle has its base on the x-axis and its upper two vertices on the parabola $y = 12 - x^2$, as shown in the diagram. What is the largest area that the rectangle can have, and what are its dimensions (i.e. length and width)?

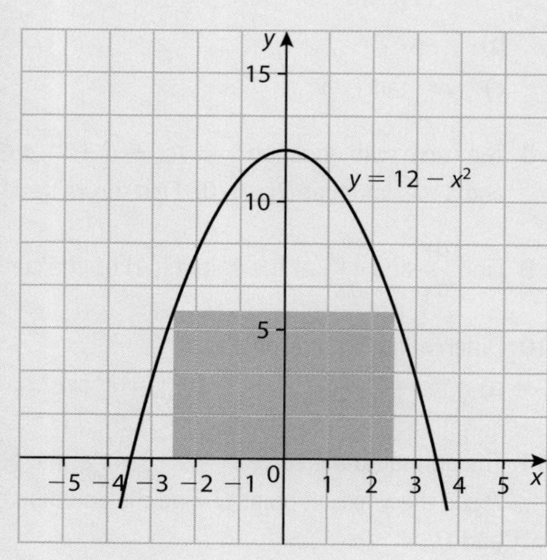

16 The figure below shows the graph of a function $y = f(x)$. At which one of the five points on the graph:

a) are $f'(x)$ and $f''(x)$ both negative?

b) is $f'(x)$ negative and $f''(x)$ positive?

c) is $f'(x)$ positive and $f''(x)$ negative?

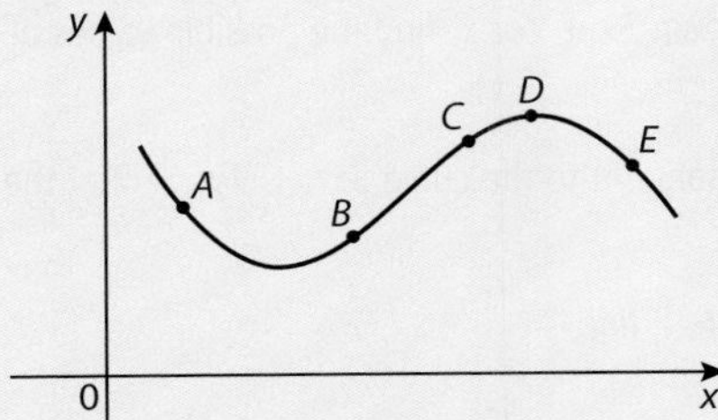

17 Find the equation of the normal to the curve with equation $y = \dfrac{2x - 1}{x + 2}$ at the point $(-3, 7)$.

18 Find the equation of a) the tangent, and b) the normal to the curve $y = \ln(4x - 3)$ at the point $(1, 0)$.

19 Consider the function $f(x) = x^2 \ln x$.

a) Find the exact coordinates of any stationary points. Indicate whether it is a maximum or minimum (and absolute or relative).

b) Find the exact coordinates of any inflexion points.

20 a) Determine the constant a such that the function $f(x) = x^2 + \dfrac{a}{x}$ has (i) a local minimum at $x = 2$ and (ii) a local minimum at $x = -3$.

c) Show that the function cannot have a local maximum for any value of a.

21 A line passes through the point $(3, 2)$ and intersects both the x-axis and the y-axis, forming a triangular region in the first quadrant bounded by the x-axis, the y-axis and the line. Find the equation of such a line that creates a triangle of minimum area.

22 Find the equation of both the tangent and normal to the curve $y = x \tan x$ at the point where $x = \dfrac{\pi}{4}$.

23 A very important function in statistics is the equation for the **standard normal curve** (mean = 0, standard deviation = 1) given by $f(x) = \dfrac{e^{-\frac{x^2}{2}}}{\sqrt{(2\pi)}}$.

a) Find the coordinates of any stationary points and of any inflexion points.

b) What happens when $x \to \infty$, and when $x \to -\infty$? Give the equation for any asymptotes.

c) Sketch a graph of $f(x)$ and indicate the location of any of the points found in part **a)**.

24 Let f be the function given by $f(x) = 2\ln(x^2 + 3) - x$.

a) Find the x-coordinate of each maximum and minimum point of f. Justify your answer(s).

b) Find the x-coordinate of each inflexion point of f. Justify your answer(s).

25 The rate at which cars on a road pass a certain point is known as the flow rate and is in units of cars per hour. The flow rate, F, of a certain road is given by $F(x) = \dfrac{2x}{18 + 0.015x^2}$ where x is the speed of the traffic in kilometres per hour. What speed will maximise the flow rate on the road?

26 If $2x^2 - 3y^2 = 2$, find the two values of $\frac{dy}{dx}$ when $x = 5$.

27 Differentiate $y = \arccos(1 - 2x^2)$ with respect to x, and simplify your answer.

28 For the function $f: x \mapsto x^2 \ln x$, $x > 0$, find the function f', the derivative of f with respect to x.

29 For the function $f: x \mapsto \frac{1}{2}\sin 2x + \cos x$, find the possible values of $\sin x$ for which $f'(x) = 0$.

30 Find the gradient of the tangent to the curve $3x^2 + 4y^2 = 7$ at the point where $x = 1$ and $y > 0$.

31 If $f(x) = \ln(2x - 1)$, $x > \frac{1}{2}$, find

a) $f'(x)$

b) the value of x where the gradient of $f(x)$ is equal to x.

32 Find the x-coordinate, between -2 and 0, of the point of inflexion on the graph of the function $f: x \mapsto x^2e^x$. Give your answer to 3 decimal places.

33 A normal to the graph of $y = \arctan(x - 1)$, for $x > 0$, has equation $y = -2x + c$, where $c \in \mathbb{R}$. Find the value of c.

34 The function f is given by $f: x \mapsto e^{1+\sin \pi x}$, $x \geqslant 0$.

a) Find $f'(x)$.

Let x_n be the value of x where the $(n + 1)$th maximum or minimum point occurs, $n \in \{N\}$ (i.e. x_0 is the value of x where the first maximum or minimum occurs, x_1 is the value of x where the second maximum or minimum occurs, etc.).

b) Find x_n in terms of n.

35 Let $f(x) = x(\sqrt[3]{(x^2 - 1)^2})$, $-1.4 \leqslant x \leqslant 1.4$.

a) **Sketch** the graph of $f(x)$. (An exact scale diagram is **not** required.)
On your graph indicate the approximate position of

(i) each zero

(ii) each maximum point

(iii) each minimum point.

b) **(i)** Find $f'(x)$, clearly stating its domain.

(ii) Find the x-coordinates of the maximum and minimum points of $f(x)$, for $-1 < x < 1$.

c) Find the x-coordinate of the point of inflexion of $f(x)$, where $x > 0$, giving your answer correct to **four** decimal places.

36 The line $y = 16x - 9$ is a tangent to the curve $y = 2x^3 + ax^2 + bx - 9$ at the point $(1, 7)$. Find the values of a and b.

37 Consider the function $y = \tan x - 8\sin x$.

a) Find $\frac{dy}{dx}$. **b)** Find the value of $\cos x$ for which $\frac{dy}{dx} = 0$.

38 Consider the tangent to the curve $y = x^3 + 4x^2 + x - 6$.

a) Find the equation of this tangent at the point where $x = -1$.

b) Find the coordinates of the point where this tangent meets the curve again.

39 Let $y = \sin(kx) - kx\cos(kx)$, where k is a constant.

Show that $\frac{dy}{dx} = k^2x\sin(kx)$.

40 A curve has equation $xy^3 + 2x^2y = 3$. Find the equation of the tangent to this curve at the point (1, 1).

41 The function f is defined by

$$f(x) = \frac{x^2 - x + 1}{x^2 + x + 1}.$$

a) **(i)** Find an expression for $f'(x)$, simplifying your answer.
(ii) The tangents to the curve of $f(x)$ at points A and B are parallel to the x-axis. Find the coordinates of A and of B.

b) **(i)** Sketch the graph of $y = f'(x)$.
(ii) Find the x-coordinates of the three points of inflexion on the graph of f.

c) Find the range of
(i) f
(ii) the composite function $f \circ f$.

42 Air is pumped into a spherical ball which expands at a rate of 8 cm^3 per second (8 cm^3 s^{-1}). Find the **exact** rate of increase of the radius of the ball when the radius is 2 cm.

43 A curve has equation $x^3y^2 = 8$. Find the equation of the normal to the curve at the point (2, 1).

44 The function f is defined by $f(x) = \frac{x^2}{2^x}$, for $x > 0$.

a) **(i)** Show that

$$f'(x) = \frac{2x - x^2 \ln 2}{2^x}.$$

(ii) Obtain an expression for $f''(x)$, simplifying your answer as far as possible.

b) **(i)** Find the exact value of x satisfying the equation $f'(x) = 0$.
(ii) Show that this value gives a maximum value for $f(x)$.

c) Find the x-coordinates of the two points of inflexion on the graph of f.

45 Consider the function $f(t) = 3\sec^2 t + 5t$.

a) Find $f'(t)$.

b) Find the **exact** values of
(i) $f(\pi)$
(ii) $f'(\pi)$.

46 Consider the equation $2xy^2 = x^2y + 3$.

a) Find y when $x = 1$ and $y < 0$.

b) Find $\frac{dy}{dx}$ when $x = 1$ and $y < 0$.

47 Let $y = e^{3x}\sin(\pi x)$.

a) Find $\frac{dy}{dx}$.

b) Find the smallest positive value of x for which $\frac{dy}{dx} = 0$.

48 An airplane is flying at a constant speed at a constant altitude of 3 km in a straight line that will take it directly over an observer at ground level. At a given instant the observer notes that the angle θ is $\frac{1}{3}\pi$ radians and is increasing at $\frac{1}{60}$ radians per second. Find the speed, in kilometres per hour, at which the airplane is moving towards the observer.

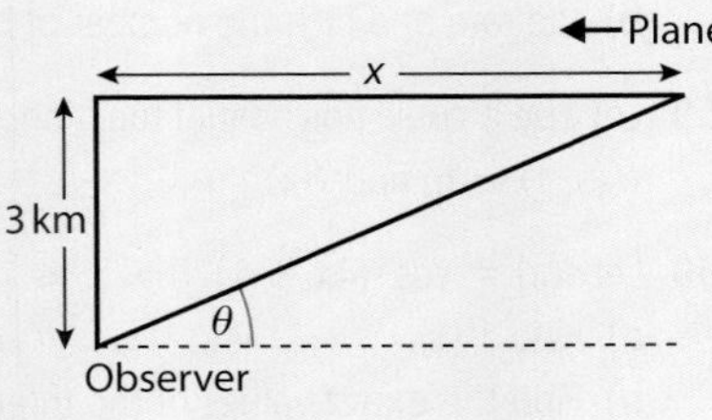

49 A curve has equation $f(x) = \frac{a}{b + e^{-cx}}$, $a \neq 0$, $b > 0$, $c > 0$.

a) Show that $f''(x) = \frac{ac^2 e^{-cx}(e^{-cx} - b)}{(b + e^{-cx})^3}$.

b) Find the coordinates of the point on the curve where $f''(x) = 0$.

c) Show that this is a point of inflexion.

50 The point $P(1, p)$, where $p > 0$, lies on the curve $2x^2y + 3y^2 = 16$.

a) Calculate the value of p.

b) Calculate the gradient of the tangent to the curve at P.

51 The function f is defined by $f: x \mapsto 3^x$.
Find the solution of the equation $f''(x) = 2$.

52 The following diagram shows an isosceles triangle ABC with $AB = 10$ cm and $AC = BC$. The vertex C is moving in a direction perpendicular to (AB) with speed 2 cm per second.

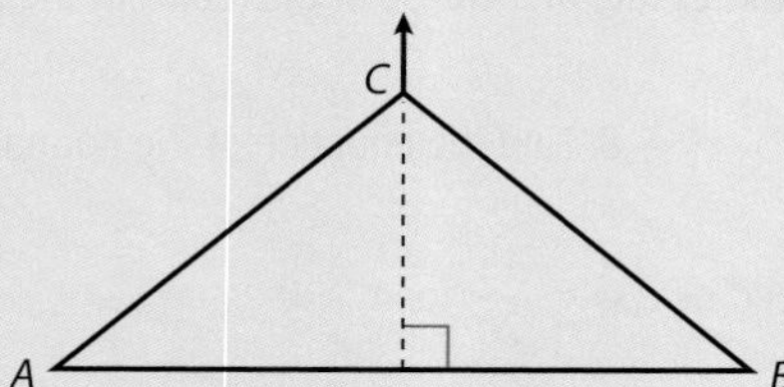

Calculate the rate of increase of the angle $C\hat{A}B$ at the moment the triangle is equilateral.

53 If $y = \ln(2x - 1)$, find $\frac{d^2y}{dx^2}$.

54 Find the equation of the normal to the curve $x^3 + y^3 - 9xy = 0$ at the point (2, 4).

55 The function f' is given by $f'(x) = 2\sin\left(5x - \frac{\pi}{2}\right)$.

a) Write down $f''(x)$.

b) Given that $f\left(\frac{\pi}{2}\right) = 1$, find $f(x)$.

56 Find the gradient of the normal to the curve $3x^2y + 2xy^2 = 2$ at the point $(1, -2)$.

57 The function f is given by $f(x) = \frac{x^5 + 2}{x}$, $x \neq 0$. There is a point of inflexion on the graph of f at the point P. Find the coordinates of P.

58 An experiment is carried out in which the number n of bacteria in a liquid is given by the formula $n = 650e^{kt}$, where t is the time in minutes after the beginning of the experiment and k is a constant. The number of bacteria doubles every 20 minutes. Find

a) the **exact** value of k

b) the rate at which the number of bacteria is increasing when $t = 90$.

59 Let f be a cubic polynomial function. Given that $f(0) = 2$, $f'(0) = -3$, $f(1) = f'(1)$ and $f''(-1) = 6$, find $f(x)$.

60 Let $f(x) = \cos^3(4x + 1)$, $0 \leqslant x \leqslant 1$.

a) Find $f'(x)$.

b) Find the **exact** values of the three roots of $f'(x) = 0$.

61 Given that $3^{x+y} = x^3 + 3y$, find $\frac{dy}{dx}$.

62 Let f be the function defined for $x > -\frac{1}{3}$ by $f(x) = \ln(3x + 1)$.

a) Find $f'(x)$.

b) Find the equation of the normal to the curve $y = f(x)$ at the point where $x = 2$.

Give your answer in the form $y = ax + b$ where $a, b \in \mathbb{R}$.

63 Let $y = x \arcsin x$, $x \in (-1, 1)$. Show that $\frac{d^2y}{dx^2} = \frac{2 - x^2}{(1 - x^2)^{\frac{3}{2}}}$.

64 Given that $e^{xy} - y^2 \ln x = e$ for $x \geqslant 1$, find $\frac{dy}{dx}$ at the point (1, 1).

65 The function f is defined by $f(x) = \frac{2x}{x^2 + 6}$ for $x \geqslant b$ where $b \in \mathbb{R}$.

a) Show that $f'(x) = \frac{12 - 2x^2}{(x^2 + 6)^2}$.

b) Hence, find the smallest exact value of b for which the inverse function f^{-1} exists. Justify your answer.

66 Consider the curve with equation $x^2 + xy + y^2 = 3$.

a) Find in terms of k, the gradient of the curve at the point $(-1, k)$.

b) Given that the tangent to the curve is parallel to the x-axis at this point, find the value of k.

67 Find the gradient of the tangent to the curve $x^3y^2 = \cos(\pi y)$ at the point $(-1, 1)$.

68 André wants to get from point A located in the sea to point Y located on a straight stretch of beach. P is the point on the beach nearest to A such that AP = 2 km and PY = 2 km. He does this by swimming in a straight line to a point Q located on the beach and then running to Y.

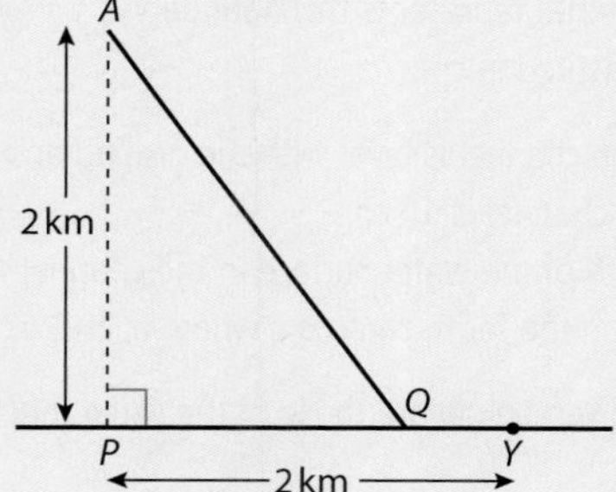

When André swims he covers 1 km in $5\sqrt{5}$ minutes. When he runs he covers 1 km in 5 minutes.

a) If PQ = x km, $0 \leqslant x \leqslant 2$, find an expression for the time T minutes taken by André to reach point Y.

b) Show that $\frac{dT}{dx} = \frac{5\sqrt{5}x}{\sqrt{x^2 + 4}} - 5$.

c) **(i)** Solve $\frac{dT}{dx} = 0$.

(ii) Use the value of x found in **part c) (i)** to determine the time, T minutes, taken for André to reach point Y.

(iii) Show that $\frac{d^2T}{dx^2} = \frac{20\sqrt{5}}{(x^2 + 4)^{\frac{3}{2}}}$ and **hence** show that the time found in **part c)** is a minimum.

69 The function f is defined by $f(x) = xe^{2x}$.

It can be shown that $f^{(n)}(x) = (2^n x + n2^{n-1})e^{2x}$ for all $n \in \mathbb{Z}^+$, where $f^{(n)}(x)$ represents the nth derivative of $f(x)$.

a) By considering $f^{(n)}(x)$ for $n = 1$ and $n = 2$, show that there is one minimum point P on the graph of f, and find the coordinates of P.

b) Show that f has a point of inflexion Q at $x = -1$.

c) Determine the intervals on the domain of f where f is

(i) concave up

(ii) concave down.

d) Sketch f', clearly showing any intercepts, asymptotes and the points P and Q.

e) Use mathematical induction to prove that $f^{(n)}(x) = (2^n x + n2^{n-1})e^{2x}$ for all $n \in \mathbb{Z}^+$, where $f^{(n)}(x)$ represents the nth derivative of $f(x)$.

70 The diagram below shows the boundary of the cross-section of a water channel.

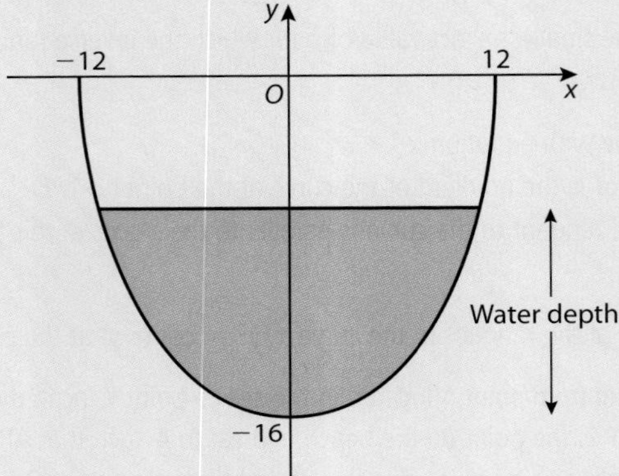

The equation that represents this boundary is $y = 16\sec\left(\frac{\pi x}{36}\right) - 32$ where x and y are both measured in cm.

The top of the channel is level with the ground and has a width of 24 cm. The maximum depth of the channel is 16 cm.
Find the width of the water surface in the channel when the water depth is 10 cm. Give your answer in the form $a \arccos b$ where $a, b \in \mathbb{R}$.

71 The graphs given below are those of the same function $y = f(x)$ for $a \leqslant x \leqslant b$.

Sketch, on the given axes, the graphs of **a)** $\frac{dy}{dx}$ and **b)** $\frac{d^2y}{dx^2}$.

Indicate clearly the positions of any asymptotes.

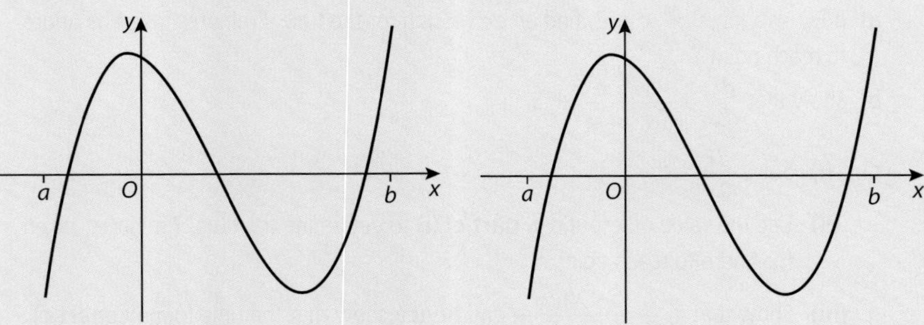

16 Integral Calculus

Assessment statements

6.4 Indefinite integration as anti-differentiation.
Indefinite integral of x^n, sin x, cos x, $1/x$ and e^x.
The composites of any of these with a linear function.

6.5 Anti-differentiation with a boundary condition to determine the constant term.
Definite integrals.
Area of the region enclosed by a curve and the x-axis or y-axis in a given interval.
Areas of regions enclosed by curves.
Volumes of revolution about the x-axis or y-axis.

6.6 Kinematic problems involving displacement s, velocity v and acceleration a.
Total distance travelled.

6.7 Further integration: integration by substitution; integration by parts.

Introduction

In Chapters 13 and 15 you learned about the process of differentiation. That is, given a function, how you can find its derivative. In this chapter, we will look at the reverse process. That is, given a function $f(x)$, how can we find a function $F(x)$ whose derivative is $f(x)$. This process is the opposite of differentiation and is therefore called **anti-differentiation.**

16.1 Anti-derivative

An **anti-derivative** of the function $f(x)$ is a function $F(x)$ such that

$$\frac{d}{dx}F(x) = F'(x) = f(x) \text{ wherever } f(x) \text{ is defined.}$$

For instance, let $f(x) = x^2$. It is not difficult to discover an anti-derivative of $f(x)$. Keep in mind that this is a power function. Since the power rule reduces the power of the function by 1, we examine the derivative of x^3: $\frac{d}{dx}(x^3) = 3x^2$.

This derivative, however, is 3 times $f(x)$. To 'compensate' for the 'extra' 3, we have to multiply by $\frac{1}{3}$, so that the anti-derivative is now $\frac{1}{3}x^3$. Now,

$$\frac{d}{dx}\left(\frac{1}{3}x^3\right) = x^2.$$

And, therefore, $\frac{1}{3}x^3$ is an anti-derivative of x^2.

Table 16.1 shows some examples of functions, each paired with one of its anti-derivatives.

Table 16.1

Function $f(x)$	Anti-derivative $F(x)$
1	x
x	$\frac{x^2}{2}$
$3x^2$	x^3
x^4	$\frac{x^5}{5}$
$\cos x$	$\sin x$
$\cos 2x$	$\frac{1}{2}\sin 2x$
e^x	e^x
$\sin x$	$-\cos x$
$2x$	x^2

The diagrams below show the relationship between the derivative and the integral as opposite operations.

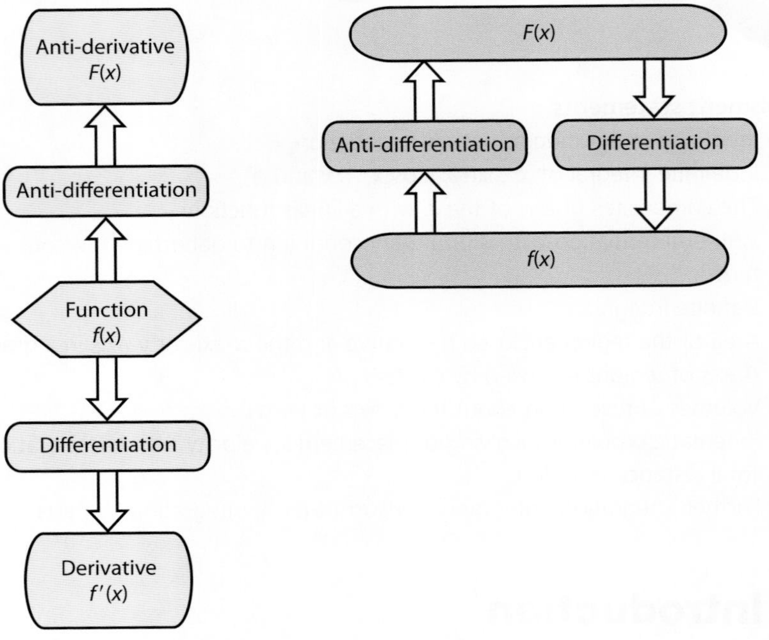

Example 1

Given the function $f(x) = 3x^2$, find an anti-derivative of $f(x)$.

Solution

$F_1(x) = x^3$ is such an anti-derivative because $\frac{d}{dx}(F_1(x)) = 3x^2$.

The following functions are also anti-derivatives because the derivative of each one of them is also $3x^2$.

$$H_1(x) = x^3 + 27,\ H_2(x) = x^3 - \pi, \text{ or } H_3(x) = x^3 + \sqrt{5}$$

Indeed, $F(x) = x^3 + c$ is an anti-derivative of $f(x) = 3x^2$ for any choice of the constant c.

This is so simply because

$$(F(x) + c)' = F'(x) + c' = F'(x) + 0 = f(x)!$$

Thus, we can say that any single function $f(x)$ has many anti-derivatives, whereas a function can have only one derivative.

> If $F(x)$ is an anti-derivative of $f(x)$, then so is $F(x) + c$ for any choice of the constant c.

Stated slightly differently, this observation says:

> If $F(x)$ is an anti-derivative of $f(x)$ over a certain interval I, then every anti-derivative of $f(x)$ on I is of the form $F(x) + c$.

This statement is an indirect conclusion of one of the results of the mean value theorem.

Two functions with the same derivative on an interval differ only by a constant on that interval.

We will state the mean value theorem here in order to establish the general rule for anti-derivatives.

Mean value theorem

A function $H(x)$, continuous over an interval $[a, b]$ and differentiable over $]a, b[$, satisfies

$$H(b) - H(a) = (b - a)H'(c) \text{ for some } c \in]a, b[.$$

Let $F(x)$ and $G(x)$ be any anti-derivatives of $f(x)$, i.e. $F'(x) = G'(x)$.

Take $H(x) = F(x) - G(x)$ and any two numbers x_1 and x_2 in the interval $[a, b]$ such that $x_1 < x_2$, then

$$\begin{aligned} H(x_2) - H(x_1) &= (x_2 - x_1)H'(c) = (x_2 - x_1)\cdot(F'(c) - G'(c)) \\ &= (x_2 - x_1)\cdot 0 = 0 \Rightarrow H(x_1) = H(x_2) \end{aligned}$$

which means that $H(x)$ is a constant function.
Hence, $H(x) = F(x) - G(x) =$ constant. That is, any two anti-derivatives of a function differ by a constant.

Notation:
The notation

$$\int f(x)\,dx = F(x) + c \qquad (1)$$

where c is an arbitrary constant, means that $F(x) + c$ is an anti-derivative of $f(x)$.

Equivalently, $F(x)$ satisfies the condition that

$$\frac{d}{dx}(F(x)) = F'(x) = f(x) \qquad (2)$$

for all x in the domain of $f(x)$.

It is important to note that (1) and (2) are just different notations to express the same fact. For example,

$$\int x^2 dx = \tfrac{1}{3}x^3 + c \text{ is equivalent to } \frac{d}{dx}\left(\tfrac{1}{3}x^3\right) = x^2.$$

Note that if we differentiate an anti-derivative of $f(x)$, we obtain $f(x)$ back again.

Thus, $\frac{d}{dx}\left(\int f(x)dx\right) = f(x)$.

The expression $\int f(x)dx$ is called an **indefinite integral** of $f(x)$. The function $f(x)$ is called the **integrand** and the constant c is called the **constant of integration**.

The integral symbol $\int$ is made like an elongated capital S. It is, in fact, a medieval S, used by Leibniz as an abbreviation for the Latin word *summa*.

We think of the combination $\int[\,]\boldsymbol{dx}$ as a single symbol; we fill in the 'blank' with the formula of the function whose anti-derivative we seek. We may regard the differential $\boldsymbol{dx}$ as specifying the independent variable $\boldsymbol{x}$ both in the function $f(x)$ and in its anti-derivatives.

If an independent variable other than $\boldsymbol{x}$ is used, say t, the notation must be adjusted appropriately.

Thus, $\frac{d}{dt}\left(\int f(t)dt\right) = f(t)$ and $\int f(t)dt = F(t) + c$ are equivalent statements.

Derivative formula	Equivalent integration formula
$\frac{d}{dx}(x^3) = 3x^2$	$\int 3x^2 dx = x^3 + c$
$\frac{d}{dx}(\sqrt{x}) = \frac{1}{2\sqrt{x}}$	$\int \frac{1}{2\sqrt{x}} dx = \sqrt{x} + c$
$\frac{d}{dt}(\tan t) = \sec^2 t$	$\int \sec^2 t\, dt = \tan t + c$
$\frac{d}{dv}(v^{\frac{3}{2}}) = \frac{3}{2}v^{\frac{1}{2}}$	$\int \frac{3}{2}v^{\frac{1}{2}} dv = v^{\frac{3}{2}} + c$

Note: The integral sign and differential serve as delimiters, adjoining the integrand on the left and right, respectively. In particular, we do not write $\int dx\, f(x)$ when we mean $\int f(x)\, dx$.

Basic integration formulae

Integration is essentially educated guesswork – given the derivative $f(x)$ of a function $F(x)$, we try to guess what the function $F(x)$ is. However, many basic integration formulae can be obtained directly from their companion differentiation formulae. Some of the most important are given in Table 16.2.

Table 16.2

	Differentiation formula	Integration formula
1	$\frac{d}{dx}(x) = 1$	$\int dx = x + c$
2	$\frac{d}{dx}(x^{n+1}) = (n+1)x^n, n \neq -1$	$\int x^n dx = \frac{x^{n+1}}{n+1} + c, n \neq -1$
3	$\frac{d}{dx}(\sin x) = \cos x$	$\int \cos x\, dx = \sin x + c$
4	$\frac{d}{dx}(\cos x) = -\sin x$	$\int \sin v\, dv = -\cos v + c$
5	$\frac{d}{dt}(\tan t) = \sec^2 t$	$\int \sec^2 t\, dt = \tan t + c$
6	$\frac{d}{dv}(e^v) = e^v$	$\int e^v dv = e^v + c$
7	$\frac{d}{dx}(\ln\lvert x\rvert) = \frac{1}{x}$	$\int \frac{1}{x} dx = \ln\lvert x\rvert + c$
8	$\frac{d}{dx}\left(\frac{a^x}{\ln a}\right) = a^x$	$\int a^x dx = \frac{1}{\ln a}a^x + c$
9	$\frac{d}{dx}(\arcsin x) = \frac{1}{\sqrt{1-x^2}}$	$\int \frac{dx}{\sqrt{1-x^2}} = \arcsin x + c$
10	$\frac{d}{dx}(\arctan x) = \frac{1}{1+x^2}$	$\int \frac{dx}{1+x^2} = \arctan x + c$

Formula (7) is a special case of the 'power' rule formula (2), but needs some modification.

If we are given the task to integrate $\frac{1}{x}$, we may attempt to do it using the power rule:

$$\int \frac{1}{x} dx = \int x^{-1} dx = \frac{1}{(-1)+1}x^{(-1)+1} + c = \frac{1}{0}x^0 + c, \text{ which is undefined.}$$

However, the solution is clearly found by observing what you learned in Chapter 15.

In Section 15.3 you learned that

$$\frac{d}{dx}(\ln x) = \frac{1}{x}, \ x > 0.$$

This implies

$$\int \frac{1}{x}\, dx = \ln x + c, x > 0.$$

However, the function $\frac{1}{x}$ is differentiable for $x < 0$ too. So, we must be able to find its integral.

The solution lies in the chain rule!

If $x < 0$, we can write $x = -u$ where $u > 0$. Then $dx = -du$, and

$$\int \frac{1}{x}\, dx = \int \frac{1}{-u}(-du) = \int \frac{1}{u}\, du = \ln u + c, u > 0.$$

But $u = -x$, therefore when $x < 0$

$\int \frac{1}{x}\, dx = \ln u + c = \ln(-x) + c$, and, combining the two results, we have

$$\int \frac{1}{x}\, dx = \ln|x| + c, x \neq 0.$$

Suppose that $f(x)$ and $g(x)$ are differentiable functions and k is a constant, then:

1. A constant factor can be moved through an integral sign, i.e.

   $\int kf(x)dx = k\int f(x)dx$

2. An anti-derivative of a sum (difference) is the sum (difference) of the anti-derivatives, i.e.

   $\int (f(x) + g(x))dx = \int f(x)dx + \int g(x)dx$, or $\int (f(x) - g(x))dx = \int f(x)dx - \int g(x)dx$

Example 2

Evaluate:

a) $\int 3\cos x\, dx$ b) $\int (x^3 + x^2)dx$

Solution

a) $\int 3\cos x\, dx = 3\int \cos x\, dx = 3\sin x + c$

b) $\int (x^3 + x^2)dx = \int x^3\, dx + \int x^2\, dx = \frac{x^4}{4} + \frac{x^3}{3} + c$

Sometimes it is useful to rewrite the integrand in a different form before performing the integration.

Example 3

Evaluate:

a) $\int \frac{t^3 - 3t^5}{t^5}\, dt$ b) $\int \frac{x + 5x^4}{x^2}\, dx$

Solution

a) $\int \frac{t^3 - 3t^5}{t^5}\, dt = \int \frac{t^3}{t^5}\, dt - \int \frac{3t^5}{t^5}\, dt = \int t^{-2}\, dt - \int 3\, dt = \frac{t^{-1}}{-1} - 3t + c$

$$= \frac{-1}{t} - 3t + c$$

b) $\int \frac{x + 5x^4}{x^2}\, dx = \int \frac{x}{x^2}\, dx + \int \frac{5x^4}{x^2}\, dx = \int \frac{1}{x}\, dx + \int 5x^2\, dx = \ln|x| + 5 \cdot \frac{x^3}{3} + c$

Integration by simple substitution

In this section, we will study a technique called substitution that can often be used to transform complicated integration problems into simpler ones.

The method of substitution depends on our understanding of the chain rule as well as the use of variables in integration. Two facts to recall:

1. When we find an anti-derivative, we established earlier that the use of x is arbitrary. We can use any other variable as you have seen in several exercises and examples so far.

   So, $\int f(u)\,du = F(u) + c$, where u is a 'dummy' variable in the sense that it can be replaced by any other variable.

2. The chain rule enables us to say
   $$\frac{d}{dx}(F(u(x))) = F'(u(x))\cdot u'(x).$$
   This can be written in integral form as
   $$\int F'(u(x))\cdot u'(x)\,dx = F(u(x)) + c$$
   or, equivalently, since $F(x)$ is an anti-derivative of $f(x)$,
   $$\int f(u(x))\cdot u'(x)\,dx = F(u(x)) + c.$$

For our purposes, it will be useful and simpler to let $u(x) = u$ and to write $\frac{du}{dx} = u'(x)$ in its 'differential' form $du = u'(x)dx$, or, simply, $du = u'dx$.

With this notation, the integral can now be written as

$$\int f(u(x))\cdot \mathbf{u'}(x)\,\mathbf{dx} = \int f(u)\,\mathbf{du} = F(u) + c.$$

The following example explains how the method works.

Example 4

Evaluate:

a) $\int (x^3 + 2)^{10}\cdot 3x^2\,dx$

b) $\int \tan x\,dx$

c) $\int \cos 5x\,dx$

d) $\int \cos x^2\cdot x\,dx$

e) $\int e^{3x+1}\,dx$

Solution

a) To integrate this function, it is simplest to make the following substitution.
Let $u = x^3 + 2$, and so $du = 3x^2\,dx$. Now the integral can be written as

$$\int (x^3 + 2)^{10}\cdot \mathbf{3x^2\,dx} = \int u^{10}\,\mathbf{du} = \frac{u^{11}}{11} + c = \frac{(x^3 + 2)^{11}}{11} + c.$$

b) This integrand has to be rewritten first and then we make the substitution.

$$\int \tan x\,dx = \int \frac{\sin x}{\cos x}\,dx = \int \frac{1}{\cos x}\cdot \sin x\,dx$$

We now let $u = \cos x \Rightarrow du = -\sin x\,dx$, and

$$\int \tan x\,dx = \int \frac{1}{\cos x}\cdot \mathbf{\sin x\,dx} = \int \frac{1}{u}\cdot(\mathbf{-du}) = -\int \frac{1}{u}\,du = -\ln|u| + c.$$

This last result can be then expressed in one of two ways:

$$\int \tan x\,dx = -\ln|\cos x| + c, \text{ or}$$

$$\int \tan x\,dx = -\ln|\cos x| + c = \ln|(\cos x)^{-1}| + c$$
$$= \ln\left|\frac{1}{(\cos x)}\right| + c = \ln|\sec x| + c$$

c) We let $u = 5x$, then $du = 5dx \Rightarrow dx = \frac{1}{5}du$, and so

$$\int \cos 5x\,dx = \int \cos u \cdot \tfrac{1}{5}du = \tfrac{1}{5}\int \cos u\,du = \tfrac{1}{5}\sin u + c$$
$$= \tfrac{1}{5}\sin 5x + c.$$

Another method can be applied here:

The substitution $u = 5x$ requires $du = 5dx$. As there is no factor of 5 in the integrand, and since 5 is a constant, we can multiply and divide by 5 so that we group the 5 and dx to form the du required by the substitution:

$$\int \cos 5x\,dx = \tfrac{1}{5}\int \cos x \cdot \mathbf{5dx} = \tfrac{1}{5}\int \cos u\,\mathbf{du} = \tfrac{1}{5}\sin u + c$$
$$= \tfrac{1}{5}\sin 5x + c$$

In integration, multiplying by a constant 'inside' the integral and 'compensating' for that with the reciprocal 'outside' the integral depends on theorem 1 (page 775). That is,

$$\int kf(x)dx = k\int f(x)dx.$$

However, you **cannot** multiply with a variable. So, you **cannot** say, for example,

$$\int \cos x^2\,dx = \frac{1}{2x}\int \cos x^2 \cdot \mathbf{2x}\,dx.$$

d) By letting $u = x^2$, $du = 2x\,dx$ and so

$$\int \cos x^2 \cdot x\,dx = \tfrac{1}{2}\int \cos x^2 \cdot 2x\,dx = \tfrac{1}{2}\int \cos u\,du = \tfrac{1}{2}\sin u + c$$
$$= \tfrac{1}{2}\sin x^2 + c.$$

e) $\int e^{3x+1}\,dx = \frac{1}{3}\int e^{3x+1}\,\mathbf{3dx} = \frac{1}{3}\int e^u\,\mathbf{du} = \frac{1}{3}e^u + c = \frac{1}{3}e^{3x+1} + c$

Note: The main challenge in using the substitution rule is to think of an appropriate substitution. You should try to select u to be a part of the integrand whose differential is also included (except for the constant). In Example 4a), we selected u to be $(x^3 + 2)$ knowing that $du = 3x^2dx$. Then we 'compensated' for the absence of 3! *Finding the right substitution is a bit of an art. You need to acquire it!* It is quite usual that your first guess may not work. Try another one!

Example 5

Evaluate each integral.

a) $\int e^{-3x}\,dx$

b) $\int \sin^2 x \cos x\,dx$

c) $\int 2\sin(3x - 5)\,dx$

d) $\int e^{mx+n}\,dx$

e) $\int x\sqrt{x}\,dx$, and $F(1) = 2$

Solution

a) Let $u = -3x$, then $du = -3dx$, and

$$\int e^{-3x}\,dx = -\tfrac{1}{3}\int e^{-3x}(\mathbf{-3dx}) = -\tfrac{1}{3}\int e^u\,\mathbf{du} = -\tfrac{1}{3}e^u + c$$
$$= -\tfrac{1}{3}e^{-3x} + c.$$

b) Let $u = \sin x$, then $du = \cos x\,dx$, and

$$\int \mathbf{\sin^2 x}\,\cos x\,dx = \int \mathbf{u^2}\,du = \tfrac{1}{3}u^3 + c = \tfrac{1}{3}\sin^3 x + c.$$

c) Let $u = 3x - 5$, then $du = 3dx$, and

$$\int 2\sin(3x - 5)\,dx = 2\cdot\frac{1}{3}\int \sin(3x - 5)3dx = \frac{2}{3}\int \sin u\,du$$
$$= -\frac{2}{3}\cos u + c = -\frac{2}{3}\cos(3x - 5) + c.$$

d) Let $u = mx + n$, then $du = m\,dx$, and

$$\int e^{mx+n}\,dx = \frac{1}{m}\int e^{mx+n}\,m\,dx = \frac{1}{m}\int e^{u}\,du$$
$$= \frac{1}{m}e^{u} + c = \frac{1}{m}e^{mx+n} + c.$$

e) $F(x) = \int x\sqrt{x}\,dx = \int x^{\frac{3}{2}}\,dx = \dfrac{x^{\frac{5}{2}}}{\left(\frac{5}{2}\right)} + c = \frac{2}{5}x^{\frac{5}{2}} + c$, but $F(1) = 2$

$F(1) = \frac{2}{5}1^{\frac{5}{2}} + c = \frac{2}{5} + c = 2 \Rightarrow c = \frac{8}{5}$

Therefore, $F(x) = \frac{2}{5}x^{\frac{5}{2}} + \frac{8}{5}$.

The previous discussion makes it clear that Table 16.2 is limited in scope, because we cannot use the integrals directly to evaluate composite integrals such as the ones in Examples 4 and 5 above. An adjusted table is therefore presented here.

Table 16.3

	Differentiation formula	Integration formula
1	$\frac{d}{dx}(u(x)) = u'(x) \Rightarrow du = u'(x)dx$	$\int du = u + c$
2	$\frac{d}{dx}\left(\frac{u^{n+1}}{(n+1)}\right) = u^n u'(x), n \neq -1 \Rightarrow d\left(\frac{u^{n+1}}{(n+1)}\right) = u^n u'(x)dx$	$\int u^n du = \frac{u^{n+1}}{n+1} + c, n \neq -1$
3	$\frac{d}{dx}(\sin(u)) = \cos(u)u'(x) \Rightarrow d(\sin(u)) = \cos(u)u'(x)dx$	$\int \cos u\,du = \sin u + c$
4	$\frac{d}{dx}(-\cos(u)) = \sin(u)u'(x) \Rightarrow d(-\cos(u)) = \sin(u)u'(x)dx$	$\int \sin u\,du = -\cos u + c$
5	$\frac{d}{dt}(\tan u) = \sec^2 u\,u'(t) \Rightarrow d(\tan u) = \sec^2 u\,u'(t)dt$	$\int \sec^2 u\,du = \tan u + c$
6	$\frac{d}{dx}(e^u) = e^u u'(x)dx \Rightarrow d(e^u) = e^u u'(x)dx$	$\int e^u du = e^u + c$
7	$\frac{d}{dx}(\ln\lvert u\rvert) = \frac{1}{u}u'(x) \Rightarrow d(\ln\lvert u\rvert) = \frac{1}{u}u'(x)dx$	$\int \frac{1}{u}du = \ln\lvert u\rvert + c$
8	$\frac{d}{dx}\left(\frac{a^u}{\ln a}\right) = a^u u'(x) \Rightarrow d\left(\frac{a^u}{\ln a}\right) = a^u u'(x)dx$	$\int a^u du = \frac{a^u}{\ln a} + c$
9	$\frac{d}{dx}(\arcsin u) = \frac{1}{\sqrt{1-u^2}}u'(x) \Rightarrow d(\arcsin u) = \frac{1}{\sqrt{1-u^2}}u'(x)dx$	$\int \frac{du}{\sqrt{1-u^2}} = \arcsin u + c$
10	$\frac{d}{dx}(\arctan u) = \frac{1}{1+u^2}u'(x) \Rightarrow d(\arctan u) = \frac{1}{1+u^2}u'(x)dx$	$\int \frac{du}{1+u^2} = \arctan u + c$

Example 6

Evaluate each integral.

a) $\int\sqrt{6x+11}\,dx$

b) $\int(5x^3+2)^8x^2dx$

c) $\displaystyle\int\frac{x^3-2}{\sqrt[5]{x^4-8x+13}}dx$

d) $\int\sin^4(3x^2)\cos(3x^2)x\,dx$

Solution

a) We let $u = 6x + 11$ and calculate du:

$$u = 6x + 11 \Rightarrow du = 6dx$$

Since du contains the factor 6, the integral is still not in the proper form $\int f(u)du$. However, here we can use of two approaches:

(i) Introduce the factor 6, as we have done before, i.e.

$$\int\sqrt{6x+11}\,dx = \tfrac{1}{6}\int\sqrt{6x+11}\,\boxed{6dx}$$
$$= \tfrac{1}{6}\int\sqrt{u}\,\boxed{du} = \tfrac{1}{6}\int u^{\frac{1}{2}}du$$
$$= \frac{1}{6}\frac{u^{\frac{3}{2}}}{\frac{3}{2}} + c = \tfrac{2}{18}u^{\frac{3}{2}} + c$$
$$= \tfrac{1}{9}(6x+11)^{\frac{3}{2}} + c$$

Or,

(ii) Since $u = 6x + 11 \Rightarrow du = 6dx \Rightarrow dx = \dfrac{du}{6}$, then

$\int\sqrt{6x+11}\,dx = \int\sqrt{u}\dfrac{du}{6} = \tfrac{1}{6}\int u^{\frac{1}{2}}du$, then we follow the same steps as before.

b) We let $u = 5x^3 + 2$, then $du = 15x^2dx$. This means that we need to introduce the factor 15 into the integrand:

$$\int(5x^3+2)^8x^2dx = \frac{1}{15}\int(5x^3+2)^8 15x^2dx$$
$$= \frac{1}{15}\int u^8du = \frac{1}{15}\frac{u^9}{9} + c$$
$$= \frac{1}{135}(5x^3+2)^9 + c$$

c) We let $u = x^4 - 8x + 13 \Rightarrow du = (4x^3 - 8)dx = 4(x^3 - 2)dx$.

$$\int\frac{x^3-2}{\sqrt[5]{x^4-8x+13}}dx = \frac{1}{4}\int\frac{4(x^3-2)dx}{\sqrt[5]{x^4-8x+13}} = \frac{1}{4}\int\frac{du}{u^{\frac{1}{5}}}$$
$$= \frac{1}{4}u^{-\frac{1}{5}}du = \frac{1}{4}\frac{u^{\frac{4}{5}}}{\frac{4}{5}} + c$$
$$= \frac{5}{16}(x^4 - 8x + 13)^{\frac{4}{5}} + c$$

d) We let $u = \sin(3x^2) \Rightarrow du = \cos(3x^2)6x\,dx$ using the chain rule!

$$\int \sin^4(3x^2)\cos(3x^2)x\,dx = \frac{1}{6}\int \sin^4(3x^2)\cos(3x^2)6x\,dx$$
$$= \frac{1}{6}\int u^4 du = \frac{1}{6}\frac{u^5}{5} + c$$
$$= \frac{1}{30}\sin^5(3x^2) + c$$

Exercise 16.1

In questions 1–15, find the most general anti-derivative of the function.

1 $f(x) = x + 2$

2 $f(t) = 3t^2 - 2t + 1$

3 $g(x) = \frac{1}{3} - \frac{2}{7}x^3$

4 $f(t) = (t - 1)(2t + 3)$

5 $g(u) = u^{\frac{2}{5}} - 4u^3$

6 $f(x) = 2\sqrt{x} - \dfrac{3}{2\sqrt{x}}$

7 $h(\theta) = 3\sin\theta + 4\cos\theta$

8 $f(t) = 3t^2 - 2\sin t$

9 $f(x) = \sqrt{x}\,(2x - 5)$

10 $g(\theta) = 3\cos\theta - 2\sec^2\theta$

11 $h(t) = e^{3t-1}$

12 $f(t) = \dfrac{2}{t}$

13 $h(t) = \dfrac{t}{3t^2 + 5}$

14 $h(\theta) = e^{\sin\theta}\cos\theta$

15 $f(x) = (3 + 2x)^2$

In questions 16–20, find f.

16 $f''(x) = 4x - 15x^2$

17 $f''(x) = 1 + 3x^2 - 4x^3$; $f'(0) = 2$, $f(1) = 2$

18 $f''(t) = 8t - \sin t$

19 $f'(x) = 12x^3 - 8x + 7$, $f(0) = 3$

20 $f'(\theta) = 2\cos\theta - \sin(2\theta)$

In questions 21–50, evaluate each integral.

21 $\int x(3x^2 + 7)^5 dx$

22 $\displaystyle\int \frac{x}{(3x^2 + 5)^4}dx$

23 $\int 2x^2\sqrt[4]{5x^3 + 2}\,dx$

24 $\displaystyle\int \frac{(3 + 2\sqrt{x})^5}{\sqrt{x}}dx$

25 $\int t^2\sqrt{2t^3 - 7}\,dt$

26 $\displaystyle\int \left(2 + \frac{3}{x}\right)^5\left(\frac{1}{x^2}\right)dx$

27 $\int \sin(7x - 3)dx$

28 $\displaystyle\int \frac{\sin(2\theta - 1)}{\cos(2\theta - 1) + 3}d\theta$

29 $\int \sec^2(5\theta - 2)d\theta$

30 $\int \cos(\pi x + 3)dx$

31 $\int \sec 2t\tan 2t\,dt$

32 $\int xe^{x^2 + 1}dx$

33 $\int \sqrt{t}\,e^{2t\sqrt{t}}dt$

34 $\displaystyle\int \frac{2}{\theta}(\ln\theta)^2 d\theta$

35 $\displaystyle\int \frac{dz}{z\ln 2z}$

36 $\int t\sqrt{3 - 5t^2}\,dt$

37 $\int \theta^2\sec^2\theta^3 d\theta$

38 $\displaystyle\int \frac{\sin\sqrt{t}}{2\sqrt{t}}dt$

39 $\int \tan^5 2t\sec^2 2t\,dt$

40 $\displaystyle\int \frac{dx}{\sqrt{x}(\sqrt{x} + 2)}$

41 $\int \sec^5 2t \tan 2t \, dt$

42 $\int \frac{x+3}{x^2+6x+7} dx$

43 $\int \frac{k^3x^3}{\sqrt{a^2-a^4x^4}} dx$

44 $\int 3x\sqrt{x-1} \, dx$

45 $\int \csc^2 \pi t \, dt$

46 $\int \sqrt{1+\cos\theta} \sin\theta d\theta$

47 $\int t^2\sqrt{1-t} \, dt$

48 $\int \frac{r^2-1}{\sqrt{2r-1}} dr$

49 $\int \frac{e^{x^2}-e^{-x^2}}{e^{x^2}+e^{-x^2}} x dx$

50 $\int \frac{t^2+2}{\sqrt{t-5}} dt$

16.2 Methods of integration: integration by parts

As far as this point, you will have noticed that while differentiation and integration are so strongly linked, finding derivatives is greatly different from finding integrals. With the derivative rules available, you are able to find the derivative of about any function you can think of. By contrast, you can compute anti-derivatives for a rather small number of functions. Thus far, we have developed a set of basic integration formulae, most of which followed directly from the related differentiation formulae that you saw in Table 16.2.

Using substitution, in some cases, helps us reduce the difficulty of evaluating some integrals by rendering them in familiar forms. However, there are far too many cases, where the simple substitution will not help. For example,

$$\int x \cos x \, dx$$

cannot be evaluated by the methods you have learned so far. We improve the situation in this section by introducing a powerful and yet simple tool called *integration by parts.*

Recall the product rule for differentiation:

$$\frac{d}{dx}(u(x)v(x)) = u'(x)v(x) + u(x)v'(x),$$

which gives rise to the differential form

$d(u(x)v(x)) = v(x)d(u(x)) + u(x)d(v(x))$, and for convenience, we will write

$$d(uv) = vdu + udv.$$

If we integrate both sides of this equation, we get

$$\int d(uv) = \int vdu + \int udv \Leftrightarrow uv = \int vdu + \int udv.$$

Solving this equation for udv, we get

$$\int udv = uv - \int vdu.$$

This rule is the **integration by parts.**

The significance of this rule is not immediately apparent. We will see its great utility in a few examples.

Brook Taylor (1685–1731) is credited with devising integration by parts. Taylor is mostly known for his contributions to power series where his 'Taylor theorem' has several very important applications in mathematics and science.

Example 7

Evaluate $\int x\cos x\,dx$.

Solution

First, observe that you cannot evaluate this as it stands, i.e. it is not one of our basic integrals and no substitution can help either.

Notice how you need to make a clever choice of u and dv so that the integral on the right side is one that will ease your work ahead. We need to choose u (to differentiate) and dv (to integrate); thus we let

$$u = x, \text{ and } dv = \cos x\,dx.$$

Then $du = dx$, and $v = \sin x$. (We will introduce c at the end of the process.)

It is usually helpful to organize your work in a table form:

$$\begin{array}{ll} u = x & du = dx \\ dv = \cos x\,dx & v = \sin x \end{array}$$

This gives us:

$$\begin{aligned} \int \underbrace{x}_{u}\ \underbrace{\cos x\,dx}_{dv} = \int u\,dv = \quad & uv - \int v\,du \\ = \ & x\sin x - \int \sin x\,dx \\ = \ & x\sin x + \cos x + c \end{aligned}$$

To verify your result, simply differentiate the right-hand side.

$$\frac{d}{dx}(x\sin x + \cos x + c) = \sin x + x\cos x - \sin x + 0 = x\cos x$$

Note: What other choices can you make?

There are three other choices of u and dv in this problem:

1 If we let

$$\left.\begin{array}{ll} u = \cos x & du = -\sin x\,dx \\ dv = dx & v = \frac{x^2}{2} \end{array}\right\} \Rightarrow \int x\cos x\,dx = \frac{x^2}{2}\cos x + \int \frac{x^2}{2}\sin x\,dx$$

This new integral is worse than the one we started with!

2 If we let

$$\left.\begin{array}{ll} u = x\cos x & du = (\cos x - x\sin x)dx \\ dv = xdx & v = x \end{array}\right\} \Rightarrow \int x\cos x\,dx = x^2\cos x - \int x(\cos x - x\sin x)dx$$

Again, this new integral is worse than the one we started with!

3 If we let

$$\begin{array}{ll} u = 1 & du = 0 \\ dv = x\cos x\,dx & v = ?? \end{array}$$

This is obviously a bad choice since we still do not know how to integrate $dv = x\cos x\,dx$.

The objective of integration by parts is to move from an integral $\int u dv$ (which we cannot see how to evaluate) to an integral $\int v du$ which we *can* integrate. So, keep in mind that integration by parts does not necessarily work all the time, and that we have to develop enough experience with such a process in order to make the 'correct' choice for u and vdu.

Example 8

Evaluate $\int xe^{-x}dx$.

Solution

We let

$$\left.\begin{array}{ll} u = x & du = dx \\ dv = e^{-x}\,dx & v = -e^{-x} \end{array}\right\} \Rightarrow \int xe^{-x}dx = -xe^{-x} + \int e^{-x}dx$$

$$= -xe^{-x} - e^{-x} + c$$

Example 9

Evaluate $\int \ln x\, dx$.

Solution

$$\left.\begin{array}{ll} u = \ln x & du = \frac{dx}{x} \\ dv = dx & v = x \end{array}\right\} \Rightarrow \int \ln x\, dx = x\ln x - \int \cancel{x}\frac{dx}{\cancel{x}}$$

$$= x\ln x - x + c$$

Example 10

Evaluate $\int x^2 \ln x\, dx$.

Solution

Since x^2 is easier to integrate than $\ln x$, and the derivative of $\ln x$ is also easier than $\ln x$ itself, we make the following substitution:

$$\left.\begin{array}{ll} u = \ln x & du = \frac{dx}{x} \\ dv = x^2\, dx & v = \frac{x^3}{3} \end{array}\right\} \Rightarrow \int x^2 \ln x\, dx = \frac{x^3}{3}\ln x - \int \frac{x^{\cancel{3}2}}{3}\frac{dx}{\cancel{x}}$$

$$= \frac{x^3}{3}\ln x - \int \tfrac{1}{3}x^2 dx$$

$$= \frac{x^3}{3}\ln x - \tfrac{1}{9}x^3 + c$$

Example 11 – Repeated use of integration by parts

Evaluate $\int x^2 \sin x\, dx$.

Solution

Since $\sin x$ is equally easy to integrate or differentiate while x^2 is easier to differentiate, we make the following substitution:

$$\left.\begin{array}{ll} u = x^2 & du = 2x\, dx \\ dv = \sin x\, dx & v = -\cos x \end{array}\right\} \Rightarrow \int x^2 \sin x\, dx = -x^2 \cos x + 2\int x \cos x\, dx$$

This first step simplified the original integral. However, the right-hand side still needs further integration. Here again, we use integration by parts.

$$\left.\begin{array}{ll} u = x^2 & du = 2x\,dx \\ dv = \cos x\,dx & v = \sin x \end{array}\right\} \Rightarrow \int 2x\cos x\,dx = 2x\sin x - 2\int \sin x\,dx$$
$$= 2x\sin x + 2\cos x + c$$

Combining the two results, we can now write

$$\int x^2 \sin x\,dx = -x^2\cos x + 2\int x\cos x\,dx$$
$$= x^2\cos x + 2x\sin x + 2\cos x + c.$$

Note: When making repeated applications of the integration by parts, you need to be careful not to change the 'nature' of the substitution in successive applications. For instance, in the previous example, the first substitution was $u = x^2$ and $dv = \sin x\,dx$. If in the second step, you had switched the substitution to $u = \cos x$ and $dv = 2x dx$, you would have obtained

$$\int x^2 \sin x\,dx = -x^2\cos x + x^2\cos x + \int x^2\sin x\,dx$$
$$= \int x^2 \sin x\,dx,$$

thus 'undoing' the previous integration and returning to the original integral.

Example 12

Evaluate $\int x^2 e^x dx$.

Solution

Since e^x is equally easy to integrate or differentiate while x^2 is easier to differentiate, we make the following substitution:

$$\left.\begin{array}{ll} u = x^2 & du = 2x\,dx \\ dv = e^x dx & v = e^x \end{array}\right\} \Rightarrow \int x^2 e^x dx = x^2 e^x - 2\int x e^x dx$$

This first step simplified the original integral. However, the right-hand side still needs further integration. Here again, we use integration by parts.

$$\left.\begin{array}{ll} u = 2x & du = 2\,dx \\ dv = e^x dx & v = e^x \end{array}\right\} \Rightarrow \int 2x e^x dx = 2xe^x - 2\int e^x dx$$
$$= 2xe^x - 2e^x + c$$

Hence,

$$\int x^2 e^x dx = x^2 e^x - \int 2x e^x dx$$
$$= x^2 e^x - 2xe^x + 2e^x + c.$$

Using integration by parts to find unknown integrals

Integrals like the one in the next example occur frequently in electricity problems. Their evaluation requires repeated applications of integration by parts followed by algebraic manipulation.

Example 13

Evaluate $\int \cos x\, e^x\, dx$.

Solution

Let

$$\left.\begin{array}{ll} u = e^x & du = e^x\,dx \\ dv = \cos x\,dx & v = \sin x \end{array}\right\} \Rightarrow \int \cos x\, e^x\, dx = e^x \sin x - \int \sin x\, e^x\, dx$$

The second integral is of the same nature, so we use integration by parts again.

$$\left.\begin{array}{ll} u = e^x & du = e^x\,dx \\ dv = \sin x\,dx & v = -\cos x \end{array}\right\} \Rightarrow \int \sin x\, e^x\, dx = -e^x \cos x + \int \cos x\, e^x\, dx$$

Hence,

$$\begin{aligned} \int \cos x\, e^x\, dx &= e^x \sin x - \int \sin x\, e^x\, dx \\ &= e^x \sin x - \left(-e^x \cos x + \int \cos x\, e^x\, dx\right) \\ &= e^x \sin x + e^x \cos x - \int \cos x\, e^x\, dx. \end{aligned}$$

Now, the unknown integral appears on both sides of the equation, thus

$$\int \cos x\, e^x\, dx + \int \cos x\, e^x\, dx = e^x \sin x + e^x \cos x$$

$$\Rightarrow 2\int \cos x\, e^x\, dx = e^x \sin x + e^x \cos x$$

$$\Rightarrow \int \cos x\, e^x\, dx = \frac{e^x \sin x + e^x \cos x}{2} + c.$$

Example 14

Evaluate $\int x \ln x\, dx$.

Solution

$$\left.\begin{array}{ll} u = \ln x & du = \dfrac{dx}{x} \\ dv = x\,dx & v = \dfrac{x^2}{2} \end{array}\right\} \Rightarrow \int x^2 \ln x\, dx = \frac{x^2}{2} \ln x - \int \frac{x^2}{2} \frac{dx}{x}$$

$$= \frac{x^2}{2} \ln x - \int \frac{x dx}{2} = \frac{x^2}{2} \ln x - \frac{x^2}{4} + c$$

Alternatively, we could have used a different substitution:

$$\left.\begin{array}{ll} u = x \ln x & du = (\ln x + 1)dx \\ dv = dx & v = x \end{array}\right\} \Rightarrow \int x \ln x\, dx = x^2 \ln x - \int x(\ln x + 1)dx$$

$$= x^2 \ln x - \int x \ln x\, dx - \int x dx$$

Adding $\int x\ln x\,dx$ to both sides and integrating $\int x dx$ we get

$$\int x\ln x\,dx + \int x\ln x\,dx = x^2\ln x - \frac{x^2}{2} + c$$

$$\Rightarrow 2\int x\ln x\,dx = x^2\ln x - \frac{x^2}{2} + c$$

$$\Rightarrow \int x\ln x\,dx = \frac{1}{2}\left(x^2\ln x - \frac{x^2}{2} + c\right) = \frac{x^2\ln x}{2} - \frac{x^2}{4} + C.$$

Note: The constant c is arbitrary, and hence it is unimportant that we use $c/2$ or C in our final answer.

Exercise 16.2

In questions 1–22, evaluate each integral.

1 $\int x^2 e^{-x^3} dx$

2 $\int x^2 e^{-x} dx$

3 $\int x^2 \cos 3x\,dx$

4 $\int x^2 \sin ax\,dx$

5 $\int \cos x \ln(\sin x)dx$

6 $\int x \ln x^2\,dx$

7 $\int x^2 \ln x\,dx$

8 $\int x^2(e^x - 1)dx$

9 $\int x \cos \pi x\,dx$

10 $\int e^{3t}\cos 2t\,dt$

11 $\int \arcsin x\,dx$

12 $\int x^3 e^x dx$

13 $\int e^{-2x}\sin 2x\,dx$

14 $\int \sin(\ln x)dx$

15 $\int \cos(\ln x)dx$

16 $\int \ln(x + x^2)dx$

17 $\int e^{kx}\sin x\,dx$

18 $\int x \sec^2 x\,dx$

19 $\int \sin x \sin 2x\,dx$

20 $\int x \arctan x\,dx$

21 $\int \frac{\ln x}{\sqrt{x}}dx$

22 $\int t \sec^2 t\,dt$

23 In one scene of the movie *Stand and Deliver*, the teacher shows his students how to evaluate $\int x^2 \sin x\,dx$ by setting up a chart similar to the following.

	$\sin x$	
x^2	$-\cos x$	$+$
$2x$	$-\sin x$	$-$
2	$\cos x$	$+$

Multiply across each row and add the result. The integral is

$$\int x^2 \sin x\,dx = -x^2\cos x + 2x\sin x + 2\cos x + c.$$

Explain why the method works for this problem.

In questions 24–26, use the result of question 23 to evaluate each integral.

24 $\int x^4 \sin x\,dx$

25 $\int x^5 \cos x\,dx$

26 $\int x^4 e^x dx$

27 Show that the method used in question 23 will not work with

$$\int x^2 \ln x \, dx.$$

28 Show that $\int x^n e^x dx = x^n e^x - n\int x^{n-1} e^x dx$, then use this *reduction formula* to show that $\int x^4 e^x dx = ax^4 e^x + bx^3 e^x + cx^2 e^x + dxe^x + fe^x + g$, where $a, b, c, \ldots, g$ are to be determined.

29 Show that $\int x^n \ln x \, dx = \frac{x^{n+1}}{n+1} \ln x - \frac{x^{n+1}}{(n+1)^2} + c.$

30 Show that $\int e^{mx} \cos nx \, dx = \frac{e^{mx}(m \cos nx + n \sin nx)}{m^2 + n^2} + c.$

31 Show that $\int e^{mx} \sin nx \, dx = \frac{e^{mx}(m \sin nx - n \cos nx)}{m^2 + n^2} + c.$

16.3 More methods of integration

In the previous section, we looked at a very powerful method for integration that has a wide range of applications. However, integration by parts does not work for all situations, and in some cases where it works, it may not be the most efficient of methods. We learned about substitution before. In this section we will consider a few trigonometric integrals and some substitutions related to trigonometric functions or their inverses.

This section is basically a set of examples that will show you how to deal with a variety of cases.

Some of the trigonometric identities you learned before will prove very helpful in this section. Key identities we will make use of are the following:

1 $\cos^2 \theta + \sin^2 \theta = 1$

2 $\sin^2 \theta = \frac{1 - \cos 2\theta}{2}$

3 $\cos^2 \theta = \frac{1 + \cos 2\theta}{2}$

4 $\sec^2 \theta = 1 + \tan^2 \theta$

Example 15

Evaluate $\int \sin^2 x \, dx$.

Solution

We can use identity (2) from the list above.

$$\int \sin^2 x \, dx = \int \frac{1 - \cos 2x}{2} dx = \frac{1}{2}\int (1 - \cos 2x) dx$$

$$= \frac{1}{2}\left(x - \frac{1}{2}\sin 2x\right) + c$$

Example 16

Evaluate $\int \cos^4 \theta d\theta$.

Solution

Identity (3) will give us the following:

$$\int \cos^4 \theta d\theta = \int \left(\frac{1+\cos 2\theta}{2}\right)^2 d\theta = \frac{1}{4}\int (1 + 2\cos 2\theta + \cos^2 2\theta) d\theta$$

$$= \frac{1}{4}\int \left(1 + 2\cos 2\theta + \frac{1+\cos 4\theta}{2}\right) d\theta$$

$$= \frac{1}{8}\left(2\theta + 2\sin 2\theta + \theta + \frac{1}{4}\sin 4\theta\right) + c$$

$$= \frac{1}{32}(12\theta + 8\sin 2\theta + \sin 4\theta) + c$$

Here is a list of a few cases and how to find the integral. There are a few more integrals that we did not list here. On exams, any non-standard cases will be accompanied by a recommended substitution.

Integral	How to find it
$\int \sin^m x \cos^n x\, dx$	If m is odd, then break $\sin^m x$ into $\sin x$ and $\sin^{m-1} x$, use the substitution $u = \cos x$ and change the integral into the form $\int \cos^p x \sin x\, dx = \int u^p du$. Similarly if n is odd.
$\int \tan^m x \sec^n x\, dx$	If m and n are odd, break off a term for $\sec x \tan x\, dx$ and express the integrand in terms of $\sec x$ since $d(\sec x) = \sec x \tan x\, dx$.
$\int \tan^n x\, dx$	Write the integrand as $\int \tan^{n-2} x \tan^2 x\, dx$, replace $\tan^2 x$ with $\sec^2 x - 1$ and then use $u = \tan x$.
$\int \sec^n x\, dx$	If n is even, factor a $\sec^2 x$ out and write the rest in terms of $\tan^2 x + 1$. If n is odd, factor a $\sec^3 x$ out. Here, integration by parts may be useful.

Example 17

Evaluate $\int \sec x\, dx$.

Solution

This integral is evaluated using a 'clever' multiplication by an atypical factor, then:

$$\int \sec x\, dx = \int \sec x \frac{\tan x + \sec x}{\tan x + \sec x} dx = \int \frac{\sec x \tan x + \sec^2 x}{\tan x + \sec x} dx$$

Now use the substitution $u = \sec x + \tan x \Rightarrow du = (\sec x \tan x + \sec^2 x)$ dx; hence,

$$\int \sec x\, dx = \int \frac{\sec x \tan x + \sec^2 x}{\tan x + \sec x} dx = \int \frac{du}{u}$$

$$= \ln|u| + c = \ln|\tan x + \sec x| + c.$$

Example 18

Evaluate $\int \sec^3 x\, dx$.

Solution

This can be evaluated using integration by parts and some of the results we have already established.

$$u = \sec x \qquad du = \sec x \tan x\, dx$$

$$dv = \sec^2 x dx \qquad v = \tan x$$

Hence,

$$\begin{aligned}\int \sec^3 x\, dx &= \sec x \tan x - \int \tan x \sec x \tan x\, dx \\ &= \sec x \tan x - \int \sec x \tan^2 x\, dx \\ &= \sec x \tan x - \int \sec x[\sec^2 x - 1]\, dx \\ &= \sec x \tan x - \int \sec^3 x\, dx + \int \sec x\, dx.\end{aligned}$$

Adding $\int \sec^3 x\, dx$ to both sides:

$$\begin{aligned}2\int \sec^3 x\, dx &= \sec x \tan x + \int \sec x\, dx \\ &= \sec x \tan x + \ln|\sec x + \tan x|\end{aligned}$$

And finally,

$$\int \sec^3 x\, dx = \frac{\sec x \tan x + \ln|\sec x + \tan x|}{2} + c.$$

Example 19

Evaluate $\int \sin^3 x \cos^3 x\, dx$.

Solution

This integral can be evaluated by separating either a cosine or a sine, then writing the rest of the expression in terms of sine or cosine.

We will separate a cosine here.

$$\begin{aligned}\int \sin^3 x \cos^3 x\, dx &= \int \sin^3 x \cos^2 x \cos x\, dx \\ &= \int \sin^3 x(1 - \sin^2 x)\cos x\, dx \\ &= \int (\sin^3 x - \sin^5 x)\cos x\, dx\end{aligned}$$

Now we let

$u = \sin x \Rightarrow du = \cos x\, dx$, and hence

$$\begin{aligned}\int \sin^3 x \cos^3 x\, dx &= \int (\sin^3 x - \sin^5 x)\cos x\, dx \\ &= \int (u^3 - u^5)\, du = \frac{u^4}{4} - \frac{u^6}{6} + c \\ &= \frac{\sin^4 x}{4} - \frac{\sin^6 x}{6} + c.\end{aligned}$$

Some useful trigonometric substitutions

Evaluating integrals that involve $(a^2 - u^2)$, $(a^2 + u^2)$ or $(u^2 - a^2)$ may be rendered simpler by using some trigonometric substitution like the ones listed below.

Expression	Substitution	Simplified	du
$a^2 - u^2$	$u = a\sin\theta$	$a^2 - u^2 = a^2 - a^2\sin^2\theta$ $= a^2(1 - \sin^2\theta) = a^2\cos^2\theta$	$a\cos\theta d\theta$
$a^2 + u^2$	$u = a\tan\theta$	$a^2 + u^2 = a^2 + a^2\tan^2\theta$ $= a^2(1 + \tan^2\theta) = a^2\sec^2\theta$	$a\sec^2\theta d\theta$
$u^2 - a^2$	$u = a\sec\theta$	$u^2 - a^2 = a^2\sec^2 4\theta - a^2$ $= a^2(\sec^2\theta - 1) = a^2\tan^2\theta$	$a\sec\theta\tan\theta d\theta$

As you notice, this substitution is not the usual form. For convenience, we express the variable of integration in terms of the new variable. For example, rather than saying let $\theta = \arcsin\frac{u}{a}$, we say $u = a\sin\theta$. This allows us to easily find an expression for du. We will clarify the use of this type of substitution with a few examples. One important aspect of the process is how to revert back to the variable of integration. We will demonstrate that in the following examples.

Example 20

Evaluate $\int\frac{dx}{\sqrt{a^2 - x^2}}$.

Solution

This integrand is of the form involving $a^2 - u^2$, where $u = x$. We use the substitution $x = a\sin\theta$.

$\Rightarrow dx = a\cos\theta d\theta$,

$$\sqrt{a^2 - x^2} = \sqrt{a^2\cos^2\theta} = a\cos\theta$$

Hence,

$$\int\frac{dx}{\sqrt{a^2 - x^2}} = \int\frac{a\cos\theta d\theta}{a\cos\theta} = \int d\theta = \theta + c.$$

Now, consider the right triangle where $x = a\sin\theta \Leftrightarrow \sin\theta = \frac{x}{a}$.

$$\int\frac{dx}{\sqrt{a^2 - x^2}} = \theta + c = \arcsin\frac{x}{a} + c.$$

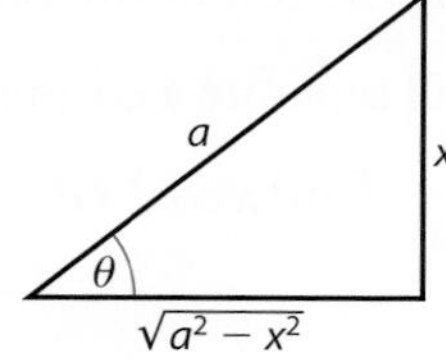

Example 21

Evaluate $\int\frac{dt}{\sqrt{a^2 - t^2}}$.

Solution

This integrand is of the form involving $a^2 - u^2$, where $u = t$. We use the substitution $t = a\sin\theta$.

$$\Rightarrow dt = a\cos\theta d\theta,$$
$$a^2 - t^2 = a^2\cos^2\theta$$

And so

$$\int \frac{dt}{a^2 - t^2} = \int \frac{a\cos\theta d\theta}{a^2\cos^2\theta} = \frac{1}{a}\int \sec\theta d\theta = \frac{1}{a}\ln|\sec\theta + \tan\theta| + c.$$

Now, in the triangle right,

$$t = a\sin\theta \Leftrightarrow \sin\theta = \frac{t}{a},$$

$$\cos\theta = \frac{\sqrt{a^2 - t^2}}{a};$$

$$\tan\theta = \frac{t}{\sqrt{a^2 - t^2}};\ \sec\theta = \frac{a}{\sqrt{a^2 - t^2}}$$

Consequently,

$$\int \frac{dt}{a^2 - t^2} = \frac{1}{a}\ln|\sec\theta + \tan\theta| + c = \frac{1}{a}\ln\left|\frac{a}{\sqrt{a^2 - t^2}} + \frac{t}{\sqrt{a^2 - t^2}}\right| + c$$

$$= \frac{1}{a}\ln\left|\frac{a + t}{\sqrt{a^2 - t^2}}\right| + c.$$

This is an acceptable answer. However, using the logarithmic properties you learned in Chapter 5, you can simplify further.

$$\int \frac{dt}{a^2 - t^2} = \frac{1}{a}\ln\left|\frac{a + t}{\sqrt{a^2 - t^2}}\right| + c = \frac{1}{a}\ln\sqrt{\frac{(a + t)^2}{a^2 - t^2}} + c$$

$$= \frac{1}{a}\ln\sqrt{\frac{(a + t)^2}{(a - t)(a + t)}} + c = \frac{1}{a}\ln\sqrt{\frac{(a + t)}{(a - t)}} + c$$

$$= \frac{1}{2a}\ln\left|\frac{(a + t)}{(a - t)}\right| + c$$

Example 22

Evaluate $\int \frac{dt}{a^2 + t^2}$.

Solution

This integrand is of the form involving $a^2 + u^2$, where $u = t$.
We use the substitution $t = a\tan\theta$.

$$\Rightarrow dt = a\sec^2\theta d\theta,$$

$$a^2 + t^2 = a^2(1 + \tan^2\theta) = a^2\sec^2\theta$$

And so

$$\int \frac{dt}{a^2 + t^2} = \int \frac{a\sec^2\theta d\theta}{a^2\sec^2\theta} = \frac{1}{a}\int d\theta = \frac{1}{a}\theta + c.$$

Since $t = a\tan\theta$, then $\tan\theta = \frac{t}{a} \Rightarrow \theta = \arctan\frac{t}{a}$.

Consequently,

$$\int \frac{dt}{a^2 + t^2} = \frac{1}{a}\theta + c = \frac{1}{a}\arctan\frac{t}{a} + c.$$

Example 23

Evaluate $\int\sqrt{x^2+5}\,dx$.

Solution

This integrand is of the form involving $a^2 + u^2$, where $u = x$. We use the substitution $x = a\tan\theta = \sqrt{5}\tan\theta$.

$$\Rightarrow dx = \sqrt{5}\sec^2\theta d\theta,$$

$$5 + x^2 = 5(1+\tan^2\theta) = 5\sec^2\theta d\theta$$

And so

$$\int\sqrt{x^2+5}\,dx = \int\sqrt{5\sec^2\theta}\sqrt{5}\sec^2\theta d\theta$$

$$= \int 5\sec^3\theta d\theta.$$

Now, earlier in Example 18, we have seen that

$$\int\sec^3 x\,dx = \frac{\sec x\tan x + \ln|\sec x+\tan x|}{2} + c.$$

And therefore

$$\int\sqrt{x^2+5}\,dx = 5\int\sec^3\theta d\theta = 5\left(\frac{\sec x\tan x+\ln|\sec x+\tan x|}{2}\right) + c.$$

Now, in the triangle right,

$$\tan\theta = \frac{x}{\sqrt{5}},$$

$$\sec\theta = \frac{\sqrt{5+x^2}}{\sqrt{5}} = \sqrt{\frac{5+x^2}{5}}, \text{ and so}$$

$\sqrt{5+x^2}$

x

θ

$\sqrt{5}$

$$\int\sqrt{x^2+5}\,dx = 5\left(\frac{\sec\theta\tan\theta+\ln|\sec\theta+\tan\theta|}{2}\right) + c$$

$$= 5\left(\frac{\sqrt{\frac{5+x^2}{5}}\cdot\frac{x}{\sqrt{5}} + \ln\left|\sqrt{\frac{5+x^2}{5}}+\frac{x}{\sqrt{5}}\right|}{2}\right) + c$$

$$= \frac{\sqrt{5}}{2}(\sqrt{5+x^2}\cdot x) + \frac{1}{2}\ln\left(\frac{\sqrt{5+x^2}+x}{\sqrt{5}}\right) + c$$

$$= \frac{\sqrt{5}}{2}(\sqrt{5+x^2}\cdot x) + \frac{1}{2}(\ln(\sqrt{5+x^2}+x) - \ln\sqrt{5}) + c$$

$$= \frac{\sqrt{5}}{2}(x\sqrt{5+x^2}) + \frac{1}{2}(\ln(\sqrt{5+x^2}+x)) + C.$$

In the last step we set $-\frac{1}{2}\ln\sqrt{5} + c = C$.

Example 24

Evaluate $\int\sqrt{25 - 4x^2}\,dx$.

Solution

This integrand is of the form involving $a^2 - u^2$, where $u = 2x$. We use the substitution $2x = 5\sin\theta$.

$$2dx = 5\cos\theta d\theta \Rightarrow dx = \frac{5}{2}\cos\theta d\theta$$

$$\sqrt{25 - 4x^2} = \sqrt{25 - 25\sin^2\theta} = 5\cos\theta$$

And so

$$\int\sqrt{25 - 4x^2}\,dx = \int 5\cos\theta\left(\frac{5}{2}\cos\theta d\theta\right) = \frac{25}{2}\int\cos^2\theta d\theta$$

$$= \frac{25}{2}\int\left(\frac{1 + \cos 2\theta}{2}\right)d\theta = \frac{25}{2}\left(\frac{\theta}{2} + \frac{1}{4}\sin 2\theta\right) + c$$

$$= \frac{25}{8}(2\theta + \sin 2\theta) + c.$$

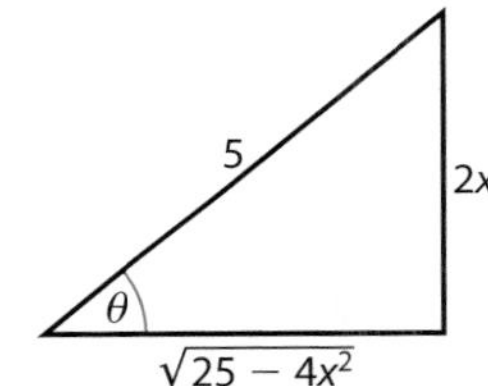

But, since $2x = 5\sin\theta$, then $\sin\theta = \frac{2x}{5} \Rightarrow \theta = \arcsin\frac{2x}{5}$, and since $\sin 2\theta = 2\sin\theta\cos\theta$, then

$$\int\sqrt{25 - 4x^2}\,dx = \frac{25}{8}(2\theta + \sin 2\theta) + c$$

$$= \frac{25}{8}\left(2\arcsin\frac{2x}{5} + 2\left(\frac{2x}{5}\right)\left(\frac{\sqrt{25 - 4x^2}}{5}\right)\right) + c$$

$$= \frac{25}{4}\arcsin\frac{2x}{5} + \frac{x\sqrt{25 - 4x^2}}{2} + c.$$

Exercise 16.3

In questions 1–44, evaluate each integral.

1 $\int\sin^3 t\cos^2 t\,dt$

2 $\int\sin^3 t\cos^3 t\,dt$

3 $\int\sin^3 3\theta\cos 3\theta\,d\theta$

4 $\int\frac{1}{t^2}\sin^5\left(\frac{1}{t}\right)\cos^2\left(\frac{1}{t}\right)dt$

5 $\int\frac{\sin^3 x}{\cos^2 x}dx$

6 $\int\tan^5 3x\sec^2 3x\,dx$

7 $\int\theta\tan^3\theta^2\sec^4\theta^2\,d\theta$

8 $\int\frac{1}{\sqrt{t}}\tan^3\sqrt{t}\sec^3\sqrt{t}\,dt$

9 $\int\tan^4(5t)dt$

10 $\int\frac{dt}{1 + \sin t}$

• **Hint:** multiply the integrand by $\frac{1 - \sin t}{1 - \sin t}$.

11 $\displaystyle\int \frac{d\theta}{1+\cos\theta}$

12 $\displaystyle\int \frac{1+\sin t}{\cos t}dt$

13 $\displaystyle\int \frac{\sin x - 5\cos x}{\sin x + \cos x}dx$

• **Hint:** find numbers a and b such that $\sin x - 5\cos x = a(\sin x + \cos x) + b(\cos x - \sin x)$.

14 $\displaystyle\int \frac{\sec\theta\tan\theta}{1+\sec^2\theta}d\theta$

15 $\displaystyle\int \frac{\arctan t}{1+t^2}dt$

16 $\displaystyle\int \frac{1}{(1+t^2)\arctan t}dt$

17 $\displaystyle\int \frac{dx}{x\sqrt{1-(\ln x)^2}}$

18 $\int \sin^3 x\, dx$

19 $\displaystyle\int \frac{\sin^3 x}{\sqrt{\cos x}}dx$

20 $\displaystyle\int \frac{\sin^3\sqrt{x}}{\sqrt{x}}dx$

21 $\int \cos t\cos^3(\sin t)dt$

22 $\displaystyle\int \frac{\cos\theta + \sin 2\theta}{\sin\theta}d\theta$

23 $\int t\sec t\tan t\, dt$

24 $\displaystyle\int \frac{\cos x}{2-\sin x}dx$

25 $\int e^{-2x}\tan(e^{-2x})\, dx$

26 $\displaystyle\int \frac{\sec(\sqrt{t})}{\sqrt{t}}dt$

27 $\displaystyle\int \frac{dt}{1+\cos 2t}$

28 $\int \sqrt{1-9x^2}\, dx$

29 $\displaystyle\int \frac{dx}{(x^2+4)^{\frac{3}{2}}}$

30 $\int \sqrt{4+t^2}\, dt$

31 $\displaystyle\int \frac{3e^t dt}{4+e^{2t}}$

32 $\displaystyle\int \frac{1}{\sqrt{9-4x^2}}dx$

33 $\displaystyle\int \frac{1}{\sqrt{4+9x^2}}dx$

34 $\displaystyle\int \frac{\cos x}{\sqrt{1+\sin^2 x}}dx$

35 $\displaystyle\int \frac{x}{\sqrt{4-x^2}}dx$

36 $\displaystyle\int \frac{x}{x^2+16}dx$

37 $\displaystyle\int \frac{\sqrt{4-x^2}}{x^2}dx$

38 $\displaystyle\int \frac{dx}{(9-x^2)^{\frac{3}{2}}}$

39 $\int x\sqrt{1+x^2}\, dx$

40 $\int e^{2x}\sqrt{1+e^{2x}}\, dx$

41 $\int e^{x}\sqrt{1-e^{2x}}\, dx$

42 $\displaystyle\int \frac{e^x dx}{\sqrt{e^{2x}+9}}$

43 $\displaystyle\int \frac{\ln x}{\sqrt{x}}dx$

44 $\displaystyle\int \frac{x^3}{(x+2)^2}dx$

45 The integral $\displaystyle\int \frac{x}{x^2+9}dx$ can be evaluated either by trigonometric substitution or by direct substitution. Do it both ways and reconcile the results.

46 The integral $\displaystyle\int \frac{x^2}{x^2+9}dx$ can be evaluated either by trigonometric substitution or by rewriting the numerator as $(x^2+9)-9$. Do it both ways and reconcile the results.

16.4 Area and definite integral

The main goal of this section is to introduce you to the following major problem of calculus.

The area problem: Given a function $f(x)$ that is continuous and non-negative on an interval $[a, b]$, find the area between the graph of $f(x)$ and the interval $[a, b]$ on the x-axis.

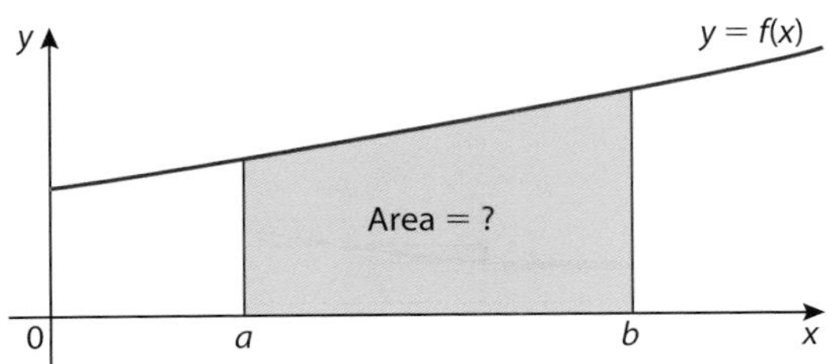

• **Hint:** This is only an expository treatment that explains to you how the definite integral is developed. You will not be required to reproduce this calculation yourself.

We divide the base interval $[a, b]$ into n equal sub-intervals, and over each sub-interval construct a rectangle that extends from the x-axis to any point on the curve $y = f(x)$ that is above the sub-interval. The particular point does not matter – it can be above the centre, above one endpoint, or any other point in the sub-interval. In Figure 16.1 it is above the centre.

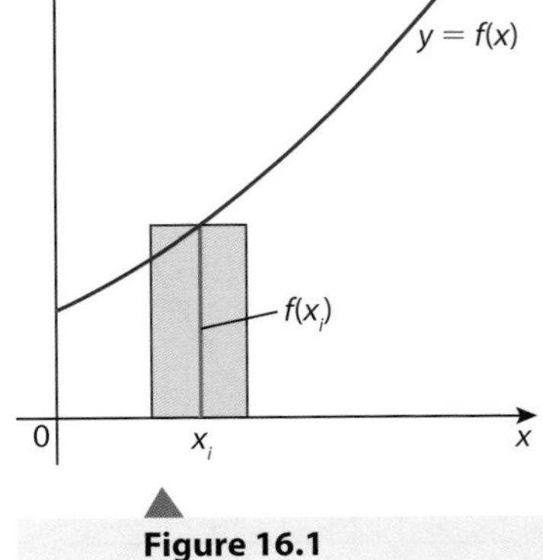

Figure 16.1

For each n, the total area of the rectangles can be viewed as an approximation to the exact area in question. Moreover, it is evident intuitively that as n increases, these approximations will get better and better and will eventually approach the exact area as a limit. See Figure 16.2.

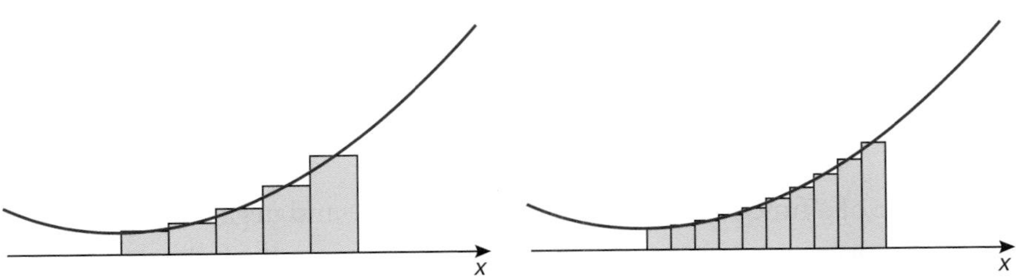

Figure 16.2

A traditional approach to this would be to study how the choice of where to erect the rectangular strip does not affect the approximation as the number of intervals increases. You can construct 'inscribed' rectangles, which, at the start, give you an underestimate of the area. On the other hand, you can construct 'circumscribed' rectangles that, at the start, overestimate the area. See Figure 16.3.

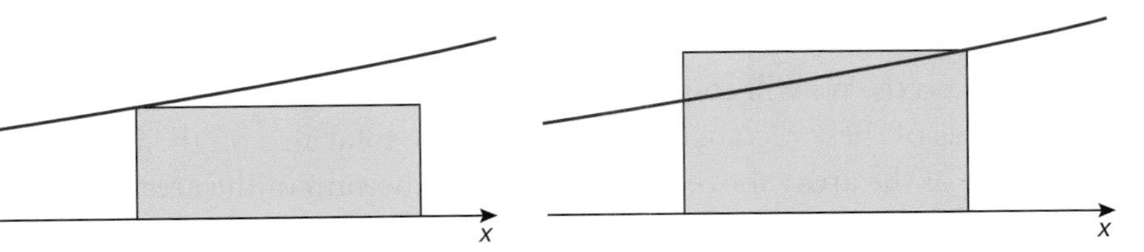

Figure 16.3

As the number of intervals increases, the difference between the overestimates and the underestimates will approach 0.

Figure 16.4

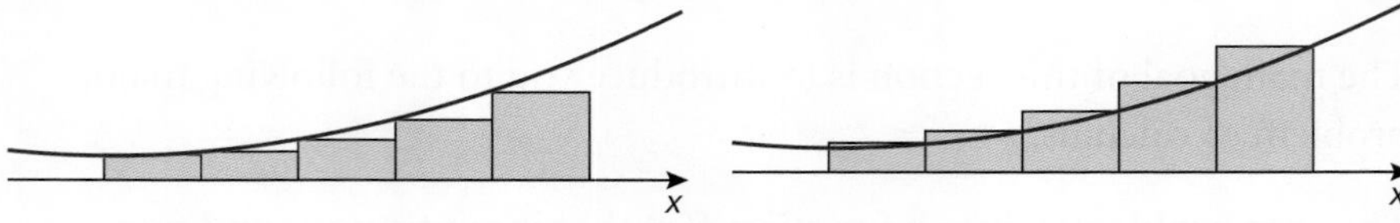

Figure 16.4 above shows n inscribed and subscribed rectangles and Figure 16.5 shows us the difference between the overestimates and the underestimates.

Figure 16.5

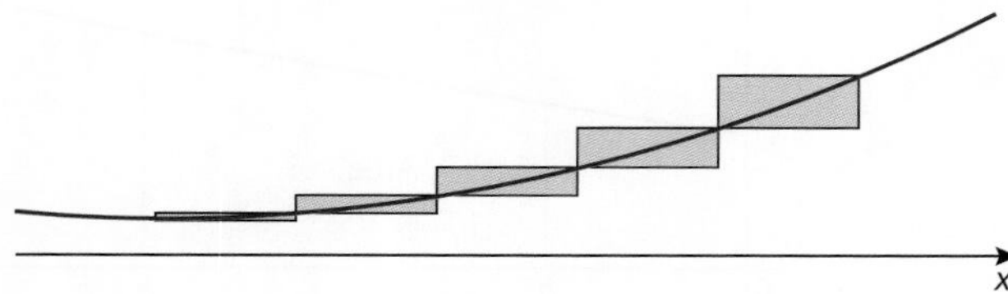

Figure 16.5 demonstrates that as the number n increases, the difference between the estimates will approach 0. Since, in Figure 16.1, we set up our rectangles by choosing a point inside the interval, the areas of the rectangles will lie between the overestimates and the underestimates, and hence as the difference between the extremes approaches zero, the rectangles we constructed will give the area of the region required.

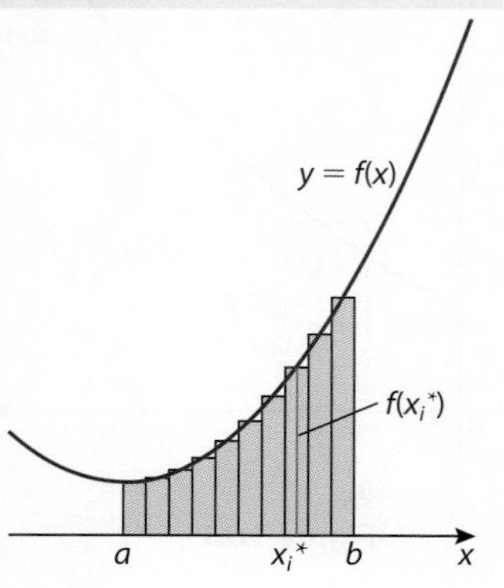

Figure 16.6

If we consider the width of each interval to be Δx, the area of any rectangle is given as

$$A_i = f(x_i^*)\Delta x.$$

The total area of the rectangles so constructed is

$$A_n = \sum_{t=0}^{n} f(x_i^*)\Delta x$$

where x_i^* is an arbitrary point within any sub-interval $[x_{i-1}, x_i]$, $x_0 = a$ and $x_n = b$.

In the case of a function $f(x)$ that has both positive and negative values on $[a, b]$, it is necessary to consider the *signs* of the areas in the following sense.

Figure 16.7

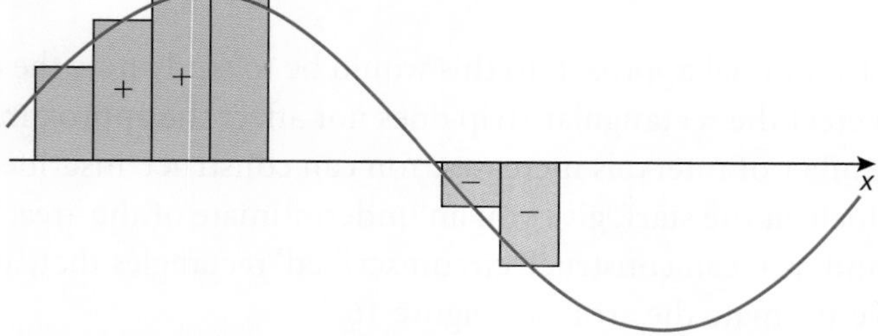

On each sub-interval, we have a rectangle with width Δx and height $f(x^*)$. If $f(x^*) > 0$, this rectangle is above the x-axis; if $f(x^*) < 0$, this rectangle is below the x-axis. We will consider the sum defined above as the sum of the signed areas of these rectangles. That means the total area on the interval is the sum of the areas above the x-axis minus the sum of the areas of the rectangles below the x-axis.

We are now ready to look at a 'loose' definition of the definite integral.

If $f(x)$ is a continuous function defined for $a \leqslant x \leqslant b$, we divide the interval $[a, b]$ into n sub-intervals of equal width $\Delta x = (b - a)/n$. We let $x_0 = a$ and $x_n = b$ and we choose $x_1^*, x_2^*, \ldots, x_n^*$ in these sub-intervals, so that x_i^* lies in the ith sub-interval $[x_{i-1}, x_i]$. Then the definite integral of $f(x)$ from a to b is

$$\int_a^b f(x)\, dx = \lim_{n \to \infty} \sum_{i=1}^{n} f(x_i^*)\Delta x.$$

For a list of recommended resources about definite integrals, visit www.pearsonhotlinks.com, enter the ISBN or title of this book and select weblink 2.

In the notation $\int_a^b f(x)\, dx$, in addition to the known integrand and differential, a and b are called the limits of integration: a is the lower limit and b is the upper limit.

Note: Because we have assumed that $f(x)$ is continuous, it can be proved that the limit definition above always exists and gives the same value no matter how we choose the points x_i^*. If we take these points at the centre, at two-thirds the distance from the lower endpoint or at the upper endpoint, the value is the same. This is why we will state the definition of the integral from now on as

$$\int_a^b f(x)\, dx = \lim_{n \to \infty} \sum_{i=1}^{n} f(x_i)\Delta x.$$

Calling the area under the function an integral is no coincidence. To make the point, let us take the following example.

Example 25(I)

Find the area, $A(x)$, between the graph of the function $f(x) = 3$ and the interval $[-1, x]$, and find the derivative $A'(x)$ of this area function.

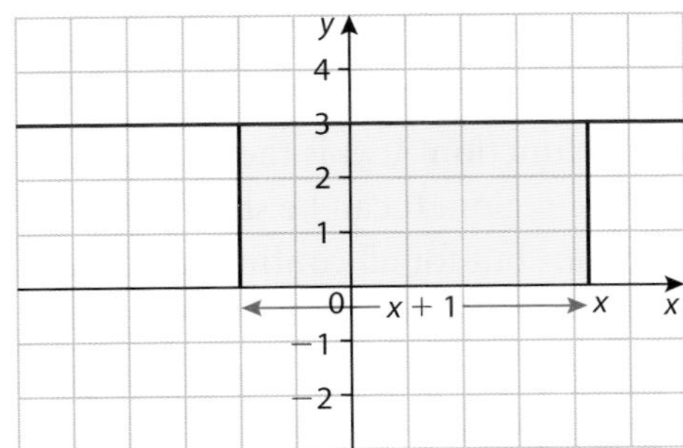

Solution

The area in question is

$$A(x) = 3(x - (-1)) = 3x + 3, \text{ and}$$

$$A'(x) = 3 = f(x).$$

Example 25(II)

Find the area, $A(x)$, between the graph of the function $f(x) = 3x + 2$ and the interval $[-2/3, x]$, and find the derivative $A'(x)$ of this area function.

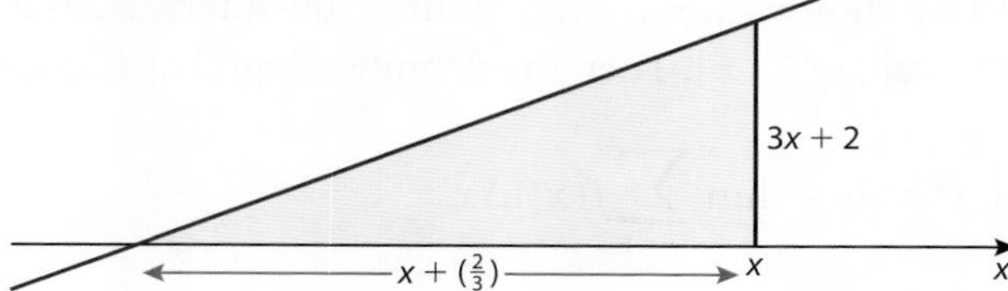

Solution

The area in question is

$A(x) = \frac{1}{2}\left(x + \frac{2}{3}\right)(3x + 2) = \frac{1}{6}(3x + 2)^2$, since this is the area of a triangle.

Hence, $A'(x) = \frac{1}{6} \times 2(3x + 2) \times 3 = 3x + 2 = f(x)$.

Example 25(III)

Find the area, $A(x)$, between the graph of the function $f(x) = x + 2$ and the interval $[-1, x]$, and find the derivative $A'(x)$ of this area function.

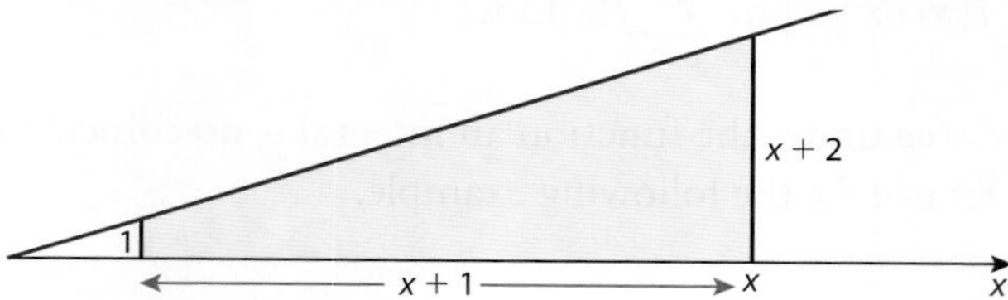

Solution

This is a trapezium, so the area is

$A(x) = \frac{1}{2}(1 + (x + 2))(x + 1) = \frac{1}{2}(x^2 + 4x + 3)$, and

$A'(x) = \frac{1}{2} \times (2x + 4) = x + 2 = f(x)$.

Note that, in every case, $A'(x) = f(x)$.

The derivative of the area function $A(x)$ is the function whose graph forms the upper boundary of the region. It can be shown that this relation is true, not only for linear functions but for all continuous functions. Thus, to find the area function $A(x)$, we can look instead for a particular function whose derivative is $f(x)$. This is, of course, the anti-derivative of $f(x)$.

So, intuitively, as we have seen above, we define the area function as

$$A(x) = \int_a^x f(t)\,dt, \text{ that is, } A'(x) = f(x).$$

This is the trigger to the **fundamental theorem of calculus** which we will introduce in the following few pages. As we stressed at the outset, our intention here is to show you that this important theorem has its solid mathematical basis. However, examinations will not include questions requiring you to repeat the steps developed here. Just enjoy the discussion!

Before we begin the discussion, it is worth looking at some of the obvious properties of the definite integral.

Properties of the definite integral

1. $\int_a^b f(x)\,dx = -\int_b^a f(x)\,dx$

   When we defined the definite integral $\int_a^b f(x)\,dx$, we implicitly assumed that $a < b$. When we reverse a and b, then Δx changes from $(b-a)/n$ to $(a-b)/n$. Therefore, the result above follows.

2. $\int_a^a f(x)\,dx = 0$

   When $a = b$, then $\Delta x = 0$ and so the result above follows.

The following are a few straightforward properties:

3. $\int_a^b c\,dx = c(b-a)$

4. $\int_a^b [f(x) \pm g(x)]\,dx = \int_a^b f(x)dx \pm \int_a^b g(x)dx$

5. $\int_a^b cf(x)dx = c\int_a^b f(x)dx$, where c is any constant

6. $\int_a^b f(x)dx = \int_a^c f(x)dx + \int_c^b f(x)dx$

   Property 6 can be demonstrated with a diagram (Figure 16.8) where the area from a to b is the sum of the two areas, i.e. $A(x) = A1 + A2$. Additionally, even if $c > b$ the relationship holds because the area from c to b in this case will be negative.

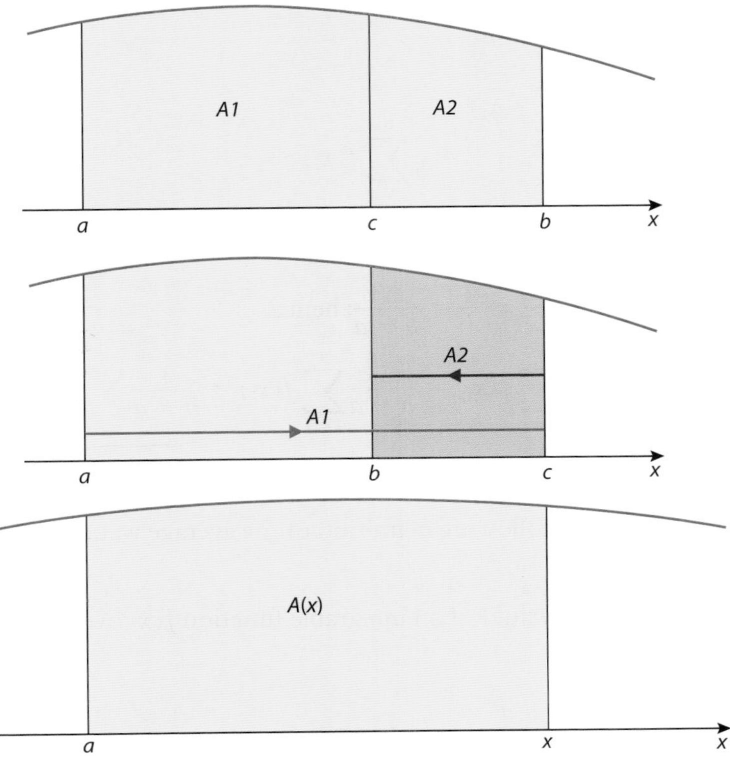

Figure 16.8

Average value of a function

As you recall from statistics, the average value of a variable is

$$\overline{X} = \frac{\sum_{i=1}^{n} X_i}{n}.$$

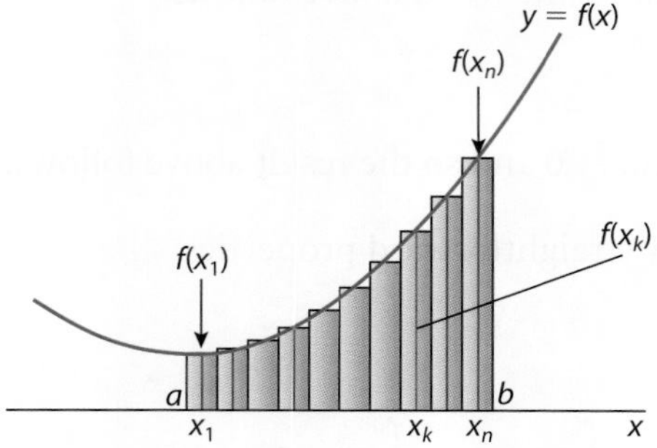

We can also think of the average value of a function in the same manner. Consider a continuous function $f(x)$ defined over a closed interval $[a, b]$. We partition this interval into n sub-intervals of equal length in a fashion similar to the previous discussion. Each interval has a length

$$\Delta x = \frac{b-a}{n}.$$

Now, the average value of $f(x)$ can be defined as

$$av(f) = \frac{f(x_1) + f(x_2) + \ldots + f(x_n)}{n}.$$

Written in sigma notation:

$$av(f) = \frac{\sum_{k=1}^{n} f(x_k)}{n} = \frac{1}{n}\sum_{k=1}^{n} f(x_k)$$

However,

$$\Delta x = \frac{b-a}{n} \Rightarrow \frac{1}{n} = \frac{\Delta x}{b-a}; \text{ hence,}$$

$$av(f) = \frac{1}{n}\sum_{k=1}^{n} f(x_k) = \frac{\Delta x}{b-a}\sum_{k=1}^{n} f(x_k) = \underbrace{\frac{1}{b-a}\sum_{k=1}^{n} f(x_k)\Delta x}_{\text{A Riemann sum for } f \text{ on } [a,\, b]}$$

This leads us to the following definition of the average value of a function $f(x)$ over an interval $[a, b]$.

The average (mean value) of an integrable function $f(x)$ over an interval $[a, b]$ is given by

$$av(f) = \frac{1}{b-a}\int_a^b f(x)\,dx.$$

Max–min inequality

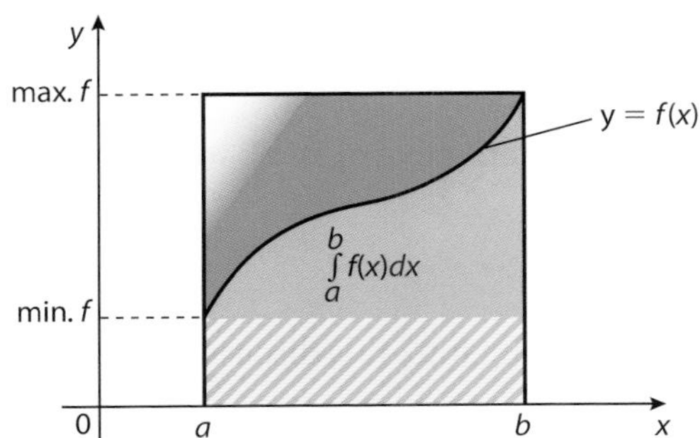

If *max. f* and *min. f* represent the maximum and minimum values of a non-negative continuous differentiable function $f(x)$ over an interval $[a, b]$, then the area under the curve lies between the area of the rectangle with base $[a, b]$ and the *min. f* as height and the rectangle with *max. f* as height.

That is,

$$(b - a)\min. f \leq \int_a^b f(x)dx \leq (b - a)\max. f.$$

With the assumption that $b > a$, this in turn is equivalent to

$$\min. f \leq \frac{1}{b - a}\int_a^b f(x)dx \leq \max. f.$$

Now using the intermediate value theorem we can ascertain that there is at least one point $c \in [a, b]$ where $f(c) = \frac{1}{b - a}\int_a^b f(x)dx$.

The value $f(c)$ in this theorem is in fact the average value of the function.

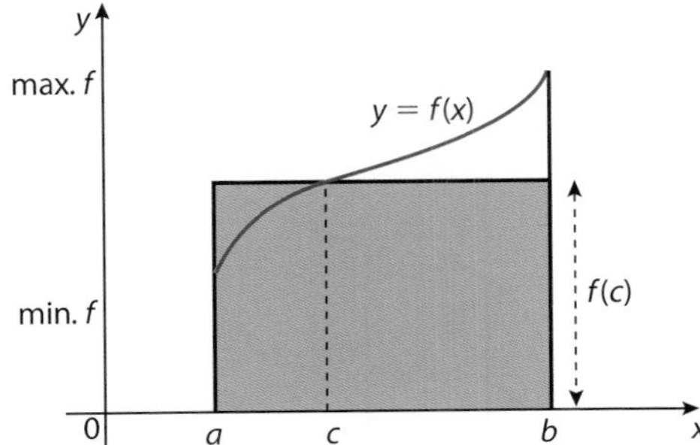

The first fundamental theorem of integral calculus

Our understanding of the definite integral as the area under the curve for $f(x)$ helps us establish the basis for the fundamental theorem of integral calculus.

In the definition of definite integral, let us make the upper limit a variable, say x. Then we will call the area between a and x, $A(x)$, i.e.

$$A(x) = \int_a^x f(t)dt.$$

Consequently,

$$A(x+h) = \int_a^{x+h} f(t)dt.$$

Now, if we want to find the derivative of $A(x)$, we evaluate.

$$\lim_{h \to 0} \frac{A(x+h) - A(x)}{h}.$$

Using the properties of definite integrals discussed earlier, we have:

$$\begin{aligned} A(x+h) - A(x) &= \int_a^{x+h} f(t)dt - \int_a^x f(t)dt \\ &= \int_x^a f(t)dt + \int_a^{x+h} f(t)dt \\ &= \int_x^{x+h} f(t)dt \end{aligned}$$

Therefore,

$$\lim_{h \to 0} \frac{A(x+h) - A(x)}{h} = \lim_{h \to 0} \frac{\int_x^{x+h} f(t)dt}{h} = \lim_{h \to 0} \frac{1}{h} \int_x^{x+h} f(t)dt.$$

Looking at this result and what we established about the average value of $f(x)$ over the interval $[x, x+h]$ we can conclude that there is a point $c \in [x, x+h]$ such that

$$f(c) = \frac{1}{h} \int_x^{x+h} f(t)dt.$$

What happens to c as h approaches 0?

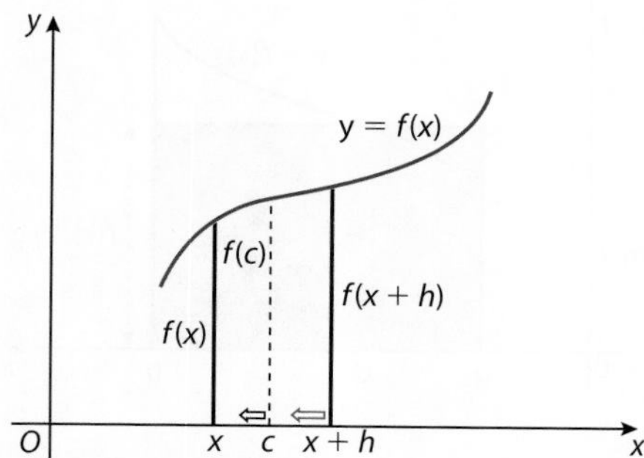

Answer: as h approaches 0, $x + h$ must approach x. This means, we are 'squeezing' c between x and a number approaching x. So, c must also approach x. That is,

$f(c) = f(x)$, and consequently

$$\lim_{h \to 0} \frac{A(x+h) - A(x)}{h} = \lim_{h \to 0} \frac{1}{h} \int_x^{x+h} f(t)dt = f(c) = f(x).$$

This last equation is stating that

$$\frac{d}{dx}(A(x)) = A'(x) = \frac{d}{dx}\left(\int_a^x f(t)\,dt\right) = f(x).$$

This very powerful statement is called the first fundamental theorem of integral calculus. In essence it says that the processes of integration and derivation are inverses of one another.

Note: It is important to remember that $\int_a^x f(t)\,dt$ is a function of x!

Example 26

Find each of the following.

a) $\frac{d}{dx}\int_{-e}^{x} \sec^2 t\,dt$

b) $\frac{d}{dx}\int_0^x \frac{dt}{1+t^4}$

c) $\frac{d}{dx}\int_x^{\pi} \frac{1}{1+t^4}\,dt$

d) $\frac{d}{dx}\int_0^{2x+x^3} \frac{1}{1+t^4}\,dt$

e) $\frac{d}{dx}\int_x^{2x+x^3} \frac{1}{1+t^4}\,dt$

Solution

a) This is a direct application of the fundamental theorem.

$$\frac{d}{dx}\int_{-e}^{x} \sec^2 t\,dt = \sec^2 x$$

b) This is also straightforward.

$$\frac{d}{dx}\int_0^x \frac{dt}{1+t^4} = \frac{1}{1+x^4}$$

c) In this exercise, we need to rewrite the expression before we perform the calculation.

$$\frac{d}{dx}\int_x^{\pi} \frac{1}{1+t^4}\,dt = \frac{d}{dx}\int_{\pi}^{x} -\frac{1}{1+t^4}\,dt = -\frac{d}{dx}\int_x^{\pi} \frac{1}{1+t^4}\,dt = \frac{-1}{1+x^4}$$

d) Remembering that this is a function of x, and that the upper limit is a function of x, which makes $\int_0^{2x+x^3} \frac{1}{1+t^4}\,dt$ a composite of $\int_0^u \frac{1}{1+t^4}\,dt$ and $u = 2x + x^3$. So, we have to resort to the chain rule!

$$\begin{aligned}\frac{d}{dx}\int_0^{2x+x^3} \frac{1}{1+t^4}\,dt &= \left(\frac{d}{du}\int_0^u \frac{1}{1+t^4}\right)\left(\frac{du}{dx}\right)\\ &= \frac{1}{1+u^4}\cdot\frac{du}{dx}\\ &= \frac{1}{1+(2x+x^3)^4}\cdot(2+3x^2)\\ &= \frac{2+3x^2}{1+(2x+x^3)^4}\end{aligned}$$

e) Again, here we need to rewrite the integral before evaluation.

$$\frac{d}{dx}\int_x^{2x+x^3}\frac{1}{1+t^4}dt = \frac{d}{dx}\left(\int_x^k \frac{1}{1+t^4}dt + \int_k^{2x+x^3}\frac{1}{1+t^4}dt\right)$$

$$= \frac{2+3x^2}{1+(2x+x^3)^4} - \frac{1}{1+x^4}$$

The second fundamental theorem of integral calculus

Recall that $A(x) = \int_a^x f(t)dt$. If $F(x)$ is any anti-derivatives of $f(x)$, then applying what we learned in earlier sections:

$$F(x) = A(x) + c \text{ where } c \text{ is an arbitrary constant.}$$

Now,

$$F(b) = A(b) + c = \int_a^b f(t)dt + c, \text{ and}$$

$$F(a) = A(a) + c = \int_a^a f(t)dt + c = 0 - c, \text{ and hence}$$

$$F(b) - F(a) = \int_a^b f(t)dt + c - c$$

$$= \int_a^b f(t)dt.$$

Second fundamental theorem of calculus

$$\int_a^b f(t)dt = F(b) - F(a)$$

The fundamental theorem is also referred to as the **evaluation theorem.** Also, since we know that $F'(x)$ is the rate of change in $F(x)$ with respect to x and that $F(b) - F(a)$ is the change in y when x changes from a to b, we can reformulate the theorem in words.

The integral of a rate of change is the **total change**.

$$\int_a^b F'(x)dx = F(b) - F(a)$$

Here are a few instances where this applies:

1. If $V'(t)$ is the rate at which a liquid flows into or out of a container at time t, then

$$\int_{t_1}^{t_2} V'(t)dt = V(t_2) - V(t_1)$$

is the change in the amount of liquid in the container between time t_1 and t_2.

2. If the rate of growth of a population is $n'(t)$, then

$$\int_{t_1}^{t_2} n'(t)dt = n(t_2) - n(t_1)$$

is the increase (decrease!) in population during the time period from t_1 to t_2.

3. Displacement situations are described separately later in the chapter.

This theorem has many other applications in calculus and several other fields. It is a very powerful tool to deal with problems of area, volume and work among other applications. In this book, we will apply it to finding areas between functions and volumes of revolution as well as displacement problems.

Notation:

We will use the following notation:

$$\int_a^b f(t)\, dt = F(x)]_a^b = F(b) - F(a)$$

Example 27

a) Evaluate the integral $\int_{-1}^{3} x^5\, dx$.

b) Evaluate the integral $\int_0^4 \sqrt{x}\, dx$.

c) Evaluate the integral $\int_{\pi}^{2\pi} \cos\theta\, d\theta$.

d) Evaluate the integral $\int_1^2 \frac{4 + u^2}{u^3} du$.

Solution

a) $\int_{-1}^{3} x^5\, dx = \frac{x^6}{6}\Big]_{-1}^{3} = \frac{3^6}{6} - \frac{1}{6} = \frac{364}{3}$

b) $\int_0^4 \sqrt{x}\, dx = \frac{2}{3}x^{\frac{3}{2}}\Big]_0^4 = \frac{2}{3}4^{\frac{3}{2}} - 0 = \frac{16}{3}$

c) $\int_{\pi}^{2\pi} \cos\theta\, d\theta = \sin\theta\Big]_{\pi}^{2\pi} = 0 - 0 = 0$

d) $\int_1^2 \frac{4 + u^2}{u^3} du = \int_1^2 \left(\frac{4}{u^3} + \frac{1}{u}\right) du = \left[4 \cdot \frac{u^{-2}}{-2} + \ln|u|\right]_1^2$

$$= \left[-2u^{-2} + \ln u\right]_1^2$$

$$= (-2 \cdot 2^{-2} + \ln 2) - (-2 \cdot 1 + \ln 1) = -\frac{1}{2} + \ln 2 + 2$$

$$= \frac{3}{2} + \ln 2$$

Using substitution with definite integral

In Section 16.1 we discussed the use of substitution to evaluate integrals in cases that are not easily recognized. We established the following rule:

$$\int f(u(x)) \cdot u'(x)dx = \int f(u)du = F(u(x)) + c = F(x) + c$$

When evaluating definite integrals by substitution, two methods are available.

1 Evaluate the indefinite integral first, revert to the original variable, and then use the fundamental theorem. For example, to evaluate

$$\int_0^{\frac{\pi}{3}} \tan^5 x \sec^2 x\, dx,$$

we find the indefinite integral

$$\int \tan^5 x \sec^2 x\, dx = \int u^5 du = \tfrac{1}{6}u^6 = \tfrac{1}{6}\tan^6 x,$$

and then we use the fundamental theorem, i.e.

$$\int_0^{\frac{\pi}{3}} \tan^5 x \sec^2 x\, dx = \tfrac{1}{6}\tan^6 x\Big]_0^{\frac{\pi}{3}} = \tfrac{1}{6}(\sqrt{3})^6 = \tfrac{27}{6} = \tfrac{9}{2}.$$

2 Use the following 'substitution rule' for definite integrals:

$$\int_a^b f(u(x))u'(x)dx = \int_{u(a)}^{u(b)} f(u)du$$

Proof:

If $F(x)$ is an anti-derivative of $f(x)$, then by the fundamental theorem

$$\int_b^a f(u(x))u'(x)dx = F(u(x))\Big]_a^b = F(u(b)) - F(u(a)).$$

Also,

$$\int_{u(a)}^{u(b)} f(u)du = F(u)\Big]_{u(a)}^{u(b)} = F(u(b)) - F(u(a)).$$

Therefore, to evaluate

$$\int_0^{\frac{\pi}{3}} \tan^5 x \sec^2 x\, dx,$$

letting $u = \tan x \Rightarrow u\left(\frac{\pi}{3}\right) = \sqrt{3}$, $u(0) = 0$, and so

$$\int_0^{\frac{\pi}{3}} \tan^5 x \sec^2 x\, dx = \int_0^{\sqrt{3}} u^5 du = \tfrac{1}{6}u^6\Big]_0^{\sqrt{3}} = \frac{9}{2}.$$

Example 28

Evaluate $\int_2^6 \sqrt{4x+1}\, dx$.

Solution

Let $u = 4x + 1$, then $du = 4dx$. The limits of integration are

$u(2) = 9$ and $u(6) = 25$, therefore

$$\int_2^6 \sqrt{4x+1}\, dx = \frac{1}{4}\int_9^{25} \sqrt{u}\, du = \frac{1}{4}\left(\frac{2}{3}u^{\frac{3}{2}}\right)\Big]_9^{25}$$

$$= \frac{1}{6}(125 - 27) = \frac{49}{3}.$$

Observe that using this method, we do not return to the original variable of integration. We simply evaluate the 'new' integral between the appropriate values of u.

Exercise 16.4

In questions 1–42, evaluate the integral.

1 $\int_{-2}^{1} (3x^2 - 4x^3)\, dx$

2 $\int_{2}^{7} 8\, dx$

3 $\int_{1}^{5} \frac{2}{t^3}\, dt$

4 $\int_{2}^{2} (\cos t - \tan t)\, dt$

5 $\int_{1}^{7} \frac{2x^2 - 3x + 5}{\sqrt{x}}\, dx$

6 $\int_{0}^{\pi} \cos \theta\, d\theta$

7 $\int_{0}^{\pi} \sin \theta\, d\theta$

8 $\int_{3}^{1} (5x^4 + 3x^2)\, dx$

9 $\int_{1}^{3} \frac{u^5 + 2}{u^2}\, du$

10 $\int_{1}^{e} \frac{2\, dx}{x}$

11 $\int_{1}^{3} \frac{2x}{x^2 + 2}\, dx$

12 $\int_{1}^{3} (2 - \sqrt{x})^2\, dx$

13 $\int_{0}^{\frac{\pi}{4}} 3 \sec^2 \theta\, d\theta$

14 $\int_{0}^{1} (8x^7 + \sqrt{\pi})\, dx$

15 a) $\int_{0}^{2} |3x| dx$ b) $\int_{-2}^{0} |3x| dx$ c) $\int_{-2}^{2} |3x| dx$

16 $\int_{0}^{\frac{\pi}{2}} \sin 2x\, dx$

17 $\int_{1}^{9} \frac{1}{\sqrt{x}} dx$

18 $\int_{-2}^{2} (e^x - e^{-x}) dx$

19 $\int_{-1}^{1} \frac{dx}{1 + x^2}$

20 $\int_{0}^{\frac{1}{2}} \frac{dx}{\sqrt{1 - x^2}}$

21 $\int_{-1}^{1} \frac{dx}{\sqrt{4 - x^2}}$

22 $\int_{-2}^{0} \frac{dx}{4 + x^2}$

23 $\int_{0}^{4} \frac{x^3\, dx}{\sqrt{x^2 + 1}}$

24 $\int_{1}^{\sqrt{e}} \frac{\sin(\pi \ln x)}{x}\, dx$

25 $\int_{e}^{e^2} \frac{dt}{t \ln t}$

26 $\int_{-1}^{2} 3x\sqrt{9 - x^2}\, dx$

27 $\int_{-\frac{\pi}{3}}^{\frac{2\pi}{3}} \frac{\sin x}{\sqrt{3 + \cos x}}\, dx$

28 $\displaystyle\int_e^{e^2} \frac{\ln x}{x}dx$

29 $\displaystyle\int_1^{\sqrt{3}} \frac{\sqrt{\arctan x}}{1+x^2}dx$

30 $\displaystyle\int_1^{\sqrt{e}} \frac{dx}{x\sqrt{1-(\ln x)^2}}$

31 $\displaystyle\int_{-\ln 2}^{\ln 2} \frac{e^{2x}}{e^{2x}+9}dx$

32 $\displaystyle\int_{\ln 2}^{\ln\left(\frac{2}{\sqrt{3}}\right)} \frac{e^{-2x}dx}{\sqrt{1-e^{-4x}}}$

33 $\displaystyle\int_0^{\frac{\pi}{4}} \sqrt{\tan x}\sec^2 x\, dx$

34 $\displaystyle\int_0^{\sqrt{\pi}} 7x\cos x^2\, dx$

35 $\displaystyle\int_{\pi^2}^{4\pi^2} \frac{\sin\sqrt{x}}{\sqrt{x}}dx$

36 $\displaystyle\int_0^1 \frac{\sqrt{3}x}{\sqrt{4-3x^4}}dx$

37 $\displaystyle\int_0^{\frac{2}{\sqrt{3}}} \frac{dx}{9+4x^2}$

38 $\displaystyle\int_1^{\sqrt{2}} \frac{x\,dx}{3+x^4}$

39 $\displaystyle\int_0^{\frac{\pi}{6}} (1-\sin 3t)\cos 3t\, dt$

40 $\displaystyle\int_0^{\frac{\pi}{4}} e^{\sin 2\theta}\cos 2\theta\, d\theta$

41 $\displaystyle\int_0^{\frac{\pi}{8}} (3+e^{\tan 2t})\sec^2 2t\, dt$

42 $\displaystyle\int_0^{\sqrt{\ln \pi}} 4te^{t^2}\sin(e^{t^2})dt$

In questions 43–47, find the average value of the given function over the given interval.

43 x^4, $[1, 2]$

44 $\cos x$, $\left[0, \frac{\pi}{2}\right]$

45 $\sec^2 x$, $\left[\frac{\pi}{6}, \frac{\pi}{4}\right]$

46 e^{-2x}, $[0, 4]$

47 $\dfrac{e^{3x}}{1+e^{6x}}$, $\left[\dfrac{-\ln 3}{6}, 0\right]$

In questions 48–55, find the indicated derivative.

48 $\displaystyle\frac{d}{dx}\int_2^x \frac{\sin t}{t}dt$

49 $\displaystyle\frac{d}{dt}\int_t^3 \frac{\sin x}{x}dx$

50 $\displaystyle\frac{d}{dx}\int_{x^2}^0 \frac{\sin t}{t}dt$

51 $\displaystyle\frac{d}{dx}\int_0^{x^2} \frac{\sin u}{u}du$

52 $\displaystyle\frac{d}{dt}\int_{-\pi}^t \frac{\cos y}{1+y^2}dy$

53 $\displaystyle\frac{d}{dx}\int_{ax}^{bx} \frac{dt}{5+t^4}$

54 $\displaystyle\frac{d}{d\theta}\int_{\sin\theta}^{\cos\theta} \frac{1}{1-x^2}dx$

55 $\displaystyle\frac{d}{dx}\int_5^{x^{\frac{1}{4}}} e^{t^4+3t^2}dt$

56 Does the function $F(x) = \displaystyle\int_0^{2x-x^2} \cos\left(\frac{1}{1+t^2}\right)dt$ have an extreme value?

57 a) Find $\displaystyle\int_0^k \frac{dx}{3x+2}$, giving your answer in terms of k.

b) Given that $\displaystyle\int_0^k \frac{dx}{3x+2} = 1$, calculate the value of k.

58 Given that $p, q \in \mathbb{N}$, show that

$$\int_0^1 x^p(1-x)^q dx = \int_0^1 x^q(1-x)^p dx.$$

Do not attempt to evaluate the integrals.

59 Given that $k \in \mathbb{N}$, evaluate the integral.

a) $\int x(1-x)^k\,dx$ b) $\int_0^1 x(1-x)^k dx$

60 Let $F(x) = \int_3^x \sqrt{5t^2+2}\,dt$. Find

a) $F(3)$ b) $F'(3)$ c) $F''(3)$

61 Show that the function

$$f(x) = \int_x^{3x} \frac{dt}{t}$$

is constant over the set of positive real numbers.

16.5 Integration by method of partial fractions (Optional)

In this section, we will see how rational functions with polynomial denominators can be integrated. For example, if we were to find the indefinite integral $\int \frac{x+1}{x^2+5x+6}dx$, we first decompose the integrand into partial fractions and then the integration process would be straightforward.

$$\frac{x+1}{x^2+5x+6} \equiv \frac{a}{x+2} + \frac{b}{x+3}$$

(See Section 3.6 for details.)

After solving for a and b we can perform the integration:

$$\int \frac{x+1}{x^2+5x+6} \equiv \int\left(\frac{-1}{x+2} + \frac{2}{x+3}\right)dx = -\ln|x+2| + 2\ln|x+3| + c$$
$$= \ln\left|\frac{(x+3)^2}{x+2}\right| + c$$

Example 29

Find the indefinite integral $\int \frac{3x-1}{x^2+4x+4}dx$.

Solution

Using partial fractions will make the work simpler than otherwise.

From Example 42 of Section 3.6 we know:

$$\frac{3x-1}{x^2+4x+4} = \frac{3}{x+2} - \frac{7}{(x+2)^2}$$

Hence, the integral can be rewritten as:

$$\int \frac{3x-1}{x^2+4x+4}dx = \int \frac{3}{x+2}dx - \int \frac{7}{(x+2)^2}dx$$

These two integrals can be found by inspection, giving:

$$\int \frac{3x-1}{x^2+4x+4}dx = 3\ln|x+2| + \frac{7}{x+2} + c$$

Example 30

Find the indefinite integral $\int \frac{2}{x^3+2x^2+2x}dx$.

Solution

Again, from Example 43 of Section 3.6, we have:

$$\frac{2}{x^3+2x^2+2x} = \frac{1}{x} - \frac{x+2}{x^2+2x+2}$$

Hence, we can write the integral as:

$$\int \frac{2}{x^3+2x^2+2x}dx = \int \frac{dx}{x} - \frac{x+2}{x^2+2x+2}dx = \int \frac{dx}{x} - \int \frac{x+1+1}{x^2+2x+2}dx$$

$$= \int \frac{dx}{x} - \frac{1}{2}\int \frac{2x+4}{x^2+2x+2}dx = \int \frac{dx}{x} - \frac{1}{2}\int \frac{2x+4}{(x+1)^2+1}dx$$

$$= \ln|x| - \frac{1}{4}\ln(x^2+2x+2) - \arctan(x+1) + c$$

Example 31

Find the indefinite integral $\int \frac{5x^2+16x+17}{2x^3+9x^2+7x-6}dx$.

Solution

Again from Example 41 of Section 3.6 we have:

$$\int \frac{5x^2+16x+17}{2x^3+9x^2+7x-6}dx = \int \frac{3}{2x-1}dx - \int \frac{1}{x+2}dx + \int \frac{2}{x+3}dx$$

$$= \frac{3}{2}\ln|2x-1| - \ln|x+2| + 2\ln|x+3| + c$$

Example 32

Evaluate $\int \frac{3x-1}{x^3+8}dx$.

Solution

We first factorize the denominator:

$$x^3+8 = (x+2)(x^2-2x+4)$$

Now, by using partial fractions we have:

$$\frac{3x-1}{x^3+8} = \frac{a}{x+2} + \frac{bx+c}{x^2-2x+4}$$

Solving for a, b, and c will yield:

$$3x-1 \equiv a(x^2-2x+4) + (bx+c)(x+2)$$
$$= (a+b)x^2 + (2b-2a+c)x + 4a + 2c$$

This implies: $\begin{cases} a + b = 0 \\ 2b - 2a + c = 3 \\ 4a + 2c = -1 \end{cases}$

Solving this system of equations will yield:

$$a = \frac{-7}{12}, b = \frac{7}{12}, c = \frac{2}{3}$$

Therefore,

$$\int \frac{3x - 1}{x^3 + 8} dx = \int \frac{-\frac{7}{12}}{x + 2} dx + \int \frac{\frac{7}{12}x + \frac{2}{3}}{x^2 - 2x + 4} dx$$

Finally, using what you learned so far you can verify the answer to be:

$$\int \frac{3x - 1}{x^3 + 8} dx = -\frac{7}{12} \ln|x+2| + \frac{7}{24} \ln (x^2 - 2x + 4) - \frac{5\sqrt{3}}{12} \arctan\left(\frac{x - 1}{\sqrt{3}}\right) + c$$

Summary of procedures

In this book we will only consider five general cases. They are outlined below.

Possible cases for partial fractions

1 **Denominator is a quadratic** that factorises into two distinct linear factors, and numerator $p(x)$ is a constant or linear.

$$\frac{p(x)}{(ax + b)(cx + d)} = \frac{A}{ax + b} + \frac{B}{cx + d}$$

2 **Denominator is a quadratic** that factorises into two repeated linear factors, and numerator $p(x)$ is a constant or linear.

$$\frac{p(x)}{(ax + b)^2} = \frac{A}{ax + b} + \frac{B}{(ax + b)^2}$$

3 **Denominator is a cubic** that factorises into three repeated linear factors, and numerator $p(x)$ is a constant, linear or quadratic.

$$\frac{p(x)}{(ax + b)^3} = \frac{A}{ax + b} + \frac{B}{(ax + b)^2} + \frac{C}{(ax + b)^3}$$

4 **Denominator is a cubic** that factorises into one linear factor and one quadratic factor (that cannot be factorised), and numerator $p(x)$ is a constant, linear or quadratic.

$$\frac{p(x)}{(ax + b)(cx^2 + dx + e)} = \frac{A}{ax + b} + \frac{Bx + C}{cx^2 + dx + e}$$

5 **Denominator is a cubic** that factorises into three distinct linear factors, and numerator $p(x)$ is a constant, linear or quadratic.

$$\frac{p(x)}{(ax + b)(cx + d)(ex + f)} = \frac{A}{ax + b} + \frac{B}{cx + d} + \frac{C}{ex + f}$$

A consequence of the Fundamental Theorem of Algebra (see margin note in Section 3.3) guarantees that any polynomial with real coefficients can only have factors that are linear or quadratic.

Exercise 16.5

Evaluate each integral.

1 $\int \frac{5x+1}{x^2+x-2}dx$

2 $\int \frac{x+4}{x^2-2}dx$

3 $\int \frac{x+2}{x^2+4x+3}dx$

4 $\int \frac{5x^2+20x+6}{x^3+2x^2+x}dx$

5 $\int \frac{2x^2+x-12}{x^3+5x^2+6x}dx$

6 $\int \frac{4x^2+2x-1}{x^3+x^2}dx$

7 $\int \frac{3}{x^2+x-2}dx$

8 $\int \frac{5-x}{2x^2+x-1}dx$

9 $\int \frac{3x+4}{(x+2)^2}dx$

10 $\int \frac{12}{x^4-x^3-2x^2}dx$

11 $\int \frac{2}{x^3+x}dx$

12 $\int \frac{x+2}{x^3+3x}dx$

13 $\int \frac{3x+2}{x^3+6x}dx$

14 $\int \frac{2x+3}{x^3+8x}dx$

15 $\int \frac{x+5}{x^3-4x^2-5x}dx$

16.6 Areas

We have seen how the area between a curve, defined by $y = f(x)$, and the x-axis can be computed by the integral $\int_a^b f(x)dx$ on an interval $[a, b]$, where $f(x) \geqslant 0$. In this section, we shall find that integration can be used to find the area of more general regions between curves.

Areas between curves of functions of the form $y = f(x)$ and the x-axis

If the function $y = f(x)$ is always above the x-axis, finding the area is a straightforward computation of the integral $\int_a^b f(x)dx$.

Example 33

Find the area under the curve $f(x) = x^3 - x + 1$ and the x-axis over the interval $[-1, 2]$.

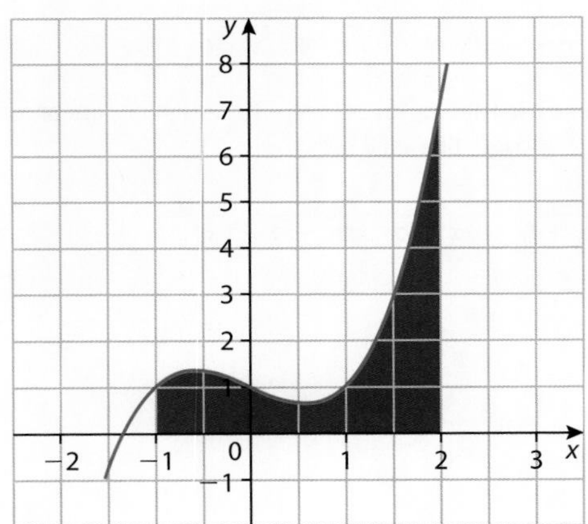

Solution

This area is simply

$$\int_{-1}^{2} (x^3 - x + 1)\, dx = \left[\frac{x^4}{4} - \frac{x^2}{2} + x\right]_{-1}^{2}$$

$$= (4 - 2 + 2) - \left(\frac{1}{4} - \frac{1}{2} - 1\right) = 5\frac{1}{4}.$$

Using your GDC, this is done by simply choosing the 'MATH' menu, then the 'fnInt' menu item.

```
fnInt(X^3-X+1,X,
-1,2)
                5.25
```

Or, you can type in your function and then go to the 'CALC' menu, where you choose '$\int f(x)\, dx$' and type in your integration limits. Here is what you see.

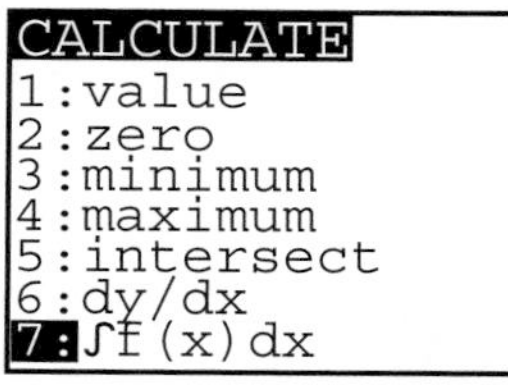

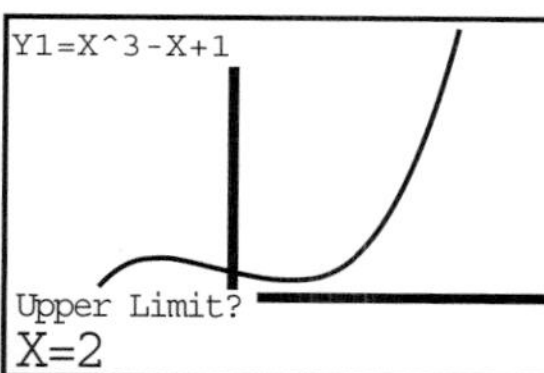

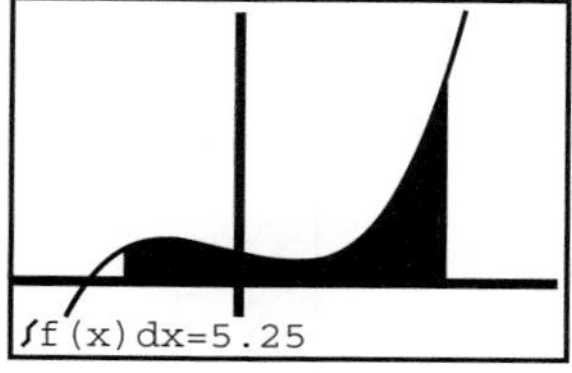

In some cases, you will have to adjust how you work. This is the case when the graph intersects the x-axis. Since you are interested in the area bounded by the curve and the interval $[a, b]$ on the x-axis, you do not want the 'signed' areas to cancel each other. This is why you have to split the process into different sub-intervals where you take the absolute values of the areas found and add them.

Example 34

Find the area under the curve $f(x) = x^3 - x - 1$ and the x-axis over the interval $[-1, 2]$.

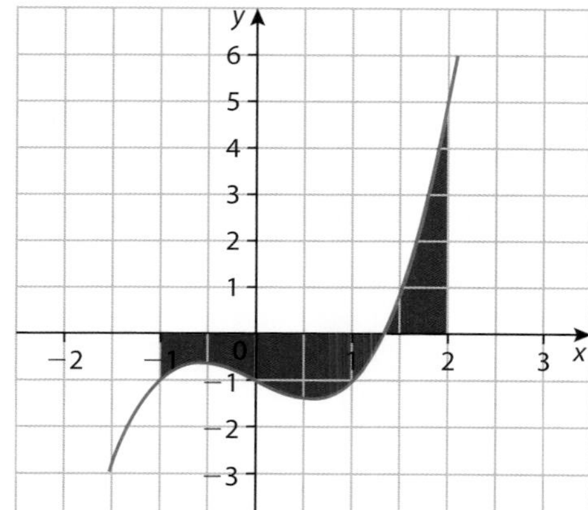

Solution

As you see from the diagram, a part of the graph is below the x-axis, and its area will be negative. If you try to integrate this function without paying attention to the intersection with the x-axis, this is what you get:

$$\int_{-1}^{2} (x^3 - x - 1)\, dx = \left[\frac{x^4}{4} - \frac{x^2}{2} - x\right]_{-1}^{2}$$

$$= (4 - 2 - 2) - \left(\frac{1}{4} - \frac{1}{2} + 1\right) = -\frac{3}{4}$$

This integration has to be split before we start. However, this is a function where you cannot find the intersection point. So, we either use our GDC to find the intersection, or we just take the absolute values of the different parts of the region. This is done by integrating the absolute value of the function:

$$\text{Area} = \int_a^b |f(x)|\,dx$$

$$\text{Hence, area} = \int_{-1}^{2} |(x^3 - x - 1)|\,dx.$$

As we said earlier, this is not easy to find given the difficulty with the x-intercept. It is best if we make use of a GDC.

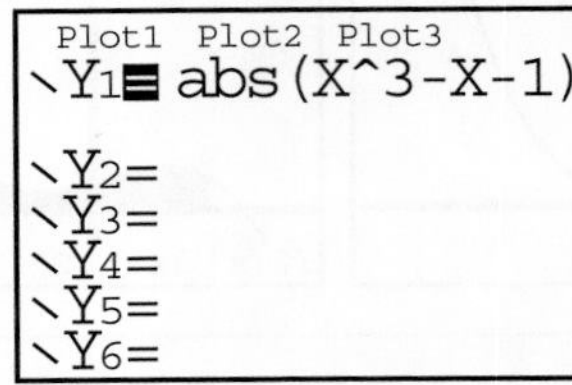

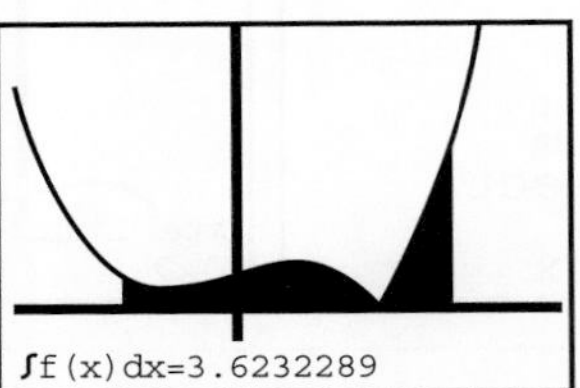

Or, using 'fnInt' directly:

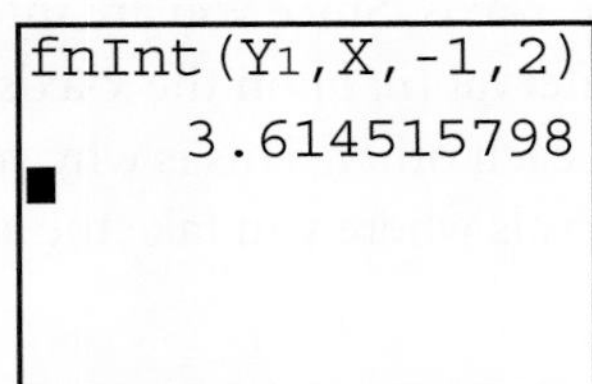

The difference between them is that the latter is more of a rough approximation than the first.

Example 35

Find the area enclosed by the graph of the function $f(x) = x^3 - 4x^2 + x + 6$ and the x-axis.

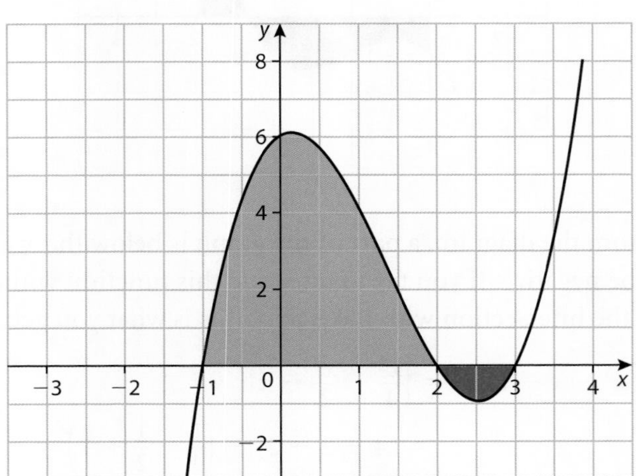

Solution

This function intersects the x-axis at three points where $x = -1, 2$ and 3. To find the area, we split it into two and then add the absolute values:

$$\text{Area} = \int_{-1}^{3} |f(x)|\,dx = \int_{-1}^{2} f(x)\,dx + \int_{2}^{3} (-f(x))\,dx$$

$$= \int_{-1}^{2} (x^3 - 4x^2 + x + 6)\,dx + \int_{2}^{3} (-x^3 + 4x^2 - x - 6)\,dx$$

$$= \left[\frac{x^4}{4} - \frac{4x^3}{3} + \frac{x^2}{2} + 6x\right]_{-1}^{2} + \left[-\frac{x^4}{4} + \frac{4x^3}{3} - \frac{x^2}{2} - 6x\right]_{2}^{3}$$

$$= \frac{45}{4} + \frac{7}{12} = \frac{71}{6}$$

Area between curves

In some practical problems, you may have to compute the area between two curves. Suppose $f(x)$ and $g(x)$ are functions such that $f(x) \geqslant g(x)$ on the interval $[a, b]$, as shown in the diagram. Note that we do not insist that both functions are non-negative, but we begin by showing that case for demonstration purposes.

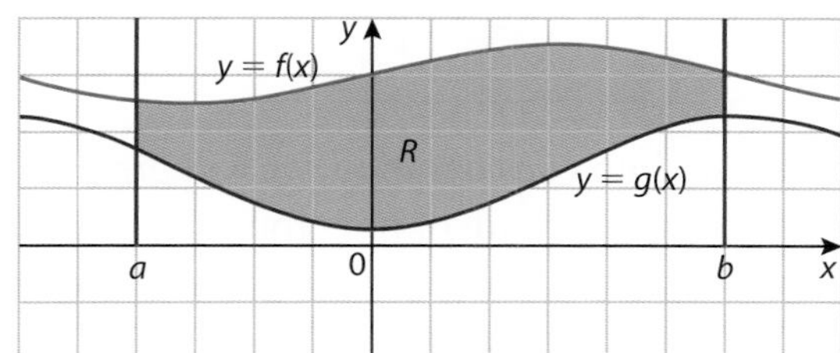

To find the area of the region R between the curves from $x = a$ to $x = b$, we subtract the area between the lower curve $g(x)$ and the x-axis from the area between the upper curve $f(x)$ and the x-axis; that is,

$$\text{Area of } R = \int_{a}^{b} f(x)\,dx - \int_{a}^{b} g(x)\,dx = \int_{a}^{b} [f(x) - g(x)]\,dx.$$

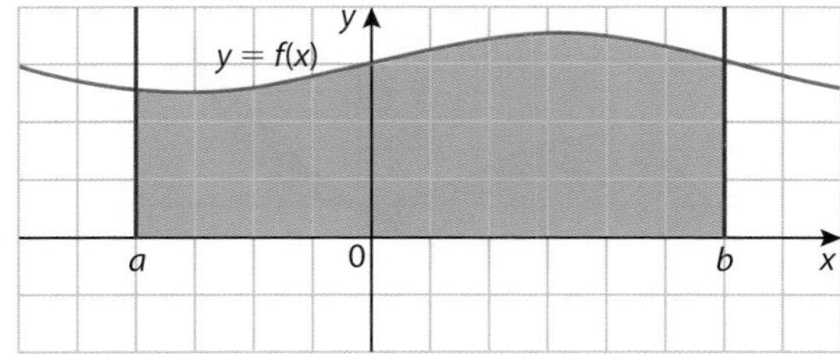

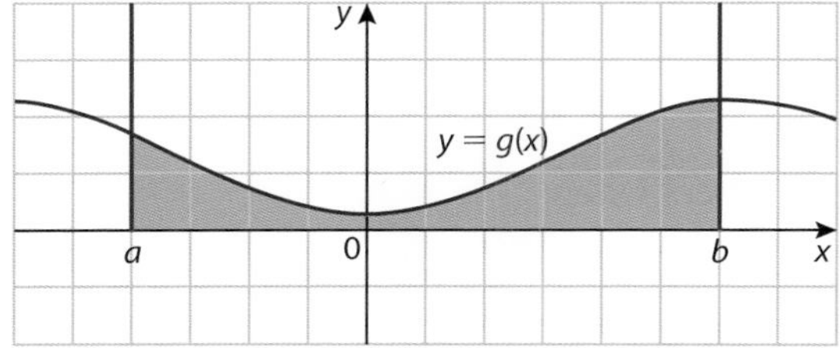

The fact just mentioned applies to all functions, not only positive functions. These facts are used to define the area between curves.

If $f(x)$ and $g(x)$ are functions such that $f(x) \geqslant g(x)$ on the interval $[a, b]$, then the area between the two curves is given by

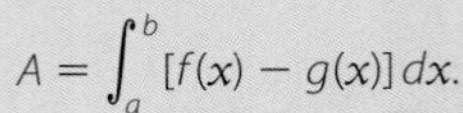

$$A = \int_a^b [f(x) - g(x)]\, dx.$$

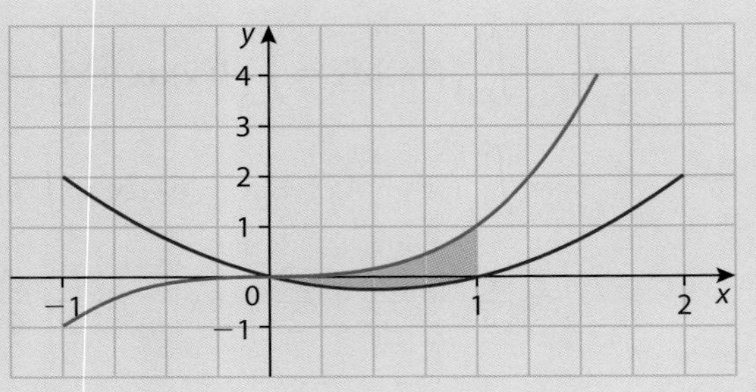

Example 36

Find the area of the region between the curves $y = x^3$ and $y = x^2 - x$ on the interval $[0, 1]$. (See diagram above.)

Solution

$y = x^3$ appears to be higher than $y = x^2 - x$ with one intersection at $x = 0$. Thus, the required area is

$$A = \int_0^1 [x^3 - (x^2 - x)]\, dx = \left[\frac{x^4}{4} - \frac{x^3}{3} + \frac{x^2}{2}\right]_0^1 = \frac{5}{12}.$$

In order to take all cases into consideration, we will present here another case where you must be very careful of how you calculate the area. This is the case where the two functions in question intersect at more than one point. We will clarify this with an example.

Example 37

Find the area of the region bounded by the curves $y = x^3 + 2x^2$ and $y = x^2 + 2x$.

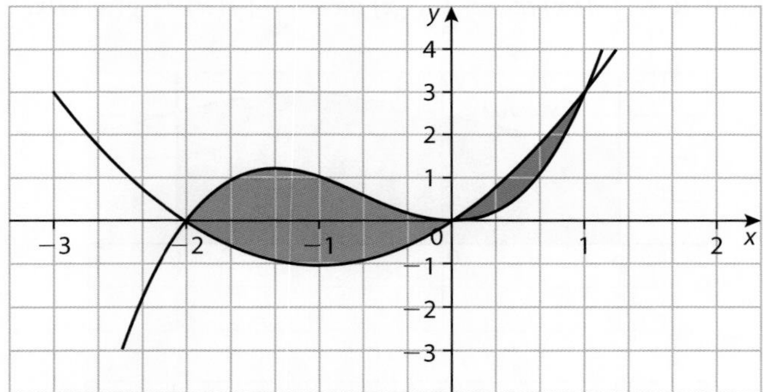

Solution

The two curves intersect when

$$x^3 + 2x^2 = x^2 + 2x \Rightarrow x^3 + x^2 - 2x = 0 \Rightarrow x(x + 2)(x - 1) = 0,$$

i.e. when $x = -2, 0$ or 1.

The area is equal to

$$A = \int_{-2}^{0} [x^3 + 2x^2 - (x^2 + 2x)]\, dx + \int_{0}^{1} [x^2 + 2x - (x^3 + 2x^2)]\, dx$$

$$= \int_{-2}^{0} [x^3 + x^2 - 2x]\, dx + \int_{0}^{1} [-x^2 + 2x - x^3]\, dx$$

$$= \left[\frac{x^4}{4} + \frac{x^3}{3} - x^2\right]_{-2}^{0} + \left[-\frac{x^4}{4} - \frac{x^3}{3} + x^2\right]_{0}^{1}$$

$$= 0 - \left[\frac{16}{4} - \frac{8}{3} - 4\right] + \left[-\frac{1}{4} - \frac{1}{3} + 1\right] - 0 = \frac{37}{12}.$$

This discussion leads us to stating the general expression you should use in evaluating areas between curves.

If $f(x)$ and $g(x)$ are continuous functions on the interval $[a, b]$, the area between the two curves is given by

$$A = \int_{a}^{b} |f(x) - g(x)|\, dx.$$

The above computation can be done with your GDC as follows:

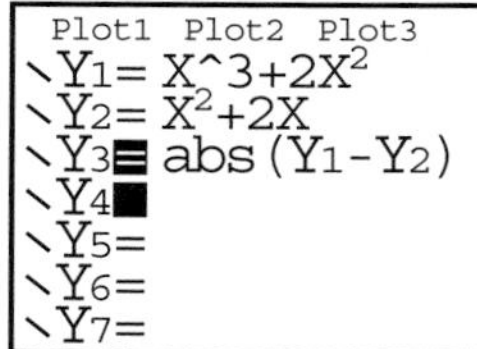

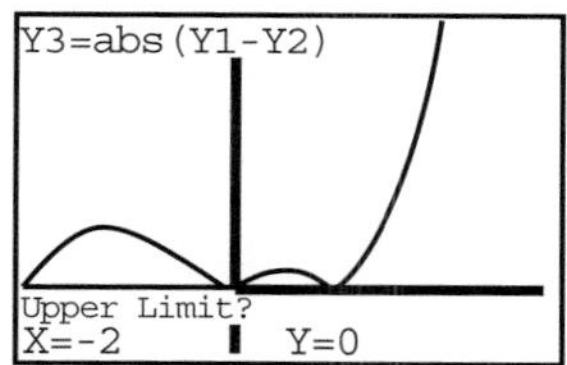

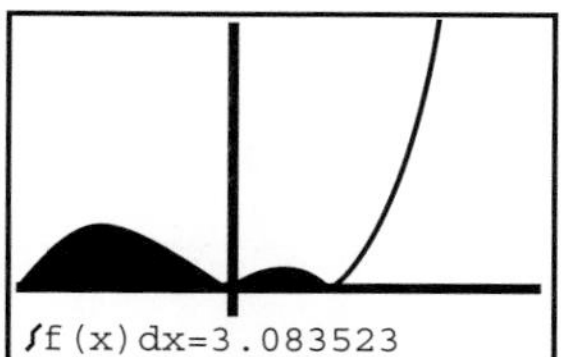

Areas along the y-axis

If we were to find the area enclosed by $y = 1 - x$ and $y^2 = x + 1$, it would be best to treat the region between them by regarding x as a function of y as you see in the graph here.

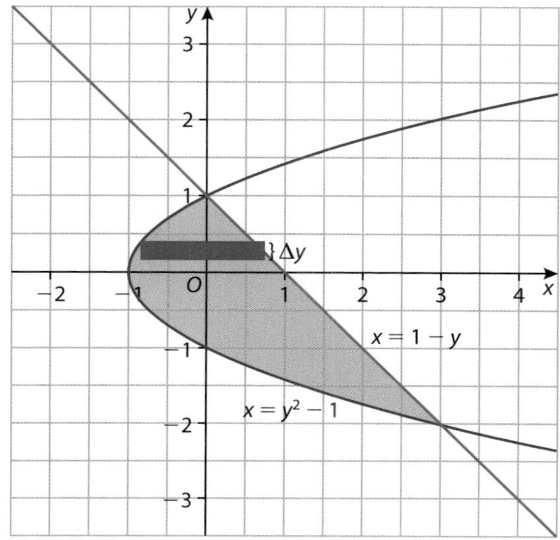

The area of the shaded region can be calculated using the following integral:

$$A(y) = \int_{-1}^{1} |(1 - y) - (y^2 - 1)| dy$$

$$= \int_{-2}^{1} |2 - y - y^2| dy = \left| 2y - \frac{y^2}{2} - \frac{y^3}{3} \right|_{-2}^{1} = \frac{9}{2}$$

If we were to use y as a function of x, then the calculation would have involved calculating the area by dividing the interval into two: $[-1, 0]$ and $[0, 3]$.

In the first part the area is enclosed between $y = \sqrt{x+1}$ and $y = -\sqrt{x+1}$, and the area in the second part is enclosed by $y = 1 - x$ and $y = -\sqrt{x+1}$:

$$2\int_{-1}^{0} \sqrt{x+1}\, dx + \int_{0}^{3} ((1-x) - (-\sqrt{x+1}))\, dx$$

(Calculation is left as an exercise.)

Exercise 16.6

In questions 1–22, sketch the region whose area you are asked for, and then compute the required area. In each question, find the area of the region bounded by the given curves.

1 $y = x + 1, y = 7 - x^2$

2 $y = \cos x, y = x - \frac{\pi}{2}, x = -\pi$

3 $y = 2x, y = x^2 - 2$

4 $y = x^3, y = x^2 - 2, x = 1$

5 $y = x^6, y = x^2$

6 $y = 5x - x^2, y = x^2$

7 $y = 2x - x^3, y = x - x^2$

8 $y = \sin x, y = 2 - \sin x$ (one period)

9 $y = \frac{x}{2}, y = \sqrt{x}, x = 9$

10 $y = \frac{x^4}{10}, y = 3x - x^3$

11 $y = \frac{1}{x}, y = \frac{1}{x^3}, x = 8$

12 $y = 2\sin x, y = \sqrt{3}\tan x, -\frac{\pi}{4} \leqslant x \leqslant \frac{\pi}{4}$

13 $y = x - 1$ and $y^2 = 2x + 6$

14 $x = 2y^2$ and $x = 4 + y^2$

15 $4x + y^2 = 12$ and $y = x$

16 $x - y = 7$ and $x = 2y^2 - y + 3$

17 $x = y^2$ and $x = 2y^2 - y - 2$

18 $y = x^3 + 2x^2, y = x^3 - 2x, x = -3$ and $x = 2$

19 $y = \sec^2 x, y = \sec x \tan x, x = -\frac{\pi}{3}$ and $x = \frac{\pi}{6}$,

20 $y = x^3 + 1$ and $y = (x + 1)^2$

21 $y = x^3 + x$ and $y = 3x^2 - x$

22 $y = 3 - \sqrt{x}$ and $y = \frac{2\sqrt{x} + 1}{2\sqrt{x}}$

23 Find the area of the shaded region.

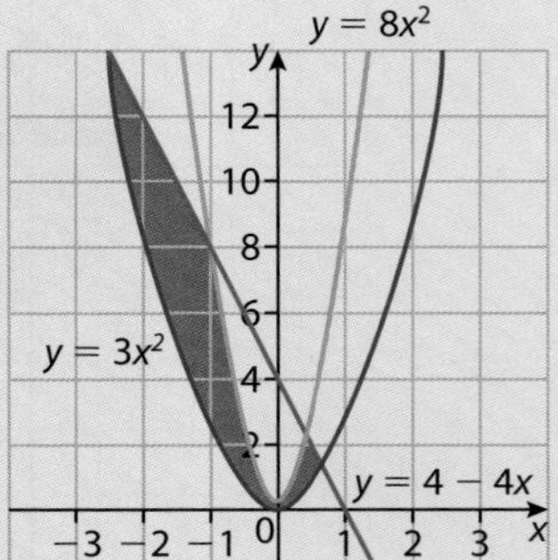

24 Find the area of the region enclosed by $y = e^x$, $x = 0$ and the tangent to $y = e^x$ at $x = 1$.

25 Find the area of the region inside the 'loop' in the graph of the curve $y^2 = x^4(x + 3)$.

26 Find the area enclosed by the curve $y^2 = 2x^2 - 4x^4$.

27 Find the area of the region enclosed by $x = 3y^2$ and $x = 12y - y^2 - 5$.

28 Find the area of the region enclosed by $y = (x - 2)^2$ and $y = x(x - 4)^2$.

29 Find a value for $m > 0$ such that the area under the graph of $y = e^{2x}$ over the interval $[0, m]$ is 3 square units.

30 Find the area of the region bounded by $y = x^3 - 4x^2 + 3x$ and the x-axis.

16.7 Volumes with integrals

Recall that the underlying principle for finding the area of a plane region is to divide the region into thin strips, approximate the area of each strip by the area of a rectangle, and then add the approximations and take the limit of the sum to produce an integral for the area. The same strategy can be used to find the volume of a solid.

• **Hint:** This is an introductory section that will not be examined. It is only used to give you an idea of why we use integrals to find volumes.

The idea is to divide the solid into thin slabs, approximate the volume of each slab, add the approximations and take the limit of the sum to produce an integral of the volume.

Given a solid whose volume is to be computed, we start by taking cross-sections perpendicular to the x-axis as shown in Figure 16.9. Each slab will be approximated by a cylindrical solid whose volume will be equal to the product of its base times its height.

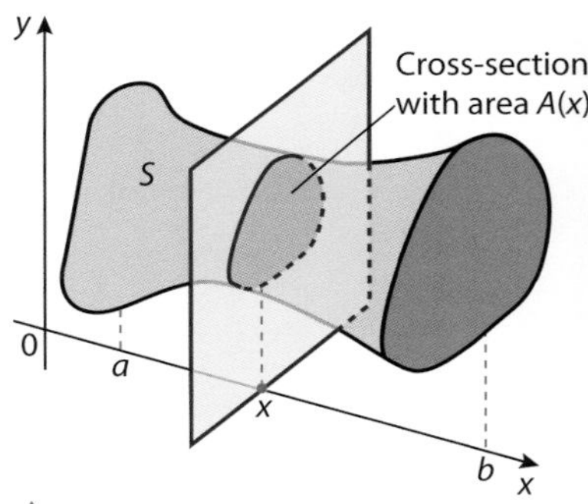

Figure 16.9

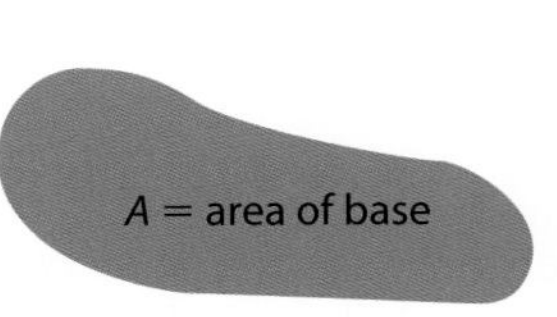

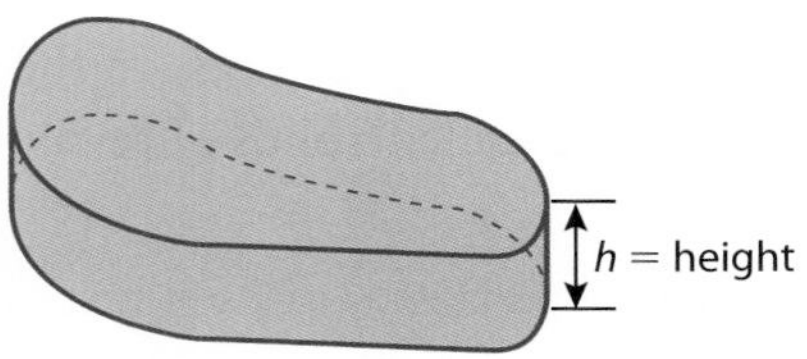

If we call the volume of the slab v_i and the area of its base $A(x)$, then

$$v_i = A(x_i) \cdot h = A(x_i) \cdot \Delta x_i.$$

Using this approximation, the volume of the whole solid can be found by

$$V \approx \sum_{i=1}^{n} A(x_i)\Delta x_i.$$

Taking the limit as n increases and the widths of the sub-intervals approach zero yields the definite integral:

$$V = \lim_{n \to \infty} \sum_{i=1}^{n} A(x_i)\Delta x_i = \int_a^b A(x)\, dx$$

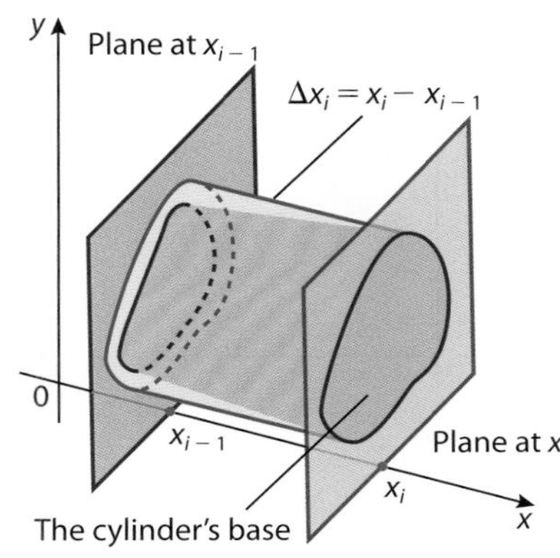

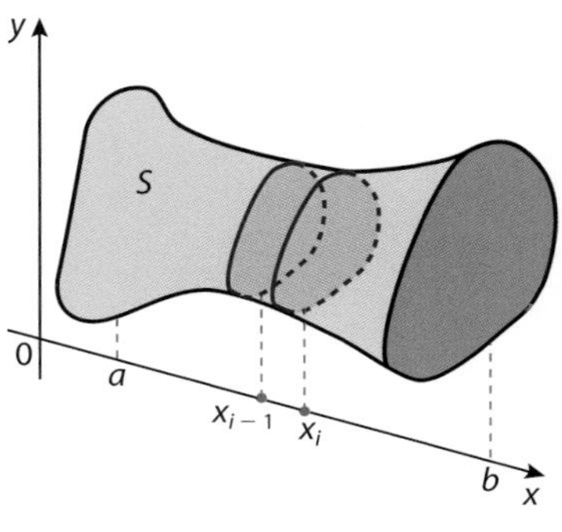

Note: If we place the solid along the y-axis and take the cross-sections perpendicular to that axis, we will arrive at a similar expression for the volume of the solid, i.e.

$$V = \lim_{n\to\infty}\sum_{i=1}^{n} A(y_i)\Delta y_i = \int_a^b A(y)dy$$

Example 38

Find the volume of the solid formed when the graph of the parabola $y = \sqrt{2x}$ over $[0, 4]$ is rotated around the x-axis through an angle of 2π radians, as shown in the diagram.

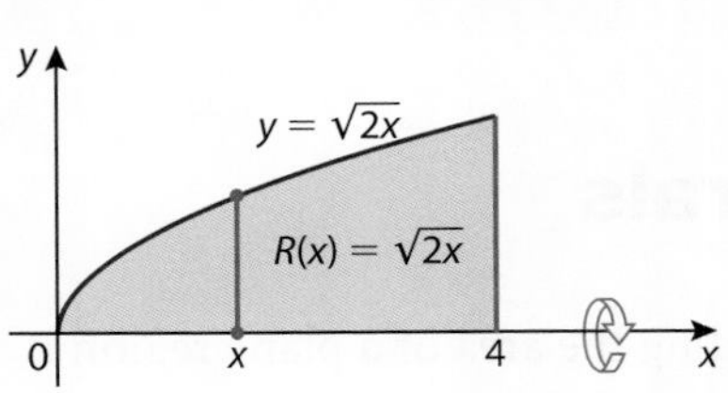

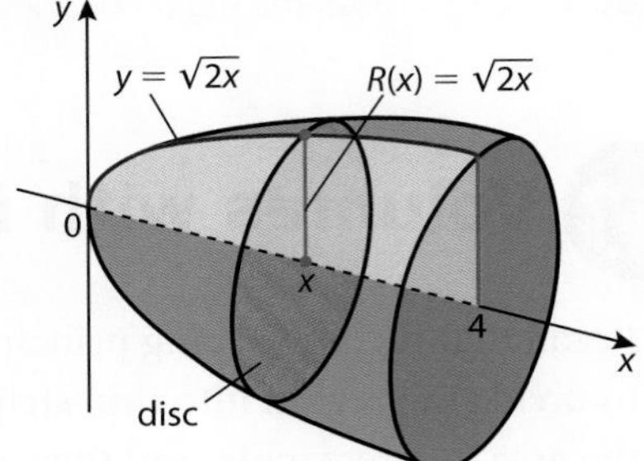

Solution

The cross-section here is a circular disc whose radius is $y = \sqrt{2x}$. Therefore,

$$A(x) = \pi R^2 = \pi(\sqrt{2x}\,)^2 = 2\pi x.$$

The volume is then

$$V = \int_0^4 A(x)\,dx = \int_0^4 2\pi x\,dx = \left[2\pi\frac{x^2}{2}\right]_0^4 = 16\pi \text{ cubic units.}$$

Example 38 above is a special case of the general process for finding volumes of the so-called 'solids of revolution'.

1 If a region is bounded by a closed interval $[a, b]$ on the x-axis and a function $f(x)$ is rotated about the x-*axis*, the volume of the resulting **solid of revolution** is given by

$$V = \int_a^b \pi(f(x))^2\,dx.$$

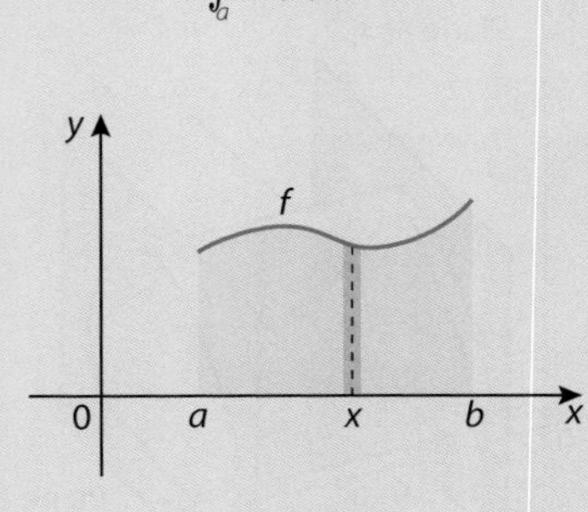

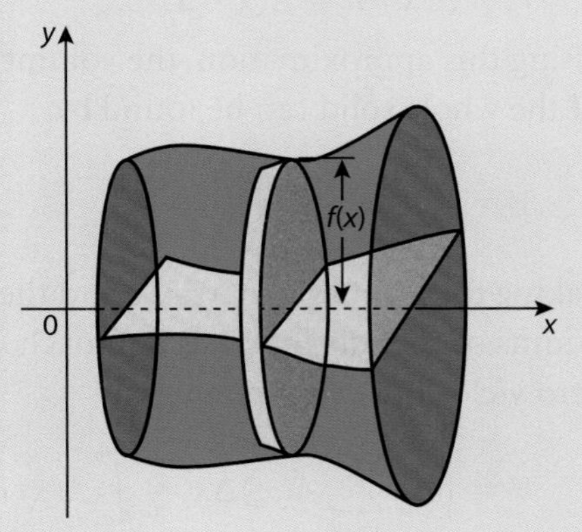

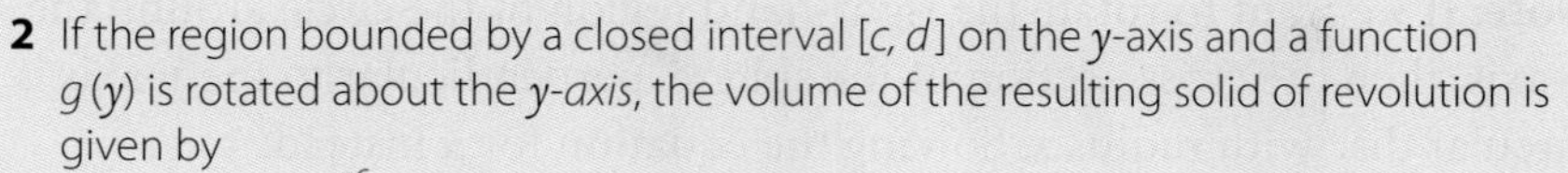

2 If the region bounded by a closed interval $[c, d]$ on the y-axis and a function $g(y)$ is rotated about the y-axis, the volume of the resulting solid of revolution is given by

$$V = \int_d^c \pi(g(y))^2 dy.$$

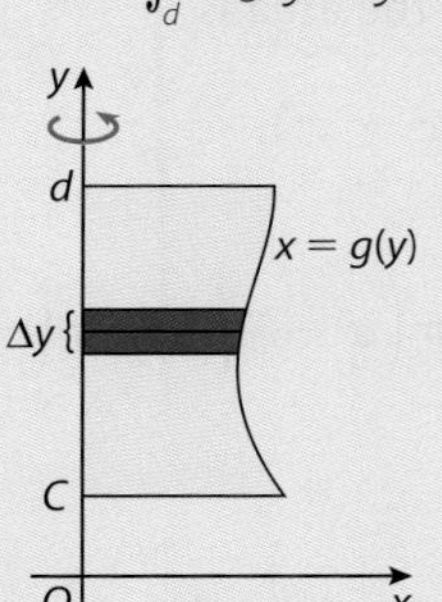

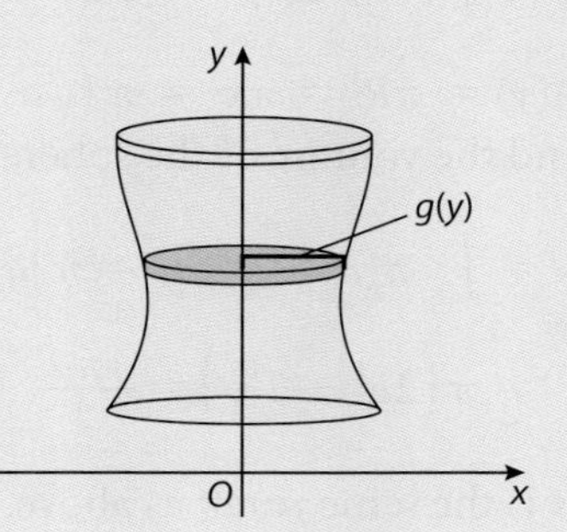

Example 39

Find the volume of a sphere with radius $R = a$.

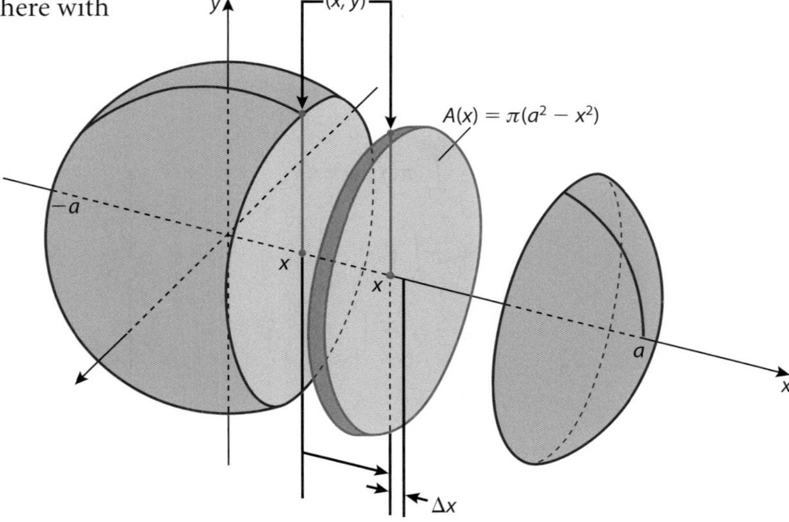

Solution

If we place the sphere with its centre at the origin, the equation of the circle will be

$$x^2 + y^2 = a^2 \Rightarrow y = \pm\sqrt{a^2 - x^2}.$$

The cross-section of the sphere, perpendicular to the x-axis, is a circular disc with radius y, so the area is

$$A(x) = \pi R^2 = \pi y^2 = \pi(\sqrt{a^2 - x^2}) = \pi(a^2 - x^2).$$

So, the volume of the sphere is

$$V = \int_{-a}^{a} \pi(a^2 - x^2)\, dx = \pi\left[a^2x - \frac{x^3}{3}\right]_{-a}^{a}$$

$$= \pi\left(a^3 - \frac{a^3}{3}\right) - \pi\left(-a^3 + \frac{a^3}{3}\right)$$

$$= \pi\left(2a^3 - 2\frac{a^3}{3}\right) = \frac{4\pi a^3}{3}.$$

Note: If we want to rotate the right-hand region of the circle around the y-axis, then the cross-section of the sphere, perpendicular to the y-axis is a circular disc with radius x. Solving the equation for x instead:

$x^2 + y^2 = a^2 \Rightarrow x = \pm\sqrt{a^2 - y^2}$, and hence the area is

$A(y) = \pi R^2 = \pi x^2 = \pi\left(\sqrt{a^2 - y^2}\right)^2 = \pi(a^2 - y^2)$,
and the volume of the sphere is

$$V = \int_{-a}^{a} \pi(a^2 - y^2)dy = \pi\left[a^2y - \frac{y^3}{3}\right]_{-a}^{a} = \pi\left(a^3 - \frac{a^3}{3}\right) - \pi\left(-a^3 + \frac{a^3}{3}\right)$$

$$= \pi\left(2a^3 - 2\frac{a^3}{3}\right) = \frac{4\pi a^3}{3}$$

This is the same result as above.

Example 40

Find the volume of the solid generated when the region enclosed by $y = \sqrt{3x}$, $x = 3$ and $y = 0$ is revolved about the x-axis.

Solution

$$V = \int_0^3 \pi(f(x))^2dx$$

$$= \pi\int_0^3 (\sqrt{3x})^2dx$$

$$= 3\pi\left[\frac{x^2}{2}\right]_0^3 = \frac{27\pi}{2}$$

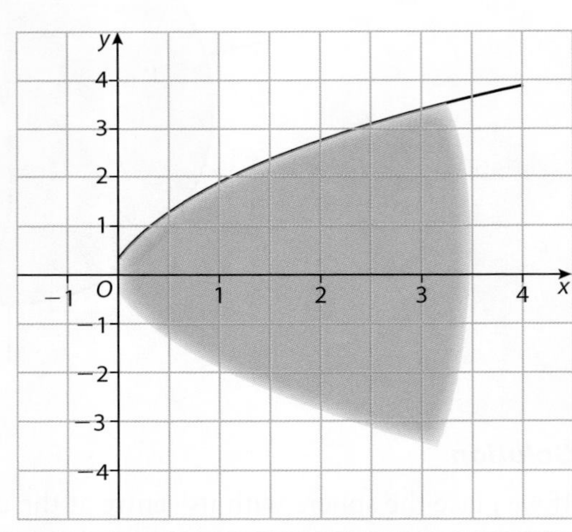

Example 41

Find the volume of the solid generated when the region enclosed by $y = \sqrt{3x}$, $y = 3$ and $x = 0$ is revolved about the y-axis.

Solution

Here, we first find x as a function of y.

$y = \sqrt{3x} \Rightarrow x = \frac{y^2}{3}$, the interval on the y-axis is $[0, 3]$

So, the volume required is

$$V = \int_0^3 \pi\left(\frac{y^2}{3}\right)^2 dy = \frac{\pi}{9}\int_0^3 y^4dy = \frac{\pi}{9}\left[\frac{y^5}{5}\right]_0^3 = \frac{27\pi}{5}.$$

Washers

Consider the region R between two curves, $y = f(x)$ and $y = g(x)$, and from $x = a$ to $x = b$ where $f(x) > g(x)$. Rotating R about the x-axis generates a solid of revolution S. How do we find the volume of S?

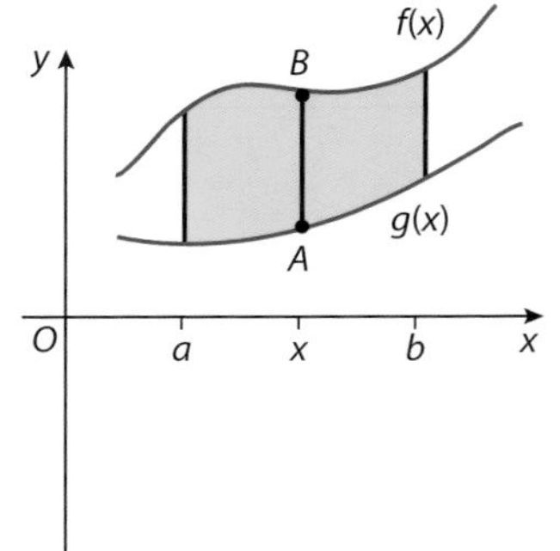

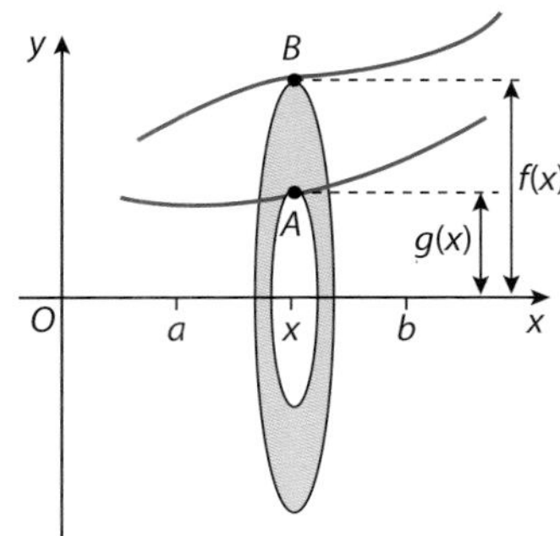

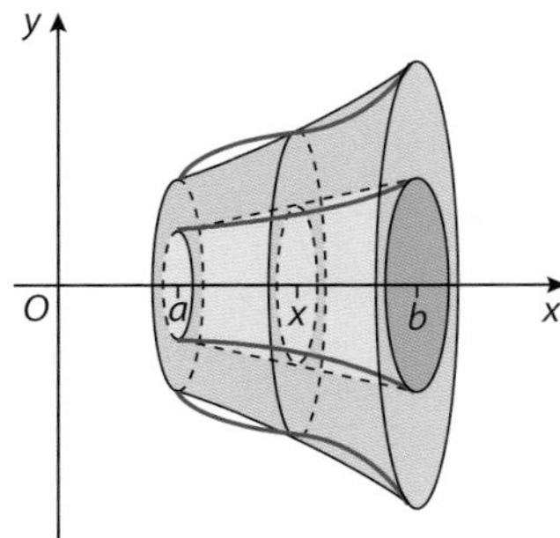

Consider an arbitrary point x in the interval $[a, \mathrm{b}]$. The segment AB represents the difference $f(x) - g(x)$. When we rotate this slice, the cross-section perpendicular to the x-axis is going to look like a 'washer' whose area is

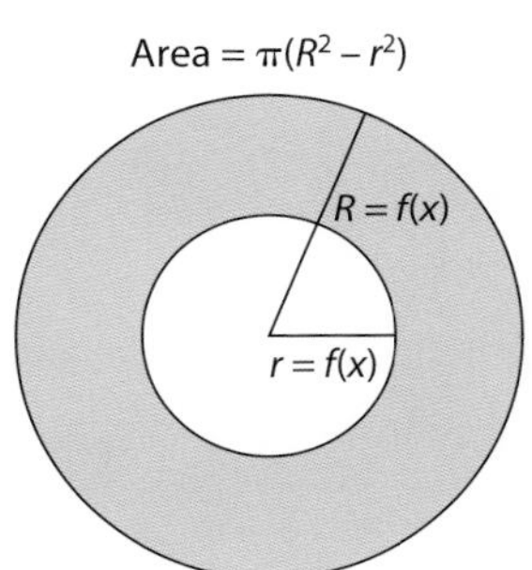

$$A = \pi(R^2 - r^2) = \pi((f(x))^2 - (g(x))^2).$$

So, the volume of S is

$$V = \int_a^b A(x)dx = \pi\int_a^b ((f(x))^2 - (g(x))^2)dx.$$

Note: If you are rotating about the y-axis, a similar formula applies.

$$V = \pi\int_c^d ((p(y))^2 - (q(y))^2)dy$$

Note: To understand the washer more, you can think of it in the following manner: Let P be the solid generated by rotating the curve $y = f(x)$ and Q be the solid generated by rotating the curve $y = g(x)$. Then S can be found by removing the solid of revolution generated by $y = g(x)$ from the solid of revolution generated by $y = f(x)$, as shown.

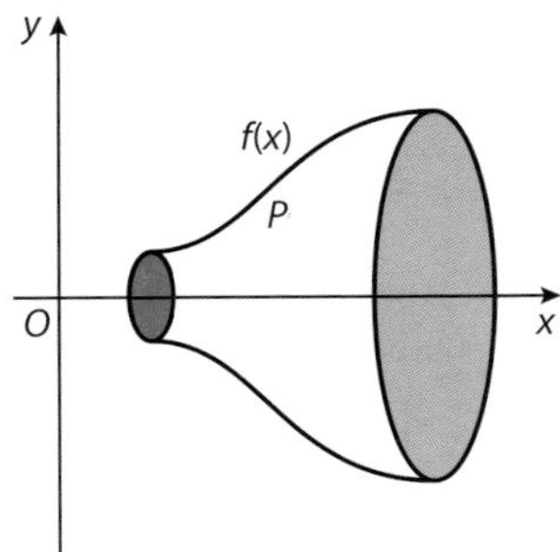

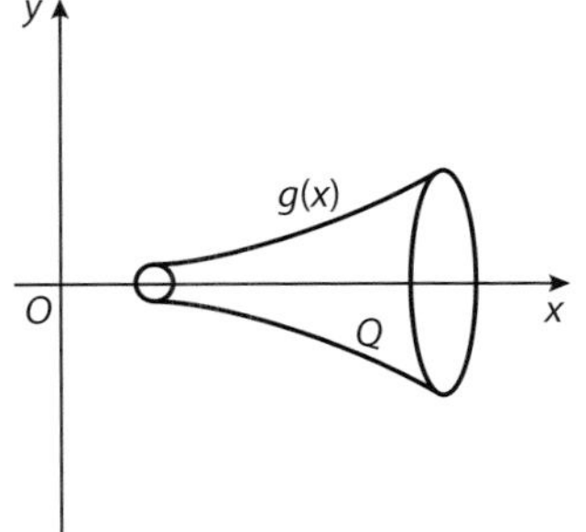

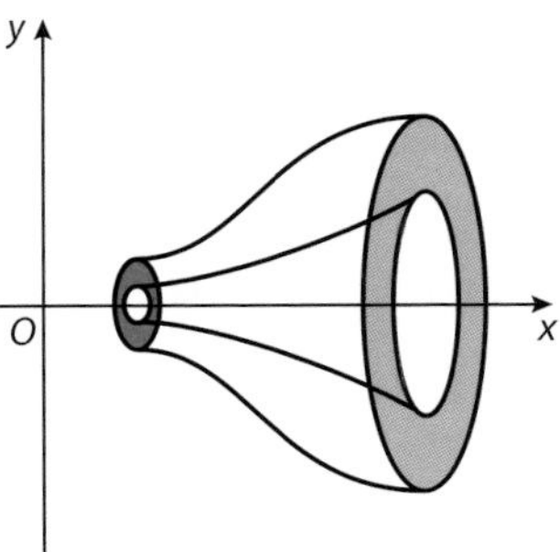

Therefore, volume of S = volume of P − volume of Q. And this justifies the formula:

$$V = \pi\int_a^b (f(x))^2dx - \pi\int_a^b (g(x))^2dx = \pi\int_a^b ((f(x))^2 - (g(x))^2)dx$$

Example 42

The region in the first quadrant between $f(x) = 6 - x^2$ and $h(x) = \frac{8}{x^2}$ is rotated about the x-axis. Find the volume of the generated solid.

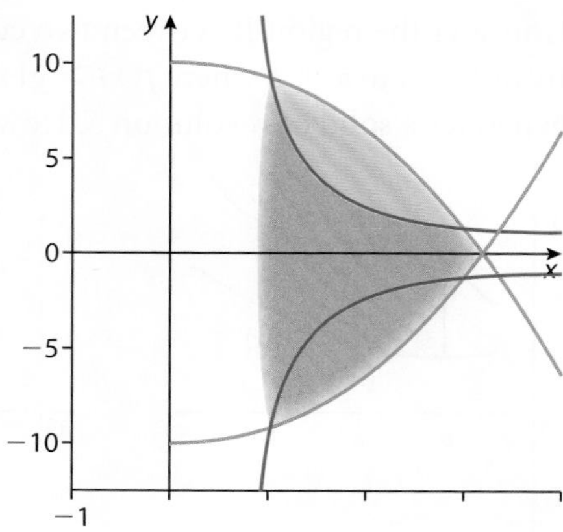

Solution

The rotated region is shown in the diagram. $f(x)$ is larger than $h(x)$in this interval. Moreover, the two curves intersect at:

$$\frac{8}{x^2} = 6 - x^2 \Rightarrow x = \sqrt{2}, x = 2$$

Hence, the volume of the solid of revolution is

$$V = \pi\int_{\sqrt{2}}^{2}\left((6 - x^2)^2 - \left(\frac{8}{x^2}\right)^2\right)dx$$

$$= \pi\int_{\sqrt{2}}^{2}\left(x^4 - 12x^2 + 36 - \frac{64}{x^4}\right)dx$$

$$= \pi\left[\frac{x^5}{5} - 4x^3 + 36x + \frac{64}{3x^3}\right]_{\sqrt{2}}^{2}$$

$$= \frac{736 - 512\sqrt{2}}{15}\pi.$$

An alternative method: Volumes by cylindrical shells

Consider the region R under the curve $y = f(x)$. Rotate R about the y-axis. We divide R into vertical strips of width Δx each as shown. When we rotate a strip around the y-axis, we generate a cylindrical shell of Δx thickness and height $f(x)$. To understand how we get the volume, we can cut the shell vertically as shown and 'unfold' it. The resulting rectangular parallelpiped has length $2\pi x$, height $f(x)$ and thickness Δx.

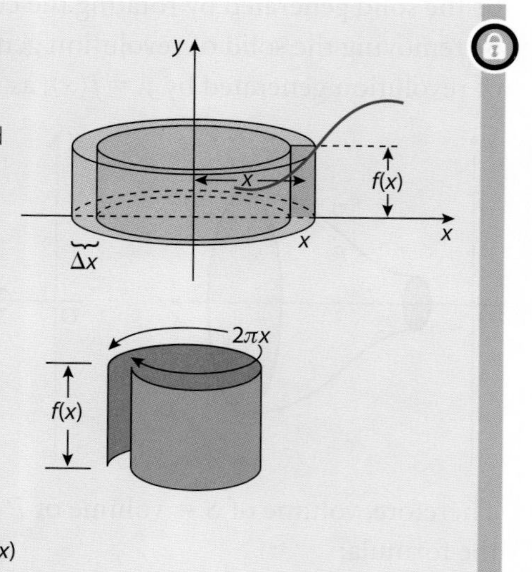

So, the volume of this shell is

$$\begin{aligned}\Delta v_i &= \text{length} \times \text{height} \times \text{thickness}\\ &= (2\pi x) \times f(x) \times \Delta x.\end{aligned}$$

The volume of the whole solid is the sum of the volumes of these shells as the number of shells increases, and consequently

$$V = \lim_{n\to\infty}\sum_{i=1}^{v}\Delta v_i = \lim_{\Delta x\to 0}\sum(2\pi x) \times f(x) \times \Delta x$$

$$= 2\pi\int_a^b xf(x)dx.$$

In many problems involving rotation about the y-axis, this would be more accessible than the disc/washer method.

Example 43

Find the volume of the solid generated when we rotate the region under $f(x) = \frac{2}{1+x^2}$, $x = 0$ and $x = 3$ around the y-axis.

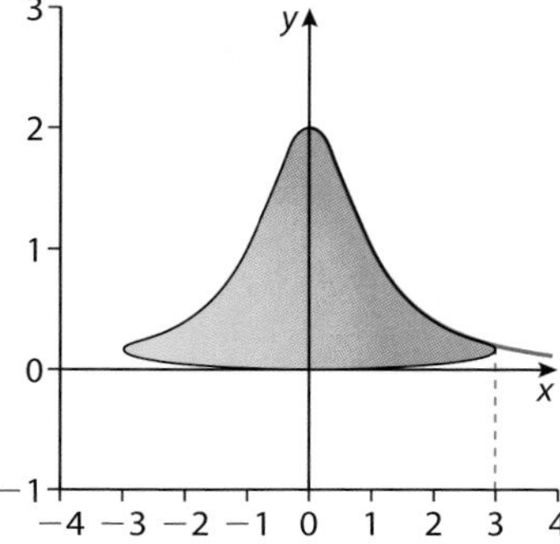

Solution

Using the shell method, we have

$$V = 2\pi\int_0^3 x \times \frac{2}{1+x^2}dx$$

$$= 2\pi\int_0^3 \frac{2x}{1+x^2}dx = 2\pi\int_1^{10}\frac{du}{u}$$

$$= 2\pi[\ln u]_1^{10} = 2\pi \ln 10.$$

Exercise 16.7

In questions 1–19, find the volume of the solid obtained by rotating the region bounded by the given curves about the x-axis. Sketch the region, the solid and a typical disc.

1 $y = 3 - \frac{x}{3}, y = 0, x = 2, x = 3$

2 $y = 2 - x^2, y = 0$

3 $y = \sqrt{16 - x^2}, y = 0, x = 1, x = 3$

4 $y = \frac{3}{x}, y = 0, x = 1, x = 3$

5 $y = 3 - x, y = 0, x = 0$

6 $y = \sqrt{\sin x}, y = 0, 0 \leqslant x \leqslant \pi$

7 $y = \sqrt{\cos x}, y = 0, -\frac{\pi}{2} \leqslant x \leqslant \frac{\pi}{3}$

8 $y = 4 - x^2, y = 0$

9 $y = x^3 + 2x + 1, y = 0, x = 1$

10 $y = -4x - x^2, y = x^2$

11 $y = \sec x, x = \frac{\pi}{4}, x = \frac{\pi}{3}, y = 0$

12 $y = 1 - x^2, y = x^3 + 1$

13 $y = \sqrt{36 - x^2}, y = 4$

14 $x = \sqrt{y}, y = 2x$

15 $y = \sin x, y = \cos x, x = \frac{\pi}{4}, x = \frac{\pi}{2}$

16 $y = 2x^2 + 4, y = x, x = 1, x = 3$

17 $y = \sqrt{x^4 + 1}, y = 0, x = 1, x = 3$

18 $y = 16 - x, y = 3x + 12, x = -1$

19 $y = \frac{1}{x}, y = \frac{5}{2} - x$

20 Find the volume resulting from a rotation of this region about

a) the x-axis

b) the y-axis.

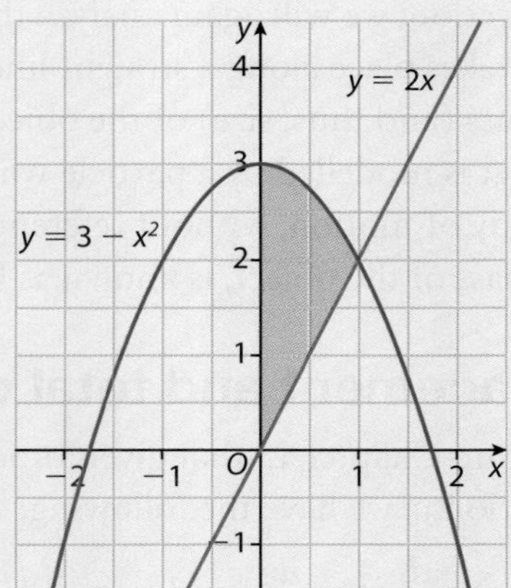

In questions 21–31, find the volume of the solid obtained by rotating the region bounded by the given curves about the y-axis. Sketch the region, the solid and a typical disc/shell.

21 $y = x^2, y = 0, x = 1, x = 3$

22 $y = x, y = \sqrt{9 - x^2}, x = 0$

23 $y = x^3 - 4x^2 + 4x, y = 0$

24 $y = \sqrt{3x}, x = 5, x = 11, y = 0$

25 $y = x^2, y = \frac{2}{1 + x^2}$

26 $y = \sqrt{x^2 + 2}, x = 3, y = 0, x = 0$

27 $y = \frac{7x}{\sqrt{x^3 + 7}}, x = 3, y = 0$

28 $y = \sin x, y = \cos x, x = \frac{\pi}{4}, x = \frac{\pi}{2}$

29 $y = 2x^2 + 4, y = x, x = 1, x = 3$

30 $y = \sin(x^2), y = 0, x = 0, x = \sqrt{\pi}$

31 $y = 5 - x^3, y = 5 - 4x$

16.8 Modelling linear motion

In previous sections of this text, we have examined problems involving displacement, velocity and acceleration of a moving object. In different sections of Chapter 13, we applied the fact that a derivative is a rate of change to express velocity and acceleration as derivatives. Even though our earlier work on motion problems involved an object moving in one, two or even three dimensions, our mathematical models considered the object's motion occurring only along a straight line. For example, projectile motion (e.g. a ball being thrown) is often modelled by a position function that simply gives the height (displacement) of the object. In that way, we are modelling the motion as if it were restricted to a vertical line.

In this section, we will again analyze the motion of an object as if its motion takes place along a straight line in space. This can only make sense if the mass (and thus, size) of the object is not taken into account. Hence, the object is modelled by a particle whose mass is considered to be zero. This study of motion, without reference either to the forces that cause it or to the mass of the object, is known as **kinematics**.

Displacement and total distance travelled

Recall from Chapter 13 that given time t, displacement s, velocity v and acceleration a, we have the following:

$$v = \frac{ds}{dt}, a = \frac{dv}{dt}, \text{ and } a = \frac{d}{dt}\left(\frac{ds}{dt}\right) = \frac{d^2s}{dt^2}$$

Let's review some of the essential terms we use to describe an object's motion.

Position, distance and displacement

- The **position s** of a particle, with respect to a chosen axis, is a measure of how far it is from a fixed point (usually the origin) *and* of its direction relative to the fixed point.
- The **distance $|s|$** of a particle is a measure of how far it is from a fixed point (usually the origin) and does *not* indicate direction. Thus, distance is the magnitude of position and is always positive.
- The **displacement** is the *change* in position. The displacement of an object may be positive, negative or zero, depending on its motion.

It is important to understand the difference between displacement and distance travelled. Consider a couple of simple examples of an object moving along the x-axis.

1. In this first example, assume that the object does not change direction during the interval $0 \leqslant t \leqslant 5$. In other words, its velocity does not change from positive to negative or from negative to positive. If the position of the object at $t = 0$ is $x = 2$ and then the object moves so that at $t = 5$ its position is $x = -3$, its displacement, or change in position, is -5 because the object changed its position by 5 units in the negative direction. This can be calculated by (final position) $-$ (initial position) $= -3 - 2 = -5$. However, the distance travelled would be the absolute value of displacement, calculated by $|\text{final position} - \text{initial position}| = |-3 - 2| = +5$.

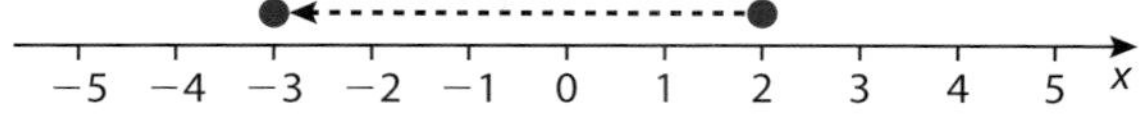

2. In this example, the object's initial and final positions are the same as in the first example – that is, at $t = 0$ its position is $x = 2$ and at $t = 5$ its position is $x = -3$. However, the object changed direction in that it first travelled to the left (negative velocity) from $x = 2$ to $x = -5$ during the interval $0 \leqslant t \leqslant 3$, and then travelled to the right (positive velocity) from $x = -5$ to $x = -3$. The object's displacement is -5 – the same as in the first example because its net change in position is just the difference between final and initial positions. However, it's clear that the object has travelled further than in the first example. But we cannot calculate it in the same way as we did in the first example. We will have to make a separate calculation for each interval where the direction changed. Hence, total distance travelled $= |-5 - 2| + |-3 - (-5)| = 7 + 2 = 9$.

$3 < t \leqslant 5$

$0 \leqslant t \leqslant 3$

−5 −4 −3 −2 −1 0 1 2 3 4 5 x

Velocity and speed

- The **velocity** $\boldsymbol{v} = \frac{ds}{dt}$ of a particle is a measure of how fast it is moving *and* of its direction of motion relative to a fixed point.
- The **speed** $|\boldsymbol{v}|$ of a particle is a measure of how fast it is moving and does *not* indicate direction. Thus, speed is the magnitude of velocity and is always positive.

Acceleration

- The **acceleration** $\boldsymbol{a} = \frac{dv}{dt}$ of a particle is a measure of how fast its velocity is changing.

There is no separate word to describe the magnitude of acceleration, $|\boldsymbol{a}|$.

The definite integral is a mathematical tool that can be used in applications to calculate net change of a quantity (e.g. Δ position → displacement) and total accumulation (e.g. Σ area → volume).

Example 44

The displacement s of a particle on the x-axis, relative to the origin, is given by the position function $s(t) = -t^2 + 6t$, where s is in centimetres and t is in seconds.

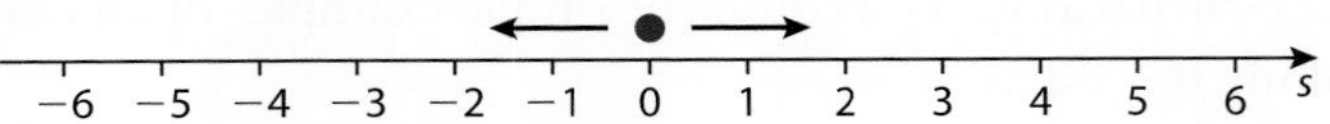

a) Find a function for the particle's velocity $v(t)$ in terms of t. Graph the functions $s(t)$ and $v(t)$ on separate axes.

b) Find the particle's position at the following times: $t = 0, 1, 3$ and 6 seconds.

c) Find the particle's displacement for the following intervals: $0 \leqslant t \leqslant 1$, $1 \leqslant t \leqslant 3$, $3 \leqslant t \leqslant 6$ and $0 \leqslant t \leqslant 6$.

d) Find the particle's total distance travelled for the following intervals: $0 \leqslant t \leqslant 1$, $1 \leqslant t \leqslant 3$, $3 \leqslant t \leqslant 6$ and $0 \leqslant t \leqslant 6$.

Solution

Position function: $s(t) = -t^2 + 6t$

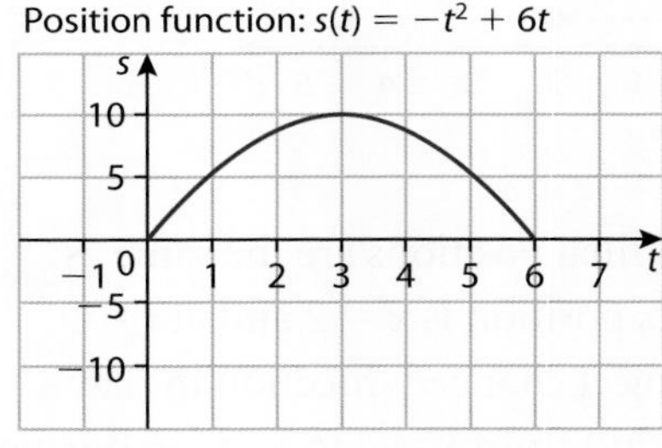

Velocity function: $v(t) = s'(t) = -2t + 6$

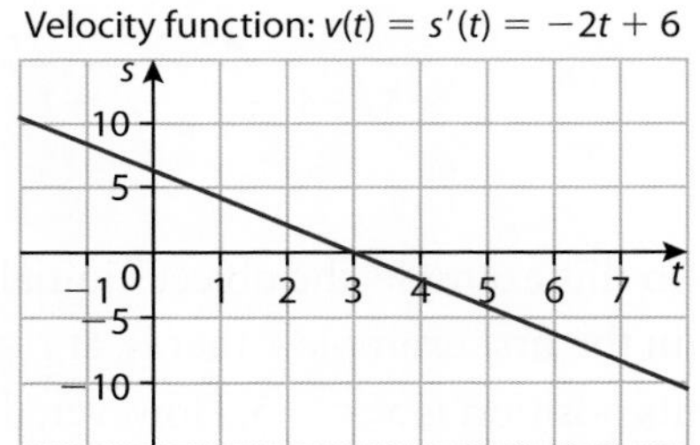

a) $v(t) = \frac{d}{dt}(-t^2 + 6t) = -2t + 6$

b) The particle's position at:
- $t = 0$ is $s(0) = -(0)^2 + 6(0) = 0$ cm
- $t = 1$ is $s(1) = -(1)^2 + 6(1) = 5$ cm
- $t = 3$ is $s(3) = -(3)^2 + 6(3) = 9$ cm
- $t = 6$ is $s(6) = -(6)^2 + 6(6) = 0$ cm

c) The particle's displacement for the interval:
- $0 \leqslant t \leqslant 1$ is Δ position $= s(1) - s(0) = 5 - 0 = 5$ cm
- $1 \leqslant t \leqslant 3$ is Δ position $= s(3) - s(1) = 9 - 5 = 4$ cm
- $3 \leqslant t \leqslant 6$ is Δ position $= s(6) - s(3) = 0 - 9 = -9$ cm
- $0 \leqslant t \leqslant 6$ is Δ position $= s(6) - s(0) = 0 - 0 = 0$ cm

This last result makes sense considering the particle moved to the right 9 cm then at $t = 3$ turned around and moved to the left 9 cm, ending where it started – thus, no change in net position.

d) The particle's total distance travelled for the interval:

- $0 \leqslant t \leqslant 1$ is $|s(1) - s(0)| = |5 - 0| = 5$ cm
- $1 \leqslant t \leqslant 3$ is $|s(3) - s(1)| = |9 - 5| = 4$ cm
- $3 \leqslant t \leqslant 6$ is $|s(6) - s(3)| = |0 - 9| = |-9| = 9$ cm
- $0 \leqslant t \leqslant 6$: The object's motion changed direction (velocity $= 0$) at $t = 3$, so total distance is $|s(3) - s(0)| + |s(6) - s(3)|$
  $= |9 - 0| + |0 - 9| = 9 + 9 = 18$ cm

Since differentiation of the position function gives the velocity function $\left(\text{i.e. } v = \frac{ds}{dt}\right)$, we expect that the inverse of differentiation, integration, will lead us in the reverse direction – that is, from velocity to position. When velocity is constant, we can find the displacement with the formula:

$$\text{displacement} = \text{velocity} \times \Delta \text{ in time}$$

If we drove a car at a constant velocity of 50 km/h for 3 hours, our displacement (same as distance travelled in this case) is 150 km. If a particle travelled to the left on the x-axis at a constant rate of -4 units/sec for 5 seconds, the particle's displacement is -20 units.

The velocity–time graph below depicts an object's motion with a constant velocity of 5 cm/s for $0 \leqslant t \leqslant 3$. Clearly, the object's displacement is 5 cm/s $\times$ 3 sec $=$ 15 cm for this interval.

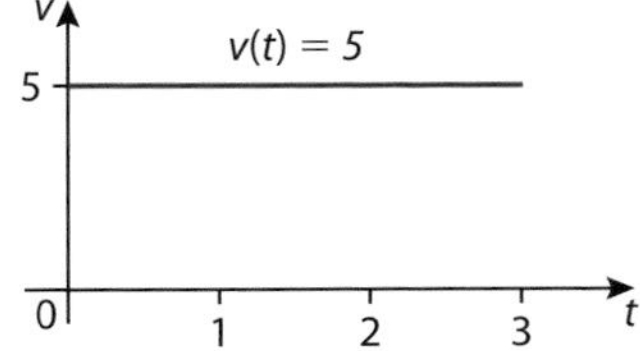

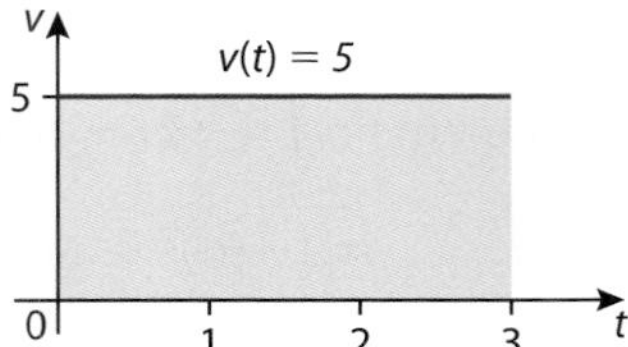

The rectangular area ($3 \times 5 = 15$) under the velocity curve is equal to the object's displacement.

Looking back at Example 44, consider the area under the graph of $v(t)$ from $t = 0$ to $t = 3$.

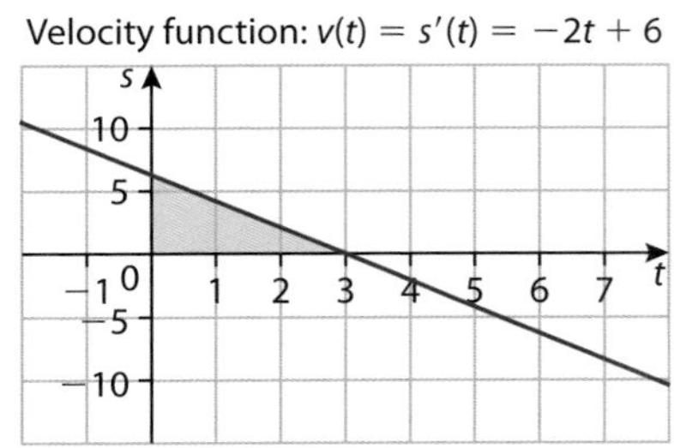

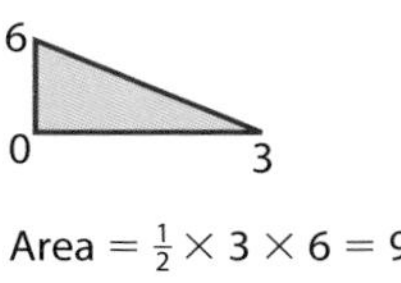

Given the discussion above, we should not be surprised to see that the area under the velocity curve for a certain interval is equal to the displacement for that interval. We can argue that just as the total area can be found by summing the areas of narrow rectangular strips, the displacement can be found by summing small displacements ($v \cdot \Delta t$). Consider:

$$\text{displacement} = \text{velocity} \times \Delta \text{ in time} \Rightarrow s = v \cdot \Delta t \Rightarrow s = v \cdot dt$$

We learned earlier in this chapter that if $f(x) \geqslant 0$ then the definite integral $\int_a^b f(x)\,dx$ gives the area between $y = f(x)$ and the x-axis from $x = a$ to $x = b$. And if $f(x) \leqslant 0$ then $\int_a^b f(x)\,dx$ gives a number that is the opposite of the area between $y = f(x)$ and the x-axis from a to b.

Using integration to find displacement and total distance travelled

Given that $v(t)$ is the velocity function for a particle moving along a line, then:

$\int_a^b v(t)\,dt$ gives the displacement from $t = a$ to $t = b$.

$\left|\int_a^b v(t)\,dt\right|$ gives the total distance travelled from $t = a$ to $t = b$ if the particle does not change direction during the interval $a < t < b$.

If a particle changes direction at some $t = c$ for $a < c < b$, the total distance travelled for the particle is given by $\left|\int_a^c v(t)\,dt\right| + \left|\int_c^b v(t)\,dt\right|$.

In general, the total distance travelled by an object from time t_0 to t_1, with many switches in direction is given by $\int_{t_0}^{t_1} |v(t)|dt$.

Let's apply integration to find the displacement and distance travelled for the two intervals $3 \leqslant t \leqslant 6$ and $0 \leqslant t \leqslant 6$ in Example 40.

- For $3 \leqslant t \leqslant 6$:

$$\text{Displacement} = \int_3^6 (-2t + 6)\,dt = \left[-t^2 + 6t\right]_3^6$$
$$= [-(6)^2 + 6(6)] - [-(3)^2 + 6(3)] = 0 - 9 = -9$$

$$\text{Distance travelled} = \left|\int_3^6 (-2t + 6)\right| dt = \left|\left[-t^2 + 6t\right]_3^6\right|$$
$$= |[-(6)^2 + 6(6)] - [-(3)^2 + 6(3)]| = |0 - 9| = 9$$

- For $0 \leqslant t \leqslant 6$:

$$\text{Displacement} = \int_0^6 (-2t + 6)\,dt = \left[-t^2 + 6t\right]_0^6$$
$$= [-(6)^2 + 6(6)] - [0] = 0$$

$$\text{Distance travelled} = \left|\int_0^3 (-2t + 6)\,dt\right| + \left|\int_3^6 (-2t + 6)\,dt\right|$$

Particle changed direction at $t = 3$.

$$= \left|\left[-t^2 + 6t\right]_3^6\right| + \left|\left[-t^2 + 6t\right]_3^6\right|$$
$$= |(-9 + 18) - 0| + |0 - (-9 + 18)|$$
$$= |9| + |-9| = 9 + 9 = 18$$

Example 45

The function $v(t) = \sin(\pi t)$ gives the velocity in m/s of a particle moving along the x-axis.

a) Determine when the particle is moving to the right, to the left, and stopped. At any time it stops, determine if it changes direction at that time.

b) Find the particle's displacement for the time interval $0 \leqslant t \leqslant 3$.

c) Find the particle's total distance travelled for the time interval $0 \leqslant t \leqslant 3$.

Solution

a) $v(t) = \sin(\pi t) = 0 \Rightarrow \sin(k \cdot \pi) = 0$ for $k \in \mathbb{Z} \Rightarrow \pi t = k\pi \Rightarrow t = k, k \in \mathbb{Z}$ for $0 \leqslant t \leqslant 3$, $t = 0, 1, 2, 3$. Therefore, the particle is stopped at $t = 0, 1, 2, 3$.

Since $t = 0$ and $t = 3$ are endpoints of the interval, the particle can only change direction at $t = 1$ or $t = 2$.

$v(\frac{1}{2}) = \sin(\pi \cdot \frac{1}{2}) = 1$; $v(\frac{3}{2}) = \sin(\pi \cdot \frac{3}{2}) = -1 \Rightarrow$ direction changes at $t = 1$

$v(\frac{3}{2}) = \sin(\pi \cdot \frac{3}{2}) = -1$; $v(\frac{5}{2}) = \sin(\pi \cdot \frac{5}{2}) = 1 \Rightarrow$ direction changes again at $t = 2$

b) Displacement $= \int_0^3 \sin(\pi t)\, dt = \left[-\frac{1}{\pi}\cos(\pi t)\right]_0^3$

$$= -\frac{1}{\pi}\cos(3\pi) - \left(-\frac{1}{\pi}\cos(0)\right) = -\frac{1}{\pi}(-1) + \frac{1}{\pi}(1) = \frac{2}{\pi} \approx 0.637 \text{ metres}$$

c) Total distance travelled $= \left|\int_0^1 \sin(\pi t)\, dt\right| + \left|\int_1^2 \sin(\pi t)\, dt\right|$

$$+ \left|\int_2^3 \sin(\pi t)\, dt\right| = \left|\left[-\frac{1}{\pi}\cos(\pi t)\right]_0^1\right|$$

$$+ \left|\left[-\frac{1}{\pi}\cos(\pi t)\right]_1^2\right| + \left|\left[-\frac{1}{\pi}\cos(\pi t)\right]_2^3\right|$$

$$= \left|\frac{2}{\pi}\right| + \left|-\frac{2}{\pi}\right| + \left|\frac{2}{\pi}\right| = \frac{6}{\pi} \approx 1.91 \text{ metres}$$

Note that, in Example 45, the position function is not known precisely. The position function can be obtained by finding the anti-derivative of the velocity function.

$$s(t) = \int v(t)\, dt = \int \sin(\pi t)\, dt = -\frac{1}{\pi}\cos(\pi t) + C$$

We can only determine the constant of integration C if we know the particle's initial position (or position at any other specific time). However, the particle's initial position will not affect displacement or distance travelled for any interval.

Position and velocity from acceleration

If we can obtain position from velocity by applying integration then we can also obtain velocity from acceleration by integrating. Consider the following example.

Example 46

The motion of a falling parachutist is modelled as linear motion by considering that the parachutist is a particle moving along a line whose positive direction is vertically downwards. The parachute is opened at $t = 0$ at which time the parachutist's position is $s = 0$. According to the model, the acceleration function for the parachutist's motion for $t > 0$ is given by:

$$a(t) = -54e^{-1.5t}$$

a) At the moment the parachute opens, the parachutist has a velocity of 42 m/s. Find the velocity function of the parachutist for $t > 0$. What does the model say about the parachutist's velocity as $t \to \infty$?

b) Find the position function of the parachutist for $t > 0$.

Solution

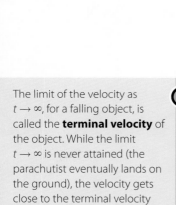

a) $v(t) = \int a(t)\,dt = \int(-54e^{-1.5t})\,dt$

$$= -54\left(\frac{1}{-1.5}\right)e^{-1.5t} + C$$

$$= 36e^{-1.5t} + C$$

Since $v = 42$ when $t = 0$, then $42 = 36e^0 + C \Rightarrow 42 = 36 + C \Rightarrow C = 6$

Therefore, after the parachute opens $(t > 0)$ the velocity function is $v(t) = 36e^{-1.5t} + 6$.

Since $\lim\limits_{t\to\infty} e^{-1.5t} = \lim\limits_{t\to\infty} \frac{1}{e^{1.5t}} = 0$, then as $t \to \infty$, $\lim\limits_{t\to\infty} v(t) = 6$ m/sec.

The limit of the velocity as $t \to \infty$, for a falling object, is called the **terminal velocity** of the object. While the limit $t \to \infty$ is never attained (the parachutist eventually lands on the ground), the velocity gets close to the terminal velocity very quickly. For example, after just 8 seconds, the velocity is $v(8) = 36e^{-1.5(8)} + 6 \approx 6.0002$ m/s.

b) $s(t) = \int v(t)\,dt = \int(36e^{-1.5t} + 6)\,dt$

$$= 36\left(\frac{1}{-1.5}\right)e^{-1.5t} + 6t + C$$

$$= -24e^{-1.5t} + 6t + C$$

Since $s = 0$ when $t = 0$, then $0 = -24e^0 + 6(0) + C$
$\Rightarrow 0 = -24 + C \Rightarrow C = 24$

Therefore, after the parachute opens $(t > 0)$ the position function is $s(t) = -24e^{-1.5t} + 6t + 24$.

Uniformly accelerated motion

Motion under the effect of gravity in the vicinity of Earth (or other planets) is an important case of rectilinear motion. This is called uniformly accelerated motion.

If a particle moves with constant acceleration along the s-axis, and if we know the initial speed and position of the particle, then it is possible to have specific formulae for the position and speed at any time t. This is how:

Assume acceleration is constant, i.e. $a(t) = a$, $v(0) = v_0$ and $s(0) = s_0$.

$v(t) = \int a dt = at + c$, we know that $v(0) = v_0$, then

$v(0) = v_0 = a(0) + c \Rightarrow c = v_0$; hence $v(t) = at + v_0$

$s(t) = \int v(t) dt = \int (at + v_0) dt = \frac{1}{2}at^2 + v_0 t + c$, but $s(0) = s_0$, then

$s(0) = s_0 = \frac{1}{2}a(0^2) + v_0(0) + c \Rightarrow c = s_0$; hence

$s(t) = \frac{1}{2}at^2 + v_0 t + s_0$

When this is applied to a free-fall model (s-axis vertical), then

$v(t) = -gt + v_0$, and

$s(t) = -\frac{1}{2}gt^2 + v_0 t + s_0$, where $g = 9.8 \text{ m/s}^2$.

Example 47

A ball is hit, from a point 2 m above the ground, directly upward with initial velocity of 45 m/s. How high will the ball travel?

Solution

$v(t) = -9.8t + 45$

$s(t) = -\frac{1}{2}(9.8)t^2 + 45t + 2 = -4.9t^2 + 45t + 2$

The ball will rise till $v(t) = 0, \Rightarrow 0 = -9.8t + 45, \Rightarrow t \approx 4.6 \text{ s}$

At this time,

$s(4.6) = -4.9(4.6)^2 + 45(4.6) + 2 \approx 105.32\text{m}.$

Example 48

Tim is running at a constant speed of 5 m/s to catch a bus that stopped at the station. The bus started as it was 11 m away with an acceleration of 1 m/s². How long will it take Tim to catch up with the bus?

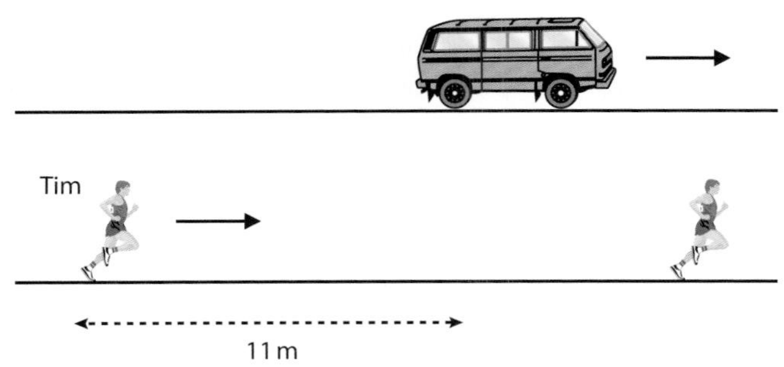

Solution

To catch the bus at some time t, Tim will have to cover a distance s_T that is equal to 11 m plus s_b travelled by the bus.

$$s_T = 5t$$

$$s_b = \tfrac{1}{2}t^2$$

But $s_T = s_b + 11 = \tfrac{1}{2}t^2 + 11$, therefore

$$5t = \tfrac{1}{2}t^2 + 11 \Rightarrow t^2 - 10t + 22 = 0$$

So, $t \approx 3.3$ s, or $t \approx 6.7$ s.

Note: The reason we have two answers is that since Tim is travelling at a constant rate he may miss the door at first, and if he continues, the bus will catch up with him 6.7 s later!

Exercise 16.8

In questions 1–6, the velocity of a particle along a rectilinear path is given by the equation $v(t)$ in m/s. Find both the net distance and the total distance it travels between the times $t = a$ and $t = b$.

1 $v(t) = t^2 - 11t + 24, a = 0, b = 10$

2 $v(t) = t - \frac{1}{t^2}, a = 0.1, b = 1$

3 $v(t) = \sin 2t, a = 0, b = \frac{\pi}{2}$

4 $v(t) = \sin t + \cos t, a = 0, b = \pi$

5 $v(t) = t^3 - 8t^2 + 15t, a = 0, b = 6$

6 $v(t) = \sin\left(\frac{\pi t}{2}\right) + \cos\left(\frac{\pi t}{2}\right), a = 0, b = 1$

In questions 7–11, the acceleration of a particle along a rectilinear path is given by the equation $a(t)$ in m/s^2, and the initial velocity v_0 m/s is also given. Find the velocity of the particle as a function of t, and both the net distance and the total distance it travels between the times $t = a$ and $t = b$.

7 $a(t) = 3, v_0 = 0, a = 0, b = 2$

8 $a(t) = 2t - 4, v_0 = 3, a = 0, b = 3$

9 $a(t) = \sin t, v_0 = 0, a = 0, b = \frac{3\pi}{2}$

10 $a(t) = \frac{-1}{\sqrt{t+1}}, v_0 = 2, a = 0, b = 4$

11 $a(t) = 6t - \frac{1}{(t+1)^3}, v_0 = 2, a = 0, b = 2$

In each question 12–15, the velocity and initial position of an object moving along a coordinate line are given. Find the position of the object at time t.

12 $v = 9.8t + 5, s(0) = 10$

13 $v = 32t - 2, s(0.5) = 4$

14 $v = \sin \pi t, s(0) = 0$

15 $v = \frac{1}{t+2}, t > -2, s(-1) = \frac{1}{2}$

In each question 16–19, the acceleration is given as well as the initial velocity and initial position of an object moving on a coordinate line. Find the position of the object at time t.

16 $a = e^t$, $v(0) = 20$, $s(0) = 5$

17 $a = 9.8$, $v(0) = -3$, $s(0) = 0$

18 $a = -4 \sin 2t$, $v(0) = 2$, $s(0) = -3$

19 $a = \frac{9}{\pi^2} \cos \frac{3t}{\pi}$, $v(0) = 0$, $s(0) = -1$

In questions 20–23, an object moves with a speed of $v(t)$ m/s along the s-axis. Find the displacement and the distance travelled by the object during the given time interval.

20 $v(t) = 2t - 4; 0 \leqslant t \leqslant 6$

21 $v(t) = |t - 3|; 0 \leqslant t \leqslant 5$

22 $v(t) = t^3 - 3t^2 + 2t; 0 \leqslant t \leqslant 3$

23 $v(t) = \sqrt{t} - 2; 0 \leqslant t \leqslant 3$

In questions 24–26, an object moves with an acceleration $a(t)$ m/s^2 along the s-axis. Find the displacement and the distance travelled by the object during the given time interval.

24 $a(t) = t - 2$, $v_0 = 0$, $1 \leqslant t \leqslant 5$

25 $a(t) = \frac{1}{\sqrt{5t + 1}}$, $v_0 = 2$, $0 \leqslant t \leqslant 3$

26 $a(t) = -2$, $v_0 = 3$, $1 \leqslant t \leqslant 4$

27 The velocity of an object moving along the s-axis is

$$v = 9.8t - 3.$$

a) Find the object's displacement between $t = 1$ and $t = 3$ given that $s(0) = 5$.

b) Find the object's displacement between $t = 1$ and $t = 3$ given that $s(0) = -2$.

c) Find the object's displacement between $t = 1$ and $t = 3$ given that $s(0) = s_0$.

28 The displacement **s** metres of a moving object from a fixed point O at time t seconds is given by $s(t) = 50t - 10t^2 + 1000$.

a) Find the velocity of the object in m s^{-1}.

b) Find its maximum displacement from O.

29 A particle moves along a line so that its speed v at time t is given by

$$v(t) = \begin{cases} 5t, & 0 \leqslant t < 1 \\ 6\sqrt{t} - \frac{1}{t}, & t \geqslant 1 \end{cases}$$

where t is in seconds and v is in cm/s. Estimate the time(s) at which the particle is 4 cm from its starting position.

30 A projectile is fired vertically upward with an initial velocity of 49 m/s from a platform 150 m high.

a) How long will it take the projectile to reach its maximum height?

b) What is the maximum height?

c) How long will it take the projectile to pass its starting point on the way down?

d) What is the velocity when it passes the starting point on the way down?

e) How long will it take the projectile to hit the ground?

f) What will its speed be at impact?

16.9 Differential equations (Optional)

This section presents only an introduction to differential equations. More on differential equations can be found in the Options part: Calculus.

A differential equation is an equation that relates an unknown function and one or more of its derivatives. A *first-order* differential equation is an equation that involves an unknown function and its *first derivative.* Examples of first-order differential equations are:

$$y' + 2xy = \sin x, \frac{dy}{dx} = y + 2x, \text{ and } \frac{dy}{dx} = -ky$$

In this part of the textbook we will consider only first-order differential equations that can be written in the form

$$\frac{dy}{dx} = f(x, y).$$

Here $f(x, y)$ is a function of two variables defined on a *region* in the xy-plane. By a solution to the differential equation, we mean the following.

Solution of a differential equation

We say that a *differentiable* function $y = y(x)$ is a solution to the differential equation

$$\frac{dy}{dx} = f(x, y)$$

on an interval of x-values (sometimes $\mathbb{R}$) when

$$\frac{d}{dx}y(x) = f(x, y(x)).$$

By $y(x)$, we mean 'y of x', i.e. y as a function of x, and not 'y times x'.

The initial condition $y(x_0) = y_0$ amounts to requiring the solution curve $y = y(x)$ to pass through the point (x_0, y_0).

Let us clarify these initial ideas by some examples.

Note: In algebra we usually seek the unknown variable values that satisfy an equation such as $3x^2 - 2x - 5 = 0$. By contrast, in solving a differential equation, we are looking for the unknown functions $y = y(x)$ for which an identity such as $y'(x) = 3x^2y(x)$ holds on some interval of real numbers. Usually, we will desire to find all solutions of the differential equation, if achievable.

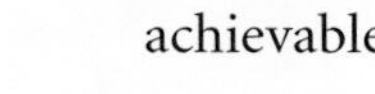

Example 49

Verify that $y(x) = Ce^{x^3}$ is a solution to the differential equation

$$\frac{dy}{dx} = 3x^2y.$$

Solution

Since C is a constant in $y(x) = Ce^{x^3}$, then

$$\frac{dy}{dx} = C(3x^2e^{x^3}) = 3x^2(Ce^{x^3}) = 3x^2y.$$

Consequently every function $y(x)$ of the form $y(x) = Ce^{x^3}$ satisfies – and thus is a solution of – the differential equation

$$\frac{dy}{dx} = 3x^2y$$

for all real x. In fact $y(x) = Ce^{x^3}$ defines an infinite family of different solutions to this differential equation, one for each choice of the arbitrary constant C.

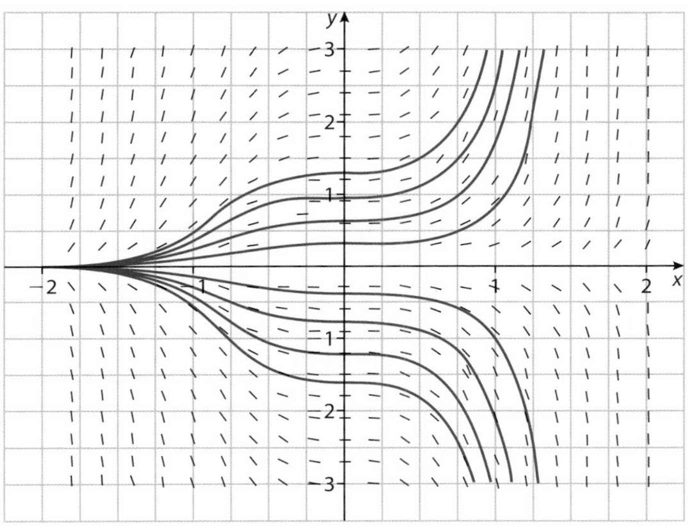

Example 50

Verify that

$$y(x) = -\frac{1}{2x^4 + 3}$$

is a solution to the differential equation

$$\frac{dy}{dx} = 8x^3y^2$$

over the interval $]-\infty, \infty[$.

Solution

Notice that the denominator in $y(x)$ is never zero and that $y(x)$ is differentiable everywhere. Furthermore, for all real numbers x,

$$\frac{d}{dx}y(x) = \frac{d}{dx}\left(-\frac{1}{2x^4 + 3}\right) = \frac{8x^3}{(2x^4 + 3)^2}$$

$$= 8x^3\left(-\frac{1}{2x^4 + 3}\right)^2 = 8x^3y^2$$

Thus,

$$y(x) = -\frac{1}{2x^4 + 3}$$

is a solution to the given differential equation.

Differential equations as mathematical models

The following examples illustrate typical cases where scientific principles are translated into differential equations.

1 **Newton's law of cooling** states that the rate of change of the temperature T of an object is proportional to the difference between T and the temperature of the surrounding medium S.

That is,

$$\frac{dT}{dt} = k(T - S)$$

where k is a constant and S is usually considered constant.

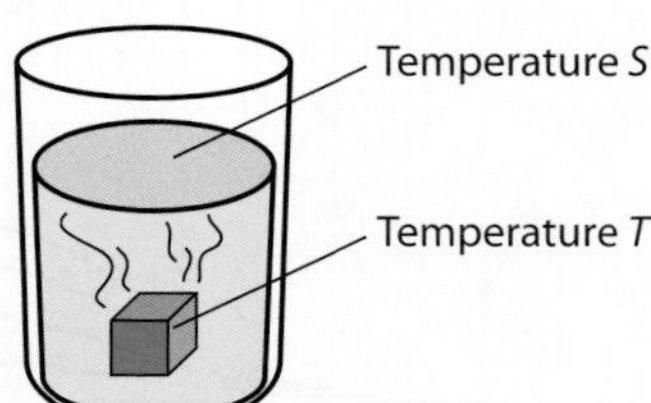

2 **Population growth rate** in cases where the birth and death rates are not variable is proportional to the size of the population. That is,

$$\frac{dP}{dt} = kP$$

where k is a constant.

Shortly, we will learn how to solve such problems.

Separable differential equations

In this section, we will limit our discussion to one basic type, the **separable differential equations**, also called **variables-separable differential equations.**

The first-order differential equation

$$\frac{dy}{dx} = f(x, y)$$

is called variable separable when the function $f(x, y)$can be factored into a product or quotient of two functions such as

$$\frac{dy}{dx} = g(x)h(y) \text{ or } \frac{dy}{dx} = \frac{p(x)}{q(y)}.$$

In such cases, the variables x and y can be separated by writing

$$\frac{dy}{h(y)} = g(x)dx \text{ or } q(y)dy = p(x)dx$$

and then simply integrating both sides with respect to x. That is,

$$\int \frac{dy}{h(y)} = \int g(x)dx + c \text{ or } \int q(y)dy = \int p(x)dx + c.$$

Note: You need to remember that $h(y)$ is a continuous function of y alone and $g(x)$ is a continuous function of x alone. The same goes for $q(y)$ and $p(x)$.

Note: We also may say that the method of solution is separation of variables.

Here are some examples of differential equations that are separable

Original differential equation	Rewritten with variables separated
$(x^2 + 4)y' = 3xy$	$\frac{dy}{y} = \frac{3x}{x^2 + 4}dx$
$\frac{3xe^y y'}{1 + e^2 y} = 5$	$\frac{3e^y}{1 + e^{2y}}dy = \frac{5}{x}dx$
$\frac{dy}{dx} = xy + 4$	Not separable!
$3x^2 + y\frac{dy}{dx} = 7$	$y\,dy = (7 - 3x^2)dx$
$x^2\frac{dy}{dx} + y^2 = xy^2$	$\frac{1}{y^2}dy = \frac{(x - 1)}{x^2}dx$
$y^2\frac{dy}{dx} + x^2 = xy^2$	Not separable!

We will end this section by looking at a few examples.

Example 51

Solve

$$y' - 9x^2y^2 = 5y^2.$$

Solution

We first factor the equation to separate the variables.

$$\frac{dy}{dx} = 5y^2 + 9x^2y^2 \Rightarrow \frac{dy}{dx} = y^2(5 + 9x^2)$$

$$\Rightarrow \frac{dy}{y^2} = (5 + 9x^2)dx$$

$$\Rightarrow -\frac{1}{y} = 5x + 3x^3 + c$$

$$\Rightarrow y = \frac{-1}{5x + 3x^3 + c}$$

This is a general solution for the differential equation. In this case we are able to express this function in explicit form.

Example 52

Solve

$$\frac{dy}{dx} = \frac{3x^2y}{1 + 4y^2}.$$

Solution

With very few steps, we can separate the variables:

$$\frac{1 + 4y^2}{y} = 3x^2dx$$

And now we can integrate both sides:

$$\int \frac{1 + 4y^2}{y} dy = \int 3x^2dx \Leftrightarrow \int \left(\frac{1}{y} + 4y\right) dy = \int 3x^2dx$$

$$\ln|y| + 2y^2 = x^3 + c$$

For every value of arbitrary constant c, this defines an exact but implicit solution $y(x)$ as it cannot be written in an explicit form $y = f(x)$.

Here are some of the solution curves for a few values of c.

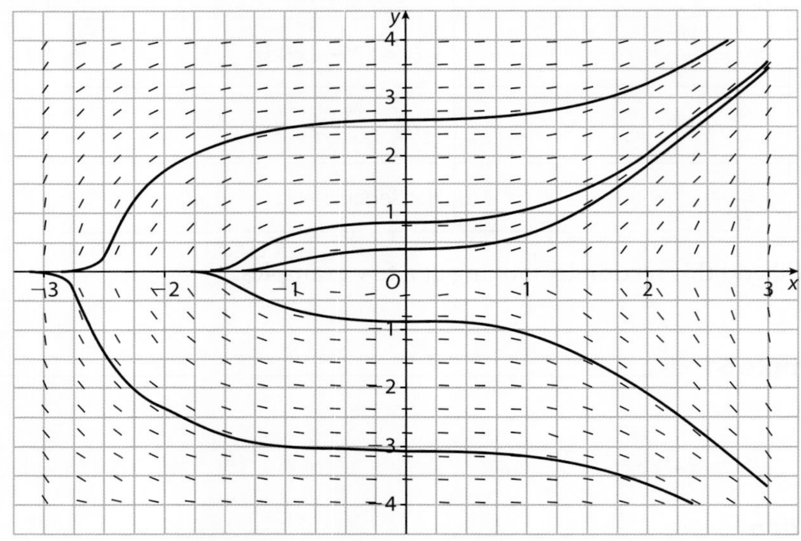

Note: Here is a summary of *solving equations by separation of variables.*

1 Write the differential equation in the standard form $\frac{dy}{dx} = f(x, y)$.

2 Can you separate the variables, i.e. is $\frac{dy}{dx} = g(x)h(y)$ or $\frac{dy}{dx} = \frac{p(x)}{q(y)}$?

3 If so, separate the variables, to get $\frac{dy}{h(y)} = g(x)dx$ or $q(y)dy = p(x)dx$.

4 Integrate both parts to get $\int \frac{dy}{h(y)} = \int g(x)dx + c$ or $\int q(y)dy$

$= \int p(x)dx + c.$

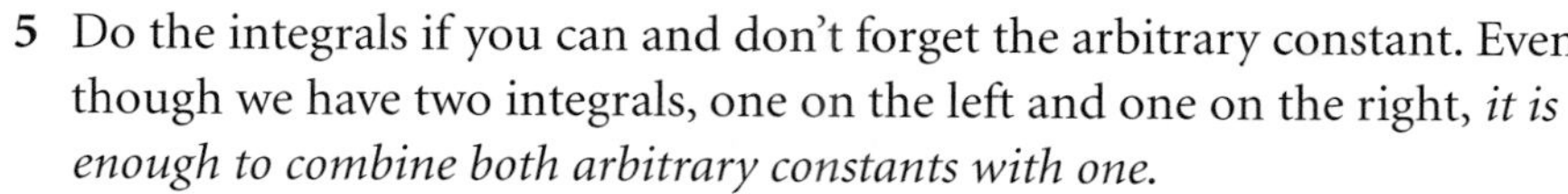

5 Do the integrals if you can and don't forget the arbitrary constant. Even though we have two integrals, one on the left and one on the right, *it is enough to combine both arbitrary constants with one.*

6 If possible, resolve the resulting equation with respect to y, to get your equation in explicit form $y = f(x)$.

Example 53

Find the general solution of the population growth model

$$\frac{dP}{dt} = kP.$$

Solution

In this problem, we can easily separate the variables.

$$\frac{dP}{P} = kt$$

Now integrate both sides to get

$$\int \frac{1}{P}\, dP = \int k\, dt$$

$$\ln|P| = kt + c$$

where c is an arbitrary constant. This last equation can be simplified to render an explicit expression for P:

$$\ln|P| = kt + c$$

$$\Rightarrow |P| = e^{kt+c} = e^{kt}e^{c} = Ae^{kt}$$

where we replaced e^c with A. Thus,

$$P = Ae^{kt} \text{ or } P = -Ae^{kt}.$$

This is the general solution and all solutions to this problem will be in this form.

If the constant k is positive, the model describes population growth; if it is negative, it is decay.

The first one corresponds to positive values of k and the second to negative values of k.

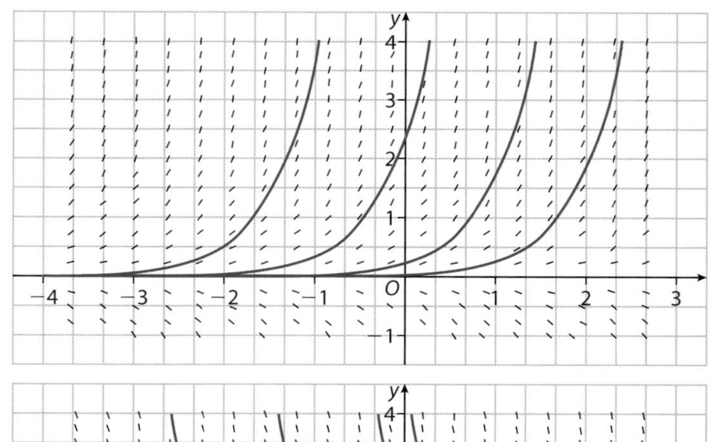

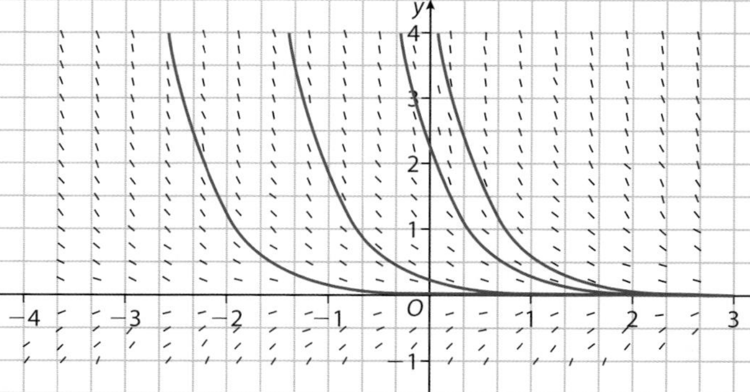

If the problem above had the additional 'initial value' that at t_0 the population is P_0, then this particular population satisfies

$$P = Ae^{kt}$$

and hence

$$P_0 = Ae^{kt_0} \Rightarrow A = \frac{P_0}{e^{kt_0}} = P_0 e^{-kt_0}$$

and the solution to the initial value problem is

$$P = Ae^{kt} = P_0 e^{-kt_0} e^{kt} = P_0 e^{k(t - t_0)}.$$

There is a very important special case when $t_0 = 0$. The solution becomes

$$P = P_0 e^{k(t - t_0)} = P_0 e^{kt}$$

which is the usual growth model which starts at time $t = 0$ with initial population P_0.

Example 54

If a cold object is placed in warmer medium that is kept at a constant temperature S, then the rate of change of the temperature $T(t)$ with respect to time t is proportional to the difference between the surrounding medium and the object and hence it satisfies

$$\frac{dT}{dt} = k(S - T) \qquad T(0) = T_0$$

where $k > 0$ and $T_0 < S$, i.e. the initial temperature is less than the temperature of the surrounding medium. Find the solution to the initial value problem.

Solution

It is immediately apparent that this is a variables separable type of differential equations as:

$$\frac{dT}{dt} = k(S - T) \Leftrightarrow \frac{dT}{S - T} = k\,dt$$

We integrate and find the general solution first.

$$\int \frac{dT}{S - T} = \int k\,dt$$

$$-\ln|S - T| = kt + c_1$$

$$\ln|S - T| = -kt - c_1$$

where c_1 is an arbitrary constant. Now since we know that the temperature T is less than the surrounding temperature, then

$$\ln|S - T| = \ln(S - T).$$

The general solution then is:

$$\ln(S - T) = -kt - c_1$$

$$S - T = e^{-kt - c_1}$$

$$T = S - e^{-kt - c_1}$$

The initial condition implies:

$$T = S - e^{-kt - c1}$$
$$T_0 = S - e^{0 - c1}$$
$$e^{-c1} = S - T_0$$
$$-c_1 = \ln(S - T_0)$$
$$c_1 = -\ln(S - T_0)$$

Therefore, substituting this value in the general solution:

$$\ln(S - T) = -kt - c_1$$
$$\ln(S - T) = -kt + \ln(S - T_0)$$
$$\ln(S - T) - \ln(S - T_0) = -kt$$
$$\ln\left(\frac{S - T}{S - T_0}\right) = -kt$$
$$\frac{S - T}{S - T_0} = e^{-kt}$$
$$S - T = (S - T_0)e^{-kt}$$
$$T = S - (S - T_0)e^{-kt}$$

This is an example of what is called 'limited growth'. This is so because the maximum value that T can achieve is S. For example, if a can of soda is left in a room with constant temperature of 21°, then the temperature of the soda will increase to reach the room temperature!

In fact, since $k > 0$ and S is a constant, then

$$T = S - (S - T_0)e^{-kt}$$
$$\frac{dT}{dt} = k(S - T_0)e^{-kt}.$$

Also, since $T_0 < S$, then

$$\frac{dT}{dt} = k(S - T_0)e^{-kt} > 0.$$

The temperature will always increase. As time passes, i.e.

$$\lim_{t \to \infty} e^{-kt} = 0$$
$$\Rightarrow \lim_{t \to \infty} T = \lim_{t \to \infty}(S - (S - T_0)e^{-kt}) = S$$

The graph shows how the temperature climbs up to 21° but does not exceed it.

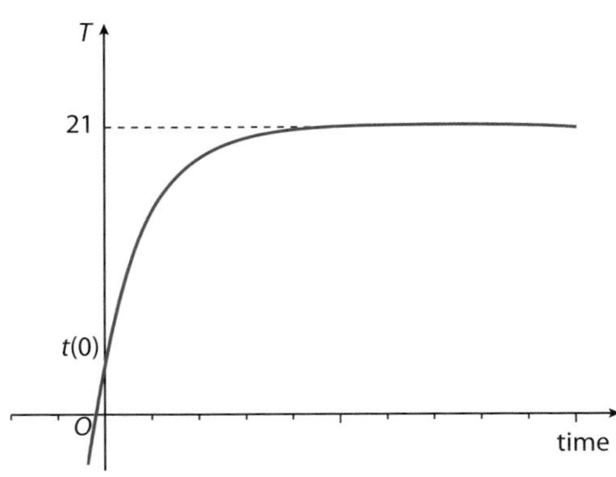

Example 55

Solve the initial value problem:

$$\frac{dy}{dx} = \frac{y}{x+1};\ y(1) = 4$$

Solution

This is a variables separable type. We will separate the variables and integrate.

$$\frac{dy}{dx} = \frac{y}{x+1}$$

$$\frac{dy}{y} = \frac{dx}{x+1}$$

$$\int \frac{dy}{y} = \int \frac{dx}{x+1}$$

$$\ln|y| = \ln|x+1| + c$$

$$|y| = e^{\ln|x+1|+c} = e^{\ln|x+1|}e^{c} = |x+1|e^{c}$$

Now, since c is an arbitrary constant, we can replace e^c with a constant C, and our solution becomes

$$|y| = C|x+1|.$$

Using the initial condition:

$4 = C|1+1| \Rightarrow C = 2$, and the particular solution

$|y| = 2|x+1|$, that is,

$y = \pm 2(x+1)$

Example 56

Solve the initial value problem:

$$\frac{dy}{dt} = e^{y-t}\frac{1+t^2}{\cos y};\ y(0) = 0$$

Solution

This problem needs some work to get it separated.

$$\frac{dy}{dt} = e^{y}e^{-t}\frac{1+t^2}{\cos y}$$

$$e^{-y}\cos y\,dy = e^{-t}(1+t^2)\,dt$$

Both sides need integration by parts (left as an exercise for you).

$$\int e^{-y}\cos y\,dy = \int e^{-t}(1+t^2)\,dt$$

$$\frac{1}{2}e^{-y}(\cos y - \sin y) = e^{-t}(t^2 + 2t + 3) + c$$

With initial conditions applied:

$$\tfrac{1}{2}e^{-y}(\cos y - \sin y) = e^{-t}(t^2 + 2t + 3) + c$$

$$\tfrac{1}{2}e^{-0}(\cos 0 - \sin 0) = e^{-0}(0^2 + 2(0) + 3) + c$$

$$\tfrac{1}{2} = 3 + c \Rightarrow c = \tfrac{5}{2}$$

Therefore, our particular solution is:

$$\tfrac{1}{2}e^{-y}(\cos y - \sin y) = e^{-t}(t^2 + 2t + 3) + \tfrac{5}{2}$$

$$e^{-y}(\cos y - \sin y) = 2e^{-t}(t^2 + 2t + 3) + 5$$

Notice here that our solution cannot be expressed explicitly. In many cases, solutions to differential equations are given in implicit form.

Exercise 16.9

In questions 1–27, solve the given differential equation.

1 $x^{-3}dy = 4y\,dx,\ y(0) = 3$

2 $\frac{dy}{dx} = xy,\ y(0) = 1$

3 $y' - xy^2 = 0,\ y(1) = 2$

4 $y' - y^2 = 0,\ y(2) = 1$

5 $\frac{dy}{dx} - e^y = 0,\ y(0) = 1$

6 $y'e^{y-x} = 1$

7 $\frac{dy}{dx} = y^{-2}x + y^{-2},\ y(0) = 1$

8 $x\,dy - y^2\,dx = -dy,\ y(0) = 1$

9 $y^2dy - x\,dx = dx - dy,\ y(0) = 3$

10 $yy' = xy^2 + x,\ y(0) = 0$

11 $\frac{dy}{dx} = y^2x + x$

12 $y' = \frac{xy - y}{y + 1},\ y(2) = 1$

13 $e^{x-y}dy = x\,dx$

14 $y' = xy^2 - x - y^2 + 1$

15 $xy\ln xy' = (y + 1)^2$

16 $\frac{dy}{dx} = \frac{1 + 2y^2}{y\sin x}$

17 $\frac{dy}{dx} = x\sqrt{\frac{1 - y^2}{1 - x^2}},\ y(0) = 0$

18 $y'(1 + e^x) = e^{x-y},\ y(1) = 0$

19 $(y + 1)dy = (x^2y - y)dx,\ y(3) = 1$

20 $\cos y\,dx + (1 + e^{-x})\sin y\,dy = 0,\ y(0) = \frac{\pi}{4}$

21 $xy' - y = 2x^2y,\ y(1) = 1$

22 $xy\,dx + e^{-x^2}(y^2 - 1)dy = 0,\ y(0) = 1$

23 $(1 + \tan y)y' = x^2 + 1$

24 $\frac{dy}{dt} = \frac{te^t}{y\sqrt{y^2 + 1}}$

25 $y\sec\theta\,dy = e^y\sin^2\theta\,d\theta$

26 $x\cos x = (2y + e^{3y})y',\ y(0) = 0$

27 $\frac{dy}{dx} = e^x - 2x,\ y(0) = 3$

28 The temperature T of a kettle in a room satisfies the differential equation

$$\frac{dT}{dt} = m(T - 21), \text{ where } t \text{ is in minutes and } m \text{ is a constant.}$$

a) Solve the differential equation showing that $T = Ce^{mt} + 21$, where C is an arbitrary constant.

b) Given that $T(0) = 99$ and $T(15) = 69$, find
   (i) the value of m and C
   (ii) t when $T = 39$.

Practice questions

1 The graph represents the function

$f: x \mapsto p\cos x, p \in \mathbb{N}$.

Find

a) the value of p

b) the area of the shaded region.

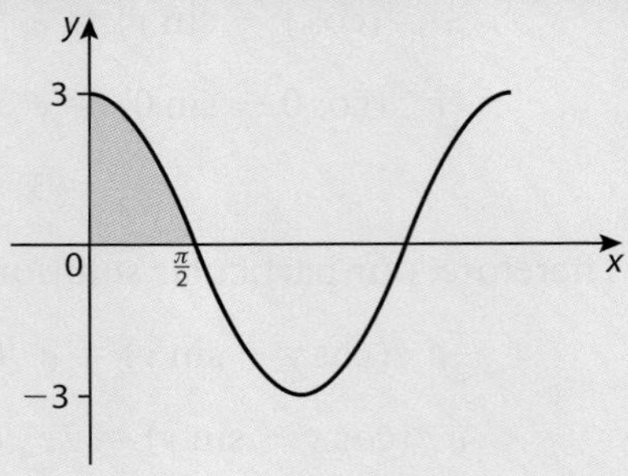

2 The diagram shows part of the graph of $y = e^{\frac{x}{2}}$.

a) Find the coordinates of the point P, where the graph meets the y-axis.

The shaded region between the graph and the x-axis, bounded by $x = 0$ and $x = \ln 2$, is rotated through 360° about the x-axis.

b) Write down an integral that represents the volume of the solid obtained.

c) Show that this volume is π cubic units.

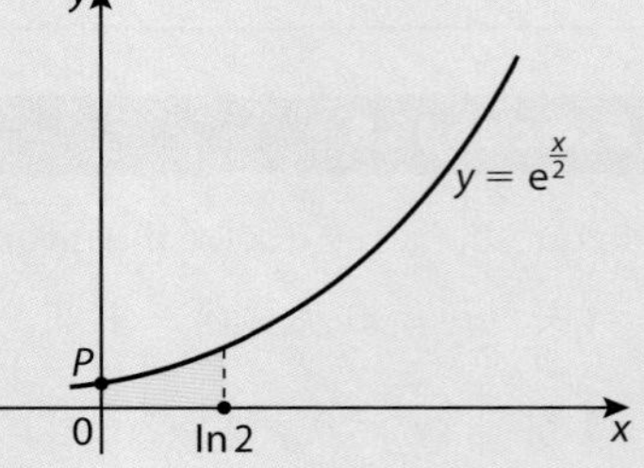

3 The diagram shows part of the graph of $y = \frac{1}{x}$. The area of the shaded region is 2 units.

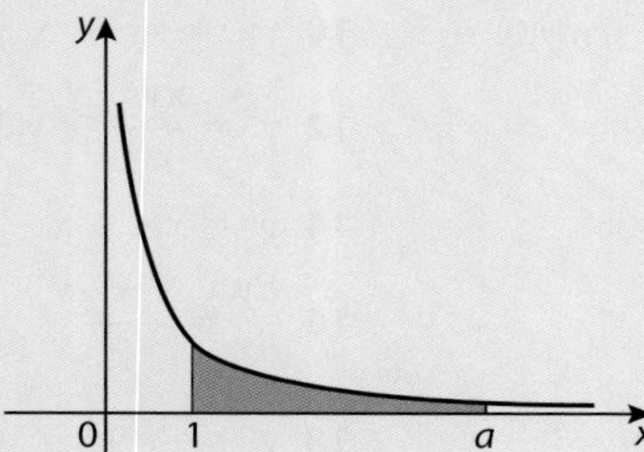

Find the exact value of a.

4 a) Find the equation of the tangent line to the curve $y = \ln x$ at the point $(e, 1)$, and verify that the origin is on this line.

b) Show that $(x \ln x - x)' = \ln x$.

c) The diagram shows the region enclosed by the curve $y = \ln x$, the tangent line in part **a)**, and the line $y = 0$.

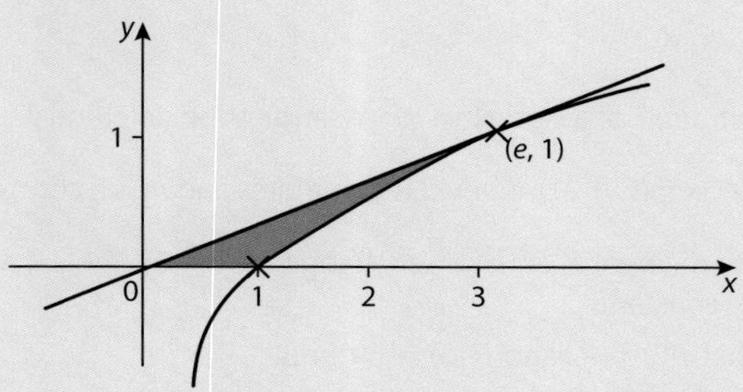

Use the result of part **b)** to show that the area of this region is $\frac{1}{2}e - 1$.

5 The main runway at Concordville airport is 2 km long. An aeroplane, landing at Concordville, touches down at point *T*, and immediately starts to slow down. The point *A* is at the southern end of the runway. A marker is located at point *P* on the runway.

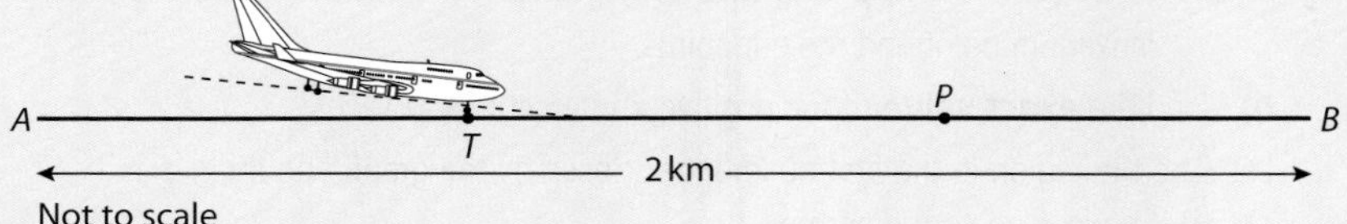

As the aeroplane slows down, its distance, *s*, from *A*, is given by

$$s = c + 100t - 4t^2$$

where *t* is the time in seconds after touchdown and *c* metres is the distance of *T* from *A*.

a) The aeroplane touches down 800 m from *A* (i.e. $c = 800$).

(i) Find the distance travelled by the aeroplane in the first 5 seconds after touchdown.

(ii) Write down an expression for the velocity of the aeroplane at time *t* seconds after touchdown, and hence find the velocity after 5 seconds.

The aeroplane passes the marker at *P* with a velocity of 36 m s^{-1}. Find

(iii) how many seconds after touchdown it passes the marker

(iv) the distance from *P* to *A*.

b) Show that if the aeroplane touches down before reaching the point *P*, it can stop before reaching the northern end, *B*, of the runway.

6 a) Sketch the graph of $y = \pi \sin x - x$, $-3 \leq x \leq 3$, on millimetre square paper, using a scale of 2 cm per unit on each axis.

Label and number both axes and indicate clearly the approximate positions of the *x*-intercepts and the local maximum and minimum points.

b) Find the solution of the equation $\pi \sin x - x = 0,\ x > 0$.

c) Find the indefinite integral

$$\int (\pi \sin x - x)\, dx$$

and hence, or otherwise, calculate the area of the region enclosed by the graph, the *x*-axis and the line $x = 1$.

7 The diagram shows the graph of the function $y = 1 + \frac{1}{x}$, $0 < x \leq 4$. Find the **exact** value of the area of the shaded region.

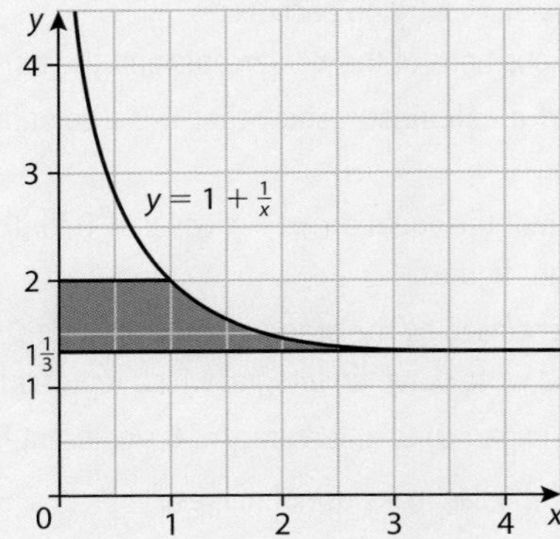

8 Note: Radians are used throughout this question.

a) **(i)** Sketch the graph of $y = x^2 \cos x$, for $0 \leqslant x \leqslant 2$, making clear the approximate positions of the positive intercept, the maximum point and the endpoints.

(ii) Write down the **approximate** coordinates of the positive x-intercept, the maximum point and the endpoints.

b) Find the **exact value** of the positive x-intercept for $0 \leqslant x \leqslant 2$.

Let R be the region in the first quadrant enclosed by the graph and the x-axis.

c) **(i)** Shade R on your diagram.

(ii) Write down an integral that represents the area of R.

d) Evaluate the integral in part **c) (ii)**, either by using a graphic display calculator, or by using the following information.

$$\frac{d}{dx}(x^2 \sin x + 2x \cos x - 2 \sin x) = x^2 \cos x$$

9 Note: Radians are used throughout this question.

The function f is given by

$$f(x) = (\sin x)^2 \cos x.$$

The diagram shows part of the graph of $y = f(x)$.

The point A is a maximum point, the point B lies on the x-axis, and the point C is a point of inflexion.

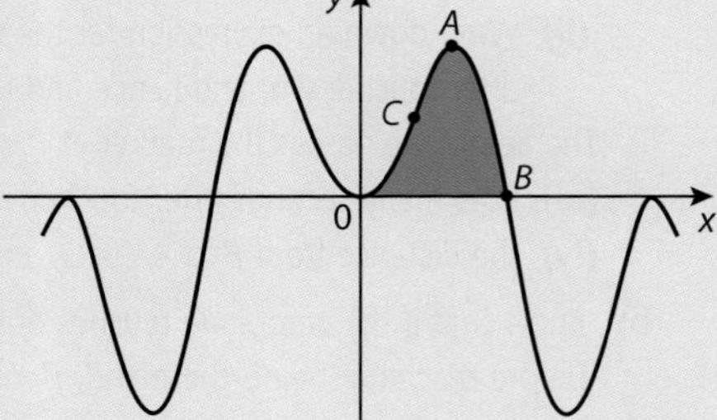

a) Give the period of f.

b) From consideration of the graph of $y = f(x)$, find, **to an accuracy of 1 significant figure**, the range of f.

c) **(i)** Find $f'(x)$.

(ii) Hence, show that at the point A $\cos x = \sqrt{\frac{1}{3}}$.

(iii) Find the exact maximum value.

d) Find the exact value of the x-coordinate at the point B.

e) **(i)** Find $\int f(x)\,dx$.

(ii) Find the area of the shaded region in the diagram.

f) Given that $f''(x) = 9(\cos x)^3 - 7 \cos x$, find the x-coordinate at the point C.

10 Note: Radians are used throughout this question.

a) Draw the graph of $y = \pi + x \cos x$, $0 \leqslant x \leqslant 5$, on millimetre square paper, using a scale of 2 cm per unit. Make clear

(i) the integer values of x and y on each axis

(ii) the approximate positions of the x-intercepts and the turning points.

b) **Without the use of a calculator**, show that π is a solution of the equation $\pi + x \cos x = 0$.

c) Find another solution of the equation $\pi + x \cos x = 0$ for $0 \leqslant x \leqslant 5$, giving your answer to 6 significant figures.

d) Let R be the region enclosed by the graph and the axes for $0 \leqslant x \leqslant \pi$. Shade R on your diagram, and write down an integral which represents the area of R.

e) Evaluate the integral in part **d)** to an accuracy of **6** significant figures. (If you consider it necessary, you can make use of the result $\frac{d}{dx}(x \sin x + \cos x) = x \cos x$.)

11 The diagram right shows the graphs of $f(x) = 1 + e^{2x}$ and $g(x) = 10x + 2, 0 \leqslant x \leqslant 1.5$.

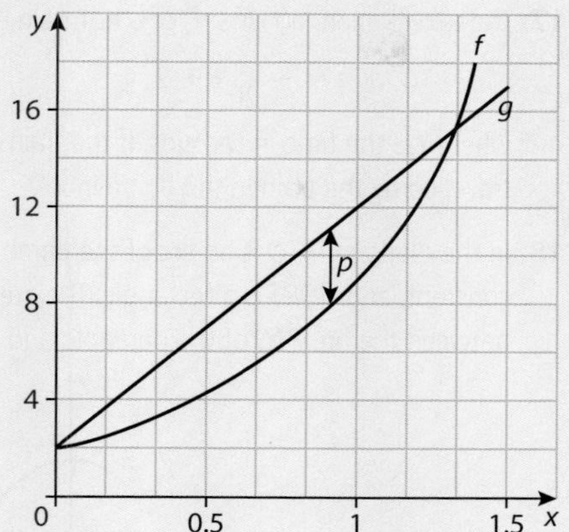

a) (i) Write down an expression for the vertical distance p between the graphs of f and g.

(ii) Given that p has a maximum value for $0 \leqslant x \leqslant 1.5$, find the value of x at which this occurs.

The graph of $y = f(x)$ only is shown in the diagram right. When $x = a, y = 5$.

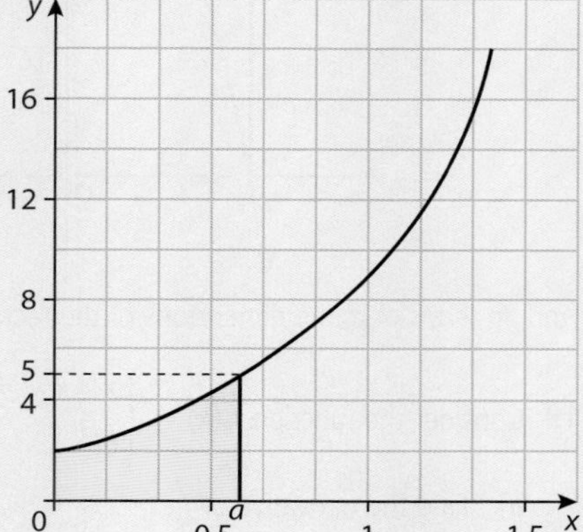

b) (i) Find $f^{-1}(x)$.

(ii) Hence, show that $a = \ln 2$.

c) The region shaded in the diagram is rotated through 360° about the x-axis. Write down an expression for the volume obtained.

12 The area of the enclosed region shown in the diagram is defined by

$y \geqslant x^2 + 2, y \leqslant ax + 2$, where $a > 0$.

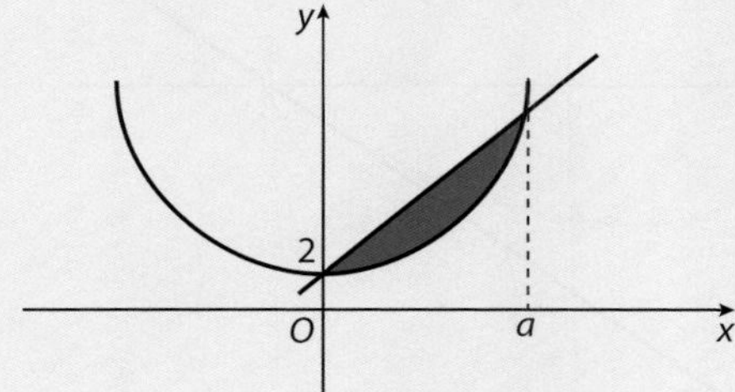

This region is rotated 360° about the x-axis to form a solid of revolution. Find, in terms of a, the volume of this solid of revolution.

13 Using the substitution $u = \frac{1}{2}x + 1$, or otherwise, find the integral $\int x\sqrt{\frac{1}{2}x + 1}\,dx$.

14 A particle moves along a straight line. When it is a distance s from a fixed point, where $s > 1$, the velocity v is given by $v = \dfrac{3s + 2}{2s - 1}$. Find the acceleration when $s = 2$.

15 The area between the graph of $y = e^x$ and the x-axis from $x = 0$ to $x = k\ (k > 0)$ is rotated through 360° about the x-axis. Find, in terms of k and e, the volume of the solid generated.

16 Find the real number $k > 1$ for which $\displaystyle\int_1^k \left(1 + \frac{1}{x^2}\right)dx = \frac{3}{2}$.

17 The acceleration, $a(t)\,\mathrm{m\,s^{-2}}$, of a fast train during the first 80 seconds of motion is given by

$$a(t) = -\frac{1}{20}t + 2$$

where t is the time in seconds. If the train starts from rest at $t = 0$, find the distance travelled by the train in the first minute.

18 In the diagram, PTQ is an arc of the parabola $y = a^2 - x^2$, where a is a positive constant, and $PQRS$ is a rectangle. The area of the rectangle $PQRS$ is equal to the area between the arc PTQ of the parabola and the x-axis.

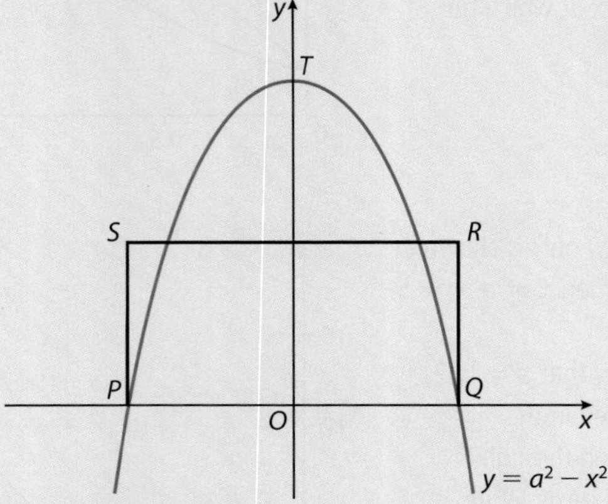

Find, in terms of a, the dimensions of the rectangle.

19 Consider the function $f_k(x) = \begin{cases} x\ln x - kx, & x > 0 \\ 0, & x = 0 \end{cases}$, where $k \in \mathbb{N}$

a) Find the derivative of $f_k(x)$, $x > 0$.

b) Find the interval over which $f(x)$ is increasing.

The graph of the function $f_k(x)$ is shown below.

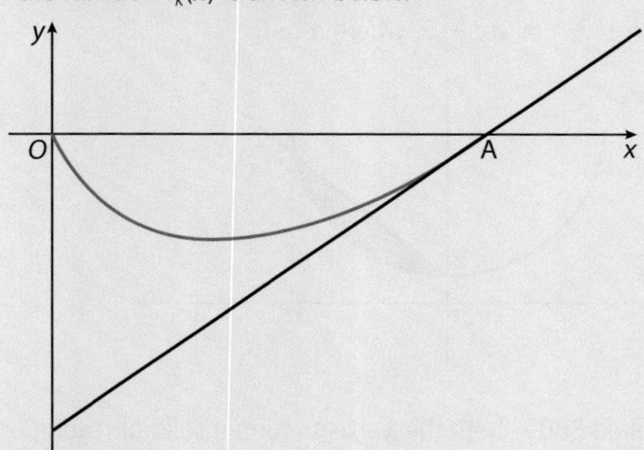

c) **(i)** Show that the stationary point of $f_k(x)$ is at $x = e^{k-1}$.

(ii) One x-intercept is at $(0, 0)$. Find the coordinates of the other x-intercept.

d) Find the area enclosed by the curve and the x-axis.

e) Find the equation of the tangent to the curve at A.

f) Show that the area of the triangular region created by the tangent and the coordinate axes is twice the area enclosed by the curve and the x-axis.

g) Show that the x-intercepts of $f_k(x)$ for consecutive values of k form a geometric sequence.

20 Solve the differential equation $\frac{dy}{dx} = 1 + y^2$ given that $y = 0$ when $x = 2$.

21 The equation of motion of a particle with mass m, subjected to a force kx can be written as $kx = mv\frac{dv}{dx}$, where x is the displacement and v is the velocity. When $x = 0$, $v = v_0$. Find v, in terms of v_0, k and m, when $x = 2$.

22 **a)** Sketch and label the graphs of $f(x) = e^{-x^2}$ and $g(x) = e^{x^2} - 1$ for $0 \leqslant x \leqslant 1$, and shade the region A which is bounded by the graphs and the y-axis.

b) Let the x-coordinate of the point of intersection of the curves $y = g(x)$ and $y = g(x)$ be p.
Without finding the value of p, show that

$$\frac{p}{2} < \text{area of region } A < p.$$

c) Find the value of p correct to four decimal places.

d) Express the area of region A as a definite integral and calculate its value.

23 Let $f(x) = x \cos 3x$.

a) Use integration by parts to show that

$$\int f(x)dx = \tfrac{1}{3}x \sin 3x + \tfrac{1}{9}\cos 3x + c.$$

b) Use your answer to part **a)** to calculate the **exact** area enclosed by $f(x)$ and the x-axis in each of the following cases. **Give your answers in terms of π.**

(i) $\frac{\pi}{6} \leqslant x \leqslant \frac{3\pi}{6}$

(ii) $\frac{3\pi}{6} \leqslant x \leqslant \frac{5\pi}{6}$

(iii) $\frac{5\pi}{6} \leqslant x \leqslant \frac{7\pi}{6}$

c) Given that the above areas are the first three terms of an arithmetic sequence, find an expression for the total area enclosed by $f(x)$ and the x-axis for

$$\frac{\pi}{6} \leqslant x \leqslant \frac{(2n+1)\pi}{6}, \text{ where } n \in \mathbb{Z}.$$

Give your answers in terms of n and π.

24 A particle is moving along a straight line so that t seconds after passing through a fixed point O on the line its velocity $v(t)\,\text{m s}^{-1}$ is given by

$$v(t) = t\sin\left(\frac{\pi}{3}t\right).$$

a) Find the values of t for which $v(t) = 0$, given that $0 \leqslant t \leqslant 6$.

b) **(i)** Write down a mathematical expression for the **total** distance travelled by the particle in the first six seconds after passing through O.

(ii) Find this distance.

25 A particle is projected along a straight-line path. After t seconds, its velocity v metres per second is given by $v = \frac{1}{2 + t^2}$.

a) Find the distance travelled in the first second.

b) Find an expression for the acceleration at time t.

26 The diagram below shows the shaded region R enclosed by the graph of $y = 2x\sqrt{1 + x^2}$, the x-axis, and the vertical line $x = k$.

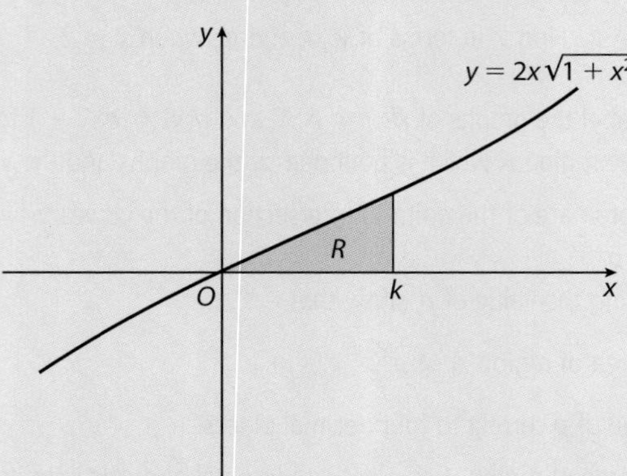

a) Find $\frac{dy}{dx}$.

b) Using the substitution $u = 1 + x^2$ or otherwise, show that

$$\int 2x\sqrt{1 + x^2}\, dx = \tfrac{2}{3}(1 + x^2)^{\frac{3}{2}} + c.$$

c) Given that the area of R equals 1, find the value of k.

27 A particle moves in a straight line with velocity, in metres per second, at time t seconds, given by

$$v(t) = 6t^2 - 6t, t \geqslant 0.$$

Calculate the total distance travelled by the particle in the first two seconds of motion.

28 A particle moves in a straight line. Its velocity $v\,\text{m s}^{-1}$ after t seconds is given by

$$v = e^{-\sqrt{t}} \sin t.$$

Find the total distance travelled in the time interval $[0, 2\pi]$.

29 The temperature T °C of an object in a room, after t minutes, satisfies the differential equation

$$\frac{dT}{dt} = k(T - 22), \text{ where } k \text{ is a constant.}$$

a) Solve the differential equation showing that $T = Te^{kt} + 22$, where A is a constant.

b) When $t = 0$, $T = 100$, and when $t = 15$, $T = 70$.

(i) Use this information to find the value of A and of k.

(ii) Hence, find the value of t when $T = 40$.

30 Solve the differential equation $x\frac{dy}{dx} - y^2 = 1$ given that $y = 0$ when $x = 2$. Give your answer in the form $y = f(x)$.

31 Use the substitution $u = x + 2$ to find $\int \frac{x^3}{(x + 2)^2} dx$.

32 a) On the same axes sketch the graphs of the functions, $f(x)$ and $g(x)$, where

$$f(x) = 4 - (1 - x)^2, \text{ for } -2 \leqslant x \leqslant 4,$$

$$g(x) = \ln(x + 3) - 2, \text{ for } -3 \leqslant x \leqslant 5.$$

b) **(i)** Write down the equation of any vertical asymptotes.

(ii) State the x-intercept and y-intercept of $g(x)$.

c) Find the values of x for which $f(x) = g(x)$.

d) Let A be the region where $f(x) \geqslant g(x)$ and $x \geqslant 0$.

(i) On your graph shade the region A.

(ii) Write down an integral that represents the area of A.

(iii) Evaluate this integral.

e) In the region A find the maximum vertical distance between $f(x)$ and $g(x)$.

33 Consider the differential equation $\frac{dy}{d\theta} = \frac{y}{e^{2\theta} + 1}$.

a) Use the substitution $x = e^{\theta}$ to show that

$$\int \frac{dy}{y} = \int \frac{dx}{x(x^2 + 1)}.$$

b) Find $\int \frac{dx}{x(x^2 + 1)}$.

c) Hence, find y in terms of θ, if $y = \sqrt{2}$ when $\theta = 0$.

17 Probability Distributions

5.5	Concept of discrete and continuous random variables and their probability distributions. Definition and use of probability density functions. Expected value (mean), mode, median, variance and standard deviation.
5.6	Binomial distribution, its mean and variance. Poisson distribution, its mean and variance.
5.7	Normal distribution. Properties of the normal distribution. Standardization of normal variables.

Introduction

Investing in securities, calculating premiums for insurance policies or overbooking policies used in the airline industry are only a few of the many applications of probability and statistics. Actuaries, for example, calculate the expected 'loss' or 'gain' that an insurance company will incur and decide on how high the premiums should be. These applications depend mainly on what we call probability distributions. A probability distribution describes the behaviour of a population in the sense that it lists the distribution of possible outcomes to an event, along with the probability of each potential outcome. This can be done by a table of values with their corresponding probabilities or by using a mathematical model.

In this chapter, you will get an understanding of the basic ideas of distributions and will study three specific ones: the binomial, Poisson and normal distributions.

17.1 Random variables

In Chapter 11, **variables** were defined as characteristics that change or vary over time and/or for different objects under consideration. A numerically valued variable x will vary or change depending on the outcome of the experiment we are performing. For example, suppose you are counting the number of mobile phones families in a certain city own. The variable of interest, x, can take any of the values 0, 1, 2, 3, etc. depending on the *random* outcome of the experiment. For this reason, we call the variable x a **random variable**.

Random variable

A **random variable** is a variable that takes on numerical values determined by the outcome of a random experiment.

When a probability experiment is performed, often we are not interested in all the details of the outcomes, but rather in the value of some numerical quantity determined by the result. For instance, in tossing two dice (used in plenty of games), often we care about their sum and not the values on the individual dice. Consider this specific experiment: A sample space for which the points are equally likely is given in Table 17.1 below. It consists of 36 ordered pairs (a, b) where a is the number on the first die and b is the number on the second die. For each sample point, we can let the *random* variable x stand for the sum of the numbers. The resulting values of x are also presented in Table 17.1.

$(1, 1); x = 2$	$(2, 1); x = 3$	$(3, 1); x = 4$	$(4, 1); x = 5$	$(5, 1); x = 6$	$(6, 1); x = 7$
$(1, 2); x = 3$	$(2, 2); x = 4$	$(3, 2); x = 5$	$(4, 2); x = 6$	$(5, 2); x = 7$	$(6, 2); x = 8$
$(1, 3); x = 4$	$(2, 3); x = 5$	$(3, 3); x = 6$	$(4, 3); x = 7$	$(5, 3); x = 8$	$(6, 3); x = 9$
$(1, 4); x = 5$	$(2, 4); x = 6$	$(3, 4); x = 7$	$(4, 4); x = 8$	$(5, 4); x = 9$	$(6, 4); x = 10$
$(1, 5); x = 6$	$(2, 5); x = 7$	$(3, 5); x = 8$	$(4, 5); x = 9$	$(5, 5); x = 10$	$(6, 5); x = 11$
$(1, 6); x = 7$	$(2, 6); x = 8$	$(3, 6); x = 9$	$(4, 6); x = 10$	$(5, 6); x = 11$	$(6, 6); x = 12$

Table 17.1 Sample space and the values of the random variable x in the two-dice experiment.

Random variables are customarily denoted by upper-case letters, such as X and Y. Lower-case letters are used to represent particular values of the random variable. That is, if X represents the numbers resulting in the throw of a die, then $x = 2$ represents the case when the outcome is 2.

Notice that events can be more accurately and concisely defined in terms of the random variable x; for example, the event of tossing a sum at least equal to 5 but less than 9 can be replaced by $5 \leqslant x < 9$.

We can think of many examples of random variables:

- $X =$ the number of calls received by a household on a Friday night.
- $X =$ the number of free beds available at hotels in a large city.
- $X =$ the number of customers a sales person contacts on a working day.
- $X =$ the length of a metal bar produced by a certain machine.
- $X =$ the weight of newborn babies in a large hospital.

As you have seen in Chapter 11, these variables are classified as **discrete** or **continuous**, according to the values that x *can* assume. In the examples above, the first three are discrete and the last two are continuous. The random variable is discrete if its set of *possible* values is isolated points on the number line, i.e. there is a *countable* number of possible values for the variable. The variable is continuous if its set of *possible* values is an entire interval on the number line, i.e. it can take any value in an interval. Consider the number of times you toss a coin until the head side appears. The possible values are $x = 1, 2, 3, \ldots$. This is a discrete variable, even though the number of times may be infinite! On the other hand, consider the time it takes a student at your school to eat/have his/her lunch. This can be anywhere between zero and 50 minutes (given that the lunch period at your school is 50 minutes).

Example 1

State whether each of the following is a discrete or a continuous random variable.

1. The number of hairs on a Scottish Terrier
2. The height of a building
3. The amount of fat in a steak
4. A high school student's grade on a maths test
5. The number of fish in the Atlantic Ocean
6. The temperature of a wooden stove

Solution

1. Even though the number of hairs is 'almost' infinite, it is countable. So, it is a discrete random variable.
2. This can be any real number. Even when you say this building is 15 m high, the number could be 15.1 or 15.02, etc. Hence, it is continuous.
3. This is continuous, as the amount of fat could be zero or anything up to the maximum amount of fat that can be held in one piece.
4. Grades are discrete. No matter how detailed a score the teacher gives, the grades are isolated points on a scale.
5. This is almost infinite, but countable, and hence discrete.
6. This is continuous, as the temperature can take any value from room temperature to 100 degrees.

Discrete probability distribution

In Chapter 11, you learned how to work with the frequency distribution and relative or percentage frequency distribution for a set of numerical measurements on a variable X. The distribution gave the following information about X:

- The value of x that occurred.
- How often each value occurred.

You also learned how to use the mean and standard deviation to measure the centre and variability of the data set.

Here is an example of the frequency distribution of 25 families in Lower Austria that were polled in a marketing survey to list the number of litres of milk consumed during a particular week, reproduced on the next page. As you will observe, the table lists the number of litres consumed along with the relative frequency with which that number is observed. As you recall from Chapter 12, one of the interpretations of probability is that it is understood to be the long-term relative frequency of the event.

Number of litres	Relative frequency
0	0.08
1	0.20
2	0.36
3	0.20
4	0.12
5	0.04

Table 17.2

A table like this, where we replace the relative frequency with probability, is called a **probability distribution** of the random variable.

The **probability distribution** for a discrete random variable is a table, graph or formula that gives the possible values of X, and the probability $P(X = x)$ associated with each value of x.
This is also called the **probability mass function (pmf)** and in many sources it is called the **probability distribution function (pdf)**.

In other words, for every possible value x of the random variable X, the probability mass function specifies the probability of observing that value when the experiment is performed.

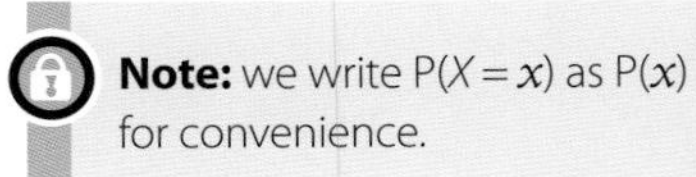

Note: we write $P(X = x)$ as $P(x)$ for convenience.

Letting x be the number of litres of milk consumed by a family above, the **probability distribution** of x would be as follows:

Table 17.3

x	0	1	2	3	4	5
$P(x)$	0.08	0.20	0.36	0.20	0.12	0.04

The other form of representing the probability distribution is with a histogram, as shown below. Every column corresponds to the probability of the associated value of x. The values of x naturally represent mutually exclusive events. Summing $P(x)$ over all values of x is equivalent to adding all probabilities of all simple events in the sample space, and hence the total is 1.

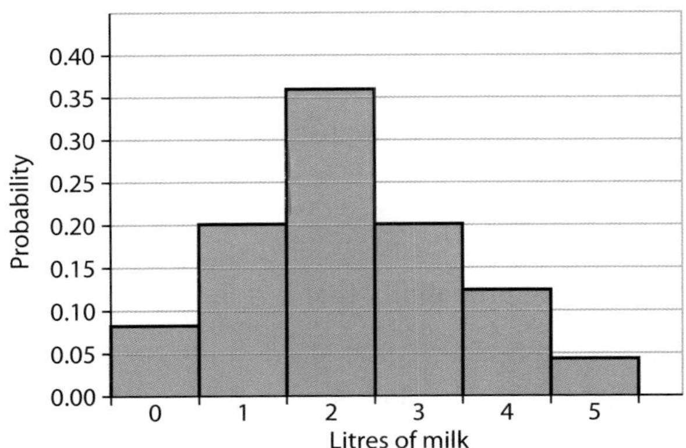

The result above can be generalized for all probability distributions:

Required properties of probability distribution functions of discrete random variables

Let X be a discrete random variable with probability distribution function, $P(x)$. Then:

- $0 \leqslant P(x) \leqslant 1$, for any value x.
- The individual probabilities sum to 1; that is, $\sum_x P(x) = 1$ where the notation indicates summation over all possible values x.

For some value x of the random variable X, we often wish to compute the probability that the observed value of X is at most x. This gives rise to the **cumulative distribution function (cdf)**.

Cumulative distribution function (cdf) (optional but very helpful)
The **cumulative distribution function** of a random variable X (also known as the 'cumulative probability function $F(x)$'), expresses the probability that X does not exceed the value x, as a function of x. That is,

$$F(x) = P(X \leqslant x) = \sum_{y:y \leqslant x} P(y)$$

The notation here indicates that summation is over all possible values of y that are less than or equal to x.

For example, in the milk consumption case, the cdf will look like the following table:

x	$F(x)$
0	0.08
1	0.28
2	0.64
3	0.84
4	0.96
5	1.00

So, $F(3) = 0.84$, stands for the probability of families that consume up to 3 litres of milk. This result of course can be achieved by adding the probabilities corresponding to $x = 0, 1, 2$ and 3.

In many cases, as we will see later, we use the cumulative distribution to find individual probabilities,

$$P(X = x) = P(X \leqslant x) - P(X < x).$$

For example, to find the probability that $x = 3$, we can use the cumulative distribution table.

$$P(x = 3) = P(x \leqslant 3) - P(x < 3) = 0.84 - 0.64 = 0.2$$

This property is of great value when studying the binomial and the Poisson distributions.

Example 2

Radon is a major cause of lung cancer. It is a radioactive gas produced by the natural decay of radium in the ground. Studies in areas rich with radium revealed that one-third of houses in these areas have dangerous levels of this gas. Suppose that two houses are randomly selected and we define the random variable X to be the number of houses with dangerous levels. Find the probability distribution of x by a table, a graph and a formula.

Solution

Since two houses are selected, the possible values of x are 0, 1 or 2. To find their probabilities, we utilize what we learned in Chapter 12. The assumption here is that we are choosing the houses randomly and independently of each other!

$P(x = 2) = P(2) = P(\text{1st house with gas } \textit{and} \text{ 2nd house with gas})$

$= P(\text{1st house with gas}) \times P(\text{2nd house with gas}) = \frac{1}{3} \times \frac{1}{3} = \frac{1}{9}$

$P(x = 0) = P(0) = P(\text{1st house without gas } \textit{and} \text{ 2nd house without gas})$

$= P(\text{1st house without gas}) \times P(\text{2nd house without gas})$
$= \frac{2}{3} \times \frac{2}{3} = \frac{4}{9}$

$P(x = 1) = 1 - [P(0) + P(2)] = 1 - \left[\frac{4}{9} + \frac{1}{9}\right] = \frac{4}{9}$

Table

x	0	1	2
P(x)	$\frac{4}{9}$	$\frac{4}{9}$	$\frac{1}{9}$

Graph

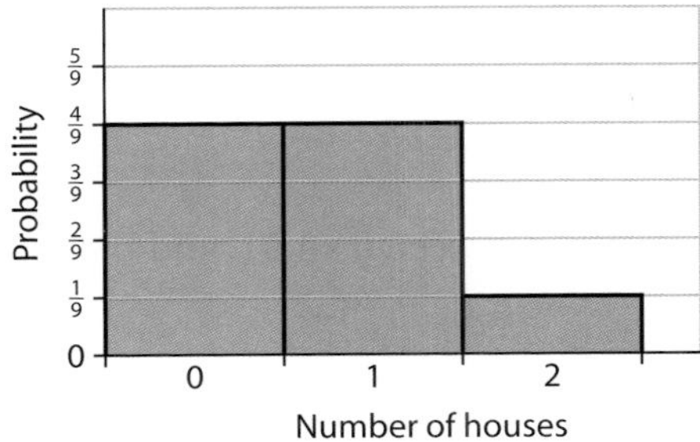

Any type of graph can be used to give the probability distribution, as long as it shows the possible values of x and the corresponding probabilities. The probability here is graphically displayed as the height of a rectangle. Moreover, the rectangle corresponding to each value of x has an area equal to the probability $P(X = x)$. The histogram is the preferred tool due to its connection to the continuous distributions discussed later in the chapter.

Formula/rule

The probability distribution of X can also be given by the following rule. Don't be concerned now with how we came up with this formula, as we will discuss it later in the chapter. The only reason we are looking at it now is to illustrate the fact that a formula/rule can sometimes be used to give the probability distribution.

$$P(x) = \binom{2}{x} \cdot \left(\frac{1}{3}\right)^x \cdot \left(\frac{2}{3}\right)^{2-x}$$

where $\binom{2}{x}$ represents the binomial coefficient you saw in Chapter 4.

Notice that when x is replaced by 0, 1 or 2 we obtain the results we are looking for:

$$P(0) = \binom{2}{0} \cdot \left(\frac{1}{3}\right)^0 \cdot \left(\frac{2}{3}\right)^{2-0} = 1 \cdot 1 \cdot \frac{4}{9} = \frac{4}{9}$$

$$P(1) = \binom{2}{1} \cdot \left(\frac{1}{3}\right)^1 \cdot \left(\frac{2}{3}\right)^{2-1} = 2 \cdot \frac{1}{3} \cdot \frac{2}{3} = \frac{4}{9}$$

$$P(2) = \binom{2}{2} \cdot \left(\frac{1}{3}\right)^2 \cdot \left(\frac{2}{3}\right)^{2-2} = 1 \cdot \frac{1}{9} \cdot 1 = \frac{1}{9}$$

Example 3

Many universities have the policy of posting the grade distributions for their courses. Several of the universities have a grade-point average that codes the grades in the following manner: A = 4, B = 3, C = 2, D = 1 and F = 0. During the spring term at a certain large university, 13% of the students in an introductory statistics course received A's, 37% B's, 45% C's, 4% received D's and 1% received F's. The experiment here is to choose a student at random and mark down his/her grade. The student's grade on the 4-point scale is a random variable X.

Here is the probability distribution of X:

x	0	1	2	3	4
P(x)	0.01	0.04	0.45	0.37	0.13

Is this a probability distribution?

Solution

Yes, it is. Each probability is between 0 and 1, and the sum of all probabilities is 1.

What is the probability that a randomly chosen student receives a B or better?

$$P(x \geqslant 3) = P(x = 3) + P(x = 4) = 0.37 + 0.13 = 0.40$$

Example 4

In the codes example in Chapter 12, we saw the probability with which people choose the first digits for the codes for their cellphones. The probability distribution is copied below for reference.

First digit	0	1	2	3	4	5	6	7	8	9
Probability	0.009	0.300	0.174	0.122	0.096	0.078	0.067	0.058	0.051	0.045

Here, X is the first digit chosen.

What is the probability that you pick a first digit and it is more than 5? Show a probability histogram for the distribution.

Solution

$$P(x > 5) = P(x = 6) + P(x = 7) + P(x = 8) + P(x = 9) = 0.221$$

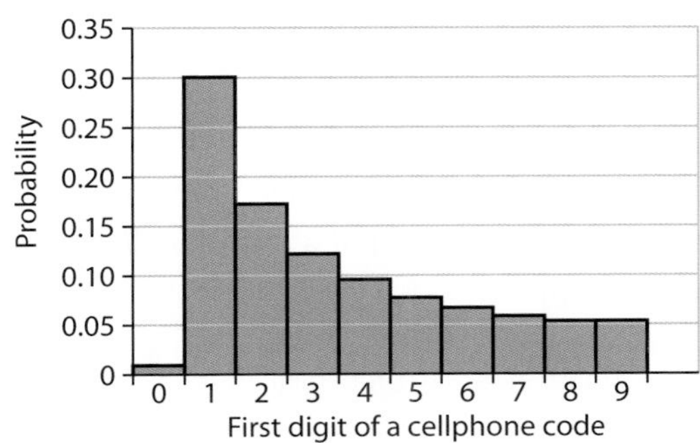

Note that the height of each bar shows the probability of the outcome at its base. The heights add up to 1, of course. The bars in this histogram have the same width, namely 1. So, the areas also display the probability assignments of the outcomes. Think of such histograms (probability histograms) as idealized pictures of the results of very many repeated trials.

Expected values

The probability distribution for a random variable looks very similar to the relative frequency distribution discussed in Chapter 11. The difference is that the relative frequency distribution describes a *sample* of measurements, whereas the probability distribution is constructed as a *model* for the *entire population.* Just as the mean and standard deviation gave you measures for the centre and spread of the sample data, you can calculate similar measures to describe the centre and spread of the population.

The population mean, which measures the average value of X in the population, is also called the **expected value** of the random variable X. It is the value that you would *expect* to observe *on average* if you repeat the experiment an infinite number of times. The formula we use to determine the expected value can be simply understood with an example.

Let's revisit the milk consumption example. Let X be the number of litres consumed. Here is the table of probabilities again:

x	0	1	2	3	4	5
P(x)	0.08	0.20	0.36	0.20	0.12	0.04

Suppose we choose a large number of families, say 100 000. Intuitively, using the relative frequency concept of probability, you would expect to observe 8000 families consuming no milk, 20 000 consuming 1 litre, and the rest similarly done: 36 000, 20 000, 12 000 and 4000.

The average (mean) value of X, as defined in Chapter 11, would then be equal to

$$\frac{\text{sum of all measurements}}{n}$$

$$= \frac{0\cdot 8000 + 1\cdot 20\,000 + 2\cdot 36\,000 + 3\cdot 20\,000 + 4\cdot 12\,000 + 5\cdot 4000}{100\,000}$$

$$= \frac{0\cdot 8000}{100\,000} + \frac{1\cdot 20\,000}{100\,000} + \frac{2\cdot 36\,000}{100\,000} + \frac{3\cdot 20\,000}{100\,000} + \frac{4\cdot 12\,000}{100\,000} + \frac{5\cdot 4000}{100\,000}$$

$$= 0\cdot 0.08 + 1\cdot 0.20 + 2\cdot 0.36 + 3\cdot 0.20 + 4\cdot 0.12 + 5\cdot 0.04$$

$$= 0\cdot \text{P}(0) + 1\cdot \text{P}(1) + 2\cdot \text{P}(2) + 3\cdot \text{P}(3) + 4\cdot \text{P}(4) + 5\cdot \text{P}(5) = 2.2$$

That is, we expect to see families, *on average,* consuming 2.2 litres of milk! This does not mean that we know what a family *will* consume, but we can say what we *expect* to happen.

Let X be a discrete random variable with probability distribution $\text{P}(x)$. The mean or **expected value** of X is given by

$$\mu = \text{E}(X) = \sum x\cdot \text{P}(x).$$

Insurance companies make extensive use of expected value calculations. Here is a simplified example.

An insurance company offers a policy that pays you €10 000 when you totally damage your car or €5000 for major damages (50%). They charge you €50 per year for this service. The question is, how can they make a profit?

To understand how they can afford this, suppose that the 'total damage' car accident rate, in any year, is 1 out of every 1000 cars, and that another 2 out of 1000 will have serious damages. Then we can display the probability model for this policy in a table like this:

Type of accident	Amount paid x	Probability $\text{P}(X = x)$
Total damage	10 000	$\frac{1}{1000}$
Major damage	5000	$\frac{2}{1000}$
Minor or no damage	0	$\frac{997}{1000}$

The expected amount the insurance company pays is given by:

$$\mu = \text{E}(X) = \sum x\text{P}(x) = €10\,000\left(\frac{1}{1000}\right) + €5000\left(\frac{2}{1000}\right)$$
$$+ €0\left(\frac{997}{1000}\right) = €20$$

This means that the insurance company *expects* to pay, on average, an amount of €20 per insured car. Since it is charging people €50 for the policy, the company *expects* to make a profit of €30 per car. Thinking about the problem in a different perspective, suppose they insure 1000 cars, then the company would expect to pay €10 000 for 1 car and €5000 to each of two cars with major damage. This is a total of €20 000 for all cars, or an average of $\frac{20\,000}{1000} = €20$ per car.

Of course, this expected value is not what actually happens to any *particular* policy. No individual policy actually costs the insurance company €20. We are dealing with random events, so a few car owners may require a payment of €10 000 or €5000, many others receive *nothing*!

Because of the need to anticipate such variability, the insurance company needs to know a measure of this variability, which is nothing but the **standard deviation**.

Variance and standard deviation

For data in Chapter 11, we calculated the variance by computing the deviation from the mean, $x - \mu$, and then squaring it. We do that with random variables as well.

We can use similar arguments to justify the formulae for the population variance σ^2 and, consequently, the population standard deviation σ. These measures describe the spread of the values of the random variable around the centre. We similarly use the idea of the 'average' or 'expected' value of the squared deviations of the x-values from the mean μ or $E(x)$.

Let X be a discrete random variable with probability distribution $P(x)$ and mean μ. The **variance of X** is given by

$$\sigma^2 = E((X - \mu)^2) = \sum (x - \mu)^2 \cdot P(x).$$

(This is sometimes called Var(X).)

Note: It can also be shown, similar to what you saw in Chapter 11, that you have another 'computation' formula for the variance:

$$\sigma^2 = \sum (x - \mu)^2 \cdot P(x) = \sum x^2 \cdot P(x) - \mu^2 = \sum x^2 \cdot P(x) - [E(x)]^2$$
$$= \sum x^2 \cdot P(x) - \left[\sum x P(x)\right]^2$$

The **standard deviation** σ of a random variable X is equal to the positive square root of its variance.

Let us go back to the milk consumption example. Recall that we calculated the expected value, mean, to be 2.2 litres. In order to calculate the variance, we can tabulate our work to make the manual calculation simple.

x	$P(x)$	Deviation $(x - \mu)$	Squared deviation $(x - \mu)^2$	$(x - \mu)^2 \cdot P(x)$
0	0.08	−2.2	4.84	0.3872
1	0.20	−1.2	1.44	0.2880
2	0.36	−0.2	0.04	0.0144
3	0.20	0.8	0.64	0.1280
4	0.12	1.8	3.24	0.3888
5	0.04	2.8	7.84	0.3136
		Total	$\sum (x - \mu)^2 \cdot P(x)$	1.52

So, the variance of the milk consumption is 1.52 litres^2, or the standard deviation is 1.233 litres.

GDC notes

The above calculations, along with the expected value calculation, can be easily done using your GDC.

First, store x and P(x) into L1 and L2.

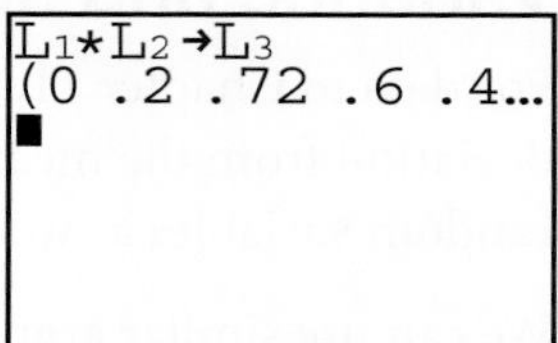

Then, to find x P(x), we multiply L1 and L2 and store the result in L3.

```
L1*L2→L3
(0 .2 .72 .6 .4…
```

To find the expected value, you simply get the sum of the entries in L3, since they correspond to $\sum x \cdot P(x)$.

```
L1*L2→L3
(0 .2 .72 .6 .4…
sum(L3)
                2.2
```

To find the variance, we need to find the deviations from the mean; so we make L4 that deviation, i.e. we store L1 − 2.2 into L4. Then, to get the squared deviations multiplied by the corresponding probability, we set up L5 to be L4 squared multiplied by L2, the probability. Now, to find the variance, just add the terms of L5.

```
L1-2.2→L4
(-2.2 -1.2 -.2…
(L4)²*L2→L5
(.3872 .228 .01…
sum(L5)
               1.52
```

Software note

In the comfort of home/class, the above calculation can be performed on a computer with a simple spreadsheet like the following one:

x	P(x)	xP(x)	$x - \mu$	$(x - \mu)^2$	$(x - \mu)^2$P(x)	
0	0.08	0	−2.2	4.84	0.3872	
1	0.2	0.2	−1.2	1.44	0.288	
2	0.36	0.72	−0.2	0.04	0.0144	
3	0.2	0.6	0.8	0.64	0.128	
4	0.12	0.48	1.8	3.24	0.3888	E4*B4
5	0.04	0.2	2.8	7.84	0.3136	
Totals	1	2.2			1.52	
				A3 − 2.2	E6^2	
	A2*B2		SUM(C2:C7)			

Example 5

A computer store sells a particular type of laptop. The daily demand for the laptops is given in the table below. X is the number of laptops in demand. They have only 4 laptops left in stock and would like to know how well they are prepared for all eventualities. Work out the expected value of the demand and the standard deviation.

x	0	1	2	3	4	5
$P(X = x)$	0.08	0.40	0.24	0.15	0.08	0.05

Solution

$$E(X) = \sum xP(x) = 0 \times 0.08 + 1 \times 0.40 + 2 \times 0.24 + 3 \times 0.15 + 4 \times 0.08 + 5 \times 0.05 = 1.90$$

$$\begin{aligned}\mathrm{Var}(X) &= \sigma^2 = \sum (x - \mu)^2 P(x)\\ &= (0 - 1.9)^2 \cdot 0.08 + (1 - 1.9)^2 \cdot 0.40 + (2 - 1.9)^2 \cdot 0.24\\ &\quad + (3 - 1.9)^2 \cdot 0.15 + (4 - 1.9)^2 \cdot 0.08 + (5 - 1.9)^2 \cdot 0.05\\ &= 1.63\end{aligned}$$

$$\sigma = 1.28$$

Spreadsheet output is also given.

x	P(x)	xP(x)	$x - \mu$	$(x - \mu)^2$	$(x - \mu)^2P(x)$
0	0.08	0	−1.9	3.61	0.2888
1	0.4	0.4	−0.9	0.81	0.324
2	0.24	0.48	0.1	0.01	0.0024
3	0.15	0.45	1.1	1.21	0.1815
4	0.08	0.32	2.1	4.41	0.3528
5	0.05	0.25	3.1	9.61	0.4805
Totals	1	1.9			1.63

The graph of the probability distribution is given below.

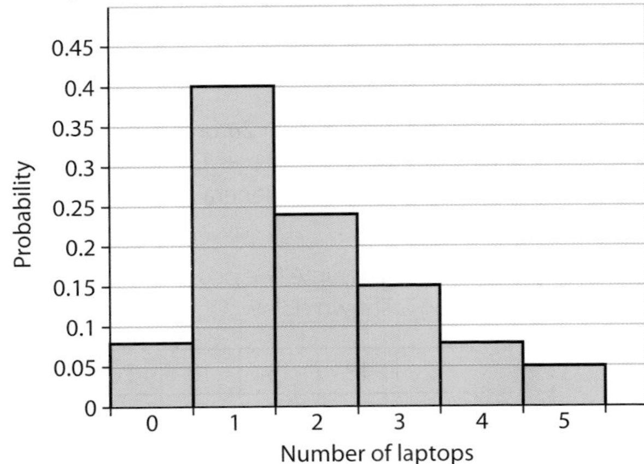

As an approximation, we can use the *empirical rule* to see where most of the demand is expected to be. Recall that the empirical rule tells us that about 95% of the values would lie within 2 standard deviations from the mean. In this case $\mu \pm 2\sigma = 1.9 \pm 2 \times 1.28 \Rightarrow (-0.66, 4.46)$. This interval does not contain the 5 units of demand. We can say that it is unlikely that 5 or more customers of this shop will want to buy a laptop today.

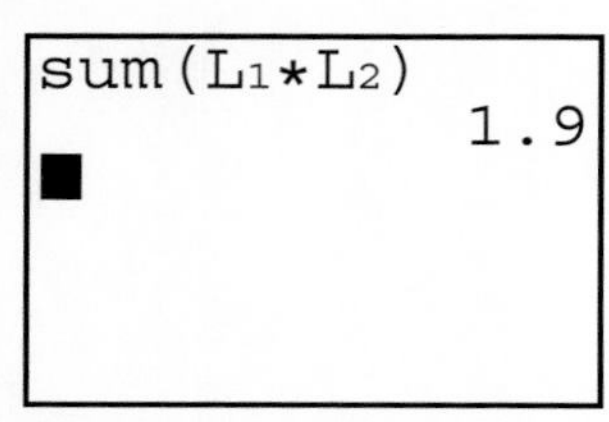


```
L1*L2→L3
(0 .4 .48 .45 ....
(L1-1.9)²*L2→L5
(.2888 .324 .00...
sum(L5)
                1.63
```

GDC

After entering the demand in L1 and the probabilities in L2, it is enough to find the sum of their product.

For the variance, we follow the same procedure as described in the previous example, see left.

Notice here that we combined several steps in one.

Exercise 17.1

1 Classify each of the following as discrete or continuous random variables.

a) The number of words spelled correctly by a student on a spelling test.
b) The amount of water flowing through the Niagara Falls per year.
c) The length of time a student is late to class.
d) The number of bacteria per cc of drinking water in Geneva.
e) The amount of CO produced per litre of unleaded gas.
f) The amount of a flu vaccine in a syringe.
g) The heart rate of a lab mouse.
h) The barometric pressure at Mount Everest.
i) The distance travelled by a taxi driver per day.
j) Total score of football teams in national leagues.
k) Height of ocean tides on the shores of Portugal.
l) Tensile breaking strength (in newtons per square metre) of a 5 cm diameter steel cable.
m) Number of overdue books in a public library.

2 A random variable Y has this probability distribution:

y	0	1	2	3	4	5
$P(y)$	0.1	0.3		0.1	0.05	0.05

a) Find P(2).
b) Construct a probability histogram for this distribution.
c) Find μ and σ.
d) Locate the interval $\mu \pm \sigma$ as well as $\mu \pm 2\sigma$ on the histogram.
e) We create another random variable $Z = b + 1$. Find μ and σ of Z.
f) Compare your results for c) and e) and generalize for $Z = Y + b$, where b is a constant.

3 A discrete random variable X can assume five possible values: 12, 13, 15, 18 and 20. Its probability distribution is shown below.

x	12	13	15	18	20
$P(x)$	0.14	0.11		0.26	0.23

a) What is P(15)?
b) What is the probability that x equals 12 or 20?
c) What is $P(x \leqslant 18)$?
d) Find E(X).
e) Find V(X).
f) Let $Y = 0.5X - 4$. Find E(Y) and V(Y).
g) Compare your results in d), e) and f) and generalize for $Y = aX + b$, where a and b are constants.

4 Medical research has shown that a certain type of chemotherapy is successful 70% of the time when used to treat skin cancer. In a study to check the validity of such a claim, researchers chose different treatment centres and chose five of their patients at random. Here is the probability distribution of the number of successful treatments for groups of five:

x	0	1	2	3	4	5
$P(x)$	0.002	0.029	0.132	0.309	0.360	0.168

a) Find the probability that at least two patients would benefit from the treatment.

b) Find the probability that the majority of the group does not benefit from the treatment.

c) Find $E(X)$ and interpret the result.

d) Show that $\sigma(X) = 1.02$.

e) Graph $P(x)$. Locate μ, $\mu \pm \sigma$ and $\mu \pm 2\sigma$ on the graph. Use the empirical rule to approximate the probability that x falls in this interval. Compare this with the actual probability.

5 The probability function of a discrete random variable X is given by

$$P(X = x) = \frac{kx}{2}, \text{ for } x = 12, 14, 16, 18.$$

Set up the table showing the probability distribution and find the value of k.

6 X has probability distribution as shown in the table.

x	5	10	15	20	25
$P(x)$	$\frac{3}{20}$	$\frac{7}{30}$	k	$\frac{3}{10}$	$\frac{13}{60}$

a) Find the value of k.

b) Find $P(x > 10)$.

c) Find $P(5 < x \leqslant 20)$.

d) Find the expected value and the standard deviation.

e) Let $Y = \frac{1}{5}X - 1$. Find $E(Y)$ and $V(Y)$.

7 The discrete random variable Y has a probability density function

$$P(Y = y) = k(16 - y^2), \text{ for } y = 0, 1, 2, 3, 4.$$

a) Find the value of the constant k.

b) Draw a histogram to illustrate the distribution.

c) Find $P(1 \leqslant y \leqslant 3)$.

d) Find the mean and variance.

8 The probability distribution of students categorized by age that visit a certain movie house on weekends is given on the right. The probabilities for 18- and 19-year-olds are missing. We know that

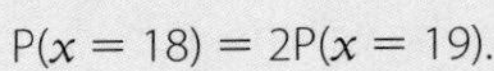

$$P(x = 18) = 2P(x = 19).$$

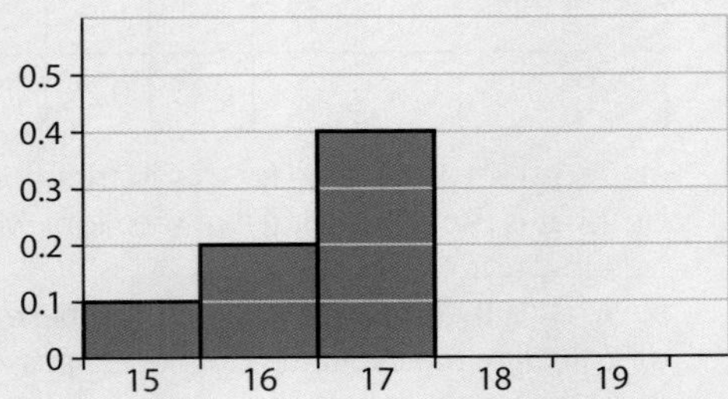

a) Complete the histogram and describe the distribution.

b) Find the expected value and the variance.

9 In a small town, a computer store sells laptops to the local residents. However, due to low demand, they like to keep their stock at a manageable level. The data they have indicate that the weekly demand for the laptops they sell follows the distribution given in the table below.

X: number of laptops bought	0	1	2	3	4	5
$P(X = x)$	0.10	0.40	0.20	0.15	0.10	0.05

a) Find the mean and standard deviation of this distribution.
b) Use the empirical rule to find the approximate number of laptops that is sold about 95% of the time.

10 The discrete random variable X has probability function given by

$$P(x) = \begin{cases} \left(\frac{1}{4}\right)^{x-1} & x = 2, 3, 4, 5, 6 \\ k & x = 7 \\ 0 & \text{otherwise} \end{cases}$$

where k is a constant. Determine the value of k and the expected value of X.

11 The following is a probability distribution for a random variable Y.

y	0	1	2	3
$P(Y = y)$	0.1	0.11	k	$(k - 1)^2$

a) Find the value of k.
b) Find the expected value.

12 A closed box contains eight red balls and four white ones. A ball is taken out at random, its colour noted, and then returned. This is done three times. Let X represent the number of red balls drawn.
a) Set up a table to show the probability distribution of X.
b) What is the expected number of red balls in this experiment?

13 A discrete random variable Y has the following probability distribution function:
$P(Y = y) = k(4 - y)$, for $y = 0, 1, 2, 3$ and 4.
a) Find the value of k.
b) Find $P(1 \leqslant y < 3)$.

14 Airlines sometimes overbook flights. Suppose for a 50-seat plane, 55 tickets were sold. Let X be the number of ticketed passengers that show up for the flight. From records, the airline has the following **pmf** for this flight.

x	45	46	47	48	49	50	51	52	53	54	55
$P(x)$	0.05	0.08	0.12	0.15	0.25	0.20	0.05	0.04	0.03	0.02	0.01

a) Construct a **cdf** table for this distribution.
b) What is the probability that the flight will accommodate all ticket holders that show up?
c) What is the probability that not all ticket holders will have a seat on the flight?
d) Calculate the expected number of passengers who will show up.
e) Calculate the standard deviation of the passengers who will show up.
f) Calculate the probability that the number of passengers showing up will be within one standard deviation of the expected number.

15 A small internet provider has 6 telephone service lines operating 24-hours daily. Defining X as the number of lines in use at any specific 10-minute period of the day, the **pmf** of X is given in the following table.

x	0	1	2	3	4	5	6
P(x)	0.08	0.15	0.22	0.27	0.20	0.05	0.03

a) Construct a **cdf** table.
b) Calculate the probability that at most three lines are in use.
c) Calculate the probability that a customer calling for service will have a free line.
d) Calculate the expected number of lines in use.
e) Calculate the standard deviation of the number of lines in use.

16 Some flashlights use one AA-type battery. The voltage in any new battery is considered acceptable if it is at least 1.3 volts. 90% of the AA batteries from a specific supplier have an acceptable voltage. Batteries are usually tested till an acceptable one is found. Then it is installed in the flashlight. Let X be the number of batteries that must be tested.
a) What is P(1), i.e. P($x = 1$)?
b) What is P(2)?
c) What is P(3)?
d) To have $x = 5$, what must be true of the fourth battery tested? of the fifth one?
e) Use your observations above to obtain a general model for P(x).

17 Repeat question 16, but now consider the flashlight as needing two batteries.

18 A biased die with four faces is used in a game. A player pays 10 counters to roll the die. The table below shows the possible scores on the die, the probability of each score and the number of counters the player receives in return for each score.

Score	1	2	3	4
Probability	$\frac{1}{2}$	$\frac{1}{5}$	$\frac{1}{5}$	$\frac{1}{10}$
Number of counters player receives	4	5	15	n

Find the value of n in order for the player to get an expected return of 9 counters per roll.

19 Two children, Alan and Belle, each throw two fair cubical dice simultaneously. The score for each child is the sum of the two numbers shown on their respective dice.
a) (i) Calculate the probability that Alan obtains a score of 9.
(ii) Calculate the probability that Alan and Belle both obtain a score of 9.
b) (i) Calculate the probability that Alan and Belle obtain the same score.
(ii) Deduce the probability that Alan's score exceeds Belle's score.
c) Let X denote the largest number shown on the four dice.
(i) Show that for $P(X \leqslant x) = \left(\frac{x}{6}\right)^4$, for $x = 1, 2, \ldots, 6$.
(ii) Copy and complete the following probability distribution table.

x	1	2	3	4	5	6
$P(X = x)$	$\frac{1}{1296}$	$\frac{15}{1296}$				$\frac{671}{1296}$

(iii) Calculate E(X).

20 Consider the 10 data items $x_1, x_2, \ldots, x_{10}$. Given that $\sum_{i=1}^{10} x^2_i = 1341$ and the standard deviation is 6.9, find the value of $\bar{x}$.

Questions 18–20, © International Baccalaureate Organization

17.2 The binomial distribution

Examples of discrete random variables are abundant in everyday situations. However, there are a few discrete probability distributions that are widely applied and serve as *models* for a great number of the applications. In this book, we will study two of them only: the **binomial distribution** and the Poisson distribution.

We will start with the basis of the binomial distribution.

Bernoulli distribution

If an experiment has two possible outcomes, '*success*' and '*failure*', and their probabilities are p and $1 - p$, respectively, then the number of successes, 0 or 1, has a **Bernoulli distribution**.

A discrete random variable X has a Bernoulli distribution if and only if it has two possible outcomes labelled by $x = 0$ and $x = 1$ in which $x = 1$ ('success') occurs with probability p and $x = 0$ ('failure') occurs with probability $1 - p$, where $0 < p < 1$. It therefore has probability function

$$p(x) = \begin{cases} 1 - p & \text{for } x = 0 \\ p & \text{for } x = 1, \end{cases}$$

which can also be written as

$$p(x) = p^x(1 - p)^{1-x}, x = 0, 1.$$

The corresponding distribution function is

$$D(x) = \begin{cases} 1 - p & \text{for } x = 0 \\ 1 & \text{for } x = 1. \end{cases}$$

A sequence of **Bernoulli trials** occurs when a Bernoulli experiment is performed several independent times so that the probability of success, p, remains the same from trial to trial. In addition, we frequently use q to denote the probability of failure, i.e. $q = 1 - p$.

The distribution of heads and tails in coin tossing is an example of a Bernoulli distribution with $p = q = \frac{1}{2}$.

The Bernoulli distribution is one of the simplest discrete distributions, and it is the basis for other more complex discrete distributions. The definitions of a few types of distributions based on sequences of independent Bernoulli trials are summarized in the following table:

Distribution[1]	Definition
Binomial distribution	number of successes in n trials
Geometric distribution	number of failures before the first success
Negative binomial distribution	number of failures before the xth success

[1]These distributions will be discussed in more detail in the Options part.

In this part of the book, we will study the binomial distribution. The other two will be discussed in the options section.

Expected value and variance

The mean of a random variable X that has a Bernoulli distribution with parameter p is:

$$\begin{aligned} \mathrm{E}(X) &= \sum_x xp(x) \\ &= 1(p) + 0(1 - p) = p \end{aligned}$$

The variance of X is:

$$\begin{aligned} Var(X) &= \mathrm{E}(X^2) - (\mathrm{E}(X))^2 \\ &= \sum_x x^2 p(x) - p^2 \\ &= 1^2 \cdot p + 0^2(1 - p) - p^2 \\ &= p - p^2 = p(1 - p) = pq \end{aligned}$$

The binomial distribution

We will start our discussion of the binomial distribution with an example.

Suppose a cereal company puts miniature figures in boxes of cornflakes to make them attractive for children and thus boost sales. The manufacturer claims that 20% of the boxes contain a figure. You buy three boxes of this cereal. What is the probability that you'll get exactly three figures?

To get three figures means that the first box contains a figure (0.20 chance), as does the second (also 0.20), and the third (0.20). You want three figures; therefore, this is the intersection of three events and the probability is simply $0.20^3 = 0.008$.

If you want to calculate the probability of getting exactly two figures, the situation becomes more complicated. A tree diagram can help you visualize it better.

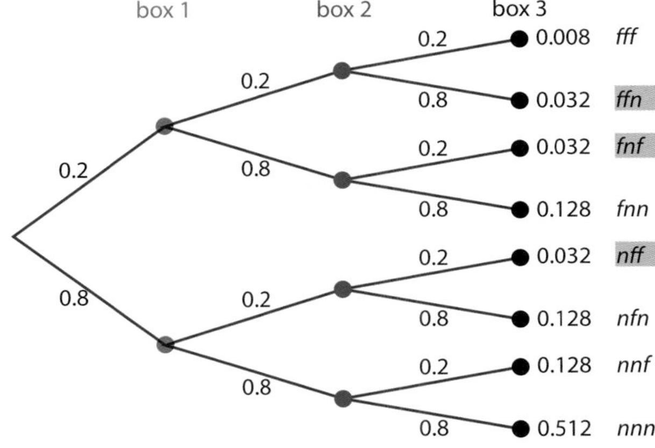

Let f stand for figure and n for no figure. There are three events of interest to us. Since we are interested in two figures, we want to see ffn, which has a probability of $0.2 \times 0.2 \times 0.8 = 0.2^2 \times 0.8 = 0.032$, and the other events of interest are *fnf* and *nff*, with probabilities $0.2 \times 0.8 \times 0.2 = 0.032$ and $0.8 \times 0.2 \times 0.2 = 0.032$.

Since the order of multiplication is not important, you see that three probabilities are the same. These three events are disjoint, as can be clearly seen from the tree diagram, and hence the probability of exactly two figures is the sum of the three numbers: $0.032 + 0.032 + 0.032$. Of course, you may realize by now that it would be much simpler if you wrote $3(0.032)$, since there are three events with the same probability.

What if you have five boxes?

The situation is similar, of course. However, a tree diagram would not be useful in this case, as there is too much information to assemble to see the solution. As you have seen above, no matter how you succeed in finding a figure, whether it is in the first box, the second or the third, it has the same probability, 0.2. So, to have two successes (finding figures) in the five boxes, you need the other three to be failures (no figures), with a probability of 0.8 for each failure. Therefore, the chance of having a case like *ffnnn* is $0.2^2 \times 0.8^3$. However, this can happen in several disjoint ways. How many? If you count them, you will find 10. This means the probability of having exactly two figures in five boxes is $10 \times 0.2^2 \times 0.8^3 = 0.2048$.
(Here are the 10 possibilities: *ffnnn*, *fnfnn*, *fnnfn*, *fnnnf*, *nffnn*, *nnffn*, *nnnff*, *nfnfn*, *nnfnf*, *nfnnf*.)

The number 10 is nothing but the *binomial* coefficient (Pascal's entry) you saw in Chapter 4. This is also the 'combination' of three events out of five.

The previous result can be written as $\binom{5}{2}0.2^2 \cdot 0.8^3$, where $\binom{5}{2}$ is the binomial coefficient.

You can find experiments like this one in many situations. Coin-tossing is only a simple example of this. Another very common example is opinion polls which are conducted before elections and used to predict voter preferences. Each sampled person can be compared to a coin – but a biased coin! A voter you sample in favour of your candidate can correspond to either a 'head' or a 'tail' on a coin. Such experiments all exhibit the typical characteristics of the **binomial experiment**.

A **binomial experiment** is one that has the following five characteristics:

1. The experiment consists of n identical trials.
2. Each trial has one of two outcomes. We call one of them success, S, and the other failure, F.
3. The probability of success on a single trial, p, is constant throughout the whole experiment. The probability of failure is $1 - p$, which is sometimes denoted by q. That is, $p + q = 1$.
4. The trials are independent.
5. We are interested in the number of successes x that are possible during the n trials. That is, $x = 0, 1, 2, \ldots, n$.

In the cereal company's example above, we started with $n = 3$ and $p = 0.2$ and asked for the probability of two successes, i.e. $x = 2$. In the second part, we have $n = 5$.

Let us imagine repeating a binomial experiment n times. If the probability of success is p, the probability of having x successes is $pppp\ldots$, x times (p^x), because the order is not important, as we saw before. However, in order to have exactly x successes, the rest, $(n - x)$ trials, must be failures, that is, with probability of $qqqq\ldots$, $(n - x)$ times (q^{n-x}). This is only one order (combination) where the successes happen the first x times and the rest are failures. In order to cater for 'all orders', we have to count the number of orders (combinations) possible. This is given by the binomial coefficient $\binom{n}{x}$.

We will state the following result without proof.

The binomial distribution

Suppose that a random experiment can result in two possible mutually exclusive and collectively exhaustive outcomes, 'success' and 'failure,' and that p is the probability of a success resulting in a single trial. If n independent trials are carried out, the distribution of the number of successes 'x' resulting is called the **binomial distribution**. Its probability distribution function for the binomial random variable X is:

$$\begin{aligned} \text{P}(x \text{ successes in } n \text{ independent trials}) &= \text{P}(x) = \binom{n}{x}p^x(1-p)^{n-x} \\ &= \binom{n}{x}p^x q^{n-x}, \text{ for } x = 0, 1, 2, \ldots, n. \end{aligned}$$

Notation:

The notation used to indicate that a variable has a binomial probability distribution with n trials and success probability of p is: $X \sim \text{B}(n, p)$.

Example 6

The computer shop orders its notebooks from a supplier, which like many suppliers has a rate of defective items of 10%. The shop usually takes a sample of 10 computers and checks them for defects. If they find two computers defective, they return the shipment. What is the probability that their random sample will contain two defective computers?

Solution

We will consider this to be a random sample and the shipment large enough to render the trials independent of each other. The probability of finding two defective computers in a sample of 10 is given by

$$\text{P}(x = 2) = \binom{10}{2}0.1^2 0.9^{10-2} = 45 \times 0.01 \times 0.43047 = 0.194.$$

Of course, it is a daunting task to do all the calculations by hand. A GDC can do this calculation for you in two different ways.

The first possibility is to let the calculator do all the calculations in the formula above: Go to the math menu, then choose PRB, then go to #3.

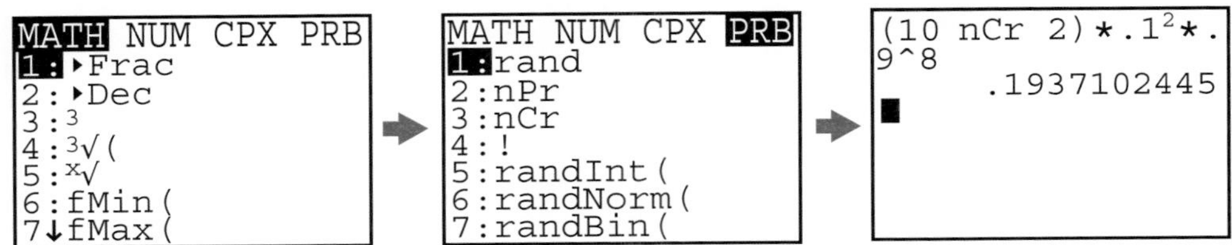

The second one is direct. We go to the 'DISTR' button, then scroll down to 'binompdf' and write down the two parameters followed by the number of successes:

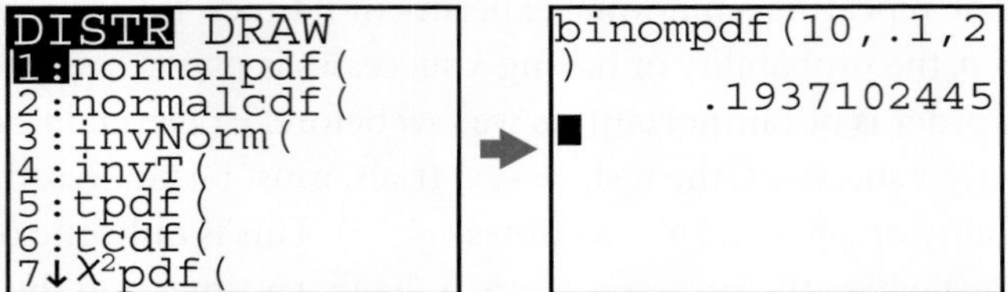

Using a spreadsheet, you can also produce this result or even a set of probabilities covering all the possible values. The command used here for Excel is (BINOMDIST(B1:G1,10,0.1,FALSE)) which produced the table below:

x	0.00	1.00	**2.00**	3.00	4.00	5.00	6.00	7.00	8.00	9.00	10.00
P(x)	0.349	0.387	**0.194**	0.057	0.011	0.001	0.000	0.000	0.000	0.000	0.000

Similarly, the GDC can also give you a list of the probabilities:

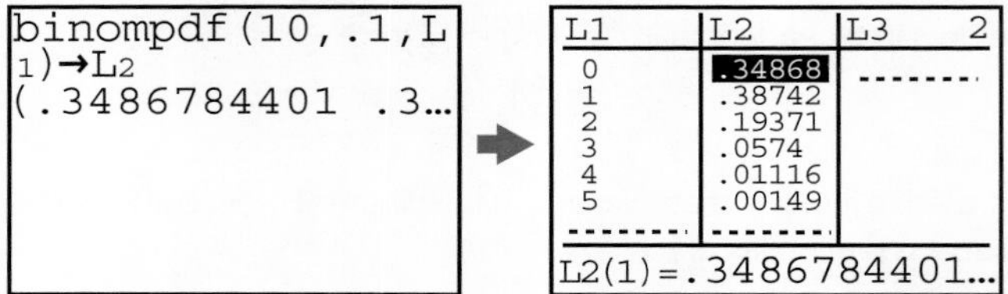

Like other distributions, when you look at the binomial distribution, you want to look at its expected value and standard deviation.

Using the formula we developed for the expected value, $\sum x\text{P}(x)$, we can of course add $x\text{P}(x)$ for all the values involved in the experiment. The process would be long and tedious for something we can intuitively know. For example, in the defective items sample, if we know that the defective rate of the computer manufacturer is 10%, it is natural to *expect* to have $10 \times 0.1 = 1$ defective computer! If we have 100 computers with a defective rate of 10%, how many would you expect to be defective? Can you think of a reason why it would not be 10?

This is so simple that many people would not even consider it. The expected value of the successes in the binomial is actually nothing but the number of trials n multiplied by the probability of success, i.e. np!

The binomial probability model

n = number of trials

p = probability of success, probability of failure $q = 1 - p$

x = number of successes in n trials

$\text{P}(x) = \binom{n}{x}p^x(1-p)^{n-x} = \binom{n}{x}p^x q^{n-x}$, for $x = 0, 1, 2, \ldots, n$

Expected value $= \mu = np$

Variance $= \sigma^2 = npq$, $\sigma = \sqrt{npq}$

So, in the defective notebooks case, the expected number of defective items in the sample of 10 is $np = 10 \times 0.1 = 1$!

And the standard deviation is $\sigma = \sqrt{npq} = \sqrt{10 \times 0.1 \times 0.9} = 0.949$.

Question:
How do we know that the binomial distribution is a probability distribution?

Answer:
We can easily verify that the binomial distribution as developed satisfies the probability distribution conditions:

1. $0 \leqslant p(x) \leqslant 1$
2. $\sum_x p(x) = 1$

1. Since $p > 0$ by definition, then $p^x > 0$, for $x = 0, 1, 2, \ldots$. Similarly, $q^{n-x} > 0$. We also know that $\binom{n}{x} > 0$. Therefore,

$$p(x) = \binom{n}{x} p^x q^{n-x} > 0.$$

$p(x) \leqslant 1$ will be a natural result of proving the second condition. If the sum of n positive parts is equal to 1, none of the parts can be greater than 1!

2. $\sum_{x=0}^{n} p(x) = \sum_{x=0}^{n} \binom{n}{x} p^x q^{n-x}$

Recalling from Chapter 4, that the binomial theorem states

$$(p + q)^n = \sum_{x=0}^{n} \binom{n}{x} p^x q^{n-x} = \sum_{x=0}^{n} p(x).$$

Since $p + q = 1$, then $(p + q)^n = 1$, and therefore

$$\sum_{x=0}^{n} p(x) = \sum_{x=0}^{n} \binom{n}{x} p^x q^{n-x} = (p + q)^n = 1.$$

Expected value of the binomial (optional)

$$\mathrm{E}(X) = \sum_{x=0}^{n} x p(x) = \sum_{x=0}^{n} x \binom{n}{x} p^x q^{n-x}$$

Notice that when $x = 0$, the first term in the summation equation is 0. Hence,

$$\mathrm{E}(X) = \sum_{x=0}^{n} x \binom{n}{x} p^x q^{n-x} = \sum_{x=1}^{n} x \binom{n}{x} p^x q^{n-x}$$

$$= \sum_{x=1}^{n} x \frac{n!}{(n-x)!x!} p^x q^{n-x} = \sum_{x=1}^{n} \frac{n!}{(n-x)!(x-1)!} p^x q^{n-x}$$

$$= \sum_{x=1}^{n} \frac{n(n-1)!}{(n-x)!(x-1)!} p p^{x-1} q^{n-x}$$

n and p are independent of x, so they can be factored out of the summation.

$$\text{E}(X) = \sum_{x=1}^{n} \frac{n(n-1)!}{(n-x)!(x-1)!} pp^{x-1}q^{n-x}$$

$$= np\sum_{x=1}^{n} \frac{(n-1)!}{(n-x)!(x-1)!} p^{x-1}q^{n-x}$$

The term in the summation expression appears to be nothing but the probability of $(x-1)$ successes among $(n-1)$ trials.

$$\sum_{x=1}^{n} \frac{(n-1)!}{(n-x)!(x-1)!} p^{x-1}q^{n-x}$$

$$= \sum_{x=1}^{n} \frac{(n-1)!}{(n-1-(x-1)!(x-1))!} p^{x-1}q^{n-1-(x-1)}$$

If you replace $x-1$ by y and $n-1$ by m, then,

$$\sum_{x=1}^{n} \frac{(n-1)!}{(n-1-(x-1)!(x-1))!} p^{x-1}q^{n-1-(x-1)}$$

$$= \sum_{y=0}^{m} \frac{m!}{(m-y)!y!} p^{y}q^{m-y} = \sum_{y} p(y) = 1$$

This is nothing but the sum of all the probabilities of the random variable $Y = X - 1$ successes in $m = n - 1$ trials, and hence it is 1. Therefore,

$$\mu = \text{E}(X) = np.$$

A slightly different manipulation of the summation rules will also be helpful to prove that

$$\sigma^2 = \text{Var}(X) = npq.$$

The proof of both is optional and we will be content by providing you with the proof of the expected value only. Some of the references cited at the end of the book will contain detailed proofs of the variance formula.

Example 7

Among the studies carried out to examine the effectiveness of advertising methods, a study reported that 4 out of 10 web surfers remember advertisement banners after they have seen them.

a) If 20 web surfers are chosen at random and shown an ad, what is the expected number of surfers that would remember the ad?

b) What is the chance that 5 of those 20 will remember the ad?

c) What is the probability that at most 1 surfer would remember the ad?

d) What is the chance that at least two surfers would remember the ad?

Solution

a) $X \sim (20, 0.4)$. The expected number is simply $20 \times 0.4 = 8$. We expect 8 of the surfers to remember the ad. Notice on the histogram below that the area in red corresponds to the expected value 8.

```
binompdf(20,.4,5
)
         .0746470195
```

b) $P(5) = \binom{20}{5} 0.4^5 (0.6)^{15} = 0.0746$, or see the output from the GDC to the right. Graphically, this area is shown on the histogram as the green area.

```
binompdf(20,.4,0
)
      3.65615844E-5
binompdf(20,.4,1
)
      4.87487792E-4
      4.87487792E-4
```

c) $P(x \leqslant 1) = P(x = 0) + P(x = 1) = 0.000\,524$

d) $P(x \geqslant 2) = 1 - P(x \leqslant 1)$
$= 1 - 0.000\,524 = 0.999\,475$

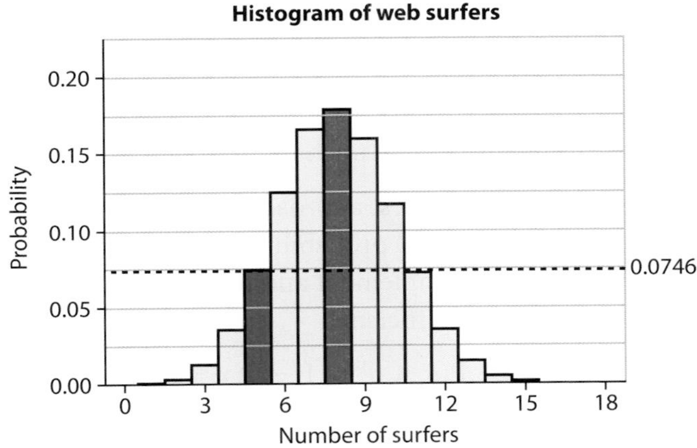

The cumulative binomial distribution function

As you have seen in Section 17.1, the cumulative distribution function $F(x)$ of a random variable X expresses the probability that X does not exceed the value x. That is,

$$F(x) = P(X \leqslant x) = \sum_{y:y \leqslant x} p(y).$$

So, for the binomial distribution, the cumulative distribution function (cdf) is given by:

$$F(x) = P(X \leqslant x) = \sum_{y:y \leqslant x} p(y)$$

$$= \sum_{y:y \leqslant x} \binom{n}{y} p^y q^{n-y}$$

The cumulative distribution is very helpful when we need to find the probability that a binomial variable assumes values over a certain interval.

Example 8

A large shipment of light bulbs contains 4% defective bulbs. In a sample of 20 randomly selected bulbs from the shipment, what is the probability that

a) there are at most three defective bulbs?

b) there are at least 6 defective bulbs?

Solution

a) This can be considered as a binomial distribution with $n = 20$ and $p = 0.04$.

We need $P(x \leqslant 3)$, which we can calculate by either finding the probabilities for $x = 0, 1, 2$ and 3, and then add them, or by using the cumulative function. In both cases, we will use a GDC to produce the answers.

```
binompdf(20,0.04
,{0,1,2,3})
{.4420024339 .3...
Ans→L1
{.4420024339 .3...
sum(L1)
         .9925870629
```

```
binomcdf(20,0.04
,3)
        .9925870629
```

As you can see, using the cdf is a much more straightforward procedure.

b) Here we need $P(x \geqslant 6)$. The first approach is not feasible at all as we need to calculate 15 individual probabilities and add them. However, setting the problem as a complement and then using the cumulative distribution is much more efficient.

$$P(x \geqslant 6) = 1 - P(x < 6) = 1 - P(x \leqslant 5)$$

```
1-binomcdf(20,0.
04,5)
  9.765401703E-5
```

Exercise 17.2

1 Consider the following binomial distribution:

$$P(x) = \binom{5}{x}(0.6)^x(0.4)^{5-x}, x = 0, 1, \ldots, 5$$

a) Make a table for this distribution.

b) Graph this distribution.

c) Find the mean and standard deviation in two ways:

(i) by formula

(ii) by using the table of values you created in part a).

d) Locate the mean μ and the two intervals $\mu \pm \sigma$ and $\mu \pm 2\sigma$ on the graph.

e) Find the actual probabilities for x to lie within each of the intervals $\mu \pm \sigma$ and $\mu \pm 2\sigma$ and compare them to the empirical rule.

2 A poll of 20 adults is taken in a large city. The purpose is to determine whether they support banning smoking in restaurants. It is known that approximately 60% of the population supports the decision. Let X represent the number of respondents in favour of the decision.

a) What is the probability that 5 respondents support the decision?

b) What is the probability that none of the 20 supports the decision?

c) What is the probability that at least 1 respondent supports the decision?

d) What is the probability that at least two respondents support the decision?

e) Find the mean and standard deviation of the distribution.

3 Consider the binomial random variable with $n = 6$ and $p = 0.3$.

a) Fill in the probabilities below.

k	0	1	2	3	4	5	6
$P(x \leqslant k)$							

b) Fill in the table below. Some cells have been filled for you to guide you.

Number of successes x	List the values of x	Write the probability statement	Explain it, if needed	Find the required probability
At most 3				
At least 3				
More than 3	4, 5, 6	$P(x > 3)$	$1 - P(x \leqslant 3)$	0.070 47
Fewer than 3				
Between 3 and 5 (inclusive)				
Exactly 3				

4 Repeat question 3 with $n = 7$ and $p = 0.4$.

5 A box contains 8 balls: 5 are green and 3 are white, red and yellow. Three balls are chosen at random without replacement and the number of green balls Y is recorded.

a) Explain why Y is not a binomial random variable.

b) Explain why, when we repeat the experiment with replacement, then Y is a binomial.

c) Give the values of n and p and display the probability distribution in tabular form.

d) What is the probability that at most 2 green balls are drawn?

e) What is the expected number of green balls drawn?

f) What is the variance of the number of balls drawn?

g) What is the probability that some green balls will be drawn?

6 On a multiple choice test, there are 10 questions, each with 5 possible answers, one of which is correct. Nick is unaware of the content of the material and guesses on all questions.

a) Find the probability that Nick does not answer any question correctly.

b) Find the probability that Nick answers at most half of the questions correctly.

c) Find the probability that Nick answers at least one question correctly.

d) How many questions should Nick expect to answer correctly?

7 Houses in a large city are equipped with alarm systems to protect them from burglary. A company claims their system to be 98% reliable. That is, it will trigger an alarm in 98% of the cases. In a certain neighbourhood, 10 houses equipped with this system experience an attempted burglary.

a) Find the probability that all the alarms work properly.

b) Find the probability that at least half of the houses trigger an alarm.

c) Find the probability that at most 8 alarms will work properly.

8 Harry Potter books are purchased by readers of all ages! 40% of Harry Potter books were purchased by readers 30 years of age or older! 15 readers are chosen at random. Find the probability that

a) at least 10 of them are 30 or older

b) 10 of them are 30 or older

c) at most 10 of them are younger than 30.

9 A factory makes computer hard disks. Over a long period, 1.5% of them are found to be defective. A random sample of 50 hard disks is tested.

a) Write down the expected number of defective hard disks in the sample.

b) Find the probability that three hard disks are defective.

c) Find the probability that more than one hard disk is defective.

10 Car colour preferences change over time and according to the area the customer lives in and the car model he/she is interested in. In a certain city, a large dealer of BMW cars noticed that 10% of the cars he sells are 'metallic grey'. Twenty of his customers are selected at random, and their car orders are checked for colour. Find the probability that

a) at least five cars are 'metallic grey'

b) at most 6 cars are 'metallic grey'

c) more than 5 are 'metallic grey'

d) between 4 and 6 are 'metallic grey'

e) more than 15 are not 'metallic grey'.

In a sample of 100 customer records, find

f) the expected number of 'metallic grey' car orders

g) the standard deviation of 'metallic grey' car orders.

According to the empirical rule, 95% of the 'metallic grey' orders are between a and b.

h) Find a and b.

11 Dogs have health insurance too! Owners of dogs in many countries buy health insurance for their dogs. 3% of all dogs have health insurance. In a random sample of 100 dogs in a large city, find

a) the expected number of dogs with health insurance

b) the probability that 5 of the dogs have health insurance

c) the probability that more than 10 dogs have health insurance.

12 A balanced coin is tossed 5 times. Let X be the number of heads observed.

a) Using a table, construct the probability distribution of X.

b) What is the probability that no heads are observed?

c) What is the probability that all tosses are heads?

d) What is the probability that at least one head is observed?

e) What is the probability that at least one tail is observed?

f) Given that the coin is unbalanced in such a way that it shows 2 heads in every 10 tosses, answer the same questions above.

13 When John throws a stone at a target, the probability that he hits the target is 0.4. He throws a stone 6 times.

a) Find the probability that he hits the target exactly 4 times.

b) Find the probability that he hits the target for the first time on his third throw.

14 On a television channel the news is shown at the same time each day. The probability that Alice watches the news on a given day is 0.4. Calculate the probability that on five consecutive days, she watches the news on at most three days.

15 A satellite relies on solar cells for its power and will operate provided that at least one of the cells is working. Cells fail independently of each other, and the probability that an individual cell fails within one year is 0.8.

a) For a satellite with ten solar cells, find the probability that all ten cells fail within one year.

b) For a satellite with ten solar cells, find the probability that the satellite is still operating at the end of one year.

c) For a satellite with n solar cells, write down the probability that the satellite is still operating at the end of one year. Hence, find the smallest number of solar cells required so that the probability of the satellite still operating at the end of one year is at least 0.95.

17.3 Poisson distribution

The Poisson distribution arises when you count a number of events across time or over an area. You should think about the Poisson distribution for any situation that involves counting events. Some examples are:

- the number of emergency visits by an infant during the first year of life,
- the number of white blood cells found in a cubic centimetre of blood
- the number of sample defects on a car
- the number of typographical errors on a page
- the number of failures in a large computer system during a given day
- the number of delivery trucks arriving at a central warehouse in an hour
- the number of customers arriving for flights during each 15-minute time interval from 3:00 p.m. to 6:00 p.m. on weekdays
- the number of customers arriving at a checkout aisle in your local grocery store during a particular time interval.

Sometimes, you will see the count represented as a rate, such as

- the number of injuries per year due to horse kicks, or
- the number of defects per square metre.

So, in general, the Poisson distribution is used when measuring the number of occurrences of 'something' (number of successes) over an interval, or time period.

Four assumptions

Information about how the data was generated can help you decide whether the Poisson distribution fits. Assume that an interval is divided into a very large number of sub-intervals so that the probability of the occurrence of an event in any sub-interval is very small. The Poisson distribution is based on four assumptions. We will use the term 'interval' to refer to either a time interval or an area, depending on the context of the problem.

1. The probability of observing a single event over a small interval is approximately proportional to the size of that interval.
2. The probability of two events occurring in the same narrow interval is negligible.
3. The probability of an event within a certain interval does not change over different intervals.
4. The probability of an event in one interval is independent of the probability of an event in any other non-overlapping interval.

Notation
If a random variable follows a Poisson distribution, we write $X \sim P_0(\mu)$.

You should examine all of these assumptions carefully, but especially the last two. If either of these last two assumptions is violated, they can lead to extra variation, sometimes referred to as overdispersion.

Mathematical details

The Poisson distribution depends on a single parameter μ. The probability that the Poisson random variable equals x is

$$P(X = x) = e^{-\mu}\frac{\mu^x}{x!}$$

where μ is the average number of events observed over the specific interval, and x is the number of 'successes' we are interested in. μ can be any positive real number, while x has to be a positive integer. We will show that the parameter is actually the expected value below.

Note: IBO uses m instead of μ for the Poisson parameter.

Only for the curious!
The Poisson probability is related to the binomial probability. Here is a justification.

Suppose we want to find the probability distribution of the number of telephone calls to the front desk of a large company over a period of one hour. Think of the one hour as being split into n sub-intervals, each of which is small enough that at most one call could arrive within it with a probability p.

This means:

$$P(\text{one call}) = p$$

$$P(\text{no call}) = 1 - p, \text{ i.e. } P(\text{ more than one call}) = 0$$

This in turn will mean that the total number of calls received within one hour is equal to the total number of sub-intervals that contain one call! Hence, if we can consider the calls to be arriving independently of each other from one interval to the other, then the distribution of the number of calls per hour is a binomial distribution.

As n increases, p will become smaller. Since the binomial distribution has its expected value $\mu = np$, and considering the probability of x successes within one hour as n increases indefinitely, that is, $n \to \infty$, we have:

$$p(x) = \binom{n}{x}p^x(1-p)^{n-x}$$

$$\Rightarrow \lim_{n\to\infty}\binom{n}{x}p^x(1-p)^{n-x} = \lim_{n\to\infty}\frac{n(n-1)(n-2)\ldots(n-x+1)}{x!}\left(\frac{\mu}{n}\right)^x\left(1-\frac{\mu}{n}\right)^{n-x}$$

This is so, because if $\mu = np$, then $p = \frac{\mu}{n}$ and $1 - p = 1 - \frac{\mu}{n}$.

$$\lim_{n\to\infty} \frac{n(n-1)(n-2)\dots(n-x+1)}{x!}\left(\frac{\mu}{n}\right)^x\left(1-\frac{\mu}{n}\right)^{n-x}$$

$$= \lim_{n\to\infty} \frac{n(n-1)(n-2)\dots(n-x+1)}{x!}\frac{\mu^x}{n^x}\left(1-\frac{\mu}{n}\right)^n\left(1-\frac{\mu}{n}\right)^{-x}$$

$$= \lim_{n\to\infty} \frac{\mu^x}{x!}\left(1-\frac{\mu}{n}\right)^n \frac{n(n-1)(n-2)\dots(n-x+1)}{n^x}\left(1-\frac{\mu}{n}\right)^{-x}$$

Now, since $\frac{\mu^x}{x!}$ is independent of n, then it can be factored out of the limit expression and we are left with:

$$\frac{\mu^x}{x!}\lim_{n\to\infty}\left(1-\frac{\mu}{n}\right)^n \frac{n(n-1)(n-2)\dots(n-x+1)}{n^x}\left(1-\frac{\mu}{n}\right)^{-x}$$

But, using the theorem that the limit of a product is the product of limits, then:

$$\lim_{n\to\infty}\left(1-\frac{\mu}{n}\right)^n \frac{n(n-1)(n-2)\dots(n-x+1)}{n^x}\left(1-\frac{\mu}{n}\right)^{-x}$$

$$= \lim_{n\to\infty}\left(1-\frac{\mu}{n}\right)^n \cdot \lim_{n\to\infty}\frac{n(n-1)(n-2)\dots(n-x+1)}{n^x} \cdot \lim_{n\to\infty}\left(1-\frac{\mu}{n}\right)^{-x}$$

Also,

$$\lim_{n\to\infty}\left(1-\frac{\mu}{n}\right)^n = e^{-\mu};$$

$$\lim_{n\to\infty}\frac{n(n-1)(n-2)\dots(n-x+1)}{n^x} = \lim_{n\to\infty}\frac{n}{n}\cdot\frac{n-1}{n}\dots\frac{n-x+1}{n} = 1\cdot 1\dots 1 = 1$$

$$\lim_{n\to\infty}\left(1-\frac{\mu}{n}\right)^{-x} = 1^{-x} = 1$$

So, finally, we have

$$p(x) = \binom{n}{x}p^x(1-p)^{n-x} = \frac{\mu^x}{x!}\cdot e^{-\mu}\cdot 1\cdot 1 = e^{-\mu}\frac{\mu^x}{x!}.$$

Expected value of the Poisson distribution

Like any discrete distribution, the Poisson distribution has an expected value that can be found using the definition of expected value developed earlier.

$$\mathrm{E}(X) = \sum_{x=0}^{\infty} xp(x) = \sum_{x=0}^{\infty} xe^{-\mu}\frac{\mu^x}{x!}$$

Notice that when $x = 0$, the first term in the summation equation is 0. Hence

$$\mathrm{E}(X) = \sum_{x=1}^{\infty} xe^{-\mu}\frac{\mu^x}{x!} = \sum_{x=1}^{\infty} e^{-\mu}\frac{\mu\cdot\mu^{x-1}}{(x-1)!} = \mu\sum_{x=1}^{\infty} e^{-\mu}\frac{\mu^{x-1}}{(x-1)!}$$

We factored μ out of the summation as it is independent of x. Now notice that the expression in the summation is nothing but the total of all probabilities of the variable $x - 1$ over its domain. So, if we replace $x - 1$ by y, the expression in the summation formula becomes

$$\sum_{x=1}^{\infty} e^{-\mu}\frac{\mu^{x-1}}{(x-1)!} = \sum_{y=0}^{\infty} e^{-\mu}\frac{\mu^y}{y!} = 1$$

(since subtracting 1 from ∞ does not change anything.)

Therefore,

$$E(X) = \mu \cdot \sum_{y=1}^{\infty} e^{-\mu} \frac{\mu^y}{y!} = \mu.$$

Also, since we considered the Poisson as a binomial when n tends to infinity, the variance can be verified to be μ in the following manner.

$$V(X) = npq = n\left(\frac{\mu}{n}\right)\left(1 - \frac{\mu}{n}\right) = \mu\left(1 - \frac{\mu}{n}\right)$$

But as $n \longrightarrow \infty$, $\frac{\mu}{\text{n}} \longrightarrow 0$; hence,

$$V(X) = \mu\left(1 - \frac{\mu}{n}\right) = \mu.$$

Note: Observe that for a Poisson distribution, $E(X) = V(X) = \mu$.

Question:
How do we know that the Poisson distribution is a probability distribution?

Answer:
We can easily verify that Poisson as developed satisfies the probability distribution conditions:

1. $0 \leqslant p(x) \leqslant 1$
2. $\sum_x p(x) = 1$

1. Since $\mu > 0$, it is obvious that $p(x) > 0$, for $x = 0, 1, 2, \ldots$.
   $p(x) \leqslant 1$ will be a natural result of proving the second condition. If the sum of n positive parts is equal to 1, none of the parts can be greater than 1!
2. $\sum_{x=0}^{\infty} p(x) = \sum_{x=0}^{\infty} e^{-\mu} \frac{\mu^x}{x!} = e^{-\mu} \sum_{x=0}^{\infty} \frac{\mu^x}{x!}$

But, $\sum_{x=0}^{\infty} \frac{\mu^x}{x!} = e^{\mu}$.

This is proved in Option 10: Series and differential equations.

This is a Taylor expansion of e^x: $e^x = 1 + x + \frac{x^2}{2!} + \frac{x^3}{3!} + \ldots$; hence,

$$\sum_{x=0}^{\infty} p(x) = e^{-\mu} e^{\mu} = 1.$$

Example 9

Police use speed cameras to record violations of speed limits. At a strategic spot in a city, the installed camera automatically turns itself on, on average once every 10 minutes. The pattern follows, approximately, a Poisson distribution.

a) Within a 10-minute interval, what is the chance that it is on
   (i) once
   (ii) twice
   (iii) at least once?

b) Within an hour, what is the chance that it is on
   (i) once
   (ii) twice
   (iii) at least once?

Solution

a) For this part the interval is 10 minutes and $\mu = 1$. Then,
   (i) $P(x = 1) = e^{-1}\frac{1^1}{1!} = e^{-1} = 0.368$
   (ii) $P(x = 2) = e^{-1}\frac{1^2}{2!} = \frac{e^{-1}}{2} = 0.184$
   (iii) $P(\text{at least once}) = 1 - P(\text{at most } 0) = 1 - P(x = 0)$
   $= 1 - e^{-1}\frac{1^0}{0!} = 1 - e^{-1} = 0.632$

b) Here the expected value $\mu = 6$.
   (i) $P(x = 1) = e^{-6}\frac{6^1}{1!} = 6e^{-6} = 0.0149$
   (ii) $P(x = 2) = e^{-6}\frac{6^2}{2!} = e^{-6}\frac{36}{2} = 0.0446$
   (iii) $P(\text{at least once}) = 1 - P(\text{at most } 0) = 1 - P(x = 0)$
   $= 1 - e^{-6}\frac{6^0}{0!} = 1 - e^{-6} = 0.998$

The results above can of course be calculated directly using your GDC.

```
poissonpdf(1,1)
      .3678794412
poissonpdf(1,2)
      .1839397206
1-poissonpdf(1,0)
      .6321205588
```

```
poissonpdf(6,1)
      .0148725131
poissonpdf(6,2)
      .0446175392
1-poissonpdf(6,0)
      .9975212478
```

Cumulative Poisson distribution function

As you have seen in Sections 17.1 and 17.2, it is more practical in several situations to start with a cumulative distribution in order to calculate probabilities of several consecutive values. The cumulative Poisson distribution function plays the same role introduced in the previous sections.

The cumulative Poisson distribution function of a Poisson random variable X expresses the probability that X does not exceed a value x.

$$P(x \leq x) = \sum_{i=0}^{x} e^{-\mu}\frac{\mu^i}{i!}$$

Example 10

The number of radioactive particles released per minute from a meteoroid after it enters the atmosphere is recorded and the average is found to be 3.5 particles per minute. Find the probability that in any one minute there are at least 5 particles released.

Solution

This appears to be a Poisson model. We first set it up to make the calculation through the cumulative distribution function.

$P(x \geqslant 5) = 1 - P(x \leqslant 4)$, and so

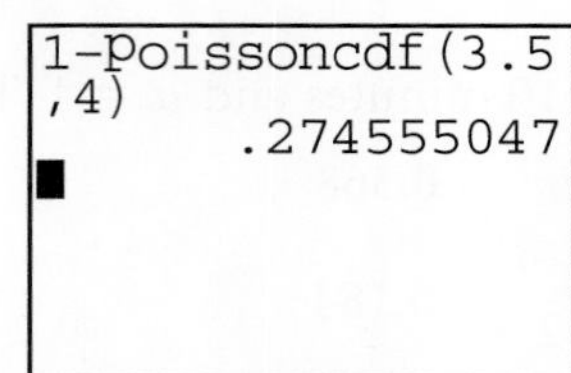
1-Poissoncdf(3.5,4)
.274555047

That is, there is a probability of 27.5% that at least 5 particles are emitted.

Example 11

Small aircraft arrive at a certain airport according to a Poisson process with rate of 10 per hour.

a) What is the probability that during a 1-hour period
   (i) 8 small aircraft arrive?
   (ii) at most 8 small aircraft arrive?
   (iii) at least 9 small aircraft arrive?

b) What is the expected value and standard deviation of the number of small aircraft that arrive during a 90-minute period?

c) What is the probability that at least 1 small aircraft arrives during a 6-minute period?

d) What is the probability that 1 small aircraft arrives during two 6-minute separate periods?

e) What is the probability that 1 small aircraft arrives during a 12-minute period?

Solution

a) (i) $Po(x = 8|\mu = 10) = e^{-10}\frac{10^8}{8!} \approx 0.113$

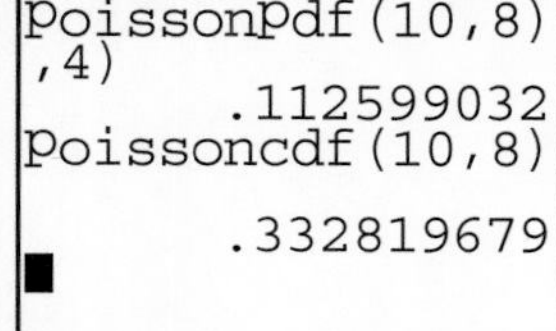
PoissonPdf(10,8),4)
.112599032
Poissoncdf(10,8)
.332819679

(ii) $Po(x \leqslant 8|\mu = 10) = \sum_{x=0}^{8} e^{-10}\frac{10^x}{x!} \approx 0.333$

(iii)

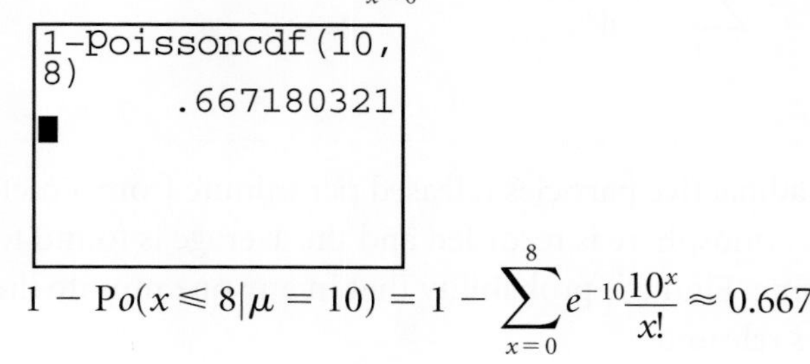
1-Poissoncdf(10,8)
.667180321

$1 - Po(x \leqslant 8|\mu = 10) = 1 - \sum_{x=0}^{8} e^{-10}\frac{10^x}{x!} \approx 0.667$

b) A 90-minute period is 1.5 hours.
So, the expected value is $1.5 \times 10 = 15$ and the standard deviation is $\sqrt{15} = 3.87$. Recall that $V(X) = \mu$.

c) During a 6-minute period, the expected value is $\lambda = \frac{\mu}{10} = 1$, and

$$\begin{aligned} Po(x \geqslant 1 | \lambda = 1) &= 1 - Po(x = 0 | \lambda = 1) = 1 - e^{-1}\frac{1^0}{0!} \\ &= 1 - e^{-1} \approx 0.632. \end{aligned}$$

d) This event consists of two simple events: either 1 plane the first period and no plane the second, or no plane the first period and 1 plane the second.
Hence, [let $P(a, b)$ be the probability of a planes first and b planes second]

$$\begin{aligned} P(1 \text{ plane in two 6-minute periods}) &= P(1, 0)P(0, 1) + P(0, 1)P(1, 0) \\ &= e^{-1}\frac{1^1}{1!} \cdot e^{-1}\frac{1^0}{0!} + e^{-1}\frac{1^0}{0!} \cdot e^{-1}\frac{1^1}{1!} \\ &= 2e^{-2} \approx 0.271. \end{aligned}$$

e) Here the expected value is 2 aircraft, and hence
$Po(x = 1 | \mu = 2) = e^{-2}\frac{2^1}{2!} \approx 0.135.$

Exercise 17.3

1 Let X denote a random variable that has a Poisson distribution with mean $\mu = 3$. Find the following probabilities, both manually and with a GDC:

a) $P(x = 5)$

b) $P(x < 5)$

c) $P(x \geqslant 5)$

d) $P(x \geqslant 5 | x \geqslant 3)$

2 Let X denote a random variable that has a Poisson distribution with mean $\mu = 5$. Find the following probabilities, both manually and with a GDC:

a) $P(x = 5)$

b) $P(x < 4)$

c) $P(x \geqslant 4)$

d) $P(x \leqslant 6 | x \geqslant 4)$

3 The number of support phone calls coming into the central switchboard of a small computer company averages 6 per minute.

a) Find the probability that no calls will arrive in a given one-minute period.

b) Find the probability that at least two calls will arrive in a given one-minute period.

c) Find the probability that at least two calls will arrive in a given 2-minute period.

4 DVDs are tested by sending them through an analyzer that measures imbalance, using accepted industry standards. A brand of DVDs is known to have an error score of 0.1 per DVD, which is within the acceptable standards.

a) Find the probability that the next inspected DVD will have no error.

b) Find the probability that the next inspected DVD will have more than one error.

c) Find the probability that neither of the next two inspected DVDs will have any error.

5 In 2000, after an extensive study of road safety, Japan decided to set a maximum speed limit on their expressways of 100 km/h. In the study, it was reported that the number of deaths and serious injuries on expressways for regular passenger vehicles was 0.024 per million vehicle–kilometres.

a) (i) Find the probability that at most 15 serious incidents happen in a given block of 10^9 vehicle–kilometres.

(ii) Find the probability that at least 20 serious incidents happen in a given block of 10^9 vehicle–kilometres.

b) The rate for light motor vehicles was 0.036.

(i) Find the probability that at most 15 serious incidents happen in a given block of 10^9 vehicle–kilometres.

(ii) Find the probability that at least 20 serious incidents happen in a given block of 10^9 vehicle–kilometres.

6 Passengers arrive at a security checkpoint in a busy airport at the rate of 8 per 10-minute period. For the time between 8:00 and 8:10 on a specific day, find the probability that

a) 8 passengers arrive

b) no more than 5 passengers arrive

c) at least 4 passengers arrive.

7 In question 6 above, find each of the following probabilities.

a) The probability that three passengers arrive between 8:00 and 8:20.

b) The probability that three passengers arrive between 8:00 and 8:10 and 9:00 and 9:10.

8 A certain internet service website receives on average 0.2 hits per second. It is known that the number of hits on this site follows a Poisson distribution.

a) Find the probability that no hits are registered during the next second.

b) Find the probability that no hits are registered for the next 3 seconds.

9 The number of faults in the knit of a certain fabric has an average of 4.4 faults per square metre. It is also assumed to have a Poisson distribution.

a) Find the probability that a 1 m^2 piece of this fabric contains at least 1 fault.

b) Find the probability that a 3 m^2 piece of this fabric contains at least 1 fault.

c) Find the probability that three 1 m^2 pieces of this fabric contain 1 fault.

10 A supplier of copper wire looks for flaws before despatching it to customers. It is known that the number of flaws follows a Poisson probability distribution with a mean of 2.3 flaws per metre.

a) Determine the probability that there are exactly 2 flaws in 1 metre of the wire.

b) Determine the probability that there is at least one flaw in 2 metres of the wire.

11 a) Patients arrive at random at an emergency room in a hospital at the rate of 15 per hour throughout the day. Find the probability that 6 patients will arrive at the emergency room between 08:00 and 08:15.

b) The emergency room switchboard has two operators. One operator answers calls for doctors and the other deals with enquiries about patients. The first operator fails to answer 1% of her calls and the second operator fails to answer 3% of his calls. On a typical day, the first and second telephone operators receive 20 and 40 calls respectively during an afternoon session. Using the Poisson distribution find the probability that, between them, the two operators fail to answer two or more calls during an afternoon session.

12 The random variable X is Poisson distributed with mean μ and satisfies $P(x = 3) = P(x = 0) + P(x = 1)$.

a) Find the value of μ, correct to four decimal places.

b) For this value of μ evaluate $P(2 \leqslant x \leqslant 4)$.

13 Give all numerical answers to this question correct to three significant figures. Two typists were given a series of tests to complete. On average, Mr Brown made 2.7 mistakes per test while Mr Smith made 2.5 mistakes per test. Assume that the number of mistakes made by any typist follows a Poisson distribution.

a) Calculate the probability that, in a particular test,

(i) Mr Brown made two mistakes

(ii) Mr Smith made three mistakes

(iii) Mr Brown made two mistakes and Mr Smith made three mistakes.

b) In another test, Mr Brown and Mr Smith made a combined total of five mistakes. Calculate the probability that Mr Brown made fewer mistakes than Mr Smith.

17.4 Continuous distributions

Continuous random variables

When a random variable X is discrete, you assign a positive probability to each value that X can take and get the probability distribution for X. The sum of all the probabilities associated with the different values of X is 1.

You have seen, in the discrete variable case, that we graphically represent the probabilities corresponding to the different values of the random variable X with a probability histogram (relative frequency histogram), where the area of each bar corresponds to the probability of the specific value it represents.

Consider now a continuous random variable X, such as height and weight, and length of life of a particular product – a TV set for example. Because it is continuous, the possible values of X are over an interval. Moreover, there are an infinite number of possible values of X. Hence, we cannot find a probability distribution function for X by listing all the possible values of X along with their probabilities, as you see in the histogram on the next page. If we try to assign probabilities to each of these uncountable values, the probabilities will no longer sum to 1, as is the case with discrete variables. Therefore, you must use a different approach to generate the probability distribution for such random variables.

Suppose that you have a set of measurements on a continuous random variable, and you create a relative frequency histogram to describe their distribution. For a small number of measurements, you can use a small number of classes, but as more and more measurements are collected, you can use more classes and reduce the **class width**.

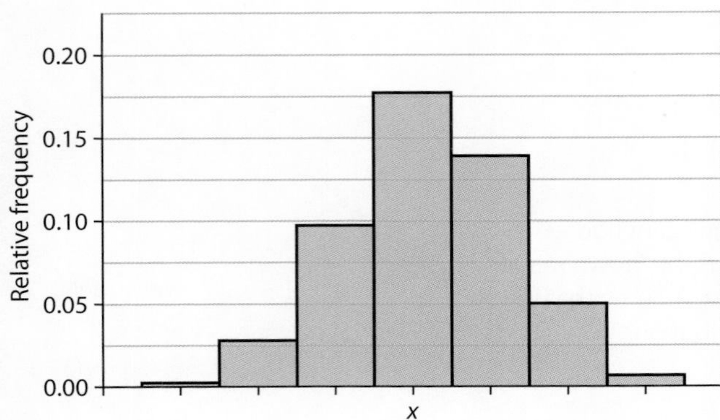

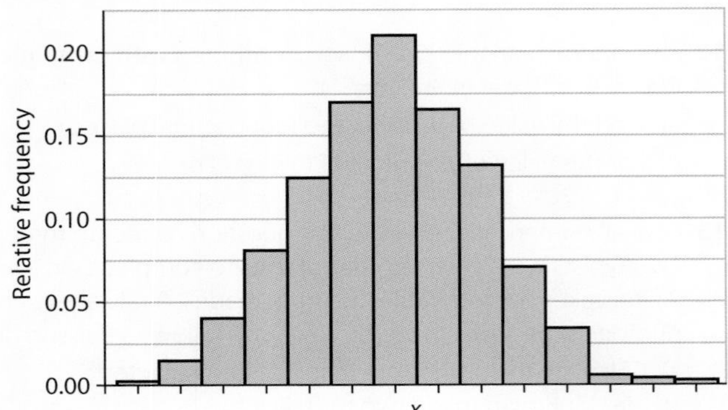

The histogram will slightly change as the class width becomes smaller and smaller, as shown in the diagrams below. As the number of measurements becomes very large and the class width becomes very narrow, the relative frequency histogram appears more and more like the smooth curve you see below. This is what happens in the continuous case, and the smooth curve describing the probability distribution of the continuous random variable becomes the **PDF** (**probability density function**) of X, represented by a curve $y = f(x)$. This curve is such that the entire area under the curve is 1 and the area between any two points is the probability that x falls between those two points.

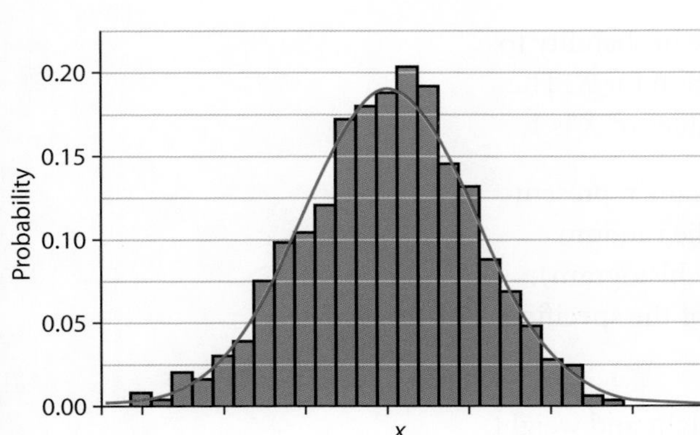

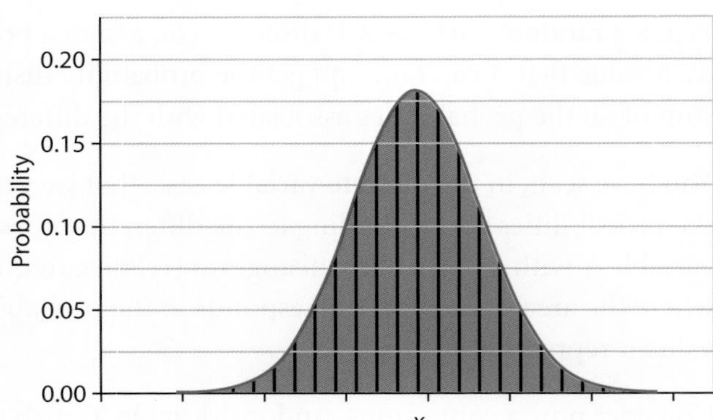

Probability density function

Let X be a continuous random variable. The probability density function, $f(x)$, of the random variable is a function with the following properties:

1. $f(x) > 0$ for all values of x.

2. The area under the probability density function $f(x)$ over all values of the random variable X is equal to 1.0, i.e. $\int_{-\infty}^{\infty} f(x)dx = 1$.

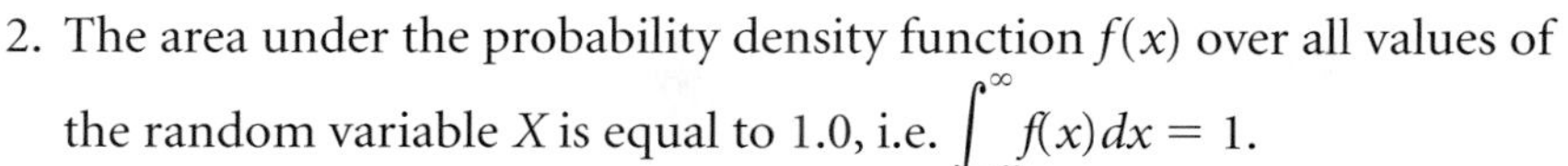

3. Suppose this density function is graphed. Let a and b be two possible values of the random variable X, with $a < b$. Then the probability that x lies between a and b $[P(a < x < b)]$ is the area under the density function between these points.

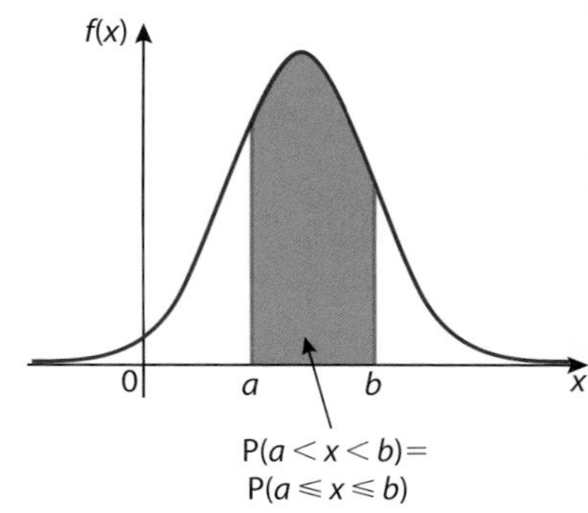

Notice that, based on this definition, the probability that x equals any point a is 0. This is so because the area above a value, say a, is a rectangle whose width is 0 or equivalently

$$P(X = a) = \int_a^a f(x)dx = 0.$$

So, for the continuous case, regardless of whether the endpoints a and b are themselves included, the area included between a and b is the same.

$$P(a < x < b) = P(a \leqslant x \leqslant b) = P(a \leqslant x < b) = P(a < x \leqslant b)$$

For example, the graph shows a model for the pdf f for a random variable X defined to be the height, in cm, of an adult female in Spain. The probability that the height of a female chosen at random from this population is between 160 and 175 is equal to the area under the curve between 160 and 175.

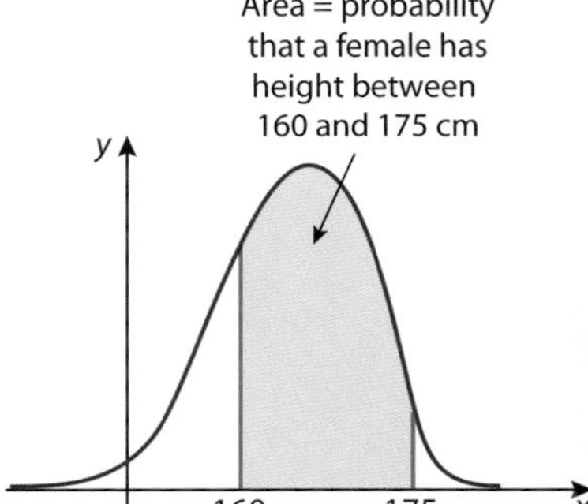

The function represented here is:

$$f(x) = \int_{160}^{175} \frac{e^{-\frac{(x-165)^2}{50}}}{5\sqrt{2\pi}} dx$$

As you know from your integral calculus class, it is not an integral you can calculate exactly. We use a GDC to approximate it.

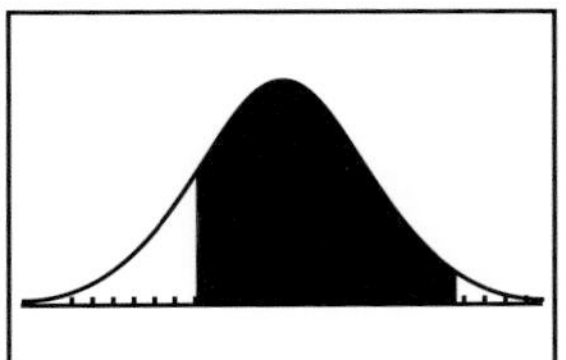

```
fnInt(Y1,X,160,1
75)
         .8185946141
```

So, the chance of choosing a female at random with a height between 160 cm and 175 cm is approximately 81.9%.

Example 12

$f(x)$ as defined below describes a random variable X.

$$f(x) = \begin{cases} \frac{1}{512}(12x^2 - x^3 - 20x) & 2 \leqslant x \leqslant 10 \\ 0 & \text{otherwise} \end{cases}$$

a) Verify that $f(x)$ is a probability density function.

b) Find $P(5 \leqslant x \leqslant 8)$.

Solution

a) For $2 \leqslant x \leqslant 10$ we have $12x^2 - x^3 - 20x \geqslant 0$, so $f(x) \geqslant 0$.

We also need to check that $\int_{-\infty}^{\infty} f(x)\,dx = 1$.

$$\begin{aligned}\int_{-\infty}^{\infty} f(x)\,dx &= \int_{2}^{10} \tfrac{1}{512}(12x^2 - x^3 - 20x)\,dx \\ &= \tfrac{1}{2048}(16x^3 - x^4 - 40x^2)\Big]_2^{10} \\ &= \tfrac{1}{2048}(16\,000 - 100\,000 - 4000 - 128 + 16 + 160) = 1\end{aligned}$$

Therefore, $f(x)$ is a pdf.

b) 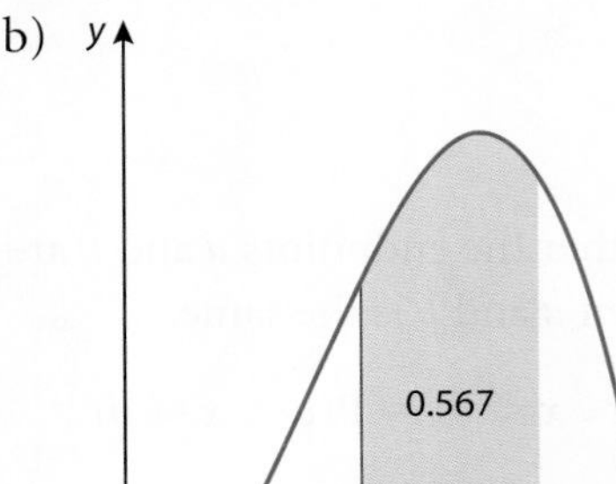

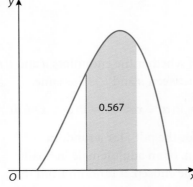

The probability that x lies between 5 and 8 is

$$\begin{aligned}\int_{5}^{8}(12x^2 - x^3 - 20x)\,dx &= \tfrac{1}{2048}(16x^3 - x^4 - 40x^2)\Big]_5^{8} \\ &= \tfrac{1}{2048}(1536 - 375) \\ &= \tfrac{1161}{2048} \approx 0.567.\end{aligned}$$

Example 13

Find the value of k such that the following represents a probability density function of a random variable X.

$$f(x) = \begin{cases} kx^2(2 - x) & 0 \leqslant x \leqslant 2 \\ 0 & \text{otherwise} \end{cases}$$

Solution

For $f(x)$ to be a pdf, we need to satisfy both conditions.

a) For $0 \leqslant x \leqslant 2$ we must have $kx^2(2 - x) \geqslant 0$, and since $x^2 \geqslant 0$, then k must be positive.

b) For $\int_{-\infty}^{\infty} f(x)\,dx = 1$, then

$\int_{-\infty}^{\infty} f(x)\,dx = \int_{0}^{2} kx^2(2 - x)\,dx = 1$, and hence

$\int_{0}^{2} kx^2(2 - x)\,dx = k\left(\frac{2}{3}x^3 - \frac{x^4}{4}\right)\Big]_0^2 = 1$, and this in turn leads to

$k\left(\frac{16}{3} - \frac{16}{4}\right) = 1 \Rightarrow k\left(\frac{4}{3}\right) = 1$, and $k = \frac{3}{4}$.

Cumulative distribution functions

You have met the idea of the cumulative distribution functions for discrete random variables in Sections 17.1 to 17.3. In the same way, and using the fact that an integral is the limit of a sum, we have the following definition.

A **cumulative distribution function**, $F(x)$, of a random variable X with a density function $f(t)$ is defined by

$F(x) = P(X \leqslant x) = \int_{-\infty}^{x} f(t)dt$, where x is a value in the domain of the function $f(t)$.

$F(x)$ gives us the proportion of the population having values smaller than x.
Note here that $F(x)$ is an anti-derivative of $f(x)$, that is, $F'(x) = f(x)$.
Any distribution function has the following properties:

1. $F(x)$ is non-decreasing.
2. $\lim_{x\to\infty} F(x) = 0$; $\lim_{x\to-\infty} F(x) = 0$.
3. Since $P(a < x < b) = \int_a^b f(x)dx$, then $P(a < x < b) = \int_a^b f(x)dx = F(b) - F(a)$.

Note: The lower limit of integration is given as $-\infty$, but in essence, it is the smallest possible value of x.

Measures of centre, position and spread of a continuous distribution

Like discrete distributions, continuous distributions have their characteristics including **mean, median, mode, variance** and the **percentiles.** Next we will discuss each of them in more detail.

Mean

Recall that for a discrete random variable

$\mathrm{E}(X) = \sum_x xp(x)$. Similarly, if we have a continuous random variable X with a *pdf* $f(x)$, then

$$\mathrm{E}(X) = \int_x xf(x)\,dx.$$

$\mathrm{E}(X)$ is called the expected value of X and it is also referred to as the mean μ.

Example 14

The function $f(x)$ is a pdf for a random variable X.

$$f(x) = \begin{cases} \frac{3}{4}x^2(2-x) & 0 \leqslant x \leqslant 2 \\ 0 & \text{otherwise} \end{cases}$$

a) Find μ.

b) Find $P(x < \mu)$.

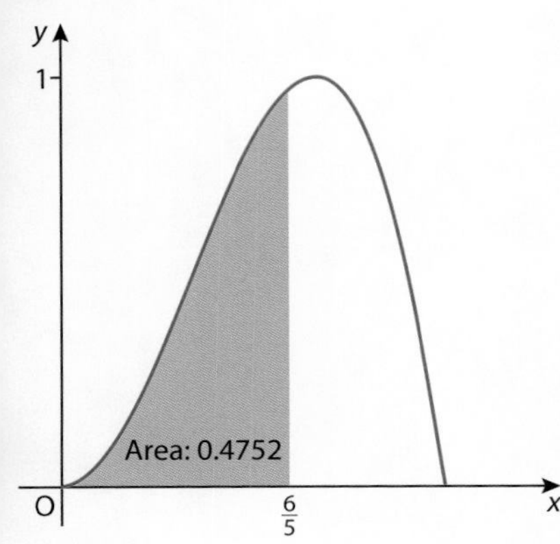

Solution

a) $\mu = \mathrm{E}(X) = \int_x xf(x)\,dx = \int_0^2 \frac{3}{4}x^3(2-x)\,dx = \frac{3}{4}\left[\frac{2x^4}{4} - \frac{x^5}{5}\right]_0^2 = \frac{3}{4}\cdot\frac{8}{5} = \frac{6}{5}$

b) $\mathrm{P}\left(x \leqslant \frac{6}{5}\right) = \int_0^{\frac{6}{5}} \frac{3}{4}x^2(2-x)\,dx = \frac{3}{4}\left[\frac{2}{3}x^3 - \frac{x^4}{4}\right]_0^{\frac{6}{5}} = \frac{3}{4}\cdot\frac{386}{624} = 0.4752$

Mode

The mode, as you know, is the value of X for which $f(x)$ is the largest in the given domain of X.

To locate the mode, you may first draw a graph of $f(x)$ and then use the first or first and second derivative tests to find the maximum. Just recall that the maximum can happen at critical points.

So, in the previous example, it appears that the mode is slightly higher than the mean.

$$f(x) = \begin{cases} \frac{3}{4}x^2(2-x) & 0 \leqslant x \leqslant 2 \\ 0 & \text{otherwise} \end{cases} \Rightarrow f'(x) = 3x - \frac{9}{4}x^2;\ f''(x) = 3 - \frac{9}{2}x$$

Now, $f'(x) = 0 \Rightarrow x = 0, x = \frac{4}{3}$, and at $x = 0, f(x) = 0$, while at $x = \frac{4}{3}, f''(x) = -3 < 0$, which means that $f(x)$ has a maximum.
So, the mode is at $x = \frac{4}{3}$.

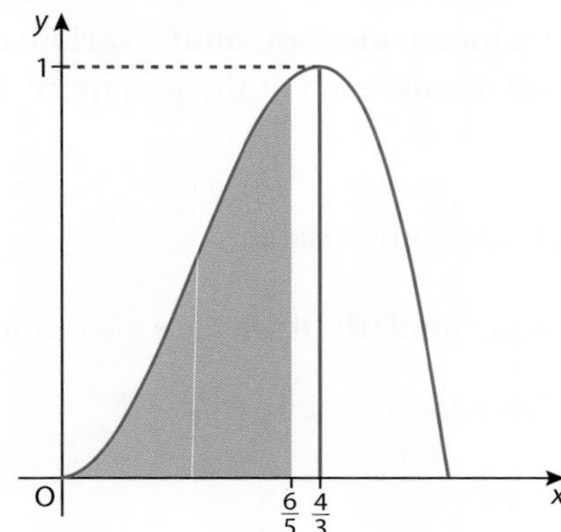

Example 15

The random variable X has a pdf defined by

$$f(x) = \begin{cases} k(-x^2 + 2x + 15); & 0 \leqslant x \leqslant 5 \\ 0 & \text{otherwise} \end{cases}$$

a) Find the value of k.

b) Determine the mean and the mode.

c) Find the value, m, of the random variable which is larger than 50% of the population.

Solution

a) The area under the curve must be equal to 1:

$$\int_{-\infty}^{\infty} f(x)dx = \int_0^5 k(-x^2 + 2x + 15)dx = 1$$

$$k\left[-\frac{x^3}{3} + x^2 + 15x\right]_0^5 = k\frac{175}{3} = 1 \Rightarrow k = \frac{3}{175}$$

b) $\mu = \int_0^5 x\left(\frac{3}{175}(-x^2 + 2x + 15)\right)dx = \frac{3}{175}\int_0^5 (-x^3 + 2x^2 + 15x)dx$

So, the mean is

$$\frac{3}{175}\left[-\frac{x^4}{4} + \frac{2x^3}{3} + \frac{15x^2}{2}\right]_0^5 = \frac{55}{28}.$$

$f'(x) = 0 \Rightarrow -2x + 2 = 0$, and so, the mode is $x = 1$.

c) The value of m can be found by finding the value where the area to the left of it under the pdf is 0.5. Hence,

$$\int_{-\infty}^{m} f(x)dx = \int_0^m \frac{3}{175}(-x^2 + 2x + 15)dx = 0.5$$

$$\Rightarrow \frac{3}{175}\left[-\frac{x^3}{3} + x^2 + 15x\right]_0^m = \frac{1}{2}$$

$$\Rightarrow x \approx 1.857$$

Note: This is the method used next to find the median of the data.

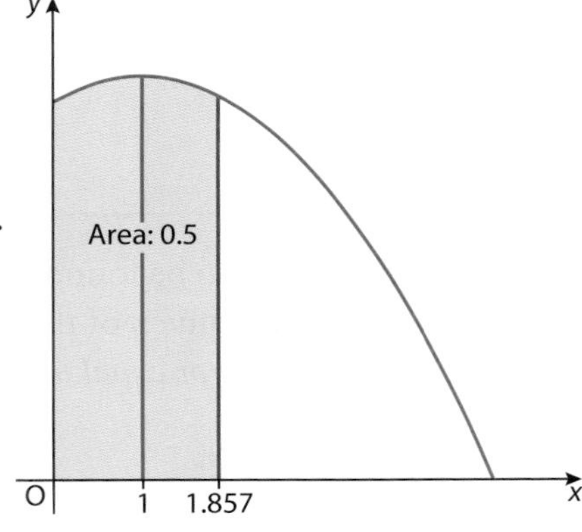

Median and percentiles

The median for a random variable X that has a pdf $f(x)$ is a value m of the random variable such that 50% of the values of X are less than or equal to m. (Similarly, 50% of the values are larger than or equal to m). Thus, the median m satisfies

$$\int_{-\infty}^{m} f(x)dx = 0.5.$$

In other words, half of the area under the pdf lies to the left of m.

Example 16

The function $f(x)$ is a pdf for a random variable X.

$$f(x) = \begin{cases} \frac{3}{4}x^2(2 - x) & 0 \leq x \leq 2 \\ 0 & \text{otherwise} \end{cases}$$

a) Find the median m.

b) Find $P(x < m)$.

Solution

a) To find the median, we set up the integral as given by the definition,
$\int_{-\infty}^{m} f(x)dx = \int_{0}^{m} \frac{3}{4}x^2(2 - x)dx = 0.5$, and then we solve for m,
$\int_{0}^{m} \frac{3}{4}x^2(2 - x)dx = \frac{3}{4}\left[\frac{2}{3}x^3 - \frac{x^4}{4}\right]_0^m = \frac{3}{4}\left(\frac{2}{3}m^3 - \frac{m^4}{4}\right) = 0.5.$
In the interval [0, 2], the solution is approx. 1.2285.

b) $P(x < 1.2285) = \int_{0}^{1.2285} \frac{3}{4}x^2(2 - x)dx = 0.4999 \approx 0.5.$

This result confirms the calculation that 1.2285 is the median. Notice how the median is to the right of the mean as the distribution is slightly skewed to the left!

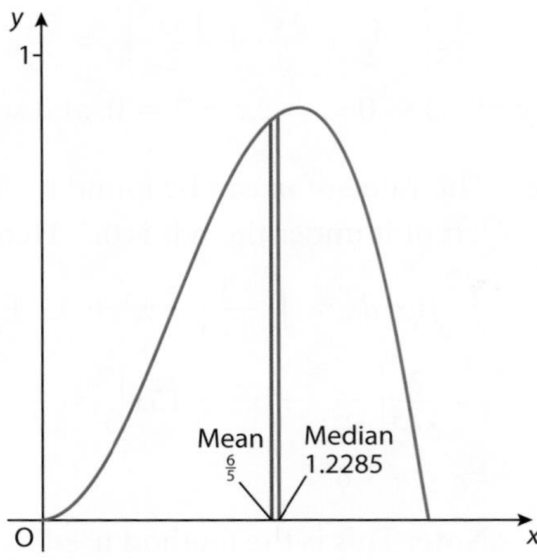

The **percentiles** can also be found in a similar manner. The kth percentile can be defined as the value n of the random variable such that *k% of the values of X are less than or equal to n.* Thus, the **kth percentile** satisfies

$$\int_{-\infty}^{n} f(x)dx = k\%.$$

Example 17

In the previous example,

a) find the first quartile

b) find the third quartile and the IQR.

Solution

a) The first quartile is the 25th percentile, and hence
$\int_{0}^{n} \frac{3}{4}x^2(2 - x)dx = \frac{3}{4}\left(\frac{2}{3}n^3 - \frac{n^4}{4}\right) = 0.25.$
In the interval [0, 2], the solution is approx. **0.913**.

b) The third quartile is the 75th percentile, and hence
$\int_{0}^{n} \frac{3}{4}x^2(2 - x)dx = \frac{3}{4}\left(\frac{2}{3}n^3 - \frac{n^4}{4}\right) = 0.75.$
In the interval [0, 2], the solution is approx. 1.514. The IQR is therefore
IQR = 1.514 − 0.913 = **0.601**.

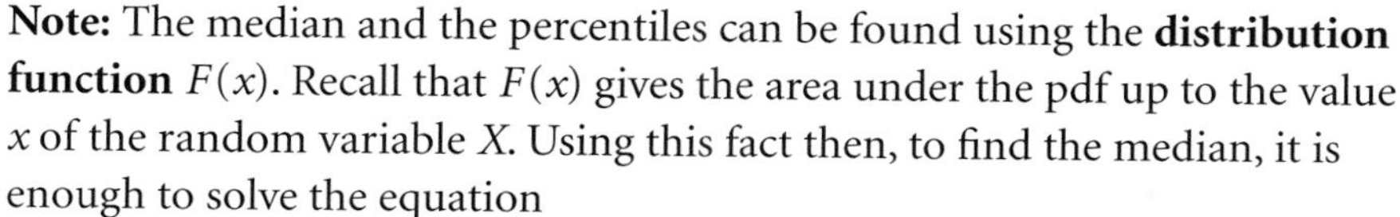

Note: The median and the percentiles can be found using the **distribution function** $F(x)$. Recall that $F(x)$ gives the area under the pdf up to the value x of the random variable X. Using this fact then, to find the median, it is enough to solve the equation

$$F(x) = 0.5.$$

For the percentiles, a similar equation can be used, i.e.

$$F(x) = k\%.$$

Example 18

Use the distribution function above to find the median, first and third quartiles.

Solution

The distribution function is given by:

$$F(x) = \int_{-\infty}^{x} f(t)\,dt = \int_{0}^{x} \frac{3}{4}t^2(2 - t)\,dt = \frac{3}{4}\left(\frac{2}{3}x^3 - \frac{x^4}{4}\right)$$

Hence, to find the median we solve

$$F(m) = 0.5, \text{ i.e. } \frac{3}{4}\left(\frac{2}{3}m^3 - \frac{m^4}{4}\right) = 0.5,$$

which gives us the median as 1.2285.

To find the quartiles, we also solve

$$F(n) = 0.25, \text{ i.e. } \frac{3}{4}\left(\frac{2}{3}n^3 - \frac{n^4}{4}\right) = 0.25,$$

which gives us the first quartile as 0.913, and

$$F(n) = 0.75, \text{ i.e. } \frac{3}{4}\left(\frac{2}{3}n^3 - \frac{n^4}{4}\right) = 0.75,$$

which gives us the third quartile as 1.514.

Notice that $P(0.913 \leqslant x \leqslant 1.514) = F(1.514) - F(0.913) = 0.5$.

Variance and standard deviation

The variance of a random variable is defined as

$$\text{Var}(X) = \text{E}(X - \mu)^2.$$

This formula can be simplified, as in the discrete case, to

$$\text{Var}(X) = \text{E}(X)^2 - \mu^2 = \text{E}(X)^2 - (\text{E}(X))^2.$$

In the continuous case, if X has a pdf function $f(x)$, then

$$\text{Var}(X) = \int_x (x - \text{E}(X))^2 f(x)\,dx = \int_x x^2 f(x)\,dx - \mu^2$$

$$= \int_x x^2 f(x)\,dx - (\text{E}(X))^2.$$

The standard deviation is of course the square root of the variance,

$$\sigma = \sqrt{\text{Var}\,(X)}.$$

Example 19

The function $f(x)$ is a *pdf* for a random variable X.

$$f(x) = \begin{cases} \frac{3}{4}x^2(2-x) & 0 \leqslant x \leqslant 2 \\ 0 & \text{otherwise} \end{cases}$$

Find the variance and the standard deviation.

Solution

$\text{Var}(X) = \int_x x^2 f(x)dx - (\text{E}(X))^2 = \int_0^2 x^2 f(x)dx - (\text{E}(X))^2$

Since we have already calculated $\text{E}(X) = 6/5$, we only need to calculate $\int_0^2 x^2 f(x)dx$.

$$\int_0^2 x^2 f(x)dx = \int_0^2 x^2\left(\frac{3}{4}x^2(2-x)\right)dx = \left[\frac{3x^5}{10} - \frac{3x^6}{24}\right]_0^2 = 1.6\text{; hence,}$$

$$\text{Var}(X) = 1.6 - (6/5)^2 = 0.16\text{, and } \sigma = \sqrt{0.16} = 0.4.$$

Exercise 17.4

1 The continuous random variable X has a pdf $f(x)$ where

$$f(x) = \begin{cases} kx^2 + \frac{3}{2} & 0 \leqslant x \leqslant 1 \\ 0 & \text{otherwise} \end{cases}$$

a) Find the value of k.
b) Find $P(x > 0.5)$.
c) Find $P(0 < x < 0.5)$.
d) Find the mean, median and standard deviation.

2 The continuous random variable X has a pdf $f(x)$ where

$$f(x) = \begin{cases} k(5-2x) & 0 \leqslant x \leqslant 2 \\ 0 & \text{otherwise} \end{cases}$$

a) Find the value of k.
b) Find $P(x > 1.5)$.
c) Find $P(0.5 < x < 1.5)$.
d) Find the mean, median and standard deviation.

3 The continuous random variable X has a pdf $f(x)$ where

$$f(x) = \begin{cases} 2x - x^3 & 0 \leqslant x \leqslant k \\ 0 & \text{otherwise} \end{cases}$$

a) Find the value of k.
b) Find $P(x > 0.5)$.
c) Find $P(0 < x < 0.5)$.
d) Find the mean, median and standard deviation.

4 The continuous random variable X has a pdf $f(x)$ where

$$f(x) = \begin{cases} k(x+1) & 0 \leqslant x \leqslant 1 \\ 2kx^2 & 1 \leqslant x \leqslant 2 \\ 0 & \text{otherwise} \end{cases}$$

a) Find the value of k.
b) Find $P(x > 0.5)$
c) Find $P(1 < x < 1.5)$
d) Find the mean, median and standard deviation.

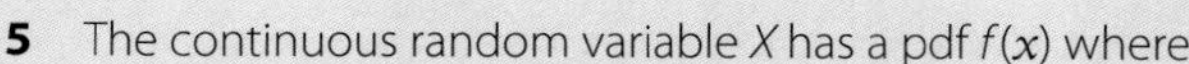

5 The continuous random variable X has a pdf $f(x)$ where

$$f(x) = \begin{cases} 2kx & 0 \leqslant x < 1 \\ 2kx^2 & 1 \leqslant x < 2 \\ k(8 - 2x) & 2 \leqslant x \leqslant 4 \\ 0 & \text{otherwise} \end{cases}$$

a) Sketch a graph for $f(x)$.
b) Find k.
c) Find the mean, median and standard deviation.
d) Find the IQR of this distribution.

6 The lifetime of flat batteries used in remote control units for one type of tv sets has a probability density function defined by

$$f(x) = \begin{cases} \frac{15}{76}(x^4 - 2x^2 + 2) & 0 \leqslant x \leqslant 1 \\ -\frac{15}{8056}(15x - 121) & 1 < x \leqslant 8\frac{1}{15} \\ 0 & \text{otherwise} \end{cases}$$

where x is measured in tens of hours.
a) Find the mean life of such batteries.
b) What is the probability that a battery will last at least 20 hours?
c) Each control unit is fitted with two batteries. The unit can only function if both batteries work. What is the probability that the unit will work for more than 20 hours?

7 The lifetime Y, in tens of hours, of light bulbs produced by a certain company has a probability density function defined by

$$f(y) = \begin{cases} \frac{3}{500}y(10 - y) & 0 \leqslant y \leqslant 10 \\ 0 & \text{otherwise} \end{cases}$$

a) Find the mean life of a light bulb.
b) Find the median life of a light bulb.
c) Find the standard deviation of the life of a light bulb.
d) Find the probability that a light bulb will last more than 80 hours.
e) A lamp set is fitted with two such bulbs. For a new set, find
 (i) the probability that both bulbs will last more than 80 hours
 (ii) the probability that at least one of the bulbs has to be replaced before 80 hours.

8 The weekly amount of oil pumped out of an oil well, in hundreds of barrels, has density function $f(y)$ defined by

$$f(y) = \begin{cases} \frac{1}{8}y^2 & 0 \leqslant y < 2 \\ \frac{y}{8}(4 - y) & 2 \leqslant y \leqslant 4 \\ 0 & \text{otherwise} \end{cases}$$

a) Sketch the graph of the pdf.
b) Find the mean production per week of this well.
c) Find the IQR for the production per week.
d) When the production falls below 10% of the weekly production, some maintenance will have to be done in terms of replacing the pumps with more specialized ones. What level of production will warrant that?

9 A continuous random variable Y has a pdf $f(y)$ given by

$$f(y) = \begin{cases} \frac{c}{(1 - y)(y - 6)} & 2 \leqslant y \leqslant 5 \\ 0 & \text{otherwise} \end{cases}$$

a) Show that $c = \frac{5}{4 \ln 2}$.
b) Calculate the mean and standard deviation of Y.

10 A random variable Y has a pdf $f(y)$ defined as follows

$$f(y) = \begin{cases} a(by - y^2) & 0 \leqslant y \leqslant 5 \\ 0 & \text{otherwise, } a, b > 0 \end{cases}$$

a) Show that $b \geqslant 5$ and that $a = \dfrac{6}{25(3b - 10)}$.

b) If the expected value of Y is 2.5, find the values of a and b.

c) Find the variance of Y.

11 A random variable X has a pdf $f(x)$ defined over an interval $[a, b]$ where $b > a > 0$. The equation that defines $f(x)$ is $f(x) = k$.

a) Find k in terms of a and b.

b) Also, in terms of a and b, find

(i) the mean

(ii) the median

(iii) the variance.

Note: This is known as the **uniform distribution**.

12 The random variable X has the probability density function

$$f(x) = \begin{cases} \dfrac{5x^4}{31} & 1 \leqslant x \leqslant 2 \\ 0 & \text{otherwise} \end{cases}$$

a) Find

(i) $P(1.2 < x < 1.7)$

(ii) the median

(iii) the value of k such that $P(x > k) = 0.25$.

b) Two independent observations from X are made. What is the probability that at least one of them is larger than 1.5?

13 (Optional) The distribution function $F(x)$ of a random variable X is

$$F(x) = \begin{cases} 0 & 0 \leqslant x < 5 \\ k(x^3 - 21x^2 + 147x - 335 & 5 \leqslant x \leqslant 7 \\ 1 & x > 7 \end{cases}$$

Find

a) the value of k

b) the probability density function $f(x)$

c) the median of X

d) $\mathrm{Var}(X)$.

14 The random variable Y has pdf

$$f(y) = \begin{cases} 4y^k & 0 \leqslant y \leqslant 1 \\ 0 & \text{otherwise, } k > 0 \end{cases}$$

a) Find the value of k.

b) Find the mean of Y.

c) Find the value a for which $P(y > a) = 0.5$.

15 The time to first failure of an engine valve (in thousands of hours) is a random variable Y with a pdf

$$f(y) = \begin{cases} 2ye^{-y^2} & 0 \leqslant y \\ 0 & \text{otherwise} \end{cases}$$

a) Show that $f(y)$ satisfies the requirements of a density function.

b) Find the probability that the valve will last at least 2000 hours before being serviced.

c) Find the mean of the random variable.

(Hint: You need to know that $\lim_{x \to \infty} \int_0^x e^{-t^2} dt = \dfrac{\sqrt{\pi}}{2}$)

d) Find the median of Y.

e) Find the IQR.

f) An engine utilizes two such valves and needs servicing as soon as any of the two valves fail. Find the probability that the engine needs servicing before 200 hours of work.

16 The length of time required by students to complete a paper 2 HL exam is a random variable with a pdf

$$f(y) = \begin{cases} \frac{1}{2}\left(cy + \frac{y^2}{3}\right) & 0 \leqslant y \leqslant 2 \\ 0 & \text{otherwise} \end{cases}$$

a) Find c.
b) Find the probability that a randomly selected student will finish in less than 1 hour.
c) In a randomly selected group of 10 students, what is the probability that 3 students will finish the exam in less than 1 hour?
d) Given that Casper, who happened to be randomly selected, needs at least 1 hour to finish the exam, what is the probability that he will require at least 90 minutes?

17 The time, in months, in excess of one year to complete a building construction project is modelled by a continuous random variable Y months with a pdf

$$f(y) = \begin{cases} ky^2(5 - y) & 0 \leqslant y \leqslant 5 \\ 0 & \text{otherwise} \end{cases}$$

a) Show that $k = \frac{12}{625}$.
b) Find the mean, median and mode of this distribution (1 decimal place accuracy).
c) What proportion of the projects is completed in less than three months of excess time?
d) Find the standard deviation of the excess time.
e) What proportion of the projects is finished within one standard deviation of the mean excess time? Does your answer contradict the 'empirical rule'?

18 The delay time T for flights of a certain airline, in hours, has the pdf

$$f(t) = \begin{cases} \frac{4}{625}(5t^3 - t^4) & 0 \leqslant t \leqslant 5 \\ 0 & \text{otherwise} \end{cases}$$

If a flight is delayed more than 5 hours, it is cancelled.
a) Find the mean and mode of the delay time.
b) Show that the median delay time is approximately 3.43 hours.
c) Find the probability that a randomly chosen flight will be delayed between 1 and 2 hours.
d) Find the standard deviation of the delay time.
e) Two flights are chosen at random. What is the probability that (assume flights delay times are independent of each other)
   (i) both are delayed more than 1 hour?
   (ii) at least one of them is delayed for more than 1 hour?
   (iii) only one of them is delayed more than 1 hour?

19 The probability density function $f(x)$ of the continuous random variable X is defined on the interval $[0, a]$ by

$$f(x) = \begin{cases} \frac{1}{8}x & 0 \leqslant x < 3 \\ \frac{27}{8x^2} & 3 < x \leqslant a \end{cases}$$

Find the value of a.

20 The probability density function $f(x)$, of a continuous random variable X is defined by

$$f(x) = \begin{cases} \frac{1}{4}x(4 - x^2) & 0 \leqslant x \leqslant 2 \\ 0 & \text{otherwise} \end{cases}$$

Calculate the median value of X.

17.5 The normal distribution

Continuous probability distributions can assume a variety of shapes. However, for reasons of staying within (with some extensions) the boundaries of the IB syllabus, we will focus on one distribution. In fact, a large number of random variables observed in our surroundings possess a frequency distribution that is approximately bell-shaped. We call that distribution the **normal probability distribution.**

The most important type of continuous random variable is the *normal* random variable. The probability density function of a normal random variable X is determined by two parameters: the mean or expected value μ and the standard deviation σ of the variable.

The normal probability density function is a bell-shaped density curve that is symmetric about the mean μ. Its variability is measured by σ. The larger the value of σ the more variability there is in the curve. That is, the higher the probability of finding values of the random variable further away from the mean. Figure 17.1 represents three different normal density functions with the same mean but different standard deviations. Note how the curves 'flatten' as σ increases. This is so because the area under the curve has to stay equal to 1.

Figure 17.1

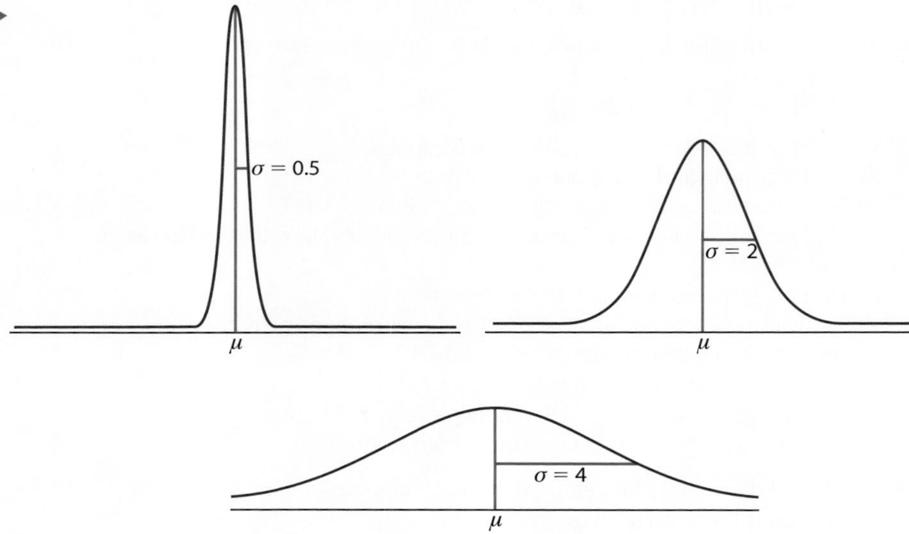

Probability density function of the normal distribution

The probability density function for a normally distributed random variable x is

$$f(x) = \frac{1}{\sigma\sqrt{2\pi}}e^{-\frac{(x-\mu)^2}{2\sigma^2}} = \frac{1}{\sigma\sqrt{2\pi}}e^{-\frac{1}{2}\left(\frac{x-\mu}{\sigma}\right)^2} \quad \text{for } -\infty < x < \infty$$

where μ and σ^2 are any number such that $-\infty < \mu < \infty$ and $0 < \sigma^2 < \infty$, and where e and π are the well-known constants $e = 2.718\,28\ldots$ and $\pi = 3.141\,59\ldots$.

Notation:
When a variable is normally distributed, we write $X \sim \mathrm{N}(\mu, \sigma^2)$.

Although we will not make direct use of the formula above, it is interesting to note its properties, because they help us understand how the normal distribution works. Notice that the equation is completely determined by the mean μ and the standard deviation σ.

The graph of a normal probability distribution is shown in Figure 17.2. As you notice, the mean or expected value locates the centre of the distribution, and the distribution is symmetric about this mean. Since the total area under the curve is 1, the symmetry of the curve implies that the area to the right of the mean and the area to the left are both equal to 0.5. The shape, or how 'flat' it is, is determined by σ, as we have seen in Figure 17.1. Large values of σ tend to reduce the height of the curve and increase the spread, and small values of σ increase the height to compensate for the narrowness of the distribution.

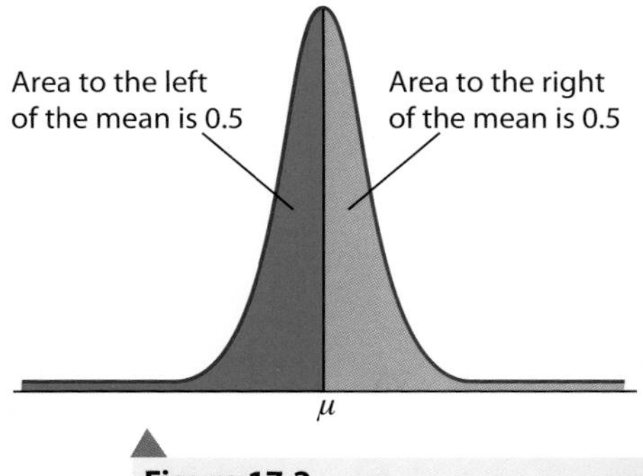

Figure 17.2

So, the normal distribution is fully determined by its mean, μ, and its standard deviation, σ. Changing μ without changing σ moves the normal curve along the horizontal axis without changing its spread. As you have seen above, the standard deviation σ controls the spread of the curve. You can also locate the standard deviation by eye on the curve. One σ to the right or left of the mean μ marks the point where the curvature of the curve changes. That is, as you move right from the mean, at the point where $x = \mu + \sigma$, the curve changes its curvature from downwards to upwards. Similarly, as you move one σ to the left from the mean the curve changes its curvature from downwards to upwards.

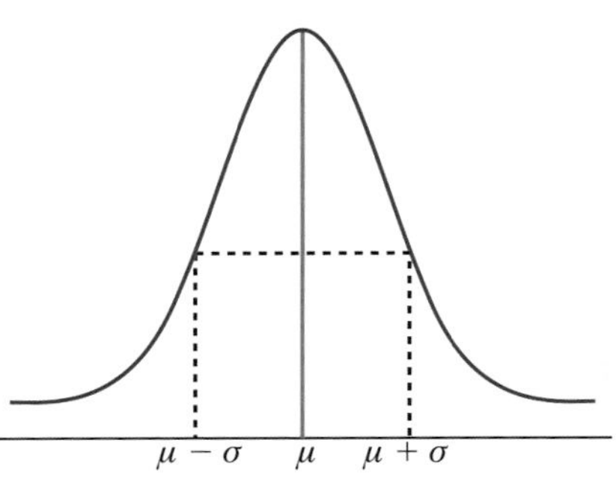

Figure 17.3

Although there are many normal curves, they all have common properties. Here is one important one that you have seen in Chapter 11:

The empirical rule – restated

In the normal distribution with mean μ and standard deviation σ:

- Approximately 68% of the observations fall within σ of the mean μ.
- Approximately 95% of the observations fall within 2σ of the mean μ.
- Approximately 99.7% of the observations fall within 3σ of the mean μ.

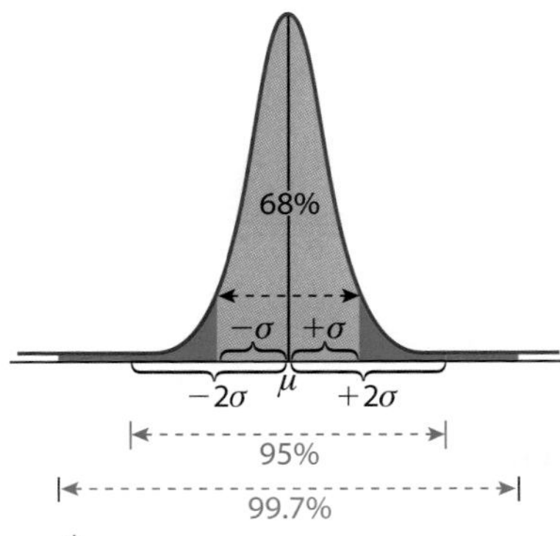

Figure 17.4

Figure 17.4 illustrates this rule. Later in this section, you will learn how to find these areas from a table or from your GDC.

Example 20

Heights of young German men between 18 and 19 years of age follow a distribution that is approximately normal, with a mean of 181 cm and a standard deviation of 8 cm (approximately). Describe this population of young men.

Solution

According to the empirical rule, we find that approximately 68% of those young men have a height between 173 cm and 189 cm, 95% of them between 165 cm and 197 cm, and 99.7% between 157 cm and 205 cm. Looking further, you can say that only 0.15% are taller than 205 cm, or shorter than 157 cm.

As the empirical rule suggests, all normal distributions are the same if we measure in units of size σ about the mean μ as centre. Changing to these units is called *standardizing*. To standardize a value, measure how far it is from the mean and express that distance in terms of σ. This is how the calculation can be done:

Standardizing

If x is a value of a normal random variable, with mean μ and standard deviation σ, the standardized value of x is

$$z = \frac{x-\mu}{\sigma}.$$

A standardized value is also called the **z-score**.

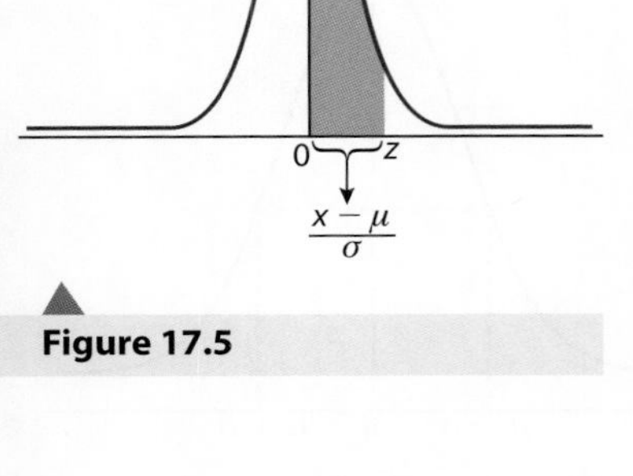

Figure 17.5

The quantity $x - \mu$ tells us how far our value is from the mean; dividing by σ then tells us how many standard deviations that distance is equal to.

The standardizing process, as you notice, is a transformation of the normal curve. For discussion purposes, assume the mean μ to be positive. The transformation $x - \mu$ shifts the graph back μ units. So, the new centre is shifted from μ back μ units. That is, the new centre is 0! Dividing by σ is going to 'scale' the distances from the mean and express everything in terms of σ. So, a point that is one standard deviation from the mean is going to be 1 unit above the new mean, i.e. it will be represented by $+1$. Now, if you look at the empirical rule we discussed earlier, points that are within one standard deviation from the mean will be within a distance of 1 in the new distribution. Instead of being at $\mu + \sigma$ and $\mu - \sigma$, they will be at $0 + 1$ and $0 - 1$ respectively, i.e. -1 and $+1$. (See Figure 17.6.)

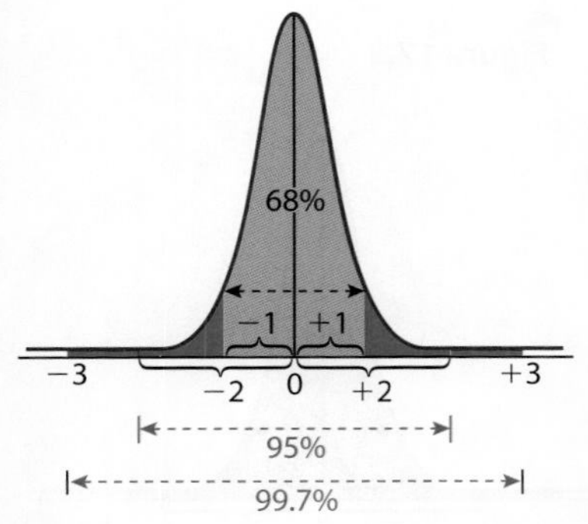

Figure 17.6

The new distribution we created by this transformation is called the **standard normal distribution**. It has a mean of 0 and a standard deviation of 1. It is a very helpful distribution because it will enable us to read the areas under any normal distribution through the standardization process, as will be demonstrated in the examples that follow.

Probability density function of the standard normal distribution

The probability density function for standard normal distribution is

$$f(z) = \frac{1}{\sqrt{2\pi}}\, e^{-\frac{1}{2}(z)^2} \text{ for } -\infty < z < \infty.$$

Since linear transformations can transform all normal functions to standard, this becomes a very convenient and efficient way of finding the area under any normal distribution.

The proof that the mean and the variance of the standard normal variable are 0 and 1 respectively is straightforward.

Let $z = \frac{x-\mu}{\sigma}$ be the standard variable corresponding to a normal variable x.

$$E(z) = E\left(\frac{x-\mu}{\sigma}\right) = E\left(\frac{1}{\sigma}(x-\mu)\right) = \frac{1}{\sigma}E(x-\mu) = \frac{1}{\sigma}(\mu-\mu) = 0$$

$$V(z) = V\left(\frac{x-\mu}{\sigma}\right) = V\left(\frac{1}{\sigma}(x-\mu)\right) = \frac{1}{\sigma^2}V(x-\mu) = \frac{1}{\sigma^2}V(x) = \frac{1}{\sigma^2}\sigma^2 = 1$$

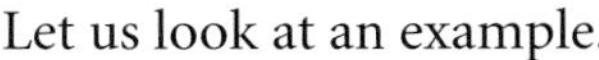

Let us look at an example.

A young German man with a height of 192 cm has a z-score of

$$z = \frac{x - \mu}{\sigma} = \frac{192 - 181}{8} = 1.375$$

or 1.375 standard deviations above the mean. Similarly, a young man with a height of 175 cm is

$$z = \frac{x - \mu}{\sigma} = \frac{175 - 181}{8} = -0.75$$

or 0.75 standard deviations below the mean.

To find the probability that a normal variable x lies in the interval a to b, we need to find the area under the normal curve $N(\mu, \sigma^2)$ between the points a and b. However, there is an infinitely large number of normal curves – one for each mean and standard deviation. (See Figure 17.7.)

A separate table of areas for each of these curves is obviously not practical. Instead, we use one table for the standard normal distribution, which gives us the required areas. When you standardize a and b, you get two standard numbers z_1 and z_2 such that the area between z_1 and z_2 is the same as the area we need.

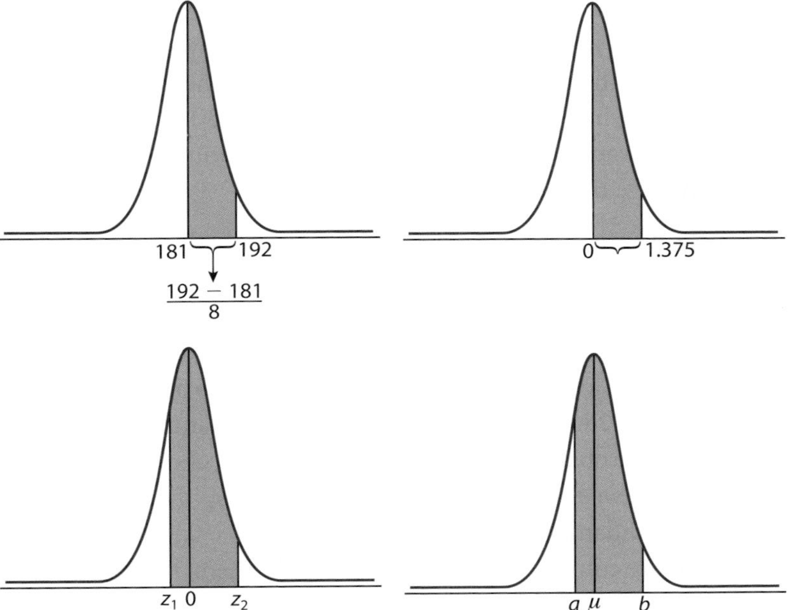

Figure 17.7

In the example above, if we are interested in the proportion of young German men whose height is between 175 cm and 192 cm, we calculate the z-scores for these numbers and then read the area from the table. (The normal distribution table is in the appendix). Here is an abbreviated version of the table and instructions on how to use it. (There are many tables of the areas under normal distributions. We will use a table constructed in a similar way to the one used on IB examinations.)

z	0.00	0.01	→	0.05	0.06	0.07	0.08	0.09
0.0	0.5000	0.5040		0.5199	0.5239	0.5279	0.5319	0.5359
0.1	0.5398	0.5438		0.5596	0.5636	0.5675	0.5714	0.5753
0.2	0.5793	0.5832		0.5987	0.6026	0.6064	0.6103	0.6141
↓			→			↓		
↓			→			↓		
↓			→			↓		
1.3	0.9032	0.9049		0.9115	0.9131	0.9147	**0.9162**	0.9177
1.4	0.9192	0.9207		0.9265	0.9279	0.9292	0.9306	0.9319
1.5	0.9332	0.9345		0.9394	0.9406	0.9418	0.9429	0.9441
1.6	0.9452	0.9463		0.9505	0.9515	0.9525	0.9535	0.9545

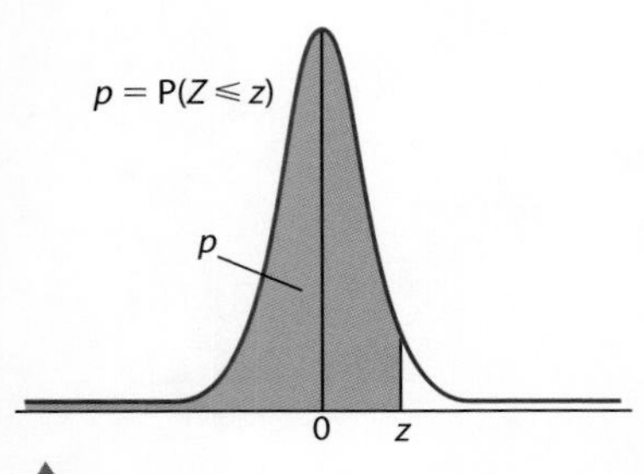

Figure 17.8

The table, as constructed, gives you the areas under the normal distribution to the left of some value z, as you see in Figure 17.8.

The table starts at 0, and gives the areas till $z = 3.9$. To read an area to the left of a number z, say 1.37, you read the first column to find the first two digits of z. So, in the first column, we stop at the cell containing 1.3. To get the area for 1.37, we look at the first row and choose the column corresponding to 0.07. Where the row at 1.3 meets the column at 0.07 is the area under the normal distribution corresponding to 1.37, namely, 0.9147. That is, the probability of at most a height with $z = 1.37$ is 0.9147. Since the table does not go to 4 decimal places, our answers will not be very precise. So, to find the probability corresponding to a height of 192 cm, we need a z of 1.375, which is not in the table. We can use 1.37, 1.38, or take an average. If we want an average, we read the neighbouring area of 0.9162, and get the average to be 0.915 45.

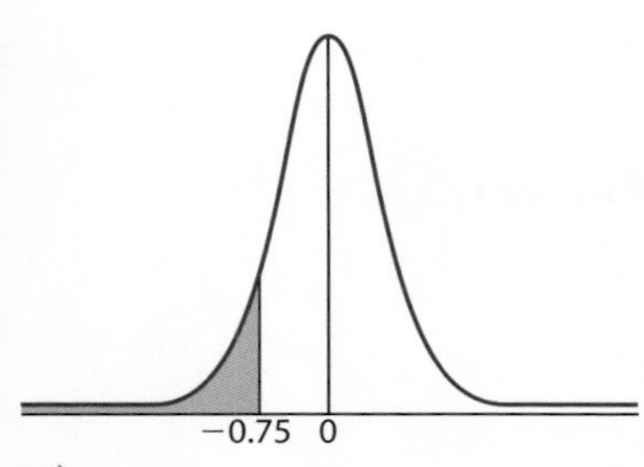

Figure 17.9

Unfortunately, due to limitations of space, this type of table does not cater for negative values of z. The good news is that, due to the symmetry of the distribution, the area to the left of a negative value of z is the same as the area to the right of its absolute value. So, if we are interested in the area to the left of -0.75, we look for the area to the right of 0.75, i.e. $1 - P(z < 0.75) = 1 - 0.7734 = 0.2266$ (see Figure 17.10). So, in the example above, if we want to know the probability of a young German man, chosen at random, having a height between 175 cm and 192 cm, we look up the corresponding area under the standard normal distribution between -0.75 and 1.375. Since these two areas are cumulative, we need to subtract them, i.e. the required area is $0.915\,45 - 0.2266 = 0.6885$.

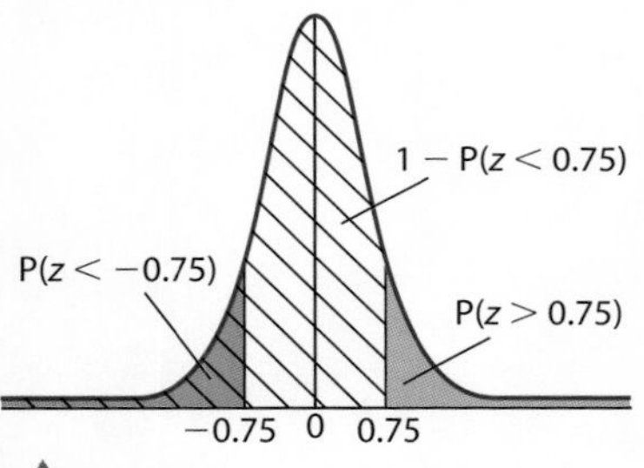

Figure 17.10
$P(z > 0.75) = 1 - P(z < 0.75)$.

Figure 17.11

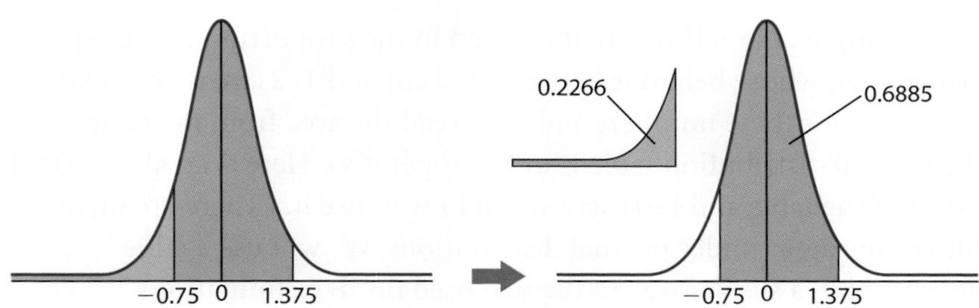

What is the chance that a young German man is taller than 175 cm?

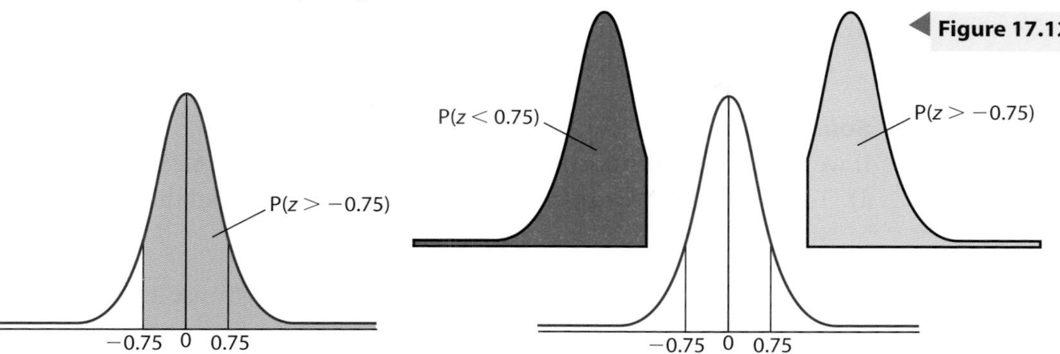

Figure 17.12

This means that we have to look at the area above -0.75. Due to symmetry, the area in question, which is to the right of -0.75, is equal to the area below 0.75, which in turn can be read directly from the table as 0.7734.

These calculations are much easier to calculate using a GDC, of course. Also, with the GDC, you do not need to standardize your variables either. However, because there are cases where you need to understand standardization and other cases where you are *required to use a table*, you need to know both methods.

Here is how your GDC can give you your answers.

You first go to the 'Distribution' menu and choose 'normalcdf'. Then you enter the numbers in the following order: *Lower limit, upper limit, mean, and standard deviation.* The result will be the area you need. See the screen images below.

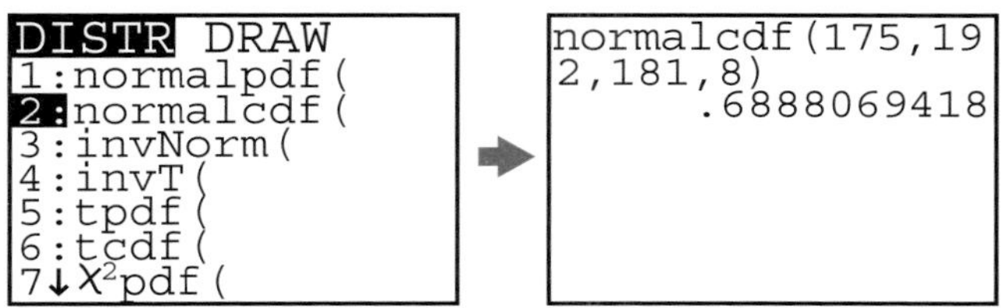

If you want to use the standard normal, your commands will be the same, but you do not need to include the mean and standard deviation. They are the default.

```
normalcdf(-.75,1
.375)
          .6888069418
```

If you need the probability that a young man is taller than 175 cm, you can also read it either by looking at the distribution with the original data or by standardizing.

Example 21

The age of graduate students in engineering programmes throughout the US is normally distributed with mean $\mu = 24.5$ and standard deviation $\sigma = 2.5$.

If a student is chosen at random,

a) what is the probability he/she is younger than 26 years old?

b) what proportion of students is older than 23.7 years?

c) what percentage of students is between 22 and 28 years old?

d) what percentage of the ages falls within 1 standard deviation of the mean? 2 standard deviations? 3 standard deviations?

Solution

If we let X = age of students, then $X \sim N(\mu = 24.5, \sigma^2 = 6.25)$.

a) To answer this, we can either standardize and then read the table for the area left of 0.6:

$$P\left(z < \frac{26 - 24.5}{2.5}\right) = P(z < 0.6) = 0.7257, \text{ or use a GDC:}$$

```
normalcdf(0,26,2
4.5,2.5)
          .7257469354
```

Notice here that we put 0 as a lower limit. You can put a number as a lower limit far enough from the mean to make sure you are receiving the correct cumulative distribution.

b) This can be done similarly:

$$P(x > 23.7) = P\left(z > \frac{23.7 - 24.5}{2.5} = -0.32\right)$$

So, by symmetry we know that

$$P(z > -0.32) = P(z < 0.32) = 0.6255.$$

With a GDC:

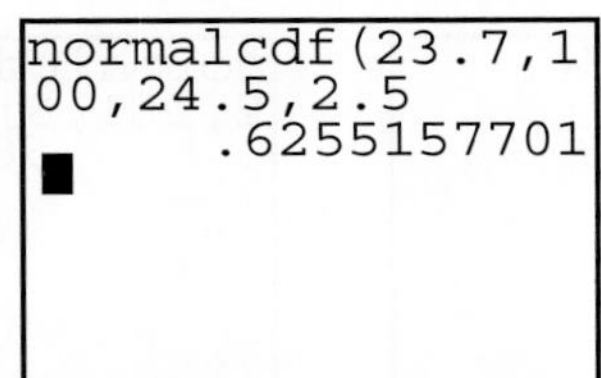

Also, notice here that we wrote 100 as an upper limit, which is an arbitrary number far enough to the right to be sure we include the whole population.

c) $P(22 < x < 28) = P\left(\frac{22 - 24.5}{2.5} < z < \frac{28 - 24.5}{2.5}\right) = P(-1 < z < 1.4)$

We find the area to the left of 1.4 and to the left of -1 and subtract them:

$$= 0.9192 - 0.1587 = 0.7606 = 76.06\%$$

With a GDC:

```
normalcdf(22,28,
24.5,2.5)
          .7605880293
```

d) This, as you know, is the empirical rule we talked about before. Let us see what percentage of the approximately normal data will lie within 1, 2 or 3 standard deviations.
We start with the traditional table:

$$P(-1 \leqslant z \leqslant 1) = P(z \leqslant 1) - P(z \leqslant -1) = 0.8413 - 0.1587 = 0.6826$$

This is the exact value corresponding to the empirical rule's 68%!

$$P(-2 \leqslant z \leqslant 2) = P(z \leqslant 2) - P(z \leqslant -2) = 0.9772 - 0.0228 = 0.9544$$

Again, this is the exact value corresponding to the empirical rule's 95%!

$$P(-3 \leqslant z \leqslant 3) = P(z \leqslant 3) - P(z \leqslant -3) = 0.9987 - 0.0013 = 0.9973$$

And again, this is the exact value corresponding to the empirical rule's 99.7%!

The inverse normal distribution

Another type of problem arises in situations similar to the one above when we are given a cumulative probability and would like to find the value in our data that has this cumulative probability. For example, what age marks the 95th percentile? That is, what age is higher or equal to 95% of the population? To answer this question, we need to reverse our steps. So far, we are given a value and then we look for the area corresponding to it. Now, we are given the area and we have to look for the number. That is why this is called the **inverse normal distribution**. Again, the approach is to find the *standard inverse normal number and then to 'de-standardize' it.* That is, to find the value from the original data that corresponds to the z-value at hand.

There is an inverse normal table available online. We will produce a part of the inverse normal table here for explanation.

p	0.000	0.001	→	0.005	0.006	0.007	0.008	0.009
0.50	0.0000	0.0025		0.0125	0.0150	0.0175	0.0201	0.0226
0.51	0.0251	0.0276		0.0376	0.0401	0.0426	0.0451	0.0476
0.52	0.0502	0.0527		0.0627	0.0652	0.0677	0.0702	0.0728
0.53	0.0753	0.0778		0.0878	0.0904	0.0929	0.0954	0.0979
↓			→					
↓			→					
0.74	0.6433	0.6464		0.6588	0.6620	0.6651	0.6682	0.6713
0.75	0.6745	0.6776		0.6903	0.6935	0.6967	0.6999	0.7031
0.76	0.7063	0.7095		0.7225	0.7257	0.7290	0.7323	0.7356
0.77	0.7388	0.7421		0.7554	0.7588	0.7621	0.7655	0.7688
0.78	0.7722	0.7756		0.7892	0.7926	0.7961	0.7995	0.8030
0.79	0.8064	0.8099		0.8239	0.8274	0.8310	0.8345	0.8381

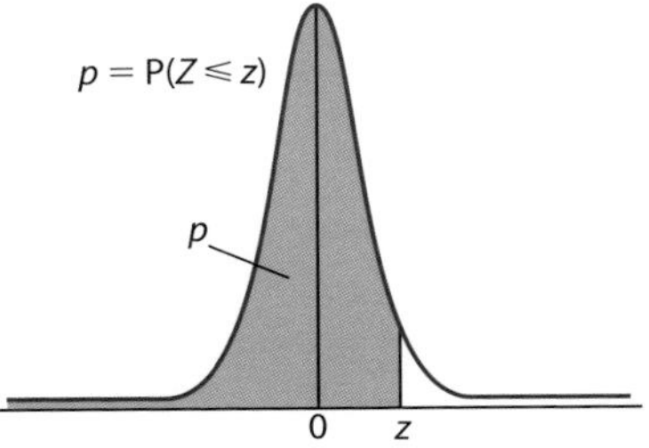

Figure 17.13

The table gives a selection of probabilities above the mean and the body of the table gives the z-value corresponding to that area. You know that 0 has a cumulative probability of 0.5. Look at the table and observe the intersection of the 0.50 row and the 0.000 column. It is 0, the mean of the standard normal distribution.

If we need to know what z-score the third quartile Q_3 is, for example, we need to look up 0.75. The z-score corresponding to Q_3 is 0.6745 as you see.

Suppose you want to find the z-score that leaves an area of 0.915 below it.

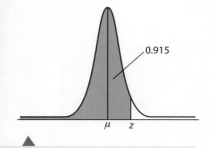

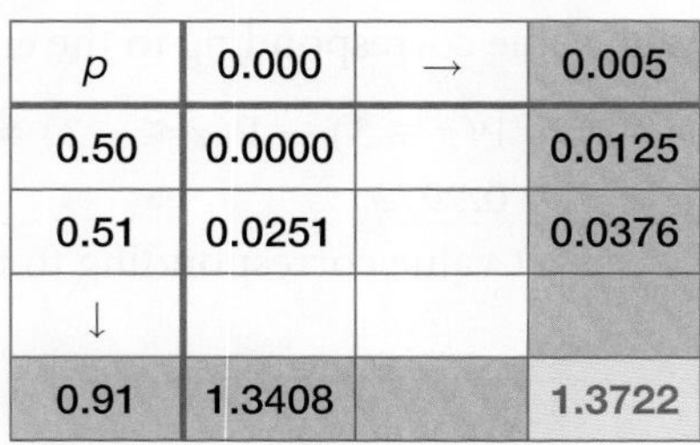
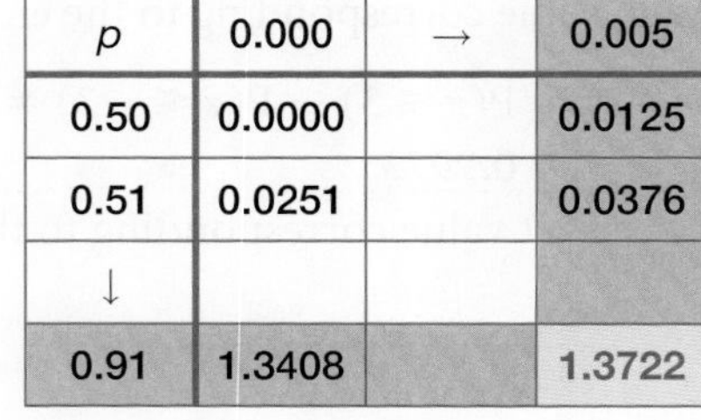

p	0.000	→	0.005
0.50	0.0000		0.0125
0.51	0.0251		0.0376
↓			
0.91	1.3408		1.3722

0.915

μ z

Figure 17.14

In the first column, we choose 0.91, then at the intersection of the row at 0.91 and the column at 0.005 the z-score corresponds to 0.915. So,

$$P(z < 1.3722) = 0.915.$$

The GDC can also be used in this case. The process is identical to the normal calculation. The difference is in choosing 'invNorm' instead.

```
invNorm(.5)
                  0
invNorm(.915)
         1.37220381
invNorm(.75)
        .6744897495
```

In the young German men example, we would like to find what height leaves 95% of the population below it.

In this case, we look up the z-score corresponding to 0.95 and we find that it is $z = 1.6449$.

$$\text{Now } z = 1.6449 = \frac{x - 181}{8} \Rightarrow x - 181 = 8 \times 1.6449$$

$$\Rightarrow x = 181 + 8 \times 1.6449 = 194.16.$$

So, 95% of the young German men are shorter than 194.16 cm.

The GDC gives you this number with less effort:

```
invNorm(.95,181,
8)
         194.158829
```

Example 22

Since November 2007, the average time it takes fast trains (Eurostar) to travel between London and Paris is 2 hours 15 minutes, with a standard deviation of 4 minutes. Assume a normal distribution.

a) What is the probability that a randomly chosen trip will take longer than 2 hours and 20 minutes?

b) What is the probability that a randomly chosen trip will take less than 2 hours and 10 minutes?

c) What is the IQR of a trip on these trains?

Solution

We will do each problem using a table and a GDC to acquaint you with both methods.

a) The mean $\mu = 2.25$ and $\sigma = 0.067$.

2 hours 20 minutes $= 2.33$

$$P(x > 2.33) = P\left(z > \frac{2.33 - 2.25}{0.067}\right) = P(z > 1.25)$$

From the table: $P(z > 1.25) = 1 - P(z < 1.25) = 1 - 0.8944 = 0.1056$

Using your GDC:

```
normalcdf(2.3333
,100,2.25,.06667
)
             .10575261
```

The number 100 is arbitrary!

b) 2 hours 10 minutes $= 2.167$

$$P(x < 2.167) = P\left(z > \frac{2.167 - 2.25}{0.067}\right) = P(z < -1.25)$$

From the table, and by symmetry, this is the same as $P(z > 1.25)$, which we found in part a) above.

GDC

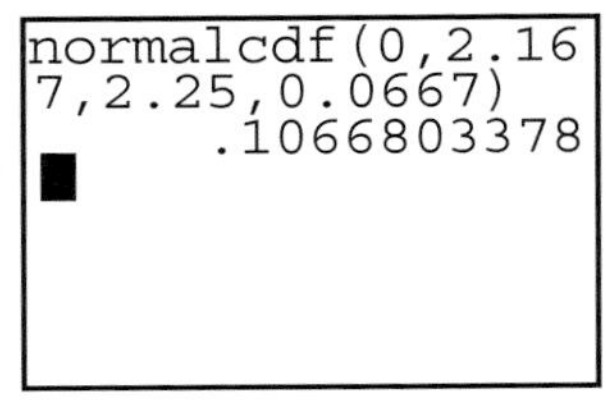

or

```
ShadeNorm(0,2.16
7,2.25,0.0667)
```

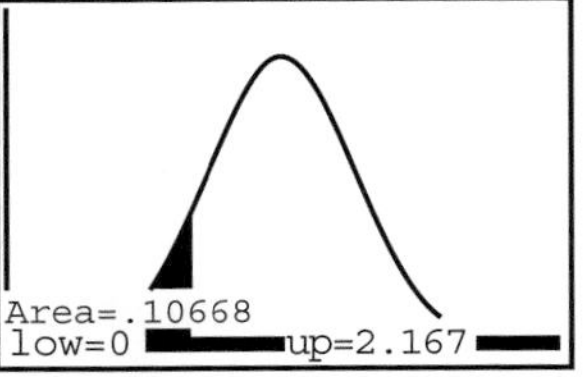

c) To find the IQR, we need to find Q_1 and Q_3.

Q_1 is the number that leaves 25% of the data before it. So, we need to find the inverse normal variable that has an area of 0.25 before it.

From the table we can only do so using symmetry. So, we find the *z*-score that corresponds to 0.25 by finding its symmetrical number, which is the z-score with 0.75. So, we only need to find $z(0.75)$. The table of standard inverse normal gives us $z = 0.6745$.

So, Q_1 corresponds to -0.6745.

$$z = -0.6745 = \frac{x - 2.25}{0.067} \Rightarrow x - 2.25 = 0.067 \times (-0.6745)$$
$$\Rightarrow x = 2.25 - 0.045 = 2.205$$

Q_3 corresponds to 0.6745.

$$z = 0.6745 = \frac{x - 2.25}{0.067} \Rightarrow x - 2.25 = 0.067 \times (0.6745)$$
$$\Rightarrow x = 2.25 + 0.045 = 2.295$$

IQR = 2.295 − 2.205 = 0.090 of an hour, i.e. 5.4 minutes.

Example 23

The age at which babies develop the ability to walk can be described by a normal model. It is known that 5% of babies learn how to walk by the age of 10 months and 25% need more than 13 months. Find the mean and standard deviation of the distribution.

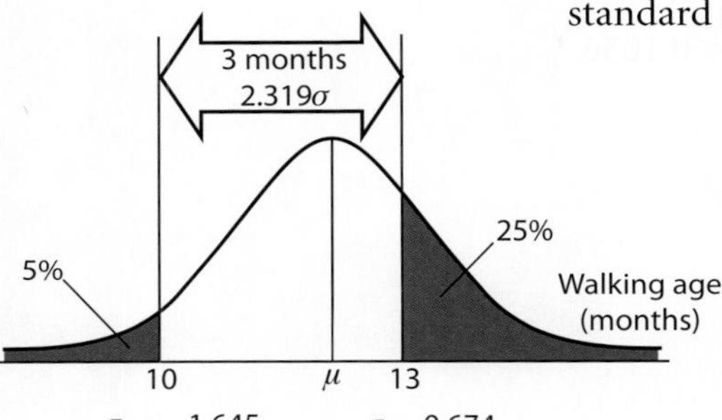

Solution

Looking at the diagram at left will help you visualize the solution. We will show two approaches to this problem.

The first approach is to consider the distance between 10 and 13 months. In our data, that distance is 3 months, but how many standard deviations does that represent? Since we know that 10 months represents the lower 5%, and 13 months represents the upper 25%, we can obtain z-scores for those two data points without knowing the mean and standard deviation. Use the inverse table or a GDC:

```
invNorm(.05)
        -1.644853626
invNorm(.75)
         .6744897495
```

Therefore, the 3-month distance is equivalent to 0.674 – (–1.64) = 2.319 standard deviations or:

$$2.319\sigma = 3 \Rightarrow \sigma = 1.294$$

And finally we use either of the data points (10 months or 13 months) in our z-score formula to find the mean:

$$z = \frac{x - \mu}{\sigma} \Rightarrow 0.674 = \frac{13 - \mu}{1.294} \Rightarrow \mu = 12.128$$

Thus the mean age that babies begin to walk is 12.1 months with a standard deviation of 1.29 months.

The second approach uses a bit of algebra instead. After obtaining the z-scores from the inverse table or GDC (as above), we begin by writing two equations using the z-score formula:

$$-1.645 = \frac{10 - \mu}{\sigma} \Rightarrow \mu - 1.645\sigma = 10,$$

$$0.674 = \frac{13 - \mu}{\sigma} \Rightarrow \mu + 0.674\sigma = 13$$

Aha! We have two linear equations with two unknowns. Solve these to obtain:

$$\mu = 12.128$$
$$\sigma = 1.294$$

Again, we conclude the mean age that babies begin to walk is 12.1 months with a standard deviation of 1.29 months.

Exercise 17.5

1 The time it takes to change the batteries of your GDC is approximately normal with mean 50 hours and standard deviation of 7.5 hours.
Find the probability that your newly equipped GDC will last

a) at least 50 hours

b) between 50 and 75 hours

c) less than 42.5 hours

d) between 42.5 and 57.5 hours

e) more than 65 hours

f) 47.5 hours.

2 Find each of the following probabilities.

a) $P(|z| < 1.2)$

b) $P(|z| > 1.4)$

c) $P(x < 3.7)$, where $X \sim N(3, 3)$

d) $P(x > -3.7)$, where $X \sim N(3, 3)$

3 A car manufacturer introduces a new model that has an in-city mileage of 11.4 litres/100 kilometres. Tests show that this model has a standard deviation of 1.26. The distribution is assumed to be normal.
A car is chosen at random from this model.

a) What is the probability that it will have a consumption less than 8.4 litres/100 kilometres?

b) What is the probability that the consumption is between 8.4 and 14.4 litres/100 kilometres?

4 Find the value of z that will be exceeded only 10% of the time.

5 Find the value of $z = z_0$ such that 95% of the values of z lie between $-z_0$ and $+z_0$.

6 The scores on a public schools examination are normally distributed with a mean of 550 and a standard deviation of 100.

a) What is the probability that a randomly chosen student from this population scores below 400?

b) What is the probability that a student will score between 450 and 650?

c) What score should you have in order to be in the 90th percentile?

d) Find the IQR of this distribution.

7 A company producing and packaging sugar for home consumption put labels on their sugar bags noting the weight to be 500 g. Their machines are known to fill the bags with weights that are normally distributed with a standard deviation of 5.7 g. A bag that contains less than 500 g is considered to be underweight and is not appreciated by consumers.

a) If the company decides to set their machines to fill the bags with a mean of 512 g, what fraction will be underweight?
b) If they wish the percentage of underweight bags to be at most 4%, what mean setting must they have?
c) If they do not want to set the mean as high as 512 g, but instead at 510 g, what standard deviation gives them at most 4% underweight bags?

8 In a large school, heights of students who are 13 years old are normally distributed with a mean of 151 cm and a standard deviation of 8 cm.
Find the probability that a randomly chosen child is
a) shorter than 166 cm
b) within 6 cm of the average.

9 The time it takes Kevin to get to school every day is normally distributed with a mean of 12 minutes and standard deviation of 2 minutes. Estimate the number of days when he takes
a) longer than 17 minutes
b) less than 10 minutes
c) between 9 and 13 minutes.
There are 180 school days in Kevin's school.

10 X has a normal distribution with mean 16. Given that the probability that x is less than 16.56 is 64%, find the standard deviation, σ, of this distribution.

11 X has a normal distribution with mean 91. Given that the probability that x is larger than 104 is 24.6%, find the standard deviation σ of this distribution.

12 X has a normal distribution with variance of 9. Given that the probability that x is more than 36.5 is 2.9%, find the mean μ of this distribution.

13 X has a normal distribution with standard deviation of 32. Given that the probability that x is more than 63 is 87.8%, find the mean μ of this distribution.

14 X has a normal distribution with variance of 25. Given that the probability that x is less than 27.5 is 0.312, find the mean μ of this distribution.

15 X has a normal distribution such that the probability that x is larger than 14.6 is 93.5% and $P(x > 29.6) = 2.2\%$. Find the mean μ and the standard deviation σ of this distribution.

16 $X \sim N(\mu, \sigma^2)$. $P(x > 19.6) = 0.16$ and $P(x < 17.6) = 0.012$. Find μ and σ.

17 $X \sim N(\mu, \sigma^2)$. $P(x > 162) = 0.122$ and $P(x < 56) = 0.0276$. Find μ and σ.

18 Wooden poles used for electricity networks in rural areas are produced and have lengths that are normally distributed.
2% of the poles are rejected because they are considered too short, and 5% are rejected because they are too long.
a) Find the mean and standard deviation of these poles if the acceptable range is between 6.3 m and 7.5 m.
b) In a randomly selected sample of 20 poles, find the probability of finding 2 rejected poles.

19 Bottles of mineral water sold by a company are advertised to contain 1 litre of water. To guarantee customer satisfaction the company actually adjusts its filling process to fill the bottles with an average of 1012 ml. The process follows a normal distribution with standard deviation of 5 ml.

a) Find the probability that a randomly chosen bottle contains more than 1010 ml.
b) Find the probability that a bottle contains less than the advertised volume.
c) In a shipment of 10 000 bottles, what is the expected number of 'underfilled' bottles?

20 Cholesterol plays a major role in a person's heart health. High blood cholesterol is a major risk factor for coronary heart disease and stroke. The level of cholesterol in the blood is measured in milligrams per decilitre of blood (mg/dl). According to the WHO, in general, less than 200 mg/dl is a desirable level, 200 to 239 is borderline high, and above 240 is a high-risk level and a person with this level has more than twice the risk of heart disease as a person with less than a 200 level. In a certain country, it is known that the average cholesterol level of their adult population is 184 mg/dl with a standard deviation of 22 mg/dl. It can be modelled by a normal distribution.

a) What percentage do you expect to be borderline high?
b) What percentage do you consider are high risk?
c) Estimate the interquartile range of the cholesterol levels in this country.
d) Above what value are the highest 2% of adults' cholesterol levels in this country?

21 A manufacturer of car tyres claims that the treadlife of its winter tyres can be described by a normal model with an average life of 52 000 km and a standard deviation of 4000 km.

a) You buy a set of tyres from this manufacturer. Is it reasonable for you to hope they last more than 64 000 km?
b) What fraction of these tyres do you expect to last less than 48 000 km?
c) What fraction of these tyres do you expect to last between 48 000 km and 56 000 km?
d) What is the IQR of the treadlife of this type of tyre?
e) The company wants to guarantee a minimum life for these tyres. That is, they will refund customers whose tyres last less than a specific distance. What should their minimum life guarantee be so that they do not end up refunding more than 2% of their customers?

22 Chicken eggs are graded by size for the purpose of sales. In Europe, modern egg sizes are defined as follows: very large has a mass of 73 g or more, large is between 63 and 73 g, medium is between 53 and 63 g, and small is less than 53 g. The small size is usually considered as undesirable by consumers.

a) Mature hens (older than 1 year) produce eggs with an average mass of 67 g. 98% of the eggs produced by mature hens are above the minimum desirable weight. What is the standard deviation if the egg production can be modelled by a normal distribution?
b) Young hens produce eggs with a mean mass of 51 g. Only 28% of their eggs exceed the desired minimum. What is the standard deviation?
c) A farmer finds that 7% of his farm's eggs are 'underweight', and 12% are very large. Estimate the mean and standard deviation of this farmer's eggs.

23 A machine produces bearings with diameters that are normally distributed with mean 3.0005 cm and standard deviation 0.0010 cm. Specifications require the bearing diameters to lie in the interval 3.000 $\pm$ 0.0020 cm. Those outside the interval are considered scrap and must be disposed of. What fraction of the production will be scrap?

24 A soft-drink machine can be regulated so that it discharges an average μ cc per bottle. If the amount of fill is normally distributed with a standard deviation 9 cc, give the setting for μ so that 237 cc bottles will overflow only 1% of the time.

25 The machine described in the previous problem has standard deviation σ that can be adjusted to the required levels when needed. What is the largest value of σ that will allow the actual amount dispensed to fall within 30 cc of the mean with probability at least 95%?

26 The speeds of cars on a main highway are approximately normal. Data collected at a certain point show that 95% of the cars travel at a speed less than 140 km/h, and 10% travel at a speed less than 90 km/h.

a) Find the average speed and the standard deviation for the cars travelling that specific stretch of the highway.

b) Find the proportion of cars that travel more at speeds exceeding 110 km/h.

27 The random variable X is normally distributed and:
$P(x \leq 10) = 0.670$
$P(x \leq 12) = 0.937$
Find $E(X)$.

28 A machine is set to produce bags of salt, whose weights are distributed normally, with a mean of 110 g and standard deviation of 1.142 g. If the weight of a bag of salt is less than 108 g, the bag is rejected. With these settings, 4% of the bags are rejected.
The settings of the machine are altered and it is found that 7% of the bags are rejected.

a) (i) If the mean has not changed, find the new standard deviation, correct to three decimal places.
The machine is adjusted to operate with this new value of the standard deviation.
(ii) Find the value, correct to two decimal places, at which the mean should be set so that only 4% of the bags are rejected.

b) With the new settings from part a), it is found that 80% of the bags of salt have a weight which lies between A g and B g, where A and B are symmetric about the mean. Find the values of A and B, giving your answers correct to two decimal places.

Practice questions

1 Residents of a small town have savings which are normally distributed with a mean of \$3000 and a standard deviation of \$500.

a) What percentage of townspeople have savings greater than \$3200?

b) Two townspeople are chosen at random. What is the probability that both of them have savings between \$2300 and \$3300?

c) The percentage of townspeople with savings less than d dollars is 74.22%. Find the value of d.

2 A box contains 35 red discs and 5 black discs. A disc is selected at random and its colour noted. The disc is then replaced in the box.

a) In eight such selections, what is the probability that a black disc is selected
(i) exactly once?
(ii) at least once?

b) The process of selecting and replacing is carried out 400 times. What is the expected number of black discs that would be drawn?

3 The graph shows a normal curve for the random variable X, with mean μ and standard deviation σ.

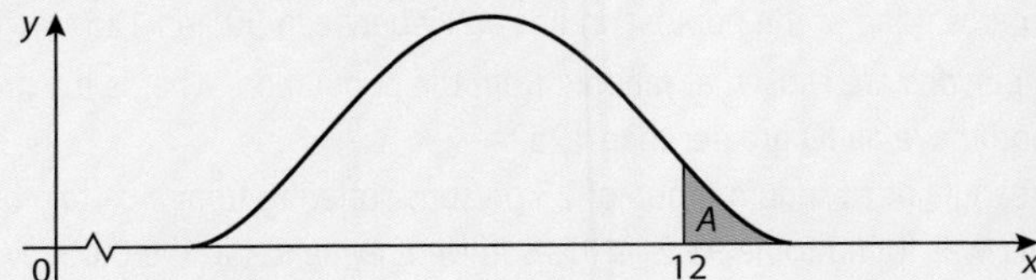

It is known that $P(x \geqslant 12) = 0.1$.

a) The shaded region A is the region under the curve where $x \geqslant 12$. Write down the area of the shaded region A.

It is also known that $P(x \leqslant 8) = 0.1$.

b) Find the value of μ, explaining your method in full.

c) Show that $\sigma = 1.56$ to an accuracy of 3 significant figures.

d) Find $P(x \leqslant 11)$.

4 A fair coin is tossed eight times. Calculate

a) the probability of obtaining exactly 4 heads

b) the probability of obtaining exactly 3 heads

c) the probability of obtaining 3, 4 or 5 heads.

5 The lifespan of a particular species of insect is normally distributed with a mean of 57 hours and a standard deviation of 4.4 hours.

The probability that the lifespan of an insect of this species lies between 55 and 60 hours is represented by the shaded area in the following diagram. This diagram represents the standard normal curve.

a) Write down the values of a and b.

b) Find the probability that the lifespan of an insect of this species is

(i) more than 55 hours

(ii) between 55 and 60 hours.

90% of the insects die after t hours.

c) **(i)** Represent this information on a standard normal curve diagram, similar to the one shown, indicating clearly the area representing 90%.

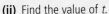

(ii) Find the value of t.

6 An urban highway has a speed limit of 50 km h^{-1}. It is known that the speeds of vehicles travelling on the highway are normally distributed, with a standard deviation of 10 km h^{-1}, and that 30% of the vehicles using the highway exceed the speed limit.

a) Show that the mean speed of the vehicles is approximately 44.8 km h^{-1}.

The police conduct a 'Safer Driving' campaign intended to encourage slower driving, and want to know whether the campaign has been effective. It is found that a sample of 25 vehicles has a mean speed of 41.3 km h^{-1}.

b) Given that the null hypothesis is
H0: the mean speed has been unaffected by the campaign state
state H1, the alternative hypothesis.

c) State whether a one-tailed or two-tailed test is appropriate for these hypotheses, and explain why.

d) Has the campaign had significant effect at the 5% level?

7 Intelligence quotient (IQ) in a certain population is normally distributed with a mean of 100 and a standard deviation of 15.

a) What percentage of the population has an IQ between 90 and 125?

b) If two persons are chosen at random from the population, what is the probability that both have an IQ greater than 125?

c) The mean IQ of a random group of 25 persons suffering from a certain brain disorder was found to be 95.2. Is this sufficient evidence, at the 0.05 level of significance, that people suffering from the disorder have, on average, a lower IQ than the entire population? State your null hypothesis and your alternative hypothesis, and explain your reasoning.

8 Bags of cement are labelled 25 kg. The bags are filled by machine and the actual weights are normally distributed with mean 25.7 kg and standard deviation 0.50 kg.

a) What is the probability a bag selected at random will weigh less than 25.0 kg?

In order to reduce the number of underweight bags (bags weighing less than 25 kg) to 2.5% of the total, the mean is increased without changing the standard deviation.

b) Show that the increased mean is 26.0 kg.

It is decided to purchase a more accurate machine for filling the bags. The requirements for this machine are that only 2.5% of bags be under 25 kg and that only 2.5% of bags be over 26 kg.

c) Calculate the mean and standard deviation that satisfy these requirements.

The cost of the new machine is $5000. Cement sells for $0.80 per kg.

d) Compared to the cost of operating with a 26 kg mean, how many bags must be filled in order to recover the cost of the new equipment?

9 The mass of packets of a breakfast cereal is normally distributed with a mean of 750 g and standard deviation of 25 g.

a) Find the probability that a packet chosen at random has mass

(i) less than 740 g

(ii) at least 780 g

(iii) between 740 g and 780 g.

b) Two packets are chosen at random. What is the probability that both packets have a mass that is less than 740 g?

c) The mass of 70% of the packets is more than x grams. Find the value of x.

10 In a country called Tallopia, the height of adults is normally distributed with a mean of 187.5 cm and a standard deviation of 9.5 cm.

a) What percentage of adults in Tallopia have a height greater than 197 cm?

b) A standard doorway in Tallopia is designed so that 99% of adults have a space of at least 17 cm over their heads when going through a doorway. Find the height of a standard doorway in Tallopia. Give your answer to the nearest cm.

11 It is claimed that the masses of a population of lions are normally distributed with a mean mass of 310 kg and a standard deviation of 30 kg.

a) Calculate the probability that a lion selected at random will have a mass of 350 kg or more.

b) The probability that the mass of a lion lies between a and b is 0.95, where a and b are symmetric about the mean. Find the values of a and b.

12 Reaction times of human beings are normally distributed with a mean of 0.76 seconds and a standard deviation of 0.06 seconds. The graph right is that of the standard normal curve. The shaded area represents the probability that the reaction time of a person chosen at random is between 0.70 and 0.79 seconds.

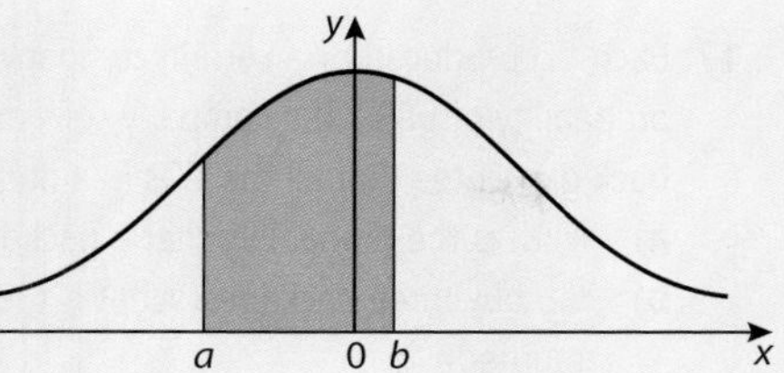

a) Write down the values of a and b.

b) Calculate the probability that the reaction time of a person chosen at random is

(i) greater than 0.70 seconds

(ii) between 0.70 and 0.79 seconds.

Three per cent (3%) of the population have a reaction time less than c seconds.

c) (i) Represent this information on a diagram similar to the one above. Indicate clearly the area representing 3%.

(ii) Find c.

13 A factory makes calculators. Over a long period, 2% of them are found to be faulty. A random sample of 100 calculators is tested.

a) Write down the expected number of faulty calculators in the sample.

b) Find the probability that three calculators are faulty.

c) Find the probability that more than one calculator is faulty.

14 The speeds of cars at a certain point on a straight road are normally distributed with mean μ and standard deviation σ. 15% of the cars travelled at speeds greater than 90 km h^{-1} and 12% of them at speeds less than 40 km h^{-1}. Find μ and σ.

15 Bag A contains 2 red balls and 3 green balls. Two balls are chosen at random from the bag without replacement. Let X denote the number of red balls chosen. The following table shows the probability distribution for X.

x	0	1	2
$P(X = x)$	$\frac{3}{10}$	$\frac{6}{10}$	$\frac{1}{10}$

a) Calculate E(X), the mean number of red balls chosen.

Bag B contains 4 red balls and 2 green balls. Two balls are chosen at random from bag B.

b) (i) Draw a tree diagram to represent the above information, including the probability of each event.

(ii) Hence, find the probability distribution for Y, where Y is the number of red balls chosen.

A standard die with six faces is rolled. If a 1 or 6 is obtained, two balls are chosen from bag A, otherwise two balls are chosen from bag B.

c) Calculate the probability that two red balls are chosen.

d) Given that two red balls are obtained, find the conditional probability that a 1 or 6 was rolled on the die.

16 Ball bearings are used in engines in large quantities. A car manufacturer buys these bearings from a factory. They agree on the following terms: The car company chooses a sample of 50 ball bearings from the shipment. If they find more than 2 defective bearings, the shipment is rejected. It is a fact that the factory produces 4% defective bearings.

a) What is the probability that the sample is clear of defects?

b) What is the probability that the shipment is accepted?

c) What is the expected number of defective bearings in the sample of 50?

17 Each CD produced by a certain company is guaranteed to function properly with a probability of 98%. The company sells these CDs in packages of 10 and offers a money-back guarantee that all the CDs in a package will function.

a) What is the probability that a package is returned?

b) You buy three packages. What is the probability that exactly 1 of them must be returned?

18 The table below shows the probability distribution of a random variable X.

x	0	1	2	3
$P(X = x)$	$2k$	$2k^2$	$k^2 + k$	$2k^2 + k$

a) Calculate the value of k.

b) Find $E(X)$.

19 It is estimated that 2.3% of the cherry tomato fruits produced on a certain farm are considered to be small and cannot be sold for commercial purposes. The farmers have to separate such fruits and use them for domestic consumption instead.

a) 12 tomatoes are randomly selected from the produce. Calculate

(i) the probability that three are not fit for selling

(ii) the probability that at least four are not fit for selling.

b) It is known that the sizes of such tomatoes are normally distributed with a mean of 3 cm and a standard deviation of 0.5 cm. Tomatoes that are categorized as large will have to be larger than 2.5 cm. What proportion of the produce is large?

20 A factory has a machine designed to produce 1 kg bags of sugar. It is found that the average weight of sugar in the bags is 1.02 kg. Assuming that the weights of the bags are normally distributed, find the standard deviation if 1.7% of the bags weigh below 1 kg. Give your answer correct to the nearest 0.1 gram.

21 The continuous random variable X has probability density function $f(x)$ where

$$f_k(x) = \begin{cases} e - ke^{kx} & 0 \leqslant x \leqslant 1 \\ 0 & \text{otherwise} \end{cases}$$

a) Show that $k = 1$.

b) What is the probability that the random variable X has a value that lies between $\frac{1}{4}$ and $\frac{1}{2}$? Give your answer in terms of e.

c) Find the mean and variance of the distribution. Give your answers *exactly* in terms of e.

The random variable X above represents the lifetime, in years, of a certain type of battery.

d) Find the probability that a battery lasts more than six months.

A calculator is fitted with three of these batteries. Each battery fails independently of the other two. Find the probability that at the end of six months

(i) none of the batteries has failed;

(ii) exactly one of the batteries has failed.

22 The lifetime of a particular component of a solar cell is Y years, where Y is a continuous random variable with probability density function

$$f(y) = \begin{cases} 0 & y < 0 \\ 0.5e^{-\frac{y}{2}} & y \geqslant 0 \end{cases}$$

a) Find the probability, correct to four significant figures, that a given component fails within six months.

Each solar cell has three components which work independently and the cell will continue to run if at least two of the components continue to work.

b) Find the probability that a solar cell fails within six months.

23 Ian and Karl have been chosen to represent their countries in the Olympic discus throw. Assume that the distance thrown by each athlete is normally distributed. The mean distance thrown by Ian in the past year was 60.33 m with a standard deviation of 1.95 m.

a) In the past year, 80% of Ian's throws have been longer than x metres.

Find x, correct to two decimal places.

b) In the past year, 80% of Karl's throws have been longer than 56.52 m. If the mean distance of his throws was 59.39 m, find the standard deviation of his throws, correct to two decimal places.

c) This year, Karl's throws have a mean of 59.50 m and a standard deviation of 3.00 m. Ian's throws still have a mean of 60.33 m and standard deviation 1.95 m. In a competition an athlete must have at least one throw of 65 m or more in the first round to qualify for the final round. Each athlete is allowed three throws in the first round.

(i) Determine which of these two athletes is more likely to qualify for the final on their first throw.

(ii) Find the probability that both athletes qualify for the final.

24 The continuous random variable X has probability density function

$$f(x) = \begin{cases} \frac{1}{6}x(1 + x^2) & 0 \leqslant x \leqslant 2 \\ 0 & \text{otherwise} \end{cases}$$

a) Sketch the graph of f for $0 \leqslant x \leqslant 2$.

b) Write down the mode of X.

c) Find the mean of X.

d) Find the median of X.

25 A company buys 44% of its stock of bolts from manufacturer A and the rest from manufacturer B. The diameters of the bolts produced by each manufacturer follow a normal distribution with a standard deviation of 0.16 mm.
The mean diameter of the bolts produced by manufacturer A is 1.56 mm.
24.2% of the bolts produced by manufacturer B have a diameter less than 1.52 mm.

a) Find the mean diameter of the bolts produced by manufacturer B.

A bolt is chosen at random from the company's stock.

b) Show that the probability that the diameter is less than 1.52 mm is 0.312, to three significant figures.

c) The diameter of the bolt is found to be less than 1.52 mm. Find the probability that the bolt was produced by manufacturer B.

d) Manufacturer B makes 8000 bolts in one day. It makes a profit of \$1.50 on each bolt sold, on condition that its diameter measures between 1.52 mm and 1.83 mm.
Bolts whose diameters measure less than 1.52 mm must be discarded at a loss of \$0.85 per bolt.
Bolts whose diameters measure over 1.83 mm are sold at a reduced profit of \$0.50 per bolt.
Find the expected profit for manufacturer B.

18 The Mathematical Exploration – Internal Assessment

'If digressions can bring knowledge of new truths, why should they trouble us? … how do we know that we shall not discover curious things that are more interesting than the answers we originally sought?'

Galileo, *Discourses and Mathematical Demonstrations Relating to Two New Sciences*, 1638

At the end of the Mathematics Higher Level course, you will take three written exams that will constitute the **External Assessment** component of the course: Paper 1 (covering the core syllabus – no GDC), Paper 2 (covering the core syllabus – GDC required), and Paper 3 (covering the option topic for which you registered). These written exams will contribute 80% to your final grade for the course that will be reported to you about six weeks after the exams finish.

Internal Assessment (IA) is another important component of the Mathematics Higher Level course and will contribute 20% to your final grade for the course. Thus, IA does comprise a significant part of the overall assessment for the course and should be taken seriously. It should also be pointed out that your work in completing the IA component differs in important ways from the written exams for the course.

- You do **not** perform IA work under strict time constraints as with written examinations.
- You have some freedom to help decide what mathematical topic you wish to explore.
- Your IA work involves writing about mathematics and not just doing mathematical procedures.
- Regular discussion with, and feedback from, your teacher will be essential.
- You should endeavour to explore a topic in which you have a genuine personal interest.
- You will be rewarded for evidence of creativity, curiosity and independent thinking.

The Mathematical Exploration

To satisfy the Internal Assessment component, you are required to write a report on a mathematical topic that you choose in consultation with your teacher. This report is formally referred to as the **Mathematical Exploration**. Throughout this chapter 'Mathematical Exploration' and 'report' refer to the same thing, i.e. the written piece of work that you submit for the Internal Assessment component of the course.

The Mathematical Exploration is aptly named because your primary objective in writing this report is to *explore* a topic in which you are genuinely interested and that is at an appropriate level for the course. Your teacher may provide you with a list of ideas (or 'stimuli') from which to choose a topic or which may help you to develop your own ideas for a topic to explore (see the **list of 200 ideas** printed later in this chapter). It is your responsibility to determine whether or not you are sufficiently interested in a particular topic – and it is your teacher's responsibility to determine if an exploration of the topic can be conducted at a level mathematically suitable for the course. Your teacher will help you determine if an exploration of a certain topic can potentially address the five assessment criteria satisfactorily. Your report should be approximately 6 to 12 pages long.

Internal Assessment Criteria

Your **Mathematical Exploration** report will be assessed by your teacher according to the following five criteria.

A <u>**Communication**</u>: This criterion assesses the **organisation** and **coherence** of the exploration. A well-organised exploration has an **introduction** and a **rationale** (which includes a brief explanation of why the topic was chosen). It describes the **aim of the exploration** and has a **conclusion**.

B <u>**Mathematical presentation**</u>: This criterion assesses to what extent you are able to:

- use appropriate mathematical language (notation, symbols and terminology);
- define key terms, where necessary;
- use multiple forms of mathematical representation such as formulae, diagrams, tables, charts, graphs and models.

C <u>**Personal engagement**</u>: This criterion assesses the extent to which you engage with the exploration, and present it in such a way that clearly shows **your own personal approach**. Personal engagement may be recognised in different attributes and skills. These include thinking independently and/or creatively, addressing personal interest, presenting mathematical ideas in your own way, using simple language to describe complex ideas, and applying unfamiliar mathematics.

D <u>Reflection</u>: This criterion assesses how well you **review, analyse** and **evaluate** the exploration. Although reflection may be seen in the conclusion to the exploration, you should also give evidence of reflective thought throughout the exploration. Reflection may be demonstrated by consideration of limitations and/or extensions and by relating mathematical ideas to your own previous knowledge.

E <u>Use of mathematics</u>: This criterion assesses to what extent and **how well you use mathematics** in your exploration. The mathematics that is explored in your report needs to be **sufficiently sophisticated.** The chosen topic should involve mathematics either in the Mathematics Higher Level syllabus, at a similar level, or beyond the level of the syllabus. Sophistication in mathematics may include understanding and use of challenging mathematical concepts, looking at a problem from different perspectives, or seeing underlying structures to link different areas of mathematics.

Your report will earn a numerical score out of a total of 20 possible marks. The five criteria do not contribute equally to the overall score for your Mathematical Exploration. For example, criterion E (Use of mathematics) is 30% of the overall score, whereas criteria B (Mathematical presentation) and D (Reflection) contribute 15% each.

It is very important that you familiarize yourself with the assessment criteria and refer to them while you are writing your report. The scoring levels for each criteria and associated descriptors are as follows.

A Communication	
0	The exploration does not reach the standard described by the descriptors below.
1	The exploration has some coherence.
2	The exploration has some coherence and shows some organisation.
3	The exploration is coherent and well organised.
4	The exploration is coherent, well organised, concise and complete.

B Mathematical presentation	
0	The exploration does not reach the standard described by the descriptors below.
1	There is some appropriate mathematical presentation.
2	The mathematical presentation is mostly appropriate.
3	The mathematical presentation is appropriate throughout.

C	Personal engagement
0	The exploration does not reach the standard described by the descriptors below.
1	There is evidence of limited or superficial personal engagement.
2	There is evidence of some personal engagement.
3	There is evidence of significant personal engagement.
4	There is abundant evidence of outstanding personal engagement.

D	Reflection
0	The exploration does not reach the standard described by the descriptors below.
1	There is evidence of limited or superficial reflection.
2	There is evidence of meaningful reflection.
3	There is substantial evidence of critical reflection.

E	Use of Mathematics
0	The exploration does not reach the standard described by the descriptors below.
1	Some relevant mathematics is used. Limited understanding is demonstrated.
2	Some relevant mathematics is used. The mathematics explored is partially correct. Some knowledge and understanding are demonstrated.
3	Relevant mathematics commensurate with the level of the course is used. The mathematics explored is correct. Good knowledge and understanding are demonstrated.
4	Relevant mathematics commensurate with the level of the course is used. The mathematics explored is correct and reflects the sophistication expected. Good knowledge and understanding are demonstrated.
5	Relevant mathematics commensurate with the level of the course is used. The mathematics explored is correct and reflects the sophistication and rigour expected. Thorough knowledge and understanding are demonstrated.
6	Relevant mathematics commensurate with the level of the course is used. The mathematics explored is precise and reflects the sophistication and rigour expected. Thorough knowledge and understanding are demonstrated.

Guidance

Conducting an in-depth individual exploration into the mathematics of a particular topic can be an interesting and very rewarding experience. It is important to take all stages of your work on the Mathematical Exploration seriously – not only because it is worth 20% of your final grade for the course, but also because of the opportunity to pursue your own personal interests without the pressure of examination conditions. The Mathematical Exploration will require a significant amount of time and energy to complete successfully. It should *not* be approached as simply an extended homework assignment. The task of writing the report will demand a considerable amount of research, analysis, reading, consultation (with your teacher only), thinking, writing, editing, mathematical work, problem solving and proofreading. Hopefully, it will also be enjoyable, thought provoking and satisfying, and give you the opportunity to gain a deeper appreciation for the beauty, power and usefulness of mathematics.

Although it is required that your Mathematical Exploration be completely your own work, you should be consulting with your teacher on a regular basis throughout the time given to you to research and write your report. Your teacher should provide support and advice during the planning and writing stages of your report. Both you and your teacher will need to sign the internal assessment coversheet verifying the authenticity of your Mathematical Exploration.

All of the work connected with the exploration must be your own. Your Mathematical Exploration must reflect intellectual honesty in research practices and must provide the reader with the exact sources of quotations, ideas and points of view with a complete and accurate bibliography. There are a number of acceptable bibliographic styles. Whatever style is chosen, it must include all relevant source information and be applied consistently. Group work is not allowed with the Mathematical Exploration. Also, if you are writing an Extended Essay for mathematics, you are not allowed to submit the same piece of work for the Mathematical Exploration – and you are strongly advised not to write about the same mathematical topic for both.

In organizing a successful Mathematical Exploration, consider the following suggestions.

1 Select a topic in which you are **genuinely interested**. Include a brief explanation in the early part of your report about why you chose your topic – including why you find it interesting.

2 Consult with your teacher that the topic is at the **appropriate level of mathematics**, i.e. that it is at the same level of mathematics in the HL syllabus, or beyond.

3 Find as much **information** about the topic as possible. Although information found on internet websites can be very helpful, try to also find information from books, journals, textbooks and other print material.

4 Prepare and organize your material into a **thorough and interesting report**. Although there is no requirement that you present your report to your class, it should be written so that your fellow classmates can follow it without trouble. Your report needs to be **logically organized** and use appropriate mathematical terminology and notation.

5 The most important aspects of your report should be about **mathematical communication and using mathematics**. Although other aspects of your topic (e.g. historical, personal, cultural etc.) can be discussed, be careful not to lose focus on the mathematical features.

6 Two of the assessment criteria – personal engagement and reflection – are about **what you think about the topic** you are exploring. Don't hesitate to pose your own relevant and insightful questions as part of your report, and then to address these questions using mathematics at a suitably sophisticated level along with sufficient written commentary.

7 Although your teacher will expect and require you to work independently, you are allowed to **consult with your teacher** – and your teacher is allowed to give you advice and feedback to a certain extent while you are working on your report. It is especially important to check with your teacher that any **mathematics in your report is correct**. Your teacher will not give mathematical answers or corrections, but can indicate where any errors have been made or where improvement is needed.

Mathematical Exploration HL – Student Checklist

Is your report written entirely by yourself – and trying to avoid simply replicating work and ideas from sources you found during your research?	☐ Yes	☐ No
Have you strived to: apply your personal interest; develop your own ideas; and use critical thinking skills during your exploration and demonstrate these in your report?	☐ Yes	☐ No
Have you referred to the five assessment criteria while writing your report?	☐ Yes	☐ No
Does your report focus on good mathematical communication – and does it read like an article for a mathematical journal?	☐ Yes	☐ No
Does your report have a clearly identified introduction and conclusion?	☐ Yes	☐ No
Have you documented all of your source material in a detailed bibliography in line with the IB academic honesty policy?	☐ Yes	☐ No
Not including the bibliography, is your report 6 to 12 pages?	☐ Yes	☐ No
Are graphs, tables and diagrams sufficiently described and labelled?	☐ Yes	☐ No
To the best of your knowledge, have you used and demonstrated mathematics that is at the same level, or above, of that studied in IB Mathematics HL?	☐ Yes	☐ No
Have you attempted to discuss mathematical ideas, and use mathematics, with a sufficient level of sophistication and rigour?	☐ Yes	☐ No
Are formulae, graphs, tables and diagrams in the main body of text? (preferably no full-page graphs; and no separate appendices)	☐ Yes	☐ No
Have you used technology – such as a GDC, spreadsheet, mathematics software, drawing and word-processing software – to enhance mathematical communication?	☐ Yes	☐ No
Have you used appropriate mathematical language (notation, symbols, terminology) and defined key terms?	☐ Yes	☐ No
Is the mathematics in your report performed precisely and accurately?	☐ Yes	☐ No
Has calculator/computer notation and terminology **not** been used? ($y = x^2$, not $y = x$^2; $\approx$, not = for approximate values; π, not pi; $\|x\|$, not abs(x); etc)	☐ Yes	☐ No
At suitable places in your report – especially in the conclusion – have you included reflective and explanatory comments about the mathematical topic being explored?	☐ Yes	☐ No

List of 200 ideas/topics for a Mathematical Exploration

The topics listed here range from fairly broad to quite narrow in scope. It is possible that some of these 200 could be the title or focus of a **Mathematical Exploration**, while others will require you to investigate further to identify a narrower focus to explore. Do not restrict yourself only to the topics listed below. This list is only the 'tip of the iceberg' with regard to potential topics for your Mathematical Exploration. Reading through this list may stimulate you to think of some other topic in which you would be interested in exploring. Many of the items listed below may be unfamiliar to you. A quick search on the internet should give you a better idea what each is about and help you determine if you're interested enough to investigate further – and see if it might be a suitable topic for your Mathematical Exploration.

Algebra and number theory		
Modular arithmetic	Goldbach's conjecture	Probabilistic number theory
Applications of complex numbers	Diophantine equations	Continued fractions
General solution of a cubic equation	Applications of logarithms	Polar equations
Patterns in Pascal's triangle	Finding prime numbers	Random numbers
Pythagorean triples	Mersenne primes	Magic squares and cubes
Loci and complex numbers	Matrices and Cramer's rule	Divisibility tests
Egyptian fractions	Complex numbers and transformations	Euler's identity: $e^{i\pi} + 1 = 0$
Chinese remainder theorem	Fermat's last theorem	Natural logarithms of complex numbers
Twin primes problem	Hypercomplex numbers	Diophantine application: Cole numbers
Odd perfect numbers	Euclidean algorithm for GCF	Palindrome numbers
Factorable sets of integers of the form $ak + b$	Algebraic congruences	Inequalities related to Fibonacci numbers
Combinatorics – art of counting	Boolean algebra	Graphical representation of roots of complex numbers
Roots of unity	Fermat's little theorem	Prime number sieves
Recurrence expressions for phi (golden ratio)		
Geometry		
Non-Euclidean geometries	Cavalieri's principle	Packing 2D and 3D shapes
Ptolemy's theorem	Hexaflexagons	Heron's formula
Geodesic domes	Proofs of Pythagorean theorem	Minimal surfaces and soap bubbles
Tesseract – a 4D cube	Map projections	Tiling the plane – tessellations
Penrose tiles	Morley's theorem	Cycloid curve

Geometry (continued)		
Symmetries of spider webs	Fractal tilings	Euler line of a triangle
Fermat point for polygons and polyhedra	Pick's theorem and lattices	Properties of a regular pentagon
Conic sections	Nine-point circle	Geometry of the catenary curve
Regular polyhedra	Euler's formula for polyhedra	Eratosthenes – measuring earth's circumference
Stacking cannon balls	Ceva's theorem for triangles	Constructing a cone from a circle
Conic sections as loci of points	Consecutive integral triangles	Area of an ellipse
Mandelbrot set and fractal shapes	Curves of constant width	Sierpinksi triangle
Squaring the circle	Polyominoes	Reuleaux triangle
Architecture and trigonometry	Spherical geometry	Gyroid – a minimal surface
Geometric structure of the universe	Rigid and non-rigid geometric structures	Tangrams
Calculus/analysis and functions		
Mean value theorem	Torricelli's trumpet (Gabriel's horn)	Integrating to infinity
Applications of power series	Newton's law of cooling	Fundamental theorem of calculus
Brachistochrone (minimum time) problem	Second order differential equations	L'Hôpital's rule and evaluating limits
Hyperbolic functions	The harmonic series	Torus – solid of revolution
Projectile motion	Why ***e*** is base of natural logarithm function	
Statistics and modelling		
Traffic flow	Logistic function and constrained growth	Modelling growth of tumours
Modelling epidemics/spread of a virus	Modelling the shape of a bird's egg	Correlation coefficients
Central limit theorem	Modelling change in record performances for a sport	Hypothesis testing
Modelling radioactive decay	Least squares regression	Modelling the carrying capacity of the earth
Regression to the mean	Modelling growth of computer power past few decades	
Probability and probability distributions		
The Monty Hall problem	Monte Carlo simulations	Random walks
Insurance and calculating risks	Poisson distribution and queues	Determination of π by probability
Lotteries	Bayes' theorem	Birthday paradox
Normal distribution and natural phenomena	Medical tests and probability	Probability and expectation

Games and game theory		
The prisoner's dilemma	Sudoku	Gambler's fallacy
Poker and other card games	Knight's tour in chess	Billiards and snooker
Zero sum games		
Topology and networks		
Knots	Steiner problem	Chinese postman problem
Travelling salesman problem	Königsberg bridge problem	Handshake problem
Möbius strip	Klein bottle	
Logic and sets		
Codes and ciphers	Set theory and different 'size' infinities	Mathematical induction (strong)
Proof by contradiction	Zeno's paradox of Achilles and the tortoise	Four colour map theorem
Numerical analysis		
Linear programming	Fixed-point iteration	Methods of approximating π
Applications of iteration	Newton's method	Estimating size of large crowds
Generating the number e	Descartes' rule of signs	Methods for solving differential equations
Physical, biological and social sciences		
Radiocarbon dating	Gravity, orbits and escape velocity	Mathematical methods in economics
Biostatistics	Genetics	Crystallography
Computing centres of mass	Elliptical orbits	Logarithmic scales – decibel, Richter, etc.
Fibonacci sequence and spirals in nature	Predicting an eclipse	Change in a person's BMI over time
Concepts of equilibrium in economics	Mathematics of the 'credit crunch'	Branching patterns of plants
Column buckling – Euler theory		
Miscellaneous		
Paper folding	Designing bridges	Mathematics of rotating gears
Mathematical card tricks	Curry's paradox – 'missing' square	Bar codes
Applications of parabolas	Music – notes, pitches, scales…	Voting systems
Flatland by Edwin Abbott	Terminal velocity	Towers of Hanoi puzzle
Photography	Art of M.C. Escher	Harmonic mean
Sundials	Navigational systems	The abacus
Construction of calendars	Slide rules	Different number systems
Mathematics of juggling	Global positioning system (GPS)	Optical illusions
Origami	Napier's bones	Celtic designs/knotwork
Design of product packaging	Mathematics of weaving	

Website support

Further guidance and information concerning Internal Assessment is available from the authors' website at www.wazir-garry-math.org. You are encouraged to register with our site. Along with a considerable amount of support for other aspects of the IB Mathematics Higher Level course, there will be a section on our website devoted specifically to the Mathematical Exploration. We will be regularly updating our site so that you will have access to thorough and useful advice, materials and updates regarding how to get the most out of your Mathematical Exploration.

19 Sample Examination Papers

Paper 1 Sample A

Paper 1 is a non-calculator paper. Your exam paper will have instructions on the first page, some of which are reproduced here.

Full marks are not necessarily awarded for a correct answer with no working. Answers must be supported by working and/or explanations. Where an answer is incorrect, some marks may be given for a correct method, provided this is shown by written working. You are therefore advised to show all working.

It is important that you remember to show work because examiners will award marks for correct work leading to the final solution. Also, if your final answer is incorrect, you will not end up losing all the marks.

Section A

Answer all the questions in the spaces provided. Working may be continued below the lines, if necessary.

Section B

Answer all the questions on the answer sheets provided. Please start each question on a new page.

Section A

1 [*Maximum mark:* 6]
Given that $\log_b a = 0.74$ and $\log_b(a - 1) = 0.65$, find the value of the following expression:

$$\log_b(a^4 - 1) - 2\log_b(a^2 + 1) + \log_b(a^3 + a) - \log_b(a + 1)$$

Give your answer to 2 decimal places.

2 [*Maximum mark:* 6]
Find the volume of the solid obtained by rotating the region under the curve

$$y = \frac{1}{x^2 + 1}$$

from 0 to $\sqrt{e^3 - 1}$ about the y-axis.

3 [*Maximum mark:* 6]
Find out where the normal line to the curve $x^2 - xy + y^2 = 3$ at the point $(-1, 1)$ intersects the curve a second time.

4 [*Maximum mark:* 8]
Flaws appear randomly in a roll of textile at an average of 2 per metre length.
   a) Find the probability that more than 3 flaws appear in a randomly chosen metre from this material.
   b) Find the probability that more than 3 flaws appear in a randomly chosen 2-metre piece from this fabric.
   c) Two pieces of 1 metre each are chosen at random. Find the probability the total number of flaws is 3.

5 [*Maximum mark:* 5]
If $\frac{3\pi}{2} < t < 2\pi$ and $\cos t = \frac{3}{\sqrt{10}}$, find the value of $\operatorname{cosec} t + \cos 2t$.

6 [*Maximum mark:* 6]
When $P(x) = ax^5 + 3x^2 - 2x + b$ is divided by $(x + 2)$ the remainder is -47, while if it is divided by $(x - 1)$ the remainder is 4. Find the values of a and b.

7 [*Maximum mark:* 4]
Simplify the following expression and write your answer in the form $a + bi$ where a and b are real numbers.

$$\frac{(2 + 3i)^2}{3 - 2i}$$

8 [*Maximum mark:* 5]
Find the value of k such that the following is a convergent geometric series whose sum to infinity is 12.

$$\sum_{i=1}^{n} 4k^{i-1}$$

9 [*Maximum mark:* 6]
The following graph is the graph of a function of the form

$$y = a\sin(b(x - c)) + d.$$

Determine the values of a, b, c and d.

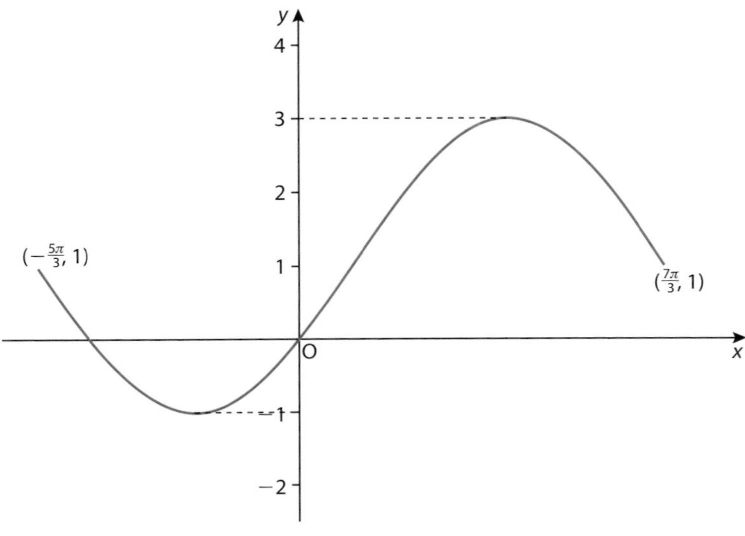

10 [*Maximum mark:* 8]
Graph the rational function

$$y = \frac{2x^2 + 3x}{x^2 + x - 2}.$$

Show clearly all x- and y-intercepts and asymptotes.

Section B

11 [*Maximum mark:* 13]
The position vectors of the points A, B and C are

$$\mathbf{a} = 4\mathbf{i} - 9\mathbf{j} - \mathbf{k}, \mathbf{b} = \mathbf{i} + 3\mathbf{j} + 5\mathbf{k} \text{ and } \mathbf{c} = 4\mathbf{i} - 7\mathbf{j} - 5\mathbf{k}.$$

a) Find the unit vector parallel to $\overrightarrow{AB}$. [*3 marks*]
b) Find the angle between the vectors $\overrightarrow{AB}$ and $\overrightarrow{AC}$. [*3 marks*]
c) Find a set of parametric equations for plane ABC. [*3 marks*]
d) Find the area of triangle ABC. [*4 marks*]

12 [*Maximum mark:* 21]
a) A and B are two events such that $P(A) = \frac{1}{3}$, $P(B) = \frac{1}{6}$, $P(B/A) = \frac{2}{5}$.
Calculate the probability that
(i) both A and B occur
(ii) either A or B occurs, but not both
(iii) A occurs, knowing that B has occurred. [*10 marks*]

b) An industrial company has f female and m male employees. The employees arrive in the morning at random, and we will assume that the probability that the first employee to arrive any day will be a female is

$$\frac{f}{f + m}.$$

Mr Guard plans to watch the employees arrive on four consecutive days. If females arrive first every day on all four days, he will conclude that $f > m$; if males arrive first every day of the four days, he will conclude that $f < m$; otherwise, he will conclude that $f = m$. Ms Reception on the other hand wants to watch for 7 days. If females arrive first on 6 or 7 days, she will conclude that $f > m$; if males arrive first on 6 or 7 days, she will conclude that $f < m$; otherwise, she will conclude that $f = m$.
(i) If $f = m$, who, if any, is more likely to be wrong?
(ii) If $f = \frac{m}{2}$, what is the probability that Ms Reception will wrongly conclude that $f = m$? [*11 marks*]

13 [*Maximum mark:* 14]
a) A sequence of real numbers $\{u_n\}$ is defined by

$$u_{n+1} = 2u_n \cos\theta - u_{n-1}; n > 1, u_0 = 1, u_1 = \cos\theta.$$

Prove by mathematical induction that $u_n = \cos n\theta$. [*8 marks*]

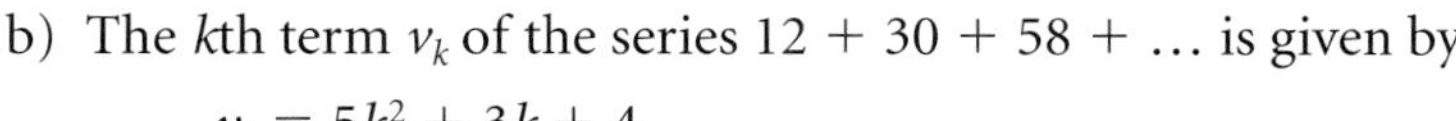

b) The kth term v_k of the series $12 + 30 + 58 + \ldots$ is given by

$$v_k = 5k^2 + 3k + 4.$$

Find $\sum_{k=1}^{n} v_k$, and express your answer in terms of n. [*6 marks*]

Hint: use the fact that $\sum_{i=1}^{n} i^2 = \frac{n(n+1)(2n+1)}{6}$.

[IBO, 1978]

14 [*Maximum mark:* 12]
From a fixed point A on a circle with centre O and radius a, a perpendicular is dropped to the tangent at P to the circle.

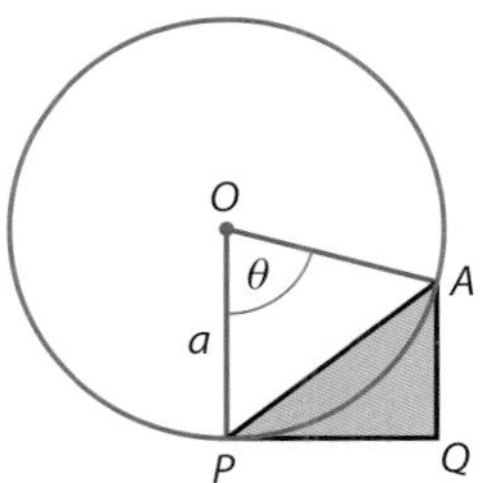

a) Given the central angle θ as shown, prove that the area of triangle APQ is

$$\tfrac{1}{2}a^2|\sin\theta|(1 - \cos\theta).$$ [*7 marks*]

b) As P moves along the circle, find the maximum value of the area of triangle APQ. [*5 marks*]

Paper 1 Sample B

Section A

1 [*Maximum mark:* 6]
12 and $-\frac{4}{9}$ are the second and fifth terms of a geometric sequence.

a) Find the sum of the first n terms of this sequence.

b) Find the sum to infinity of this sequence.

2 [*Maximum mark:* 4]
The data $\{0, 11, 12, 12, 14, 16, 17, 18, 19, 20, 21\}$ are represented by the box-plot below.

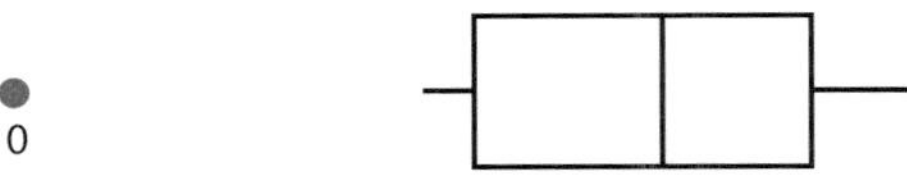

0 is considered an outlier because it is more than 1.5 IQR (interquartile range) below the first quartile. Show that this is true.

3 [*Maximum mark:* 8]
The diagram below shows the shaded region A, in the first quadrant that is enclosed by the curve $y^2 = 8(2 - x)$. Find the ratio of the volume of the solid formed when A is rotated through 2π radians around the x-axis to the volume of the solid formed when A is rotated through 2π radians around the y-axis.

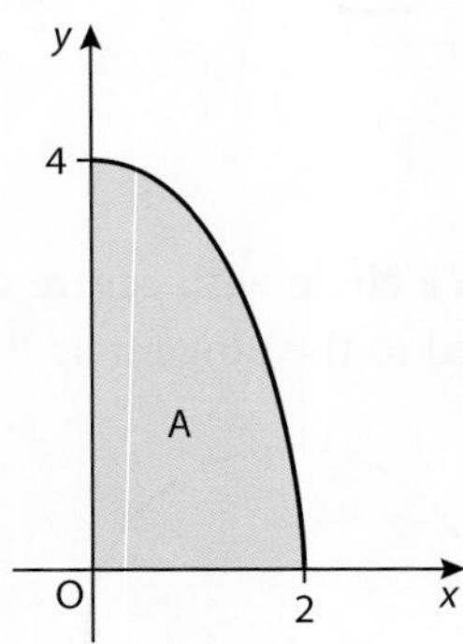

4 [*Maximum mark:* 6]
$P(x) = x^3 + bx^2 + x + c$ is divisible by $(x - 2)$ but leaves a remainder of -35 when divided by $(x + 3)$. Find b and c.

5 [*Maximum mark:* 6]
Consider the function

$$f(x) = e^{2x-x^2}.$$

a) Find the maximum value of this function.

b) Find the x-coordinate of the points of inflexion.

6 [*Maximum mark:* 6]
The diagram shows the graph of a function $y = f(x)$, which passes through the points $A(-3, 0)$, $B(-1, -2)$, $C(1, 0)$ and $D(4, 0)$.

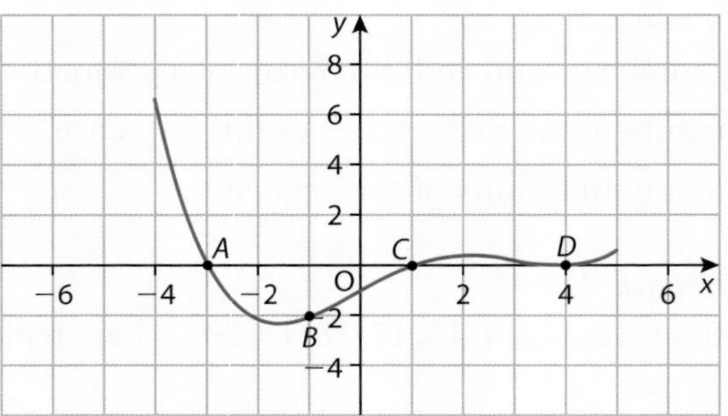

a) Graph the function $g(x) = f(x + 2)$ clearly indicating the coordinates of the images of A, B, C and D.

b) Graph the function $h(x) = f(2x + 2)$ clearly indicating the coordinates of the images of A, B, C and D.

7 [*Maximum mark:* 6]

The lines L_1 and L_2 have the following equations:

$$L_1: r = \begin{pmatrix} 2 \\ -1 \\ -4 \end{pmatrix} + \lambda \begin{pmatrix} 1 \\ 3 \\ -2 \end{pmatrix}; \ L_2: \frac{x-5}{2} = \frac{y+1}{-3} = z+5$$

a) Find the point of intersection between these lines.
b) Find the cosine of the acute angle between the lines.

8 [*Maximum mark:* 6]

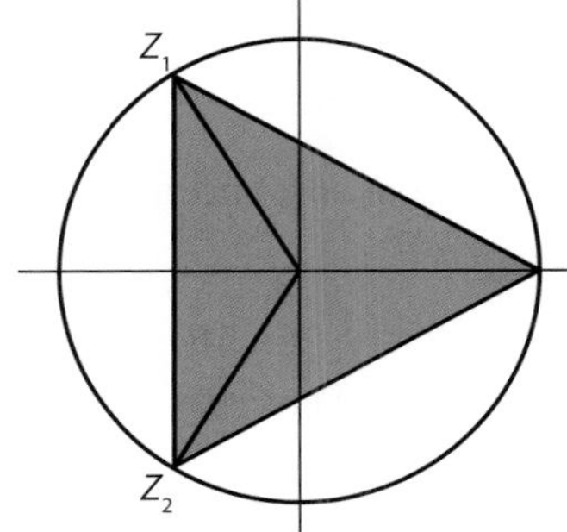

The cube roots of 1 are 1, $z_1 = a + bi$, and $z_2 = a - bi$.
a) Find the values of a and b.
b) In the Argand diagram, join the points corresponding to the roots. Find the area of the triangle formed by the points as vertices.

9 [*Maximum mark:* 6]
Find all solutions to the equation

$$\cos 4\theta + \sin^2 2\theta = \frac{1}{4}$$

in the interval $[0, 2\pi]$.

10 [*Maximum mark:* 6]

The sum of the first n terms of a sequence is $\frac{3n^2 + 4n}{2}$, $n \in \mathbb{Z}^+$.

a) Find the first three terms of the sequence.
b) Find an expression for the nth term of the sequence, giving your answer in terms of n.

Section B

11 [*Maximum mark:* 20]
Consider the plane P with equation $2x + 3y - z = 11$ and the line L with equation

$$\frac{x-2}{-1} = \frac{y-1}{2} = \frac{z-3}{5}.$$

a) Show that the point $A(1, 3, 8)$ lies on the line L. [*3 marks*]
b) Find the coordinates of point B, the intersection between line L and plane P. [*3 marks*]

c) Find an equation of a line M containing A and perpendicular to P. [4 marks]

d) Find the coordinates of the point C, the intersection of line M and P. [2 marks]

e) Hence or otherwise, find the point D symmetric to A about plane P. [4 marks]

f) Find a set of parametric equations of the line through B and D. [4 marks]

12 [*Maximum mark:* 20]

Consider the complex number $z = \cos\theta + i\sin\theta$.

a) Show that $z + \frac{1}{z} = 2\cos\theta$ and $z - \frac{1}{z} = 2i\sin\theta$. [2 marks]

b) Show that $z^n + \frac{1}{z^n} = 2\cos(n\theta)$, and find a similar expression for $z^n - \frac{1}{z^n}$. [4 marks]

c) Hence, show that $\sin^5\theta = \frac{1}{16}(\sin 5\theta - 5\sin 3\theta + 10\sin\theta)$, and find a similar expression for $\cos^5\theta$. [8 marks]

d) Hence, find $\int\left(4\sin^2\left(\frac{3\theta}{2}\right) - 32\sin^5\theta\right)d\theta$. [6 marks]

13 [*Maximum mark: 20*]

A function f is defined by

$$f(x) = \frac{x^2 + x - 2}{x^2 - 2x - 3}.$$

a) What is the largest possible domain of f? Find its derivative $f'(x)$. Show that $f'(x)$ has a constant sign over its domain. [6 marks]

b) Write down the equations of the asymptotes of the curve $y = f(x)$. [3 marks]

c) Use the information developed so far to sketch the graph of $f(x)$. [3 marks]

d) Find the real numbers P, Q and R such that the following is true for all values of x in the domain of f:

$$f(x) = P + \frac{Q}{x+1} + \frac{R}{x-3}.$$ [4 marks]

e) Use the expression above to find

$$\int f(x)dx.$$ [4 marks]

Paper 1 Sample C

Section B

1 [*Maximum mark:* 5]

When the function $f(x) = 6x^4 + 11x^3 - 22x^2 + ax + 6$ is divided by $(x + 1)$, the remainder is -20.

Find the value of a.

2 [*Maximum mark:* 5]

A bag contains 2 red balls, 3 blue balls and 4 green balls. A ball is chosen at random from the bag and is not replaced. A second ball is chosen. Find the probability of choosing one green ball and one blue ball in any order.

3 [*Maximum mark:* 6]

The 80 applicants for a Sports Science course were required to run 800 metres and their times were recorded. The results were used to produce the following cumulative frequency graph.

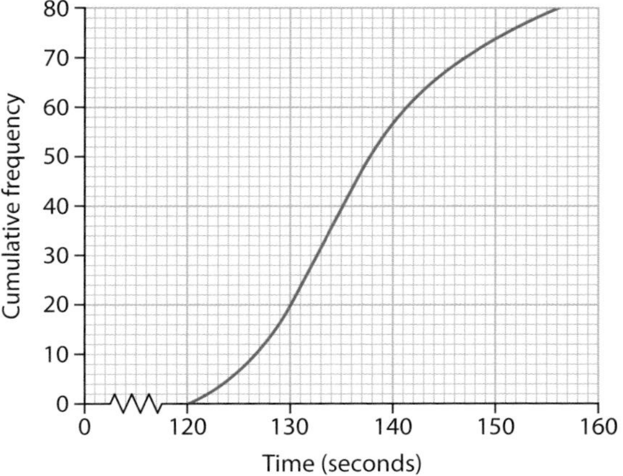

Estimate

a) the median [2 *marks*]

b) the interquartile range. [4 *marks*]

4 [*Maximum mark:* 6]

Find the coordinates of the point where the line with the vector equation $\boldsymbol{r} = \begin{pmatrix} 4 \\ -2 \\ 2 \end{pmatrix} + \gamma \begin{pmatrix} 2 \\ -1 \\ 3 \end{pmatrix}$ intersects the plane with the equation $2x + 3y - z = 2$.

5 [*Maximum mark:* 7]

a) Express the complex number $8i$ in polar form. [3 *marks*]

b) The cube root of $8i$ which lies in the first quadrant is denoted by z. Express z

(i) in polar form [2 *marks*]

(ii) in Cartesian form. [2 *marks*]

6 [*Maximum mark:* 7]

Find the equation of the line that is tangent to the curve $3x^2 + 4y^2 = 7$ where $x = 1$ and $y > 0$.

7 [*Maximum mark:* 6]

Find the value of x satisfying the equation

$(3^x)(4^{2x+1}) = 6^{x+2}$

Give your answer in the form $\frac{\ln a}{\ln b}$ where $a, b \in \mathbb{Z}$.

8 [*Maximum mark:* 6]

a) The independent events A and B are such that $P(A) = 0.4$ and $P(A \cup B) = 0.88$. Find $P(B)$. [4 *marks*]

b) Find the probability that either A occurs or B occurs, but **not** both. [2 *marks*]

9 [*Maximum mark:* 6]

The area of the enclosed region shown in the diagram is defined by

$y \geqslant x^2 + 2$, $y \leqslant ax + 2$, where $a > 0$

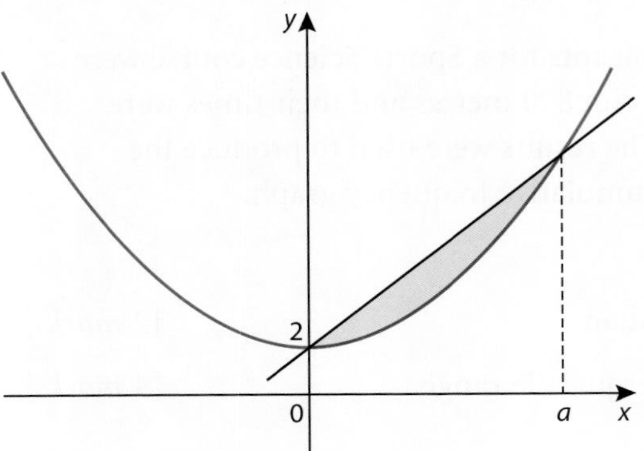

The region is rotated 360° about the x-axis to form a solid of revolution. Find, in terms of a, the volume of this solid of revolution.

10 [*Maximum mark:* 6]

The diagram below shows the graph of equation $y_1 = f(x)$, $0 \leqslant x \leqslant 4$.

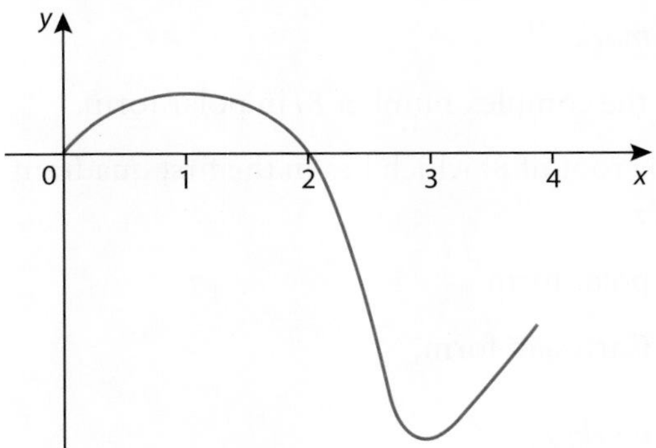

Copy the axes below and sketch the graph of $y_2 = \int_0^x f(t)\, dt$, marking clearly the points of inflexion.

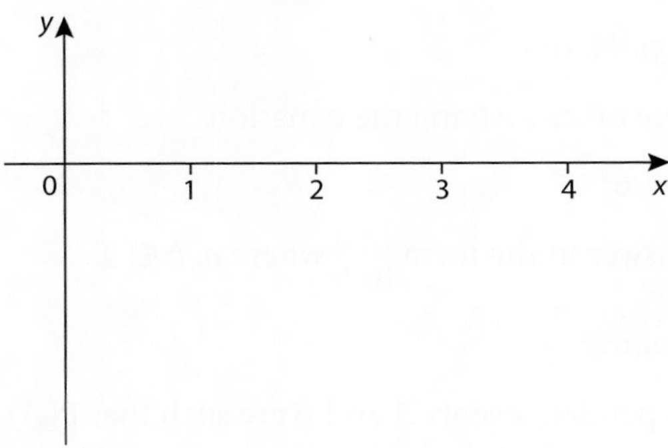

Section B

11 [*Maximum mark:* 17]

The probability density function of the random variable X is given by

$$f(x) = \begin{cases} \dfrac{k}{\sqrt{4-x^2}}, & \text{for } 0 \leqslant x \leqslant 1 \\ 0 & \text{otherwise} \end{cases}$$

a) Find the value of the constant k. [5 *marks*]

b) Show that $E(X) = \dfrac{6(2-\sqrt{3})}{\pi}$ [7 *marks*]

c) Find the median of X. [5 *marks*]

12 [*Maximum mark:* 16]

a) Find the root of the equation $e^{2-2x} = 2e^{-x}$ giving the answer as a logarithm. [4 *marks*]

b) The curve $y = e^{2-2x} - 2e^{-x}$ has a minimum point. Find the coordinates of this minimum. [7 *marks*]

c) The curve $y = e^{2-2x} - 2e^{-x}$ is shown below.

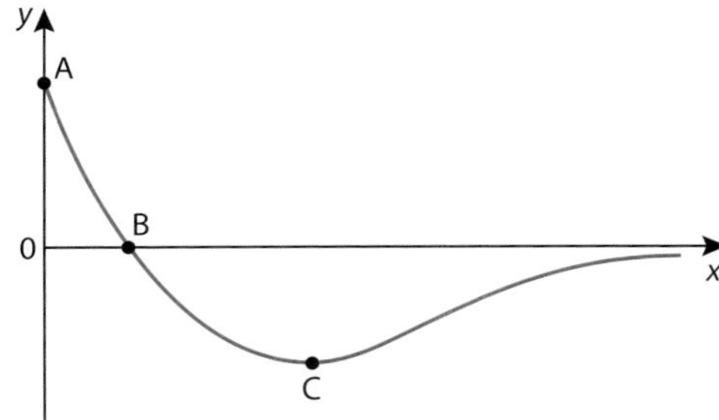

Write down the coordinates of the points A, B and C. [3 *marks*]

d) Hence state the set of values of k for which the equation $e^{2-2x} - 2e^{-x} = k$ has two distinct roots. [2 *marks*]

13 [*Maximum mark:* 13]

a) Show that the following system of equations will have a unique solution when $a \neq -1$.

$$\begin{aligned} x + 3y - z &= 0 \\ 3x + 5y - z &= 0 \\ x - 5y + (2-a)z &= 9 - a^2 \end{aligned}$$

[5 *marks*]

b) Given that $a \neq -1$, state the solution in terms of a. [6 *marks*]

c) Hence, solve

$$\begin{aligned} x + 3y - z &= 0 \\ 3x + 5y - z &= 0 \\ x - 5y + z &= 8 \end{aligned}$$

[2 *marks*]

14 [*Maximum mark:* 14]

a) Using mathematical induction, prove that

$$\sum_{r=1}^{n}(r+1)2^{r-1} = n(2^n).$$ [7 *marks*]

b) The first three terms of a geometric sequence are also the first, eleventh and sixteenth terms of an arithmetic sequence.

The terms of the geometric sequence are all different.

The sum to infinity of the geometric sequence is 18.

(i) Find the common ratio of the geometric sequence, clearly showing all working. [4 *marks*]

(ii) Find the common difference of the arithmetic sequence. [3 *marks*]

Paper 2 Sample A

Paper 2 is a GDC paper. Your exam paper will have some instructions on the first page, some of which are reproduced here.

Full marks are not necessarily awarded for a correct answer with no working. Answers must be supported by working and/or explanations. Where an answer is incorrect, some marks may be given for a correct method, provided this is shown by written working. You are therefore advised to show all working.

It is important that you remember to show work because examiners will award marks for correct work leading to the final solution. Also, if your final answer is incorrect, you will not end up losing all the marks.

Specific to GDC papers: If you use a GDC to arrive at your conclusion, you need to show work leading to what you entered into your GDC. For example, if you are to find the area of a certain region under a curve between two points a and b, then you set up the integral leading to the solution but not necessarily the symbolic manipulation required.

Example
Find the area enclosed by the curve $f(x) = 2x^3 - 9x^2 + x + 12$ and the x-axis.

Suggested answer
To find the area of this region, we observe that the function intersects the x-axis at three different points: at $x = -1$, $x = 1.5$ and $x = 4$.

Therefore, the area of the region is $\int_{-1}^{4}|2x^3 - 9x^2 + x + 12|dx \approx 39.1$.

Section A

1 [*Maximum mark:* 5]
For what values of x is the following inequation true?

$$-7x^2 - 27x + 4 \geqslant 0$$

2 [*Maximum mark:* 6]
In triangle ABC, $BC = 6$, $AC = 7$ and $\angle A = 30°$. Find all possible values of AB.

3 [*Maximum mark:* 6]
An experiment can result in one or both of events A or B with the following probabilities:

	A	A'
B	0.34	0.46
B'	0.15	0.05

Find:
a) $P(A \cup B)$ b) $P(A|B)$.
c) Are A and B independent? Justify.

4 [*Maximum mark:* 4]
In a binomial experiment with n trials, the probability of success $p = 0.6$ and $P(x < 2) = 0.1792$. Find the value of n.

5 [*Maximum mark:* 7]
Consider the function $f(x) = \dfrac{2e^x}{1 + 3e^x}$.
a) Find $f^{-1}(x)$.
b) Find the exact domain of $f^{-1}(x)$.

6 [*Maximum mark:* 5]
A part of a track is shown in the diagram. The radius of the inner circle is 60 m and the width of the track is 3 m. The length of the inner arc is 20π and the outer arc is 21π. Find the area of the track.

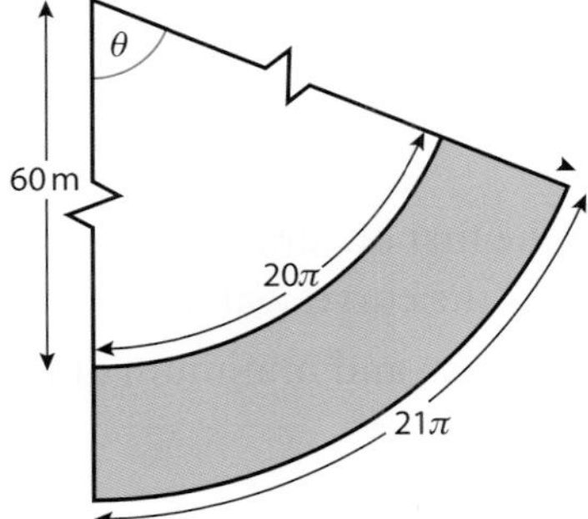

7 [*Maximum mark:* 6]
Consider the complex number $z = 5\sqrt{3} - 5i$.
a) Express z in the form $re^{i\theta}$, presenting your answer in exact form.
b) Find the fifth roots of the complex number and sketch them in an Argand diagram.

8 [*Maximum mark:* 7]
The figure below is that of the function

$$f(x) = \cos x \ln x, 0 \leqslant x \leqslant 2\pi.$$

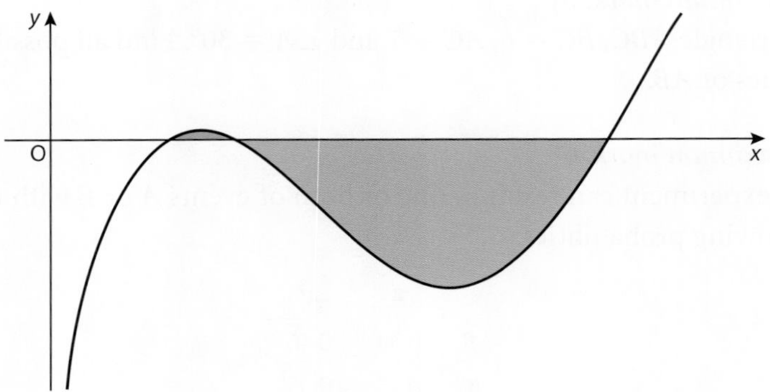

Find the ratio of the shaded area below the x-axis to the shaded area above the x-axis.

9 [*Maximum mark:* 9]
The continuous random variable has the following pdf:

$$f(x) = \begin{cases} mx\sqrt{4 - x^2} & x \in [0, 2] \\ 0 & \text{otherwise} \end{cases}$$

a) Find the value of m.
b) Find the ratio of the area between the mean and median to that between the mean and mode.

10 [*Maximum mark:* 5]
Solve the initial value problem

$$e^x \sin 2y \frac{dy}{dx} = \cos y(e^{2x} - x), y(0) = 0.$$

Section B

11 [*Maximum mark:* 18]
A function is defined by

$$f(x) = \tfrac{1}{3}x^3 - x^2 - 3x + 9, x \in \{\mathrm{R}\}.$$

a) Find the points where the graph of this function intersects the x-axis. [*3 marks*]
b) Find the point in the first quadrant where the normal to the curve at $(0, 9)$ meets the curve again. [*4 marks*]
c) Find the local maximum and minimum of the function in the interval $[-2, 5]$. [*3 marks*]
d) Sketch the graph of the function. [*3 marks*]
e) Find the area enclosed by the function and the line connecting the maximum to the minimum points. [*5 marks*]

12 [*Maximum mark:* 20]
a) Write down the expanded form of $(1 + x)^n$ using binomial coefficients. Include the term containing x^r where $0 \leq r \leq n$. [*4 marks*]
b) Calculate the coefficient of x^5 in the expansion of $(1 + x)^7(1 + x)^{11}$. [*4 marks*]

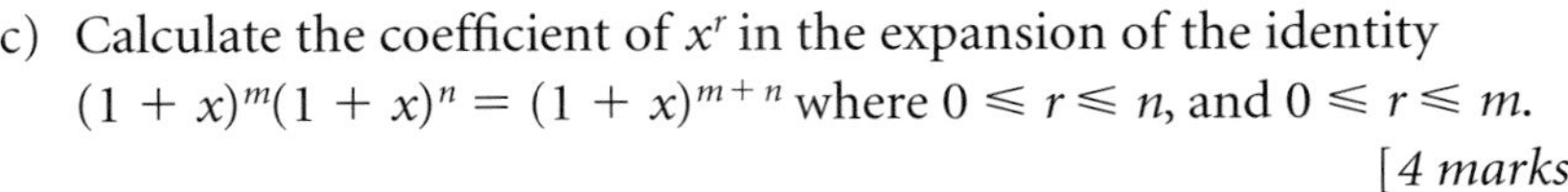

c) Calculate the coefficient of x^r in the expansion of the identity $(1 + x)^m(1 + x)^n = (1 + x)^{m+n}$ where $0 \leqslant r \leqslant n$, and $0 \leqslant r \leqslant m$. *[4 marks]*

d) Hence, show that

$$\binom{m}{0}\binom{n}{r} + \binom{m}{1}\binom{n}{r-1} + \binom{m}{2}\binom{n}{r-2} + \ldots + \binom{m}{r}\binom{n}{0} = \binom{m+n}{r}.$$

[4 marks]

e) By considering $n = m = r$, show that

$$\binom{n}{0}^2 + \binom{n}{1}^2 + \ldots + \binom{n}{n}^2 = \binom{2n}{n}.$$

[4 marks]

13 [*Maximum mark:* 22]

The amount of salt extracted in a large salt mine is modelled by a normal distribution with a mean of 1.5 cubic metres per hour of production time, and a standard deviation of 0.375 cubic metres.

a) (i) Find the probability that in a randomly chosen hour of production, the output is between 1.2 cubic metres and 1.875 cubic metres. *[3 marks]*

(ii) In 10% of the production hours, the output is considered low. How many cubic metres are considered low? *[3 marks]*

b) The production process can be adjusted to meet production demand. So, the mean and the standard deviation can be altered. The management would like to see that the production exceeds 2 cubic metres at most 10% of the time, and falls short of 0.5 cubic metres 5% of the time. Find the values of the required mean and standard deviation. *[8 marks]*

Because of the hard nature of the extraction process, the machines used in the process occasionally stop and have to be restarted. The number of stoppages per hour of production is modelled by a Poisson distribution with a mean of 3 stoppages.

c) (i) Find the probability that the machines stop at least 4 times in each of three successive hours of production. *[5 marks]*

(ii) Find the probability that the machines stop 20 times during a randomly chosen 8-hour shift. *[3 marks]*

Paper 2 Sample B

Section A

1 [*Maximum mark:* 7]

A pizza producer packs half-baked pizzas in boxes, freezes them and distributes them to consumers. For a pizza to fit in a box, the diameter must not exceed 30 centimetres. All pizzas with diameter larger than 30 cm have to be re-done. To comply with the label, pizzas must not be smaller than 27 cm in diameter. It is found that 4% of the pizzas are too large while 1% are too small. Assuming the diameters of these pizzas to be normally distributed, find the mean and standard deviation.

2 [*Maximum mark:* 5]
An infinite geometric series converges to 24. The sum of the first three terms is 208/9.
Find the sum of the first 6 terms.

3 [*Maximum mark:* 6]
Consider the function

$$f(x) = 3^{3x-x^2}.$$

a) Find the maximum value of this function.

b) Find the coordinates of the points of inflexion.

4 [*Maximum mark:* 5]
Solve the following inequation:

$$\frac{|x-1|+3}{|x+1|-2} < 2$$

5 [*Maximum mark:* 6]
Consider the function

$$f(x) = \sin(\sqrt{4-x^2}).$$

a) Find the domain and range of the function.

b) For what values of x does this function have an extreme value?

6 [*Maximum mark:* 7]
The probability density function of a random variable is

$$f(x) = \begin{cases} k(2-\log_3(4x^2+1)) & -a \leqslant x \leqslant a \\ 0 & \text{otherwise} \end{cases}$$

a) Show that $a = \sqrt{2}$.
b) Find the value of k correct to 3 decimal places.

7 [*Maximum mark:* 7]
The number of defects per square metre of fabric is known to follow a Poisson distribution. It is discovered that

$$P(x \leqslant 3) = 0.2381.$$

a) Find the average number of defects per square metre, to the nearest integer.
b) You randomly pick 1 square metre of this fabric for inspection. Find the probability of observing at least 3 defects.

8 [*Maximum mark:* 6]
You invest an amount of \$1000 at an interest rate of 6% compounded semi-annually.
How much money will you have in 20 years?
If you were offered to invest the money at continuous compounding, how long will it take you to earn the same amount?

9 [*Maximum mark:* 5]

Find the equation of the tangent line to the curve defined by

$$\ln(xy) = 2x$$

at the point $(1, e^2)$.

10 [*Maximum mark:* 6]

Solve the differential equation

$$\frac{dy}{dx} = \frac{e^{(y^2 + \sin x)}}{y \sec x};\ y(0) = \sqrt{\ln 2}.$$

Section B

11 [*Maximum mark:* 21]

a) $5^x = e^{kx}$ for all real numbers x. Find the value of k. [*4 marks*]

b) Use the value of k found in a) to find the derivative of $f(x) = 5^x$. [*3 marks*]

c) A random variable X has the following probability density function:

$$f(x) = \begin{cases} 5^x & 0 \leqslant x \leqslant a \\ 0 & \text{otherwise} \end{cases}$$

(i) Find the value of a. [*4 marks*]

(ii) Find the expected value of X. [*4 marks*]

d) Three values of this random variable are chosen at random. What is the probability that:

(i) at least one of the values is larger than 0.5? [*3 marks*]

(ii) at most two of the values are less than 0.5? [*3 marks*]

12 [*Maximum mark:* 18]

Let A be the matrix $\begin{pmatrix} -2 & 2 & 3 \\ -4 & 5 & 5 \\ 2 & 1 & -3 \end{pmatrix}$ and I be the identity matrix of order 3.

a) Show that: $\det(A - kI) = -k^3 + 22k - 6$. [*4 marks*]

b) With an appropriate choice of k, find the determinant of A. [*3 marks*]

c) You are given that the matrix A satisfies the equation

$$-6I + 22A - A^3 = 0.$$

(i) Express the matrix A^{-1} in terms of A. [*5 marks*]

(ii) Hence, show that the three planes

$$20x - 9y + 5z = 2,$$
$$2x + 2z = 3,$$
$$14x - 6y + 2z = 5$$

intersect at one point. Find the coordinates of that point. [*6 marks*]

13 [*Maximum mark:* 21]

Consider the complex number $z = \cos\theta + i\sin\theta$.

a) Use DeMoivre's theorem to find z^5. [3 *marks*]

b) Hence, show that $\cos 5\theta = 11\cos^5\theta - 10\cos^3\theta + 5\sin^4\theta\cos\theta$, and $\sin 5\theta = 15\sin\theta\cos^4\theta - 10\sin\theta\cos^2\theta + \sin^5\theta$. [6 *marks*]

c) Hence, show that $\tan 5\theta = \dfrac{5t - 10t^3 + t^5}{1 - 10t^2 + 5t^4}$ where $t = \tan\theta$. [6 *marks*]

d) Hence, find the solutions to the equation
$t^5 - 5t^4 - 10t^3 + 10t^2 + 5t - 1 = 0$,
expressing your answer correct to 3 d.p. [6 *marks*]

Paper 2 Sample C

Section A

1 [*Maximum mark:* 6]

Triangle ABC has $\hat{C} = 42°$, BC = 1.74 cm, and area 1.19 cm^2.

a) Find AC. [2 *marks*]

b) Find AB. [4 *marks*]

2 [*Maximum mark:* 5]

Find the values of a and b, where a and b are real, given that $(a + bi)(2 - i) = 5 - i$.

3 [*Maximum mark:* 6]

The function f is defined as $f(x) = \dfrac{3x - 4}{x + 2}, x \neq -2$.

a) Find an expression for $f^{-1}(x)$. [5 *marks*]

b) Write down the domain of f^{-1}. [1 *marks*]

4 [*Maximum mark:* 6]

The function f is defined as $f(x) = \sin x \ln x$ for $x \in [0.5, 3.5]$.

a) Write down the x-intercepts. [2 *marks*]

b) The area above the x-axis is A and the **total** area below the x-axis is B.

If $A = kB$, find k. [4 *marks*]

5 [*Maximum mark:* 6]

The weights in grams of bread loaves sold at a supermarket are normally distributed with mean 200 grams. The weights of 88% of the loaves are less than 200 grams. Find the standard deviation.

6 [*Maximum mark:* 6]

Find $\int e^{2x} \sin x \, dx$.

7 [*Maximum mark:* 6]

The number of car accidents occurring per day on a highway follows a Poisson distribution with mean 1.5.

a) Find the probability that more than two accidents will occur on a given day. [*2 marks*]

b) Given that at least one accident occurs on another day, find the probability that more than two accidents occur on that day. [*4 marks*]

8 [*Maximum mark:* 6]

There are 10 seats in a row in a waiting room. There are six people in the room.

a) In how many different ways can they be seated? [*2 marks*]

b) In the group of six people, there are three sisters who must sit next to each other.
In how many different ways can the group be seated? [*4 marks*]

9 [*Maximum mark:* 6]

Solve the differential equation given that $y = 1$ when $x = -1$.

$$(x + 2)^2 \frac{dy}{dx} = 4xy \quad (x > -2)$$

10 [*Maximum mark:* 6]

The radius and height of a cylinder are both equal to x cm. The curved surface area of the cylinder is increasing at a constant rate of 10 cm^2/sec. When $x = 2$, find the rate of change of

a) the radius of the cylinder [*4 marks*]

b) the volume of the cylinder. [*2 marks*]

Section B

11 [*Maximum mark:* 12]

A machine is set to produce bags of salt, whose weights are distributed normally, with a mean of 110 grams and standard deviation of 1.142 grams. If the weight of a bag of salt is less than 108 grams, the bag is rejected. With these settings, 4% of the bags are rejected.

The settings of the machine are altered and it is found that 7% of the bags are rejected.

a) (i) If the mean has not changed, find the new standard deviation, **correct to three decimal places.** [*4 marks*]

The mean is adjusted to operate with this new value of the standard deviation.

(ii) Find the value, **correct to two decimal places**, at which the mean should be set so that only 4% of the bags are rejected. *[4 marks]*

b) With the new settings from part (a), it is found that 80% of the bags of salt have a weight which lies between A grams and B grams, where A and B are symmetric about the mean.

Find the values of A and B, giving your answers **correct to two decimal places.** *[4 marks]*

12 [*Total mark:* 22]

Part A [*Maximum mark:* 12]

A bag contains a very large number of ribbons. One quarter of the ribbons are yellow and the rest are blue. Ten ribbons are selected at random from the bag.

a) Find the expected number of yellow ribbons selected. *[2 marks]*

b) Find the probability that exactly six of these ribbons are yellow. *[2 marks]*

c) Find the probability that at least two of these ribbons are yellow. *[3 marks]*

d) Find the most likely number of yellow ribbons selected. *[4 marks]*

e) What assumption have you made about the probability of selecting a yellow ribbon? *[1 mark]*

Part B [*Maximum mark:* 10]

The continuous random variable X has probability density function

$$f(x) = \begin{cases} \dfrac{x}{1+x^2}, & \text{for } 0 \leqslant x \leqslant k \\ 0, & \text{otherwise} \end{cases}$$

a) Find the exact value of k. *[5 marks]*

b) Find the mode of X. *[2 marks]*

c) Calculate $P(1 \leqslant X \leqslant 2)$. *[3 marks]*

13 [*Total mark:* 26]

Part A [*Maximum mark:* 14]

a) The line L_1 passes through the point A(0, 1, 2) and is perpendicular to the plane $x - 4y - 3z = 0$. Find a Cartesian equation of L_1. *[2 marks]*

b) The line L_2 is parallel to L_1 and passes through the point P(3, −8, −11). Find the vector equation of the line L_2. *[2 marks]*

c) (i) The point Q is on the line L_1 such that $\overrightarrow{PQ}$ is perpendicular to L_1 and L_2.

Find the coordinates of Q.

(ii) Hence find the distance between L_1 and L_2. [*10 marks*]

Part B [*Maximum mark:* 12]

Consider this system of equations.

$$x + 2y + kz = 0$$
$$x + 3y + z = 3$$
$$kx + 8y + 5z = 6$$

a) Find the set of values of k for which this system of equations has a **unique** solution. [*6 marks*]

b) For each value of k that results in a non-unique solution, find the solution set. [*6 marks*]

20 Theory of Knowledge

What is TOK?

Theory of knowledge is concerned with how we know what we claim to know. As an IB diploma student you take classes in a number of areas of knowledge corresponding to the IB hexagon. While we call what we learn in each of these subjects 'knowledge', each seems to go about the process of getting this knowledge in a different way. Theory of knowledge examines these different ways of knowing and asks a number of questions about what sort of things can be considered facts, knowledge, good evidence and truth in each of the IB subjects.

Mathematics is rather puzzling as an area of knowledge. Most other subjects that we study in the IB base their knowledge claims upon observations of the world. Mathematics does not. Yet mathematics has profoundly practical applications in the world. How can this be? Knowledge claims in the sciences – while often fairly secure – are nonetheless provisional in some sense. Science allows the possibility that it is wrong – that some new observation or discovery will overturn previously held beliefs. The statements of mathematics, on the other hand, are certain. $1 + 1 = 2$ is not just probably true. It is certain. **It cannot be otherwise.** This is because $1 + 1 = 2$ can be proved. These features give mathematics a special place in TOK.

Explain why probability theory is certain even though it deals in probabilities.

Think of your favourite topic in the HL course. In this topic, can you identify (1) a mathematical transformation and (2) an invariant under this transformation? If you get stuck, ask your maths teacher. (Hint: when studying a function $f(x)$ defined on the real numbers, the function itself is a transformation of the whole real number line, and the set of points that are unmoved by the function, i.e. for which $f(x) = x$ are the invariants. This set is the fixed point set of the function. These points would be represented graphically as the points where the graph of $y = f(x)$ intersected the 45 degree line $y = x$.)

What is mathematics?

It is remarkably difficult to pin down exactly what mathematics is about. A first attempt might be: 'mathematics is the study of numbers'. Certainly, much of our school mathematics is concerned with operations on numbers and the relations between them. This is what is called **arithmetic**. But there is much more to mathematics than numbers, and mathematicians do not take kindly to being thought of as simply good at adding up the bill in the restaurant (actually many of them are not). One of the oldest fields of mathematical thought is **geometry**. When we study geometrical objects such as points, lines, planes, triangles, circles and ellipses we are not studying numbers as such. Rather we are studying the structure of space itself – in particular those aspects that stay unchanged under various types of geometrical transformation. These aspects we call **invariants**. Modern mathematics takes this idea further and studies structures, which are far removed from numbers or even our everyday intuitions about space and time. We could do far worse than define mathematics as the study of transformations and invariants.

What are the foundations of mathematics?

Sets

Modern mathematicians build up the raw materials of their subject from quite humble beginnings. Let us look at how they do this. They start off with some basic concepts about sets. A set, as you know, is just a collection of elements placed inside curly brackets. For example, we could consider a set $A = \{1, 2, 3, 4\}$. We can say that 1 belongs to A: $1 \in A$, but that 5 does not belong to A: $5 \notin A$. The notions of what it is to be a set and to belong to a set are **primitive**. This means that they cannot be explained in terms of more simple notions. If you keep on asking the question 'why?' (as some small children do), the questions stop when you get to a primitive concept (you find yourself answering: 'it just is'). Aristotle was aware of this when he stated that any explanation has to end somewhere. We can now answer him that explanations end in primitive concepts.

Think about an explanation in one of your IB subjects. Keep on asking the question 'why?' until you can go no further. What you are left with is a primitive notion. Are the primitive notions in physics different from those in history?

Mappings

We also need the idea of a mapping between sets. A mapping from A to B is a rule that assigns an element of B to each element of A. The functions that you study in your Higher Level Mathematics course are examples of mappings between the set of real numbers and itself.

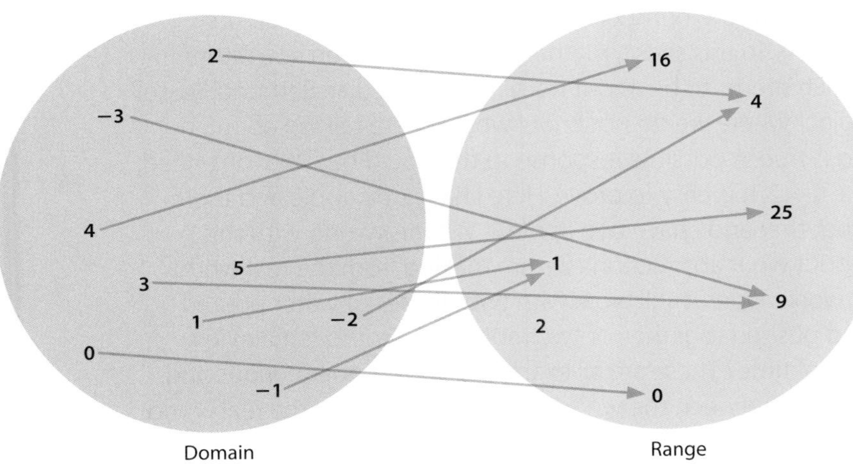

An example of a mapping between two sets.

Notice that for a mapping to be well defined, every member of the domain set has to have an arrow (and only one arrow) pointing from it. But some members of the range set can have more than one arrow pointing to them (and the number 2 has no arrow pointing to it). This is an example of a many-to-one mapping. What is the mapping represented here?

Bertrand Russell and A. N. Whitehead, in their monumental book *Principia Mathematica* (1913), reduced the whole of mathematics to these simple notions. With a bit of work and a great deal of care and patience we can establish basic truths of arithmetic, such as $1 + 1 = 2$. (Proving this takes about four pages of quite sophisticated mathematical argument; this is a surprise to many students who think that 2 is *defined* to be $1 + 1$.)

Because we can build the whole of mathematics out of these primitive ideas of sets and mappings, does this mean that this is what mathematics is about?

Russell's paradox[1]

In constructing mathematics from set theory, we must be careful that we do not allow sets to be members of themselves. Consider the collection $D = \{d$: d is a set **and** d contains more than 1 element$\}$. By this definition, D actually belongs to itself, since D contains more than one element. There is something rather strange about this, which might make us suspicious. The self-reference involved in thinking about sets that are members of themselves leads to a famous paradox discovered by the English philosopher and mathematician Bertrand Russell. He considered the set that is defined as follows: $S = \{s$: s is a set **and** s does not belong to itself$\}$. The question he then asked was: Does S belong to itself or not? If the answer is yes – S does belong to itself – then, by the definition of S, S does not belong to itself. If S does not belong to itself, then, by the definition of S, S does belong to itself. Either way we get a contradiction. Russell realized that certain large collections (such as that of all sets) were actually too big to be a set. A collection like this is called a **proper class**.

[1]Bertrand Russell *The Principles of Mathematics* (1903) Cambridge

The barber of Seville

Russell's paradox is similar to the story of the barber of Seville. There was a man who lived in Seville who was a barber. He had a monopoly on the shaving industry in Seville. He shaved every man in the town who did not shave himself. What is contradictory about this?

Mathematics and the real world

1 + 1 = 2?

The objects of mathematics, such as the number systems that we use, are built up from elementary ideas about sets. In this sense, mathematics can be seen as a rather elaborate abstract game, which seems to be about nothing in particular. Bertrand Russell wrote: 'Mathematics is a subject where we do not know what we are talking about, nor whether what we are saying is true.' A possible response to this could be: 'We don't need to establish formally that $1 + 1 = 2$. It is easy to prove. Here I have one apple and there another apple. I put them together and I have two apples!' What is wrong with this approach? Think carefully about what abstractions we are making from the real world in order that this argument works. Does it still work with two glasses of water poured together, or two piles of sand pushed together, or two rabbits (male and female) left together for a suitable length of time? These are all examples of the rather curious and sometimes awkward connection between the world of mathematics and the real world.

The Platonist view of mathematics

Plato was aware of the tension between the world of perfect geometrical objects – points with no area, lines with no width, perfect circles and triangles – and the messy physical world. There are no perfectly thin lines, infinitely small points and perfect circles in the real world. But he thought there was a world of perfect mathematical objects underlying the imperfect physical world of our everyday experiences. This mathematical world existed independently of human beings. There would still be nine planets in the solar system long after human beings have ceased to exist on Earth (well, eight actually!). Plato's thinking can help explain the usefulness of mathematics. After all, mathematics is often described as the language of the natural sciences – it is almost impossible to do biology, chemistry or physics without it. But increasingly, mathematics is becoming the *lingua franca* of the social sciences. For example, cutting-edge research in economics is highly mathematical. Governments use highly complex mathematical models to make predictions about future inflation, unemployment and growth rates. This makes a lot of sense, if we grant that mathematics is 'out there' as part of the structure of our physical and social world, as Plato thought it was. That would explain why mathematical methods are so effective in solving real world problems. This is called the **Platonist** view of mathematics.

Formalist and constructivist mathematics

There are two responses to Plato's view that mathematics is 'out there'. One emphasizes the game-like nature of mathematics and the other the fact that it is played by human beings. The **formalist** approach treats mathematics as an abstract formal game. The game proceeds using an agreed set of rules from agreed starting points, or **axioms**. The individual symbolic statements of mathematics mean nothing outside the game, just as 'checkmate' is meaningless outside chess and 'fifteen-love' is meaningless outside the game of tennis.

The formalist must concede that any use mathematics has in the outside world is largely a coincidence. Is this a point against the formalist view of mathematics?

The **constructivist** sees mathematics as a human activity. To this way of thinking, when there are no more humans there will be no more mathematics. Mathematics is produced by individuals or societies in much the same way as literature and other cultural artefacts. Again, the constructivist has the problem of explaining the success of mathematics in describing, understanding and predicting the outside world. How can a man-made system fit the non-human world so well?

There is another problem with the constructivist view of mathematics. It seems that we are accountable to the truths of mathematics. Mathematicians speak of mathematics as having an independent existence – maths is there to be discovered rather than being man-made. It is certainly true, as we shall see later, that maths can throw up quite unexpected results. Is this compatible with the description of mathematics as being built up out of a few basic and abstract raw materials? In order to try to answer this question, we need to look a little more closely at what constitutes mathematical truth.

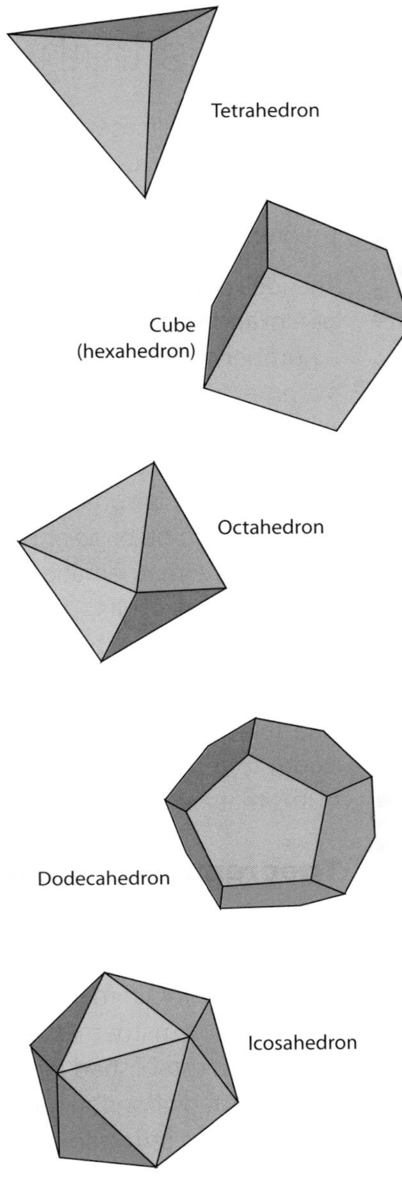

▲ Plato thought that underlying the messy real world was the perfect world of mathematics. The five regular polyhedra shown are often called the Platonic solids.

Think about the question of whether mathematics is 'out there' in the world or whether it is an invention of human beings. Does this question occur in other areas of knowledge? Does it make sense to ask if English literature is out there in the world? Does it make sense to ask if chemistry is invented by humans?

What is truth in mathematics?

Let us look again at what we mean by **mathematical truth**. Mathematical statements are true if they can be proved. Before it is proved, a mathematical statement is called a **conjecture**. Once it is proved it is called a **theorem**. So, theorems are mathematical truths.

The idea of proof in mathematics is very old. In around 300 BC, the geometer Euclid of Alexandria formalized the notion of proof in his book *The Elements*. He proved a number of truths about geometrical figures. A proof is a list of statements. Each statement is derived from the preceding statement in the list using only the rules of logic. This is called **chain reasoning**. But what starts the chain in the first place? The first statement in the chain must be, in some sense, either true by definition or self-evident in some way. These self-evident truths are called **axioms**. They are considered to be basic or primitive mathematical truths. By definition, they cannot be proved. A mathematical proof builds a chain of reasoning from the axioms to final mathematical results – theorems.

They are very special from a TOK perspective because it seems that a theorem is an example of knowledge that is certain. A mathematical theorem is not just probably true. It is true in the sense that, given the definitions of the terms it uses and the axiom system used to prove it, **it cannot be otherwise**. In TOK, we rarely meet truths that are certain in this absolute sense.

Theorem, theory and proof

Be careful that you do not confuse the word 'theorem' with the similar-sounding 'theory'. In mathematics, a theory is an established piece of mathematical work that might contain many theorems. In other words, mathematical theories are pieces of true mathematics. In science, the word is more problematic. It might apply to an established piece of science that has been tested and found to yield accurate predictions and to give good explanations of phenomena in the physical world. But the term can also refer to a more tentative idea that has not yet been thoroughly tested. It is a common mistake in TOK essays to make a statement such as: 'The theory of evolution is only a theory so it cannot be considered knowledge'. Evolution theory belongs to the first type of theory – it is as well supported by evidence as the fact that water is H_2O – but the essay treats it as belonging to the second.

A word of warning is also needed about the word **proof**. Strictly speaking, proof is the mathematical process outlined above – where a mathematical statement is derived from axioms in a step-by-step manner. Proof implies absolute certainty. Be careful applying this word outside mathematics.

Part of a manuscript by the French mathematician Evariste Galois.

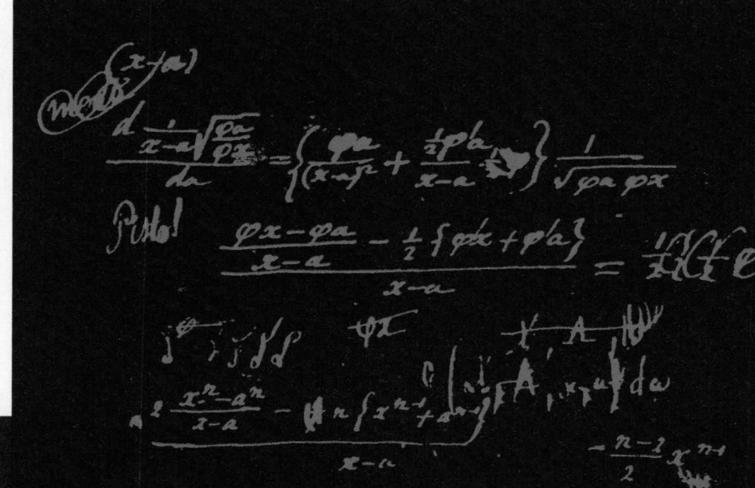

Absolute certainty is generally not achievable in science for a number of reasons that you may have discovered in your TOK course. Scientific results are not proved in this strict sense, it is better to describe them as being 'secure' or 'well supported by the evidence'.

To see how mathematical proof works, let us prove a simple theorem.

Theorem: Let x and y be odd integers. Then $x + y$ is an even integer.

Proof: The definition of an odd number is that it is an even number plus 1. An even number is a number in the 2× table.

So, write $x = 2m + 1$ for some integer m. In a similar fashion, $y = 2n + 1$ for some integer n.

$x + y = (2m + 1) + (2n + 1) = 2m + 2n + 2$

We can take out the common factor of 2 to give: $x + y = 2(m + n + 1)$

Since m, n and 1 are integers, it follows that $m + n + 1$ is also an integer.[2]

Hence, $x + y$ is 2× an integer and so must be even. QED

We write QED (*Quad Erat Demonstrandum* – meaning 'which was to be shown') at the end, to show that the proof is finished.

Let us take a closer look at some features of this method of proof. First notice that we have in effect proved an infinite sequence of statements including: 3 + 5 is even, 3 + 7 is even, 5 + 7 is even, and so on. We could have attempted to do a sort of mathematical experiment by checking whether the result holds for some randomly chosen odd numbers: 3 + 5 = 8, which is even; 3 + 7 = 10, which is even; 5 + 7 = 12, which is even; and so on. But this is not a proof. There is always an infinite number of examples that we have not tried and for which the result might not hold. This is what mathematicians call a **semantic** method. But, as you have probably learned from studying the natural sciences in TOK, it takes a single counter-example to disprove a conjecture. The same is true in mathematics. Why not try Moser's circle problem (shown right) to see what we mean?

A proof is a **syntactic** method. It does not look at particular examples of odd numbers but rather depends on features that all odd numbers have in common (namely their oddness!). We have been able to do this by using algebra. We have substituted letters for numbers to allow us to talk generally about odd numbers rather than specific examples. This is typical of a mathematical

Moser's circle problem illustrates the difference between an experimental approach to mathematics – a semantic method (trying out a conjecture to see if it works) – and proving it – a syntactic method.

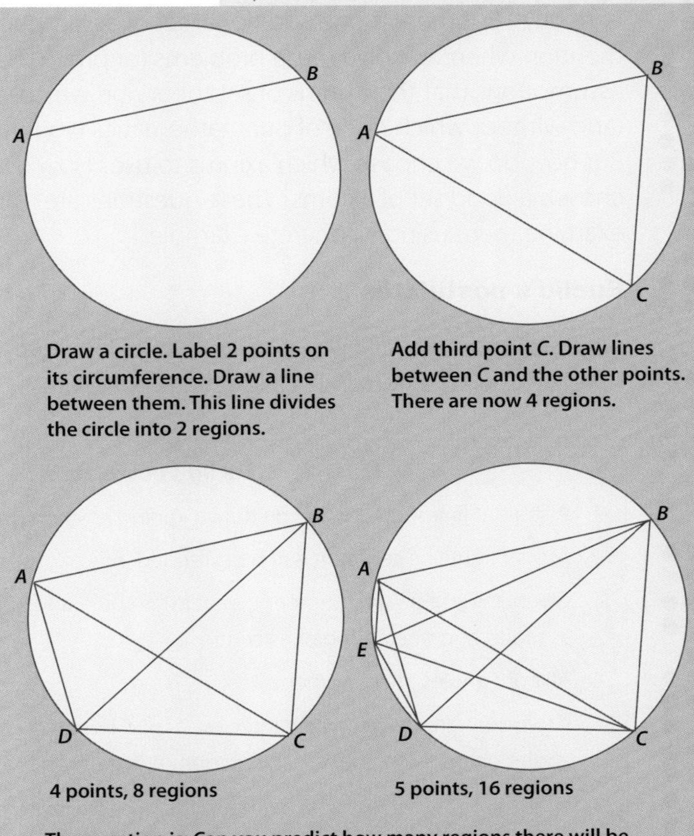

Draw a circle. Label 2 points on its circumference. Draw a line between them. This line divides the circle into 2 regions.

Add third point C. Draw lines between C and the other points. There are now 4 regions.

4 points, 8 regions

5 points, 16 regions

The question is: Can you predict how many regions there will be when you add a sixth point? Can you prove why this is so?

[2] There is a further subtlety here in the statement that $n + m + 1$ is an integer because m and n are. This is because the integers are closed under addition because $\mathbb{Z}^+$ is a group – closure is a property of groups. Groups are structures that underlie most mathematics.

proof. Once the conjecture is proved we can state categorically that it is true, now and for all time. It does not depend on culture, nationality, personal points of view, language or gender. It does not matter who proved it. It could be a university professor of mathematics or it could be an eight-year-old. It simply does not matter. In mathematics, proof means truth and that is the end of the story.

Axioms

A mathematical statement is true if it could be derived from the basic axioms of set theory by using only the rules of logic. In the HL course, the rules of logic are packaged in a convenient way to help us solve problems. We call this package 'the rules of algebra'. These are rules such as: You can add the same number to both sides of an equality and it remains equal (if $y = x$ then $y + 5 = x + 5$).

We can use these rules of algebra to solve mathematical problems. Each problem we solve is a little theorem. An example is: If $x + 5 = 10$ then $x = 5$. This is rather a simple theorem, but it is a theorem nevertheless. If you write any of your standard maths problems in the form '**If** … (problem to be solved) **then** (solution)' you get a theorem. (This assumes that you have solved the problem correctly!) But there is one additional set of assumptions that we do not explicitly mention when we solve these problems (or prove these theorems). That is, the assumptions that the axioms of set theory on which we base all our mathematics (and without which none of our mathematics would mean anything) are true. But how do we choose which axioms to use? How do we know that we have chosen a good set of axioms? These questions are not easy to answer. We shall examine them using a concrete example.

Euclid's postulates

What axioms did Euclid propose for doing plane geometry?
Here are Euclid's axioms. He called them 'postulates'.

Euclid's postulates

1 A straight line segment can be drawn joining any two points.
2 Any straight line segment can be extended indefinitely in a straight line.
3 Given any straight line segment, a circle can be drawn having the given line segment as radius and one endpoint as centre.
4 All right angles are congruent.
5 If two lines are drawn, which intersect a third in such a way that the sum of the inner angles on one side is less than two right angles, then the two lines inevitably must intersect each other on that side, if extended far enough.

In some sense, Euclid's axioms express mathematical intuitions about the nature of geometrical objects. What is clear in any case is that they are not established using observation of the external world. Objects such as points, lines, circles and planes do not exist in the real world with the perfect qualities they possess in mathematics.

Euclidian geometry

Let us try to use Euclid's postulates to do some geometry (see right).

Let us now examine the construction and see which postulates were used.

Step 1: drawing the arcs is allowed by postulate 3 (twice).

Step 2: drawing the line segments *AC* and *BC* is allowed by postulate 1 (twice).

It follows from step 1 that the line segments *AC* and *BC* are both equal to *AB*. Therefore, they must be equal (this is sometimes quoted as a separate axiom – that two line segments equal to the same line segment must be of equal length).

Problem: To construct an equilateral triangle on a given line segment using Euclid's axioms.

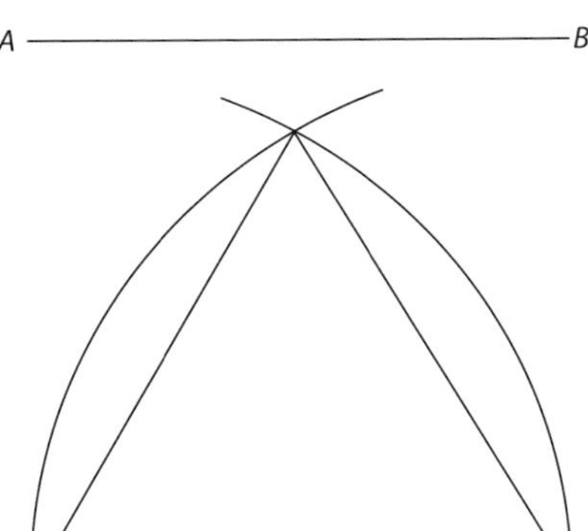

Non-Euclidian geometry

Take a look at Euclid's postulate 5. This cannot be proved as a theorem from the other axioms (that this is impossible can itself be proved!) although many people have attempted this. Euclid himself only used the first four axioms in the first 28 propositions of the *Elements*, but he was forced to use the fifth axiom, so-called 'parallel postulate', in the 29th proposition. The independence of the parallel postulate means that we can choose whether we accept it or not. If we accept it, parallel lines do not meet. The geometry we get is the familiar geometry of the plane. This is the geometry that we can use to construct buildings and other physical objects. In 1823, Janos Bolyai and Nicolai Lobachevsky independently realized that entirely self-consistent non-Euclidian geometries could be envisaged in which the fifth axiom did not hold. There are two quite different geometries in this case: those in which parallel lines meet at some point – elliptical geometry – and those in which parallel lines diverge – hyperbolic geometry. An example of elliptical geometry is the geometry of long distance travel on the Earth's surface. The shortest path between two points (say the most efficient route of a jet airliner) is a curve called a great circle.[3] The parallel lines of longitude are great circles that meet at the poles. If parallel lines diverge, we get so-called hyperbolic geometry. An example of doing hyperbolic geometry would be to draw lines on a saddle.

Why not try out a construction yourself, using the postulates of Euclidean geometry? Extend the arcs below the segment *AB* to meet again at *D*. Join *CD* with a line segment. The task is to prove that *CD* is the perpendicular bisector of *AB* using Euclid's postulates. (The perpendicular bisector of *AB* is a line segment *CD* that cuts *AB* exactly in half and the angle it makes with *AB* is exactly a right angle.)

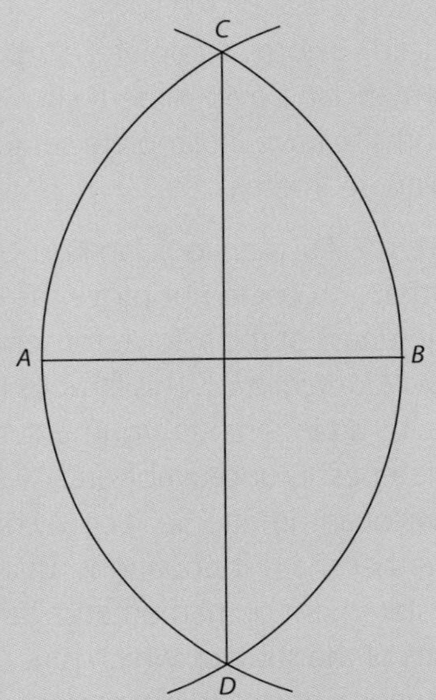

[3] A 'great circle' is the largest circle that can be drawn on a sphere, and is the intersection between the surface of a sphere and a plane passing through the centre of the sphere. The shortest path between two points on a sphere follows a great circle.

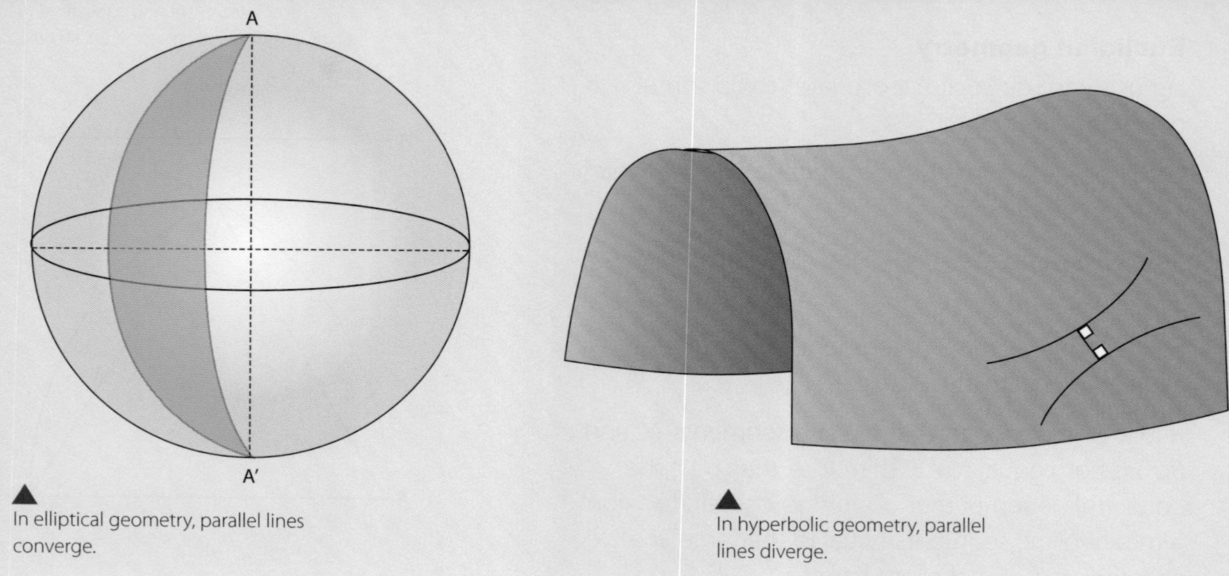

In elliptical geometry, parallel lines converge.

In hyperbolic geometry, parallel lines diverge.

Consistency and completeness

We saw that postulate 5 is independent of the other four - that it could not be derived from them. More generally, there are two questions that can be asked of any set of axioms:
(1) Is the set **consistent**? In other words, is it impossible to derive a contradiction from them (to derive both the statements 'P is true' and 'not P is true')?
(2) Is the set **complete**? That is, any (semantically) true statement can be derived from them.

Are Euclid's axioms complete? Surprisingly, the German mathematician David Hilbert[4] showed that Euclid needs another 15 axioms to have a complete set to do what we now call Euclidian geometry.

[4]David Hilbert *Foundations of Geometry* (1902) Gottingen

In 1931, the Austrian logician Kurt Gödel[5] proved the devastating result that you could not prove the consistency and the completeness of the axioms for set theory that were used in *Principia Mathematica*. This famous incompleteness theorem proves, by an ingenious argument, that consistency and completeness is unprovable in any system rich enough to include the laws of arithmetic. So, it could be that mathematics is based upon rather shaky foundations. This might mean that there is a true statement of mathematics lurking somewhere in the recesses of the subject, which is not provable within the system. More serious, from a mathematical point of view, is the possibility that we can derive a contradiction of the form 'P is true' and 'not P is true' from the axioms using the rules of logic. Producing a contradiction means instant death for any area of knowledge. If you believe that 'P is true' and that 'not P is true' then one of your beliefs has to be false. This makes the combined belief 'P is

[5]Über formal unentscheidbare Sätze der Principia Mathematica und verwandter Systeme, I. *Monatshefte für Mathematik und Physik* 38, 173-98 (1931)

Do you hold any contradictory beliefs? If so, what are the implications for what you consider to be knowledge?

true and not P is true' false under all circumstances. So, if an area of knowledge throws up a contradiction, it simply cannot ever be true. It is condemned to being false whatever the actual state of the world. Since knowledge can be thought of (at least as a first approximation) as justified true belief, a statement that is forever false cannot be knowledge.

To what extent is mathematics really a language?

Beautiful equations

Einstein suggested that the most incomprehensible thing about the universe was that it was comprehensible. From a TOK point of view, the most incomprehensible thing about the universe is that it is comprehensible in the language of mathematics. Galileo wrote: 'Philosophy is written in this grand book, the universe … It is written in the language of mathematics, and its characters are triangles, circles, and other geometric figures….'[6]

[6]Galileo, *Il Saggiatore* (1623) Rome

What is perhaps most puzzling is not just that we can describe the universe in mathematical terms, but the mathematics we need to do this is mostly simple, elegant and even beautiful.

To illustrate this, let us look at some of the famous equations of physics. Most of you will be familiar with at least some of the following:

Relation between force and acceleration: $F = ma$ (more generally this is $F = \frac{d}{dt}(mv)$)

Gravitational force between two bodies: $F = \frac{Gm_1m_2}{r^2}$

Energy of rest mass: $E = mc^2$

Kinetic energy of a moving body: $E = \frac{1}{2}mv^2$

Electrostatic force between two charges: $F = \frac{kq_1q_2}{r^2}$

Maxwell's equations: $\nabla \times \mathbf{B} - \frac{d\mathbf{E}}{dt} = 4\pi\mathbf{J} \quad \nabla \times \mathbf{E} + \frac{d\mathbf{B}}{dt} = 0 \quad \nabla \cdot \mathbf{B} = 0 \quad \nabla \cdot \mathbf{E} = 4\pi\rho$

Einstein's field equation for general relativity: $R_{\mu\nu} - \frac{1}{2}g_{\mu\nu} = 8\pi T_{\mu\nu}$

Similar equations can be found in the other natural sciences. Can you think of any?

Is there a sense in which these equations are elegant or beautiful?

I must admit that I find it perplexing that the whole crazy complex universe can be described by such simple, elegant and even beautiful equations. It seems that our mathematics fits the universe rather well. It is difficult to believe that maths is just a mind game that we humans have invented.

But the argument for simplicity and beauty goes further. Symmetry in the underlying algebra led mathematical physicists to propose the existence of new fundamental particles, which were subsequently discovered. In some cases, beauty and elegance of the mathematical description have even been used as evidence of its truth. The physicist Paul Dirac said: 'It seems that if one is working from the point of view of getting beauty in one's equations, and if one has really a sound insight, one is on a sure line of progress.'

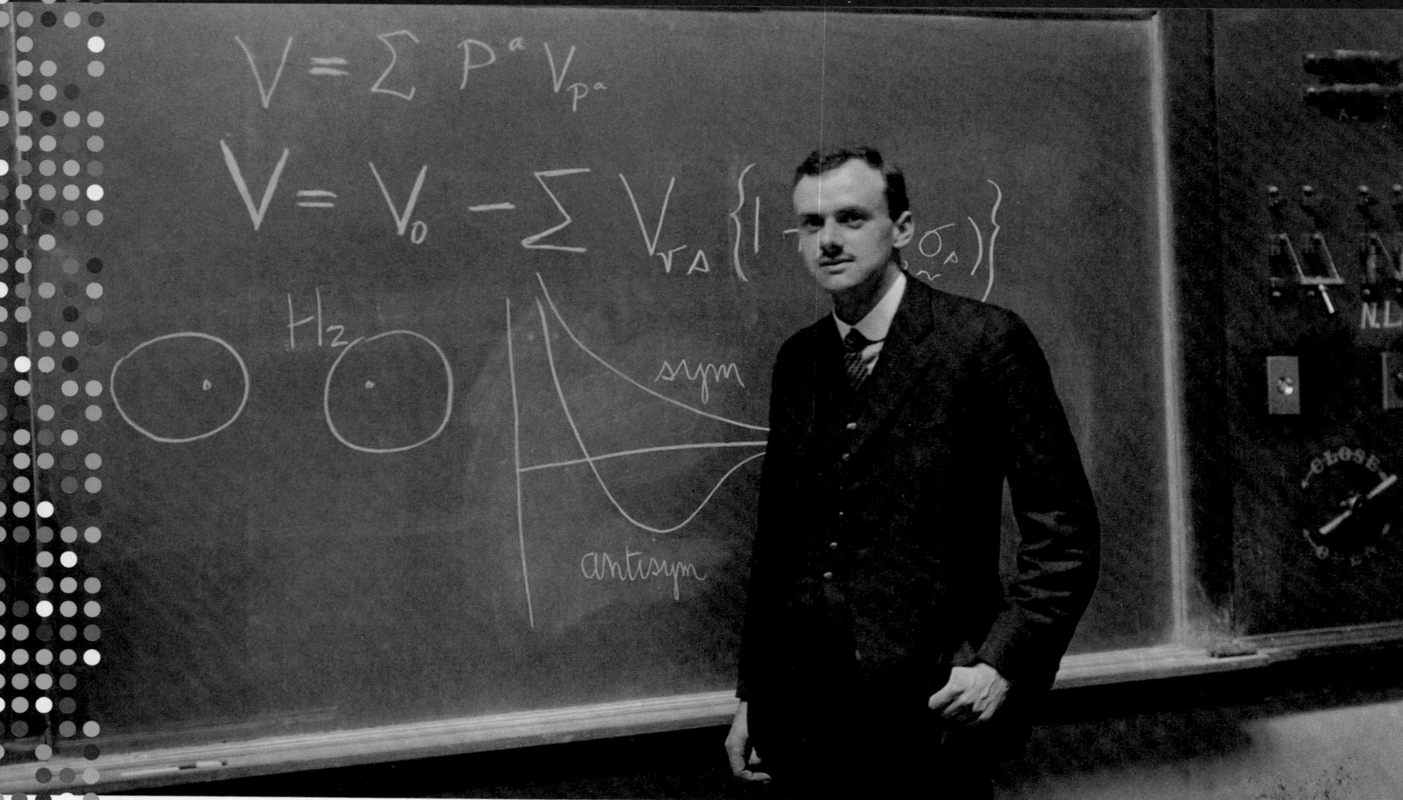

'God used beautiful mathematics in creating the world.'
Paul Dirac

Dirac's own equation for the electron must qualify for being one of the most profoundly beautiful of all. Its beauty lies in the extraordinary neatness of the underlying mathematics – it all seems to fit so perfectly together:

$$\left(\beta mc^2 + \sum_{k=1}^{3} \alpha_k p_k c\right) \psi(x, t) = ih\frac{d\psi}{dt}(x, t)$$

The physicist and mathematician Palle Jorgensen[7] has written: '[Dirac] … liked to use his equation for the electron as an example, stressing that he was led to it by paying attention to the beauty of the math, more than to the physics experiments.'

[7]Palle Jorgensen *Operator Commutation Relations* (1984) New York

I shall leave the last word on this subject to Dirac himself, writing in *Scientific American* in 1963:

'I think that there is a moral to this story, namely that it is more important to have beauty in one's equations than to have them fit experiment.'

By any standards, this is an extraordinary statement for a mathematical physicist to make.

How good are your mathematical intuitions?

Mathematics can sometimes surprise us. Our mathematical intuitions can sometimes let us down, badly. In this section, we shall try out two basic scenarios upon our unsuspecting intuitions and see how they fare.

Scenario 1: The rare genetic disease

Consider the following. There is a very rare genetic disease amongst the population. Very few people have the disease. As a precaution, a test has been developed to check in a particular case whether a person has the disease or not. Although the test is quite good, it is not perfect – it is only 99% accurate. A person X takes the test and it shows positive. The question for your intuition: What is the probability that X actually has the disease?

Think about this for a moment before we go on with the analysis.

Many of the students and teachers that I have worked with in the past have given the same answer: The probability that X actually has the disease, given a positive test result, is around 99%. Did you say the same?

If you did, your mathematical intuition let you down – very badly.

Let us put some numbers into this problem. For the sake of simplicity, let us assume that the country in which the test takes place has a population of 10 million. We are told that the disease is very rare. Let us assume that only 100 people have it in the whole country.

We are told that the test is 99% accurate so that of the 100 cases of the disease the test would show positive in 99 cases and negative in 1 case. So far, so good.

Now let us look at the 9 999 900 people who do not have the disease. In 99% of these cases the test does its job and records a negative result. But in 1% of these cases the test records a positive result. 1% of 9 999 900 is 99 999. This means that if the whole population were tested 99 999 + 99 = 100 098 test results would be positive. Of these, only 99 people have the disease. Therefore, the probability of having the disease, given a positive test result, is 99/100 098 = 0.0989% – in other words, about a tenth of a per cent or one in one thousand. This is a bit different from the 99% that most people guess. How well did you do?

What went wrong with the intuition here?

The important factor in this problem is not just the accuracy of the test *but the accuracy of the test relative to the incidence of the disease.* Because the disease is so rare, the actual number of people with the disease is overwhelmed by the false positive results of the test – the 1% or so of the population who do not have the disease, but the test shows positive anyway. If more people had the disease and if the test was more accurate, the test scoring positive would be a better predictor of X actually having the disease.

Try this problem out with some other numbers to check how the test could be made more useful.

Scenario 2: The Monty Hall game

The second scenario is also based on probability theory. The problem refers to a TV game show, which is loosely based on the actual show *Let's Make a Deal*[8]. A contestant in the show is shown three doors and told (truthfully), by the game show host Monty Hall, that behind one of the doors is a luxury sports car and behind the other two doors are goats. The contestant is told that she must pick a door. She will be allowed to take home whatever is behind the door she picks. We shall assume at this stage that she prefers to win the car. So she goes ahead and picks a door. At this point, Monty Hall opens another door to reveal a goat. (Whenever this game is played, Monty Hall chooses a door concealing a goat.) He then asks the competitor: 'Do you want to switch to the other closed door?'

[8]A widely known statement of the problem was published in Marilyn vos Savant's *Ask Marilyn* column in *Parade* (1990).

What does your intuition tell you? Should the contestant switch or should she stick to her original choice?

The Monty Hall problem: should the contestant switch?

Take a little time to think this through. You might like to try this game with a friend to see experimentally what the best strategy is.

Clearly, because there is one car and two goats, the probability of picking the car if the competitor does not switch doors is 1/3.

If she does switch, what is the probability of winning the car? Let us ask a related question. If she does switch, under what circumstances can she lose the car? Clearly, the only way she can lose the car is if her original choice was right. In other words, she has a 1/3 probability of losing. This must mean that by switching, her probability of winning the car is 2/3.

In other words, by switching she doubles her probability of winning.

Does this make sense? Even after this explanation many of the students and teachers that attend my workshops are not convinced. They argue that they cannot see how an asymmetry has been introduced into the situation.

The crucial point is that Monty Hall knows where the car is. He always opens a door to reveal a goat. It is this act that produces the required asymmetry.

Consider an extreme version of the Monty Hall problem. Imagine 100 closed doors containing 1 car and 99 goats. Let us suppose, for the sake of the argument, that our contestant chooses door number 1. Monty Hall then opens 98 doors to reveal goats. The contestant would be foolish not to switch to the one remaining door (and multiply her probability of winning by a factor of 99).

Try this problem out on your friends and relatives. Are their mathematical intuitions letting them down?

Is the fact that mathematics can surprise us and go against our intuitions evidence that mathematics exists independently of us?

[9]John R. Searle *The Construction of Social Reality* (1995) London

What is a social fact?

The philosopher John Searle[9] points out that many of the facts in our lives are actually socially constructed. He uses money as his central example. Money is money because we believe it to be money. There is something rather strange about this. Normally speaking, when we define a term X, we do not expect the definition to refer to X. Did we not learn in TOK that it was bad to define X in terms of itself? Was this not the reason why our TOK teacher advised us to keep clear of dictionary definitions: 'knowledge – that which is known'. Searle thinks that this sort of circularity is characteristic of what he calls a **socially constructed fact**. He asserts that the social agreements that we make collectively that something should be money makes it such. So 'X functions as money in society S' is a socially constructed fact. As such, statements about it are objective and capable of being evaluated as true or false. Our socially constructed reality includes the concepts of wife, girlfriend, driving licence, bank account, traffic lights, rules of etiquette, nationality, legality, country, nationality, debt, honour, and so on. Many of the physical objects around us are defined in terms of their function, and hence in terms of our intentions, and hence are socially constructed. The concept of a chair or a knife is socially constructed. This is what makes them so difficult to define without using the words 'function' or 'intention'.

Try to define a chair without making reference to human intentions.

Is mathematics a social fact?

Reuben Hersch[10] argues that numbers (and any other mathematical entities) are social constructions. If we acknowledge that they are not just out there in the world independent of human beings and they are not just thoughts in people's heads (our intuitions can be wrong after all) then what are they? There is a third possibility. Mathematics is a construction of human society.

Hersch proposes that mathematics is itself a whole interconnected web of socially constructed reality. Here he is in an interview with John Brockman on the Edge website:[11]

[10]Reuben Hersch *What Is Mathematics, Really?* (1997) Oxford

[11] http://www.edge.org/3rd_culture/hersh/hersh_p1.html (accessed Feb 2008)

'Mathematics is neither physical nor mental, it's social. It's part of culture, it's part of history, it's like law, like religion, like money, like all those very real things, which are real only as part of collective human consciousness. Being part of society and culture, it's both internal and external. Internal to society and culture as a whole, external to the individual, who has to learn it from books and in school. That's what math is.'

When asked what he called his theory of mathematics, Hersch replied that he calls it humanism 'because it's saying that math is something human. There's no math without people. Many people think that ellipses and numbers and so on are there whether or not any people know about them; I think that's a confusion.'

Hersch points out that we do use numbers to describe physical reality and that this seems to contradict the idea that numbers are a social construction. It is important to note here that we use numbers in two distinct ways: as nouns and adjectives. When we say nine apples, nine is an adjective. 'If it's an objective fact that there are nine apples on the table, that's just as objective as the fact that the apples are red, or that they're ripe, or anything else about them, that's a fact'. The problem occurs when we make a subconscious switch to 'nine' as an abstract noun in the sort of problems we deal with in maths class. Hersch thinks that this is not really the same nine. They are connected, but the number nine is an abstract object as part of a number system. It is a human creation.

Politics and maths learning

Hersch sees both a political and a pedagogic dimension to his thinking about mathematics. He thinks that a humanistic vision of mathematics chimes in with more progressive politics. How can politics enter mathematics? As soon as we think of mathematics as a social construction then the exact arrangements by which this construction comes about – the institutions that build and maintain it – become important. These arrangements are political. Particularly interesting for us here is how a different view of maths can bring about changes in maths teaching and learning. Let us return to Hersch:

'Let me state three possible philosophical attitudes towards mathematics. Platonism says mathematics is about some abstract entities, which are independent of humanity. Formalism says mathematics is nothing but calculations. There's no meaning to it at all. You just come out with the right answer by following the rules. Humanism sees mathematics as part of human culture and human history. It's hard to come to rigorous conclusions about this kind of thing, but I feel it's almost obvious that Platonism and Formalism are anti-educational, and interfere with understanding, and Humanism at least doesn't hurt and could be beneficial. Formalism is connected with rote, the traditional method, which is still common in many parts of the world. Here's an algorithm; practise it for a while; now here's another one. That's certainly what makes a lot of people hate mathematics. (I don't mean that mathematicians who are formalists advocate teaching by rote. But the formalist conception of mathematics fits naturally with the rote method of instruction.) There are various kinds of Platonists. Some are good teachers, some are bad. But the Platonist idea, that, as my friend Phil Davis puts it, Pi is in the sky, helps to make mathematics intimidating and remote. It can be an excuse for a pupil's failure to learn, or for a teacher's saying, "Some people just don't get it." The humanistic philosophy brings mathematics down to earth, makes it accessible psychologically, and increases the likelihood that someone can learn it, because it's just one of the things that people do.'

Do you agree with Reuben Hersch's humanist picture of mathematics – that mathematics is a social construction? Do you think he is right in his association of formalism with rote learning of maths and Platonism with the idea of maths being something remote that some people simply 'do not get'?

Are you really only intelligent if you can do maths?

There is a possibility that the arguments explored in this section might cast light on an aspect of mathematics learning which has seemed puzzling – why it is that mathematical ability is seen to be closely correlated with a certain type of intelligence. Mathematics has, moreover, seemed to polarize society into two distinct groups: those that can do it and those that cannot. Those that cannot do it often feel the stigma of failure. Is Hersch right in attributing this to a formalistic or platonic view? Is he right to suggest that if maths is just a meaningless set of formal exercises, then it will not be valued in the main by society? If maths is out there to be discovered, it does seem reasonable to imagine that a particular individual who does not make the discovery might experience a sense of failure. The interesting question in this case is: What practical consequences in the classroom would follow from a humanist view of mathematics?

The golden ratio

There are some intriguing links between mathematics and the arts. One link that seems to fascinate many students of mathematics is the ancient idea of the **golden ratio**. Consider a line segment *AB*. The Greek mathematicians were interested in dividing *AB* by placing a point *X* in such a way that the ratio of the smaller piece to the longer piece was equal to the ratio of the longer piece to the whole line.

In other words: $XB/AX = AX/AB$

A ——————— X ———— B

Let us rescale our units so that $AB = 1$ unit. Let $AX = x$. Then $XB = 1 - x$.

The equation above gives us: $\frac{1-x}{x} = \frac{x}{1}$

Rearranging gives us: $1 - x = x^2$

This gives the quadratic equation: $x^2 + x - 1 = 0$

Solving this equation using the quadratic formula gives:

$x = \frac{-1+\sqrt{5}}{2}$ and $x = \frac{-1-\sqrt{5}}{2}$ or $x = 0.618\,033\,988\,75\ldots$ or $-1.618\,033\,988\,75\ldots$

The first of these solutions is known as the **golden ratio**. Because of the special symmetry of the relationship between the different parts of the line segment above, this ratio was thought to be special or perfect in some way. Rectangles in which the ratio of the shorter to the longer side is equal to the golden ratio were thought to be especially beautiful. Try this out yourself in the rectangle beauty contest. Choose the rectangle that is most pleasing to you. Measure the sides and calculate the ratio between the shorter and the longer side. How close are you to the golden ratio?

A4 paper has dimensions of 210 mm × 297 mm. 210/297 = 0.707, which is a little high. A4 paper is a little too 'fat' to be a golden rectangle.

Measure some rectangles in your school or home environment – for example, credit cards, postcards, books, tables. How close are they to golden rectangles?

Rectangle beauty contest

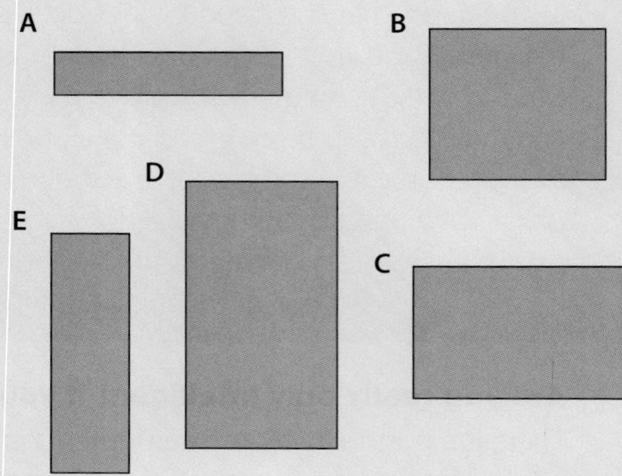

Which rectangle do you find the most pleasing?

A2, A3 and A5 paper are also all a little too fat to be golden rectangles. Why is this?

The golden ratio and the arts

There are many studies of the occurrence of the golden ratio in the natural and human worlds. It occurs in nature in connection with spirals and the Fibonacci sequence. The golden ratio has also been exploited by human beings in art, architecture and music. For example, the golden ratio was exploited by the ancient Greeks in their designs for temples and other buildings. The Parthenon in Athens is constructed using the golden section at key points.

The Greek letter ϕ is often used for the golden section.

Golden ratios have been consciously used in the structure of some musical compositions. The French composer Debussy is known to have used this ratio in his orchestral piece *La Mer,* for example. The 55 bar introduction to 'Dialogue du vent et de la mer' breaks down into five sections of 21, 8, 8, 5 and 13 bars in length, which are numbers in the Fibonacci sequence. The golden ratio point of bar 34 in this passage is signalled by the entry of the trombones and percussion. More generally, we can ask ourselves how many pieces of music (or films or plays or dance performances) have some sort of structurally significant event roughly two-thirds of the way through the piece?

Think of a film you have seen recently. At what point in the film did the moment of highest tension occur? How far into the film did this happen? Calculate this as a proportion. Is it close to 62%?

The Fibonacci sequence

The golden ratio is linked closely to the Fibonacci sequence:

1, 1, 2, 3, 5, 8, 13, 21, 34, 55, 89, …

What are the next two terms in the sequence?

If we divide successive terms in the sequence:

$\frac{1}{1} = 1,\ \frac{1}{2} = 0.5,\ \frac{2}{3} = 0.6667,\ \frac{3}{5} = 0.6,\ \frac{5}{8} = 0.625,$
$\frac{8}{13} = 0.6154,\ \frac{13}{21} = 0.6190,\ \frac{21}{34} = 0.6176, \ldots$

What is going on here?

Much has been written about how this sequence occurs in nature. It is naturally associated with certain types of growth. Ian Stewart, in his book *Nature's Numbers,* describes how these numbers are naturally associated with the spiral growth of many types of shell, for example. There is nothing mystical about this link. But it is tempting to think again about the Platonists and their view of mathematics as somehow embedded in the outside world.

The golden ratio suggests a strong link between mathematics and the arts. In theory of knowledge, it also raises a set of interesting questions about the nature of beauty. If we find certain rectangles pleasing because of the golden ratio, we might also find certain faces beautiful because of the ratios between the features, and find certain paintings or pieces of music beautiful, because of their proportions. Beauty would not be entirely in the eye of the beholder – it would be in the mathematics.

Answers

Chapter 1

Exercise 1.1

1 $\mathbb{Z} \subset \mathbb{Q}$ **2** $\mathbb{N} \subset \mathbb{Q}$ **3** $\mathbb{R} \subset \mathbb{C}$

4 $\mathbb{N} \subset \mathbb{Z}$ **5** $\mathbb{Z}^+ \subset \mathbb{Z}$ **6** $\mathbb{N} \subset \mathbb{R}$

7 $\frac{71}{33}$ **8** $\frac{1787}{150}$ **9** $\frac{61}{7}$

10 $\{1, 3, 5, 7\}$ **11** $\{1, 2, 3, 4, 5, 6, 7, 8, 9\}$

12 $\varnothing$ (empty set) **13** $\{1, 2, 3, 4, 5, 6, 7, 8\}$

14 $\{2, 4, 6\}$ **15** $\{1, 2, 3, 4, 5, 6, 7, 8, 9\}$

16 a) A is set of all even numbers between 2 and 10 inclusive; B is set of all odd numbers between 1 and 9 inclusive; C is set of all multiples of 3 between 3 and 9 inclusive

b) (i) $\varnothing$ (ii) U (iii) B
(iv) A (v) $\{6\}$ (vi) $\{3, 9\}$
(vii) $\{9\}$ (viii) $\{2, 4, 5, 8, 10\}$ (ix) $\{2, 4, 8, 10\}$

c)

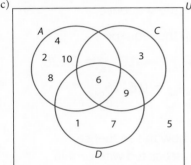

17 a) $\{x \mid 0 < x \leqslant 1\}$ b) $\{x \mid -2 \leqslant x < 3\}$
c) $\{x \mid 1 < x \leqslant 4\}$ d) $\{x \mid -2 \leqslant x \leqslant 0,\ 3 \leqslant x \leqslant 4\}$
e) $\{x \mid 3 \leqslant x \leqslant 4\}$ f) $\{x \mid -2 \leqslant x \leqslant 0,\ 1 < x \leqslant 4\}$

18 $x > -10$ **19** $x \leqslant -3$ **20** $x > \frac{10}{3}$

21 $x > -\frac{3}{8}$ **22** $\frac{3}{2} \leqslant x < \frac{7}{4}$ **23** $-3 \leqslant x < 1$

24 False, $x = -1$ **25** True **26** False, $x = 0$

27 False, $x = \frac{1}{2}$ **28** True **29** True

30 False, $x = -1$ **31** False, $x = \frac{1}{2}$ **32** 14.5

33 9 **34** 8.2 **35** $\pi - 3$

36 $\frac{11\pi}{3}$ **37** $\frac{832}{77}$

38 $-5 \leqslant x \leqslant 3$, closed, bounded

39 $-10 < x \leqslant -2$, half-open, bounded

40 $x \geqslant 1$, half-open, unbounded

41 $x < 4$, open, unbounded

42 $0 \leqslant x < 2\pi$, half-open, bounded

43 $a \leqslant x \leqslant b$, closed, bounded

44 $]-3, \infty[$ **45** $]-4, 6[$ **46** $]-\infty, 10]$

47 $[0, 12[$ **48** $]-\infty, \pi[$ **49** $[-3, 3]$

50 $x \geqslant 6$ $[6, \infty[$ **51** $4 \leqslant x < 10$ $[4, 10[$

52 $x < 0$ $]-\infty, 0[$ **53** $0 < x < 25$ $]0, 25[$

54 $|x| < 6$ **55** $|x| \geqslant 4$ **56** $|x| \leqslant \pi$ **57** $|x| > 1$

58 13 **59** 4 **60** -25 **61** -5

62 $3 - \sqrt{3}$ **63** -1

64 $x = -5$ or $x = 5$ **65** $x = -1$ or $x = 7$

66 $x = -4$ or $x = 16$ **67** No solution; false for all x

68 $x = -2$ or $x = -\frac{4}{3}$ **69** $x = \frac{32}{3}$ or $x = -\frac{28}{3}$

70 $x = -\frac{42}{5}$ or $x = \frac{72}{5}$ **71** $x = 0$ or $x = -4$

72 a) $x = 2, y = -2$ b) $x = 2, y = -2$

Exercise 1.2

1 h^2 **2** 3 **3** $6\sqrt{5}$ **4** $\frac{2\sqrt{7}}{7}$

5 4 **6** $\frac{\sqrt{3}}{2}$ **7** $3\sqrt{5} + 20$ **8** -2

9 $7\sqrt{2}$ **10** $40\sqrt{10}$ **11** $2\sqrt[3]{6}$ **12** $2|xy|\sqrt{3xy}$

13 m **14** $\frac{3\sqrt{2}}{2}$ **15** $x^8|1+x|$ **16** $3\sqrt{7}$

17 $6\sqrt{2} + 4\sqrt{3}$ **18** $17\sqrt{5}$ **19** $\frac{\sqrt{5}}{5}$ **20** $\frac{\sqrt{2}}{5}$

21 $2\sqrt{21}$ **22** $\frac{\sqrt{2}}{2}$ **23** $\frac{\sqrt{5} - 1}{2}$ **24** $\frac{2\sqrt{5} - 3}{11}$

25 $3 + 2\sqrt{3}$ **26** $\frac{4\sqrt{5} - 4\sqrt{2}}{3}$ **27** $\sqrt{x} - \sqrt{y}$ **28** $5 + 3\sqrt{3}$

29 $\frac{\sqrt{1 - x^2}}{x}$ **30** $\sqrt{x + h} + \sqrt{x}$ **31** $\frac{1}{\sqrt{a} + 3}$ **32** $\frac{1}{\sqrt{x} + \sqrt{y}}$

33 $\frac{m - 7}{(7 - x)(\sqrt{m} + \sqrt{7})}$

Exercise 1.3

1 2 **2** 27 **3** 16 **4** 16

5 8 **6** 8 **7** $\frac{4}{9}$ **8** $\frac{3}{4}$

9 $\frac{125}{8}$ **10** $\frac{1}{9}$ **11** 1 **12** $\frac{16}{3}$

13 $\frac{-64}{27}$ **14** x^2y^6 **15** $-x^2y^6$ **16** $-8x^3y^9$

17 $\frac{32y^7}{x}$ **18** $\frac{1}{64m^6}$ **19** $\frac{p^2}{3k^3}$ **20** -8

21 25 **22** $x^{\frac{7}{6}}$ **23** $\frac{1}{4a^2}$ **24** x

25 $2(a - b)$ **26** $\frac{(x + 4y)^{\frac{3}{2}}}{2}$ **27** $\sqrt{p^2 + q^2}$ **28** 5^{3x-1}

29 $\frac{1}{x^{\frac{1}{6}}} + \frac{1}{x^{\frac{1}{4}}}$ **30** $\frac{26}{9}(3^n)$ **31** 16 **32** $2\sqrt{3}x^2y^4$

33 $\sqrt{1 + n^2}$ **34** $\sqrt{x}$

Exercise 1.4

1 2.54×10^2 **2** 7.81×10^{-3} **3** 7.41×10^6

4 1.04×10^{-6} **5** 4.98 **6** 1.99×10^{-3}

7 1.49×10^8 **8** 8.99×10^{-5} **9** 1.50×10^8

10 9.11×10^{-31} **11** 0.0027 **12** 50 000 000

13 0.000 000 090 35 **14** 4 180 000 000 000 **15** 2.5×10^3

16 2×10^4 **17** 8.2×10^{-5} **18** 5.6×10^{18}

19 1.8×10^5 **20** 5×10^1 **21** 8.2×10^{-5}
22 5.56×10^1

Exercise 1.5

1 x^2+x-20 **2** $6h^2-11h+3$ **3** y^2-81
4 $16x^2+16x+4$ **5** $4n^2-10n+25$
6 $4y^2-20y+25$ **7** $36a^2-49b^2$
8 $4x^2+12x+9-y^2$ **9** $a^3x^3+3a^2bx^2+3ab^2x+b^3$
10 $a^4x^4+4a^3bx^3+6a^2b^2x^2+4ab^3x+b^4$
11 $4-5x^2$ **12** $8x^3-1$
13 $x^2+y^2+z^2+2xy+2xz+2yz$ **14** x^2+y^2
15 $9-m^2$ **16** $x^2-2\sqrt{x^2+1}+2$
17 $12(x+2)(x-2)$ **18** $x^2(x-6)$
19 $(x+4)(x-3)$ **20** $-(m-1)(m+7)$
21 $(x-8)(x-2)$ **22** $(y+1)(y+6)$
23 $3(n-5)(n-2)$ **24** $2x(x+1)(x+9)$
25 $(a+4)(a-4)$ **26** $(3y+1)(y-5)$
27 $(5n^2+2)(5n^2-2)$ **28** $a(x+3)^2$
29 $(m+1)^2(2n-1)$ **30** $(x+1)(x-1)(x^2+1)$
31 $y(6-y)$ **32** $2y^2(2y^2-5y-48)$
33 $(2x-5)^2$ **34** $(2x+3)^{-3}(4x+3)=\dfrac{4x+3}{(2x+3)^3}$
35 $(n-2)^3(1-n)$ **36** $m\left(m-\dfrac{2}{3}\right)^2$ **37** $\dfrac{1}{x+1}$
38 $\dfrac{1}{2n}$ **39** $\dfrac{a+b}{5}$ **40** $x+2$
41 -1 **42** $4x$ **43** $\dfrac{3x+2}{x+1}$
44 $\dfrac{3y-1}{y+2}$ **45** -1 **46** $\dfrac{(2x-1)(x-1)}{x(x-2)}$
47 $\dfrac{1-n}{n}$ **48** $\dfrac{6-8x}{2x-1}$ **49** $\dfrac{-2x+5}{15}$
50 $\dfrac{b-a}{ab}$ **51** $\dfrac{10-3x}{(x-3)^2}$ **52** $\dfrac{x^2+x+3}{x^2+3x}$
53 $\dfrac{2x}{x^2-y^2}$ **54** $\dfrac{-2}{x-2}$ **55** 6
56 $\dfrac{2}{7x-21}$ **57** $\dfrac{1}{ab-b^2}$ **58** $-\dfrac{5}{2}x(x+1)$
59 $\dfrac{3y-10}{(y+2)(y-5)}$ **60** $\dfrac{(x-3)(x+2)}{23x^2}$ **61** $\dfrac{x+\sqrt{2}}{x^2-2}$
62 $\dfrac{10-5x\sqrt{3}}{4-3x^2}$ **63** $\dfrac{x+2\sqrt{xy}+y}{x-y}$ **64** $\dfrac{\sqrt{x+h}-\sqrt{x}}{h}$

Exercise 1.6

1 $x=h-\dfrac{n}{m}$ **2** $a=\dfrac{v^2+t}{b}$ **3** $b_1=\dfrac{2A}{h}-b_2$
4 $r=\sqrt{\dfrac{2A}{\theta}}$ **5** $k=\dfrac{gh}{f}$ **6** $t=\dfrac{x}{a+b}$
7 $r=\sqrt[3]{\dfrac{3V}{\pi h}}$ **8** $k=\dfrac{g}{F(m_1+m_2)}$ **9** $y=-\dfrac{2}{3}x-5$
10 $y=-4$ **11** $y=\dfrac{5}{4}x+6$ **12** $x=\dfrac{7}{3}$
13 $y=-4x+11$ **14** $y=-\dfrac{5}{2}x-7$
15 a) 17 b) $\left(0,\dfrac{5}{2}\right)$ **16** a) $\sqrt{40}$ b) $(2,3)$
17 a) $\dfrac{\sqrt{82}}{3}$ b) $\left(-1,\dfrac{7}{6}\right)$ **18** a) $\sqrt{533}$ b) $\left(1,\dfrac{11}{2}\right)$
19 $k=1$ or 9 **20** $k=-11$ or -3
21 $(\sqrt{5})^2+(\sqrt{45})^2=(\sqrt{50})^2$ **22** Sides are: $\sqrt{29}$, $\sqrt{29}$, $\sqrt{58}$
23 Sides are: $\sqrt{45}$, $\sqrt{10}$, $\sqrt{45}$, $\sqrt{10}$ **24** $(5,1)$
25 $\left(4,\dfrac{1}{2}\right)$ **26** $(3,-4)$ **27** $(3.8,-1.6)$
28 No solution **29** $(-1,2)$ **30** $(-1,3)$
31 $(-3,-8)$
32 Lines are coincident; solution set is all points on the line $y=-\dfrac{1}{4}x-\dfrac{3}{4}$
33 $\left(\dfrac{20}{3},\dfrac{40}{3}\right)$ **34** $\left(\dfrac{1}{2},3\right)$ **35** $(-5,10)$
36 $(5,-3)$ **37** $(14.1, 10.4)$ **38** $\left(\dfrac{11}{19},-\dfrac{18}{19}\right)$

Chapter 2

Exercise 2.1

1 G **2** L **3** H **4** K
5 J **6** C **7** A **8** I
9 F **10** $A=\dfrac{C^2}{4\pi}$ **11** $A=\dfrac{l^2\sqrt{3}}{4}$ **12** $A=4x^2+60x$
13 $h=x\sqrt{2}$
14 a) 9.4 b) $V=\dfrac{3525}{P}$
15 a) $F=kx$ b) 6.25 c) 37.5 N
16 $\{-6.2,-1.5,0.7,3.2,3.8\}$ **17** $r>0$
18 $\mathbb{R}$ **19** $\mathbb{R}$ **20** $t\leqslant3$ **21** $\mathbb{R}$
22 $x\neq\pm3$ **23** $-1\leqslant x\leqslant1$ and $x\neq0$
24 No, $x=c$ is a vertical line
25 a) (i) $\sqrt{17}$ (ii) 7 (iii) 0
b) $x<4$ c) Domain: $x\geqslant4$, range: $h(x)\geqslant0$
26 a) Domain $\{x: x\in\mathbb{R},\ x\neq5\}$, range $\{y: y\in\mathbb{R},\ y\neq0\}$
b) y-intercept $\left(0,-\dfrac{1}{5}\right)$, vertical asymptote $x=5$, horizontal asymptote $y=0$

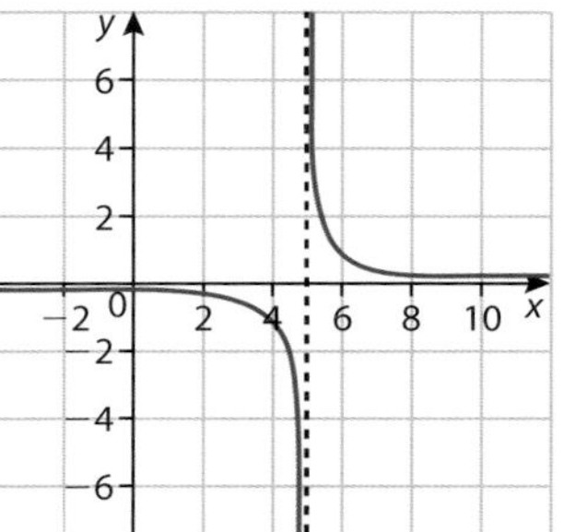

27 a) Domain $\{x: x<-3,\ x>3\}$, range $\{y: y>0\}$
b) Vertical asymptotes $x=-3$ and $x=3$

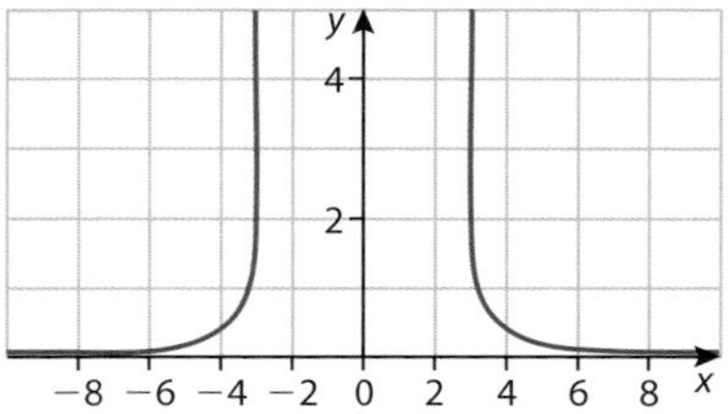

28 a) Domain $\{x: x\in\mathbb{R},\ x\neq-2\}$, range $\{y: y\in\mathbb{R},\ y\neq2\}$

b) y-intercept $\left(0, -\frac{1}{2}\right)$, vertical asymptote $x = -2$, horizontal asymptote $y = 2$

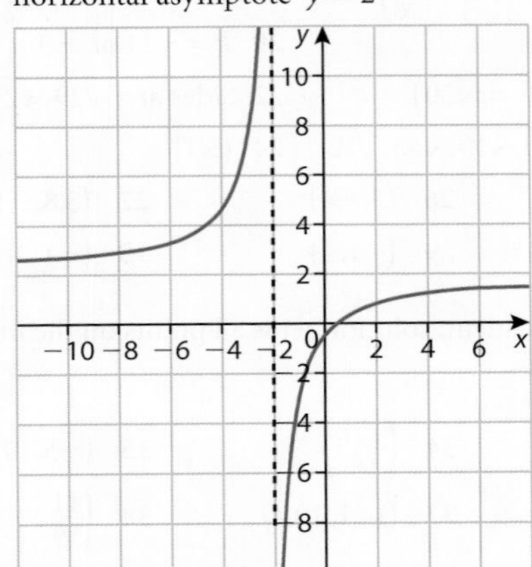

29 a) Domain $\left\{x: -\frac{\sqrt{10}}{2} \leqslant x \leqslant \frac{\sqrt{10}}{2}\right\}$, range $\{y: 0 \leqslant y \leqslant \sqrt{5}\}$

b) y-intercept $(0, \sqrt{5})$, x-intercepts $\left(-\frac{\sqrt{10}}{2}, 0\right)$ and $\left(\frac{\sqrt{10}}{2}, 0\right)$

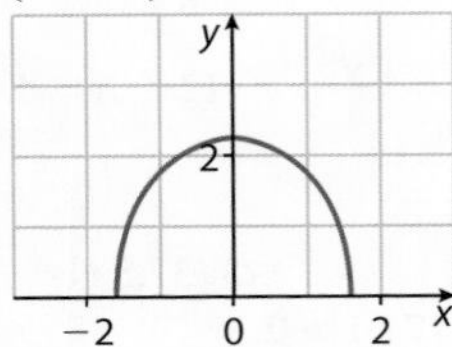

30 a) Domain $\{x: x \in \mathbb{R},\ x \neq 0\}$, range $\{y: y \in \mathbb{R},\ y \neq -4\}$

b) Vertical asymptote $x = 0$, horizontal asymptote $y = -4$

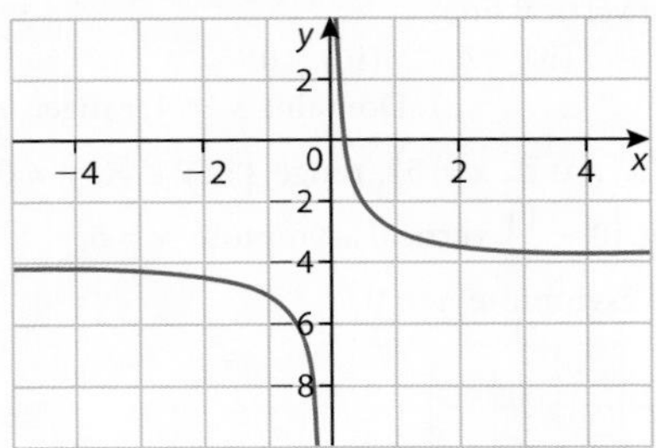

Exercise 2.2

1 a) $(f \circ g)(5) = 1$, $(g \circ f)(5) = \frac{1}{7}$

b) $(f \circ g)(x) = \frac{2}{x-3}$, $(g \circ f)(x) = \frac{1}{2x-3}$

2 a) 1 b) -7 c) 7 d) -47 e) -1 f) -79 g) $1 - 2x^2$ h) $-4x^2 + 12x - 7$ i) $4x - 9$ j) $-x^4 + 4x^2 - 2$

3 $(f \circ g)(x) = 12x + 7$, domain: $x \in \mathbb{R}$; $(g \circ f)(x) = 12x - 1$, domain: $x \in \mathbb{R}$

4 $(f \circ g)(x) = 4x^2 + 1$, domain: $x \in \mathbb{R}$; $(g \circ f)(x) = -2x^2 - 2$, domain: $x \in \mathbb{R}$

5 $(f \circ g)(x) = \sqrt{x^2 + 2}$, domain: $x \in \mathbb{R}$; $(g \circ f)(x) = x + 2$, domain: $x \geqslant -1$

6 $(f \circ g)(x) = \frac{2}{x+3}$, domain: $x \in \mathbb{R}$, $x \neq -3$; $(g \circ f)(x) = -\frac{x+2}{x+4}$, domain: $x \in \mathbb{R}$, $x \neq -4$

7 $(f \circ g)(x) = x$, domain: $x \in \mathbb{R}$; $(g \circ f)(x) = x$, domain: $x \in \mathbb{R}$

8 $(f \circ g)(x) = x^4$, domain: $x \in \mathbb{R}$; $(g \circ f)(x) = -x^4 + 4x^3 - 6x^2 + 4x$, domain: $x \in \mathbb{R}$

9 $(f \circ g)(x) = \frac{2}{4x^2 - 1}$, domain: $x \neq 0, x \neq \pm\frac{1}{2}$; $(g \circ f)(x) = \frac{(4-x)^2}{4x^2}$, domain: $x \neq 0, x \neq 4$

10 $(f \circ g)(x) = 1 + x^2$, domain: $-1 \leqslant x \leqslant 1$; $(g \circ f)(x) = \sqrt[3]{-x^6 + 4x^3 - 3}$, domain: $x \in \mathbb{R}$

11 $(f \circ g)(x) = x$, domain: $x \neq -3$; $(g \circ f)(x) = x$, domain: $x \neq -3$

12 $(f \circ g)(x) = \frac{x^2 - 1}{x^2 - 2}$, domain: $x \neq \pm\sqrt{2}$; $(g \circ f)(x) = \frac{2x - 1}{(x-1)^2}$, domain: $x \neq 1$

13 a) $(g \circ h)(x) = \sqrt{9 - x^2}$, domain: $-3 \leqslant x \leqslant 3$, range: $y \geqslant 0$

b) $(h \circ g)(x) = -x + 11$, domain: $x \geqslant 1$, range: $y \leqslant 10$

14 a) $(f \circ g)(x) = \frac{1}{10 - x^2}$, domain: $x \neq \pm\sqrt{10}$, range: $y \neq 0$

b) $(g \circ f)(x) = 10 - \frac{1}{x^2}$, domain: $x \neq 0$, range: $y < 10$

15 $h(x) = x + 3$, $g(x) = x^2$

16 $h(x) = x - 5$, $g(x) = \sqrt{x}$

17 $h(x) = \sqrt{x}$, $g(x) = 7 - x$

18 $h(x) = x + 3$, $g(x) = \frac{1}{x}$

19 $h(x) = x + 1$, $g(x) = 10^x$

20 $h(x) = x - 9$, $g(x) = \sqrt[3]{x}$

21 $h(x) = x^2 - 9$, $g(x) = |x|$

22 $h(x) = \sqrt{x - 5}$, $g(x) = \frac{1}{x}$

23 a) Domain of f: $x \geqslant 0$ b) Domain of g: $x \in \mathbb{R}$

c) $(f \circ g)(x) = \sqrt{x^2 + 1}$, domain $x \in \mathbb{R}$

24 a) Domain of f: $x \neq 0$ b) Domain of g: $x \in \mathbb{R}$

c) $(f \circ g)(x) = \frac{1}{x+3}$, domain $x \neq -3$

25 a) Domain of f: $x \neq \pm 1$ b) Domain of g: $x \in \mathbb{R}$

c) $(f \circ g)(x) = \frac{3}{x^2 + 2x}$, domain $x \neq 0, -3$

26 a) Domain of f: $x \in \mathbb{R}$ b) Domain of g: $x \in \mathbb{R}$

c) $(f \circ g)(x) = x + 3$, domain $x \in \mathbb{R}$

Exercise 2.3

1 a) 2 b) 6

2 a) -1 b) b

3 4

4 6

5

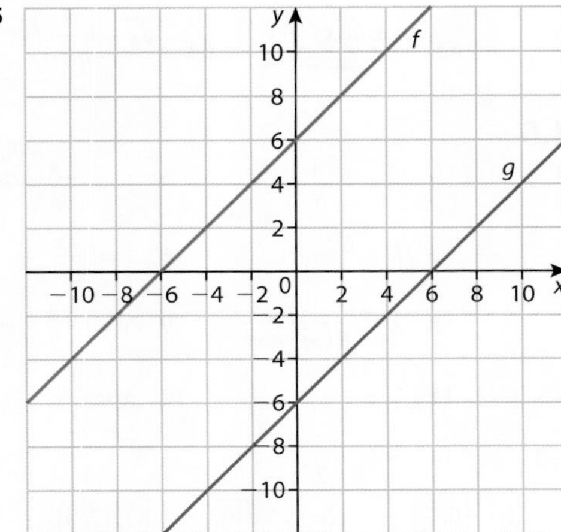

6

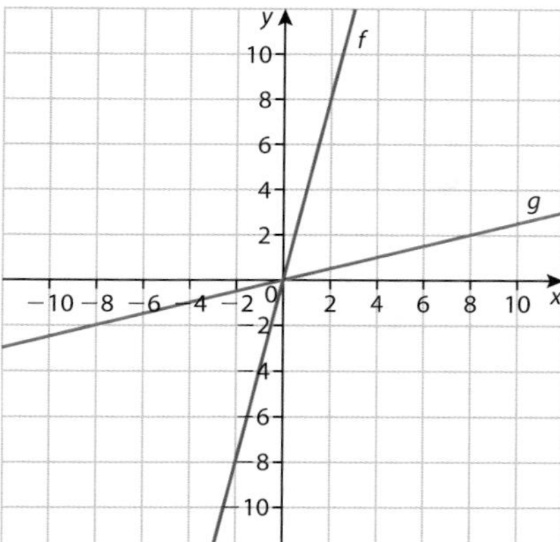

7

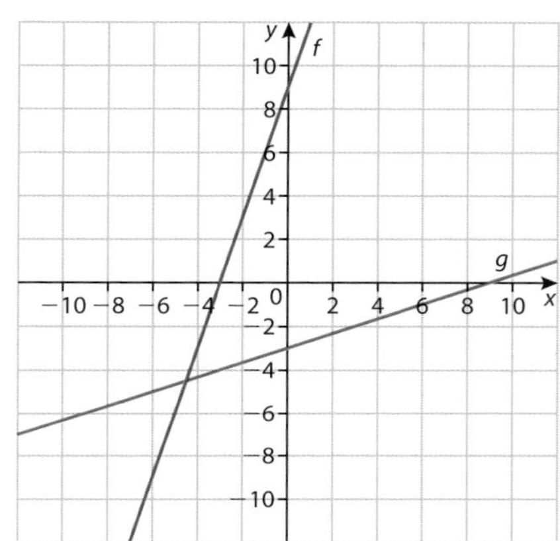

8

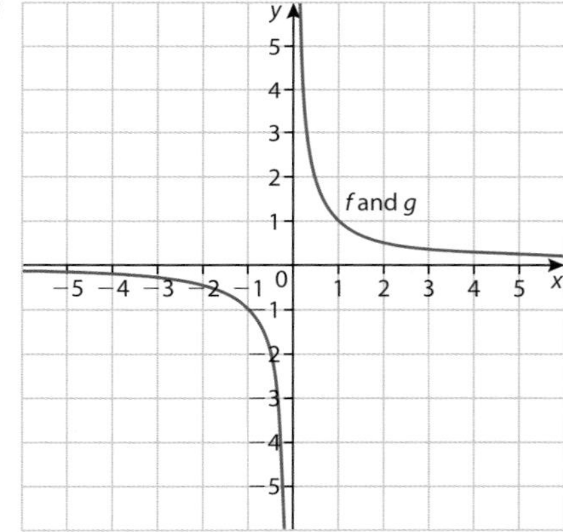

9

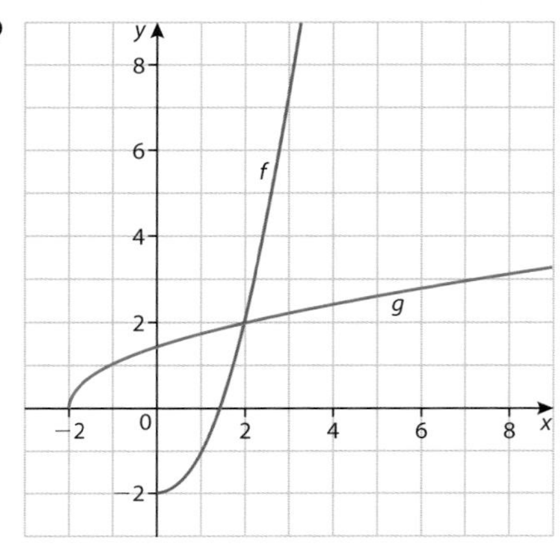

10

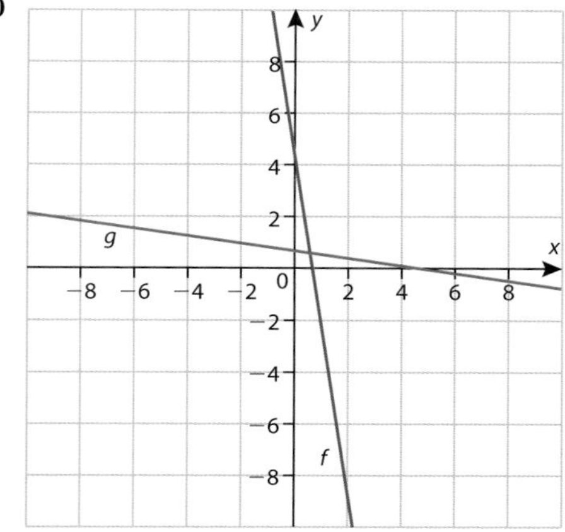

11

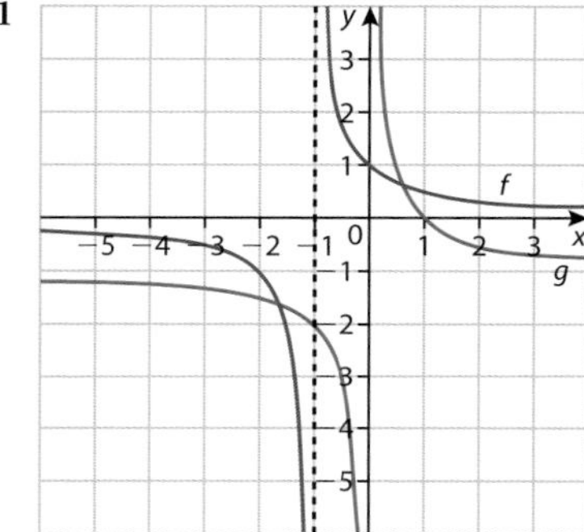

12

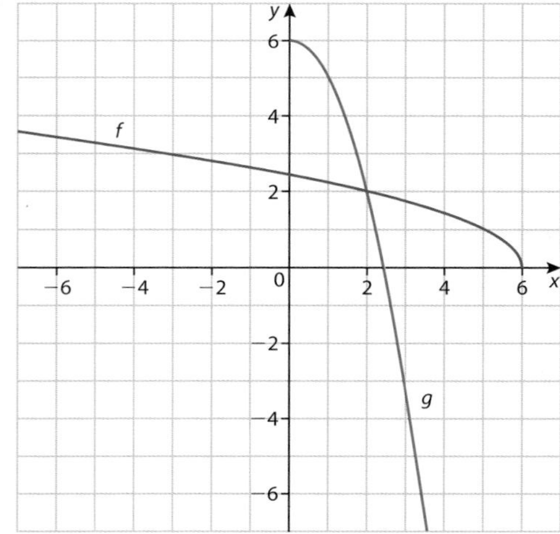

13

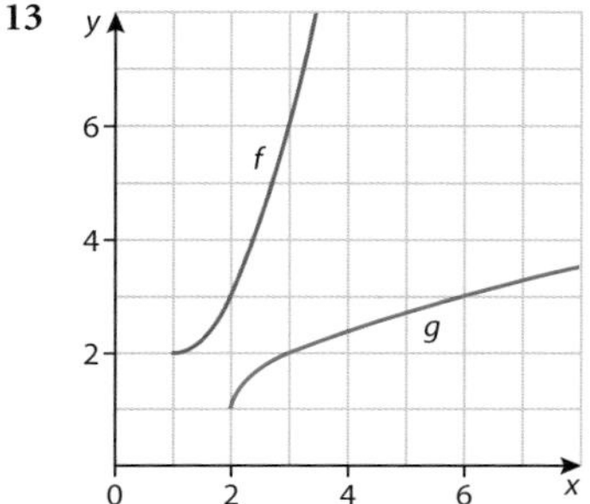

14

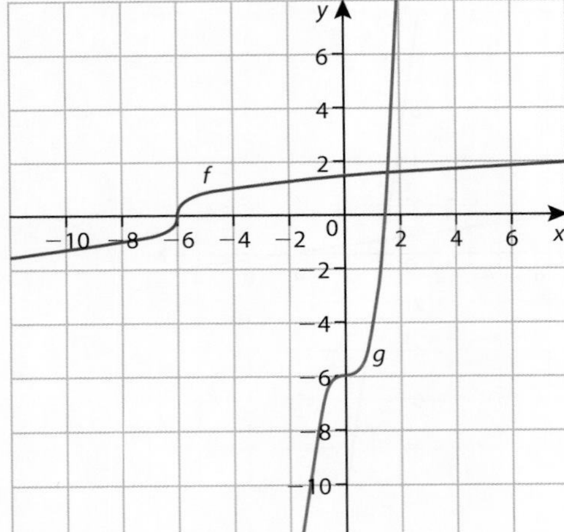

15 $f^{-1}(x) = \frac{1}{2}x + \frac{3}{2}, x \in \mathbb{R}$

16 $f^{-1}(x) = 4x - 7, x \in \mathbb{R}$

17 $f^{-1}(x) = x^2, x \geqslant 0$

18 $f^{-1}(x) = \frac{1}{x} - 2, x \in \mathbb{R}, x \neq 0$

19 $f^{-1}(x) = \sqrt{4 - x}, x \leqslant 4$

20 $f^{-1}(x) = x^2 + 5, x \geqslant 0$

21 $f^{-1}(x) = \frac{1}{a}x - \frac{b}{a}, x \in \mathbb{R}$

22 $f^{-1}(x) = 1 + \sqrt{x + 1}, x \geqslant -1$

23 $f^{-1}(x) = \sqrt{\frac{1 + x}{1 - x}}, -1 \leqslant x \leqslant 1$

24 $f^{-1}(x) = \sqrt[3]{x - 1}, x \in \mathbb{R}$

25 $f^{-1}(x) = \frac{x + 3}{x - 2}$

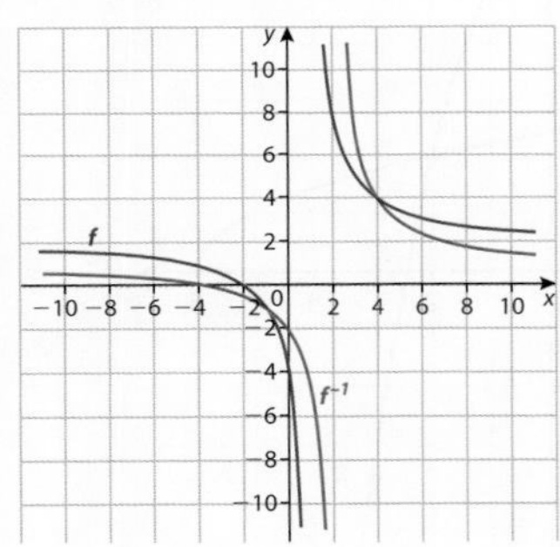

26 $x \geqslant 2,\ f^{-1}(x) = \sqrt{x} + 2$

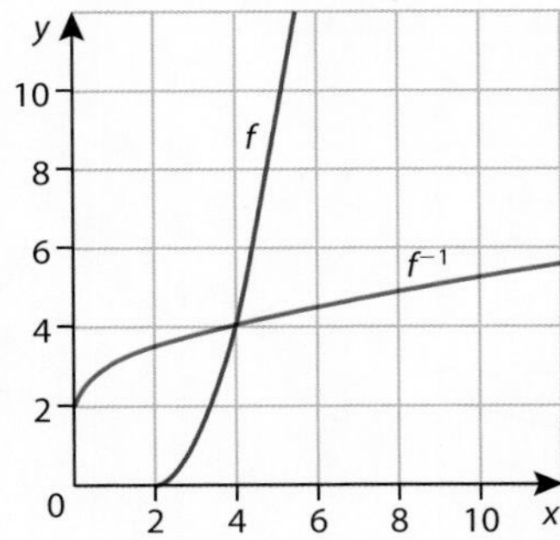

27 $x > 0,\ f^{-1}(x) = \sqrt{\frac{1}{x}}$

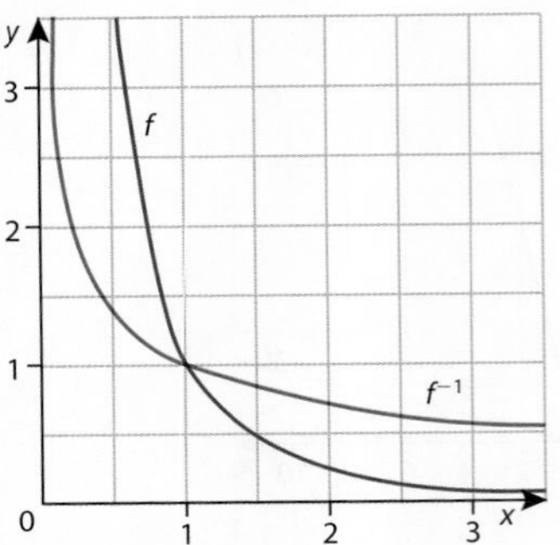

28 $x > 0,\ f^{-1}(x) = \sqrt[4]{2 - x}$

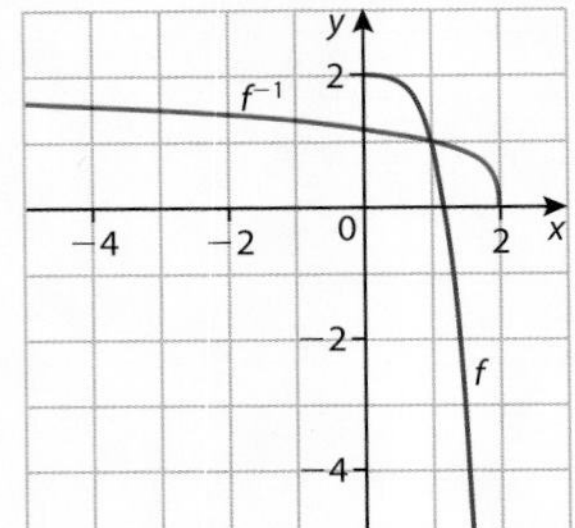

29 $x < -1,\ -1 \leqslant x \leqslant 1,\ x > 1$

30 $\frac{3}{2}$

31 5

32 -4

33 $\frac{7}{2}$

34 $g^{-1} \circ h^{-1} = \frac{1}{2}x - 3$

35 $h^{-1} \circ g^{-1} = \frac{1}{2}x - \frac{3}{2}$

36 $(g \circ h)^{-1} = \frac{1}{2}x + \frac{1}{2}$

37 $(h \circ g)^{-1} = 2x + 2$

38 $f(f(x)) = f\left(\frac{a}{x+b} - b\right) = \frac{a}{\frac{a}{x+b} - b + b} - b$

$$= \frac{a}{\frac{a}{x+b}} - b = \frac{a}{1} \cdot \frac{x+b}{a} - b = x + b - b = x$$

Since $f(f(x)) = x$, then the function f is its own inverse.

Exercise 2.4

1

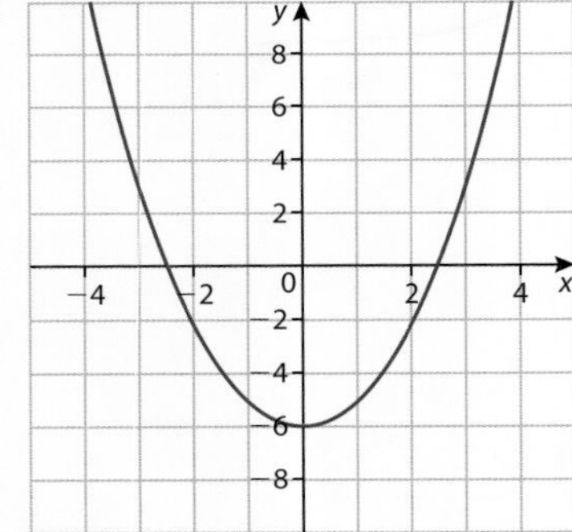

2

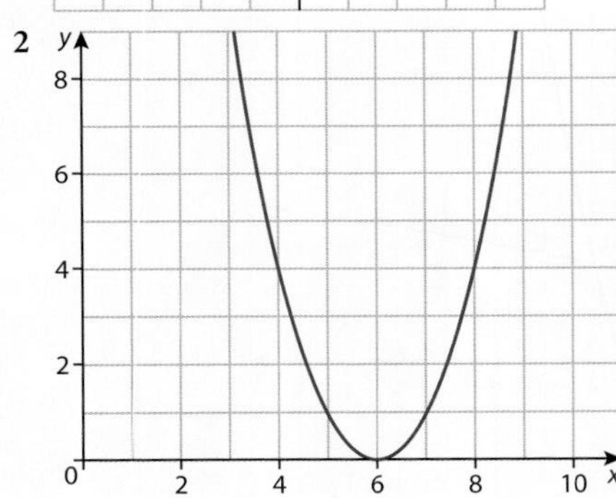

3

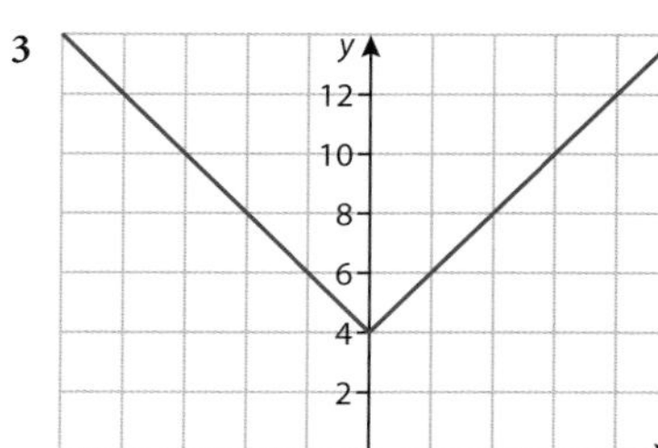

4

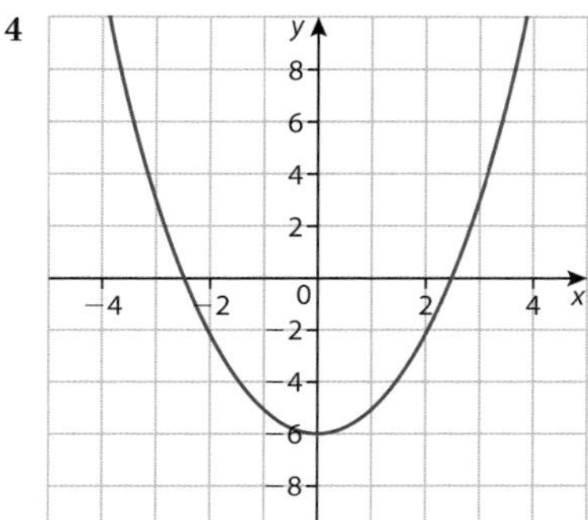

5

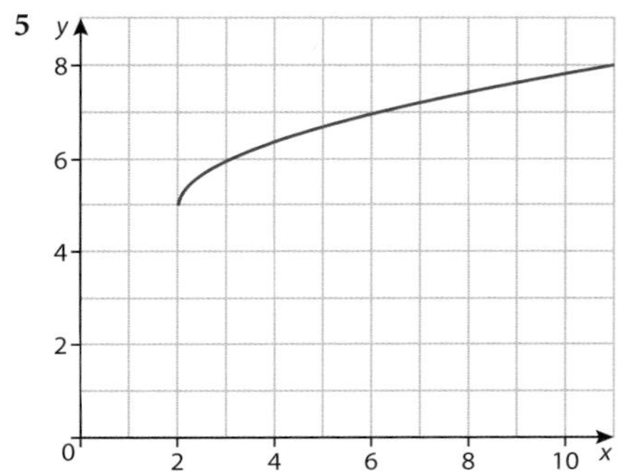

6

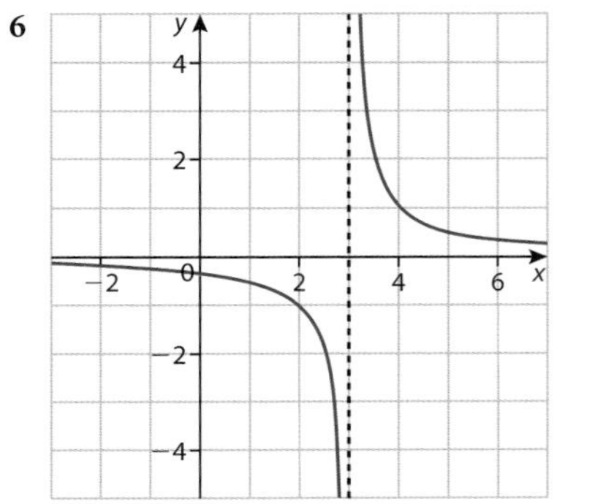

7

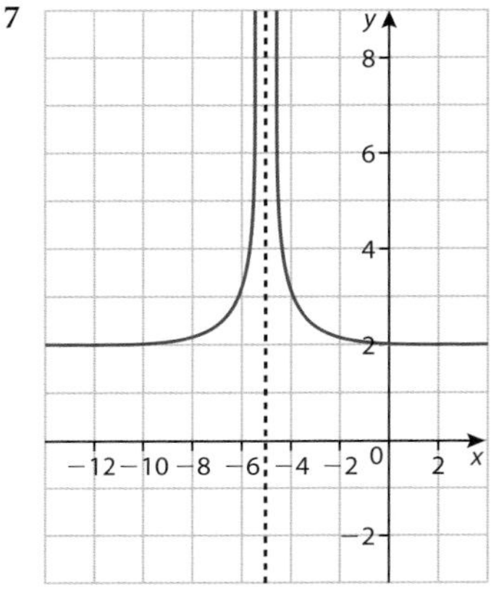

8

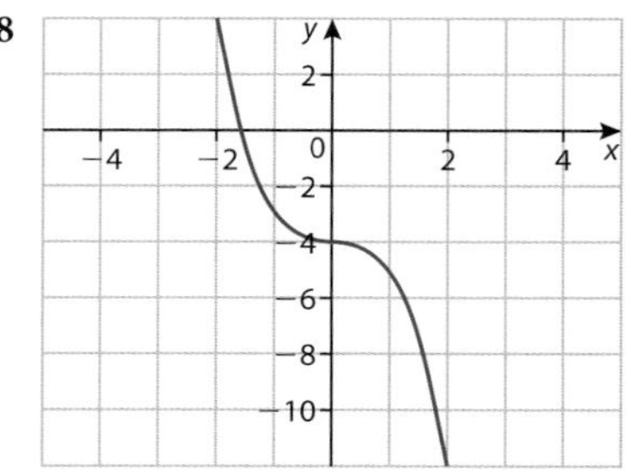

9

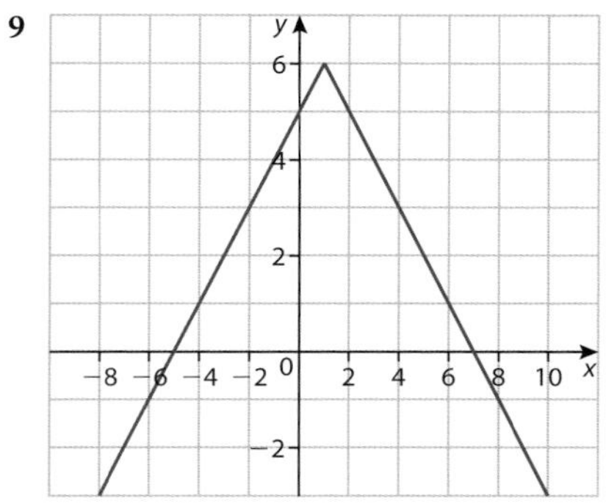

10

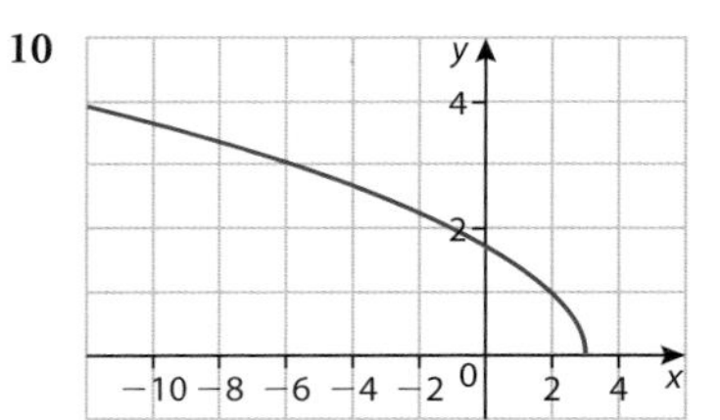

11

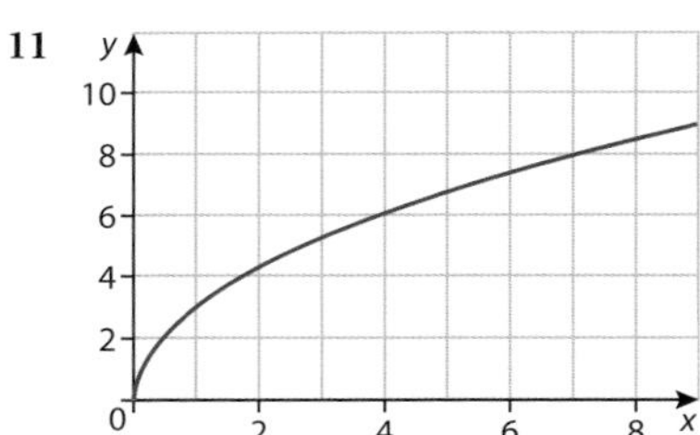

12

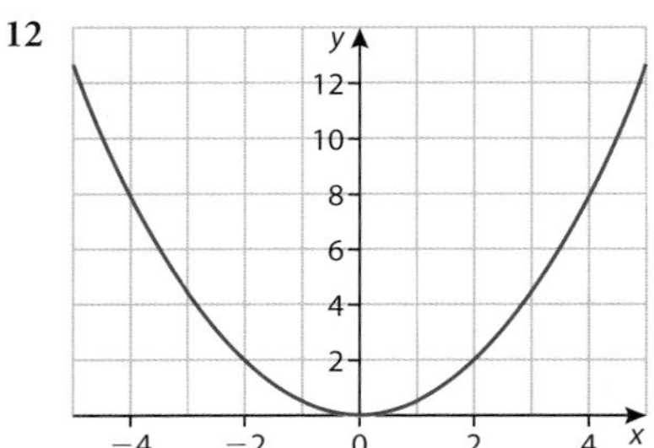

13

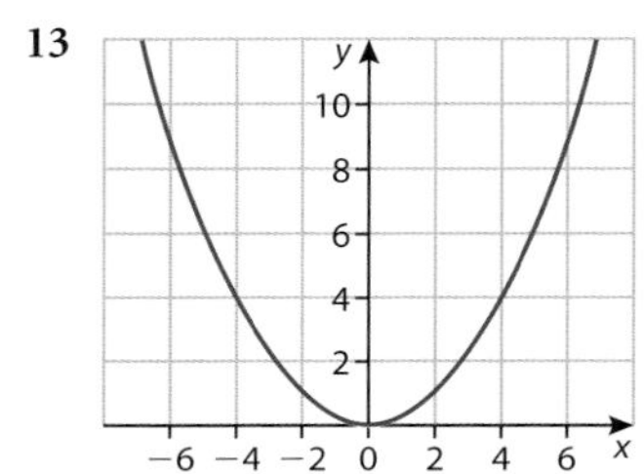

14

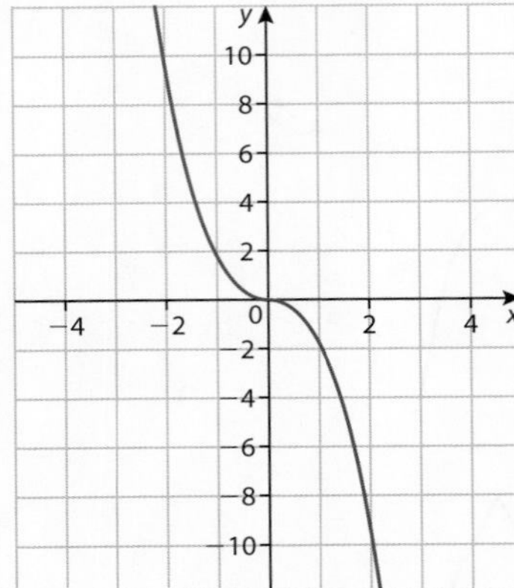

15 $y = -x^2 + 5$

16 $y = \sqrt{-x}$

17 $y = -|x + 1|$

18 $y = \dfrac{1}{x-2} - 2$

19 a)

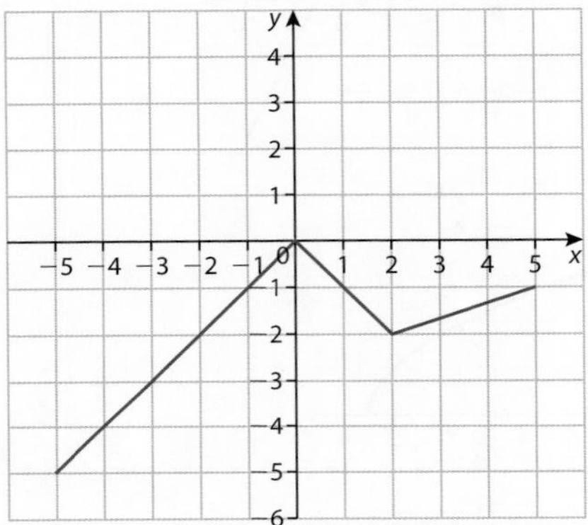

b)

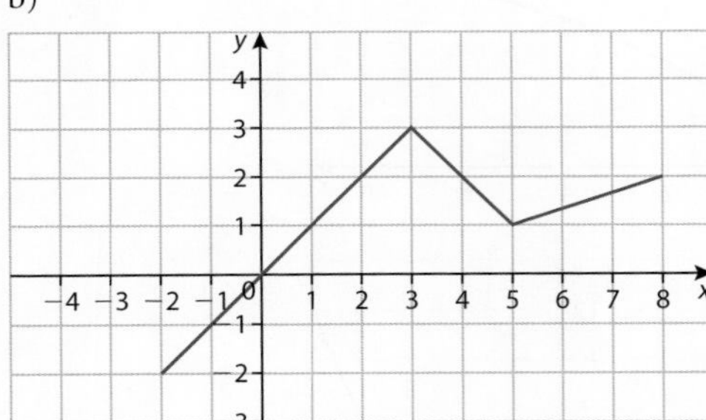

c)

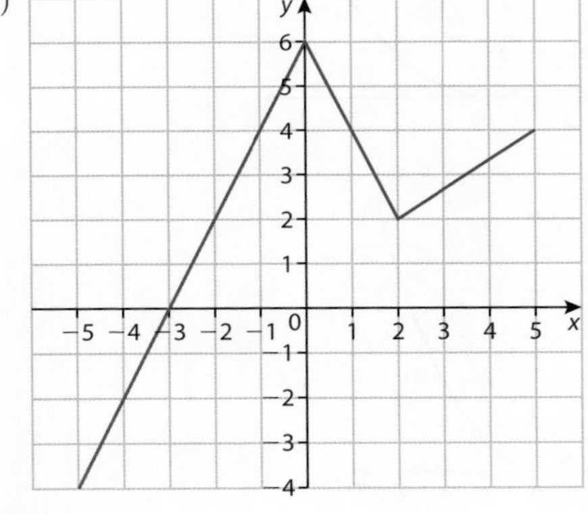

d)

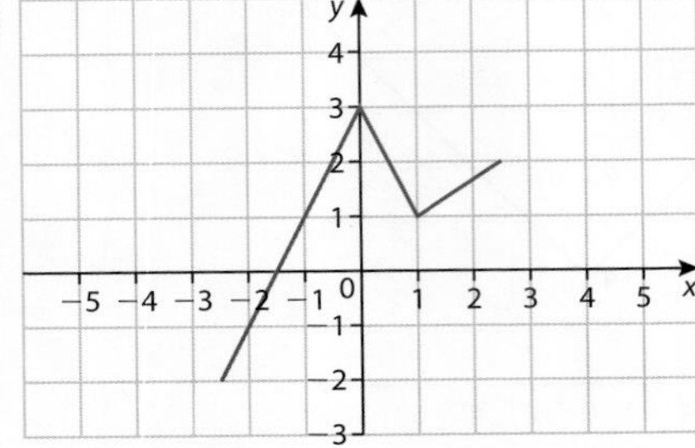

e)

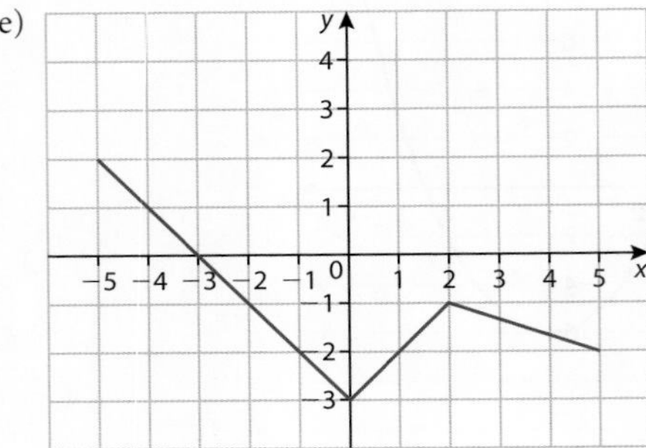

f)

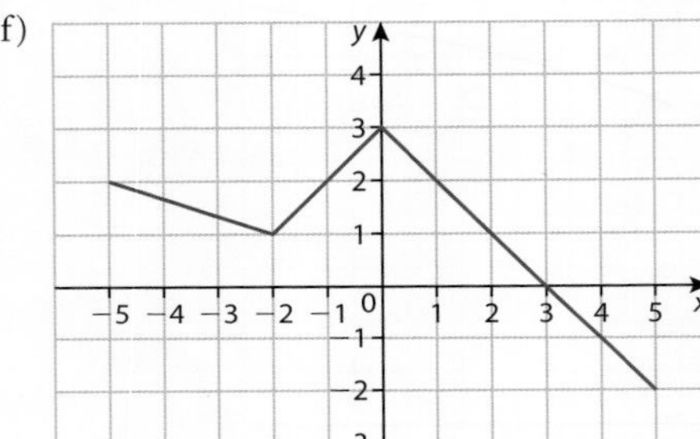

g)

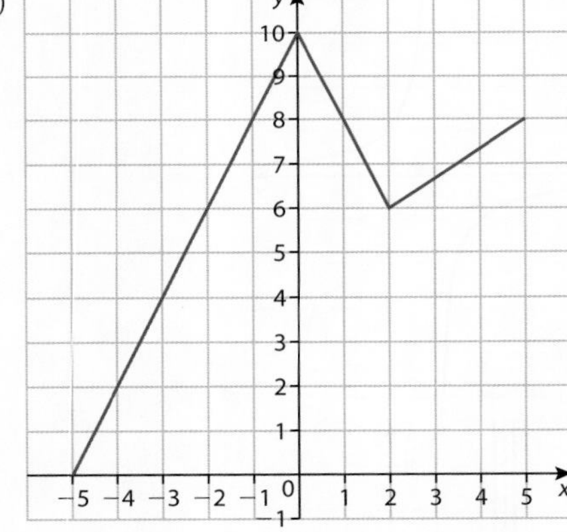

20 Horizontal translation 3 units right; vertical translation 5 units up (or reverse order).

21 Reflect over the x-axis; vertical translation 2 units up (or reverse order).

22 Horizontal translation 4 units left; vertical shrink by factor $\frac{1}{2}$ (or reverse order).

23 Horizontal translation 1 unit right; horizontal shrink by factor $\frac{1}{3}$; vertical translation 6 units down.

24 a)

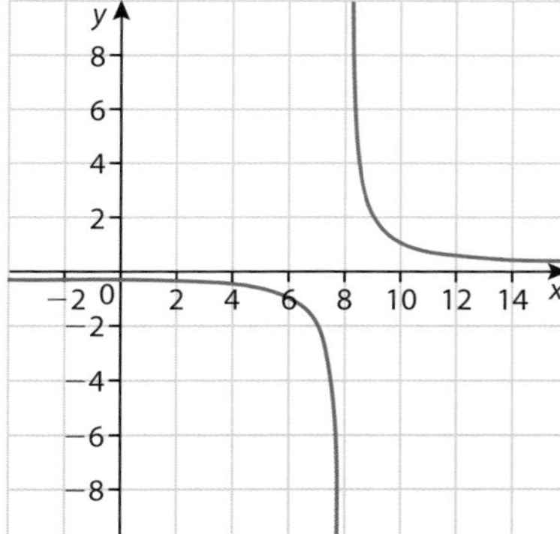

b)

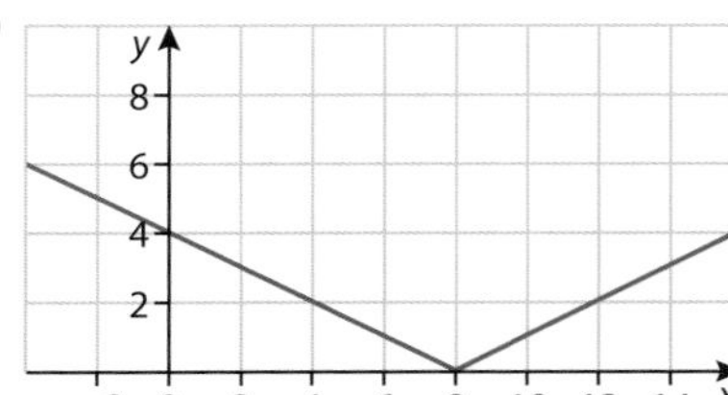

c)

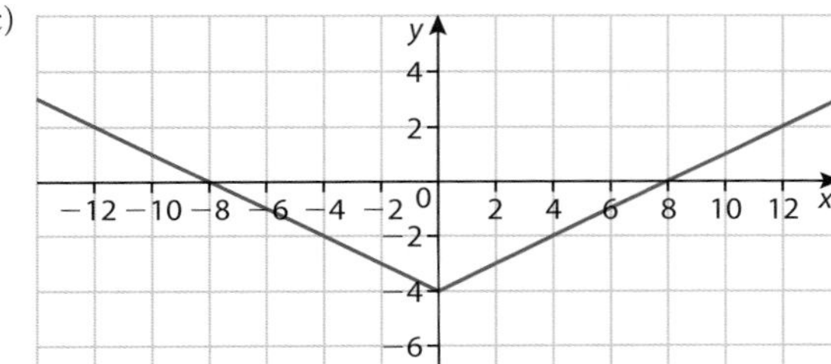

25 a)

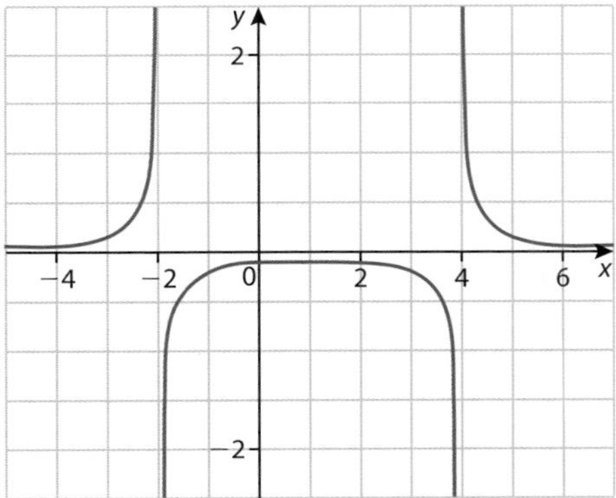

b)

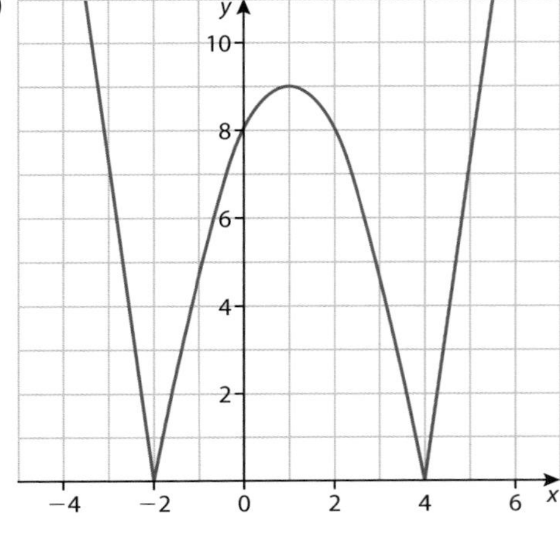

c)

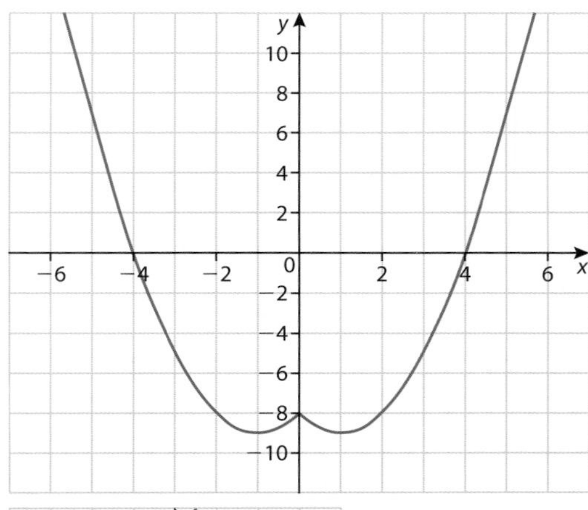

26 a)

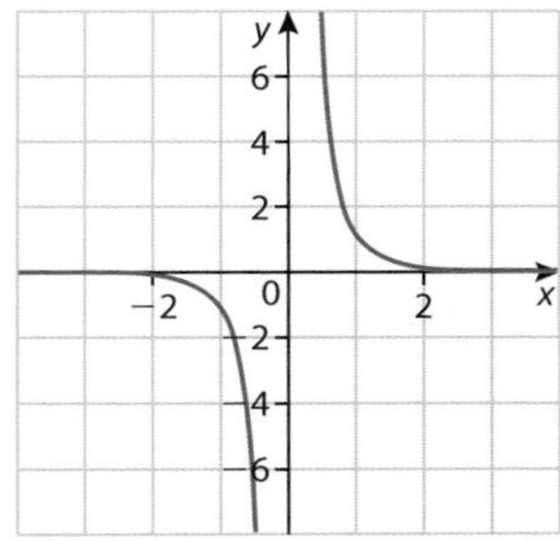

b)

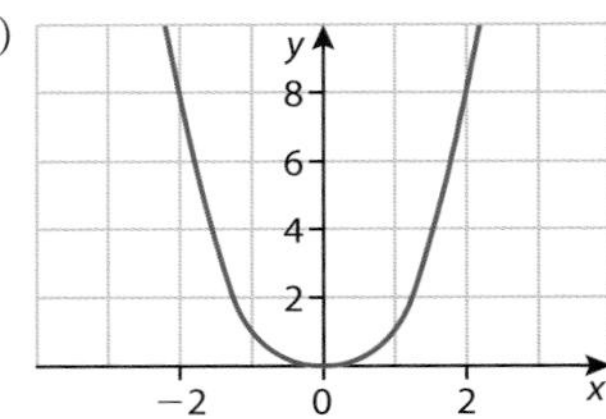

c)

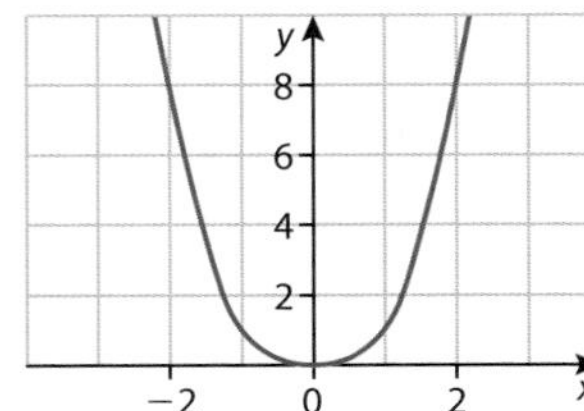

Practice questions

1 a) $a = -3,\ b = 1$ b) range: $y \geqslant 0$

2 a) 5 b) 3

3 a) $g^{-1}(x) = -3x + 4$ b) $x = \frac{2}{3}$

4 a) $(g \circ h)(x) = 2x - 3$ b) 24

5 a)

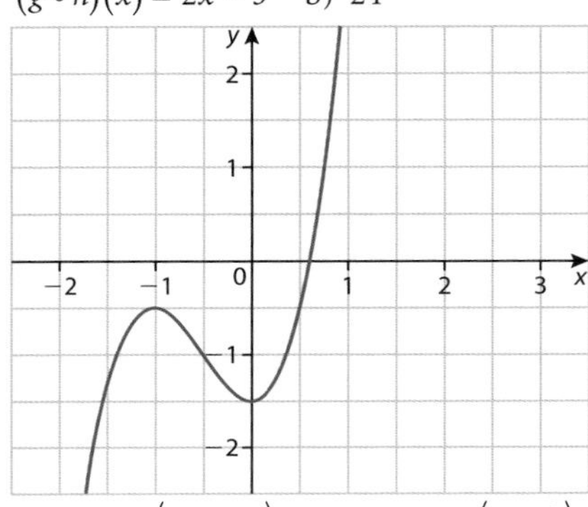

maximum at $\left(-1, -\frac{1}{2}\right)$; minimum at $\left(0, -\frac{3}{2}\right)$

6 a) $k = \frac{1}{2}$ b) $p = -5$ c) $q = 3$

7 a)

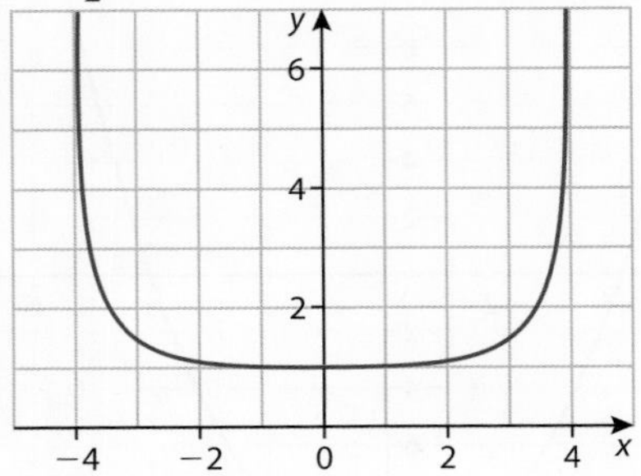

b) $x = 4, x = -4$ c) range: $y \geqslant 1$

8 a)

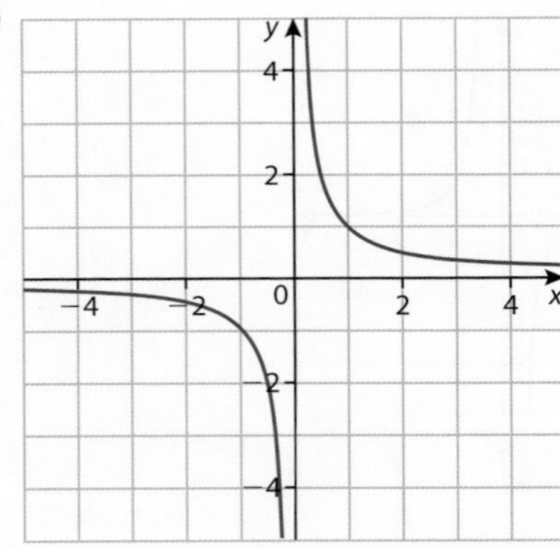

b) $h(x) = \frac{1}{x+4} - 2$

c) (i) x-intercept: $\left(-\frac{7}{2}, 0\right)$; y-intercept: $\left(0, -\frac{7}{4}\right)$

(ii) Vertical asymptote: $x = -4$; horizontal asymptote: $y = -2$

(iii)

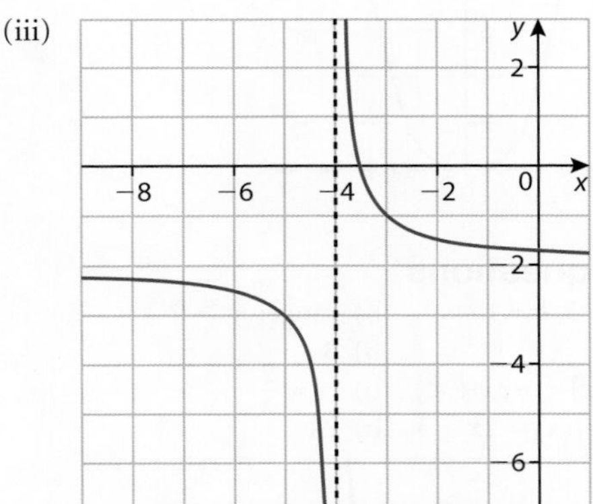

9 a) (i) $\sqrt{11}$ (ii) 7 (iii) 0
b) $x < -3$
c) $(g \circ f)(x) = x - 2$

10 a) 4 b) $(g^{-1} \circ h)(x) = 2x^2 + 6$ c) $x = \pm 2\sqrt{2}$

11 a) $f^{-1}(x) = \frac{1}{3}x + \frac{1}{3}$
b) $(f \circ g)(x) = \frac{12}{x} - 1$
c) $(f \circ g)^{-1}(x) = \frac{12}{x+1}$
d) $(g \circ g)(x) = x$

12 a) (i) $a = 8$ (ii) $b = -3$
b) Reflection over x-axis

13 a)

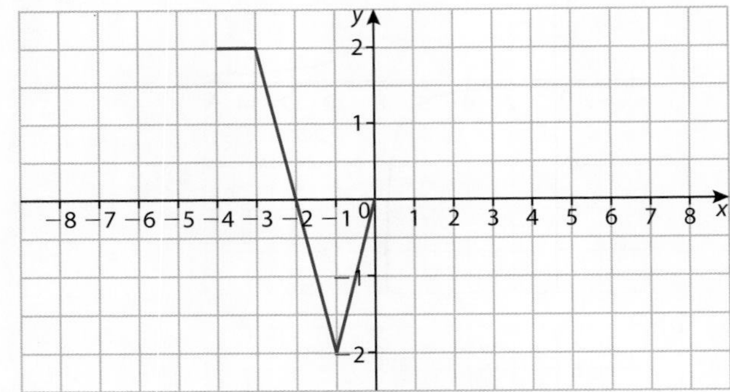

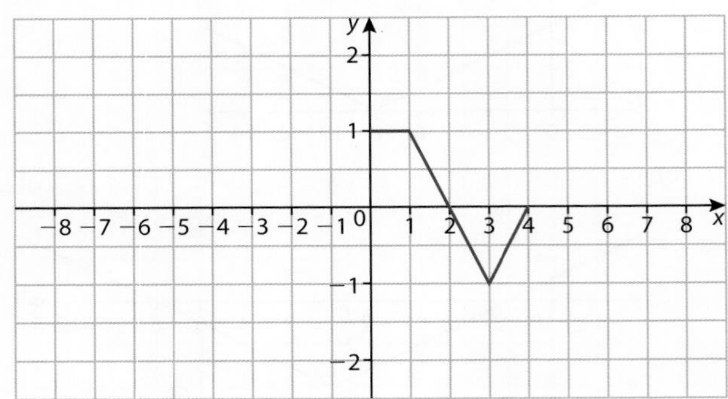

b) $A'(-3, -2)$

14

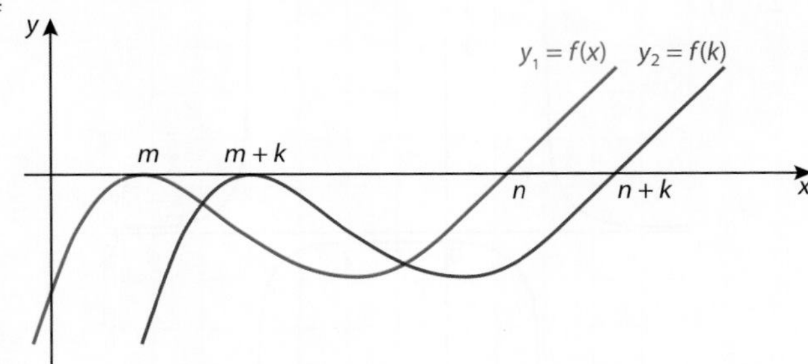

15 $(f \circ g^{-1})(x) = \sqrt[3]{x} + 1$

16 a) $g(x) = \frac{x}{2x+1}$ b) $\frac{2}{9}$

17 a) $-\frac{\sqrt{2}}{2} \leqslant x \leqslant \frac{\sqrt{2}}{2}, x \neq 0$
b) $f(x) \geqslant 0$

18 $f^{-1}(x) = \frac{x+1}{x-2}, x \neq 2$

19 a) $-\frac{1}{2} < A < 2$ b) $f^{-1}(x) = \frac{-2x-1}{x-2}$

20 a) $g(x) = \sqrt[3]{x+1}$ b) $g(x) = \sqrt[3]{x} + 1$

21 a) $S = \{x : -\sqrt{3} < x < \sqrt{3}\}$ b) $f(x) \geqslant \frac{\sqrt{3}}{3}$

22 $\frac{x}{4}$

23 a) $A(1, 25), B(4, 0), C(7, -35), D(10, 0)$
b) $A(-1, -25), B(0, 0), C(1, 35), D(2, 0)$

Chapter 3

Exercise 3.1

1 $-8; -8$ **2** $0; 33$
3 $29; 2375$ **4** $0; -3c+6$
5 $k=2$ **6** $k=2$

7 a) $-16, 2, 2, -4, -4, 14, 62$
b) 3
c)

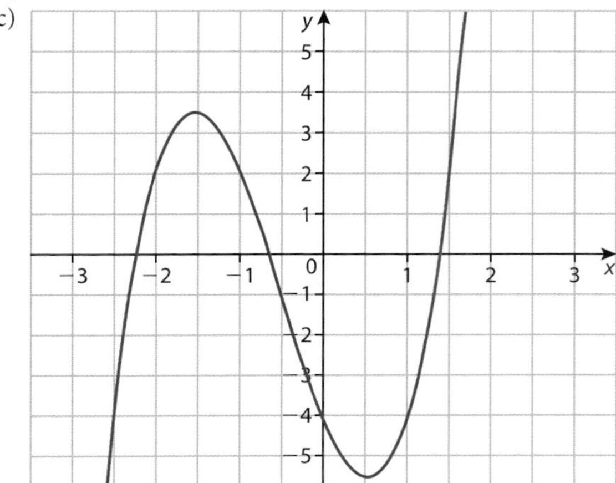

8 a) $52, 5, 0, 1, -4, -3, 40$
b) 4
c)

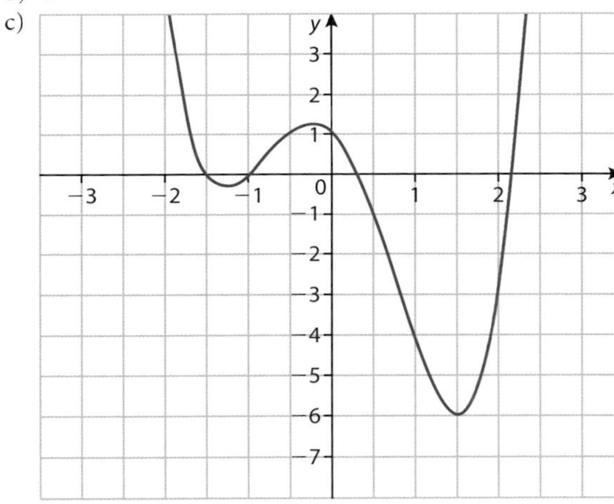

9 $a = \frac{12}{11}$

10 $b = -\sqrt{3}$

11 a) (i) $(\nwarrow, \nearrow)$ (ii) $(\swarrow, \searrow)$
(iii) $(\nwarrow, \searrow)$ (iv) $(\swarrow, \nearrow)$
(v) $(\nwarrow, \nearrow)$ (vi) $(\swarrow, \searrow)$
(vii) $(\swarrow, \nearrow)$ (viii) $(\nwarrow, \searrow)$

b) If leading term has positive coefficient and even exponent, then $(\nwarrow, \nearrow)$.
If leading term has negative coefficient and even exponent, then $(\swarrow, \searrow)$.
If leading term has positive coefficient and odd exponent, then $(\swarrow, \nearrow)$.
If leading term has negative coefficient and odd exponent, then $(\nwarrow, \searrow)$.

Exercise 3.2

1 a) $f(x) = (x-5)^2 + 7$
b) Horizontal translation 5 units right; vertical translation 7 units up.
c) Minimum: $(5, 7)$

2 a) $f(x) = (x+3)^2 - 1$
b) Horizontal translation 3 units left; vertical translation 1 unit down.
c) Minimum: $(-3, -1)$

3 a) $f(x) = -2(x+1)^2 + 12$
b) Horizontal translation 1 unit left; reflection over x-axis; vertical stretch by factor 2; vertical translation 12 units up.
c) Maximum: $(-1, 12)$

4 a) $f(x) = 4\left(x - \frac{1}{2}\right)^2 + 8$
b) Horizontal translation $\frac{1}{2}$ unit right; vertical stretch by factor 4; vertical translation 8 units up.
c) Maximum: $\left(\frac{1}{2}, 8\right)$

5 a) $f(x) = \frac{1}{2}(x+7)^2 + \frac{3}{2}$
b) Horizontal translation 7 units left; vertical shrink by factor $\frac{1}{2}$; vertical translation $\frac{3}{2}$ units up.
c) Minimum: $\left(-7, \frac{3}{2}\right)$

6 $x=2, x=-4$ **7** $x=5, x=-2$

8 $x=\frac{3}{2}, x=0$ **9** $x=6, x=-1$

10 $x=3$ **11** $x=\frac{1}{3}, x=-4$

12 $x=3, x=2$ **13** $x=2, x=\frac{1}{4}$

14 $x=-2\pm\sqrt{7}$ **15** $x=5, x=-1$

16 No real solution **17** $x=-4\pm\sqrt{13}$

18 $x=2, x=-4$ **19** $x=\frac{2\pm\sqrt{22}}{2}$

20 a) $x = 2 \pm \sqrt{5}$
b) Axis of symmetry: $x=2$
c) Minimum value of f is -5

21 Two real solutions **22** No real solutions

23 Two real solutions **24** No real solutions

25 $p = \pm 2\sqrt{2}$ **26** $k<4$

27 $k<-1,\ k>1$ **28** $m<-3,\ m>3$

29 $k>12$

30 $x - 2 - x^2 \Rightarrow -(x^2 - x + 2) \Rightarrow -\left(x^2 - x + \frac{1}{4}\right) - \frac{7}{4}$
$\Rightarrow -\left(x - \frac{1}{2}\right)^2 - \frac{7}{4} \leqslant -\frac{7}{4}$ for all x

31 $y = -2x^2 + 6x + 8$ **32** $y = x^2 - \frac{7}{2}x - \frac{1}{2}$

33 $-1<k<15$ **34** $m < -2\sqrt{10}$ or $m > 2\sqrt{10}$

35 $f(x) = 3x^2 + 5x - 2$ **36** $f(2) = 8$

37 $x<1$ or $x>3$

38 $\Delta = (2-t)^2 - 4(2)(t^2+3) > 0 \Rightarrow -7t^2 - 4t - 20 > 0$; because $\Delta_1 = -544$ for $-7t^2 - 4t - 20$ and leading coefficient is negative, then graph of $y = -7t^2 - 4t - 20$ is a parabola opening down and always below x-axis; hence, Δ for original equation is always negative; thus, no real roots

39 $x = \frac{-(-a^2-1) \pm \sqrt{(-a^2-1)^2 - 4a(a)}}{2a} = \frac{a^2+1 \pm \sqrt{a^4 - 2a^2 + 1}}{2a}$
$= \frac{a^2+1 \pm \sqrt{(a^2-1)^2}}{2a} = \frac{a^2+1 \pm (a^2-1)}{2a} \Rightarrow x = \frac{2a^2}{2a}$
$= a$ or $x = \frac{2}{2a} = \frac{1}{a}$

40 a) sum $= -3$, product $= -\frac{5}{2}$

b) sum $= -3$, product $= -1$

c) sum $= 0$, product $= -\frac{3}{2}$

d) sum $= a$, product $= -2a$

e) sum $= 6$, product $= -4$

f) sum $= \frac{1}{3}$, product $= -\frac{2}{3}$

41 $4x^2 + 5x + 4 = 0$

42 a) $\frac{1}{9}$ b) $\frac{1}{12}$ c) $\frac{55}{27}$

43 a) -2 and -6 b) $k = 12$

44 a) $-\frac{1}{4}$ b) $4x^2 + x + 1 = 0$

45 a) $x^2 - 19x + 25 = 0$ b) $25x^2 + 72x - 5 = 0$

Exercise 3.3

1 $3x^2 + 5x - 5 = (x+3)(3x-4) + 7$

2 $3x^4 - 8x^3 + 9x + 5 = (x-2)(3x^3 - 2x^2 - 4x + 1) + 7$

3 $x^3 - 5x^2 + 3x - 7 = (x-4)(x^2 - x - 1) - 11$

4 $9x^3 + 12x^2 - 5x + 1 = (3x-1)(3x^2 + 5x) + 1$

5 $x^5 + x^4 - 8x^3 + x + 2 = (x^2 + x - 7)(x^3 - x + 1) + (-7x + 9)$

6 $(x-7)(x-1)(2x-1)$ **7** $(x-2)(2x+1)(3x+2)$

8 $(x-2)^2(x+4)(3x+2)$ **9** $Q(x) = x - 2,\ R = -2$

10 $Q(x) = x^2 + 2,\ R = -3$ **11** $Q(x) = 3,\ R(x) = 20x + 5$

12 $Q(x) = x^4 + x^3 + 4x^2 + 4x + 4,\ R = -2$

13 $P(2) = 5$ **14** $P(-1) = -17$

15 $P(-7) = -483$ **16** $P\left(\frac{1}{4}\right) = \frac{49}{64}$

17 $x = 2 + i$ or $x = 2 - i$ **18** $x = \frac{1+\sqrt{5}}{2}$ or $x = \frac{1-\sqrt{5}}{2}$

19 $k = \sqrt{1-x}\sqrt{3}$ or $k = -\sqrt{1-x}\sqrt{3}$

20 $a = 5,\ b = 12$

21 $x^3 - 3x^2 - 6x + 8$ **22** $x^4 - 3x^3 - 7x^2 + 15x + 18$

23 $x^3 - 6x^2 + 12x - 8$ **24** $x^3 - x^2 + 2$

25 $x^4 + 2x^3 + x^2 + 18x - 72$ **26** $x^4 - 8x^3 + 27x^2 - 50x + 50$

27 $x = 2 + 3i,\ x = 3$

28 a) $a = -1,\ b = -2$ b) $3x + 2$

29 $a = \frac{4}{3},\ b = \frac{1}{3}$

30 $x = 3,\ x = -1,\ x = -\frac{1}{4} + \frac{\sqrt{3}}{4}i,\ x = -\frac{1}{4} - \frac{\sqrt{3}}{4}i$

31 $a = -1,\ b = -4,\ c = 4$ **32** $p = -5,\ q = 23,\ r = -51$

33 $a = -5$ **34** $m = -2,\ n = -6$

35 $b = 18$ **36** b) $R = 3$

37 a) sum $= \frac{2}{3}$, product $= 5$ b) sum $= 1$, product $= 7$

c) sum $= \frac{1}{3}$, product $= -\frac{1}{2}$

39 $-9, 3, 6$

40 $2, -4, 8$

41 $3 + 2i, 2 + i, 2 - i$

42 $k = 3$

43 $k = -8$

Exercise 3.4

1

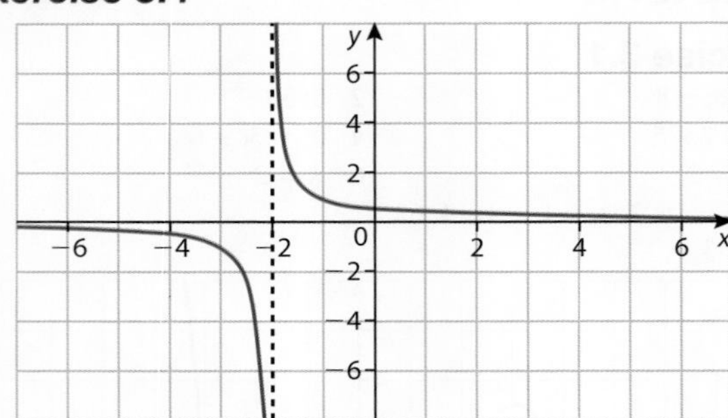

vertical asymptote: $x = -2$
horizontal asymptote: $y = 0$

2

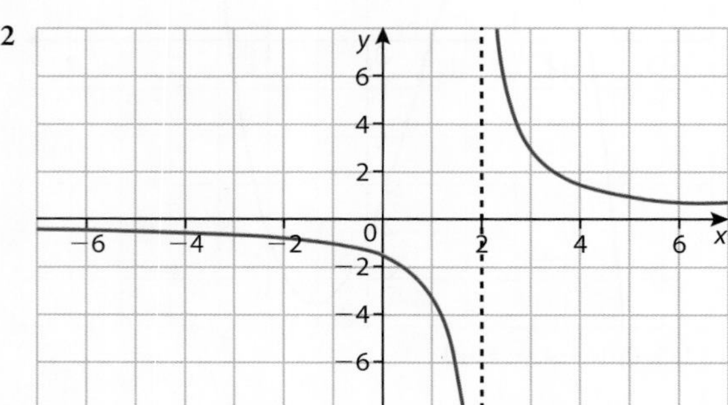

vertical asymptote: $x = 2$
horizontal asymptote: $y = 0$

3

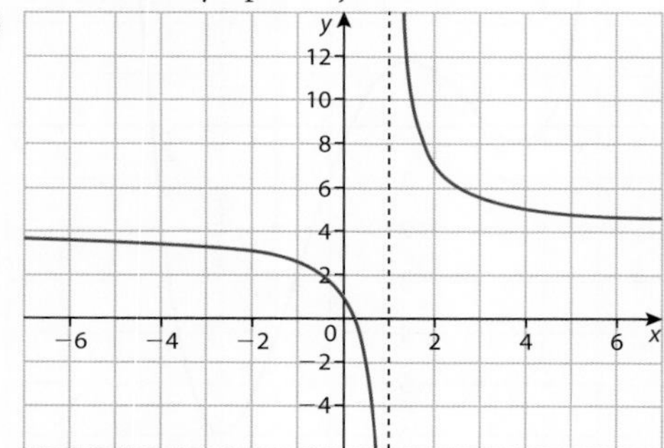

x-intercept: $\left(\frac{1}{4}, 0\right)$, y-intercept: $(0, 1)$
vertical asymptote: $x = 1$ horizontal asymptote: $y = 4$

4

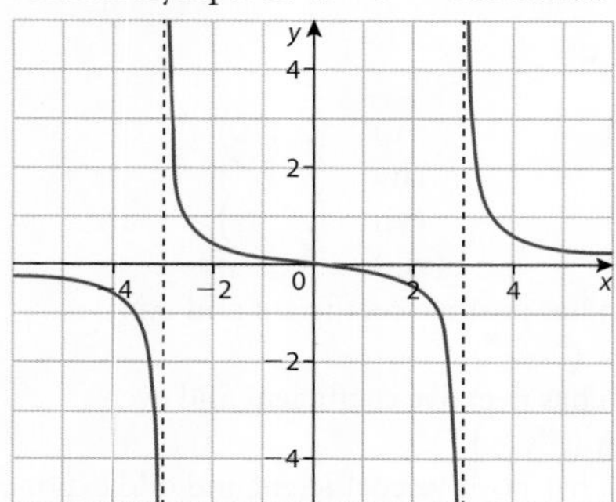

x- and y-intercept: $(0, 0)$
vertical asymptotes: $x = -3, x = 3$
horizontal asymptote: $y = 0$

5 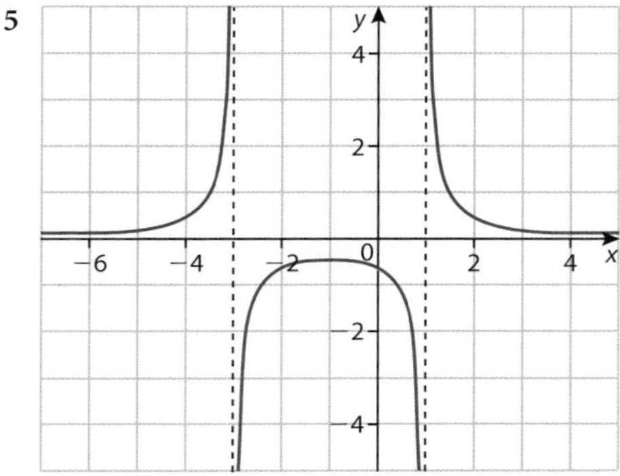

x-intercept: none, y-intercept: $\left(0, -\frac{2}{3}\right)$
vertical asymptotes: $x = -3$, $x = 1$
horizontal asymptote: $y = 0$

6 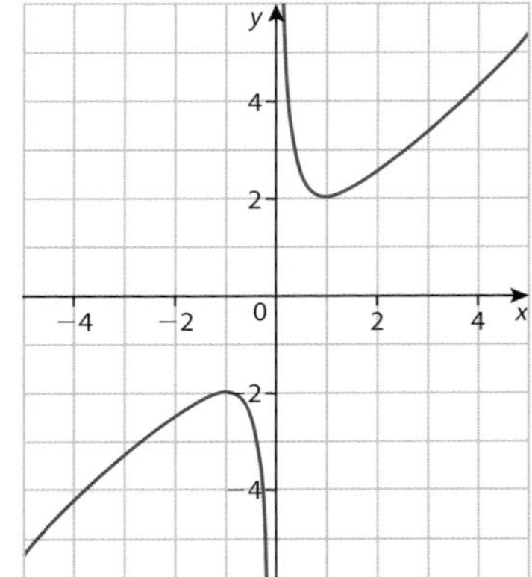

oblique asymptote: $y = x$ vertical asymptote: $x = 0$

7 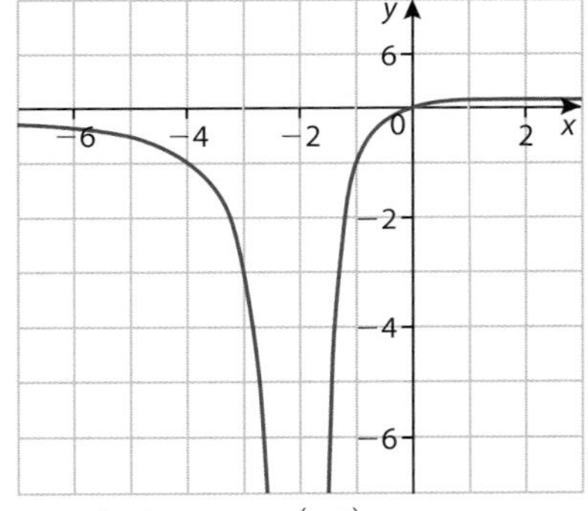

x- and y-intercept: $(0, 0)$
vertical asymptote: $x = -2$ horizontal asymptote: $y = 0$

8 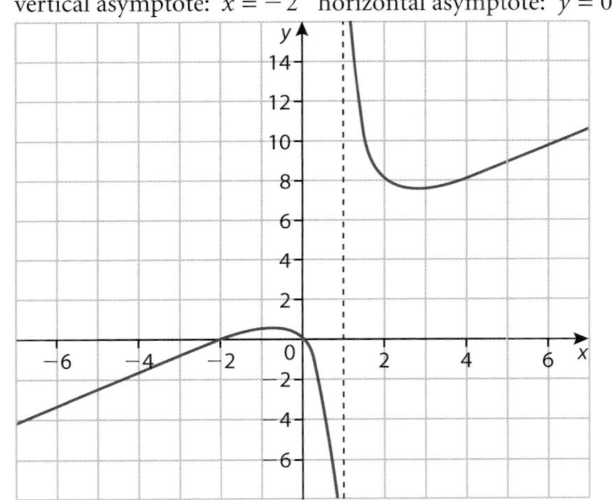

x- and y-intercept: $(0, 0)$
vertical asymptote: $x = 1$ oblique asymptote: $y = x + 3$

9 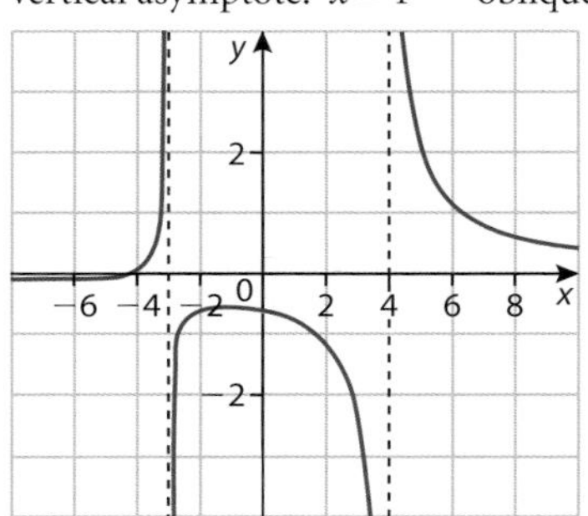

x-intercept: $(-4, 0)$ y-intercept: $\left(0, -\frac{2}{3}\right)$
vertical asymptotes: $x = -3$ and $x = 4$
horizontal asymptote: $y = 0$

10 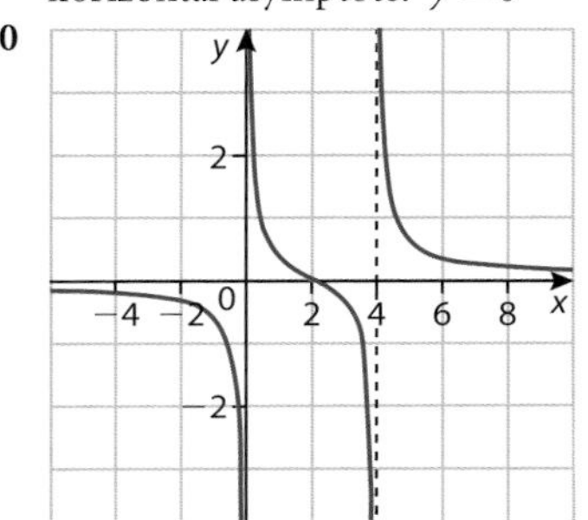

x-intercept: $(2, 0)$ y-intercept: none
vertical asymptotes: $x = 0$ and $x = 4$
horizontal asymptote: $y = 0$

11 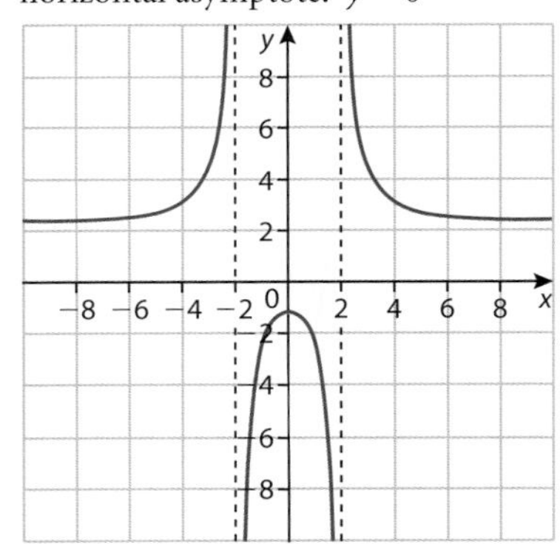

domain $\{x : x \in \mathbb{R}, x \neq \pm 2\}$ range $\left\{y : y \leqslant -\frac{5}{4} \text{ or } y > 2\right\}$

12 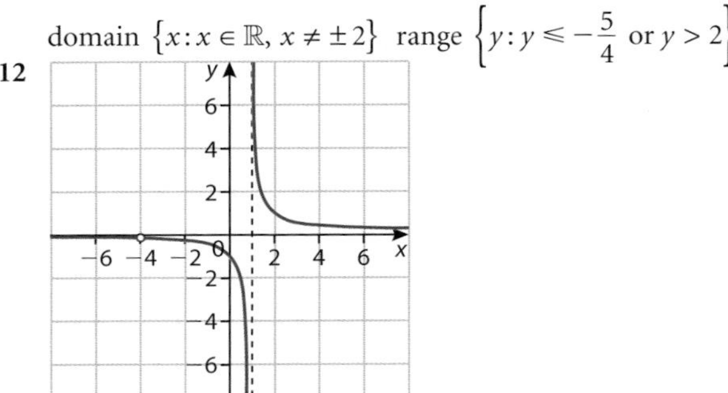

domain $\{x : x \in \mathbb{R}, x \neq -4, 1\}$ range $\{y : y \in \mathbb{R}, y \neq 0\}$

13

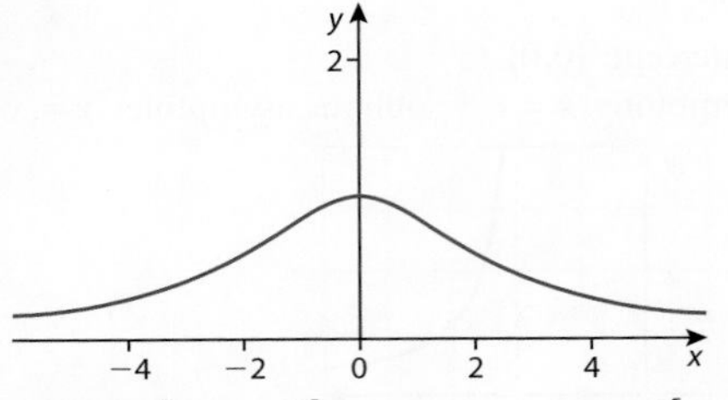

domain $\{x : x \in \mathbb{R}\}$ range $\{y : 0 < y \leq 1\}$

14

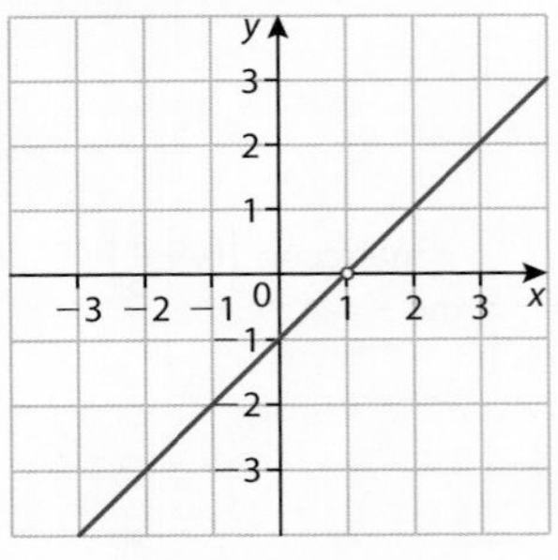

domain $\{x : x \in \mathbb{R}, x \neq 1\}$
range $\{y : y \in \mathbb{R}, y \neq 0\}$

15

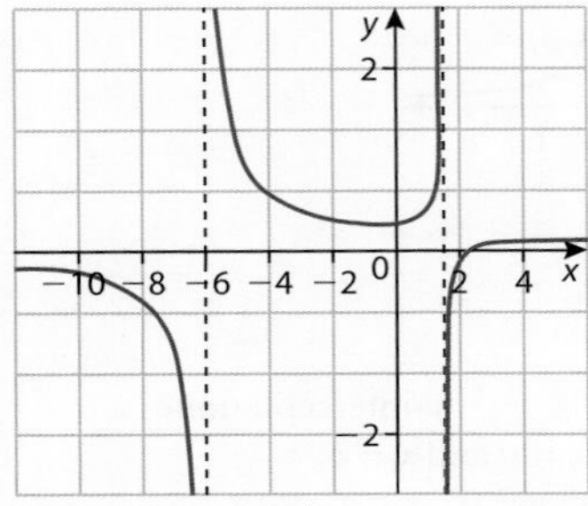

x-intercept: $\left(\frac{5}{2}, 0\right)$ y-intercept: $\left(0, \frac{5}{18}\right)$
vertical asymptotes: $x = -6$ and $x = \frac{3}{2}$
horizontal asymptote: $y = 0$

16

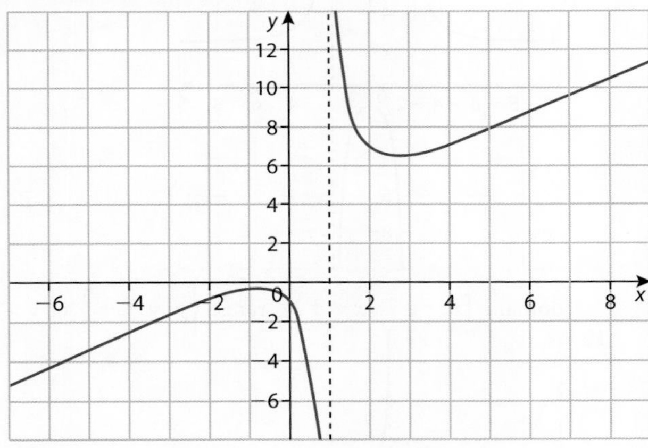

x-intercept: none
y-intercept: $(0, -1)$
vertical asymptote: $x = 1$
oblique asymptote: $y = x + 2$

17

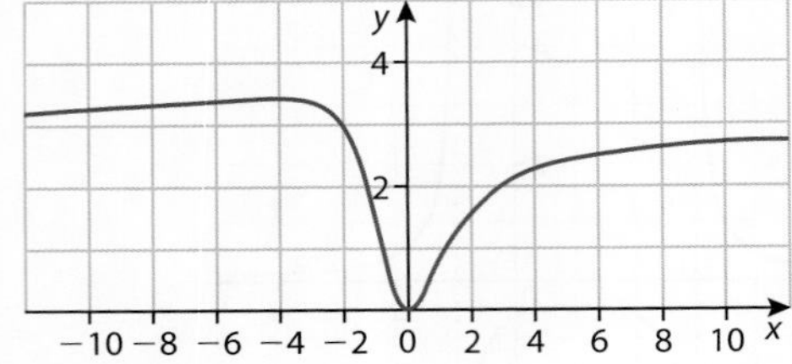

x- and y-intercept: $(0, 0)$
horizontal asymptote: $y = 3$

18

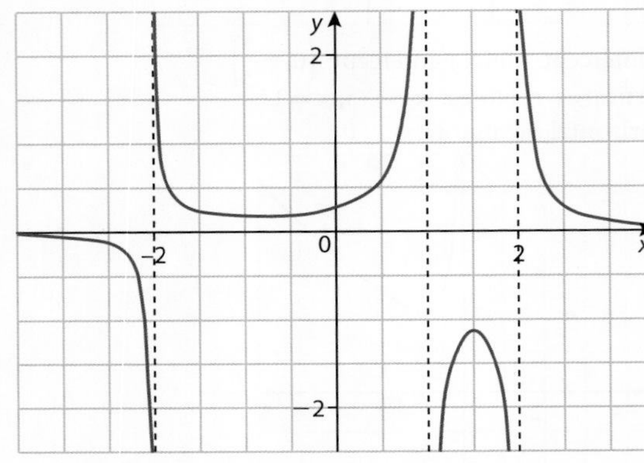

x-intercept: none y-intercept: $\left(0, \frac{1}{4}\right)$
vertical asymptotes: $x = -2$, $x = 1$ and $x = 2$
horizontal asymptote: $y = 0$

19 a)

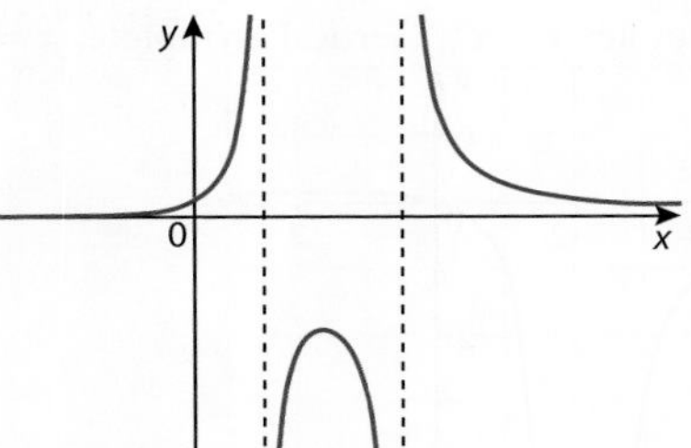

b)

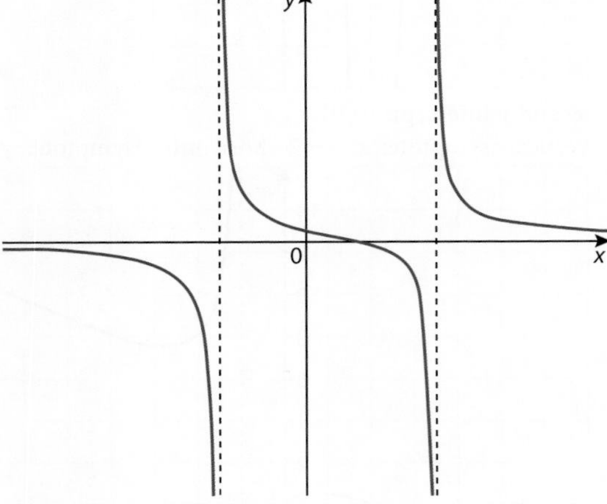

c)

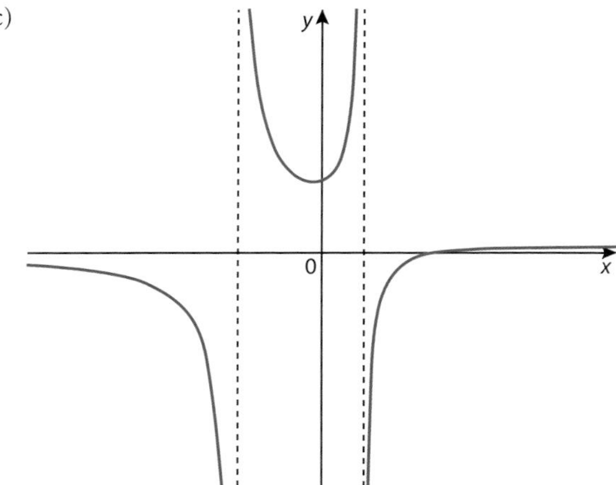

20 a)

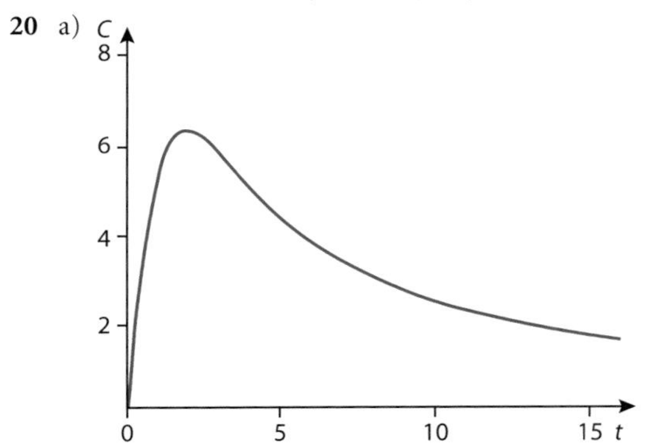

b) At $t = 2$ minutes, concentration is 6.25 mg/l.

c) It continues to decrease and approaches zero as amount of time increases.

d) 50 minutes (49 minutes 55 seconds)

Exercise 3.5

1 $x = 3$ | **2** $x = 9$

3 $x = 5$ or $x = -2$ | **4** $x = 11$ or $x = 3$

5 $x = -5$ | **6** $x = 1$ or $x = -2$

7 $x = 2$ or $x = -2$ | **8** $x = \frac{1}{2}$

9 $x = \pm\sqrt{5}$ | **10** $x = 27$ or $x = -\frac{125}{8}$

11 $x = 3$ or $x = 2$ | **12** $x = \frac{1 \pm \sqrt{41}}{4}$

13 $x = \frac{4}{3}$ or $x = -4$ | **14** $x = 15$ or $x = \frac{9}{2}$

15 No solution | **16** $x = 2$ or $x = -1$

17 $x = 2$ or $x = \frac{1}{2}$ | **18** $x = 9$

19 $x = -5$ | **20** $x = \frac{\pm 2\sqrt{5}}{5}$ or $x = \pm 1$

21 $x = \frac{1 + \sqrt{41}}{2}$ | **22** $x = \frac{49}{4}$ or $x = \frac{64}{9}$

23 $-\frac{2}{3} < x < 2$ | **24** $x < -2, x \geqslant 3$

25 $-10 \leqslant x \leqslant 6$ | **26** $x < \frac{3}{2}, x > 2$

27 $x > \frac{17}{2}$ | **28** $-4 \leqslant x \leqslant -1, 1 \leqslant x \leqslant 4$

29 $x < -1, x > 2$ | **30** $x < -1, -\frac{2}{3} < x < 3, x > 4$

31 a) $p = \frac{9}{4}$ b) $p < \frac{9}{4}$ c) $p > \frac{9}{4}$

32 $x < -1, x > \frac{1}{3}$

33 a) $m + \frac{1}{n} > 2 \Rightarrow mn + 1 > 2n \Rightarrow mn - 2n + 1 > 0$; since $m > n \Rightarrow mn > n^2$ it follows that $mn - 2n + 1 > n^2 - 2n + 1$ and since $n^2 - 2n + 1 = (n - 1)^2 > 0$ then $mn - 2n + 1 > 0$ $\Rightarrow m + \frac{1}{n} > 2$

b) $(m + n)\left(\frac{1}{m} + \frac{1}{n}\right) > 4 \Rightarrow (m + n)\left(\frac{1}{m} + \frac{1}{n}\right)mn > 4mn \Rightarrow$ $(m + n)(n + m) > 4mn \Rightarrow m^2 + 2mn + n^2 > 4mn \Rightarrow$ $m^2 - 2mn + n^2 > 0 \Rightarrow (m - n)^2 > 0$ which is true for all x and is equivalent to original inequality – thus, $(m + n)\left(\frac{1}{m} + \frac{1}{n}\right) > 4$ is true for all x.

34 $x = \frac{-1 \pm \sqrt{13}}{2}$, $x = 1$ or $x = -2$

35 $(a + b + c)^2 < 3\,(a^2 + b^2 + c^2)$
$\Rightarrow a^2 + b^2 + c^2 + 2ab + 2ac + 2bc < 3a^2 + 3b^2 + 3c^2$
$\Rightarrow 0 < 2a^2 + 2b^2 + 2c^2 - 2ab - 2ac - 2bc$
$\Rightarrow a^2 - 2ab + b^2 + b^2 - 2bc + c^2 + a^2 - 2ac + c^2 > 0$
$\Rightarrow (a - b)^2 + (b - c)^2 + (c - a)^2 > 0.$
Since all the numbers are unequal, the squares of their differences are strictly larger than zero therefore their sum too is strictly larger than zero.

36 a) $1 < x < 3$ b) $x < -2, -1 < x < 1, x > 3$

37 If a and b have the same sign, then $|a + b| = |a| + |b|$; and if a and b are of opposite sign, then $|a + b| < |a| + |b|$.

Practice questions

1 $x = a$ or $x = 3b$ | **2** $x \leqslant 4$

3 $c = 5$ | **4** $a = -\frac{1}{2}, b = 4, c = -2$

5 $\omega = -2, p = 2, q = -8$

6 a) $m > -2$ b) $-2 < m < 0$

7 $a = 2, b = -1, c = -2$

8 $x < 5, x > \frac{15}{2}$

9 $-1 < k < 15$

10 a) $f(x) = 2 - \frac{3}{(x + 2)^2 + 1}$

b) (i) $\lim_{x \to +\infty} f(x) = 2$ (ii) $\lim_{x \to -\infty} f(x) = 2$

c) $(-2, -1)$

11 $k \in \mathbb{R}$ | **12** $a = -1$

13 $a = \frac{7}{4}, b = -\frac{1}{4}$ | **14** $a = -6$

15 $a = 4$ | **16** $a = -2, b = 6$

17 $a = 1$ | **18** $k = 6$

19 $k = 6$ | **20** $-2.80 < k < 0.803$ (3 s.f.)

21 $-3 \leqslant k \leqslant 4.5$ | **22** $-4 \leqslant m \leqslant 0$

23 $1 \leqslant x \leqslant 3$ | **24** $-2.30 < x < 0$ or $1 < x < 1.30$

25 $-3 \leqslant x \leqslant \frac{1}{3}$ | **26** $x < -1$ or $4 < x \leqslant 14$

27 $x \leqslant 3$ or $x \geqslant 27$ | **28** $x = 2 - i$ and $x = 2$

29 $x < \frac{1}{3}$

Chapter 4

Exercise 4.1

1 $-1, 1, 3, 5, 7$ | **2** $-1, 1, 5, 13, 29$

3 $\frac{3}{2}, \frac{3}{4}, \frac{3}{8}, \frac{3}{16}, \frac{3}{32}$ | **4** $5, 8, 11, 14, 17$

5 $1, 7, -5, 19, -29$ | **6** $3, 7, 13, 21, 31$

7 $-1, 1, 3, 5, 7, 97$

8 $2, 6, 18, 54, 162, 4.786 \times 10^{23}$

9 $\frac{2}{3}, -\frac{2}{3}, \frac{6}{11}, -\frac{4}{9}, \frac{10}{27}, \frac{50}{1251}$
10 1, 2, 9, 64, 625, 1.776×10^{83}
11 3, 11, 27, 59, 123, 4.50×10^{15}
12 0, 3, $\frac{3}{7}, \frac{21}{13}, \frac{39}{55}$, approx. 1
13 2, 6, 18, 54, 162, 4.786×10^{23}
14 −1, 1, 3, 5, 7, 97 **15** $u_n = \frac{1}{4}u_{n-1}, u_1 = \frac{1}{3}$
16 $u_n = \frac{4a^2}{3}u_{n-1}, u_1 = \frac{1}{2}a$ **17** $u_n = u_{n-1} + a - k, u_1 = a - 5k$
18 $u_n = n^2 + 3$ **19** $u_n = 3n - 1$
20 $u_n = \frac{2n-1}{n^2}$ **21** $u_n = \frac{2n-1}{n+3}$
22 a) $1, 2, \frac{3}{2}, \frac{5}{3}, \frac{8}{5}, \frac{13}{8}, \frac{21}{13}, \frac{34}{21}, \frac{55}{34}, \frac{89}{55}$
23 a) 1, 1, 2, 3, 5, 8, 13, 21, 34, 55, 89, 144

Exercise 4.2

1 $3, \frac{19}{5}, \frac{23}{5}, \frac{27}{5}, \frac{31}{5}, 7$
2 a) Arithmetic, $d = 2, a_{50} = 97$
b) Arithmetic, $d = 1, a_{50} = 52$
c) Arithmetic, $d = 2, a_{50} = 97$
d) Not arithmetic, *no common difference*
e) Not arithmetic, *no common difference*
f) Arithmetic, $d = -7, a_{50} = -341$
3 a) 26
b) $a_n = -2 + 4(n-1)$
c) $a_1 = -2, a_n = a_{n-1} + 4$ for $n > 1$
4 a) 1
b) $a_n = 29 - 4(n-1)$
c) $a_1 = 29, a_n = a_{n-1} - 4$ for $n > 1$
5 a) 57
b) $a_n = -6 + 9(n-1)$
c) $a_1 = -6, a_n = a_{n-1} + 9$ for $n > 1$
6 a) 9.23
b) $a_n = 10.07 - 0.12(n-1)$
c) $a_1 = 10.07, a_n = a_{n-1} - 0.12$ for $n > 1$
7 a) 79
b) $a_n = 100 - 3(n-1)$
c) $a_1 = 100, a_n = a_{n-1} - 3$ for $n > 1$
8 a) $-\frac{27}{4}$
b) $a_n = 2 - \frac{5}{4}(n-1)$
c) $a_1 = 2, a_n = a_{n-1} - \frac{5}{4}$ for $n > 1$
9 13, 7, 1, −5, −11, −17, −23
10 299, $299\frac{1}{4}$, $299\frac{1}{2}$, $299\frac{3}{4}$, 300
11 $a_n = -10 + 4(n-1) = 4n - 14$
12 $a_n = -\frac{142}{3} + \frac{11}{3}(n-1) = -51 + \frac{11}{3}n$
13 88 **14** 36
15 11 **16** 16
17 11 **18** 9, 3, −3, −9, −15
19 99.25, 99.50, 99.75 **20** $a_n = 4n - 1$
21 $a_n = \frac{19n - 277}{3}$ **22** $a_n = 4n + 27$
23 Yes, 3271th term **24** Yes, 1385th term
25 No

Exercise 4.3

1 Geometric, $r = 3^a, g_{10} = 3^{9a+1}$
2 Arithmetic, $d = 3, a_{10} = 27$
3 Geometric, $r = 2, b_{10} = 4096$
4 Neither, not geometric, $r = 2, c_{10} = -1534$
5 Geometric, $r = 3, u_{10} = 78\,732$
6 Geometric, $r = 2.5, a_{10} = 7629.394\,531\,25$
7 Geometric, $r = -2.5, a_{10} = -7629.394\,531\,25$
8 Arithmetic, $d = 0.75, a_{10} = 8.75$
9 Geometric, $r = -\frac{2}{3}, a_{10} = -\frac{1024}{2187}$
10 Arithmetic, $d = 3$ **11** Geometric, $r = -3$
12 Geometric, $r = 2$ **13** Neither
14 Neither **15** Arithmetic, $d = 1.3$
16 a) 32 b) $-3 + 5(n-1)$
c) $a_1 = -3, a_n = a_{n-1} + 5$ for $n > 1$
17 a) −9 b) $19 - 4(n-1)$
c) $a_1 = 19, a_n = a_{n-1} - 4$ for $n > 1$
18 a) 69 b) $-8 + 11(n-1)$
c) $a_1 = -8, a_n = a_{n-1} + 11$ for $n > 1$
19 a) 9.35 b) $10.05 - 0.1(n-1)$
c) $a_1 = 10.05, a_n = a_{n-1} - 0.1$ for $n > 1$
20 a) 93 b) $100 - (n-1)$
c) $a_1 = 100, a_n = a_{n-1} - 1$ for $n > 1$
21 a) $-\frac{17}{2}$ b) $2 - 1.5(n-1)$
c) $a_1 = 2, a_n = a_{n-1} - 1.5$ for $n > 1$
22 a) 384 b) $3 \times 2^{n-1}$
c) $a_1 = 3, a_n = 2a_{n-1}$ for $n > 1$
23 a) 8748 b) $4 \times 3^{n-1}$
c) $a_1 = 4, a_n = 3a_{n-1}$ for $n > 1$
24 a) −5 b) $5 \times (-1)^{n-1}$
c) $a_1 = 5, a_n = -a_{n-1}$ for $n > 1$
25 a) −384 b) $3 \times (-2)^{n-1}$
c) $a_1 = 3, a_n = -2a_{n-1}$ for $n > 1$
26 a) $-\frac{4}{9}$ b) $972 \times (-\frac{1}{3})^{n-1}$
c) $a_1 = 972, a_n = (-\frac{1}{3})a_{n-1}$ for $n > 1$
27 a) $\frac{2187}{64}$ b) $a_n = -2\left(-\frac{3}{2}\right)^{n-1}$
c) $a_1 = -2, a_n = -\frac{3}{2}a_{n-1}, n > 1$
28 a) $\frac{390\,625}{117\,649}$ b) $a_n = 35\left(\frac{5}{7}\right)^{n-1}$
c) $a_1 = 35, a_n = \frac{5}{7}a_{n-1}, n > 1$
29 a) $-\frac{3}{64}$ b) $a_n = -6\left(\frac{1}{2}\right)^{n-1}$
c) $a_n = -6, a_n = \frac{1}{2}a_{n-1}, n > 1$
30 a) 1216 b) $9.5 \times 2^{n-1}$
c) $a_1 = 9.5, a_n = 2a_{n-1}, n > 1$
31 a) $69.833\,729\,609\,375 = \frac{893\,871\,739}{12\,800\,000}$
b) $a_n = 100\left(\frac{19}{20}\right)^{n-1}$
c) $a_1 = 100, a_n = \frac{19}{20}a_{n-1}, n > 1$
32 a) $0.002\,085\,685\,73 = \frac{2187}{1\,048\,576}$
b) $a_n = 2\left(\frac{3}{8}\right)^{n-1}$ c) $a_1 = 2, a_n = \frac{3}{8}a_{n-1}, n > 1$
33 6, 12, 24, 48 **34** 35, 175, 875
35 36 **36** 21, 63, 189, 567
37 −24, 24 **38** 1.5, $a_n = 24\left(\frac{1}{2}\right)^{n-1}$
39 $a_4 = \pm 3, r = \pm\frac{1}{2}, a_n = 24\left(\pm\frac{1}{2}\right)^{n-1}$ **40** $\frac{49}{3}$
41 10th term **42** Yes, 10th term
43 Yes, 10th term **44** 2228.92
45 £945.23 **46** €2968.79
47 7745 thousands **48** $\frac{98}{9}$

49 10th term **50** €3714.87

51 £2921.16

Exercise 4.4

1 11 280 **2** $-\frac{10\,5469}{1024}$ **3** 0.7

4 $\frac{10}{7}$ **5** $\frac{16 + 4\sqrt{3}}{39}$

6 a) $\frac{52}{99}$ b) $\frac{449}{990}$ c) $\frac{7459}{2475}$

7 13 026.135 (£13 026.14)

8 940 **9** 6578

10 42 625 **11** $\frac{n(7 + 3n)}{2}$

12 17 terms **13** 85 terms

14 $d = 4$ **15** a) 250, 125 250, b) 83 501

16 $a = 1, d = 5$ **17** 2890

18 0.290 **19** −2.065

20 11 400 **21** 1.191

22 49.2 **23** $\frac{6}{5}$

24 $\frac{3 + \sqrt{6}}{2}$ **25** $3, \frac{18}{5}, \frac{93}{25}, \frac{468}{125}, \frac{15}{4}\left(1 - \frac{1}{5^n}\right)$

26 $\frac{1}{6}, \frac{1}{4}, \frac{3}{10}, \frac{1}{3}; \frac{n}{2n + 4}$

27 $\sqrt{2} - 1, \sqrt{3} - 1, 1, \sqrt{5} - 1; \sqrt{n + 1} - 1$

28 1.945, 152.42 **29** 127, 128

30 $\frac{819}{128}, \frac{32}{5}$ **31** 11 866

32 763 517 **33** 14 348 906

34 ≈150

Exercise 4.5

1 a) 120 b) 120 c) 20 d) 336

2 a) 1 b) 1 c) 120 d) 120

3 a) 70 b) 70 c) 330 d) 330

4 a) 0 b) 39 916 800 c) 0 d) 10

5 a) F b) F c) T

6 24

7 72 **8** 312

9 16 777 216 **10** 262 144

11 1 757 600 000 **12** 81 000

13 a) 40 320 b) 384

14 a) 40 320 b) 720

15 JANE, JAEN, JNAE, JNEA, JEAN, JENA, AJNE, AJEN, ANJE, ANEJ, AEJN, AENJ, NJAE, NJEA, NEJA, NEAJ, NAJE, NAEJ, EJAN, EJNA, EAJN, EANJ, ENJA, ENAJ

16 Mag, Mga, Mai, …(60 of them)

17 a) 175 760 000 b) 174 790 000

18 a) 4080 b) 1680 c) 1050 d) 1980 e) 3150

19 a) 296 b) 1460 c) 504

20 a) 125 000 b) 117 600 c) 61 250 d) 176 400

21 768

22 a) 36 b) 256

23 a) 5985 b) 2376 c) 2475

24 a) 2280 b) 748 c) 770

25 a) 1 192 052 400 b) 4560, 0.000 38% c) 265 004 096, 22.2%

26 a) 74 613 b) 7560

27 54 867 456 000

Exercise 4.6

1 a) $x^5 + 10x^4y + 40x^3y^2 + 80x^2y^3 + 80xy^4 + 32y^5$

b) $a^4 - 4a^3b + 6a^2b^2 - 4ab^3 + b^4$

c) $x^6 - 18x^5 + 135x^4 - 540x^3 + 1215x^2 - 1458x + 729$

d) $16 - 32x^3 + 24x^6 - 8x^9 + x^{12}$

e) $x^7 - 21bx^6 + 189b^2x^5 - 945b^3x^4 + 2835b^4x^3 - 5103b^5x^2 + 5103b^6x - 2187b^7$

f) $64n^6 + 192n^3 + 240 + \frac{160}{n^3} + \frac{60}{n^6} + \frac{12}{n^9} + \frac{1}{n^{12}}$

g) $\frac{81}{x^4} - \frac{216}{x^2\sqrt{x}} + \frac{216}{x} - 96\sqrt{x} + 16x^2$

2 a) 56 b) 0 c) 1225 d) 32 e) 0

3 a) $x^7 + 14x^6y + 84x^5y^2 + 280x^4y^3 + 560x^3y^4 + 672x^2y^5 + 448xy^6 + 128y^7$

b) $a^6 - 6a^5b + 15a^4b^2 - 20a^3b^3 + 15a^2b^4 - 6ab^5 + b^6$

c) $x^5 - 15x^4 + 90x^3 - 270x^2 + 405x - 243$

d) $x^{18} - 12x^{15} + 60x^{12} - 160x^9 + 240x^6 - 192x^3 + 64$

e) $x^7 - 21bx^6 + 189b^2x^5 - 945b^3x^4 + 2835b^4x^3 - 5103b^5x^2 + 5103b^6x - 2187b^7$

f) $64n^6 + 192n^3 + 240 + \frac{160}{n^3} + \frac{60}{n^6} + \frac{12}{n^9} + \frac{1}{n^{12}}$

g) $\frac{81}{x^4} - \frac{216}{x^2\sqrt{x}} + \frac{216}{x} - 96\sqrt{x} + 16x^2$

h) 112 i) $1792\sqrt{3}$

j) 16 k) $-23 + 10i\sqrt{2}$

4 a) $x^{45} - 90x^{43} + 3960x^{41}$

b) Does not exist as the powers of x decrease by 2's starting at 45. There is no chance for any expression to have zero exponent.

c) $\binom{45}{43}x^2\left(\frac{-2}{x}\right)^{43} + \binom{45}{44}x\left(\frac{-2}{x}\right)^{44} + \left(\frac{-2}{x}\right)^{45} = -\binom{45}{43}\frac{2^{43}}{x^{41}} + \binom{45}{44}\frac{2^{44}}{x^{43}} - \frac{2^{45}}{x^{45}}$

d) $\binom{45}{21}x^{24}\left(\frac{-2}{x}\right)^{21} = -\binom{45}{21}\cdot 2^{21}x^3$

5 $\binom{n}{k} = \frac{n!}{k!(n-k)!} = \frac{n!}{(n-k)!k!} = \frac{n!}{(n-k)!(n-(n-k))!} = \binom{n}{n-k}$

6 $(1 + 1)^n = \binom{n}{0} + \binom{n}{1} + \binom{n}{2} + \dots + \binom{n}{n}$

$2^n = 1 + \binom{n}{1} + \binom{n}{2} + \dots + \binom{n}{n} \Rightarrow 2^n - 1 = \binom{n}{1} + \binom{n}{2} + \dots \binom{n}{n}$

7 Answers vary **8** $\left(\frac{1}{3} + \frac{2}{3}\right)^6 = 1$

9 $\left(\frac{2}{5} + \frac{3}{5}\right)^8 = 1$ **10** $\left(\frac{1}{7} + \frac{6}{7}\right)^n = 1$

11 15 **12** 90 720

13 16 128 **14** 1.1045, 0.9045

15 Proof

16 a) $\frac{7}{9}$ b) $\frac{38}{110}$ c) $\frac{31\,808}{9900}$

17 −145 152 **18** $35a^3$ **19** 96 096

20 $243n^5 - 810n^4m + 1080n^3m^2 - 720n^2m^3 + 240nm^4 - 32m^5$

21 7 838 208

Exercise 4.7

1 $2 + 4 + 6 + \dots + 2n = n(n + 1)$

2–20 All proofs

Practice questions

1 $D = 5, n = 20$

2 €2098.63

3 a) Nick: 20 Charlotte: 17.6
b) Nick: 390 Charlotte: 381.3
c) Charlotte will exceed the 40 hours during week 14.
d) In week 12 Charlotte will catch up with Nick and exceed him.

4 a) Loss for the second month = 1060 g
Loss for the third month = 1123.6 g
b) Plan A loss = 1880 g
Plan B loss = 1898.3 g
c) (i) Loss due to plan A in all 12 months = 17 280 g
(ii) Loss due to Plan B in all 12 months = 16 869.9 g

5 a) €895.42 b) €6985.82

6 a) 142.5 b) 19 003.5

7 $1, \sqrt[3]{7}, 1, 1, \sqrt[3]{7}, 1, \ldots; 2, 0, 2, 0, 2, \ldots$

8 a) On the 37th day b) 407 km

9 a) 1.5 b) 207 595
c) 2009 d) 619 583
e) Market saturation

10 $-4, 3006$

11 a) $\sqrt{\frac{1}{4}+\frac{1}{4}}=\frac{\sqrt{2}}{2}$ b) $\frac{1}{2}$
c) (i) $\frac{1}{4}$ (ii) $\frac{1}{2}$ d) (i) $\frac{1}{512}$ (ii) 2

12 a) 1220 b) 36 920

13 a) Area A = 1, Area B = $\frac{1}{9}$ b) $\frac{1}{81}$
c) $1+\frac{8}{9}, 1+\frac{8}{9}+\left(\frac{8}{9}\right)^2$ d) 0

14 a) Neither, geometric converging, arithmetic, geometric diverging
b) 6

15 a) (i) Kell: 18 400, 18 800; YBO: 18 190, 19 463.3
(ii) Kell: 198 000; YBO: 234 879.62
(iii) Kell: 21 600; YBO: 31 253.81
b) (i) After the second year
(ii) 4th year

16 a) 62 b) 936

17 a) $7000(1 + 0.0525)^t$ b) 7 years
c) Yes, since 10 084.7 > 10 015.0

18 a) 11 b) 2 c) 15

19 $15, -8$ **20** $-2, -7$ **21** 10 300

22 Proof

23 a) $a_n = 8n - 3$ b) 50

24 2 099 520

25 $6n - 5$ **26** 72 **27** 559

28 $-3, 3$ **29** 9 **30** 62

31 $-\frac{36}{5}$

32 a) 4 b) $16(4^n - 1)$

33 a) $|x| < 1.5$ b) 5

34 3168

35 a) $\frac{n(3n+1)}{2}$ b) 30

36 -7

37 $1275 \ln 2$

38 a) 4, 8, 16
b) (i) $u_n = 2^n$ (ii) proof

39 a) $\frac{2}{3}$ b) 9

40 $2, -3$ **41** 55 **42** $-2, 4$

43 $\frac{\theta}{1-\cos\theta}$

44 a) 1, 5, 9 b) $4n - 3$

45 a) $32 + 80x + 80x^2 + 40x^3 + 10x^4 + x^5$
a) 32.808 040 1001

46 a) $5000(1.063)^n$ b) 6786.35
c) (i) $5000(1.063)^n > 1000$ (ii) 12

47 Proof **48** 7

Chapter 5

Exercise 5.1 and 5.2

1 a) $y = b^x$
b) Domain $\{x : x \in \mathbb{R}\}$, range $\{y : y > 0\}$
c) (i) 

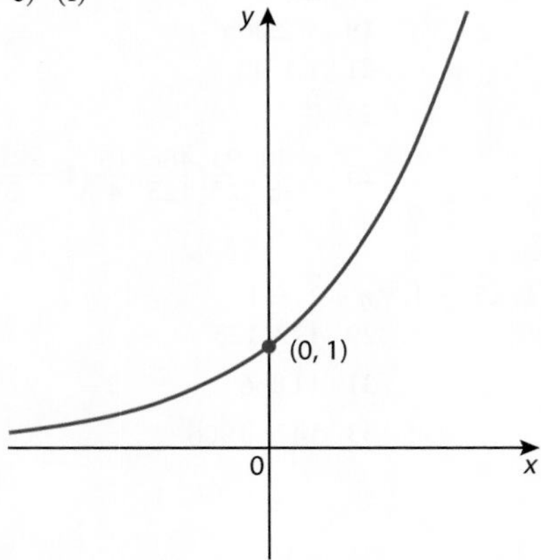

(ii)

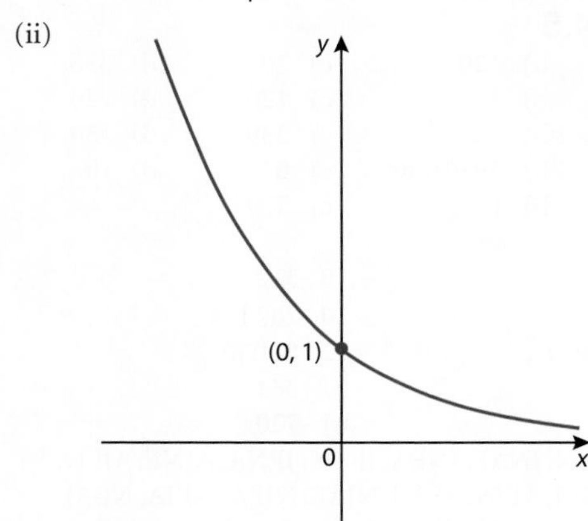

2

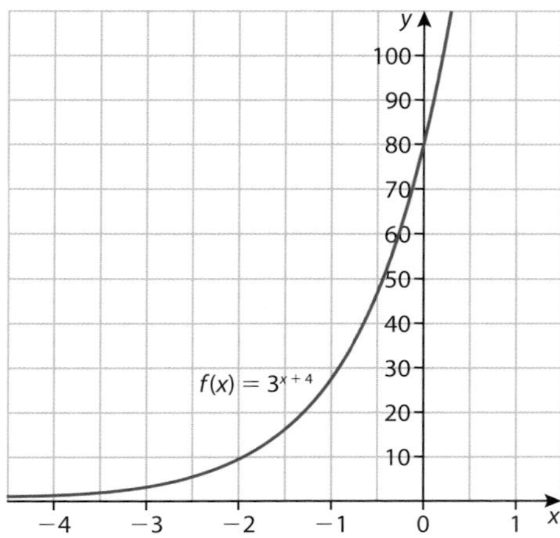

domain: $x \in \mathbb{R}$

range: $y > 0$

y-intercept: $(0, 81)$

horizontal asymptote: $y = 0$ (x-axis)

3

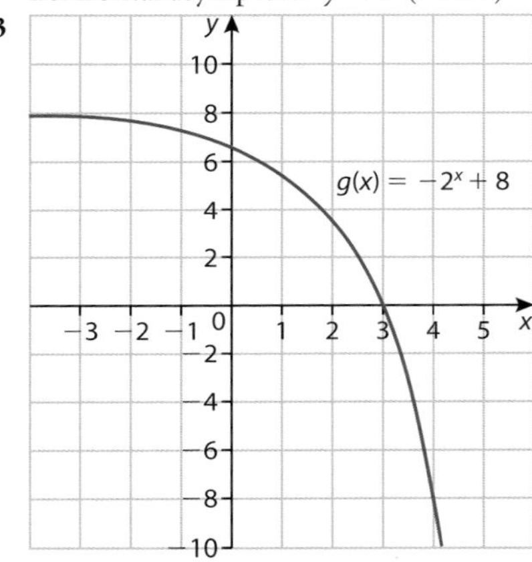

domain: $x \in \mathbb{R}$ range: $y < 8$

y-intercept: $(0, 7)$ horizontal asymptote: $y = 8$

4

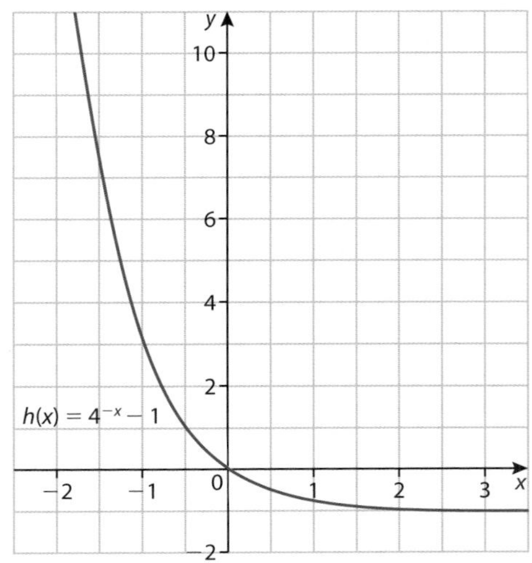

domain: $x \in \mathbb{R}$ range: $y > -1$

y-intercept: $(0, 0)$ horizontal asymptote: $y = -1$

5

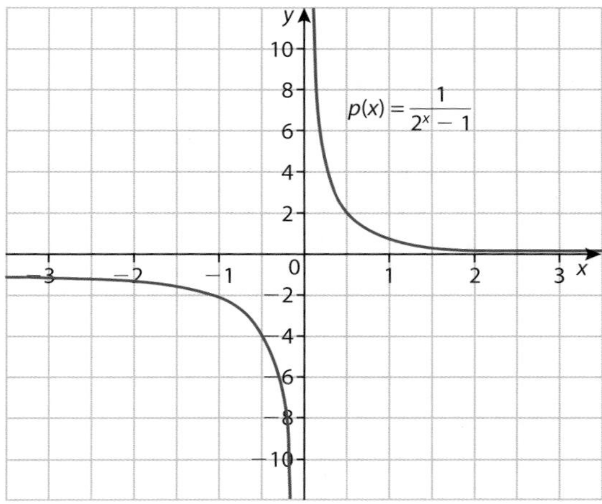

domain: $x \in \mathbb{R}$, $x \neq 0$ range: $y < -1$ or $y > 0$

y-intercept: none

horizontal asymptotes: $y = 0$ and $y = -1$

6

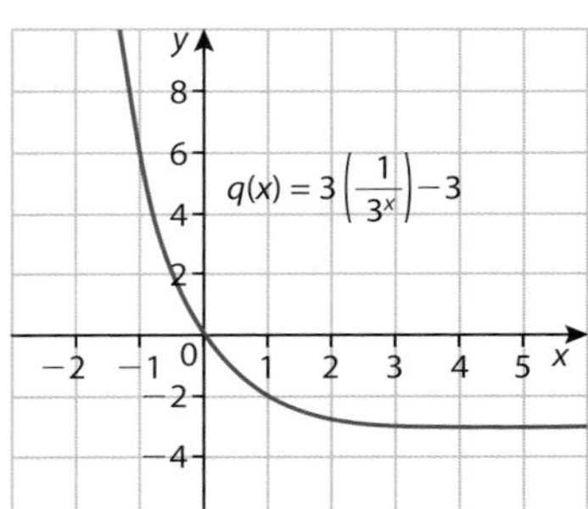

domain: $x \in \mathbb{R}$ range: $y > -3$

y-intercept: $(0, 0)$ horizontal asymptote: $y = -3$

7

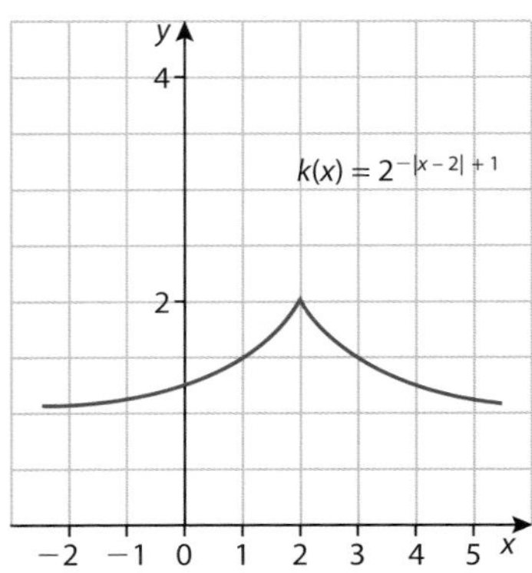

domain: $x \in \mathbb{R}$ range: $y > 1$

y-intercept: $\left(0, \frac{5}{4}\right)$ horizontal asymptote: $y = 1$

8 Domain: $x \in \mathbb{R}$

range: if $a > 0 \Rightarrow y > d$, if $a < 0 \Rightarrow y < d$

y-intercept: $\left(0, a(b)^{-c} + d\right)$ horizontal asymptote: $y = d$

9

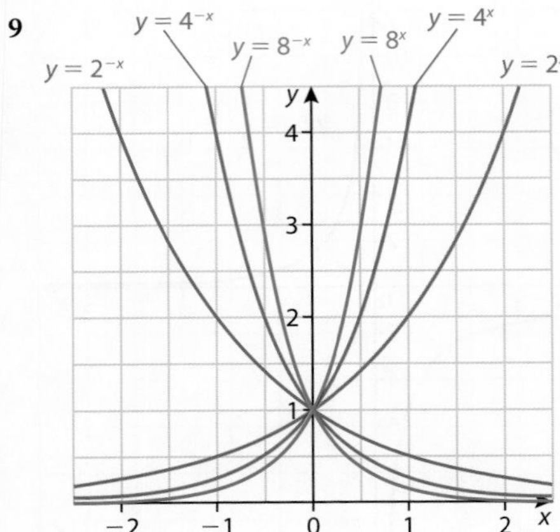

10 a) $y = \left(\frac{1}{2}\right)^x$ b) $y = \left(\frac{1}{4}\right)^x$ c) $y = \left(\frac{1}{8}\right)^x$

11 $y = b^x$ is steeper

12 $P(t) = 100\,000(3)^{\frac{t}{25}}$ where t is number of years
a) 900 000 b) 2 167 402 c) 8 100 000

13 $N(t) = 10^4(2)^{\frac{t}{3}}$
a) 20 000 b) 80 000
c) 5 120 000 d) 10 485 760 000

14 a) $A(t) = A_0(2)^{\frac{t}{10}}$ b) 7.18%

15 a) \$17 204.28 b) \$29 598.74
c) \$50 922.51

16 a) $A(t) = 5000\left(1 + \frac{.09}{12}\right)^{12t}$

b)

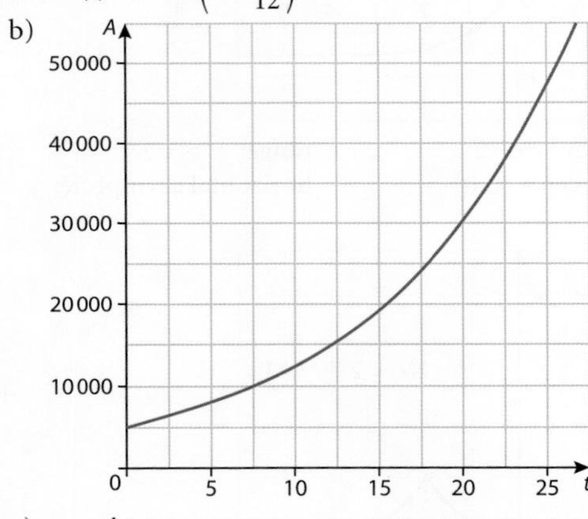

c)

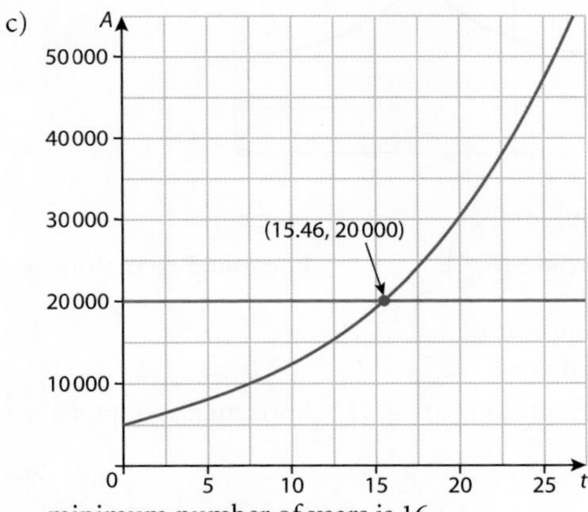

minimum number of years is 16

17 a) \$16 850.58 b) \$17 289.16
c) \$17 331.09 d) \$17 332.47

18 a) \$2 b) \$2.61 c) \$2.71 d) \$2.72 e) \$2.72

19 a) 240 310 b) 192 759

20 8.90%

21 $0.0992A_0$ (or 9.92% of A_0 remains)

22 a) $A(w) = 1000(0.7)^w$ b) About 20 weeks

23 $b > 0$ because if $b = 0$ then the result is always zero, and if $b < 0$ then b^x gives a positive result when x is an even integer and a negative result when x is an odd integer.

24 Payment plan I: \$465; payment plan II: \$10 737 418.23

25 a) $a = 2, k = 3$ b) $a = \frac{1}{3}, k = 2$
c) $a = 3, k = -4$ d) $a = 10, k = \frac{3}{2}$

Exercise 5.3

1

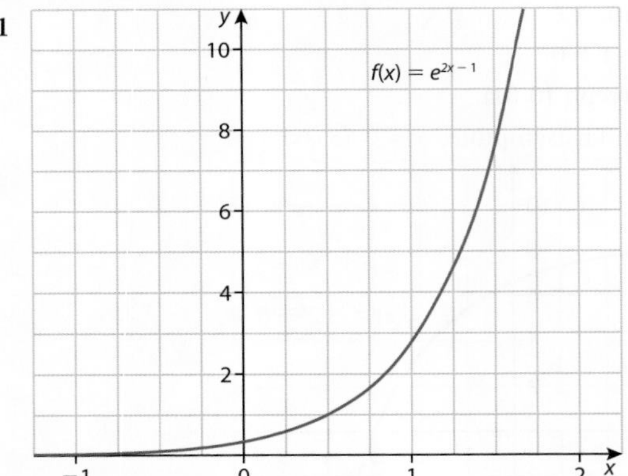

a) Domain: $x \in \mathbb{R}$, range: $y > 0$

b) x-intercept: none, y-intercept: $\left(0, \frac{1}{e}\right)$

c) Horizontal asymptote: $y = 0$

2

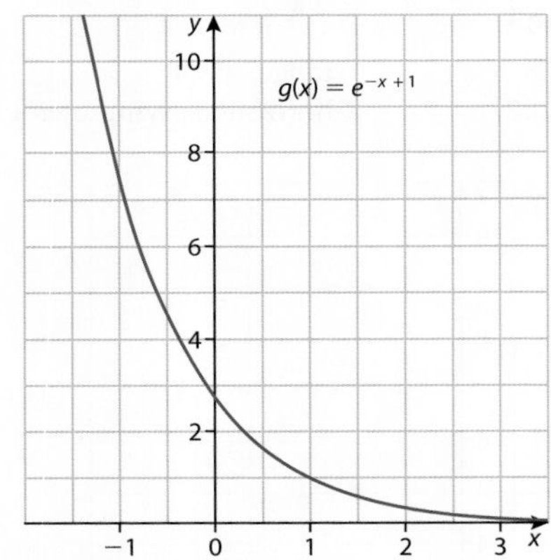

a) Domain: $x \in \mathbb{R}$, range: $y < 0$

b) x-intercept: none, y-intercept: $(0, e)$

c) Horizontal asymptote: $y = 0$

3

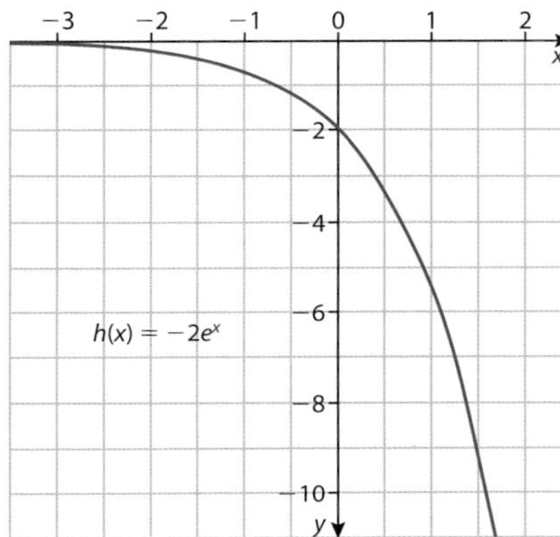

a) Domain: $x \in \mathbb{R}$, range: $y < 0$
b) x-intercept: none, y-intercept: $(0, -2)$
c) Horizontal asymptote: $y = 0$

4

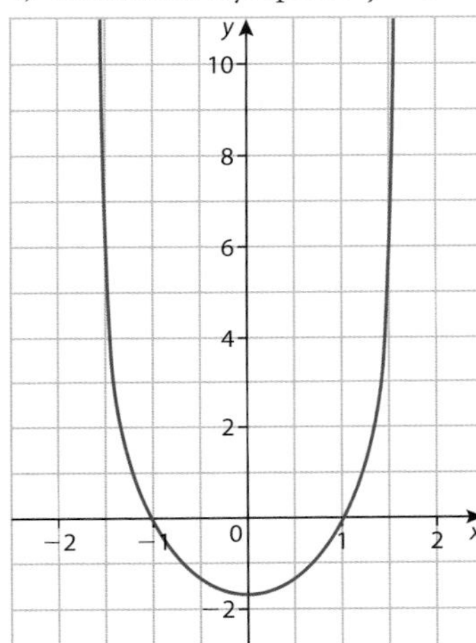

a) Domain: $x \in \mathbb{R}$, range: $y \geqslant 1 - e$
b) x-intercept: $(-1, 0)$ and $(1, 0)$, y-intercept: $(0, 1 - e)$
c) No asymptotes

5

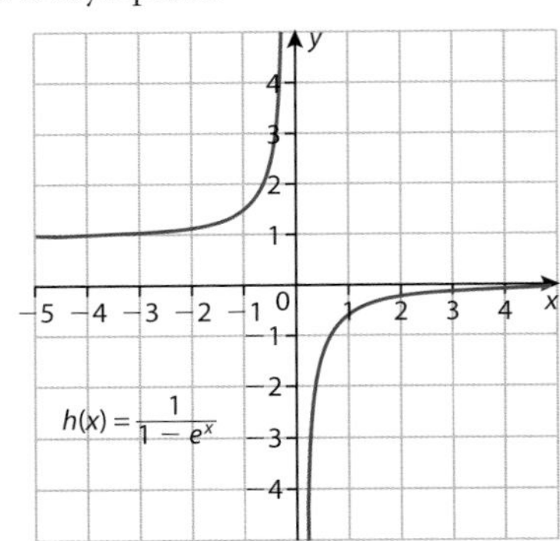

a) Domain: $x \in \mathbb{R}, x \neq 0$, range: $y < 0, y > 1$
b) x-intercept: none, y-intercept: none
c) Horizontal asymptotes: $y = 0$ and $y = 0$

6

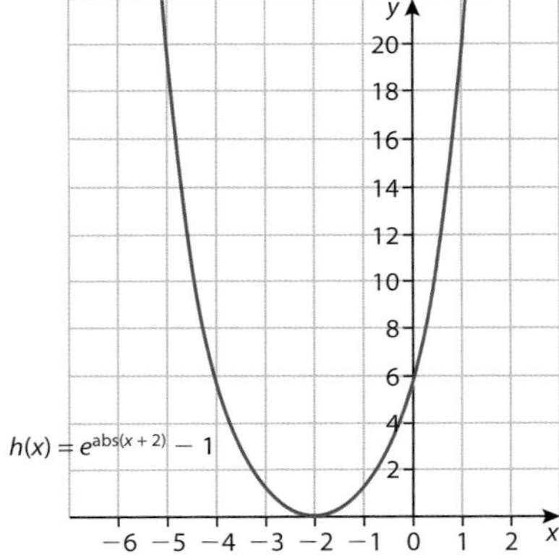

a) Domain: $x \in \mathbb{R}$, range: $y \geqslant 0$
b) x-intercept: $(-2, 0)$, y-intercept: $(0, e^2 - 1)$
c) No asymptotes

7 a) $e = \lim_{x \to \infty} \left(1 + \frac{1}{n}\right)^n$
b) 0.366 032 3413, 0.367 861 0464, 0.367 879 2572
c) 0.367 88; reciprocal of e, $\frac{1}{e} \approx 0.367\,879\,4412$

8 $y = \left(x + \frac{1}{x}\right)^x$ will not intersect $y = 2.72$ because

$\lim_{x \to \infty} \left(x + \frac{1}{x}\right)^x = e \approx 2.718\,281\,828\ldots < 2.72$

9 Bank A: earn 608.79 euros in interest.
Bank B: earn 609.16 euros in interest.
Bank B account earns 0.37 euros more in interest.

10 Blue Star has greater total of \$1358.42 which is \$11.93 more than the Red Star.

11 a) 0.976 kg b) 0.787 kg c) 0.0916 kg d) 0.002 54 kg

12 a) 5 kg b) 71.7%
c)

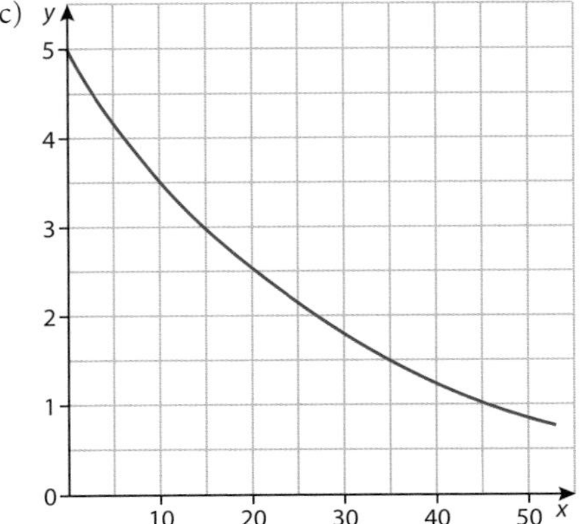

d) 20 days

13 a) $8\frac{1}{2}$% compounded semi-annually is the better investment.

14 a) $r \approx 1.070\,37$ (6 s.f.) b) 7.037% (4 s.f.)

15 a) Less than 1 b) Less than 1
c) Greater than 1 d) Greater than 1

16 a) £1568.31, £2459.60
b) 15.4 years
c) 15.4 years
d) Same; doubling time is independent of initial amount

Exercise 5.4

1 $2^4 = 16$ **2** $e^0 = 1$ **3** $10^2 = 100$
4 $10^{-2} = 0.01$ **5** $7^3 = 343$ **6** $e^{-1} = \frac{1}{e}$
7 $10^y = 50$ **8** $e^{12} = x$ **9** $e^3 = x + 2$
10 $\log_2 1024 = 10$ **11** $\log_{10} 0.0001 = -4$ **12** $\log_4\left(\frac{1}{2}\right) = -\frac{1}{2}$
13 $\log_3 81 = 4$ **14** $\log_{10} 1 = 0$ **15** $\ln 5 = x$
16 $\log_2 0.125 = -3$ **17** $\ln y = 4$
18 $\log_{10} y = x + 1$ **19** 6 **20** 3
21 -3 **22** 5 **23** $\frac{3}{4}$ **24** $\frac{1}{3}$
25 -3 **26** 13 **27** 0 **28** 6
29 -3 **30** $\sqrt{2}$ **31** 3 **32** $\frac{1}{2}$
33 -2 **34** 88 **35** $\frac{1}{2}$ **36** 18
37 $\frac{1}{3}$ **38** π **39** 1.6990 **40** 0.2386
41 3.912 **42** 0.5493 **43** 1.398 **44** 0.2090
45 4.605 **46** 13.82 **47** $x > 2$ **48** $x \in \mathbb{R}$
49 $x > 0$ **50** $x < \frac{8}{5}$ **51** $-2 \leqslant x < 3$ **52** $x < 0$
53 Domain $\{x : x > 0,\ x \neq 1\}$, range $\{y : y \in \mathbb{R},\ y \neq 0\}$
54 Domain $\{x : x > 1\}$, range $\{y : y \geqslant 0\}$
55 Domain: $x > 0, x \neq 1$, range: $y < 0$
56 $f(x) = \log_4 x$ **57** $f(x) = \log_2 x$
58 $f(x) = \log_{10} x$ **59** $f(x) = \log_3 x$
60 $\log_2 2 + \log_2 m = 1 + \log_2 m$ **61** $\log 9 - \log x$
62 $\frac{1}{5}\ln x$ **63** $\log a + 3\log b$
64 $\log 10x + \log(1+r)^t = \log 10 + \log x + t\log(1+r)$
65 $3\ln m - \ln n$ **66** $\log_b p + \log_b q + \log_b r$
67 $2\log_b p + 3\log_b q - \log_b r$ **68** $\frac{\log_b p}{4} + \frac{\log_b q}{4}$
69 $\frac{\log_b q}{2} + \frac{\log_b r}{2} - \frac{\log_b p}{2}$ **70** $\log_b p + \frac{1}{2}\log_b q - \log_b r$
71 $3\log_b p + 3\log_b q - \frac{1}{2}\log_b r$ **72** $\log x$
73 $\log_3 72$ **74** $\ln\left(\frac{y^4}{4}\right)$ **75** $\log_b 4$ **76** $\log\left(\frac{p}{qr}\right)$
77 $\ln\left(\frac{36}{e}\right)$ **78** 9.97 **79** -5.32 **80** 2.06
81 -0.179 **82** 4.32 **83** 1.86
84 $\log_b a = \frac{\log_a a}{\log_a b} = \frac{1}{\log_a b}$ **85** $\log e = \frac{\ln e}{\ln 10} = \frac{1}{\ln 10}$
86 $dB = 10\log\left(\frac{I}{10^{-16}}\right) = 10(\log I - \log 10^{-16}) = 10(\log I + 16)$
$= 10\log 10^{-4} + 160 = 10(-4) + 160 = 120$ decibels

Exercise 5.5

1 0.699 **2** 2.5 **3** 7.99 **4** 3.64
5 -1.92 **6** 2.71 **7** 0.434 **8** 2.12
9 4.42 **10** 0.225 **11** 0.642 **12** 22.0
13 3 **14** 0 or -1 **15** $\frac{\ln\left(\frac{3}{2}\right)}{\ln 6}$ or $\frac{\ln\left(\frac{4}{3}\right)}{\ln 6}$
16 1 or -1
17 a) \$6248.58 b) $9\frac{1}{4}$ years
18 12.9 years
19 20 hours (≈ 19.93)
20 a) 24 years (≈ 23.45) b) 12 years (≈ 11.9)
c) 9 years (≈ 8.04)
21 6 years
22 a) 99.7% b) 139 000 years
23 a) 37 dogs b) 9 years
24 a) 458 litres b) 8.89 minutes $\approx$ 8 min. 53 seconds
c) 39 minutes
25 a) 5 kg b) 17.7 days
26 $x = \frac{20}{3}$ **27** $x = 104$ **28** $x = \frac{1}{e^3}$
29 $x = 4$ **30** $x = 98$ **31** $x = \pm\sqrt{e^{16}} \approx \pm 2980.96$
32 $x = 2$ or $x = 4$ **33** $x = 9$ **34** $x = \frac{13}{5}$
35 $x = 3$ **36** $x = 1$ or $x = 100$
37 $x > \frac{1}{\sqrt[5]{100}}$ **38** $x < 2$ **39** $0 < x < \ln 6$
40 $0.161 < x < 1.14$ (approx. to 3 s.f.)

Practice questions

1 a) $(8, 0)$ b) $(0, 2)$ c) $\left(-\frac{2}{3}, 3\right)$
2 a) 183 g (3 s.f.) b) 154 years (3 s.f.)
3 a) $a_n = \ln(y^n)$, $S_n = \frac{n(n+1)}{2}\ln y$
b) $a_n = \ln(xy^n)$, $S_n = n\ln x + \frac{n(n+1)}{2}\ln y$
4 $x = 2$ **5** $y = 16$ **6** $x = 0, \ln\left(\frac{1}{2}\right)$ or $-\ln$
7 $x = e^{-4e}$ or e^{2e}
8 a) $x = 3$ b) $x = 6$
9 a) $\log\left(\frac{a^2b^3}{c}\right)$ b) $\ln\left(\frac{ex^3}{\sqrt{y}}\right)$
10 1900 years
11 $c = 22$
12 a)

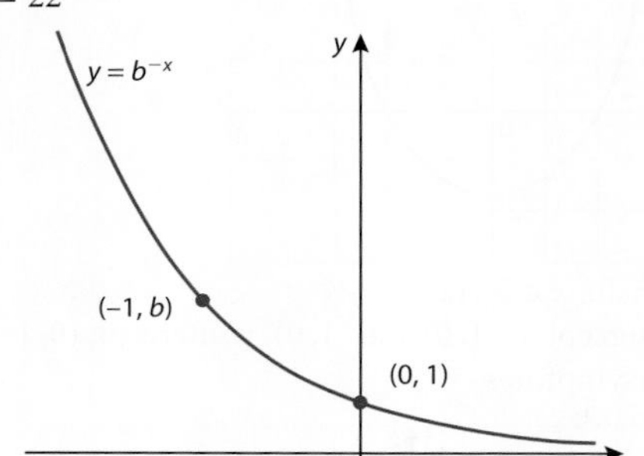

b)

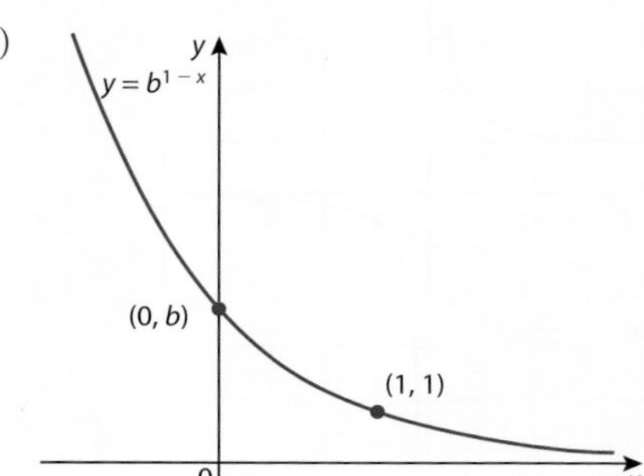

13 a) $k \approx 0.000\,4332$ b) 17.7% (3 s.f.)
14 $x \approx 1.28$
15 $1.52 < x < 1.79 \cup 17.6 < x < 19.1$
16 $-1 < x < -0.800 \cup x > 1$
17 a) $x = -\frac{1}{2}$ or $x = 0$
b) $x = \frac{1}{\ln a - 2}$ or $x = \frac{\log_a e}{1 - 2\log_a e}$
c) $a = e^2$
18 $a = -2, b = 3$

19 $x = \sqrt{e}, x = e$

20 a) $V = \$265.33$ b) 235 months

21 $x = 5^{\frac{5}{3}}$ or $x = 5^{\frac{-5}{3}}$

22 $x = e - 3$ or $x = \frac{1}{e} - 3$

23 $x = -2.50, -1.51$ or 0.440 (3 s.f.)

24 $k = \frac{\ln 2}{20}$

25 a) $f(x) = \ln\left(\frac{x}{x+2}\right)$ b) $f^{-1}(x) = -\frac{2e^x}{e^x - 1}$ or $\frac{2e^x}{1 - e^x}$

26 a) (i) Minimum value of f is 0.
(ii) From part (i) $f(x) \geqslant 0 \Rightarrow e^x - 1 - x \geqslant 0 \Rightarrow e^x \geqslant 1 + x$

d) $n > e^{100}$

Chapter 6

Exercise 6.1 and 6.2

1 a) (i) $\begin{pmatrix} x-1 & x-3 \\ y+3 & y+1 \end{pmatrix}$ (ii) $\begin{pmatrix} -x-7 & 3x+3 \\ 3y-7 & 11-y \end{pmatrix}$

b) $x = -3, y = 5$ c) $x = 3, y = -3$

d) $AB = \begin{pmatrix} 2x-2 & xy-2x+6 \\ xy-x+y+11 & -3 \end{pmatrix}$;

$BA = \begin{pmatrix} -2x-3y+1 & x^2+x-9 \\ y^2-3y-6 & 4x+3y-6 \end{pmatrix}$

2 a) $x = 2, y = -10$ b) $p = 2, q = -4$

3 a) $\begin{bmatrix} 0&1&0&0&1&2&0 \\ 1&0&1&1&1&1&0 \\ 0&1&0&2&0&0&2 \\ 0&1&2&0&1&0&0 \\ 1&1&0&1&0&1&0 \\ 2&1&0&0&1&0&0 \\ 0&0&2&0&0&0&0 \end{bmatrix}$ b) $\begin{bmatrix} 6&3&1&2&3&2&0 \\ 3&5&2&3&3&3&2 \\ 1&2&9&1&3&1&0 \\ 2&3&1&6&1&2&4 \\ 3&3&3&1&4&3&0 \\ 2&3&1&2&3&6&0 \\ 0&2&0&4&0&0&4 \end{bmatrix}$

Matrix signifies the number of routes between each pair that go via one other city.

4 a) $A + C = \begin{pmatrix} x+1 & 10 & y+1 \\ 0 & -x-3 & y+3 \\ 2x+y+7 & x-3y & -x+2y-1 \end{pmatrix}$

b) $\begin{pmatrix} 17m+2 & -6 \\ 4-9m & 9 \\ 7m-2 & -17 \end{pmatrix}$

c) Not possible d) $x = 3, y = 1$

e) Not possible f) $m = 3$

5 $a = -3, b = 3, c = 2$

6 $x = 4, y = -3$

7 $m = 2, n = 3$

8 Shop A: €18.77

9 a) $\begin{pmatrix} 2 & 4 \\ -2 & 12 \end{pmatrix}$ b) associative

c) $\begin{pmatrix} -22 & 16 \\ 60 & -7 \end{pmatrix}$ d) associative

10 $AB = [88\ 142]$, which represents total profit.

11 $r = 3, s = -2$

12 a) (i) $\begin{pmatrix} 1 & 2 \\ 0 & 1 \end{pmatrix}$ (ii) $\begin{pmatrix} 1 & 3 \\ 0 & 1 \end{pmatrix}$

(iii) $\begin{pmatrix} 1 & 4 \\ 0 & 1 \end{pmatrix}$ (iv) $\begin{pmatrix} 1^n & n \\ 0 & 3^n \end{pmatrix}$

b) (i) $\begin{pmatrix} 9 & 18 \\ 0 & 9 \end{pmatrix}$ (ii) $\begin{pmatrix} 27 & 81 \\ 0 & 27 \end{pmatrix}$

(iii) $\begin{pmatrix} 81 & 324 \\ 0 & 81 \end{pmatrix}$ (iv) $\begin{pmatrix} 3^n & 3^{n+1} \\ 0 & 3^n \end{pmatrix}$

13 $\left(\frac{11}{3}, \frac{8}{3}\right)$ **14** $(1, -4)$

15 5 **16** $(5, 1)$

Exercise 6.3

1 a) $\begin{pmatrix} -9 & -7 \\ 4 & 3 \end{pmatrix}$ b) $M = \begin{pmatrix} -9 & -7 \\ 4 & 3 \end{pmatrix}\begin{pmatrix} 2 & 1 \\ 3 & 5 \end{pmatrix}$

c) $\begin{pmatrix} -39 & -44 \\ 17 & 19 \end{pmatrix}$

d) (i) $N = \begin{pmatrix} 2 & 1 \\ 3 & 5 \end{pmatrix}\begin{pmatrix} -9 & -7 \\ 4 & 3 \end{pmatrix}$ (ii) $N = \begin{pmatrix} -14 & -11 \\ -7 & -6 \end{pmatrix}$

e) If $AB = C$ then $B = A^{-1}C$, while if $BA = C$, then $B = CA^{-1}$. Also, $A^{-1}C \neq CA^{-1}$.

2 $\begin{pmatrix} 1 & -\frac{3}{5} \\ 0 & 0 \end{pmatrix}$

3 a) $|A| = -5 \neq 0$ b) $\begin{pmatrix} \frac{9}{5} & \frac{11}{5} & -\frac{8}{5} \\ \frac{6}{5} & \frac{9}{5} & -\frac{7}{5} \\ 1 & 1 & -1 \end{pmatrix}$ c) $\begin{pmatrix} \frac{1}{2} \\ -1 \\ \frac{1}{5} \end{pmatrix}$

4 a) $\begin{pmatrix} \frac{\sqrt{3}}{2} & \frac{1}{2} \\ -\frac{1}{2} & \frac{\sqrt{3}}{2} \end{pmatrix}$ b) $\begin{pmatrix} \frac{3}{a}+1 & -1 \\ -a-2 & a \end{pmatrix}$

5 $x = 2$ or $x = 3$

6 $n = 0.5$

7 a) $X = \begin{pmatrix} \frac{1}{2} & 0 \\ \frac{3}{4} & -\frac{7}{6} \end{pmatrix}$ b) $Y = \begin{pmatrix} 1 & \frac{13}{12} \\ -1 & -\frac{5}{3} \end{pmatrix}$

c) $X \neq Y$ – not commutative

8 a) $PQ = \begin{pmatrix} 5 & -4 & 3 \\ 33 & 5 & -1 \\ 2 & -3 & 2 \end{pmatrix}, QP = \begin{pmatrix} 4 & -5 & -8 \\ 8 & 0 & -4 \\ 7 & 10 & 8 \end{pmatrix}$

b) $P^{-1} = \begin{pmatrix} 1 & 0 & -1 \\ -\frac{7}{5} & \frac{1}{5} & \frac{11}{5} \\ 1 & 0 & -2 \end{pmatrix}, Q^{-1} = \begin{pmatrix} 0 & \frac{1}{4} & 0 \\ 1 & -1 & 1 \\ 2 & -\frac{7}{4} & 1 \end{pmatrix}$

$P^{-1}Q^{-1} = \begin{pmatrix} -2 & 2 & -1 \\ \frac{23}{5} & -\frac{22}{5} & \frac{12}{5} \\ -4 & \frac{15}{4} & -2 \end{pmatrix}$

$Q^{-1}P^{-1} = \begin{pmatrix} -\frac{7}{20} & \frac{1}{20} & \frac{11}{20} \\ \frac{17}{5} & -\frac{1}{5} & -\frac{26}{5} \\ \frac{109}{20} & -\frac{7}{20} & -\frac{157}{20} \end{pmatrix}$

$(PQ)^{-1} = \begin{pmatrix} -\frac{7}{20} & \frac{1}{20} & \frac{11}{20} \\ \frac{17}{5} & -\frac{1}{5} & -\frac{26}{5} \\ \frac{109}{20} & -\frac{7}{20} & -\frac{157}{20} \end{pmatrix}$

$(QP)^{-1} = \begin{pmatrix} -2 & 2 & -1 \\ \frac{23}{5} & -\frac{22}{5} & \frac{12}{5} \\ -4 & \frac{15}{4} & -2 \end{pmatrix}$

9 a) $\begin{pmatrix} -7 \\ 3 \\ -2 \end{pmatrix}$ b) $\begin{pmatrix} -7 \\ 3 \\ -2 \end{pmatrix}$

10 $x = -1$ **11** $x = 1, y = 2$

12 $(0, 1)$ **13** $(-3, -29), (0, 1)$

14 $17x - 8y + 37 = 0; y + 2 = 0; x + 5 = 0$ **15** 165; 80; 136

16 $x = \frac{89}{2}$ or $x = \frac{129}{8}$; $x = -4$ or $x = -2$ or $x = -3 \pm\sqrt{21}$

17 $-3; 3$

18 a) -25
b) $x^2 - 7x - 25$, constant $= \det(A)$
c) $-(a + d)$
d) $f(A) = 0$
e) $ad - bc$; $x^2 - (a + d)x + (ad - bc)$, constant $= \det(A)$; $f(A) = 0$

19 a) -22
b) $x^3 - x^2 - 22x + 22$, constant $= -\det(A)$
c) Opposite of the sum of the main diagonal
d) $f(A) = 0$

Exercise 6.4

1 $\mathbf{m} = 2$ or $\mathbf{m} = 3$

2 a) $a = 7, b = 2$ b) $(-1, 2, -1)$

3 $\mathbf{m} = 2$

4 a) $(-1, 3, 2)$ b) $(5, 8, -2)$
c) $\left(\frac{13}{16} + \frac{5}{16}t, \frac{11}{16} + \frac{19}{16}t, t\right)$ d) $(-7, 3, -2)$
e) $(-1 + 2t, 2 - 3t, t)$ f) inconsistent
g) $(-2, 4, 3)$ h) $(4, -2, 1)$

5 a) $k \neq \frac{-1 \pm \sqrt{33}}{4}$ b) $k = 1$

c) $\begin{pmatrix} 1 & 0 & 0 & -2 & -3 & 1 \\ 0 & 1 & 0 & 3 & 3 & -1 \\ 0 & 0 & 1 & -2 & -4 & 1 \end{pmatrix}$

6 a) $\frac{71 \pm i\sqrt{251}}{42}$ b) $k = 2$

c) $\begin{pmatrix} 1 & 0 & 0 & \frac{3}{5} & \frac{1}{5} & \frac{2}{5} \\ 0 & 1 & 0 & \frac{2}{5} & \frac{4}{5} & \frac{-3}{5} \\ 0 & 0 & 1 & \frac{3}{5} & \frac{6}{5} & -1 \end{pmatrix}$

7 $\begin{pmatrix} 1 & 0 & 0 & \frac{1}{2} & -1 & -\frac{1}{2} \\ 0 & 1 & 0 & \frac{1}{2} & -\frac{2}{3} & -\frac{5}{6} \\ 0 & 0 & 1 & 0 & \frac{2}{3} & \frac{1}{3} \end{pmatrix}; \begin{pmatrix} 1 & 0 & 0 & 2 & \frac{-16}{13} & \frac{-19}{13} \\ 0 & 1 & 0 & 1 & \frac{-11}{13} & \frac{-9}{13} \\ 0 & 0 & 1 & -1 & \frac{12}{13} & \frac{11}{13} \end{pmatrix}$

B is the inverse of A

8 a) $f(x) = 4x^2 - 6x - 5$
b) $f(x) = \frac{1}{2}(m - 27)x^2 + \frac{3}{2}(17 - m)x + m, m \in \mathbb{R}$
c) $f(x) = 3x^3 - 2x^2 - 7x + 3$
d) $f(x) = \frac{1}{6}(4 - m)x^3 + \frac{1}{3}(4 - m)x^2 - \frac{5}{6}(4 - m)x + m, m \in \mathbb{R}$

9 $m = 2, \begin{pmatrix} -t - \frac{3}{5} \\ -t - \frac{19}{5} \\ 5t \end{pmatrix}$ **10** $m = -1, \begin{pmatrix} 7t - \frac{9}{5} \\ \frac{3}{5} - 11t \\ 5t \end{pmatrix}$

11 a) 3 b) $\begin{pmatrix} 3 & -4 & -6 \\ 0 & -2 & -3 \\ 0 & 0 & -\frac{1}{2} \end{pmatrix}$ c) 3

d) -1672 e) $\begin{pmatrix} 2 & 1 & -3 & 5 \\ 0 & 1 & 2 & -16 \\ 0 & 0 & 36 & -184 \\ 0 & 0 & 0 & -\frac{209}{9} \end{pmatrix}$ f) -1672

Practice questions

1 $x = -7$ or $x = 1$

2 a) $\begin{pmatrix} a^2 + 4 & 2a - 2 \\ 2a - 2 & 5 \end{pmatrix}$
b) $a = -1$; $\binom{x}{y} = \binom{1}{-1}$

3 $B = \begin{pmatrix} 1 & 3 \\ 4 & 12 \end{pmatrix}$

4 $a = \frac{28}{33}; b = \frac{59}{33}; c = \frac{20}{33}; d = \frac{28}{33}$

5 a) $A^{-1} = \begin{pmatrix} \frac{1}{19} & \frac{2}{19} \\ \frac{-7}{19} & \frac{5}{19} \end{pmatrix}$
b) (i) $X = (C - B)A^{-1}$ (ii) $X = \begin{pmatrix} 2 & -3 \\ -4 & 1 \end{pmatrix}$

6 a) $A + B = \begin{pmatrix} a + 1 & b + 2 \\ c + d & 1 + c \end{pmatrix}$
b) $AB = \begin{pmatrix} a + bd & 2a + bc \\ c + d & 3c \end{pmatrix}$

7 a) $\begin{pmatrix} 0.1 & 0.4 & 0.1 \\ -0.7 & 0.2 & 0.3 \\ -1.2 & 0.2 & 0.8 \end{pmatrix}$
b) $x = 1.2, y = 0.6, z = 1.6$

8 a) $Q = \begin{pmatrix} -3 & 2 \\ 1 & \frac{14 - a}{3} \end{pmatrix}$
b) $CD = \begin{pmatrix} -14 & -4 + 4a \\ -2 & 2 + 7a \end{pmatrix}$
c) $D^{-1} = \frac{1}{5a + 2}\begin{pmatrix} a & -2 \\ 1 & 5 \end{pmatrix}$

9 a) $(7, 2)$ b) $(-1, 2, -1)$

10 a) $B = A^{-1}C$ b) $DA = \begin{pmatrix} 1 & 0 & 0 \\ 0 & 1 & 0 \\ 0 & 0 & 1 \end{pmatrix}, B = \begin{pmatrix} 1 \\ -1 \\ 2 \end{pmatrix}$
c) $(1, -1, 2)$

11 a) Det = 0 b) $\lambda = 5$ c) $(2 - 3t, 1 + t, t)$

12 No answer required – proof

Chapter 7

Exercise 7.1

1 $\frac{\pi}{3}$ **2** $\frac{5\pi}{6}$ **3** $-\frac{3\pi}{2}$ **4** $\frac{\pi}{5}$

5 $\frac{3\pi}{4}$ **6** $\frac{5\pi}{18}$ **7** $-\frac{\pi}{4}$ **8** $\frac{20\pi}{9}$

9 $-\frac{8\pi}{3}$

10 135° **11** -630° **12** 115° **13** 210°

14 -143° **15** 300° **16** 115° **17** $89.95^\circ \approx 90^\circ$

18 480°　**19** 390°, −330°　**20** $\frac{7\pi}{2}, -\frac{\pi}{2}$　**21** 535°, −185°

22 $\frac{11\pi}{6}, -\frac{13\pi}{6}$　**23** $\frac{11\pi}{3}, -\frac{\pi}{3}$

24 $3.25 + 2\pi \approx 9.5$, $3.25 - 2\pi \approx -3.03$

25 12.6 cm　**26** 14.7 cm

27 1.5 radians, or approx. 85.9°　**28** $r \approx 7.16$

29 Area ≈ 13.96 ≈ 14.0 cm²　**30** Area ≈ 131 cm²

31 $\alpha = 3$ (radian measure), or $\alpha = 172°$　**32** 32 cm

33 6.77 cm

34 a) 3π radians/second　b) 11.9 km/hr

35 19.8 radians/second　**36** $v = \omega \times r$

37 28.3 cm　**38** 20 944 sq metres

39 a) $r \approx 30.6$ cm　b) $r \approx 0.0771$ cm

40 $150\sqrt{3}$ cm²　**41** Area of circle $= \left(\frac{\pi - 2}{4\pi}\right)A$

Exercise 7.2

1 a) $t = \frac{\pi}{6} : \left(\frac{\sqrt{3}}{2}, \frac{1}{2}\right)$; $t = \frac{\pi}{3} : \left(\frac{1}{2}, \frac{\sqrt{3}}{2}\right)$

2 0.6　**3** 1.0　**4** 0.5　**5** 0.5

6 2.7　**7** 0.1　**8** 0.3　**9** 1.6

10 a) I　b) $\left(\frac{\sqrt{3}}{2}, \frac{1}{2}\right)$

11 a) IV　b) $\left(\frac{1}{2}, -\frac{\sqrt{3}}{2}\right)$

12 a) IV　b) $\left(\frac{\sqrt{2}}{2}, -\frac{\sqrt{2}}{2}\right)$

13 a) Negative x-axis　b) $(0, -1)$

14 a) II　b) $(-0.416, 0.909)$

15 a) I　b) $\left(\frac{\sqrt{2}}{2}, \frac{\sqrt{2}}{2}\right)$

16 a) IV　b) $(0.540, 0.841)$

17 a) II　b) $\left(-\frac{\sqrt{2}}{2}, \frac{\sqrt{2}}{2}\right)$

18 a) III　b) $(-0.929, -0.369)$

19 $\sin\frac{\pi}{3} = \frac{\sqrt{3}}{2}$, $\cos\frac{\pi}{3} = \frac{1}{2}$, $\tan\frac{\pi}{3} = \sqrt{3}$

20 $\sin\frac{5\pi}{6} = \frac{1}{2}$, $\cos\frac{5\pi}{6} = -\frac{\sqrt{3}}{2}$, $\tan\frac{5\pi}{6} = -\frac{\sqrt{3}}{3}$

21 $\sin\left(-\frac{3\pi}{4}\right) = -\frac{\sqrt{2}}{2}$, $\cos\left(-\frac{3\pi}{4}\right) = -\frac{\sqrt{2}}{2}$, $\tan\left(-\frac{3\pi}{4}\right) = 1$

22 $\sin\frac{\pi}{2} = \frac{1}{2}$, $\cos\frac{\pi}{2} = 0$, $\tan\frac{\pi}{2}$ is undefined

23 $\sin\left(-\frac{4\pi}{3}\right) = \frac{\sqrt{3}}{2}$, $\cos\left(-\frac{4\pi}{3}\right) = -\frac{1}{2}$, $\tan\left(-\frac{4\pi}{3}\right) = -\sqrt{3}$

24 $\sin 3\pi = 0$, $\cos 3\pi = -1$, $\tan 3\pi = 0$

25 $\sin\frac{3\pi}{2} = -1$, $\cos\frac{3\pi}{2} = 0$, $\tan\frac{3\pi}{2}$ is undefined

26 $\sin\left(-\frac{7\pi}{6}\right) = \frac{1}{2}$, $\cos\left(-\frac{7\pi}{6}\right) = -\frac{\sqrt{3}}{2}$, $\tan\left(-\frac{7\pi}{6}\right) = -\frac{\sqrt{3}}{3}$

27 $\sin(1.25\pi) = -\frac{\sqrt{2}}{2}$, $\cos(1.25\pi) = -\frac{\sqrt{2}}{2}$, $\tan(1.25\pi) = 1$

28 $\sin\frac{13\pi}{6} = \sin\frac{\pi}{6} = \frac{1}{2}$; $\cos\frac{13\pi}{6} = \cos\frac{\pi}{6} = \frac{\sqrt{3}}{2}$

29 $\sin\frac{10\pi}{3} = \sin\frac{4\pi}{3} = -\frac{\sqrt{3}}{2}$; $\cos\frac{10\pi}{3} = \cos\frac{4\pi}{3} = -\frac{1}{2}$

30 $\sin\frac{15\pi}{4} = \sin\frac{7\pi}{4} = -\frac{\sqrt{2}}{2}$; $\cos\frac{15\pi}{4} = \cos\frac{7\pi}{4} = \frac{\sqrt{2}}{2}$

31 $\sin\frac{17\pi}{6} = \sin\frac{5\pi}{6} = \frac{1}{2}$; $\cos\frac{17\pi}{6} = \cos\frac{5\pi}{6} = -\frac{\sqrt{3}}{2}$

32 a) $-\frac{\sqrt{3}}{2}$　b) $-\frac{\sqrt{2}}{2}$　c) undefined

d) 2　e) $-\frac{2\sqrt{3}}{3}$

33 a) 0.598　b) $-\frac{\sqrt{3}}{3}$　c) $\frac{1}{2}$　d) 1.04　e) 0

34 I, II　**35** II

36 III　**37** II

38 I, IV　**39** I

40 IV　**41** II, IV

Exercise 7.3

1

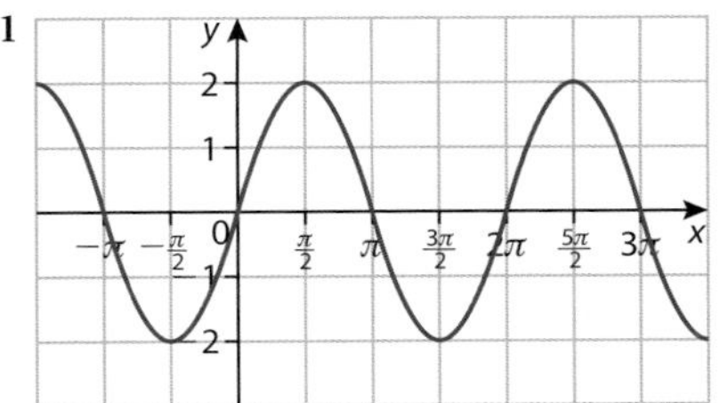

2

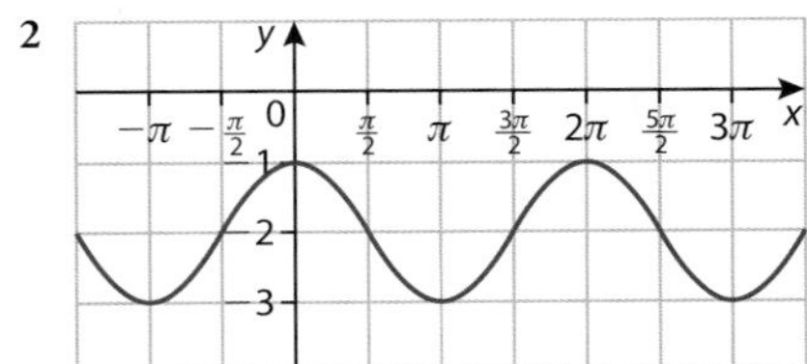

3

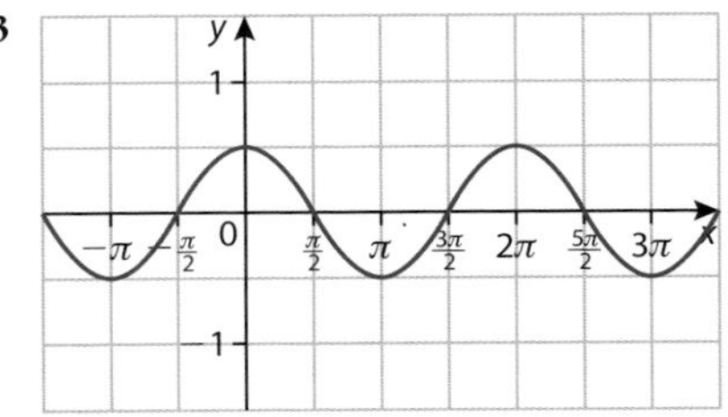

4

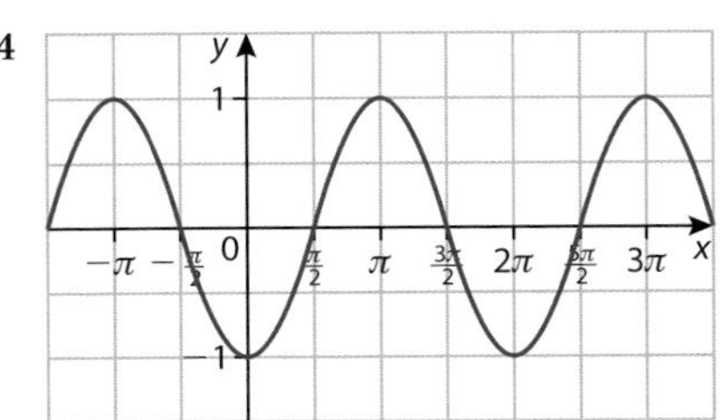

5

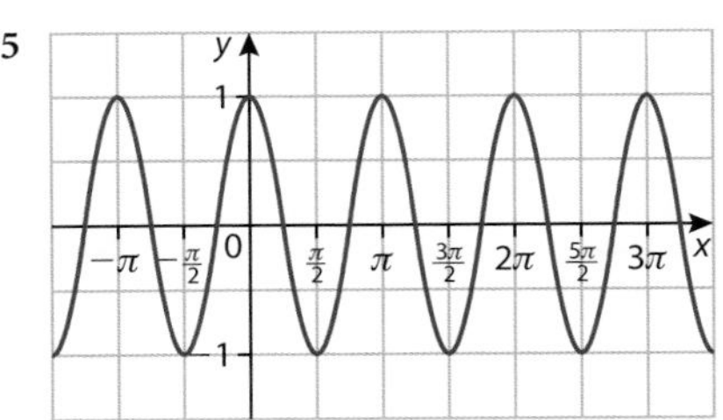

6

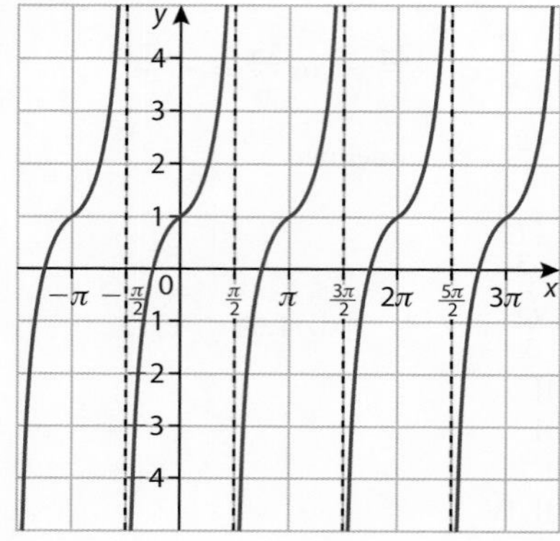

7

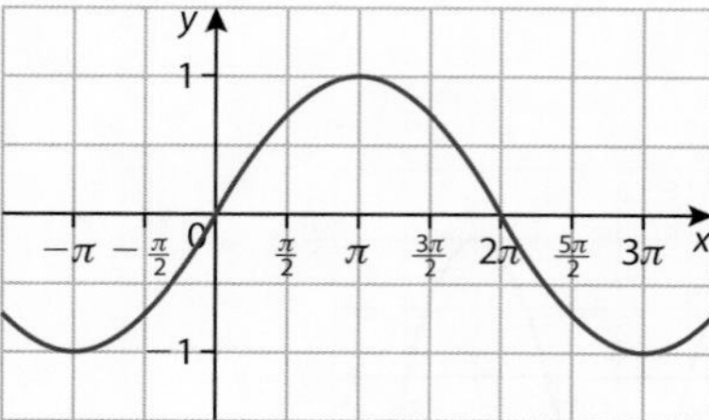

8

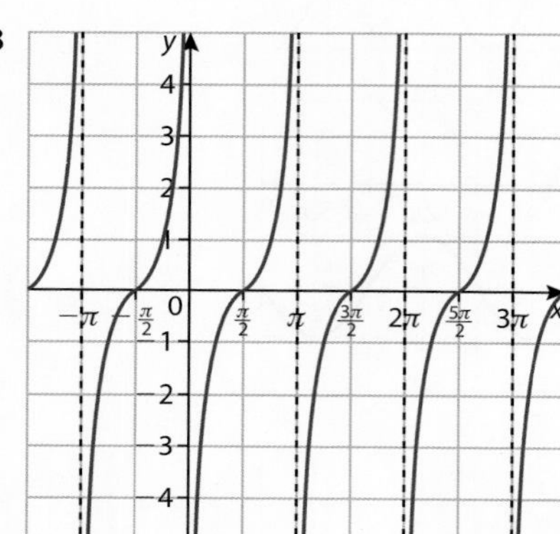

9

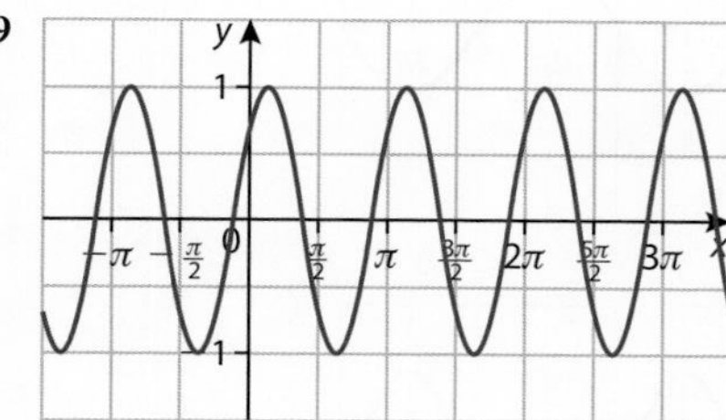

10 a)

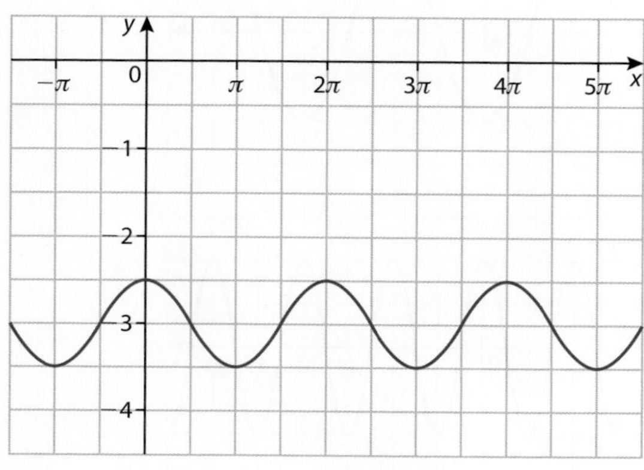

amplitude $= \frac{1}{2}$, period $= 2\pi$

b) Domain: $x \in \mathbb{R}$, range: $-3.5 \leqslant y \leqslant -2.$

11 a)

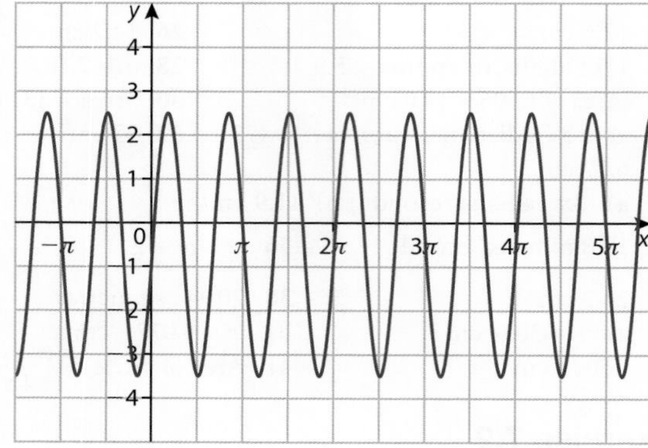

amplitude $= 3$, period $= \frac{2\pi}{3}$

b) Domain: $x \in \mathbb{R}$, range: $-3.5 \leqslant y \leqslant 2.5$

12 a)

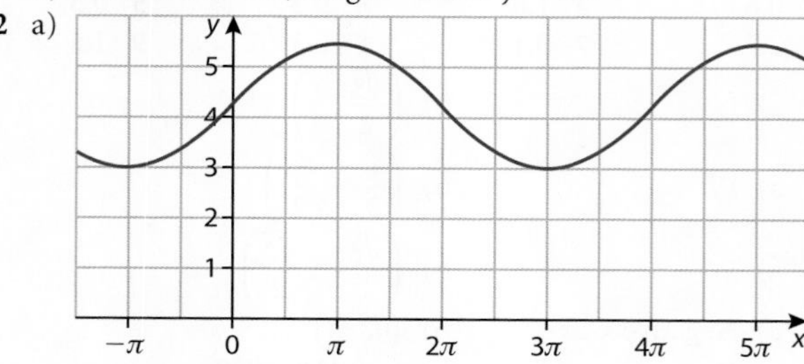

amplitude 1.2, period $= 4\pi$

b) Domain: $x \in \mathbb{R}$, range: $3.1 \leqslant y \leqslant 5.5$

13 $A = 3$, $B = 7$

14 $A = 2.7$, $B = 5.9$

15 $A = 1.9$, $B = 4.3$

16 a) $p = 8$ b) $q = 6$

17 a)

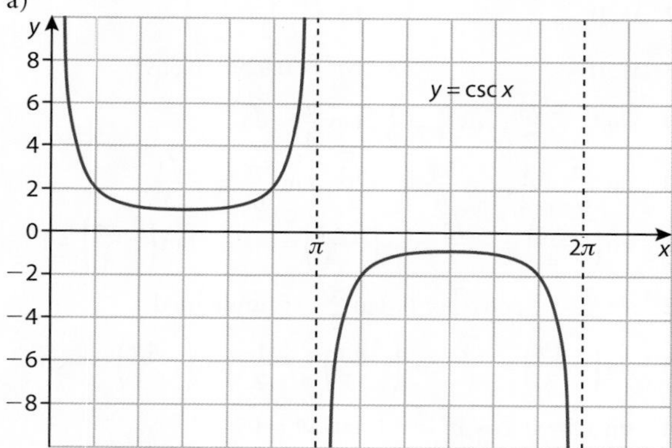

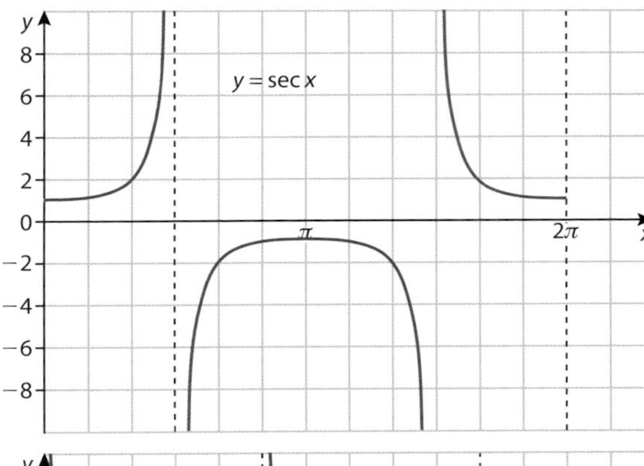

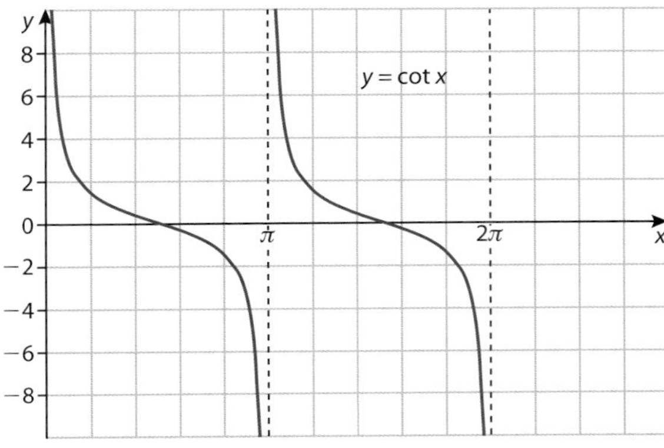

b) $y = \sec x$, range: $y \geqslant 1$, $y \leqslant -1$;
$y = \csc x$, range: $y \geqslant 1$, $y \leqslant -1$; $y = \cot x$, range: $y \in \mathbb{R}$

18 a) $a = 2, b = 3, c = -1$ b) $\frac{5\pi}{18}$

19 $a = 3, b = -\frac{\pi}{4}, c = -1$

Exercise 7.4

1 $x = \frac{\pi}{3}, \frac{5\pi}{3}$ **2** $x = \frac{7\pi}{6}, \frac{11\pi}{6}$

3 $x = \frac{\pi}{4}, \frac{5\pi}{4}$ **4** $x = \frac{\pi}{3}, \frac{2\pi}{3}$

5 $x = \frac{\pi}{4}, \frac{3\pi}{4}, \frac{5\pi}{4}, \frac{7\pi}{4}$ **6** $x = \frac{\pi}{6}, \frac{5\pi}{6}, \frac{7\pi}{6}, \frac{11\pi}{6}$

7 $x = \frac{\pi}{4}, \frac{3\pi}{4}, \frac{5\pi}{4}, \frac{7\pi}{4}$ **8** $x = \frac{\pi}{3}, \frac{2\pi}{3}, \frac{4\pi}{3}, \frac{5\pi}{3}$

9 $x = 0, \frac{3\pi}{4}, \pi, \frac{7\pi}{4}, 2\pi$ **10** $x = 0, \frac{\pi}{2}, \pi, \frac{3\pi}{2}, 2\pi$

11 $x = \frac{\pi}{3}, \frac{5\pi}{3}$ **12** $x = \frac{\pi}{4}, \frac{3\pi}{4}, \frac{5\pi}{4}, \frac{7\pi}{4}$

13 $x \approx 0.412, 2.73$ **14** $x \approx 1.91, 4.37$

15 $x \approx 1.11, 4.25$ **16** $x \approx 5.64, 3.78, 2.50, 0.639$

17 $x \approx 2.96, 5.32$ **18** $x = \frac{\pi}{6}, \frac{5\pi}{6}, \frac{7\pi}{6}, \frac{11\pi}{6}$

19 $x \approx 5.85, 5.01, 2.71, 1.86$ **20** $x \approx 3.43, 0.291, 2.71, 1.86$

21 $\frac{5\pi}{2}, \frac{3\pi}{2}, \frac{\pi}{2}, -\frac{\pi}{2}, -\frac{3\pi}{2}, -\frac{5\pi}{2}$

22 $\frac{\pi}{6}, -\frac{11\pi}{6}$ **23** $\frac{7\pi}{12}, \frac{19\pi}{12}$

24 $0, \frac{\pi}{4}, \frac{\pi}{2}, \frac{3\pi}{4}, \pi, \frac{5\pi}{4}, \frac{3\pi}{2}, \frac{7\pi}{4}, 2\pi$

25 $x = \frac{5\pi}{6}, \frac{3\pi}{2}$ **26** $\theta = -\frac{3\pi}{4}, \frac{\pi}{4}$

27 $x = 30°, 60°, 210°, 240°$ **28** $\alpha = -\frac{\pi}{6}, \frac{\pi}{6}$

29 $\theta = \frac{2\pi}{3}, \frac{4\pi}{3}$ **30** $x = \frac{\pi}{6}, \frac{5\pi}{6}$

31 $x = 225°, 315°$ **32** $\theta = \frac{\pi}{6}, \frac{5\pi}{6}$

33 $t \approx 1.5$ hours

34 a) 80th day (March 21) and approximately 263rd day (September 20)
b) 105th day (April 15) and approximately 238th day (August 26)
c) 94 days – from 125th day to 218th day

35 $x = \frac{\pi}{2}, \frac{2\pi}{3}, \frac{3\pi}{2}, \frac{4\pi}{3}$ **36** $\theta = \frac{\pi}{2}, \frac{7\pi}{6}, \frac{11\pi}{6}$

37 $x = -45°, 63.4°$ **38** $x \approx -1.87, 1.87$

39 $x \approx 56.3°$ **40** $x = \frac{\pi}{4}, \frac{3\pi}{4}$

41 No solution **42** $x \approx 0°, 71.6°, 180°, 252°$

Exercise 7.5

1 $\frac{\sqrt{2}-\sqrt{6}}{4}$ **2** $\frac{\sqrt{6}-\sqrt{2}}{4}$ **3** $2-\sqrt{3}$

4 $\frac{-\sqrt{6}-\sqrt{2}}{4}$ **5** $\frac{\sqrt{2}-\sqrt{6}}{4}$ **6** $2-\sqrt{3}$

7 a) $\frac{\sqrt{6}+\sqrt{2}}{4}$ b) $\sqrt{\frac{\sqrt{6}+\sqrt{2}+4}{8}}$

8 $\tan\left(\frac{\pi}{2}-\theta\right) = \frac{\sin\left(\frac{\pi}{2}-\theta\right)}{\cos\left(\frac{\pi}{2}-\theta\right)} = \frac{\sin\frac{\pi}{2}\cos\theta - \cos\frac{\pi}{2}\sin\theta}{\cos\frac{\pi}{2}\cos\theta + \sin\frac{\pi}{2}\sin\theta} = \frac{\cos\theta}{\sin\theta} = \cot\theta$

9 $\sin\left(\frac{\pi}{2}-\theta\right) = \sin\frac{\pi}{2}\cos\theta - \cos\frac{\pi}{2}\sin\theta = \cos\theta$

10 $\csc\left(\frac{\pi}{2}-\theta\right) = \frac{1}{\sin\left(\frac{\pi}{2}-\theta\right)} = \frac{1}{\sin\frac{\pi}{2}\cos\theta - \cos\frac{\pi}{2}\sin\theta} = \frac{1}{\cos\theta} = \sec\theta$

11 a) $\frac{4}{5}$ b) $\frac{7}{25}$ c) $\frac{24}{25}$

12 a) $\frac{\sqrt{5}}{3}$ b) $-\frac{4\sqrt{5}}{9}$ c) $-\frac{1}{9}$

13 $\sin 2\theta = -\frac{4\sqrt{5}}{9}$, $\cos 2\theta = \frac{1}{9}$, $\tan 2\theta = -4\sqrt{5}$

14 $\sin 2\theta = \frac{24}{25}$, $\cos 2\theta = \frac{7}{25}$, $\tan 2\theta = \frac{24}{7}$

15 $\sin 2\theta = \frac{4}{5}$, $\cos 2\theta = -\frac{3}{5}$, $\tan 2\theta = -\frac{4}{3}$

16 $\sin 2\theta = -\frac{2\sqrt{15}}{16}\left(\text{or } \sin 2\theta = -\frac{\sqrt{15}}{8}\right)$, $\cos 2\theta = -\frac{7}{8}$, $\tan 2\theta = \frac{\sqrt{15}}{7}$

17 $-\cos x$ **18** $-\cos x$ **19** $\tan x$

20 $-\sin x$ **21** $\frac{1+\sin\theta\cos\theta}{\cos\theta}$ **22** $\frac{1}{\sin^3\theta}$

23 $\frac{\sin\theta+\cos\theta}{\sin 2\theta}$ **24** $\frac{1+\sin^2\theta}{\cos^2 x}$ **25** $\cos^3\theta$

26 1 **27** $\cos^2\theta$ **28** $2\tan^2\theta$

29 $2\sin\alpha\cos\beta$ **30** $\cos^2 A$ **31** $2\cos\alpha\cos\beta$

32 1 **33–46** No answers required (proofs)

47 $\tan\theta = \frac{5x}{x^2+14}$ **48** $x = \frac{\pi}{3}, \pi, \frac{5\pi}{3}$

49 $x = \frac{\pi}{3}, \frac{5\pi}{3}$ **50** $x = 90°$ and $-90°$

51 $x \approx 0.375, 2.77$ **52** $x \approx 0.615, 2.53, 3.76, 5.67$

53 $x = \frac{3\pi}{4}, \frac{7\pi}{4}$ **54** $x = \frac{\pi}{3}, \frac{2\pi}{3}$

55 $x = 0, \frac{\pi}{4}, \pi, \frac{5\pi}{4}$ **56** $x = 0, \frac{\pi}{3}, \frac{2\pi}{3}, \pi, \frac{4\pi}{3}, \frac{5\pi}{3}$

57 $x = 30°, 90°, 105°, 150°, 165°$ **58** $3\sin x - 4\sin^3 x$

59 b) $x = \frac{\pi}{4}, \frac{3\pi}{4}, \frac{5\pi}{4}, \frac{7\pi}{4}$

Exercise 7.6

1 $\frac{\pi}{2}$ **2** $\frac{\pi}{4}$ **3** $-\frac{\pi}{3}$ **4** $\frac{2\pi}{3}$

5 0 **6** $-\frac{\pi}{3}$ **7** $\frac{\pi}{3}$ **8** $\frac{3}{2}$

9 12 **10** Not possible **11** $\frac{\pi}{4}$ **12** Not possible

13 $\frac{3}{5}$ **14** $\frac{24}{25}$ **15** Not possible **16** $\frac{\pi}{3}$

17 $\frac{2\sqrt{5}}{5}$ **18** $\frac{4}{5}$ **19** $\frac{63}{65}$

20 $\frac{2\sqrt{20}-3\sqrt{10}}{30}\left(\text{or } \frac{4\sqrt{5}-3\sqrt{10}}{30}\right)$

21 $\sqrt{1-x^2}$ **22** $\frac{\sqrt{1-x^2}}{x}$ **23** $\frac{1}{\sqrt{x^2+1}}$

24 $2x\sqrt{1-x^2}$ **25** $\sqrt{\frac{1-x}{1+x}}$

26 $\frac{-x^3+x+2x\sqrt{1-x^2}}{x^2+1}$

27 $\cos\left(\arcsin\frac{4}{5}+\arcsin\frac{5}{13}\right)=\cos\left(\arccos\frac{16}{65}\right)$

$\cos\left(\arcsin\frac{4}{5}\right)\cos\left(\arcsin\frac{5}{13}\right)-\sin\left(\arcsin\frac{4}{5}\right)\sin\left(\arcsin\frac{5}{13}\right)=\frac{16}{65}$

$\frac{3}{5}\cdot\frac{12}{13}-\frac{4}{5}\cdot\frac{5}{13}=\frac{36}{65}-\frac{20}{65}=\frac{16}{65}$ Q.E.D

28 $\sin\left(\arctan\frac{1}{2}+\arcsin\frac{1}{3}\right)=\sin\left(\frac{\pi}{4}\right)$

$\sin\left(\arctan\frac{1}{2}\right)\cos\left(\arctan\frac{1}{3}\right)+\cos\left(\arctan\frac{1}{2}\right)\sin\left(\arctan\frac{1}{3}\right)=\frac{\sqrt{2}}{2}$

$\frac{\sqrt{5}}{5}\cdot\frac{3\sqrt{10}}{10}+\frac{2\sqrt{5}}{5}\cdot\frac{\sqrt{10}}{10}=\frac{3\sqrt{50}}{50}+\frac{2\sqrt{50}}{50}=\frac{25\sqrt{2}}{50}=\frac{\sqrt{2}}{2}$ Q.E.D

29 $x=\frac{1}{2}$ **30** $x\approx 0.580, 2.56$

31 $x\approx 2.21$ **32** $x\approx 1.11, 4.25$

33 $x=\frac{\pi}{4},\frac{5\pi}{4}$; $x\approx 2.82, 5.96$ **34** $x=\frac{\pi}{4}$; $x\approx 0.464$

35 $x\approx 1.37, 4.91$

36 $x=\pi, 2\pi$; $x\approx 0.912, 2.23, 4.05, 5.37$

37 $x=0,\pi$; $x\approx 1.89, 5.03$ **38** $\theta=\arctan\left(\frac{2}{d}\right)$

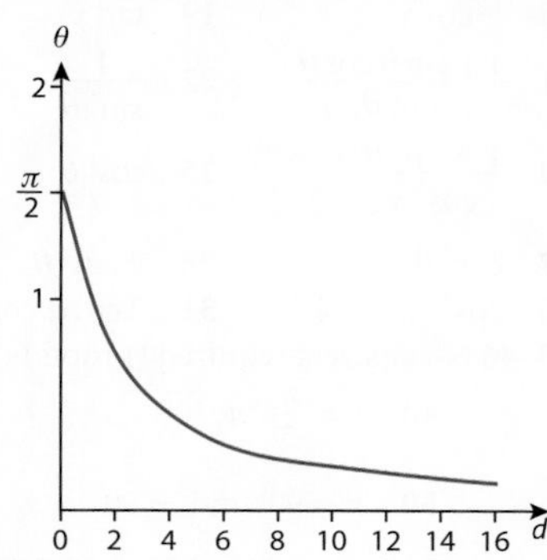

39 a) (ii) $\theta=\arctan\left(\frac{7x}{x^2+15.84}\right)$

(iii)

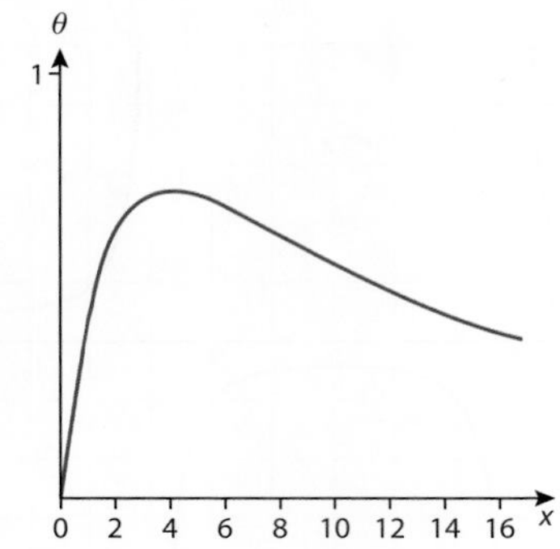

(iv) 3.98 m; sit in the 2nd row

b) (ii) $\theta=\arctan\left(\frac{7\left(x\cos\frac{\pi}{9}+2.5\right)}{\left(x\cos\frac{\pi}{9}+2.5\right)^2+\left(8.8-x\sin\frac{\pi}{9}\right)\left(1.8-x\sin\frac{\pi}{9}\right)}\right)$

$\left[\text{note: } 20^\circ=\frac{\pi}{9}\right]$

(iii)

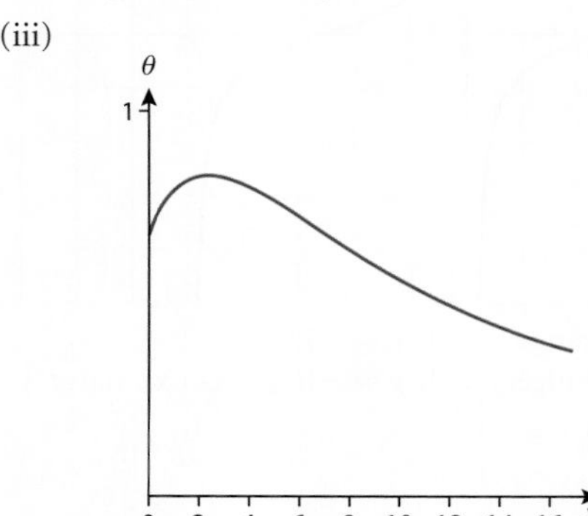

(iv) 2.5 m; sit in the 3rd row

Practice questions

1 a) 135 cm b) 85 cm
c) $t=0.5$ sec d) 1 sec

2 $x=0, 2\pi$

3 $\theta\approx 2.12$ (radian measure)

4 a) (i) -1 (ii) 4π
b) four

5 a) $p=35$ b) $q=29$ c) $m=\frac{1}{2}$

6 $x=0, 1.06, 2.05$

7 a) $x=\frac{2\pi}{3},\frac{4\pi}{3}$ b) $x=\frac{\pi}{6},\frac{\pi}{2},\frac{5\pi}{6},\frac{3\pi}{2}$

8 a) $\sin x=\frac{1}{3}$ b) $\cos 2x=\frac{7}{9}$
c) $\sin 2x=-\frac{4\sqrt{2}}{9}$

9 a) $1.6\sin\left(\frac{2\pi}{11}\left(x-\frac{9}{4}\right)\right)+4.2$
b) Approximately 3.15 metres
c) Approximately 12:27 p.m. to 7:33 p.m.

10 $x\approx 0.785, 1.89$

11 a) 15 cm
b) area ≈ 239 cm^2

12 $k > 2.5, k < -2.5$

13 $k = 1, a = -2$

14 $\sec\theta = -\frac{3}{2}$

15 a) $\frac{84}{85}$ b) $-\frac{13}{85}$ c) $-\frac{84}{13}$

16 $\sin 2p^\circ = \frac{4}{5}, \sin 3p^\circ = \frac{11\sqrt{5}}{25}$

17 a) $-\frac{5}{13}$ b) $\frac{12}{13}$ c) $-\frac{120}{169}$ d) $\frac{119}{169}$

18 $\tan\theta = \frac{1}{3}$ or -3

19 $\tan x = \frac{-(k+1)}{k-1}\tan\alpha\left(\text{or } \tan x = \frac{\tan\alpha(k+1)}{1-k}\right)$

20 $\theta = \pm\frac{3\pi}{8}, \pm\frac{\pi}{8}$

21 b) $x \approx 0.412$
c) $\cos(2) \leqslant g(x) \leqslant 1$

22 24.1°

23 $\frac{72}{\pi}\arccos\frac{8}{13}$ cm

Chapter 8

Exercise 8.1

1 b) $\cos\theta = \frac{4}{5}, \tan\theta = \frac{3}{4}, \cot\theta = \frac{4}{3}, \sec\theta = \frac{5}{4}, \csc\theta = \frac{5}{3}$
c) $\theta \approx 36.9^\circ; 53.1^\circ$

2 b) $\sin\theta = \frac{\sqrt{39}}{8}, \tan\theta = \frac{\sqrt{39}}{5}, \cot\theta = \frac{5\sqrt{39}}{39}, \sec\theta = \frac{8}{5},$
$\csc\theta = \frac{8\sqrt{39}}{39}$
c) $\theta \approx 51.3^\circ; 38.7^\circ$

3 b) $\sin\theta = \frac{2\sqrt{5}}{5}, \cos\theta = \frac{\sqrt{5}}{5}, \cot\theta = \frac{1}{2}, \sec\theta = \sqrt{5},$
$\csc\theta = \frac{\sqrt{5}}{2}$
c) $\theta \approx 63.4^\circ; 26.6^\circ$

4 b) $\sin\theta = \frac{\sqrt{51}}{10}, \tan\theta = \frac{\sqrt{51}}{7}, \cot\theta = \frac{7\sqrt{51}}{51}, \sec\theta = \frac{10}{7},$
$\csc\theta = \frac{10\sqrt{51}}{51}$
c) $\theta \approx 45.6^\circ; 44.4^\circ$

5 b) $\sin\theta = \frac{3\sqrt{10}}{10}, \cos\theta = \frac{\sqrt{10}}{10}, \tan\theta = 3, \sec\theta = \sqrt{10},$
$\csc\theta = \frac{\sqrt{10}}{3}$
c) $\theta \approx 71.6^\circ; 18.4^\circ$

6 b) $\cos\theta = \frac{3}{4}, \tan\theta = \frac{\sqrt{7}}{3}, \cot\theta = \frac{3\sqrt{7}}{7}, \sec\theta = \frac{4}{3},$
$\csc\theta = \frac{4\sqrt{7}}{7}$
c) $\theta \approx 41.4^\circ; 48.6^\circ$

7 b) $\sin\theta = \frac{\sqrt{60}}{11}, \cos\theta = \frac{\sqrt{61}}{11}, \tan\theta = \frac{2\sqrt{915}}{61},$
$\cot\theta = \frac{\sqrt{915}}{30}, \csc\theta = \frac{11\sqrt{60}}{60}$
c) $\theta \approx 44.8^\circ; 45.2^\circ$

8 b) $\sin\theta = \frac{9\sqrt{181}}{181}, \cos\theta = \frac{10\sqrt{181}}{181}, \cot\theta = \frac{10}{9}, \sec\theta = \frac{\sqrt{181}}{10},$
$\csc\theta = \frac{\sqrt{181}}{9}$
c) $\theta \approx 42.0^\circ; 48.0^\circ$

9 b) $\sin\theta = \frac{7\sqrt{65}}{65}, \tan\theta = \frac{7}{4}, \cot\theta = \frac{4}{7}, \sec\theta = \frac{\sqrt{65}}{4},$
$\csc\theta = \frac{\sqrt{65}}{7}$
c) $\theta \approx 60.3^\circ; 29.7^\circ$

10 $\theta = 60^\circ, \frac{\pi}{3}$

11 $\theta = 45^\circ, \frac{\pi}{4}$

12 $\theta = 60^\circ, \frac{\pi}{3}$

13 $\theta = 60^\circ, \frac{\pi}{3}$

14 $\theta = 45^\circ, \frac{\pi}{4}$

15 $\theta = 30^\circ, \frac{\pi}{6}$

16 $x \approx 86.6$

17 $x \approx 8.60$

18 $x \approx 20.6$

19 $x \approx 374$

20 $x = 18$

21 $x = 200$

22 $\alpha = 30^\circ, \beta = 60^\circ$

23 $\alpha \approx 67.4^\circ, \beta \approx 22.6^\circ$

24 $\alpha = 30^\circ, \beta = 60^\circ$

25 $\alpha \approx 20.0^\circ, \beta \approx 70.0^\circ$

26 114 metres

27 67.4°

28 4.05 metres

29 4105 m

30 $44^\circ, 68^\circ, 68^\circ$

31 5.76 km/hr

32 69.5 m

33 28.7 m

34 151 m

35 59.2 m

36 $3\sqrt{5}$

37 -0.6

38 $\frac{|ap + bq + c|}{\sqrt{a^2 + b^2}}$

39 Verify

40 14°

Exercise 8.2

1 $\sin\theta = \frac{3}{5}, \cos\theta = \frac{4}{5}, \tan\theta = \frac{3}{4}$

2 $\sin\theta = \frac{12}{37}, \cos\theta = -\frac{35}{37}, \tan\theta = -\frac{12}{35}$

3 $\sin\theta = -\frac{\sqrt{2}}{2}, \cos\theta = \frac{\sqrt{2}}{2}, \tan\theta = -1$

4 $\sin\theta = -\frac{1}{2}, \cos\theta = \frac{\sqrt{3}}{2}, \tan\theta = \frac{\sqrt{3}}{3}$

5 a) $\sin 120^\circ = \frac{\sqrt{3}}{2}, \cos 120^\circ = -\frac{1}{2}, \tan 120^\circ = -\sqrt{3}, \cot 120^\circ = -\frac{\sqrt{3}}{3}, \sec 120^\circ = -2, \csc 120^\circ = \frac{2\sqrt{3}}{3}$

b) $\sin 135^\circ = \frac{\sqrt{2}}{2}, \cos 135^\circ = -\frac{\sqrt{2}}{2}, \tan 135^\circ = -1, \cot 135^\circ = -1, \sec 135^\circ = -\sqrt{2}, \csc 135^\circ = \sqrt{2}$

c) $\sin 330^\circ = -\frac{1}{2}, \cos 330^\circ = \frac{\sqrt{3}}{2}, \tan 330^\circ = -\frac{1}{2}, \cot 330^\circ = -2, \sec 330^\circ = \frac{2\sqrt{3}}{3}, \csc 330^\circ = -2$

d) $\sin 270^\circ = -1, \cos 270^\circ = 0, \tan 270^\circ = \text{undef.}, \cot 270^\circ = 0, \sec 270^\circ = \text{undef.}, \csc 270^\circ = -1$

e) $\sin 240^\circ = -\frac{\sqrt{3}}{2}, \cos 240^\circ = -\frac{1}{2}, \tan 240^\circ = \sqrt{3}, \cot 240^\circ = \frac{\sqrt{3}}{3}, \sec 240^\circ = -2, \csc 240^\circ = -\frac{2\sqrt{3}}{3}$

f) $\sin\frac{5\pi}{4} = -\frac{\sqrt{2}}{2}, \cos\frac{5\pi}{4} = -\frac{\sqrt{2}}{2}, \tan\frac{5\pi}{4} = 1, \cot\frac{5\pi}{4} = 1, \sec\frac{5\pi}{4} = -\sqrt{2}, \csc\frac{5\pi}{4} = -\sqrt{2}$

g) $\sin\left(-\frac{\pi}{6}\right) = -\frac{1}{2}, \cos\left(-\frac{\pi}{6}\right) = \frac{\sqrt{3}}{2}, \tan\left(-\frac{\pi}{6}\right) = -\frac{\sqrt{3}}{3}, \cot\left(-\frac{\pi}{6}\right) = -\sqrt{3}, \sec\left(-\frac{\pi}{6}\right) = \frac{2\sqrt{3}}{3}, \csc\left(-\frac{\pi}{6}\right) = -2$

h) $\sin\left(\frac{7\pi}{6}\right) = \frac{\sqrt{3}}{2}, \cos\left(\frac{7\pi}{6}\right) = -\frac{1}{2}, \tan\left(\frac{7\pi}{6}\right) = -\sqrt{3}, \cot\left(\frac{7\pi}{6}\right) = -\frac{\sqrt{3}}{3}, \sec\left(\frac{7\pi}{6}\right) = -2, \csc\left(\frac{7\pi}{6}\right) = \frac{2\sqrt{3}}{3}$

i) $\sin(-60^\circ) = -\frac{\sqrt{3}}{2}, \cos(-60^\circ) = \frac{1}{2}, \tan(-60^\circ) = -\sqrt{3}, \cot(-60^\circ) = -\frac{\sqrt{3}}{3}, \sec(-60^\circ) = 2, \csc(-60^\circ) = -\frac{2\sqrt{3}}{3}$

j) $\sin\left(-\frac{3\pi}{2}\right) = 1, \cos\left(-\frac{3\pi}{2}\right) = 0, \tan\left(-\frac{3\pi}{2}\right) = \text{undef.}, \cot\left(-\frac{3\pi}{2}\right) = 0, \sec\left(-\frac{3\pi}{2}\right) = \text{undef.}, \csc\left(-\frac{\pi}{6}\right) = 1$

k) $\sin\left(\frac{5\pi}{3}\right) = \frac{1}{2}, \cos\left(\frac{5\pi}{3}\right) = -\frac{\sqrt{3}}{2}, \tan\left(\frac{5\pi}{3}\right) = -\frac{\sqrt{3}}{3}, \cot\left(\frac{5\pi}{3}\right) = -\sqrt{3}, \sec\left(\frac{5\pi}{3}\right) = -\frac{2\sqrt{3}}{3}, \csc\left(\frac{5\pi}{3}\right) = 2$

l) $\sin(-210^\circ) = -\frac{1}{2}, \cos(-210^\circ) = -\frac{\sqrt{3}}{2}, \tan(-210^\circ) = \frac{\sqrt{3}}{3}, \cot(-210^\circ) = \sqrt{3}, \sec(-210^\circ) = -\frac{2\sqrt{3}}{3}, \csc(-210^\circ) = -2$

m) $\sin\left(-\frac{\pi}{4}\right) = -\frac{\sqrt{2}}{2}, \cos\left(-\frac{\pi}{4}\right) = \frac{\sqrt{2}}{2}, \tan\left(-\frac{\pi}{4}\right) = -1, \cot\left(-\frac{\pi}{4}\right) = -1, \sec\left(-\frac{\pi}{4}\right) = \sqrt{2}, \csc\left(-\frac{\pi}{4}\right) = -\sqrt{2}$

n) $\sin\pi = 0, \cos\pi = -1, \tan\pi = 0, \cot\pi = \text{undef.}, \sec\pi = -1, \csc\pi = \text{undef.}$

o) $\sin 4.25\pi = \frac{\sqrt{2}}{2}, \cos 4.25\pi = \frac{\sqrt{2}}{2}, \tan 4.25\pi = 1, \cot 4.25\pi = 1, \sec 4.25\pi = \sqrt{2}, \csc 4.25\pi = \sqrt{2}$

6 $\sin\theta = \frac{15}{17}, \tan\theta = \frac{15}{8}, \cot\theta = \frac{8}{15}, \sec\theta = \frac{17}{8}, \csc\theta = \frac{17}{15}$

7 $\sin\theta = -\frac{6\sqrt{61}}{61}, \cos\theta = \frac{5\sqrt{61}}{61}$

8 $\cos\theta = -1, \tan\theta = 0, \cot\theta = \text{undef.}, \sec\theta = -1, \csc\theta = \text{undef.}$

9 $\sin\theta = -\frac{\sqrt{3}}{2}, \cos\theta = \frac{1}{2}, \tan\theta = -\sqrt{3}, \cot\theta = -\frac{\sqrt{3}}{3}, \csc\theta = -\frac{2\sqrt{3}}{3}$

10 a) (i) 30° (ii) 85°
b) (i) 45° (ii) 7°
c) (i) 60° (ii) 20°

11 a) $6\sqrt{3}$ b) 87.5 c) $675\sqrt{2}$

12 28.5°

13 a) 236 b) 97.4

14 a) 9.06 b) 119

15 $ab\sin\theta$

16 $x\sqrt{3}$ 17 $\frac{2hf\cos\theta}{h+f}$ 18 Verify

19 a) $A(x) = 24\sin x$ b) $0^\circ < x < 180^\circ$

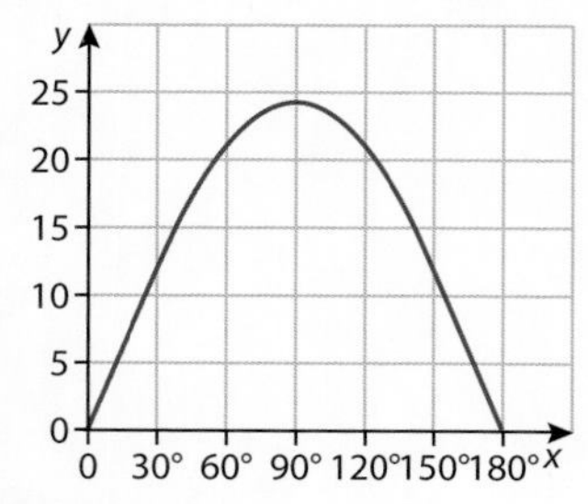

c) $(90^\circ, 24)$

20 c) 7.02 m

21 1740 km

22 a) $\sec\theta = \frac{1}{\sqrt{1-x^2}}, 0 \leqslant x < \frac{\pi}{2}$

b) $\sin\beta = \frac{y\sqrt{1+y^2}}{1+y^2}$

23 $\cos\theta = OA, \tan\theta = PB, \cot\theta = CP, \sec\theta = OB, \csc\theta = OC$

Exercise 8.3 and 8.4

1 Infinite triangles 2 One triangle 3 One triangle
4 One triangle 5 Two triangles 6 One triangle

7 $BC \approx 17.9, AC \approx 27.0, A\hat{C}B = 115^\circ$

8 $AB \approx 18.1, BC \approx 22.5, B\hat{A}C = 65^\circ$

9 $AB \approx 3.91, BC \approx 1.56, A\hat{B}C = 111^\circ$

10 $AB \approx 326, AC \approx 149, B\hat{A}C = 43^\circ$

11 $AB \approx 74.1, B\hat{A}C \approx 60.2^\circ, A\hat{B}C \approx 48.8^\circ$

12 $B\hat{A}C \approx 75.5^\circ, A\hat{B}C \approx 57.9^\circ, A\hat{C}B \approx 46.6^\circ$

13 $B\hat{A}C \approx 81.6^\circ, A\hat{B}C \approx 60.6^\circ, A\hat{C}B \approx 37.8^\circ$

14 Two possible triangles:
(1) $B\hat{A}C \approx 55.9^\circ, A\hat{C}B \approx 81.1^\circ, AB \approx 40.6$
(2) $B\hat{A}C \approx 124.1^\circ, A\hat{C}B \approx 12.9^\circ, AB \approx 9.17$

15 Two possible triangles:
(1) $A\hat{B}C \approx 72.2^\circ$, $A\hat{C}B \approx 45.8^\circ$, $AB \approx 0.414$
(2) $A\hat{B}C \approx 107.8^\circ$, $A\hat{C}B \approx 10.2^\circ$, $AB \approx 0.102$

16 10.8 cm and 30.4 cm **17** $51.3^\circ, 51.3^\circ, 77.4^\circ$

18 71.6° or 22.4° **19** Distance ≈ 743 metres

20 20.7° **21** Area ≈ 151.2 cm^2

22 a) $BC = 5\sin 36^\circ$ or $BC \geqslant 5$ b) $5\sin 36^\circ < BC < 5$
c) $BC < 5\sin 36^\circ$

23 a) $BC = 5\sqrt{3}$ or $BC \geqslant 10$ b) $5\sqrt{3} < BC < 10$
c) $BC < 5\sqrt{3}$

24 $x \approx 64.9$ m, $y \approx 56.9$ m

25 a) $x = 5$ c) $\frac{15\sqrt{3}}{14}$

26 $\frac{21\sqrt{15}}{4}$

27 a) Obtuse triangle b) acute triangle

28 21.1

29 a) 14 b) $\cos\theta = \frac{3}{5}$, $WY = 2\sqrt{65}$
c) $2\sqrt{5}$ d) 13.9°

30 51.3° **31–32** Verify

Exercise 8.5

1 a) $\tan 70^\circ \approx 2.75$ b) $y = x\tan 70^\circ$

2 a) $\tan(-20^\circ) \approx -0.364$ b) $y = x\tan(-20^\circ)$

3 a) 1 b) $y = -x + 2$

4 a) $\tan 22^\circ \approx 0.404$ b) $y = x\tan 22^\circ - \frac{3}{2}$

5 45° **6** 33.7° **7** 60.3° **8** 71.6° **9** 45°

10 a) $y = \frac{\sqrt{3}}{3}x$ b) 56.6°

11 $AB \approx 19.3$ cm

12 $P\hat{R}O \approx 71.8^\circ$, $S\hat{R}O \approx 51.3^\circ$, area ≈ 20.9 cm^2

13 406.1 metres **14** 2.70 metres

15 a) 1291.8 km b) 42.8°

16 59.5 cm

17 $\Delta ABC = 72\text{ cm}^2$, $\Delta ABD = 24\sqrt{3} \approx 41.6\text{ cm}^2$,
$\Delta BCD \approx 34.6\text{ cm}^2$, $\Delta ACD \approx 69.3\text{ cm}^2$

18 $D\hat{E}F \approx 41.9^\circ$ **19** 43.0 metres **20** 95.9°

Practice questions

1 $\sin A\hat{O}B = \frac{24}{25}$ **2** $\sin 2\theta = \frac{21}{29}$, $\cos 2\theta = \frac{20}{29}$

3 101.5° **4** $\sin 2A = \frac{120}{169}$

5 a) 29.1 m b) 41.9 m

6 $C\hat{A}B \approx 86.4^\circ$

7 a) 38.2° b) 17.3 cm^2

8 a) $A\hat{C}B \approx 116^\circ$ b) 155 cm^2

9 78.5 km **10** $J\hat{K}L \approx 31^\circ$

11 a) 3.26 cm b) 7.07 cm^2

12 70.5°

13 a) 91 m b) $1690\sqrt{3}$
c) (ii) $A_2 = 26x$ (iii) $x = 40\sqrt{3}$
d) (i) Supplementary angles have equal sines.

14 a) $2\sqrt{2} + 4$ b) $2\sqrt{6} + 3\sqrt{3} + 2\sqrt{2} + 3$

15 Proof

16 a) $0 < \theta < 120^\circ$ b) verify c) 60°

17 a) 120 cm^2 b) 2.16 c) 161 cm^2

18 Verify

19 $\cos\theta = \frac{b}{2a}$

Chapter 9

Exercise 9.1 and 9.2

1

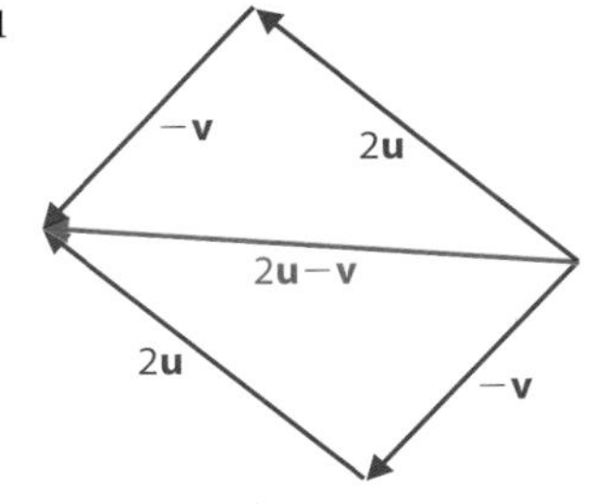

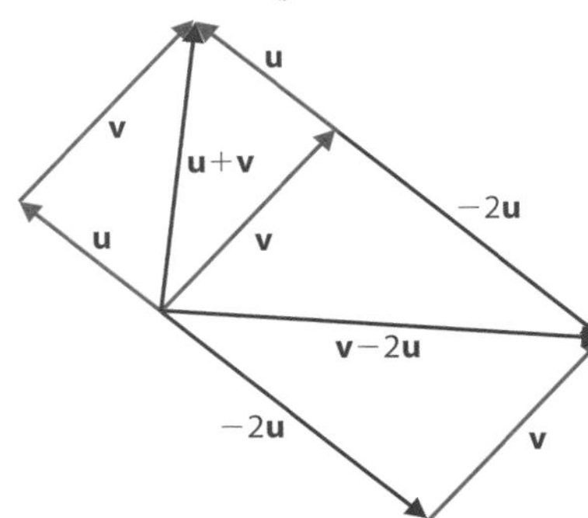

2 a) $\sqrt{41}$ b) $\mathbf{u} = (4, -5)$
c) $\mathbf{v} = \left(\frac{4}{\sqrt{41}}, \frac{-5}{\sqrt{41}}\right)$ d) 1

3 a) $\sqrt{53}$ b) $\mathbf{u} = (7, -2)$
c) $\mathbf{v} = \left(\frac{7}{\sqrt{53}}, \frac{-2}{\sqrt{53}}\right)$ d) 1

4 a) 3 b) $(-3, 0)$ c) $(-1, 0)$ d) 1

5 a) 5 b) $(0, 5)$ c) $(0, 1)$ d) 1

6 a) $\overrightarrow{PQ} = (5, -6)$ b) $\sqrt{61}$ d) $(4, -5)$

7 a) $\overrightarrow{PQ} = (4, 6)$ b) $2\sqrt{13}$ d) $(3, 7)$

8 a) $\overrightarrow{PQ} = (5, 5)$ b) $5\sqrt{2}$ d) $(4, 6)$

9 a) $\overrightarrow{PQ} = (4, 6)$ b) $2\sqrt{13}$ d) $(3, 7)$

10 a, c

11 $(1, -1)$ **12** $(8, -1)$

13 $(4, 8)$ **14** $(-5, -5)$

15 a) $\mathbf{u} + \mathbf{v} = 2\mathbf{i} + 2\mathbf{j}$, $\mathbf{u} - \mathbf{v} = 4\mathbf{i} - 4\mathbf{j}$, $2\mathbf{u} + 3\mathbf{v} = 3\mathbf{i} + 7\mathbf{j}$,
$2\mathbf{u} - 3\mathbf{v} = 9\mathbf{i} - 11\mathbf{j}$
b) $|\mathbf{u} + \mathbf{v}| = 2\sqrt{2}$, $|\mathbf{u} - \mathbf{v}| = 4\sqrt{2}$, $|\mathbf{u}| + |\mathbf{v}| = 2\sqrt{10}$,
$|\mathbf{u}| - |\mathbf{v}| = 0$
c) $|2\mathbf{u} + 3\mathbf{v}| = \sqrt{58}$, $|2\mathbf{u} - 3\mathbf{v}| = \sqrt{202}$, $2|\mathbf{u}| + 3|\mathbf{v}| = 5\sqrt{10}$,
$2|\mathbf{u}| - 3|\mathbf{v}| = -\sqrt{10}$

16 $\left(\frac{11}{8}, -\frac{1}{4}\right)$

17 $\mathbf{u} = \frac{8}{5}\mathbf{i} - \frac{7}{5}\mathbf{j}$; $\mathbf{v} = -\frac{1}{5}\mathbf{i} + \frac{4}{5}\mathbf{j}$

18 $\sqrt{13}, \sqrt{17}$

19 a) $\mathbf{v} + \mathbf{u}$ b) $\mathbf{v} + 0.5\mathbf{u}$ c) $\mathbf{v} - \mathbf{u}$ d) $0.5(\mathbf{v} - \mathbf{u})$

20 $(6, 8)$

21 $x = 3, y = 5$

22 $(6, 2)$

23 $\frac{5}{2}(2, 3) - \frac{1}{2}(2, 1)$

24 $r(1, -1) + (r - 5)(-1, 1)$

25 $2(2, 5) - 5(3, 2)$

26 $\binom{x}{y} = \left(\frac{x + y}{2}\right)\binom{1}{1} + \left(\frac{y - x}{2}\right)\binom{-1}{1}$

Exercise 9.3

1 a) $0°$ b) $90°$ c) $180°$ d) $56.31°$ e) $135°$

2 a) $\sqrt{13}, 33.69°$ b) $\sqrt{13}, 213.69°$
c) $2\sqrt{13}, 33.69°$ d) $3\sqrt{13}, 213.69°$
e) $5\sqrt{13}, 213.69°$ f) $\sqrt{13}, 33.69°$

3 a) $\sqrt{65}, \tan^{-1}\left(-\frac{7}{4}\right) + \pi$ b) $\sqrt{29}, \tan^{-1}\left(\frac{5}{2}\right)$
c) $3\sqrt{65}, \tan^{-1}\left(-\frac{7}{4}\right) + \pi$ d) $2\sqrt{29}, \tan^{-1}\left(\frac{5}{2}\right) + \pi$
e) $5\sqrt{41}, \tan^{-1}\left(-\frac{31}{8}\right) + \pi$ f) $2\sqrt{10}, \tan^{-1}\left(-\frac{1}{3}\right) + \pi$

4 a) $(145.54, 273.71)$ b) $(40.70, 14.49)$
c) $(-6\sqrt{2}, 6\sqrt{2})$ d) $(120, -120\sqrt{3})$

5 $(0, 4)$

6 a) $\left(\frac{3}{5}, \frac{4}{5}\right)$ b) $\frac{2}{\sqrt{29}}\mathbf{i} - \frac{5}{\sqrt{29}}\mathbf{j}$

7 $\left(-\frac{\sqrt{3}}{2}, \frac{1}{2}\right); \left(\frac{\sqrt{2}}{2}, -\frac{\sqrt{2}}{2}\right)$

8 $\frac{21}{5}\mathbf{i} - \frac{28}{5}\mathbf{j}$

9 $\pm\frac{3}{\sqrt{13}}(2\mathbf{i} + 3\mathbf{j})$

10 $\pm\frac{7}{5}(4\mathbf{i} + 3\mathbf{j})$

11 $\pm\frac{3}{\sqrt{13}}(3\mathbf{i} - 2\mathbf{j})$

12 a) $\vec{P} = (840\cos 80°, -840\sin 80°)$;
$\vec{W} = (60\cos 30°, -60\sin 30°)$
b) $\vec{V} = (840\cos 80° + 60\cos 30°, -840\sin 80° - 60\sin 30°)$
$= (197.83, -857.24)$
c) Speed = 879.77 km/h, bearing 167°

13 a) $\vec{P} = (520\cos 110)\vec{i} + (520\sin 110)\vec{j}$
$= -177.85\vec{i} + 488.64\vec{j}$
$\vec{W} = (64\cos 160)\vec{i} + (64\sin 160)\vec{j} = -60.14\vec{i} + 21.89\vec{j}$
b) Speed = 580.6 km/h, bearing 337.8°

14 24.15, 6.47

15 200 m east of the initial point.

16 Force = 8176.152 N at an angle of $-10.85°$ to the x-axis.

17 Water = 12.36, boat = 38.04

18 T = 35.89, S = 41.57

19 35.9 km/h at N 12.88° W

20 At N 11.54° W

21 P = (10, 6)

22 N 11.54° E, 293.9 km/h

23 a) $(4, 6)$ b) $(0, -2)$ and $(20, 6)$

24 No answer required – proof

25 No answer required – proof

26 No answer required – proof

27 a) 50 m b) 5 minutes
c) N 19.47° W, 5.3 minutes

28 a) $\mathbf{p} = (220, 200\sqrt{3})$ b) speed = 410.37, N 32.42° E

29 66.6 N, S 28.5° E (or N 151.5° E)

Exercise 9.4

1 a) $0, 90°$ b) $13, 54°$ c) $11, 42°$ d) $2\sqrt{3}, 30°$
e) $4, 90°$ f) $3\sqrt{3}, 30°$ g) $-12\sqrt{3}, 150°$ h) $-16, 180°$

2 a) -1 b) -1 c) $(57, -38)$
d) $(-12, -15)$ e) -6 f) 3
g) Scalar multiplication is distributive over addition of vectors. Multiplication is not associative.

3 Neither, perpendicular, perpendicular

4 a) 2000 b) 6450 c) 155

5 a) 26.6, 63.4, 90 b) 41.4, 74.5, 64.1
c) 41.6, 116.6, 21.8

6 a) $(5t, -3t)$ b) $(3t, 2t)$

7 a) $(x-1)(x-3) + (y-2)(y-4) = 0$
b) $(x-3)(x+1) + (y-4)(y+7) = 0$

8 No

9 $t = \frac{21}{5}$

10 $b = \sqrt{6}$ or $b = -\sqrt{6}$

11 $\left(\frac{4\sqrt{3}+3}{10}, \frac{4-3\sqrt{3}}{10}\right)$ or $\left(\frac{3-4\sqrt{3}}{10}, \frac{4+3\sqrt{3}}{10}\right)$

12 $t = 0$

13 Sides of rhombus: $\vec{a}$ and $\vec{b}$ with $|\vec{a}| = |\vec{b}|$, diagonals are $\vec{a} + \vec{b}$ and $\vec{a} - \vec{b} \Rightarrow (\vec{a} + \vec{b})(\vec{a} - \vec{b}) = (\vec{a})^2 - \vec{a}\vec{b} + \vec{a}\vec{b} - (\vec{b})^2 = 0$

14 a) 5.6 b) $\frac{8}{\sqrt{17}}$

15 $440\sqrt{2}$

16 a) 1 b) 0 c) $\frac{21}{\sqrt{34}}$

17 No answer required – proof

18 $\frac{48 \pm 25\sqrt{3}}{39}$

19 $\alpha = 63.4°, \beta = 71.6°, \theta = 45°$

20 No answer required – proof

Practice questions

1 a) $\mathbf{v} - \mathbf{u}$
b) $(\frac{1}{2})(\mathbf{v} - \mathbf{u})$
c) $(\frac{1}{2})(\mathbf{u} + \mathbf{v})$
d) $(\frac{3}{2})\mathbf{v} - (\frac{1}{2})\mathbf{u}$

2 a) $(6, -1)$ b) $\frac{6}{\sqrt{37}}(6, -1)$

3 a) $OR = 15$ b) $\begin{pmatrix} -5 \\ 5\sqrt{5} \end{pmatrix}$ c) $\frac{1}{\sqrt{6}}$ d) $75\sqrt{5}$

4 a) $\overrightarrow{MR} = \begin{pmatrix} 11 \\ 4 \end{pmatrix}, \overrightarrow{AC} = \begin{pmatrix} -3 \\ 6 \end{pmatrix}$
b) 83.4°
c) $\mathbf{u} = \frac{1}{2}\overrightarrow{MR}, \mathbf{v} = -\frac{1}{2}\overrightarrow{MR} \Rightarrow \mathbf{u} \parallel \mathbf{v}$ and $|\mathbf{u}| = |\mathbf{v}|$

5 $m = \frac{63}{46}, n = \frac{37}{46}$

6 a) 15 km/h, 19.7 km/h b) $\begin{pmatrix} 4.5 \\ 6 \end{pmatrix}; \begin{pmatrix} 9 \\ -4 \end{pmatrix}$
c) 11.4 km d) At 8 a.m.
e) 12.2 km f) 54 minutes

7 a)

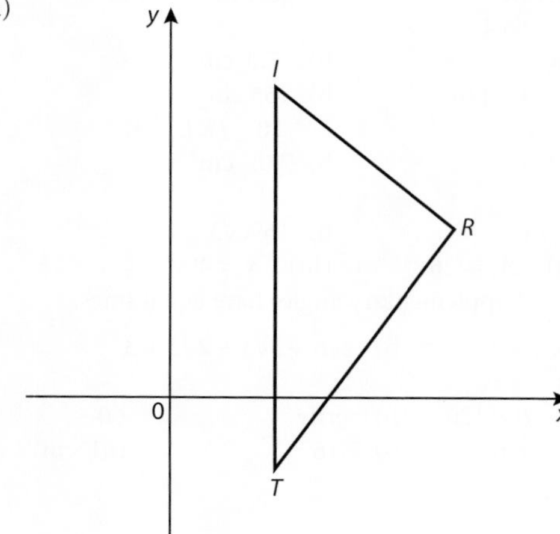

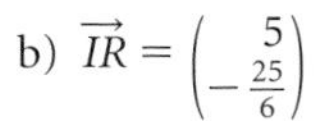

b) $\overrightarrow{IR} = \begin{pmatrix} 5 \\ -\frac{25}{6} \end{pmatrix}$

8 a) $\begin{pmatrix} 745 \\ 1000 \end{pmatrix}$ b) 600 km/h c) at 1.5 hrs

d) $\begin{pmatrix} 325 \\ 940 \end{pmatrix}$ e) 451 km

9 $2n^2 - n + 12 = 0$ does not have real solutions, so it is not possible.

10 $\alpha = \frac{\pi}{2} - 2\theta$ **11** 0

Chapter 10

Exercise 10.1

1 $5 + 2i$ **2** $7 - \sqrt{7}i$ **3** $-6 + 0i$

4 $-7 + 0i$ **5** $0 + 9i$ **6** $0 - \frac{5}{4}i$

7 $-1 - i$ **8** $-5 + 9i$ **9** $14 + 23i$

10 $-2 + 7i$ **11** $34 - 13i$ **12** $5 - i$

13 $\frac{16}{29} - \frac{11}{29}i$ **14** $\frac{4}{13} - \frac{7}{13}i$ **15** 1

16 $\frac{25}{36}$ **17** $-\frac{1}{13} - \frac{18}{13}i$ **18** $8 - i$

19 $-7 - 3i$ **20** $4 + 10i$ **21** $\frac{5}{13} + \frac{12}{13}i$

22 $\frac{48}{25} + \frac{36}{25}i$ **23** $2 + 9i$ **24** 68

25 $\frac{8}{13} - \frac{63}{26}i$ **26** $\frac{7}{65} + \frac{4}{65}i$ **27** $\frac{5}{169} + \frac{12}{169}i$

28 $\frac{12}{25} + \frac{8}{25}i$ **29** $\frac{498}{169} + \frac{553}{169}i$ **30** $-\frac{33}{25} - \frac{56}{25}i$

31 $\frac{17}{13} - \frac{19}{13}i$ **32** $x = -\frac{1}{2}, y = -2$; and $x = 1, y = 1$

33 a) -8 c) 2^{48}

34 a) $-4i$ c) 2^{46}

35 $x^2 + y^2 = 4$ **36** $\frac{9 - \sqrt{2}}{3} + \frac{2}{3}i$

37 $x = -\frac{2}{65}, y = \frac{29}{65}$ **38** $\frac{1}{2}(1 + i)$

39 $5 + 12i$ **40** $(x, y) = (2, -1)$ or $(x, y) = (-2, 1)$

41 a) $(x, y) = (1, 3)$ or $(x, y) = (-1, -3)$

b) $2i, -1 - i$

42 $\left\{-3i, \frac{3(\sqrt{3} + i)}{2}, \frac{3(-\sqrt{3} + i)}{2}\right\}$

43 $\frac{1}{2} - 2i, 3$ **44** $f(x) = 2x^4 - 11x^3 + 15x^2 + 17x - 11$

45 $f(x) = x^4 + 2x^3 + 8x + 16$

46 $5 - 2i, -3$ **47** $1 + i\sqrt{3}, -\frac{2}{3}$ **48** Verify

49 a) $k = 0 \pm 1$ b) $k = \pm\sqrt{3 \pm 2\sqrt{2}}$

50 $z_1 = 1 + i, z_2 = 2 - i$ **51** $z_1 = \frac{7 - 4i}{3}, z_2 = \frac{1 + 6i}{3}$

Exercise 10.2

1 $2\sqrt{2}\operatorname{cis}\left(\frac{\pi}{4}\right)$ **2** $2\operatorname{cis}\left(\frac{\pi}{6}\right)$ **3** $2\sqrt{2}\operatorname{cis}\left(\frac{7\pi}{4}\right)$

4 $2\sqrt{2}\operatorname{cis}\left(\frac{11\pi}{6}\right)$ **5** $4\operatorname{cis}\frac{5\pi}{3}$ **6** $3\sqrt{2}\operatorname{cis}\left(\frac{3\pi}{4}\right)$

7 $4\operatorname{cis}\left(\frac{\pi}{2}\right)$ **8** $6\operatorname{cis}\left(\frac{7\pi}{6}\right)$ **9** $\sqrt{2}\operatorname{cis}\left(\frac{\pi}{4}\right)$

10 $15\operatorname{cis}\pi$ **11** $\frac{1}{5}\operatorname{cis}(5.64)$ **12** $3\sqrt{2}\operatorname{cis}\left(\frac{3\pi}{4}\right)$

13 $\pi\operatorname{cis}(0)$ **14** $e\operatorname{cis}\left(\frac{\pi}{2}\right)$ **15** $\frac{-\sqrt{3}}{2} + \frac{i}{2}, \frac{\sqrt{3}}{2} + \frac{i}{2}$

16 $1, \frac{1}{2} - \frac{\sqrt{3}i}{2}$ **17** $\frac{-\sqrt{3}}{2} + \frac{i}{2}, -i$ **18** $-i, \frac{-1}{2} + \frac{\sqrt{3}i}{2}$

19 $\frac{\sqrt{6} + \sqrt{2}}{2} + i\frac{\sqrt{6} - \sqrt{2}}{2}, \frac{9(-\sqrt{6} + \sqrt{2})}{8} - i\frac{9(\sqrt{6} + \sqrt{2})}{8}$

20 $-3\sqrt{3} - 3 + i(3\sqrt{3} - 3), \frac{3\sqrt{3} - 3}{4} - \frac{i(3\sqrt{3} + 3)}{4}$

21 $\frac{-\sqrt{2}}{2}(1 + i), \frac{\sqrt{2}}{2}(1 + i)$

22 $6, \frac{-3}{4} - \frac{3\sqrt{3}i}{4}$

23 $\frac{5\sqrt{6} - 15\sqrt{2}}{48} - i\frac{5\sqrt{6} + 15\sqrt{2}}{48}, \frac{-5\sqrt{6} - 15\sqrt{2}}{64} + i\frac{5\sqrt{6} - 15\sqrt{2}}{64}$

24 $-3\sqrt{3} + 3 + i(3\sqrt{3} + 3), \frac{3\sqrt{3} + 3}{4} + \frac{i(3\sqrt{3} - 3)}{4}$

25 $z_1 = 2\operatorname{cis}\frac{\pi}{6}, z_2 = 4\operatorname{cis}\frac{-\pi}{3}, \frac{1}{z_1} = \frac{1}{2}\operatorname{cis} -\frac{\pi}{6}, \frac{1}{z_2} = \frac{1}{4}\operatorname{cis}\frac{\pi}{3},$

$z_1z_2 = 8\operatorname{cis}\frac{-\pi}{6}, \frac{z_1}{z_2} = \frac{1}{2}\operatorname{cis}\frac{\pi}{2}$

26 $z_1 = 2\sqrt{2}\operatorname{cis}\frac{\pi}{6}, z_2 = 4\sqrt{3}\operatorname{cis}\frac{-\pi}{3}, \frac{1}{z_1} = \frac{\sqrt{2}}{4}\operatorname{cis}\frac{\pi}{6}, \frac{1}{z_2} = \frac{\sqrt{3}}{12}\operatorname{cis}\frac{\pi}{3}$

,

$z_1z_2 = 8\sqrt{6}\operatorname{cis}\frac{-\pi}{6}, \frac{z_1}{z_2} = \frac{\sqrt{6}}{6}\operatorname{cis}\frac{\pi}{2}$

27 $z_1 = 8\operatorname{cis}\frac{\pi}{6}, z_2 = 3\sqrt{2}\operatorname{cis}\frac{-3\pi}{4}, \frac{1}{z_1} = \frac{1}{8}\operatorname{cis}\frac{-\pi}{6}, \frac{1}{z_2} = \frac{\sqrt{2}}{6}\operatorname{cis}\frac{3\pi}{4},$

$z_1z_2 = 24\sqrt{2}\operatorname{cis}\frac{-7\pi}{12}, \frac{z_1}{z_2} = \frac{4\sqrt{2}}{3}\operatorname{cis}\frac{11\pi}{12}$

28 $z_1 = \sqrt{3}\operatorname{cis}\frac{\pi}{2}, z_2 = 2\sqrt{2}\operatorname{cis}\frac{-2\pi}{3}, \frac{1}{z_1} = \frac{\sqrt{3}}{3}\operatorname{cis}\frac{-\pi}{2},$

$\frac{1}{z_2} = \frac{\sqrt{2}}{8}\operatorname{cis}\frac{2\pi}{3}, z_1z_2 = 2\sqrt{6}\operatorname{cis}\frac{-\pi}{6}, \frac{z_1}{z_2} = \frac{\sqrt{6}}{4}\operatorname{cis}\frac{-5\pi}{6}$

29 $z_1 = \sqrt{10}\operatorname{cis}\frac{\pi}{4}, z_2 = 2\sqrt{2}\operatorname{cis}\frac{\pi}{2}, \frac{1}{z_1} = \frac{\sqrt{10}}{10}\operatorname{cis}\frac{\pi}{4}, \frac{1}{z_2} = \frac{\sqrt{2}}{8}\operatorname{cis}\frac{\pi}{2}$

$, z_1z_2 = 4\sqrt{5}\operatorname{cis}\frac{3\pi}{4}, \frac{z_1}{z_2} = \frac{\sqrt{5}}{2}\operatorname{cis}\frac{-\pi}{4}$

30 $z_1 = 2\operatorname{cis}\frac{\pi}{3}, z_2 = 2\sqrt{3}\operatorname{cis}0, \frac{1}{z_1} = \frac{1}{2}\operatorname{cis}\frac{-\pi}{3}, \frac{1}{z_2} = \frac{\sqrt{3}}{6}\operatorname{cis}0,$

$z_1z_2 = 4\sqrt{3}\operatorname{cis}\frac{\pi}{3}, \frac{z_1}{z_2} = \frac{\sqrt{3}}{3}\operatorname{cis}\frac{\pi}{3}$

31 b) (i)

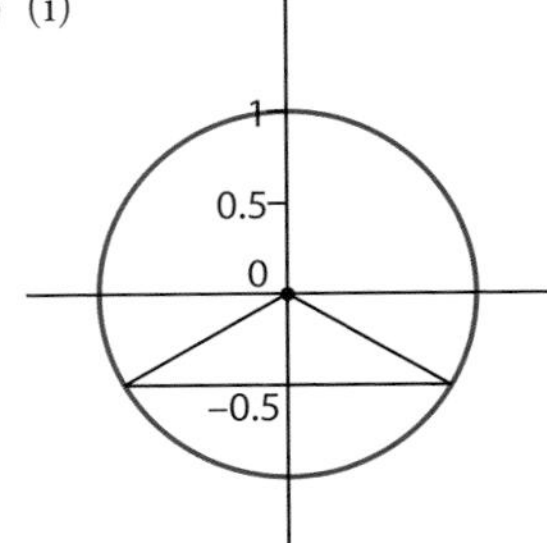

(ii) $\arg(z_1) = \frac{-\pi}{6}, \arg(z_2) = \frac{-5\pi}{6}$

32 Verify

33 a) $\frac{\sqrt{3}}{2} - \frac{3i}{2}$ b) $\frac{-2\sqrt{3}}{3}$ c) $\sqrt{3}i$

34 $|z_1| = 4, \arg(z_1) = \frac{-\pi}{6}, |z_2| = 2\sqrt{2}, \arg(z_2) = \frac{\pi}{4}, |z_3| = 8\sqrt{2},$

$\arg(z_3) = \frac{\pi}{12}$

35 $22 - 2\sqrt{3} \approx 18.5$

36 a) $\{(x, y): x^2 + y^2 = 9\}$, the circle centre (0, 0) radius 3

b) $\{(x, y): x = 0\}$, the y-axis

c) $\{(x, y): x = 4\}$, the line $x = 4$

d) $\{(x, y): (x - 3)^2 + y^2 = 4\}$, the circle centre (3, 0) radius 2

e) $\{(x, y): 1 - x + 3$ and $y = 0\}$, the line segment between (1, 0) and (3, 0)

37 a) $\{(x, y): x^2 + y^2 \leqslant 9\}$, the disk centre $(0, 0)$ radius 3

b) $\{(x, y): x^2\}\ (y + 3)^2 - 4|$, all points excluding the interior of the disk centre $(0, 3)$ radius 2

Exercise 10.3

1 $-2-2i\sqrt{3}$ 2 3 3 $3i$

4 $\sqrt{2}-\sqrt{6}+i\left(\sqrt{2}+\sqrt{6}\right)$ 5 $\frac{13}{2}+\frac{13i\sqrt{3}}{2}$

6 $\frac{3e}{2}+\frac{3ei\sqrt{3}}{2}$ 7 $2\sqrt{2}e^{i\frac{\pi}{4}}$ 8 $2e^{i\frac{\pi}{6}}$

9 $2\sqrt{2}e^{-i\frac{\pi}{6}}$ 10 $4e^{-i\frac{\pi}{3}}$ 11 $3\sqrt{2}e^{i\frac{3\pi}{4}}$

12 $4e^{i\frac{\pi}{2}}$ 13 $6e^{i\frac{7\pi}{6}}$ 14 $3\sqrt{2}e^{i\frac{3\pi}{4}}$

15 $\pi e^{2\pi i}$ or simply π 16 $e^{1+i\frac{\pi}{2}}$ 17 $32i$

18 -64 19 $-10\,077\,696$ 20 $-262\,144$

21 1296 22 $17\,496(-1-i)$ 23 $\frac{1}{1296}$

24 $\frac{1}{559\,872}\left(\sqrt{3}-i\right)$

25 $-128\sqrt{3}-128i$

26 $\sqrt{6}+i\sqrt{2}\,, -\sqrt{6}-i\sqrt{2}$

27 $2e^{i\frac{\pi}{9}}; 2e^{i\frac{7\pi}{9}}; 2e^{i\frac{13\pi}{9}}$

28 $\pm\frac{\sqrt{2}}{2}\pm i\frac{\sqrt{2}}{2}$

29 $\left(-\frac{\sqrt{6}}{4}-\frac{\sqrt{2}}{4}\right)+i\left(\frac{\sqrt{2}}{4}-\frac{\sqrt{6}}{4}\right); \left(-\frac{\sqrt{6}}{4}+\frac{\sqrt{2}}{4}\right)+i\left(-\frac{\sqrt{2}}{4}-\frac{\sqrt{6}}{4}\right)$;

$\left(\frac{\sqrt{6}}{4}-\frac{\sqrt{2}}{4}\right)+i\left(\frac{\sqrt{2}}{4}+\frac{\sqrt{6}}{4}\right); \left(\frac{\sqrt{6}}{4}+\frac{\sqrt{2}}{4}\right)+i\left(\frac{\sqrt{6}}{4}-\frac{\sqrt{2}}{4}\right);$

$-\frac{\sqrt{2}}{2}+i\frac{\sqrt{2}}{2}; \frac{\sqrt{2}}{2}-i\frac{\sqrt{2}}{2}$

30 $\sqrt[5]{18}e^{i\frac{4\pi}{15}}; \sqrt[5]{18}e^{i\frac{10\pi}{15}}; \sqrt[5]{18}e^{i\frac{16\pi}{15}}; \sqrt[5]{18}e^{i\frac{22\pi}{15}}; \sqrt[5]{18}e^{i\frac{28\pi}{15}}$

31 $2; 2e^{i\frac{2\pi}{5}}; 2e^{i\frac{4\pi}{5}}; 2e^{i\frac{6\pi}{5}}; 2e^{i\frac{8\pi}{5}}$

32 $e^{i\left(-\frac{\pi}{16}\right)}; e^{i\frac{3\pi}{16}}; e^{i\frac{7\pi}{16}}; e^{i\frac{11\pi}{16}}; \ldots; e^{i\frac{27\pi}{16}}$

33 $2e^{i\frac{5\pi}{18}}; 2e^{i\frac{17\pi}{18}}; 2e^{i\frac{29\pi}{18}}$

34 $\pm 2, \pm 2i$

35 $\sqrt{8}e^{i\left(\frac{3\pi}{20}\right)}; \sqrt{8}e^{i\left(\frac{11\pi}{20}\right)}; \ldots; \sqrt{8}e^{i\left(\frac{35\pi}{20}\right)}$

36 $\left(-\frac{\sqrt{6}}{2}-\frac{\sqrt{2}}{2}\right)+i\left(\frac{\sqrt{2}}{2}-\frac{\sqrt{6}}{2}\right); \left(-\frac{\sqrt{6}}{2}+\frac{\sqrt{2}}{2}\right)+i\left(-\frac{\sqrt{2}}{2}-\frac{\sqrt{6}}{2}\right)$;

$\left(\frac{\sqrt{6}}{2}-\frac{\sqrt{2}}{2}\right)+i\left(\frac{\sqrt{6}}{2}+\frac{\sqrt{2}}{2}\right); -\sqrt{2}+i\sqrt{2}; \sqrt{2}-i\sqrt{2}$

37 $\cos(4\beta)+i\sin(4\beta)$ 38 $\cos(7\beta)+i\sin(7\beta)$

39 $\cos(3\beta)+i\sin(3\beta)$ 40 $\cos(2\beta)+i\sin(2\beta)$

41 Proof 42–43 Verify

45 b) $2\cos 2n\alpha = z^n + \frac{1}{z^n}; 2i\sin 2n\alpha = z^n - \frac{1}{z^n}$

46 7 47 b) $1-i$ 48 b) $\frac{3+\sqrt{5}}{2}$

49 524 288 50 $\frac{3}{2}$

Practice questions

1 $x = 2, y = -1$

2 a) 0 b) $x^2 + y^2 - xy$

3 a) $2i$ c) 65 536

4 a) $z_1 = \sqrt{2}\operatorname{cis}\left(-\frac{\pi}{6}\right); z_2 = \sqrt{2}\operatorname{cis}\left(-\frac{\pi}{4}\right)$

c) $\frac{z_1}{z_2} = \frac{\sqrt{6}+\sqrt{2}}{4} + i\frac{\sqrt{6}-\sqrt{2}}{4}; \cos\frac{\pi}{12} = \frac{\sqrt{6}+\sqrt{2}}{4};$ $\sin\frac{\pi}{12} = \frac{\sqrt{6}-\sqrt{2}}{4}$

5 $\left(\frac{z_1}{z_2}\right)^3 = \left(\frac{a^3\sqrt{2}}{2b^3}\right) - i\left(\frac{a^3\sqrt{2}}{2b^3}\right)$

6 $|z| = 4$ 7 $a = \frac{11}{5}, b = \frac{3}{5}$

8 $b = \sqrt{3}$ 9 $a = 0, b = -1$

10 a) $z^5 - 1 = (z-1)\left(z^4 + z^3 + z^2 + z + 1\right)$

b) $\operatorname{cis}\left(\pm\frac{2\pi}{5}\right); \operatorname{cis}\left(\pm\frac{4\pi}{5}\right)$

c) $\left(z^2 - \left(2\cos\frac{2\pi}{5}\right)z + 1\right)\left(z^2 - \left(2\cos\frac{4\pi}{5}\right)z + 1\right)$

11 a) $8i = 8\operatorname{cis}\frac{\pi}{2}$

b) (i) $z = 2\operatorname{cis}\frac{\pi}{6}$

(ii) $z = \sqrt{3} + i$

12 a) $|z| = 1; \arg(z) = \frac{2\pi}{3}$

c) $\frac{3}{2} + \frac{3\sqrt{3}}{2}i$

13 $-5 - 12i$

14 c) $z^5 + 5z^3 + 10z + \frac{10}{z} + \frac{5}{z^3} + \frac{1}{z^5}$

15 $p = -\frac{2}{5}; q = \frac{6}{5}$

16 (ii) $z_1^2 = 4\operatorname{cis}\frac{4\pi}{5}; z_1^3 = 8\operatorname{cis}\frac{6\pi}{5}; z_1^4 = 16\operatorname{cis}\frac{\pi}{5}; z_1^5 = 32$

(iii)

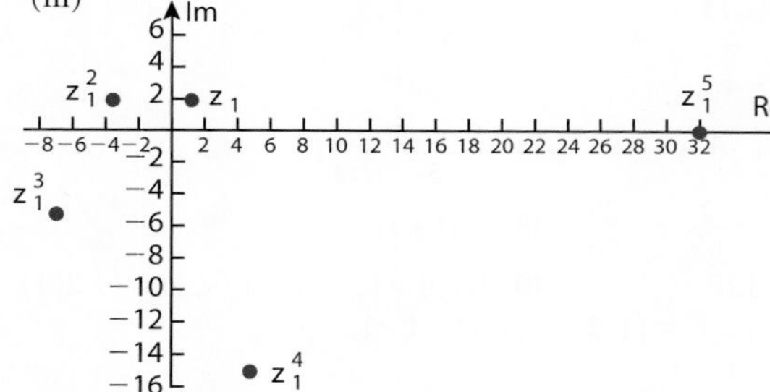

(iv) Enlargement scale factor of 2 with $(0, 0)$ as centre, and a rotation of $\frac{2\pi}{5}$.

17 b) (i)

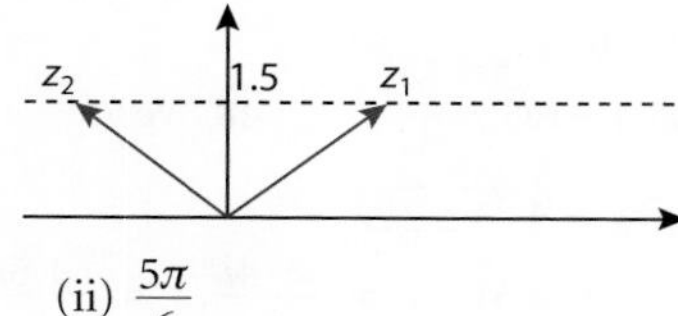

(ii) $\frac{5\pi}{6}$

c) $k = 4$

18 $a = 3, b = 1$

19 No answers required – proofs

20 a) $z = \frac{1}{2}e^{i\theta}$

c) $S_\infty = \frac{e^{i\theta}}{1 - \frac{1}{2}e^{i\theta}}$

d) (i) $S_\infty = \frac{\cos\theta + i\sin\theta}{1 - \frac{1}{2}(\cos\theta + i\sin\theta)}$

21 $a = 8; b = 25; c = 26$

22 $z = 2 + 4i$

23 $z_1 = 1 + 4i; z_2 = \frac{7}{2} - \frac{1}{2}i$

24 a) $z_1 = 2 + 2i; z_2 = 2 - 2i$ b) $\frac{z_1^4}{z_2^2} = -8i$

d) 0 e) $n = 4k$, where $k \in \mathbb{Z}$

Chapter 11

Exercise 11.1

Note: Some answers may differ from one person to another due to different graph accuracies.

1 a) Student, all students in a community, random sample of few students, qualitative

b) Exam, 10th-grade students in a country, a sample from a few schools, quantitative

c) Newborns, heights of newborns in a city, sample from a few hospitals, quantitative

d) Children, eye colour of children in a city, sample of children at schools, qualitative

e) Working persons, commuters in a city, sample of few districts, quantitative

f) Country leaders, sample of few presidents, qualitative

g) Students, origin countries of a group of international school students, qualitative

2 Answers are not unique!

a) Skewed to the right as few players score very high

b) Symmetric

c) Skewed to the right

d) Unimodal, or bi-modal, symmetric or skewed, etc.

3 a) b) Quantitative

c) d) Qualitative

4 a) Discrete b) Continuous

c) Continuous d) Discrete

e) Continuous f) Discrete (debatable!)

5

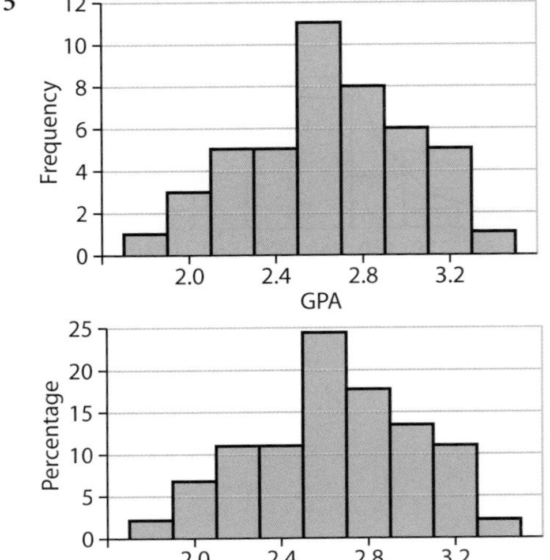

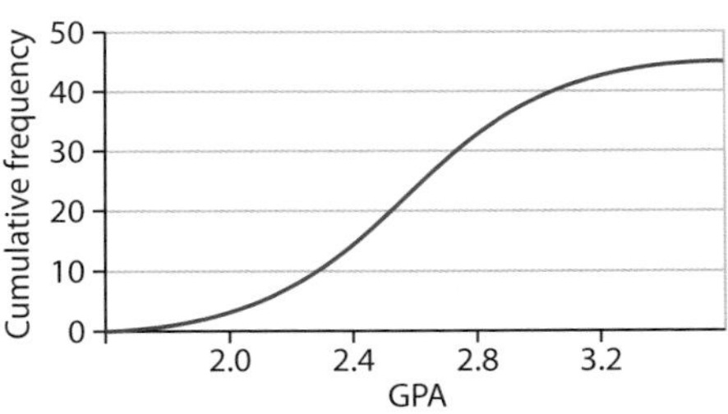

Relatively symmetric; no outliers

6

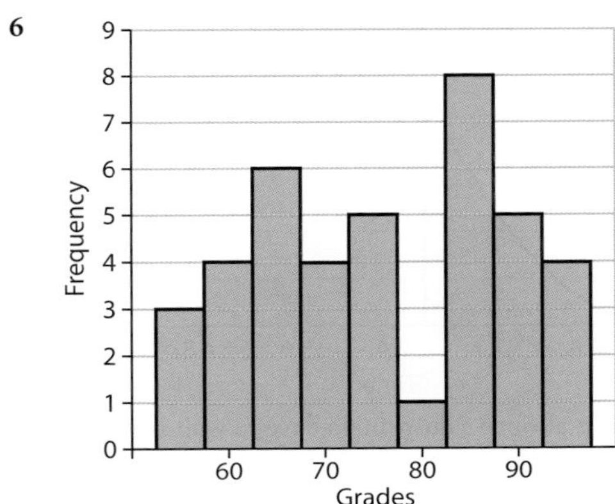

The grades appear to be divided into two groups, one with mode around 65 and the other around 85. No outliers are detected.

7 a)

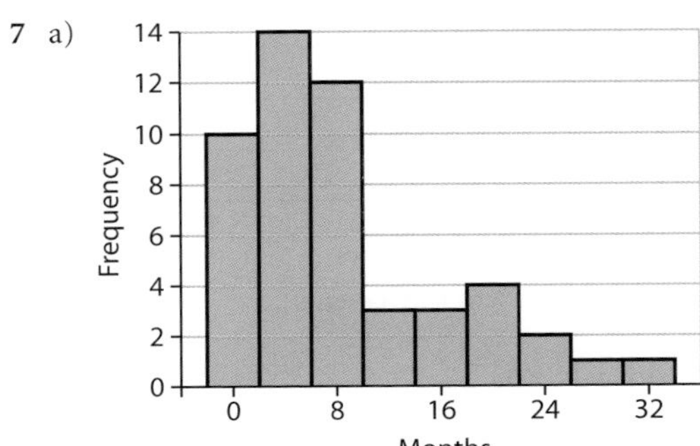

b) The data is skewed to the right.

c)

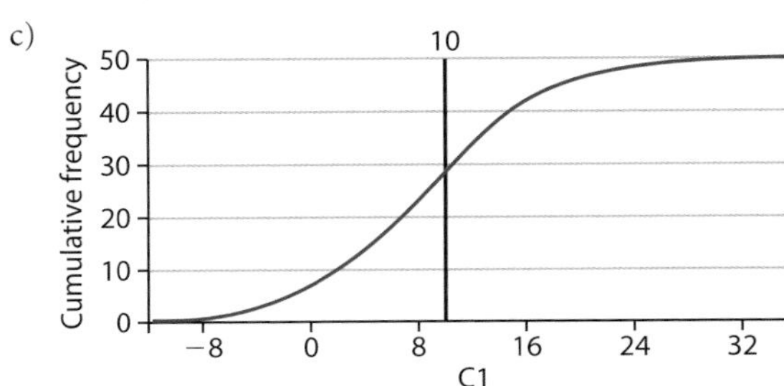

Apparently, more than 35 out of the 50 will lose the licence, about 70%.

8 a)

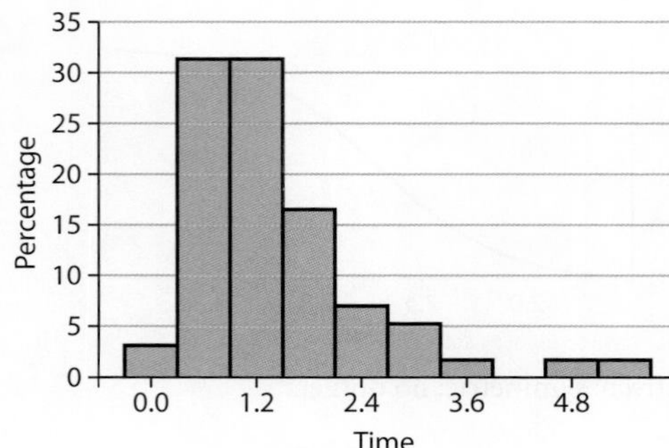

b)

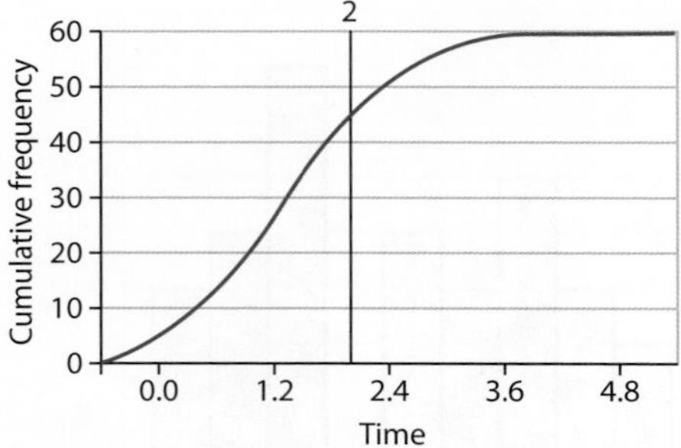

Apparently, about 15 customers have to wait more than 2 minutes.

9 a) Skewed to the right, there is a mode at about 7 days stay, and a few extremes that stayed more than 20 days. A good proportion stayed for about 3 days.

b)

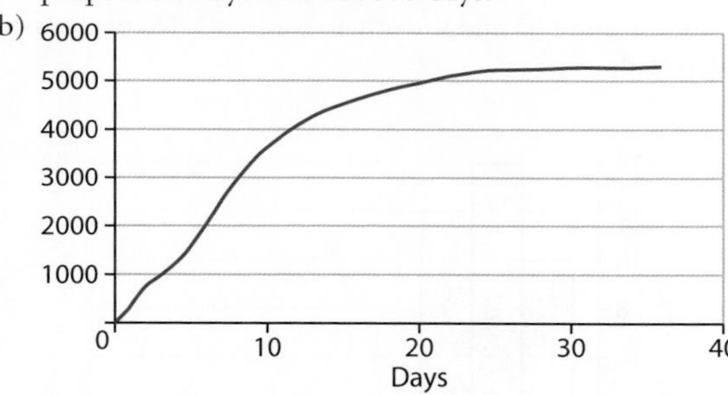

c) Approximately 35% of the patients

10 a) 40 minutes

b) Approximately 30%

c)

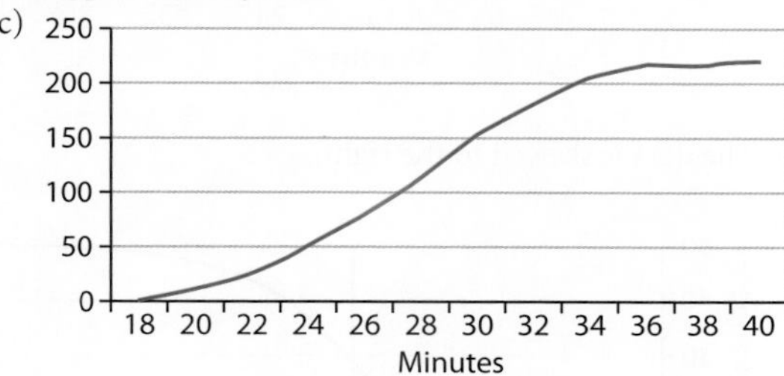

11 a)

Speed	Frequency
60–75	20
75–90	70
90–105	110
105–120	150
120–135	40
135– …	10

b) c)

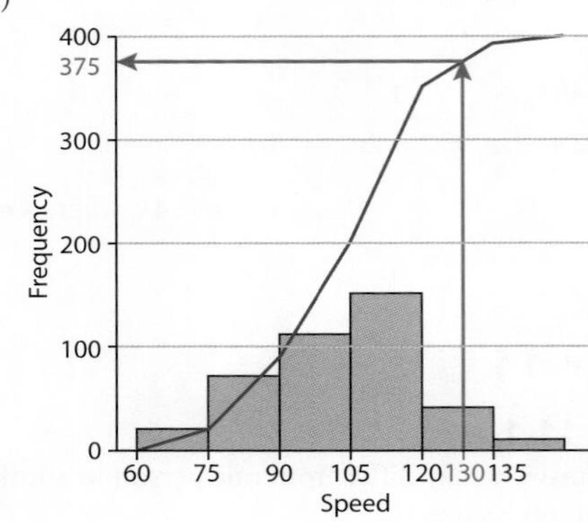

d) As you see from the diagram, $\frac{25}{400} = 6.25\%$.

12 a)

Histogram of C1

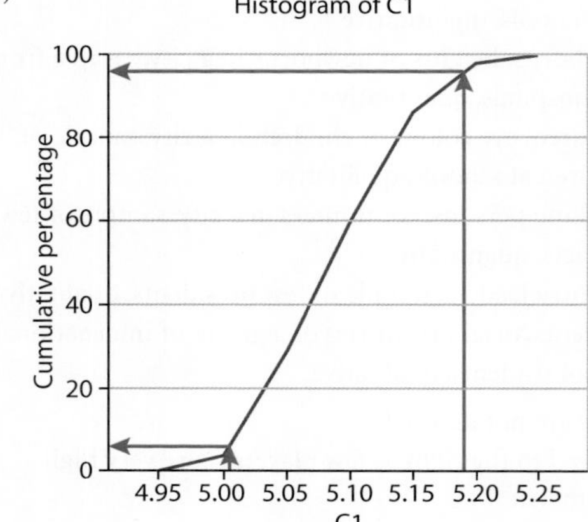

b) About 5% at the lower end and also about 5% at the upper end.

13 a) b)

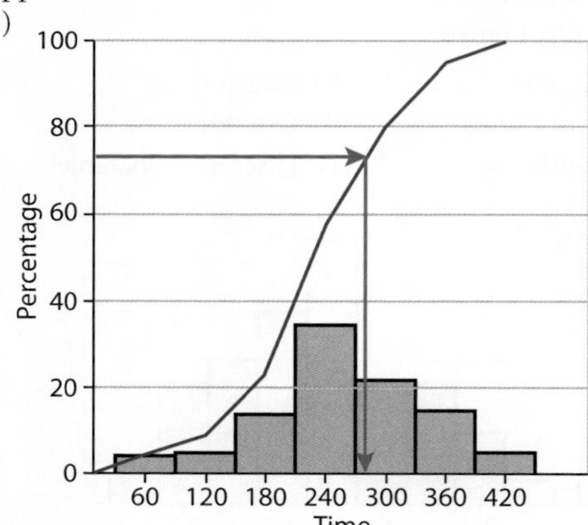

c) As you see from the diagram, about 250 seconds.

14

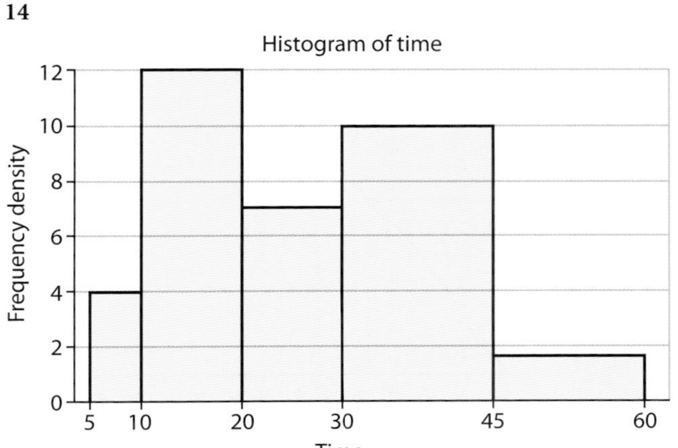

15

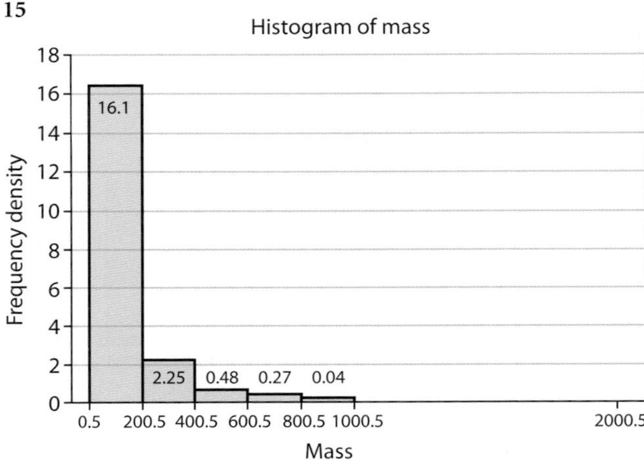

16

x	$0 \leqslant x < 10$	$10 \leqslant x < 20$	$20 \leqslant x < 30$	$30 \leqslant x < 60$	$60 \leqslant x < 80$	$80 \leqslant x < 120$
Frequency	120	80	60	60	40	40

17

x	$0 \leqslant x < 1$	$1 \leqslant x < 3$	$3 \leqslant x < 6$	$6 \leqslant x < 10$	$10 \leqslant x < 15$	$15 \leqslant x < 20$	$20 \leqslant x < 30$
Frequency	6	10	20	30	50	20	18

18

x	$0 \leqslant x < 3$	$3 \leqslant x < 8$	$8 \leqslant x < 12$	$12 \leqslant x < 16$	$16 \leqslant x < 24$	$24 \leqslant x < 30$	$30 \leqslant x < 36$
Frequency	10	10	18	20	30	20	10

19 $m = 10$, $n = 45$, $p = 2.5$, $q = 33$

Exercise 11.2

1 a) 6 b) 6
c) It appears to be symmetric as the mean and median are the same. A histogram supports this view.

2 a) 7.8 b) 7.5 c) 7 or 8

3 Average = 1.16, median = 1. Median is more appropriate as the data is skewed to the right.

4 Mean = 7494.7, median = 837.5. There are extreme values and hence the median is more appropriate.

5 Mean = median = 430. It appears to be symmetric and hence either measure would be fine.

6 a) 49.56 b) 49.93

7 2.052

8 a) 29.96

b)

1	89
2	0223344
2	5666777
3	34
3	568
4	022
4	66

Median is 27

c)

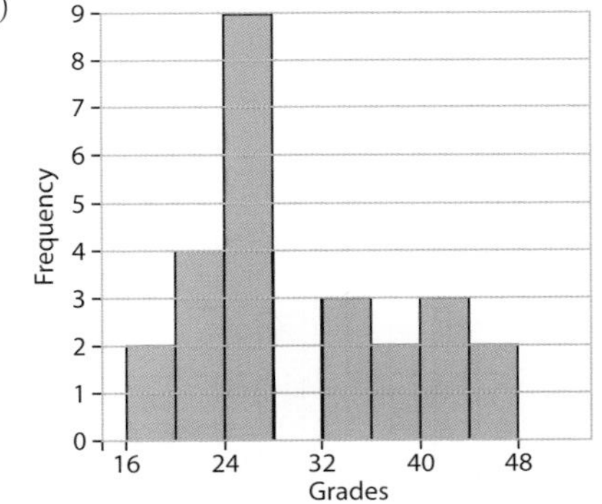

d)

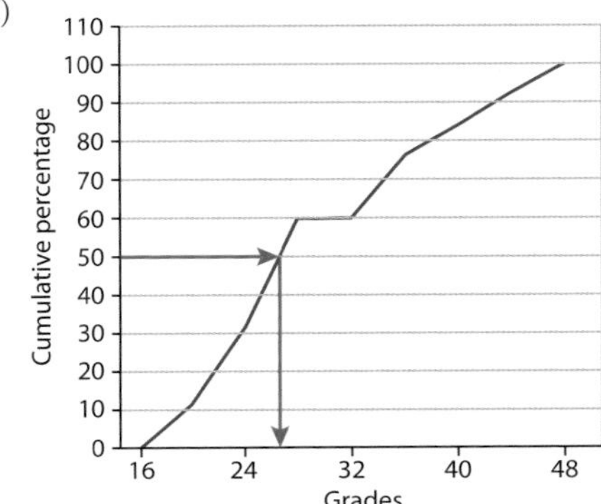

The median ≈ 27

9 a)

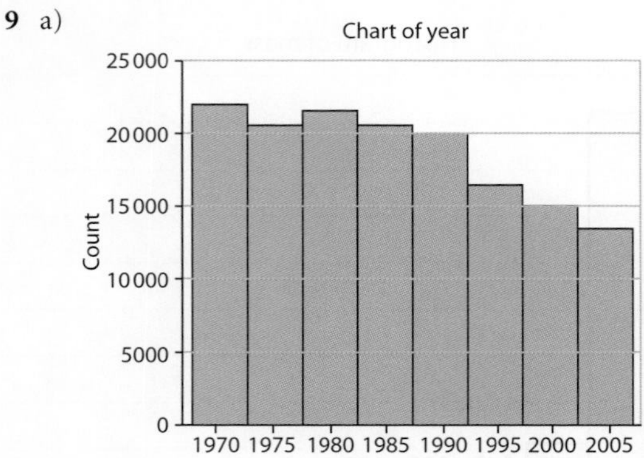

There appears to be a decline in the total number of injuries.

b)

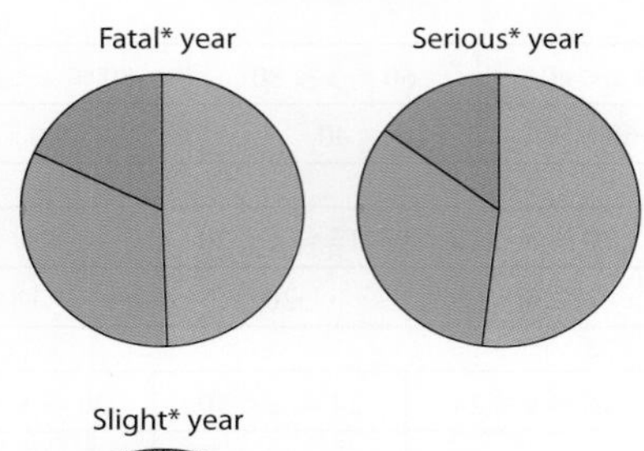

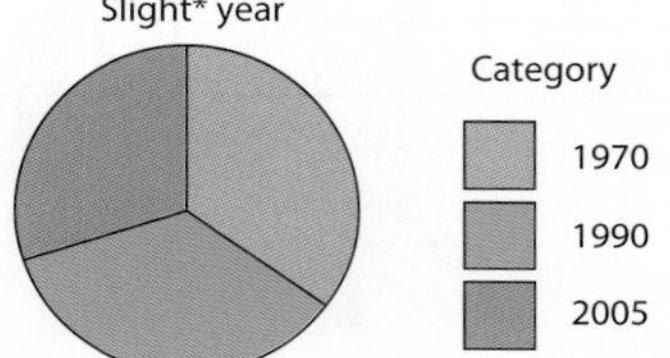

10 a)

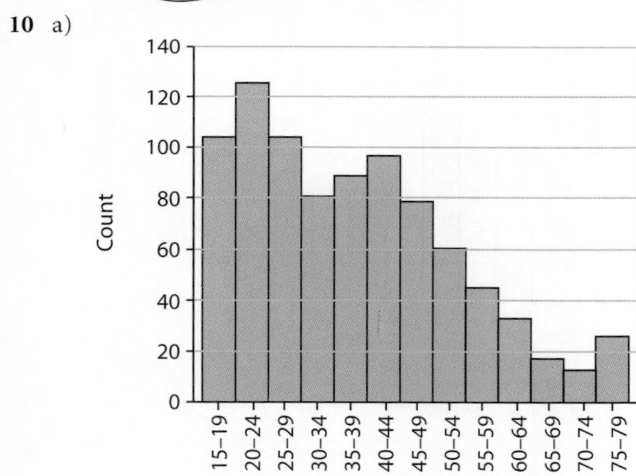

b) 37.6

c)

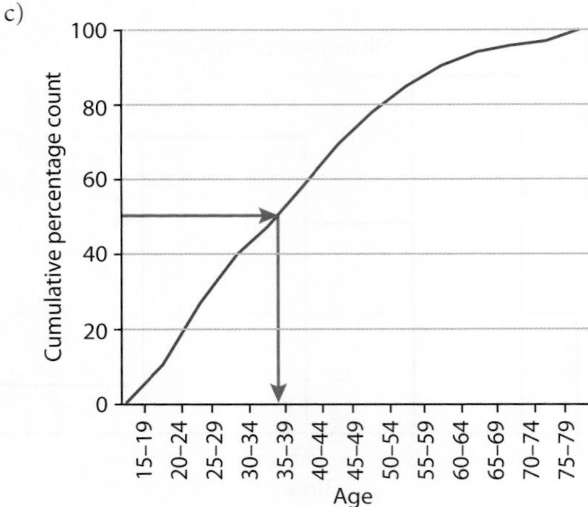

From the graph, the median is approx. at 36.

11 Median ≈ 8 days; mean = 9.5 days

12 Median ≈ 28 minutes; mean = 28.7 minutes

13 Median ≈ 105; mean = 103 km/h

14 Median ≈ 5.075; mean = 5.09

15 Median ≈ 210; mean = 228.6

16 a) 41.6

b) 61.6

17 a) 61.4

b) 63.8

Exercise 11.3

Where S_{n-1} is given, please multiply the answer by $\sqrt{\frac{n-1}{n}}$ to get the answer to S_n.

1 a) Mean = 71.47, $S_{n-1} = 7.29$

b)

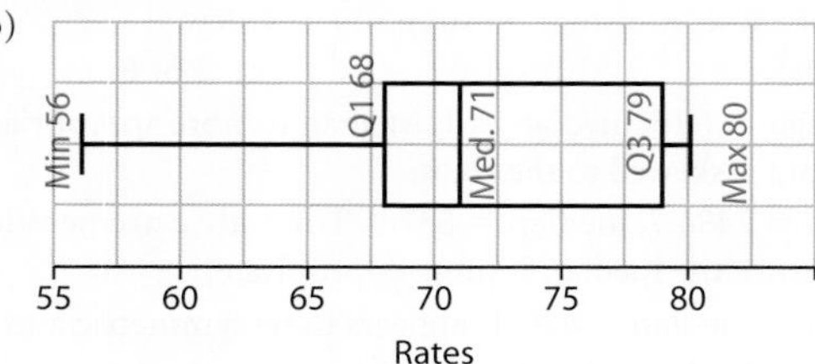

c) No outliers

2 a) Mean = 162.6, $S_{n-1} = 23.35$

b)

Stem	Leaf
11	79
12	567
13	089
14	123679
15	033445689
16	02334568
17	1344789
18	02255779
19	8
20	9
21	08

Median = 162.5

c)

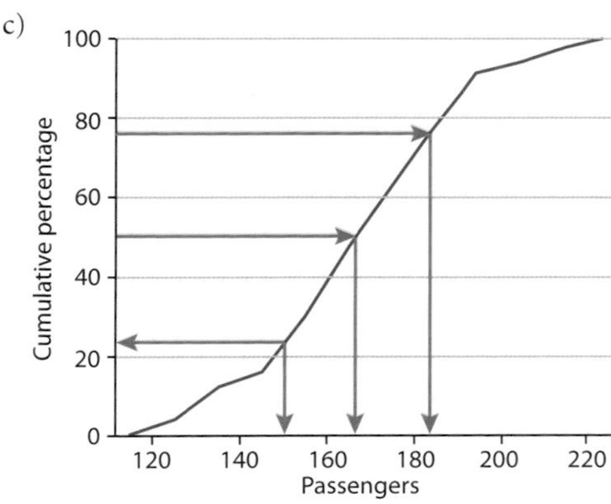

Q1 ≈150, median ≈ 165, Q3 ≈ 182

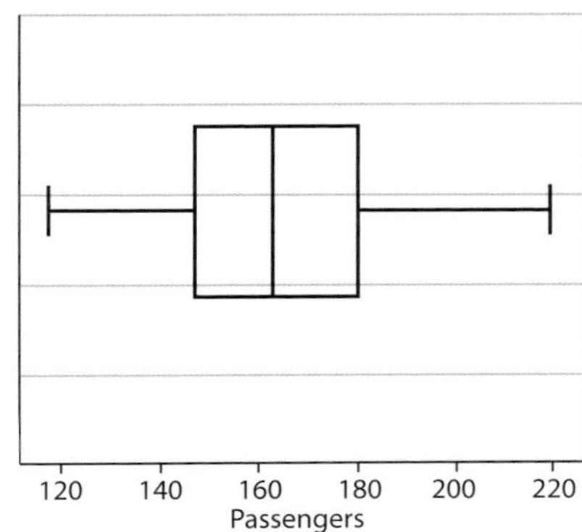

d) Real Q1 = 146.75, Q3 = 179.25, IQR = 32.5. No outliers

e) $\bar{x} \pm 3s_{n-1} = (92.55, 232.65)$. No outliers

3 a) and b)

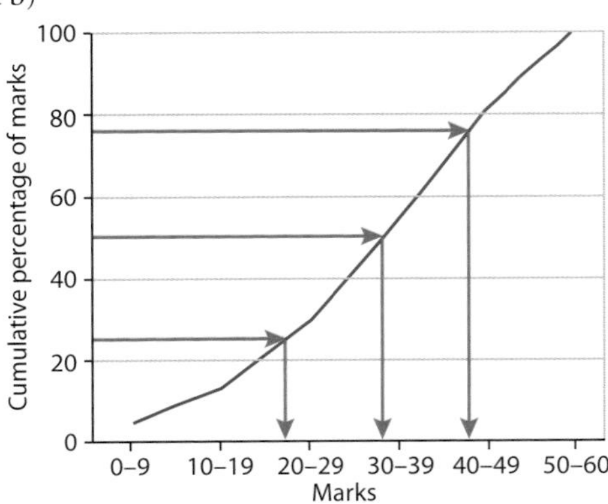

Percentage within all data.

Q1 ≈ 18, median ≈ 29, Q3 ≈ 39

4 a)

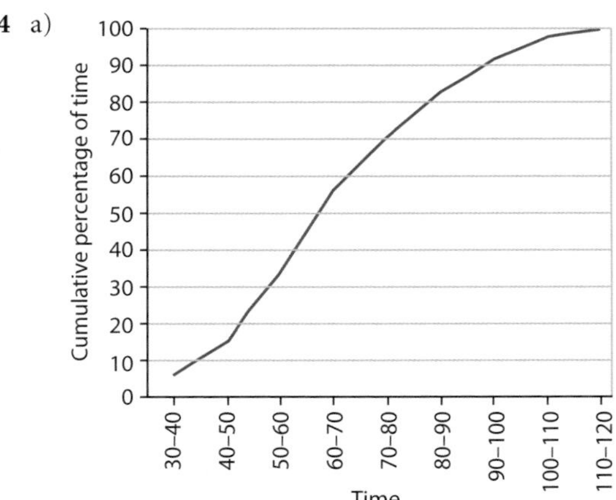

Percentage within all data.

b) Median = 63, IQR = 27

c) About 68

5 29.6

6 a) Mean = 72.1, $S_{n-1} = 6.1$

b) New mean = 85.1, S will not change.

7 a)

$x \leqslant 10$	$x \leqslant 20$	$x \leqslant 30$	$x \leqslant 40$	$x \leqslant 50$
15	65	165	335	595

$x \leqslant 60$	$x \leqslant 70$	$x \leqslant 80$	$x \leqslant 90$	$x \leqslant 100$
815	905	950	980	1000

b)

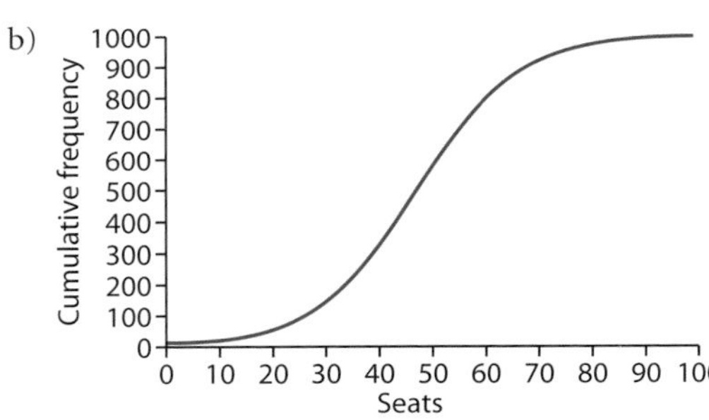

c) (i) Around 50

(ii) Q1 = 40, Q3 = 60, IQR = 20

(iii) About 170 days

(iv) Approximately 70 seats

8 a)

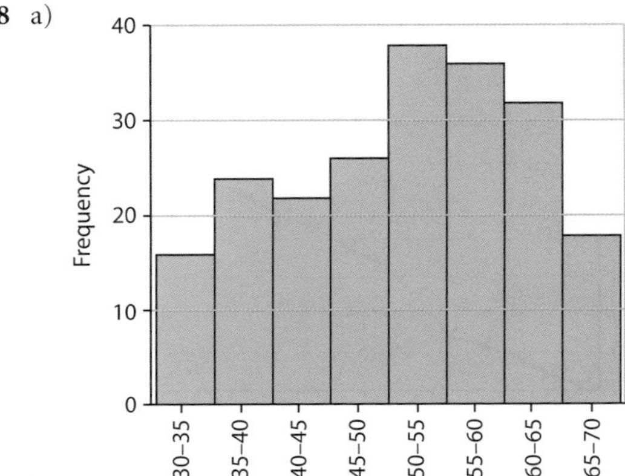

b) 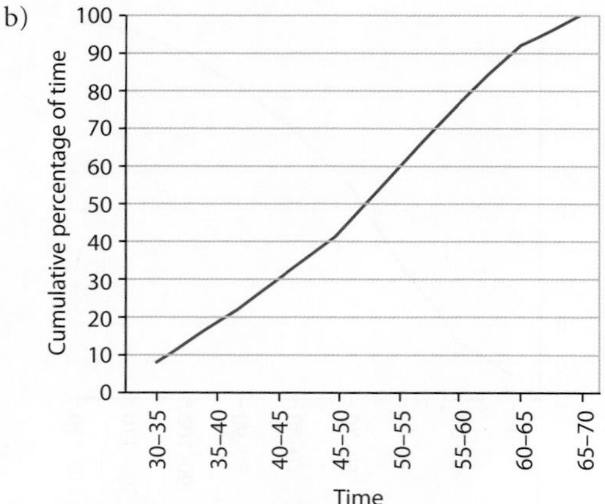

Percentage within all data.

Median = 53, IQR = 15

c) Mean = 51.3 and $S_{n-1} = 34.8$

9 a) Q1 = 165.1, median = 167.64, Q3 = 177.8, minimum = 152, maximum = 193

b) 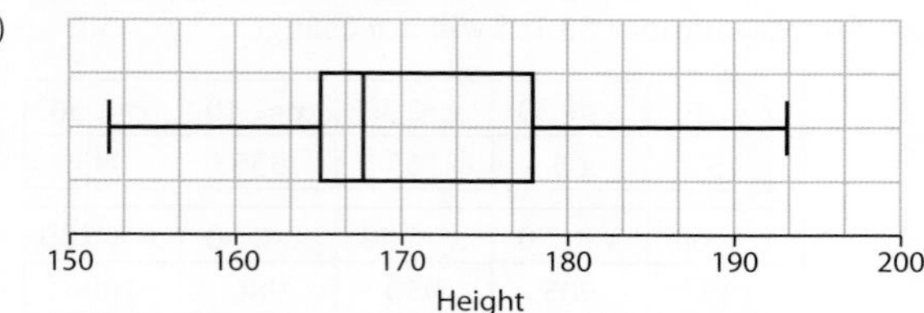

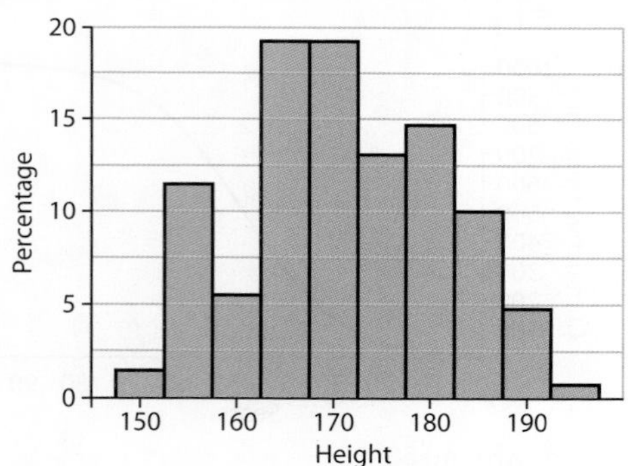

c) Mean = 170.5, standard deviation = 9.61

d) The heights are widely spread from very short to very tall players. Heights are slightly skewed to the right, bimodal at 165 and 170, no apparent outliers. The heights between the first quartile and the median are closer together than the rest of the data.

e) 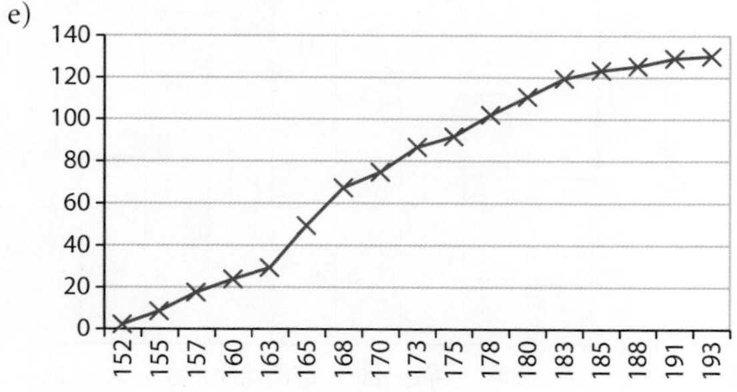

Approximately 183 cm tall

f) 171.3

10 a) 12 b) 12 c) 111

11 a) 31 b) Increase

12 36.7

13 $x = 6, y = 11$

14 Mean = 11.12, variance = 24.6 (calculating $\sigma^2 = 23.6$)

15 Standard deviation = 6.1, IQR ≈ 6

16 Standard deviation ≈ 4.5, IQR ≈ 6

17 Standard deviation ≈ 16.7, IQR ≈ 15

18 Standard deviation ≈ 0.056, IQR ≈ 0.05

19 Standard deviation ≈ 82.3, IQR ≈ 60

Practice questions

1 a) 12 b) $\sqrt{30.83}$

2 4

3 a)

Time	1.6	2.1	2.6	3.1	3.6	4.1	4.6	5.1	5.6	6.1	6.6
Frequency	2	2	6	4	11	10	5	5	3	2	0

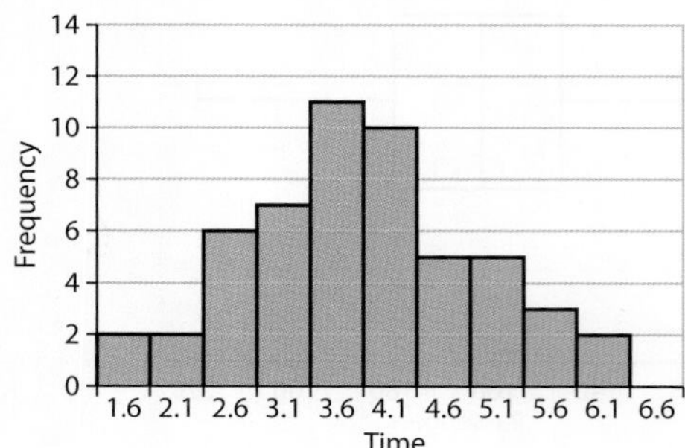

b) 86% c) approx. 4 d) 3.86, 1.1

e) 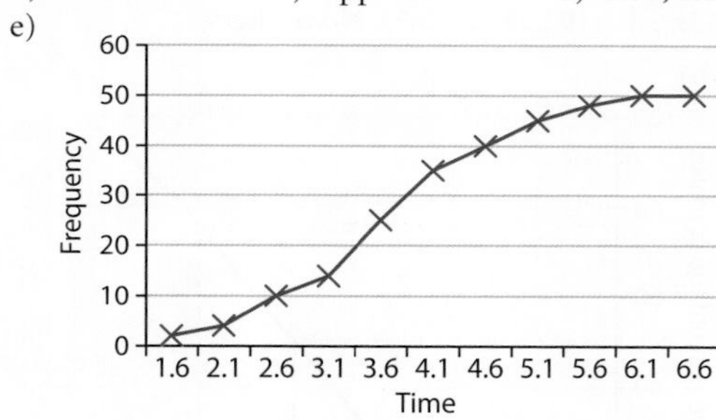

f) Minimum = 1.6, Q1 = 3, median = 4, Q3 = 4.5, maximum = 6.2

4 a) Median and IQR as the data is skewed with outliers.

b) Mean = 682.6, standard deviation = 536.2

c) 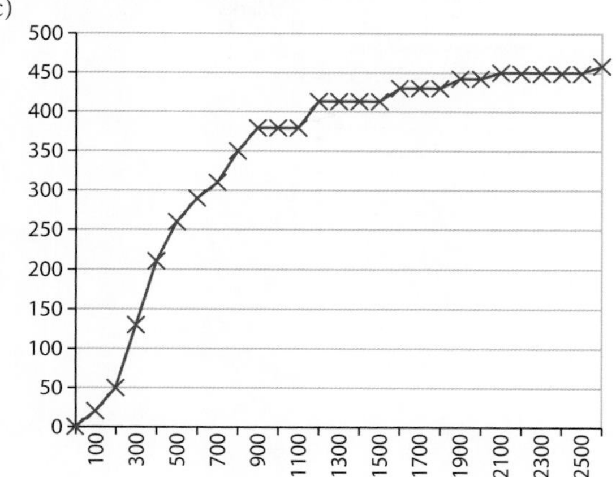

d) Q1 = 300, median = 500, Q3 = 800, IQR = 500
e) There are a few outliers on the right side. Outliers lie above Q3 + 1.5IQR = 1550.
f) Data is skewed to the right, with several outliers from 1600 onwards. It is bimodal at 300–400.

5 a) Spain, Spain b) France
c) On average, it appears that France produces the more expensive wines as 50% of its wines are more expensive than most of the wines from the other countries. Italy's prices seem to be symmetric while France's prices are skewed to the left. Spain has the widest range of prices.

6 a) Mean = 52.65, standard deviation = 7.66
b) Median = 51.34, IQR = 2.65
c) Apparently, the data is skewed to the right with a clear outlier of 112.72! This outlier pulled the value of the mean to the right and increased the spread of the data. The median and IQR are not influenced by the extreme value.

7 a) The distribution does not appear to be symmetric as the mean is less than the median, the lower whisker is longer than the upper one and the distance between Q1 and the median is larger than the distance between the median and Q3. Left skewed.
b) There are no outliers as Q1 − 1.5IQR = 37 < 42 and Q3 + 1.5IQR = 99 > 86.
c)

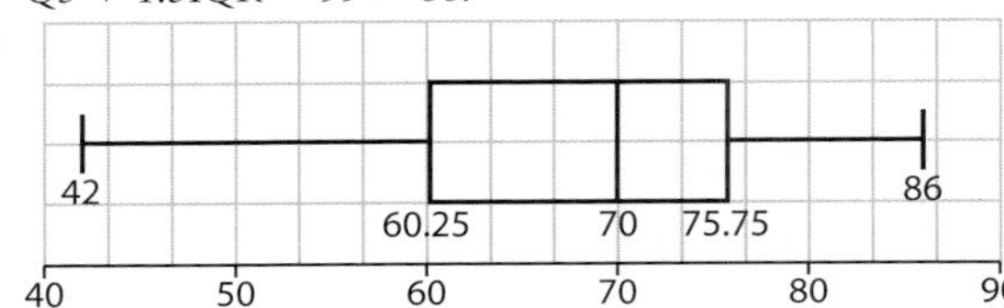

d) See a)

8 a) 225
b) Q1=205, Q3 = 255, 90th percentile=300, 10th percentile = 190
c) IQR = 50, since Q1 − 1.5IQR = 130 > minimum and Q3 + 1.5IQR = 330 < 400 then there are outliers on both sides.
d)

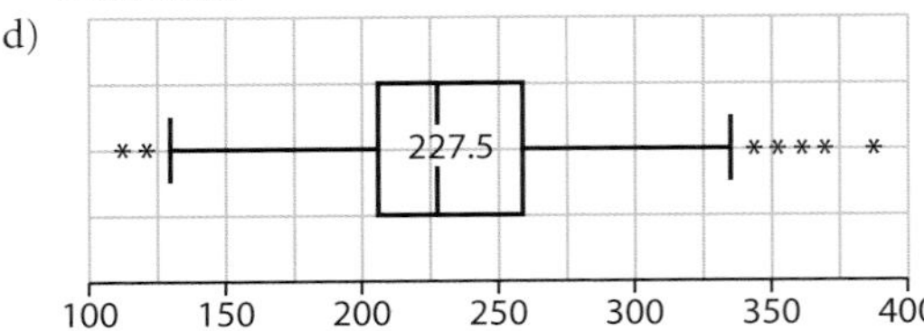

e) The distribution has many outliers. Apparently skewed to the right with more outliers there. The middle 50% seem to be very close together while the whiskers appear to be quite spread.

9 a)

Speed	Frequency
26–30	8
31–34	15
35–38	31
39–42	24
43–46	10
47–50	10
51–54	2

b)

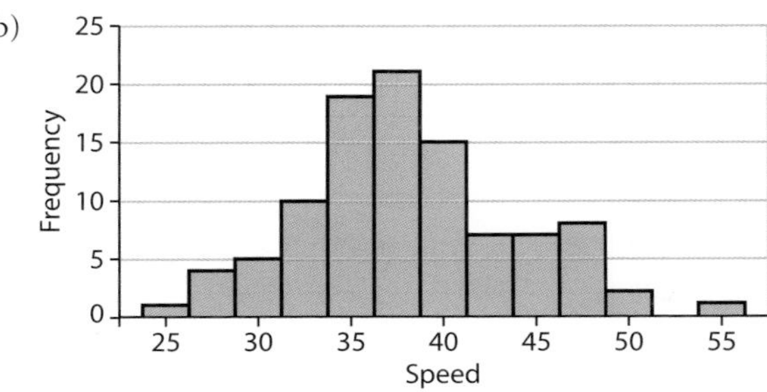

Data is relatively symmetric with possible outlier at 55. The mode is approximately 37.
Histogram created from table:

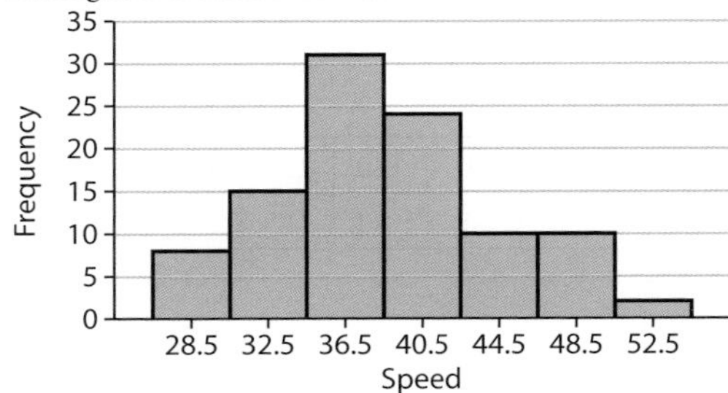

c) Mean = 38.2, standard deviation = 5.7
d)

Speed	Cu. frequency
26–30	8
31–34	23
35–38	54
39–42	78
43–46	88
47–50	98
51–54	100

e) Median = 37.6, Q1 = 34.5, Q3 = 41.3, IQR = 6.8
f) There are outliers on the right since Q3 + 1.5IQR = 51.5 < maximum = 54.

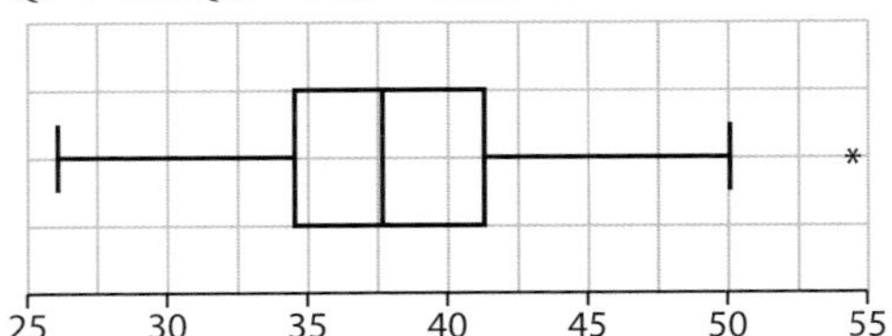

10 a) Mean = 1846.9, median = 1898.6, standard deviation = 233.8, Q1 = 1711.8, Q3 = 2031.3, IQR = 319.5
b) Q1 − 1.5IQR = 1232.55 > minimum, so there is an outlier on the left.
c)

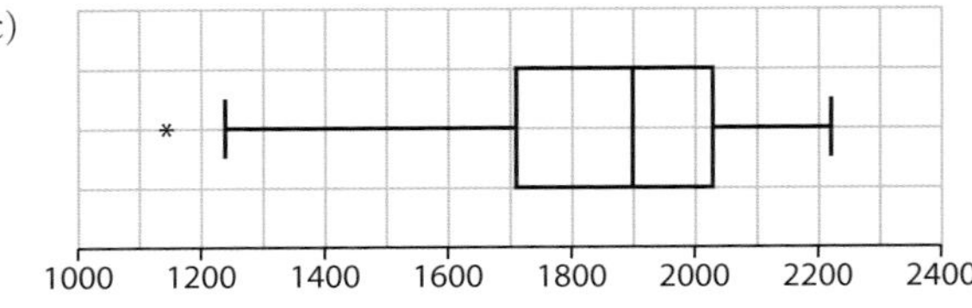

d)]1613, 2081[
e) The mean and standard deviation will get larger. The rest will not change much.

11 a) 49.6 minutes b) 48.9 minutes

12 a)

≤10	≤20	≤30	≤40	≤50	≤60	≤70	≤80	≤90	≤100
30	130	330	670	1190	1630	1810	1900	1960	2000

b)

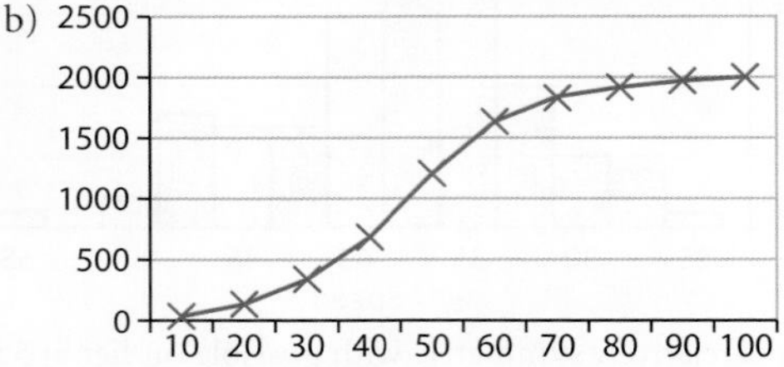

c) (i) 47 (ii) About 500 (iii) Above 60

13 1.74

14 a) $m = 12$ b) Standard deviation = 5

15 $k = 4$

16 a) 97.2

b)

30	60	90	120	150	180	210	240
5	20	53	74	85	92	97	100

c)

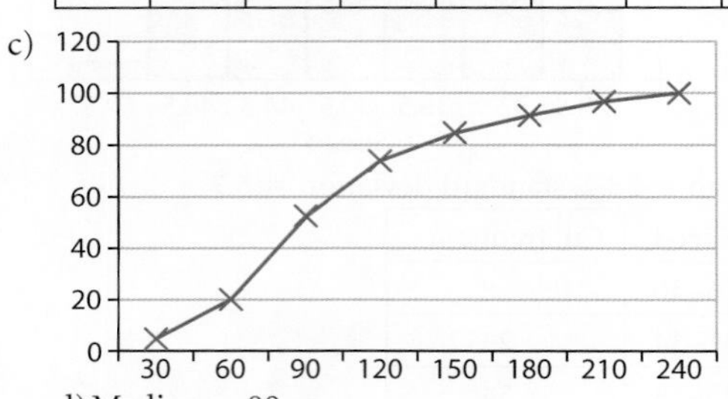

d) Median = 88
Q1 = 66
Q3 = 124

17 a) (i) 10 (ii) 24
b) Mean = 63, standard deviation = 20.5
c) Skew to the left d) 65

18 a) 7.41

b)

Weight	Number of packets
$w \leq 85$	5
$w \leq 90$	15
$w \leq 95$	30
$w \leq 100$	56
$w \leq 105$	69
$w \leq 110$	76
$w \leq 115$	80

c) (i) Median = 97 (ii) Q3 = 101 d) 0 e) 0.282

19 a) 98.2
b) (i) $a = 165, b = 275$
(ii)

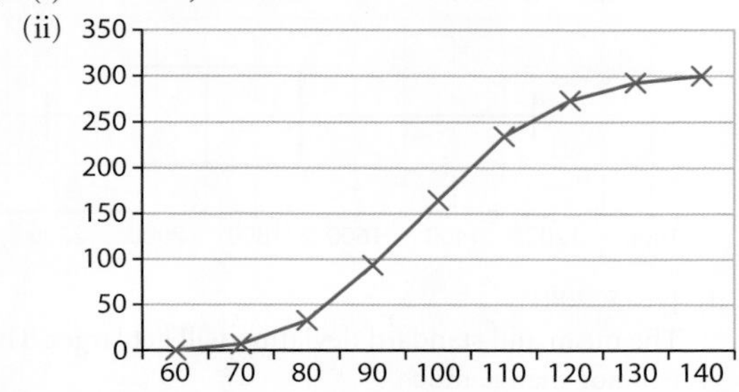

c) (i) 34% (ii) 115

20 a) (i) 24 (ii) 158 b) 40 c) 7%

21 $a = 3$

22 a)

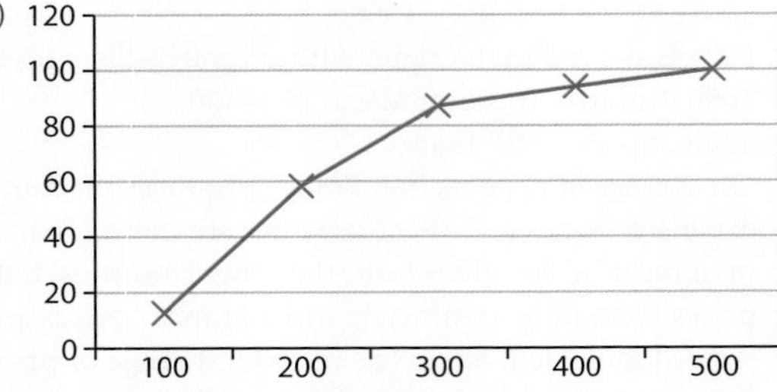

b) IQR = 110 c) $a = 7, b = 6$
d) 199 e) (i) 9 (ii) $\frac{15}{28}$

23 a) (i) 20 (ii) 24 b) 10

24 a)

Mark	[0, 20[	[20, 40[	[40, 60[	[60, 80[	[80, 100[
Number of students	22	50	66	42	20

b) Pass mark = 43%

25 a) 183 b) 14

26 $a = 3, b = 7, c = 11, d = 11$

27 a) 100 b) $a = 55, b = 75$

28 $x = 4, y = 10$

Chapter 12

Exercise 12.1 and 12.2

Note: Some answers may differ from one person to another due to different graph accuracies.

1 a) {left-handed, right-handed}
b) All real numbers from (say) 50 cm to 210 cm.
c) All real numbers from 0 to 720 (say).

2 {(1, h), (2, h), …, (1, t), …, (6, t)}

3 a) {(1, hearts), …, (king, hearts), (1, spades), …}
b) {[(1, hearts), (king, diamonds)], …,[(1, spades), (10, diamonds)],…}
c) a: 52, b: 1326

4 a) 0.47 b) Anywhere from 0 to 20
c) 10 000

5 a) {(1, 1), (1, 2), …, (4, 4)} b) {3, 4, …, 9}

6 a) {(b, b), (b, g), (b, y), (g, b), (g, g), (g, y), (y, b), (y, g), (y, y)}
b) {(y, y), (y, b), (y, g)}
c) {(b, b), (g, g), (y, y)}

7 a) {(b, g), (b, y), (g, b), (g, y), (y, b), (y, g)}
b) {(y, b), (y, g)} c) $\varnothing$

8 a) {(t, t, t), (t, t, h), (t, h, t), (h, t, t), (h, t, h), (h, h, t), (t, h, h), (h, h, h)}
b) {(h, t, h), (h, h, t), (t, h, h), (h, h, h)}

9 {(I, fly), (I, dr), (I, tr), (H, dr), (H, b)}
{(I, fly)}

10 a) {(1, g), (1, f), …, (0, c)}
b) {(0, c), (0, s)}
c) {(1, g), (1, f), (0, g), (0, f)}
d) {(1, g), (1, f), (1, s), (1, c)}

11 a) $\{(G_1, K_1, M_1), (G_1, K_2, M_1), (G_1, K_1, M_2), \ldots\}$
b) A = all triplets containing G_2; B = all triplets not containing K_1; C = all triplets containing M_2.

c) $A \cup B$ = all males or persons who drink; $A \cap C$ = all single males; C' = all non–single persons; $A \cap B \cap C$ = all single males who drink; $A' \cap B$ = all females who drink.

12 a) {(R, L, L, S), (L, R, L, R), …}, 81
b) {(R, R, R, R), (L, L, L, L), (S, S, S, S)}
c) {(R, R, L, L), (R, L, R, S), …}
d) {(R, L, R, S), (S, S, R, L), …}

13 a) {(*T*, *SY*, *O*), (*C*, *SN*, *O*), …}
b) {(*T*, *SY*, *O*), (*T*, *SY*, *F*), (*B*, *SY*, *O*), …}
c) {(*C*, *SY*, *O*), (*C*, *SN*, *O*), (*C*, *SY*, *F*), …}
d) $C \cap SY$ = {(*C*, *SY*, *O*), (*C*, *SY*, *F*)}
C' = {(*T*, …, …), (*B*, …, …)}
$C \cup SY$ = all triplets containing *C* or *SY*.

14 a) {(1, 1, 1), (1, 1, 0), (0, 1, 0), …}
b) X = {(1, 1, 0), (1, 0, 1), (0, 1, 1)}
c) Y = {(1, 1, 1), (1, 1, 0), (1, 0, 1), (0, 1, 1)}
d) Z = {(1, 1, 1), (1, 1, 0), (1, 0, 1)}
e) Z' = {(0, 1, 1), (0, 1, 0), (0, 0, 1), (0, 0, 0), (1, 0, 0)}
$X \cup Z$ = {(1, 1, 0), (1, 0, 1), (0, 1, 1), (1, 1, 1)}
$X \cap Z$ = { (1, 1, 0), (1, 0, 1)}
$Y \cup Z$ = {(1, 1, 0), (1, 0, 1), (0, 1, 1), (1, 1, 1)}
$Y \cap Z$ = { (1, 1, 1), (1, 1, 0), (1, 0, 1)}

15 a) {1, 2, 31, 32, 41, 42, 51, 52, 341, 342, …, 3452}
b) {31, 32, 41, 42, 51, 52}
c) All except {1, 2}
d) {1, 31, 41, 51, 341, 351, 431, 451, 531, 541, 3451, 4351, …}

Exercise 12.3

1 a) $\frac{3}{10}$ b) $\frac{3}{4}$
2 a) 0.63 b) 1
3 a) (i) $\frac{1}{52}$ (ii) $\frac{7}{26}$ (iii) $\frac{4}{13}$ (iv) $\frac{10}{13}$
b) (i) $\frac{1}{51}$ (ii) $\frac{13}{17}$
c) (i) $\frac{1}{52}$ (ii) $\frac{10}{13}$
4 a) $\frac{4}{5}$ b) $\frac{11}{30}$ c) 1
5 a) $\frac{1}{2}$ b) $\frac{1}{12}$
6 a) $\frac{1}{7}$ b) $\frac{4}{7}$
7 a) (i) {(1, 1), (1, 2), …, (6, 6)}
(ii) $\frac{1}{6}$ (iii) $\frac{2}{9}$ (iv) $\frac{5}{6}$
b) (i) 0 (ii) $\frac{1}{9}$ (iii) $\frac{5}{36}$ (iv) 0
8 a) 0.04 b) 0.55 c) 0.1548
d) 0.060 372 e) 0.104 022
9 a) Yes b) no c) no
10 a) 0.06 b) 0.42 c) 0.3364 d) 0.412
11 a) 0.183 b) 0.69
12 $x > \frac{n-1}{2}$
13 a) $n = 20$ b) $n = 12$
14 a) $\frac{5}{18}$ b) $\frac{1}{3}$ c) $\frac{1}{4}$
15 a) $\frac{1}{3}$ b) $\frac{5}{14}$
16 a) $\frac{3243}{10\,829} \approx 0.299$ b) $\frac{143}{39\,984} \approx 0.0036$
17 a) $\frac{144}{3553} \approx 0.0405$ b) $\frac{943}{2261} \approx 0.417$
18 a) $\frac{1}{91}$ b) $\frac{4}{91}$
19 a) $\frac{78}{253} \approx 0.308$ b) $\frac{576}{1265} \approx 0.455$
20 a) 593 775 b) $\frac{608}{2639} \approx 0.230$
c) $\frac{2426}{7917} \approx 0.306$ d) $\frac{1045}{1131} \approx 0.924$
21 a) $\frac{10\,005}{29\,900\,492} \approx 0.000\,33$ b) $\frac{269\,265}{777\,412\,792} \approx 0.000\,35$
c) $\frac{777\,143\,527}{777\,412\,792} \approx 0.9997$ d) $\frac{85\,266\,221}{777\,412\,792} \approx 0.1097$
22 a) $\frac{3}{28} \approx 0.107$ b) $\frac{17}{190} \approx 0.0895$
c) $\frac{153}{190} \approx 0.805$
23 $\frac{7}{16} \approx 0.4375$
24 $\frac{111}{400} \approx 0.2775$
25 a) $\frac{264}{1885} \approx 0.140$ b) $\frac{166}{84\,825} \approx 0.001\,96$
c) $\frac{2584}{395\,85} \approx 0.0653$
26 a) 0.096 b) 0.008 c) 0.512

Exercise 12.4

1 $\frac{7}{20}$
2 a) $\frac{5}{10}$ b) $\frac{4}{10}$ c) $\frac{2}{10}$ d) $\frac{1}{10}$ e) $\frac{2}{5}$
3 $P(A \cap B) = \frac{1}{9} \neq 0 \neq P(A)P(B)$
4 $\frac{29}{35}$
5 0.90
6 a) 92%
b) (i) 0.64% (ii) 15.36% (iii) 14.72%
c) 48.68%
7 a) 10 000 b) $\frac{9}{10}$ c) 0.3439 d) $\frac{1000}{3439}$
8 a) $\frac{15}{16}$ b) $\frac{4}{5}$ c) $\frac{1}{5}$
9 a) {(1, 1), (1, 2), …, (6, 6)}
b)

x	2	3	4	5	6	7	8	9	10	11	12
P(x)	$\frac{1}{36}$	$\frac{1}{18}$	$\frac{1}{12}$	$\frac{1}{9}$	$\frac{5}{36}$	$\frac{1}{6}$	$\frac{5}{36}$	$\frac{1}{9}$	$\frac{1}{12}$	$\frac{1}{18}$	$\frac{1}{36}$

c) $\frac{11}{36}$ d) $\frac{11}{12}$ e) $\frac{1}{3}$ f) $\frac{2}{3}$
10 a) $\frac{7}{15}$ b) $\frac{11}{75}$ c) $\frac{9}{35}$
d) $\frac{46}{75}$ e) $\frac{11}{20}$
f) No: P(female) ≠ P(female/grade 12) – for example
11 a) (i) 0.56 (ii) 0.15
b) $\frac{15}{56}$ c) no
12

P(A)	P(B)	Conditions for events A and B	$P(A \cap B)$	$P(A \cup B)$	$P(A \mid B)$
0.3	0.4	Mutually exclusive	0.00	0.7	0.00
0.3	0.4	Independent	0.12	0.58	0.30
0.1	0.5	Mutually exclusive	0.00	0.60	0.00
0.2	0.5	Independent	0.10	0.60	0.20

13 a) 0.30 b) yes
14 a) 65% b) 35% c) 52%
15 a) 0.56 b) 0.10
16 a) $\frac{1}{216}$ b) $\frac{91}{216}$ c) $\frac{75}{216}$
17 a) 0.21 b) 0.441 c) 0.657
18 a) $\frac{23}{144}$ b) $\frac{11}{144}$ c) $\frac{15}{144}\left(\text{or } \frac{5}{48}\right)$ d) $\frac{9}{23}$
19 a) $A \cap B = \{(10,5),(10,4),...,(10,1),(1,10),...,(5,10)\}$,

P = 0.069

b) $A \cup B = \{(1,12), ..., (1,1), (2,12), ..., (3,12), ..., (4,11), ..., (5,10), ...\}$, P = 0.778

c) list, P = 0.931 d) list, P = 0.222

e) same as c) f) same as d)

g) This is $(A \cup B) - (A \cap B)$; P = 0.709

20 Proof

21 a) $\frac{1}{36}$ b) $\frac{1}{28}$ c) $\frac{91}{216}$ d) 0.5

22 $\frac{2}{3}$

23 a) 0.103 b) 0.0887 c) 0.537

24 a) 0.10 b) 0.00001

25 a) $\frac{21}{22} \approx 0.955$ b) $\frac{23}{66} \approx 0.348$ c) $\frac{13}{63} \approx 0.206$

26 a) 0.36 b) 0.64 c) 0.75 d) 0.17
e) 0.0455 f) 0.682

27 a) 0.8805 b) 0.0471

Exercise 12.5

1 a)

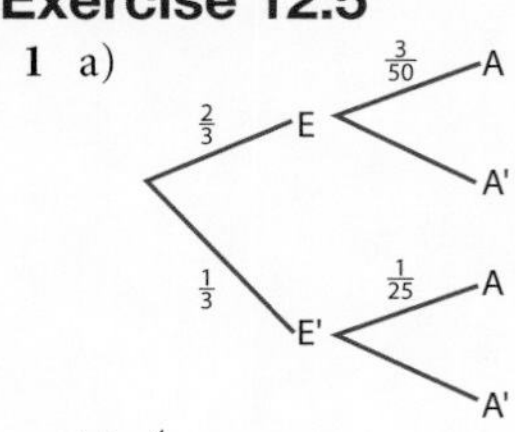

b) $\frac{4}{75}$ c) $\frac{3}{4}$

2 a) 0.060 1872 b) 0.003 37

3 a) 0.52 b) 0.692

4 a) $\frac{1}{3}$ b) $\frac{1}{2}$

5 a) 0.055 b) 0.444

6 0.875

7 a) 85.5% b) 10.5%

8 a) 0.64 b) 0.703

9 a) 0.7 b) 0.50

10 0.915

11 a) 3.6% b) 66.7% c) 5.37% d) 4.24%

12 0.382 **13** Antonio

14 a) 0.1445 b) 0.000 58

15 a) 0.425 b) 0.176

16 a) 0.93 b) 0.108

17 a) P(F) = 0.4, P(F ∩ T) = 0.224, P(F ∪ A') = 0.944, P(F'|A) = 0.35

b) M is mutually exclusive with F. T is independent as P(F ∩ T) = P(T) P(F), …

c) (i) 0.716 (ii) 0.704

18 a) 0.012 b) 0.030 c) 0.40

Practice questions

1 a) 0.30 b) 0.72 c) 0.70

2 a) 0.0004 b) 0.9996 c) 0.0004

3 0.999 98

4 a) (i) 0.85 (ii) 0.80 (iii) 0.15 b) 0.083

5 a) (i) 0.3405 (ii) 0.0108 (iii) 0.9622 (iv) 0.30
b) Yes

6 a) 0.63 b) 0.971

7 a) 0.60 b) yes, $P(B|A) = P(B) = 0.60$ c) 0.42

8 a)

	Boys	Girls
Passed the ski test	32	16
Failed the ski test	14	12
Training, but did not take the test yet	20	16
Too young to take the test	4	6

b) (i) 0.6167 (ii) 0.56 (iii) 0.1463

9 a) $\frac{3}{32}$ b) $\frac{3}{4}$ c) $\frac{5}{32}$

10 a) 0.02 b) 0.64

11 a) 0.4 b) 0.6

12 a) 0.38 b) 0.283

13 b) $\frac{11}{36}$

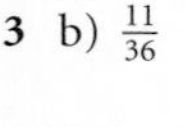
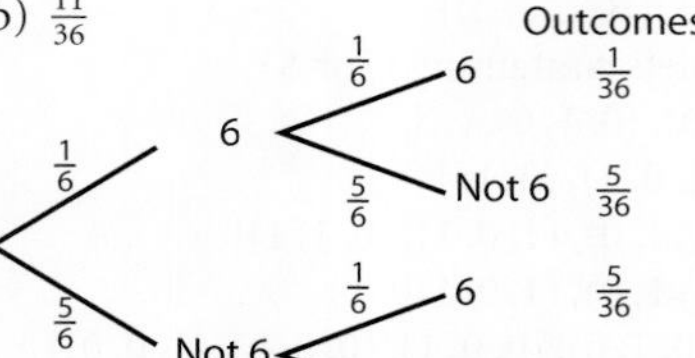

14 a)

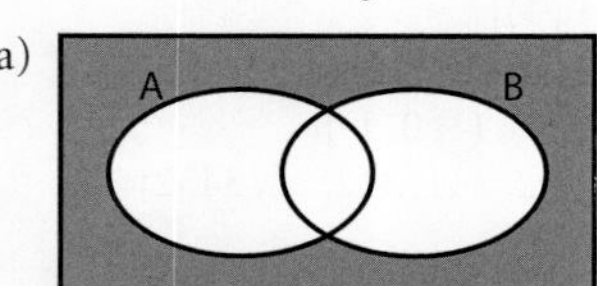

b) (i) 2 (ii) $\frac{1}{18}$ c) $n(A \cap B) \neq 0$

15 a)

	Male	Female	Total
Unemployed	20	40	60
Employed	90	50	140
Total	110	90	200

b) (i) $\frac{1}{5}$ (ii) $\frac{9}{14}$

16 $\frac{44}{65}$

17 a)

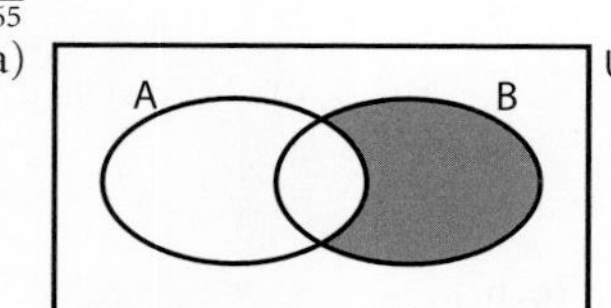

b) 35 c) 0.35

18 a)

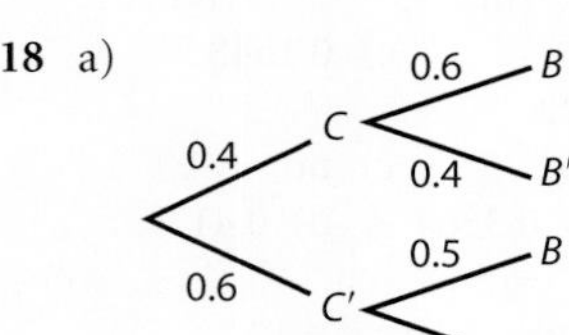

b) 0.54 c) 0.444

19 a) $\frac{7}{12}$ b) $\frac{11}{36}$ c) $\frac{1}{3}$

20 a) $\frac{1}{11}$ b) $\frac{12}{121}$

21 a) Independent b) M c) N

22 a) $a = 21, b = 11, c = 17$
b) (i) $\frac{1}{8}$ (ii) $\frac{21}{32}$
c) (i) 0.253 (ii) 0.747

23 $\frac{31}{66}$

24 a)

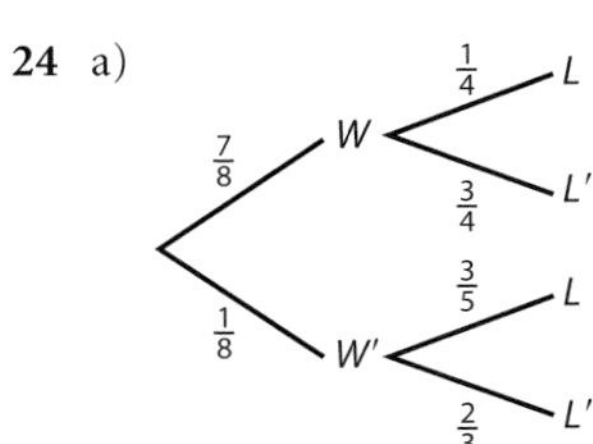

b) $\frac{47}{160}$ c) $\frac{35}{47}$

25 a) $\frac{1}{3}$ b) $\frac{7}{12}$ c) $\frac{3}{7}$

26 a)

Red: 0.4 — Grows 0.9; Does not grow 0.1
Yellow: 0.6 — Grows 0.8; Does not grow 0.2

b) (i) 0.36 (ii) 0.84 (iii) 0.429

27 a) $\frac{1}{6}$ b) $\frac{1}{12}$ c) $\frac{2}{9}$

28 a) (i) $\frac{8}{21}$ (ii) $\frac{1}{6}$ (iii) no, $P(A \cap B) \neq P(A)P(B)$

b) $\frac{10}{17}$ c) $\frac{200}{399}$

29 $\frac{1}{3}$ 30 $\frac{4}{5}$ 31 0.001 98 32 $\frac{19}{30}$

33 0.80 34 $\frac{10}{19}$ 35 a) $\frac{13}{20}$ b) $\frac{11}{15}$

36 $\frac{2}{5}$

37 a) (i) $\frac{5}{36}$ (ii) $\frac{25}{216}$ (iii) $\frac{1}{6}\left(\frac{5}{6}\right)^{2n-2}$

b) No answer required – proof

c) $\frac{5}{11}$ d) 0.432

38 a) 0.957 b) 0.301

39 $\frac{1}{9}$

40 a) 0.25 b) 0.083

41 a) 0.80 b) 0.56

42 a) 0.732 b) $\frac{11}{61}$

43 a) $\frac{2}{3}$ b) $\frac{2}{9}$ c) $\frac{3}{4}$

44 a) $\frac{1}{10}$ b) proof

c) $\frac{11}{90}$ d) $\frac{3}{11}$

45 $\frac{3}{7}$

Chapter 13

Exercise 13.1

1 4 2 $3x^2$ 3 $2x$ 4 6

5 0 6 $\frac{5}{2}$ 7 d.n.e. (increases without bound)

8 $\frac{1}{8}$ 9 $\frac{3}{2}$ 10 $\frac{\sqrt{2}}{4}$ 11 $\frac{1}{4}$

12 1 13 3 14 $\frac{1}{e}$

15 $\frac{d}{dx}[\log_b x] = \frac{1}{x \ln b}$

16 As $x \to a$, $g(x) \to +\infty$

17 a) Horizontal: $y = 3$; vertical: $x = -1$
b) Horizontal: $y = 0$; vertical: $x = 2$
c) Horizontal: $y = b$; vertical: $x = a$
d) Horizontal: $y = 2$; vertical: $x = \pm 3$
e) Horizontal: $y = 0$; vertical: $x = 0$, $x = 5$
f) Horizontal: none; vertical: $x = 4$

18 $\frac{1}{3}$ 19 4

Exercise 13.2

1 $f'(x) = -2x$ 2 $g'(x) = 3x^2$

3 $h'(x) = \frac{1}{2\sqrt{x}}$ 4 $r'(x) = -\frac{2}{x^3}$

5 (i)

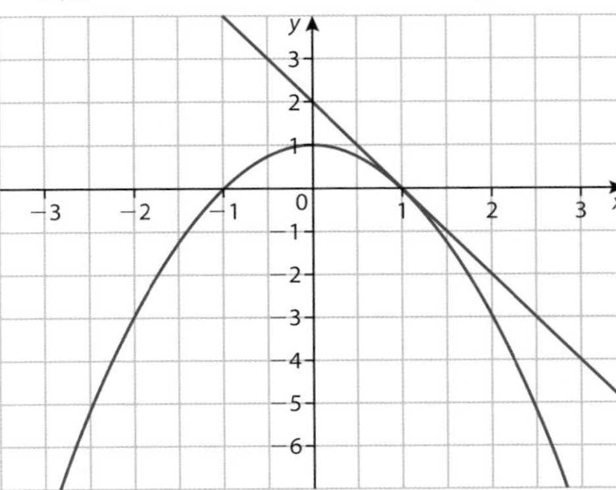

(ii)

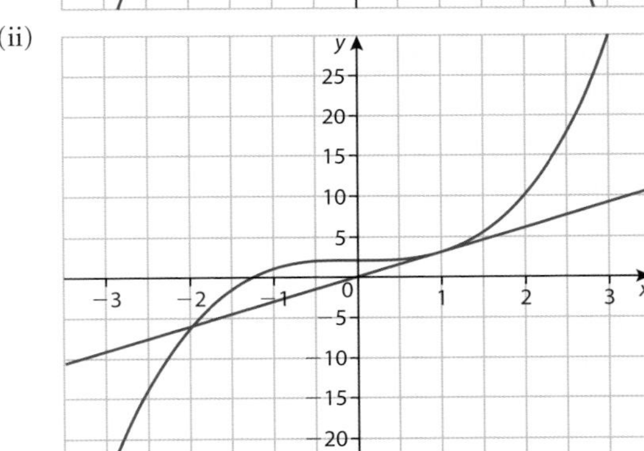

(iii)

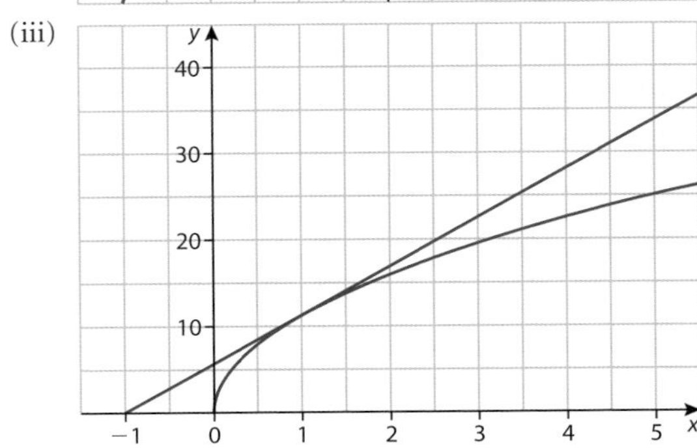

(iv)

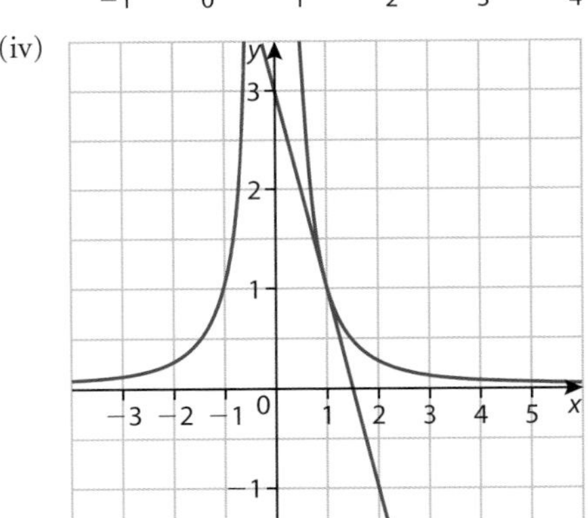

6 a) $y' = 6x - 4$ b) -4

7 a) $y' = -2x - 6$ b) 0

8 a) $y' = -\dfrac{6}{x^4}$ b) -6

9 a) $y' = 5x^4 - 3x^2 - 1$ b) 1

10 a) $y' = 2x - 4$ b) 0

11 a) $y' = 2 - \dfrac{1}{x^2} + \dfrac{9}{x^4}$ b) 10

12 a) $y' = 1 - \dfrac{2}{x^3}$ b) 3

13 $a = -5, b = 2$ **14** $(0,0)$

15 $(2,8)$ and $(-2,-8)$ **16** $\left(\dfrac{5}{2}, -\dfrac{21}{4}\right)$

17 $(1,-2)$

18 a) Between A and B
b) Rate of change is positive at A, B and F; rate of change is negative at D and E; rate of change is zero at C
c) Pair B and D, and pair E and F

19 $a = 1, b = 5$ **20** $a = 1$ **21** $(3,6)$

22 a) 12.61 b) 12 **23** $f'(x) = 2ax + b$

24 a) $4.\dot{6}$ degrees Celsius per hour
b) $C'(t) = 3\sqrt{t}$
c) $t = \dfrac{196}{81} \approx 2.42$ hours

25–26 Proof

27 $\dfrac{1}{2\sqrt{x}}$ **28** $-\dfrac{1}{x^2}$

29 $\dfrac{5}{(3-x)^2}$ $\left[\text{or } \dfrac{5}{(x-3)^2}\right]$ **30** $-\dfrac{1}{2\sqrt{(x+2)^3}}$

Exercise 13.3

1 $(1,-7)$ **2** $\left(-\dfrac{3}{2}, 8\right)$ **3** $(3,2)$

4 a) $y' = 2x - 5$ b) increasing for $x > \dfrac{5}{2}$
c) decreasing for $x < \dfrac{5}{2}$

5 a) $y' = -6x - 4$ b) increasing for $x < -\dfrac{2}{3}$
c) decreasing for $x > -\dfrac{2}{3}$

6 a) $y' = x^2 - 1$ b) increasing for $x > 1, x < -1$
c) decreasing for $-1 < x < 1$

7 a) $y' = 4x^3 - 12x^2$ b) increasing for $x > 3$
c) decreasing for $x < 0, 0 < x < 3$

8 a) $(3,-130), (-4,213)$
b) $(3,-130)$ minimum because 2nd derivative is positive at $x = 3$
$(-4,213)$ maximum because 2nd derivative is negative at $x = -4$
c)

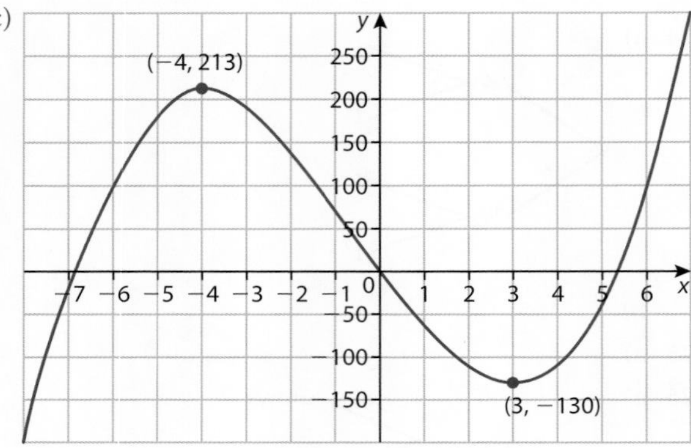

9 a) $(0,-5)$
b) Stationary point is neither a maximum nor minimum because 1st derivative is always positive.
c)

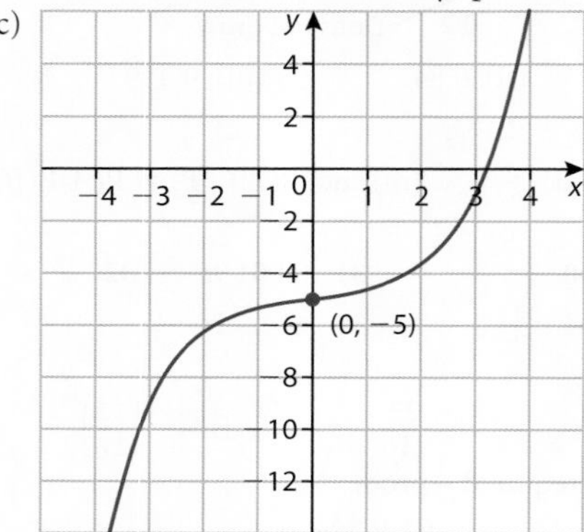

10 a) $(1,4), (3,0)$
b) $(1,4)$ maximum because 2nd derivative is negative at $x = 1$
$(3,0)$ minimum because 2nd derivative is positive at $x = 3$
c)

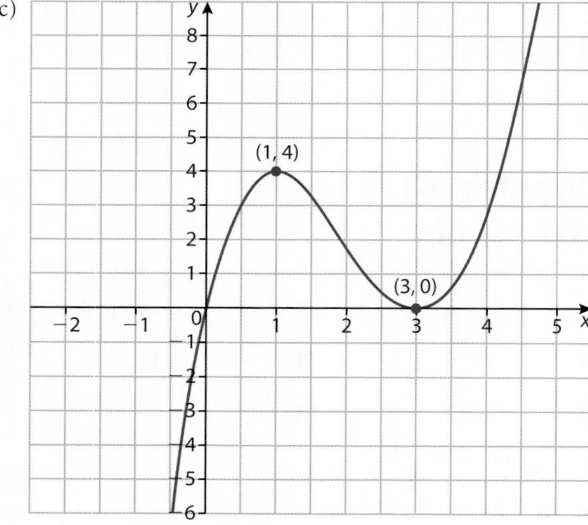

11 a) $(-1,4), (0,6), \left(\dfrac{5}{2}, -\dfrac{279}{16}\right)$
b) $(-1,4)$ minimum because 2nd derivative is positive at $x = -1$
$(0,6)$ maximum because 2nd derivative is negative at $x = 0$

at $\left(\frac{5}{2}, -\frac{279}{16}\right)$ minimum because 2nd derivative is positive at $x = \frac{5}{2}$

c)

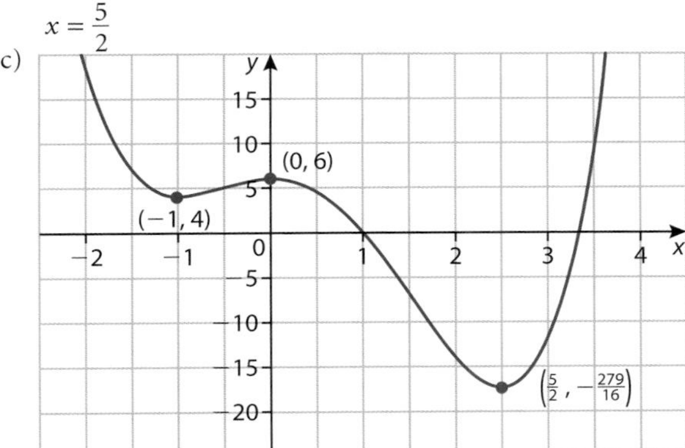

12 a) $(-1, 14), \left(\frac{7}{3}, -\frac{122}{27}\right)$

b) $(-1, 14)$ maximum because 2nd derivative is negative at $x = -1$

$\left(\frac{7}{3}, -\frac{122}{27}\right)$ minimum because 2nd derivative is positive at $x = \frac{7}{3}$

c)

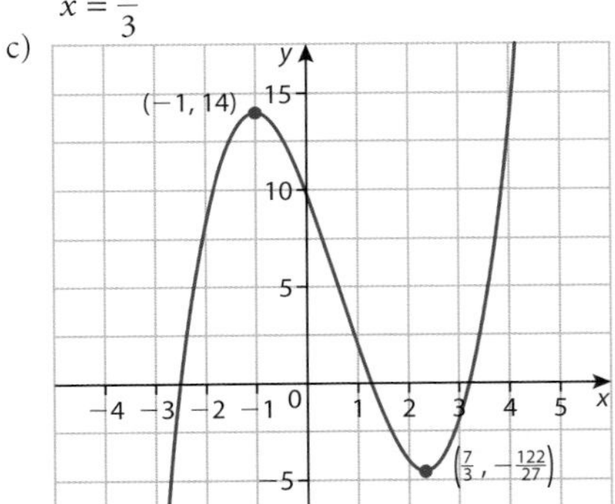

13 a) $\left(\frac{1}{4}, -\frac{1}{4}\right)$

b) $\left(\frac{1}{4}, -\frac{1}{4}\right)$ minimum because 2nd derivative is positive at $x = \frac{1}{4}$

c)

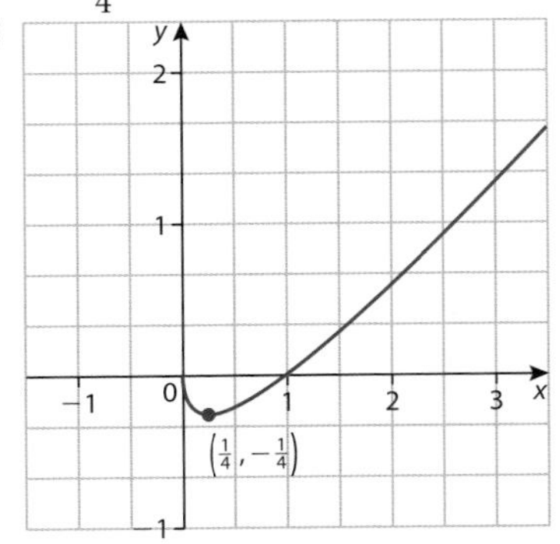

14 a) $v(t) = 3t^2 - 8t + 1$; $a(t) = 6t - 8$

b)

Displacement function:
$s(t) = t^3 - 4t^2 + t$

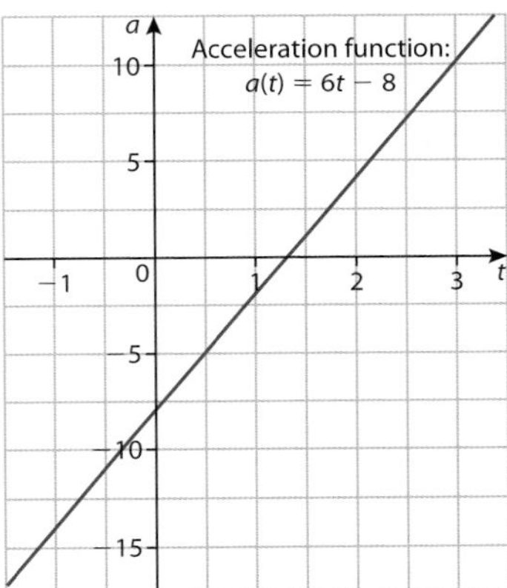

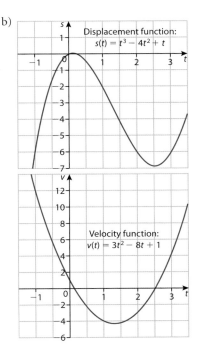

Acceleration function:
$a(t) = 6t - 8$

c) $t \approx 0.131$, displacement ≈ 0.0646

d) $t = 1.\overline{3}$, displacement $= -4.\overline{3}$

e) Object moves right at a decreasing velocity then turns left with increasing velocity then slowing down and turning right with increasing velocity.

15 Relative maximum at $(-2, 16)$; relative minimum at $(2, 16)$; inflexion point at $(0, 0)$

16 Absolute minima at $(-2, -4)$ and $(2, -4)$; relative maximum at $(0, 0)$; inflexion points at $\left(-\frac{2\sqrt{3}}{3}, -\frac{20}{9}\right)$ and $\left(\frac{2\sqrt{3}}{3}, -\frac{20}{9}\right)$

17 Relative maximum at $(-2, -4)$; relative minimum at $(2, 4)$; no inflexion points

18 Relative minimum at $\left(-\frac{\sqrt[3]{4}}{2}, \frac{3\sqrt[3]{2}}{2}\right)$; inflexion point at $(1, 0)$

19 Relative minimum at $(-1, -2)$; relative maximum at $(1, 2)$; inflexion points at $\left(-\frac{\sqrt{2}}{2}, -\frac{7\sqrt{2}}{8}\right)$, $(0, 0)$ and $\left(\frac{\sqrt{2}}{2}, \frac{7\sqrt{2}}{8}\right)$

20 Relative minimum at $(-1, 0)$; absolute minimum at $(2, -27)$; relative maximum at $(0, 5)$; inflexion points at $(1.22, -13.4)$ and $(-0.549, 2.32)$

21 a) $v(0) = 27 \text{ m s}^{-1}$, $a(0) = -66 \text{ m s}^{-2}$

b) $v(3) = 45 \text{ m s}^{-1}$, $a(3) = 78 \text{ m s}^{-2}$

c) $t = \frac{1}{2}$ and $t = 2\frac{1}{4}$; where displacement has a relative maximum or minimum

d) $t = \frac{11}{8} = 1.375$; where acceleration is zero

22 $x \approx 5.77$ tonnes; $D \approx 34.6$ (\$34 600); this cost is a minimum because cost decreases to this value then increases

23 $a - 3$, $b = 4$, $c = -2$

24 Relative maximum at $\left(-2, -\frac{15}{4}\right)$, stationary inflexion point at $(1, 3)$

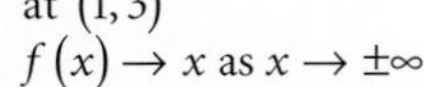

$f(x) \to x$ as $x \to \pm\infty$

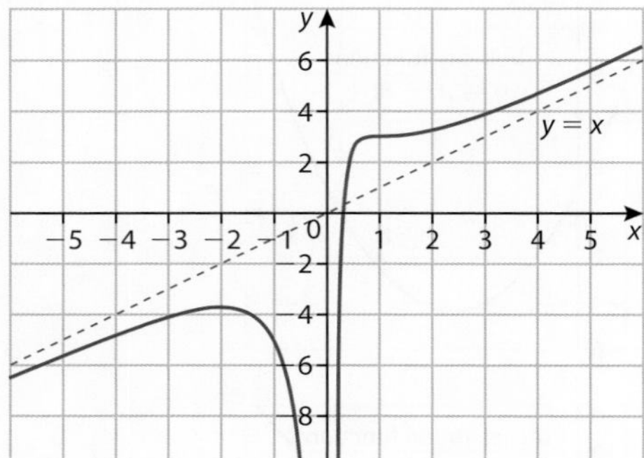

25 a)

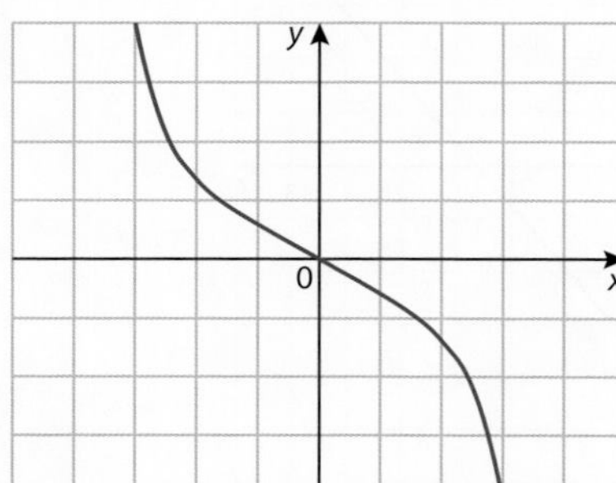

b)

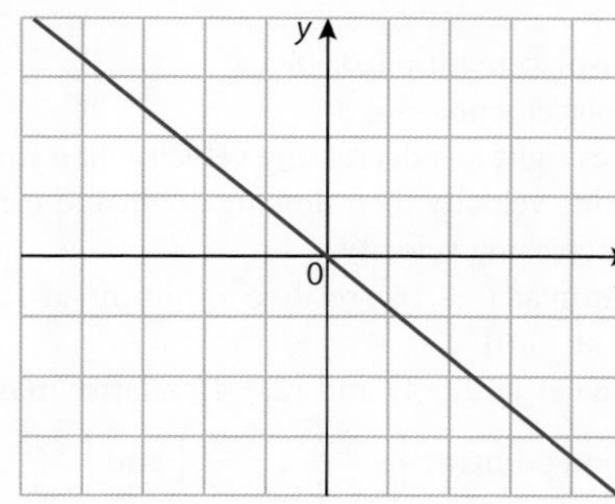

c)

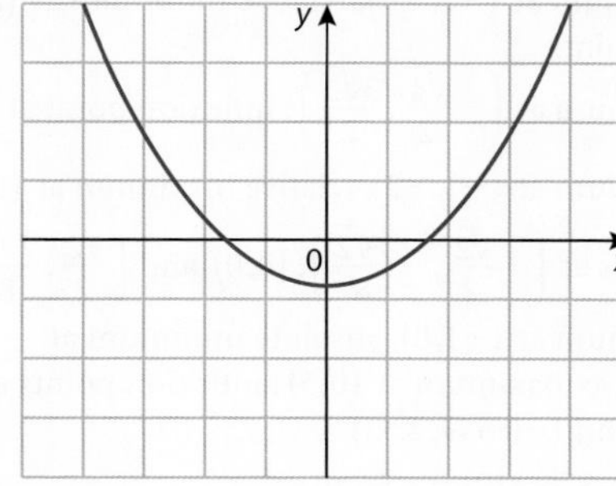

d)

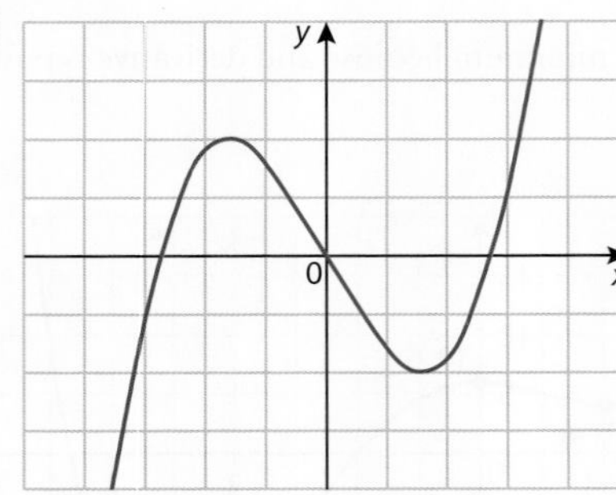

e)

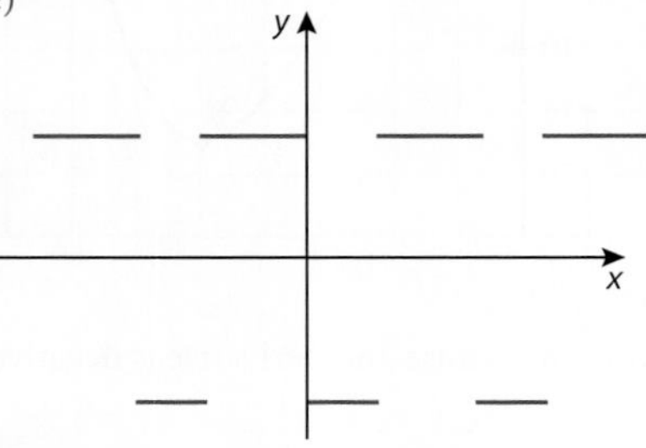

26 a) Increasing on $1 < x < 5$; decreasing on $x < 1$, $x > 5$

b) Minimum at $x = 1$; maximum at $x = 5$

27 a) Increasing on $0 \leqslant x < 1$, $3 < x < 5$; decreasing on $1 < x < 3$, $x > 5$

b) Minimum at $x = 3$; maximum at $x = 1$ and $x = 5$

28 $x \approx 0.5$ and $x \approx 7.5$

29

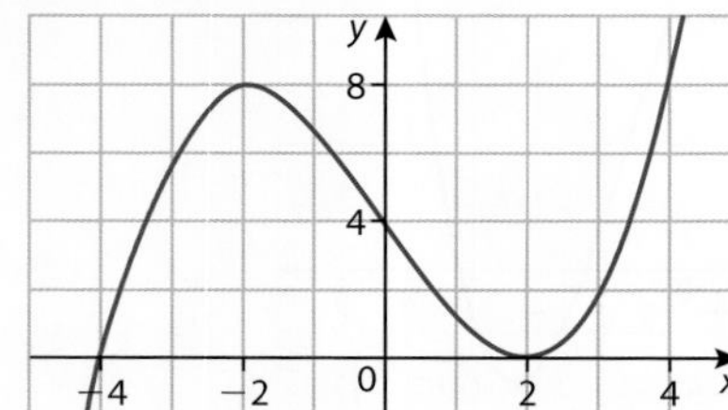

30 a) Right $1 < t < 4$; left $t < 1$, $t > 4$

b) $v_0 = -24$, $a_0 = 30$

c) $d_{max} = 16$ at $t = 4$, $v_{max} = 13.5$ at $t = 2.5$

d) Velocity is maximum at $t = 2.5$

31 a) Maximum at $x \approx 6.50$, minimum at $x \approx -0.215$

b) Maximum is $\frac{7\pi}{4} + 1$, minimum is $\frac{\pi}{4} - 1$

Exercise 13.4

1 a) $y = -4x - 8$ b) $y = \frac{4}{27}$

c) $y = -x + 1$ d) $y = -2x + 4$

2 a) $y = \frac{1}{4}x + \frac{19}{4}$ b) $x = -\frac{2}{3}$

c) $y = x + 1$ d) $y = \frac{1}{2}x + \frac{11}{4}$

3 At $(0, 0)$: $y = 2x$; at $(1, 0)$: $y = -x + 1$; at $(2, 0)$: $y = 2x - 4$

4 $y = -2x$

5 a) $x = 1$

b) For $y = x^2 - 6x + 20$, eq. of tangent is $y = -4x + 19$
For $y = x^3 - 3x^2 - x$, eq. of tangent is $y = -4x + 1$

6 Normal: $y = \frac{1}{2}x - \frac{7}{2}$; intersection pt: $\left(-\frac{1}{2}, -\frac{15}{4}\right)$

7 Eq. of tangent: $y = -3x + 3$; eq. of normal: $y = \frac{1}{3}x - \frac{1}{3}$

8 $a = 4$, $b = -7$

9 a) $y = 2x + \frac{5}{2}$ b) $\left(\frac{2}{3}, \frac{41}{27}\right)$

10 Eq. of tangent: $y = -\frac{3}{4}x + 1$; eq. of normal: $y = \frac{4}{3}x - \frac{22}{3}$

11

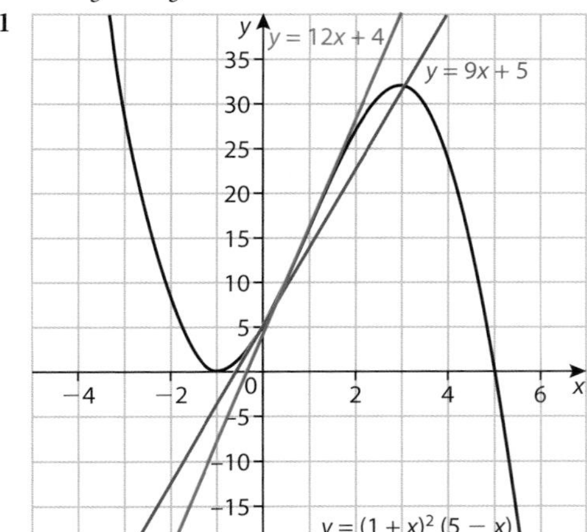

12 $y = 11x - 25$ and $y = -x - 1$

13 $y = \left(2\sqrt{2} - 2\right)x$ and $y = -\left(2\sqrt{2} + 2\right)x$

14 a) $y = \frac{1}{12}x + \frac{4}{3}$ b) $\sqrt[3]{9} \approx 2.08$

15 $y = -\frac{1}{2\sqrt{a^3}}x + \frac{3}{2\sqrt{a}}$

16 $x_Q = -2x_P$, $y_Q = -8y_P$

Practice questions

1 a) Gradient = 3 b) $y = 3x - \frac{9}{4}$

c)

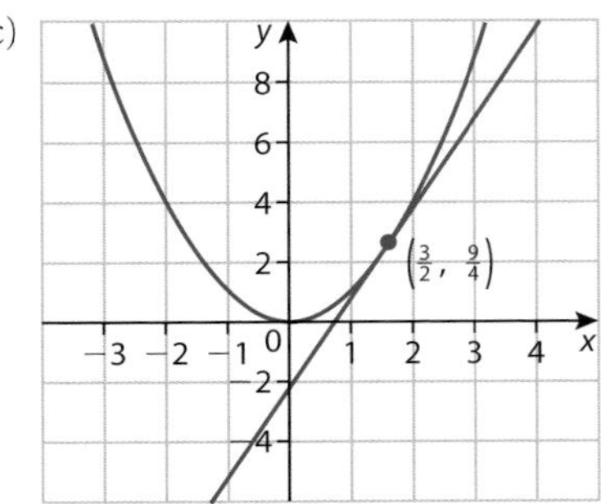

d) $Q\left(\frac{3}{4}, 0\right)$, $R\left(0, -\frac{9}{4}\right)$

f) $y = 2ax - a^2$

g) $T\left(\frac{a}{2}, 0\right)$, $U\left(0, -a^2\right)$

h) x-coord.: $\frac{a+0}{2} = \frac{a}{2}$; y-coord.: $\frac{a^2 - a^2}{2} = 0$

2 $A = 1$, $B = 2$, $C = 1$

3 a) $4x - 15x^4$

b) $-\frac{1}{x^2}$

4 a) $x = 2$ or -2; $f'(1) = -6 < 0$ (decreasing) and $f'(3) = \frac{10}{9} > 0$ (increasing) $\therefore f(2)$ is a turning point

b) vertical asymptote: $x = 0$ (y-axis); oblique asymptote: $y = 2x$

5 $\left(\frac{1}{2}, 3\right)$

6 $a = 1$

7 a) $y = 5x - 7$

b) $y = -\frac{1}{5}x + \frac{17}{5}$

8 a) $x = 1$

b) $-3 < x < -2$, $1 < x < 3$

c) $x = -\frac{1}{2}$

d)

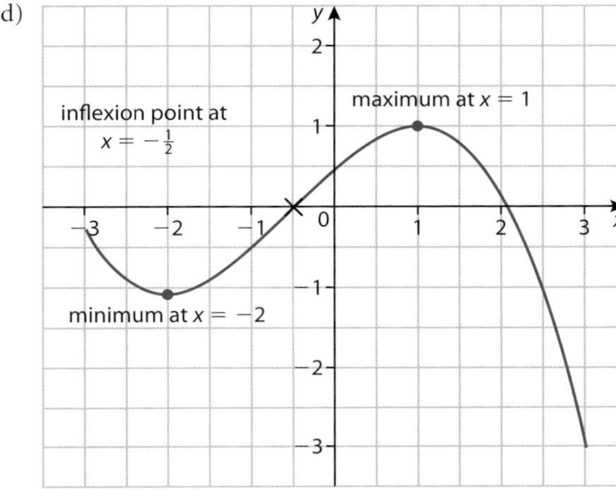

9 $b = 2$, $c = 3$

10

function	diagram
f_1	d
f_2	e
f_3	b
f_4	a

11 a) $\frac{2}{\pi}$ b) $\frac{\sqrt{2}}{2}$ c) $x \approx 0.881$

12 a) (i) $x = 0$ (ii) $y = 3$

b) $\frac{dy}{dx} = \frac{2}{x^2}$

c) Increasing for all x, except $x = 0$

d) No stationary points because $\frac{dy}{dx} = \frac{2}{x^2} \neq 0$

13 Maximum at $(-1, 1)$, minimum at $(0, 0)$, maximum at $(1, 1)$

14 $a = \frac{8}{3}$, $b = \frac{16}{5}$

15 a) $10\ \text{m s}^{-1}$ b) 10 sec c) 50 metres

16 a) $v = 14 - 9.8t$

b) $t \approx 1.43$ sec

c) Velocity = 0, acceleration = $-9.8\ \text{m s}^{-2}$

17 $(-4, 120)$

18 a) $y = \left(-\sqrt{3}\right)x + \frac{\pi\sqrt{3}}{3} - 2$

b) $y = \left(\frac{\sqrt{3}}{3}\right)x - \frac{\pi\sqrt{3}}{9} - 2$

19 a) $h = \frac{27 - r^2}{r}$; $V = \pi r\left(27 - r^2\right)$

b) $r = 3$

20 $a=-2, b=8, c=10$

21 a) $y=-7x+1$

b) $y=\frac{x}{7}+\frac{107}{7}$

22 a) Absolute minimum at $\left(\frac{3}{4},-\frac{27}{256}\right)$

b) Domain: $x \in \mathbb{R}$, range: $y \geqslant -\frac{27}{256}$

c) Inflexion points at $(0,0)$ and $\left(\frac{1}{2},-\frac{1}{16}\right)$

d)

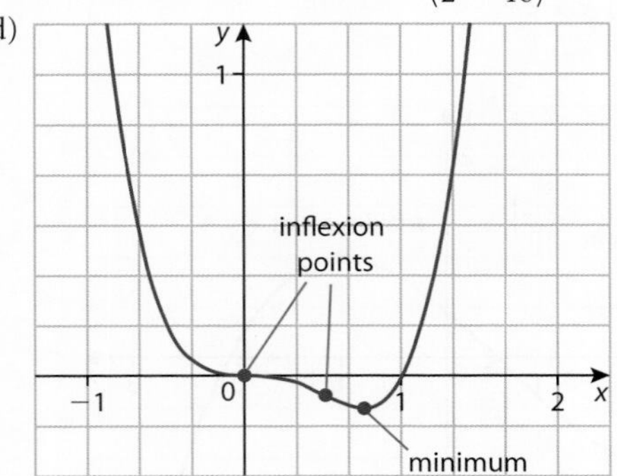

23 a) $-\frac{5}{3}$ b) $\frac{1}{4}$ c) 3 d) $\frac{1}{2\sqrt{x+2}}$

24 a) $f'(x)=\frac{3x-4}{2\sqrt{x}}$

b) $f'(x)=3x^2-3\cos x$

c) $f'(x)=-\frac{1}{x^2}+\frac{1}{2}$

d) $f'(x)=-\frac{91}{3x^{14}}$

25 3 solutions: $\left(\frac{11}{2},\frac{1105}{8}\right)$, $(2,-15)$, and $(-2,5)$

26 $\frac{17}{2}$

27 $\left(2,\frac{2}{3}\right),\left(-2,-\frac{2}{3}\right)$

28 $(-1,-2)$

30 $(2,20),(4,16)$

31 a) particle does not change direction for $0 \leqslant t \leqslant 2\pi$

b) $v=1+\cos t \geqslant 0$ for $0 \leqslant t \leqslant 2\pi$

c) $t=0, \pi, 2\pi$

d) Maximum value of s is 2π

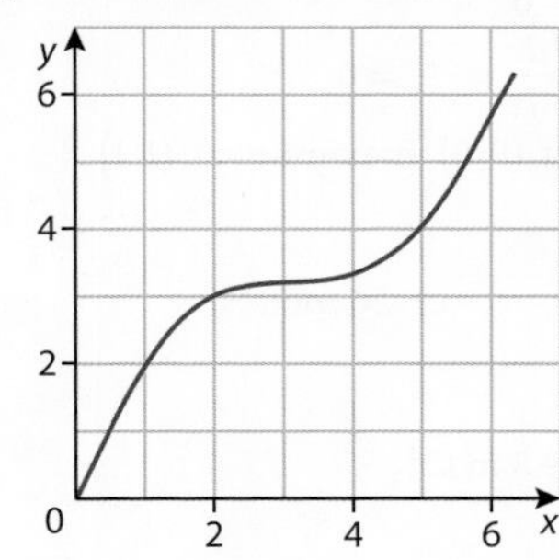

32 $a=\frac{1}{4}, b=\frac{3}{4}, c=-6, d=-\frac{5}{2}$; y-coord. is $-\frac{19}{2}$

33 Absolute minimum points at $\left(-2,-\frac{1}{8}\right)$ and $\left(2,-\frac{1}{8}\right)$

34 a) $y=-x+2$ b) $y=-x+\frac{\pi}{2}$

35 b) $y=2x$, $(1,2)$

36 a) $v=50-20t$ b) $s=1062.5$ m

37

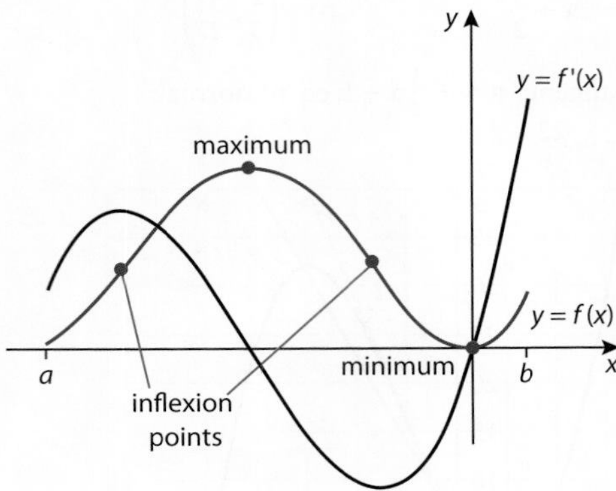

Chapter 14

Exercise 14.1

1 a) $\left(\frac{5}{2},-2,0\right)$

b) $(3,2\sqrt{3},0)$

c) $(-1,2,-2)$

d) $(a,-4a,-a)$

2 a) $Q\left(-\frac{1}{2},-3,2\right)$

b) $P\left(\frac{5}{2},-2,0\right)$

c) $Q(0,-4a,3a)$

3 a) $(x,y,z)=(t,t,5-5t)$, or $(x,y,z)=(1+t,1+t,-5t)$

b) $(x,y,z)=(-1+4t,5t,1-3t)$

c) $(x,y,z)=(2-4t,3-6t,4+t)$

4 a) $C(7,-8,-1)$

b) $C\left(-1,\frac{11}{2},\frac{29}{3}\right)$

c) $C(2-a,4-2a,-b-2)$

5 a) $\left(-\frac{1}{3},1,\frac{1}{3}\right)$

b) $\left(1,-\frac{5}{3},-1\right)$

c) $\left(\frac{a+b+c}{3},\frac{2a+2b+2c}{3},a+b+c\right)$

6 a) $D(-1,1,-6)$

b) $D(-2\sqrt{2},2\sqrt{3},1-4\sqrt{5})$

c) $D\left(\frac{5}{2},-\frac{2}{3},-4\right)$

7 $m=5, n=1$

8 a) $\mathbf{v}=\frac{2}{3}\mathbf{i}+\frac{2}{3}\mathbf{j}-\frac{1}{3}\mathbf{k}$

b) $\mathbf{v}=\frac{3}{\sqrt{14}}\mathbf{i}-\frac{2}{\sqrt{14}}\mathbf{j}+\frac{1}{\sqrt{14}}\mathbf{k}$

c) $\mathbf{v}=\frac{2}{3}\mathbf{i}-\frac{1}{3}\mathbf{j}-\frac{2}{3}\mathbf{k}$

9 a) $\frac{2}{3}(2\mathbf{i}+2\mathbf{j}-\mathbf{k})$

b) $\frac{2}{\sqrt{14}}(6\mathbf{i}-4\mathbf{j}+2\mathbf{k})$

c) $\frac{5}{3}(2\mathbf{i}-\mathbf{j}-2\mathbf{k})$

10 a) $|\mathbf{u}+\mathbf{v}|=\sqrt{29}$

b) $|\mathbf{u}|+|\mathbf{v}|=\sqrt{14}+\sqrt{5}$

c) $|-3\mathbf{u}|+|3\mathbf{v}|=3\sqrt{14}+3\sqrt{5}$

d) $\frac{1}{|\mathbf{u}|}\mathbf{u}=\frac{\mathbf{i}}{\sqrt{14}}+\frac{3\mathbf{j}}{\sqrt{14}}-\frac{2\mathbf{k}}{\sqrt{14}}$

e) $\left|\frac{1}{|\mathbf{u}|}\mathbf{u}\right|=1$

11 a) $(3,4,-5)$ b) $(0,-2,5)$

12 a) $(1, -\frac{4}{3})$ b) $\sqrt{6}(4\mathbf{i} + 2\mathbf{j} - 2\mathbf{k})$ c) $-\frac{2}{3}\mathbf{i} + \frac{8}{3}\mathbf{j} - 2\mathbf{k}$

13 0 **14** $\pm\frac{\sqrt{14}}{14}$ **15** None **16** None

17 a) $\mathbf{a} = (8, 0, 0), \mathbf{b} = (8, 8, 0), \mathbf{c} = (0, 8, 0), \mathbf{d} = (0, 0, 8),$ $\mathbf{e} = (8, 0, 8), \mathbf{f} = (8, 8, 8)$
b) $\mathbf{l} = (8, 4, 8), \mathbf{m} = (4, 8, 8), \mathbf{n} = (8, 8, 4)$
c) proof

18 a) $\mathbf{c} = (8, 0, 12), \mathbf{d} = (0, 10, 12)$
b) $\mathbf{f} = (4, 5, 0), \mathbf{g} = (4, 5, 12)$
c) $\overrightarrow{AG} = (-4, 5, 12) = \overrightarrow{FD}$

19 $\pm\frac{\sqrt{6}}{3}$

20 $(\alpha,\beta,\mu) = \left(\frac{31}{7}, -\frac{15}{7}, \frac{6}{7}\right)$ **21** $(\alpha,\beta,\mu) = (2, -1, 3)$

22 Not possible **23** Rectangle

24 $T_1 = 125(\sqrt{3} - 1)$ N; $T_2 = 175\left(\frac{3\sqrt{2} - \sqrt{6}}{2}\right)$ N

25 $T_1 = 150$ N; $T_2 = 150\sqrt{3}$ N

Exercise 14.2

1 a) $-16, 117.65°$ b) $-20, 64.68°$ c) $13, 40.24°$
d) $-15, 151.74°$ e) $6, 60°$ f) $-6, 120°$

2 a) Orthogonal b) acute c) orthogonal

3 a) $\mathbf{v} \cdot \mathbf{u} = 0 = \mathbf{wu}$ b) $\frac{3}{\sqrt{13}}\mathbf{i} + \frac{2}{\sqrt{13}}\mathbf{j}, \frac{-3}{\sqrt{13}}\mathbf{i} - \frac{2}{\sqrt{13}}\mathbf{j}$

4 a) (i) $\cos\alpha = \frac{2}{\sqrt{14}}, \cos\beta = \frac{-3}{\sqrt{14}}, \cos\gamma = \frac{1}{\sqrt{14}}$
(ii) $\cos^2\alpha + \cos^2\beta + \cos^2\gamma = \frac{2^2}{14} + \frac{(-3)^2}{14} + \frac{1^2}{14} = 1$
(iii) $\alpha \approx 58°, \beta \approx 143°, \gamma \approx 74°$

b) (i) $\cos a = \frac{1}{\sqrt{6}}, \cos\beta = \frac{-2}{\sqrt{6}}, \cos\gamma = \frac{1}{\sqrt{6}}$
(ii) $\cos^2\alpha + \cos^2\beta + \cos^2\gamma = \frac{1^2}{6} + \frac{2^2}{6} + \frac{1^2}{6} = 1$
(iii) $\alpha \approx 66°, \beta \approx 145°, \gamma \approx 66°$

c) (i) $\cos\alpha = \frac{3}{\sqrt{14}}, \cos\beta = \frac{-2}{\sqrt{14}}, \cos\gamma = \frac{1}{\sqrt{14}}$
(ii) $\cos^2\alpha + \cos^2\beta + \cos^2\gamma = \frac{3^2}{14} + \frac{(-2)^2}{14} + \frac{1^2}{14}$ $= 1$
(iii) $\alpha \approx 37°, \beta \approx 122°, \gamma \approx 74°$

d) (i) $\cos\alpha = \frac{3}{5}, \cos\beta = 0, \cos\gamma = \frac{-4}{5}$
(ii) $\cos^2\alpha + \cos^2\beta + \cos^2\gamma = \frac{3^2}{25} + \frac{0^2}{25} + \frac{4^2}{25} = 1$
(iii) $\alpha \approx 53°, \beta \approx 90°, \gamma \approx 143°$

5 $\begin{pmatrix} \frac{1}{2} \\ \frac{\sqrt{2}}{2} \\ -\frac{1}{2} \end{pmatrix}$ **6** $\begin{pmatrix} \frac{3\sqrt{2}}{2} \\ \frac{3\sqrt{2}}{2} \\ 0 \end{pmatrix}$

7 a) $m = -\frac{9}{8}$ b) $m = 1$ or $-\frac{1}{4}$

8 $m = -14$

9 a) 127° b) 63° c) 73°

10 a) $m = \frac{1}{3}$ b) $m = -\frac{1}{4}$

11 m_A: $\mathbf{r} = (4, -2, -1) + m(-1, 0, \frac{3}{2})$;
m_B: $\mathbf{r} = (3, -5, -1) + n(\frac{1}{2}, \frac{9}{2}, \frac{3}{2})$
m_C: $\mathbf{r} = (3, 1, 2) + k(\frac{1}{2}, -\frac{9}{2}, -3)$; centroid $(\frac{10}{3}, -2, 0)$

12 90, 90, 82, 74, 60, 54, 53, 52, 47, 43, 38, 37

13 68.22

14 103.3°, 133.5°, 46.5°

15 0

16 $k = 2$

17 $k = 0$ or $k = 4$

18 $x = -20, y = -14$

19 $x = 5$

20 117°, $\overrightarrow{AC} = \begin{pmatrix} 0 \\ 6 \\ 3 \end{pmatrix}$, 33°

21 a) $b = -\frac{1}{2}$ b) $b = 0$ or $b = \frac{1}{2}$
c) $b = \frac{5}{2}$ or $b = 3$ d) $b = \pm 4$

22 a) $b = -\frac{1}{2}$ b) $b = \frac{1}{2}$

23 $(-140.8, 140.8, 18)$ **24** $t = 2$

25 $t = -\frac{1}{2}$ **26** $t = 0$ or $t = \frac{1}{2}$

27 90° or $\cos^{-1}\left(\frac{2}{\sqrt{6}}\right)$ **28** Proof

29 $m = \frac{7}{4}, n = -\frac{1}{4}$ **30** Proof

31 $\frac{\pi}{3}, -\frac{2\pi}{3}$ **32** $\cos^{-1}\left(\pm\frac{\sqrt{3}}{3}\right)$

33 $\pi - \alpha, \pi - \beta, \pi - y$ **34** $\mathbf{k}(8\mathbf{i} + \mathbf{j} - 10\mathbf{k})$

Exercise 14.3

1 a) $\mathbf{k} - \mathbf{j}$ b) same

2 a) $\mathbf{i} - \mathbf{k}$ b) same

3 a) $\mathbf{j} - \mathbf{i}$ b) same

4 Proof **5** (13, 0, 13) **6** $6\mathbf{i} - 8\mathbf{j} - 8\mathbf{k}$

7 $\begin{pmatrix} -5 \\ -1 \\ -7 \end{pmatrix}$ **8** $\mathbf{i} + \mathbf{j} - 3\mathbf{k}$

9 a) $-2m^2 + 9m - 11$ b) $-2m^2 + 9m - 11$
c) $-2m^2 + 9m - 11$

10 a) $(-40, -115, 30)$ b) $(-150, 60, 0)$
c) $(-80, -160, -640)$
d) $(80, 160, 640)$ e) $(-40, -115, 30)$
f) $(-150, 60, 0)$

11 $\frac{\sqrt{1774}}{1774}\begin{pmatrix} 19 \\ 33 \\ -18 \end{pmatrix}$ **12** $\frac{\sqrt{6}}{6}\begin{pmatrix} 2 \\ 1 \\ 1 \end{pmatrix}$ **13** $\sqrt{209}$

14 $\sqrt{139}$ **15** $2\sqrt{43}$ **16** Proof

17 $m = 1$ or $m = \frac{11}{4}$ **18** $\frac{\sqrt{374}}{2}$ **19** $5\sqrt{29}$

20 128 **21** 21 **22** 1 **23** 78

24 63 **25** No **26** Yes **27** $-2, \frac{6}{5}$

28 Not possible

29 a) 49 b) $7\sqrt{5}$ c) $\frac{7\sqrt{5}}{5}$ d) $\cos^{-1}\left(\frac{7\sqrt{10}}{30}\right)$

30 a) $\frac{49}{3}$, V(tetrahedron) $= \frac{1}{3}$ V(parallelepiped) b) $\frac{4}{3}$

31 45° **32** Proof **33** Proof

34 a) $\sqrt{\frac{564}{29}}$ b) $\frac{6\sqrt{5}}{5}$ c) $\sqrt{\frac{3}{2}}$

35 $2(\mathbf{u} \times \mathbf{v})$ **36** $23(\mathbf{u} \times \mathbf{v})$ **37** $(mp + nq)(\mathbf{u} \times \mathbf{v})$

38 a) $o = \frac{1}{2}\left(\sqrt{(ab)^2 + (ac)^2 + (bc)^2}\right)$
b) $a = \frac{1}{2}ab; b = \frac{1}{2}bc; c = \frac{1}{2}ac$
c) result obvious

39 $\begin{pmatrix} 5t - \frac{1}{3} \\ -t + \frac{2}{3} \\ 3t \end{pmatrix}$ **40** Not possible

Exercise 14.4

1 a) $\mathbf{r} = \begin{pmatrix} -1 \\ 0 \\ 2 \end{pmatrix} + t\begin{pmatrix} 1 \\ 5 \\ -4 \end{pmatrix}$ $\begin{pmatrix} x \\ y \\ z \end{pmatrix} = \begin{pmatrix} -1+t \\ 5t \\ 2-4t \end{pmatrix}$

b) $\mathbf{r} = \begin{pmatrix} 3 \\ -1 \\ 2 \end{pmatrix} + t\begin{pmatrix} 2 \\ 5 \\ -1 \end{pmatrix}$ $\begin{pmatrix} x \\ y \\ z \end{pmatrix} = \begin{pmatrix} 3+2t \\ -1+5t \\ 2-t \end{pmatrix}$

c) $\mathbf{r} = \begin{pmatrix} 1 \\ -2 \\ 6 \end{pmatrix} + t\begin{pmatrix} 3 \\ 5 \\ -11 \end{pmatrix}$ $\begin{pmatrix} x \\ y \\ z \end{pmatrix} = \begin{pmatrix} 1+3t \\ -2+5t \\ 6-11t \end{pmatrix}$

2 a) $\mathbf{r} = \begin{pmatrix} -1 \\ 4 \\ 2 \end{pmatrix} + t\begin{pmatrix} 8 \\ 1 \\ -2 \end{pmatrix}$ b) $\mathbf{r} = \begin{pmatrix} 4 \\ 2 \\ -3 \end{pmatrix} + t\begin{pmatrix} -4 \\ -4 \\ 4 \end{pmatrix}$

c) $\mathbf{r} = \begin{pmatrix} 1 \\ 3 \\ -3 \end{pmatrix} + t\begin{pmatrix} 4 \\ -2 \\ 5 \end{pmatrix}$

3 a) $\mathbf{r} = \begin{pmatrix} 3 \\ -2 \end{pmatrix} + t\begin{pmatrix} 2 \\ 3 \end{pmatrix}$ b) $\mathbf{r} = \begin{pmatrix} 0 \\ -2 \end{pmatrix} + t\begin{pmatrix} 5 \\ 2 \end{pmatrix}$

4 $2x + 3y = 7$

5 $\mathbf{r} = 2\mathbf{i} - 3\mathbf{j} + \lambda(4\mathbf{i} - 3\mathbf{j})$

6 $\mathbf{r} = (-2, 1, 4) + t(3, -4, 7)$

7 a) $(1, -1, 2)$ b) $(-17, -1, 1)$
c) No d) No

8 a) $\mathbf{r} = (2, -1) + t(1, 3)$ $\begin{pmatrix} x \\ y \end{pmatrix} = \begin{pmatrix} 2+t \\ -1+3t \end{pmatrix}$

b) $\mathbf{r} = (2, -1) + t(-3, 7)$ $\begin{pmatrix} x \\ y \end{pmatrix} = \begin{pmatrix} 2-3t \\ -1+7t \end{pmatrix}$

c) $\mathbf{r} = (2, -1) + t(7, 3)$ $\begin{pmatrix} x \\ y \end{pmatrix} = \begin{pmatrix} 2+7t \\ -1+3t \end{pmatrix}$

d) $\mathbf{r} = (0, 2) + t(2, -4)$ $\begin{pmatrix} x \\ y \end{pmatrix} = \begin{pmatrix} 2t \\ 2-4t \end{pmatrix}$

9 a) $t = \frac{3}{2}$ b) no c) $m = \frac{7}{2}$

10 a) (i) $(3, -4)$ (ii) $(7, 24)$ (iii) 25
b) (i) $(-3, 1)$ (ii) $(5, -12)$ (iii) 13
c) (i) $(5, -2)$ (ii) $(24, -7)$ (iii) 25

11 a) $(-96, 128)$ b) $\left(\frac{2040}{13}, -\frac{850}{13}\right)$

12 a) $(24, 18)$
b) $\mathbf{r} = (3, 2) + t(24, 18)$
c) In 10 minutes

13 a) $a = -3, b = -5$
b) $-\frac{\sqrt{21}}{6}$
c) $\frac{\sqrt{15}}{6}, \frac{\sqrt{35}}{2}$

14 a) 146.8° b) 3.87
c) (i) L_1: $\mathbf{r} = (2, -1, 0) + t(0, 1, 2)$; L_2: $\mathbf{r} = (-1, 1, 1) + t(1, -3, -2)$

15 a) $(x, y, z) = (1 + t, 3 - 2t, -17 + 5t)$
b) $(4, -3, -2)$

16 a) $\mathbf{r} = \left(\frac{p}{m}, 0\right) + t(n, -m)$
b) (i) $bx - ay = bx_0 - ay_0$ (ii) slope $= \frac{b}{a}$

17 (i) $\mathbf{r} = (t, t, 3t), 0 \leqslant t \leqslant 1$
(ii) $\mathbf{r} = (2t - 1, t, 1 - 3t), 0 \leqslant t \leqslant 1$
(iii) $\mathbf{r} = (1 - t, 3t, t - 1), 0 \leqslant t \leqslant 1$

18 $\mathbf{r} = (2\mathbf{j} + 3\mathbf{k}) + 2t\mathbf{k}$
$\begin{cases} x = 0 \\ y = 2 \\ z = 3 + 2t \end{cases}$

19 $\mathbf{r} = (\mathbf{i} + 2\mathbf{j} - \mathbf{k}) + t(2\mathbf{i} - 3\mathbf{j} + \mathbf{k})$
$\begin{cases} 1 + 2t \\ 2 - 3t \\ -1 + t \end{cases}$

20 $\mathbf{r} = t(x_0\mathbf{i} + y_0\mathbf{j} + z_0\mathbf{k})$
$\begin{cases} tx_0 \\ ty_0 \\ tz_0 \end{cases}$

21 a) $\mathbf{r} = (3\mathbf{i} + 2\mathbf{j} - 3\mathbf{k}) + t\mathbf{j}$
$\begin{cases} 3 \\ 2 + t \\ -3 \end{cases}$
b) $\mathbf{r} = (3\mathbf{i} + 2\mathbf{j} - 3\mathbf{k}) + t\mathbf{i}$
$\begin{cases} 3 + t \\ 2 \\ -3 \end{cases}$

22 $\frac{x - x_0}{x_0} = \frac{y - y_0}{y_0} = \frac{z - z_0}{z_0}$

23 Intersect at $(1, 3, 1)$

24 Parallel **25** Skew lines

26 Skew lines **27** Parallel

28 Skew lines **29** $(4, -4, 8)$

30 $\left(\frac{16}{11}, \frac{35}{11}, \frac{13}{11}\right)$ **31** $\left(\frac{17}{11}, -\frac{7}{11}, \frac{72}{11}\right)$ **32** $\left(\frac{43}{11}, \frac{58}{11}, -\frac{1}{11}\right)$

Exercise 14.5

1 B and C

2 A

3 $\begin{pmatrix} 2 \\ -4 \\ 3 \end{pmatrix} \begin{pmatrix} x \\ y \\ z \end{pmatrix} = 26$; $2x - 4y + 3z - 26 = 0$

4 $\begin{pmatrix} 2 \\ 0 \\ 3 \end{pmatrix} \begin{pmatrix} x \\ y \\ z \end{pmatrix} = -3$; $2x + 3z + 3 = 0$

5 $\begin{pmatrix} 0 \\ 0 \\ 3 \end{pmatrix} \begin{pmatrix} x \\ y \\ z \end{pmatrix} = 3$; $3z - 3 = 0$; $\mathbf{r} = \begin{pmatrix} 0 \\ 3 \\ 1 \end{pmatrix} + t\begin{pmatrix} 2 \\ -1 \\ 0 \end{pmatrix} + s\begin{pmatrix} 1 \\ 1 \\ 0 \end{pmatrix}$

6 $\begin{pmatrix} 5 \\ 1 \\ -2 \end{pmatrix} \begin{pmatrix} x \\ y \\ z \end{pmatrix} = 5$; $5x + y - 2z - 5 = 0$

7 $\begin{pmatrix} 0 \\ 1 \\ -2 \end{pmatrix} \begin{pmatrix} x \\ y \\ z \end{pmatrix} = -2$; $y - 2z + 2 = 0$

8 $\begin{pmatrix}1\\-6\\2\end{pmatrix}\begin{pmatrix}x\\y\\z\end{pmatrix} = 23;\ r = \begin{pmatrix}3\\-2\\4\end{pmatrix} + \lambda\begin{pmatrix}2\\1\\2\end{pmatrix} + \mu\begin{pmatrix}2\\0\\-1\end{pmatrix}$

9 $\begin{pmatrix}-2\\2\\1\end{pmatrix}\begin{pmatrix}x\\y\\z\end{pmatrix} = -1;\ -2x + 2y + z = -1$

10 $\begin{pmatrix}18\\-3\\-11\end{pmatrix}\begin{pmatrix}x\\y\\z\end{pmatrix} = 5;\ 18x - 3y - 11z = 5$

11 $\begin{pmatrix}p\\q\\r\end{pmatrix}\begin{pmatrix}x\\y\\z\end{pmatrix} = p^2 + q^2 + r^2;\ px + qy + rz = p^2 + q^2 + r^2$

12 $4x - 2y + 7z = 14;\ \begin{pmatrix}4\\-2\\7\end{pmatrix}\begin{pmatrix}x\\y\\z\end{pmatrix} = 14;$

$\mathbf{r} = \begin{pmatrix}1\\2\\2\end{pmatrix} + m\begin{pmatrix}2\\-3\\-2\end{pmatrix} + n\begin{pmatrix}4\\1\\-2\end{pmatrix}$

13 $8x + 17y - 5z + 8 = 0;\ \begin{pmatrix}8\\17\\-5\end{pmatrix}\begin{pmatrix}x\\y\\z\end{pmatrix} = -8;$

$\mathbf{r} = \begin{pmatrix}2\\-2\\-2\end{pmatrix} + s\begin{pmatrix}1\\1\\5\end{pmatrix} + t\begin{pmatrix}-3\\2\\2\end{pmatrix}$

14 $\begin{pmatrix}1\\-1\\0\end{pmatrix}\begin{pmatrix}x\\y\\z\end{pmatrix} = 3;\ x - y = 3$

15 $\begin{pmatrix}30\\1\\-23\end{pmatrix}\begin{pmatrix}x\\y\\z\end{pmatrix} = -86;\ 30x + y - 23z + 86 = 0$

16 $\begin{pmatrix}1\\0\\-1\end{pmatrix}\begin{pmatrix}x\\y\\z\end{pmatrix} = 1;\ x - z = 1;\ r = \begin{pmatrix}1\\1\\0\end{pmatrix} + m\begin{pmatrix}1\\0\\1\end{pmatrix} + n\begin{pmatrix}1\\-1\\1\end{pmatrix}$

Note: All answers for 17–22 are to the nearest degree.

17 64° **18** 90° **19** 45° **20** 50°

21 24° **22** 55° **23** $(3, 6, -10)$ **24** $(2, -2, 6)$

25 No intersection **26** Plane contains line

27 $\mathbf{r} = \begin{pmatrix}10\\-7\\0\end{pmatrix} + t\begin{pmatrix}0\\-1\\1\end{pmatrix}$ **28** $\mathbf{r} = \begin{pmatrix}3\\1\\0\end{pmatrix} + t\begin{pmatrix}0\\1\\1\end{pmatrix}$

29 No intersection

30 $\mathbf{r} = \begin{pmatrix}\ \\ \ \\ \ \end{pmatrix} + t\begin{pmatrix}0\\-1\\1\end{pmatrix}$ **31** $\begin{pmatrix}1\\-1\\1\end{pmatrix}\begin{pmatrix}x\\y\\z\end{pmatrix} = 0$

32 $x + 6y + z = 16$ **33** $\left(\frac{31}{21}, \frac{37}{21}, \frac{85}{21}\right)$

34 $\begin{pmatrix}10\\1\\-8\end{pmatrix}\begin{pmatrix}x\\y\\z\end{pmatrix} = -32;\ 10x + y - 8z + 32 = 0$

35 $\begin{pmatrix}4\\-3\\2\end{pmatrix}\begin{pmatrix}x\\y\\z\end{pmatrix} = 5;\ 4x - 3y + 2z - 5 = 0$

36 $(BC)x + (AC)y + (AB)z = ABC$

37 $\mathbf{r} = \begin{pmatrix}4\\-3\\-1\end{pmatrix} + \mathbf{r}\begin{pmatrix}2\\-3\\4\end{pmatrix} + \mathrm{s}\begin{pmatrix}4\\0\\-3\end{pmatrix}$

38 $\mathbf{r} = \begin{pmatrix}2\\3\\0\end{pmatrix} + m\begin{pmatrix}2\\-3\\4\end{pmatrix} + n\begin{pmatrix}1\\-2\\1\end{pmatrix}$

Practice questions

1 a) $\overrightarrow{OD} - \overrightarrow{OC}$ b) $\frac{1}{2}(\overrightarrow{OD} - \overrightarrow{OC})$ c) $\frac{1}{2}(\overrightarrow{OD} + \overrightarrow{OC})$

2 a) $5\mathbf{i} + 12\mathbf{j}$ b) $10\mathbf{i} + 24\mathbf{j}$

3 a) $|\overrightarrow{OA}| = |\overrightarrow{OB}| = |\overrightarrow{OC}| = 6$

b) $\overrightarrow{AC} = \begin{pmatrix}-1\\\sqrt{11}\end{pmatrix}$ c) $\frac{1}{\sqrt{12}}$ d) $6\sqrt{11}$

4 a) $(10, 5)$ b) $(-3, 6)$; 90°

5 $a = 2, b = 8$

6 $\mathbf{r} = (3, -1) + t(4, -5)$

7 a) 39.4 b) (i) $(9, 12), (18, -8)$ (ii) $\sqrt{481}$

c) 7 a.m. d) 24.4 km e) 54 minutes

8 $\mathbf{r} = t(2\mathbf{i} + 3\mathbf{j})$

9 b) $(2, 3.25)$

10 c) 90°

d) (i) $12x - 5y = 301$ (ii) $(28, 7)$

11 117°

12 $2x + 3y = 5$

13 a) $(6, 20)$ b) (i) $(6, -8)$ (ii) 10

c) $4x + 3y = 84$ d) collide at 15:00

f) 26 km

14 72°

15 a) 3.94 m b) 1.22 m/s

c) $x - 0.7y = 2$ d) $\left(\frac{170}{29}, \frac{160}{29}\right)$

e) Speed = 1.24 m/s

16 $\begin{pmatrix}x\\y\end{pmatrix} = \begin{pmatrix}1\\3\end{pmatrix} + t\begin{pmatrix}5\\2\end{pmatrix}$

17 $2x^2 + 7x - 15 = 0, x = \frac{3}{2}, x = -5$

18 a) (ii) $(288, 84)$ (iii) 50 minutes b) 20.6°

c) (i) $(99, 168)$ (iii) $XY = 75$ d) 180 km

19 $3x + 2y = 7$

20 a) $\overrightarrow{ST} = \begin{pmatrix}9\\9\end{pmatrix}$, $V(-4, 6)$ b) $\mathbf{r} = (-4, 6) + \lambda(1, 1)$

c) $\lambda = 5$ d) (i) $a = 5$ (ii) 157°

21 81.9°

22 a) 13 b) $\frac{1}{5}(3\mathbf{i} + 4\mathbf{j})$ c) $\frac{56}{65}$

23 $(2, 3)$

24 a) $(3, -2)$ c) (iii) 23 square units

25 a) $\overrightarrow{OB} = \begin{pmatrix}-1\\7\end{pmatrix}; \overrightarrow{OC} = \begin{pmatrix}8\\9\end{pmatrix}$ b) $d = 11$

c) $\overrightarrow{BD} = \begin{pmatrix}12\\-3\end{pmatrix}$ d) (i) $\begin{pmatrix}x\\y\end{pmatrix} = \begin{pmatrix}-1\\7\end{pmatrix} + t\begin{pmatrix}12\\-3\end{pmatrix}$ (ii) $t = 0$

26 a) (i) $\overrightarrow{AB} = \begin{pmatrix}-5\\1\end{pmatrix}$ (ii) $AB = \sqrt{26}$ b) $\overrightarrow{AD} = \begin{pmatrix}d - 2\\25\end{pmatrix}$

c) (ii) $\overrightarrow{OD} = \begin{pmatrix}7\\23\end{pmatrix}$ d) $\overrightarrow{OC} = \begin{pmatrix}2\\24\end{pmatrix}$ e) 130

27 a) (i) $\overrightarrow{BC} = -6\mathbf{i} - 2\mathbf{j}$ (ii) $\overrightarrow{OD} = -2\mathbf{i}$ b) 82.9°

c) $\mathbf{r} = \mathbf{i} - 3\mathbf{j} + t(2\mathbf{i} + 7\mathbf{j})$ d) $15\mathbf{i} + 46\mathbf{j}$

28 a) $(5, 5, -5)$ b) $(-5, 0, 5)$ c) $(5, 5, -5)$

29 b) (i) $(49, 32, 0)$ (ii) 54 km/h

c) (i) $\frac{5}{6}$ hours (ii) $(9, 12, 5)$

30 a) (i) $\overrightarrow{AB} = \begin{pmatrix}800\\600\end{pmatrix}$

b) (ii) $\begin{pmatrix}-400\\-50\end{pmatrix}$ (iii) at 16:00 hours

c) 27.8 km

31 a) $c = 1$ b) $3\mathbf{i} + 3\mathbf{k}$

c) $\mathbf{r} = 3(1 - t)\mathbf{i} + (3 - t)\mathbf{j} + (5 + 3t)\mathbf{k}$

d) $9x - 15y + 4z - 2 = 0$ e) $\frac{15}{\sqrt{322}}$

32 a) $\overrightarrow{AB} = -\mathbf{i} - 3\mathbf{j} + \mathbf{k}; \overrightarrow{BC} = \mathbf{i} + \mathbf{j}$

b) $-\mathbf{i} + \mathbf{j} + 2\mathbf{k}$ c) $\frac{\sqrt{6}}{2}$

d) $-x + y + 2z = 3$

e) $\begin{cases} 2-t \\ -1+t \\ -6+2t \end{cases}$ f) $3\sqrt{6}$

g) $\frac{1}{\sqrt{6}}(-\mathbf{i} + \mathbf{j} + 2\mathbf{k})$ h) E$(-4, 5, 6)$

33 Proof

34 a) P$(4, 0, -3)$, Q$(3, 3, 0)$, R$(3, 1, 1)$, S$(5, 2, 1)$

b) $3x + 2y + 4z = 0$

c) 0

35 a) 147° b) 2.29

c) (i) $L_1: \begin{cases} 2 \\ -1+\lambda \\ 2\lambda \end{cases}; L_2: \begin{cases} -1+\mu \\ 1-3\mu \\ 1-2\mu \end{cases}$ (ii) no solution

d) $\frac{9}{\sqrt{21}}$

36 a) $(1, -1, 2)$

b) $11\mathbf{i} - 7\mathbf{j} - 5\mathbf{k}$

c) $\mathbf{v}.\mathbf{u} = 0$

d) $\mathbf{r} = \begin{pmatrix} 1 \\ -1 \\ 2 \end{pmatrix} + t\begin{pmatrix} 6 \\ 13 \\ -5 \end{pmatrix}$

37 a) (i) $-5\mathbf{i} + 3\mathbf{j} + \mathbf{k}$ (ii) $\frac{\sqrt{35}}{2}$

b) (i) $-5x + 3y + z = 5$

(ii) $\frac{x-5}{-5} = \frac{y+2}{3} = z - 1$

c) $(0, 1, 2)$ d) $\sqrt{35}$

38 a) $x - 2 = y - 5 = z + 1$ b) $\left(\frac{1}{3}, \frac{10}{3}, -\frac{8}{3}\right)$

c) $A'\left(-\frac{4}{3}, \frac{5}{3}, -\frac{13}{3}\right)$ d) $\frac{\sqrt{654}}{3}$

39 a) $3x - 4y + z = 6$

b) (ii) $\mathbf{r} = \begin{pmatrix} 1 \\ 2 \\ 11 \end{pmatrix} + t\begin{pmatrix} 1 \\ 4 \\ 13 \end{pmatrix}$ c) 53.7°

40 a) $(3\mu - 2, \mu, 9 - 2\mu)$

b) (i) $\mathbf{r} = \begin{pmatrix} 4 \\ 0 \\ -3 \end{pmatrix} + \lambda\begin{pmatrix} 3 \\ 1 \\ -2 \end{pmatrix}$

(ii) $\overrightarrow{PM} = \begin{pmatrix} 3\mu - 6 \\ \mu \\ 12 - 2\mu \end{pmatrix}$

c) (i) $\mu = 3$ (ii) $3\sqrt{6}$

d) $2x - 4y + z = 5$ e) verify

41 a) $(1, -1, 2)$

b) $2x - y + z = 5$

c) $(3, 1, 3)$ and $(1, 2, 2)$

42 a) (i) $\lambda = \mu$

(ii) $\mathbf{r} = \begin{pmatrix} 2 \\ 1 \\ 1 \end{pmatrix} + t\begin{pmatrix} -1 \\ -2 \\ -1 \end{pmatrix}$

b) $3x - 2y + z = 5$

c) $\mathbf{r} = \begin{pmatrix} 2 \\ 1 \\ 1 \end{pmatrix} + t\begin{pmatrix} -1 \\ -2 \\ -1 \end{pmatrix}$

Chapter 15

Exercise 15.1

1 a) $y' = 12(3x-8)^3$ b) $y' = -\frac{1}{2\sqrt{1-x}}$

c) $y' = \cos^2 x - \sin^2 x$

d) $y' = \cos\left(\frac{x}{2}\right)$ e) $y' = -\frac{4x}{(x^2+4)^3}$

f) $y' = \frac{-2}{(x-1)^2}$

g) $y' = \frac{-1}{2\sqrt{(x+2)^3}}$ $\left[\text{or } \frac{-1}{(2x+4)\sqrt{x+2}}\right]$

h) $y' = -2\sin x\cos x$

i) $y' = \frac{-x+2}{2\sqrt{(1-x)^3}}$ $\left[\text{or } \frac{-x+2}{(2-2x)\sqrt{1-x}}\right]$

j) $y' = \frac{-6x+5}{(3x^2-5x+7)^2}$ k) $y' = \frac{2}{3\sqrt[3]{(2x+5)^2}}$

l) $y' = 2(2x-1)^2(7x^4 - 2x^3 + 3)$

2 a) $y = -12x - 11$ b) $y = \frac{9}{5}x - \frac{2}{5}$

c) $y = 2x - 2\pi$ d) $y = \frac{1}{2}x + \frac{1}{2}$

3 a) $v(t) = -2t\sin(t^2 - 1)$ b) velocity $= 0$

c) $t = \sqrt{\pi + 1} \approx 2.04$, $t = 1$

d) Accelerating to the right then slowing down, turning around, accelerating to the left, slowing down, turning around again, then accelerating to the right.

4 a) $y = -12x + 38$ b) $y = \frac{1}{12}x + \frac{7}{4}$

5 a) $y = \frac{2}{3}x + \frac{5}{3}$ b) $y = -\frac{3}{2}x + 6$

6 a) $y = \frac{1}{4}x + \frac{1}{4}$ b) $y = -4x + \frac{9}{2}$

7 a) $\frac{dy}{dx} = 2\sin(2x)$; $\frac{d^2y}{dx^2} = 4\cos(2x)$

b) $\left(\frac{\pi}{4}, 0\right)$ and $\left(\frac{3\pi}{4}, 0\right)$

8 a) (i) $(0,0)$ and $(4,0)$ (ii) $\left(\frac{4}{3}, \frac{256}{27}\right)$ (iii) $\left(\frac{8}{3}, \frac{128}{27}\right)$

b)

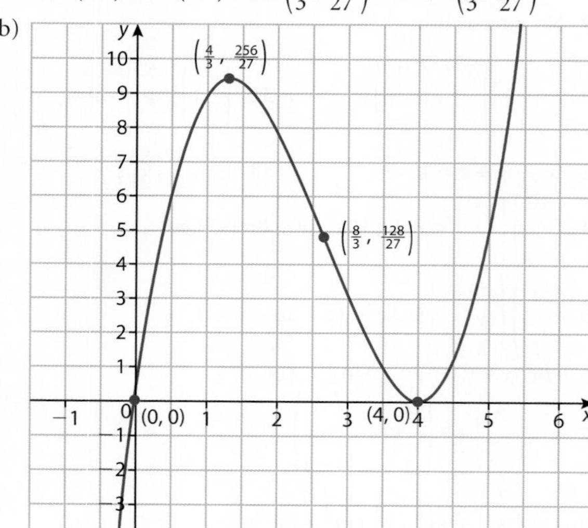

9 c) $f''(3.8) = 0$ and $f''(3) = \frac{1}{3} > 0$, $f''(4) = -\frac{2}{625} < 0$, therefore graph of f changes concavity from up to down at $x = 3.8$ verifying that graph of f does have an inflexion point at $x = 3.8$

10 $\frac{dy}{dx} = \frac{2a}{(x+a)^2}$; $\frac{d^2y}{dx^2} = \frac{-4a}{(x+a)^3}$

11 $\frac{d^n y}{dx^n} = \frac{(-1)^{n+1} n!}{(x-1)^{n+1}} \left(\text{or } \frac{n!}{(1-x)^{n+1}} \right)$

12 a) Max. at $(0, 2)$; inflexion pts at or $\left(-\frac{2\sqrt{3}}{3}, \frac{3}{2}\right)$ and $\left(\frac{2\sqrt{3}}{3}, \frac{3}{2}\right)$

b) (i) None (ii) none (iii) all $x \in$ °

c) (i) $\lim_{x\to\infty} g(x) = 0$ (ii) $\lim_{x\to-\infty} g(x) = 0$

d)

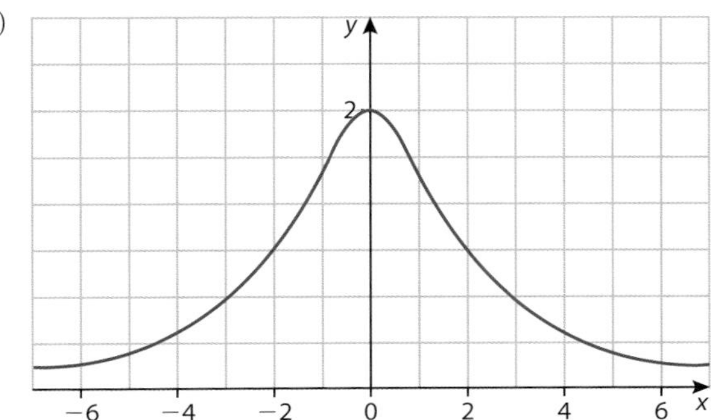

13 $\frac{d}{dx}(c \cdot f(x)) = \frac{d}{dx}(c) \cdot f(x) + c \cdot \frac{d}{dx}(f(x))$

$= 0 \cdot f(x) + c \cdot \frac{d}{dx}(f(x)) = c \cdot \frac{d}{dx}(f(x))$

14 $y = x^2(x^2 - 6) = 0$ when $x = 0$ and $x = \pm\sqrt{6}$;

$y\left(\frac{1}{2}\right) = -\frac{23}{16} < 0$, so $y < 0$ for $0 < x < 1$

$\frac{dy}{dx} = 4x(x^2 - 3) = 0$ when $x = 0$, $x = \pm\sqrt{3}$; when $x = \frac{1}{2}$, $\frac{dy}{dx} = -\frac{11}{2} < 0$, so $\frac{dy}{dx} < 0$ for $0 < x < 1$

$\frac{d^2y}{dx^2} = 12(x^2 - 1) = 0$ when $x = 0$, $x = \pm 1$; when $x = \frac{1}{2}$, $\frac{d^2y}{dx^2} = -9 < 0$, so $\frac{d^2y}{dx^2} < 0$ for $0 < x < 1$

$\frac{d^3y}{dx^3} = 24x > 0$ for $0 < x < 1$

Exercise 15.2

1 a) $y' = x^2e^x + 2xe^x$ b) $y' = 8^x \ln 8$

c) $y' = e^x \sec^2(e^x)$ d) $y' = \frac{\cos x + x\sin x + 1}{(1+\cos x)^2}$

e) $y' = \frac{xe^x - e^x}{x^2}$ f) $y' = 2\tan^3(2x)\sec(2x)$

g) $y' = \left(\frac{1}{4}\right)^x \ln\left(\frac{1}{4}\right)$ h) $y' = \cos x$

i) $y' = \frac{-xe^x + e^x - 1}{(e^x - 1)^2}$ j) $y' = -12\cos(3x)\sin(\sin(3x))$

k) $y' = 2\ln 2(2^x)$ l) $y' = \frac{\cos^3 x - \sin^3 x}{(\cos x - \sin x)^2}$

2 a) $y = \frac{1}{2}x + \frac{3\sqrt{3} - \text{p}}{6}$

b) $y = 2x + 1$

c) $y = 16x + 4 - 2\text{p}$

3 a) $x = \frac{\text{p}}{6}$, $x = \frac{5\text{p}}{6}$

b) Maximum at $\frac{\text{p}}{6}$, minimum at $\frac{5\text{p}}{6}$

4 $(0, -1)$ is an absolute maximum

5 a) Maximum at $\left(\frac{\text{p}}{2}, 5\right)$; minimum at $\left(\frac{3\text{p}}{2}, -3\right)$

b) Minimum at $\left(\frac{3\text{p}}{4}, -1\right)$ and $\left(\frac{7\text{p}}{4}, -1\right)$

6 $x = \frac{\text{p}}{2}$

7 a) $f'(x) = e^x - 3x^2$; $f''(x) = e^x - 6x$

b) $x \approx 3.73$ or $x \approx 0.910$ or $x \approx -0.459$

c) Decreasing on $(-\infty, -0.459)$ and $(0.910, 3.73)$; increasing on $(-0.459, 0.910)$ and $(3.73, \infty)$

d) $x \approx -0.459$ (minimum); $x \approx 0.910$ (maximum); $x \approx 3.73$ (minimum)

e) $x \approx 0.204$ or $x \approx 2.83$

f) Concave up on $(-\infty, 0.204)$ and $(2.83, \infty)$; concave down on $(0.204, 2.83)$

8 The two functions intersect for all x such that $\cos x = 1$, i.e. $x = k \cdot 2\text{p}$, $k \in$ ¢. The derivatives for the two functions are $y' = -e^{-x}$ and $y' = -e^{-x}(\cos x + \sin x)$. The derivatives are equal whenever $x = k \cdot 2\text{p}$, $k \in$ ¢. Therefore, the functions are tangent at all of the intersection points.

9 a) $8\,\text{m}\,\text{s}^{-2}$ b) $2.09\,\text{m}\,\text{s}^{-1}$

10 $y = ex$

11 a) $f'(x) = 2^x \ln 2$

b) $y = x\ln 2 + 1$

c) $f'(x) = 2^x \ln 2 \neq 0$ for any x

12 a) $(-1, -2e)$ and $\left(3, \frac{6}{e^3}\right)$

b) $(-1, -2e)$ is a minimum; $\left(3, \frac{6}{e^3}\right)$ is a maximum

c) (i) $\lim_{x\to\infty} h(x) = 0$

(ii) as $x \to -\infty$, $h(x)$ increases without bound

d) Horizontal asymptote $y = 0$

e)

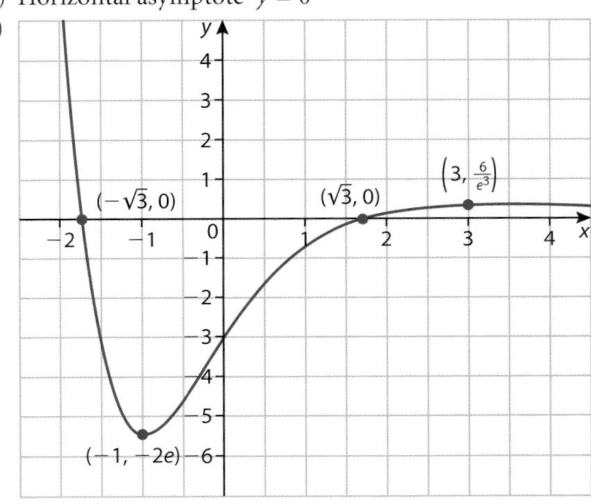

13 a) $a = \frac{\pi}{2}, b = \pi, c = \frac{3\pi}{2}$

b) $\frac{d^{(n)}}{dx^{(n)}}(\sin x) = \sin\left(x + n \cdot \frac{\pi}{2}\right), \quad n \in \mathbb{Z}^+$

14 a) $\frac{d}{dx}\left(xe^x\right) = xe^x + e^x$; $\frac{d^2}{dx^2}\left(xe^x\right) = xe^x + 2e^x$;

$\frac{d^3}{dx^3}\left(xe^x\right) = xe^x + 3e^x$

b) $\frac{d^{(n)}}{dx^{(n)}}\left(xe^x\right) = xe^x + ne^x$

Exercise 15.3

1 $\frac{dy}{dx} = -\frac{x}{y}$

2 $\frac{dy}{dx} = \frac{-2xy - y^2}{x^2 + 2xy}$

3 $\frac{dy}{dx} = \cos^2 y \quad \left[\text{or } \frac{dy}{dx} = \frac{1}{1+x^2}\right]$

4 $\frac{dy}{dx} = \frac{-2x + 3y^2 - y^3}{-6xy + 3xy^2 - 2y}$

5 $\frac{dy}{dx} = \frac{x^2y + y^3}{x^3 + xy^2}$

6 $\frac{dy}{dx} = \frac{-2xy - 2y^2 - xy}{2x^2 + 2xy + xy}$

7 $\frac{dy}{dx} = \frac{y - 1}{\cos y - x}$

8 $\frac{dy}{dx} = \frac{4x^3 - 2xy^3}{3x^2y^2 + 4y^3}$

9 $\frac{dy}{dx} = \frac{-y}{x + e^y}$

10 $\frac{dy}{dx} = \frac{x+2}{y+3}$

11 $\frac{dy}{dx} = -\sin^2(x+y) \quad \left[\text{or } \frac{dy}{dx} = -\frac{x^2}{x^2+1}\right]$

12 $\frac{dy}{dx} = \frac{18x^2\sqrt{xy} - y}{x + 2\sqrt{xy}}$

13 $y = -\frac{7}{5}x + \frac{4}{5}$; $y = \frac{5}{7}x - \frac{24}{7}$

14 $y = -2x + 4$; $y = \frac{1}{2}x + \frac{3}{2}$

15 $y = -\frac{\pi}{2}x + \pi$; $y = \frac{2}{\pi}x + \frac{\pi^2 - 4}{2\pi}$

16 $y = -\frac{352}{23}x - \frac{32}{23}$; $y = \frac{23}{352}x - \frac{5655}{176}$

17 $x^2 + y^2 = r^2 \Rightarrow \frac{dy}{dx} = -\frac{x}{y}$; at point (x_1, y_1), $m = -\frac{x_1}{y_1}$; centre of circle is $(0,0)$; slope of line through (x_1, y_1) and $(0,0)$ is $\frac{y_1}{x_1}$; because $-\frac{x_1}{y_1} \times \frac{y_1}{x_1} = -1$, the tangent to the circle at (x_1, y_1) and the line through (x_1, y_1) and $(0,0)$ are perpendicular

18 a) $(\sqrt{7}, 0)$, $(-\sqrt{7}, 0)$; $\frac{dy}{dx} = \frac{-2x - y}{x + 2y}$, at both points $\frac{dy}{dx} = -2$

b) $\left(\sqrt{\frac{7}{3}}, -2\sqrt{\frac{7}{3}}\right)$ and $\left(-\sqrt{\frac{7}{3}}, 2\sqrt{\frac{7}{3}}\right)$

c) $\left(2\sqrt{\frac{7}{3}}, -\sqrt{\frac{7}{3}}\right)$ and $\left(-2\sqrt{\frac{7}{3}}, \sqrt{\frac{7}{3}}\right)$

19 $(0,0)$

20 $\frac{dy}{dx} = -\frac{4x}{9y}$, $\frac{d^2y}{dx^2} = \frac{-36y^2 - 16x^2}{81y^3}$

21 $\frac{dy}{dx} = \frac{2-y}{(x+3)^2}$, $\frac{d^2y}{dx^2} = \frac{2y-4}{(x+3)^2}$

22 a) $\frac{dy}{dx} = \frac{-1}{3x^{\frac{4}{3}}}$, $\frac{d^2y}{dx^2} = \frac{4}{9x^{\frac{7}{3}}}$

b) $\frac{dy}{dx} = -\frac{y}{3x}$, $\frac{d^2y}{dx^2} = \frac{4y}{9x^2}$

23 $y = x + \frac{1}{2}$

24 $\frac{dy}{dx} = \frac{3x^2}{x^3 + 1}$

25 $\frac{dy}{dx} = \cot x$

26 $\frac{dy}{dx} = \frac{x}{(x^2 - 1)\ln 5}$

27 $\frac{dy}{dx} = \frac{-1}{x^2 - 1}$

28 $\frac{dy}{dx} = \frac{1}{2x \ln 10 \sqrt{\log x}}$

29 $\frac{dy}{dx} = \frac{2a}{x^2 - a}$

30 $\frac{dy}{dx} = -\sin x$

31 $\frac{dy}{dx} = \frac{-1}{x \ln 3 (\log_3 x)^2}$

32 $\frac{dy}{dx} = \ln x$

33 0

34 $y = \left(\frac{1}{8 \ln 2}\right)x - \frac{1}{\ln 2} + 3$

35 Verify

36 $x = \frac{1}{e^{\frac{3}{2}}}$

37 a) $g'(x) = \frac{1 - \ln x}{x^2}$, $g''(x) = \frac{-3 + 2\ln x}{x^3}$

b) $g'(x) = 0$ only at $x = e$; $g''(e) = -\frac{1}{e^3} < 0$, $\therefore$ abs. max. at $x = e$, max. value of g is $\frac{1}{e}$

38 $\frac{dy}{dx} = \frac{1}{x^2 + 2x + 2}$

39 $\frac{dy}{dx} = \frac{1}{x^2 + 1}$

40 $\frac{dy}{dx} = \frac{6}{x\sqrt{x^4 - 9}}$

41 $\frac{dy}{dx} = \left(\tan^{-1} x + \frac{x}{x^2 + 1}\right)e^{x \tan^{-1} x}$

42 $f'(x) = 0$; the graph of $f(x)$ is horizontal

43 Verify

44 $y = \left(\frac{\pi + 4}{2}\right)x + \frac{\pi - 4}{4}$

45 a) For $0 \leqslant x < \pi$, $f'(x) = -1$, therefore $f(x)$ is linear

b) $y = -x + \frac{\pi}{2}$

46 $\sqrt{10} \approx 3.16$ m

47 a) $\frac{1}{4}\text{m s}^{-1}, \frac{1}{20}\text{m s}^{-1}$

b) $-\frac{1}{4}\text{m s}^{-2}, -\frac{13}{800}\text{m s}^{-2}$

c) The particle initially is moving very fast to the right and then gradually slows down while continuing to move to the right.

d) $\lim_{t \to \infty} s(t) = \frac{\pi}{2}$ m

Exercise 15.4

1 a) −18.1 cm/min b) −6.79 cm/min

2 a) 0.298 cm/sec b) 0.439 cm/sec

3 a) 2π cm/hr b) 8π cm/hr

4 $\frac{d\theta}{dt} = \frac{3}{34} \approx 0.0882$ radians/min

5 26.4 m/sec

6 2 ft/sec

7 69.6 km/hr

8 $\frac{dy}{dt} = \frac{12}{\sqrt{10}} \approx 3.79$

9 0.01 m/sec

10 30 mm^3/sec

11 45 km/hr

12 $\frac{8\sqrt{3}}{3} \approx 4.62$ cm/sec

13 1.5 units/sec

14 $222.\bar{2}$ m/sec = 800 km/hr

15 a) 115 degrees/sec b) 57 degrees/sec

16 −485 km/hr

Exercise 15.5

1 $\sqrt{2}$ by $\frac{\sqrt{2}}{2}$

2 $13\frac{1}{3}$ cm by $6\frac{2}{3}$ cm

3 $\frac{\sqrt{5}}{2}$

4 b) $S = 4x^2 + \frac{3000}{x}$ c) 7.21 cm × 14.4 cm × 9.61 cm

5 $x = 5\sqrt{2\pi} \approx 12.5$ cm 6 $x \approx 3.62$ m

7 Longest ladder ≈ 7.02 m 8 $d \approx 2.64$ km

9 $\frac{8}{5}$ units^2 10 6 nautical miles

11 $h = R\sqrt{2}$, $r = \frac{R\sqrt{2}}{2}$

12 Distance of point P from point X is $\frac{ac}{\sqrt{r^2 - c^2}}$

13 $x \approx 51.3$ cm, maximum volume ≈ 403 cm^3

Practice questions

1

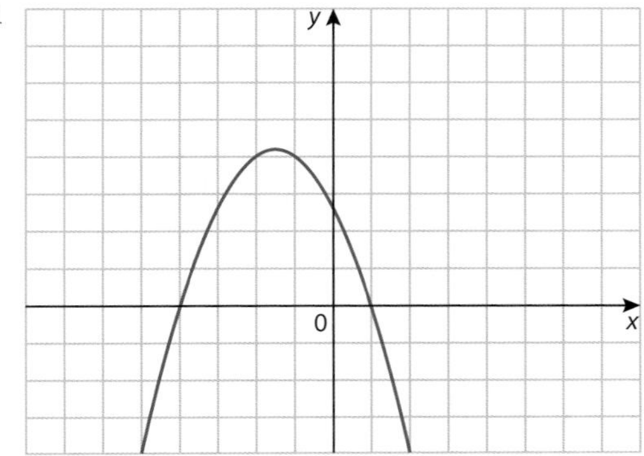

2 a) (i) $a = -4$ (ii) $b = 2$

(i) $f'(x) = -3x^2 - 4x + 8$

(ii) $\frac{-2+2\sqrt{7}}{3}$, $\frac{-2-2\sqrt{7}}{3}$

(iii) $f(1) = 5$

c) (i) $y = 8x$ (ii) $x = -2$

3 a) (i) $v(0) = 0$ (ii) $v(10) \approx 51.3$

b) (i) $a(t) = 0.99e^{-0.15t}$ (ii) $a(0) = 0.99$

c) (i) 66 (ii) 0

(iii) As object falls it approaches terminal velocity

4 a) $\left(-\frac{2}{3}, -\frac{149}{27}\right)$ is a minimum, $(-4, 13)$ is a maximum

b) $\left(-\frac{7}{3}, \frac{101}{27}\right)$ is an inflexion point

5 a) (i) $g'(x) = -\frac{3}{e^{3x}}$

(ii) $e^{3x} > 0$ for all x, hence $-\frac{3}{e^{3x}} < 0$ for all x; therefore, $f(x)$ is decreasing for all x

b) (i) $e + 2$

(ii) $g'(-\frac{1}{3}) = -3e$

c) $y = -3ex + 2$

6 b) $f'(3) = 0$ and $f''(3) > 0 \Rightarrow$ stationary point at $x = 3$ and graph of f is concave up at $x = 3$, so $f(3)$ is a minimum

c) $(4, 0)$

7 a) $-\frac{4}{(2x+3)^3}$

b) $5\cos(5x)e^{\sin(5x)}$

8 $A = 1$, $B = 2$, $C = 1$

9 $\frac{dy}{dx} = -1$, $\frac{d^2y}{dx^2} = -4$

10 a) $\frac{dy}{dx} = \frac{-xe^x + e^x - 1}{(e^x - 1)^2}$

b) $\frac{dy}{dx} = 2e^x\cos(2x) + e^x\sin(2x)$

c) $\frac{dy}{dx} = 2x\ln x + 2x\ln 3 + x - \frac{1}{x}$

11 $y = -\frac{1}{2}x - \frac{3}{2}$, $P(-3, 0)$, $Q\left(0, -\frac{3}{2}\right)$

12 a) $x = 3$; sign of $h''(x)$ changes from negative (concave down) to positive (concave up) at $x = 3$

b) $x = 1$; $h'(x)$ changes from positive (h increasing) to negative (h decreasing) at $x = 1$

13 $y = \frac{5}{7}x + \frac{11}{7}$

14 $h = 8$ cm, $r = 4$ cm

15 Maximum area is 32 square units; dimensions are 4 by 8

16 a) E b) A c) C

17 $y = -\frac{1}{5}x + \frac{32}{5}$

18 a) $y = 4x - 4$

b) $y = -\frac{1}{4}x + \frac{1}{4}$

19 a) Absolute minimum at $\left(\frac{1}{\sqrt{e}}, -\frac{1}{2e}\right)$

b) Inflexion point at $\left(\frac{1}{\sqrt{e^3}}, -\frac{3}{2e^3}\right)$

20 a) (i) $a = 16$ (ii) $a = 54$

b) $f'(x) = 2x - \frac{a}{x^2} = 0 \Rightarrow x = \sqrt[3]{\frac{a}{2}}$;

$f''(x) = 2 + \frac{2a}{x^3} \Rightarrow f''\left(\sqrt[3]{\frac{a}{2}}\right) = 4 > 0$; hence, f is concave up at any critical point, so it cannot be a maximum

21 $y = -\frac{2}{3}x + 4$

22 $y = \left(\frac{\pi+2}{2}\right)x - \frac{\pi^2}{8}$; $y = \left(\frac{-2}{\pi+2}\right)x + \frac{\pi}{2\pi+4} + \frac{\pi}{4}$

23 a) Maximum at $\left(0, \frac{1}{\sqrt{2\pi}}\right)$, inflexion points at $\left(-1, \frac{1}{\sqrt{2e\pi}}\right)$ and $\left(1, \frac{1}{\sqrt{2e\pi}}\right)$

b) $\lim_{x \to \pm\infty} f(x) = 0$; $y = 0$ (x-axis) is a horizontal asymptote

c)

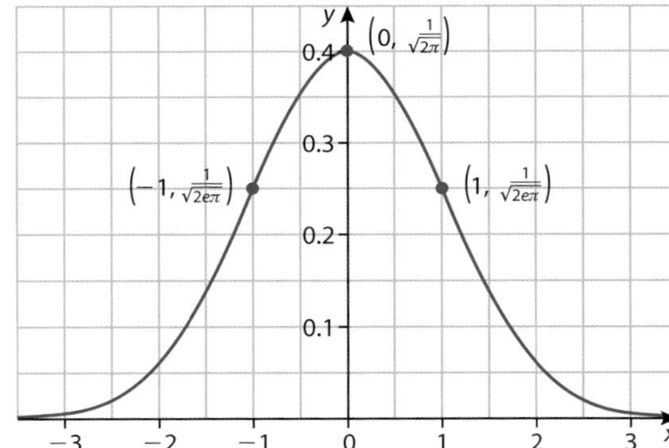

24 a) Min. at $x = 1$ because $f''(1) = \frac{1}{2} > 0$; max. at $x = 3$ because $f''(3) = -\frac{1}{6} < 0$

b) Inflexion points at $x = -\sqrt{3}$ and $x = \sqrt{3}$ because $f''(x)$ changes sign at both values

25 $x = 20\sqrt{3} \approx 34.6$ km/hr

26 $\frac{dy}{dx} = \frac{5}{6}$ or $\frac{dy}{dx} = -\frac{5}{6}$

27 $\frac{dy}{dx} = \frac{-2x}{2x^4 - 2x^2 + 1}$

28 $\frac{dy}{dx} = 2x\ln x + x$

29 $\sin x = \frac{1}{2}$, $\sin x = -1$

30 $-\frac{3}{4}$

31 a) $f'(x) = \frac{2}{2x-1}$ b) $x = \frac{1+\sqrt{17}}{4}$

32 $x \approx -0.586$

33 $c = 4 + \frac{\pi}{4}$

34 a) $f'(x) = \pi\cos(\pi x)e^{1+\sin\pi x}$ b) $x_n = \frac{2n+1}{2}$

35 a)

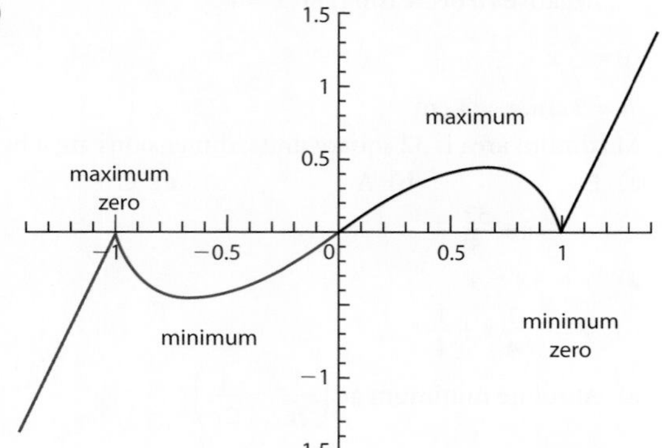

b) (i) $f'(x) = \frac{7x^2 - 3}{3(x^2-1)^{\frac{1}{3}}}$, domain: $-1.4 \leqslant x \leqslant 1.4, x \neq \pm 1$

(ii) Maximum at $x = \sqrt{\frac{3}{7}}$, minimum at $x = -\sqrt{\frac{3}{7}}$

c) $x \approx 1.1339$

36 $a = -4, b = 18$

37 a) $\frac{dy}{dx} = \sec^2 x - 8\cos x$ b) $\cos x = \frac{1}{2}$

38 a) $y = -4x - 8$ b) $(-2, 0)$

39 Proof

40 $y = -x + 2$

41 a) (i) $f'(x) = \frac{2(x^2-1)}{(x^2+x+1)^2}$

(ii) $A\left(1, \frac{1}{3}\right)$, $B(-1, 3)$ $\left(\text{or } A(-1,3), B\left(1, \frac{1}{3}\right)\right)$

b) (i)

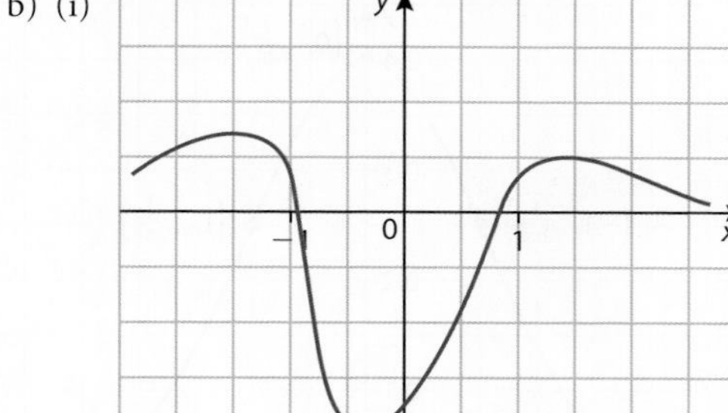

(ii) $x \approx -0.347, 1.53, 1.88$

c) (i) Range of f: $\left[\frac{1}{3}, 3\right]$ (ii) range of $f \circ f$: $\left[\frac{1}{3}, \frac{7}{13}\right]$

42 $\frac{1}{2\pi}$ cm/s

43 $y = \frac{4}{3}x - \frac{5}{3}$

44 a) (ii) $f''(x) = \frac{x^2(\ln 2)^2 - 4x\ln 2 + 2}{2^x}$

b) (i) $x = \frac{2}{\ln 2}$

(ii) $f''\left(\frac{2}{\ln 2}\right) < 0$; therefore, a maximum

c) $x = \frac{2+\sqrt{2}}{\ln 2} \approx 4.93$, $x = \frac{2-\sqrt{2}}{\ln 2} \approx 0.845$

45 a) $f'(t) = 6\sec^2 t\tan t + 5$ $\left[\text{or } f'(t) = \frac{6\sin t}{\cos^3 t} + 5\right]$

b) (i) $3 + 5\pi$ (ii) 5

46 a) $y = -1$ b) $\frac{dy}{dx} = \frac{4}{5}$

47 a) $\frac{dy}{dx} = 3e^{3x}\sin(\pi x) + \pi e^{3x}\cos(\pi x)$ b) $x \approx 0.743$

48 240 km/hr

49 b) $\left(-\frac{1}{c}\ln b, \frac{a}{2b}\right)$

50 a) $p = 2$ b) $-\frac{4}{7}$

51 $x \approx 0.460$

52 $\frac{1}{10}$ radians/sec

53 $\frac{d^2y}{dx^2} = \frac{-4}{(2x-1)^2}$

54 $y = -\frac{5}{4}x + \frac{13}{2}$

55 a) $f''(x) = 10\cos\left(5x - \frac{\pi}{2}\right)$

b) $f(x) = -\frac{2}{5}\cos\left(5x - \frac{\pi}{2}\right) + \frac{7}{5}$

56 $\frac{5}{4}$

57 $(-0.803, -2.08)$

58 a) $k = \frac{\ln 2}{20}$ b) 510 bacteria per minute

59 $f(x) = -\frac{1}{5}x^3 + \frac{12}{5}x^2 - 3x + 2$

60 a) $f'(x) = -12\cos^2(4x+1)\sin(4x+1)$

b) $x = \frac{\pi - 2}{8}$, $x = \frac{3\pi - 2}{8}$, $x = \frac{\pi - 1}{4}$

61 $\frac{dy}{dx} = \frac{3x^2 - (\ln 3)3^{x+y}}{(\ln 3)3^{x+y} - 3}$

62 a) $f'(x) = \frac{3}{3x+1}$ b) $y = -\frac{7}{3}x + \frac{14}{3} + \ln 7$

63 Verify

64 $\frac{dy}{dx} = \frac{1-e}{e}$

65 b) $b = \sqrt{6}$

66 a) $\frac{dy}{dx} = \frac{2-k}{2k-1}$ b) $k = 2$

67 $\frac{3}{2}$

68 a) $5\sqrt{5}\sqrt{x^2+4} + 5(2-x)$ minutes

c) (i) $x = 1$ (ii) 30 minutes

(iii) $\frac{d^2T}{dx^2} > 0$ for $x = 1$; therefore, it's a minimum

69 a) $P\left(-\frac{1}{2}, -\frac{1}{2e}\right)$

b) $f''(x) = 4x + 4 = 0$ at $x = -1$, and $f''(x)$ changes sign at $x = -1$

c) (i) Concave up for $x > -1$

(ii) Concave down for $x < -1$

d)

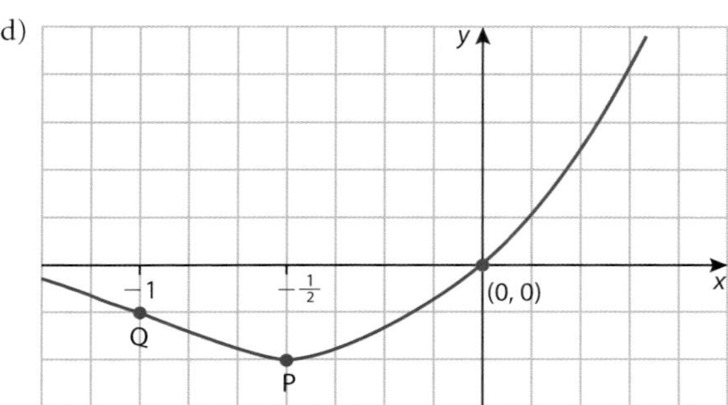

e) Show true for $n = 1$:

$f'(x) = e^{2x} + 2xe^{2x}$
$= e^{2x}(1 + 2x) = (2x + 2^0)\,e^{2x}$

Assume true for $n = k$, i.e. $f^{(k)}(x)$
$= (2^k x + k \times 2^{k-1})\,e^{2x},\ k \geqslant 1$

Consider $n = k + 1$, *i.e.* an attempt to find $\frac{d}{dx}(f^{(k)}(x))$

$f^{(k+1)}(x) = 2^k e^{2x} + 2e^{2x}(2^k x + k \times 2^{k-1})$
$= (2^k + 2(2^k x + k \times 2^{k-1}))\,e^{2x}$
$= (2 \times 2^k x + 2^k + k \times 2 \times 2^{k-1})\,e^{2x}$
$= (2^{k+1} x + 2^k + k \times 2^k)\,e^{2x}$
$= (2^{k+1} x + (k+1)2^k)\,e^{2x}$

$P(n)$ is true for $k \Rightarrow P(n)$ is true for $k + 1$, and since true for $n = 1$, result proved by mathematical induction $\forall n \in \mathbb{Z}^+$

70 $\frac{72}{\pi}\arccos\frac{8}{13}$ cm

71 a)

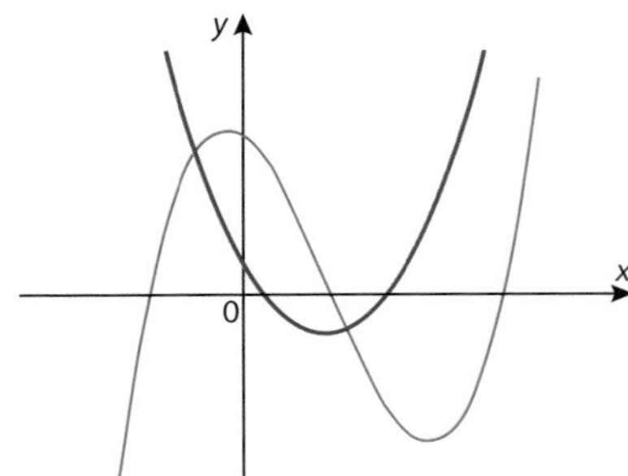

b)

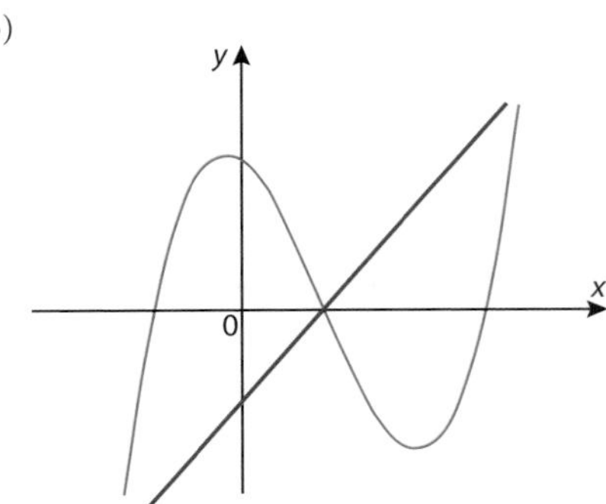

Chapter 16

Exercise 16.1

1 $\frac{x^2}{2} + 2x + c$

2 $t^3 - t^2 + t + c$

3 $\frac{x}{3} - \frac{x^4}{14} + c$

4 $\frac{2t^3}{3} + \frac{t^2}{2} - 3t + c$

5 $\frac{5u^{\frac{7}{5}}}{7} - u^4 + c$

6 $\frac{4x\sqrt{x}}{3} - 3\sqrt{x} + c$

7 $-3\cos\theta + 4\sin\theta + c$

8 $t^3 + 2\cos t + c$

9 $\frac{4x^2\sqrt{x}}{5} - \frac{10x\sqrt{x}}{3} + c$

10 $3\sin\theta - 2\tan\theta + c$

11 $\frac{1}{3}e^{3t-1} + c$

12 $2\ln|t| + c$

13 $\frac{1}{6}\ln(3t^2 + 5) + c$

14 $e^{\sin\theta} + c$

15 $\frac{(2x+3)^3}{6} + c$

16 $-\frac{5x^4}{4} + \frac{2x^3}{3} + cx + k$

17 $-\frac{x^5}{5} + \frac{x^4}{4} + \frac{x^2}{2} + 2x - \frac{11}{20}$

18 $\frac{4t^3}{3} + \sin t + ct + k$

19 $3x^4 - 4x^2 + 7x + 3$

20 $2\sin\theta + \frac{1}{2}\cos 2\theta + c$

21 $\frac{(3x^2+7)^6}{36} + c$

22 $-\frac{1}{18(3x^2+5)^3} + c$

23 $\frac{8\sqrt[4]{(5x^3+2)^5}}{75} + c$

24 $\frac{(2\sqrt{x}+3)^6}{6} + c$

25 $\frac{\sqrt{(2t^3-7)^3}}{9} + c$

26 $-\frac{(2x+3)^6}{18x^6} + c$

27 $-\frac{\cos(7x-3)}{7} + c$

28 $-\frac{1}{2}\ln(\cos(2\theta - 1) + 3) + c$

29 $\frac{1}{5}\tan(5\theta - 2) + c$

30 $\frac{1}{\pi}\sin(\pi x + 3) + c$

31 $\frac{1}{2}\sec 2t + c$

32 $\frac{1}{2}e^{x^2+1} + c$

33 $\frac{1}{3}e^{2t\sqrt{t}} + c$

34 $\frac{2}{3}(\ln\theta)^3 + c$

35 $\ln|\ln 2z| + c$

36 $-\frac{1}{15}\sqrt{(3-5t^2)^3} + c$

37 $\frac{1}{3}\tan\theta^3 + c$

38 $-\cos\sqrt{t} + c$

39 $\frac{1}{12}\tan^6 2t + c$

40 $2\ln(\sqrt{x} + 2) + c$

41 $\frac{1}{10}\sec^5 2t + c$

42 $\frac{1}{2}\ln|x^2 + 6x + 7| + c$

43 $-\frac{k^3}{2a^4}\sqrt{a^2 - a^4x^4} + c = -\frac{k^3}{2|a|^3}\sqrt{1 - a^2x^4} + c$

44 $\frac{2}{5}(3x^2 - x - 2)\sqrt{x-1} + c$

45 $-\frac{1}{\pi}\cot\pi t + c$

46 $-\frac{2}{3}\sqrt{(1+\cos\theta)^3} + c$

47 $\frac{2}{105}(15t^3 - 3t^2 - 4t - 8)\sqrt{1-t} + c$

48 $\frac{1}{15}(3r^2 + 2r - 13)\sqrt{2r-1} + c$

49 $\frac{1}{2}\ln\left(e^{x^2} + e^{-x^2}\right) + c$

50 $\frac{2}{15}(3t^2 + 20t + 230)\sqrt{t-5} + c$

Exercise 16.2

1 $-\frac{1}{3}e^{-x^3} + c$

2 $-e^{-x}(x^2 + 2x + 2) + c$

3 $\frac{2}{9}x\cos 3x - \frac{2}{27}\sin 3x + \frac{1}{3}x^2\sin 3x + c$

4 $\frac{1}{a^3}(2\cos ax - a^2x^2\cos ax + 2ax\sin ax) + c$

5 $\sin x(\ln(\sin x) - 1) + c$

6 $\frac{1}{2}x^2(\ln x^2 - 1) + c$

7 $\frac{1}{3}x^3\ln x - \frac{1}{9}x^3 + c$

8 $2e^x + x^2e^x - 2xe^x - \frac{1}{3}x^3 + c$

9 $\frac{1}{\pi^2}(\cos\pi x + \pi x\sin\pi x) + c$

10 $\frac{3}{13}\cos 2t\,e^{3t} + \frac{2}{13}e^{3t}\sin 2t + c$

11 $\sqrt{1-x^2} + x\arcsin x + c$ 12 $e^x(x^3 - 3x^2 + 6x - 6) + c$
13 $-\frac{1}{4}e^{-2x}(\cos 2x + \sin 2x) + c$
14 $\frac{1}{2}x\left(\sin(\ln x) - \cos(\ln x)\right) + c$
15 $\frac{1}{2}x\left(\sin(\ln x) + \cos(\ln x)\right) + c$
16 $\ln|x+1| - 2x + x\ln|x^2 + x| + c$
17 $\dfrac{e^{kx}(k\sin x - \cos x) + c}{k^2+1}$ 18 $x\tan x + \ln|\cos x|$
19 $\frac{2}{3}\sin^3 x$ 20 $\frac{1}{2}\arctan x(1+x^2) - \frac{1}{2}x + c$
21 $2\sqrt{x}\left(\ln x - 2\right) + c$ 22 $t\tan t + \ln|\cos x| + c$
23 Verification
24 $-x^4\cos x + 4x^3\sin x + 12x^2\cos x - 24x\sin x - 24\cos x + c$
25 $x^5\sin x + 5x^4\cos x - 20x^3\sin x - 60x^2\cos x + 120x\sin x + 120\cos x + c$
26 $e^x\left(x^4 - 4x^3 + 12x^2 - 24x + 24\right) + c$
27 Proof 28 Proof 29 Proof
30 Proof 31 Proof

Exercise 16.3

1 $\frac{1}{80}\cos 5t - \frac{1}{48}\cos 3t - \frac{1}{8}\cos t$; $c\,\dfrac{\cos^5 t}{5} - \dfrac{\cos^3 t}{3} + c$
2 $\dfrac{\cos^6 t}{6} - \dfrac{\cos^4 t}{4} + c$
3 $\dfrac{\cos^4 3\theta}{12} + c$
4 $\frac{1}{3}\cos^3\left(\frac{1}{t}\right) - \frac{2}{5}\cos^5\left(\frac{1}{t}\right) + \frac{1}{7}\cos^7\left(\frac{1}{t}\right) + c$
5 $\sec x + \cos x + c$ 6 $\frac{1}{18}\tan^6 3x + c$
7 $\frac{1}{24}\left(3\tan^4\theta^2 + 2\tan^6\theta^2\right) + c$
8 $\frac{2}{5}\sec^5\sqrt{t} - \frac{2}{3}\sec^3\sqrt{t} + c$
9 $\frac{1}{15}\left(\tan^3 5t - 3\tan 5t + 15t\right) + c$
10 $\tan t - \sec t + c$ 11 $\csc t - \cot t + c$
12 $-\ln|1 - \sin t| + c$ 13 $-2x - 3\ln|\sin x + \cos x| + c$
14 $\arctan\left(\sec\theta\right) + c$ 15 $\frac{1}{2}\left(\arctan t\right)^2 + c$
16 $\ln|\arctan t| + c$ 17 $\arcsin\left(\ln x\right) + c$
18 $\dfrac{-\cos x}{3}\left(\sin^2 x + 2\right) + c$ 19 $\frac{2}{5}\left(\cos^2 x\sqrt{\cos x} - 5\sqrt{\cos x}\right) + c$
20 $\dfrac{-\cos\sqrt{x}}{3}\left(2\sin^2\sqrt{x} + 4\right) + c$
21 $\dfrac{\sin\left(\sin t\right)}{3}\left(\cos^2\left(\sin t\right) + 2\right) + c$
22 $\ln|\sin\theta| + 2\sin\theta + c$ 23 $t\sec t - \ln|\sec t + \tan t| + c$
24 $-\ln\left(2 - \sin x\right) + c$ 25 $\frac{1}{2}\ln\left|\cos\left(e^{-2x}\right)\right| + c$
26 $2\ln\left|\sec\sqrt{x} + \tan\sqrt{x}\right| + c$ 27 $\frac{1}{2}\tan x + c$
28 $\frac{1}{6}\left(\arcsin 3x + 3x\sqrt{1 - 9x^2}\right) + c$
29 $\dfrac{x}{4\sqrt{x^2+4}} + c$
30 $2\ln\left|t + \sqrt{t^2+4}\right| + \frac{1}{2}t\sqrt{t^2+4} + c$
31 $\frac{3}{2}\arctan\left(\frac{1}{2}e^t\right) + c$ 32 $\frac{1}{2}\arcsin\left(\frac{2}{3}x\right) + c$
33 $\frac{1}{3}\ln\left|\frac{3}{2}x + \frac{1}{2}\sqrt{9x^2+4}\right| + c$ 34 $\ln\left|\sqrt{1+\sin 2x} + \sin x\right| + c$
35 $-\sqrt{4-x^2} + c$ 36 $\frac{1}{2}\ln\left(x^2+16\right) + c$

37 $-\arcsin\left(\dfrac{x}{2}\right) - \dfrac{\sqrt{4-x^2}}{x} + c$ 38 $\dfrac{1}{9}\dfrac{x}{\sqrt{9-x^2}} + c$
39 $\dfrac{(x^2+1)^{\left(\frac{3}{2}\right)}}{3} + c$ 40 $\dfrac{(e^{2x}+1)^{\left(\frac{3}{2}\right)}}{3} + c$
41 $\frac{1}{2}\left(\arcsin\left(e^x\right) + e^x\sqrt{1-e^{2x}}\right) + c$ 42 $\ln\left(\frac{1}{3}e^x + \frac{1}{3}\sqrt{e^{2x}+9}\right) + c$
43 $2\sqrt{x}\left(\ln x - 2\right) + c$
44 $12\ln\left(x+2\right) + \dfrac{8}{x+2} + \dfrac{x^2}{2} - 4x + c$
45 $\frac{1}{2}\ln\left(x^2+9\right) + c_1$; $x = 3\tan\theta$ yields $\ln\left(\dfrac{\sqrt{x^2+9}}{3}\right) + c_2$; they differ by a constant
46 $x - 3\arctan\left(\frac{x}{3}\right) + c_1$; $x = 3\tan\theta$ yields $3(\tan\theta - \theta) + c_2 = 3\left(\frac{x}{3} - \arctan\frac{x}{3}\right) + c_2$

Exercise 16.4

1 24 2 40
3 $\frac{24}{25}$ 4 0
5 $\dfrac{176\sqrt{7} - 44}{5}$ 6 0
7 2 8 -268
9 $\frac{64}{3}$ 10 2
11 $\ln\left(\frac{11}{3}\right)$ 12 $\frac{44}{3} - 8\sqrt{3}$
13 3 14 $\sqrt{\pi} + 1$
15 a) 6 b) 6 c) 12 16 1
17 4 18 0
19 $\frac{\pi}{2}$ 20 $\frac{\pi}{6}$
21 $\frac{\pi}{3}$ 22 $\frac{\pi}{8}$
23 $\dfrac{14\sqrt{17} + 2}{3}$ 24 $\frac{1}{\pi}$
25 In(2) 26 $16\sqrt{2} - 5\sqrt{5}$
27 $\sqrt{14} - \sqrt{10}$ 28 $\frac{3}{2}$
29 $\pi^{\frac{3}{2}}\left(\dfrac{2\sqrt{3}}{27} - \dfrac{1}{12}\right)$ 30 $\frac{\pi}{6}$
31 $-\frac{1}{2}\ln\left(\frac{37}{52}\right)$
32 $-\arctan\left(\dfrac{\sqrt{15} - \sqrt{7}}{4}\right)$ or $\frac{1}{2}\left(\arcsin\left(\frac{1}{4}\right) - \arcsin\left(\frac{3}{4}\right)\right)$
33 $\frac{2}{3}$ 34 0
35 -4 36 $\frac{\pi}{6}$
37 $\frac{1}{6}\arctan\left(\dfrac{4\sqrt{3}}{9}\right)$ 38 $\dfrac{\pi\sqrt{3} - 3\sqrt{3}\arctan\left(\frac{\sqrt{3}}{2}\right)}{18}$
39 $\frac{1}{6}$ 40 $\dfrac{e-1}{2}$
41 $1 + \dfrac{e}{2}$ 42 $2\cos(1) + 2$
43 $\frac{31}{5}$ 44 $\frac{2}{\pi}$
45 $\dfrac{12 - 4\sqrt{3}}{\pi}$ 46 $\dfrac{e^8 - 1}{8e^8}$
47 $\dfrac{\pi}{6\ln 3}$ 48 $\dfrac{\sin x}{x}$

49 $-\dfrac{\sin t}{t}$

50 $-2x\dfrac{\sin x^2}{x^2}$

51 $2x\dfrac{\sin x^2}{x^2}$

52 $\dfrac{\cos t}{1+t^2}$

53 $\dfrac{b-a}{5+x^4}$

54 $-\csc\theta-\sec\theta$

55 $-\dfrac{1}{4x^{\frac{3}{4}}}\left(e^{x+3x^{\frac{1}{2}}}\right)$

56 Yes

57 a) $\dfrac{1}{3}\ln\left(\dfrac{3k+2}{2}\right)$ b) $k=\dfrac{2\left(e^3-1\right)}{3}$

58 Proof

59 $-(1-x)^{k+1}\left(\dfrac{1}{k+1}+\dfrac{1-x}{k+2}\right)$

60 a) 0 b) $\sqrt{47}$ c) $\dfrac{15\sqrt{47}}{47}$

61 Proof

Exercise 16.5

1 $\frac{1}{2}((1+2\sqrt{2})\ln|x-\sqrt{2}|+(1-2\sqrt{2})\ln|x+\sqrt{2}|)$

2 $3\ln|x-2|-2\ln|x|+c$

3 $\frac{1}{2}\ln|x^2+4x+3|+c$

4 $-\ln|x+1|+6\ln|x|-\dfrac{9}{x+1}+c$

5 $\ln|x+3|+3\ln|x+2|-2\ln|x|+c$

6 $\ln|x+1|+3\ln|x|+\dfrac{1}{x}+c$

7 $-\ln|x+2|+\ln|x-1|+c$

8 $\dfrac{3\ln|2x-1|}{2}-2\ln|x+1|+c$

9 $3\ln|x+2|+\dfrac{2}{x+2}+c$

10 $\ln|x-2|-4\ln|x+1|+3\ln|x|+\dfrac{6}{x}+c$

11 $-\ln|x^2+1|+2\ln|x|+c$

12 $\dfrac{\sqrt{3}}{3}\arctan\left(\dfrac{\sqrt{3}x}{3}\right)-\dfrac{\ln|x^2+3|}{3}+\dfrac{2\ln|x|}{3}+c$

13 $\dfrac{\sqrt{3}}{2}\arctan\left(\dfrac{x}{\sqrt{6}}\right)-\dfrac{\ln|x^2+6|}{6}+\dfrac{\ln|x|}{3}+c$

14 $\dfrac{\sqrt{2}}{2}\arctan\left(\dfrac{\sqrt{2}x}{4}\right)-\dfrac{3}{16}\ln|x^2+8|+\dfrac{3}{8}\ln|x|+c$

15 $\dfrac{\ln|x-5|}{3}+\dfrac{2\ln|x+1|}{3}-\ln|x|+c$

Exercise 16.6

1 $\dfrac{125}{6}$

2 $\dfrac{9\pi^2}{8}+1$

3 $4\sqrt{3}$

4 $\frac{10}{3}$

5 $\frac{8}{21}$

6 $\frac{125}{24}$

7 $\frac{13}{12}$

8 4π

9 $\frac{59}{12}$

10 Approx. 361.95 (4 points of intersection!)

11 $3\ln 2-\dfrac{63}{128}$

12 Between $-\dfrac{\pi}{6}$ and $\dfrac{\pi}{6}$, $\sqrt{3}\ln\left(\dfrac{3}{4}\right)-2\sqrt{3}+4$

13 18

14 $\frac{32}{3}$

15 $\frac{64}{3}$

16 9

17 $\frac{9}{2}$

18 19

19 $\dfrac{2\sqrt{3}}{3}+2$

20 $\frac{37}{12}$

21 $\frac{1}{2}$

22 $\dfrac{2\sqrt{2}}{3}$

23 $\frac{269}{54}$

24 $\dfrac{e}{2}-1$

25 $\dfrac{288\sqrt{3}}{35}$

26 $\dfrac{2\sqrt{2}}{3}$

27 $\dfrac{16}{3}$

28 25.36

29 $m=0.973$

30 $\dfrac{37}{12}$

Exercise 16.7

1 $\dfrac{127\pi}{27}$

2 $\dfrac{64\sqrt{2}\pi}{15}$

3 $\dfrac{70\pi}{3}$

4 6π

5 9π

6 2π

7 $\left(\dfrac{\sqrt{3}}{2}+1\right)\pi$

8 $\dfrac{512\pi}{15}$

9 Approx. 5.937π

10 $\dfrac{32\pi}{3}$

11 $\pi\left(\sqrt{3}-1\right)$

12 $\dfrac{23\pi}{210}$

13 $288\pi-\dfrac{160\pi\sqrt{5}}{3}$

14 $\dfrac{64}{15}\pi$

15 $\pi\left(\dfrac{1}{2}-\dfrac{1}{4}\sqrt{3}\right)$

16 $\dfrac{1778}{5}\pi$

17 $\dfrac{252}{5}\pi$

18 1419π

19 $\dfrac{9}{8}\pi$

20 a) $\dfrac{88}{15}\pi$ b) $\dfrac{7}{6}\pi$

21 40π

22 $9\pi\left(2-\sqrt{2}\right)$

23 $\dfrac{32}{15}\pi$

24 $\dfrac{4}{5}\pi\left(121\sqrt{33}-25\sqrt{15}\right)$

25 $2\pi\left(\ln 2-\dfrac{1}{4}\right)$

26 $2\pi\left(\dfrac{11}{3}\sqrt{11}-\dfrac{2}{3}\sqrt{2}\right)$

27 $\dfrac{28}{3}\pi\left(\sqrt{34}-\sqrt{7}\right)$

28 $\pi\left(\dfrac{1}{2}\sqrt{2}\pi-\pi+2\right)$

29 $\dfrac{284}{3}\pi$

30 2π

31 $\dfrac{256}{15}\pi$

Exercise 16.8

1 $\frac{70}{3}$ m, 65 m

2 8.5 m to the left, 8.5 m

3 1 m, 1 m

4 2 m, $2\sqrt{2}$ m

5 18 m, 28.67 m

6 $\dfrac{4}{\pi}$ m, $\dfrac{4}{\pi}$ m

7 $3t$, 6 m, 6 m

8 t^2-4t+3, 0, 2.67 m

9 $1-\cos t$, $\left(\dfrac{3\pi}{2}+1\right)$m, $\left(\dfrac{3\pi}{2}+1\right)$m

10 $4-2\sqrt{t+1}$, 2.43 m, 2.91 m

11 $3t^2+\dfrac{1}{2(1+t)^2}+\dfrac{3}{2}$, 11.3 m, 11.3 m

12 $4.9t^2+5t+10$

13 $16t^2-2t+1$

14 $\dfrac{1}{\pi}-\dfrac{\cos\pi t}{\pi}$

15 $\ln(t+2)+\dfrac{1}{2}$

16 $e^t+19t+4$

17 $4.9t^2-3t$

18 $\sin(2t)-3$

19 $-\cos\left(\dfrac{3t}{\pi}\right)$

20 12; 20

21 $\dfrac{13}{2};\dfrac{13}{2}$

22 $\dfrac{9}{4};\dfrac{11}{4}$

23 $2\sqrt{3}-6$; $6-2\sqrt{3}$

24 $-\dfrac{10}{3};\dfrac{17}{3}$

25 $\dfrac{204}{25}$

26 $-6;\dfrac{13}{2}$

27 $\dfrac{166}{5};\dfrac{166}{5};\dfrac{166}{5}$

28 a) $50-20t$ b) 1062.5

29 1.27 s

30 a) 5 s b) 272.5 m c) 10 s d) -49 m/s e) 12.46 s f) -73.08 m/s

Exercise 16.9

1 $y = \pm 10e^{x^4}$

2 $y = \pm e^{\frac{1}{2}x^2}$

3 $y = \dfrac{2}{2-x^2}$

4 $y = \dfrac{1}{3-x}$

5 $y = \ln\left(\dfrac{e}{1-ex}\right)$

6 $y = \ln\left(e^x - C\right)$

7 $y^3 = \dfrac{3(x+1)^2}{2} - \dfrac{1}{2}$

8 $y = \dfrac{1}{\ln|x+1|+1}$

9 $2y^3 + 6y = 3x^2 + 6x + 72$

10 $y^2 = e^{x^2} - 1$

11 $\arctan y = \dfrac{x^2}{2} + c$

12 $y + \ln|y| = \dfrac{x^2}{2} - x + 1$

13 $x + \ln\dfrac{1}{x + Ce^x + 1}$

14 $\dfrac{y-1}{y+1} = e^{(x-1)^2} + c$

15 $(y+1)\ln|y+1| + 1 = (y+1)(\ln|\ln x|) + c$

16 $1 + 2y^2 = c\tan^4\dfrac{x}{2}$

17 $\arcsin y = 1 - \sqrt{1-x^2}$

18 $y = \ln\left(\ln\dfrac{e(e^x+1)}{1+e}\right)$

19 $y + |\ln y| = \dfrac{x^3}{3} - x - 5$

20 $\cos y = \dfrac{\sqrt{2}}{4}(e^x + 1)$

21 $|y| = |x|e^{x^2-1}$

22 $2\ln|y| - y^2 = e^{x^2} - 2$

23 $y + \ln|\sec y| = \dfrac{1}{3}x^3 + x + c$

24 $\sqrt{(y^2+1)^3} = 3e^t(t-1) + c$

25 $e^{-y}(y+1) = -\dfrac{1}{3}\sin^3\theta + c$

26 $e^{3y} + 3y^2 = 3(\cos x + x\sin x) - 2$

27 $y = e^x - x^2 + 2$

28 b) $C = 78; m = \dfrac{1}{15}\ln\dfrac{8}{13}$; 45.3 minutes

Practice questions

1 a) $p = 3$ b) 3 square units

2 a) (0, 1) b) $V = \displaystyle\int_0^{\ln 2} (e^{\frac{x}{2}})^2 dx$

3 $a = e^2$

4 a) $y = \dfrac{x}{e}$

5 a) (i) 400 m (ii) $v = 100 - 8t$, 60 m/s
(iii) 8 s (iv) 1344 m
b) Distance needed 625

6 b) 2.31 c) $-\pi\cos x - \dfrac{x^2}{2} + c$; 0.944

7 ln 3

8 a) (ii) (1.57, 0); (1.1, 0.55); (0, 0), (2, −1.66)
b) $x = \dfrac{\pi}{2}$ c) (ii) $\displaystyle\int_0^{\frac{\pi}{2}} x^2\cos x\, dx$ d) $\dfrac{\pi^2}{2} - 2$

9 a) 2π
b) Range: $\{y \mid -0.4 < y < 0.4\}$
c) (i) $-3\sin^3 x + 2\sin x$ (iii) $\dfrac{2\sqrt{3}}{9}$
d) $\dfrac{\pi}{2}$
e) (i) $\frac{1}{3}\sin^3 x + c$ (ii) $\frac{1}{3}$
f) $\arccos\dfrac{\sqrt{7}}{3} \approx 0.491$

10 c) 3.696 72 d) $\displaystyle\int_0^{\pi} (\pi + x\cos x)dx$
e) $\pi^2 - 2 \approx 7.869\,60$

11 a) (i) $10x + 1 - e^{2x}$ (ii) $\dfrac{\ln 5}{2} \approx 0.805$
b) (i) $f^{-1}(x) = \dfrac{\ln(x-1)}{2}$
c) $v = \pi\displaystyle\int_0^{\ln 2} (1 + e^{2x})^2\, dx$

12 $\pi\left(\dfrac{2}{15}a^5 + \dfrac{2}{3}a^3\right)$

13 $4\left(\dfrac{2}{5}\left(\dfrac{1}{2}x+1\right)^{\frac{5}{2}} - \dfrac{2}{3}\left(\dfrac{1}{2}x+1\right)^{\frac{3}{2}}\right) + c$

14 $a = -\dfrac{56}{27}$

15 $\dfrac{\pi}{2}\left(e^{2k} - 1\right)$

16 $k = 2$

17 1800 m

18 $2a$ by $\dfrac{2}{3}a^2$

19 a) $\ln x + 1 - k$ b) $x > \dfrac{1}{e}$
c) (ii) $\left(e^k, 0\right)$ d) $\dfrac{e^{2k}}{4}$
e) $y = x - e^k$ f) Verify
g) Common ratio = e

20 $x^2 - 4y^2 = 4$

21 $v = \sqrt{v_0^2 + \dfrac{4k}{m}}$

22 a)

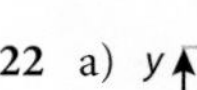

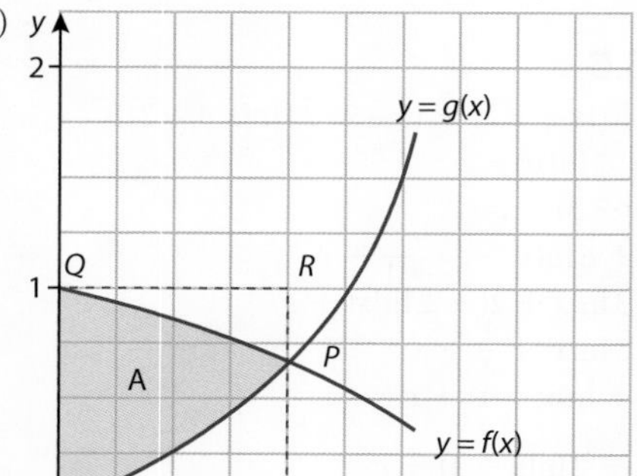

b) Proof c) 0.6937
d) $\int_0^p \left(e^{-x^2} - \left(e^{-x^2} - 1\right)\right) dx \approx 0.467$

23 a) Verify
b) $\dfrac{2\pi}{9}; \dfrac{4\pi}{9}; \dfrac{6\pi}{9}$
c) $\dfrac{n\pi}{9}(n+1)$

24 a) $t = 0, 3$, or 6
b) (i) $\left|\int_0^6 t\sin\left(\dfrac{\pi}{3}\right)t\,dt\right|$ (ii) 11.5 m

25 a) 0.435 b) $\dfrac{-2t}{\left(2+t^2\right)^2}$

26 a) $\dfrac{dy}{dx} = \dfrac{2x^2}{\sqrt{1+x^2}} + 2\sqrt{1+x^2}$
b) Verify c) $k = 0.918$

27 6 m

28 0.852

29 a) Verify
a) (i) $A = 78$; $k = \dfrac{1}{15}\ln\dfrac{48}{78}$ (ii) 45.3

30 $y = \tan\left(\ln\dfrac{x}{2}\right)$

31 $\dfrac{(x+2)^2}{2} - 6(x+2) + 12\ln|x+2| + \dfrac{8}{x+2} + c$

32 a)

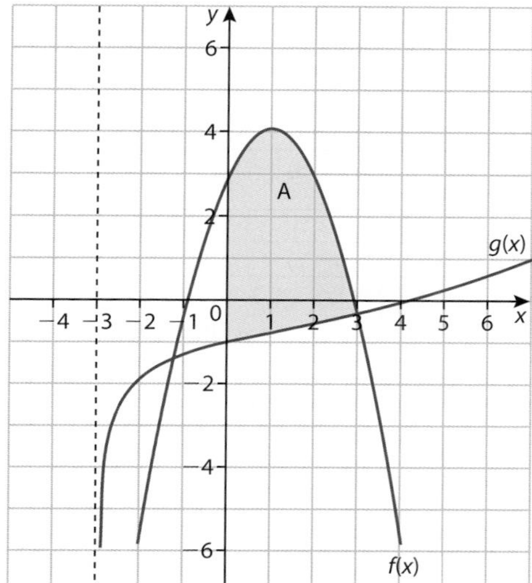

b) (i) $x = -3$;
(ii) $x - \text{int} = e^2 - 3$; $y - \text{int} = \ln 3 - 2$

c) -1.34; 3.05

d) (ii) $\int_0^{3.05} \left(4 - (1-x)^2 - \left(\ln(x+3) - 2\right)\right) dx$
(iii) 10.6

e) 4.63

33 a) Verify
b) $\ln x - \frac{1}{2}\ln\left(x^2+1\right) + c$
c) $y = \dfrac{2e^{\theta}}{\sqrt{e^{2\theta}+1}}$

Chapter 17

Exercise 17.1

1 a) Discrete b) Continuous c) Continuous
d) Discrete e) Continuous f) Continuous
g) Discrete h) Continuous i) Continuous
j) Discrete k) Continuous l) Continuous
m) Discrete

2 a) 0.4
b)

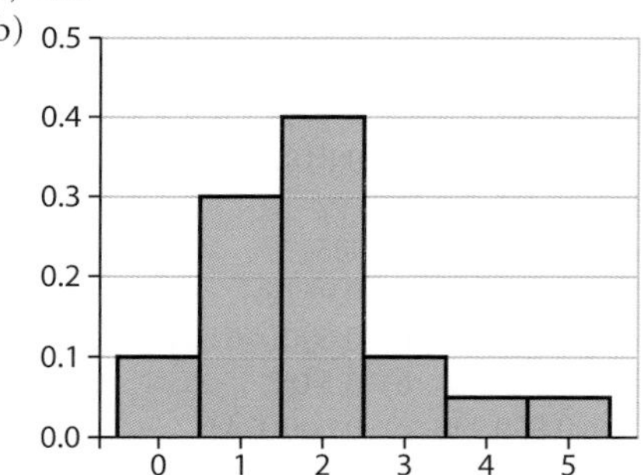

c) 1.85, 1.19 e) 2.85, 1.19
f) $E(X + b) = E(X) + b$; $V(X + b) = V(X)$

3 a) 0.26 b) 0.37 c) 0.77
d) 16.29 e) 8.126 f) 4.125; 2.013 25
g) $E(aX + b) = aE(X) + b$; $V(aX + b) = a^2V(X)$

4 a) 0.969 b) 0.163 c) 3.5

5 $k = \frac{1}{30}$

x	12	14	16	18
$P(X = x)$	$6k$	$7k$	$8k$	$9k$

6 a) $k = \frac{1}{10}$ b) $\frac{37}{60}$ c) $\frac{19}{30}$
d) $E(X) = 16$, SD = 7 e) $E(Y) = \frac{11}{5}$; SD $= \frac{7}{5}$

7 a) $\frac{1}{50}$
b)

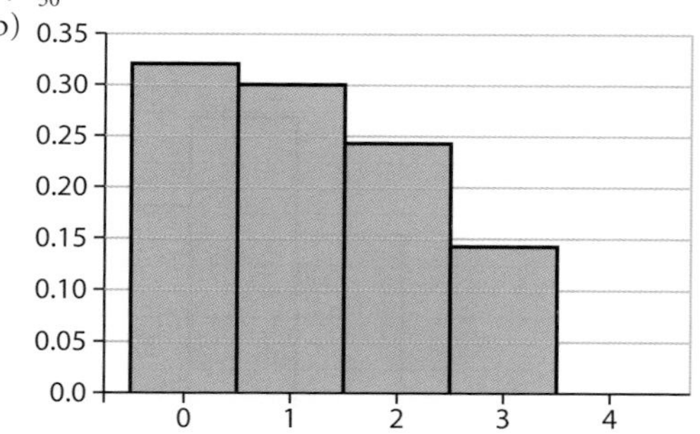

c) $\frac{17}{25}$
d) $\mu = 1.2$; var = 0.9

8 a) $P(x = 18) = 0.2$, $P(x = 19) = 0.1$, symmetric distribution.
b) $\mu = 17$, SD = 1.095

9 a) $\mu = 1.9$, SD = 1.34
b) Between 0 and 5

10 $k = 0.667$, $E(X) = 5.44$

11 a) $k = 0.3$ or 0.7
b) for $k = 0.3$: $E(X) = 2.18$; for $k = 0.7$: $E(X) = 1.78$

12 a)

y	0	1	2	3
$P(Y = y)$	$\frac{1}{27}$	$\frac{2}{9}$	$\frac{4}{9}$	$\frac{8}{27}$

b) 2

13 a) $k = \frac{1}{10}$ b) $\frac{1}{2}$

14 a) <See table below>
b) 0.85 c) 0.15 d) 48.87
e) 2.057 f) 0.77

x	45	46	47	48	49	50	51	52	53	54	55
CDF	0.05	0.13	0.25	0.4	0.65	0.85	0.9	0.94	0.97	0.99	1

15 a)

x	0	1	2	3	4	5	6
CDF	0.08	0.23	0.45	0.72	0.92	0.97	1

b) 0.72 c) 0.97
d) 2.63 e) 1.44

16 a) 0.90 b) 0.09 c) 0.009
d) Unacceptable, acceptable e) $p(x) = (0.1^{x-1}) \times 0.90$

17 a) 0 b) 0.81 c) 0.162
d) Either acceptable or unacceptable, acceptable
e) $(x - 1)\,(0.1^{x-2}) \times 0.90^2$, $x > 1$

18 $n = 30$

19 a) (i) $\frac{1}{9}$ (ii) $\frac{1}{81}$
b) (i) $\frac{73}{648}$ (ii) $\frac{575}{1296}$

c) (ii)

x	1	2	3	4	5	6
$P(X=x)$	$\frac{1}{1296}$	$\frac{15}{1296}$	$\frac{65}{1296}$	$\frac{175}{1296}$	$\frac{369}{1296}$	$\frac{671}{1296}$

(iii) $\frac{6797}{1296}$

20 9.3

Exercise 17.2

1 a)

x	0	1	2	3	4	5
$P(X=x)$	0.010 24	0.0768	0.2304	0.3456	0.2592	0.077 76

b)

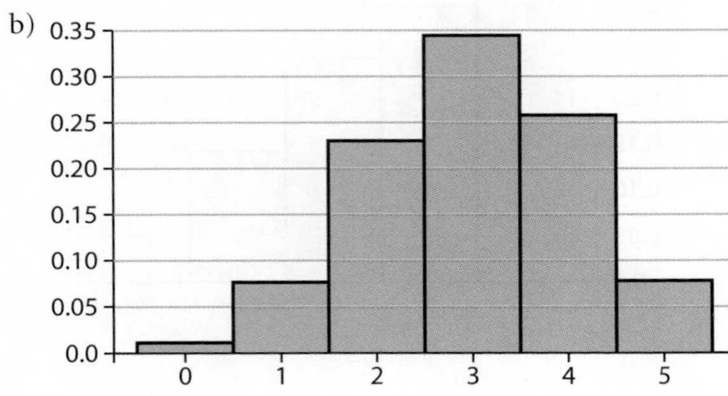

c) (i) Mean = 3, SD = 1.095
(ii) Mean = 3, SD = 1.095

d) Between 2 and 4, and between 1 and 5

e) 0.8352, 0.990. Slightly more than the empirical rule.

2 a) 0.001 294 494 b) 0.000 000 011
c) 0.999 999 99 d) 0.999 999 66
e) Mean = 12, SD = 2.19

3 a)

k	0	1	2	3	4	5	6
$P(x \leq k)$	0.117 65	0.420 17	0.744 31	0.929 53	0.989 07	0.999 27	1

b)

Number of successes x	List the values of x	Write the probability statement	Explain it, if needed	Find the required probability
At most 3	0, 1, 2, 3	$P(x \leq 3)$	$P(x \leq 3)$	0.929 53
At least 3	3, 4, 5, 6	$P(x \geq 3)$	$1 - P(x \leq 2)$	0.255 69
More than 3	4, 5, 6	$P(x > 3)$	$1 - P(x \leq 3)$	0.070 47
Fewer than 3	0, 1, 2	$P(x \leq 2)$	$P(x \leq 2)$	0.744 31
Between 3 and 5 (inclusive)	3, 4, 5	$P(3 \leq x \leq 5)$	$P(x \leq 5) - P(x \leq 2)$	0.254 96
Exactly 3	3	$P(x = 3)$	$P(x = 3)$	0.185 22

4 a)

k	0	1	2	3	4	5	6	7
$P(x \leq k)$	0.027 99	0.158 63	0.419 90	0.710 21	0.903 74	0.981 16	0.998 36	1

b)

Number of successes x	List the values of x	Write the probability statement	Explain it, if needed	Find the required probability
At most 3	0, 1, 2, 3	$P(x \leq 3)$	$P(x \leq 3)$	0.710 21
At least 3	3, 4, 5, 6, 7	$P(x \geq 3)$	$1 - P(x \leq 2)$	0.580 10
More than 3	4, 5, 6, 7	$P(x > 3)$	$1 - P(x \leq 3)$	0.289 79
Fewer than 3	0, 1, 2	$P(x \leq 2)$	$P(x \leq 2)$	0.419 90
Between 3 and 5 (inclusive)	3, 4, 5	$P(3 \leq x \leq 5)$	$P(x \leq 5) - P(x \leq 2)$	0.561 26
Exactly 3	3	$P(x = 3)$	$P(x = 3)$	0.290 304

5 a) p is not constant, trials are not independent.
b) p becomes constant.
c) $n = 3, p = \frac{5}{8}$

y	0	1	2	3
$P(Y=y)$	0.052 73	0.263 672	0.439 453	0.244 141

d) 0.755 86 e) 1.875 f) 0.703 125 g) 0.947 27

6 a) 0.107 374 b) 0.993 63 c) 0.892 63 d) 2

7 a) 0.817 073 b) 1 c) 0.016 1776

8 a) 0.033 833 b) 0.024 486 c) 0.782 722

9 a) 0.75 b) 0.032 5112 c) 0.172 678

10 a) 0.043 1745 b) 0.997 614 c) 0.011 2531
d) 0.130 567 e) 0.956 826 f) 10
g) 3 h) 4, 16

11 a) 3 b) 0.101 308 c) 0.000 214 925

12 a)

x	0	1	2	3	4	5
$P(x)$	0.031 25	0.156 25	0.312 50	0.312 50	0.156 25	0.031 25

b) 0.031 25 c) 0.031 25 d) 0.968 75 e) 0.968 75
f) a)

x	0	1	2	3	4	5
$P(x)$	0.327 68	0.409 60	0.204 80	0.051 20	0.006 40	0.000 32

b) 0.327 68 c) 0.000 32 d) 0.672 32 e) 0.999 68

13 a) 0.138 b) 0.144

14 0.912 96

15 a) 0.107 b) 0.893 c) $n = 14$

Exercise 17.3

Note: most answers are rounded.

1 a) 0.100 82 b) 0.8153 c) 0.1847 d) 0.3203

2 a) 0.1755 b) 0.2650 c) 0.7350 d) 0.6764

3 a) 0.0025 b) 0.9826 c) 0.9999

4 a) 0.9048 b) 0.0047 c) 0.8187

5 a) (i) 0.0344 (ii) 0.8197
b) (i) 0.0001 (ii) 0.9986

6 a) 0.1396 b) 0.1912 c) 0.9576

7 a) 0.000 0768 b) 0.000 076 824

8 a) 0.8187 b) 0.5488

9 a) 0.9877 b) 0.999 998 c) 0.000 0244

10 a) 0.265 b) 0.990

11 a) 0.0908 b) 0.408

12 a) 2.8473 b) 0.617

13 a) 0.245, 0.214, 0.0524 b) 0.464

Exercise 17.4

1 a) $k = -\frac{3}{2}$ b) 0.3125 c) 0.6875
d) 0.375, 0.3473, 0.2437

2 a) $\frac{1}{6}$ b) $\frac{1}{8}$ c) $\frac{1}{2}$
d) $\frac{7}{9}$, 0.697, 0.533

3 a) $k = \sqrt{2}$ b) 0.766 c) 0.234
d) 0.754, 0.765, 0.3127

4 a) $\frac{6}{37}$ b) $\frac{133}{148}$ c) $\frac{19}{74}$
d) $\frac{50}{37}$, 1.5, 0.528

5 a)

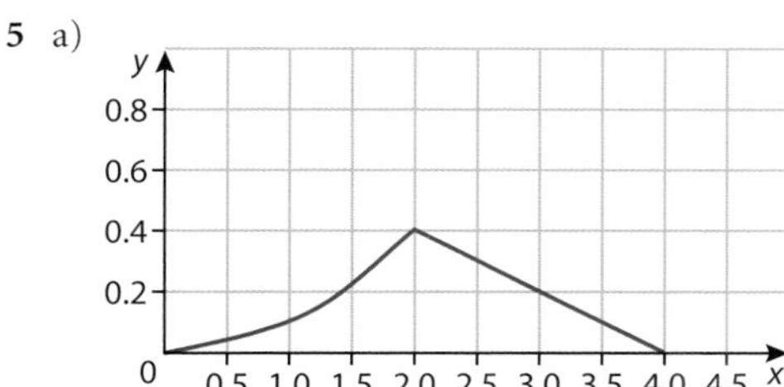

b) $\frac{3}{29}$ c) $\frac{113}{58}$, 1.89, 0.757 d) 0.983

6 a) 24.7 hours b) 0.514 c) 0.264

7 a) 50 hours b) 50 hours c) 22.4 hours
d) 0.104 e) (i) 0.010 82 (ii) 0.9892

8 a)

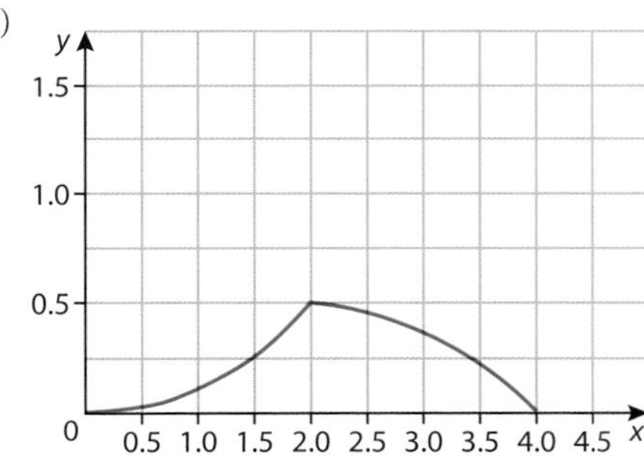

b) $\frac{7}{3}$ c) 0.694 d) 134 barrels

9 b) $\frac{7}{2}$, 0.916

10 b) $a = \frac{6}{125}$; $b = 5$ c) 1.25

11 a) $k = \frac{1}{(b-a)}$
b) mean = median = $\frac{(a+b)}{2}$; variance = $\frac{(a-b)^2}{12}$

12 a) (i) 0.378 (ii) 1.752 (iii) 1.892
b) 0.955

13 a) $\frac{1}{8}$
b) $f(x) = \begin{cases} 0 & 0 \leqslant x < 5 \\ \frac{3(x-7)^2}{8} & 5 \leqslant x \leqslant 7 \\ 0 & x > 7 \end{cases}$
c) 5.4126 d) 0.15

14 a) $k = 3$ b) $\frac{4}{5}$ c) 0.8409

15 b) 0.0183 c) $\frac{\sqrt{\pi}}{2}$ d) 0.8326 e) 0.641 f) 0.0769

16 a) $\frac{5}{9}$ b) 0.1944 c) 0.1941 d) 0.6207

17 b) 3, 3.1, 3.3 c) 0.475 d) 1 e) 0.64, no

18 a) $\frac{10}{3}, \frac{15}{4}$ c) 0.0803 d) 0.891
e) (i) 0.987 (ii) 0.9999 (iii) 0.9996

19 $\frac{54}{11}$ **20** 1.08

Exercise 17.5

Note: some answers are rounded.

1 a) 0.5 b) 0.499 571 c) 0.158 655
d) 0.682 690 e) 0.022 750 f) 0

2 a) 0.769 86 b) 0.161 514 c) 0.656 947
d) 0.999 944

3 a) 0.008 634 b) 0.982 732

4 1.28

5 1.96

6 a) 0.066 807 b) 0.682 69 c) 678.16
d) 134.898

7 a) 1.8% b) 509.975 c) 5.71

8 a) 0.9696 b) 0.546 746

9 a) 1 day b) 29 days c) 112 days

10 1.56 **11** 18.95

12 30.81 **13** 100.28

14 29.95

15 $\mu = 21.037$, $\sigma = 4.252$

16 $\mu = 18.988$, $\sigma = 0.615$

17 $\mu = 121.936$, $\sigma = 34.39$

18 a) $\mu = 6.966$, $\sigma = 0.324$ b) 0.252

19 a) 0.655 422 b) 0.008 198 c) 82 bottles

20 a) 0.227 319 b) 0.55% c) 29.678
d) 229.182

21 a) Not likely: chance is 0.14% b) 15.87%
c) 68.27% d) 5396 e) 43 785

22 a) 6.817 b) 3.4315
c) $\mu = 64.14$, $\sigma = 7.545$

23 7.3% **24** 216.06 **25** 15.31

26 a) $\mu = 111.89$, $\sigma = 17.9$ **27** 0.919

28 a) (i) $\sigma = 1.355$ (ii) $\mu = 110.37$
b) A = 108.63; B = 112.11

Practice questions

1 a) 34.5% b) 0.416 c) 3325

2 a) (i) 0.393 (ii) 0.656 b) 50

3 a) 0.1 b) 10 d) 0.739

4 a) $\frac{35}{128}$ b) $\frac{7}{32}$ c) $\frac{91}{128}$

5 a) $a = -0.455$, $b = 0.682$
b) (i) 0.675
(ii) 0.428
c) (ii) $t = 62.6$

6 a) $\mu = 50 - 10(0.522\,44) \approx 44.8$
b) HI: the mean speed has been affected by the campaign.
c) One-tailed test, as we are interested in a decrease in the mean only (not also an increase).

7 a) 70.1% b) 0.002 26 c) p-value = 5.48%

8 a) 0.0808 c) $\mu = 25.5$, $\sigma = 0.255$ d) 12 500

9 a) (i) 0.345 (ii) 0.115 (iii) 0.540
b) 0.119 c) 737

10 a) 15.9% b) 210

11 a) 0.0912 b) $a = 251, b = 369.$

12 a) $a = -1, b = 0.5$ b) (i) 0.841 (ii) 0.533

c) (i)

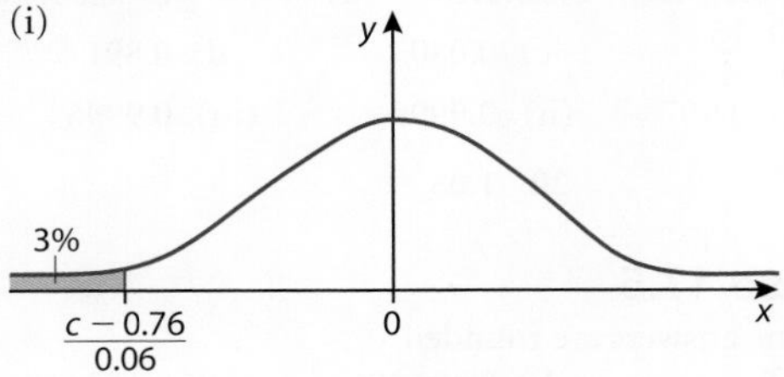

(ii) 0.647

13 a) 2 b) 0.182 c) 0.597

14 $\mu = 66.6, \sigma = 22.6$

15 a) 0.8

b) (i)

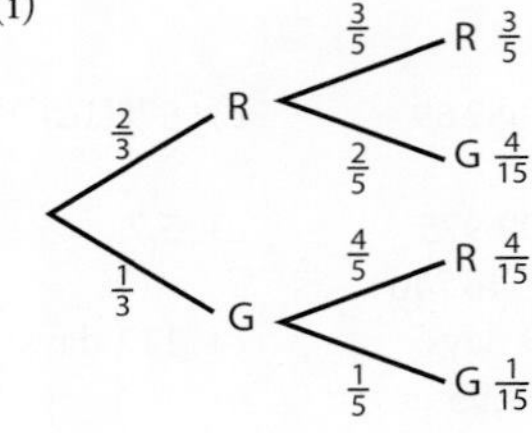

(ii)

x	0	1	2
$P(X = x)$	$\frac{1}{15}$	$\frac{8}{15}$	$\frac{2}{5}$

c) $\frac{3}{10}$ d) $\frac{1}{9}$

16 a) 0.129 886 b) 0.676 714 c) 2

17 a) 0.1829 b) 0.3664

18 a) $\frac{1}{5}$ b) $\frac{7}{5}$

19 a) (i) 0.217% (ii) 0.012%

b) 84.13%

20 $\sigma = 0.009\,43$ kg ≈ 9.4 g

21 b) $\frac{e}{4} - \sqrt{e} + \sqrt[4]{e}$

c) $\mu = \frac{e}{2} - 1$; $\sigma^2 = 1 + \frac{e}{3} - \frac{e^2}{4}$

d) $\sqrt{e} - \frac{e}{2}$

e) $\left(\sqrt{e} - \frac{e}{2}\right)^3$

f) $\binom{3}{2}\left(1 - \sqrt{e} + \frac{e}{2}\right)\left(\sqrt{e} - \frac{e}{2}\right)^2$

22 a) 0.2212 b) 0.125

23 a) $x = 58.69$ b) $\sigma = 3.41$

c) (i) Karl

(ii) 0.002 39

24 a)

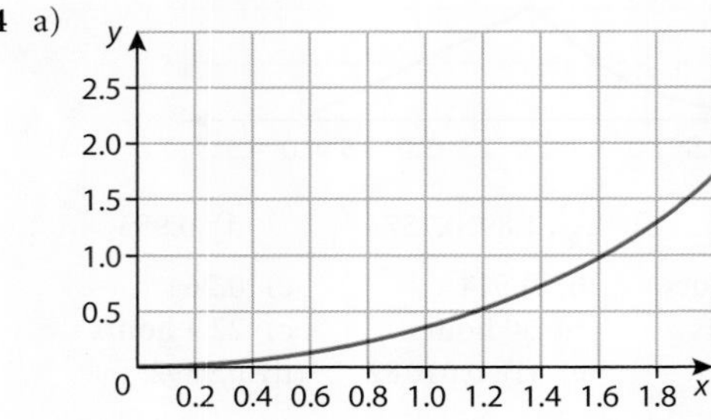

b) 2 c) 1.51 d) 1.61

25 a) $\mu = 1.63$ c) 0.434 d) $6605.28

Index

Page numbers in *italics* refer to information boxes and hint boxes.

E

H

I

O

P

S

T